The Auditory Cortex

Jeffery A. Winer · Christoph E. Schreiner
Editors

The Auditory Cortex

Springer

Editors
Jeffery A. Winer
Department of Molecular & Cell Biology
University of California
Berkeley, CA 94720-3200, USA

Christoph E. Schreiner
Coleman Memorial Laboratory
Department of Otolaryngology
University of California
San Francisco, CA 94143-0732, USA
chris@phy.ucsf.edu

ISBN 978-1-4419-0073-9 e-ISBN 978-1-4419-0074-6
DOI 10.1007/978-1-4419-0074-6
Springer New York Dordrecht Heidelberg London

Printed on acid-free paper

Springer is part of Springer Science+Business Media (www.springer.com)

Photograph courtesy of Jane A. Winer

Jeffery A. Winer (1945–2008)
Scholar, scientist, colleague, mentor, friend

Photograph courtesy of Jane A. Winer

Preface

This volume is a summary and synthesis of the current state of auditory forebrain organization. We think it a timely contribution in view of the growing interest in this network as the arbiter for hearing, as a key element in the larger communications network that spans and links the parietal, temporal, and frontal cortices, and as a candidate for clinical intervention, whether through cochlear implants or more exotic upstream prostheses that, one day, may involve the forebrain more directly.

The present account differs from the available efforts (Aitkin 1990; König et al. 2005) in two significant ways. First, the medial geniculate body is included as a full partner since it has cooperative, reciprocal, and robust relations with the auditory cortex that suggest a partnership in which the exclusion of either structure detracts from a functional portrait of their interactions. Second, our aim has been systematic and synoptic, including as it does a wide range of species, methods, subsystems, physiological perspectives, and functional architectures. We look back on 100 years of the discipline of auditory forebrain studies with a view to framing a future agenda. As new methods emerge and as older approaches exhaust their potential, it seems appropriate to attempt a summing up and to forge a prospectus for future work. We cannot present a full theory of auditory forebrain organization since the field is still so new as a discipline; that task we must leave to a later, more mature volume that recognizes the distributed nature of forebrain operations in a more refined way than is now possible. Our goal is to provide an experimental foundation and a conceptual framework for the auditory forebrain useful to the discipline as a whole, and which one might consult as both a summary of work in progress and an invitation to explore further. This formidable task could not have been accomplished without the contribution of an expert cohort of collaborators on whose efforts this enterprise rests.

Several methodological and conceptual insights have converged to create the present, congenial atmosphere for this effort. The emergence of new functional approaches such as the tissue slice and its varieties has enabled the exploration of new neurochemical and synaptic vistas (Metherate and Hsieh 2004) and allowed a more formal and anatomical–physiological characterization of identified neurons (Verbny et al. 2006). Related advances include the important insights gleaned from large silicon electrodes that span the full cortical depth and reveal critical facets of interneuronal and laminar organization invisible to a single extracellular pipet (Atencio and Schreiner 2008). Such local circuits in the medial geniculate body and auditory cortex are the functional building blocks upon which the large-scale operations of spectral analysis, aurality, and frequency modulation are arrayed. How these several subsystems interact cooperatively as a network is among the most challenging questions for the future. Other powerful insights flowed from the ability to record from synaptically joined pairs of cells (Miller et al. 2001) contributing to a new perspective on the thalamocortical transformation (Winer et al. 2005). Understanding such transformations—tectothalamic, thalamocortical, corticocortical, and corticofugal—remains an enterprise for the future.

A second wave of insight arose from the neuroimaging domain, where positron emission tomography, functional magnetic resonance imaging, and magnetoencephalography each provided powerful documentation of the locus and density of activity in the living brain during specific tasks or after particular pathologies. This work not only defined the site of activation, but related measures such as 2-deoxygluose provided the first full perspective on the limits of auditory-responsive cortex (Poremba et al. 2003).

Neuroanatomical and immunocytochemical approaches have provided credible maps of connectivity in the thalamocortical and corticocortical systems (Huang and Winer 2000; de La Mothe et al. 2006), documenting a vast web of forebrain long- and short-range circuits. The implementation of studies of lamina-specific interneuronal properties has provided valuable insights into these dynamic systems (Atencio et al. 2009). The corticothalamic and other corticofugal systems likewise are now construed as prospective dynamic players in regulating auditory cortical excitability rather than as feedback pathways (Winer et al. 2001). Combined physiological-connectional studies established the existence of specific pathways for sound localization and object identification (Rauschecker and Tian 2000).

The dramatic demonstration and ensuing exploration of widespread auditory forebrain plasticity (Kilgard and Merzenich 1998; Weinberger 1998) was a watershed and its implementation in the descending systems (Zhang et al. 2005) suggested a role for the corticofugal systems very different from earlier accounts that emphasized feedback. The auditory cortex now appears to be as concerned with the control of inferior colliculus excitability and plasticity and information processing as it is in the analysis of sound parameters and categorical perceptual analyses. Such findings were a linchpin in larger efforts to characterize the distributed auditory cortex as an entity that represents hearing in its largest and most inclusive sense (Winer and Lee 2007). The present volume can be construed as a multidisciplinary effort to further implement and instantiate that perspective.

Berkeley, CA Jeffery A. Winer
San Francisco, CA Christoph E. Schreiner

References

Aitkin L (1990) *The Auditory Cortex: Structural and Functional Bases of Auditory Perception*. Chapman and Hall, London.

Atencio CA and Schreiner CE (2008) Spectrotemporal processing differences between auditory cortical fast-spiking and regular-spiking neurons. *Journal of Neuroscience* **28**: 3897–3910.

Atencio CA, Sharpee TO, and Schreiner CE (2009) Hierarchical computation in the canonical auditory cortical circuit. *Proceedings of the National Academy of Sciences of the United States of America* **106**:21894–21899.

de La Mothe L, Blumell S, Kajikawa Y, and Hackett TA (2006) Thalamic connections of the auditory cortex in marmoset monkeys: core and medial belt regions. *Journal of Comparative Neurology* **496**:72–96.

Huang CL and Winer JA (2000) Auditory thalamocortical projections in the cat: laminar and areal patterns of input. *Journal of Comparative Neurology* **427**:302–331.

Kilgard MP and Merzenich MM (1998) Cortical map reorganization enabled by nucleus basalis activity. *Science* **279**:1714–1718.

König R, Heil P, Budinger E, and Scheich H (eds.) (2005) *The Auditory Cortex. A Synthesis of Human and Animal Research*, Lawrence Erlbaum Associates, Mahwah.

Metherate R and Hsieh CY (2004) Synaptic mechanisms and cholinergic regulation in auditory cortex. *Progress in Brain Research* **145**:143–156.

Miller LM, Escabí MA, and Schreiner CE (2001) Feature selectivity and interneuronal cooperation in the thalamocortical system. *Journal of Neuroscience* **21**:8136–8144.

Poremba A, Saunders RC, Crane AM, Cook M, Sokoloff L, and Mishkin M (2003) Functional mapping of the primate auditory system. *Science* **299**:568–572.

Rauschecker JP and Tian B (2000) Mechanisms and streams for processing of "what" and "where" in auditory cortex. *Proceedings of the National Academy of Sciences of the United States of America* **97**:11800–11806.

Verbny YL, Erdélyi F, Szabó G, and Banks MI (2006) Properties of a population of GABAergic cells in murine auditory cortex weakly excited by thalamic stimulation. *Journal of Neurophysiology* **96**:3194–3208.

Weinberger NM (1998) Tuning the brain by learning and by stimulation of the nucleus basalis. *Trends in Cognitive Sciences* **2**:271–273.

Winer JA, Diehl JJ, and Larue DT (2001) Projections of auditory cortex to the medial geniculate body of the cat. *Journal of Comparative Neurology* **430**:27–55.

Winer JA and Lee CC (2007) The distributed auditory cortex. *Hearing Research* **229**:3–13.

Winer JA, Miller LM, Lee CC, and Schreiner CE (2005) Auditory thalamocortical transformation: structure and function. *Trends in Neurosciences* **28**:255–263.

Zhang Y, Hakes JJ, Bonfield SP, and Yan J (2005) Corticofugal feedback for auditory midbrain plasticity elicited by tones and electrical stimulation of basal forebrain in mice. *European Journal of Neuroscience* **22**:871–879.

Acknowledgments

Our colleagues, collaborators, and students were instrumental in expressing the need for and inspiring the preparation of this book. In lieu of a list that might inadvertently omit some worthy of mention, we simply thank the many whose encouragement and labors made this volume possible. Interactions with them were a delight and their ingenuity and creativity was a source of much pleasure. We thank the chapter authors for striving to capture the essence of a dynamic field within the space constraints.

On a more individual basis we are especially grateful to our editor, Ann Avouris of Springer New York for her unfailing efforts to help us realize our vision. She was the ideal editor, providing timely and balanced counsel and reminding us with courtesy of the shortest path to our goal. David T. Larue used his graphics expertise to organize and perfect the many figures that are at the heart of this volume.

Each of us were beneficiaries of the aid and encouragement of our families, Jane M. Winer, Carol R. Galbraith, and Eileen G. Winer for Jeff, and Marcia Raggio and Christina Schreiner for Christoph. We are most thankful for their unfailing and loving support.

Berkeley, CA — Jeffery A. Winer
San Francisco, CA — Christoph E. Schreiner

Contents

Contributors

Ana Amador Department of Organismal Biology and Anatomy, University of Chicago, Chicago, IL 60637, USA, ana.amador@gmail.com

Craig A. Atencio Coleman Memorial Laboratory, Department of Otolaryngology, University of California, School of Medicine, San Francisco, CA 94143-0732, USA, craig@phy.ucsf.edu

Matthew I. Banks Department of Anesthesiology, University of Wisconsin, Madison, WI 53706, USA, mibanks@facstaff.wisc.edu

John F. Brugge Department of Physiology, University of Wisconsin, Madison, WI 53706, USA, brugge@physiology.wisc.edu

Mike B. Calford Faculty of Health, The University of Newcastle Australia, Callaghan, NSW 2308, Australia, mike.calford@newcastle.edu.au

Hubert R. Dinse Department of Theoretical Biology, Institute for Neuroinformatics, Ruhr-University Bochum, 44780 Bochum, Germany, hubert.dinse@neuroinformatik.ruhr-uni-bochum.de

Jean-Marc Edeline Centre de Neurosciences de Paris-Sud (CNPS) UMR 8195, CNRS & Université Paris-Sud, 91405 Orsay, Cedex, France, jean-marc.edeline@u-psud.fr

Jos J. Eggermont Department of Psychology, University of Calgary, Calgary, AB, Canada T2N 1N4, eggermon@ucalgary.ca

Günter Ehret Institute of Neurobiology, University of Ulm, D-89069, Ulm, Germany, guenter.ehret@uni-ulm.de

Taffeta M. Elliott Department of Psychology, University of California, Berkeley, CA 94720-1650, USA, taffeta@berkeley.edu

Robert C. Froemke Skirball Institute for Biomolecular Medicine and Departments of Otolaryngology, Physiology and Neuroscience, New York University School of Medicine, New York, NY 10016, USA, robert.froemke@med.nyu.edu

Troy A. Hackett Department of Hearing and Speech Sciences, Vanderbilt University, Nashville, TN 37203, USA, troy.a.hackett@vanderbilt.edu

Amee J. McMillan Cerebral Systems Laboratory, Department of Physiology and Pharmacology, Department of Psychology, The University of Western Ontario, London, ON N6A 5C1, Canada, ahall59@uwo.ca

Jufang He Department of Rehabilitation Sciences, The Hong Kong Polytechnic University, Kowloon, Hong Kong, rsjufang@polyu.edu.hk

Junsei Horikawa Department of Knowledge-Based Information Engineering, Toyohashi University of Technology, Toyohashi-shi 441–8580, Japan, horikawa@tutkie.tut.ac.jp

Edward G. Jones Center for Neuroscience, University of California, Davis, CA 95616, USA, ejones@ucdavis.edu

Jon H. Kaas Department of Psychology, Vanderbilt University, Nashville 37240, TN, USA, jon.h.kaas@vanderbilt.edu

Jagmeet S. Kanwal Department of Physiology and Biophysics, Georgetown University Medical Center, Washington, DC 20057-1460, USA, kanwalj@georgetown.edu

Andrew J. King Department of Physiology, Anatomy and Genetics, University of Oxford, Oxford, OX1 3PT, UK, andrew.king@dpag.ox.ac.uk

Andrej Kral Institute of Audioneurotechnology and Department of Experimental Otology, ENT Clinics, Medical University Hannover, Feodor-Lynen-Str. 35, D-30625 Hanover, Germany, a.kral@uke.de

Charles C. Lee Department of Neurobiology, The University of Chicago, Chicago, IL 60637, USA, clee@bsd.uchicago.edu

Catherine Liégeois-Chauvel Laboratoire de Neurophysiologie et Neuropsychologie, Université de la Méditerranée, INSERM EPI 9926 13005, Marseille, France, catherine.liegeois@medecine.univ-mrs.fr

Stephen G. Lomber Cerebral Systems Laboratory, Department of Physiology and Pharmacology, Department of Psychology, The University of Western Ontario, N6A 5C1, London, ON, Canada, steve.lomber@uwo.ca

Bernd Lütkenhöner Department of Otorhinolaryngology, University Hospital Münster, 48129, Münster, Germany, lutkenh@uni-muenster.de

Manuel S. Malmierca Auditory Neurophysiology Unit, Laboratory for the Neurobiology of Hearing, Faculty of Medicine, Institute for Neuroscience of Castilla y León, and Dept Cell Biology and Pathology, Med School, University of Salamanca, Salamanca, Spain, msm@usal.es

Daniel Margoliash Department of Organismal Biology and Anatomy, Department of Psychology, University of Chicago, Chicago, IL 60637, USA, dan@bigbird.uchicago.edu

Julie R. Mendelson Research, Toronto Rehabilitation Institute, Toronto, ON, M5G 2A2, Canada, j.mendelson@utoronto.ca

Raju Metherate Department of Neurobiology and Behavior, Center for Hearing Research, University of California, Irvine, CA 92697, USA, rmethera@uci.edu

John C. Middlebrooks Department of Otolaryngology, University of California at Irvine, Irvine, CA 92697-5310, USA, j.midd@uci.edu

Israel Nelken Department of Neurobiology, The Edmond and Lily Safra Center for Brain Sciences and the Silberman Institute of Life Sciences, Hebrew University, Jerusalem 91904, Israel, israel@cc.huji.ac.il

Frank W. Ohl Abteilung für Neuroscience, Leibniz-Institute for Neurobiology, D-39118 Magdeburg, Germany, frank.ohl@ifn-magdeburg.de

Hisayuki Ojima Tokyo Medical and Dental University, Tokyo 113-8549, Japan, yojima.cnb@tmd.ac.jp

Sarah L. Pallas Neuroscience Institute, Georgia State University, Atlanta, GA 30302-5030, USA, spallas@gsu.edu

Dennis P. Phillips Department of Psychology, Dalhousie University, Halifax, NS B3H 4J1, Canada

David Poeppel Department of Psychology, New York University, New York, NY 10003, USA, david.poeppel@nyu.edu

Ramesh Rajan Department of Physiology, Monash University, Clayton, VIC 3168, Australia, ramesh.rajan@med.monash.edu.au

Josef P. Rauschecker Department of Physiology and Biophysics, Georgetown University School of Medicine, Washington, DC 20007-2197, USA, josef@helix.nih.gov

Lizabeth M. Romanski Department of Neurobiology and Anatomy, University Rochester Medical Center, Rochester, NY 14642, USA, liz_romanski@urmc.rochester.edu

Eric M. Rouiller Institute de Physiologie, 1700 Fribourg, Switzerland, eric.rouiller@unifr.ch

David K. Ryugo Program for Neuroscience, Garvan Institute of Medical Research, Darlinghurst, NSW 2010, Australia, d.ryugo@garvan.org.au

Henning Scheich Leibniz Institute for Neurobiology 39118, Magdeburg, Germany, henning.scheich@ifn-magdeburg.de

Marc Schönwiesner International Laboratory for Brain, Music and Sound Research, Université de Montréal, Montréal, QC, Canada H2V 4P3, marc.schoenwiesner@umontreal.ca

Christoph E. Schreiner Coleman Memorial Laboratory, Department of Otolaryngology, W.M. Keck Center for Integrative Neuroscience, University of California, School of Medicine, San Francisco, CA 94143-0732, USA, chris@phy.ucsf.edu

Shihab A. Shamma Department of Electrical Engineering, University of Maryland, College Park, MD 20742, USA, sas@eng.umd.edu

Philip H. Smith Department of Anatomy, University Wisconsin Medical School, Madison, WI 53706, USA, smith@neurophys.wisc.edu

Mitchell Steinschneider Kennedy Center, Albert Einstein Medical College, Bronx, NY 10461, USA, steinsch@aecom.yu.edu

Nobuo Suga Department of Biology, Washington University, St. Louis, MO 63130, USA, suga@biology.wustl.edu

Mitchell L. Sutter Center for Neuroscience, University of California, Davis, CA 95618, USA, mlsutter@ucdavis.edu

Frédéric E. Theunissen Department of Psychology, University of California, Berkeley, CA 94720-1650, USA, theunissen@berkeley.edu

Mark N. Wallace Medical Research Council Institute of Hearing Research, University Park, Nottingham NG7 2RD, UK, markw@ihr.mrc.ac.uk

Xiaoqin Wang Department of Biomedical Engineering, Johns Hopkins University, Baltimore, MD 21205, USA, xwang@bme.jhu.edu

Michael Wehr Department of Psychology, Institute of Neuroscience, University of Oregon, Eugene, OR 97403-1227, USA, wehr@uoneuro.uoregon.edu

Norman M. Weinberger Qureshey Research Laboratory, Department of Neurobiology and Behavior, Center of Neurobiology Learning and Memory, University of California at Irvine, Irvine, CA 92697-3800, USA, nmweinbe@uci.edu

Jeffery A. Winer Division of Neurobiology, Department of Molecular and Cell Biology, University of California at Berkeley, Berkeley, CA 94720-3200, USA, chris@phy.ucsf.edu

Robert J. Zatorre Department of Neuropsychology, Montreal Neurological Institute and Hospital, McGill University, Montréal, QC, Canada H3A 2B4; International Laboratory for Brain, Music and Sound Research, Université de Montréal, Montréal, QC, Canada H2V 4P3, robert.zatorre@mcgill.ca

Chapter 1

The Historical Development of Ideas About the Auditory Cortex

Edward G. Jones

1 Introduction: Early Theories of Brain and the Perception of Sound

The realization that auditory perception depended upon the cerebral hemispheres came to science and medicine rather later than that of the other principal sensations. Thomas Willis (1664, 1681) while recognizing the dependence of auditory perception upon the forebrain, felt that some aspects of the hearing sense, especially the appreciation of music, depended upon the cerebellum, a view that was to persist well into the eighteenth century and even beyond. Willis, knowing that the auditory nerves (his seventh pair of nerves) were concerned with hearing, and tracing them to the vicinity of the cerebellum, considered that "the impression of the sound or the Species admitted to the Ears ...[is] carried inwardly towards the Cerebel and *sensorium commune*," that is, to both the cerebellum and higher levels of the brain. Of the latter, he felt that the corpus striatum was the eventual arrival place. "Ideas of sounds conveyed also to the Cerebel; which forming there footsteps or tracts, impress a remembrance of themselves, from whence when afterwards the Species there laid up are drawn forth by the help of the vocal process, voices, like the sounds before admitted, and breaking forth in a certain ordained series, come to be made." That is, the cerebellum maintains the beat and tempo of a series of sounds and permits them to be reproduced later, in this case mediated by the outgoing facial nerve, which Willis also saw as part of the auditory nerve arising from the vicinity of the cerebellum. "Hence it is usual, that musick or melody is soon learnt by some men, which afterwards they bring forth with exact Symphony ... the Spirits moving within the Cerebel [being] disposed into peculiar Schemes; to which when they flow on both sides into the vocal process of the auditory Nerve, they render as it were with a certain spontaneous voice, and like a Machine or Clock with the succession of Species, the measures or Tunes of the Instrument which they had drunk in at the ears."

The realization that the cerebral cortex formed the substratum for sensation and motion starts to become implicit in many of the numerous anatomical studies devoted to charting the cerebral sulci and gyri in the latter part of the eighteenth and early part of the nineteenth century. By the time that Ecker (1869) wrote his *Die Hirnwindung des Menschen*, he could begin by saying "That the cortex of the cerebrum, the undoubted material substratum of our mental operations, is not a single organ, which is brought into play as a whole in the exercise of each and every psychical function, but consists rather of a multitude of mental organs, each of which is subservient to certain intellectual processes, is a conviction which forces itself upon us almost with the necessity of a claim of reason" (Translation by John Galton 1873). No friend to phrenology, then in its dying days, Ecker considered that uncovering the localization of "psychical functions" in the cortex of the cerebral hemisphere was destined to become one of the most important problems for anatomy and physiology and destined to bring about a revolution in psychology. Ecker's work, which summed up in brief format the knowledge that had accumulated about the human cerebral sulci and gyri and provided a systematic nomenclature that remains in use today, came at a time when experimental studies that were to reveal the localization of the sensory and motor areas of the cerebral cortex were about to begin. A curious feature of his work, however, is his unusually superficial description of the gyri of the insula and temporal operculum. This stands in marked contrast to his detailed descriptions of the other gyri and sulci of the hemisphere. The discovery of Heschl's gyrus had to wait until 1877 (Heschl 1877). Heschl's gyrus is in fact two gyri, which Heschl himself called the anterior and posterior transverse temporal gyri.

E.G. Jones (✉)
Center for Neuroscience, University of California, Davis, CA 95616, USA
e-mail: ejones@ucdavis.edu

J.A. Winer, C.E. Schreiner (eds.), *The Auditory Cortex*,
DOI 10.1007/978-1-4419-0074-6_1,

2 First Experimental Studies in Monkeys: David Ferrier

If Ecker was disdainful of phrenology, it was nevertheless an interest in cerebral localization derived from phrenology that induced Sir James Crichton Browne, the Director of the West Riding Lunatic Asylum to invite a young Scottish neurologist with time on his hands in London to come to Yorkshire and commence investigations of the cerebrum in animals using lesions and electrical stimulation. David Ferrier commenced his investigations by confirming and extending the studies of Fritsch and Hitzig (1870; Hitzig 1874) that had led to the identification of the motor cortex. By using Faradic rather than Galvanic stimulation, Ferrier (1873, 1876) was able to obtain a far more precise localization of the motor cortex than had Fritsch and Hitzig, and in a series of experiments on monkeys, dogs, cats, jackals, rabbits, guinea pigs, and rats, he demonstrated regions from which movements of comparable parts of the limbs could be obtained, and thus confirmed the presence of a similar motor map. All his experiments were carried out under ether or chloroform anesthesia and the level of current used was that which elicited a tingling sensation when the electrodes were applied to the tongue of the investigator! He also identified regions from which stimulation evoked movements, such as eye and head turning, that he considered to be reflex responses to sensory experiences, and thus to betoken the presence of specific sensory areas. These observations led him to ablate various gyri or surface regions of the hemisphere in up to 25 monkeys (Ferrier 1875) in the search for specific losses of sensation. It was from these experiments that there emerged the first intimations of the existence of an auditory cortex.

Ferrier localizations were not always correct. He located the visual cortex, for example, in the angular gyrus since destruction of that gyrus led to "blindness in the opposite eye." As discussed below, a deeply penetrating lesion, compounded by secondary infection, undoubtedly led him into this error, his lesions having severed the optic radiation. He was closer to the truth with his identification of the auditory region. He had located it in the first instance in the upper part of the superior temporal gyrus (called by him the superior temporo-sphenoidal convolution) by noting that electrical stimulation of that region caused monkeys to prick up the opposite ear and turn the head and eyes to the opposite side. Following bilateral lesions of the superior temporal gyri (Fig. 1.1), "the animal, though fully conscious and on the alert to everything attracting sight, failed to respond to auditory stimuli usually exciting active reaction and attention." After unilateral lesions, "the animal continued to respond to auditory stimuli, turning its head if called to; ... reactions, however, which did not ensue when the ear on the same side as the lesion was securely stopped with cotton wool." Ferrier's account of how he determined that his animals with bilateral lesions were indeed deaf bears quoting in full: "In

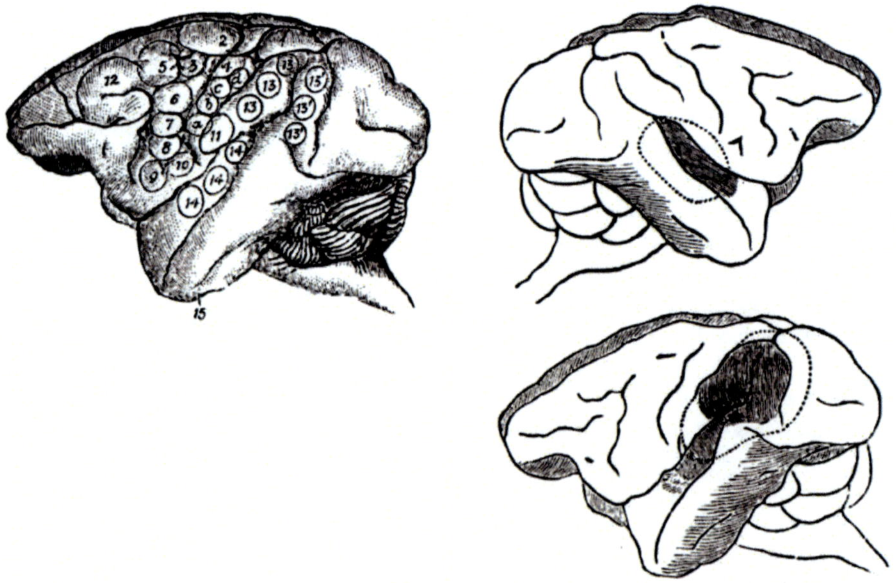

Fig. 1.1 Figures from Ferrier (1876) illustrating his stimulation and ablation experiments in monkeys; from these, he located the auditory cortex in the superior temporal gyrus. *Left*: the locations of regions which when stimulated electrically gave rise to movements of different parts of the body. From regions labeled 14 he reported "pricking of the opposite ear, head and eyes turn to the opposite side, pupils dilate widely." *Right*: the locations of bilateral lesions that led to "loss of hearing in both ears, and loss of sight in the right eye." The *dotted lines* indicate the extent of brain surface exposed by removal of part of the skull

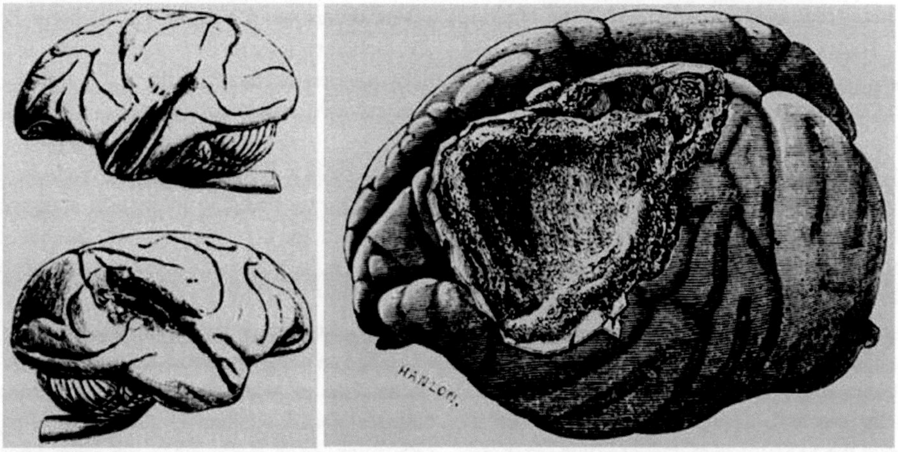

Fig. 1.2 *Left*: the extent of bilateral superior temporal lesions in one of Ferrier's monkeys, demonstrated at the *International Medical Congress* in 1881 and found to be profoundly deaf. From Ferrier (1886). ***Right***: the extent of the lesion in the second, hemiplegic monkey. From Ferrier et al. (1881)

order to avoid attracting its attention by sight, I retired behind the door and watched the animal through a chink, while it sat comfortably before the fire. When all was still I called loudly, whistled, knocked, &c., without attracting the animal's attention to the source of the sound, though it was sitting perfectly awake and looking around. On my cautiously approaching it, it remained unaware of my proximity until I came within the field of vision, when it started suddenly and made grimaces as if in terror or alarm. On repeating these tests when the monkey was sitting quietly along with a companion monkey whose powers of hearing were unquestionable, the companion invariably became startled at the sounds, and came peering curiously to ascertain their origin, while the other remained quite still. Ten hours subsequently I again repeated these various tests with the same results—results which justified the conclusion that whether the animal heard or not, it certainly gave no signs of hearing that which, in another animal, excited lively curiosity. Beyond this, without personal testimony from the subject of experiment, it is impossible to go, but I think that when the two sets of experiments are taken together,—*viz.*, the positive reactions to electric stimulation, and the absence of reaction to usual forms of auditory stimuli when the superior temporo-sphenoidal convolutions were destroyed,—the evidence of the localization of the centre of hearing in this region amounts to positive demonstration."

All of Ferrier's lesions in his first and largest series of monkeys were deeply penetrating and apparently heavily compromised by infection. They undoubtedly undercut the region on the supratemporal plane where we now know the primary auditory cortex to be located; and the presence of severe infection seems to have led Ferrier to kill the animals after relatively short survivals so that recovery of function could not be tested. By the time that he demonstrated his monkeys at the 1881 International Medical Congress in London (Ferrier et al. 1881), he had obtained the assistance of the surgeon, Gerald Yeo, who made lesions in a group of monkeys by the newly introduced antiseptic method, with the result that animals could survive postoperatively for long periods free of infection. One of the monkeys that demonstrated by Ferrier at the Congress had survived for 6 weeks subsequent to a bilateral ablation of the superior temporal gyri (Fig. 1.2). The other was a monkey with hemiplegia as the result of a lesion of the pre- and postcentral gyri carried out some 7 months before. The brain of a third monkey that had been blind as the result of ablation of the angular gyri and, significantly, of the occipital lobes, was also shown. When the monkeys were demonstrated at King's College on August 3rd, Yeo confessed that he had had some earlier skepticism about cerebral localization but now admitted to being completely won over. As has often been described, the condition of the hemiplegic monkey led the French neurologist, Charcot, to exclaim: "It is a patient." The second monkey is described in the *Proceedings of the Congress* as "the one which had had the region of the superior temporo-sphenoidal convolution destroyed in both hemispheres 10 weeks previously. The animal was seen to be active and vigorous without the slightest sign of motor paralysis in any part of the body. Its vision was evidently perfect, the animal snatching eagerly at pieces of food offered it. That it was deaf, however, was demonstrated most clearly. While the two monkeys were on the floor together before the audience, Dr. Ferrier snapped a percussion cap in their immediate proximity, whereupon the hemiplegic monkey started with the most lively signs of surprise, whereas the other exhibited not the slightest indication whatever of hearing. This experiment was repeated several times with the same result. The animal was admitted to be perfectly deaf, and no other deficiency could be detected." Ferrier had thus demonstrated the general location of the auditory cortex in the region of the upper part of the superior temporal gyrus. That success and Charcot's earlier poking fun at the English paradox of promoting fox hunting while enacting the most stringent of anti-vivisection laws, did not ensure Ferrier's protection from the hounds of righteousness and, within days, he was hauled off to Bow Street Magistrates Court and charged with cruelty to animals. Fortunately, it was proven that Yeo had performed the operations under general anesthesia and that he had the appropriate license, so Ferrier was acquitted, but not before he had been pilloried in the popular press and made a martyr by the medical establishment.

Ferrier seems not to have performed any further lesion experiments on the cerebral cortex, although the three that he had described at the 1881 International Medical Congress were written up in Ferrier and Yeo (1884) and presented in the second edition of his *Functions of the Brain* (1886). He was not, however, to escape from controversy over his localization of the auditory cortex. He was able to dismiss the observations of Luciani and Tamburini (Luciani 1884), who had described an early loss but considerable recovery of auditory function subsequent to bilateral lesions of the perisylvian regions, on the grounds that they had been made on dogs rather than monkeys and the testing of auditory function was crude. His own method wasn't much better. A more serious attack came from quarters closer to home, when Edward Schäfer, in a series of papers published between 1887 and 1888, described monkeys operated on with aseptic techniques and with extensive bilateral lesions of the temporal lobes (Fig. 1.3) and that displayed no evidence of severe or sustained hearing loss (Schäfer 1888a, b; Horsley and Schäfer 1888; Brown and Schäfer 1888). The dispute between Ferrier and Schäfer was perhaps as bitter as any between two rather correct Victorian gentlemen (Fig. 1.4) could be, Ferrier responding with a further review of his own work in which he presented additional data from his and Yeo's experimental notebooks, along with a discussion of cases in the human literature associated with bilateral strokes affecting both superior temporal gyri. He concluded that Schäfer's lesions were too small and superficial. They probably were too superficial, unlike Ferrier's, not penetrating deeply enough to undercut the auditory cortex. Perhaps Victor Horsley, who made most of the lesions, had a lighter neurosurgical hand than Gerald Yeo. Schäfer wasn't silenced and fought back along much the same lines as had Ferrier. Eventually however, Ferrier seems to have won the day and most neurology texts subsequent to this era localized the auditory cortex in the superior temporal gyrus. Even Schäfer (1900) seems to have retreated. By 1905, Campbell (1905) was in little doubt that Schäfer's lesions did not penetrate deeply enough to undercut the transverse temporal gyri.

3 The Clinical Experience: Localization of Human Auditory Cortex in the Superior Temporal Gyrus

A typical illustration from the later nineteenth century (Fig. 1.5) shows where many neurologists believed the

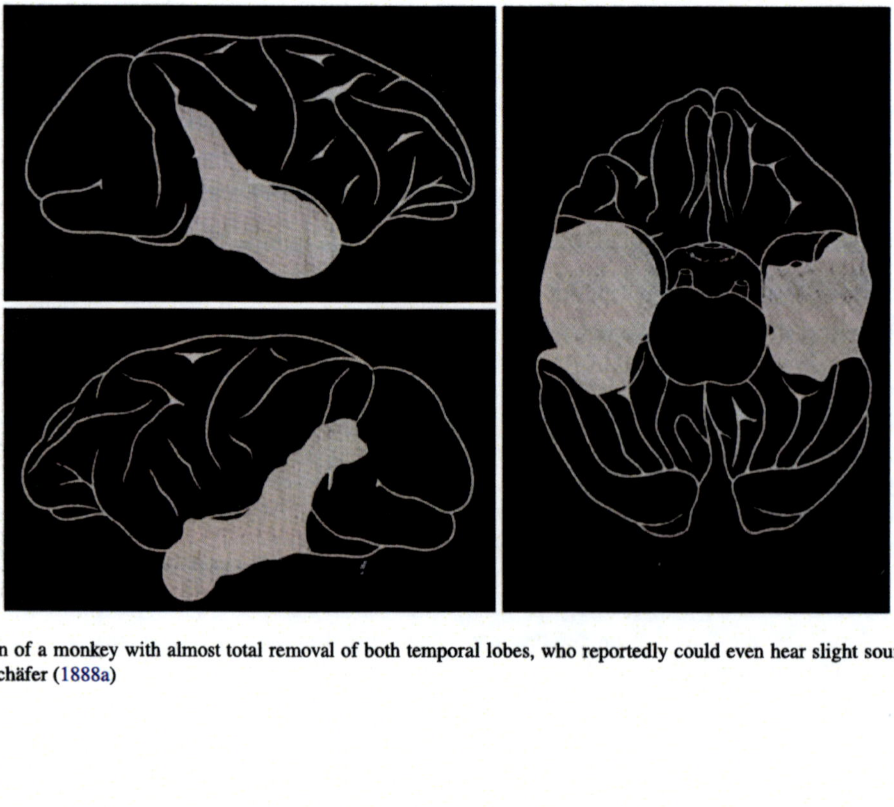

Fig. 1.3 The brain of a monkey with almost total removal of both temporal lobes, who reportedly could even hear slight sounds soon after the operation. From Schäfer (1888a)

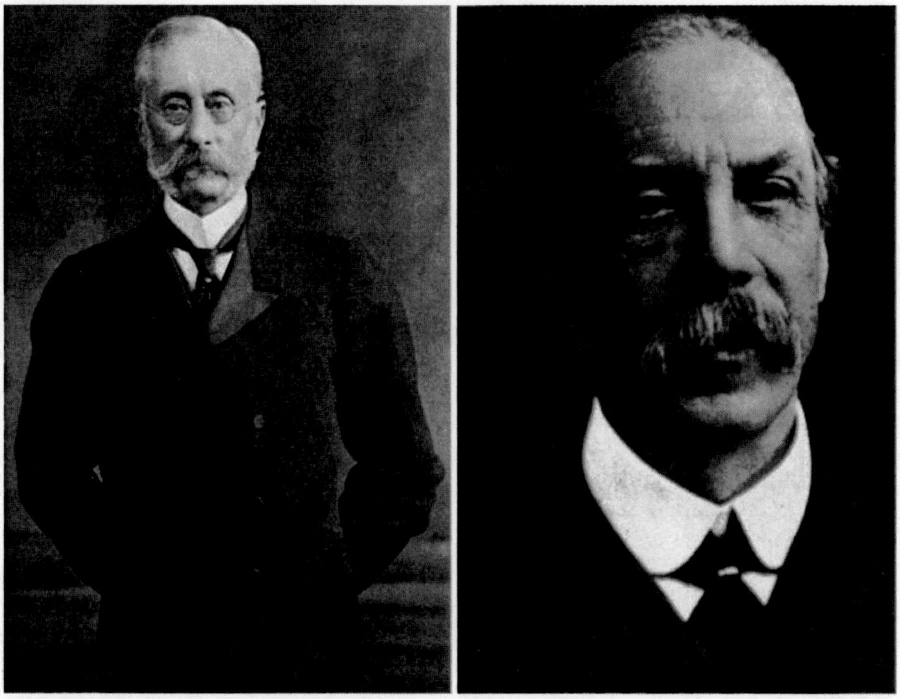

Fig. 1.4 Sir David Ferrier (1843–1928) and Sir Edward A. Schäfer (later Sharpey-Schäfer) (1850–1935), who fought bitterly over the location of the auditory cortex in the superior temporal gyrus. From *Biographical Memoirs of the Fellows of the Royal Society* (*left*) and from the *Quarterly Journal of Experimental Physiology* (*right*)

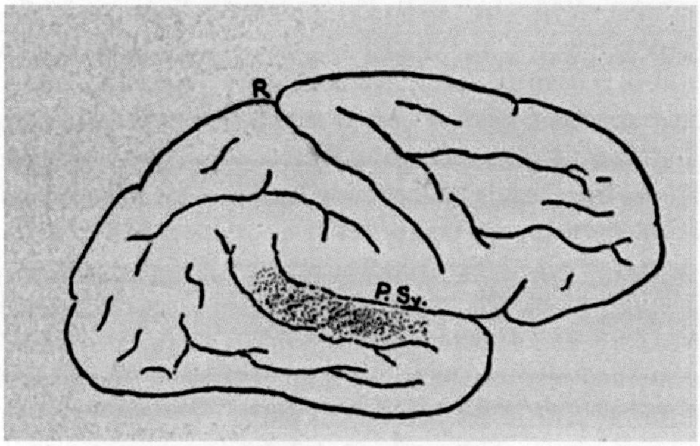

Fig. 1.5 The location of the auditory cortex in the human superior temporal gyrus, as commonly understood by neurologists, from about 1880 to about 1900. From Gowers (1885)

human auditory cortex to be located (Wernicke and Friedlander 1883; Gowers 1885; Mills 1891; Dejerine and Dejerine-Klumpke 1895). Examination of postmortem brains from patients who had suffered from strokes that impaired hearing invariably revealed large lesions of the upper part of the superior temporal gyrus; these findings were taken to be confirmatory of Ferrier's original observations. Bilaterally symmetrical lesions were rare but, when reported, were usually associated with total deafness. In the case reported by Wernicke and Friedlander (1883), there were bilateral gummata (abscess-like lesions of tertiary syphilis) in the upper parts of the superior temporal gyri. A case reported by Sérieux and Mignot (1901) had bilateral hydatid cysts; others such as those of Pick (1892), Anton (1899) and Mills (1891) were stroke cases with bilateral softening. Most neurologists, however, noted that lesions that impaired audition were also accompanied by alterations in the comprehension of language and extended into the temporal and parietal opercula and onto the insula.

4 Experimental Studies in Other Animals: Cats, Dogs, Rabbits

Studies on dogs have already been mentioned. In them Fritsch and Hitzig had first localized the motor cortex and in these and other carnivores Ferrier had obtained what he thought was evidence of an auditory area which, when stimulated, elicited movements suggestive of the animal attending to a sound (Fig. 1.6). The area in the dog, cat, and jackal and the equivalent area in a rabbit are labeled "14" in the figure. With the success of his work on monkeys, Ferrier was to turn away from dogs and other non-primates for his later experimental studies and, in general, to discount recovery of

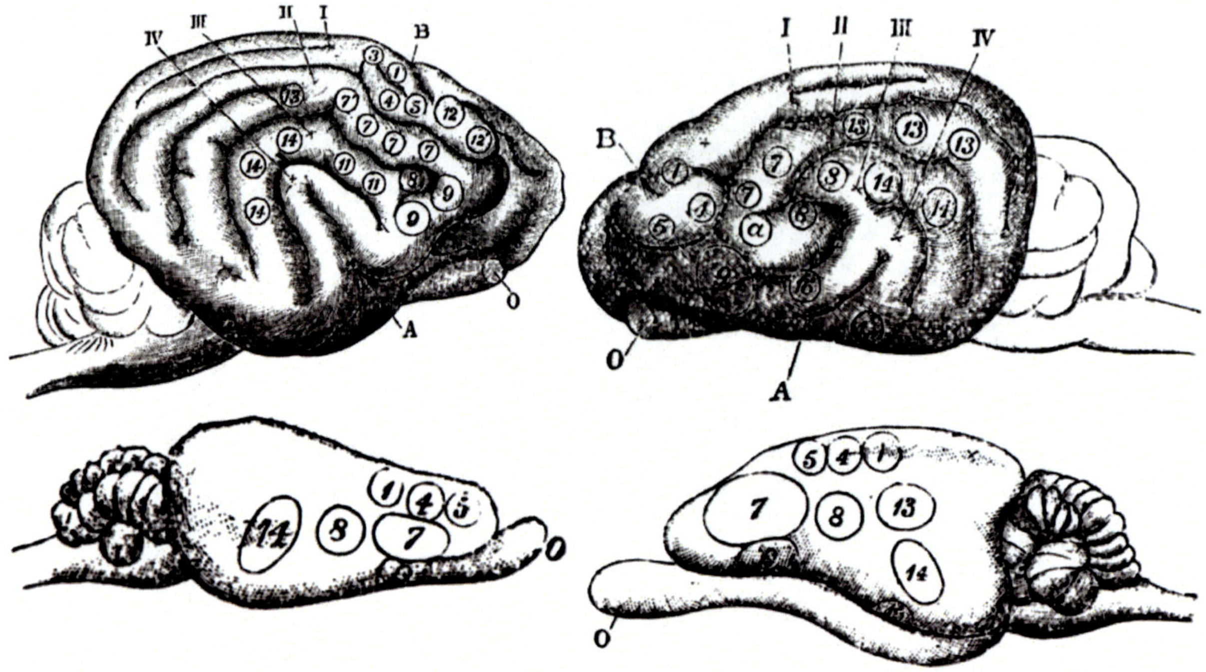

Fig. 1.6 Results of Ferrier's electrical stimulation experiments in a jackal (*top left*), cat (*top right*), rat (*lower left*) and guinea pig (*lower right*). In each case, stimulation of the area labeled 14 resulted in pricking of the ears and turning of the head to the opposite side. This area was therefore identified as the auditory cortex. From Ferrier (1876)

auditory function after lesions of the cerebral cortex in dogs and cats. Luciani and Tamburini (1879) and Luciani (1884) had bilaterally extirpated large regions of the sylvian and ectosylvian gyri in dogs and had observed an initial deafness and subsequently a significant loss of hearing acuity. Their method of testing hearing was to throw pieces of meat onto a tin plate. The German investigator, Hermann Munk (1881), had also observed a loss of hearing in dogs with bilateral lesions of the perisylvian region, localizing the principal focus at the ventral ends of the posterior ectosylvian and posterior suprasylvian gyri (Fig. 1.7) but noting that, to prevent any recovery of auditory function, it was essential to ablate much wider and deeper territories, perhaps extending to the hippocampus. To Munk, the key region in the posterior ectosylvian gyrus was a center for the comprehension of the meaning of sounds, likening the effects of its removal to something resembling psychic blindness. Clearly, these early investigators with their rather primitive means of testing auditory function were coming up against the capacity of animals to discriminate some aspects of sound without a cortex, so long as the inferior colliculus remains intact something that was demonstrated much later. For example, the auditory cortex is not essential for frequency discrimination, cats being able to perform this if the auditory midbrain is intact (Goldberg and Neff 1961). The cortex is, however, necessary for most aspects of sound localization (Whitfield et al. 1972; Heffner 1978). The ability of dogs to discriminate musical notes of different pitch after perisylvian region lesions was examined in a preliminary way by Munk who felt that anterior lesions were associated with deficits in the discrimination of higher pitched sounds and posterior lesions with deficits in the discrimination of lower pitched sounds. A Russian student of Bechterew, Larionow (1899), followed up this observation by testing the ability of dogs to discriminate tones after small but penetrating lesions in the ventral ends of the posterior sylvian, posterior ectosylvian and posterior suprasylvian gyri (Fig. 1.8). The resultant map was a remarkable facsimile of the cochlea, with higher tones represented anteriorly and ventrally and lower tones represented posteriorly and dorsally. Perhaps Larionow's lesions penetrated different parts of the auditory radiation and thus interfered with the tonotopically organized thalamocortical projection to the primary auditory areas that we now know to be located more dorsally in the middle ectosylvian regions (see below). Larionow was an inventive man and tried to record responses to tuning forks of different pitches by placing a galvanometer on the surface of the dog's cortex. He may have been at least partially successful.

5 Anatomical Identification of Auditory Cortex

By this time, of course, terminations of the ascending auditory pathways in the medial geniculate nuclei had been demonstrated with the Marchi technique by Ferrier and

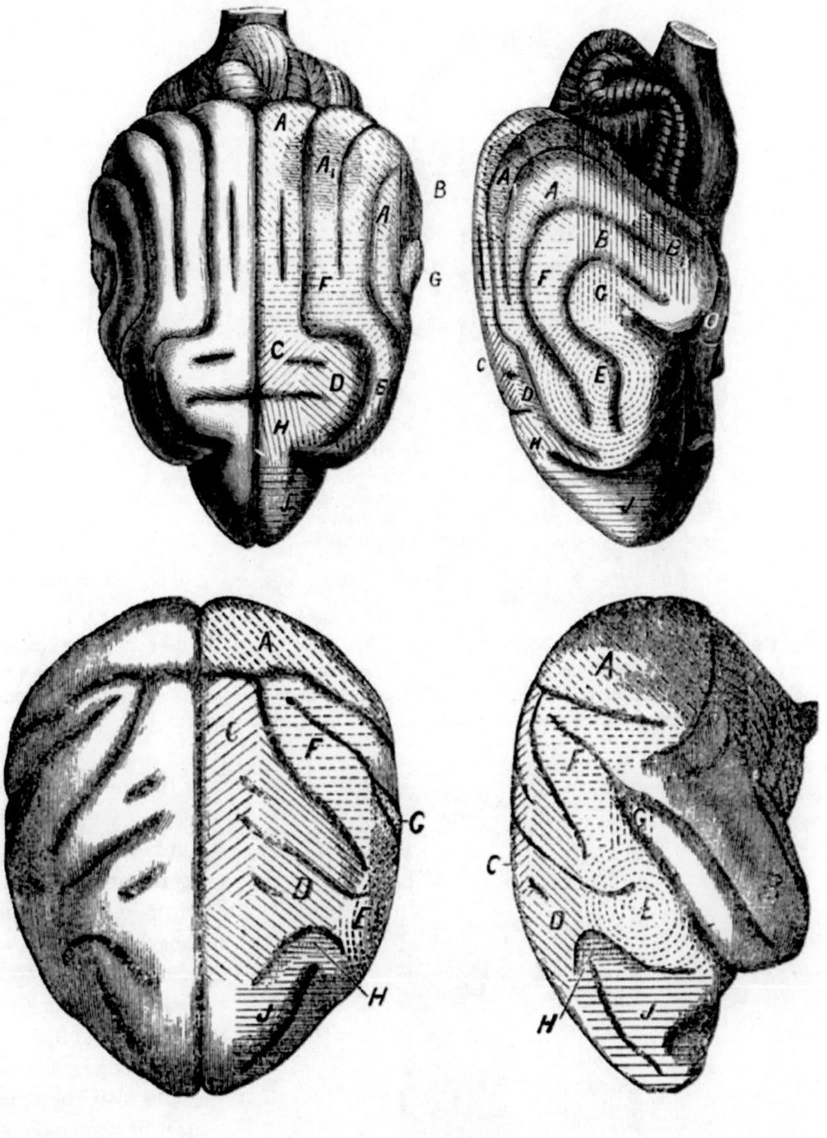

Fig. 1.7 Functional areas of the cerebral cortex of the dog (*upper*) and monkey (*lower*) as located by Munk on the basis of experimental lesions. (*A*) Sehsphäre or visual cortex. (*B*) Hörsphäre or auditory cortex. (*C–J*), Fühlsphäre or somatic sensory cortex (*C*, hindlimb region, *D*; forelimb region; *E*, head region; *F*, eye region; *G*, ear region; *H*, neck region). From Munk (1881)

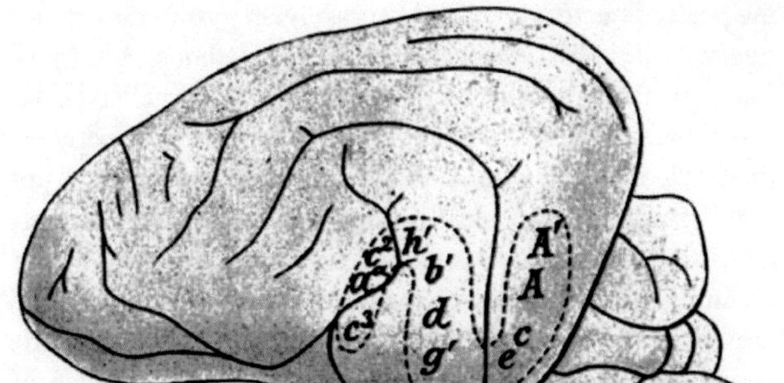

Fig. 1.8 Results obtained by Larionow (1899) showing the distribution of tone centers in the brain of the dog. Lesions located at different points along the S-shaped trajectory result in a failure to respond to tones of different frequencies. *Lower tones* are represented posteriorly and higher tones anteriorly. From Bechterew (1911)

Turner (1894) and by Monakow (1895). A next step in localizing the auditory cortex was, therefore, to identify the region that received its thalamic input from the medial geniculate body. The first major studies with the retrograde degeneration technique of Gudden were carried out by Constantine von Monakow (1895) who found atrophy and fiber loss in the cat medial geniculate body after lesions of the posterior ectosylvian regions (Fig. 1.9). This was apparently a confirmation of the localization of the auditory cortex as identified by Munk.

With the beginnings of higher resolution studies of cortical histology by Santiago Ramón y Cajal, the first efforts at identifying a structural correlate of the auditory area were focused, not without reason, on the superior temporal gyrus. In the human brain (Fig. 1.10), Cajal (1900b) identified a region on the "anterior half of the first sphenoidal

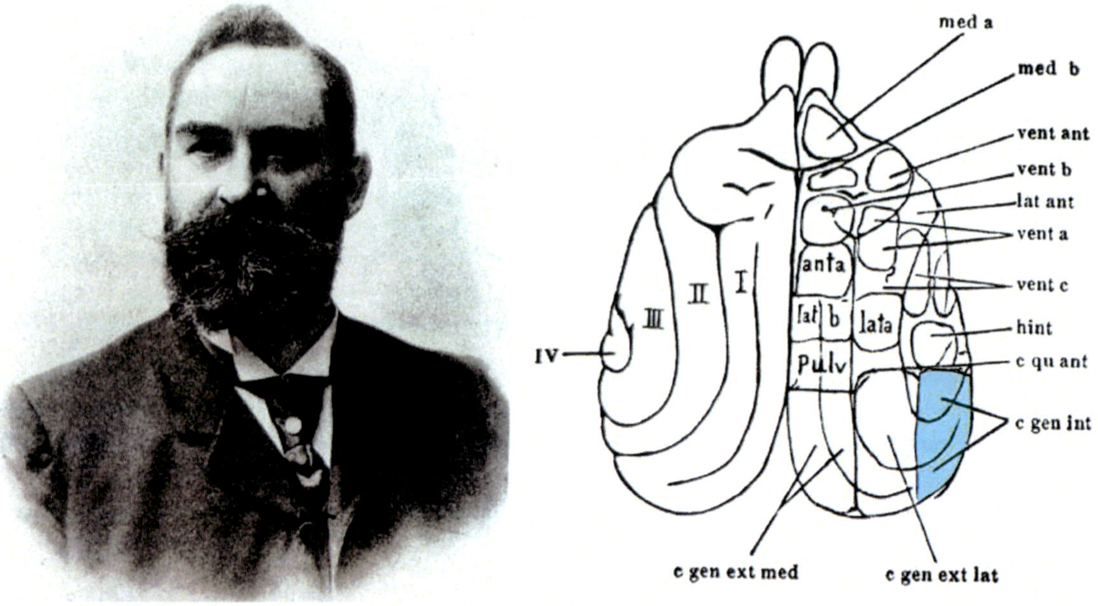

Fig. 1.9 *Left*: Constantine von Monakow (1853–1930). *Right*: summary of Monakow's experiments in cats in which he identified the thalamic nuclei projecting to different areas of the cerebral cortex on the basis of the retrograde atrophy that ensued from localized lesions of the cortex. The *colored* region labeled c gen int was identified as the projection field of the medial geniculate body and was thus equated with the auditory cortex. From Monakow (1895)

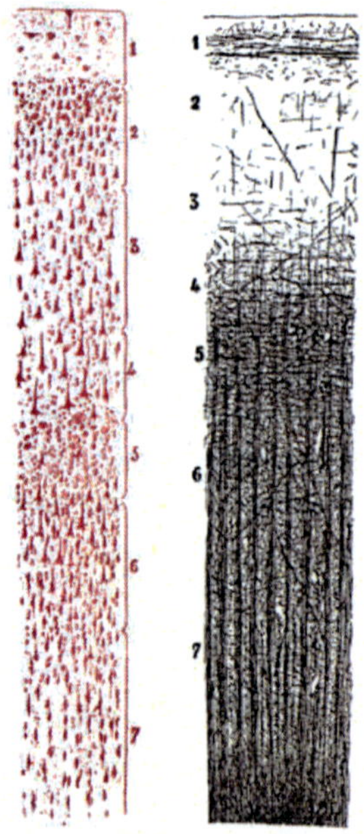

Fig. 1.10 The structure of the human anterior sphenoidal (superior temporal) gyrus as seen in Nissl (*left*) and Weigert (*right*) stains by Santiago Ramón y Cajal. From Cajal (1904, *left*, 1900b, *right*)

[superior temporal] gyrus" that he equated with the auditory region from his readings of Munk, Luciani, Ferrier and Monakow. Although repeatedly referring to the anterior part of the gyrus, Cajal's description makes it clear that by this he meant the upper part of the gyrus where neurologists had localized the auditory area. Cajal's reading of Munk in particular left him in no doubt that the corresponding region of the canine cortex was a center for auditory perception. Other, more ventral areas of the dog's posterior ectosylvian regions were, according to Cajal, concerned with the comprehension of the significance of sounds, an interpretation of Munk's experiments that had involved deeply penetrating lesions. Monakow's findings on the degeneration of the medial geniculate body subsequent to temporal cortical lesions also impressed Cajal. Cajal studied both the human superior temporal cortex and that of the "central regions" of the posterior ectosylvian and suprasylvian gyri in the cat and dog with the Nissl, Weigert, and Golgi methods. After giving a detailed description, layer by layer and cell type by cell type, he sums up by saying that comparisons between the cortices of the human and of the two carnivores are not easy because of the great differences in cellular morphology present. But he stresses that what he interpreted as the auditory cortex in all three species was characterized by the presence of a distinct layer of granule cells containing a variety of types of cells with short axons, by the existence of cells resembling pyramidal cells and having a long axon projecting out of the cortex but devoid of an apical dendrite, and an excessive development of the deeper cortical layers. Noteworthy features were the greater abundance of cells with short axons and especially those of the tufted (i.e., double bouquet) type in the human and some special giant cells not found in other areas were described in all layers. Cajal's auditory cortex in the human had 7 layers: plexiform or layer 1; layer of small pyramids or layer 2; layer of medium pyramids or layer 3; layer of giant pyramids or layer 4; layer of diminutive or granule cells or layer 5; layer of deep medium pyramids or layer 6; layer of fusiform cells or layer 7. Apart from the specific features mentioned above, this cortex had many similarities to the cortex of the postcentral gyrus but differed in its layering pattern from that of the precentral gyrus and of the visual cortex (Cajal 1899a, b, 1900a).

It is difficult to trace who first directed attention about the auditory cortex away from the superior temporal gyrus per se and onto the supratemporal plane. It may have been Flechsig (1898) who had recognized that a region corresponding mainly to the transverse temporal gyri (Heschl's gyrus) was the endpoint of fibers radiating into the hemisphere from the medial geniculate body and which, along with the visual and somatic sensory radiations, was the first to myelinate in the human embryo (Fig. 1.11). In the first of the great cytoarchitectonic studies of the human cerebral cortex, A. Walter Campbell (1905) and Korbinian Brodmann (1909) (Fig. 1.12), both located an area of granular cortex on the transverse temporal gyri that they thought corresponded to the region that Flechsig (1898) had identified.

Campbell found the fiber architecture of the transverse temporal gyri to be a more distinctive feature of the cortex

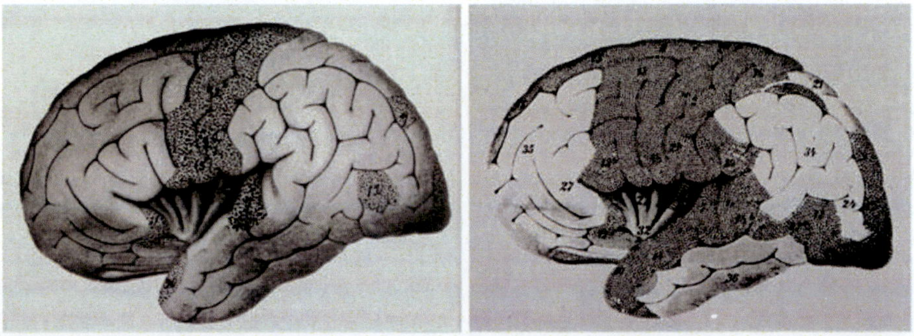

Fig. 1.11 Paul Flechsig's location of the cortical areas of the human brain that show myelination before birth (*left*) and postnatally (*right*). An early myelinating field located on the anterior and anterior half of the posterior transverse temporal gyri represents the primary auditory cortex. From Flechsig (1904)

Fig. 1.12 Alfred Walter Campbell (1868–1937, *left*) and Korbinian Brodmann (1868–1918, *right*). From the *Medical Journal of Australia* (*left*) and *World Neurology* (*right*)

here than the cytoarchitecture, noting that the outstanding features of this type of cortex are the many large fibers entering in a radial fashion from what he took to be the auditory radiation, the existence of a pronounced stria of Kaes which is the transverse band of myelinated fibers in our layer II, and a general wealth of fibers in all layers. It is interesting to read Campbell's description of his approach to the study of cellular lamination: "It is possible to distinguish three types, but the topical variations in cell lamination are not equivalent in degree to the differences in fiber-arrangement, also the intervening gradations are by no means abrupt: hence the extent and limits of these types of lamination are by no means easy to define; however, I may say that the following description has been built up on a particularly full and careful examination of the lobe, and above all things I would mention that judgments concerning the size, number, and general dis-position of cells in various parts have been based not upon mere microscopic inspection, but upon the comparative results given by a great number of camera lucida drawings made at various magnifications. This statement is necessary because experience gained in this work has proved to me over and over again that the eye cannot be trusted to make reliable comparisons, especially when the matter concerns the relative magnitude, or the number of given cells in different sections: accordingly when any doubt has existed on these points I have always settled the matter by making a drawing; and tedious as this procedure undoubtedly is, it is a very necessary, indeed an essential, safeguard in work of this description" (Campbell 1905). The chief distinguishing features of the cellular architecture of the cortex on the transverse temporal gyri were "a general rich supply of cells," prominent giant pyramidal cells in the external pyramidal layer (our layer III), and a thick stellate cell layer (our layer IV), the cells being divided into columns by radial fasciculi of fibers. The extent of this cortex "corresponds exactly with the area mapped out by fiber-arrangement." In some brains the area "is completely concealed, in others it is found peeping over the lip of the [Sylvian] fissure on to the free surface of the first temporal convolution" (Fig. 1.13). Campbell, in reviewing the neurological literature, was convinced that clinical distinctions could be made between cases of deafness, pure word deafness, amusia, and psychic deafness but he could find little pathological data to support the localization of causative lesions other than broadly in the upper part of the superior temporal gyrus and surrounding regions. The lesions in reported cases were simply too extensive. In trying to identify the primary auditory cortex, he was more impressed by the myelogenetic studies of Flechsig (1898) and of Cécile and Oskar Vogt (1902) which had clearly shown the auditory radiation emanating from the

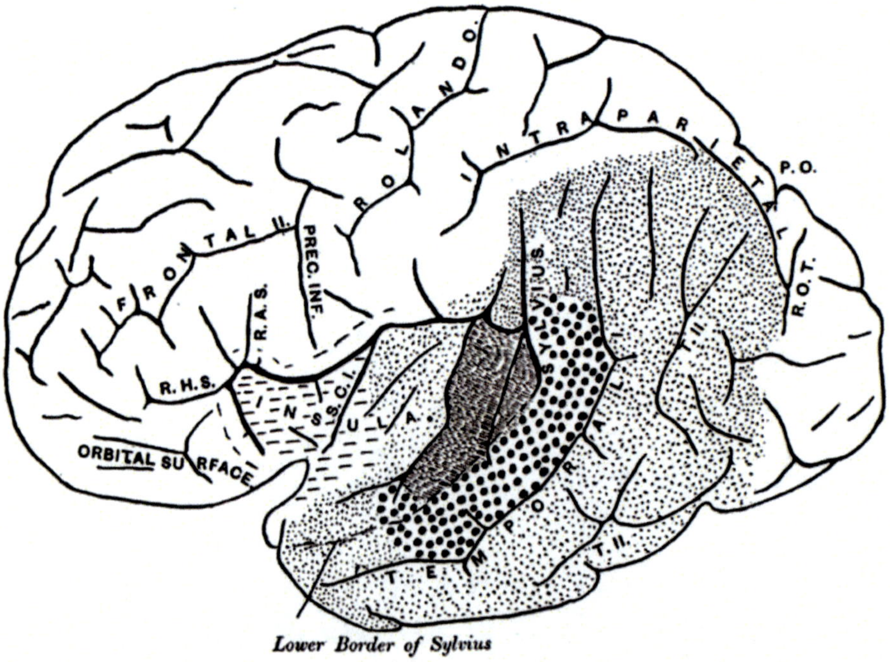

Fig. 1.13 Campbell's drawing of the human cerebral hemisphere with the Sylvian fissure opened out to reveal the audito-sensory area (*shaded*) confined to the transverse temporal gyri, the audito-psychic area (*large dots*) on the exposed surface of the upper two thirds of the superior temporal gyrus, and the common temporal cortex (*small dots*). From Campbell (1905)

medial geniculate body of human fetuses and its termination beneath the two transverse temporal gyri, particularly the anterior gyrus. Marchi-based studies of degenerating fibers or studies of the secondary atrophy following cortical and subcortical lesions in stroke patients also helped to relate the outflow tract of the medial geniculate body to the transverse temporal gyri (Monakow 1895; Dejerine and Dejerine-Klumpke 1901). Campbell briefly describes atrophy more or less restricted to the transverse temporal gyri in the brain of a 40-year old man who had been deaf from birth. In comparing the architecture of the area that he had described on the transverse temporal gyri with that of the visual cortex, Campbell concluded that this "restricted transverse temporal area is the part of the temporal lobe on which auditory stimuli first impinge." He named the area the audito-sensory area and, in a typically Edwardian railway analogy, he called it "the arrival platform of auditory stimuli." Its bilateral destruction should lead, therefore, to total deafness. He pointed out, however, that this could not be definitively determined from the case studies that had been described because the lesions were far too large to have the requisite localizing value. Because unilateral lesions involving the transverse temporal gyri were usually reported to be accompanied by "a dulling of the sharpness of hearing" rather than unilateral deafness, he was inclined to believe that the auditory pathways providing the input to the audito-sensory area were bilateral.

Following on from his identification of "psychic areas" around the primary visual and somatosensory cortical areas, Campbell was led to search for a comparable "audito-psychic" area adjoining the audito-sensory area. This he identified in a "skirt area" located lateral to the audito-sensory area on the exposed surface of the superior temporal gyrus (Fig. 1.13) and possessing many structural similarities to the audito-sensory area. It differed, however, in that the large, deeply placed oblique fibers and the giant cells were less prominent, and the radial fasciculi were more prominent (Figs. 1.14 and 1.15).

Brodmann (1909), also influenced by Flechsig, outlined in the human an area 41 which he called "the internal (anterior) transverse temporal area," corresponding "approximately but not precisely to the anterior transverse gyrus." It extended into the circular sulcus medially where it was sharply demarcated from a parainsular area that he numbered area 52 (Fig. 1.16). A second area, area 42, called "the external (posterior) transverse temporal area," formed an arc around area 41 posteromedially, posteriorly and laterally, coming to the surface of the superior temporal gyrus lateral to area 41. Brodmann was never very explicit about the details of the cytoarchitecture of these three areas, although his context makes clear that area 41 was the area of smallest cells and highest granularity while area 42 was somewhat less granular. Areas 41, 42, and 52 were later named areas TC, TB, and TD respectively, by Economo and Koskinas (1925) (Fig. 1.17).

Campbell went on to locate an "audito-sensory area" and a surrounding "audito-psychic area" in comparable locations in the brains of a chimpanzee and an orangutan, but he was uncertain about the location of a homologous primary auditory sensory area in cats, dogs, or pigs (Figs. 1.18 and 1.19). The apes were the species that Grünbaum and Sherrington (1902, 1903) had used for mapping the motor cortex by electrical stimulation. Campbell was then working at the Rainhill Asylum, on the outskirts of Liverpool, where Sherrington had his laboratory in Liverpool University. He was undoubtedly influenced by the ablation studies of Munk (1881) although his area "ectosylvian a," which he thought might correspond to the human auditory sensory area, is located on one or both of the two sylvian gyri anterior to where Munk had located the center of his auditory cortex (Fig. 1.19). Brodmann (1905), impressed by a lack of anything resembling typical koniocortex in the temporal cortex of monkeys and many other species (Figs. 1.20 and 1.21), considered either that animals lacked a specialized auditory cortex or that the human auditory fields (his areas 41, 42, and 52) represented specialization for functions additional to audition. Perhaps this is not surprising, given that the human auditory cortex when visualized in Nissl stains (Fig. 1.22) lacks the intense granularity of the postcentral somatic sensory and the primary visual areas. As Brodmann put it: "...the cell and fiber architecture, so very characteristic of both transverse gyri in man, is lacking in all other animals. To put it anther way, a human structural zone in which Flechsig locates the cortical end-station of the auditory pathway, the auditory [cortex], is completely absent in animals, even in monkeys that otherwise possess a very similar cortical structure to man."

6 New Experimental Studies in Animals

Brodmann's view held for many years, and when A. Earl Walker (1937) discovered an area of granular cortex connected with the medial geniculate nucleus and located on the posterior part of the supratemporal plane in the macaque monkey (Fig. 1.23), there is a note of mild surprise in his description, even though Stephan Polyak (1932) had already traced degenerating fibers with the Marchi technique from the posterior part of the thalamus to the supratemporal plane (Fig. 1.24). Once Walker (1937) had recognized the location of what he thought must be the primary auditory cortex in the region to which Polyak had traced the putative auditory radiation, he could lesion it and observe not only the occurrence of retrograde degeneration in the heart of the

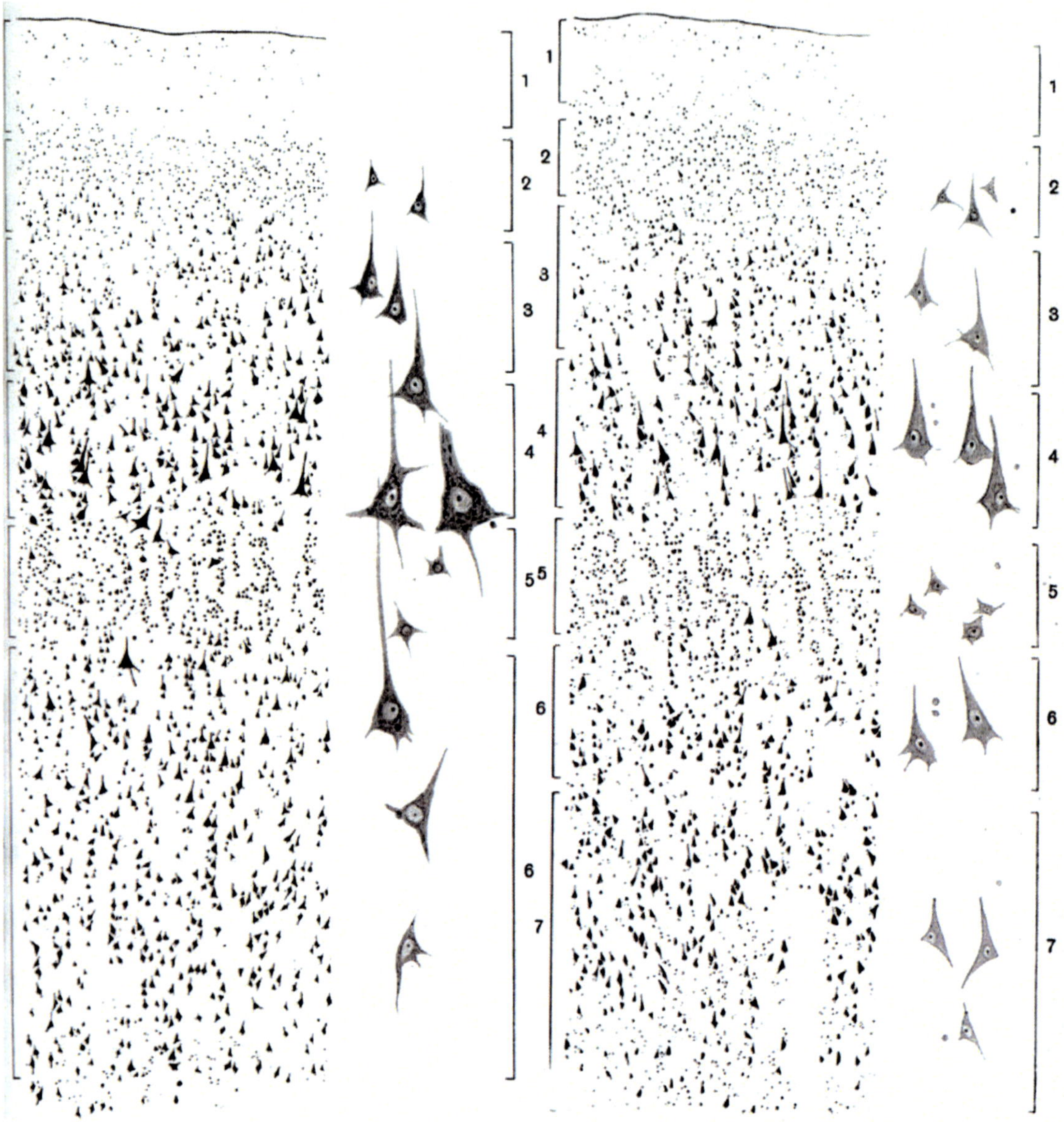

Fig. 1.14 Campbell's drawings of cells and cell lamination in the audito-sensory area (*left*) and the audito-psychic area (*right*) of the human brain. From Campbell (1905)

medial geniculate body, but also the systematic movement of the locus of degeneration with lesions in different locations of the auditory cortex, implying a topographically ordered geniculo-cortical projection. Wilfrid Le Gros Clark (1936) had similar, although less detailed, findings. Later, Ades and Felder (1942), in delineating the cortical region from which evoked potentials could be recorded in response to click stimuli, confirmed Walker's location of the primary auditory cortex on the supratemporal plane of the monkey but found that it fell within a much wider area that was activated by click stimuli (Fig. 1.25).

In cats, as noted above, Monakow had observed retrograde degeneration in the medial geniculate body after ablations of the posterior sylvian and ectosylvian regions, and Mettler (1932) later observed retrograde degeneration in the medial geniculate nucleus after dorsal ectosylvian lesions as well, while Woollard and Harpman (1939) traced Marchi-stained degenerating fibers to middle ectosylvian and sylvian gyri after lesions of the medial geniculate body. Retrograde degeneration was also described in the medial geniculate complex after temporal lesions in rats (Waller 1934; Waller and Barris 1937). The first investigation of the area of the

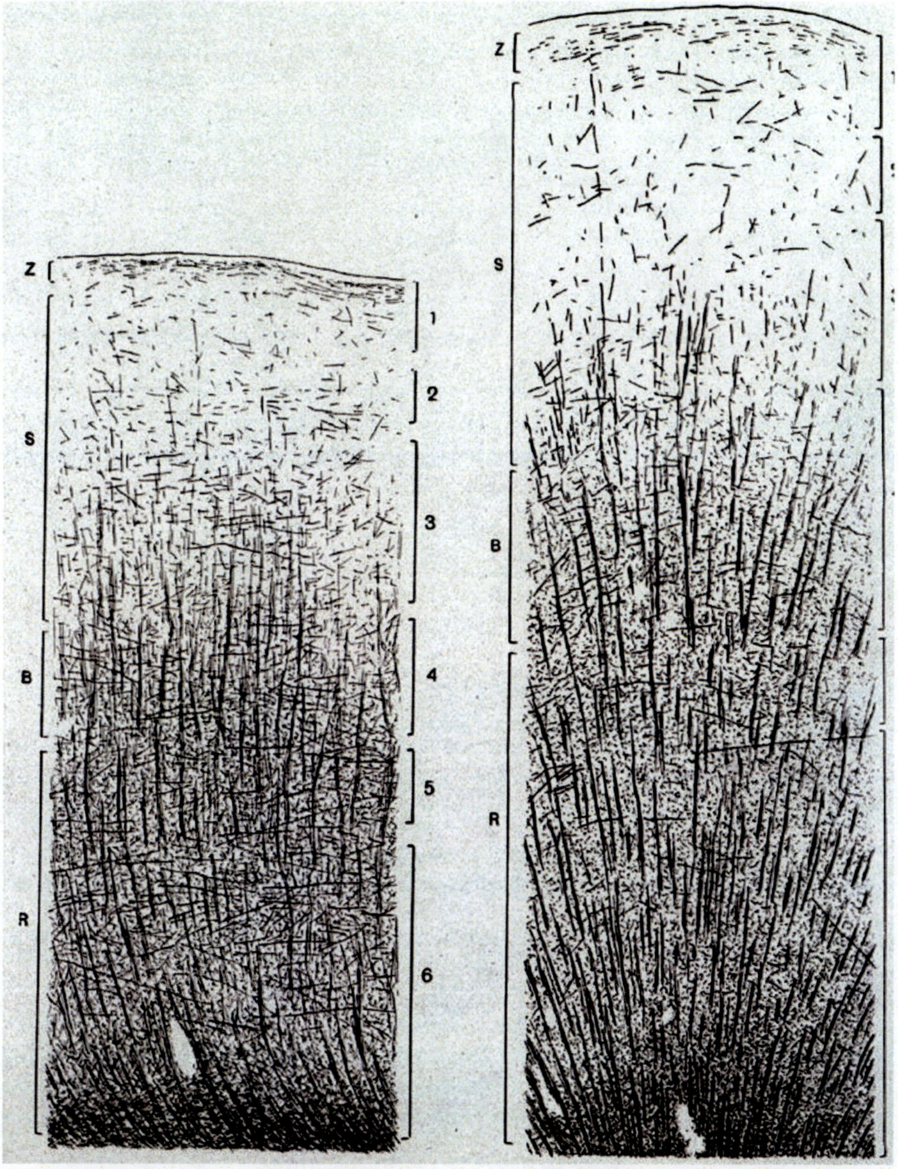

Fig. 1.15 Campbell's drawings of fiber patterning in the audito-sensory (*left*) and audito-psychic (*right*) areas of the human cortex. From Campbell (1905)

cat cortex in which evoked potentials could be recorded in response to auditory stimuli was made by Kornmüller in 1937 and was found anterior and dorsal to the region delineated by Munk as that which was essential for auditory perception (Fig. 1.7). Bremer and Dow (1939), in applying the click-evoked potential method for the first time in the cortex (Fig. 1.26), mapped out a larger region, the dorsal part of which, equivalent more or less to what we now call area AI, had a cytoarchitecture that was granular and typical of sensory cortex. An area virtually identical to the granular area of Bremer and Dow was also mapped using click-evoked potentials by Ades (1941). Much earlier, the same region had been delineated by Cécile Vogt as the zone of earliest and heaviest myelination in the developing cat brain (Fig. 1.27). Waller (1934) made relatively small lesions in the region defined by Bremer and Dow and in certain areas around it and examined the distribution of retrograde degeneration in the thalamus (Fig. 1.28). With lesions largely restricted to the granular area of Bremer and Dow, he observed retrograde degeneration in what we would now regard as the ventral nucleus of the medial geniculate nucleus. With lesions located ventrally and posteriorly, degeneration was mainly

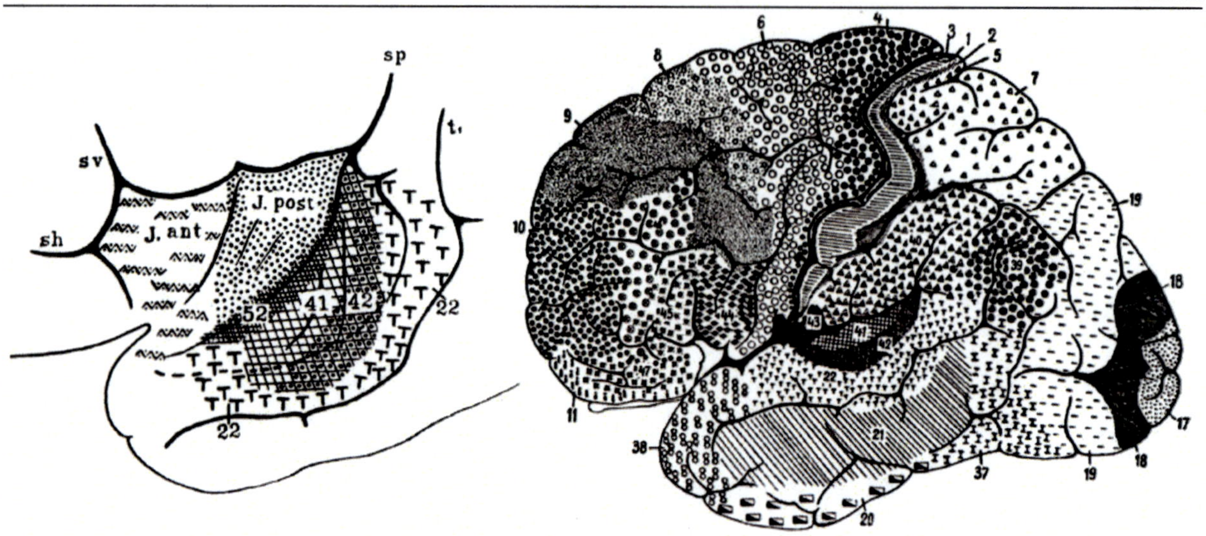

Fig. 1.16 Brodmann's drawings of the insular region and upper surface of the superior temporal gyrus (*left*) and of the lateral aspect of the human cerebral hemisphere (*right*) showing areas 41 and 42, which are called the internal or anterior and the external or posterior transverse temporal areas, respectively. Area 52 is the parainsular area and area 22 the cortex on the exposed surface of the superior temporal gyrus. From Brodmann (1909)

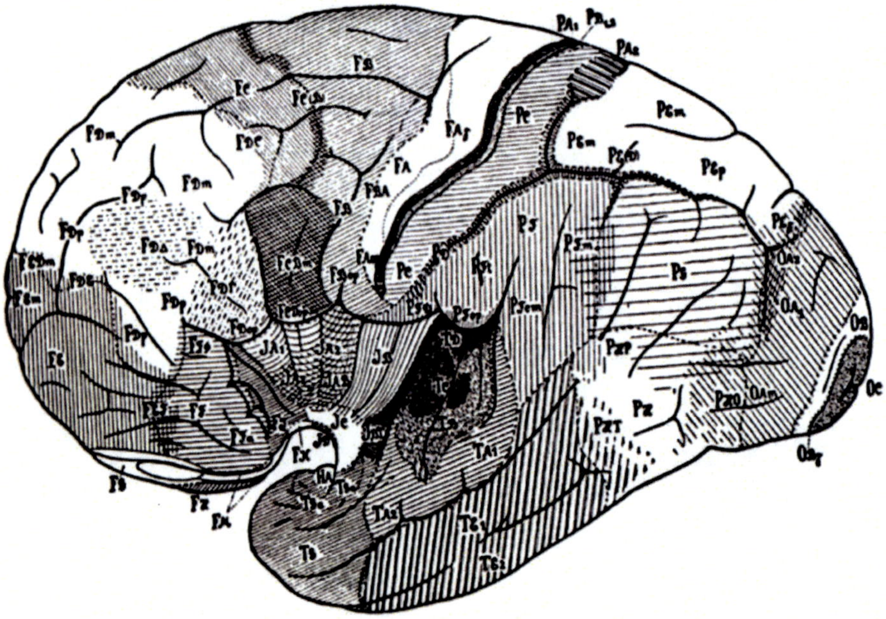

Fig. 1.17 Map of the cytoarchitectonic areas of the human cerebral cortex by Economo and Koskinas (1925). Te and Ts are the two transverse temporal areas of Brodmann and Campbell. From Economo and Koskinas (1925)

in what we would now call the dorsal and medial nuclei. These studies were forerunners of the concerted investigation made by Rose and Woolsey on the plan of organization of the cat auditory cortex and its thalamic connections (Rose 1949; Rose and Woolsey 1949, 1958; Woolsey 1961). The in-depth studies of Rose and Woolsey (Fig. 1.29) were to provide the first detailed parcellations of the auditory cortex.

Woolsey and Walzl (1942; Walzl and Woolsey 1946; Woolsey 1971a, b) had extended the studies of Bremer and Dow and Ades in the cat by recording surface evoked potentials in response to electrical stimulation of small bundles of nerve fibers leaving different parts of the cochlea. By this means, they could demonstrate that, within an area corresponding approximately to the granular area outlined by Bremer and Dow, the apex of the cochlea, and thus low

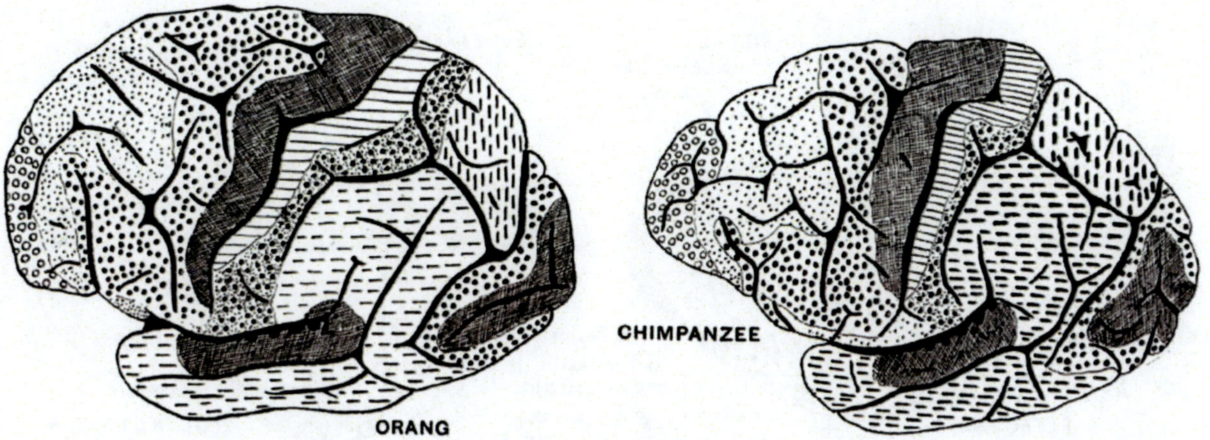

Fig. 1.18 Campbell's maps of the orangutan and chimpanzee cortex. The audito-sensory area is represented by a row of *large dots*. The *hatched* region lateral to it is the audito-psychic area. From Campbell (1905)

tones, was represented posteriorly and the base, and thus high tones, was represented anteriorly. This area they called the first auditory area or AI. They also noted that with increased intensity of stimulation, a much wider area could be activated and that the part of this wider area lying ventral to the primary area showed a cochleotopic representation that was a mirror image of that in the AI area. This area they called the second auditory area or AII (Fig. 1.30). Almost contemporaneously, Ades (1943) showed a similar region of click-evoked responses extending over the middle and posterior ectosylvian gyri. Responses in the posterior ectosylvian area could also be evoked by applying strychnine to the surface of the middle ectosylvian "primary area," so the auditory responsiveness of the "secondary area" was thought to be dependent on corticocortical projections from the primary area.

7 Entering the Modern Era: Multiple Cortical Fields, Tonotopicity, and Thalamocortical Projections

The next steps in the delineation of the cat's auditory cortex came in the combined anatomical and physiological studies of Rose and Woolsey. In 1949, Rose subdivided the cat auditory cortex regions into a central, moderately granular area, coincident with the anterior half to two-thirds of the first auditory or AI field as delineated with the evoked potential method by Woolsey and Walzl (1942), and several surrounding areas with different cytoarchitectonic characteristics (Fig. 1.31). The AII area was reduced by Rose to only the anterior half of the original AII field, the posterior parts of both it and the old AI now being subsumed into a posterior ectosylvian or EP field virtually identical to the secondary auditory field of Ades (1943). The surrounding areas were later mapped with the evoked potential method by Woolsey and his co-workers, further subdivided, and most of these areas were demonstrated to contain complete and independent representations of the cochlea (Fig. 1.32) (Woolsey 1958, 1964). Rose and Woolsey (1949) were able to show that destruction of the AI area resulted in retrograde degeneration in the anterior part of the medial geniculate complex, in a region corresponding to what we now call the ventral nucleus. They also found that lesions of different parts of AI led to degeneration in different parts of the nucleus in a manner that implied a cochlear representation within it, and a cochleotopic projection on the cortex (Fig. 1.33). They further confirmed this by showing that electrical stimulation of fibers from the apex or base of the cochlea in the cat resulted in evoked potentials in lateral or medial aspects of the ventral nucleus, respectively (Rose and Woolsey 1958), and that electrical stimulation at progressively more medial sites in the ventral nucleus led to evoked potentials at progressively more anterior sites in AI (Woolsey 1964).

The first single unit responses to pure tone stimuli were recorded from cat AI area by Erulkar et al. (1956), the locations of high and low tone-responding units being located anteriorly and posteriorly, respectively, confirming the electrical stimulation results of Woolsey and Walzl. Woolsey and his colleagues continued to map tonotopically organized evoked potentials in cat auditory regions (Woolsey 1959, 1960, 1964, 1971a). The culmination of several years of intensive work first appeared in the 1961 map (Fig. 1.32), where the AI field is embraced anteriorly, dorsally and posterodorsally by a "suprasylvian fringe" area, the apical cochlear representation of which has been taken from the old AII field which has become restricted to a region between the anterior and posterior ectosylvian sulci, and the posterior

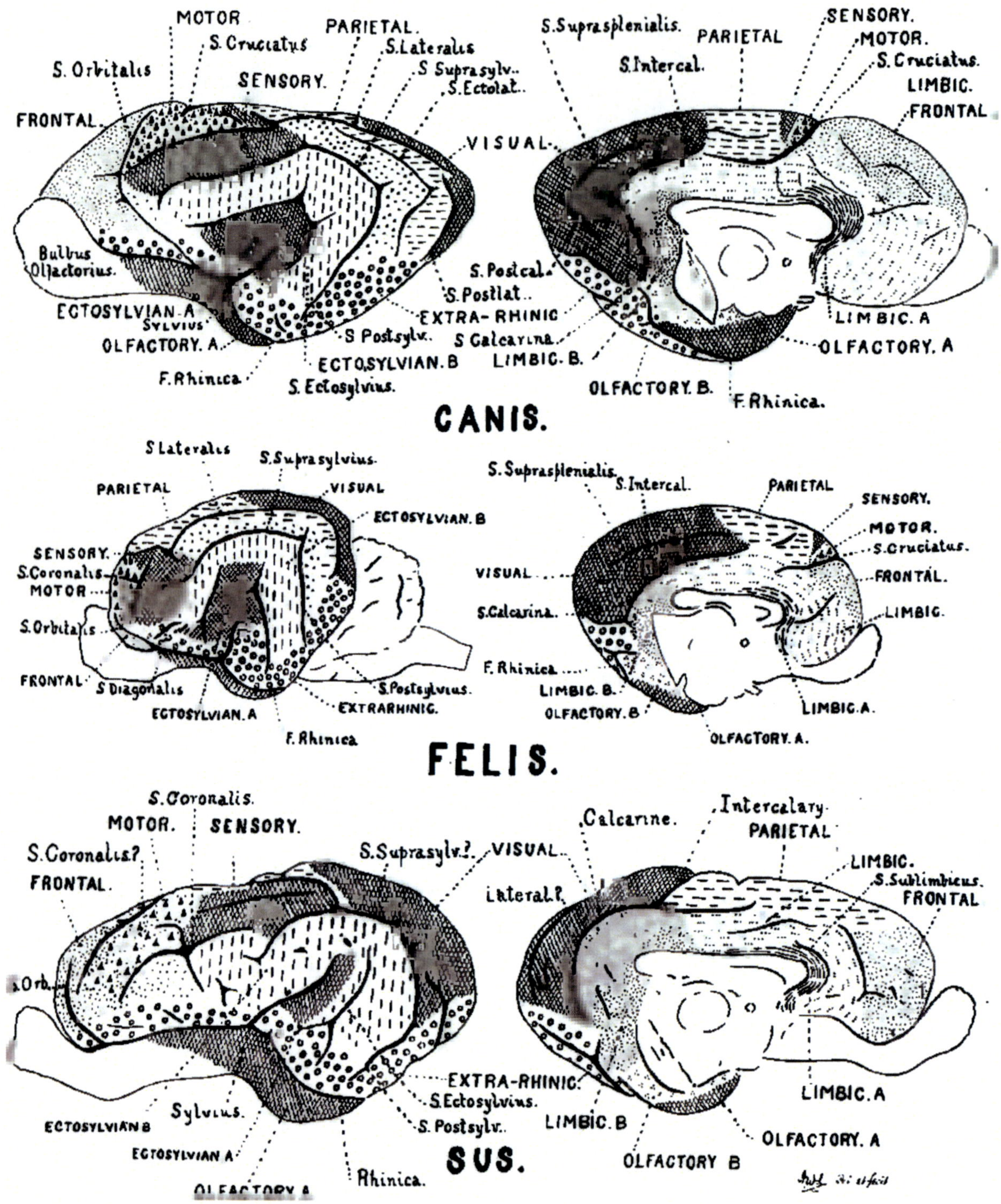

Fig. 1.19 Campbell's maps of the dog, cat, and pig cortex. He considered that the area labeled ectosylvian A might represent the auditory sensory area. From Campbell (1905)

ectosylvian area (Ep) has been displaced a little ventrally. By 1962, Sindberg and Thompson had observed auditory responses extending towards the tip of the posterior Sylvian gyrus and this region is identified as the temporal area (Te). There is a small additional, tonotopically organized field in the insular region (Ins). A further non-tonotopic area labeled

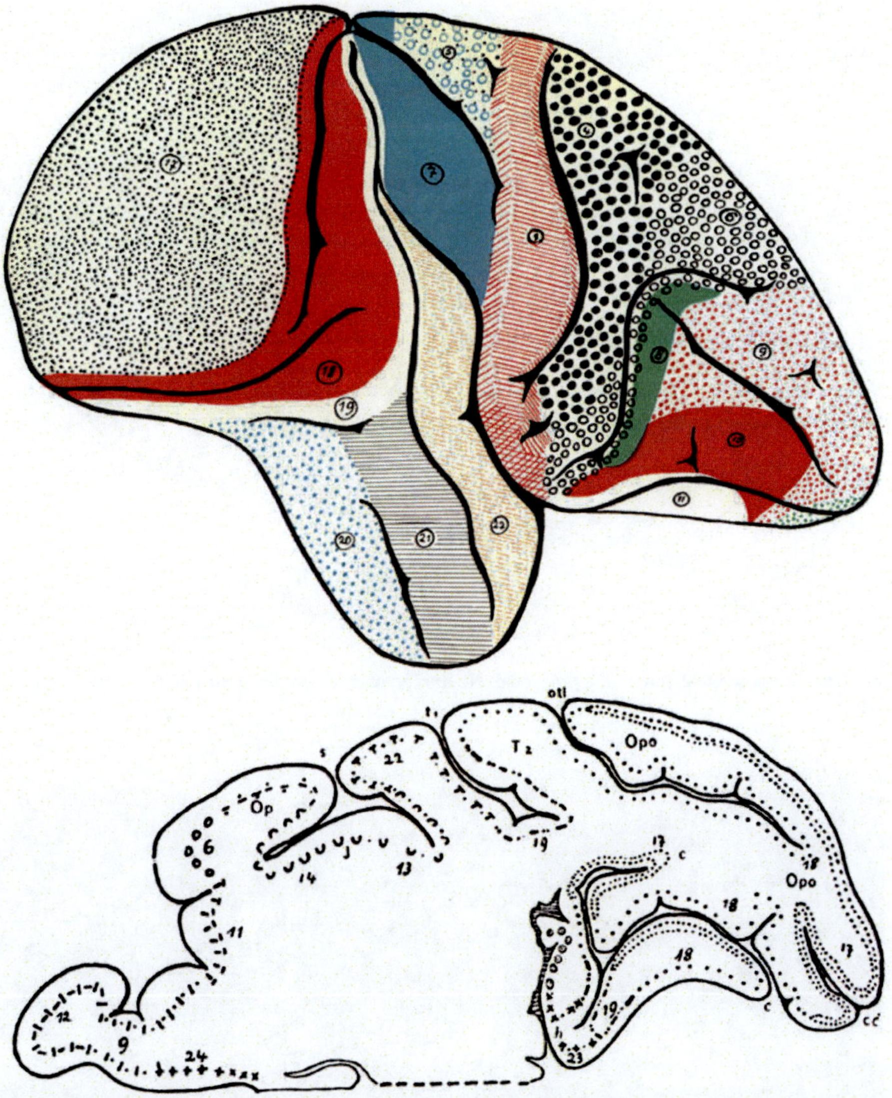

Fig. 1.20 Brodmann's map of the cortex of a cercopithecine monkey (*upper*) and one of his drawings of a horizontal section (*lower*) through the insula and adjoining superior temporal gyrus of the same animal (anterior and posterior are reversed in the two figures). He was unable to identify a separate auditory sensory area. From Brodmann (1905)

AIII had been described by Tunturi (1945) in the anterior ectosylvian gyrus. The region labeled "Association area" is a region of long latency responses without tonotopic order; late responses, according to Woolsey, can sometimes also be recorded in the second visual area (VII) and in the primary motor area (MI).

8 Later Studies in Cats, Monkeys, and Other Species

Mapping the cat auditory cortex for multiunit responses to pure tone stimuli represented a refinement of the evoked potential method and led to re-parcellation of the fields originally delineated by Woolsey (Merzenich et al. 1975; Reale and Imig 1980; Schreiner and Cynader 1984) and shown in Fig. 1.34. The identity of the fields mainly depends upon the representation of the full range of audible frequencies, that is, a representation of the complete cochlear partition. Borders are customarily identified by a reversal in a progression of best frequency responses recorded as microelectrode penetrations traverse the cortex. In some instances, especially for AI, a unique chemoarchitectonic profile (Wallace et al. 1991) enables the areal borders to be identified histologically. The consensus view, based on multiunit mapping, is that four of the later delineated fields contain cells with sharp

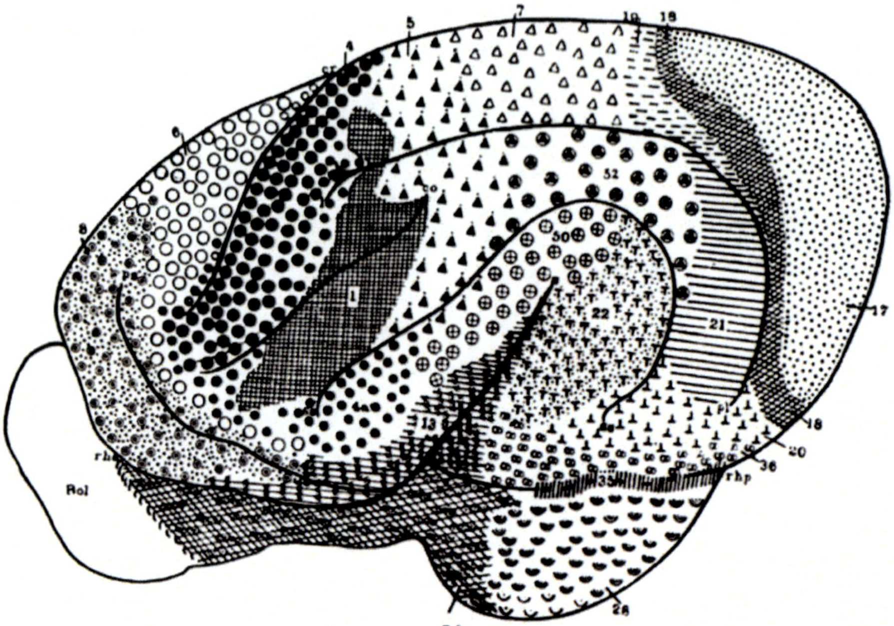

Fig. 1.21 Brodmann's map of the cerebral cortex of a procyonid. No area comparable to the human auditory sensory area could be identified. From Brodmann (1909)

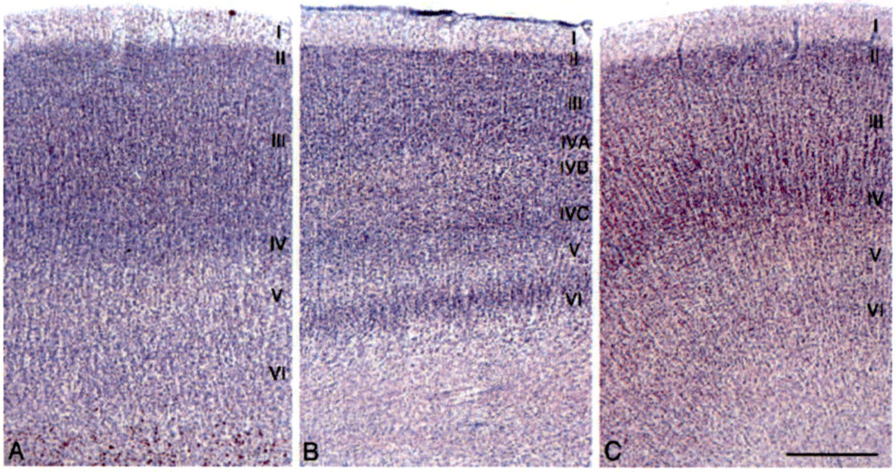

Fig. 1.22 Photomicrographs of Nissl-stained sections through the human somatic sensory (**a**), visual (**b**) and auditory (**c**) areas. The auditory area is distinctly less granular and more radially disposed than the other two. Bar 500 μm

frequency tuning and independent tonotopic representations (Phillips and Irvine 1981). Two of these fields, separated out of Woolsey's old AI, are AI and the anterior auditory field (AAF). The other two, located mainly in the banks of the posterior ectosylvian sulcus and derived from Woolsey's old posterior ectosylvian field (Ep), are termed the posterior (PAF) and ventral posterior (VPAF) auditory fields. At variance with Woolsey's studies, the old AII and temporal areas are now reported to be non-tonotopically organized. The suprasylvian fringe has lost its name but sometimes receives passing attention as a region of ill-defined auditory responses dorsal or posterodorsal to AI. Its thalamic input comes from the posterior complex of nuclei (Heath and Jones 1971a, b, Fig. 1.35). The insular region is usually ignored as an auditory area. Its input comes from the suprageniculate nucleus and not from the medial geniculate complex (Jones and Leavitt 1973; Winer et al. 1977; Bowman and Olson 1988; Clascá et al. 1997) (Fig. 1.35).

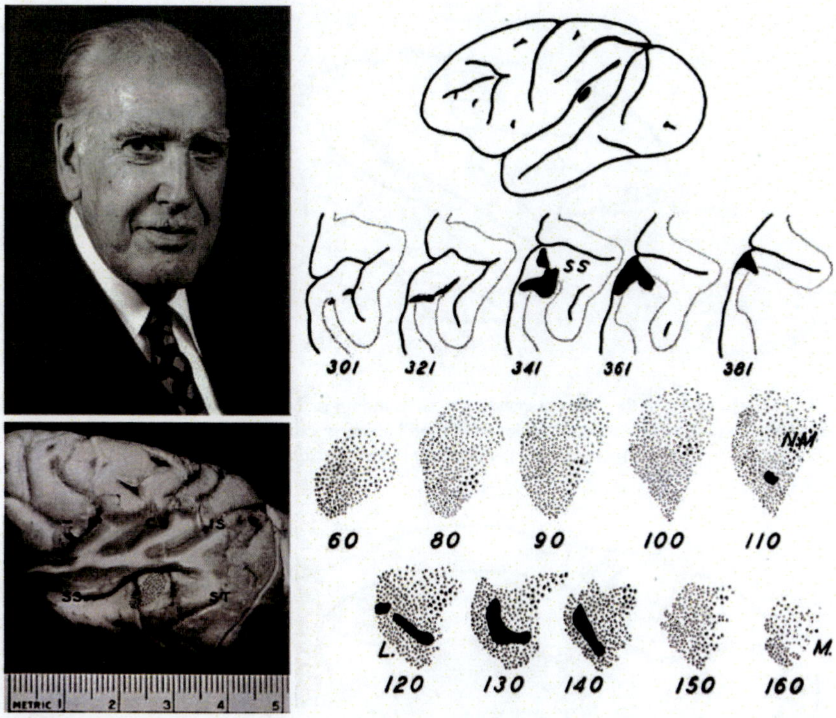

Fig. 1.23 A Earl Walker (1907–1994) and his illustration of the location of the primary auditory cortex in a rhesus monkey and his illustration of the location of a lesion and the ensuing retrograde degeneration in the medial geniculate nucleus. From Walker (1938)

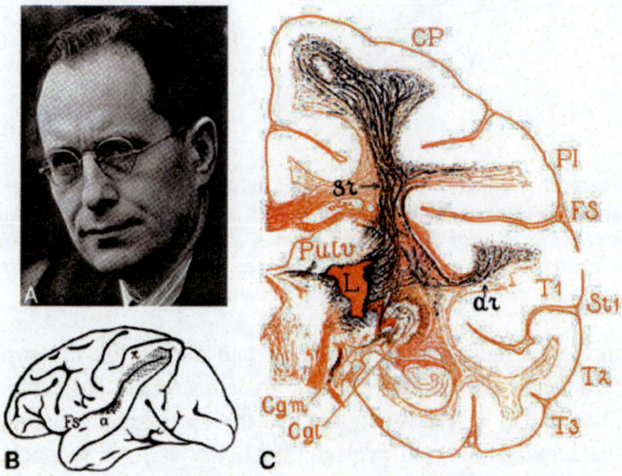

Fig. 1.24 Stephan Polyak (1889–1955) **a** and his depiction of the auditory radiation (ar), as stained by the Marchi method after a lesion interrupting the outflow from the medial geniculate nucleus (**b**, **c**). **b** and **c** from Poliak (1932)

In Rose and Woolsey's hands, isolated destruction of the auditory fields outside AI resulted in very little retrograde degeneration in the medial geniculate complex, and thus it was uncertain how acoustic input reached them. Destruction of two or more fields often led to unmistakable retrograde degeneration, implying that cells in the affected part of the nucleus had branched axons to each of the two fields and that the presence of the collateral axon "sustained" the cell when the cortical terminations of its fellow were destroyed. For example, destruction of the AI or more ventral auditory fields alone elicited little or no retrograde degeneration in the magnocellular medial geniculate nucleus but the degeneration grew more severe as fields additional to AI, including those such as the second somatic sensory area outside the auditory regions, became involved. This, Rose and Woolsey felt, was evidence for widespread, probably collateral, projections from the magnocellular nucleus.

Diamond et al. (1958) were the first to demonstrate an independent projection from the medial geniculate nucleus to an area outside AI when they showed that destruction of a temporal field (Te; Figs. 1.32, 1.36, and 1.37) lying ventral to AII led to retrograde degeneration at the posterior pole of the medial geniculate body, a part of what is now called the dorsal nucleus (Jones 2007). The degeneration became especially severe if the insular field and the AII field were destroyed as well. They interpreted their results to indicate direct projections from the posterior pole of the medial geniculate complex to the temporal region and collateral projections to the insular and AII fields. Unlike all the other cortical fields related to the medial geniculate nucleus, the temporal field has never been shown to be responsive to cochlear stimulation.

Anterograde axonal degeneration studies based on the method of lesioning the medial geniculate complex and studying the distribution of axonal degeneration in the cortex

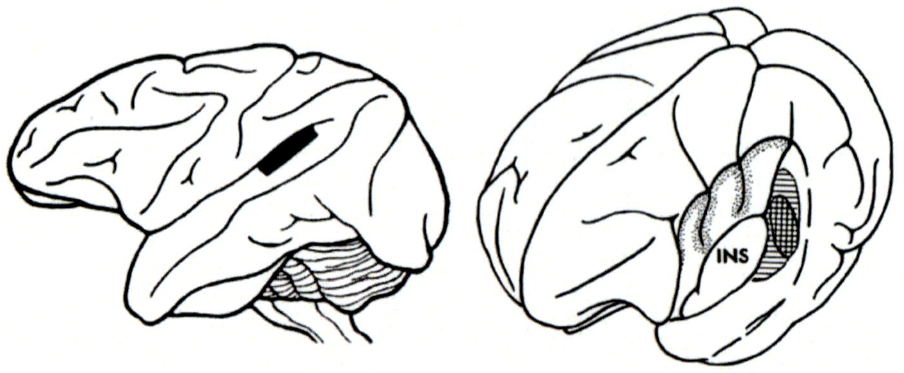

Fig. 1.25 Location of the cortex responding to click stimuli in the rhesus monkey (*left*, from Ades and Felder 1942) and the larger extent of the responsive region (*right, transverse hatching*) in comparison with the projection target of the medial geniculate nucleus as mapped by Walker (*cross hatching*) (from Ades 1959)

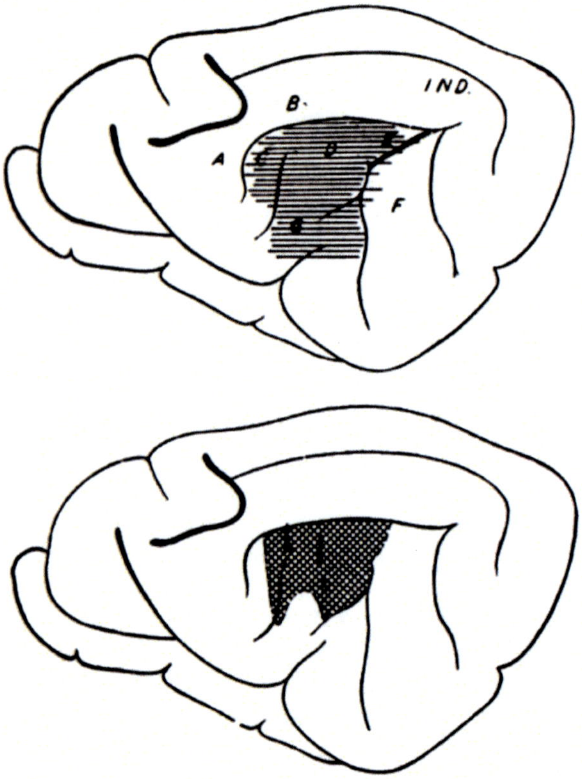

Fig. 1.26 Drawings from Bremer and Dow (1939) showing the extent of the area of cortex activated by click stimuli in the cat (*upper*) and the smaller extent of granular cortex in the same animal

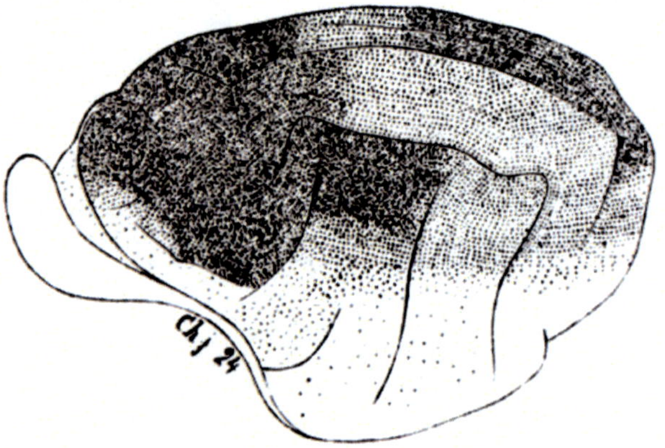

Fig. 1.27 Dense myelination outlining the auditory, visual, somatosensory and motor areas of the cerebral cortex in a 12-day old cat. Note the anterior location of the auditory region in comparison with its location marked in Figs. 1.5, 1.6, 1.7 and 1.9. From Vogt and Vogt (1919) after a rare work of Cécile Vogt published in Paris in 1900

did not lead to results that could help resolve the uncertainties about the origins of thalamic inputs to all the auditory cortical fields (Wilson and Cragg 1969; Heath and Jones 1971a, b; Niimi and Naito 1974), although Sousa-Pinto (1973) argued for a projection from each nucleus to an independent field. From the differential distribution of corticothalamic fibers in the medial geniculate complex and making the assumption that these reciprocated the thalamocortical projection, Diamond et al. (1969) argued that the ventral medial geniculate nucleus would project to AI, the various subnuclei of the dorsal nucleus to independent fields around AI, and the magnocellular nucleus to all fields (Fig. 1.38). With the introduction of more sensitive anterograde and retrograde tracing techniques, studies in the cat, tree shrew, and monkey reported that the ventral nucleus projected only to AI and favored the view that the various subdivisions of the dorsal nucleus projected independently to separate cortical fields around AI (Burton and Jones 1976; Casseday et al. 1976; Oliver and Hall 1978). The magnocellular nucleus was considered to project widely and diffusely. Others reported that the subnuclei of the dorsal nucleus in the cat projected upon more than one cortical field (Winer et al. 1977). After injections of tracer aimed at the individual auditory cortical fields, identified from maps such as those shown in Figs. 1.32

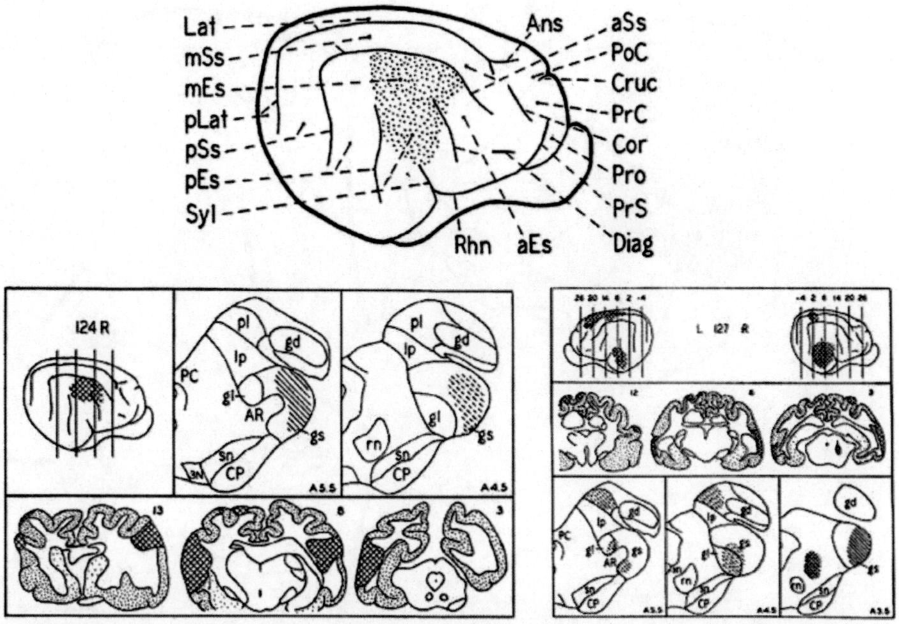

Fig. 1.28 Waller's reproduction of the click-evoked activity map of Bremer and Dow (*upper*) and the location of retrograde degeneration in the medial geniculate nuclei of cats in which this area (*lower left*) and adjoining areas (*lower right*) were lesioned. From Waller (1940)

Fig. 1.29 Jerzy E. Rose (1909–1992) (*left*) and Clinton N. Woolsey (1904–1993) (*right*). From Jones (2007)

and 1.34, retrogradely labeled cells were found in more than one nucleus of the medial geniculate complex. The labeling of cells in the magnocellular nucleus was consistent with other reports of widespread thalamocortical projections from this nucleus. But it was felt that similarly widespread thalamocortical projections emanated from the subdivisions of the dorsal nucleus and even from the ventral nucleus. When these results are examined closely, it is evident that an injection apparently centered in one or other of the auditory fields invariably led to a major focus of retrograde labeling in a single nucleus other than the magnocellular, with fewer, less concentrated cells being labeled in other nuclei. In view of what follows below, it would be easy to reinterpret these results to indicate that the ventral nucleus of the medial geniculate complex projects primarily to AI; the deep dorsal nucleus projects primarily to AII and to the anterior auditory field; the posterior pole of the dorsal nucleus projects to the temporal field; and the other dorsal nucleus components

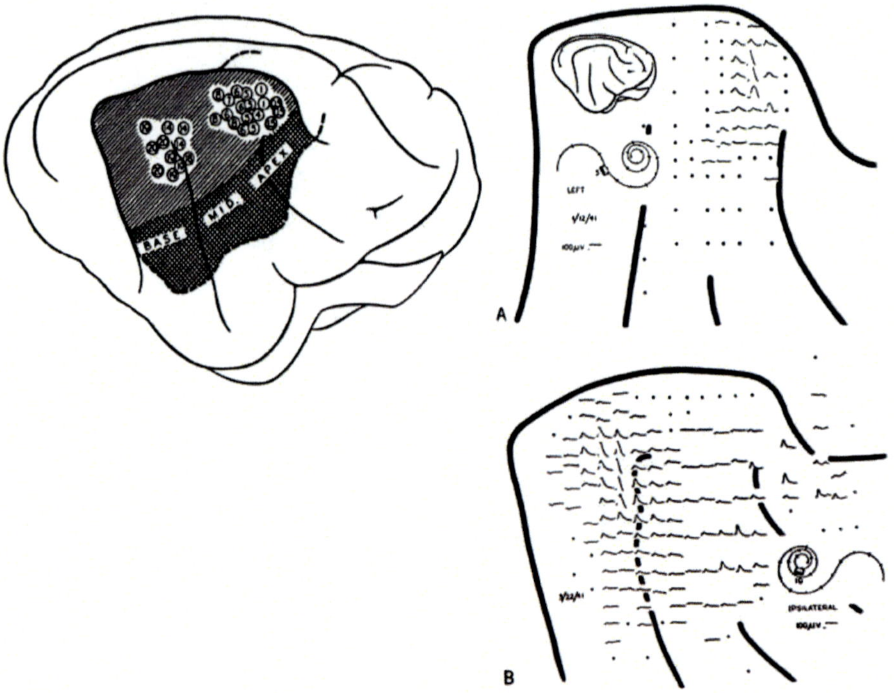

Fig. 1.30 Distribution of responses to electrical stimulation of small bundles of cochlear nerve filaments entering the osseous spiral lamina at different levels. The numbers in *circles* are within the main auditory area and indicate centers of maximal response to stimulation of fibers at distances from the basal end of the cochlea. The *cross hatched* area showed a reversed order of cochlear representation determined by the same methods. The *smaller figures* show the distributions of evoked responses resulting from electrical stimulation at points 5 mm from the basal end (**a**) or at the apex (**b**) of the cochlea. From Woolsey and Walzl (1942)

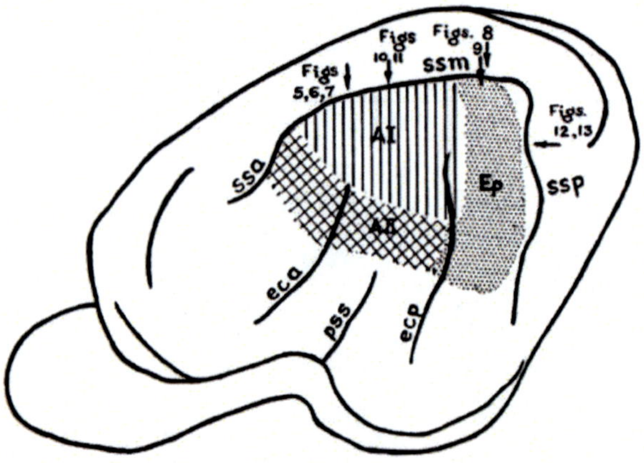

Fig. 1.31 Rose's (1949) cytoarchitectonic map of the auditory regions in the cortex of the cat. Within the *upper* evoked potential field of Woolsey and Walzl (Fig. 1.30), Rose delineated a first auditory field (AI) and part of a second posterior ectosylvian field (Ep). The *lower* representation was also divided architectonically into an anterior second auditory area (AII) and the remainder of the Ep field. Contemporary work by Rose and Woolsey identified AI as the essential target of the principal division of the medial geniculate body. From Rose (1949)

project to the two posterior auditory (posterior ectosylvian) fields. The few cells labeled in nuclei outside the confines of that containing the major concentration of labeled cells may have merely represented inadvertent spread of an injection to cortical fields other than the one aimed at, or they might have represented a true, sparse projection of the kind described below.

In later studies in the cat, Morel and Imig (1987) placed moderate sized injections of retrogradely transported tracers aimed at different parts of the tonotopic representations in fields AAF (called A by Morel and Imig), AI, PAF (called P) and VPAF (called VP). They presented a case for the four auditory fields each receiving a major (quantitatively larger) projection from cells located in spatially separated parts of the ventral medial geniculate nucleus, in the lateral division of the posterior complex or in different divisions of the dorsal medial geniculate nucleus, with minor (quantitatively smaller) projections from other parts of the same nuclei or from different nuclei, including large- and small-celled divisions of the magnocellular nucleus. For the lateral division of the posterior complex and the ventral medial geniculate nucleus, the cell populations that form the major projections are shown in Fig. 1.39a. Those projecting from the ventral nucleus to fields beyond AI field are posterior to those projecting to AI and close to the posterior pole of the medial geniculate complex, so they are more likely to be in the dorsal than in the ventral nucleus. Given that Morel and Imig's scheme of nuclear borders may not match exactly the

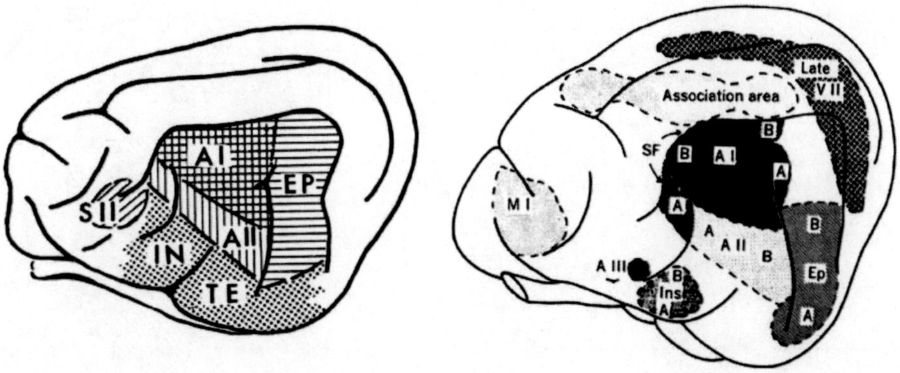

Fig. 1.32 *Left*: summary by Harlow Ades (1959) of "all areas of the cat brain showing auditory function," plus the second somatic sensory area (SII). IN is the insular area and TE the temporal area. ***Right***: Woolsey's 1961 map of all fields in which auditory evoked responses could be elicited in the cat. SF is the suprasylvian fringe; *A* and *B* indicate the representations of the cochlear apex and base, respectively. AIII is the third auditory area of Tunturi. Under chloralose anesthesia, association cortex (Assoc) and primary motor area (MI) responded to click stimuli at 15 ms latencies and the second visual area (VII) at 100 ms latencies. From Woolsey (1961)

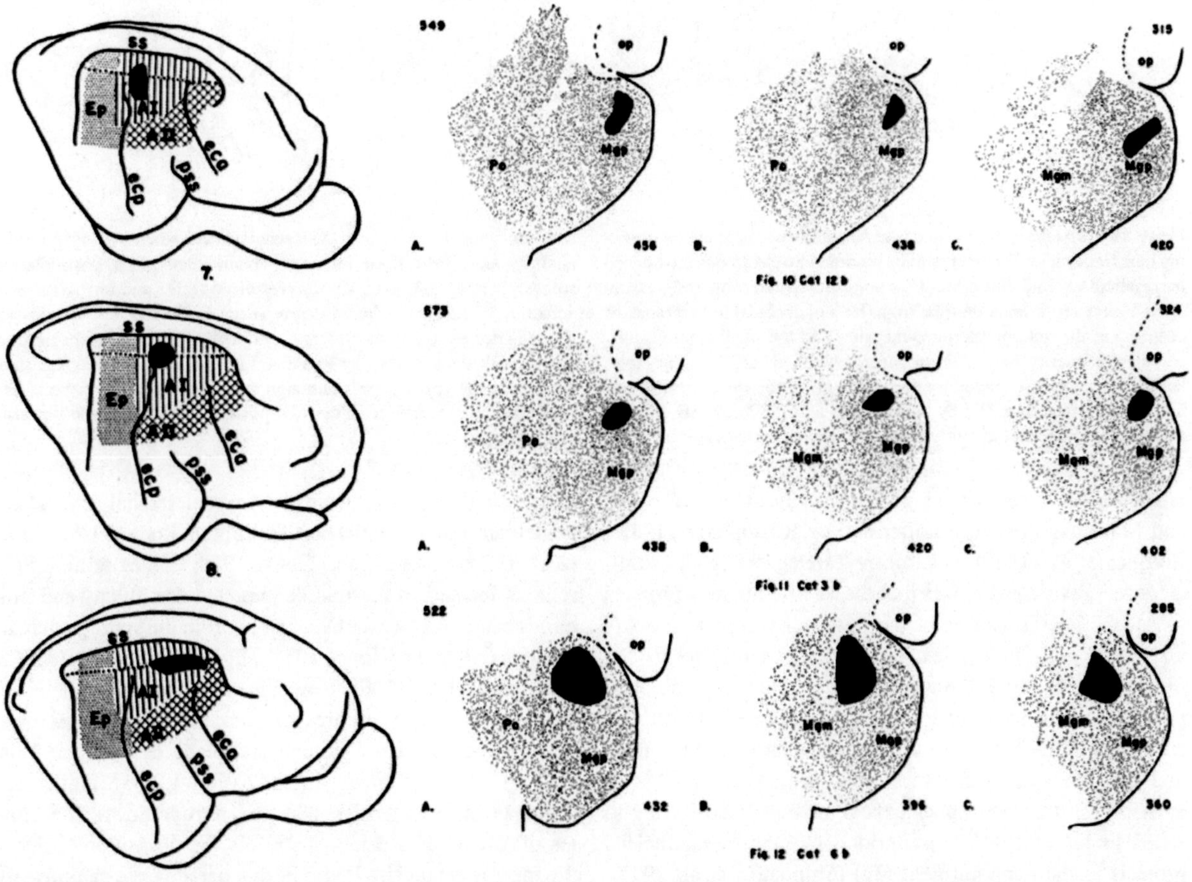

Fig. 1.33 Three experiments in which lesions located in different parts of the cochlear representation in AI of the cat led to retrograde degeneration in different parts of the principal (ventral) nucleus of the medial geniculate nucleus (Mgp). From Rose and Woolsey (1949)

cytoarchitectonic borders, the complex pattern of thalamocortical projections presented by them is not greatly different from that of earlier authors. All cells projecting to PAF and VPAF in Morel and Imig's experiments are located in the dorsal nucleus or at the posterior pole of the ventral nucleus, where the dorsal nucleus begins to expand at the expense of the ventral. Where the border between the ventral and dorsal nuclei is placed can, therefore, affect where one locates

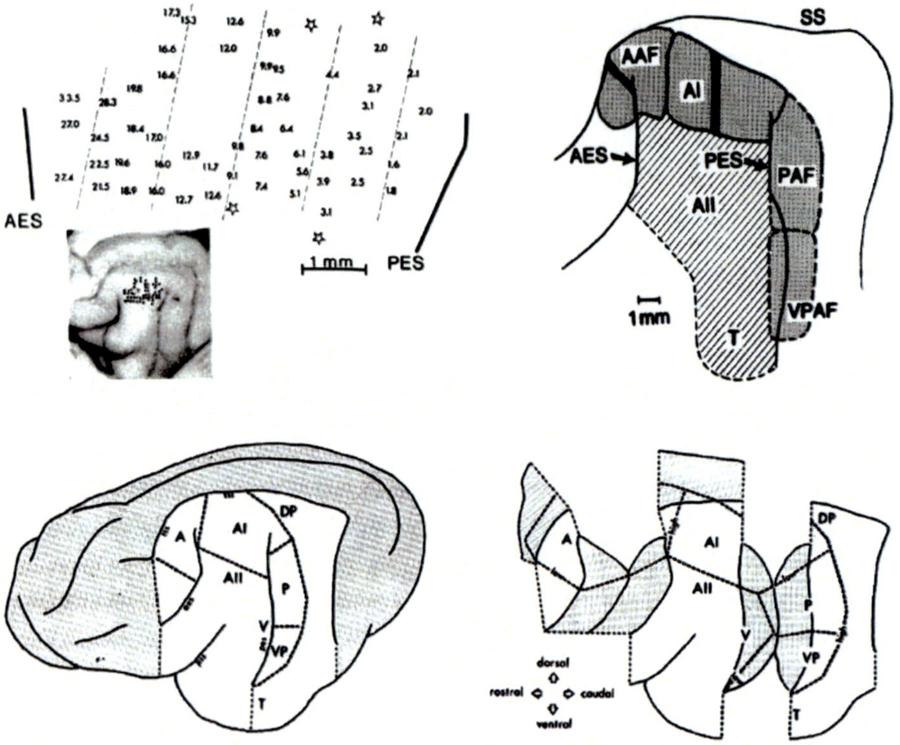

Fig. 1.34 *Top left*: surface map of the AI auditory area of the cat showing best frequencies (in kHz) of units recorded by electrodes introduced perpendicularly into the cortex at the *points* indicated. Note progression of frequency representation from high (base of cochlea) to low (apex of cochlea) in the anteroposterior dimension and the mediolaterally oriented isofrequency bands. From Merzenich et al. (1975). *Top right*: later delineation of the auditory cortical fields of the cat as determined by multiunit recording (D) by Andersen et al. (1980b). AAF, anterior auditory field; A, first auditory area; AII, second auditory area; AES, anterior ectosylvian sulcus; SS, suprasylvian sulcus; PAF, posterior auditory area; PES, posterior suprasylvian sulcus; PLs, posterolateral sulcus; T, temporal area; VPAF, ventral posterior auditory area. *Bars* across AAF and AI indicate isofrequency lines. AII and T areas are not regarded as tonotopically organized and are indistinguishable from one another. *Lower*: map by Reale and Imig (1977) showing the same areas as in the *upper panels* although with different abbreviations, and indicating the extension of the two posterior fields into the posterior ectosylvian sulcus

labeled cells. When taken together with the studies of Niimi and Matsuoka (1979), Rouiller and de Ribaupierre (1985), Rouiller et al. (1989) and Rodrigues-Dagaeff et al. (1989) a conservative viewpoint is that the ventral nucleus projects to AI, the lateral nucleus of the posterior complex to AAF, and the dorsal nucleus, excluding the part that forms the most posterior cap of the whole medial geniculate complex, projects to PAF and VPAF (Fig. 1.39c). There may well be an incipient division of the dorsal nucleus revealed in Morel and Imig's results, since most cells projecting to PAF lie anteriorly in the dorsal nucleus and most cells projecting to VPAF lie posteriorly. The posterior cap of the dorsal nucleus projects to the temporal field (Te) (Shinonaga et al. 1994). Thus, the posterior pole may represent another subdivision of the dorsal nucleus, a feature that is also found in monkeys (see below).

The fringe areas around the major auditory fields of the cat tend to receive their inputs from thalamic nuclei other than the medial geniculate body. Parts of the anterior ectosylvian gyrus anterior and ventral to AAF and AI receive their inputs predominantly from the ventral medial and suprageniculate nuclei (Reinoso-Suárez and Roda 1983; Norita et al. 1986; Bowman and Olson 1988; Clascá et al. 1997); regions located in the middle suprasylvian sulcus and posterodorsal to AI receive their inputs from the suprageniculate nucleus (Heath and Jones 1971a, b; Jones and Leavitt 1973; Winer et al. 1977, 2001; Bowman and Olson 1988).

As noted above, an area of click-evoked responses was first outlined on the supratemporal plane of macaque monkeys by Ades and Felder in 1942 (Fig. 1.25). Significantly, this was more extensive than the retrograde tracing studies of Walker had demonstrated as the projection target of the medial geniculate body; in this paradox we can now see parallels with the once uncertain state of the medial geniculocortical projection as reported from retrograde degeneration studies in the cat. In his review of 1959, Ades noted how Poliak (1932) in his Marchi study of projections from the medial geniculate body of the monkey had defined an area larger than Walker's as the terminus of fibers emanating from that nucleus. As early as 1942, Licklider and Kryter had

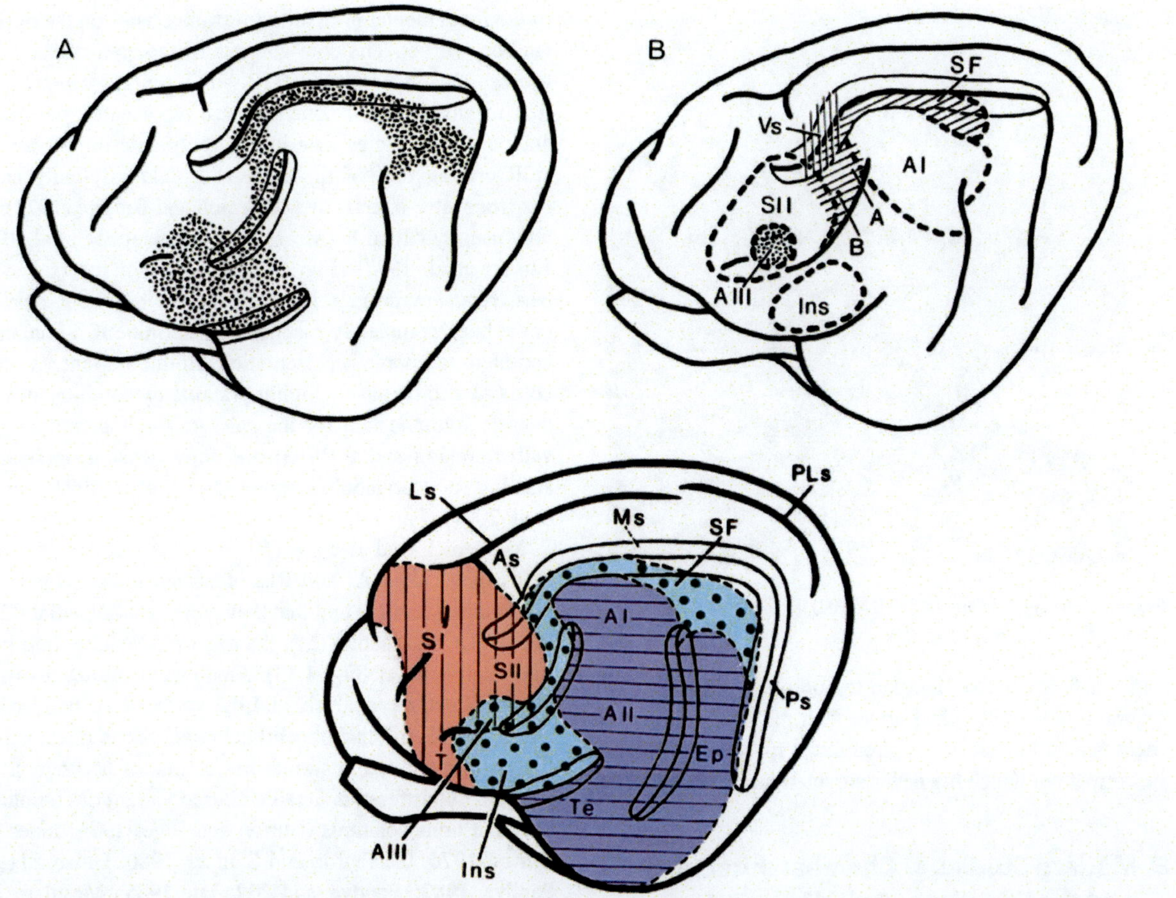

Fig. 1.35 Upper continuous projection of the posterior group of thalamic nuclei to the insular, anterior auditory and suprasylvian fringe areas, as denoted in (**b**). *A* and *B* in **b** are the *A* and *B* areas of Carreras and Andersson (1963); Vs, vestibular area. From Heath and Jones (1971a). Lower comparison of the projection areas of the ventral posterior nucleus (*red*), posterior group (*green*) and medial geniculate complex (*blue*) in the cat cerebral cortex. From Heath and Jones (1971b)

found some degree of tonotopic organization in the general region of the area defined by Ades and Felder by recording responses to short bursts of pure tone stimuli. Pribram et al. (1954) recorded responses to click stimuli over a large area of the supratemporal plane, insula, and parietal operculum and attempted to correlate parts of this region with the subdivisions of the auditory cortex made in the cat by Rose and Woolsey, mainly on the basis of response latencies. The shortest latency region, considered equivalent to AI, was located posteriorly; an anteriorly located region was felt to be equivalent to area EP (Fig. 1.40). In 1961, Neff reported a tonotopic progression of auditory evoked potentials in the region identified by Ades and Felder, with high tones represented posteriorly and low tones represented anteriorly. Woolsey by 1964 had confirmed this by the electrical stimulation of small cochlear nerve bundles, calling the field so identified AI, and he had provisionally identified a second, AII, field with a possibly reversed representation, lying medial to it (Fig. 1.41). In New World and Old World monkeys, the downward and forward growth of the temporal lobe has led to a rotation of the auditory cortical fields in comparison with those of the cat (Fig. 1.42). The representation of the basal turn of the cochlea in the AI field thus lies posteromedially instead of anteriorly as in the cat, and the representation of the apical turn lies anterolaterally instead of posteriorly (Woolsey 1964; Merzenich and Brugge 1973). The apparent equivalent of the AII field, thus, was expected to lie medial rather than ventral to AI (Woolsey 1964). Moreover, any equivalents of the several other auditory fields of the cat would then have to be sought posterior, lateral, and anterior instead of anterior, dorsal, and posterior to AI. In a short and now little-known paper, Woolsey (1971b) described the results of a number of evoked potential mapping studies carried out in monkeys in which he and his colleagues had obtained some evidence for other fields containing tonotopic representations located around his previously identified AI

Fig. 1.36 Irving T. Diamond (1922–2004). From Jones (2007)

and AII fields. These, he tried to homologize (Fig. 1.42) as Pribram et al. (1954) had done, with the fields that he had identified in the cat. It is a thoughtful speculation that may be correct but which has never attracted much attention.

9 Modern Studies of Chemoarchitecture and the Functional Parcellation of Auditory Cortex

Multiple auditory fields, each with relatively strong evidence of tonotopicity, were documented in the first comprehensive microelectrode mapping study carried out on rhesus monkeys by Merzenich and Brugge (1973) (Fig. 1.43). Their parcellation of the macaque auditory cortex forms the basis of most recent studies of the auditory cortex and its connections in monkeys. Where modified, on the basis of more extensive multiunit mapping (Imig et al. 1977; Aitkin et al. 1986, 1988; Luethke et al. 1989; Morel and Kaas 1992; Morel et al. 1993; Kosaki et al. 1997), and by immunocytochemical (Jones et al. 1995), or histochemical (Hackett et al. 2001) investigations, allied with connectional tracing, it has mainly been the extent of areas and the placement of areal borders that have been changed.

The delineations of auditory fields on and adjacent to the supratemporal plane of the macaque monkey could be related to a cytoarchitectonic plan originally proposed by Pandya and Sanides (1973) (Fig. 1.44) which defined a primary auditory core area with a distinctive cytoarchitecture, and a surrounding belt of secondary areas. This idea did not immediately catch on but it now forms one of the fundamental principles informing work on the primate auditory cortex. The core consists of the area called AI by Merzenich and Brugge (1973), with a tonotopic representation in which high frequencies are represented posteriorly and low frequencies anteriorly. At its anterior border and still forming part of the core, is a field originally called RL (for rostro-lateral) by Merzenich and Brugge (1973) and now usually called R (Morel et al. 1993; Jones et al. 1995; Kosaki et al. 1997) (Fig. 1.45). There is a reversal in the best frequency progression at the border but the R field has never been completely mapped to determine if it contains a complete tonotopic representation. Both components of the core (AI and R) have a highly granular cytoarchitecture and densely immunostain for the calcium binding protein, parvalbumin (Jones et al. 1995), and show dense histochemical staining for acetylcholinesterase (Morel et al. 1993; Hackett et al. 2001) (Fig. 1.45).

Merzenich and Brugge (1973), in addition to the core fields they called AI and RL, identified three fields in the surrounding cortical belt that they called caudomedial (CM), lateral (L), and medial (M), the names indicating their position relative to AI (Fig. 1.43). Anatomical studies involving cytoarchitectonic analysis and the study of corticocortical, commissural, and thalamocortical connections made parcellations that had many similarities to that of Merzenich and Brugge, but differed in detail and used variant or completely different nomenclatures (Burton and Jones 1976; Jones and Burton 1976; Galaburda and Sanides 1980; Galaburda and Pandya 1983; Pandya and Yeterian 1985; Mesulam and Mufson 1985). Later mapping studies in New World monkeys (Aitkin et al. 1986; Luethke et al. 1989; Morel and Kaas 1992), and investigations of greater or lesser completeness in Old World monkeys (Morel et al. 1993; Rauschecker et al. 1995; Kosaki et al. 1997; Hackett et al. 1998a, b, 2001) still recognize the AI and R fields as the core, primary auditory cortex. Around these areas, Morel and Kaas (1992), Morel et al. (1993) and Hackett et al. (1998a, b, 2001) recognized a rostromedial field lying medial to the primary core, a rostrotemporal field anterior to the core, two fields (anterolateral and posterolateral) lying lateral to the core, and both caudal and caudomedial fields lying behind the core (Figs. 1.45 and 1.46). Rauschecker et al. (1995) showed reversals in tonotopic representations along the lateral aspect of the supratemporal plane that suggested the presence of three auditory fields which they called anterolateral, middle lateral, and caudolateral. They probably correspond to the anterolateral, posterolateral, and caudal fields of Morel et al. (1993).

Kosaki et al. (1997) delineated the auditory cortical fields by immunocytochemical staining for two common calcium binding proteins, parvalbumin and 28 kDa calbindin. Parvalbumin immunostaining is particularly useful as a marker to delineate the core and belt regions of the

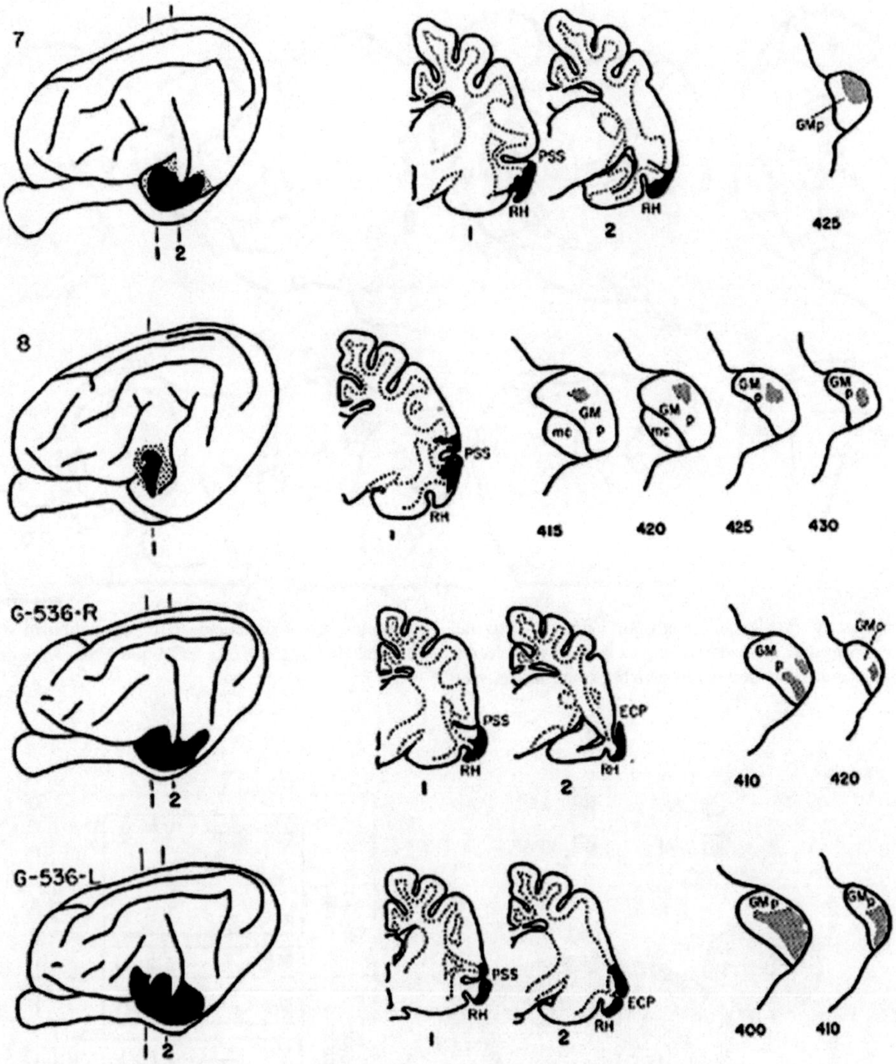

Fig. 1.37 Experiments in four cats in which lesions affecting the temporal field of the cortex resulted in retrograde degeneration of the caudal polar region of the medial geniculate complex. From Diamond et al. (1958)

monkey auditory cortex and to identify subareas within them (Jones et al. 1995; Molinari et al. 1995). There are major differences in the intensity of immunoreactive staining of fiber plexuses in layers III, IV, and VI that permit distinct chemoarchitectonic areas to be defined (Fig. 1.47). The stained plexuses represent mainly thalamocortical fibers and their different density reflects the relative proportions of parvalbumin immunoreactive cells in the medial geniculate complex projecting to each area. Four principal zones are distinguishable on the supratemporal plane: a central region of densest immunostaining coincides with the auditory koniocortex and is more or less coextensive with the cortex of the small annectant gyrus commonly found on the posteromedial aspect of the supratemporal plane in the larger macaques (Fig. 1.45). Surrounding the core is an inner belt of moderately dense immunostaining which is heavier anteromedial and posteromedial to the core zone and somewhat lighter lateral to the core zone. Outside the inner belt is an outer belt of much weaker parvalbumin immunostaining extending out onto the surface of the superior temporal gyrus and around the anterior aspect of the inner belt. The outer belt is embraced in turn by a fourth zone or outermost belt in which parvalbumin immunostaining of fiber plexuses is essentially absent (Fig. 1.47). It occupies the anterior end of the temporal operculum and extends to the temporal pole and over the surface of the middle and inferior temporal gyri. As parvalbumin immunoreactivity declines in intensity, immunoreactivity for another calcium binding protein, calbindin, increases (Fig. 1.47). A core zone of dense acetylcholinesterase activity, surrounded by a belt region of lesser

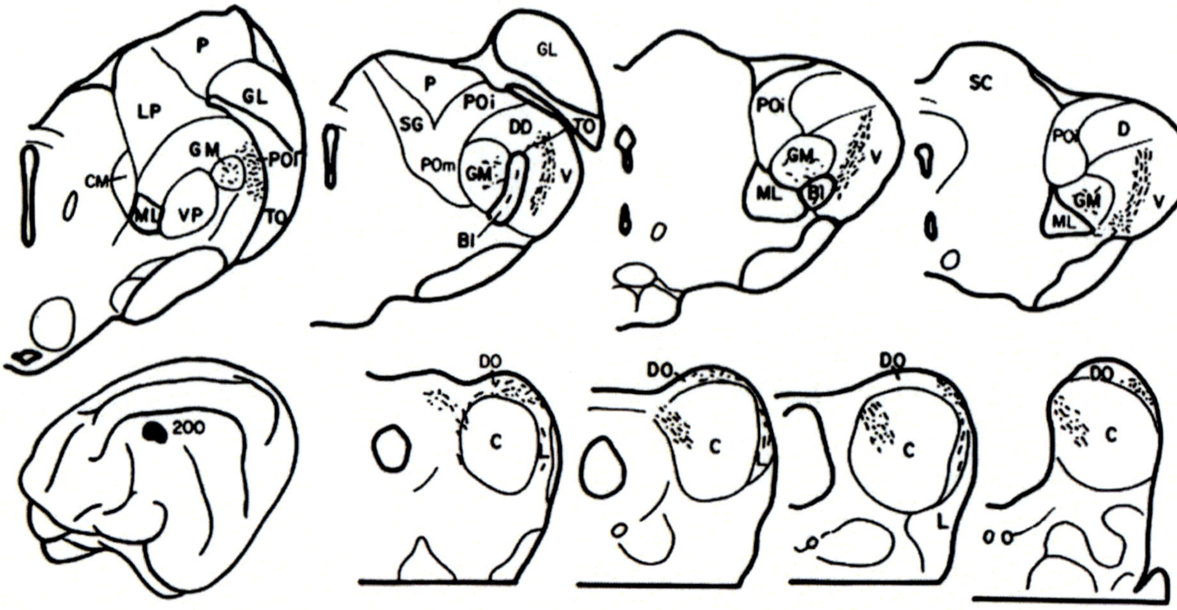

Fig. 1.38 Topographically organized corticothalamic projection to the ventral nucleus of the medial geniculate complex and to the central nucleus of the inferior colliculus in a cat in which a small lesion was placed in part of the cochlear representation in AI and degenerating fibers labeled by the Nauta technique. From Diamond et al. (1969)

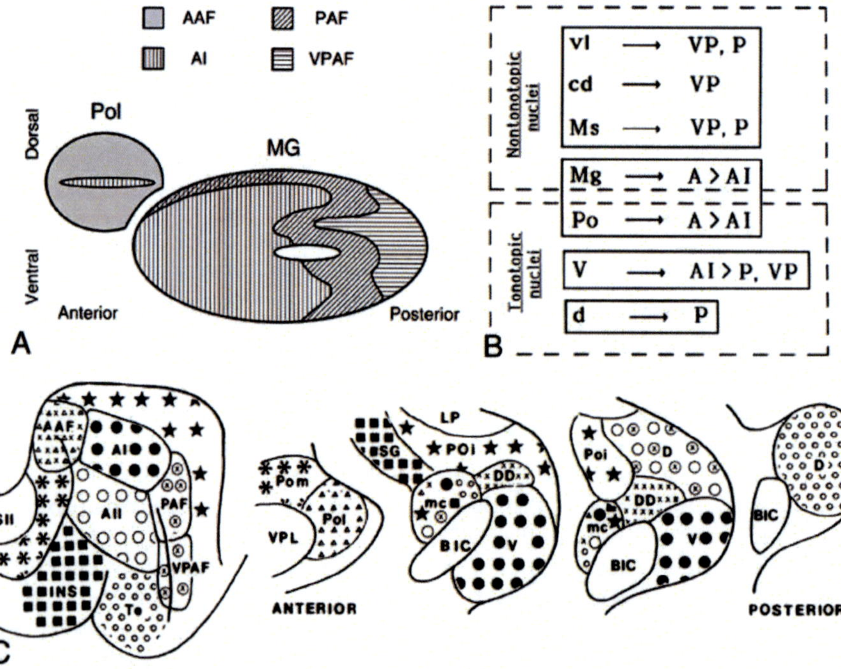

Fig. 1.39 **a** The major projections of the lateral division of the posterior group (Pol) and of the ventral medial geniculate nucleus according to Morel and Imig (1987). **b** The heaviest projections from the nuclei of the medial geniculate complex and lateral division of the posterior group. From Morel and Imig (1987). **c** Schematic representation of the major thalamocortical projections of the subnuclei of the medial geniculate complex and adjacent nuclei in the cat. The *same symbol* indicates the cortical field and the thalamic nucleus from which it receives its predominant projection. This figure represents a consensus view of reports that do not always agree on all counts except for AI and MGv. Modified from Jones (2007)

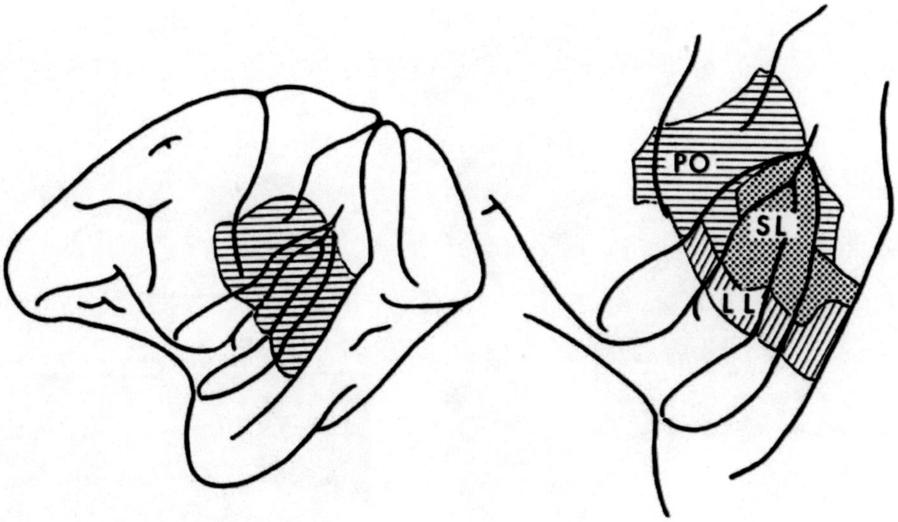

Fig. 1.40 Total extent of cerebral cortex responsive to click stimuli in the rhesus monkey. Shortest latency responses are found in a central core area labeled SL. From Pribram et al. (1954)

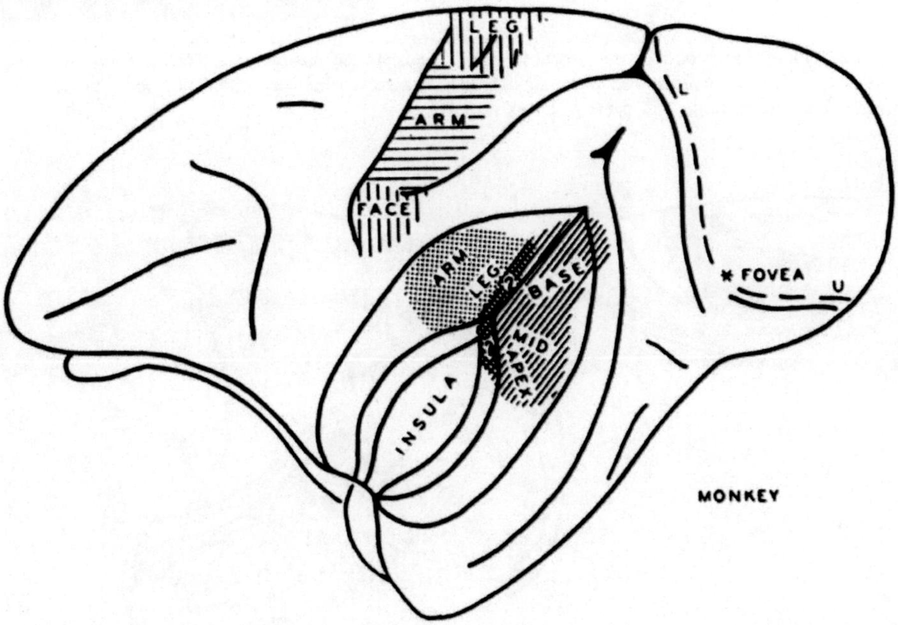

Fig. 1.41 Locations of the first and second auditory areas in the macaque monkey as mapped by evoked potentials. BASE, MID and APEX indicate cochlear representation in the AI field. Area AII is designated by 2. ARM, LEG and FACE indicate body representation in first and second somatic sensory areas. From Woolsey (1971)

activity, characterizes comparable regions of the macaque, chimpanzee, and human supratemporal plane (Hackett et al. 2001).

Kosaki et al. (1997) mapped the parvalbumin immunostained regions on the basis of best frequency responses to pure tone stimuli and demonstrated that the central core zone of densest parvalbumin immunoreactivity contained two auditory frequency range representations (Fig. 1.48). Frequencies >20 kHz are represented posteromedially in the core, as in the AI field of Woolsey (1971) and Merzenich and Brugge (1973). On moving anteriorly in the core, there is a gradual reduction in the frequencies represented, then a reversal and a progressive increase until, at the anterior border of the core, neurons respond again to frequencies up

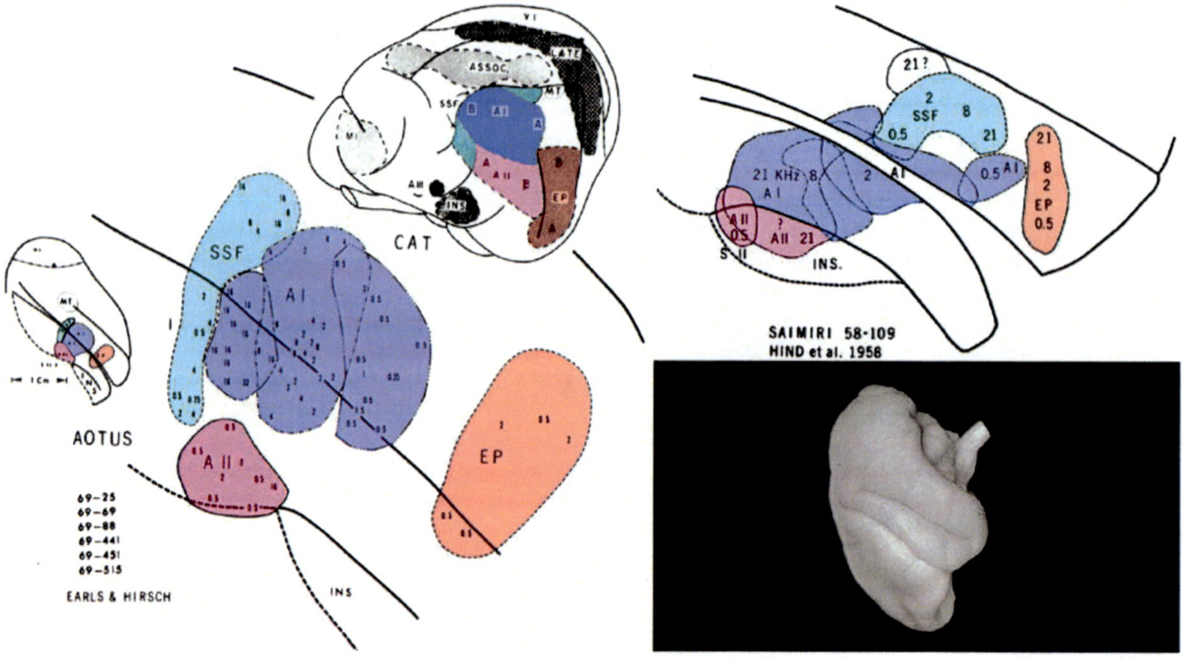

Fig. 1.42 Woolsey's (1971) homologies of auditory cortical fields of the owl monkey (*Aotus*, *left*) and squirrel monkey (*Saimiri*, *right*) with those of the cat (*inset*). Inset above *Aotus* and panel at *lower right* indicate the rotation of the owl and squirrel monkey brains required to achieve symmetry with the fields of the cat

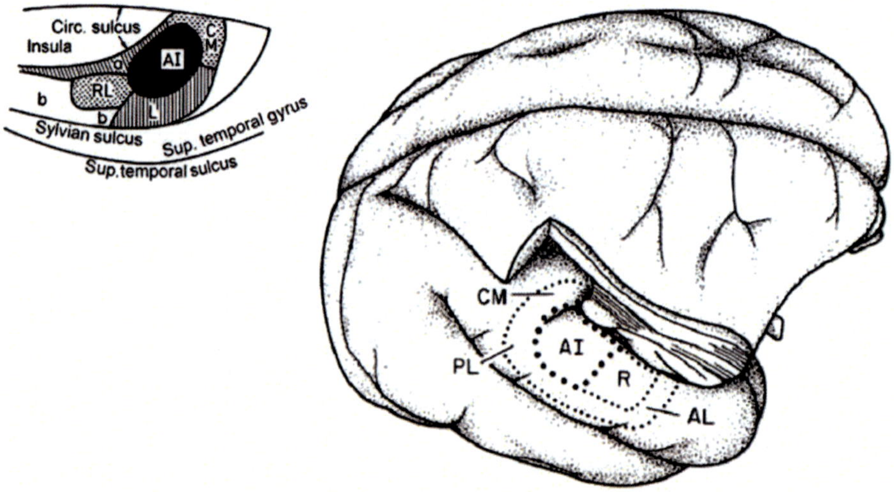

Fig. 1.43 *Left*: auditory fields of the rhesus monkey, as delineated by tonotopic maps derived from multiunit recordings. Adapted from Merzenich and Brugge (1973). *Right*: map of the supratemporal plane in the rhesus monkey brain with the locations of the fields on the *left* shown in situ. From Merzenich and Brugge (1973)

to 20 kHz. This reversal in the representation enables fields AI and R to be delineated within the primary core region. Because the central core curves following the long axis of the supratemporal plane, the high-to-low frequency representation runs from posteromedial to anterolateral in AI and from posterolateral to anteromedial in R.

Reversals of frequency representation in the lateral division of the inner belt surrounding the central core delineates two fields, probably corresponding to the anterolateral and posterolateral fields of Morel et al. (1993) and to the anterior and middle fields of Rauschecker et al (1995). A posteromedial region with some indications of a tonotopic reversal in its

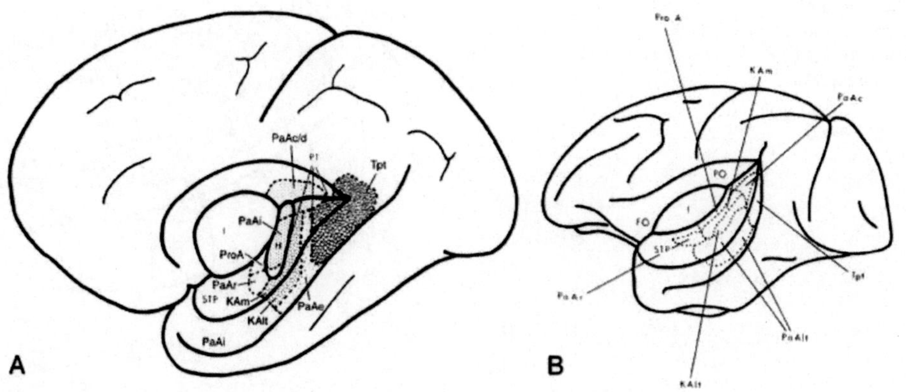

Fig. 1.44 The auditory and adjacent cortical areas of the human (**a**) and rhesus monkey (**b**) brains, as delineated cytoarchitectonically by Galaburda and Sanides (1980) and Pandya and Sanides (1973). The core koniocortical area (KAm plus KAl) represents the primary auditory cortex which is surrounded by a belt of para-auditory areas (PA, Pa) beyond which lie temporo-parietal (Tpt, tpt) and other fields of association cortex (ProA). I, insula; STP, supratemporal plane

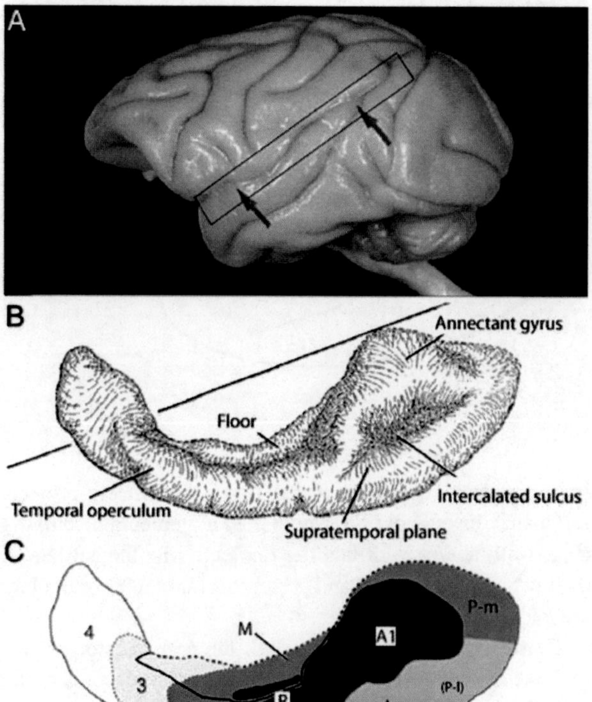

Fig. 1.45 **a** Lateral view of the brain of a macaque monkey, showing the lateral sulcus (*boxed area*) and the location of the auditory cortical areas (*arrows*). **b** Drawing of the lower bank of the lateral sulcus, showing the supratemporal plane and the annectant gyrus on which the primary auditory cortex is located. **c** Reconstruction of the same region as (**b**), showing the locations of the core (primary) auditory area and the surrounding belt areas, based on relative intensities of parvalbumin immunostaining. Based on Jones et al. (1995)

middle probably contains the posterior field of Rauschecker et al., and the caudal and caudomedial fields of Morel et al. (1993).

A medial (M) field containing a tonotopic representation, with low frequencies represented posteriorly and high frequencies represented anteriorly, is confined to the floor and adjacent lateral bank of the inferior limiting sulcus of the insula and seems to be equivalent to the medial field of Merzenich and Brugge (1973), but it is much thinner than the rostromedial field of Morel et al. (1993) which extends for a considerable distance on to the insula. The anterior, high frequency representation in field M is separated from the high frequency representation at the anterior end of the anterolateral field by a part of the outer belt in which tonotopic order is indistinct. This is the anteromedial (A–M) field, which may correspond to the rostrotemporal (RT) field of Morel et al. (1993).

Neurons in the fields lying lateral to the core fields are only weakly responsive to pure tone stimuli but respond to band-passed noise around a center frequency, permitting a type of tonotopic map to be constructed (Rauschecker et al. 1995). Neurons in these areas seem to prefer more complex sounds, including species-specific calls. Kosaki et al. (1997) found that neurons in all three of the lateral fields (A–L, P–L, and P–M) had broader tuning curves than neurons in the AI and R core regions (Fig. 1.48). Neurons in the medial and anteromedial fields are either very broadly tuned or not tuned. Outside the inner belt, in the outer and outermost belts of weak or absent parvalbumin immunoreactivity, neurons cannot be driven by tonal stimuli and, if they respond at all, only do so to white noise. Studies in awake monkeys performing an auditory discrimination task generally confirm the observations made in anesthetized monkeys (Recanzone et al. 1999).

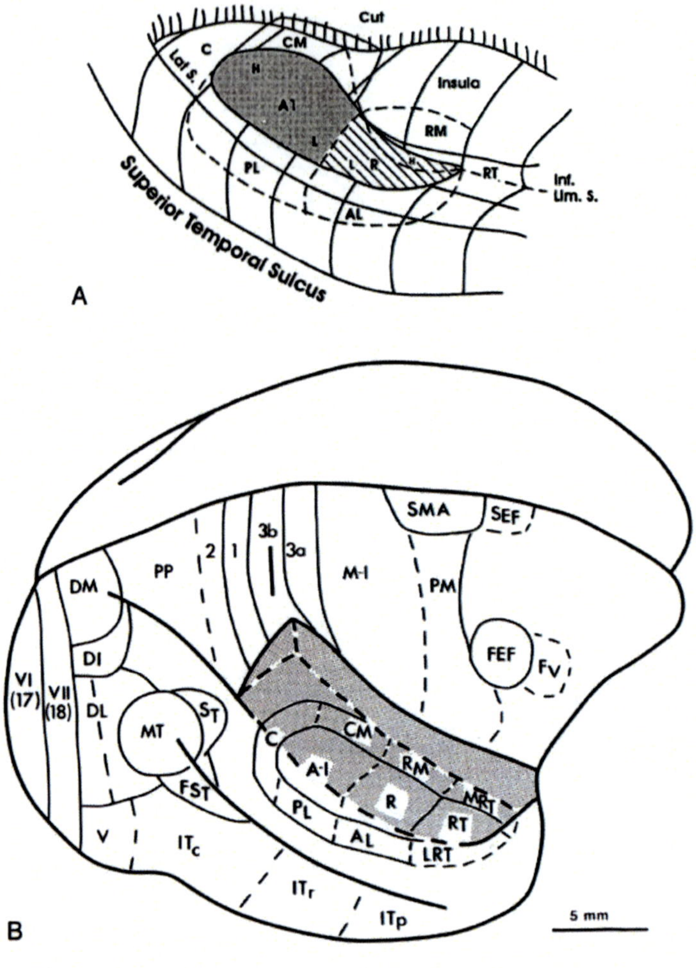

Fig. 1.46 Schematic views of areas containing representations of the auditory frequency range in **a** Old World and **b** New World monkeys. From Morel et al. (1993) and Morel and Kaas (1992)

Early connectional studies in monkeys showed that the ventral medial geniculate nucleus provided thalamic input to AI and suggested that the surrounding fields were the targets of other medial geniculate nuclei (Mesulam and Pandya 1973; Burton and Jones 1976; FitzPatrick and Imig 1978). Later studies demonstrated that the predominant input to fields AI and R arises in the ventral nucleus of the medial geniculate complex (Aitkin et al. 1988; Luethke et al. 1989; Morel and Kaas 1992; Morel et al. 1993; Molinari et al. 1995; Hackett et al. 1998b) (Fig. 1.49). The major inputs to the fields of the belt regions are from the dorsal nucleus. The anterodorsal nucleus projects most posteriorly in the belt, the greater part of the posterodorsal nucleus projects to middle and anterior fields of the belt, and the extreme posterior pole of the posterodorsal nucleus projects the most anteriorly (Molinari et al. 1995). Minor projections from the dorsal nuclei to AI and R and from the ventral nucleus to the belt can be attributed to the diffusely, projecting, calbindin immunoreactive population that forms a diffusely projecting matrix throughout the whole medial geniculate complex (Fig. 1.49) and/or to difficulties in identifying the borders of the ventral, anterodorsal and posterodorsal medial geniculate nuclei.

Corticocortical connections link the core auditory fields to those of the inner belt and further projections connect the inner belt to the outer belt and to more distant areas of cortex (Pandya et al. 1969; Pandya and Sanides 1973; Aitkin et al. 1988; Luethke et al. 1989; Morel et al. 1993; Jones et al. 1995; Hackett et al 1998a). Three streams of corticocortical connections can be traced from the inner belt towards areas of association and limbic cortex (Pandya and Yeterian 1985; Romanski et al. 1999a, b). One stream arises from the anterolateral areas and extends towards orbital regions of the prefrontal cortex and to anterior entorhinal areas; a second stream arises from lateral areas and passes into lateral prefrontal and posterior entorhinal areas; the third stream arises from posterolateral areas and passes into lateral

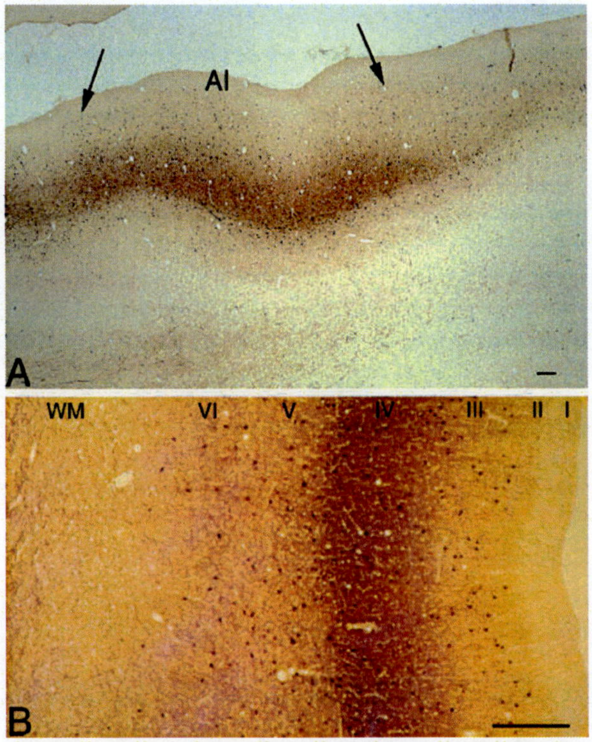

Fig. 1.47 **a** Photomicrograph of parvalbumin immunostaining in the core (AI) and belt (*to left* and *right*) regions of the macaque auditory cortex. **b** Higher magnification view of the immunostaining of parvalbumin immunoreactive thalamocortical fibers and cortical GABAergic neurons in the AI area. Bars: 100 μm

prefrontal, posterior parietal, and posterior cingulate cortex (Fig. 1.50). Functional imaging studies suggest that these different streams are engaged in processing different aspects of auditory perception and different components of auditory behavior (Jones 2003).

The intensity of parvalbumin fiber staining in layer IV and deep layer III of the areas on the monkey supratemporal plane is directly correlated with the relative concentrations of parvalbumin immunoreactive cells in the ventral and dorsal nuclei of the medial geniculate complex, and inversely correlated with the relative concentrations of cells immunoreactive for 28 kDa calbindin in the same nuclei (Molinari et al. 1995) (Fig. 1.49). These differences among the nuclei are also correlated with the relative proportions of cells projecting to middle and superficial layers of the auditory cortex. The densest middle layer fiber plexus is in the two core areas, AI and R, and this correlates with the very high proportion of parvalbumin cells in the ventral nucleus of the medial geniculate complex. The vast majority of relay cells in the ventral medial geniculate nucleus are parvalbumin immunoreactive (Hashikawa et al. 1991; Molinari et al. 1995). The ventral nucleus contains very few, very widely dispersed calbindin immunoreactive cells. The anterodorsal nucleus of the complex also contains a majority of parvalbumin cells but more calbindin cells are present. The posterodorsal nucleus contains approximately equal numbers of parvalbumin and calbindin cells anteriorly but the posterior part of the nucleus, which forms the posterior cap of the medial geniculate complex, contains mainly calbindin immunoreactive cells (Fig. 1.49). The magnocellular nucleus contains islands of cells in which parvalbumin or calbindin immunoreactive types predominate. With the exception of the ventral nucleus, the density of calbindin cells does not change greatly throughout the nuclei of the medial geniculate complex and the calbindin cells form a diffuse matrix very similar to that found in the ventral posterior and ventral lateral nuclei of the thalamus (Rausell and Jones 1991; Rausell et al. 1992; Jones 1998a–c).

Parvalbumin cells located in the monkey dorsal thalamus project only to the middle layers, and calbindin cells to the superficial layers, of the cerebral cortex (Hashikawa et al. 1991; Molinari et al. 1995; Rausell and Jones 1991; Rausell et al. 1992; Jones 1998a–c). Injections of retrogradely transported tracers that affect all layers of the dense parvalbumin immunoreactive core fields of the auditory cortex, AI and/or R, invariably label a focus of parvalbumin positive cells in the ventral nucleus, with a few scattered calbindin cells in the ventral and dorsal nuclei and a few cells of both types in the magnocellular nucleus. Injections in the fields of the moderate-to-densely parvalbumin immunoreactive inner belt around AI and R label a focus of parvalbumin cells in the anterodorsal nucleus when posteriorly placed and label a focus in anterior and middle parts of the posterodorsal nucleus when anteriorly placed, plus scattered calbindin cells in these nuclei and a few parvalbumin and calbindin cells in the magnocellular nucleus. Injections in the parvalbumin-weak fields beyond the inner belt invariably label parvalbumin and calbindin cells in the magnocellular nucleus but only injections in anterior regions near the temporal pole label cells elsewhere in the medial geniculate complex. These labeled cells are located at the posterior pole of the posterodorsal nucleus and all are calbindin positive.

From these data, we can conclude that parvalbumin cells form the basis of a topographically organized projection from the ventral nucleus of the medial geniculate complex to middle layers of AI and R and from the anterodorsal and most of the posterodorsal nuclei to one or more of the immediately surrounding fields of the inner belt. The calbindin cells, by contrast, form a diffusely projecting system whose axons terminate in superficial layers and are unconstrained by the borders of cortical fields. The magnocellular nucleus, although projecting to more than one field, also possesses this organization since its parvalbumin cells project to layer IV of single fields while its calbindin cells project to layer I of multiple fields (Hashikawa et al. 1995). The posterior, polar part of the posterodorsal nucleus forms a unique,

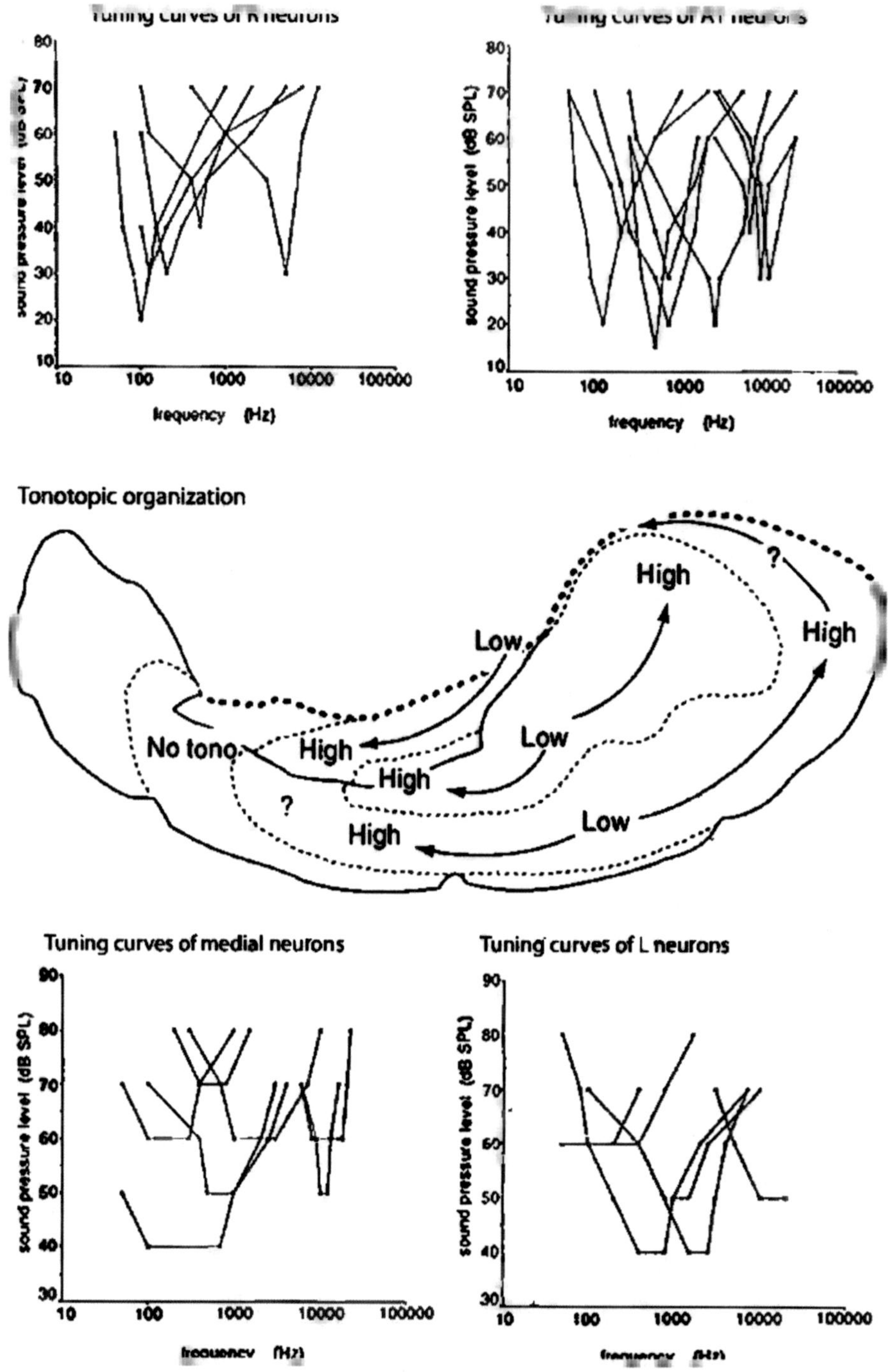

Fig. 1.48 Tonotopic organization of the core and belt areas of the auditory cortex, based on multiunit mapping of responses to pure tone stimuli (*middle figure*), and the tuning curves of individual neurons in the two core areas and in two of the surrounding belt areas. Based on Kosaki et al. (1997)

calbindin population of cells projecting to anterior field(s) of the outer belt region.

The dominant parvalbumin positive input to core fields of the auditory cortex and to the fields of the inner belt is a reflection of the predominance of parvalbumin immunoreactivity in the most direct brain stem auditory pathways leading to the medial geniculate complex. The ventral cochlear nuclei, the trapezoid body and lateral lemniscus and their

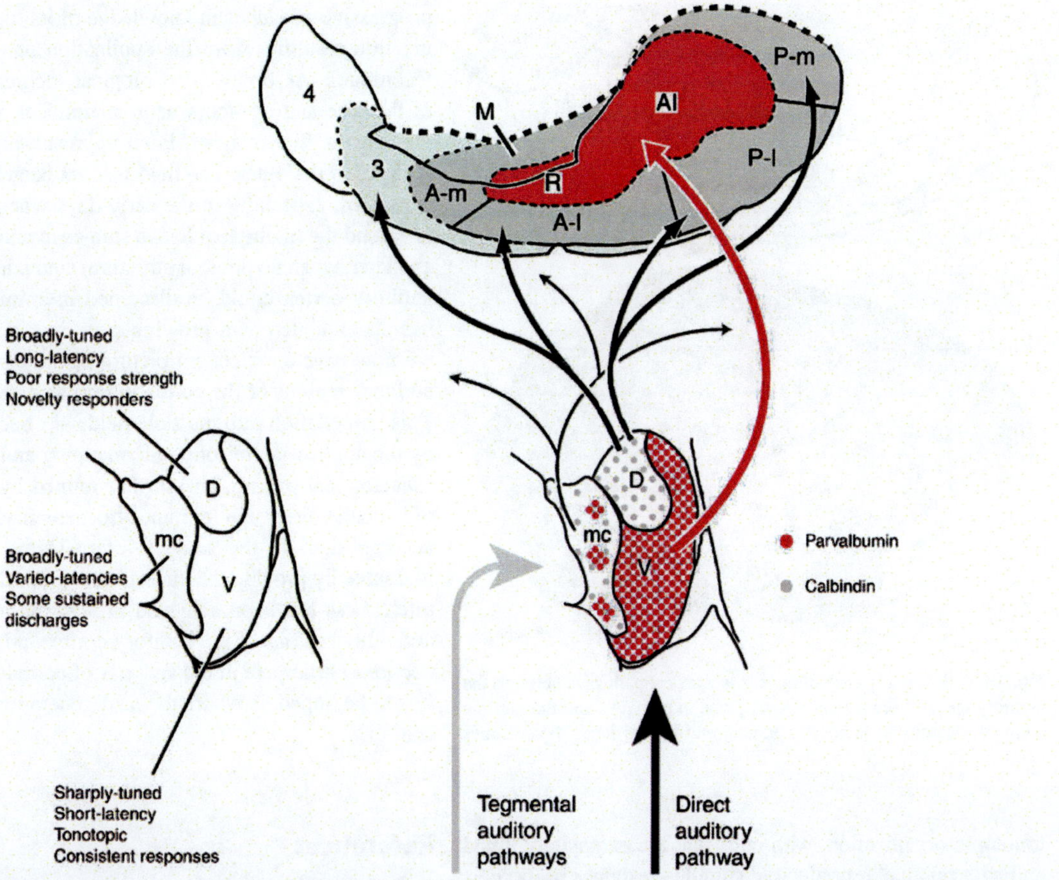

Fig. 1.49 *Left*: predicted typical responses to auditory stimuli of neurons located in the dorsal (D), ventral (V) and magnocellular (mc) nuclei of the monkey medial geniculate complex. Based on work by Calford (1983) in the cat. *Right*: differential expression of the calcium binding proteins, parvalbumin and calbindin, in the nuclei of the medial geniculate complex, their innervation by brain stem pathways expressing the same proteins, and their projections to the auditory cortex. Based on Molinari et al. (1995) and Jones (2007)

associated nuclei, the central nucleus of the inferior colliculus and the brachium of the inferior colliculus in the monkey brain stem all show heavy parvalbumin immunostaining for cells and/or fibers (Fig. 1.49) (Jones 2003). The parvalbumin immunoreactive fibers of the brachium of the inferior colliculus enter the ventral and anterodorsal medial geniculate nuclei and form a dense parvalbumin positive neuropil around the predominant, parvalbumin immunoreactive cell population in these nuclei (Fig. 1.49). The posterodorsal nucleus, by contrast, contains very weak neuropil staining, implying that few parvalbumin positive fibers arising in the ventral nucleus of the inferior colliculus terminate there. Instead, the posterodorsal nucleus contains a weakly calbindin immunoreactive neuropil, especially at its posterior pole; this is formed by the terminations of calbindin positive fibers entering from the lateral midbrain tegmentum and probably arising from calbindin cells that predominate in the pericentral and external nuclei of the inferior colliculus (Fig. 1.49). The magnocellular nucleus receives local concentrations of parvalbumin or calbindin rich fibers derived from both ascending pathways.

The monkey subcortical auditory pathways and their continuations in the auditory thalamocortical projection seem to reflect their organization as two, chemically distinct parallel streams (Fig. 1.49). One, characterized by parvalbumin immunoreactivity, leads through the tonotopically organized nuclei to the primary, core areas of auditory cortex and to the fields of the inner belt with which they are most closely connected. The other, characterized by calbindin immunoreactivity, is diffusely projected onto the core and inner belt areas, as well as to wider areas of cortex forming the outer and outermost belts. Distinct parvalbumin- and calbindin-immunoreactive auditory pathways have also been described in other species such as the chinchilla (Kelley et al. 1992) and in bats (Zettel et al. 1991; Vater and Braun 1994), with calbindin predominating in the centers that are afferent to the dorsal medial geniculate nucleus.

The functional properties of the two parallel auditory pathways through the medial geniculate complex, so far as they are known, mainly from investigations in the cat (Calford 1983; Calford and Aitkin 1983), imply that the parvalbumin pathway is characterized by a high degree of

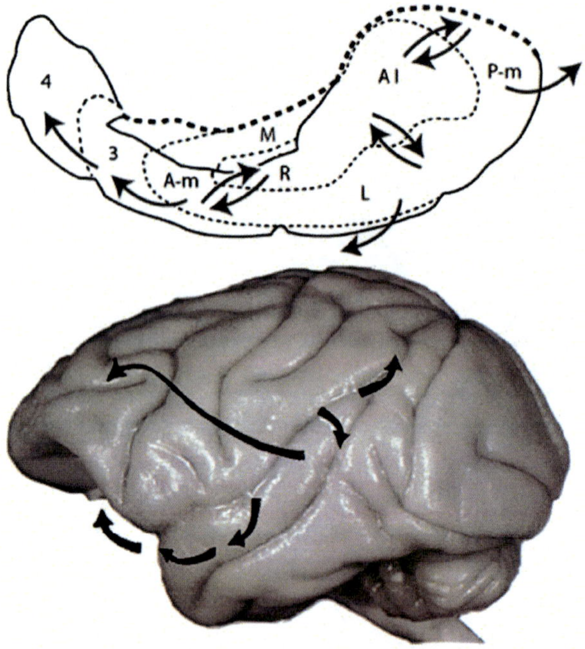

Fig. 1.50 Schematic view of the corticocortical pathways between the auditory areas of the supratemporal plane in monkeys and the outflow pathways leading to temporal, parietal and frontal lobes. From Jones (2003)

tonotopic organization, with cells that are exquisitely tuned and with reliable, reproducible stimulus-response properties. The calbindin pathway, by contrast appears to be characterized by absent or weak tonotopicity, weakly tuned or un-tuned cells with unreliable and fatigable responses, responding best to novel and complex stimuli rather than to pure tones (Fig. 1.49). It is perhaps the most interesting pathway from the perspective of an infrastructure that enables the nervous system to perceive complex sounds; the more thoroughly studied parvalbumin pathway through the ventral nucleus of the medial geniculate complex seemingly underlies mainly the perception of pitch (Jones 2003). At the cortical level, the two pathways converge in the auditory belt region in particular and from there parallel corticocortical pathways convey information relevant to the perception of species-specific vocalizations, moving sounds and other complex aspects of auditory appreciation into the parietal, frontal and temporal lobes (Romanski et al. 1999b; Jones 2003) (Fig. 1.50).

10 Summary

The historical development of knowledge about the auditory cortex, as with that about most other functional areas of the cerebral cortex, has been characterized by progressive advances in knowledge of its location and organization resulting from the application of new and refined techniques. At times, physiological techniques have been to the fore and at others neuroanatomical approaches have yielded the first insights. Each approach has informed and stimulated the other. The field has not been without its controversies, especially in the early days when it was difficult to extend the findings of lesion studies in animals to humans, and later, when no obvious structural equivalent of the human auditory cortex could be discerned in animals. Knowledge has accumulated at a growing pace in the recent past and we now have an excellent picture of the parcellation of the auditory regions of the cortex in primates and non-primates. This parcellation into multiple fields has been accomplished by the application of tonotopic mapping and correlated neuroanatomical tracing, and further refined by the revelations of histochemistry and immunocytochemistry. Much has been accomplished in the analysis of pathways and areas that undoubtedly provide the underpinnings for the perception of pitch. Less has been achieved in learning about how these and other regions of the auditory cortex and their input connections participate in the analysis of complex sounds. That, it is to be hoped, is where the next generation of studies will lead us.

References

Ades HW (1941) Connections of the medial geniculate body of the cat. *Archives of Neurology and Psychiatry* **45**:138–144.

Ades HW (1959) Central auditory mechanisms. In: Field J, Magoun HW, and Hall VE (eds). *Handbook of Physiology. Section 1: Neurophysiology*, Volume I. American Physiological Society, Washington DC, pp. 585–613.

Ades HW and Felder R (1942) The acoustic area of the monkey (*Macaca mulatta*). *Journal of Neurophysiology* **5**:49–54.

Ades HW (1943) A secondary acoustic area in the cerebral cortex of the cat. *Journal of Neurophysiology* **6**:59–63.

Aitkin LM, Merzenich MM, Irvine DRF, Clarey JC, and Nelson JE (1986) Frequency representation in auditory cortex of the common marmoset (*Callithrix jacchus jacchus*). *Journal of Comparative Neurology* **252**:175–185.

Aitkin LM, Kudo M, and Irvine DRF (1988) Connections of the primary auditory cortex in the common marmoset, *Callithrix jacchus jacchus*. *Journal of Comparative Neurology* **269**: 235–248.

Anton G (1899) Ueber die Selbstwahrnehmung der Herderkrankungen des Gehirns bei Rindenblindheit und Rindentaubheit. *Archiv für Psychiatrie und Nervenkrankheiten* **32**:86–127.

Bechterew W von (1911) *Die Funktionen der Nervencentra*. German translation by Weinberg R. Volume III. Fischer, Jena, pp. 1840–1882.

Bowman EM and Olson CR (1988) Visual and auditory association areas of the cat's posterior ectosylvian gyrus: thalamic afferents. *Journal of Comparative Neurology* **272**:15–29.

Bremer F and Dow RS (1939) The cerebral acoustic area of the cat. A combined oscillographic and cytoarchitectonic study. *Journal of Neurophysiology* **2**:308–318.

Brodmann K (1905) Beiträge zur histologischen Lokalisation des Grosshinrinde dritte Mitteilung: die Rindenfelder der niederen Affen. *Journal für Psychologie und Neurologie* **4**:177–226.

Brodmann K (1909) *Vergleichende Lokalisationslehre der Grosshirnrinde in ihren Prinzipien dargestellt auf Grund des Zellenbaues*. Barth, Leipzig.

Brown S and Schäfer EA (1888) An investigation into the functions of the occipital and temporal lobes of the monkey's brain. *Philosophical Transactions of the Royal Society of London, series B, Biological Sciences* **179**:303–327.

Burton H and Jones EG (1976) The posterior thalamic region and its cortical projection in New World and Old World monkeys. *Journal of Comparative Neurology* **168**:249–301.

Cajal S Ramón y (1899a) Estudios sobre la corteza cerebral humana. I. Corteza visual. *Revista trimestral Micrográfica* **4**:1–63.

Cajal S Ramón y (1899b) Estudios sobre la corteza cerebral humana. II. Estructura de la corteza motriz del hombre y mamíferos superiores. *Revista trimestral Micrográfica* **4**:117–200.

Cajal S Ramón y (1900a) Estudios sobre la corteza cerebral humana. II. Corteza motriz (conclusion). *Revista trimestral Micrográfica* **5**: 1–11.

Cajal S Ramón y (1900b) Estudios sobre la corteza cerebral humana. III. Corteza acústica. *Revista trimestral Micrográfica* **5**:129–183.

Cajal S Ramón y (1904) *Textura Del Sistema Nervioso del Hombre y de Los Vertebrados Volume II, second part*. N Moya, Madrid. (2002) English translation by Pasik T and Pasik P. *Texture of the Nervous System of Man and the Vertebrates*, Volume III. Springer, Wien, New York, Barcelona.

Calford MB and Aitkin LM (1983) Ascending projections to the medial geniculate body of the cat: evidence for multiple, parallel auditory pathways through thalamus. *Journal of Neuroscience* **3**:2365–2380.

Calford MB (1983) The parcellation of the medial geniculate body of the cat defined by the auditory response of single units. *Journal of Neuroscience* **3**:2350–2364.

Campbell AW (1905) *Histological Studies on the Localisation of Cerebral Function*. Cambridge University Press, Cambridge.

Casseday JH, Diamond IT, and Harting JK (1976) Auditory pathways to the cortex in *Tupaia glis*. *Journal of Comparative Neurology* **166**:303–340.

Clascá F, Llamas A and Reinoso-Suárez F (1997) Insular cortex and neighboring fields in the cat: a redefinition based on cortical microarchitecture and connections with the thalamus. *Journal of Comparative Neurology* **384**:456–482.

Dejerine J and Dejerine-Klumpke M (1895) *Anatomie des Centres Nerveux, Volume 1*. Rueff, Paris.

Dejerine J and Dejerine-Klumpke M (1901) *Anatomie des Centres Nerveux, Volume 2, Part 1*. Rueff, Paris.

Diamond IT, Chow KL, and Neff WD (1958) Degeneration of caudal medial geniculate body following cortical lesions ventral to auditory area II in the cat. *Journal of Comparative Neurology* **109**:349–362.

Diamond IT, Jones EG, and Powell TPS (1969) The projection of the auditory cortex upon the diencephalon and brain stem in the cat. *Brain Research* **15**:305–340.

Ecker A (1869) *Die Hirnwindungen des Menschen*. Vieweg, Braunschweig.

Ecker A (1873) *On the Convolutions of the Human Brain*. Translated by Galton JC. Smith Elder, London.

Economo C von and Koskinas GN (1925) *Die Cytoarchitektonik der Hirnrinde des Erwachsenen Menschen*. Springer, Vienna.

Erulkar SD, Rose JE, and Davies PW (1956) Single unit activity in the auditory cortex of the cat. *Bulletin of the Johns Hopkins Hospital* **99**:55–86.

Ferrier D (1873) Experimental researches in cerebral physiology and pathology. *West Riding Lunatic Asylum Medical Reports* **3**: 1–50.

Ferrier D (1876) *Functions of the Brain*. Smith Elder, London.

Ferrier D (1875) The Croonian lecture: experiments on the brain of monkeys (second series). *Philosophical Transactions of the Royal Society of London* **165**:433–488.

Ferrier D, Golz F, Cros A, and Yeo G (1881) Discussion on the localization of function in the cortex cerebri. In: MacCormac W (ed). *Transactions of the International Medical Congress, Seventh Session*, Volume I. Kolckmann, London, pp. 228–242.

Ferrier D and Yeo GF (1884) XIX. A record of experiments on the effects of lesion of different regions of the cerebral hemispheres. *Philosophical Transactions of the Royal Society of London, series B, Biological Sciences* **175**:479–564.

Ferrier D (1886) *The Functions of the Brain*. Smith Elder, London.

Ferrier D and Turner WA (1894) A record of experiments illustrative of the symptomatology and degenerations following lesions of the cerebellum and its peduncles and related structures in monkeys. *Philosophical Transactions of the Royal Society of London, series B, Biological Sciences***185**:719–778.

FitzPatrick K and Imig TJ (1978) Projections of auditory cortex upon the thalamus and midbrain in the owl monkey. *Journal of Comparative Neurology* 177:437–556.

Flechsig P (1898) Neue Untersuchungen über die Markbildung in den menschlichen Grosshirnlappen. *Neurologisches Centralblatt* **17**:977–996.

Fritsch GT and Hitzig E (1870) Über die elektrische Erregbarkeit des Grosshirns. *Archiv für Anatomie und Physiologie* **37**:300–332.

Galaburda A and Sanides F (1980) Cytoarchitectonic organization of the human auditory cortex. *Journal of Comparative Neurology* **190**:597–610.

Galaburda AM and Pandya DN (1983) The intrinsic architectonic and connectional organization of the superior temporal region of the rhesus monkey. *Journal of Comparative Neurology* **221**: 169–184.

Goldberg JM and Neff WD (1961) Frequency discrimination after bilateral section of the brachium of the inferior colliculus. *Journal of Comparative Neurology* **116**:265–289.

Gowers WR (1885) *Epilepsy*. Wood, London.

Grünbaum ASF and Sherrington CS (1902) Observations on the physiology of the cerebral cortex of some of the highest apes preliminary communication. *Proceedings of the Royal Society of London* **69**:206–209.

Grünbaum ASF, Sherrington CS (1903) observations on the physiology of the cerebral cortex of the anthropoid apes. *Proceedings of the Royal Society of London* **72**:152–155.

Hackett TA, Stepniewska I, and Kaas JH (1998a) Subdivisions of auditory cortex and ipsilateral cortical connections of the parabelt auditory cortex in macaque monkeys. *Journal of Comparative Neurology* **394**:475–495.

Hackett TA, Stepniewska I, and Kaas JH (1998b) Thalamocortical connections of the parabelt auditory cortex in macaque monkeys. *Journal of Comparative Neurology* **400**:271–286.

Hackett TA, Preuss TM, and Kaas JH (2001) Architectonic identification of the core region in auditory cortex of macaques, chimpanzees, and humans. *Journal of Comparative Neurology* **441**:197–222.

Hashikawa T, Rausell E, Molinari M, Jones EG (1991) Parvalbumin- and calbindin-containing neurons in the monkey medial geniculate complex: differential distribution and cortical layer specific projections. *Brain Research* **544**:335–341.

Hashikawa T, Molinari M, Rausell E, and Jones EG (1995) Patchy and laminar terminations of medial geniculate axons in monkey auditory cortex. *Journal of Comparative Neurology* **362**:195–208.

Heath CJ and Jones EG (1971a) An experimental study of ascending connections from the posterior group of thalamic nuclei in the cat. *Journal of Comparative Neurology* **141**:397–426.

Heath CJ, Jones EG (1971b) The anatomical organization of the suprasylvian gyrus of the cat. *Ergebnisse für Anatomie und Entwicklungsgeschichte* **45**:3–64.

Heffner H (1978) Effect of auditory cortex ablation on localization and discrimination of brief sounds. *Journal of Neurophysiology* **41**: 963–976.

Heschl RL (1877) Tiefenwindungen des menschlichen Grosshirns und die Überbrückung der Zentralfurche. *Wien Medizinische Wochenschrift* **27**:985–988.

Hitzig E (1874) *Untersuchungen über das Gehirn*. Hirschwald, Berlin.

Horsley V and Schäfer EA (1888) A record of experiments upon the functions of the cerebral cortex. *Philosophical Transactions of the Royal Society of London, series B, Biological Sciences* **179**: 1–45.

Imig TJ, Ruggero MA, Kitzes LM, Javel E, and Brugge JF (1977) Organization of auditory cortex in the owl monkey (*Aotus trivirgatus*). *Journal of Comparative Neurology* **171**: 111–128.

Jones EG and Leavitt RY (1973) Demonstration of thalamo-cortical connectivity in the cat somato-sensory system by retrograde axonal transport of horseradish peroxidase. *Brain Research* **63**: 414–418.

Jones EG and Burton H (1976) Areal differences in the laminar distribution of thalamic afferents in cortical fields of the insular, parietal and temporal regions of primates. *Journal of Comparative Neurology* **168**:197–248.

Jones EG, Dell'Anna ME, Molinari M, Rausell E, and Hashikawa T (1995) Subdivisions of macaque monkey auditory cortex revealed by calcium-binding protein immunoreactivity. *Journal of Comparative Neurology* **362**:153–170.

Jones EG (1998a) A new view of specific and non-specific thalamocortical connections. In: Jasper HH, Descarries L, Castelluci VF, and Rossignol S (eds). *Consciousness: At the Frontiers of Neuroscience, Advances in Neurology*, Volume 77. Wiley-Liss, New York, pp. 49–73.

Jones EG (1998b) The thalamus of primates. In: Bloom FE, Björklund A, and Hökfelt T (eds). *The Primate Nervous System, part II*. Elsevier, Amsterdam, pp. 1–298.

Jones EG (1998c) Viewpoint: the core and matrix of thalamic organization. *Neuroscience* **85**:331–345.

Jones EG (2003) Chemically defined parallel pathways in the monkey auditory system. *Annals of the New York Academy of Sciences* **999**:218–233.

Jones EG (2007) *The Thalamus*. Cambridge University Press, Cambridge.

Kelley PE, Frisina RD, Zettel ML, and Walton JP (1992) Differential calbindin-like immunoreactivity in the brain stem auditory system of the chinchilla. *Journal of Comparative Neurology* **319**:196–212.

Kornmüller AE (1937) *Bioelektrische Erscheinungen der Hirnrindenfelder*. Thieme, Leipzig.

Kosaki H, Hashikawa T, He J, and Jones EG (1997) Tonotopic organization of auditory cortical fields delineated by parvalbumin immunoreactivity in macaque monkeys. *Journal of Comparative Neurology* **386**:304–316.

Larionow W (1899) Ueber die musikalischen Centren des Gehirns. *Pflügers Archiv European Journal of Physiology* **76**:608–625.

Le Gros Clark WE (1936) The thalamic connections of the temporal lobe of the brain in the monkey. *Journal of Anatomy (London)* **70**:447–464.

Licklider JCR and Kryter KD (1942) frequency localization in the auditory cortex of the monkey. *Federation Proceedings of the American Society of Experimental Biology* **1**:51 (abstract).

Luciani L and Tamburini A (1879) *Sulle Funzioni del Cervello*. Reggio-Emilia: S. Calderini.

Luciani L (1884) On the sensorial localisations in the cortex cerebri. *Brain* **7**:145–160.

Luethke LE, Krubitzer LA, and Kaas JH (1989) Connections of primary auditory cortex in the New World monkey, *Saguinus*. *Journal of Comparative Neurology* **285**:487–513.

Merzenich MM and Brugge JF (1973) Representation of the cochlear partition on the superior temporal plane of the macaque monkey. *Brain Research* **50**:275–296.

Merzenich MM, Knight PL and Roth GL (1975) Representation of the cochlea within the primary auditory cortex in the cat. *Journal of Neurophysiology* **38**:231–249.

Mesulam MM and Pandya DN (1973) The projection of the medial geniculate complex within the Sylvian fissure of the rhesus monkey. *Brain Research* **60**:315–353.

Mesulam M-M and Mufson EJ (1985) The insula of Reil in man and monkey: architectonics, connectivity, and function. In: Peters A and Jones EG (eds). *Cerebral Cortex*, Volume 4, *Auditory and Association Cortices*. Plenum, New York, pp. 179–228.

Mettler FA (1932) Connections of the auditory cortex of the cat. *Journal of Comparative Neurology* **55**:139–183.

Mills CK (1891) On the localization of the auditory centre. *Brain* **14**:465–472.

Molinari M, Dell'Anna ME, Rausell E, Leggio MG, Hashikawa T, and Jones EG (1995) Auditory thalamocortical pathways defined in monkeys by calcium-binding protein immunoreactivity. *Journal of Comparative Neurology* **362**:171–194.

Monakow C von (1895) Experimentelle und pathologisch-anatomische Untersuchungen über die Haubenregion, den Sehhügel und die Regio subthalamica, nebst Beiträgen zur Kentniss früh erworbener Gross- und Kleinhirndefekte. *Archiv für Psychiatrie und Nervenkranke* **27**:1–128, 386–478.

Morel A and Imig TJ (1987) Thalamic projections to fields A, AI, P, and VP in the cat auditory cortex. *Journal of Comparative Neurology* **265**:119–144.

Morel A and Kaas JH (1992) Subdivisions and connections of auditory cortex in owl monkeys. *Journal of Comparative Neurology* **318**: 27–63.

Morel A, Garraghty PE, and Kaas JH (1993) Tonotopic organization, architectonic fields, and connections of auditory cortex in macaque monkeys. *Journal of Comparative Neurology* **335**:437–459.

Munk H (1881) *Über die Funktionen der Grosshirnrinde*. Hirschwald, Berlin.

Neff WD (1961) Neural mechanisms of auditory discrimination. In: Rosenblith WA (ed). *Principles of Sensory Communication*. MIT Press, Cambridge, pp. 259–278.

Niimi K and Naito F (1974) Cortical projections of the medial geniculate body in the cat. *Experimental Brain Research* **19**: 326–342.

Niimi K and Matsuoka H (1979) Thalamocortical organization of the auditory system in the cat studied by retrograde axonal transport of horseradish peroxidase. *Advances in Anatomy, Embryology and Cell Biology* **57**:1–56.

Norita M, Mucke L, Benedek G, Albowitz B, Katoh Y, and Creutzfeldt OD (1986) Connections of the anterior ectosylvian area (AEV). *Experimental Brain Research* **62**:225–240.

Oliver DL and Hall WC (1978) The medial geniculate body of the tree shrew, *Tupaia glis*. II. connections with the neocortex. *Journal of Comparative Neurology* **182**:459–493.

Pandya DN, Hallett M, and Mukherjee SK (1969) Intra- and interhemispheric connections of the neocortical auditory system in the rhesus monkey. *Brain Research* **13**:13–36.

Pandya DN, Sanides F (1973) Architectonic parcellation of the temporal operculum in rhesus monkey and its projection pattern. *Zeitschrift für Anatomie und Entwicklungsgeschichte* **139**: 127–161.

Pandya DP, Yeterian EH (1985) Architecture and connections of cortical association areas. In: Peters A and Jones EG (eds). *Cerebral Cortex, Volume 4. Association and Auditory Cortices*. Plenum, New York, pp. 3–62.

Phillips DP and Irvine DRF (1981) Responses of single neurons in physiologically defined primary auditory cortex (AI) of the cat: frequency tuning and responses to intensity. *Journal of Neurophysiology* **45**:48–58.

Pick A (1892) Beiträge zur Lehre von den Störungen der Sprache. *Archiv für Psychiatrie* **23**:896–918.

Polyak S (1932) *The Main Afferent Fiber Systems of the Cerebral Cortex in Primates*. University of California Press, Berkeley.

Pribram KH, Rosner BS, and Rosenblith WA (1954) Electrical responses to acoustic clicks in monkey: extent of neocortex activated. *Journal of Neurophysiology* **17**:336–344.

Rausell E and Jones EG (1991) Chemically distinct compartments of the thalamic VPM nucleus in monkeys relay principal and spinal trigeminal pathways to different layers of the somatosensory cortex. *Journal of Neuroscience* **11**:226–237.

Rausell E, Bae CS, Viñuela A, Huntley GW, and Jones EG (1992) Calbindin and parvalbumin cells in monkey VPL thalamic nucleus: distribution, laminar cortical projections, and relations to spinothalamic terminations. *Journal of Neuroscience* **12**: 4088–4111.

Rauschecker JP, Tian B, and Hauser M (1995) Processing of complex sounds in the macaque nonprimary auditory cortex. *Science* **268**:111–114.

Reale RA and Imig TJ (1980) Tonotopic organization in auditory cortex of the cat. *Journal of Comparative Neurology* **192**:265–292.

Recanzone GH, Schreiner CE, Sutter ML, Beitel RE, and Merzenich MM (1999) Functional organization of spectral receptive fields in the primary auditory cortex of the owl monkey. *Journal of Comparative Neurology* **415**:460–481.

Reinoso-Suárez F and Roda JM (1983) Topographical organization of the thalamic projections to the cortex of the anterior ectosylvian sulcus in the cat. *Experimental Brain Research* **49**:131–139.

Rodrigues-Dagaeff C, Simm G, de Ribaupierre Y, Villa A, de Ribaupierre F, and Rouiller EM (1989) Functional organization of the ventral division of the medial geniculate body of the cat: evidence for a rostro-caudal gradient of response properties and cortical projections. *Hearing Research* **39**:103–125.

Romanski LM, Tian B, Fritz J, Mishkin M, Goldman-Rakic PS, and Rauschecker JP (1999) Dual streams of auditory afferents target multiple domains in the primate prefrontal cortex. *Nature Neuroscience* **2**:1131–1136.

Romanski LM, Bates JF, and Goldman-Rakic PS (1999) Auditory belt and parabelt projections to the prefrontal cortex in the rhesus monkey. *Journal of Comparative Neurology* **403**:141–157.

Rose JE (1949) The cellular structure of the auditory region of the cat. *Journal of Comparative Neurology* **91**:409–440.

Rose JE and Woolsey CN (1949) The relations of thalamic connections, cellular structure and evocable electrical activity in the auditory region of the cat. *Journal of Comparative Neurology* **91**: 441–466.

Rose JE and Woolsey CN (1958) Cortical connections and functional organization of the thalamic auditory system of the cat. In: Harlow HF and Woolsey CN (eds). *Biological and Biochemical Bases of Behavior*. University Wisconsin Press, Madison, pp. 127–150.

Rouiller EM and de Ribaupierre F (1985) Origin of afferents to physiologically defined regions of the medial geniculate body of the cat: ventral and dorsal divisions. *Hearing Research* **19**:97–114.

Rouiller EM, Rodrigues-Dagaeff C, Simm G, de Ribaupierre Y, Villa A, and de Ribaupierre (1989) Functional organization of the medial division of the medial geniculate body of the cat: tonotopic organization, spatial distribution of response properties and cortical connections. *Hearing Research* **39**:127–142.

Schäfer EA (1888a) Experiments on special sense localisations in the cortex cerebri. *Brain* **10**:362–380.

Schäfer EA (1888b) On the functions of the temporal and occipital lobes: a reply to Dr Ferrier. *Brain* **11**:145–166.

Schäfer EA (1900) The cerebral cortex. In: Schäfer EA (ed).*Text-Book of Physiology*. Young Pentland, Edinburgh, pp. 697–782.

Schreiner CE and Cynader MS (1984) Basic functional organization of second auditory cortical field (AII) of the cat. *Journal of Neurophysiology* **51**:1284–1305.

Sérieux P and Mignot PV (1901) Surdité corticale avec paralexie et hallucinations de l'ouvië due à des kystes hydatides du cerveau. Nouvelle Iconographie de la Salpêtrière. *L'Année Biologique: Comptes Rendus Annuels des Travaux de Biologie*, p. 560.

Shinonaga Y, Takada M, and Mizuno N (1994) Direct projections from the non-laminated divisions of the medial geniculate nucleus to the temporal polar cortex and amygdala in the cat. *Journal of Comparative Neurology* **340**:405–426.

Sindberg RM and Thompson RF (1962) Auditory response fields in ventral temporal and insular cortex of cat. *Journal of Neurophysiology* **25**:21–28.

Sousa-Pinto A (1973) Cortical projections of the medial geniculate body in the cat. *Advances in Anatomy, Embryology and Cell Biology* **48**:1–42.

Tunturi AR (1945) Further afferent connections to the acoustic cortex of the dog. *American Journal of Physiology* **144**:389–394.

Vater M and Braun K (1994) Parvalbumin, calbindin D-28 k, and calretinin immunoreactivity in the ascending auditory pathway of horseshoe bats. *Journal of Comparative Neurology* **341**: 534–558.

Vogt C and Vogt O (1919) Allgemeinere Ergebnisse unserer Hirnforschung. *Journal für Psychologie und Neurologie* **25**: 279–462.

Vogt O and Vogt C (1902) *Neurobiologische Arbeiten. Beiträge der Hirnfaserlehre*. Fischer, Leipzig.

Walker AE (1937) A note on the thalamic nuclei of *Macaca mulatta*. *Journal of Comparative Neurology* **66**:145–155.

Walker AE (1937) The projection of the medial geniculate body to the cerebral cortex in the macaque monkey. *Journal of Anatomy (London)* **71**:319–331.

Walker AE (1938) *The Primate Thalamus*. University of Chicago Press, Chicago.

Wallace MN, Kitzes LM, and Jones EG (1991) Chemoarchitectonic organization of the cat primary auditory cortex. *Experimental Brain Research* **86**:518–526.

Waller WH (1934) Topographical relations of cortical lesions to thalamic nuclei in the albino rat. *Journal of Comparative Neurology* **60**:237–269.

Waller WH and Barris W (1937) Relationship of thalamic nuclei to the cerebral cortex in the cat. *Journal of Comparative Neurology* **67**:317–341.

Waller WH (1940) Thalamic degeneration induced by temporal lesions in the cat. *Journal of Anatomy (London)* **74**:528–536.

Walzl EM and Woolsey CN (1946) Effects of cochlear lesions on click responses in the auditory cortex of the cat. *Bulletin of the Johns Hopkins Hospital* **79**:309–319.

Wernicke C and Friedlander C (1883) Ein Fall von Taubheit in Folge von doppelseitiger Läsion des Schläfenlappens. *Fortschritte der Medizin* **1**:177–185.

Whitfield IC, Cranford J, Ravizza R, and Diamond IT (1972) Effects of unilateral ablation of auditory cortex in cat on complex sound localization. *Journal of Neurophysiology* **35**:718–731.

Willis T (1664) *Cerebri Anatome: Cui Accessit Nervorum Descriptio et Usus*. Martyn and Allestry, London.

Willis T (1681) *The Remaining Medical Works of that Famous and Renowned Physician, Dr. Thomas Willis*. Dring, Harper & Leigh, London.

Wilson ME and Cragg BG (1969) Projections from the medial geniculate body to the cerebral cortex in the cat. *Brain Research* **13**:462–475.

Winer JA, Diamond IT, and Raczkowski D (1977) Subdivisions of the auditory cortex of the cat: The retrograde transport of horseradish peroxidase to the medial geniculate body and posterior thalamic nuclei. *Journal of Comparative Neurology* **176**: 387–418.

Winer JA, Diehl JJ, and Larue DT (2001) Projections of auditory cortex to the medial geniculate body of the cat. *Journal of Comparative Neurology* **430**:27–55.

Woollard HH and Harpman A (1939) The cortical projection of the medial geniculate body. *Journal of Neurology and Psychiatry* **2**: 35–44.

Woolsey CN and Walzl EM (1942) Topical projection of nerve fibers from local regions of the cochlea to the cerebral cortex of the cat. *Bulletin of the Johns Hopkins Hospital* **71**:315–344.

Woolsey CN (1958) Organization of somatic sensory and motor areas of the cerebral cortex. In: Harlow H and Woolsey CN (eds). *Biological and Biochemical Bases of Behavior*. University Wisconsin Press, Madison, pp. 63–81.

Woolsey CN (1959) Organization of cortical auditory system. In: Rosenblith WA (ed). *Principles of Sensory Communication* MIT Press, Cambridge, pp. 235–257.

Woolsey CN (1960) Some observations on brain fissuration in relation to cortical localization of function. In: Tower DB and Schadé JPS (eds). *Structure and Function of the Cerebral Cortex*. Elsevier, Amsterdam, pp. 64–68.

Woolsey CN (1961) Organization of cortical auditory system: review and a synthesis. In: Rasmussen GL and Windle WF (eds). *Neural Mechanisms of the Auditory and Vestibular Systems* Charles C. Thomas, Springfield, pp. 165–180.

Woolsey CN (1964) Electrophysiological studies on thalamocortical relations in the auditory system. In: Abrams A, Garner HH, and Toman JEP (eds). *Unfinished Tasks in the Behavioral Sciences*. Williams and Wilkins, Baltimore, pp. 45–57.

Woolsey CN (1971a) Tonotopic organization of the auditory cortex. In: Sachs MB (ed). *Physiology of the Auditory System*. National Educational Consultants, Baltimore, pp. 271–282.

Woolsey CN (1971b) Dedication of the workshop to the memory of Edward McColgan Walzl. In: Sachs MB (ed). *Physiology of the Auditory System*. National Educational Consultants, Baltimore, pp. 5–20.

Zettel ML, Carr CE, and O'Neill WE (1991) Calbindin-like immunoreactivity in the central auditory system of the mustached bat, *Pteronotus parnelli*. *Journal of Comparative Neurology* **313**: 1–16.

Chapter 2

A Profile of Auditory Forebrain Connections and Circuits

Jeffery A. Winer

Abbreviations

AA	amygdala, anterior nucleus
AAF	anterior auditory field
ABm	basomedial nucleus of the amygdala
ACe	central nucleus of the amygdala
AD	anterior part of the DCN
AES	anterior ectosylvian area
aes	anterior ectosylvian sulcus
AI	primary auditory cortex
AlP	anterolateral periolivary nucleus
APt	anterior pretectum
AII	second auditory cortex
AM	anterior medial nucleus
AV	anterior ventral thalamic nucleus
AV	anteroventral cochlear nucleus
Ava	anteroventral cochlear nucleus, anterior part
BB	broadband
BM	amygdala, basomedial nucleus
BIC	brachium of the inferior colliculus
Bl	amygdala, basolateral nucleus
C	caudal *or* C layer of the lateral geniculate body
Ca	caudate nucleus
CBM	cerebellum
CC	caudal cortex of the inferior colliculus
CC	corpus callosum
CF,CF-CF	constant frequency cortical area
CF	characteristic frequency
Cl	claustrum
DF	dorsal fringe auditory cortical area
DI–DIV	layers of dorsal cortex of inferior colliculus
DM	dorsomedial auditory cortical area
CG	central gray
CIC	commissure of the inferior colliculus
CM	central medial nucleus
CM	central medial/caudomedial auditory cortical area
CN	central nucleus of the inferior colliculus
cNB	central narrowband module in AI
CP	cerebral peduncle
Cu	cuneiform nucleus
D	dorsal nucleus of the MGB *or* dorsal
d1–d4	narrowband modules in AI
DD	deep dorsal nucleus of the MGB
DC	dorsal cortex of the inferior colliculus
DCa	caudal pole of the inferior colliculus
DCN	dorsal cochlear nucleus
DF	dorsal cochlear nucleus, fusiform cell layer
DL	dorsal nucleus of the lateral lemniscus
DlP	dorsolateral periolivary nucleus
DM	dorsal cochlear nucleus, molecular layer
DmP	dorsomedial periolivary nucleus
DP	dorsoposterior auditory area of cat
DS	dorsal superficial nucleus of the MGB
DSCF	Doppler-shifted constant frequency region
DZ	dorsal auditory zone
ED	posterior ectosylvian gyrus, dorsal part
EE	excitatory-excitatory binaural interaction
EI	posterior ectosylvian gyrus, intermediate part *or* excitatory-inhibitory binaural interaction
En	entopeduncular nucleus
EP	posterior ectosylvian gyrus
EV	posterior ectosylvian gyrus, ventral part
EW	Edinger-Westphal nucleus
FF	fields of Forel
FM	frequency modulated
FM-FM	frequency-modulated auditory cortical area
FM_1-FM	first harmonic frequency-modulated auditory cortical area
FM_1-FM_2	first harmonic, second harmonic frequency-modulated auditory cortical area
FM_1-FM_3	first harmonic, third harmonic frequency-modulated auditory cortical area

J.A. Winer (✉)
Division of Neurobiology, Department of Molecular and Cell Biology, University of California at Berkeley, Berkeley, CA 94720-3200, USA
e-mail: chris@phy.ucsf.edu

J.A. Winer, C.E. Schreiner (eds.), *The Auditory Cortex*,
DOI 10.1007/978-1-4419-0074-6_2, © Springer Science+Business Media, LLC 2011

FM_1-FM_4	first harmonic, fourth harmonic frequency-modulated auditory cortical area
GP	globus pallidus
Ha	habenula
Hip	hippocampus
HiT	habenulointerpeduncular tract
IC	inferior colliculus
ICa	internal capsule
IcT	intercollicular tegmentum
IL	intermediate nucleus of the lateral lemniscus
In	insular cortex
IlN	intralaminar thalamic nuclei
Int	thalamic intralaminar nuclei
IT	intercollicular tegmentum
La	lateral nucleus of the amygdala
LC	lateral cortex of the inferior colliculus
LD	lateral dorsal nucleus
LGB,LGBd	lateral geniculate body, dorsal part
LGBv	lateral geniculate body, ventral part
LM	lateral medial nucleus
LMN	lateral mesencephalic nucleus
LN	lateral nucleus of the inferior colliculus
LP	lateral posterior nucleus
LS	lateral superior olive
LT	lateral nucleus of the trapezoid body
M	medial division of the MGB
MCP	middle cerebellar peduncle
MGB	medial geniculate body
ML	medial lemniscus
MLF	medial longitudinal fasciculus
MR	mesencephalic reticular formation
MRF	mesencephalic reticular formation
MS	medial superior olive
MT	medial nucleus of the trapezoid body
Mv	medioventral thalamic nucleus
MZ	marginal zone of MGB
NB	narrowband
NBIC	nucleus of the brachium of the inferior colliculus
NRTP	reticular tegmental nucleus of the pons
OT	optic tract
OR	optic radiation
Ov	*pars ovoidea* of the ventral division of the MGB
P	posterior auditory field
PC	posterior commissure
Pd	posterodorsal division of the DCN
PeN	periolivary nuclei
PHy	posterior hypothalamus
PL	posterior limitans nucleus
PL	posterior lateral auditory area of bushbaby
PLSS	posterior lateral suprasylvian area
Pl	paralemniscal zone
PN	pontine nuclei
Pol	rostral pole of the MGB
Pom	medial part of the posterior group
Pt	pretectum
Pu	pulvinar nucleus
Pv	posteroventral cochlear nucleus
PvO	posteroventral cochlear nucleus, octopus cell layer
Py	pyramidal tract
R	rostral
R	rostral auditory area of squirrel *or* monkey
Ra	raphe
RF	reticular formation
Rh	rhomboid nucleus
RL	rostral lateral auditory are in monkey
RN	red nucleus
RP	rostral pole nucleus of the inferior colliculus *or* MGB
Sa	nucleus sagulum
SC	superior colliculus
SCP	superior cerebellar peduncle
SCPX	decussation of the superior cerebellar peduncle
SF	suprasylvian fringe area
SF/daz	suprasylvian fringe/dorsal auditory zone
SGS	superficial gray layer of superior colliculus
SGi	intermediate gray layer of superior colliculus
SGP	deep layer of superior colliculus
Sg	suprageniculate nucleus
Sgl/Sl	suprageniculate nucleus, lateral part
Sgm/S	suprageniculate nucleus, medial part
SN	substantia nigra
SNc	substantia nigra, *pars compacta*
SNL	substantia nigra, *pars lateralis*
SNR,SNr	substantia nigra, *pars reticulata*
Spf	subparafascicular nucleus
SpN	suprapeduncular nucleus
TA	temporal auditory area of squirrel
TE1	primary auditory cortex of rat
TE2	second auditory cortex of rat
TE3	third auditory cortex of rat
Te	temporal cortex
TL	lateral nucleus of the trapezoid body
TM	medial nucleus of the trapezoid body
TRN	thalamic reticular nucleus
Tr	trochlear nerve
TV	ventral nucleus of the trapezoid body
V	*pars lateralis* of the ventral division *or* ventral *or* ventral auditory area
VA	ventroanterior auditory cortical area
Vb	ventrobasal complex
Ve	ventral auditory area
VF	ventral fringe auditory cortical area
VL	ventral nucleus of the lateral lemniscus
VLa	ventral lateral thalamic nucleus
Vl	ventrolateral nucleus of the MGB
VM	ventral medial thalamic nucleus

Vm	mesencephalic nucleus of the trigeminal
VmP	ventromedial periolivary nucleus
VP	ventral posterior auditory area
Vpl	ventral posterolateral nucleus
Vpm	ventral posteromedial nucleus
VT	ventral nucleus of the trapezoid body
wm	white matter
ZI	zona incerta
I-IV	layers of the dorsal cortex of the inferior colliculus
I-VI	layers of cerebral cortex
α	layer IVCα in primary visual cortex
β	layer IVCβ in primary visual cortex
c	layer IVc in primary visual cortex
35/36	perirhinal cortex

1 Profiling the Auditory Forebrain

Establishing rules for auditory information processing requires knowledge of the physiology of the neurons, their connections, and of how local circuits shape signals. When available, as in the cochlear nucleus (Cant and Benson 2003), such profiles underlie plausible models of receptive field (RF) genesis (Davis and Young 2000), serial information transfer (Smith et al. 1993), and feature detection (Nelken 2002). Progress in this endeavor in the medial geniculate body (MGB) and auditory cortex (AC) since 1990 is the subject of this review, and it is prerequisite to understanding how auditory thalamic (Senatorov and Hu 2002), cortical (de Ribaupierre 1997; Rouiller and Welker 2000), and subcortical sites (Winer 2006) interact. A second theme is the function of massive, focal, and precise corticocortical (Lee and Winer 2005) and corticofugal (Winer 2006) projections. The emerging picture of multiple ascending and descending pathways with intricate convergence and divergence patterns (Smith and Spirou 2002) and robust interneuronal substrates for modulation (Huang et al. 1999) is at odds with more serial models of information flow (Brandner and Redies 1990). Each section summarizes views prevailing circa 1990, then assesses subsequent studies in cat, rodents, bats, and primates. For areas with little change earlier accounts are available (Clarey et al. 1992; Winer 1992).

2 Medial Geniculate Body Organization

The MGB is part of a neuronal network extending from the cochlear nucleus to the cerebral cortex (Winer and Schreiner 2005) (Fig. 2.1). As such, it is no more independent of the midbrain or cortex than the inferior colliculus is from the cochlear nucleus. A principled analysis of MGB function must therefore integrate the architecture of its neurons, their physiological responses, the main extrinsic influences, the primary neurochemical components, and, when available, comparative functional adaptations.

2.1 Ventral Division

The number of MGB neurons relative to the IC is species specific, ranging from a 1:5 ratio in rat (Kulesza et al. 2002) to structures of more equal volume in the cat (Berman and Jones 1982). Assumptions that a nucleus (or area) is analogous or homologous require caution because of species-specific internal differences of unknown significance (Winer 1984b; Morest and Winer 1986; Winer and Larue 1996).

Many contemporary studies recognize three MGB territories: a large ventral division, which constitutes the principal part, and smaller dorsal and medial divisions; each division differs in size and internal architecture (Winer 1992) (Fig. 2.1). The ventral division can be construed as the thalamic target and representative of the disc-shaped neurons of the central nucleus of the IC, and it contains bushy tufted cells with a characteristic fibrodendritic arrangement. Bushy cells target (McMullen et al. 2005) the nonpyramidal (and other) AC neurons (Winer 1984a; Smith and Populin 2001). Bushy cells share with disc-shaped neurons a morphologically polarized dendritic axis (Oliver 2005), and the tectothalamic axon plexus terminates in ordered arrays (Wenstrup and Leroy 2001) that embody IC tonotopy (Merzenich and Reid 1974) and critical bands (Schreiner and Langner 1997), while enabling novel or enhanced representations of particular physiological features such as species-specific elements of the echolocation call (Wenstrup 1999). This is consistent with bushy cell variability within (Cetas et al. 2003) and between (Winer and Wenstrup 1994) species. Laminar regularity in many species is clearest in the low-frequency (lateral) part of the ventral division, and less so dorsomedially, where passing fibers complicate the neuropil (Morest 1965a) and affect local fibrodendritic patterns.

Ventral division neurons respond to stimulation of the brachium of the IC with single spikes and at the shortest latencies in the MGB (Hu 1995); their membrane properties and response properties contrast with those in the dorsal division (see below). One map of characteristic frequency is present in the cat ventral division, and it spans ~9 octaves (Imig and Morel 1985b); neuron types in barbiturate anesthetized animals respond with onset, onset with inhibition, offset, on–off, sustained, and complex responses, of which onset cells were most common and on–off rarest (Cetas et al. 2002). The ventral division contains a further

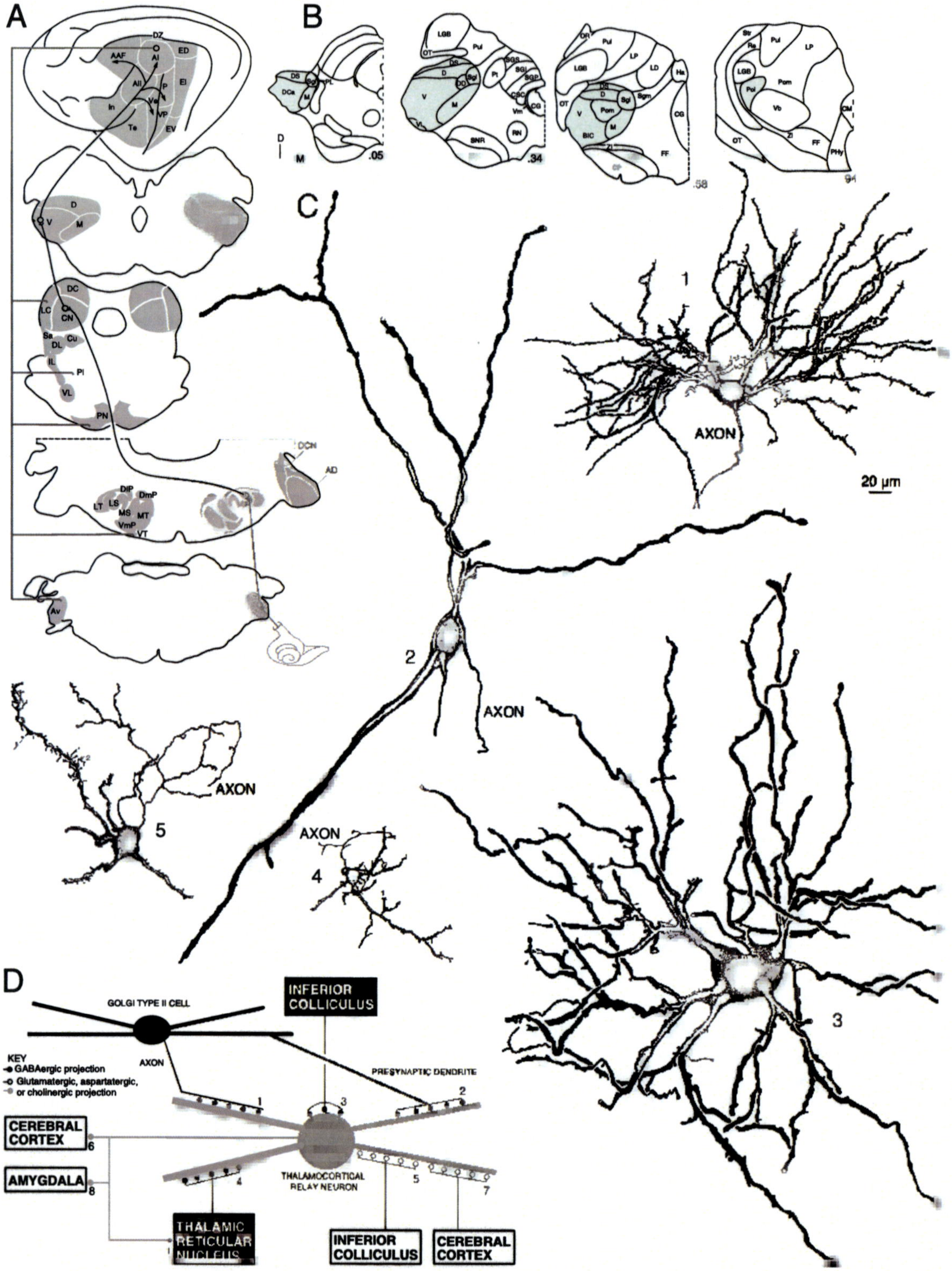

Fig. 2.1 Overview of the cat central auditory system. **a** Ascending (*right side*) and descending (*left side*) pathways. Only a small part of the ascending system, involving the interaural intensity stream passing through the lateral superior olive (LS; for abbreviations, see the list), is shown. The descending system is the largest in the medial geniculate body (MGB), has diverse targets, and reaches the cochlear nucleus (Winer 2006). **b** Medial geniculate body subdivisions drawn from Nissl preparations show that the ventral division (V) is the largest, the dorsal

caudo-rostral gradient of sharpness of tuning (Rouiller et al. 1989) whose significance is unknown. The classes of monaural and binaural responsive MGB neurons resemble those in AC (Samson et al. 2000) and there is evidence for binaural modules (Middlebrooks and Zook 1983) related to AC.

Damage to the ventral division and related thalamic regions has species-specific consequences, leaving rodent sound localization intact (Kelly and Judge 1985), while severely impairing human localization and attention (Wester et al. 2001).

2.2 Dorsal Division

The dorsal division differs from the ventral division: it has more subdivisions, greater neuronal diversity, stellate rather than bushy cells predominate, it contains at least two varieties of interneuron, the principal inputs are from IC and AC subdivisions with less regular tonotopic organization, its cells have broader tuning curves and temporally extended responses to tonal stimuli, damage to it does not affect sound localization, and the neurons are implicated in complex perceptual behaviors.

The dorsal division consists of several nuclei that comprise the caudal, dorsal, dorsomedial, and anterodorsal facets of the MGB (Winer and Morest 1983b). A cardinal feature is the neuronal diversity, including highly tufted cells among which neurons with radiate dendritic fields mingle; this disrupts any laminar pattern. Such heterogeneity extends to the suprageniculate nucleus, where only radiate neurons conspicuously larger are found, and the posterior limitans nucleus, whose major type of principal cell has long, sparsely spinous dendrites. Golgi type II cells are plentiful, with evidence for a small, and a larger, much rarer, class (Huang et al. 1999). Ultrastructural arrangements feature γ-aminobutyric acid-positive (GABAergic) axons presynaptic to immunonegative dendrites, GABAergic profiles postsynaptic to GABA-negative input, and presynaptic dendrites (Coomes et al. 2002), each reminiscent of other thalamic nuclei (Sherman and Guillery 2000).

There is physiological diversity to dorsal division function consonant with its structural complexity. Major features are a tonotopic organization favoring high frequencies (Bordi and LeDoux 1994a), wide tuning curves (Calford 1983), protracted excitatory–inhibitory oscillations (He 2003), and a bursting rather than the single-spiking firing mode (He and Hu 2002) prevalent in the ventral division (Hu 1995). The intrinsic membrane profile of rat ventral and dorsal division neurons is similar except for a depolarizing sag potential in ventral division cells, while suprathreshold excitatory responses are confined to tufted cells (Bartlett and Smith 1999). This implies continuity among MGB physiological parameters irrespective of the cell's cortical target (Winer et al. 1999b). Moreover, pairs of dorsal division neurons show the most independence of discharge in the MGB (Kvasnak et al. 2000b), suggesting that they may not encode a singular or continuous sensory domain.

2.3 Medial Division

The distinction between lemniscal and extralemniscal streams is embodied by the medial division (Morest 1965b). Auditory input is only one of several modalities that converge upon it, it has no map of characteristic frequency, nor architectonic subdivisions, and it projects to many cortical areas including auditory and non-auditory fields, and subcortically as well.

The medial division comprises the ventromedial aspect of the MGB and extends from nearly the caudal pole almost to the rostral pole. Its neurons are the most diverse in the MGB and comprise a wide range differing in size, shape, and dendritic complexity and which are embedded in heterogeneous axons (Winer and Morest 1983a). Dendritic fields radiate widely or have tufts polarized axially.

Medial division cells in anesthetized preparations respond to pure tones with extended bursts and symmetric interval histograms (Kvasnak et al. 2000a). Single cell tuning is typically broad (Calford 1983), often multisensory and convergent, and potentiated by polymodal activation (Bordi and LeDoux 1994b). Finally, they show more RF plasticity than other MGB cells (Lennartz and Weinberger 1992) and are implicated in autonomic learning paradigms using acoustic cues (McEchron et al. 1996).

Fig. 2.1 (continued) division (DS, D, etc.) has the most nuclei, and the medial division (M) is a single nucleus with several cell types (Huang et al. 1999). **c** Representative MGB neurons. *1*, A thalamocortical (TC) neuron in the ventral division, with highly polarized dendritic tufts. *2*, An elongated cell in the posterior limitans nucleus, with smooth dendrites. *3*, A suprageniculate TC neuron, with sparse appendages and a soma exceeded only by medial division magnocellular neurons. *4*, A Golgi type II cell with a local axon. *5*, A rare, much larger type II neuron. Rapid Golgi method, planapochromat, N.A. 1.32, ×2000. **d** Canonical MGB circuitry showing multiple, convergent, chemically specific inputs to a typical bushy neuron, including interneuronal contributions (Winer et al. 1996)

3 Auditory Thalamic Neurotransmitter Profile

The discovery of thalamic Golgi type II interneurons (Jacobson 1975) was a watershed in understanding sensory information sent to neocortex is modulated by local circuits (Scheibel and Scheibel 1966). Characterizing these neurons in the MGB (Morest 1971) was a vital step in studying thalamocortical (TC) relations (Sherman and Guillery 1996) and clarified the genesis and control of thalamic oscillations (Jones 2002).

3.1 Excitatory Amino Acids

A major tectothalamic component is glutamatergic axons acting on *N*-methyl-D-aspartate (NMDA) and non-NMDA receptors, in both lemniscal and non-lemniscal streams (Hu et al. 1994).

3.2 Gamma-Aminobutyric Acid

GABA is the principal compound implicated in auditory thalamic inhibition and disinhibition. Immunocytochemical analysis finds a subdivision-specific concentration of GABAergic neurons and axon terminals (puncta), with the cat ventral division having 33% such cells, the dorsal division 26%, and the medial division 18%; ventral division puncta were dense and primarily medium sized, those in the dorsal division were variable, ranging from small to giant, while medial division endings were sparser and heterogeneous (Figs. 2.2 and 2.3). The GABAergic neurons are $\sim$10 μm in diameter and correspond to neurons in Golgi preparations that have small somata and long, slender dendrites with stringy appendages (Huang et al. 1999); a second, larger and much rarer type II cell is also been recognized (Winer and Morest 1983b).

Three sources of GABA are known: the (two varieties of) intrinsic neurons, the thalamic reticular nucleus projection (Crabtree 1998), and robust feedforward projections from all inferior colliculus (IC) subdivisions (Winer et al. 1996), each contributing to MGB neuropil (Morest 1975). The specific physiological impact of each GABAergic source to function remains unclear. A clue to this complexity is the broad afferent tuning of dorsal (Aitkin and Dunlop 1968) and medial division (Aitkin 1973) neurons, processes which can hardly reflect inhibitory sharpening despite the many GABAergic neurons and extrinsic sources of GABA.

Possible parallels to intrinsic circuit functions in an analogous structure come from the lateral geniculate body, where dendrodendritic synapses between type II cells and principal neurons modulate TC transmission via metabotropic and ionotropic receptors (Cox and Sherman 2000) and glutamatergic processes (Cox et al. 1998). Analogous mechanisms may operate in the auditory thalamus (Kudoh et al. 2002). A modality-specific role in attentional control is proposed for the thalamic reticular nucleus (McAlonan et al. 2000) that could reflect dynamic shifts in its discharge behavior (Bazhenov et al. 2000). IC GABAergic projections to the MGB evoke a $GABA_A$-related IPSP/EPSP (inhibitory/excitatory postsynaptic potential) sequence followed by a $GABA_B$ IPSP, suggesting a rapid monosynaptic IC influence on TC transmission. The differential effects of such activation may reflect the several types of GABA-positive IC neurons (Oliver et al. 1994). These observations are compatible with models for bat temporal processing which incorporate parallel $GABA_A$/$GABA_B$ streams (Llano and Feng 2000).

Lateral geniculate body interneurons studied in vitro show short action potentials, can produce action potentials >500 Hz without robust adaptation of output, and exhibit a regenerative, low-threshold response extending from depolarization below threshold to multiple spikes. They are depolarized by glutamate, kainate, quisqualate, and NMDA, whereas GABA blocked action potentials and baclofen (a $GABA_B$ agonist) hyperpolarized membranes weakly and blocked spontaneous discharge. Acetylcholine hyperpolarized membrane potentials and serotonin affected a subset by enhancing spontaneous discharge, and other compounds have no effect on membrane behavior at rest or spontaneous rate (adenosine, nor adrenaline) or elicit minute, protracted depolarization (histamine). Suggested roles for these interneurons include local inhibitory influences driven by influences arising in the retina, cortex, and brain stem (Pape and McCormick 1995). In the rat ventral division muscarinic agonists induce extended membrane depolarization that block burst responses (Mooney et al. 1995) in a nucleus with few Golgi type II cells (Winer and Larue 1988).

4 Medial Geniculate Body Connections

At least five connectional roles for the MGB can be identified. First, it is the target of IC inputs via convergence and divergence of chemically specific inputs, which it then modifies with intrinsic networks. A second role is redistribution of information to cortical and subcortical targets. Third, brain stem extralemniscal input creates parallel streams. Fourth, the corticothalamic pathways shape MGB representations and may modify ongoing processing. Finally, a small thalamotectal system gives MGB neurons direct access to the IC. These several pathways suggest a view of thalamic

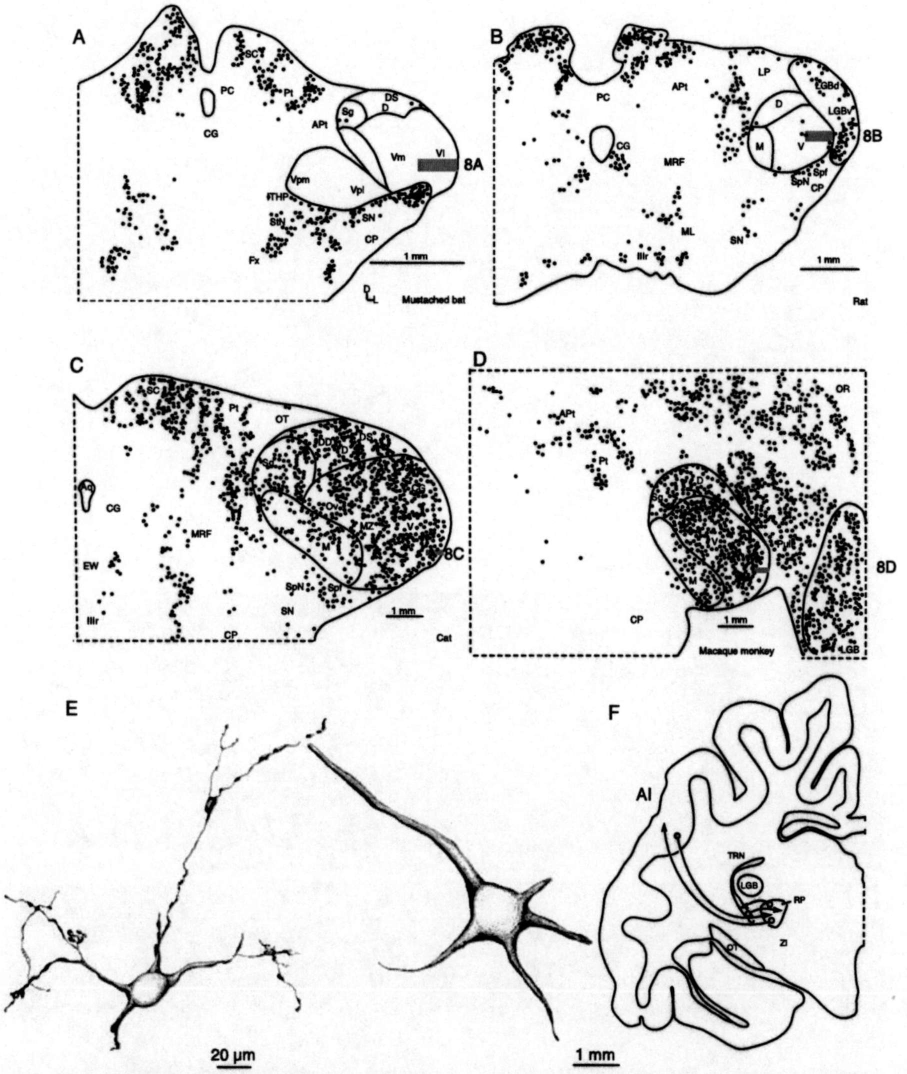

Fig. 2.2 γ-aminobutyric acid-containing (GABAergic) cells in the MGB. **a** In the mustached bat, the few GABAergic neurons present concentrate in the dorsal division, though there are abundant puncta. **b** In the rat, MGB GABAergic neurons are sparse, ~1% of the total, and relatively fewer than in the LGBd. **c** Thalamic GABAergic cells are abundant in the cat, representing ~25% of MGB cells. **d** In the rhesus, the proportion of GABAergic cells is at least equal to that in the cat **c**. **e** Representative GABAergic neurons, showing the more numerous small variety (*left*) corresponding to the classic Golgi type II cell (Fig. 2.1c:*4*). The second, much larger type (*right*) is far less common and its somatodendritic profile corresponds to the equally rare large Golgi type II cell (Fig. 2.1c:*5*). **f** Possible interconnections between the MGB and thalamic reticular nucleus, and TC and corticothalamic circuits. The chemical sign of the specific connections is shown in Fig. 2.1d; TC and corticofugal connections are thought to be excitatory. Plots made from sections 30 μm thick and immunostained with antibodies to glutamic acid decarboxylase (Winer and Wenstrup 1994)

processing incompatible with the concept of the MGB as a relay nucleus.

4.1 Tectothalamic Pathway

This principal ascending input to the MGB is massive, arising from all IC subdivisions and reaching all MGB subdivisions. Projections are origin and target specific, and there is not a direct relation between a particular IC and MGB division that does not involve other subdivisions of each. Moreover, the tectothalamic projections are bilateral and asymmetrical in density. There are also parallel, chemically specific IC inputs (Wenstrup 2005).

The lemniscal pathway is represented by a massive topographic input from the IC central nucleus that reaches all tonotopic (laminated) parts of the MGB (Calford and Aitkin 1983), where the axons form sheets contributing to restricted

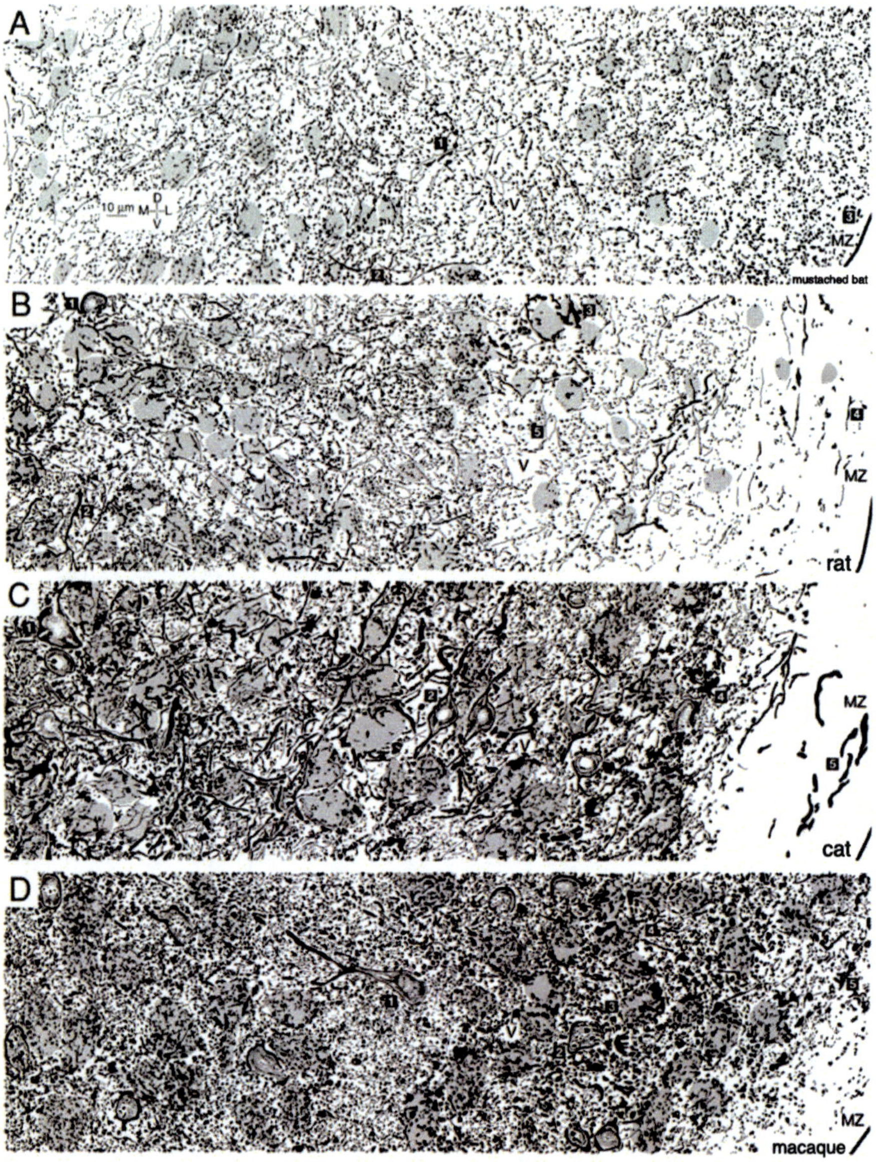

Fig. 2.3 GABAergic axon terminals *fine dots* immunostained neurons (*dark stippled* profiles), and immunonegative MGB neurons in the ventral division in four species; cf. Fig. 2.2 *insets* for loci. Planapochromat, N.A. 1.32, ×2000. (**a**) The mustached bat ventral division is virtually devoid of GABAergic neurons and contains a fine plexus of puncta. *1*, Beaded axons. *2*, Axons projecting laterally. *3*, The bat marginal zone (MZ) has more puncta than do the other species (**b–d**). **b** The rat has a few GABAergic neurons (*1*), some thick, possibly ascending preterminal fibers (*2*), some coarse terminals (*3*), just a few MZ endings, and many extremely fine (~0.5 µm in diameter) fibers (*5*). **c** There is a marked increase in neuropil density in the cat, with an unusually large somatodendritic profile (*1*) possibly corresponding to a large type II cell (Figs. 2.1c:*5* and 2.2e, *right*). Many GABAergic neurons lie parallel to fibrodendritic laminae (*2*), some thick axons are present (*3*) as well as giant ones (*4*), and the marginal zone has much thicker preterminal fibers (*5*) than does the rat (**b**). **d** The rhesus ventral division has an even denser GABAergic neuropil organization than the cat c. Some GABAergic cells have immunostaining to their secondary dendrites (*1*), and puncta range from granular to far larger and clustered (*2*), often virtually engulfing GABA-negative somata (*3*), and with complex terminal axosomatic architectures (*4*)

synaptic arrays (McMullen et al. 2005), and other regions (medial division) whose representation of characteristic frequency is far less ordered (Rouiller et al. 1989). Projections from IC subdivisions (lateral cortex and dorsal cortex) with non-auditory affiliations (Syka et al. 2000) and from lateral tegmental regions (Morest 1965b) reach other MGB subdivisions beside those with a tonotopic representation (Fig. 2.4).

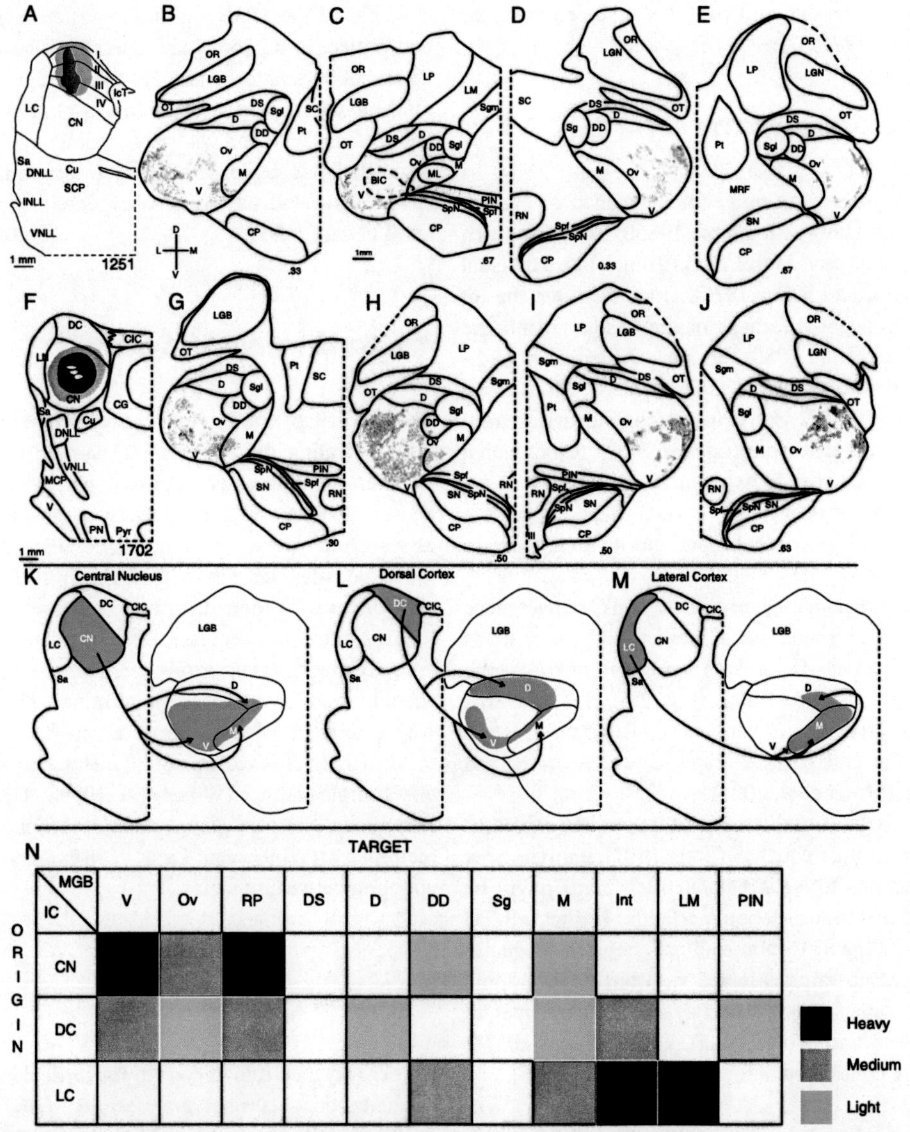

Fig. 2.4 Some inferior colliculus (**IC**) inputs to the cat **MGB. a–e** A biotinylated dextran amines (**BDA**) deposit in the IC dorsal cortex (**I–IV**) labels a broad, ipsilateral swath of the ventral division frequency representation in the **MGB** (Imig and Morel 1985b) (**V, Ov**) and a smaller, less continuous, and overlapping crossed projection; note the ipsilateral intralaminar (Spf, **PIN**) input. **f–j** A **BDA** injection in the central nucleus (CN), extending to its high frequency border (Merzenich and Reid 1974), labels nearly the entire ventral division (**h**) exclusively, with appreciable contralateral input (**j**). **k–m** Schematics of tectothalamic input showing that each IC targets more than one MGB subdivision and that an **MGB** target receives convergent input from more than one **IC** source. **n** A schematic of cat tectothalamic anterograde (present results) and retrograde (Calford and Aitkin 1983) results, showing extensive convergence and divergence suggesting that the tectothalamic transformation is as individuated as the TC system (Winer et al. 2005)

Input from the IC targets principal thalamocortical neurons as well as Golgi type II cells (Morest 1975), a convergence that may well synchronize their temporal discharge behavior for the propagation of thalamic information to the neocortex. The type II neuron's axon and the dendrites are presynaptic to the thalamocortical cells and might gate transmission as in the visual system (Cox and Sherman 2000). The cholinergic system (Caballero-Bleda et al. 1991) is associated with RF plasticity (Edeline and Weinberger 1992) and the central adjustments ensuing from peripheral trauma (Kamke et al. 2003).

4.2 Extracollicular Projections

A monosynaptic projection from the small cell cap of the dorsal cochlear nucleus targets the MGB medial division (Malmierca et al. 2002). It may have a role in polymodal and visuomotor processing and was seen in the chimpanzee (Strominger et al. 1977).

4.3 Thalamocortical System

The MGB may contain a single, complete map of characteristic frequency (Imig and Morel 1985b) or a few such representations (Imig and Morel 1985a), much like the visual thalamus (Malpeli and Baker 1975). How, then, are the (at least) five independent AC maps of the basilar membrane created (Reale and Imig 1980)? In areas AI (primary auditory cortex) and AAF (anterior auditory field), MGB input to the corresponding AC isofrequency contour arises from nearby thalamocortical neurons (Edeline 2003), few of which project to both fields even in experiments designed to maximize this possibility (Lee et al. 2004a). This suggests a conservation of thalamocortical input and its divergence to many areal targets.

As in the tectothalamic system, there is TC convergence and divergence, with single nuclei projecting to many fields and each field receiving input from more than one thalamic nucleus (Huang and Winer 2000) (Fig. 2.5). Such patterns likely contribute to the conservation of thalamic information in AC and to the emergence of new RF architectures and representations (Miller et al. 2001).

Few of the many AC fields have a topographic representation of characteristic frequency (Ehret 1997), implying that thalamic projections to areas without such maps might be less ordered than those to tonotopic fields. In fact, all TC projections to AC (and to non-auditory areas) are equally ordered and specific when assessed with three topographic metrics. Such topographic rules may have developmental implications (Lee and Winer 2005). Other aspects of TC organization are considered below (Section 6.1).

4.4 Thalamoamygdaloid System

Neurons from the MGB dorsal and medial divisions project to the lateral amygdaloid nucleus and nearby polar temporal cortex (Shinonaga et al. 1994). This permits of thalamofugal and corticofugal convergence in the lateral amygdala and confirms the different amygdaloid targets of specific AC subdivisions (Romanski and LeDoux 1993).

4.5 Thalamotectal System

This pathway permits thalamofugal access to midbrain targets. Neurons dispersed widely in the dorsal and medial divisions of the MGB and in the posterior intralaminar system project to IC dorsal and lateral cortices (Kuwabara and Zook 2000; Senatorov and Hu 2002), regions implicated in multimodal convergence (Aitkin et al. 1994) and attention (Jane et al. 1965) and whose ascending projections reach the same thalamic regions that target non-primary AC (Winer et al. 2001) and the corticoamygdaloid stream (Romanski and LeDoux 1993).

4.6 Corticothalamic System

This is perhaps the largest cortical projection (Winer et al. 2001), rivaling the corticospinal system, and itself part of an even more massive corticofugal system that reaches nearly all levels of the auditory system (Winer 2006). It arises from every AC subdivision and follows many of the same rules that govern thalamocortical connectivity: an area projects to more than one thalamic target, and a thalamic subdivision receives input from more than one AC area. Corticothalamic axons are, likewise, origin and target specific, and comprise giant terminals (Bajo et al. 1995) which might enable thalamic access to cortex (Diamond et al. 1992) and which could interact with equally large GABAergic endings (Winer et al. 1999a). Despite significant exceptions, there is global thalamocortical-corticothalamic reciprocity (Winer and Larue 1987) suggesting powerful coupling between systems.

5 Auditory Cortex

Cortical layers are analogous to subcortical nuclei: each layer has a unique neuronal architecture, individuated connections, a specific neurochemical organization, and a particular functional arrangement. Given the many auditory areas in different species—three in rat (Shi and Cassell 1997), six in gerbil (Thomas and Lopez 2003), thirteen in cat (Lee and Winer 2005), and twelve in monkey (Hackett et al. 1998)—the prospective complexity of forebrain connectional relations is impressive (Fig. 2.6). Amplified as these relations are by sublaminar organization, neurotransmitter receptor diversity and synaptic plasticity, the task of dissecting AC functionally is formidable. It leaves open the question of species differences.

5.1 Supragranular Layers

Layer I has few neurons, >90% are GABAergic, and these are primarily in layer Ib and unexpectedly diverse

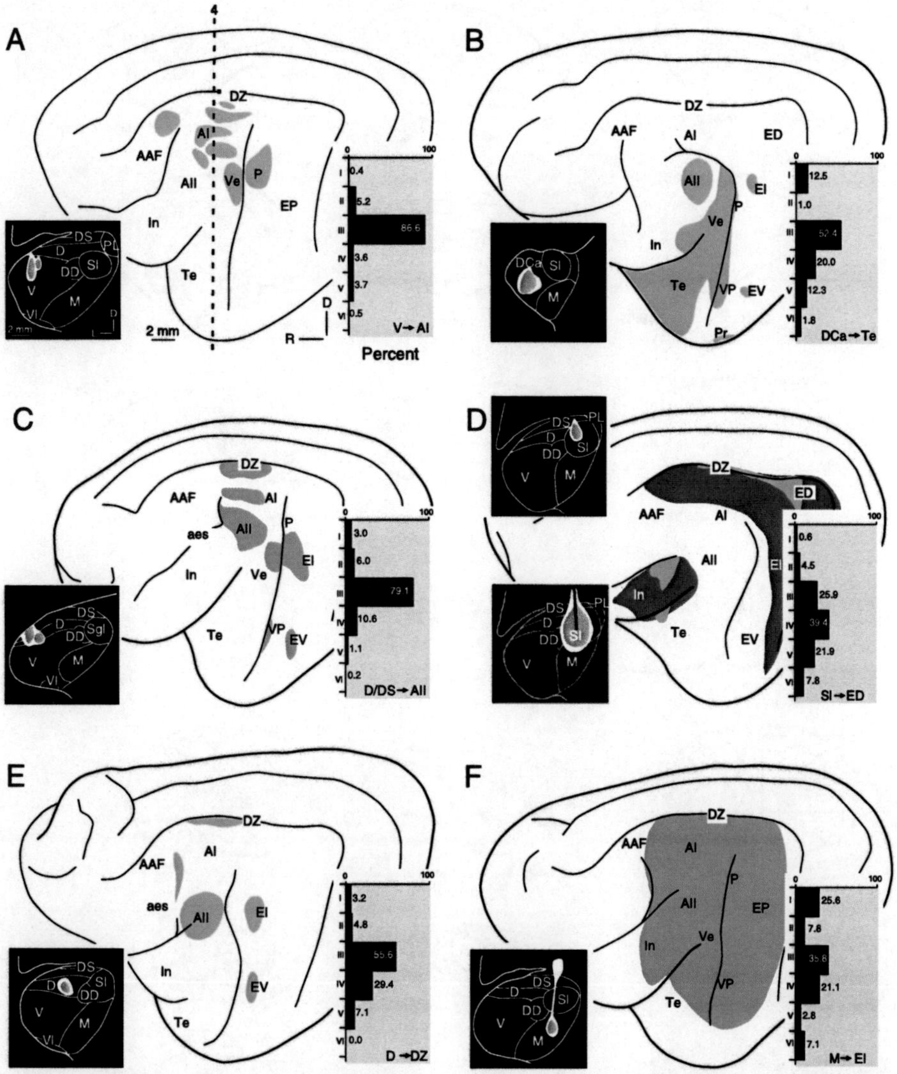

Fig. 2.5 Areal and laminar distribution of TC projections with BDA. **a** A deposit in the central part of the ventral division (*black inset*; V) labels multiple patches in AI and has a more continuous distribution in three other primary fields. *Right inset* on this and subsequent panels: distribution of boutons (cf. Fig. 2.9b) by layers, with >80% in layer III. **b** The caudal dorsal nucleus (DCa) targets primary (Ve) non-primary (EV) and limbic-related (Te) areas, and has a much more variable laminar pattern that the ventral division to AI projection (**a**). **c** The dorsal/dorsal superficial nucleus has as precise, and a similar, laminar organization as the ventral division to AI projection (**a**), but terminates only in non-primary, extralimbic fields. **d** Two experiments with deposits in the lateral part of the suprageniculate nucleus (*black inset*; SI) showing the consistency of TC labeling, its specificity (e.g., in area In but not Te, though both are limbic-related) and unique laminar distribution, with the heaviest input to layer IV in this small series, and significant labeling in 5/6 layers. **e** A deposit in the medial part of the dorsal nucleus (*black inset*; **d**) had a similar labeling pattern as in the lateral dorsal nucleus (**c**) except for AAF involvement, and a more equal layer III/IV distribution. **f** The medial division (*black inset*; M) has the broadest areal suite of projections, and a diverse laminar profile involving all layers in area EI. Modified from prior work (Huang and Winer 2000)

(Winer and Larue 1989). Layer Ia contains the apical dendrites of deep-lying pyramidal cells (Sousa-Pinto 1973) and the largest thalamocortical axons (Huang and Winer 1997).

Layer II is comparatively cell rich, has many pyramidal cells and some unique neurons (Winer 1985), and projects chiefly in the corticocortical system (Fig. 2.7).

5.2 Granular Layers

Layer III is eclipsed in size only by layer V, and pyramidal cells dominate layer IIIa (Winer 1984d) and non-pyramidal cells layer IIIb (Winer 1984c), where MGB input is near-maximal (Hashikawa et al. 1995; Huang and Winer 2000; Linke and Schwegler 2000; Kimura et al. 2003). It contains

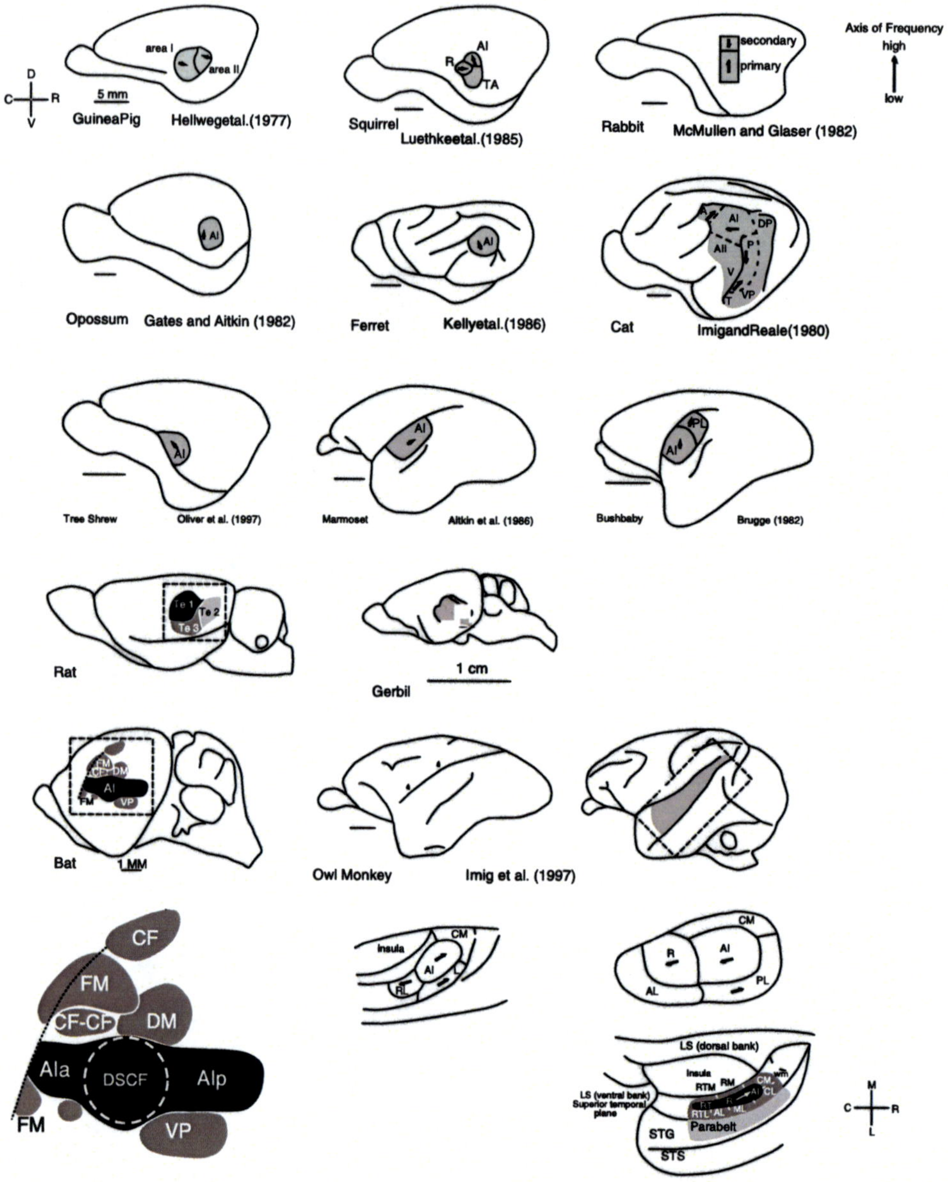

Fig. 2.6 Comparative physiological organization of mammalian AC. Arrows indicate the orientation of the tonotopic organization axis (increase of CF_s)

corticocortical feedforward (Thomas and Lopez 2003) and commissural projection cells and axons (Imig and Brugge 1978).

The major features of layer IV are its non-pyramidal neuronal population (Winer 1984a) (Fig. 2.8) which receives a dense MGB input (Smith and Populin 2001) (Fig. 2.9), and whose primary projection is local (Mitani et al. 1985).

5.3 Infragranular Layers

Layer V is the thickest AI layer, with three sublaminae: Va and Vc contain corticothalamic neurons, layer Vb has corticocollicular projection cells (Winer and Prieto 2001). The neurons are diverse and the proportion of GABAergic cells is among the lowest in AC (Prieto et al. 1994b).

Layers VI and I are the only AC layers with horizontal cells (Radnikow et al. 2002), layer VI has the lowest proportion of GABAergic cells, and its chief subcortical target is the MGB (Prieto and Winer 1999; Winer et al. 2001).

6 Auditory Cortex Connectivity

The partition of extrinsic connections to AC is approximately 15% from the thalamus, 15% from the contralateral

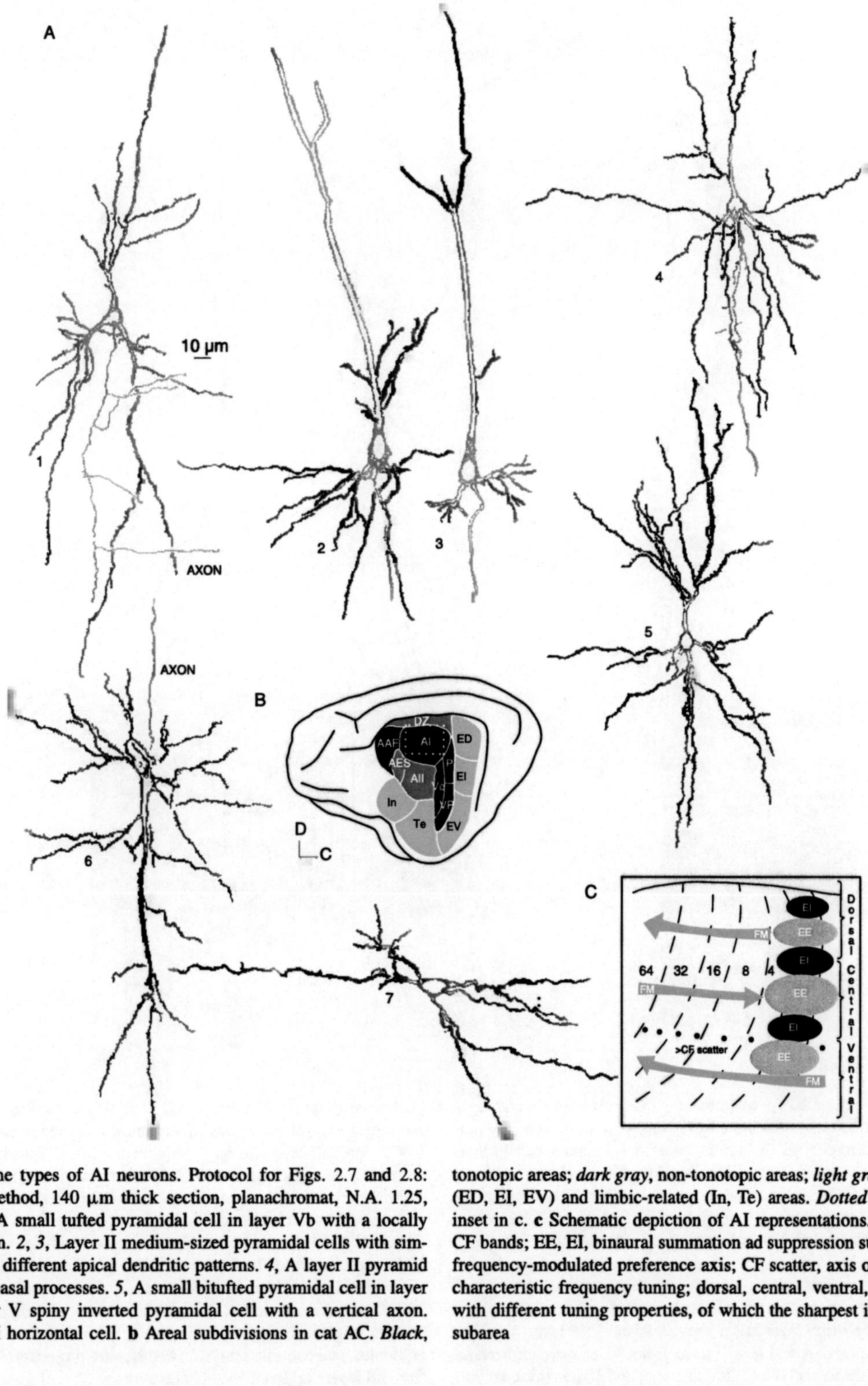

Fig. 2.7 Some types of AI neurons. Protocol for Figs. 2.7 and 2.8: Golgi-Cox method, 140 μm thick section, planachromat, N.A. 1.25, ×1000. **a** *1*, A small tufted pyramidal cell in layer Vb with a locally branched axon. *2*, *3*, Layer II medium-sized pyramidal cells with similar basal and different apical dendritic patterns. *4*, A layer II pyramid with stellate basal processes. *5*, A small bitufted pyramidal cell in layer V. *6*, A layer V spiny inverted pyramidal cell with a vertical axon. *7*, A layer VI horizontal cell. **b** Areal subdivisions in cat AC. *Black*, tonotopic areas; *dark gray*, non-tonotopic areas; *light gray*, multimodal (ED, EI, EV) and limbic-related (In, Te) areas. *Dotted lines*, locus of inset in c. **c** Schematic depiction of AI representations. *Dashed lines*, CF bands; EE, EI, binaural summation ad suppression subregions; FM, frequency-modulated preference axis; CF scatter, axis of dispersion of characteristic frequency tuning; dorsal, central, ventral, AI subregions with different tuning properties, of which the sharpest is in the central subarea

hemisphere, and 70% from ipsilateral cortex (Lee et al. 2004a; Lee et al. 2004b), and each source has a similar topography of precision irrespective of the degree of tonotopic representation in origin or target (Lee and Winer 2005). The corticofugal pathways represent, with the corticocortical projections, the largest projection system (Winer

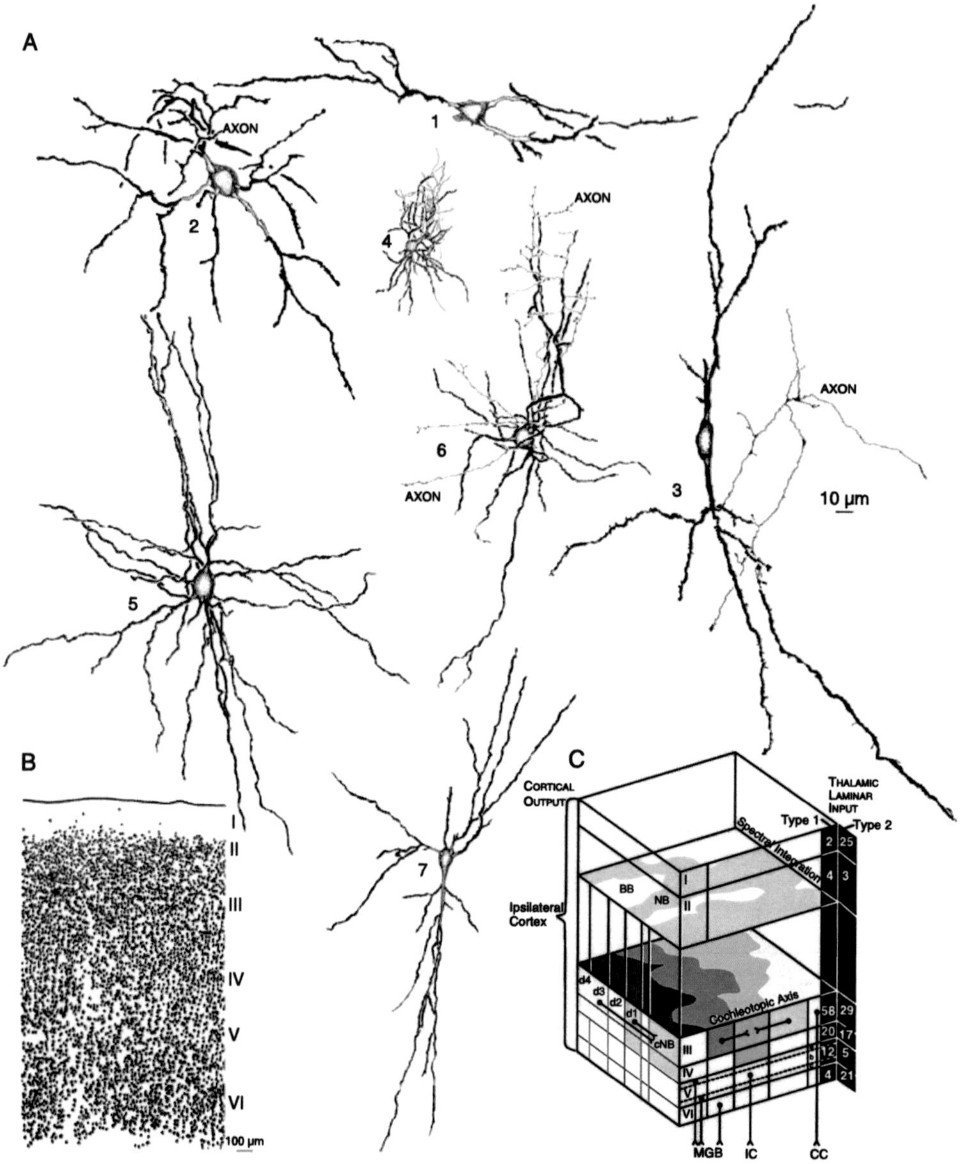

Fig. 2.8 Non-pyramidal AI neurons and AC architectonic and laminar organization. Most, if not all, of these types are GABAergic (Prieto et al. 1994b). Protocol as in Fig. 2.7. **a** *1*, A horizontal cell in layer Ib. 2, A sparsely spinous layer I multipolar cell. *3*, A layer II bipolar neuron with a local branched axon. *4*, A layer III neurogliaform cell whose delicate processes ramify locally. *5*, A layer II large multipolar cell with smooth vertical and lateral dendritic fascicles. *6*, A layer IV basket cell with aspiny dendrites and a branched local axon. *7*, A layer V smooth inverted pyramidal cell. **b** AI cytoarchitecture in a Nissl preparation showing a thick layer I, a small cell-dominated layer II, pyramidal cell-rich layer III, a slender layer IV, a layer V with a wide range of neurons, and a cell-dense layer VI in which horizontal cells dominate the lower half. Celloidin embedded 30 µm thick section, planapochromat, N.A. 0.65, ×500. **c** A schematic view of a cube of AI integrating some features of structural and functional organization. I–VI, cortical layers. ***Light gray shapes***, the spectral integration domains, which include broad- (BB) and narrowband (NB) subregions. cNB, the central narrowband domain; d1–d4, flanking narrowband subregions whose corticocortical projections converge onto the cNB region (Read et al. 2001). Gray-white contours, the cochleotopic axis. Types 1, 2, percentages of MGB boutons in specific thalamic layers after anterograde tracer deposits in the auditory thalamus (Huang and Winer 2000) show three patterns (cf. Fig. 2.11). MGB, IC, and CC refer to the laminar origins of input to, respectively, the medial geniculate, inferior colliculus, and contralateral AC. Feedforward projections (Output) arise from all layers (Unpublished observations)

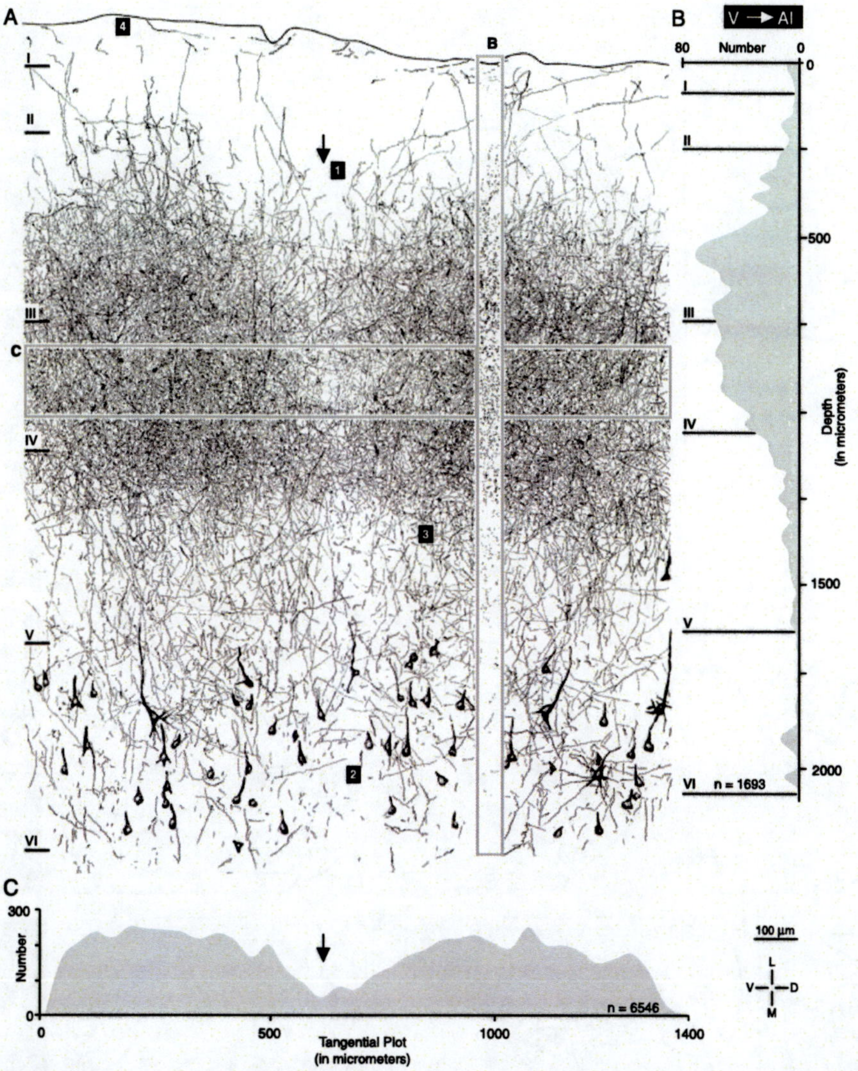

Fig. 2.9 TCC and corticothalamic labeling in AI after a deposit in the ventral division of the MGB (cf. Fig. 2.5a). **a** The TC axonal plexus forms clusters separated by bouton-sparse regions (*arrow*; *1*). Some lateral, preterminal fibers are present (2), and the terminal plexus is heaviest at the base of layer IV; *some thick, horizontal fibers* (*4*) are present in layer I. **b** A *vertical traverse* (*gray lines* in **a** denote the locus) in which only boutons were counted shows a peak at the layer IIIa-b border. The corticothalamic neurons are unlikely to have contributed to the bouton profile since their axons were rarely filled. **c** A horizontal traverse confirms the bouton clusters (*arrow*) and denotes a relatively constant lateral plexus. Modified from prior work (Huang and Winer 2000)

2006). Knowledge of these connectivities is approximately the inverse of their size.

6.1 Thalamic Areal and Laminar Input

All MGB divisions project to AC, and the TC pathway is both nucleus and area specific, with a representative subdivision projecting to half of the fourteen areas identified as auditory (Huang and Winer 2000) (Figs. 2.10 and 2.11). Ventral division input to primary auditory cortex (AI) is clustered in the cat, rabbit (McMullen and de Venecia 1993), and monkey (Hashikawa et al. 1995), with ~85% of the TC boutons in layers III–IV, and lesser involvement of many other layers; layer II receives only minute input and is thus remote from thalamic influence except via polysynaptic intracortical processes (Fig. 2.9). The dorsal division also concentrates its input to the granular layers, but has more laminar dispersion than the ventral division; the medial division pattern involves layers I and VI in some areas and the granular layers in others. TC axons are diverse and

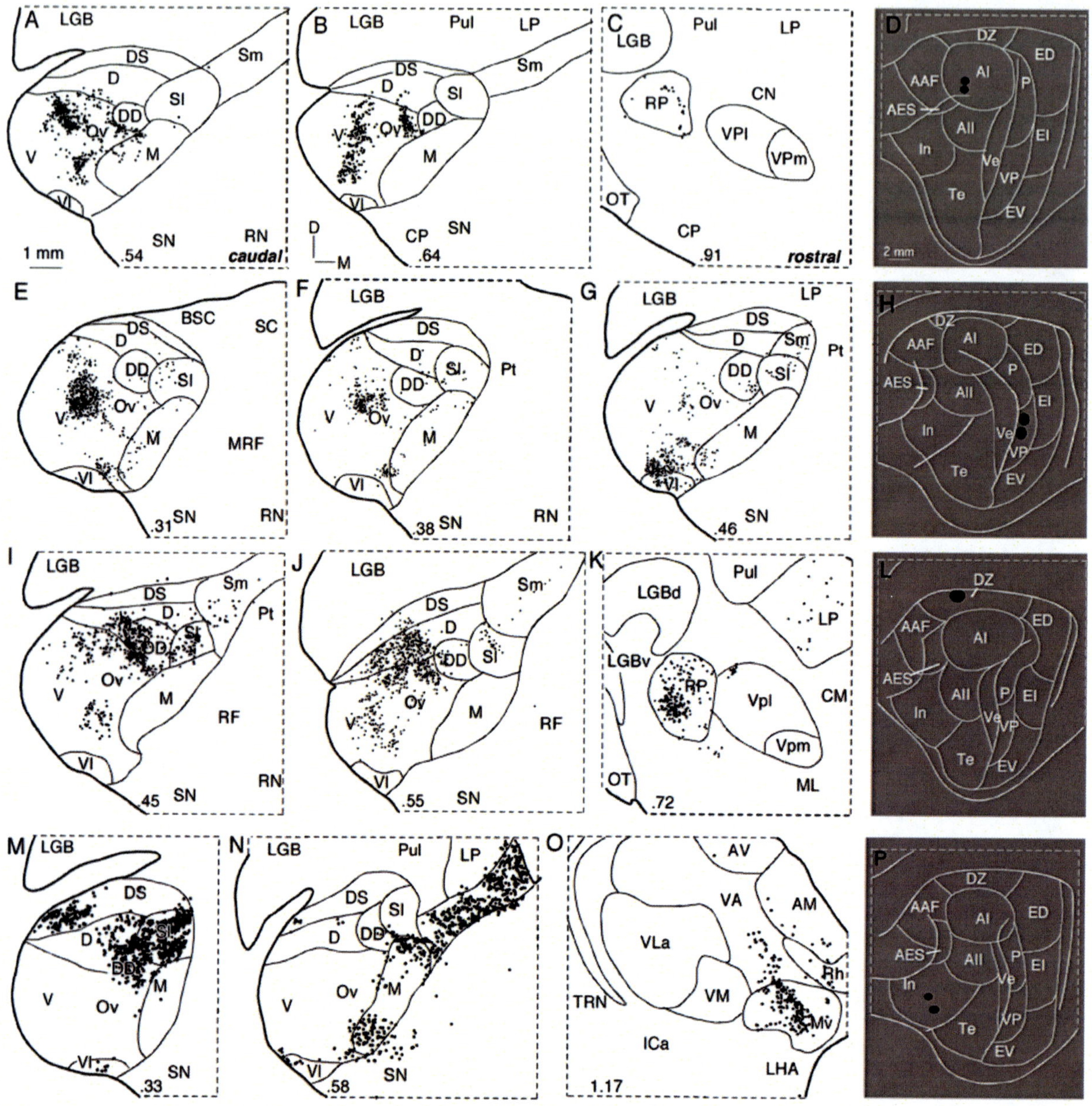

Fig. 2.10 TC input to four AC areas (D,H,L,P) demonstrated with cholera toxin β fragment, and each showing that more than one thalamic nucleus converges to an area and that the projection is topographic. **a–d** Deposits in mid-to-high frequency AI (D) labeled few cells outside the ventral division ad rostral pole. **e–h** Injections in area VP label ventral division neurons and suprageniculate complex (Sl, Sm) cells. **i–l** Tracer in the dorsal auditory zone (DAZ) labels mainly dorsal division (DD, Sg) and rostral pole cells, and lateral posterior nucleus (LP) cells. **m–p** Input to In cortex is from nuclei almost wholly outside the lemniscal pathway, yet equally focal and topographic (see **a**)

layer-specific in form, with pronounced lateral plexuses in layers I, IV, and VI, and surprisingly massive fibers in layer I (Huang and Winer 2000), a feature consistent with a projection (Mitani et al. 1984) from magnocellular medial division neurons (Winer and Morest 1983a). Posterior thalamic (suprageniculate, posterior intralaminar, and peripeduncular nucleus) projections to rat AC also target layer Ia, and the lower part of layer III and layer IV in ectorhinal and visceral areas (Linke and Schwegler 2000). Perhaps layer I intralaminar influences modulate the corticofugal system (Vogt 1991; Cauller 1995) via the distal dendrites of projection neurons (Prieto and Winer 1999; Winer and Prieto 2001). Layer II/III cells have a lower discharge threshold than layer V neurons, permitting the latter neurons

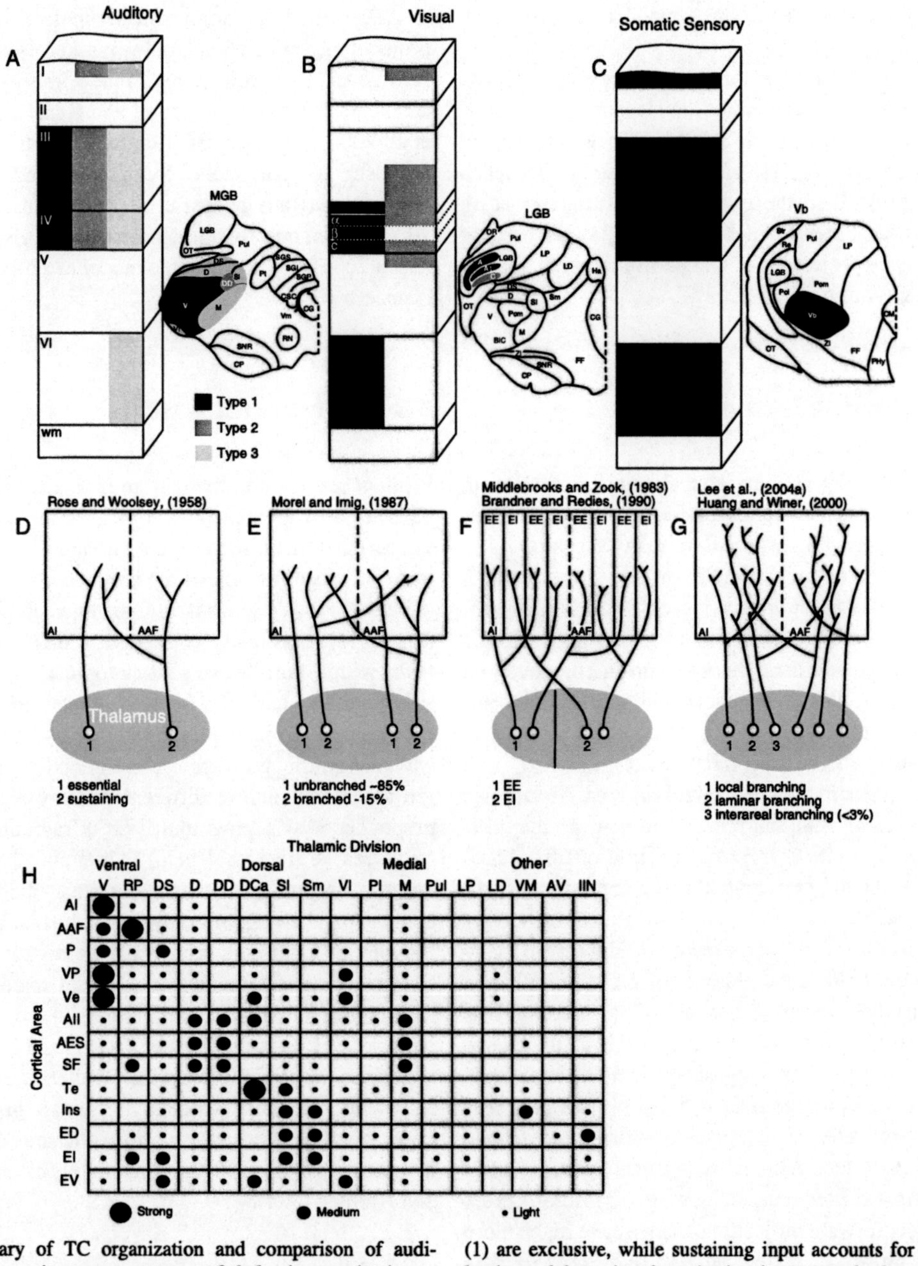

Fig. 2.11 Summary of TC organization and comparison of auditory, visual, and somatic sensory patterns of thalamic organization. **a** Auditory TC input has three arrangements related to the functional affiliations of the parallel streams. Lemniscal input (*black*) targets layers III–IV, dorsal division lemniscal-adjunct (Winer and Morest 1983a) neurons target layers III, IV, and I, and projections from polysensory MGB subdivisions end in all but layer V, many of whose cells receive thalamic input to their distal dendrites (Mitani et al. 1984). **b** In contrast, lateral geniculate projections have only two patterns: those from the a layers end in layers IV and VI and those from the c layer overlap in the *upper* and *lower parts* of layer IV with those from the a layers, but occupy parts of layers I, III, and the *upper part* of layer V (Humphrey et al. 1985). **c** Thalamic input to somatic sensory cortex terminates in all layers except Ib, II, and V (Landry and Deschênes 1981). **d–g** Four models of TC connectivity. **d** Essential projections (1) are exclusive, while sustaining input accounts for why some thalamic nuclei survive decortication in retrograde degeneration studies (Rose and Woolsey 1958). **e** Most TC axons end in one area (1), and some (2) terminate in more than one (Morel and Imig 1987). **f** Binaural modules in AI receive TC input from MGB neurons with like aural features relating to suppression (EI) or summation (EE) (Middlebrooks and Zook 1983), and TC input is point-to-point (Brandner and Redies 1990). **g** Studies with CTβ and CTβG (CTβ conjugated to gold) and BDA show that few (<2%) of TC cells have branched projections, that almost all layers receive such input, and that the channels to areas AI and AAF are almost entirely parallel (Huang and Winer 2000; Lee et al. 2004b). **h** Families of auditory TC relations, with lemniscal input the heaviest and most restricted, lemniscal-adjunct lighter and more variable, and multimodal associative projections yet lighter and diverging more

to restrict intracortical processing spatially (Atzori et al. 2004).

Some MGB subdivisions project to area outside AC (Fig. 2.11). The suprageniculate nucleus of the dorsal division (Winer and Morest 1983b), whose neurons respond to acoustic, somatic sensory, and visual stimuli (Benedek et al. 1997), projects to the frontal lobe (Kurokawa et al. 1990) and insular cortex (Winer et al. 1977); other parts of the dorsal division project to AC and perirhinal areas, mainly in layers III/IV (Kimura et al. 2003).

6.2 Corticocortical System

The arrival of parallel streams of thalamic information in AC triggers an immense series of feedforward and feedback intracortical systems which have many essential roles. Among these is the coordination of intrinsic processes with layers and modules in an area; a second facet is the propagation of corticocortical information to areas higher in the presumed sequence and their participation in emergent and parallel processing; finally, these local and remote computations must each converge upon the corticofugal systems to affect subsequent subcortical computations.

Studies with sensitive tracers reveal an area AI intrinsic convergence system with sharpness of tuning as the critical metric, and modules segregated within an isofrequency contour and projecting preferentially to a central, narrowly tuned band (Read et al. 2001; Read et al. 2002). Larger deposits find massive convergent feedback input from many areas, e.g., some fields are the target of all other AC area (Fig. 2.12). The three largest projections often contribute less than half the total input. Tonotopic fields tend to receive input from tonotopic fields, non-tonotopic fields from non-tonotopic fields, and limbic-related fields project preferentially to limbic-related areas; there are significant exceptions, e.g., non-tonotopic area AII (Schreiner and Cynader 1984) projects to limbic-related temporal, but not to insular, cortex (Ch. 7). The feedforward projections likewise are much more elaborate than those seen in degeneration studies (Kawamura 1973) and suggest principles analogous to those noted for the feedback relations. Thus, an area projects to at least three, and up to ten, other fields; feedforward projections have functional affiliative preferences, e.g., tonotopic to tonotopic, etc., but with exceptions; non-primary areas have more divergent, and functionally heterogeneous, targets than tonotopic fields; the projections of adjacent tonotopic fields can be almost entirely independent; and the intricacy and selectivity of the corticocortical projections is consonant with the existence of many, as opposed to a few, AC areas (Fig. 2.13).

In macaque AC, where the areal equivalence the cat includes several tonotopic fields but no obvious equivalent to AII, there is a comparable richness in the feedforward projections, and an equally marked tendency for the "crossover" of projections from primary to non-primary areas and the converse (Morel and Kaas 1992; de la Mothe et al. 2006). One model for these patterns in primates envisages streams for sound localization and auditory object recognition (Rauschecker et al. 1997; Rauschecker and Tian 2000) analogous to those proposed in visual cortex (Van Essen 2005). Whether such a model extends to other species is uncertain.

6.3 Commissural System

Without exception, the cat interhemispheric connections have an area to area homotopy that sets them apart from the corticocortical system; the strongest single commissural projection is a fraction of the corticocortical input and there are always fewer commissural than ipsilateral inputs (Ch. 7) (Fig. 2.14). Secondary commissural input to an area arises from within families (e.g., tonotopic to tonotopic) or outside (auditory and visual association to limbic areas). The commissural projection originates in layers III and V almost exclusively, though there is no relation between laminar origin and functionality: different primary areas may have a preponderantly supragranular or infragranular origin, and only area Te (temporal field) is bilaminar. In macaque, the commissural projections to non-primary belt cortex arise in belt and parabelt regions only (Hackett et al. 1999), divorcing the commissural core from the belt except via polysynaptic corticocortical pathways. Callosal axonal terminations are modular in AI with considerable anterograde–retrograde reciprocity (Code and Winer 1986), and projection bands vary from 200 to 800 μm wide (Wallace and Harper 1997). Magnetic resonance studies of human primary AC find a symmetrical tonotopic map (Formisano et al. 2003). In macaques damage reveals a functionally asymmetry (Heffner and Heffner 1989).

6.4 Corticothalamic System

This is largest of the corticofugal system and among the largest in the brain, with each AC locus projecting to more thalamic venues than project to it (Winer and Larue 1987; Deschênes et al. 1998). In AI it arises from varieties of layer Va, Vc, and VI pyramidal cells (Winer 1992). The projection is topographic irrespective of the tonotopic status of its AC origin or the MGB target (Winer et al. 2001), suggesting an intrinsic order independent of physiologic topography and analogous to other forebrain connectional systems (Lee and

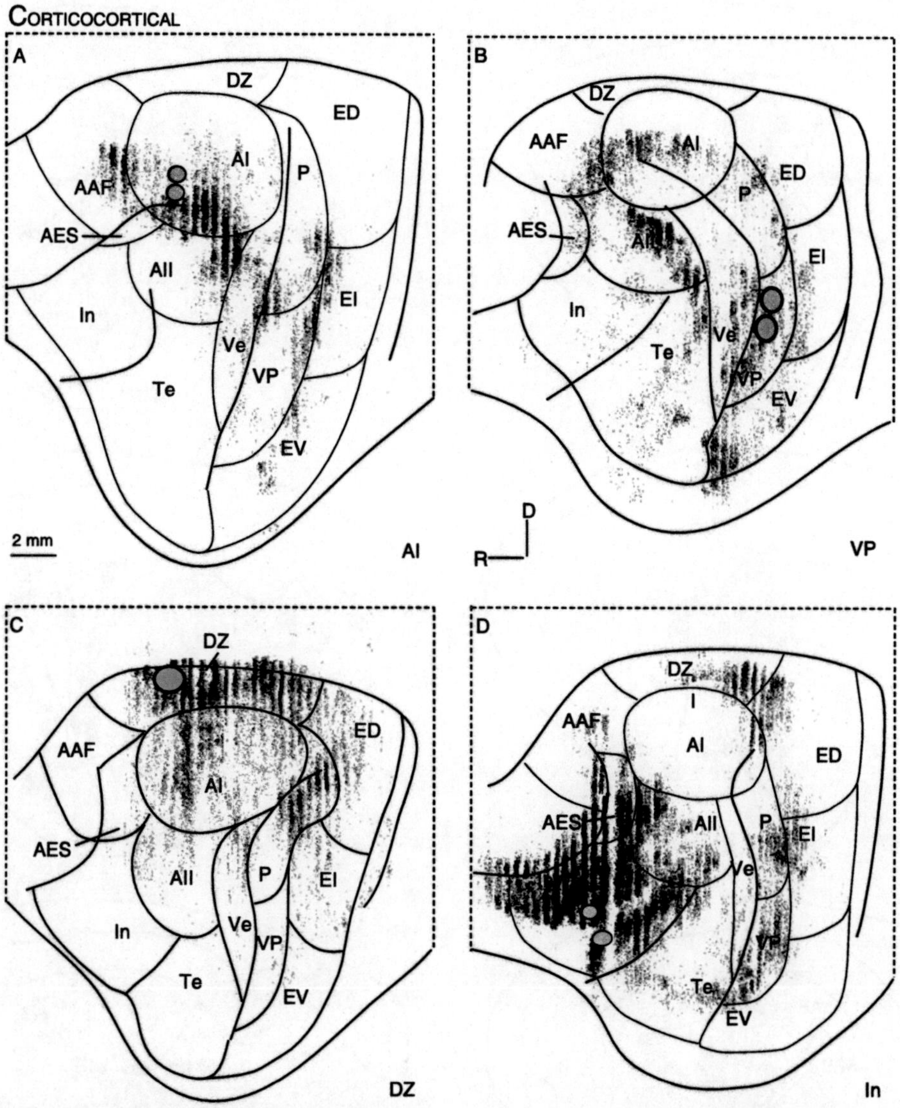

Fig. 2.12 Corticocortical cells of origin revealed with CTβ. The vertical banding of labeling is an artifact of collapsing three dimensions onto two. **a** Convergent input to AI arises chiefly from other fields with a tonotopic map (areas AAF, P, VP, Ve), and weak input from the posterior ectosylvian gyrus (EI) and a limbic-related area Te. **b** VP deposits involve label all AC areas significantly, including both limbic fields (Te, In). **c** Projections to the dorsal auditory zone (DZ) are heaviest from area P, and involve vast expanses of auditory and adjacent suprasylvian territories. **d** Insular cortex is remarkable for having no AI or area P projection, and massive AII and ED projections, attesting to non-tonotopic and multisensory and affiliations

Winer 2005) (Fig. 2.15). The five tonotopic fields in the cat each have fewer MGB targets than non-tonotopic and limbic-related areas, which average three times as many, and these are often discontinuous and widely distributed. Perhaps the corticothalamic system contains parallel streams to specific MGB subdivisions. This is further supported by the existence of small and large corticofugal boutons in cat (Bajo et al. 1995), and monkey (Rouiller and Durif 2004) and other species (Rouiller and Welker 2000); the large endings are among the biggest MGB terminals and may complement a system of equally large GABAergic endings prominent in cat (Winer et al. 1999a) and sparse in rat (Winer and Larue 1988) MGB, consistent with species-specific neurochemical patterns (Winer and Larue 1996).

6.5 Corticocollicular System

Auditory corticocollicular arrangements differ in almost every respect from their corticothalamic counterparts

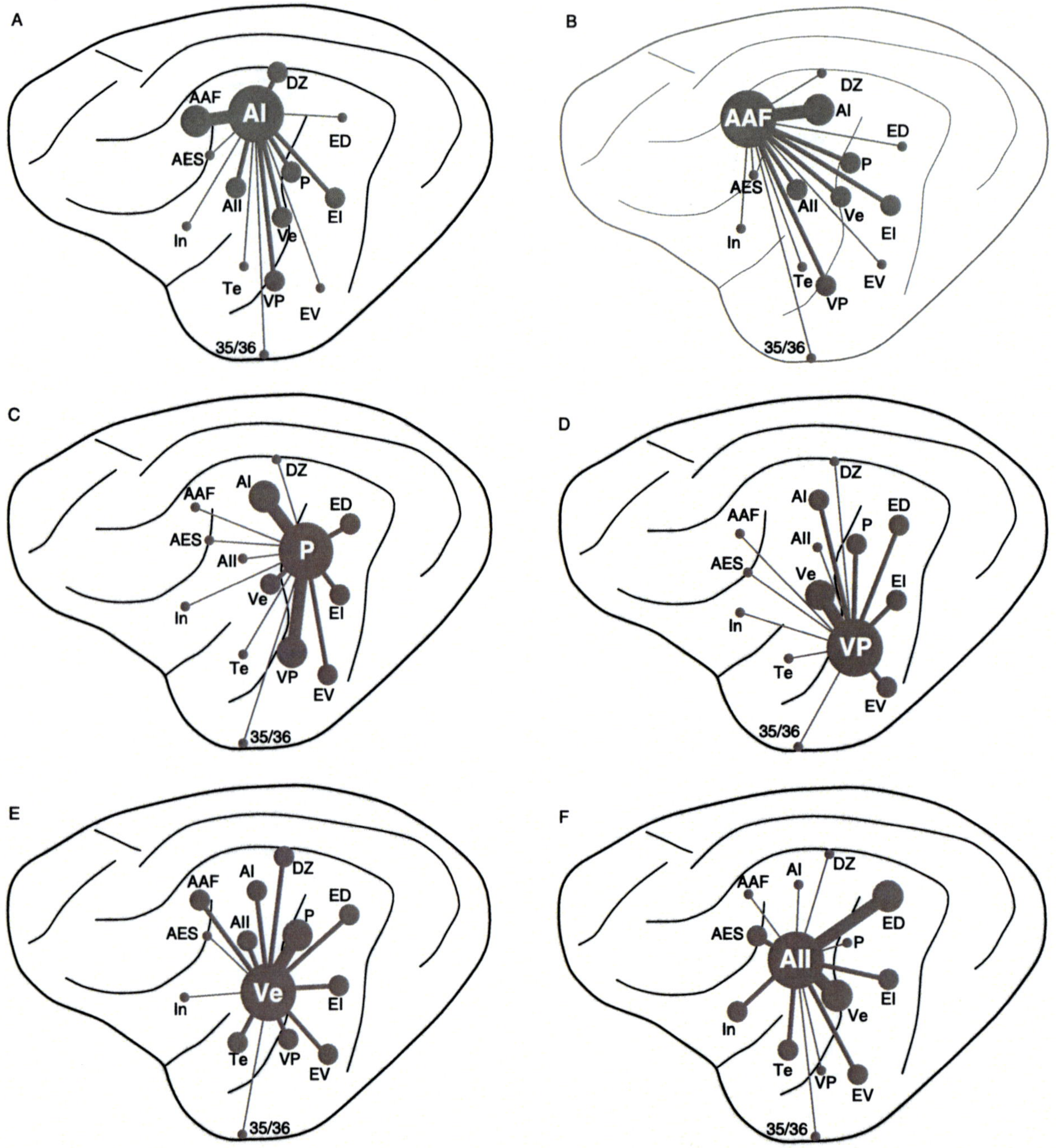

Fig. 2.13 Summaries of corticocortical convergence patterns in Fig. 2.12 converted to density plots in which line thickness is proportional to input strength. The principles are that each area receives projections from many (sometimes all) AC areas and that AI is no more a redistributive hub (Winer 1992) than is any other area. **a** AI resembles other areas in having within-group (i.e., areas with tonotopic affiliations) as its main input. **b** Likewise, AAF is dominated by projections from tonotopic area, but receives substantial input from multimodal posterior ectosylvian regions. **c** Area P receives a smaller constellation of input than AI, with the heaviest projections from tonotopic areas, consistent with a hierarchical organization. **d** Area VP has a special relationship with area Ve, with second tier input from multiple primary and perivisual fields. **e** Area Ve has many second tier projections, and a differential limbic-related input. **f** AII is hardly a target of AI, and receives area-specific primary input (from Ve but not AAF) and substantial posterior ectosylvian projections (ED, EI, EV)

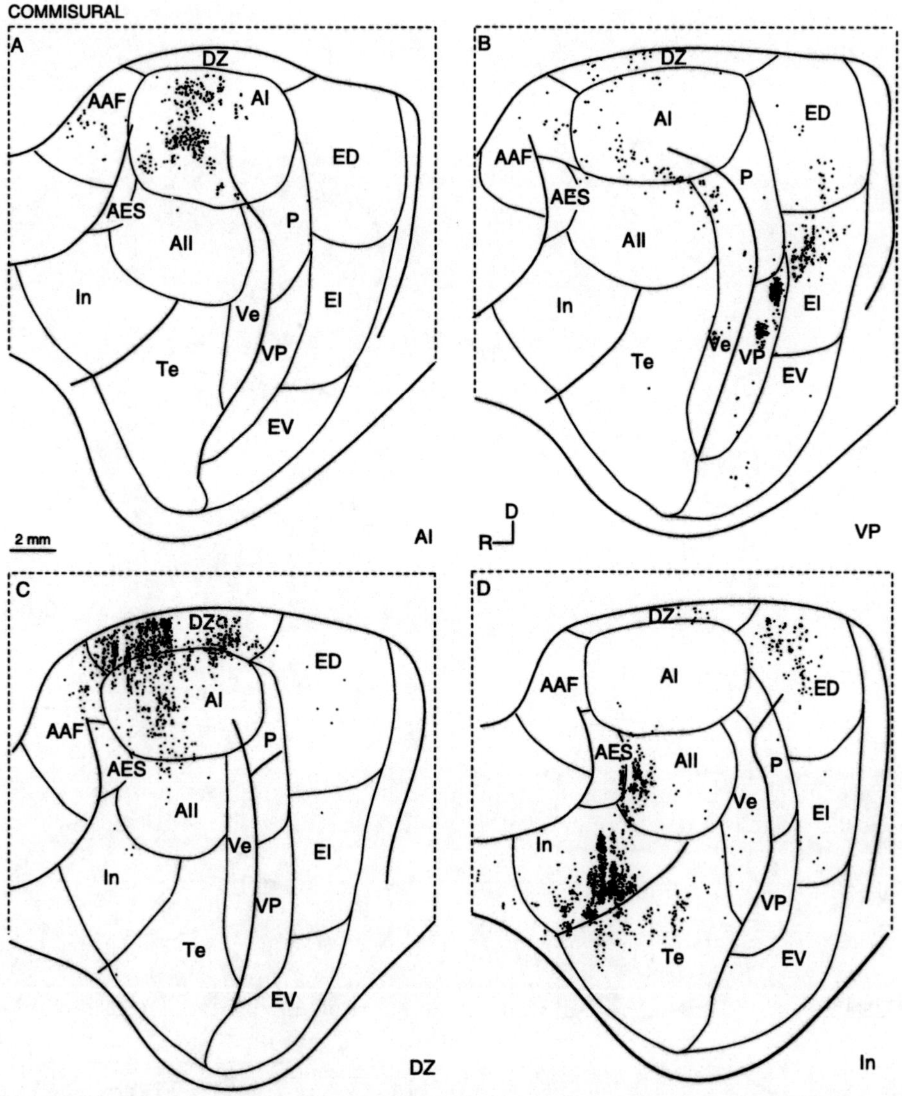

Fig. 2.14 Commissural projections from the experiments in Fig. 2.12. These differ from the corticocortical projections is being smaller, more reciprocal, and within-family (e.g., limbic to limbic) though selectively (only some tonotopic areas project). **a** Input to AI is the most symmetrical in this series, with little involvement of areas P, VP, and VE, and moderate AAF projections. **b** Area VP receives significant input from AI but does not project to it (**a**), and unexpected projections from periauditory posterior ectosylvian fields. **c** Area DZ receives input from nearly all of contralateral DZ and substantial, topographic AI projections. **d** Area In is a commissural target of areas ED, Te, and AII/AES, suggesting a widespread integrative role

(Fig. 2.16). Their laminar origins are limited to layer Vb (Winer and Prieto 2001), and few corticofugal cells project to both the MGB and the inferior colliculus (IC) (Wong and Kelly 1981), a pattern consistent with the virtual absence of branched forebrain auditory projections (Lee et al. 2004a, b). The input to the IC central nucleus is a mere fraction of that to the MGB ventral division. The chief targets of the tonotopic fields are, rather, the dorsal and caudal cortices, and regarded as outside the primary auditory pathway (Winer 2005). In contrast, non-tonotopic areas project only weakly to these venues, and moderately to the central nucleus, implying extralemniscal influences on the lemniscal stream; AC areas with perivisual affiliations project to the superior colliculus. The limbic-related auditory association fields have a unique projection pattern which includes non-tonotopic IC subdivisions and motor affiliated extracollicular targets perhaps related to vocalization and other behaviors (Jürgens 2002). Finally, corticocollicular axon morphology differs from that of corticogeniculate axons, suggesting different spatiotemporal dynamics (Winer et al. 1998; Winer et al. 2001; Winer 2005).

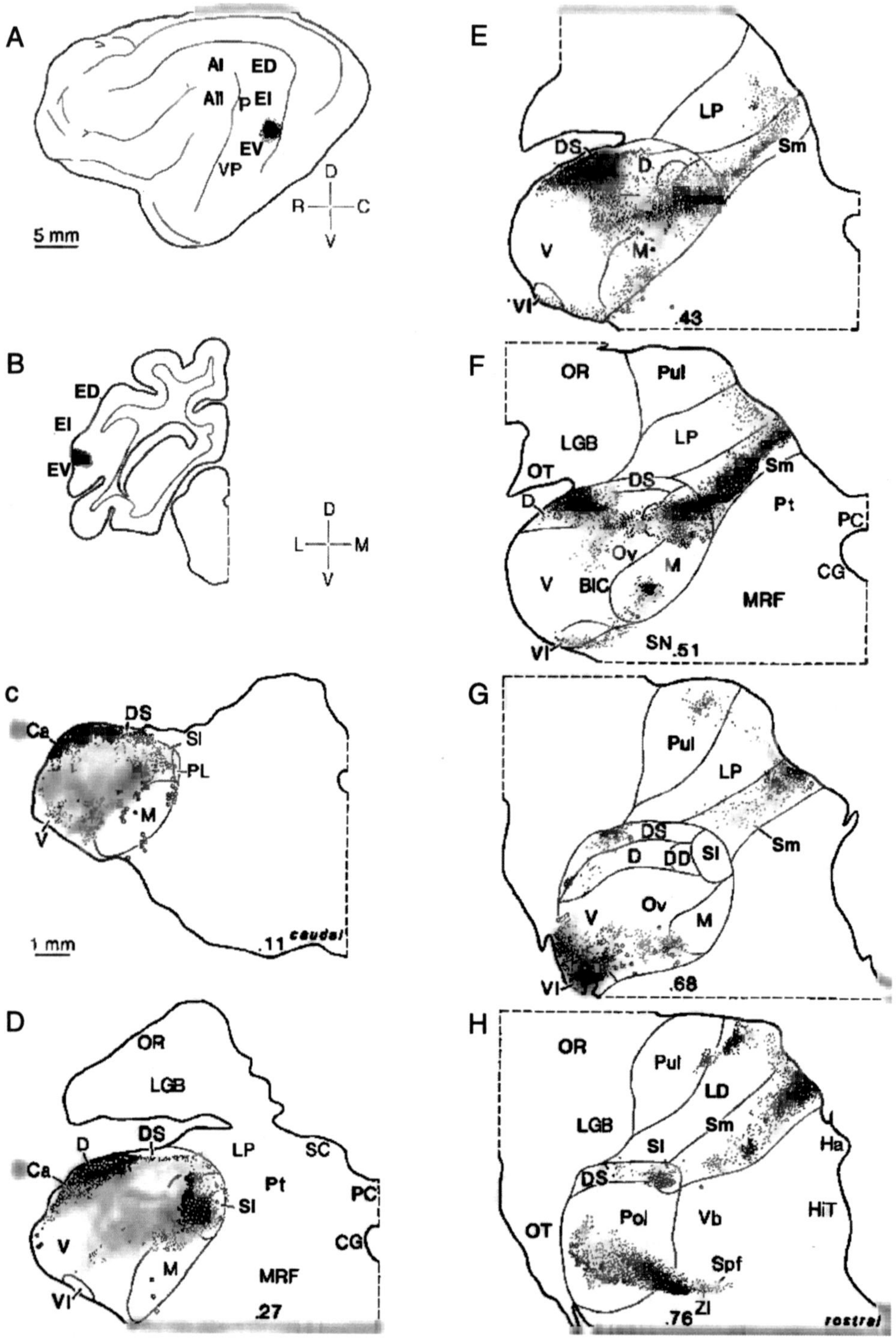

Fig. 2.15 AC projections to MGB subdivisions studied with a tracer transported bidirectionally (wheat germ agglutinin conjugated to horseradish peroxidase; WGA). *Fine dots*: corticofugal axon terminals; *circles*, TC neurons. **a, b** The deposit was at the EI-EV border and did not enter the white matter. **c** Non-tonotopic MGB divisions were principally involved, especially the dorsal nuclei. **d** Independent projection foci target the dorsal division (D, DCa) and the suprageniculate (SI) nuclei. This implies multimodal-to-auditory crosstalk in the descending system. **e** Extraauditory targets include the lateral posterior (LP) and suprageniculate nucleus medial part (Sm). **f** The medial division receives AC input as specific and focal as that to the ventral division in other experiments (not shown). **g** Even a small part of the dorsal division, the ventrolateral nucleus (Vl) has a massive and reciprocal projection. **h** The labeling extends nearly to the rostral tip of the MGB and terminates in the posterior intralaminar system (Spf) (Winer et al. 1988)

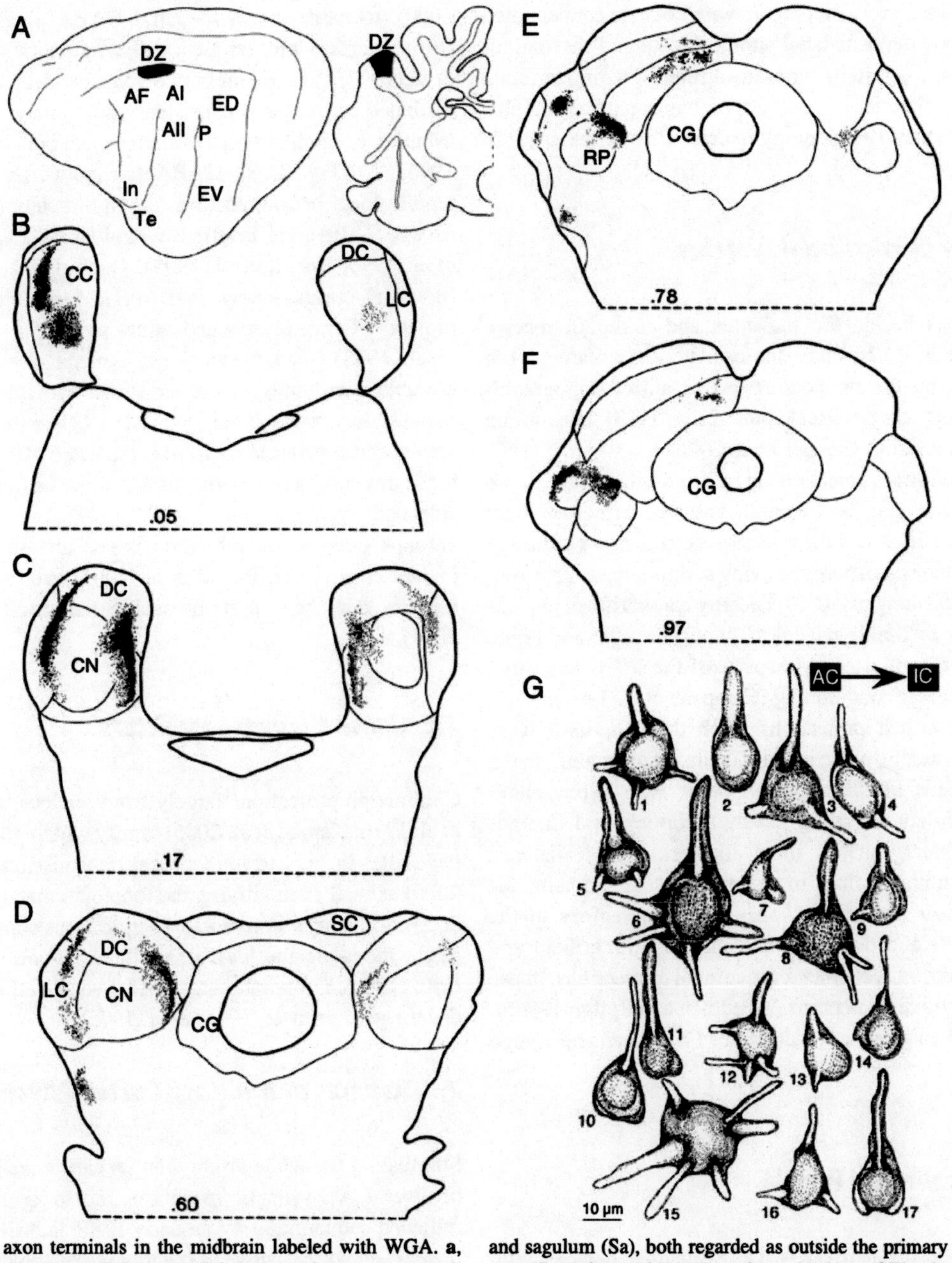

Fig. 2.16 AC axon terminals in the midbrain labeled with WGA. **a**, **b** A deposit on the crest of the ectosylvian gyrus does not encroach into medial perivisual areas. **c** The caudal inferior colliculus cortex receives the bulk of input. **d** The projection is bilateral, and symmetrical, and largely outside the central nucleus. **e** The lateral nucleus (La) and sagulum (Sa), both regarded as outside the primary auditory pathway (Casseday et al. 2005), each receive input. **f** The deep layers of the superior colliculus (SCx) receive bilateral projections. **g** Axons extend from virtually the caudal tip to the inferior colliculus rostral pole (RP), and are focal and dense

6.6 Corticopontine System

Like the corticocollicular pathway, the corticopontine system arises from layer V cells that do not project to the cochlear nucleus or superior olivary complex (Doucet et al. 2003). This projection comes from all AC areas and, like the corticogeniculate (Winer et al. 2001) and corticocollicular systems (Winer et al. 1998), is topographic and reaches most pontine subdivisions. A given AC locus, which labels a discrete IC target, may span much larger pontine territories, suggesting an area- and target-specific pattern of corticofugal divergence. Corticopontine axons have an architecture entirely different from the other corticofugal axons, forming narrow sheets 200–300 µm long whose preterminal segments fill a narrow subdomain with boutons concentrated focally. Single deposits label strongly focal and discontinuous domains, suggesting a mosaic of interdigitating terminal architectures (Perales et al. 2006) perhaps contributing ultimately to fractured somatotopy in cerebellar cortex (Arends 1997).

6.7 Other Corticofugal Systems

Other regions beside the thalamus and midbrain receive AC input (Fig. 2.17). Thus, the caudate and putamen both are targets and the projections are organized topographically, forming sheets (Reale and Imig 1983) resembling corticopontine axons (Perales et al. 2006).

Corticoclaustral projections arise from all AC areas, are reciprocal, and may be bilateral, with the principal input from insular cortex and the posterior ectosylvian gyrus, and the intermediate claustrum receiving auditory input preferentially and projecting to AC. The nearby endopiriform nucleus is dominated by limbic-related AC input (Beneyto and Prieto 2001) and is implicated, with parts of the MGB and other structures, in fear conditioning (Campeau et al. 1997).

Temporal cortex projections reach the superior olivary complex bilaterally, preferentially ipsilaterally, mainly in the ventral nucleus of the trapezoid body, with axons either branching widely or ending focally (Schofield and Coomes 2004). Cochlear nucleus input likewise was ipsilateral-dominant, ending mainly in the granule cell domain, the dorsal cochlear nucleus, and magnocellular regions of the posteroventral and dorsal cochlear nucleus (Schofield and Coomes 2005). Like other corticofugal projections, these arise from pyramidal neurons (Weedman and Ryugo 1996b), and they end on granule cell dendrites (Weedman and Ryugo 1996a).

7 Neurochemical Profile

In the visual cortex there is a vast array of chemically specific circuits and cell types (Lund et al. 1995; Gonchar et al. 2002) contributing to local microcircuitry (Callaway 1998). Such a profile is not available for AC.

7.1 Gamma-Aminobutyric Acid

GABAergic neurons and axons in AC influence RF architecture (Foeller et al. 2001), intensity tuning (Sutter and Loftus 2003), excitatory–inhibitory interactions (Volkov and Galazjuk 1991), and tuning curve shape (Chen and Jen 2000), to name just a few, and GABA antagonists selectively interfere with frequency discrimination (Riquimaroux et al. 1992). The richness and robustness of these GABA-mediated responses is consistent with the layer-specific distribution of immunopositive neurons and puncta (Prieto et al. 1994a, b) (Fig. 2.18). GABAergic networks are probably largely local or limited to a few millimeters (Winer 1986) except GABAergic inverted pyramidal cells (Bueno-López et al. 1990; Reblet et al. 1992) which may project farther. In visual cortex, many GABAergic local circuit neurons project ~1 mm or less and target pyramidal cells (Freund et al. 1983). GABAergic arrangements elsewhere in AC are unknown, though there are marked regional patterns in non-auditory areas (Hendry et al. 1987) and evidence for area-specific patterns of GABA-mediated effects, e.g., posterior auditory area neurons have a more intricate inhibitory sideband structure than AI cells (Loftus and Sutter 2001). Perhaps GABAergic inhibition is site and species specific (Winer et al. 1995; Pollak et al. 2003) and even TC transmission may have system-specific features (Atzori et al. 2001).

7.2 Other Neurotransmitters

Cholinergic projections largely from nucleus basalis (Jones et al. 1976; Kamke et al. 2005) ramify widely in AC and may have roles in TC and intracortical transmission (Hsieh et al. 2000) as well as modifying the tonotopic map (Kamke et al. 2003). Influences from noradrenergic locus ceruleus neurons might modulate the level of vigilance (Foote et al. 1983). Both GABA and noradrenalin affect vocalization related AC discharge in primate (Foote et al. 1975).

7.3 Aspects of Auditory Cortex Physiology

Stimulating the white or gray matter can evoke a wide range of diverse AC synaptic responses, including a rapid EPSP followed sequentially by an early IPSP, late EPSP, and late IPSP; each was sensitive to specific pharmacologic blockade (e.g., a quisqualate/kainate receptor antagonist abolished the early EPSP, while the late EPSP was affected by an NMDA receptor antagonist, etc.) (Cox et al. 1992).

EPSPs evoked in young mouse AC slices by thalamic stimulation are reliable and have little temporal variation, with both regular- and fast-spiking cells receiving TC input; the EPSPs can also summate to elicit multisynaptic activity modulated by NMDA receptors (Rose and Metherate 2005). This is in accord with observations that the postsynaptic targets of TC neurons are heterogeneous (Smith and Populin 2001), a pattern of diversity that could subserve the several motifs of TC transformation (Winer et al. 2005). Tracing

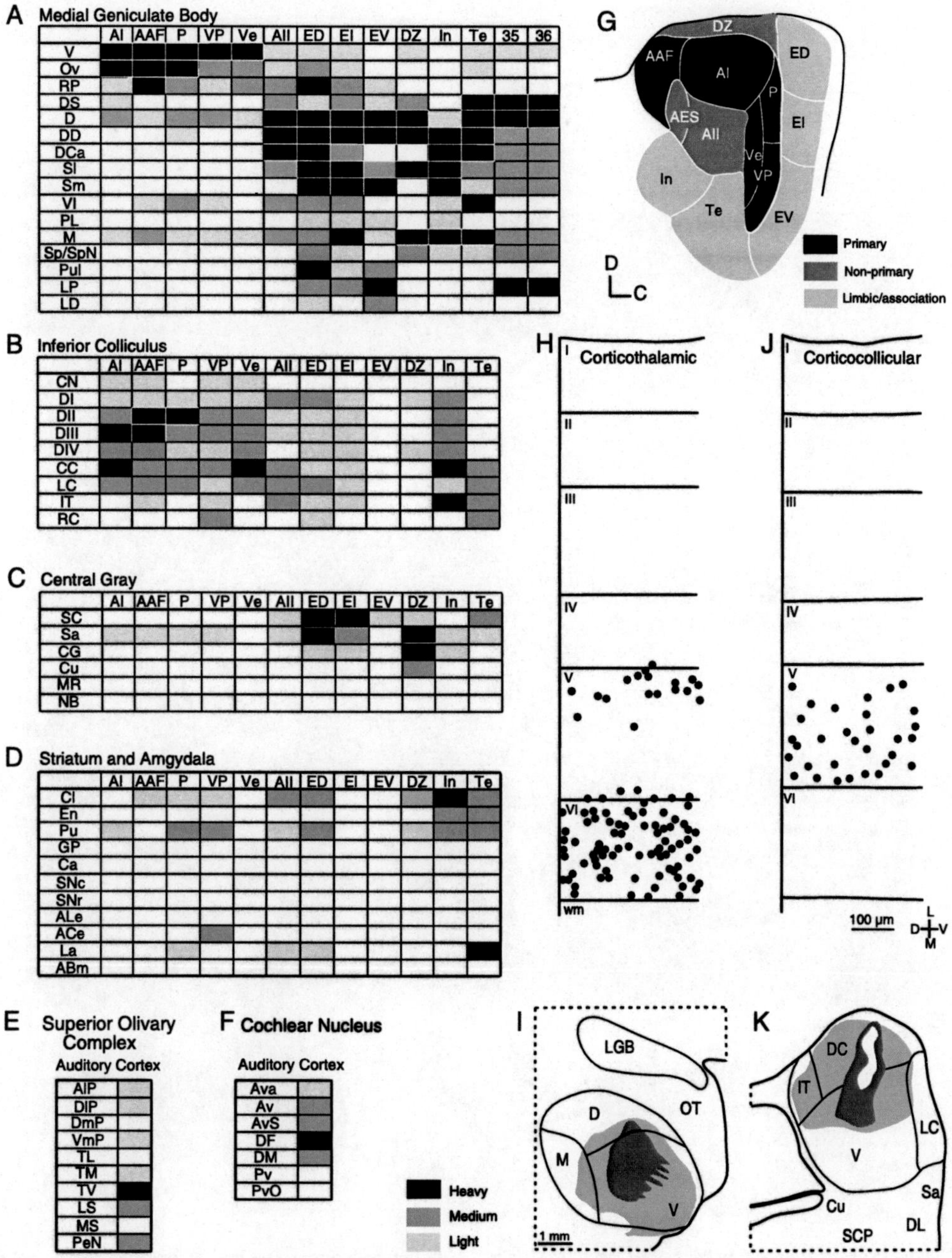

Fig. 2.17 Targets of the auditory corticofugal system. The main points are that the system is highly divergent, that all levels of the central auditory pathway receive some descending projections, that the projection patterns are areally and target specific, and that the axons (not shown) are also related to their targets. **a** Corticofugal projections from 14 cortical areas to the MGB and adjoining thalamic nuclei. Primary areas have the most limited projections, non-primary (e.g., AII) the most diverse, and limbic-related (e.g., In) areas have projections as focal as those of primary areas but little nuclear overlap. This suggests area-specific parallel corticothalamic pathways (Winer et al. 2001). **b** Corticocollicular projections are principally to targets outside the central nucleus, few are as heavy as those to the MGB, and the divergence is comparable to that of the corticogeniculate system. **c** Central *gray input* is restricted and supports a functional distinction between limbic-related areas (In, Te) and specificity in their corticofugal role (Winer et al. 1998). **d** Corticolimbic and corticostriatal projections are more widespread than those to the *central gray*, and involve different areas (EV and DZ in the latter and not in the former), again supporting a functional disjunction among areas whose roles remain uncertain. **e** Near all rodent superior olivary subdivisions receive AC input (Schofield and Coomes 2004), as does the (**f**) cochlear nucleus with the conspicuous exception of the posteroventral (Pv) and octopus cell (PvO) regions, which suggests differential corticofugal influence. **g** Principal AC areas recognized in the cat. **h, i** Laminar distribution of corticothalamic projections to the ventral division (**i**), showing that layers Va, Vc, and VI project. **j, k** The contrasting corticocollicular projection arises from different parts of layer V after a central nucleus deposit (**k**) (Winer et al. 2001) and has few targets (**b**)

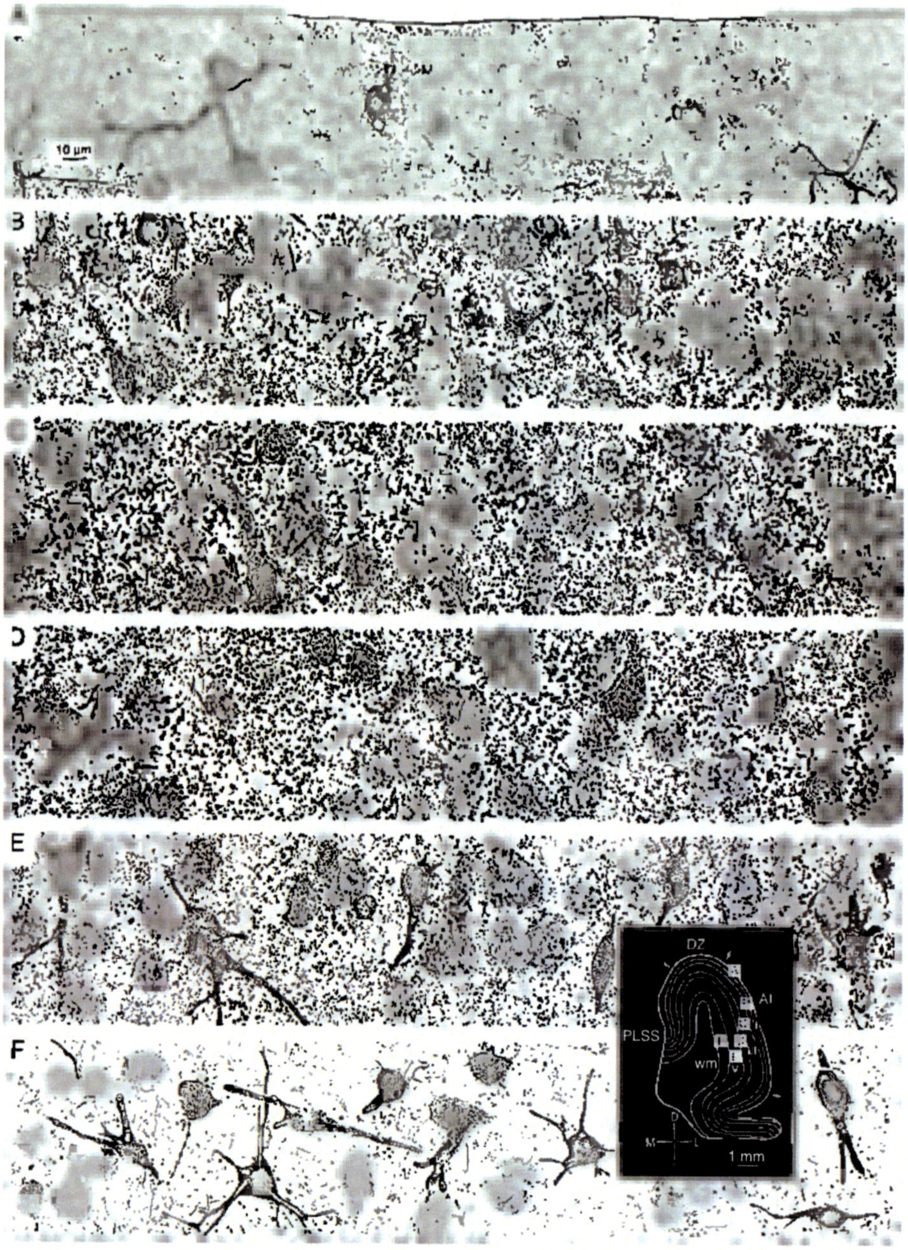

Fig. 2.18 Glutamic acid decarboxylase-positive AI neurons (*heavy stipple*) and puncta (*fine dots*) and immunonegative somata (*gray stipple*). Frozen section, 30 μm thick, planapochromat, N.A. 1.32, ×2000. Each layer has a unique organization, a pattern consistent with the proposition that cortical layers have as specific an organization as subcortical nuclei. **a** Layer I puncta are finer than in any other layer, with few GABAergic layer Ia cells. **b** Layer II puncta are coarser and target both immunopositive and immunonegative somata. **c** Layer III puncta cover triangular, immunonegative somata and their apical processes. **d** Layer IV neurons are almost entirely non-pyramidal and the puncta are larger than those in layer II. **e** Layer V puncta are finer than those inlayer IV and sparser, and cluster on all somata. **f** Layer VI terminals are as fine and delicate as those in layer I, and often less numerous on immunonegative somata (Prieto et al. 1994a)

the further intracortical distribution of these signals using current source density analysis reveals strong early sinks in layers III and IV, while stimuli far (3 octaves from CF) preferentially activated primarily infragranular sinks, and later overlap presumed to reflect lemniscal and non-lemniscal influence. MGB stimulation in vitro in rat elicits horizontal dispersion of excitation and sinks in layers III and IV (Kaur et al. 2005). Such dispersion is consistent with estimates of the size of TC axonal plexuses (Huang and Winer 2000; McMullen et al. 2005). In anesthetized rats muscimol greatly attenuated local field potentials and RF size, but did not affect threshold at characteristic frequency or the timing of the onset response, consistent with an effect on intracortical rather than TC processes. Subthreshold EPSPs and local field potentials were unexpectedly broad (Kaur et al. 2004), consistent with models of TC input emphasizing intracortical dispersion (Lee et al. 2004b). Other evidence suggests coactivation of primary and non-primary areas in tandem from different MGB subdivisions (Barth et al. 1995).

AI neurons have heterogeneous tuning curves which may embody a continuous rather than a categorical distribution of response types, ranging from classic V- to U-shaped to sloped or slanted RF profiles, many showing pronounced sharpening (Sutter 2000) and a range of responses consistent with diverse roles for GABAergic neurons (Prieto et al. 1994b). Convergent excitatory–inhibitory interactions find an exclusively early-stage inhibitory contribution to excitatory intensity tuning (Sutter and Loftus 2003) consistent with the specificity of intracortical inhibition (Prieto et al. 1994a). In awake marmosets AC cells exposed to preferred stimuli can fire for longer than expected periods, consistent with a model in which neural ensembles have extended representational-computational roles (Bendor and Wang 2005).

More complex stimulus representations are also found in AC, with primate vocalizations eliciting considerable response specificity. Thus, responses concentrate in supragranular layers and neuron pairs discharge independently, with phrase-specific temporal release, and excitatory–inhibitory events reflect sound frequency and energy (Eliades and Wang 2005). The dispersion of TC projections in cat (Huang and Winer 2000; Lee et al. 2004b) is in accord with the distributed spatial arrangement.

8 Toward a Theory of Auditory Forebrain Operations

The premise explored here is that the auditory forebrain (and perhaps the auditory system) is comprised of several streams which have evolved interdependently and which interact cooperatively. The concept of an "auditory system" is thus a synthetic construct whose value is mainly heuristic and whose objective correlative may not exist.

8.1 Forebrain Auditory System

A classical view of the auditory forebrain views it as an extension of systems arising in the cochlear nucleus and olivary complex and which are exquisitely adapted for analyzing interaural time and intensity differences. Signals arising in the cochlea are propagated to hindbrain, midbrain, thalamic and cortical levels, presumably for further analysis of complex features, before corticocortical feedforward input to higher areas for more global processing, to provide descending output for ongoing activity, and for perceptual purposes (Bregman 1990). While this view might be pertinent to the function of end bulbs of Held, bushy cells in the anteroventral cochlear nucleus, the superior and medial superior olives, and disc shaped or bushy cells in the central nucleus of the IC, the ventral nucleus of the MGB, and the several tonotopic areas of AC, it says little about the explicit roles of much (perhaps most) of the midbrain and forebrain. Areas omitted include five of six IC subdivisions (dorsal cortex, lateral cortex, caudal cortex, intercollicular tegmentum, and rostral pole nucleus), vast MGB regions (dorsal division, caudal cortex, medial division, and rostral pole), and all of AC except the five tonotopic areas (omitting nine others connected with the MGB and IC). This view is insufferably parochial.

8.2 Multimodal Interactions

Trigeminal somatic sensory information reaches the granule cell domain of the cochlear nucleus (Haenggeli et al. 2005), and eye position signals influence RFs in primate IC (Groh et al. 2001). In the IC extensive non-auditory connections establish multimodal representations (Syka et al. 2000), and convey auditory input to other systems (Linke 1999; Harting and Van Lieshout 2000) and this pattern is elaborated in the MGB (Wepsic 1966).

8.3 Auditory-Motor Relations

Motor activity influenced by audition includes somatic, visceral, vocal behavior, and movement planning components. Acoustic startle and its inhibition are shaped by sensory input that requires integration across these four domains (Fig. 2.19a). Substantia nigra projections to the IC (Olazábal and Moore 1989) may coordinate motor orientation to sounds, while AC input to the basal ganglia (Reale and Imig 1983) could provide a premotor signal essential for cross-modal calibration (Fig. 2.19a). Analogous corticofugal circuits (Feliciano et al. 1995) might subserve auditory influence on vocalization (Schuller et al. 1997).

8.4 Auditory-Limbic Interactions

Circuits linking the MGB and the amygdala (LeDoux et al. 1985) bidirectionally (Marsh et al. 2002) could

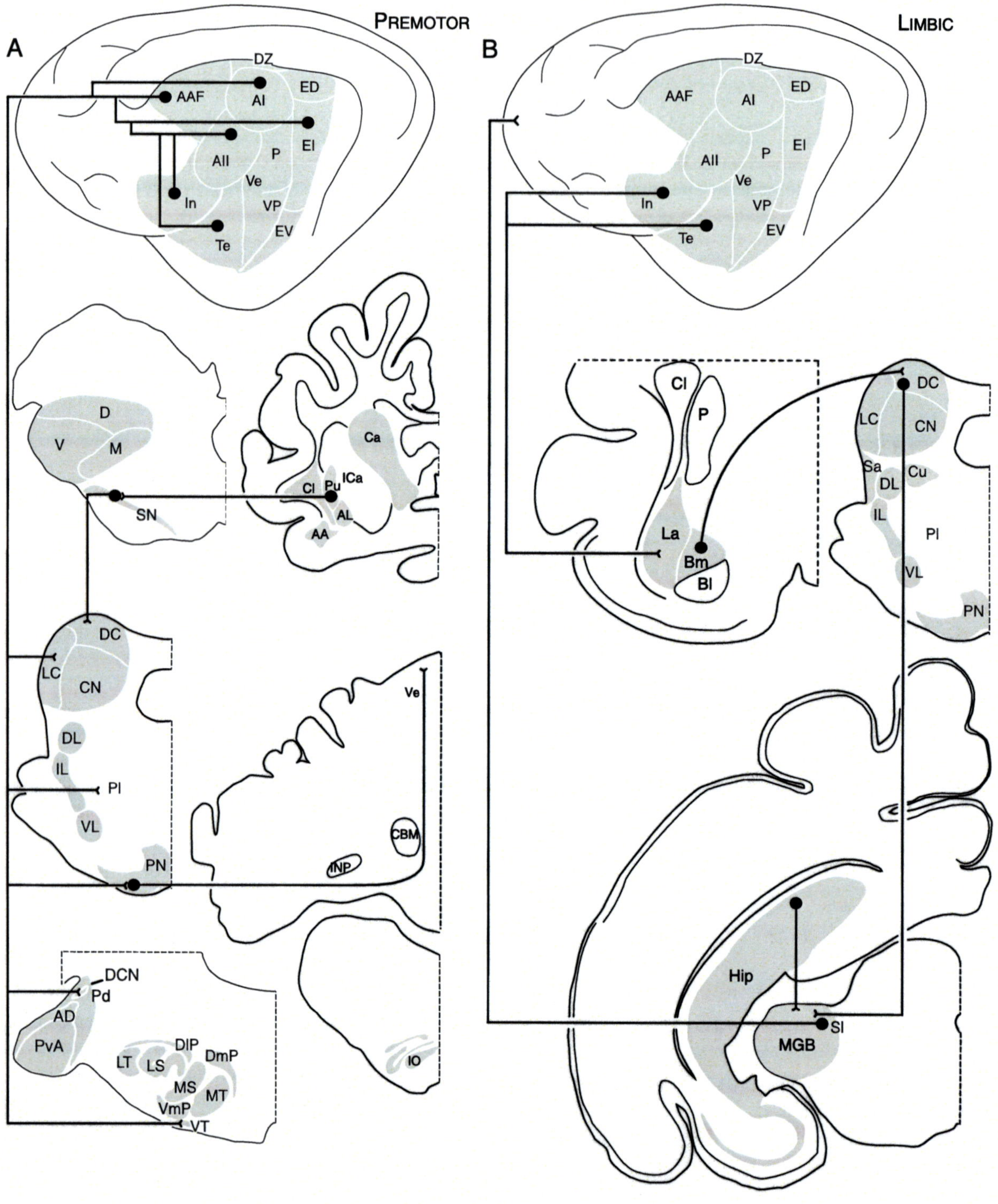

Fig. 2.19 Reciprocal AC connection to premotor and limbic structures. **a** The premotor relations with nigral, striatal, and paralemniscal areas might coordinate skeletal (Olazábal and Moore 1989) and smooth muscle (Winer 2006) and vocalization-related pathways (Feliciano et al. 1995). AC input to the putamen arises from primary, non-primary, multisensory, and limbic related fields (Beneyto and Prieto 2001) and might affect motor set and cognitive aspects of movement planning. Corticocollicular projections target IC subdivisions (Winer et al. 1998) with robust substantia nigra input (Olazábal and Moore 1989), and corticofugal AC axons end in the adjoining intralaminar nuclei (Winer et al. 2001), which modulate global TC excitability and vigilance. AC input to the paralemniscal zone can affect bat vocalizations (Schuller et al. 1997). Corticopontine projections arise from all AC regions, consistent with the view that AC tonotopic, non-tonotopic, multisensory, and limbic areas each influence premotor control (Perales et al. 2006). **b** The AC input to the amygdala (La) allows access to many extraauditory

shape autonomic responses to sound (LeDoux et al. 1986) (Fig. 2.19b). A massive AC projection to the amygdala (Romanski et al. 1993; Shi and Cassell 1997) testifies to this strong relation.

9 Directions for Future Research

It will suffice here to point out a few directions in which future research may be most helpful in providing a stronger underpinning for the development of a theory of auditory forebrain function. Further thoughts are expressed in Chapters 10 and 32.

1. What rules (construction, combination, etc.) govern the tectothalamic transformation?
2. Are the principles for information transfer similar among MGB subdivisions?
3. How does reversible inactivation of MGB subdivisions affect behavioral performance?
4. Are different MGB and AC subdivisions equally plastic from a physiologic perspective?
5. How is forebrain plasticity related to synaptic processes of facilitation and depression?
6. What do interneurons do in the MGB and AC? Is their performance nucleus and layer specific, or do they serve more general processes?
7. What is the significance—physiologically and behaviorally—of the several-fold species differences in the proportion of interneurons in the MGB and AC?

Acknowledgment Thanks to Christoph E. Schreiner for helpful comments. This work was supported by United States Public Health Service grant R01 DC02319-26.

References

Aitkin L, Tran L, and Syka J (1994) The responses of neurons in subdivisions of the inferior colliculus of cats to tonal, noise and vocal stimuli. *Experimental Brain Research* **98**:53–64.

Aitkin LM (1973) Medial geniculate body of the cat: responses to tonal stimuli of neurons in medial division. *Journal of Neurophysiology* **36**:275–283.

Aitkin LM and Dunlop CW (1968) Interplay of excitation and inhibition in the cat medial geniculate body. *Journal of Neurophysiology* **31**:44–61.

Arends JJ (1997) Sensory representation in the cerebellum and control circuits of motion. *European Journal of Morphology* **35**: 234–245.

Atzori M, Flores Hernández J, and Pineda JC (2004) Interlaminar differences of spike activation threshold in the auditory cortex of the rat. *Hear Res* **189**:101–106.

Atzori M, Lei S, Evans DI, Kanold PO, Phillips-Tansey E, McIntyre O, and McBain CJ (2001) Differential synaptic processing separates stationary from transient inputs to the auditory cortex. *Nature Neuroscience* **4**:1230–1237.

Bajo VM, Rouiller EM, Welker E, Clarke S, Villa AEP, de Ribaupierre Y, and de Ribaupierre F (1995) Morphology and spatial distribution of corticothalamic terminals originating from the cat auditory cortex. *Hearing Research* **83**:161–174.

Barth DS, Goldberg N, Brett B, and Di S (1995) The spatiotemporal organization of auditory, visual, and auditory-visual evoked potentials in rat cortex. *Brain Research* **678**:177–190.

Bartlett EL and Smith PH (1999) Anatomic, intrinsic, and synaptic properties of dorsal and ventral division neurons in rat medial geniculate body. *Journal of Neurophysiology* **81**:1999–2016.

Bazhenov M, Timofeev I, Steriade M, and Sejnowski T (2000) Spiking-bursting activity in the thalamic reticular nucleus initiates sequences of spindle oscillations in thalamic networks. *Journal of Neurophysiology* **84**:1076–1087.

Bendor D and Wang X (2005) The neuronal representation of pitch in primate auditory cortex. *Nature* **436**:1161–1165.

Benedek G, Perény J, Kovács G, Fischer-Szátmári L, and Katoh YY (1997) Visual, somatosensory, auditory and nociceptive modality properties in the feline suprageniculate nucleus. *Neuroscience* **78**:179–189.

Beneyto M and Prieto JJ (2001) Connections of the auditory cortex with the claustrum and endopiriform nucleus in the cat. *Brain Research Bulletin* **54**:485–498.

Berman AL and Jones EG (1982) *The Thalamus and Basal Telencephalon of the Cat: A Cytoarchitectonic Atlas with Stereotaxic Coordinates.* University of Wisconsin Press, Madison.

Bordi F and LeDoux JE (1994a) Response properties of single units in areas of rat auditory thalamus that project the amygdala. I. Acoustic discharge patterns and frequency receptive fields. *Experimental Brain Research* **98**:261–274.

Bordi F and LeDoux JE (1994b) Response properties of single units in areas of rat auditory thalamus that project the amygdala. II. Cells receiving convergent auditory and somatosensory inputs and cells antidromically activated by amygdala stimulation. *Experimental Brain Research* **98**:275–286.

Brandner S and Redies H (1990) The projection of the medial geniculate body to field AI: organization in the isofrequency dimension. *Journal of Neuroscience* **10**:50–61.

Bregman AS (1990) *Auditory Scene Analysis. The Perceptual Organization of Sound.* MIT Press, Cambridge.

Bueno-López JL, Reblet C, López-Medina A, Gómez-Urquijo SM, Grandes P, Gondra J, and Hennequet L (1990) Targets and laminar distribution of projection neurons with 'inverted' morphology in rabbit cortex. *European Journal of Neuroscience* **3**: 415–430.

Caballero-Bleda M, Fernandez B, and Puelles L (1991) Acetylcholinesterase and NADH-diaphorase chemoarchitectonic

Fig. 2.19 (continued) sites (Clascá et al. 2000). Robust corticoamygdaloid input may mediate visceromotor and appetitive behavior (Romanski and LeDoux 1993). These pathways also have reciprocal and extended connections with the distributed AC. Corticogeniculate projections target MGB divisions receiving input from parahippocampal areas that have strong reciprocal amygdaloid connections (Shinonaga et al. 1994). This enables monosynaptic auditory corticoamygdaloid pathways and disynaptic and reciprocal corticothalamoamygdaloid and corticoamygdalothalamic streams. (Adapted from Winer and Lee 2007) subdivisions in the rabbit medial geniculate body. *Journal of Chemical Neuroanatomy* **4**:271–280.

Calford MB (1983) The parcellation of the medial geniculate body of the cat defined by the auditory response properties of single units. *Journal of Neuroscience* **3**:2350–2364.

Calford MB and Aitkin LM (1983) Ascending projections to the medial geniculate body of the cat: evidence for multiple, parallel auditory pathways through the thalamus. *Journal of Neuroscience* **3**: 2365–2380.

Callaway EM (1998) Local circuits in primary visual cortex of the macaque monkey. *Annual Review of Neuroscience* **21**:47–74.

Campeau S, Falls WA, Cullinan WE, Helmreich DL, Davis M, and Watson SJ (1997) Elicitation and reduction of fear: behavioural and neuroendocrine indices and brain induction of the immediate early gene c-*fos*. *Neuroscience* **78**:1087–1104.

Cant NB and Benson CG (2003) Parallel auditory pathways: projection patterns of the different neuronal populations in the dorsal and ventral cochlear nuclei. *Brain Research Bulletin* **60**: 457–474.

Casseday JH, Schreiner CE, and Winer JA (2005) The inferior colliculus: past, present, and future. In: Winer JA and Schreiner CE (eds). *The Inferior Colliculus*. Springer, New York, pp. 626–640.

Cauller L (1995) Layer I of primary sensory neocortex: where top-down converges upon bottom-up. *Behavioural Brain Research* **71**: 163–170.

Cetas JS, Price RO, Crowe J, Velenovsky DS, and McMullen NT (2003) Dendritic orientation and laminar architecture in the rabbit auditory thalamus. *Journal of Comparative Neurology* **458**: 307–317.

Cetas JS, Price RO, Velenovsky DS, Crowe JJ, Sinex DG, and McMullen NT (2002) Cell types and response properties of neurons in the ventral division of the medial geniculate body of the rabbit. *Journal of Comparative Neurology* **445**:78–96.

Chen QC and Jen PH (2000) Bicuculline application affects discharge patterns, rate-intensity functions, and frequency tuning characteristics of bat auditory cortical neurons. *Hearing Research* **150**: 161–174.

Clarey JC, Barone P, and Imig TJ (1992) Physiology of thalamus and cortex. In: Popper AN and Fay RR (eds). *Springer Handbook of Auditory Research*, volume 2, *The Mammalian Auditory Pathway: Neurophysiology*. Springer, New York, pp. 232–334.

Clasca F, Llamas A, and Reinoso-Suarez F (1997) Insular cortex and neighboring fields in the cat: a redefinition based on cortical microarchitecture and connections with the thalamus. *Journal of Comparative Neurology* **384**:456–482.

Code RA and Winer JA (1986) Columnar organization and reciprocity of commissural connections in cat primary auditory cortex (AI). *Hearing Research* **23**:205–222.

Coomes DL, Bickford ME, and Schofield BR (2002) GABAergic circuitry in the dorsal division of the cat medial geniculate nucleus. *Journal of Comparative Neurology* **453**:45–56.

Cox CL, Metherate R, Weinberger NM, and Ashe JH (1992) Synaptic potentials and effects of amino acid antagonists in the auditory cortex. *Brain Research Bulletin* **28**:401–410.

Cox CL and Sherman SM (2000) Control of dendritic outputs of inhibitory interneurons in the lateral geniculate nucleus. *Neuron* **27**:597–610.

Cox CL, Zhou Q, and Sherman SM (1998) Glutamate locally activates dendritic outputs of thalamic interneurons. *Nature* **394**:478–482.

Crabtree JW (1998) Organization in the auditory sector of the cat's thalamic reticular nucleus. *Journal of Comparative Neurology* **390**:167–182.

Davis KA and Young ED (2000) Pharmacological evidence of inhibitory and disinhibitory neuronal circuits in dorsal cochlear nucleus. *Journal of Neurophysiology* **83**:926–940.

de la Mothe L, Blumell S, Kajikawa Y, and Hackett TA (2006) Cortical connections of the auditory cortex in marmoset monkeys: core and medial belt regions. *Journal of Comparative Neurology* **496**:27–71.

de Ribaupierre F (1997) Acoustic information processing in the auditory thalamus and cerebral cortex. In: Ehret G (ed). *The Central Auditory System*. Oxford University Press, New York, pp. 317–397.

Deschênes M, Veinante P, and Zhang Z-W (1998) The organization of corticothalamic projections: reciprocity versus parity. *Brain Research Reviews* **28**:286–308.

Diamond ME, Armstrong-James M, Budway MJ, and Ebner FF (1992) Somatic sensory responses in the rostral sector of the posterior group (POm) and in the ventral posterior medial nucleus (VPM) of the rat thalamus: dependence on the barrel field cortex. *Journal of Comparative Neurology* **319**:66–84.

Doucet JR, Molavi DL, and Ryugo DK (2003) The source of corticocollicular and corticobulbar projections in area Te1 of the rat. *Experimental Brain Research* **153**:477–485.

Edeline J-M and Weinberger NM (1992) Associative retuning in the thalamic source of input to the amygdala and auditory cortex: receptive field plasticity in the medial division of the medial geniculate body. *Behavioral Neuroscience* **106**:81–105.

Edeline JM (2003) The thalamo-cortical auditory receptive fields: regulation by the states of vigilance, learning and the neuromodulatory systems. *Experimental Brain Research* **153**:554–572.

Ehret G (1997) The auditory cortex. *Journal of Comparative Physiology A* **181**:547–557.

Eliades SJ and Wang X (2005) Dynamics of auditory-vocal interaction in monkey auditory cortex. *Cerebral Cortex* **15**:1510–1523.

Feliciano M, Saldaña E, and Mugnaini E (1995) Direct projections from the rat primary auditory neocortex to nucleus sagulum, paralemniscal regions, superior olivary complex and cochlear nuclei. *Auditory Neuroscience* **1**:287–308.

Foeller E, Vater M, and Kössl M (2001) Laminar analysis of inhibition in the gerbil primary auditory cortex. *Journal of the Association for Research in Otolaryngology* **2**:279–296.

Foote SL, Bloom FE, and Aston-Jones G (1983) Nucleus locus ceruleus: new evidence of anatomical and physiological specificity. *Physiological Reviews* **63**:844–914.

Foote SL, Freedman R, and Oliver AP (1975) Effects of putative transmitters on neuronal activity in monkey auditory cortex. *Brain Research* **86**:229–242.

Formisano E, Kim DS, Di Salle F, van de Moortele PF, Ugurbil K, and Goebel R (2003) Mirror-symmetric tonotopic maps in human primary auditory cortex. *Neuron* **40**:859–869.

Freund TF, Martin KAC, Smith AD, and Somogyi P (1983) Glutamate decarboxylase-immunoreactive terminals of Golgi-impregnated axo-axonic cells and of presumed basket cells in synaptic contact with pyramidal neurons of the cat's visual cortex. *Journal of Comparative Neurology* **221**:263–278.

Gonchar Y, Turney S, Price JL, and Burkhalter A (2002) Axo-axonic synapses formed by somatostatin-expressing GABAergic neurons in rat and monkey visual cortex. *Journal of Comparative Neurology* **443**:1–14.

Groh JM, Trause AS, Underhill AM, Clark KR, and Inati S (2001) Eye position influences auditory responses in primate inferior colliculus. *Neuron* **29**:509–518.

Hackett TA, Stepniewska I, and Kaas JH (1998) Subdivisions of auditory cortex and ipsilateral cortical connections of the parabelt auditory cortex in macaque monkeys. *Journal of Comparative Neurology* **394**:475–495.

Hackett TA, Stepniewska I, and Kaas JH (1999) Callosal connections of the parabelt auditory cortex in macaque monkeys. *European Journal of Neuroscience* **11**:856–866.

Haenggeli CA, Pongstaporn T, Doucet JR, and Ryugo DK (2005) Projections from the spinal trigeminal nucleus to the cochlear nucleus in the rat. *Journal of Comparative Neurology* **484**: 191–205.

Harting JK and Van Lieshout DP (2000) Projections from the rostral pole of the inferior colliculus to the cat superior colliculus. *Brain Research* **881**:244–247.

Hashikawa T, Molinari M, Rausell E, and Jones EG (1995) Patchy and laminar terminations of medial geniculate axons in monkey auditory cortex. *Journal of Comparative Neurology* **362**:195–208.

He J (2003) Slow oscillation in non-lemniscal auditory thalamus. *Journal of Neuroscience* **23**:8281–8290.

He J and Hu B (2002) Differential distribution of burst and single-spike responses in auditory thalamus. *Journal of Neurophysiology* **88**:2152–2156.

Heffner HE and Heffner RS (1989) Unilateral auditory cortex ablation in macaques results in a contralateral hearing loss. *Journal of Neurophysiology* **62**:789–801.

Hendry SHC, Schwark HD, Jones EG, and Yan J (1987) Numbers and proportions of GABA-immunoreactive neurons in different areas of monkey cerebral cortex. *Journal of Neuroscience* **7**:1503–1519.

Hsieh CY, Cruikshank SJ, and Metherate R (2000) Differential modulation of auditory thalamocortical and intracortical synaptic transmission by cholinergic agonist. *Brain Research* **880**:51–64.

Hu B (1995) Cellular basis of temporal synaptic signalling: an in vitro electrophysiological study in rat auditory thalamus. *Journal of Physiology (London)* **483**:167–182.

Hu B, Senatorov V, and Mooney D (1994) Lemniscal and non-lemniscal synaptic transmission in rat auditory thalamus. *Journal of Physiology (London)* **479**:217–231.

Huang CL, Larue DT, and Winer JA (1999) GABAergic organization of the cat medial geniculate body. *Journal of Comparative Neurology* **415**:368–392.

Huang CL and Winer JA (1997) Areal and laminar distribution of cat auditory thalamocortical projections. *Proceedings of the Society for Neuroscience* **24**:185.

Huang CL and Winer JA (2000) Auditory thalamocortical projections in the cat: laminar and areal patterns of input. *Journal of Comparative Neurology* **427**:302–331.

Humphrey AL, Sur M, Uhlrich DJ, and Sherman SM (1985) Projection patterns of individual X- and Y-cell axons from the lateral geniculate nucleus to cortical area 17 in the cat. *Journal of Comparative Neurology* **233**:159–189.

Imig TJ and Brugge JF (1978) Sources and terminations of callosal axons related to binaural and frequency maps in primary auditory cortex of the cat. *Journal of Comparative Neurology* **182**:637–660.

Imig TJ and Morel A (1985a) Tonotopic organization in lateral part of posterior group of thalamic nuclei in the cat. *Journal of Neurophysiology* **53**:836–851.

Imig TJ and Morel A (1985b) Tonotopic organization in ventral nucleus of medial geniculate body in the cat. *Journal of Neurophysiology* **53**:309–340.

Jacobson M (1975) Development and evolution of type II neurons: conjectures a century after Golgi. In: Santini M (ed). *Golgi Centennial Symposium: Perspectives in Neurobiology*. Raven Press, New York, pp. 147–160.

Jane JA, Masterton RB, and Diamond IT (1965) The function of the tectum for attention to auditory stimuli in the cat. *Journal of Comparative Neurology* **125**:165–192.

Jones EG (2002) Thalamic circuitry and thalamocortical synchrony. *Philosophical Transactions of the Royal Society of London, series B, Biological Sciences* **357**:1659–1673.

Jones EG, Burton H, Saper CB, and Swanson LW (1976) Midbrain, diencephalic and cortical relationships of the basal nucleus of Meynert and associated structures in primates. *Journal of Comparative Neurology* **167**:385–420.

Jürgens U (2002) Neural pathways underlying vocal control. *Neuroscience and Biobehavioral Reviews* **26**:235–258.

Kamke MR, Brown M, and Irvine DR (2003) Plasticity in the tonotopic organization of the medial geniculate body in adult cats following restricted unilateral cochlear lesions. *Journal of Comparative Neurology* **459**:355–367.

Kamke MR, Brown M, and Irvine DRF (2005) Origin and immunolesioning of cholinergic basal forebrain innervation of cat primary auditory cortex. *Hearing Research* **206**:89–106.

Kaur S, Lazar R, and Metherate R (2004) Intracortical pathways determine breadth of subthreshold frequency receptive fields in primary auditory cortex. *Journal of Neurophysiology* **91**: 2551–2567.

Kaur S, Rose HJ, Lazar R, Liang K, and Metherate R (2005) Spectral integration in primary auditory cortex: laminar processing of afferent input, in vivo and in vitro. *Neuroscience* **134**: 1033–1045.

Kawamura K (1973) Corticocortical fiber connections of the cat cerebrum. I. The temporal region. *Brain Research* **51**:1–21.

Kelly JB and Judge PW (1985) Effects of medial geniculate lesions on sound localization by the rat. *Journal of Neurophysiology* **53**: 361–372.

Kimura A, Donishi T, Sakoda T, Hazama M, and Tamai Y (2003) Auditory thalamic nuclei projections to the temporal cortex in the rat. *Neuroscience* **117**:1003–1016.

Kudoh M, Sakai M, and Shibuki K (2002) Differential dependence of LTD on glutamate receptors in the auditory cortical synapses of cortical and thalamic inputs. *Journal of Neurophysiology* **88**: 3167–3174.

Kulesza RJ, Viñuela A, Saldaña E, and Berrebi AS (2002) Unbiased stereological estimates of neuron number in subcortical auditory nuclei of the rat. *Hearing Research* **168**:12–24.

Kurokawa T, Yoshida K, Yamamoto T, and Oka H (1990) Frontal cortical projections from the suprageniculate nucleus in the rat, as demonstrated by the PHA-L method. *Neuroscience Letters* **120**:259–262.

Kuwabara N and Zook JM (2000) Geniculo-collicular descending projections in the gerbil. *Brain Research* **878**:79–87.

Kvasnak E, Popelar J, and Syka J (2000a) Discharge properties of neurons in subdivisions of the medial geniculate body of the guinea pig. *Physiological Research* **49**:369–378.

Kvasnak E, Suta D, Popelar J, and Syka J (2000b) Neuronal connections in the medial geniculate body of the guinea-pig. *Experimental Brain Research* **132**:87–102.

Landry P and Deschênes M (1981) Intracortical arborizations and receptive fields of identified ventrobasal thalamocortical afferents to the primary somatic sensory cortex in the cat. *Journal of Comparative Neurology* **199**:345–372.

LeDoux JE, Ruggiero DA, and Reis DJ (1985) Projections to the subcortical forebrain from anatomically defined regions of the medial geniculate body in the rat. *Journal of Comparative Neurology* **242**:182–213.

LeDoux JE, Sakaguchi A, Iwata J, and Reis DJ (1986) Interruption of projections from the medial geniculate body to an archi-neostriatal field disrupts the classical conditioning of emotional responses to acoustic stimuli. *Neuroscience* **17**:615–627.

Lee CC, Imaizumi K, Schreiner CE, and Winer JA (2004a) Concurrent tonotopic processing streams in auditory cortex. *Cerebral Cortex* **14**:441–451.

Lee CC, Schreiner CE, Imaizumi K, and Winer JA (2004b) Tonotopic and heterotopic projection systems in physiologically defined auditory cortex. *Neuroscience* **128**:871–887.

Lee CC and Winer JA (2005) Principles governing auditory forebrain connections. *Cerebral Cortex* **15**:1804–1814.

Lennartz RC and Weinberger NM (1992) Frequency-specific receptive field plasticity in the medial geniculate body induced by Pavlovian fear conditioning is expressed in the anesthetized brain. *Behavioral Neuroscience* **106**:484–497.

Linke R (1999) Differential projection patterns of superior and inferior collicular neurons onto posterior paralaminar nuclei of the thalamus surrounding the medial geniculate body in the rat. *European Journal of Neuroscience* **11**:187–203.

Linke R and Schwegler H (2000) Convergent and complementary projections of the caudal paralaminar thalamic nuclei to rat temporal and insular cortex. *Cerebral Cortex* **10**:753–771.

Llano DA and Feng AS (2000) Computational models of temporal processing in the auditory thalamus. *Biological Cybernetics* **83**:419–433.

Loftus WC and Sutter ML (2001) Spectrotemporal organization of excitatory and inhibitory receptive fields of cat posterior auditory field neurons. *Journal of Neurophysiology* **86**:475–491.

Lund JS, Wu Q, Hadingham PT, and Levitt JB (1995) Cells and circuits contributing to functional properties in area V1 of macaque monkey cerebral cortex: bases for neuroanatomically realistic models. *Journal of Anatomy (London)* **187**:563–581.

Malmierca MS, Merchán MA, Henkel CK, and Oliver DL (2002) Direct projections from cochlear nuclear complex to auditory thalamus in the rat. *Journal of Neuroscience* **22**:10891–10897.

Malpeli JG and Baker FH (1975) The representation of the visual field in the lateral geniculate nucleus of *Macaca mulatta*. *Journal of Comparative Neurology* **161**:569–594.

Marsh RA, Fuzessery ZM, Grose CD, and Wenstrup JJ (2002) Projection to the inferior colliculus from the basal nucleus of the amygdala. *Journal of Neuroscience* **22**:10449–10460.

McAlonan K, Brown VJ, and Bowman EM (2000) Thalamic reticular nucleus activation reflects attentional gating during classical conditioning. *Journal of Neuroscience* **20**:8897–8901.

McEchron MD, Green EJ, Winters RW, Nolen TG, Schneiderman N, and McCabe PM (1996) Changes of synaptic efficacy in the medial geniculate nucleus as a result of auditory classical conditioning. *Journal of Neuroscience* **16**:1273–1283.

McMullen NT and de Venecia RK (1993) Thalamocortical patches in auditory neocortex. *Brain Research* **620**:317–322.

McMullen NT, Velenovsky DS, and Holmes MG (2005) Auditory thalamic organization: cellular slabs, dendritic arbors and tectothalamic axons underlying the frequency map. *Neuroscience* **136**: 927–943.

Merzenich MM and Reid MD (1974) Representation of the cochlea within the inferior colliculus of the cat. *Brain Research* **77**: 397–415.

Middlebrooks JC and Zook JM (1983) Intrinsic organization of the cat's medial geniculate body identified by projections to binaural response-specific bands in the primary auditory cortex. *Journal of Neuroscience* **3**:203–225.

Miller LM, Escabí MA, Read HL, and Schreiner CE (2001) Functional convergence of response properties in the auditory thalamocortical system. *Neuron* **32**:151–160.

Mitani A, Itoh K, Nomura S, Kudo M, Kaneko T, and Mizuno N (1984) Thalamocortical projections to layer I of the primary auditory cortex in the cat: a horseradish peroxidase study. *Brain Research* **310**: 347–350.

Mitani A, Shimokouchi M, Itoh K, Nomura S, Kudo M, and Mizuno N (1985) Morphology and laminar organization of electrophysiologically identified neurons in primary auditory cortex in the cat. *Journal of Comparative Neurology* **235**:430–447.

Mooney DM, Hu B, and Senatorov VV (1995) Muscarine induces an anomalous inhibition of synaptic transmission in rat auditory thalamic neurons in vitro. *Journal of Pharmacology and Experimental Therapeutics* **275**:838–844.

Morel A and Imig TJ (1987) Thalamic projections to fields A, AI, P ,and VP in the cat auditory cortex. *Journal of Comparative Neurology* **265**:119–144.

Morel A and Kaas JH (1992) Subdivisions and connections of auditory cortex in owl monkeys. *Journal of Comparative Neurology* **318**: 27–63.

Morest DK (1965a) The laminar structure of the medial geniculate body of the cat. *Journal of Anatomy (London)* **99**:143–160.

Morest DK (1965b) The lateral tegmental system of the midbrain and the medial geniculate body: study with Golgi and Nauta methods in cat. *Journal of Anatomy (London)* **99**:611–634.

Morest DK (1971) Dendrodendritic synapses of cells that have axons: the fine structure of the Golgi type II cell in the medial geniculate body of the cat. *Zeitschrift für Anatomie und Entwicklungsgeschichte* **133**:216–246.

Morest DK (1975) Synaptic relationships of Golgi type II cells in the medial geniculate body of the cat. *Journal of Comparative Neurology* **162**:157–194.

Morest DK and Winer JA (1986) The comparative anatomy of neurons: homologous neurons in the medial geniculate body of the opossum and the cat. *Advances in Anatomy, Embryology and Cell Biology* **97**:1–96.

Nelken I (2002) Feature detection by the auditory cortex. In: Oertel D, Popper AN and Fay RR (eds). *Springer Handbook of Auditory Research*, volume 15, *Integrative Functions in the Mammalian Auditory Pathway*. Springer, New York, pp. 359–416.

Olazábal UE and Moore JK (1989) Nigrotectal projection to the inferior colliculus: horseradish peroxidase transport and tyrosine hydroxylase immunohistochemical studies in rats, cats, and bats. *Journal of Comparative Neurology* **282**:98–118.

Oliver DL (2005) Neuronal organization of the inferior colliculus. In: Winer JA and Schreiner CE (eds). *The Inferior Colliculus*. Springer, New York, pp. 69–114.

Oliver DL, Winer JA, Beckius GE, and Saint Marie RL (1994) Morphology of GABAergic cells and axon terminals in the cat inferior colliculus. *Journal of Comparative Neurology* **340**:27–42.

Pape HC and McCormick DA (1995) Electrophysiological and pharmacological properties of interneurons in the cat dorsal lateral geniculate nucleus. *Neuroscience* **68**:1105–1125.

Perales M, Winer JA, and Prieto JJ (2006) Focal projections of cat auditory cortex to the pontine nuclei. *Journal of Comparative Neurology* **497**:959–980.

Pollak GD, Burger RM, and Klug A (2003) Dissecting the circuitry of the auditory system. *Trends in Neurosciences* **26**:33–39.

Prieto JJ, Peterson BA, and Winer JA (1994a) Laminar distribution and neuronal targets of GABAergic axon terminals in cat primary auditory cortex (AI). *Journal of Comparative Neurology* **344**: 383–402.

Prieto JJ, Peterson BA, and Winer JA (1994b) Morphology and spatial distribution of GABAergic neurons in cat primary auditory cortex (AI). *Journal of Comparative Neurology* **344**:349–382.

Prieto JJ and Winer JA (1999) Layer VI in cat primary auditory cortex (AI): Golgi study and sublaminar origins of projection neurons. *Journal of Comparative Neurology* **404**:332–358.

Radnikow G, Feldmeyer D, and Lübke J (2002) Axonal projection, input and output synapses, and synaptic physiology of Cajal-Retzius cells in the developing rat neocortex. *Journal of Neuroscience* **22**:6908–6919.

Rauschecker JP and Tian B (2000) Mechanisms and streams for processing of "what" and "where" in auditory cortex. *Proceedings of the National Academy of Sciences of the United States of America* **97**:11800–11806.

Rauschecker JP, Tian B, Pons T, and Mishkin M (1997) Serial and parallel processing in rhesus monkey auditory cortex. *Journal of Comparative Neurology* **382**:89–103.

Read HL, Winer JA, and Schreiner CE (2001) Modular organization of intrinsic connections associated with spectral tuning in cat auditory cortex. *Proceedings of the National Academy of Sciences of the United States of America* **98**:8042–8047.

Read HL, Winer JA, and Schreiner CE (2002) Functional architecture of auditory cortex. *Current Opinion in Neurobiology* **12**:433–440.

Reale RA and Imig TJ (1980) Tonotopic organization in auditory cortex of the cat. *Journal of Comparative Neurology* **192**:265–291.

Reale RA and Imig TJ (1983) Auditory cortical field projections to the basal ganglia of the cat. *Neuroscience* **8**:67–86.

Reblet C, López-Medina A, Gómez-Urquijo SM, and Bueno-López JL (1992) Widespread horizontal connections arising from layer 5/6 border inverted cells in rabbit visual cortex. *European Journal of Neuroscience* **4**:221–234.

Riquimaroux H, Gaioni SJ, and Suga N (1992) Inactivation of the DSCF area of the auditory cortex with muscimol disrupts frequency discrimination in the mustached bat. *Journal of Neurophysiology* **68**:1613–1623.

Romanski LM, Clugnet M, Bordi F, and LeDoux JE (1993) Somatosensory and auditory convergence in the lateral nucleus of the amygdala. *Behavioral Neuroscience* **107**:444–450.

Romanski LM and LeDoux JE (1993) Information cascade from primary auditory cortex to the amygdala: corticocortical and corticoamygdaloid projections of temporal cortex in the rat. *Cerebral Cortex* **3**:515–532.

Rose HJ and Metherate R (2005) Auditory thalamocortical transmission is reliable and temporally precise. *Journal of Neurophysiology* **94**:2019–2030.

Rose JE and Woolsey CN (1958) Cortical connections and functional organization of thalamic auditory system of cat. In: Harlow HF and Woolsey CN (eds). *Biological and Biochemical Bases of Behavior*. University of Wisconsin Press, Madison, pp. 127–150.

Rouiller EM and Durif C (2004) The dual pattern of corticothalamic projection of the primary auditory cortex in macaque monkey. *Neuroscience Letters* **358**:49–52.

Rouiller EM, Rodrigues-Dagaeff C, Simm GM, de Ribaupierre Y, Villa AEP, and de Ribaupierre F (1989) Functional organization of the medial division of the medial geniculate body of the cat: tonotopic organization, spatial distribution of response properties and cortical connections. *Hearing Research* **39**:127–146.

Rouiller EM and Welker E (2000) A comparative analysis of the morphology of corticothalamic projections in mammals. *Brain Research Bulletin* **53**:727–741.

Samson FK, Barone P, Irons WA, Clarey JC, Poirier P, and Imig TJ (2000) Directionality derived from differential sensitivity to monaural and binaural cues in the cat's medial geniculate body. *Journal of Neurophysiology* **84**:1330–1345.

Scheibel ME and Scheibel AB (1966) Patterns of organization in specific and nonspecific thalamic fields. In: Purpura DP and Yahr MD (eds). *The Thalamus*. Columbia University Press, New York, pp. 13–46.

Schofield BR and Coomes DL (2004) Projections from the auditory cortex to the superior olivary complex in guinea pigs. *European Journal of Neuroscience* **19**:2188–2200.

Schofield BR and Coomes DL (2005) Auditory cortical projections to the cochlear nucleus in guinea pigs. *Hearing Research* **199**:89–102.

Schreiner CE and Cynader MS (1984) Basic functional organization of second auditory cortical field (AII) of the cat. *Journal of Neurophysiology* **51**:1284–1305.

Schreiner CE and Langner G (1997) Laminar fine structure of frequency organization in auditory midbrain. *Nature* **388**:383–386.

Schuller G, Fischer S, and Schweizer H (1997) Significance of the paralemniscal tegmental area for audio-motor control in the moustached bat, *Pteronotus p. parnellii*: the afferent and efferent connections of the paralemniscal area. *European Journal of Neuroscience* **9**:342–355.

Senatorov VV and Hu B (2002) Extracortical descending projections to the rat inferior colliculus. *Neuroscience* **115**:243–250.

Sherman SM and Guillery RW (1996) Functional organization of thalamocortical relays. *Journal of Neurophysiology* **76**:1367–1395.

Sherman SM and Guillery RW (2000) *Exploring the Thalamus*. Academic Press, Orlando.

Shi C-J and Cassell MD (1997) Cortical, thalamic, and amygdaloid projections of rat temporal cortex. *Journal of Comparative Neurology* **382**:153–175.

Shinonaga Y, Takada M, and Mizuno N (1994) Direct projections from the non-laminated divisions of the medial geniculate nucleus to the temporal polar cortex and amygdala in the cat. *Journal of Comparative Neurology* **340**:405–426.

Smith PH, Joris PX, and Yin TCT (1993) Projections of physiologically characterized spherical bushy cell axons from the cochlear nucleus of the cat: evidence for delay lines to the medial superior olive. *Journal of Comparative Neurology* **331**:245–260.

Smith PH and Populin LC (2001) Fundamental differences between the thalamocortical recipient layers of the cat auditory and visual cortices. *Journal of Comparative Neurology* **436**:508–519.

Smith PH and Spirou GA (2002) From the cochlea to the cortex and back. In: Oertel D, Fay RR and Popper AN (eds). *Springer Handbook of Auditory Research*, volume 15, *Integrative Functions in the Mammalian Auditory Pathway*. Springer, New York, pp. 6–71.

Sousa-Pinto A (1973) The structure of the first auditory cortex (A I) in the cat. I. — Light microscopic observations on its structure. *Archives Italiennes de Biologie* **111**:112–137.

Strominger NL, Nelson LR, and Dougherty WJ (1977) Second order auditory pathways in the chimpanzee. *Journal of Comparative Neurology* **172**:349–366.

Sutter ML (2000) Shapes and level tolerances of frequency tuning curves in primary auditory cortex: quantitative measures and population codes. *Journal of Neurophysiology* **84**:1012–1025.

Sutter ML and Loftus WC (2003) Excitatory and inhibitory intensity tuning in auditory cortex: evidence for multiple inhibitory mechanisms. *Journal of Neurophysiology* **90**:2629–2647.

Syka J, Popelar J, Kvasnak E, and Astl J (2000) Response properties of neurons in the central nucleus and external and dorsal cortices of the inferior colliculus in guinea pig. *Experimental Brain Research* **133**:254–266.

Thomas H and Lopez V (2003) Comparative study of inter- and intrahemispheric cortico-cortical connections in gerbil auditory cortex. *Biological Research* **36**:155–169.

Van Essen DC (2005) Corticocortical and thalamocortical information flow in the primate visual system. *Progress in Brain Research* **69**:215–237.

Vogt BA (1991) The role of layer I in cortical function. In: Peters A (ed). *Cerebral Cortex*. Plenum Press, New York, pp. 49–80.

Volkov IO and Galazjuk AV (1991) Formation of spike response to sound tones in cat auditory cortex neurons: interaction of excitatory and inhibitory effects. *Neuroscience* **43**:307–321.

Wallace MN and Harper MS (1997) Callosal connections of the ferret primary auditory cortex. *Experimental Brain Research* **116**:367–374.

Weedman DL and Ryugo DK (1996a) Projections from auditory cortex to the cochlear nucleus in rats: synapses on granule cell dendrites. *Journal of Comparative Neurology* **371**:311–324.

Weedman DL and Ryugo DK (1996b) Pyramidal cells in primary auditory cortex project to cochlear nucleus in rat. *Brain Research* **706**:97–102.

Wenstrup JJ (1999) Frequency organization and responses to complex sounds in the medial geniculate body of the mustached bat. *Journal of Neurophysiology* **82**:2528–2544.

Wenstrup JJ (2005) The tectothalamic system. In: Winer JA and Schreiner CE (ed). *The Inferior Colliculus*. Springer, New York, pp. 200–230.

Wenstrup JJ and Leroy SA (2001) Spectral integration in the inferior colliculus: role of glycinergic inhibition in response facilitation. *Journal of Neuroscience* **21**:RC124 (121–126).

Wepsic JG (1966) Multimodal sensory activation of cells in the magnocellular medial geniculate nucleus. *Experimental Neurology* **15**:299–318.

Wester K, Irvine DRF, and Hugdahl K (2001) Auditory laterality and attentional deficits after thalamic haemorrhage. *Journal of Neurology* **248**:676–683.

Winer JA (1984a) Anatomy of layer IV in cat primary auditory cortex (AI). *Journal of Comparative Neurology* **224**:535–567.
Winer JA (1984b) The human medial geniculate body. *Hearing Research* **15**:225–247.
Winer JA (1984c) The non-pyramidal neurons in layer III of cat primary auditory cortex (AI). *Journal of Comparative Neurology* **229**: 512–530.
Winer JA (1984d) The pyramidal cells in layer III of cat primary auditory cortex (AI). *Journal of Comparative Neurology* **229**:476–496.
Winer JA (1985) Structure of layer II in cat primary auditory cortex (AI). *Journal of Comparative Neurology* **238**:10–37.
Winer JA (1986) Neurons accumulating [3H]gamma-aminobutyric acid (GABA) in supragranular layers of cat primary auditory cortex (AI). *Neuroscience* **19**:771–793.
Winer JA (1992) The functional architecture of the medial geniculate body and the primary auditory cortex. In: Webster DB, Popper AN, and Fay RR (eds). *Springer Handbook of Auditory Research, volume 1, The Mammalian Auditory Pathway: Neuroanatomy*. Springer, New York, pp. 222–409.
Winer JA (2005) Three systems of descending projections to the inferior colliculus. In: Winer JA and Schreiner CE (ed). *The Inferior Colliculus*. Springer, New York, pp. 231–247.
Winer JA (2006) Decoding the auditory corticofugal systems. *Hearing Research* **212**:1–8.
Winer JA, Diamond IT, and Raczkowski D (1977) Subdivisions of the auditory cortex of the cat: the retrograde transport of horseradish peroxidase to the medial geniculate body and posterior thalamic nuclei. *Journal of Comparative Neurology* **176**:387–418.
Winer JA, Diehl JJ, and Larue DT (2001) Projections of auditory cortex to the medial geniculate body of the cat. *Journal of Comparative Neurology* **430**:27–55.
Winer JA and Larue DT (1987) Patterns of reciprocity in auditory thalamocortical and corticothalamic connections: study with horseradish peroxidase and autoradiographic methods in the rat medial geniculate body. *Journal of Comparative Neurology* **257**:282–315.
Winer JA and Larue DT (1988) Anatomy of glutamic acid decarboxylase (GAD) immunoreactive neurons and axons in the rat medial geniculate body. *Journal of Comparative Neurology* **278**:47–68.
Winer JA and Larue DT (1989) Populations of GABAergic neurons and axons in layer I of rat auditory cortex. *Neuroscience* **33**: 499–515.
Winer JA and Larue DT (1996) Evolution of GABAergic circuitry in the mammalian medial geniculate body. *Proceedings of the National Academy of Sciences of the United States of America* **93**: 3083–3087.
Winer JA, Larue DT, Diehl JJ, and Hefti BJ (1998) Auditory cortical projections to the cat inferior colliculus. *Journal of Comparative Neurology* **400**:147–174.
Winer JA, Larue DT, and Huang CL (1999a) Two systems of giant axon terminals in the cat medial geniculate body: convergence of cortical and GABAergic inputs. *Journal of Comparative Neurology* **413**:181–197.
Winer JA, Larue DT, and Pollak GD (1995) GABA and glycine in the central auditory system of the mustache bat: structural substrates for inhibitory neuronal organization. *Journal of Comparative Neurology* **355**:317–353.
Winer JA and Lee CC (2007) The distributed auditory cortex. *Hearing Research* **229**:3–13.
Winer JA, Miller LM, Lee CC, and Schreiner CE (2005) Auditory thalamocortical transformation: structure and function. *Trends in Neurosciences* **28**:255–263.
Winer JA and Morest DK (1983a) The medial division of the medial geniculate body of the cat: implications for thalamic organization. *Journal of Neuroscience* **3**:2629–2651.
Winer JA and Morest DK (1983b) The neuronal architecture of the dorsal division of the medial geniculate body of the cat. A study with the rapid Golgi method. *Journal of Comparative Neurology* **221**:1–30.
Winer JA, Morest DK, and Diamond IT (1988) A cytoarchitectonic atlas of the medial geniculate body of the opossum, *Didelphys virginiana*, with a comment on the posterior intralaminar nuclei of the thalamus. *Journal of Comparative Neurology* **274**:422–448.
Winer JA and Prieto JJ (2001) Layer V in cat primary auditory cortex (AI): cellular architecture and identification of projection neurons. *Journal of Comparative Neurology* **434**:379–412.
Winer JA, Saint Marie RL, Larue DT, and Oliver DL (1996) GABAergic feedforward projections from the inferior colliculus to the medial geniculate body. *Proceedings of the National Academy of Sciences of the United States of America* **93**:8005–8010.
Winer JA, Sally SL, Larue DT, and Kelly JB (1999b) Origins of medial geniculate body projections to physiologically defined regions of rat auditory cortex. *Hearing Research* **130**:42–61.
Winer JA and Schreiner CE (2005) The central auditory system: a functional analysis. In: Winer JA and Schreiner CE (eds). *The Inferior Colliculus*. Springer, New York, pp. 1–68.
Winer JA and Wenstrup JJ (1994) The neurons of the medial geniculate body in the mustached bat *(Pteronotus parnellii)*. *Journal of Comparative Neurology* **346**:183–206.
Wong D and Kelly JP (1981) Differentially projecting cells in individual layers of the auditory cortex: a double-labeling study. *Brain Research* **230**:362–366.

Chapter 3

Thalamocortical Relations

Matthew I. Banks and Philip H. Smith

Abbreviations

AC	auditory cortex
AI	primary auditory cortex
AM	amplitude modulated
AMPA	α-amino-3-hydroxy-4-isoxazoepropionic acid
CB	calbindin
CC	corticocortical
RF	receptive field
CF	characteristic frequency
CM	caudomedial cortical area
DC	dorsal cortex
EC	external nucleus
ECT	ectorhinal
EE	bilateral summation
EI	bilateral suppression
EO	bilateral occlusion
EPSP	excitatory postsynaptic potentials
FS	fast spiking
GABA	g-aminobutyric acid
IB	intrinsically bursting
IC	inferior colliculus
ICC	central nucleus of the inferior colliculus
IPSP	inhibitory postsynaptic potential
MGad	anterior subdivision of the dorsal division of the medial geniculate body
MGB	medial geniculate body
MGD	dorsal nucleus of the medial geniculate body
MGM	medial division of the medial geniculate body
MGpd	posterior subdivision of the dorsal division of the medial geniculate body
MGV	ventral nucleus of the medial geniculate body
NMDA	*N*-methyl-D-aspartate
OS	on-spiking
P	perirhinal
PD	posterodorsal region
PI	posterior interlaminar
PP	peripeduncular
PV	parvalbumin
RECT	rectifying
RF	receptive field
RM	rostromedial cortical area
RS	regular-spiking
SG	suprageniculate
STRF	spectrotemporal receptive field
TC	thalamocortical
TRN	thalamic reticular nucleus

P.H. Smith (✉)
Department of Anatomy, University Wisconsin Medical School, Madison, WI 53706, USA
e-mail: smith@neurophys.wisc.edu

1 The Auditory Thalamus

Since about 1995, a new view of the sensory thalamus and its influence on the cortex has emphasized that thalamocortical (TC) cells are not simple relay neurons whose sole function is to transfer sensory information, without modification, from the animal's environment to the sensory cortex. Rather, TC cells are now viewed as highly sensitive modifiers of this information that base the nature and degree of their modification upon the state of the organism (Edeline 2003; Winer et al. 2005). The effect of these state changes is brought about by the other indirect non-environmental inputs which make up more than 80% of the synapses impinging on a thalamocortical cell. Equally striking is that some TC cells may not be activated primarily by their direct inputs from the sensory environment but instead may be more strongly driven by the internal environment, that is, by neural sources such as the cerebral and cerebellar cortex that lie outside the ascending sensory pathway (Bender 1983; Diamond et al. 1992; Guillery 1995; Rouiller and Welker 2000). Thus, the information reaching the cortex by the TC system is in a continuous state-dependent flux.

To add to this complexity, the axons that constitute the thalamic projections onto sensory cortex project to multiple

J.A. Winer, C.E. Schreiner (eds.), *The Auditory Cortex*,
DOI 10.1007/978-1-4419-0074-6_3, © Springer Science+Business Media, LLC 2011

layers and terminate at different locations on a wide variety of cell types. The TC synapse and the response of the cortical cell postsynaptic to that input can also be modified by other projections, only some of which are modality specific (Edeline 2003; Metherate et al. 2005). The cortical circuitry on which auditory TC synapses act differs in significant ways from that found in other sensory cortical areas (Miller et al. 2001a; Smith and Populin 2001). In this chapter, we shall provide a picture of what is known about TC cells and how they influence their postsynaptic auditory cortical targets.

1.1 Auditory Thalamic Subdivisions

In species studied thus far (cat: Morest 1964; rat: Clerici and Coleman 1990, monkey: de la Mothe et al. 2006; human: Winer 1984), the auditory thalamus or medial geniculate body (MGB) has been subdivided into ventral, dorsal, and medial divisions.

1.1.1 Ventral Division

The ventral division of the MGB (MGV) is usually subdivided into a tonotopically organized *pars lateralis*, where the cells are organized into a series of parallel laminae, and a more medial *pars ovoidea*, whose cells have a less tonotopic organization and a coiled arrangement. MGV is part of the tonotopically organized primary or lemniscal pathway and virtually all of its projections are thalamocortical. In cat, a second tonotopic auditory nucleus, the lateral part of the posterior thalamic nucleus, is rostral to MGV (Clarey et al. 1992). Anatomically, TC MGV cells are classified as bushy or tufted by their highly confined and well-branched bipolar dendritic tree (Morest 1964; Winer 1985; Clerici et al. 1990).

The input to these cells comes from the tonotopically organized central nucleus of the inferior colliculus (IC) and arises from both its major cell types, disc shaped and stellate neurons (Fig. 3.1) (Andersen et al. 1980; Calford and Aitkin 1983; Oliver 1984). Their terminals form excitatory or inhibitory synapses on the dendrites and use glutamate

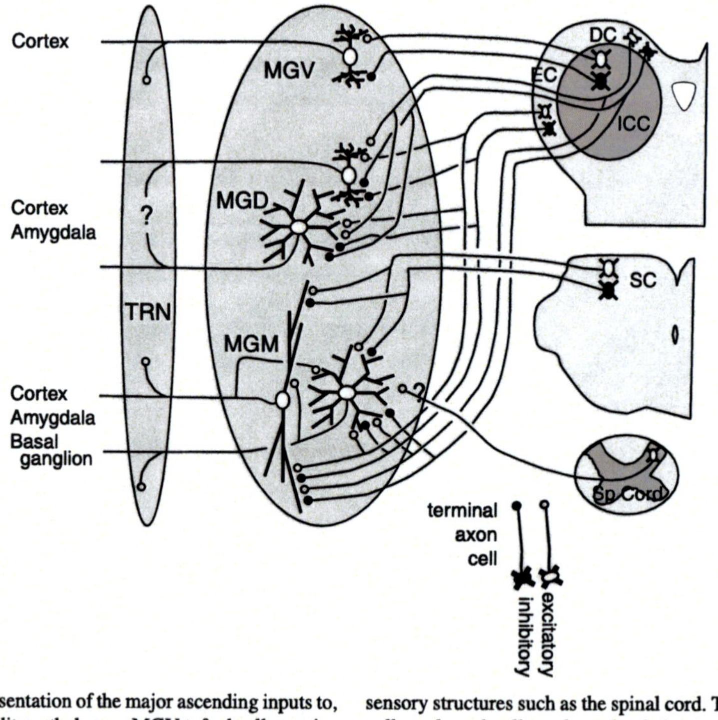

Fig. 3.1 Diagrammatic representation of the major ascending inputs to, and outputs of, cells in the auditory thalamus. MGV tufted cells receive ascending inputs primarily from the central nucleus of the inferior colliculus and project to primary auditory cortex, giving off collaterals to the thalamic reticular nucleus. MGD tufted and stellate cells receive ascending inputs from cortical areas of the inferior colliculus and can project to secondary cortical areas or the amygdala. MGM stellate or elongate cells can receive ascending inputs from cortical subdivisions of the inferior colliculus as well as the superior colliculus and somatic sensory structures such as the spinal cord. Their axons can give off local collaterals and collaterals to the thalamic reticular nucleus before projecting to secondary auditory cortical areas, amygdala, or basal ganglia. Abbreviations: DC, dorsal cortex of IC; EC, external cortex of the IC; ICC, central nucleus of the inferior colliculus; MGD, dorsal division of the medial geniculate body; MGM, medial division of the medial geniculate body; MGV, ventral division of the medial geniculate body; SC, superior colliculus; Sp Cord, spinal cord; TRN, thalamic reticular nucleus

and GABA (gamma-aminobutyric acid) as their neurotransmitters (Peruzzi et al. 1997; Bartlett and Smith 2000). The glutamatergic inputs act on AMPA (α-amino-3-hydroxy-4-isoxazoepropionic acid) and NMDA (*N*-methyl-D-aspartate) receptors, while GABAergic inputs act on $GABA_A$ and $GABA_B$ receptors (Fig. 3.2a, bottom two panels). The axons of TC cells in MGV leave MGB without local collaterals (Winer 1985; Bartlett and Smith 1999) but in *Galago* (a prosimian primate) provide collaterals to the GABAergic cells of the thalamic reticular nucleus (TRN) (Fig. 3.1) (Conley et al. 1991) en route to the cortex.

1.1.2 Dorsal Division

The dorsal division (MGD) is a component of the extra- or nonlemniscal auditory thalamus and in some species has been divided into caudodorsal, deep dorsal, superficial

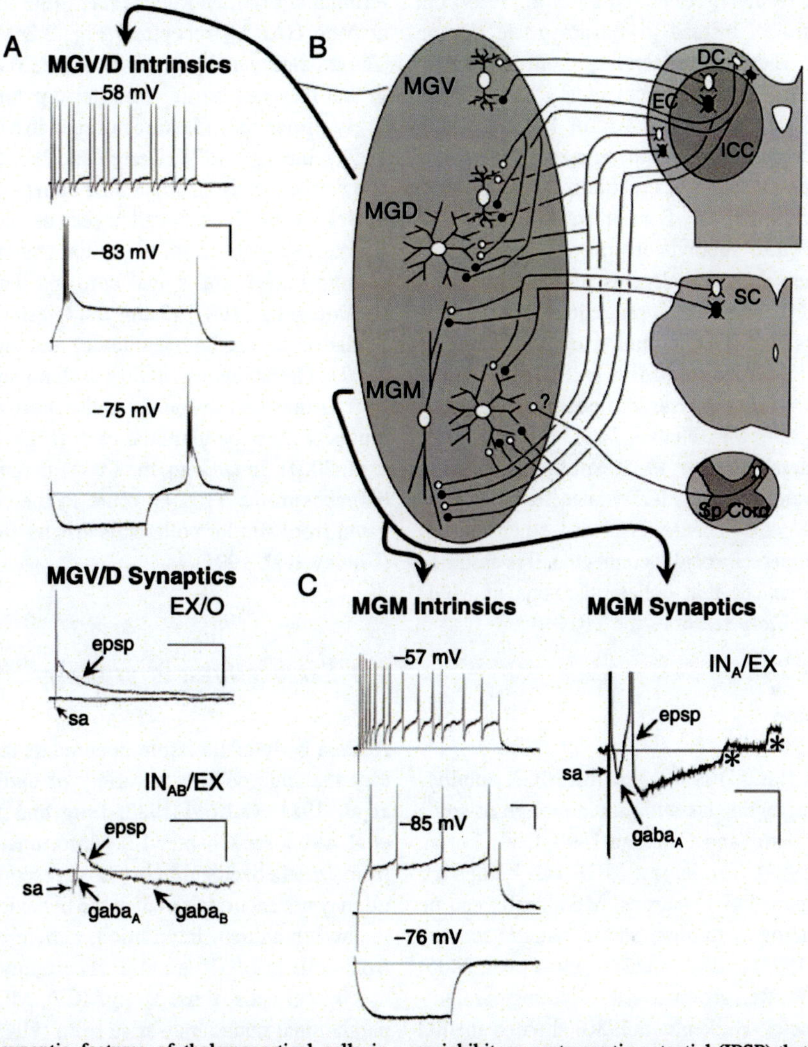

Fig. 3.2 Intrinsic and synaptic features of thalamocortical cells in MGB subdivisions. **a** MGV/D intrinsic organization. Typical response of a MGV or MGD cell to a depolarizing current pulse as the cell membrane potential is held at a depolarized (*top*) or a hyperpolarized (*middle*) level. When a hyperpolarizing pulse is applied at the appropriate membrane potential (*bottom*), the cell generates a rebound burst of spikes. MGV/D synaptic organization. Typical synaptic response of cells in the MGD or MGV to shocking the input from the IC. *Top*: in some tufted cells, shocking the IC input elicited a large amplitude excitatory postsynaptic potential (EPSPs) that could be suprathreshold at the cells resting membrane potential. *Bottom*: shocking the IC input to some tufted or stellate cells elicited a smaller amplitude EPSP as well as an inhibitory postsynaptic potential (IPSP) that had both $GABA_A$ and $GABA_B$ components. **b** Diagrammatic representation of the MGB cell types and their inputs. **c** MGM intrinsic organization. Typical response of an MGM cell to a depolarizing current pulse as the cell membrane potential is held at a depolarized (*top*) or hyperpolarized (*middle*) level. Note the lack of a calcium burst at either membrane potential. When a hyperpolarizing pulse is applied, no rebound calcium burst is activated at the pulse offset at any membrane potential (*bottom*). MGM synaptic organization: shocking either the IC or SC inputs to MGM cells could elicit both an EPSP and an IPSP. The IPSP typically had only a $GABA_A$ component. Sa, shock artifact. Other abbreviations as in Fig. 3.1

dorsal, and dorsal subnuclei (Clerici and Coleman 1990; Clerici et al. 1990; Winer et al. 1999). In monkeys only anterior and posterior dorsal divisions are recognized (de la Mothe et al. 2006). The adjacent suprageniculate nucleus is usually grouped with the paralaminar nuclei (see below) or is considered as part of MGD. Many MGD cells are tufted or stellate and they do not have a precise tonotopic arrangement (Clarey et al. 1992). Both types can be thalamocortical (Arnault and Roger 1990; Roger and Arnault 1989; Clerici and Coleman 1990; McMullen and de Venecia 1993; Kharazia and Weinberg 1994; Winer et al. 1999) but many MGD cells project instead to the amygdala (Doron and LeDoux 2000). The major ascending input to cells in MGD arises from the IC external nucleus (EC) and dorsal cortex (DC) (Fig. 3.1) (Calford and Aitkin 1983; LeDoux et al. 1985) and the midbrain lateral tegmental system, a region medial and ventral to the IC brachium (Morest 1965), although the relative amounts of input from different IC areas to the various MGD subdivisions may be species specific. The MGD deep dorsal nucleus is unusual in that it may receive most of its collicular input from ICC. Like the MGV, the IC inputs to MGD terminate on the dendrites and are glutamatergic, acting on AMPA and NMDA receptors, or GABAergic (γ-aminobutyric acid-positive), acting on $GABA_A$ or $GABA_B$ receptors. Neither MGD cell type gives off local axon collaterals (Winer 1985; Smith et al. 2006). Because the TRN projects to the MGD in cat (Rouiller et al. 1985; Crabtree 1998) and because TRN and other sensory thalamic areas are often reciprocally connected, it is assumed (although not documented) that collateral axons of MGD cells project to the auditory sector of the TRN.

1.1.3 Medial Division

The medial division (MGM) is the smallest MGB subdivision yet the most heterogeneous with a wide range of soma size (Morest 1964: Clerici and Coleman 1990; Clerici et al. 1990; Winer et al. 1999). The largest MGB cell bodies are found here, the magnocellular neurons. MGM cells can be multisensory, responding to auditory and to somatic sensory (Bordi and LeDoux 1994) and/or visual (Komura et al. 2005) stimuli. Thus MGM is often grouped with adjoining paralaminar multisensory nuclei (Herkenham 1980) situated medial and ventromedial to the MGB. This group includes the posterior interlaminar (PI), suprageniculate (SG), and peripeduncular (PP) nuclei (Winer and Morest 1983; LeDoux et al. 1987; Winer and Larue 1988). In MGM and paralaminar nuclei, several cell types, including magnocellular cells, are thalamocortical, but a substantial proportion instead project to the amygdala and striatum (LeDoux et al. 1985). Some cells have stellate morphology like MGD cells, but with fewer, less branched dendrites. Other MGM cells have long, sparsely branching elongate dendritic fields that encompass a much larger area. Both the stellate and elongate dendritic trees may be smooth but, unlike MGV and MGD cells, some MGM cells have many spines (Smith et al. 2006). The major ascending auditory input to the paralaminar nuclei is from the external nucleus and dorsal cortex with minor inputs from ICC (Fig. 3.1). The DC and EC input to each paralaminar nucleus arise from different DC and EC layers (Linke 1999). Slice experiments show that, like MGV and MGD, MGM cells receive both excitatory and inhibitory IC inputs terminating on dendrites, except that inhibitory inputs only activate $GABA_A$ receptors (Fig. 3.2c) (Smith et al. 2007). The excitatory auditory inputs to MGM cells are also unusual in that they can display long-term potentiation when repetitively stimulated, a feature not seen in MGV or MGD (Ryugo and Weinberger 1978; Gerren and Weinberger 1983). Visual and somatic sensory input also converge on the paralaminar nuclei, primarily from all layers of the superior colliculus (Linke 1999) and from the spinal trigeminal nucleus, dorsal column nuclei, and spinal cord via the spinothalamic tract (LeDoux et al. 1987; Zhang and Giesler 2005). Superior collicular inputs are both excitatory and inhibitory (Smith et al. 2007). The axons of MGM/paralaminar cells can have collaterals that branch locally and appear to provide excitatory inputs to other paralaminar cells (Figs. 3.1 and 3.2c) (Smith et al. 2007). In *Galago*, the MGM thalamocortical cells send branches to the TRN en route to the cortex that are segregated from similar collaterals arising from MGV cell axons (Conley et al. 1991).

1.2 Extracellular Physiology

A large body of literature documents the responses of auditory thalamic cells to a variety of auditory stimuli (Aitkin et al. 1981; Calford 1983; Imig and Morel 1988; Clarey et al. 1992; see Chapter 12). Unfortunately, most reports are from anesthetized animals whose responses to auditory stimuli may not accurately reflect the behavior of the same cells in the awake animal. In addition, although responses recorded from cells in MGV are almost certainly coming from thalamocortical cells, those from MGD, MGM and surrounding paralaminar nuclei may arise from cells whose axons project elsewhere.

1.2.1 Ventral Division

Despite these caveats, findings from such studies indicate that MGV cells are arranged tonotopically and that most are binaural and driven by both ears. They tend to be sharply tuned and have a reliable short latency transient discharge (onset response) to repeated tonal stimuli at their characteristic frequency (CF).

1.2.2 Dorsal Division

About 10% of MGD units have responses resembling those typical of MGV cells. However, MGD is the least auditory-responsive thalamic nucleus. Responses resemble those of MGV cells in that they are binaural, excited by both ears and respond transiently to repetitive tonal stimuli; they differ in that they are broadly tuned, have longer first spike latencies, greater spike timing variability, and low reliability even at 1 Hz repetition rates. The area between MGV and MGD, sometimes known as the deep dorsal nucleus, may be a transition zone with responses that are intermediate to these extremes.

1.2.3 Medial Division

Like cells in MGV and MGD, most cells in MGM and the paralaminar nuclei are binaural and excited by both ears. Many MGM and paralaminar cells are similar in having onset responses, though 20–30% show sustained responses to auditory stimuli. The CF distribution favors high frequencies (Aitkin 1973; Calford 1983; Bordi and LeDoux 1994), though some MGM cells with low CFs show unusually short click latencies and appreciable phase locking (Wallace et al. 2007), possibly a result of their direct input from the cochlear nucleus (Malmierca et al. 2002; Anderson et al. 2006). Thus, MGM cell responses are heterogeneous, like their anatomical features. Besides their auditory responses, many MGM/paralaminar cells respond to somatic sensory input (Bordi and LeDoux 1994) primarily from nociceptors (Zhang and Giesler 2005). Cells respond to both somatic sensory and auditory stimuli or, those responding only to auditory stimuli show enhanced responsiveness to concurrent somatic sensory input. Visual input can also influence these cells. MGB cells in the paralaminar nuclei respond to either visual or auditory stimuli presented separately or their auditory response is significantly enhanced by concurrent visual stimuli (Komura et al. 2005). Stimulation of the superior colliculus (SC), a probable source of the visual input, evoked spikes in MGM cells only (McEchron et al. 1996).

1.3 Intracellular Physiology

A major finding in the thalamus since 1980 was the discovery that many TC neurons can respond to the same input stimulus in fundamentally different ways depending upon their membrane potential. At a resting potential of about –70 mV, depolarization activates a Ca^{2+} conductance, resulting in a large, rapid depolarization that can elicit 3–5 high frequency (>200 Hz) spikes (Fig. 3.2a). If depolarized to –50 or –60 mV, the Ca^{2+} conductance is inactivated and that same depolarization will, instead, elicit a shorter latency, well-timed spike (Jahnsen and Llinás 1984). Thus, the temporal information conveyed to the cortex is closely related to membrane potential (Sherman 2001). The important observation is exemplified by finding thalamocortical synapses on layer IV GABAergic interneurons in the visual cortex, in whom the first spike in a burst is twice as likely to elicit a postsynaptic spike as subsequent arrivals (Swadlow and Gusev 2001). This indicates that bursting provides a more reliable, powerful signaling mechanism for the cortex.

Intracellular recordings in brain slices from rat MGV and MGD show that, like thalamocortical cells elsewhere in the lemniscal sensory thalamus, MGV tufted TC cells and tufted or stellate MGD TC cells respond to depolarization with a tonic or a burst response contingent on membrane potential (Hu 1995; Peruzzi et al. 1997; Tennigkeit et al. 1997; Bartlett and Smith 1999). The membrane potential is labile over time and can be modified by the activation of metabotropic glutamatergic receptors at cortical synapses (Bartlett and Smith 1999), by brain stem muscarinic cholinergic inputs (Mooney et al. 2003) and probably by other modulatory inputs as well. The few in vivo intracellular recordings from MGV cells (Yu et al. 2004) confirm a similar membrane-dependent change in response pattern in vivo. In contrast, some MGM cells show bursting behavior, although many cells there and in the adjoining paralaminar nuclei have a reduced or complete absence of the low threshold, voltage-sensitive Ca^{2+} conductance that reduces or eliminates the voltage-dependent burst response (Fig. 3.2c) (Smith et al. 2006).

1.4 Medial Geniculate Body Responses in Unanesthetized Animals

Cells in the auditory thalamus and cortex are affected more dramatically by anesthetics than are the responses of cells at lower levels in the pathway including the IC (Zurita et al. 1994; Ter Mikaelian et al. 2007). An accurate record of the responses of TC cells and the possible transformations ensuing as information flows from thalamus to cortex (see below) rests on the results from awake animals. Moreover, the responses of TC cells to a given input vary considerably depending on the state of the animal when the input arrives. Some variability reflects the change in the relative percentages of spikes occurring in bursts versus single spikes, changes likely reflecting shifts in membrane potential that alter the low threshold calcium conductance described above. Recordings from visual thalamus (lateral geniculate nucleus, LGN) show that the fraction of spikes in bursts and the number of bursts per minute reflect behavioral state: in awake monkeys (Ruiz et al. 2006) and cats (Weyand et al. 2001) burst rates were lowest, whereas during sleep and under anesthesia they increased several-fold (Weyand et al. 2001; Denning and Reinagel 2005).

There have been a few studies in the auditory system in which responses in awake conditions were recorded. Responses from barbiturate-anesthetized and awake (galamine-paralyzed) cats were compared and cells in the awake animal had higher spontaneous spike rates and they responded in either single spike or burst mode, although the MGB recording loci were not specified (Aitkin et al. 1966). A small population of MGB cells in awake cats anatomically verified to be in the three major subdivisions also showed higher spontaneous spike activity and these spikes occurred more often as irregular trains than the "high frequency bursts" seen in the anesthetized preparation. Although their sample was small, sustained responses to tones were much more prevalent than in anesthetized preparations, where phasic onset responses were the norm (Aitkin and Prain 1974).

Similar results are seen in more recent studies which have considered MGB response features in unanesthetized, awake animals and the effects of alertness and anesthesia. In awake macaque monkey, MGV cell activity was primarily single spikes with burst responses comprising <10% of spikes (Ramcharan et al. 2005). In all MGB subdivisions, the proportion of spontaneous or tone-evoked spikes in bursts was lowest in the awake and highest in barbiturate-anesthetized guinea-pigs, and spontaneous activity was higher in awake animals (Massaux and Edeline 2003). Spontaneous and evoked bursting activity was highest in the awake MGV compared to sleeping and anesthetized states and bursting was lowest in awake guinea-pigs. Even though spikes in bursts were minor response component, they appeared preferentially at frequencies near the cells' CF and, when present, response latency and variability decreased (Massaux et al. 2004). MGB activity also varies as a function of sleep/wake states and these changes were consistent in all subdivisions. In slow wave sleep spontaneous and evoked firing rates, onset latency and receptive field size decreased, while threshold and number of spontaneous and tone-evoked spikes in bursts increased relative to the awake state. Similar changes were noted in most of these parameters when activity awake was compared with that in paradoxical sleep (PS). However, in PS the spontaneous activity was higher and the number of spontaneous spikes in bursts was no different, while the number during tones was lower (Edeline et al. 2000). Thus, there are significant state-dependent changes in MGB cell responses.

2 Thalamocortical Projections

Auditory TC neurons project primarily to the temporal cortex. Because they are the primary thalamic input to this region and because sound is the stimulus preferred by these cells, temporal cortex is designated as auditory. Although TC projections show some regional specializations that might be a basis for subdividing temporal cortex, other physiological and anatomical features must support any such parcellation. This is challenging as one moves from primary core areas to secondary and higher order processing centers because response and connectional features differ or increase in complexity and subdivisions remain in flux (Polley et al. 2007). Designating a cortical area in one species as homologous to that in another, even for areas considered core or primary areas, is not always easy. Thus, any description of the thalamocortical projection patterns and their significance for the cortical parcellation schemes must reflect that we are still in the early stages of this endeavor (Kaas 2005).

2.1 Thalamic Projections to the Thalamic Reticular Nucleus

The thalamic reticular nucleus (TRN) is a thin sheet of GABAergic thalamic neurons that forms a shell-shaped nucleus around the rostral and lateral surface of the dorsal thalamus. Its cells provide powerful inhibitory feedback to the thalamus and are implicated in several important functions including normal and abnormal thalamocortical oscillations (Pinault 2004). TRN receives its major excitatory inputs from cortical layer VI pyramidal cell axons that emit corticoreticular collaterals as they project to the thalamus. The second major excitatory input is from collaterals of TC cell axons en route to the cortex. In other sensory systems, both terminal types produce strong excitatory postsynaptic potentials (EPSPs) in TRN neurons (Steriade 1997). From 40 to 75% of TC axons have TRN collaterals (Yen and Jones 1983; Yamamoto et al. 1985; Harris 1987).

Because the cortical and thalamic representatives of each sensory modality converge onto a specific part of TRN, functionally distinct sectors exist within it. TRN cells in each sector in turn project back to the part of the thalamic nucleus that innervates them. The auditory sector in rats and cats is in the caudoventral TRN (Shosaku and Sumitomo 1983; Simm et al. 1990; Villa 1990; Crabtree 1998). Cells in the auditory TRN project back to the MGB topographically: those projecting to the laminar part of MGV and MGD are intermingled but separated from those cells projecting to MGM and the *pars ovoidea* portion of MGV (Crabtree 1998). The synaptology of MGB TC axons on TRN cells in the auditory

TRN sector is unknown; in other sensory systems TC axons form small (compared to cortical terminals) asymmetric glutamatergic synapses on the dendrites. IC stimulation elicits short latency bursts of spikes in cells in the rat TRN auditory sector, presumably from the activation of the MGB TC cell inputs, quickly followed by 2–6 spike bursts (Shosaku and Sumitomo 1983).

In anesthetized rats and cats, there is not a strict tonotopic organization in the auditory TRN cells since most cells are driven preferentially by broadband stimuli (Shosaku and Sumitomo 1983; Simm et al. 1990). Cells likely at the edge of the auditory sector can be bimodal, responding to auditory and somatic sensory stimulation (Sugitani 1979). Unimodal cells are typically binaural and the response patterns are variable from cell to cell, with much of the spike activity showing bursts, more so then that seen from MGB recordings. TRN cells often respond to pure tones or shocks applied to the IC with latencies 1–1.5 ms longer than MGB cells. TRN frequency tuning is typically several octaves wider than MGV cells and rate-level functions that are more often non-monotonic. Little is known about the specifics of the TRN projection onto the MGB TC cells but TRN terminals contain pleomorphic vesicles and form symmetric synaptic contacts on MGV TC cell bodies (Montero 1983). Physiological studies in vivo show that TRN electrical stimulation can suppress MGB spontaneous or sound-evoked responses (Shosaku and Sumitomo 1983). It is unclear whether, under normal conditions, the TRN input would provide on-CF or sideband inhibition as suggested in other sensory systems (Pinault and Deschênes 1998). In brain slices electrical stimulation of TRN axons projecting to MGB generates GABAergic IPSPs (inhibitory postsynaptic potentials) in tufted cells in MGV and tufted or stellate cells in MGD, with $GABA_A$ and $GABA_B$ components (Bartlett and Smith 1999).

2.2 Thalamic Projections to the Cortex: The Core and Matrix Theory

TC projections do not appear to respect divisional boundaries based on MGB cytoarchitecture alone. An alternative organizational scheme for the cat and monkey proposes two populations of thalamocortical cells based on the cortical layer(s) in which they terminate (Jones 1998a, b, 2001). Cells with smaller somata, concentrated in MGD and MGM but scattered in all major MGB subdivisions and the associated paralaminar nuclei, project to layer I of the cat primary (core) and non-primary (belt/parabelt) auditory cortical areas (Mitani et al. 1984, 1987; Niimi et al. 1984). Layer IV inputs to these same cortical areas arise from larger cells mostly in MGV and MGD and from MGM as well. These two sets of cells are distinguished in some species by their immunoreactivity to the calcium binding proteins parvalbumin (PV) or calbindin (CB). Injections of retrograde tracers into layer I or IV of the monkey or rabbit primary auditory cortex label calbindin-positive (CB+) or parvalbumin-positive (PV+) MGB cells, respectively (Hashikawa et al. 1991; de Venecia et al. 1995, 1998). In rat the distribution of CB staining is similar but there appear to be no PV-immunoreactive cells (Celio 1990; Friauf 1994).

These observations underlie the core and matrix theory based on the origin and projection of the MGB cells that express these calcium binding proteins (Hashikawa et al. 1991 1995; Molinari et al. 1995; Jones 1998a). The PV-labeled cells (core) are confined to certain MGB divisions and are highly concentrated in MGV, where CB-labeled cells are largely absent. The core cells receive afferent input from the tonotopically organized lemniscal ascending auditory pathway (i.e., ICC) whose cells have well-defined receptive fields. Axons of PV-positive core cells are thick (i.e., fast-conducting) and project to one or a few AC fields in a precise tonotopic fashion. CB-labeled cells form a background thalamic matrix which is not confined by MGB divisional borders and is widely distributed in all MGB divisions and the associated paralaminar nuclei. The inputs to these cells are more diffuse and arise primarily from non-lemniscal sources. The axons are thinner (i.e., slower conducting) and end in AC layer I of auditory cortex, which itself is diffuse and independent of regional cortical borders. At the risk of oversimplification, this model is a valuable framework for understanding the two TC projections onto AC. It implies that the thalamic core relay cells have projections focused to middle cortical layers which provide parametric information. Matrix cell projections, in contrast, diverge over large areas and convey contextual information synchronously to multiple independent content-activated core sites. This arrangement promotes the synchronous activation of separated sites thus binding the dispersed activity into a single unique event.

2.3 Core, Belt, and Parabelt Cortical Regions

In mammalian species studied so far there are usually multiple core AC areas with primary or primary-like features which include robust responses to tones at short latencies and narrow, well-defined frequency response areas. Cells in each of these areas form a tonotopic map of frequency made up of sequential narrow isofrequency strips. Anatomical features include a koniocellular (well-defined granular layer IV) arrangement, a robust cholinergic organization, high levels of PB, and a high level of cytochrome oxidase, the endogenous mitochondrial enzyme that is a sensitive indicator of neural activity. Each primary AC area is highly interconnected with other primary areas and with non-primary areas immediately abutting each core region (Hackett et al. 2001).

These adjoining areas are often designated as belt region containing different fields which are considered a second stage in the hierarchical auditory processing from core to belt to parabelt AC regions (Kaas and Hackett 1999, 2000). Belt areas are highly interconnected and receive major input from immediately adjacent core areas and from MGD and MGM (see below). Ablation of part of the core cortex projecting to a belt cortex subdivision reduced belt cell sensitivity to tone stimuli but not to more complex stimuli, indicating that MGD and core AC inputs to this belt area have a role in shaping belt cell responses (Rauschecker et al. 1997). Adjacent to the lateral belt areas are the parabelt regions that form the next processing stage. They receive considerable input from adjacent belt regions, from MGD and MGM, and sparser projections from core AC or MGV. These parabelt regions (and to a lesser extent the belt areas) are the AC output pathway to higher temporal, frontal, and parietal cortex processing centers.

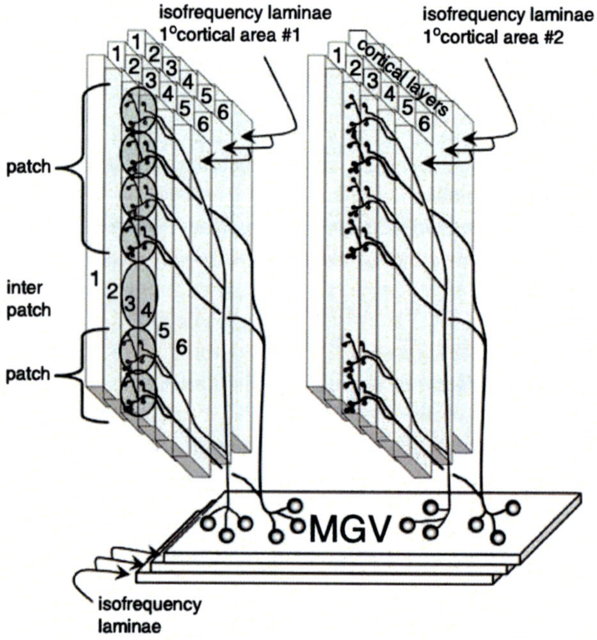

Fig. 3.3 Projection patterns of thalamocortical cell axons from an isofrequency lamina in MGV to primary auditory cortical areas. MGV axon collaterals and their terminals form patches in layers III and IV interspersed with terminal-free patches. The MGV cell axons may contribute terminals to one or more of the patches. Axons of other cells in the same isofrequency lamina may project to a second primary cortical area and form similar patches

2.4 Medial Geniculate Body Cortical Projection: Ventral Division

A further morphological criterion for a cortical area to be designated as a core region is a thalamic input arising primarily from the MGV. Individual MGV cells do not project uniformly to core cortical areas. Rather, the axons of PV+ MGV cells form multiple terminal patches along isofrequency contours in layers III and IV of the primary AC and are interspersed with terminal-sparse regions (Fig. 3.3) (McMullen and de Venecia 1993; Hashikawa et al. 1995; Kimura et al. 2003). Such thalamocortical patches correspond to the patchy parvalbumin staining in layer IV. Single MGV axons send collateral branches and terminals to one or more of these patches in layers III and IV (Hashikawa et al. 1995; Cetas et al. 1999). The patch/interpatch boundary may represent the transition between distinct binaural domains for bilateral summation (EE), suppression (EI), or occlusion (EO) (Middlebrooks and Zook 1983; Velenovsky et al. 2003). In contrast to the strict division of synaptic targets outlined in the core/matrix model above, some of the axons forming the patches in layers III and IV also send collaterals to unknown synaptic targets in layer I (Velenovsky et al. 2003).

There is little information on the postsynaptic targets of MGV axons. In layers III and IV in rat primary AC, >90% of these terminals contact dendritic spines, indicating that the postsynaptic cells are excitatory (Fig. 3.4) (Kharazia and Weinberg 1994). However, there is no information on the identity of these cells and the spines may be from layer III or IV pyramidal cells, layer IV spiny stellate cells, or layer V pyramidal cells; the identity of these targets has different implications for AC processing. GABAergic layers III and IV interneurons also receive direct MGV input (Fig. 3.4) (Cipolloni and Keller 1989; Kharazia and Weinberg 1994), though their identity and local synaptic targets are unknown.

2.5 Medial Geniculate Body Cortical Projection: Dorsal Division

In monkey, cat and rodent only general statements can be made about MGD projections. MGD cells target layers III and IV of the belt non-tonotopic areas that are near the core regions. MGD input to layers III and IV is more dispersed than that from MGV and does not form the patchy terminal distributions seen in the MGV projections (Huang and Winer 2000). Weaker projections are seen in layers III and IV of primary/core cortical areas. Terminals are also observed in layers I and VI of these core and belt areas (Fig. 3.4) (Kimura et al. 2003; de la Mothe et al. 2006) and some terminals may be present in all layers (Winer and Lee 2007). In all species

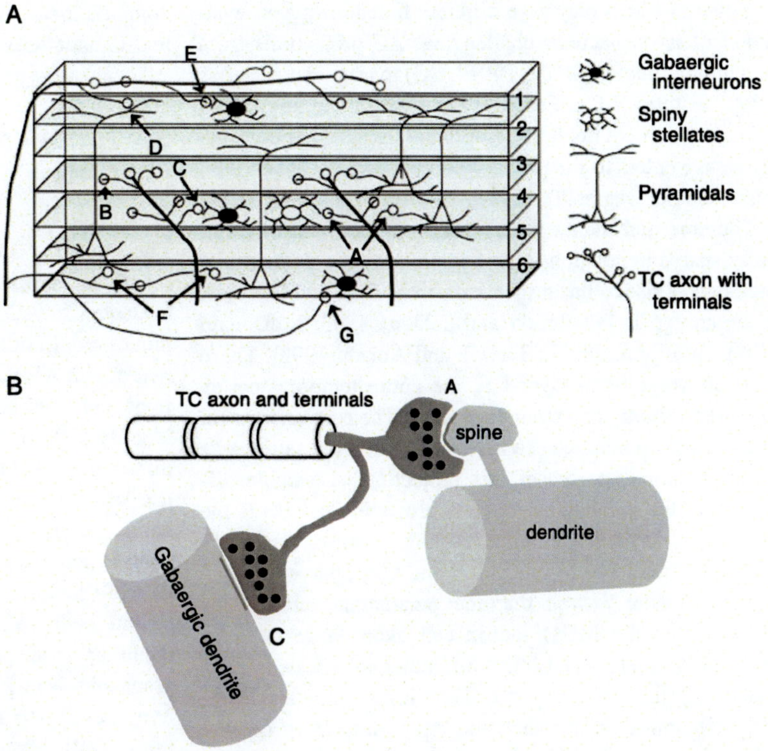

Fig. 3.4 Schematic representation of thalamocortical inputs to auditory cortical layers. **a** Potential cellular targets of TC axon terminals in layers I, III/IV, and VI. Layer III/IV terminals could be ending on the dendrites of pyramidal cells or spiny stellate cells (of which the latter are rare) in the same (*A*) or deeper layers (*B*), or on the GABAergic interneurons in the same layers (*C*). Layer I terminals could end on the dendrites of pyramidal cells in deeper layers (*D*) or on GABAergic cells within the layer (*E*). Layer VI terminals could target pyramidal cells in the same layer, on dendrites of pyramidal cells in higher layers (*F*) or on interneurons (*G*). **b** Documented termination sites of TC axons in layer IV showing terminals on the shafts of GABAergic dendrites and the spines, attached to presumptively glutamatergic dendrites

studied, MGD cells project predominantly to layers I, III, and IV and perhaps VI in secondary AC areas and may have lighter input than similar MGV projections to primary AC. In species where PV+ core and CB+ matrix cells are found in MGB it is assumed that the sparser MGD PV+ core cells are the source of the layer III/IV terminals, while the MGD CB+ matrix cells form the layer I terminals and perhaps those inputs to the lightly innervated layers. These assumptions could be tested in single cell/axon projection pattern studies.

The specifics of MGD projections are further complicated by other anatomical features. First, in the cat there are several MGD subdivisions and these may have unique TC projection patterns (Huang and Winer 2000). Even small injections into a region targeted by MGD may label cells and their axons from multiple MGD subdivisions and also label cells and axons of the PV+ core and CB+ matrix (Molinari et al. 1995). The unique projection patterns may also be due in part to the different distribution of PV+ core and CB+ matrix cell types in MGD. In the monkey, where only two MGD subdivisions, anterior (MGad) and posterior (MGpd) are present, PV+ cells are predominant while CB+ cells are rarer in the anterior division, and this ratio reverses in the posterior division. Injections of retrograde tracer into one region of the belt cortex designated caudomedial (CM) labeled mostly PV+ cells that presumably project to layers III/IV in MGad (Jones, 1998a, b, 2001), while injections into another cortical belt region (rostromedial; RM) labeled mainly CB+ MGpd cells (that presumably project to layer I (de la Mothe et al. 2006)). Does CM receive primarily layer I terminals while RM receives primarily layer III/IV terminals? Single axon studies will be required to address this question.

In the rat, MGD injections labeled some terminals in layer I and VI in both primary and non-primary AC (Kimura et al. 2003) and many more in layers III and IV of two specific regions of non-primary AC. One region (PD) lies posterodorsal to the primary cortical areas (TE1) and its cell prefer frequencies >15 kHz. Projections from this area may be involved in spatial perception and directed attention (Kimura et al. 2004, 2007). The other region receiving strong layer III/IV projections from MGD is area V, ventral to TE1 and responsive to frequencies <15 kHz; area V projections to other cortical areas suggest a role in processing the emotive content of auditory stimuli (Donishi et al. 2006; Kimura et al. 2007).

2.6 Medial Geniculate Body Cortical Projection: Medial Division

MGM is a small nucleus bordered by a group of equally small paralaminar nuclei. Dissecting the projections of these small closely abutting individual nuclei using injection methods is difficult. Moreover, the diversity of resident cell types (see above) which may have different functions makes evaluation of its projections challenging. AC projections from paralaminar nuclei (MGM, PP, PI, SG) overlap but each has unique features. All project to lower layer III and layer IV of secondary cortical fields, the ectorhinal (ECT) and perirhinal (P) areas ventral to the primary cortical areas (Linke 1999; Linke and Schwegler 2000).

The core and matrix theory would predict that the relatively sparse population of core neurons in the paralaminar nuclei gives rise to this projection. Areas ECT and P project to the amygdala (Romanski and LeDoux 1993; McDonald 1998) and hippocampus (Burwell and Amaral 1998). These areas converge upon layer I of the entire temporal cortex including primary and secondary areas. The core and matrix theory would predict an extensive population of matrix cells in the thalamic sources of this projection. A notable difference in the paralaminar nuclear projection pattern is the fairly heavy layer VI terminal labeling in primary and secondary AC after MGM injections, and very sparse terminal labeling in layer VI from the other paralaminar nuclei. This suggests that the MGM matrix cell axons have collaterals to both layers I and VI, while other matrix cells have branches primarily in layer I. Thus layer I has a convergence of non-lemniscal thalamic inputs related to context rather than content of sensory stimuli, feedback connections from higher cortical areas (Rockland and Drash 1996), extensive innervation by subcortical neurotransmitter systems (Lysakowski et al. 1989) whose activity is state-dependent (Lucas-Meunier et al. 2003), and long-range horizontal connections by its GABAergic cells (Verbny et al. 2006). Each of these features suggests a role for layer I in coordinating activity across cortical columns in a behaviorally dependent fashion.

3 Auditory Thalamocortical Physiology

A large literature describes the acoustic response properties of neurons in AI and in MGB. However, how stimulus representation changes across the thalamocortical synapse and any unique role that primary AC plays in encoding stimulus features remains unclear. Two critical issues are (1) methodological variability, including species, anesthetic state and stimuli, and (2) the few studies available on the basic anatomy and physiology of the thalamocortical circuit, and how these synaptic, cellular, and network properties relate to response properties in vivo. Since 2000, the in vitro brain slice and in vivo recording techniques have begun to reveal common elements among findings obtained with the many in vivo techniques. Studies on the auditory TC system can now specify how it contributes to forebrain processing of sound information in vivo, and suggest how these computational architectures differ from those of the visual and somatic sensory systems.

3.1 Intrinsic Membrane Properties in Auditory Cortex

The impact of thalamocortical synaptic inputs reflects the membrane properties of recipient cells, which have been evaluated in AC using in vitro brain slice recording techniques. As in other primary sensory areas, layer III and IV neurons in AI receive synaptic input from thalamic projections and from other cortical cells (Winer et al. 2005). The identity of the thalamorecipient cells in auditory cortex likely differs from that of primary visual and somatic sensory cortex, where many thalamic inputs synapse on non-pyramidal spiny stellate cells whose dendrites are confined largely to layer IV and whose axons project to more superficial pyramidal cells and interneurons (Lund 1984). However, AC spiny stellate cells are rare and layer IV pyramidal cells, whose dendrites extend into more superficial layers, may have assumed this task (McMullen and Glaser 1984; Winer 1984; Meyer et al. 1989; Smith and Populin 2001). Layer V pyramidal cells receive monosynaptic thalamic inputs (Verbny et al. 2006), and nonlemniscal thalamic input reaches as yet unidentified layers I and VI cells.

3.1.1 Intracellular Properties of Auditory Cortex Cells

Whole-cell recordings from rat and cat AC slices reveal that most thalamorecipient cells (except layer I) are best classified physiologically as regular-spiking (RS); from10 to 20% are fast-spiking (FS), intrinsically bursting (IB), and, at least in developing tissue, rectifying (RECT), or on-spiking (OS) (Metherate and Aramakis 1999; Hefti and Smith 2000). RS, FS, and IB cells in AI resemble comparable types elsewhere in cortex (Connors and Gutnick 1990). Many RS and IB cells are pyramidal in shape, but IB cells, found almost exclusively in layer V, are significantly larger, and their long apical dendrites extend into layer I, where they form tufts (Hefti and Smith 2000). As in other cortical areas, FS cells are non-pyramidal and are presumed to be GABAergic (Metherate and Aramakis 1999). GABAergic cells have heterogeneous firing properties, however, and represent a continuum that includes RS and other categories (Verbny et al. 2006). RECT and OS cells can have pyramidal or non-pyramidal morphology, and fire transiently to depolarizing currents. For RECT cells, the transient response is a train of spikes that typically last <250 ms. OS cells fire only one or two action potentials in response to depolarization, and do so with very short onset latencies (Metherate and Aramakis 1999).

3.2 Thalamocortical Synaptic Physiology

Understanding the physiology of the thalamocortical synapse that terminates on these cell types with their varying intrinsic membrane properties is critical to elucidating how information is transformed between MGB and AI. As examples (1) short-term plasticity at the TC synapse contributes to the increased incidence of phasic responses in AC cells compared to their thalamic inputs (Wehr and Zador 2005), and (2) the reduced efficacy of TC synapses onto interneurons versus pyramidal cells likely explains the relative absence of feedforward inhibition in AI compared to other sensory cortices (Rose and Metherate 2005; Verbny et al. 2006).

3.2.1 Thalamocortical Synapses in Slice Preparations

In the murine auditory TC slice preparation (Cruikshank et al. 2002), the connections between auditory thalamus and auditory cortex are preserved (Fig. 3.5), and two experimental slices are popular: a primary slice contains largely MGV and primary AC and in which lemniscal inputs can be activated by thalamic stimulation, and a shell slice containing mostly nonprimary thalamic nuclei and secondary AC in which stimulating the thalamus may activate non-lemniscal TC inputs to layer I and other layers.

Stimulation of thalamic afferents in this preparation evokes monosynaptic excitatory postsynaptic responses in AC cells mediated by AMPA/kainate and NMDA glutamate receptors (Cruikshank et al. 2002). Efficacy at a particular synapse is the product of the number of release sites, release probability, and the quantal amplitude. The latter two parameters likely differ between release sites and may vary dynamically depending on the recent history of the synapse.

Pyramidal cells receive thousands of synaptic inputs, with excitatory synapses arising from both TC and corticocortical (CC) afferents. Although estimates of the percentage of TC excitatory synapses onto pyramidal cells varies in different areas from <10% (Ahmed et al. 1994; Latawiec et al. 2000) to ~33% (Kubota et al. 2007), these are vastly outnumbered by CC synapses. In other sensory cortices, however, TC synapses have far greater efficacy, with more release sites per fiber (~5 to 10 for TC fibers) and higher release probability than CC synapses (Gil et al. 1999; Amitai 2001). Although comparable AC data is not available, the precision and reliability of lemniscal, thalamically evoked responses in pyramidal cells and interneurons in layers III and IV of primary slices suggest that release probabilities at AI TC synapses are high as well. This reliability, with low jitter in response latency, may contribute to the precision of cortical

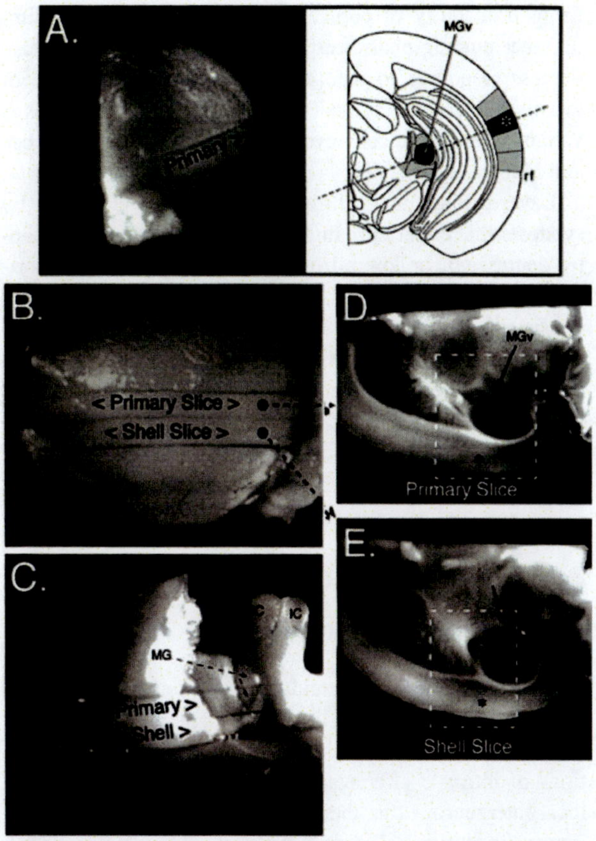

Fig. 3.5 Thalamocortical slice preparation in the mouse. **a** A cut angled 15 degrees to the horizontal plane preserves connections between MGv and AI ('primary slice') or nonprimary thalamic nuclei and secondary auditory cortical areas ('shell slice'). **b** Same hemisphere as in (**a**) but viewed from lateral side. Orientation: anterior (*left*); dorsal (*top*). **c** Same as (**b**) except with the cortex removed, revealing midbrain and thalamic structures (SC, superior colliculus; IC, inferior colliculus). (**d, e**) The primary and shell slices viewed from the dorsal side. Orientation: anterior (*left*); lateral (*bottom*). From the original source (Cruikshank et al. 2002)

responses to envelope transients in acoustic stimuli (Rose and Metherate 2005).

Because of the stochastic nature of neurotransmitter release and its dependence on interacting dynamic intracellular processes, synaptic efficacy is itself dynamic and depends on the recent history of activity at the synapse. Short-term synaptic plasticity, i.e., changes in synaptic efficacy on the scale of milliseconds to seconds, plays an important role in shaping the information that can cross the TC synapse. As is typical for synapses with high release probability, efficacy at TC synapses onto many AI layer III/IV cells rapidly depresses over tens of milliseconds for repetitive stimuli, and requires >1 s to recover (Rose and Metherate 2005), as has been observed in other cortical areas (Stratford et al. 1996; Gil et al. 1997). Depression at the TC synapse reduces the firing probability of cortical cells during prolonged stimuli, accentuating phasic responses in them, and would also depress responses to repetitive stimuli. The magnitude of synaptic depression at the TC synapse in pyramidal cells is variable, and some cells even show mild facilitation (Rose and Metherate 2005).

Layer II pyramidal cells are connected to one another by synapses that either have high release probability and synaptic depression, or low release probability and show short-term facilitation (Atzori et al. 2001). These observations are consistent with the range of response properties observed in AI, with some cells exhibiting sustained responses and others phasic responses to prolonged stimuli, though phasic responses dominate in AC layers III and IV in anesthetized cats (Volkov and Galazjuk 1991).

3.3 Intracortical Feedforward and Feedback Connections

As part of a network, pyramidal cell responses in AI reflect intrinsic membrane properties, TC synaptic physiology, the response properties of thalamic inputs, as well as contributions of intrinsic afferents from other pyramidal cells and local interneurons that themselves may be activated by TC inputs. Repetitive TC stimuli evoke excitatory polysynaptic responses in layer III/IV pyramidal cells in AI, presumably from the activation of excitatory corticocortical connections (Rose and Metherate 2005). Since most AI TC synapses depress rapidly, lateral excitation within AI may explain the prevalence of sustained responses to stimuli seen in the awake monkey (Wang et al. 2005). Lateral excitation may also broaden AI receptive fields, as suggested by in vivo pharmacological studies (Kaur et al. 2004; Metherate et al. 2005).

3.3.1 Cortical Inhibitory Networks

Local inhibition also plays a critical role in shaping responses of AI pyramidal cells (Wang et al. 2000; Chen and Jen 2000; Foeller et al. 2001). Such inhibition could arise from direct thalamic excitation of cortical inhibitory cells (feedforward), or from activation of inhibitory cells by thalamically activated excitatory cells in cortex (feedback/lateral). In somatic sensory cortex, feedforward inhibition is prominent: excitatory thalamic inputs to fast spiking (FS) cells are more efficacious than in pyramidal cells and often suprathreshold, such that thalamic stimulation typically evokes disynaptic inhibitory currents (Gibson et al. 1999; Gabernet et al. 2005; Cruikshank et al. 2007). In contrast, there is little direct evidence for strong feedforward inhibition in AI. Here, thalamic stimulation evokes excitatory responses in interneurons whose responses are of comparable amplitude (Rose and Metherate 2005) or far smaller (Verbny et al. 2006) than thalamic-evoked responses in pyramidal cells, depending on the type of interneuron, and are rarely superthreshold (Fig. 3.6).

Furthermore, disynaptic inhibitory inputs are rarely seen in layer IV pyramidal cells upon thalamic stimulation (Rose and Metherate 2005; Verbny et al. 2006), though in layer V pyramidal cells they are observed in about one-third of cells (Verbny et al. 2006) (Fig. 3.7). The latter study focused on TC slices prepared from transgenic mice in which a subset of GABAergic interneurons express green fluorescent protein (Lopez-Bendito et al. 2004). This study showed that thalamic afferents contact interneurons throughout layers I–V of AI (Fig. 3.6). In this interneuronal population, there was no difference in response amplitude based on the laminar position of the interneuron: excitation was uniformly weak throughout AI. Thus, this population of cells does not appear to mediate feedforward inhibition of pyramidal cells, and may require coincident excitation, from other cortical cells or from subcortical neuromodulators, to reach spike threshold.

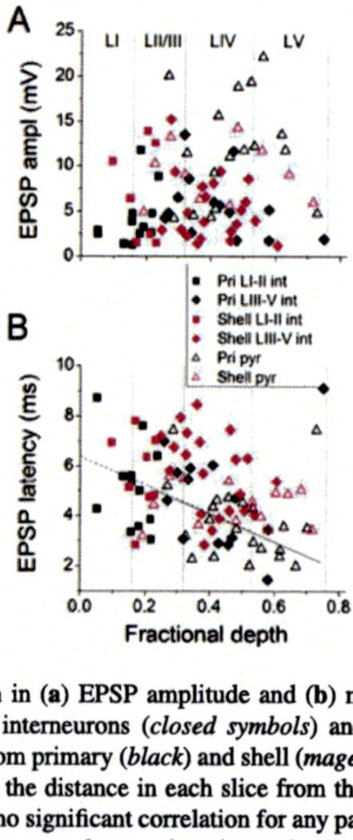

Fig. 3.6 Variation in (**a**) EPSP amplitude and (**b**) mean latency with cortical depth for interneurons (*closed symbols*) and pyramidal cells (*open triangles*) from primary (*black*) and shell (*magenta*) slices. Depth was normalized to the distance in each slice from the pia to the white matter. There was no significant correlation for any parameter for either cell type except for mean latency in primary interneurons (**b**, *dashed line*; r = –0.504, $P<0.05$) and primary pyramidal cells (**b**, *solid line*; r = –0.522, $P<0.05$). *Vertical dotted lines*, the boundaries of the cortical layers, as indicated in (**a**). From the original source (Verbny et al. 2006)

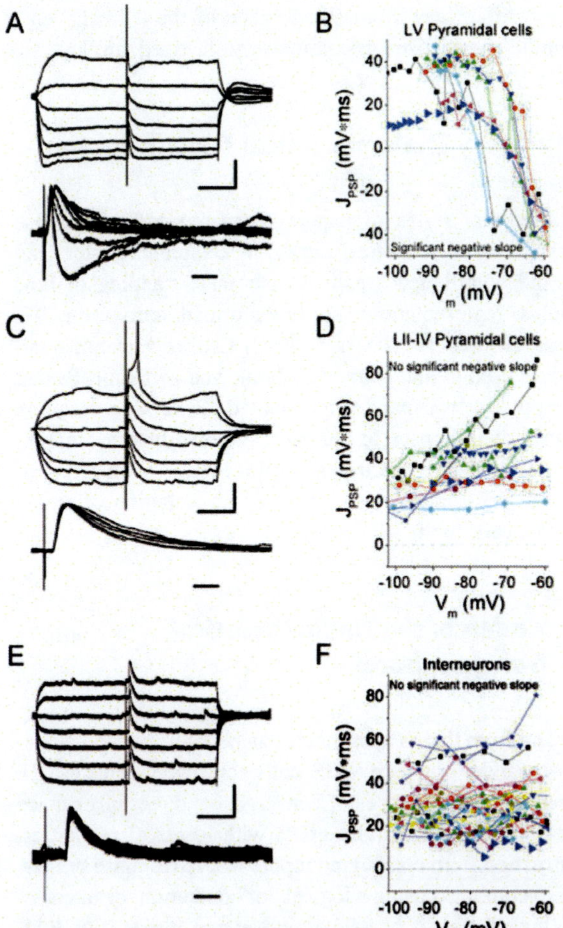

Fig. 3.7 Thalamic stimulation evokes feedforward inhibition in layer V pyramidal cells (**a**, **b**), but not in layers II–IV pyramidal cells (**c**, **d**) or in interneurons (**e**, **f**). Cells were hyper- and depolarized about the resting potential (ERest) with 250 ms current steps, and thalamic afferents stimulated at constant intensity (**a**, **c**, **e**, *top traces*). Responses were then normalized and aligned (**a**, **c**, **e**, *bottom traces*) and the time integral of the response computed (JPSP; **b**, **d**, **f**, *large symbols*). This analysis was repeated for all cells in which at least three membrane potentials were tested (**b**, **d**, **f**, *small symbols*). For pyramidal cells in layer V, inhibition was apparent in the responses and JPSP(Vm) had significant negative slope (**b**). For pyramidal cells in layers II–IV (**d**) and interneurons (**f**), no inhibition was apparent and JPSP(Vm) did not have a significant negative slope. Resting potentials were: –78.2 mV (**a**), –75.1 (**c**), –67.9 mV (**e**). Scale bars: 100 ms, 10 mV (**a**, **c**, *upper traces*); 100 ms, 6 mV (**e**, *upper traces*); 10 ms (**a**, **c**, **e**, *lower traces*). From the original source (Verbny et al. 2006)

3.3.2 Comparing Auditory Cortex with Other Cortical Areas

It is not yet clear whether the inhibitory circuitry of AI is fundamentally different from that of primary somatic sensory cortex, but because short latency inhibitory currents are evoked by acoustic stimuli in AI neurons in vivo (Zhang et al. 2003; Wehr and Zador 2003), some cellular mechanism for the rapid inhibition of cortical responses must operate. The evidence so far suggests that this inhibition arises from feedback/lateral connections within AI, rather than from direct thalamic excitation. Further investigations are necessary to test this hypothesis.

3.3.3 Cortical Interneurons and Thalamic Projections

Layer I interneurons may also play a role in TC processing in AI. These cells likely receive excitatory TC lemniscal input from axons with collaterals extending into this layer and non-lemniscal axons whose primary target is layer I (Cetas et al. 1999; Huang and Winer 2000). Many layer I interneurons have long axonal horizontal projections across cortical columns, possibly beyond AI (Verbny et al. 2006). This morphology is consistent with what has been seen in other neocortical regions, and has led to the suggestion that layer I interneurons may coordinate activity patterns across multiple columns or even disparate areas of neocortex (Vogt 1991; Nieuwenhuys 1994; Hestrin and Armstrong 1996; Zhou and Hablitz 1996). AI layer I interneurons form networks connected by both GABAergic synapses and gap junctions (Merriam et al. 2005), similar to those in somatic sensory cortex (Chu et al. 2003). Reciprocal inhibitory and gap junction synapses connecting these cells cause phase-locking of their action potentials during prolonged stimulation, with enhancement of synchronous spiking when the cells fire at high frequencies. Thus, thalamocortical activation of layer I inhibitory networks may trigger synchronous inhibition that could, through interneuron axon collaterals, target layer II and layer V pyramidal cells both locally and across cortical columns. Further examination of AI cell types and of their synaptic interactions are needed to dissect the cellular and synaptic factors subserving early cortical processing of sound.

3.4 Comparison of Thalamic and Cortical Response Properties In Vivo

A central issue in TC research is the unique roles that AC plays in processing sensory information. As one proceeds from core to belt to parabelt and higher cortical areas, identifying the TC role becomes more difficult. However, if we restrict ourselves to primary AC areas directly excited by MGB, then the question becomes, what transformations occur between thalamus and cortex? We approach this question by comparing MGB and AC organizational features. For example, a postulated role of the thalamocortical system is to create new tonotopic maps that may be critical for parallel cortical processing of auditory information (Winer and Lee 2007). Thus two cat MGB tonotopic maps give rise to at least four tonotopic maps in AC (Imig and Morel 1985). Anatomical–physiological studies indicate that these maps arise from interspersed but largely independent cell MGB populations (Lee et al. 2004a, b). This expansion may form part of the structural basis for the more complex stimulus representations in auditory cortex compared to the brain stem (Micheyl et al. 2007). Similarly, in AI topographies beyond that of best frequency appear to emerge in the form of binaural interaction (Brugge and Merzenich 1973; Imig and Adrián 1977; Middlebrooks et al. 1980; Reale and Kettner 1986; see also Reser et al. 2000), response latency (Mendelson et al. 1997), and intensity tuning (Schreiner et al. 1992; Heil et al. 1994), each of which is organized topographically perpendicular to the cochleotopic axis (Linden and Schreiner 2003).

3.5 Comparing Auditory Cortex and Midbrain Models of Information Processing

If AC processes information in a manner fundamentally different from IC circuits, then we might expect to see dramatic transformations of stimulus representations at the TC synapse. Several studies have compared the response properties of AC cells with those in MGB or IC, but there have been few studies in which comparisons between AC and MGB responses were made in (putatively) synaptically connected pairs of cells (Creutzfeldt et al. 1980; Miller et al. 2001a, b, 2002). Other studies have compared the properties of populations of MGB and AC cells (Clarey et al. 1995; Barone et al. 1996; Chechik et al. 2006; Coffey et al. 2006; Bartlett and Wang 2007), more peripheral structures and AC (Heil and Irvine 1997; Ter Mikaelian et al. 2007), or between intracortically recorded multiunit responses and putative thalamic afferent responses (Steinschneider et al. 1993, 1994). All such studies have focused exclusively on the lemniscal auditory pathway, i.e., the MGB ventral division and AI. Insights into stimulus coding transformations across the TC synapse are constrained by the variety of species (mice, rats, cats, marmosets, and rhesus and squirrel monkeys) and by differences in anesthetic state of the animals (e.g., ketamine-anesthetized, barbiturate-anesthetized, awake).

4 Auditory Thalamocortical Transformation

An important series of papers compared MGB response properties to their cortical targets in ketamine-anesthetized cats and contributed greatly to our understanding of how stimulus representations are transformed across the TC synapse (Miller et al. 2001a, b, 2002). Cross correlation analysis was used to find pairs of thalamic and cortical cells that were putatively synaptically coupled. Reverse correlation analysis of responses to complex stimuli (dynamic ripple) was used to derive spectrotemporal receptive fields (RFs) of pre- and postsynaptic cells, and the contribution of single thalamic cells to the RFs of their AI targets.

4.1 Models of the Thalamocortical Transformation

The data from these experiments can be interpreted in a conceptual model in which MGB cells contribute to the AI RF properties via three types of convergence: direct inheritance from converging thalamic inputs with identical properties, construction from converging inputs whose response properties are combined using a logical "or" operation, or intersection (logical "and") of RFs of converging inputs (Fig. 3.8). Indeed, a TC cell pair might exhibit characteristics of two or three types of convergence, presumably reflecting an optimal underlying mechanism for transforming stimulus features specific to that pair.

4.2 Phasic Responses

Because of the importance of temporal features in complex stimuli such as communication sounds, several studies have

Fig. 3.8 (continued) two units, normalized to firing rate. The bar plot is the cross correlogram under the ripple-driven condition, and the line plot (truncated for clarity) is under the spontaneous condition. The brief, short-latency peak, with cortical spike lagging thalamic (2 ms), is indicative of a monosynaptic-like functional connection. The *cyan* line is the mean, and the *red* lines are the 99% confidence intervals for the stimulus-driven correlogram. **c** Expanded views of the excitatory peaks of the STRFs in (**a**). Superimposed on the cortical STRF are contours circumscribing the high-energy region of the cortical (*green*) and thalamic (*black*) STRFs. The thalamic contour has been shifted in time by the peak correlogram delay. In this case, the cortical cell appears to inherit its excitatory features from the thalamic input. (**d–f**) Mixed constructive/ensemble convergence. **d** STRFs. **e** Thalamocortical cross correlograms. **f** Expanded views of the excitatory peaks. In this case, a thalamic cell with smaller receptive field helps construct a larger, composite cortical STRF. **g–i** Ensemble convergence. **g** STRFs. **h** Thalamocortical cross correlograms. **i** Expanded views of the excitatory peaks. A thalamic cell with much larger excitatory receptive field is reduced to contribute to a smaller cortical STRF. This logically demands the participation of an ensemble of other inputs, acting in concert. From the original source (Miller et al. 2001a)

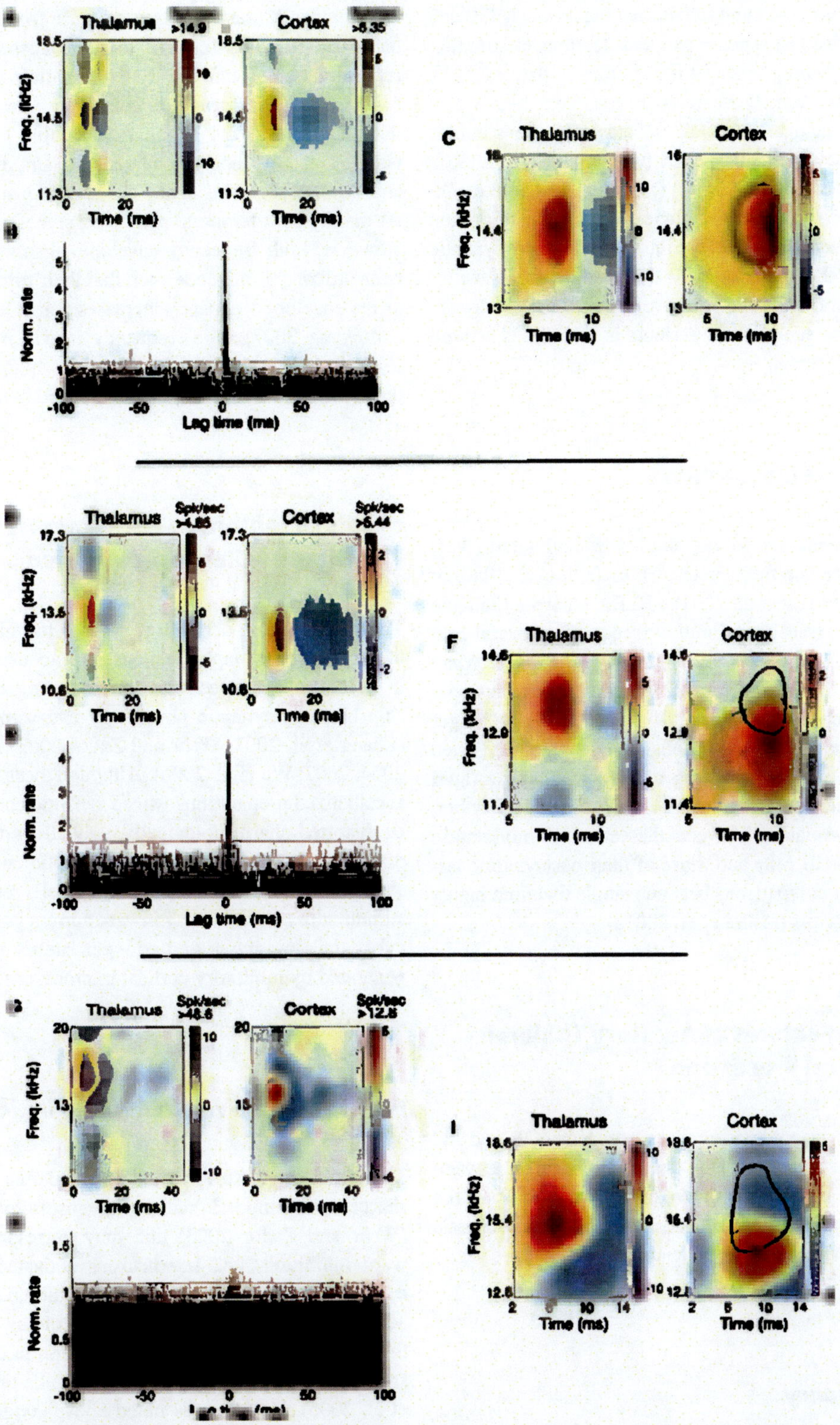

Fig. 3.8 Contribution of thalamic receptive fields to cortical response properties via three types of convergence: Inheritance (**a**–**c**), constructive/ensemble convergence (**d**–**f**), ensemble convergence (**g**–**i**). **a** Spectrotemporal receptive fields (STRFs) for a simultaneously recorded thalamic and cortical cell pair. The STRFs are depicted with time-preceding-spike on the abscissa, and frequency on the ordinate. *Warm* and *cool* colors indicate an excitatory or inhibitory effect, respectively, that the stimulus induced in a particular spectrotemporal region. The values on the color bar are thus differential rates, in spikes/s, relative to the mean rate. **b** Cross correlograms between the

focused on the precision and reliability of spikes in MGB and AC neurons driven by specific temporal features of stimuli.

A common finding among these studies is that AC cells are more phasic than those in thalamus, i.e., they fire at stimulus onset and at temporal transients in the stimulus envelope and are then suppressed (Creutzfeldt et al. 1980; Steinschneider et al. 1993. 1994) (Fig. 3.9). Possible mechanisms for this suppression include intracortical inhibition (Calford and Semple 1995; Tan et al. 2004) and synaptic depression (Wehr and Zador 2005). Despite their preference for temporal transients, however, AC cells typically are unable to follow successive transients in a stimulus as well as their MGB inputs.

4.3 Amplitude Modulation

Thus few AC cells can follow amplitude modulated (AM) stimuli at modulation frequencies (Creutzfeldt et al. 1980) or click frequencies (Lu et al. 2001) >50 Hz, whereas thalamic cells can follow transient stimuli at frequencies several-fold higher values (Creutzfeldt et al. 1980; Bartlett and Wang 2007) (Fig. 3.10). This shift in AC temporal following capability is unlikely to be the result of substantially longer membrane time constants. AC cells can fire precisely when driven directly with depolarizing currents containing rapid temporal transients (Mainen and Sejnowski 1995). Further, there is no correlation between the temporal modulation properties of MGB cells and those of their postsynaptic targets, at least for comparisons between single thalamic inputs and their target cells (Miller et al. 2001a).

5 Common Features of Auditory Thalamic and Cortical Processing

Despite some systematic differences between MGB and AC responses, what is most striking are the similarities between responses pre- and postsynaptic to the TC synapse. Thus, the information content of spike trains in response to acoustic stimulation in MGB and AC cells are similar, and are vastly different from those in the IC.

5.1 Redundancy

Similarly, information redundancy across cell populations is high in the IC, and much lower and virtually indistinguishable in AI and MGB (Chechik et al. 2006). A reduction in redundancy may represent a shift from representations based on stimulus features to more abstract object-based representations (Schnupp 2006), and thus may represent a fundamental transformation in encoding strategy that occurs not in AI but in the MGB. An example is low frequency (<100 Hz) envelope fluctuations in amplitude modulated sounds or click trains. They are encoded in both MGB and AI mainly by a temporal code with spikes at each transient. However, high frequency envelope signals are encoded by cells utilizing a rate code such that their firing rates increase as the envelope frequency increases (Lu et al. 2001; Bartlett and Wang 2007). The frequency at which the population shifts from a temporal to rate code is higher in MGB, but the overall coding strategy is established in thalamus.

6 Role of Inhibition and Synaptic Plasticity in Shaping Responses In Vivo

The use of whole-cell voltage clamp techniques to recording single-cell synaptic responses to acoustic stimuli in AI in vivo has enhanced our understanding of the relationship between synaptic physiology and response properties (Zhang et al. 2003; Wehr and Zador 2003, 2005; Tan et al. 2004, 2007; Wu et al. 2006). Although these data come from anesthetized preparations whose synaptic conductances may be distorted pharmacologically, they allow direct evaluation of the excitatory and inhibitory synaptic conductances that shape the cells' firing properties. AC cells respond to simple stimuli with a short latency inward current generated by excitatory thalamic input, and a longer latency outward current generated by inhibitory cortical interneurons.

6.1 Thalamic-Evoked Cortical Inhibition

The inhibitory circuit activated by thalamic afferents to AI sharpens the temporal precision of sensory-evoked AC spikes (Wehr and Zador 2003) and may correspond to feedforward inhibition, i.e., interneurons activated by TC afferents directly, or feedback/lateral inhibition, i.e., interneurons activated by AC pyramidal cells receiving superthreshold thalamic stimuli. Feedforward inhibition in rat somatic sensory cortex enhances pyramidal cell spike timing (Gabernet et al. 2005). However, as noted above, evidence for strong AC feedforward inhibition from in vitro preparations is lacking (Rose and Metherate 2005; Verbny et al. 2006) and a comparison of extracellular recordings from AC neurons between awake and anesthetized animals indicates that this

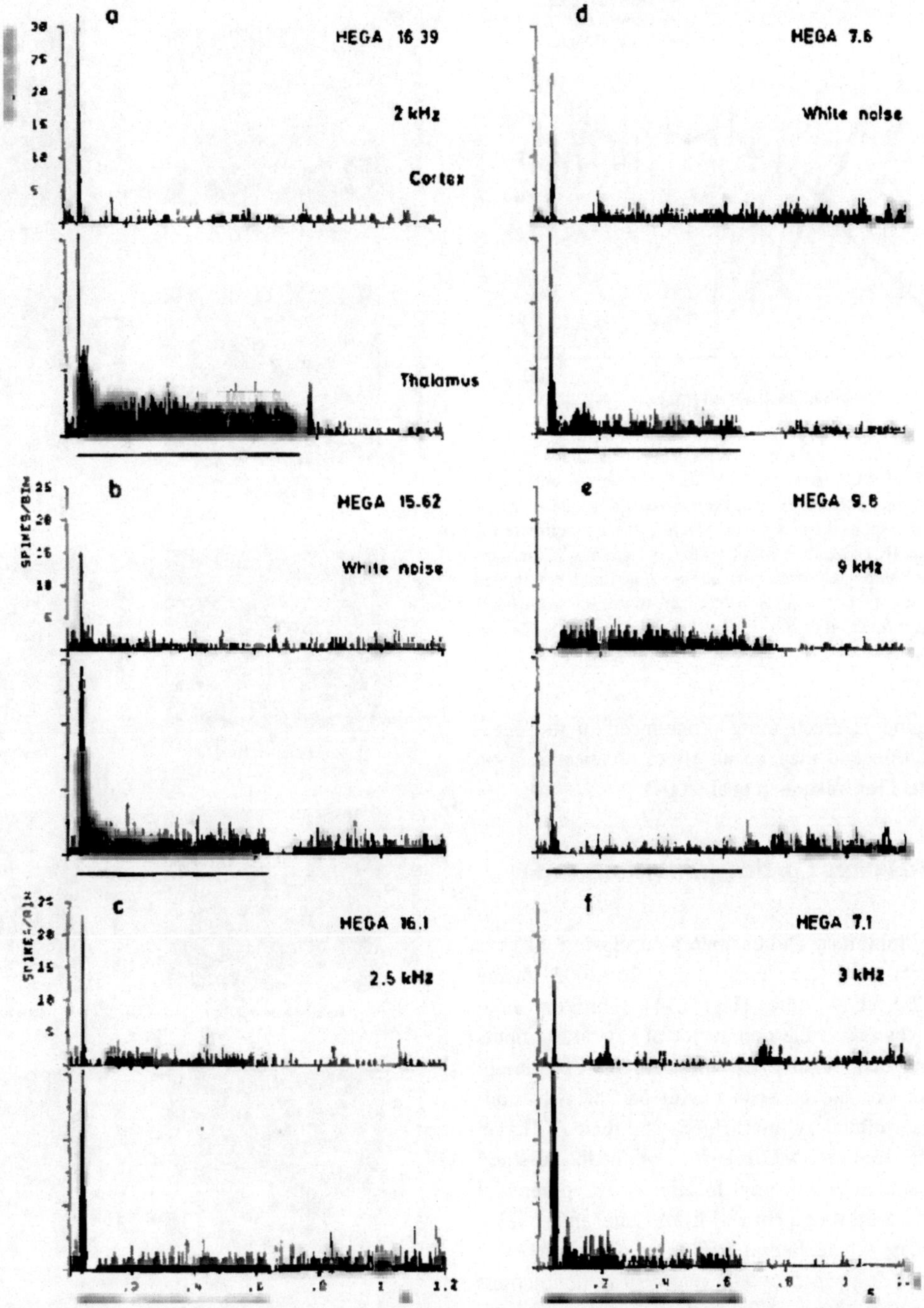

Fig. 3.9 Responses of various response types of thalamocortical neuron pairs to pure tones at CF (**a**, **c**, **e**, **f**) or to white noise (**b**, **d**). In each neuron pair, the record on top is from the cortical cell, and at bottom from the thalamic neuron. The neuron pair in (**a**) and (**b**) showed a positive cross correlation of the spontaneous activity, the spontaneous activities of the neurons in (**c**, **e**, **f**) were not correlated. In the neuron pair (**d**), cross correlation was not analyzed. PSTHs from 20 stimuli, bin width 2 ms. The duration of the tone or noise stimulus is indicated as a line under the record. From the original source (Creutzfeldt et al. 1980)

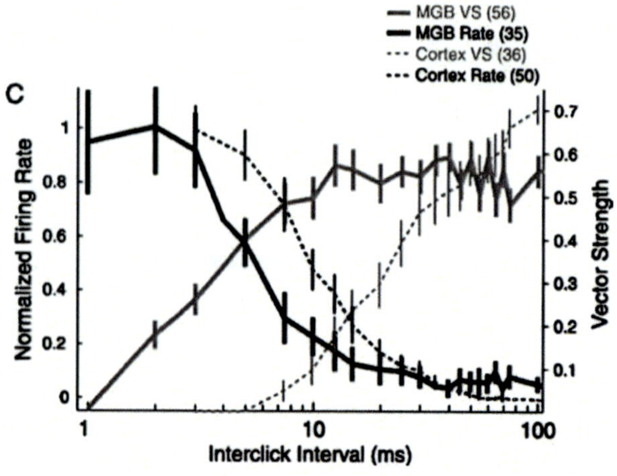

Fig. 3.10 Changes in rate and timing codes between thalamus and cortex. Mean ± SE of vector strength of MGB synchronized and mixed responses (*thick gray solid lines*) and normalized firing rate of MGB nonsynchronized responses (*thick black solid lines*) as a function of the interclick interval (ICI). Mean ± SE of vector strength of auditory cortex synchronized responses (*thin gray dashed lines*) and normalized firing rate of auditory cortex nonsynchronized responses (*thin black dashed lines*) as a function of ICI. From the original sources (Bartlett and Wang 2007; cortex data from Lu et al. 2001)

onset sharpening is much more pronounced in the anesthetized condition and may be an effect of anesthesia on synaptic events (Ter Mikaelian et al. 2007).

6.2 Sound-Evoked Cortical Inhibition

Sound-evoked inhibitory and excitatory currents in AI have almost identical tonal RFs (Wehr and Zador 2003; Zhang et al. 2003; Tan et al. 2004) (Fig. 3.11). Inhibitory sidebands in the RFs reflect the dominance of excitatory inputs for on-CF, and their relative weakness for off-CF stimuli. These observations are consistent with one of two models: feedforward inhibitory interneurons and their AC target cells receiving common MGB inputs, or feedback/lateral inhibitory interneurons activated locally by AC pyramidal cells. Either model distinguishes AI from visual and somatic sensory cortices, where lateral inhibition of responses is thought to arise from feedforward interneuron populations tuned more broadly than principal cells (Ferster and Miller 2000; Swadlow 2003).

6.3 Transient Cortical Responses

Many AC neurons have transient responses to acoustic stimuli in anesthetized and non-anaesthetized animals (Evans and Whitfield 1964; Creutzfeldt et al. 1980; DeWeese et al.

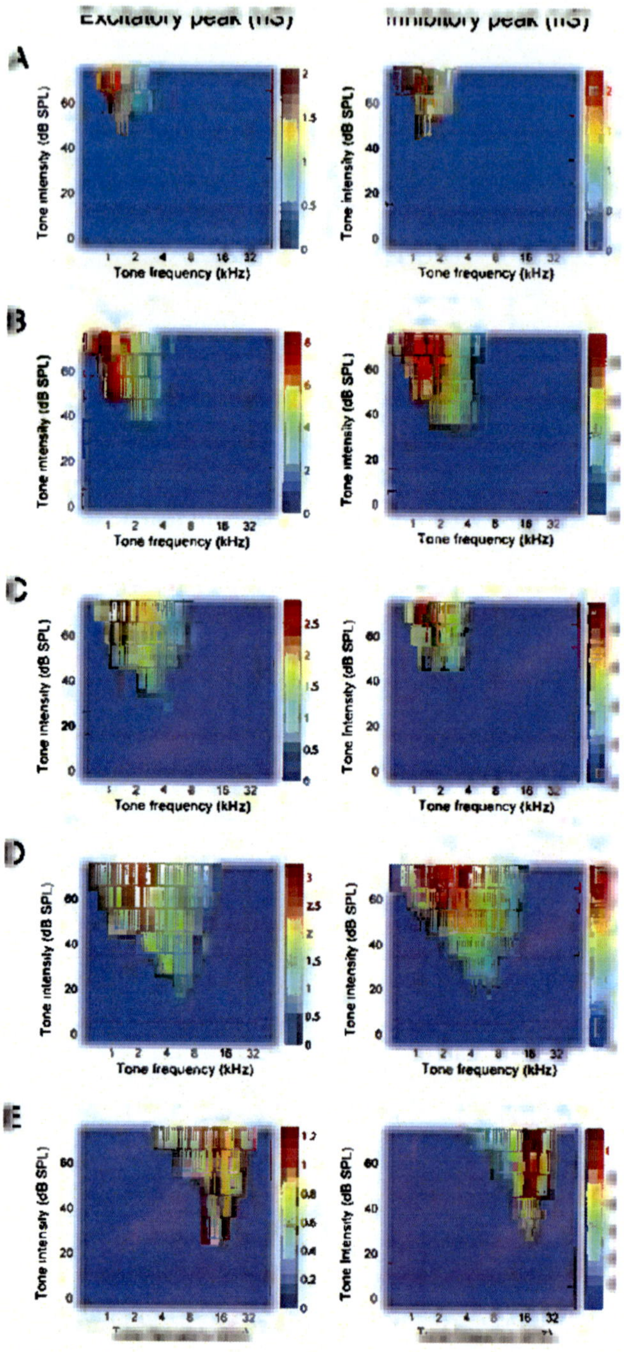

Fig. 3.11 Excitatory and inhibitory synaptic currents have similar tuning in AI. *Each row* represents data from one neuron. *Column 1*: Tonal receptive field (TRF) of excitatory conductance peak amplitude. *Column 2*: TRF of inhibitory conductance peak amplitude. From the original source (Tan et al. 2004)

2003; also Wang et al. 2005), and in paired recordings, AC responses are more phasic than their thalamic inputs (Creutzfeldt et al. 1980). The basis for this TC transformation likely involves synaptic, cellular, and network mechanisms.

Nearly all pyramidal cells exhibit spike frequency adaptation, which would raise spike thresholds during sustained stimuli, albeit to varying degrees (Metherate and Aramakis 1999; Hefti and Smith 2000; Verbny et al. 2006). Inhibitory synaptic inputs are seen in AC pyramidal cells after early MGB excitation, and the concomitant reduction in excitability matches the length of the $GABA_A$ receptor-mediated synaptic conductance (de Ribaupierre et al. 1972; Volkov and Galazjuk 1991). However, these conductances are too brief to account fully for the strong forward suppression effect in many AI cells, and thus short-term TC synaptic depression onto pyramidal cells may curtail sustained AI responses (Wehr and Zador 2005). Thus, short-term synaptic plasticity, intrinsic membrane properties, and intracortical inhibition may all play a role in converting sustained MGB responses to phasic AC responses.

6.4 Spectrotemporal Receptive Fields

Additional evidence for the importance of intracortical inhibition for shaping responses arises from paired recordings in vivo which mapped the spectrotemporal receptive fields (STRFs) of MGB cells and their AI synaptic targets (Miller et al. 2001a, b, 2002). Inhibitory sidebands in STRFs are prominent in AC cells, but are often dissimilar in connected thalamic/cortical cell pairs (Miller et al. 2001a). Thus inhibitory regions of STRFs are largely recreated in AC cells via intracortical circuitry. Because the relationship between excitatory and inhibitory zones in STRFs determines the preferences of AC cells for spectral and temporal modulation rates, these data suggest a role for intracortical inhibition in shaping temporal spike patterns.

Acknowledgment This work was supported by NIH Grant R01 DC006013 to Matthew Banks and NIH Grant R01 DC006212 to Philip Smith.

References

Ahmed B, Anderson JC, Douglas RJ, Martin KAC, and Nelson JC (1994) Polyneuronal innervation of spiny stellate neurons in cat visual cortex. *Journal of Comparative Neurology* **341**:39–49.

Aitkin LM, Calford MB, Kenyon CE, and Webster WR (1981) Some facets of organization of the principal division of the cat medial geniculate body. In: Syka J and Aitkin LM (eds). *Neural Mechanisms of Hearing*. Plenum Press, New York, pp. 163–181.

Aitkin LM, Dunlop CW, and Webster WR (1966) Click-evoked response patterns of single units in the medial geniculate body of the cat. *Journal of Neurophysiology* **29**:109–123.

Aitkin LM (1973) Medial geniculate body of the cat: responses to tonal stimuli of neurons in medial division. *Journal of Neurophysiology* **36**:275–283.

Aitkin LM and Prain SM (1974) Medial geniculate body: unit responses in the awake cat. *Journal of Neurophysiology* **37**:512–521.

Amitai Y (2001) Thalamocortical synaptic connections: efficacy, modulation, inhibition and plasticity. *Review of Neuroscience* **12**: 159–173.

Anderson LA, Malmierca MS, Wallace MN, and Palmer AR (2006) Evidence for a direct, short latency projection from the dorsal cochlear nucleus to the auditory thalamus in the guinea pig. *European Journal of Neuroscience* **24**:491–498.

Andersen RA, Roth GL, Aitkin LM and Merzenich MM (1980) The efferent projections of the central nucleus and the pericentral nucleus of the inferior colliculus in the cat. *Journal of Comparative Neurology* **194**:649–662.

Arnault P and Roger M (1990) Ventral temporal cortex in the rat: connections of secondary auditory areas Te2 and Te3. *Journal of Comparative Neurology* **302**:111–123.

Atzori M, Lei S, Evans DI, Kanold PO, Phillips-Tansey E, McIntyre O, and McBain CJ (2001) Differential synaptic processing separates stationary from transient inputs to the auditory cortex. *Nature Neuroscience* **4**:1230–1237.

Barone P, Clarey JC, Irons WA, and Imig TJ (1996) Cortical synthesis of azimuth-sensitive single-unit responses with nonmonotonic level tuning: a thalamocortical comparison in the cat. *Journal of Neurophysiology* **75**:1206–1220.

Bartlett, EL and Smith PH (2000) An ultrastructural comparison of axon terminals in the rat MGB. *Neuroscience* **100**:811–828.

Bartlett EL and Smith PH (1999) Anatomical, intrinsic, and synaptic properties of dorsal and ventral division neurons in the rat medial geniculate body. *Journal of Neurophysiology* **81**: 1999–2016.

Bartlett EL and Wang X (2007) Neural representations of temporally modulated signals in the auditory thalamus of awake primates. *Journal of Neurophysiology* **97**:1005–1017.

Bender DB (1983) Visual activation of the pulvinar depends on cortex but not colliculus. *Brain Research* **279**:258–261.

Bordi F and LeDoux JE (1994) Response properties of single units in areas of rat auditory thalamus that project to the amygdala. I. Acoustic discharge patterns and frequency receptive fields. *Experimental Brain Research* **98**:261–274.

Brugge JF and Merzenich MM (1973) Responses of neurons in auditory cortex of the macaque monkey to monaural and binaural stimulation. *Journal of Neurophysiology* **36**:1138–1158.

Burwell RD and Amaral DG (1998) Perirhinal and postrhinal cortices of the rat: interconnectivity and connections with the entorhinal cortex. *Journal of Comparative Neurology* **391**:293–321.

Calford MB (1983) The parcellation of the medial geniculate body of the cat defined by the auditory response properties of single units. *Journal of Neuroscience* **3**:2350–2364.

Calford MB and Aitkin LM (1983) Ascending projections to the medial geniculate body of the cat: evidence for multiple parallel pathways through thalamus. *Journal of Neuroscience* **3**:395–401.

Calford MB and Semple MN (1995) Monaural inhibition in cat auditory cortex. *Journal of Neurophysiology* **73**:1876–1891.

Celio MR (1990) Calbindin D-28 k and parvalbumin in the rat nervous system. *Neuroscience* **35**:375–475.

Cetas JS, de Venecia RK, and McMullen NT (1999) Thalamocortical afferents of Lorente de No: medial geniculate axons that project to primary auditory cortex have collateral branches to layer I. *Brain Research* **830**:203–208.

Chechik G, Anderson MJ, Bar-Yosef O, Young ED, Tishby N, and Nelken I (2006) Reduction of information redundancy in the ascending auditory pathway. *Neuron* **51**:359–368.

Chen QC and Jen PH-S (2000) Bicuculline application affects discharge patterns, rate-intensity functions, and frequency tuning characteristics of bat auditory cortical neurons. *Hearing Research* **150**: 161–174.

Chu Z, Galarreta M, and Hestrin S (2003) Synaptic interactions of late-spiking neocortical neurons in layer 1. *Journal of Neuroscience* **23**:96–102.

Cipolloni PB and Keller A (1989) Thalamocortical synapses with identified neurons in monkey primary auditory cortex: a combined Golgi/EM and GABA/peptide immunocytochemistry study. *Brain Research* **492**:347–355.

Clarey JC, Barone P, Irons WA, Samson FK, and Imig TJ (1995) Comparison of noise and tone azimuth tuning of neurons in cat primary auditory cortex and medical geniculate body. *Journal of Neurophysiology* **74**:961–980.

Clarey JC, Barone P, and Imig TJ (1992) Physiology of auditory thalamus. In: Popper AN and Fay RR (eds). *Springer Handbook of Auditory Research*, Volume 2, *The Mammalian Auditory Pathway: Neurophysiology*. Springer, New York, pp. 232–334.

Clerici WJ and Coleman JR (1990) Anatomy of the rat medial geniculate body: I. Cytoarchitecture, myeloarchitecture, and neocortical connectivity. *Journal of Comparative Neurology* **297**:14–31.

Clerici WJ, McDonald AJ, Thompson R, and Coleman JR (1990) Anatomy of the medial geniculate body: II. Dendritic morphology. *Journal of Comparative Neurology* **297**:32–54.

Coffey CS, Ebert J, Marshall AF, Skaggs JD, Falk SE, Crocker WD, Pearson JM, and Fitzpatrick DC (2006) Detection of interaural correlation by neurons in the superior olivary complex, inferior colliculus and auditory cortex of the unanesthetized rabbit. *Hearing Research* **221**:1–16.

Conley M, Kupersmith AC, and Diamond IT (1991) The organization of projections from subdivisions of the auditory cortex and thalamus to the auditory sector of the thalamic reticular nucleus in *Galago*. *European Journal of Neuroscience* **3**:1089–1103.

Connors BW and Gutnick MJ (1990) Intrinsic firing patterns of diverse neocortical neurons. *Trends in Neurosciences* **13**:99–104.

Crabtree JW (1998) Organization in the auditory sector of the cat's thalamic reticular nucleus. *Journal of Comparative Neurology* **390**:167–182.

Creutzfeldt O, Hellweg F-C, and Schreiner C (1980) Thalamocortical transformation of responses to complex auditory stimuli. *Experimental Brain Research* **39**:87–104.

Cruikshank SJ, Lewis TJ, and Connors BW (2007) Synaptic basis for intense thalamocortical activation of feedforward inhibitory cells in neocortex. *Nature Neuroscience* **10**:462–468.

Cruikshank SJ, Rose HJ, and Metherate R (2002) Auditory thalamocortical synaptic transmission in vitro. *Journal of Neurophysiology* **87**:361–384.

de la Mothe, Blumell S, Kajikawa Y, and Hackett TA (2006) Thalamic connections of the auditory cortex in marmoset monkeys: core and medial belt regions. *Journal of Comparative Neurology* **496**:72–96.

Denning KS and Reinagel P (2005) Visual control of burst priming in the anesthetized lateral geniculate nucleus. *Journal of Neuroscience* **25**:3531–3538.

de Ribaupierre F, Goldstein MH and Yeni-Komshian G (1972) Intracellular study of the cat's primary auditory cortex. *Brain Research* **48**:185–204.

de Venecia RK, Smelser CB, Lossman SD and McMullen NT (1995) Complementary expression of parvalbumin and calbindin D-28 k delineates subdivisions of the rabbit medial geniculate body. *Journal of Comparative Neurology* **359**:595–612.

de Venecia RK, Smelser CB, and McMullen NT (1998) Parvalbumin is expressed in a reciprocal circuit linking the medial geniculate body and auditory neocortex in the rabbit. *Journal of Comparative Neurology* **400**:349–362.

DeWeese MR, Wehr M, and Zador AM (2003) Binary spiking in auditory cortex. *Journal of Neuroscience* **23**:7940–7949.

Diamond ME, Armstrong-James M, Budway MJ, and Ebner FF (1992) Somatic sensory responses in the rostral sector of the posterior group (POm) and the ventral posterior medial nucleus (VPM) of the rat thalamus. Dependence on the barrel field cortex. *Journal of Comparative Neurology* **319**:66–84.

Donishi D, Kimura, A, Okamoto, K, and Tamai Y (2006) "Ventral" area in the rat auditory cortex: A major auditory field connected with the dorsal division of the medial geniculate body. *Neuroscience* **141**:1553–1567.

Doron NN and LeDoux JE (2000) Cells in the posterior thalamus project to both amygdala and temporal cortex: a quantitative retrograde double-labeling study in the rat. *Journal of Comparative Neurology* **425**:257–274.

Edeline JM, Manunta Y, and Hennevin E (2000) Auditory thalamus neurons during sleep: changes in frequency selectivity, threshold, and receptive field size. *Journal of Neurophysiology* **84**: 934–952.

Edeline J-M (2003) The thalamo-cortical auditory receptive fields: regulation by the states of vigilance, learning and the neuromodulatory systems. *Experimental Brain Research* **153**:554–572.

Evans EF and Whitfield IC (1964) Classification of unit responses in the auditory cortex of the unanesthetized and unrestrained cat. *Journal of Physiology (London)* **171**:476–493.

Ferster D and Miller KD (2000) Neural mechanisms of orientation selectivity in the visual cortex. *Annual Review of Neuroscience* **23**:441–471.

Foeller E, Vater M, and Kössl M (2001) Laminar analysis of inhibition in the gerbil primary auditory cortex. *Journal of the Association for Research in Otolaryngology* **2**:279–296.

Friauf E (1994) Distribution of calcium-binding protein calbindin-D28k in the auditory system of adult and developing rats. *Journal of Comparative Neurology* **349**:193–211.

Gabernet L, Jadhav SP, Feldman DE, Carandini M, and Scanziani M (2005) Somatosensory integration controlled by dynamic thalamocortical feed-forward inhibition. *Neuron* **48**:315–327.

Gerren RA and Weinberger NM (1983) Long term potentiation in the magnocellular medial geniculate nucleus of the anesthetized cat. *Brain Research* **265**:138–142.

Gibson JR, Beierlein M, and Connors BW (1999) Two networks of electrically coupled inhibitory neurons in neocortex. *Nature* **402**: 75–79.

Gil Z, Connors BW, and Amitai Y (1997) Differential regulation of neocortical synapses by neuromodulators and activity. *Neuron* **19**:679–686.

Gil Z, Connors BW, and Amitai Y (1999) Efficacy of thalamocortical and intracortical synaptic connections: quanta, innervation, and reliability. *Neuron* **23**:385–397.

Guillery RW (1995) Anatomical role concerning the role of the thalamus in corticocortical communication: a brief review. *Journal of Anatomy (London)* **187**:583–592.

Hackett TA, Preuss TM, and Kaas JH (2001) Architectonic identification of the core region in auditory cortex of macaques, chimpanzees, and humans. *Journal of Comparative Neurology* **441**:197–222.

Harris RM (1987) Axon collaterals in the thalamic reticular nucleus from thalamocortical neurons of the rat ventrobasal thalamus. *Journal of Comparative Neurology* **258**:397–406

Hashikawa T, Molinari M, Rausell E, and Jones EG (1995) Patchy and laminar terminations of medial geniculate axons in monkey auditory cortex. *Journal of Comparative Neurology* **362**:195–208.

Hashikawa T, Rausell E, Molinari M, and Jones EG (1991) Parvalbumin- and calbindin containing neurons in the monkey medial geniculate complex: differential distribution and cortical layer specific projections. *Brain Research* **544**:335–341.

Hefti BJ and Smith PH (2000) Anatomy, physiology, and synaptic responses of rat layer V auditory cortical cells and effects of intracellular $GABA_A$ blockade. *Journal of Neurophysiology* **83**:2626–2638.

Heil P and Rajan R, and Irvine DRF (1994) Topographic representation of tone intensity along the isofrequency axis of cat primary auditory cortex. *Hearing Research* **76**:188–202.

Heil P and Irvine DRF (1997) First-spike timing of auditory-nerve fibers and comparison with auditory cortex. *Journal of Neurophysiology* **78**:2438–2454.

Herkenham M (1980) Laminar organization of thalamic projections to the rat neocortex. *Science* **207**:532–535.

Hestrin S and Armstrong WE (1996) Morphology and physiology of cortical neurons in layer I. *Journal of Neuroscience* **16**: 5290–5300.

Hu B (1995) Cellular basis of temporal synaptic signaling: an in vitro electrophysiological study in rat auditory thalamus. *Journal of Physiology(London)* **483**:167–182.

Huang CL and Winer JA (2000) Auditory thalamocortical projections in the cat: laminar and areal patterns of input. *Journal of Comparative Neurology* **427**:302–331.

Imig TJ and Adrián HO (1977) Binaural columns in the primary field (A1) of cat auditory cortex. *Brain Research* **138**:241–257.

Imig TJ and Morel A (1985) Tonotopic organization in lateral part of posterior group of thalamic nuclei in the cat. *Journal of Neurophysiology* **53**:309–340.

Imig TJ and Morel A (1988) Organization of the cat's auditory thalamus. In: Edelman GM, Gall WE and Cowen WM (eds) *Auditory Function. Neurobiological Bases of Hearing*. Wiley, New York, pp. 457–485.

Jahnsen H and Llinás R (1984) Electrophysiological properties of guinea-pig thalamic neurones: an in vitro study. *Journal of Physiology (London)* **349**:205–226.

Jones EG (1998a) A new view of specific and nonspecific thalamocortical connections. *Advances in Neurology* **77**:49–71.

Jones EG (1998b) Viewpoint: the core and matrix of thalamic organization. *Neuroscience* **85**:331–345.

Jones EG (2001) The thalamic matrix and thalamocortical synchrony. *Trends in Neurosciences* **24**:595–601.

Kaas JH (2005) The future of mapping sensory cortex in primates: three of the many remaining issues. *Philosophical Transactions of the Royal Society of London. series B, Biological Sciences* **360**:653–664.

Kaas JH and Hackett TA (1999) 'What' and 'where' processing in auditory cortex. *Nature Neuroscience* **2**:1045–1047.

Kaas JH and Hackett TA (2000) Subdivisions of auditory cortex and processing streams in primates. *Proceedings of the National Academy of Science of the United States of America* **97**: 11793–11799.

Kaur S, Lazar R, and Metherate R (2004) Intracortical pathways determine breadth of subthreshold frequency receptive fields in primary auditory cortex. *Journal of Neurophysiology* **91**:2551–2567.

Kharazia VN and Weinberg RJ (1994) Glutamate in thalamic fibers terminating in layer IV of primary sensory cortex. *Journal of Neuroscience* **14**:6021–6032.

Kimura A, Donishi T, Okamoto K, and Tamai Y (2004) Efferent connections of "posterodorsal" auditory area in the rat cortex: implications for auditory spatial processing. *Neuroscience* **128**: 399–419.

Kimura A, Donishi T, Okamoto K, Imbe H, and Tamai Y (2007) Efferent connections of the ventral auditory area in the cortex: Implications for auditory processing related to emotion. *European Journal of Neuroscience* **25**:2819–2834.

Kimura A, Donishi T, Sakoda T, Hazama M, and Tamai Y (2003) Auditory thalamic nuclei projections to the temporal cortex in the rat. *Neuroscience* **117**:1003–1016.

Komura Y, Tamara R, Uwano T, Nishijo H, and Ono T (2005) Auditory thalamus integrates visual inputs into behavioral gains. *Nature Neuroscience* **8**:1203–1209.

Kubota Y, Hatada S, Kondo S, Karube F, and Kawaguchi Y (2007) Neocortical inhibitory terminals innervate dendritic spines targeted by thalamocortical afferents. *Journal of Neuroscience* **27**: 1139–1150.

LeDoux JE, Ruggiero DA, and Reis DJ (1985) Projections to the subcortical forebrain from anatomically defined regions of the medial geniculate body in the rat. *Journal of Comparative Neurology* **242**:182–213.

LeDoux JE, Ruggiero DA, Forest R, Stornetta R, and Reis DJ (1987) Topographic organization of convergent projections to the thalamus from the inferior colliculus and spinal cord in the rat. *Journal of Comparative Neurology* **264**:123–146.

Latawiec D, Martin KAC, and Meskenaite V (2000) Termination of the geniculocortical projection in the striate cortex of macaque monkey: a quantitative immunoelectron microscopic study. *Journal of Comparative Neurology* **419**:306–319.

Lee CC, Imaizumi K, Schreiner CE, and Winer JA (2004a) Concurrent tonotopic processing streams in auditory cortex. *Cerebral Cortex* **14**:441–451.

Lee CC, Schreiner CE, Imaizumi K, and Winer JA (2004b) Tonotopic and heterotopic projection systems in physiologically defined auditory cortex. *Neuroscience* **128**:871–887.

Linden JF and Schreiner CE (2003) Columnar transformations in auditory cortex? A comparison to visual and somatosensory cortices. *Cerebral Cortex* **13**:83–89.

Linke R. (1999) Differential projection patterns of superior and inferior collicular neurons onto posterior paralaminar neurons of the thalamus surrounding the medial geniculate body in the rat. *European Journal of Neuroscience* **11**:187–203.

Linke R and Schwegler H (2000) Convergent and complimentary projections of the caudal paralaminar thalamic nuclei to the rat temporal and insular cortex. *Cerebral Cortex* **10**:753–771.

Lopez-Bendito G, Sturgess K, Erdelyi F, Szabó G, Molnár Z, and Paulsen O (2004) Preferential origin and layer destination of GAD65-GFP cortical interneurons. *Cerebral Cortex* **14**: 1122–1133.

Lu T, Liang L, and Wang XQ (2001) Temporal and rate representations of time-varying signals in the auditory cortex of awake primates. *Nature Neuroscience* **4**:1131–1138.

Lucas-Meunier E, Fossier P, Baux G, and Amar M (2003) Cholinergic modulation of the cortical neural network. *Pflügers Archiv – European Journal of Physiology* **446**:17–29.

Lund JS (1984) Spiny stellate cells. In: Peters A and Jones EG (eds) *Cerebral Cortex*, Volume 1, *Cellular Components of the Cerebral Cortex*. Plenum Press, New York, pp. 255–308.

Lysakowski A, Wainer BH, Bruce G, and Hersh LB (1989) An atlas of regional and laminar distribution of choline acetyltransferase immunoreactivity in rat cortex. *Neuroscience* **28**:291–336.

Mainen ZF and Sejnowski TJ (1995) Reliability of spike timing in neocortical neurons. *Science* **268**:1503–1506.

Malmierca MS, Merchán MA, Henkel CK, and Oliver DL (2002) Direct projections from cochlear nuclear complex to auditory thalamus in the rat. *Journal of Neuroscience* **22**:10891–10897.

Massaux A, Dutrieux G, Cotillon-Williams N, Manunta Y, and Edeline J-M (2004) Auditory thalamus bursts in anesthetized and non-anesthetized states: contribution to functional properties. *Journal of Neurophysiology* **91**:2117–2134.

Massaux A and Edeline J-M (2003) Bursts in the medial geniculate body: a comparison between anesthetized and unanesthetized states in guinea pig. *Experimental Brain Research* **153**: 573–578.

McDonald AJ (1998) Cortical pathways to the mammalian amygdale. *Progress in Neurobiology* **55**:257–332.

McEchron MD, Green EJ, Winters RW, Nolen TG, Schneiderman N, and McCabe PM (1996) Changes in synaptic efficacy in the medial geniculate nucleus as a result of auditory classical conditioning. *Journal of Neuroscience* **16**:1273–1283.

McMullen NT and Glaser EM (1984) Morphology and laminar distribution of nonpyramidal neurons in the auditory cortex of the rabbit. *Journal of Comparative Neurology.* **208**:85–106.

McMullen NT and de Venecia RK (1993) Thalamocortical patches in auditory neocortex. *Brain Research* **620**:317–322.

Mendelson JR, Schreiner CE, and Sutter ML (1997) Functional topography of cat primary auditory cortex: response latencies. *Journal of Comparative Physiology A* **181**:615–633.

Merriam EB, Netoff TI, and Banks MI (2005) Bistable network behavior of layer I interneurons in auditory cortex. *Journal of Neuroscience* **25**:6175–6186.

Metherate R and Aramakis VB (1999) Intrinsic electrophysiology of neurons in thalamorecipient layers of developing rat auditory cortex. *Brain Research Developmental Brain Research* **115**:131–144.

Metherate R, Kaur S, Kawai H, Lazar R, Liang K, and Rose HJ (2005) Spectral integration in auditory cortex: mechanisms and modulation. *Hearing Research* **206**:146–158.

Meyer G, González-Hernández TH, and Ferres-Torres R (1989) The spiny stellate neurons in layer IV of the human auditory cortex. A Golgi study. *Neuroscience* **33**:489–498.

Micheyl C, Carlyon RP, Gutschalk A, Melcher JR, Oxenham AJ, Rauschecker JP, Tian B, and Courtenay WE (2007) The role of auditory cortex in the formation of auditory streams. *Hearing Research* **229**:116–131.

Middlebrooks JC, Dykes RW, and Merzenich MM (1980) Binaural response-specific bands in primary auditory cortex (AI) of the cat: topographical organization orthogonal to isofrequency contours. *Brain Research* **181**:31–48.

Middlebrooks JC and Zook JM (1983) Intrinsic organization of the cat's medial geniculate body identified by projections to binaural response-specific bands in the primary auditory cortex. *Journal of Neuroscience* **3**:203–224.

Miller LM, Escabí MA, Read HL, and Schreiner CE (2002) Spectrotemporal receptive fields in the lemniscal auditory thalamus and cortex. *Journal of Neurophysiology* **87**:516–527.

Miller LM, Escabí MA, Read HL, and Schreiner CE (2001a) Functional convergence of response properties in the auditory thalamocortical system. *Neuron* **32**:151–160.

Miller LM, Escabí MA. and Schreiner CE (2001b) Feature selectivity and interneuronal cooperation in the thalamocortical system. *Journal of Neuroscience* **21**:8136–8144.

Mitani A, Itoh K, Nomura S, Kudo M, Kaneko T, and Mizuno N (1984) Thalamocortical projections to layer I of the primary auditory cortex in the cat: a horseradish peroxidase study. *Brain Research* **310**: 347–350.

Mitani A, Itoh K, and Mizuno N (1987) Distribution and size of thalamic neurons projecting to layer I of the auditory cortical fields of the cat compared to those projecting to layer IV. *Journal of Comparative Neurology* **257**:105–121.

Molinari M, Dell'Anna ME, Russell E, Leggio MG, Hashikawa T, and Jones EG (1995) Auditory thalamocortical pathways defined in monkeys by calcium-binding protein immunoreactivity. *Journal of Comparative Neurology* **362**:171–194.

Montero VM (1983) Ultrastructural identification of axon terminals from the thalamic reticular nucleus in the medial geniculate body of the rat: an EM autoradiographic study. *Experimental Brain Research* **51**:338–342.

Mooney DM, Zhang L, Basile C, Senatorov VV, Ngsee J, Omar A, and Hu B (2003) Distinct forms of cholinergic modulation in parallel thalamic sensory pathways. *Proceedings of the National Academy of Science of the United States of America* **101**:320–324.

Morest, DK (1964) The neuronal architecture of the medial geniculate body of the cat. *Journal of Anatomy (London)* **98**:611–630.

Morest, DK (1965) The lateral tegmental system of the midbrain and the medial geniculate body: study with Golgi and Nauta methods in cat. *Journal of Anatomy (London)* **99**:611–634.

Nieuwenhuys R (1994) The neocortex. An overview of its evolutionary development, structural organization and synaptology. *Anatomy & Embryology* **190**:307–337.

Niimi K, Ono K, and Kusunose M (1984) Projections of the medial geniculate nucleus to layer 1 of the auditory cortex in the cat traced with horseradish peroxidase. *Neuroscience Letters* **45**:223–288.

Oliver DL (1984) Neuron types in the central nucleus of the inferior colliculus that project to the medial geniculate body. *Neuroscience* **11**:409–424

Peruzzi D, Bartlett E, Smith PH and Oliver DL (1997) A monosynaptic GABAergic input from the inferior colliculus to the medial geniculate body in the rat. *Journal of Neuroscience* **17**:3766–3777.

Pinault D (2004) The thalamic reticular nucleus: structure, function and concept. *Brain Research Brain Research Review* **46**:1–31.

Pinault D and Deschênes M (1998) Projection and innervation pattern of individual thalamicreticular axons in the thalamus of the adult rat: a three-dimensional, graphic, and morphometric analysis. *Journal of Comparative Neurology* **391**:180–203.

Polley DB, Read HL, Storace DA, and Merzenich MM (2007) Multiparametric auditory field organization across five cortical fields in the albino rat. *Journal of Neurophysiology* **97**: 3621–3638.

Ramcharan EJ, Gnadt JW, and Sherman SM (2005) Higher-order thalamic relays burst more than first-order relays. *Proceedings of the National Academy of Science of the United States of America* **102**:12236–12241.

Rauschecker JF, Tian B, Pons T, and Mishkin M (1997) Serial and parallel processing in rhesus monkey auditory cortex. *Journal of Comparative Neurology* **382**:89–103.

Reale RA and Kettner RE (1986) Topography of binaural organization in primary auditory cortex of the cat: effects of changing interaural intensity. *Journal of Neurophysiology* **56**:663–682.

Reser DH, Fishman YI, Arezzo JC, and Steinschneider M (2000) Binaural interactions in primary auditory cortex of the awake macaque. *Cerebral Cortex* **10**:574–584.

Rockland KS and Drash GW (1996) Collateralized divergent feedback connections that target multiple cortical areas. *Journal of Comparative Neurology* **373**:529–548.

Roger M and Arnault P (1989) Anatomical study of the connections of the primary auditory area in the rat. *Journal of Comparative Neurology* **287**:339–356.

Romanski LM and LeDoux JE (1993) Organization of rodent auditory cortex: anterograde transport of PHA-L from MGV to temporal cortex. *Cerebral Cortex* **3**:499–514.

Rose HJ and Metherate R (2005) Auditory thalamocortical transmission is reliable and temporally precise. *Journal of Neurophysiology* **94**:2019–2030.

Rouiller EM, Colomb E, Capt M, and de Ribaupierre F (1985) Projections of the reticular complex of the thalamus onto physiologically characterized regions of the medial geniculate body. *Neuroscience Letters* **53**:227–232.

Rouiller EM and Welker E (2000) A comparative analysis of the morphology of corticothalamic projections in mammals. *Brain Research Bulletin* **53**:727–741.

Ruiz O, Royal D, Sary G, Chen X, Schall JD, and Casagrande VA (2006) Low threshold CA^{2+} associated bursts are rare events in the LGN of the awake behaving monkey. *Journal of Neurophysiology* **95**:3401–3413.

Ryugo DK and Weinberger NM (1978) Differential plasticity of morphologically distinct neuron populations in the medical geniculate body of the cat during classical conditioning. *Behavioral Biology* **22**:275–301.

Schnupp JWN (2006) Auditory filters, features, and redundant representations. *Neuron* **51**:278–280.

Schreiner CE, Mendelson JR, and Sutter ML (1992) Functional topography of cat primary auditory cortex: representation of tone intensity. *Experimental Brain Research* **92**:105–122.

Sherman SM (2001) Thalamic relay functions. *Progress in Brain Research* **134**:51–69.

Shosaku A and Sumitomo I (1983) Auditory neurons in the rat thalamic reticular nucleus. *Experimental Brain Research* **49**:432–442.

Simm GM, de Ribaupierre F, de Ribaupierre Y, and Rouiller EM (1990) Discharge properties of single units in auditory part of reticular nucleus of thalamic in cat. *Journal of Neurophysiology* **63**:1010–1021.

Smith PH and Populin LC (2001) Fundamental differences between the thalamocortical recipient layers of the cat auditory and visual cortices. *Journal of Comparative Neurology* **436**:508–519.

Smith PH, Bartlett EL, and Kowalkowski A (2006) Unique combination of anatomy and physiology in cells of the rat paralaminar thalamic nuclei adjacent to the medial geniculate body. *Journal of Comparative Neurology* **496**:314–334.

Smith PH, Bartlett EL, and Kowalkowski A (2007) Cortical, superior and inferior collicular inputs to cells in the rat paralaminar thalamic nuclei adjacent to the medial geniculate body. *Journal of Neurophysiology* **98**:681–695.

Steinschneider M, Schroeder CE, Arezzo JC, and Vaughan HG Jr (1993) Temporal encoding of phonetic features in auditory cortex. *Annals of the New York Academy of Science* **682**:415–417.

Steinschneider M, Schroeder CE, Arezzo JC, and Vaughan HG Jr (1994) Speech-evoked activity in primary auditory cortex: effects of voice onset time. *Electroencephalography and Clinical Neurophysiology* **92**:30–43.

Stratford KJ, Tarczy-Hornoch K, Martin KAC, Bannister NJ, and Jack JJB (1996) Excitatory synaptic inputs to spiny stellate cells in cat visual cortex. *Nature* **382**:258–261.

Steriade M (1997) Synchronized activities of coupled oscillators in the cerebral cortex and thalamus at different levels of vigilance. *Cerebral Cortex* **7**:583–604.

Sugitani M (1979) Electrophysiological and sensory properties of the thalamic reticular neurones related to somatic sensation in rats. *Journal of Physiology* **290**:79–95.

Swadlow HA (2003) Fast-spike interneurons and feedforward inhibition in awake sensory neocortex. *Cerebral Cortex* **13**:25–32.

Swadlow HA and Gusev AG (2001) The impact of 'bursting' thalamic impulses at a neocortical synapse. *Nature Neuroscience* **4**:402–408.

Tan AY, Zhang LI, Merzenich MM, and Schreiner CE (2004) Tone-evoked excitatory and inhibitory synaptic conductances of primary auditory cortex neurons. *Journal of Neurophysiology* **92**: 630–643.

Tan AYY, Atencio CA, Polley DB, Merzenich MM, and Schreiner CE (2007) Unbalanced synaptic inhibition can create intensity-tuned auditory cortex neurons. *Neuroscience* **146**:449–462.

Tennigkeit F, Puil E, and Schwarz DWF (1997) Firing modes and membrane properties in lemniscal auditory thalamus. *Acta Oto-Laryngologica* **117**:254–257.

Ter Mikaelian M, Sanes DH, and Semple MN (2007) Transformation of temporal properties between auditory midbrain and cortex in the awake Mongolian gerbil. *Journal of Neuroscience* **27**:6091–6102.

Velenovsky DS, Cetas JS, Price RO, Sinex DG, and McMullen NT (2003) Functional subregions in primary auditory cortex defined by thalamocortical terminal arbors: an electrophysiological and anterograde labeling study. *Journal of Neuroscience* **23**: 308–316.

Verbny YI, Erdélyi F, Szabó G, and Banks MI (2006) Properties of a population of GABAergic cells in murine auditory cortex weakly excited by thalamic stimulation. *Journal of Neurophysiology* **96**:3194–3208.

Villa AEP (1990) Physiological differentiation within the auditory part of the thalamic reticular nucleus of the cat. *Brain Research Review* **15**:25–40.

Vogt BA (1991) The role of layer I in cortical function. In: Peters A and Jones EG (eds). *Cerebral Cortex*, Volume 9, *Normal and Altered States of Function*. Plenum Press, New York, pp. 49–79.

Volkov IO and Galazjuk AV (1991) Formation of spike response to sound tones in cat auditory cortex neurons: interaction of excitatory and inhibitory effects. *Neuroscience* **43**:307–321.

Wallace MN, Anderson LA, and Palmer AR (2007) Phase-locked responses to pure tones in auditory thalamus. *Journal of Neurophysiology* **98**:1941–1952.

Wang J, Caspary D, and Salvi RJ (2000) GABA-A antagonist causes dramatic expansion of tuning in primary auditory cortex. *Neuroreport* **11**:1137–1140.

Wang X, Lu T, Snider RK, and Liang L (2005) Sustained firing in auditory cortex evoked by preferred stimuli. *Nature* **435**:341–346.

Wehr M and Zador AM (2003) Balanced inhibition underlies tuning and sharpens spike timing in auditory cortex. *Nature* **426**:442–446.

Wehr M and Zador AM (2005) Synaptic mechanisms of forward suppression in rat auditory cortex. *Neuron* **47**:437–445.

Weyand TG, Boudreaux M, and Guido W (2001) Burst and tonic response modes in thalamic neurons during sleep and wakefulness. *Journal of Neurophysiology* **85**:1107–1118

Winer JA (1984) The human medial geniculate body. *Hearing Research* **15**:225–247.

Winer JA (1985) The medial geniculate body of the cat. *Advances in Anatomy, Embryology and Cell Biology* **86**:1–97.

Winer JA, Kelly JB, and Larue DT (1999) Neural architecture of the rat medial geniculate body *Hearing Research* **130**:42–61.

Winer JA and Larue DT (1988) Anatomy of glutamic acid decarboxylase immunoreactive neurons and axons in rat medial geniculate body. *Journal of Comparative Neurology* **274**:422–448.

Winer JA and Lee CC (2007) The distributed auditory cortex. *Hearing Research* **229**:3–13.

Winer JA, Miller LM, Lee CC, and Schreiner CE (2005) Auditory thalamocortical transformation: structure and function. *Trends in Neurosciences* **28**:255–263.

Winer JA and Morest DK (1983) The neuronal architecture of the dorsal division of the medial geniculate body of the cat. A study with the rapid Golgi method. *Journal of Comparative Neurology* **221**: 1–30.

Winer JA, Sally SL, Larue DT, and Kelly JB (1999b) Origins of medial geniculate body projections to physiologically defined zones of rat primary auditory cortex. *Hearing Research* **130**:42–61.

Wu GK, Li P, Tao HW, and Zhang LI (2006) Nonmonotonic synaptic excitation and imbalanced inhibition underlying cortical intensity tuning. *Neuron* **52**:705–715.

Yamamoto T, Noda T, Samejima A, and Oka H (1985) A morphological investigation of thalamic neurons by intracellular HRP staining in cats. *Journal of Comparative Neurology* **236**:331–347.

Yen CT and Jones EG (1983) Intracellular staining of physiologically identified neurons and axons in the somatosensory thalamus of cat. *Brain Research* **280**:148–154.

Yu YQ, Xiong Y, Chan YS, and He J (2004) In vivo intracellular responses of the medial geniculate neurones to acoustic stimuli in anaesthetized guinea pigs. *Journal of Physiology (London)* **560**:191–205.

Zhang X and Giesler GJ Jr (2005) Response characteristics of spinothalamic tract neurons that project to the posterior thalamus in rats. *Journal of Neurophysiology* **93**:2552–2564.

Zhang LI, Tan AYY, Schreiner CE, and Merzenich MM (2003) Topography and synaptic shaping of direction selectivity in primary auditory cortex. *Nature* **424**:201–205.

Zhou FM and Hablitz JJ (1996) Morphological properties of intracellularly labeled layer I neurons in rat neocortex. *Journal of Comparative Neurology* **376**:198–213.

Zurita P, Villa AEP, de Ribaupierre Y, de Ribaupierre F, and Rouiller EM (1994) Changes of single unit activity in the cat's auditory thalamus and cortex associated with different anesthetic conditions. *Neuroscience Research* **19**:303–316.

Chapter 4

Auditory Cortical Organization: Evidence for Functional Streams

Josef P. Rauschecker and Lizabeth M. Romanski

Abbreviations

AI	primary auditory cortex
AL	anterior region of the lateral belt
BBW	best bandwidth
BPN	band pass noise
CL	caudolateral area
CM	caudal medial area
CPB	caudal parabelt region
DLPFC	dorsolateral prefrontal cortex
FM	frequency modulation
fMRI	functional magnetic resonance imaging
Ig	granular insula
L	lateral region
LB	lateral belt
Lim	limitans
MB	medial belt
MC	monkey call
MCPI	monkey call preference index
MGad	anterodorsal division of the medial geniculate complex
MGC	medial geniculate complex
MGd	dorsal division of the medial geniculate complex
MGm	magnocellular division of the medial geniculate complex
MGpd	dorsoposterior division of the medial geniculate complex
MGv	ventral division of the medial geniculate complex
ML	mediolateral area
PET	positron emission tomography
PFC	prefrontal cortex
PM	medial pulvinar
Po	posterior nucleus of the multisensory thalamic complex
R	rostral field
Ri	retroinsular cortex
RM	rostromedial belt region
RPB	rostral parabelt region
RT	rostral temporal field
RTL	lateral rostrotemporal belt region
RTM	medial rostrotemporal belt region
Sg	suprageniculate
STG	superior temporal gyrus
STP	superior temporal plane
STS	superior temporal sulcus
TAa	anterior temporal area
TE	inferior temporal lobe region
TEO	inferior temporal lobe region
TPO	temporal polysensory area
Tpt	temporo-parietal area
TS1,2	rostral areas of the superior temporal gyrus
VI	primary visual cortex
VLPFC	ventro-lateral prefrontal cortex

J.P. Rauschecker (✉)
Department of Physiology and Biophysics, Georgetown University School of Medicine, Washington, DC 20007-2197, USA
e-mail: josef@helix.nih.gov

1 Introduction

Understanding auditory cortex functional organization lags far behind the current understanding of visual cortex. One reason may be that auditory research has traditionally taken a bottom-up approach dealing first with cochlear and brain stem mechanisms of auditory coding. However, to understand how complex sounds are processed, stored, and recognized, we must understand how auditory cortex functions. In the cortex, the primary auditory cortex (AI) has long received the most attention. Physiological and anatomical studies in cats and monkeys find that multiple auditory areas surround AI, just as multiple representations of the visual world surround primary visual cortex (VI). It is reasonable to propose that these multiple fields support unique specialized functions in the complex behavioral repertoire of higher mammals.

J.A. Winer, C.E. Schreiner (eds.), *The Auditory Cortex*,
DOI 10.1007/978-1-4419-0074-6_4, © Springer Science+Business Media, LLC 2011

In the visual system, behavioral functions comprise at least two major categories (Ungerleider and Mishkin 1982): one function of vision is the identification of objects and is based on the detection of object boundaries, contrast or color, texture gradients, spatial frequency content, and analysis of other feature domains. A second major visual function supports spatial behavior and includes the analysis of spatial motion and depth. The functions of hearing are analogous, and auditory perception has pursued similar goals: identification of objects on the basis of their sounds or, more generally, identification of auditory objects, i.e., sounds with meaning or behavioral significance; and guidance of spatial behavior based on sound localization, including sound source motion in space. A task for auditory neuroscientists is to determine if these major auditory functions are implemented by separate functional brain entities, especially in auditory cortex, where multiple fields are connected to form functional streams.

The hypothesis of segregated functional streams in auditory cortical processing was first proposed a decade ago (Rauschecker 1997, 1998a, b; Romanski et al. 1999b), and this chapter will consider the evidence for it that has accumulated since. We will first describe the neuroanatomy of nonprimary auditory cortex in nonhuman primates, its cytoarchitectonic organization, and its thalamocortical and corticocortical projections. We then consider neurophysiological findings in nonprimary auditory cortex and prefrontal cortex. Finally, we discuss results from human brain physiology in functional neuroimaging, which can speak to the organization of auditory processing streams directly.

2 Anatomy of Nonprimary Auditory Cortex

2.1 Cytoarchitectonic Organization

Primate auditory cortex contains a core of three primary areas surrounded by a belt region of secondary areas, which are in turn bounded by parabelt auditory cortex. Even at the earliest stages of cortical auditory processing, there are likely divergent streams emanating from the core areas. The centrally located core region contains three subdivisions: a primary auditory area (AI), a rostral field (R) and, an even more rostral temporal field (RT) (Hackett et al. 1998a, 1999; Morel et al. 1993; Kosaki et al. 1997). The three core areas respond robustly at short latencies to pure tones. Each has a well-developed layer IV granule cell architecture. The middle layers also stain densely for parvalbumin, cytochrome oxidase, and acetylcholinesterase (Hackett et al. 1998a). Differences in the tonotopic map and in the distribution for these three markers have been helpful in differentiating areas AI, R, and RT. The core region projects to, and is surrounded by a cortical belt with eight subdivisions, and a lateral parabelt region of at least two fields.

The auditory belt can be differentiated from the core using both anatomical and physiological methods. In coronal sections, the belt cortex stains less densely for parvalbumin, acetylcholinesterase, and myelin. Two caudal belt regions, the caudomedial (CM) area and the mediolateral (ML) area, have a slightly more intense staining pattern for parvalbumin, while the rostromedial (RM) and medial rostrotemporal (RTM) medial belt regions stain more lightly for parvalbumin than the caudomedial or the lateral belt areas (Hackett et al. 1998). Subdivisions in the auditory belt, especially the lateral belt, are best demonstrated by connectional and physiological criteria (Fig. 4.1).

The parabelt is adjacent and lateral to the auditory lateral belt regions and on the dorsal gyral surface of the superior temporal gyrus. Staining for parvalbumin in it results in paler staining than in the adjacent belt. Parabelt parvalbumin staining is densest in layer IV just as in the core and belt regions, and the caudal parabelt stain slightly more than the rostral parabelt. As with the lateral belt, the parabelt and higher auditory areas are best demonstrated by physiological and connectional methods.

2.2 Thalamocortical Projections of Nonprimary Auditory Cortex

Connections of non-primary auditory cortex arise from a wide region of the posterior thalamus. For example, tracer injections in the parabelt label many nuclei, including dorsal (MGd) and magnocellular (MGm) divisions of the medial geniculate complex (MGC), suprageniculate (Sg), limitans (Lim), and medial pulvinar (PM) nuclei (Hackett et al. 1998b). A rostro-caudal topography exists such that rostral superior temporal gyrus (STG) receives input from posterior MGd, while the caudal STG is targeted by anterior MGd. Injections in the lateral belt label the primary relay nucleus, MGv, prominently. The principal inputs of CM are MGad (anterodorsal), MGv, and MGm, with secondary inputs from multisensory posterior (Po), suprageniculate (Sg), limitans (Lim), and medial pulvinar (PM) nuclei. The main inputs of belt caudolateral area (CL) are Po and MGpd, with secondary inputs from MGad, MGm, and multisensory nuclei (Hackett et al. 2007). The thalamic input to the tempero-parietal region (Tpt) resembles that to CL with less input from the MGC. Connections of the superior temporal sulcus (STS) and anterior parts of the superior temporal gyrus also include more input from multisensory nuclei and less from the MGC.

2.3 Corticocortical Connections

The three core cortical regions function in parallel and lesions in one do not abolish pure tone responses in the

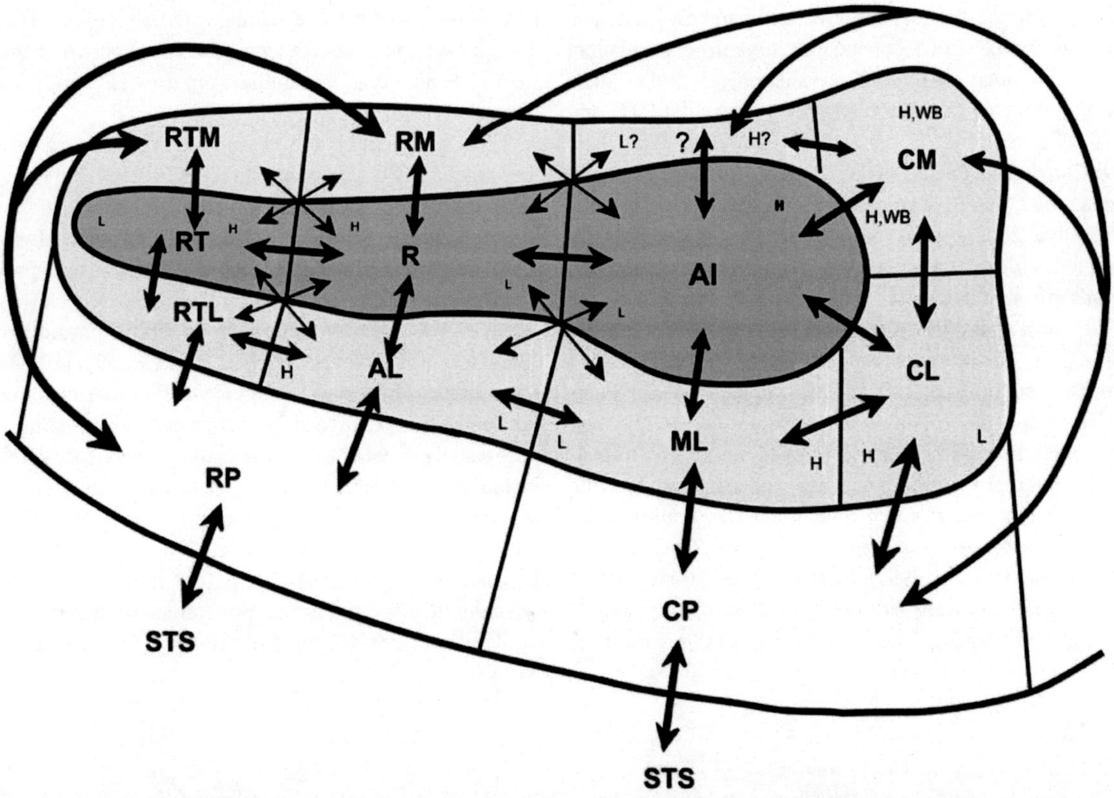

Fig. 4.1 Schematic of the macaque auditory cortex organization and connections. The core regions (AI, R, and RT) are bounded medially by the medial belt regions (CM, RM, and RTM). The lateral belt regions (CL, ML, AL, and RTL) form the lateral boundary of the core. Just lateral to the lateral belt is the parabelt, divided into rostral (RP) and caudal regions (CP). Reprinted with permission from Hackett et al. (1998)

others, indicating that separate, parallel thalamic inputs drive each core region independently. The auditory cortex core regions are densely connected with each other (Hackett et al. 1998; Kaas and Hackett 1998) (Fig. 4.1). To distinguish the belt cortex from the core regions, lesions placed in AI abolished pure tone responses in belt area CM. This study confirmed that auditory information flows from area AI to CM serially and that these connections are distinct from those of rostral core area R (Rauschecker et al. 1997). Previous studies found that each of the core regions is most densely connected with adjacent belt regions (Morel et al. 1993; Hackett et al. 1998a) (Fig. 4.1). Additional studies have confirmed connections of the rostral core with the anterior regions of the lateral belt (areas AL and ML) and caudal core with caudal belt regions. Thus, information leaves the auditory core in parallel streams and a topographic manner that suggests dorso-caudal and rostro-ventral streams.

The auditory belt has reciprocal connections with the core but also projects to the parabelt, the third stage of auditory cortical processing, and beyond the auditory cortex as well. Within the medial belt, area CM has connections with the caudal auditory core and with somatic sensory areas in retroinsular cortex (area Ri) and granular insula (Ig) as well as multisensory temporal pole areas Tpt and TPO (Hackett et al. 2007). Medial belt area RM is connected with rostral and caudal areas of the parabelt. The lateral belt projects to the parabelt topographically, with caudal parabelt regions connected most densely to the caudal belt, and rostral parabelt to rostral belt (Hackett et al. 1998) (Fig. 4.1). Beyond the belt and parabelt, auditory afferents target fields in the rostral STG and in the superior temporal sulcus (STS). The upper and lower banks of the STS receive dense parabelt projections (Hackett et al. 1999). The connections are mainly with the polysensory STS areas in the rostral half of the gyrus and also with the anterior temporal area TAa. The rostral STG areas TS1 and TS2 receive projections from the rostral parabelt while caudal parabelt projects to area Tpt in the superior temporal gyrus, which may be multisensory and also has parietal cortex connections (Hackett et al. 2007). Thus, connections arise topographically with the rostral parabelt more densely connected to the regions in the anterior superior temporal gyrus and the caudal parabelt projecting to caudal areas and area Tpt, with some projections to the caudal STS as well.

The connections of the auditory cortex with regions outside the temporal lobe have been examined in the most detail with regard to the frontal lobe. Belt and parabelt each have prefrontal cortex (PFC) connections organized as distinct rostral and caudal streams. Early anatomical studies indicated that a rostro-caudal topography with caudal STG and caudal PFC reciprocally connected (Pandya and Kuypers 1969; Chavis and Pandya 1976; Barbas 1992; Petrides and Pandya 1988, 2002; Romanski et al. 1999a, b; Petrides and Pandya 2002), while the rostral STG is reciprocally connected with rostral principalis (rostral areas 46 and 10) and orbitofrontal areas 11 and 12 (Pandya and Kuypers 1969; Pandya et al. 1969; Chavis and Pandya 1976). Further studies have characterized temporo-prefrontal connections to include auditory belt (Romanski et al. 1999b), parabelt (Hackett et al. 1998a), and prefrontal cortex (Hackett et al. 1999; Romanski et al. 1999a) to refine the rostral-caudal topography previously noted and showing that the frontal pole and anterior principal sulcus are densely connected with anterior belt and parabelt regions (Hackett et al. 1999; Romanski et al. 1999a). Moreover, the caudal parabelt and belt are reciprocally connected with dorsal prearcuate and caudal principalis. The ventrolateral PFC connections involve both rostral and caudal parabelt regions (Hackett et al. 1999) and it also receives a dense input from the dorsal STS bank including multisensory area TPO and area TAa (Romanski et al. 1999a).

These anatomical studies suggest that auditory information reaches PFC, and more direct evidence from combined anatomical and physiological methods confirms it. The lateral belt auditory areas AL, ML and CL, were physiologically identified (Romanski et al. 1999b), as in previous work (Rauschecker et al. 1995) then injected with anterograde and retrograde anatomical tracers in physiologically characterized belt areas AL, ML, and CL (Fig. 4.2a–b) and anterograde fibers and retrograde cells were examined in prefrontal cortex. Five frontal lobe regions contained both retrogradely labeled cells and anterogradely labeled fibers, including the frontal pole, the principal sulcus, the lateral inferior convexity, the lateral orbital cortex, and the dorsal periarcuate region (Fig. 4.2c). These connections were organized topographically such that projections from AL typically involved the frontal pole (area 10), the rostral principal sulcus (area 46), ventrolateral prefrontal cortex (areas

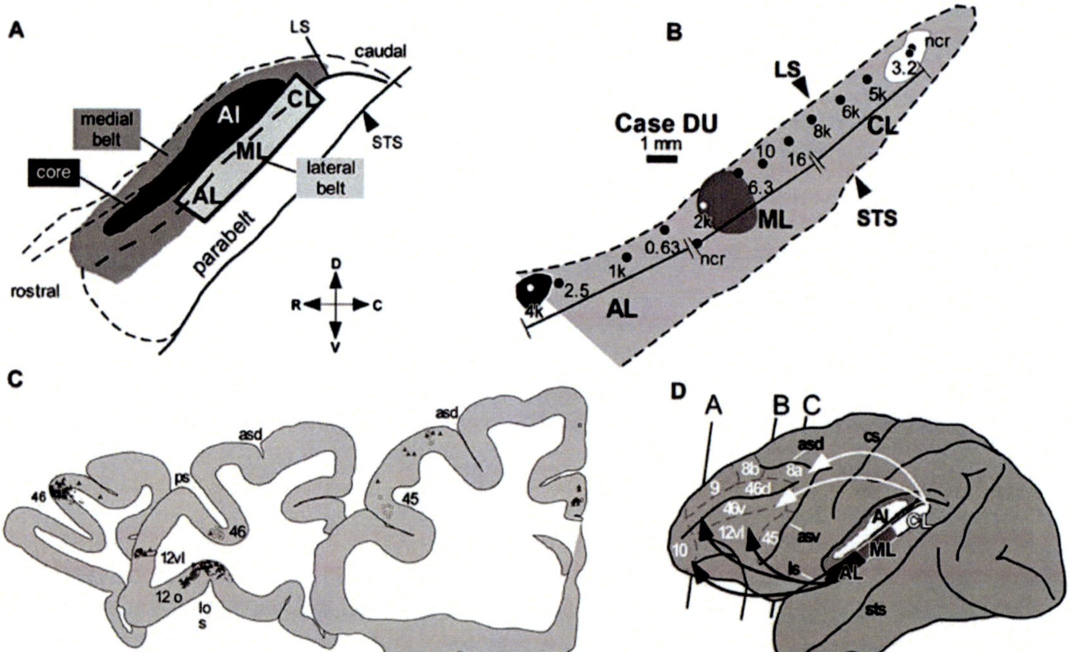

Fig. 4.2 Auditory parabelt projections target specific prefrontal regions of the cortex. Adapted with permission from Romanski et al. (1999b). Areas AL, ML, and CL comprise the lateral auditory belt cortex (*black rectangle*) (**a**), surrounding the core auditory cortex (A1) in *black*. (**b**) Tracer injections s into the auditory belt areas AL, ML, and CL are shown. Recording tracks are depicted by *black dots* and the average characteristic frequency of recorded neurons in that track is denoted beside each track. Injections were made in the 4 kHz region in AL, ML, and CL. (**c**) Coronal sections through the prefrontal cortex (anterior to posterior; *left* to *right*) with anterograde and retrograde cells and fibers in colors that match the injections sites. The injection in AL (*black circle* in **b**) labeled cells and fibers (*black squares* and *lines*) in the most rostral prefrontal regions, including the frontal pole, rostral principal sulcus, and the inferior convexity (areas 12vl and 12o). Injections into CL (*white circle* in **b**) labeled fibers and cells (*white lines* and *while squares*) in the caudal principal sulcus, frontal eyefields, and the most caudal edge of area 45 on the inferior convexity. ML injections labeled a combination of AL and ML (*grey triangles* and *lines*). A topographic summary (**d**) shows that anterior auditory cortex projects to rostroventral prefrontal areas, while caudal auditory cortex projects to the caudodorsal frontal lobe, confirming the dorsal and ventral auditory streams

12 vl and 45) and the lateral orbital cortex (areas 11, 12o). In contrast, area CL projections targeted the dorsal periarcuate cortex (area 8a, frontal eye fields), caudal principal sulcus (area 46), the caudal inferior convexity (area 45) and, in two cases, premotor cortex (area 6d). The frontal pole (area 10) and the lateral orbital cortices (areas 11 and 12) were devoid of anterograde labeling from caudal auditory injections. Conversely, the frontal eye fields did not receive projections from anterior auditory area AL. Inputs to area ML were a combination of the connections of AL and CL. These specific rostrocaudal topographical connections suggest that separate streams of auditory information target distinct frontal lobe domains. A pathway originating in CL targets caudal dorsolateral prefrontal cortex (DLPFC); another, arising in AL, targets rostral and ventral prefrontal areas (Fig. 4.2d). Visual system studies find spatial and nonspatial visual streams that target dorsal-spatial and ventral-object prefrontal regions (Ungerleider and Mishkin 1982; Wilson et al. 1993). Perhaps pathways originating from anterior and posterior auditory belt and parabelt cortices are analogous to the "what" and "where" visual system streams. These dual anatomical streams further support the physiological distinctions between anterior and posterior auditory cortex in non-human primates and in the human brain.

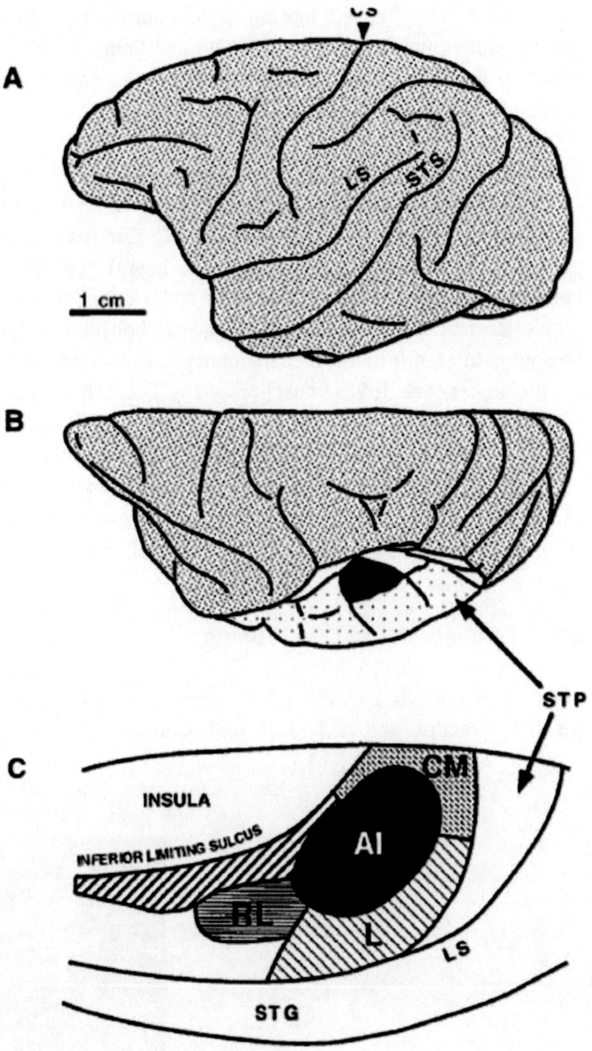

Fig. 4.3 Original discovery of multiple representations in rhesus monkey auditory cortex (Merzenich and Brugge 1973). **a** Lateral view of rhesus brain (LS, lateral sulcus; STS, superior temporal sulcus). **b** Dorsal view of left hemisphere with parietal cortex partly removed; *black area*, primary auditory cortex (AI); stippled region, surrounding on the superior temporal fields (STP). **c** Schematic view of auditory cortical fields. In the lateral field (L), a tonotopic map could not be established as its neurons did not respond to pure tones. CM, caudomedial field; L, lateral field; RL, rostrolateral field

3 Physiology of Nonprimary Auditory Cortex

Initial microelectrode analysis of rhesus monkey nonprimary auditory cortex used pure tone stimuli (Brugge and Merzenich 1973; Merzenich and Brugge 1973). Besides AI two other supratemporal plane (STP) areas were identified, both tonotopically organized: one (RL) was rostrolateral to AI and shared its low-frequency border; a caudomedial area (CM) shared its high-frequency border with AI. A lateral region (L) had neurons responsive to sounds but which were not driven consistently with pure tones (Fig. 4.3a–c). Later analysis posited a hierarchical model of auditory processing based on the idea that neurons in higher cortical areas would have larger receptive fields that integrate information over a wider area of the receptor surface. This led to studies with broader spectrum sounds.

3.1 Neurophysiology of Lateral Belt

3.1.1 Bandpass Tuning

Rauschecker and colleagues (1995) elicited reliable responses from most lateral belt (LB) neurons by broadening the bandwidth of frequency-centered sound bursts. Unexpectedly, most neurons did not simply increase their response monotonically with increasing bandwidth, i.e., white noise was not normally the best stimulus; instead, they preferred specific bandwidths of band-passed noise (BPN) Most LB neurons showed level tolerance in terms of bandwidth preference, i.e., their bandwidth tuning was independent of sound intensity, which makes these cells suitable for auditory pattern recognition (Rauschecker et al. 1995).

Bandwidth tuning or best bandwidth (BBW) varied systematically across the LB along an axis orthogonal to the cochleotopic organization of best center frequencies, which was confirmed in more extensive recordings (Rauschecker and Tian 2004). Related findings are reported for other species (Ehret and Schreiner 1997; Ohl and Scheich 1997). BBW in the LB is equally distributed over the bandwidth spectrum, whereas AI neurons prefer pure tones to BPN (Fig. 4.4). BBW decreases medially direction towards the core regions, and increases in the lateral direction.

Finding robust auditory responses in lateral belt neurons to BPN stimuli was of great practical value because it permitted the systematic mapping of the lateral belt. BPN bursts have a defined center frequency and a defined bandwidth. Rostrocaudal mapping of the lateral belt revealed a smooth gradient for best center frequency with two reversals (Rauschecker et al. 1995; Rauschecker and Tian 2004). This demonstrated three cochleotopically organized LB areas, named the anterolateral, middle lateral, and caudolateral areas (AL, ML, and CL). They are partly on the free surface of the superior temporal gyrus, shifted laterally and parallel to areas R, AI, and CM (Morel et al. 1993) (Fig. 4.3d).

3.1.2 Frequency-Modulated Tuning

Another prominent feature of LB neurons is their tuning for FM direction and rate (Tian and Rauschecker 2004), as seen in cat nonprimary auditory cortex areas (Tian and Rauschecker 1994, 1998). Although FM selectivity is found throughout the auditory pathway, it is pronounced in the LB, with >90% of cells responding strongly and selectively to FM stimuli.

Preferred FM rate differs among cortical areas, and this has provided clues about functional specificity of LB areas in processing of some types of complex sounds (Rauschecker 1997, 1998b). AL cells, for instance, prefer FM rates (<64 Hz/ms), which match those contained in most rhesus-specific communication calls ranging mostly between 8 and 50 Hz/ms (Hauser 1996; Rauschecker 1998b). By contrast, neurons in CL respond best to median FM rates 160 Hz/ms, which are better suited for sound localization (Tian and Rauschecker 2004).

3.1.3 Visual Analogies

Both types of selectivity for sounds of intermediate complexity have visual system analogs. Bandwidth selectivity in auditory belt cortex is reminiscent of size selectivity in monkey visual area V4, where cells prefer light bars of specific width and length (Desimone and Schein 1987). The visual

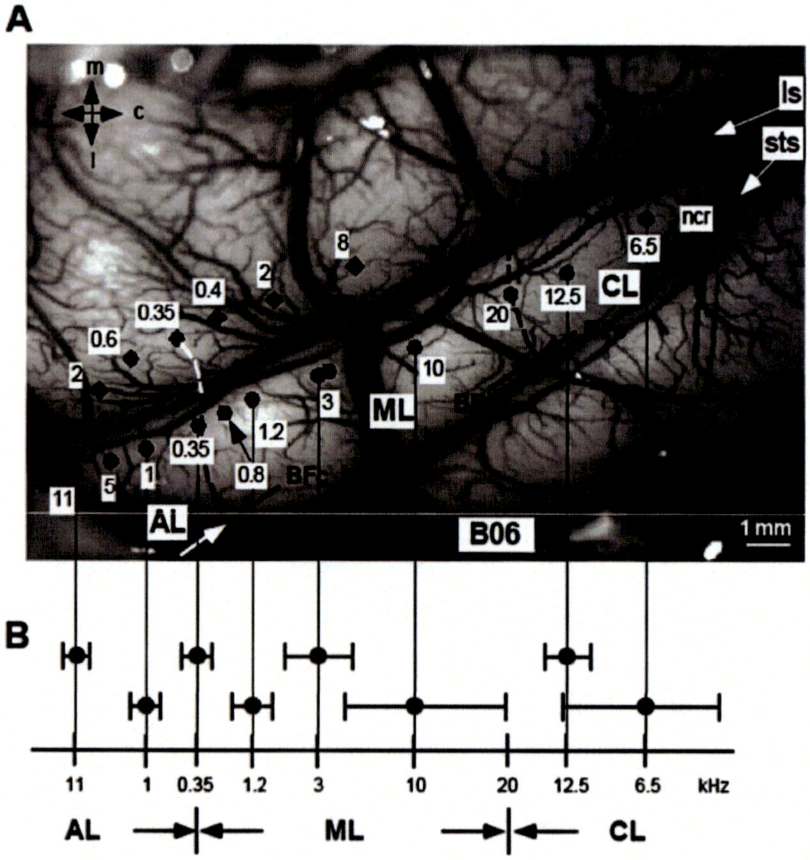

Fig. 4.4 Mapping of lateral auditory belt cortex in a rhesus monkey using band passed noise (BPN) bursts. **a** Lateral belt neurons in the respond briskly to BPN bursts of a specific bandwidth and center frequency, making it possible to map lateral belt and to establish areas AL, ML, and CL (anterolateral, middle lateral, and caudolateral fields, respectively), on the basis of best-frequency reversals, besides the cochleotopically organized areas R, A1, and CM. R, rostral field, identical with RL. From the original source (Rauschecker et al. 1995). **b** Best center frequencies in each electrode track were determined by taking the arithmetic mean of at least three recordings in that track. Used with permission from Rauschecker and Tian (2004)

analog for auditory FM selectivity is tuning for movement direction and velocity (Hubel and Wiesel 1962; Movshon and Newsome 1996), which is common in visual cortex. It seems, therefore, that fundamentally similar neurophysiological computations are performed in both sensory cortices. These computations may use similar local circuitry.

3.2 Neurophysiology of Medial Belt

Medial belt (MB) is less studied than the lateral belt, partly because it is more difficult to access. Medial belt cells are more responsive to BPN stimuli than to pure tones (Kusmierek and Rauschecker 2006, 2007). Neurons have various selectivities to complex sounds depending on their rostro-caudal location. Nothing is known of their FM selectivity. Comparison between lateral and medial belt areas will be conceptually important for determining whether they are functionally distinct entities supporting different behavioral goals, or whether rostral and caudal medial and lateral belt each is a functionally homogenous system.

Systematic studies of parabelt cortex are sparse. Initial results are compatible with the notion of a hierarchical system, neurons becoming increasingly selective for more specific features with distance from core areas (Kikuchi et al. 2007).

3.3 Neurophysiology of Rostral and Caudal Superior Temporal Gyrus

Early findings of differences in FM-rate tuning in LB areas led to more direct studies of functional differences in rostral and caudal STG. Assuming that the neural substrates for the processing of auditory objects, or the identity of sounds, including species-specific communication sounds, and for auditory spatial processing are at least partially segregated, several studies tested neurons across the auditory belt for space and object quality selectivity (Rauschecker and Tian 2000; Tian et al. 2001) (Fig. 4.5).

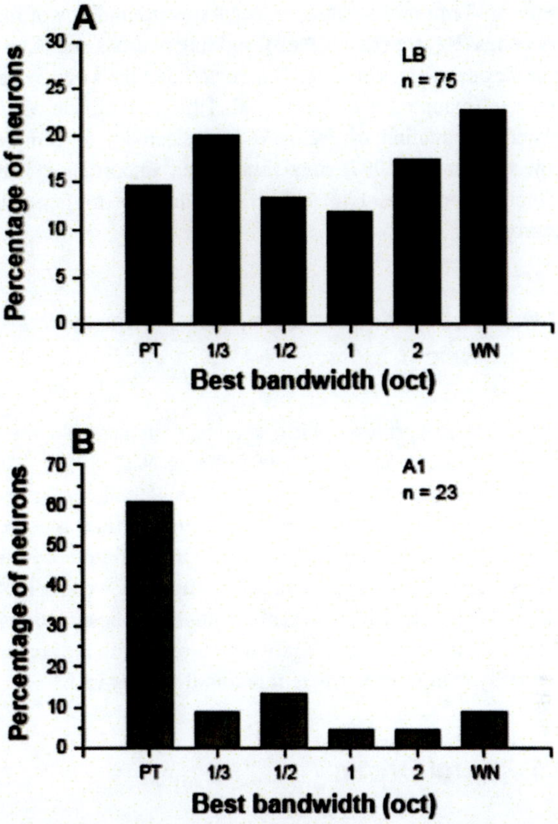

Fig. 4.5 Bandwidth preferences across lateral belt and primary auditory cortex. Best bandwidth was approximately evenly distributed in the lateral belt (LB) **a**. By contrast, core area A1 cells in preferred pure tones to sounds with wider bandwidth

3.3.1 Responses to Species-Specific Communication Calls

Moderately complex sounds discussed above, BPN bursts and FM sweeps, are ubiquitous components of communication sounds in many species. Prior work on squirrel monkey auditory cortex (Winter and Funkenstein 1973) was extended to macaque LB cells for complete species-specific vocalizations available in digitized form from a library of wild calls (Hauser 1996). Many neurons responded vigorously to monkey calls or their components. Unexpectedly, the neurons showed selectivity for different types of calls which did not reflect frequency tuning. Often, frequencies outside the cell's pure-tone tuning range, which were ineffective alone, elicited response facilitation when combined with frequencies inside the tuning range (Rauschecker et al. 1995; Rauschecker 1998b). Sometimes two complex sounds evoked a response only in a certain temporal order (Rauschecker 1997).

3.3.2 Combination Sensitivity

In the bat and songbird auditory system, this facilitation in both the spectral and temporal domain has been termed "combination sensitivity" (Suga et al. 1978; Margoliash and Fortune 1992). Nonlinear summation may be the main mechanism creating such selectivity in monkey auditory cortex, although suppression is also seen. Spectral summation involves convergence of input from more narrowly tuned neurons. Temporal summation occurs over hundreds of milliseconds (Rauschecker 1998b), yielding neurons selective to complex sequences of sounds characterizing their own vocalizations (Doupe 1997; Esser et al. 1997). Nonlinear visual system summation mechanisms for selectivity to complex objects (Tanaka 1997) may thus be an important general principle for generating feature specificity in higher-order neurons.

3.4 Lateral Belt Versus Primary Auditory Cortex

Responses to species-specific vocalizations are also found in AI (Steinschneider et al. 1995; Wang et al. 1995). This is not necessarily the same, however, as genuine call selectivity, where the response to calls is greater than that to a frequency band that falls within the neurons' frequency tuning range. In the macaque, call selectivity is far weaker in AI than in the lateral belt (Fritz et al., unpublished observations). Perhaps combination sensitivity is created later in the cortical hierarchy than in bats, where it is found already in AI.

3.5 Rostral Stream

Whether the lateral belt areas are the final communication call processing stage in monkeys is doubtful, but they constitute an important station in this complicated process. Neuroanatomical studies (Jones et al. 1995; Hackett et al. 1998) show robust feedforward projections toward anterior and lateral parts of the STG and the prefrontal cortex. If progressively more call-selective neurons are found in these areas, as seems to be the case (Kikuchi et al. 2004), this would be compatible with a hierarchical organization of auditory cortical processing. It would also be comparable to the visual concentration of face-selective cells in the rostral inferior temporal cortex (Desimone 1991). Such neurophysiological findings in nonhuman primates are in accord with human neuroimaging data, which suggest phonemic processing in the anterior superior temporal region.

3.6 Spatial Selectivity in the Caudal Belt

Some earlier studies of area CM cells found preferential tuning to the free field spatial location of a sound (Rauschecker et al. 1997) and such preferences occur in posterior STG (Leinonen et al. 1980). Spatially tuned neurons are present in AI, but they are more common in caudal belt areas CM and CL (Recanzone 2000; Tian et al. 2001). Work on alert monkeys trained in an auditory localization task found that the firing rate of CM cells correlates more with behavioral performance than that of AI cells, indicating that the caudal belt is important in sound localization (Recanzone et al. 2000).

Area CM receives direct input from the MGd which in turn receives most of its input from the dorsal cochlear nucleus via the external nucleus of the inferior colliculus, structures which are each implicated in auditory spatial functions (Knudsen 1983; Young et al. 1992). The spatial selectivity of caudal belt neurons may be set up in earlier processing stations.

4 "What" and "Where" in Nonprimary Auditory Cortex

To compare the response selectivity of the rostral and caudal lateral belt neurons directly in the same animals, auditory belt cells were examined for selectivity to both space and monkey calls. Broad-band species-specific communication calls were presented from different locations (Tian et al. 2001).

To quantify the selectivity for different monkey calls (MC), a monkey call preference index (MCPI) was calculated for the number of calls the cell responds to. Usually, a standard battery of seven common calls was used so that an MCPI of 7 means that the cell responded to all the calls presented. An MCPI of 3 or less indicated responsiveness to less than half of the calls, termed MC-selective, whereas those responsive to 5 or more of the calls are termed MC-nonselective. The LB areas differed in MC selectivity, with area AL having the most selective neurons (MCPI$\leq$2), then ML, and CL had the fewest highly selective neurons. For the most non-selective neurons (MCPI$\geq$6), the opposite was found: CL had the most non-selective neurons, AL the least, with ML intermediate.

Lateral belt spatial tuning was measured in these animals and showed the opposite areal distribution: highest selectivity in CL and lowest in AL. This finding supports the claim of functional specialization in monkey nonprimary auditory cortex. With the neuroanatomical findings (cf. 2.), this has led to the hypothesis that these areas, at the poles of lateral belt along its rostro-caudal extent, are the source of dual processing streams for auditory object and space information (Rauschecker and Tian 2000; Tian et al. 2001). The anterior "what"-stream may extend all the way to the temporal pole, which was demonstrated to be responsive acoustically using 2-deoxyglucose (Poremba et al. 2003). At that level a hemispheric difference for species-specific communication sounds also emerges (Poremba et al. 2004). The anterior "what"-stream extends further into ventrolateral PFC, as shown in connectional (Romanski et al. 1999a, b) and physiological (Romanski and Goldman-Rakic 2002) studies, whereas the posterior stream projects to parietal cortex and DLPFC (Fig. 4.6).

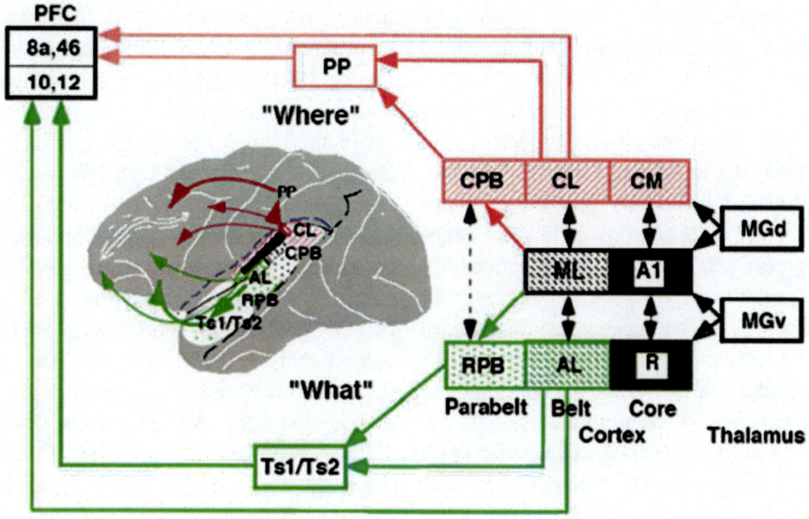

Fig. 4.6 Dual processing streams for "what" and "where" in the macaque auditory cortex. Modified from the original source and reproduced with permission (Rauschecker 1998b; Rauschecker and Tian 2000). Discrete thalamic input to the two pathways is provided from different medial geniculate (MG) nuclei: the ventral part (MGv) projects only to the core fields A1 and R, whereas its dorsal part (MGd) projects to A1 and CM (Rauschecker et al. 1997). Likewise, feedforward projections from AL and CL are largely separated and target the rostral parabelt (RPB) and caudal parabelt (CPB) regions, respectively (Hackett et al. 1998). Prefrontal cortex projections (PFC) are segregated in Brodmann areas 10 and 12 versus 8a and 46, respectively (Romanski et al. 1999b)

Primate functional magnetic resonance imaging (fMRI) studies confirm the subdivision of auditory cortex on the supratemporal plane into core and belt by their responses to tones and BPN (Petkov et al. 2006) and they have also provided clues about functional streams that attest to hierarchical complex sound processing network involving a higher-order, specialized voice region in the anterior superior temporal plane of the rhesus monkey (Kikuchi et al. 2008; Petkov et al. 2008). Combined fMRI and microelectrode recordings in nonhuman primates may be instrumental in further searches for functional specialization.

Behavioral studies on cats with reversible lesions of the anterior and posterior auditory belt induced by cooling the respective areas find a double dissociation of pattern and spatial functions that confirms the dual stream idea (Lomber and Malhotra 2008).

5 Beyond Classical Auditory Cortex

5.1 Auditory Processing in the Superior Temporal Sulcus

Few studies have examined auditory responses in areas beyond classical auditory cortex. Two promising areas are the cortex on the dorsal bank of the STS and the prefrontal cortex. Electrophysiological recordings find responses to complex sounds and multisensory neurons responsive to audiovisual stimuli in both.

While auditory afferents project to the STS, there have been few analyses of physiological responses to auditory stimuli in awake, behaving animals. Three studies recorded responses to auditory and other modality sensory stimuli in anesthetized animals (Benevento et al. 1977; Bruce et al. 1981; Hikosaka et al. 1988). These studies targeted the posterior two-thirds of the dorsal and ventral banks of the STS. Thus recordings were mainly confined to polymodal area TPO as well as visual processing regions. The auditory stimuli tested consisted of pure tones and clicks (Benevento et al. 1977), as well as complex sounds which included monkey calls, human vocalizations and environmental sounds (Bruce et al. 1981; Hikosaka et al. 1988). In Benevento et al. (1977) unimodal auditory responses, although sparse ($N = 14/107$ cells), were observed and included onset and offset excitatory responses and onset inhibitory responses. Hikosaka et al. (1988) also reported unimodal auditory responses in the caudal STS polysensory region, where the responses were all broadly tuned with little stimulus specificity. Responses to auditory stimuli in Bruce et al. (1981) were only obtained in combination with visual stimuli. The most comprehensive study Baylis et al. (1987) recorded more than 2600 neurons in alert rhesus macaques across a wide region of the STS including the ventral STG and area TAa. This study demonstrated responses to complex auditory stimuli in area TAa and in area TPO. Approximately 50% of the neurons in TAa were responsive to auditory stimuli while only 8% of the recorded neurons in TPO were auditory responsive. Earlier evidence for auditory spatial processing in caudal STS showed that Tpt neurons were most active when animals actively performed an auditory localization task, though sharp tuning was not described (Vaadia et al. 1986).

Other work carefully examined STS responses in alert non-human primates to complex sounds and actions including monkey's vocalizing, lip-smacking, paper ripping, etc. (Barraclough et al. 2005). The primary goal of the study was to investigate whether STS neurons coding the sight of actions also integrated the sound of those actions. In this study pictures or movies of an action were presented separately or combined with the accompanying sound. Approximately 32% of cells in the anterior STS responded to the auditory component of the audio-visual stimulus. However, most of these cells were multimodal, and the auditory responses were not tested further. Ghazanfar et al. (2008) have also evaluated responses to vocalizations in the STS by recording local field potentials and found responses to the area was responsive to voice, face and combinations of face–voice stimuli (Ghazanfar et al. 2008).

Although responses to auditory stimuli can be evoked, most neurons within the dorsal bank of the STS, specifically in TPO, are polysensory and may prefer audiovisual stimuli. The few auditory physiology studies performed in the STS make it difficult to conclusively define these responses as related to object or spatial processing. Only one study (Hikosaka et al. 1988) tested directional preference in caudal STS auditory cells and it found that most had large spatial receptive fields and a contralateral preference, as expected, and with little sharp spatial tuning. Furthermore, auditory and visual receptive fields of multimodal neurons overlapped (Hikosaka et al. 1988). A better candidate for the continuation of the rostral stream of auditory information may be area TAa, which in all studies had more auditory responsive cells. Comparing the responses in area TAa and caudal STS regions shows more auditory responsive cells in TAa and more multisensory and visual responses in caudal STS (Baylis et al. 1987). Thus, anterior parts of the STS and area TAa participate in ventral stream object processing. Middle STS regions integrate audio-visual information, and caudal areas, such as Tpt, are active during auditory spatial processing, a dorsal stream attribute.

5.2 Auditory Processing in the Prefrontal Cortex

5.2.1 Prefrontal Auditory Object Processing: The Ventral Stream

The prefrontal cortex receives auditory inputs from stations as early as the caudal belt (Romanski et al. 1999b) with more robust parabelt and STS projections. For more than a century, the inferior frontal gyrus, including Broca's area, has been linked with speech and language processes (Broca 1861). Recent positron emission tomography (PET) and functional magnetic resonance imaging (fMRI) studies of the human brain implicate ventrolateral frontal lobe Brodmann's areas 44, 45, and 47 in auditory working memory, phonological processing, comprehension, and semantic judgment (Buckner et al. 1995; Demb et al. 1995; Fiez et al. 1996; Stromswold et al. 1996; Zatorre et al. 1996; Gabrieli et al. 1998; Friederici et al. 2003). Although the cellular analysis of frontal lobe auditory neurons lagged behind that of the visual processes, a few studies noted sparse auditory responsive neurons across the frontal lobe in Old and New World primates (Newman and Lindsley 1976; Azuma and Suzuki 1984; Tanila et al. 1992, 1993; Watanabe 1992; Bodner et al. 1996). Furthermore, deficits in auditory processing followed lesions of lateral frontal cortical areas in non-human primates. Electrophysiological and behavioral studies in non-human primates provide evidence for dorsal-spatial and ventral-object auditory streams that form separate frontal lobe domains with auditory spatial analysis concentrated in dorsal prefrontal cortex (DLPFC) and auditory-object related processing in ventral prefrontal cortex (VLPFC). This extends the what and where streams of the visual system into the auditory system reaching the frontal lobes.

For auditory object processing, several studies have demonstrated a role for VLPFC in auditory discrimination, or acoustic feature processing. Large lateral frontal cortex lesions (some including ventral prefrontal regions) in non-human primates disrupt auditory discrimination task performance (Gross and Weiskrantz 1962; Gross 1963; Goldman and Rosvold 1970; Iversen and Mishkin 1970), though only one physiological study had observed ventral prefrontal auditory responses (Benevento et al. 1977). Years later, an auditory responsive region was localized to the primate VLPFC (Romanski and Goldman-Rakic 2002). VLPFC cells in area 12/47 respond to complex stimuli including species-specific vocalizations. These cells are adjacent to face responsive cells ventral to the principal sulcus (Wilson et al. 1993; O'Scalaidhe et al. 1997, 1999) and receive acoustic input from ventral stream auditory neurons in the anterior belt and parabelt (Hackett et al. 1999; Romanski et al. 1999a, b).

If the VLPFC is the terminus of a ventral auditory object pathway, then one would expect a role in auditory-object related functions. Although the precise function of primate VLPFC is unknown, physiological studies suggest roles in the processing and discrimination of complex sounds and sound features. VLPFC auditory cells do not respond readily to simple acoustic stimuli such as pure tones (Romanski and Goldman-Rakic 2002) but robustly respond to complex sounds that include species-specific vocalizations, human vocalizations as well as other complex features (Averbeck and Romanski 2004; Gifford et al. 2005; Romanski et al. 2005; Russ et al. 2008). The feature(s) of complex sounds encoded by VLPFC auditory neurons likely include combinations of complex acoustic and behavioral or referential factors. In Romanski et al. (2005) VLPFC neurons were tested with an array of exemplars from 10 vocalization categories which included food and non-food calls, agonistic calls and affiliative calls. An MCPI preference index for each prefrontal auditory cell was calculated in a manner similar to Tian et al. (2001). Prefrontal cell selectivity resembles that in the lateral belt area AL with most cells selective for ~2 vocalizations when tested with 10 call types. A cluster analysis of VLPFC responses shows that cells prefer vocalizations with similar acoustic morphology (Fig. 4.7) (Romanski et al. 2005), suggesting that VLPFC neurons encode complex acoustic features. Since the repertoire of rhesus macaque vocalizations is divided into calls both by behavioral context and acoustic features, VLPFC neurons are likely capable of categorical discrimination of vocalizations using complex feature analysis to encode category membership. Averbeck and Romanski (2006) utilized a probabilistic approach in their analysis of feature encoding in prefrontal cortex, which assumes that VLPFC is involved in the discrimination of vocalizations and their acoustic categorical boundaries. Using these probabilities to characterize acoustic category membership, it has been shown that VLPFC neurons can be described as linear functions of the probabilities that individual calls belong to each of the acoustic categories that have been described (Averbeck and Romanski 2006). This hidden Markov model demonstrates that prefrontal neurons respond similarly to vocalizations from the same or acoustically similar categories (Averbeck and Romanski 2006). In this model categorical response depends on complex feature encoding and not responses to simple features that do not adequately capture prefrontal neuronal responses (Averbeck and Romanski 2006). Other studies de-emphasize the complex feature processing in categorical discrimination and suggest that prefrontal neurons reflect contextual semantic categorical rather than acoustic changes (Gifford et al. 2005). Nonetheless, evidence suggests a VLPFC role in the analysis of the combinations of complex features (acoustic, semantic, etc.) of auditory objects in recognition and communication. An object-based auditory stream is further supported by human neuroimaging studies that confirm ventral prefrontal activity during complex auditory processing (Belin et al. 2000; Scott et al. 2000; Binder et al. 2004; Zatorre et al. 2004), complementing work on speech and language processing in the human inferior frontal gyrus.

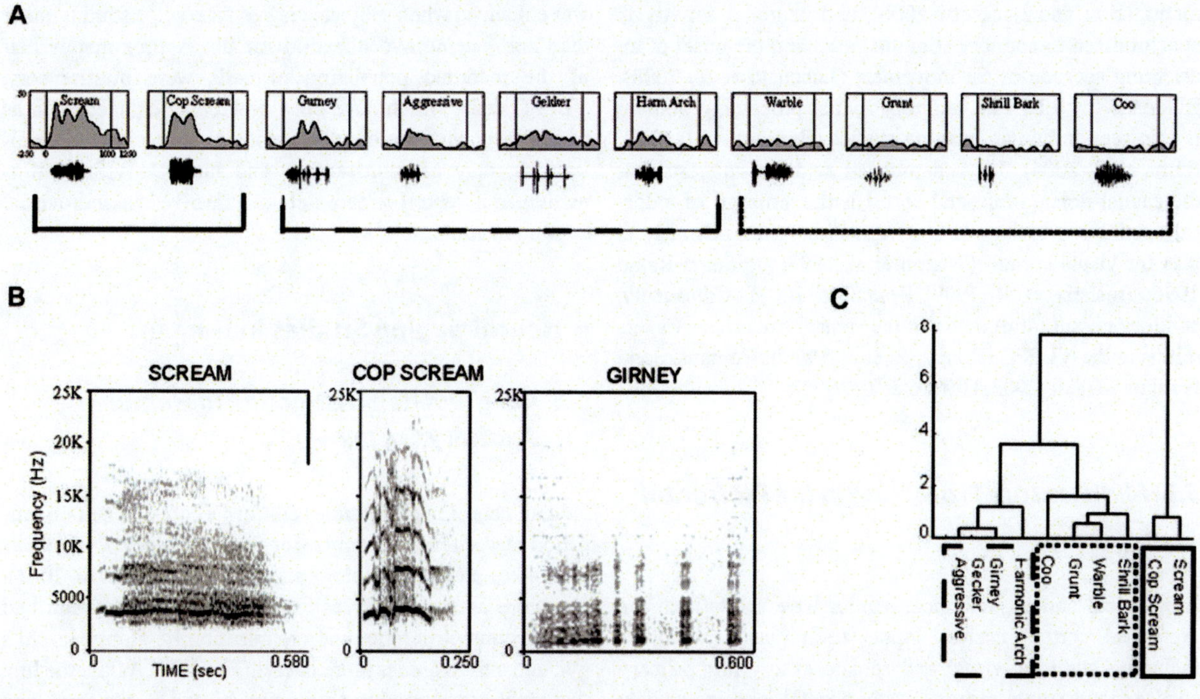

Fig. 4.7 A prefrontal auditory response to species-specific vocalizations. **a** The response of a single cell to 10 exemplars from 10 different call types with the vocalization waveform beneath each raster/spike density plot. The maximal response increase in was to the Agonistic Scream (Scream) and to the Copulation Scream. The girney also evoked a response significantly above background firing rate. The spectrograms of the three sounds that evoked a response **b**. A cluster analysis of the neural response shows that sounds that evoked a similar response cluster and that sounds with similar acoustic features likely elicit similar neuronal responses

5.2.2 Prefrontal Auditory Spatial Processing: The Dorsal Stream

The dorsal prefrontal cortex (DLPFC) has long been associated with visuospatial processing. If auditory ventral and dorsal streams parallel the visual system, there might be auditory spatial processing in the DLPFC. Physiological studies in non-human primates support a caudal and dorsal prefrontal cortex role in auditory processing. Cells responsive to simple auditory stimuli in the periarcuate region (Ito 1982; Azuma and Suzuki 1984; Vaadia et al. 1986) and lesions there impair primate auditory discrimination (Gross and Weiskrantz 1962; Petrides 1986). Importantly, periarcuate cells that respond to auditory stimuli are affected by the sound source location (Azuma and Suzuki 1984; Russo and Bruce 1989; Kikuchi-Yorioka and Sawaguchi 2000) and affected by changes in gaze (Russo and Bruce 1989). DLPFC houses the frontal eye fields which play a vital role in controlling saccadic eye movements to salient stimuli including auditory targets (Russo and Bruce 1994). An enhanced response of periarcuate cells is seen when non-human primates engage in an auditory localization task rather than a passive listening task (Vaadia et al. 1986). Therefore, the evidence suggests that neurons in the DLPFC respond to acoustic stimuli and are sensitive to sound location. Furthermore, stimulation in area 8B, dorsal to the principal sulcus, elicits ear and eye movements (Bon and Lucchetti 2006), and neuronal activity is synchronized to auditory stimulus onset and the onset of an orienting saccade or ear movement (Lucchetti et al. 2008). Evidence for a DLPFC auditory spatial processing domain is also seen in human imaging studies (Bushara et al. 1999; Rama et al. 2004). The accumulated evidence suggests that the caudal-dorsal prefrontal cortex is the terminus of a dorsal spatial processing stream for auditory stimuli much as it is for visual stimuli (Wilson et al. 1993; Goldman-Rakic 1996; Bushara et al. 1999; Romanski 2007). Conversely, auditory object information in the ventral stream has its terminus in the VLPFC (Romanski et al. 1999b; Romanski and Goldman-Rakic 2002; Gifford et al. 2003).

5.3 Multisensory Processing in the Prefrontal Cortex

Anatomical and physiological studies have shown that the prefrontal cortex receives inputs from many modalities including auditory, visual and somatosensory non-primary cortices, and single neurons in the frontal lobe respond to each of these modalities (O'Scalaidhe et al. 1997, 1999; Romo et al. 1999; Romanski et al. 2005). These features make it an excellent candidate for integrating complex sensory signals. It is hypothesized that dorsal and ventral PFC participate in spatial and object processing, respectively, but that each region received inputs from all modalities which could be processed independently or integrated (Goldman-Rakic 1996; Romanski 2004). Auditory afferents from caudal auditory association cortex and visual input from parietal and caudal temporal lobe both target similar DLPFC regions, suggesting the possibility of integration. Populations of cells responding to auditory and visual spatial information overlap (Kikuchi-Yorioka and Sawaguchi 2000) and may integrate this information during spatial localization.

In the object domain, single VLPFC neurons, which receive converging auditory and visual afferents from the temporal lobe, integrate audio-visual stimuli. Auditory projections emanate from anterior lateral belt, parabelt, and rostral temporal lobe (cf. 2.), while ventral stream visual projections to VLPFC originate in inferior temporal lobe regions TE and TEO, and the STS, a multisensory processing area (Barraclough et al. 2005; Ghazanfar et al. 2008), which is a significant source of already-integrated audiovisual projections to the VLPFC. Physiological recordings in VLPFC showed that single cells were multisensory and responded to facial gestures and their corresponding vocalizations (Sugihara et al. 2006). VLPFC cells exhibited multisensory enhancement and multisensory suppression in their response to simultaneously presented audio-visual vocalization stimuli (Fig. 4.8). The probability of a multisensory response was enhanced when subjects viewed face/voice stimuli rather than non-face/non-vocalization stimuli. Approximately half of the recorded population of cells were multisensory. VLPFC cells, which have been previously characterized as purely responsive to complex sounds and especially to vocalizations, may be multisensory and integrate visual information in a ventral stream network involved in audiovisual communication.

6 Neuroimaging Studies in Humans

6.1 Core and Belt Distinction in Human Auditory Cortex

Human neuroimaging studies confirm a core and belt organization of auditory cortex by using the same types of stimuli as in monkey electrophysiology studies (Wessinger et al. 2001). Two core areas robustly activated by pure-tone stimuli and mirror-symmetric tonotopic organization lie along Heschl's gyr, and a third area is more lateral (Fig. 4.9a). While the first two areas correspond to core areas A1 and R, the third area may be homologous to area ML, which, like the core areas AI and R (Rauschecker et al. 1997), receives direct input from the MGv (Morel et al. 1993; Kaas and Hackett 2000). These three pure-tone responsive areas were surrounded by belt

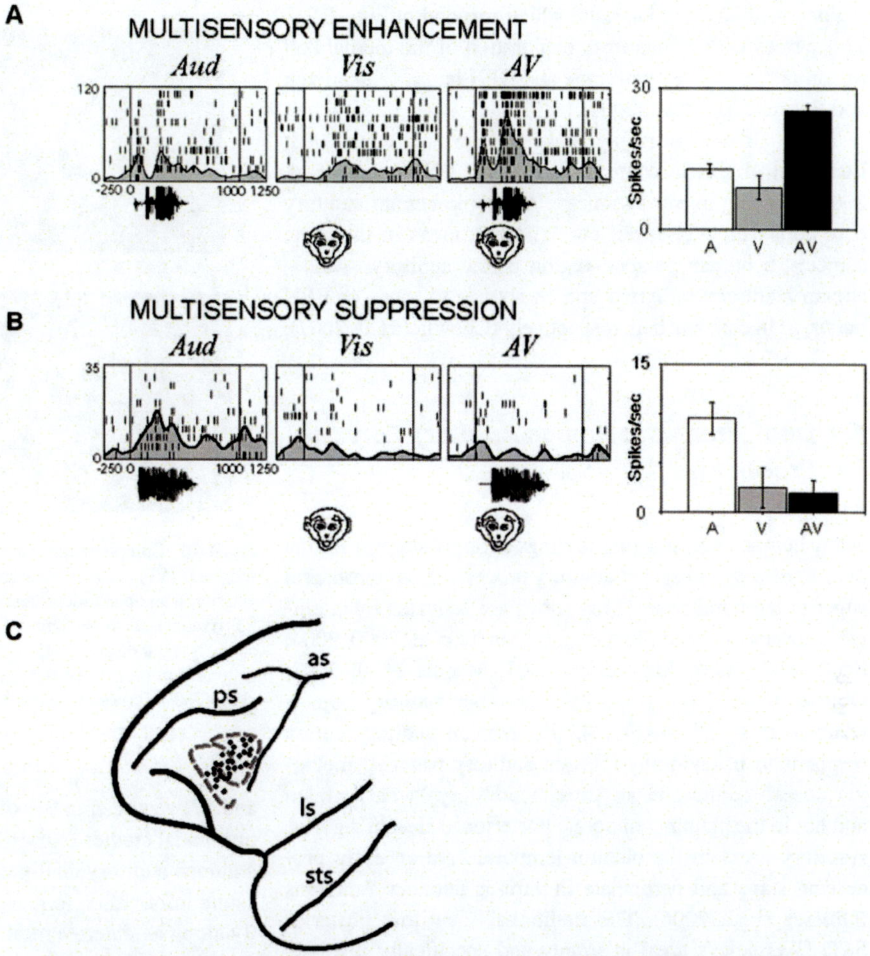

Fig. 4.8 Prefrontal cortex multisensory responses to faces and vocalizations. Different single unit responses to auditory (Aud), visual (Vis) and combined (AV) stimuli shown as raster/spike density responses and as bar graphs of the mean response. **a** The cell responded moderately to the auditory vocalization, weakly to the visual face stimulus, and more strongly to the combined audio-visual vocalization movie. This is an example of multisensory enhancement. **b** In multisensory suppression, the response to the unimodal stimulus (A) is suppressed when the auditory and visual stimuli are combined (AV). **c** The locations of multisensory cells in a lateral schematic of macaque prefrontal cortex. *Black dots*, cells within the boundaries of previously characterized unimodal auditory and visual areas (*gray dotted lines*). This suggests that many neuronal responses to auditory and visual stimuli, thought to be unimodal, may be multisensory

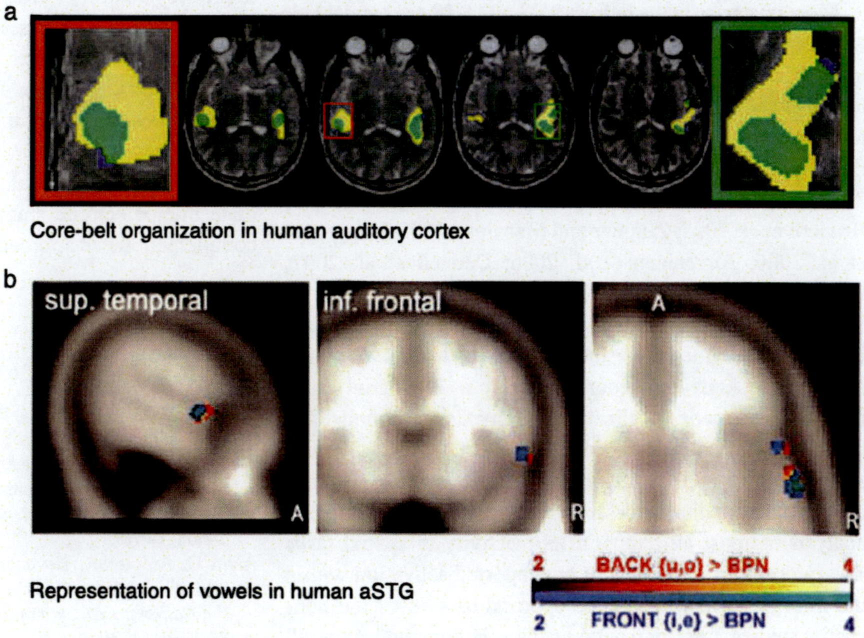

Fig. 4.9 Hierarchical organization in human auditory cortex revealed with functional imaging techniques. **a** Core-belt organization shown with tone bursts and BPN bursts, as in prior monkey studies. *Yellow regions* (belt) are activated only by BPN, *green regions* (core) by both tone and BPN bursts. Modified from the original (Wessinger et al. 2001). **b** Speech sounds consistently activate a region in the anterior superior temporal cortex (aST) more strongly than band passed noise. Within this region, different types of vowels activate discrete but overlapping clusters (Obleser et al. 2006)

regions medially and laterally, which respond only to BPN. This agrees with results from exploration of the medial belt region in the monkey with tone and BPN bursts (Kusmierek and Rauschecker 2006, 2007).

High-field MRI scanning studies (Formisano et al. 2003; Leaver et al. 2007) confirm prior work (Wessinger et al. 2001) showing mirror-symmetric tonotopic human auditory core cortex areas. Besides core and belt areas in the same subjects, a human parabelt region in the auditory "what"-stream hierarchy preferred speech sounds to tones or BPN bursts, as in core and belt, respectively (Chevillet et al. 2007).

6.2 Dual Streams in Human Auditory Cortical Processing

Many human neuroimaging findings strongly support for the dual-stream hypothesis of auditory processing. Anterolateral superior temporal areas of the cortex are activated by intelligible speech or speech-like sounds (Binder et al. 2000, 2004; Scott et al. 2000; Alain et al. 2001; Maeder et al. 2001; Obleser et al. 2005) (Fig. 4.9b) or other auditory objects (Zatorre et al. 2004) (Fig. 4.10a). Human auditory cortex responds to behaviorally relevant auditory patterns, including speech sounds, in an anterior auditory "what"-stream and not in the planum temporale posterior to Heschl's gyrus. Auditory areas in the planum temporale are an early processing stage and participate in various auditory functions (Obleser et al. 2006). The traditional view that posterior STG (Wernicke's area) is simply and specifically involved in speech perception is, therefore, in question.

More posterior in the human STG and STS are caudal belt and parabelt regions projecting up dorsally into inferior posterior parietal cortex and that are active in auditory spatial tasks, such as spatial discrimination or tasks involving auditory motion in space (Maeder et al. 2001; Warren et al. 2002; Zatorre et al. 2002; Arnott et al. 2004; Jääskeläinen et al. 2004; Krumbholz et al. 2005a, b; Tata and Ward 2005a, b; Brunetti et al. 2005; Zimmer and Macaluso 2005; Degerman et al. 2006; Ahveninen et al. 2006; Deouell et al. 2007; Rauschecker 2007) (Fig. 4.10b).

A meta-analysis, of evidence from fMRI and PET studies examined evidence for the auditory dual-pathway model in humans. Activation coordinates from 11 spatial studies (listeners made localization judgments on sounds that could occur at perceptually different positions) and 27 nonspatial studies (listeners completed tasks involving sounds presented from the same location) were entered into the analysis. Almost all temporal lobe activity in spatial tasks was in posterior areas; all but one reported activation within the inferior parietal lobule, as opposed to 41% of the nonspatial studies. Inferior frontal activity (Brodmann's areas 45 and 47) was seen in 9% of the spatial studies, and in 56% of nonspatial studies (Arnott et al. 2004). These results support a human auditory dual-pathway model s in which nonspatial sound information (e.g., sound identity) is processed primarily along an anteroventral stream, whereas sound location is processed along a posterodorsal stream, i.e., in areas caudal to primary auditory cortex.

As in the visual system, work on nonhuman primates can guide human studies, and human imaging findings can provide useful guidance for nonhuman primate microelectrode studies to permit analyses at much higher spatial and temporal resolution than in most human studies, with some exceptions (Howard et al. 1996, 2000). Imaging-guided primate microelectrode studies (Tsao et al. 2006) can further complement these approaches.

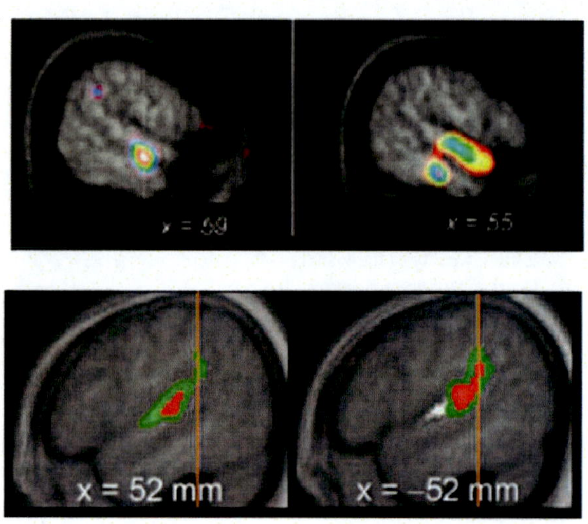

Fig. 4.10 Complementary activation of anterior and posterior superior temporal (ST) cortex by different auditory tasks. **a** Anterior ST cortex activity in an object identification task (Zatorre et al. 2004). **b** Posterior ST cortex is activated by a sound localization task and, even more strongly, by a discrimination task involving motion in space. From the original source with permission (Krumbholz et al. 2005b); see Warren et al. (2002) for a similar result

References

Ahveninen J, Jääskeläinen IP, Raij T, Bonmassar G, Devore S, Hamalainen M, Levanen S, Lin FH, Sams M, Shinn-Cunningham BG, Witzel T, and Belliveau JW (2006) Task-modulated "what" and "where" pathways in human auditory cortex. *Proceedings of the National Academy of Sciences of the United States of America* **103**: 14608–14613.

Alain C, Arnott SR, Hevenor S, Graham S, and Grady CL (2001) "What" and "where" in the human auditory system. *Proceedings of the National Academy of Sciences of the United States of America* **98**:12301–12306.

Arnott SR, Binns MA, Grady CL, and Alain C (2004) Assessing the auditory dual-pathway model in humans. *Neuroimage* **22**:401–408.

Averbeck BB and Romanski LM (2006) Probabilistic encoding of vocalizations in macaque ventral lateral prefrontal cortex. *Journal of Neuroscience* **26**:11023–11033.

Averbeck BB and Romanski LM (2004) Principal and independent components of macaque vocalizations: constructing stimuli to probe high-level sensory processing. *Journal of Neurophysiology* **91**:2897–2909.

Azuma M and Suzuki H (1984) Properties and distribution of auditory neurons in the dorsolateral prefrontal cortex of the alert monkey. *Brain Research* **298**:343–346.

Barbas H (1992) Architecture and cortical connections of the prefrontal cortex in the rhesus monkey. *Advances in Neurology* **57**:91–115.

Barraclough NE, Xiao D, Baker CI, Oram MW, and Perrett DI (2005) Integration of visual and auditory information by superior temporal sulcus neurons responsive to the sight of actions. *Journal of Cognitive Neuroscience* **17**:377–391.

Baylis GC, Rolls ET, and Leonard CM (1987) Functional subdivisions of the temporal lobe neocortex. *Journal of Neuroscience* **7**:330–342.

Belin P, Zatorre RJ, Lafaille P, Ahad P, and Pike B (2000) Voice-selective areas in human auditory cortex. *Nature* **403**:309–312.

Benevento LA, Fallon J, Davis BJ, and Rezak M (1977) Auditory-visual interaction in single cells in the cortex of the superior temporal sulcus and the orbital frontal cortex of the macaque monkey. *Experimental Neurology* **57**:849–872.

Binder JR, Frost JA, Hammeke TA, Bellgowan PS, Springer JA, Kaufman JN, and Possing ET (2000) Human temporal lobe activation by speech and nonspeech sounds. *Cerebral Cortex* **10**: 512–528.

Binder JR, Liebenthal E, Possing ET, Medler DA, and Ward BD (2004) Neural correlates of sensory and decision processes in auditory object identification. *Nature Neuroscience* **7**:295–301.

Bodner M, Kroger J, and Fuster JM (1996) Auditory memory cells in dorsolateral prefrontal cortex. *Neuroreport* **7**:1905–1908.

Bon L and Lucchetti C (2006) Auditory environmental cells and visual fixation effect in area 8B of macaque monkey. *Experimental Brain Research* **168**:441–449.

Broca P (1861) Remarques su le siege defaulte de langage articule suivies d'une observation d'aphemie (perte de la parole). *Bulletin de la Societe' d'Anthropologie* **2**:330–337.

Bruce C, Desimone R, and Gross CG (1981) Visual properties of neurons in a polysensory area in superior temporal sulcus of the macaque. *Journal of Neurophysiology* **46**:369–384.

Brugge JF and Merzenich MM (1973) Responses of neurons in auditory cortex of the macaque monkey to monaural and binaural stimulation. *Journal of Neurophysiology* **36**:1138–1158.

Brunetti M, Belardinelli P, Caulo M, Del Gratta C, Della Penna S, Ferretti A, Lucci G, Moretti A, Pizzella V, Tartaro A, Torquati K, Olivetti Belardinelli M, and Romani GL (2005) Human brain activation during passive listening to sounds from different locations: an fMRI and MEG study. *Human Brain Mapping* **26**:251–261.

Buckner RL, Raichle ME, and Petersen SE (1995) Dissociation of human prefrontal cortical areas across different speech production tasks and gender groups. *Journal of Neurophysiology* **74**: 2163–2173.

Bushara KO, Weeks RA, Ishii K, Catalan MJ, Tian B, Rauschecker JP, and Hallett M (1999) Modality-specific frontal and parietal areas for auditory and visual spatial localization in humans. *Nature Neuroscience* **2**:759–766.

Chavis DA and Pandya DN (1976) Further observations on corticofrontal connections in the rhesus monkey. *Brain Research* **117**:369–386.

Chevillet MA, Riesenhuber M, and Rauschecker JP (2007) Functional localization of the auditory "what" stream hierarchy. *Society for Neuroscience Abstracts* **33**:174.9.

Degerman A, Rinne T, Salmi J, Salonen O, and Alho K (2006) Selective attention to sound location or pitch studied with fMRI. *Brain Research* **1**:123–134

Demb JB, Desmond JE, Wagner AD, Vaidya CJ, Glover GH, and Gabrieli JD (1995) Semantic encoding and retrieval in the left inferior prefrontal cortex: a functional MRI study of task difficulty and process specificity. *Journal of Neuroscience* **15**: 5870–5878.

Deouell LY, Heller AS, Malach R, D'Esposito M, and Knight RT (2007) Cerebral responses to change in spatial location of unattended sounds. *Neuron* **55**:985–996.

Desimone R (1991) Face-selective cells in the temporal cortex of monkeys. *Journal of Cognitive Neuroscience* **3**:1–8.

Desimone R and Schein SJ (1987) Visual properties of neurons in area V4 of the macaque: sensitivity to stimulus form. *Journal of Neurophysiology* **57**:835–868.

Doupe AJ (1997) Song- and order-selective neurons in the songbird anterior forebrain and their emergence during vocal development. *Journal of Neuroscience* **17**:1147–1167.

Ehret G and Schreiner CE (1997) Frequency resolution and spectral integration (critical band analysis) in single units of the cat primary auditory cortex. *Journal of Comparative Physiology A* **181**: 635–650.

Esser K-H, Condon CJ, Suga N, and Kanwal JS (1997) Syntax processing by auditory cortical neurons in the FM-FM area of the mustached bat *Pteronotus parnellii*. *Proceedings of the National Academy of Sciences of the United States of America* **94**: 14019–14024.

Fecteau S, Armony JL, Joanette Y, and Belin P (2005) Sensitivity to voice in human prefrontal cortex. *Journal of Neurophysiology* **94**:2251–2254.

Fiez JA, Raife EA, Balota DA, Schwarz JP, Raichle ME, and Petersen SE (1996) A positron emission tomography study of the short-term maintenance of verbal information. *Journal of Neuroscience* **16**:808–822.

Formisano E, Kim DS, Di Salle F, van de Moortele PF, Ugurbil K, and Goebel R (2003) Mirror-symmetric tonotopic maps in human primary auditory cortex. *Neuron* **40**:859–869.

Friederici AD, Ruschemeyer SA, Hahne A, and Fiebach CJ (2003) The role of left inferior frontal and superior temporal cortex in sentence comprehension: localizing syntactic and semantic processes. *Cerebral Cortex* **13**:170–177.

Gabrieli JDE, Poldrack RA, and Desmond JE (1998) The role of left prefrontal cortex in language and memory. *Proceedings of the National Academy of Sciences of the United States of America* **95**:906–913.

Ghazanfar AA, Chandrasekaran C, and Logothetis NK (2008) Interactions between the superior temporal sulcus and auditory cortex mediate dynamic face/voice integration in rhesus monkeys. *Journal of Neuroscience* **28**:4457–4469.

Gifford GW, III, Hauser MD, and Cohen YE (2003) Discrimination of functionally referential calls by laboratory-housed rhesus macaques: Implications for neuroethological studies. *Brain Behavior and Evolution* **61**:213–224.

Goldman-Rakic, PS (1996). The prefrontal landscape: implications of functional architecture for understanding human mentation and the central executive. *Philosophical Transactions of the Royal Society, London. B Biological Sciences* **351**:1445–1453.

Goldman PS and Rosvold HE (1970) Localization of function within the dorsolateral prefrontal cortex of the rhesus monkey. *Experimental Neurology* **27**:291–304.

Gross CG (1963) A comparison of the effects of partial and total lateral frontal lesions on test performance by monkeys. *Journal of Comparative Physiological Psychology* **56**:41–47.

Gross CG and Weiskrantz L (1962) Evidence for dissociation of impairment on auditory discrimination and delayed response following

lateral frontal lesions in monkeys. *Experimental Neurology* **5**: 453–476.
Hackett TA, De La Mothe LA, Ulbert I, Karmos G, Smiley J, and Schroeder CE (2007) Multisensory convergence in auditory cortex, II. Thalamocortical connections of the caudal superior temporal plane. *Journal of Comparative Neurology* **502**:924–952.
Hackett TA, Stepniewska I, and Kaas JH (1999) Prefrontal connections of the parabelt auditory cortex in macaque monkeys. *Brain Research* **817**:45–58.
Hackett TA, Stepniewska I, and Kaas JH (1998a) Subdivisions of auditory cortex and ipsilateral cortical connections of the parabelt auditory cortex in macaque monkeys. *Journal of Comparative Neurology* **394**:475–495.
Hackett TA, Stepniewska I, and Kaas JH (1998b) Thalamocortical connections of the parabelt auditory cortex in macaque monkeys. *Journal of Comparative Neurology*, **400**:271–286.
Hauser MD (1996) *The Evolution of Communication*. MIT Press, Cambridge.
Hikosaka K, Iwai E, Saito H, and Tanaka K (1988) Polysensory properties of neurons in the anterior bank of the caudal superior temporal sulcus of the macaque monkey. *Journal of Neurophysiology* **60**:1615–1637.
Howard MA, Volkov IO, Abbas PJ, Damasio H, Ollendieck MC, and Granner M (1996) A chronic microelectrode investigation of the tonotopic organization of human auditory cortex. *Brain Research* **724**:260–264.
Howard MA, Volkov IO, Mirsky R, Garell PC, Noh MD, Granner M, Damasio H, Steinschneider M, Reale RA, Hind JE, and Brugge JF (2000) Auditory cortex on the human posterior superior temporal gyrus. *Journal of Comparative Neurology* **416**: 79–92.
Hubel DH and Wiesel TN (1962) Receptive fields, binocular interaction and functional architecture in the cat's visual cortex. *Journal of Physiology (London)* **160**:106–154.
Gifford GGW, Maclean KA, Hauser MD, and Cohen YE (2005) The neurophysiology of functionally meaningful categories: macaque ventrolateral prefrontal cortex plays a critical role in spontaneous categorization of species-specific vocalizations. *Journal of Cognitive Neuroscience* **17**:1471–1482.
Ito SI (1982) Prefrontal unit activity of macaque monkeys during auditory and visual reaction time tasks. *Brain Research* **247**: 39–47.
Iversen SD and Mishkin M (1970) Perseverative interference in monkeys following selective lesions of the inferior prefrontal convexity. *Experimental Brain Research* **11**:376–386.
Jääskeläinen IP, Ahveninen J, Bonmassar G, Dale AM, Ilmoniemi RJ, Levänen S, Lin FH, May P, Melcher J, Stufflebeam S, Tiitinen H, and Belliveau JW (2004) Human posterior auditory cortex gates novel sounds to consciousness. *Proceedings of the National Academy of Sciences of the United States of America* **101**: 6809–6814.
Jones EG, Dell'Anna ME, Molinari M, Rausell E, and Hashikawa T (1995) Subdivisions of macaque monkey auditory cortex revealed by calcium-binding protein immunoreactivity. *Journal of Comparative Neurology* **362**:153–170.
Kaas JH and Hackett TA (2000) Subdivisions of auditory cortex and processing streams in primates. *Proceedings of the National Academy of Sciences of the United States of America* **97**: 11793–11799.
Kaas JH and Hackett TA (1998) Subdivisions of auditory cortex and levels of processing in primates. *Audiology & Neuro-Otology* **3**: 73–85.
Kikuchi Y, Horwitz B, and Mishkin M (2007) Auditory response properties in the rostral and caudal stations of the auditory stimulus processing stream of the macaque superior temporal cortex. *Society for Neuroscience Abstracts* **33**:278.16.
Kikuchi Y, Horwitz B, and Mishkin M (2004) A patch of neurons in the monkey's rostral superior temporal gyrus are activated by conspecific calls. *Society for Neuroscience Abstracts* **30**:650:10.
Kikuchi Y, Rauschecker JP, Mishkin M, Augath M, Logothetis NK, and Petkov CI (2008) Voice region connectivity in the monkey assessed with microstimulation and functional imaging. *Society for Neuroscience Abstracts* **34**:850.2.
Kikuchi-Yorioka Y and Sawaguchi T (2000) Parallel visuospatial and audiospatial working memory processes in the monkey dorsolateral prefrontal cortex. *Nature Neuroscience* **3**: 1075–1076.
Knudsen EI (1983) Subdivisions of the inferior colliculus in the barn owl (*Tyto alba*). *Journal of Comparative Neurology* **218**: 174–186.
Kosaki H, Hashikawa T, He J, and Jones EG (1997) Tonotopic organization of auditory cortical fields delineated by parvalbumin immunoreactivity in macaque monkeys. *Journal of Comparative Neurology* **386**:304–316.
Krumbholz K, Schonwiesner M, Rübsamen R, Zilles K, Fink GR, and von Cramon DY (2005a) Hierarchical processing of sound location and motion in the human brainstem and planum temporale. *The European Journal of Neuroscience* **21**:230–238.
Krumbholz K, Schonwiesner M, von Cramon DY, Rübsamen R, Shah NJ, Zilles K, and Fink GR (2005b) Representation of interaural temporal information from left and right auditory space in the human planum temporale and inferior parietal lobe. *Cerebral Cortex* **15**:317–324.
Kusmierek P and Rauschecker JP (2006) Selectivity for environmental sounds and conspecific vocalizations in the anterolateral auditory belt cortex of the awake rhesus monkey. *Society for Neuroscience Abstract* **32**:798.10.
Kusmierek P and Rauschecker JP (2007) Response properties of medial belt neurons in rhesus monkey auditory cortex. *Society for Neuroscience Abstracts* **33**:174.8.
Leaver A, Chevillet MA, Renier L, Purcell JJ, and Rauschecker JP (2007) Visualization of multiple tonotopic fields in human auditory cortex. *Society for Neuroscience Abstracts* **33**:174.1.
Leinonen L, Hyvärinen J, and Sovijarvi AR (1980) Functional properties of neurons in the temporo-parietal association cortex of awake monkey. *Experimental Brain Research* **39**:203–215.
Lomber SG and Malhotra S (2008) Double dissociation of 'what' and 'where' processing in auditory cortex. *Nature Neuroscience* **11**: 609–616.
Lucchetti C, Lanzilotto M, and Bon L (2008) Auditory-motor and cognitive aspects in area 8B of macaque monkey's frontal cortex: a premotor ear-eye field (PEEF). *Experimental Brain Research* **186**:131–141.
Maeder PP, Meuli RA, Adriani M, Bellmann A, Fornari E, Thiran JP, Pittet A, and Clarke S (2001) Distinct pathways involved in sound recognition and localization: a human fMRI study. *Neuroimage* **14**:802–816.
Margoliash D and Fortune ES (1992) Temporal and harmonic combination-sensitive neurons in the zebra finch's HVc. *Journal of Neuroscience*, **12**:4309–4326.
Merzenich MM and Brugge JF (1973) Representation of the cochlear partition of the superior temporal plane of the macaque monkey. *Brain Research* **50**:275–296.
Morel A, Garraghty PE, and Kaas JH (1993) Tonotopic organization, architectonic fields, and connections of auditory cortex in macaque monkeys. *Journal of Comparative Neurology* **335**:437–459.
Movshon JA and Newsome WT (1996) Visual response properties of striate cortical neurons projecting to area MT in macaque monkeys. *Journal of Neuroscience* **16**:7733–7741.
Newman JD and Lindsley DF (1976) Single unit analysis of auditory processing in squirrel monkey frontal cortex. *Experimental Brain Research* **25**:169–181.

Obleser J, Boecker H, Drzezga A, Haslinger B, Hennenlotter A, Roettinger M, Eulitz C, and Rauschecker JP (2006) Vowel sound extraction in anterior superior temporal cortex. *Human Brain Mapping* **27**:562–571.

Obleser J, Boecker H, Drzezga A, Haslinger B, Hennenlotter A, Roettinger M, Eulitz C, and Rauschecker JP (2005) Vowel sound extraction in anterior superior temporal cortex. *Human Brain Mapping* **7**:562–571.

Ohl FW and Scheich H (1997) Orderly cortical representation of vowels based on formant interaction. *Proceedings of the National Academy of Sciences of the United States of America* **94**:9440–9444.

O'Scalaidhe SP, Wilson FA, and Goldman-Rakic PS (1997) Areal segregation of face-processing neurons in prefrontal cortex. *Science* **278**:1135–1138.

O'Scalaidhe SP, Wilson FA, and Goldman-Rakic PG (1999) Face-selective neurons during passive viewing and working memory performance of rhesus monkeys: evidence for intrinsic specialization of neuronal coding. *Cerebral Cortex* **9**:459–475.

Pandya DN, Hallett M, and Mukherjee SK (1969) Intra- and interhemispheric connections of the neocortical auditory system in the rhesus monkey. *Brain Research* **14**:49–65.

Pandya DN and Kuypers HGJM (1969) Cortico-cortical connections in the rhesus monkey. *Brain Research* **13**:13–36.

Petkov CI, Kayser C, Augath M, and Logothetis NK (2006) Functional imaging reveals numerous fields in the monkey auditory cortex. *Public Library of Science Biology* **4**:e215.

Petkov CI, Kayser C, Steudel T, Whittingstall K, Augath M, and Logothetis NK (2008) A voice region in the monkey brain. *Nature Neuroscience* **11**:367–374.

Petrides M (1986) The effect of periarcuate lesions in the monkey on the performance of symmetrically and asymmetrically reinforced visual and auditory go, no-go tasks. *Journal of Neuroscience* **6**: 2054–2063.

Petrides M and Pandya DN (1988) Association fiber pathways to the frontal cortex from the superior temporal region in the rhesus monkey. *Journal of Comparative Neurology* **273**:52–66.

Petrides M and Pandya DN (2002) Comparative cytoarchitectonic analysis of the human and the macaque ventrolateral prefrontal cortex and corticocortical connection patterns in the monkey. *European Journal of Neuroscience* **16**:291–310.

Poremba A, Malloy M, Saunders RC, Carson RE, Herscovitch P, and Mishkin M (2004) Species-specific calls evoke asymmetric activity in the monkey's temporal poles. *Nature* **427**:448–451.

Poremba A, Saunders RC, Crane AM, Cook M, Sokoloff L, and Mishkin M (2003) Functional mapping of the primate auditory system. *Science* **299**:568–572.

Rama P, Poremba A, Sala JB, Yee L, Malloy M, Mishkin M, and Courtney SM (2004) Dissociable functional cortical topographies for working memory maintenance of voice identity and location. *Cerebral Cortex* **14**:768–780.

Rauschecker JP (2007) Cortical processing of auditory space: pathways and plasticity. In: Mast F and Jäncke L (eds). *Spatial Processing in Navigation, Imagery, and Perception*. Springer-Verlag, New York, pp. 389–410.

Rauschecker JP (1998a) Cortical processing of complex sounds. *Current Opinion in Neurobiology* **8**:516–521.

Rauschecker JP (1998b) Parallel processing in the auditory cortex of primates. *Audiology & Neuro-Otology* **3**:86–103.

Rauschecker JP (1997) Processing of complex sounds in the auditory cortex of cat, monkey, and man. *Acta Oto-Laryngologica Supplement* **532**:34–38.

Rauschecker JP and Tian B (2004) Processing of band-passed noise in the lateral auditory belt cortex of the rhesus monkey. *Journal of Neurophysiology* **91**:2578–2589.

Rauschecker JP and Tian B (2000) Mechanisms and streams for processing of "what" and "where" in auditory cortex. *Proceedings of the National Academy of Sciences of the United States of America* **97**:11800–11806.

Rauschecker JP, Tian B, and Hauser M (1995) Processing of complex sounds in the macaque nonprimary auditory cortex. *Science* **268**:111–114.

Rauschecker JP, Tian B, Pons T, and Mishkin M (1997) Serial and parallel processing in rhesus monkey auditory cortex. *Journal of Comparative Neurology* **382**:89–103.

Recanzone GH (2000) Spatial processing in the auditory cortex of the macaque monkey. *Proceedings of the National Academy of Sciences of the United States of America* **97**:11829–11835.

Recanzone GH, Guard DC, Phan ML, and Su TK (2000) Correlation between the activity of single auditory cortical neurons and sound-localization behavior in the macaque monkey. *Journal of Neurophysiology* **83**:2723–2739.

Romanski LM (2004) Domain specificity in the primate prefrontal cortex. *Cognitive, Affective & Behavioral Neuroscience* **4**: 421–429.

Romanski LM (2003) Anatomy and physiology of auditory-prefrontal interactions in non-human primates. In: Ghazanfar AA (ed). *Primate Audition: Ethology and Neurobiology*. CRC Press, New York, pp. 259–278.

Romanski LM (2007) Representation and Integration of Communication Stimuli by the Primate Prefrontal Cortex. *Cerebral Cortex* **17**:61–69.

Romanski LM, Averbeck BB, and Diltz M (2005) Neural representation of vocalizations in the primate ventrolateral prefrontal cortex. *Journal of Neurophysiology* **93**:734–747.

Romanski LM and Goldman-Rakic PS (2002) An auditory domain in primate prefrontal cortex. *Nature Neuroscience* **5**:15–16.

Romanski LM, Bates JF, and Goldman-Rakic PS (1999a) Auditory belt and parabelt projections to the prefrontal cortex in the rhesus monkey. *Journal of Comparative Neurology* **403**:141–157.

Romanski LM, Tian B, Fritz J, Mishkin M, Goldman-Rakic PS, and Rauschecker JP (1999b) Dual streams of auditory afferents target multiple domains in the primate prefrontal cortex. *Nature Neuroscience* **2**:1131–1136.

Romo R, Brody CD, Hernandez A, and Lemus L (1999) Neuronal correlates of parametric working memory in the prefrontal cortex. *Nature* **399**:470–473.

Russ BE, Ackelson AL, Baker AE, and Cohen YE (2008) Coding of auditory-stimulus identity in the auditory non-spatial processing stream. *Journal of Neuroscience* **99**:87–95.

Russo GS and Bruce CJ (1989) Auditory receptive fields of neurons in frontal cortex of rhesus monkey shift with direction of gaze. *Society for Neuroscience Abstracts* **15**:120.4.

Russo GS and Bruce CJ (1994) Frontal eye field activity preceding aurally guided saccades. *Journal of Neurophysiology* **71**: 1250–1253.

Scott SK, Blank CC, Rosen S, and Wise RJ (2000) Identification of a pathway for intelligible speech in the left temporal lobe. *Brain: A Journal of Neurology* **12**:2400–2406.

Steinschneider M, Reser D, Schroeder CE, and Arezzo JC (1995) Tonotopic organization of responses reflecting stop consonant place of articulation in primary auditory cortex (A1) of the monkey. *Brain Research* **674**:147–152.

Stromswold K, Caplan D, Alpert N, and Rauch S (1996) Localization of syntactic comprehension by positron emission tomography. *Brain & Language* **52**:452–473.

Suga N, O'Neill WE, and Manabe T (1978) Cortical neurons sensitive to combinations of information-bearing elements of biosonar signals in the mustache bat. *Science* **200**:778–781.

Sugihara T, Diltz MD, Averbeck BB, and Romanski LM (2006) Integration of auditory and visual communication information in the primate ventrolateral prefrontal cortex. *Journal of Neuroscience* **26**:11138–11147.

Tanaka K (1997) Mechanisms of visual object recognition: monkey and human studies. *Current Opinion in Neurobiology* **7**:523–529.

Tanila H, Carlson S, Linnankoski I, and Kahila H (1993) Regional distribution of functions in dorsolateral prefrontal cortex of the monkey. *Behavioural Brain Research* **53**:63–71.

Tanila H, Carlson S, Linnankoski I, Lindroos F, and Kahila H (1992) Functional properties of dorsolateral prefrontal cortical neurons in awake monkey. *Behavioral Brain Research* **47**:169–180.

Tata MS and Ward LM (2005a) Early phase of spatial mismatch negativity is localized to a posterior "where" auditory pathway. *Experimental Brain Research* **167**:481–486.

Tata MS and Ward LM (2005b) Spatial attention modulates activity in a posterior "where" auditory pathway. *Neuropsychologia* **43**:509–516.

Tian B and Rauschecker JP (2004) Processing of frequency-modulated sounds in the lateral auditory belt cortex of the rhesus monkey. *Journal of Neurophysiology* **92**:2993–3013.

Tian B and Rauschecker JP (1998) Processing of frequency-modulated sounds in the cat's posterior auditory field. *Journal of Neurophysiology* **79**:2629–2642.

Tian B and Rauschecker JP (1994) Processing of frequency-modulated sounds in the cat's anterior auditory field. *Journal of Neurophysiology* **71**:1959–1975.

Tian B, Reser D, Durham A, Kustov A, and Rauschecker JP (2001) Functional specialization in rhesus monkey auditory cortex. *Science* **292**:290–293.

Tsao DY, Freiwald WA, Tootell RB, and Livingstone MS (2006) A cortical region consisting entirely of face-selective cells. *Science* **311**:670–674.

Ungerleider LG and Mishkin M (1982) Two cortical visual systems. In: Ingle DJ, Goodale MA, Mansfield RJW (eds). *Analysis of Visual Behaviour*. MIT Press, Cambridge, pp. 549–586.

Vaadia E, Benson DA, Hienz RD, and H. Goldstein MH Jr (1986) Unit study of monkey frontal cortex: active localization of auditory and of visual stimuli. *Journal of Neurophysiology* **56**:934–952.

Wang X, Merzenich MM, Beitel R, and Schreiner CE (1995) Representation of a species-specific vocalization in the primary auditory cortex of the common marmoset: temporal and spectral characteristics. *Journal of Neurophysiology* **74**:2685–2706.

Warren JD, Zielinski BA, Green GGR, Rauschecker JP, and Griffiths TD (2002) Analysis of sound source motion by the human brain. *Neuron* **34**:1–20.

Watanabe M (1992) Frontal units of the monkey coding the associative significance of visual and auditory stimuli. *Experimental Brain Research* **89**:233–247.

Wessinger CM, VanMeter J, Tian B, Van Lare J, Pekar J, and Rauschecker JP (2001) Hierarchical organization of the human auditory cortex revealed by functional magnetic resonance imaging. *Journal of Cognitive Neuroscience* **13**:1–7.

Wilson FA, O'Scalaidhe SP, and Goldman-Rakic PS (1993) Dissociation of object and spatial processing domains in primate prefrontal cortex. *Science* **260**:1955–1958.

Winter P and Funkenstein HH (1973) The effect of species-specific vocalization on the discharge of auditory cortical cells in the awake squirrel monkey. (Saimiri sciureus). *Experimental Brain Research* **18**:489–504.

Young ED, Spirou GA, Rice JJ, Voigt HF, and Rees A (1992) Neural organization and responses to complex stimuli in the dorsal cochlear nucleus [and discussion]. *Philosophical Transactions of the Royal Society of London, series B, Biological Sciences* **336**: 407–413.

Zatorre RJ, Bouffard M, Ahad P, and Belin P (2002) Where is 'where' in the human auditory cortex? *Nature Neuroscience* **5**: 905–909.

Zatorre RJ, Bouffard M, and Belin P (2004) Sensitivity to auditory object features in human temporal neocortex. *Journal of Neuroscience* **24**:3637–3642.

Zatorre RJ, Meyer E, Gjedde A, and Evans AC (1996) PET studies of phonetic processing of speech: review, replication, and reanalysis. *Cerebral Cortex* **6**:21–30.

Zimmer U and Macaluso E (2005) High binaural coherence determines successful sound localization and increased activity in posterior auditory areas. *Neuron* **47**:893–905.

Chapter 5
The Commissural Auditory System

Troy A. Hackett and Dennis P. Phillips

Abbreviations

AAF	anterior auditory field (non-primate cortex)
APD	auditory processing disorder
AI	auditory area 1 (primary auditory cortex)
AII	auditory area 2 (non-primate cortex)
CF	characteristic frequency
dB	decibel
EE	excitatory (contralateral)—excitatory (ipsilateral)
EI	excitatory (contralateral)—inhibitory (ipsilateral)
EO	excitatory (contralateral)—no effect (ipsilateral)
GABA	gamma-aminobutyric acid
HRP	horseradish peroxidase
ILD	interaural level difference
ITD	interaural time difference
MGC	medial geniculate complex (thalamus)
PAF	posterior auditory field (non-primate cortex)
PB	predominately binaural
PV	parietal ventral area (primate somatic sensory cortex)
SI	somatic sensory area 1 (non-primate cortex)
VI	visual area 1 (area 17)
VII	visual area 2 (area 18)
VPAF	ventral posterior auditory field (non-primate cortex)

1 Introduction

We consider the anatomy and the function of the forebrain auditory callosal system. We begin with an anatomical description of callosal organization, drawing on comparative evidence, present evidence for common principles and area-specific departures from these in audition and other modalities, and consider this system in its own right. We also consider experience-dependent development of commissural connectivity and how it is perturbed by experience and disease. We then explore the functional correlates of this anatomical organization with particular attention to the empirical link between callosal and intrahemispheric connectivity on the one hand, and binaural processing on the other. We conclude by exploring the hypothesis that callosal connectivity supports continuity of sensation across the midline to create perceptual unity.

T.A. Hackett (✉)
Department of Hearing and Speech Sciences, Vanderbilt University, Nashville, TN 37203, USA
e-mail: troy.a.hackett@vanderbilt.edu

2 Structure of the Forebrain Auditory Callosal System

2.1 Callosal Organization Is Species Specific

The corpus callosum is the major pathway for transfer of sensory and motor information between the cerebral hemispheres in eutherian (placental) mammals. Monotremes and marsupials have no corpus callosum and their interhemispheric connections are made entirely via a relatively enormous anterior commissure, mediating a rather limited subset of connections between temporal cortices in eutherian species (Kaas 1995). Variability in interhemispheric connectional patterns is surprisingly high, within taxa and across species. This may reflect differences in brain size and an increased number of cortical areas in species with larger brains, since these have an arguably more elaborate pattern of callosal connections (CC) linking more areas (Cusick and Kaas 1986). Second, there are differences in CC among the cortical areas that are common to all mammals. Thus, in species with smaller cortices (rodents) most of area 17 (VI, primary visual cortex) is broadly interconnected between hemispheres, whereas Old World primates are nearly devoid of such connections (Kennedy and Dehay 1988; Olavarria and Abel 1996). By comparison, in prosimian primates and

J.A. Winer, C.E. Schreiner (eds.), *The Auditory Cortex*,
DOI 10.1007/978-1-4419-0074-6_5, © Springer Science+Business Media, LLC 2011

tree shrews, area VI has an intermediate degree of callosal connectivity (Cusick et al. 1984, 1985). These examples demonstrate that callosal organization embodies species-specific features of brain organization. While these differences constrain the generalizations across taxa, several other features of interhemispheric organization are conserved, and from these patterns some general conclusions follow.

2.2 Callosal Organization Is Modality Specific

The primary auditory (AI), somatic sensory (SI), and visual (V1) cortices differ in their CC (Fig. 5.1). In primary somatic sensory (SI) cortex of cats (Caminiti et al. 1979; Ebner and Myers 1965) in cats and primates (areas 3b, 1, and 2), the distal limb representations have few interhemispheric connections, while the proximal limbs and remainder of the body surface are densely interconnected (Jones et al. 1975, 1979; Killackey et al. 1983). Similarly, in visual cortex, most V1 neurons do not project commissurally, but are concentrated along the area V1 and V1 (area 17/18) boundary at the representation of the zero (vertical) meridian (Hubel and Wiesel 1965; Zeki 1970; Berlucchi 1972; Sanides 1978; Newsome and Allman 1980; Van Essen et al. 1982; Cusick et al. 1984; Lewis and Olavarria 1995; Olavarria 1996). While VI CC are evenly distributed in some species, they are most dense at the V1/V2 border in all species studied. Therefore, the CC of areas SI and V1 favor the representations of the somatic sensory visual and midlines, respectively, at the expense of peripheral representations and have been proposed to mediate midline perceptual continuity (Choudhury et al. 1965; Berlucchi 1972; Manzoni et al. 1989).

AI differs from VI and SI in that the cochlear frequency representation anchors the functional topography, while auditory space may be encoded in a non-topographic way. The CC of cat AI are distributed across the entire area (Code and Winer 1985, 1986) and they link most strongly the topographically corresponding best-frequency representations, whereas weaker connections link heterotopic AI loci (Imig et al. 1977; Brugge and Imig 1978; Imig and Brugge 1978; Imig and Reale 1980; Code and Winer 1985; Rouiller et al. 1991). Despite local irregularities (e.g., patchiness and related forms of modularity) in the CC within AI (Diamond et al. 1968; Imig and Brugge 1978), evidence from studies in cats argues against a gradient or gaps in the field. In primates, some early studies found denser CC in medial AI (deep in the lateral fissure) (Pandya et al. 1969; Pandya and Sanides 1973), but this pattern has not been consistently observed (Pandya and Rosene 1993) and requires further study. Thus, it appears that the interhemispheric connections of auditory cortex are widely distributed, while larger discontinuities characterize portions of the primary somatic sensory and visual cortices.

2.3 Callosal Organization Varies by Cortical Area

While most areas of the cerebral cortex are connected with one or more areas in the opposite hemisphere, projection patterns vary between areas in ways that provide clues about

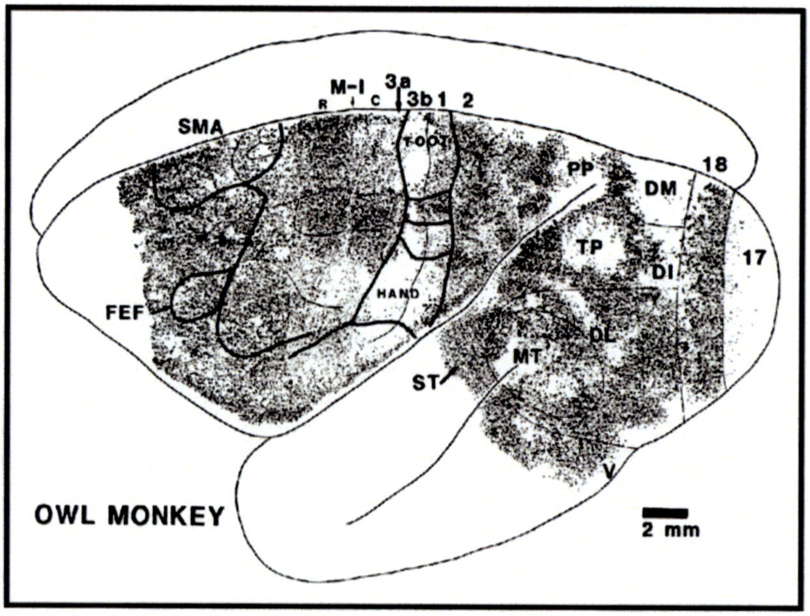

Fig. 5.1 Callosally projection neurons in owl monkey sensory and motor cortex. *Black dots*, cells of origin. Note: absence of neurons in primary visual area 17 and distal limb representations in primary somatosensory areas 3b, 1, and 2. Adapted from the original source (Cusick and Kaas 1986)

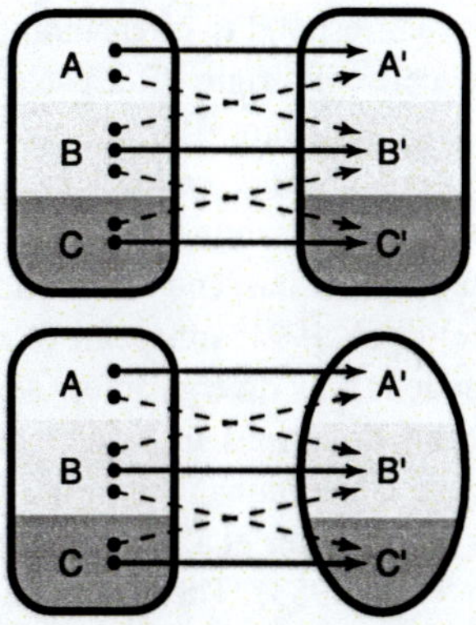

Fig. 5.2 Schematic representation of callosal connection patterns. *Top*, homotopic connections link corresponding areas in each hemisphere (*boxes*). *Bottom*, heterotopic connections link non-homologous areas (*box* and *oval*). Connections between matching receptive fields of both homotopic and heterotopic areas (*solid arrows*) are stronger than connections between dissimilar receptive fields (*dashed arrows*)

the flow of information. Two principal types of CC are typically recognized for sensory and motor areas. Homotopic connections link corresponding areas in both hemispheres. Heterotopic connections link non-homologous areas. The majority of both homotopic and heterotopic connections are between matched receptive fields in both hemispheres, while connections between non-matched receptive fields are relatively weaker (Fig. 5.2). After injection with a neuroanatomical tracer at a single location in cortex, labeled cells and terminals in the contralateral hemisphere are usually distributed over an area larger than the size of the tracer injection, and often in multiple areas, indicating that CC are both homotopic and heterotopic. Therefore, CC may or may not link matched and mismatched receptive fields within and across cortical areas. Smaller injections (<1 mm in diameter), however, have been reported to be strictly point-to-point, linking only homotopic domains (Jones et al. 1979).

The density and spatial distribution of both homotopic and heterotopic projection also varies between areas, ranging from very dense to absent, and from uniformly distributed to patchy (Ebner and Myers 1965). In most species, significant portions of the primary fields lack CC, while other regions are densely interconnected. In visual cortex, it was noted above that the interhemispheric connections of most of area VI are rather sparse, especially in higher primates. This contrasts with relatively dense connections at the borders between areas, e.g., V1 and V2 where reversals in the visual field representation occur. The density of CC also differs in areas of motor (Pandya and Vignolo 1971; Künzle 1976; Jenny 1979) and somatic sensory cortex (Shanks et al. 1975; Jones et al. 1975, 1978, 1979; Killackey et al. 1983). They are weakest in the distal limb representations (e.g., primate area 3b), and the most dense in area 2 (Killackey et al. 1983). Areas SII (second somatic sensory cortex), PV, and 5 are also densely interconnected in primates (Qi et al. 2002) and in cats (Caminiti et al. 1979; Barbaresi et al. 1989, 1994). Sectioning the corpus callosum eliminates responses to ipsilateral stimulation of bilateral neurons in these non-primary areas (Iwamura et al. 1994).

In AC, the density of CC appears to vary between areas, but homotopic projections are the strongest overall (Lee and Winer 2008; Rouiller et al. 1991). In cats (Fig. 5.3) AI is connected preferentially with the contralateral AI, then the anterior auditory field (AAF), with weaker input from posterior (PAF) and ventral posterior (VPAF) auditory fields and the second auditory area (AII). The CC of the anterior auditory field (area AAF) are similar to those of AI, with the strongest projections to AAF and AI, while weaker

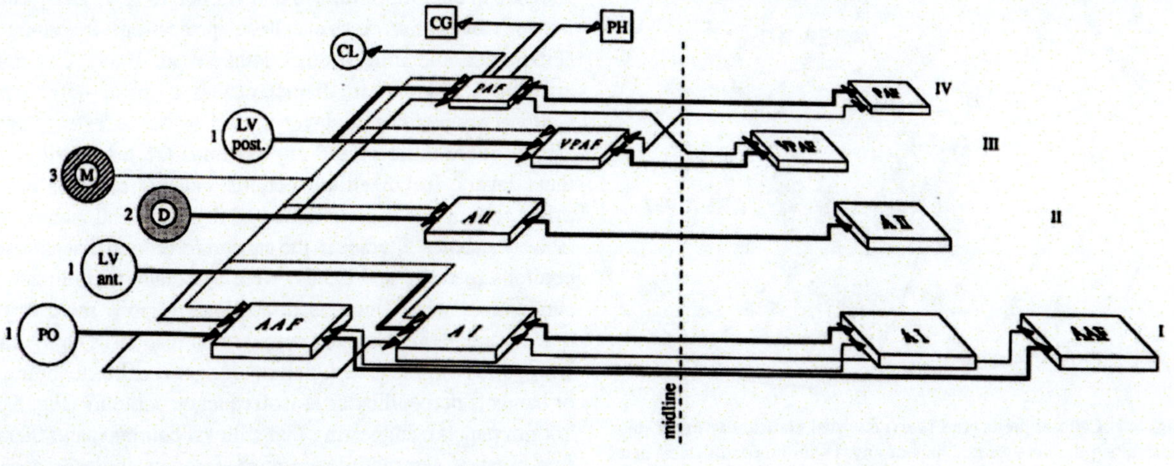

Fig. 5.3 Thalamocortical and interhemispheric connections of auditory cortex in cats. *Bold arrows* denote strong reciprocal connections. *Thin arrows* denote weaker connections. The strongest connections link homotopic areas across the midline. Adapted from the original source (Rouiller et al. 1991)

CC link AAF with PAF, VPAF, and AII. Just as AI and AAF are preferentially connected between hemispheres, PAF and VPAF are strongly linked, with few heterotopic inputs from AI and AAF. AII, by contrast, has the weakest heterotopic connections. Less is known about the organization of CC in primates, but the data available indicate that strong homotopic connections prevail (Pandya et al. 1969; Forbes and Moskowitz 1977; Fitzpatrick and Imig 1980; Aitkin et al. 1988; Cipolloni and Pandya 1989; Luethke et al. 1989; Hackett et al. 1999; Morel and Kaas 1992; Morel et al. 1993; Pandya and Rosene 1993; de la Mothe et al. 2006a).

2.4 Callosal Projections Are Discontinuous

Patchy terminations at the VI/V2 border form clusters coextensive with small patches (i.e., blobs) that stain intensely for the metabolic enzyme, cytochrome oxidase (CO) (Cusick et al. 1984). On the V2 margin (Fig. 5.4), callosal bands or stripes extend several millimeters from the border linking mirror-symmetrical V2 regions (Cusick and Kaas 1986; Innocenti 1986; Olavarria and Abel 1996; Abel et al. 2000). These bands overlie strips of dense CO staining which have strong connections with the superior colliculus and magnocellular pathway of the lateral geniculate nucleus (Olavarria and Abel 1996; Abel et al. 1997). Since CC and thalamocortical inputs to VI are matched for orientation, the callosal

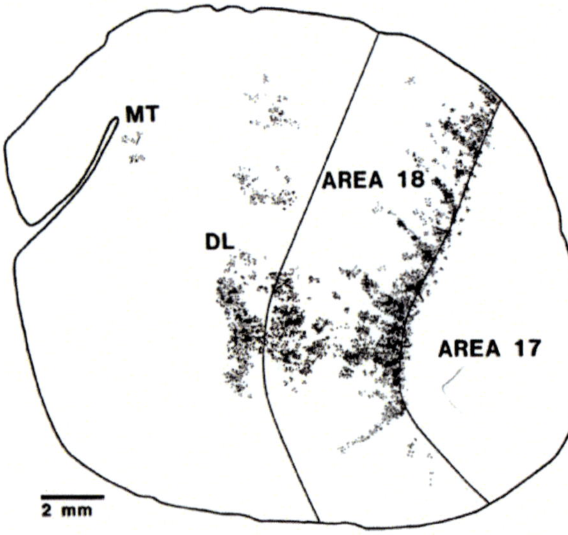

Fig. 5.4 Callosal projections (*dots*) in visual cortex after tracer injections along the dorsoventral axis of area 18 in the contralateral hemisphere are clustered at the areas 17 and 18 border with patchy labeling in area 17 and band-like extensions in area 18. Adapted from the original source (Cusick and Kaas 1986)

terminations may correspond to orientation columns matching that of the cells of origin (Berlucchi and Rizzolatti 1968; Leporé and Guillemot 1982; Leporé et al. 1992). At the V1–V2 border, some callosal axons extend 100–3,500 μm. Some callosal axons branch along infragranular and/or supragranular columns, and others end in neighboring columns (Houzel et al. 1994; Innocenti et al. 1995). Thus, CC axon terminations may diverge across several columns.

Tracer injections or lesions in SI or motor cortex label patchy anterograde CC terminations in the form of callosal columns (Jenny 1979; Jones et al. 1975, 1979; Shanks et al. 1975; Killackey et al. 1983). These columns extend across all layers, ranging from 500 to 1,100 μm in diameter, and overlap with groups of neurons from which reciprocal projections originate. Columns are separated by smaller gaps in which somatic and terminal labeling is weak or absent. In sections parallel to the pia, callosal columns form irregular strips or bands whose orientation, size, and shape vary across areas, animal subjects, and between studies, so that it cannot be determined if there is a consistent pattern of organization. In some areas (e.g., area 4 of primates), however, their alignment and orientation is more regular than in areas such as SI (Jones et al. 1979). This may reflect that the homotopic ipsi- and contralateral projections tend to overlap more in SI, while there is greater pathway segregation in areas 4, 5, and SII. Similarly, thalamocortical inputs to SI are distributed evenly, while callosal inputs are patchier. Thus, some columns may be dominated by thalamic input, while others are also influenced by callosal input (Jones and Burton 1976).

Irregularities in CC spatial distribution are also characteristic of AC (Fig. 5.5, top). Terminal labeling in contralateral AI forms radial aggregates across laminae, comprising callosal columns (Jacobson and Trojanowski 1974; Imig and Brugge 1978; Kelly and Wong 1981; Code and Winer 1986; Luethke et al. 1988; Wallace and Harper 1997). Patchy connections also characterize callosal projections in primates (Fitzpatrick and Imig 1980; Luethke et al. 1989). Columns are 300–1,200 μm in diameter with a mean ~500 μm. Callosal projections to layer I, and to some extent layer VI, are more uniform and any columns are less obvious in these layers. In tangential sections, callosal columns comprise continuous bands separated by acallosal zones, or isolated patches. Studies in the cat and ferret find that callosal columns (patches and bands) were associated with binaural summation, while neurons in acallosal regions more often had contralateral dominant suppression responses (Imig and Brugge 1978). The orientation of elongated callosal columns or bands is perpendicular to isofrequency contours (Fig. 5.5, bottom panels), suggesting that callosal columns, and therefore binaural summation, are represented periodically along the isofrequency axis (Imig and Adrián 1977; Middlebrooks et al. 1980; Middlebrooks and Zook 1983; Kelly and Judge

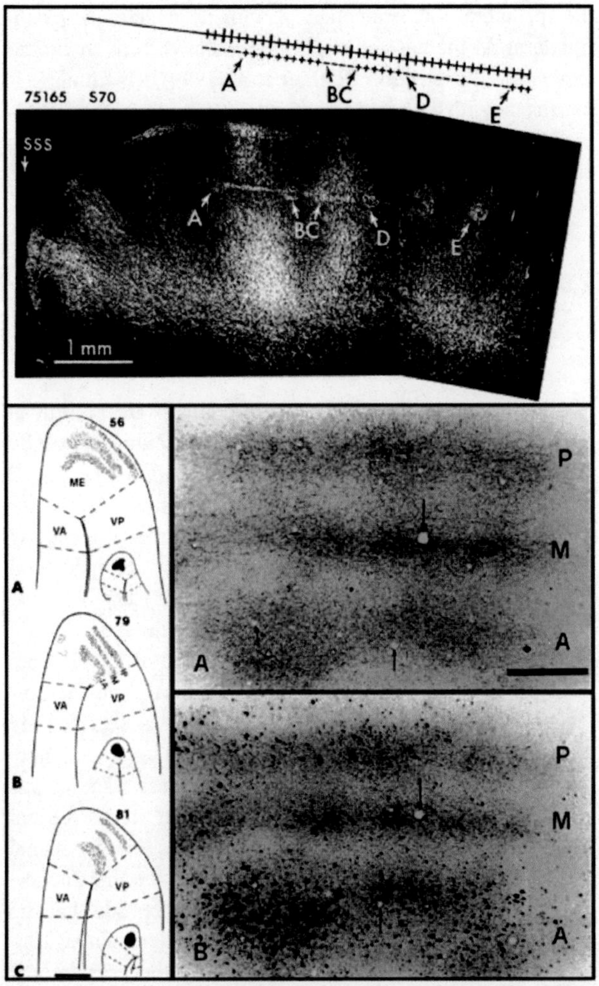

Fig. 5.5 Callosal projections of AI. *Top*, coronal section in the cat after injections of tritiated proline into the opposite hemisphere label callosal columns. Microelectrode recordings in the diagram above the *top panel* show alternating regions of contralateral dominant suppression (–) and summation (+) along an electrode track. The locations of lesion markers are indicated by letters (*A–E*). Adapted from the original source (Imig and Brugge 1978). *Bottom*, labeled terminals and cell soma in primary auditory cortex (ME) of the ferret after tracer injections in the contralateral hemisphere. *Bottom left*, bands of labeling in three experiments. *Insets*, injection sites. *Bottom right*, photomicrograph of cells and terminals in tangential sections through layer II (*top*) and layer III (*bottom*). Adapted from the original source (Wallace and Harper 1997)

1994; Wallace and Harper 1997). Callosal columns, however, do not clearly relate to patches of intrinsic (ipsilateral) connections, since single injections in AI may label groups of neurons with the same or higher CF (Imig and Reale 1981; Matsubara and Phillips 1988). Callosal columns also seem independent of patchy thalamocortical inputs to AI (Hashikawa et al. 1995; McMullen and de Venecia 1993) and may interdigitate with them (Pandya and Rosene 1993).

2.5 Callosal Projections Are Layer Specific

Commissural projections have laminar connection patterns that are modality and area specific. In primary sensory areas, callosal terminations and cells of origin concentrate in the supragranular layers, which is also the main intrahemispheric origin (Jacobson and Trojanowski 1974; Swadlow et al. 1978; Jones et al. 1979; Kelly and Wong 1981; Killackey et al. 1983; Conti et al. 1986; Kennedy and Dehay 1988; Innocenti et al. 1995; Budinger et al. 2000; de la Mothe et al. 2006a). Thus, CC projections from primary sensory areas resemble feedforward intrahemispheric connections, with layer III neurons projecting to layers III and IV of areas presumed to be higher in the cortical sequence of processing. Most callosal axons originate from, and synapse on, pyramidal neurons, especially in the deep half of layer III, and to a lesser extent layer VI (Kelly and Wong 1981; Tigges et al. 1981; Van Essen et al. 1982; Cusick et al. 1984; Voigt et al. 1988; Buhl and Singer 1989). In primary sensory cortex, the infragranular layers represent less of the total callosal projection, no more than 10% (Tigges et al. 1981; Van Essen et al. 1982). Outside the primary fields, however, the density of infragranular connections is greater, reflecting a different interhemispheric circuitry (Innocenti 1986; Manzoni 1997).

Similar proportions of supragranular and infragranular projections have been estimated for AC (Fig. 5.6) (Imig and Brugge 1978; Kelly and Wong 1981; Wong and Kelly 1981; Imig et al. 1982; Code and Winer 1985, 1986; Cipolloni and Pandya 1989; Rouiller et al. 1991; Hackett et al. 1999; Budinger et al. 2000; de la Mothe et al. 2006a), though there is evidence for area-specific laminar origins, with more ventral areas in cats having a preponderance of cells in layer V and more dorsal areas having a concentration in layers

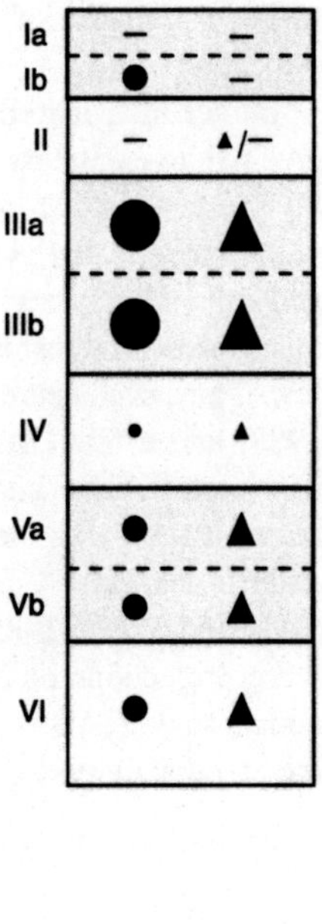

Fig. 5.6 Relative density of commissural connections of primary auditory cortex by layer. *Circles*, density of axonal terminals (anterograde); *triangles*, cell somata (retrograde); *dashes*, no connections. Symbol size denotes increased connection magnitude. The strongest projections arise from, and terminate in, layer III in auditory, somatic sensory and visual cortex. Connections with other laminae appear to vary by modality, but are generally weakest in layers II and IV

III and V (Lee and Winer 2008). CC arises mainly from layer III pyramidal cells with lesser involvement of layers V and VI. Layer I neurons in do not have callosal projections, but callosal terminations are present in the lower half of layer I and approximate those in layers V and VI (Code and Winer 1986). The weakest input target layers II and IV. A few layer IV non-pyramidal neurons in project in the CC, with no layer II callosal neurons (Code and Winer 1985). The callosal layer III pyramidal cells concentrate in its lower half, and are the largest layer III pyramidal neurons (de la Mothe et al. 2006a). Dispersed throughout layer III are several non-pyramidal varieties that project contralaterally. Nevertheless, some pyramidal and non-pyramidal neurons are not labeled commissurally, indicating that some cells in callosal columns do not have interhemispheric connections (Code and Winer 1985). Moreover, there is incomplete reciprocity between the cells and terminals, with the zone of retrogradely labeled cells sometimes larger than the area of labeled terminals (Kelly and Wong 1981), and there are discontinuities between labeled cells and terminals (Imig et al. 1982; Code and Winer 1986) implying that some callosally projecting neurons do not receive homotopic commissural input, or that they receive heterotopic input.

In aggregate, there is a degree of specificity in the callosal system that remains to be defined, especially concerning the postsynaptic relationships of the different neuronal populations. The synapses made on pyramidal neurons are usually asymmetric, targeting dendritic spines rather than the shafts or somata (Cipolloni and Peters 1983; Vaughan and Peters 1985; Voigt et al. 1988; Buhl and Singer 1989; White and Czeiger 1991). Non-pyramidal neurons, chiefly spiny stellate cells, are a far smaller proportion of callosally projecting neurons (Jacobson and Trojanowski 1974; Lund et al. 1975; Code and Winer 1985; Hughes and Peters 1990). Only 4.6% of supragranular synapses are callosal in origin, testifying to the dominance of intrahemispheric and thalamocortical projections.

Intracellular recordings show that most of these neurons give rise to excitatory monosynaptic CC projections, probably via fast-conducting myelinated fibers, whereas neurons with unmyelinated axons may play a neuromodulatory role, and are not likely excitatory. Glutamate is the probable commissural neurotransmitter (Conti and Manzoni 1994) with few, if any, such neurons releasing GABA in AC (Voigt et al. 1988) and a few GABAergic cells in rodent somatic sensory cortex (Fabri and Manzoni 2004). These observations are consistent with the morphological profile of the pyramidal neurons with CC projections (Cipolloni and Peters 1983; Code and Winer 1985). On the other hand, the impact of these projections on AC activity may not be simply excitatory, since GABA antagonists injected into gerbil AI of increased activity-dependent (fluro-deoxyglucose) uptake in the ipsilateral AC dramatically, with no concomitant contralateral AI increase as might be predicted from an excitatory callosal projection (Richter et al. 1999). This finding is consistent with the effects of corpus callosum electrical stimulation, which suppressed responses in ferret AC s (Kitzes and Doherty 1994).

2.6 Development of Callosal Projection Patterns Is Experience Dependent

Mature primate callosal projection patterns are present at birth, though some laminar adjustment occurs postnatally (Dehay et al. 1986; Beck and Kaas 1994), while in cats and ferrets it is less mature at birth. Ablation of the ferret superior and inferior colliculus shortly after birth evokes a re-routing of otherwise contralateral retinal axons into the ipsilateral medial geniculate complex (Sur et al. 1988) and AI cells develop retinotopic and orientation-selective response properties (Roe et al. 1990, 1992). Thus, callosal axons from AI were substantially reduced in much of AI, and the normal commissural banding pattern (Fig. 5.5) was significantly disrupted (Pallas et al. 1999). By contrast, the CC projections in animals deafened by bilateral cochlear ablation was immature pattern and unusually widespread across AI and relatively diffuse, with fewer distinct patches than in normal animals (Feng and Brugge 1983). Because the thalamocortical connections remained intact, it was proposed that the callosal development reflects activity-dependent factors contributing to cortical circuit formation. In the cross-modal animals, converging thalamic auditory and visual input from the thalamus may reduce the callosal projections to the affected portion of AI, thus ameliorating perceptual errors. In the deafened animals, the missing auditory input eliminated the experience-dependent pruning of connections that would normally culminate in callosal banding pattern. Thus, the thalamocortical ipsilateral input can modify the homotopic callosal projection contralaterally.

3 Function of Auditory Callosal and Intrahemispheric Systems

3.1 Binaural and Spatial Responses of Auditory Cortical Neurons

The callosal and ipsilateral corticocortical connectivity of cat AI has been established using retrograde (horseradish peroxidase, HRP; cholera toxin beta fragment, CTb) and/or anterograde (tritiated proline) tracing methods (Imig and Brugge 1978; Imig and Reale 1981; Lee and Winer 2008). AC areas participating in these of connections have patchy, discontinuous distributions, and these may be topographically complementary (Fig. 5.5). Certain studies derive special significance by the documented association between callosal connectivity and the binaural interaction patterns of these neurons (Imig and Brugge 1978; Imig and Reale 1981). This relationship suggests that a feature distinguishing intra- from interhemispheric processing is binaural interactions, which are essential for auditory spatial processing.

3.2 Patterns of Binaural Input and Interaction in Auditory Cortex

Almost all primary AC neurons are influenced binaurally (Imig and Adrían 1977; Phillips and Irvine 1983), and this likely extends to other AC fields (Phillips and Irvine 1982; Orman and Phillips 1984; Lee and Winer 2008). Classification schemes for binaural properties usually incorporate some description of the responses to monaural stimulation, and of the interaction between the monaural inputs seen by comparison of responses to binaural stimulation with the stronger monaural response. The contralateral ear usually provides a short-latency excitatory input, and the ipsilateral ear may elicit a short-latency excitatory input (EE) or not (EO) cells. When the ears are stimulated concurrently, the binaural response may evoke stronger (facilitative or summative) or weaker (suppressive) interactions. Suppressive interactions are particularly common among EO cells, suggesting that the ipsilateral input is in fact a short-latency inhibitory one (EI cells). EE cells often show summative interactions, as do some EO cells. A few AC cells are predominantly binaural (PB) (Kitzes et al. 1980), with very weak monaural responses vigorous response to dichotic stimuli with interaural disparities in time (ITD) or intensity (IID) near zero.

When the contralateral stimulus is held constant, and the level of the ipsilateral stimulus is varied, EO cells which display summative interactions for interaural disparities favoring the contralateral ear (or near zero dB) may display suppressive interactions when the interaural level difference (ILD) significantly favors the ipsilateral ear (Phillips and Irvine 1981). This suggests the existence of a mixed excitatory–inhibitory ipsilateral input, and much the same inference can be drawn from studies of sensitivity to interaural phase differences (Brugge et al. 1969, 1970). Among EI cells, the relative sensitivities of the two monaural inputs may vary, but spike output is usually a sigmoidal function of ILD, with spike rates maximal for ILDs favoring the contralateral, excitatory ear (Phillips and Irvine 1981; Phillips and Brugge 1985). EE cells tend to be less sensitive to ILDs, and PB cells often have steep, nonmonotonic relations of spike output to ILD, with response rates greatest for zero or near-zero disparities.

3.3 Spatial Receptive Fields of Cortical Auditory Neurons

In principle, the free-field spatial receptive fields (RFs) of AC neurons are understandable in terms of the binaural properties seen in dichotic studies. Variations in sound source azimuth cause variations in ILD and ITD, and a cell's spike rate is an orderly function of the sign and size of the ILD and ITD. Some AC cells respond in a relatively undifferentiated way across auditory azimuth (and/or elevation), and have been termed omnidirectional (Middlebrooks and Pettigrew 1981) and likely are of the EE class because only such cells are excited over wide ranges of ILD sign and magnitude. Hemifield units (again, after Middlebrooks and Pettigrew 1981) are usually excited by sources in the contralateral auditory hemifield and have RF boundaries within about 30° of the midline (Imig et al. 1990; Rajan et al. 1990; Eggermont and Mossop 1998; Middlebrooks et al. 1998). These are often EI cells (Clarey et al. 1995); the proximity of the RF boundary to the midline may depend on the relative sensitivities of the inputs from the two ears.

In cats, some high-frequency neurons have focal, circumscribed spatial RFs located on the acoustical axis of the contralateral pinna (axial units; Middlebrooks and Pettigrew 1981; see also Brugge et al. 1996). The cat's pinna enhances its directional selectivity at high frequencies (Phillips et al. 1982), and it is likely that the shape and position of axial RFs in high-frequency neurons in part reflects this directionality. Perhaps simple binaural interactions determine RF shape, while the distribution of spike rates for a given source located at different points within the RF reflects pinna directionality through the proximity of the source to the pinna axis and, thus, the strength of the excitatory (or inhibitory) drive on the neuron. In cats (Clarey et al. 1995) and primates (Recanzone et al. 2000), some neurons are maximally driven by sources near 30° of the midline, and their selectivity seems to depend on facilitative binaural interactions (Clarey et al. 1995), suggesting that they may be PB cells.

3.4 Binaural Organization of Primary Auditory Cortex

In the primary auditory cortex of cats (Imig and Adrián 1977; Imig and Brugge 1978; Middlebrooks et al. 1980) and other species (Kelly and Judge 1994; Liu and Suga 1997), local territories are dominated by cells with specific binaural properties. An organization with local regions dominated by suppressive or summative binaural interactions in cats, and a topographic organization according to binaural responses (EI and EE) (Imig and Brugge 1978) was confirmed (Middlebrooks et al. 1980). The criteria for one or other classification system have been ad hoc, and often the recordings subject to those classifications have been from neuron clusters, rather than single units, rendering the classifications somewhat ambiguous, with EE and summative designations including EE cells, PB cells, and EO cells with facilitative interactions. Nevertheless, there is some segregation of binaural cell types in AC. This sometimes manifested as elongated bands of cells with the same general binaural responsiveness, oriented orthogonal to lines of cells with the same characteristic frequencies (CFs), and spanning a significant fraction of the high-frequency portion of the tonotopic map. The binaural maps are usually based on recordings from the middle cortical layers, although there is evidence that the binaural response-specificity may extend to all cortical layers (Imig and Adrián 1977).

3.5 Binaural Responses and Patterns of Callosal and Intrahemispheric Connectivity

In cat primary AC, callosal axons arise from cells in columns of tissue often characterized by summative binaural interactions (Imig and Brugge 1978; see Fig. 5.5). The callosal terminals have a columnar distribution which also is associated with summative binaural interactions in recordings from the same tissue. The correspondence between physiologically and anatomically characterized territories is imperfect, but it can be striking in individual cases. The anatomically defined columns average about half a millimeter in width, and, on occasion, are elongated parallel to the tonotopic axis. The match between the columns giving rise to callosal axons, and those receiving them is good, but also imperfect (Code and Winer 1986). Thus, there are some callosal recipient zones without callosal axons, and vice versa. For the present purposes, however, the characterization of callosal neurons as summative in binaural interactions potentially includes EE, PB, and some EO cells, since each can exhibit summative interactions. Perhaps the apparent mismatches in the correspondence of callosal columns with summative binaural ones in part stems from inappropriate physiological characterization of the territories. Constraint of callosal connectivity to the EE region of the primary auditory cortex extends to bats (Liu and Suga 1997). In ferrets (Wallace and Harper 1997), owl monkeys (FitzPatrick and Imig 1980, 1982), and tamarins (Luethke et al. 1989), callosal connectivity of the primary AC is again patchy, but it is unknown whether this patchiness is associated with patterns of binaural responses.

The binaural properties of callosally projecting auditory neurons have also been revealed in recordings from callosal axons (Poirier et al. 1995). Almost 80% of these were EE and, of those, half had binaural interactions which could be described as summative or facilitative. Of the remaining 20% (EO or OE cells), half again had summative interactions. Of callosal axons tuned to ITDs over the behaviorally relevant range (about ±450 μs in cats), almost half were tuned to near-zero ITDs. Other neurons had an ITD preference associated with one or other acoustic hemifield, or were unclassifiable (Poirier et al. 1995).

The AI territories participating in intrahemispheric corticocortical connectivity are also patchy. In cats, the connections most commonly described are those with other cortical AC fields. Ipsilateral corticocortical connectivity is associated with suppressive binaural interactions in AI (Imig and Reale 1981). These neurons are likely EI cells and/or EO cells with mixed binaural interactions. The patchiness of the participating territories extends to primates (FitzPatrick and Imig 1980, 1982; Cipolloni and Pandya 1989) and squirrels (Luethke et al. 1988), but it is unknown whether the patchiness in those species has binaural physiological correlates.

Binaural interactions are initially created by the convergence of monaural inputs in the superior olivary complex, and are modified through circuits involving, e.g., the dorsal nucleus of the lateral lemniscus (Phillips 2001). There is further afferent convergence in the thalamocortical projection, as the sheet-like array of thalamic neurons representing any given cochlear site collapses and converges onto a strip-like assembly of neurons in the AI tonotopic map. AC neurons are generally binaural by virtue of their subcortical afferent inputs, not because of their callosal connectivity. However, there is evidence for de novo binaural interactions in the ferret AC and its mediation by callosal connectivity (Kitzes and Doherty 1994). Thus, acoustical stimulation of one ear and electrical stimulation of the opposite AI evokes an excitatory response in some PB cells, while the acoustic stimulation alone does not. Most commonly, however, callosal activation suppresses acoustically evoked activity, and does so with latencies of 2–4 ms (Kitzes and Doherty 1994). Thus, while the ascending input to AI cells might already be binaural, that binaurality could be modified by independent callosal inputs.

The foregoing has important implications for the spatial properties of processing executed within and across the cerebral hemispheres. Only a subset of AC cells has restricted spatial RFs, or binaural interactions which would otherwise support restricted RFs. Most of these neurons are maximally responsive to, and/or are most differentially sensitive to, spatial cues for the contralateral auditory hemifield or the midline (Phillips and Brugge 1985). That is, the AC of each hemisphere may be independently capable of processing sound sources in the contralateral hemifield. This conclusion is supported by studies of the effects of unilateral cortical ablations on sound localization behavior. Most commonly, such lesions result in profound deficits in sound localization performance only for sources in the hemifield contralateral to the ablation in animals (Jenkins and Masterton 1982; Thompson and Cortez 1983; Kavanagh and Kelly 1987; Heffner 1997) and humans (Sanchez-Longo and Forster 1958; Poirier et al. 1994).

Now, since the auditory spatial information provided to higher stations in the same hemisphere may be derived somewhat selectively from the subset of AC neurons with hemifield spatial sensitivity, it follows that those higher centers also will process spatial information only for the contralateral hemifield. This appears to be true for the human parietal cortex, since unilateral damage to the posterior parietal cortex can result in an auditory neglect or inattention or extinction which is restricted to the hemifield contralateral to the damage (Heilman and Valenstein 1972). Frontal cortex cells with auditory input often are spatially selective for sources in the contralateral hemifield (Azuma and Suzuki 1984; Vaadia et al. 1986). Interestingly, this preservation by the frontal cortex of the extent of spatial tuning seen in the AC is in contrast to the development of dramatically different other features. One of the more striking of these is the expression of responses only when the animal is actively localizing the source (Vaadia et al. 1986). What seems clear, however, is that there is a hemifield-specificity of spatial information available in each side of the forebrain. Such lateralization might support the hemifield tuning of human perceptual channels for auditory azimuth (Boehnke and Phillips 1999; Phillips et al. 2003; Phillips and Hall 2005; Stecker et al. 2005; Dingle et al. 2010). The great acuity listeners have for localizing sources near the midline likely reflects the fact that the azimuthal processing channels have their most differentiated outputs for sources within about 30° of the midline, i.e., the edges of their spatial RFs (Stecker et al. 2005), which in part reflects that the stimulus information available for encoding source azimuth is itself most informative for near-midline eccentricities (Phillips and Brugge 1985).

In contrast, the binaural classifications of callosal cells are more likely to be EE or summative. EE cells are perhaps least likely to transmit information about source location. Some cells with summative interactions are associated with cues for near-midline azimuths (PB cells). Lesion of the corpus callosum in cats have little effect on sound lateralization (discriminating left from right) (Moore et al. 1974), but the effects on sound localization (discrimination of source location within an auditory hemifield) have not been explored in detail. In humans, callosal agenesis, or early callosotomy, has little or no effect on the localization of stationary white noise sources, but may affect on the precision with which moving targets can be localized (Lessard et al. 2002). These human data are hard to interpret, because a life-long callosal agenesis might have elicit the development of any of a number of compensatory circuits to mediate normal or near-normal localization ability.

4 The Continuity of Sensation Across the Midline: The Midline Fusion Hypothesis

4.1 Three Sensory Systems

In both the primary visual and primary somatic sensory cortices, cells participating in callosal connectivity have restricted distributions within the sensory representation (see Figs. 5.3, 5.4, and 5.5). The visual cortex neural population projecting to the callosum usually has RFs centered very close to the vertical meridian. In the somatosensory cortex, callosally projecting cells have RFs with axial locations, or para-axial locations and medial borders at the midline. In other (i.e., nonprimary) sensory cortices, receptive fields can be quite large, and the extension often is far from the midline. It is thus because of the involvement of the other areas that the callosum as a whole can convey information from complete sensory hemi-worlds to the opposite cerebral hemisphere. Nevertheless, the highly uneven distributions of callosal cells in the primary sensory areas have led to the hypothesis that callosal connectivity is important for continuity of sensation across the midline (the midline fusion hypothesis) (Berlucchi 1972; Berlucchi et al. 1986; Leporé et al. 1986; Guillemot et al. 1992).

The primary cortical representation of the retina or the body surface is inherently spatial. It is direct in that it requires no computations, save for the preservation of receptive field size, and the topography of the point-to-point projections from the sensory epithelium to the cortex. In the AC, the most fundamental mapping is the topographic arrangement of cells by their CFs (tonotopic organization). Any representation of auditory space is necessarily computational, insofar as it embodies interaural disparities. The information about source locations is biased heavily towards locations in the contralateral hemifield, or near the midline. This is expressed in two ways: cells with spatial RFs centered on the midline, or cells with RFs in the lateral hemispheres, but whose medial borders are near the midline. The remaining cells (likely EE ones) have very broad RFs, spanning the midline and encompassing both acoustic hemifields. In this regard, the finding that a disproportionate number of callosal axons (Poirier et al. 1995) are tuned to ITDs associated with near-midline azimuths fits well with the findings from the primary visual and somatic sensory cortices. One could construe that ITD tuning as relatively broad because the selectivity was often expressed over very broad ITD ranges. However, naturally occurring ITDs are almost always associated with ILDs favoring the same ear, and the conjunction of those two stimulus parameters—which would occur in the free-field—would likely only enhance the azimuthal tuning of the callosal cells. Note that the compatibility of the AC data with the midline fusion hypothesis was a foregone conclusion: the spatial RFs of almost all AI cells either encompass the midline or abut it. The only exceptions are seen in animals with highly directional pinnae whose high-frequency cells can have RFs on the pinna axis (see above).

It is clear that the neuroanatomical and neurophysiological data, that the RFs of callosally projecting cells span or abut the midline, and that at least some (Berlucchi 1972) of the bilaterality of midline cortical representation is attributable to callosal inputs. How does this relate to the midline fusion hypothesis and the "continuity of sensation across the midline?" We offer two approaches. The first is a simple conceptualization of cortical wiring; the second is an examination of sensory saltation.

One can conceptualize AC RFs as being elaborations of those provided by individual thalamic inputs through their convergence and/or by local intracortical connectivity. Cells with RF centers very close to the midline may require contributions from intracortical commissural sources. This provides a continuity of cortical representation (see also Hubel and Wiesel 1967). However, to understand the role of the callosum in continuity of sensation across the midline, one needs access to behavioral data.

4.2 Sensory Saltation

Sensory saltation is an illusion of perceived motion for transient stimuli presented repetitively at successive sites with perfectly regular interstimulus intervals (Geldard and Sherrick 1986). In the somatic sensory system, repetitive taps to skin at successive sites along the forearm led to the percept of stimulation not only at these anchor points, but also at points between them. The effect occurs in vision (Geldard 1976) and in audition (Phillips and Hall 2001; Phillips et al. 2002; Boehnke and Phillips 2005). Auditory saltation is usually studied with dichotic stimuli: a brief train of identical dichotic clicks, lateralized by an ITD, is followed in perfect temporal cadence by an identical train of clicks lateralized to the opposite side. If the interclick intervals are less than about 120 ms, the ensuing percept is of clicks emanating not only from the anchor points, but from an orderly spacing of points between them in intracranial space. The effect remains robust if the stimulus lateralization is achieved with ILDs rather than ITDs, or even if the cue identity is switched from ITD to ILD in mid-train (Phillips et al. 2002). The effect is also robust in the auditory free-field, even when the ITD and ILD cue sizes are near zero and unchanging between anchor points, e.g., the vertical midsagittal plane (Boehnke and Phillips 2005). The independence of the effect from the stimulus cue information available to localize the anchor points suggests that the illusion arises from a high-level spatial representation (Phillips et al. 2002). The illusion likely reflects the fact that the sensory system uses stimulus information sampled across a temporal window a few hundred milliseconds wide in the process of constructing the percept (Phillips and Hall 2001; Boehnke and Phillips 2005). That is, the time required for fabricating the conscious percept for one event is long enough that the percept can be updated by (influenced by) information from a subsequent event occurring within that temporal window.

The saltation illusion allows exploration of the continuity of processing across the auditory midline. Despite callosal connectivity across the representations of the vertical meridian in vision and axial body part representations in somatic sensation, the saltation illusion does not span the midline is either modality (Geldard 1976; Geldard and Sherrick 1986; but see Eimer et al. 2005). In the auditory system, the illusion does span the midline (Phillips and Hall 2001; Phillips et al. 2002) suggesting that callosal connectivity is not necessary for midline continuity. Perhaps the midline effects works in the auditory case because each hemisphere contains cells with omnidirectional receptive fields, i.e., saltation is only supported under conditions in which the actual stimulus locations fall within the RF boundaries of individual neurons, i.e., most likely for cells with very large RFs (Phillips and Hall 2001).

Thus, the corpus callosum might indeed be important for the continuity of sensory representations across the midline, but with fewer data on its role in the continuity of sensation across the midline. The apparently seamless continuity of sensation across the midline must partly reflect callosal contributions, since it mediates what would otherwise be the local, intracortical connectivity that exists as a continuous, intrahemispheric network for spatial locations off the midline. Data from studies of sensory saltation in normal subjects suggest that continuity of saltation across the midline might be independent of the existence of callosal connectivity across midline representations. The saltation illusion may thus distinguish continuity of representation from continuity of sensation.

5 Conclusions and Directions for Future Research

The auditory component of the corpus callosum is a highly organized array of axons which arises relatively from particular morphological cell types within only some layers within cortical columns, and whose laminar origin is area-specific. Those cortical columns contain neurons with particular patterns of binaural interactions and, therefore, particular patterns of spatial selectivity. Further work is needed to establish in more detail the form-function relationships that characterize the cells participating in interhemispheric connectivity. Neurophysiological studies in animals and behavioral studies in humans provide some evidence on how the callosum establishes a fusion or continuity of sensory representation across the two hemiworlds. The rules governing perceptual corollaries of this continuity remain to be elaborated.

Sensory representation and its perceptual corollaries might, however, be construed as very primitive operations. Of further interest, then, is the role of auditory callosal connectivity in mediating the performance of higher-level tasks that require coordination and cooperation between the two cerebral hemispheres, as has been demonstrated for visual and attentional realms in cognitive science (Banich 1998a, b; Santhouse et al. 2002). Deviations in interhemispheric connectivity in schizophrenia, and its relation to interhemispheric transfer of (visual) verbal information (Highley et al. 1999; Endrass et al. 2002) could be extended to the auditory sense relatively easily. A temporal correlations study of left- and right-hemisphere electrical responses evoked by dichotic words in a set of twins, one with an auditory processing disorder (APD) and who showed disturbed patterns of hemispheric correlations and reduced callosal myelin integrity (Jerger et al. 2002, 2004). These model studies might inspire future in their use of quantitative, objective approaches to the evaluation of auditory callosal structure and function in high-level perceptual and cognitive operations.

Ongoing study of human callosal sexual dimorphisms leave open the question of gender-related anatomical differences in callosal shape (Allen et al. 1991) rather than cross-sectional area or volume (Oka et al. 1999; Mitchell et al. 2003). Callosal structural changes in aging (Inzitari 2000; Peters and Sethares 2002), Alzheimer's disease (Janowsky et al. 1996; Thompson et al. 1998), and dyslexia (Hynd et al. 1995) are documented. The cognitive operations whose impairment marks aging, dementia or dyslexia are mediated by neural networks distributed widely across the cerebral hemispheres and which often include auditory and/or phonological processing. The widely distributed callosal origins (Fig. 5.1) suggest that the corpus callosum may have massive perceptual roles. The task is to specify precisely the white matter deficits that contribute to such functional declines.

Acknowledgments Some of the work described in this chapter was supported by grants from the Natural Sciences and Engineering Research Council of Canada, the Canadian Language and Literacy Research Network, and the Killam Trust (D.P.P.)

References

Abel PL, O'Brien BJ, Lia B, and Olavarria JF (1997) Distribution of neurons projecting to the superior colliculus correlates with thick cytochrome oxidase stripes in macaque visual area V2. *Journal of Comparative Neurology* **377**:313–323.

Abel PL, O'Brien BJ, and Olavarria JF (2000) Organization of callosal linkages in visual area V2 of macaque monkey. *Journal of Comparative Neurology* **428**:278–293.

Aitkin LM, Kudo M, and Irvine DRF (1988) Connections of the primary auditory cortex in the common marmoset, *Callithrix jacchus jacchus*. *Journal of Comparative Neurology* **269**:235–248.

Allen LS, Richey MF, Chai YM, and Gorski RA (1991) Sex differences in the corpus callosum of the living human being. *Journal of Neuroscience* **11**:933–942.

Azuma M and Suzuki H (1984) Properties and distribution of auditory neurons in the dorsolateral prefrontal cortex of the alert monkey. *Brain Research* **298**:343–346.

Banich MT (1998a) Integration of information between the cerebral hemispheres. *Current Directions in Psychological Science* **7**:32–37.

Banich MT (1998b) The missing link: the role of interhemispheric interaction in attentional processing. *Brain and Cognition* **36**:128–157.

Barbaresi P, Bernardi S, and Manzoni T (1989) Callosal connections of the somatic sensory areas II and IV in the cat. *Journal of Comparative Neurology* **283**:355–373.

Barbaresi P, Minelli A, and Manzoni T (1994) Topographical relations between ipsilateral cortical afferents and callosal neurons in the second somatic sensory area of cats. *Journal of Comparative Neurology* **343**:582–596.

Beck PD and Kaas JH (1994) Interhemispheric connections in neonatal owl monkeys (*Aotus trivirgatus*) and galagos (*Galago crassicaudatus*). *Brain Research* 651:57–75.

Berlucchi G (1972) Anatomical and physiological aspects of visual functions of corpus callosum. *Brain Research* **37**:371–392.

Berlucchi G and Rizzolatti G (1968) Binocularly driven neurons in visual cortex of split-chiasm cats. *Science* **159**:308–310.

Berlucchi G, Tassinari G, and Antonini A (1986) The organization of the callosal connections according to Sperry's principle of supplemental complementarity. In: Leporé F, Ptito M, and Jasper HH (eds). *Two Hemispheres – One Brain: Functions of the Corpus Callosum.* Alan R Liss, New York, pp 171–188.

Boehnke SE and Phillips DP (1999) Azimuthal tuning of human perceptual channels for sound location. *Journal of the Acoustical Society of America* **106**:1948–1955.

Boehnke SE and Phillips DP (2005) Auditory saltation in the vertical, midsagittal plane. *Perception* **34**:371–377.

Brugge JF, Anderson DJ, and Aitkin LM (1970) Responses of neurons in the dorsal nucleus of the lateral lemniscus of cat to binaural tonal stimulation. *Journal of Neurophysiology* **33**:441–458

Brugge JF, Dubrovsky NA, Aitkin LM, and Anderson DJ (1969) Sensitivity of single neurons in auditory cortex of cat to binaural tonal stimulation; effects of varying interaural time and intensity. *Journal of Neurophysiology* **32**:1005–1024.

Brugge JF and Imig TJ (1978) Some relationships of binaural response patterns of single neurons to cortical columns and interhemispheric connections of auditory area AI of cat cerebral cortex. In: Fernández C and Naunton RF (eds). *Evoked Electrical Activity in the Auditory Nervous System.* Academic Press, New York, pp. 487–504.

Brugge JF, Reale RA, and Hind JE (1996) The structure of spatial receptive fields of neurons in primary auditory cortex of the cat. *Journal of Neuroscience* **16**:4420–4437.

Budinger E, Heil P, Scheich H (2000) Functional organization of auditory cortex in the Mongolian gerbil (*Meriones unguiculatus*). III. Anatomical subdivisions and corticocortical connections. *European Journal of Neuroscience* **12**:2425–2451.

Buhl EH and Singer W (1989) The callosal projection in cat visual cortex as revealed by a combination of retrograde tracing and intracellular injection. *Experimental Brain Research* **75**:470–476.

Caminiti R, Innocenti GM, and Manzoni T 1979 The anatomical substrate of callosal messages from SI and SII in the cat. *Experimental Brain Research* **35**:295–314.

Choudhury BP, Whitteridge D, and Wilson ME (1965) The function of the callosal connections of the visual cortex. *Quarterly Journal Experimental Physiology and Cognate Medical Sciences* **50**: 214–219.

Cipolloni PB and Pandya DN (1989) Connectional analysis of the ipsilateral and contralateral afferent neurons of the superior temporal region in the rhesus monkey. *Journal of Comparative Neurology* **281**:567–585.

Cipolloni PB and Peters A (1983) The termination of callosal fibres in the auditory cortex of the rat. A combined Golgi-electron microscope and degeneration study. *Journal of Neurocytology* **12**: 713–726.

Clarey JC, Barone P, Irons WA, Samson FK. and Imig TJ (1995) Comparison of noise and tone azimuth tuning of neurons in cat primary auditory cortex and medial geniculate body. *Journal of Neurophysiology* **74**:961–980.

Code RA and Winer JA (1985) Commissural neurons in layer III of cat primary auditory cortex (AI): pyramidal and non-pyramidal cell input. *Journal of Comparative Neurology* **242**:485–510.

Code RA and Winer JA (1986) Columnar organization and reciprocity of commissural connections in cat primary auditory cortex (AI). *Hearing Research* **23**:205–222.

Conti F, Fabri M, and Manzoni T (1986) Bilateral receptive fields and callosal connectivity of the body midline representation in the first somatosensory area of primates. *Somatosensory and Motor Research* **3**:273–289.

Conti F and Manzoni T (1994) The neurotransmitters and postsynaptic actions of callosally projecting neurons. *Behavioral Brain Research* **64**:37–53.

Cusick CG, Gould HJ 3rd, and Kaas JH (1984) Interhemispheric connections of visual cortex of owl monkeys (*Aotus trivirgatus*), marmosets (*Callithrix jacchus*), and galagos (*Galago crassicaudatus*). *Journal of Comparative Neurology* **230**:311–336.

Cusick CG and Kaas JH (1986) Interhemispheric connections of cortical sensory and motor representations in primates. In: Leporé F, Ptito M, and Jasper HH (eds). *Two Hemispheres – One Brain: Functions of the Corpus Callosum.* Alan R. Liss, New York, pp. 83–102.

Cusick CG, MacAvoy MG, and Kaas JH (1985) Interhemispheric connections of cortical sensory areas in tree shrews. *Journal of Comparative Neurology* **235**:111–128.

Dehay C, Kennedy H, and Bullier J (1986) Callosal connectivity of areas V1 and V2 in the newborn monkey. *Journal of Comparative Neurology* **254**:20–33.

de la Mothe L, Blumell S, Kajikawa Y, and Hackett TA (2006a) Cortical connections of auditory cortex in marmoset monkeys: core and medial belt regions. *Journal of Comparative Neurology* **496**:27–71.

Diamond IT, Jones EG, and Powell TPS (1968) Interhemispheric fiber connections of the auditory cortex of the cat. *Brain Research* **11**:177–193.

Dingle RN, Hall SE, and Phillips DP (2010) A midline azimuthal channel in human spatial hearing. *Hearing Research* **268**:67–74.

Ebner FF and Myers RE (1965) Distribution of corpus callosum and anterior commissure in cat and raccoon. *Journal of Comparative Neurology* **124**:353–365.

Eggermont JJ and Mossop JE (1998) Azimuth coding in primary auditory cortex of the cat. I. Spike synchrony versus spike count representations. *Journal of Neurophysiology* **80**: 2133–2150.

Eimer M, Fortser B, and Vibell J (2005) Cutaneous saltation within and across arms: A new measure of the saltation illusion in somatosensation. *Perception and Psychophysics* **67**:458–468.

Endrass T, Mohr B, and Rockstroh B (2002) Reduced interhemispheric transmission in schizophrenia patients: evidence from event-related potentials. *Neuroscience Letters* **320**:57–60.

Fabri M and Manzoni T (2004) Glutamic acid decarboxylase immunoreactivity in callosal projecting neurons of cat and rat somatic sensory areas. *Neuroscience* **123**:557–566.

Feng JZ and Brugge JF (1983) Postnatal development of auditory callosal connections in the kitten. *Journal of Comparative Neurology* **214**:416–426.

Fitzpatrick KA and Imig TJ (1980) Auditory cortico-cortical connections in the owl monkey. *Journal of Comparative Neurology* **192**:589–610.

FitzPatrick KA and Imig TJ (1982) Organization of auditory connections. The primate auditory cortex. In: Woolsey CN (ed) *Cortical Sensory Organization*, Volume 3, *Multiple Auditory Areas*. Humana Press, Clifton, NJ, pp 71–109.

Forbes BF and Moskowitz N (1977) Cortico-cortical connections of the superior temporal gyrus in the squirrel monkey. *Brain Research* **136**:547–552.

Geldard FA (1976) The auditory saltation effect in vision. *Sensory Processes* **1**:77–86.

Geldard FA and Sherrick CE (1986) Space, time and touch. *Scientific American* **255**:90–95.

Guillemot J-P, Richer L, Ptito M, Guilbert M, and Leporé F (1992) Somatosensory receptive field properties of corpus callosum fibres in the raccoon. *Journal of Comparative Neurology* **321**: 124–132.

Hackett TA, Stepniewska I, and Kaas JH (1999) Callosal connections of the parabelt auditory cortex in macaque monkeys. *European Journal of Neuroscience* **11**:856–866.

Hashikawa T, Molinari M, Rausell E, and Jones EG (1995) Patchy and laminar terminations of medial geniculate axons in monkey auditory cortex. *Journal of Comparative Neurology* **362**:195–208.

Heffner HE (1997) The role of macaque auditory cortex in sound localization. *Acta Otolaryngology (Stockholm) Supplement* **532**: 22–27.

Heilman KM and Valenstein E (1972) Auditory neglect in man. *Archives of Neurology* **26**:32–35.

Highley JR, Esiri MM, McDonald B, Roberts HC, Wakler MA, and Crow TJ (1999) The size and fiber composition of the anterior commissure with respect to gender and schizophrenia. *Biological Psychiatry* **45**:1120–1127.

Houzel JC, Milleret C, and Innocenti G (1994) Morphology of callosal axons interconnecting areas 17 and 18 of the cat. *European Journal of Neuroscience* **6**:898–917.

Hubel DH and Wiesel TN (1965) Receptive fields and functional architecture in two nonstriate visual areas (18 and 19) of the cat. *Journal of Neurophysiology* **28**:229–289.

Hubel DH and Wiesel TN (1967) Cortical and callosal connections concerned with the vertical meridian of visual fields in the cat. *Journal of Neurophysiology* **30**:1561–1573.

Hughes CM and Peters A (1990) Morphological evidence for callosally projecting nonpyramidal neurons in rat visual cortex. *Anatomy and Embryology (Berlin)* **182**:591–603.

Hynd GW, Hall J, Novey ES, Eliopulos D, Black K, Gonzales JJ, Edmonds JE, Riccio C, Cohen M (1995) Dyslexia and corpus callosum morphology. *Archives of Neurology* **52**:32–38.

Imig TJ and Adrián HO (1977) Binaural columns in the primary field (A1) of cat auditory cortex. *Brain Research* **138**:241–257.

Imig TJ and Brugge JF (1978) Sources and terminations of callosal axons related to binaural and frequency maps in primary auditory cortex of the cat. *Journal of Comparative Neurology* **182**:637–660.

Imig TJ, Irons WA, and Samson FR (1990) Single-unit selectivity to azimuthal direction and sound pressure level of noise bursts in cat high-frequency primary auditory cortex. *Journal of Neurophysiology* **63**:1448–1466.

Imig TJ, Morel A, and Kauer CD (1982) Covariation of distributions of callosal cell bodies and callosal axon terminals in layer III of cat primary auditory cortex. *Brain Research* **251**:157–159.

Imig TJ and Reale RA (1980) Patterns of cortico-cortical connections related to tonotopic maps in cat auditory cortex. *Journal of Comparative Neurology* **192**:293–332.

Imig TJ and Reale RA 1981 Ipsilateral corticocortical projections related to binaural columns in cat primary auditory cortex. *Journal of Comparative Neurology* **203**:1–14.

Imig TJ, Reale RA, and Brugge JF (1977) The origin and targets of primary field efferents related to tonotopic and binaural maps of cat auditory cortex. *Proceedings of the International Union of Physiology* **12**:706.

Innocenti GM (1986) General organization of commissural connections in the cerebral cortex. In: Jones EG and Peters A (eds). *Cerebral Cortex*, volume 5, *Sensory-Motor Areas and Aspects of Cortical Connectivity*. Plenum Press, New York, pp. 291–353.

Innocenti GM, Aggoun-Zouaoui D, and Lehmann P (1995) Cellular aspects of callosal connections and their development. *Neuropsychologia* **33**:961–987.

Inzitari D (2000) Age-related white matter changes and cognitive impairment. *Annals of Neurology* **47**:141–143.

Iwamura Y, Iriki A, and Tanaka M (1994) Bilateral hand representation in the postcentral somatosensory cortex. *Nature* **369**:554–556.

Jacobson S and Trojanowski JQ (1974) The cells of origin of the corpus callosum in rat, cat and rhesus monkey. *Brain Research* **74**: 149–155.

Janowsky JS, Kaye JA, and Carper RA (1996) Atrophy of the corpus callosum in Alzheimer's Disease versus healthy aging. *Journal of the American Geriatric Society* **44**:798–803.

Jenkins WM and Masterton RB (1982) Sound localization: effects of unilateral lesions in the central auditory system. *Journal of Neurophysiology* **47**:987–1016.

Jenny AB (1979) Commissural projections of the cortical hand motor area in monkeys. *Journal of Comparative Neurology* **188**:137–145.

Jerger J, Martin J, and McColl R (2004) Interaural cross correlations of event-related potentials and diffusion tensor imaging in the evaluation of auditory processing disorder: a case study. *Journal of the American Academy of Audiology* **15**:79–87.

Jerger J, Thibodeau L, Martin J, Mehta J, Tillman G, Greenwald R, Britt L, Scott J, and Overson G (2002) Behavioral and electrophysiological evidence of auditory processing disorder: a twin study. *Journal of the American Academy of Audiology* **13**: 438–460.

Jones EG and Burton H (1976) Areal differences in the laminar distribution of thalamic afferents in cortical fields of the insular, parietal and temporal regions of primates. *Journal of Comparative Neurology* **168**:197–248.

Jones EG, Burton H, and Porter R (1975) Commissural and cortico-cortical "columns" in the somatic sensory cortex of primates. *Science* **190**:572–574.

Jones EG, Coulter JD, and Hendry SHC (1978) Intracortical connectivity of architectonic fields in the somatic sensory, motor and parietal cortex of monkeys. *Journal of Comparative Neurology* **181**:291–347.

Jones EG, Coulter JD, and Wise SP (1979) Commissural columns in the sensory-motor cortex of monkeys. *Journal of Comparative Neurology* **188**:113–135.

Kaas JH (1995) The organization of callosal connections in primates. In: Reeves AG and Roberts DW (eds) *Epilepsy and the Corpus Callosum II*. New York: Plenum Press. pp 15–27.

Kavanagh GL and Kelly JB (1987) Contribution of auditory cortex to sound localization by the ferret (*Mustela putorius*). *Journal of Neurophysiology* **57**:1746–1766.

Kelly JB and Judge PW (1994) Binaural organization of primary auditory cortex in the ferret (*Mustela putorius*). *Journal of Neurophysiology* **71**:904–913.

Kelly JP and Wong D (1981) Laminar connections of the cat's auditory cortex. *Brain Research* **212**:1–15.

Kennedy H and Dehay C (1988) Functional implications of the anatomical organization of the callosal projections of visual areas VI and V2 in the macaque monkey. *Behavioral Brain Research* **29**:225–236.

Killackey HP, Gould HJ 3rd, Cusick CG, Pons TP, and Kaas JH (1983) The relation of corpus callosum connections to architectonic fields and body surface maps in sensorimotor cortex of new and old world monkeys. *Journal of Comparative Neurology* **219**: 384–419.

Kitzes LM and Doherty D (1994) Influence of callosal activity on units in the auditory cortex of ferret (*Mustela putorius*). *Journal of Neurophysiology* **71**:1740–1751.

Kitzes LM, Wrege KS, and Cassady JM (1980) Patterns of responses of cortical cells to binaural stimulation. *Journal of Comparative Neurology* **192**:455–472.

Künzle H (1976) Alternating afferent zones of high and low axon terminal density within the macaque motor cortex. *Brain Research* **106**:365–370.

Lee CC and Winer JA (2008) Connections of cat auditory cortex: II. Commissural system. *Journal of Comparative Neurology* **507**: 1901–1919.

Leporé F and Guillemot J-P (1982) Visual receptive field properties of cells innervated through the corpus callosum in the cat. *Experimental Brain Research* **46**:413–424.

Leporé F, Ptito M, and Guillemot J-P (1986) The role of the corpus callosum in midline fusion. In: Leporé F, Ptito M, and Jasper HH (eds). *Two Hemispheres-One Brain: Functions of the Corpus Callosum*. Alan R Liss, New York, pp. 211–229.

Lessard N, Leporé F, Villemagne J, and Lassonde M (2002) Sound localization in callosal agenesis and early callosotomy subjects: brain reorganization and/or compensatory strategies. *Brain* **125**:1039–1053.

Leporé F, Samson A, Paradis MC, Ptito M, and Guillemot J-P (1992) Binocular interaction and disparity coding at the 17–18 border: contribution of the corpus callosum. *Experimental Brain Research* **90**:129–140.

Lewis JW and Olavarria JF (1995) Two rules for callosal connectivity in striate cortex of the rat. *Journal of Comparative Neurology* **361**: 119–137.

Luethke LE, Krubitzer LA, and Kaas JH (1988) Cortical connections of electrophysiologically and architectonically defined subdivisions of auditory cortex in squirrels. *Journal of Comparative Neurology* **268**:181–203.

Luethke LE, Krubitzer LA, and Kaas JH (1989) Connections of primary auditory cortex in the New World monkey, *Saguinus*. *Journal of Comparative Neurology* **285**:487–513.

Liu W and Suga N (1997) Binaural and commissural organization of the primary auditory cortex of the mustached bat. *Journal of Comparative Physiology A* **181**:599–605.

Lund JS, Lund RD, Hendrickson AE, Bunt AH, and Fuchs AF (1975) The origin of efferent pathways from the primary visual cortex, area 17, of the macaque monkey as shown by retrograde transport of horseradish peroxidase. *Journal of Comparative Neurology* **164**:287–303.

Manzoni T (1997) The callosal connections of the hierarchically organized somatosensory areas of primates. *Journal of Neurosurgical Sciences* **41**:1–22.

Manzoni T, Barbaresi P, Conti F, and Fabri M (1989) The callosal connections of the primary somatosensory cortex and the neural bases of midline fusion. *Experimental Brain Research* **76**: 251–266.

Matsubara JA and Phillips DP (1988) Intracortical connections and their physiological correlates in the primary auditory cortex (AI) of the cat. *Journal of Comparative Neurology* **268**:38–48.

McMullen NT and de Venecia RK (1993) Thalamocortical patches in auditory neocortex. *Brain Research* **620**:317–322.

Middlebrooks JC, Dykes RW, and Merzenich MM (1980) Binaural response-specific bands in primary auditory cortex (AI) of the cat: topographical organization orthogonal to isofrequency contours. *Brain Research* **181**:31–48.

Middlebrooks JC and Pettigrew JD (1981) Functional classes of neurons in primary auditory cortex of the cat distinguished by sensitivity to sound location. *Journal of Neuroscience* **1**:107–120.

Middlebrooks JC, Xu L, Eddins AC, and Green DM (1998) Codes for sound source location in nontonotopic auditory cortex. *Journal of Neurophysiology* **80**:863–881.

Middlebrooks JC and Zook JM (1983) Intrinsic organization of the cat's medial geniculate body identified by projections to binaural response-specific bands in the primary auditory cortex. *Journal of Neuroscience* **3**:203–224.

Mitchell TN, Free SL, Merschhemke M, Lemieux L, Sisodiya SM, and Shorvon SD (2003) Reliable callosal measurement: population normative data confirm sex-related differences. *American Journal of Neuroradiology* **24**:410–418.

Morel A, Garraghty PE, and Kaas JH (1993) Tonotopic organization, architectonic fields, and connections of auditory cortex in macaque monkeys. *Journal of Comparative Neurology* **335**:437–459.

Moore CN, Casseday JH, and Neff WD (1974) Sound localization: the role of the commissural pathways of the auditory system of the cat. *Brain Research* **82**:13–26.

Morel A and Kaas JH (1992) Subdivisions and connections of auditory cortex in owl monkeys. *Journal of Comparative Neurology* **318**: 27–63.

Newsome WT and Allman JM (1980) Interhemispheric connections of visual cortex in the owl monkey, *Aotus trivirgatus*, and the bushbaby, *Galago senegalensis*. *Journal of Comparative Neurology* **194**: 209–233.

Oka S, Miyamoto O, Janjua NA, Honjo-Fujiwara N, Ohkawa M, Nagao S, Kondo H, Minami T, Toyoshima T, and Itano T (1999) Re-evaluation of sexual dimorphism in human corpus callosum. *NeuroReport* **10**:937–940.

Olavarria JF (1996) Non-mirror-symmetric patterns of callosal linkages in areas 17 and 18 in cat visual cortex. *Journal of Comparative Neurology* **366**:643–655.

Olavarria JF and Abel PL (1996) The distribution of CC correlates with the pattern of cytochrome oxidase stripes in visual area V2 of macaque monkeys. *Cerebral Cortex* **6**:631–639.

Orman SS and Phillips DP (1984) Binaural interactions of single neurons in posterior field of cat auditory cortex. *Journal of Neurophysiology* **51**:1028–1039.

Pallas SL, Littman T, and Moore DRF (1999) Cross-modal reorganization of callosal connectivity without altering thalamocortical projections. *Proceedings of the National Academy of Sciences of the United States of America* **96**:8751–8756.

Pandya DN, Hallett M, and Mukherjee SK (1969) Intra- and interhemispheric connections of the neocortical auditory system in the rhesus monkey. *Brain Research* **14**:49–65.

Pandya DN and Rosene DL (1993) Laminar termination patterns of thalamic, callosal, and association afferents in the primary auditory area of the rhesus monkey. *Experimental Neurology* **119**:220–234.

Pandya DN and Sanides (1973) Architectonic parcellation of the temporal operculum in rhesus monkey and its projection pattern. *Zeitschrift für Anatomie und Entwicklungsgeschichte* **139**: 127–161.

Pandya DN and Vignolo LA (1971) Intra- and interhemispheric projections of the precentral, premotor and arcuate areas in the rhesus monkey. *Brain Research* **26**:217–233.

Peters A and Sethares C (2002) Aging and the myelinated fibers in prefrontal cortex and corpus callosum of the monkey. *Journal of Comparative Neurology* **442**:277–291.

Phillips DP (2001) Introduction to the central auditory nervous system. In: Jahn AF and Santos-Sacchi JR (eds). *Physiology of the Ear*. Singular, San Diego, pp. 613–638.

Phillips DP and Brugge JF (1985) Progress in neurophysiology of sound localization. *Annual Review of Psychology* **36**:245–274.

Phillips DP and Hall SE (2001) Spatial and temporal factors in auditory saltation. *Journal of the Acoustical Society of America* **110**: 1539–1547.

Phillips DP and Hall SE (2005) Psychophysical evidence for adaptation of central auditory processors for interaural differences in time and level. *Hearing Research* **202**:188–199.

Phillips DP, Hall SE, Boehnke SE, and Rutherford LED (2002) Spatial stimulus cue information supplying auditory saltation. *Perception* **31**: 875–885.

Phillips DP and Irvine DRF (1981) Responses of single neurons in physiologically defined area AI of cat cerebral cortex: sensitivity to interaural intensity differences. *Hearing Research* **4**:299–307.

Phillips DP and Irvine DRF (1982) Properties of single neurons in the anterior auditory field (AAF) of cat cerebral cortex. *Brain Research* **248**:237–244.

Phillips DP and Irvine DRF (1983) Some features of binaural input to single neurons in physiologically defined area AI of cat cerebral cortex. *Journal of Neurophysiology* **49**:383–395.

Phillips DP, Calford MB, Pettigrew JD, Aitkin LM, and Semple MN (1982) Directionality of sound pressure transformation at the cat's pinna. *Hearing Research* **8**:13–28.

Phillips DP, Vigneault-MacLean BK, Boehnke SE, and Hall SE (2003) Acoustic hemifields in the spatial release from masking of speech by noise. *Journal of the American Academy of Audiology* **14**:518–524.

Poirier P, Lassonde M, Villemure J-G, Geoffroy G, and Leporé F (1994) Sound localization in hemispherectomized patients. *Neuropsychologia* **32**:541–553.

Poirier P, Leporé F, Provencal C, Ptito M, and Guillemot J-P (1995) Binaural noise stimulation of auditory callosal fibers of the cat: responses to interaural time delays. *Experimental Brain Research* **104**:30–40.

Qi HX, Lyon DC, and Kaas JH (2002) Cortical and thalamic connections of the parietal ventral somatosensory area in marmoset monkeys (*Callithrix jacchus*). *Journal of Comparative Neurology* **443**:168–182.

Rajan R, Aitkin LM, Irvine DRF, and McKay J (1990) Azimuthal sensitivity of neurons in primary auditory cortex of cats. I. Types of sensitivity and effects of variations in stimulus parameters. *Journal of Neurophysiology* **64**:872–887.

Recanzone GH, Guard DC, Phan ML, and Su T-IK (2000) Correlation between the activity of single auditory cortical neurons and sound-localization behavior in the macaque monkey. *Journal of Neurophysiology* **83**:2723–2739.

Richter K, Hess A, and Scheich H (1999) Functional mapping of transsynaptic effects of local manipulation of inhibition in gerbil auditory cortex. *Brain Research* **831**:184–199.

Roe AW, Pallas SL, Hahm JO, and Sur M (1990). A map of visual space induced in primary auditory cortex. *Science* **250**:818–820.

Roe AW, Pallas SL, Kwon YH, and Sur M (1992) Visual projections routed to the auditory pathway in ferrets: receptive fields of visual neurons in primary auditory cortex. *Journal of Neuroscience* **12**:3651–3664.

Rouiller EM, Simm GM, Villa AE, de Ribaupierre Y, and de Ribaupierre F (1991) Auditory corticocortical interconnections in the cat: evidence for parallel and hierarchical arrangement of the auditory cortical areas. *Experimental Brain Research* **86**:483–505.

Sanchez-Longo LP and Forster FM (1958) Clinical significance of impairment of sound localization. *Neurology* **8**:119–125.

Sanides D (1978) The retinotopic distribution of visual callosal projections in the suprasylvian visual areas compared to the classical visual areas (17, 18, 19) in the cat. *Experimental Brain Research* **33**:435–443.

Santhouse AM, Ffytche DH, Howard RJ, Williams SCR, Rifkin L, and Murray RM (2002) Functional imaging of the mechanisms underlying the bilateral field advantage. *NeuroImage* **17**:680–687.

Shanks MF, Rockel AJ, and Powell TPS (1975) The commissural fibre connections of the primary somatic sensory cortex. *Brain Research* **98**:166–171.

Stecker GC, Harrington IA, and Middlebrooks JC (2005) Location coding by opponent neural populations in the auditory cortex. *Public Library of Science Biology* **3**:520–528 (e78).

Sur M, Garraghty PE, and Roe AW (1988) Experimentally induced visual projections into auditory thalamus and cortex. *Science* **242**:1437–1441.

Swadlow HA, Weyand TG, and Waxman SG (1978) The cells of origin of the corpus callosum in rabbit visual cortex. *Brain Research* **156**:129–134.

Thompson GC and Cortez AM (1983) The inability of squirrel monkeys to localize sound after unilateral ablation of auditory cortex. *Behavioral Brain Research* **8**:211–216.

Thompson PM, Moussai J, Zohoori S, Goldkorn A, Khan AA, Mega MS, Small GW, Cummings JL, and Toga AW (1998) Cortical variability and asymmetry in normal aging and Alzheimer's disease. *Cerebral Cortex* **8**:492–509.

Tigges J, Tigges M, Anschel S, Cross NA, Letbetter WD, and McBride RL (1981) Areal and laminar distribution of neurons interconnecting the central visual cortical areas 17, 18, 19, and MT in squirrel monkey (*Saimiri*). *Journal of Comparative Neurology* **202**: 539–560.

Vaadia E, Benson DA, Hienz RD, and Goldstein MH Jr (1986) Unit study of monkey frontal cortex: active localization of auditory and of visual stimuli. *Journal of Neurophysiology* **56**: 934–952.

Van Essen DC, Newsome WT, and Bixby JL (1982) The pattern of interhemispheric connections and its relationship to extrastriate visual areas in the macaque monkey. *Journal of Neuroscience* **2**: 265–283.

Vaughan DW and Peters A (1985) Proliferation of thalamic afferents in cerebral cortex altered by callosal deafferentation. *Journal of Neurocytology* **14**:705–716.

Voigt T, LeVay S, and Stamnes MA (1988) Morphological and immunocytochemical observations on the visual callosal projections in the cat. *Journal of Comparative Neurology* **272**: 450–460.

Wallace MN and Harper MS (1997) Callosal connections of the ferret primary auditory cortex. *Experimental Brain Research* **116**: 367–374.

White EL and Czeiger D (1991) Synapses made by axons of callosal projection neurons in mouse somatosensory cortex: emphasis on intrinsic connections. *Journal of Comparative Neurology* **303**: 233–244.

Wong D and Kelly JP (1981) Differentially projecting cells in individual layers of the auditory cortex: a double-labeling study. *Brain Research* **230**:362–366.

Zeki SM (1970) Interhemispheric connections of prestriate cortex in monkey. *Brain Research* **19**:63–75.

Chapter 6

Intrinsic Connections of the Auditory Cortex

Mark N. Wallace and Jufang He

Abbreviations

AC	auditory cortex
AI	primary auditory cortex
AII	second auditory cortical area
BDNF	brain-derived neurotrophic factor
BP	bipolar cell
BTC	bitufted cell
CB	calbindin
CB1	cannabinoid receptor Type 1
CHC	chandelier cell
DBC	double bouquet cell
EVMC	extraverted multipolar cell
FGF	fibroblast growth factor
FS	fast-spiking
GABA	γ-aminobutyric acid
HBTC	horizontal bitufted cell
HC	horizontal cell
IB	intrinsic bursting
LBC	large basket cell
LMC	large multipolar cell
LS	late-spiking
LTS	low-threshold spiking
MAC	Martinotti cell
MBC	medium basket cell
MGB	medial geniculate body
MMC	medium multipolar cell
NBC	nest basket cells
NG	neuroglia
NGC	neurogliaform cell
NMDA	N-methyl-D-aspartate
NWAA	neuron with axonal arcade
PC	pyramidal cell
PV	parvalbumin
RS	regular spiking
SBC	small basket cell
SMC	small multipolar cell
SST	somatostatin
STG	superior temporal gyrus

M.N. Wallace (✉)
Medical Research Council Institute of Hearing Research, University Park, Nottingham NG7 2RD, UK
e-mail: markw@ihr.mrc.ac.uk

1 Introduction

The patterns of connectivity between cortical neurons define and constrain the basic functional organization of the neocortex. Understanding the local and long distance connections is a prerequisite for developing an integrative description of cortical structure and function. The circuits to which such cells contribute likely control perception and action.

The traditional wiring diagram for sensory neocortex, including the primary auditory area (AI), involves specific thalamic afferents terminating on spiny stellate cells in layer IV, which then project onto cells in layer III (Mitani et al. 1985). These cells in turn contribute to a series of interconnections which integrate neural activity within a vertical module. If this module spans all six layers, and if the cells within it have a narrow range of latency differences in their excitatory responses and show prominent lateral inhibition, then the module is defined as a column (Mountcastle 1997; Jones 2000). Columns, dependent on thalamic afferents, may be present in the rabbit auditory cortex (McMullen and de Venecia 1993) and perhaps in other species. However, auditory cortex columns are less pronounced and regular than those in the visual or somatic sensory cortex (Linden and Schreiner 2003).

An early cortical wiring diagram (Mitani et al. 1985) must incorporate more recent findings which emphasize the role of clusters of superficial pyramidal cells (Douglas and Martin 2004), which receive direct/indirect thalamic input, and input from other cortical areas. Any realistic circuit diagram must consider the many different cell types—there are at least 15 types of cell in layer VI alone (Prieto and Winer 1999)—and it must include new data on their connections. The total number of distinct neuronal types, even in AI, is unknown. Although there may be 20 basic cortical neuron types in an

J.A. Winer, C.E. Schreiner (eds.), *The Auditory Cortex*,
DOI 10.1007/978-1-4419-0074-6_6,

area (Binzegger et al. 2004), there may be even more when variation in their laminar origin and specific synaptic target are included as criteria (Kozloski et al. 2001). There may also be species differences in the evolution of new neuronal types (DeFelipe et al. 2002b), a question awaiting further study.

A diagram summarizing all the intrinsic connections within the auditory cortex does not exist. It may be more fruitful to think of auditory cortex as functioning like an "enchanted loom" (Sherrington 1933), where patterns of connections are continually being formed and then dissolving in a three-dimensional matrix. At one moment a particular circuit may dominate the output of a group of related pyramidal cells, whereas at the next instant a different circuit emerges. Although the dendritic structure of auditory cortex cells has been comprehensively described (Prieto et al. 1994a; DeFelipe 2002a), the intrinsic axonal projection for each cell type is still incomplete and the specificity of synaptic connections is only beginning to emerge (Watts and Thomson 2005). The prospects for identifying the patterns woven on the cortical loom are indeed challenging (Margrie et al. 2003).

A problem in working on auditory cortex is that species from at least six phyla are used: primates, carnivores, rodents, lagomorphs, odontocetes, and chiroptera, as well as avians. All are valuable models, but even identifying AI can be challenging (Ehret 1997) and the differences between the auditory cortex regions of different species are unclear. Until homologies are established, we cannot assess critically the parallels between auditory areas other than for AI. This analysis of intrinsic connections is limited mainly to cat AI and concentrates on work after 1990 as earlier studies have been summarized (Winer 1992). The four sections below outline the main types of auditory cortex cells and their connections.

2 Local Connections of Excitatory Interneurons

Some proportion of the vertical integration within a column is mediated by excitatory interneurons. These generally have spiny dendrites and axons that ramify locally and may enter the white matter for a short distance. These cells likely form less than 5% of cortical neurons. In visual and somatic sensory areas they largely comprise the layer IV spiny stellate (granule) cells. However, it has been suggested for the auditory cortex, that non-pyramidal neurons containing excitatory neurotransmitter occur in every layer and they seem to be more evenly distributed than in other sensory areas (Winer 1992).

The human (Meyer et al. 1989), bat (Fitzpatrick and Henson 1994), and cat (Mitani et al. 1985; Smith and Populin 2001) auditory cortex have comparatively few spiny stellate cells. The function of these cells may have been subsumed by many small pyramidal cells at the base of layer III and in layer IV or by other non-pyramidal cells. The layer IV spiny interneurons are diverse, both in dendritic morphology and axonal targets. Most axonal branches appear to terminate in, or near, layer IV, with input to layers II and III and some deeper branches. The layer VI spiny interneurons (Mitani et al. 1985) include bipolar, inverted pyramidal, and spiny stellate cells. All such spinous interneurons may be excitatory, though the intracellular data documenting this assertion is sparse. The axons of most of these interneurons terminate locally, within their layer of origin, and do not appear to contribute to vertical integration. Their function may be to selectively amplify or suppress certain inputs.

3 Local Connections of Inhibitory Interneurons

Virtually all inhibitory neurons in the mammalian neocortex contain gamma aminobutyric acid (GABA). These GABAergic cells compose between 12 and 25% of auditory cortex neurons (Hendry and Jones 1991; Prieto et al. 1994a) and they are concentrated in the supragranular layers. Most auditory cortex inhibition has an intrinsic origin as few afferents provide an inhibitory input (Winer 1992). Much of the rapid auditory cortex inhibition reflects the $GABA_A$ receptor (linked to chloride channels), while slower-acting metabotropic $GABA_B$ receptors are linked to potassium channels (Eder et al. 2001).

GABAergic neurons are present in all auditory cortex layers, including the white matter, and have a variety of morphologies. They have smooth, or sparsely spinous, dendrites and have a wide range in the size and form of their axonal tree. Based on their laminar position, somatic size and dendritic morphology, they form more than 30 different subtypes in cat AI (Fig. 6.1) (Prieto et al. 1994a), with different proportions in each layer. Layer I has the highest proportion (94%), layer VI the lowest (16%). Similar cell types have been described in other species and other cortical areas (Prieto et al. 1994a). Auditory cortex cells are classified by four criteria: (1) dendritic morphology, (2) axonal morphology, (3) presence of chemical markers, and (4) discharge characteristics after intracellular current injection. Ideally, a comprehensive description of an inhibitory interneuron should include details from all four (and even more) criteria (Markram et al. 2004), but large gaps in knowledge impede this process.

Virtually all inhibitory interneurons have axonal branches which terminate close to their soma; the axon is often restricted locally to an area matching the dendritic arbor. Other inhibitory interneurons, such as the large basket cells (LBC), have longer-range axons with a more lateral distribution and these may be involved in lateral inhibition. A third axonal group has a more vertical arrangement which crosses

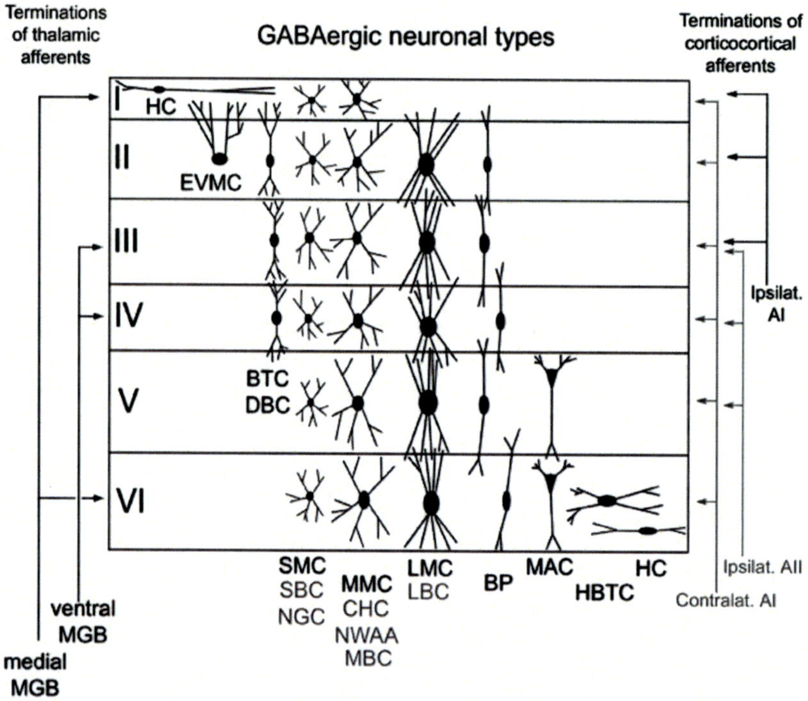

Fig. 6.1 Schematic summary of GABAergic neuronal populations in cat AI (Prieto et al. 1994a). The six layers of the cortex are indicated by the Roman numerals. The laminar distribution of nine different types of cell is shown. This classification is based on dendritic and axonal morphology and the types are abbreviated in *black* type. Some of these types can be subdivided on the basis of their axonal morphology and the names of these are abbreviated in *gray* type. Certain types are only found in one layer (e.g., EVMC), while others are present in all layers (e.g., SMC). The laminar input for various afferents is indicated by *arrows* showing afferents from the medial geniculate body (MGB), ipsilateral AI, the ipsilateral second auditory area (AII), and the contralateral AI. Their distribution indicates that cells of one type may be receiving afferent inputs from different extrinsic sources depending upon their laminar position. Key (*left* to *right*): HC, horizontal cell; EVMC, extraverted multipolar cell; BTC, bitufted cell; DBC, double bouquet cell; SMC, small multipolar cell; SBC, small basket cell; NGC neurogliaform cell; MMC, medium multipolar cell; CHC, chandelier cell; NWAA, neuron with axonal arcade; MBC, medium basket cell; LMC, large multipolar cell; LBC, large basket cell; BP, bipolar cell; MAC, Martinotti cell; HBTC, horizontal bitufted cell

multiple layers (Fig. 6.2), including Martinotti cells (MAC), neurons with axonal arcades (NWAA), double bouquet cells (DBC), and bipolar cells (BP). This vertical arrangement suggests a role in the (currently unknown) transformations associated with columnar microcircuits (DeFelipe et al. 2002b), whose diverse functions have been described (Markram et al. 2004).

3.1 Interneurons Synapsing Near the Soma

It is estimated (Peters 2002) that a representative pyramidal cell receives input from about 75 inhibitory interneurons, and excitatory input from up to 1,000 neurons. Significant synaptic input, much of it inhibitory in origin, targets the soma, where it can maximally affect initiation of the action potential at the axon hillock. Indeed, one type of interneuron, the chandelier cell (Fig. 6.2: CHC) specifically targets the axon hillock and forms strings of synapses almost exclusively along the axon initial segment of pyramidal cells in the rat and monkey visual cortex (Peters 2002). It is also present in auditory cortex (Prieto et al. 1994b) though perhaps less common. Its activity could check the firing of nearby pyramidal cells.

Much of the GABAergic input to the pyramidal cell soma arises from basket cells, multipolar neurons found in layers II–V (Prieto et al. 1994b). They form basket-like terminal endings around the somata of target neurons (Figs. 6.1 and 6.2). About 70% of the somatic synapses on pyramidal cells are inhibitory and most arise from basket cells (Peters 2002). In the somatic sensory cortex basket cells constitute ~50% of inhibitory cells and have large, small, and nest varieties (Markram et al. 2004), each with a specific dendritic morphology and different neurochemical markers. Large basket cell (LBC) axons project widely between layers and to local and more distant columns. The small and medium basket cells (MBC) have much smaller axonal trees, and more synaptic swellings and their targets seldom extend beyond

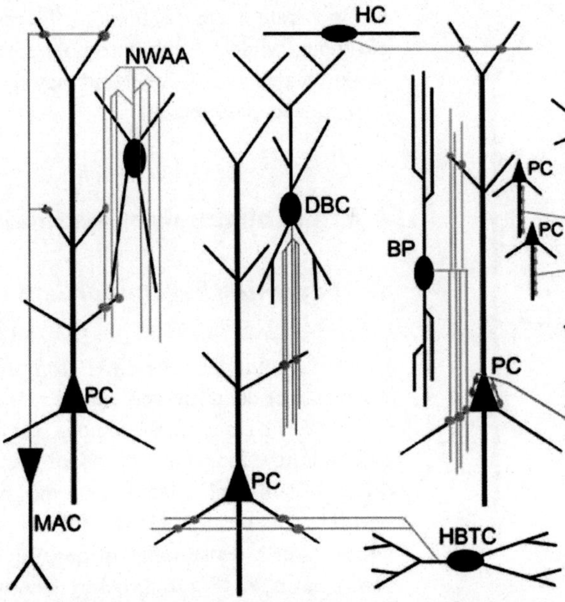

Fig. 6.2 Diagram showing the eight main types of inhibitory interneurons classified according to their axonal morphology. The interneurons have axons (*gray*) and synaptic boutons, while the pyramidal cells (PC) have short sections of axon (*black*). Three of the interneurons have horizontal axons restricted to one lamina: the small or medium basket cell (MBC), the horizontal cell in layer I (HC), and the horizontal bitufted cell (HBTC). Four of the interneurons have axons with a prominent vertical arrangement: Martinotti cell (MAC), bipolar cell (BP), double bouquet cell (DBC), and neuron with axonal arcade (NWAA). The final type (chandelier cell) has axo-axonal endings resembling a chandelier (CHC) and these may be arranged over two or three layers

their layer or column of origin. The nest basket cells (NBC) have not been studied in detail in auditory cortex.

3.2 Interneurons Synapsing on Distal Dendrites

Horizontal cells (HC) in layer I make small synapses on the distal dendrites of the many pyramidal cells whose apical dendrites branch and ramify widely (Fig. 6.2). Martinotti cells (MAC) in layers V and VI also project to layer I, presumably to similar targets (Figs. 6.2 and 6.3). Both HC and MAC appear to avoid layer VI corticothalamic pyramidal cells whose apical dendrites do not enter layer I. Cells in layers I and VI receive non-specific thalamic afferents from the medial division of the medial geniculate body (Fig. 6.1) and they may provide feed-forward inhibition onto pyramidal cell distal dendrites (Prieto et al. 1994b).

Double bouquet cells (DBC) are more common in the primate auditory cortex than other species. Their axons target the pyramidal cell distal dendrites as vertical bundles of descending collaterals resembling a horse tail (Fig. 6.2). Bipolar cells appear to be more common in rodents than primates and resemble the resemble the DBC. Bipolar cells have vertically arranged dendrites and their axon forms a loose plexus of ascending and descending branches. Bitufted cells (BTC) are similar to DBCs but have a more limited axonal interlaminar range and a wider intracolumnar dispersion (Markram et al. 2004).

Neurogliaform cells (NGC) are small multipolar cells, with somata in layers II, III and IV, with 7–9 delicate dendrites that branch profusely (Prieto et al. 1994b), and axons that project locally, but over an area twice that of the dendrites. Other inhibitory neurons have (Fig. 6.1) axonal arcades (NWAA) (DeFelipe 2002a), and the extraverted multipolar cell (EVMC) is unique to layer II (Prieto et al. 1994a). Martinotti cells (MAC) also form synapses on dendrites of layer II–IV cells as do the various basket cells. The variety of inhibitory interneurons likely contribute to different types of inhibition in adjacent layers and even among nearby cells (Foeller et al. 2001).

3.3 Networks of Electrically Coupled Interneurons

Inhibitory interneurons also contact each other and, in some cases, they form synchronized nets which are coupled electrotonically via gap junctions. This discovery extends and refines the classical neuron doctrine, which stipulates that neurons can form chemical synapses or electrical contacts, but not both. At least six sets of network-forming inhibitory cells (Fig. 6.3) are known (Hestrin and Galarreta 2005). These classes have different morphological,

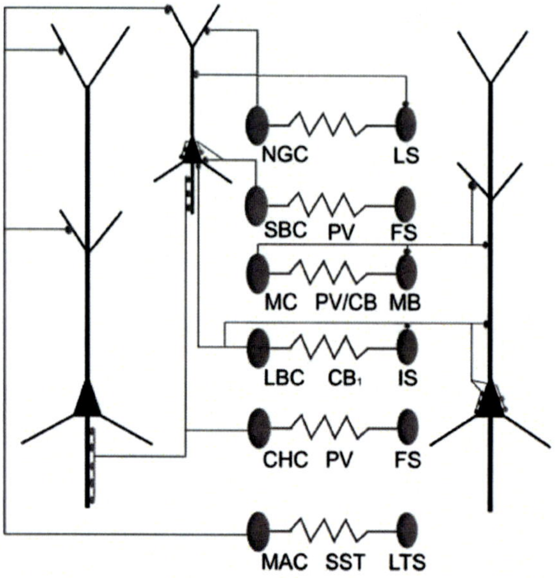

Fig. 6.3 Diagram showing the inhibitory interneurons which may form networks of electrotonically coupled cells. The inhibitory cells (*gray ovals*) are joined by an electrical junction to another cell in their network and with axons (*gray*) forming synaptic swellings on their target cells. The letters beside each interneuron denote morphological type (*left*), chemical identity (*center*), and spike generation type (*right*). The neurogliaform cell (NGC) has a late-spiking (LS) pattern following current injection. The small basket cell (SBC) may contain parvalbumin (PV) and shows fast-spiking (FS) behavior. The multipolar cell (MC) may contain either parvalbumin or calbindin (CB) and has multipolar-bursting (MB) characteristics. The large basket cell (LBC) has cannabinoid type 1 (CB_1) receptors and shows irregular-spiking (IS). The chandelier cell (CHC) may be parvalbumin-positive and is fast-spiking. The Martinotti cell (MAC) may contain somatostatin (SST) and shows low-threshold spiking (LTS)

chemical and functional properties and include: (1) late-spiking (LS) neurogliaform cells (NGC); (2) fast-spiking (FS), basket (SBC) parvalbumin-positive cells (PV); (3) multipolar-bursting cells (MB) whose axon mainly targets dendrites and expresses parvalbumin or calbindin (CB); (4) irregularly spiking (IS) large basket cells (LBC) expressing cannabinoid receptors (CB_1); (5) fast-spiking chandelier cells (CHC) containing parvalbumin; (6) low-threshold spiking (LTS) Martinotti cells (MAC) containing somatostatin (SST). Electrical junctions often link the same types of inhibitory cells and are rare between those of different types. Besides these direct electrical contacts, some inhibitory, fast-spiking cells and multipolar-bursting cells have inhibitory GABAergic synapses. These independent electrotonically coupled networks of inhibitory cells may synchronize the activity of pyramidal cell populations through temporally precise bursts of inhibition. The fast-spiking cells receive strong thalamocortical input and may participate in rapid feed-forward inhibition. The multipolar-bursting neurons can generate theta oscillations in response to cholinergic input, while the low-threshold spiking neurons have a weak thalamocortical input and may function independently (Hestrin and Galarreta 2005).

4 Axon Collaterals of Pyramidal Cells

4.1 Layer II/III Pyramidal Cells

Layer II pyramidal cells differ from those in layer III and have distinct dendritic and axonal morphologies. The layer II pyramids have sparsely spinous dendrites, while layer III pyramids have longer apical dendrites and more spinous dendrites (Mitani et al. 1985). Both types have axon collaterals which branch locally in layers I–IV (Ojima et al. 1991) and which form a dense input in deep layer V and the upper part of layer VI (Fig. 6.4) before entering the white matter. The layer II pyramidal cell axon extends several millimeters in layers III–V with patches of terminal labeling in layers I–III. Layer III pyramids also have long axon branches (Fig. 6.4) which terminate in up to eight vertical patches in a 5 mm-wide expanse (Wallace et al. 1991). These patches concentrate in layers I–III and some extend into layers IV–V.

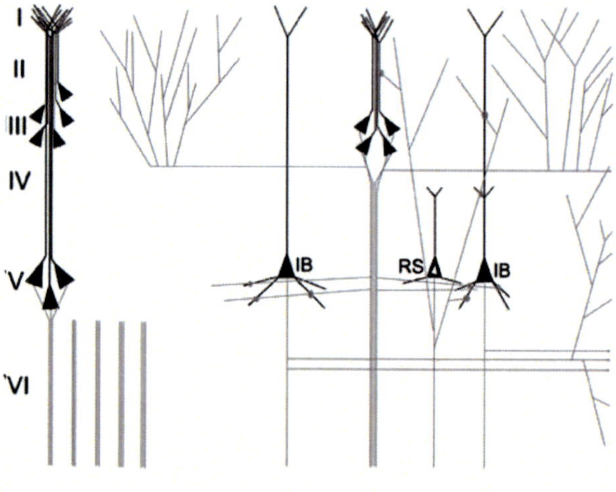

Fig. 6.4 Diagram of some pyramidal cell axonal and dendritic arrangements (*black triangles*). Cortical laminae are denoted by Roman numerals (*left*). Pyramidal cells in layers II–V have large bundles of apical dendrites which terminate with tufts in layer I. Some of these cells also have axons arranged in descending bundles forming a regular pattern. Pyramidal cells in layers II/III have axon collaterals forming periodic patches of dense terminals in the *upper* layers and collaterals in layer V that synapse (*small gray oval*) on the large pyramidal cells with intrinsic bursting (IB) properties, while avoiding the smaller, regular spiking (RS) pyramids in the same layer. The large pyramidal cells have few or no local collateral branches but do form patches of terminal fibers remotely. The regular spiking pyramids have local collaterals ascending towards layer II

4.2 Layer V/VI Pyramidal Cells

Layer V contains six types of pyramidal cells, but only the large pyramidal cell apical dendrites reach layer I (Winer and Prieto 2001). These are the intrinsic bursting cells (Fig. 6.4:IB), which project to the thalamus and various midbrain and brain stem targets (Hefti and Smith 2000). Most common are the medium-sized, regular spiking pyramids whose dendrites reach layer II and whose axon projects to other auditory cortex areas and the caudatoputamen. Small pyramidal cell apical dendrites sometimes do not exit layer V. Star pyramidal cell apical dendrites enter layer IV, fusiform pyramids have a modest apical dendrite, and inverted pyramidal cells have a dendrite which can reach the white matter. Each of these cell types has many dendritic appendages and the cells are thought to be glutamatergic (Winer and Prieto 2001).

The large intrinsic bursting cells are unique to layer V and the deep part of layer IV (Hefti and Smith 2000) and do not form local axonal endings, and their long horizontal collaterals project in many directions and have abundant boutons (Ojima et al. 1992). These projections are asymmetric and can end in all layers (Wallace et al. 1991). By contrast the medium-sized, regular spiking pyramids correspond to cells with a dense network of recurrent collaterals embedded in the dendritic tree and extending into layers II and III, with few if any horizontal collaterals (Ojima et al. 1992).

Eight types of pyramidal cells are found in layer VI of cat AI, none with apical dendrites extending beyond layer III and some of which are confined to layer VI (Prieto and Winer 1999). These include small, medium, large, star, fusiform vertical, fusiform horizontal, tangential, and inverted pyramids. As with layer V pyramids the pattern of axon collaterals has not been studied in detail, but two patterns are noted: (1) cells projecting towards the thalamus and with a dense network of local collaterals overlapping the layers III–VI dendritic field, and (2) cells projecting to other areas or the claustrum and with sparse local collaterals in layers V and VI and relatively long horizontal branches (Ojima et al. 1992).

Different pyramidal cell types can have unique axonal endings and functional properties. Thus, the layer V bursting cells may participate in corticothalamic oscillation/synchronization. Layer V neurons projecting to the non-lemniscal medial geniculate body (MGB) have giant terminals, while layer VI neurons terminating in the ventral MGB and medial MGB have small terminals (Rouiller and de Ribaupierre 1990; Winer et al. 1999). Non-lemniscal MGB neurons have more bursting responses and may oscillate (He and Hu 2002; He 2003), processes implicated in switching from waking to sleep states.

5 Specificity of Interlaminar Connections

Some quantitative models of intrinsic cortical circuitry assume that synapses arising within a layer (other than those of chandelier cells) randomly target all postsynaptic elements (Binzegger et al. 2004). However, the picture emerging is that even pyramidal cell collaterals may show considerable specificity in their postsynaptic targets (Kozloski et al. 2001; Thomson and Morris 2002; White 2002; Watts and Thomson 2005) in non-auditory areas, a principle likely to pertain in auditory cortex.

5.1 Layer IV Connections

The main output of layer IV excitatory cells is to layer III; these axons diverge to a local region at least three times the area of the dendritic tree (Mitani et al. 1985). In other fields, this output is primarily to the basal dendrites of layer III pyramidal cells (Watts and Thomson 2005), but in auditory cortex such cells likely receive monosynaptic thalamic input (Winer 1992; Smith and Populin 2001). Inhibitory layer IV cells also project to layer III but seldom beyond.

5.2 Connections from Cells in Layers I–III

Layer I cells mainly project within layer I among the apical dendritic branches of pyramidal cells, and it is unknown whether they have specific targets. By contrast, layer III pyramidal cells axons specifically target dendrites in layers II/III and in layer V (Fig. 6.4). In layer V in various areas in the rat neocortex the descending axons from layer III pyramids target the large burst-firing layer V pyramids and rarely contact the smaller regular-spiking pyramids (Thomson and Bannister 1998). Thus, the probability of a connection between a layer III pyramid and a large layer V pyramid is higher than that for any other neocortical pyramidal cell circuit. Layer III pyramids also target layer IV inhibitory interneurons of various types, but do not project to the excitatory layer IV cells (Watts and Thomson, 2005). There is little evidence for specific connections of layer II pyramids.

5.3 Connections Produced by Cells in Layers V/VI

The smaller, regular firing layer V pyramidal cells project to layers II and III, unlike the larger, intrinsic bursting cells, which have no local branches (Ojima et al. 1992). Such recurrent layer V axons are highly specific and, in the rat neocortex, almost never contact layer III pyramidal cells; rather, they preferentially contact the distal dendrites of other layer V pyramids or certain interneuronal subclasses (Thomson and Bannister 2003).

AC layer VI corticothalamic cells have robust recurrent branches in layers III–VI (Ojima et al. 1992). In the rat somatic sensory cortex this input to layer IV excitatory cells constitutes a massive 45% of asymmetric synapses (Zhang and Deschênes 1997) and dwarfs the thalamocortical synapses which are only ~6% of the total synapses. Neocortical layer VI corticothalamic axon collaterals target interneurons in layers IV–VI, while collaterals of corticocortical cells mainly innervate other pyramidal cells in layers V and VI (Watts and Thomson 2005).

6 Columnar Arrangement of Intrinsic Connections

6.1 Columnar Arrangement of Somata and Dendrites

A fundamental concept of neocortical architecture is the idea that an area contains many repeating units, each with perhaps a few hundred cells, forming a cylinder across all the cortical depth (Mountcastle 1997; Jones 2000; Buxhoeveden and Casanova 2002). These modules, microcolumns, or minicolumns contain heterogeneous cell types and are thought to be the building blocks for larger macrocolumns, which might reflect the spatial dispersion of a single suite of thalamic afferents (McMullen and de Venecia 1993), and could be conserved across cortical areas (Mountcastle 2003). The microcolumns might arise from clonally related pyramidal cells derived from the same progenitor and which have migrated along radially oriented glia into the cortical plate (Rakic 1995). As the pyramidal cells form radial cylinders, the interneurons migrate tangentially in the marginal zone from different, independent proliferative sources (Ang et al. 2003). In humans they may also migrate in the subventricular zone or may arise locally (Letinic et al. 2002). After reaching their target area, the marginal zone interneurons descend into the underlying cortex beside the pyramidal cells born at similar times. Different morphological subgroups of interneurons may arise in different proliferative zones (Xu et al. 2004), implying that the various cortical areas may have specific proportions or types of interneuron in a layer (Chiry et al. 2003).

In adults, these microcolumns are evident in the primate temporal cortex as long, radial strings of somata (Jones 2000; Hackett et al. 2001). The regular bundles of apical dendrites formed by layer V pyramidal cells may also contribute to them in cat AI (Winer and Prieto 2001), where they are separated by 50–70 μm intervals (Feldman and Peters 1974), and they occur elsewhere in temporal cortex (Viebahn 1990). The layer V dendritic bundles are often joined by the dendrites of layer II and III pyramidal cells and other cell types (Peters and Sethares 1997). In human auditory cortex, some pyramidal cells contain acetylcholinesterase and their apical dendrites form bundles (Fig. 6.5a). The dendritic columns contain close membrane appositions, without evidence of adult gap junctions (Rockland and Ichinohe 2004). About one-third of layer V pyramids are not part of the dendritic bundles; likewise, the apical dendrites of layer VI pyramids lie in the intrabundle space.

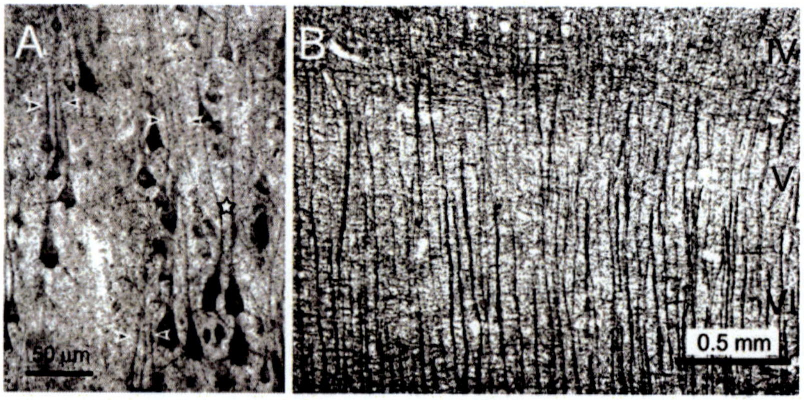

Fig. 6.5 Sections from the human auditory region (lateral posterior area of Heschl's sulcus) showing evidence of apical dendritic bundles and bundles of pyramidal cell output axons. Details of the tissue preparation are available in the original source (Wallace et al., 2002). (**a**) Pyramidal cells stained for acetylcholinesterase. The apical dendrites often form bundles in close proximity (three pairs of *arrowheads*). Adjacent pyramidal cells apical dendrites sometimes diverge from each other (*star*). (**b**) Regular bundles of myelinated fibers. These bundles are most prominent in layers V and VI and have a center-to-center distance of about 45 μm. The bundles are less prominent in AI

6.2 Columnar Arrangement of Axonal Connections

A striking feature in the auditory neocortex is the regular vertical bundles of thick myelinated fibers in various areas, particularly in the primate brain (Hackett et al. 2001). These may represent the efferent axons of pyramidal cells and contribute to microcolumns. They are particularly evident in some human auditory cortex areas (Fig. 6.5b), and less prominent in AI.

The long-range collaterals of pyramidal cells in layers II and III end in regular dense terminal patches, particularly in the upper layers (Fig. 6.4), and these may contribute to macrocolumns 0.5–1 mm in diameter. Layer V pyramidal cells also have remote collaterals that may contribute to a columnar organization (Wallace et al. 1991; Ojima et al. 1992).

Another case of a columnar structure is the vertical bundles of double bouquet cell axons in the primate brain (Fig. 6.2). Their tightly interwoven bundles span the lower half of layer II to the upper half of layer V (DeFelipe et al. 1990; del Rio and DeFelipe 1995) and contain many synaptic swellings that form inhibitory synapses upon the spines and shafts of the basal dendrites or oblique dendritic branches of pyramidal cells, while avoiding the bundles of apical dendrites running parallel to them. These bundles are present in layer III and may form a regular interdigitating pattern with the apical dendritic bundles (Peters and Sethares 1997). Other interneuron types provide inhibition within a vertical cylinder that crosses several layers (Fig. 6.2). Although there is abundant structural evidence for a modular, and even columnar auditory cortex organization, its functional significance is unclear (Horton and Adams 2005).

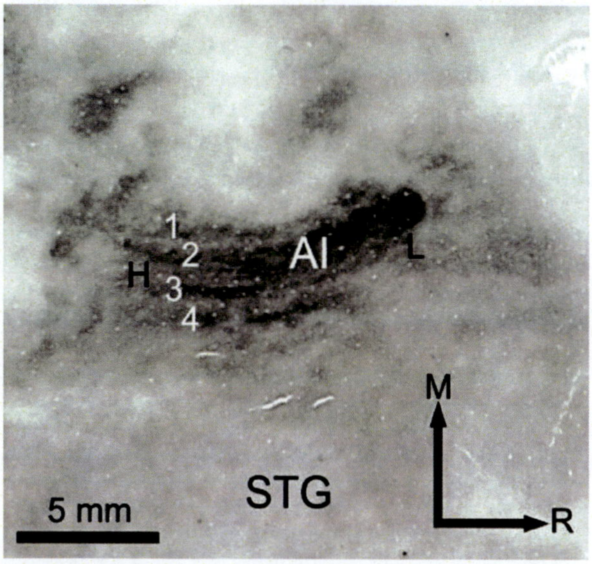

Fig. 6.6 Cytochrome oxidase staining in a section through the *upper* layers of a flatmount of the right neocortex from a macaque monkey (Sincich et al. 2003) and including the ventral bank of the lateral fissure and the superior temporal gyrus (STG). AI has relatively high levels of enzyme activity and best frequency maps correlated with an area of high enzyme activity (Morel et al. 1993), with the frequency gradient running from high (H) to low (L) in a caudal to rostral (R) direction. Four narrow bands of intense enzyme activity are present at the high-frequency end of AI and bands 2 and 3 fuse at the low-frequency end to produce a broader band. Flattening the cortex before sectioning creates complex changes in the orientation of the cortical sheet; the approximate direction of the midline (M) is indicated (*arrow*)

7 Interspecies Comparisons of Connections in AI

7.1 Comparison of Metabolic Activity Bands in Primate Species

When functional modules formed by masses of interlinked neurons have a higher average level of synaptic activity than cells outside the modules, this is reflected in the concentration of enzymes involved in oxidative metabolism such as cytochrome oxidase, whose activity is readily demonstrated in fixed tissue. It has been used to demonstrate a periodic internal structure represented by blobs or bands in five primate visual cortex areas (Horton and Adams 2005). A regular pattern of cytochrome oxidase staining was also observed in macaque monkey AI consisting of 4–5 parallel bands of dense enzyme activity, each about 8 mm long and 0.5 mm wide (Fig. 6.6), and separated by pale strips of lower enzyme activity (Sincich et al. 2003). Similar bands of high and low cytochrome oxidase activity ~0.5 mm wide occur in layers III–IV of human AI (Clarke and Rivier 1998). It was suggested that the bands might correspond to binaural or amplitopic domains because they appeared to be oriented at right angles to the human isofrequency axes. The bands of high cytochrome oxidase activity in the macaque AI also appear to be at right angles to the isofrequency lines from mapping experiments (Morel et al. 1993).

7.2 Comparison of Primate and Non-primate Species

In the monkey auditory cortex, interneurons with smaller somata are numerous in the upper layers, whereas larger interneurons dominate in the lower layers. The opposite is found in the rat where the largest interneurons are in layers II and III (Valcanis and Tan 2003). Double bouquet cells may have an important role in the cat and primate temporal cortex, where their strict geometrical arrangement is striking and, although they have been described in rat frontal cortex (Kawaguchi and Kubota 1997), they appear to be sparser in rodents where the geometrical regularity of axonal arrangement seen in the human is also absent (DeFelipe 2002a). Indeed, it has been suggested that double bouquet cells with horse-tail axons are numerous in primates, rarer in carnivores, and absent in lagomorphs or rodents (Yanez et al. 2005). Rodents have numerous bipolar cells in layers II–VI, whereas bipolar cells are less common in primate cortex and are located mainly in layers II–III (DeFelipe 2002a).

In cat AI, layer III pyramidal cells have extensive collateral branches that form up to eight patches in layers I–V (Ojima et al. 1991; Wallace et al. 1991). Similar patches are also found in humans (Galuske et al. 2000), owl (Morel and Kaas 1992) and macaque monkeys (Morel et al. 1993), and in ferret AI (Wallace and Bajwa 1991; Gao and Pallas 1999), but there is little evidence for multiple patches in other species.

8 Comparison of Connections in Core and Belt Areas

There are few studies of the intrinsic connections of any auditory cortex area other than AI, and studies of temporal cortex have usually not been defined. A detailed study of interneuron types in human AI and the auditory belt areas of the cortex (Chiry et al. 2003) found significant differences between AI and the belt areas in the distribution of three calcium binding proteins, which are markers for different populations of GABAergic interneurons (see Chapter 10).

In the cat there was a higher proportion of GABAergic cells in layers II and III in AII than in AI (Clemo et al. 2003). There were also differences in the proportion of interneurons containing the calcium binding proteins in these two areas. The numbers of inhibitory neurons may not vary greatly in AI and the belt areas, but the differences in protein markers imply a unique distribution of neuronal types and suggest that the belt areas perform different intrinsic operations.

9 Comparison with Connections in Other Cortical Areas

Intrinsic connections have been studied in more detail in primate visual and rodent somatic sensory areas than in auditory cortex. The basic circuits are thought to be similar in all areas (DeFelipe et al. 2002b), especially when studied with the same methods in a species (Clemo et al. 2003). Less is known about the primate auditory cortex, making comparisons with the visual cortex problematic (Callaway 1998; Douglas and Martin 2004). Nevertheless, differences exist between the main sensory areas, especially in layer IV. In the somatic sensory cortex layer IV has discrete, repeating representations of the whiskers (barrels) or specialized modules for the digits (Mountcastle 1997; Horton and Adams 2005). Layers III and IV of the primate visual cortex have more cells than other areas (Callaway 1998), whereas auditory cortex layer IV has few spiny stellate cells (Smith and Populin 2001). The ionic channels associated with GABA receptors also differ in auditory cortex, with faster time constants than in other sensory areas (Hefti and Smith 2003), and which may reflect the importance of auditory temporal processing.

10 Development of Intrinsic Connections

Molecular signals help to specify the characteristics of neocortical areas. An example is a member of the fibroblast growth factor family (FGF8). Introducing supplemental FGF8 into embryonic somatic sensory cortex induced a duplication of part of the whisker barrel field (Fukuchi-Shimogori and Grove 2001). The duplicated field had a mirror image of the whisker barrels compared to the normal field, but otherwise was normal. The ordered appearance of the whisker barrels suggests that their specialized intrinsic connections were intact.

It has been suggested (Thomson and Morris 2002) that, during development, chandelier cells seek specific postsynaptic targets, while excitatory cells seek out different sources of input. Two mechanisms may be involved in forming specific connections: one is the molecules that guide growing axons; the second is the activity-dependent refinement of initially exuberant connections (Price et al. 2006).

10.1 Molecular Mechanisms Specifying Intrinsic Connections

The genesis of specific connections may involve the families of nerve cell-adhesion molecules such as the cadherins, or protein groups produced by alternative splicing from one gene such as the neurexins (Cline 2003) and neuroligins. The postsynaptic neuroligins and presynaptic neurexins bind one another and induce functional synapses (Dean and Dresbach 2006). It is uncertain if there are individual interneuron types with unique surface recognition molecules. Diffusible or membrane-associated molecules (e.g. neurotrophin-3) are present in specific auditory cortex layers and guide axons to specific layers (Castellani and Bolz 1999).

Molecular signals may participate in the tangential migration of cortical interneurons. Thus, brain-derived neurotrophic factor (BDNF) and neurotrophin-4 have dramatic effects on the tangential migration of inhibitory interneurons in developing cortex (Polleux et al. 2002; Woo and Lu 2006). BDNF may act in the formation of cortical columns (Alcantara et al. 2006). Agonists acting on cannabinoid receptors can also affect the migration of cholecystokinin-expressing interneurons (Berghuis et al. 2005).

10.2 Role of Functional Activity in Specifying Intrinsic Connections

Molecular signals affect the early formation of cortical circuits, whereas activity dependent mechanisms refine the initially imprecise connections (Price et al. 2006). One approach to studying this rewires the ascending auditory input such that the medial geniculate body instead receives its main afferent input from retino-geniculate axons after midbrain ablation (Angelucci et al. 1998). Auditory cortex neurons which should have been auditory now respond to visual line orientation, movement direction, and velocity much like cells in primary visual cortex (Roe et al. 1992).

11 Intrinsic Connections and Functional Plasticity

Functional plasticity may be associated with axonal pruning, or with modifying dendritic branching or spines (Maravall et al. 2004). Spines participate in synaptic plasticity, and long-term potentiation is correlated with spine enlargement and addition of new spines, while long-term depression is correlated with spine shrinkage (Sur and Rubenstein 2005). In the rat whisker barrel cortex, early postnatal sensory deprivation reduces the secondary branching of the dendrites of layer II/III pyramidal cells (Maravall et al. 2004). Similar processes presumably occur in auditory cortex and would adjust spiny cells and their excitatory inputs. However, many intrinsic neurons are inhibitory and the mechanisms for their maturation are less well documented than those for excitatory projections. Refinement of excitatory connections may involve Hebbian processes where a correlation between the pre- and postsynaptic action potentials strengthens some synapses and a lack of correlation elicits synaptic pruning (Sur et al. 1999). Inhibitory synapses, especially perisomatic ones, correlate negatively with postsynaptic firing and they may have unique regulatory mechanisms. Inhibitory cells may, early in development, have an excitatory function because of changes in the postsynaptic chloride reversal potential (Issa 2003) or because inhibitory neurons transiently release glutamate and GABA and thus activate postsynaptic NMDA receptors (Gillespie et al. 2005). GABA can have excitatory roles in maturity at axoaxonic synapses because the postsynaptic chloride reversal potential is different in the axon than the soma (Szabadics et al. 2006).

12 Functional Role of Intrinsic Connections

12.1 Connections Within the Isofrequency Domain

Layer II/III pyramidal cells have long intrinsic axons which can form regular patches in an isofrequency band (Wallace et al. 1991). These feed-forward terminal patches preferentially connect groups of cells with related functional properties such as sensitivity to spectral bandwidth, e.g., clusters of cells with narrow bandwidth tuning (Read et al. 2001). Other functional parameters, such as binaural, temporal or intensity aspects, may also be associated with the intrinsic patches.

12.2 Purposes of Intrinsic Connections

In the visual cortex emergent properties such as line orientation sensitivity likely arise from intrinsic processing of thalamic input. A similar form of constructive convergence may also be present in auditory cortex involving the summation of a group of simple receptive fields, present in the thalamic input, to create a more complex auditory cortical receptive field (Miller et al. 2001). Constructive convergence in AI may contribute to the analysis of the velocity and vector of frequency modulation (Zhang et al. 2003). Ensemble convergence also occurs in AI to extract features common to different thalamic inputs, while suppressing features present only in one input (Miller et al. 2001), enabling auditory cortex cells to extract specific signals from the acoustic environment while suppressing competing sounds (Nelken 2004). In both cases convergence is modified by local inhibition.

13 Future Directions

Future studies of cortical microcircuits need to define classes of cells, estimate their prevalence, identify their inputs, and determine the extent and specificity of their axonal distribution. It should be possible to correlate these morphologic types with cytoplasmic or cell surface markers which would quantify areal and cross-species comparisons. There is still not a single cell type in AI for which such features have been adequately characterized.

Laminar interactions require investigations with a multielectrode array, oriented vertically, and an intracellular electrode, containing a tracer. In vitro preparations provide a window on the specificity of local connections. Optical approaches now permit visualization of individual cells in vivo in the upper auditory cortex layers using two-photon scanning microscopy and bulk loading of calcium-sensitive dyes. Two-photon scanning microscopy should enable recording in vivo from one or more target neurons that contain a particular molecular marker or form part of a microcircuit (Margrie et al. 2003). This should allow a much closer correlation of structure and function than is now available.

Acknowledgments We thank Alan Palmer for helpful comments.

References

Alcantara S, Pozas E, Ibanez CF, and Soriano E (2006) BDNF-modulated spatial organization of Cajal-Retzius and GABAergic neurons in the marginal zone plays a role in the development of cortical organization. *Cerebral Cortex* **16**:487–499.

Ang Jr ESBC, Haydar TF, Gluncic V, and Rakic P (2003) Four-dimensional migratory coordinates of GABAergic interneurons in the developing mouse cortex. *Journal of Neuroscience* **23**: 5805–5815.

Angelucci A, Clasca F, and Sur M (1998) Brainstem inputs to the ferret medial geniculate nucleus and the effect of early deafferentation on novel retinal projections to the auditory thalamus. *Journal of Comparative Neurology* **400**:417–439.

Berghuis P, Dobszay MB, Wang XY, Spano S, Ledda F, Sousa KM, Schulte G, Ernfors P, Mackie K, Paratcha G, Hurd YL, and Harkany T (2005) Endocannabinoids regulate interneuron migration and morphogenesis by transactivating the TrkB receptor. *Proceedings of the National Academy of Sciences of the United States of America* **102**:19115–19120.

Binzegger T, Douglas RJ, and Martin KAC (2004) A quantitative map of the circuit of cat primary visual cortex. *Journal of Neuroscience* **24**:8441–8453.

Buxhoeveden DP and Casanova MF (2002) The minicolumn hypothesis in neuroscience. *Brain* **125**:935–951.

Callaway EM (1998) Local circuits in primary visual cortex of the macaque monkey. *Annual Review of Neuroscience* **21**:47–74.

Castellani V and Bolz J (1999) Opposing roles for neurotrophin-3 in targeting and collateral formation of distinct sets of developing cortical neurons. *Development* **126**:3335–3345.

Chiry O. Tardif E, Magistretti PJ, and Clarke S (2003) Patterns of calcium-binding proteins support parallel and hierarchical organization of human auditory areas. *European Journal of Neuroscience* **17**:397–410.

Clarke S and Rivier F (1998) Compartments within human primary auditory cortex: evidence from cytochrome oxidase and acetylcholinesterase staining. *European Journal of Neuroscience* **10**: 741–745.

Clemo HR, Keniston L, and Meredith MA (2003) A comparison of the distribution of GABA-ergic neurons in cortices representing different sensory modalities. *Journal of Chemical Neuroanatomy* **26**:51–63.

Cline H (2003) Sperry and Hebb: oil and vinegar? *Trends in Neurosciences* **26**:655–661.

Dean C and Dresbach T (2006) Neuroligins and neurexins: linking cell adhesion, synapse formation and cognitive function. *Trends in Neurosciences* **29**:21–29.

DeFelipe J (2002a) Cortical interneurons: from Cajal to 2001. *Progress in Brain Research* **136**:215–238.

DeFelipe J, Alonso-Nanclares L, and Arellano JI (2002b) Microstructure of the neocortex: comparative aspects. *Journal of Neurocytology* **31**:299–316.

DeFelipe J, Hendry SHC, Hashikawa T, Molinari M, and Jones EG (1990) A microcolumnar structure of monkey cerebral cortex revealed by immunocytochemical studies of double bouquet cell axons. *Neuroscience* **37**:655–673.

del Rio MR and DeFelipe J (1995) A light and electron microscopic study of calbindin D-28 k immunoreactive double bouquet cells in the human temporal cortex. *Brain Research* **690**:133–140.

Douglas RJ and Martin KAC (2004) Neuronal circuits of the neocortex. *Annual Review of Neuroscience* **27**:419–451.

Eder M, Rammes G, Zieglgansberger W, and Dodt H-U (2001) $GABA_A$ and $GABA_B$ receptors on neocortical neurons are differentially distributed. *European Journal of Neuroscience* **13**:1065–1069.

Ehret G (1997) The auditory cortex. *Journal of Comparative Physiology (A)* **181**:547–557.

Feldman ML and Peters A (1974) A study of barrels and pyramidal dendritic clusters in the cerebral cortex. *Brain Research* **77**:55–76.

Fitzpatrick DC and Henson OW Jr (1994) Cell types in the mustached bat auditory cortex. *Brain Behavior and Evolution* **43**:79–91.

Foeller E, Vater M, and Kössl M (2001) Laminar analysis of inhibition in the gerbil primary auditory cortex. *Journal of the Association for Research in Otolaryngology* **2**:279–296.

Fukuchi-Shimogori T and Grove EA (2001) Neocortex patterning by the secreted signaling molecule FGF8. *Science* **294**:1071–1074.

Galuske RAW, Schlote W, Bratzke H, and Singer W (2000) Interhemispheric asymmetries of the modular structure in human temporal cortex. *Science* **289**:1946–1949.

Gao W-J and Pallas SL (1999) Cross-modal reorganization of horizontal connectivity in auditory cortex without altering thalamocortical projections. *Journal of Neuroscience* **19**:7940–7950.

Gillespie DC, Kim G, and Kandler K (2005) Inhibitory synapses in the developing auditory system are glutamatergic. *Nature Neuroscience* **8**:332–338.

Hackett TA, Preuss TM, and Kaas JH (2001) Architectonic identification of the core region in auditory cortex of macaques, chimpanzees, and humans. *Journal of Comparative Neurology* **441**:197–222.

He J (2003) Slow oscillation in non-lemniscal auditory thalamus. *Journal of Neuroscience* **23**:8281–8290.

He JF and Hu B (2002) Differential distribution of burst and single-spike responses in auditory thalamus. *Journal of Neurophysiology* **88**:2152–2156.

Hefti BJ and Smith PH (2000) Anatomy, physiology, and synaptic responses of rat layer V auditory cortical cells and effects of intracellular $GABA_A$ blockade. *Journal of Neurophysiology* **83**: 2626–2638.

Hefti BJ and Smith PH (2003) Distribution and kinetic properties of GABAergic inputs to layer V pyramidal cells in rat auditory cortex. *Journal of the Association for Research in Otolaryngology* **4**: 106–121.

Hendry SHC and Jones EG (1991) GABA neuronal subpopulation in cat primary auditory cortex: co-localization with calcium binding proteins. *Brain Research* **543**:45–55.

Hestrin S and Galarreta M (2005) Electrical synapses define networks of neocortical GABAergic neurons. *Trends in Neurosciences* **28**: 304–309.

Horton JC and Adams DL (2005) The cortical column: a structure without a function. *Philosophical Transactions of the Royal Society B* **360**:837–862.

Issa NP (2003) Inhibitory circuits in sensory maps develop through excitation. *Trends in Neurosciences* **26**:456–458.

Jones EG (2000) Microcolumns in the cerebral cortex. *Proceedings of the National Academy of Sciences of the United States of America* **97**:5019–5021.

Kawaguchi Y and Kubota Y (1997) GABAergic cell subtypes and their synaptic connections in rat frontal cortex. *Cerebral Cortex* **7**: 476–486.

Kozloski J, Hamzei-Sichani F, and Yuste R (2001) Stereotyped position of local synaptic targets in neocortex. *Science* **293**:868–872.

Letinic K, Zoncu R, and Rakic P (2002) Origin of GABAergic neurons in the human neocortex. *Nature* **417**:645–649.

Linden JF and Schreiner CE.(2003) Columnar transformations in auditory cortex? A comparison to visual and somatosensory cortices. *Cerebral Cortex* **13**:83–89.

Maravall M, Koh IYY, Lindquist WB, and Svoboda K (2004) Experience-dependent changes in basal dendritic branching of layer 2/3 pyramidal neurons during a critical period for developmental plasticity in rat barrel cortex. *Cerebral Cortex* **14**: 655–664.

Margrie TW, Meyer AH, Caputi A, Monyer H, Hasan MT, Schaefer AT, Denk W, and Brecht M (2003) Targeted whole-cell recordings in the mammalian brain in vivo. *Neuron* **39**:911–918.

Markram H, Toledo-Rodriguez M, Wang Y, Gupta A, Silberberg G, and Wu C (2004) Interneurons of the neocortical inhibitory system. *Nature Reviews Neuroscience* **5**:793–807.

McMullen NT and de Venecia RK (1993) Thalamocortical patches in auditory neocortex. *Brain Research* **620**:317–322.

Meyer G, Gonzalez-Hernandez TH, and Ferres-Torres R (1989) The spiny stellate neurons in layer IV of the human auditory cortex. A Golgi study. *Neuroscience* **33**:489–498.

Miller LM, Escabí MA, Read HL, and Schreiner CE (2001) Functional convergence of response properties in the auditory thalamocortical system. *Neuron* **32**:151–160.

Mitani A, Shimokouchi M, Itoh K, Nomura S, Kudo M, and Mizuno N (1985) Morphology and laminar organization of electrophysiologically identified neurons in the primary auditory cortex in the cat. *Journal of Comparative Neurology* **235**:430–447.

Morel A, Garraghty PE, and Kaas JH (1993) Tonotopic organization, architectonic fields, and connections of auditory cortex in macaque monkeys. *Journal of Comparative Neurology* **335**: 437–459.

Morel A and Kaas JH (1992) Subdivisions and connections of auditory cortex in owl monkeys. *Journal of Comparative Neurology* **318**: 27–63.

Mountcastle VB (1997) The columnar organization of the neocortex. *Brain* **120**:701–722.

Mountcastle VB (2003) Introduction. *Cerebral Cortex* **13**:2–4.

Nelken I (2004) Processing of complex stimuli and natural scenes in the auditory cortex. *Current Opinion in Neurobiology* **14**:474–480.

Ojima H, Honda CN, and Jones EG (1991) Patterns of axon collateralization of identified supragranular pyramidal neurons in the cat auditory cortex. *Cerebral Cortex* **1**:80–94.

Ojima H, Honda CN, and Jones EG (1992) Characteristics of intracellularly injected infragranular pyramidal neurons in cat primary auditory cortex. *Cerebral Cortex* **2**:197–216.

Peters A (2002) Examining neocortical circuits: some background and facts. *Journal of Neurocytology* **31**:183–193.

Peters A and Sethares C (1997) The organization of double bouquet cells in monkey striate cortex. *Journal of Neurocytology* **26**: 779–797.

Polleux F, Whitford KL, Dijkhuizen PA, Vitalis T, and Ghosh A (2002) Control of cortical interneuron migration by neurotrophins and P13-kinase signaling. *Development* **129**:3147–3160.

Price DJ, Kennedy H, Dehay C, Zhou L, Mercier M, Jossin Y, Goffinet AM, Tissir F, Blakey D, and Molnar Z (2006) The development of cortical connections. *European Journal of Neuroscience* **23**: 910–920.

Prieto JJ, Peterson BA, and Winer JA (1994a) Morphology and spatial distribution of GABAergic neurons in cat primary auditory cortex (AI). *Journal of Comparative Neurology* **344**:349–382.

Prieto JJ, Peterson BA, and Winer JA (1994b) Laminar distribution and neuronal targets of GABAergic axon terminals in cat primary auditory cortex (AI). *Journal of Comparative Neurology* **344**: 383–402.

Prieto JJ and Winer JA (1999) Layer VI in cat primary auditory cortex: Golgi study and sublaminar origins of projection neurons. *Journal of Comparative Neurology* **404**:332–358.

Rakic P (1995) A small step for the cell, a giant leap for mankind: a hypothesis of neocortical expansion during evolution. *Trends in Neurosciences* **18**:383–388.

Read HL, Winer JA, and Schreiner CE (2001) Modular organization of intrinsic connections associated with spectral tuning in cat auditory cortex. *Proceedings of the National Academy of Sciences of the United States of America* **98**:8042–8047.

Rockland KS and Ichinohe N (2004) Some thoughts on cortical minicolumns. *Experimental Brain Research* **158**:265–277.

Roe AW, Pallas SL, Kwon YH, and Sur M (1992) Visual projections routed to the auditory pathway in ferrets: receptive fields of visual neurons in primary auditory cortex. *Journal of Neuroscience* **12**:3651–3664.

Rouiller EM and de Ribaupierre F (1990) Arborization of corticothalamic axons in the auditory thalamus of the cat: a PHA-L tracing study. *Neuroscience Letters* **108**:29–35.

Sherrington CS (1933) *The Brain and Its Mechanism*. Cambridge University Press, London.

Sincich LC, Adams DL, and Horton JC (2003) Complete flatmounting of the macaque cerebral cortex. *Visual Neuroscience* **20**: 663–686.

Smith PH and Populin LC (2001) Fundamental differences between the thalamocortical recipient layers of the cat auditory and visual cortices. *Journal of Comparative Neurology* **436**:508–519.

Sur M, Angelucci A, and Sharma J (1999) Rewiring cortex: the role of patterned activity in development and plasticity of neocortical circuits. *Journal of Neurobiology* **41**:33–43.

Sur M and Rubenstein JLR (2005) Patterning and plasticity of the cerebral cortex. *Science* **310**:805–810.

Szabadics J, Varga C, Molnar G, Olah S, Barzo P, and Tamas G (2006) Excitatory effect of GABAergic axo-axonic cells in cortical microcircuits *Science* **311**:233–235.

Thomson AM and Bannister AP (1998) Postsynaptic pyramidal target selection by descending layer III pyramidal axons: dual intracellular recordings and biocytin filling in slices of rat neocortex. *Neuroscience* **84**:669–683.

Thomson AM and Bannister AP (2003) Interlaminar connections in the neocortex. *Cerebral Cortex* **13**:5–14.

Thomson AM and Morris OT (2002) Selectivity in the interlaminar connections made by neocortical neurones. *Journal of Neurocytology* **31**:239–246.

Valcanis H and Tan S-S (2003) Layer specification of transplanted interneurons in developing mouse neocortex. *Journal of Neuroscience* **23**:5113–5122.

Viebahn C (1990) Correlation between differences in the structure of dendrite bundles and cytoarchitectonic patterns in the cerebral cortex of the rabbit. *Journal für Hirnforschung* **31**:645–652.

Wallace, M.N. and Bajwa, S. (1991) Patchy intrinsic connections of the ferret primary auditory cortex. *Neuroreport* **2**:417–420.

Wallace MN, Johnston PW, and Palmer AR (2002) Histochemical identification of cortical areas in the auditory region of the human brain. *Experimental Brain Research* **143**:499–508.

Wallace MN, Kitzes LM, and Jones EG (1991) Intrinsic inter- and intralaminar connections and their relationship to the tonotopic map in cat primary auditory cortex. *Experimental Brain Research* **86**:527–544.

Watts J and Thomson AM (2005) Excitatory and inhibitory connections show selectivity in the neocortex. *Journal of Physiology* **562**:89–97.

White EL (2002) Specificity of cortical synaptic connectivity: emphasis on perspectives gained from quantitative electron microscopy. *Journal of Neurocytology* **31**:195–202.

Winer JA (1992) The functional architecture of the medial geniculate body and the primary auditory cortex. In: Webster DB, Popper AN, and Fay RR (eds). *Springer Handbook of Auditory Research*, volume 1, *The Mammalian Auditory Pathway: Neuroanatomy*. Springer, New York, pp. 222–409.

Winer JA, Larue DT, and Huang CL (1999) Two systems of giant axon terminals in the cat medial geniculate body: convergence of cortical and GABAergic inputs. *Journal of Comparative Neurology* **413**:181–197.

Winer JA and Prieto JJ (2001) Layer V in cat primary auditory cortex (AI): cellular architecture and identification of projection neurons. *Journal of Comparative Neurology* **434**:379–412.

Woo NH and Lu B (2006) Regulation of cortical interneurons by neurotrophins: from development to cognitive disorders. *Neuroscientist* **12**:43–56.

Xu Q, Cobos I, de la Cruz E, Rubenstein JL, and Anderson SA (2004) Origins of cortical interneuron subtypes. *Journal of Neuroscience* **24**:2612–2622.

Yanez IB, Munoz A, Contreras J, Gonzalez J, Rodriguez-Veiga E, and DeFelipe J (2005) Double bouquet cell in the human cerebral cortex and a comparison with other mammals. *Journal of Comparative Neurology* **486**:344–360.

Zhang LI, Tan AYY, Schreiner CE, and Merzenich MM (2003) Topography and synaptic shaping of direction selectivity in primary auditory cortex. *Nature* **424**:201–205.

Zhang ZW and Deschênes M (1997) Intracortical axonal projections of lamina VI cells of the primary somatosensory cortex in the rat: a single-cell labeling study. *Journal of Neuroscience* **17**: 6365–6379.

Chapter 7

A Synthesis of Auditory Cortical Connections: Thalamocortical, Commissural and Corticocortical Systems

Charles C. Lee and Jeffery A. Winer

Abbreviations

AAF	anterior auditory field
AC	auditory cortex
AES	anterior ectosylvian field
AI	primary auditory cortex
AII	secondary auditory cortex
AL	anterior lateral auditory belt, macaque
AM	anterior medial thalamic nucleus
APt	anterior pretectum
AS	arcuate sulcus
AV	anterior ventral thalamic nucleus
BIC	brachium of the inferior colliculus
BSC	brachium of the superior colliculus
CB	cerebellum
CC	corticocortical
CF	constant frequency region or characteristic frequency
CF–CF	constant frequency–constant frequency region, bat
CG	central gray
CL	caudal lateral auditory belt, macaque
CM	caudomedial auditory belt, macaque
CMN	centromedial nucleus
CO	commissural or contralateral
CP	cerebral peduncle, or caudal parabelt, macaque
CS	central sulcus
CTb	cholera toxin beta subunit
CTbG	cholera toxin beta subunit, gold-conjugate
D	dorsal nucleus of the medial geniculate body *or* dorsal
DD	deep dorsal nucleus
DCa	caudal dorsal nucleus of the medial geniculate body
DF	dorsal fringe area, bat
DlF	dorsal lateral fringe area, bat
DM	dorsomedial area, bat
DSCF	Doppler-shifted constant frequency region, bat
DD	deep dorsal nucleus of the medial geniculate body
DS	dorsal superficial nucleus of the medial geniculate body
DZ	dorsal auditory zone
ED	posterior ectosylvian gyrus, dorsal part
EI	posterior ectosylvian gyrus, intermediate part
EN	entopeduncular
EPP	posterior ectosylvian gyrus, caudal par
EV	posterior ectosylvian gyrus, ventral part
FM	frequency modulated area, bat
FM–FM	FM–FM area, bat
GABA	gamma aminobutyric acid
Ha	habenula
HiT	habenulointerpeduncular tract
III	oculomotor nucleus
In	insular cortex
L	lateral
LD	lateral dorsal thalamic nucleus
LGB	lateral geniculate nucleus
LGBd	lateral geniculate body, dorsal nucleus
LGBv	lateral geniculate body, ventral nucleus
LLS	lateral visual association area, lateral part
LP	lateral posterior nucleus
LOS	lateral orbital sulcus
LS	lateral sulcus
LS	lateral suprasylvian visual association area
LuS	lunate sulcus
M	medial division of the medial geniculate body *or* medial
MD	mediodorsal nucleus
MeV	mesencephalic nucleus of the trigeminal
MGB	medial geniculate body
ML	middle lateral auditory belt, macaque
MLS	middle lateral suprasylvian visual association area
MRF	mesencephalic reticular formation
OT	optic tract

C.C. Lee (✉)
Department of Neurobiology
The University of Chicago, Chicago, IL 60637, USA
e-mail: clee@bsd.uchicago.edu

J.A. Winer, C.E. Schreiner (eds.), *The Auditory Cortex*,
DOI 10.1007/978-1-4419-0074-6_7, © Springer Science+Business Media, LLC 2011

Ov	ovoid part of the medial geniculate body
P	posterior auditory cortex
PAC	paracentral thalamic nucleus
PC	posterior commissure
PFC	prefrontal cortex
PHyp	posterior hypothalamus
Ps	principal sulcus
Ps	posterior sylvian gyrus
Pt	pretectum
Pul	pulvinar
R	rostral, *or* rostral auditory area, macaque
RM	rostromedial region, macaque
RN	red nucleus
RP	rostral pole nucleus of the medial geniculate body
RT	rostrotemporal area, macaque
RTL	lateral rostrotemporal auditory belt, macaque
RTM	medial rostrotemporal auditory belt, macaque
RP	rostral pole division of the medial geniculate body, *or* rostral parabelt, macaque
SC	superior colliculus
SG	suprageniculate nucleus
Sgl	suprageniculate nucleus, lateral part
Sl	suprageniculate nucleus, lateral part
Sm	suprageniculate nucleus, medial part
SN	substantia nigra
Spf	subparafascicular nucleus
STG	superior temporal gyrus
STS	superior temporal sulcus
TC	thalamocortical
Te1	temporal area, rat
TRN	thalamic reticular nucleus
V	ventral division of the medial geniculate body *or* ventral
VA	ventral anterior thalamic nucleus, *or* ventroanterior area, bat
Vb	ventrobasal complex
Ve	ventral auditory area
VF	ventral fringe, bat
VL	ventral lateral thalamic nucleus
Vl	ventrolateral nucleus of the medial geniculate body
VP	ventral posterior auditory area, *or* ventroposterior area, bat
Vpl	ventral posterior nucleus, lateral part
Vpm	ventral posterior nucleus, medial part
Vpmpc	ventral posteromedial nucleus, parvocellular part
wm	white matter
7	parietal area 7
20	posterior sylvian visual association area 20
21b	posterior sylvian visual association area 21b
35	perirhinal area 35
36	perirhinal area 36

1 Introduction

The study of mammalian auditory cortex (AC) began with the delineation of a few areas which appeared to have a limited set of connections from other AC fields (Brugge and Reale 1985) and from a small number of nuclei in the medial geniculate body (MGB) (Rose and Woolsey 1958). Subsequent work with more sensitive methods revealed many more areas in a wider temporal lobe expanse (Schreiner and Cynader 1984; Bowman and Olson 1988; Clarey and Irvine 1990; Shinonaga et al. 1994; Clascá et al. 1997) and, within some regions, multiple subdivisions specialized physiologically for sound processing or communication signals (Ehret 1997; Winer and Lee 2007).

In the cat auditory cortex, at least 13 areas each receive unique patterns of convergent thalamic (Lee and Winer 2008a), ipsilateral cortical (Lee and Winer 2008c), and commissural (Lee and Winer 2008b) projections (Fig. 7.1c). These many inputs contribute to the diverse physiological properties (Reale and Imig 1980) and the differential functional role (Lomber et al. 2007) in each area. A systematic representation of characteristic frequency (CF) is the chief organizing feature in five areas only (Imig and Reale 1980; Reale and Imig 1980), which leaves open the question of how the others are organized and their role(s). In eight non-tonotopic AC areas, defined by criteria other than tonotopy, neurons respond to acoustic stimuli (Woolsey 1960; Schreiner and Cynader 1984; Clarey and Irvine 1990; He et al. 1997), receive area- and nucleus-specific input from the medial geniculate body (MGB) (Lee and Winer 2008a) and AC areas (Lee and Winer 2008c), and have diverse roles in pattern recognition (Eggermont 1998), sound localization (Middlebrooks et al. 1980; Stecker et al. 2005), and in multimodal (Bowman and Olson 1988; Meredith et al. 2006) and limbic (Clascá et al. 1997) interactions (Fig. 7.1c).

Efforts to characterize auditory forebrain connections have largely concentrated on areas AI (primary AC) and AAF (the anterior auditory field), both with an ordered (though different) organization of characteristic frequency (CF) (Merzenich et al. 1975; Knight 1977; Imaizumi et al. 2004a); however, even in these areas no current profile with contemporary tracers is available which includes the convergent thalamic, ipsilateral, and contralateral cortical projection systems. Since these pathways operate in tandem, it would seem important to compare these multiple systems as a basis for principled hypotheses of forebrain operations. Despite the functional areal diversity, the extrinsic connections subserving receptive field (RF) formation have rarely been attempted in regions other than AI (Miller et al. 2001, 2002). The present synthesis is an early step toward a connectionist framework. Such a profile of extrinsic input can

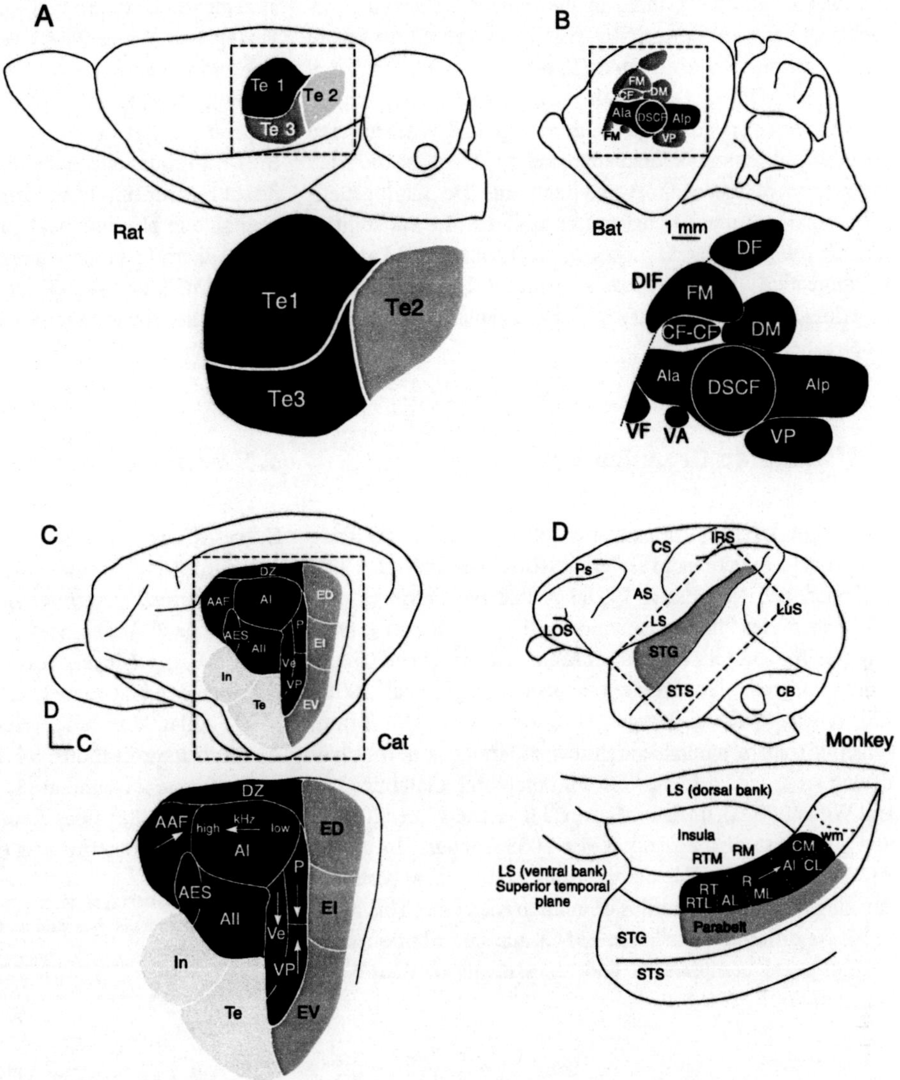

Fig. 7.1 Comparative auditory cortical organization in rat, bat, cat, and monkey. Tonotopic (*black*), non-tonotopic (*dark gray*), multisensory (*medium gray*), and limbic (*light gray*) areas vary in number, size, distribution, and connectivity. **a** The rat auditory cortex (AC) contains a tonotopic region (Te1), a non-tonotopic area (Te3), and a multisensory area (Te2) (Roger and Arnault 1989). **b** The mustached bat has a specialized Doppler-shifted constant frequency tonotopic region (DSCF) embedded in AI containing an overrepresentation at the third harmonic of echolocation frequencies and is surrounded by several non-tonotopic areas (CF, FM-FM, DIF, DF, DM, VA, VF, VP) (Fitzpatrick et al. 1998). **c** The cat AC has an expansive representation with five tonotopic (AI, AAF, P, VP, Ve), two non-tonotopic (AII, DZ, AES), four multisensory (AES, ED, EI, EV) and two limbic areas (Te, In) (Lee and Winer 2008c). **d**: Monkey auditory cortex is composed of a core tonotopic region (AI, R, RT) surrounded by a belt of non-tonotopic or weakly-tonotopic areas (CL, CM, RM, RTM, ML, AL, RTL), which is flanked by a parabelt multisensory/limbic region (Hackett et al. 1998)

encourage structure–function experiments and serve as a framework for exploring intrinsic connectivity (Read et al. 2001).

2 Connectional Models

One view of forebrain processing is that serial corticocortical connections (Felleman and Van Essen 1991; Zeki 1993) from lower hierarchical levels combine input to create more complex RFs with emergent properties. Such models can minimize or relegate thalamic and commissural inputs to a modulatory role in processes such as attention (Olshausen et al. 1993) or largely omit them in view of their small absolute magnitude. This view does not stipulate whether the relative thalamic and commissural contributions differ within any cortical hierarchy, or whether different areas each have unique structural adaptations for specialized processing. If such contributions are weighted equally in all areas, does this imply a more distributed web of parallel processing? Is CF a primitive or a derived feature of AC organization? Is there one connectionist hierarchy in the thalamocortical, corticocortical, and commissural systems?

3 Comparative Framework

Another fundamental question is the role of multiple auditory areas. There are two robustly tonotopic areas in the mouse

(Stiebler et al. 1997), three in the monkey (Hackett et al. 1998) and five in the cat (Reale and Imig 1980), in whom AC operations are distributed (Eggermont 1998; Imaizumi et al. 2004a) (Fig. 7.1). Does this comparative areal diversity reflect species-specific computational repertoires or subserve generalizable computational approaches to the dissection and integration of the auditory environment? Do phylogenetically emergent areas transform or augment the capabilities of their progenitor areas or operate independently? Parallels in connectivity between areas may reveal their ontogenetic, functional, and evolutionary patterns and clarify differences among them.

4 Distributed Organization

Models of forebrain organization often focus on thalamic or cortical projections in isolation (Rose and Dobson 1985; Felleman and Van Essen 1991). A thalamocentric or corticocentric view limits an understanding of the operations required for auditory computation, and a more refined picture would integrate all convergent inputs (Budinger et al. 2008). This view of the convergent connectivity would require sensitive tracers, a reliable architectonic framework for identifying areas, and a framework for comparing thalamic (Lee and Winer 2008a), ipsilateral cortical (Lee and Winer 2008c), and commissural (Lee and Winer 2008b) inputs. In the 13 AC areas in cat, such comparisons could be a basis for deriving quantitative profiles of input to each area. This highlights the differential thalamic and cortical contributions, and enables areal comparisons and assessments of distributed connectivity.

5 Auditory Cortex Connections

The multiple convergent inputs received by each cortical area from thalamic, ipsilateral cortical, and contralateral cortical sources define their familial affiliations, i.e., tonotopic, non-tonotopic, multisensory, or limbic (Fig. 7.1c). This parcellation is provisional and provides a functional framework within which to examine connectional relations. Here we present a quantitative summary of the sources and anatomical magnitude of each projection as a stimulus for deriving basic principles governing auditory cortex organization.

Some caveats pertain to interpreting the quantitative patterns of anatomical connectivity described here. First, the conclusions are based on retrograde labeling studies entirely; no correlations with the synaptic strength of any input are implied or available, since an entirely different set of methods and analytical procedures would be required for such statements (Davis and Sterling 1979; Humphrey et al. 1985; Huang and Winer 2000). Moreover, anatomical projection size and synaptic strength often are negatively correlated (see below). Second, the validity of the analysis of group relations depends in part on methods such as Nissl preparations and SMI-32 immunohistochemistry that, while useful, have no direct relationship to any functional parameter such as those available in physiological experiments (Lee et al. 2004a, b). While there is broad agreement for the principal divisions of cat MGB and AC (Winer 1992), independent confirmatory studies using other methods would be essential for the non-primary auditory forebrain regions.

5.1 Intrinsic Projections

Every AC area receives a massive input from intrinsic cortical sources that originate from within the area. Anatomically and numerically, these provide ~50% of the total input to an area (Table 7.1). These intrinsic projections are densely clustered across all layers (excluding layer I) and isotropically distributed within a ~2 mm radius (Fig. 7.2a) (Imaizumi et al. 2004a; Lee and Winer 2008c). Exceptions are the AI intrinsic projections, which cluster anisotropically along the isofrequency contour (Matsubara and Phillips 1988; Lee and Winer 2008c), consistent with the modular physiological organization along this axis (Read et al. 2001).

Table 7.1 Percentages of intrinsic and extrinsic input from thalamic and cortical sources. Intrinsic and extrinsic sources each provide 50% of the projections

Area	Intrinsic	Extrinsic[a]
AI	60.2	39.8
AAF	61.1	38.9
P	50.6	49.4
VP	45.2	54.8
Ve	36.9	63.1
AII	42.0	58.0
AES	70.5	29.5
DZ	41.0	59.0
Te	36.3	63.7
In	52.4	47.6
ED	43.0	57.0
EI	48.4	51.6
EV	50.3	49.7
Average[b]	49.1	50.9
Std. Dev.	10.1	10.1

[a] Sum of percentage of labeling in thalamus, extrinsic ipsilateral cortex, and contralateral hemisphere.
[b] $p>0.05$, paired t-test.

The intrinsic projections are the major input to each area, with a nearly twofold range (Fig. 7.3 and Table 7.1). The multisensory area AES (Clarey and Irvine 1990) receives the largest intrinsic input (71%), while areas Te (Shinonaga

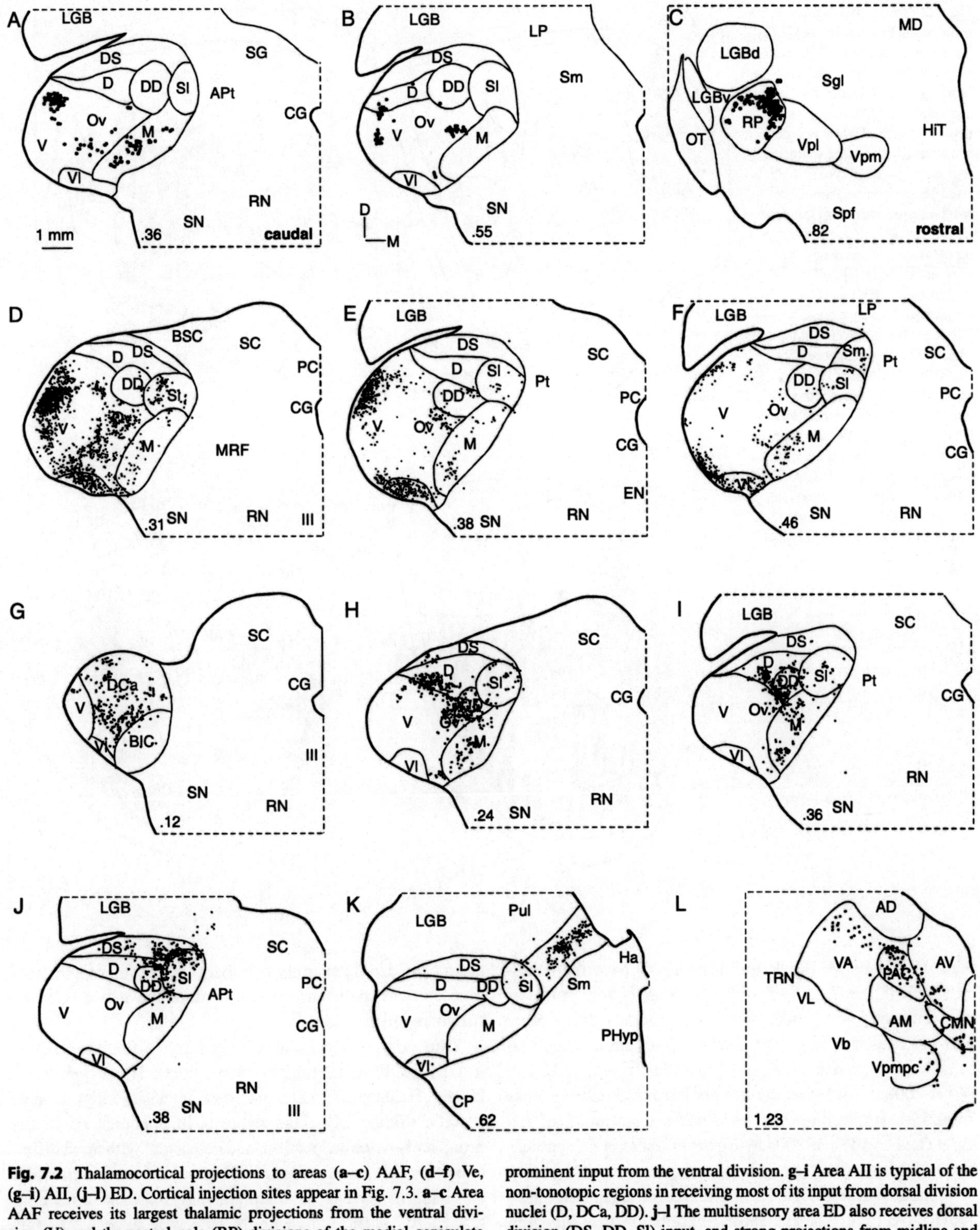

Fig. 7.2 Thalamocortical projections to areas (**a–c**) AAF, (**d–f**) Ve, (**g–i**) AII, (**j–l**) ED. Cortical injection sites appear in Fig. 7.3. **a–c** Area AAF receives its largest thalamic projections from the ventral division (V) and the rostral pole (RP) divisions of the medial geniculate body (MGB). **d–f** Area Ve, another tonotopic region, also receives a prominent input from the ventral division. **g–i** Area AII is typical of the non-tonotopic regions in receiving most of its input from dorsal division nuclei (D, DCa, DD). **j–l** The multisensory area ED also receives dorsal division (DS, DD, SI) input, and strong projections from midline and intralaminar nuclei (CMN, PAC, VA)

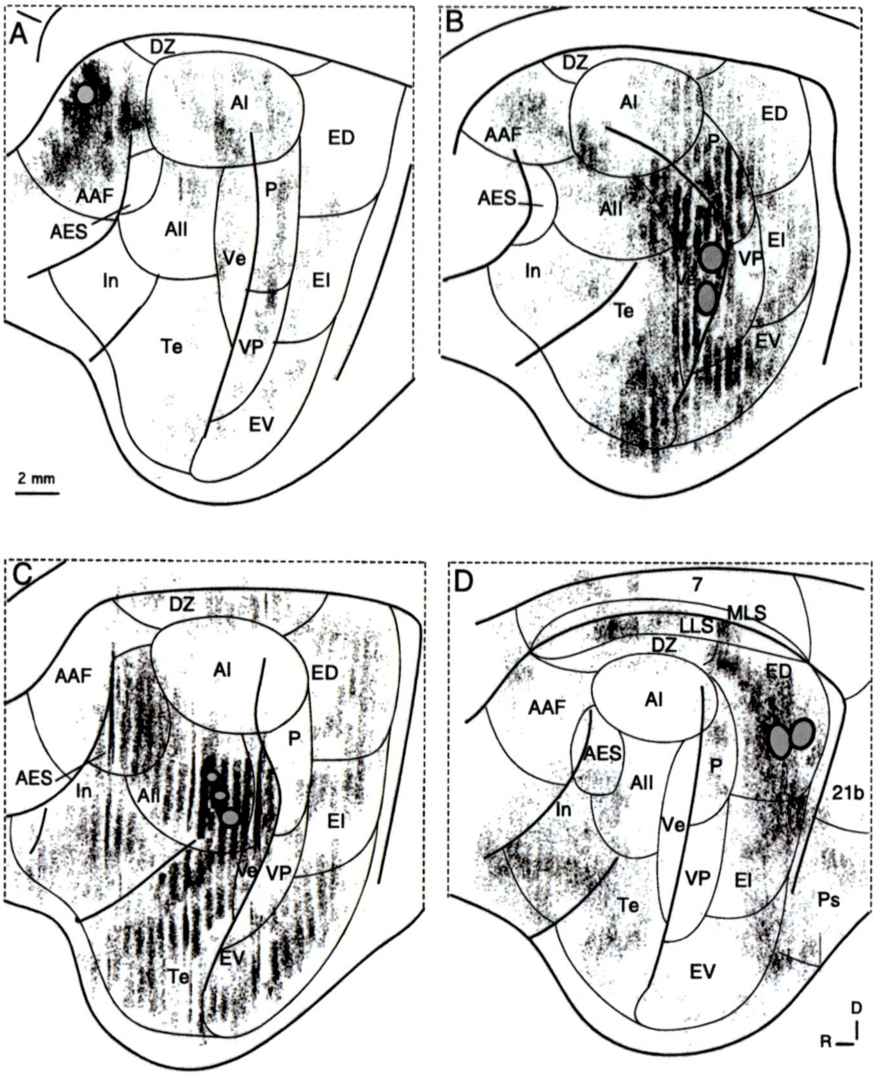

Fig. 7.3 Ipsilateral cortical projections to areas (**a**) AAF, (**b**) Ve, (**c**) AII, and (**d**) ED. Injection sites correspond to thalamic labeling (Fig. 7.2) and contralateral AC labeling (Fig. 7.4). **a** AAF receives most of its input from other tonotopic regions, e.g., AI, P, Ve, and VP. **b** Area Ve, by comparison, also receives predominantly tonotopic inputs, but also receives strong inputs from surrounding non-tonotopic (AII) and multisensory (EV) areas. **c** Projections to non-tonotopic area AII arise in other non-tonotopic (AES, DZ), multisensory (ED, EI, EV), and limbic (Te, In) areas. However, projections from tonotopic areas are largely absent. **d** Similarly, the multisensory area ED receives strong inputs from other non-tonotopic (DZ, AES), multisensory (EI, EV), and limbic (Te, In) areas. It also receives strong perivisual inputs (LLS, 7, Ps, 21b)

et al. 1994) and Ve (Reale and Imig 1980) have the smallest (37%) (Table 7.1). Such differences are counterbalanced by correspondingly fewer (or greater) inputs from other sources. Thus, the larger AES intrinsic proportion is counterbalanced by fewer extrinsic corticocortical inputs (Lee and Winer 2008c). Likewise, the smaller intrinsic inputs to areas Te and Ve are accompanied by a wider distribution of AC input (Lee and Winer 2008c). However, there is no segregation of intrinsic input strength reflecting anatomical location or functional type (Table 7.5). A tonotopic region can have above (AAF; Fig. 7.3a and Table 7.1) or below (Ve; Fig. 7.3b and Table 7.1) average intrinsic input. Moreover, functionally similar areas, such as Te and In, have different intrinsic projection strengths (Table 7.1). This supports the idea that areas considered as family members (limbic, tonotopic, non-tonotopic, or multisensory) are distinct and it is a criterion for distinguishing them.

Thus, the bulk of the connectivity within an area is usually intrinsic (Schreiner et al. 2000; Binzegger et al. 2004). Further, the percentage of intrinsic input is area specific (Table 7.1). This range is independent of family group and anatomic location, and consistent with areal differences for intrinsic processing. The variability also provides an anatomical basis for distinguishing intrinsic projections among AC areas, which might otherwise appear homogeneous. This is the case except in AI, some of whose projections are markedly anisotropic and modular (Middlebrooks et al. 1980; Huang and Winer 2000; Read et al. 2001).

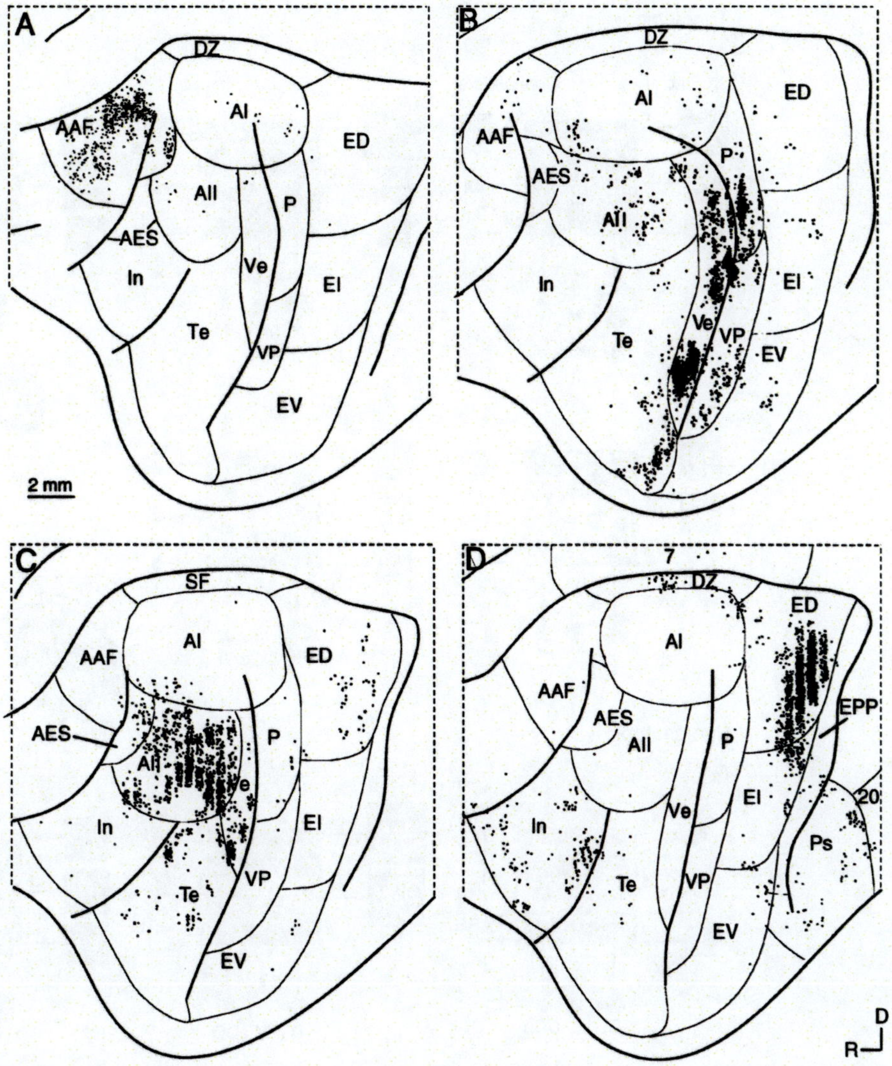

Fig. 7.4 Contralateral cortical projections to areas **a** AAF, **b** Ve, **c** AII, and **d** ED. Injection sites appear in Fig. 7.3. Orientation of the contralateral hemisphere is mirrored to match the ipsilateral hemisphere. All AC regions receive their strongest projection from the homolateral cortical area, but also weaker projections from areas and location similar to that of the ipsilateral cortical projections. **a** Area AAF receives its strongest projections from contralateral AAF. **b** Area Ve receives its strongest projection from contralateral Ve, and also from other tonotopic areas, such as P and VP. **c** Area AII receives its strongest input from contralateral AII, and weaker inputs from Ve and Te. **d** Likewise, ED is heavily connected with its contralateral counterpart, as well as weakly with DZ, In, and PS

These intrinsic differences might entail a concomitant rebalancing of their respective extrinsic inputs, so that area AES (29%) has a smaller percentage than Te and Ve (63%); therefore, the relative impact of intrinsic and extrinsic connections also may differ among areas. The implications of such findings can be examined directly in immunoconnectional studies to determine how differential concentrations in the proportion of local circuit gamma-aminobutyric acid-containing (GABAergic) neurons also vary and whether their connections are area specific (Clemo et al. 2003).

If there is parity with other convergent inputs, the functional impact of individual intrinsic neurons should be weaker relative to these numerically sparser projections from each extrinsic source. Binzegger et al. (2004) estimated the synaptic weights to cat visual cortex layer IV neurons, and found that the many intrinsic cortical circuits have comparatively weaker synapses, with stronger connections arising from smaller extrinsic sources such as the thalamus. Such an inverse functional weighting is corroborated by physiological studies on the efficacy of TC and CC synapses (Stratford et al. 1996; Gil et al. 1999). An inverse correlation between anatomical projection size and functional weight may be characteristic of the thalamus, e.g., the functionally potent retinogeniculate input is anatomically much smaller (~6%) than modulatory (e.g., aminergic) inputs to the lateral geniculate nucleus (Sherman and Guillery 2002). This rule predicts that smaller intrinsic contributions, such as in areas Te and Ve, would be more salient synaptically than the larger intrinsic contributions in AES, and it implies a prospectively potent role for branched auditory forebrain axons, which are usually <2% of the total projection (Kishan et al. 2008). However, further physiological experiments will be required to properly assign the synaptic weights to the corresponding anatomical projection.

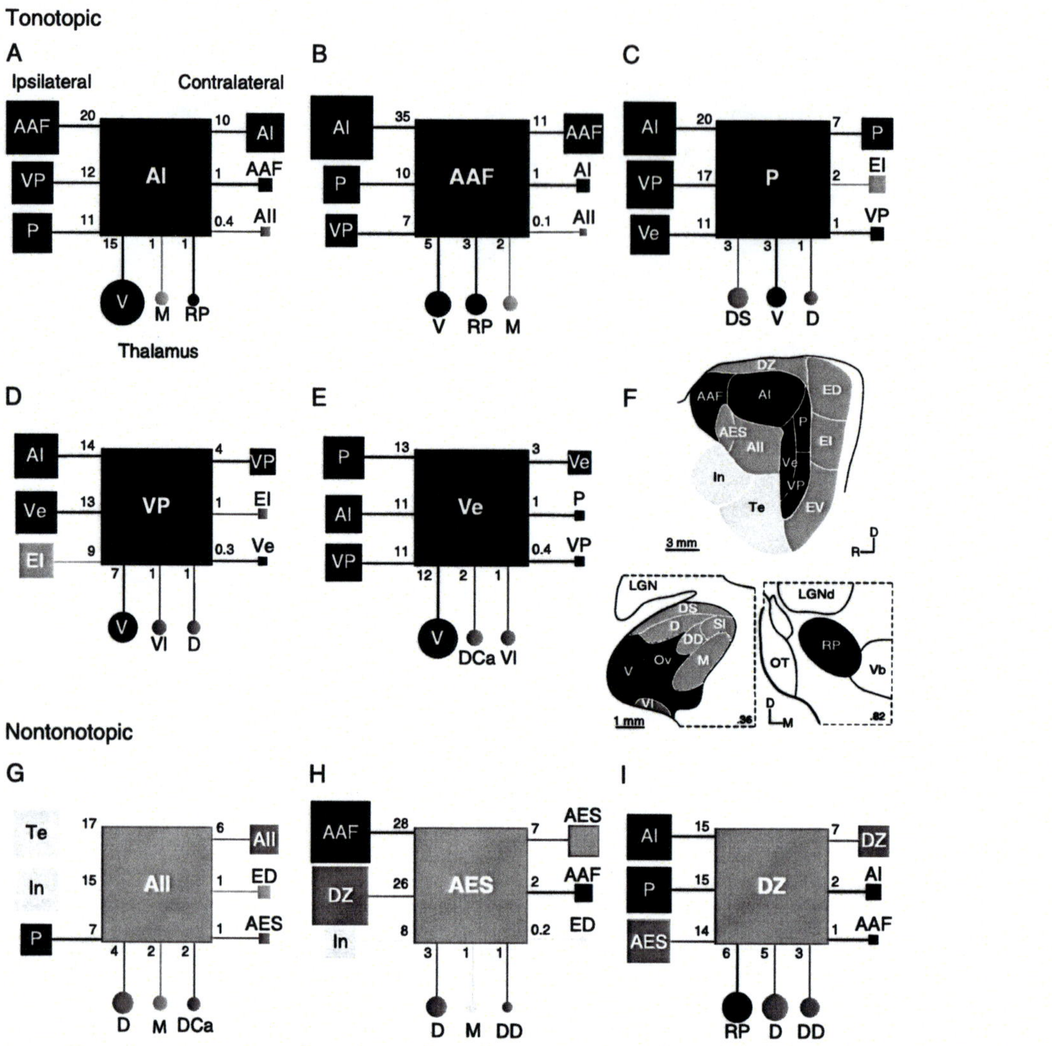

Fig. 7.5 Summaries of convergence of MGB and AC projections in tonotopic and non-tonotopic areas. The three largest inputs to tonotopic and non-tonotopic areas from the MGB, and ipsi- and contralateral AC. *Large filled squares*, projection targets from MGB (*circles*), ipsilateral AC (*left-side squares*), and contralateral AC (*right-side squares*) sites in tonotopic (*black*), non-tonotopic (*dark gray*), multisensory (*medium gray*), and limbic (*pale gray*) regions. Size of *small boxes* and *circles* is proportional to input percentages (numbers). **a–e** Tonotopic areas (*large black boxes*) receive their main input from other tonotopic nuclei and areas. In each, the ventral division is among the strongest MGB input, while the homotopic AC is the major contralateral input. Ipsilateral AC input is predominantly from other tonotopic sources. **f–h** Nontonotopic areas (*large dark gray boxes*) receive strong MGB dorsal division input, with the homotopic area again the main contralateral AC input. The ipsilateral AC input arises largely from adjacent tonotopic, non-tonotopic, multisensory, and limbic areas (adapted from Lee and Winer 2010)

5.2 Extrinsic Input

Extrinsic projections form the remaining half of the total convergent input (Table 7.1). The ipsilateral AC projections dominate and contribute ~75% (or ~37.5% of the total combined input), while the thalamocortical (~15% extrinsic; ~7.5% total) and commissural (~10% extrinsic; ~5% total) inputs are nearly an order of magnitude less. At this population level, convergent MGB and AC inputs are independent of functional type (Table 7.5), and no AC area connects only with either the thalamus or cortex, even in regions as remote functionally as AI (Fig. 7.5a) and area In (Fig. 7.6e). This conserved distribution of global extrinsic inputs suggests shared functional and developmental constraints for all AC areas (Kaas 1995; Catalano and Shatz 1998), and may also conserve topography between areas (Lee and Winer 2005). Ontogenetic (Pallas and Sur 1993) and experience-dependent (Catalano and Shatz 1998; Rutkowski and Weinberger 2006)

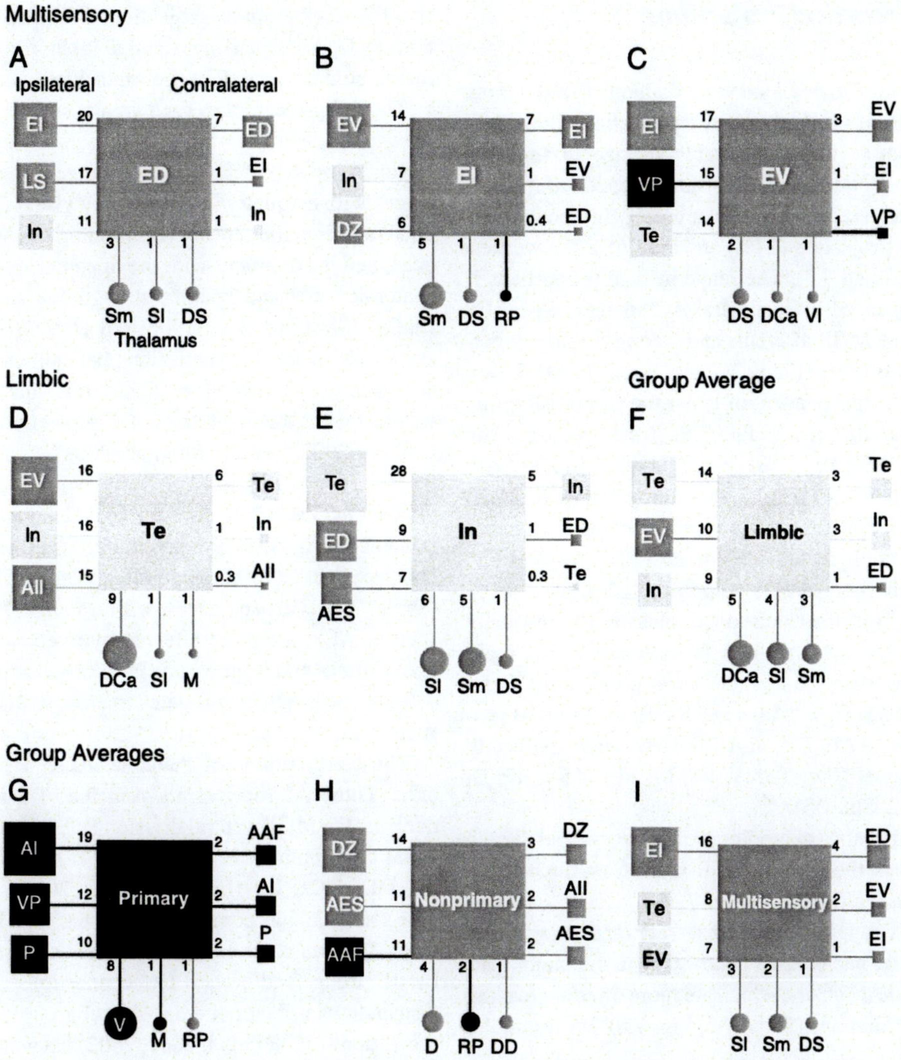

Fig. 7.6 Summaries of convergence of MGB and AC projections in multisensory and limbic areas, and group averages. The three largest inputs to multisensory and limbic areas and to each areal group from the MGB and AC. *Large filled squares*, MGB projection targets (*filled circles*) and ipsilateral AC (*left-side squares*) and contralateral AC (*right-side squares*) sites in tonotopic (*black*), non-tonotopic (*dark gray*), multisensory (*medium gray*), and limbic (*pale gray*) regions. Sizes of *small boxes* and *circles* reflect the input percentages (numbers). **a–e** Multisensory (*medium gray boxes*) and limbic areas (*large light gray boxes*) receive their primary MGB input from shell nuclei (DCa, DS, Sm). The contralateral projections are chiefly from the homotopic area, and ipsilateral AC input is from other multisensory and limbic areas. **f–i** Projection averages for each group capture the self-connected nature of each areal family. Thus, **g** tonotopic, **h** non-tonotopic, **i** multisensory, and **f** limbic groups all share strong input from MGB and AC sources within the same family

processes may constrain the number and type of each extrinsic input across areas.

At a finer scale, the local variation of nuclear and areal inputs and GABAergic organization likely endows each area with a specific identity, and supports the idea of connectional families based on common inputs and shared functions (Lee and Winer 2008a, b, c). Each family (tonotopic, non-tonotopic, multisensory and limbic) has shared cortical and thalamic connections. Tonotopic regions receive their largest input from tonotopic nuclei and areas, while non-tonotopic areas and nuclei are also connected preferentially (Fig. 7.5 and Table 7.3), perhaps to coordinate the shared physiologies among family members. However, smaller projections between family groups may be more salient (Binzegger et al. 2004), as with the intrinsic projections, and link groups. Smaller connections can link otherwise functionally unrelated areas, such as AI and In (Figs. 7.5a and 7.6e and Table 7.3). Each thalamic and cortical input may have variable influence in an area, and the range of these potential roles is explored below.

5.3 Thalamocortical Projections

Every auditory cortex area receives a thalamic (MGB) input averaging ~15% of the total extrinsic projection (Table 7.2). The MGB input is thus comparable in size to the total commissural projection (see below) and a fraction of the ipsilateral projection (see below). Insular cortex (In) receives the strongest MGB projection (24%) and area AES the weakest (4%) (Tables 7.2 and 7.3). The range in such proportions is often counterbalanced by ipsilateral AC sources, e.g., AES has relatively few MGB afferents and among the largest corticocortical inputs (86%) (Table 7.2 and Fig. 7.7). MGB projection strength is independent of either the functional group or anatomical arrangement (Table 7.5). Tonotopic areas can have large (AI: 18%; Ve: 17%) (Fig. 7.2d–e) or small (AAF: 11%, VP: 11%; P: 9%) (Fig. 7.2a–c) proportions of MGB input (Table 7.2). Similarly, anatomically remote (i.e., unconnected) and functionally unrelated fields (AES: 5% and EV: 7%) might receive comparable proportions of input.

The MGB projections are often distributed unequally among nuclei. Areas AI, Ve, and Te receive thalamic input four times larger than the next strongest projection (V and DCa, respectively) (Figs. 7.5a, e and 7.6d), while MGB projections to areas AAF, DZ, and ED arise more uniformly from several nuclei (Fig. 7.5b:V, RP, M; Fig. 7.5 h:RP, D, DD; Fig. 7.6b:Sl, Sm, DS).

The tonotopic, non-tonotopic, multisensory, and limbic areas each receive their major MGB input from nuclei with similar functional affiliations (Fig. 7.6f–h). Each area also receives a wider range of (usually lesser) input from functionally dissimilar nuclei. This contrasts with the major ipsi- and contralateral AC inputs, which are more often segregated by functional affiliations (Table 7.3, Fig. 7.8). For example, area DZ receives major projections from non-tonotopic MGB nuclei (D, DD), and substantial input from the tonotopic rostral pole nucleus (RP) (Lee and Winer 2008a).

Table 7.2 Percentages of input from the thalamus, ipsilateral, and contralateral cortex. Ipsilateral input dominates (75%); thalamic (15%) and contralateral (10%) sources are far smaller

Area	Thalamus	Ipsilateral	Contralateral
AI	18.1	70.2	11.7
AAF	10.9	76.4	12.7
P	9.4	77.6	13.0
VP	11.3	84.2	4.5
Ve	17.1	77.8	5.1
AII	13.1	77.8	9.1
AES	4.5	86.1	9.4
DZ	20.2	70.5	9.3
Te	14.7	78.5	6.8
In	23.8	69.9	6.3
ED	12.0	78.8	9.2
EI	14.8	75.6	9.6
EV	6.7	87.7	5.6
Average	13.6	77.8	8.7
Std. Dev.	5.4	5.7	2.8

The TC projection in each area is smaller than the intrinsic and ipsilateral cortical input, comprising ~15% of the extrinsic input (~7% of the convergent input). This value agrees with estimates in the primary visual (Binzegger et al. 2004) and somatic sensory (Benshalom and White 1986) cortices, and is congruent with the apparent mismatch between anatomical size and synaptic strength found in physiological studies (Stratford et al. 1996; Gil et al. 1999). Indeed, the small size of the TC projections belies their potential functional impact on auditory areas, one that may match or exceed that of the much larger CC projections (Sherman and Guillery 2002; Lee and Sherman 2008).

The TC contribution also varies widely, from 4% (AES) to 24% (In; Table 7.2), confirming that the relative functional influence is similarly broad. Smaller TC projections are often counterbalanced by correspondingly larger CC and CO input (Table 7.2). Perhaps areas with smaller TC connections, such as AES, are more affected by selective thalamic damage (Carrera and Bogousslavsky 2006), leading to functional deficits that might be mitigated in areas with larger TC input, such as In.

The contributions of thalamic nuclei are also area specific. Thus, AI receives its principal TC input from one nucleus, the MGB ventral division (V), whereas area AII has equal contributions from many MGB dorsal division nuclei (D, DD, DCa, M) (Lee and Winer 2008a). This suggests that some unique computation occurs in AI and in V and that AI serves some larger role for all other tonotopic fields by its strong feedforward CC connections (Fig. 7.8). Such distributions reflect potential TC subgroups; direct thalamic projections (Miller et al. 2001, 2002) from one nucleus to an area (e.g. V to AI), while input from many nuclei (e.g., DS, V, and D to area P) represents convergent and/or extraauditory streams (Aitkin and Dunlop 1968; Bordi and LeDoux 1994). Despite their smaller size, these distributed TC projections may have a functional salience inversely correlated with numerical size (Winer et al. 2005).

Thalamocortical projections also differ with respect to their areal and laminar terminations in AC (Huang and Winer 2000). Comparing thalamocortical inputs from both retrograde and anterograde data illustrates three general patterns (Fig. 7.9), e.g., nuclear input proportional to termination strength (Fig. 7.9b), nuclear input greater in proportion to termination strength suggesting (Fig. 7.9d), or nuclear input lesser in proportion to termination strength (Fig. 7.9c). Each pattern implies a difference in functional salience. For instance, balanced contributions (Fig. 7.9b) could support the reliable transmission of essential information, while weaker inputs (Fig. 7.9c) may modulate cortical activity. Thus, fully determining the functional effects of AC thalamic input must

Table 7.3 Average percentage of extrinsic input to each auditory area originating from each thalamic nucleus (top rows), ipsilateral cortical area (middle rows), and contralateral cortical area (bottom rows)

	AI	AAF	P	VP	Ve	AII	AES	DZ	Te	In	ED	EI	EV
V	15	5	3	7	12	1	1	2	1	0.2	0	0.1	2
RP	1	3	0.3	0.4	0.2	0.2	0.1	6	0	0	0.2	1	0
DS	0.1	0.4	3	0.3	0.4	1	0.1	1	1	1	1	1	2
D	1	1	1	1	0	4	3	5	0.2	1	0.4	0	0.2
DD	0.2	1	0	0	0	1	1	3	0	1	0.2	0	0.1
DCa	0	0	0.3	0.1	2	1	0	0	9	1	0	0	1
Sl	0.1	0.2	1	0.3	1	2	0.1	1	1	6	1	1	0.4
Sm	0	0	0.2	0.1	0.2	1	0	1	0.4	5	3	5	0
Vl	0	0	1	1	1	0.2	0	0	0.3	0.2	0.1	0.1	1
M	1	2	0.1	1	1	2	1	0.3	1	1	1	1	1
Other	0.1	0	0.2	0.1	0.2	1	0.3	1	1	7	5	7	0
AI		35	20	14	11	2	1	15	0	0.1	0	0	0
AAF	20		2	3	3	1	28	3	0	0	0	0	0.3
P	11	10		5	13	1	0.2	15	0	0.1	1	1	4
VP	12	7	17		11	2	3	1	2	0.2	0	0	15
Ve	8	5	11	13		6	6	3	12	0.3	0	0	1
AII	5	5	0.2	5	5		2	4	15	3	0.2	0.1	3
AES	0.1	1	2	3	0.4	8		14	0	7	2	2	1
DZ	6	2	2	6	11	1	26		0	2	7	6	0
Te	1	1	3	4	4	17	1	2		28	3	2	14
In	1	0	4	1	1	15	8	3	16		11	7	5
ED	1	1	6	7	1	7	2	6	1	9			3
EI	3	3	5	9	1	5	4	2	13	7	20	13	17
EV	1	3	5	9	3	3	1	1	16	4	6	14	
Other	1	3	1	5	14	9	3	1	4	10	28	30	24
AI	10	1	0.4	0.1	0.1	0.2	0.1	2	0	0	0.2	0	0
AAF	1	11	0	0.1	0	0	2	1	0	0	0	0	0
P	0.2	0	7	0.3	1	0.2	0	0	0	0	0	0	0
VP	0	0	1	4	0.4	0	0	0	0	0	0	0	1
Ve	0.1	0	1	0.3	3	0.1	0	0	0.2	0	0	0	0.2
AII	0.4	0.1	0	0.1	0.2	6	0.2	0	0.3	0.2	0	0	0
AES	0	0	0	0	0	1	7	0	0	0.1	0	0	0
DZ	0.2	0.1	0	0	0	0.1	0.3	7	0	0.1	0.3	0.1	0
Te	0	0	0	0	0.4	0.3	0	0	6	0.3	0	0	0.2
In	0	0	0.4	0	0	0.4	0	0	1	5	1	1	0
ED	0	0	1	0.2	0	1	0.2	0.1	0	1	7	0.4	0.1
EI	0	0	2	1	0	0.1	0	0	0.1	0.1	1	7	1
EV	0	0	0.1	0.3	0	0	0	0	0.1	0	0.1	1	3
Other	0	0	0	0	0	0	0	0	0	0.1	1	0.4	1

also consider the wide ranging proportion of cells, their relative termination strength, and their laminar and postsynaptic targets.

5.4 Corticocortical Inputs

Corticocortical projections comprise ~75% of the total AC projection (Table 7.2). EV has the largest input (88%), area In the smallest (70%), and there is a narrower range for these values than in the thalamic or commissural projections (Table 7.2). As with the thalamic inputs, these overall weights are independent of known functional or anatomical arrangements (Table 7.2 and Fig. 7.10). Non-tonotopic regions can have above (AES: 86%) or below (DZ: 70%) average percentages of corticocortical input.

Nearly half the total ipsilateral input arises from the three principal sources (Figs. 7.5 and 7.6: ipsilateral). The major ipsilateral inputs to 8/13 areas are ~15% each. Functionally unrelated areas such as VP (14%: tonotopic), DZ (15%: non-tonotopic), and Te (16%: limbic) each receive similar weights of input from their three main ipsilateral AC sources (Figs. 7.5c, h and 7.6a). A homogenous weighting of input suggests widely dispersed but quantitatively similar distributed projections among areas. In areas AAF

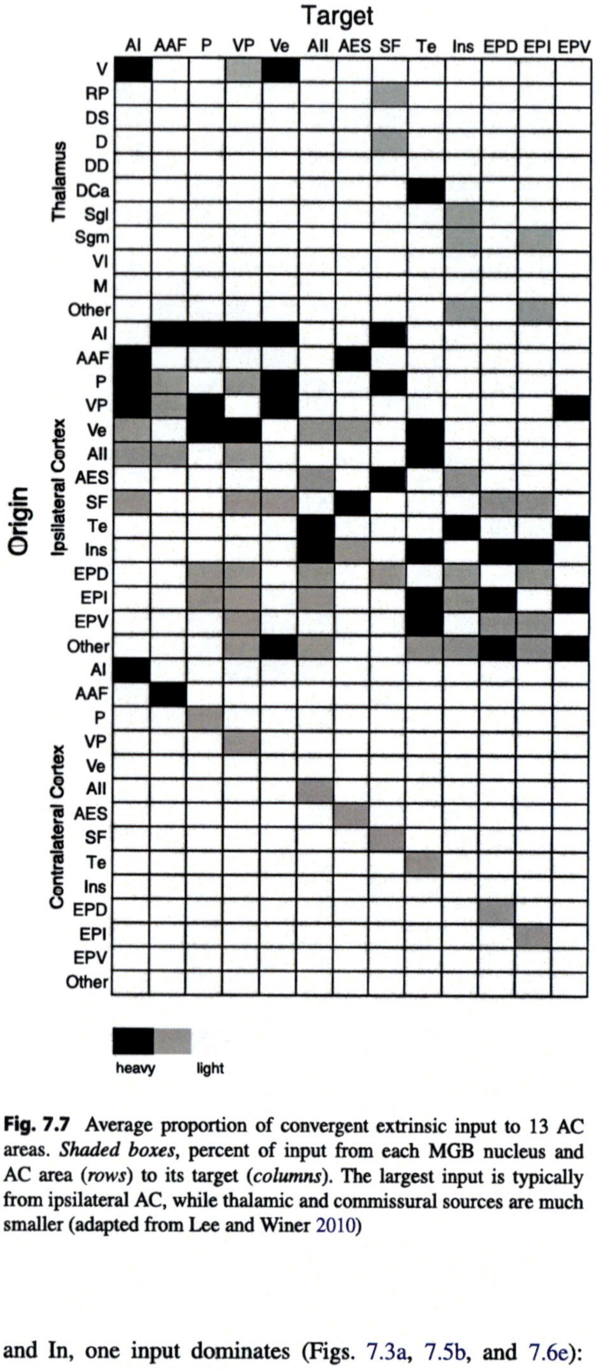

Fig. 7.7 Average proportion of convergent extrinsic input to 13 AC areas. *Shaded boxes*, percent of input from each MGB nucleus and AC area (*rows*) to its target (*columns*). The largest input is typically from ipsilateral AC, while thalamic and commissural sources are much smaller (adapted from Lee and Winer 2010)

and In, one input dominates (Figs. 7.3a, 7.5b, and 7.6e): area In receives ~28% of its extrinsic input from Te, and <10% from the next strongest sources (ED: 9%; AES: 7%) (Fig. 7.6e).

The largest AC input to an area often arises from within the same class (Table 7.3), i.e., tonotopic, non-tonotopic, etc. Tonotopic areas are the most stereotyped, with all but VP receiving their three principal inputs from other tonotopic regions (Fig. 7.5a–e). Limbic and multisensory regions have more variable input, with major non-tonotopic projections (Fig. 7.6a–e). Non-tonotopic areas receive the widest range of tonotopic, multisensory, and limbic input (Fig. 7.5f–h), suggesting that they link these three systems. This is more marked for the chief projections to areas within a group, with non-tonotopic areas collectively receiving input from non-primary and primary regions (Fig. 7.6f–h).

Like the other extrinsic inputs, each area receives variable CC contributions, with EV receiving the largest (88%) and In the smallest (70%). Because of the larger proportion and number of CC connections, this range is smaller than that in the TC and CO projections (Table 7.2). Most areas receive nearly equivalent inputs (~15%) from their three top sources (Figs. 7.5 and 7.6), implying that many ipsilateral projections follow a common plan (Winer and Lee 2007), and it might explain why selective cortical lesions can have a negligible influence on activity (Kitzes and Hollrigel 1996) and why inactivation of whole areas is required to elicit global deficits (Malhotra et al. 2004).

The reciprocity between areas can be quantified as an average of reciprocal (+1.00), neutral (0), and non-reciprocal (–1.00) connections (Table 7.4). With this metric, all areas are reciprocally connected (indices >0.00), and none have dominant non-reciprocal connections (indices <0.00). Reciprocal AC projections are similar in size (Table 7.4), except in a few cases, where they are not reciprocal, e.g., EI to VP (8.9 vs. 0.1%; Figs. 7.5d and 7.6b). Such patterns are favorable candidates for inactivation studies (Lomber et al. 2007), since they represent strong examples of serial processing. Thus, the tonotopic, multisensory, and limbic areas are asymmetrically weighted, and limbic areas send more non-reciprocal input to tonotopic areas (Table 7.3). Some areas have stronger reciprocal connections (AI, AES, In) and others weaker reciprocity (P, AII, EI) (Table 7.4). Such differences are independent of functional affiliations (tonotopic, non-tonotopic, etc.) (Table 7.5) and support the familial affiliations proposed here, where a group member should have common as well as unique relations with its neighbors. Non-reciprocal projections are rare, and they reflect inputs between functional groups, e.g., the EI–VP projection, which flows heavily from EI to VP, but is not reciprocated (Figs. 7.5d and 7.6b).

Many ipsilateral cortical connections are reciprocal (Table 7.4), suggesting that interconnected areas influence each other to a similar degree. It is unclear whether such reciprocity is also maintained on a laminar or cellular basis, a question requiring anterograde studies. Non-reciprocal mismatches suggest an ordinal area-to-area sequence of information flow and distinguish the corticocortical system from the corticothalamic stream, which is far more constrained connectionally and reciprocally (Winer and Larue 1987; Deschênes et al. 1998). Interestingly, both EI and VP have similar sized projections with EV. Thus, the effects of

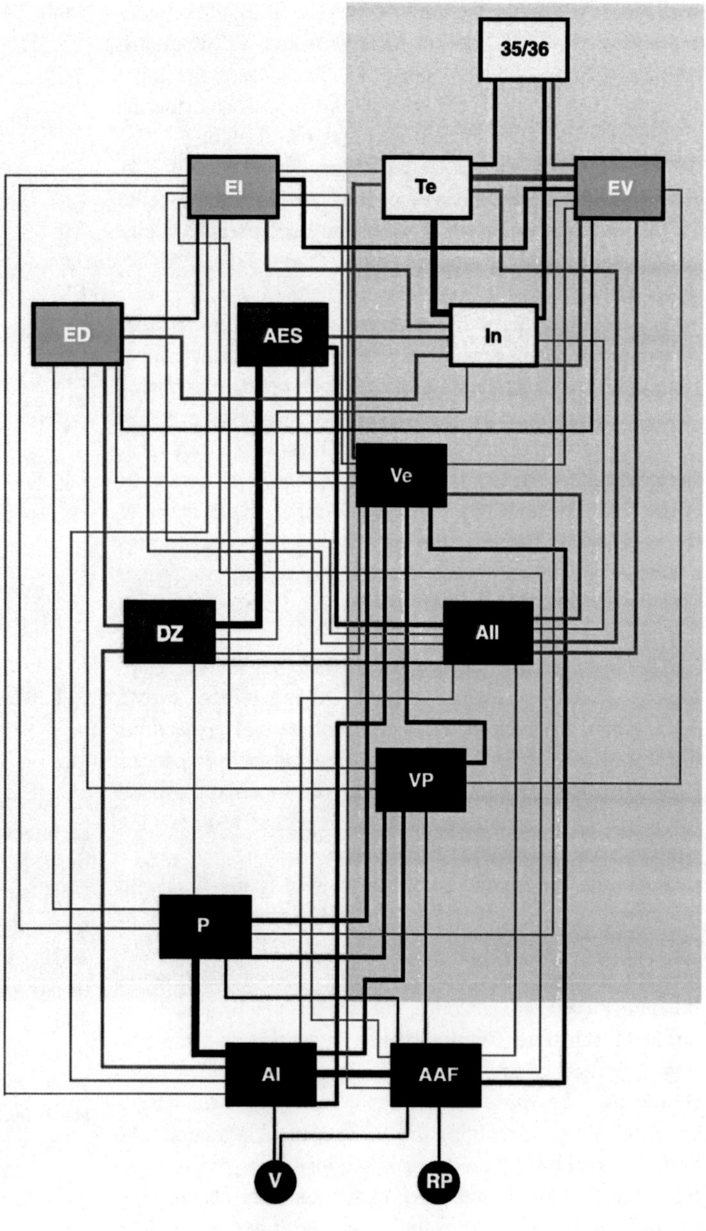

Fig. 7.8 Hierarchical connections of AC areas ordered by the projection laminar origins (Lee and Winer 2008c) from lowest to highest, with AI and AAF at the base and areas 35/36 at the top (Felleman and Van Essen, 1991). Connectional strength is indicated by line thickness: strong (*thick lines*), medium (*bold lines*), and weak (*thin lines*). When connectional strengths are non-reciprocal, the average strength is indicated. *Dark black lines* connect areas within the same family, while *gray lines* connect areas in different families. Laminar origins of projections are indicated by the line origins on the area boxes: supragranular (*top of box*), bilaminar (*side*), infragranular (*bottom*). Areas can also be grouped according to putative functional relations: dorsal areas (*light gray background shading*) may analyze auditory space, while ventral areas (*gray background shading*) may be primarily responsible for spectral analysis (adapted from Lee and Winer 2010)

mismatches could be mitigated by compensatory connections via an intermediary area.

Finally, the widespread CC connectivity provides many more opportunities for information transfer between families than the more constrained TC and CO systems. While the major inputs to each area are within-group, many other, smaller connections arise outside the group (Table 7.3). Thus, a key CC role may be to link otherwise separate functional groups. Non-tonotopic areas receive the most varied inputs (Fig. 7.3c and Table 7.3), suggesting a vital role in linking groups.

5.5 Commissural Projections

The commissural input is the smallest and most stereotyped projection (Fig. 7.4), <10% of the extrinsic system (Table 7.2). Area P has the largest (13%) and area VP the smallest (4.5%) commissural input (Fig. 7.10). The contribution is independent of functional group (Table 7.5), with tonotopic (9.4±4.2%), non-tonotopic (9.3±0.2%), multisensory (8.1±2.2%), and limbic (6.6±0.4%) areas each receiving similar proportions (Fig. 7.4). Unlike the ipsilateral AC inputs, the major contralateral input is homotypic (Fig. 7.4 and Table 7.3), irrespective of functional affiliation (Figs. 7.5, 7.6, and 7.4). The homotypic input is more than half of the total commissural contribution (Lee and Winer 2008b), and averages 6% (range: 3–11%) of the total extrinsic projection (Figs. 7.5, 7.6, and 7.4). However, the homotypic projections are half the size of the major ipsilateral cortical inputs, and comparable in size to the main thalamic inputs (Table 7.2 and Figs. 7.5, 7.6, and 7.4).

The residual heterotypic commissural sources are an order of magnitude weaker, often <1% of the total extrinsic input (Table 7.3); this implies a more point-to-point relationship in commissural than in ipsilateral operations. However, the main heterotypic sources in 9 of 13 areas arise from functionally related areas, i.e., tonotopic, non-tonotopic, etc. (Figs. 7.5, 7.6, and 7.4). Heterotypic inputs have many of the same areal sources as the main ipsilateral cortical inputs (Table 7.3), with similar topographic origins (Lee and Winer 2008b), e.g., area Te receives ipsilateral and contralateral AII and In input (Fig. 7.6a).

The commissural projections are also organized reciprocally (Table 7.4), with 8/13 areas having indices >0.50. As a group, the reciprocity of the contralateral projections does not differ from that of the ipsilateral AC projections (0.54±0.13 vs. 0.47±0.22), nor between functional groups (Table 7.5). Reciprocal projections in either system may enable strong coactivation among areas.

A clue to the unique commissural role is the dominant homotypic input in all areas, which is more than half of the commissural input (~6% of the total extrinsic). Since one commissural function is to unify sensory information bilaterally (Gazzaniga 2000), the homotypic projection is an ideal substrate for combining interhemispheric auditory influence at an areal level. Thus, this smaller input source may be more salient synaptically than the larger CC projections, resembling the TC system. Contralateral AI deficits might have more impact than ipsilateral projections, e.g., area P to AI or area AAF to AI, and may account for the behavioral deficits in sound localization in contralateral inactivation studies (Malhotra et al. 2004; Lomber et al. 2007).

Other heterotypic input sources are often reciprocal (Table 7.4). These sources vary in size across areas, but are usually <1% of the total extrinsic input. Although heterotypic sources are nearly an order of magnitude smaller than the homotypic projection, they could trigger widespread physiological interhemispheric activation (Bozhko and Slepchenko 1988), and perhaps align commissural and ipsilateral operations, which are mirrored topographically (Lee and Winer 2005). The wide numerical range in projection strength in each system may allow smaller, more agile inputs to act as functional fulcrums on large neuronal populations.

Table 7.4 Reciprocity index. Ipsilateral and contralateral projections are reciprocally organized in all areas (indices >0.0)

Area	Ipsilateral[a]	Contralateral[a]
AI	0.62	0.62
AAF	0.38	0.69
P	0.23	0.46
VP	0.54	0.46
Ve	0.31	0.62
AII	0.23	0.54
AES	0.69	0.38
DZ	0.46	0.54
Te	0.69	0.77
In	0.77	0.61
ED	0.38	0.31
EI	0.07	0.38
EV	0.69	0.62
Average	0.47	0.54
Std. Dev.	0.22	0.13

[a]Reciprocity index: +1.00 (symmetric) to –1.00 (asymmetric).

6 Organizational Features

6.1 Modularity

The apparent absence of connectional modularity except in AI is striking and suggests a unique anatomical-functional architecture. Many areas have clusters of focal projections, but these are particularly marked in AI. The fine-grained local AI organization is evident in experiments with small tracer deposits (Matsubara and Phillips 1988; Read et al. 2001). Further mapping and tracer studies may reveal a degree of modularity in areas such as AAF (Imaizumi et al. 2004b). A future challenge, especially in non-primary areas, is to dissect their intrinsic connections. If AI alone has such modularity, then it stands apart and has the closet AC resemblance to visual area 17 (Lund et al. 1995) and somatic sensory area 3 (Jones and Porter 1980).

6.2 Topography

A major organizing feature of the primary AC is the topographic representation of best frequency (Merzenich et al. 1975). In cat AC, five areas are tonotopically organized. Such tonotopic maps are absent or highly reduced in most other areas (Schreiner and Winer 2007). Despite the lack of functional topography in these areas, connectional studies find that all auditory areas receive equally precise projections from thalamic and cortical sources (Lee and Winer 2005). These provide an ideal anatomical substrate for communication between the families of areas by establishing

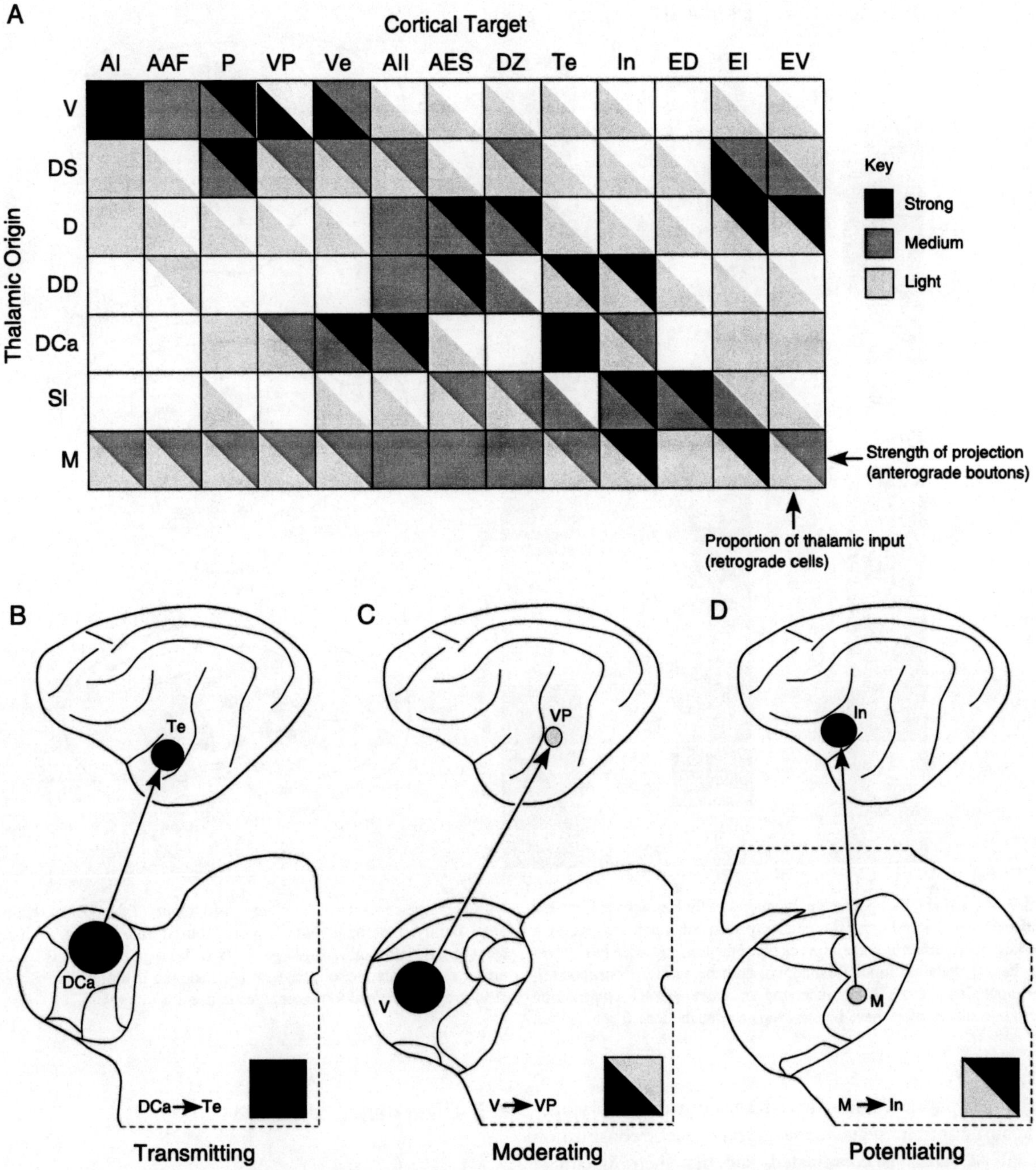

Fig. 7.9 Synthesis of retrograde and anterograde TC studies. **a** A comparison of the proportion of thalamocortical cells from each nucleus (*lower left triangles*) with the number of anterograde boutons in each cortical area (*upper right triangles*) (Huang and Winer, 2000). Each area receives a unique combination of input from multiple thalamic sources, and their strength is area specific. Thus, AI and VP both receive much of their input (>60%) from V, but VP has fewer anterograde boutons than AI. **b–d** Three thalamocortical patterns. **b** The number of anterograde boutons is proportional to the retrograde input from a thalamic nucleus. Individual neurons may contribute equally to the projection, enabling direct patterns of information transfer. **c** Other areas receive input with relatively fewer anterograde boutons, suggesting that a neuron receives weaker projections to moderate activity more pluralistically. **d** Few cells in some nuclei project to an area, but they have dense terminal arrays. Such cells may provide stronger input than in the previous cases, and serve as potent drivers of global cortical activity

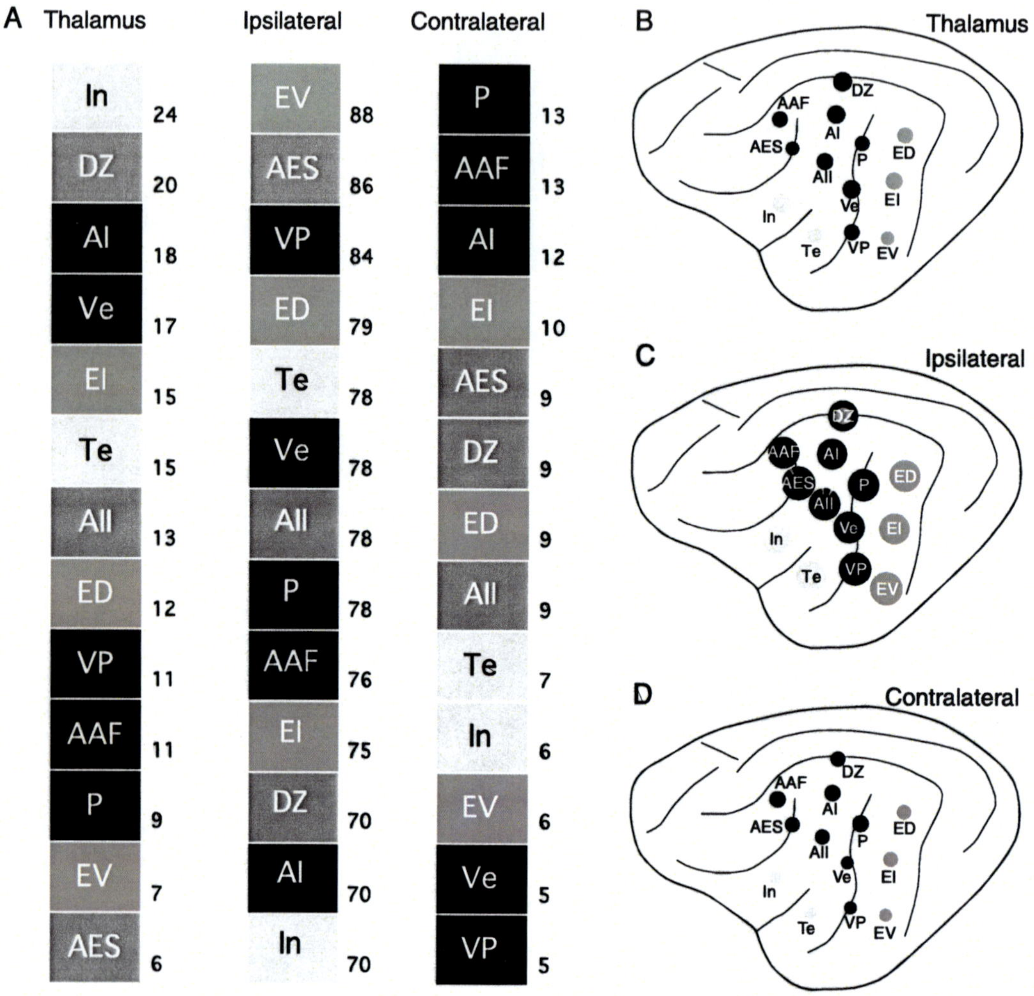

Fig. 7.10 Areal ranking by the percentages of thalamocortical, ipsilateral, and contralateral cortical inputs. **a** In each column, areas receiving the largest contribution from a source are at the *top*, e.g., area In receives the largest thalamic input (Thalamus), and the smallest corticocortical input (Ipsilateral). Areas receiving an above-average contribution from one source often have below-average contributions from another. Numbers, percentage of total extrinsic input (Table 7.2); *shading*, areal family: tonotopic (*black*), non-tonotopic (*dark gray*), multisensory (*gray*), and limbic (*light gray*). **b–d** Average total inputs to each area from **b** thalamic, **c** ipsilateral cortical, and **d** limbic sources. *Dot sizes* are proportional to percentages in **panel a**

a common physical metric for interareal signaling. This topographic uniformity suggests that the ontogenetic construction of all AC areas is coordinated, and that their functional identity arises principally through their differential input, network interactions, and local circuitry.

The proportions of intrinsic and extrinsic inputs by functional groups (Table 7.5) are also statistically similar. Thus, the balance of local and remote connections is area specific (Table 7.2) but not family specific (Table 7.5), while the areal patterns of connectivity are family specific (Lee and Winer 2008c). Such topographies could coordinate cortical and perhaps corticofugal operations across different scales of resolution.

6.3 Divergent Projections

Divergent projections in different AC areas can contribute to multiple feature-specific maps or synchronize processing in remote targets. Such projections could account for the expansion of the two thalamic tonotopic maps to the five cortical maps. However, studies of axonal divergence using dual retrograde tracers (Kishan et al. 2008) find that the proportion of double-labeled cells in the MGB and AC is ~2%, even when tracers are deposited at physiologically matched sites in different areas in a strategy that should maximize any double labeling (Lee et al. 2004a). Thus, massive single cell divergence does not appear to be a major auditory forebrain organizing feature, unlike the subcortical auditory system, where branched axons can be far more numerous (Irvine 1986). Despite their paucity, the few branched projections may have a functional impact inversely proportional to their anatomical size (Binzegger et al. 2004). In addition, the auditory forebrain branching is much less than in the visual (Bullier et al. 1984) and somatic sensory (Spreafico et al. 1987) forebrain.

7 Comparative Organization of Auditory Cortex

From a comparative standpoint, the evolution of mammalian AC areas is an intriguing functional and developmental issue. The number and organization of areas varies widely across species (Reale and Imig 1980; Stiebler et al. 1997; Hackett et al. 1998; Budinger et al. 2000; Bizley et al. 2005) (Figs. 7.1, 7.11, and 7.12). Nuclear and areal physiologies have adapted to specific ecological niches (Xiao and Suga 2004). Can homologous areas be identified, and do the various AC areas support unique behavioral requirements?

Convergent thalamic and cortical inputs are independent of functional type (Fig. 7.12 and Table 7.5). No AC area exclusively connects with either the thalamus or cortex, even in regions as remote functionally as AI and In (Fig. 7.7). However, on a finer scale, an area's specific connectivity with different nuclear and areal groups imbues it with a functional identity. This suggests that the parcellation into tonotopic, non-tonotopic, multisensory, and limbic-related areas is valid and may capture the functional axes in AC. A highly conserved, locally differentiated connectivity pattern suggests an ontogenetic and evolutionary mechanism for creating new auditory areas (Fig. 7.11).

The classic evolutionary view of new biological structures posits an initial duplication of pre-existing structures, then their subsequent adaptation to serve new functional requirements (Hall 2003). Auditory areas so established by descent with modification may thus distribute computational processes across many areas, which otherwise collapse or are absent in AC in other species (Frost et al. 2000).

The ontogenetic program specifying thalamic and cortical connectivity of structures may extend to new areas (Diamond and Hall 1969), with subsequent specialization being cued by both particular developmental programs and experience (Kral and Eggermont 2007). Cortical areas are ontogenetically pluripotent (Pallas and Sur 1993), with their functional capabilities assigned by their specific connections and activity (Catalano and Shatz 1998) and refined by experience-dependent (Rutkowski and Weinberger 2006) processes (see Chapters 21 and 22).

A subsequent fractionation of connections across nuclei and areas may reflect this later specialization, with subtle differences defining their specialized roles. Cortical areal duplication may expand processing capabilities, allowing additional areas to be recruited as processing demands emerge ecologically (Eggermont 1999). Areas share many similar common processing motifs (neuron types, laminar architecture, topography, local circuit organization), while local differences allow an expanded and expansive functional repertoire (multiple tonotopic maps, area-specific effects of lesions, differential roles in localization) (Lee et al. 2004a). Modular cortical connections allow multiple degenerate combinations of areas and nuclei to converge upon similar computational outcomes.

8 Synthesis of Auditory Cortical Connections

8.1 Hierarchical Models

Hierarchical models begin with the receptor epithelium, and imply serial processing among successive areal groups. This idea has been most fully developed in the monkey visual system (Felleman and Van Essen 1991). The increased

Table 7.5 Comparison of inputs by functional group. Percentages of extrinsic and intrinsic input are similar and independent of group (ANOVA, $p>0.05$), as is reciprocity (ANOVA, $p>0.05$)

Projection	Tonotopic	Non-tonotopic	Multisensory	Limbic	F	p
Intrinsic	50.8 ± 10.2; 104.8[a]	51.2 ± 16.8; 280.6	47.2 ± 3.8; 14.4	44.3 ± 11.4; 129.6	0.22[b]	0.88
Extrinsic (total)	49.2 ± 10.2; 104.8	48.8 ± 16.8; 280.6	52.8 ± 3.8; 14.4	55.7 ± 11.4; 129.6	0.22	0.88
Extrinsic (individual)						
Thalamocortical	13.4 ± 4.0; 15.6	12.6 ± 7.9; 61.8	11.2 ± 4.1; 16.9	19.2 ± 6.4; 41.4	0.98	0.44
Corticocortical	77.2 ± 5.0; 24.8	78.1 ± 7.8; 60.9	80.7 ± 6.3; 39.3	74.2 ± 6.1; 37.0	0.47	0.71
Commissural	9.4 ± 4.2; 17.9	9.3 ± 0.2; 0.02	8.1 ± 2.2; 4.9	6.6 ± 0.4; 0.1	0.50	0.69
Reciprocity						
Ipsilateral	0.42 ± 0.16; 0.02	0.46 ± 0.23; 0.05	0.38 ± 0.31; 0.31	0.73 ± 0.06; 0.01	1.29	0.34
Contralateral	0.57 ± 0.10; 0.01	0.49 ± 0.09; 0.01	0.43 ± 0.16; 0.03	0.69 ± 0.11; 0.01	2.13	0.17

[a]Values: Mean percentages ± std. dev.; variance.
[b]ANOVA, single factor, df=2.

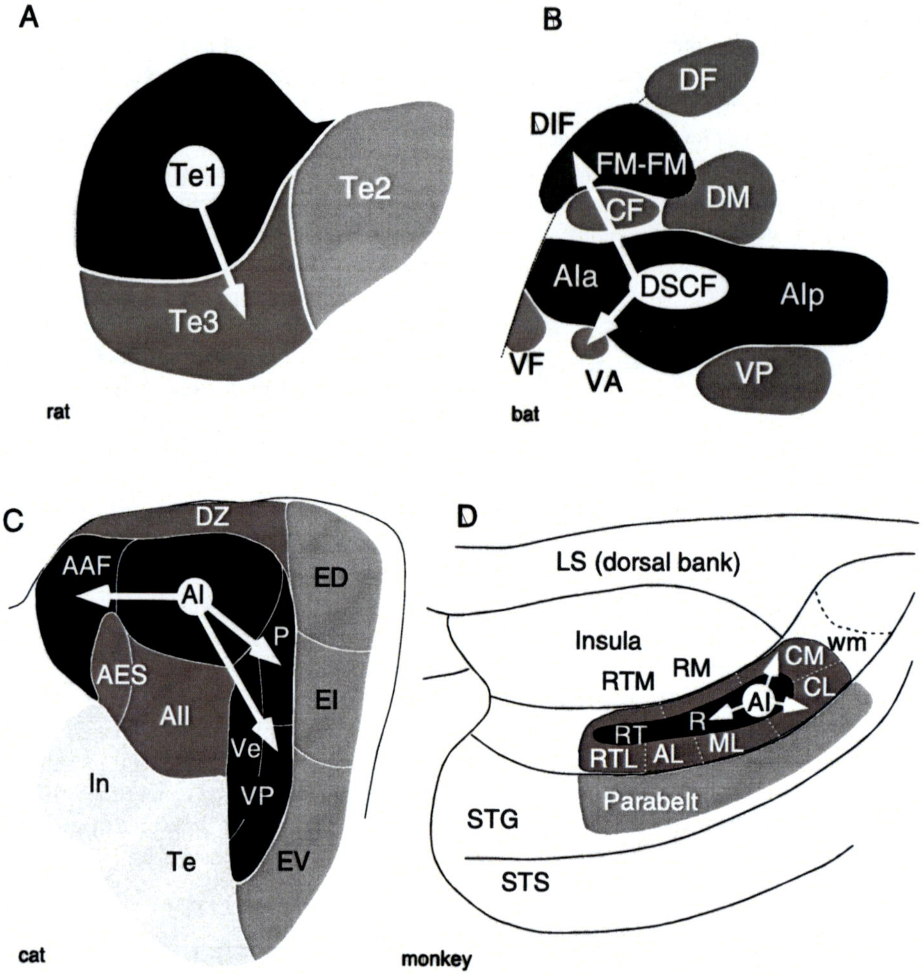

Fig. 7.11 Principal outputs of primary auditory cortex in rat, bat, cat, and monkey. **a** The primary area Te in rat outputs principally to the non-tonotopic region Te3 (Roger and Arnault 1989). **b** The Doppler-shifted constant frequency region in the bat has strong outputs to VA and DIF (Fitzpatrick et al. 1998). **c** In the cat, primary AC is principally connected with other tonotopic areas, AAF, P, and VP (Lee and Winer 2008c). **d** In the monkey, AI connects with both tonotopic core (R) and non-tonotopic (CM, CL) belt areas (Hackett et al. 1998)

complexity and specialization of receptive fields beyond the striate cortex (Desimone et al. 1985) reflects the convergence of simpler input from lower levels of processing. Thus visual areas MT (motion) and V4 (color) (Desimone and Schein 1987; Priebe et al. 2003) process more complex features over larger spatial scales than V1 cells (Angelucci et al. 2002). This may reflect summation of serial input from lower stations to create neurons which embody unique specificities (Gross 2002), and pose a conundrum for distributed processing models.

Hierarchical models have influenced many different theories of auditory cortical organization, from tonotopic processing in the cat (Rouiller et al. 1991) to the areal parcellations into core, belt, and parabelt regions in the monkey (Hackett et al. 1998; Kaas and Hackett 1998) (Fig. 7.11d). Hierarchical models are also compatible with the auditory parallel pathways analogous to those in the visual system (Romanski et al. 1999a, b; Kaas and Hackett 2000).

Analysis of the laminar origins of cat AC projections (Lee and Winer 2008c) using this framework results in an eight-stage hierarchy, with the primary auditory cortex (AI) and the anterior auditory field (AAF) at the base and the parahippocampal regions (35/36) at the apex (Fig. 7.8), respectively (Lee and Winer 2008c). This model is consistent with known physiologies. Thus, the complex tuning curves in AII (Schreiner and Cynader 1984) suggest a higher ordinal position than AI or AAF, whose frequency tuning is sharper (Phillips and Irvine 1981, 1982). Multimodal responses in areas such as AES (Clarey and Irvine 1990) suggest a yet higher categorical assignment. At the highest levels

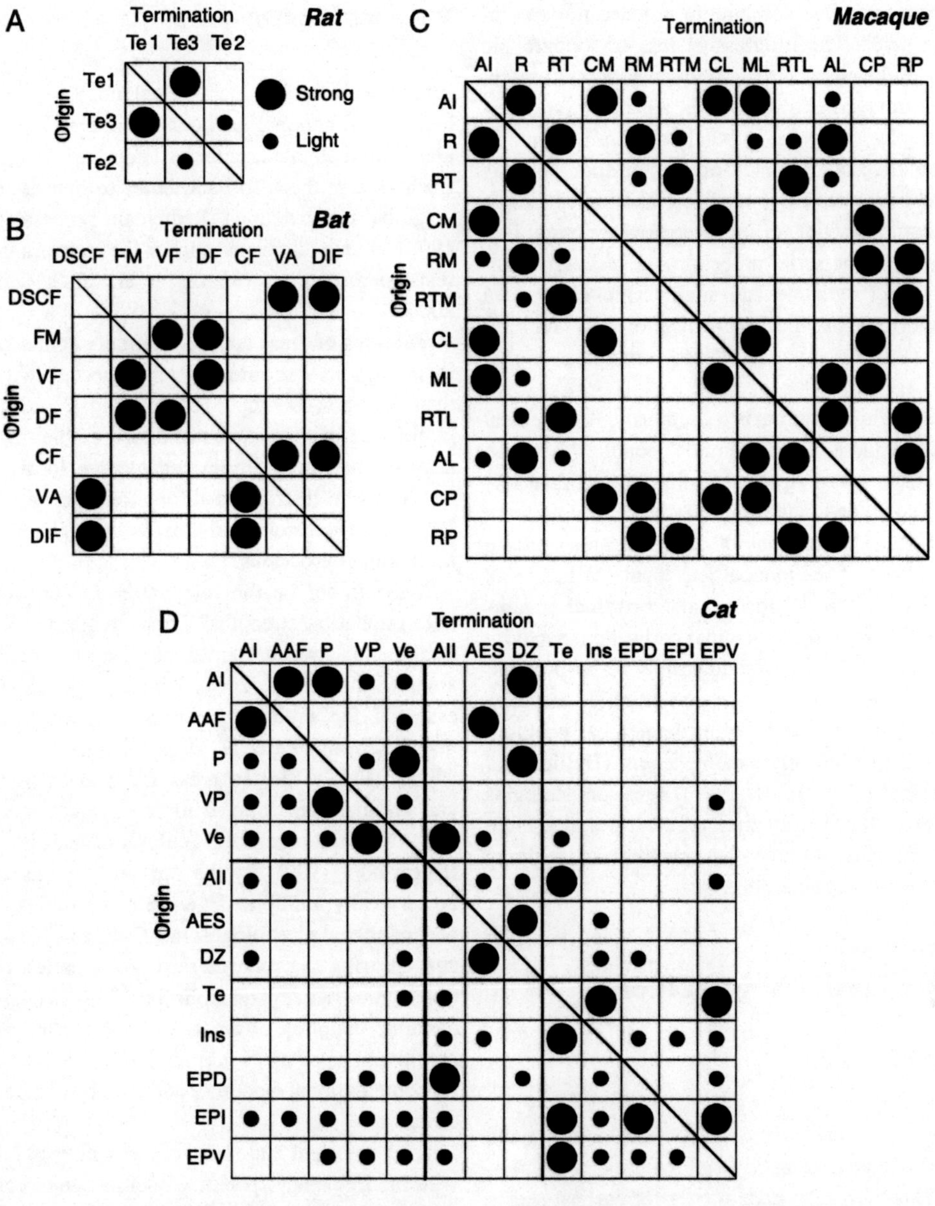

Fig. 7.12 Graphical summary of the ipsilateral corticocortical inputs in the rat, bat, cat, and monkey. Projection size is indicated by dot size. **a** Connections of rat AC (Roger and Arnault 1989). **b** Bat AC connections (Fitzpatrick et al. 1998). **c** Macaque AC (Hackett et al. 1998). **d** Convergent AC input in the cat (Lee and Winer 2008c)

are limbic and parahippocampal areas with roles in cognitive, affective, and memory processes (Squire et al. 2004).

A simple feedforward model uses frequency information from the ventral nucleus of the medial geniculate body to activate early tonotopic areas AI and AAF (Fig. 7.8). The results of these computations next reach higher tonotopic areas such as P, where the complex frequency tuning suggests even more refined or locally specialized processing (Loftus and Sutter 2001). From these higher areas, the now-modified tonotopic information is sent to non-tonotopic areas (DZ, AII), where local processing regimes establish segregated spatial processing in the dorsal stream (DZ) (Stecker et al. 2005) (Fig. 7.8: light gray background) or spectral information in the ventral stream (AII) (Schreiner and Cynader 1984) (Fig. 7.8: gray background), analogous to the primate model (Romanski et al. 1999b). At higher levels, information would reach multimodal areas such as AES and ED (Bowman and Olson 1988; Clarey and Irvine 1990) and then be further integrated at subsequent stations and stored in memory-related centers. Analogous patterns of divergent projections from each area combine to produce more complex receptive fields. The progressive loss of topographic representation in the characteristic frequency domain might reflect that this analysis is already well advanced (and perhaps even completed) in the subthalamic brain stem, in the MGB, and in primary fields of AC; further analysis of this information would not require elaborate tonotopic maps, which could account for the decreased tonotopy in a hierarchical model. The serial processing at successive stations should reflect dynamic and site-specific changes, a prediction confirmed by the emergence of new regimes (larger receptive fields, broader afferent tuning) at higher levels.

This model should be approached cautiously. Hierarchical groupings may be more indicative of the coupling between functionally related areas. Areas at similar hierarchical levels could form computational ensembles, in accord with the many corticocortical projections of each AC area and the physiological model of thalamocortical input (Miller et al. 2001; Winer et al. 2005). Multiple thalamocortical streams (Lee and Winer 2008a) and intralaminar cortical connectivity between levels constrain and violate strictly serial processing models. In the visual cortex, average stimulus response latencies often do not correlate with hierarchical position, or do not differ significantly among levels (Bullier and Nowak 1995; Nowak and Bullier 1997). Areal physiologies may partly derive from hierarchical processes, and a more principled model must include the spectrum of regional connections (Budinger et al. 2008).

9 Distributed Cortical Organization

9.1 *Beyond Hierarchy*

A tenet of hierarchical processing is that serial convergence creates complex RF arrangements (e.g., feature detectors) in higher areas. This view may pertain only for earliest stages of AC processing. Each area receives multiple convergent projections that vary in numerical strength (Fig. 7.7), but with unknown salience, as inactivating one projection may have little, if any, effect on activity (Kitzes and Hollrigel 1996). Thus, even robust AC connections may not elicit a response, but their coincident action could shape receptive fields both selectively and integratively (Angelucci et al. 2002). All AC areas receive thalamic and commissural input, and each provides ~15% of the extrinsic areal input (Figs. 7.5, 7.6, and 7.7 and Table 7.3) (Lee and Winer 2008a, b). Combined with massive corticofugal projections (Winer et al. 2001) this offers alternative avenues for information flow outside a corticocentric framework (Sherman and Guillery 2002; Lee and Winer 2008a, b, c).

9.2 *Areal Ensembles*

A complement to the hierarchical model views areas as members of a functional network formed by the conjunction of areal and thalamic ensembles (Figs. 7.5, 7.6, 7.7, and 7.10). Each area in the web contributes to one or several aspects of global computations. Redundant processing (Eggermont 1998) among elements allows the network as a whole to resist perturbations (Sporns et al. 2002; Izhikevich et al. 2004).

Families of areas are preferentially connected, e.g., tonotopic regions are more interconnected with each other than with non-tonotopic regions (Fig. 7.12d). Connectional families define one feature of the ensemble space, while connections between families enable the local computational results to be distributed throughout the network (Lee and Winer 2010). Specialized processing might thus coexist with more global processes.

What might be the role of an AC network with variable projection strengths? Such architectures are ideal for establishing transient ensembles of functionally connected areas (Sporns et al. 2002; Izhikevich et al. 2004) perhaps essential for auditory scene decomposition and recombination (Hromádka et al. 2008). Thus, AII may transiently couple (Fig. 7.12d) to areas ED and In to bind auditory and visual information, while an alternative union with AI and AAF may permit the analysis of auditory spectral cues (Eggermont 1999). By the same token, this analysis could be served by multiple, degenerate ensembles, taking advantage of shared physiologies in AC regions (Eggermont 1998) and ensuring the robustness to perturbation noted in ablation studies (Kitzes and Hollrigel 1996) through a distributed dynamic coupling of areal ensembles. Multiple degenerate combinations of areas and nuclei may converge upon similar computational outcomes and thereby construct perceptual unity.

Such transient and transitory coupling is fully consistent with the frequency-specific plasticity seen in awake animals (Kilgard and Merzenich 1998) and the broader afferent tuning in these preparations (Whitfield and Evans 1965). Such an ensemble scheme is also consistent with the immense, continuous, and dynamic demands imposed by auditory stream analysis (Bregman et al. 2001) and would likely be essential for rapidly and efficiently unifying the otherwise independent "what" and "where" streams (Romanski et al. 1999b) into a continuous perceptual entity (Winer and Lee 2007).

9.3 *Laminar Ensembles*

The diverse laminar origins of convergent projections represent an additional axis along which AC areas may distribute and segregate information in the corticocortical (Lee and Winer 2008c), commissural (Lee and Winer 2008b), and thalamocortical (Huang and Winer 2000) systems. Thus, each layer is a potential source of segregated processing, even in projections from the same area. Infragranular corticofugal projections to the midbrain (Winer et al. 1998) and thalamus (Winer and Prieto 2001) may thus influence their own activity via top-down corticothalamocortical interactions (Sherman and Guillery 2002). By contrast, the supragranular layers have a more cortical bias, connecting almost exclusively with ipsilateral (Lee and Winer 2008c) and commissural (Lee and Winer 2008b) AC. This diversity of laminar origins, while observed in other studies (Imig and Reale 1980; Rouiller et al. 1991) has not typically been interpreted in the context of segregated processing systems.

If areal ensembles arise transiently for the global analysis of stimulus parameters, separate laminar sources might be recruited independently to form specialized ensembles. Predominantly infragranular projections could contribute to a network ensemble to recruit subcortical pathways, while supragranular ensembles would be segregated from subcortical streams. Areal laminar sources would specify and limit the ensembles available for prospective recruitment. Projections between areas arise from bilaminar and infragranular sources, with no area receiving predominantly supragranular input (Lee and Winer 2008c). Perhaps few laminar ensembles are exclusively cortical, and most cortical processes inform subcortical streams.

10 Future Directions

The convergent auditory forebrain connections underscore the global quality of telencephalic computations. It suggests why inactivating one area may have comparatively little effect on the physiology of another (Kitzes and Hollrigel 1996; Malhotra et al. 2004), since functional computations are likely distributed across many degenerate ensembles. It also captures the experimental challenge to understanding, uncoupling, and reconstructing these global operations. A next step is to correlate anatomical and functional weight with their laminar and cellular targets. Concurrent recording from many nuclear and areal sites in vivo and in vitro will reveal the focal cellular and widespread network interactions of coupled auditory regions. Models of forebrain acoustic organization will evolve from simple linear frameworks to a global view of the distributed auditory cortex.

Acknowledgments Supported by the National Institutes of Health grant R01 DC2319-29. David Larue's help was essential in constructing the figures.

References

Aitkin LM and Dunlop CW (1968) Interplay of excitation and inhibition in the cat medial geniculate body. *Journal of Neurophysiology* **31**:44–61.

Angelucci A, Levitt JB, Walton EJS, Hupé J-M, Bullier J, and Lund JS (2002) Circuits for local and global signal integration in primary visual cortex. *Journal of Neuroscience* **22**:8633–8646.

Benshalom G and White EL (1986) Quantification of thalamocortical synapses with spiny stellate neurons in layer IV of mouse somatosensory cortex. *Journal of Comparative Neurology* **253**: 303–314.

Binzegger T, Douglas RJ, and Martin KA (2004) A quantitative map of the circuit of cat primary visual cortex. *Journal of Neuroscience* **24**:8441–8453.

Bizley JK, Nodal FR, Nelken I, and King AJ (2005) Functional organization of ferret auditory cortex. *Cerebral Cortex* **15**:1637–1653.

Bordi F and LeDoux JE (1994) Response properties of single units in areas of rat auditory thalamus that project the amygdala. I. Acoustic discharge patterns and frequency receptive fields. *Experimental Brain Research* **98**:261–274.

Bowman EM and Olson CR (1988) Visual and auditory association areas of the cat's posterior ectosylvian gyrus: cortical afferents. *Journal of Comparative Neurology* **272**:30–42.

Bozhko GT and Slepchenko AF (1988) Functional organization of the callosal connections of the cat auditory cortex. *Neuroscience and Behavioral Physiology* **18**:323–330.

Bregman AS, Ahad PA, and Van Loon C (2001) Stream segregation of narrow-band noise bursts. *Perception and Psychophysics* **63**: 790–797.

Brugge JF and Reale RA (1985) Auditory cortex. In: Peters A and Jones EG (eds). *Cerebral Cortex*, volume 4, *Association and Auditory Cortices*. Plenum Press, New York, pp. 229–271.

Budinger E, Heil P, and Scheich H (2000) Functional organization of auditory cortex in the Mongolian gerbil (*Meriones unguiculatus*). IV. Connections with anatomically characterized subcortical structures. *European Journal of Neuroscience* **12**:2452–2474.

Budinger E, Laszcz A, Lison H, Scheich H, and Ohl FW (2008) Non-sensory cortical and subcortical connections of the primary auditory cortex in Mongolian gerbils: bottom-up and top-down processing of neuronal information via field AI. *Brain Research* **1220**: 70–80.

Bullier J, Kennedy H, and Salinger W (1984) Bifurcation of subcortical afferents to visual areas 17, 18, and 19 in the cat cortex. *Journal of Comparative Neurology* **228**:308–328.

Bullier J and Nowak JG (1995) Parallel versus serial processing: new vistas on the distributed organization of the visual system. *Current Opinion in Neurobiology* **5**:497–503.

Carrera E and Bogousslavsky J (2006) The thalamus and behavior: effects of anatomically distinct strokes. *Neurology* **66**: 1817–1823.

Catalano SM and Shatz CJ (1998) Activity-dependent cortical target selection by thalamic axons. *Science* **281**:559–562.

Clarey JC and Irvine DRF (1990) The anterior ectosylvian sulcal auditory field in the cat: I. An electrophysiological study of its relation to surrounding auditory cortical fields. *Journal of Comparative Neurology* **301**:289–303.

Clascá F, Llamas A, and Reinoso-Suárez F (1997) Insular cortex and neighboring fields in the cat: a redefinition based on cortical microarchitecture and connections with the thalamus. *Journal of Comparative Neurology* **384**:456–482.

Clemo HR, Keniston L, and Meredith MA (2003) A comparison of the distribution of GABA-ergic neurons in cortices representing different sensory modalities. *Journal of Chemical Neuroanatomy* **26**:51–63.

Davis TL and Sterling P (1979) Microcircuitry of cat visual cortex: classification of neurons in layer IV of area 17, and identification of the patterns of lateral geniculate input. *Journal of Comparative Neurology* **188**:599–628.

Deschênes M, Veinante P, and Zhang Z-W (1998) The organization of corticothalamic projections: reciprocity versus parity. *Brain Research Reviews* **28**:286–308.

Desimone R and Schein SJ (1987) Visual properties of neurons in area V4 of the macaque: sensitivity to stimulus form. *Journal of Neurophysiology* **57**:835–868.

Desimone R, Schein SJ, Moran J, and Ungerleider LG (1985) Contour, color and shape analysis beyond the striate cortex. *Vision Research* **25**:441–452.

Diamond IT and Hall WC (1969) Evolution of neocortex. *Science* **164**:251–262.

Eggermont JJ (1998) Representation of spectral and temporal sound features in three cortical fields of the cat. Similarities outweigh differences. *Journal of Neurophysiology* **80**: 2743–2764.

Eggermont JJ (1999) Neural correlates of gap detection in three auditory cortical fields in the cat. *Journal of Neurophysiology* **81**: 2570–2581.

Ehret G (1997) The auditory cortex. *Journal of Comparative Physiology A* **181**:547–557.

Felleman DJ and Van Essen DC (1991) Distributed hierarchical processing in the primate cerebral cortex. *Cerebral Cortex* **1**:1–47.

Fitzpatrick DC, Olsen JF, and Suga N (1998) Connections among functional areas in the mustached bat auditory cortex. *Journal of Comparative Neurology* **391**:366–396.

Frost SB, Milliken GW, Plautz EJ, Masterton RB, and Nudo RJ (2000) Somatosensory and motor representation in cerebral cortex of a primitive mammal (Monodelphis domestica): a window in the early evolution of sensorimotor cortex. *Journal of Comparative Neurology* **421**:29–51.

Gazzaniga MS (2000) Cerebral specialization and interhemispheric communication: does the corpus callosum enable the human condition? *Brain* **123**:1293–1326.

Gil Z, Connors BW, and Amitai Y (1999) Efficacy of thalamocortical synaptic connections: quanta, innervation, and reliability. *Neuron* **23**:385–397.

Gross CG (2002) Genealogy of the "grandmother cell". *Neuroscientist* **8**:512–518.

Hackett TA, Stepniewska I, and Kaas JH (1998) Subdivisions of auditory cortex and ipsilateral cortical connections of the parabelt auditory cortex in macaque monkeys. *Journal of Comparative Neurology* **394**:475–495.

Hall BK (2003) Descent with modification: the unity underlying homology and homoplasy as seen through an analysis of development and evolution. *Biological Reviews of the Cambridge Philosophical Society* **78**:409–433.

He J, Hashikawa T, Ojima H, and Kinouchi Y (1997) Temporal integration and duration tuning in the dorsal zone of cat auditory cortex. *Journal of Neuroscience* **17**:2615–2625.

Hromádka T, DeWeese MR, and Zador AM (2008) Sparse representation of sounds in the unanesthetized auditory cortex. *Public Library of Science Biology* **6**:e16.

Huang CL and Winer JA (2000) Auditory thalamocortical projections in the cat: laminar and areal patterns of input. *Journal of Comparative Neurology* **427**:302–331.

Humphrey AL, Sur M, Uhlrich DJ, and Sherman SM (1985) Termination patterns of individual X- and Y-cell axons in the visual cortex of the cat: projections to area 18, to the 17/18 border region, and to both areas 17 and 18. *Journal of Comparative Neurology* **233**:190–212.

Imaizumi K, Lee CC, Linden JF, Winer JA, and Schreiner CE (2004a) The anterior auditory field of auditory cortex: neurophysiological and neuroanatomical organization. In: König R, Heil P, Budinger E, and Scheich H (eds). *Auditory Cortex: Neurophysiological and Neuroanatomical Organization*. Lawrence Erlbaum Associates, Mahwah, pp. 95–110.

Imaizumi K, Priebe NJ, Crum PAC, Bedenbaugh PH, Cheung SW, and Schreiner CE (2004b) Modular functional organization of cat anterior auditory field. *Journal of Neurophysiology* **92**: 444–457.

Imig TJ and Reale RA (1980) Patterns of cortico-cortical connections related to tonotopic maps in cat auditory cortex. *Journal of Comparative Neurology* **192**:293–332.

Irvine DRF (1986) The Auditory Brainstem. A Review of the Structure and Function of Auditory Brainstem Processing Mechanisms. In: Autrum H, Ottoson D, Perl ER, Schmidt RF, Shimazu H, and Willis WD (eds). *Progress in Sensory Physiology*. Springer, New York, pp. 1–279.

Izhikevich EM, Gally JA, and Edelman GM (2004) Spike-timing dynamics of neuronal groups. *Cerebral Cortex* **14**:933–944.

Jones EG and Porter R (1980) What is area 3a? *Brain Research Reviews* **2**:1–43.

Kaas JH (1995) The evolution of isocortex. *Brain, Behaviour and Evolution* **46**:187–196.

Kaas JH and Hackett TA (1998) Subdivisions of auditory cortex and levels of processing in primates. *Audiology and Neurotology* **3**: 73–85.

Kaas JH and Hackett TA (2000) Subdivisions of auditory cortex and processing streams in primates. *Proceedings of the National Academy of Sciences of the United States of America* **97**: 11793–11799.

Kilgard MP and Merzenich MM (1998) Cortical map reorganization enabled by nucleus basalis activity. *Science* **279**:1714–1718.

Kishan AU, Lee CC, and Winer JA (2008) Branched projections in the auditory thalamocortical and corticocortical systems. *Neuroscience* **154**:283–293.

Kitzes LM and Hollrigel GS (1996) Response properties of units in the posterior auditory field deprived of input from the ipsilateral primary auditory cortex. *Hearing Research* **100**:120–130.

Knight PL (1977) Representation of the cochlea within the anterior auditory field (AAF) of the cat. *Brain Research* **130**:447–467.

Kral A and Eggermont JJ (2007) What's to lose and what's to learn: development under auditory deprivation, cochlear implants and limits of cortical plasticity. *Brain Research Reviews* **56**: 256–269.

Lee CC, Imaizumi K, Schreiner CE, and Winer JA (2004a) Concurrent tonotopic processing streams in auditory cortex. *Cerebral Cortex* **14**:441–451.

Lee CC, Schreiner CE, Imaizumi K, and Winer JA (2004b) Tonotopic and heterotopic projection systems in physiologically defined auditory cortex. *Neuroscience* **128**:871–887.

Lee CC and Sherman SM (2008) Synaptic properties of thalamic and intracortical inputs to layer 4 of the first- and higher-order cortical areas in the auditory and somatosensory systems. *Journal of Neurophysiology* **100**:317–326.

Lee CC and Winer JA (2005) Principles governing auditory cortex connections. *Cerebral Cortex* **15**:1804–1818.

Lee CC and Winer JA (2008a) Connections of cat auditory cortex. I. Thalamocortical system. *Journal of Comparative Neurology* **507**:1879–1900.

Lee CC and Winer JA (2008b) Connections of cat auditory cortex. II. Commissural system. *Journal of Comparative Neurology* **507**: 1901–1919.

Lee CC and Winer JA (2008c) Connections of cat auditory cortex. III. Corticocortical system. *Journal of Comparative Neurology* **507**:1920–1943.

Lee CC and Winer JA (2010) Convergence of thalamic and cortical pathways. *Hearing Research* doi:10.1016.

Loftus WC and Sutter ML (2001) Spectrotemporal organization of excitatory and inhibitory receptive fields of cat posterior auditory field neurons. *Journal of Neurophysiology* **86**:475–491.

Lomber SG, Malhotra S, and Hall AJ (2007) Functional specialization in non-primary auditory cortex of the cat: areal and laminar contributions to sound localization. *Hearing Research* **229**: 31–45.

Lund JS, Wu Q, Hadingham PT, and Levitt JB (1995) Cells and circuits contributing to functional properties in area V1 of macaque monkey cerebral cortex: bases for neuroanatomically realistic models. *Journal of Anatomy (London)* **187**:563–581.

Malhotra S, Hall AJ, and Lomber SG (2004) Cortical control of sound localization in the cat: unilateral cooling deactivation of 19 cerebral areas. *Journal of Neurophysiology* **92**:1625–1643.

Matsubara JA and Phillips DP (1988) Intracortical connections and their physiological correlates in the primary auditory cortex (AI) of the cat. *Journal of Comparative Neurology* **268**:38–48.

Meredith MA, Keniston LR, Dehner LR, and Clemo HR (2006) Crossmodal projections from somatosensory area SIV to the anterior ectosylvian sulcus (FAES) in cat: further evidence for subthreshold forms of multisensory processing. *Experimental Brain Research* **172**:472–484.

Merzenich MM, Knight PL, and Roth GL (1975) Representation of cochlea within primary auditory cortex in the cat. *Journal of Neurophysiology* **38**:231–249.

Middlebrooks JC, Dykes RW, and Merzenich MM (1980) Binaural response-specific bands in primary auditory cortex (AI) of the cat: topographic organization orthogonal to isofrequency contours. *Brain Research* **181**:31–48.

Miller LM, Escabí MA, Read HL, and Schreiner CE (2001) Functional convergence of response properties in the auditory thalamocortical system. *Neuron* **32**:151–160.

Miller LM, Escabí MA, Read HL, and Schreiner CE (2002) Spectrotemporal receptive fields in the lemniscal auditory thalamus and cortex. *Journal of Neurophysiology* **87**:516–527.

Nowak JG and Bullier J (1997) The timing of information transfer in the visual system. In: Rockland KS, Kaas JH, and Peters A (eds). *Cerebral Cortex*, volume 12, *Extrastriate Cortex in Primates*. Plenum Press, New York, pp. 205–241.

Olshausen BA, Anderson CH, and Van Essen DC (1993) A neurobiological model of visual attention and invariant pattern recognition based on dynamic routing of information. *Journal of Neuroscience* **16**:1180–1192.

Pallas SL and Sur M (1993) Visual projections induced into the auditory pathway of ferrets: II. corticocortical connections of primary auditory cortex. *Journal of Comparative Neurology* **1993**: 317–333.

Phillips DP and Irvine DRF (1981) Responses of single neurons in physiologically defined primary auditory cortex (AI) of the cat: frequency tuning and responses to intensity. *Journal of Neurophysiology* **45**:48–58.

Phillips DP and Irvine DRF (1982) Properties of single neurons in the anterior auditory field (AAF) of the cat cerebral cortex. *Brain Research* **248**:237–244.

Priebe NJ, Cassanello CR, and Lisberger SG (2003) The neural representation of speed in macaque area MT/V5. *Journal of Neuroscience* **23**:5650–5661.

Read HL, Winer JA, and Schreiner CE (2001) Modular organization of intrinsic connections associated with spectral tuning in cat auditory cortex. *Proceedings of the National Academy of Sciences of the United States of America* **98**:8042–8047.

Reale RA and Imig TJ (1980) Tonotopic organization in auditory cortex of the cat. *Journal of Comparative Neurology* **182**:265–291.

Roger M and Arnault P (1989) Anatomical study of the connections of the primary auditory area in the rat. *Journal of Comparative Neurology* **287**:339–356.

Romanski LM, Bates JF, and Goldman-Rakic PS (1999a) Auditory belt and parabelt projections to the prefrontal cortex in the rhesus monkey. *Journal of Comparative Neurology* **403**:141–157.

Romanski LM, Tian B, Fritz J, Mishkin M, Goldman-Rakic PS, and Rauschecker JP (1999b) Dual streams of auditory afferents target multiple domains in the primate prefrontal cortex. *Nature Neuroscience* **2**:1131–1136.

Rose D and Dobson VG (eds) (1985) *Models of the Visual Cortex*, Wiley, Chichester.

Rose JE and Woolsey CN (1958) Cortical connections and functional organization of the thalamic auditory system of the cat. In: Harlow HF, and Woolsey CN (eds). *Biological and Biochemical Bases of Behavior*. University of Wisconsin Press, Madison, pp. 127–150.

Rouiller EM, Simm GM, Villa AEP, de Ribaupierre Y, and de Ribaupierre F (1991) Auditory corticocortical interconnections in the cat: evidence for parallel and hierarchical arrangement of the auditory cortical areas. *Experimental Brain Research* **86**:483–505.

Rutkowski RG and Weinberger NM (2006) Encoding of learned importance of sound by magnitude of representational area in primary auditory cortex. *Proceedings of the National Academy of Sciences of the United States of America* **102**:13664–13669.

Schreiner CE and Cynader MS (1984) Basic functional organization of second auditory cortical field (AII) of the cat. *Journal of Neurophysiology* **51**:1284–1305.

Schreiner CE, Read HL, and Sutter ML (2000) Modular organization of frequency integration in primary auditory cortex. *Annual Reviews of Neuroscience* **23**:501–529.

Schreiner CE and Winer JA (2007) Auditory cortex mapmaking: principles, projections, and plasticity. *Neuron* **56**:356–365.

Sherman SM and Guillery RW (2002) The role of the thalamus in the flow of information to the cortex. *Philosophical Transactions of the Royal Society London, series B, Biological Sciences* **357**: 1695–1708.

Shinonaga Y, Takada M, and Mizuno N (1994) Direct projections from the non-laminated divisions of the medial geniculate nucleus to the temporal polar cortex and amygdala in the cat. *Journal of Comparative Neurology* **340**:405–426.

Sporns O, Tononi G, and Edelman GM (2002) Theoretical neuroanatomy and the connectivity of the cerebral cortex. *Behavioural Brain Research* **135**:69–74.

Spreafico R, Barbaresi P, Weinberg RJ, and Rustioni A (1987) SII-projecting neurons in the rat thalamus: a single- and double-retrograde-tracing study. *Somatosensory Research* **4**:359–375.

Squire LR, Stark CE, and Clark RE (2004) The medial temporal lobe. *Annual Reviews of Neuroscience* **27**:279–306.

Stecker GC, Harrington IA, Macpherson EA, and Middlebrooks JC (2005) Spatial sensitivity in the dorsal zone (area DZ) of cat auditory cortex. *Journal of Neurophysiology* **94**:1267–1280.

Stiebler I, Neulist R, Fichtel I, and Ehret G (1997) The auditory cortex of the house mouse: left–right differences, tonotopic organization and quantitative analysis of frequency representation. *Journal of Comparative Physiology A* **181**:559–571.

Stratford KJ, Tarczy-Hornoch K, Martin KAC, Bannister NJ, and Jack JJB (1996) Excitatory synaptic inputs to spiny stellate cells in cat visual cortex. *Nature* **382**:258–261.

Whitfield IC and Evans EF (1965) Responses of auditory cortical neurons to stimuli of changing frequency. *Journal of Neurophysiology* **28**:655–672.

Winer JA (1992) The functional architecture of the medial geniculate body and the primary auditory cortex. In: Webster DB, Popper AN, and Fay RR (eds). *Springer Handbook of Auditory Research*, volume 1, *The Mammalian Auditory Pathway: Neuroanatomy*. Springer, New York, pp. 222–409.

Winer JA, Diehl JJ, and Larue DT (2001) Projections of auditory cortex to the medial geniculate body of the cat. *Journal of Comparative Neurology* **430**:27–55.

Winer JA and Larue DT (1987) Patterns of reciprocity in auditory thalamocortical and corticothalamic connections: study with horseradish peroxidase and autoradiographic methods in the rat medial geniculate body. *Journal of Comparative Neurology* **257**: 282–315.

Winer JA, Larue DT, Diehl JJ, and Hefti BJ (1998) Auditory cortical projections to the cat inferior colliculus. *Journal of Comparative Neurology* **400**:147–174.

Winer JA and Lee CC (2007) The distributed auditory cortex. *Hearing Research* **229**:3–13.

Winer JA, Miller LM, Lee CC, and Schreiner CE (2005) Auditory thalamocortical transformation: structure and function. *Trends in Neurosciences* **28**:255–263.

Winer JA and Prieto JJ (2001) Layer V in cat primary auditory cortex (AI): cellular architecture and identification of projection neurons. *Journal of Comparative Neurology* **434**:379–412.

Woolsey CN (1960) Organization of cortical auditory system: a review and synthesis. In: Rasmussen GL and Windle WF (eds). *Neural Mechanisms of the Auditory and Vestibular Systems*. Charles C Thomas, Springfield, pp. 165–180.

Xiao Z and Suga N (2004) Reorganization of the auditory cortex specialized for echo-delay processing in the mustached bat. *Proceedings of the National Academy of Sciences of the United States of America* **101**:1769–1774.

Zeki S (1993) The visual association cortex. *Current Opinion in Neurobiology* **3**:155–159.

Chapter 8

Auditory Cortical Projections to the Medial Geniculate Body

Hisayuki Ojima and Eric M. Rouiller

Abbreviations

AC	auditory cortex
AI	primary auditory cortex
AMPA	α-amino-3-hydroxy-5-methylisoxazole-4-priopionic acid receptor
BDA	biotinylated dextran amine
BF	best frequency
CT	corticothalamic
dMGB	dorsal division of MGB
EPSC	excitatory postsynaptic conductance *or* current
EPSP	excitatory postsynaptic potential
FO	first order
GABA	γ-aminobutyric acid
HO	higher order
HRP	horseradish peroxidase
IC	inferior colliculus
IPSP	inhibitory postsynaptic potential
dLGN	dorsal lateral geniculate nucleus
LPl	lateral division of LPN
LPN	lateral posterior nucleus
MGB	medial geniculate body
mMGB	medial division of MGB
mGluR	metabotropic glutamate receptor
NMDA	*N*-methyl-D-aspartate
P	pulvinar nucleus
PHA-L	*Phaseolus vulgaris* leucoagglutinin
Po	posterior nuclear complex
Pol	lateral part of Po
Pom	medial part of Po
RTN	thalamic reticular nucleus
S1	primary somatosensory cortex
TC	thalamocortical
vMGB	ventral division of MGB
VB	ventrobasal nuclear complex
VP, VPN	ventral posterior nucleus
V1	primary visual cortex
WGA	wheat germ agglutinin

E.M. Rouiller (✉)
Institute de Physiologie, 1700 Fribourg, Switzerland
e-mail: eric.rouiller@unifr.ch

H. Ojima (✉)
Tokyo Medical and Dental University, Tokyo 113-8549, Japan
e-mail: yojima.cnb@tmd.ac.jp

1 General Features of the Corticothalamic System

Although the thalamus is traditionally regarded as a simple relay station for sensory information reaching cerebral cortex from the periphery, growing evidence supports a new view: that it actively regulates the passage of sensory information and modulates sensory signals (Jones 2002; Alitto and Usrey 2003). These modulations are enabled by inhibitory circuits, ascending inputs from brain stem nuclei, and corticofugal projections (Scannell et al. 1999). Ascending and descending inputs to the thalamus act as "drivers" and "modulators" (Guillery and Sherman 2002; Sherman and Guillery 2006). Thus, the ascending drivers to the medial geniculate body (MGB) represent projections coming from the inferior colliculus (IC, mainly its central nucleus), whereas other brain stem inputs are regarded as modulators.

Descending inputs to the thalamus, mainly of cortical origin, terminate on principal neurons as well as on local thalamic interneurons, and on GABAergic projection neurons in the reticular nucleus of the thalamus (RTN; Fig. 8.1). The number of GABAergic thalamic interneurons varies among nuclei and animal species. There are only a few MGB interneurons in the rat/mouse MGB (<1%) or in the guinea pig MGB (1%), as previously reported (Thompson et al. 1985; Winer and Larue 1996; Arcelli et al. 1997). The rat ventrobasal/ventroposterior nuclear complex contains no or very few intrinsic interneurons (Barbaresi et al. 1986; Harris and Hendrickson 1987). The cerebral cortex thus can exert direct influences on thalamic principal neuron responsiveness and indirect influences via inhibitory neurons.

J.A. Winer, C.E. Schreiner (eds.), *The Auditory Cortex*,
DOI 10.1007/978-1-4419-0074-6_8, © Springer Science+Business Media, LLC 2011

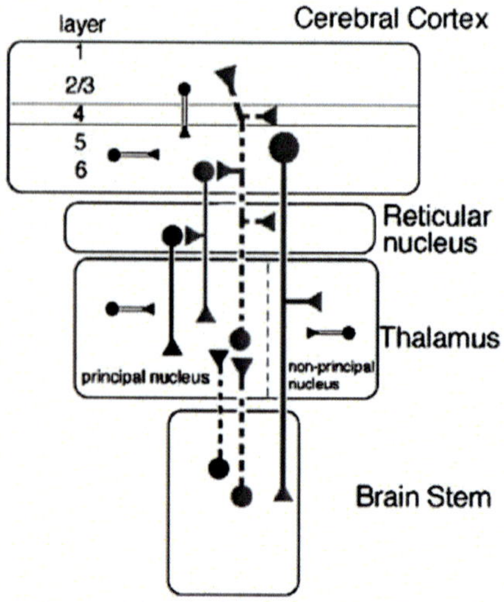

Fig. 8.1 Schematic of auditory connectivity between the cerebral cortex, thalamic reticular nucleus, thalamus, and brain stem. *Circles*, cell bodies; *triangles*, terminals; *dotted lines*, ascending pathways; *solid lines*, descending pathways; *double-lines*, local connectivity; *black*, inhibitory; *gray*, excitatory. The scheme may be extended to other sensory systems except for the ascending inhibitory inferior colliculus projection

These cortical influences were classically considered to derive from layer 6 pyramidal neurons. Modern anatomical tracing studies reveal a more complex scheme for the corticothalamic (CT) projection (Hoogland et al. 1987), consistent with a conceptual division between drivers and modulators. It is now proposed that the CT system has dual origins (Kelly and Wong 1981; Landry and Dykes 1985; Pandya et al. 1994) and termination modes and terminal morphologies (Rouiller and de Ribaupierre 1990; Rouiller and Welker 1991; Ojima 1994; Bajo et al. 1995; Steriade et al. 1997; Deschênes et al. 1998; Winer et al. 1999; Rouiller and Welker 2000; Rouiller and Durif 2004; Winer 2006).

Anterograde tracer injections in specific auditory cortex (AC) layers provide direct evidence for the relationships between the laminar origin and terminal morphology of the CT projection (cat primary AC, area AI) (Ojima 1994). Labeling and tracing of single axons provided a morphological profile of the CT fibers, showing a sublaminar location of cells of origin projecting to distinct thalamic targets (rat, Bourassa and Deschênes 1995; Bourassa et al. 1995; Veinante et al. 2000). A dual CT system from layers 5 and 6 forms large (or giant) and small terminals, respectively, in thalamic nuclei. As described below, functional characterization suggests that the layer 6-derived system is a modulator, while the layer 5-derived system is a driver (Sherman and Guillery 2006). This morphological dualism serves for the auditory, visual, and somatic sensory thalamic nuclei (Rouiller and Welker 2000), in anterior and medial thalamic nuclei (Schwartz et al. 1991; Kuroda et al. 1993; Negyéssy et al. 1998) and for motor thalamic nuclei (Rouiller et al. 1991, 1998; Kakei et al. 2001). It agrees with an earlier finding that small and large CT terminals degenerate following lesions of the cortical input to the monkey pulvinar (Mathers 1972).

The CT projections forming small and large terminals have been demonstrated for all the AC fields with different projection strengths (Winer et al. 2001). However, the CT projection targets and the pattern of each terminal type reflect the AC fields of origin (Bajo et al. 1995; Rouiller and Welker 2000; Winer et al. 2001). Collateral axons to the RTN are a general feature of the CT system (Jones 1975) and may derive from the layer 6 CT cells (Rouiller and de Ribaupierre 1990; Ojima 1994; Deschênes et al. 1994; Bourassa and Deschênes 1995; Bourassa et al. 1995). The cerebral cortex also projects to brain stem nuclei distal to the thalamus (see Chapter 2) including the inferior colliculus (IC) (Diamond et al. 1969; Winer et al. 1998), whose axon terminals resemble those in the thalamus (Ojima 1994) and derive from large and medium sized layer 5 pyramidal neurons (Winer and Prieto 2001). Separate layer 5 pyramidal neurons project to the MGB and IC (Wong and Kelly 1981; Winer and Prieto 2001), although collaterals with clusters of large boutons in the dMGB also belong to axons ending in the IC (Ojima 1994).

The CT projection is largely ipsilateral and bilateral for some regions (rat, Negyéssy et al. 1998; cat, Rinvik 1968; Molinari et al. 1985; monkey, Preuss and Goldman-Rakic 1987; Velayos 1997). The AI CT projection is ipsilateral in cat (Winer et al. 2001).

2 Quantitative Comparison of Corticofugal and Afferent Input

How strong is the CT projection relative to other inputs to the MGB? A retrograde tracer, horseradish peroxidase, was injected in restricted MGB loci in the cat (Rouiller and de Ribaupierre 1985). RTN inputs were quantitatively relatively modest, less than 10% of the total number of retrogradely labeled neurons in the auditory cortex, midbrain, and brain stem. However, many RTN neurons were labeled. About 90% of labeled cells were found in the IC and AC, divided equally between them (Rouiller and de Ribaupierre 1985).

Retrograde labeling of CT projection neurons in the cat AI finds 50% of layer 6 pyramidal neurons project to the MGB ventral division (vMGB, Kelly and Wong 1981), much like the cat area 17 CT projection to the dorsal geniculate nucleus (dLGN; Gilbert and Kelly 1975). Thus, the CT projection is likely the largest input to thalamic relay neurons, with their axons outnumbering TC axons by a factor of two (Guillery 1967). CT synapses form 40–45% of all synapses on LGN principal neurons, while retinal ganglion cells represent 10–20% of the total. Numerical estimates of the CT synapses on cat A-laminae principal neurons (Budd 2004) suggest one order of magnitude more CT synapses (12,000–15,000) than prior estimates (1,200–1,500). A comparable quantification of synapses on principal neurons in the MGB has not been made.

3 Organization of the Medial Geniculate Complex

Consideration of the CT auditory projections requires a brief survey of the auditory thalamic parcellation, including the MGB, the auditory sector of the RTN (Jones 1975; Shosaku and Sumitomo 1983; Rouiller et al. 1985; Simm et al. 1990; Villa 1990) and the lateral part of the posterior nuclear complex (Pol; Diamond et al. 1969; Imig and Morel 1985).

3.1 *Medial Geniculate Body Parcellation*

Morest (1964, 1965) subdivided the cat MGB into ventral (vMGB), medial (magnocellular, mMGB), and dorsal (dMGB) divisions. The vMGB consists of the *pars lateralis* and the *pars ovoidea* based on myeloarchitectonic and Golgi preparations and midbrain connections. The densely packed *pars lateralis* has a laminar arrangement (Morest 1965) aligned to its tonotopic organization (Aitkin and Webster 1972). The dMGB has three nuclei: (superficial) dorsal nucleus, deep dorsal nucleus, and suprageniculate nuclei, none having a tonotopic organization. mMGB cells have large somata and low packing density, receive auditory, vestibular, somatosensory, and probably visual inputs, and have broad auditory tuning. The mMGB has a tonotopic map in its anterior sector that is less ordered than in vMGB (Aitkin 1973; Rouiller et al. 1989). This MGB parcellation is consistent with subsequent architectonic (Winer and Morest 1983, 1984) and functional (Calford 1983) studies.

The posterior nuclear complex (Po) is an ill-defined transitional zone anterior to the MGB. Rostrally, it adjoins the lateral ventral posterior nucleus, ventrally the medial MGB, and dorsally the lateral posterior-pulvinar (LP-P) complex. Po has lateral (Pol), medial (Pom), and intermediate parts (Jones 1985). Pol neuron response properties resemble those in vMGB, with sharp tuning and a tonotopic organization (Imig and Morel 1985).

3.2 *Descending Projections to the Auditory Thalamus*

Cat AC contains 4–5 tonotopic and several non-tonotopic fields (Reale and Imig 1980). Areas AI, anterior auditory field (AAF), posterior auditory field (PAF), ventroposterior auditory field (VPAF) and, probably, the ventral auditory area (VAF) are organized tonotopically. AII, the secondary auditory field is weakly tonotopic (Schreiner and Cynader 1984), with limited frequency selectivity and broad neural tuning. Other non-tonotopic fields include the posterior ectosylvian gyrus, the temporal area, dorsoposterior field, and the insular cortex (Imig and Reale 1980; Reale and Imig 1980; Schreiner and Cynader 1984; He 1997; Winer et al. 2001). Anterior ectosylvian sulcus neurons respond to pure tones and visual stimuli, or to auditory, visual, or somatosensory stimuli. This multimodal region is distinct from nearby AAF or AII (Clarey and Irvine 1986, 1990; Kimura and Tamai 1992; Kimura et al. 1996).

Each AC region projects upon different MGB nuclei with a distinct pattern and termination strength (Diamond et al. 1969; Pontes et al. 1975; Sousa-Pinto and Reis 1975; Andersen et al. 1980; Kelly and Wong 1981; Wong and Kelly 1981; Rouiller and de Ribaupierre 1985; Pandya et al. 1994; Bajo et al. 1995; Winer et al. 2001). Regionally, the projection from each AC field to the MGB is divergent. Cat AI CT input ends principally in vMGB, with additional input to mMGB, dMGB, and Pol (Andersen et al. 1980). A comparable pattern of CT projection is seen for AAF. AII, in contrast, targets principally dMGB, with weaker input to vMGB and mMGB. At a finer level, the CT projection is also divergent, with single domains in an AC field projecting to many thalamic nuclei (Andersen et al. 1980; Ojima 1994; Bajo et al. 1995). Within the major target, on the contrary, the distribution of the CT projection follows a regular orientation along the mediolateral vMGB axis consistent with the tonotopic vMGB map. It is unknown whether each axon has collaterals also to the dMGB and mMGB as in the somatic sensory cortex (Deschênes et al. 1998).

The CT projection also displays some regional convergence, and each MGB division receives CT input from multiple AC fields (Andersen et al. 1980). The convergence, however, depends on the combination of the AC areas. Areas AAF and AI project differentially to the same MGB nuclei. Thus, dMGB, mMGB, and Pol receive prominent convergent projections from both, whereas the vMGB receives far less (Andersen et al. 1980; Lee et al. 2004a, b). The AI and AII CT projections terminate in largely segregated MGB nuclei, suggesting that each receives little convergence from these areas. Since there are other tonotopic fields in the cat AC, the full CT convergence pattern will need further study.

4 Dual Termination Patterns

The AC laminar sources of the CT projections are layers 5 or 6 (Fig. 8.2). The dendritic morphology, local collateralization, and intrinsic membrane properties of cells in these two AC origins differ and specific axon morphology, projection targets, and excitatory postsynaptic potentials (EPSPs)

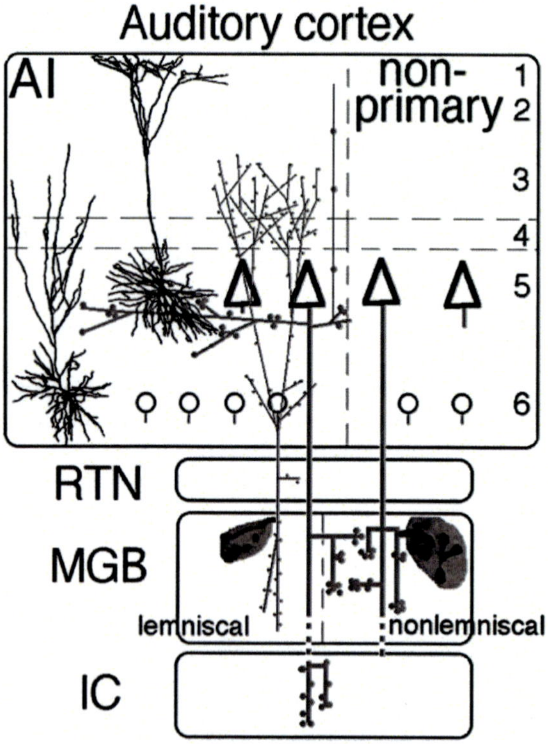

Fig. 8.2 A scheme for the laminar origin of the dual corticothalamic (CT) systems, morphological features of the cells of origin, and patterns of the CT projections to the auditory lemniscal and non-lemniscal thalamic nuclei. *Black triangles*, layer 5 pyramidal cell bodies; *black circles*, layer 6 pyramidal neuron somata; *gray lines* with small *or* large *dots*, axons with terminal small or large varicosities, respectively. Photomicrographs of the small and large medial geniculate body (MGB) terminal boutons. Typical dendritic arborizations of auditory cortex (AC) layer 5 and 6 CT pyramidal neurons

induced at their MGB synaptic targets vary. Layer 5 (Ojima 1994) and 6 (Rouiller and de Ribaupierre 1990) CT fibers also have different terminal field patterns.

The dual CT projections with distinct termination modes are derived from both tonotopic and non-tonotopic AC fields (Bajo et al. 1995; Winer et al. 2001) and often end in separate thalamic nuclei, suggesting heterogeneous or complex influences on acoustic signal processing.

5 Dual Cortical Origins

5.1 Laminar Corticothalamic Origins

The dual CT origins were originally demonstrated in AI by retrograde tracer injection in the MGB (Kelly and Wong 1981). The pyramidal cells labeled formed two populations, one in superficial layer 5 and the other throughout layer 6. Later studies with an anterograde tracer (biocytin) gave direct evidence for the source and target relationships between AC and MGB by injecting sublayers. Layer 6 injections labeled small axon terminals in vMGB, mMGB and Pol, while layer 5 deposits labeled axon terminals almost twice as large in dMGB (both superficial and deep nuclei) and, less so, likely in the mMGB (Ojima et al. 1992). Such CT dual origins occur in visual, somatic sensory, and motor cortex (Rouiller and Welker 2000), and in non-primary areas, including prefrontal and cingulate cortices (Velayos 1997).

5.2 Morphology of Auditory Corticothalamic Cells

Each AC layer has neurons with distinct morphologies and intrinsic membrane properties. One view of layer 5 in cat AI recognizes three sublayers: a superficial part with many large pyramidal neurons, a deep portion dominated by small pyramidal neurons, and a cell sparse belt in between (Rose 1949). Other schemes have also been proposed (Winer and Prieto 2001). Layer 6 has diverse pyramidal neurons forming upper and lower portions (VIa, VIb). In this scheme, pyramidal neurons that are the sources of the layer 5 and layer 6 CT systems differ in somatic morphology, dendritic arbors, and local horizontal axon branching.

Tracing the projection of single neurons intracellularly filled in vivo to the MGB reveals the morphology of one population of layer 5 CT neurons (Fig. 8.2; Ojima et al. 1992). They have a large soma and typical pyramid shape, a thick apical dendrite, and many basal dendrites. The apical dendrite ascends to layer 1 and forms 2–3 branches in upper layer 3 and a terminal tuft in layer 1a (Winer and Prieto 2001), reminiscent of pyramidal neurons with a thick apical dendritic morphology seen in vitro (Chagnac-Amitai et al. 1990; Larkman and Mason 1990; Mason and Larkman 1990; Hefti and Smith 2000). Their apical dendrite and arborization pattern are shown by retrograde tracer injections in cat visual system (Lund et al. 1975) and by intracellular filling after physiological identification of motor cortex projections to a thalamic target (Na et al. 1997). The range of morphological diversity of the layer 5 CT neurons is unknown.

Cat AI layer 5 large pyramidal neurons are a source of the CT projection, with axon collaterals projecting laterally in layer 5 and, less so, layer 6, with a few terminal arbors in layers 1 and 2. Synaptic varicosities up to 2 μm in diameter cluster and form regular horizontal collaterals (Fig. 8.2).

Nearly half of neurons in layer 6 sublayers are labeled by MGB retrograde tracer injections targeting the ventral division (Prieto and Winer 1999). Most CT neurons are pyramidal, with a few fusiform cells (Kelly and Wong 1981; Wong and Kelly 1981). Perhaps pyramidal neurons with different dendritic patterns project to distinct thalamic loci.

One type of the layer 6 CT pyramidal cell had an apical dendrite that branches and ascended to layer 3. Local axons are thin and, in contrast to those of layer 5 CT neurons, mainly vertical. Their terminals are ~1 μm in diameter and prominent in layer 4 and lower layer 3 and less so in layers 5 and 6 (Ojima et al. 1992). Juxtacellular labeling (Pinault 1996) of a few cells and single axon tracing in other modalities finds unique CT projection patterns from different sublaminar sources (Deschênes et al. 1998). Rat SI barrel cortex superficial layer 6 projections target the medial division of the VP nucleus (VPm), and the deep layer 6 projection ends in the medial part of the posterior nuclear complex (Pom) and the VPm (Bourassa et al. 1995). Upper layer 6 in primary visual cortex projects to the dLGN, and lower layer 6 to the lateral part of the LP nucleus and the dLGN (Bourassa and Deschênes 1995). The relationship between cortical sublayers and thalamic nuclei for higher order sensory connections is unknown (Levesque et al. 1996).

6 Corticothalamic Collaterals in the Thalamic Reticular Nucleus

6.1 Reticular Thalamic Sensory Subdivisions

The RTN receives cortical CT collaterals and TC collaterals (Fig. 8.1). The sensory RTN contains distinct modality-specific cortical and thalamic sectors (Jones 1975). The auditory sector occupies the caudoventral RTN (Shosaku and Sumitomo 1983; Rouiller et al. 1985; Simm et al. 1990; Villa 1990; Conley et al. 1991) and is closely connected to the MGB; the rostrodorsal portion is connected with the LP nucleus, (cat: Sakoda et al. 2004), and also has an auditory responsive region.

RTN subsectors have been defined for each MGB nuclei or cortical field. Thus, in the RTN auditory sector of the prosimian *Galago* (bushbaby), mMGB and Pol inputs target the auditory sector border region, while vMGB projections end in the central region. AII injections labeled the border subsectors, and the entire auditory sector was targeted by AI (Conley et al. 1991). A similar local pattern of connectivity is seen in the cat (Crabtree 1998). The connectivity is constrained by the cortical or thalamic map such that the RTN labeling is distributed systematically. Single RTN cells project to more than one MGB nucleus (Crabtree 1998).

Three types of axon terminals dominate the RTN (for other types, see Steriade et al. 1997). Small endings with round vesicles likely arise from CT axons, large terminals with round vesicles are from TC axons, and GABAergic endings with flattened vesicles are from RTN axon collaterals. In rat somatic sensory RTN these three terminals target proximal dendritic segments in proportions of 50, 30–40, and 10–25%, respectively, and distal dendrites at 60–65, 20, and 15%, respectively (Liu and Jones 1999). Comparable auditory sector data for the RTN are not available.

6.2 Laminar Origin of Corticoreticular Projections

The direct influence from the reciprocal TC and CT projections implies that RTN neurons play some role in generating TC oscillations via its inhibitory effect on thalamic transmission (Warren et al. 1994). Cortico-RTN cells of origin in rat visual (Bourassa and Deschênes 1995) and somatosensory cortex (Bourassa et al. 1995) are layer 6 pyramidal neurons. Layer 6 CT axon terminals in RTN are small and comparable to those found in the vMGB, though small and large terminals occur in monkey prefrontal-to-RTN projection (but see Zikopoulos and Barbas 2006). It is unknown whether all CT axons have RTN collaterals.

7 Corticothalamic Axon Terminal Morphology

As expected from differences in dendritic arborization, the distribution of local axons, and the size of their cortical axon terminals, the CT axon terminals in MGB may differ for layers 5 and 6 neurons. Rat layer 6 CT pyramidal neurons have small axon terminals ~1 μm in diameter, while giant CT terminals of layer 5 pyramidal neurons are 5–10 μm in diameter (Rouiller and Welker, 1991). As in the rat, cat layer 6 CT terminals are 1–2 μm spherical endings distinct from giant grape-like bouton clusters with 2–3 μm boutons (Ojima 1994; Bajo et al. 1995; Winer et al. 1999). A similar dual auditory CT pattern of terminals was seen in monkeys, but with smaller giant endings (2–6 μm) (Fig. 8.3; Rouiller and Durif 2004); smaller giant endings also arise from rat and monkey motor cortex (Rouiller et al. 1991, 1998). In all species noted, both small and giant endings form *boutons en passant* or *boutons terminaux*. There is little size overlap between the small and giant CT endings in the monkey motor thalamus (Rouiller et al. 1998). A dual CT projection pattern exists in rat AI (Rouiller and Welker 1991) and monkey (Rouiller and Durif 2004). In the cat this pattern is seen in many areas, including AI, AAF, PAF, and AII (Bajo et al. 1995; Winer et al. 1999).

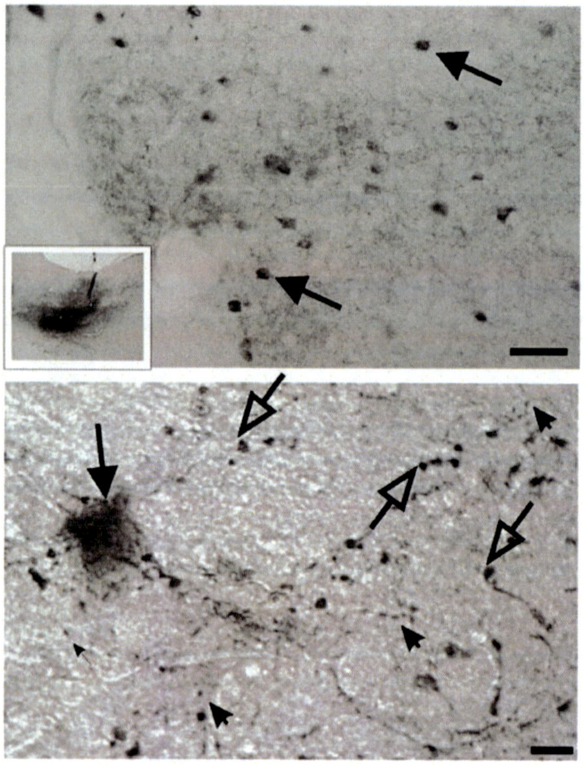

Fig. 8.3 A BDA injection in the deep layers of monkey AI (*inset, left* of *top panel*), and anterograde and retrograde MGB labeling (*top panel*). The halo represents anterogradely labeled terminal fields of small endings (*top panel*). Some retrogradely labeled neurons are present (*arrow, top panel*). Some anterogradely labeled terminal fields have giant CT endings (*bottom panel, open arrows*). *Bottom panel, arrow*, a retrogradely labeled neuron; *thin arrows*, a few small CT endings. Scale bars: 100 and 20 μm, *top* and *bottom panels*, respectively

8 Spatial Distribution of Corticothalamic Axon Terminals

8.1 Nuclear Distribution

Layer 6 AC CT axons target mainly the corresponding nucleus which project to AC and, to a lesser extent, other thalamic nuclei. Anterograde tracer deposits in AI label terminal fields of small endings concentrated in vMGB (rat: Rouiller and Welker 1991; Hazama et al. 2004; cat: Bajo et al. 1995, Winer et al. 1999). Smaller and sparser fields of small terminals end in Pol, dMGB, and mMGB. Injections in AAF and PAF label small endings and reciprocate the TC projection pattern in vMGB and in rostral vMGB (Bajo et al. 1995). Again, smaller and sparser terminal sets of small endings target Pol, dMGB, and mMGB. In sharp contrast, AII deposits label small endings mainly in dMGB (Bajo et al. 1995), reciprocating the dMGB TC projection. The overlap of small CT terminals and TC neurons is discussed below.

The giant CT endings of layer 5 pyramidal neurons do not coincide in MGB with the cell bodies of principal relay neurons. AI, AAF, and PAF tracer deposits label giant endings in dMGB of the rat (Rouiller and Welker 1991; Bartlett et al. 2000; Hazama et al. 2004) and cat (Bajo et al. 1995; Winer et al. 1999). In the rat AI and AAF deposits also labeled large boutons in Pol. The giant AI CT endings in monkey were in the posterior nucleus and between vMGB and dMGB (Rouiller and Durif 2004). As in the cat, in contrast with the tonotopic areas (AI, AAF and PAF), the giant layer 5 pyramidal endings from AII preserved the reciprocal TC projection in dMGB, where many small CT endings and the TC cells of origin to AII are found. CT giant endings from AII are in the dorsal and deep dorsal nuclei (Bajo et al. 1995). The pattern from AI, AAF, PAF, and AII (Bajo et al. 1995) includes areas EP, Ins, and TE (Winer et al. 1999), where CT giant endings covary with other giant GABAergic axon terminals of unknown origin (Winer et al. 1999).

In rat, cat, and monkey, the CT giant terminal fields from all areas are restricted spatially and sparser than the terminal fields of small endings. Giant endings overlap with the small CT endings. The functional significance of this topography is considered below.

Single CT axons may have both small and giant *boutons en passant*. Thus, layer 5 CT neurons send giant and small endings to the MGB, while the layer 6 neurons have small endings only. How many nuclei one CT neuron projects to is unknown.

8.2 Corticothalamic and Thalamocortical Reciprocity

To determine whether the CT terminal fields in the MGB overlap the TC neurons (Winer and Larue 1987), anterograde CT labeling was combined in the same experiment with retrograde TC labeling (Colwell 1975). Despite global CT-TC overlap, there were small zones of mismatch with CT terminal fields larger than TC clusters (Winer and Larue 1987). A similar situation exists in the motor thalamus (Rouiller et al. 1998). Analysis of the rat somatic sensory system finds that the upper layer 6 CT projection is more reciprocal than that from lower layer 6 (Deschênes et al. 1998; see also Land et al. 1995). Visual thalamic CT projections largely reciprocate the TC projection though non-reciprocal regions contained giant endings (van Horn and Sherman 2004). Caution is required when interpreting mixed anterograde and retrograde tracers (Winer and Larue 1987), since it is difficult to equate injection sizes even when tracers are ejected from the same pipette and damage can label passing fibers and neurons far from the deposit site (Paré and Smith 1996).

8.3 Layer 6 Corticothalamic Topography

All AC fields have CT projections (Winer et al. 2001; Winer 2006). Projections of the first order (FO) thalamic nucleus (vMGB) to the tonotopic AC are topographic and tonotopic, matching thalamic source and cortical target frequency preference (Imig and Morel 1984, 1985). The layer 6 AI CT system is likewise topographic in the auditory FO relay nucleus as a sheet-like plexus of axon terminals. Mapping studies of best frequencies (BFs) in AI and vMGB support the topographic organization of plexuses of small-terminal CT input to vMGB. Injections of anterograde tracers in different AI frequency domains label sheet-like plexuses that shift along the mediolateral (frequency) axis from lower (lateral) frequency representation lamellae to medial (higher) frequencies (cat: Andersen et al. 1980; Imig and Morel 1984, 1985; Takayanagi and Ojima 2006).

Other topographic relations of AI and vMGB neuron clusters exist. Based on response classifications related to spatial cues and binaural interaction properties (Zhang et al. 2004), it is unclear whether binaural bands (binaurally facilitated: EE, or inhibited: EI) have topographic relations in MGB (Middlebrooks and Zook 1983; Brandner and Redies 1990). The layer 6 AI CT projection targets MGB domains whose binaural interaction class could match their AC source (Imig and Adrián 1977).

An RTN (sub)sector receives input from reciprocally connected cortical and thalamic areas (Jones 1975). However, this topography is not point-to-point because of extensive dendritic RTN overlap.

Studies of rat AI and other AC input to RTN and vMGB show that tonotopically comparable subfields of both project to the same part of the vMGB and to different RTN regions (Kimura et al. 2005). Perhaps the direct CT projection from domains in different AC areas with the same frequency selectivity converges on corresponding thalamic domains. Alternatively, an indirect CT pathway via inhibitory RTN neurons allows different frequency-matched AC domains to exert divergent inhibitory effects on MGB domains at non-corresponding frequencies.

8.4 Layer 5 Corticothalamic Projection

AI is linked to non-lemniscal MGB nuclei, especially dMGB, via the layer 5 CT system. The topographical relationship may be weaker or coarser than that for layer 6. However, deposits in rat rostral area Te1 (comparable to the AI high best frequency domain) labeled large layer 5 terminals concentrated in the most rostromedial dMGB (Hazama et al. 2004; Kimura et al. 2005).

Small anterograde tracer deposits in cat AI label clusters of large terminals at ~15 dMGB loci. Single fibers in the dMGB branched within the internal capsule and form 7–10 small clusters of large terminals dispersed in the dMGB (Ojima 1994). The dMGB may be non-tonotopic though the deep dorsal nucleus has neurons with relatively sharp frequency selectivity, perhaps comparable to that of vMGB neurons (Calford and Webster 1981; Calford 1983; Imig and Morel 1985). Interestingly, this division receives most layer 5 AI CT projection, suggesting an organization related to dMGB frequency selectivity.

Large-terminal clusters in layer 5-derived CT dMGB projections occur at multiple loci. This sparse distribution complicates interpretation with respect to frequency. Injections of two different anterograde tracers in neighboring frequency axis domains in mapped cat AI reveal the spatial relationships of pair-labeling of large terminals in the dMGB in the deep and superficial dorsal nuclei, where clusters of different origins partly overlap (Takayanagi and Ojima 2006). Some clusters are horizontal, others vertical, suggesting that the AI layer 5 CT input may link AC and MGB domains of like frequency selectivity, though the target is not tonotopic.

More large terminals were found in cat dMGB originating from AII (Bajo et al. 1995; Winer et al. 2001), and the AII CT projections target the caudal nucleus of the dMGB, where response properties are unlike those of vMGB. However, the nature of non-tonotopic AC CT projections to non-tonotopic MGB is unknown.

9 Comparison with Other Modalities

Ultrastructural studies described many kinds of CT synapses (Jones and Powell 1969a; Jones 1985; Steriade et al. 1997), though a distinction between the two types of CT endings emerged only with anterograde tracers (PHA-L, biocytin and BDA) that provided cellular resolution since degeneration, autoradiography with [^{3}H]amino acids, or WGA-HRP tracing, though useful for topographic purposes, have insufficient resolution. Dual CT endings were seen first in the mouse somatic sensory system after PHA-L injection in the barrel cortex (Hoogland et al. 1987, 1988; Welker et al. 1988; Wouterlood et al. 1990). Later work found dual CT projections to visual, auditory, and motor systems in rodents, cats, and monkeys (Fig. 8.4) (Rouiller and Welker 1991; Rouiller et al. 1991; Ojima 1994; Bajo

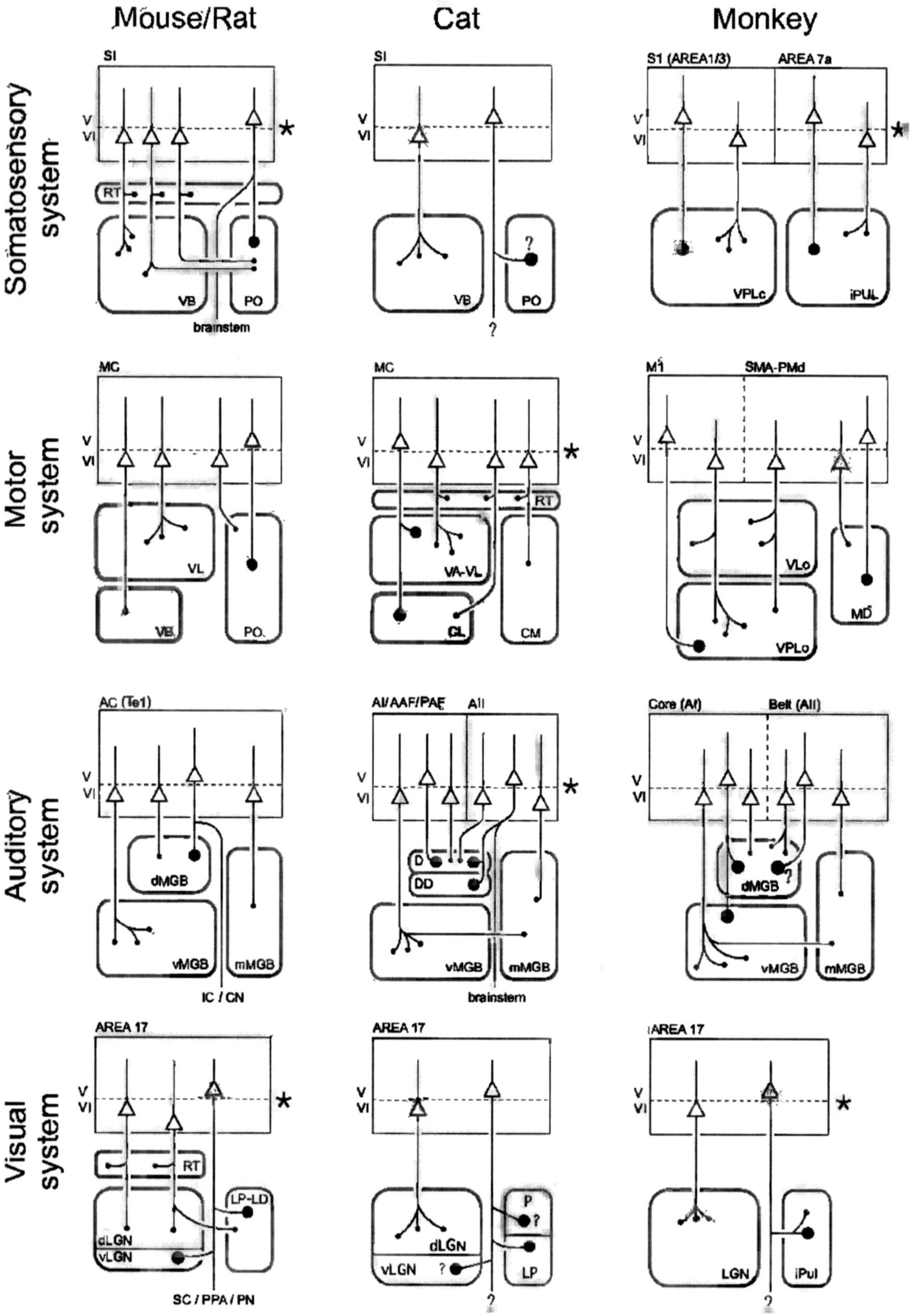

Fig. 8.4 (continued)

et al. 1995; Guillery 1995; Bourassa and Deschênes 1995; Bourassa et al. 1995; Rockland 1996; Ojima et al. 1996; Deschênes et al. 1998; Feig and Harting 1998; Rouiller et al. 1998; Darian-Smith et al. 1999; Winer et al. 1999; Rouiller and Welker 2000; Kakei et al. 2001; Guillery et al. 2001; Rouiller et al. 2003; Rouiller and Durif 2004; van Horn and Sherman 2004; Huppé-Gourges et al. 2006). A dual CT projection arises from prefrontal cortex (Schwartz et al. 1991; Negyéssy et al. 1998), the presubiculum (Oda 1997), prelimbic (Levesque and Parent 1998) and piriform cortex (Kuroda et al., 1992), and monkey posterior parietal cortex area 5 and the medial bank of the intraparietal sulcus (Cappe et al. 2007). Moreover, inferior parietal lobule area PFG and area Opt between parietal and occipitotemporal cortex give rise to such endings (Taktakishvili et al. 2002).

The dual CT projection might support a general principle of information processing. Thus, some thalamic first order (FO) nuclei receive driving ascending input (in audition mainly from the IC central nucleus) and project to low hierarchical-level primary cortical areas (Sherman and Guillery 2006). The auditory FO nucleus is vMGB, which drives AI/AAF, while AI/AAF send massive layer 6 CT input to vMGB, whose small axon terminals provide feedback modulation. In contrast, dMGB is a high order (HO) nucleus receiving driving layer 5 input from giant terminals. HO denotes that dMGB receives messages that have already reached AC. The main TC projection of dMGB to AII thus provides a rapid, indirect transthalamic link between AI and AII. Such indirect feed-forward projections are consistent with the large axons of layer 5 pyramidal neurons and their giant endings, which may ensure secure synaptic transmission (Bartlett et al. 2000; see also Miller 1996). This indirect connection via the thalamus between two cortical areas is quantitatively minor relative to the direct corticocortical connections. However, such corticothalamocortical route originating from layer 5 might drive AC, whereas the direct corticocortical pathways are modulators (Sherman and Guillery 2002, 2006). The feed-forward layer 5 CT projection is the first step in a proposed transthalamic corticocortical communication route (Rouiller and Welker 1991).

The distinction between FO and HO thalamic nuclei is not, however, strict: thus, dMGB is a mixed nucleus (HO and FO), relaying ascending input from the dorsal cortex of the IC to area AII, which then sends a modulatory layer 6 CT projection to dMGB. The output from cortex via the thalamus to another area also reaches brain stem or spinal cord targets, including motor centers. Corticocortical communication alone could not affect such remote processes.

10 Ultrastructure of Corticothalamic Terminals

Limited data as to the features of the CT terminal ultrastructure in the auditory system exist. The descriptions below are thus primarily from other modalities.

10.1 Layer 6 Corticothalamic Terminal Ultrastructure

In rat (Ohara and Lieberman 1981; Li et al. 2003c) and cat (Wilson et al. 1984; Montero 1991) dLGN, small terminals containing round vesicles forming asymmetrical synaptic thickenings on the shafts of the vMGB relay neuron dendrites (Majorossy and Réthelyi 1968; Morest 1975; Bartlett et al. 2000) likely represent layer 6 CT axon terminals. They contact cat VP nucleus distal segments and intermediate segments, and rarely end on relay neuron proximal dendrites (Liu et al. 1995a). Quantitatively, the CT terminals on cat LGN X and Y neurons (Wilson et al. 1984; Montero 1991) or cat VP cells (Liu et al. 1995a) form nearly 50% of the synapses, but the proportion may vary as on cat lateral geniculate body (LGN) W neurons (Wilson and Forestner 1995).

Small CT terminals also contact thalamic intrinsic neurons, mainly their dendrites, in all modalities (Jones and Powell 1969a, b; Morest 1975; Vidnyanszky and Hámori 1994; Barbaresi and Manzoni 2003; Li et al. 2003c). Their relative proportion varies among species.

In the RTN, small layer 6 CT terminals resemble those in the principal thalamic relay nucleus (Ohara and Lieberman 1981; Williamson et al. 1993; Liu and Jones 2003) and have equal somatic and dendritic input to RTN neurons (Liu and

Fig. 8.4 (continued) Schematic of corticothalamic (CT) topography in rodent, cat, and macaque monkey somatosensory, motor, auditory, and visual cortices (*four rows*, top to bottom). *Black rectangles*, cortical areas; *gray boxes*, thalamic nuclei. See list for abbreviations. *Open triangles* in cortex, pyramidal CT neurons in layers 5 and 6. *Asterisk, right side of rectangle* (cortical area) denotes layer 5 giant endings and layer 6 small endings. *Small filled thalamic circles*, small CT endings; *large filled circles*, giant endings. The thalamic reticular nucleus (RTN) is shown only when single cell data confirm that layer 6 neurons send collaterals to it; layer 5 neurons do not. If the collateral difference is unknown, RTN was omitted. *Question mark*, indicates that giant endings have not been confirmed in monkey auditory system. See text for citations (modified from Rouiller and Welker 2000)

Jones 2003). All rat RTN neurons are GABAergic (Houser et al. 1980) and presynaptic to vMGB relay neuron dendrites and somata (Montero 1983). They may synapse on cat interneuron dendrites as well (Liu et al. 1995b). It is unknown if these patterns exist in the MGB.

10.2 Layer 5 Corticothalamic Terminal Ultrastructure

Layer 5 CT large terminals have been studied in the rat dMGB (Bartlett et al., 2000) and confirm prior work on mouse somatic sensory Po nucleus system (Hoogland et al. 1991) and rat LP (lateral posterior) (Li et al. 2003c) and cat LP-P (lateral posterior nucleus-pulvinar) complex (Paré and Smith 1996; Vidnyanszky et al. 1996; Feig and Harting 1998; van Horn and Sherman 2004). These large, oval or irregular terminals have round synaptic vesicles and target either thick dendrites with a large apposition and asymmetric postsynaptic densities at multiple sites, or GABAergic interneuronal dendritic appendages.

In the cat dMGB, these two postsynaptic elements, and large CT terminals, rarely form the typical glomerular arrangement originally found in the FO thalamic nuclei (Guillery 1969; Jones and Powell 1969b). Dendritic spines receiving large CT terminals were not always presynaptic to the relay neuron dendrite postsynaptic to the same CT terminal (Fig. 8.5) (Ojima and Murakami unpublished observations). This suggests diverse CT arrangements which may be nucleus-specific.

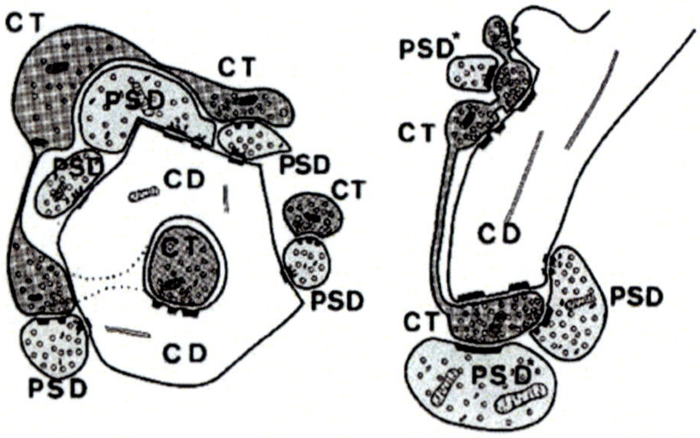

Fig. 8.5 Schematic of larger CT terminals near a thick, perhaps thalamocortical, dendrite, and vesicle-filled structures, likely thalamic interneuronal dendritic appendages in the MGB dorsal nucleus. Reconstructed from serial ultrathin sections containing a large-terminal cluster labeled following after anterograde tracer deposits in cat AI. Terminals (*dark gray*) of cortical origin (CT). The presynaptic dendrite (PSD, light gray) postsynaptic to a labeled CT terminal is in turn presynaptic to the CT in full reconstruction.

The proportion of visual system non-GABAergic/GABAergic postsynaptic targets of large CT terminals reflects animal species and/or methods. In rat area 17-to-LP CT system, almost all targets are large GABA-negative dendrites (Li et al. 2003c). In contrast, in the cat area 17-to-LPl (lateral part of lateral posterior nucleus) and the monkey area 17-to-inferior pulvinar projections, nearly 40% are also presynaptic to GABAergic interneuron dendritic appendages and almost all large CT terminals end on thalamic neurons' dendrites (Feig and Harting 1998; Vidnyanszky et al. 1996).

11 Corticothalamic Neurotransmitter Receptors

CT synapses on thalamic relay neurons are glutamatergic (McCormick and von Krosigk 1992) and activation of vMGB neurons is mediated via NMDA, AMPA, and metabotropic glutamate receptors (mGluR) (Bartlett and Smith 1999, 2002), as in other sensory systems (Salt and Eaton 1995; Eaton and Salt 1996; Liu 1997; Godwin et al. 1996a, b; Vidnyanszky et al. 1996). It is unknown if the same glutamatergic receptor subtypes participate in CT synaptic transmission for GABAergic interneurons. Interestingly, the CT terminal itself has presynaptic mGluRs and these receptors (Liu et al. 1998; Turner and Salt 1999) may reduce CT EPSCs (excitatory postsynaptic conductances) induced in ferret dLGN neurons after high-frequency activation of CT fibers (Alexander and Godwin 2005).

CT terminals ending on RTN neurons are also glutamatergic (de Curtis et al. 1989) and are activated by NMDA, AMPA (rat and cat; Liu 1997; Golshani et al. 2001; Alexander et al. 2006), and probably mGluRs (Martin et al. 1992). Similarly, RTN GABAergic axon terminals have presynaptic mGluRs in rat VB (Salt and Eaton 1995). In rat AC, inhibitory transmission from RTN axons to vMGB and dMGB neurons activates $GABA_A$ and $GABA_B$ receptors (Bartlett and Smith 1999). dMGB relay neurons, a major target of rat AC layer 5 large CT terminals, have NMDA and non-NMDA synaptic components, though cells of origin were not confirmed (Bartlett and Smith 1999). A similar CT transmission mode exists in other modalities in mice (medial part of the posterior group, Pom) (Reichhova and Sherman 2004).

12 Corticothalamic Modulations in Slice Preparations

As in the ultrastructural features of the CT terminals, limited data are available for the auditory system. Most accounts are from sensory systems other than auditory.

12.1 Thalamic Responses Following Corticothalamic Activation In Vitro

Most studies of CT influences on thalamic neurons record from principal relay nuclei (Watanabe et al. 1966), which receive small layer 6 terminals from FO thalamic nuclei. Responses to a single shock in layer 6 of the CT pathway in vitro elicit in rat vMGB (Bartlett and Smith 2002) and dLGN (McCormick and von Krosigk 1992; von Krosigk et al. 1999; Castro-Alamancos and Calcagnotto 1999; Li et al. 2003a, b) a small-amplitude EPSP, graded EPSPs with increasing stimulus intensity, and facilitation in paired or repetitive pulse stimulation. The growth of EPSP amplitude with stimulus intensity indicates that CT fibers converging on a cell are recruited progressively.

Studies of rat visual HO (Li et al. 2003a, b) and mouse somatic sensory (Reichhova and Sherman 2004) thalamic neurons show that single CT shocks elicit all-or-none responses in regions where CT large terminals alone target visual LP and somatic sensory Pom neurons. These responses likely reflect layer 5 pyramidal neurons' synaptic input. Efforts to correlate the synaptic transmission with CT terminal morphologies in the auditory system did not find marked vMGB and dMGB differences (Bartlett and Smith 1999).

12.2 Indirect Corticofugal Circuitry

The CT projection can be viewed as part of the thalamocortical-reticulothalamic loop. CT, reticulothalamic and TC projections generate oscillatory membrane potential changes during EEG spindling (Steriade et al. 1997) and characterized by 7–14 Hz alternating de- and hyperpolarizing waves of membrane potential and lasting 1–3 s with a 0.1–0.5 Hz repetition rate. The RTN may pace this oscillation.

CT activation of direct and indirect influences might modulate thalamic relay neuron excitability. CT fibers can strongly excite RTN GABAergic neurons (Zhang and Jones 2004), which then inhibit glutamatergic thalamic relay neurons. As these neurons recover from this RTN-mediated inhibition, the firing subsequently activates postsynaptic RTN cell targets, eliciting thalamocortical oscillations.

CT layer 6 input reaches both the thalamic principal nucleus and RTN. The balance between direct CT influence (glutamatergic) and indirect transreticular CT influence (GABAergic) should affect thalamic excitability. Comparing the synaptic efficacy among these three inputs reveals that CT and corticoreticular input differentially induce EPSCs in their respective targets, with EPSCs 2–3 times larger in RTN than in the FO VPN (ventral posterior nucleus) neurons. This might reflect different AMPA receptor subunit compositions such as GluR4-type subunit with more on mouse RTN synapses than on those of the principal neurons (Golshani et al. 2001). Alternatively, when a stimulus train is used, larger facilitation occurs in the ferret visual FO dLGN than that in the RTN (800% vs. 200% at maximum facilitation) (Alexander et al. 2006). In the somatic sensory system, the corticoreticular EPSP is ~33% of the amplitude, and decays slightly faster, than that of the FO thalamoreticular transmission (Gentet and Ulrich 2003, 2004).

12.3 Transthalamic Control of Higher Order Areas

Perhaps the indirect transthalamic activation of HO cortical fields could affect latencies of the circuits involved. Voltage-sensitive dye optical imaging in guinea pig (Horikawa et al. 2001) found a ~9 ms activation difference between tonotopic and non-tonotopic AC after pure tone stimulation. One latency value reflects HO thalamic activation after stimulation of layer 5 CT neurons, the other the non-primary cortical neurons following the HO thalamic stimulation. The interval in rat dMGB between non-lemniscal thalamic activation and CT stimulation at its midpoint (though not at layer 5) is estimated at 3.3 ms (Bartlett and Smith 1999) and at 1.6 ms in rat LP (Li et al. 2003a, b). The gap between activation in non-primary mouse AC and that for non-lemniscal thalamic nuclei is 6 ms (Cruikshank et al. 2002). Summing these latency values yields a total delay comparable to the optically measured interval. This is consistent with an HO thalamic involvement in sequential activation of hierarchically connected AC fields via transthalamic activation.

13 Corticofugal Modulation of Medial Geniculate Body Neurons

13.1 Auditory Cortex Activation

AC effects on MGB neuron responses in vivo are excitatory, inhibitory, mixed, or absent (Aitkin and Dunlop 1969; Ryugo and Weinberger 1976; Orman and Humphrey 1981; Villa et al. 1991). A plausible interpretation to such response heterogeneity is that the excitatory–inhibitory balance is variable and includes spatial and temporal factors.

These influences reflect stimulus repetition rate in vitro (McCormick and von Krosigk 1992), and effective spatial extent and stimulus strength. Electrically stimulating an AI site whose BF matches that of a vMGB recording site elicits mainly facilitation of pure tone responses. If the stimulated site is frequency unmatched, a suppressive modulation results (He 1997). In the cat this may be mediated either by intrinsic MGB interneurons (Rinvik et al. 1987; Rouiller et al. 1990; Huang et al. 1999), RTN inhibition, ascending GABAergic IC input (Winer et al. 1996; Peruzzi et al. 1997) or combinations of these. AC cells of origin for these modulation effects are unknown. If the inhibitory influence is mediated via IC neurons, layer 5 input to the IC must play a substantial role. The AC locus for facilitatory or suppressive effects is ~1 mm in diameter in cat (He 1997) and guinea pig (He et al. 2002).

From the anatomical relationship between the AC stimulation and vMGB recording sites, such in vivo modulatory effects may be mediated via layer 6 CT fibers. In contrast to the effects on the lemniscal MGB (vMGB), AC stimulation also affects the non-lemniscal mMGB and dMGB, with a strong effect in guinea pig dMGB of suppressive modulation, especially at the ON phase of firing, while facilitation is induced at the OFF phase (He 2003). The paired-pulse depression induced in layer 5 CT synapses on rat thalamic non-principal neurons in vitro (Li et al. 2003a, b) suggests that this suppressive effect is mediated via the RTN input to the thalamic neurons or via synaptic depression after activating layer 5 CT synapses.

Mechanisms underlying the modulation are depolarizing and hyperpolarizing synaptic membrane potentials induced in guinea pig MGB thalamic neurons and lasting 125–210 ms, after 5–20 pulses at 50–200 Hz (Yu et al. 2004). Which membrane potential is induced is nucleus-specific in the guinea pig (He et al. 2002; He 2003), perhaps reflecting the different density of inhibitory interneurons in each nucleus (Arcelli et al. 1997).

13.2 Auditory Cortex Inactivation and Thalamic Response Properties

Inactivating AC while recording in the MGB (Ryugo and Weinberger 1976; Orman and Humphrey 1981; Villa et al. 1991) shows that CT effects are complex, including inhibitory (Amato et al. 1969), excitatory–inhibitory (Ryugo and Weinberger 1976), or mainly excitatory (Orman and Humphrey 1981) effects. A facilitatory CT role is supported by the marked decrease of MGB spontaneous discharge rate when AC was reversibly cooled (Villa et al. 1991). For the responses evoked by acoustic stimulation, this inactivation modified some unit tuning properties and/or the ratio between peak and the spontaneous firing rates, but such changes were subtle (Villa et al. 1991) and MGB and RTN receptive fields units were not modified dramatically, as an inactivation of the modulatory layer 6 CT projection might predict given this substantial projection and massive AC inactivation. To assess the effect of inactivating the driver layer 5 CT projection, it would be necessary to record selectively from the spatially restricted territories (mainly in dMGB) with giant terminals, which was not done in this cooling experiment (Villa et al. 1991).

Visual system experiments that ablate or silence cortex likewise elicit modest changes on thalamic discharge properties (Kalil and Chase 1970; Baker and Malpeli 1977; Sillito and Jones 2002). These data suggest that the layer 6 CT projection has a modulatory role. In contrast, removing the somatosensory or visual cortex abolished thalamic receptive fields in an HO nucleus receiving layer 5 CT inputs, but did not affect those of FO neurons receiving layer 6 CT input (Bender 1983; Diamond et al. 1992). This implies that layer 5 CT inputs are indeed drivers, establishing HO neuron receptive field properties (Sherman and Guillery 2006).

14 Concluding Remarks

Despite parallels in CT projections and actions in different modalities and species, its functional significance remains uncertain. Most studies emphasize its modulatory effects on principal thalamic relay neuron response properties following cortical manipulation. Most such influences arise in layer 6 CT neurons. In contrast, the layer 5 CT system, though far smaller and sparser, may strongly affect large EPSP amplitudes, secure all-or-none transmission, and spike-frequency dependent synaptic plasticity. Layer 5 may have roles beyond the efficacy of sensory processing or its gain control. Its input

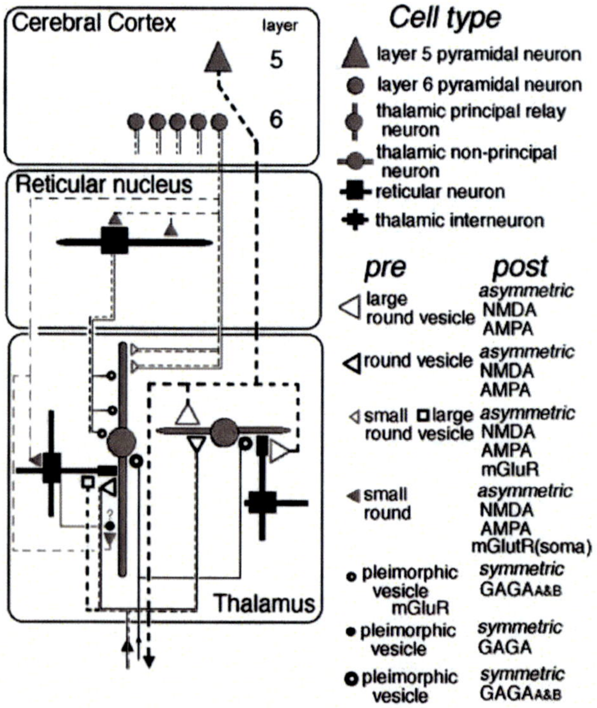

Fig. 8.6 A synthesis of synaptic connections and their receptor types. *Dotted lines*, data from other sensory systems; *solid lines*, auditory system data; pre, presynaptic; post, postsynaptic

to the HO thalamic nuclei may support functions other than modulation. In the visual system, a major non-principal layer 5 CT thalamic target, the pulvinar nucleus, participates in the attentional selection of visual information (LaBerge and Buchsbaum 1990). A similar mechanism may also exist in the auditory system (Woldorff et al. 1993).

The layer 5 CT system may route information indirectly between different AC fields via the transthalamic pathways (Sherman and Guillery 2006). For example, AI receives sensory activation from the FO thalamic nucleus (vMGB), then projects to the HO thalamic nucleus (dMGB) via its descending layer 5 CT system, perhaps activating other AC fields (AII) via the TC dMGB ascending projections. This pathway might recruit HO cortical fields when it functions in tandem with the direct corticocortical pathways. Such transthalamic activation of AC seems plausible, but its confirmation requires further evaluation of this indirect connectivity to decipher the dynamic interplay between cortex and thalamus (Fig. 8.6).

References

Aitkin LM and Dunlop CW (1969) Inhibition in the medial geniculate body of the cat. *Experimental Brain Research***7**:68–83.

Aitkin LM and Webster WR (1972) Medial geniculate body of the cat: organization and responses to tonal stimuli of neurons in ventral division. *Journal of Neurophysiology* **35**:365–380.

Aitkin LM (1973) Medial geniculate body of the cat: responses to tonal stimuli of neurons in medial division. *Journal of Neurophysiology* **36**:275–283.

Alexander GM, Fisher TL, and Godwin DW (2006) Differential response dynamics of corticothalamic glutamatergic synapses in the lateral geniculate nucleus and thalamic reticular nucleus. *Neuroscience* **137**:367–372.

Alexander GM and Godwin DW (2005) Presynaptic inhibition of corticothalamic feedback by metabotropic glutamate receptors. *Journal of Neurophysiology* **94**:163–175.

Amato G, Lagrutta V, and Enia F (1969) The control exerted by the auditory cortex on the activity of the medial geniculate body and inferior colliculus. *Archives Italiennes de Biologie* **53**: 291–313.

Alitto HJ and Usrey WM (2003) Corticothalamic feedback and sensory processing. *Current Opinion in Neurobiology* **13**:440–445.

Andersen RA, Knight PL, and Merzenich MM (1980) The thalamocortical and corticothalamic connections of AI, AII, and the anterior auditory field (AAF) in the cat: evidence for two largely segregated systems of connections. *Journal of Comparative Neurology* **194**:663–701.

Arcelli P, Frassoni C, Regondi MC, De Biasi S, and Spreafico R (1997) GABAergic neurons in mammalian thalamus: a marker of thalamic complexity? *Brain Research Bulletin* **42**:27–37.

Bajo VM, Rouiller EM, Welker E, Clarke S, Villa AEP, de Ribaupierre Y, and de Ribaupierre F (1995) Morphology and spatial distribution of corticothalamic terminals originating from the cat auditory cortex. *Hearing Research* **83**:161–174.

Baker FH and Malpeli JG (1977) Effects of cryogenic blockade of visual cortex on the responses of lateral geniculate neurons in the monkey. *Experimental Brain Research* **29**:433–444.

Barbaresi P and Manzoni T (2003) Direct synaptic connections between corticothalamic axon terminals and interneurons in the cat ventrobasal complex. *Neuroscience Letters* **339**:211–214.

Barbaresi P, Spreafico R, Frassoni C, and Rustioni A (1986) GABAergic neurons are present in the dorsal column nuclei but not in the ventroposterior complex of rats. *Brain Research* **382**: 305–326.

Bartlett EL and Smith PH (1999) Anatomic, intrinsic, and synaptic properties of dorsal and ventral division neurons in rat medial geniculate body. *Journal of Neurophysiology* **81**:1999–2016.

Bartlett EL and Smith PH (2002) Effects of paired-pulse and repetitive stimulation on neurons in the rat medial geniculate body. *Neuroscience* **113**:957–974.

Bartlett EL, Stark JM, Guillery RW, and Smith PH (2000) Comparison of the fine structure of cortical and collicular terminals in the rat medial geniculate body. *Neuroscience* **100**:811–828.

Bender DB (1983) Visual activation of neurons in the primate pulvinar depends on cortex but not colliculus. *Brain Research* **279**: 258–261.

Bourassa J and Deschênes M (1995) Corticothalamic projections from the primary visual cortex in rats: a single fiber study using biocytin as an anterograde tracer. *Neuroscience* **66**:253–263.

Bourassa J, Pinault D, and Deschênes M (1995) Corticothalamic projections from the cortical barrel field to the somatosensory thalamus in rats: a single-fibre study using biocytin as an anterograde tracer. *European Journal of Neuroscience* **7**:19–30.

Brandner S and Redies H (1990) The projection from medial geniculate to field AI in cat: organization in the isofrequency dimension. *Journal of Neuroscience* **10**:50–61.

Budd JM (2004) How much feedback from visual cortex to lateral geniculate nucleus in cat: a perspective. *Visual Neuroscience* **21**:487–500.

Calford MB (1983) The parcellation of the medial geniculate body of the cat defined by the auditory response properties of single units. *Journal of Neuroscience* **3**:2350–2364.

Calford MB and Webster WR (1981) Auditory representation within principal division of cat medial geniculate body: an electrophysiology study. *Journal of Neurophysiology* **45**: 1013–1028.

Cappe C, Morel A, and Rouiller EM (2007) Thalamocortical and the dual pattern of corticothalamic projections of the posterior parietal cortex in macaque monkeys. *Neuroscience* **146**:1371–1387.

Castro-Alamancos MA and Calcagnotto ME (1999) Presynaptic long-term potentiation in corticothalamic synapses. *Journal of Neuroscience* **19**:9090–9097.

Chagnac-Amitai Y, Luhmann HJ, and Prince DA (1990) Burst generating and regular spiking layer 5 pyramidal neurons of rat neocortex have different morphological features. *Journal of Comparative Neurology* **296**:598–613.

Clarey JC and Irvine DRF (1986) Auditory response properties of neurons in the anterior ectosylvian sulcus of the cat. *Brain Research* **386**:12–19.

Clarey JC and Irvine DRF (1990) The anterior ectosylvian sulcal auditory field in the cat: II. A horseradish peroxidase study of its thalamic and cortical connections. *Journal of Comparative Neurology* **301**:304–324.

Colwell SA (1975) Thalamocortical-corticothalamic reciprocity: a combined anterograde-retrograde tracer technique. *Brain Research* **92**:443–449.

Conley M, Kupersmith AC, and Diamond IT (1991) The organization of projections from subdivisions of the auditory cortex and thalamus to the auditory sector of the thalamic reticular nucleus in Galago. *European Journal of Neuroscience* **3**:1089–1103.

Crabtree JW (1998) Organization in the auditory sector of the cat's thalamic reticular nucleus. *Journal of Comparative Neurology* **390**:167–182.

Cruikshank SJ, Rose HJ, and Metherate R (2002) Auditory thalamocortical synaptic transmission in vitro. *Journal of Neurophysiology* **87**:361–384.

Darian-Smith C, Tan A, and Edwards S (1999) Comparing thalamocortical and corticothalamic microstructure and spatial reciprocity in the macaque ventral posterolateral nucleus (VPLc) and medial pulvinar. *Journal of Comparative Neurology* **410**:211–234.

Deschênes M, Bourassa J, and Pinault D (1994) Corticothalamic projections from layer V cells in rat are collaterals of long-range corticofugal axons. *Brain Research* **664**:215–219.

Deschênes M, Veinante P, and Zhang ZW (1998) The organization of corticothalamic projections: reciprocity versus parity. *Brain Research Brain Research Reviews* **28**:286–308.

de Curtis M, Spreafico R, and Avanzini G (1989) Excitatory amino acids mediate responses elicited in vitro by stimulation of cortical afferents to reticularis thalami neurons of the rat. *Neuroscience* **33**:275–283.

Diamond IT, Jones EG, and Powell TPS (1969) The projection of the auditory cortex upon the diencephalon and brain stem in the cat. *Brain Research* **15**:305–340.

Diamond ME, Armstrong-James M, Budway MJ, and Ebner FF (1992) Somatic sensory responses in the rostral sector of the posterior group (POm) and in the ventral posterior medial nucleus (VPM) of the rat thalamus: dependence on the barrel field cortex. *Journal of Comparative Neurology* **319**:66–84.

Eaton SA and Salt TE (1996) Role of *N*-methyl-D-aspartate and metabotropic glutamate receptors in corticothalamic excitatory postsynaptic potentials in vivo. *Neuroscience* **73**:1–5.

Feig S and Harting JK (1998) Corticocortical communication via the thalamus: ultrastructural studies of corticothalamic projections from area 17 to the lateral posterior nucleus of the cat and inferior pulvinar nucleus of the owl monkey. *Journal of Comparative Neurology* **395**:281–295.

Gentet LJ and Ulrich D (2003) Strong, reliable and precise synaptic connections between thalamic relay cells and neurones of the nucleus reticularis in juvenile rats. *Journal of Physiology (London)* **546**:801–811.

Gentet LJ and Ulrich D (2004) Electrophysiological characterization of synaptic connections between layer VI cortical cells and neurons of the nucleus reticularis thalami in juvenile rats. *European Journal of Neuroscience* **19**:625–633.

Gilbert CD and Kelly JP (1975) The projections of cells in different layers of the cat's visual cortex. *Journal of Comparative Neurology* **163**:81–105.

Golshani P, Liu X-B, and Jones EG (2001) Differences in quantal amplitude reflect GluR4-subunit number at corticothalamic synapses on two populations of thalamic neurons. *Proceedings of the National Academy of Sciences of the United States of America* **98**:4172–4177.

Godwin DW, Van Horn SC, Erisçir A, Sesma M, Romano C, and Sherman SM (1996a) Ultrastructural localization suggests that retinal and cortical inputs access different metabotropic glutamate receptors in the lateral geniculate nucleus. *Journal of Neuroscience* **16**:8181–8192.

Godwin DW, Vaughan JW, and Sherman SM (1996b) Metabotropic glutamate receptors switch visual response mode of lateral geniculate nucleus cells from burst to tonic. *Journal of Neurophysiology* **76**:1800–1816.

Guillery RW (1967) Patterns of fiber degeneration in the dorsal lateral geniculate nucleus of the cat following lesions in the visual cortex. *Journal of Comparative Neurology* **130**:197–221.

Guillery RW (1969) The organization of synaptic interconnections in the laminae of the dorsal lateral geniculate nucleus of the cat. *Zeitschrift für Zellforschung und Mikroskopische Anatomie* **96**: 1–38.

Guillery RW (1995) Anatomical evidence concerning the role of the thalamus in corticocortical communication: a brief review. *Journal of Anatomy (London)* **187**:583–592.

Guillery RW, Feig SL, and Van Lieshout DP (2001) Connections of higher order visual relays in the thalamus: a study of corticothalamic pathways in cats. *Journal of Comparative Neurology* **438**:66–85.

Guillery RW and Sherman SM (2002) Review. Thalamic relay functions and their role in corticocortical communication: generalizations from the visual system. *Neuron* **33**:163–175.

Harris RM and Hendrickson AE (1987) Local circuit neurons in the rat ventrobasal thalamus – A GABA immunocytochemical study. *Neuroscience* **21**:229–236.

Hazama M, Kimura A, Donishi T, Sakoda T, and Tamai Y (2004) Topography of corticothalamic projections from the auditory cortex of the rat. *Neuroscience* **124**:655–667.

He J (1997) Modulatory effects of regional cortical activation on the onset responses of the cat medial geniculate neurons. *Journal of Neurophysiology* **77**:896–908.

He J, Yu Y-Q, Xiong Y, Hashikawa T, and Chan U-S (2002) Modulatory effect of cortical activation on the lemniscal auditory thalamus of the guinea pig. *Journal of Neurophysiology* **88**:1040–1050.

He J (2003) Corticofugal modulation on both on and off responses in the nonlemniscal auditory thalamus of the guinea pig. *Journal of Neurophysiology* **89**:367–381.

Hefti BJ and Smith PH (2000) Anatomy, physiology, and synaptic responses of rat layer V auditory cortical cells and effects of intracellular $GABA_A$ blockade. *Journal of Neurophysiology* **83**:2626–2638.

Hoogland PV, Welker E, and van der Loos H (1987) Organization of the projections from barrel cortex to thalamus in mice studied with *Phaseolus vulgaris*-leucoagglutinin and HRP. *Experimental Brain Research* **68**:73–87.

Hoogland PV, Welker E, van der Loos H, and Wouterlood FG (1988) The organization and structure of the thalamic afferents from the barrel cortex in the mouse; a PHA-L study. In: Bentivoglio M and Spreafico R (eds). *Cellular Thalamic Mechanisms*. Elsevier Science Publishers BV, Amserdam, pp. 151–162.

Hoogland PV, Wouterlood FG, Welker E, and van der Loos H (1991) Ultrastructure of giant and small thalamic terminals of cortical origin: a study of the projections from the barrel cortex in mice using *Phaseolus vulgaris* leuco-agglutinin (PHA-L). *Experimental Brain Research* **87**:159–172.

Horikawa J, Hess A, Nasu M, Hosokawa Y, Scheich H, and Taniguchi I (2001) Optical imaging of neural activity in multiple auditory cortical fields of guinea pigs. *Neuroreport* **12**: 3335–3339.

Houser CR, Vaughn JE, Barber RP, and Roberts E (1980) GABA neurons are the major cell type of the nucleus reticularis thalami. *Brain Research* **200**:341–354.

Huang CL, Larue DT, and Winer JA (1999) GABAergic organization of the cat medial geniculate body. *Journal of Comparative Neurology* **415**:368–392.

Huppé-Gourges F, Bickford ME, Boire D, Ptito M, and Casonova M (2006) Distribution, morphology, and synaptic targets of corticothalamic terminals in the cat lateral posterior-pulvinar complex that originate from the posteromedial lateral suprasylvian cortex. *Journal of Comparative Neurology* **497**:847–863.

Imig TJ and Adrián HO (1977) Binaural columns in the primary field (A1) of cat auditory cortex. *Brain Research* **138**:241–257.

Imig TJ and Morel A (1984) Topographic and cytoarchitectonic organization of thalamic neurons related to their targets in low-, middle-, and high-frequency representations in cat auditory cortex. *Journal of Comparative Neurology* **227**:511–539.

Imig TJ and Morel A (1985) Tonotopic organization in ventral nucleus of medial geniculate body in the cat. *Journal of Neurophysiology* **53**:309–340.

Imig TJ and Reale RA (1980) Patterns of cortico-cortical connections related to tonotopic maps in cat auditory cortex. *Journal of Comparative Neurology* **192**:293–332.

Jones EG (1975) Some aspects of the organization of the thalamic reticular complex. *Journal of Comparative Neurology* **162**:285–308.

Jones EG (1985) *The Thalamus*. Plenum Press, New York, London.

Jones EG (2002) Thalamic circuitry and thalamocortical synchrony. *Philosophical Transactions of the Royal Society of London* **357**:1659–1673.

Jones EG and Powell TPS (1969a) An electron microscopic study of the mode of termination of cortico-thalamic fibres within the sensory relay nuclei of the thalamus. *Proceedings of the Royal Society of London, series B, Biological Sciences* **172**:173–185.

Jones EG and Powell TPS (1969b) Electron microscopy of synaptic glomeruli in the thalamic relay nuclei of the cat. *Proceedings of the Royal Society of London, series B, Biological Sciences* **172**: 153–171.

Kakei S, Na J, and Shinoda Y (2001) Thalamic terminal morphology and distribution of single corticothalamic axons originating from layers 5 and 6 of the cat motor cortex. *Journal of Comparative Neurology* **437**:170–185.

Kalil RE and Chase R (1970) Corticofugal influence on activity of lateral geniculate neurons in the cat. *Journal of Neurophysiology* **33**:459–474.

Kelly JP and Wong D (1981) Laminar connections of the cat's auditory cortex. *Brain Research* **212**:1–15.

Kimura A, Donishi T, Okamoto K, and Tamai Y (2005) Topography of projections from the primary and non-primary auditory cortical areas to the medial geniculate body and thalamic reticular nucleus in the rat. *Neuroscience* **135**:1325–1342.

Kimura A, Hamada Y, Kawai Y, and Tamai Y (1996) Sensory response properties of cortical neurons in the anterior ectosylvian sulcus of cats: intracellular recording and labeling. *Neuroscience Research* **26**:357–367.

Kimura A and Tamai Y (1992) Sensory response of cortical neurons in the anterior ectosylvian sulcus, including the area evoking eye movement. *Brain Research* **575**:181–186.

Kuroda M, Murakami K, Kishi K, and Price JL (1992) Distribution of the piriform cortical terminals to cells in the central segment of the mediodorsal thalamic nucleus of the rat. *Brain Research* **595**: 159–163.

Kuroda M, Sugiura T, Shinkai M, Murakami K, Oda S, and Kishi K (1993) Synaptic organization and prefrontal corticothalamic termination in the mediodorsal thalamic nucleus of the cat. *Journal für Hirnforschung* **34**:417–430.

LaBerge D and Buchsbaum MS (1990) Positron emission tomographic measurements of pulvinar activity during an attention task. *Journal of Neuroscience* **10**:613–619.

Land PW, Buffer SA Jr, and Yaskosky JD (1995) Barreloids in adult rat thalamus: three-dimensional architecture and relationship to somatosensory cortical barrels. *Journal of Comparative Neurology* **355**:573–588.

Landry P and Dykes RW (1985) Identification of two populations of corticothalamic neurons in cat primary somatosensory cortex. *Experimental Brain Research* **60**:289–298.

Larkman A and Mason A (1990) Correlations between morphology and electrophysiology of pyramidal neurons in slices of rat visual cortex. I. Establishment of cell classes. *Journal of Neuroscience* **10**: 1407–1414.

Lee CC, Imaizumi K, Schreiner CE, and Winer JA (2004a) Concurrent tonotopic processing streams in auditory cortex. *Cerebral Cortex* **14**:441–451.

Lee CC, Schreiner CE, Imaizumi K, and Winer JA (2004b) Tonotopic and heterotopic projection systems in physiologically defined auditory cortex. *Neuroscience* **128**:871–887.

Levesque M, Gagnon S, Parent A, and Deschênes M (1996) Axonal arborizations of corticostriatal and corticothalamic fibers arising from the second somatosensory area in the rat. *Cerebral Cortex* **6**:759–770.

Levesque M and Parent A (1998) Axonal arborization of corticostriatal and corticothalamic fibers arising from prelimbic cortex in the rat. *Cerebral Cortex* **8**:602–613.

Li J, Guido W, and Bickford ME (2003a) Two distinct types of corticothalamic EPSPs and their contribution to short-term synaptic plasticity. *Journal of Neurophysiology* **90**:3429–3440.

Li J, Bickford ME, and Guido W (2003b) Distinct firing properties of higher order thalamic relay neurons. *Journal of Neurophysiology* **90**:291–299.

Li J, Wang S, and Bickford ME (2003c) Comparison of the ultrastructure of cortical and retinal terminals in the rat dorsal lateral geniculate and lateral posterior nuclei. *Journal of Comparative Neurology* **460**:394–409.

Liu X-B (1997) Subcellular distribution of AMPA and NMDA receptor subunit immunoreactivity in ventral posterior and reticular nuclei of rat and cat thalamus. *Journal of Comparative Neurology* **388**: 587–602.

Liu X-B, Honda CN, and Jones EG (1995a) Distribution of four types of synapse on physiologically identified relay neurons in the ventral posterior thalamic nucleus of the cat. *Journal of Comparative Neurology* **352**:69–91.

Liu X-B, Warren RA, and Jones EG (1995b) Synaptic distribution of afferents from reticular nucleus in ventroposterior nucleus of cat thalamus. *Journal of Comparative Neurology* **352**: 187–202.

Liu X-B, Munoz A, and Jones EG (1998) Changes in subcellular localization of metabotropic glutamate receptor subtypes during postnatal development of mouse thalamus. *Journal of Comparative Neurology* **388**:450–465.

Liu X-B and Jones EG (1999) Predominance of corticothalamic synaptic inputs to thalamic reticular nucleus neurons in the rat. *Journal of Comparative Neurology* **414**:67–79.

Liu X-B and Jones EG (2003) Fine structural localization of connexin-36 immunoreactivity in mouse cerebral cortex and thalamus. *Journal of Comparative Neurology* **466**:457–467.

Lund JS, Lund RD, Hendrickson AE, Bunt AH, and Fuchs AF (1975) The origin of efferent pathways from the primary visual cortex, area 17, of the macaque monkey as shown by retrograde transport of horseradish peroxidase. *Journal of Comparative Neurology* **164**:287–303.

Majorossy K and Réthelyi M (1968) Synaptic architecture in the medial geniculate body (ventral division). *Experimental Brain Research* **6**:306–323.

Martin LJ, Blackstone CD, Huganir RL, and Price DL (1992) Cellular localization of a metabotropic glutamate receptor in rat brain. *Neuron* **9**:259–270.

Mason A and Larkman A (1990) Correlations between morphology and electrophysiology of pyramidal neurons in slices of rat visual cortex. II. Electrophysiology. *Journal of Neuroscience* **10**: 1415–1428.

Mathers LH (1972) The synaptic organization of the cortical projection to the pulvinar of the squirrel monkey. *Journal of Comparative Neurology* **146**:43–60.

McCormick DA and von Krosigk M (1992) Corticothalamic activation modulates thalamic firing though glutamate "metabotropic" receptors. *Proceedings of the National Academy of Sciences of the United States of America* **89**:2774–2778.

Middlebrooks JC and Zook JM (1983) Intrinsic organization of the cat's medial geniculate body identified by projections to binaural response-specific bands in the primary auditory cortex. *Journal of Neuroscience* **3**:203–224.

Miller R (1996) Cortico-thalamic interplay and the security of operation of neural assemblies and temporal chains in the cerebral cortex. *Biological Cybernetics* **75**:263–275.

Molinari M, Minciacchi D, Bentivoglio M, and Macchi G (1985) Efferent fibers from the motor cortex terminate bilaterally in the thalamus of rats and cats. *Experimental Brain Research* **57**: 305–312.

Montero VM (1983) Ultrastructural identification of axon terminals from the thalamic reticular nucleus in the medial geniculate body in the rat: an EM autoradiographic study. *Experimental Brain Research* **51**:338–342.

Montero VM (1991) A quantitative study of synaptic contacts on interneurons and relay cells of the cat lateral geniculate nucleus. *Experimental Brain Research* **86**:257–270.

Morest DK (1964) The neuronal architecture of the medial geniculate body of the cat. *Journal of Anatomy (London)* **98**:611–630.

Morest DK (1965) The laminar structure of the medial geniculate body of the cat. *Journal of Anatomy (London)* **99**:143–160.

Morest DK (1975) Synaptic relationships of Golgi type II cells in the medial geniculate body of the cat. *Journal of Comparative Neurology* **162**:157–194.

Na J, Kakei S, and Shinoda Y (1997) Cerebellar input to corticothalamic neurons in layers V and VI in the motor cortex. *Neuroscience Research* **28**:77–91.

Negyéssy L, Hámori J, and Bentivoglio M (1998) Contralateral cortical projection to the mediodorsal thalamic nucleus: origin and synaptic organization in the rat. *Neuroscience* **84**:741–753.

Ohara PT and Lieberman AR (1981) Thalamic reticular nucleus: anatomical evidence that cortico-reticular axons establish monosynaptic contact with reticulo-geniculate projection cells. *Brain Research* **207**:153–156.

Oda S (1997) Ultrastructure and distribution of corticothalamic fiber terminals from the posterior cingulate cortex and the presubiculum to the anteroventral thalamic nucleus of the rat. *Brain Research Bulletin* **42**:485–491.

Ojima H, Honda CN, and Jones EG (1992) Characteristics of intracellularly injected infragranular pyramidal neurons in cat primary auditory cortex. *Cerebral Cortex* **2**:197–216

Ojima H (1994) Terminal morphology and distribution of corticothalamic fibers originating from layers 5 and 6 of cat primary auditory cortex. *Cerebral Cortex* **6**:646–663.

Ojima H, Murakami K, and Kishi K (1996) Dual termination modes of corticothalamic fibers originating from pyramids of layers 5 and 6 in cat visual cortical area 17. *Neuroscience Letters* **208**:57–60.

Orman SS and Humphrey GL (1981) Effects of changes in cortical arousal and of auditory cortex cooling on neuronal activity in the medial geniculate body. *Experimental Brain Research* **42**: 475–482.

Pandya DN, Rosene D, and Doolittle AM (1994) Corticothalamic connections of auditory-related areas of the temporal lobe in the rhesus monkey. *Journal of Comparative Neurology* **345**:447–471.

Paré D and Smith Y (1996) Thalamic collaterals of corticostriatal axons: their termination field and synaptic targets in cats. *Journal of Comparative Neurology* **372**:551–567.

Peruzzi D, Bartlett E, Smith PH, and Oliver DL (1997) A monosynaptic GABAergic input from the inferior colliculus to the medial geniculate body in rat. *Journal of Neuroscience* **17**:3766–3777.

Pinault D (1996) A novel single-cell staining procedure performed in vivo under electrophysiological control: morpho-functional features of juxtacellularly labeled thalamic cells and other central neurons with biocytin or Neurobiotin. *Journal of Neuroscience Methods* **65**:113–136.

Pontes C, Reis FF, and Sousa-Pinto A (1975) The auditory cortical projections onto the medial geniculate body in the cat. An experimental anatomical study with silver and autoradiographic methods. *Brain Research* **91**:43–63.

Preuss TM and Goldman-Rakic PS (1987) Crossed corticothalamic and thalamocortical connections of macaque prefrontal cortex. *Journal of Comparative Neurology* **257**:269–281.

Prieto JJ and Winer JA (1999) Layer VI in cat primary auditory cortex: Golgi study and sublaminar origins of projection neurons. *Journal of Comparative Neurology* **404**:332–358.

Reale RA and Imig TJ (1980) Tonotopic organization in auditory cortex of the cat. *Journal of Comparative Neurology* **192**: 265–291.

Rinvik E (1968) The corticothalamic projection from the pericruciate and coronal gyri in the cat. An experimental study with silver-impregnation methods. *Brain Research* **10**:79–119.

Rinvik E, Ottersen OP, and Storm-Mathisen J (1987) Gamma-aminobutyrate-like immunoreactivity in the thalamus of the cat. *Neuroscience* **21**:781–805.

Reichhova I and Sherman SM (2004) Somatosensory corticothalamic projections: distinguishing drivers from modulators. *Journal of Neurophysiology* **92**:2185–2197.

Rockland KS (1996) Two types of corticopulvinar terminations: round (type 2) and elongate (type 1). *Journal of Comparative Neurology* **368**:57–87.

Rose JE (1949) The cellular structure of the auditory region of the cat. *Journal of Comparative Neurology* **91**:409–440.

Rouiller EM, Capt M, Hornung JP, and Streit P (1990) Correlation between regional changes in the distributions of GABA-containing neurons and unit response properties in the medial geniculate body of the cat. *Hear Res.* **49**:249–258.

Rouiller EM, Colomb E, Capt M, and de Ribaupierre F (1985) Projections of the reticular complex of the thalamus onto physiologically characterized regions of the medial geniculate body. *Neuroscience Letters* **53**:227–232.

Rouiller EM and de Ribaupierre F (1985) Origin of afferents to physiologically defined regions of the medial geniculate body of the cat: ventral and dorsal divisions. *Hearing Research* **19**:97–114.

Rouiller EM, Rodrigues-Dagaeff C, Simm G, de Ribaupierre Y, Villa A, and de Ribaupierre F (1989) Functional organization of the medial division of the medial geniculate body of the cat: tonotopic organization, spatial distribution of response properties and cortical connections. *Hearing Research* **39**:127–142.

Rouiller EM and de Ribaupierre F (1990) Arborization of corticothalamic axons in the auditory thalamus of the cat: a PHA-L tracing study. *Neuroscience Letters* **108**:29–35.

Rouiller EM and Durif C (2004) The dual pattern of corticothalamic projection of the primary auditory cortex in macaque monkey. *Neuroscience Letters* **358**:49–52.

Rouiller EM, Liang FY, Moret V, and Wiesendanger M (1991) Patterns of corticothalamic terminations following injection of *Phaseolus vulgaris* leucoagglutinin (PHA-L) in the sensorimotor cortex of the rat. *Neuroscience Letters* **125**:93–97.

Rouiller EM, Tanne J, Moret V, Kermadi I, Boussaoud D, and Welker E (1998) Dual morphology and topography of the corticothalamic terminals originating from the primary, supplementary motor, and dorsal premotor cortical areas in macaque monkeys. *Journal of Comparative Neurology* **396**:169–185.

Rouiller EM, Wannier T, and Morel A (2003) The dual pattern of corticothalamic projection of the premotor cortex in macaque monkeys. *Thalamus and Related Systems* **2**:189–197.

Rouiller EM and Welker E (1991) Morphology of corticothalamic terminals arising from the auditory cortex of the rat: a *Phaseolus vulgaris*-leucoagglutinin (PHA-L) tracing study. *Hearing Research* **56**:179–190.

Rouiller EM and Welker E (2000) A comparative analysis of the morphology of corticothalamic projections in mammals. *Brain Research Bulletin* **53**:727–741.

Ryugo DK and Weinberger NM (1976) Corticofugal modulation of the medial geniculate body. *Experimental Neurology* **51**: 377–391.

Sakoda T, Kimura A, Donishi T, Kitano H, and Tamai Y (2004) Presence and connections of auditory neurons in the rostrodorsal and rostrolateral parts of the thalamic reticular nucleus. *Acta Otolaryngologica* **Supplement 553**:36–42.

Salt TE and Eaton SA (1995) Modulation of sensory neurone excitatory and inhibitory responses in the ventrobasal thalamus by activation of metabotropic excitatory amino acid receptors. *Neuropharmacology* **34**:1043–1051.

Scannell JW, Burns GA, Hilgetag CC, O'Neil MA, and Young MP (1999) The connectional organization of the cortico-thalamic system of the cat. *Cerebral Cortex* **9**:277–299.

Schreiner CE and Cynader MS (1984) Basic functional organization of second auditory cortical field (AII) of the cat. *Journal of Neurophysiology* **51**:1284–1305.

Schwartz M, Dekker JJ, and Goldman-Rakic PS (1991) Dual mode of corticothalamic synaptic termination in the mediodorsal nucleus of the rhesus monkey. *Journal of Comparative Neurology* **309**: 289–304.

Shosaku A and Sumitomo I (1983) Auditory neurons in the rat thalamic reticular nucleus. *Experimental Brain Research* **49**: 432–442.

Sherman SM and Guillery RW (2002) The role of the thalamus in the flow of information to the cortex. *Philosophical Transactions of the Royal Society of London, series B, Biological Sciences* **357**: 1695–1708.

Sherman SM and Guillery RW (2006) *Exploring the Thalamus and its Role in Cortical Function*. MIT Press, Cambridge.

Sillito AM and Jones HE (2002) Corticothalamic interactions in the transfer of visual information. *Philosophical Transactions of the Royal Society of London, series B, Biological Sciences* **357**: 1739–1752.

Simm GM, de Ribaupierre F, de Ribaupierre Y, and Rouiller EM (1990) Discharge properties of single units in auditory part of reticular nucleus of thalamus in cat. *Journal of Neurophysiology* **63**:1010–1021.

Sousa-Pinto A and Reis FF (1975) Selective uptake of (^{3}H)leucine by projection neurons of the cat auditory cortex. *Brain Research* **85**:331–336.

Steriade M, Jones EG, and McCormick DA (eds) (1997) *Thalamus*, volume 1, *Organization and Function*. Elsevier Scientific Publishing Company BV, Amsterdam.

Takayanagi M and Ojima H (2006) Microtopography of the dual corticothalamic projections originating from domains along the frequency axis of the cat primary auditory cortex. *Neuroscience* **142**:769–780.

Taktakishvili O, Sivan-Loukianova E, Kultas-Ilinsky K, and Ilinsky IA (2002) Posterior parietal cortex projections to the ventral lateral and some association thalamic nuclei in *Macaca mulatta*. *Brain Research Bulletin* **59**:135–150.

Thompson GC, Cortez AM, and Lam DM-K (1985) Localization of GABA immunoreactivity in the auditory brainstem of guinea pigs. *Brain Research* **339**:119–122.

Turner JP and Salt TE (1999) Group III metabotropic glutamate receptors control corticothalamic synaptic transmission in the rat thalamus in vitro. *Journal of Physiology (London)* **519**: 481–491.

Van Horn SC and Sherman SM (2004) Differences in projection patterns between large and small corticothalamic terminals. *Journal of Comparative Neurology* **475**:406–415.

Veinante P, Lavallee P, and Deschênes M (2000) Corticothalamic projections from layer 5 of the vibrissal barrel cortex in the rat. *Journal of Comparative Neurology* **424**:197–204.

Velayos JL (1997) Laminar and areal distribution of corticothalamic neurons with contralateral projection. *European Journal of Anatomy* **1**:105–116.

Vidnyanszky Z and Hámori J (1994) Quantitative electron microscopic analysis of synaptic input from cortical areas 17 and 18 to the dorsal lateral geniculate nucleus in cats. *Journal of Comparative Neurology* **349**:259–268.

Vidnyanszky Z, Gorcs TJ, Negyéssy L, Borostyankio Z, Knopfel T, and Hámori J (1996) Immunocytochemical visualization of the mGluR1a metabotropic glutamate receptor at synapses of corticothalamic terminals originating from area 17 of the rat. *European Journal of Neuroscience* **8**:1061–1071.

Villa AEP (1990) Physiological differentiation within the auditory part of the thalamic reticular nucleus of the cat. *Brain Research Brain Research Reviews* **15**:25–40.

Villa AEP, Rouiller EM, Simm GM, Zurita P, de Ribaupierre Y, and de Ribaupierre F (1991) Corticofugal modulation of the information processing in the auditory thalamus of the cat. *Experimental Brain Research* **86**:506–517.

von Krosigk M, Monckton JE, Reiner PB, and McCormick DA (1999) Dynamic properties of corticothalamic excitatory postsynaptic potentials and thalamic reticular inhibitory postsynaptic potentials in thalamocortical neurons of the guinea-pig dorsal lateral geniculate nucleus. *Neuroscience* **91**:7–20.

Warren RA, Agmon A, and Jones EG (1994) Oscillatory synaptic interactions between ventroposterior and reticular neurons in mouse thalamus in vitro. *Journal of Neurophysiology* **72**:1993–2003.

Watanabe T, Yanagisawa K, Kanzaki J, and Katsuki Y (1966) Cortical efferent flow influencing unit responses of medial geniculate body to sound stimulation. *Experimental Brain Research* **2**:302–317.

Welker E, Hoogland PV, and van der Loos H (1988) Organization of feedback and feedforward projections of the barrel cortex: a PHA-L study in the mouse. *Experimental Brain Research* **73**:411–435.

Williamson AM, Ohara PT, and Ralston HJ III (1993) Electron microscopic evidence that cortical terminals make direct contact onto cells of the thalamic reticular nucleus in the monkey. *Brain Research* **631**:175–179.

Wilson JR, Friedlander MJ, and Sherman SM (1984) Fine structural morphology of identified X- and Y-cells in the cat's lateral geniculate nucleus. *Proceedings of the Royal Society of London, series B, Biological Sciences* **221**:411–436.

Wilson JR and Forestner DM (1995) Synaptic inputs to single neurons in the lateral geniculate nuclei of normal and monocularly deprived squirrel monkeys. *Journal of Comparative Neurology* **362**: 468–488.

Winer JA (2006) Decoding the auditory corticofugal systems. *Hearing Research* **212**:1–8.

Winer JA, Diehl JJ, and Larue DT (2001) Projections of auditory cortex to the medial geniculate body of the cat. *Journal of Comparative Neurology* **430**:27–55.

Winer JA and Larue DT (1996) Evolution of GABAergic circuitry in the mammalian medial geniculate body. *Proceedings of the National Academy of Sciences of the United States of America* **93**: 3083–3087.

Winer JA, Larue DT, and Huang CL (1999) Two systems of giant axon terminals in the cat medial geniculate body: convergence of cortical and GABAergic inputs. *Journal of Comparative Neurology* **413**:181–197.

Winer JA, Larue DT, Diehl JJ, and Hefti BJ (1998) Auditory cortical projections to the cat inferior colliculus. *Journal of Comparative Neurology* **400**:147–174.

Winer JA and Larue DT (1987) Patterns of reciprocity in auditory thalamocortical and corticothalamic connections: study with horseradish peroxidase and autoradiographic methods in the rat medial geniculate body. *Journal of Comparative Neurology* **257**:282–315.

Winer JA and Morest DK (1983) The medial division of the medial geniculate body of the cat: implications for thalamic organization. *Journal of Neuroscience* **3**:2629–2651.

Winer JA and Morest DK (1984) Axons of the dorsal division of the medial geniculate body of the cat: a study with the rapid Golgi method *Journal of Comparative Neurology* **224**:344–370.

Winer JA and Prieto JJ (2001) Layer V in cat primary auditory cortex (AI): cellular architecture and identification of projection neurons. *Journal of Comparative Neurology* **434**:379–412.

Winer JA, Saint Marie RL, Larue DT, and Oliver DL (1996) GABAergic feedforward projections from the inferior colliculus to the medial geniculate body. *Proceedings of the National Academy of Sciences of the United States of America* **93**:8005–8010.

Woldorff MG, Gallen CC, Hampson SA, Hillyard SA, Pantev C, Sobel D, and Bloom FE (1993). Modulation of early sensory processing in human auditory cortex during auditory selective attention. *Proceedings of the National Academy of Sciences of the United States of America* **90**:8722–8726.

Wong D and Kelly JP (1981) Differentially projecting cells in individual layers of the auditory cortex: a double-labeling study. *Brain Research* **230**:362–366.

Wouterlood FG, Goede PH, and Groenewegen HJ (1990) The in situ detectability of the neuroanatomical tracer *Phaseolus vulgaris*-leucoagglutinin (PHA-L). *Journal of Chemical Neuroanatomy* **3**:11–18.

Yu YQ, Xiong Y, Chan YS, and He J (2004) Corticofugal gating of auditory information in the thalamus: an in vivo intracellular recording study. *Journal of Neuroscience* **24**:3060–3069.

Zhang J, Nakamoto KT, and Kitzes LM (2004) Binaural interaction revisited in the cat primary auditory cortex. *Journal of Neurophysiology* **91**:101–117.

Zhang L and Jones EG (2004) Corticothalamic inhibition in the thalamic reticular nucleus. *Journal of Neurophysiology* **91**: 759–766.

Zikopoulos B and Barbas H (2006) Prefrontal projections to the thalamic reticular nucleus form a unique circuit for attentional mechanisms. *Journal of Neuroscience* **26**:7348–7361.

Chapter 9

Descending Connections of Auditory Cortex to the Midbrain and Brain Stem

Manuel S. Malmierca and David K. Ryugo

Abbreviations

AI	primary auditory cortex
AC	auditory cortex
BDA	biotinylated dextran amines
BF	best frequency
CNC	cochlear nuclear complex
CNIC	central of the inferior colliculus
DCN	dorsal cochlear nucleus
DCIC	dorsal cortex of the inferior colliculus
DNLL	dorsal nucleus of the lateral lemniscus
DSCF	Doppler-shifted constant frequency
ECIC	external cortex of the inferior colliculus
ES	electrical stimulation
GABA	g-aminobutyric acid
IC	inferior colliculus
ICH	inner hair cell
LOC	lateral olivocochlear bundle
LSO	lateral superior olive
MGB	medial geniculate body
MOC	medial olivocochlear bundle
NB	nucleus basalis
NLL	nuclei of the lateral lemniscus
OHC	outer hair cell
PHA-L	*Phaseolus vulgaris*-leucoagglutinin
PN	pontine nuclei
SOC	superior olivary complex
SPO	superior paraolivary nucleus
VCN	ventral cochlear nucleus
VNLL	ventral nucleus of the lateral lemniscus
VNTB	ventral nucleus of the trapezoid body

D.K. Ryugo (✉)
Program for Neuroscience, Garvan Institute of Medical Research, Darlinghurst, NSW 2010, Australia
e-mail: d.ryugo@garvan.org.au

1 Introduction

Descending pathways in the brain have been known, since the end of the nineteenth century (Held 1891) but their significance was unappreciated due to the focus on ascending pathways and the unsuitability of the tract tracing methods then available to reveal these projections. Renewed interest was triggered by the discovery of the olivocochlear bundle (Rasmussen 1946, 1953), and interest surged as the magnitude of the descending pathways emerged (Bourassa et al. 1997; Winer 2006). The auditory cortex (AC) projects to a wide range of subcortical targets in the auditory pathway (Winer 2005, 2006; Winer and Lee 2007). By far, the projections to the auditory thalamus and midbrain are the largest and the projections to subcollicular nuclei such as nucleus sagulum, the paraleminscal regions, superior olivary complex (SOC), cochlear nuclear complex (CNC), and pontine nuclei (PN) were not appreciated until recently (Feliciano and Potashner 1995; Weedman and Ryugo 1996; Doucet et al. 2002; Doucet et al. 2003; Meltzer and Ryugo 2006; Perales et al. 2006). The AC also projects to subcortical forebrain structures such as the amygdala (Romanski and LeDoux 1993), the basal ganglia, and premotor structures including the striatum (Beneyto and Prieto 2001), superior colliculus (Paula-Barbosa and Sousa-Pinto 1973), and central gray (Winer et al. 1998), suggesting that the AC has an important role not only in sensory processing of audition, but also in motor behavior, autonomic function, and state dependent changes (Winer 2005, 2006).

Currently, the descending auditory system (Fig. 9.1) is viewed as a series of regional feedback loops and as a descending chain, since both arrangements coexist (Spangler and Warr 1991). Feedback loops comprise cortical input to subcortical nuclei that project back to cortex, directly or indirectly, allowing the cortex to modulate input to it from lower centers. The impact of this projection is also influenced by ascending fibers, whereby feedback loops of various sizes and complexities are established. There is physiological evidence for inhibitory and facilitatory actions (Watanabe et al.

J.A. Winer, C.E. Schreiner (eds.), *The Auditory Cortex*,
DOI 10.1007/978-1-4419-0074-6_9, © Springer Science+Business Media, LLC 2011

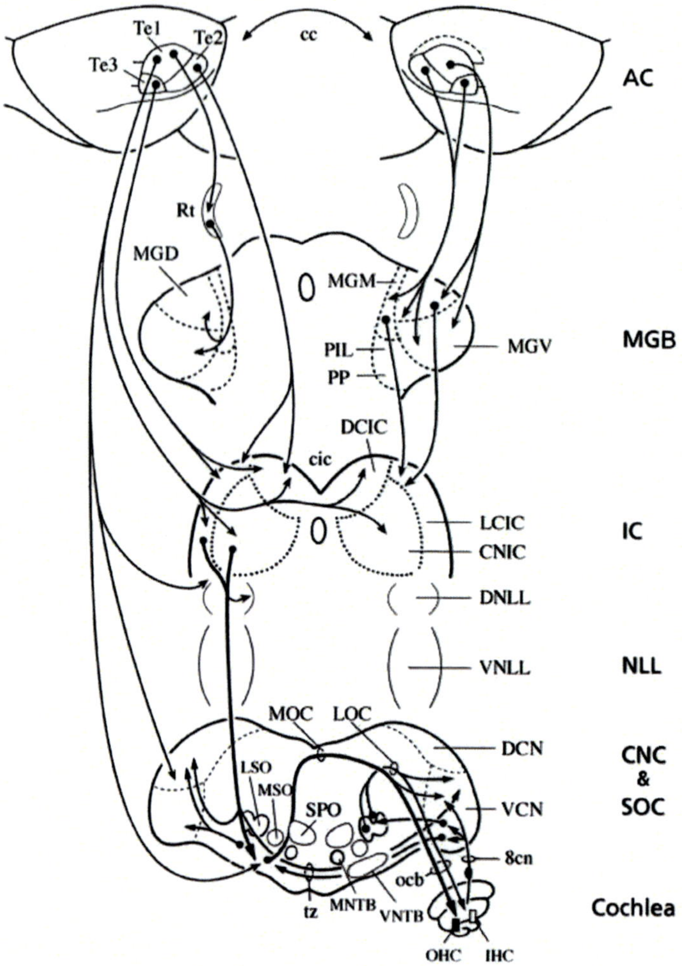

Fig. 9.1 Diagrammatic and simplified representation of the auditory corticofugal system. For clarity, some pathways are omitted or shown only unilaterally. Right auditory cortex (AC) shows the corticothalamic projections; right medial geniculate body (MGB) shows thalamotectal projections. Left inferior colliculus (IC) shows the cortiococollicular and corticobulbar projections. Left cochlea shows the olivocochlear system. Modified from the original source (Malmierca and Merchán 2004)

1966; Zhang and Suga 1997; Zhang et al. 1997; He et al. 2002; He 2003a, b; Wu and Yan 2007), but the role of these pathways in audition is still poorly understood (Warr et al. 1986; Pickles 1988; Huffman and Henson 1990; Spangler and Warr 1991; Warr 1992).

In contrast, descending chains involve projections that contact neurons that then project to still lower auditory centers (Fig. 9.1). The first link of the chain emerges from AC. A second link is fibers from the IC (Spangler and Warr 1991; Warr 1992; Malmierca 2003). These pathways are colliculoolivary and colliculo-cochlear nuclear projections, respectively. The colliculoolivary fibers arise from the external cortex of the inferior colliculus (ECIC) and the ventral part of the central nucleus of the inferior colliculus (CNIC) (Faye-Lund 1985; Caicedo and Herbert 1993; Vetter et al. 1993; Malmierca et al. 1996) and terminate on the medial olivocochlear cells (MOC), which innervate the outer hair cells (Fig. 9.1). Thus, this circuit may constitute a three-neuron pathway from the AC to the receptors (Mulders and Robertson 2000). The lateral olivocochlear cells (LOC) which project to the inner hair cells (Fig. 9.1) also are directly influenced from higher auditory centers (Feliciano and Potashner 1995).

A direct corticoolivary projection has effects upon LOC neurons. Electrical stimulation of the IC produces novel cochlear effects attributable to LOC activation, which was a long-lasting (5–20 min) enhancement or suppression of compound action potentials without concomitant changes in otoacoustic emissions and cochlear microphonics that would be attributable to the MOC system (Groff and Liberman 2003). These efferent neurons, therefore, may be controlled by multiple neuronal loops that commence in AC and the IC. The third link, the olivocochlear system, constitutes the efferent innervation of the cochlea (Rasmussen 1946; Warr 1992).

We will focus on the description and analysis of the structural and functional organization of the descending corticofugal network of projections that originate from the AC. The descending projections from the IC and SOC are outside the scope of this review (Malmierca and Merchán 2004;

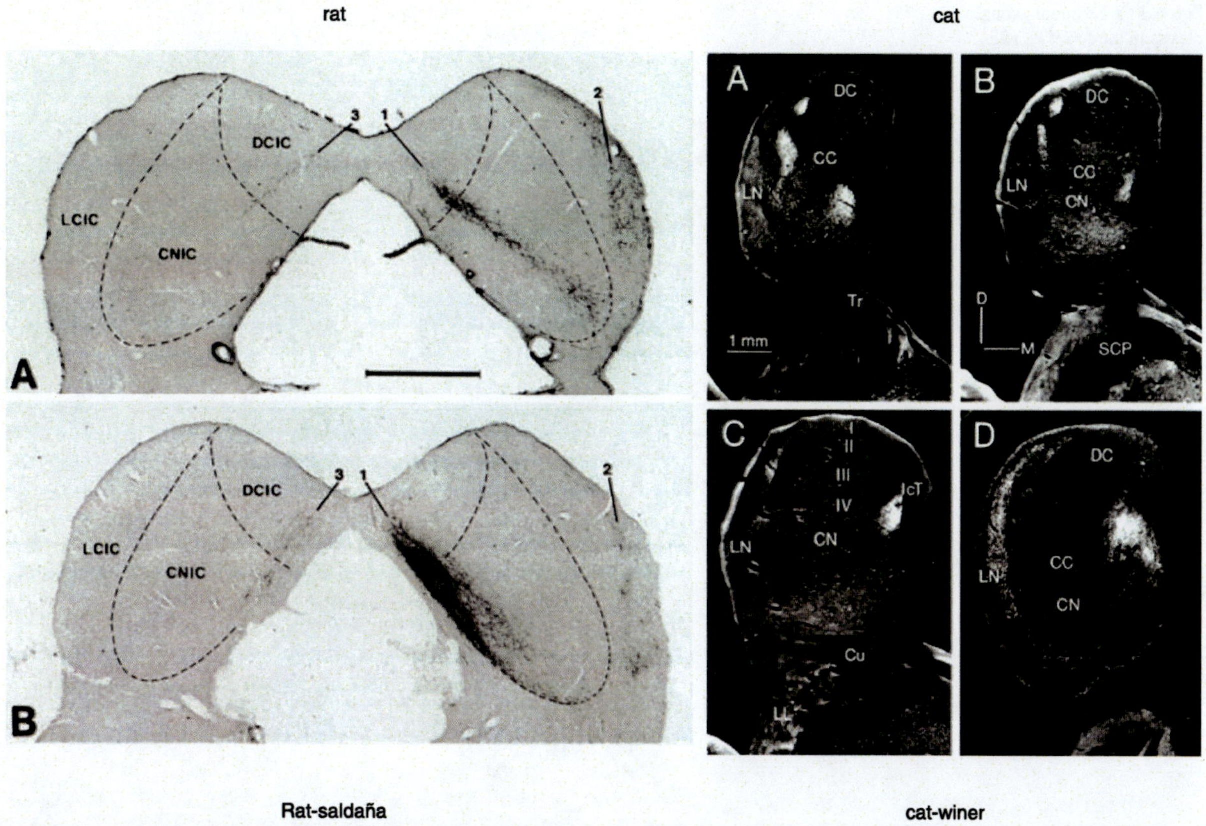

Fig. 9.2 *Left panel*, Topography and distribution of the terminal axonal plexus in the IC after an injections of (**a**) *Phaseolus vulgaris*-leucoagglutinin (PHA-L) or (**b**) a biotinylated dextran amines (BDA) deposit in different frequency regions of the rat left AC. Reproduced from the original source (Saldaña et al. 1996). *Right panel*, Terminal axonal plexus in the IC after injections in the cat left AC. The terminal fields in the rat extend into the central nucleus of the IC (CNIC), whereas those in the cat are more pronounced in the IC dorsal cortex (DCIC). Reproduced from the original source (Winer et al. 1998)

Thompson 2005). The network of descending projections from AC gives off four major tracts: (1) corticothalamic (Figs. 9.1), (2) corticocollicular (Figs. 9.2, 9.3, 9.4, 9.8, and 9.9), (3) corticobulbar (Figs. 9.5, 9.6, 9.7, 9.8, and 9.9), and (4) corticopontine projections.

The corticothalamic system is among the largest projections in the brain, rivaled only by the corticospinal tract (Winer et al. 2001; Winer 2006). Because the corticothalamic system has already been considered in some depth (see Chapter 8 by Ojima and Rouiller), we consider it only briefly. The thalamus has reciprocal connections with the cortex and there is a large-scale topographical overlap in the spatial territories of thalamocortical cells and corticothalamic axonal terminals (Winer 2006; see also Llano and Sherman 2008). Most auditory corticofugal boutons in the medial geniculate body (MGB) are small (approximately 0.5 μm^2 in diameter) and likely arise from the pyramidal cells of layer VI (Romanski and LeDoux 1993; Romanski et al. 1993; Bartlett et al. 2000; Winer 2006). A few are larger boutons (>2 μm^2). These terminals usually form complexes with the dendrites partially surrounded by astrocytic processes (Bartlett et al. 2000; Winer 2006). These endings are thought to originate from neurons in layer V (Rouiller and Welker 1991; Bajo et al. 1995; Shi and Cassell 1997; Bartlett et al. 2000; Winer 2006).

The major corticofugal projections are likely glutamatergic (Potashner et al. 1988) suggesting an excitatory postsynaptic effect. The AC projects to the auditory sector of the reticular thalamic nucleus, which in turns projects to the MGB (Rouiller and Welker 1991; Bartlett and Smith 1999; Bartlett et al. 2000), thus providing the MGB with an inhibitory influence (Montero 1983; Bartlett et al. 2000; Yu et al. 2009). The corticofugal projection, therefore, can modulate the MGB responses to sound through a direct excitatory pathway and/or an indirect inhibitory pathway, consistent with physiological results in the MGB based on electrical stimulation of the AC (Watanabe et al. 1966; Ryugo and

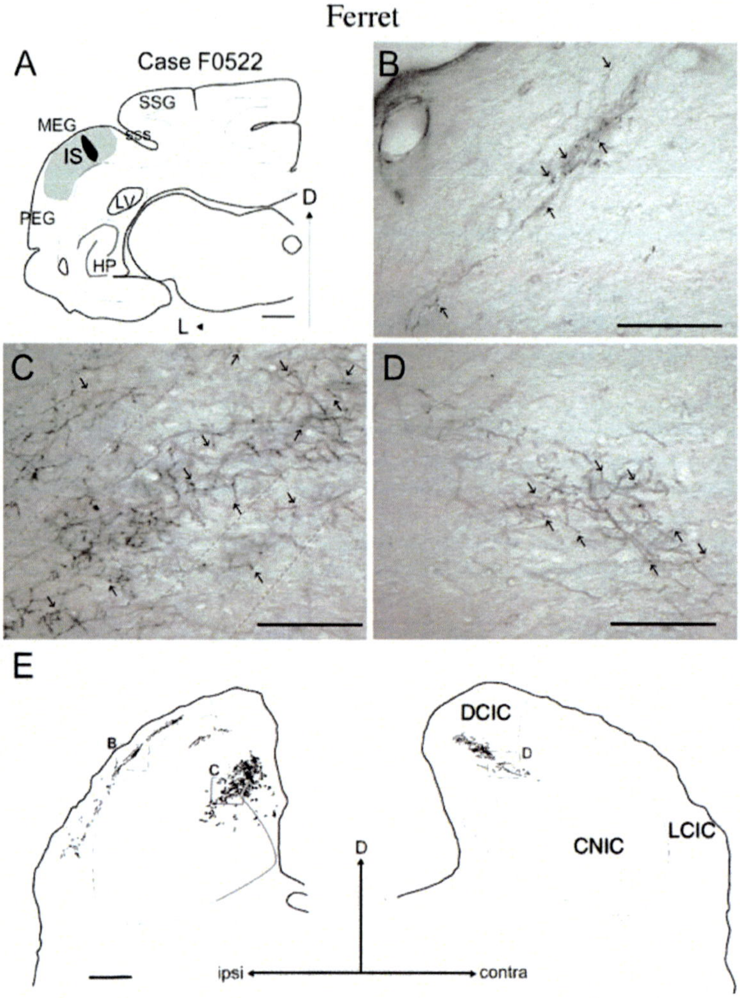

Fig. 9.3 **a** Terminal axonal plexus in the ferret IC after Fluororuby injections in the left AC. **b–d** Details of terminals shown in **e**. The terminal fields extend into the DCIC, LCIC, and CNIC. **e** Overview of the terminals. Reproduced from the original source (Bajo et al. 2007)

Weinberger 1976; Zhang and Suga 1997; Zhang et al. 1997; He et al. 2002; He 2003a, b; Yu et al. 2004).

2 The Corticocollicular System

Corticocollicular projections arise in all AC subdivisions, bypass the MGB and terminate in the IC (Faye-Lund 1985; Herbert et al. 1991; Saldaña et al. 1996; Budinger et al. 2000; Doucet et al. 2002; Bajo and Moore 2005, Bajo et al. 2007, 2010). Since an early report of a temporal cortex projection to the primate corpora quadrigemina (Thompson 1900), many studies have documented this pathway in several species (Massopust and Ordy 1962; Andersen et al. 1980; Morest and Oliver 1984; Faye-Lund 1985; Games and Winer 1988; Feliciano and Potashner 1995; Saldaña et al. 1996; Winer et al. 1998; Budinger et al. 2000; Bajo and Moore 2005; Coomes et al. 2005; Bajo et al. 2007). Most studies find a topographic (tonotopic) organization of these projections arising from the primary auditory cortex (AI), such that the low frequency AI regions project to the dorsolateral IC and the high frequency part projects to the ventromedial IC (Fig. 9.2).

Species differences are marked in AC, where the number of areas identified range from 5 to 6 in mice and rats, 6–9 in cats and ferrets, 10–12 in primates, and 30 or more in some studies of humans. Species differences include the number of areas, their relative position and arrangement, cell density, connections, and tonotopic organization. However, a common theme is that a central primary region, or core, designated here as AI, is surrounded by a variable number of secondary, or belt, areas. Descending projections originate bilaterally in multiple cortical areas (Winer 2006).

As a rule, the projections from AI are the heaviest (Fig. 9.2). But projections to the IC also originate from non-primary area and these inputs are more variable and lighter than those from AI (Herbert et al. 1991; Budinger et al.

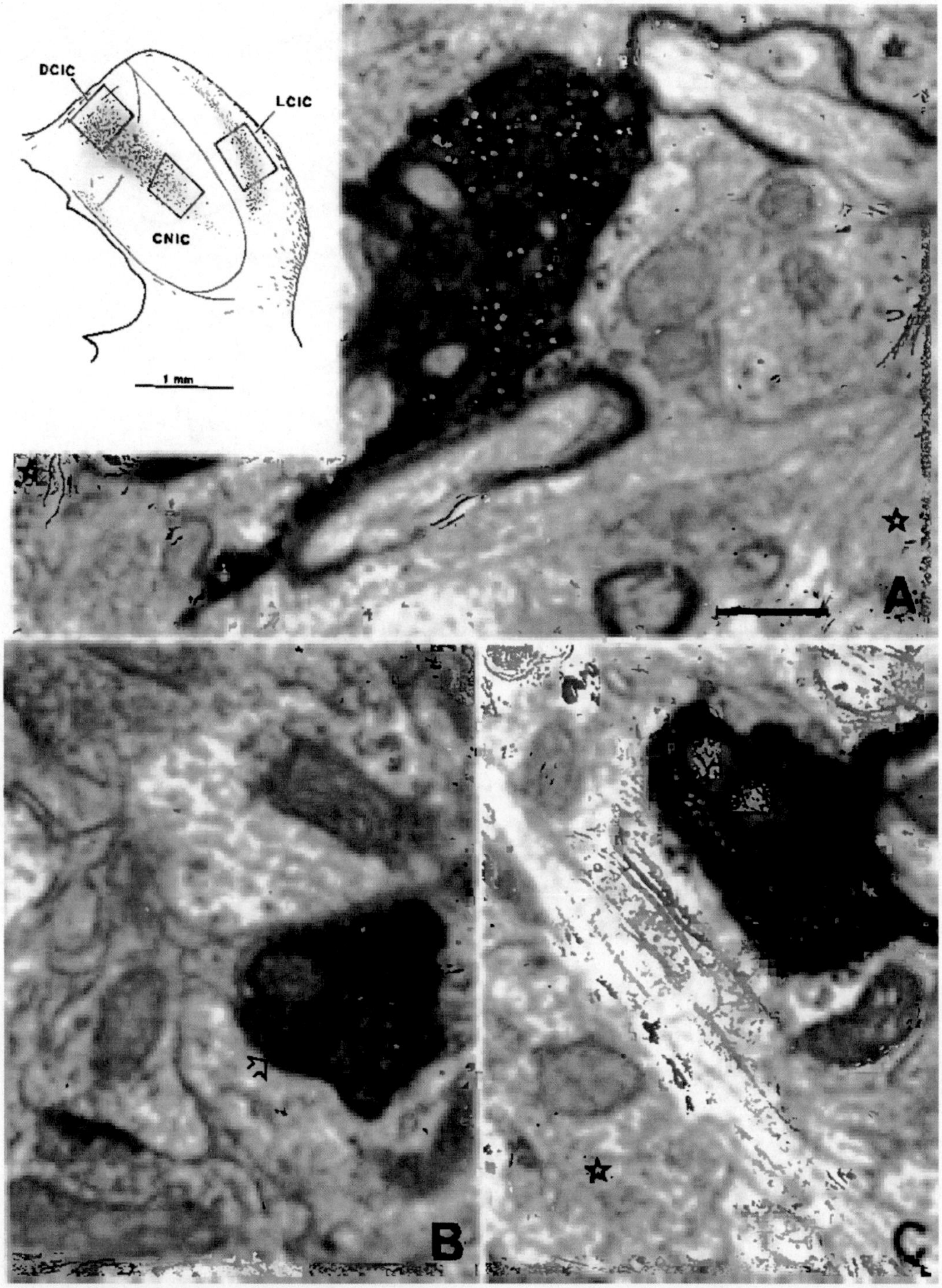

Fig. 9.4 Electron micrographs of auditory corticocollicular endings labeled in the (**a**) CNIC, (**b**) DCIC and (**c**) LCIC after a large injection of biotinylated dextran amines into the ipsilateral AC in rat (Tel). All labeled endings contain round synaptic vesicles and make asymmetric synaptic junctions (*arrows*). **a** A terminal bouton synapsing on a dendritic spine. **b** An ending contacting a thin dendritic shaft or spine. **c** An ending makes two synapses, one with a dendritic shaft containing a mitochondrion (*small open arrow*), the other with a spine or a dendritic branchlet (*large open arrow*). *Stars*, (**a**) unlabeled terminal boutons with pleomorphic synaptic vesicles. *Inset*: camera lucida drawing of a representative flat embedded transverse section illustrating the distribution of the corticocollicular terminal fields. Scale bar: 0.4 μm. Reproduced from the original source (Saldaña et al. 1996)

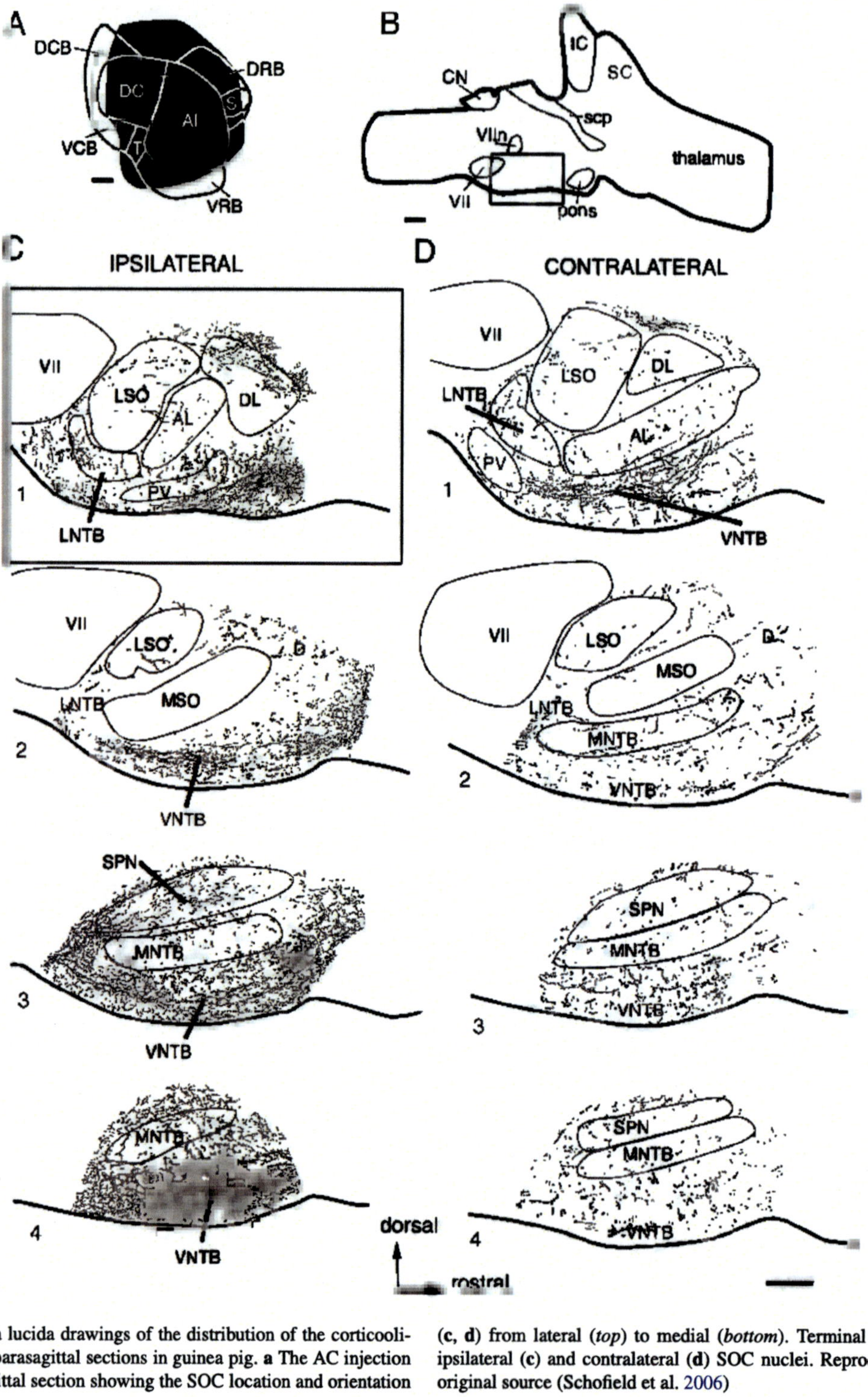

Fig. 9.5 Camera lucida drawings of the distribution of the corticoolivary boutons in parasagittal sections in guinea pig. **a** The AC injection site. **b** A parasagittal section showing the SOC location and orientation (**c, d**) from lateral (*top*) to medial (*bottom*). Terminal boutons in the ipsilateral (**c**) and contralateral (**d**) SOC nuclei. Reproduced from the original source (Schofield et al. 2006)

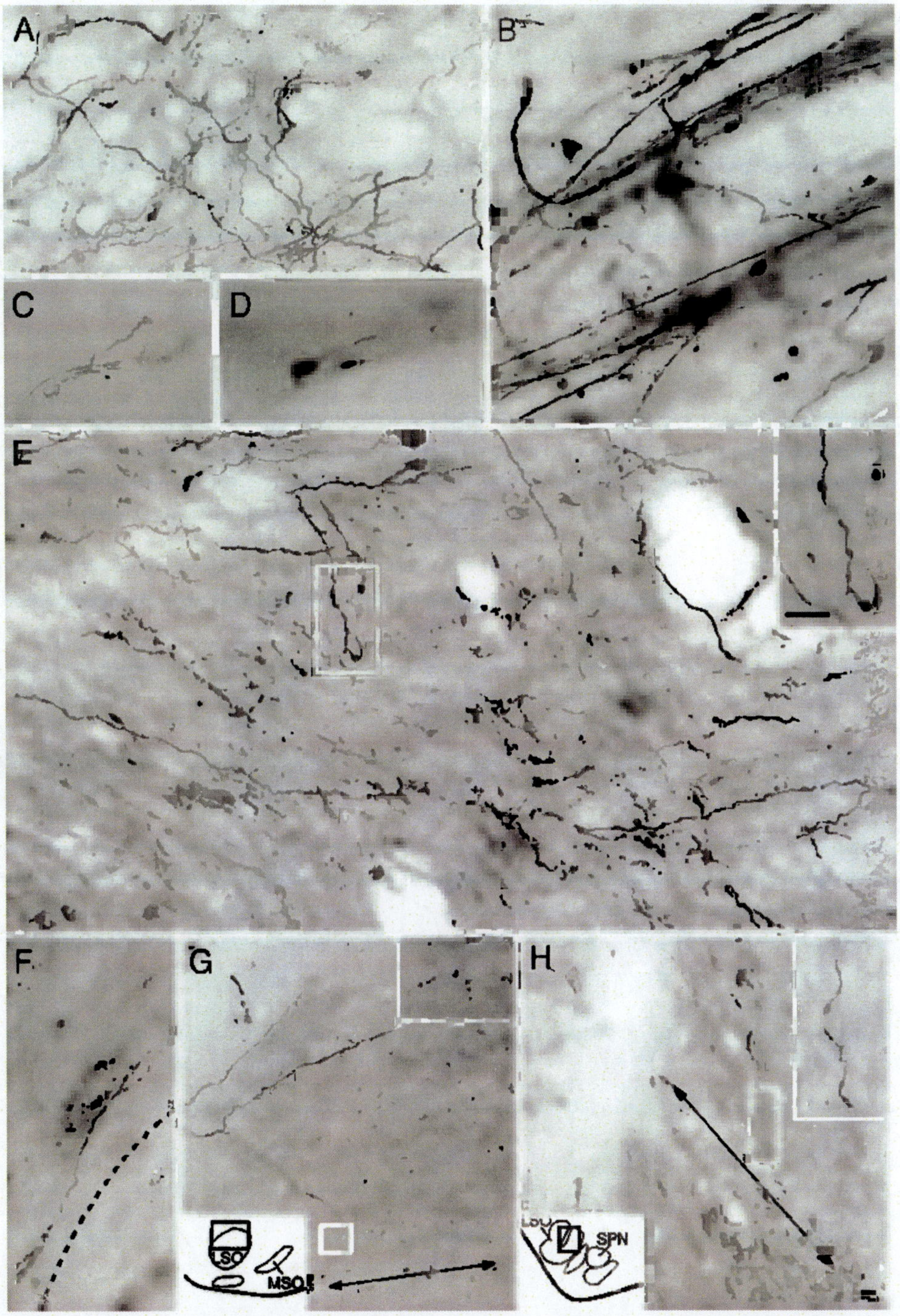

Fig. 9.6 Corticoolivary axon morphology. **a–d** A variety of axon sizes and orientations in the ventral nucleus of the trapezoid body. **e** Axons in the dorsolateral periolivary nucleus. **f** Axons along the margin of the lateral superior olive (LSO). **(g, h)** Axons within the LSO. Reproduced from the original source (Schofield et al. (2004)

2000; Bajo et al. 2007). The AI projections target the inferior colliculus dorsal and external (lateral and rostral) cortices (DCIC and ECIC) bilaterally, with the ipsilateral projection densest (Figs. 9.2 and 9.3).

However, the degree to which the CNIC is also a target remains controversial (Saldaña et al. 1996; Winer et al. 1998) and studies have tended to emphasized the traditional view that AC targets the IC cortices only (Massopust and Ordy

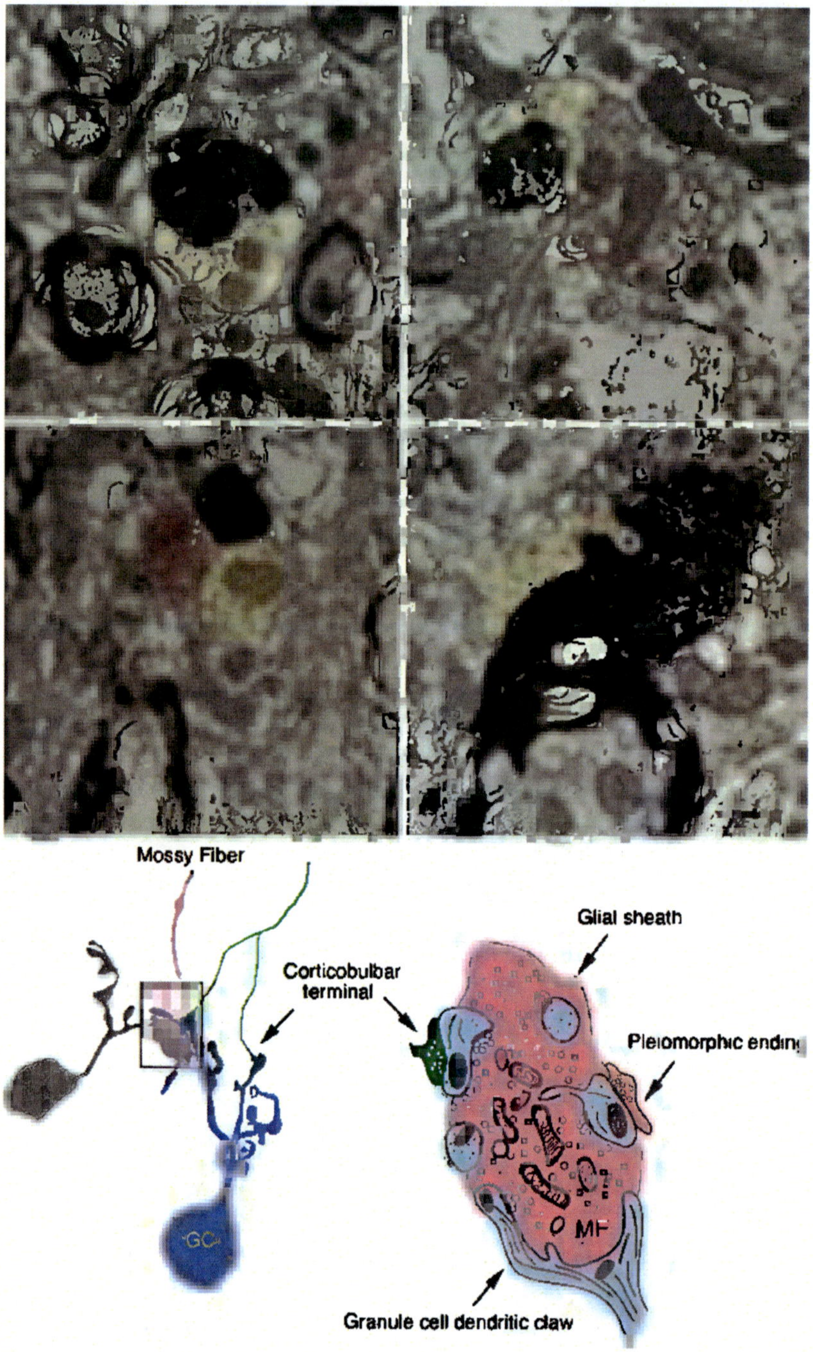

Fig. 9.7 Electron micrographs of labeled cortical terminals in the ipsilateral (**a**, **b**) and contralateral (**c**, **d**) cochlear nucleus granule cell domain. The labeled endings contain round synaptic vesicles and form asymmetric contacts (*arrows*) with thin dendrites (*yellow*). These dendrites often have hair-like protrusions (*asterisk*) that penetrate the afferent ending. Often, these dendrites are also contacted by terminals containing pleomorphic synaptic vesicles (*pink*). The features of these postsynaptic dendrites are typical of granule cells. Scale bar: 0.5 μm. *Bottom*: Summary diagram of the synaptic glomerulus consisting of mossy fibers, granule cell dendrites, and corticobulbar endings. The small size and remote location of the cortical terminals suggest that a modulatory postsynaptic effect. Reproduced from the original source (Meltzer and Ryugo 2006)

1962; Beyerl 1978; Fitzpatrick and Imig 1978; Casseday et al. 1979; Andersen et al. 1980; Faye-Lund 1985; Coleman and Clerici 1987; Herbert et al. 1991). The differences in results, however, may reflect technical limitations and variations of the tracers used, cytoarchitectonic criteria, and/or species differences (Bajo and Moore 2005). Larger injections of stable tracers such as *Phaseolus vulgaris*-leucoagglutinin (PHA-L) or biotinylated dextran amine (BDA) (Saldaña et al. 1996; Winer et al. 1998; Haas et al., 2003; Bajo and Moore 2005; Bajo et al. 2007; Budinger et al. 2000) label CNIC terminal fields, whereas smaller injections and/or more rapidly metabolized tracers produced weak or no terminal fields (Herbert et al. 1991; Budinger et al. 2000). The terminal boutons density is always lower in the CNIC, and the morphology of the terminal boutons that innervate the CNIC and the IC cortices differs. Projections to the CNIC have thinner axons and smaller boutons than those in the IC cortices (Fig. 9.3).

The projection to the IC from the secondary AC ends primarily in the superficial layers of the DCIC and CIC and the rostral ECIC or rostral cortex (Herbert et al. 1991; Budinger et al. 2000; Bajo and Moore 2005; Coomes et al. 2005; Bajo et al. 2007). The projections arise primarily in layer V with some layer VI cells also contributing (Wong and Kelly 1981; Games and Winer 1988; Bajo et al. 1995; Winer and Prieto 2001; Doucet et al. 2003; Bajo and Moore 2005). Layer V neurons include pyramidal cells (Figs. 9.8 and 9.9), and those from layer VI are described as small cells deep in layer VI (Bajo and Moore 2005; Bajo et al. 2010). The largest population of these pyramidal neurons projects to the ipsilateral IC and a smaller population project to the contralateral IC or bilaterally to both ICs.

At least two types of layer V pyramidal neurons with different morphologies participate in the AI corticocollicular pathway (Figs. 9.8 and 9.9); these are tufted and nontufted pyramidal neurons. The morphological range of terminal fields in the IC after anterograde injections in AI also supports the notion of two separate populations (Bajo and Moore 2005). Two populations of projecting neurons in layer V in rat AC have been described (Hefti and Smith 2000). Regular spiking neurons resemble the nontufted type and the intrinsic bursting neurons correspond to tufted neurons noted in gerbil (Bajo and Moore 2005). It has been suggested that only the intrinsic bursting neurons project to the IC, whereas the regular spiking neurons project to other cortical areas and to the putamen (Games and Winer 1988; Moriizumi and Hattori 1991; Ojima et al. 1992; Hefti and Smith 2000; Bajo and Moore 2005).

Cortical projections to the IC contact cells that project to many ascending and descending targets (Games and Winer 1988; Moriizumi and Hattori 1991; Ojima et al. 1992; Hefti and Smith 2000; Malmierca 2003; Bajo and Moore 2005). IC neurons project to the thalamus (Malmierca et al. 1997; Oliver et al. 1999; Peruzzi et al. 1997) and they are also the source of the colliculo-lemniscal, colliculoolivary and colliculo-cochlear nucleus projections (Caicedo and Herbert 1993; Vetter et al. 1993; Malmierca et al. 1996; Schofield and Coomes 2006). The tectothalamic neurons receive not only ascending lemniscal fibers (Oliver et al. 1999) but also AC projections to ipsilateral IC neurons with ascending projections to the ipsi- and contralateral MGB (Coomes-Peterson and Schofield 2007).

The colliculo-lemniscal projections are largely confined to the dorsal nucleus, the sagulum, the horizontal cell group, and the perilemniscal zone (Herbert et al. 1991; Feliciano and Potashner 1995). The sagulum and perilemniscal zones receive the input from the ECIC (lateral and rostral) and DCIC. The colliculoolivary projections originate in the CNIC and ECIC and end in a terminal band that extends to the ventral nucleus of the trapezoid body (VNTB) Herbert et al. 1991; Malmierca et al. 1996; Vetter et al. 1993) and it is also topographic, such that the dorsolateral low-frequency IC projects to the lateral part of the VNTB and the ventromedial high-frequency IC targets the medial VNTB. The terminal fibers overlap the origin of the medial olivocochlear (MOC) system (White and Warr 1983), as shown by double labeling (Vetter et al. 1993). PHA-L-labeled fibers from the IC were in close apposition to retrogradely labeled MOC neurons; though it is unknown whether they make synapses with MOC neurons, the observations strongly suggest that the IC may modulate cochlear responses (Vetter et al. 1993). This idea is supported by the electrophysiological studies showing that electrical stimulation of the IC produces an increase in the latency and a reduction in the amplitude of the auditory whole-nerve response (Dolan and Nuttall 1998), similar to the efffects elicited by electrical stimulation of the MOC (Rajan 1990). Finer dissection of the responses to electrical stimulation of the IC, however, reveals complex actions that involve both the LOC and the MOC (Groff and Liberman 2003; Darrow et al. 2006).

The colliculo-cochlear nucleus projection originates in the CNIC and ECIC and targets the dorsal cochlear nucleus (DCN) and ventral cochlear nucleus (VCN), including the latter's granule cell domain (Caicedo and Herbert 1993; Malmierca et al. 1996). The DCN projections were bilateral and topographic and suggest that neurons with a certain frequency preference project to those with similar, though not identical, tuning. Alternatively, the descending projection may be somewhat mismatched, perhaps underlying sideband influence. There was a projection to the granule cell domain above VCN but its topography is uncertain. The IC projection overlaps with that from the AC (Weedman and Ryugo 1996). The question remains whether these projections target the same neurons and/or the same dendritic shaft or spine.

IC neurons also project to the pontine and mesencephalic reticular formation (Caicedo and Herbert 1993). Targets include the pontine nuclei, the lateral paragigantocellular nucleus, gigantocellular reticular nucleus, the ventrolateral tegmental nucleus, and caudal pontine reticular nucleus. Presumably the IC neurons targeting these nuclei are also under the AC influence, but this remains to be shown (but see below the corticopontine system).

The glutamatergic nature of the guinea pig corticocollicular projection was inferred by IC glutamate decrease after AC ablation (Feliciano and Potashner 1995), consistent with electron microscopic observations showing that in

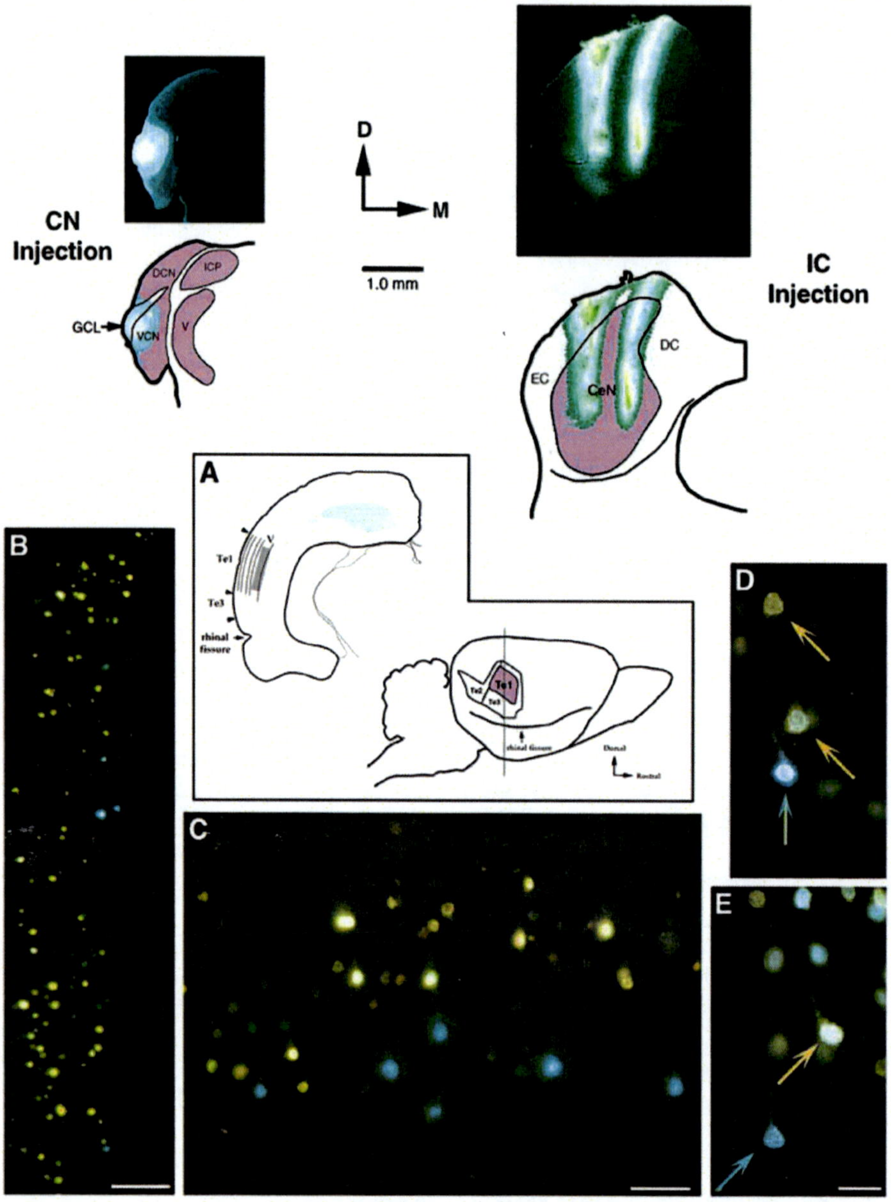

Fig. 9.8 Retrograde labeling in rat area Te1 after a Fast *blue* (FB) injection in the cochlear nucleus (CN) and Diamidino *yellow* (DiY) deposit in the IC. *Top*, a photomicrograph and drawing of a coronal section through each injection site (CeN central nucleus; DCN dorsal cochlear nucleus; DC dorsal cortex; EC external cortex; GCL granule cell lamina; ICP inferior cerebellar peduncle; V spinal tract of the trigeminal nerve). **a** The locus of labeled cortical cells (**b–e**). In the sagittal view, the *gray line* through auditory cortex (temporal areas Te1, Te2, and Te3) indicates the approximate position of the cells along the rostral/caudal axis. The *gray rectangle* in the coronal section ipsilateral to the injection sites denotes the region shown in the subsequent panels. **b** Photomontage of layer V. The pia is to the *left*. Few FB-labeled cells are in deep layer V, whereas the DiY-labeled neurons are distributed more broadly. Scale bar: 100 μm. **c** Higher magnification view of the laminar organization of layer V cells projecting to the CN (*blue*) versus those targeting the IC (*yellow*). The cortical surface is towards the *top* of the figure. Scale: 50 μm. **d**, **e** Labeled cortical cells, most with FB or DiY (*blue* or *yellow arrows*). Cortical cells with both (**d**, *blue* and *yellow arrow*) were far rarer. Reproduced from the original source (Doucet et al. 2003)

all rat IC subdivisions these axons formed small terminal boutons with round synaptic vesicles and made asymmetric synapses onto thin dendritic shafts and spines (Fig. 9.4) (Saldaña et al. 1996). Although these studies suggest a purely excitatory corticollicular projection, electrical stimulation of cat AC elicits excitatory as well as inhibitory and complex interactions in IC neurons (Mitani et al. 1983). In the rat, using tetrodotoxin to block the effect of AC stimulation on

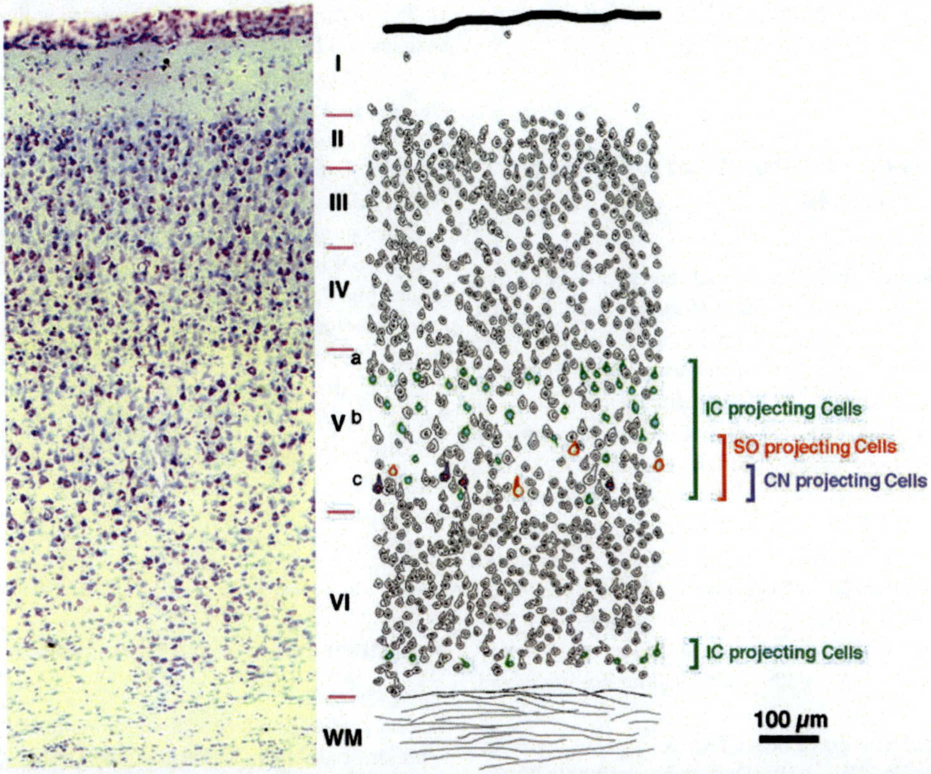

Fig. 9.9 Summary of findings with respect to the laminar organization of AC cells projecting to the inferior colliculus (IC), superior olivary complex (SOC), and cochlear nucleus (CN). *Left*, a photomicrograph of a Nissl preparation in area rat Te1. The cortical surface is towards the *top*. *Right*, a schematic of layers V and VI showing the neurons that project to the IC, SOC, and CN. A few corticocollicular cells are near the border of layer VI and the white matter (WM), but most are in layer V. All three distributions overlap. However, the cortical cells projecting to more distant targets are more narrowly distributed and deeper in layer V. Reproduced from the original source (Doucet et al. 2003)

the IC revealed enhancement or suppression of the neuronal activity and extended the first spike latency (Popelár et al. 2001). The AC may modulate the IC via direct activation of excitatory or inhibitory IC circuits, and/or act on ascending circuits from lower centers.

3 The Corticobulbar System

Early studies suggested a direct neocortical projection to auditory brain stem nuclei but the data were difficult to interpret due to technical limitations (Mettler 1935; Kuypers and Lawrence 1967). Later analysis with PHA-L (Saldaña et al. 1996) demonstrated that AI projects to regions near the lateral lemniscal nuclei (NLL), including ipsilaterally to the nucleus sagulum; and bilaterally to the superior olivary complex (SOC) and CNC (Figs. 9.1, 9.5, 9.6, and 9.7), to the ventral nucleus of the trapezoid body (VNTB); to the lateral superior olive (LSO); to a narrow region above the SOC (Fig. 9.5); to the dorsal cochlear nucleus (DCN) and to the VCN granule cells. Since this descending input terminates in the LSO and the LOC (at least to the shell neurons), it is probably involved in local feedback loops and in modulating afferent responses from inner hair cells (Groff and Liberman 2003).

3.1 Auditory Cortex Projections to the Lateral Lemniscal Nuclei

These projections have been described in most detail in rats, gerbils, and cats (Feliciano et al. 1995; Beneyto et al. 1998; Budinger et al. 2000). In rats, terminations target areas near the NLL, in a paralemniscal area medial to the ventral nucleus of the lateral lemniscus (VNLL), in the horizontal cell group between the VNLL and dorsal nucleus of the lateral lemniscus (DNLL), and in the sagulum. The projections are ipsilateral and originate in AI. In gerbils, terminations in the DNLL and adjacent areas, including the cuneiform nucleus (Fig. 9.1), arise from the anterior auditory field (but not from AI) and from other nonprimary areas (Budinger et al. 2000). The laminar identity and morphology of the cells of origin are unknown in either species.

3.2 Auditory Cortex Projections to Superior Olivary Complex

A direct projection from auditory cortex to the SOC has been suggested by numerous authors (Kuypers and Lawrence 1967; Feliciano et al. 1995; Coomes and Schofield 2004; Doucet et al. 2002, 2003). This pathway was demonstrated from AI in rats to several SOC termination zones (Feliciano et al. 1995) and in cats and guinea pigs (Kuypers and Lawrence 1967; Coomes and Schofield 2004). The AC axons end bilaterally in several SOC regions (Figs. 9.1, 9.5 and 9.6), including the VNTB, LSO, the periolivary region above the LSO, and the superior paraolivary nucleus (SPO) (Feliciano et al. 1995). Most terminations occur in the ipsilateral VNTB, where the terminals have a topographic pattern that presumably reflects VNTB tonotopic organization. It is unclear whether AC projections to other SOC regions are organized tonotopically.

The principal neurons of origin to SOC neurons are layer V pyramidal cells (Fig. 9.9) (Doucet et al. 2002, 2003). Corticoolivary axons and boutons are more numerous ipsilaterally. The majority of boutons were located in the VNTB and the SPO, bilaterally. Boutons are present in LSO and in the other periolivary nuclei.

AC projections to the SOC may contact cells with many targets including the ipsilateral MOC cells which project to the opposite cochlea as well as bilateral SOC cells that project to the CNC ipsi-, contra- or bilaterally (Mulders and Robertson 2000; Schofield et al. 2006), or ipsilaterally or contralaterally to the IC (Schofield and Coomes 2005; Schofield and Coomes 2006; Coomes-Peterson and Schofield 2007).

3.3 Auditory Cortex Projections to Cochlear Nuclear Complex

Cortical projections to the CNC originate primarily in the ipsilateral AI (Weedman and Ryugo 1996; Feliciano et al., 1995; Meltzer and Ryugo 2006), but studies in guinea pig have also found nonprimary AC projections (Jacomme et al. 2003; Schofield and Coomes 2005; Schofield and Coomes 2006). The principal neurons of origin are pyramidal cells from layer V (Figs. 9.8 and 9.9). The guinea pig AC projects bilaterally and symmetrically to the CN. Axons end in the DCN and the granule cell area. Electron microscopic tracing studies in rat and mouse found boutons with round synaptic vesicles and forming asymmetric junctions typical of excitatory synapses on granule cell dendrites (Fig. 9.7) (Weedman and Ryugo 1996; Meltzer and Ryugo 2006). These boutons converge on the mossy fiber-dendritic complex and accentuate the complexity of the granule cell synaptic neuropil region (Meltzer and Ryugo 2006).

Projections to the DCN fusiform cell layer, and to much of the VCN including the small cell cap, have been found in guinea pig. AC boutons outside the granule cell areas, however, are closely apposed to giant DCN fusiform neurons, and to VCN multipolar neurons. The multipolar cells project mostly contralaterally and also ipsilaterally or bilaterally to the IC (Schofield and Coomes 2005). Species differences raise important questions about a global plan for descending auditory pathways, since behavioral specialization and ecologic niche must play roles in brain organization.

4 Auditory Corticopontine System

Several studies find a neocortical input to PN (Brodal 1972; Schuller et al. 1991; Kawamura and Chiba 1979; Schofield and Coomes; 2005; Perales et al. 2006) from AI and secondary AC. As in AC projections to the IC and brain stem, these are topographically and tonotopically organized, although the corticopontine projection is topographically but not tonotopically organized (Perales et al. 2006). The PN terminal plexuses differ from those targeting the cochlear nucleus as many axon terminal fields are widespread and diffuse, with sparse ramifications extending in several axes. PN neurons target the cerebellum and the cochlear nucleus granule cell domain (Ohlrogge et al. 2001). The pontine projection is primarily to the contralateral granule cell domain, on granule cell distal dendrites, as are endings of AC origin. Both terminal types contain round synaptic vesicles making asymmetric contacts (Weedman and Ryugo 1996; Ohlrogge et al. 2001).

5 Neuronal Source of the Descending Connections to the Midbrain and Brain Stem: Collateral Projections

We have described the corticothalamic, corticocollicular, corticobulbar, and corticopontine systems. The corticothalamic projections arise mostly from layer VI and a few from layer V pyramidal neurons (Figs. 9.8 and 9.9), whereas corticocollicular and corticobulbar projections arise from layer V pyramidal neurons located save for a few layer VI nonpyramidal neurons. Thus, the main corticocollicular and corticobulbar projection cell in AI is the layer V pyramidal neuron. Since these neurons represent different classes on grounds of cell size and dendritic morphology in rat and cat (Games and Winer 1988; Winer and Prieto 2001), we propose that different pyramidal cell classes are distinguished by their projection target. Layer V in cats has been partitioned into three sublaminae (Va, Vb, and Vc) on the basis of cytoarchitecture and connectional data (Winer and Prieto 2001). A similar scheme was used to summarize the pyramidal cell distribution of s projecting to the rat IC and brain stem (Doucet et al. 2003). Superficial layer V (Va) cells project to contralateral AC, and a few neurons project to the IC (Games and Winer 1988). Most corticocollicular projections are from middle (Vb) and deep (Vc) layer V neurons. Layer Vc has corticobulbar projections to both the CNC and SOC, but differences in their distributions suggest even finer distinctions within layer V. The laminar distribution of corticostriatal and corticopontine pyramidal cells has not been described.

A general important issue is the extent of collateral projections. Given the many auditory targets in the descending system, there may be many such opportunities. Such a study found a surprisingly small population of cells that project to the thalamus and to the IC (Wong and Kelly 1981). Subsequent studies of AC projections to the IC, SOC, and CNC are in accord with these results (Doucet et al. 2002, 2003) and the most common pattern of projection is for a pyramidal cell to have one midbrain or brain stem target (Figs. 9.8 and 9.9). Differences in these projection patterns are seen in guinea pigs (Coomes et al. 2005; Schofield and Coomes 2005; Schofield et al. 2006). About 5% of layer V corticocollicular cells project to both ICs suggesting that a small but significant population projects bilaterally (Coomes et al. 2005).

Corticofugal projections to the ipsilateral IC and one or the other parts of the CNC are seen in guinea pigs (Coomes et al. 2005; Schofield et al. 2006). Fewer neurons projected to the CNC than to the IC, supporting the conclusion that neurons with collateral projections were rare. A neuron might have axon collaterals to three targets: both ICs and one CNC. The IC projections to the ipsi- and contralateral CNC arise from different cell groups, i.e., without collaterals (Schofield 2002). It was proposed that layer V projections have two patterns: projections to one target (most common), and rarer projections to more targets. A minority of cells with collateral (divergent) projections may exert broader effects and serve a different function than cells with single projection targets.

Some of the principles governing the organization of corticofugal auditory pathways structures are emerging. The set of AC cells projecting to proximal targets is larger and the more distal-projecting neurons are more evenly distributed between the two hemispheres and centered in deeper layer V. Most layer V pyramidal neurons project to one brain stem target.

6 Functional Significance of the Descending Connections to the Midbrain and Brain Stem

The large and diverse sets of descending projections to subcortical auditory nuclei imply variable functional roles in auditory signal processing. The data suggest that the effects of the corticofugal system must be an important function of the AC.

Despite early physiological experiments showing inhibitory and facilitatory actions (Watanabe et al. 1966; Ryugo and Weinberger 1976; Mitani et al. 1983), the role of these loops and chain of descending connections in audition is not well understood, and most functional studies on the descending auditory pathway have focused on the corticothalamic projection (Watanabe et al. 1966; Ryugo and Weinberger 1976; Zhang and Suga 1997; Zhang et al. 1997; He et al. 2002; He 2003a, b).

The roles of the corticocollicular pathway and of the corticobulbar pathway have been explored in most detail in echolocating bats. Descending projections with excitatory and/or inhibitory effects on IC neurons can sharpen and amplify ascending inputs at the same best frequency as the AC activation site (Zhang et al. 1997; Gao and Suga 1998, 2000; Suga et al. 2000, 2002; Zhang and Suga 1997, 2000, 2005; Suga 2008). Thus, AC influences IC spectral processing (Figs. 9.10 and 9.11). IC neuronal activity was recorded before and after AC inactivation with lidocaine, or stimulation with acoustic or electrical stimuli, or both experimental manipulations (Fig. 9.11). The results of these inactivation and stimulation experiments were complementary and show that the IC BF sharpened its tuning when it is matched in frequency with that of the stimulated AC site but does not shift the IC response curve in frequency. When a recorded cell BF is matched to that of the stimulated AC cell, the response of the former is augmented at its BF and is inhibited at frequencies above and below, enhancing its frequency tuning. AC activation shifts IC tuning curves when the AC and IC BFs are mismatched (Fig. 9.11). This dual effect is defined as egocentric selection (Zhang and Suga 1997) and it is seen in gerbil (Sakai and Suga 2001, 2002) and mouse (Yan and Ehret 2002; Yan and Zhang 2005; Yan et al. 2005). While the shape of frequency response areas or tuning curves is not altered in the bat experiments, they do change in mice (Yan et al. 2005). Thus, some changes associated with egocentric selection may be species specific (Fig. 9.10).

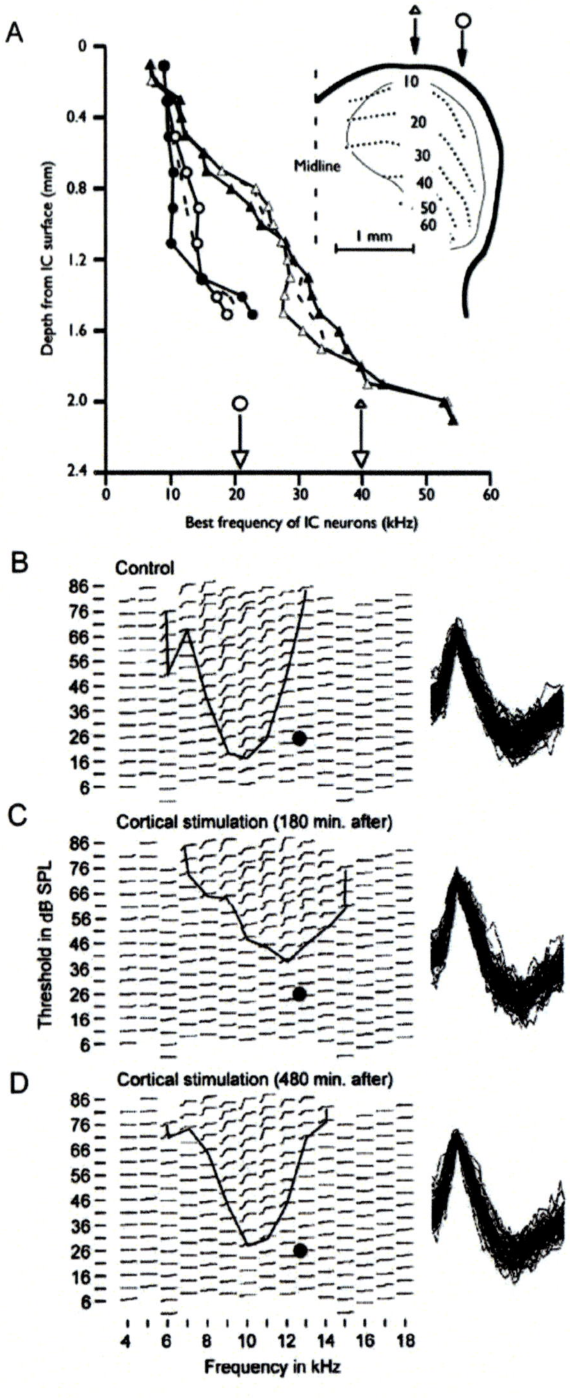

Fig. 9.10 *Top panel*, Best frequency (BF) of IC neurons recorded in two separate CNIC penetrations. IC BF increases systematically with depth (*circles* and *triangles*). When the AC is stimulated electrically, the BF shifts to that of the AC site (*open arrows* and *opened circles* and *triangles*, 3 h after cortical stimulation). At 8 h poststimulation (*dashed lines*) the responses recovers to the original value. Reproduced from the original source (Yan and Ehret 2001). *Bottom panels*, A single frequency response area (FRA) of a well-isolated CNIC neuron from the before (**a**) and after AC stimulation (**b**, 180 min after; **c**, 480 min after). **Fig. 9.10** (continued) The IC BF was 10 kHz and threshold was ~25 dB SPL. After 12.5 kHz AC stimulation (*dots*), the IC BF and threshold shifted (B), and recovers 8 h later (**c**). The unit spike waveform was unchanged over 8 h. The excitatory FRA was significantly changed after AC stimulation. Reproduced from the original source (Yan et al. 2005)

In the mustached bat, AC neurons mediate a highly focused positive feedback associated with widespread bilateral (symmetric) inhibition (Zhang et al. 1997). Compared to control conditions, AC inactivation shifts these curves towards the BF of the AC cell, suggesting that, in control conditions, AC neurons enhance the neural contrast of auditory information. By contrast, egocentric selection is asymmetric in the big brown bat (Yan and Suga 1998) so that BF shifts of IC cells away from the AC BF are seen only in those IC cells with BFs below the AC BF. For neurons above BF, the IC cells shift BF towards the AC BF. In mice, IC tuning shifts are always towards the AC BF (Yan and Ehret 2001). Comparisons with psychophysical measures suggested that the mouse corticofugal adjustments were related to critical bandwidths in the frequency resolution domain (Yan et al. 2005). These BF shifts alter the IC frequency or tonotopic map (Yan and Ehret 2001) and can be induced by repeated moderate intensity sound bursts.

A study in the guinea pig showed that the corticocollicular pathway also plays a critical role in the processing of sound localization cues. Using the cryoloop technique allowing non-focal IC deactivation by cooling the AC radically altered sensitivity to interaural level differences. Interaural level differences are created by the head shadow such that the ear nearer to the sound source receives a louder signal than the opposite ear. This observation is important because, as described above, experimentally induced frequency shifts are seen after AC focal inactivation or stimulation (Nakamoto et al. 2008). When the entire cortex was inactivated or strongly electrically stimulated, however, no IC frequency shifts were found (Suga et al. 2000). These data show that global cortical inactivation has a profound effect on other response features in IC neurons.

Because the major descending projection to subcortical nuclei is ispsilateral, it is not surprising that the contralateral BF shift is somewhat smaller but otherwise similar to the ipsilateral BF shift (Ma and Suga 2004). However, differences between ipsi- and contralateral corticofugal modulation are seen and contralateral BF shift differs from the ipsilateral when the electrical stimulation is made in the mustached bat Doppler-shifted constant frequency (DSCF) area, which is implicated in fine frequency analysis and echo processing. Here, the BF shift depends on the stimulation site (Xiao and Suga 2005). Thus, the corticofugal system may adjust subcortical sensory maps in response to sensory

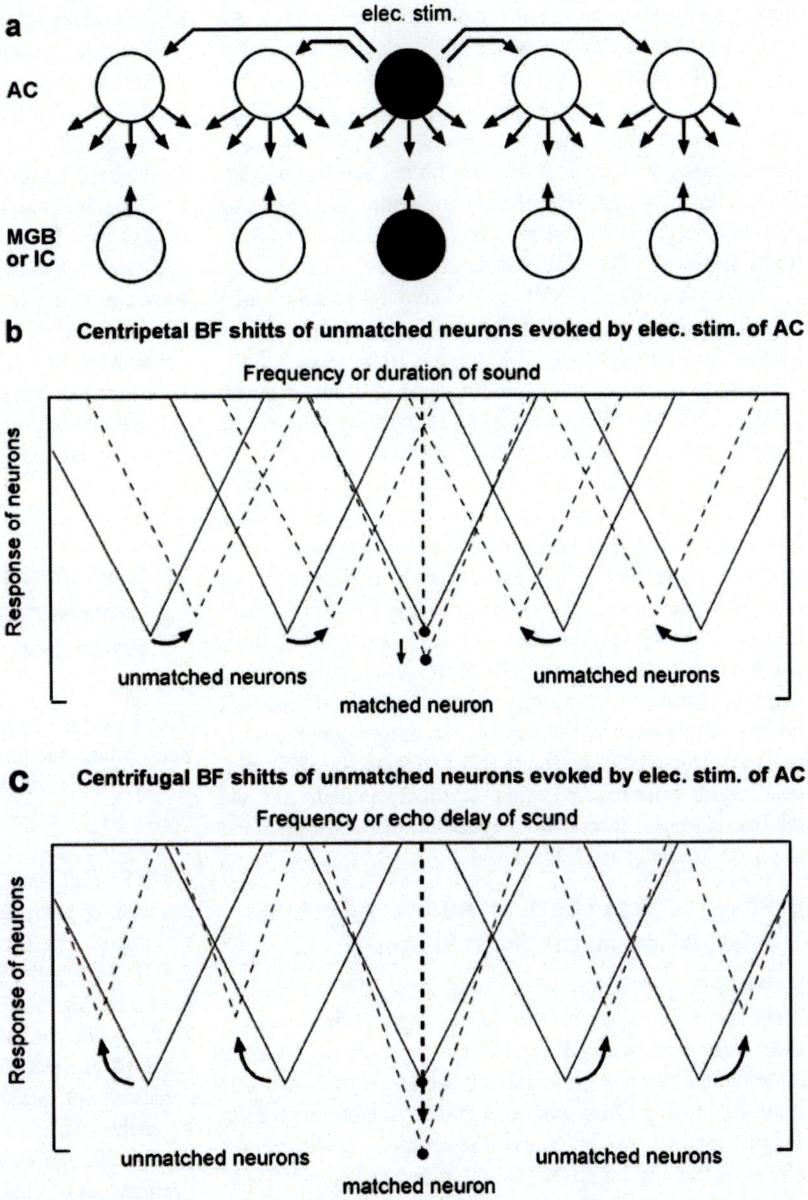

Fig. 9.11 Model of the facilitation of matched neurons and centripetal or centrifugal BF shift of unmatched IC or MGB neurons after electrical AC stimulation. Facilitation of matched neurons and centripetal or centrifugal best frequency-shifts of unmatched neurons evoked by electrical stimulation of cortical neurons. **a** Arrays of neurons in the AC, IC and MGB tuned to different frequencies. Electrical stimulation of an AC neuron (*upper filled circle*) evokes different changes in matched (*lower filled circle*) and unmatched (*lower open circles*) cells. Electrical stimulation of AC neurons evokes facilitation, inhibition, and best frequency (BF) shifts in the AC and subcortical auditory nuclei. There are two types of BF shifts: centripetal (**b**) and centrifugal (**c**). The discontinuous and continuous *triangular curves* represent the FRA in the control and shifted conditions, respectively. Reproduced from the original source (Suga 2008)

experience, and the nature of the modification may depend on the functional system (Yan and Suga 1998).

7 Interpretative Constraints and the Corticofugal System

Several limitations affect interpretation of the anatomical and physiological data. Most physiological studies on the modulatory effect of the corticocollicular pathway have been carried out in bats; we are not aware of any anatomical study of AC-to-IC projections in this specialized animal model. Furthermore, most studies after cortical inactivation or stimulation have recorded the changes observed in the neuronal activity of units from the IC central nucleus, where the AC projections are weakest.

8 Possible Functional Significance

Human speech and animal communication sounds are complex time-varying stimuli whose parameters include frequency, amplitude, duration, interval between sounds, etc. Corticofugal modulation is multiparametric and occurs in different types of subcortical neurons (Suga and Ma 2003). The behavioral changes related to the changes evoked by the corticofugal system, however, remain to be explored. AC focal electrical stimulation evokes highly specific changes in subcortical neurons in the frequency, amplitude, and time domains (Yan and Suga 1996; Ma and Suga 2001, 2007; Yan and Ehret 2002). Auditory learning or conditioning also evokes changes in cortical and subcortical neurons that are specific to the parameters of the conditioned sound (Bakin and Weinberger 1990; Recanzone et al. 1993; Gao and Suga 1998; Bao et al. 2004; Polley et al. 2004).

The nucleus basalis (NB) of the cholinergic basal forebrain is proposed as an essential neural substrate for learning-induced auditory plasticity (Weinberger 1998; Suga and Ma 2003). A tone paired with NB electrical stimulation (tone-ES_{NB}) shifts the frequency tuning of cortical (Bakin and Weinberger 1996; Kilgard and Merzenich 1998; Ma and Suga 2003; Yan and Zhang 2005) and subcortical (Ma and Suga 2003) neurons towards the frequency of the paired tone. Since there is no evidence that the NB projects to the auditory midbrain, these findings suggest that corticofugal projections may influence experience-dependent neural plasticity of subcortical auditory neurons (Gao and Suga 1998, 2000; Suga and Ma 2003; Yan and Zhang 2005). IC plasticity elicited by auditory fear conditioning may reflect corticofugal feedback because application of the $GABA_A$ receptor agonist muscimol to AC blocked IC plasticity (Ma and Suga 2004). Similar conclusions were reached using electrical stimulation of the cholinergic basal forebrain to evoke IC plasticity (Ma and Suga 2003; Zhang and Suga 2005).

9 Physiological Effects of Auditory Cortex Stimulation on the Brain Stem

Electrical stimulation of the AC at high rates evokes a short-term centrifugal BF shift of the contralateral cochlear microphonic receptor potential, indicating a shift in hair cell frequency tuning (Xiao and Suga 2002). Anatomical results suggest that the AC influences the cochlea via the medial olivocochlear system (MOC) (Mulders and Robertson 2000). The MOC projects to the contralateral cochlea and synapses on the outer hair cells (OHCs). The somatic electromotility (Mulders and Robertson 2000; Zheng et al. 2000) of OHCs may be the anatomical substrate for the active cochlear micromechanics supporting fine frequency selectivity of the normal ear (Brownell et al. 1985). The AC influence on the SOC modulates OHC electromotility through the MOC to adjust afferent signals early in peripheral auditory processing (Xiao and Suga 2002; Perrot et al. 2006).

Gentamicin abolishes efferent cochlear effects (Mulders and Robertson 2006) and efferent effects in the IC (Seluakumaran et al. 2008), suggesting that MOC-induced changes in monaural responses primarily reflect the actions of efferent terminals in the cochlea. In addition to learning-based plasticity (described above) that clearly utilizes descending pathways as its substrate, a form of lesion-based plasticity is seen after damage in a restricted part of the cochlea eliminates a range of frequencies (Irvine et al. 2001, 2003; Irvine and Wright 2005; Kamke et al. 2003, 2005). When AI was examined weeks later, the AC region deprived of input was not silent but was now occupied by frequencies at the perimeter of the cochlear lesion. The exaggerated frequency expansion and changes in thresholds and other response characteristics suggest that the changes were not a passive consequence of the lesion but true plastic alterations. Equivalent changes were observed in the ventral division of the MGB, but not in the inferior colliculus, nor was there plasticity in the DCN frequency map, emphasizing that such plasticity was a forebrain phenomenon (Rajan and Irvine 1998a, b).

10 Summary and General Principles Governing Auditory Corticofugal Projections

The ascending sensory pathways carry topographic information that underlies various aspects of sensory processing and map construction. However, the descending systems may be ten times larger at the thalamus (Deschênes et al. 1998; Jones 2002) and corticofugal projections are significant at lower levels (Towe and Jabbur 1961; Dewson 1968). These descending paths involving sight, touch and hearing appear to have a core projection with a light halo (Winer et al. 2001). This pattern suggests a facilitative effect for interconnected topographic regions (matched) (Monconduit et al. 2006), while the halo of projections would represent a mismatched substrate of lateral inhibition for possible response enhancement (Malmierca and Núñez 1998). These descending pathways, however, are more complex since some targets are specific relay structures, whereas others are nonspecific (Veinante et al. 2000). These widespread descending connections link brain structures via direct and indirect connections to AC. These pathways might mediate specific sensory features in the topographic maps that can be modified by injury, sensory deprivation, and experience. We have reviewed data that suggest different roles for corticofugal feedback in modulating these changes in subcortical structures to adjust and enhance the extraction of biologically significant signals from noise.

Acknowledgments Supported by grants from the Spanish MEC (BFU2009-07286), EU (EUI2009-04083) and JCYL-UE (GR221) to MSM to Manuel S. Malmierca, and NIH grants DC00232 and DC04395 and a Life Sciences Research Award from the Office for Science and Medicine, NSW, to David K. Ryugo. We are grateful to our colleagues whose work made this review possible.

References

Andersen RA, Snyder RL, and Merzenich MM (1980) The topographic organization of corticocollicular projections from physiologically identified loci in the AI, AII, and anterior auditory cortical fields of the cat. *Journal of Comparative Neurology* **191**: 479–494.

Bajo VM and Moore DR (2005) Descending projections from the auditory cortex to the inferior colliculus in the gerbil, *Meriones unguiculatus*. *Journal of Comparative Neurology* **486**: 101–116.

Bajo VM, Nodal FR, Bizley JK, Moore DR, and King AJ (2007) The ferret auditory cortex: descending projections to the inferior colliculus. *Cerebral Cortex* **17**:475–491.

Bajo VM, Rouiller EM, Welker E, Clarke S, Villa AEP, de Ribaupierre Y, and de Ribaupierre F (1995) Morphology and spatial distribution of corticothalamic terminals originating from the cat auditory cortex. *Hearing Research* **83**:161–174.

Bajo VM, Nodal FR, Moore DR, King AJ (2010) The descending corticocollicular pathway mediates learning-induced auditory plasticity. *Nature Neuroscience* **13**:253–260.

Bakin JS and Weinberger NM (1990) Classical conditioning induces CS-specific receptive field plasticity in the auditory cortex of the guinea pig. *Brain Research* **536**:271–286.

Bakin JS, Weinberger NM (1996) Induction of a physiological memory in the cerebral cortex by stimulation of the nucleus basalis. *Proceedings of the National Academy of Sciences of the United States of America* **93**:11219–11224.

Bao S, Chang EF, Woods J, and Merzenich MM (2004) Temporal plasticity in the primary auditory cortex induced by operant perceptual learning. *Nature Neuroscience* **7**:974–981.

Bartlett EL and Smith PH (1999) Anatomic, intrinsic, and synaptic properties of dorsal and ventral division neurons in rat medial geniculate body. *Journal of Neurophysiology* **81**:1999–2016.

Bartlett EL, Stark JM, Guillery RW, and Smith PH (2000) Comparison of the fine structure of cortical and collicular terminals in the rat medial geniculate body. *Neuroscience* **100**:811–828.

Beneyto M and Prieto JJ (2001) Connections of the auditory cortex with the claustrum and the endopiriform nucleus in the cat. *Brain Research Bulletin* **54**:485–498.

Beneyto M, Winer JA, Larue DT, and Prieto JJ (1998) Auditory connections and neurochemistry of the sagulum. *Journal of Comparative Neurology* **401**:329–351.

Beyerl BD (1978) Afferent projections to the central nucleus of the inferior colliculus in the rat. *Brain Research* **145**:209–223.

Bourassa J, Pinault D, and Deschênes M (1997) Corticothalamic projections from the cortical barrel field to the somatosensory thalamus in rats: a single-fibre study using biocytin as an anterograde tracer. *European Journal of Neuroscience* **7**:19–30.

Brodal P (1972) The corticopontine projection in the cat. The projection from the auditory cortex. *Archives Italianes de Biologie* **110**: 119–144.

Brownell WE, Bader CR, Bertrand D, and de Ribaupierre Y (1985) Evoked mechanical responses of isolated cochlear outer hair cells. *Science* **227**:194–196.

Budinger E, Heil P, and Scheich H (2000) Functional organization of auditory cortex in the Mongolian gerbil (*Meriones unguiculatus*). IV. Connections with anatomically characterized subcortical structures. *European Journal of Neuroscience* **12**:2452–2474.

Caicedo A and Herbert H (1993) Topography of descending projections from the inferior colliculus to auditory brainstem nuclei in the rat. *Journal of Comparative Neurology* **328**:377–392.

Casseday JH, Jones DR, and Diamond IT (1979) Projections from cortex to tectum in the tree shrew, *Tupaia glis*. *Journal of Comparative Neurology* **185**:253–291.

Coleman JR and Clerici WJ (1987) Sources of projections to subdivisions of the inferior colliculus in the rat. *Journal of Comparative Neurology* **262**:215–226.

Coomes DL and Schofield BR (2004) Projections from the auditory cortex to the superior olivary complex in guinea pigs. *European Journal of Neuroscience* **19**:2188–2200.

Coomes DL, Schofield RM, and Schofield BR (2005) Unilateral and bilateral projections from cortical cells to the inferior colliculus in guinea pigs. *Brain Research* **1042**:62–72.

Coomes-Peterson D and Schofield BR (2007) Projections from auditory cortex contact ascending pathways that originate in the superior olive and inferior colliculus. *Hearing Research* **232**:67–77.

Darrow KN, Maison SF, and Liberman MC (2006) Cochlear efferent feedback balances interaural sensitivity. *Nature Neuroscience* **9**:1474–1476.

Deschênes MM, Veinante P, and Zhang ZW (1998) The organization of corticothalamic projections: reciprocity versus parity. *Brain Research Brain Research Review* **28**:286–308.

Dewson JH, III (1968) Some relationships to stimulus discrimination in noise. *Journal of Neurophysiology* **31**:122–130.

Dolan DF and Nuttall AL (1988) Masked cochlear whole-nerve response intensity functions altered by electrical stimulation of the crossed olivocochlear bundle. *Journal of the Acoustical Society of America* **83**:1081–1086.

Doucet JR, Molavi DL, and Ryugo DK (2003) The source of corticocollicular and corticobulbar projections in area Te1 of the rat. *Experimental Brain Research* **153**:461–466.

Doucet JR, Rose L, and Ryugo DK (2002) The cellular origin of corticofugal projections to the superior olivary complex in the rat. *Brain Research* **925**:28–41.

Faye-Lund H (1985) The neocortical projection to the inferior colliculus in the albino rat. *Anatomy & Embryology (Berlin)* **173**:53–70.

Feliciano M and Potashner SJ (1995) Evidence for a glutamatergic pathway from the guinea pig auditory cortex to the inferior colliculus. *Journal of Neurochemistry* **65**:1348–1357.

Feliciano M, Saldaña E, and Mugnaini E (1995). Direct projections from the rat primary auditory neocortex to nucleus sagulum, paralemniscal regions, superior olivary complex and cochlear nuclei. *Auditory Neuroscience* **1**:287–308.

Fitzpatrick KA and Imig TJ (1978) Projections of auditory cortex upon the thalamus and midbrain in the owl monkey. *Journal of Comparative Neurology* **177**:537–556.

Games KD and Winer JA (1988) Layer V in rat auditory cortex: projections to the inferior colliculus and contralateral cortex. *Hearing Research* **34**:1–25.

Gao E and Suga N (1998) Experience-dependent corticofugal adjustment of midbrain frequency map in bat auditory system. *Proceedings of the National Academy of Sciences of the United States of America*. **95**:12663–12670.

Gao E and Suga N (2000) Experience-dependent plasticity in the auditory cortex and the inferior colliculus of bats: role of the corticofugal system. *Proceedings of the National Academy of Sciences of the United States of America* **97**:8081–8086.

Groff JA and Liberman MC (2003) Modulation of cochlear afferent response by the lateral olivocochlear system: activation via electrical stimulation of the inferior colliculus. *Journal of Neurophysiology* **90**:3178–3200.

Haas M, Coomes D, Kuwabara N, and Schofield BR (2003) Laminar distribution and projection patterns of corticocollicular cells in guinea pig auditory cortex. *Association for Research in Otolaryngology Abstracts* **26**:817.

He J (2003a) Corticofugal modulation on both ON and OFF responses in the nonlemniscal auditory thalamus of the guinea pig. *Journal of Neurophysiology* **89**:367–381.

He J (2003b) Corticofugal modulation of the auditory thalamus. *Experimental Brain Research* **153**:579–590.

He J, Yu YQ, Xiong Y, Hashikawa T, and Chan YS (2002) Modulatory effect of cortical activation on the lemniscal auditory thalamus of the Guinea pig. *Journal of Neurophysiology* **88**:1040–1050.

Held H (1891) Die centralen Bahnen des Nervus acusticus bei der Katze. *Archiv für Anatomie und Physiologie. Anatomische Abteilung* **15**:271–290.

Hefti BJ and Smith PH (2000) Anatomy, physiology, and synaptic responses of rat layer V auditory cortical cells and effects of intracellular $GABA_A$ blockade. *Journal of Neurophysiology* **83**:2626–2638.

Herbert H, Aschoff A, and Ostwald J (1991) Topography of projections from the auditory cortex to the inferior colliculus in the rat. *Journal of Comparative Neurology* **304**:103–122.

Huffman RF and Henson OW J. (1990). The descending auditory pathway and acousticomotor systems, connections with the inferior colliculus. *Brain Research Brain Research Reviews* **15**:295–323.

Irvine DRF, Rajan R, and Brown M (2001) Injury- and use-related plasticity in adult auditory cortex. *Audiology & Neurootology* **6**:192–195.

Irvine DRF, Rajan R, and Smith S (2003) Effects of restricted cochlear lesions in adult cats on the frequency organization of the inferior colliculus. *Journal of Comparative Neurology* **467**:354–374.

Irvine DRF and Wright BA (2005) Plasticity of spectral processing. *International Review of Neurobiology* **70**:435–472.

Jacomme AV, Nodal FR, Bajo VM Manunta Y, Edeline J-M, Babalian A, and Rouiller EM (2003) The projection from auditory cortex to cochlear nucleus in guinea pigs: an in vivo anatomical and in vitro electrophysiological study. *Experimental Brain Research* **153**: 467–476.

Jones EG (2002) Thalamic organization and function after Cajal. *Progress in Brain Research* **136**:333–357.

Kamke MR, Brown M, and Irvine DRF (2005) Basal forebrain cholinergic input is not essential for lesion-induced plasticity in mature auditory cortex. *Neuron* **48**:675–686.

Kamke MR, Brown M, and Irvine DRF (2003) Plasticity in the tonotopic organization of the medial geniculate body in adult cats following restricted unilateral cochlear lesions. *Journal of Comparative Neurology* **459**:355–367.

Kawamura K and Chiba M (1979) Cortical neurons projecting to the pontine nuclei in the cat. An experimental study with the horseradish peroxidase technique. *Experimental Brain Research* **35**:269–285.

Kilgard MP and Merzenich MM (1998) Plasticity of temporal information processing in the primary auditory cortex. *Nature Neuroscience* **1**:727–731.

Kuypers HGJM and Lawrence DG (1967) Cortical projections to the red nucleus and the brain stem in the Rhesus monkey. *Brain Research* **4**:151–188.

Llano DA and Sherman SM (2008) Evidence for nonreciprocal organization of the mouse auditory thalamocortical-corticothalamic projection systems. *Journal of Comparative Neurology* **507**: 1209–1227.

Ma X and Suga N (2001) Corticofugal modulation of duration-tuned neurons in the midbrain auditory nucleus in bats. *Proceedings of the National Academy of Sciences of the United States of America* **98**:14060–14065

Ma X and Suga N (2003) Augmentation of plasticity of the central auditory system by the basal forebrain and/or somatosensory cortex. *Journal of Neurophysiology* **89**:90–103.

Ma X and Suga N (2004) Lateral inhibition for center-surround reorganization of the frequency map of bat auditory cortex. *Journal of Neurophysiology* **92**:3192–3199.

Ma X and Suga N (2007) Multiparametric corticofugal modulation of collicular duration-tuned neurons: modulation in the amplitude domain. *Journal of Neurophysiology* **97**:3722–3730.

Malmierca E and Núñez A (1998) Corticofugal action on somatosensory response properties of rat nucleus gracilis cells. *Brain Research* **810**:172–180.

Malmierca MS (2003) The structure and physiology of the rat auditory system: an overview. *International Review of Neurobiology* **56**: 147–211.

Malmierca MS and Merchán M (2004) The auditory system. In: Paxinos G (ed) *The Rat Nervous System*. Academic Press, San Diego, pp. 997–1082.

Malmierca MS, Le Beau FEN, and Rees A (1996) The topographical organization of descending projections from the central nucleus of the inferior colliculus in guinea pig. *Hearing Research* **93**: 167–180.

Malmierca, MS, Rees A, and Le Beau FEN (1997). Ascending projections to the medial geniculate body from physiologically identified loci in the inferior colliculus. In Syka J (ed). *Acoustic Signal Processing in the Central Auditory System*. Plenum Press, New York, pp. 295–302.

Massopust LC Jr and Ordy JM (1962) Auditory organization of the inferior colliculi in the cat. *Experimental Neurology* **6**:465–477.

Meltzer NE and Ryugo DK (2006) Projections from auditory cortex to cochlear nucleus: a comparative analysis of rat and mouse. *Anatomical Record Advances in Integrative Anatomy and Evolutionary Biology* **288**:397–408.

Mettler FA (1935) Corticofugal fiber connections of the cortex of *Macaca mulatta*. The temporal region. *Journal of Comparative Neurology* **63**:25–47.

Mitani A, Shimokouchi M, and Nomura S (1983) Effects of stimulation of the primary auditory cortex upon colliculogeniculate neurons in the inferior colliculus of the cat. *Neuroscience Letters* **42**:185–189.

Monconduit L, López-Ávila A, Molat JL, Chalus M, and Villanueva L (2006) Corticofugal output from the primary somatosensory cortex selectively modulates innocuous and noxious inputs in the rat spinothalamic system. *Journal of Neuroscience* **26**: 8441–8450.

Montero VM (1983). Ultrastructural identification of axon terminals from the thalamic reticular nucleus in the medial geniculate body of the rat: an EM autoradiographic study. *Experimental Brain Research* **51**:338–342.

Morest DK and Oliver DL (1984) The neuronal architecture of the inferior colliculus in the cat: defining the functional anatomy of the auditory midbrain. *Journal of Comparative Neurology* **222**: 209–236.

Moriizumi T and Hattori T (1991) Pyramidal cells in rat temporoauditory cortex project to both striatum and inferior colliculus. *Brain Research Bulletin* **27**:141–144.

Mulders WHAM and Robertson D (2000) Evidence for direct cortical innervation of medial olivocochlear neurones in rats. *Hearing Research* **144**:65–72.

Mulders WHAM and Robertson D (2006) Gentamicin abolishes all cochlear effects of electrical stimulation of the inferior colliculus. *Experimental Brain Research* **174**:35–44.

Nakamoto KT, Jones SJ, and Palmer AR (2008) Descending projections from auditory cortex modulate sensitivity in the midbrain to cues for spatial position. *Journal of Neurophysiology* **99**:2347–2356.

Ohlrogge M, Doucet JR, and Ryugo DK (2001) Projections of the pontine nuclei to the cochlear nucleus in rats. *Journal of Comparative Neurology* **436**:290–303.

Ojima H, Honda CN, and Jones EG (1992) Characteristics of intracellularly injected infragranular pyramidal neurons in cat primary auditory cortex. *Cerebral Cortex* **2**:197–216.

Oliver DL, Ostapoff E-M, and Beckius GE (1999) Direct innervation of identified tectothalamic neurons in the inferior colliculus by axons from the cochlear nucleus. *Neuroscience* **93**:643–658.

Paula-Barbosa MM and Sousa-Pinto A (1973) Auditory cortical projections to the superior colliculus in the cat. *Brain Research* **50**:47–61.

Perales M, Winer JA, and Prieto JJ (2006) Focal projections of cat auditory cortex to the pontine nuclei. *Journal of Comparative Neurology* **497**:959–980.

Perrot X, Ryvlin P, Isnard J, Guénot M, Catenoix H, Fischer C, Mauguière F, and Collet L (2006) Evidence for corticofugal modulation of peripheral auditory activity in humans. *Cerebral Cortex* **16**:941–948.

Peruzzi D, Bartlett E, Smith PH, and Oliver DL (1997) A monosynaptic GABAergic input from the inferior colliculus to the medial geniculate body in rat. *Journal of Neuroscience* **17**:3766–3777.

Pickles, JO (1988) *An Introduction to the Physiology of Hearing*. Academic Press, Orlando.

Popelár J, Erre JP, Syka J, and Aran JM (2001) Effects of contralateral acoustical stimulation on three measures of cochlear function in the guinea pig. *Hearing Research* **152**:128–138.

Polley DB, Heiser MA, Blake DT, Schreiner CE, Merzenich MM (2004) Associative learning shapes the neural code for stimulus magnitude in primary auditory cortex. *Proceedings of the National Academy of Sciences of the United States of America* **101**: 16351–16356.

Potashner SJ, Dymczyk L, and Deangelis MM (1988) D-aspartate uptake and release in the guinea pig spinal cord after partial ablation of the cerebral cortex. *Journal of Neurochemistry* **50**: 103–111.

Rajan R (1990) Electrical stimulation of the inferior colliculus at low rates protects the cochlea from auditory desensitization. *Brain Research* **506**:192–204.

Rajan R and Irvine DRF (1998a) Absence of plasticity of the frequency map in dorsal cochlear nucleus of adult cats after unilateral partial cochlear lesions. *Journal of Comparative Neurology* **399**:35–46.

Rajan R and Irvine DRF (1998b) Neuronal responses across cortical field A1 in plasticity induced by peripheral auditory organ damage. *Audiology & Neurootology* **3**:123–144.

Rasmussen GL (1946) The olivary peduncle and other fiber projections of the superior olivary complex. *Journal of Comparative Neurology* **99**:61–74.

Rasmussen GL (1953) Further observations of the efferent cochlear bundle. *Journal of Comparative Neurology* **99**:61–74.

Recanzone GH, Schreiner CE, and Merzenich MM (1993) Plasticity in the frequency representation of primary auditory cortex following discrimination training in adult owl monkeys. *Journal of Neuroscience* **13**:87–103.

Romanski LM, Clugnet MC, Bordi F, and LeDoux JE (1993) Somatosensory and auditory convergence in the lateral nucleus of the amygdala. *Behavioral Neuroscience* **107**:444–450.

Romanski LM and LeDoux JE (1993) Information cascade from primary auditory cortex to the amygdala: corticocortical and corticoamygdaloid projections of temporal cortex in the rat. *Cerebral Cortex* **3**:515–532.

Rouiller EM and Welker E (1991) Morphology of corticothalamic terminals arising from the auditory cortex of the rat: a *Phaseolus vulgaris*-leucoagglutinin (PHA-L) tracing study. *Hearing Research* **56**:179–190.

Ryugo DK and Weinberger NM (1976) Corticofugal modulation of the medial geniculate body. *Experimental Neurology* **51**:377–391.

Sakai M and Suga N (2001) Plasticity of the cochleotopic (frequency) map in specialized and nonspecialized auditory cortices. *Proceedings of the National Academy of Sciences of the United States of America* **98**:3507–3512.

Sakai M and Suga N (2002) Centripetal and centrifugal reorganizations of frequency map of auditory cortex in gerbils. *Proceedings of the National Academy of Sciences of the United States of America* **99**:7108–7112.

Saldaña E, Feliciano M, and Mugnaini E (1996) Distribution of descending projections from primary auditory neocortex to inferior colliculus mimics the topography of intracollicular projections. *Journal of Comparative Neurology* **371**:15–40.

Schofield BR (2002) Ascending and descending projections from the superior olivary complex in guinea pigs: different cells project to the cochlear nucleus and the inferior colliculus. *Journal of Comparative Neurology* **453**:217–225.

Schofield BR and Coomes DL (2005) Auditory cortical projections to the cochlear nucleus in guinea pigs. *Hearing Research* **199**:89–102.

Schofield BR and Coomes DL (2006) Pathways from auditory cortex to the cochlear nucleus in guinea pigs. *Hearing Research* **216–217**: 81–89.

Schofield BR, Coomes DL, and Schofield RM (2006) Cells in auditory cortex that project to the cochlear nucleus in guinea pigs. *Journal of the Association for Research in Otolaryngology* **7**:95–109.

Schuller G, Covey E, and Casseday JH (1991) Auditory pontine grey: connections and response properties in the horseshoe bat. *European Journal of Neuroscience* **3**:648–662.

Seluakumaran K, Mulders WH, and Robertson D (2008) Effects of medial olivocochlear efferent stimulation on the activity of neurons in the auditory midbrain. *Experimental Brain Research* **186**: 161–174.

Shi CJ and Cassell MD (1997) Cortical, thalamic, and amygdaloid projections of rat temporal cortex. *Journal of Comparative Neurology* **382**:153–175.

Spangler KM and Warr WB (1991). The descending auditory system. In: Altschuler R, Hoffman DW, Bobbin RP and Clopton BM (eds) *The Neurobiology of Hearing*. Volume. II, Raven Press, New York. pp. 27–45.

Suga N (2008) Role of corticofugal feedback in hearing. *Journal of Comparative Physiology A* **194**:169–183.

Suga N, Gao E, Zhang Y, Ma X, and Olsen JF (2000) The corticofugal system for hearing: recent progress. *Proceedings of the National Academy of Sciences of the United States of America* **97**:11807–11814.

Suga N and Ma X (2003) Multiparametric corticofugal modulation and plasticity in the auditory system. *Nature Reviews Neuroscience* **4**:783–794.

Suga N, Xiao Z, Ma X, and Ji W (2002) Plasticity and corticofugal modulation for hearing in adult animals. *Neuron* **36**:9–18.

Thompson AM (2005) Descending connections of the auditory midbrain. In: Winer JA and Schreiner CE (eds), *The Inferior Colliculus*. Springer, New York. pp. 182–199.

Thompson WH (1900) Degenerations resulting from lesions of the cortex of the temporal lobe. *Journal of Anatomy and Physiology* **35**:147–165.

Towe AL and Jabbur SJ (1961) Cortical inhibition of neurons in dorsal column nuclei of cat. *Journal of Neurophysiology* **24**:488–498.

Veinante P, Jacquin MF, and Deschênes M (2000) Thalamic projections from the whisker-sensitive regions of the spinal trigeminal complex in the rat. *Journal of Comparative Neurology* **420**:233–243.

Vetter DE, Saldaña E, and Mugnaini E (1993) Input from the inferior colliculus to medial olivocochlear neurons in the rat: a double label study with PHA-L and cholera toxin. *Hearing Research* **70**: 173–186.

Warr WB (1992) Organization of olivocochlear efferent systems in mammals. In: Webster DB, Popper AN, and Fay RR (eds). *The Mammalian Auditory Pathway, Neuroanatomy*. Volume 2, *Springer Handbook of Auditory Research*. Springer, Berlin, pp. 410–448.

Warr WB, Guinan JJ Jr, and White JS (1986) Organization of the efferent fibers: the lateral and medial olivocochlear systems. In: Altschuler RA, Hoffman DW, and Bobbin RP (eds). *Neurobiology of Hearing, The Cochlea*. Raven Press, New York, pp. 333–348.

Watanabe T, Yanagisawa K, Kanzaki J, and Katsuki Y (1966) Cortical efferent flow influencing unit responses of medial geniculate body to sound stimulation. *Experimental Brain Research* **2**: 302–317.

Weedman DL and Ryugo DK (1996) Projections from auditory cortex to the cochlear nucleus in rats: synapses on granule cell dendrites. *Journal of Comparative Neurology* **371**:311–324.

Weinberger NM (1998) Physiological memory in primary auditory cortex: characteristics and mechanisms. *Neurobiology of Learning and Memory* **70**:226–251.

White JS and Warr WB (1983) The dual origins of the olivocochlear bundle in the albino rat. *Journal of Comparative Neurology* **219**:203–214.

Winer JA (2005) Three systems of descending projections to the inferior colliculus. In: Winer JA and Schreiner CE (eds). *The Inferior Colliculus*, Springer, New York, pp. 231–247.

Winer JA (2006) Decoding the auditory corticofugal systems. *Hearing Research* **212**:1–8.

Winer JA, Diehl JJ, and Larue DT (2001) Projections of auditory cortex to the medial geniculate body of the cat. *Journal of Comparative Neurology* **430**:27–55.

Winer JA, Larue DT, Diehl JJ, and Hefti BJ (1998) Auditory cortical projections to the cat inferior colliculus. *Journal of Comparative Neurology* **400**:147–174.

Winer JA and Lee CC (2007) The distributed auditory cortex. *Hearing Research* **229**:3–13.

Winer JA and Prieto JJ (2001) Layer V in cat primary auditory cortex (AI): cellular architecture and identification of projection neurons. *Journal of Comparative Neurology* **434**:379–412.

Wong D and Kelly JP (1981) Differentially projecting cells in individual layers of the auditory cortex: a double-labeling study. *Brain Research* **230**:362–366.

Wu Y and Yan J (2007) Modulation of the receptive fields of midbrain neurons elicited by thalamic electrical stimulation through corticofugal feedback. *Journal of Neuroscience* **27**:10651–10658.

Xiao Z and Suga N (2002) Modulation of cochlear hair cells by the auditory cortex in the mustached bat. *Nature Neuroscience* **5**: 57–63.

Xiao Z and Suga N (2005) Asymmetry in corticofugal modulation of frequency-tuning in mustached bat auditory system. *Proceedings of the National Academy of Sciences of the United States of America* **102**:19162–19167.

Yan J and Ehret G (2001) Corticofugal reorganization of the midbrain tonotopic map in mice. *Neuroreport* **12**:3313–3316.

Yan J and Ehret G (2002) Corticofugal modulation of midbrain sound processing in the house mouse. *European Journal of Neuroscience* **16**:119–128.

Yan J and Suga N (1996) Corticofugal modulation of time-domain processing of biosonar information in bats. *Science* **273**:1100–1103.

Yan J and Zhang Y (2005) Sound-guided shaping of the receptive field in the mouse auditory cortex by basal forebrain activation. *European Journal of Neuroscience* **21**:563–576.

Yan J, Zhang Y, and Ehret G (2005) Corticofugal shaping of frequency tuning curves in the central nucleus of the inferior colliculus of mice. *Journal of Neurophysiology* **93**:71–83.

Yan W and Suga N (1998) Corticofugal modulation of the midbrain frequency map in the bat auditory system. *Nature Neuroscience* **1**:54–58.

Yu YQ, Xiong Y, Chan YS, and He J (2004) Corticofugal gating of auditory information in the thalamus: an in vivo intracellular recording study. *Journal of Neuroscience* **24**:3060–3069.

Yu XJ, Xu XX, He S, and He J (2009) Change detection by thalamic reticular neurons. *Nature Neuroscience* **12**:1165–1170.

Zhang Y and Suga N (1997) Corticofugal amplification of subcortical responses to single tone stimuli in the mustached bat. *Journal of Neurophysiology* **78**:3489–3492.

Zhang Y and Suga N (2000) Modulation of responses and frequency tuning of thalamic and collicular neurons by cortical activation in mustached bats. *Journal of Neurophysiology* **84**:325–333.

Zhang Y and Suga N (2005) Corticofugal feedback for collicular plasticity evoked by electric stimulation of the inferior colliculus. *Journal of Neurophysiology* **94**:2676–2682.

Zhang Y, Suga N, and Yan J (1997) Corticofugal modulation of frequency processing in bat auditory system. *Nature* **387**:900–903.

Zheng J, Shen W, He DZ, Long KB, Madison LD, and Dallos P (2000) Prestin is the motor protein of cochlear outer hair cells. *Nature* **405**:149–155.

Chapter 10

Neurochemical Organization of the Medial Geniculate Body and Auditory Cortex

Jeffery A. Winer

Abbreviations

AAF	anterior auditory field
AI	primary auditory cortex
AII	second auditory cortex area
APt	anterior pretectum
Aq	cerebral aqueduct
BIC	brachium of the inferior colliculus
BSC	brachium of the superior colliculus
C	caudal
c	contralateral
CG	central gray
CN	central nucleus of the inferior colliculus
CP	cerebral peduncle
D	dorsal nucleus of the medial geniculate body *or* dorsal
DD	deep dorsal nucleus of the medial geniculate body
DC	dorsal cortex of the inferior colliculus
DS	dorsal superficial nucleus of the medial geniculate body
DZ	dorsal auditory zone
ED	posterior ectosylvian gyrus, intermediate area
EP	posterior ectosylvian gyrus
EPSP	excitatory postsynaptic potential
EV	posterior ectosylvian gyrus, ventral part
EW	Edinger-Westphal nucleus
FSU	fast-spiking unit
GABA	gamma aminobutyric acid
GAD	glutamic acid decarboxylase
ICc	central nucleus of the inferior colliculus
ICp	caudal cortex of the inferior colliculus
I	Golgi type I cell
II	Golgi type II cell
III	oculomotor nucleus
In	insular cortex
LC	lateral cortex of the inferior colliculus
LD	lateral dorsal nucleus
LGB	lateral geniculate body
LMN	lateral mesencephalic nucleus
LP	lateral posterior nucleus
M	medial division of the medial geniculate body *or* medial
ML	medial lemniscus
MRF	mesencephalic reticular formation
MZ	marginal zone of medial geniculate body
NBIC	nucleus of the brachium of the inferior colliculus
NMDA	N-methyl-D-aspartate
OR	optic radiation
OT	optic tract
Ov	*pars ovoidea* of the ventral division of the medial geniculate body
PKC	protein kinase c
PLSS	posterior lateral suprasylvian area
Pol	rostral pole of the medial geniculate body
Pom	medial part of the posterior group
Pt	pretectum
Pul	pulvinar nucleus
Pv	parvalbumin
Re	thalamic reticular nucleus
RF	reticular formation
RN	red nucleus
RP	rostral pole nucleus of the inferior colliculus
RSU	regular-spiking unit
SC	superior colliculus
SCi	intermediate gray layer of superior colliculus
SCp	deep layer of superior colliculus
SCs	superficial gray layer of superior colliculus
Sl	suprageniculate nucleus, lateral part
Sm	suprageniculate nucleus, medial part
SN	substantia nigra
SNR, SNr	substantia nigra, *pars reticulata*
Spf	subparafascicular nucleus
SpN	suprapeduncular nucleus
Te	temporal cortex

J.A. Winer (✉)
Division of Neurobiology, Department of Molecular and Cell Biology, University of California at Berkeley, Berkeley, CA 94720-3200, USA
e-mail: chris@phy.ucsf.edu

J.A. Winer, C.E. Schreiner (eds.), *The Auditory Cortex*,
DOI 10.1007/978-1-4419-0074-6_10, © Springer Science+Business Media, LLC 2011

V	*pars lateralis* of the ventral division *or* ventral
Vb	ventrobasal complex
Vl,vl	ventrolateral nucleus of the medial geniculate body
VP	ventral posterior area
VTA	ventral tegmental area
WAHG	wheatgerm apo-horseradish gold
wm	white matter
ZI	zona incerta
I–VI,1–6	layers of cerebral cortex

1 Chemical Neuroanatomy of the Auditory System

The auditory, visual, and somatic sensory systems each have a topographic receptor epithelium, multiple central representations of the peripheral mosaic, parallel neural streams serving epicritic processing, vast networks of central connectivity, and a web of intrinsic circuits at all levels of processing. Marked neurochemical differences distinguish the central auditory system from its visual and somatic sensory counterparts, which each largely conserves the parallel contribution from specific classes of retinal, cutaneous, or neuromuscular receptors; moreover, sight and touch have, relative to audition, fewer synapses in the ascending pathways to neocortex (Dykes 1983; Stone 1983). Perhaps the auditory system has more opportunities for inhibitory and other local interactions.

Each sensory system has abundant substrates for inhibitory/disinhibitory local processing via Golgi type II local circuit interneurons (Mugnaini and Oertel 1985). However, the auditory system is unique in several ways. First, interneurons in it are numerous, comprising up to one-third of medial geniculate body neurons (Huang et al. 1999); they are fewer in the analogous visual and somatic sensory thalamic nuclei (Jones 1985). Second, some auditory nuclei contain almost exclusively cells that use γ-aminobutyric acid and are GABAergic (dorsal nucleus of the lateral lemniscus) or glycine (medial nucleus of the trapezoid body); neither specialization has a visual or somatic sensory counterpart. Third, the auditory system contains many GABAergic and glycinergic neurons (the latter up to the level of the medulla), while the visual system has both transmitters only in the retina (after which GABA alone is present) and the somatic sensory system has only GABAergic cells (Emson 1983; Ottersen and Storm-Mathisen 1990; Winer et al. 1995). Fourth, many auditory brain stem neurons receive a convergent GABAergic and glycinergic input, whereas in the visual system the interaction between these transmitters except in the retina (Marc and Cameron 2001) is weaker. Why two such Cl^--mediated and interdependent pathways exist is unknown. Finally, auditory system GABAergic or glycinergic neurons (Osen et al. 1990; Oliver et al. 1991) below the inferior colliculus are as likely to project remotely (Koch and Grothe 2000) as to participate in local circuits (Kuwabara et al. 1991), or both, whereas the GABAergic visual and somatic sensory neurons seem to be confined to local projections (Spreafico et al. 1994) with some significant exceptions (Cucchiaro et al. 1993; Fabri and Manzoni 2004; Higo et al. 2007).

These differences suggest that one plan for inhibitory organization does not prevail in all modalities. This account explores the organization of GABAergic and related circuitry in the medial geniculate body and auditory cortex in pursuit of functional hypotheses about the role of these many neurons. A further question pertinent to any hierarchical contribution of intrinsic circuitry is why these substrates are so abundant in the forebrain when their brain stem representation is already robust.

2 Tectothalamic System

The inferior colliculus is vital in hearing, since it is among the earliest synaptic stations where most (Aitkin and Phillips 1984; Cant 2005; Schofield 2005) though not all (Anderson et al. 2007) brain stem input converges, and it is influenced strongly by auditory cortex projections (Winer et al. 1998) and by a smaller descending input from the medial geniculate body (Kuwabara and Zook 2000), whose role is unknown. Tectothalamic signals likely incorporate ascending and descending influences and imply that intrinsic midbrain operations modify, and do not merely relay, ascending information.

2.1 Glutamate and Calcium Binding Proteins: Differential Distribution

The primary tectothalamic transmitter candidate(s) are likely glutamate/aspartate (Hu et al. 1994; Webber et al. 1999). This pathway has parallel excitatory channels, since inferior colliculus stimulation in the rat evoked either a pure excitatory or a mixed excitatory–inhibitory response with effects far briefer than those elicited by corticofugal stimulation (Bartlett and Smith 2002). The distribution of calcium binding proteins (parvalbumin and calbindin D-28k) largely parallels the patterns of tectothalamic input, with parvalbumin concentrating in the rabbit ventral and medial divisions and calbindin-positive cells in the dorsal division (de Venecia et al. 1995); in primates, the ventral division is more associated with parvalbumin than the dorsal division (Molinari et al. 1995). Parvalbumin was concentrated preferentially in thalamic subdivisions with input from the inferior colliculus central nucleus, whereas calbindin dominated nuclei whose input arose from outside the inferior colliculus central nucleus and in non-auditory brain stem regions. In the mouse auditory thalamus, parvalbumin largely defines core (primary; lemniscal) auditory centers, and calbindin is confined to shell (nonprimary; belt; extralemniscal) regions (Cruikshank et al. 2001; Jones 2003).

2.2 Gamma-Aminobutyric Acid: Local Circuitry and Extrinsic Influences

The diversity of tectothalamic operations has a morphological basis and there is a wide range of medial geniculate body responses to auditory and nonauditory stimuli (Clarey et al. 1992; Hu 1995). The largest inferior colliculus neurons are GABAergic (Oliver et al. 1994), whereas Golgi type II cells elsewhere usually have smaller somata and thinner dendrites than principal cells, and a thin, locally projecting, unmyelinated axon (Morest 1975). GABA was colocalized subsequently in tectothalamic projection cells (Winer et al. 1996) and large caliber GABAergic axons were seen in the brachium of the inferior colliculus (Saint Marie et al. 1997). Physiological work found that some GABAergic signals reached the medial geniculate body before the excitatory midbrain projections (Peruzzi et al. 1997). This implies parallel GABAergic and glutamatergic/aspartatergic tectothalamic systems whose interaction in the auditory thalamus is unknown.

2.3 Cholinergic and Other Subsystems

Several neurochemically specific projections reach the medial geniculate body, though their origin is not extensively documented nor is their function known. Thus, the ventral division receives far less cholinergic input than the retinorecipient layers of the lateral geniculate body, and serotonin and tyrosine hydroxylase are more prominent in the medial geniculate body and ventrobasal complex than in the lateral geniculate body (Fitzpatrick et al. 1989). The rat also has a differential pattern of acetylcholinesterase distribution, with the dorsal and medial divisions stained much more heavily than the ventral division; a differential pattern of c-fos activity distinguishes the ventral division, with the lateral part conspicuously darker (Olucha-Bordonau et al. 2004). Pharmacological studies find that muscarinic actions in mouse slice preparations can alter the impact of thalamocortical and intracortical activity selectively and specifically, suppressing local circuits and thereby amplifying the effects of thalamocortical transmission (Hsieh et al. 2000).

C-kinase α (PICK1) may regulate PCKα and Glu2R receptors and thereby influence synaptic activity and its modification. PICK1 is widely distributed in the auditory pathway, and concentrates in the rat medial geniculate ventral and medial divisions, and in the suprageniculate nucleus especially, though not elsewhere in the dorsal division (McInvale et al. 2002). In rat different protein kinase C (PKC) isoforms yield unique patterns of medial geniculate immunoreactivity, with PKC βI staining many terminals in the ventral division and few elsewhere. In contrast, PKC δ immunostains many cells, especially in the ventral division (Garcia and Harlan 1997).

Likewise, the dorsal division contains enkephalinergic neurons (Coveñas et al. 1986) which may extend the oscillatory discharge dynamics that are so marked in the dorsal division (Aitkin and Dunlop 1968). Extended dorsal division reverberatory sequences are prevalent (Hu et al. 1994) and might occur in nonprimary auditory cortical areas to which its cells project (Hall 2005).

3 Thalamotectal System

This small thalamofugal projection is present in bat, rat, cat, and monkey (Senatorov and Hu 2002; Winer et al. 2002); its neurotransmitter and synaptic organization is unknown. It arises from cells scattered in the medial and dorsal divisions of the medial geniculate body and adjoining intralaminar thalamic nuclei and targets inferior colliculus regions largely outside the central nucleus. There is no analogous pathway in the visual and somatic sensory thalamus.

4 Thalamocortical System

The principal, presumptively glutamatergic/aspartatergic ventral division thalamocortical neurons have their long dendritic axis largely confined to and parallel with the isofrequency domains imposed by inferior colliculus afferents (McMullen et al. 2005) and corticofugal axons (Morest 1975). Midbrain projections to the medial geniculate body do not follow a point-to-point projection pattern. Rather, most cells in an inferior colliculus division target one medial geniculate body division, though some project more divergently (Wenstrup et al. 1994). Each auditory thalamic subdivision receives convergent input from several midbrain origins (Calford and Aitkin 1983), suggesting that information transfer involves patterns of convergence and divergence analogous to those in the thalamocortical system (Miller et al. 2001). In the rat somatic sensory thalamus, $GABA_A$ antagonists enhance whisker-evoked response probability in the receptive field center or alter the responsiveness of the surround (peripheral) receptive field (Lee et al. 1994). In a mouse brain slice, $GABA_A$-mediated control of spike timing was found (Bright et al. 2007). The idea of the thalamus as a relay nucleus has given way to the view that local circuitry influences signal processing, without being able to specify how such interactions transform signals passing through the thalamus (Sherman and Guillery 2006).

4.1 Gamma-Aminobutyric Acid in the Auditory Thalamus

GABAergic neurons and axon terminals are plentiful and have a differential, nucleus-specific distribution in each auditory thalamic subdivision. In the cat ventral division, 33% of cells are GABAergic, with 26% in the dorsal division, and 18% in the medial division (Huang et al. 1999) (Figs. 10.1d–f and 10.2). This implies a subdivision-specific distribution and concentration of GABAergic effects within the auditory thalamus. In the ventral division GABAergic neurons have a caudal-to-rostral gradient (Rouiller et al. 1990) perhaps associated with local differences in inhibition. The GABAergic Golgi type II cells are interspersed among the thalamocortical neurons, to which they are presynaptic, forming both axodendritic and dendrodendritic synapses (Morest 1971). The type II cells are smaller, with thin, sparsely spinous dendrites, and a fine, unmyelinated axon whose projection appear confined to the division of origin (Fig. 10.1a: Type II). There is a small subpopulation of larger GABAergic neurons (Huang et al. 1999). Axodendritic synaptic arrangements are prevalent and dendrodendritic synapses are seen (Sherman 2004). The role of presynaptic dendrites is unknown (Fig. 10.3h).

Two further sources of GABAergic influence are present in the auditory thalamus. About 20% of inferior colliculus cells projecting to the medial geniculate body are GABAergic (Winer et al. 1996) (Fig. 10.3a–g), and the thalamic reticular nucleus has reciprocal projections with the medial geniculate body (Rouiller et al. 1985; Crabtree 1998). The functional impact of these multiple sources of GABAergic convergence is a major issue in understanding the tectothalamic transformation (Wenstrup 2005).

The proportion of GABAergic medial geniculate cells is species specific (Fig. 10.4). Mice have none or a few (Arcelli et al. 1997), mustached bats <1% (Winer et al. 1992), rats ~1%, cats 18–33%, and macaques many (Winer and Larue 1996). Even in species with few such cells, a subdivision-specific concentration of GABAergic puncta (boutons) likely arising from the inferior colliculus and the thalamic reticular nucleus implies that such projections have particular functional roles. In the mustached bat this does not reflect an overall or species-specific decline of GABAergic cells, since these are almost as abundant in the rat inferior colliculus (Merchán et al. 2005) and auditory cortex (Winer 1992) as they are in the cat inferior colliculus (Oliver et al. 1994) and auditory cortex (Prieto et al. 1994b). Perhaps the medial geniculate body and the somatic sensory ventrobasal complex, which share global patterns of organization (Arcelli et al. 1997), represent the differential evolution of interneuronal circuitry.

The ventral division ultrastructural profile (Fig. 10.3h) suggests that principal cells receive excitatory axodendritic input of cortical and midbrain origin, and interneuronal dendrodendritic and axodendritic synapses (Morest 1971, 1974). This is in accord with the distribution of glutamic acid decarboxylase-positive axon terminals in medial geniculate body divisions (Fig. 10.2) and with evidence that thalamocortical cells are glutamatergic, as their auditory cortex synaptic terminals are enriched for glutamate (Weinberg and Kharazia 1996). Dorsal division GABAergic axon terminals terminate on principal cell dendrites, while GABAergic postsynaptic elements receive nonGABAergic axons (Coomes et al. 2002). These synaptic arrangements are characteristic in the dorsal thalamus (Jones 2007).

Fig. 10.1 (continued) Medial geniculate body local circuit and thalamocortical neurons. **a** A glutamic acid decarboxylase (GAD) immunostained local circuit neuron in (**b**) the ventral division of the rat. Characteristic features are long, poorly branched dendrites arising irregularly from a spindle shaped soma, sparse and slender dendritic appendages (*arrowheads*), and dendritic arbors with an axoniform configuration (*lower left*). Modified from the original source (Winer and Larue 1988). **a**, **c** Rapid Golgi method, planapochromat, N.A. 1.32, ×1250. **c** A cat type II cell in the ventral division of the medial geniculate body with many local axonal branches confined to a narrow venue approximately the dendritic width of the type I cell's arbor; the soma size is about half that of the thalamocortical (type I) cell. Modified from the original source (Winer 1992). **d** A *horizontal* section showing midbrain and diencephalic GABAergic somata (*dots*). The ventral division has by far the densest contribution of immunoreactive cells. **d–f** Modified from the original source (Huang et al. 1999). Planapochromat, N.A. 0.65, ×500. **e** The caudal dorsal medial geniculate body (DCa) has far fewer GABAergic cells than the ventral division of the medial geniculate body (f:V) or the dorsal superficial nucleus (f:DS), suggesting quantitative intranuclear differences in GABAergic circuitry. For abbreviations see the list

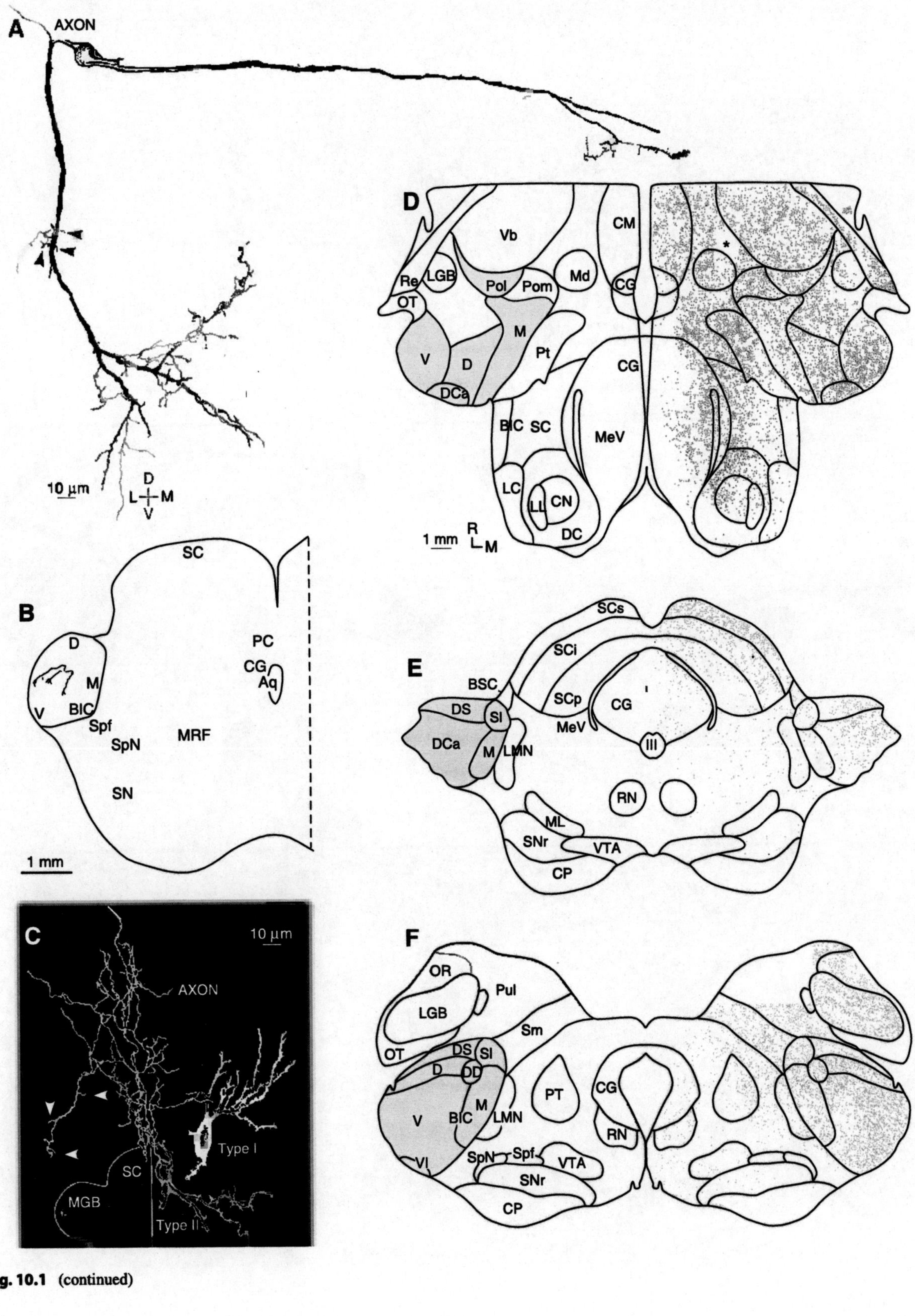

Fig. 10.1 (continued)

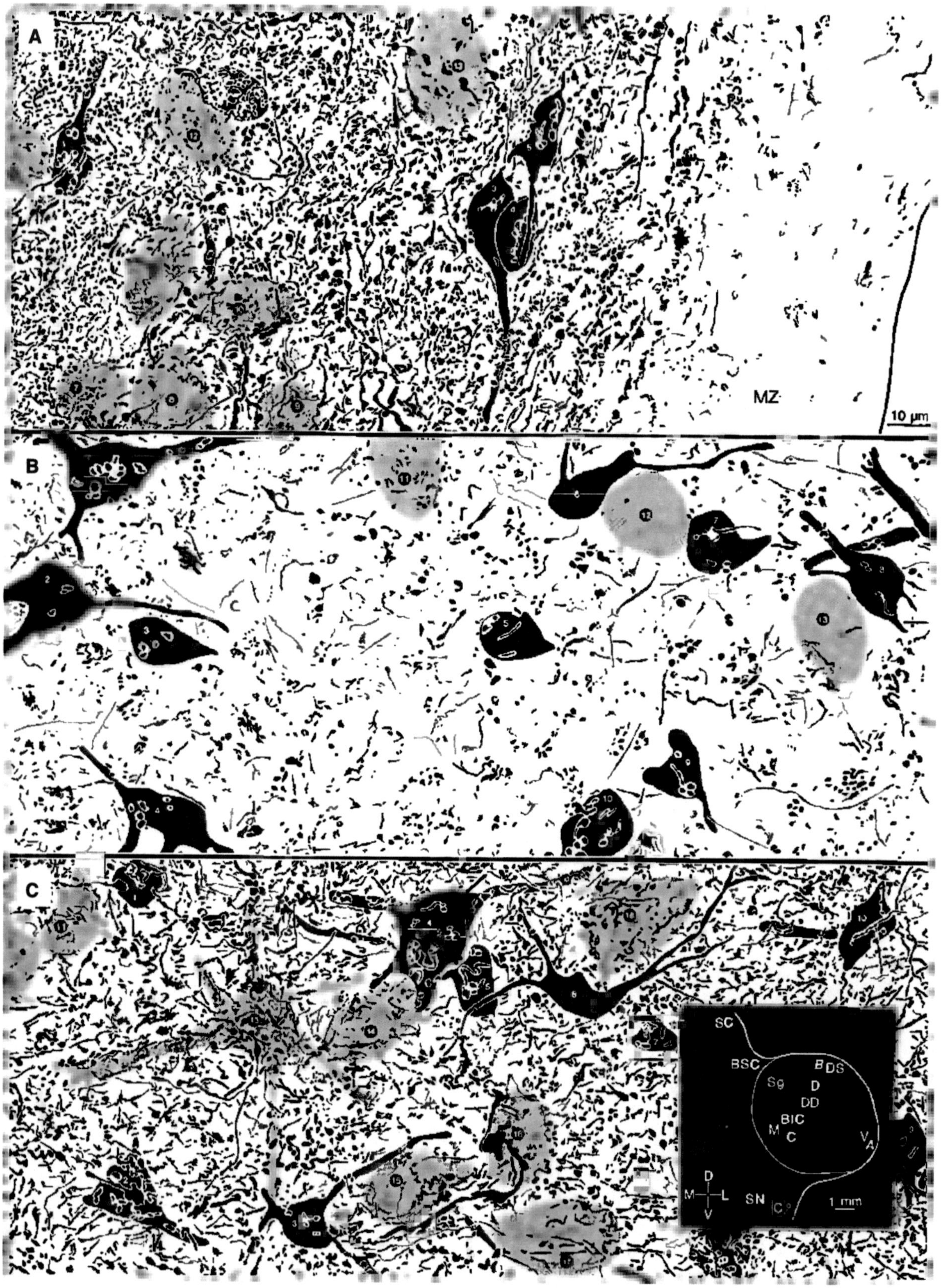

Fig. 10.2 (continued)

4.2 Auditory Thalamocortical Relations

The thalamocortical projection, like the tectothalamic system, has both convergent and divergent connections: each medial geniculate division projects to about five noncontiguous cortical areas, and each area receives input from several medial geniculate sources (Huang and Winer 2000). Single thalamic cells rarely project to more than one area, and neurons near one another in a division can project to the same or to different areas (Lee et al. 2004a). As in the tectothalamic system, a convergent–divergent model is more appropriate (Lee et al. 2004b) than a pure point-to-point pattern (Brandner and Redies 1990). Likewise, laminar thalamocortical terminal patterns to an area often involve several layers; these axons are diverse in form, with the largest, unexpectedly, in layer I (Huang and Winer 2000). The chief target of input to layer IV is nonpyramidal cells (Smith and Populin 2001) of which there are several varieties (Winer 1984a). Parvalbumin-positive axons form clusters ~500 μm wide in rabbit layers III–IV interspersed with pale zones; the clusters often overlie immunoreactive layer V cells (de Venecia et al. 1998). In the rat PKC δ-positive axons of medial geniculate origin are proposed to contact a wide variety of postsynaptic cells, including multipolar, spiny stellate, and pyramidal cells in different layers (Garcia and Harlan 1997) extrapolating from models of rat visual cortex circuitry (Sefton and Dreher 1995). This suggests that thalamocortical input drives, synchronizes, and modulates several physiologically and neurochemically specific operations in tandem (Sherman and Guillery 1998).

This model of thalamocortical connectivity and function entails extensive lateral areal input and the multilaminar transfer of information, rather than point-to-point topographic precision. It postulates widespread convergence, with neurons from different thalamic origins targeting nearby cortical loci (Lee et al. 2004b; Read et al. 2008) and some divergent projections to more than one area (Kishan et al. 2008); each input pattern may influence local circuitry. The model predicts, and may even require, the concurrent coactivation of other large systems—corticocortical and corticofugal—for calibration within a frame of reference such as the thalamocortical to corticothalamic transformation (White and Hersch 1982). This mechanism can also help coordinate multisensory operations, as between the auditory and visual spatial coordinate systems (Groh et al. 2001) or somatic sensory and auditory interactions (Rodgers et al. 2008). The neurochemical substrate for each operation is unknown, and their connectional interactions are obscure.

5 Thalamic Reticular Nucleus

The thalamic reticular nucleus is unlike any other thalamic nucleus, and it is unrelated to the brain stem reticular formation; the name reflects the web-like texture of reticular nucleus fibrodendritic organization, which forms a shell along the rostrolateral margin of the dorsal thalamus. No other thalamic nuclei contain exclusively GABAergic cells (Houser et al. 1980), none have GABAergic cells that project extrinsically, and few receive input from, or projects to, other thalamic nuclei; it is among the few thalamic nuclei that do not project to the cortex (Jones 2007). Each feature suggests that a special thalamic reticular nucleus role in modulating thalamocortical and corticothalamic networks. While the thalamic reticular nucleus appears cytologically homogeneous, it has separate though overlapping auditory, visual, and somatic sensory sectors, with the auditory part most caudoventral.

There is a systematic arrangement between medial geniculate body subdivisions and the reticular nucleus, with each major division sending slab-like projections oriented dorsoventrally and in an oblique caudorostral axis to the reticular nucleus, with some overlap in their terminal fields. Some reticular cells, in turn, project to more than one medial geniculate subdivision. All reticulothalamic auditory projections are organized topographically (Crabtree 1998).

Another major extrinsic input to the reticular nucleus arises from auditory cortex layer VI cells, many also targeting the medial geniculate complex (Ojima et al. 1996). However, the synaptic arrangement of afferents onto single

Fig. 10.2 (continued) The GAD-immunostained neuropil in medial geniculate body subdivisions. **a** The type II cells (*black*) are aligned *vertically*, parallel to the dendrites of type I cells (*gray stipple*) (Morest 1975), whose somata receive variable numbers of axosomatic puncta (boutons). The type II cells receive few GABAergic axosomatic synaptic boutons, while the type I cell somata receive many. *Inset* in **c**: locus of observation for each panel. The borders between divisions (MZ, marginal zone) are sharp. All panels: 25 μm thick GAD immunostained frozen section. Planapochromat, N.A. 1.32, ×1250. **b** The dorsal division has far fewer and smaller GABAergic boutons (puncta) than the ventral division (**a**) and the somatic orientation is mainly mediolateral, parallel to the principal cells dendrites. GABAergic cells receive a few clusters of axosomatic endings. **c** Medial division type I neurons are far more homogeneous, and include large radiate (*13*), magnocellular (*18*) and smaller (*12*) subtypes. The GABAergic cells have a diverse orientation, from lateral (*6*) to *vertical* (*10*) to radiate (*3*) and some (*7*) are among the smallest medial geniculate body GABAergic cells. Puncta are large, coarser than those in the ventral and dorsal divisions, uniform in density, and form prominent clusters on (*4*) or avoid type II somata (*6*)

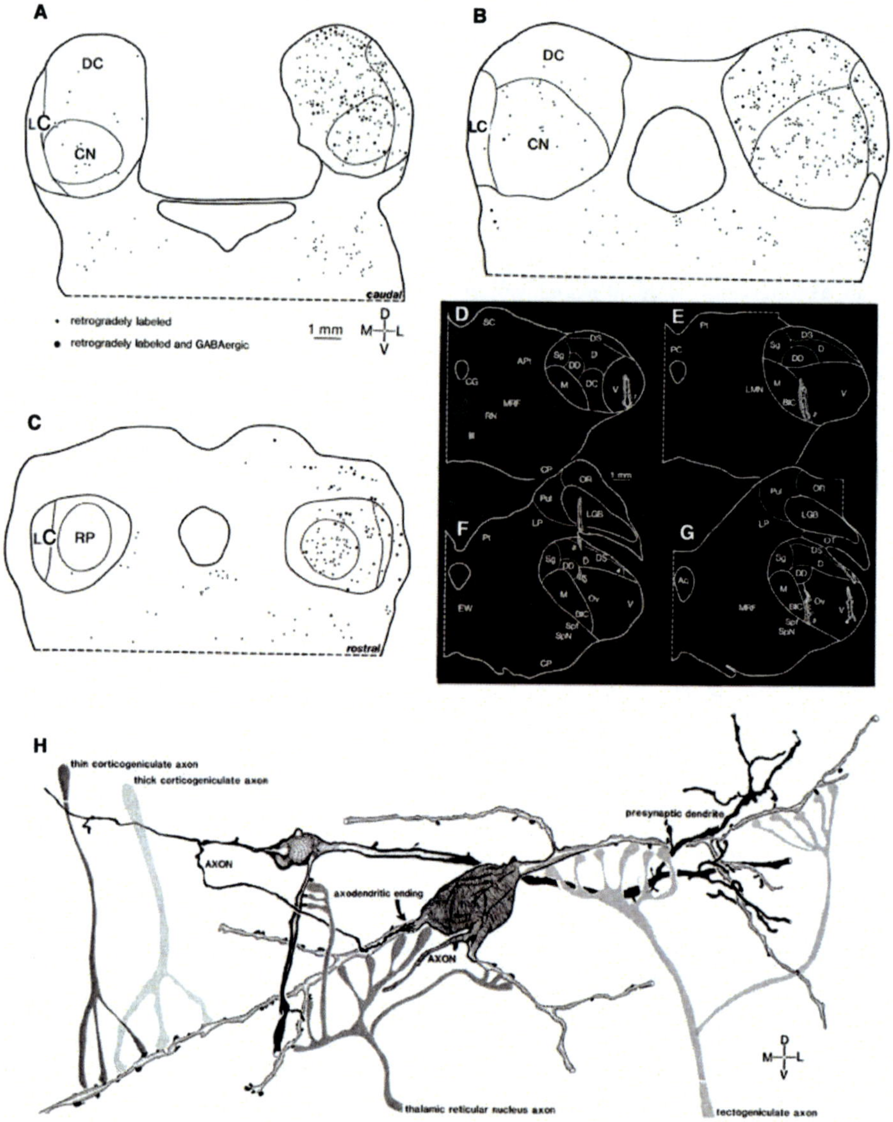

Fig. 10.3 Inferior colliculus projections from GABAergic cells to the medial geniculate body and a profile of intrinsic circuitry in the ventral division. **a** The inferior colliculus contains ~20% GABAergic cells (Oliver et al. 1994). Deposits of a retrograde tracer in the medial geniculate body (**d–g**: *light gray stippling*) label many inferior colliculus cells retrogradely (**a–c**: *small dots*), and ~20% of these are also GABAergic (Winer et al. 1996). **b** The double labeling is bilateral. **c** Inferior colliculus projections arise from adjoining parts of the subcollicular midbrain. **d–g** The two deposits (**g**) were centered in the ventral division and included the poles of the tonotopic representation (Imig and Morel 1985). **h** A profile of medial geniculate body circuitry in the ventral division. A thalamocortical tufted cell (*dark lines*) receives an array of afferents on its dendrites, which are often more spinous (Morest 1965) than in this example. The axon is myelinated after the initial segment. Afferents are segregated spatially, with those from GABAergic Golgi type II cells (*stippled outline*) providing proximal axodendritic and distal dendrodendritic synapses (Morest 1975). Tectogeniculate axons favor the proximal and intermediate dendrites, and the two types of corticogeniculate fibers target the distal dendritic segments. Thalamic reticular nucleus axons are a third source of synaptic input to dendrites. For illustrative purposes, the cells chosen are simplified. Rapid Golgi method, planapochromat, N.A. 1.32, ×1250

reticular nucleus cells is complex: proximal dendrites receive large thalamocortical collaterals and cholinergic afferents, intermediate dendrites are targets of converging serotoninergic and noradrenergic axons, and the distal, spinous dendrites receive thalamocortical collateral and intrinsic GABAergic endings. The thalamic (Kharazia and Weinberg 1994; Warren and Jones 1994) and cortical (Kharazia et al. 1996) endings are likely glutamatergic and may underlie the precise

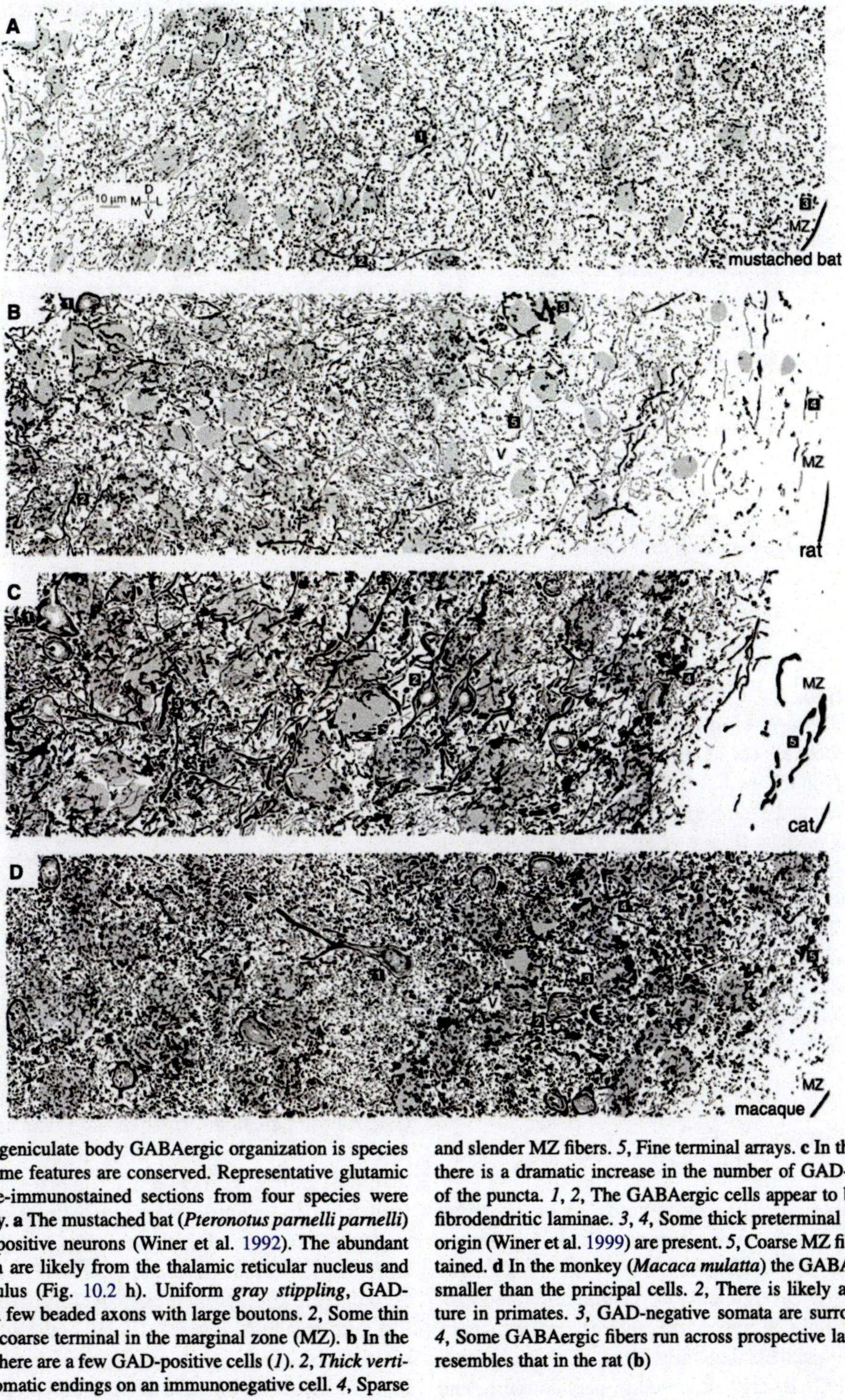

Fig. 10.4 Medial geniculate body GABAergic organization is species specific, though some features are conserved. Representative glutamic acid decarboxylase-immunostained sections from four species were prepared identically. **a** The mustached bat (*Pteronotus parnelli parnelli*) has ~0.5% GAD-positive neurons (Winer et al. 1992). The abundant GABAergic puncta are likely from the thalamic reticular nucleus and the inferior colliculus (Fig. 10.2 h). Uniform *gray stippling*, GAD-negative cells. *1*, A few beaded axons with large boutons. 2, Some thin lateral fibers. *3*, A coarse terminal in the marginal zone (MZ). **b** In the rat (*Rattus rattus*) there are a few GAD-positive cells (*1*). *2*, *Thick vertical* fibers. *3*, Axosomatic endings on an immunonegative cell. *4*, Sparse and slender MZ fibers. *5*, Fine terminal arrays. **c** In the cat (*Felis catus*) there is a dramatic increase in the number of GAD-positive cells and of the puncta. *1*, *2*, The GABAergic cells appear to be associated with fibrodendritic laminae. *3*, *4*, Some thick preterminal fibers of unknown origin (Winer et al. 1999) are present. *5*, Coarse MZ fibers are immunostained. **d** In the monkey (*Macaca mulatta*) the GABAergic cells are far smaller than the principal cells. *2*, There is likely a laminar architecture in primates. *3*, GAD-negative somata are surrounded by puncta. *4*, Some GABAergic fibers run across prospective laminae. *5*, The MZ resembles that in the rat (**b**)

transfer of information (Sherman and Guillery 2006), the modulatory aminergic inputs could contribute to a shift to bursting mode of discharge (Pape and McCormick 1989), and the topographic reticulothalamic relations assure coordination between thalamic divisions. Reticular nucleus cells have presynaptic dendrites with an unusual arrangement:

one cell's terminal dendrites contact the intermediate dendrites of the targeted neuron (Steriade et al. 1997), suggesting asymmetry in the information flow. In other thalamic nuclei such dendrodendritic arrangements are optimal for generating brief inhibitory postsynaptic potentials (Paré et al. 1991). The precise rhythmic interplay between reticular and thalamic principal cells implies control of temporal discharge cadence (Warren et al. 1994) with possible roles in vigilance and state-dependent oscillations (Destexhe and Sejnowski 2001) and attentional mechanisms (Crick 1984).

6 Thalamoamygdaloid System

There is a substantial projection from the dorsal and medial divisions of the medial geniculate body to the amygdala (Shinonaga et al. 1994). The transmitter may be glutamate (Hu et al. 1994) and synaptic endings in the rat lateral amygdaloid nucleus target the dendrites of inhibitory postsynaptic cells, perhaps to increase sound salience (Woodson et al. 2000). The NR2B subtype of NMDA (*N*-methyl-D-aspartate) receptor is concentrated in thalamoamygdaloid postsynaptic spines associated with fear conditioning (Radley et al. 2007), and $GABA_A$ and $GABA_B$ receptors are associated with shorter- and longer-term modulation of glutamatergic excitability, respectively (Li et al. 1996). Perhaps the physiological plasticity in the medial division of the medial geniculate body (Gerren and Weinberger 1983) and nearby posterior intralaminar nucleus (Linke et al. 2000) interacts with amygdaloid operations (Apergis-Schoute et al. 2005) and with multisensory corticoamygdaloid influences (Romanski et al. 1993). Little is known about the neurochemical identity of parahippocampal input to these same, nontonotopic medial geniculate subdivisions (Witter and Groenewegen 1986) and to limbic-affiliated auditory cortex (Lee and Winer 2008b).

7 Auditory Cortex

The data available are primarily for area AI in the rabbit (McMullen et al. 1994), monkey (Jones et al. 1995; Molinari et al. 1995) and cat (Hendry and Jones 1991), with some observations on GABAergic cells or neurons immunoreactive for parvalbumin in secondary auditory cortex areas. There are close parallels between areas in their laminar concentration of GABAergic, calbindin-, parvalbumin-, and calretinin-positive cells (Clemo et al. 2003). Physiological studies find areal differences in the regional concentration of complex inhibitory sidebands (Loftus and Sutter 2001), with implications for local processing regimes (Clemo et al. 2003).

7.1 Neurochemical Convergence in Auditory Cortex

The principal extrinsic afferents to neocortex are glutamatergic (Conti and Minelli 1996; Weinberg and Kharazia 1996), with complementary cholinergic afferents from the nucleus basalis (Kamke et al. 2005) and norepinephrine (Edeline 1995) of brain stem origin (Foote et al. 1983). Chemically specific markers include a parvalbumin-specific pathway in rabbit primary auditory cortex consists of bands of immunoreactivity in layers III/IV and in the dorsal half of layer VI in rabbit. Focal clusters of parvalbumin immunostaining resemble foci of thalamocortical axons (de Venecia et al. 1998), and the layer VI involvement suggests a relation between thalamocortical and corticothalamic pathways consistent with prior work in rat (Winer and Larue 1987) and with models predicting corticofugal influences on subcortical plasticity (King 1997; Winer 2006).

7.2 Layer I: Intrinsic Matrix

Layer I differs from other layers in almost all respects. It consists chiefly of neuropil (Fig. 10.5a: I), it has a predominantly lateral organization, it has an exclusively nonpyramidal and therefore almost entirely GABAergic neuronal population (Fig. 10.5c: I), there are few extrinsic connections, it has a highly species- and area-dependent organization, and it is the main synaptic input to pyramidal cell distal dendrites.

Some 94% of layer I cells were GABAergic (Fig. 10.6a) and these included horizontal and two varieties of multipolar cells (Prieto et al. 1994b); both multipolar types have a substantial lateral dendritic component to their arbors (Fig. 10.7). A similar proportion and comparable types of layer I cells are seen in the rat (Winer and Larue 1989). GABAergic puncta are smaller and twice as dense as those in other layers. Moreover, layer I alone has a significant (twofold) difference between its upper and lower halves, with layer Ia having almost seven times as many puncta as layer VI; these differences were comparable for glutamic acid decarboxylase and GABA (Prieto et al. 1994a). Serotonergic afferents in primate AI form axosomatic basket-like terminals in AI layers I–III and in the white matter, with most baskets in layers I and II; there is also evidence for multiple serotonin-positive axonal subtypes, not all of which have classical synaptic arrangements (DeFelipe et al. 1991).

Layer I receives a slow-conducting monosynaptic input from the medial geniculate body (Mitani and Shimokouchi 1985). EPSPs were evoked in a layer I horizontal cell by stimulation of the medial geniculate body, area AII (second auditory cortex), the posterior ectosylvian gyrus, and the contralateral AI (primary auditory cortex); the axon of the layer I cell projects to layer II (Mitani et al. 1985). Physiologically characterized, intracellularly labeled mouse layer I cells have within-layer axonal projections of ~750 μm (Fig. 10.8f) and the axon forms ~23 boutons/100 μm^3, whereas those in layers II/III have ~33/100 μm^3 and layer IV ~23/100 μm^3 (Verbny et al. 2006). Considering possible species differences, the ~50% increment in layer I GABAergic puncta suggests that other intracortical GABAergic projections to layer I must account for the doubling of its puncta (Prieto et al. 1994a). Horizontal and small multipolar layer I cells have wide axonal arrays in layer Ia, and some horizontal cells project to layer II (Verbny et al. 2006). Fast-spiking neurons were <5% of auditory cortex cells in the most superficial 600 μm (Atencio and Schreiner 2008), corresponding approximately to the layer II–III border (Winer 1984b, 1985) (Fig. 10.5a), and are present in much larger numbers in layers III–VI, suggesting a distinction in the spiking behavior of layers I–II GABAergic cells and those in other layers (Fig. 10.9c). Diversity in the laminar distribution of GABAergic operations is supported by a corresponding physiological range in AI inhibitory response areas, where only 38% of cells had inhibitory flanks with two simple lateral suppression bands and, in the more broadly tuned dorsal part of AI, 16% of cells had such flanks (Sutter et al. 1999).

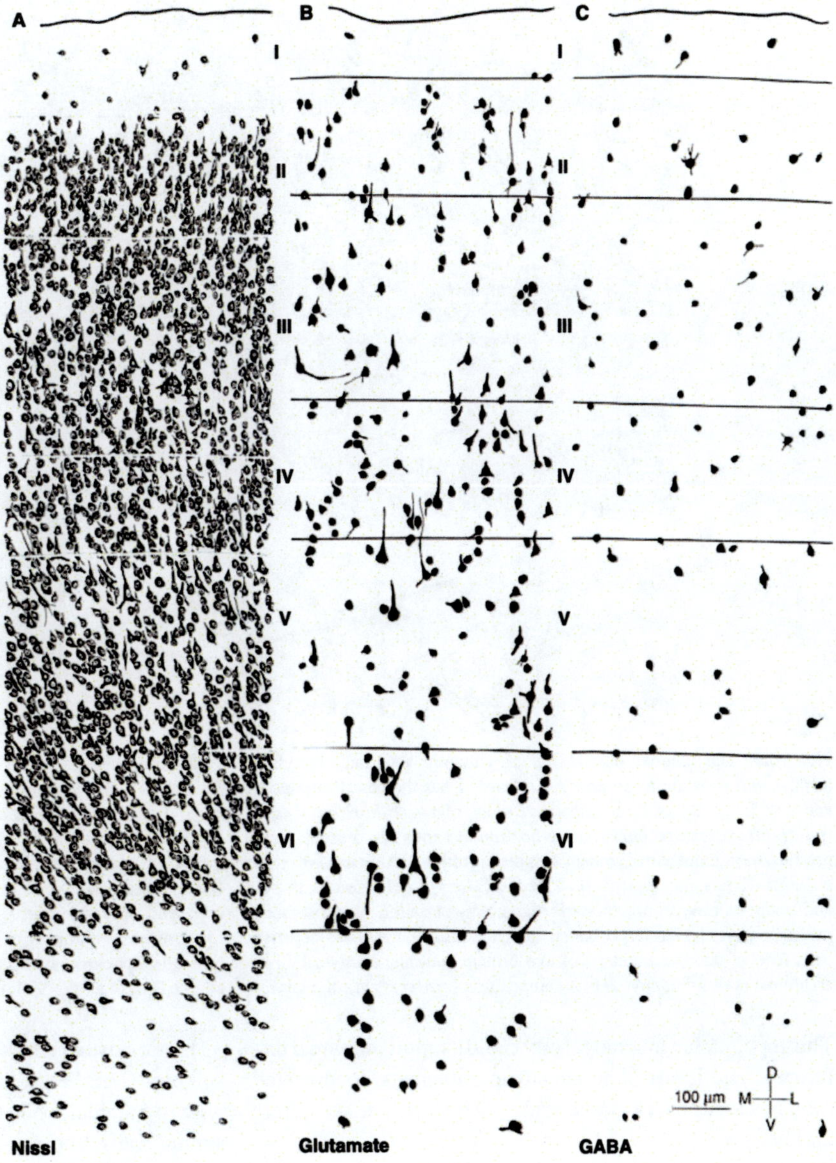

Fig. 10.5 Cytoarchitecture and neurochemistry of primary auditory cortex (AI). **a** The typical features of AI in Nissl preparations are a cell-sparse layer I, a layer II dominated by small pyramidal cells, a thick layer III with medium-sized pyramidal cells, a slender layer IV with few pyramidal cells, a broad layer V with a few large superficial pyramids and many deep ones, a rich plexus of axons in the upper half, and a layer VI with small pyramids in the upper tier and *horizontal* cells below. Celloidin embedded 25 μm thick section. For **a–c**: Planapochromat, N.A. 0.65, ×500. **b** In material prepared for glutamate, the pyramidal cell size distribution is marked, with the largest cells in layer V and the smallest in layer II; there is a substantial population in layer IV and even some in layer I. **c** In Vibratomed material, GABAergic cells dominate layer I, are diverse in layers II–IV, are rare in the fiber rich outer half of layer V (**a**), notably sparser in layer VI, and extend into the white matter

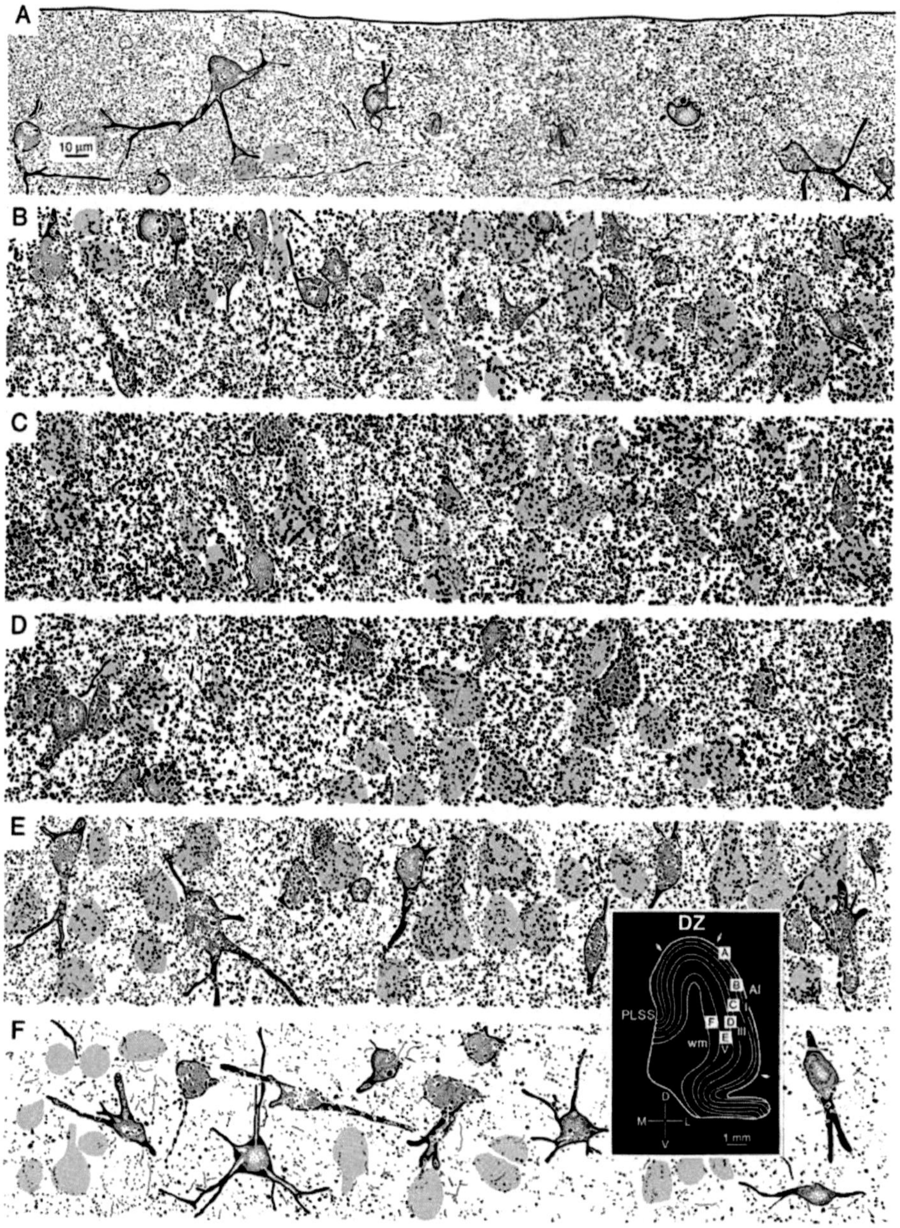

Fig. 10.6 The laminar distribution of glutamic acid decarboxylase (GAD) immunostaining in area AI. **a** Layer I has the smallest puncta, and a wide variety of immunopositive cells, of which *horizontal* cells (*far right*) are conspicuous. A few *horizontal* axons are present. Gray profiles, immunonegative somata. Stippled profiles, GAD-positive cells. **b** Layer II has the typical 20–25% GABAergic cells found in AI, and many receive an unexpectedly large concentration of axosomatic puncta. **c** Layer III has larger puncta than layer II, and the GAD-negative cells receive massive numbers of axosomatic boutons. **d** Layer IV has an almost entirely nonpyramidal composition (*center right*), the largest GABAergic cells in AI, and a preponderance of coarse puncta. **e** In layer V there is a sharp decline in the number of boutons as the proportion of GABAergic cells reaches the second highest level after layer I. This implies that the axons of layer V GABAergic cells either project to other (granular and supragranular) layers or that their local branches have fewer boutons. **f** Layer VI has a wide morphologic range of GABAergic cells and only fine and sparse puncta relative to other layers. *Inset*, locus of observations. GAD immunostained, 25 μm thick frozen section. Planapochromat, N.A. 1.32, ×1250. Modified from the original source (Prieto et al. 1994a)

This topographic inhomogeneity could support intraareal differences and levels of architectonic refinement documented for a few areas only (Clascá et al. 1997; Schreiner and Winer 2007).

In rat visual and somatic sensory cortical slices, late-spiking GABAergic cells can make chemical synapses with pyramidal cell apical dendrites and with non-late-spiking cells, and can be coupled by gap junctions to other

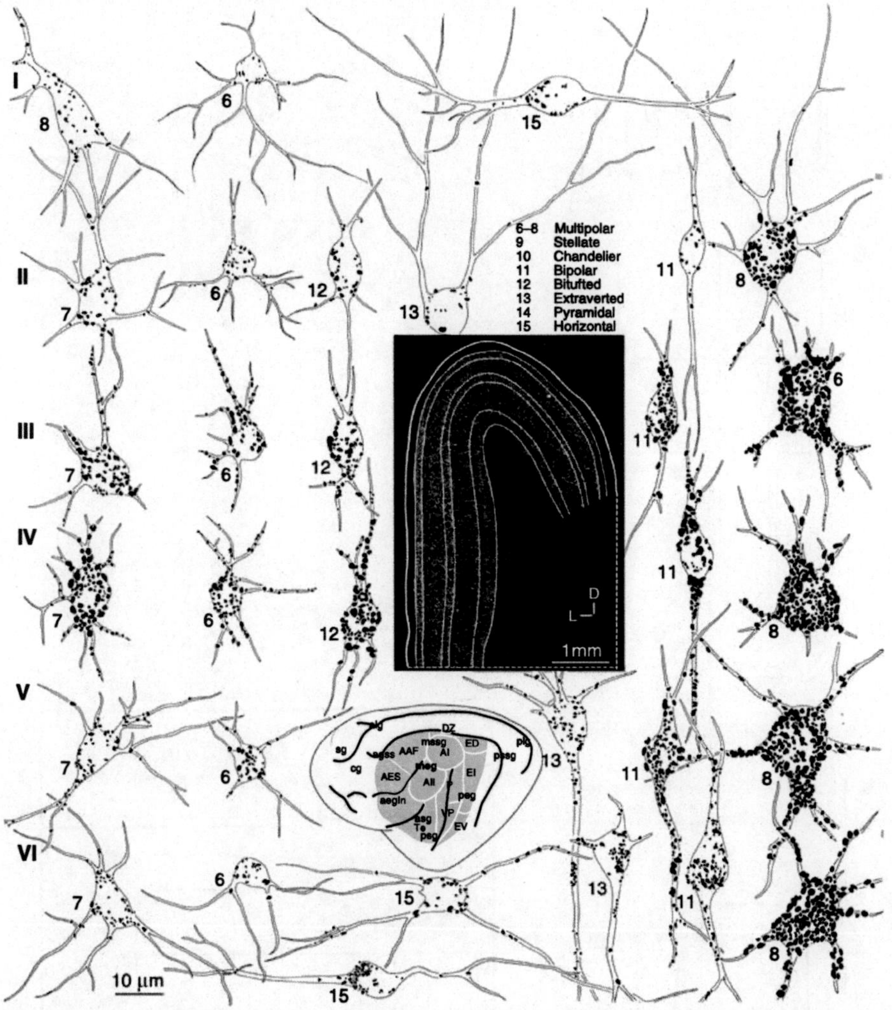

Fig. 10.7 Types of glutamic acid decarboxylase (GAD) immunostained area AI cells and the patterns of axosomatic puncta. Each layer has a unique pattern. In some, the axosomatic endings are sparse despite their prominence in the neuropil (Fig. 10.6a). In no layer are the puncta uniformly present or absent without reference to specific types of cells. The puncta are more variable on a laminar basis than with regard to a type of cell: all large multipolar cells (*8*) receive substantial axosomatic puncta in layers II–VI, whereas other types of cells in a layer (e.g., V: *6*, *7*, *13*) receive variable numbers of such boutons. Cells with the fewest axosomatic boutons tended to be unique to a layer, e.g., layer I and VI *horizontal* cells (*15*), or layer II extraverted multipolar cells (*13*). Moreover, even in a layer with fewer GAD-positive puncta, large multipolar cells received many coarse axosomatic terminals, suggesting that these might arise from another source. GAD immunostained, 25 µm thick frozen section. Planapochromat, N.A. 1.32, ×1250. Modified from the original source (Prieto et al. 1994a)

late-spiking cells (Chu et al. 2003). This suggests within- and between-class interactions, much as basket cells project to one another and can be presynaptic to pyramidal cells (Kisvárday et al. 1993).

Models of auditory cortex layer I local connectivity emphasize thalamic projections from extralemniscal sources (Mitani et al. 1984), including the largest thalamocortical axons reported (Huang and Winer 2000), intrinsic projections from layer II cells, and output to layer II cells which project to corticofugal infragranular cells (Mitani et al. 1985).

7.3 Layer II: Intracortical Refinement

Layer II has perhaps the most enigmatic role in cortical processing, and even in visual cortex its connections and impact on receptive field organization are not well understood (Thomson and Bannister 2003). Its limited auditory cortex connections largely segregate it from corticocortical or commissural influence and the thalamic input reaching it does so largely by polysynaptic routes (Mitani et al. 1985); it has modest long-range projections (Winguth and

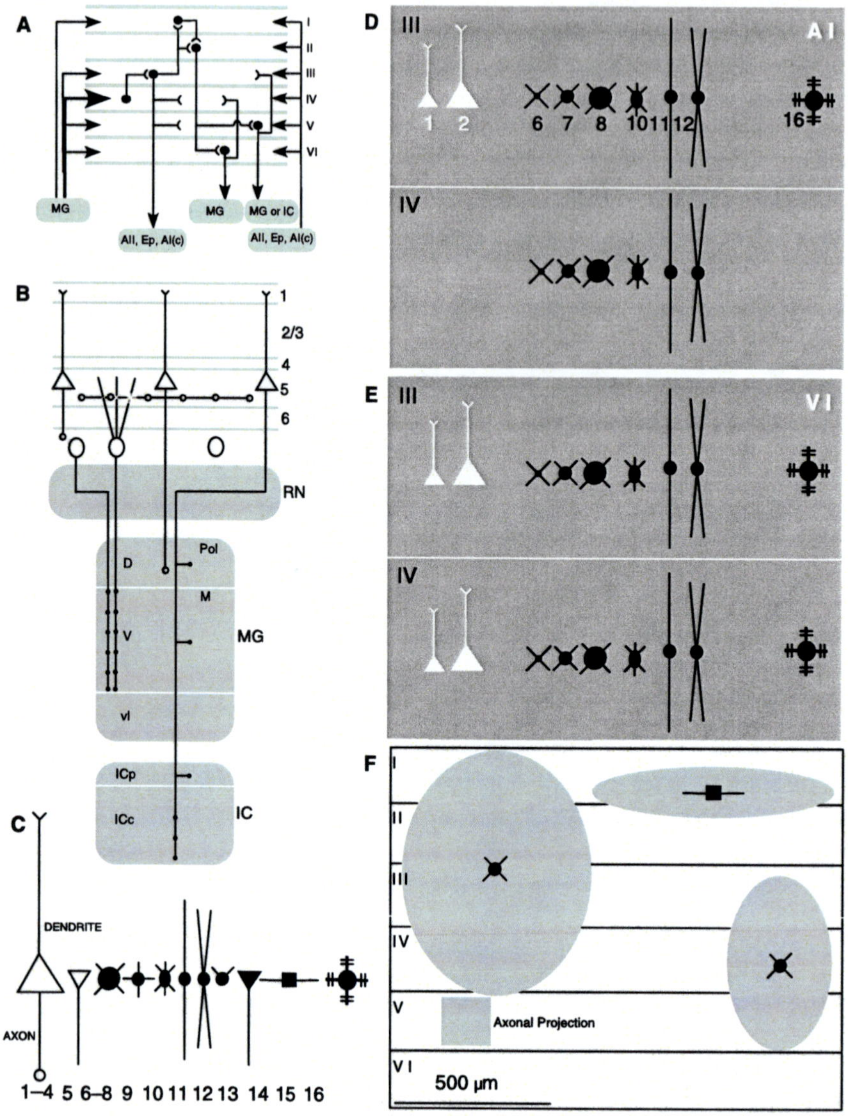

Fig. 10.8 Aspects of circuit organization in area AI. **a** Basic connectionist model of AI, with medial geniculate body (MG) afferents and their laminar targets (*left side*) concentrated in layer IV and with weaker projections to all layers. The postsynaptic cells are sources of intralaminar connections, interlaminar feedforward projections to supragranular layers, and feedforward ipsilateral corticocortical input to areas AII and the posterior ectosylvian gyrus (Ep), and to the contralateral AI. Interlaminar connections reach intracortical cells that contact corticofugal neurons projecting to the MG or inferior colliculus (IC); *right side*, extensive ipsi- (AII, Ep) and contralateral (AI(c)) projections also reach all AI layers. Modified from the original (Mitani et al. 1985). **b** Basic elements in the auditory corticofugal systems. Layer 5 pyramidal cells have extensive lateral intralaminar connections in AI; they project to the inferior colliculus (IC) and the medial geniculate body (MG) and their intralaminar axons may affect the dendrites of layer 6 neurons, whose axons are the primary input to the MG. Modified from the original (Ojima 1994). **c** Some of the more than 15 main types of cortical glutamatergic (*white*) or GABAergic (*black*) defined on the basis of morphological and immunocytochemical studies (Winer 1992; Prieto et al. 1994a, b). **d**, **e** Contrasting patterns of auditory (**d**; AI) and visual (**e**; VI) cortex organization. The principal differences are the virtual absence of layer IV spiny stellate cells in morphologic (Winer 1984a) and physiological studies combined with intracellular injection (Smith and Populin 2001), and the near-absence of layer IV pyramidal cells. Common features are the several types of nonpyramidal cells (*6–8*, *11*, *12*). Other differences are that the visual cortex (VI) layer III pyramidal cells in the foveal region do not participate in the commissural system (Fisken et al. 1975), whereas AI commissural projections are widespread and do not involve layer IV (Lee and Winer 2008a). **f** The intracortical sphere of axonal (presumptive synaptic) influence (*gray perimeters*) for three classes of GABAergic interneurons (*black outlines*) filled intracellularly in mouse auditory cortex. Projections usually involve more than one layer and can be 500 μm or more. Modified and interpreted from the original source (Verbny et al. 2006)

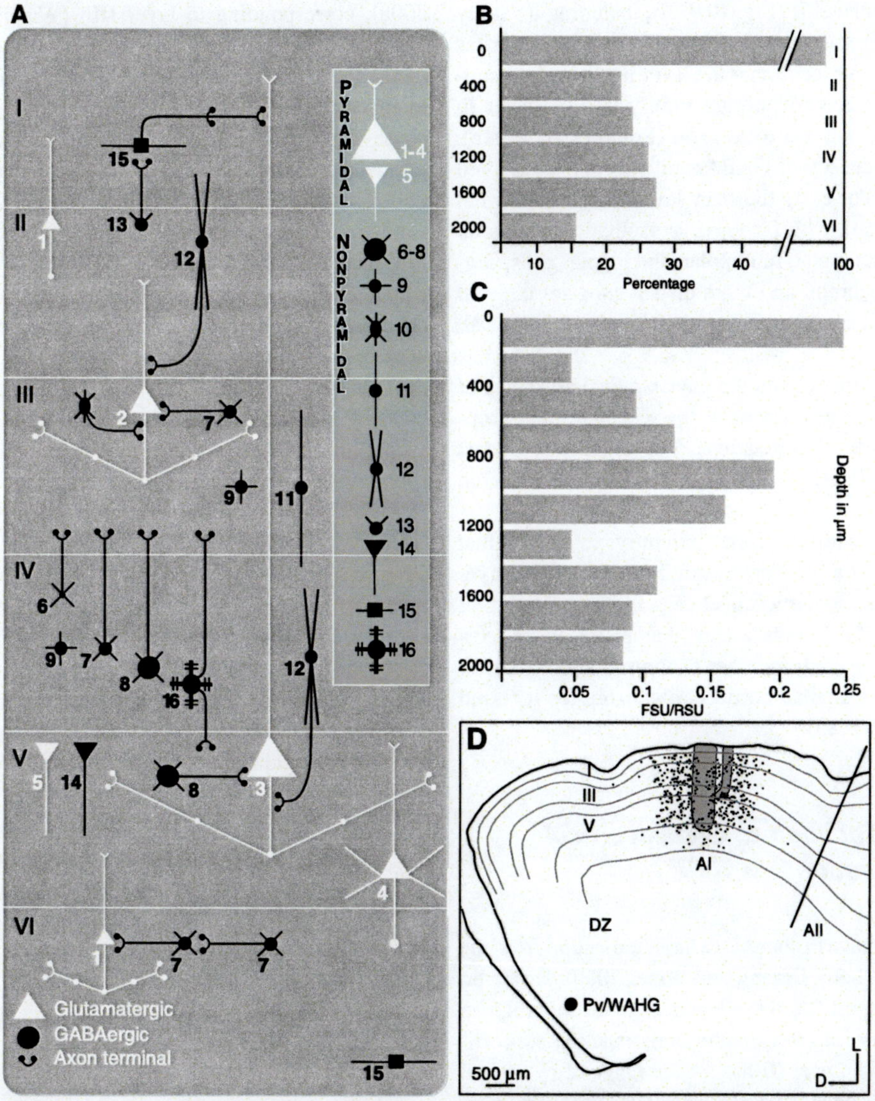

Fig. 10.9 Elements of area AI circuit organization. **a** Some aspects of local and intralaminar connectivity. An example of local connectivity is layer I *horizontal* cells, whose projections appear to be confined to layer I (see **d**). Some cells integrate input from many layers, while the small dendritic arbors of most GABAergic cells (*6–8*) would intercept input from more limited spatial domains and redistribute it far more widely (Verbny et al. 2006). The lateral integrative span of the *horizontal* cells in layers I and VI (*15*) is complemented by the *vertical* domain of bipolar (*11*) and bitufted (*12*) cells. Pyramidal cells represent an excitatory interneuronal system in all but layer I as well as corticofugal and feed-forward projections. **b** Proportions of GABAergic neurons in AI. The largest differences are in the layers that are the sources of the corticofugal systems, with layer V having the largest proportion of GABAergic cells and the second-lowest puncta contribution (Fig. 10.6e). Layer VI has the fewest GABAergic cells and the lowest number of puncta (Fig. 10.6f). Modified and interpreted from the original source (Prieto et al. 1994b). **c** Proportions of fast- and regular-spiking cells recorded from a silicon microprobe and showing striking laminar differences in fast-spiking dominance, including a precipitous decline at the layer V border, where medial geniculate body input abruptly falls (Huang and Winer 2000). Modified from the original (Atencio and Schreiner 2008). **d** Spatial and laminar origins of intracortical parvalbumin (Pv) positive projection neurons labeled by a deposit of wheatgerm apo-horseradish peroxidase conjugated to colloidal gold (WAHG) in AI. About 90% of such cells were within ±1,500 µm of the deposit perimeters, in close accord with estimates of the axonal domain of intracellularly filled GABAergic mouse auditory cortex cells (Fig. 10.8f). Layer I cells were unlabeled. Modified from the original source (Yuan et al. 2010)

Winer 1986), and some layer II pyramidal cell axons reach layer V and may have intracortical branches (Mitani et al. 1985).

About 24% of layer II cells are GABAergic, a value comparable to that in layers III–V; these cells are similar in size to those in layer III. Most of the same types resident in layers

III–IV are seen in layer II (Fig. 10.9: II), including a variety of multipolar cells with smooth or sparsely spinous dendrites and bipolar cells; an exception is the layer II extraverted multipolar cell whose apical dendrites ramify into the layer Ib neuropil, and which has a local axon (Prieto et al. 1994b). The number and density of GABAergic puncta is conserved in layers II–IV; however, those in layer II are somewhat smaller (Fig. 10.6b), with far fewer axosomatic endings on the extraverted multipolar, multipolar, and bipolar cells, suggesting that the neuropil and more distant sites are targeted by these cells and by neurons in other layers (Prieto et al. 1994a).

Little data are available on the physiological attributes of layer II GABAergic neurons or of the receptive field properties of other layer II populations. The proportion of fast spiking layer II cells was ~20% of that in layer IV (Atencio and Schreiner 2008).

One model of auditory cortex circuitry postulates that layer II cells act as a hub for integrating output from layer III and layer I cells to corticofugal cells in layers V and VI (Mitani et al. 1985). In contrast, models of the visual cortex regard it and layer III together (Thomson and Bannister 2003) or suggest a limited corticocortical role for it (Lund et al. 1979; Peters 1985).

7.4 Layer III: Thalamic to Corticocortical Transformation

Layers III and IV have in common a large projection from the medial geniculate body (Huang and Winer 2000). However, other features suggest that these layers have distinct roles in processing sensory information and in modulating subcortical projection cells. Layer III has a wide range of pyramidal and nonpyramidal cells, whereas layer IV is dominated by the latter; layer III receives abundant thalamic, corticocortical, and commissural projections and projects in the latter system, whereas layer IV does not appear to contribute to commissural processing; the subtle architectonic distinction between the upper and lower halves is sharper in layer III and glutamatergic cells are more plentiful in it (Fig. 10.5b).

Layer III GABAergic cells are almost as numerous as those in layer IV (24% versus 26%) but they are ~25% smaller. GABAergic layer III cells include a range of multipolar subtypes, including sparsely spinous varieties, and bitufted, bipolar and neurogliaform types (Prieto et al. 1994b). Puncta were lighter in layer IIIa and increased continuously from layer IIIb through layer IV; they were slightly smaller than layer IV puncta and targeted both pyramidal and GABAergic somata (Prieto et al. 1994a).

Fast-spiking, presumptively GABAergic cells are concentrated 800–1,000 μm deep in AI (Atencio and Schreiner 2008), corresponding to layer IIIb (Winer 1984b). A factor analysis whose main axes were response mode, spectral modulation, and temporal modulation found significant differences between fast- and regular-spiking cells in their rate, phase-locking index, spectral best modulation frequency and for many other functional indices (Atencio and Schreiner 2008); such differences may ultimately covary with cell type and laminar distribution.

Layer II–III GABAergic cells in mouse auditory cortex, which are driven weakly by thalamic input, also have a multipolar configuration, with axons projecting locally in 550 μm wide by ~600 μm high domains, and they have even more boutons than in layer IV (33 vs. 22/100 μm^3) (Verbny et al. 2006). Differences in bouton concentration between mouse and cat may reflect schemes of laminar division or actual concentrations in cell-specific laminar projections.

A model of cat AI intracortical connectivity postulates a layer IV projection to layer III cells which, in turn, have direct projections to layer I, II, IV and V, and polysynaptic access to corticofugal cells in layers V and VI via layer II interneurons (Mitani et al. 1985). Whether these intralaminar connections arise from GABAergic neurons is unknown, as is the chemical specificity of these pathways.

Likewise, models of visual cortex intralaminar information flow recognize that layer III pyramidal cells have reciprocal projections with GABAergic cells in the same layer, that layer IV GABAergic cells project to layer III pyramids, and that layer IV pyramidal cells project to layer II pyramids; layer III pyramidal cells also project to large and tufted layer V pyramidal cells (Thomson and Bannister 2003).

7.5 Layer IV: Thalamic to Intrinsic Cortical Transformation

Nonpyramidal cells dominate layer IV (Prieto et al. 1994b), with few pyramidal cells (Winer 1984a; Smith and Populin 2001). Layer IV receives massive medial geniculate body glutamatergic (Cruikshank et al. 2002) input through its full depth (Huang and Winer 2000). Layer IV somata cluster between the columns of apical dendrites from layer V and VI pyramidal cells (Sousa-Pinto 1973).

Layer IV contains a diverse population of GABAergic neurons, with multipolar, bipolar, and bitufted types predominating (Fig. 10.6d: IV). The proportion of GABAergic cells, 26%, is the highest in auditory cortex except in layer I (94%) (Prieto et al. 1994b). Many layer IV cells receive a dense concentration of unusually large axosomatic puncta, relatively more than in other layers; the only exception is the large multipolar (basket) cells, which receive such endings in all but layer I and, in this respect only, are comparable to pyramidal cells (Fig. 10.7: IV). The layer IV neuropil had larger and more dense puncta than did other layers (Prieto et al. 1994a), suggesting a layer- and size-specific concentration of putative inhibitory/disinhibitory synaptic endings and a possible role for such endings along the pyramidal cell processes traversing layer IV, including those of deep-lying layer IIIb cells (Winer 1984b).

Intracellularly recorded and morphologically characterized layer IV cells in mouse thalamocortical brain slices have multipolar or vertical dendritic arrangements (Verbny et al. 2006) that correlate well with two of the main types of local circuit cells in Golgi (Winer 1984a) and immunocytochemical (Prieto et al. 1994b) work. The axons of the filled, physiologically characterized cells were ~430 μm wide and 505 μm tall; scaling the murine layer IV cells to the dimensions of a cat, a layer IV cell's axon might overlap much of an isofrequency representation (Reale and Imig 1980) and/or any of several modular arrays within AI (Ehret 1997). Intracellular filled layer IV cat spiny stellate cells project to layers III and V and have lateral branches that extend 1 mm (Mitani et al. 1985); these cells were driven by thalamic, corticocortical, and commissural stimulation, suggesting considerable convergence onto their small dendritic arbors. The widespread local distribution of AI intrinsic axons implies a spatial diversity of GABAergic actions from a single neuron, as does the relatively high density of boutons (~22/100 μm^3). Interestingly, these cells were driven only weakly by thalamic stimulation (Verbny et al. 2006), consistent with the idea that a major input to them is of intracortical origin based on the massive investment of axosomatic puncta (Prieto et al. 1994a). Perhaps a principal role of thalamic input to auditory cortex GABAergic layer IV cells is temporal coordination of thalamic and intracortical processes, such as shifts from unbalanced (inhibitory dominant) to balanced (inhibitory–excitatory equality) in intensity tuning (Tan et al. 2007). Both kainate (alpha-amino-3-hydroxy-5-methylisoxazole-4-proprionic acid) and NMDA receptors are implicated in thalamocortical processing (Cruikshank et al. 2002).

Despite some parallels in layer IV organization between modalities, there is reason to believe that there are major functional differences between systems and species. Thus, in cat AI there are few layer IV pyramidal cells and almost no spiny stellate cells, while both are abundant in cat and monkey primary visual cortex (Smith and Populin 2001). A related issue is that the pyramidal-to-pyramidal cell intralaminar layer IV visual cortex circuit thus must be reduced or absent in AI, as must be the intralaminar nonpyramidal-to-nonpyramidal cell pathway in the visual cortex (Thomson and Bannister 2003). Finally, the case for sublaminar arrangements in layer IV is clearest in primate visual cortex (Lund and Yoshioka 1991) and more difficult to discern in auditory cortex. In auditory cortex gap junctions between layer IV cells in primate (Smith and Moskowitz 1979) may enable large-scale network or lateral interactions (Galarreta and Hestrin 2001). The apparent postsynaptic connectional divergence of layer IV GABAergic cells in visual and auditory cortex may have a shared feature (Smith and Populin 2001). The reported diminution in mouse auditory cortex of feedforward intracortical inhibition (Verbny et al. 2006) contrasts with its more robust expression in somatic sensory cortex (Agmon and Connors 1992).

7.6 Layer V: Corticocortical to Corticofugal Transformation

Layer V has massive extrinsic projections to the midbrain (Winer and Prieto 2001), thalamus (Winer 1992) and pons (Perales et al. 2006) to name just a few of its targets (Doucet et al. 2003) and a slightly higher proportion of GABAergic cells than layers II–IV. It represents the principal GABAergic influence on auditory cortex output to extrathalamic sources (Winer 2006). From a molecular perspective, layer V cells in nonauditory cortex have wide range of subtypes (Molnár and Cheung 2006). In auditory cortex the distinctions among layer V cells have been made on structural and connectional grounds in AI only (Winer and Prieto 2001) and with regard to their differential spiking and bursting behavior in rat auditory cortex (Hefti and Smith 2000). The kinetics of inhibitory postsynaptic currents and spontaneous inhibitory currents in rat auditory cortex are more rapid than in other areas, regular spiking cells have larger inhibitory postsynaptic currents with more rapid growth and decline than bursting cells, and bursting cells have more such events; differences in the intrinsic filters for these currents (Hefti and Smith 2003) are consistent with a corresponding diversity of pyramidal cell connectivity and architecture.

Some subtypes of layer V nonpyramidal cells are smooth multipolar and bipolar neurons; they resemble their counterparts in other auditory cortex layers and are similar in size to those in layers I–III and slightly smaller than those in layer IV. Unique types in layer V are the inverted pyramidal cells which themselves have chemically specific subtypes (Winer and Prieto 2001) and neurogliaform cells are rare. The 27% proportion of GABAergic cells is the highest in auditory cortex (Prieto et al. 1994b) and approximates that in the ventral division of the medial geniculate body (Huang et al. 1999). The GABAergic puncta differ from their counterparts in layers II–IV: they are smaller, ~40% less numerous and, except for large multipolar cells, appear to make fewer axosomatic contacts (Fig. 10.7: V) (Prieto et al. 1994a). An implication is that the putative synaptic targets of layer V (and perhaps layer VI; cf. below) GABAergic cells lie outside the layer in which their somata resides. A convergence of GABAergic boutons in granular and supragranular layers may be at the expense of local circuits in the corticofugal layers.

Two-thirds of mouse layer V pyramidal cells show feedforward inhibition after medial geniculate body stimulation. Possible sources, other than intrinsic layer V cells, include layer III and IV cells in primary slices and layer IV cells in shell slices (Verbny et al. 2006). Layer V corticogeniculate or corticocollicular cells have collaterals to unknown cells in layer III (Mitani et al. 1985).

In primary visual cortex layer V pyramidal cells have reciprocal connections with GABAergic layer V cells, and the latter are driven by layer II pyramids; the layer V pyramidal cells have nonreciprocal connections with one another, and feedforward connections to layer VI pyramidal cells (Thomson and Bannister 2003).

7.7 Layer VI: Corticocortical to Corticothalamic Transformation

From a connectionist perspective, layer VI has more restricted (though equally dense) efferent connections than layer V, projecting chiefly to the medial geniculate body (Winer et al. 2001); with regard to its input, it also differs since it receives significant projections from nonlemniscal thalamic nuclei (Huang and Winer 2000). Finally, its neuronal organization is very different from layer V: its pyramidal cells are smaller, the deep half of layer VI is dominated by horizontal cells, these are more numerous than those elsewhere in AI, and they do not appear to constitute a specific sublayer (Prieto and Winer 1999).

The layer VI GABAergic arrangements are as layer specific as those in other AI layers. However, it has only 16% GABAergic neurons; the lowest proportion in AI and about two-thirds the layer II value, the next-sparsest layer. The low relative proportion is in contrast to the morphological range of GABAergic neurons, which include large, medium-sized and small multipolar cells, inverted pyramids, bipolar, and horizontal cells, the latter dominating layer VIb (Prieto et al. 1994b). The overall low level of layer VI GABAergic immunoreactivity is conserved in AII (the second auditory cortex) and SII (second somatic sensory cortex), and each area has only modest layer VI calbindin, calretinin, or parvalbumin immunoreactivity (Clemo et al. 2003). In contrast to other AI layers, layer VI contains far fewer puncta, and many are small. Thus, layer VI has ~40% of the layer IV value and, with the conspicuous exception of the large multipolar cell, few puncta are axosomatic (Prieto et al. 1994a). Fast-spiking units (<0.2 ms) were about half as numerous in layer VI as in layer IV, and regular-spiking layer VI cells were ~80% as numerous as those in layer IV (Atencio and Schreiner 2008).

Layer VI AI cells show evoked responses to stimulation in medial geniculate body, AII and EP (posterior ectosylvian) areas. The local cells have modest intrinsic axonal collateral systems, sometimes extending laterally or sending branches to layer IV. Layer I cells project to layer VI (and V) corticofugal cells (Mitani et al. 1985).

Visual cortex layer VIa cells have long vertical axonal arbors in a column-like configuration, and corticocortical projections as well as projections to the thalamic reticular nucleus; in contrast, layer VIb cells have modest, mainly vertical axonal projections beside their corticofugal axons in the parvocellular lateral geniculate body (Thomson and Bannister 2003).

7.8 Nucleus Basalis and the Cholinergic Forebrain

The nucleus basalis/substantia innominata is a group of diverse nuclei situated in a fiber-rich basal forebrain region with a complex cytoarchitecture. It has a wide range of connections resembling those of the amygdala. Inputs include the bed nucleus of the stria terminalis, preoptic and lateral hypothalamic nuclei, parabrachial nucleus, dorsal raphe, peripeduncular nuclei, and caudoputamen and subthalamic nuclei (Grove 1988b). Targets are the ventral tegmental areas, substantia nigra and peripeduncular area, the amygdala and hypothalamus, bed nucleus of the stria terminalis, the thalamic subparafascicular, gustatory and midline nuclei, and widespread cortical input to piriform, infralimbic, prelimbic, anterior cingulate, granular and agranular insular cortex, and perirhinal and entorhinal areas as well as the olfactory bulb (Grove 1988a). Despite its overall small size, the nucleus basalis has access to an array of structures whose common task is motivational, affective, and cognitive behaviors (Wenk 1997; Froemke et al. 2007).

Activation of the cholinergic stream via the nucleus basalis (Kilgard and Merzenich 1998) or dopaminergic ventral tegmental system (Bao et al. 2001) elicits massive functional changes in the frequency specific map in primary auditory cortex. How these forms of plasticity differ, and the neural mechanisms initiating and supporting them, is unknown (Weinberger 2004).

7.9 Other Chemically Specific Cortical Subsystems

Further specific neurochemical inputs arise from noradrenergic (Descarries et al. 1977) and serotoninergic (Cransac et al. 1998) sources and terminate widely in areal and laminar domains. Noradrenergic effects include a reduction of receptive field size (Manunta and Edeline 1999) without altering their spiking behavior (Manunta and Edeline 2000). Serotonin receptor-specific iontophoresis can enhance, attenuate or reverse the change in long-term best-frequency shifts in a conditioning paradigm in the big brown bat (Ji and Suga 2007).

Besides the morphological diversity of neocortical neuron types, there is a parallel form of neurochemical heterogeneity in which antibodies to parvalbumin label basket cells preferentially and calbindin is selective for double bouquet cells, both of which also are GABAergic (Morino-Wannier et al. 1992). Combined with the differential expression of pre- and postsynaptic receptors on specific types of cells, this suggests massive diversity in the substrates for intracortical function.

A parallel cholinergic pathway to cortex has a differential laminar distribution: choline acetyltransferase immunoreactivity in rodent AI is maximal in layers I and V, lower in layer IV, and weakest in layer VI; this largely agrees with the acetylcholinesterase patterns. Nonprimary areas have a different pattern consistent with corresponding areas in other modalities, with heavy layer I and V immunostaining, medium-to-heavy staining in layers II and III, a layer V sublaminar distribution, and weaker layer I immunoreactivity (Lysakowski et al. 1989). Sources of cholinergic input to AI are diverse and include the caudate, internal capsule, putamen, globus pallidus, among others (Kamke et al. 2005). There is pharmacological evidence for cholinergic–GABAergic interactions on pyramidal cells in rat layer II/III slice preparations (Bandrowski et al. 2001). Some non-pyramidal cells might also receive serotoninergic input (Mulligan and Törk 1988).

8 Intralaminar System

In the primary visual cortex, there are extensive connections between the layers (Lübke et al. 2000) for which chemically specific substrates have been described (Gonchar et al. 2002). Serotonin receptor agonists or antagonists have opposite effects on layer V pyramidal cells (Xiang and Prince 2003). GABAergic circuits affect specific aspects of receptive field organization such as response duration in a subset of cells (Liu et al. 2007). Such analyses remain to be performed for specific types of auditory cortex cells other than layer V pyramids (Hefti and Smith 2003).

Only a sketch of a prospective auditory cortex scenario is possible. The data for AI indicate that thalamic input to layer IV (Huang and Winer 2000) reaches a diverse group of aspiny nonpyramidal cells (Winer 1984a) that project toward populations of layer III neurons with intralaminar projections to layers I and II; these neurons then project to corticofugal cells in layers V and VI (Fig. 10.8a) (Mitani et al. 1985). The corticofugal layer V cells have lateral intralaminar branches that may reach the dendrites of layer VI corticofugal cells projecting to the medial geniculate body (Fig. 10.8b) (Ojima 1994) (cf. 7.7.).

Anesthetized rat auditory cortex receptive fields and local field potential are similar in their characteristic frequency and tuning breadth, and excitatory postsynaptic and local field potentials also were closely aligned. The subthreshold tuning of both of these was often unexpectedly broad, up to five octaves. Muscimol ejected iontophoretically decreased the size of the local field potentials and narrowed receptive field bandwidth, without changing onset latency or the response to the characteristic frequency. The enhancement in local selectivity was attributed in part to GABAergic mechanisms (Kaur et al. 2004). In a mouse slice preparation, lemniscal thalamocortical activation at a characteristic frequency elicits current sinks in layers III and IV, and stimulation ± 3 octaves evoke changes in infragranular layers. A noteworthy feature of the activation was its lateral dispersion within cortex (Kaur et al. 2005). Some 13.4% of labeled, morphologically characterized neurons were considered as intracortical, putatively inhibitory cells, and their axons projected from 2 to 7 mm in all layers and consisted of multiple subtypes (Clarke et al. 1993). This broad horizontal arrangement is consistent with the distribution of thalamocortical afferents in layers III and IV (Huang and Winer 2000), while the vertical interactions may be driven by the clustered thalamocortical axons (Velenovsky et al. 2003), and the effects of activation can extend across much of an isofrequency contour (Song et al. 2006). A more fine grained adjustment of receptive field dynamics could use cholinergic mechanisms to alter aspects of the tuning of single cells (Metherate et al. 2005) and to reduce near-threshold auditory evoked responses by altering excitability along myelinated thalamocortical fibers (Kawai et al. 2007). Other GABAergic influences include distinct early and late components for intensity tuning, where rapid inhibition is specific to the excitatory intensity tuning (Sutter and Loftus 2003). Such findings are compatible with patterns of thalamocortical convergence, inheritance and construction (Miller et al. 2001) and congruent with local neurochemical specificity since most GABAergic intralaminar projections to a ~400 μm-wide column in AI are within ~2 mm (Fig. 10.9d) (Yuan et al. 2010), which corresponds well with the lateral range of visual cortex basket cell axons (Kisvárday and Eysel 1993). The latter observations support the idea of a layer-specific organization of GABAergic circuits, with common and unique features in each layer (Fig. 10.9b, c) (Atencio and Schreiner 2008). Analyses of the spatial distribution of activity with voltage sensitive dye studies reveal that the spatiotemporal spread of horizontal excitation and its modulation by GABA differ significantly in rat somatic sensory (barrel) and insular areas, the latter having a more columnar configuration (Sato et al. 2008).

9 Corticocortical System

A detailed profile of corticocortical connectivity is available for visual cortex (Felleman and Van Essen 1991; Conway et al. 2000) and in rodents the excitatory amino acids are strongly implicated in such pathways, including the thalamocortical system (Johnson and Burkhalter 1992, 1994).

10 Commissural System

The laminar origins and projection neurons in this system are known in AI and involve principally layers III and V and a variety of neurons, including a few non-pyramidal cells (Code and Winer 1985). In the visual system, glutamatergic, aspartatergic, and possible GABAergic projection neurons have been identified (Conti and Manzoni 1994). Rare (~1%) GABAergic commissural projections are present in rat and cat somatic sensory cortex (Fabri and Manzoni 2004) and nonpyramidal callosal cells are found in rat visual cortex (Peters et al. 1990). How these pathways interact with GABAergic cells is unknown.

11 Corticofugal Systems

Glutamate is strongly implicated as a corticogeniculate and corticotectal neurotransmitter in the visual pathways (Lund-Karlsen and Fonnum 1978; Baughman and Gilbert 1981; Conti and Hicks 1996). Many auditory corticocollicular projection neurons in the guinea pig are glutamatergic (Feliciano and Potashner 1995). The physiological relations of these neurons with GABAergic cells in the medial geniculate body and inferior colliculus are unknown.

12 Dissecting Cortical Circuitry

A complete cellular profile that includes the transmitters of local corticocortical feedforward and feedback circuits is not yet available even for the visual cortex (Callaway 1998; Douglas and Martin 2004). In cat (Higo et al. 2007) and monkey (Tomioka and Rockland 2007) auditory and temporal cortex, GABAergic projections from >2 mm away are as rare as those in rat somatic sensory cortex (Fabri and Manzoni 1996), suggesting that many synapses have a proximal origin. How might such pathways be disentangled?

One strategy is to use the types of neurons and their resident layers as a template for exploring hypotheses. This would entail creating a profile for each major type of cell, with the main inputs to it and the projections from it specified anatomically, and a concomitant neurochemical and electrophysiological profile. Finally, the pattern of convergence onto morphologically identified cells would be derived, as has been done to some degree in the cochlear nucleus with respect to GABAergic cells (Saint Marie et al. 1989). Such profiles are now emerging for pyramidal cells (Spruston 2008) and considerable data exist for visual cortex spiny stellate cells (Saint-Marie and Peters 1985). A case in point is the chandelier cell (Howard et al. 2005), which is an ideal candidate given its stereotyped morphology (Somogyi 1977). It has an axon presynaptic to the pyramidal cell's axon initial segment (Szentágothai 1979), and it prefers pyramidal cells projecting preferentially in the corticocortical rather than corticofugal system (Fariñas and DeFelipe 1991). Their axon avoids other axoaxonal cells (Somogyi et al. 1982) and constitutes >90% of axoaxonal synapses (Kawaguchi and Kubota 1998; Howard et al. 2005). There is spatial segregation of ion channels on the axon initial segment suggesting that Kv1.2 channels associated with a specific adhesion molecule concentrate in the distal initial segment, whereas Na^+-associated channels prefer the proximal part (Inda et al. 2005). Some axoaxonic cells can elicit pyramidal cell excitation because of the depolarized reversal potential in contrast to axosomatic inhibitory input (Szabadics et al. 2006). Chandelier cells have larger receptive fields and poorer acuity than other fast-spiking interneurons and are proposed to check unusual excitatory activity in their targets (Zhu et al. 2004). No comparably detailed profile exists for any GABAergic cell in auditory cortex or the medial geniculate body.

13 Concluding Comments

Understanding interneuron performance presents conceptual challenges and practical difficulties. All cortical cells with local axonal branches and remote projections to other areas, layers, or specific circuits can have proximal roles, as do pyramidal cell axon collaterals, which are present on most subtypes and whose role is known in a few cases only (Somogyi et al. 1979; Lübke et al. 1996), or interneuronal self-innervation (Tamás et al. 1997). The conceptual problems that classic Golgi type II cells entail are many. Why are these cells so variable morphologically? Do similar types of neurons found at different sites (e.g., basket cells in layers II and IV) have similar roles? How are interneuronal receptive fields related to those of their targets, and to those of the cells that target them? Are interneurons capable of plasticity? If so, is this matched, greater than, or anticorrelated with the cells to which they project? Are any operations of principal cells outside the ambit or influence of interneurons? Are any interneuronal operations independent of those of principal cells? Does the molecular heterogeneity of pyramidal cells have an interneuronal correlate? Why do interneurons appear to have such extremes of species specific distribution? Challenging as these conceptual issues might seem, the concomitant practical difficulties will require an unprecedented degree of technical refinement. What do interneurons do in awake animals (Swadlow 2003)? How can interneuronal axons, many of which are ~0.4 μm thick, conduct impulses? In what polysynaptic circuits do they project (Swadlow 2002)? How does the cholinergic system impinge upon them? Do they have true axons and dendrites (Winer and Larue 1988)? What is their role in learning and synaptic plasticity? The answers to these questions could constitute the basis for an empirical and predictive theory of interneuronal function.

14 Future Questions

- What do different types of thalamic and cortical interneurons do?
- How do cholinergic, GABAergic, and glutamatergic/aspartatergic inputs target, and interact differentially with, specific types of cells?
- Does axosomatic and axoaxonal GABAergic input to corticothalamic, corticocollicular, and corticopontine cells follow the same principles?
- What is the neurochemical basis for the thalamocortical-to-corticocortical and other transformations?
- What are the roles of interneurons in broadly tuned nuclei and areas or regions with large receptive fields and coarsely articulated sensory maps?
- Do interneurons participate in interactions between modalities?

Acknowledgments Thanks to Ms. Katie D. Dorsch for conscientious preparation of the figures. This work was supported by United States Public Health Service grant R01 DC02319-28.

References

Agmon A and Connors BW (1992) Correlation between intrinsic firing patterns and thalamocortical synaptic responses of neurons in mouse barrel cortex. *Journal of Neuroscience* **12**:319–329.

Aitkin LM and Dunlop CW (1968) Interplay of excitation and inhibition in the cat medial geniculate body. *Journal of Neurophysiology* **31**:44–61.

Aitkin LM and Phillips SC (1984) Is the inferior colliculus an obligatory relay in the cat auditory system? *Neuroscience Letters* **44**:259–264.

Anderson LA, Malmierca MS, Wallace MN, and Palmer AR (2007) Evidence for a direct, short latency projection from the dorsal cochlear nucleus to the auditory thalamus in the guinea pig. *European Journal of Neuroscience* **24**:491–498.

Apergis-Schoute AM, Debiec J, Doyère V, LeDoux JE, and Schafe GE (2005) Auditory fear conditioning and long-term potentiation in the lateral amygdala require ERK/MAP kinase signaling in the auditory thalamus: a role for presynaptic plasticity in the fear system. *Journal of Neuroscience* **25**:5730–5739.

Arcelli P, Frassoni C, Regondi MC, De Biasi S, and Spreafico R (1997) GABAergic neurons in mammalian thalamus: a marker of thalamic complexity? *Brain Research Bulletin* **42**:27–37.

Atencio CA and Schreiner CE (2008) Spectrotemporal processing differences between auditory cortical fast-spiking and regular-spiking neurons. *Journal of Neuroscience* **28**:3897–3910.

Bandrowski AE, Moore SL, and Ashe JH (2001) Cholinergic synaptic potentials in the supragranular layers of auditory cortex. *Synapse* **41**:118–130.

Bao S, Chan VT, and Merzenich MM (2001) Cortical remodelling induced by activity of ventral tegmental dopamine neurons. *Nature* **412**:79–83.

Bartlett EL and Smith PH (2002) Effects of paired-pulse and repetitive stimulation on neurons in the rat medial geniculate body. *Neuroscience* **113**:957–974.

Baughman RW and Gilbert CD (1981) Aspartate and glutamate as possible neurotransmitters in the visual cortex. *Journal of Neuroscience* **1**:427–438.

Brandner S and Redies H (1990) The projection of the medial geniculate body to field AI: organization in the isofrequency dimension. *Journal of Neuroscience* **10**:50–61.

Bright DP, Aller MI, and Brickley SG (2007) Synaptic release generates a tonic $GABA_A$ receptor-mediated conductance that modulates burst precision in thalamic relay neurons. *Journal of Neuroscience* **27**:2560–2569.

Calford MB and Aitkin LM (1983) Ascending projections to the medial geniculate body of the cat: evidence for multiple, parallel auditory pathways through the thalamus. *Journal of Neuroscience* **3**: 2365–2380.

Callaway EM (1998) Local circuits in primary visual cortex of the macaque monkey. *Annual Review of Neuroscience* **21**:47–74.

Cant NB (2005) Projections from the cochlear nuclear complex to the inferior colliculus. In: Winer JA and Schreiner CE (eds.). *The Inferior Colliculus*. Springer-Verlag, New York, pp. 115–131.

Chu Z, Galaretta M, and Hestrin S (2003) Synaptic interactions of late-spiking neocortical neurons in layer 1. *Journal of Neuroscience* **23**:96–102.

Clarey JC, Barone P, and Imig TJ (1992) Physiology of thalamus and cortex. In: Popper AN and Fay RR (eds.). *Springer Handbook of Auditory Research*, volume 2, *The Mammalian Auditory Pathway: Neurophysiology*. Springer-Verlag, New York, pp. 232–334.

Clarke S, de Ribaupierre F, Rouiller EM, and de Ribaupierre Y (1993) Several neuronal and axonal types form long intrinsic connections in the cat primary auditory cortical field (AI). *Anatomy and Embryology* **188**:117–138.

Clascá F, Llamas A, and Reinoso-Suárez F (1997) Insular cortex and neighboring fields in the cat: a redefinition based on cortical microarchitecture and connections with the thalamus. *Journal of Comparative Neurology* **384**:456–482.

Clemo HR, Keniston L, and Meredith MA (2003) A comparison of the distribution of GABA-ergic neurons in cortices representing different sensory modalities. *Journal of Chemical Neuroanatomy* **26**:51–63.

Code RA and Winer JA (1985) Commissural neurons in layer III of cat primary auditory cortex (AI): pyramidal and non-pyramidal cell input. *Journal of Comparative Neurology* **242**:485–510.

Conti F and Hicks TP (eds.) (1996) *Excitatory Amino Acids and the Cerebral Cortex*, MIT Press, Cambridge.

Conti F and Manzoni T (1994) The neurotransmitters and postsynaptic actions of callosally projecting neurons. *Behavioural Brain Research* **64**:37–53.

Conti F and Minelli A (1996) The anatomy of glutamatergic transmission in the cerebral cortex. In: Conti F and Hicks TP (eds). *Excitatory Amino Acids and the Cerebral Cortex.* MIT Press, Cambridge, pp. 81–98.

Conway B, Boyd JD, Stewart TH, and Matsubara JA (2000) The projection from V1 to extrastriate area 21a: a second patchy efferent pathway that colocalizes with the CO blob columns in cat visual cortex. *Cerebral Cortex* **10**:149–159.

Coomes DL, Bickford ME, and Schofield BR (2002) GABAergic circuitry in the dorsal division of the cat medial geniculate nucleus. *Journal of Comparative Neurology* **453**:45–56.

Coveñas R, Romo R, Cheramy A, Cesselin F, and Conrath M (1986) Immunocytochemical study of enkephalin-like cell bodies in the thalamus of the cat. *Brain Research* **377**: 355–361.

Crabtree JW (1998) Organization in the auditory sector of the cat's thalamic reticular nucleus. *Journal of Comparative Neurology* **390**:167–182.

Cransac H, Cottet-Emard JM, Hellstrom S, and Peyrin L (1998) Specific sound-induced noradrenergic and serotonergic activation in central auditory structures. *Hearing Research* **118**:151–156.

Crick F (1984) Function of the thalamic reticular nucleus: the searchlight hypothesis. *Proceedings of the National Academy of Sciences of the United States of America* **81**:4586–4590.

Cruikshank SJ, Killackey HP, and Metherate R (2001) Parvalbumin and calbindin are differentially distributed within primary and secondary subregions of the mouse auditory forebrain. *Neuroscience* **105**: 553–569.

Cruikshank SJ, Rose HJ, and Metherate R (2002) Auditory thalamocortical synaptic transmission in vitro. *Journal of Neurophysiology* **87**:361–384.

Cucchiaro JB, Uhlrich DJ, and Sherman SM (1993) Ultrastructure of synapses from the pretectum in the A-laminae of the cat's lateral geniculate nucleus. *Journal of Comparative Neurology* **334**: 618–630.

de Venecia RK, Smelser CB, Lossman SD, and McMullen NT (1995) Complementary expression of parvalbumin and calbindin D-28 k delineate subdivisions of the rabbit medial geniculate body. *Journal of Comparative Neurology* **359**:595–612.

de Venecia RK, Smelser CB, and McMullen NT (1998) Parvalbumin is expressed in a reciprocal circuit linking the medial geniculate body and auditory neocortex in the rabbit. *Journal of Comparative Neurology* **400**:349–362.

DeFelipe J, Hendry SH, Hashikawa T, and Jones EG (1991) Synaptic relationships of serotonin-immunoreactive terminal baskets on GABA neurons in the cat auditory cortex. *Cerebral Cortex* **1**: 117–133.

Descarries L, Watkins KC, and Lapierre Y (1977) Noradrenergic axon terminals in the cerebral cortex of rat: topometric ultrastructural analysis. *Brain Research* **133**:197–222.

Destexhe A and Sejnowski TJ (2001) *Thalamocortical Assemblies. How Ion Channels, Single Neurons and Large-Scale Networks Organize Sleep Oscillations. Monographs of the Physiological Society*, volume 49. Oxford University Press, New York.

Doucet JR, Molavi DL, and Ryugo DK (2003) The source of corticocollicular and corticobulbar projections in area Te1 of the rat. *Experimental Brain Research* **153**:477–485.

Douglas RJ and Martin KAC (2004) Neuronal circuits of the neocortex. *Annual Review of Neuroscience* **29**:419–451.

Dykes RW (1983) Parallel processing of somatosensory information: a theory. *Brain Research Reviews* **6**:47–115.

Edeline J-M (1995) The α2-antagonist idazoxan enhances the frequency selectivity and increases the threshold of auditory cortex neurons. *Experimental Brain Research* **107**:221–240.

Ehret G (1997) The auditory cortex. *Journal of Comparative Physiology A* **181**:547–557.

Emson PC (ed.) (1983) *Chemical Neuroanatomy*, Raven Press, New York.

Fabri M and Manzoni T (1996) Glutamate decarboxylase immunoreactivity in corticocortical projecting neurons of rat somatic sensory cortex. *Neuroscience* **72**:435–448.

Fabri M and Manzoni T (2004) Glutamic acid decarboxylase immunoreactivity in callosal projecting neurons of cat and rat somatic sensory areas. *Neuroscience* **123**:557–566.

Fariñas I and DeFelipe J (1991) Patterns of synaptic input on corticocortical and corticothalamic cells in the cat visual cortex. II. The axon initial segment. *Journal of Comparative Neurology* **304**: 70–77.

Feliciano M and Potashner SJ (1995) Evidence for a glutamatergic pathway from the guinea pig auditory cortex to the inferior colliculus. *Journal of Neurochemistry* **65**:1348–1357.

Felleman DJ and Van Essen DC (1991) Distributed hierarchical processing in the primate cerebral cortex. *Cerebral Cortex* **1**:1–47.

Fisken RA, Garey LJ, and Powell TPS (1975) The intrinsic, association and commissural connections of area 17 of the visual cortex. *Philosophical Transactions of the Royal Society of London, series B, Biological Sciences* **272**:487–536.

Fitzpatrick D, Diamond IT, and Raczkowski D (1989) Cholinergic and monoaminergic innervation of the cat's thalamus: comparison of the lateral geniculate nucleus with other principal sensory nuclei. *Journal of Comparative Neurology* **288**:647–675.

Foote SL, Bloom FE, and Aston-Jones G (1983) Nucleus locus ceruleus: new evidence of anatomical and physiological specificity. *Physiological Reviews* **63**:844–914.

Froemke RC, Merzenich MM, and Schreiner CE (2007) A synaptic memory trace in auditory cortex. *Nature* **450**:425–429.

Galarreta M and Hestrin S (2001) Spike transmission and synchrony detection in networks of GABAergic interneurons. *Science* **292**:2295–2299.

Garcia MM and Harlan RE (1997) Protein kinase C in central auditory pathways of the rat. *Journal of Comparative Neurology* **385**:1–25.

Gerren RA and Weinberger NM (1983) Long term potentiation in the magnocellular medial geniculate nucleus of the anesthetized cat. *Brain Research* **265**:138–142.

Gonchar Y, Turney S, Price JL, and Burkhalter A (2002) Axo-axonic synapses formed by somatostatin-expressing GABAergic neurons in rat and monkey visual cortex. *Journal of Comparative Neurology* **443**:1–14.

Groh JM, Trause AS, Underhill AM, Clark KR, and Inati S (2001) Eye position influences auditory responses in primate inferior colliculus. *Neuron* **29**:509–518.

Grove EA (1988a) Efferent connections of the substantia innominata in the rat. *Journal of Comparative Neurology* **277**:347–364.

Grove EA (1988b) Neural associations of the substantia innominata in the rat: afferent connections. *Journal of Comparative Neurology* **277**:315–346.

Hall DA (2005) Sensitivity to spectral and temporal properties of sound in human non-primary auditory cortex. In: König R, Heil P, Budinger E, and Scheich H (eds). *The Auditory Cortex. A Synthesis of Human and Animal Research.* Lawrence Erlbaum Associates, Mahwah, pp. 51–76.

Hefti BJ and Smith PH (2000) Anatomy, physiology, and synaptic responses of rat layer V auditory cortical cells and effects of intracellular $GABA_A$ blockade. *Journal of Neurophysiology* **83**:2626–2638.

Hefti BJ and Smith PH (2003) Distribution and kinetic properties of GABAergic inputs to layer V pyramidal cells. *Journal of the Association for Research in Otolaryngology* **4**:106–121.

Hendry SHC and Jones EG (1991) GABA neuronal subpopulations in cat primary auditory cortex—co-localization with calcium binding proteins. *Brain Research* **543**:45–55.

Higo S, Udaka N, and Tamamaki N (2007) Long-range GABAergic projection neurons in the cat neocortex. *Journal of Comparative Neurology* **503**:421–431.

Houser CR, Vaughn JE, Barber RP, and Roberts E (1980) GABA neurons are the major cell type in the nucleus reticularis thalami. *Brain Research* **200**:341–354.

Howard A, Tamas G, and Soltesz I (2005) Lighting the chandelier: new vistas for axo-axonic cells. *Trends in Neurosciences* **28**: 310–316.

Hsieh CY, Cruikshank SJ, and Metherate R (2000) Differential modulation of auditory thalamocortical and intracortical synaptic transmission by cholinergic agonist. *Brain Research* **880**:51–64.

Hu B (1995) Cellular basis of temporal synaptic signalling: an in vitro electrophysiological study in rat auditory thalamus. *Journal of Physiology (London)* **483**:167–182.

Hu B, Senatorov V, and Mooney D (1994) Lemniscal and non-lemniscal synaptic transmission in rat auditory thalamus. *Journal of Physiology (London)* **479**:217–231.

Huang CL, Larue DT, and Winer JA (1999) GABAergic organization of the cat medial geniculate body. *Journal of Comparative Neurology* **415**:368–392.

Huang CL and Winer JA (2000) Auditory thalamocortical projections in the cat: laminar and areal patterns of input. *Journal of Comparative Neurology* **427**:302–331.

Imig TJ and Morel A (1985) Tonotopic organization in ventral nucleus of medial geniculate body in the cat. *Journal of Neurophysiology* **53**:309–340.

Inda MC, DeFelipe J, and Muñoz A (2005) Voltage-gated ion channels in the axon initial segment of human cortical pyramidal cells and their relationship with chandelier cells. *Proceedings of the National Academy of Sciences of the United States of America* **103**:2920–2925.

Ji W and Suga N (2007) Serotoninergic modulation of plasticity of the auditory cortex elicited by fear conditioning. *Journal of Neuroscience* **27**:4910–4918.

Johnson RR and Burkhalter A (1992) Evidence for excitatory amino acid neurotransmitters in the geniculo-cortical pathway and local projections within rat primary visual cortex. *Experimental Brain Research* **89**:20–30.

Johnson RR and Burkhalter A (1994) Evidence for excitatory amino acid neurotransmitters in forward and feedback corticocortical connections within rat visual cortex. *European Journal of Neuroscience* **6**:272–286.

Jones EG (1985) *The Thalamus*. Plenum Press, New York.

Jones EG (2003) Chemically defined parallel pathways in the monkey auditory system. *Annals of the New York Academy of Sciences* **999**:218–233.

Jones EG (2007) *The Thalamus*. Cambridge University Press, New York.

Jones EG, Dell'Anna ME, Molinari M, Rausell E, and Hashikawa T (1995) Subdivisions of macaque monkey auditory cortex revealed by calcium-binding protein immunoreactivity. *Journal of Comparative Neurology* **362**:153–170.

Kamke MR, Brown M, and Irvine DRF (2005) Origin and immunolesioning of cholinergic basal forebrain innervation of cat primary auditory cortex. *Hearing Research* **206**:89–106.

Kaur S, Lazar R, and Metherate R (2004) Intracortical pathways determine breadth of subthreshold frequency receptive fields in primary auditory cortex. *Journal of Neurophysiology* **91**: 2551–2567.

Kaur S, Rose HJ, Lazar R, Liang K, and Metherate R (2005) Spectral integration in primary auditory cortex: laminar processing of afferent input, in vivo and in vitro. *Neuroscience* **134**:1033–1045.

Kawaguchi Y and Kubota Y (1998) Neurochemical features and synaptic connections of large physiologically-identified GABAergic cells in the rat frontal cortex. *Neuroscience* **85**:677–701.

Kawai H, Lazar R, and Metherate R (2007) Nicotinic control of axon excitability regulates thalamocortical transmission. *Nature Neuroscience* **10**:1168–1175.

Kharazia VN, Phend KD, Weinberg RJ, and Rustioni A (1996) Excitatory amino acids in corticofugal projections: microscopic evidence. In: Conti F and Hicks TP (eds). *Excitatory Amino Acids and the Cerebral Cortex*. MIT Press, Cambridge, pp. 127–136.

Kharazia VN and Weinberg RJ (1994) Glutamate in thalamic fibers terminating in layer IV of primary sensory cortex. *Journal of Neuroscience* **14**:6021–6032.

Kilgard MP and Merzenich MM (1998) Cortical map reorganization enabled by nucleus basalis activity. *Science* **279**:1714–1718.

King AJ (1997) Signal selection by cortical feedback. *Current Biology* **7**:R85-R88.

Kishan AU, Lee CC, and Winer JA (2008) Branched projections in the auditory thalamocortical and corticocortical systems. *Neuroscience* **154**:283–293.

Kisvárday Z, Beaulieu C, and Eysel U (1993) Network of GABAergic large basket cells in cat visual cortex (area 18): implication for lateral disinhibition. *Journal of Comparative Neurology* **327**:398–415.

Kisvárday ZF and Eysel UT (1993) Functional and structural topography of horizontal inhibitory connections in cat visual cortex. *European Journal of Neuroscience* **5**:1558–1572.

Koch U and Grothe B (2000) Interdependence of spatial and temporal coding in the auditory midbrain. *Journal of Neurophysiology* **83**:2300–2314.

Kuwabara N, DiCaprio RA, and Zook JM (1991) Afferents to the medial nucleus of the trapezoid body and their collateral projections. *Journal of Comparative Neurology* **314**:684–706.

Kuwabara N and Zook JM (2000) Geniculo-collicular descending projections in the gerbil. *Brain Research* **878**:79–87.

Lee CC, Imaizumi K, Schreiner CE, and Winer JA (2004a) Concurrent tonotopic processing streams in auditory cortex. *Cerebral Cortex* **14**:441–451.

Lee CC, Schreiner CE, Imaizumi K, and Winer JA (2004b) Tonotopic and heterotopic projection systems in physiologically defined auditory cortex. *Neuroscience* **128**:871–887.

Lee CC and Winer JA (2008a) Connections of cat auditory cortex: II. Commissural system. *Journal of Comparative Neurology* **507**: 1901–1919.

Lee CC and Winer JA (2008b) Connections of cat auditory cortex: III. Corticocortical system. *Journal of Comparative Neurology* **507**:1920–1943.

Lee SM, Friedberg MH, and Ebner FF (1994) The role of GABA-mediated inhibition in the rat ventral posterior medial thalamus. II. Differential effects of $GABA_A$ and $GABA_B$ receptor antagonists on responses of VPM neurons. *Journal of Neurophysiology* **71**:1716–1726.

Li XF, Armony JL, and LeDoux JE (1996) $GABA_A$ and $GABA_B$ receptors differentially regulate synaptic transmission in the auditory thalamo-amygdala pathway: an in vitro microiontophoretic study and a model. *Synapse* **24**:115–124.

Linke R, Braune G, and Schwegler H (2000) Differential projection of the posterior paralaminar thalamic nuclei to the amygdaloid complex in the rat. *Experiment Brain Research* **134**:520–532.

Liu S, Liu Y-J, and Li B (2007) Spatiotemporal structure of complex cell receptive fields and influence of GABAergic inhibition. *NeuroReport* **18**:1577–1581.

Loftus WC and Sutter ML (2001) Spectrotemporal organization of excitatory and inhibitory receptive fields of cat posterior auditory field neurons. *Journal of Neurophysiology* **86**:475–491.

Lübke J, Egger V, Sakmann B, and Feldmeyer D (2000) Columnar organization of dendrites and axons of single and synaptically coupled excitatory spiny neurons in layer 4 of the rat barrel cortex. *Journal of Neuroscience* **20**:5300–5311.

Lübke J, Markram H, Frotscher M, and Sakmann B (1996) Frequency and dendritic distribution of autapses established by layer 5 pyramidal neurons in the developing rat neocortex: comparison with synaptic innervation of adjacent neurons of the same class. *Journal of Neuroscience* **16**:3209–3218.

Lund JS, Henry GH, MacQueen CL, and Harvey AR (1979) Anatomical organization of the primary visual cortex (area 17) of the cat. A comparison with area 17 of the monkey. *Journal of Comparative Neurology* **184**:599–618.

Lund JS and Yoshioka T (1991) Local circuit neurons of macaque monkeys striate cortex: III Neurons of laminae 4B, 4A, and 3B. *Journal of Comparative Neurology* **311**:234–258.

Lund-Karlsen R and Fonnum F (1978) Evidence for glutamate as a neurotransmitter in the corticofugal fibres to the dorsal lateral geniculate body and the superior colliculus in rats. *Brain Research* **151**:457–468.

Lysakowski A, Wainer BH, Bruce G, and Hersh LB (1989) An atlas of the regional and laminar distribution of choline acetyltransferase immunoreactivity in rat cerebral cortex. *Neuroscience* **28**:291–336.

Manunta Y and Edeline J-M (1999) Effects of noradrenaline on frequency tuning of auditory cortex neurons during wakefulness and slow-wave sleep. *European Journal of Neuroscience* **11**: 2134–2150.

Manunta Y and Edeline JM (2000) Noradrenaline does not change the mode of discharge of auditory cortex neurons. *NeuroReport* **11**: 23–26.

Marc RE and Cameron D (2001) A molecular phenotype atlas of the zebrafish retina. *Journal of Neurocytology* **30**:593–674.

McInvale AC, Staudinger J, Harlan RE, and Garcia MM (2002) Immunolocalization of PICK1 in the ascending auditory pathways of the adult rat. *Journal of Comparative Neurology* **450**:382–394.

McMullen NT, Smelser CB, and de Venecia RK (1994) A quantitative analysis of parvalbumin neurons in rabbit auditory neocortex. *Journal of Comparative Neurology* **349**:493–511.

McMullen NT, Velenovsky DS, and Holmes MG (2005) Auditory thalamic organization: cellular slabs, dendritic arbors and tectothalamic axons underlying the frequency map. *Neuroscience* **136**:927–943.

Merchán MA, Aguilar LA, Lopez-Poveda EA, and Malmierca MS (2005) The inferior colliculus of the rat: quantitative immunocytochemical study of GABA and glycine. *Neuroscience* **136**:907–925.

Metherate R, Kaur S, Kawai H, Lazar R, Liang K, and Rose HJ (2005) Spectral integration in auditory cortex: mechanisms and modulation. *Hearing Research* **206**:146–158.

Miller LM, Escabí MA, Read HL, and Schreiner CE (2001) Functional convergence of response properties in the auditory thalamocortical system. *Neuron* **32**:151–160.

Mitani A, Itoh K, Nomura S, Kudo M, Kaneko T, and Mizuno N (1984) Thalamocortical projections to layer I of the primary auditory cortex in the cat: a horseradish peroxidase study. *Brain Research* **310**: 347–350.

Mitani A and Shimokouchi M (1985) Neuronal connections in the primary auditory cortex: an electrophysiological study in the cat. *Journal of Comparative Neurology* **235**:417–429.

Mitani A, Shimokouchi M, Itoh K, Nomura S, Kudo M, and Mizuno N (1985) Morphology and laminar organization of electrophysiologically identified neurons in primary auditory cortex in the cat. *Journal of Comparative Neurology* **235**:430–447.

Molinari M, Dell'Anna ME, Rausell E, Leggio MG, Hashikawa T, and Jones EG (1995) Auditory thalamocortical pathways defined in monkeys by calcium-binding protein immunoreactivity. *Journal of Comparative Neurology* **362**:171–194.

Molnár Z and Cheung AFP (2006) Towards the classification of subpopulations of layer V pyramidal neurons. *Neuroscience Research* **55**:105–115.

Morest DK (1965) The laminar structure of the medial geniculate body of the cat. *Journal of Anatomy (London)* **99**:143–160.

Morest DK (1971) Dendrodendritic synapses of cells that have axons: the fine structure of the Golgi type II cell in the medial geniculate body of the cat. *Zeitschrift für Anatomie und Entwicklungsgeschichte* **133**:216–246.

Morest DK (1974) LCN's in the medial geniculate body of the cat. *Neurosciences Research Program Bulletin* **13**:367–377.

Morest DK (1975) Synaptic relationships of Golgi type II cells in the medial geniculate body of the cat. *Journal of Comparative Neurology* **162**:157–194.

Morino-Wannier P, Fujita SC, and Jones EG (1992) GABAergic neuronal populations in monkey primary auditory cortex defined by co-localized calcium binding proteins and surface antigens. *Experimental Brain Research* **88**:422–432.

Mugnaini E and Oertel WH (1985) An atlas of the distribution of GABAergic neurons and terminals in the rat CNS as revealed by GAD immunocytochemistry. In: Björklund A and Hökfelt T (eds). *Handbook of Chemical Neuroanatomy*, volume 4*: GABA and Neuropeptides in the CNS, part 1*. Elsevier Scientific Publishers BV, Amsterdam, pp. 436–608.

Mulligan KA and Törk I (1988) Serotoninergic innervation of the cat cerebral cortex. *Journal of Comparative Neurology* **270**:86–110.

Ojima H (1994) Terminal morphology and distribution of corticothalamic fibers originating from layers 5 and 6 of cat primary auditory cortex. *Cerebral Cortex* **6**:646–663.

Ojima H, Murakami K, and Kishi K (1996) Dual termination modes of corticothalamic fibers originating from pyramids of layers 5 and 6 in cat visual cortical area 17. *Neuroscience Letters* **208**:57–60.

Oliver DL, Kuwada S, Yin TCT, Haberly LB, and Henkel CK (1991) Dendritic and axonal morphology of HRP-injected neurons in the inferior colliculus of the cat. *Journal of Comparative Neurology* **303**:75–100.

Oliver DL, Winer JA, Beckius GE, and Saint Marie RL (1994) Morphology of GABAergic cells and axon terminals in the cat inferior colliculus. *Journal of Comparative Neurology* **340**: 27–42.

Olucha-Bordonau FE, Pérez-Villalba A, Teruel-Martí V, and Ruiz-Torner A (2004) Chemical divisions in the medial geniculate body and surrounding paralaminar nuclei of the rat: quantitative comparison of cell density, NADPH diaphorase, acetyl cholin esterase and basal expression of c-fos. *Journal of Chemical Neuroanatomy* **28**:147–162.

Osen KK, Ottersen OP, and Storm-Mathisen J (1990) Colocalization of glycine-like and GABA-like immunoreactivities: a semiquantitative study in the dorsal cochlear nucleus of the cat. In: Ottersen OP and Storm-Mathisen J (eds). *Glycine Neurotransmission*. John Wiley & Sons, Chichester, pp. 417–451.

Ottersen OP and Storm-Mathisen J (eds) (1990) *Glycine Neurotransmission*, John Wiley & Sons, Chichester.

Pape HC and McCormick DA (1989) Norepinephrine and serotonin selectively modulate thalamic burst firing by enhancing a hyperpolarization-activated cation current. *Nature* **340**: 715–718.

Paré D, Curró Dossi R, and Steriade M (1991) Three types of inhibitory postsynaptic potentials generated by interneurons in the anterior thalamic complex of cat. *Journal of Neurophysiology* **66**:1190–1204.

Perales M, Winer JA, and Prieto JJ (2006) Focal projections of cat auditory cortex to the pontine nuclei. *Journal of Comparative Neurology* **497**:959–980.

Peruzzi D, Bartlett E, Smith PH, and Oliver DL (1997) A monosynaptic GABAergic input from the inferior colliculus to the medial geniculate body in rat. *Journal of Neuroscience* **17**:3766–3777.

Peters A (1985) Neuronal composition and circuitry of rat visual cortex. In: Rose D and Dobson V (eds). *Models of the Visual Cortex*. John Wiley & Sons, Chichester, pp. 492–503.

Peters A, Payne BR, and Josephson K (1990) Transcallosal nonpyramidal cell projections from visual cortex in the cat. *Journal of Comparative Neurology* **302**:124–142.

Prieto JJ, Peterson BA, and Winer JA (1994a) Laminar distribution and neuronal targets of GABAergic axon terminals in cat primary auditory cortex (AI). *Journal of Comparative Neurology* **344**: 383–402.

Prieto JJ, Peterson BA, and Winer JA (1994b) Morphology and spatial distribution of GABAergic neurons in cat primary auditory cortex (AI). *Journal of Comparative Neurology* **344**:349–382.

Prieto JJ and Winer JA (1999) Layer VI in cat primary auditory cortex (AI): Golgi study and sublaminar origins of projection neurons. *Journal of Comparative Neurology* **404**:332–358.

Radley JJ, Farb CR, He Y, Janssen WG, Rodrigues SM, Hof PR, LeDoux JE, and Morrison JH (2007) Distribution of NMDA and AMPA receptor subunits at thalamo-amygdaloid dendritic spines. *Brain Research* **1134**:87–94.

Read HL, Miller LM, Schreiner CE, and Winer JA (2008) Two thalamic pathways to primary auditory cortex. *Neuroscience* **152**:151–159.

Reale RA and Imig TJ (1980) Tonotopic organization in auditory cortex of the cat. *Journal of Comparative Neurology* **192**:265–291.

Rodgers KM, Benison AM, Klein A, and Barth DS (2008) Auditory, somatosensory, and multisensory insular cortex in the rat. *Cerebral Cortex* **18**:2941–2951.

Romanski LM, Clugnet M, Bordi F, and LeDoux JE (1993) Somatosensory and auditory convergence in the lateral nucleus of the amygdala. *Behavioral Neuroscience* **107**:444–450.

Rouiller EM, Capt M, Hornung JP, and Streit P (1990) Correlation between regional changes in the distributions of GABA-containing neurons and unit response properties in the medial geniculate body of the cat. *Hearing Research* **49**:249–258.

Rouiller EM, Colomb E, Capt M, and de Ribaupierre F (1985) Projections of the reticular complex of the thalamus onto physiologically characterized regions of the medial geniculate body. *Neuroscience Letters* **53**:227–232.

Saint Marie RL, Morest DK, and Brandon CJ (1989) The form and distribution of GABAergic synapses on the principal cell types of the ventral cochlear nucleus of the cat. *Hearing Research* **42**:97–112.

Saint Marie RL, Stanforth DA, and Jubelier EM (1997) Substrate for rapid feedforward inhibition of the auditory forebrain. *Brain Research* **765**:173–176.

Saint-Marie RL and Peters A (1985) The morphology and synaptic connections of the spiny stellate neurons in monkey visual cortex (area 17). *Journal of Comparative Neurology* **223**:213–235.

Sato H, Shimanuki Y, Saito M, Toyoda H, Nokubi T, Maeda Y, Yamamoto T, and Kang Y (2008) Differential columnar processing in local circuits of barrel and insular cortices. *Journal of Neuroscience* **28**:3076–3089.

Schofield BR (2005) Superior olivary complex and lateral lemniscal connections of the auditory midbrain. In: Winer JA and Schreiner CE (eds). *The Inferior Colliculus*. Springer-Verlag, New York, pp. 132–154.

Schreiner CE and Winer JA (2007) Auditory cortex mapmaking: principles, projections, and plasticity. *Neuron* **56**:356–365.

Sefton AJ and Dreher B (1995) Visual system. In: Paxinos G (ed). *The Rat Nervous System*. Academic Press, San Diego, pp. 833–898.

Senatorov VV and Hu B (2002) Extracortical descending projections to the rat inferior colliculus. *Neuroscience* **115**:243–250.

Sherman SM (2004) Interneurons and triadic circuitry of the thalamus. *Trends in Neurosciences* **27**:670–675.

Sherman SM and Guillery RW (1998) On the actions that one nerve cell can have on another: distinguishing "drivers" from "modulators". *Proceedings of the National Academy of Sciences of the United States of America* **95**:7121–7126.

Sherman SM and Guillery RW (2006) *Exploring the Thalamus and Its Role in Cortical Function*. MIT Press, Cambridge.

Shinonaga Y, Takada M, and Mizuno N (1994) Direct projections from the non-laminated divisions of the medial geniculate nucleus to the temporal polar cortex and amygdala in the cat. *Journal of Comparative Neurology* **340**:405–426.

Smith DE and Moskowitz N (1979) Ultrastructure of layer IV of the primary auditory cortex of the squirrel monkey. *Neuroscience* **4**:349–359.

Smith PH and Populin LC (2001) Fundamental differences between the thalamocortical recipient layers of the cat auditory and visual cortices. *Journal of Comparative Neurology* **436**:508–519.

Somogyi P (1977) A specific "axo-axonal" interneuron in the visual cortex of the rat. *Brain Research* **136**:345–350.

Somogyi P, Freund TF, and Cowey A (1982) The axo-axonic interneuron in the cerebral cortex of the rat, cat and monkey. *Neuroscience* **7**:2577–2607.

Somogyi P, Hodgson AJ, and Smith AD (1979) An approach to tracing neuron networks in the cerebral cortex and basal ganglia: combination of Golgi staining, retrograde transport of horseradish peroxidase and anterograde degeneration of synaptic boutons in the same material. *Neuroscience* **4**:1805–1852.

Song W-J, Kawaguchi H, Totoki S, Inoue Y, Katura T, Maeda S, Inagaki S, Shirasawa H, and Nishimura M (2006) Cortical intrinsic circuits can support activity propagation through an isofrequency strip of the guinea pig primary auditory cortex. *Cerebral Cortex* **16**:718–729.

Sousa-Pinto A (1973) The structure of the first auditory cortex (A I) in the cat. I. — Light microscopic observations on its structure. *Archives Italiennes de Biologie* **111**:112–137.

Spreafico R, Frassoni C, Arcelli P, and De Biasi S (1994) GABAergic interneurons in the somatosensory thalamus of the guinea-pig: a light and ultrastructural immunocytochemical investigation. *Neuroscience* **59**:961–973.

Spruston N (2008) Pyramidal neurons: dendritic structure and synaptic integration. *Nature Reviews Neuroscience* **9**:206–221.

Steriade M, Jones EG, and McCormick DA (1997) *Thalamus*, volume I, *Organisation and Function*. Elsevier Science, Amsterdam.

Stone J (1983) *Parallel Processing in the Visual System. The Classification of Retinal Ganglion Cells and Its Impact on the Neurobiology of Vision*. Plenum Press, New York and London.

Sutter ML and Loftus WC (2003) Excitatory and inhibitory intensity tuning in auditory cortex: evidence for multiple inhibitory mechanisms. *Journal of Neurophysiology* **90**:2629–2647.

Sutter ML, Schreiner CE, McLean M, O'Connor KN, and Loftus WC (1999) Organization of inhibitory frequency receptive fields in cat primary auditory cortex. *Journal of Neurophysiology* **82**:2358–2371.

Swadlow HA (2002) Thalamocortical control of feed-forward inhibition in awake somatosensory 'barrel' cortex. *Proceedings of the Royal Society of London* **375**:1717–1727.

Swadlow HA (2003) Fast-spike interneurons and feedforward inhibition in awake sensory neocortex. *Cerebral Cortex* **13**:25–32.

Szabadics J, Varga C, Molnár G, Oláh S, Barzó P, and Tamás G (2006) Excitatory effect of GABAergic axo-axonic cells in cortical microcircuits. *Science* **311**:233–235.

Szentágothai J (1979) Local neuron circuits of the neocortex. In: Schmitt FO and Worden FG (eds). *The Neurosciences: Fourth Study Program*. MIT Press, Cambridge, pp. 399–415.

Tamás G, Buhl E, and Somogyi P (1997) Massive autaptic self-innervation of GABAergic neurons in cat visual cortex. *Journal of Neuroscience* **17**:6352–6364.

Tan AYY, Atencio CA, Polley DB, Merzenich MM, and Schreiner CE (2007) Unbalanced synaptic inhibition can create intensity-tuned auditory cortex neurons. *Neuroscience* **146**:449–462.

Thomson AM and Bannister AP (2003) Intralaminar connections in the neocortex. *Cerebral Cortex* **13**:5–14.

Tomioka R and Rockland KS (2007) Long-distance corticocortical GABAergic neurons in the adult monkey white and gray matter. *Journal of Comparative Neurology* **505**:526–538.

Velenovsky DS, Cetas JS, Price RO, Sinex DG, and McMullen NT (2003) Functional subregions in primary auditory cortex defined by thalamocortical terminal arbors: an electrophysiological and anterograde labeling study. *Journal of Neuroscience* **23**:308–316.

Verbny YL, Erdélyi F, Szabó G, and Banks MI (2006) Properties of a population of GABAergic cells in murine auditory cortex weakly excited by thalamic stimulation. *Journal of Neurophysiology* **96**:3194–3208.

Warren RA, Agmon A, and Jones EG (1994) Oscillatory synaptic interactions between ventroposterior and reticular neurons in mouse thalamus in vitro. *Journal of Neurophysiology* **72**:1993–2003.

Warren RA and Jones EG (1994) Glutamate activation of cat thalamic reticular nucleus: effects on response properties of ventroposterior neurons. *Experimental Brain Research* **100**:215–226.

Webber TJ, Green EJ, Winters RW, Schneiderman N, and McCabe PM (1999) Contribution of NMDA and non-NMDA receptors to synaptic transmission from the brachium of the inferior colliculus to the medial subdivision of the medial geniculate nucleus in the rabbit. *Experimental Brain Research* **124**:295–303.

Weinberg RJ and Kharazia VN (1996) Excitatory amino acids in thalamocortical transmission. In: Conti F and Hicks TP (eds.). *Excitatory Amino Acids and the Cerebral Cortex*. MIT Press, Cambridge, pp. 109–117.

Weinberger NM (2004) Specific long-term memory traces in primary auditory cortex. *Nature Reviews Neuroscience* **5**:279–290.

Wenk GL (1997) The nucleus basalis magnocellularis cholinergic system: one hundred years of progress. *Neurobiology of Learning and Memory* **67**:85–95.

Wenstrup JJ (2005) The tectothalamic system. In: Winer JA and Schreiner CE (eds). *The Inferior Colliculus*. Springer-Verlag, New York, pp. 200–230.

Wenstrup JJ, Larue DT, and Winer JA (1994) Projections of physiologically defined subdivisions of the inferior colliculus in the mustached bat: targets in the medial geniculate body and extrathalamic nuclei. *Journal of Comparative Neurology* **346**:207–236.

White EL and Hersch SM (1982) A quantitative study of thalamocortical and other synapses involving the apical dendrites of corticothalamic projection cells in mouse SmI cortex. *Journal of Neurocytology* **11**:137–158.

Winer JA (1984a) Anatomy of layer IV in cat primary auditory cortex (AI). *Journal of Comparative Neurology* **224**:535–567.

Winer JA (1984b) The pyramidal cells in layer III of cat primary auditory cortex (AI). *Journal of Comparative Neurology* **229**:476–496.

Winer JA (1985) Structure of layer II in cat primary auditory cortex (AI). *Journal of Comparative Neurology* **238**:10–37.

Winer JA (1992) The functional architecture of the medial geniculate body and the primary auditory cortex. In: Webster DB, Popper AN, and Fay RR (eds). *Springer Handbook of Auditory Research*, volume 1, *The Mammalian Auditory Pathway: Neuroanatomy*. Springer-Verlag, New York, pp. 222–409.

Winer JA (2006) Decoding the auditory corticofugal systems. *Hearing Research* **212**:1–8.

Winer JA, Chernock ML, Larue DT, and Cheung SW (2002) Descending projections to the inferior colliculus from the posterior thalamus and the auditory cortex in rat, cat, and monkey. *Hearing Research* **168**:181–195.

Winer JA, Diehl JJ, and Larue DT (2001) Projections of auditory cortex to the medial geniculate body of the cat. *Journal of Comparative Neurology* **430**:27–55.

Winer JA and Larue DT (1987) Patterns of reciprocity in auditory thalamocortical and corticothalamic connections: study with horseradish peroxidase and autoradiographic methods in the rat medial geniculate body. *Journal of Comparative Neurology* **257**:282–315.

Winer JA and Larue DT (1988) Anatomy of glutamic acid decarboxylase (GAD) immunoreactive neurons and axons in the rat medial geniculate body. *Journal of Comparative Neurology* **278**:47–68.

Winer JA and Larue DT (1989) Populations of GABAergic neurons and axons in layer I of rat auditory cortex. *Neuroscience* **33**:499–515.

Winer JA and Larue DT (1996) Evolution of GABAergic circuitry in the mammalian medial geniculate body. *Proceedings of the National Academy of Sciences of the United States of America* **93**:3083–3087.

Winer JA, Larue DT, Diehl JJ, and Hefti BJ (1998) Auditory cortical projections to the cat inferior colliculus. *Journal of Comparative Neurology* **400**:147–174.

Winer JA, Larue DT, and Huang CL (1999) Two systems of giant axon terminals in the cat medial geniculate body: convergence of cortical and GABAergic inputs. *Journal of Comparative Neurology* **413**:181–197.

Winer JA, Larue DT, and Pollak GD (1995) GABA and glycine in the central auditory system of the mustache bat: structural substrates for inhibitory neuronal organization. *Journal of Comparative Neurology* **355**:317–353.

Winer JA and Prieto JJ (2001) Layer V in cat primary auditory cortex (AI): cellular architecture and identification of projection neurons. *Journal of Comparative Neurology* **434**:379–412.

Winer JA, Saint Marie RL, Larue DT, and Oliver DL (1996) GABAergic feedforward projections from the inferior colliculus to the medial geniculate body. *Proceedings of the National Academy of Sciences of the United States of America* **93**:8005–8010.

Winer JA, Wenstrup JJ, and Larue DT (1992) Patterns of GABAergic immunoreactivity define subdivisions of the mustached bat's medial geniculate body. *Journal of Comparative Neurology* **319**:172–190.

Winguth SD and Winer JA (1986) Corticocortical connections of cat primary auditory cortex (AI): laminar organization and identification of supragranular neurons projecting to area AII. *Journal of Comparative Neurology* **248**:36–56.

Witter MP and Groenewegen HJ (1986) Connections of the parahippocampal cortex in the cat. III. Cortical and thalamic efferents. *Journal of Comparative Neurology* **252**:1–31.

Woodson W, Farb CR, and LeDoux JE (2000) Afferents from the auditory thalamus synapse on inhibitory interneurons in the lateral nucleus of the amygdala. *Synapse* **38**:124–137.

Xiang Z and Prince DA (2003) Heterogeneous actions of serotonin on interneurons in rat visual cortex. *Journal of Neurophysiology* **89**:1278–1287.

Yuan K, Fink KL, Schreiner CE, and Winer JA (2008) Local connection patterns of parvalbumin-positive inhibitory interneurons in rat primary auditory cortex. *Hearing Research* doi:10.1016

Zhu Y, Stornetta RL, and Zhu JJ (2004) Chandelier cells control excessive cortical excitation: characteristics of whisker-evoked synaptic responses of layer 2/3 nonpyramidal and pyramidal neurons. *Journal of Neuroscience* **24**:5101–5108.

Chapter 11

Synaptic Integration in Auditory Cortex

Michael Wehr and Raju Metherate

Abbreviations

ACh	acetylcholine
AI	primary auditory cortex
AHP	afterhyperpolarization potential
CF	characteristic frequency
EPSP	excitatory postsynaptic potentials
FM	frequency modulation
FS	fast-spiking
GABA	γ-aminobutyric acid
LFP	local field potential
mAChR	muscarinic acetylcholine receptor
MGv	ventral division of the medial geniculate body
nAChR	nicotinic acetylcholine receptor
NMDA	*N*-methyl-d-aspartate
PET	positron emission tomography
SPL	sound pressure level
STRF	spectrotemporal receptive field

1 Introduction: What Can We Learn from Synaptic Mechanisms?

What does the auditory cortex do? Most would agree that it processes auditory information, but few would assert that we understand just what computations are performed by auditory cortical neurons. If we describe computation as the transformation of information from one representation to another, then which transformations are accomplished by the auditory cortex remains an open question at the heart of the discipline.

At the level of individual cortical neurons, representation has to be understood in terms of receptive fields and stimulus selectivity. The question of what the auditory cortex does then becomes: which receptive field properties are synthesized de novo in cortex, which are enhanced by cortical processing, and which are passively inherited from subcortical inputs? For example, cortical neurons are tuned for frequency, but this tuning is inherited, at least partly, from thalamic inputs (and ultimately from the tuning of the basilar membrane and all synapses interposed between it and the auditory cortex). Nevertheless, the frequency tuning of cortical neurons may be transformed by cortical circuitry into sharper, broader, or even multipeaked tuning in different neurons. Which synaptic mechanisms and cortical circuits are involved in these various forms of spectral integration remains a focus of research, as we will see below.

If a computation is performed in the auditory cortex, then we should be able to observe the synaptic mechanisms that underlie it. Likewise, understanding the synaptic interactions will help reveal the computations being performed by the cortex. Here we use this guiding principle to explore what is known about the synaptic mechanisms underlying a range of receptive field properties. Much has been inferred about synaptic processing from extracellular single-unit recordings, but conclusive demonstrations of cellular and synaptic mechanisms generally rely on the gold standard of whole cell and sharp intracellular recording techniques, both in vivo and with in vitro methods such as the thalamocortical slice preparation. Indeed, these approaches have led to some surprising conclusions about the role of synaptic inhibition in such receptive field properties as surround suppression, forward masking, and intensity tuning, which challenge existing models of auditory processing based on extracellular recording studies (Oswald et al. 2006).

2 Spectral Processing

The most prominent feature of primary auditory cortex (AI) is its tonotopic organization, which reflects the topographically organized input of ascending auditory pathways. AI

R. Metherate (✉)
Department of Neurobiology and Behavior, Center for Hearing Research, University of California, Irvine, CA 92697, USA
e-mail: rmethera@uci.edu

J.A. Winer, C.E. Schreiner (eds.), *The Auditory Cortex*,
DOI 10.1007/978-1-4419-0074-6_11, © Springer Science+Business Media, LLC 2011

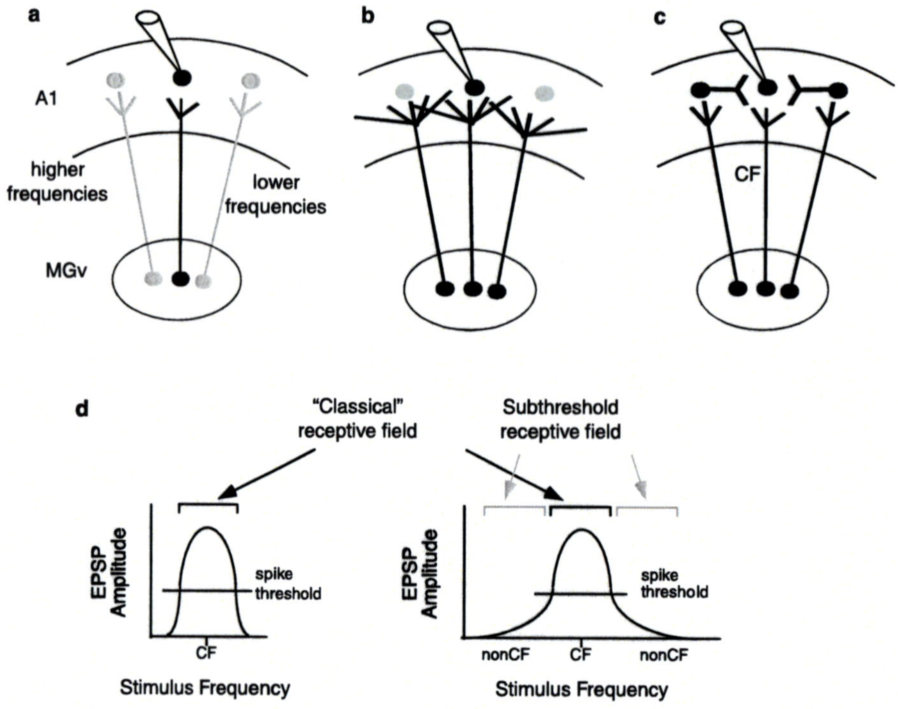

Fig. 11.1 Alternative scenarios for synaptic circuitry and the resulting (**d**) acoustic-evoked EPSP amplitudes underlying spectral integration in AI. **a** The schematic depicts restricted terminal fields of thalamocortical axons that might relay to auditory cortex information about classical (suprathreshold) receptive fields with little subthreshold information (**d**, *left*). **b**, **c** Broader divergence of thalamocortical information relay that could produce significant subthreshold receptive fields (**d**, *right*), with (**c**) or without (**b**) the contribution of long-range intracortical projections. EPSP amplitudes (**d**) are as would be recorded by the intracellular electrode depicted in **a**–**c**

neurons have narrow frequency receptive fields, which are similar in breadth to those throughout the lemniscal auditory pathway (Calford et al. 1983), suggesting, on the one hand, that cortical receptive fields are passively inherited from the auditory thalamus (Fig. 11.1a). However, other evidence suggests that cortical receptive fields actually reflect substantial intracortical processing (Fig. 11.1c). If so, then studying the cellular and synaptic bases of cortical receptive fields may reveal mechanisms for spectral processing unique to the cortex, and perhaps unique in the auditory system.

To address this issue, two basic questions arise: (1) to what extent do cortical neurons integrate inputs across sound frequencies, and (2) by what mechanism is this done? The first question has been addressed traditionally by determining the breadth (bandwidth) of frequency receptive fields or response areas using pure tone stimuli and extracellular single-unit recordings. Receptive fields so determined are similar in breadth throughout the lemniscal (primary) auditory pathway, including AI (Calford et al. 1983). These data might be interpreted as suggesting that spectral integration is performed early in the auditory pathway and the outcome simply relayed to higher centers (Fig. 11.1a). However, conventional measures of receptive field breadth only reflect the strongest inputs to a neuron, i.e., those capable of eliciting a spike (Fig. 11.1d). We will not consider here inhibitory subfields that can be detected, for example, with paired stimuli or reverse correlation techniques (Miller et al. 2002; Sutter and Loftus 2003). Receptive fields thus delineated are suprathreshold, or classical, receptive fields, but do not reveal the presence or absence of subthreshold inputs. Subthreshold receptive fields (sometimes referred to as subliminal, surround, or nonclassical receptive fields) require other techniques to be detected; most directly, intracellular recording. When subthreshold receptive fields are considered, AI spectral integration is substantially broader than previously thought (Fig. 11.1d, right).

2.1 Extent of Spectral Integration for AI Neurons

The narrow bandwidth of classical receptive fields (Calford et al. 1983) clearly underestimates the spectral breadth of inputs to cortical neurons, as evidenced by several different experimental approaches. Administration of the $GABA_A$ receptor antagonist, bicuculline, results in an expansion of the classical receptive field (Müller and Scheich 1988); Foeller et al. 2001; Wang et al. 2000, 2002), suggesting the presence of excitatory postsynaptic potentials (EPSPs) that normally are subthreshold due to inhibition. Bicuculline likely has nonspecific excitatory actions (Debarbieux et al. 1998; Kurt et al. 2006); however, such actions still should reveal only pre-existing subthreshold inputs. Direct measures of synaptic inputs by intracellular recordings reveal subthreshold receptive fields that extend well beyond the boundaries of classical receptive fields (de Ribaupierre et al. 1972; Volkov and Galazjuk 1991; Wehr and Zador 2003; Kaur et al. 2004). Optical imaging (intrinsic signal imaging or voltage-sensitive dyes) of activity shows that, while peaks of tone-evoked activity are organized tonotopically, the full extent of activity evoked by a single tone can encompass all of auditory cortex (Bakin et al. 1996; Hess and Scheich 1996; Horikawa et al. 1996). Finally, studies using tone-evoked local field potentials (LFPs), which reflect synchronous synaptic potentials in a local group of neurons, demonstrate receptive field bandwidths of several octaves at moderate intensities, considerably broader than the classical receptive fields of single-unit or multi-unit activity recorded with the same electrode (Eggermont 1996; Galvan et al. 2001; Norena and Eggermont 2002). A comparison of intracellular (synaptic) receptive fields with subsequently determined LFP receptive fields from the same cortical site shows matching bandwidths that can span five octaves or more at moderate intensities (>40 dB SPL), i.e., much broader than classical receptive fields (Kaur et al. 2004).

The breadth of intracellular and LFP receptive fields raises the intriguing possibility that some AI neurons may integrate information over much of the audible spectrum. This possibility was first raised by an extracellular, multiunit study of AI demonstrating responses to stimuli completely outside the classical receptive field (Schulze and Langner 1999). Amplitude modulated stimuli with carrier frequencies several octaves above the classical receptive field were effective in eliciting responses. Similarly, frequency modulated stimuli with spectra completely outside a neuron's classical receptive field can elicit responses (Whitfield and Evans 1965). These findings show that spectral integration in AI is more extensive than the classical receptive field might suggest. Moreover, there is evidence that spectral integration is more extensive than in subcortical auditory nuclei, since blockade of local inhibition in AI expands frequency receptive fields (see above), and similar manipulations in the cochlear nucleus or inferior colliculus produce lesser or no expansion (Palombi and Caspary 1992; Caspary et al. 1994; LeBeau et al. 2001). However, this result may only indicate that cortical and subcortical receptive fields are not similarly modulated by inhibition; in fact, intracellular recordings from inferior colliculus neurons in the awake bat demonstrate subthreshold receptive fields 1.5–2 octaves wide. Although this value is twice the breadth of classical receptive fields in the same animal, and indicates subthreshold spectral integration (Xie et al. 2007), it is still considerably narrower than in rodent AI (above). Direct comparison of subthreshold receptive fields at cortical and subcortical (especially thalamic) levels will be needed to resolve this issue.

In other sensory systems, similar differences between inhibitory control of cortical versus subcortical (thalamic) receptive fields have been observed (Sillito 1975; Dykes et al. 1984; Hicks et al. 1986; Lee et al. 1994), and intracellular recordings find extensive subthreshold cortical receptive fields (Smits et al. 1991; Bakin et al. 1996; Bringuier et al. 1999). These data imply similar integrative functions that may distinguish sensory cortex in each modality from corresponding subcortical relays.

2.2 Mechanisms of Spectral Integration: Role of Thalamocortical Input

To understand spectral integration mechanisms in AI, we must know how spectral information converges onto single neurons, i.e., which anatomical pathways contribute to a neuron's receptive field. The inputs that could contribute to the underlying circuitry can be divided into thalamocortical and intracortical pathways (the latter referring to long-distance horizontal, rather than local, pathways). To simplify examination of this problem, we can focus on the receptive field's center and edge—i.e., comparing how information about characteristic frequency (CF) and spectrally distant non-CF stimuli converge on single neurons (empirically, non-CF stimuli can be defined as ~3 octaves below CF). This large spectral distance should maximize any differences in the underlying circuitry activated by CF and non-CF stimuli. However, in reality most stimuli will activate both pathways to varying degrees. The relevant concepts and pathways are shown schematically (Fig. 11.1).

The thalamocortical contribution pathway is relatively straightforward to predict, and involves relaying information predominantly for stimuli around CF, since lemniscal thalamic neurons have narrow tuning. This discussion is focused on short-latency responses of AI neurons, since activity beyond the thalamocortical synapse obviously involves intracortical circuits. Lemniscal input to AI comes from the ventral division of the medial geniculate body (MGv) (Romanski and LeDoux 1993; Kimura et al. 2003). Combined physiological and anatomical studies show that thalamocortical projections link MGv and AI neurons with similar CFs (Imig and Morel 1984; Winer et al. 1999; Budinger et al. 2000). Thalamocortical arbors labeled by injections of tracer into small portions of MGv cover regions of cortex containing neurons with similar CFs (Velenovsky et al. 2003). Similarly, paired recordings of neurons in MGv and AI show that cells with correlated discharge have CFs within one-third of an octave (Miller et al. 2001). If, as these data imply, thalamocortical axons do not diverge to contact neurons throughout AI (shown schematically in Fig. 11.1b), but rather project to restricted portions of AI (Fig. 11.1c), then direct thalamocortical projections are not solely responsible for the broad spectral integration observed in functional studies. Rather, it seems that thalamocortical inputs directly mediate only cortical responses to CF and near-CF stimuli.

2.3 Spectral Processing by Intracortical Pathways

A more difficult problem is to understand how AI neurons respond to spectrally distant (non-CF) stimuli. We consider three possibilities. First, it is unlikely that the same thalamocortical inputs carry complete information about CF and non-CF stimuli, since classical MGv receptive fields are narrower than subthreshold AI receptive fields (Calford et al. 1983; Kaur et al. 2004). Only classical—suprathreshold—MGv receptive fields are relevant, since an MGv spike is required to produce an EPSP in AI. Moreover, stimuli several octaves from CF are likely to elicit a spike in MGv neurons only at long latencies. There are no published data for MGv, but examples from AI of single-unit responses to CF and non-CF stimuli show approximate increases in minimum spike latency of 12–60 ms/octave (Brugge et al. 1969; Phillips and Hall 1992; Eggermont 1996). In contrast, AI synaptic onset latencies increase with spectral distance from CF at the rate of only a few ms/octave (Kaur et al. 2004). Thus, the earliest synaptic responses in AI to non-CF stimuli occur before direct-projecting MGv neurons spike in response to the same stimulus.

A second possibility is that thalamocortical neurons have very broad terminal arbors (shown schematically in Fig. 11.1b), endowing AI neurons with a broad range of convergent thalamocortical CFs from MGv neurons. This possibility is unlikely given the physiological and anatomical studies described above and the additional findings described next.

The third and most likely possibility is that AI spectral integration involves intracortical horizontal pathways (Fig. 11.1c). Intracortical injection of the $GABA_A$ receptor agonist, muscimol, inhibited cortical neurons (but not thalamocortical axons), and the effect on cortical responses to CF and non-CF stimuli was assessed (Kaur et al. 2004). For CF stimuli, muscimol partly suppressed initial response components (the first ~10 ms) and fully suppressed longer-latency response components, consistent with inhibition of cortical neurons but not thalamocortical inputs. In contrast, for non-CF stimuli, muscimol sometimes completely suppressed both initial and longer-latency response components (reducing receptive field bandwidth by ~2 octaves at 20 dB above CF threshold), suggesting a major involvement of intracortical pathways. In a second manipulation, local antagonism of inhibition at the recording electrode (via drug ejected from the recording pipet) preferentially reversed inhibition of responses to CF stimuli over responses to non-CF stimuli. Thus, it appears that direct thalamocortical inputs preferentially contribute to the response to CF stimuli, whereas responses to non-CF stimuli mainly involve intracortical horizontal projections from neurons with spectrally distant CFs.

The schematic (Fig. 11.1c), which implies a clear dissociation between pathways relaying information about CF versus non-CF stimuli, is simplistic. Such a complete dissociation is unlikely, and the figure suggests a concept rather than specific circuitry.

2.4 Spectral Integration: Possible Mechanisms and Functions

The anatomical pathways described in the previous sections enable spectral information to converge upon single AI neurons. How that information is used reflects cellular and synaptic processes, such as the spatial and temporal integration of excitatory and inhibitory inputs in dendrites. This section describes cellular mechanisms that likely contribute to spectral processing.

The auditory system is capable of precisely timed responses to acoustic stimuli. Although most attention has been on specialized auditory brain stem synapses (Oertel 1999; Trussell 1999), cells in higher auditory centers also respond to stimuli with precision. AI neurons respond strongly at stimulus onset, and to spectral transitions in complex stimuli (Phillips and Hall 1990; Heil and Irvine 1997; Elhilali et al. 2004). AI spikes can be as tightly locked to stimulus onset as spikes in cochlear nerve fibers, with spike latency jitter often <1 ms (Phillips and Hall 1990; Heil and Irvine 1997). Even responses to spectrotemporal transitions within complex stimuli often have <10 ms jitter (Elhilali et al. 2004). Cortical neurons respond with great reliability and precision to thalamic inputs, due, at least in part, to specializations at the thalamocortical synapse (Rose and Metherate 2005). Moreover, thalamocortical EPSPs in inhibitory fast-spiking (FS) cells have faster rise times and shorter peak latencies than thalamocortical EPSPs in excitatory cells (Rose and Metherate 2005), suggesting that some inhibitory neurons mediate fast, disynaptic inhibition to limit thalamocortical excitation. Other inhibitory neurons are not driven strongly by thalamocortical inputs (Verbny et al. 2006). Thus, fast, feed-forward inhibition could enhance the precision of initial spike timing (see next section) (Bao et al. 2003; Wehr and Zador 2003; Tan et al. 2004).

Precise timing of synaptic onsets in cortex may be important for spectral integration. For a given cell, the synaptic response latency to a pure tone increases with spectral distance from CF (Kaur et al. 2004). Although a similar phenomenon is seen with spike latency throughout the auditory pathway (Hind et al. 1963; Brugge et al. 1969; Kitzes et al. 1978; Phillips and Hall 1992; Eggermont 1996), the changes in cochlear nucleus spike latency with spectral distance from CF (~8–15 ms/octave at 50–60 dB above threshold) (Kitzes et al., 1978) are greater than changes in AI synaptic latencies, as reflected in LFPs or intracellular recordings (~1 or 4 ms/octave, respectively, for 50 or 10 dB above threshold) (Kaur et al. 2004). Thus, as noted above, the earliest cortical responses to non-CF stimuli occur before direct-projecting MGv neurons spike in response to the same stimulus.

The precise synaptic delays in AI raise the question as to their purpose. Perhaps the circuitry and mechanisms that establish systematic changes in onset latency determine optimal spectral integration in AI neurons. More specifically, summation of EPSPs elicited by tones of different frequency may be enhanced when those tones are presented asynchronously. Maximal summation would occur with optimal staggering of tone onsets, approaching, in the limit, a frequency-modulated (FM) sweep. If so, then the optimal stimulus would be an FM sweep moving toward the CF, from either higher or lower non-CFs, at the specific sweep rate that produces maximal summation of EPSPs. This potential mechanism for sensitivity to optimal FM stimuli has been proposed in extracellular single-unit studies (Phillips et al. 1985; Heil et al. 1992). Similar requirements for optimal integration may underlie the finding that some AI neurons respond more strongly to FM stimuli than to pure tones (Whitfield and Evans 1965; Tian and Rauschecker2004), or prefer specific kinds of FM stimuli with faster or slower rates, or with rising or falling frequencies (Mendelson and Cynader 1985; Nelken and Versnel 2000; Ohl et al. 2000; Zhang et al. 2003). Preferences of individual neurons for particular stimulus configurations may also be dictated by local inhibitory circuits (Zhang et al. 2003), as well as intrinsic (both active and passive) membrane mechanisms.

The mechanism described here for spectral integration could apply equally to any stimulus feature mapped across AI that is integrated in single neurons via converging, intracortical pathways (Schreiner et al. 2000). Such notions remain largely untested.

3 Temporal Processing

Time is inherently important in the processing and perception of sound. Although temporal coding and phase locking are seen in the periphery, these aspects of temporal processing are progressively transformed along the auditory hierarchy. In AI, at least three related aspects of temporal processing appear to be transformed compared to subcortical sites. For one tone or click, responses are typically transient, consisting of one or a few precisely timed and reliable spikes. For pairs of tones, the response to the second tone is usually suppressed, but can also be facilitated. Finally, for multiple tones presented in a train, cortical neurons can phase lock up only to ~5–15 Hz. In each case the temporal response properties differ from those at thalamic and lower levels. Synaptic inhibition, depression, and facilitation each appear to play distinct roles in these phenomena.

In anesthetized animals, AI neurons respond transiently only at tone onset, and for some neurons, also at the offset (Brugge et al. 1969; DeWeese et al. 2003). In awake animals, a subset of neurons show sustained or phasic-tonic firing patterns for optimal stimuli, but a substantial proportion of neurons only respond transiently (Chimoto et al. 2002; Wang et al. 2005; Ter-Mikaelian et al. 2007). The transient responses typically consist of one or a few spikes, show high temporal precision, and can be highly reliable across trials (DeWeese et al. 2003). In contrast, subcortical cells in the auditory nerve or inferior colliculus can show sustained firing throughout a stimulus, even in the anesthetized animal (Pickles 1988). Investigations of the cellular and synaptic mechanisms underlying these temporal response properties have been limited to anesthetized animals and the slice preparation, and have therefore focused mainly on transient onset responses. Using these methods (see Box 11.1), the synaptic mechanisms for these transient, temporally precise, reliable spiking responses can now be understood.

Box 11.1 Disentangling excitation and inhibition

In an ideal world, pure excitation and pure inhibition could be represented as depolarizations and hyperpolarizations in the membrane potential. Unfortunately, excitation and inhibition almost invariably occur together, and the resulting sum is a voltage-dependent mixture of the two. A method often used to dissect excitation and inhibition is whole cell recording in voltage clamp mode, where voltage clamping a cell to different holding potentials alters the driving force of excitatory and inhibitory synaptic channels. When a neuron is clamped to 0 mV (the reversal potential for excitatory synaptic currents), no excitatory synaptic current flows, and any synaptic currents will be exclusively inhibitory (*gray trace*). Likewise, when a neuron is clamped to the reversal potential for inhibition, only excitatory synaptic currents will be measured (*black trace*). This method for disentangling excitation and inhibition in vivo was pioneered in visual cortex (Borg-Graham et al. 1998) and has been extended to somatic sensory and auditory cortex. In practice, the decomposition is improved by using 3–5 holding potentials, and linear regression to extract excitatory and inhibitory conductances. Unlike the currents shown here, conductances (measured in nS) are always non-negative, so tone-evoked excitatory and inhibitory conductances are both positive deflections (see Fig. 11.2). The isolation of synaptic currents can be improved by blocking voltage-dependent currents in the recorded neuron. A major limitation is that the method depends on assumptions about linearity and isopotentiality, which are clearly violated by cortical neurons. Fortunately, the errors introduced by violating these assumptions are well understood, and the compounds used to block voltage dependent channels render cortical neurons surprisingly linear. Most such errors can be avoided by designing experiments to compare excitation and inhibition evoked by different stimuli in a given neuron, such that any errors apply equally across responses.

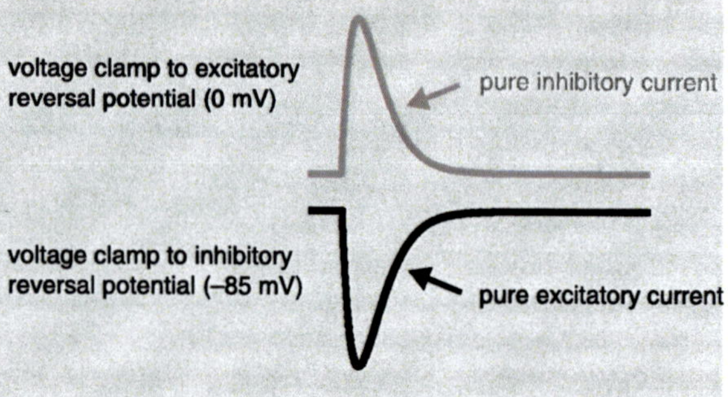

The main contribution to transient firing in auditory cortex is a brief delay between excitatory and inhibitory inputs (Fig. 11.2). Tones and noise bursts evoke a stereotyped sequence of synaptic inputs: an initial volley of excitatory inputs is followed after a brief delay by a precisely timed volley of inhibition (Wehr and Zador 2003; Tan et al. 2004). This ~3 ms average delay forms a precise window in which pure excitation can elicit one or at most a few action potentials, before the spiking response is quenched by the wave of inhibition. The synaptic conductances of both excitation and inhibition typically last 50–100 ms, far outlasting the spiking response, but the inhibition prevents firing after the first few milliseconds. Because of the tonotopic organization of auditory cortex, tones evoke a similar sequence in most neighboring neurons, suggesting that the local circuit will experience a relatively synchronized firing event. Thus a substantial proportion of synapses will simultaneously enter synaptic depression. This event is followed by a silent period (enforced by inhibition) in which many synapses will emerge from synaptic depression together. The network will therefore be poised to respond synchronously if another stimulus should arrive 200–300 ms after the previous tone. Evidence for such brief synchronous volleys can be seen in the highly non-Gaussian membrane potential dynamics of auditory cortical neurons, which consist of occasional large excursions (bumps) separated by quiescent periods (DeWeese and Zador 2006). This synchronous response of a large proportion of neurons in a recurrently connected network might enhance the temporal precision of the spiking response. Thus the transient and temporally precise spiking responses of auditory cortical neurons are shaped not only by the sequence of excitation and inhibition, but also by the consequences of that sequence for synaptic depression in the surrounding circuits.

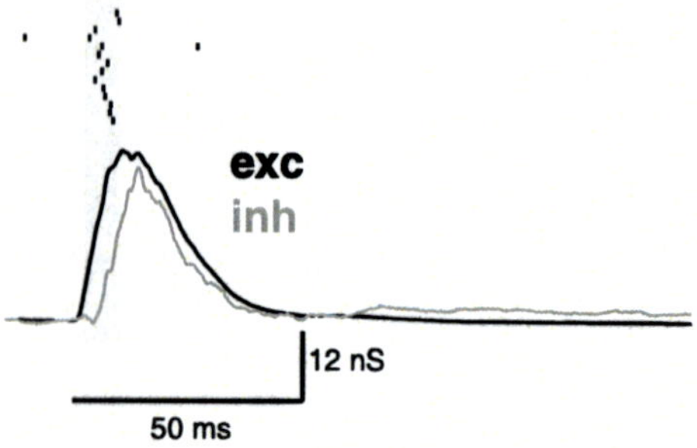

Fig. 11.2 Delayed inhibition enforces precise, transient spiking responses. Tone-evoked excitatory (*black*) and inhibitory (*gray*) conductances in an auditory cortical neuron. This neuron reliably fired one spike on each trial, which coincided with the brief window of pure excitation

When two tones are presented, depending on the temporal separation, the first tone can have a profound impact on the response to the second. This interaction can be facilitative or suppressive. A reduced response to the second tone is forward suppression, or forward masking (by analogy with psychoacoustics). Forward suppression is more pronounced and longer lasting in AI than in the MGv (Wehr and Zador 2005), suggesting a contribution by some cortical mechanism. Long-lasting synaptic inhibition has been proposed as the mechanism likely underlying forward suppression (Fig. 11.3a). Synaptic depression, however, might also influence forward suppression (Eggermont 1999; Denham and Denham 2001). In this model, excitatory and inhibitory conductances decay together with a similar time course (Fig. 11.3a), but synapses remain depressed long after the conductances have decayed. Using whole cell methods to tease apart excitation and inhibition can distinguish

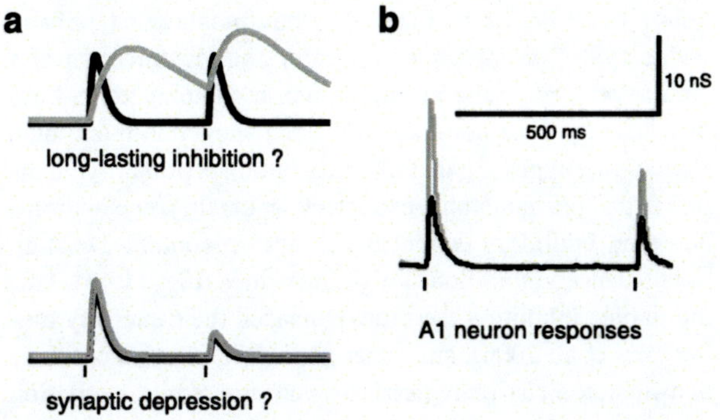

Fig. 11.3 Forward suppression is not due to long-lasting synaptic inhibition. **a** Two models for the synaptic mechanisms underlying forward suppression. *Top*: Long-lasting inhibition (*gray trace*) could overlap with the excitation (*black trace*) evoked by a second sound, and suppress the spiking response. *Bottom*: Alternatively, excitation and inhibition could decay with the same time course, but synaptic depression would reduce both conductances evoked by the second sound. **b** Excitatory and inhibitory conductances in an AI neuron, evoked by a pair of clicks separated by 512 ms (clicks indicated by *black tick marks*). The data are not consistent with long-lasting inhibition accounting for forward suppression

between these two possibilities. These methods revealed that inhibitory conductances rarely last longer than 50–100 ms, whereas spiking responses and synaptic currents remain suppressed for several hundred milliseconds (Fig. 11.3b). Thus, synaptic inhibition does not contribute forward suppression beyond 50–100 ms after stimulus onset. Synaptic depression is the likely mechanism for suppression at intervals beyond 100 ms (Wehr and Zador 2005). Although this has not been demonstrated directly in auditory cortex, synaptic depression underlies similar forms of adaptation in somatic sensory (Chung et al. 2002) and visual cortex (Carandini et al. 2002; Freeman et al. 2002). At short intervals, forward suppression reflects a mixture of several mechanisms (synaptic inhibition, synaptic depression, and inherited subcortical suppression), whereas for longer intervals synaptic depression is likely to dominate.

Facilitation is also seen for pair of tones (Brosch et al. 1999; Brosch and Schreiner 2000; Wehr and Zador 2005), but the synaptic mechanisms have not been conclusively demonstrated. Synaptic facilitation likely plays a major role, though other circuit mechanisms may also contribute. For example, auditory cortex inhibitory synapses are relatively stronger than excitatory synapses (Metherate and Ashe 1994), causing a net increase in excitation relative to inhibition for the second tone, increasing the evoked response. Indeed, some cells have an increased ratio of excitation to inhibition for the second tone (Wehr et al. 2005), suggesting that this mechanism may contribute to facilitation. Intrinsic properties such as voltage dependent channels, e.g., *N*-methyl-d-aspartate (NMDA) receptors, could enhance the facilitation of spiking responses. Facilitation is probably a combination of some or all of these cellular, synaptic, and circuit mechanisms.

The mechanisms underlying the responses to trains of sounds can be largely understood from those involved in the responses to single tones or pairs of tones. Cortical neurons can follow temporal modulations or tone trains with repetition rates up to ~5–15 Hz (Eggermont 1999), far slower than seen in the thalamus (up to 100 Hz; Creutzfeldt et al. 1980) or inferior colliculus (up to several hundred Hz; Langner and Schreiner 1988; Ter-Mikaelian et al. 2007). This cortical cutoff rate is well matched to the time courses of the mechanisms underlying forward suppression: synaptic inhibition (which lasts 50–100 ms, preventing neurons from following beyond 10–20 Hz) and synaptic depression (lasting several hundred ms, reducing steady-state response amplitudes for rates as slow as 1–2 Hz (Ulanovsky et al. 2004)). The role of synaptic facilitation in steady-state responses to tone or click trains is uncertain. A subset of neurons or recording sites in cat and rat AI shows enhanced steady-state responses in the 8–16 Hz range (Eggermont 1999; Kilgard and Merzenich 1999), consistent with the enhancement of synaptic currents in a subset of neurons for click pairs separated by ~100 ms (Wehr and Zador 2005). Synaptic facilitation, intrinsic properties, or preferential depression of inhibitory synapses could explain this enhancement, although this remains to be demonstrated.

Others find, however, that steady-state responses to tone trains were invariably suppressed in cat AI (Phillips et al. 1989). Perhaps this discrepancy reflects that neurons facilitated to a pair of tones are expected to show an enhanced response for the first few tones in a train, an enhancement that could rapidly give way to suppression. Whether they show facilitation or suppression would therefore depend critically on which tone responses were used to measure steady state. That is, the third response in a train might provide a different view of steady state than later responses. Because synaptic facilitation and depression are each complex phenomena with multiple mechanisms, the rapid components of synaptic facilitation may initially prevail over depression at these synapses, but would eventually be overwhelmed by slower but stronger components of depression such as vesicle depletion.

4 Intensity Processing

Sound intensity can provide important information about the identity and location of sound sources, and conveys prosodic information in speech. Representation of sound intensity may involve intensity-tuned neurons, whose firing rates vary non-monotonically with intensity. Because auditory nerve fibers have exclusively monotonic rate-level functions, it has long been argued that intensity tuning must be created by central inhibitory circuitry. But because both monotonic and non-monotonic neurons are found throughout the central auditory system, including the cochlear nuclei (Rhode and Smith 1986), inferior colliculus (Sivaramakrishnan et al. 2004), thalamus (Aitkin and Webster 1972), and cortex (Sutter and Schreiner 1995), it is unclear whether intensity tuning is created at an early stage and then passively inherited by higher structures, or whether successive structures actively enhance or even synthesize intensity tuning de novo. Several studies have addressed this issue using whole cell voltage clamp methods in auditory cortex, finding that all three possibilities can occur.

In monotonic cells, the magnitudes of both excitatory and inhibitory conductances are monotonic, and well correlated with each other (Wehr and Zador 2003; Tan et al. 2004). In contrast, three different patterns of excitation and inhibition have been seen in non-monotonic neurons (Fig. 11.4). In some, the magnitudes of excitatory and inhibitory conductances are non-monotonic themselves, suggesting that these neurons passively inherit their intensity tuning from presynaptic inputs (Fig. 11.4a) (Tan et al. 2007; Wehr and Zador 2003; Wu et al. 2006). These non-monotonic presynaptic inputs could be thalamocortical afferents (from non-monotonic thalamic neurons), or they could be intracortical inputs from other non-monotonic cortical neurons.

Different patterns of excitation and inhibition have also been demonstrated in non-monotonic neurons, in which unbalanced inhibition enhances or even creates intensity tuning (Wu et al. 2006; Tan et al. 2007). A major obstacle in obtaining whole cell recordings from non-monotonic cells is their sparseness in rat primary auditory cortex. In cat AI, ~25% of cells are strongly non-monotonic, whereas in rat AI only 5% are (Sutter and Schreiner 1995; Wu et al. 2006). In cat PAF, 80% of cells are non-monotonic (Heil and Irvine 1998), prompting a search for and identification of a similar posterior zone in rat, the non-monotonic zone, 80% of whose cells are non-monotonic and show non-monotonic excitation, suggesting that intensity tuning is primarily inherited from non-monotonic excitatory inputs. In some neurons, however, inhibitory conductances were monotonic, leading to unbalanced inhibition at high intensities (Fig. 11.4b). This monotonic inhibition therefore enhanced their intensity tuning (Wu et al. 2006), and synaptic inhibition enhances their non-monotonicity compared to non-monotonic excitation alone.

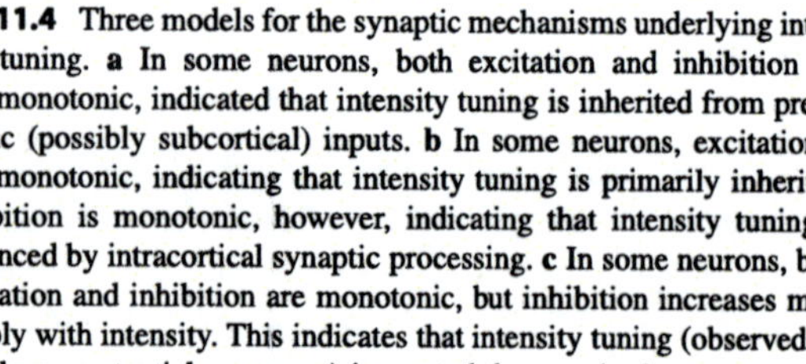

Fig. 11.4 Three models for the synaptic mechanisms underlying intensity tuning. **a** In some neurons, both excitation and inhibition are non-monotonic, indicated that intensity tuning is inherited from presynaptic (possibly subcortical) inputs. **b** In some neurons, excitation is non-monotonic, indicating that intensity tuning is primarily inherited. Inhibition is monotonic, however, indicating that intensity tuning is enhanced by intracortical synaptic processing. **c** In some neurons, both excitation and inhibition are monotonic, but inhibition increases more steeply with intensity. This indicates that intensity tuning (observed for membrane potential responses) is created de novo in these neurons by cortical synaptic interactions

Spiking responses are shaped not just by the magnitudes of excitation and inhibition, but by their relative timing. In monotonic neurons, the spiking response is limited by the brief delay between excitation and inhibition, which is typically stable across frequencies and intensities. In non-monotonic rat cells, however, this delay was intensity-tuned, so that at higher intensities, inhibition arrived earlier. Both the monotonically increasing magnitude and the earlier arrival of inhibition for loud intensities enhance the intensity tuning of these neurons (Wu et al. 2006).

A different approach to overcoming the sparseness of non-monotonic cells in AI (Tan et al. 2007) uses a paradigm in which rats were trained on a fine intensity discrimination task to increase the number of intensity-tuned cells (Polley et al. 2004, 2006). In whole cell recordings from these rats, non-monotonic cells were found (Tan et al. 2007), as in other studies (Wu et al. 2006), and monotonic inhibition enhances the intensity tuning in neurons which inherit non-monotonic excitation. However, the present study also observed non-monotonic neurons in which both excitation and inhibition were monotonic. In these cells, although both excitatory and inhibitory magnitudes rose monotonically with intensity, the magnitude of inhibition rose faster (Fig. 11.4c). At the highest intensities, inhibition was unbalanced, creating intensity tuning anew in these cells (Tan et al. 2007). Intensity tuning in cortical cells can be passively inherited from non-monotonic inputs, enhanced by the tuning and timing of cortical inhibition, or even synthesized de novo from monotonic inputs. Such neurons were found in both trained and untrained animals, suggesting that they are not an artifact of training. The increase of intensity-tuned neurons in trained animals, moreover, suggests that an alteration of the balance of excitation and inhibition in cortical neurons may subserve plasticity in fine intensity-discrimination tasks (Tan et al. 2007).

5 Spectrotemporal Interactions

We have discussed spectral and temporal processing separately, tacitly assuming that the complete spectrotemporal response properties of cortical neurons can be understood simply by combining these two properties. While this assumption is true to some extent, it remains a major challenge to characterize how and under what conditions it breaks down. The synaptic mechanisms underlying spectrotemporal interactions pose a special challenge given the vast combinatorial space of spectrotemporal stimuli. Nevertheless, some key advances have been made towards understanding the synaptic mechanisms underlying simple spectrotemporal stimuli such as two-tone combinations and frequency modulated (FM) sweeps.

Spectrotemporal response properties can be characterized by the spectrotemporal receptive field (STRF), which is estimated using a reverse correlation approach. The STRF provides a linear estimate of the spectrotemporal response properties of a neuron. Most cortical STRFs are spectrotemporally separable, which means that their spectral and temporal response properties can be considered independently without a loss of predictive value (Linden et al. 2003). However, a large minority of STRFs is spectrotemporally inseparable, for example, those with a slanted spectrotemporal structure which correspond to FM sweep direction selectivity. STRFs computed from the subthreshold membrane potential of cortical neurons are qualitatively similar to those computed from spiking responses (Machens et al. 2004). However, a given neuron responding to different stimulus sets can generate different STRFs, and for either spikes or membrane potential responses, these STRFs typically poorly predict the responses to spectrotemporally complex stimuli such as natural sounds (Linden et al. 2003; Machens et al. 2004). This implies a substantial degree of nonlinear spectrotemporal interaction in cortical neurons.

Perhaps the simplest and most widely used method to assess spectrotemporal interactions is the use of two successive tones. These elicit various forms of side band suppression and facilitation of spiking responses (Calford and Semple 1995; Brosch and Schreiner 1997, 2000; Sutter et al. 1999). The synaptic mechanisms of these two-tone interactions have not yet been investigated in detail, but are likely similar to those underlying forward suppression (which have been studied with pairs of clicks, but not tones of different frequencies). Sideband suppression is often described in terms of the inhibitory receptive field, or as lateral inhibition, perhaps as synaptic lateral inhibition from inhibitory interneurons (Calford and Semple 1995; Ojima and Murakami 2002; Zhang et al. 2003; Tan et al. 2004). However, the finding that tone-evoked excitatory and inhibitory inputs are co-tuned for frequency (Wehr and Zador 2003) suggests that synaptic lateral inhibition is unlikely to account for side-band suppression. Instead, tones outside the spiking receptive field (but within the suppressive sideband) evoke a balanced mixture of subthreshold excitation and inhibition. Even though these responses are subthreshold, they still result from presynaptic release events, which will engage synaptic depression lasting hundreds of milliseconds, and which would also be expected to contribute to suppression for 50–100 ms. In this view, the prime cause for the amount of sideband suppression from cortical circuitry is the degree of overlap in the set of synapses activated by two different tones. Additional sideband suppression may be inherited from subcortical response properties.

FM sweeps can be thought of as a continuous extension of two-tone stimuli. Many auditory cortical neurons show FM sweep selectivity, preferring upward or downward sweeps. In rats this may be topographic, such that low-CF neurons prefer upward sweeps, and high-CF neurons prefer downward sweeps (Zhang et al. 2003). Whole cell recordings show sweep selectivity in both excitatory and inhibitory inputs, suggesting that it is inherited from presynaptic (possible thalamocortical) inputs (Zhang et al. 2003). Indeed, thalamic neurons possess sweep selectivity (Lui and Mendelson 2003). However, synaptic interactions do appear to enhance sweep selectivity in cortical neurons (Zhang et al. 2003). The synaptic mechanisms underlying this enhancement can be interpreted in two ways. First, inhibition is delayed relative to excitation, as previously mentioned. Second, suppressive side bands are often asymmetric. In rats this asymmetry is topographic, such that suppressive sidebands of low-CF neurons tend to be on the high frequency side of their receptive fields, whereas for high-CF neurons they are on the low frequency side. For a low-CF neuron, an upward sweep will first produce excitation, which can evoke spikes before the onset of inhibition. In contrast, a downward sweep will first traverse the suppressive sideband, producing lasting forward suppression that prevents the spikes that would otherwise be evoked as the sweep continues into the receptive field center. The neuron thus spikes in response to upward, but not downward, sweeps. Synaptic inhibition and synaptic depression can again be invoked, just as in forward suppression. In addition, tonal receptive fields are asymmetric, such that low frequencies evoked stronger currents in low-CF neurons than higher frequencies (and vice versa for high-CF neurons). This contribution of synaptic inhibition to direction selectivity can be seen directly from the relative timing of excitation and inhibition evoked by sweeps: for preferred sweeps, strong excitation leads inhibition and evokes spiking, but for non-preferred sweeps, weak inhibition overlaps with excitation and prevents spiking (Zhang et al. 2003).

6 Neuromodulation of Cortical Processing

AI processing is regulated dynamically and changes in behavioral state (e.g., sleep versus waking versus attention) alter receptive field shape and the magnitude of evoked responses (Edeline et al. 2001). Behavioral training can alter receptive fields (Weinberger 2004), and single-unit recordings demonstrate rapid and reversible receptive field changes during learned behavior (Fritz et al. 2003). Rapid changes in receptive fields are likely to be caused by physiological mechanisms, such as transient neuromodulation, which may contribute to longer-lasting plasticity underlying large scale tonotopic map reorganization (Recanzone et al. 1993; Kilgard and Merzenich 1998). Thus, spectral integration (as reflected in receptive fields) is regulated by neuromodulation, perhaps continuously, and on both fast and long-lasting time scales.

Mechanisms of neuromodulation in AI are diverse, but poorly understood, with most attention on the neuromodulators, acetylcholine (ACh), and norepinephrine. We focus on ACh, a neuromodulator with roles in arousal, attention, and learning (Hasselmo 1999; Sarter et al. 2001; Weinberger 2004). Cholinergic actions at both major ACh receptors—nicotinic (nAChR) and muscarinic (mAChR)—influence AI spectral processing, including specific and differential actions on thalamocortical and intracortical transmission, which relates to earlier points (Fig. 11.1). The roles of norepinephrine, dopamine, and serotonin in auditory cortex are considered elsewhere (Stark and Scheich 1997; Bao et al. 2003; Edeline 2003; Atzori et al. 2005; Ji and Suga 2007).

Many in vivo studies of cholinergic modulation in sensory cortex found that mAChRs enhance responsiveness to afferent inputs (Sillito and Kemp 1983; Metherate et al. 1988; McKenna et al. 1989). This results, in part, from increased postsynaptic membrane resistance due to decreased conductance of several K^+ channels (Krnjevic et al. 1971; Halliwell and Adams 1982; McCormick and Prince 1986; Madison et al. 1987) and/or activation of a nonselective cation current (Haj-Dahmane and Andrade 1996). Activation of intrinsic cholinergic synapses in brain slices containing auditory cortex also increases postsynaptic excitability via slow depolarization, increased membrane input resistance, and decreased afterhyperpolarization potentials (AHPs) (Metherate et al. 1992; Cox et al. 1994). Electrical or chemical stimulation in vivo of cortically projecting basal forebrain (nucleus basalis) neurons enhances auditory cortex EPSPs evoked by electrical MGv stimulation (Metherate et al. 1992, Metherate and Ashe 1993). Basal forebrain stimulation similarly enhances responses to acoustic stimuli (Edeline et al. 1994). All of these actions involve mAChRs and demonstrate functional implications of cellular muscarinic actions.

Muscarinic modulation in the auditory thalamocortical slice allows selective activation of thalamocortical versus long-distance intracortical pathways (Hsieh et al. 2000). The cholinergic agonist, carbachol, suppressed intracortical EPSPs with lesser or no effect on thalamic-evoked EPSPs. Atropine blocked the effect implicating mAChRs. These effects are seen at the cell's resting membrane potential, so that voltage-dependent postsynaptic actions of ACh are minimal. The reduced EPSP amplitudes likely reflect reduction of neurotransmitter release by presynaptic mAChRs (Hounsgaard 1978; Valentino and Dingledine 1981; Segal 1982, 1989), and the differential effects on intracortical versus thalamocortical EPSPs may reflect the preferential distribution of presynaptic mAChRs on intracortical, but not thalamocortical, synapses (Sahin et al. 1992). The effects of mAChR activation on auditory cortex intracortical versus thalamocortical pathways resemble those reported for intrinsic versus extrinsic afferents to entorhinal and perirhinal cortex, and hippocampus (Hasselmo 2006).

ACh also acts at nAChRs, which are widely distributed in the auditory system (Morley and Happe 2000). Two main functions of nAChRs in sensory cortex are proposed: regulation of thalamocortical transmission and postsynaptic excitation of GABAergic interneurons (Metherate 2004). We relate cholinergic actions to mechanisms of spectral integration (Fig. 11.1c) by considering nicotinic regulation of thalamocortical transmission.

The distribution of nAChRs in sensory cortex varies among species, but supports the widely held hypothesis that nAChRs regulate thalamocortical transmission (Clarke 2004). Cat and rat show dense, high-affinity [^{3}H]nicotine binding (nicotine binding) in layers 3–4 where thalamocortical input terminates (and also layers 1 and 5/6) (Clarke et al. 1984, 1985; Prusky et al. 1987; Parkinson et al. 1988; Sahin et al. 1992; Lavine et al. 1997). High-affinity nicotine binding reflects nAChRs containing α4 and β2 subunits (Couturier et al. 1990; Séguéla et al. 1993). The middle layer nicotine binding is reduced by thalamic lesions, but not by excitotoxic cortical lesions that spare axons and terminals. Thus, it appears that α4- and β2-containing nAChRs in cat and rat sensory cortex are associated with thalamocortical terminals. In mouse, however, middle layer nAChRs are less prominent, but are dense in layers 1 and 5/6, and in sensory thalamus (Rogers et al. 1998; Zoli et al. 1998). Moreover, immunostaining for β2-containing nAChRs is found in the mouse thalamocortical pathway (unpublished observations), reinforcing recent positron emission tomography (PET) and radioligand binding studies in humans, primates and rats showing nicotinic ligand binding along thalamocortical pathways (Ding et al. 2004; Chattopadhyay et al. 2005; Easwaramoorthy et al. 2007). Thalamocortical transmission in cat and rat may be regulated by nAChRs located at (presynaptic) or near (preterminal) thalamocortical terminals, a mechanism supported by functional studies (Gil et al. 1997; Lambe et al. 2003; Clarke 2004). In mouse, rat, primates, and humans nAChRs associated with thalamocortical axons suggest an additional mechanism that may involve regulation of axon excitability, since activation of nAChRs in the mouse auditory thalamocortical pathway, in vitro and

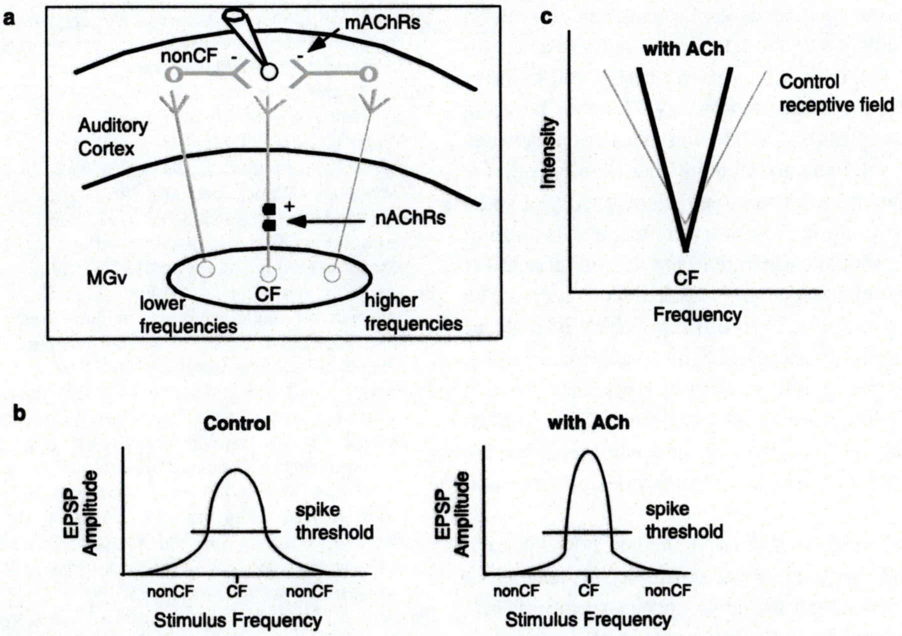

Fig. 11.5 Hypothesized effect of ACh on frequency receptive fields due to (**a**): (i) muscarinic enhancement of postsynaptic excitability, (ii) muscarinic presynaptic reduction of intracortical transmission, and (iii) nicotinic enhancement of thalamocortical transmission. The resulting combined effects on tone-evoked EPSPs (**b**) would reduce receptive field breadth, lower CF threshold, and enhance responses to stimuli within the sharpened receptive field (**c**)

in vivo, enhances thalamocortical transmission (Kawai et al. 2007).

These studies suggest that mAChR activation would enhance postsynaptic excitability, while decreasing intracortical transmission via presynaptic receptors, whereas nAChR activation of would enhance thalamocortical transmission. Thus, the combined action of ACh at mAChRs and nAChRs might enhance thalamocortical transmission, while suppressing intracortical excitatory transmission. Given the hypothesized contribution of thalamocortical and intracortical inputs to frequency receptive fields (Fig. 11.1c), the integrated actions of ACh would produce (Fig. 11.5): muscarinic suppression of responses to non-CF stimuli (mediated by intracortical horizontal inputs), nicotinic facilitation of responses to CF stimuli (mediated by thalamocortical inputs), and muscarinic enhancement of responsiveness to remaining inputs (from enhanced postsynaptic excitability). The net effect could be to reduce receptive field breadth, lower the threshold to CF stimuli, and enhance responses to stimuli within the narrowed receptive field. These actions of ACh have been modeled quantitatively (Soto et al. 2006) and shown to be consistent with rapid changes in receptive fields observed during attentive behavior (Fritz et al. 2003). Similar narrowing of frequency tuning in auditory cortex occurs during selective attention in humans (Okamoto et al. 2007). As these examples suggest, understanding the cellular actions of ACh can promote a deeper understanding of sensory processing in AI during specific behaviors.

7 Future Directions

The studies reviewed suggest that many basic synaptic and cellular mechanisms in auditory cortex have been identified. For the most part, these mechanisms resemble those in other cortical areas, with little evidence of the striking synaptic and cellular specializations that characterize the lower parts of the auditory pathways (Oertel 1999; Trussell 2002). Even for mechanisms first identified in auditory cortex (Atzori et al. 2001; Kawai et al. 2007), there is little to suggest that they are unique to auditory cortex and there is evidence to the contrary. Future studies increasingly will turn to understanding how basic cellular and synaptic mechanisms are engaged, potentially in novel ways, in the service of processing acoustic information in auditory cortex.

With more studies turning to vivo intracellular approaches, the thalamocortical slice preparation, techniques for complex stimulus generation and analysis, and physiological recordings in behaving animals, other basic mechanisms will likely emerge. However, these approaches have so far been limited to fairly simplistic distinctions such as categorizing synaptic inputs as excitatory or inhibitory; this important advance ignores the impressive diversity of both excitatory and inhibitory auditory cortex cell types as defined by firing properties, morphology, laminar position, and molecular expression. While the distinction between thalamocortical and intracortical inputs has been useful, it is an early step towards understanding the complexity of cortical circuitry. The promise of new developments in molecular genetic methods should contribute to the functional analysis of molecularly defined auditory cortical cell types. The ability to silence, activate, or record the activity of molecularly defined classes of neurons might reveal, for example, whether the synaptic inhibition underlying intensity tuning arises from parvalbumin-positive chandelier cells or another type of inhibitory interneuron. The discipline is therefore poised on the brink of a new era of unprecedented detail and refinement.

Important issues to be addressed include how synaptic mechanisms give rise to neuronal sensitivity to, and selectivity for, complex stimuli including species-specific vocalizations such as speech. Early studies must be extended to more complex stimuli and to situations involving changes in behavioral state. The latter are part of a general trend towards understanding regulation of neural responsiveness during behavior, e.g., the rapid and reversible changes in spectrotemporal receptive fields during specific behaviors, and longer lasting changes underlying the learned significance of specific stimuli. Specific neuromodulators will play a key role in such changes, and their roles, interactions, and regulation of cellular mechanisms are essential for understanding of auditory cortex function. An integrated understanding of auditory cortex function, from synapses to systems to behavior, will entail such an approach.

References

Aitkin LM and Webster WR (1972) Medial geniculate body of the cat: organization and responses to tonal stimuli of neurons in ventral division. *Journal of Neurophysiology* **35**:365–380.

Atzori M, Kanold PO, Pineda JC, Flores-Hernandez J, and Paz RD (2005) Dopamine prevents muscarinic-induced decrease of glutamate release in the auditory cortex. *Neuroscience* **134**:1153–1165.

Atzori M, Lei S, Evans DI, Kanold PO, Phillips-Tansey E, McIntyre O, and McBain CJ (2001) Differential synaptic processing separates stationary from transient inputs to the auditory cortex. *Nature Neuroscience* **4**:1230–1237.

Bakin JS, Kwon MC, Masino SA, Weinberger NM, and Frostig RD (1996) Suprathreshold auditory cortex activation visualized by intrinsic signal optical imaging. *Cerebral Cortex* **6**:120–130.

Bao S, Chan VT, Zhang LI, and Merzenich MM (2003) Suppression of cortical representation through backward conditioning. *Proceedings of the National Academy of Sciences of the United States of America* **100**:1405–1408.

Borg-Graham LJ, Monier C, and Fregnac Y (1998) Visual input evokes transient and strong shunting inhibition in visual cortical neurons. *Nature* **393**:369–373.

Bringuier V, Chavane F, Glaeser L, and Fregnac Y (1999) Horizontal propagation of visual activity in the synaptic integration field of area 17 neurons. *Science* **283**:695–699.

Brosch M and Schreiner CE (1997) Time course of forward masking tuning curves in cat primary auditory cortex. *Journal of Neurophysiology* **77**:923–943.

Brosch M and Schreiner CE (2000) Sequence sensitivity of neurons in cat primary auditory cortex. *Cerebral Cortex* **10**:1155–1167.

Brosch M, Schulz A, and Scheich H (1999) Processing of sound sequences in macaque auditory cortex: response enhancement. *Journal of Neurophysiology* **82**:1542–1559.

Brugge JF, Dubrovsky NA, Aitkin LM, and Anderson DJ (1969) Sensitivity of single neurons in auditory cortex of cat to binaural tonal stimulation; effects of varying interaural time and intensity. *Journal of Neurophysiology* **32**:1005–1024.

Budinger E, Heil P, and Scheich H (2000) Functional organization of auditory cortex in the Mongolian gerbil (*Meriones unguiculatus*). IV. Connections with anatomically characterized subcortical structures. *European Journal of Neuroscience* **12**:2452–2474.

Calford MB and Semple MN (1995) Monaural inhibition in cat auditory cortex. *Journal of Neurophysiology* **73**:1876–1891.

Calford MB, Webster WR, and Semple MM (1983) Measurement of frequency selectivity of single neurons in the central auditory pathway. *Hearing Research* **11**:395–401.

Carandini M, Heeger DJ, and Senn W (2002) A synaptic explanation of suppression in visual cortex. *Journal of Neuroscience* **22**:10053–10065.

Caspary DM, Backoff PM, Finlayson PG, and Palombi PS (1994) Inhibitory inputs modulate discharge rate within frequency receptive fields of anteroventral cochlear nucleus neurons. *Journal of Neurophysiology* **72**:2124–2133.

Chattopadhyay S, Xue B, Collins D, Pichika R, Bagnera R, Leslie FM, Christian BT, Shi B, Narayanan TK, Potkin SG, and Mukherjee J (2005) Synthesis and evaluation of nicotine alpha4beta2 receptor radioligand, 5-(3′-18F-fluoropropyl)-3-(2-(S)-pyrrolidinylmethoxy)pyridine, in rodents and PET in nonhuman primate. *Journal of Nuclear Medicine* **46**:130–140.

Chimoto S, Kitama T, Qin L, Sakayori S, and Sato Y (2002) Tonal response patterns of primary auditory cortex neurons in alert cats. *Brain Research* **934**:34–42.

Chung S, Li X, and Nelson SB (2002) Short-term depression at thalamocortical synapses contributes to rapid adaptation of cortical sensory responses in vivo. *Neuron* **34**:437–446.

Clarke PB (2004) Nicotinic modulation of thalamocortical neurotransmission. *Progress in Brain Research* **145**:253–260.

Clarke PB, Pert CB, and Pert A (1984) Autoradiographic distribution of nicotine receptors in rat brain. *Brain Research* **323**:390–395.

Clarke PB, Schwartz RD, Paul SM, Pert CB, and Pert A (1985) Nicotinic binding in rat brain: autoradiographic comparison of [3H]acetylcholine, [3H]nicotine, and [125I]-alpha-bungarotoxin. *Journal of Neuroscience* **5**:1307–1315.

Couturier S, Bertrand D, Matter JM, Hernandez MC, Bertrand S, Millar N, Valera S, Barkas T, and Ballivet M (1990) A neuronal nicotinic acetylcholine receptor subunit (alpha 7) is developmentally regulated and forms a homo-oligomeric channel blocked by a-BTX. *Neuron* **5**:847–856.

Cox CL, Metherate R, and Ashe JH (1994) Modulation of cellular excitability in neocortex: muscarinic receptor and second messenger-mediated actions of acetylcholine. *Synapse* **16**: 123–136.

Creutzfeldt O, Hellweg FC, and Schreiner C (1980) Thalamocortical transformation of responses to complex auditory stimuli. *Experimental Brain Research* **39**:87–104.

Debarbieux F, Brunton J, and Charpak S (1998) Effect of bicuculline on thalamic activity: a direct blockade of IAHP in reticularis neurons. *Journal of Neurophysiology* **79**:2911–2918.

Denham SL and Denham MJ (2001). An investigation into the role of cortical synaptic depression in auditory processing. In: Wermter S, Austin JL, and Willshaw D (eds). *Lecture Notes in Artificial Intelligence*. Springer, New York, pp. 494–506.

DeWeese MR and Zador AM (2006) Non-Gaussian membrane potential dynamics imply sparse, synchronous activity in auditory cortex. *Journal of Neuroscience* **26**:12206–12218.

DeWeese MR, Wehr M, and Zador AM (2003) Binary spiking in auditory cortex. *Journal of Neuroscience* **23**:7940–7949.

Ding YS, Fowler JS, Logan J, Wang GJ, Telang F, Garza V, Biegon A, Pareto D, Rooney W, Shea C, Alexoff D, Volkow ND, and Vocci F (2004) 6-[18F]Fluoro-A-85380, a new PET tracer for the nicotinic acetylcholine receptor: studies in the human brain and in vivo demonstration of specific binding in white matter. *Synapse* **53**:184–189.

Dykes RW, Landry P, Metherate R, and Hicks TP (1984) Functional role for GABA in primary somatosensory cortex: shaping receptive fields of cortical neurons. *Journal of Neurophysiology* **52**: 1066–1093.

Easwaramoorthy B, Pichika R, Collins D, Potkin SG, Leslie FM, and Mukherjee J (2007) Effect of acetylcholinesterase inhibitors on the binding of nicotinic alpha4beta2 receptor PET radiotracer, (18)F-nifene: A measure of acetylcholine competition. *Synapse* **61**: 29–36.

Edeline JM, Hars B, Maho C, and Hennevin E (1994) Transient and prolonged facilitation of tone-evoked responses induced by basal forebrain stimulations in the rat auditory cortex. *Experimental Brain Research* **97**:373–386.

Edeline JM (2003) The thalamo-cortical auditory receptive fields: regulation by the states of vigilance, learning and the neuromodulatory systems. *Experimental Brain Research* **153**: 554–572.

Edeline JM, Dutrieux G, Manunta Y, and Hennevin E (2001) Diversity of receptive field changes in auditory cortex during natural sleep. *European Journal of Neuroscience* **14**:1865–1880.

Eggermont JJ (1996) How homogeneous is cat primary auditory cortex? Evidence from simultaneous single-unit recordings. *Auditory Neuroscience* **2**:79–96.

Eggermont JJ (1999) The magnitude and phase of temporal modulation transfer functions in cat auditory cortex. *Journal of Neuroscience* **19**:2780–2788.

Elhilali M, Fritz JB, Klein DJ, Simon JZ, and Shamma SA (2004) Dynamics of precise spike timing in primary auditory cortex. *Journal of Neuroscience* **24**:1159–1172.

Foeller E, Vater M, and Kössl M (2001) Laminar analysis of inhibition in the gerbil primary auditory cortex. *Journal of the Association for Research in Otolaryngology* **2**:279–296.

Freeman TC, Durand S, Kiper DC, and Carandini M (2002) Suppression without inhibition in visual cortex. *Neuron* **35**: 759–771.

Fritz J, Shamma S, Elhilali M, and Klein D (2003) Rapid task-related plasticity of spectrotemporal receptive fields in primary auditory cortex. *Nature Neuroscience* **6**:1216–1223.

Galvan VV, Chen J, and Weinberger NM (2001) Long-term frequency tuning of local field potentials in the auditory cortex of the waking guinea pig. *Journal of the Association for Research in Otolaryngology* **2**:199–215.

Gil Z, Connors BW, and Amitai Y (1997) Differential regulation of neocortical synapses by neuromodulators and activity. *Neuron* **19**:679–686.

Haj-Dahmane S and Andrade R (1996) Muscarinic activation of a voltage-dependent cation nonselective current in rat association cortex. *Journal of Neuroscience* **16**:3848–3861.

Halliwell JV and Adams PR (1982) Voltage-clamp analysis of muscarinic excitation in hippocampal neurons. *Brain Research* **250**: 71–92.

Hasselmo ME (1999) Neuromodulation: acetylcholine and memory consolidation. *Trends in Cognitive Science* **3**:351–359.

Hasselmo ME (2006) The role of acetylcholine in learning and memory. *Current Opinion in Neurobiology* **16**:710–715.

Heil P and Irvine DRF (1997) First-spike timing of auditory-nerve fibers and comparison with auditory cortex. *Journal of Neurophysiology* **78**:2438–2454.

Heil P and Irvine DRF (1998) The posterior field P of cat auditory cortex: coding of envelope transients. *Cerebral Cortex* **8**:125–141.

Heil P, Langner G, and Scheich H (1992) Processing of frequency-modulated stimuli in the chick auditory cortex analogue: evidence for topographic representations and possible mechanisms of rate and directional sensitivity. *Journal of Comparative Physiology A* **171**:583–600.

Hess A and Scheich H (1996) Optical and FDG mapping of frequency-specific activity in auditory cortex. *Neuroreport* **7**:2643–2647.

Hicks TP, Metherate R, Landry P, and Dykes RW (1986) Bicuculline-induced alterations of response properties in functionally identified ventroposterior thalamic neurones. *Experimental Brain Research* **63**:248–264.

Hind JE, Goldberg JM, Greenwood DG, and Rose JE (1963) Some discharge characteristics of single neurons in the inferior colliculus of the cat. II. Timing of the discharges and observations on binaural stimulation. *Journal of Neurophysiology* **26**:321–341.

Horikawa J, Hosokawa Y, Kubota M, Nasu M, and Taniguchi I (1996) Optical imaging of spatiotemporal patterns of glutamatergic excitation and GABAergic inhibition in the guinea-pig auditory cortex in vivo. *Journal of Physiology (London)* **497**:629–638.

Hounsgaard J (1978) Presynaptic inhibitory action of acetylcholine in area CA1 of the hippocampus. *Experimental Neurology* **62**: 787–797.

Hsieh CY, Cruikshank SJ, and Metherate R (2000) Differential modulation of auditory thalamocortical and intracortical synaptic transmission by cholinergic agonist. *Brain Research* **880**:51–64.

Imig TJ and Morel A (1984) Topographic and cytoarchitectonic organization of thalamic neurons related to their targets in low-, middle-, and high-frequency representations in cat auditory cortex. *Journal of Comparative Neurology* **227**:511–539.

Ji W and Suga N (2007) Serotonergic modulation of plasticity of the auditory cortex elicited by fear conditioning. *Journal of Neuroscience* **27**:4910–4918.

Kaur S, Lazar R, and Metherate R (2004) Intracortical pathways determine breadth of subthreshold frequency receptive fields in primary auditory cortex. *Journal of Neurophysiology* **91**:2551–2567.

Kawai H, Lazar R, and Metherate R (2007) Nicotinic control of axon excitability regulates thalamocortical transmission. *Nature Neuroscience* **10**:1168–1175.

Kilgard MP and Merzenich MM (1998) Cortical map reorganization enabled by nucleus basalis activity. *Science* **279**:1714–1718.

Kilgard MP and Merzenich MM (1999) Distributed representation of spectral and temporal information in rat primary auditory cortex. *Hearing Research* **134**:16–28.

Kimura A, Donishi T, Sakoda T, Hazama M, and Tamai Y (2003) Auditory thalamic nuclei projections to the temporal cortex in the rat. *Neuroscience* **117**:1003–1016.

Kitzes LM, Gibson MM, Rose JE, and Hind JE (1978) Initial discharge latency and threshold considerations for some neurons in cochlear nuclear complex of the cat. *Journal of Neurophysiology* **41**: 1165–1182.

Krnjevic K, Pumain R, and Renaud L (1971) The mechanism of excitation by acetylcholine in the cerebral cortex. *Journal of Physiology (London)* **215**:247–268.

Kurt S, Crook JM, Ohl FW, Scheich H, and Schulze H (2006) Differential effects of iontophoretic in vivo application of the $GABA_A$-antagonists bicuculline and gabazine in sensory cortex. *Hearing Research* **212**:224–235.

Lambe EK, Picciotto MR, and Aghajanian GK (2003) Nicotine induces glutamate release from thalamocortical terminals in prefrontal cortex. *Neuropsychopharmacology* **28**:216–225.

Langner G and Schreiner CE (1988) Periodicity coding in the inferior colliculus of the cat. I. Neuronal mechanisms. *Journal of Neurophysiology* **60**:1799–1822.

Lavine N, Reuben M, and Clarke PBS (1997) A population of nicotine receptors is associated with thalamocortical afferents in the adult rat: laminar and areal analysis. *Journal of Comparative Neurology* **380**:175–190.

LeBeau FEN, Malmierca MS, and Rees A (2001) Iontophoresis in vivo demonstrates a key role for $GABA_A$ and glycinergic inhibition in shaping frequency response areas in the inferior colliculus of guinea pig. *Journal of Neuroscience* **21**:7303–7312.

Lee SM, Friedberg MH, and Ebner FF (1994) The role of GABA-mediated inhibition in the rat ventral posterior medial thalamus. II. Differential effects of $GABA_A$ and $GABA_B$ receptor antagonists on responses of VPM neurons. *Journal of Neurophysiology* **71**:1716–1726.

Linden JF, Liu RC, Sahani M, Schreiner CE, and Merzenich MM (2003) Spectrotemporal structure of receptive fields in areas AI and AAF of mouse auditory cortex. *Journal of Neurophysiology* **90**: 2660–2675.

Lui B and Mendelson JR (2003) Frequency modulated sweep responses in the medial geniculate nucleus. *Experimental Brain Research* **153**:550–553.

Machens CK, Wehr MS, and Zador AM (2004) Linearity of cortical receptive fields measured with natural sounds. *Journal of Neuroscience* **24**:1089–1100.

Madison DV, Lancaster B, and Nicoll RA (1987) Voltage clamp analysis of cholinergic action in the hippocampus. *Journal of Neuroscience* **7**:733–741.

McCormick DA and Prince DA (1986) Mechanism of action of acetylcholine in the guinea-pig cerebral cortex in vitro. *Journal of Physiology (London)* **375**:169–194.

McKenna TM, Ashe JH, and Weinberger NM (1989) Cholinergic modulation of frequency receptive fields in auditory cortex: I. Frequency-specific effects of muscarinic agonists. *Synapse* **4**: 30–43.

Mendelson JR and Cynader MS (1985) Sensitivity of cat primary auditory cortex (AI) neurons to the direction and rate of frequency modulation. *Brain Research* **327**:331–335.

Metherate R (2004) Nicotinic acetylcholine receptors in sensory cortex. *Learning & Memory* **11**:50–59.

Metherate R and Ashe JH (1993) Ionic flux contributions to neocortical slow waves and nucleus basalis-mediated activation: whole-cell recordings in vivo. *Journal of Neuroscience* **13**:5312–5323.

Metherate R and Ashe JH (1994) Facilitation of an NMDA receptor-mediated EPSP by paired-pulse stimulation in rat neocortex via depression of GABAergic IPSPs. *Journal of Physiology (London)* **481**:331–348.

Metherate R, Tremblay N, and Dykes RW (1988) The effects of acetylcholine on response properties of cat somatosensory cortical neurons. *Journal of Neurophysiology* **59**:1231–1251.

Metherate R, Cox CL, and Ashe JH (1992) Cellular bases of neocortical activation: modulation of neural oscillations by the nucleus basalis and endogenous acetylcholine. *Journal of Neuroscience* **12**:4701–4711.

Miller LM, Escabí MA, and Schreiner CE (2001) Feature selectivity and interneuronal cooperation in the thalamocortical system. *Journal of Neuroscience* **21**:8136–8144.

Miller LM, Escabi MA, Read HL, and Schreiner CE (2002) Spectrotemporal receptive fields in the lemniscal auditory thalamus and cortex. *Journal of Neurophysiology* **87**:516–527.

Morley BJ and Happe HK (2000) Cholinergic receptors: dual roles in transduction and plasticity. *Hearing Research* **147**: 104–112.

Müller CM and Scheich H (1988) Contribution of GABAergic inhibition to the response characteristics of auditory units in the avian forebrain. *Journal of Neurophysiology* **59**:1673–1689.

Nelken I and Versnel H (2000) Responses to linear and logarithmic frequency-modulated sweeps in ferret primary auditory cortex. *European Journal of Neuroscience* **12**:549–562.

Norena A and Eggermont JJ (2002) Comparison between local field potentials and unit cluster activity in primary auditory cortex and anterior auditory field in the cat. *Hearing Research* **166**:202–213.

Oertel D (1999) The role of timing in the brain stem auditory nuclei of vertebrates. *Annual Review of Physiology* **61**:497–519.

Ohl FW, Schulze H, Scheich H, and Freeman WJ (2000) Spatial representation of frequency-modulated tones in gerbil auditory cortex revealed by epidural electrocorticography. *Journal of Physiology (Paris)* **94**:549–554.

Ojima H and Murakami K (2002) Intracellular characterization of suppressive responses in supragranular pyramidal neurons of cat primary auditory cortex in vivo. *Cerebral Cortex* **12**:1079–1091.

Okamoto H, Stracke H, Wolters CH, Schmael F, and Pantev C (2007) Attention improves population-level frequency tuning in human auditory cortex. *Journal of Neuroscience* **27**:10383–10390.

Oswald AM, Schiff ML, and Reyes AD (2006) Synaptic mechanisms underlying auditory processing. *Current Opinion in Neurobiology* **16**:371–376.

Palombi PS and Caspary DM (1992) GABAA receptor antagonist bicuculline alters response properties of posteroventral cochlear nucleus neurons. *Journal of Neurophysiology* **67**:738–746.

Parkinson D, Kratz KE, and Daw NW (1988) Evidence for a nicotinic component to the actions of acetylcholine in cat visual cortex. *Experimental Brain Research* **73**:553–568.

Phillips DP and Hall SE (1990) Response timing constraints on the cortical representation of sound time structure. *Journal of the Acoustical Society of America* **88**:1403–1411.

Phillips DP and Hall SE (1992) Multiplicity of inputs in the afferent path to cat auditory cortex neurons revealed by tone-on-tone masking. *Cerebral Cortex* **2**:425–433.

Phillips DP, Hall SE, and Hollett JL (1989) Repetition rate and signal level effects on neuronal responses to brief tone pulses in cat auditory cortex. *Journal of the Acoustical Society of America* **85**:2537–2549.

Phillips DP, Mendelson JR, Cynader MS, and Douglas RM (1985) Responses of single neurones in cat auditory cortex to time-varying stimuli: frequency-modulated tones of narrow excursion. *Experimental Brain Research* **58**:443–454.

Pickles JO (1988) *An Introduction to the Physiology of Hearing*. Academic Press, London, San Diego.

Polley DB, Steinberg EE, and Merzenich MM (2006) Perceptual learning directs auditory cortical map reorganization through top-down influences. *Journal of Neuroscience* **26**:4970–4982.

Polley DB, Heiser MA, Blake DT, Schreiner CE, and Merzenich MM (2004) Associative learning shapes the neural code for stimulus magnitude in primary auditory cortex. *Proceedings of the National Academy of Sciences of the United States of America* **101**:16351–16356.

Prusky GT, Shaw C, and Cynader MS (1987) Nicotine receptors are located on lateral geniculate nucleus terminals in cat visual cortex. *Brain Research* **412**:131–138.

Recanzone GH, Schreiner CE, and Merzenich MM (1993) Plasticity in the frequency representation of primary auditory cortex following discrimination training in adult owl monkeys. *Journal of Neuroscience* **13**:87–103.

Rhode WS and Smith PH (1986) Encoding timing and intensity in the ventral cochlear nucleus of the cat. *Journal of Neurophysiology* **56**:261–286.

de Ribaupierre F, Goldstein MH, and Yeni-Komshian G (1972) Intracellular study of the cat's primary auditory cortex. *Brain Research* **48**:185–204.

Rogers SW, Gahring LC, Collins AC, and Marks M (1998) Age-related changes in neuronal nicotinic acetylcholine receptor subunit α4 expression are modified by long-term nicotine administration. *Journal of Neuroscience* **18**:4825–4832.

Romanski LM and LeDoux JE (1993) Organization of rodent auditory cortex: anterograde transport of PHA-L from MGv to temporal neocortex. *Cerebral Cortex* **3**:499–514.

Rose HJ and Metherate R (2005) Auditory thalamocortical transmission is reliable and temporally precise. *Journal of Neurophysiology* **94**:2019–2030.

Sahin M, Bowen WD, and Donoghue JP (1992) Location of nicotinic and muscarinic cholinergic and μ-opiate receptors in rat cerebral neocortex: evidence from thalamic and cortical lesions. *Brain Research* **579**:135–147.

Sarter M, Givens B, and Bruno JP (2001) The cognitive neuroscience of sustained attention: where top-down meets bottom-up. *Brain Research Reviews* **35**:146–160.

Schreiner CE, Read HL, and Sutter ML (2000) Modular organization of frequency integration in primary auditory cortex. *Annual Review of Neuroscience* **23**:501–529.

Schulze H and Langner G (1999) Auditory cortical responses to amplitude modulations with spectra above frequency receptive fields: evidence for wide spectral integration. *Journal of Comparative Physiology [A]* **185**:493–508.

Segal M (1982) Multiple action of acetylcholine at a muscarinic receptor studied in the rat hippocampal slice. *Brain Research* **246**:77–87.

Segal M (1989) Presynaptic cholinergic inhibition in hippocampal cultures. *Synapse* **4**:305–312.

Séguéla P, Wadiche J, Dineley-Miller K, Dani JA, and Patrick JW (1993) Molecular cloning, functional properties, and distribution of rat brain α_7: a nicotinic cation channel highly permeable to calcium. *Journal of Neuroscience* **13**:596–604.

Sillito AM (1975) The contribution of inhibitory mechanisms to the receptive field properties of neurones in the striate cortex of the cat. *Journal of Physiology (London)* **250**:305–329.

Sillito AM and Kemp JA (1983) Cholinergic modulation of the functional organization of the cat visual cortex. *Brain Research* **289**:143–155.

Sivaramakrishnan S, Sterbing-D'Angelo SJ, Filipovic B, D'Angelo WR, Oliver DL, and Kuwada S (2004) $GABA_A$ synapses shape neuronal responses to sound intensity in the inferior colliculus. *Journal of Neuroscience* **24**:5031–5043.

Smits E, Gordon DC, Witte S, Rasmusson DD, and Zarzecki P (1991) Synaptic potentials evoked by convergent somatosensory and corticocortical inputs in raccoon somatosensory cortex: substrates for plasticity. *Journal of Neurophysiology* **66**:688–695.

Soto G, Kopell N, and Sen K (2006) Network architecture, receptive fields, and neuromodulation: computational and functional implications of cholinergic modulation in primary auditory cortex. *Journal of Neurophysiology* **96**:2972–2983.

Stark H and Scheich H (1997) Dopaminergic and serotonergic neurotransmission systems are differentially involved in auditory cortex learning: a long-term microdialysis study of metabolites. *Journal of Neurochemistry* **68**:691–697.

Sutter ML and Schreiner CE (1995) Topography of intensity tuning in cat primary auditory cortex: single-neuron versus multiple-neuron recordings. *Journal of Neurophysiology* **73**:190–204.

Sutter ML and Loftus WC (2003) Excitatory and inhibitory intensity tuning in auditory cortex: evidence for multiple inhibitory mechanisms. *Journal of Neurophysiology* **90**:2629–2947.

Sutter ML, Schreiner CE, McLean M, O'Connor KN, and Loftus WC (1999) Organization of inhibitory frequency receptive fields in cat primary auditory cortex. *Journal of Neurophysiology* **82**:2358–2371.

Tan AY, Zhang LI, Merzenich MM, and Schreiner CE (2004) Tone-evoked excitatory and inhibitory synaptic conductances of primary auditory cortex neurons. *Journal of Neurophysiology* **92**:630–643.

Tan AYY, Atencio CA, Polley DB, Merzenich MM, and Schreiner CE (2007) Unbalanced synaptic inhibition can create intensity-tuned auditory cortex neurons. *Neuroscience* **146**:449–462.

Ter-Mikaelian M, Sanes DH, and Semple MN (2007) Transformation of temporal properties between auditory midbrain and cortex in the awake Mongolian gerbil. *Journal of Neuroscience* **27**:6091–6102.

Tian B and Rauschecker JP (2004) Processing of frequency-modulated sounds in the lateral auditory belt cortex of the rhesus monkey. *Journal of Neurophysiology* **92**:2993–3013.

Trussell LO (1999) Synaptic mechanisms for coding timing in auditory neurons. *Annual Review of Physiology* **61**:477–496.

Trussell LO (2002) Modulation of transmitter release at giant synapses of the auditory system. *Current Opinion in Neurobiology* **12**:400–404.

Ulanovsky N, Las L, Farkas D, and Nelken I (2004) Multiple time scales of adaptation in auditory cortex neurons. *Journal of Neuroscience* **24**:10440–10453.

Valentino RJ and Dingledine R (1981) Presynaptic inhibitory effect of acetylcholine in the hippocampus. *Journal of Neuroscience* **1**:784–792.

Velenovsky DS, Cetas JS, Price RO, Sinex DG, and McMullen NT (2003) Functional subregions in primary auditory cortex defined by thalamocortical terminal arbors: an electrophysiological and anterograde labeling study. *Journal of Neuroscience* **23**:308–316.

Verbny YI, Erdelyi F, Szabo G, and Banks MI (2006) Properties of a population of GABAergic cells in murine auditory cortex weakly excited by thalamic stimulation. *Journal of Neurophysiology* **96**:3194–3208.

Volkov IO and Galazjuk AV (1991) Formation of spike response to sound tones in cat auditory cortex neurons: interaction of excitatory and inhibitory effects. *Neuroscience* **43**:307–321.

Wang J, Caspary D, and Salvi RJ (2000) GABA-A antagonist causes dramatic expansion of tuning in primary auditory cortex. *Neuroreport* **11**:1137–1140.

Wang J, McFadden SL, Caspary D, and Salvi R (2002) Gamma-aminobutyric acid circuits shape response properties of auditory cortex neurons. *Brain Research* **944**:219–231.

Wang X, Lu T, Snider RK, and Liang L (2005) Sustained firing in auditory cortex evoked by preferred stimuli. *Nature* **435**:341–346.

Wehr M and Zador AM (2003) Balanced inhibition underlies tuning and sharpens spike timing in auditory cortex. *Nature* **426**:442–446.

Wehr M and Zador AM (2005) Synaptic mechanisms of forward suppression in rat auditory cortex. *Neuron* **47**:437–445.

Weinberger NM (2004) Specific long-term memory traces in primary auditory cortex. *Nature Reviews Neuroscience* **5**:279–290.

Whitfield IC and Evans EF (1965) Responses of auditory cortical neurons to stimuli of changing frequency. *Journal of Neurophysiology* **28**:655–672.

Winer JA, Sally SL, Larue DT, and Kelly JB (1999) Origins of medial geniculate body projections to physiologically defined zones of rat primary auditory cortex. *Hearing Research* **130**:42–61.

Wu GK, Li P, Tao HW, and Zhang LI (2006) Nonmonotonic synaptic excitation and imbalanced inhibition underlying cortical intensity tuning. *Neuron* **52**:705–715.

Xie R, Gittelman JX, and Pollak GD (2007) Rethinking tuning: in vivo whole-cell recordings of the inferior colliculus in awake bats. *Journal of Neuroscience* **27**:9469–9481.

Zhang LI, Tan AY, Schreiner CE, and Merzenich MM (2003) Topography and synaptic shaping of direction selectivity in primary auditory cortex. *Nature* **424**:201–205.

Zoli M, Léna C, Picciotto MR, and Changeux J-P (1998) Identification of four classes of brain nicotinic receptors using β2 mutant mice. *Journal of Neuroscience* **18**:4461–4472.

Chapter 12
Physiological Properties of Neurons in the Medial Geniculate Body

Jean-Marc Edeline

Abbreviations

AI	primary auditory cortex
AHP	action potential hyperpolarization
AMPA	a-amino-3-hydroxyl-5-methyl-4-isoxazole-propionate
AP	action potential
APV	aminophosphonovalerate
AS	azimuth sensitivity
ATP	adenosine triphosphate
BIC	brachium of the inferior colliculus
CB	calbindin
CF	characteristic frequency
CNIC	central nucleus of the inferior colliculus
CNQX	6-cyano-7-nitroquinoxaline-2,3-dione
CR	conditioned response
CS	conditioned stimulus
E	excitatory
EEG	electroencephalogram
EPSP	excitatory postsynaptic potential
F	facilitatory
GABA	gamma-aminobutyric acid
GAD	glutamic acid decarboxylase
HRTF	head-related transfer function
IC	inferior colliculus
IID	interaural intensity difference
I	inhibitory
IPD	interaural phase difference
IPSP	inhibitory postsynaptic potential
ITD	interaural time difference
LSO	lateral nucleus of the superior olivary complex
LTS	low threshold spikes
MAP	microtubule associated protein
MD	monaural direction
MG	medial geniculate body
MGd	dorsal division of the medial geniculate body
MGm	medial division of the medial geniculate body
MGv	ventral division of the medial geniculate body
MSO	medial nucleus of the superior olivary complex
NMDA	*N*-methyl-D-aspartate
PIN	posterior intralaminar nucleus
Pol	lateral region of the posterior nucleus
PV	parvalbumin
PS	paradoxical sleep
RF	receptive field
SC	superior colliculus
SG	suprageniculate nucleus
SPL	sound pressure level
STRF	spectrotemporal receptive field
SWS	slow-wave sleep
TC	thalamocortical
TRN	thalamic reticular nucleus
TWIN	two-way intensity network
W	waking

1 Introduction

An important synthesis of thalamic organization noted that: '...the thalamus has not had good press in the recent past' (Sherman and Guillery 1996). The functional role of the auditory thalamus (medial geniculate body; MG) has been eclipsed by the enormous effort aimed at dissecting the properties of cortical neurons. In contrast to the visual and somatic sensory somatic sensory systems, the MG differs conspicuously between the richness of the analysis already performed in the various brain stem stations, and the exquisite range of physiological properties embodied by auditory cortex cells. Is there a specific role for the MG or is it a passive relay between brain stem and cortex?

This account defends the proposition that the auditory thalamus has a critical impact on the abilities of auditory cortex neurons to process and integrate acoustic information. Knowledge of the morphological and intrinsic properties of medial geniculate body (MG) cells has grown since

J.-M. Edeline (✉)
Centre de Neurosciences de Paris-Sud (CNPS) UMR 8195, CNRS & Université Paris-Sud, 91405 Orsay, Cedex, France
e-mail: jean-marc.edeline@u-psud.fr

J.A. Winer, C.E. Schreiner (eds.), *The Auditory Cortex*,
DOI 10.1007/978-1-4419-0074-6_12,

1990. A major goal was to provide a cellular substrate to the well-known partition of the MG into three major divisions. Intracellular in vitro and in vivo labeling studies and biochemical and pharmacological analyses substantiate the notion that the lemniscal and non-lemniscal systems have structural, cellular, connectional, and functional substrates. However, the differences and similarities between the intrinsic and synaptic properties observed in the MG divisions are still a matter of debate. New lines of research have also considered the impact of auditory cortex on the functional properties of MG neurons (He 2003b) and have begun to dissect the responses of MG cells to natural stimuli such as conspecific vocalizations.

Many advances have been made in understanding the role of auditory thalamus in integrative functions. In contrast with the dogma that the cortex has the unique ability to reorganize after peripheral injury and in accord with data from the somatic sensory thalamus (Rasmusson 1996) the MG has massive potential for reorganization after specific hearing loss. MG neurons also have receptive fields involved in learning, implying strong links between sensory plasticity and cognitive processes (Edeline 2003). Besides its auditory role in the elaboration of cortical representations, the MG has direct interactions with limbic structures, such as the amygdala, which suggests a pivotal role in integrating acoustic stimuli with the emotional and cognitive context of any perception. Each of these features imply a multifaceted role for the auditory thalamus.

2 Cellular Bases of Auditory Thalamic Functions

In most sensory modalities, thalamocortical communication employs two largely segregated channels, the lemniscal and non-lemniscal pathways (Graybiel 1972; Sherman and Guillery 1996). In the auditory system, anatomical (Morest 1965b, 1975; LeDoux et al. 1987; Winer 1991, 1992) and physiological (Aitkin and Webster 1972; Imig and Morel 1985a, b) studies have firmly established that, at the thalamic level, the lemniscal component is represented by the ventral, tonotopically organized, division (MGv). The non-lemniscal component is divided into dorsal (MGd) and medial (MGm) nuclei, with the addition of the posterior intralaminar nucleus (PIN), the suprageniculate nucleus (SG), and the lateral region of the posterior nucleus (Pol). Many electrophysiological studies have confirmed that acoustic responses differ fundamentally in the lemniscal and non-lemniscal MG (Aitkin and Webster 1972; Aitkin 1973; Calford 1983; Bordi and LeDoux 1994a, b; Edeline et al. 1999): MGv neurons have shorter latency responses, sharper frequency-tuning curves, and shorter duration responses than MGd and MGm neurons.

The challenge is now to understand how the morphological, anatomical, cellular and molecular characteristics of neurons are related with the functional properties assessed in anesthetized and non-anesthetized animals, and the nature of species differences. What are the anatomical and cellular characteristics of lemniscal and non-lemniscal neurons underlying their diverse functional properties such as frequency tuning, rate-level functions, binaural properties, spectrotemporal receptive field, communication call preferences that exist in the MG?

2.1 Connectivity of the Lemniscal and Non-lemniscal Auditory Thalamus

Evidence for dissociation between the MGv and the other MG divisions comes from its afferent and efferent connections. These connections are only briefly summarized here; more detailed analyses are found elsewhere (Chapters 2 and 3). It is classically considered that the principal input to MGv comes from the central nucleus of the ipsilateral inferior colliculus (CNIC; Andersen et al. 1980; Rouiller and de Ribaupierre 1985; Cant and Benson 2007). Several types of CNIC neurons project to the MGv (Oliver 1984) and all such projections are topographically organized. For example, in the cat, neurons representing low frequencies and located dorsally in the CNIC project to MGv neurons representing low frequencies and located laterally (Rouiller and de Ribaupierre 1985). A more complex picture, suggesting that parallel pathways arising from the CNIC remain separate in MGv, has been proposed in the gerbil: the part of the CNIC that receives ascending input from the main nuclei of the superior olivary complex (LSO and MSO) as well as from the nuclei of the lateral lemniscus and cochlear nuclei projects to rostral parts of the MGv, whereas the part of the CNIC that do not receives ascending inputs from MSO and LSO projects to more caudal parts of the MGv (Cant and Benson 2007).

The brain stem connections of MGd are more diverse than those of MGv and the strength of the different projections reflects this variability (Winer 1991, 1992). In cat, the MG dorsal and deep dorsal nuclei receive strong input from the dorsomedial part and from the dorsal cortex of the IC (Calford and Aitkin 1983). In addition, a substantial input from the lateral tegmental system of the midbrain terminates in the deep dorsal and suprageniculate nuclei (Morest 1965a). It is unknown whether there is a point-to-point topographic relationship between the IC (or other brain stem nuclei) and the MGd. Other non-auditory brain stem sites (nucleus sagulum and the superior colliculus) project to the dorsal division and suprageniculate (Calford and Aitkin 1983; Morest and Winer 1986).

The projections targeting MGm have diverse origins. Nearly every IC subdivision projects to MGm (Kudo and Niimi 1980). Other inputs come from the vicinity of the superior olivary complex (Henkel 1983), the ventral lateral lemniscus (Whitley and Henkel 1984), and the dorsal cochlear nucleus (Malmierca et al. 2002; Anderson et al. 2006). Non-auditory projections originate from the spinal cord (Jones and Burton 1974; Winer and Morest 1983; LeDoux et al. 1987), the vestibular nuclei, and the superior colliculus (Graham 1977).

The role of the multisensory PIN may be crucial in behavioral tasks. This area lies beneath MGm and medial to the deepest part of MGv. The inputs to it are as diverse as those of the MGm. The output of PIN to limbic areas such as the lateral amygdala (Doron and LeDoux 1999, 2000) is more numerous as those from MGm; PIN thus seems to be in a key position for processing acoustic stimuli during emotional situations, as in fear conditioning. In addition, the widespread projections of PIN to all cortical auditory areas, and to non-auditory areas as well, suggest that it can influence or even synchronize the activity of large cortical territories in awake behaving animals.

In all species, axons from MGv neurons terminate in layer III–IV of the primary auditory cortex (AI; Huang and Winer 2000; Smith and Populin 2001), whereas MGd neurons essentially project outside AI (also reaching layers III/IV). The projections from MGm and PIN neurons reach all auditory cortical fields and, more rostrally, the somatic sensory somatic sensory cortex (Linke and Schwegler 2000). The MGm and the suprageniculate have descending projections to the IC (Senatorov and Hu 2002). All parts of the MG receive cortical input, and there is evidence in several species that the projections between MGv and the auditory cortex are topographic and reciprocal (Diamond et al. 1969; Andersen et al. 1980; Winer and Larue 1987; Winer et al. 2001).

The interconnections between the MG and the thalamic reticular nucleus (TRN) are unique in several ways. Except in its rostral pole, the TRN is a thin layer (200–300 μm) of GABAergic cells surrounding the dorsal thalamus laterally and ventrally. It receives sensory information from collaterals of thalamocortical fibers and from corticothalamic fibers (Pinault 2004). The TRN projects to the dorsal thalamus, not to the cortex, and it includes separate visual, somatic sensory and auditory sectors. Anatomical studies indicate that, within a given sensory modality, the topographic projections are preserved. In the auditory part, MGv and MGd/MGm project to different regions of the auditory TRN (Conley et al. 1991), but TRN cells project either to one or two MG divisions, e.g., to MGv and MGm (Crabtree 1998). TRN neurons are sometimes considered as local interneurons displaced outside the sensory relays, but this might not be valid since TRN is a developmental derivative of the ventral thalamus (Jones 1985). In rat, TRN neurons and IC neurons are the main GABAergic inputs onto MG cells, whereas in cat and primate an additional source of inhibitory input comes from local interneurons.

2.2 Cell Types in Medial Geniculate Body Divisions

Based on Golgi material, only a comparatively few cell types have been identified in MGv whereas more diverse cell types are present in MGd and MGm. In the MGv of all species, the most prominent cell type is the bushy tufted neuron (Morest 1964; Winer 1985; Clerici and Coleman 1990) with primary dendrites confined to zones 30–100 μm wide, and whose long axis follows the somatic orientation. The marked orientation of the dendritic trees confers a laminar structure upon MGv because long rows of these neurons form conspicuous fibrodendritic laminae whose arrangement conserves, and reflects, the patterns of inputs arising from CNIC. The thick myelinated axons of tufted cells project to layer III/IV of auditory cortex. In cat, a second cell type, the small stellate cells, represents ~30% of MGv neurons. Their flask-shaped somata are about two-third the size of the tufted neurons and their thin dendrites have a stellate configuration and thin, unmyelinated axons which terminate among the tufted neurons; their terminal axonal fields are probably confined to relatively few fibrodendritic laminae (Morest 1965b, 1971). In each cat MG division, cells immunoreactive for glutamic acid decarboxylase (GAD) were found in the same proportion as small stellate cells (Huang et al. 1999). In all the rat MG divisions, in contrast, GAD-positive neurons represent only 1% of neurons (Winer and Larue 1988) and small stellate cells are also only sparsely observed in MGv (Clerici et al. 1990; Winer et al. 1999). Therefore, it is likely that, in cat and in rat, small stellate cells found in MGv represent GABAergic, inhibitory interneurons.

Two main cell types are present in MGd. The bushy tufted neurons are still present, but less prominent than in MGv. They do not have a laminar organization, and their dendritic branches may be less tufted than those in MGv (Clerici et al. 1990). In the rat, small stellate cells resembling their ventral counterpart were also described and also represent ~1% of the population (Winer et al. 1999). However, stellate cells (or radiate cells) with medium- to large-sized somata and extensive dendritic arbors are numerous and are characteristic of MGd (Clerici et al. 1990; Winer et al. 1999). In rat, these stellate/radiate cells are probably not GABAergic interneurons because (i) they are far more numerous than that of GAD-positive neurons and (ii) their axons project to cortex. Minor cell types found in the dorsal division include bitufted neurons present in the deep dorsal nucleus where their dendrites are horizontally arranged parallel to the midgeniculate bundle (Clerici et al. 1990).

The MGm has the most diverse MG cell types. The most conspicuous neurons, especially caudally, are the magnocellular neurons noted in Nissl material. Their 25–35 μm in diameter somata have a few thick dendrites that branch sparsely and can extend in a stellate arrangement. Medium-sized stellate cells with extensive arbors are also frequently observed, and again, in rat, they are probably not local GABAergic interneurons. Some bushy cells are also present but they have less tufted and longer arbors than those of MGd and MGv neurons (Clerici et al. 1990; Winer et al. 1999).

Intracellular labeling studies in rat during electrophysiological recordings confirmed the dominant cell types of the main MG divisions. MGv neurons have a tufted morphology characterized by shorter dendrites that branch profusely within 50 μm of the cell body with varying dendritic orientations. The MGd neurons display either a tufted or a stellate morphology with long dendrites that divide moderately in all directions and have distal dendritic branch points (Bartlett and Smith 1999). The morphological properties of MGm cells differ fundamentally from those of MGd and MGv cells (Smith et al. 2006). First, the somatic size range in MGm is wider (118–452 μm^2) than in MGd/MGv (122–226 μm^2). Second, significant differences were found between the dendritic branching of MGm cells and that of the MGd/MGv cells. MGm cells have elongated and oriented arbors with a few sparsely branching dendrites that could extend for >1 mm. In ~20% of MGm cells many spines were seen along the dendrites, and other cells were sparsely spiny. In contrast to MGd/MGv cells, some MGm cells have one or more local axon collaterals that arise hundreds of micrometers from the axon origin. The local collaterals can project toward the paralaminar nuclei but not to other MG divisions (Smith et al. 2006).

2.3 Intrinsic Electrophysiological Properties of Medial Geniculate Body Cells

The electrophysiological properties of thalamic relay cells have been investigated in vivo and in vitro in various thalamic regions in several species (Steriade et al. 1997). There is a striking similarity in electrophysiological properties between thalamocortical neurons in different dorsal thalamic regions.

The current versus voltage (I–V) plots of thalamocortical cells always indicate that they show rectification in both the depolarizing and hyperpolarizing ranges leading to the typical sigmoid shape of the I–V plots with the maximal slope near the resting membrane potential (Jahnsen and Llinás 1984a, b; Crunelli et al. 1987).

Thalamocortical relay cells have two basic modes of action potential (AP) generation: burst firing and single-spike activity. In vitro, burst firing involves high-frequency (200–400 Hz) bursts of 2–8 APs, while single-spike activity is the generation of trains of APs with the number of spikes being determined by the intensity and duration of the current pulse (Steriade and Llinás 1988). In vivo extracellular and intracellular recordings show that burst firing is more prevalent during anesthesia and slow-wave sleep, while single-spike activity is more common during wakefulness (Steriade and Deschênes 1984; McCormick and Bal 1997).

In vitro and in vivo intracellular recordings confirm that the two modes of AP generation in thalamocortical (TC) cells exist at different membrane potentials (Steriade et al. 1997). When TC cells are at membrane potentials positive to –65 mV, a suprathreshold depolarizing current injection elicits one or more APs. In contrast, when TC cells are at membrane potentials below –65 mV, the suprathreshold depolarizing injection current activates slow depolarizing potentials on top of which a burst of APs is emitted rather than a train of APs. This depolarizing potential is generated by a specialized Ca^{2+} current, the low-threshold Ca^{2+} current or the T current (I_t for transient Ca^{2+} current) (Jahnsen and Llinás 1984a, b).

Two views have been proposed to account for the physiological properties of lemniscal and non-lemniscal MG cells: one stresses the differences, the other the similarities, between the intrinsic properties of ventral and the dorsal division neurons.

An in vitro explant preparation containing both the MG and the brachium of the inferior colliculus (BIC) was used to show that BIC stimulation triggers single- or dual-spike responses in 70% of the MGv neurons, whereas burst responses of 3–8 spikes were more common in MGd (Hu 1995). The first spike latency of the MGd burst response ranged from 10 to 70 ms, and the single-spike response was 5–15 ms. MGd neurons have a more negative resting membrane potential than MGv neurons (–61 mV vs. –72 mV), perhaps because the hyperpolarization-activated inward current I_h (seen as a depolarizing sag in current clamp recordings) was present in MGv neurons only. Two further mechanisms might contribute to a lower burst proportion in MGv: first, the EPSP evoked in MGv is usually smaller and shorter than that in MGd and, second, BIC stimulation elicited a prolonged EPSP in MGv which was curtailed by prominent IPSP, whereas no IPSPs were seen in MGd (Hu 1995). Whole-cell recording studies found that MGd neurons express enhanced activity of Na^+-K^+-ATPase relative to MGv neurons, a difference attributed to differential membrane pump densities (Senatorov and Hu 1997; Senatorov et al. 1997).

Other recording studies from morphologically identified neurons in MG slices find only minor differences between

MGd and MGv cells (Bartlett and Smith 1999). Bursts were seen in MGv and MGd after the offset of hyperpolarizing current, and the resting membrane potential in both was similar. The neuronal intrinsic properties were also comparable, except that a sag current was found in fewer MGd neurons. In anatomically identified MGd cells with a sag, half had a stellate morphology and half were tufted. The presence of bursts in rat MGv neurons with a tufted morphology was confirmed in whole-cell recordings in slices (Tennigkeit et al. 1996). Low threshold spikes (LTSs) were seen after release from hyperpolarization or when depolarizing current pulses were applied at hyperpolarized membrane potentials. In half of MGv neurons, bursts of 2–7 AP were seen, and for the others only one or two AP were noted on the LTS hump (Tennigkeit et al. 1996, 1997). Methodological differences may account for the discrepancies between the results from different laboratories. First, the age of the animal, 2–3 weeks (Bartlett and Smith 1999) versus older animals (Hu 1995) may underlie differences in intrinsic properties. Second, the diencephalic in vitro explant (Hu 1995) could include TRN neurons which might provide additional IPSPs onto MG cells.

The physiological properties of MGm cells differ from those in the MGd/MGv (Smith et al. 2006). First, most MGm cells showed little or no calcium burst. Some paralaminar nucleus and MGm cells fire in a regular sustained mode at all membrane potentials in response to a suprathreshold current pulse; others showed persistent firing with some spike frequency adaptation. Their action potential amplitude was larger than that of the MGd/MGv cells, and 72% had a biphasic action potential hyperpolarization (bAHP), which was never observed in MGd/MGv cells, but is seen in thalamic interneurons (Pape and McCormick 1995). The input resistance of the bAHP cells was higher than the MGd/MGv cells and higher than that of the paralaminar and MGm cells with monophasic AHP (Smith et al. 2006).

While these studies were done in vitro, some in vivo intracellular recordings partly confirmed their results: MG neurons respond to sound presentation with one AP at resting membrane potential (–66 mV), whereas they have a LTS spike burst when hyperpolarized below –77 mV (Yu et al. 2004b). It is unknown whether MG nuclei differ in their propensity to respond with LTS spike bursts under in vivo conditions.

2.4 Synaptic Properties of Medial Geniculate Body Cells

BIC stimulation in explant preparations evokes monosynaptic EPSPs mediated by glutamate acting on both on *N*-methyl-D-Aspartate (NMDA) and non-NMDA receptors (Hu et al. 1994). Here, too, differences were seen between MGv and MGd. In MGv, AMPA receptor blockers (e.g., CNQX) completely abolished the BIC-evoked responses. In contrast, in MGd, AMPA receptor blockers and NMDA receptor blockers (e.g., aminophosphonovalerate, APV) only partly suppressed the BIC-evoked responses. This suggests that MGv cells only have non-NMDA responses, whereas MGd cells show NMDA and non-NMDA responses.

In contrast, in slice preparation, no differences were seen between the synaptic responses of MGv and MGd cells (Bartlett and Smith 1999). Stimulating axons of IC origin led to different patterns of responses depending on the presence of a $GABA_A$ IPSP and its latency relative to the EPSP latency. The two dominant responses were EX/O and IN/EX. The EX/O pattern occurs only in the presence of EPSPs, whereas the IN/EX pattern was contingent on an EPSP preceded by an IPSP. Neurons with different patterns were found in similar proportion in MGd and MGv, but there was more EX/IN in MGd than in MGv (23% vs. 7%). In both MGd and MGv, EPSP latencies to EX/O inputs were significantly shorter than the latency of IN/EX EPSPs (2.3 ms vs. 3.8 ms in MGv). In MGd but not MGv, $GABA_A$ IPSPs in EX/IN cells were significantly longer than all other $GABA_A$ latencies, whereas the EPSP latencies for these neurons were shorter than the IN/EX EPSP latencies. This suggests that excitatory and inhibitory input arriving at MG cells has short- and long-latency components.

Neurons with the EX/O patterns have strong paired-pulse depression of their large, short-latency EPSPs, whereas those with the IN/EX pattern have much weaker paired-pulse depression or even paired-pulse facilitation of their smaller, long-latency EPSPs (Bartlett and Smith 2002). After BIC stimulation, EPSPs reaching MG cells have AMPA or NMDA components; the IPSPs involved only a $GABA_A$component. Stimulating the thalamic radiation to excite corticothalamic and thalamic reticular nucleus fibers triggers a more uniform response than the IC stimulation, eliciting a $GABA_A$ IPSP/EPSP or $GABA_B$ IPSP sequence in 74% of tufted and stellate neurons in all MG divisions (Bartlett and Smith 1999).

2.5 Pharmacological and Biochemical Markers of Parallel Pathways in Medial Geniculate Body

Application of cholinergic, catecholaminergic or cholecystokininergic receptor antagonists to in vitro explant preparation did not affect the monosynaptic excitatory potentials either in the MGv or MGd (Hu et al. 1994). However, in MGv, muscarinic agonists block burst but not single-spike responses, an effect attributed to membrane depolarization per se, since it was mimicked by K^+-induced membrane depolarizations (Mooney et al. 1995). There are differential MGd and MGv neuron responses to muscarine application. In MGv, muscarine induced a sustained membrane depolarization and tonic firing by closing a linear K^+ conductance, while in MGd it evoked a membrane hyperpolarization by opening a voltage-independent K^+ conductance. Immunohistochemistry and western blot techniques indicate that MGd neurons predominantly expressed M2 muscarinic receptors, whereas MGv neurons expressed more M1 receptors (Mooney et al. 2004).

The distribution of calcium-binding proteins suggests parallel pathways in the auditory thalamus and forebrain. Studies on the MG of old world monkeys find that every nucleus had at least two populations of relay cells: large cells immunoreactive for the calcium-binding protein, parvalbumin (PV), and smaller cells immunoreactive for another calcium-binding protein, calbindin-D28k (CB); some cells are also immunoreactive for calretinin-29 k (Hashikawa et al. 1991; Molinari et al. 1995). PV cells are numerous in MGv and correspond to cells projecting to layers III and IV, whereas CB cells are concentrated outside MGv and correspond to cells projecting to superficial layers including layer I (Hashikawa et al. 1991). Related patterns are seen in other species. The rabbit MG has segregation between a parvalbumin-rich MGv and a calbindin-rich MGd (De Venecia et al. 1995). In mouse, similar immunostaining was seen in MGv, whereas the surrounding areas had strong calbindin and lighter parvalbumin reactivity (Cruikshank et al. 2001). Comparative studies find that PV demarcates the lemniscal auditory thalamus and cortex, and that the non-lemniscal system is rich in CB. The functional role of these calcium-binding proteins in integrative functions is still unclear but several hypotheses have been proposed (Cruikshank et al. 2001).

3 Functional Aspects of Auditory Thalamic Sensory Processing

Early descriptions of MG neuronal responses captured the most prominent response properties (Galambos 1952; Galambos et al. 1952). Many MG cells show stronger responses and shorter latency as sound intensity increases; however, a substantial minority has a non-monotonic response. Despite the predominance of cells with stronger discharges to contralateral sound, their preference for the ipsilateral or the contralateral sound might reflect their MG location. The diversity of frequency response areas, the existence of 'on' and 'off' responses, and the suppressive effect of pure tones on click-evoked responses were noteworthy.

3.1 Frequency Tuning and Tonotopic Organization

3.1.1 Frequency Tuning

Most single unit studies have been performed on cat and the exploration of other species is more recent. In MGv, the sharpness of tuning in a population of cells does not suggest a progressive improvement from the brain stem tuning (Aitkin and Webster 1972; Calford and Webster 1981; Calford 1983). In cat, rat, and guinea-pig response latency and breadth of tuning differ between the lemniscal and non-lemniscal divisions. In cat, the mean latency is at least 5 ms longer in MGv than in MGd/MGm; the mean breadth of tuning increases twofold in MGd/MGm compared with MGv (Calford 1983). A relationship between response latency and breadth of tuning also exists: the shorter the latency the sharper the tuning (Calford 1983; Bordi and LeDoux 1994a; Edeline et al. 1999). The distributions of the breadth of tuning and that of the response latency largely overlap anatomical divisions: some MGm cells have a latency as short and/or tuning curves as sharp as those in MGv (Aitkin 1973; Calford 1983; Edeline et al. 1999).

3.1.2 Tonotopic Organization

MGv tonotopic organization has been mainly studied in barbiturate anesthetized cat: neurons with high characteristic frequency (CF) were found medially and neurons with low CF laterally (Aitkin and Webster 1971, 1972; Calford and Webster 1981; Morel et al. 1987) (Fig. 12.1). A more complex three dimensional tonotopic arrangement is proposed from multiunit recordings: while the lateral MGv contains only low frequency CFs, the medial part has a concentric organization with the low frequency CF region surrounded by middle frequency CFs, which themselves are partly embedded in a region of high CF (Imig and Morel 1985b). The rabbit MGv has a steep tonotopic gradient along the dorsoventral axis, with low CFs dorsally and higher CFs more ventral. In addition, the tonotopic gradient was weaker along the anteroposterior than across the mediolateral axes (Cetas et al. 2001). In guinea-pig, retrograde tracers injected in auditory cortex label a concentric projection with high frequencies in the central MGv core surrounded by the middle frequencies which are themselves surrounded by lower frequencies (Redies et al. 1989), a finding partly confirmed in physiological studies (Redies and Brandner 1991).

Tonotopic organization elsewhere in the auditory thalamus has not been fully analyzed. In cat, a tonotopic sequence was found in the lateral part of the posterior nucleus:

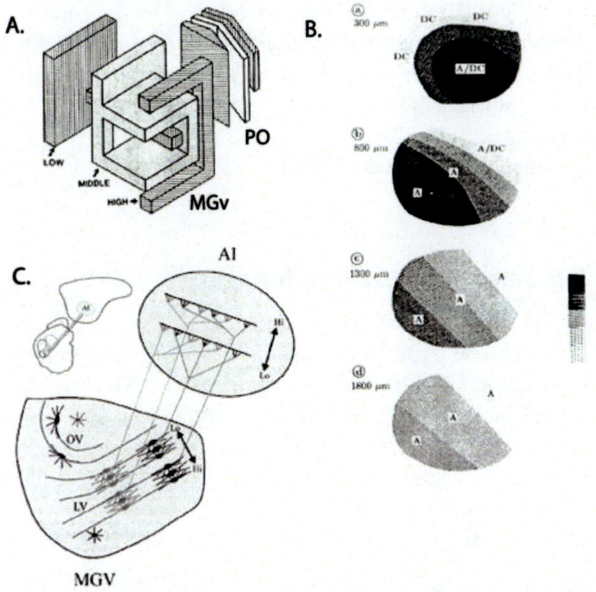

Fig. 12.1 Tonotopic organization in the MG of (**a**) cat, (**b**) guinea-pig, and (**c**) rabbit. **a** This block model depicts the tonotopic organization in the cat ventral MG (MGv) and posterior nucleus (PO). In MGv the lateral aspect has low-frequency CFs, whereas the medial part has a concentric structure where the center represents low-frequency CFs surrounded by middle-frequency CFs, themselves surrounded by high-frequency CFs. A simpler organization is present in PO with a progression from low-to-high frequencies rostrocaudally. Modified from the original source (Imig and Morel 1985a). **b** In the guinea-pig MG, the CFs of ventral MG neurons are shown in rostrocaudal sections 300 μm (*a*) to 1,800 μm (*d*) from the MGv rostral pole. The CFs are labeled in gray levels (scale at right). 'A' and 'DC' indicate that the neurons s project either to the cortical areas A or DC. Reproduced from the original (Redies et al. 1989). **c** Rabbit MGv projection. Adjacent cells, in the same frequency domain but with different binaural properties, project to separate loci in the same cortical frequency strip. Reproduced from the original (Velenovsky et al. 2003)

isofrequency contours progress from low to high frequencies in the rostrocaudal axis (Imig and Morel 1985a). In MGm, a tonotopic gradient finds lower CFs lateroventrally and higher CFs dorsomedially (Rouiller et al. 1989). The tonotopic factor (Morel et al. 1987) was lower in MGm than in MGv, but in both there was a more precise tonotopic gradient in the rostral than in the caudal MG. In the cat MG, a rostrocaudal gradient was detected for several parameters of MG cells, including inhibitory patterns, non-monotonic intensity functions, and response latency and variability; each of these decreases along the caudorostral axis and it was hypothesized that the modulation of these functional properties might be related with the density of GABAergic cells (Rouiller et al. 1990). This rostrocaudal gradient might not have been present in species in whom the MG is virtually devoid of GABAergic interneurons (Winer and Larue 1996; Huang et al 1999).

3.1.3 Intensity Tuning

Few studies have evaluated the intensity-function of MG neurons, a measure of the effect of sound intensity on evoked firing rate. Non-monotonic cells were three times as numerous as monotonic cells (Rouiller et al. 1983). This ratio was similar in the different MG subdivisions, but the proportions of monotonic units increased from 30% caudally to 70% rostrally (Rouiller et al. 1989).

3.1.4 Comparing Awake and Anesthetized Conditions

How neural functional properties and tonotopic maps are affected by anesthesia remains a matter of debate. Compared with the findings on deeply anesthetized animals, studies of lightly anaesthetized ones found a less ordered mediaolateral frequency gradient and a modulation of the tonotopy along the rostrocaudal axis and a more precise tonotopic arrangement anteriorly (Morel et al 1987; Rodrigues-Dagaeff et al. 1989). Relative to anesthetized animals, the breadth of tuning seen in the awake MGv is twice as wide, the mean latency is longer, and the threshold values are ~20 dB higher (Allon et al. 1981; Edeline et al. 1999). A high proportion of units were inhibited by tones or had a narrow excitatory area surrounded by wide-band inhibitory regions (Whitfield and Purser 1972). Inhibition is rarer in deeply anesthetized animals perhaps because of the sparse spontaneous activity. In awake animals, the high level of spontaneous activity can impair response measures of threshold and latency.

3.2 Temporal Aspects of Neuronal Responses

The temporal dynamics of neuronal discharges occur on different time-scales. These range from the response patterns at presentation of a brief tone to rhythmic activities over long epochs.

3.2.1 Discharge Pattern

The most common response of MG cells is a transient response at the onset or offset of acoustic stimuli. Cells with 'off' responses are spatially organized in clusters in MGv at the boundaries between it and other divisions (He 2001). The 'off' cells are duration sensitive, with responses at tone offset stronger when the tones last 100–400 ms. For 'on–off' neurons, the CF and the tuning curve shape differ for the 'on' and 'off' responses, and the 'off' threshold is often higher (He 2002). 'Off' responses could be generated by strong inhibitory inputs coming either from the IC or from the reticular nucleus.

Sustained responses are more common in the non-lemniscal divisions in anesthetized (Aitkin and Webster 1972; Calford 1983) and non-anesthetized animals (Allon et al. 1981; Edeline et al. 1999). To determine if sustained responses can be phase-locked to low-frequency stimuli, a large database of MG cells was analyzed: out of the 10% of the sustained responding cells, only 20% exhibit phased-locked responses (2% of the population). A good phase-locking index (vector strength R >0.5) is present for 20% of cells in the most lateral part of MGv and is far lower elsewhere (Rouiller et al. 1979); this can be elicited only by pure tones <500 Hz. When phase-locking was assessed with click trains, among 'locker' cells (discharges time-locked to individual clicks in the train), one-third had a strong preference for a particular rate (10–100 Hz), and in some cases complete unresponsiveness at low presentation rates (Rouiller and de Ribaupierre 1982).

3.2.2 Evoked Oscillations

Under several anesthetics, and with low rates of stimulation (0.1–0.2 Hz), MG cells have rhythmic 7–14 Hz discharges ~400–1,500 ms long (Galambos 1952; Aitkin et al. 1966) that likely result from neuronal interactions between MG and the auditory sector of the reticular nucleus and display the features of evoked spindles: they occur during cortical inactivation and disappear after reticular nucleus inactivation (Cotillon and Edeline 2000; Cotillon et al. 2000). In multiunit activity collected from the same sites first in awake animals then under several anesthetics, these oscillations were never observed during waking and were seen sometimes during slow-wave sleep (Cotillon-Williams and Edeline 2003). During sleep, the oscillations were non-stimulus locked (detectable on autocorrelograms), whereas under anesthesia half were stimulus locked (detectable from peri-stimulus-time histograms). Under some anesthetics, a slow oscillation (0.03–0.25 Hz) is seen in non-lemniscal neurons (He 2003a). Unknown is whether this slow oscillation relies on mechanisms similar to the slow oscillations in other cortical areas (Steriade et al. 1993a, b, c, d). This slow rhythm, present in spontaneous activity, strongly affects the responsiveness of MG neurons: every 10–20 s the evoked discharges shifted from a one-spike response to a robust 10-spike response. Whether such oscillations control the thalamic neural responsiveness in awake animals remains an open question.

Spontaneous and evoked gamma auditory cortex oscillations occur in anesthetized (Franowicz and Barth 1995) and in vitro (Metherate and Cruikshank 1999) preparations. Although MG lesions do not impair spontaneous gamma oscillations (Brett et al. 1996), activation of the MG division modulate differentially these oscillations: they are inhibited by MGd and MGv stimulation, and evoked by stimulation of the adjacent posterior intralaminar nucleus (Barth and MacDonald 1996; Sukov and Barth 2001) and of the reticular nucleus (MacDonald et al. 1998).

3.3 Sensitivity to Directional Cues

Most binaural thalamocortical cells are sensitive to interaural disparities such as the interaural intensity difference (IID) and the interaural phase differences (IPD). The initial MG dichotic studies classified the cells in a few categories (see below), and some subsequent studies have described their directional sensitivity.

3.3.1 Initial Classification of Binaural Interactions

MG cells are sensitive to IIDs which in free-field conditions ensue from head shadowing and pinna amplification (Adrian et al. 1966; Aitkin and Webster 1972). As in other auditory nuclei, binaural interactions are defined as excitatory (E), inhibitory (I) or no (O) events evoked by stimulation of each ear. An EI unit is excited by contralateral stimuli and inhibited by ipsilateral stimuli; an EO(I) unit shows no ipsilateral effect alone and an inhibition of contralateral excitation by a simultaneous ipsilateral stimulus. Most cells show binaural inhibitory or facilitatory interactions and 10–20% respond only to stimulation of one ear, with contralateral cells being far more common (Adrian et al. 1966; Aitkin and Webster 1972; Calford and Webster 1981).

3.3.2 Sensitivity to Interaural Intensity Differences

MG EI cells have IID-sensitivity that is relatively independent of binaural intensity (Ivarsson et al. 1988). Based on binaural intensity response fields which represent the changes in both IID and overall sound pressure level (SPL), MG cells form four categories: lateralized, centered, bilateral, and monaural-like. These categories can be integrated in a more commonly used classification, since the cells have either inhibitory or facilitatory binaural interactions (Clarey et al. 1992; Irvine 1992). The inhibitory or facilitatory binaural interactions are based on IID functions created by maintaining the average binaural intensity and changing the IID such that an SPL increase at one ear decreases the SPL at the other ear. In MG and auditory cortex, most cells with inhibitory binaural interactions correspond to EI and EO(I), and few correspond to the reverse pattern (IE) and respond maximally to IIDs favoring the ipsilateral ear. On the other hand, the IID sensitivity of MG cells, classified by dichotic studies as EE(F) or EO(F), is the product of facilitatory binaural interactions: cells weakly responsive to monaural stimulation of either ear can be strongly responsive to binaural stimulation at a particular IID value, thus showing facilitation. When the optimum IID value is near zero, this type of cell corresponds to the centered cells (Ivarsson et al. 1988). The proportion of monaural MG cells is lower than in the IC, suggesting that the projections from the midbrain to the thalamus provide additional binaural convergence.

3.3.3 Sensitivity Revealed by Free-Field Stimulation

Analyses of the azimuth tuning in MG and primary auditory cortex (AI) of barbiturate-anesthetized cats in free-field conditions quantified azimuth tuning for a range of intensities, thus providing azimuth-level response areas (Barone et al. 1996; Clarey et al. 1995; Imig et al. 1997; Samson et al. 2000) (Fig. 12.2).

Although the best azimuth to noise and pure tone usually matched, many cells had a more selective azimuth function to noise than to pure tones. Azimuth sensitivity does not differ between MG divisions but several differences were found between thalamus and cortex (Barone et al. 1996). Mean azimuth sensitivity was greater in AI than in MG (82% vs. 75%), and the proportion of azimuth sensitive (AS) units was higher in MG (69% vs. 57%). Non-monotonicity strength was greater in AI (35% vs. 23%), indicating decreased responsiveness with increasing SPL; the number of non-monotonic AS units was higher in cortex (18.5% vs. 7.4%). There was a significant relationship between breadth of azimuth tuning and breadth of level tuning in AI, but not in MG. The non-monotonic AS cortical cells in these free-field conditions might correspond to the two-way intensity network (TWIN) cells (Semple and Kitzes 1993a, b) seen when manipulating the IID; this response was interpreted as the result of cortical processing. These results suggest that the distribution of AI azimuth preference largely reflects that of the MG. However, the higher proportion of AS and non-monotonic AI cells indicates that there may be further AS synthesis in AI cells.

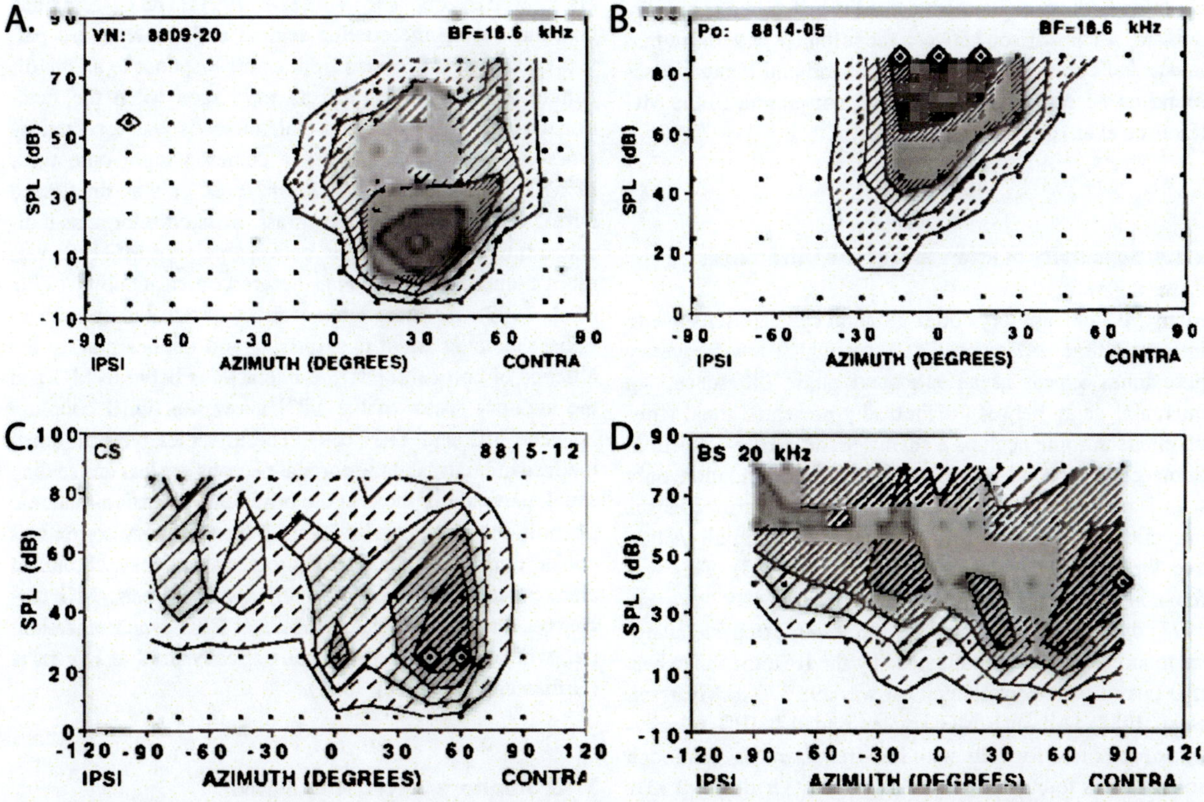

Fig. 12.2 Azimuth-level response areas of a cat MG cell. **a**, **b** Units were studied with various SPLs for a set of frontal hemifield azimuths (±90°). The areas display the response magnitude as a joint function of azimuth and SPL. Interpolated isoresponse *contour lines* and *shadings* delineate azimuth-level combinations that produced >5, 25, 50, and 75% responses relative to the maximum. Modified from the original source (Barone et al. 1996). **c**, **d** Azimuth-level response area of an MG cell to monaural contralateral stimulation with (**c**) broad band noise and (**d**) to binaural sound at the CF (20 kHz) of the cell. In **c**, the cell responded optimally at 45–60° azimuth, in **d** to each azimuth. Modified from the original source (Imig et al. 1997)

Although most directional sensitivity is presumed to result from binaural interactions, unilaterally deaf humans can localize broadband high-frequency sound (but not tonal stimuli). To what extent can monaural cues provide sufficient information to generate directional selectivity of auditory neurons? Unilateral ear occlusion revealed that a significant number of cells show directional sensitivity based on monaural cues. These monaural direction (MD) cells were first described in auditory cortex (Samson et al. 1993), then in the MG (Imig et al. 1997), dorsal cochlear nucleus (Imig et al. 2000) and inferior colliculus (Poirier et al. 2003). Perhaps this property arises from pinna-derived spectral cues and from the head-related transfer function (HRTF): broadband noises from different locations are affected differentially by the HRTF and therefore generate greater sound pressure at certain frequencies (Musicant and Butler 1984; Musicant et al. 1990). In the MG, the frequency-responses areas of MD cells have excitatory and inhibitory domains, and their excitatory domains are narrower than those of binaural direction-sensitive cells (Imig et al. 1997). Furthermore, the thalamic MD cells were direction-sensitive both in azimuth and elevation to spectral stimuli that engage both excitatory and inhibitory domains, and they were insensitive to stimulus direction with spectral components limited to one excitatory domain. A comparison between the response patterns in thalamus and cortex sensitivity to monaural and binaural cues found no AI response type that was not present in the MG (Samson et al. 2000).

3.3.4 Sensitivity of Interaural Phase Differences

Many MG and auditory cortex binaural cells are sensitive to interaural time difference (ITD), for which low-frequency pure tones appear as an interaural phase difference. An interaural delay between dichotically presented tonal stimuli elicits in some neurons a characteristic periodic function. Some cells respond strongly to binaural stimulation only when a particular delay of a few hundreds of microseconds is introduced between the ipsi- and contralateral stimuli (Aitkin and Webster 1972; Aitkin 1973; Calford 1983). This was found in MGv and in MGm (Aitkin and Webster 1972; Aitkin 1973) or only in MGv (Calford 1983). An extensive analysis found 28% of cells influenced by the IPD; the mean best IPD favoring the contralateral ear was 29±170 μs (Ivarsson et al. 1988). All such MG studies find that IPD selectivity is prominent for cells with low-frequency CF. Although some studies found all delay-sensitive units had BF <1 kHz (Calford, 1983), others reported a more progressive cut-off: e.g., 52% of the cells with BF <3 kHz were sensitive to IPD, but only 25% of those with BF >3 kHz were (Ivarsson et al. 1988).

3.4 Corticofugal Influence

Two techniques can probe how cortical activity affects the functional properties of thalamic neurons: cortical inactivation and cortical electrical stimulation. Each has advantages and constraints. Inactivation removes the cortex from the thalamocortical loop but leaves unclear when and how cortical neurons influence thalamic cells. Electrical stimulation enables the dissection of the circuits and their timing, but they produce highly and non-physiologically synchronized activation of corticothalamic fibers.

3.4.1 Inactivation of Auditory Cortex

When MG activity was recorded during auditory cortex cooling, dissociation was found between onset and late reverberatory responses. Phasic onset responses were unaffected, but reverberatory responses systematically disappeared (Ryugo and Weinberger 1976). Subsequent studies of cortical cooling on MG performance produced mixed results. Cortical cooling decreased spontaneous activity for 60% of MG cells, and some cells had enhanced frequency tuning whereas others showed the opposite (Villa et al. 1991). The CF remained constant during the cooling regime and the response pattern could be transformed from an off-response to an on–off pattern. Heterogeneous effects were seen when the functional connectivity between thalamic cells was assessed by cross-correlograms indicating a common input onto pairs of MG cells, which either disappeared or was unaffected during cortical cooling. In contrast, cross-correlograms indicate a functional connection between pairs of MG cells was more numerous during than before cortical cooling (Villa et al. 1999). As there are no direct intrathalamic connections, this latter result is surprising and implies that, in the absence of cortical input, the interactions between MG and the auditory sector of the TRN allow functional coupling between MG and TRN cells. These studies were in anesthetized animals, and the cortical neurons are less active than usual, particularly those at the origin of the cortico-thalamic fibers in the infragranular layers. When auditory cortex was cooled during periods of synchronized and desynchronized electroencephalogram (EEG) in awake animals, MG discharges diminished during periods of EEG synchronization, i.e., when cortical and MG cells already fired at low rates (Orman and Humphrey 1981).

3.4.2 Auditory Cortex Stimulation

Few studies have evaluated how activation of auditory cortex neurons modifies the responsiveness and functional properties of MG cells (He 2003a). In cat, auditory cortex electrical stimulation increased excitatory tone-evoked responses in most MGv cells (77%). This facilitation developed when there was a good match between the cortical stimulation site CF and the MG neuron (He 1997); the effective facilitatory sites formed patches ~1 mm aligned along isofrequency contours and separated by more than a millimeter. Similar effects were seen in the guinea-pig MGv but a given cortical site facilitated much wider MGv regions (He et al. 2002). A more complex picture was seen in the non-lemniscal MG where cortical stimulation largely attenuated (sometimes totally inhibited) the 'On' phasic responses in the medial and shell nucleus, whereas neurons with 'Off' or 'On–Off' patterns of responses expressed facilitation (He 2003c). The dichotomy between the lemniscal and non-lemniscal MG was confirmed by intracellular recordings. In most cases, cortical stimulation depolarized lemniscal neurons and hyperpolarized non-lemniscal neurons (Yu et al. 2004a). It was proposed that the EPSP-mediated facilitation results from a direct action of the corticofugal fibers onto thalamic relay cells, whereas IPSP-mediated inhibition ensues from the activation of TRN cells which provide a powerful and long-lasting control of thalamic relay cells through $GABA_B$ receptors. An unresolved question is why the direct corticofugal excitatory effect overcomes the inhibition provided by the TRN neurons in the lemniscal MG, whereas this inhibition dominates the non-lemniscal MG. As these experiments used pulse trains to activate the cortex, it is difficult to evaluate how cortical neurons influence thalamic cells in normal physiological conditions, but changing the number of stimulation pulses (e.g., 1–20) sometimes elicited only quantitative differences in IPSP amplitude and duration (Yu et al. 2004b).

3.5 Responses to Natural Stimuli

The responses of cortical neurons to natural stimuli such as conspecific or heterospecific vocalizations raise important issues about serial processing. A minority of cortical neurons behave as call detectors, i.e., respond exclusively to one or two naturalistic calls from a sample (Funkenstein and Winter 1973; Winter and Funkenstein 1973). Nevertheless, MG cells respond to a wide range of natural calls, and to more such components of a call than do cortical cells (Creutzfeldt et al. 1980), even when the thalamic and cortical cells display functional interactions (assessed by peaks in the cross-correlograms). This property may simply reflect higher values of MG best amplitude modulation (Joris et al. 2004). Other differences are that cortical responses are poorly predictable by linear analyses such as the spectrotemporal receptive field (STRF) derived from complex stimuli that included animal vocalizations (Machens et al. 2004), whereas the responses of MG cells to natural calls seem easier to predict (Yeshurun et al. 1989, 1985). Moreover, based upon their discharge rates, cortical neurons apparently have species-specific specializations, whereas MG neurons do not. In marmoset auditory cortex, most cells respond more strongly to marmoset calls than to their time-reversed version, which is not the case for cat neurons (Wang and Kadia 2001). In contrast, guinea-pig MG neurons respond similarly to normal and time-reversed guinea-pig calls and no differences were noted between guinea-pig and rat MG cells (Philibert et al. 2005). However, studies quantifying the information transmitted by spikes trains emitted at presentation of vocalizations find that spike-timing precision allows discrimination between normal and time-reversed calls (Fig. 12.3).

4 Auditory Thalamus and Integrative Function

4.1 State Dependent Changes

The thalamus is the first station where sensory messages are attenuated during sleep (Steriade 1984, 1989; Steriade et al. 1997; but see Morales-Cobas et al. 1995; Pena et al. 1992); Cairns et al. 1996; Soja et al. 1996). This conclusion reflects a limited set of results from the visual thalamus based on juxtacellular recordings (Coenen and Vendrick 1972) which estimated that the transfer ratio between EPSPs and spikes fell from 0.9–1.0 in waking (W) to 0.4–0.5 in slow-wave sleep (SWS) preparations. Visual thalamic neurons were hyperpolarized by ~4 mV when the animal shifted from quiet waking to SWS, whereas in paradoxical sleep (PS) neurons were depolarized by ~10 mV relative to SWS (Hirsch et al. 1983).

Extracellular recordings from the auditory system of awake (non-sleep-deprived) animals confirmed that most MG cells decrease spontaneous and evoked activity when the animal shifted from W to SWS; decreases or no change in spontaneous and evoked activity were seen when comparing W and PS states (Fig. 12.4; Edeline et al. 2000). As a consequence of these changes, the frequency-tuning curve of 70% of neurons narrowed in SWS compared with W, and their frequency response area decreased (Fig. 12.4). During PS, 60% of cells had narrower frequency tuning than in SWS whereas 40% had tuning like that in W. Acoustic threshold was increased in SWS and was even higher in PS, though the shape of the rate-level functions was rarely affected by the state of vigilance. The state-dependent modulation at the cortex differs from that in MG: most cortical cells show reliable modification of their response from W to SWS and to PS, but the diversity of the effects from one cell to the next produce on average no significant modification of the parameters quantifying the frequency response area (Edeline et al. 2001).

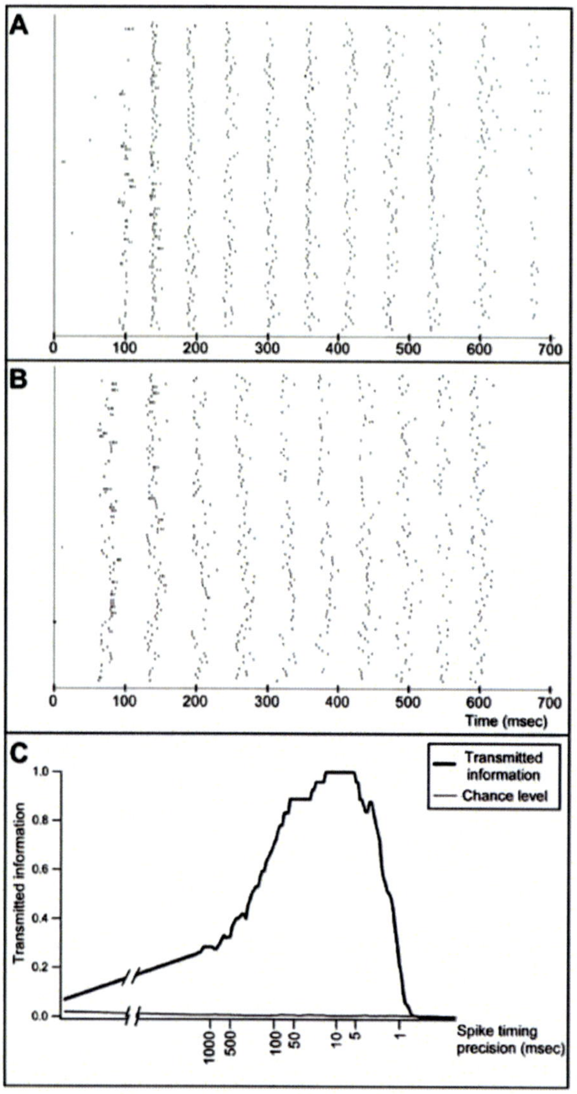

Fig. 12.3 The rate of transmitted information increases when spike-timing is considered. **a**, **b** Raster plots of MG neuron responses to 100 presentations of a guinea-pig vocalization ('purr') in (natural, **a**) forward and backward (artificial, **b**) versions. The stimulus began at time 0 and was 700 ms long. **c** Transmitted information (*upper curve*) and chance level (*lower curve*) as a function of the spike-timing precision computed with the metric-space analysis (Victor and Purpura 1996, 1997). The *upper curve* shows that almost no transmitted information is based on spike count (i.e., for a temporal precision equal to zero) and that the maximum transmitted information is reached for spike-timing values ~10 ms long. Modified from the original source (Huetz and Edeline 2006)

Several temporal characteristics of the MG cell signals are also modified by the state of vigilance. The proportion of high-frequency (>200 Hz) bursts, long viewed as the EEG signature of synchronized states such as SWS, increased from ~5% in W to ~10% in SWS (Massaux and Edeline 2003; Massaux et al. 2004). Higher burst proportions (15–30%) were seen in various anesthetic regimes (urethane, pentobarbital, ketamine/xylazine). As the burst mode never dominated the cell discharge mode, the view that the tonic mode of discharge corresponds to W and the burst mode to SWS should be reconsidered. In fact, the notion of 'mode of discharge,' though relevant to describe the results from in vitro conditions, may be inadequate to characterize the thalamic neuron discharge patterns in awake animals. The evoked neuronal rhythmic activities triggered by sensory stimuli differ in W and in EEG synchronized states, respectively. The temporal profile of multiunit responses revealed that acoustic stimuli often trigger 7–15 Hz oscillations in the anesthetized state. These oscillations can also be detected (but less often) during natural SWS, but were never seen in W and PS (Cotillon-Williams and Edeline 2003).

It is noteworthy that the effects seen at the shift from W to SWS contradict the EEG changes in anesthetized preparations (for review see Hennevin et al. 2007). For example, in anesthetized animals, the receptive fields of visual cortex neurons are wider during synchronized EEG epochs (Wörgötter et al. 1998) and the responses of barrel cortex neurons are suppressed at arousal evoked by stimulating the brain stem reticular formation (Castro-Alamancos 2002; Castro-Alamancos and Oldford 2002). Collecting data in anesthetized animals is much easier and less time-consuming than in awake animals, and one can easily succumb to the temptation of considering that the changes observed under anesthesia mimic changes observed during sleep. This oversimplification can only contribute to generate confusions in a domain where data are rare.

4.2 Post-injury Plasticity

It has been often shown that adult sensory cortex reorganizes after peripheral injury (Kaas et al. 1983; Jenkins et al. 1990; Calford 2002). In the auditory modality, Robertson and Irvine (1989) were the first to show reorganizations after restricted cochlear lesions in adult animals. One month after the lesion, the frequency representation of the contralateral auditory cortex is reorganized in such a way that the cortical region normally receiving inputs from the lesioned cochlea is entirely or partly occupied by an expanded representation of the frequencies represented at the edges of the cochlear lesion. Subcortical reorganization also occurs and 1–3 months after restricted unilateral cochlear lesions, the ventral MG tonotopic organization reorganized much as did the cortical map (Kamke et al. 2003). Both in cortex and MG, the normal thresholds from the regions of the expanded representation argue against an interpretation in terms of 'residues of prelesion responses': in both, a dynamic process reorganized the tonotopic map. In contrast, the cochlear nucleus tonotopic map does not reorganize after cochlear lesions (Rajan and Irvine 1998) and only modest reorganization is seen in CNIC, where half the electrode penetrations show changes in frequency organization explicable as residues of prelesion responses and the remainder were interpretable as unmasking of normally inhibited inputs or as dynamic reorganization (Irvine et al. 2003). Discrete zones of reorganization may exist in the CNIC, although most of the observed changes in it are passive consequences of the lesion. Thus, the MG is the first site where massive tonotopic reorganisations are expressed after restricted cochlear lesions, suggesting a bottom-up process as proposed in other modalities (Pons et al. 1991; Jones 2000).

4.3 Learning-Induced Plasticity

4.3.1 Findings Obtained During Behavioral Training

Sensory responses can be modified when awake animals learn (Weinberger and Diamond 1987). Multiunit or single unit recordings demonstrate that MGm neurons exhibit discharge plasticity in aversive (Gabriel et al. 1975; Ryugo and Weinberger 1978; Supple and Kapp 1989, Edeline 1990; Edeline et al. 1990a, b) and appetitive tasks (Disterhoft and Olds 1972; Birt and Olds 1981; Maho and Hennevin 2002). MGm plasticity is long lasting (Edeline et al. 1988) and robust enough to survive extinction trials (Supple and Kapp 1989). In a trace conditioning protocol, relationships between MGm responses and the behavioral conditioned responses (CR) were found: MGm neuron activity on fast CR trials was higher and occurred earlier than on slow CR trials (O'Connors et al. 1997). To evaluate if changes in MGm synaptic efficacy underlie the plasticity in a learning task, the responses of MG neurons were tested before and after conditioning, to BIC stimulation and to superior colliculus (SC) stimulation. After training, BIC but not SC stimulation evoked larger responses at shorter latency (McEchron et al. 1996). Although it is challenging to assess synaptic efficacy from extracellular recordings, this study suggested prospective intracellular experiments.

MG neurons can code more than the significance of a particular sound. MGm neuron responses analyzed at presentation of an acoustic conditioned stimulus (CS) signaling the occurrence of rewards of different magnitudes at various delays after the CS show that the late responses gradually increased, peaked before the reward presentation, and were correlated with reward intensity (Komura et al. 2001). The early and the late response components were also influenced by significant visual cues (Komura et al. 2005), supporting a multimodal effect on acoustic stimuli. Moreover, the non-lemniscal MG may provide a shortcut to the cortical network for multimodal integration in goal-directed behaviors.

An interesting characteristic of the learning-induced plasticity occurring in MGm is its potency to be transfered to other behavioral states such as PS. In several experiments, the increased responses obtained at the CS presentation were detected during phases of PS following conditioning with no sign of behavioral awaking (Hennevin et al. 1993, 1998; Maho and Hennevin 2002). In contrast, the plasticity of MG neurons poorly transfers to SWS (Hennevin and Maho 2005).

4.3.2 Modification of Medial Geniculate Body Frequency Tuning: A Bottom-Up Process for Cortical Plasticity?

Thalamocortical neurons show selective functional modification during behavioral training (Edeline 1999, 2003; Weinberger 2004; Fritz et al. 2005a) as studied in three basic designs. A 'Pre-Post' task compares neural tuning before and after a brief conditioning session and shows that auditory cortex neurons can be retuned to the frequency of a significant sound (Weinberger 2004). A second set of studies involved a training extending over several weeks or months during which a frequency predicted the occurrence of an appetitive reward. Cortical maps quantified under general anesthesia were compared with those from controls. Two such studies found cortical map reorganization after extensive training (Recanzone et al. 1993; Rutkowski and Weinberger 2005) and another found no map reorganization (Brown et al. 2004). The third type of investigation assesses the neurons' functional properties in awake animals during a behavioral task. An initial study showed a particular type of selective effect where the CS+ was in a "valley" surrounded by two increases (Ohl and Scheich 1996, 1997). Using dynamic noise to construct the neurons' spectrotemporal receptive field, active listening for a particular sound frequency reinforced the excitatory areas (and/or decrease the inhibitory area) at or near this frequency (Fritz et al. 2003, 2005b).

The MG data from 'Pre–Post' designs show selective re-tuning favoring the CS frequency after a brief conditioning session. The selective effects ranged from ~50% in MGd and MGm to 29% in MGv (Edeline and Weinberger 1991a, b, 1992); this plasticity lasted for more than an hour in MGd/MGm but dissipated quickly in MGv. Perhaps the narrow tuning curves of MGv neurons are maintained by GABAergic projections which allow short-term, but not long-term, plasticity. This was confirmed in MGm: the narrower the tuning curve, the briefer the re-tuning after conditioning (Edeline and Weinberger 1992). Is thalamic plasticity responsible for cortical plasticity, or do corticofugal inputs control the emergence of thalamic plasticity? The question might not be relevant because of the tight links between thalamic and cortical networks; a co-emergence of plastic changes in both structures is highly probable (see Chapter 22).

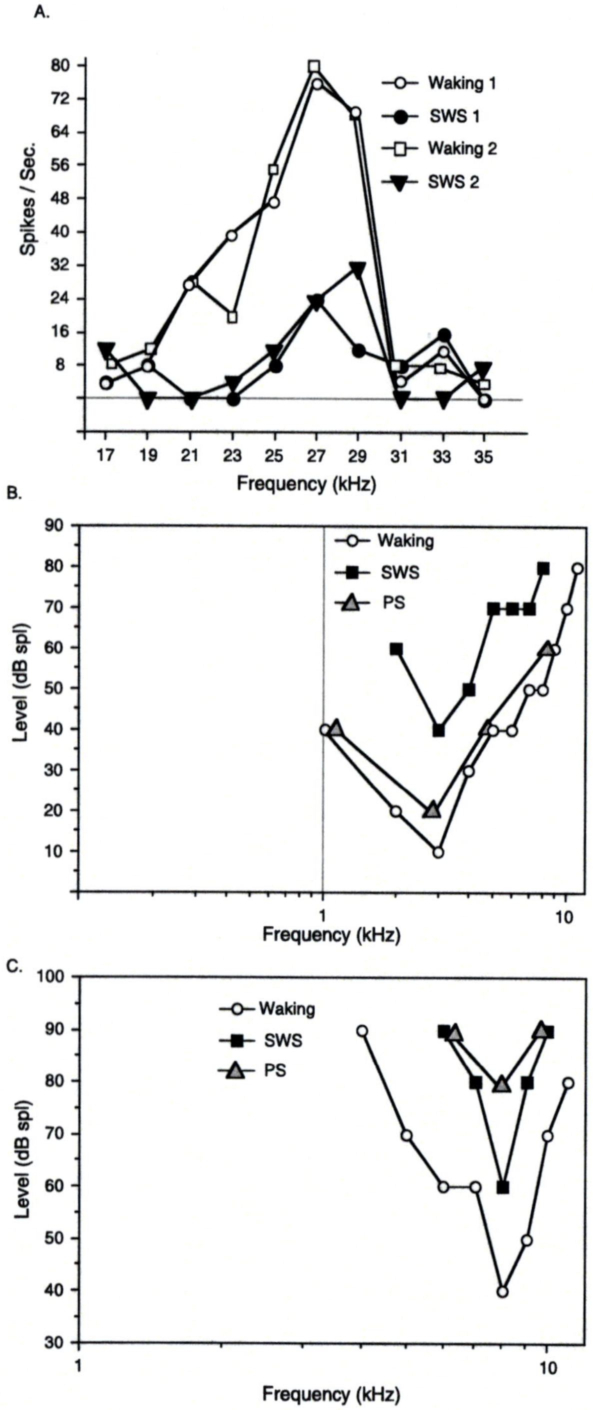

Fig. 12.4 Modulation of MGv frequency tuning by the state of vigilance. **a** The cell was tested at 70 dB twice during periods of wakefulness and twice in slow-wave sleep (SWS) episodes of 120 s. The responses were largely decreased in SWS compared to waking, and were reliable within a given state of vigilance. **b, c** Threshold tuning curves in waking, SWS and paradoxical sleep (PS). **b** The tuning curve was smaller in SWS than in waking, and in PS it resembled that in waking. **c** The tuning curve, reduced in SWS, was still smaller in PS. Modified from the original source (Edeline et al. 2000)

4.3.3 Plasticity of Medial Geniculate Body Neurons in Fear Conditioning: Relation with Limbic Plasticity

Classical fear conditioning has emerged as an assay for potential cellular learning and memory mechanisms (Fanselow and LeDoux 1999; Schafe et al. 2001; McGaugh 2004). In many learning situations, plastic changes were detected simultaneously in the thalamocortical auditory system and in non-sensory, limbic structures (Olds et al. 1972; Disterhoft and Stuart 1976; Edeline et al. 1990a, b; Maho et al. 1995; Quirk et al. 1997). The amygdala exhibits conditioned responses and is essential for fear conditioning and it has been proposed as a primary site for gating plasticity in the auditory cortex and thalamus (Fanselow and LeDoux 1999; Maren and Quirk 2004). However, as the learning-induced changes occurring in auditory cortex and MG are highly selective for the CS, they most likely result from afferent-specific plasticity and are not easily explained by the influence of non-auditory structures (note that there is no direct pathway from the amygdala to MG). For example, the tuning shifts to the CS frequency in MG likely reflect changes in synaptic efficacy between IC terminals and thalamic relay cells: the efficacy of the synapses conveying the CS information is reinforced and that of the synapses conveying the initial best frequency information is decreased. That the amygdala influence produces such opposite changes in synaptic efficacy is hardly plausible. The MGm plasticity impaired by muscimol inactivation of the amygdala (Maren et al. 2001; Poremba and Gabriel 2001) might support this scenario, although the large volumes of injected muscimol could invade the MG and confound the interpretation. MG plasticity can be expressed without amygdala plasticity during PS after appetitive conditioning (Maho and Hennevin 2002); this is also the case when a microtubule associated protein (MAP) kinase inhibitor blocks amygdaloid plasticity without preventing MGm plasticity (Schafe et al. 2005).

Surprisingly, lesions of different MG divisions produce heterogeneous results. Work in rabbits found that MGm lesions attenuated conditioned bradycardia to an acoustic CS (Jarrell et al. 1986a, b; McCabe et al. 1993). However, MGm lesions in rats did not prevent fear conditioning as assessed by freezing and arterial pressure (Romanski and LeDoux 1992): only when these lesions were combined with auditory cortex lesions was there a fear conditioning deficit. Lesions of MGm/PIN do not prevent fear-potentiated startle, whereas MGv/MGd lesions impaired auditory (but not visual) fear-potentiated startle, an impairment ameliorated by retraining (Campeau and Davis 1995). MG lesions alone can block corticosterone release normally induced by noise (Campeau et al. 1997). The fear conditioning results contrast with the effects of auditory forebrain lesions on instrumental tasks ((Ravizza and Belmore 1978; Whitfield 1979). Auditory forebrain lesions do not impair the analysis of acoustic parameters such as frequency and intensity, but do impair spatial localization and the temporal analysis of acoustic stimuli (Phillips and Farmer 1990). Human MG lesions can produce transient auditory illusions (Fukutake and Hattori 1998) and auditory cortex damage impairs the comprehension of verbal material and/or attentional features (Wester et al. 2001).

5 Conclusions and Future Directions

5.1 Special Features of Medial Geniculate Body Compared with the Visual and Somatic Sensory Thalamus

5.1.1 Unique Features of Medial Geniculate Body Inhibition

The diversity of inhibition converging onto MG cells sets the auditory thalamus apart from the visual and somatic sensory thalamus. The IC inhibitory projections target MG relay cells. In cat and monkey, other IPSPs come from local GABAergic (probably small stellate) cells, whereas in rodent the few GABAergic cells suggest that local inhibitory circuits do not play this role. The massive TRN input to MG cells triggers further inhibition as in visual and somatic sensory systems. Few studies have examined TRN cells effects on thalamic neuron functional properties. In the somatic sensory thalamus, the receptive fields (RFs) of ventral posterior neurons grew larger after somatic sensory TRN excitotoxic lesions (Lee et al. 1994), with little effect on RF properties after glutamatergic activation of the somatic sensory TRN (Warren and Jones 1994). In the MG, altering TRN activity by glutamate or GABA iontophoresis affects evoked discharge, tuning breadth and acoustic threshold only if the distance between the CF in the TRN and the MG cells is <0.25 octave (Fig. 12.5), suggesting that functional interactions between TRN cells and MG cells reflect topographic connections (Cotillon-Williams et al. 2008).

The sparse electrophysiological data impede understanding how TRN affects MG relay cells. Anatomical data suggest that TRN and MG cells form open loop connections and that TRN may create lateral inhibition in the thalamus (Pinault and Deschênes, 1998). However, this scheme may hold in the somatic sensory thalamus and not in other modalities. In the visual system (Fig. 12.6b), inhibitory interneurons receiving retinal inputs suggests alternative forms of thalamic integration. The auditory system differs since the TC cells receive inhibitory and excitatory IC input (Fig. 12.6c). How these several inhibitory inputs to MG cells shape lateral inhibition and/or to feedforward and feedback inhibition is unknown.

5.1.2 Temporal Constraints

Many models of thalamic functional organization are based on the visual system (Usrey 2002; Alitto and Usrey 2003). However, the temporal resolution for feature extraction differs between modalities. Several auditory time scales encode simultaneously the temporal envelope (hundreds of ms) and the fine structure (a few ms) of acoustic signals (Elhilali et al. 2004). Thus, the transient (phasic) responses in the auditory TC system are likely precise and synchronized across neurons. Analyses of evoked response strength (excluding spike-timing) that succeed in the visual system often fail in the auditory system. For example, a receiver operating characteristic analysis applied to thalamic spike trains indicated that bursts of visual thalamus neurons promote a better detection of the stimuli than single spikes (Guido et al. 1995), whereas bursts of auditory thalamus neurons do not (Massaux et al. 2004).

5.2 Future Directions: Toward Understanding Medial Geniculate Body Function

5.2.1 Neuromodulators

Neuromodulators affect the excitability and functional properties of auditory cortex neurons. However, neuromodulatory effects of acetylcholine, noradrenalin, and serotonin on MG cell function are unknown. Iontophoretic application or stimulation of the source nuclei (peripeduncular tegmentum, locus ceruleus, dorsal raphe) requires study in anesthetized and awake animals.

5.2.2 Studying the Thalamocortical System as a Whole

Work on the plasticity of auditory cortex neurons is often done without determining whether such properties exist in the MG (Bao et al. 2004; Polley et al. 2004). MG cells have many features like those of cortical cells whereas IC neurons do not (Las et al. 2005). It is vital to understand how MG functional properties are transferred to auditory cortex cells (Miller et al. 2001).

5.2.3 Using Biologically Relevant Stimuli and Natural Conditions

Auditory neurons are usually studied with standardized artificial stimuli: pure tones and AM or FM sounds. Complex dynamic signals assess and sample functional properties of auditory neurons in a different way (Klein et al. 2000; Escabi and Schreiner 2002). The complex stimuli used to map the

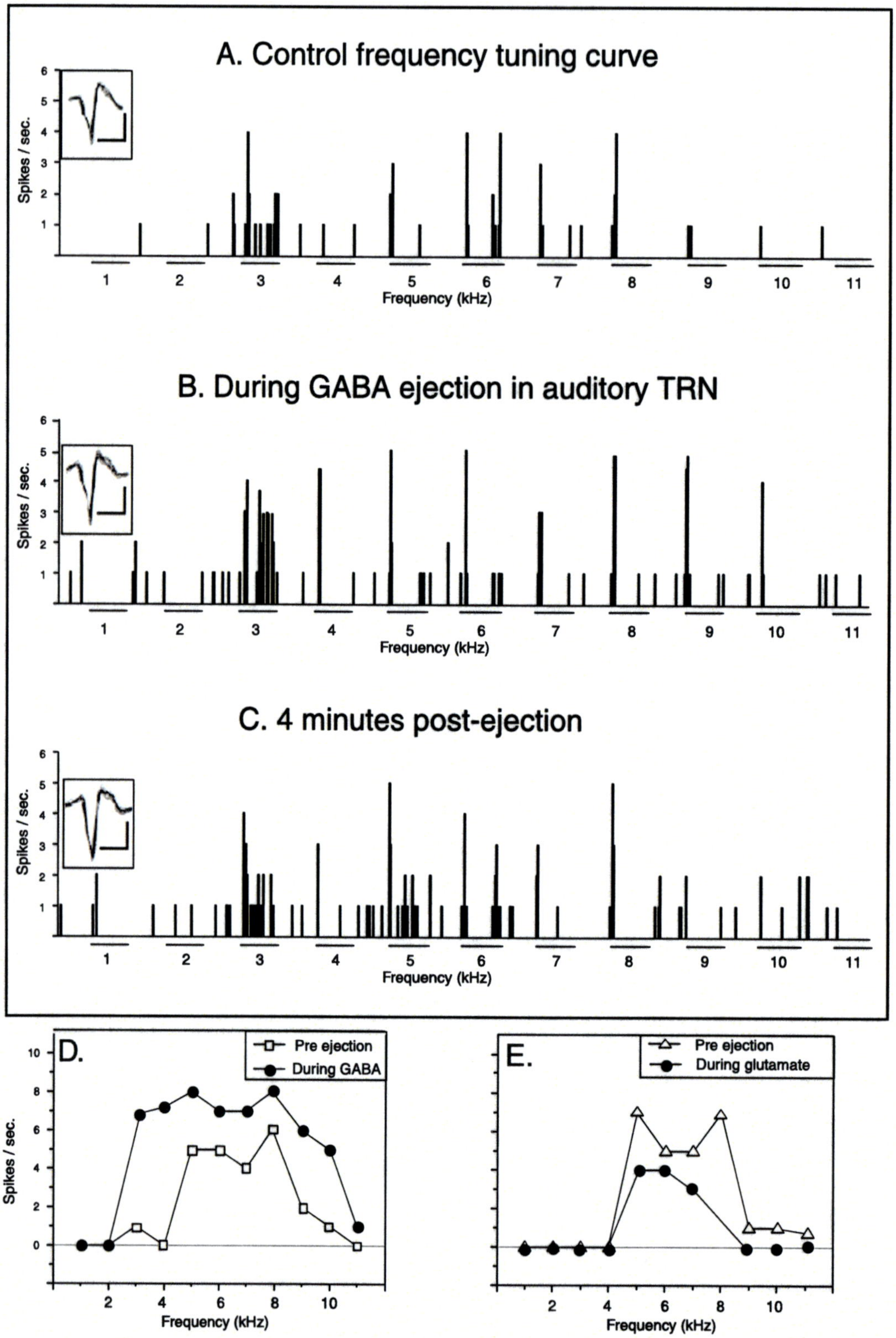

Fig. 12.5 Tuning curve expansion of a ventral MG during GABA ejection (250 nA) and reduction after glutamate ejection (260 nA) in the auditory TRN. **a** In the control tuning curve, this cell (8 kHz CF at 40 dB) responded from 3 to 8 kHz. GABA was ejected in the auditory TRN at a site with a 9 kHz CF. **b, d** During GABA ejection, responses at all frequencies were facilitated and new responses emerged at 9 and 10 kHz. **c** One minute after the GABA application ended, some responses were still facilitated but 4 min later the tuning curve resembled the control. **e** In contrast, during glutamate ejection, the responses were attenuated for almost all the frequencies. Modified from the original source (Cotillon-Williams et al. 2008)

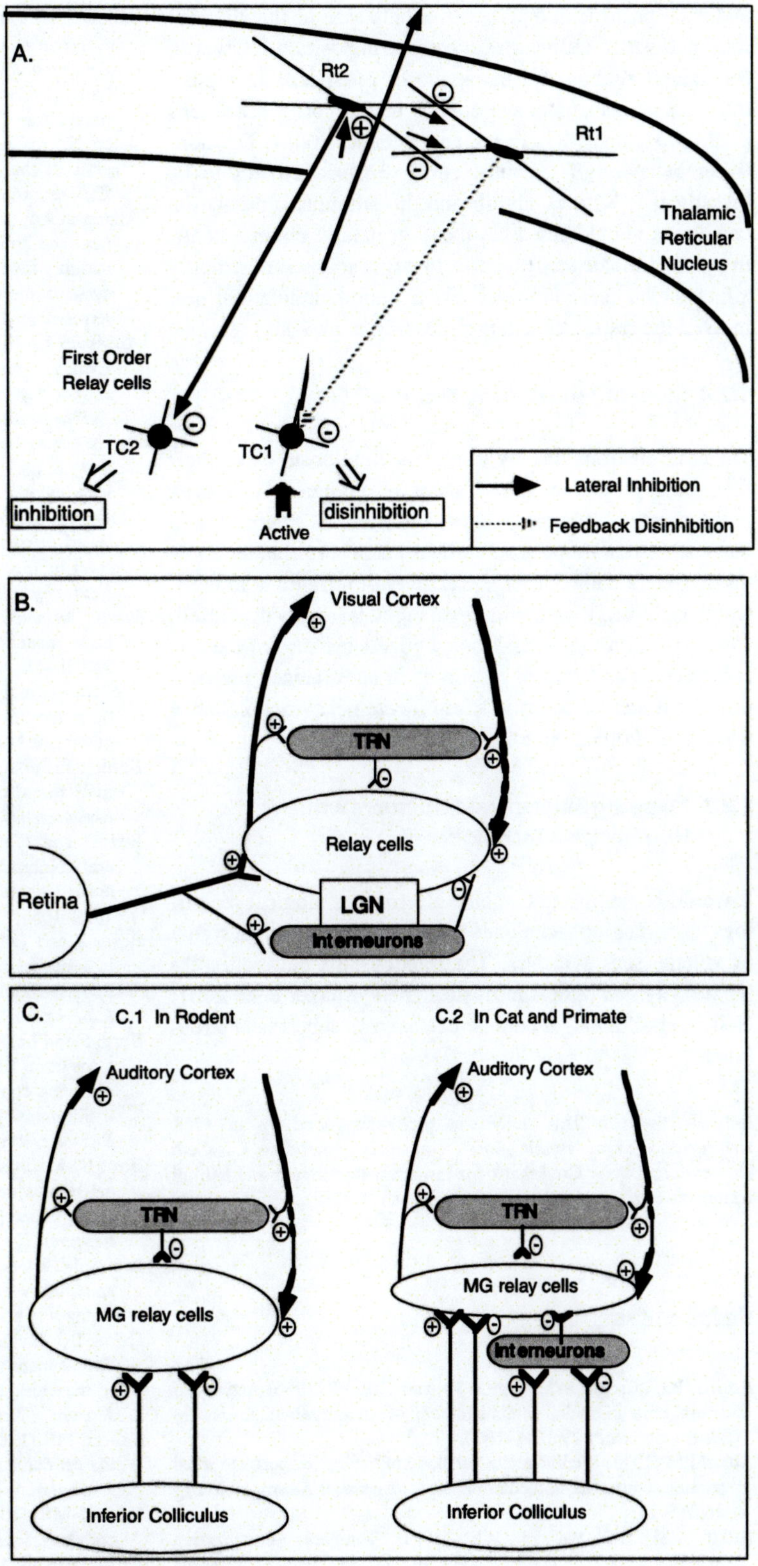

Fig. 12.6 Schematic of the MG excitatory–inhibitory interplay. **a** In the somatic sensory thalamus, the inhibitory input comes from TNR neurons and the interaction between reticular (Rt) and thalamocortical (TC) cells might underlie lateral inhibition. Modified from the original source (Pinault and Deschênes 1998). **b** In the visual thalamus, the local GABAergic interneurons provide feedforward inhibition and relay cells receive feedback inhibition from TRN neurons. **c** In the MG, the relay cells receive inhibition and excitation, with two variations. *1*. In rodent, TRN neurons are the only source of feedback inhibition. *2*. In cat and primate, TC cells receive mixed excitation and inhibition, with the IC and auditory cortex the primary excitatory sources, and local interneurons and TRN providing feedforward and feedback inhibitory input to relay cells, respectively

auditory spectrotemporal receptive field assume that the neurons perform a multiscale linear decomposition. However, for natural sounds, the linear model often fails to predict the responses to unknown sounds at the cortex (Machens et al. 2004) and MG (Huetz and Edeline, 2006). The nonlinear behavior of thalamocortical neurons remains to be investigated. Natural stimuli and the stimulus presentation conditions should resemble those in natural environments. In awake animals the responses to target sounds are strongly affected by noise, and speech-like temporal modulations best masked the responses to targets (Martin et al. 2004).

5.2.4 Studying the Waking Brain

Thalamic neurons display important modifications of their functional properties across states of vigilance, and even more drastic differences exist between anesthetized and unanesthetized animals. In cortical inactivation under anesthesia (see Section 3.4.1.), the activity of infragranular cells, which is the origin of corticofugal input, is largely depressed. Cortical cooling may mask many effects because of the low spontaneous and evoked responses in anesthetized preparations, and natural acoustic stimuli have behavioral salience (Fritz et al. 2003, 2005b).

5.2.5 Mapping Subcortical Structures with Non-invasive Techniques

It would be helpful to combine electrophysiological recordings with imaging techniques such as intrinsic cortical signal or voltage sensitive dyes. The relationships between spiking activity and optical intrinsic signals (Spitzer et al. 2001; Nelken et al. 2004) should be extended to subcortical structures.

Acknowledgments This study was partly supported by an ANR Neuroscience grant 'Hearing loss'. I am very grateful to Elizabeth Hennevin and Scott Cruikshank for insightful discussions and helpful comments on earlier versions of this chapter.

References

Adrian HO, Lifschitz WM, Tavitas RJ, and Galli FP (1966) Activity of neural units in medial geniculate body of cat and rabbit. *Journal of Neurophysiology* **29**:1046–1060.

Aitkin LM (1973) Medial geniculate body of the cat: response to tonal stimuli of neurons in medial division. *Journal of Neurophysiology* **36**:275–283.

Aitkin LM and Webster WR (1971) Tonotopic organization in the medial geniculate body of the cat. *Brain Research* **26**:402–405.

Aitkin LM and Webster WR (1972) Medial geniculate body of the cat: organization and responses to tonal stimuli of neurons in ventral division. *Journal of Neurophysiology* **35**:365–380.

Aitkin LM, Dunlop CW, and Webster WR (1966) Click-evoked response patterns of single units in the medial geniculate body of the cat. *Journal of Neurophysiology* **29**:109–123.

Alitto HJ and Usrey WM (2003) Corticothalamic feedback and sensory processing. *Current Opinion in Neurobiology* **13**:440–445.

Allon N, Yeshurun Y, and Wollberg Z (1981) Responses of single cells in the medial geniculate body of awake squirrel monkey. *Experimental Brain Research* **41**:222–232.

Andersen RA, Knight PL, and Merzenich MM (1980) The thalamocortical and corticothalamic connections of AI, AII, and the anterior auditory field (AAF) in the cat: evidence for two largely segregated systems of connections. *Journal of Comparative and Physiological Psychology* **194**:663–701.

Anderson LA, Malmierca MS, Wallace MN, and Palmer AR (2006) Evidence for a direct, short latency projection from the dorsal cochlear nucleus to the auditory thalamus in the guinea pig. *European Journal of Neuroscience* **24**:491–498.

Bao S, Chang EF, Woods J, and Merzenich MM (2004) Temporal plasticity in the primary auditory cortex induced by operant perceptual learning. *Nature Neuroscience* **7**:974–981.

Barone P, Clarey JC, Irons WA, and Imig TJ (1996) Cortical synthesis of azimuth-sensitive single-unit responses with nonmonotonic level tuning: a thalamocortical comparison in the cat. *Journal of Neurophysiology* **75**:1206–1220.

Barth DS and MacDonald KD (1996) Thalamic modulation of high-frequency oscillating potentials in auditory cortex. *Nature* **383**:78–81.

Bartlett EL and Smith PH (1999) Anatomic, intrinsic, and synaptic properties of dorsal and ventral division neurons in rat medial geniculate body. *Journal of Neurophysiology* **81**:1999–2016.

Bartlett EL and Smith PH (2002) Effects of paired-pulse and repetitive stimulation on neurons in the rat medial geniculate body. *Neuroscience* **113**:957–974.

Birt D and Olds M (1981) Associative response changes in lateral midbrain tegmentum and medial geniculate during differential appetitive conditioning. *Journal of Neurophysiology* **46**: 1039–1055.

Bordi F and LeDoux J (1994a) Response properties of single units in areas of rat auditory thalamus that project to the amygdala. I. Acoustic discharge patterns and frequency receptive fields. *Experimental Brain Research* **98**:261–274.

Bordi F and LeDoux JE (1994b) Response properties of single units in areas of rat auditory thalamus that project to the amygdala. II. Cells receiving convergent auditory and somatosensory inputs and cells antidromically activated by amygdala stimulation. *Experimental Brain Research* **98**:275–286.

Brett B, Krishnan G, and Barth DS (1996) The effects of subcortical lesions on evoked potentials and spontaneous high frequency (gamma-band) oscillating potentials in rat auditory cortex. *Brain Research* **721**:155–166.

Brown M, Irvine DRF, and Park VN (2004) Perceptual learning on an auditory frequency discrimination task by cats: association with changes in primary auditory cortex. *Cerebral Cortex* **14**:952–965.

Cairns BE, McFarlane SA, Fragoso MC, Jia WG, and Soja PJ (1996) Spontaneous discharge and peripherally evoked orofacial responses of trigemino-thalamic tract neurons during wakefulness and sleep. *Journal of Neuroscience* **16**:8149–8159.

Calford MB (1983) The parcellation of the medial geniculate body of the cat defined by the auditory response properties of single units. *Journal of Neuroscience* **3**:2350–2365.

Calford MB (2002) Dynamic representational plasticity in sensory cortex. *Neuroscience* **111**:709–738.

Calford MB and Webster WR (1981) The auditory representation within the principal division of cat medial geniculate: an electrophysiological study. *Journal of Neurophysiology* **45**:1013–1028.

Calford MB and Aitkin LM (1983) Ascending projections to the medial geniculate body of the cat: evidence for multiple

parallel auditory pathways through thalamus. *Journal of Neuroscience* **11**:2365–2380.
Campeau S and Davis M (1995) Involvement of subcortical and cortical afferents to the lateral nucleus of the amygdala in fear conditioning measured with fear-potentiated startle in rats trained concurrently with auditory and visual conditioned stimuli. *Journal of Neuroscience* **15**:2312–2327.
Campeau S, Falls WA, Cullinan WE, Helmreich DL, Davis M, and Watson SJ (1997) Elicitation and reduction of fear: behavioural and neuroendocrine indices and brain induction of the immediate-early gene c-fos. *Neuroscience* **78**:1087–1104.
Cant NB and Benson CG (2007) Multiple topographically organized projections connect the central nucleus of the inferior colliculus to the ventral division of the medial geniculate nucleus in the gerbil, Meriones unguiculatus. *Journal of Comparative Neurology* **503**:432–453.
Castro-Alamancos MA (2002) Role of thalamocortical sensory suppression during arousal: focusing sensory inputs in neocortex. *Journal of Neuroscience* **22**:9651–9655.
Castro-Alamancos MA and Oldford E (2002) Cortical sensory suppression during arousal is due to the activity-dependent depression of thalamocortical synapses. *Journal of Physiology (London)* **541**: 319–331.
Cetas JS, Price RO, Velenovsky DS, Sinex DG, and McMullen NT (2001) Frequency organization and cellular lamination in the medial geniculate body of the rabbit. *Hearing Research* **155**:113–123.
Clarey JC, Barone P, and Imig TJ (1992) Physiology of thalamus and cortex. In: Popper AN and Fay RR (eds). *Springer Handbook of Auditory Research. The Mammalian Auditory Pathway*, volume 2, *Neurophysiology*. Springer, New York, pp. 232–334.
Clarey JC, Barone P, Irons WA, Samson FK, and Imig T (1995) Comparison of noise and tone azimuth tuning of neurons in cat primary auditory cortex and medial geniculate body. *Journal of Neurophysiology* **74**:961–980.
Clerici WJ and Coleman JR (1990) Anatomy of the rat medial geniculate body: I. Cytoarchitecture, myeloarchitecture, and neocortical connectivity. *Journal of Comparative Neurology* **297**:14–31.
Clerici WJ, McDonald AJ, Thompson R, and Coleman JR (1990) Anatomy of the rat medial geniculate body: II. Dendritic morphology. *Journal of Comparative Neurology* **297**:32–54.
Coenen ALM and Vendrick AJH (1972) Determination of the transfer ratio of cat's geniculate neurons through quasi-intracellular recordings and the relation with the level of alertness. *Experimental Brain Research* **14**:227–242.
Conley M, Kupersmith AC, and Diamond IT (1991) The organization of projections from subdivisions of the auditory cortex and thalamus to the auditory sector of the thalamic reticular nucleus in Galago. *European Journal of Neuroscience* **3**:1089–1103.
Cotillon N and Edeline JM (2000) Tone-evoked oscillations in the rat auditory cortex result from interactions between the thalamus and reticular nucleus. *European Journal of Neuroscience* **12**:3637–3650.
Cotillon N, Nafati M, and Edeline JM (2000) Characteristics of reliable tone-evoked oscillations in the rat thalamo-cortical auditory system. *Hearing Research* **142**:113–130.
Cotillon-Williams N, and Edeline JM (2003) Evoked oscillations in the thalamo-cortical auditory system are present in anesthetized but not in unanesthetized rats. *Journal of Neurophysiology* **89**:1968–1984.
Cotillon-Williams N, Huetz C, Hennevin E, and Edeline J-M (2008) Topographic control of auditory thalamus frequency tuning by reticular thalamic neurons. *Journal of Neurophysiology* **99**:1137–1151.
Crabtree JW (1998) Organization in the auditory sector of the cat's thalamic reticular nucleus. *Journal of Comparative Neurology* **390**:167–182.
Creutzfeldt O, Hellweg FC, and Schreiner C (1980) Thalamocortical transformation of responses to complex auditory stimuli. *Experimental Brain Research* **39**:87–104.
Cruikshank SJ, Killackey HP, and Metherate R (2001) Parvalbumin and calbindin are differentially distributed within primary and secondary subregions of the mouse auditory forebrain. *Neuroscience* **105**: 553–569.
Crunelli V, Leresche N, and Parnavelas JG (1987) Membrane properties of morphologically identified X and Y cells in the lateral geniculate nucleus of the cat in vitro. *Journal of Physiology (London)* **390**: 243–256.
de Venecia RK, Smelser CB, Lossman SD, and McMullen NT (1995) Complementary expression of parvalbumin and calbindin D-28 k delineates subdivisions of the rabbit medial geniculate body. *Journal of Comparative Neurology* **359**:595–612.
Diamond IT, Jones EG, and Powell TPS (1969) The projection of the auditory cortex upon the diencephalon and brain stem in the cat. *Brain Research* **15**:305–340.
Disterhoft JF and Olds J (1972) Differentials development of conditioned unit changes in thalamus and cortex of rat. *Journal of Neurophysiology* **35**:665–679.
Disterhoft JF and Stuart DK (1976) Trial sequence of changed unit activity in auditory system of alert rat during conditioned response acquisition and extinction. *Journal of Neurophysiology* 39: 266–281.
Doron NN and LeDoux JE (1999) Organization of projections to the lateral amygdala from auditory and visual areas of the thalamus in the rat. *Journal of Comparative Neurology* **412**:383–409.
Doron NN and LeDoux JE (2000) Cells in the posterior thalamus project to both amygdala and temporal cortex: a quantitative retrograde double-labeling study in the rat. *Journal of Comparative Neurology* **425**:257–274.
Edeline JM (1990) Frequency-specific plasticity of single unit discharges in the rat medial geniculate body. *Brain Research* **529**: 109–119.
Edeline JM (1999) Learning-induced physiological plasticity in the thalamo-cortical sensory systems: a critical evaluation of receptive field plasticity, map changes and their potential mechanisms. *Progress in Neurobiology* **57**:165–224.
Edeline JM (2003) The thalamo-cortical auditory receptive fields: regulation by the states of vigilance, learning and the neuromodulatory systems. *Experimental Brain Research* **153**:554–572.
Edeline JM and Weinberger NM (1991a) Subcortical adaptive filtering in the auditory system: associative receptive field plasticity in the dorsal medial geniculate body. *Behavioral Neuroscience* **105**: 154–175.
Edeline JM and Weinberger NM (1991b) Thalamic short term plasticity in the auditory system: associative retuning of receptive fields in the ventral medial geniculate body. *Behavioral Neuroscience* **105**: 618–639.
Edeline JM and Weinberger NM (1992) Associative retuning in the thalamic source of input to the amygdala and auditory cortex: receptive field plasticity in the medial division of the medial geniculate body. *Behavioral Neuroscience* **106**:81–105.
Edeline JM, Dutrieux G, Manunta Y, and Hennevin E (2001) Diversity of receptive field changes in auditory cortex during natural sleep. *European Journal of Neuroscience* **14**:1865–1880.
Edeline JM, Dutrieux G, and Neuenschwander-El Massioui N (1988) Multiunit changes in hippocampus and medial geniculate body in free-behaving rats during acquisition and retention of a conditioned response to a tone. *Behavioral and Neural Biology* **50**: 61–79.
Edeline JM, Manunta Y and Hennevin E (2000) Auditory thalamus neurons during sleep: changes in frequency selectivity, threshold and receptive field size. *Journal of Neurophysiology* **84**:934–952.
Edeline JM, Manunta Y, Nodal FR, and Bajo V (1999) Do auditory responses recorded from awake animals reflect the anatomical parcellation of the auditory thalamus? *Hearing Research* **131**: 135–152.

Edeline JM, Neuenschwander-El Massioui N, and Dutrieux G (1990a) Discriminative long-term retention of rapidly induced multiunit changes in the hippocampus, medial geniculate and auditory cortex. *Behavioural Brain Research* **39**:145–155.

Edeline JM, Neuenschwander-El Massioui N, and Dutrieux G (1990b) Frequency-specific cellular changes in the auditory system during acquisition and reversal of discriminative conditioning. *Psychobiology* **18**:382–393.

Elhilali M, Fritz JB, Klein DJ, Simon JZ, and Shamma SA (2004) Dynamics of precise spike timing in primary auditory cortex. *Journal of Neuroscience* **24**:1159–1172.

Escabi MA and Schreiner CE (2002) Nonlinear spectrotemporal sound analysis by neurons in the auditory midbrain. *Journal of Neuroscience* **22**:4114–4131.

Fanselow MS and LeDoux JE (1999) Why we think plasticity underlying Pavlovian fear conditioning occurs in the basolateral amygdala. *Neuron* **23**:229–232.

Franowicz MN and Barth DS (1995) Comparison of evoked potentials and high-frequency (gamma-band) oscillating potentials in rat auditory cortex. *Journal of Neurophysiology* **74**:96–112.

Fritz JB, Shamma S, Elhilali M, and Klein D (2003) Rapid task-related plasticity of spectrotemporal receptive fields in primary auditory cortex. *Nature Neuroscience* **6**:1216–1223.

Fritz JB, Elhilali M, and Shamma S (2005a) Active listening: task-dependent plasticity of spectrotemporal receptive fields in primary auditory cortex. *Hearing Research* **206**:159–176.

Fritz JB, Elhilali M, and Shamma SA (2005b) Differential dynamic plasticity of A1 receptive fields during multiple spectral tasks. *Journal of Neuroscience* **25**:7623–7635.

Fukutake T and Hattori T (1998) Auditory illusions caused by a small lesion in the right medial geniculate body. *Neurology* **51**: 1469–1471.

Funkenstein HH and Winter P (1973) Responses to acoustic stimuli of units in the auditory cortex of awake squirrel monkeys. *Experimental Brain Research* **18**:464–488.

Gabriel M, Saltwick SE, and Miller JD (1975) Conditioning and reversal of short-latency multiple-unit responses in the rabbit medial geniculate nucleus. *Science* **189**:1108–1109.

Galambos R (1952) Microelectrode studies on medial geniculate body of cat. III. Response to pure tones. *Journal of Neurophysiology* **15**:381–400.

Galambos R, Rose JE, Bromiley RB, and Hughes JR (1952) Microelectrode studies on medial geniculate body of cat. II. Response to clicks. *Journal of Neurophysiology* **15**:359–380.

Graham J (1977) An autoradiographic study of the efferent connections of the superior colliculus in the cat. *Journal of Comparative Neurology* **173**:629–654.

Graybiel AM (1972) Some fiber pathways related to the posterior thalamic region in the cat. *Brain Behavior and Evolution* **6**:363–393.

Guido W, Lu SM, Vaughan JW, Godwin DW, and Sherman SM (1995) Receiver operating characteristic (ROC) analysis of neurons in the cat's lateral geniculate nucleus during tonic and burst response mode. *Visual Neuroscience* **12**:723–741.

Hashikawa T, Rausell E, Molinari M, and Jones EG (1991) Parvalbumin- and calbindin-containing neurons in the monkey medial geniculate complex: differential distribution and cortical layer specific projections. *Brain Research* **544**:335–341.

He J (1997) Modulatory effects of regional cortical activation on the onset responses of the cat medial geniculate neurons. *Journal of Neurophysiology* **77**:896–908.

He J (2001) On and off pathways segregated at the auditory thalamus of the guinea pig. *Journal of Neuroscience* **21**:8672–8679.

He J (2002) OFF Responses in the auditory thalamus of the guinea pig. *Journal of Neurophysiology* **88**:2377–2386.

He J (2003a) Slow oscillation in non-lemniscal auditory thalamus. *Journal of Neuroscience* **23**:8281–8290.

He J (2003b) Corticofugal modulation of the auditory thalamus. *Experimental Brain Research* **153**:579–590.

He J (2003c) Corticofugal modulation on both ON and OFF responses in the nonlemniscal auditory thalamus of the guinea pig. *Journal of Neurophysiology* **89**:367–381.

He J, Yu YQ, Xiong Y, Hashikawa T, and Chan YS (2002) Modulatory effect of cortical activation on the lemniscal auditory thalamus of the Guinea pig. *Journal of Neurophysiology* **88**:1040–1050.

Henkel CK (1983) Evidence of sub-collicular auditory projections to the medial geniculate nucleus in the cat: an autoradiographic and horseradish peroxidase study. *Brain Research* **259**:21–30.

Hennevin E and Maho C (2005) Fear conditioning-induced plasticity in auditory thalamus and cortex: to what extent is it expressed during slow-wave sleep? *Behavioral Neuroscience* **119**:1277–1289.

Hennevin E, Huetz C, and Edeline JM. (2007) Neural representations during sleep: from sensory processing to memory traces. *Neurobiology of Learning and Memory* **87**:416–440.

Hennevin E, Maho C, and Hars B (1998) Neuronal plasticity induced by fear conditioning is expressed during paradoxical sleep: evidence from simultaneous recordings in the lateral amygdala and the medial geniculate in rats. *Behavioral Neuroscience* **112**:839–862.

Hennevin E, Maho C, Hars B, and Dutrieux G (1993) Learning-induced plasticity in the medial geniculate nucleus is expressed during paradoxical sleep. *Behavioral Neuroscience* **107**:1018–1030.

Hirsch JC, Fourment A, and Marc ME (1983) Sleep-related variations of membrane potential in the lateral geniculate body relay neurons of the cat. *Brain Research* **259**:308–312.

Hu B (1995) Cellular basis of temporal synaptic signalling: an in vitro electrophysiological study in rat auditory thalamus. *Journal of Physiology (London)* **483**:167–182.

Hu B, Senatorov V, and Mooney D (1994) Lemniscal and non-lemniscal synaptic transmission in rat auditory thalamus. *Journal of Physiology (London)* **479**:217–231.

Huang CL and Winer JA (2000) Auditory thalamocortical projections in the cat: laminar and areal patterns of input. *Journal of Comparative Neurology* **427**:302–331.

Huang CL, Larue DT, and Winer JA (1999) GABAergic organization of the cat medial geniculate body. *Journal of Comparative Neurology* **415**:368–392.

Huetz C and Edeline J-M (2006) From receptive field dynamics to the rate of transmitted information: some facets of the thalamocortical system. *Neuroembryology and Aging* **3**:230–238.

Imig TJ and Morel A (1985a) Tonotopic organization in the lateral part of posterior group of thalamic nuclei in the cat. *Journal of Neurophysiology* **53**:836–851.

Imig TJ and Morel A (1985b) Tonotopic organization in ventral nucleus of medial geniculate body in the cat. *Journal of Neurophysiology* **53**:309–340.

Imig TJ, Poirier P, Irons WA, and Samson FK (1997) Monaural spectral contrast mechanism for neural sensitivity to sound direction in the medial geniculate body of the cat. *Journal of Neurophysiology* **78**:2754–2771.

Imig TJ, Bibikov NG, Poirier P, and Samson FK (2000) Directionality derived from pinna-cue spectral notches in cat dorsal cochlear nucleus. *Journal of Neurophysiology* **83**:907–925.

Irvine DRF (1992) Physiology of the auditory brainstem. In: Popper AN and Fay RR (eds). *Springer Handbook of Auditory Research*, volume 2, *The Mammalian Auditory Pathway: Neurophysiology*. Springer, New York, pp. 153–231.

Irvine DRF, Rajan R, and Smith S (2003) Effects of restricted cochlear lesions in adult cats on the frequency organization of the inferior colliculus. *Journal of Comparative Neurology* **467**: 354–374.

Ivarsson C, de Ribaupierre Y, and de Ribaupierre F (1988) Influence of auditory localization cues on neuronal activity in the auditory thalamus of the cat. *Journal of Neurophysiology* **59**:586–606.

Jahnsen H and Llinás R (1984a) Electrophysiological properties of guinea-pig thalamic neurones: an in vitro study. *Journal of Physiology (London)* **349**:205–226.

Jahnsen H and Llinás R (1984b) Ionic basis for the electroresponsiveness and oscillatory properties of guinea-pig thalamic neurones in vitro. *Journal of Physiology (London)* **349**:227–247.

Jarrell TW, Gentile CG, McCabe PM, and Schneiderman N (1986a) The role of the medial geniculate region in differential Pavlovian conditioning of bradycardia in rabbits. *Brain Research* **374**: 126–136.

Jarrell TW, Romanski LW, Gentile CG, McCabe PM, and Schneiderman N (1986b) Ibotenic acid lesions in the medial geniculate region prevent the acquisition of differential Pavlovian conditioning of bradycardia to acoustic stimuli in rabbits. *Brain Research* **382**: 199–203.

Jenkins WM, Merzenich MM, and Recanzone G (1990) Neocortical representational dynamics in adult primates: implications for neuropsychology. *Neuropsychologia* **28**:573–584.

Jones EG (1985) *The Thalamus*. Plenum Press, New York.

Jones EG (2000) Cortical and subcortical contributions to activity-dependent plasticity in primate somatosensory cortex. *Annual Review of Neuroscience* **23**:1–37.

Jones EG and Burton H (1974) Cytoarchitecture and somatic connectivity of thalamic nuclei other than the ventrobasal complex in the cat. *Journal of Comparative Neurology* **154**:395–432.

Joris PX, Schreiner CE, and Rees A (2004) Neural processing of amplitude-modulated sounds. *Physiological Reviews* **84**: 541–577.

Kaas JH, Merzenich MM, and Killackey HP (1983) The reorganization of somatosensory cortex following peripheral nerve damage in adult and developing mammals. *Annual Review of Neuroscience* **6**: 325–356.

Kamke MR, Brown M, and Irvine DRF (2003) Plasticity in the tonotopic organization of the medial geniculate body in adult cats following restricted unilateral cochlear lesions. *Journal of Comparative Neurology* **459**:355–367.

Klein DJ, Depireux DA, Simon JZ, and Shamma SA (2000) Robust spectrotemporal reverse correlation for the auditory system: optimizing stimulus design. *Journal of Computational Neuroscience* **9**:85–111.

Komura Y, Tamura R, Uwano T, Nishijo H, and Ono T (2005) Auditory thalamus integrates visual inputs into behavioral gains. *Nature Neuroscience* **8**:1203–1209.

Komura Y, Tamura R, Uwano T, Nishijo H, Kaga K, and Ono T (2001) Retrospective and prospective coding for predicted reward in the sensory thalamus. *Nature* **412**:546–549.

Kudo M and Niimi K (1980) Ascending projection of the inferior colliculus in the cat: an autoradiographic study. *Journal of Comparative Neurology* **191**:545–556.

Las L, Stern EA, and Nelken I (2005) Representation of tone in fluctuating maskers in the ascending auditory system. *Journal of Neuroscience* **25**:1503–1513.

LeDoux JE, Ruggiero DA, Forest R, Stornetta R, and Reis DJ (1987) Topographic organization of convergent projections to the thalamus from the inferior colliculus and spinal cord in the rat. *Journal of Comparative Neurology* **264**:123–146.

Lee SM, Friedberg MH, and Ebner FF (1994) The role of GABA-mediated inhibition in the rat ventral posterior medial thalamus. I. Assessment of receptive field changes following thalamic reticular nucleus lesions. *Journal of Neurophysiology* **71**:1702–1715.

Linke R and Schwegler H (2000) Convergent and complementary projections of the caudal paralaminar thalamic nuclei to rat temporal and insular cortex. *Cerebral Cortex* **10**:753–771.

MacDonald KD, Fifková E, Jones MS, and Barth DS (1998) Focal stimulation of the thalamic reticular nucleus induces focal gamma waves in cortex. *Journal of Neurophysiology* **79**:474–477.

Machens CK, Wehr MS, and Zador AM (2004) Linearity of cortical receptive fields measured with natural sounds. *Journal of Neuroscience* **24**:1089–1100.

Maho C and Hennevin E (2002) Appetitive conditioning-induced plasticity is expressed during paradoxical sleep in the medial geniculate, but not in the lateral amygdala. *Behavioral Neuroscience* **116**: 807–823.

Maho C, Hars B, Edeline JM, and Hennevin E (1995) Conditioned changes in the basal forebrain: relations with learning-induced cortical plasticity. *Psychobiology* **23**:10–25.

Malmierca MS, Merchán MA, Henkel CK, and Oliver DL (2002) Direct projections from cochlear nuclear complex to auditory thalamus in the rat. *Journal of Neuroscience* **22**:10891–10897.

Maren S and Quirk GJ (2004) Neuronal signalling of fear memory. *Nature Reviews Neuroscience* **5**:844–852.

Maren S, Yap SA, and Goosens KA (2001) The amygdala is essential for the development of neuronal plasticity in the medial geniculate nucleus during auditory fear conditioning in rats. *Journal of Neuroscience* **21**:RC135.

Martin EM, West MF, and Bedenbaugh PH (2004) Masking and scrambling in the auditory thalamus of awake rats by Gaussian and modulated noises. *Proceedings of the National Academy of Sciences of the United States of America* **101**:14961–14965.

Massaux A and Edeline JM (2003) Bursts in the medial geniculate body: a comparison between anesthetized and unanesthetized states in guinea pig. *Experimental Brain Research* **153**: 573–578.

Massaux A, Dutrieux G, Cotillon-Williams N, Manunta Y, and Edeline JM (2004) Auditory thalamus bursts in anesthetized and non-anesthetized states: contribution to functional properties. *Journal of Neurophysiology* **91**:2117–2134.

McCabe PM, McEchron MD, Green EJ, and Schneiderman N (1993) Electrolytic and ibotenic acid lesions of the medial subnucleus of the medial geniculate prevent the acquisition of classically conditioned heart rate to a single acoustic stimulus in rabbits. *Brain Research* **619**:291–298.

McCormick DA and Bal T (1997) Sleep and arousal: thalamocortical mechanisms. *Annual Review of Neuroscience* **20**:185–215.

McEchron MD, Green EJ, Winters RW, Nolen TG, Schneiderman N, and McCabe PM (1996) Changes of synaptic efficacy in the medial geniculate nucleus as a result of auditory classical conditioning. *Journal of Neuroscience* **16**:1273–1283.

McGaugh JL (2004) The amygdala modulates the consolidation of memories of emotionally arousing experiences. *Annual Review of Neuroscience* **27**:1–28.

Metherate R and Cruikshank SJ (1999) Thalamocortical inputs trigger a propagating envelope of gamma-band activity in auditory cortex in vitro. *Experimental Brain Research* **126**:160–174.

Miller LM, Escabí MA, Read HL, and Schreiner CE (2001) Functional convergence of response properties in the auditory thalamocortical system. *Neuron* **32**:151–160.

Molinari M, Dell'Anna ME, Rausell E, Leggio MG, Hashikawa T, and Jones EG (1995) Auditory thalamocortical pathways defined in monkeys by calcium-binding protein immunoreactivity. *Journal of Comparative Neurology* **362**:171–194.

Mooney DM, Hu B, and Senatorov VV (1995) Muscarine induces an anomalous inhibition of synaptic transmission in rat auditory thalamic neurons in vitro. *Journal of Pharmacology and Experimental Therapeutics* **275**:838–844.

Mooney DM, Zhang L, Basile C, Senatorov VV, Ngsee J, Omar A, and Hu B (2004) Distinct forms of cholinergic modulation in parallel thalamic sensory pathways. *Proceedings of the National Academy of Sciences of the United States of America* **101**:320–324.

Morales-Cobas G, Ferreira MI, and Velluti RA (1995) Firing of inferior colliculus neurons in response to low-frequency sound stimulation during sleep and waking. *Journal of Sleep Research* **4**:242–251.

Morel A, Rouiller E, de Ribaupierre Y, and de Ribaupierre F (1987) Tonotopic organization in the medial geniculate body (MGB) of lightly anesthetized cats. *Experimental Brain Research* **69**: 24–42.

Morest DK (1964) The neuronal architecture of the medial geniculate body of the cat. *Journal of Anatomy (London)* **98**:611–630.

Morest DK (1965a) The lateral tegmental system of the midbrain and the medial geniculate body: study with Golgi and Nauta methods in cat. *Journal of Anatomy (London)* **99**:611–634.

Morest DK (1965b) The laminar structure of the medial geniculate body of the cat. *Journal of Anatomy (London)* **99**:143–160.

Morest DK (1971) Dendrodendritic synapses of cells that have axons: the fine structure of the Golgi type II cell in the medial geniculate body of the cat. *Zeitschrift für Anatomie und Entwicklungsgeschichte* **133**:216–246.

Morest DK (1975) Synaptic relationships of Golgi type II cells in the medial geniculate body of the cat. *Journal of Comparative Neurology* **162**:157–193.

Morest DK and Winer JA (1986) The comparative anatomy of neurons: homologous neurons in the medial geniculate body of the opossum and the cat. *Advances in Anatomy, Embryology and Cell Biology* **97**:1–96.

Musicant AD and Butler RA (1984) The influence of pinnae-based spectral cues on sound localization. *Journal of the Acoustic Society of America* **75**:1195–1200.

Musicant AD, Chan JC, and Hind JE (1990) Direction-dependent spectral properties of cat external ear: new data and cross-species comparisons. *Journal of the Acoustical Society of America* **87**: 757–781.

Nelken I, Bizley JK, Nodal FR, Ahmed B, Schnupp JWH, and King AJ (2004) Large-scale organization of ferret auditory cortex revealed using continuous acquisition of intrinsic optical signals. *Journal of Neurophysiology* **92**:2574–2588.

O'Connors KN, Allison TL, Rosenfield ME, and Moore JW (1997) Neural activity in the medial geniculate nucleus during auditory trace conditioning. *Experimental Brain Research* **113**:534–556.

Ohl F and Scheich H (1996) Differential frequency conditioning enhances spectral contrast sensitivity of units in the auditory cortex (field AI) of the alert Mongolian gerbil. *European Journal of Neuroscience* **8**:1001–1017.

Ohl FW and Scheich H (1997) Learning-induced dynamic receptive field changes in primary auditory cortex of the unanaesthetized Mongolian gerbil. *Journal of Comparative Physiology* [A] **181**: 685–696.

Olds J, Disterhoft JF, Segal M, Kornblith DL, and Hirch R (1972) Learning center of rat brain mapped by measuring latencies of conditioned unit responses. *Journal of Neurophysiology* **35**:202–219.

Oliver DL (1984) Neuron types in the central nucleus of the inferior colliculus that project to the medial geniculate body. *Neuroscience* **11**:409–424.

Orman SS and Humphrey GL (1981) Effects of changes in cortical arousal and of auditory cortex cooling on neuronal activity in the medial geniculate body. *Experimental Brain Research* **42**: 475–482.

Pape HC and McCormick DA (1995) Electrophysiological and pharmacological properties of interneurons in the cat dorsal lateral geniculate nucleus. *Neuroscience* **68**:1105–1125.

Pena JL, Pedemonte M, Ribeiro MF, and Velluti R (1992) Single unit activity in the guinea-pig cochlear nucleus during sleep and wakefulness. *Archives Italiennes de Biologie* **130**:179–189.

Philibert B, Laudanski J, and Edeline JM (2005) Auditory thalamus responses to guinea-pig vocalizations: a comparison between rat and guinea-pig. *Hearing Research* **209**:97–103.

Phillips DP and Farmer ME (1990) Acquired word deafness, and the temporal grain of sound representation in the primary auditory cortex. *Behavioral Brain Research* **40**:85–94.

Pinault D. (2004) The thalamic reticular nucleus: structure, function and concept. *Brain Research Review* **46**:1–31.

Pinault D and Deschênes M (1998) Anatomical evidence for a mechanism of lateral inhibition in the rat thalamus. *European Journal of Neuroscience* **10**:3462–3469.

Poirier P, Samson FK, and Imig TJ (2003) Spectral shape sensitivity contributes to the azimuth tuning of neurons in the cat's inferior colliculus. *Journal of Neurophysiology* **89**:2760–2777.

Polley DB, Heiser MA, Blake DT, Schreiner CE, and Merzenich MM (2004) Associative learning shapes the neural code for stimulus magnitude in primary auditory cortex. *Proceedings of the National Academy of Sciences of the United States of America* **101**:16351–16356.

Pons TP, Garraghty PE, Ommaya AK, Kaas JH, Taub E, and Mishkin M (1991) Massive cortical reorganization after sensory deafferentation in adult macaques. *Science* **252**:1857–1860.

Poremba A and Gabriel M (2001) Amygdalar efferents initiate auditory thalamic discriminative training-induced neuronal activity. *Journal of Neuroscience* **21**:270–278.

Quirk GJ, Armony JL, and LeDoux JE (1997) Fear conditioning enhances different temporal components of tone-evoked spike trains in auditory cortex and lateral amygdala. *Neuron* **19**: 613–624.

Rajan R and Irvine DRF (1998) Absence of plasticity of the frequency map in dorsal cochlear nucleus of adult cats after unilateral partial cochlear lesions. *Journal of Comparative Neurology* **399**:35–46.

Rasmusson D (1996) Changes in the organization of the ventroposterior lateral thalamic nucleus after digit removal in adult raccoon. *Journal of Comparative Neurology* **364**:92–103.

Ravizza RJ and Belmore SM (1978) Auditory forebrain: evidence from anatomical and behavioral experiments involving human and animal subjects. In: Masterton RB (ed). *Handbook of Behavioral Neurobiology*. Plenum Press, New York, pp. 459–501.

Recanzone GH, Schreiner CE, and Merzenich MM (1993) Plasticity in the frequency representation of primary auditory cortex following discrimination training in adult owl monkeys. *Journal of Neuroscience* **13**:87–103.

Redies H and Brandner S (1991) Functional organization of the auditory thalamus in the guinea pig. *Experimental Brain Research* **86**:384–392.

Redies H, Brandner S, and Creutzfeldt OD (1989) Anatomy of the auditory thalamocortical system of the guinea pig. *Journal of Comparative Neurology* **282**:489–511.

Robertson D and Irvine DRF (1989) Plasticity of frequency organization in auditory cortex of guinea pigs with partial unilateral deafness. *Journal of Comparative Neurology* **282**:456–471.

Rodrigues-Dagaeff C, Simm G, de Ribaupierre Y, Villa A, de Ribaupierre F, and Rouiller EM (1989) Functional organization of the ventral division of the medial geniculate body of the cat: evidence for a rostro-caudal gradient of response properties and cortical projections. *Hearing Research* **39**:103–126.

Romanski LM and LeDoux JE (1992) Equipotentiality of thalamoamygdaloid and thalamocorticoamygdaloid circuits in auditory fear conditioning. *Journal of Neuroscience* **12**:4501–4509.

Rouiller E and de Ribaupierre F (1982) Neurons sensitive to narrow ranges of repetitive acoustic transients in the medial geniculate body of the cat. *Experimental Brain Research* **48**:323–326.

Rouiller E, de Ribaupierre Y, and de Ribaupierre F (1979) Phase-locked responses to low-frequency tones in the medial geniculate body. *Hearing Research* **1**:213–226.

Rouiller E, de Ribaupierre Y, Morel A, and de Ribaupierre F (1983) Intensity functions of single unit responses to tone in the medial geniculate body of cat. *Hearing Research* **11**:235–247.

Rouiller EM and de Ribaupierre F (1985) Origin of afferents to physiologically defined regions of the medial geniculate body of the cat: ventral and dorsal divisions. *Hearing Research* **19**:97–114.

Rouiller EM, Capt M, Hornung JP, and Streit P (1990) Correlation between regional changes in the distributions of GABA-containing neurons and unit response properties in the medial geniculate body of the cat. *Hearing Research* **49**:249–258.

Rouiller EM, Rodrigues-Dagaeff C, Simm G, de Ribaupierre Y, Villa AEP, and de Ribaupierre F (1989) Functional organization of the medial division of the medial geniculate body of the cat: tonotopic organization, spatial distribution of response properties and cortical connections. *Hearing Research* **39**:127–142.

Rutkowski RG and Weinberger NM (2005) Encoding of learned importance of sound by magnitude of representational area in primary auditory cortex. *Proceedings of the National Academy of Sciences of the United States of America* **102**:13664–13669.

Ryugo DK and Weinberger NM (1976) Corticofugal modulation of the medial geniculate body. *Experimental Neurology* **51**:377–391.

Ryugo DK and Weinberger NM (1978) Differential plasticity of morphologically distinct neuron populations in the medial geniculate body of the cat during classical conditioning. *Behavioral Biology* **22**:275–301.

Samson FK, Clarey JC, Barone P, and Imig TJ (1993) Effects of ear plugging on single-unit azimuth sensitivity in cat primary auditory cortex. I. Evidence for monaural directional cues. *Journal of Neurophysiology* **70**:492–511.

Samson FK, Barone P, Irons WA, Clarey JC, Poirier P, and Imig TJ (2000) Directionality derived from differential sensitivity to monaural and binaural cues in the cat's medial geniculate body. *Journal of Neurophysiology* **84**:1330–1345.

Schafe GE, Doyere V, and LeDoux JE (2005) Tracking the fear engram: the lateral amygdala is an essential locus of fear memory storage. *Journal of Neuroscience* **25**:10010–10014.

Schafe GE, Nader K, Blair HT and LeDoux JE (2001) Memory consolidation of Pavlovian fear conditioning: a cellular and molecular perspective. *Trends in Neurosciences* **24**:540–546.

Semple MN and Kitzes LM (1993a) Binaural processing of sound pressure level in cat primary auditory cortex - evidence for a representation based on absolute levels rather than interaural level differences. *Journal of Neurophysiology* **69**:449–461.

Semple MN and Kitzes LM (1993b) Focal selectivity for binaural sound pressure level in cat primary auditory cortex - 2-way intensity network tuning. *Journal of Neurophysiology* **69**: 462–473.

Senatorov VV and Hu B (1997) Differential Na(+)-K(+)-ATPase activity in rat lemniscal and non-lemniscal auditory thalami. *Journal of Physiology (London)* **502**:387–395.

Senatorov VV and Hu B (2002) Extracortical descending projections to the rat inferior colliculus. *Neuroscience* **115**:243–250.

Senatorov VV, Mooney D, and Hu B (1997) The electrogenic effects of Na(+)-K(+)-ATPase in rat auditory thalamus. *Journal of Physiology (London)* **502**:375–385.

Sherman SM and Guillery RW (1996) Functional organization of thalamocortical relays. *Journal of Neurophysiology* **76**:1367–1395.

Smith PH and Populin LC (2001) Fundamental differences between the thalamocortical recipient layers of the cat auditory and visual cortices. *Journal of Comparative Neurology* **436**:508–519.

Smith PH, Bartlett EL, and Kowalkowski A (2006) Unique combination of anatomy and physiology in cells of the rat paralaminar thalamic nuclei adjacent to the medial geniculate body. *Journal of Comparative Neurology* **496**:314–334.

Soja PJ, Fragoso MC, Cairns BE, and Jia WG (1996) Dorsal spinocerebellar tract neurons in the chronic intact cat during wakefulness and sleep: analysis of spontaneous spike activity. *Journal of Neuroscience* **16**:1260–1272.

Spitzer MW, Calford MB, Clarey JC, Pettigrew JD, and Roe AW (2001) Spontaneous and stimulus-evoked intrinsic optical signals in primary auditory cortex of the cat. *Journal of Neurophysiology* **85**:1283–1298.

Steriade M (1984) The excitatory-inhibitory response sequence in thalamic and neocortical cells: state-related changes and regulatory systems. In: Edelman GM, Gall WE, and Cowan WM (eds). *Dynamic Aspects of Neocortical Functions*. Wiley, New York, pp. 107–157.

Steriade M (1989) Brain electrical activity and sensory processing during waking and sleep states. In: Kryger MH, Roth T, and Dement WC (eds). *Principles and Practices of Sleep Medicine*. W.B. Saunders Company, Philadelphia, pp. 86–103.

Steriade M and Deschênes M (1984) The thalamus as a neuronal oscillator. *Brain Research Reviews* **8**:1–63.

Steriade M and Llinás R (1988) The functional states of the thalamus and the associated neuronal interplay. *Physiological Reviews* **68**:649–742.

Steriade M, Contreras D, Curro Dossi R, and Nunez A (1993a) The slow (<1 Hz) oscillation in reticular thalamic and thalamocortical neurons: scenario of sleep rhythm generation in interacting thalamic and neocortical networks. *Journal of Neuroscience* **13**: 3284–3299.

Steriade M, Curro Dossi R, and Contreras D (1993b) Electrophysiological properties of intralaminar thalamocortical cells discharging rhythmic (approximately 40 Hz) spike-bursts at approximately 1000 Hz during waking and rapid eye movement sleep. *Neuroscience* **56**:1–9.

Steriade M, Nunez A, and Amzica F (1993c) A novel slow (<1 Hz) oscillation of neocortical neurons in vivo: depolarizing and hyperpolarizing components. *Journal of Neuroscience* **13**: 3252–3265.

Steriade M, Nunez A and Amzica F (1993d) intracellular analysis of relations between the slow (<1 Hz) neocortical oscillation and other sleep rhythms of the electroencephalogram. *Journal of Neuroscience* **13**:3266–3283.

Steriade M, Jones EG, and McCormick DA (1997) *Thalamus, volume 1*. Elsevier, Amsterdam.

Sukov W and Barth DS (2001) Cellular mechanisms of thalamically evoked gamma oscillations in auditory cortex. *Journal of Neurophysiology* **85**:1235–1245.

Supple WF and Kapp BS (1989) Response characteristics of neurons in the medial component of the medial geniculate nucleus during Pavlovian differential fear conditioning in rabbits. *Behavioral Neuroscience* **103**:1276–1286.

Tennigkeit F, Schwarz DWF, and Puil E (1996) Mechanisms for signal transformation in lemniscal auditory thalamus. *Journal of Neurophysiology* **76**:3597–3608.

Tennigkeit F, Puil E, and Schwarz DWF (1997) Firing modes and membrane properties in lemniscal auditory thalamus. *Acta Otolaryngologica* **117**:254–257.

Usrey WM (2002) The role of spike timing for thalamocortical processing. *Current Opinion in Neurobiology* **12**:411–417.

Velenovsky DS, Cetas JS, Price RO, Sinex DG, and McMullen NT (2003) Functional subregions in primary auditory cortex defined by thalamocortical terminal arbors: an electrophysiological and anterograde labeling study. *Journal of Neuroscience* **23**: 308–316.

Victor JD and Purpura KP (1996) Nature and precision of temporal coding in visual cortex: a metric-space analysis. *Journal of Neurophysiology* **76**:1310–1326.

Victor, JD and Purpura K (1997) Metric-space analysis of spike trains: theory, algorithms, and application. *Network* **8**:127–164.

Villa AE, Tetko IV, Dutoit P, de Ribaupierre Y and, de Ribaupierre F (1999) Corticofugal modulation of functional connectivity within the auditory thalamus of rat, guinea pig and cat revealed by cooling deactivation. *Journal of Neuroscience Methods* **86**: 161–178.

Villa AEP, Rouiller EM, Simm GM, Zurita P, de Ribaupierre Y, and de Ribaupierre F (1991) Corticofugal modulation of the information

processing in the auditory thalamus of the cat. *Experimental Brain Research* **86**:506–517.
Wang X and Kadia SC (2001) Differential representation of species-specific primate vocalizations in the auditory cortices of marmoset and cat. *Journal of Neurophysiology* **86**:2616–2620.
Warren RA and Jones EG (1994) Glutamate activation of cat thalamic reticular nucleus: effects on response properties of ventroposterior neurons. *Experimental Brain Research* **100**:215–226.
Weinberger NM (2004) Specific long-term memory traces in primary auditory cortex. *Nature Reviews Neuroscience* **5**:279–290.
Weinberger NM and Diamond DM (1987) Physiological plasticity in auditory cortex: rapid induction by learning. *Progress in Neurobiology* **29**:1–55.
Wester K, Irvine DRF, and Hugdahl K (2001) Auditory laterality and attentional deficits after thalamic haemorrhage. *Journal of Neurology* **248**:676–683.
Whitfield (1979) The object of the sensory cortex. *Brain Behavior and Evolution* **16**:129–154.
Whitfield IC and Purser D (1972) Microelectrode study of the medial geniculate body in unanaesthetized free-moving cats. *Brain Behavior and Evolution* **6**:311–322.
Whitley JM and Henkel CK (1984) Topographical organization of the inferior collicular projection and other connections of the ventral nucleus of the lateral lemniscus in the cat. *Journal of Comparative Neurology* **229**:257–270.
Winer JA (1985) The medial geniculate body of the cat. *Advances in Anatomy, Embryology and Cell Biology* **86**:1–98.
Winer JA (1991) Anatomy of the medial geniculate body. In: Altschuler RA, Bobbins RP, Clopton BM and Hoffman DW (eds). *Neurobiology of Hearing: The Central Auditory System*. Raven Press, New York, pp. 293–333.
Winer JA (1992) The functional architecture of the medial geniculate body and primary auditory cortex. In: Webster DB, Popper AN, and Fay RR (eds). *Springer Handbook of Auditory Research*, volume 1, *The Mammalian Auditory Pathway: Neuroanatomy*. Springer, New York, pp. 222–409.
Winer JA, Diehl JJ, and Larue DT (2001) Projections of auditory cortex to the medial geniculate body of the cat. *Journal of Comparative Neurology* **430**:27–55.
Winer JA and Morest DK (1983) The medial division of the medial geniculate body of the cat: implications for thalamic organization. *Journal of Neuroscience* **3**:2629–2651.
Winer JA and Larue DT (1987) Patterns of reciprocity in auditory thalamocortical and corticothalamic connections: study with horseradish peroxidase and autoradiographic methods in the rat medial geniculate body. *Journal of Comparative Neurology* **257**: 282–315.
Winer JA and Larue DT (1996) Evolution of GABAergic circuitry in the mammalian medial geniculate body. *Proceedings of the National Academy of Sciences of the United States of America* **93**: 3083–3087.
Winer JA, Kelly JB, and Larue DT (1999) Neural architecture of the rat medial geniculate body. *Hearing Research* **130**:19–41.
Winer JE and Larue DT (1988) Anatomy of glutamic acid decarboxylase immunoreactive neurons and axons in the rat medial geniculate body. *Journal of Comparative Neurology* **278**:47–68.
Winter P and Funkenstein HH (1973) The effect of species-specific vocalization on the discharge of auditory cortical cells in the awake squirrel monkey. (Saimiri sciureus). *Experimental Brain Research* **18**:489–504.
Wörgötter F, Suder K, Zhao Y, Kerscher N, Eysel UT, and Funke K (1998) State-dependent receptive-field restructuring in the visual cortex. *Nature* **396**:165–168.
Yeshurun Y, Wollberg Z, and Dyn N (1989) Prediction of linear and non-linear responses of MGB neurons by system identification methods. *Bulletin of Mathematical Biology* **51**: 337–346.
Yeshurun Y, Wollberg Z, Dyn N, and Allon N (1985) Identification of MGB cells by Volterra kernels. I. Prediction of responses to species specific vocalizations. *Biological Cybernetics* **51**: 383–390.
Yu YQ, Xiong Y, Chan YS, and He J (2004a) Corticofugal gating of auditory information in the thalamus: an in vivo intracellular recording study. *Journal of Neuroscience* **24**:3060–3069.
Yu YQ, Xiong Y, Chan YS, and He J (2004b) In vivo intracellular responses of the medial geniculate neurones to acoustic stimuli in anaesthetized guinea pigs. *Journal of Physiology (London)* **560**:191–205.

Chapter 13
Spectral Processing in Auditory Cortex

Christoph E. Schreiner, Robert C. Froemke, and Craig A. Atencio

Abbreviations

AAF	anterior auditory field
ADF	anterior dorsal field
AI	primary auditory cortex
AII	second auditory cortical field
AL	antero-lateral field
BF	best frequency
BW	bandwidth
CF	characteristic frequency
CL	caudolateral field
CM	caudal medial field
DC	dorsal-caudal field
DCB	dorsocaudal belt
DRB	dorsorostral belt
DZ	dorsal zone
EP	ectosylvian fields
FRA	frequency response area
FSU	fast-spiking unit
FTC	frequency tuning curve
GABA	gamma-aminobutyric acid
MGB	medial geniculate body
MI	mutual information
MID	maximally informative dimension
ML	medial-lateral field
MM	middle medial field
MTF	modulation transfer function
P	postnatal day
PAF	posterior auditory field
PDF	posterior dorsal field
PPF	posterior pseudosylvian field
PSF	posterior suprasylvian field
Q	quality factor
R	rostral field
RF	receptive field
RM	rostro-medial field
RSS	random spectral stimulus
RSU	regular-spiking unit
RT	rostro-temporal field
SRAF	suprarhinal auditory field
STA	spike-triggered average
STRF	spectro-temporal receptive field
TORC	temporally orthogonal ripple combinations
VCB	ventrocaudal belt
VPAF	ventroposterior auditory field
VRB	ventrorostral belt

1 Introduction

Historically, the main purpose of the auditory system has been interpreted as a frequency analyzer (Ohm 1843; von Helmholtz 1863) that provides a faithful spectral representation of the received acoustic waveform. Analysis and characterization of spectral processing, beginning with the principle of parallel signal processing in narrow, partially overlapping frequency channels in the cochlea, has provided a framework for all subsequent stages of computation, information extraction and encoding in the auditory system, including the auditory cortex. This still evolving bottom-up characterization around the concept of a set of parallel frequency filters has been significantly enhanced by including temporal or dynamic and nonlinear aspects of spectral processing. Quantitative and rigorous systems and information analysis approaches have resulted in more complete characterizations of spectral encoding and decoding abilities throughout the auditory system.

However, the view of the ear as a mere frequency analyzer, even a nonlinear, dynamic one, is an incomplete characterization of the auditory system, especially when it comes to more central stations, including the auditory cortex. Firstly, the ability to process complex, natural acoustic environments, including transmission of communication sounds in the presence of background noise or competing signals in a

C.E. Schreiner (✉)
Coleman Memorial Laboratory, Department of Otolaryngology, W.M. Keck Center for Integrative Neuroscience, University of California, School of Medicine, San Francisco, CA 94143–0732, USA
e-mail: chris@phy.ucsf.edu

J.A. Winer, C.E. Schreiner (eds.), *The Auditory Cortex*,
DOI 10.1007/978-1-4419-0074-6_13, © Springer Science+Business Media, LLC 2011

complex or reverberant auditory environment, is likely to require special mechanisms that may not be apparent using simple spectral analysis methods.

Secondly, and perhaps more importantly, our experience of the world around us is not simply an accurate reflection of its physical features. Determining the meaning of stimuli, and generating behaviors that lead us to successfully and efficiently achieve our immediate and long-term goals, is an inherent aspect of sensory processing. Consequently, sensory stimuli often need to be grouped according to their category membership in behaviorally equivalent classes of sounds. For interpretational purposes, sound classes require a grouping process into categories along various dimensions that can be perceptual, in that stimuli share perceivable attributes, or interpretational, in that stimuli share a behavioral response. A purely spectrally based solution to this problem seems unlikely.

Conceptually, cortical stimulus representations must employ mechanisms to compensate for natural variations in stimuli, such as intensity, timing, vocal tract length, noise interference and speed of presentation, that otherwise may hamper if not preclude efficient and robust sound classification and categorization tasks essential for speech perception (King and Nelken 2009; Winkler et al. 2009). Potential auditory cortical stimulus encoding principles that differ from subcortical stations have been proposed: (i) shifts from temporal coding to rate-coding (Wang et al. 2008); (ii) non-isomorphic transformations of acoustic features (Barbour and Wang 2003b; Wang 2007); (iii) emphasis of natural sound statistics (David et al. 2009; Mesgarani et al. 2009; Nagel and Doupe 2008; Sen et al. 2001; Theunissen and Shaevitz 2006; Woolley et al. 2006); (iv) creation of feature combinations toward an "object"-based representation (Bar-Yosef and Nelken 2007; King and Nelken 2009); and (v) creation of representational invariances, e.g., for intensity (Billimoria et al. 2008; Sadagopan and Wang 2009), background noise robustness (Mesgarani et al. 2009; Nagarajan et al. 2002), or sound source properties (Grana et al. 2009; Margoliash and Fortune 1992; Theunissen and Shaevitz 2006). These cortical processes may include stimulus transformations into internal representations that may no longer be faithful to their physical structure (Wang 2007) and have to reflect influences from behavioral states, such as attention and vigilance, in the context of optimal behavioral task performance (Edeline 2003; Fritz et al. 2007a, b).

While spectral analysis aspects alone may seem inadequate in addressing these issues, new estimation methods of dynamic spectral processing (Atencio et al. 2008, 2009; Bruno and Simons 2002) indicate that emergent processing aspects do exist in auditory cortex and that they may contribute to some of these proposed encoding principles of auditory stimuli.

The types and spatial distribution of physiological response properties have provided crucial information for deciphering principles and mechanisms underlying processing in cat and primate visual cortex (Callaway 1998; Henry 1991; Hirsch 2003; Lund 1990). Similarly, in auditory cortex, non-uniform spatial distributions of functional properties have been found for many basic response properties reflecting regional specializations.

Expansion of the central auditory representation of a given frequency from a point in the cochlea to many neurons tuned to the same frequency in cortex introduces the ability to treat many different aspects of required multiple analyses in parallel. This is further reflected in a reduction of redundancy between different stations: cortical neurons are less redundant than subcortical neurons suggesting that different cortical neurons, even when tuned to the same frequency, can convey different perceptual or interpretational aspects of stimuli (Chechik et al. 2006; Nelken and Bar-Yosef 2008).

Spectral processing in the auditory forebrain appears to undergo major transformations relative to the initial coding of acoustic information in the cochlea and compared to various principles that shape brainstem processing. However, our knowledge of the nature, purpose and mechanisms of these cortical transformations, especially in light of the dual purpose of stimulus representation and stimulus interpretation, is still rather rudimentary. The need for profound changes in the way spectral information must be processed becomes evident from the very diverse roles that auditory cortex has to play. In the following sections, we review some of the emerging and emergent properties of auditory cortical processing following a largely historical development in the sophistication of the employed spectral analysis methods. The focus is on more recent accomplishments. Several recent reviews (Escabí and Read 2003; Escabí and Read 2005; Schreiner et al. 2000; Sutter 2005; Young 2008) and other chapters in this book complement and often expand on aspects of spectral auditory cortical processing. If data are available, we consider spectral processing at different structural levels of cortical organization, such as cell types, cortical layers, and cortical fields and subfields, especially within the framework of general divisions such as primary and non-primary areas or auditory core, belt and parabelt areas – connectionally differentiated by thalamic input sources and cortico-cortical projection patterns (Hackett 2008; Hackett and Schroeder 2009; Kaas and Hackett 2000).

2 Spectral Analysis of Tonal Stimuli

2.1 Frequency Specificity

The most basic approach to characterize the excitatory spectral response of auditory neurons has been to present single tones of different frequencies and intensities to the ear and record the evoked neuronal responses. From the responses, different response profiles, such as the frequency tuning curve (FTC), frequency response area (FRA), or iso-intensity frequency profile of excitatory responses can be reconstructed. Two main aspects of response characterization commonly have been extracted. The first is the frequency preference or sensitivity of a neuron as captured by the characteristic frequency (CF), the tone that produces a response at the lowest intensity of any tested frequency, or the best frequency (BF), the tone that produces the strongest response for a given sound intensity. The second is the frequency selectivity or sharpness of tuning, often expressed as the bandwidth (BW) or range of frequencies, at a given sound intensity, that produce an excitatory response. Alternatively, a relative measure of sharpness of tuning, the Q-factor, is used which is defined as CF/BW and stated for a given sound intensity above minimum response threshold, such as Q10, Q20, or Q40.

Areal Organization: Many neurons in early auditory cortical stations, such as primary auditory cortex (AI), appear to have fairly simple, often V-shaped FRAs (Fig. 13.1), especially in various anesthetized preparations (e.g., rats: Gaese and Ostwald 2001; Sally and Kelly 1988; cats: Brugge and Reale 1985; Phillips and Irvine 1981; monkeys: Merzenich and Brugge 1973; Recanzone et al. 2000). Frequency specificity of cortical neurons, i.e., the presence of frequency-specific channels, is reflected in a wide range of CFs for many cortical fields and is largely independent of the particular cell type such as excitatory pyramidal cells or inhibitory interneurons (Atencio and Schreiner 2008). For many cortical areas, the full range of CFs, corresponding to the species-specific cochlear frequency extent, is present.

Convergent frequency information from the two ears is usually matched in auditory cortex, resulting in similar CFs for the two inputs. CFs derived from contralateral stimulation can be, on average, slightly higher (0.06 octave;

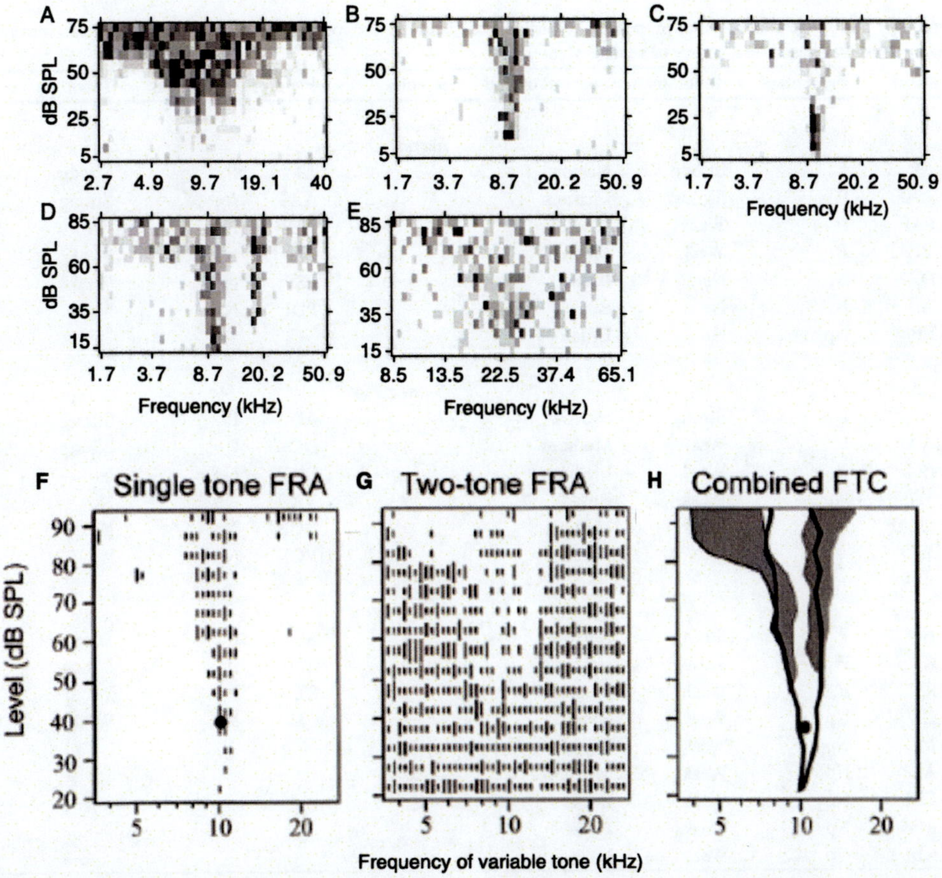

Fig. 13.1 Examples of cortical frequency response areas (FRAs; cat AI). The firing rate during the presentation of tones of different frequency and intensity are displayed. **a** Broadly tuned V-shaped FRA. **b** Narrowly tuned, I-shaped FRA. **c** Non-monotonic, O-shaped FRA. **d** Multipeaked FRA. **e** Diffuse FRA. **f** Single-tone FRA. **g, h** Two-tone FRAs. One tone is varied in frequency and intensity, similar to the single-tone FRAs in **a–e**. A second, constant tone at CF and at moderate to low levels (*black dot*) is presented conjointly with the varying tone to create an increase in baseline activity. This allows distinction of excitatory regions (firing rate above baseline) and suppressive regions (firing rate below baseline, *gray area* in H). Adapted from Sutter et al. (1999)

squirrel monkey (Cheung et al. 2009)). The significance of interaural CF asymmetry in normal hearing animals, however, is unlikely to be physiologically meaningful. Aurally asymmetric hearing loss can result in mismatch of convergent frequency information in cortical neurons with potential perceptual consequences (Cheung et al. 2009).

Stimulus information is distributed across a wide range of cortical neuron types, laminae, and areas. Knowledge of the spatial layout of information processing is important because it can provide crucial insights into the local functional tasks and algorithms (Eggermont 2001; Schreiner and Winer 2007). In primary/core cortical areas, neighboring neurons often have similar CF values. Spatial analysis of cortical frequency distributions obtained with extracellular, action potential-based mapping reveals that local clustering of similar functional properties, i.e., exceeding the expectations from random parameter distributions, is a general feature of many response and receptive field parameters (Schreiner and Winer 2007). Only few parameters, however, show a systematic spatial gradient across an entire cortical field. For CF, such functional gradients have been shown for many auditory cortical areas across many different species (e.g., Table 13.1). For classifying the degree of local clustering and global CF gradients, little quantitative information is available although precise measures have been used (Imaizumi et al. 2010; Bandyopadhyay et al. 2010; Rothschild et al. 2010). A coarse classification can, however, be derived for some of the more completely assessed animal models based on general descriptions of their CF organization in primary/core and non-primary/belt areas. Fields can be considered to have "strong" tonotopy if they show both local and global frequency organization, commensurate with a smooth CF gradient across most of the cochlear frequency range (e.g., cat AI, Merzenich et al. 1975; Fig. 13.2). "Weak" tonotopic fields are considered to have considerable variability in local clustering and neighborhood relationships but show evidence of a global gradient (e.g., cat PAF; Loftus and Sutter 2001; Reale and Imig 1980). "Non-tonotopic" areas may still

Table 13.1 Tonotopy and spectral bandwidth properties across cortical fields in six species. Classification of the fields into primary/core and non-primary/belt regions was based on a survey of several studies

Species	Field	Field class	Tonotopy	Spectral tuning	Species	Field	Field class	Tonotopy	Spectral tuning
Carnivores									
Cat					*Ferret*				
	AI	P	Strong	Narrow		AI	P	Strong	Narrow
	AAF	P	Strong	Medium		AAF	*Np*	Strong	Medium
	PAF	*P*	Weak	Medium		ADF	Np	No	Medium
	VPAF	*P*	Weak	Medium		PPF	Np	Weak	Medium
	DZ	Np	No	Broad		PSF	Np	Weak	Medium
	AII	Np	No	Broad		PDF	Np	No	Medium
	EP	Np	No	Broad					
Rodents									
Rat					*Guinea Pig*				
	AI	P	Strong	Narrow		AI	P	Strong	Narrow
	AAF	P	Weak	Medium		DC	P	Strong	Narrow
	PAF	*P*	Weak	Medium		DRB	Np	No	Medium
	VAF	Np	Weak	Broad		VRB	Np	No	Broad
	SRAF	Np	Weak	Medium		DCB	Np	No	Broad
						VCB	Np	No	Broad
Primates									
Macaque					*Marmoset*				
	AI	P	Strong	Narrow		AI	P	Strong	Narrow
	R	P	Strong	Narrow		R	P	Strong	Narrow
	RT	P	Weak	Medium		RT	P	Strong	Narrow
	CL	Np	Weak	Broad		CM	Np	Weak	Medium
	ML	Np	Weak	Broad					
	AL	Np	Weak	Medium					
	CM	Np	Weak	Broad					
	MM	Np	Weak	Broad					
	RM	Np	Weak	Medium					

Areas with some uncertainty regarding this classification are indicated in italic. Classification of tonotopy and spectral tuning was based largely on verbal description of these properties, since uniform quantitative measures (see text) are rare beyond primary fields. Among the studies that were surveyed are: Bendor and Wang (2008), Bizley et al. (2005), Imaizumi et al. (2004), Hackett et al. (1998), Hackett (2008, 2010), Kajikawa et al. (2008), Kowalski et al. (1995), Kusmierek and Rauschecker (2009), Loftus and Sutter (2001), Merzenich and Brugge (1973), Nishimura et al. (2007), Polley et al. (2007), Rauschecker and Tian (2004), Recanzone (2000, 2008), Rutkowski et al. (2002), Sally and Kelly (1988), Schreiner and Cynader (1984), and Tian and Rauschecker (2004).

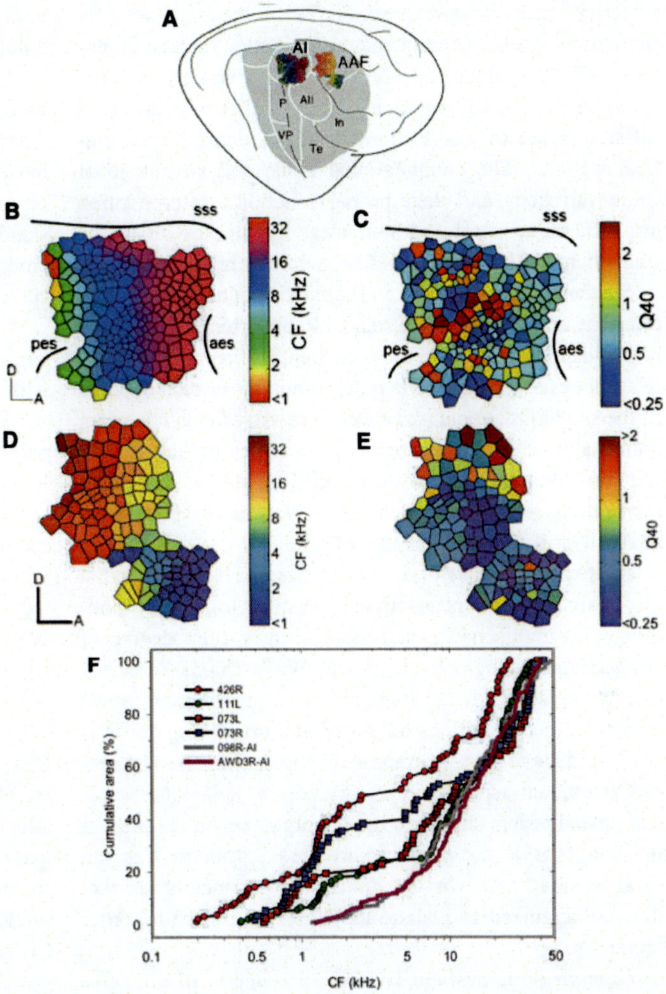

Fig. 13.2 Spatial distribution of CF and sharpness of tuning (Q) across cat AI and AAF. **a** Schematic view of cat auditory cortex. Recording locations of example maps are superimposed on AI and AAF, respectively. **b** Cat AI CF map. **c** Cat AI Q40 map. **d** Cat AAF CF map. **e** Cat AAF Q40 map. **f** Cumulative cortical area as a function of CF. *Solid lines* (*gray*, *purple*; no data symbols) are two cat AI cumulative area functions for two AI maps. The area functions with data symbols are from four cat AAF maps. Adapted from Imaizumi et al. (2004)

contain some local CF clustering but show no indication of a single spatial gradient covering significant portions of the cochlear frequency range (e.g., cat AII; Reale and Imig 1980; Schreiner and Cynader 1984). For two primates, two carnivores, and two rodents, with fairly advanced characterizations of several cortical fields, 11 of 16 (70%, Table 13.1) primary/core fields exhibit strong tonotopy while only 1 of 21 (5%) non-primary/belt areas show this trait. Conversely, none of the primary/core fields lacks tonotopy whereas 9 of 21 (43%, Table 13.1) non-primary/belt areas are non-tonotopic. Differences in map structure may reflect differences in underlying intracortical circuits, related to differences in input statistics, local algorithms, or in behavioral tasks requirements (Chklovskii and Koulakov 2004; Schreiner and Winer 2007).

Even in primary/core areas, the frequency representation of sounds, as reflected in the distribution of CFs, is not a faithful replica of the cochlear frequency map. Fine-grain electrophysiological cortical frequency mapping usually shows a clear CF gradient in cat AI (Fig.13.2). The mean gradient changes as a function of CF with the steepest slope below 5 kHz and differs from the cochlear frequency gradient. The steep section corresponds to a smaller magnification factor and a relative under-representation of those frequencies (Merzenich et al. 1975). However, the AI tonotopic gradient is relatively smooth compared to that in other primary fields, such as the anterior auditory field (AAF). Cat, gerbil, and ferret AAF all express gross local distortions and apparent omissions in their CF representations that appear to be unique to each individual animal and species (Bizley et al. 2005; Imaizumi et al. 2004; Thomas et al. 1993; Fig. 13.2). The functional implications of these uneven frequency representations remain unclear but likely reflect specific environmental or task-specific adaptations of cortical or subcortical processing that benefit from non-uniform spectral emphasis.

A further reduction or even elimination of tonotopy is often connected to a loss of neuronal frequency selectivity near response threshold. This is the main cause for the virtual absence of frequency organization in cat auditory field AII (Schreiner and Cynader 1984) and ferret anterior dorsal field (Bizley et al. 2005) and is suggestive of different types of information transformation between cortical stations. The computational goals and advantage of these variations and their proper functional interpretation are difficult to assess without clear hypotheses about the implemented perceptually and behaviorally relevant tasks.

Systematic degradations in the fidelity of auditory cortex tonotopy across areas seem related to other hierarchical area classifications, such as in the core, belt, and parabelt scheme (Rauschecker 1998). However, tonotopy alone cannot serve as the single functional parameter to classify fields regarding their status as primary/core of non-primary or belt, which requires additional information based on source and target connectivity of its projections (e.g., Hackett 2010; Hackett et al. 1998; Kaas and Hackett 1999).

Anatomical studies of auditory cortex have revealed that all extrinsic areal connections, whether tonotopic, non-tonotopic, multisensory, or limbic, show a high degree of connectional topography (Lee and Winer 2005; Schreiner and Winer 2007). Local topographies in convergent inputs create distinct conditions for functional processing and it is not surprising to see topographic principles expressed by several functional aspects in auditory cortex. It is conceivable that spatial orders similar to the CF organization are present in areas outside the primary/core areas although it is not clear, at this time, what the functional parameters are that may be organized and where they fall along a spatial order hierarchy.

Laminar Organization: Evidence in support of a precise anatomical lamination of auditory cortex is manifold and compelling (Kelly and Wong 1981; Mitani and Shimokouchi 1985; Mitani et al. 1985; Winer 1984a, c; Winguth and Winer 1986). Laminar borders, defined by cell structure, connections, or chemical anatomy, are precise to within a few micrometers, as is the spatial segregation of afferents (Winer 1992). Each layer differs in its neuronal architecture and cytoarchitecture, GABAergic organization, thalamic input, commissural input and output, cortico-cortical input and output, and corticofugal projections to the telencephalon and brainstem (Winer 1992; see Chapter 2).

In AI, cells are vertically arranged in a more conspicuous manner than in other sensory systems (Jones 2000; Winer 1984b). This vertical arrangement is accompanied by highly specific interlaminar connections (Barbour and Callaway 2008; Mitani and Shimokouchi 1985; Mitani et al. 1985; Wallace et al. 1991). This vertical microcircuitry has been considered a key element of cortical processing (Mountcastle 1997). Thus, the connections between layers follow a precise and characteristic pattern that offers the opportunity to compare the function of specific components of the cortical microcircuit (Martinez et al. 2005). Functionally defined columns may not be a fundamental (canonical) building block or provide a transcendent principle given their variability in presence and appearance in some species (Horton and Adams 2005). However, the vertical circuit – influenced by horizontal inputs and feedbacks – does provide a more robust organizational principle that may contain the key to understanding the local transformations and output patterns that emerge from every point in the horizontal sheet of cortical cells (Atencio and Schreiner 2010a,b; Atencio et al. 2009).

A basic feature of sensory cortex is that certain response parameters are conserved across cortical depth, especially with regard to the location of the receptor surface (Linden and Schreiner 2003). In auditory cortex, the evidence is compelling that this is also the case for frequency sensitivity. Vertical electrode penetrations across all cortical layers often show a clear and moderately tight alignment and correspondence of CFs, supporting a strong columnar organization principle, at least in primary/core areas (Abeles and Goldstein 1970; Phillips and Irvine 1981; Shen et al. 1999; Wallace and Palmer 2008; Atencio and Schreiner 2010a,b; Atencio et al. 2009). Similar studies in non-primary/belt areas are still lacking. In some subregions of cat AI, e.g., in the central narrowly tuned section, the average deviation of CFs in an orthogonal penetration across all cortical layers is only 0.1–0.2 octaves (Fig. 13.3) (Abeles and Goldstein 1970; Atencio and Schreiner 2010a,b). CF variations of similar magnitude across depth have been observed in unanesthesized mice (Shen et al. 1999). Other regions in cat AI proper, such as near the ventral or dorsal borders, can show a larger CF scatter across layers with some CFs within a penetration deviating by as much as 1 octave (Abeles and Goldstein 1970; Schreiner and Sutter 1992; Phillips and Irvine 1981; Atencio and Schreiner 2010a). This indicates that a strict columnar frequency organization, preserving close functional neighborhood relations across different layers, may be common, especially in cortical core areas, but is not a universal principle of auditory cortex organization. In fact, recent studies of the fidelity of the tonotopic organization in mouse AI, using two-photon calcium imaging techniques, have revealed evidence for a highly fractured local frequency organization in the horizontal domain of the upper cortical layers (Bandyopadhyay et al. 2010; Rothschild et al. 2010). At a fine spatial scale, local CFs differed by up to an octave creating a highly diffuse local frequency organization, while maintaining a rather coarse tonotopic gradient on a global scale. By contrast, mapping in the thalamic input layers has demonstrated a reasonably strong tonotopic organization in mouse AI (Stiebler et al. 1997). These discrepancies in the observed fine and global frequency organization, such as tight alignment across layers in some cases and large local CF scatter within a cortical layer in other cases, require further attention because it has profound consequences on

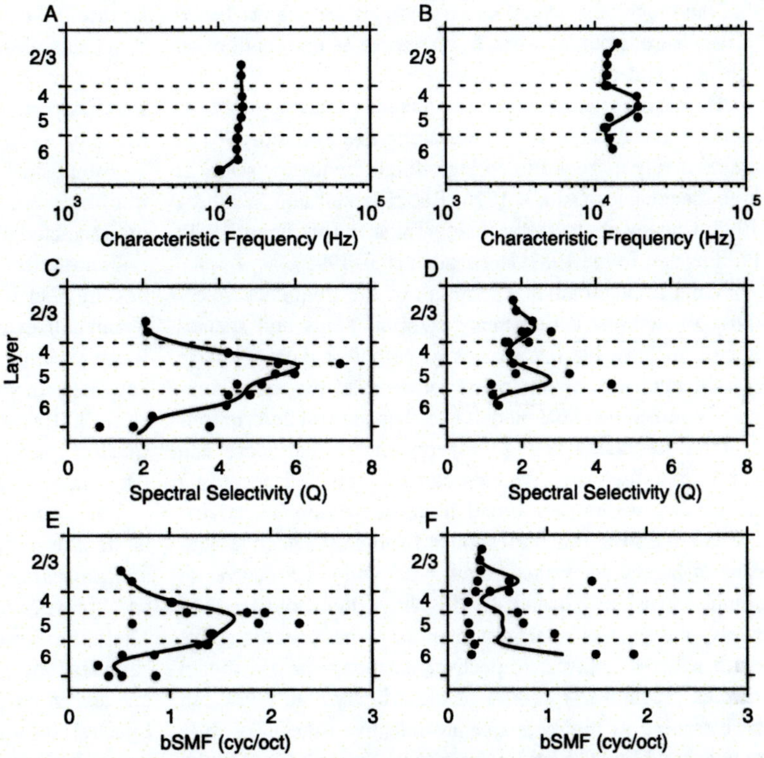

Fig. 13.3 Laminar distribution of spectral response properties. **a, b** *Vertical* electrode penetrations in cat AI with low and high variability in the CF distribution. **c,d** Depth profile of spectral tuning width (Q; high values correspond to more narrow tuning) for the same penetrations shown in (**a**) and (**b**). **e, f** Laminar profile of best spectral modulation frequency distribution for the same penetrations shown in (**a, c**) and (**b, d**), respectively. Data based on Atencio and Schreiner (2010b) and unpublished observations by Atencio and Schreiner

our understanding of cortical processing principles. Issues that certainly play a significant role in accounting for these differences are related to methodologically determined selectivity biases toward cell types, spatial integration, and anesthesia influences. Further biases arise from uncertainties regarding developmental stage, environmental properties and demands, and species-specific organization and processing principles. Species-, areal-, laminar-, and cell-specific computational tasks are not stereotypic but likely involve many different algorithms and serve different goals. The main limitation in interpreting any of the auditory cortex organizational features is, for many species, a lack in understanding the purposes served by individual processing steps. Together, these points emphasize the need for thorough comparative studies and highlight the limits of interpretational generalizations.

2.2 Frequency Selectivity

For tonotopy, quite precise maps can be derived from near-threshold pure-tone responses, especially in primary/core areas, but the validity of an interpretation of the frequency sensitivity for suprathreshold stimuli is limited without considering other aspects of stimulus parameter covariations, such as the spread of excitation across the receptor surface with sound intensity and systematic changes in filter bandwidth in the cochlea and in subcortical processing stations, as well as behavioral task relevance. Thus, frequency specificity does not reflect the actual frequency selectivity of neurons and, consequently, is a poor substrate for understanding spectral processing, especially of broad-band sounds.

Areal Organization: Excitatory bandwidths of neurons have generally been assessed by varying pure-tone stimuli over a large range of frequencies and intensities (Fig. 13.1). For many cortical neurons this results in a single, circumscribed frequency/intensity region of elevated activity. The differences in upper and lower frequency limits of the excitatory region serve as a measure of excitatory bandwidth, although one has to take into consideration that the range can strongly depend on sound intensity. As a consequence, frequency selectivity measures are often expressed with reference to a specific stimulus intensity, such as 10 or 40 dB above minimum response threshold. In primary/core areas, many neurons show a fairly narrow excitatory range, especially in the anesthetized preparation. Other fields show low frequency selectivity for all neurons and across all stimulus intensities. Even in primary/core areas, the range of Q and BW values can span 1–1.5 orders of magnitude (Phillips and Irvine 1981; Schreiner and Sutter 1992; Cheung et al. 2001a; Kowalski et al. 1995; Recanzone et al. 1999). This means that the range of potential spectral integration – as reflected in the neurons output – can be as narrow as one tenth of an octave or wider than five octaves.

At least in primary/core fields, there is a tendency for Q to increase as a function of CF indicating that excitatory FRAs are relatively narrower (on a logarithmic frequency scale) at high frequencies (Aitkin 1976; Batzri-Izraeli and Wollberg 1992; Cheung et al. 2001a; Pelleg-Toiba and Wollberg 1989; Phillips and Irvine 1981; Recanzone et al. 1999).

While a quantitative description of the frequency selectivity of neurons across many auditory fields and species is still not possible due to lack of sufficient data, a coarse classification can be attempted for some of the more completely assessed animal models based on general descriptions of their frequency tuning properties in primary/core and non-primary/belt areas. Fields can be classified according to narrow, medium, or broad frequency tuning/selectivity. For our purpose, this corresponds for highly selective neurons to bandwidth values below ~0.5 and ~1.5 octaves at sound intensities 10 and 40 dB above threshold, respectively, and for low selectivity neurons to bandwidths above ~1.5 and ~4 octaves, respectively. Among the six model species (Table 13.1), ~60% of the primary/core fields can be classified as highly frequency selective (narrow), while none were found to have low selectivity (broad). Conversely, 52% of the non-primary/belt areas have low frequency selectivity (broad) and none were classified as highly selective. Similar to the classification based on tonotopy, frequency selectivity alone does not provide a functional differentiation of fields that accurately corresponds to that based on anatomical/connectivity aspects. Relating frequency selectivity to tonotopy estimates strengthens the global field classification. All ten narrowly tuned fields (Table 13.1) are primary/core areas and have strong tonotopy. Conversely, all eleven broadly tuned areas are non-primary/belt and 55% of these show no evidence of a tonotopic gradient. Eleven areas with weak tonotopy and medium frequency selectivity split nearly evenly between primary/core and non-primary/belt regions underscoring that basic frequency processing aspects alone cannot align functional and anatomical cortical field classifications.

In some primary/core fields, clusters of neurons sharply or broadly tuned to frequency are segregated along the iso-frequency axis of the tonotopic map. Cortico-cortical connectivity in cat AI finds that broad or narrow spectral bandwidth clusters predominantly are connected with other clusters of the same property (Imaizumi et al. 2004; Read et al. 2001), thus creating a functional and connectional mosaic of interconnected, interleaved modules of different spectral integration. This topographic arrangement can be interpreted as an iterated map of spectral integration (Schreiner et al. 2000) that is independent of, or orthogonal to, the frequency decomposition domain of the receptor surface. A clear functional, task-directed interpretation of these modules is still elusive but they may enhance processing of spectral shape as in the determination of vocal tract properties (Calhoun and Schreiner 1998; Versnel and Shamma 1998). Functional significance, however, needs to be established related to particular steps in a sequence of transformations and integrations rather than as an isolated, disassociated phenomenon.

Non-uniform distributions of spectral integration properties are also seen in other primary fields, such as cat AAF (Imaizumi et al. 2004), and in other species, such as the ferret (Shamma et al. 1993; Bizley et al. 2005), owl monkey (Recanzone et al. 1999), and squirrel monkey (Cheung et al. 2001a). However, in awake preparations, evidence of spectral integration topography has not been unambiguous (Recanzone et al. 2000).

The systematic change in spectral selectivity across AI is significant for understanding the cortical representation and processing of spectrally complex signals, like species-specific vocalizations, speech, music, and ambient noise. These topographies suggest that any incoming signal is simultaneously processed through many filters with different center frequencies and a broad range of bandwidths. Spectral information in AI is extracted and represented by multiple modules for frequency resolution along the iso-frequency domain, and the center frequency of each bandwidth module is aligned to the "frequency decomposition" or tonotopic axis. Parallel analysis by multiple bandwidths results in an iterative, multi-resolution representation of information within each iso-frequency domain differentially weighted by filter width. This parallel analysis may aid in the extraction and evaluation of complex spectral shapes, e.g., formant structure of vowels, and establish multiple, parallel output streams for further processing (Mesgarani et al. 2008; Schreiner and Calhoun 1994; Shamma et al. 1993; Sutter 2005; Wang and Shamma 1995)

The heterogeneity of spectral integration properties across primary and non-primary fields is in contrast to psychophysically determined spectral integration that is relatively constant at a "critical bandwidth" of ~1/3 octave throughout the cat hearing range (Ehret and Schreiner 1997; Nienhuys and Clark 1979; Pickles 1975). The module-like spatial organization of Q values across CFs in AI and AAF may be related to peripheral and thalamocortical mechanisms as well as to the RF construction in auditory cortex (Miller et al. 2001; Suga 1995; Sutter et al. 1999; Cheung et al. 2001a). Spectral bandwidth is already influenced by cochlear tuning properties (Liberman 1978; Narayan et al. 1998) and is reflected in subsequent processing stations. However, spectral integration differences in different frequency regions likely reflect higher-order processing principles, perhaps reflecting specific behavioral tasks (e.g., Razak et al. 2007; Suga 1995) or neuroanatomical arrangements (Prieto et al. 1994a, b; Read et al. 2002).

Anesthesia strongly affects the responses of neurons in the central auditory pathway, from the dorsal cochlear nucleus (Young and Brownell 1976) to the auditory cortex (Gaese and Ostwald 2001; Sally and Kelly 1988; Schreiner and Sutter 1992; Sutter and Schreiner 1991, 1995). In particular, the frequency selectivity in barbiturate- or higher dose isoflurane-anesthetized animals (Sutter and Schreiner 1991, 1995; Cheung et al. 2001b) appears to be generally narrower than in other anesthetic regimens, such as under halothane (Moshitch et al. 2006), or in awake animals. For example, awake rats and cats can show a 3–4 times wider bandwidth of excitatory tuning curves than under barbiturate (Gaese and Ostwald 2001; Qin et al. 2003). However, similarly highly frequency selective and unselective neurons can be encountered in both awake and anesthetized models (e.g., Abeles and Goldstein 1970; Moshitch et al. 2006; Schreiner and Sutter 1992, 1995; Kadia and Wang 2003). The shift toward higher frequency selectivity under certain anesthetic regimens may be due to an increase in the effectiveness of inhibition in the cortex. The consequences of bandwidth differences due to anesthesia for the emergence and functional interpretation of the wide range of spectral integration properties in auditory cortex and their relationship to behavior remain to be fully evaluated.

Laminar Organization: Laminar differences in frequency tuning bandwidths have been seen in several studies of cat, bat, and rodent auditory cortex (Dear et al. 1993; Eggermont 1996; Norena and Eggermont 2002; Sugimoto et al. 1997; Wallace and Palmer 2008; Atencio and Schreiner 2010a, b). The tuning bandwidth was generally broader for single neurons in the deep layers (IV to VI) compared to layers I to III of the Guinea pig (Wallace and Palmer 2008) and was sharpest for layers III and IV in the Mongolian gerbil (Sugimoto et al. 1997). In AI of ketamine-anesthetized cats, layer-specific frequency selectivity was also present; however, sites with fairly constant BW values across depth were also encountered (Fig. 13.1d). On average, the cat data also reflect a lower frequency selectivity for infragranular layers (Atencio and Schreiner 2010a, b). This indicates that strict columnar invariance in frequency selectivity is not the rule. In addition to layer-specific differences, pyramidal cells appear to have slightly higher frequency selectivity than putative inhibitory interneurons when they are recorded from within the same layer (Atencio and Schreiner 2008).

Overall, auditory cortex shows a wide range of frequency specificity and selectivity. However, to adequately appreciate this broad and varied repertoire of frequency filters and its impact for signal analysis, other aspects of cortical signal encoding need to be taken into consideration (see below) and, foremost, a better understanding of local and global processing goals and algorithms has to be developed (e.g., Griffiths et al. 2004; King and Nelken 2009).

2.3 Shape of Frequency Response Areas

In primary auditory cortical fields, most extensively observed in AI of various species, many frequency/intensity response areas have a rather uniform V-shape under anesthesia, i.e., the frequency selectivity decreases with increasing intensity (Brugge and Reale 1985; Sally and Kelly 1988; Phillips and Irvine 1981). However, a substantial proportion of neurons have quite different FRA shapes, including intensity-independent frequency tuning (I-shape), and circumscribed FRAs with no or substantially reduced responses at higher sound intensities (O-shape) (e.g., Abeles and Goldstein 1972; deCharms et al. 1998; Goldstein and Abeles 1975; Pelleg-Toiba and Wollberg 1989). Some neurons have multiple, non-continuous response areas (multipeaked) (Abeles and Goldstein 1972; Sutter and Schreiner 1991, Kadia and Wang 2003; He and Hashikawa 1998) or diffuse/patchy response areas composed out of many local intensity/frequency combinations without a clear, joint appearance that fits into standard classification schemes (Moshitch et al. 2006; Sadagopan and Wang 2009). A higher incident of complexly shaped response patterns can be found in unanesthetized and halothane preparations (Abeles and Goldstein 1972; deCharms et al. 1998; Pelleg-Toiba and Wollberg 1989; Kadia and Wang 2003; Moshitch et al. 2006; Sadagopan and Wang 2009).

A large diversity of FRA shapes, including some with very broad frequency tuning and some with multiple distinct excitatory frequency ranges, are also seen in other cortical fields, especially in non-primary areas (e.g., cat PAF; Loftus and Sutter 2001; Horseshoe bat; Radtke-Schuller and Schuller 1995). However, more quantitative studies of non-primary FRAs are needed to fully assess systematic filter-shape differences between most cortical fields.

Under anesthesia, most AI neurons have a single peaked FRA (Phillips and Irvine 1981), i.e., they have a single region of low-intensity responses centered at the CF. However, multipeaked tuning curves with two or three distinct low-threshold peaks have been described (Abeles and Goldstein 1972; Sutter and Schreiner 1991, Kadia and Wang 2003; Fitzpatrick et al. 1998; He and Hashikawa 1998; Oonishi and Katsuki 1965; Wenstrup and Grose 1995). In AI of awake marmosets, 20% of neurons have multipeaked FRAs. In both cats and marmosets, the excitatory spectral peaks in the multipeaked FRAs are often harmonically related (Kadia and Wang 2003; Sutter and Schreiner 1991). Stimuli presented at the spectral peaks of the multipeaked FRA can result in a facilitated response compared to either component presented in isolation. This suggests that sounds containing multiple, prominent spectral components may be processed by different classes of neurons (Kadia and Wang 2003).

Relating the position of single neurons with multipeaked tuning curves to the excitatory bandwidth distribution in cat AI reveals a distinct spatial distribution of these neurons (Sutter and Schreiner 1991). Multipeaked tuning curves are primarily found in the dorsal third of AI, whereas the rest of AI shows little evidence of single neurons with multiple FRAs. Multipeaked tuning curves are also characteristic for the Dorsal Zone, a non-primary area located adjacent and dorsal to AI (He and Hashikawa 1998). This subpopulation of cortical neurons may be sensitive to specific spectro-temporal combinations in the acoustic input (Sutter and Schreiner 1991; He et al. 1997). The spatial clustering of these specialized multipeaked neurons implies a functional segregation. Spatial and functional segregation of spectral analysis appears to be a general organizing principle of AI.

In the auditory cortex of awake animals, a substantial number of neurons do not respond to pure tones (Sadagopan and Wang 2009; Bandyopadhyay et al. 2010; Rothschild et al. 2010). At least some of these "unresponsive" neurons are likely to be selective for complex sound features with highly nonlinear combination-sensitive responses (Sadagopan and Wang 2009). Specific combinations of several tones with appropriate spectral and timing relationships can elicit strong responses whereas each component alone fails to produce an excitatory response (Sadagopan and Wang 2009), highly reminiscent of combination-sensitive neurons in echolocating bats (Suga 1984). Characterizing cortical neurons with more complex, broad-band spectra, including naturally occurring sounds, may reveal more appropriate response classifications that transcend the diversity of pure-tone FRA shapes.

2.4 Temporal Dependence of Pure-Tone Tuning

Frequency specificity (e.g., BF) and frequency selectivity (e.g., BW) are usually determined by integrating spikes over the entire duration of a tone stimulus for the construction of FRAs. This procedure masks three potential changes in frequency tuning during the time course of the response: (i) response latency differences for different intensities and for frequencies near the margins of the FRA, (ii) response duration differences, such as phasic versus sustained responses, and (iii) occurrence of "off" responses, i.e., excitatory activity following the end of tones.

Neurons with phasic response profiles predominate in anesthetized animals, and account for up to 50% of responses in awake animals (DeWeese et al. 2003; Evans and Whitfield 1964; Wang et al. 2005). In these neurons, frequency specificity strongly depends on the time relative to the stimulus onset (Schreiner et al. 2006). Early, short latency responses account for the high-intensity, broadly tuned region of most V-shaped FRAs (Fig. 13.4). Slightly longer latency responses provide lower intensity, near BF regions of the FRA. The longest latency phasic responses supply the off-CF regions of the FRA margins. Therefore, frequency specificity, including sensitivity and selectivity, for single neurons and for the neuronal population evolves rapidly over the course of the first ~40 ms after stimulus onset.

This is also the case for the phasic portion of neurons with sustained responses, however, the impact on the global tuning is diminished by the sustained portion of the activity. Yet, the frequency specificity of sustained neurons also undergoes a clear temporal evolution. FRAs of phasic (<30 ms) and early-sustained responses (<100 ms) were found to be highly similar, with BF differences of < 1/4 octaves (awake macaque; Fishman and Steinschneider 2009). In contrast, FRAs based on phasic and late-sustained (>100 ms) response portions differed considerably (BF differences: 2/3 octaves).

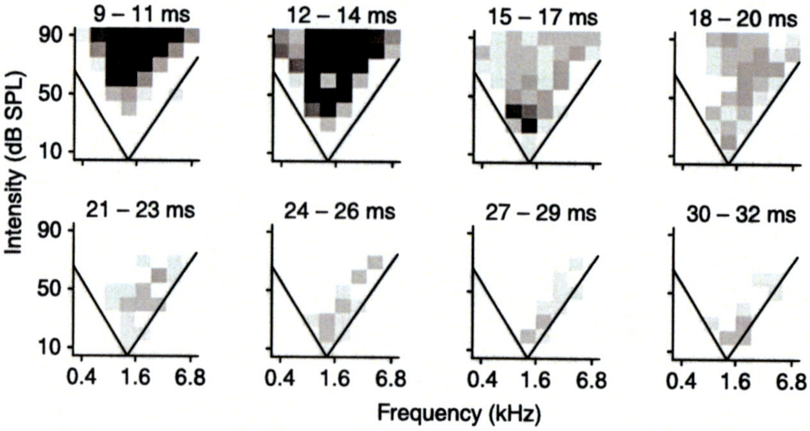

Fig. 13.4 Tuning curve shape as a function of time. The shape of a pure-tone FRA is plotted in 2 ms time intervals relative to tone-onset. *Shaded boxes* correspond to firing rate strength for different frequency-intensity combinations (*darker squares* correspond to higher firing rates). The *solid lines* indicate the lowest threshold across all time intervals, indicating the traditional frequency tuning curve that is customarily integrated across the total stimulus duration. Modified from Schreiner et al. (2006)

Many neurons with strong phasic or phasic/sustained response profiles also exhibit offset responses, especially in awake preparations (Fishman and Steinschneider 2009; Qin et al. 2007; Recanzone 2000). Prevalence of off-responsive neurons are about 30% in awake monkey (Recanzone 2000) and ketamine- or halothane-anesthetized cat (Volkov and Galazjuk 1991; Moshitch et al. 2006), and roughly 60% in awake cat (Qin et al. 2007). The frequency-filtering property of the off-responses differs from that of the phasic and sustained portions. Off-response FRAs usually are non-overlapping with or inversely related to that of the on-responses. Frequency tuning of off-responses is often ~1–2 octaves above that of on-responses in the awake macaque (Fishman and Steinschneider 2009; Pelleg-Toiba and Wollberg 1989). However, in awake cats, a similarly consistent relationship was not found (Qin et al. 2007). The different frequency tuning and excitatory–inhibitory compositions underlying on- and off-responses strongly suggest that they are driven by largely non-overlapping sets of synapses (Scholl et al. 2010). Frequency tuning of population responses may vary considerably over the course of the response to a tone, demonstrating a strong temporal dependence of the cortical spectral representation of sounds (Fishman and Steinschneider 2009).

2.5 Inhibitory Response Areas

Processing properties of cortical neurons are shaped by the convergence and interaction of excitatory thalamocortical and cortico-cortical inputs and inhibitory projections (see Section 13.6 and Chapter 2). Stimulus components outside of the excitatory FRA can exert strong suppressive effects on responses. If sufficient spontaneous activity is present, as is often the case in awake animals, suppressive effects from single tones can be observed (Qin and Sato 2004). Phasic neurons in awake cats showed that tone-evoked suppression and excitation temporally alternated and spectrally co-occurred, restricting excitatory spike-responses within narrow temporal limits but not setting the spectral limits. By contrast, sustained neurons showed that the suppression and excitation spectrally alternated and temporally co-occurred, restricting excitatory frequency tuning but not setting the time limits (Qin and Sato 2004). These observations hint at complex interactions of excitatory and inhibitory forces.

Many neurons, especially in anesthetized preparations, do not have sufficient spontaneous activity to observe suppressive effects at the level of extracellular recordings of spiking activity. By eliciting a mildly excitatory response, for example by a soft CF tone, suppressive effects of an additional tone can be observed. Application of this two-tone interaction paradigm has revealed a high incidence of neurons (>90%) with suppressive response regions outside the excitatory (one tone) FRA. A wide variety in the structure of these "inhibitory bands" has been observed ranging from a single V- or I-shaped band to more than four distinct suppressive regions (e.g., Sutter et al. 1999; Loftus and Sutter 2001) (Fig. 13.1). The most common arrangement of suppressive bands (~35%) in the anesthetized cat, ferret, and gerbil AI is a single suppressive band on either side of the excitatory FRA (Loftus and Sutter 2001; Sutter et al. 1999; Shamma et al. 1993; Foeller et al. 2001). Regional differences in the distribution of suppressive regions across AI have also been reported (Loftus and Sutter 2001; Kowalski et al. 1995). In cat dorsal AI, only 16% of the neurons had one suppressive band on either side of the FRA whereas 50% of ventral AI neurons had this organization. Regional organizational differences, thus, are also present when considering suppressive areas of the spectral filters that may be part of functionally distinct auditory cortical processing streams (Sutter et al. 1999). No laminar differences in strength of inhibition were observed (Foeller et al. 2001), although the distribution and density of different interneuron classes varies across lamina (Prieto et al. 1994a,b).

Suppressive interactions can also play a role in shaping the response magnitude within the excitatory FRA such as in the generation of O-shaped, circumscribed FRAs (Fig. 13.1). In extracellular (Sutter and Loftus 2003) and intracellular recordings (Tan et al. 2007), the intensity tuning of excitatory and inhibitory/suppressive components can be negatively correlated, supporting the hypothesis that cortical inhibition can contribute to intensity tuning within the excitatory domain.

Most studies of inhibitory cortical properties in the auditory system have been limited to AI. Studies in cat PAF revealed a higher incident of complexly shaped inhibitory FRAs, such as with more than 2 suppressive regions (Loftus and Sutter 2001). It is likely that more complex suppressive frequency bands indicate an analysis of greater spectral complexity. However, detailed studies at the synaptic level are needed to clearly establish the role of inhibitory/excitatory interactions in the shaping of spectral filter properties and the generation of excitatory and suppressive FRA regions throughout auditory cortex (see Section 13.6).

3 Cortical Frequency Channels

Psychophysical experiments in humans and animals have demonstrated that auditory processing makes use of a set of frequency channels with well-defined bandwidth for the processing and resolution of complex stimuli. The components of such a filter bank with intensity-tolerant and frequency-dependent bandwidth are known as critical bands (e.g., Greenwood 1974).

Speech recognition in humans requires relatively coarse spectral information, provided sufficiently resolved temporal information is available (Shannon et al. 1998, 2004). As little as four independent frequency channels may suffice for some basic speech identification. More channels, 16–64, can provide sufficient clues for nearly full speech perception, even in noise (Shannon 2005; Shannon et al. 1998, 2004; Smith et al. 2002). Music processing requires even higher spectral resolution (Shannon 2005). In addition to integration across relatively narrow frequency bands, for example for loudness formation and discrimination between different frequency components in a complex sound, integration across wider frequency regions is also perceptually utilized such as in comodulation masking release and profile analysis (Bregman 1990; Hall and Grose 1988).

Although the spectral RFs of auditory cortical neurons derived from tones are useful for estimating properties of spectral integration, a more direct measure of the effective auditory filter bandwidth is necessary to establish the relationship between psychophysics and neuronal behavior. Methods analogous to psychophysical measurements of critical bands applied to single neuron responses, such as suppression of a tone response by noises of different bandwidths or by flanking noise-bands at different frequency separations, are useful to establish a neural-perceptual correspondence (Ehret and Merzenich 1985, 1988; Ehret and Schreiner 1997; Fishman and Steinschneider 2006). By repeating this measurement for different tone intensities, the level dependence of the neural critical bandwidth can be assessed.

A majority of neurons in anesthetized cat AI show spectral integration properties that remain relatively constant across intensity. However, the critical bandwidth of many intensity-tolerant neurons is broader than predicted from behavioral measurements of the critical band. Neurons that are intensity tolerant and have critical bandwidths similar to the behaviorally known values for cats (Pickles 1975; Nienhuys and Clark 1979) are less common but cluster in the central, narrow-band region of cat AI (Ehret and Schreiner 1997). Only in a subgroup of neurons does the spectral integration width estimated from pure-tone responses match that derived from noise masking with clear discrepancies between the two measures in the remaining neurons (Ehret and Schreiner 1997). Consequently, the actual spectral integration properties depend on the specific stimulus conditions and pure-tone excitatory measures are not sufficient to fully explain broad-band spectral integration behavior (Schreiner et al. 2000).

Using a two-noise masking paradigm, the spectral resolution of neural populations in AI of awake macaques also was found to parallel results of psychoacoustic studies in both monkeys and humans. The best fit of auditory filter shapes in psychoacoustic and these neural studies of frequency resolution was found in cortical layers IV and lower layer III compared to lower quality fits for more superficial cortical layers (Fishman and Steinschneider 2006). Evidence for physiological correlates of perceptual critical bands was also found in human auditory cortex using magneto-encephalographic measures (Soeta and Nakagawa 2006). These studies indicate that a cortical representation of perceptual frequency resolution is available, at least at the level of AI.

Evidence of a correspondence between psychophysical and neural spectral integration properties in non-primary/belt areas is still lacking. Broader pure-tone tuning in many non-primary fields may indicate that such a correspondence may be less likely than for narrowly tuned cortical fields and wider frequency integration may be emphasized at those later levels of analysis. Neurons in non-primary areas, especially in awake preparations, have been shown to respond often better to noise than to tonal stimuli (Recanzone 2000; Rauschecker and Tian 2004). However, the consequences of such observations for the formation of perceptual attributes, in particular for spectral integration and resolution, remain unclear.

4 Static Spectral Profile Analysis

Given that naturally occurring sounds are usually neither tone- nor noise-like, the discrepancies between spectral response characterization between pure-tone and noise stimuli indicated in the previous section become even more relevant. Spectral profiles of environmental sounds, and in particular of communication sounds, typically are composed of distinct spectral peaks and troughs distributed over a wide frequency range. Examples are the formant structure of vowels, a fundamental spectral feature of the vocal tract expressed in speech and animal vocalization sounds, and the spectral notches and peaks introduced by head shadows and outer-ear resonance utilized for sound localization processing.

Sensitivity and selectivity of neurons for more natural, complex spectral profiles can be assessed with broad-band stimuli using various methods. Random spectrum stimulus (RSS) sets, i.e., time-invariant broad-band stimuli with complex spectral envelopes, have been used to estimate the spectral weighting function that a neuron applies to sound energy across frequency. A linear frequency weighting function can be deduced by presenting stimuli with many different predetermined spectral shapes, by recording the observed discharge rates, and by subsequent superposition of the profiles proportional to their evoked activity. The resulting function is a rate-code based, normalized and

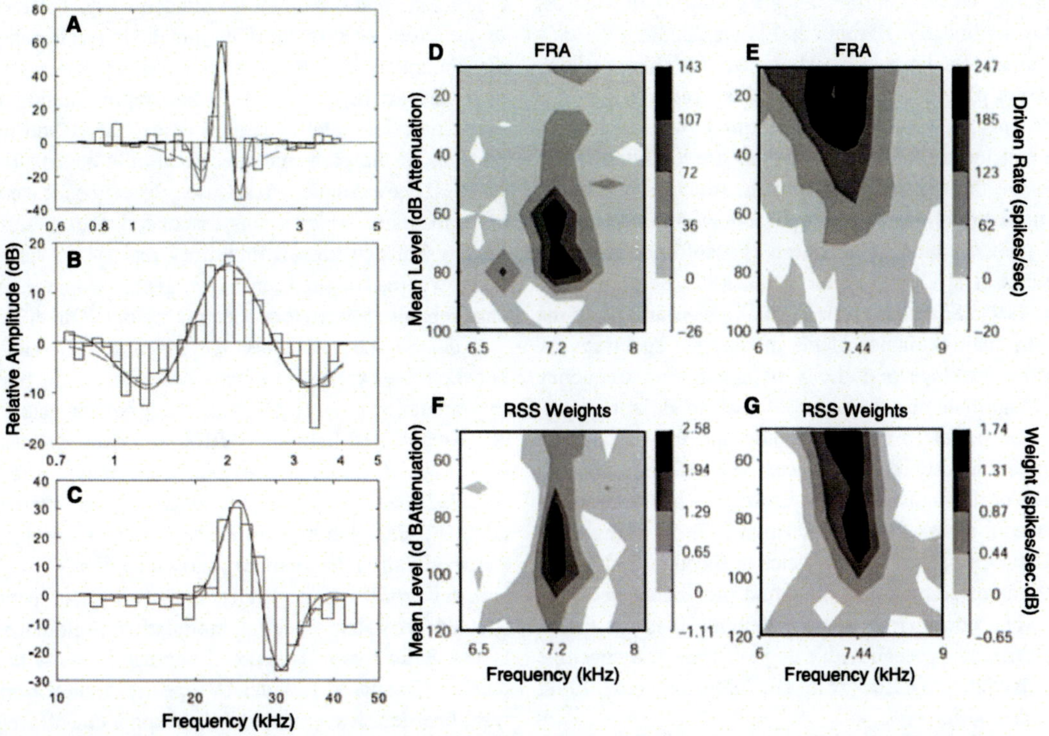

Fig. 13.5 Static spectral profiles determined with random sequence stimuli (RSS). **a–c** Three examples of spectral profiles indicating excitatory regions (activity above mean firing rate) and suppressive regions (activity below mean firing rate) (macaque monkey; adapted from O'Connor et al. 2005). **d, e** Two frequency response areas determined with pure tones (marmoset monkey; adapted from Barbour and Wang 2003a). **f, g** Two frequency response areas reconstructed from RSS obtained at different mean intensity levels for the same neurons as shown in **(d, e)**

weighted average spectral profile and corresponds to the spectral receptive field (Fig. 13.5; Barbour and Wang 2003a; Yu and Young 2000). Function values above the mean correspond to frequencies at which stimulus energy addition increases the driven rate of a neuron. Values below the mean are frequencies at which energy elimination increases the driven rate (Fig. 13.5a–c).

Similar estimates of neuronal spectral profile preference can be derived with adaptive stimulus optimization (Nelken et al. 1994a; O'Connor et al. 2005) by using variations of a static spectral stimulus profile to iteratively reach a maximum in the response rate. The resulting preferred stimulus profile also is a robust estimate of the neuron's actual spectral tuning, effectively representing properties found in natural sounds. While spectral profile estimations are not identical between the different methods, similarities exist revealing linear and nonlinear aspects of spectral integration properties (Sutter 2005).

RSS produced significant firing rate changes in 60–80% of neurons encountered in AI of awake marmoset and rhesus monkeys (Barbour and Wang 2003a; O'Connor et al. 2005) most of them showing sustained spiking. The resulting shapes of preferred spectral profiles (Fig. 13.5) showed a range of appearances with narrow or broad excitatory maxima and various suppressive/inhibitory troughs on either side, described as circumscribed, multi-lobed antagonistic structures (O'Connor et al. 2005). When obtained for a range of different mean stimulus intensities, the shape of the estimate function closely resembled two-tone FRAs (Fig. 13.5d–g). In contrast to typical V-shaped FRAs, they remained relatively constant throughout the stimulus interval and across the stimulus properties of mean sound level (Fig. 13.5f, g), spectral density, and spectral contrast (Barbour and Wang 2003b; O'Connor et al. 2005). Similarities to FRAs include the occurrence of multiple excitatory bands, their shape and bandwidth, and the position of suppressive sidebands. However, it is highly likely that many auditory cortex neurons behave in a substantially nonlinear manner in response to complex spectral input (Barbour and Wang 2003a; Calhoun and Schreiner 1998; Linden et al. 2003; Machens et al. 2004; Nelken et al. 1994b; Sahani and Linden 2003). This should result in distinct differences between narrow- and broad-band estimates of spectral processing. This is emphasized by the observation that even linear predictions of rate responses from preferred spectral profiles for other RSSs yielded poor results, again implying that auditory cortex neurons integrate information across frequency nonlinearly (Barbour and Wang 2003a).

Several other techniques have been used for characterizing the structure of auditory receptive fields. Auditory gratings or ripple spectra, i.e., broad-band stimuli with sinusoidal spectral envelopes (linear spacing along the logarithmic frequency axis) that resemble the formant structure of vowels, can be used to obtain the spectral modulation spectrum or spectral gain function of a neuron (Escabí and Schreiner 2002; Klein et al. 2000; Miller et al. 2002; Schreiner and Calhoun 1994; Shamma et al. 1995; Versnel and Shamma 1998). The main variables of the modulation spectrum are the spectral envelope periodicity or modulation frequency and the magnitude and the phase of each modulation component. The preferred spectral profile and the modulation spectrum are directly related and can be translated into each other via Fourier transform. The usefulness of the modulation spectrum approach as a descriptor is in its straightforward parametric space. The relevance of spectral modulation information for communication sound processing becomes evident when considering how challenging it is for listeners to discriminate speech with a degraded spectral envelope (Dreisbach et al. 2005; Leek et al. 1987; Shannon et al. 1998; Smith et al. 2002).

Cat and ferret cortical neurons respond preferentially to a limited range of spectral envelope frequencies (Calhoun and Schreiner 1998; Klein et al. 2000; Kowalski et al. 1996a, b; Schreiner and Calhoun 1994; Shamma et al. 1995). For these static ripple stimuli, preferred ripple frequencies for AI range between 0.2 and 4 cycles/octave (Schreiner and Calhoun 1994; Keeling et al. 2008; Shamma et al. 1995) with mean frequencies of ~1.0 cycles/octave. This range corresponds well to the best sine-profile frequencies that can be fit to the preferred spectral profiles obtained with RSS which range between 0.2 and 3 cycles/octave with a mean of 1.17 in the awake rhesus monkey (O'Connor et al. 2005). As with preferred spectral profiles, the relative response to different spectral modulation/ripple frequencies remains fairly constant with changes in the intensity and the spectral density of the broad-band carrier signal. However, variations of spectral modulation depth or contrast can result in nonlinear behaviors of the spectral modulation spectrum (Calhoun and Schreiner 1998). There is only sparse experimental evidence for a spatial organization or clustering of ripple transfer functions (Shamma et al. 1995; Kowalski et al. 1996a, b).

Studies in ferret AI find that ripple responses allow reasonable predictions of responses to pure tones and to spectrally complex natural sounds (Shamma et al. 1995; Versnel and Shamma 1998; Klein et al. 2000; David et al. 2009), suggesting that AI neurons analyze the shape of acoustic spectra in a substantially linear manner.

Details of the spectral shape of natural broad-band sounds, such as sharpness of formants or attributes of spectral edges, contribute to the perceived sound quality. Different types of preferred spectral profiles and their relationship to the distribution of excitatory and inhibitory subregions in AI neurons can help in an effective representation of these properties. The relative pattern of excitatory and inhibitory portions of the preferred spectral profile contributes to this process. For example, a response preference for steep slopes of formants or edges seems associated with a shift of processing balance toward inhibitory regions of the receptive field, whereas a preference for gentle slopes emphasizes engagement of excitatory spectral regions (Qin et al. 2004).

Laminar Organization: Significant differences exist between the expression of spectral modulation preferences in granular, supragranular, and infragranular neurons in cat AI (Atencio and Schreiner 2010a, b). Simultaneous recordings from 8 to 20 single neurons across cortical layers revealed that CFs show only small laminar variations. By contrast, clear laminar differences were evident for spectral modulation preferences, and equivalently, of preferred spectral profiles (Fig. 13.3f). Only ~30% of penetrations showed consistent spectral modulation preferences across layers, indicative of functional laminar diversity or specialization. Compared to layer IV, spectral modulation spectra were broader on average, and their upper cut-off frequencies higher, in layers V and VI. This suggests a higher representational fidelity of sharp edges in the spectral profile in the infragranular layers. Ensembles of auditory neurons that are tuned to different auditory features enhance the acoustic differences between classes of natural sounds and their distribution may reflect high informational regions in the environmental sound statistics (Woolley et al. 2005). Functional layer differences, reflecting different pre-processing for their respective projection targets, suggest then specific sensitivities to spectral profiles that need to be understood based on the goals and algorithms at each point in the circuit.

5 Dynamic Spectro-Temporal Profile Analysis

5.1 Spectro-Temporal Receptive Fields

When the sensory functions and response characteristics of a neuron are unknown, it is preferable to make few assumptions and to explore a large set of stimulus attributes in an unbiased way. Reverse correlation or spike-triggered average (STA) techniques embody this principle. Synthetic, spectro-temporally complex stimuli, such as dynamic chord stimuli, dynamic ripples, ripple noise, and temporally orthogonal ripple combinations (TORCs) (Escabí and Schreiner 2002; Blake and Merzenich 2002; Klein et al. 2000), share many properties with natural sounds and satisfy formal requirements for deriving spectro-temporal receptive fields (STRFs) through the STA. The STRF is a linear, time-frequency representation of neural stimulus preferences as shown by the excitatory and inhibitory STRF subregions (Aertsen and Johannesma 1981; Depireux et al. 2001; Eggermont et al. 1983; Gill et al. 2006). The two-dimensional Fourier transform of an STRF yields modulation transfer functions (MTFs) that characterize the neurons preferred spectral and temporal stimulus envelopes. The STRF and its relatives remain among the richest unbiased, linear descriptors of neuronal function. Compared to static spectral profiles, STRFs add a temporal axis that characterizes the temporal evolution or dynamics of the spectral influences. While informative, STRFs may be biased by stimulus correlations, may reflect nonlinear behavior in a very limited way, and do not characterize neural sensitivity to multiple stimulus dimensions. STRFs provide a versatile and integrated, spectral and temporal, functional characterization of neural responses (Klein et al. 2000, 2006). STRFs express a single feature dimension that captures the time-dependent behavior of stimulus envelope processing in auditory neurons. This combined spectro-temporal processing is advantageous for encoding of natural sounds which are rarely static. It enables – at least partially – the basic reconstruction of the input signal (David et al. 2009; Mesgarani et al. 2009).

To extract additional feature dimensions and to account for nonlinear response rules, an alternative approach can be used that is based on maximizing the mutual information (MI) between the stimulus and the evoked spike train of a neuron (Atencio et al. 2008; Clifford et al. 2007; Sharpee 2007; Sharpee et al. 2006, 2008). The resulting maximally informative dimension (MID) can share many aspects with STRFs obtained through reverse correlation (Fig. 13.6) and has additional advantages, such as suitability for derivation with non-Gaussian signals and elimination of effects from stimulus correlations (Sharpee et al. 2004a, b).

Both MIDs and STAs can provide the linear component in a linear–nonlinear neuron model (Sharpee et al. 2008; Schwartz et al. 2006). In this model, spectro-temporal stimulus features, or linear filters, are combined with a static nonlinearity to compactly represent neural processing. This

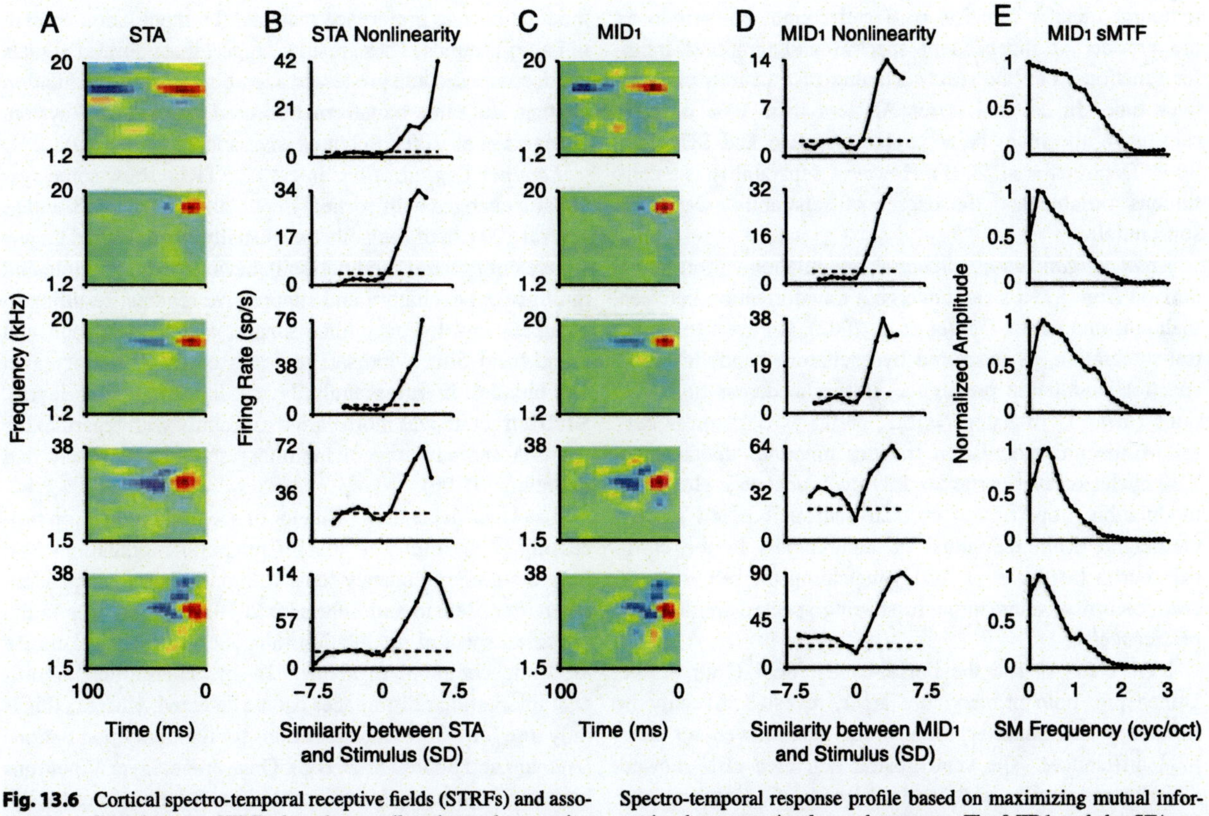

Fig. 13.6 Cortical spectro-temporal receptive fields (STRFs) and associated nonlinearities. **a** STRFs based on spike-triggered averaging (STA) in response to a 20 min dynamic moving ripple stimulus (*red*: excitatory regions; *blue* suppressive/inhibitory regions). **b** Associated nonlinearities. Nonlinearities express the firing rate as a function of the similarity between the stimulus and the STA or MID. **c** MID1. Spectro-temporal response profile based on maximizing mutual information between stimulus and response. The MID1 and the STA are highly correlated (mean r=0.8) indicating that the two estimates of spectro-temporal processing are capturing highly similar aspects. **d** MID1 nonlinearity. MID1 and STA nonlinearity are highly correlated. **e** Spectral modulation transfer functions based on the MID1

linear–nonlinear model can account for features in stimulus space that best capture the variability in neuronal responses. The nonlinear input–output function, or nonlinearity, describes the firing probability of a neuron as the similarity, or correlation, between the stimulus and the STRF/MID changes (Fig. 13.6) and forms a fundamental component in linear/nonlinear cascade models of neuronal function (Chichilnisky 2001; Schwartz et al. 2006). Most STRF/MID nonlinearities in ketamine-anesthetized cat AI are asymmetric and sigmoidal in shape, representative of a thresholding and smoothing operation. Parameter ranges, such as slope and position of inflection point, of asymmetric nonlinearities provide a rich substrate for response differences in neurons with similar STRFs.

The main feature of STRFs is that they can capture temporal dependencies of spectral processing. Many neurons in cat and ferret AI have STRFs with "sloped" response maxima or minima, indicating that the frequency position of excitatory and/or inhibitory regions shift with time (Atencio and Schreiner 2008; Depireux et al. 2001). This means that spectral and temporal processes can interact and cannot be considered in isolation. This inseparability of spectral and temporal processing implies that the combined spectro-temporal transfer function of a cell cannot be written as the product of independent spectral and temporal transfer functions; i.e., the spectral tuning of a neuron changes over time. In cat and ferret AI, less than 10% of neurons were shown to be separable (Atencio and Schreiner 2008; Depireux et al. 2001). However, separability is a continuous variable and the degree of separability can vary substantially.

Areal Organization: Spectral modulation information derived from STRFs can undergo a transformation between thalamus and cortex (Miller et al. 2002). On average, spectral integration, as measured by excitatory bandwidth and spectral modulation preference, is similar across both stations (mean Q: thalamus = 5.8, cortex = 5.4; upper cut-off of spectral modulation transfer function: thalamus = 1.30 cycles/octave, cortex = 1.37 cycles/octave). However, modulation properties of cortical neurons are not strictly predictable from individual thalamic inputs to the cortical neuron (Miller et al. 2002) indicating the relevance of cortico-cortical interactions in shaping spectral modulation preferences.

STRFs in AI and the dorsal-caudal field (DC) of the guinea pig, both primary/core areas, revealed diversity in excitatory and inhibitory bandwidths but showed no clear field differences. The ventrorostral belt area also showed STRF types similar to those in AI and DC. However, the proportions of STRF types were significantly different, suggesting a difference in spectro-temporal processing between the ventrorostral belt and the core areas (Rutkowski et al. 2002).

Spectral properties of AI and AAF receptive fields in mice were largely similar, although STRF bandwidths were slightly broader in AI than in AAF. In both, AI and AAF, only a small proportion of STRFs were spectro-temporally inseparable, e.g., revealing slanted STRFs. This suggests still a fairly independent processing of temporal and spectral aspects in these core areas (Linden et al. 2003). In cat PAF, a higher hierarchy core area, about half of the neurons have non-separable STRFs (Loftus and Sutter 2001) indicating a potential increase in spectral–temporal interactions at later stages of the cortical pathways.

Attempts to derive STRFs in prefrontal cortex of awake macaque monkeys (Averbeck and Romanski 2006; Cohen et al. 2007) did not reveal significant internal structures despite the fact that neurons responded strongly to acoustic stimuli, especially if they were complex in structure, such as vocalizations. A faithful time-frequency representation appears to be less useful at this stage and other processing aspects, such as time-probability representations, may play a larger role (Romanski and Averbeck 2009).

Noninvasive imaging methods showed selective tuning to combined spectro-temporal modulations in the primary and secondary auditory cortex in humans. The overall low-pass modulation rate preference matched the modulation content of natural sounds. These results suggest that complex signals are decomposed and processed according to their modulation content, the same transformation used by the visual system (Langers et al. 2003; Schönwiesner and Zatorre 2009).

Laminar Organization: In cat AI STRFs show some systematic changes with cortical depth, although STRFs within several 100 μm of each other are usually quite similar. Layer-dependent spectral modulation behavior includes single and multipeaked excitatory and suppressive regions, resulting in bandpass and lowpass filter shapes, and narrow-band and broad-band filter widths. The width of the excitatory area was broadest in infragranular layers. In infragranular layers, STRF structure was more varied especially with regard to the position and structure of inhibitory subfields (Atencio and Schreiner 2010b).

The layer-dependent behavior of spectral modulation processing is dissimilar to that of temporal modulations that have a stronger tendency for a columnar, layer-independent behavior (Atencio and Schreiner 2010b). Differences in the preferred spectral modulation range across cortical laminae are quite common. In about 70% of penetrations, significant interlaminar differences can be detected, whereas this is only true for ~30% of penetrations for temporal modulations (Atencio and Schreiner 2010b). On average, layer V neurons have the highest preferred spectral modulation frequencies. Compared to layer IV, spectral MTFs are broader and their upper cut-off frequency higher in layers V and VI. This filter broadening and increase in preferred spectral modulation frequencies in infragranular layers can be accounted for by the shift of the strength and location of inhibitory sidebands. Spectral integration appears to increase in infragranular layers (Wallace and Palmer 2008; Volkov and Galaziuk 1989; Atencio and Schreiner 2010b). Responsiveness of infragranular layers to higher modulations than in granular layers clearly requires additional inputs not provided by a simple columnar feedforward stream from the thalamo-recipient layers.

In cat AI, STRFs are less separable in supra- and infragranular layers, indicating that spectral and temporal processing aspects become more interdependent compared to the main thalamic input layer. In granular layers, the STRF nonlinearities were most asymmetric, revealing that in these layers responses are greatest for stimuli that are highly matched to the STRF. On average, the STRF nonlinearity of supragranular neurons showed the same degree of asymmetry as granular layer neurons. Infragranular neurons, however, had a clearly reduced asymmetry, suggestive of a processing manner less sensitive to the phase, or polarity, of the spectro-temporal envelope.

Receptive fields in the cortical input layers may be predominantly created via three general schemes: inheritance from thalamic inputs, constructive convergence of different narrow thalamic and cortical inputs, and/or by assembly convergence of combined, broader thalamic and cortical inputs (Miller et al. 2001). After this initial integration stage, further transformations occur related to the primary interlaminar flow of information from the thalamocortical input layers to the supragranular and then to infragranular output layers, by intralaminar cortico-cortical inputs as well as cortico-cortical feedforward and feedback connections (Wallace et al. 1991; Mitani and Shimokouchi 1985; Mitani et al. 1985; Winer 2006). The direction of STRF changes, however, is not strictly linked to a simple interlaminar flow pattern from thalamic input layers to supra- and infragranular output layers. Changes in modulation properties captured in STRFs make it feasible to dissect laminar-specific, module-specific, and field-specific variations in the cortical processing regime and can help to determine whether common functional patterns pertain to cortical or subcortical inputs, and how they reflect local, lamina-specific circuitry (Atencio and Schreiner 2010a, b).

5.2 STRF Differences Between Cell Classes

Excitatory pyramidal neurons and inhibitory interneurons constitute the main elements of the cortical circuitry and have distinctive morphologic and electrophysiological properties. Functional differences between these different neuronal classes have been found in mammalian cortex (Bartho et al. 2004; Bruno and Simons 2002; Hirsch 2003; Simons and Carvell 1989; Swadlow and Gusev 2002; Zhang and Alloway 2004). Differences in spike duration and amplitude ratios are associated with specific classes of cortical neurons. "Regular-spiking" neurons (RSUs) have slow action potentials (initial negative wave >300 μs) and are presumably excitatory pyramidal cells, though some inhibitory interneurons also show this spike waveform (Bruno and Simons 2002; Kawaguchi and Kubota 1993; Swadlow 2003; Simons and Carvell 1989). "Fast-spiking" or "thin-spike" neurons (FSUs) have shorter action potentials (initial wave <200 μs) and are associated with inhibitory interneurons, although some excitatory neurons also show this spike waveform (Connors and Gutnick 1990; McCormick et al. 1985).

Excitatory sharpness of frequency tuning among simultaneously recorded fast-spiking and regular-spiking neurons differs despite the similarity of layer and local CF. Fast-spiking cells have slightly broader spectral tuning than RSUs. At a given intensity, fast-spiking inhibitory neurons exhibit less-selective frequency tuning than nearby excitatory neurons (Atencio and Schreiner 2008; Wu et al. 2008). A possible consequence of the wider FSU bandwidth is that the spike-based tuning of RSUs, the potential synaptic target of FSUs, is narrower than their synaptic inputs (Tan et al. 2004; Wu et al. 2008). No significant differences were found between FSUs and RSUs in relation to best spectral modulation frequency and spectral MTF width. Although the range of preferred spectral modulations values does not differ for the two cell distributions, the manner in which FSUs and RSUs respond to spectral and temporal envelope modulations does differ. A slightly higher proportion of RSUs show band-pass spectral modulation transfer functions (25%) as compared to FSUs (15%). Response latency was shorter for FSUs versus RSUs within a given cortical layer (Atencio and Schreiner 2008). This could enable them to transmit feedforward inhibition to nearby cells.

STRF structure differs between FSUs and RSUs. FSU STRFs are more separable, thus dissociating more fully spectral and temporal processing, since they can be approximated as the product of two independent functions. Whether this reflects different cortical connection patterns and/or different distributions and kinetic properties of GABAergic inputs to RSUs (Hefti and Smith 2003) is unknown, since detailed accounts of cortico-cortical inputs to inhibitory neurons are not yet available. The nonlinearities associated with the two cell classes revealed a stronger asymmetry for FSUs indicative of higher feature selectivity.

These global functional differences between RSUs and FSUs suggest clear distinctions between putative excitatory and inhibitory neurons that shape auditory cortical processing. FSUs have response characteristics more closely related to thalamic input properties than RSUs. Connected thalamocortical neuron pairs usually differ in most of their modulation properties (Miller et al. 2001). Intracortical recurrent excitation appears to amplify the thalamocortical inputs to determine stimulus selectivity of cortical neurons (Wu et al. 2008). Cortical modulation likely is also shaped by local inhibitory mechanisms. The precise role of inhibition in determining modulation preferences is still unclear (Kurt et al. 2006) and contributing factors, such as convergence of different modulation ranges and synaptic depression/facilitation, play major roles in the modulation of cortical responses (Eggermont 2002; Wehr and Zador 2005).

Differences in STRFs of RSUs and FSUs provide a useful first step in the analysis of local circuits and laminar functional diversity and segregations within an auditory context. The extension of this approach to nonlinear, multi-feature receptive fields is required and will further delineate systematic processing differences between cell types.

5.3 Multi-filter Spectral Analysis

One of the advantages of the linear-nonlinear STRF/MID characterization is that it provides a rigorous framework to predict neuronal response behavior to novel sounds. Some success in STRF-based response prediction and stimulus reconstruction has been reported for auditory cortical neurons (e.g., Kowalski et al. 1996b; Versnel and Shamma 1998; Mesgarani et al. 2009). However, other studies have fallen short of successfully predicting responses to complex sounds, especially when test stimuli differ in their statistical properties from those sounds used to derive STRFs (e.g., Machens et al. 2004; Sahani and Linden 2003; Theunissen et al. 2000). One possible cause for low predictive power may be that standard STRFs/MIDs represent a single stimulus dimension that influences a neuron's response. In visual cortex it has been shown that an additional stimulus dimension may be necessary to provide a more complete depiction of the effective stimulus configurations (Rust et al. 2005). An extension of the information-based MID method has demonstrated that auditory cortical neurons as well are better characterized by at least two independent but interacting spectro-temporal filters (Atencio et al. 2008). In this method, two parametrically independent but jointly operating filters are iteratively adjusted until the mutual information between stimulus and response is maximized, resulting in two (or more) MIDs, and their associated nonlinearities. The first MID (MID1) maximizes the MI with respect to one STRF and the second MID (MID2) is an additional STRF that further maximizes the MI. The concurrent operation of these two MIDs in combination with their nonlinearities can capture a substantially larger proportion of the mutual information of cortical neurons than the STA or a single MID alone (Sharpee et al. 2004a, b; 2008).

The main observations (Atencio et al. 2008, 2009) from this approach include: (1) All neurons in cat AI with an STA/MID1 also have a significant although slightly less informative MID2, i.e., each neuron can be modeled as a combination of at least two stimulus dimensions (Fig. 13.7). The contribution of the MID2 to the combined mutual information is in the range of 20–40%. (2) MID1 and STA-based STRF and their nonlinearities are highly correlated, thus validating the use of spike-triggered averaging in previous studies to identify the strongest contributing filter (Fig. 13.7). (3) The nonlinearities of the two MIDs differ in character. The nonlinearity of the first STA/MID is asymmetric and sigmoidal, while the nonlinearity of the second MID is usually symmetrical. The asymmetric nonlinearity is typical for a feature detector. The symmetric nature of the MID2 nonlinearity shows that for this dimension the neuron has an increased probability of firing when a stimulus is either correlated or anti-correlated with the filter (Fig. 13.7). This type of nonlinearity is often seen in visual neurons that are envelope-phase insensitive or shift-invariant (Emerson et al. 1992; Dellen et al. 2009). The difference in the shape of the nonlinearities implies that a given AI neuron in this extended model contains functional subunits that both threshold (MID1) and square (MID2) the outputs of the individual filters. (4) Best frequencies of the two filters are usually closely matched. However, the shape of the two MIDs (i.e., the distribution and relationship of excitatory and inhibitory subregions) differs, reflecting their orthogonality and providing different constellations of spectral modulation preferences. The preferred spectral envelope modulation frequencies of a population of AI neurons span an equally wide range for both MIDs but are uncorrelated. As a consequence, spectral processing properties of cortical neurons reflect at least two differently tuned spectral filters (Fig. 13.7e). (5) The two MIDs cooperate in a nonlinear fashion, creating combination-sensitive, and sometimes synergistic, processing. On average, the combined applied filters account for 28% more information than the sum of each filter in isolation. This type of nonlinear combination-sensitivity differs from previously described combinations-sensitivity in subcortical and cortical auditory stations. It requires two interacting filters and cannot be explained by the shape and properties of a single, one-dimensional nonlinearity as is the case for combination-sensitivity described for tone-on-tone interactions, for example, in bats (Suga 1984), or in awake marmosets (Sadagopan and Wang 2009).

Of relevance is that the contributions of MID2s in subcortical stations, such as the central nucleus of the inferior colliculus or the ventral nucleus of the medial geniculate body, seem to be absent or significantly smaller than in AI (Atencio, Shih, Schreiner unpublished observation). This suggests that the generation of multiple STRFs/MIDs expressed in a single neuron is an emergent property of

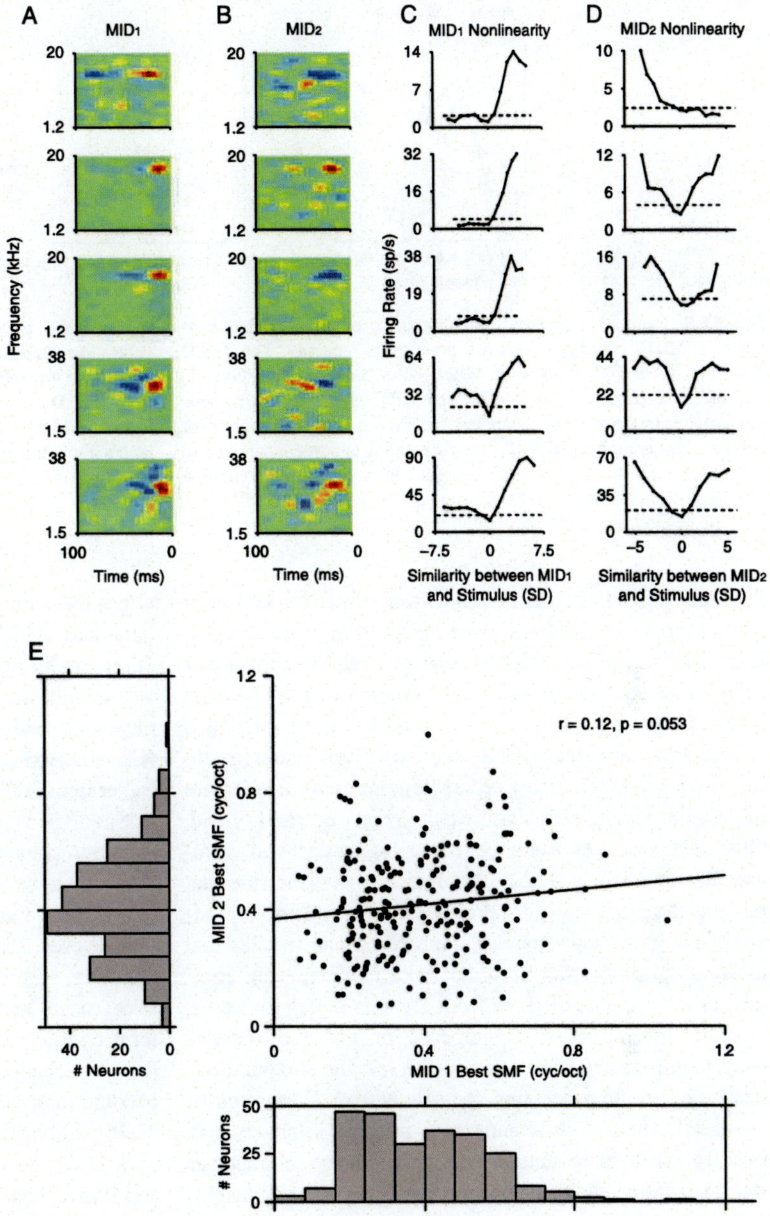

Fig. 13.7 Auditory cortical responses are more fully characterized by two filters and their associated nonlinearities. **a** MID1 s of five single AI neurons (*red*: excitatory regions, *blue* suppressive/inhibitory regions). Same neurons as in Fig. 13.6. **b** MID2 for the same neurons. Note similar CFs but different distributions of excitatory and inhibitory subregions. **c** MID1 nonlinearities. MID1 nonlinearities are typically asymmetric, i.e., positive stimulus/MID correlation can result in increased firing rates, whereas negative correlations have little effect on firing rate. **d** MID2 nonlinearities. Note that the nonlinearities are mostly symmetric, i.e., either positive or negative stimulus/MID2 correlations can increase firing rate. **e** Distribution of the best spectral modulation frequency of MID1 and MID2 with marginal distribution histograms. The preferred spectral modulation frequency of the two filters is essentially uncorrelated

auditory cortex, similar – but not identical – to processing principles emerging in visual cortex, such as simple and complex cells (Hubel and Wiesel 1962; Movshon et al. 1978). This finding suggests that there may be general principles in cortical processing and hierarchical computation across different sensory modalities. As of yet, it is unknown whether higher cortical areas also have multiple STRF/MID filters.

Laminar Organization: For STRFs and MIDs in cat AI, a sequential evolution within the interlaminar columnar microcircuit is evident (Atencio and Schreiner 2010a, b; Atencio et al. 2009). Processing in all AI layers is more completely captured with a two-filter MID characterization. In granular layers, the MID1 is most dominant, with a high degree of feature selectivity and separability (Fig. 13.8). A MID2 is found in all layers although its contribution is smaller in granular layers (Fig. 13.8). The two MIDs, and their nonlinearities, differ in shape, and show different properties with cortical depth. In supra- and infragranular layers, the MID1 contribution is reduced, and the synergy or positive interactions between the filters increases (Atencio et al. 2010a) (Fig. 13.8).

The sequential information processing across the different AI layers is progressive and becomes more complex, and synergistic, as the auditory signal moves from

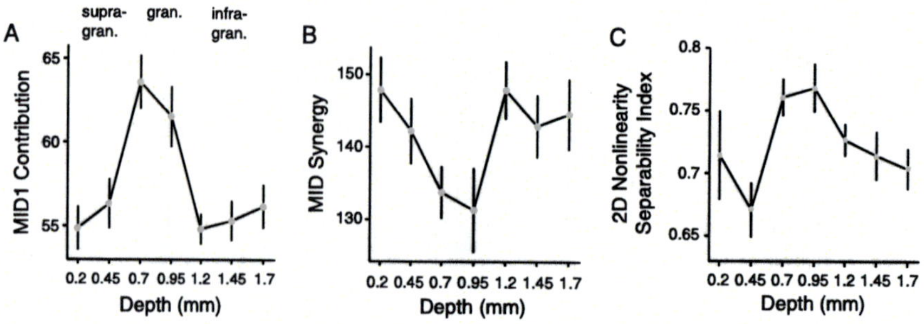

Fig. 13.8 Laminar differences exist for basic multi-MID characteristics. **a** MID1 contribution across the cortical laminae. The MID1 contribution (in %) quantifies the MID1 mutual information relative to the information from the joint MID1 and MID2 processing. *Gray area* corresponds to the granular layers IIIb and IV. Supra- and infragranular layer regions are indicated. MID1 provides the strongest contribution to granular layers. In the supra- and infragranular layers, the contribution of MID2 is, on average, nearly as strong as that of MID1. **b** MID synergy (in %) expresses the cooperativity between the two MIDs. Joint application of the two filters often results in higher mutual information values than the sum of the two MIDs applied in isolation. Highest cooperativity is typically found outside the granular layers. **c** Separability of the joint 2D nonlinearities of the two MIDFs. High values correspond to reduced interactions between the two nonlinearities. Supragranular layers show the least separability suggestive of nonlinear interactions between the two MIDs and their individual nonlinearities. Adapted from Atencio et al. (2009)

thalamic input to cortical output layers. Spectral and temporal processing becomes more complex in structure, less linear in interaction and response generation, and potentially more abstract and stimulus-variation tolerant. All AI neurons exhibit some degree of inseparability of their two-dimensional nonlinearity, i.e., the two filters cooperate to various degrees. The most separable joint nonlinearities are in granular layers, with significantly lower separability and, thus, increased cooperativity, in supragranular and infragranular layers (Fig. 13.8). This indicates that the rule that governs the joint, two-filter processing is not a simple product of two one-dimensional nonlinearities, and implies that information processing becomes more nonlinear and complex as the synaptic distance from granular layers increases. The relationship of the emergence of multiple spectro-temporal filters in auditory cortex with specific computations and task-specific processing remains elusive. Formation of enhanced stimulus invariance may indicate improvements in foreground/background separation and noise tolerance as well as in perceptual and conceptual category formation.

5.4 Receptive Fields: Constancy Versus Malleability

5.4.1 Short-Term Changes of Receptive Fields

Receptive field properties are measured at certain points in time, after presentation of a specific stimulus set. Thus, the empirically determined receptive fields of cortical neurons are thought to be approximations of their "true," intrinsic functional characteristics. However, many aspects can affect the outcome of receptive field estimations. Neuronal parameter sensitivity and selectivity may depend, among other conditions, on stimulus statistics, response adaptation, task conditions, context, and attention, consistent with complex, nonlinear and recurrent processing in neural assemblies (Christianson et al. 2008; Fritz et al. 2007c; Pienkowski and Eggermont 2009).

Spontaneous variations of STRF parameters in repeated estimations have been shown to usually be quite small, suggesting that neuronal properties can be stable over hours and days (Blake and Merzenich 2002; Elhilali et al. 2007).

However, state-dependencies, such as arousal, alertness, attention, stimulus statistics dependencies – including variance, mean, and skewness of the distributions – and behavioral context and task-dependencies can induce temporary RF perturbations that, under certain circumstance, may become long-lasting changes, usually referred to as reorganizational plasticity.

A main utility of STRFs is their versatility in capturing and classifying the large range of cortical processing properties. However, a significant problem is that responses are nonlinear, adaptive, and sensitive to biased stimuli. With nonlinear processing, STRFs inevitably become stimulus and context dependent, e.g., altering polarity, shape and extent of STRFs (Christianson et al. 2008). Especially when applying non-Gaussian, natural stimulus statistics, STA methods may produce biased STRFs leading to features that are shifted away from the most relevant dimensions (David et al. 2009; Machens et al. 2004; Nagel and Doupe 2006; Rotman et al. 2001). STRFs computed for natural stimuli in a nonlinear MID model have been shown to be significantly different from those computed with a linear STA model, and usually show a better description of the neuronal responses (Sharpee 2007). A number of potential causes for nonlinear responses have been proposed, including short-term depression (David et al. 2009), divisive surround inhibition (Carandini et al. 1998), or thresholding of spiking output (Qiu et al. 2003), although definitive links between cellular and synaptic mechanisms and the model nonlinearities remain to be fully established.

STRF perturbations have been described for a number of stimulus parameters, such as the density and bandwidth of random cord and ripple stimuli (Blake and Merzenich 2002; Gourevitch et al. 2009; Norena et al. 2008), and for stimuli with more natural parameter statistics, such as speech and vocalizations (David et al. 2009; Theunissen and Shaevitz 2006). For random chord stimuli with different sound densities, STRFs often develop larger inhibitory fields and narrower spectral tuning (Valentine and Eggermont 2004; Blake and Merzenich 2002). Comparing STRFs obtained for dynamic ripple stimuli (composed of a single pair of spectral/temporal properties at any given time) and ripple noise stimuli (composed of multiple spectral and temporal features at any given time) also revealed some differences. Cells with low firing rates often respond better to the dynamic ripple than to the ripple noise, a highly nonlinear behavior (Escabí and Schreiner 2002). More natural stimulus statistics, as compared to Gaussian distributions, also have large effects on the estimated filters and nonlinearities, and seem to increase precision of temporal coding and emphasize the most informative features of natural sounds (Theunissen and Shaevitz 2006; Woolley et al. 2006).

Accommodation of the neural response to an ongoing stimulus is called adaptation. Input–output functions for intensity, temporal preferences, or spectral receptive fields are shifted or altered (Gourevitch and Eggermont 2008; Ohl and Scheich 1996; Pienkowski and Eggermont 2009; Ulanovsky et al. 2004). Consequences of adaptation are thought to rearrange the neural response sensitivity of neurons to optimize their information transmission. This can be achieved by providing a better match of the statistical distribution in the ongoing stimulus and the response preferences.

Attention is essential for performing auditory tasks (see Chapter 29). Neural correlates of this perceptual ability have been demonstrated in STRFs of AI in behaving ferrets (Fritz et al. 2003; Fritz et al. 2005, 2007c) during the detection of a target tone embedded in noise. Compared with responses in the passive state, the gain of STRFs decreased in most cells and STRF shape changes were specific to the stimuli in the task, and were strongest in cells with best frequencies near the target tone. These adaptations accentuate the spectro-temporal representation of the target tone relative to the noise (Atiani et al. 2009).

The non-static properties of spectral integration can also be seen with changes in behavioral states such as sleep versus wakefulness (Edeline et al. 2001; Edeline 2003; Issa and Wang 2008; Pena et al. 1999). During slow-wave-sleep, as compared to waking animals, the receptive field size – and implicitly the spectral integration behavior – varied as a function of the changes in evoked responses: it was reduced for cells whose responses were decreased, and enlarged for the cells whose responses were increased (Edeline et al. 2001).

5.4.2 Long-Term Plasticity of Spectral Modulation Filters

Cortical representations of signal dimensions have been shown to be alterable over extended periods of time when behavioral significance is attached to parts of those dimensions (Allard et al. 1985; Calford and Tweedle 1988; Gilbert et al. 2009; Recanzone 1998; Recanzone et al. 1992, 1993). Animals that learn to distinguish between certain spectral or temporal properties of sensory stimuli show an expanded and/or more refined cortical representation of relevant stimulus features and concomitant changes in perceptual ability (Jenkins et al. 1990; Recanzone et al. 1993).

Many studies have been undertaken which demonstrate plasticity in the receptive field of auditory cortical neurons during classical conditioning (e.g., Diamond and Weinberger 1986, 1989; Edeline 1998; see Chapter 22). Significant changes in discharge activity in auditory cortical cells follow the associative pairing of an acoustic conditioned stimulus with an unconditioned stimulus. Since the extent of these physiological changes does not occur during the sensitization and extinguishing phases of the training session, it becomes clear that the associative process plays the most salient role in discharge plasticity. Plasticity in auditory cortical neurons and the spatial distribution of receptive field properties have been demonstrated for a number of other learning conditions, e.g., operant detection and discrimination training and exposure to altered sensory inputs (e.g., Diamond and Weinberger 1986; Harrison et al. 1991; Rajan 2001; Robertson and Irvine 1989). For example, the distribution of the CF of AI neurons can be altered by frequency discriminative training (Recanzone et al. 1993). The representation of the frequency domain over which animals were trained expanded, and the excitatory bandwidth of cortical neurons was sharpened with training, reflecting task-dependent demands on sound processing.

Changes in spectral bandwidth properties of auditory cortical receptive fields occur during and after certain forms of perceptual learning. Prolonged exposure to a spectral profile with a fixed spectral periodicity (e.g., 1 ripple/octave) embedded into a perceptual training task influences the distribution of neuronal ripple transfer functions and pure-tone tuning curves (Keeling et al. 2008). The animals had to discriminate between stimuli that contained equally spaced formants but differed in their frequency positions. Following discrimination training, the preferred ripple density shifted toward the spectral spacing in the training stimuli (Fig. 13.9).

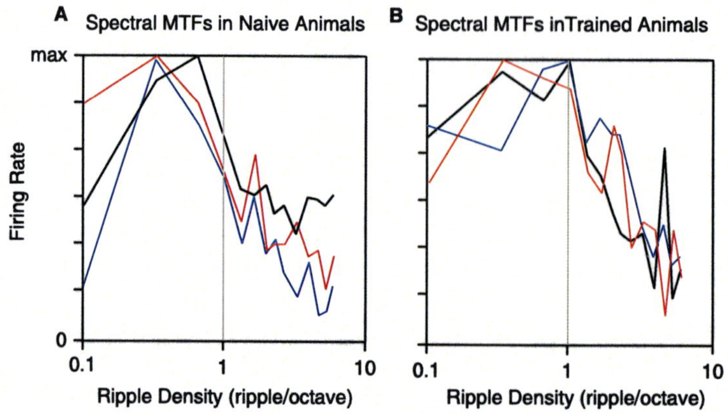

Fig. 13.9 Effects of behavioral training on spectral modulation transfer functions (sMTFs). **a** Population sMTFs for three untrained cats (Keeling et al. 2008). **b** population sMTFs for cats that were trained to perform a spectral envelope discrimination task. The training stimulus was a three-octave wide ripple sound with a sinusoidal spectral envelope of 1 ripple/octave (indicated by the *vertical gray line*). Animals were required to discriminate between stimuli with shifted positions of the spectral peaks (envelope phase) but with constant peak spacing. Note the shift in the preferred ripple frequency toward the trained ripple spacing and the relative increase in firing rate at the trained ripple density. Adapted from Keeling et al. (2008)

This is equivalent to an expansion of cortical space for the most task-relevant stimulus feature and increases stimulus sensitivity. In addition, the bandwidth of ripple transfer functions, a direct measure of the selectivity of neurons to specific formant spacings, became significantly narrower (Keeling et al. 2008) in conjunction with a narrowing of the bandwidth of pure-tone tuning curves. This change corresponds to an increase in selectivity.

Exposure to stimuli without overt behavioral consequence or explicit learning task can also have a long-term effect on the properties of cortical receptive fields (Gourevitch et al. 2009).

These observations indicate that the rules of short- and long-term cortical plasticity alike can operate on elemental stimulus features independent or in conjunction with others. The effect is governed by the stimulus statistics and their relationship to associative tasks. The cortex seems to use these features to guide several forms of receptive field reorganization, including reorganization of feature maps, plasticity of spectral and temporal specificity and selectivity, emphasis of relevant parameter ranges and combinations, and altered strength of evoked responses.

6 Synaptic Mechanisms of Spectral Processing

6.1 Synaptic Frequency Tuning

Most studies of cortical receptive fields have relied on extracellular recordings of spike output. However, recent advances in understanding the organization and dynamics of cortical circuits have been obtained using intracellular techniques such as in vivo whole-cell voltage-clamp recording. This is primarily for two reasons: first, spiking receptive fields necessarily are subsets of the underlying synaptic receptive fields; and second, excitatory responses are strongly governed by the inhibitory inputs received by a given neuron. Extracellular and optical approaches cannot at present directly measure these subthreshold inhibitory responses. Thus in vivo whole-cell recording experiments have provided the highest resolution descriptions of cortical tuning curves and receptive field properties, particularly for responses to pure tones and frequency modulation sweeps.

In terms of spectral tuning in adult cat, rat, and mouse AI, a major feature of synaptic receptive fields is that the relative strengths of excitatory and inhibitory inputs are proportional across tone frequency, i.e., synaptic excitation and inhibition are essentially balanced in mature AI (Froemke et al. 2007; Tan and Wehr 2009; Tan et al. 2004; Volkov and Galazjuk 1991; Wehr and Zador 2003; Zhang et al. 2003). Excitatory and inhibitory responses are balanced in the sense that they are usually co-tuned, i.e., sharing best frequencies and having correlated response magnitudes across other frequencies (Fig. 13.10a, c, d). However, although the relative amplitudes of inhibitory responses scale with the size of excitatory responses for a given stimulus, the onset of inhibition is delayed by a few milliseconds (Wehr and Zador 2003). As a consequence, there is a brief window in which excitatory responses can sum together and generate action potentials. This phase lag for inhibition is likely due to the architecture of thalamocortical circuitry in that there are few if any direct inhibitory projections from the MGB to AI (Winer 1992), leading to a short disynaptic delay between the onset of excitation and the onset of inhibition.

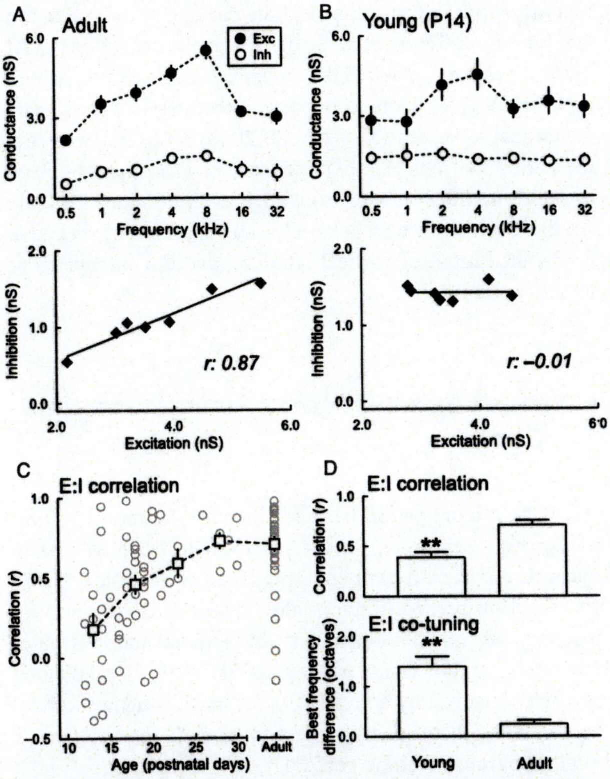

Fig. 13.10 Spectral tuning of synaptic excitation and inhibition in adult and developing rat AI. **a** Balanced tone-evoked excitation and inhibition in adult AI. Whole-cell recording from an adult (3-month old) rat. *Top*, frequency tuning of excitatory (*filled symbols*) and inhibitory (*open symbols*) conductances. *Bottom*, correlation between excitation and inhibition. Error bars represent s.e.m. **b** Imbalanced excitatory and inhibitory frequency tuning early in development. Whole-cell recording from AI of a young (P14) rat. **c** Increase of excitatory–inhibitory balance during the AI critical period. At the end of the second postnatal week, excitation and inhibition were uncorrelated. By the end of the third week, the correlation rapidly improved, and by the end of the first month, the excitation–inhibition correlation was similar to that measured in adult animals. **d** Summary of changes to excitatory–inhibitory balance during development. *Top*, mean correlation between excitation and inhibition in young (P12-21) and adult animals. *Bottom*, mean difference in excitatory and inhibitory best frequencies in young and adult animals

While on average, synaptic frequency tuning of AI neurons is balanced, sensory-evoked excitatory and inhibitory responses are not always so closely matched. For some cells in adult AI, excitation and inhibition are uncorrelated or even anti-correlated (Fig. 13.10c). There is additional evidence for untuned or cross-tuned inhibitory inputs from intracellular recording studies in visual cortex (Douglas et al. 1991; Ferster 1986; Monier et al. 2003; Pei et al. 1991; Schummers et al. 2002). Recordings from interneurons in both auditory and visual cortex indicate that inhibitory cells are frequently less tuned than excitatory neurons (Atencio and Schreiner 2008; Liu et al. 2009; Niell and Stryker 2008; Sohya et al. 2007; Wu et al. 2008). Also, depending on the position of a neuron within the AI frequency map, there may be asymmetrical sidebands of inhibitory inputs within an octave or so, helping to selectively shape the responses to up or down frequency sweeps (Zhang et al. 2003). Likewise, other receptive field properties, such as intensity tuning, may be regulated by focally imbalanced inhibition (Tan et al. 2007; Wu et al. 2006). In general, diversity in the synaptic organization of cortical receptive fields might be important for detection and discrimination of different classes of auditory stimuli, and theoretical models suggest that both balanced inhibition and relatively broad lateral inhibition schemes are required to explain the range of spiking responses observed in extracellular recordings in vivo (de la Rocha et al. 2008). One challenge for future studies will be to determine how the various types of interneurons, such as basket cells and Martinotti cells (Petilla International Nomenclature Group 2008), might be activated by specific patterns of auditory stimulation and differentially affect synaptic receptive fields.

While the exact sources of intracortical inhibition remain unknown, it is also still unclear to what degree thalamic or intracortical excitatory inputs contribute to the net excitation evoked by tones or other stimuli. Kaur and colleagues (2004) reported that intracortical injections of muscimol, a $GABA_A$ receptor agonist, reduced the bandwidth of frequency-intensity receptive fields, but left characteristic frequency responses relatively intact. These results suggest that intracortical inputs help define the width of excitatory receptive fields, broadening frequency tuning curves beyond the extent determined by more sharply tuned thalamic input. However, a study by a different group attempted to isolate thalamic inputs using muscimol in combination with SCH50911 (a $GABA_B$ receptor antagonist), to prevent reduction of presynaptic transmitter release at thalamocortical afferents while simultaneously reducing intracortical excitation. They found that tuning curve width was unaffected by this pharmacological treatment (Liu et al. 2007), suggesting that the range of thalamic input alone may set the width of subthreshold frequency tuning. Regardless of the anatomical basis of synaptic receptive fields, the relative connection strengths of thalamic and intracortical inputs can be changed by various forms of experience, with intracortical synapses seemingly expressing a higher degree of plasticity than thalamic inputs (Diamond et al. 1994; Froemke et al. 2007).

6.2 Development of Synaptic Frequency Tuning

Although cortical synapses can be modified all throughout life, receptive fields are especially plastic during developmental critical periods, epochs during which cortical circuits are particularly susceptible to changes in sensory input (Buonomano and Merzenich 1998; Hensch 2005; Katz and Shatz 1996). Auditory cortical critical periods usually last for a few days or weeks, beginning just after the start of hearing, and possibly are overlapping or staggered for different components of the auditory system or different receptive field properties.

In rodent AI, representations of sound frequency and intensity can be profoundly altered if young animals are exposed to pulsed pure tones for a brief period immediately after hearing onset, between postnatal day (P) 11 and 13. This form of patterned exposure was found to both rapidly alter tonotopic map structure and close the cortical critical period for frequency tuning (de Villers-Sidani et al. 2007; Dorrn et al. 2010). Conversely, exposure to pulsed white noise stimuli early in life was found to degrade the tonotopic organization of rodent AI (Zhou and Merzenich 2007). Therefore, exposure to either pulsed pure tones or white noise bursts has opposing effects on AI feature selectivity. In both cases, however, receptive fields are remodeled to match the statistics of the sensory environment.

Exposure to continual white noise, rather than periodic bursts of noise, has also been found to degrade cortical receptive fields. However, continual noise exposure prolongs the extent of the critical period into adulthood (Chang and Merzenich 2003). Thus while the spectral structure of acoustic stimuli controls the formation of AI frequency tuning profiles, the temporal pattern of sensory input regulates the overall duration of the AI critical period. Continual stimuli keep the critical period open, perhaps because of the strong neuronal adaptation driven by tonic input, while pulsed or phasic input precociously close the critical period, probably because of the increase in correlated or coincident neuronal activity that should drive long-term synaptic modifications throughout the cortical network (Dorrn et al. 2010).

These forms of receptive field plasticity can also be observed at the synaptic level. In rodent AI, synaptic maturation occurs between P12-21 (Dorrn et al. 2010; Oswald and Reyes 2008). Excitatory inputs seem to mature first, and are tuned for sound frequency by approximately P14 (de Villers-Sidani et al. 2007; Dorrn et al. 2010; Sun et al. 2010). However, inhibitory inputs are potentially equally as strong in young versus adult AI, but exhibit little to no frequency tuning after the second postnatal week, resulting in imbalanced excitation and inhibition and erratic receptive field organization (Fig. 13.10b). After three postnatal weeks of relatively normal acoustic experience, though, cortical inhibition progressively becomes tuned to sound frequency, matching and balancing excitatory inputs (Fig. 13.10c, d). This experience-dependent process of inhibitory maturation can be affected in a similar manner to tonotopic maps: continual white noise exposure delays maturation, while repetitive tonal exposure accelerates balancing of excitation and inhibition (Dorrn et al. 2010). Furthermore, studies in AI brain slices have revealed that postnatal hearing loss, even to a partial degree, leads to persistent changes in the efficacy of cortical synapses (Kotak et al. 2008). Thus early in life, the patterns of acoustic experience – or lack thereof – lead to rapid modifications of excitatory and inhibitory synaptic strength, which in turn govern the organization of receptive fields, the output of cortical circuitry, and the perception of auditory stimuli.

6.3 Plasticity of Frequency Tuning in the Adult Cortex

After the critical period has ended, patterned auditory stimulation by itself is no longer sufficient to drive long-term synaptic modifications or enduring changes to cortical receptive field properties. Rather, adult receptive field plasticity also depends on stimulus history and internal state variables such as arousal level and motivation. This behavioral context is often conveyed by activation of subcortical neuromodulatory systems that directly project to AI, e.g., the cholinergic nucleus basalis (Weinberger 2007; see Chapter 22).

Acetylcholine release is essential for learning and memory, and is believed to be involved in arousal and attentional modulation of cortical responses (Froemke et al. 2007; Parikh et al. 2007). Classic studies using extracellular recordings have shown that pairing pure tones of a specific frequency with electrical stimulation of nucleus basalis induces large, long-lasting enhancements of spontaneous and tone-evoked spiking (Bakin and Weinberger 1996; Kilgard and Merzenich 1998; Rasmusson and Dykes 1988). Although electrical stimulation of nucleus basalis should activate a heterogeneous population of projection neurons, including those that release acetylcholine, glutamate, GABA, and various peptides (Henny and Jones 2008; Lin and Nicolelis 2008), pharmacological evidence indicates that cortical muscarinic acetylcholine receptors are specifically required for the long-term effects on AI receptive fields of this pairing procedure. Acetylcholine has a wide range of effects on cortical neurons, but a consistent observation is increased excitability (Woody and Gruen 1987) and suppression of intracortical synaptic transmission (Metherate et al. 2005; Sarter and Parikh 2005; Xiang et al. 1998).

Intracellular recordings in vivo revealed the mechanisms by which stimulation of the nucleus basalis neuromodulatory system activates cortical networks (Metherate and Ashe 1993; Metherate et al. 1992) and enables receptive field plasticity (Froemke et al. 2007). In these latter experiments, whole-cell voltage-clamp recordings from individual neurons were obtained in anesthetized adult rat

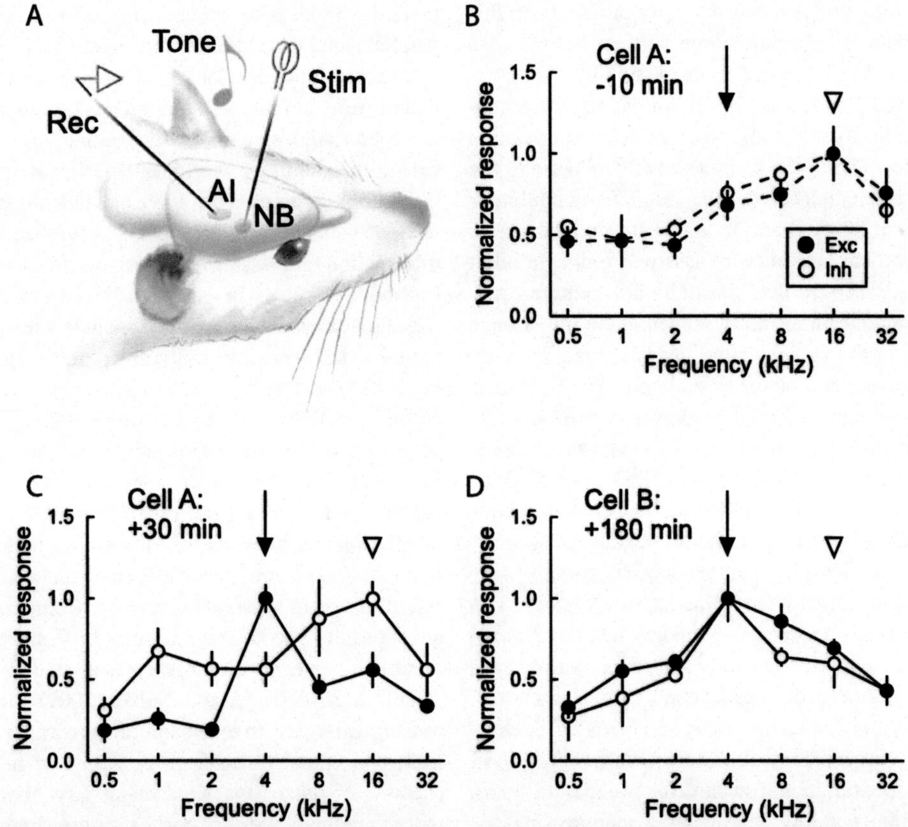

Fig. 13.11 Temporal dynamics of progressive synaptic receptive field plasticity induced by nucleus basalis pairing. **a** Experimental configuration. The stimulation electrode was acutely implanted in the right nucleus basalis, and whole-cell recordings were obtained from neurons in right AI. Pure tones of various frequencies were played to the contralateral ear, and synaptic responses were recorded in voltage-clamp. **b** Frequency tuning of synaptic excitation (*filled*) and inhibition (*open*) for the first cell 10 min prior to nucleus basalis pairing. Note the initial balance of excitation and inhibition across frequencies (linear correlation coefficient *r*: 0.9). *Arrow* indicates the paired frequency (4 kHz). *Arrowhead* indicates the original best frequency (16 kHz) for this region of AI. Error bars represent s.e.m. **c** Frequency tuning of the same cell in (**b**), recorded 30 min after nucleus basalis pairing. The paired frequency had become the best frequency for excitatory tuning but not inhibitory tuning because of the enhancement of excitation and suppression of inhibition, leading to a decrease in excitatory–inhibitory balance (*r*: 0.3). **d** Another cell from same region of AI, recorded 180 min after nucleus basalis pairing. The paired frequency was now the best frequency for both excitation and inhibition, and excitatory–inhibitory balance across all frequencies was restored (*r*: 0.9). Adapted from Froemke et al. (2007)

AI (Fig. 13.11a), and excitatory and inhibitory synaptic frequency tuning was initially determined (Fig.13.11b). Afterwards, tones of a specific non-preferred frequency were paired with electrical stimulation of nucleus basalis. Several seconds after the start of pairing, there was a large suppression of inhibitory events evoked by the paired tone, followed by a more gradual enhancement of tone-evoked excitation. These changes were long-lasting, persisting at least 20 min or more after the end of the pairing procedure. While nucleus basalis stimulation has immediate effects on both thalamocortical and intracortical transmission, longer-term synaptic modifications appear to be specific to intracortical connections and not to the primary thalamic input to AI (Metherate and Ashe 1993; Froemke et al. 2007).

Due to the cooperative effects of suppression of inhibition and enhancement of excitation, nucleus basalis pairing disrupted excitatory–inhibitory balance in adult AI (Fig. 13.11c). However, over a longer time period (several hours), synaptic modifications continually evolved, with inhibition progressively increasing to a higher level than before, eventually re-balancing the persistent increase of excitation at the paired frequency (Fig. 13.11d). These results indicate that the dynamics of inhibitory transmission could serve as a synaptic memory trace of the brief pairing event (Froemke et al. 2007). The duration of input-selective disinhibition may permit self-reorganization of AI receptive fields to emphasize the new preference for paired stimuli, in a manner independent of further evoked neuromodulator release. Under natural conditions, this memory trace could represent episodes or events that have acquired new behavioral meaning, or might be similar to the sorts of cortical changes that occur during perceptual learning, especially for those tasks requiring focal attention and sensory discrimination. In this way, neuromodulatory systems allow cortical networks to selectively respond to important or novel stimuli.

Transient, focal suppression of inhibition may be a general mechanism for induction of receptive field modification in the adult cortex. During developmental critical periods, the high level of plasticity may be due to a less-refined inhibitory tone (Chang et al. 2005; Dorrn et al. 2010), permissive for alterations of cortical networks by passive stimuli. In adult cortex, however, receptive field plasticity also requires activation of neuromodulator systems, reflecting the importance of behavioral context in associative learning and memory provided by subcortical systems (Weinberger 2007). This is further demonstrated by a series of studies from Fritz and colleagues (Fritz et al. 2003, 2005), using single-unit recordings in AI of head-restrained behaving ferrets. Receptive fields of AI neurons were powerfully modified after behavioral conditioning. Excitatory and suppressive subregions of spectro-temporal receptive fields evoked by certain stimuli were altered when those stimuli were followed by tail-shock. The predominant changes to spectro-temporal receptive fields were increases of excitatory regions and reductions of suppressive regions around the conditioned tone (Fritz et al. 2003), strikingly similar to the synaptic effects of nucleus basalis pairing (Froemke et al. 2007). These changes in receptive field structure could endure for minutes to hours after conditioning, possibly serving as a memory in sensory cortex for the contingencies of behavioral training and reinforcement.

Intracellular recordings have been essential for describing cortical organization and dynamics at the synaptic level. During development, perturbations in the sensory environment drive changes in synaptic strength, functioning to model cortical receptive fields around the statistics of sensory inputs. In the adult brain, receptive field plasticity is controlled by behavioral context and motivational state, acting through neuromodulators to gate long-term changes in excitatory and inhibitory synaptic receptive fields. It remains an open question how distinct elements of cortical networks and subcortical neuromodulatory systems are recruited by various forms of sensation, experience, and internal drive for the control of synaptic modifications, circuit dynamics, perception, and cognition.

7 Conclusions and Future Directions

Elucidating stimulus-centered complex coding principles and placing them into a functional and behavioral context remains a primary goal of future studies of the auditory cortex. Without that information, hypotheses about local and global tasks and mechanisms as well as the properties of potential processing streams in higher cortical areas and parallels among modalities remain speculative or untestable.

Linking functional organization and structural substrates that govern complex sound processing in auditory cortex is an essential step in understanding how the brain represents the auditory world and performs specific auditory tasks. Similar approaches in visual and somatic sensory cortices of cats and primates revealed fundamentally different information processing mechanisms from subcortical processing strategies. In early visual cortex, locally created stimulus-based representations include substrates for binocularity, orientation selectivity, and motion selectivity (Bishop et al. 1973; Henry et al. 1974; Hubel and Livingstone 1990; Hubel and Wiesel 1970; Movshon 1975). In somatic sensory cortex, the segregation of slowly and rapidly adapting peripheral mechanoreceptors (Mountcastle 1957), single-to-multiple whisker integration (Mirabella et al. 2001), and integration mechanisms for vibrotactile frequency information (Luna et al. 2005) each offer essential clues as to how the brain interprets sensory experience. Comparable and emergent stimulus processing attributes have not yet been clearly identified for early auditory cortical stations. Instead, it is often assumed (King and Nelken 2009) that cortical processing is largely an extension of subcortical processes with little conceptual changes in content ("what") and manner ("how") of processing. One major impediment to progress is that uniquely auditory cortical processing principles have not been unambiguously identified. The observation of emergent, multi-dimensional spectro-temporal feature processing in AI (Atencio et al. 2008) may hold the key to an advancement in stimulus-centered cortical processing attributes.

The observation of an ordinal laminar progression of *how* information is processed – as opposed to *what* stimulus content is processed – represents a departure from traditional models of auditory cortical stimulus feature extraction and representation (Atencio et al. 2009). The additional informative dimensions express further relevant spectro-temporal aspects. Their interactions with the traditional, feature-selective filter (Atencio et al. 2008) are reminiscent of the notion of combination-sensitivity epitomized in the processing of biosonar signal (Portfors and Felix 2005; Suga 1984; Yan and Suga 1996). However, differences in the filter nonlinearity and the synergistic cooperation of the filters introduce new processing dimensions beyond the combination of highly defined stimulus features that is already present in subcortical stations (Gans et al. 2009; Olsen and Suga 1991; Peterson et al. 2008; Portfors and Felix 2005). Further investigations along these lines, especially in non-primary/belt areas may provide a key step in our understanding of laminar RF transitions and the evolution toward increasingly more complex, nonlinear, robust, stimulus invariant, categorical and/or abstract processing principles.

Cortical microcircuits should be understood according to their different tasks, requirements of the auditory system, and how cortical connection patterns subserve these operations. Simple stereotypical columnar maps repeated across the spatial extent of auditory cortex can be excluded as a dominant computational principle. However, it is conceivable that the main functions of auditory cortex circuits may remain hidden when applying simple, stimulus-based parameter analyses. For the processing rules to emerge fully, a more task-dependent analysis, including determining more complete and higher-order receptive field properties, may have to be performed (Ahissar et al. 2009; Fritz et al. 2003; King and Nelken 2009). The manner in which stimulus information is processed may be a more relevant organizing principle for auditory cortex than the encoding of acoustic content itself. In this framework, increased nonlinear dynamics may emerge as information moves from input to output layers (Ahmed et al. 2006) analogous to the different nonlinearities inherent in simple and complex cell processing in the cat primary visual cortex (Hubel and Wiesel 1962; Linden and Schreiner 2003; Martinez and Alonso 2003).

While much is known about how the brain processes and encodes basic sensory features such as color, orientation, or motion direction in vision and frequency, intensity, and sound source location in audition, much less is known about how the brain acquires and represents the behavioral relevance of stimuli. The neuronal encoding of meaning, as expressed in the creation of sound categories, must involve something beyond the neuronal encoding observed for basic stimulus features. The gradual emergence of these coding aspects, or at least initial steps toward such goals, and their redistribution via extensive feedback connections (Winer 2006) likely renders most stations that have been traditionally considered purely sensory as substrates for combined sensory and cognitive processes.

An array of new methods, including optical methods to record from hundreds of neurons simultaneously, optogenetic methods to manipulate activity in specific cell classes, and computational approaches to dissect and model neuronal ensemble activity across multiple stations during behaviors, are being increasingly exploited to address fundamental issues of spectral and spectral–temporal coding in auditory cortex. It is clear that the focus of research has to shift from single neurons to neuron assemblies, from early cortical regions to later cortical regions, from stimulus-based to cognition-based aspects, and from animal-based to human-based studies in order to fully appreciate and understand the complexity of auditory cortical processing.

Acknowledgements Work supported by NIH grants DC 02260 and MH 077970 (C.E.S.) and DC 09635 (R.C.F.). Fruitful collaborations and always lively, illuminating and stimulating interactions with Dr. Jeffery Winer over the last 20 years have influenced much of the thinking presented here.

References

Abeles M and Goldstein MH, Jr. (1970) Functional architecture in cat primary auditory cortex: columnar organization and organization according to depth. *Journal of Neurophysiology* **33**:172–187.

Abeles M and Goldstein MH Jr (1972) Responses of single units in the primary auditory cortex of the cat to tones and to tone pairs. *Brain Research* **42**:337–352.

Aertsen AM and Johannesma PI (1981) The spectro-temporal receptive field. A functional characteristic of auditory neurons. *Biological Cybernetics* **42**:133–143.

Ahissar M, Nahum M, Nelken I, and Hochstein S (2009) Reverse hierarchies and sensory learning. *Philosophical Transactions of the Royal Society of London, series B, Biological Sciences* **364**: 285–299.

Ahmed B, Garcia-Lazaro JA, and Schnupp JW (2006) Response linearity in primary auditory cortex of the ferret. *Journal of Physiology* **572**:763–773.

Aitkin LM (1976) Tonotopic organization at higher levels of the auditory pathway. In: Porter R (ed). *International Review of Physiology, Neurophysiology II*. University Park Press, Baltimore, pp. 249–279.

Allard TT, Clark SA, Jenkins WM, and Merzenich MM (1985) Syndactyly results in the emergence of double-digit receptive fields in somatosensory cortex in adult owl monkeys. *Proceedings of the Society for Neuroscience* **11**:965.

Atencio CA and Schreiner CE (2010a) Columnar connectivity and laminar processing in cat primary auditory cortex. *Public Library of Science One* **5**:e9521.

Atencio CA and Schreiner CE (2010b) Laminar diversity of dynamic sound processing in cat primary auditory cortex. *Journal of Neurophysiology* **103**:192–205.

Atencio CA and Schreiner CE (2008) Spectrotemporal processing differences between auditory cortical fast-spiking and regular-spiking neurons. *Journal of Neuroscience* **28**:3897–3910.

Atencio CA, Sharpee TO, and Schreiner CE (2008) Cooperative nonlinearities in auditory cortical neurons. *Neuron* **58**:956–966.

Atencio CA, Sharpee TO, and Schreiner CE (2009) Hierarchical computation in the canonical auditory cortical circuit. *Proceedings of the National Academy of Sciences of the United States of America* **106**:21894–21899.

Atiani S, Elhilali M, David SV, Fritz JB, and Shamma SA (2009) Task difficulty and performance induce diverse adaptive patterns in gain and shape of primary auditory cortical receptive fields. *Neuron* **61**:467–480.

Averbeck BB and Romanski LM (2006) Probabilistic encoding of vocalizations in macaque ventral lateral prefrontal cortex. *Journal of Neuroscience* **26**:11023–11033.

Bakin JS and Weinberger NM (1996) Induction of a physiological memory in the cerebral cortex by stimulation of the nucleus basalis. *Proceedings of the National Academy of Sciences of the United States of America* **93**:11219–11224.

Bandyopadhyay S, Shamma SA, and Kanold PO (2010) Dichotomy of functional organization in the mouse auditory cortex. *Nature Neuroscience* **13**:361–368.

Bar-Yosef O and Nelken I (2007) The effects of background noise on the neural responses to natural sounds in cat primary auditory cortex. *Frontiers in Computational Neuroscience* **1**:1:3, doi: 10.3389/neuro.3310/3003.2007.

Barbour DL and Callaway EM (2008) Excitatory local connections of superficial neurons in rat auditory cortex. *Journal of Neuroscience* **28**:11174–11185.

Barbour DL and Wang X (2003a) Auditory cortical responses elicited in awake primates by random spectrum stimuli. *Journal of Neuroscience* **23**:7194–7206.

Barbour DL and Wang X (2003b) Contrast tuning in auditory cortex. *Science* **299**:1073–1075.

Bartho P, Hirase H, Monconduit L, Zugaro M, Harris KD, and Buzsaki G (2004) Characterization of neocortical principal cells and interneurons by network interactions and extracellular features. *Journal of Neurophysiology* **92**:600–608.

Batzri-Izraeli R and Wollberg Z (1992) Auditory cortex of the long-eared hedgehog (Hemiechinus auritus): II. Tuning properties. *Brain Behavior and Evolution* **39**:143–152.

Bendor D and Wang X (2008) Neural properties of primary, rostral, and rostrotemporal core fields in the auditory cortex of marmoset monkeys. *Journal of Neurophysiology* **100**:888–906.

Billimoria CP, Kraus BJ, Narayan R, Maddox RK, and Sen K (2008) Invariance and sensitivity to intensity in neural discrimination of natural sounds. *Journal of Neuroscience* **28**:6304–6308.

Bishop PO, Coombs JS, and Henry GH (1973) Receptive fields of simple cells in the cat striate cortex. *Journal of Physiology (London)* **231**:31–60.

Bizley JK, Nodal FR, Nelken I, and King AJ (2005) Functional organization of ferret auditory cortex. *Cerebral Cortex* **15**:1637–1653.

Blake DT and Merzenich MM (2002) Changes of AI receptive fields with sound density. *Journal of Neurophysiology* **88**:3409–3420.

Brandner S and Redies H (1990) The projection of the medial geniculate body to field AI: organization in the isofrequency dimension. *Journal of Neuroscience* **10**:50–61.

Bregman AS (1990) *Auditory Scene Analysis. The Perceptual Organization of Sound.* MIT Press, Cambridge.

Brugge JF and Reale RA (1985) Auditory cortex. In: Peters A and Jones EG (eds). *Cerebral Cortex*, volume 4, *Association and Auditory Cortices*. Plenum Press, New York, pp. 229–271.

Bruno RM and Simons DJ (2002) Feedforward mechanisms of excitatory and inhibitory cortical receptive fields. *Journal of Neuroscience* **22**:10966–10975.

Buonomano DV and Merzenich MM (1998) Cortical plasticity: from synapses to maps. *Annual Review of Neuroscience* **21**:149–186.

Calford MB and Tweedle R (1988) Immediate and chronic changes in responses of somatosensory cortex in adult flying-fox after induced sensory deficits. *Nature* **332**:446–448.

Calhoun BM and Schreiner CE (1998) Spectral envelope coding in cat primary auditory cortex: linear and non-linear effects of stimulus characteristics. *European Journal of Neuroscience* **10**: 926–940.

Callaway EM (1998) Local circuits in primary visual cortex of the macaque monkey. *Annual Review of Neuroscience* **21**:47–74.

Carandini M, Movshon JA, and Ferster D (1998) Pattern adaptation and cross-orientation interactions in the primary visual cortex. *Neuropharmacology* **37**:501–511.

Chang EF, Bao S, Imaizumi K, Schreiner CE, and Merzenich MM (2005) Development of spectral and temporal response selectivity in the auditory cortex. *Proceedings of the National Academy of Sciences of the United States of America* **102**:16460–16465.

Chang EF and Merzenich MM (2003) Environmental noise retards auditory cortical development. *Science* **300**:498–502.

Chechik G, Anderson MJ, Bar-Yosef O, Young ED, Tishby N, and Nelken I (2006) Reduction of information redundancy in the ascending auditory pathway. *Neuron* **51**:359–368.

Cheung SW, Bedenbaugh PH, Nagarajan SS, and Schreiner CE (2001a) Functional organization of squirrel monkey primary auditory cortex: responses to pure tones. *Journal of Neurophysiology* **85**:1732–1749.

Cheung SW, Bonham BH, Schreiner CE, Godey B, and Copenhaver DA (2009) Realignment of interaural cortical maps in asymmetric hearing loss. *Journal of Neuroscience* **29**:7065–7078.

Cheung SW, Nagarajan SS, Bedenbaugh PH, Schreiner CE, Wang X, and Wong A (2001b) Auditory cortical neuron response differences under isoflurane versus pentobarbital anesthesia. *Hearing Research* **156**:115–127.

Chichilnisky EJ (2001) A simple white noise analysis of neuronal light responses. *Network* **12**:199–213.

Chklovskii DB and Koulakov AA (2004) Maps in the brain: what can we learn from them? *Annual Review of Neuroscience* **27**:369–392.

Christianson GB, Sahani M, and Linden JF (2008) The consequences of response nonlinearities for interpretation of spectrotemporal receptive fields. *Journal of Neuroscience* **28**:446–455.

Clifford CW, Webster MA, Stanley GB, Stocker AA, Kohn A, Sharpee TO, and Schwartz O (2007) Visual adaptation: neural, psychological and computational aspects. *Vision Research* **47**:3125–3131.

Cohen YE, Theunissen F, Russ BE, and Gill P (2007) Acoustic features of rhesus vocalizations and their representation in the ventrolateral prefrontal cortex. *Journal of Neurophysiology* **97**:1470–1484.

Connors BW and Gutnick MJ (1990) Intrinsic firing patterns of diverse neocortical neurons. *Trends in Neurosciences* **13**:99–104.

David SV, Mesgarani N, Fritz JB, and Shamma SA (2009) Rapid synaptic depression explains nonlinear modulation of spectro-temporal tuning in primary auditory cortex by natural stimuli. *Journal of Neuroscience* **29**:3374–3386.

de la Rocha J, Marchetti C, Schiff M, and Reyes AD (2008) Linking the response properties of cells in auditory cortex with network architecture: cotuning versus lateral inhibition. *Journal of Neuroscience* **28**:9151–9163.

Dellen BK, Clark JW, and Wessel R (2009) Contextual interaction in a generalized model of complex cells. *Spatial Vision* **22**:301–324.

de Villers-Sidani E, Chang EF, Bao S, and Merzenich MM (2007) Critical period window for spectral tuning defined in the primary auditory cortex (A1) in the rat. *Journal of Neuroscience* **27**: 180–189.

Dear SP, Fritz J, Haresign T, Ferragamo M, and Simmons JA (1993) Tonotopic and functional organization in the auditory cortex of the big brown bat, *Eptesicus fuscus*. *Journal of Neurophysiology* **70**:1988–2009.

deCharms RC, Blake DT, and Merzenich MM (1998) Optimizing sound features for cortical neurons. *Science* **280**:1439–1443.

Depireux DA, Simon JZ, Klein DJ, and Shamma SA (2001) Spectro-temporal response field characterization with dynamic ripples in ferret primary auditory cortex. *Journal of Neurophysiology* **85**: 1220–1234.

DeWeese MR, Wehr M, and Zador AM (2003) Binary spiking in auditory cortex. *Journal of Neuroscience* **23**:7940–7949.

Diamond DM and Weinberger NM (1986) Classical conditioning rapidly induces specific changes in frequency receptive fields of single neurons in secondary and ventral ectosylvian auditory cortical fields. *Brain Research* **372**:357–360.

Diamond DM and Weinberger NM (1989) Role of context in the expression of learning-induced plasticity of single neurons in auditory cortex. *Behavioral Neuroscience* **103**:471–494.

Diamond ME, Huang W, and Ebner FF (1994) Laminar comparison of somatosensory cortical plasticity. *Science* **265**:1885–1888.

Dorrn AJ, Yuan K, Barker AJ, Schreiner CE, and Froemke RC (2010) Developmental sensory experience balances cortical excitation and inhibition. *Nature* **465**:932–936.

Douglas RJ, Martin KAC, and Whitteridge D (1991) An intracellular analysis of the visual responses of neurones in cat visual cortex. *Journal of Physiology (London)* **440**:659–696.

Dreisbach LE, Leek MR, and Lentz JJ (2005) Perception of spectral contrast by hearing-impaired listeners. *Journal of Speech Language and Hearing Research* **48**:910–921.

Edeline J, Dutrieux G, Manunta Y, and Hennevin E (2001) Diversity of receptive field changes in auditory cortex during natural sleep. *European Journal of Neuroscience* **14**:1865–1880.

Edeline J-M (1998) Learning-induced physiological plasticity in the thalamo-cortical sensory systems: a critical evaluation of receptive field plasticity, map changes and their potential mechanisms. *Progress in Neurobiology* **57**:165–224.

Edeline JM (2003) The thalamo-cortical auditory receptive fields: regulation by the states of vigilance, learning and the neuromodulatory systems. *Experimental Brain Research* **153**:554–572.

Eggermont JJ (1996) How homogeneous is cat primary auditory cortex? Evidence from simultaneous single-unit recordings. *Auditory Neuroscience* **2**:79–96.

Eggermont JJ (2001) Between sound and perception: reviewing the search for a neural code. *Hearing Research* **157**:1–42.

Eggermont JJ (2002) Temporal modulation transfer functions in cat primary auditory cortex: separating stimulus effects from neural mechanisms. *Journal of Neurophysiology* **87**:305–321.

Eggermont JJ, Johannesma PM, and Aertsen AM (1983) Reverse-correlation methods in auditory research. *Quarterly Reviews in Biophysics* **16**:341–414.

Ehret G and Merzenich MM (1985) Auditory midbrain responses parallel spectral integration phenomena. *Science* **227**: 1245–1247.

Ehret G and Merzenich MM (1988) Complex sound analysis (frequency resolution, filtering and spectral integration) by single units of the inferior colliculus of the cat. *Brain Research* **472**:139–163.

Ehret G and Schreiner CE (1997) Frequency resolution and spectral integration (critical band analysis) in single units of the cat primary auditory cortex. *Journal of Comparative Physiology A* **181**: 635–650.

Elhilali M, Fritz JB, Chi T-S, and Shamma SA (2007) Auditory cortical receptive fields: stable entities with plastic abilities. *Journal of Neuroscience* **27**:10372–10382.

Emerson RC, Korenberg MJ, and Citron MC (1992) Identification of complex-cell intensive nonlinearities in a cascade model of cat visal cortex. *Biological Cybernetics* **66**:291–300.

Escabí MA and Read HL (2003) Representation of spectrotemporal sound information in the ascending auditory pathway. *Biological Cybernetics* **89**:350–362.

Escabí MA and Read HL (2005) Neural mechanisms for spectral analysis in the auditory midbrain, thalamus, and cortex. *International Review of Neurobiology* **70**:207–252.

Escabí MA and Schreiner CE (2002) Nonlinear spectrotemporal sound analysis by neurons in the auditory midbrain. *Journal of Neuroscience* **22**:4114–4131.

Evans EF and Whitfield IC (1964) Classification of unit responses in the auditory cortex of unanaesthetized and unrestrained cat. *Journal of Physiology (London)* **171**:476–493.

Ferster D (1986) Orientation selectivity of synaptic potentials in neurons of cat primary visual cortex. *Journal of Neuroscience* **6**: 1284–1301.

Fishman YI and Steinschneider M (2006) Spectral resolution of monkey primary auditory cortex (A1) revealed with two-noise masking. *Journal of Neurophysiology* **96**:1105–1115.

Fishman YI and Steinschneider M (2009) Temporally dynamic frequency tuning of population responses in monkey primary auditory cortex. *Hearing Research* **254**:64–76.

Fitzpatrick DC, Suga N, and Olsen JF (1998) Distribution of response types across entire hemispheres of the mustached bat's auditory cortex. *Journal of Comparative Neurology* **391**:353–365.

Foeller E, Vater M, and Kössl M (2001) Laminar analysis of inhibition in the gerbil primary auditory cortex. *Journal of the Association for Research in Otolaryngology* **2**:279–296.

Fritz J, Shamma S, Elhilali M, and Klein D (2003) Rapid task-related plasticity of spectrotemporal receptive fields in primary auditory cortex. *Nature Neuroscience* **6**:1216–1223.

Fritz JB, Elhilali M, David SV, and Shamma S (2007a) Does attention play a role in dynamic receptive field adaptation to changing acoustic salience in A1? *Hearing Research* **229**:186–203.

Fritz JB, Elhilali M, David SV, and Shamma SA (2007b) Auditory attention--focusing the searchlight on sound. *Current Opinions in Neurobiology* **17**:437–455.

Fritz JB, Elhilali M, and Shamma SA (2005) Differential dynamic plasticity of A1 receptive fields during multiple spectral tasks. *Journal of Neuroscience* **25**:7623–7635.

Fritz JB, Elhilali M, and Shamma SA (2007c) Adaptive changes in cortical receptive fields induced by attention to complex sounds. *Journal of Neurophysiology* **98**:2337–2346.

Froemke RC, Merzenich MM, and Schreiner CE (2007) A synaptic memory trace in auditory cortex. *Nature* **450**:425–429.

Gaese BH and Ostwald J (2001) Anesthesia changes frequency tuning of neurons in the rat primary auditory cortex. *Journal of Neurophysiology* **86**:1062–1066.

Gans D, Sheykholeslami K, Peterson DC, and Wenstrup J (2009) Temporal features of spectral integration in the inferior colliculus: effects of stimulus duration and rise time. *Journal of Neurophysiology* **102**:167–180.

Gilbert CD, Li W, and Piech V (2009) Perceptual learning and adult cortical plasticity. *Journal of Physiology* **587**:2743–2751.

Gill P, Zhang J, Woolley SM, Fremouw T, and Theunissen FE (2006) Sound representation methods for spectro-temporal receptive field estimation. *Journal of Computational Neuroscience* **21**:5–20.

Goldstein MH, Jr. and Abeles M (1975) Single unit activity of the auditory cortex. In: Keidel WD and Neff WD (eds). *Handbook of Sensory Physiology*, volume V, part 2, *Auditory System Anatomy, Physiology (Ear)*. Springer-Verlag, Berlin, pp. 199–218.

Gourevitch B and Eggermont JJ (2008) Spectro-temporal sound density-dependent long-term adaptation in cat primary auditory cortex. *European Journal of Neuroscience* **27**:3310–3321.

Gourevitch B, Norena A, Shaw G, and Eggermont JJ (2009) Spectrotemporal receptive fields in anesthetized cat primary auditory cortex are context dependent. *Cerebral Cortex* **19**: 1448–1461.

Grana GD, Billimoria CP, and Sen K (2009) Analyzing variability in neural responses to complex natural sounds in the awake songbird. *Journal of Neurophysiology* **101**:3147–3157.

Greenwood DD (1974) Critical bandwidth in man and in some other species in relation to the traveling wave envelope. In: Moskowitz HR, Scharf B, and Stevens JC (eds). *Sensation and Measurement*. Dordecht, Reidel, pp. 231–239.

Griffiths TD, Warren JD, Scott SK, Nelken I, and King AJ (2004) Cortical processing of complex sound: a way forward? *Trends in Neurosciences* **27**:181–185.

Hackett TA (2010) Information flow in the auditory cortical network. *Hearing Research* doi:10.1016/j.heares.2010.01.011.

Hackett TA (2008) Anatomical organization of the auditory cortex. *Journal of the American Academy of Audiology* **19**:774–779.

Hackett TA and Schroeder CE (2009) Multisensory integration in auditory and auditory-related areas of cortex. *Hearing Research* **258**:1–3.

Hackett TA, Stepniewska I, and Kaas JH (1998) Subdivisions of auditory cortex and ipsilateral cortical connections of the parabelt auditory cortex in macaque monkeys. *Journal of Comparative Neurology* **394**:475–495.

Hall JW, 3rd and Grose JH (1988) Comodulation masking release: evidence for multiple cues. *Journal of the Acoustical Society of America* **84**:1669–1675.

Harrison RV, Nagasawa A, Smith DW, Stanton S, and Mount RJ (1991) Reorganization of auditory cortex after neonatal high frequency cochlear hearing loss. *Hearing Research* **54**:11–19.

He J and Hashikawa T (1998) Connections of the dorsal zone of cat auditory cortex. *Journal of Comparative Neurology* **400**:334–348.

He J, Hashikawa T, Ojima H, and Kinouchi Y (1997) Temporal integration and duration tuning in the dorsal zone of cat auditory cortex. *Journal of Neuroscience* **17**:2615–2625.

Hefti BJ and Smith PH (2003) Distribution and kinetic properties of GABAergic inputs to layer V pyramidal cells. *Journal of the Association for Research in Otolaryngology* **4**:106–121.

Henny P and Jones BE (2008) Projections from basal forebrain to prefrontal cortex comprise cholinergic, GABAergic and glutamatergic inputs to pyramidal cells or interneurons. *European Journal of Neuroscience* **27**:654–670.

Henry GH (1991) Afferent inputs, receptive field properties and morphologial cell types in different layers. In: Leventhal AG (ed). *Vision and Visual Dysfunction*. Macmillan Press, London, pp. 223–240.

Henry GH, Dreher B, and Bishop PO (1974) Orientation specificity of cells in cat striate cortex. *Journal of Neurophysiology* **37**: 1394–1409.

Hensch TK (2005) Critical period plasticity in local cortical circuits. *Nature Reviews Neuroscience* **6**:877–888.

Hirsch JA (2003) Synaptic physiology and receptive field structure in the early visual pathway of the cat. *Cerebral Cortex* **13**:63–69.

Horton JC and Adams DL (2005) The cortical column: a structure without a function. *Philosophical Transactions of the Royal Society of London, series B, Biological Sciences* **360**:837–862.

Hubel DH and Livingstone MS (1990) Color and contrast sensitivity in the lateral geniculate body and primary visual cortex of the macaque monkey. *Journal of Neuroscience* **10**:2223–2237.

Hubel DH and Wiesel TN (1962) Receptive fields, binocular interaction and functional architecture in the cat's visual cortex. *Journal of Physiology (London)* **160**:106–154.

Hubel DH and Wiesel TN (1970) Cells sensitive to binocular depth in area 18 of the macaque monkey cortex. *Nature* **225**:41–42.

Imaizumi K, Priebe NJ, Crum PAC, Bedenbaugh PH, Cheung SW, and Schreiner CE (2004) Modular functional organization of cat anterior auditory field. *Journal of Neurophysiology* **90**:444–457.

Imaizumi K, Priebe NJ, Sharpee TO, Cheung SW, and Schreiner CE (2010) Encoding of temporal information rate, timing, and place incat auditory cortex. *Public Library of Science One* (in press).

Issa EB and Wang X (2008) Sensory responses during sleep in primate primary and secondary auditory cortex. *Journal of Neuroscience* **28**:14467–14480.

Jenkins WM, Merzenich MM, Ochs MT, Allard T, and Guíc-Robles E (1990) Functional reorganization of primary somatosensory cortex in adult owl monkeys after behaviorally controlled tactile stimulation. *Journal of Neurophysiology* **63**:82–104.

Jones EG (2000) Microcolumns in the cerebral cortex. *Proceedings of the National Academy of Sciences of the United States of America* **97**:5019–5021.

Kaas JH and Hackett TA (1999) 'What' and 'where' processing in auditory cortex. *Nature Neuroscience* **2**:1045–1047.

Kaas JH and Hackett TA (2000) Subdivisions of auditory cortex and processing streams in primates. *Proceedings of the National Academy of Sciences of the United States of America* **97**: 11793–11799.

Kadia SC and Wang X (2003) Spectral integration in A1 of awake primates: neurons with single and multipeaked tuning characteristics. *Journal of Neurophysiology* **89**:1603–1622.

Kajikawa Y, de la Mothe LA, Blumell S, Sterbing-D'Angelo SJ, D'Angelo W, Camaier CR, and Hackett TA (2008) Coding of FM sweep trains an twitter calls in area CM of marmoset auditory cortex. *Hearing Research* **239**:107–125.

Katz LC and Shatz CJ (1996) Synaptic activity and the construction of cortical circuits. *Science* **274**:1133–1138.

Kaur S, Lazar R, and Metherate R (2004) Intracortical pathways determine breadth of subthreshold frequency receptive fields in primary auditory cortex. *Journal of Neurophysiology* **91**:2551–2567.

Kawaguchi Y and Kubota Y (1993) Correlation of physiological subgroupings of nonpyramidal cells with paravalbumin- and calbindin D28k-immunoreactive neurons in layer V of rat frontal cortex. *Journal of Neurophysiology* **70**:387–396.

Keeling MD, Calhoun BM, Krueger K, Polley DB, and Schreiner CE (2008) Spectral integration plasticity in cat auditory cortex induced by perceptual training. *Experimental Brain Research* **184**:493–509.

Kelly JP and Wong D (1981) Laminar connections of the cat's auditory cortex. *Brain Research* **212**:1–15.

Kilgard MP and Merzenich MM (1998) Cortical map reorganization enabled by nucleus basalis activity. *Science* **279**:1714–1718.

King AJ and Nelken I (2009) Unraveling the principles of auditory cortical processing: can we learn from the visual system? *Nature Neuroscience* **12**:698–701.

Klein DJ, Depireux DA, Simon JZ, and Shamma SA (2000) Robust spectrotemporal reverse correlation for the auditory system: optimizing stimulus design. *Journal of Computational Neuroscience* **9**:85–111.

Klein DJ, Simon JZ, Depireux DA, and Shamma SA (2006) Stimulus-invariant processing and spectrotemporal reverse correlation in primary auditory cortex. *Journal of Computational Neuroscience* **20**:111–136.

Kotak VC, Takesian AE, and Sanes DH (2008) Hearing loss prevents the maturation of GABAergic transmission in the auditory cortex. *Cerebral Cortex* **18**:2098–2108.

Kowalski N, Depireux DA, and Shamma SA (1996a) Analysis of dynamic spectra in ferret primary auditory cortex. I. Characteristics of single-unit responses to moving ripple spectra. *Journal of Neurophysiology* **76**:3503–3523.

Kowalski N, Depireux DA, and Shamma SA (1996b) Analysis of dynamic spectra in ferret primary auditory cortex. II. Prediction of unit responses to arbitrary dynamic spectra. *Journal of Neurophysiology* **76**:3524–3534.

Kowalski N, Versnel H, and Shamma SA (1995) Comparison of responses in the anterior and primary auditory fields of the ferret cortex. *Journal of Neurophysiology* **73**: 1513–1523.

Kurt S, Crook JM, Ohl FW, Scheich H, and Schulze H (2006) Differential effects of iontophoretic in vivo application of $GABA_A$-antagonists bicuculline and gabazine in sensory cortex. *Hearing Research* **212**:224–235.

Kusmierek P and Rauschecker JP (2009) Functional specialization of medial auditory belt cortex in the alert rhesus monkey. *Journal of Neurophysiology* **102**:1606–1622.

Langers DR, Backes WH, and van Dijk P (2003) Spectrotemporal features of the auditory cortex: the activation in response to dynamic ripples. *Neuroimage* **20**:265–275.

Lee CC and Winer JA (2005) Principles governing auditory forebrain connections. *Cerebral Cortex* **15**:1804–1814.

Leek MR, Dorman MF, and Summerfield Q (1987) Minimum spectral contrast for vowel identification by normal-hearing and hearing-impaired listeners. *Journal of the Acoustical Society of America* **81**:148–154.

Liberman MC (1978) Auditory-nerve response from cats raised in a low-noise chamber. *Journal of the Acoustical Society of America* **63**:442–455.

Lin SC and Nicolelis MA (2008) Neuronal ensemble bursting in the basal forebrain encodes salience irrespective of valence. *Neuron* **59**:138–149.

Linden JF and Schreiner CE (2003) Columnar transformations in auditory cortex? A comparison to visual and somatosensory cortices. *Cerebral Cortex* **13**:83–89.

Linden JS, Liu RC, Sahani M, Schreiner CE, and Merzenich MM (2003) Spectrotemporal structure of receptive fields in areas AI and AAF of mouse auditory cortex. *Journal of Neurophysiology* **90**:2660–2675.

Liu BH, Li P, Li YT, Sun YJ, Yanagawa Y, Obata K, Zhang LI, and Tao HW (2009) Visual receptive field structure of cortical inhibitory neurons revealed by two-photon imaging guided recording. *Journal of Neuroscience* **29**:10520–10532.

Liu BH, Wu GK, Arbuckle R, Tao HW, and Zhang LI (2007) Defining cortical frequency tuning with recurrent excitatory circuitry. *Nature Neuroscience* **10**:1594–1600.

Loftus WC and Sutter ML (2001) Spectrotemporal organization of excitatory and inhibitory receptive fields of cat posterior auditory field neurons. *Journal of Neurophysiology* **86**:475–491.

Luna R, Hernandez A, Brody CD, and Romo R (2005) Neural codes for perceptual discrimination in primary somatosensory cortex. *Nature Neuroscience* **8**:1210–1219.

Lund JS (1990) Mapping strategies of monkey primary visual cortex. In: Lund JS (ed). *Sensory Processing in the Mammalian Brain: Neural Substrates and Experimental Strategies*. Oxford Press, New York, pp. 209–225.

Machens CK, Wehr MS, and Zador AM (2004) Linearity of cortical receptive fields measured with natural sounds. *Journal of Neuroscience* **24**:1089–1100.

Margoliash D and Fortune ES (1992) Temporal and harmonic combination-sensitive neurons in the zebra finch's HVc. *Journal of Neuroscience* **12**:4309–4326.

Martinez LM and Alonso JM (2003) Complex receptive fields in primary visual cortex. *Neuroscientist* **9**:317–331.

Martinez LM, Wang Q, Reid RC, Pillai C, Alonso J-M, Sommer FT, and Hirsch JA (2005) Receptive field varies with layer in the primary visual cortex. *Nature Neuroscience* **8**:372–379.

McCormick DA, Connors BW, Lighthall JW, and Prince DA (1985) Comparative electrophysiology of pyramidal and sparsely spiny stellate neurons of the neocortex. *Journal of Neurophysiology* **54**:782–806.

Merzenich MM and Brugge JF (1973) Representation of the cochlear partition on the superior temporal plane of the macaque monkey. *Brain Research* **50**:275–296.

Merzenich MM, Knight PL, and Roth GL (1975) Representation of cochlea within primary auditory cortex in the cat. *Journal of Neurophysiology* **38**:231–249.

Mesgarani N, David SV, Fritz JB, and Shamma SA (2008) Phoneme representation and classification in primary auditory cortex. *Journal of the Acoustical Society of America* **123**:899–909.

Mesgarani N, David SV, Fritz JB, and Shamma SA (2009) Influence of context and behavior on stimulus reconstruction from neural activity in primary auditory cortex. *Journal of Neurophysiology* **102**: 3329–3339.

Metherate R and Ashe JH (1993) Nucleus basalis stimulation facilitates thalamocortical synaptic transmission in the rat auditory cortex. *Synapse* **14**:132–143.

Metherate R, Cox CL, and Ashe JH (1992) Cellular bases of neocortical activation: modulation of neural oscillations by the nucleus basalis and endogenous acetylcholine. *Journal of Neuroscience* **12**: 4701–4711.

Metherate R, Kaur S, Kawai H, Lazar R, Liang K, and Rose HJ (2005) Spectral integration in auditory cortex: mechanisms and modulation. *Hearing Research* **206**:146–158.

Miller LM, Escabi MA, Read HL, and Schreiner CE (2001) Functional convergence of response properties in the auditory thalamocortical system. *Neuron* **32**:151–160.

Miller LM, Escabí MA, Read HL, and Schreiner CE (2002) Spectrotemporal receptive fields in the lemniscal auditory thalamus and cortex. *Journal of Neurophysiology* **87**:516–527.

Mirabella G, Battiston S, and Diamond ME (2001) Integration of multiple-whisker inputs in rat somatosensory cortex. *Cerebral Cortex* **11**:164–170.

Mitani A and Shimokouchi M (1985) Neuronal connections in the primary auditory cortex: an electrophysiological study in the cat. *Journal of Comparative Neurology* **235**:417–429.

Mitani A, Shimokouchi M, Itoh K, Nomura S, Kudo M, and Mizuno N (1985) Morphology and laminar organization of electrophysiologically identified neurons in primary auditory cortex in the cat. *Journal of Comparative Neurology* **235**:430–447.

Monier C, Chavane F, Baudot P, Graham LJ, and Fregnac Y (2003) Orientation and direction selectivity of synaptic inputs in visual cortical neurons: a diversity of combinations produces spike tuning. *Neuron* **37**:663–680.

Moshitch D, Las L, Ulanovsky N, Bar-Yosef O, and Nelken I (2006) Responses of neurons in primary auditory cortex (A1) to pure tones in the halothane-anesthetized cat. *Journal of Neurophysiology* **95**:3756–3769.

Mountcastle VB (1957) Modality and topographic properties of single neurons of cat's somatic sensory cortex. *Journal of Neurophysiology* **20**:408–434.

Mountcastle VB (1997) The columnar organization of the neocortex. *Brain* **120**:701–722.

Movshon JA (1975) The velocity tuning of single units in cat striate cortex. *Journal of Physiology (London)* **249**:445–468.

Movshon JA, Thompson ID, and Tolhurst DJ (1978) Receptive field organization of complex cells in the cat's striate cortex. *Journal of Physiology (London)* **283**:79–99.

Nagarajan SS, Cheung SW, Bedenbaugh P, Beitel RE, Schreiner CE, and Merzenich MM (2002) Representation of spectral and temporal envelope of twitter vocalizations in common marmoset primary auditory cortex. *Journal of Neurophysiology* **87**: 1723–1737.

Nagel KI and Doupe AJ (2006) Temporal processing and adaptation in the songbird auditory forebrain. *Neuron* **51**:845–859.

Nagel KI and Doupe AJ (2008) Organizing principles of spectro-temporal encoding in the avian primary auditory area field L. *Neuron* **58**:938–955.

Narayan SS, Temchin AN, Recio A, and Ruggero MA (1998) Frequency tuning of basilar membrane and auditory nerve fibers in the same cochleae. *Science* **282**:1882–1884.

Nelken I and Bar-Yosef O (2008) Neurons and objects: the case of auditory cortex. *Frontiers in Neuroscience* **2**:107–113.

Nelken I, Prut Y, Vaadia E, and Abeles M (1994a) In search of the best stimulus: an optimization procedure for finding efficient stimuli in the cat auditory cortex. *Hearing Research* **72**:237–253.

Nelken I, Prut Y, Vaadia E, and Abeles M (1994b) Population responses to multifrequency sounds in the cat auditory cortex: four-tone complexes. *Hearing Research* **72**:223–236.

Niell CM and Stryker MP (2008) Highly selective receptive fields in mouse visual cortex. *Journal of Neuroscience* **28**:7520–7536.

Nienhuys TGW and Clark GM (1979) Critical bands following the selective destruction of cochlear inner and outer hair cells. *Acta Oto-Laryngologica* **88**:350–358.

Nishimura M, Shirasawa H, Kaizo H, and Sonf WJ (2007) New field with tonotopic organization in guinea pig auditory cortex. *Journal of Neurophysiology* **97**:927–932.

Norena A and Eggermont JJ (2002) Comparison between local field potentials and unit cluster activity on primary auditory cortex and anterior auditory field in the cat. *Hearing Research* **166**: 202–213.

Norena AJ, Gourevitch B, Pienkowski M, Shaw G, and Eggermont JJ (2008) Increasing spectrotemporal sound density reveals an octave-based organization in cat primary auditory cortex. *Journal of Neuroscience* **28**:8885–8896.

O'Connor K N, Petkov CI, and Sutter ML (2005) Adaptive stimulus optimization for auditory cortical neurons. *Journal of Neurophysiology* **94**:4051–4067.

Ohl FW and Scheich H (1996) Differential frequency conditioning enhances spectral contrast sensitivity of units in auditory cortex (field AI) of the alert Mongolian gerbil. *European Journal of Neuroscience* **8**:1001–1017.

Ohm GS (1843) Über die Definition des Tones, nebst daran geknüpfter Theorie der Sirene und ähnlicher Tonbildender Vorrichtungen. *Annalen der Physik* **59**:497–565.

Olsen JF and Suga N (1991) Combination-sensitive neurons in the medial geniculate body of the mustached bat: encoding of target range information. *Journal of Neurophysiology* **65**:1275–1296.

Oonishi S and Katsuki Y (1965) Functional organization and integrative mechanism on the auditory cortex of the cat. *Japanese Journal of Physiology* **15**:342–365.

Oswald AM and Reyes AD (2008) Maturation of intrinsic and synaptic properties of layer 2/3 pyramidal neurons in mouse auditory cortex. *Journal of Neurophysiology* **99**:2998–3008.

Parikh V, Kozak R, Martinez V, and Sarter M (2007) Prefrontal acetylcholine release controls cue detection on multiple timescales. *Neuron* **56**:141–154.

Pei X, Volgushev M, Vidyasagar TR, and Creutzfeldt OD (1991) Whole cell recording and conductance measurements in cat visual cortex in-vivo. *NeuroReport* **2**:485–488.

Pelleg-Toiba R and Wollberg Z (1989) Tuning properties of auditory cortex cells in the awake squirrel monkey. *Experimental Brain Research* **74**:353–364.

Pena JL, Perez-Perera L, Bouvier M, and Velluti RA (1999) Sleep and wakefulness modulation of the neuronal firing in the auditory cortex of the guinea pig. *Brain Research* **816**:463–470.

Peterson DC, Voytenko S, Gans D, Galazyuk A, and Wenstrup J (2008) Intracellular recordings from combination-sensitive neurons in the inferior colliculus. *Journal of Neurophysiology* **100**:629–645.

Petilla International Nomenclature Group (2008) Petilla terminology: nomenclature of features of GABAergic interneurons of the cerebral cortex. *Nature Reviews Neuroscience* **9**:557–568.

Phillips DP and Irvine DRF (1981) Responses of single neurons in physiologically defined primary auditory cortex (AI) of the cat: frequency tuning and responses to intensity. *Journal of Neurophysiology* **45**:48–58.

Pickles JO (1975) Normal critical bands in the cat. *Acta Otolaryngologica* **80**:245–254.

Pienkowski M and Eggermont JJ (2009) Effects of adaptation on spectrotemporal receptive fields in primary auditory cortex. *NeuroReport* **20**:1198–1203.

Polley DB, Read HL, Storace DA, and Merzenich MM (2007) Multiparametric auditory receptive field organization across five cortical fields in the albino rat. *Journal of Neurophysiology* **97**:3621–3638.

Portfors CV and Felix RA, 2nd (2005) Spectral integration in the inferior colliculus of the CBA/CaJ mouse. *Neuroscience* **136**: 1159–1170.

Prieto JJ, Peterson BA, and Winer JA (1994a) Laminar distribution and neuronal targets of GABAergic axon terminals in cat primary auditory cortex (AI). *Journal of Comparative Neurology* **344**:383–402.

Prieto JJ, Peterson BA, and Winer JA (1994b) Morphology and spatial distribution of GABAergic neurons in cat primary auditory cortex (AI). *Journal of Comparative Neurology* **344**:349–382.

Qin L, Chimoto S, Sakai M, and Sato Y (2004) Spectral-shape preference of primary auditory cortex neurons in awake cats. *Brain Research* **1024**:167–175.

Qin L, Chimoto S, Sakai M, Wang J, and Sato Y (2007) Comparison between offset and onset responses of primary auditory cortex ON-OFF neurons in awake cats. *Journal of Neurophysiology* **97**: 3421–3431.

Qin L, Kitama T, Chimoto S, Sakayori S, and Sato Y (2003) Time course of tonal frequency-response-area of primary auditory cortex neurons in alert cats. *Neuroscience Research* **46**:145–152.

Qin L and Sato Y (2004) Suppression of auditory cortical activities in awake cats by pure tone stimuli. *Neuroscience Letters* **365**:190–194.

Qiu A, Schreiner CE, and Escabi MA (2003) Gabor analysis of auditory midbrain receptive fields: spectro-temporal and binaural composition. *Journal of Neurophysiology* **90**:456–476.

Radtke-Schuller S and Schuller G (1995) Auditory cortex of the rufous horseshoe bat: 1. Physiological response properties to acoustic stimuli and vocalizations and the topographical distribution of neurons. *European Journal of Neuroscience* **7**: 570–591.

Rajan R (2001) Plasticity of excitation and inhibition in the receptive field of primary auditory cortical neurons after limited receptor organ damage. *Cerebral Cortex* **11**:171–182.

Rasmusson DD and Dykes RW (1988) Long-term enhancement of evoked potentials in cat somatosensory cortex produced by co-activation of the basal forebrain and cutaneous receptors. *Experimental Brain Research* **70**:276–286.

Rauschecker JP (1998) Parallel processing in the auditory cortex of primates. *Audiology & Neuro-Otology* **3**:86–103.

Rauschecker JP and Tian B (2004) Processing of band-passed noise in the lateral auditory belt cortex of the rhesus monkey. *Journal of Neurophysiology* **91**:2578–2589.

Razak KA, Shen W, Zumsteg T, and Fuzessery ZM (2007) Parallel thalamocortical pathways for echolocation and passive sound localization in a gleaning bat, Antrozous pallidus. *Journal of Comparative Neurology* **500**:322–338.

Read HL, Winer JA, and Schreiner CE (2001) Modular organization of intrinsic connections associated with spectral tuning in cat auditory cortex. *Proceedings of the National Academy of Sciences of the United States of America* **98**:8042–8047.

Read HL, Winer JA, and Schreiner CE (2002) Functional architecture of auditory cortex. *Current Opinion in Neurobiology* **12**:433–440.

Reale RA and Imig TJ (1980) Tonotopic organization in auditory cortex of the cat. *Journal of Comparative Neurology* **192**:265–291.

Recanzone GH (1998) Rapidly induced auditory plasticity: the ventriloquism aftereffect. *Proceedings of the National Academy of Sciences of the United States of America* **95**:869–875.

Recanzone GH (2000) Response profiles of auditory cortical neurons to tones and noise in behaving macaque monkeys. *Hearing Research* **150**:104–118.

Recanzone GH (2008) Representation of con-specific vocalizations in the core and belt areas of the auditory cortex in the alert macaque monkey. *Journal of Neuroscience* **28**:13184–13193.

Recanzone GH, Guard DC, and Phan ML (2000) Frequency and intensity response properties of single neurons in the auditory cortex of the behaving macaque monkey. *Journal of Neurophysiology* **83**:2315–2331.

Recanzone GH, Merzenich MM, and Schreiner CE (1992) Changes in the distributed temporal responses of SI cortical neurons reflect improvements in performance on a temporally based tactile discrimination task. *Journal of Neurophysiology* **67**.

Recanzone GH, Schreiner CE, and Merzenich MM (1993) Plasticity in the frequency representation of primary auditory cortex following discrimination training in adult owl monkeys. *Journal of Neuroscience* **13**:87–103.

Recanzone GH, Schreiner CE, Sutter ML, Beitel RE, and Merzenich MM (1999) Functional organization of spectral receptive fields in the primary auditory cortex of the owl monkey. *Journal of Comparative Neurology* **415**:460–481.

Robertson D and Irvine DRF (1989) Plasticity of frequency organization in auditory cortex of guinea pigs with partial unilateral deafness. *Journal of Comparative Neurology* **282**:456–471.

Romanski LM and Averbeck BB (2009) The primate cortical auditory system and neural representation of conspecific vocalizations. *Annual Review of Neuroscience* **32**:315–346.

Rothschild G, Nelken I, and Mizrahi A (2010) Functional organization and population dynamics in the mouse primary auditory cortex. *Nature Neuroscience* **13**:353–360.

Rotman Y, Bar-Yosef O, and Nelken I (2001) Relating cluster and population responses to natural sounds and tonal stimuli in cat primary auditory cortex. *Hearing Research* **152**:110–127.

Rust NC, Schwartz O, Movshon JA, and Simoncelli EP (2005) Spatiotemporal elements in macaque v1 receptive fields. *Neuron* **46**: 945–956.

Rutkowski RG, Shackleton TM, Schnupp JW, Wallace MN, and Palmer AM (2002) Spectrotemporal receptive field properties of single units in the primary, dorsocausal and ventrorostral auditory cortex in the guinea pig. *Audiology and Neurootology* **7**:214–227.

Sadagopan S and Wang X (2009) Nonlinear spectrotemporal interactions underlying selectivity for complex sounds in auditory cortex. *Journal of Neuroscience* **29**:11192–11202.

Sahani M and Linden JF (2003) How linear are auditory cortical responses? In: Becker S, Thrun S, and Obermayer K (eds). *Advances in Neural Information Processing Systems*, Volume 15. MIT Press, Cambridge, pp. 109–116.

Sally SL and Kelly JB (1988) Organization of auditory cortex in the albino rat: sound frequency. *Journal of Neurophysiology* **59**: 1627–1638.

Sarter M and Parikh V (2005) Choline transporters, cholinergic transmission and cognition. *Nature Reviews Neuroscience* **6**:48–56.

Scholl B, Gao X, and Wehr M (2010) Nonoverlapping sets of synapses drive on responses and off responses in auditory cortex. *Neuron* **65**:412–421.

Schoenwiesner M and Zatorre RJ (2009) Spectro-temporal modulation transfer function of single voxels in the human auditory cortex measured with high-resolution fMRI. *Proceedings of the National Academy of Sciences of the United States of America* **106**: 14611–14616.

Schreiner CE and Calhoun BM (1994) Spectral envelope coding in cat primary auditory cortex: properties of ripple transfer functions. *Auditory Neuroscience* **1**:39–61.

Schreiner CE and Cynader MS (1984) Basic functional organization of second auditory cortical field (AII) of the cat. *Journal of Neurophysiology* **51**:1284–1305.

Schreiner CE, Read HL, and Sutter ML (2000) Modular organization of frequency integration in primary auditory cortex. *Annual Review of Neuroscience* **23**:501–529.

Schreiner CE and Sutter ML (1992) Topography of excitatory bandwidth in cat primary auditory cortex: single-neuron versus multiple-neuron recordings. *Journal of Neurophysiology* **68**:1487–1502.

Schreiner CE and Winer JA (2007) Auditory cortex mapmaking: principles, projections, and plasticity. *Neuron* **56**:356–365, PMC51590.

Schreiner CE, Wong SW, and Dinse HR (2006) Temporal processing in cat primary auditory cortex: dynamic frequency tuning and spectrotemporal representation of speech sounds . In: Greenberg S and Ainsworth WA (eds). *Listening to Speech: An Auditory Perspective*. Lawrence Erlbaum Assoc., Mahwah, pp. 129–141.

Schummers J, Marino J, and Sur M (2002) Synaptic integration by V1 neurons depends on location within the orientation map. *Neuron* **36**:969–978.

Schwartz O, Pillow JW, Rust NC, and Simoncelli EP (2006) Spike-triggered neural characterization. *Journal of Vision* **6**:484–507.

Sen K, Theunissen FE, and Doupe AJ (2001) Feature analysis of natural sounds in the songbird auditory forebrain. *Journal of Neurophysiology* **86**:1445–1458.

Shamma SA, Fleshman JW, Wiser PR, and Versnel H (1993) Organization of response areas in ferret primary auditory cortex. *Journal of Neurophysiology* **69**:367–383.

Shamma SA, Versnel H, and Kowalski N (1995) Ripple analysis in ferret primary auditory cortex. I. Response characteristics of single units to sinusoidal rippled spectra. *Auditory Neuroscience* **1**:233–254.

Shannon RV (2005) Speech and music have different requirements for spectral resolution. *International Review of Neurobiology* **70**: 121–134.

Shannon RV, Fu QJ, and Galvin J, 3rd (2004) The number of spectral channels required for speech recognition depends on the difficulty of the listening situation. *Acta Otolaryngologica Supplement* **552**: 50–54.

Shannon RV, Zeng FG, and Wygonski J (1998) Speech recognition with altered spectral distribution of envelope cues. *Journal of the Acoustical Society of America* **104**:2467–2476.

Sharpee T, Rust NC, and Bialek W (2004a) Analyzing neural responses to natural signals: maximally informative dimensions. *Neural Computation* **16**:223–250.

Sharpee T, Sugihara H, Kurgansky AV, Rebrik S, Stryker MP, and Miller KD (2004b) Probing feature selectivity of neurons in primary visual cortex with natural stimuli. *Proceedings of the Society for Photo and Optical Instrumentation Engineering* **5467**:212–222.

Sharpee TO (2007) Comparison of information and variance maximization strategies for characterizing neural feature selectivity. *Statistical Medicine* **26**:4009–4031.

Sharpee TO, Miller KD, and Stryker MP (2008) On the importance of static nonlinearity in estimating spatiotemporal neural filters with natural stimuli. *Journal of Neurophysiology* **99**: 2496–2509.

Sharpee TO, Sugihara H, Kurgansky AV, Rebrik SP, Stryker MP, and Miller KD (2006) Adaptive filtering enhances information transmission in visual cortex. *Nature* **439**:936–942.

Shen J-X, Xu Z-M, and Yao Y-D (1999) Evidence for columnar organization in the auditory cortex of the mouse. *Hearing Research* **137**:174–177.

Simons DJ and Carvell GE (1989) Thalamocortical response transformation in the rat vibrissa/barrel system. *Journal of Neurophysiology* **61**:311–330.

Smith ZM, Delgutte B, and Oxenham AJ (2002) Chimaeric sounds reveal dichotomies in auditory perception. *Nature* **416**:87–90.

Soeta Y and Nakagawa S (2006) Complex tone processing and critical band in the human auditory cortex. *Hearing Research* **222**:125–132.

Sohya K, Kameyama K, Yanagawa Y, Obata K, and Tsumoto T (2007) GABAergic neurons are less selective to stimulus orientation than excitatory neurons in layer II/III of visual cortex, as revealed by in vivo functional Ca2+ imaging in transgenic mice. *Journal of Neuroscience* **27**:2145–2149.

Stiebler I, Neulist R, Fichtel I, and Ehret G (1997) The auditory cortex of the house mouse: left-right differences, tonotopic organization and quantitative analysis of frequency representation. *Journal of Comparative Physiology A* **181**:559–571.

Suga N (1984) The extent to which biosonar information is represented in the bat auditory cortex. In: Edelman GM, Gall WE, and Cowan WM (eds). *Dynamic Aspects of Neocortical Function*. John Wiley & Sons, New York, New York, pp. 315–373.

Suga N (1995) Sharpening of frequency tuning by inhibition in the central auditory system: tribute to Yasuji Katsuki. *Neuroscience Research* **21**:287–299.

Sugimoto S, Sakurada M, Horikawa J, and Taniguchi I (1997) The columnar and layer-specific response properties of neurons in the primary auditory cortex of Mongolian gerbils. *Hearing Research* **112**:175–185.

Sun YJ, Wu GK, Liu BH, Li P, Zhou M, Xiao Z, Tao HW, and Zhang LI (2010) Fine-tuning of pre-balanced excitation and inhibition during auditory cortical development. *Nature* **465**:927–931.

Sutter ML (2005) Spectral processing in auditory cortex. *International Review of Neurobiology* **70**:253–298.

Sutter ML and Loftus WC (2003) Excitatory and inhibitory intensity tuning in auditory cortex: evidence for multiple inhibitory mechanisms. *Journal of Neurophysiology* **90**:2629–2647.

Sutter ML and Schreiner CE (1991) Physiology and topography of neurons with multipeaked tuning curves in cat primary auditory cortex. *Journal of Neurophysiology* **65**:1207–1226.

Sutter ML and Schreiner CE (1995) Topography of intensity tuning in cat primary auditory cortex: single-neuron versus multiple-neuron recordings. *Journal of Neurophysiology* **73**:190–204.

Sutter ML, Schreiner CE, McLean M, O'Connor KN, and Loftus WC (1999) Organization of inhibitory frequency receptive fields in cat primary auditory cortex. *Journal of Neurophysiology* **82**:2358–2371.

Swadlow HA (2003) Fast-spike interneurons and feedforward inhibition in awake sensory neocortex. *Cerebral Cortex* **13**:25–32.

Swadlow HA and Gusev AG (2002) Receptive-field construction in cortical inhibitory interneurons. *Nature Neuroscience* **5**:403–404.

Tan AY and Wehr M (2009) Balanced tone-evoked synaptic excitation and inhibition in mouse auditory cortex. *Neuroscience* **163**: 1302–1315.

Tan AY, Zhang LI, Merzenich MM, and Schreiner CE (2004) Tone-evoked excitatory and inhibitory synaptic conductances of primary auditory cortex neurons. *Journal of Neurophysiology* **92**:630–634.

Tan AY, Atencio CA, Polley DB, Merzenich MM, and Schreiner CE (2007) Unbalanced synaptic inhibition can create intensity-tuned auditory cortex neurons. *Neuroscience* **146**:449–462.

Theunissen FE, Sen K, and Doupe AJ (2000) Spectral-temporal receptive fields of nonlinear auditory neurons obtained using natural sounds. *Journal of Neuroscience* **20**:2315–2331.

Theunissen FE and Shaevitz SS (2006) Auditory processing of vocal sounds in birds. *Current Opinion in Neurobiology* **16**:400–407.

Thomas H, Tillein J, Heil P, and Scheich H (1993) Functional organization of auditory cortex in the mongolian gerbil (*Meriones unguiculatus*). I. Electrophysiological mapping of frequency representation and distinction of fields. *European Journal of Neuroscience* **5**: 882–897.

Tian B and Rauschecker JP (2004) processing of frequency-modulated sounds in the lateral belt cortex of the rhesus monkey. *Journal of Neurophysiology* **92**:2993–3013.

Ulanovsky N, Las L, Farkas D, and Nelken I (2004) Multiple time scales of adaptation in auditory cortex neurons. *Journal of Neuroscience* **24**:10440–10453.

Valentine PA and Eggermont JJ (2004) Stimulus dependence of spectro-temporal receptive fields in cat primary auditory cortex. *Hear Res* **196**:119–133.

Versnel H and Shamma SA (1998) Spectral-ripple representation of steady-state vowels in primary auditory cortex. *Journal of the Acoustical Society of America* **103**:2502–2514.

Volkov IO and Galaziuk AV (1989) Reactions of tonic-type neurons in the cat auditory cortex to tones of various frequency and intensity. *Neirofiziologiia* **21**:498–506.

Volkov IO and Galazjuk AV (1991) Formation of spike response to sound tones in cat auditory cortex neurons: interaction of excitatory and inhibitory effects. *Neuroscience* **43**:307–321.

von Helmholtz HLF (1863) *Die Lehre von den Tonempfindungen als physiologische Grundlage für die Theorie der Musik*. Vieweg und Sohn, Braunschweig.

Wallace MN, Kitzes LM, and Jones EG (1991) Intrinsic inter- and intralaminar connections and their relationship to the tonotopic map in cat primary auditory cortex. *Experimental Brain Research* **86**:527–544.

Wallace MN and Palmer AR (2008) Laminar differences in the response properties of cells in the primary auditory cortex. *Experimental Brain Research* **184**:179–191.

Wang K and Shamma S (1995) Representation of spectral profiles in primary auditory cortex. *IEEE Transactions on Speech and Audio Processing* **3**:382–395.

Wang X (2007) A sharper view from the top. *Nature Neuroscience* **10**:1509–1511.

Wang X, Lu T, Bendor D, and Bartlett E (2008) Neural coding of temporal information in auditory thalamus and cortex. *Neuroscience* **157**:484–494.

Wang X, Lu T, Snider RK, and Liang L (2005) Sustained firing in auditory cortex evoked by preferred stimuli. *Nature* **435**:341–346.

Wehr M and Zador AM (2003) Balanced inhibition underlies tuning and sharpens spike timing in auditory cortex. *Nature* **426**:442–446.

Wehr M and Zador AM (2005) Synaptic mechanisms of forward suppression in rat auditory cortex. *Neuron* **47**:437–445.

Weinberger NM (2007) Auditory associative memory and representational plasticity in the primary auditory cortex. *Hearing Research* **229**:54–68.

Wenstrup JJ and Grose CD (1995) Inputs to combination-sensitive neurons in the medial geniculate body of the mustached bat: the missing fundamental. *Journal of Neuroscience* **15**:4693–4711.

Winer JA (1984a) Anatomy of layer IV in cat primary auditory cortex (AI). *Journal of Comparative Neurology* **224**:535–567.

Winer JA (1984b) The pyramidal cells in layer III of cat primary auditory cortex (AI). *Journal of Comparative Neurology* **229**:476–496.

Winer JA (1992) The functional architecture of the medial geniculate body and the primary auditory cortex. In: Webster DB, Popper AN, and Fay RR (eds). *Springer Handbook of Auditory Research*, volume 1, *The Mammalian Auditory Pathway: Neuroanatomy*. Springer-Verlag, New York, pp. 222–409.

Winer JA (2006) Decoding the auditory corticofugal systems. *Hearing Research* **212**:1–8.

Winguth SD and Winer JA (1986) Corticocortical connections of cat primary auditory cortex (AI): laminar organization and identification of supragranular neurons projecting to area AII. *Journal of Comparative Neurology* **248**:36–56.

Winkler I, Denham SL, and Nelken I (2009) Modeling the auditory scene: predictive regularity representations and perceptual objects. *Trends in Cognitive Sciences* **13**:532–540.

Woody CD and Gruen E (1987) Acetylcholine reduces net outward currents measured in vivo with single electrode voltage clamp techniques in neurons of the motor cortex of cats. *Brain Research* **424**:193–198.

Woolley SM, Fremouw TE, Hsu A, and Theunissen FE (2005) Tuning for spectro-temporal modulations as a mechanism for auditory discrimination of natural sounds. *Nature Neuroscience* **8**:1371–1379.

Woolley SM, Gill PR, and Theunissen FE (2006) Stimulus-dependent auditory tuning results in synchronous population coding of vocalizations in the songbird midbrain. *Journal of Neuroscience* **26**: 2499–2512.

Wu GK, Arbuckle R, Liu B-h, Tao HW, and Zhang LI (2008) Lateral sharpening of cortical frequency tuning by approximately balanced inhibition. *Neuron* **58**:132–143.

Wu GK, Li P, Tao HW, and Zhang LI (2006) Nonmonotonic synaptic excitation and imbalanced inhibition underlying cortical intensity tuning. *Neuron* **52**:705–715.

Xiang Z, Huguenard JR, and Prince DA (1998) Cholinergic switching within neocortical inhibitory networks. *Science* **281**:985–988.

Yan J and Suga N (1996) The midbrain creates and the thalamus sharpens echo-delay tuning for the cortical representation of target-distance information in the mustached bat. *Hearing Research* **93**:102–110.

Young ED (2008) Neural representation of spectral and temporal information in speech. *Philosophical Transactions of the Royal Society of London, series B, Biological Sciences* **363**: 923–945.

Young ED and Brownell WE (1976) response to tones and noise of single cells in dorsal cochlear nucleus of unanesthetized cats. *Journal of Neurophysiology* **39**:282–300.

Yu JJ and Young ED (2000) Linear and nonlinear pathways of spectral information transmission in the cochlear nucleus. *Proceedings of the National Academy of Sciences of the United States of America* **97**:11780–11786.

Zhang LI, Tan AYY, Schreiner CE, and Merzenich MM (2003) Topography and synaptic shaping of direction selectivity in primary auditory cortex. *Nature* **424**:201–205.

Zhang M and Alloway KD (2004) Stimulus-induced intercolumnar synchronization of neuronal activity in rat barrel cortex: a laminar analysis. *Journal of Neurophysiology* **92**:1464–1478.

Zhou X and Merzenich MM (2007) Intensive training in adults refines A1 representations degraded in an early postnatal critical period. *Proceedings of the National Academy of Sciences of the United States of America* **104**: 15935–15940.

Chapter 14
Temporal Coding in Auditory Cortex

Jos J. Eggermont and Xiaoqin Wang

Abbreviations

AAF	Tanterior auditory field
AI	primary auditory cortex
AII	second auditory field
AL	anterior lateral field
AM	amplitude modulation
BMF	best modulation frequency
CF	characteristic frequency
CL	caudo-lateral field
EEG	electroencephalography
ERBP	event-related band power
FFR	frequency following response
FM	frequency modulation
fMRI	functional magnetic resonance imaging
FM	frequency modulation
HG	Heschl's gyrus
IBE	information bearing elements
IBP	information bearing parameters
ICEP	intracortical evoked potential
ICI	inter-click interval
ISI	inter-spike interval
LFP	local field potential
m	depth of modulation
MEG	magnetoencephalography
MF	modulation frequency
MGB	medial geniculate body
ML	middle lateral field
MU	multi-unit
PAF	posterior auditory field
PP	planum polare
PST/PSTH	post-stimulus time histogram
PT	planum temporale
R	rostral field
RL	rostrolateral field
RT	rostrotemporal field
rMTF	rate modulation transfer function
tMTF	temporal modulation transfer function
SAM	sinusoidal amplitude modulation
STG	supra temporal gyrus
STS	supra temporal sulcus
SU	single unit
VOT	voice onset time
VPAF	ventro-posterior auditory field
VS	vector strength

J.J. Eggermont (✉)
Department of Psychology, University of Calgary, Calgary, AB, Canada T2N 1N4
e-mail: eggermon@ucalgary.ca

1 Introduction

Sounds in general, and human speech and animal vocalizations in particular, are characterized by their intricate temporal structure and, often, strong harmonic content. These characteristics have been named information bearing elements (IBEs) and include steady-state harmonically related frequencies, frequency modulations, and noise bursts. The interrelationship between these IBEs is reflected in information bearing parameters (IBPs) such as onsets, slow amplitude modulations, and silent gaps (Suga 1989, 1992). The IBEs are part of the sound's texture (fine structure) and the IBPs reflect the sound contours or envelopes (Eggermont 1998, 2001). The representation of these temporal IBPs in cortical neural activity is the main topic of this Chapter.

Neural coding has become the generic characterization of those aspects of neural activity that seem to be specifically suited to signal the presence of particular stimuli. A *neural code* is best considered as a vocabulary of the firings on which perceptual discrimination is based. This vocabulary, a multi-dimensional representation depending on the size of the number of participating neurons, contains all the information needed for the perceptual decision process. Examples of such vocabularies are those based on, for instance, first-spike latencies, firing rates, or inter-spike intervals in the participating neurons (Eggermont 2001). As our definition of neural coding implies, we are considering here a population of neurons and this suggests that the term neural code does not

J.A. Winer, C.E. Schreiner (eds.), *The Auditory Cortex*,
DOI 10.1007/978-1-4419-0074-6_14, © Springer Science+Business Media, LLC 2011

apply to the activity of a single neuron; in that case we have to use *neural representation.*

Just as the vocabulary for a neural code in general, that for a *temporal* code is not uniquely defined. It has for instance been narrowly defined as the coding of stimuli by the very precise temporal structure of a spike train, i.e., by specific firing patterns that do not have to bear a one-to-one relationship to the temporal structure of the stimulus (Borst and Theunissen 1999). However, in auditory neurophysiology the use of the phrase "temporal coding" generally implies the representation of the temporal structure of sound by the *modulated* firing rate of individual neurons or neuron populations (Young and Sachs 1979; Schreiner and Urbas 1988; Eggermont 1991; Mueller-Preuss et al. 1994; Cariani and Delgutte 1996; Joris and Yin 1998; Lu and Wang 2000). This particular interpretation runs into potential difficulties as an exclusive code for the higher modulation frequencies. These are only represented in the temporal properties of auditory nerve fibers or brainstem neurons but not in the temporally modulated firing patterns of thalamic and cortical neurons. At the level of auditory nerve fibers there is a dual representation, both by modulated firing rate (more or less independent of the fibers characteristic frequency (CF)) and by a rate-place (CF-dependent) code. The emphasis at the level of the auditory nerve and brainstem nuclei is on a temporal representation but is likely to shift for all but the lowest modulation frequencies towards a rate-place code at cortical levels (Joris et al. 2004). It is the aim of this chapter to explore the evidence for the various ways the temporal structure of sound can be represented in the firing activity of individual cortical neurons and populations thereof.

1.1 Temporal Structure of Sound

1.1.1 Human Speech and Species-Specific Animal Vocalizations

Human brains appear specifically adapted to produce and represent speech, and conversely animal brains are likely better at representing species-specific vocalizations than those of other animals. As human speech processing is generally lateralized to the left hemisphere in right-handed individuals, one can ask whether a similar bias is present in handedness and cortical asymmetry in other mammals. Some reports suggest that the left hemisphere in rodents is specialized for communication sounds (Ehret 1987) and temporal processing (Fitch et al. 1993; Glass and Wollberg 1979), similarly to that in humans (Zatorre and Belin 2001). However, extensive studies by Steinschneider et al. (1980, 1982, 1994, 1998) found no difference in the representation of voice pitch and voice onset time between left and right primary auditory cortex in monkeys. In songbirds, evidence for lateralization of song producing and perceiving structures is also largely negative (Doupe and Kuhl 1999). Thus, lateralization of auditory functions may be limited to humans.

A number of studies have revealed complex temporal structure in species-specific vocalizations of various mammalian species. Moelk (1944) gave a phonetic description of the many calls in cats and their functional use in social interaction. A spectrographic description and some maturational aspects of cat vocalizations were described by Brown et al. (1978); its harmonic content and slow frequency modulation (5–10 Hz/ms) are most characteristic. Gehr et al. (2000) showed spectrograms of a natural kitten call characterized by harmonic content with about six components and a fundamental frequency of 500 Hz that is slowly amplitude and frequency modulated. Mouse vocalizations have been shown to have temporal structure in both short and long time scales (Liu et al. 2003; Holy and Guo 2005). Kanwal et al. (1994) quantitatively analyzed communication sounds emitted by the mustached bat and showed that they have a variety of amplitude and frequency modulations.

Non-human primates generally exhibit rich vocal repertories in natural environment, some of which can be observed in captive conditions, in particular those of New World monkey species. Vocalizations of both adult and developing common marmosets (*callithrix jacchus*), a New World primate, have been quantitatively analyzed and modeled (DiMattina and Wang 2006; Pistorio et al. 2006). Marmoset vocalizations have a wide range of amplitude and frequency modulations, with modulation frequencies ranging from a few hertz to tens and hundreds of hertz. Their vocalizations are also rich in harmonic structure. Similar acoustic structure has been found in other non-human primates, for example, tamarins (Moody and Menzel 1976), squirrel monkeys (Winter 1969), and macaque monkeys (Gouzoules et al. 1984; Hauser 1996).

A rich set of literature has been published on acoustic structure of bird songs (Ball and Hulse 1998). A recent study by Singh and Theunissen (2003) quantitatively compared modulation spectra of natural sounds, animal vocalizations, and human speech. They found that animal vocalizations and human speech are characterized by low temporal modulation.

1.1.2 Music

Music, the artificial extension of vocalizing behavior with instruments, shares characteristics of vocalizations. Speech and music thus have much in common, although they appear to engage potentially different cortical structures for their representation or interpretation (Peretz et al. 2002; Zatorre et al. 2002). This difference has been assigned mainly because the left hemisphere processes temporal cues

(Johnsrude et al. 1997) and the right hemisphere mainly spectral cues (Tervaniemi et al. 1999, 2000; Johnsrude et al. 2000). However, timbre, reflecting tonal color and textures as well as temporal aspects related to the attack of the sounds, was shown to involve fMRI activation in both left and right hemispheres (Menon et al. 2002). Pitch also showed bilateral activation (Griffiths et al. 1999) and so did human voices (Belin et al. 2000). Peretz and Zatorre (2005) in a review on music processing, come to the conclusion that musical pitch is dominantly processed by the spectral processing capacity in the right hemisphere, whereas timing relations in music draw on temporal processes in both hemispheres.

1.2 Perception: Rhythm, Roughness, and Pitch

Sounds of very different origin may have the same pitch, largely determined by a similar periodic envelope, but they may have a different timbre, as determined by the sound's texture, i.e., the range and strength of harmonics present as well as its attack property, a temporal characteristic. Spectral or pure tone pitch is represented in tonotopic maps all along the auditory pathway. Recent studies have begun to point out where temporal or residue pitch is represented in the primate brain (see review by Bendor and Wang 2006).

The frequency range where the cortical neural firing rate is strongly modulated by the temporal envelope of sound extends up to about 20 Hz and these sounds are perceived as having rhythm and fluctuation strength. At repetition rates of ~10 to 45 Hz, we perceive flutter (Miller and Taylor 1948; Besser 1967). The perception of pitch starts at around 30 Hz (Krumbholz et al. 2000). Roughness is perceived at higher modulation frequencies (Zwicker and Fastl 1990). The percept of roughness may be caused by the emerging dual representation, i.e., both by firing rate and by envelope synchrony for higher modulation frequencies (Schulze and Langner 1997b; Fishman et al. 2000a). The sensation of roughness disappears around 300 Hz and is strongest at about 70 Hz where the temporal coding of modulation is still present and the firing rate representation becomes more important. Modulation frequencies above about 100 Hz and up to about 2.5 kHz evoke a specific pitch sensation. This upper level of pitch perception corresponds to those modulation frequencies that are still coded in a temporal way at the level of the auditory nerve and lower brainstem (Joris et al. 2004).

Thus, a comprehensive coverage of the coding of temporal aspects of sound in cortex needs to take into account the conversion of modulated firing rates as a code for envelope periodicity into a place or rate code whereby firing rate and/or place of activation represents changes in stimulus periodicity. This transformation from a temporal representation to a rate-place representation for pitch occurs somewhere along the auditory pathway and potentially in the inferior colliculus (Schreiner and Langner 1988, 1997; Langner and Schreiner 1988; Langner 1992). A rate representation of modulation frequencies above 50 Hz by a distinct population of single neurons in primary auditory cortex as described by Lu et al. (2001b) may run into uniqueness problems because firing rate also changes with stimulus level or background noise. In the form of a rate-place code, this could work and thereby extend the representation of the modulation frequency range in the cortex dramatically. Another way to resolve the ambiguity of a monotonic rate code is by having two neuronal populations, each with an increasing or decreasing firing rate, respectively, as modulation (or repetition) frequency increases (Lu et al. 2001b).

2 Coding of Stimulus Periodicity and Envelope by Single Neurons

Periodicity information can in principle be recovered from the frequency distance between sidebands and carrier frequency and for that interactions between different parts of the tonotopic map are needed to play a role in pitch perception. Horizontal fiber interactions could play a role here that preferably tend to branch in cat auditory cortex at intervals of ~0.75 mm (Wallace et al. 1991). This periodic branching distance points to interactions over octave intervals and could serve to extract residue pitch, i.e., fundamentals of harmonic complexes. However, periodicity pitch also results from amplitude modulation (AM) of broadband noise or from periodic click trains, and in these cases purely temporal aspects of the sound carry the pitch information that can be extracted by the auditory system, putatively on the basis of an all-order inter-spike-interval representation (Cariani and Delgutte 1996) or only first-order inter-spike intervals (Kaernbach and Demany 1998).

2.1 Phase-Locked Responses

The temporal structure of sound comprises that of the carrier (determining the fine structure or texture of the sound) and that of the stimulus envelope (the contour of the sound). Temporal representations of carrier period exist for most of the auditory pathway alongside rate-place representations. In auditory nerve fibers, the timing of action potentials is locked to both the period of the carrier frequency—up to 5 kHz (Johnson 1980; Weiss and Rose 1988)—and to the sound envelope but then only time-locking up to about 2.2 kHz (Joris et al. 2004).

The upper limit of this phase locking of action potentials to the sound envelope decreases along the auditory pathway (see Eggermont 2001; Joris et al. 2004; Wang et al. 2008). The highest locking to periodic click trains found in single neurons in primary auditory cortex in paralyzed cats was around 100 Hz (Ribaupierre et al. 1972). Local field potentials (LFP) appear to lock up to 200 Hz in awake animals (Goldstein et al. 1959) and maybe even up to 300 Hz (Steinschneider et al. 1980). LFPs, however, reflect the input to the cortical cells, in the form of synchronized postsynaptic potentials, and thus in fact represent the envelope synchronization of action potentials in the thalamus where the upper limit of the phase locking is substantially higher than that of the auditory cortex (Joris et al. 2004), especially when measured in awake animals (Bartlett and Wang 2007).

As already eluded to, the representation of temporal aspects of sound in auditory cortex may be in the form of both phase-locked responses and firing rate. The neural populations that respond in a phasic or tonic way to amplitude-modulated stimuli are potentially mutually exclusive (Lu and Wang 2004) and the tonically responding population may be largely silent in anesthetized animals (Lu and Wang 2000).

2.1.1 Quantitative Measurement of Phase-Locked Responses

There are basically three ways in which the relationship between neuronal firings and sound carrier or envelope can be characterized (Fig. 14.1). The first is by the average

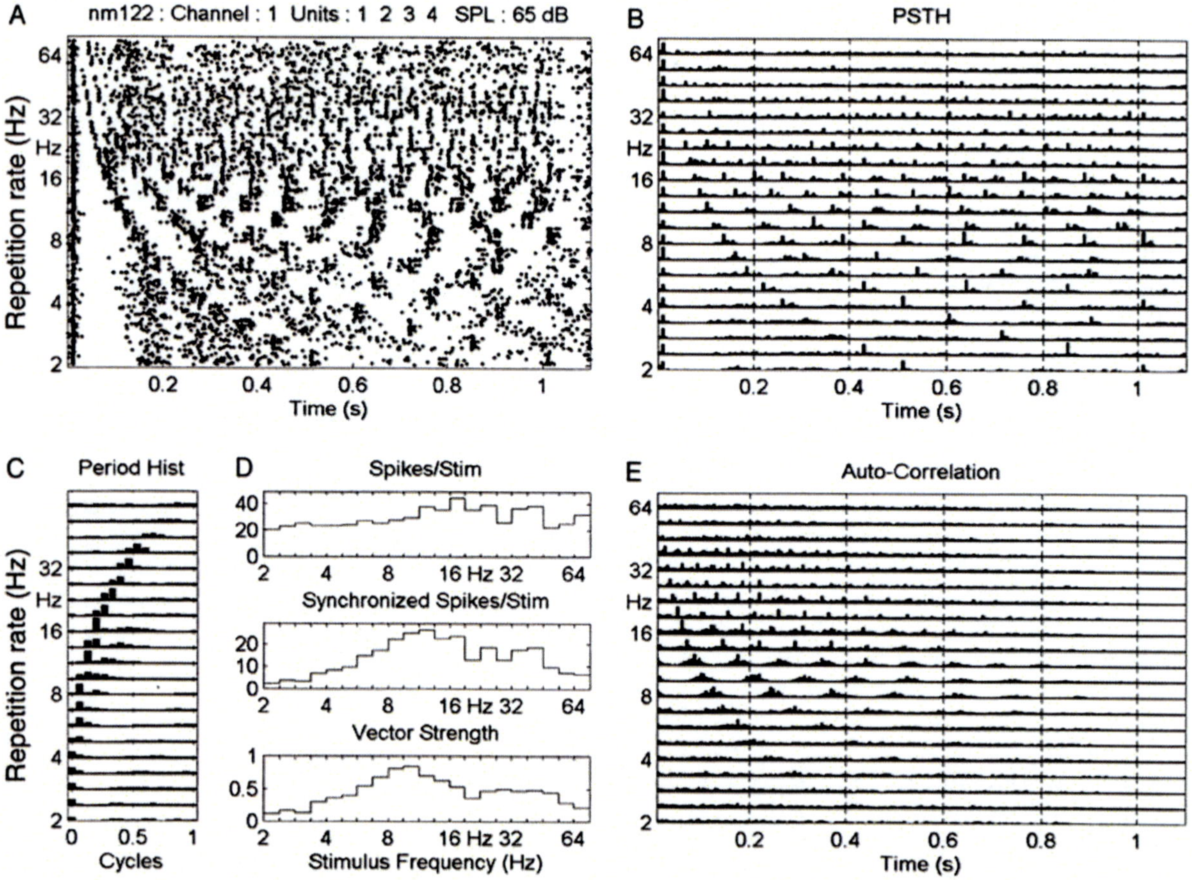

Fig. 14.1 Basic method of analysis for temporal response properties. *Top left*: a multi-unit (comprising 3 single units) *raster dot* display for periodic click train stimulation. The click repetition rate is shown vertically and ranges between 2 and 64 Hz. The time since click train onset is on the horizontal axis; the click trains last 1 s. For click rates that can be expressed in natural numbers, a click is present at the 1-s mark. *Top right*: the corresponding post-stimulus time (PST) histogram. *Bottom right*: the population autocorrelogram. The axes are the same as for the *dot* display and the PST histogram. The autocorrelation function is computed for each single unit individually, and the normalized autocorrelograms are added. The sharp peaks are the result of 1 single unit with very little spike jitter; it is the same unit that produces the short latency responses in the PSTH and *dot* display. The set of 4 small graphs at the *bottom left* represent the period histogram, the rate modulation transfer function (rMTF) (number of spikes per click train), the temporal modulation transfer function (tMTF, number of synchronized spikes per click train), and the vector strength (VS, the ratio of tMTF and rMTF). Adapted from Eggermont (2002)

firing rate as a function of modulation frequency; this is often called the rate modulation transfer function (rMTF). The second is by the ratio of the strength of the Fourier component at the modulation frequency (MF) divided by the average firing rate. This ratio is usually called the vector strength, VS (Goldberg and Brown 1968), for which the significance can be evaluated using the Rayleigh test (Fisher 1993; Mardia 1972). The third is by a ratio of modulation depth of firing rate (m_r) and the modulation depth of the stimulus envelope (m_s), this can be defined as a modulation gain (G) and is often expressed in dB (Møller 1972) and is related to the VS: $G = 20 \log_{10} (2\times \text{VS}/m_s)$. Both m_r and m_s are defined by the maximum and minimum magnitude of firing rate or stimulus amplitude, respectively: (max–min)/(max+min). The strength of the Fourier component at the MF and the VS, as a function of the MF, have both been called the temporal modulation transfer function (tMTF), although they only have the same shape if the rMTF is independent of the MF.

Limiting modulation rates have been defined as the 50% point of the tMTF at the high-frequency side (Eggermont 1991), but also as the highest modulation frequency at which the first peak in the autocorrelation function is found at a time commensurate with the period of the modulation or where the second peak in the post-stimulus-time histogram (PSTH) is commensurate with the time of the second modulation peak in the stimulus (Eggermont 2002). These definitions need not result in the same limiting rate value. It is quite often the case that for higher modulation frequencies the response to the second or even the third modulation period is skipped but that phase-locked responses resume for subsequent later peaks. In this case, the limiting rate determined from the autocorrelation function is higher than that based on the PSTH. Alternatively, if the neuron does not always fire at each modulation period, the limiting rate derived from the PSTH will be higher than that based on the autocorrelation function. The 50% point will typically be lower for tMTFs that are sharply peaked and have high VS, than for low VS, shallow peaked tMTFs. Consequently, it is required that the significance of the VS be tested at the limiting rate reported. This is usually based on the Rayleigh test looking at a circular uniform distribution of $2n\times \text{VS}^2$, where n is the average number of spikes per modulation period. One can also define the limiting rate as the highest MF at which the VS is still significantly different from zero (Lu and Wang 2000), this value is generally higher than that based upon the 50% point of the tMTF or the PSTH or autocorrelation measure.

The phase of the response relative to the modulation period can be derived from the phase spectrum of the Fourier transform of the period histogram (Eggermont 1999). The slope of the phase as a function of modulation frequency results in the group delay. The group delay can differ substantially from the average latency of the response to the first stimulus period and is caused by the additional delay introduced by the synaptic filter. This is a functional synaptic filter that under certain conditions can be equated with the interface between thalamic afferents and cortical neurons (Eggermont 2002). For large differences between the latency in response to the first stimulus (period) and group delay, indicative of a narrowly tuned synaptic filter, the extra delay was only found for the spikes and not for the LFPs. This suggests that the cause for these long delays resides in the spike-generating mechanism and in the spike-induced suppression. The units with long group delays for spikes showed slightly higher limiting rates in agreement with the enhanced response for modulation frequencies above 8 Hz. It is noted that the low-frequency slope of the tMTF is not dependent on the group delay and is thus determined by the pre-filter envelope synchronization (Eggermont 2002). The increased response strength was interpreted as the result of an amplifying filter tuned around 12–14 Hz (Eggermont 1999).

The measurement of neuronal tMTFs does not have to be based on periodic stimulation and alternative stimuli could include Poisson-distributed clicks or low-pass noise-modulated carriers as used in Møller's (1973) pioneering studies in the cochlear nucleus. For these stimuli, one can derive the tMTF by dividing the spectrum of the spike train phase-locked to the stimulus or stimulus envelope (obtained by Fourier transformation of the cross-correlation function between stimulus (envelope) and spike train, i.e., the cross spectrum) by the amplitude spectrum of the stimulus or stimulus envelope. At the level of the auditory nerve and cochlear nucleus (Møller 1973) and the auditory midbrain (Epping and Eggermont 1986a, b), the results did not differ substantially from those obtained by periodic stimuli. Data for auditory cortex are still scarce but also suggest similarity to the results obtained with periodic stimuli (Eggermont and Smith 1995). However, predictions are only satisfactory if second order non-linearities are taken into account (Pienkowski et al. 2009).

2.1.2 Phase Locking to Carrier

Phase locking of action potentials to the period of pure tones is best in the auditory periphery where it can be demonstrated up to at least 5 kHz depending on the species (Weiss and Rose 1988). In the auditory cortex, the upper limit for unit activity seems to be around 300 Hz but has only been demonstrated so far in awake monkey auditory cortex (Steinschneider et al. 1980) and in the ketamine-anesthetized guinea pig primary auditory cortex (Wallace et al. 2002) and its cortical belt areas (Wallace et al. 2000). The upper limit in the guinea pig encompasses the fundamental frequency of their vocalizations. In cats, fundamental frequencies are higher and one would need phase locking up to about 500 Hz or more to code for this in temporal fashion; phase locking to pure tones in cat auditory cortex has not been described.

2.1.3 Phase Locking to Stimulus Envelope

A thorough review of envelope locking covering the entire auditory system (Joris et al. 2004) also describes most of the individual studies on envelope coding by modulated firing rate in auditory cortex. From this compilation of data for tMTFs it follows that in primary (core) cortical areas the best modulation frequency (BMF) for single or multi-unit activity is typically in the 8–14 Hz range, the average limiting rate is between 30–50 Hz and the upper limit of phase locking is approximately 100 Hz. The upper limit of phase locking appears to be higher in awake animals than in anesthetized animals (Liang et al. 2002; Malone et al. 2007). One thing to caution in comparing data from anesthetized and awake animals is the laminar difference. In anesthetized animals data are collected primarily from middle (thalamic recipient) layers, whereas in awake animals data are often recorded from upper cortical layers. The actual difference between anesthetized and awake animals could be larger than what has been reported in the literature if laminar difference were taken into account. In the ketamine-anesthetized cat, and based on dynamic ripple noise stimulation (Miller et al. 2001), mean BMFs (32.4 Hz) and limiting rates (62.9 Hz) for the ventral medial geniculate body (MGBv) in the cat were less than a factor 2 higher than those recorded simultaneously in AI (BMF = 16.6 Hz, limiting rate = 37.4 Hz). The thalamus data should provide an estimate for the upper level of envelope locking for cortically recorded LFPs under ketamine anesthesia.

There is however a substantial difference of phase-locking ability in primary auditory cortex for different types of stimuli. Eggermont (2002), extending initial studies (Eggermont, 1994b), compared the tMTFs obtained in cat AI for six different modulated stimuli. The stimuli comprised three periodic ones based on one-second duration trains of transient stimuli such as clicks, gamma tone pips and time-reversed gamma tone pips separated by two seconds of silence. In addition, three modulated stimuli of one-second duration: exponential sine AM tones and AM noise and sinusoidally frequency-modulated (FM) tones were used. The tone pips and modulated tones were typically presented with a carrier frequency equal or close to the CF of the neurons.

The rMTFs were nearly always, and for every stimulus type, largely independent of the modulation frequency and either slightly increasing (for AM noise) or slightly decreasing (for most other stimuli) with increasing modulation frequency. This contrasts the rMTFs reported for instance by Schreiner and Urbas (1988) and Schreiner and Raggio (1996) that were clearly band-pass. A potential cause for this difference may be the type of anesthesia used, suggesting that pentobarbital results in band-pass rMTFs and ketamine in flat or slightly low-pass rMTFs. The higher spontaneous firing rates recorded under ketamine anesthesia may contribute substantially to the firing rate for long repetition periods, thereby elevating the low-frequency portion of the rMTF and converting a band-pass response into a low-pass one. However, in awake squirrel monkey at most 10% of the units in AI showed a flat rMTF, and in 30% the rMTF was basically similar to the tMTF and both showing a band-pass shape (Bieser and Mueller-Preuss 1996). In awake marmoset monkey, about 70% of AI neurons (from mostly upper layers) showed band-pass rMTFs when tested with SAM stimuli (Liang et al. 2002). Thus, the absence of band-pass rMTFs in ketamine-anesthetized cats may rest on the ability of ketamine to increase spontaneous firing rates and/or induce rebound responses after a post-activation suppression. For higher repetition rates, the inability to lock the firings to the stimulus envelope would basically result in the absence of suppression and allowing the spontaneous firings again to lift up the response for frequencies above the BMF; in combination, this results in an all pass rMTF.

The tMTFs found were typically band-pass for the transient stimuli, because of rebound responses in the long periods between stimuli at low rates and the spontaneous firings for high rates that reduce the VS. The tMTFs were typically low-pass for the AM and FM stimuli. The average peak VS was highest for periodic clicks (0.81) and lowest for the time-reversed gamma tone pips (0.37). The average BMFs followed the same trend, being highest for clicks (8.8 Hz) and again lowest for time-reversed gamma tone pips (4.7 Hz). The average limiting rates (defined as the highest significant value from either the PSTH or autocorrelation function) in contrast were highest for AM noise (19.1 Hz), followed by AM tones (16.2 Hz) and clicks (15.8 Hz), and again lowest for time-reversed gamma tone pips (11.9 Hz). The limiting rate definitions used here are more conservative than using the 50% point of the tMTF (as used in Eggermont 1998). A quantitative neuron-by-neuron comparison between responses to sinusoidally modulated AM and FM stimuli was made in the study of awake marmoset AI by Liang et al. (2002). They reported similar population median tBMFs (AM: 10.8 Hz, FM: 10.9 Hz), rBMFs (AM: 20.7 Hz, FM: 16.2 Hz) and upper limit of phase locking (AM: 47.1 Hz, FM: 40.1 Hz) for the two types of stimuli.

While auditory researchers have traditionally valued (perhaps overly valued) the phase-locking ability of neurons, how important the phase-locked firing is in underlying auditory perception is not clear. A recent study of human subjects correlated a subject's speech perception with the degree of phase locking in its core auditory cortex to speech envelope and concluded that such capacity in itself is not a limiting factor for speech comprehension (Nourski et al. 2009).

2.1.4 Gap Detection and Voice Onset Time Representation

Gap detection refers to the ability to discriminate small interruptions in a noise or constant frequency tone. For gaps inserted in the middle of a one-second duration noise the minimum gap identified on basis of single- and multi-unit recordings was 5 ms, the shortest gap duration tested (Eggermont 1999b). Gaps in sound also play an important role in speech perception. The categorical perception of a /ba/ from a /pa/ phoneme depends to a large extent on the duration of the silent gap, around 20–40 ms, voice onset time (VOT), between the short initial noise burst and the onset of voicing. The VOT is well represented in the activity patterns of cortical neurons by a double-on response; one to the leading noise burst and one to the onset of voicing (Eggermont 1995).

As already suggested on basis of the difference between the minimum detectable "late gap" in noise and the "early gap" in the phonemes, the temporal position of the gap in a noise burst makes a large difference for the minimum gap threshold determined on basis of the neural response in cat AI (Fig. 14.2). If the gap was embedded after a leading burst of at least 200 ms duration, the cortical neurons responded with an on-response to the trailing burst for gap durations as low as 5 ms (see also the similar findings by Buchhfellner et al. (1989) in the forebrain of the starling). However, if the gap was embedded after only 5 ms of noise (comparable to the duration of the leading noise burst for a /pa/), the minimum gap threshold was around 40 ms similar to that found for a /ba/-/pa/ VOT continuum (Eggermont 1999b). A systematic effect of the leading gap duration was found with longer leading bursts resulting in smaller minimum gap thresholds (Eggermont 2000). Interestingly the time after leading burst onset at which the on-response to the trailing burst occurred was constant over a large range. This near constancy can be modeled on basis of a long lasting inhibitory effect produced by the onset of the leading noise burst combined with synaptic depression.

Whether the perception of VOT depends on neurons that respond to both the noise burst and the vowel or to a comparison between the responses of populations of neurons that

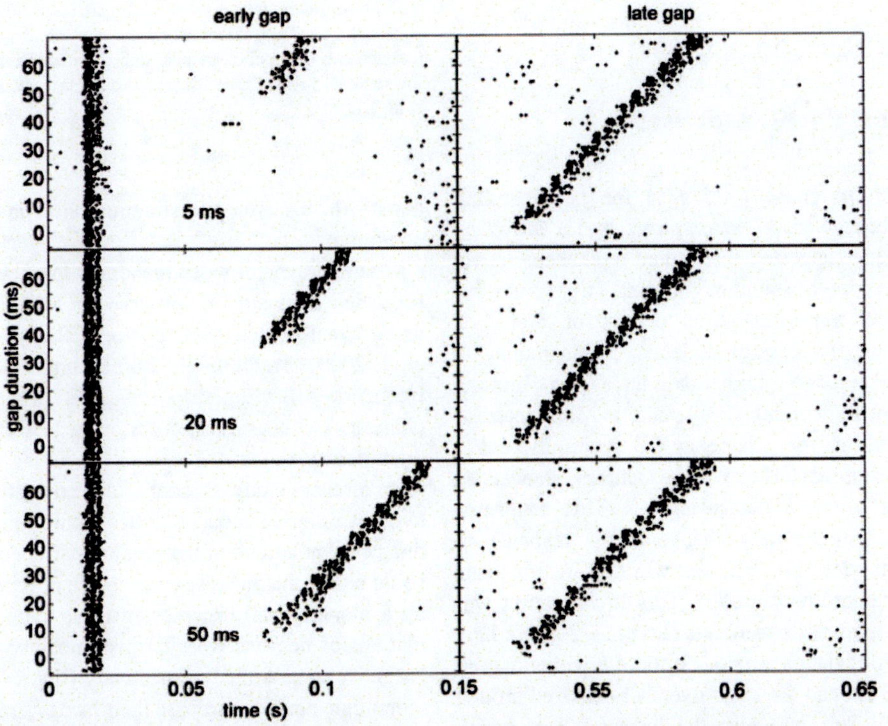

Fig. 14.2 The effect of temporal position of the gap in a noise burst on the minimum gap threshold determined from AI responses. Two gaps are inserted in the 1-s noise burst: one early in the noise burst (*left*) and another one 500 ms after noise burst onset (*right*). The duration of the burst preceding the early gap was either 5 ms (*top left*), 20 ms (*middle left*), or 50 ms (*bottom left*). The multi-unit record consisted of 2 well-separated single units with identical properties. Along the *horizontal axis*, time since noise burst onset is presented. *Left*: time runs from 0 to 150 ms; *right*: time runs from 500 to 650 ms. The *vertical axis* represents gap duration. This was advanced in 5-ms steps, which is reflected in the step-wise response change to the trailing burst. Every gap condition was presented 15 times. One observes (*left*) that the time between the onset response to the leading burst and that for the lowest gap to the trailing burst is approximately constant. When no onset response to the trailing burst is present a rebound response occurs ~130 ms after the onset response. Adapted from Eggermont (2000)

respond to either the noise (higher CF neurons) or the vowel (lower CF neurons) is not clear. However, psychoacoustical studies (Phillips et al. 1997) provide evidence for within-frequency channel mechanisms to function with an acuity comparable to that for the "late gaps" and for "between channel" mechanisms that are aimed to detect VOTs. It is surprising that the "early gap"-in-noise results in AI neurons (clearly a within channel situation) are so similar to those for VOT (Eggermont 2000).

Studies on the representation of speech sounds in primary auditory cortex of awake monkeys spanning more than two decades (Steinschneider et al.1982, 1990, 1994, 2003) revealed two important findings. First, phase locking to the fundamental frequency of voicing (100 Hz) occurs in about 40% of multi-unit (MU) activity of thalamocortical fibers and in about 25% of lower lamina III, thalamo-recipient cortical neurons. Second, voice onset times of 40 ms and longer are detectable as a double-on response in MU activity in about 25% of lower lamina III cortical cells and only in 8% of thalamocortical fibers (where it is restricted to units with CFs below 1 kHz). Thus, the phasic nature of cortical responses facilitates representation of gaps of around 30 ms by a double on-response.

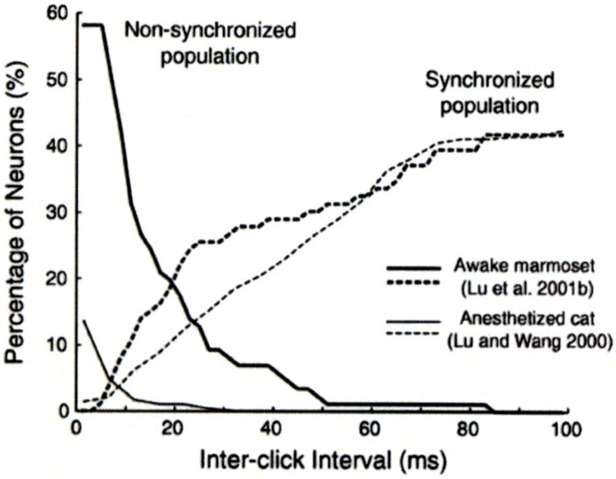

Fig. 14.3 Dual mechanisms for encoding slow and fast repetition rates by two populations of auditory cortex neurons. A combination of temporal and rate representations can encode a wide range of ICIs. The *dashed line* shows the percentage of neurons with synchronization boundaries less than or equal to a given ICI. The *solid line* shows the percentage of neurons with rate-response boundaries greater than or equal to a given ICI. *Thick curves* are based on the data of two neuronal populations recorded from A1 of awake marmosets, one with stimulus-synchronized discharges (*N*_36, *dashed curve*) and the other with non-synchronized discharges (*N*_50, *solid curve*), respectively (Lu et al. 2001b). Thin curves show the data obtained from A1 of anesthetized cats using click train stimuli (Lu and Wang 2000), analyzed in the same way as the data from awake marmosets. Adapted from Wang et al. (2008)

2.2 Tonic (Sustained) Responses

The upper limit of phase locking in the pentobarbital/ketamine-anesthetized cat AI was explored by Lu and Wang (2000). This study suggested that a dual representation for the temporal structure of sound might exist. They found that the upper limit of significant phase locking for inter click intervals (ICI) was on average 39.8 ms, corresponding to about 25 Hz and somewhat higher than the average limiting rate reported by Eggermont (1998, 2002). More interestingly, they found that for ICIs below 6.3 ms (equivalent to about 150 Hz) the response of some neurons showed an increased firing rate with decreasing ICI. This confirmed earlier findings that the only change in the response to high modulation rates (128–512 Hz) was that in firing rate (Bieser and Mueller-Preuss 1996). This investigation was extended to AI in awake marmosets (Lu et al. 2001b); here the rate-coding population was much more obvious than in the anesthetized cat and the cross over ICI occurred around 20 ms (50 Hz). This was also the lowest ICI at which significant synchronized responses were observed on average (Fig. 14.3). Another important difference between the data reported in anesthetized cats (Lu and Wang 2000) and awake marmosets (Lu et al. 2001b) is that the firing of neurons at short ICI was sustained throughout the entire stimulus duration in the awake condition, but only lasted for about 100 ms after the stimulus onset in the anesthetized condition.

The non-synchronized responses to click trains with high repetition rates are also observed in human auditory cortex using event-related band power (ERBP) analysis (Brugge et al. 2009). Traditionally, human brain recordings rely on frequency following response (FFR) that can only detect phase-locked neural responses evoked by click trains of low repetition rates.

In a recent study, Bendor and Wang (2007) showed that, over the range of acoustic flutter (approximately 10–45 Hz), distinct populations of neurons in marmoset auditory cortex could signal the increase or decrease of repetition frequency by a monotonically increasing firing rate. These two populations of neurons were termed "positive monotonic" and "negative monotonic," respectively (Fig. 14.4). The positive monotonic neurons increase their firing rate as repetition rate increases, in a manner similar to non-synchronized neurons encoding repetition rates higher than flutter as reported in Lu et al. (2001b). The negative monotonic neurons increase their firing rate as repetition rate decreases in the flutter range. Studies of the somatosensory cortex have demonstrated that such a coding scheme exists for encode tactile flutter (Romo et al. 2003).

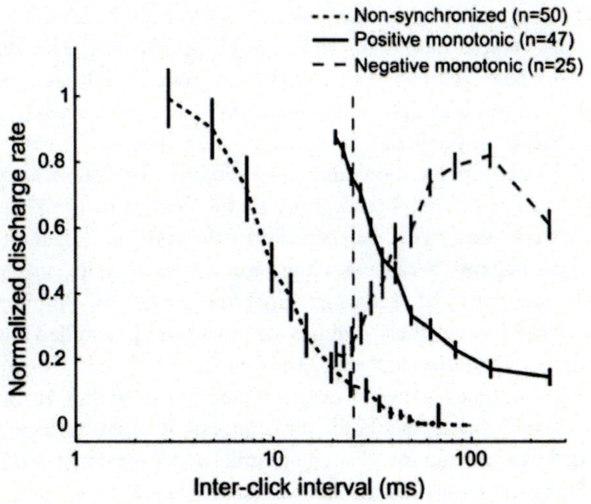

Fig. 14.4 Positive and negative monotonic tuning in firing rate of auditory cortex neurons. Comparison between non-synchronized (*dotted curve*), positive monotonic (*solid curve*), and negative monotonic (*dashed curve*) neurons. Normalized discharge rates are plotted as a function of inter-click interval (ICI). The *vertical dashed line* marks the perceptual boundary between flutter and pitch at 40 Hz (25 ms). The *upper* limit of flutter is 45 Hz and the *lower* limit of pitch is 30 Hz. Data of non-synchronized neurons are from Lu et al. (2001b). Data of positive and negative monotonic neurons are from Bendor and Wang (2007). *Error bars* represent standard error of the mean (S.E.M.). Adapted from Wang et al. (2008)

2.3 Temporal Envelope Asymmetry

One of the first studies to record responses to animal vocalizations in fentanyl anesthetized and muscle relaxed cat AI was by Sovijärvi (1975) who found that time reversal of a kitten meow had no dramatic effect on the response. No cells were encountered that responded only to cat meows. The meows largely excited cells with on–off responses and inhibited the activity in between or had no effect at all. Sovijärvi (1975) also recorded responses to a variety of other natural vocalizations that were more periodic in their structure such as songs of nightingales, barn swallows, willow warblers, golden orioles, and chaffinches. In 68% the responses to the calls were reasonably predictable on basis of the response to pure tones. The periodic song of the willow warbler elicited good responses locked to the rhythm of the song, which represents a strong mix of fast FM sweeps alternating at a rate of about 5 Hz. Gehr et al. (2000) showed that 40% of the neurons in AI of the ketamine-anesthetized cat responded to peaks in the meow envelope and to other transient changes such as the FM parts and 60% only showed an on-response. The responses occurred synchronously at various sites in AI and could be based on firing rate alone but more likely on coincident firings related to salient aspects of the sound contour. Stronger onset responses were found to the natural meow compared to the time-reversed call, whereas there was no difference in the response to the sustained part of the call albeit that its firing rates were still significantly higher than spontaneous firing rate.

Human listeners can discriminate between exponentially ramped and damped noisebursts with half-life durations between 0.5 and 64 ms (Akeroyd and Patterson 1995). This appears to be qualitatively matched—up to a half-life of 32 ms—by the neural firing rate in AI of the awake marmoset (Lu et al. 2001a). A substantial portion of neurons showed an asymmetry in their firing rate, and for a subset also in synchronization. In light of recent findings (Ahissar and Arieli 2001; VanRullen et al. 2005) that first-spike latency may matter more than an evaluation of the firing rate over an entire stimulus, temporal aspects need also to be evaluated. Consideration of the first-spike latencies alone may also have allowed this discrimination (Fig. 1 in Lu et al. 2001a). In cat AI, gamma tone pips and their time reversals show vastly different rise-times, but the firing rate did not discriminate between them although their synchronized firings did (Eggermont 2002).

The marmoset twitter call has a highly stereotyped periodicity and a shorter rise than fall time for the individual elements (Wang et al. 1995). In barbiturate-anesthetized cat AI the responses (in terms of average firing rate across the response) were not very different for a time reversal of the calls, whereas in barbiturate-anesthetized marmoset responses were much stronger for the natural twitter call. This was interpreted as evidence for either experience-dependency (e.g., Nakahara et al. 2004) or for species-specific mechanisms of processing communication sounds (Wang and Kadia 2001). In cat, if the same reasoning applies, one would expect the response to normal and time-reversed calls to be different. They are not different as far as firing rate is concerned, but they are when envelope locked spikes are taken into account (Gehr et al. 2000). The results of Eggermont (2002) showed that although the firing rates in cat were the same for normal and time-reversed gamma tone pips (somewhat similar in envelope to the elements of a twitter call, but lacking the high-to-low FM through the call), the tMTFs were clearly different. Thus, experience dependence or species-specific mechanisms either only apply to the marmoset and not to the cat or do not provide the explanation for this finding.

Two studies in barbiturate-anesthetized marmosets describe individual and population responses to the normal twitter call, compressed and expanded versions, time reversals (Wang et al. 1995; Wang 2000). In addition, the number of frequency bands was reduced to mimic a low-number-channel vocoder, and the temporal envelope was filtered, and tMTFs obtained from sinusoidally amplitude-modulated (SAM) CF tones and periodic click trains were compared (Nagarajan et al. 2002). A subpopulation of neurons in AI responded more strongly and selectively to normal twitter calls: about two third of neurons belonged to that category and showed a marked reduction in response to time-reversed, time-contracted or time-dilated calls. Spectral degradation was less effective than temporal degradation in affecting the responses, which was similar to findings by Theunissen and Doupe (1998) regarding the representation of the bird's own song in field HVc of Zebra finches. Neurons in AI phase-locked to the low-frequency modulations can distinguish individual calls but do not respond to the fast FM in the call elements. Major features were reflected in a population response created by averaging PSTHs as a function of neurons CFs. The responding neurons were evenly distributed through the entire AI and the firing rate for calls was highly correlated between units with similar or often very different CF. Combined, these studies suggest that the response to a twitter call is spectrally distributed and synchronized across AI. Although the mean periodicity in the twitter call was 7.7 Hz, similar to the BMF to SAM tones and periodic click trains, the tMTF calculated from SAM tones and periodic click trains only explained 22% or 13%, respectively, of the variance in the call response. In this light, it is not too surprising that low-pass filtering the call envelope below 10 Hz nearly abolished the temporal envelope locking in AI neurons, as did high-pass filtering above 60 Hz (Nagarajan et al. 2002). What this suggests is that low-pass filtering blurs rapid changes in the individual element's envelope that are important for the response. It would likely also blur the distinction between the normal and time-reversed call (Wang et al. 1995; Nagarajan et al. 2002).

The natural twitter call in squirrel monkeys has a repetition frequency of about 12 Hz for its highly frequency-modulated elements. Bieser (1998) compared the responses for neurons in the insula, primary auditory field and rostral auditory field to the natural call and synthesized FM calls that contained only one frequency and were modulated continuously over 500 ms. The phase locking to these sinusoidally FM tones ceased abruptly above 16 Hz. The natural call produced a stronger response than the synthesized FM calls in all cortical fields, suggesting that the present amplitude modulations that divide the natural call into individual elements also have a large contribution. Synchronization between firings in AI and insula was limited to responses for low FM rates.

3 Population Coding of Temporal Sound Structures

What is population coding? Is it similar to pooling data across neurons in one animal or even across animals? Is pooling data from sequentially recorded activity across animals' representative of what a single animal might be able to do based on its simultaneous accessible population activity? Is pooling data from sequentially recorded activity in one animal predictive of what that animal might be able to do based on its simultaneous accessible population activity? This, if true, would imply that responses of individual neurons are largely independent of those of other neurons. Alternatively, because auditory cortical neurons do not fire independently of each other (Eggermont 1992, 1994a), population coding likely requires simultaneously recorded data for its evaluation. In that case one needs to explore the potential difference between near-coincident, say within 10 ms, firings of the units and the remaining non-coincident firings in the representation of sound (Tomita and Eggermont 2005). Do single units have the same response properties as the MU recordings they were derived from? Are local field potentials (LFP) sampling from more than the multi-unit activity recorded on the same electrode and are their tMTFs comparable to those of the MU or SU? An added complication is that the LFPs represent the compound input to cortical cells, whereas MU activity represents the compound output of (a subset) of those cells. We will address these questions in the following sections.

3.1 *Comparison Between Single-Unit, Multi-Unit, and Field Potential Measurements*

Frequency-tuning curves derived from LFPs and multi-unit activity recorded from the same micro-electrode have the same CF (Eggermont 1996; Eggermont 1998; Noreña and Eggermont 2002) but a much larger bandwidth reflecting the converging inputs from a wide CF-range of thalamic neurons. LFPs are considered often as synchronized theta or spindle waves (7–12 Hz) as they have the same periodicity. A detailed comparison of the LFPs and spikes for periodic click trains (Eggermont and Smith 1995) showed that destructive interference between LFPs occurs whenever the depth-negative wave to the (n+1)th click superimposes the depth-positive wave in response to the nth click. When that occurs, spikes are also missing in the response. In contrast, constructive interference occurs when the timing of the clicks corresponds to the periodicity of the LFPs, typically in the range of 80–150 ms. This suggests that the ongoing oscillatory network activity interferes with and affects the LFPs and the action-potential firings to periodic stimuli. In the Eggermont and Smith (1995) and Eggermont (1996) studies, the differences between the tMTF for clicks and LFPs were minor. In a follow-up study encompassing recordings in three cortical fields (Eggermont 1998) the

BMFs and limiting rates for periodic click trains were about 50% higher for LFPs compared to SU or MU recordings in AI, AAF, and AII. In contrast, for AM noise and AM tones the limiting rates were nearly the same in AI and AII, and again about 50% higher for LFPs compared to MU in AAF. A comparison between SU and MU recordings was only made for periodic click trains and revealed no systematic differences, suggesting that the SU that make up the MU recording have the same temporal response properties.

3.2 Inter-spike-Interval Coding

Eggermont (1998) constructed population autocorrelation functions and inter-spike-interval histograms for all responsive units recorded in three cortical fields. These functions were then converted to a population MTF for interval coding. A comparison with the pooled tMTFs across neurons based on period histograms showed only small differences. A distinction has to be made here between the average values across a population and the value obtained by constructing pooled histograms. In the latter case the neurons with highest firing rate will dominate in the ISI population histograms, and these neurons typically have higher limiting rates (Eggermont 1999). We will focus on the limiting rates as defined here by the 50% point on the high-frequency slope of the tMTF. For primary auditory cortex, the ISI-based limiting rates for MU activity were consistently higher for AM tone and noise stimuli compared to clicks. Whereas the latter were typically around 13 Hz, they were about 30 Hz for AM stimuli. For standard tMTFs the limiting rates obtained were similar to those based on ISIs for clicks and somewhat higher at 40 Hz for AM stimuli. This similarity is expected based on the information theoretic analysis for coding based on ISI and period histogram (Lu and Wang, 2004).

If neurons fire with strong periodicity to periodic stimuli, they will also show near coincident firings across a population as already suggested by a comparison of ISI histograms for single units and the population as a whole. As the temporal response properties are mostly independent of the CF, such coincident activity could bind neurons with different CFs in different cortical areas that respond to distinct aspects of the sound's texture.

3.3 Place Coding

Intrinsic optical imaging of the activity produced by amplitude-modulated tones in gerbil primary auditory cortex suggested the existence of a horseshoe shaped periodotopic map partially overlapping with the tonotopic map (Schulze et al. 2002). The horseshoe shape was reminiscent of the pinwheel geometry of iso-orientation domains in visual cortex. This could reflect periodicity coding along the iso-frequency contours. Multi-unit recordings were also done in identified parts of the periodotopic map and confirmed the optical imaging data. These findings seem to solidify previous data form Schulze and Langner (1997, 1999) who used harmonic complexes to create different fundamental frequencies. However, a study by Fishman et al. (2000b) in awake monkey primary auditory cortex suggests that what is mapped is likely not the fundamental frequency but is potentially more related to critical band filtering effects as the fundamental covaries with the separation and thus with the resolvability of the harmonics.

4 Stimulus Dependence/Invariance of Temporal Coding

Temporal coding is at the very least determined by cortical synaptic mechanisms and by the way the stimulus determines the input to this thalamocortical synapse (Eggermont 2002). For instance, different modulation envelopes (e.g., sinusoidal vs. rectangular) will potentially produce very different inputs to the synaptic filter and thus the output, the PSPs, will differ. This in turn will lead to different spiking behavior in the cortical cell and so produce different levels of post-activation suppression. All these factors will determine the spike-based tMTF. In addition, overall stimulus level and modulation depth will potentially contribute some non-linear effects.

The contribution of the stimulus waveform will be reflected in the response to the first click, tone pip or modulation period. The effect of the synaptic filter can then be evaluated by deconvolving the tMTF (here based on VS) with the VS to the first modulation or stimulus period. This reveals the low-pass filter characteristics of the thalamocortical synapses with a cut-off frequency (–6 dB) of 14 Hz and a slope of –6 dB/oct. with the –12 dB point at about 40 Hz. This filter includes the properties of the spike-generating process (Fig. 14.5). The filter estimated on the basis of LFPs was also band-pass but had a much wider frequency range with the cut-off frequency at 40 Hz and with the –12 dB point at 100 Hz. This is commensurate with the idea that LFPs reflect the output of the synaptic filter prior to the cortical spike-generating process. On could conceive the differences of the "LFP filter" and the "spike filter" as largely produced by those post-activation suppression mechanisms, after-hyperpolarization, feed-back inhibition, and rebound processes, that are triggered by the spikes (Eggermont 1992, 2000).

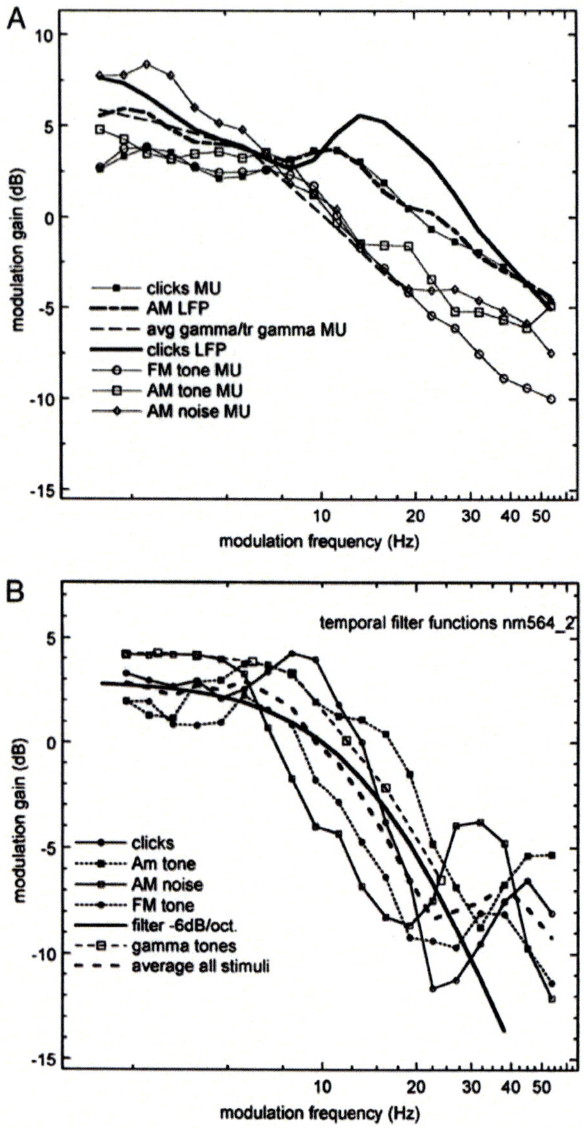

Fig. 14.5 Average modulation gain for MU responses and local field potential (LFP) triggers to the 6 stimuli used in this study compared with model results. **Part a** shows the tMTFs for the entire stimulus. **Part b** shows the temporal filter functions estimated by deconvolution of the functions in **Part a** with the response to the first click, modulation period etc. in the stimulus. The average curves for clicks, AM tones, AM noise, and FM tones are shown in thin *drawn lines* and *symbols*. The average curves for gamma tone and time-reversed gamma tones are shown as one here (*long thin dashed line*). The average functions obtained for LFP triggers are shown in *fat*, *full*, or *dashed*, *lines*. The model results are shown as a *dotted line* for the membrane filter and as *full line* for the combination of synaptic depression and membrane filter. Adapted from Eggermont (2002)

5 Temporal Coding in Different Cortical Areas

Local field potential based tMTFs showed limiting rates that were basically the same in cat AI, AAF and AII, but lower for clicks (17–21 Hz) than for AM stimuli (42–50 Hz). Spike based tMTFs showed lower limiting rates compared to LFPs: for clicks 11–14 Hz in all areas, and for AM stimuli 37–45 Hz in AI and AII and 28–29 Hz in AAF. The only interesting differences in area responses were the lower limiting rate for AM stimuli in AAF and the near absence of envelope locking for AM tones in AII. This difference was even more emphasized when considering the tMTFs based on the pooled ISI histograms. For this measure, the AM tones produced sustained AM following only in AAF and with a low limiting rate of about 13 Hz (Eggermont 1998). In awake squirrel monkey (Bieser and Mueller-Preuss 1996) observed best phase locking for MFs in the range of 2–64 Hz in fields AI, PI, and T1. In fields Pa (AII homologue), R (PAF homologue), AL, Rpi and Insula large numbers of units did not respond in a phase-locked manner and those who did hardly ever followed up to 16 Hz. Previously, Schreiner and Urbas (1988), in the paralyzed and barbiturate-anesthetized cat, had shown for AM tones that the BMFs for rMTFs were highest in AAF (31.1 Hz), lower in AI (14.2 Hz) and lowest in AII (7.0 Hz), PAF (6.8 Hz) and VPAF (5.2 Hz), suggesting large differences between fields. Comparison with Eggermont (1998)'s limiting rate data based on the tMTFs suggests substantial differences for their findings in AAF.

In monkeys, the core auditory cortex is divided into AI, R, and RT fields. Bendor and Wang (2008) found that neurons in fields RT and R have longer minimum latencies than those of field AI. They also reported that fields RT and R have poorer stimulus synchronization than field AI to AM tones.

Studies using linear FM sweeps provide another way of probing area differences in temporal processing. At low sweep rates (10–100 Hz/ms) these linear FM sweeps mimic those occurring in animal vocalizations (Tian and Rauschecker 2004) and in formant transitions in speech. Consistent differences were found between primary cortical areas, including AI (Heil et al. 1992) and AAF (Tian and Rauschecker 1994), in the cat where the majority of neurons (90–96%) showed a preference for fast changes in frequency (> 200 Hz/ms), compared to field PAF where only 22% of the neurons showed this preference and the vast majority preferred rates < 100–200 Hz/ms (Heil and Irvine 1998a, b; Tian and Rauschecker 1998). This suggests that PAF is suited for the analysis of communication sounds where formant changes are relatively slow. Both AAF and PAF showed high sensitivity for FM direction with more than half of the neurons showing a preference, this could be useful in distinguishing up- or downward formant transitions or equivalents in vocalizations. Three areas in the lateral belt of rhesus monkey were also studied using linear FM sweeps (Tian and Rauschecker 2004) and a dichotomy was found between field AL which responded to low sweep rates (< 200 Hz/ms) and field CL which preferred high sweep rates (> 200 Hz/ms). Field ML responded to the entire sweep range (6.25–640 Hz/ms). This led the authors to suggest that this reflected the presumed role of these areas in the "where" (CL) and the "what" (AL) auditory pathway.

6 Temporal Processing in Human Auditory Cortex

6.1 Representation of Amplitude and Frequency Modulations

In human auditory cortex, AM or FM sounds appear to generate more activity compared to unmodulated sounds as observed by fMRI (Hall et al. 2002; Hart et al. 2003) or MEG (Makela et al. 1987) recordings. Both types of study found that the most responsive areas for AM and FM were overlapping. Different locations for maximal activation in fMRI were found for the planum temporale (PT) and dorsal supra temporal gyrus (STG) bilaterally for FM tones compared to noise, whereas the supra temporal sulcus (STS) bilaterally was more activated by speech-like sounds than by FM tones (Binder et al. 2000). The speech-like sounds included normal and pseudo words as well as time-reversed speech and suggest that the acoustic aspects of such sounds determine the difference to FM tones, not the linguistic aspects.

AM sounds were used in an fMRI study to investigate BMFs in the auditory pathway (Giraud et al. 2000; Griffiths et al. 2001) and suggested a general decrease from lower brainstem (256 Hz), inferior colliculus (32–256 Hz), MGB (16 Hz), primary auditory cortex (8 Hz) to secondary cortical areas (4–8 Hz). These results obtained from awake humans are in general agreement with studies in anesthetized animals (Joris et al. 2004). Interestingly, evidence was found for a different coding of modulation frequencies below 16 Hz and those above 128 Hz by not mutually exclusive areas. This is reminiscent of the envelope synchrony coding and rate coding found in awake marmoset cortex (Lu et al. 2001b). Evidence for a correlation between the perceptual resolution of low rate noise bursts (1–2/s) and the one perceptual event presented by high rate noise bursts (35/s) was found in the cortical activity recorded by fMRI but not in that originating from the inferior colliculus (Harms and Melcher 2002).

6.2 Voice Onset Time

Intracortical evoked potential (ICEP) recordings from Heschl's gyrus (HG), planum temporale (PT), and posterior supra temporal gyrus (STG, Area 22) in humans (Liegeois-Chauvel et al. 1999) showed a distinct preference for the representation of voiced (/ba/, /da/, /ga/) and voiceless (/pa/, /ta/, /ka/) phonemes in the left but not in the right HG and PT. Only in the left hemisphere, ICEPs to the different components of the phoneme were present (i.e., a double on-response in the voiceless phonemes). This was also found for non-verbal sounds that mimicked the temporal structure of the phoneme. No differences were found in the representation of voiced and voiceless phonemes in the left or right Area 22. ICEPs in response to sinusoidally AM tones showed a low-pass dependence on modulation frequency in primary auditory cortex with cut-off frequency at 16 Hz. In the posterior part of the right STG and in the right secondary auditory cortex, the ICEP-amplitude MTFs were band pass with a BMF = 8 Hz, whereas in these structures in the left hemisphere the MTFs were low-pass (BMF=4 Hz). In Brodman Area 22 the MTFs were low-pass (cut-off frequency 4–8 Hz) in both the left and right hemisphere (Liegeois-Chauvel et al. 2004).

Intracranial recordings in awake humans revealed no difference between monkey and human local field potentials in the representation of VOT, which was equally well represented in core and belt areas of auditory cortex in humans (Steinschneider et al.1999). Human scalp recorded N100 EPs in response to a /da/-/ta/ continuum (Sharma and Dorman 1999) and a /ga/-/ka/ and /ba/-/pa/ continuum (Sharma et al. 2000) clearly suggested that the double-on response can be recorded when VOT values exceed 30 ms. However, the occurrence of this double-on response is not related to the categorical perception boundary, which differed for the three consonant–vowel continua used. This is not surprising as the same VOT threshold is also found in chinchillas (Kuhl and Miller 1975), cats (Eggermont 1995, 1998), and monkeys (Steinschneider et al. 1982; 2005).

6.3 Pitch and Melody

Pitch of the (missing) fundamental may be processed in the medial part of HG as a MEG study by Krumbholtz et al. (2003) claims. However, this may not be the end point in a putative hierarchy of pitch and melody perception. Patterson et al. (2002) in an fMRI study suggested that the lateral half of HG is particularly activated by pitch, which overlaps with the medial part identified by Krumbholtz et al. (2003), but when pitch changed to melody specific activations in STS and planum polare (PP) were found. Human imaging studies by Penagos et al. (2004), Schneider et al. (2005) and Puschmann et al. (2010) also provided supporting evidence for a pitch-processing area in lateral HG. Other cortical areas may also be involved in pitch processing (Hall and Plack 2009).

Pitch of the missing fundamental has been claimed to be represented as a map, in the same direction as the tonotopic map, in AI on the basis of MEG recordings and equivalent dipole reconstruction (Pantev et al. 1989). However, detailed recordings in awake monkey AI have failed to find any representation of the missing fundamental along the direction of the tonotopic gradient (Fishman et al., 1998). Langner et al. (1997), also on basis of MEG recordings, claimed that the periodotopic map of the fundamental frequency of harmonic complexes in human auditory cortex was orthogonal to the tonotopic map. There are inherently problems with localization of source dipoles (Luetkenhoner et al. 2003) based on MEG (and even more when based on EEG), so these claims have to be corroborated by more direct imaging methods before hard conclusions can be drawn. Nevertheless, the result seems to be compatible with a periodotopic map found by Schulze et al. (2002) in the gerbil auditory cortex. Studies by Bendor and Wang (2005, 2010) identified a region of auditory cortex near the low-frequency border of fields AI and R where a large proportion of low-frequency tuning neurons were found to respond selectively to missing fundamental pitch. This cortical area corresponds to the lateral HG evoked by pitch stimuli identified in several human imaging studies (Bendor and Wang 2006).

7 Synaptic Mechanisms and Modifiability of Temporal Coding

7.1 Role of Synaptic Depression, Integration and Inhibition

What determines the upper limit of envelope locking in a neuron? At the auditory nerve-fiber level it is the combination of frequency-tuning curve bandwidth and the low-pass properties of the hair-cell neuron synapse (Joris et al. 2004). If the lower or upper side-bands of sinusoidal AM sounds fall outside the frequency-tuning curve there is no phase locking of the firings. For high CFs, the width of the tuning curves is sufficiently large that the synaptic low-pass filter limits the envelope-locking capacity of the neurons.

At more central levels, the limiting factor could be the cumulative effect of more intervening synapses albeit that longer membrane time constants and other intrinsic properties of the neurons come into the picture as well (Carandini et al. 1996). At thalamic and cortical levels after-hyperpolarization (Eggermont 2000) or other forms of post-activation suppression (Brosch and Schreiner 2000) as well as synaptic depression (Tsodyks and Markram 1997; Eggermont 1999) start playing a larger role. Thalamocortical network properties may also interfere, for instance, in anesthetized animals the spindle oscillations effectively determine the BMF (Eggermont 1992; Horikawa et al. 1994; Kenmochi and Eggermont 1997) and affect the upper limit of phase locking. The depth of amplitude modulation also plays a role, since it affects the strength of the post-activation suppression. Low modulation depths result in lower VS (Fastl et al. 1986; Eggermont 1994a) because of the less punctuate response to such stimuli. Lower modulation depths do produce higher limiting rates (Middlebrooks, personal communication), but there is no effect on BMF (Eggermont 1994b). A clear correlation between the upper limit of envelope-locking and the minimum latency of AI neurons can be demonstrated (e.g., Fig. 14 in Liang et al. 2002; Schreiner and Raggio 1996). Neurons with shorter minimum latencies tend to have higher upper limit of envelop locking, which suggests that it is the temporal integration window of cortical neurons that ultimately constrain the upper limit of envelop locking. Wang et al. (2003) suggest that cortical processing of sound streams operates on a "segment-by-segment" basis with a temporal integration window on the order of 20–30 ms in AI. Bendor and Wang (2008) showed that the temporal integration window is longer in cortical fields R and RT in marmosets and proposed a temporal processing pathway in primate auditory cortex along the rostral direction (Fig. 14.6).

Other mechanisms that could shape temporal processing in auditory cortex include forward masking (Brosch and Schreiner 1997, 1999, 2000; Brosch et al. 1999). In AI of awake animals, it has been shown that a stimulus could generate long-lasting facilitatory or inhibitory influences succeeding stimuli for hundreds of milliseconds or even over 1 s (Bartlett and Wang 2005).

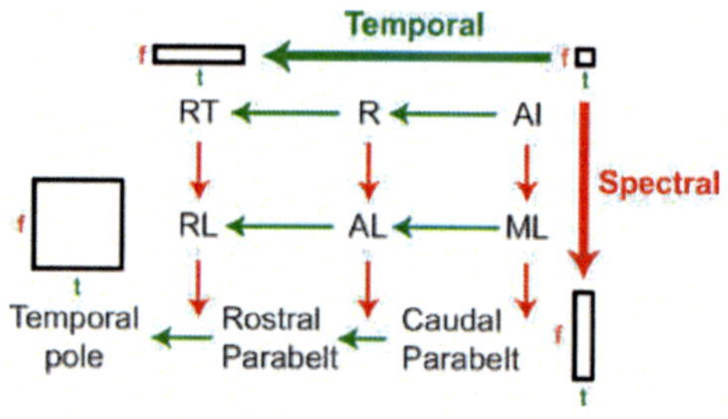

Fig. 14.6 A model of spectral and temporal processing pathways in primate auditory cortex. In the proposed model, the temporal processing pathway is in the caudal-to-rostral axis, where AI has the smallest temporal integration window and this temporal integration window increases in R and RT. The spectral processing pathway is in the medial-to-lateral axis, where AI and the other core fields have the smallest spectral integration window, and belt and parabelt areas have larger spectral integration windows. AI, primary auditory cortex; R, rostral field; RT, rostrotemporal field; ML, middle lateral field; AL, anterolateral field; RL, rostrolateral field; *f*, frequency, *t*, time. Adapted from Bendor and Wang (2008)

7.2 Modeling Temporal Processing

A model describing the dependence on modulation frequency for temporal processing based on a combination of synaptic depression and post-activation suppression was introduced 30 years ago to account for adaptation in auditory nerve fibers (Eggermont 1975, 1985) and recently modified and extended to account for temporal processing in cortical neurons (Eggermont 1999, 2000). This model is similar in spirit to that proposed by Markram and Tsodyks (1996).

The model results relate to those for the low-pass synaptic filter is described in Eggermont (2002). This low-pass filter had a high-frequency slope of –6 dB/oct, i.e., a first-order low-pass filter and a cut-off frequency of around 14 Hz. The model to fit the tMTF was based on a first-order differential equation describing the click-rate or modulation frequency dependent release of transmitter in the thalamo-cortical synapse (Eggermont 1999). It resulted in an exponentially decaying amount of inactivated immediate-release sites that after several modulation periods settled at its MF-dependent steady-state value. These steady-state values were used to generate a theoretical tMTF and could be fitted to become nearly identical to the experimentally observed ones. The model produced an exponentially decaying impulse response suggesting, under the assumption of linearity, a first-order low-pass filter. The estimated time constant τ for this impulse response was about 20 ms, indicating a high cut-off frequency $f_c = (2\pi\tau)^{-1}$ of 8 Hz about an octave lower than the results in Eggermont (2002). The problem with the model was that it needed some facilitation to account for the shape of the tMTF for the units with high firing rates (and long group delays). Facilitation does occur in auditory cortex but only in those neurons that have a low probability of transmitter release (Atzori et al. 2001) and those are likely not the ones that produce high onset-firing rates.

An alternative version of this model (Eggermont, unpublished data) suggests the use a stimulus repetition rate-dependent recovery from synaptic depression as suggested previously by Wang and Kaczmarek (1998). This use-dependent recovery model was able to produce a good fit to the PSTHs for MFs between 2 and 64 Hz and resulted in a time constant for the impulse response that was proportional to the modulation frequency $\tau_{0*}(\mathrm{MF})^{-0.5}$. Here τ_0 was equal to 0.145 s and represents the largest time constant for the impulse response (or recovery from previous stimulation) that can account for the findings at MF = 1 Hz. Because the model assumes that increased repetition rates produce higher release of transmitter, this model would only apply to units with high firing rates. These were the ones with large group delays and required the extra amplifying filter.

In the light of potentially use-dependent recovery from depression one can also describe the differences between the two groups as a result of the changing impulse response time-constant. The impulse response time constants needed to describe the neurons with high limiting rates and without assuming facilitation was 10 ms. This corresponded to a cut-off frequency for the synaptic filter of 16 Hz in good agreement with the findings in Eggermont (2002). The model fits suggest that the tMTFs for high firing rate neurons in cat AI show evidence of use-dependent recovery from synaptic depression.

An important part of the model, in addition to the low-pass synaptic filter, was the incorporation of a post-activation suppression (Eggermont 2000). This suppression needed a recovery time constant of about 55 ms and could be produced by either after-hyperpolarization or feed-forward inhibition. This suppression was needed to describe the relative larger reduction, or even absence, of the response to the second click or modulation period relative to responses to subsequent clicks or modulation periods. The model is simple, as it incorporates only an MF-dependent recovery time constant and an exponential recovery from post-activation suppression. However, it does not describe all intricacies found in the response obtained in ketamine-anesthetized cats such as the strong rebound responses after the initial post-activation suppression for lower MFs and their effect on the observed results. It is possible that in awake animals the response pattern is simpler because of the absence of strong hyperpolarizations and subsequent rebound responses (Wang et al. 2005; Sadagopan and Wang 2010).

7.3 Effects of Basal Forebrain Stimulation and Learning

There is a great deal of learning involved in the acquisition of new songs in certain passerine birds (Doupe and Kuhl 1999), and is sometimes accompanied by growth of new neurons (Goldman and Nottebohm 1983). Is learning also involved in the perception of the temporal aspects of sound in mammals, and if so does this extent beyond the short maturation period? What does it mean that we can change the representation of temporal aspects of sound in cortex? Is this of survival value? Pairing basal forebrain stimulation with periodic click stimuli with a rate above the BMF of control animals showed an increase in overall firing rate for repetition rates above the BMF, such that the low-frequency slope of the rMTF was unaffected (Kilgard et al. 2001). As a result the BMF and limiting rate shifted to higher repetition rates, not unlike the type of change described for units with short and long group delays (Eggermont 1999). In fact, the increased synchronization to higher click-repetition rates induced by basal forebrain stimulation paired with high click rate stimuli mimicked the expected increase in the VS in units with higher firing rates than normally observed in anesthetized animals. So do we have plasticity here in terms of changing (use-dependent) synaptic depression or just effects of increases in firing rate (e.g., by lower thresholds or higher polarization levels)? Are these plastic changes permanent after conditioning or learning?

7.4 Cortical versus Subcortical Contributions

Human imaging studies emphasize the role of "higher" auditory cortical areas, i.e., those equivalent to the belt and parabelt regions in monkeys, in the representation of speech and music (Griffiths et al. 1998; Zatorre and Belin 2001; Scott and Johnsrude 2003). These areas are furthermore not limited to posterior cortical areas but also feature the frontal supplementary motor areas, cingulate cortex, and others. One may wonder if cortex is needed to detect or discriminate species-specific vocalizations. A study by Liegeois-Chauvel et al. (1998), in 65 patients who underwent unilateral temporal lobe resection for incurable epilepsy, showed that discrimination of melodies based on interval information was affected by surgical removal of the posterior part of the STG on either side, but that melody discrimination based on contour information was only affected by removal of the posterior part of the STG in the right hemisphere. A review by Phillips (1993) suggests that vowel discrimination is preserved but consonant–vowel discrimination that relies on VOT is not following cortical lesions. Are there subcortical representations that can be accessed?

8 Future Perspectives

It is clear that at present only a limited picture can be drawn of the representation of the temporal structure of sound in auditory cortex. There is a reasonable coverage of core cortical areas in a representative number of species both in awake and anesthetized states. There is some information on the more limited capacity to code for temporal aspects in belt-like areas such as AII and PAF in the cat but for PAF, this is based largely on rate of change sensitivity for FM stimuli. More data of temporal processing are needed in awake animals especially from non-core auditory cortical areas.

A number of critical questions remain largely unexplored. Are "higher" cortical areas specialized to process behaviorally meaningful temporal aspects of sound? What do tMTFs tell us about the coding of species-specific vocalizations and the processing of behaviorally relevant aspects of the acoustic biotope?

Why do some studies find band-pass rMTFs and others don't? This relates to whether only spike count matters in the encoding of natural sounds regardless of phase locking to the modulation period.

The relationships between onset firing rate, minimum latency, group delay, stimulus type, and stimulus level on the tMTF properties have to be further evaluated. The role of first-spike latency in coding for MF needs to be explored.

Are tMTFs, obtained from periodic stimuli that last at most 1 s and that are separated from each other by several seconds, representative of those that would represent more steady-state conditions (e.g., for Poisson-distributed click trains or low-frequency noise-modulated sound) or of those short-duration modulated stimuli embedded in an acoustic biotope of competing sounds?

Is coding of temporal sound properties cortical-layer dependent? One could expect higher limiting rates in input layers and lower in supragranular layers. However, evidence is still lacking, especially in awake animals.

References

Ahissar E and Arieli A (2001) Figuring space by time. *Neuron* **32**: 185–201.

Akeroyd MA and Patterson RD (1995) Discrimination of wideband noises modulated by a temporally asymmetric function. *Journal of the Acoustical Society of America* **98**:2466–2474.

Atzori M, Lei S, Evans DI, Kanold PO, Phillips-Tansey E, McIntyre O, and McBain CJ (2001) Differential synaptic processing separates stationary from transient inputs to the auditory cortex. *Nature Neuroscience* **4**:1230–1237.

Ball GF and Hulse SH (1998) Birdsong. *American Psychologist* **53**: 37–58

Bartlett E L and Wang X (2005) Long-lasting modulation by stimulus context in primate auditory cortex. *Journal of Neurophysiology* **94**:83–104.

Bartlett EL and Wang X (2007) Neural representations of temporally-modulated signals in the auditory thalamus of awake primates. *Journal of Neurophysiology* **97**:1005–1017.

Belin P, Zatorre RJ, Lafaille P, Ahad P, and Pike B (2000) Voice-selective areas in human auditory cortex. *Nature* **403**:309–312.

Bendor D and Wang X (2005) The neuronal representation of pitch in primate auditory cortex. *Nature* **436**:1161–1165.

Bendor D and Wang X (2006) Cortical representations of pitch in monkeys and humans. *Current Opinion in Neurobiology* **16**:391–399.

Bendor DA and Wang X (2007) Differential neural coding of acoustic flutter within primate auditory cortex. *Nature Neuroscience* **10**: 763–771.

Bendor D and Wang X (2008) Neural response properties of primary, rostral, and rostrotemporal core fields in the auditory cortex of marmoset monkeys. *Journal of Neurophysiology* **100**: 888–906.

Bendor D and Wang X (2010) Neural coding of periodicity in marmoset auditory cortex. *Journal of Neurophysiology* **103**:1809–1822.

Besser GM (1967) Some physiological characteristics of auditory flutter fusion in man. *Nature* **214**:17–19.

Bieser A (1998) Processing of twitter-call fundamental frequencies in insula and auditory cortex of squirrel monkeys. *Experimental Brain Research* **122**:139–148.

Bieser A and Mueller-Preuss P (1996) Auditory responsive cortex in the squirrel monkey: neural responses to amplitude-modulated sounds. *Experimental Brain Research* **108**:273–284.

Binder JR, Frost JA, Hammeke TA, Bellgowan PS, Springer JA, Kaufman JN, and Possing ET (2000) Human temporal lobe

activation by speech and nonspeech sounds. *Cerebral Cortex* **10**: 512–528.
Borst A and Theunissen FE (1999) Information theory and neural coding. *Nature Neuroscience* **2**:947–957.
Brosch M and Schreiner CE (1999) Correlations between neural discharges are related to receptive field properties in cat primary auditory cortex. *European Journal of Neuroscience* **11**:3517–3530.
Brosch M and Schreiner CE (1997) Time course of forward masking tuning curves in cat primary auditory cortex. *Journal of Neurophysiology* **77**:923–943.
Brosch M and Schreiner CE (2000) Sequence sensitivity of neurons in cat primary auditory cortex. *Cerebral Cortex* **10**:1155–1167.
Brosch M, Schulz A, and Scheich H (1999) Processing of sound sequences in macaque auditory cortex: response enhancement. *Journal of Neurophysiology* **82**:1542–1559.
Brown KA, Buchwald JS, Johnson JR, and Mikolich DJ (1978) Vocalization in the cat and kitten. *Developmental Psychobiology* **11**:559–570.
Brugge JF, Nourski KV, Oya H, Reale RA, Kawasaki H, Steinschneider M, and Howard MA 3rd (2009) Coding of repetitive transients by auditory cortex on Heschl's gyrus. *Journal of Neurophysiology* **102**:2358–2374
Buchhfellner E, Leppelsack H-J, Klum GM, and Häusler U (1989) Gap detection in the starling (Sturnus vulgaris). *Journal of Comparative Physiology A* **164**:539–549.
Carandini M, Mechler F, Leonard CS, and Movshon JA (1996) Spike train encoding by regular-spiking cells of the visual cortex. *Journal of Neurophysiology* **76**:3425–3441.
Cariani PA and Delgutte B (1996) Neural correlates of the pitch of complex tones. I. Pitch and pitch salience. *Journal of Neurophysiology*76:1698–1716.
DiMattina C and Wang X (2006) Virtual vocalization stimuli for investigating neural representations of species-specific vocalizations. *Journal of Neurophysiology* **95**:1244–1262.
Doupe AJ and Kuhl PK (1999) Birdsong and human speech: common themes and mechanisms. *Annual Reviews of Neuroscience* **122**: 567–631.
Eggermont JJ (1975) Cochlear adaptation: A theoretical description. *Biological Cybernetics* **19**:181–190.
Eggermont JJ (1985) Peripheral auditory adaptation and fatigue: a model oriented review. *Hearing Research* **18**:57–71
Eggermont JJ (1991) Rate and synchronization measures of periodicity coding in cat primary auditory cortex. *Hearing Research* **56**: 153–167.
Eggermont JJ (1992) Neural interaction in cat primary auditory cortex. Dependence on recording depth, electrode separation and age. *Journal of Neurophysiology* **68**:1216–1228.
Eggermont JJ (1994a) Neural interaction in cat primary auditory cortex II. Effects of sound stimulation. *Journal of Neurophysiology* **71**:246–270.
Eggermont JJ (1994b) Temporal modulation transfer functions for AM and FM stiumuli in cat auditory cortex. Effects of carrier type, modulating waveform and intensity. *Hearing Research* **74**: 51–66.
Eggermont JJ (1995) Representation of a voice onset time continuum in primary auditory cortex of the cat. *Journal of the Acoustical Society of America* **98**:911–920.
Eggermont JJ (1996) How homogeneous is cat primary auditory cortex? Evidence from simultaneous single-unit recordings. *Auditory Neuroscience* **2**:76–96.
Eggermont JJ (1998) Representation of spectral and temporal sound features in three cortical fields of the cat. Similarities outweigh differences. *Journal of Neurophysiology* **80**:2743–2764.
Eggermont JJ (1999a) The magnitude and phase of temporal modulation transfer functions in cat auditory cortex. *Journal of Neuroscience* **19**:2780–2788.
Eggermont JJ (1999b) Neural correlates of gap detection in three auditory cortical fields in the cat. *Journal of Neurophysiology* **81**:2570–2581.
Eggermont JJ (2000) Neural responses in primary auditory cortex mimic psychophysical, across frequency-channel, gap-detection thresholds. *Journal of Neurophysiology* **84**:1453–1463.
Eggermont JJ (2001) Between sound and perception: reviewing the search for a neural code. *Hearing Research* **157**:1–42.
Eggermont JJ (2002) Temporal modulation transfer functions in cat primary auditory cortex: separating stimulus effects from neural mechanisms. *Journal of Neurophysiology* **87**:305–321.
Eggermont JJ and Smith GM (1995) Synchrony between single-unit activity and local field potentials in relation to periodicity coding in primary auditory cortex. *Journal of Neurophysiology* **73**: 227–245.
Ehret G (1987) Left hemisphere advantage in the mouse brain for recognizing ultrasonic communication calls. *Nature* **325**: 249–251.
Epping WJM and Eggermont JJ (1986a) Sensitivity of neurons in the auditory midbrain of the grassfrog to temporal characteristics of sound. I. Stimulation with acoustic clicks. *Hearing Research* **24**:37–54.
Epping WJM and Eggermont JJ (1986b) Sensitivity of neurons in the auditory midbrain of the grassfrog to temporal characteristics of sound. II. Stimulation with amplitude modulated sound. *Hearing Research* **24**:55–72.
Fastl H, Hesse A, Schorer E, Urbas J, and Mueller-Preuss P (1986) Searching for neural correlates of the hearing sensation fluctuation strength in the auditory cortex of squirrel monkeys. *Hearing Research* **23**:199–203.
Fishman YI, Reser DH, Arewzzo JC, and Steinschneider M (1998) Pitch vs. spectral encoding of harmonic complexes in primary auditory cortex of the awake monkey. *Brain Research* **786**:18–30.
Fishman YI, Reser DH, Arezzo JC, and Steinschneider M (2000a) Complex tone processing in primary auditory cortex of the awake monkey. I. Neural ensemble correlates of roughness. *Journal of the Acoustical Society of America* **108**:235–246.
Fishman YI, Reser DH, Arezzo JC, and Steinschneider M (2000b) Complex tone processing in primary auditory cortex of the awake monkey. II. Pitch versus critical band representation. *Journal of the Acoustical Society of America* **108**:247–262.
Fisher NI (1993) *Statistical analysis of circular data.* Cambridge University Press, Cambridge.
Fitch RH, Brown CP, O'Connor K, and Tallal P (1993) Functional lateralization for auditory temporal processing in male and female rats. *Behavoral Neuroscience* **107**:844–850.
Gehr DD, Komiya H, and Eggermont JJ (2000) Neuronal responses of cat primary auditory cortex to natural and altered species-specific calls. *Hearing Research* **150**: 27–42.
Glass I and Wollberg Z (1979) Lability in the responses of cells in the auditory cortex of squirrel monkeys to species-specific vocalizations. *Experimental Brain Research* **34**:489–498.
Giraud AL, Lorenzi C, Ashburner J, Wable J, Johnsrude I, Frackowiak R, and Kleinschmidt A (2000) Representation of the temporal envelope of sounds in the human brain. *Journal of Neurophysiology* **84**:1588–1598.
Goldberg JM and Brown PB (1968) Functional organization of the dog superior olivary complex: an anatomical and electrophysiological study. *Journal of Neurophysiology* **31**:639–656.
Goldman SA and Nottebohm F (1983) Neuronal production, migration, and differentiation in a vocal control nucleus of the adult female canary brain. *Proceedings of the National Academy of Sciences of the United States of America* **80**:2390–2394.
Goldstein MH, Kiang NYS, and Brown RM (1959) Responses of the auditory cortex to repetitive acoustic stimuli. *Journal of the Acoustical Society of America* **31**:356–364.

Gouzoules S, Gouzoules H, and Marler P (1984) Rhesus monkey (*Macaca mulatta*) screams: representational signaling in the recruitment of agonistic aid. *Animal Behaviour* **32**:182–193.

Griffiths TD, Büchel C, Frackowiak RS, and Patterson RD (1998) Analysis of temporal structure in sound by the human brain. *Nature Neuroscience* **1**:422–427.

Griffiths R, Johnsrude I, Dean J, and Green G (1999) A common neural substrate for the analysis of pitch and duration pattern in segmented sound? *NeuroReport* **10**:3825–3830.

Griffiths TD, Uppenkamp S, Johnsrude I, Josephs O, and Patterson RD (2001) Encoding of the temporal regularity of sound in the human brainstem. *Nature Neuroscience* **4**:633–637.

Hall DA, Johnsrude IS, Haggard MP, Palmer AR, Akeroyd MA, and Summerfield AQ (2002) Spectral and temporal processing in human auditory cortex. *Cerebral Cortex* **12**:140–149.

Hall DA and Plack CJ (2009) Pitch processing sites in the human auditory brain. *Cerebral Cortex* **19**:576–85

Harms MP and Melcher JR (2002) Sound repetition rate in the human auditory pathway: representations in the waveshape and amplitude of fMRI activation. *Journal of Neurophysiology* **88**:1433–1450.

Hart HC, Palmer AR, and Hall DA (2003) Amplitude and frequency-modulated stimuli activate common regions of human auditory cortex. *Cerebral Cortex* **13**:773–81.

Hauser MD (1996) *The Evolution of Communication*, MIT Press, Cambridge.

Heil P and Irvine DR (1998a) Functional specialization in auditory cortex: responses to frequency-modulated stimuli in the cat's posterior auditory field. *Journal of Neurophysiology* **79**:3041–3059.

Heil P and Irvine DR (1998b) The posterior field P of cat auditory cortex: coding of envelope transients. *Cerebral Cortex* **8**:125–141.

Heil P, Rajan R, and Irvine D (1992). Sensitivity of neurons in cat primary auditory cortex to tones and frequency-modulated stimuli. II: organization of response properties along the 'isofrequency' dimension. *Hearing Research* **63**:135–156.

Holy TE, Guo Z (2005) Ultrasonic songs of male mice. *PLoS Biology* **3**:e386.

Horikawa J, Tanahashi A, and Suga N (1994) After-discharges in the auditory cortex of the mustached bat: no oscillatory discharges for binding auditory information. *Hearing Research* **76**:45–52.

Johnson DH (1980) The relationship between spike rate and synchrony in responses of auditory-nerve fibers to single tones. *Journal of the Acoustical Society of America* **68**:1115–1122.

Johnsrude IS, Penhune VB, and Zatorre RJ (2000) Functional specificity in the right human auditory cortex for perceiving pitch direction. *Brain* **123**:155–163.

Johnsrude IS, Zatorre RJ, Milner BA, and Evans AC (1997) Left-hemisphere specialization for the processing of acoustic transients. *NeuroReport* **8**:1761–1765

Joris PX, Schreiner CE, and Rees A (2004) Neural processing of amplitude-modulated sounds. *Physiological Reviews* **84**:541–577.

Joris PX and Yin T (1998) Envelope coding in the lateral superior olive. III. Comparison with afferent pathways. *Journal of Neurophysiology* **79**:253–269.

Kanwal JS, Matsumura S, Ohlemiller K, and Suga N (1994). Analysis of acoustic elements and syntax in communication sounds emitted by mustached bats. *Journal of the Acoustical Society of America* **96**:1229–1254.

Kenmochi M and Eggermont JJ (1997) Autonomous cortical rhythms affect temporal modulation transfer functions. *NeuroReport* **8**: 1589–1593.

Kaernbach C and Demany L (1998) Psychophysical evidence against the autocorrelation theory of auditory temporal processing. *Journal of the Acoustical Society of America* **104**: 2298–2306.

Kilgard MP, Pandya PK, Vazquez J, Gehi A, Schreiner CE, and Merzenich MM (2001) Sensory input directs spatial and temporal plasticity in primary auditory cortex. *Journal of Neurophysiology* **86**:326–338.

Krumbholz K, Patterson RD, Seither-Preisler A, Lammertmann C and Lütkenhöner B (2003) Neuromagnetic evidence for a pitch processing center in Heschl's gyrus. *Cerebral Cortex* **13**:765–772.

Krumbholz K, Patterson RD, and Pressnitzer D (2000) The lower limit of pitch as determined by rate discrimination. *Journal of the Acoustical Society of America* **108**:1170–1180.

Kuhl PK and Miller JD (1975) Speech perception by the chinchilla: voiced-voiceless distinction in alveolar plosive consonants. *Science* **190**:69–72.

Moelk M (1944) Vocalization in the house cat; a phonetic and functional study. *Journal of Psychology* **57**:184–205.

Langner G (1992) Periodicity coding in the auditory system. *Hearing Research* **60**:115–142.

Langner G, Sams M, Heil P, and Schulze H (1997) Frequency and periodicity are represented in orthogonal maps in the human auditory cortex: evidence from magnetoencephalography. *Journal of Comparative Physiology A* **181**:665–676.

Langner G and Schreiner CE (1988) Periodicity coding in the inferior colliculus of the cat. I. Neuronal mechanisms. *Journal of Neurophysiology* **60**:1799–1822.

Liang L, Lu T, and Wang X (2002) Neural representations of sinusoidal amplitude and frequency modulations in the primary auditory cortex of awake primates. *Journal of Neurophysiology* **87**:2237–2261.

Liegeois-Chauvel C, de Graaf JB, Laguitton V, and Chauvel P (1999) Specialization of left auditory cortex for speech perception in man depends on temporal coding. *Cerebral Cortex* **9**:484–496.

Liegeois-Chauvel C, Lorenzi C, Trebuchon A, Regis J, and Chauvel P (2004) Temporal envelope processing in the human left and right auditory cortices. *Cerebral Cortex* **14**:731–740.

Liegeois-Chauvel C, Peretz I, Babai M, Laguitton V, and Chauvel P (1998) Contribution of different cortical areas in the temporal lobes to music processing. *Brain* **121**:1853–1867.

Liu RC, Miller KD, Merzenich MM, and Schreiner CE (2003) Acoustic variability and distinguishability among mouse ultrasound vocalizations. *Journal of the Acoustical Society of America* **114**:3412–322.

Lu T and Wang X (2000) Temporal discharge patterns evoked by rapid sequences of wide- and narrowband clicks in the primary auditory cortex of cat. *Journal of Neurophysiology* **84**:236–246.

Lu T and Wang X (2004) Information content of auditory cortical responses to time-varying acoustic stimuli. *Journal of Neurophysiology* **91**:301–313.

Lu T, Liang L, and Wang X (2001a) Neural representations of temporally asymmetric stimuli in the auditory cortex of awake primates. *Journal of Neurophysiology* **85**:2364–2380.

Lu T, Liang L, and Wang X (2001b) Temporal and rate representations of time-varying signals in the auditory cortex of awake primates. *Nature Neuroscience* **4**:1131–1138.

Luetkenhoner B, Krumbholz K, and Seither-Preisler A (2003) Studies of tonotopy based on wave N100 of the auditory evoked field are problematic. *Neuroimage* **19**:935–49.

Makela JP, Hari R, and Linnankivi A (1987). Different analysis of frequency and amplitude modulations of a continuous tone in the human auditory cortex: a neuromagnetic study. *Hearing Research* **27**:257–264.

Malone BJ, Scott BH, and Semple MN (2007) Dynamic amplitude coding in the auditory cortex of awake rhesus macaques. *Journal of Neurophysiology* **98**:1451–1474.

Mardia KV (1972) *Statistics of directional data*. Academic Press, London.

Menon V, Levitin DJ, Smith BK, Lembke A, Krasnow BD, Glazer D, Glover GH, and McAdams S (2002) Neural correlates of timbre change in harmonic sounds. *Neuroimage* **17**:1742–1754.

Miller GA and Taylor WG (1948) The perception of repeated bursts of noise. *Journal of the Acoustical Society of America* **20**:171–182.

Miller LM, Escabi MA, Read HL, and Schreiner CE (2001) Functional convergence of response properties in the auditory thalamocortical system. *Neuron* **32**:151–160.

Møller AR (1972). Coding of amplitude and frequency modulated sounds in the cochlear nucleus of the rat. *Acta Physiologica Scandinavica* **86**:223–238.

Møller AR (1973) Statistical evaluation of the dynamic properties of cochlear nucleus units using stimuli modulated with pseudorandom noise. *Brain Research* **57**:443–456.

Moody MI and Menzel EW Jr (1976) Vocalizations and their behavioral contexts in the tamarin Saguinus fuscicollis. *Folia Primatologica* **25**:73–94.

Mueller-Preuss P, Flachskamm C, and Bieser A (1994) Neural encoding of amplitude modulation within the auditory midbrain of squirrel monkeys. *Hearing Research* **80**:197–208.

Nagarajan SS, Cheung SW, Bedenbaugh P, Beitel RE, Schreiner CE, and Merzenich MM (2002) Representation of spectral and temporal envelope of twitter vocalizations in common marmoset primary auditory cortex. *Journal of Neurophysiology* **87**:1723–1737.

Nakahara H, Zhang LI, and Merzenich MM (2004) Specialization of primary auditory cortex processing by sound exposure in the "critical period". *Proceedings of the National Academy of Sciences of the United States of America* **101**:7170–7174.

Noreña A and Eggermont JJ (2002) Comparison between local field potentials and unit cluster activity in primary auditory cortex and anterior auditory field in the cat. *Hearing Research* **166**:202–213.

Nourski KV, Reale RA, Oya H, Kawasaki H, Kovach CK, Chen H, Howard MA 3rd, and Brugge JF (2009) Temporal envelope of time-compressed speech represented in the human auditory cortex. *Journal of Neuroscience* **29**:15564–15574.

Patterson RD, Uppenkamp S, Johnsrude IS, and Griffiths TD (2002) The processing of temporal pitch and melody information in auditory cortex. *Neuron* **36**:767–776.

Pantev C, Hoke M, Luetkenhohner B, and Lehnertz K (1989) Tonotopic organization and the representation of virtual pitch in the human auditory cortex. *Science* **246**:486–488.

Penagos H, Melcher JR, and Oxenham AJ (2004) A neural representation of pitch salience in nonprimary human auditory cortex revealed with functional magnetic resonance imaging. *Journal of Neuroscience* **24**:6810–6815.

Peretz I, Ayotte J, Zatorre RJ, Mehler J, Ahad P, Penhune VB, and Jutras B (2002) Congenital amusia: a disorder of fine-grained pitch discrimination. *Neuron* **33**:185–191.

Peretz I and Zatorre RJ (2005) Brain Organization for Music Processing. *Annual Reviews of Psychology* **56**:89–114.

Phillips DP (1993) Representation of acoustic events in the primary auditory cortex. *Journal of Experimental Psychology: Human Perception and Performance* **19**:203–216.

Phillips DP, Taylor TL, Hall SE, Carr MM, and Mossop JE (1997) Detection of silent intervals between noises activating different perceptual channels: some properties of "central" auditory gap detection. *Journal of the Acoustical Society of America* **101**: 3694–3705.

Pienkowski M, Shaw G, and Eggermont JJ (2009) Wiener-Volterra Characterization of Neurons in Primary Auditory Cortex Using Poisson-Distributed Impulse Train Inputs. *Journal of Neurophysiology* **101**:3031–3041.

Pistorio AL, Vintch B, and Wang X (2006) Acoustical analysis of vocal development in a New World primate, the common marmoset (*Callithrix jacchus*). *Journal of the Acoustical Society of America* **120**:1655–1670.

Puschmann S, Uppenkamp S, Kollmeier B, and Thiel CM (2010) Dichotic pitch activates pitch processing centre in Heschl's gyrus. *Neuroimage* **49**:1641–1649.

Ribaupierre F de, Goldstein MH Jr, and Yeni-Komshian G (1972) Cortical coding of repetitive acoustic pulses. *Brain Research* **48**:205–225.

Romo R, Hernández A, Zainos A, and Salinas E (2003) Correlated neuronal discharges that increase coding efficiency during perceptual discrimination. *Neuron* **38**:649–657.

Sadagopan S and Wang X (2010) Contribution of inhibition to stimulus selectivity in primary auditory cortex of awake primates. *Journal of Neuroscience* **30**:7314–7325.

Schneider P, Sluming V, Roberts N, Scherg M, Goebel R, Specht HJ, Dosch HG, Bleeck S, Stippich C, and Rupp A (2005) Structural and functional asymmetry of lateral Heschl's gyrus reflects pitch perception preference. *Nature Neuroscience* **8**: 1241–1247.

Schreiner CE and Langner G (1988) Periodicity coding in the inferior colliculus of the cat. II. Topographical organization. *Journal of Neurophysiology* **60**:1823–1840.

Schreiner CE and Langner G (1997) Laminar fine structure of frequency organization in auditory midbrain. *Nature* **388**:383–386.

Schreiner CE and Raggio MW (1996) Neuronal responses in cat primary auditory cortex to electrical cochlear stimulation. II. Repetition rate coding. *Journal of Neurophysiology* **75**: 1283–1300.

Schreiner CE and Urbas JV (1988) Representation of amplitude modulation in the auditory cortex of the cat. II. Comparison between cortical fields. *Hearing Research* **32**:49–63.

Schulze H and Langner G (1997a) Representation of periodicity pitch in the primary auditory cortex of the Mongolian gerbil. *Acta Otolaryngologica Supplement* **532**:89–95.

Schulze H and Langner G (1997b) Periodicity coding in the primary auditory cortex of the Mongolian gerbil (*Meriones unguiculatus*): two different coding strategies for pitch and rhythm? *Journal of Comparative Physiology A* **181**:651–663.

Schulze H and Langner G (1999) Auditory cortical responses to amplitude modulations with spectra above frequency receptive fields: evidence for wide spectral integration. *Journal of Comparative Physiology A* **185**:493–508.

Schulze H, Hess A, Ohl FW, and Scheich H (2002) Superposition of horseshoe-like periodicity and linear tonotopic maps in auditory cortex of the Mongolian gerbil. *European Journal of Neuroscience* **15**:1077–1084.

Scott SK and Johnsrude IS (2003) The neuroanatomical and functional organization of speech perception. *Trends in Neurosciences* **26**: 100–107.

Sharma A and Dorman MF (1999) Cortical auditory evoked potential correlates of categorical perception of voice-onset time. *Journal of the Acoustical Society of America* **106**:1078–1083.

Sharma A, Marsh CM, and Dorman MF (2000). Relationship between N1 evoked potential morphology and the perception of voicing. *Journal of the Acoustical Society of America* **108**:3030–3035.

Singh NC and Theunissen FE (2003) Modulation spectra of natural sounds and ethological theories of auditory processing. *Journal of the Acoustical Society of America* **114**:3394–3411.

Sovijärvi ARA (1975) Detection of natural complex sounds by cells in the primary auditory cortex of the cat. *Acta Physiologica Scandinavica* **93**:318–335.

Steinschneider M, Arezzo J, and Vaughan HG Jr (1980) Phase-locked cortical responses to a human speech sound and low-frequency tones in the monkey. *Brain Research* **198**:75–84.

Steinschneider M, Arezzo JC, and Vaughan HG (1982) Speech evoked activity in the auditory radiations and cortex of awake monkey. *Brain Research* **252**:353–365.

Steinschneider M, Arezzo JC, and Vaughan HG (1990) Tonotopic features of speech-evoked activity in primate auditory cortex. *Brain Research* **519**:158–168.

Steinschneider M, Arezzo JC, Schroeder CE, and Vaughan HG (1994) Speech-evoked activity in primary auditory cortex: effects of voice-onset time. *Electroencephalography and Clinical Neurophysiology* **92**:30–43.

Steinschneider M, Fishman YI, and Arezzo JC (2003) Representation of the voice onset time (VOT) speech parameter in population responses within primary auditory cortex of the awake monkey. *Journal of the Acoustical Society of America* **114**:307–321.

Steinschneider M, Schroeder CE, Arezzo JC, and Vaughan HG Jr (1994) Speech-evoked activity in primary auditory cortex: effects of voice onset time. *Electroencephalography and Clinical Neurophysiology* **92**:30–43.

Steinschneider M, Reser DH, Fishman YI, Schroeder CE, and Arezzo JC (1998) Click train encoding in primary auditory cortex of the awake monkey: evidence for two mechanisms subserving pitch perception. *Journal of the Acoustical Society of America* **104**: 2935–2955.

Steinschneider M, Volkov, IO, Noh MD, Garell PC, and Howard MA 3rd (1999) Temporal encoding of voice onset time phonetic parameter by field potentials recorded directly from human auditory cortex. *Journal of Neurophysiology* **82**:2346–2357.

Steinschneider M, Volkov IO, Fishman YI, Oya H, Arezzo JC, and Howard MA 3rd (2005) Intracortical Responses in Human and Monkey Primary Auditory Cortex Support a Temporal Processing Mechanism for Encoding of the Voice Onset Time Phonetic Parameter. *Cerebral Cortex* **15**:170–186.

Suga N (1989) Principles of auditory information-processing derived from neuroethology. *Journal of Experimental Biology* **146**:277–286.

Suga N (1992) Philosophy and stimulus design for neuroethology of complex-sound processing. *Philosophical Transactions of the Royal Society London B: Biological Sciences* **336**:423–428.

Tervaniemi M, Kujala A, Alho K, Virtanen J, Ilmoniemi RJ, and Naatanen R (1999). Functional specialization of the human auditory cortex in processing phonetic and musical sounds: a magnetoencephalographic (MEG) study. *Neuroimage* **9**:330–336.

Tervaniemi M, Medvedev SV, Alho K, Pakhomov SV, Roudas MS, Van Zuijen TL, and Naatanen R (2000) Lateralized automatic auditory processing of phonetic versus musical information: a PET study. *Human Brain Mapping* **10**:74–79.

Theunissen FE and Doupe AJ (1998) Temporal and spectral sensitivity of complex auditory neurons in the nucleus HVc of male zebra finches. *Journal of Neuroscience* **18**:3786–3802.

Tian B and Rauschecker JP (1994) Processing of frequency-modulated sounds in the cat's anterior auditory field. *Journal of Neurophysiology* **71**:1959–1975.

Tian B and Rauschecker JP (1998) Processing of frequency-modulated sounds in the cat's posterior auditory field. *Journal of Neurophysiology* **79**:2629–2642.

Tian B and Rauschecker JP (2004) Processing of frequency-modulated sounds in the lateral auditory belt cortex of the rhesus monkey. *Journal of Neurophysiology* **92**:2993–3013.

Tomita M and Eggermont JJ. (2005) Cross-correlation and joint spectro-temporal receptive field properties in auditory cortex. *Journal of Neurophysiology* **93**:378–392.

Tsodyks MV and Markram H (1997) The neural code between neocortical pyramidal neurons depends on neurotransmitter release probability. *Proceedings of the National Academy of Sciences of the United States of America* **94**:719–723.

VanRullen R, Guyonneau R, and Thorpe SJ (2005) Spike times make sense. *Trends in Neuroscience* **28**:1–4.

Wallace MN, Kitzes LM, and Jones EG (1991) Intrinsic inter- and intralaminar connections and their relationship to the tonotopic map in cat primary auditory cortex. *Experimental Brain Research* **86**:527–44.

Wallace MN, Rutkowski RG, and Shackleton TM, Palmer AR (2000) Phase-locked responses to pure tones in guinea pig auditory cortex. *NeuroReport* **11**:3989–3993.

Wallace MN, Shackleton TM, and Palmer AR (2002) Phase-locked responses to pure tones in the primary auditory cortex. *Hearing Research* 172:160–171.

Wang LY and Kaczmarek LK (1998) High-frequency firing helps replenish the readily releasable pool of synaptic vesicles. *Nature* **394**:384–388.

Wang X (2000) On cortical coding of vocal communication sounds in primates. *Proceedings of the National Academy of Sciences of the United States of America* **97**:11843–11849.

Wang X and Kadia SC (2001) Differential representation of species-specific primate vocalizations in the auditory cortices of marmoset and cat. *Journal of Neurophysiology* **86**:2616–2620.

Wang X, Lu T, and Liang L (2003) Cortical Processing of Temporal Modulations. *Speech Communication* **41**:107–121.

Wang X, T Lu, Snider RK, and Liang L (2005) Sustained firing in auditory cortex evoked by preferred stimuli. *Nature* **435**: 341–346.

Wang X, Lu T, Bendor D and Bartlett EL (2008) Neural Coding of Temporal Information in Auditory Thalamus and Cortex. *Neuroscience* **157**: 484–493.

Wang X, Merzenich MM, Beitel R, and Schreiner CE (1995) Representation of a species-specific vocalization in the primary auditory cortex of the common marmoset: temporal and spectral characteristics. *Journal of Neurophysiology* **74**:2685–2706.

Weiss TF and Rose C (1988) A comparison of synchronization filters in different auditory receptor organs. *Hearing Research* **33**: 175–179.

Winter P (1969) The variability of peep and twit calls in captive squirrel monkeys (*Saimiri sciureus*). *Folia Primatologica* **10**: 204–215.

Young ED and Sachs MB (1979) Representation of steady-state vowels in the temporal aspects of the discharge patterns of populations of auditory-nerve fibers. *Journal of the Acoustical Society of America* **66**:1381–1403.

Zwicker E and Fastl H (1990) *Psychoacoustics: Facts and Models.* Springer, Berlin.

Zatorre RJ and Belin P (2001) Spectral and temporal processing in human auditory cortex. *Cerebral Cortex* **11**:946–953.

Zatorre RJ, Belin P, and Penhune VB (2002) Structure and function of auditory cortex: music and speech. *Trends in Cognitive Sciences* **6**:37–46.

Chapter 15

Cortical Representation of Auditory Space

Andrew J. King and John C. Middlebrooks

Abbreviations

AAF	anterior auditory field
AI	primary auditory cortex
AII	secondary auditory cortex
AES	anterior ectosylvian sulcus
DZ	dorsal zone
IC	inferior colliculus
ILD	interaural level difference
ITD	interaural time difference
PAF	posterior auditory field
SC	superior colliculus

1 Introduction

It has been known for many years that an intact auditory cortex is necessary for the normal ability of carnivores and primates, including humans, to localize sound sources. As such, the auditory cortex plays an essential part in one of the most important functions of hearing, which is critical to the way in which these species perceive and interact with their environments. For example, the ability to determine the direction of sound-producing objects or events is often used to find potential mates or prey or to avoid and escape from approaching predators. Sound localization also contributes in important ways to the process by which different sound sources are segregated from one another and therefore aids source identification.

J.C. Middlebrooks (✉)
Department of Otolaryngology, University of California at Irvine, Irvine, CA 92697-5310, USA
e-mail: j.midd@uci.edu

A.J. King (✉)
Department of Physiology, Anatomy and Genetics, University of Oxford, Oxford, OX1 3PT, UK
e-mail: andrew.king@dpag.ox.ac.uk

Information about the direction of a sound source is provided in the form of physical cues that are generated by the way in which incoming sounds interact with the head and external ears. These cues comprise differences in the time of sound arrival and amplitude level between the two ears, together with spectral shape cues produced by the filter properties of these structures. In mammals, binaural cues are utilized for localizing sounds within the horizontal plane, with interaural time differences (ITDs) dominating at low frequencies and interaural level differences (ILDs) at high frequencies, whereas spectral cues enable listeners to localize sounds in elevation and to distinguish between front and back (Wightman and Kistler 1993). These acoustical cues are encoded in the patterns of activity in each auditory nerve and then extracted by neurons in specific brainstem nuclei (Yin 2002; Young and Davis 2002). The outputs from these nuclei converge within the inferior colliculus (IC) in the midbrain (Winer and Schreiner 2005), where neurons are typically sensitive to multiple localization cues (Chase and Young 2006). The major output of the IC is toward forebrain targets. In addition to the forebrain projection, however, a pathway to the superior colliculus (SC) within the midbrain gives rise to a point-to-point map of auditory space (King and Palmer 1983; Middlebrooks and Knudsen 1984; King and Hutchings 1987), which, together with visual and somatosensory inputs to this structure, is used to direct orienting movements towards specific spatial locations (King 2005).

The existence of a map of auditory space in the SC indicates that substantial processing of spatial information takes place subcortically. Moreover, certain aspects of auditory spatial perception can, in principle, be accounted for by the tuning properties of neurons in the IC (Shackleton et al. 2003). It could therefore be argued that the process of sound localization is largely complete at the level of the midbrain. Nevertheless, given the impaired localization abilities that result when the auditory cortex is no longer functioning, it is clear that a spatially coded signal must be transmitted to the forebrain to support spatial perception and behavior and likely that further essential processing takes place at the cortical level.

J.A. Winer, C.E. Schreiner (eds.), *The Auditory Cortex*,
DOI 10.1007/978-1-4419-0074-6_15, © Springer Science+Business Media, LLC 2011

We first review the behavioral consequences of ablating or inactivating particular auditory cortical areas and then consider how well these findings can be reconciled with the spatial sensitivity of neurons in those areas. In particular, we focus on how the location of a sound source is encoded by the firing patterns of cortical neurons and how that information might be decoded. Finally, we examine the possible role of descending corticofugal projections in sound localization and the role of auditory cortex in the plasticity of spatial hearing.

2 Inactivation of Auditory Cortex Induces Sound Localization Deficits

Evidence that an intact auditory cortex is required for normal sound localization behavior has been provided by a number of studies showing that removal of the cortex in one hemisphere in carnivores and primates results in an impaired ability to approach, discriminate or even orient toward sound sources in the contralateral hemifield, whereas localization performance on the ipsilateral side is largely unaffected (e.g., Jenkins and Masterton 1982; Jenkins and Merzenich 1984; Kavanagh and Kelly 1987; Heffner and Heffner 1990; Beitel and Kaas 1993). In fact, a contralateral deficit in localization behavior is the most obvious change observed following unilateral removal of the auditory cortex. If the cortex is ablated bilaterally, cats, dogs, ferrets, and monkeys perform poorly in both lateral hemifields, although they generally still show some ability to distinguish between sound sources located in one hemifield from the other (Neff et al. 1956; Heffner and Masterton 1975; Heffner 1978; Kavanagh and Kelly 1987; Heffner and Heffner 1990; Heffner 1997; Nodal et al. 2010).

Although the magnitude of the reported deficits varies with the size of the lesions and the methods used for measuring localization performance, these studies strongly suggest that the auditory cortex in each hemisphere of these species is primarily responsible for localization behavior in the opposite hemifield, with regions near the midline likely to be represented bilaterally. Although an impaired ability to localize sound is found following restricted lesions focused on the primary auditory cortex (AI), several authors have noted that more profound deficits are observed following lesions that extend beyond AI (Heffner and Masterton 1975; Heffner 1978; Kavanagh and Kelly 1987; Bizley et al. 2007; Nodal et al. 2010). This suggests that other cortical fields contribute further to the processing of spatial information.

The use of aspiration lesions for probing the role of auditory cortex in sound localization and other sound-related behaviors has now largely been superseded by cryogenic (Malhotra et al. 2004, 2008; Malhotra and Lomber 2007; Lomber and Malhotra 2008) or pharmacological inactivation techniques (Smith et al. 2004; Bizley et al. 2007; Nodal et al. 2010), which allow neurons in specific regions of the brain to be silenced reversibly. As expected from the lesion studies, these experiments have shown that unilateral inactivation of AI results in contralateral deficits, whereas bilateral inactivation leads to increased localization errors at all positions tested within the horizontal plane (Malhotra et al. 2004; Smith et al. 2004), as well as a reduced ability to discriminate sound sources located on the midsagittal plane (Bizley et al. 2007) (Fig. 15.1).

The deficits observed following temporary inactivation tend to be smaller than those produced by large cortical lesions, with the animals typically still able to orient toward the side on which the sounds are presented, but unable to localize them as accurately as before the cortex was inactivated. This difference is likely due to a combination of factors. First, removal of the cortex causes neuronal degeneration in brain areas, such as the thalamus, to which the affected cortical area is connected. Second, the temporary inactivation experiments have been aimed at specific cortical fields previously identified using physiological and anatomical criteria. Indeed, cooling studies in cats (Malhotra et al. 2004, 2008; Malhotra and Lomber 2007; Lomber and Malhotra 2008) have shown that, in addition to the well-established effects of silencing AI, deficits in spatial hearing result from inactivation of the posterior auditory field (PAF), anterior ectosylvian sulcus (AES), or dorsal zone (DZ), but not when other areas, such as the secondary auditory cortex (AII) or anterior auditory field (AAF) are targeted (Fig. 15.2). These findings imply that a division of labor may exist within auditory cortex, with different areas responsible for the processing of spatial and non-spatial information. There is no one "space region," however, as multiple auditory cortical fields, each with distinct sources of input (Morel and Imig 1987; Huang and Winer 2000), are necessary for normal localization behavior, with certain areas, particularly PAF and AES, appearing to contribute more than others.

Studies in humans have confirmed that damage to the auditory cortex, which can occur as a result of a stroke or following surgery to remove a tumor, results in impaired sound localization (Zatorre and Penhune 2001; Adriani et al. 2003), as well as raised ITD and ILD discrimination thresholds (Yamada et al. 1996). Difficulties in defining the precise locus of the damage, which varies between individuals both in its extent and in the age at which it occurs, inevitably limit the comparisons that can be drawn with the animal studies. However, in contrast to the contralateral representation of auditory space emphasized in other species, humans appear to show a clear right-hemisphere dominance for sound localization (Zatorre and Penhune 2001). Thus, right-sided lesions in humans often result in bilateral localization deficits, and bilateral localization is sometimes spared following a left-sided lesion.

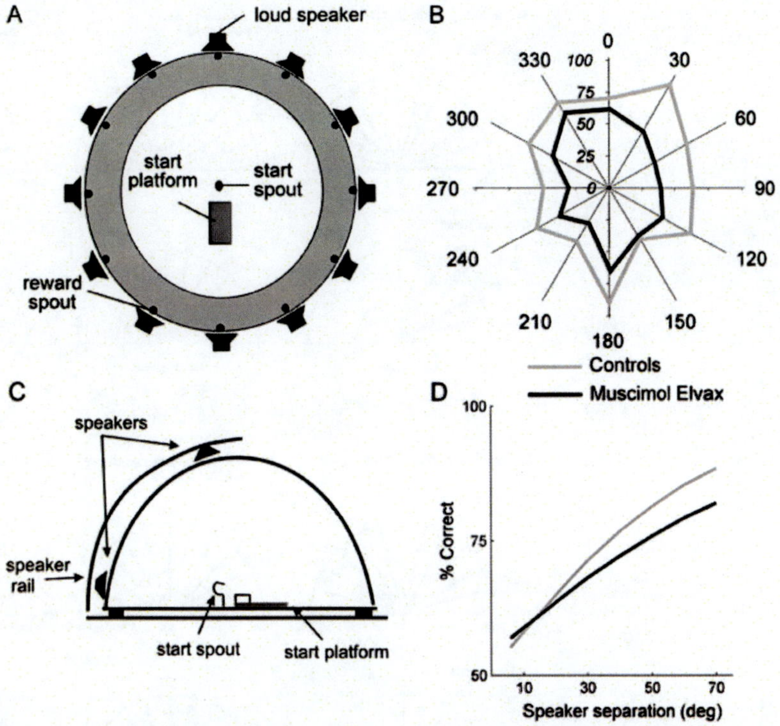

Fig. 15.1 The auditory cortex is needed for normal sound localization. **a** Setup used for measuring localization in the horizontal plane. Ferrets were trained to stand on the start platform and initiate a trial by licking the start spout. Each trial consisted of a broadband noise burst of variable duration and level presented randomly from 1 of 12 speakers positioned at 30° intervals in the horizontal plane. The animals were rewarded for approaching and licking the reward spout associated with the speaker that had been triggered. **b** The polar plot shows the mean percentage scores achieved when localizing 40-ms noise bursts by a group of 4 control ferrets and 4 animals in which A1 had been inactivated bilaterally by placing sheets of a slow-release polymer containing the $GABA_A$ agonist muscimol on the cortex. These animals achieved lower scores than the normal controls at all stimulus angles. From Smith et al. (2004). **c** Setup used for measuring localization in the *vertical* plane. The animals had to discriminate between stimuli presented from one of two speakers positioned in the midsagittal plane. Because it was not possible for the animals to approach the sound source directly, they were rewarded for responding at a reward spout to their right (+90°) when the sound was presented from the upper speaker, and at a spout to their left (–90°) when sound was presented from the lower speaker. **d** Psychometric functions fitted to the data from the same ferrets before (control) and after inactivating AI bilaterally with muscimol-Elvax. In 5 (out of 6) animals contributing to these data, AI inactivation produced a significant drop in performance. From Bizley et al. (2007)

3 Representation of Auditory Space in the Cortex

The role established by lesion-behavior studies for the auditory cortex in spatial hearing raises the question of how sound-source location is represented there. This has been addressed by either mapping out the spatial receptive fields of individual cortical neurons or by measuring their sensitivity to acoustic localization cues. As in the behavioral experiments, receptive field mapping studies typically involve recording the spiking activity of the neurons in response to sounds delivered from loudspeakers positioned around the animals' head in the free-field (e.g., Middlebrooks and Pettigrew 1981; Imig et al. 1990; Rajan et al. 1990a, b; Stecker et al. 2005a, b; Woods et al. 2006; Harrington et al. 2008; Werner-Reiss and Groh 2008). Alternatively, stimuli can be presented over headphones in virtual acoustic space, an approach that enables rapid mapping of spatial sensitivity across a broad range of stimulus directions, as well as manipulation of the localization cue values provided (Brugge et al. 1994, 1996; Mrsic-Flogel et al. 2001, 2003, 2005; Las et al. 2008).

3.1 Spatial Receptive Fields in Primary Auditory Cortex

Like the lesion and inactivation studies, early recording experiments focused on AI, while more recent studies have explored the spatial sensitivity of neurons in other cortical areas. We first review the general properties of and acoustical basis for the spatial receptive fields in the primary auditory cortex, which have been determined by recording neuronal responses from both anesthetized and awake animals, and

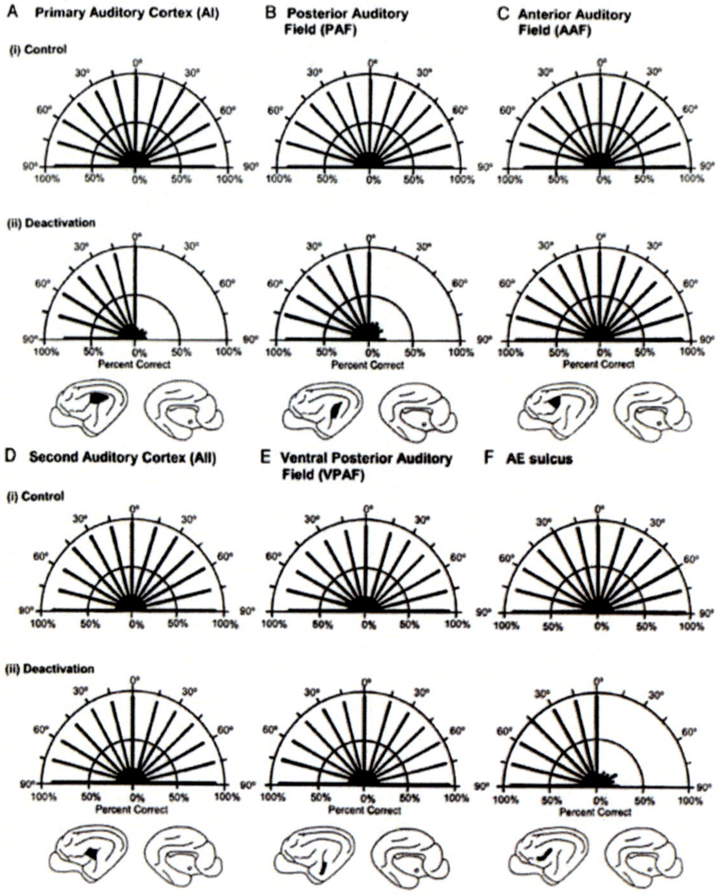

Fig. 15.2 Localization responses to sounds presented in the frontal hemifield before and after (*i*) and during (*ii*) unilateral cooling deactivation of 6 areas of auditory cortex in the cat: AI (**a**), PAF (**b**), AAF (**c**), AII (**d**), VPAF (**e**), and AES (**f**). The length of each *radial line* indicates the mean percentage correct score for that sound direction. The site and extent of deactivation are shown below each plot by the *black regions* on the medial and lateral views of the cat brain. Inactivation of AI, PAF, and AES each resulted in a contralateral, but not an ipsilateral, localization deficit. Adapted from Malhotra et al. (2004)

then, in the next section, consider the extent to which these properties vary among different cortical areas.

Cortical receptive fields vary in size, from a minority of neurons that show a clear preference for restricted regions of space to those that respond throughout an entire hemifield or beyond. Generally, receptive fields expand with increasing sound level and also vary in size according to the bandwidth of the stimulus and with other properties of the neuron in question. In keeping with the behavioral deficits produced by unilateral lesions or inactivation, most cortical neurons respond best to sounds presented on the contralateral side of the animal, although some prefer sound sources near the frontal midline or on the ipsilateral side (Fig. 15.3).

The differences in spatial receptive field properties among cortical neurons can be attributed to their tuning to monaural and binaural localization cues. As in subcortical nuclei, low-frequency cortical neurons are sensitive to ITDs (Malone et al. 2002; Scott et al. 2009), whereas high-frequency neurons rely more on ILDs (Imig and Adrian 1977; Middlebrooks et al. 1980; Irvine et al. 1996; Rutkowski et al. 2000; Zhang et al. 2004; Campbell et al. 2006). In both cats (Irvine et al. 1996; Zhang et al. 2004) and ferrets (Campbell et al. 2006), ILD sensitivity ranges from a minority of neurons showing ipsilateral dominance or tuning to values close to zero, corresponding to sound sources located in front of the animal, to the majority that respond most strongly to values that would be produced by sound sources on the contralateral side of space.

Although this continuum of ILD sensitivity matches the distribution of spatial receptive fields in auditory cortex, binaural interactions alone are insufficient to account for the representation of auditory space in the cortex. At near-threshold sound levels, high-frequency AI neurons in cat (Middlebrooks and Pettigrew 1981; Rajan et al. 1990; Brugge et al. 1994) and ferret (Mrsic-Flogel et al. 2003, 2005) tend to have "axial" receptive fields that are centered on the acoustical axis of the contralateral external ear. This is the region in which the acoustical gain of the external ear is at its maximum, therefore suggesting that, at these low sound levels, the receptive fields of the neurons are shaped by pinna directionality. Moreover, using virtual acoustic space stimuli, it has been shown that a linear combination of the frequency sensitivity to stimulation of each ear and the directional properties of the auditory periphery can account

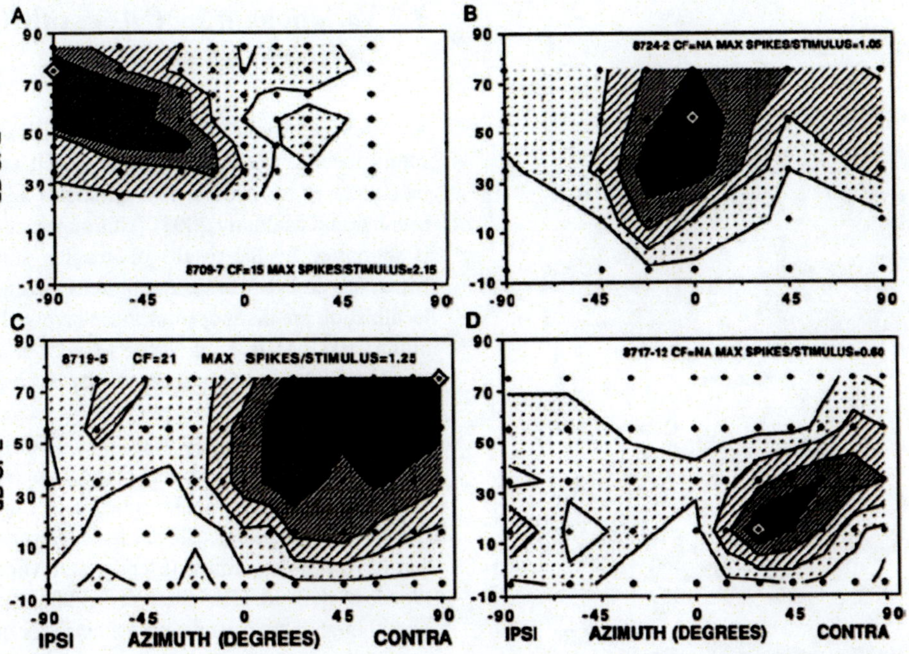

Fig. 15.3 Spatial sensitivity of four neurons in cat area AI. In each panel, contours represent normalized spike rates as a function of sound-source azimuth (*horizontal axes*) and sound level (*vertical axes*). Contours are drawn at 5, 25, 50, and 75% of maximum spike rates. The grids of *small diamonds* indicate stimulus locations and levels that were tested. Adapted from Imig et al.(1990)

for the location and shape of the spatial receptive fields of many high-frequency neurons in ferret AI (Schnupp et al. 2001; Mrsic-Flogel et al. 2005; Fig. 15.4). Changes in spatial sensitivity with increasing sound level can be explained by this linear model (Schnupp et al. 2001; Mrsic-Flogel et al. 2005), which also predicts the observed sharpening of spatial receptive fields with age as the head and ears grow (Mrsic-Flogel et al. 2003). Mrsic-Flogel et al. (2005) found that the linear model works best for neurons that receive predominantly excitatory input from the contralateral ear and inhibitory input from the ipsilateral ear, and which are therefore sensitive to ILDs, but less well for neurons that receive excitatory inputs from both ears and which are likely to be sensitive to ITDs. A similar linear estimation procedure based on the neurons' frequency selectivity and the external ear acoustics can also account for the elevation sensitivity of neurons in the primary fields AI and AAF of the cat cortex (Macpherson et al. 2004).

Several studies have observed that neurons tuned to particular regions of space are found in clusters (Middlebrooks and Pettigrew 1981; Imig et al. 1990; Rajan et al. 1990b), as is also the case for the binaural properties of cortical neurons (e.g., Imig and Adrian 1977; Middlebrooks et al. 1980; Rutkowski et al. 2000; Nakamoto et al. 2004). Although this indicates a degree of local order, there is, in most species, no evidence for a map of auditory space equivalent to that found in the SC or to the spatiotopic maps that characterize the cortices of other sensory modalities. Similarly, optical imaging of intrinsic signals in ferrets has failed to provide evidence for a systematic variation in sensitivity to ILDs across the cortical surface (Nelken et al. 2008). The only exception to this seems to be in the region of the pallid bat auditory cortex responsible for passive sound localization, where a topographic representation of ILD sensitivity has been described (Razak and Fuzessery 2002).

In addition to changes in firing rate across different loudspeaker locations, variations in the latency of the response can also signal sound-source direction. This has been observed in a number of studies in both anesthetized cats (Middlebrooks et al. 1994, 1998; Brugge et al. 1996; Jenison 2000; Furukawa and Middlebrooks 2002; Reale et al. 2003; Stecker and Middlebrooks 2003) and ferrets (Mrsic-Flogel et al. 2005; Nelken et al. 2005). Although first-spike latencies tend to vary inversely with spike counts, with sounds at preferred locations evoking more spikes with shorter latencies, spike timing can be modulated across the receptive field even at levels at which neurons respond relatively uniformly to all tested locations. Indeed, spike timing can carry as much or more information about sound-source location than spike rate (Brugge et al. 1996; Eggermont 1998; Furukawa and Middlebrooks, 2002; Stecker and Middlebrooks 2003; Nelken et al. 2005).

The proportion of location-related information carried by spike timing is somewhat lower in recordings in unanesthetized conditions (Mickey and Middlebrooks 2003; Woods et al. 2006; Werner-Reiss and Groh 2008). This is likely

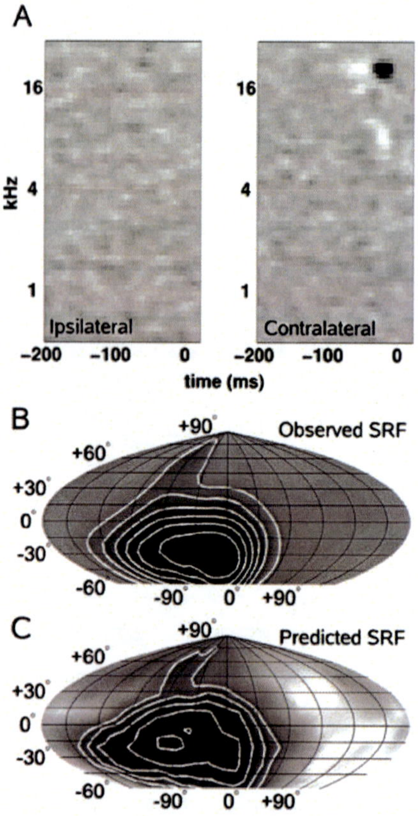

Fig. 15.4 Predicting spatial responses from the frequency tuning of neurons in AI. Examples of frequency-time response fields (FTRFs) for each ear (**a**), together with the observed (**b**) and predicted (**c**) spatial receptive fields (SRFs) of a neuron recorded in AI of an anesthetized ferret. The FTRFs were measured by reverse correlation to random chord stimuli presented to each ear. The observed SRFs were generated by presenting noise bursts from 224 virtual sound directions, covering 360° in azimuth and from –60° to +90° in elevation. The predicted SRFs were generated by convolving the FTRFs with the energy spectrum vectors of the VAS stimuli for each ear and each position in space. From Schnupp et al. (2001)

to be due to the fact that cortical neurons tend to be more active in the awake condition, providing greater potential for modulation of spike counts by sound-source location, including suppression of spontaneous activity away from the excitatory region of the receptive field. Aside from the deeper stimulus-related modulation of spike rates, spatial sensitivity in unanesthetized conditions is largely similar to that recorded under anesthesia. As in anesthetized conditions, cortical receptive fields recorded in awake animals often span a hemifield in width (Mickey and Middlebrooks 2003; Woods et al. 2006; King et al. 2007), and there is no indication of a point-to-point map of auditory space. One notable difference is that spatial sensitivity is less vulnerable to increases in stimulus level in awake conditions than in the anesthetized state (Mickey and Middlebrooks 2003).

3.2 Variations in Spatial Sensitivity Across Different Cortical Areas

As discussed above, the impact of cortical inactivation on sound localization depends upon which areas are silenced (Malhotra et al. 2004, 2008; Malhotra and Lomber 2007; Lomber and Malhotra 2008). This apparent division of labor is supported by the results of imaging studies in humans, which suggest that the cortical areas engaged during sound localization are distinct from those involved in sound recognition tasks (Alain et al. 2001; Maeder et al. 2001; Barrett and Hall 2006). A more recent study has reported, however, that widespread cortical areas may be activated during auditory spatial processing (Lewald et al. 2008). The distinction among spatial and non-spatial cortical areas also is less clear cut at the level of neuronal responses, as some degree of sensitivity to sound-source location is a property of all areas that have been examined (Stecker and Middlebrooks 2003; Woods et al. 2006; Harrington et al. 2008; Bizley et al. 2009).

Although there is as yet no evidence for qualitative differences in spatial sensitivity among cortical areas, recording studies have shown that certain cortical areas show quantitatively enhanced spatial sensitivity compared to others. In monkeys, for example, neurons in caudal auditory cortical fields are more sharply tuned for sound-source location than those in core or rostral fields (Recanzone et al. 2000; Tian et al. 2001; Woods et al. 2006; Miller and Recanzone 2009) (Fig. 15.5), which is broadly consistent with most of the imaging data in humans. Similar findings have been obtained in cats, the species in which the representation of auditory space in different cortical fields has been explored most extensively. The spike counts and first-spike latencies of neurons in PAF and DZ show greater modulation with changes in stimulus location and transmit more spatial information, particularly in the timing of their spike discharges, than those in AI, AII, or AAF (Stecker et al. 2003, 2005a; Harrington et al. 2008) (Fig. 15.6c, d). Furthermore, the receptive fields of PAF and DZ neurons are more tolerant to changes in stimulus level than those in other cortical fields.

Although these differences are fairly modest, the distinction between PAF and AAF in cats is supported by the effects of cortical cooling, which results in deficits in sound localization and in sound pattern recognition, respectively (Lomber and Malhotra 2008). Neurons in posterior AES also show greater spatial selectivity compared to those in the AI (Las et al. 2008). Again, this fits with the behavioral-inactivation evidence that AES, which is the only auditory cortical area to project heavily to the SC (Meredith and Clemo 1989), plays an important role in spatial hearing (Malhotra et al. 2004; Malhotra and Lomber 2007). By contrast, the consequences of inactivation of AI are greater than might be expected given the relatively poor spatial sensitivity of its neurons.

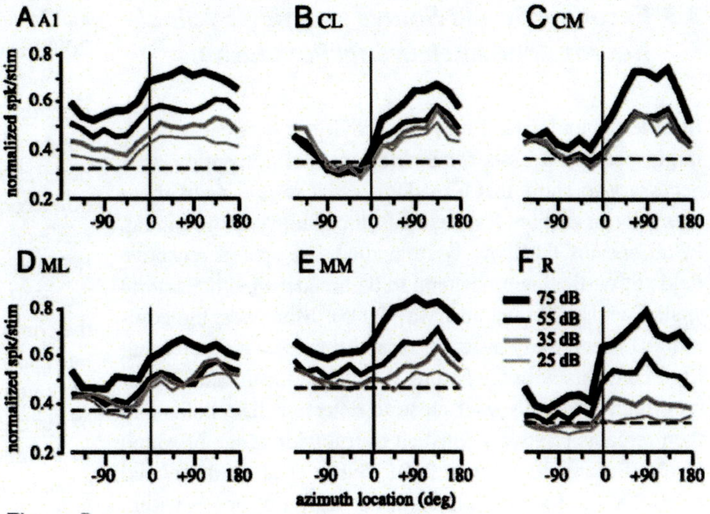

Fig. 15.5 Normalized distribution of activity as a function of stimulus level and azimuth recorded in different areas of the monkey auditory cortex. *Line thickness* and *shading* corresponds to the different levels (see *inset* in **f**). The *horizontal dashed line* is the normalized spontaneous activity. Overall the activity increased with increasing stimulus levels and was more sharply tuned for the caudal belt fields. From Woods et al. (2006)

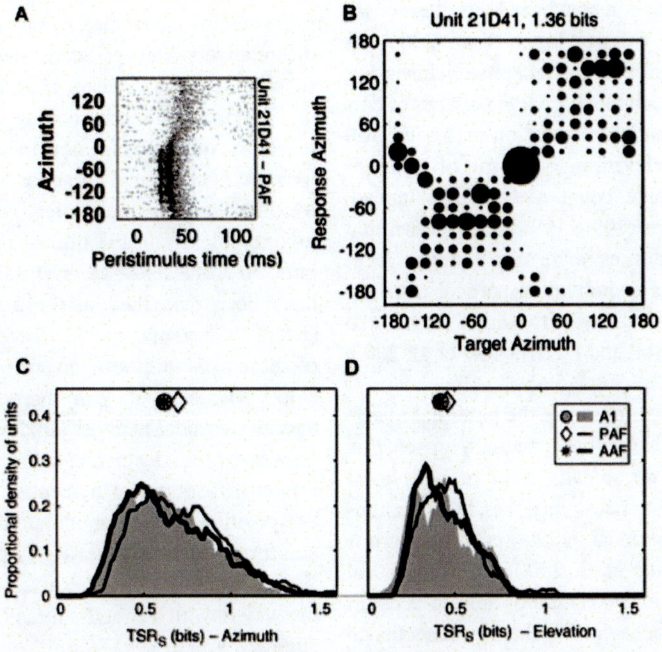

Fig. 15.6 Estimating spatial information carried by neural spike patterns. A statistical pattern recognition algorithm (see Stecker and Middlebrooks, 2003 for details) classifies each neural response according to the most likely eliciting stimulus location. **a** Peristimulus times (*x*-axis) of spikes elicited by stimuli varying in location (*y*-axis) for a neurons recorded in cat area PAF. **b** Algorithm performance for this neuron, represented by a joint stimulus-response matrix (confusion matrix). Proportions of responses at each combination of target (*x*-axis) and response (*y*-axis) location are indicated by the diameter of the *circles* inside the figure. In this case, classification is highly accurate between hemifields: contralateral targets (negative azimuths) are almost never misclassified to ipsilateral locations or vice versa. Targets on the *midline* are accurately localized, although discrimination of front from back is poor. Mutual information of target and response gives an estimate of the total stimulus-related information contained in the neural response; in this case, 1.36 bits. Distributions of total stimulus-related information (TSR) transmitted by neural responses for azimuth (**c**) and elevation (**d**) when full spike patterns were used are shown for neurons recorded in cat AI, AAF, and PAF. *Symbols* represent the median of each distribution. Overall, units in PAF transmit significantly more information than units in AI or AAF. Adapted from Stecker et al. (2003) and Harrington et al. (2008)

The magnitude of these deficits may therefore have less to do with the physiological properties of the neurons in AI than with their projections to other areas, such as PAF (Rouiller et al. 1991). A related possibility is that the responses of AI neurons might provide a temporal reference for comparison with the more pronounced location-related modulations of spike latency in PAF (Stecker and Middlebrooks 2003).

3.3 Encoding Sound-Source Location by Single Neurons and by Neuronal Populations

In order to understand how the activity of cortical neurons might provide a basis for auditory spatial perception, it is necessary to show that a readout of the responses of those neurons can account for the localization ability of the animal. In all species that have been studied, the spatial receptive fields of cortical neurons tend to be broader than behavioral spatial acuity (Brown and May 2005). Moreover, the commonly observed expansion of receptive fields with increasing level contrasts with the finding that sound localization accuracy improves with level close to detection thresholds and then remains relatively constant over a wide range of sound levels (Su and Recanzone 2001; Sabin et al. 2005; Nodal et al. 2008). However, although the region of space within which a stimulus can drive the neurons generally increases, the amount of spatial information conveyed by the responses stays effectively the same (Mrsic-Flogel et al. 2003). A potential advantage of omnidirectional receptive fields is that they provide a means by which the discharge patterns of cortical neurons can convey spatial information across the full range of sound azimuth or elevation, as a result of location-dependent variations in spike count and timing. Indeed, Middlebrooks and colleagues (1994, 1998) have shown that computer-based classifiers can estimate sound-source location from the firing patterns of individual cortical neurons, and that, as expected, the accuracy with which they do so in cats is greatest in areas PAF and DZ (Stecker et al. 2003, 2005a; Harrington et al. 2008) (Fig. 15.6).

Although some cortical neurons have the potential to signal sound-source location throughout auditory space, the accuracy with which they do so falls short of behavioral performance. Similarly, neurometric analyses have demonstrated that the tuning of individual monkey cortical neurons to sound location (Recanzone et al. 2000) or to interaural phase differences (Scott et al. 2009) is not able to account for the acuity measured in behavioral tasks. A consequence of broad tuning is that sounds emanating from a particular direction will activate many neurons distributed throughout the auditory cortex. Several studies have now emphasized the importance of population coding schemes, based either on the full spike discharge patterns (Furukawa et al. 2000; Stecker et al. 2003) or, more specifically, on the spike firing rates (Miller and Recanzone 2009) or latencies (Jenison 2000; Reale et al. 2003) of ensembles of cortical neurons. These population models provide a better fit to the behavioral data. With most receptive fields lying off the midline, the steepest—and therefore most informative—spatial gradients of the neurons' spike counts or latencies lie on or close to the midline (Stecker et al. 2005b; Campbell et al. 2006), which is where localization is most accurate (Makous and Middlebrooks 1990; May and Huang 1996; Nodal et al. 2008) and spatial discrimination most acute (Mills 1958).

One way in which sound-source location might be represented by the pooled activity of neurons is through an "opponent process," based on the relative activity of two populations of neurons, one tuned ipsilaterally and the other contralaterally (Stecker et al. 2005b). This notion has received support from studies of ITD coding in the brainstem (McAlpine and Grothe 2003) and from psychophysical studies of binaural adaptation in humans (Phillips 2008) where the comparison is thought to be made between activity in the left and right hemispheres. While most cortical neurons do respond preferentially to sounds located on the opposite side of the body, the notion that localization judgments are based on a comparison of activity in the two hemispheres is inconsistent with the contralateral deficits produced in animals by unilateral cortical damage or inactivation (see Section 2). It is possible, however, that an opponent model of sound localization could be based on the contralaterally tuned majority and the relatively few ipsilateral neurons that are found within each hemisphere (Stecker et al. 2005b).

The mode of spatial coding in the auditory cortex raises important questions for how information about sound-source location is combined and coordinated with signals provided by other sensory modalities—which are often represented topographically in the brain—or translated into motor outputs. Neurons sensitive to visual or somatic sensory stimuli have been described in the auditory cortex of numerous species (Ghazanfar and Schroeder 2006). While the function of these non-auditory sensory responses is not fully understood, visual inputs can sharpen the spatial sensitivity of auditory cortical neurons (Bizley and King 2008), and could therefore provide a neural substrate for the many crossmodal influences on spatial perception (King 2009). In monkeys, eye position can also modulate the activity of neurons in the auditory cortex (Werner-Reiss et al. 2003; Fu et al. 2004; Woods et al. 2006). These factors will therefore influence the way in which sound-source location is represented in the auditory cortex.

4 Representation of Multiple Sound Sources

The great majority of behavioral and physiological studies have focused on the localization of single, usually stationary sound sources. While this is the simplest situation to investigate, it is important to remember that real auditory objects are often encountered in reverberant environments and in the presence of other, competing sound sources. Adding diffuse background noise reduces the effective level of the stimuli used to map the responses of cortical neurons and reduces the size of their receptive fields (Brugge et al. 1998;

Furukawa and Middlebrooks 2001). By contrast, background noise originating from a specific direction in space can alter both the size and location of receptive fields (Furukawa and Middlebrooks 2001).

When brief sounds are presented from two different locations, the resulting percept can change if a delay is introduced between them. For delays of less than about 1 ms, human listeners report hearing a single stimulus that originates from a region intermediate between the two source locations, a phenomenon which is therefore known as "summing localization." If the interstimulus delay is extended out to 5 ms, a single sound is still heard, but its perceived location is dominated by the actual location of the leading source. In other words, the percept of the lagging sound is suppressed. This is the "precedence effect," which plays an important role in reducing the influence of room echoes (Litovsky et al. 1999). A neural correlate of these spatial illusions has been observed in the auditory cortex of cats (Reale and Brugge 2000; Mickey and Middlebrooks 2005) and rabbits (Fitzpatrick et al. 1999), although neuronal responses to the lagging sound tend to be suppressed out to much longer interstimulus delays than the precedence effect lasts for in humans.

Ongoing studies are exploring a cortical correlate of "spatial stream segregation," in which sequences of sounds originating from distinct locations are perceived as corresponding to distinct auditory objects. In the cortical work (Middlebrooks et al. 2009), interleaved trains of brief noise bursts are presented from sources at two locations. A spatial separation of as little as 10° can result in the time-locked response of a cortical neuron being captured by one or the other sound source. That spatial acuity is substantially greater than that which has been observed in the responses of cortical neurons mapped with single sound sources.

5 Dynamic Coding of Auditory Space

As with their other response properties, the spatial sensitivity of cortical neurons is not fixed, but depends on the animal's behavioral state and on the neurons' history of stimulation. Dependence on history of stimulation has been demonstrated for sensitivity to interaural phase differences (Malone et al. 2002) and to virtual sound locations (Jenison et al. 2001). These context-dependent effects may enhance the representation of certain stimulus values or confer sensitivity to moving sounds. Ongoing studies of the effects of behavioral state show that the spatial sensitivity of cortical neurons can sharpen markedly under conditions in which an animal is required to localize sounds (Lee et al. 2008).

Over longer time scales, changes in cortical response properties have been shown to accompany improvements in performance during perceptual learning (reviewed by Dahmen and King 2007). Although plasticity has yet to be demonstrated for spatial sensitivity at the neuronal level, auditory-evoked potential measurements in humans suggest that training-induced improvements in ITD discrimination may be associated with refinements in the cortical population response (Spierer et al. 2007). Auditory cortical plasticity may also enable adult animals to adapt to changes in the balance of inputs between the two ears. Provided that they are given appropriate auditory training, adult ferrets can rapidly adjust to the altered spatial cues produced by occluding one ear and learn to localize accurately again (Kacelnik et al. 2006). The capacity of the animals to compensate for these changes in binaural cues is impaired if different regions of the auditory cortex, including AI, are reversibly inactivated (King et al. 2007) (Fig. 15.7). Sound localization plasticity is also disrupted if a substantial portion of the descending projection from the auditory cortex to the inferior colliculus is removed using a targeted neuronal degeneration technique (King et al. 2007; Bajo et al. 2010) (Fig. 15.7). This finding is consistent with the changes in ILD sensitivity of IC neurons that have been reported in anesthetized guinea pigs following cortical cooling (Nakamoto et al. 2008), and suggests that one function played by the auditory cortex in spatial hearing is to provide signals that are transmitted via descending cortical pathways to bring about experience-driven changes in localization abilities.

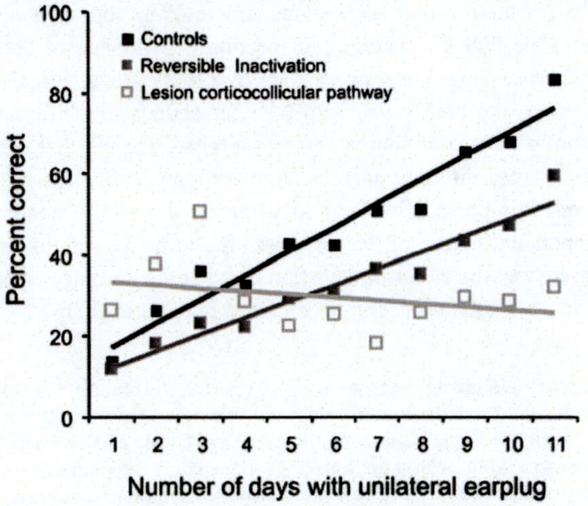

Fig. 15.7 Plasticity of auditory localization depends on the auditory cortex. Change in performance (averaged across all speaker locations, using the setup shown in Fig. 15.1) over time in 3 groups of ferrets that received daily training with unilateral earplugs. Compared to the rapid and near complete recovery in localization accuracy observed in control animals ($n = 3$; *black symbols* and *regression line*), a significantly slower improvement was observed in animals in which AI had been reversibly inactivated using muscimol-Elvax implants ($n = 4$; *dark gray symbols* and *regression line*). Moreover, no improvement in performance was observed in ferrets in which targeted apoptotic degeneration of corticocollicular neurons had been induced using a photoactivation technique ($n = 3$; *open symbols* and *light gray regression line*). Adapted from (King et al. 2007)

6 Concluding Remarks and Future Directions

That the auditory cortex plays an essential role in the ability of many species, including humans, to localize sound is beyond any doubt, but the nature of that role has yet to be fully established. Recording studies have shown that space is represented by neurons possessing very large receptive fields that most often are centered within the contralateral hemifield. The regions of greatest spatial acuity, near the frontal midline, correspond to the edges of many of these large receptive fields. Although sound-source location can be signaled by both the timing and the number of spikes evoked by individual cortical neurons, pooling of this information across populations of neurons appears to be required in order to account for behavioral performance. As with other aspects of auditory perception, further insights into the neural coding strategies used to extract spatial information will only come if recordings are made from cortical neurons, while animals perform localization tasks, so that trial-by-trial correlations can be made between the physiology and the behavior.

While the contribution of different cortical fields to spatial hearing is clearly not the same, with some areas, such as PAF and DZ in cats and the caudal fields in monkeys, showing greater and more level-tolerant spatial sensitivity than others, neurons in all cortical areas convey at least some information about sound-source location. This might simply reflect the processing that takes place subcortically, but it is also possible that the widespread location dependence of cortical processing is just one aspect of a higher-level function, such as the ability to group together sounds that originate from a particular source and to segregate sounds that originate from different sources. Approaching cortical function from this perspective, and focusing on the highly context-dependent nature of the responses found there, should help to answer the enduring question of what the auditory cortex adds to spatial processing performed in the brainstem.

Acknowledgments Andrew King's research is funded by the Wellcome Trust, the Biotechnology and Biological Sciences Research Council, the Royal National Institute for Deaf People and by Deafness Research UK. John Middlebrooks's research is supported by the National Institute on Deafness and Other Communication Disorders.

References

Adriani M, Maeder P, Meuli R, Thiran AB, Frischknecht R, Villemure JG, Mayer J, Annoni JM, Bogousslavsky J, Fornari E, Thiran JP, and Clarke S (2003) Sound recognition and localization in man: specialized cortical networks and effects of acute circumscribed lesions. *Experimental Brain Research* **153**:591–604.

Alain C, Arnott SR, Hevenor S, Graham S, and Grady CL (2001) "What" and "where" in the human auditory system. *Proceedings of the National Academy of Sciences of the United States of America* **98**:12301–12306.

Bajo VM, Nodal FR, Moore DR, and King AJ (2010) The descending corticocollicular pathway mediates learning-induced auditory plasticity. *Nature Neuroscience* **13**:253–260.

Barrett DJ and Hall DA (2006) Response preferences for "what" and "where" in human non-primary auditory cortex. *Neuroimage* **32**:968–977.

Beitel RE and Kaas JH (1993) Effects of bilateral and unilateral ablation of auditory cortex in cats on the unconditioned head orienting response to acoustic stimuli. *Journal of Neurophysiology* **70**:351–369.

Bizley JK and King AJ (2008) Visual-auditory spatial processing in auditory cortical neurons. *Brain Research* **1242**:24–36.

Bizley JK, Nodal FR, Parsons CH, and King AJ (2007) Role of auditory cortex in sound localization in the midsagittal plane. *Journal of Neurophysiology* **98**:1763–1774.

Bizley JK, Walker KM, Silverman BW, King AJ, and Schnupp JWH (2009) Interdependent encoding of pitch, timbre, and spatial location in auditory cortex. *Journal of Neuroscience* **29**: 2064–2075.

Brown CH and May BJ (2005) Comparative mammalian sound localization. In: Popper AN and Fay RR (eds). *Springer Handbook of Auditory Research*, volume 25, *Sound Source Localization*. Springer, New York, pp. 124–178.

Brugge JF, Reale RA, and Hind JE (1996) The structure of spatial receptive fields of neurons in primary auditory cortex of the cat. *Journal of Neuroscience* **16**:4420–4437.

Brugge JF, Reale RA, and Hind JE (1998) Spatial receptive fields of primary auditory cortical neurons in quiet and in the presence of continuous background noise. *Journal of Neurophysiology* **80**: 2417–2432.

Brugge JF, Reale RA, Hind JE, Chan JC, Musicant AD, and Poon PW (1994) Simulation of free-field sound sources and its application to studies of cortical mechanisms of sound localization in the cat. *Hearing Research* **73**:67–84.

Campbell RAA, Schnupp JWH, Shial A, and King AJ (2006) Binaural-level functions in ferret auditory cortex: evidence for a continuous distribution of response properties. *Journal of Neuroscience* **95**:3742–3755.

Chase SM and Young ED (2006) Spike-timing codes enhance the representation of multiple simultaneous sound-localization cues in the inferior colliculus. *Journal of Neuroscience* **26**:3889–3898.

Dahmen JC and King AJ (2007) Learning to hear: plasticity of auditory cortical processing. *Current Opinion in Neurobiology* **17**: 456–464.

Eggermont JJ (1998) Azimuth coding in primary auditory cortex of the cat. II. Relative latency and interspike interval representation. *Journal of Neurophysiology* **80**:2151–2161.

Fitzpatrick DC, Kuwada S, Kim DO, Parham K, and Batra R (1999) Responses of neurons to click-pairs as simulated echoes: auditory nerve to auditory cortex. *Journal of the Acoustical Society of America* **106**:3460–3472.

Fu KM, Shah AS, O'Connell MN, McGinnis T, Eckholdt H, Lakatos P, Smiley J, and Schroeder CE (2004) Timing and laminar profile of eye-position effects on auditory responses in primate auditory cortex. *Journal of Neurophysiology* **92**:3522–3531.

Furukawa S and Middlebrooks JC (2001) Sensitivity of auditory cortical neurons to locations of signals and competing noise sources. *Journal of Neurophysiology* **86**:226–240.

Furukawa S and Middlebrooks JC (2002) Cortical representation of auditory space: information-bearing features of spike patterns. *Journal of Neurophysiology* **87**:1749–1762.

Furukawa S, Xu L, and Middlebrooks JC (2000) Coding of sound-source location by ensembles of cortical neurons. *Journal of Neuroscience* **20**:1216–1228.

Ghazanfar AA and Schroeder CE (2006) Is neocortex essentially multisensory? *Trends in Cognitive Sciences* **10**:278–285.

Harrington IA, Stecker GC, Macpherson EA, and Middlebrooks JC (2008) Spatial sensitivity of neurons in the anterior, posterior, and primary fields of cat auditory cortex. *Hearing Research* **240**: 22–41.

Heffner H (1978) Effect of auditory cortex ablation on localization and discrimination of brief sounds. *Journal of Neurophysiology* **41**: 963–976.

Heffner H (1997) The role of macaque auditory cortex in sound localization. *Acta Otolaryngologia Supplement* **532**:22–27.

Heffner HE and Heffner RS (1990) Effect of bilateral auditory cortex lesions on sound localization in Japanese macaques. *Journal of Neurophysiology* **64**:915–931.

Heffner H and Masterton B (1975) Contribution of auditory cortex to sound localization in the monkey (*Macaca mulatta*). *Journal of Neurophysiology* **38**:1340–1358.

Huang CL and Winer JA (2000) Auditory thalamocortical projections in the cat: laminar and areal patterns of input. *Journal of Comparative Neurology* **427**:302–331.

Imig TJ and Adrian HO (1977) Binaural columns in the primary field (A1) of cat auditory cortex. *Brain Research* **138**:241–257.

Imig TJ, Irons WA, and Samson FR (1990) Single-unit selectivity to azimuthal direction and sound pressure level of noise bursts in cat high-frequency primary auditory cortex. *Journal of Neurophysiology* **63**:1448–1466.

Irvine DRF, Rajan R, and Aitkin LM (1996) Sensitivity to interaural intensity differences of neurons in primary auditory cortex of the cat. I. types of sensitivity and effects of variations in sound pressure level. *Journal of Neurophysiology* **75**:75–96.

Jenison RL (2000) Correlated cortical populations can enhance sound localization performance. *Journal of the Acoustical Society of America* **107**:414–421.

Jenison RL, Schnupp JWH, Reale RA, and Brugge JF (2001) Auditory space-time receptive field dynamics revealed by spherical white-noise analysis. *Journal of Neuroscience* **21**:4408–4415.

Jenkins WM and Masterton RB (1982) Sound localization: effects of unilateral lesions in central auditory system. *Journal of Neurophysiology* **47**:987–1016.

Jenkins WM and Merzenich MM (1984) Role of cat primary auditory cortex for sound-localization behavior. *Journal of Neurophysiology* **52**:819–847.

Kacelnik O, Nodal FR, Parsons CH, and King AJ (2006) Training-induced plasticity of auditory localization in adult mammals. *Public Library of Science Biology* **4**:627–638.

Kavanagh GL and Kelly JB (1987) Contributions of auditory cortex to sound localization in the ferret (*Mustela putorius*). *Journal of Neurophysiology* **57**:1746–1766.

King AJ (2005) The superior colliculus. *Current Biology* **14**: R335-R338.

King AJ (2009) Visual influences on auditory spatial learning. *Philosophical Transactions of the Royal Society B Biological Sciences* **364**:331–339.

King AJ, Bajo VM, Bizley JK, Campbell RAA, Nodal FR, Schulz AL, and Schnupp JWH (2007) Physiological and behavioral studies of spatial coding in the auditory cortex. *Hearing Research* **229**: 106–115.

King AJ and Hutching ME (1987) Spatial response properties of acoustically responsive neurons in the superior colliculus of the ferret: a map of auditory space. *Journal of Neurophysiology* **57**:596–624.

King AJ and Palmer AR (1983) Cells responsive to free-field auditory stimuli in guinea-pig superior colliculus: distribution and response properties. *Journal of Physiology* **342**:361–381.

Las L, Shapira AH, and Nelken I (2008) Functional gradients of auditory sensitivity along the anterior ectosylvian sulcus of the cat. *Journal of Neuroscience* **28**:3657–3667.

Lee C-C, Macpherson EA, and Middlebrooks JC (2008) Task-dependence of spatial sensitivity in cat auditory cortex: area A1 vs DZ. *Association of Research in Otolaryngology Abstracts* **31**:1204.

Lewald J, Riederer KA, Lentz T, and Meister IG (2008) Processing of sound location in human cortex. *European Journal of Neuroscience* **27**:1261–1270.

Litovsky RY, Colburn HS, Yost WA, and Guzman SJ (1999) The precedence effect. *Journal of the Acoustical Society of America* **106**:1633–1654.

Lomber SG and Malhotra S (2008) Double dissociation of 'what' and 'where' processing in auditory cortex. *Nature Neuroscience* **11**: 609–616.

MacPherson EA, Stecker GC, Harrington IA, and Middlebrooks JC (2004) Nonlinear processing of spectral cues for sound localization in areas DZ and PAF of cat auditory cortex. *Society for Neuroscience Abstracts*.

Maeder PP, Meuli RA, Adriani M, Bellmann A, Fornari E, Thiran JP, Pittet A, and Clarke S (2001) Distinct pathways involved in sound recognition and localization: a human fMRI study. *Neuroimage* **14**:802–816.

Makous JC, Middlebrooks JC (1990) Two-dimensional sound localization by human listeners. *Journal of the Acoustical Society of America* **87**:2188–2200.

Malhotra S and Lomber SG (2007) Sound localization during homotopic and heterotopic bilateral cooling deactivation of primary and nonprimary auditory cortical areas in the cat. *Journal of Neurophysiology* **97**:26–43.

Malhotra S, Hall AJ, and Lomber SG (2004) Cortical control of sound localization in the cat: unilateral cooling deactivation of 19 cerebral areas. *Journal of Neurophysiology* **92**:1625–1643.

Malhotra S, Stecker GC, Middlebrooks JC, and Lomber SG (2008) Sound localization deficits during reversible deactivation of primary auditory cortex and/or the dorsal zone. *Journal of Neurophysiology* **99**:1628–1642.

Malone BJ, Scott BH, and Semple MN (2002) Context-dependent adaptive coding of interaural phase disparity in the auditory cortex of awake macaques. *Journal of Neuroscience* **22**:4625–4638.

May BJ and Huang AY (1996) Sound orientation behavior in cats. I. Localization of broadband noise. *Journal of the Acoustical Society of America* **100**:1059–1069.

McAlpine D and Grothe B (2003) Sound localization and delay lines--do mammals fit the model? *Trends in Neurosciences* **26**: 347–350.

Meredith MA and Clemo HR (1989) Auditory cortical projection from the anterior ectosylvian sulcus (Field AES) to the superior colliculus in the cat: an anatomical and electrophysiological study. *Journal of Comparative Neurology* **289**:687–707.

Mickey BJ and Middlebrooks JC (2003) Representation of auditory space by cortical neurons in awake cats. *Journal of Neuroscience* **23**:8649–8663.

Mickey BJ and Middlebrooks JC (2005) Sensitivity of auditory cortical neurons to the locations of leading and lagging sounds. *Journal of Neuroscience* **94**:979–989.

Middlebrooks JC and Knudsen EI (1984) A neural code for auditory space in the cat's superior colliculus. *Journal of Neuroscience* **4**:2621–2634.

Middlebrooks JC and Pettigrew JD (1981) Functional classes of neurons in primary auditory cortex of the cat distinguished by sensitivity to sound location. *Journal of Neuroscience* **1**:107–120.

Middlebrooks JC, Clock AE, Xu L, and Green DM (1994) A panoramic code for sound location by cortical neurons. *Science* **264**:842–844.

Middlebrooks JC, Dykes RW, and Merzenich MM (1980) Binaural response-specific bands in primary auditory cortex (AI) of the cat: topographical organization orthogonal to isofrequency contours. *Brain Research* **181**:31–48.

Middlebrooks JC, Lee C-C, and Macpherson EA (2009) Some brain mechanisms for auditory scene analysis. *Journal of the Acoustical Society of America* **125**:2491.

Middlebrooks JC, Xu L, Eddins AC, and Green DM (1998) Codes for soundsource location in nontonotopic auditory cortex. *Journal of Neurophysiology* **80**:863–881.

Miller LM and Recanzone GH (2009) Populations of auditory cortical neurons can accurately encode acoustic space across stimulus intensity. *Proceedings of the National Academy Sciences of the United States of America* **106**:5931–5935.

Mills AW (1958) On the minimum audible angle. *Journal of the Acoustical Society of America* **30**:237–246.

Morel A and Imig TJ (1987) Thalamic projections to fields A, AI, P, and VP in the cat auditory cortex. *Journal of Comparative Neurology* **265**:119–144.

Mrsic-Flogel TD, King AJ, Jenison RL, and Schnupp JWH (2001) Listening through different ears alters spatial response fields in ferret primary auditory cortex. *Journal of Neurophysiology* **86**: 1043–1046.

Mrsic-Flogel TD, King AJ, and Schnupp JWH (2005) Encoding of virtual acoustic space stimuli by neurons in ferret primary auditory cortex. *Journal of Neurophysiology* **93**:3489–3503.

Mrsic-Flogel TD, Schnupp JWH, and King AJ (2003) Acoustic factors govern developmental sharpening of spatial tuning in the auditory cortex. *Nature Neuroscience* **6**:981–988.

Nakamoto KT, Jones SJ, and Palmer AR (2008) Descending projections from auditory cortex modulate sensitivity in the midbrain to cues for spatial position. *Journal of Neurophysiology* **99**:2347–2356.

Nakamoto KT, Zhang J, and Kitzes LM (2004) Response patterns along an isofrequency contour in cat primary auditory cortex (AI) to stimuli varying in average and interaural levels. *Journal of Neurophysiology* **91**:118–135.

Neff WD, Diamond IT, Fisher JF, and Yela M (1956) Role of auditory cortex in discrimination requiring localization of sound in space. *Journal of Neurophysiology* **19**:500–512.

Nelken I, Bizley JK, Nodal FR, Ahmed B, King AJ, and Schnupp JWH (2008) Responses of auditory cortex to complex stimuli: functional organization revealed using intrinsic optical signals. *Journal of Neurophysiology* **99**:1928–1941.

Nelken I, Chechik G, Mrsic-Flogel TD, King AJ, and Schnupp JWH (2005) Encoding stimulus information by spike numbers and mean response time in primary auditory cortex. *Journal of Computational Neuroscience* **19**:199–221.

Nodal FR, Bajo VM, and King AJ (2010) Role of auditory cortex in acoustic orientation and approach-to-target responses. In: Lopez-Poveda EA, Palmer AR, and Meddis R (eds). *Advances in Auditory Physiology, Psychophysics and Models*. Springer, New York, pp. 581–596.

Nodal FR, Bajo VM, Parsons CH, Schnupp JWH, and King AJ (2008) Sound localization behavior in ferrets: comparison of acoustic orientation and approach-to-target responses. *Neuroscience* **154**: 397–408.

Phillips DP (2008) A perceptual architecture for sound lateralization in man. *Hearing Research* **238**:124–132.

Rajan R, Aitkin LM, Irvine DRF, and McKay J (1990a) Azimuthal sensitivity of neurons in primary auditory cortex of cats. I. Types of sensitivity and the effects of variations in stimulus parameters. *Journal of Neurophysiology* **64**:872–887.

Rajan R, Aitkin LM, and Irvine DRF (1990b) Azimuthal sensitivity of neurons in primary auditory cortex of cats. II. Organization along frequency-band strips. *Journal of Neurophysiology* **64**:888–902.

Razak KA and Fuzessery ZM (2002) Functional organization of the pallid bat auditory cortex: emphasis on binaural organization. *Journal of Neurophysiology* **87**:72–86.

Reale RA and Brugge JF (2000) Directional sensitivity of neurons in the primary auditory (AI) cortex of the cat to successive sounds ordered in time and space. *Journal of Neurophysiology* **84**:435–450.

Reale RA, Jenison RL, and Brugge JF (2003) Directional sensitivity of neurons in the primary auditory (AI) cortex: effects of sound-source intensity level. *Journal of Neurophysiology* **89**:1024–1038.

Recanzone GH, Guard DC, Phan ML, and Su TK (2000) Correlation between the activity of single auditory cortical neurons and sound-localization behavior in the macaque monkey. *Journal of Neurophysiology* **83**:2723–2739.

Rouiller EM, Simm GM, Villa AE, de Ribaupierre Y, and de Ribaupierre F (1991) Auditory corticocortical interconnections in the cat: evidence for parallel and hierarchical arrangement of the auditory cortical areas. *Experimental Brain Research* **86**:483–505.

Rutkowski RG, Wallace MN, Shackleton TM, and Palmer AR (2000) Organisation of binaural interactions in the primary and dorsocaudal fields of the guinea pig auditory cortex. *Hearing Research* **145**: 177–189.

Sabin AT, Macpherson EA, and Middlebrooks JC (2005) Human sound localization at near-threshold levels. *Hearing Research* **199**: 124–134.

Scott BH, Malone BJ, and Semple MN (2009) Representation of dynamic interaural phase difference in auditory cortex of awake rhesus macaques. *Journal of Neurophysiology* **101**:1781–1799.

Shackleton TM, Skottun BC, Arnott RH, and Palmer AR (2003) Interaural time difference discrimination thresholds for single neurons in the inferior colliculus of guinea pigs. *Journal of Neuroscience* **23**:716–724.

Schnupp JWH, Mrsic-Flogel TD, and King AJ (2001) Linear processing of spatial cues in primary auditory cortex. *Nature* **414**:200–204.

Smith AL, Parsons CH, Lanyon RG, Bizley JK, Akerman CJ, Baker GE, Dempster AC, Thompson ID, and King AJ (2004) An investigation of the role of auditory cortex in sound localization using muscimol-releasing Elvax. *European Journal of Neuroscience* **19**: 3059–3072.

Spierer L, Tardif E, Sperdin H, Murray MM, and Clarke S (2007) Learning-induced plasticity in auditory spatial representations revealed by electrical neuroimaging. *Journal of Neuroscience* **27**:5474–5483.

Stecker GC and Middlebrooks JC (2003) Distributed coding of sound locations in the auditory cortex. *Biological Cybernetics* **89**:341–349.

Stecker GC, Harrington IA, Macpherson EA, and Middlebrooks JC (2005a) Spatial sensitivity in the dorsal zone (area DZ) of cat auditory cortex. *Journal of Neurophysiology* **94**:1267–1280.

Stecker GC, Harrington IA, and Middlebrooks JC (2005b) Location coding by opponent neural populations in the auditory cortex. *Public Library of Science Biology* **3**:e78.

Stecker GC, Mickey BJ, Macpherson EA, and Middlebrooks JC (2003) Spatial sensitivity in field PAF of cat auditory cortex. *Journal of Neurophysiology* **89**:2889–2903.

Su TI and Recanzone GH (2001) Differential effect of near-threshold stimulus intensities on sound localization performance in azimuth and elevation in normal human subjects. *Journal of the Association for Research in Otolaryngology* **2**:246–256.

Tian B, Reser D, Durham A, Kustov A, and Rauschecker JP (2001) Functional specialization in rhesus monkey auditory cortex. *Science* **292**:290–293.

Werner-Reiss and Groh JM (2008) A rate code for sound azimuth in monkey auditory cortex: implications for human neuroimaging studies. *Journal of Neuroscience* **28**:3747–3758.

Werner-Reiss U, Kelly KA, Trause AS, Underhill AM, and Groh JM (2003) Eye position affects activity in primary auditory cortex of primates. *Current Biology* **13**:554–562.

Wightman FL and Kistler DJ (1993) Sound localization. In: Yost WA, Popper AN, and Fay RR (eds). *Springer Handbook of Auditory Research*, volume 3, *Human Psychophysics*. Springer, New York, pp. 155–192.

Winer JA and Schreiner CE (2005) *The Inferior Colliculus*. Springer, New York.

Woods TM, Lopez SE, Long JH, Rahman JE, and Recanzone GH (2006) Effects of stimulus azimuth and intensity on the single-neuron activity in the auditory cortex of the alert macaque monkey. *Journal of Neurophysiology* **96**:3323–3337.

Yamada K, Kaga K, Uno A, and Shindo M (1996) Sound lateralization in patients with lesions including the auditory cortex: comparison of interaural time difference (ITD) discrimination and interaural intensity difference (IID) discrimination. *Hearing Research* **101**:173–180.

Yin TCT (2002) Neural mechanisms of encoding binaural localization cues in the auditory brainstem. In: Oertel D, Fay RR, and Popper AN (eds). *Springer Handbook of Auditory Research*, volume 15, *Integrative Functions in the Mammalian Auditory Pathway*. Springer, New York, pp. 99–159.

Young ED and Davis KA (2002) Circuitry and function of the dorsal cochlear nucleus. In: Oertel D, Fay RR, and Popper AN (eds). *Springer Handbook of Auditory Research*, volume 15, *Integrative Functions in the Mammalian Auditory Pathway*. Springer, New York, pp. 160–206.

Zatorre RJ and Penhune VB (2001) Spatial localization after excision of human auditory cortex. *Journal of Neuroscience* **21**:6321–6328.

Zhang J, Nakamoto KT, and Kitzes LM (2004) Binaural interaction revisited in the cat primary auditory cortex. *Journal of Neurophysiology* **91**:101–117.

Chapter 16

Communication Sounds and their Cortical Representation

Jagmeet S. Kanwal and Günter Ehret

Abbreviations

AC	auditory cortex
AI	primary auditory cortex
AIp	posterior pole of primary auditory cortex
AII	second auditory field
AL	anterior lateral belt area
AM	amplitude modulation
AAF	anterior auditory field
CF	constant frequency
DF	dorsal fringe
DSCF	Doppler-shifted constant frequency area
FM	frequency modulation/frequency sweep
fMRI	functional magnetic resonance imaging
IBE	information-bearing element
IBP	information-bearing parameter
LFP	local field potential
ML	medial lateral belt area
NB	noise burst
PAF	posterior auditory field
PET	positron emission tomography
R	rostral field
UF	ultrasonic field
VA	ventral anterior area
VOT	voice onset time
VSD	voltage sensitive dye

J.S. Kanwal (✉)
Department of Physiology and Biophysics, Georgetown University Medical Center, Washington, DC 20057-1460, USA
e-mail: kanwalj@georgetown.edu

G. Ehret (✉)
Institute of Neurobiology, University of Ulm, D-89069 Ulm, Germany
e-mail: guenter.ehret@uni-ulm.de

1 Introduction

1.1 Audiovocal Communication and the Structure of Communication Sounds

Evolutionary selection pressures and mechanisms have shaped the acoustic structure of vocalizations of senders and the perceptual abilities of receivers to allow audiovocal communication between conspecifics and different species. Such mechanisms have likely shaped the adaptations that allow humans to use sounds in speech and music (Sussman 1989; Sussman et al. 1991). Within-species communication often involves social interactions between two or a few animals, as in nursing, reproduction, or agonistic behaviors. Furthermore, within-species communication plays an important role in group interactions and the audience effect (Evans and Marler 1991). Across-species communication typically conveys other information of prospective biologic significance such as danger. Social communication between humans and their pets or between domesticated species raised together is special in this context.

Mammalian communication sounds consist of generic or basic acoustic patterns or sound elements and include constant frequency (CF) tones, frequency-modulated tones or frequency sweeps (FM), and noise bursts (NB). When a basic acoustic pattern carries behaviorally relevant information about the sender and/or the context, it is defined as an information-bearing element or IBE (Suga and Schlegel 1972; Suga 1978, 1988). IBEs contain information-bearing parameters (IBPs) (Suga 1988, 1994b, 1996) that specify sound duration, frequency and amplitude, fundamental frequency, number and frequency ranges of the predominant harmonics or formants, bandwidth, slope and central frequency in FMs, the depth and duty cycle of amplitude modulation (AM), and the noisiness perceived as roughness or harshness. IBPs may further define spectral and/or temporal relationships between IBEs, such as the duration

J.A. Winer, C.E. Schreiner (eds.), *The Auditory Cortex*,
DOI 10.1007/978-1-4419-0074-6_16, © Springer Science+Business Media, LLC 2011

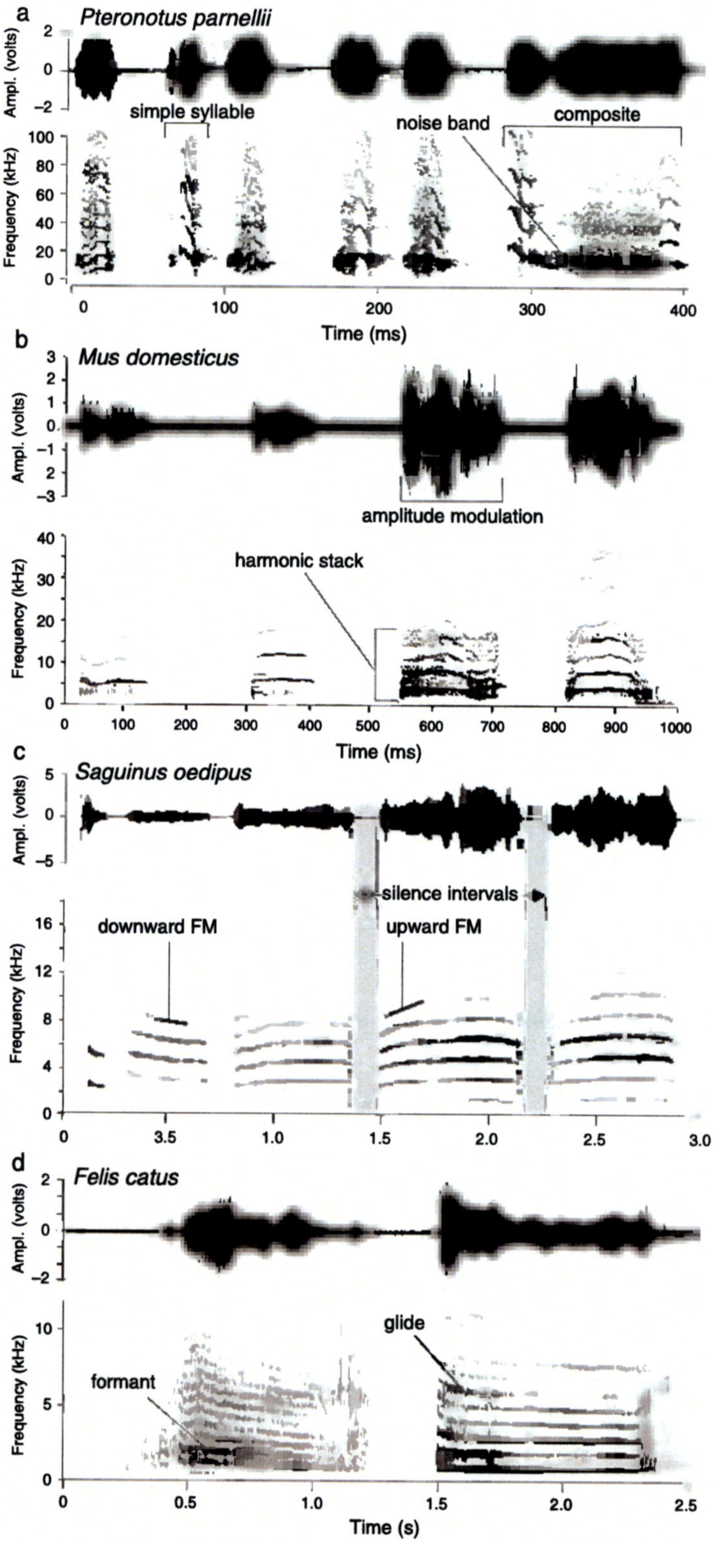

Fig. 16.1 Information-bearing elements (IBEs) and information-bearing parameters (IBPs) in vocalizations of **a** mustached bat, **b** house mouse, **c** monkey, and **d** cat. **a** Vocal exchange between a male and a female bat housed together with another male in a cage. **b** Series of wriggling calls of a 5-day-old mouse pup. Adapted from the original source (Ehret 1975; Ehret and Riecke 2002). The vocalizations release maternal behavior only if they are perceived as a series. **c** Cotton-top tamarin calls. A representative combination long call consisting of a series of whistles following a chirp (adapted from Ghazanfar and Hauser 2001). The oscillogram (*upper part*) shows the temporal structure with the amplitude modulations in the calls. The spectrogram (*lower part*) shows the frequency structure. Several harmonics, rapid amplitude modulations (side bands to the harmonics), and noise components are visible

of silent intervals between syllables. Species-specific IBEs and IBPs allow species to coexist and communicate among conspecifics in an acoustically cluttered environment. By hearing and perceiving the vocal interaction between group members, an individual and its motivational/emotional state may be identified by specific IBEs and IBPs. Examples of mammalian vocalizations from bats, mice and monkeys and cats with some IBEs and IBPs illustrate this (Fig. 16.1).

The common mechanisms that generate mammalian IBEs characterized by their IBPs are vocal cord vibrations. Therefore, CF and FM tonal components in communication sounds have a common acoustical structure with a fundamental frequency, the cord vibration frequency, and many overtones or harmonics. Some harmonics are amplified by selective resonances from the shape of the laryngeal, pharyngeal, and nasal cavities. Such amplified harmonics or groups of harmonics are vocalization formants. Thus, pitch (fundamental frequency), a formant structure, and the timbre quality result from frequency and amplitude interactions of harmonics and are important acoustic characteristics of vocalizations. Moreover, all frequency components in a vocalization are influenced by a vocal amplitude modulation and AM can produce pitch percepts identical to the frequency of AM (Langner 1992). In rapid and irregular AM, as when vocal cord air pressure is high, turbulence is created so that formant structure and vocalization timbre are superimposed by roughness or noisiness and, in the extreme case, may be lost. Such vocalizations are perceived as harsh. Brief noise bursts may be produced as clicks, by sudden openings or closings of the air pathway, longer noise bursts by passing air through the vocal tract without vocal cord vibrations. Besides vocalizing single calls or a series of calls of one type, different call types can be combined to generate composite calls (composites) (Kanwal et al. 1994). Acoustically, composites resemble phonemes, and a series of composites or call phrases are comparable to words in human speech. Recognition of composites and call phrases requires the analysis of syntax in a sound series.

We will examine our knowledge of species-specific communication sound representation in the auditory cortex (AC). We will show that complex sounds need not be represented as wholes. Rather, a coding strategy can represent both the identity of a complex sound and the sender, and its emotional state and possibly the gender of the source, simultaneously. This is accomplished by representing simply the IBEs and sometimes creating maps of IBPs. What are these IBEs and how does the auditory system extract the relevant parameters from complex sounds, and how is this reflected in the AC cells' response properties and AC organization? We begin by examining sound properties of potential importance for characterizing an auditory percept.

1.2 Methods for Characterizing the Auditory Cortex

1.2.1 In Vivo Electrophysiology

Most studies on neural responses to communication sounds (calls) use electrophysiological in vivo recordings of single and multiunit activity and local field potentials (LFPs). Simple tone bursts were first used to identify the AC fields and their frequency representation, the tonotopy (Fig. 16.2). Pure tones and their potential IBPs are insufficient, however, to understand communication sound processing. Sounds of intermediate complexity, such as FMs, and natural calls are now used to study the AC organization and neuronal receptive fields. LFPs can illuminate the role of local neural populations in encoding sounds by their coordinated firing, leading to oscillations that may represent a transient binding of neural activity for the extraction of meaning.

Most mammalian electrophysiological studies are from anesthetized animals. These AC neural responses may differ from those of the awake animal since anesthetics disturb the balance between excitation and inhibition, which can significantly change neuronal responses, especially to complex sounds (Nelken 2002; Plourde et al. 2006). Even studies in awake, restrained, and/or head-fixed animals that cannot respond to the perceived sounds, need to be interpreted cautiously. Since AC neurons are sensitive to somatic sensory input (Lakatos et al. 2007), initiation of motor output (Warren et al. 2005), and attention to sound for the preparation of a motor response (Fritz et al. 2003), any restraint may change their responses to relevant sounds.

1.2.2 Metabolic Activity Mapping and the Blood Oxygen Level Signal

An open question is how AC local and global activity patterns relate to auditory perception. This question is relevant, especially for the perception of calls, because only some sound properties may be apparent in the response properties of neurons at a given locus. To address this question, global AC activation patterns are obtained in autoradiographs of local sugar consumption (Gonzalez-Lima and Scheich 1986; Scheich et al. 1993; Ohl and Scheich 1997) or analyses of stimulus-induced gene expression (Fichtel and Ehret 1999; Wan et al. 2001; Geissler and Ehret 2004). In awake and attentive humans, positron emission tomography (PET) and functional magnetic resonance imaging (fMRI) are used to study the neural substrate for the perception of speech (Mazoyer et al. 1993; Binder et al. 2000). How do data on speech representation in the human AC compare to communication call representation in the AC of other mammals? An answer could contribute to our understanding of the

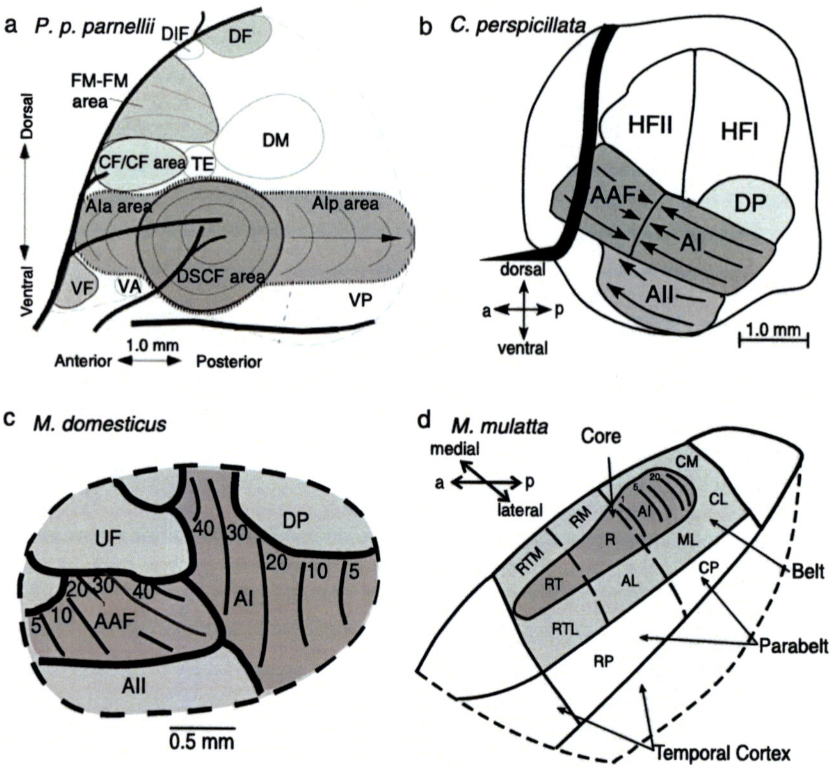

Fig. 16.2 View on the left-side auditory cortex (AC) of the (**a**) mustached bat (*Pteronotus parnellii parnellii*), (**b**) a fruit bat (*Carollia perspicillata*), (**c**) house mouse and (**d**) macaque monkey. **a** Dorsolateral view of the AC in left cerebral hemisphere of the mustached bat showing its functional organization into 12 areas specialized for processing echolocation signals. The AC consists of several nonprimary areas (CF/CF FM-FM, DF VF, DM, VA, and TE). Adapted from the original (Suga 1984). Areas A1a, A1p, and DSCF, respectively, belong to the primary AC. Four areas are implicated in communication sound processing (see text). Branching lines, segments of the medial cerebral artery. The longest branch is on the sulcus or fossa. **b** The fruit bat AC is based on tonotopic mapping of multiunit activity, adapted from the original (Esser and Eiermann 1999). **c** The mouse AC is modified from the original (Stiebler et al. 1997) and contains two primary (core) fields, the primary field (AI) and the anterior auditory field (AAF), an ultrasonic field (UF) with representation of frequencies >45 kHz, a secondary field (AII), a dorsoposterior field (DP), and two small unnamed areas. The isofrequency strips (numbers are frequencies in kHz) in AI and AAF indicate their tonotopy. The other fields have no regular tonotopy. **d** The macaque AC modified from the original source (Kaas et al. 1999) contains three tonotopically organized core fields: the primary cortex (AI), the rostral core field (R), and the rostrotemporal field (RT). Around the core is the belt, consisting of the caudomedial (CM) and caudolateral (CL), middle lateral (ML), anterolateral (AL), lateral rostrotemporal (RTL), medial rostrotemporal (RTM), and rostromedial (RM) areas. Lateral to the belt area are the rostral (RP) and the caudal (CP) parabelt areas

functional parceling of the AC for processing of IBEs, IBPs, auditory objects, syntax, and meaning, and could reveal which parts of the AC are specialized for species-specific communication.

1.2.3 Optical Imaging

Optical imaging using an intrinsic signal or fluorescence-induced change in a voltage-sensitive dye (VSD) has been used to monitor spatiotemporal activity of cortical neurons (Taniguchi et al. 1992; Wu et al. 2001; Jin et al. 2002; Kalatsky et al. 2005). These methods have not been fully exploited for studying auditory representations in vivo because of the cardiac and respiratory artifacts and cytotoxicity of voltage-sensitive dyes (Wu et al. 1998; Francois et al. 2000), though new dyes and analytic methodologies are promising (Lippert et al. 2007). VSD imaging allowed the direct visualization of the activity of populations of neurons to tone bursts and complex stimuli, including a species-specific call (Horikawa et al. 1998; Horikawa et al. 2001; Horikawa et al. 2006).

2 Auditory Cortical Fields in Representative Species

Mammalian AC is divided into primary and higher-order fields by its cytoarchitecture, its connectivity with different medial geniculate body regions and other thalamic nuclei, and AC fields (Winer et al. 1992; de Ribaupierre 1997). Eutherian mammals have at least two primary fields with a tonotopic organization resembling the frequency gradient along the cochlear basilar membrane. Three or more higher-order fields have also been characterized by their connectivity and neuronal response patterns (Fig. 16.2). Especially in primate AC, primary fields are defined as the core areas and higher-order fields as the belt and parabelt areas (Hackett et al. 1998; Rauschecker 1998a; Hackett et al. 2001).

We focus on bats as represented by the mustached bat (*Pteronotus parnellii*), rodents represented by the house mouse (*Mus domesticus*), and primates as represented by the common marmoset (*Callithrix jacchus*) and rhesus monkey (*Macaca mulatta*), which have served as model species for auditory cortical research. The mustached bat depends on audition for its survival and its auditory system is well studied, allowing for critical evaluation of intracortical versus subcortical contributions to AC responses in echolocation (Suga 1994b, 1996) and audiovocal social communication (Ohlemiller et al. 1994; Esser et al. 1997; Kanwal 1999, 2006; Medvedev and Kanwal 2004b, 2008). Research on mice has great potential for understanding the genetic bases of hearing and its disorders (Willott 2001; Delmaghani et al. 2006; Liu 2006), and instinctive differences in the perception and communicative significance of calls between individuals in different reproductive states are reflected in different neural responses in AC fields (Fichtel and Ehret 1999; Geissler and Ehret 2004). Monkeys have large vocal repertoires (Seyfarth et al. 1980; Newman et al. 1983), a suggested close resemblance to human AC organization, and a plasticity of acoustic representations amenable to study task-related analyses of single neuron responses in chronic preparations (Schwarz and Tomlinson 1990; Rauschecker et al. 1995; Fishman et al. 2001a; Nagarajan et al. 2002; Eliades and Wang 2003; Bendor and Wang 2005; Alain et al. 2007).

2.1 Mustached Bat

Electrophysiological studies in lightly anesthetized or awake, head-fixed animals have been used to characterize mustached bat AC. The AC contains several fields that are organized to represent maps of combination-sensitivity with regard to different IBEs and the IBPs relevant for echolocation. AC functional organization is the most detailed profile available for a mammal (Fig. 16.2a) (Suga 1994b, 1996). In the FM bat *Carollia persipicallata*, the AC fields identified on the basis of characteristic frequencies of neurons differ from those of the mustached bat (Fig. 16.2b) (Esser and Eiermann 1999), suggesting that AC functional organization varies even in species in the same order.

In the mustached bat, several maps derived from information processing for echolocation have been found. The primary AC (AI) tonotopy is enlarged for the representation of frequencies corresponding to the echolocation pulse second harmonic (CF_2). This resting CF_2, so-called because it is measured in stationary or resting bats, varies from 58 to 63 kHz with the subspecies. This auditory fovea in AI is the Doppler-shifted constant frequency area (DSCF) and corresponds to the frequencies over-represented in the bat's cochlea (Kössl and Vater 1985). Besides their tuning to an echo frequency, the DSCF neurons topographically represent specific echo amplitudes, creating a local response optimum correlated with the size and angle of a moving prey target. Second, CF/CF area cells are combination-sensitive to the CF component of the echolocation pulse fundamental frequency and the CF component of a higher echo harmonic, and to a Doppler-shifted echo frequency due to a velocity difference between the bat and its prey. This combination-sensitivity leads to optimum responses at multiple locations representing the instantaneous relative speed between the bat and its prey. Third, neurons in the non-primary FM–FM and dorsal fringe (DF) areas are combination-sensitive to the fundamental frequency FM component in the echolocation pulse and the FM component of a higher harmonic in the echo and to a specific delay between the echolocation pulse and echo. This combination-sensitivity creates a detailed target image (Dear et al. 1993b), including an instantaneous measure of target distance. Single neuron tuning to more than one dimension or IBE in the echolocation signal leads to multiparametric integration and will be discussed later.

2.2 House Mouse

The AC fields in the house mouse have been characterized by electrophysiological mapping of characteristic frequencies and by studying the activation patterns via *c-Fos* immunocytochemistry after the presentation of calls in natural settings (Stiebler et al. 1997; Fichtel and Ehret 1999; Geissler and Ehret 2004). The AC has five major fields and two small areas (Fig. 16.2c). The cells in the two tonotopically organized core fields, AI and anterior auditory field (AAF), have characteristic frequencies <45 kHz which is less than the range of cochlear nerve fibers (Taberner and Liberman 2005). Neurons with CFs to $\sim$70 kHz are in a non-tonotopic ultrasonic field (UF). At least part of UF is primary AC and can be regarded as a specialized part of AI, resembling the mustached bat's DSCF area. Perhaps UF contains neurons combination-sensitive to mouse ultrasonic calls that are part of the communication between mouse pups and their mothers and in sexual interactions (Ehret 1975; Liu et al. 2003; Ehret 2005).

2.3 Monkeys

Rhesus monkeys', marmosets', and owl monkeys' auditory cortical fields are well studied (Merzenich and Brugge 1973; Imig et al. 1977; Aitkin et al. 1986; Morel and Kaas 1992; Morel et al. 1993; Petkov et al. 2006). Initially, primate AC was divided into three cytoarchitectonic fields: Brodmann areas 41, 42, and 22 (or TC, TB, and TA, respectively; Bonin and Bailey 1947). Later the primary AC (areas 41 or TC) was subdivided into two to three core areas on histochemical and neurophysiological grounds (Pandya and Sanides 1973) (Fig. 16.2d). Similarly, the secondary AC, surrounding the core both laterally and medially, was subdivided into seven belt areas (Rauschecker et al. 1995; Kaas and Hackett 2000; Rauschecker and Tian 2000). Beyond the belt areas, a third level of processing is the parabelt, whose organization is less well understood but contains at least rostral (anterior) and caudal (posterior) subdivisions (Hackett et al. 1998). The rostral and caudal belt and parabelt subdivisions may initiate functional processing streams for the identification of complex auditory patterns or objects (what), and spatial perception of sound, including motion in space (where and how), respectively (Rauschecker et al. 1997; Rauschecker 1998b; Rauschecker and Tian 2000).

The division of the human AC by its anatomy, connectivity, and tonotopy is consistent with that in monkeys (Hackett et al. 2001; Formisano et al. 2003; Sweet et al. 2005), including a rostral stream for sound identification and a caudal one for the processing of spatial relationships (Alain et al. 2001). A comparison of functional specializations of fields for sound processing is difficult, if not impossible because there is little information about call processing in the belt and parabelt fields of primates and in higher-order fields of the AC of other mammals.

2.4 Other Mammals

Multiple subdivisions of the AC, tonotopy, and functional maps are found in cat (Merzenich and Brugge 1973; Imig and Reale 1980), ferret (Bizley et al. 2005), rat (Hosokawa et al. 1998; Doron et al. 2002; Polley et al. 2007), Mongolian gerbil (Budinger et al. 2000), guinea pig (Wallace et al. 2000), and some bats (Ostwald 1984; Dear et al. 1993a; Radtke-Schuller and Schuller 1995; Esser and Eiermann 1999). In the pallid bat AI low-frequency range (~8–20 kHz), there is a concentric map of the sound object (prey) azimuth angle (Razak et al. 1999; Razak and Fuzessery 2002). The dorsal low-frequency region of the gerbil AI has a horseshoe-like map selective for pitch derived from rapid amplitude modulation (periodicity pitch) (Schulze et al. 2002). Across the whole frequency range formed by the isofrequency stripes in squirrel monkey AI, there is a ~4 ms latency gradient of responses to tones (higher best frequency cells had longer latencies) (Cheung et al. 2001). Neurons preferentially responding to brief sounds (few ms) or long sounds (<100 ms) form two stripes across the AI frequency map in the little brown bat (Galazyuk and Feng 1997).

3 Complex Sound Representation and Processing

With the exception of mustached bats, the information about AC functional organization is largely from studies of tonotopy or the representation of simple acoustic parameters in AI. Homologies between fields in the mammalian AC derived from anatomy, physiology, or perception are unclear (Kaas 2005), even in the primary AC. In most studies, one of the two or three tonotopically organized core fields was defined as AI, based on the strong, non-habituating, and sharply tuned frequency responses. Since the rostrocaudal frequency gradients in each field show a frequency reversal at their borders, AI can differ in positions, size, and frequency gradients in different mammals. Across-species comparisons of communication sound processing in fields denominated as AI, the source for most data available, can include fields that may not be anatomical and connectional homologues, making generalization of the results difficult and understanding the role of AI in complex sound perception problematic.

3.1 Representation Within Primary Auditory Cortex

Acoustic parameters and neural response parameters that may code for response patterns represented in AI are available in bat (Suga 1988; Suga 1990; Ohlemiller et al. 1996), ferret (Bizley et al. 2005) and cat (Merzenich and Brugge 1973; Imig and Reale 1980). As in bats, there is systematic clustering and gradients of neural response characteristics besides tonotopy in cat AI (Schreiner 1992, 1995; Schreiner et al. 2000; Schreiner and Winer 2007) summarized in Fig. 16.3.

Gradients along and across isofrequency stripes and the clustering of neurons with specific preferences along isofrequency stripes, or in a certain area of AI do not directly inform us about the representation of calls in AI. Two exceptions are the mustached bat, where acoustic parameters (IBEs and IBPs) of importance in echolocation are mapped, and the house mouse, in which the representation of ultrasonic calls of pups and adults (Ehret 1975), was mainly in the UF

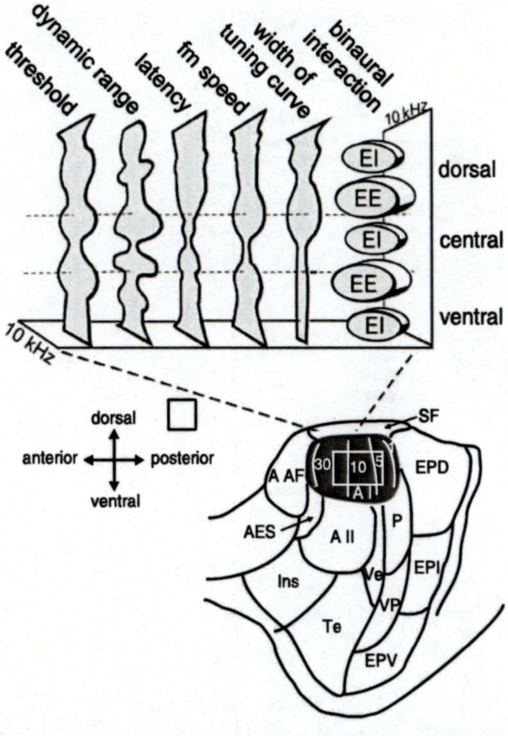

Fig. 16.3 View of the left side of cat AC (*lower part*) modified from the original source (Lee and Winer 2005). The 10 kHz isofrequency strip of the primary field (AI) is enlarged (*upper part*) and the representation of various neural response properties indicated along the dorsoventral axis; modified from the original source (Ehret 1997). Topographies are shown for tone response threshold, dynamic range, response latency, preferred rate of frequency modulation (FM rate), width of frequency tuning curve, and binaural interaction clusters of neurons (EE, excited by both ears; EI, excited by the contralateral and inhibited by the ipsilateral ear). These topographies are superimposed on the isofrequency strip; for clarity they are shown side by side. The necks of the topographies indicate small values (low thresholds, small dynamic range, short latency, low FM speed, and narrow tuning curve). All five topographies have a neck near the center of the strip, and the values increase towards the dorsal and ventral border of AI but may have more necks between the center and these borders. The superposition on these topographies of neural response properties creates local patches of neurons with unique combinations of responsiveness to sound features. Besides AI, the cat has four other tonotopic fields: AAF (anterior auditory field), P (posterior field), Ve (ventral field), and VP (ventral posterior field). Other ectosylvian gyrus areas (EPD, EPI, EPV), suprasylvian fringe area (SF; also known as the dorsal auditory zone), secondary AC (AII), temporal (Te), and insular (Ins) cortex, and an area in the anterior ectosylvian sulcus (AES) are devoid of tonotopy. The sounds used in the mapping studies were regular and amplitude modulated constant frequency (CF) tones or tone complexes, frequency-modulated (FM) sweeps, noises of various bandwidths, and tone-noise combinations (Ehret and Schreiner 1997; Ehret and Schreiner 2000). The stimuli were presented to one or both ears. Beyond the tonotopic gradient represented by CFs, neurons in an isofrequency strip (Fig. 16.3) show response properties according to their relative place in it. In the strip center in central AI, (i) tone-response thresholds are lowest, (ii) frequency tuning is sharpest and spectral resolution is highest and most compatible with psychophysical measures, (iii) dynamic ranges of rate-intensity functions are smallest, i.e., rate-intensity functions are peaked

(see above). Extrapolating these results supports two general conclusions.

First, specialized areas exist in AI for a certain frequency range, as in mouse UF (Fig. 16.2c), or amplitude in mustached bats (Fig. 16.2a), azimuth angle representation in pallid bats (Razak and Fuzessery 2002), or pitch in the gerbil (Schulze et al. 2002). These areas suggest that the represented sound property must be vital for the species, for locating or identifying prey (bats) or for sound communication (mouse). The significance of gerbil pitch representation is unknown. Second, a patchy distribution of topographies of neuronal response properties (Fig. 16.3) is a general feature of mammalian AI. Accordingly, every locus in AI can be characterized by a set of values describing a generic sound pattern. The result is a non-arbitrary, local specificity of neuronal responsiveness to a complex sound according to the IBEs and IBPs within it, such as frequency composition, intensity of frequency components (spectral shape), frequency modulation and FM sweeps, pitch by amplitude modulation, duration, and sound source location in the horizontal plane. Each locus in AI may respond to its optimal stimulus, i.e., the sound that elicits the best peak response magnitude or shortest latency. Further, this representation may be dynamic and plastic.

Different stimulation techniques reveal the optimal stimulus for a given AI neuron in monkeys (Nelken et al. 1994; deCharms et al. 1998; O'Connor et al. 2005). A linear summation of response probabilities at a given locus of AI as derived from the knowledge of patchy distributions of neural response properties to IBEs and IBPs is unlikely. These responses are expected to interact in a non-linear manner to increase the signal-to-noise ratio for the preferred stimulus. Excitatory, facilitatory, and inhibitory tuning curves in the AI of different species (Fig. 16.4) reveal that the local responsiveness is shaped by spectral and temporal inhibition and facilitation (Horikawa et al. 1996; Kanwal et al. 1999; Sutter et al. 1999; Kadia and Wang 2003; Wehr and Zador 2003; Bartlett and Wang 2005; Metherate et al. 2005; Horikawa et al. 2006). Further, a combination of different IBEs and

Fig. 16.3 (continued) (non-monotonic), (iv) tone-response latencies are shortest, and (v) neurons prefer downward FM sweeps at slow FM rates. Lateral to the center of an isofrequency strip, i.e., in the cat AI dorsal and ventral subareas, patches of neurons have higher tone-response thresholds, broader tuning, larger dynamic ranges, longer tone-response latencies, and may prefer upward FM sweeps at higher FM rates. Response properties at more lateral positions may change again. The loci of sweep direction preferences in the squirrel monkey have been mapped (Godey et al. (2005). Simple sounds, such as a pure tone burst at a certain frequency or an FM at a certain rate and level, elicit an optimal response only at one or a few loci at the corresponding AI isofrequency strip

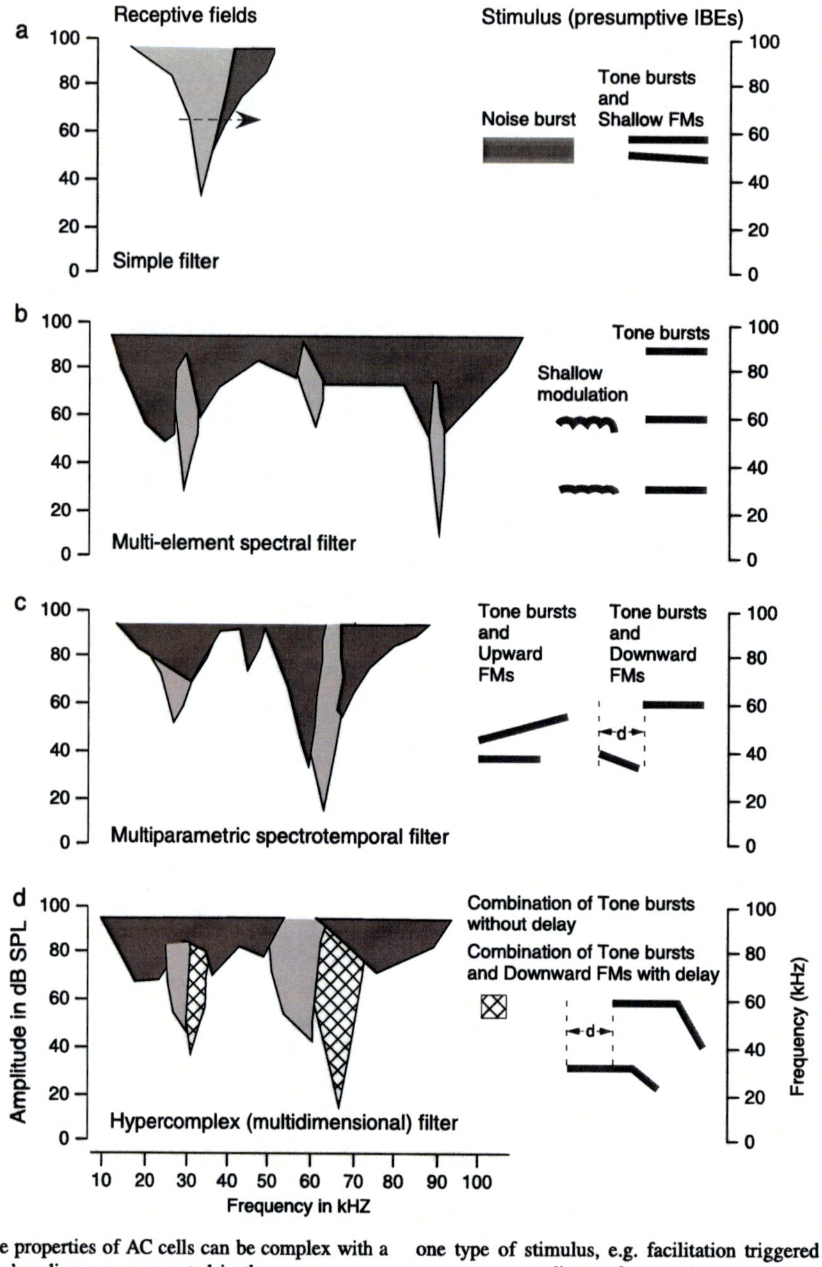

Fig. 16.4 The response properties of AC cells can be complex with a large part of the species' audiogram represented in the response profile of a single neuron. The *left panels* (**a–d**) show simple to complex filter properties of AC neurons embedded in their two-dimensional, frequency-amplitude inhibitory (*dark gray*) and excitatory/facilitatory (*light gray*) response areas. Cross-hatches (**d**) represent specialized excitatory response areas responsive only to FM sounds in hypercomplex filters. A stimulus combination traversing both excitatory response areas of a cell leads to a facilitated or combination-sensitive response. *Arrow*, the preferred direction of an FM sound. The shape of inhibitory response areas and their overlap with excitatory response areas may make the cell's response level-tolerant or modify its preference for a stimulus presented at a particular amplitude. *Right panel*, the spectrotemporal structure of the presumptive IBEs, which are extracted up to the AC. Even a relatively simple filter (**a**), with adjacent excitatory and inhibitory response areas, can respond equally well to more than one type of stimulus, e.g. facilitation triggered by a combination of tones corresponding to frequencies at the two edges of the excitatory response area, or a narrowband noise burst (NB), or an FM signal in a particular direction that traverses the excitatory and inhibitory response areas. The preferred FM directionality may be determined by an asymmetry in the organization of the inhibitory response areas abutting the excitatory response area. For a multiparametric spectrotemporal filter (**c**), the neural filter properties are designed to allow the cell to respond to more than one type of combination of sound patterns or presumptive IBEs. These can include combinations of CF, FM, and NB types of sound patterns. **d** Hypercomplex (multidimensional) filters are organized to extract one or more higher-order parameters, namely a time delay "d" between two FM sounds, but not necessarily for two CF sounds. All response area schematics are based on published data from the AC in mustached bats (e.g. Ohlemiller et al. 1996; Kanwal et al. 1999)

IBPs may be provided by specific sounds that dominate the responses at different AI loci and extracted by simple to hypercomplex filters within the AC (Suga 1994a; Ohlemiller et al. 1996; Kanwal 1999).

3.2 Representation outside Primary Auditory Cortex

Most studies of the AC are restricted to AI. However, other fields which may be as or more important than AI in processing and representing the biological significance of vocalizations (Figs. 16.2 and 16.3).

3.2.1 Fields Anterior and Ventral to Primary Auditory Cortex

Besides AI, the anterior auditory field (AAF) or the rostral field (R) (Figs. 16.2 and 16.3) belong to the primary or core AC fields. Hence, these fields are tonotopically organized and the neural responses to simple stimuli representing IBEs and IBPs, such as tones and frequency and amplitude modulated tones and noise, are comparable with AI neurons (Schreiner and Urbas 1986; Eggermont 1998; Linden et al. 2003; Imaizumi et al. 2004; Bizley et al. 2005; Polley et al. 2007). However, there are also systematic differences: in tonotopy, with frequencies in the central part of the cat, ferret, and gerbil hearing ranges possibly underrepresented in AAF (Thomas et al. 1993; Imaizumi et al. 2004; Bizley et al. 2005). AAF neurons have shorter response latencies (mouse, rat, cat, ferret) than AI neurons (Schreiner and Urbas 1986; Eggermont 1998; Linden et al. 2003; Rutkowski et al. 2003; Imaizumi et al. 2004; Bizley et al. 2005), and can follow higher rates of frequency and amplitude modulation (cat, mouse) than AI neurons (Schreiner and Urbas 1988; Linden et al. 2003; Imaizumi et al. 2004). Studies in marmosets on coding of rapid amplitude modulations (flutter) found that many AI neurons synchronized their responses to the modulation frequency, while most neurons in area R (probably corresponding to AAF) did not, and the latter encoded flutter by response rate changes (Bendor and Wang 2007). However, AAF (R) sound processing has a patchy topographic representation of acoustic parameters in neural responses, with local combination-sensitivities like those in AI (Imaizumi et al. 2004; Polley et al. 2007). A main difference between AI and more rostral primary (core) AC is in spatial information processing. Sound localization in cats remains intact after AAF, but not AI, deactivation (Malhotra and Lomber 2007). This supports the distinction between two AC processing streams of information, for object identification (what) and for sound localization (where) (Rauschecker 1998a,b). AAF (R) may belong to the anterior/ventral stream for sound identification and the extraction of meaning.

Ventral to cat, mouse, and FM bat AI is the second auditory field (AII), whose role in sound processing is understood poorly. The absence of clear tonotopy in the cat and mouse, broad frequency tuning, lability and habituation an area of to tone bursts suggests that this is of responses higher order than the primary AC (de Ribaupierre 1997; Rouiller 1997; Stiebler et al. 1997). Since AII deactivation in cat does not influence sound localization (Malhotra and Lomber 2007), it may belong with AAF to the anterior/ventral stream of information processing for identification of acoustic objects and meaning. If AII is analogous to monkey belt AC (Fig. 16.2), one would predict selectivity of neuronal responses to species-specific and other complex sounds, as in the monkey belt areas (Rauschecker et al. 1995). In fact, mouse AII neurons (Fichtel and Ehret 1999; Geissler and Ehret 2004) respond selectively to species-specific vocalizations based on their acoustic structure, behavioral setting, or both. Behaviorally relevant calls in a familiar environment (the typical communication situation) or calls of no behavioral relevance lead to only a few activated neurons as measured by *c-Fos* labeling. In the first case, the calls carried no new information. In the second case, they carried no behaviorally relevant information. Acoustically adequate calls in an unfamiliar situation or acoustically inadequate calls in a familiar situation label many more *c-Fos* positive neurons. In this scenario, a mismatch between calls and context elicits a larger response in more AII cells. A similar novelty response was seen in a rat AC area corresponding to mouse AII (Wan et al. 2001). AII cells integrate acoustic patterns together with perceptual inputs from other modalities and experience in a behavioral context. Perhaps responses in higher-order areas of the AC are not only driven by sound properties, but also by perceptual context. Where and how this contextual information reaches AII is unclear. It remains to be shown how this context- or task-related responsiveness is expressed in AI neuron responses and in higher auditory areas (Fritz et al. 2003, 2005; Scheich et al. 2005, 2007).

3.2.2 Fields Posterior to AI

Posterior to AI, a tonotopic posterior auditory field (PAF) has been described in cat (de Ribaupierre 1997). This area (Phillips and Orman 1984; Schreiner and Urbas 1988; Phillips et al. 1995; Tian and Rauschecker 1998; Loftus and Sutter 2001) and comparable fields in the ferret (PSF) (Bizley et al. 2005) and rat (PAF) (Polley et al. 2007) have cells with broader frequency tuning, longer tone response latencies, lower following rates of acoustic frequency and amplitude modulations, more non-monotonic rate-level functions and sharper intensity tuning than AI cells. Responsiveness of neurons to complex sounds is less predictable. Deactivation of cat PAF strongly impairs sound localization (Malhotra and Lomber 2007), suggesting that PAF may be involved in higher-level auditory processing for sound localization in the "where" pathway.

3.3 Representation of Communication Sounds: Beyond Tonotopy

The data on acoustic representation suggest that the AC contains several fields evolved for creating combination-sensitivity to different acoustic properties of importance in complex sounds. While the idea that IBEs and IBPs combine to represent information crucial for bat AC echolocation is evident, the nature of combination-sensitivity and the cortical fields required for its processing and for representation of communication calls is less obvious. What is this information in the calls and where does it reside in the acoustic design of a call?

3.3.1 Motivation-Structure Hypothesis

The motivation-structure hypothesis (Morton 1977; August and Anderson 1987) presents a framework in which the IBEs important to many species are defined. It delineates common rules that describe and explain from a comparative and evolutionary perspective the relationship between the physical structure of a sound and the motivational/emotional context expressed in the acoustic properties. Thus, friendly, appeasing or fearful tendencies are expressed in tonal, high-frequency sounds and aggressive or hostile motivations/emotions by harsh, noisy, low-frequency sounds. At low sound levels, the latter can express friendly motivations/emotions when the noisy elements are repeated rhythmically. How are the potential meanings that are transported in these different acoustic patterns represented in AC activity in order to eventually be perceived?

3.3.2 Rules for Call Perception

Building on Morton's rules (Morton 1977; August and Anderson 1987) and behavioral studies of sound perception in mice and other species, six rules for communication sound perception have been proposed (Ehret 2006a). These do not necessarily constrain potential IBP representations in AC, but they relate perceptions and behavior to the acoustic structure of calls. The following inferences can be made about communication sound representation in mammalian ACs.

First, IBPs are coded in AC such to accentuate just meaningful differences in the species-specific calls, not just noticeable differences. Local combination sensitivity in AI and AAF is adjusted by evolutionary forces to encode the meaning of calls in the species-specific repertoire, and by experience to distinguish learned meanings of environmental sounds including conspecific vocalizations. Attentional demands may increase AC spectrotemporal acuity in an IBP range for any IBE domain. AC evolutionary plasticity is mediated by hormones and neurotransmitter systems (dopaminergic, serotonergic, cholinergic). The perception of emotional meaning requires input from structures such as the amygdala (Naumann et al. 2006).

A second conclusion is that IBP variations describing the IBEs perception of acoustically expressed degrees of arousal, which may contribute to the meaning of a call or to perceptual processes affecting behavior. Call roughness or noisiness and duration and repetition rates can express the state of arousal and emotive and motivational states of the emitter. Sensitivity to spectrotemporal variations in basic call patterns is expected to be important for AC neurons and is supported by call responses in mustached bat DSCF areas (Kanwal and Rauschecker 2007).

Third, although acoustic parameters may change along a continuum, behavioral responses to sound are categorical and a response is elicited or not, or a sound source is approached or avoided. Auditory percepts should be categorical if behavioral decisions important for individual survival are based on them. Categorical perception of acoustic continua such as voice onset time (VOT) occurs at human speech phonetic boundaries used for semantic classification. Therefore, we should find correlates of categorization in the spectral and temporal domain of AC responses.

Based on the perception of calls in mice (Ehret 2006a), a model shows how the parameter space of sounds is partitioned for the perception the three basic meanings, "attraction" to a vocalizing conspecific animal, "cohesion" of animals in a group, and "aversion" of a vocalizing conspecific. Attraction uses primarily high-frequency tonal sounds, cohesion low-frequency sounds with rhythmic amplitude modulations, and aversion broadband, loud sounds with noisy components and rough, dissonant sounding amplitude modulations. AAF neuron special response properties are well suited for a discrimination of these three basic meanings. An underrepresentation of neurons in the central hearing range may separate cells preferring high- or low-frequency sounds corresponding to attraction or cohesion. Such coding would be enhanced by a temporally precise rhythm representation in AAF (R). Aversion coding of would require activity of low- and high-frequency AAF (R) neurons, and is enhanced by the temporal coupling from precise phase-locking of neuronal responses to rapid amplitude modulations representing sound roughness. Further studies can show whether this approach to encode basic meanings of sounds in the anterior/ventral AC (Ehret 2006a) can be substantiated.

3.4 Plasticity in Representation

Plasticity in sound representation is an inherent AC property (Section 3.3.2) that emerges in development and aging, from changes in the input to the AC after cochlear pathology, learning, and task-specific responsiveness (König et al. 2005; Syka and Merzenich 2005).

Plasticity of AC sound representation may be induced by pregnancy or estrous (menstruation) cycles and by copulation in males, including humans. Such changes affect human hearing (Parlee 1983; Elkind-Hirsch et al. 1992) and house mouse call perception (Ehret and Koch 1989; Ehret and Buckenmaier 1994; Ehret and Schmid 2009). Hormonal plasticity may underlie ultrasonic processing differences in AI and UF (Fig. 16.2c) in mothers compared to virgin females. Mothers respond to a mouse pup ultrasound series with maternal behavior, but virgin females without prior pup contact do not (Ehret et al. 1987; Ehret and Buckenmaier 1994). Maternal AC neurons also respond more precisely and with higher rates to ultrasound series (Liu et al. 2006; Liu and Schreiner 2007). These differences do not occur to sounds without biological significance and the results suggest that the development of meaning may reflect specializations enhancing IBP representation after a hormone-induced change in the animal's internal state.

4 Presumptive Information-Bearing Elements within Communication Sounds

We noted that some IBEs and IBPs may carry information about sender identity, affective state, and proximity. We now emphasize the representation of (i) harmonics which define many vocalizations in the spectral domain, (ii) rhythm, roughness/flutter, pitch and timbral features characterizing individual voices and streams of vocalizations, (iii) human speech voice onset time (VOT) that may be equated to intersyllable intervals in calls, (iv) formant transitions in phonemes equivalent to call FMs, and (v) phonetic-like syntax as a feature in the perception of meaning. We will also examine the receptive fields (tuning curves) of neurons indicating how these IBEs may be extracted and represented in the AC (Fig. 16.4).

4.1 Harmonic Complexity and Timbre

Vocalization formant structure may carry semantic content as vowels do in phonemes of human speech. Other than ultrasonic bat and rodent calls (Sales and Pye 1974; Whitney and Nyby 1983; Ehret 2005) and primate isolation peeps (Winter et al. 1966), most call types contain several harmonics and formants that determine the call spectral character. The presence and frequencies of harmonics are important for the perception of prospective sound meaning. Much as human vowel perception reflects the formant number and frequency position, adult mice require at least three resolved low-frequency harmonics to perceive a multiharmonic communication call (wriggling call of pups) (Ehret and Riecke 2002).

Early studies of species-specific calls in the squirrel monkey AC (Wollberg and Newman 1972; Newman and Wollberg 1973; Winter and Funkenstein 1973) could not determine whether responses to harmonically structured vocalizations reflected sound spectral or temporal properties. Later work found population differences in neurons with regard to spectral and temporal sound properties for response generation. Some AI neurons preferred a harmonic spectrum based on the shape of their excitatory and inhibitory receptive fields relative to the frequency spectrum (Schwarz and Tomlinson 1990; Sutter et al. 1999; Gehr et al. 2000; Kadia and Wang 2003). Response facilitation could occur for harmonics with integer frequency ratios. Neurons preferred resolved harmonics, i.e., when their frequencies differed at least by one critical band (Ehret and Schreiner 1997; Fishman et al. 2000a), which is a perceptual measure of frequency resolution (Scharf 1970).

The responsiveness of most neurons depends, by their position in the tonotopic sequence, the shapes of their excitatory, inhibitory, and facilitatory receptive fields as well as on the call power spectrum. However, they respond also to modulations of frequency and amplitude. Hence, AI neuron responses, coding the spectral contents of call types, may not provide a direct basis for perceiving different call types.

A related spectral measure, harmonic complexity, a presumptive higher order IBP, can influence oscillatory activity in local field potentials of AI neurons (Medvedev and Kanwal 2004a). Harmonic complexity increases with the harmonic number and the spectral width of the harmonic stack in a call type. It reflects the value of fundamental frequency, the absolute and/or relative value of the predominant frequency, spectral discreteness, and temporal stability of the harmonic components.

4.2 Amplitude Modulation

Amplitude modulation is a determinant of the perceived sound quality in the time domain. Amplitude modulations introduce the perception of rhythm and pitch, and influence the perceived timbre (Roederer 1975). In serial sounds with intersound intervals >2,000 ms, every sound is perceived as one event; serial sounds with intervals of ~100–2,000 ms (frequency of amplitude modulation 0.5–10 Hz) elicit a perception of rhythm; 20–100 ms intervals lead to a roughness or flutter percept (modulation frequency of 10–50 Hz); shorter intervals (higher modulation frequencies) lead to a pitch percept (Miller and Taylor 1948; Besser 1967; Terhardt 1974a,b; Schulze and Langner 1997; Krumbholz et al. 2000; Zanto et al. 2006).

4.2.1 Rhythm

Rhythms with the above interval durations are perceptually important communication features of bats (Ohlemiller et al. 1996; Kanwal 1999), mice (Gaub and Ehret 2005), monkeys (Ramus et al. 2000), and humans (speech syllable rhythm) (van Dommelen 1990; Sansavini et al. 1997; Dolata et al. 2008). Most AI neurons encode rhythm by discharge peaks to the amplitude modulations. Rhythm coding is highly susceptible to changes in the stimulus sequence, with changes of the frequency of one tone burst or omitting one in a series of otherwise identical bursts can strongly affect the response to the next burst, often enhancing it (Weinberger and McKenna 1988; Ulanovsky et al. 2003). That is, AI neurons respond to rhythmic novelty. Such changes may partly reflect release from adaptation (Ulanovsky et al. 2004). Whether novelty effects occur in responses to modified species-specific call sequences remains to be seen.

4.2.2 Roughness

Roughness or flutter may have a temporal code, since many AI neurons in bats, gerbils, and monkeys phase-lock their response to the amplitude-modulated temporal envelope of complex sounds, producing a roughness or flutter percept in humans (Bieser and Müller-Preuß 1996; Schulze and Langner 1997; Steinschneider et al. 1998; Fishman et al. 2000b; Bendor and Wang 2007). Phase-locking to amplitude fluctuations may underlie encoding dissonant compared to consonant tone complexes in AI. Monkey and human AI show a high degree of phase-locking in the modulation range of roughness (see above) to dissonant compared to consonant chords due to the much higher degree of amplitude fluctuations in the former (Fishman et al. 2001b).

4.2.3 Pitch

Pitch perception is present in cat (Heffner and Whitfield 1976) and monkey (Tomlinson and Schwarz 1988) and may be present in bat (Preisler and Schmidt 1995), mouse (Ehret and Riecke 2002), and gerbil (Deutscher et al. 2006). As noted, pitch sensitive neurons are found in the dorsal part of the low-frequency map in gerbil AI (Schulze et al. 2002). In other mammals, AI cells tuned to the missing fundamental (pitch) of a harmonic complex (the individual frequencies do not stimulate the cell) or to the repetition rate (pitch) of a carrier frequency (the carrier frequency does not stimulate the cell) have not been found (Bendor and Wang 2005). Pitch sensitive neurons in marmoset AC are in a special field between rostral AI, the rostral field (R) and lateral belt areas (AL and ML) (Bendor and Wang 2005) (Fig. 16.2d) and perhaps in area AIp (posterior pole of AI) in mustached bats (Medvedev et al. 2002). AC pitch sensitivity is expressed by an average rate code not by a time code (phase-locking) as for sounds leading to a rhythm or roughness percept; temporal fluctuations and periodicities >50 Hz have been transformed from temporal IBPs to a neural code determined by the location of active AC neurons. We suggest that phase-locking to rapid sound amplitude modulations has been abandoned at higher auditory levels (especially in the AC) because it may disturb the temporally coordinated cell discharges necessary for binding the local activities of combination-sensitive neurons for the perception of acoustic objects and auditory scenes. Integrating pitch-sensitive neurons into the distributed neural representation of sound properties permits different pitches to be analyzed concurrently, to process pitches with other locally represented individual acoustic properties, and to synchronize and bind the activity of each group. Thus, the vocalizations of many individuals can be processed simultaneously in parallel by activity in overlapping AC neuronal ensembles. A winner-take-all mechanism of pitch processing in gerbil AI (Kurt et al. 2008) can enhance the perception of the voice with the most salient pitch in a group of vocalizing individuals or in noisy environments.

4.3 Sound Duration

Duration-tuned neurons were first discovered in the mustached bat inferior colliculus (Casseday et al. 1994) and have since been shown in the AC as well (Ma and Suga 2001; Wang et al. 2005). Principal components analysis reveals that duration does not correlate with other acoustic parameters within calls (Kanwal et al. 1994); thus, duration may be an IBE important both as an independent parameter and as one that can be combined with other IBEs to extract IBPs unique to a particular call type, as in early- and late-high coo calls in monkeys (Moody and Le Prell 1997). In bat and guinea pig, the mechanism for computation of sound duration requires subthreshold inhibition to produce a temporal frame in which rebound excitation can occur from cells responding to sound offset (Yin et al. 2008). At the level of the AC, duration tuning may be accomplished via sustained neural firing seen in the marmoset (Wang et al. 2005).

4.4 Silent Intervals

In communication calls, both silence and the absence of it are important IBEs. FM–FM area cells are most sensitive to these IBEs within calls (Ohlemiller et al. 1996; Esser et al. 1997) and in pulse-echo pairs used for extracting range information in echolocation for prey capture (O'Neill and Suga 1979; Suga and O'Neill 1979). In speech sounds, the VOT phonetic parameter is the interval between a noise burst at the vocalization onset and the ensuing harmonic spectrum from vocal cord vibrations. In human speech perception, VOT is a cue for discriminating pairs of stop consonants (b/p, d/t, and g/k) in consonant/vowel combinations (Pisoni and Lazarus 1974). The shortest speech VOT boundary is at ~25 ms, separating the perception of /ba/ (VOT <25 ms) from /pa/ (VOT >25 ms). A 25 ms boundary has been found for chinchilla VOT discrimination (Kuhl and Miller 1978), for the detection of gaps in noise (Penner 1975) and other sounds (Stevens and Klatt 1974), for the perception of temporal order (Hirsh 1959) and for the categorization of bat (Zimmer et al. 1998) and mouse (Ehret 1992a) species-specific calls. This boundary is common for perception in the time domain in many species. Evoked potential recordings from human AI to synthetic syllables differing in VOT and multiunit responses from monkey AI to two tones separated by various intervals showed a 20–30 ms boundary for resolving onset differences between syllables or two tones (Steinschneider et al. 2005). A similar result for coding synthetic VOT stimuli was found in young cats (Eggermont 1995). These results and those on the change from a temporal to an average-rate code at 20 ms time intervals, or 50 Hz repetition rate transition from rhythm/roughness perception to pitch, suggest that the shortest time interval between two acoustic events represented by a temporal code in AI is ~20 ms, which corresponds to a repetition rate of 50 Hz. This mammalian boundary may define and separate many perceptual qualities such as VOT-dependent consonants and roughness from pitch.

4.5 Frequency-Modulated Slope and Direction

FMs represent the most ubiquitous sound patterns present for most bat (Kanwal et al. 1994; Clement et al. 2006; Ma et al. 2006), monkey (Hauser 1991; Dimattina and Wang 2006), rat (Boinski and Mitchell 1995; Brudzynski 2005), and whale (Payne and McVay 1971) calls. Although FMs are well represented in AC, their role in call representation in AI has been directly tested only in bats (Washington and Kanwal, unpublished observations). At least five parameters of an FM signal (slope, bandwidth, central frequency, amplitude, and direction) are important determinants for central auditory neural responses (Erulkar et al. 1968).

Analyses of FMs in AI have focused on different FM classes: linear (Heil et al. 1992a, b, c; Nelken and Versnel 2000; Washington and Kanwal 2008), logarithmic (Mendelson et al. 1993; Zhang et al. 2003), and sinusoidal FMs (Suga et al. 1983; Liang et al. 2002). FM representations were mapped by determining FM rate (or slope) and direction that elicited peak responses from unit clusters and single cells. In cats, 95% of the multiunits had a directional preference and over 50% preferred downward FMs (Mendelson et al. 1993). In ferret AI, cells preferred upward FMs (Shamma et al. 1993; Nelken and Versnel 2000). Linear FMs tended to elicit more direction preference (Nelken and Versnel 2000). FM direction preference is thus dependent upon the experimental paradigm and the species.

4.5.1 Transitions and Glides

Fast frequency modulations in the few millisecond range mark speech sound formant transitions, such as in "*ba*", "*da*" and "*ga*" phonemes (Liberman et al. 1967). Similar transitions and rapid modulations occur in animal calls, albeit across different time scales. Slower modulations over tens to a few hundred milliseconds are labeled as FM glides. Both transitions and glides are presumptive IBEs differing largely by their IBP corresponding to the rate of frequency change. Glides are prominent, for example, in the meow of a cat (Fig. 16.1d), but are not encoded uniquely in AI in a special way beyond the harmonic content (Gehr et al. 2000). Cat AI neurons are sensitive to tonal contour, i.e., their responses differed depending on ascending, descending, or non-monotonic tone sequences (Weinberger and McKenna 1988). Squirrel monkey AC response selectivity was studied with natural, reversed or spectrotemporally destructured species-specific vocalizations (Glass and Wollberg 1983). No effect of call reversal was found (Glass and Wollberg 1983) and it was not possible to define the acoustic features determining a cell's response (Wollberg and Newman 1972; Winter and Funkenstein 1973). Call responsive units were classified as generalists, specialists, or in-between by the number of vocalizations to which they responded (Newman and Wollberg 1973; Winter and Funkenstein 1973). This was also found for cells in AIp area in mustached bat AI (Kanwal 2006). In the right hemispheric DSCF area, however, most cells prefer shallow FMs, representing glides, rather than rapid FMs (Washington and Kanwal 2008); however, in the left hemisphere, both glides and rapid FM transitions (>1 kHz/ms) are represented equally well (Washington and

Kanwal 2007). Thus, FM slope and duration are important IBPs in AC and enable distinctions between different call types.

4.5.2 Pitch or Tonal Contours

Pitch contours can be considered as a sequence of FM transitions and/or glides typical of longer duration sounds whose modulation direction changes more than once irregularly (aperiodically) (Streeter 1978; Trehub et al. 1984). Both pitch and pitch contour are independent IBEs extracted by the auditory system and represented in the AC (Streeter 1978; Chandrasekaran et al. 2007). Human singing, baby, and other nonverbal sounds often contain pitch contours. The right human AC may be specialized to extract pitch contours by integrating sounds over longer intervals and hence responds best to melodies and musical sounds and to speech prosody (Sloboda 1978; Stewart et al. 2008). Few mustached bat AC cells respond equally to upward and downward FMs, but not well to CF tones, making them well suited to represent tonal contours (Washington and Kanwal, unpublished observations).

4.6 Higher Order Constructs: Syntax

Perceptual features once thought to be speech-specific, such as categorical perception (Ehret and Haack 1981; May et al. 1989), perceptual constancy despite acoustic variability, formant structure perception in multitone complexes (Ehret 1992b), and phoneme perception (Kuhl and Padden 1982), are emerging as pre-adaptations for the communication sound analysis and recognition in mammals, including humans (Ehret 1992b). Perhaps, as in mustached bat AC, human combination-sensitive neurons are also involved in the perception of parameters of speech constructs such as syntax.

From a linguistic perspective (Umiker-Sebeok and Sebeok 1980; Snowdon 1982; Hauser et al. 2002) syntax denotes a rule system for production of an infinite variety of words and sentences from a few phonemes (vowels and consonants). Besides human speech, rule systems for sequencing species-specific vocalizations have been found in birds (Balaban 1988; Marler and Peters 1988) and many non-human mammals (Kanwal et al. 1994). Syntax is any system of rules that allows the prediction of a communication signal sequence (Snowdon 1982). The neuronal processing of acoustic sequences in the AC has been studied mainly in a few species of bats (O'Neill 1995). In the mustached bat AC, FM–FM cells (Fig. 16.2a) respond facilitatively to iso-syllabic pairs and composites (Ohlemiller et al. 1996; Esser et al. 1997). This response to communication calls mediates acoustic communication besides a role in echolocation (O'Neill and Suga 1979). Facilitated responses of FM–FM cells to composites show increased integration time compared to that for signals mimicking bat biosonar (Ohlemiller et al. 1996). That ≈50% of FM–FM neurons showed complex intersyllable interactions in the time domain (facilitation or suppression) supports the hypothesis that syntax processing occurs in non-primary AC.

5 Functional Specializations

5.1 Distributed Representations and the Number of Auditory Cortex Fields

Information from complex and hypercomplex acoustic patterns, such as in whole calls and call phrases, changes simultaneously in many dimensions. A multidimensional shift between calls imparts robustness to parametric differences that can be more distinctively represented by a patchy spatiotemporal AC activity pattern. Accordingly, neural activity in response to calls need not show spatial continuity across different AC regions and one or more calls can elicit a robust response in neurons in many AC areas (Fig. 16.5a). Such distributed representation may clarify how animals discriminate call types without feature detectors. No region in the mustached bat AC seems critical for the identification, localization, pursuit as well as capture of prey (Riquimaroux et al. 1991). Similarly, there may be no single locus in the mammalian AC for optimal neuronal representation of all of the auditory features within a given call type. The direction from which a call type is received may also impart meaning within a behavioral context; thus, the call type (what) and the spatial relationship between source and receiver (where) must be integrated to trigger an adaptive response. Perhaps this integration occurs in the frontal cortex, which is the target of both the "what" and "where" streams passing through and leaving the AC in different directions (Rauschecker 1998b; Kaas and Hackett 2000; Rauschecker and Tian 2000). Processing in the mustached bat AC (Riquimaroux et al. 1991) indicates that even the auditory object representation, the what, has no defined AC locus. If so, the representation must be distributed, and the question then becomes how and in which AC fields?

The number of AC fields in a mammal might reflect the various IBE-IBP combinations important for its communication system and acoustic ecology. Highly vocal species such as primates with many different calls in complex social interactions (Pola and Snowdon 1975) may have more fields and perhaps areas of more specific combination-sensitivities

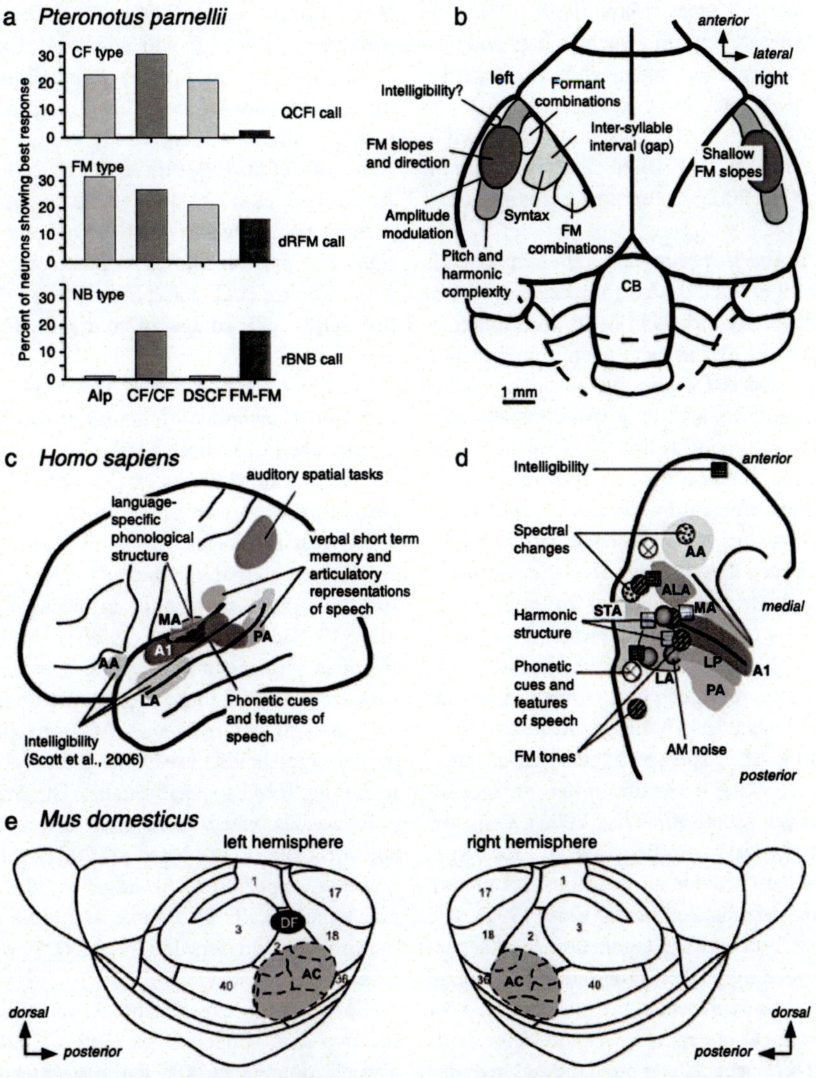

Fig. 16.5 **a** *Bar graphs* showing the relative number of neurons in the mustached bat DSCF, FM-FM, CF/CF and AIp areas preferring certain call types, as indicated for each plot. A call is labeled as a preferred call if its response is within 50% of the response to the best calls. CF/CF neurons are the most selective, whereas AIp neurons show the poorest selectivity. The CF and FM call types are well represented in all four areas and noise burst (NB) call types are represented in two areas. **b** Dorsal view of a mustached bat's brain showing the areas where presumptive IBPs characterizing IBEs are extracted and represented within the separate AC areas. These include A1 and nonprimary regions both ventral and dorsal to A1. Sound features such as amplitude modulations, formant combinations, and FM slope characteristics are extracted within the AC (see text). Intelligibility may be extracted in the tiny AV area located anterior and ventral to the A1 axis, where multiharmonic, composite pulse-echo echolocation signals are required to activate the neurons. CB, cerebellum. **c** Lateral view of the representation of speech processing fields in the left human temporal lobe. *Shaded regions* (*gray*) indicate broadly the different phonetic cues and speech features to which each area in the plane of the temporal lobe responds. Adapted from the original source (Scott 2005). **d** Dorsal view of the flattened temporal lobe showing the distribution of the same areas extracting specific phonetic and acoustic features within speech sounds. These data are based on the results of a meta-analysis of the peaks of activation seen in functional imaging studies that have considered speech and non-speech auditory processing in the temporal cortex (courtesy of S.K. Scott from metadata analysis in Scott and Johnsrude 2003). A1, primary AC; AA, anterior area; ALA, anterior lateral area; LA, lateral area; LPA, lateroposterior area; MA, medial area; PA, posterior area. **e** Lateral view on the adult mouse cerebral cortex, modified from the original source (Ehret 2006b). The AC with its fields (compare Fig. 16.2c) is indicated in *gray*. On the left hemisphere, the AC area activated by mouse pup wriggling calls (Fig. 16.1b) is ~20% larger than on the *right side*. Moreover, only on the *left side*, communication sounds activate neurons in a dorsal field (DF; *dark gray*) that is reciprocally connected with the AC (Hofstetter and Ehret 1992) and, from its position between auditory, somatic sensory (area 3) and visual (area 18) fields, likely processes multimodal information. It may be part of the "where" pathway from the auditory to the frontal cortex as defined in primates (Kaas and Hackett 2000; Rauschecker and Tian 2000)

than rodents with fewer call types (Ehret 1980). Mammals that rely on hearing for catching prey such as bats and cats should have a more diverse field organization than herbivorous rodents. Rodent AC has five (mouse) Fig. 16.2c) to seven fields (Mongolian gerbil) (Thomas et al. 1993), while in bat, cat (Fig. 16.3), monkey (Fig. 16.2d), and human (Scott and Johnsrude 2003; Scott 2005) ten or more fields are found (Fig. 16.5b–d).

For the cat AI (Fig. 16.3), spectrotemporal complex calls (Fig. 16.1) are expected to elicit a strong response at several loci in different isofrequency stripes in fields with tonotopy and in other fields defined by the local neural combination sensitivity to the IBEs and IBPs. The hypothesis of coding a complex (communication) sound by a specific spatiotemporal pattern of local "hot spots" (Ehret 1997) is supported by the twitter-call representation in marmoset monkey AI (Wang et al. 1995). Recordings from the mustached bat AC show that various social calls are represented within a consensus map in the AC such that different acoustic parameter combinations in the calls trigger activity of different sets of neurons mapped by their response selectivity in the specialized AC areas (Kanwal 2006). Optical imaging in rat AI also finds a distributed, sparse representation of various sounds (Hromádka et al. 2008). Data from human speech sound perception studied with modified speech suggest that different IBEs within sounds are extracted within different, specialized temporal lobe areas (Scott and Wise 2003). Only the simultaneous, time-coordinated profile of the AC local spatial excitation pattern may provide an instantaneous picture of the acoustical Gestalt of what was heard (Ehret 1997). The specificity of representation between combinations of acoustic properties, corresponding to their communication sound IBEs and IBPs, and their representation in a specific spatiotemporal pattern of local hot spots in AI and other fields of possible species-specific specialization in the AC, remains to be confirmed.

5.2 Lateralization of Processing and Representations

Perhaps the most emphasized specialization in humans is the right–left asymmetry for processing speech sounds and spoken language production. Since the discovery of left-hemisphere dominance of human speech production and perception (Broca 1861; Wernicke (1874), many studies have shown lateralization of auditory perceptual functions besides those for semantics and language syntax (Mateer 1983; Ojemann 1983; Benson 1986) and including music (Zatorre et al. 2002), laughing and crying (Sander and Scheich 2005), frequency-modulated tones (Poeppel et al. 2004), or tone bursts (Devlin et al. 2003). The left AC showed a higher activation for sounds with important temporal cues such as duration, rhythm and profile of amplitude modulations, formant transitions of importance combined with VOT, and auditory stream timing at the cortical (Fitch et al. 1997; Johnsrude et al. 1997; Binder et al. 2000; Deike et al. 2004; Ackermann et al. 2005; Brechmann and Scheich 2005) and perhaps subcortical levels (Kanwal and Gordon 1999). The right AC preferred pitch contours as in melodies (Wong 2002; Zatorre et al. 2002; Stewart et al. 2008). Most humans had larger left compared to right ACs (Shapleske et al. 1999).

Auditory functional lateralization is not unique to humans. Functional mapping of neural responses to tones in the anesthetized marmoset AI showed left–right individual but not population differences (Philibert et al. 2005). The AC fields (mainly higher order) in chimpanzees (Gannon et al. 1998) and mice (Stiebler et al. 1997; Geissler and Ehret 2004) were larger on the left side (Fig. 16.5e). There is left hemisphere dominance in macaque monkeys (Petersen et al. 1978; Beecher et al. 1979) and mice (Ehret 1987) for phonetic information in species-specific calls pertaining to meaningful social interactions. Anesthetized macaque monkeys had higher right-side AC activation in primary and higher-order fields in response to complex sound sequences including species-specific calls. The left anterior temporal pole was activated more intensely, but only for species-specific calls (Poremba et al. 2004). All mouse AC areas, however, appeared to be larger on the left (Fig. 16.2b) if activated by calls of high or low biological significance; a multimodal field dorsal to the AC (DF) was unique to the left (Fig. 16.5b).

Conditioning experiments in rats find a left dominance for two-tone sequence perception (Fitch et al. 1993) or a right dominance for the discrimination of rising and falling frequency sweeps in gerbils (Wetzel et al. 1998). Monkeys (Hauser and Andersson 1994), mice (Ehret 1987), and California sea lions (Böye et al. 2005) show a left-hemispheric advantage for species-specific calls that may be induced by social experience or hormonal changes via a priming process. Single cell studies in mustached bat AC show a left hemisphere (right ear) call processing advantage (Kanwal and Suga 1995). Left hemispheric neurons respond with greater relative (to tones) response to different call types. Right hemispheric neurons respond largely to shallow FM slopes (Fig. 16.5b), whereas left side neurons prefer more varied slopes (Washington and Kanwal 2007). Since many mustached bat calls contain various FM patterns, this slope tuning difference may account for the observed left AC call preference. What aspects of auditory processing and perception are lateralized in AC? Although this question cannot be answered yet, the left hemisphere prevails in time-critical sound processing and perception, categorizing stimuli, forming associative memories, and directing selective attention to stimuli (Ehret 2006b). The right hemisphere controls states of global arousal and attention. Thus, the general mechanism of lateralized AC function is dynamic and plastic, with neuromodulators such as hormones or the dopaminergic system regulating lateralization.

5.3 Auditory Objects: Extraction of Meaning

As noted earlier (Section 3.3.2), the AC representation of three basic meanings of call types may sort stimuli by the frequency spectrum and temporal characteristics of the calls in the anterior/ventral stream of processing in area AAF. Call types may act as independent auditory objects (Husain et al. 2004; Nelken 2004) comparable to visual objects. To create the perception of an acoustic object, a scene or Gestalt elicited by specific sounds, dynamic neural activities in specialized areas and at distributed AI hot spots have to be bound together (Section 4.2.3). One way to accomplish this is by convergence of inputs from different lower-order AC regions onto higher-order neurons in the nonprimary AC or frontal cortex. In bat, a ventral and anterior (VA) area may have such a role (Tsuzuki and Suga 1988), and a similar area has been found in the anterior pole of the human temporal lobe (Scott and Johnsrude 2003). A more dynamic possibility establishes a transient functional connectivity (Gruber et al. 2008). Temporal binding of coherent neural activities over a few millimeters has been shown in AI by correlated neural firing at different loci (Eggermont 1994; deCharms and Merzenich 1996; Brosch and Schreiner 1999; Gehr et al. 2000; Read et al. 2001), especially for cells with similar binaural interactions (EE or EI) or frequency selectivity (Fig. 16.3). Binding over <200 μm may occur as gamma oscillations in local field potentials (Medvedev and Kanwal 2008). The gamma-band traditionally refers to activity at 30 to ~150 Hz with oscillations >70 Hz considered as high gamma (Sinai et al. 2005). The correlative dynamics pose some temporal limitations that make oscillations in the gamma-band range within a small neural population a most suitable carrier to reliably and rapidly express correlated neural activity (König and Schillen 1991). Highly correlated, phase-locked activity in several locations of the brain underlies the Gestalt of a cohesive and coherent object and may indicate a recognition event (Crick 1984; von der Malsburg 1986).

A call recognition event might occur when transient synchronization of neuronal firing patterns ensues by the temporal convergence of synaptic currents in a subset of neurons activated by the call (Hopfield and Brody 2001). This transient neuronal synchrony is accompanied by gamma-band oscillations within the population activity. From this model, the population transient synchrony is highly sensitive to the specific sound temporal structure. Time reversal of a familiar sound changes its temporal structure disrupting population synchrony, abolishing the gamma-band response, and impairing sound recognition. Gamma-band sensitivity to call time reversal in bats is consistent with this model's predictions and provides experimental data that such activity may be a signature or an outcome of a call recognition event in a neural network (Martinovic et al. 2008; Medvedev and Kanwal 2008). For some call types, the large variation in the gamma-band activity change for forward versus reversed call types suggests that such activity is likely influenced by additional parameters, such as AM modulations and sound onset rise times and their interaction with sound duration (Kaiser et al. 2007). In humans, learning temporally modulated tone trains increases the power of high gamma band activity in the inferior frontal cortex, a potential neural substrate for top-down modulation of learning-induced AC plasticity (van Wassenhove and Nagarajan 2007). Accordingly, gamma-band oscillations have been implicated in human speech sound perception (Crone et al. 2001).

6 Conclusions

Communication sounds or calls in most mammals have common acoustic properties that derive from constant frequency tones, frequency modulations, and noise busts. Acoustic properties in calls convey behaviorally relevant information via a specified range of IBPs derived from IBEs. The mammalian AC extracts the values of the IBPs. When it is important to extract precise values within a biologically significant range of values of a given parameter, these parameters are represented in the AC maps that contain continuous scales of the biologically important IBP range. A different IBP map is needed for each IBE. The model of mammalian AC functional organization determined by IBEs and IBPs arose from and is exemplified by the mustached bat, which was studied from the viewpoint of echolocation, which requires precise and successive readouts of activity from parametric maps to support object identification and prey capture. The goal now is to determine the extent to which IBEs and IBPs in communication sounds are processed according to similar principles in the AC of bats and other species.

Unlike the continuous and systematic variations in few IBPs in echolocation sounds, calls vary inherently in many parameters. A high variability may occur in every parameter within a call utterance and between different call type utterances by the same and/or different individuals. IBPs may change with behavioral context, gender, and social status of the call sender and receiver, making AC maps representing continuous scales of all IBPs for call discrimination and recognition highly unlikely. Nevertheless, the extraction of

IBEs and IBPs is as relevant for communication calls as for echolocation. Studies in mammals, including humans, find that presumptive IBEs are indeed extracted and generic stimulus parameters may be represented in gradients across AC space and in cell clusters with similar responses to a given parameter. A picture emerges for AI and AAF that communication sounds are represented by the activity of several local patches of neurons with combination-sensitivity and/or tuning to IBPs characterizing sound IBEs. The IBEs include rate or slope, and FM directions underlying pitch contours and call terminations, harmonic complexity embodied in power spectrum peaks, amplitude modulations of noise and harmonic complexes, sound duration, syllable composition, duration of silent intervals marking syllable repetition rates (rhythm), and syntax or sequencing of syllables. For relatively simple signals such as rodent ultrasounds or messages that convey attractive or aversive meanings, few fields of the AC and limited clusters of activated neurons may underlie perception. Vocalizations from a large species-specific repertoire with many IBEs and IBPs as in bats, primates and human speech would require complex binding of temporally coherent activities in many fields and patches of neurons of different combination-sensitivity to be perceived. Besides combination-sensitivity, high gamma-band oscillations across AC may likewise bind neurons at several loci to enhance the signal-to-noise ratio for reliable perception.

Finally, the AC in each hemisphere likely extracts and represents different IBPs. The left AC integrates IBPs over brief time intervals for high temporal resolution and the right hemisphere is specialized for integrating IBPs over longer time intervals yielding a greater spectral resolution (Zatorre et al. 2002; Poeppel 2003). This translates into a right ear-left hemisphere advantage for communication sound processing in mice, monkeys, bats, and humans, particularly males, and a human left ear-right hemisphere advantage for processing musical melodies and prosody, and perhaps in bats for pulse-echo sequences used for navigation and ranging.

7 Future Directions

Communication mainly involves interindividual, species-specific interactions. Therefore, to understand the principles underlying the cortical representation of communication sounds, neurophysiological studies must use multiple species and a neuroethological approach. Further, cortical representation of different call types now needs to be tested more directly in awake behaving animals. We need to establish the close relationship between the presumptive IBEs in a call and the local zones of neural activity dispersed and distributed in the AC. The spatiotemporal patterns of cortical activation by call sequences also need to be established together with the spatial dynamics of these patterns. Finally, we must measure neuronal specificity to particular IBEs within calls. Optical imaging and/or multisite recordings of spiking and/or local field potential activity can demonstrate that the coordinated and sequential firing of a small population of cortical neurons imparts the ultimate specificity to a call's response and hence its identity.

References

Ackermann H, Hertrich I, Lutzenberger W, and Mathiak K (2005) Cerebral organization of speech sound perception: hemispheric lateralization effects at the level of the supratemporal plane, the inferior dorsolateral frontal lobe, and the cerebellum. In: König R, Heil P, Budinger E, and Scheich H (eds). *The Auditory Cortex. A Synthesis of Human and Animal Research*. Lawrence Erlbaum, Mahwa, pp. 181–200.

Aitkin LM, Merzenich MM, Irvine DRF, Clarey JC, and Nelson JE (1986) Frequency representation in auditory cortex of the common marmoset (*Callithrix jacchus jacchus*). *Journal of Comparative Neurology* **252**:175–185.

Alain C, Arnott SR, and Picton TW (2001) Bottom-up and top-down influences on auditory scene analysis: evidence from event-related brain potentials. *Journal of Experimental Psychology: Human Perception and Performance* **27**:1072–1089.

Alain C, Snyder JS, He Y, and Reinke KS (2007) Changes in auditory cortex parallel rapid perceptual learning. *Cerebral Cortex* **17**: 1074–1084.

August PV and Anderson JGT (1987) Mammal sounds and motivation-structural rules: a test of the hypothesis. *Journal of Mammalogy* **68**:1–9.

Balaban E (1988) Bird song syntax: learned intraspecific variation is meaningful. *Proceedings of the National Academy of Sciences of the United States of America* **85**:3657–3660.

Bartlett EL and Wang X (2005) Long-lasting modulation by stimulus context in primate auditory cortex. *Journal of Neurophysiology* **94**:83–104.

Beecher MD, Petersen MR, Zoloth SR, Moody DB, and Stebbins WC (1979) Perception of conspecific vocalizations by Japanese macaques. Evidence for selective attention and neural lateralization. *Brain, Behavior and Evolution* **16**:443–460.

Bendor D and Wang X (2005) The neuronal representation of pitch in primate auditory cortex. *Nature* **436**:1161–1165.

Bendor D and Wang X (2007) Differential neural coding of acoustic flutter within primate auditory cortex. *Nature Neuroscience* **10**: 763–771.

Benson DF (1986) Aphasia and the lateralization of language. *Cortex* **22**:71–86.

Besser GM (1967) Some physiological characteristics of auditory flutter fusion in man. *Nature* **214**:17–19.

Bieser A and Müller-Preuß P (1996) Auditory responsive cortex in the squirrel monkey: neural responses to amplitude-modulated sounds. *Experimental Brain Research* **108**:273–284.

Binder JR, Frost JA, Hammeke TA, Bellgowan PS, Springer JA, Kaufman JN, and Possing ET (2000) Human temporal lobe activation by speech and nonspeech sounds. *Cerebral Cortex* **10**:512–528.

Bizley JK, Nodal FR, Nelken I, and King AJ (2005) Functional organization of ferret auditory cortex. *Cerebral Cortex* **15**:1637–1653.

Boinski S and Mitchell CL (1995) Wild squirrel monkey (*Saimiri sciureus*) "caregiver" calls: contexts and acoustic structure. *American Journal of Primatology* **35**:129–137.

Bonin GV and Bailey P (1947) *The Neocortex of Macaca mulatta*. University of Illinois Press, Urbana.

Böye M, Güntürkün O, and Vauclair J (2005) Right ear advantage for conspecific calls in adults and subadults, but not infants, California sea lions (*Zalophus californianus*): hemispheric specialization for communication? *European Journal of Neuroscience* **21**:1727–1732.

Brechmann A and Scheich H (2005) Hemispheric shifts of sound representation in auditory cortex with conceptual listening. *Cerebral Cortex* **15**:578–587.

Broca P (1861) Remarques sur la siege de la faculté du langage articule suivies d'une observation d'aphemie. *Bulletin de la Société d'Anatomie* **6**:398–407.

Brosch M and Schreiner CE (1999) Correlations between neural discharges are related to receptive field properties in cat primary auditory cortex. *European Journal of Neuroscience* **11**:3517–3530.

Brudzynski SM (2005) Principles of rat communication: quantitative parameters of ultrasonic calls in rats. *Behavior Genetics* **35**:85–92.

Budinger E, Heil P, and Scheich H (2000) Functional organization of auditory cortex in the Mongolian gerbil (*Meriones unguiculatus*). III. Anatomical subdivisions and corticocortical connections. *European Journal of Neuroscience* **12**:2425–2451.

Casseday JH, Ehrlich D, and Covey E (1994) Neural tuning for sound duration: role of inhibitory mechanisms in the inferior colliculus. *Science* **264**:847–850.

Chandrasekaran B, Gandour JT, and Krishnan A (2007) Neuroplasticity in the processing of pitch dimensions: a multidimensional scaling analysis of the mismatch negativity. *Restorative Neurology and Neuroscience* **25**:195–210.

Cheung SW, Nagarajan SS, Bedenbaugh PH, Schreiner CE, Wang X, and Wong A (2001) Auditory cortical neuron response differences under isoflurane versus pentobarbital anesthesia. *Hearing Research* **156**:115–127.

Clement MJ, Dietz N, Gupta P, and Kanwal JS (2006) Audiovocal communication and social behavior in mustached bats. In: Kanwal JS and Ehret G (eds). *Behavior and Neurodynamics for Auditory Communication*. Cambridge University Press, Cambridge, pp. 57–84.

Crick F (1984) Function of the thalamic reticular complex: the searchlight hypothesis. *Proceedings of the National Academy of Sciences of the United States of America* **81**:4586–4590.

Crone NE, Hao L, Hart Jr J, Boatman D, Lesser RP, Irizarry R, and Gordon B (2001) Electrocorticographic gamma activity during word production in spoken and sign language. *Neurology* **57**:2045–2053.

de Ribaupierre F (1997) Acoustical information processing in the auditory thalamus and cerebral cortex. In: Ehret G and Romand R (eds). *The Central Auditory System*. Oxford University Press, New York, pp. 317–397.

Dear SP, Fritz J, Haresign T, Ferragamo M, and Simmons JA (1993a) Tonotopic and functional organization in the auditory cortex of the big brown bat, *Eptesicus fuscus*. *Journal of Neurophysiology* **70**:1988–2009.

Dear SP, Simmons JA, and Fritz J (1993b) A possible neuronal basis for representation of acoustic scenes in auditory cortex of the big brown bat. *Nature* **364**:620–623.

deCharms RC, Blake DT, and Merzenich MM (1998) Optimizing sound features for cortical neurons. *Science* **280**:1439–1443.

deCharms RC and Merzenich MM (1996) Primary cortical representation of sounds by the coordination of action-potential timing. *Nature* **381**:610–613.

Deike S, Gaschler-Markefski B, Brechmann A, and Scheich H (2004) Auditory stream segregation relying on timbre involves left auditory cortex. *Neuroreport* **15**:1511–1514.

Delmaghani S, del Castillo FJ, Michel V, Leibovici M, Aghaie A, Ron U, Van Laer L, Ben-Tal N, Van Camp G, Weil D, Langa F, Lathrop M, Avan P, and Petit C (2006) Mutations in the gene encoding pejvakin, a newly identified protein of the afferent auditory pathway, cause DFNB59 auditory neuropathy. *Nature Genetics* **38**: 770–778.

Deutscher A, Kurt S, Scheich H, and Schulze H (2006) Cortical and subcortical sides of auditory rhythms and pitches. *Neuroreport* **17**:853–856.

Devlin JT, Raley J, Tunbridge E, Lanary K, Floyer-Lea A, Narain C, Cohen I, Behrens T, Jezzard P, Matthews PM, and Moore DR (2003) Functional asymmetry for auditory processing in human primary auditory cortex. *Journal of Neuroscience* **23**:11516–11522.

Dimattina C and Wang X (2006) Virtual vocalization stimuli for investigating neural representations of species-specific vocalizations. *Journal of Neurophysiology* **95**:1244–1262.

Dolata JK, Davis BL, and Macneilage PF (2008) Characteristics of the rhythmic organization of vocal babbling: implications for an amodal linguistic rhythm. *Infant Behavioral Development*, **31**:422–431.

Doron NN, LeDoux JE, and Semple MN (2002) Redefining the tonotopic core of rat auditory cortex: physiological evidence for a posterior field. *Journal of Comparative Neurology* **453**:345–360.

Eggermont JJ (1994) Neural interaction in cat primary auditory cortex II. Effects of sound stimulation. *Journal of Neurophysiology* **71**:246–270.

Eggermont JJ (1995) Representation of a voice onset time continuum in primary auditory cortex of the cat. *Journal of the Acoustical Society of America* **98**:911–920.

Eggermont JJ (1998) Representation of spectral and temporal sound features in three cortical fields of the cat. Similarities outweigh differences. *Journal of Neurophysiology* **80**:2743–2764.

Ehret G (1975) Schallsignale der Hausmaus (*Mus musculus*). *Behaviour* **52**:38–56.

Ehret G (1980) Development of sound communication in mammals. *Advances in the Study of Behavior* **11**:179–225.

Ehret G (1987) Left hemisphere advantage in the mouse brain for recognizing ultrasonic communication calls. *Nature* **325**:249–251.

Ehret G (1992a) Categorical perception of mouse-pup ultrasounds in the temporal domain. *Animal Behavior* **43**:409–416.

Ehret G (1992b) Preadaptations in the auditory system of mammals for phoneme perception: from sounds to words. In: Schouten MEH (ed). *The Auditory Processing of Speech*. Mouton de Gruyter, Berlin, pp. 99–112.

Ehret G (1997) The auditory cortex. *Journal of Comparative Physiology A* **181**:547–557.

Ehret G (2005) Infant rodent ultrasounds – a gate to the understanding of sound communication. *Behavior Genetics* **35**:19–29.

Ehret G (2006a) Common rules of communication sound perception. In: Kanwal J and Ehret G (eds). *Behavior and Neurodynamics for Auditory Communication*. Cambridge University Press, Cambridge, pp. 85–114.

Ehret G (2006b) Hemisphere dominance of brain function-which functions are lateralized and why? In: van Hemmen J and Sejnowski T (eds). *23 Problems in Systems Neuroscience*. Oxford University Press, New York, pp. 44–61.

Ehret G and Buckenmaier J (1994) Estrogen-receptor occurrence in the female mouse brain: effects of maternal experience, ovariectomy, estrogen and anosmia. *Journal of Physiology (Paris)* **88**:315–329.

Ehret G and Haack B (1981) Categorical perception of mouse pup ultrasound by lactating females. *Naturwissenschaften* **68**:208–209.

Ehret G and Koch M (1989) Ultrasound-induced parental behaviour in house mice is controlled by female sex hormones and parental experience. *Ethology* **80**:81–93.

Ehret G, Koch M, Haack B, and Markl H (1987) Sex and parental experience determine the onset of an instinctive behavior in mice. *Naturwissenschaften* **74**:47.

Ehret G and Riecke S (2002) Mice and humans perceive multiharmonic communication sounds in the same way. *Proceedings of the National Academy of Sciences of the United States of America* **99**:479–482.

Ehret G and Schmid C (2009) Reproductive cycle-dependent plasticity of perception of acoustic meaning in mice. *Physiology & Behavior* **96**:428–433.

Ehret G and Schreiner CE (1997) Frequency resolution and spectral integration (critical band analysis) in single units of the cat primary auditory cortex. *Journal of Comparative Physiology A* **181**:635–650.

Ehret G and Schreiner CE (2000) Regional variations of noise-induced changes in operating range in cat AI. *Hearing Research* **141**: 107–116.

Eliades SJ and Wang X (2003) Sensory-motor interaction in the primate auditory cortex during self-initiated vocalizations. *Journal of Neurophysiology* **89**:2194–2207.

Elkind-Hirsch KE, Stoner WR, Stach BA, and Jerger JF (1992) Estrogen influences auditory brainstem responses during the normal menstrual cycle. *Hearing Research* **60**:143–148.

Erulkar SD, Nelson PG, and Bryan JS (1968) Experimental and theoretical approaches to neural processing in the central auditory pathway. *Contributions to Sensory Physiology* **3**:149–189.

Esser K-H, Condon CJ, Suga N, and Kanwal JS (1997) Syntax processing by auditory cortical neurons in the FM-FM area of the mustached bat *Pteronotus parnellii*. *Proceedings of the National Academy of Sciences of the United States of America* **94**: 14019–14024.

Esser K-H and Eiermann A (1999) Tonotopic organization and parcellation of auditory cortex in the FM-bat *Carollia perspicillata*. *European Journal of Neuroscience* **11**:3669–3682.

Evans CS and Marler P (1991) On the use of video images as social stimuli in birds: audience effects on alarm calling. *Animal Behavior* **41**:17–26.

Fichtel I and Ehret G (1999) Perception and recognition discriminated in the mouse auditory cortex by c-Fos labeling. *Neuroreport* **10**:2341–2345.

Fishman YI, Reser DH, Arezzo JC, and Steinschneider M (2000a) Complex tone processing in primary auditory cortex of the awake monkey. I. Neural ensemble correlates of roughness. *Journal of the Acoustical Society of America* **108**:235–246.

Fishman YI, Reser DH, Arezzo JC, and Steinschneider M (2000b) Complex tone processing in primary auditory cortex of the awake monkey. II. Pitch versus critical band representation. *Journal of the Acoustical Society of America* **108**:247–262.

Fishman YI, Reser DH, Arezzo JC, and Steinschneider M (2001a) Neural correlates of auditory stream segregation in primary auditory cortex of the awake monkey. *Hearing Research* **151**: 167–187.

Fishman YI, Volkov IO, Noh MD, Garell PC, Bakken H, Arezzo JC, Howard MA, and Steinschneider M (2001b) Consonance and dissonance of musical chords: neural correlates in auditory cortex of monkeys and humans. *Journal of Neurophysiology* **86**:2761–2788.

Fitch RH, Brown CP, O'Connor K, and Tallal P (1993) Functional lateralization for auditory temporal processing in male and female rats. *Behavioral Neuroscience* **107**:844–850.

Fitch RH, Miller S, and Tallal P (1997) Neurobiology of speech perception. *Annual Review of Neuroscience* **20**:331–353.

Formisano E, Kim DS, Di Salle F, van de Moortele PF, Ugurbil K, and Goebel R (2003) Mirror-symmetric tonotopic maps in human primary auditory cortex. *Neuron* **40**:859–869.

Francois O, Abdallahi LM, Horikawa J, Taniguchi I, and Herve T (2000) Statistical procedures for spatiotemporal neuronal data with applications to optical recording of the auditory cortex. *Neural Computation* **12**:1821–1838.

Fritz J, Elhilali M, and Shamma S (2005) Active listening: task-dependent plasticity of spectrotemporal receptive fields in primary auditory cortex. *Hearing Research* **206**:159–176.

Fritz J, Shamma S, Elhilali M, and Klein D (2003) Rapid task-related plasticity of spectrotemporal receptive fields in primary auditory cortex. *Nature Neuroscience* **6**:1216–1223.

Galazyuk AV and Feng AS (1997) Encoding of sound duration by neurons in the auditory cortex of the little brown bat, *Myotis lucifugus*. *Journal of Comparative Physiology A* **180**:301–311.

Gannon PJ, Holloway RL, Broadfield DC, and Braun AR (1998) Asymmetry of chimpanzee planum temporale: humanlike pattern of Wernicke's brain language area homolog. *Science* **279**:220–222.

Gaub S and Ehret G (2005) Grouping in auditory temporal perception and vocal production is mutually adapted: the case of wriggling calls of mice. *Journal of Comparative Physiology A* **191**:1131–1135.

Gehr DD, Komiya H, and Eggermont JJ (2000) Neuronal responses in cat primary auditory cortex to natural and altered species-specific calls. *Hearing Research* **150**:27–42.

Geissler DB and Ehret G (2004) Auditory perception vs. recognition: representation of complex communication sounds in the mouse auditory cortical fields. *European Journal of Neuroscience* **19**: 1027–1040.

Ghazanfar AA and Hauser MD (2001) The auditory behaviour of primates: a neuroethological perspective. *Current Opinion in Neurobiology* **11**:712–720.

Glass I and Wollberg Z (1983) Responses of cells in the auditory cortex of awake squirrel monkeys to normal and reversed species-specific vocalizations. *Hearing Research* **9**:27–33.

Godey B, Atencio CA, Bonham BH, Schreiner CE, and Cheung SW (2005) Functional organization of squirrel monkey primary auditory cortex: responses to frequency-modulated sweeps. *Journal of Neurophysiology* **94**:1299–1311.

Gonzalez-Lima F and Scheich H (1986) Neural substrates for tone-conditioned bradycardia demonstrated with 2-deoxyglucose. II. Auditory cortex plasticity. *Behavioural Brain Research* **20**: 281–293.

Gruber T, Maess B, Trujillo-Barreto NJ, and Muller MM (2008) Sources of synchronized induced gamma-band responses during a simple object recognition task: a replication study in human MEG. *Brain Research* **1196**:74–84.

Hackett TA, Preuss TM, and Kaas JH (2001) Architectonic identification of the core region in auditory cortex of macaques, chimpanzees, and humans. *Journal of Comparative Neurology* **441**:197–222.

Hackett TA, Stepniewska I, and Kaas JH (1998) Subdivisions of auditory cortex and ipsilateral cortical connections of the parabelt auditory cortex in macaque monkeys. *Journal of Comparative Neurology* **394**:475–495.

Hauser MD (1991) Sources of acoustic variation in rhesus macaque vocalization. *Ethology* **89**:29–46.

Hauser MD and Andersson K (1994) Left hemisphere dominance for processing vocalizations in adult, but not infant, rhesus monkeys: field experiments. *Proceedings of the National Academy of Sciences of the United States of America* **91**:3946–3948.

Hauser MD, Chomsky N, and Fitch WT (2002) The faculty of language: what is it, who has it, and how did it evolve? *Science* **298**: 1569–1579.

Heffner H and Whitfield IC (1976) Perception of the missing fundamental by cats. *Journal of the Acoustical Society of America* **59**:915–919.

Heil P, Langner G, and Scheich H (1992a) Processing of frequency-modulated stimuli in the chick auditory cortex analogue: evidence for topographic representations and possible mechanisms of rate and directional sensitivity. *Journal of Comparative Physiology A* **171**:583–600.

Heil P, Rajan R, and Irvine DRF (1992b) Sensitivity of neurons in cat primary auditory cortex to tones and frequency-modulated stimuli. I: Effects of variation of stimulus parameters. *Hearing Research* **63**:108–134.

Heil P, Rajan R, and Irvine DRF (1992c) Sensitivity of neurons in cat primary auditory cortex to tones and frequency-modulated stimuli. II: Organization of response properties along the 'isofrequency' dimension. *Hearing Research* **63**:135–156.

Hirsh IJ (1959) Auditory perception of temporal order. *Journal of the Acoustical Society of America* **31**:759–767.

Hofstetter KM and Ehret G (1992) The auditory cortex of the mouse: connections of the ultrasonic field. *Journal of Comparative Neurology* **323**:370–386.

Hopfield JJ and Brody CD (2001) What is a moment? Transient synchrony as a collective mechanism for spatiotemporal integration. *Proceedings of the National Academy of Sciences of the United States of America* **98**:1282–1287.

Horikawa J, Hess A, Hosokawa Y, and Taniguchi I (2006) Spatiotemporal processing in the guinea pig auditory cortex In: Kanwal JS and Ehret G (eds). *Behavior and Neurodynamics for Auditory Communication*. Cambridge University Press, Cambridge, pp. 189–204.

Horikawa J, Hess A, Nasu M, Hosokawa Y, Scheich H, and Taniguchi I (2001) Optical imaging of neural activity in multiple auditory cortical fields of guinea pigs. *Neuroreport* **12**:3335–3339.

Horikawa J, Hosokawa Y, Kubota M, Nasu M, and Taniguchi I (1996) Optical imaging of spatiotemporal patterns of glutamatergic excitation and GABAergic inhibition in the guinea-pig auditory cortex in vivo. *Journal of Physiology (London)* **497**:629–638.

Horikawa J, Nasu M, and Taniguchi I (1998) Optical recording of responses to frequency-modulated sounds in the auditory cortex. *Neuroreport* **9**:799–802.

Hosokawa Y, Horikawa J, Nasu M, Sugimoto S, and Taniguchi I (1998) Anisotropic neural interaction in the primary auditory cortex of guinea pigs with sound stimulation. *Neuroreport* **9**: 3421–3425.

Hromádka T, DeWeese MR, and Zador AM (2008) Sparse representation of sounds in the unanesthetized auditory cortex. *Public Library of Science Biology* **6**:e16.

Husain FT, Tagamets MA, Fromm SJ, Braun AR, and Horwitz B (2004) Relating neuronal dynamics for auditory object processing to neuroimaging activity: a computational modeling and an fMRI study. *Neuroimage* **21**:1701–1720.

Imaizumi K, Priebe NJ, Crum PAC, Bedenbaugh PH, Cheung SW, and Schreiner CE (2004) Modular functional organization of cat anterior auditory field. *Journal of Neurophysiology* **92**:444–457.

Imig TJ and Reale RA (1980) Patterns of cortico-cortical connections related to tonotopic maps in cat auditory cortex. *Journal of Comparative Neurology* **192**:293–332.

Imig TJ, Ruggero MA, Kitzes LM, Javel E, and Brugge JF (1977) Organization of auditory cortex in the owl monkey (*Aotus trivirgatus*). *Journal of Comparative Neurology* **171**:111–128.

Jin W, Zhang RJ, and Wu JY (2002) Voltage-sensitive dye imaging of population neuronal activity in cortical tissue. *Journal of Neuroscience Methods* **115**:13–27.

Johnsrude IS, Zatorre RJ, Milner BA, and Evans AC (1997) Left-hemisphere specialization for the processing of acoustic transients. *Neuroreport* **8**:1761–1765.

Kaas JH (2005) The future of mapping sensory cortex in primates: three of many remaining issues. *Philosophical Transactions of the Royal Society of London, series B: Biological Sciences* **360**:653–664.

Kaas JH and Hackett TA (2000) Subdivisions of auditory cortex and processing streams in primates. *Proceedings of the National Academy of Sciences of the United States of America* **97**: 11793–11799.

Kaas JH, Hackett TA, and Tramo MJ (1999) Auditory processing in primate cerebral cortex. *Current Opinion in Neurobiology* **9**: 164–170.

Kadia SC and Wang X (2003) Spectral integration in A1 of awake primates: neurons with single- and multipeaked tuning characteristics. *Journal of Neurophysiology* **89**:1603–1622.

Kaiser J, Leiberg S, Rust H, and Lutzenberger W (2007) Prefrontal gamma-band activity distinguishes between sound durations. *Brain Research* **1139**:153–162.

Kalatsky VA, Polley DB, Merzenich MM, Schreiner CE, and Stryker MP (2005) Fine functional organization of auditory cortex revealed by Fourier optical imaging. *Proceedings of the National Academy of Sciences of the United States of America* **102**:13325–13330.

Kanwal JS (1999) Processing species-specific calls by combination-sensitive neurons in an echolocating bat. In: Hauser MD and Konishi M (eds). *The Design of Animal Communication*. The MIT Press, Cambridge, pp. 135–157.

Kanwal JS (2006) A distributed cortical representation of social calls. In: Kanwal JS and Ehret G (eds). *Behavior and Neurodynamics for Auditory Communication*. Cambridge University Press, Cambridge, pp. 163–188.

Kanwal JS, Fitzpatrick DC, and Suga N (1999) Facilitatory and inhibitory frequency tuning of combination-sensitive neurons in the primary auditory cortex of mustached bats. *Journal of Neurophysiology* **82**:2327–2345.

Kanwal JS and Gordon M (1999) Segregating auditory information along two subcortical streams. *Journal of Cognitive Neuroscience*, Supplement:102.

Kanwal JS, Matsumura S, Ohlemiller K, and Suga N (1994) Analysis of acoustic elements and syntax in communication sounds emitted by mustached bats. *Journal of the Acoustical Society of America* **96**:1229–1254.

Kanwal JS and Rauschecker JP (2007) Auditory cortex of bats and primates: managing species-specific calls for social communication. *Frontiers in Bioscience* **12**:4621–4640.

Kanwal JS and Suga N (1995) Hemispheric asymmetry in the processing of calls in the auditory cortex of the mustached bat. *Association for Research in Otolaryngology Abstracts* **18**:104.

König P and Schillen TB (1991) Stimulus-dependent assembly formation of oscillatory responses: I. Synchronisation. *Neural Computation* **3**:155–166.

König R, Heil P, Budinger E, and Scheich H (2005) *The Auditory Cortex. A Synthesis of Human and Animal Research*. Lawrence Erlbaum, Mahwah.

Kössl M and Vater M (1985) The cochlear frequency map of the mustache bat, *Pteronotus parnellii*. *Journal of Comparative Physiology A* **157**:687–697.

Krumbholz K, Patterson RD, and Pressnitzer D (2000) The lower limit of pitch as determined by rate discrimination. *Journal of the Acoustical Society of America* **108**:1170–1180.

Kuhl PK and Miller JD (1978) Speech perception by the chinchilla: identification function for synthetic VOT stimuli. *Journal of the Acoustical Society of America* **63**:905–917.

Kuhl PK and Padden DM (1982) Enhanced discriminability at the phonetic boundaries for the voicing feature in macaques. *Perception and Psychophysics* **32**:542–550.

Kurt S, Deutscher A, Crook JM, Ohl FW, Budinger E, Moeller CK, Scheich H, and Schulze H (2008) Auditory cortical contrast enhancing by global winner-take-all inhibitory interactions. *Public Library of Science ONE* **3**:e1735.

Lakatos P, Chen CM, O'Connell MN, Mills A, and Schroeder CE (2007) Neuronal oscillations and multisensory interaction in primary auditory cortex. *Neuron* **53**:279–292.

Langner G (1992) Periodicity coding in the auditory system. *Hearing Research* **60**:115–142.

Lee CC and Winer JA (2005) Principles governing auditory cortex connections. *Cerebral Cortex* **15**:1804–1818.

Liang L, Lu T, and Wang X (2002) Neural representations of sinusoidal amplitude and frequency modulations in the primary auditory cortex of awake primates. *Journal of Neurophysiology* **87**:2237–2261.

Liberman AM, Cooper FS, Shankweiller DP, and Studdert-Kennedy M (1967) Perception of the speech-code. *Psychological Review* **74**:431–461.

Linden J, Liu RC, Sahani M, Schreiner CE, and Merzenich MM (2003) Spectrotemporal structure of receptive fields in areas AI and

AAF of mouse auditory cortex. *Journal of Neurophysiology* **96**: 2660–2975.
Lippert MT, Takagaki K, Xu W, Huang X, and Wu JY (2007) Methods for voltage-sensitive dye imaging of rat cortical activity with high signal-to-noise ratio. *Journal of Neurophysiology* **98**: 502–512.
Liu RC (2006) Prospective contributions of transgenic mouse models to central auditory research. *Brain Research* **1091**:217–223.
Liu RC, Linden JF, and Schreiner CE (2006) Improved cortical entrainment to infant communication calls in mothers compared with virgin mice. *European Journal of Neuroscience* **23**:3087–3097.
Liu RC, Miller KD, Merzenich MM, and Schreiner CE (2003) Acoustic variability and distinguishability among mouse ultrasound vocalizations. *Journal of the Acoustical Society of America* **114**: 3412–3422.
Liu RC and Schreiner CE (2007) Auditory cortical detection and discrimination correlates with communicative significance. *Public Library of Science Biology* **5**:e173.
Loftus WC and Sutter ML (2001) Spectrotemporal organization of excitatory and inhibitory receptive fields of cat posterior auditory field neurons. *Journal of Neurophysiology* **86**:475–491.
Ma J, Kobayasi K, Zhang S, and Metzner W (2006) Vocal communication in adult greater horseshoe bats, *Rhinolophus ferrumequinum*. *Journal of Comparative Physiology A* **192**:535–550.
Ma X and Suga N (2001) Corticofugal modulation of duration-tuned neurons in the midbrain auditory nucleus in bats. *Proceedings of the National Academy of Sciences of the United States of America* **98**:14060–14065.
Malhotra S and Lomber SG (2007) Sound localization during homotopic and heterotopic bilateral cooling deactivation of primary and nonprimary auditory cortical areas in the cat. *Journal of Neurophysiology* **97**:26–43.
Marler P and Peters S (1988) The role of song phonology and syntax in vocal learning preferences in the song sparrow, *Melospiza melodia*. *Ethology* **77**.
Martinovic J, Gruber T, Hantsch A, and Muller MM (2008) Induced gamma-band activity is related to the time point of object identification. *Brain Research* **1198**:93–106.
Mateer C (1983) Motor and perceptual functions of the left hemisphere and their interaction. In: Segalowitz S (ed). *Language Functions and Brain Organization*. Academic Press, New York, pp. 145–170.
May B, Moody DB, and Stebbins WC (1989) Categorical perception of conspecific communication sounds by Japanese macaques, *Macaca fuscata*. *Journal of the Acoustical Society of America* **85**: 837–847.
Mazoyer B, Tzourio N, Frak V, Syrota A, Murayama N, Levrier O, Salamon G, Dehaene S, Cohen L, and Mehler J (1993) The cortical representation of speech. *Journal of Cognitive Neuroscience* **5**: 467–479.
Medvedev AV, Chiao F, and Kanwal JS (2002) Modeling complex tone perception: grouping harmonics with combination-sensitive neurons. *Biological Cybernetics* **86**:497–505.
Medvedev AV and Kanwal JS (2004a) Local field potentials and spiking activity in the primary auditory cortex in response to social calls. *Journal of Neurophysiology* **92**:52–65.
Medvedev AV and Kanwal JS (2004b) Simple acoustic patterns do not predict responses to natural calls in the frontal auditory field of the mustached bat, *Pteronotus parnellii*. *Association for Research in Otolaryngology Abstracts* **27**:196.
Medvedev AV and Kanwal JS (2008) Communication call-evoked gamma-band activity in the auditory cortex of awake bats is modified by complex acoustic features. *Brain Research* **1188**:76–86.
Mendelson JR, Schreiner CE, Sutter ML, and Grasse KL (1993) Functional topography of cat primary auditory cortex: responses to frequency-modulated sweeps. *Experimental Brain Research* **94**: 65–87.
Merzenich MM and Brugge JF (1973) Representation of the cochlear partition of the superior temporal plane of the macaque monkey. *Brain Research* **50**:275–296.
Metherate R, Kaur S, Kawai H, Lazar R, Liang K, and Rose HJ (2005) Spectral integration in auditory cortex: mechanisms and modulation. *Hearing Research* **206**:146–158.
Miller G and Taylor W (1948) The perception of repeated bursts of noise. *Journal of the Acoustical Society of America* **20**:171–182.
Moody DB and Le Prell CG (1997) Perceptual salience of acoustic features of Japanese monkey coo calls. *Journal of Comparative Psychology* **111**:261–274.
Morel A, Garraghty PE, and Kaas JH (1993) Tonotopic organization, architectonic fields, and connections of auditory cortex in macaque monkeys. *Journal of Comparative Neurology* **335**:437–459.
Morel A and Kaas JH (1992) Subdivisions and connections of auditory cortex in owl monkeys. *Journal of Comparative Neurology* **318**: 27–63.
Morton ES (1977) On the occurrence and significance of motivation-structural rules in some bird and mammal sounds. *American Naturalist* **111**:855–869.
Nagarajan SS, Cheung SW, Bedenbaugh PH, Beitel RE, Schreiner CE, and Merzenich MM (2002) Representation of spectral and temporal envelope of twitter vocalizations in common marmoset primary auditory cortex. *Journal of Neurophysiology* **87**:1723–1737.
Naumann T, Washington S, and Kanwal J (2006) Auditory responses in the basolateral amygdala of the mustached bat, *Pteronotus parnellii*. *Association for Research in Otolaryngology Abstracts* **29**:577.
Nelken I (2002) Feature detection by the auditory cortex. In: Oertel D, Fay RR, and Popper AN (eds). *Springer Handbook of Auditory Research*, volume 15, *Integrative Functions in the Mammalian Auditory Pathway*. Springer, New York, pp. 358–416.
Nelken I (2004) Processing of complex stimuli and natural scenes in the auditory cortex. *Current Opinion in Neurobiology* **14**:474–480.
Nelken I, Prut Y, Vaadia E, and Abeles M (1994) Population responses to multifrequency sounds in the cat auditory cortex: one- and two-parameter families of sounds. *Hearing Research* **72**:206–222.
Nelken I and Versnel H (2000) Responses to linear and logarithmic frequency-modulated sweeps in ferret primary auditory cortex. *European Journal of Neuroscience* **12**:549–562.
Newman JD, Smith HJ, and Talmage-Riggs G (1983) Structural variability in primate vocalizations and its functional significance: an analysis of squirrel monkey chuck calls. *Folia Primatologica* **40**:114–124.
Newman JD and Wollberg Z (1973) Responses of single neurons in the auditory cortex of squirrel monkeys to variants of a single call type. *Experimental Neurology* **40**:821–824.
O'Connor KN, Petkov CI, and Sutter ML (2005) Adaptive stimulus optimization for auditory cortical neurons. *Journal of Neurophysiology* **94**:4051–4067.
O'Neill WE (1995) The bat auditory cortex. In: Popper AN and Fay RR (eds). *Springer Handbook of Auditory Research*, volume 5, *Hearing by Bats*. Springer, New York, pp. 416–480.
O'Neill WE and Suga N (1979) Target range-sensitive neurons in the auditory cortex of the mustache bat. *Science* **203**:69–73.
Ohl FW and Scheich H (1997) Orderly cortical representation of vowels based on formant interaction. *Proceedings of the National Academy of Sciences of the United States of America* **94**:9440–9444.
Ohlemiller K, Kanwal JS, Butman JA, and Suga N (1994) Stimulus design for auditory neuroethology: synthesis and manipulation of complex communication sounds. *Auditory Neuroscience* **1**:19–37.
Ohlemiller KK, Kanwal JS, and Suga N (1996) Facilitative responses to species-specific calls in cortical FM-FM neurons of the mustached bat. *Neuroreport* **7**:1749–1755.
Ojemann G (1983) Brain organization for language from the perspective of electrical stimulation mapping. *Behavioral and Brain Sciences* **2**:189–230.

Ostwald J (1984) Tonotopical organization and pure tone response characteristics of single units in the auditory cortex of the greater horseshoe bat. *Journal of Comparative Physiology A* **155**:821–834.

Pandya DN and Sanides F (1973) Architectonic parcellation of the temporal operculum in rhesus monkey and its projection pattern. *Zeitschrift für Anatomie und Entwicklungsgeschichte* **139**:127–161.

Parlee MB (1983) Menstrual rhythms in sensory processes: a review of fluctuations in vision, olfaction, audition, taste, and touch. *Psychological Bulletin* **93**:539–548.

Payne RS and McVay S (1971) Songs of humpback whales. *Science* **173**:585–597.

Penner MJ (1975) Persistence and integration: two consequences of a sliding integrator. *Perception and Psychophysics* **18**:114–120.

Petersen MR, Beecher MD, Zoloth SR, Moody DB, and Stebbins WC (1978) Neural lateralization of species-specific vocalizations by Japanese macaques (*Macaca fuscata*). *Science* **202**:324–327.

Petkov CI, Kayser C, Augath M, and Logothetis NK (2006) Functional imaging reveals numerous fields in the monkey auditory cortex. *Public Library of Science Biology* **4**:e215.

Philibert B, Beitel RE, Nagarajan SS, Bonham BH, Schreiner CE, and Cheung SW (2005) Functional organization and hemispheric comparison of primary auditory cortex in the common marmoset (*Callithrix jacchus*). *Journal of Comparative Neurology* **487**: 391–406.

Phillips DP and Orman SS (1984) Responses of single neurons in posterior field of cat auditory cortex to tonal stimulation. *Journal of Neurophysiology* **51**:147–163.

Phillips DP, Semple MN, and Kitzes LM (1995) Factors shaping the tone level sensitivity of single neurons in posterior field of cat auditory cortex. *Journal of Neurophysiology* **73**:674–686.

Pisoni DB and Lazarus JH (1974) Categorical and noncategorical modes of speech perception along the voicing continuum. *Journal of the Acoustical Society of America* **55**:328–333.

Plourde G, Belin P, Chartrand D, Fist P, Backman S, Guoming X, and Zatorre R (2006) Cortical processing of complex auditory stimuli during alterations of consciousness with the general anesthetic propofol. *Anesthesiology* **104**:448–457.

Poeppel D (2003) The analysis of speech in different temporal integration windows: cerebral lateralization as 'asymmetric sampling in time'. *Speech Communication* **41**:245–255.

Poeppel D, Guillemin A, Thompson J, Fritz J, Bavelier D, and Braun AR (2004) Auditory lexical decision, categorical perception, and FM direction discrimination differentially engage left and right auditory cortex. *Neuropsychologia* **42**:183–200.

Pola YV and Snowdon CT (1975) The vocalizations of pygmy marmosets (*Cebuella pygmaea*). *Animal Behaviour* **23**:826–842.

Polley DB, Read HL, Storace DA, and Merzenich MM (2007) Multiparametric auditory receptive field organization across five cortical fields in the albino rat. *Journal of Neurophysiology* **97**:3621–3638.

Poremba A, Malloy M, Saunders RC, Carson RE, Herscovitch P, and Mishkin M (2004) Species-specific calls evoke asymmetric activity in the monkey's temporal poles. *Nature* **427**:448–451.

Preisler A and Schmidt S (1995) Virtual pitch formation in the ultrasonic range. *Naturwissenschaften* **82**:45–47.

Radtke-Schuller S and Schuller G (1995) Auditory cortex of the rufous horseshoe bat: 1. Physiological response properties to acoustic stimuli and vocalizations and the topographical distribution of neurons. *European Journal of Neuroscience* **7**:570–591.

Ramus F, Hauser MD, Miller C, Morris D, and Mehler J (2000) Language discrimination by human newborns and by cotton-top tamarin monkeys. *Science* **288**:349–351.

Rauschecker JP (1998a) Cortical processing of complex sounds. *Current Opinion in Neurobiology* **8**:516–521.

Rauschecker JP (1998b) Parallel processing in the auditory cortex of primates. *Audiology and Neuro-Otology* **3**:86–103.

Rauschecker JP and Tian B (2000) Mechanisms and streams for processing of "what" and "where" in auditory cortex. *Proceedings of the National Academy of Sciences of the United States of America* **97**:11800–11806.

Rauschecker JP, Tian B, and Hauser M (1995) Processing of complex sounds in the macaque nonprimary auditory cortex. *Science* **268**:111–114.

Rauschecker JP, Tian B, Pons T, and Mishkin M (1997) Serial and parallel processing in rhesus monkey auditory cortex. *Journal of Comparative Neurology* **382**:89–103.

Razak KA and Fuzessery ZM (2002) Functional organization of the pallid bat auditory cortex: emphasis on binaural organization. *Journal of Neurophysiology* **87**:72–86.

Razak KA, Fuzessery ZM, and Lohuis TD (1999) Single cortical neurons serve both echolocation and passive sound localization. *Journal of Neurophysiology* **81**:1438–1442.

Read HL, Winer JA, and Schreiner CE (2001) Modular organization of intrinsic connections associated with spectral tuning in cat auditory cortex. *Proceedings of the National Academy of Sciences of the United States of America* **98**:8042–8047.

Riquimaroux H, Gaioni SJ, and Suga N (1991) Cortical computational maps control auditory perception. *Science* **251**:565–568.

Roederer JG (1975) Introduction to the Physics and Psychophysics of Music. Springer, New York.

Rouiller EM (1997) Functional organization of the auditory pathways. In: Ehret G and Romand R (eds). *The Central Auditory System*. Oxford University Press, New York, pp. 3–96.

Rutkowski RG, Miasnikov AA, and Weinberger NM (2003) Characterisation of multiple physiological fields within the anatomical core of rat auditory cortex. *Hearing Research* **181**: 116–130.

Sales G and Pye D (1974) *Ultrasonic Communication by Animals*. Chapman and Hall, London.

Sander K and Scheich H (2005) Left auditory cortex and amygdala, but right insula dominance for human laughing and crying. *Journal of Cognitive Neuroscience* **17**:1519–1531.

Sansavini A, Bertoncini J, and Giovanelli G (1997) Newborns discriminate the rhythm of multisyllabic stressed words. *Developmental Psychology* **33**:3–11.

Scharf B (1970) Critical band. In: Tobias JV (ed). *Foundations of Modern Auditory Theory*, volume 1. Academic Press, New York, pp. 159–202.

Scheich H, Brechmann A, Brosch M, Budinger E, and Ohl FW (2007) The cognitive auditory cortex: task-specificity of stimulus representations. *Hearing Research* **229**:213–224.

Scheich H, Ohl FW, Schulze H, Hess A, and Brechmann A (2005) What is reflected in auditory cortex activity: properties of sound stimuli or what the brain does with them? In: König R, Heil P, Budinger E, and Scheich H (eds). *The Auditory Cortex. A Synthesis of Human and Animal Research*. Lawrence Erlbaum Associates, Mahwah, pp. 383–407.

Scheich H, Simonis C, Ohl F, Tillein J, and Thomas H (1993) Functional organization and learning-related plasticity in auditory cortex of the Mongolian gerbil. *Progress in Brain Research* **97**: 135–143.

Schreiner CE (1992) Functional organization of the auditory cortex: maps and mechanisms. *Current Opinion in Neurobiology* **2**: 516–521.

Schreiner CE (1995) Order and disorder in auditory cortical maps. *Current Opinion in Neurobiology*:489–496.

Schreiner CE, Read HL, and Sutter ML (2000) Modular organization of frequency integration in primary auditory cortex. *Annual Review of Neuroscience* **23**:501–529.

Schreiner CE and Urbas JV (1986) Representation of amplitude modulation in the auditory cortex of the cat. I. The anterior auditory field (AAF). *Hearing Research* **21**:227–241.

Schreiner CE and Urbas JV (1988) Representation of amplitude modulation in the auditory cortex of the cat. II. Comparison between cortical fields. *Hearing Research* **32**:49–63.

Schreiner CE and Winer JA (2007) Auditory cortex mapmaking: principles, projections, and plasticity. *Neuron* **56**:356–365.

Schulze H, Hess A, Ohl FW, and Scheich H (2002) Superposition of horseshoe-like periodicity and linear tonotopic maps in auditory cortex of the Mongolian gerbil. *European Journal of Neuroscience* **15**:1077–1084.

Schulze H and Langner G (1997) Periodicity coding in the primary auditory cortex of the Mongolian gerbil (*Meriones unguiculatus*): two different coding strategies for pitch and rhythm? *Journal of Comparative Physiology A* **181**:651–663.

Schwarz DW and Tomlinson RW (1990) Spectral response patterns of auditory cortex neurons to harmonic complex tones in alert monkey (*Macaca mulatta*). *Journal of Neurophysiology* **64**: 282–298.

Scott SK (2005) Auditory processing-speech, space and auditory objects. *Current Opinion in Neurobiology* **15**:197–201.

Scott SK and Johnsrude IS (2003) The neuroanatomical and functional organization of speech perception. *Trends in Neurosciences* **26**: 100–107.

Scott SK and Wise RJS (2003) PET and fMRI studies of the neural basis of speech perception. *Speech Communication* **41**:23–34.

Seyfarth RM, Cheney DL, and Marler P (1980) Vervet monkey alarm calls: semantic communication in a free-ranging primate. *Animal Behavior* **28**:1070–1094.

Shamma SA, Fleshman JW, Wiser PR, and Versnel H (1993) Organization of response areas in ferret primary auditory cortex. *Journal of Neurophysiology* **69**:367–383.

Shapleske J, Rossell SL, Woodruff PW, and David AS (1999) The planum temporale: a systematic, quantitative review of its structural, functional and clinical significance. *Brain Research Reviews* **29**:26–49.

Sinai A, Bowers CW, Crainiceanu CM, Boatman D, Gordon B, Lesser RP, Lenz FA, and Crone NE (2005) Electrocorticographic high gamma activity versus electrical cortical stimulation mapping of naming. *Brain* **128**:1556–1570.

Sloboda JA (1978) Perception of contour in music reading. *Perception* **7**:323–331.

Snowdon CT (1982) Linguistic and psycholinguistic approaches to primate communication. In: Snowdon CT, Brown CH, and Petersen MR (eds). *Primate Communication*. Cambridge University Press, Cambridge, pp. 212–238.

Steinschneider M, Reser DH, Fishman YI, Schroeder CE, and Arezzo JC (1998) Click train encoding in primary auditory cortex of the awake monkey: evidence for two mechanisms subserving pitch perception. *Journal of the Acoustical Society of America* **104**: 2935–2955.

Steinschneider M, Volkov IO, Fishman YI, Oya H, Arezzo JC, and Howard MA 3rd (2005) Intracortical responses in human and monkey primary auditory cortex support a temporal processing mechanism for encoding of the voice onset time phonetic parameter. *Cerebral Cortex* **15**:170–186.

Stevens KN and Klatt DH (1974) Role of formant transitions in the voiced-voiceless distinction for stops. *Journal of the Acoustical Society of America* **55**:653–659.

Stewart L, Overath T, Warren JD, Foxton JM, and Griffiths TD (2008) fMRI evidence for a cortical hierarchy of pitch pattern processing. *Public Library of Science ONE* **3**:e1470.

Stiebler I, Neulist R, Fichtel I and Ehret G (1997) The auditory cortex of the house mouse: left-right differences, tonotopic organization and quantitative analysis of frequency representation. *Journal of Comparative Physiology A* **181**:559–571.

Streeter LA (1978) Acoustic determinants of phrase boundary perception. *Journal of the Acoustical Society of America* **64**:1582–1592.

Suga N (1978) Specialization of the auditory system for reception and processing of species-specific sounds. *Federation Proceedings* **37**:2342–2354.

Suga N (1984) The extent to which biosonar information is represented in the bat auditory cortex. In: Edelman GM, Gall WE, and Cowan WM (eds). *Dynamic Aspects of Neocortical Function*. John Wiley & Sons, New York, pp. 315–373.

Suga N (1988) What does single-unit analysis in the auditory cortex tell us about information processing in the auditory system? In: Rakic P and Singer W (eds). *Neurobiology of Neocortex*. John Wiley & Sons, Chichester, pp. 331–349.

Suga N (1990) Biosonar and neural computation in bats. *Scientific American* **262**:60–68.

Suga N (1994a) Multi-function theory for cortical processing of auditory information: implications of single-unit and lesion data for future research. *Journal of Comparative Physiology A* **175**:135–144.

Suga N (1994b) Processing of auditory information carried by species-specific complex sounds. In: Gazzaniga MS (ed). *The Cognitive Neurosciences*. The MIT Press, Cambridge, pp. 295–313.

Suga N (1996) Basic acoustic patterns and neural mechanisms shared by humans and animals for auditory perception: a neuroethologist's view. In: Ainsworth W and Greenberg S (eds). *Auditory Basis of Speech Perception*. ISCA Archive, Keele University, Keele, pp. 31–38.

Suga N and O'Neill WE (1979) Neural axis representing target range in the auditory cortex of the mustache bat. *Science* **206**:351–353.

Suga N, O'Neill WE, Kujirai K, and Manabe T (1983) Specificity of combination-sensitive neurons for processing of complex biosonar signals in auditory cortex of the mustached bat. *Journal of Neurophysiology* **49**:1573–1626.

Suga N and Schlegel P (1972) Neural attenuation of responses to emitted sounds in echolocating rats. *Science* **177**:82–84.

Sussman HM (1989) Neural coding of relational invariance in speech: human language analogs to the barn owl. *Psychological Review* **96**:631–642.

Sussman HM, McCaffrey HA, and Matthews S (1991) An investigation of locus equations as a source of relational invariance for stop place categorization. *Journal of the Acoustical Society of America* **90**:1309–1325.

Sutter ML, Schreiner CE, McLean M, O'Connor KN, and Loftus WC (1999) Organization of inhibitory frequency receptive fields in cat primary auditory cortex. *Journal of Neurophysiology* **82**: 2358–2371.

Sweet RA, Dorph-Petersen KA, and Lewis DA (2005) Mapping auditory core, lateral belt, and parabelt cortices in the human superior temporal gyrus. *Journal of Comparative Neurology* **491**:270–289.

Syka J and Merzenich MM (eds) (2005) *Plasticity and Signal Representation in the Auditory System*. Springer, New York. .

Taberner AM and Liberman MC (2005) Response properties of single auditory nerve fibers in the mouse. *Journal of Neurophysiology* **93**:557–569.

Taniguchi I, Horikawa J, Moriyama T, and Nasu M (1992) Spatio-temporal pattern of frequency representation in the auditory cortex of guinea pigs. *Neuroscience Letters* **146**:37–40.

Terhardt E (1974a) On the perception of periodic sound fluctuations (roughness). *Acustica* **30**:201–213.

Terhardt E (1974b) Pitch, consonance, and harmony. *Journal of the Acoustical Society of America* **55**:1061–1069.

Thomas H, Tillein J, Heil P, and Scheich H (1993) Functional organization of auditory cortex in the Mongolian gerbil (*Meriones unguiculatus*). I. Electrophysiological mapping of frequency representation and distinction of fields. *European Journal of Neuroscience* **5**: 882–897.

Tian B and Rauschecker JP (1998) Processing of frequency-modulated sounds in the cat's posterior auditory field. *Journal of Neurophysiology* **79**:2629–2642.

Tomlinson RW and Schwarz DWF (1988) Perception of the missing fundamental in nonhuman primates. *Journal of the Acoustical Society of America* **84**:560–565.

Trehub SE, Bull D, and Thorpe LA (1984) Infants' perception of melodies: the role of melodic contour. *Child Development* **55**: 821–830.

Tsuzuki K and Suga N (1988) Combination-sensitive neurons in the ventroanterior area of the auditory cortex of the mustached bat. *Journal of Neurophysiology* **60**:1908–1923.

Ulanovsky N, Las L, Farkas D, and Nelken I (2004) Multiple time scales of adaptation in auditory cortex neurons. *Journal of Neuroscience* **24**:10440–10453.

Ulanovsky N, Las L, and Nelken I (2003) Processing of low-probability sounds by cortical neurons. *Nature Neuroscience* **6**:391–398.

Umiker-Sebeok J and Sebeok TA (1980) Introduction: questioning apes. In: Sebeok TA and Umiker-Sebeok J (eds). *Speaking of Apes. A Critical Anthology of Two-Way Communication With Man*. Plenum, New York, pp. 1–59.

van Dommelen WA (1990) Acoustic parameters in human speaker recognition. *Language and Speech* **33**:259–272.

van Wassenhove V and Nagarajan SS (2007) Auditory cortical plasticity in learning to discriminate modulation rate. *Journal of Neuroscience* **27**:2663–2672.

von der Malsburg C (1986) Am I thinking assemblies? In: Palm G and Aertsen A (eds). *Brain Theory*. Springer, Berlin Heidelberg, pp. 161–176.

Wallace MN, Rutkowski RG, and Palmer AR (2000) Identification and localisation of auditory areas in guinea pig cortex. *Experimental Brain Research* **132**:445–456.

Wan H, Warburton EC, Kusmierek P, Aggleton JP, Kowalska DM, and Brown MW (2001) Fos imaging reveals differential neuronal activation of areas of rat temporal cortex by novel and familiar sounds. *European Journal of Neuroscience* **14**:118–124.

Wang X, Lu T, Snider RK, and Liang L (2005) Sustained firing in auditory cortex evoked by preferred stimuli. *Nature* **435**:341–346.

Wang X, Merzenich MM, Beitel R, and Schreiner CE (1995) Representation of a species-specific vocalization in the primary auditory cortex of the common marmoset: temporal and spectral characteristics. *Journal of Neurophysiology* **74**:2685–2706.

Warren JE, Wise RJ, and Warren JD (2005) Sounds do-able: auditory-motor transformations and the posterior temporal plane. *Trends in Neurosciences* **28**:636–643.

Washington SD and Kanwal JS (2007) Right versus left hemispheric tuning to frequency modulated sweeps in primary auditory cortex. *Society for Neuroscience Abstracts* **37**:389.9.

Washington SD and Kanwal JS (2008) DSCF neurons within the primary auditory cortex of the mustached bat process frequency modulations present within social calls. *Journal of Neurophysiology* **100**:3285–3304.

Wehr M and Zador AM (2003) Balanced inhibition underlies tuning and sharpens spike timing in auditory cortex. *Nature* **426**: 442–446.

Weinberger NM and McKenna TM (1988) Sensitivity of single neurons in auditory cortex to contour: toward a neurophysiology of music perception. *Music Perception* **5**:355–390.

Wernicke C (1874) Der Aphasische Symptomenkomplex: Eine Psychologische Studie auf Anatomischer Basis. Max Cohn und Weigert, Breslau.

Wetzel W, Ohl FW, Wagner T and Scheich H (1998) Right auditory cortex lesion in Mongolian gerbils impairs discrimination of rising and falling frequency-modulated tones. *Neuroscience Letters* **252**:115–118.

Whitney G and Nyby JG (1983) Sound communication among adults. In: Willott JF (ed). *The Auditory Psychobiology of the Mouse*. Charles C Thomas, Springfield, pp. 98–129.

Willott JF (2001) *Handbook of Mouse Auditory Research*. CRC Press, Boca Raton.

Winer JA, Wenstrup JJ, and Larue DT (1992) Patterns of GABAergic immunoreactivity define subdivisions of the mustached bat's medial geniculate body. *Journal of Comparative Neurology* **319**:172–190.

Winter P and Funkenstein HH (1973) The effect of species-specific vocalization on the discharge of auditory cortical cells in the awake squirrel monkey (*Saimiri sciureus*). *Experimental Brain Research* **18**:489–504.

Winter P, Ploog D, and Latta J (1966) Vocal repertoire of the squirrel monkey (*Saimiri sciureus*), its analysis and significance. *Experimental Brain Research* **1**:359–384.

Wollberg Z and Newman JD (1972) Auditory cortex of squirrel monkey: response patterns of single cells to species-specific vocalizations. *Science* **175**:212–214.

Wong PC (2002) Hemispheric specialization of linguistic pitch patterns. *Brain Research Bulletin* **59**:83–95.

Wu JY, Guan L, Bai L, and Yang Q (2001) Spatiotemporal properties of an evoked population activity in rat sensory cortical slices. *Journal of Neurophysiology* **86**:2461–2474.

Wu JY, Lam YW, Falk CX, Cohen LB, Fang J, Loew L, Prechtl JC, Kleinfeld D, and Tsau Y (1998) Voltage-sensitive dyes for monitoring multineuronal activity in the intact central nervous system. *Histochemical Journal* **30**:169–187.

Yin S, Chen Z, Yu D, Feng Y, and Wang J (2008) Local inhibition shapes duration tuning in the inferior colliculus of guinea pigs. *Hearing Research* **237**:32–48.

Zanto TP, Snyder JS, and Large EW (2006) Neural correlates of rhythmic expectancy. *Advances in Cognitive Psychology* **2**:221–231.

Zatorre RJ, Belin P, and Penhune VB (2002) Structure and function of auditory cortex: music and speech. *Trends in Cognitive Sciences* **6**:37–46.

Zhang LI, Tan AYY, Schreiner CE, and Merzenich MM (2003) Topography and synaptic shaping of direction selectivity in primary auditory cortex. *Nature* **424**:201–205.

Zimmer U, Ehret G, and Esser K-H (1998) Categorical perception of sounds mimicking species-specific communication calls in the FM-bat *Phyllostomus discolor*. In: *Göttingen Neurobiology Report*, Thieme, Göttingen, p. 326.

Chapter 17

A Semantic Concept of Auditory Cortex Function and Learning

Henning Scheich and Frank W. Ohl

Abbreviations

AC	auditory cortex
AI	primary auditory cortex
AM	amplitude modulation
BF	best frequency
CM	caudo-medial field
EEG	electroencephalogram
FM	frequency modulation
fMRI	functional magnetic resonance imaging
PET	positron emission tomography

1 Introduction

Auditory cortex (AC) differs from subcortical auditory nuclei due to its prominent learning-related plasticity (e.g., Suga and Ma 2003; Weinberger 2004a; Irvine and Wright 2005; Ohl and Scheich 2005; Scheich et al. 2007) and the dominance of descending inputs from other cortical areas (Budinger et al. 2000a,b, 2006, 2008; Scheich et al. 2007). Thus, AC can be considered a bottom-up/top-down interface with its function comprising more than simply refining identification and discrimination of auditory stimulus properties for use in cognitive and behavioral tasks elsewhere. AC appears to be an active participant in such tasks (Scheich et al. 2006; Scheich et al. 2010). This role would require various learning strategies. While it is known that AC processing is modifiable by learning (see Chapters 22 and 24), the role of learning-induced changes in a behavioral and cognitive context is still uncertain. Considerations of the special nature of environmental sounds as information source in shaping auditory system evolution lead us to propose a semantic hypothesis of auditory cortex function and learning: AC processing and learning promotes behaviorally meaningful interpretations of sounds. It is not the sound event itself, but its presumed cause and/or behavioral consequence toward which the concern of the listener is ultimately directed. The semantics of any sound may not be explicitly available in AC. Rather, auditory cortical stimulus representations and their plasticity reflect implicit rules and cognitive strategies in auditory tasks through interactions with other cortical areas that attach meaning to sounds and predict behavioral performance. Similar processes may occur in other sensory modalities, but we propose that attributing behavioral meaning to a sound requires learning phenomena that are particularly salient in auditory cortex. In that sense, our proposal is an extension of the concept of auditory scene analysis (Bregman 1990) toward the semantic nature of auditory processing.

Since the early findings of associative learning-induced physiological plasticity in AC (Kraus and Disterhoft 1982; Weinberger et al. 1984; Gonzalez-Lima and Scheich 1986), various neuronal stimulus representations have been described in animals and humans that likely were shaped by learning processes that affected stimulus meaning for the individual (for review, see Suga and Ma 2003; Weinberger 2004; Irvine and Wright 2005; Ohl and Scheich 2005). These phenomena are evident for various analysis methods (receptive field measurement of neurons, local field potentials, EEG, 2-deoxyglucose, immediate early genes, PET, fMRI) and at microscopic, mesoscopic and macroscopic scales of functional organization. Such wide range of plasticity expression requires explanation. Yet causes and mechanistic underpinnings are not fully understood, and a comprehensive scheme is required to assess what it is that auditory cortex allows individuals to extract from the sound environment. Here we attempt an integration of several perspectives on the role of task-specificity in stimulus representation (Ohl and Scheich 2005; Scheich et al. 2007; Scheich et al. 2010).

In the following, we consider the special nature of sounds and their role for individual behavior, i.e., which learning

F.W. Ohl (✉)
Abteilung für Neuroscience, Leibniz-Institute for Neurobiology, D-39118 Magdeburg, Germany
e-mail: frank.ohl@ifn-magdeburg.de

J.A. Winer, C.E. Schreiner (eds.), *The Auditory Cortex*,
DOI 10.1007/978-1-4419-0074-6_17,

strategies must have been developed by auditory systems to derive individually meaningful information from environmental sound patterns. Central are identification, discrimination and categorization of auditory stimuli. Specific auditory processing strategies are contrasted with those in the visual system.

Then we shall describe the main organizational principles of AC functional maps. Long standing mapping attempts to discover prominent gradient maps for the separation of various complex acoustic dimensions (similar to visual cortex maps and to specialized auditory cortex maps in echolocating bats) in different non-specialized species have failed to date. Instead a mosaic-like organization with spatially dispersed arrangements of diverse neuronal sensitivities and selectivities has emerged. Implications for the concept of stimulus contrast and forms of discrimination and discrimination learning in the auditory domain are discussed.

Finally, we shall conceptualize AC functional organization as an opportunistic principle for flexible representation of sound stimuli in a task-dependent fashion. Any given sound can be represented by different activation patterns across maps depending on the task- and interpretation-related selection of neuron assemblies. Task-dependence of neuronal representation is one form of including semantic stimulus aspects into their processing. Beside stimulus identification and discrimination, the formation of sound categories is fundamental to the interpretation of sounds, as categories are equivalence classes of meaning that can be derived by learning from multiple exemplars. For instance, some types of noisy sounds are characteristic of rain, while others are characteristic of the friction of small objects like dry leaves. We shall discuss evidence that opportunistic maps are ideal to represent multiple categories and other semantic aspects of stimulus processing.

2 The Nature of Sounds

A naïve listener typically cannot develop any reasonable behavioral strategy when perceiving a novel sound except searching for its cause or waiting for potential behavioral consequences, neither of which can result from the sound itself but only from its source or causal events involving the source.

The implications of this tenet become more apparent by imagining the following scene: In a windy dark night you find yourself in a forest and you are relatively inexperienced with the sounds in this environment. Above the ever-changing rush of wind in the trees you occasionally hear rustles, cracks, murmurs, bangs, screeches, squeals, whimpers, claps, plops, and some tonal or even voice-like events. Some sounds may form rapid sequences with major changes of sound quality. In the dark, this all alerts your senses but only leads to associations and speculations as to the origin of the sounds and to possible consequences of theses causes for your well-being. This scene demonstrates that the many potential attributes of natural sounds are making the process of learning their meaning very demanding but also suggests some strategies how these interpretational challenges may be solved.

2.1 *Sounds vs. Sound-Generating Objects and Events*

Sounds are emitted from material objects under the influence of variable forces acting upon these objects and make them vibrate. When a sound occurs, something must have moved in the environment and with the concomitant uncertainty what it was, this presents a case of enormous biological significance. Yet, depending on the causal events, numerous very different sounds can be obtained from the same object and similar sounds from unrelated objects. Thus, natural sounds are signals that, in principle, contain information about both the properties of objects and the actions leading to the sounds, although with a high degree of ambiguity. Without evolutionary knowledge or prior experience, a listener cannot disambiguate these two aspects. Sounds, per se, do not provide object information such as size, shape, material and dynamics in the sense of "images" of the sources, information that could be used for a reasonable behavioral attitude towards a novel sound. This is in contrast to the visual modality where perceived novel objects can be directly addressed and behaviorally explored. Therefore, it is primarily the source object and the causal events leading to the sound signal that a listener is attempting to infer to instruct his/her own behavior.

These considerations suggest that sounds can only be interpreted by a learning process that uses additional associative information, e.g., from visual and other sensory references to the sources (Cahill et al. 1996; Brosch et al. 2005; Budinger et al. 2006; Ghazanfar and Schroeder 2006) and from behavior of the sources related to the sounds, e.g., if there is a growl there might be a dog that subsequently bites. The interpretation of novel sounds – in contrast to novel visual objects – requires at least one additional associative learning step. By accumulated experience, a perceived sound may acquire a symbol-like character for the material object or event that has produced it.

2.2 *Source Identification in Auditory Scenes*

Highly variable sound signals can obscure the source identity and, sound changes can affect interpretation, e.g., implying a different object or causal event. These ambiguities combined with frequent sound changes require highly efficient discrimination and learning mechanisms. This may not differ from demands in other sensory modalities; however, differences emerge when considering two discriminative strategies: simultaneous and sequential contrast formation. Simultaneous sounds do not necessarily conceal each other or a background sound, unlike occlusion for visual objects. Sound waves simply sum up. Consequently, simultaneous contrasts of spectral elements – in analogy to spatial visual elements that provide enhanced contours – cannot be reliably used to discriminate simultaneous sounds or sounds from backgrounds. Rather, the auditory system exploits the temporal comodulation of spectral elements that emanate from a given sound source to distinguish simultaneous sounds (e.g., Moore 1990; Ernst and Verhey 2008). This strategy points to a prominent interpretational role of the source identity based on the implicit assumption that comodulated sound elements stem from the same source.

Nevertheless, since most natural sounds are spectrally wideband, the simultaneous contrast of different spectral elements within a given sound is an important target of contrast mechanisms (Peterson and Barney 1952; Sachs and Young 1979; Ohl and Scheich 1997b) as we will discuss later.

Due to the character of sound generation, natural sounds are typically nonstationary and contain fast transients. Thus, time-domain analysis is of paramount importance to characterize sounds and to derive their meaning. Natural sounds are often composed of sequences of virtually discrete, often rapidly changing, events with unpredictable time intervals. This is unlike the properties of most material objects that occur in natural environments as perceived by other senses. Such sound sequences, generated by single or multiple sources, are highly informative, generated by inanimate sources as well as by vocalizing animals, and are a common ingredient of speech and music. Their analysis involves a wide repertoire of mechanisms including sequential contrast formation by forward and backward masking or facilitation, streaming and other spectro-temporal integration mechanisms (Bregman 1990; Plack and Viemeister 1993; Nelken et al. 1999; Brosch and Schreiner 2000; Altmann et al. 2007; Brosch and Scheich 2008; Happel et al. 2010).

Onset asynchronies between partially overlapping sounds are also a source of perceptual sound distinction. Even delaying a single partial of a harmonic tone by more than 30 ms after stimulus-onset can lead to perception of two distinct tonal entities (Bregman and Rudnicky 1975; Rasch 1978). This corresponds to the implicit processing assumption that co-occurring sounds starting at different times must stem from different sources.

2.3 Statistics of Sound Properties

Considerations on natural sound properties must include assumption about signal statistics, including reliabilities and predictabilities, for explaining neuronal sensitivities and selectivities (Eggermont 1998a; Nelken et al. 1999; Ulanovsky et al. 2003; Machens et al. 2004; Smith and Lewicki 2006). In the following, we develop a rationale for how such response properties serve the semantic demands of auditory cortical processing and learning in identification, discrimination and categorization.

2.4 Categorization of Sound-Generating Objects and Events

Sound ambiguity with respect to sources and events as well as problems of distinguishing biologically meaningful from accidental variations can be overcome by experience and learning with formation of categories. Categories can be defined as equivalence classes of meaning and allow extrapolations on the meaning of new sounds (Ohl and Scheich 2001; Scheich et al. 2007). Sound categories can be formed by reference to similar object properties, or similar sound-causing events. Sound categories can acquire a symbolic meaning for objects and/or events. For example, in the inanimate world splashing or dripping sounds refer to liquids in different quantities, while cracking sounds usually refer to the breaking of hard material objects but without identifying the sources in both cases. The inherently symbolic usefulness of sound categories seems to be exploited in animal and human communication processing. Animal vocalizations typically refer to classes of emotional and motivational states of the callers, a strategy which is maintained in emotional prosodies of human speech. In linguistics, speech nouns and verbs refer to objects and events, respectively, in a categorical fashion.

2.5 Memory-Dependence of Sound Interpretation

Natural sounds from inanimate and animate sources are typically fleeting, i.e., short-lived or have a time course with fast varying parameters (transients). Many sounds have already vanished before they are interpreted. Thus short sounds must be kept in short-term memory for identification, discrimination and categorization and to compare them with stored references (long-term memory). This short-term memory requirement is not typical of visual scenes where most objects remain visible over sufficient time. Ultrashort-term "sensory memory" of visual objects is an order of magnitude shorter (200–400 ms; Sperling 1960) than the corresponding "echo memory" for auditory items (2–4 s; Darwin et al. 1972). Mechanisms of short-term memory are highly developed in auditory cortex, e.g., the mismatch negativity in echo memory (Näätänen and Winkler 1999; Ulanovsky et al. 2003), and the neuronal activity maintenance in working memory (Winter and Stich 2005; Brechmann et al. 2007).

2.6 Natural vs. Laboratory Situations

The semantic requirements of auditory processing and learning in principle do not change when an auditory task is brought into the laboratory situation – unless they are analyzed in anaesthetized subjects. The only difference regarding semantic uncertainty is that all sounds come from loudspeakers, i.e., cannot be directly associated with natural sources. In this case the emphasis of a listener is placed on learning to interpret sounds with regard to other concomitant information, i.e., to the context and/or the ensuing behavioral consequences. This is predominantly the situation for auditory conditioning training of animals.

In animal training paradigms, the behavioral rules providing the meaning of the sound are derived by the animals themselves. In human auditory psychophysics the behavioral rules and, consequently, the semantics are often provided explicitly by instructions. Such detection, discrimination or categorization tasks are less natural or semantically more abstract in attributing meaning to sounds. This underscores that arbitrary sounds can be made meaningful with respect to any behavioral or cognitive context, as is evident in language.

2.7 "Sound Objects" Revisited

Sounds are sometimes considered as "objects" in an auditory analogy to the definition that "objects are material things that can be seen and touched" (Oxford Dictionary). From the preceding considerations, it is apparent that this term does not capture the nature of sounds which is signals only sometimes emanating from objects. "Object" could be used in a transferred sense of an item of specific interest or reasoning. For example, (1) sound waves as physical entities can be addressed as acoustic objects in order to perform physical measurements; (2) at the perceptual level one could follow the fundamental Platonian and Aristotelian line of reasoning that distinction ("krinein") is the basic process of obtaining insight into what exists. Then any two distinguishable sounds could be considered as two objects. (3) A recent proposal was to consider "the perceptual entities (produced by sounds) as auditory objects that can, e.g., be categorized as either a particular voice or particular vowel" (Griffiths and Warren 2004). This definition is a perceptual and semantic blend that was explicitly derived from George Berkeley's rather select type of philosophical epistemology. His stance went beyond the moderate Kantian tenet that there can be no knowledge about the world independent of our percepts, by assuming that nothing in the world exists independent of our percepts ("esse est percipi"). This essentially constructivist view does not help to explain the mechanistic strategies of the auditory system both in evolution and individual learning under the constraints of the described special nature of sounds. It appears to us that any use of the term "object" for sounds requires extensive definitions without providing considerable advantages and also obscures the difference to visual and somatosensory information and their processing.

2.8 Adequacy of Visual Metaphors for Audition

The difference between the typical demands of processing environmental sound information and visual information (cf. Handel 2006) could be derived from a transformation of visual scenes applying rules of sound behavior from sonogram-like depictions of acoustic scenes. Assuming that elements of visual scenes would generally behave similar to sounds, some unusual properties would emerge. First, all objects in a visual scene would be transparent, such that foreground objects would not conceal background objects, hampering delineation of object contours and shapes. Furthermore, scenes would not be static. However, objects would not simply move in space but would rapidly change, among other properties, in size, shape and texture across an equally variable visual background. Many visual objects would vary their properties in a rather abrupt or even discontinuous fashion. They would appear and vanish, without obvious occlusion effects, and if they pop up again, they would often show considerably morphed shapes that would make it uncertain whether it is still the same object. These clearly are not typical properties of natural visual environments as we experience them and they do not allow deriving meaning from natural visual scenes as we have learned to interpret them.

This sketch therefore emphasizes some modality differences that are often neglected in attempts to use visual metaphors for auditory processing. Thus, auditory and visual cortex must have evolved under the selection pressure of different demands imposed by the environment.

3 Representation of Sounds in Auditory Cortex

3.1 Sensitivities and Selectivities of Auditory Cortical Representations

The implications of the special attributes of sounds with respect to meaningful interpretation and learning seem to be reflected by several traits of AC organization and functional properties of neurons. In all electrophysiologically well-characterized auditory cortices of animals (monkey, cat, gerbil, guinea pig, rat, mouse, bat, ferret) additional fields have been identified besides the primary field AI. These fields are still commonly distinguished by the spatial layout of their tonotopic gradient or by a lack of tonotopy (Kaas and Hackett 2000). The reason is that many neurons in a field prefer some tone frequencies, whereas response preferences to more complex acoustic dimensions remain ambiguous. Even in cortical fields of the best-characterized species (cat and monkey) no exclusive representation or dominant spatial response gradient of a complex sound dimension has been identified that would unequivocally delineate cortical fields. Tested sound parameters include frequency combinations (Sutter et al. 1999; Kadia and Wang 2003), and various amplitude modulations (AM), frequency modulations (FM), noise band or harmonicity conditions (e.g., Urbas and Schreiner 1988; Eggermont 1998b; Fishbach et al. 2001; Nelken 2002, 2004; Read et al. 2002; Kadia and Wang 2003; Linden and Schreiner 2003; Griffiths et al. 2004; Joris et al. 2004; Rauschecker and Tian 2004; Tian and Rauschecker 2004; for review see Schreiner and Winer 2007). This is in notable contrast to selective representations or exquisite spatial gradients for orientation selectivity, moving contrasts, color and other stimulus dimensions in visual cortex of the same species (cf. Nelson and Bower 1990; Chklovskii and Koulakov 2004; Basole et al. 2006).

Even the characterization of single neurons in terms of specific preference for complex stimuli can pose problems by often showing responses to seemingly unrelated parameter regions, such as different AM and FM modulation ranges, tonal and broad-band stimuli among others. It appears that neurons with highly dominant complex feature preference are rare. Perhaps the optimal stimulus could not be identified due to practical combinatorial limits in an experiment. From a population perspective, however, it is clear that exclusive response specialization must play a minor role in identification of stimuli due to the vast ensembles of neurons simultaneously activated by a given stimulus in any cortical map, parts of which also responding to very different stimuli.

We will use here operational terms of response specialization to classify neurons under the aspect what they contribute by their activity to these processes: Neurons that are responsive to stimuli can show sensitivity to systematic variation of a given stimulus parameter, and thereby are sensitive to what distinguishes two or more stimuli with respect to this parameter. Sensitivity to parameter variation also implies that such units will exhibit a preference for a particular value or set of values of this parameter. Other neurons, also responsive to these stimuli, may be largely insensitive to this particular variation. In this vein, selectivity means that neurons may maintain their discrimination of parameters of interest even if other stimulus properties are varied.

3.2 Maps in Auditory Cortical Representations

Mixed populations of neurons in a field can still show some spatial order. A principle of the AI organization seems to be that neuronal sensitivities and selectivities for various complex dimensions are distributed in patches along isofrequency contours of the field but with little spatial predictability especially in a comparison of different individuals (Middlebrooks et al. 1980; Imig et al. 1990; Sutter and Schreiner 1991; Heil et al. 1992; Mendelson et al. 1993; Shamma et al. 1993; Phillips et al. 1994; Rauschecker et al. 1997; Schreiner 1995, 1998; Read et al. 2002; Nelken et al. 2003). Some of these specializations in AI may form weak gradients, e.g., for intensity (Heil et al. 1994) or for bandwidth (Schreiner 1998), which can be used for discrimination of vowels (Ohl and Scheich 1997b). But different gradients are not spatially separated; they are rather superimposed or overlap such that the organization may be conceived as a multidimensional interdigitating mosaic of numerous neuronal sensitivities and selectivities. This mosaic-like or modular principle as an alternative to spatial gradient maps has previously been identified in the bird auditory cortex analogue, field L (Scheich et al. 1979; Langner et al. 1981, Hose et al. 1987; Scheich 1991). "Mosaic-like maps" (in contrast to separation of complex dimensions in different orderly "spatial gradient maps") may contain sufficiently diverse mechanisms to characterize any complex stimulus by ensemble representation across neurons with various sensitivities and selectivities and to recall this stimulus upon reoccurrence by the same representation. Therefore they may be considered as "opportunistic maps" here referring to an organization that provides a rich repertoire of complex neuronal sensitivities and selectivities, but with few specific assumptions about the natural prevalence of any complex acoustic dimensions or their parametric variations. Thus, beside identification of sounds, any discrimination or categorization based on selection of some of the properties of sounds can be performed. The ensemble representation of an actual stimulus in such a map consists of a spatiotemporal activation pattern of all neurons involved, i.e., a state of activation (cf. Takagaki et al. 2008). In light of the necessity for interpreting sounds one can assume that in addition to the physical parameters of a sound, aspects of its specific semantic interpretation will co-determine its neuronal representation. This hypothesis can be tested by demonstrating changes in the neuronal representation of physically identical sounds that are due to changes in their interpretation and meaning, e.g., by varying the nature of context or behavioral task (e.g., Brechmann et al. 2007).

4 Contrasts and Stimulus Representations

Cortical discrimination mechanisms play a fundamental role not only for distinguishing stimuli and for forming categorical boundaries but also for identifying relevant stimuli against backgrounds. Classically, sensory stimulus discrimination has been related to the concept of simultaneous contrast (classical work of Mach, Hering, von Helmholtz, von Békésy as discussed in Lange-Malecki et al. 1990). Subsequently this was referred in the visual system to mechanisms of lateral inhibition (reciprocal, inhibitory interaction between neighboring neurons) in gradient maps (Hartline 1949) and neuronal receptive fields with excitatory centers and inhibitory surrounds that flexibly enhance contrasts (Kuffler 1952; Hubel and Wiesel 1962). There is also evidence form auditory cortex (Ohl and Scheich 1997b) and visual cortex (Priebe and Ferster 2008) that contrast mechanisms that previously appeared to require lateral inhibition can in fact be better explained by nonlinearities of the feed-forward gains. As described, naturally varying sounds do not form predictable and stationary contrasts with sound backgrounds or interfering sounds. This is quite different from the visual domain where simultaneous spatial contrasts are omni-present and some complex dimensions such as contours, shapes and colors are prone to simultaneous contrast formation with visual cortex showing corresponding topographic organizations. If simultaneous contrast cannot be used in a predictable and systematic fashion to delineate sounds from backgrounds or from competing sounds there is no evolutionary advantage in representing complex sound dimensions separately and systematically in gradient maps that would allow any local lateral inhibition along these dimensions. Therefore, in view of the general importance of foreground–background decomposition in audition and the description of a specialized secondary area for this task in human auditory cortex (Scheich et al. 1998) relevant contrast mechanisms must be independent of spectral sound properties and cannot be stable over time using true simultaneous contrasts. Rather, as mentioned earlier, sounds from different sources can be distinguished by spectral components of different comodulation and/or onset asynchronies, i.e., by temporal cues.

Relevant spectral contrasts do exist between different components within wideband sounds, such as formants in vocalizations and spectral modes in musical sounds. They exhibit characteristic spectral energy profiles with subregions of high or low power. It remains to be shown that such contrasts form the basis for distinguishing natural sounds. Inhibitory sidebands of neurons and lateral inhibition in tonotopic maps seem to be instrumental in resolving such spectral profiles (Ohl and Scheich 1997b; Barbour and Wang 2003).

Besides simple spectral contrasts there appears little utility for other systematic intra-sound contrast enhancement mechanisms. Due to the mechanics of sound generation there is usually only one frequency modulation or amplitude modulation present at a time in a given sound (comodulation, Hall et al. 1984; Verhey et al. 2003). An exception may be "spectral motions" in formant transitions of speech phonemes (Thivard et al. 2000). However, these converging or diverging shifts of energy maxima across a harmonic spectrum cannot be considered simultaneous frequency modulations. Multiple spectral correlates of modulations in terms of stimulus carriers and sidebands may be chiefly used for the formation of spectral contrast. Tonotopic maps have the necessary properties to achieve this. Different modulation frequencies or modulation speeds usually do not occur simultaneously and systematic topographic representation in separate gradient maps with lateral inhibition may not be required. This is not trivial as the situation is quite different in visual cortex. Simultaneous contrasts between differently oriented visual contours and colors are common for textured objects (e.g., plants) and different directions of motion are informative features of visual scenes. This rich repertoire of distinctive features may, at least in part, explain the presence of various gradient maps in visual cortex.

A rare example of complex gradient maps in auditory cortex, the systematic layout of complex echo-dimensions in bats supports the above reasoning on predictabilities, contrasts and interpretation of sounds (Suga and Jen 1976; Suga 1990; Riquimaroux et al. 1991). Echo sounds have predictable relationships with the emitted biosonar signal. Echos can provide images of objects from which they are reflected. These images form contrasts with the echos from the surrounding environment. Such sound properties are unlike those directly emitted from these objects (Dear et al. 1993). Small changes of echo parameters contain highly reliable information about spatial and material object properties. Furthermore, many overlapping biosonar echos occur in large colonies of bats in caves, requiring mechanisms to resolve simultaneous contrasts of complex parameters. Considering all these aspects, it seems plausible that complex echo maps are organized analogously to visual maps.

While the competitive principle of local lateral inhibition in gradient maps seems less useful for contrast enhancement of complex sounds, except in a tonotopic (spectral) framework, there may be other competitive mechanisms that are facilitated by certain map types. In gerbil auditory cortex, a map has been described that represents modulation frequencies of amplitude modulated sounds (periodicity map, Schulze et al. 2002) with a gradient that has quasi-circular geometry. This geometry may support a winner-take-all mechanism in the context of multiple competing sounds, e.g., in a cocktail-party situation. The underlying principle may be an "economy of intrinsic connectivities" proposed as a rationale for the formation of gradient maps (Chklovskii and Kulakov 2003). In a circular map, this principle allows to enhance in an opportunistic fashion the contrast between a modulated carrier and any other competing sounds with different modulations (Kurt et al. 2008). Competition of sounds also includes the auditory streaming phenomenon (Bregman 1990) where several neuronal correlates have been described (Deike et al. 2004; Fishman et al. 2004; Snyder et al. 2006) that would suggest special competitive mechanisms (Carlyon 2004; Bee and Klump 2004, 2005) including sequential contrasts (Brosch et al. 1999; Brosch and Schreiner 2000).

Despite the dearth of separate cortical gradient maps for complex dimensions, inhibitory mechanisms in complex receptive fields of individual neurons are pronounced. These individual inhibitory mechanisms may be a substrate of discrimination ability. Neurons with various combinations of excitatory and inhibitory frequency-bands are already found at the level of the cochlear nucleus (Rohde and Greenberg 1992) and are also present in auditory cortex, although with an increasing complexity of the inhibitory sidebands (e.g., Phillips and Cynader 1985; Wehr and Zador 2003; Tan et al. 2004). These observations of complex inhibitory mechanisms have led to only few systematic experiments on contrast formation and feature discrimination (Ohl and Scheich 1997b; Barbour and Wang 2003). Initially, studies focused on tonal sound-level coding (monotonic and nonmonotonic responses) and noise masking (for review: Clarey et al. 1992). More recently, inhibitory sidebands have been related to spectral edge sensitivity for white band sounds (Qin et al. 2004, 2005). Furthermore, in the domain of sequential contrasts and inhibition, some rules for frequency relationships and time separation between successive tones have been described for AI neurons that lead to suppression of the second tone (Brosch et al. 1999, 2004; Brosch and Schreiner 2000).

Selectivities of auditory cortical neurons for complex sound dimensions have been characterized mainly with band-pass filter concepts, i.e., by a "best response" attributed to the center condition of the filter. The role of inhibition in shaping such best responses and for sequential discrimination of similar sounds of the same dimension has not been systematically studied. In discrimination learning, inhibitory components of receptive fields are likely to be responsible for much of the plasticity that leads to an improved discrimination performance.

5 Discrimination Learning Through Local Contrast Enhancement

Various neuronal and, especially, cortical mechanisms can be modified by learning as outlined above (cf. Beitel et al. 2003; Fritz et al. 2003, 2005; Ohl and Scheich 2005; Froemke et al. 2007). Thereby sounds can acquire or change a specific behavioral meaning. In auditory cortex, two types of learning-induced changes in receptive field tuning have been described and discussed in the context of spectral contrast enhancement (Fig. 17.1). The first type is a learning-induced shift of the best frequency of cortical neurons ("BF shift") towards a frequency that has gained special significance in a one-tone classical conditioning task (for review see Weinberger 2004a) (Fig. 17.1c). The second type is a learning-induced slope increase of the tuning curve near the conditioned frequency (Ohl and Scheich 1996, 1997a, 2005; Witte and Kipke 2005). Depending on the sign of the slopes (rising or falling) in the low and high frequency neighborhoods of the conditioned frequency, this type of retuning will position the conditioned frequency either on a steeper gradient or on a local minimum of the post-training tuning curve (Fig. 17.1d). This second type of retuning was found in a conditioning paradigm involving multiple non-reinforced frequencies in addition to the reinforced tone frequency. This emphasizes the discrimination of the reinforced from the non-reinforced frequencies. There has been some discussion whether conditioning-induced plasticity in auditory cortex always reflects the tagging of a relevant stimulus or whether different, task-specific plasticity phenomena are observed (Weinberger 2004a, 2007a–c; Ohl and Scheich 2004, 2005; Scheich et al. 2007; Scheich et al. 2010).

For the BF shift plasticity, contrast enhancement is accomplished by recruiting highly responsive neurons to the tonal conditioned stimulus at the expense of other frequencies (Weinberger 2007a). A similar argument is used (Weinberger 2007a, 2007b) to interpret the functional relevance of increased map representations of a trained frequency in perceptual learning (Recanzone et al. 1993). This would, however, reduce or eliminate fine discrimination of the conditioned frequency from nearby frequencies (Ohl and Scheich 1996, 1997a) since decrease of firing for nearby frequencies could be interpreted as a higher or a lower frequency than the retuned best frequency (Fig. 17.1c). Conversely, steeper slopes on the tuning curves or minima in these slopes would increase fine discrimination of conditioned frequencies from nearby frequencies (Butts and Goldmann 2006). The relevance of local slopes in neuronal tuning curves for

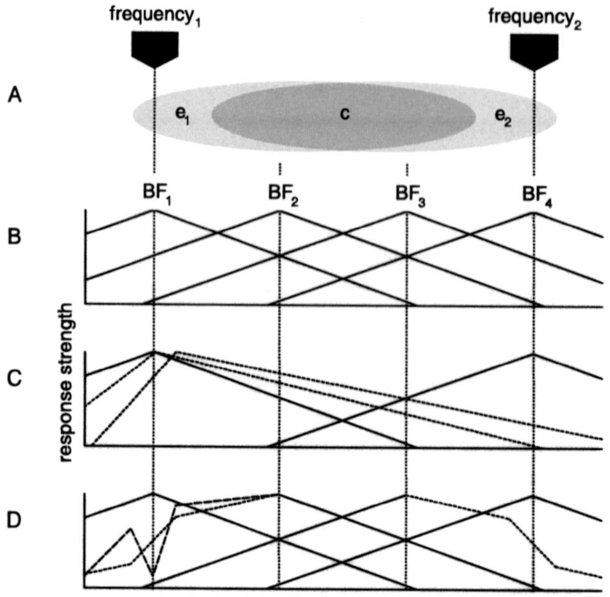

Fig. 17.1 Conceptualization of proposed receptive field plasticity of neurons in a tonotopic map for stimulus identification (detection learning) and discrimination learning and their interrelationships. **a** and **b**. Situation before learning. Four exemplary, largely overlapping, receptive fields in (**b**) are chosen in such a way that stimulus frequency 1 and frequency 2 can activate two receptive fields at their best frequency, BF1 and BF2, respectively. The two receptive fields in the middle are activated by either stimulus frequency but not at their BFs. This configuration with a larger and more dense population of neurons is represented in (**a**) by two overlapping ellipses of stimulus influence. They cover the ranges from BF1 to BF3 and from BF2 to BF4, respectively. This results in three domains of frequency influences on receptive fields. (1) The exclusive domain e1 where frequency 1 activates neurons at or close to BF1 as well as different fractions of receptive fields in the range of BF2 and BF3 but has no influence on the receptive field with BF4, (2) The exclusive domain e2 where frequency 2 activates receptive fields in a way mirror-imaged to (1), and (3) the common frequency domain c where all frequencies that are not frequency 1 or frequency 2, but are in between, activate all four receptive fields. The logic of these domains of influence on receptive fields sets the stage for the types of plastic changes that have been observed for frequency conditioning in AI. Note that in a population view of neurons in a tonotopic map consequences of receptive field plasticity for stimulus identification become inseparable from consequences for discrimination of these and other stimuli. (**c**) Effect of training-induced BF shifts. When a single frequency, e.g., frequency 1, acquires a specific associative meaning sensitive neurons excited adjacent to their best frequency (BF2 and BF3) may shift their BF towards frequency 1. In this way, more neurons would respond more strongly to the trained frequency rendering it more salient in the tonotopic representation. While this is useful for identification of that particular frequency (detection task), the crowding of BFs at or close to the trained frequency has several adverse effects for the fine discrimination of other frequencies from the trained frequency. Neurons which fully shift their BF to the trained frequency lose their previous sensitivity to small frequency changes around the trained frequency because tuning has zero slope at its peak and, in addition, become insensitive to the direction of frequency change (to higher or lower frequencies). Neurons which only partially shift their BF toward the trained frequency retain their sensitivity to direction of frequency change but become more ambiguous with respect to the magnitude of the frequency change (a smaller and a larger deviation from the training frequency might evoke identical firing rates from the two sides of the tuning curve). In the far range of the trained frequency, the shifts of BFs of the sensitive neurons must lead to a reduced density of representation of these frequencies by BFs. This reduces detectability and discriminability of frequencies in that range. (**d**) Illustration of the effects of local slope enhancement of receptive fields and local minima of tuning curves at the conditioned frequency. When a number of different frequencies are used during association training such that one of them acquires a specific associative meaning and needs to be distinguished from the others during acquisition and retrieval the most prominent and reliable phenomena are a local slope enhancement of the receptive field around the relevant frequency or a local minimum of the tuning curve at that frequency. These forms of retuning engage neurons which are sensitive to the particular frequency but do not have their BF at that frequency. Thus, they may belong to the same population of neurons that shift their BF in the simple detection task in (**c**). In the near range of the conditioned frequency the locally steeper slope will increase the discriminability of the conditioned frequency. Such neurons retain their sensitivity to the direction of frequency change from that frequency and remain unambiguous with respect to the magnitudes of frequency change. In a population view, neurons that develop a local minimum on the slope of the tuning curve at the conditioned frequency by their reduced firing could enhance the contrast to the neighboring neurons that are maximally excited by the conditioned frequency at their BF (BF1) and thereby facilitate the detection of the conditioned frequency. In the far range of the training frequency there is no loss in density of the frequency representation by BFs. The scheme also shows that multiple local slope enhancements can hypothetically serve discrimination learning (e.g., frequency 1 from frequency 2) without changing the distribution of BFs in a map

discrimination has also been demonstrated for sound localization processing in the mid-brain (Harper and McAlpine 2004). In the visual system, orientation-discrimination training has led to local sharpening of orientation tuning curves in V1 (Schoups et al. 2001) and V4 (Raiguel et al. 2006) in accordance with theoretical predictions (Ghose et al. 2001, Ghose 2004).

BF shifts appear as a suitable mechanism for tagging a single important frequency in a detection task, i.e., for emphasizing its uniqueness in the current behavioral context. The learning-induced slope enhancement better serves the discrimination of a relevant frequency from similar neighboring frequencies when the likelihood of occurrence of these similar frequencies is high. Thus, the two described forms of leaning-induced receptive field plasticity can be parsimoniously conceptualized on the basis of the different semantic contexts in which the reinforced tones were presented in the two conditioning experiments.

6 Problems of the Map Concept for Understanding Category Formation

Gradient maps of stimulus dimensions are suitable for discrimination learning and may be ideal if the competitive mechanism of lateral interaction can be used to sharpen local contrast. Category formation and categorical stimulus distinction require different mechanisms. Stimulus categories are equivalence classes of meaning for different stimuli. Categories are formed by learning and, simplified, use a selection of some shared stimulus properties irrespective of other properties of these stimuli. Category-relevant properties typically allow a range of variation determined by inclusion or exclusion criteria (positive and negative equivalence constraints). The selection of the range either coincides with naturally occurring dimensions of stimuli, e.g., rising and falling direction of FM sweeps independent of the actual frequency range (natural categories) or they are set along a dimension by behavioral relevance criteria (categories of choice), e.g., cultural differences of speech with specific ranges for different vowels within the two-dimensional formant space (Peterson and Barney 1952; Ohl and Scheich 1997b). Categories can also be formed by multiple different stimulus dimensions if a common meaning can be derived. This is particularly relevant in audition since a given object may produce very different sounds and very different objects may produce similar sounds. It is the objects and their behavior that are naturally meaningful, not the sounds. These higher-order categories are illustrated, for example, by the very different sounds of natural gaits of horses during walk, trot, and gallop which all identify a horse. There are other aspects of category formation and corresponding models from prototype-based to rule-based forms (Handel 1989; Komatsu 1992; Estes 1996), but they all have in common a classification of information provided by different stimuli according to behavioral meaning for the individual thus allowing hypotheses regarding category membership of novel stimuli.

Considering these principles of categories it is obvious that the simple mechanisms of local contrast enhancement cannot account for their formation and the inclusion or exclusion of exemplars. Due to the highly variable and complex nature of sounds emanating from natural objects and the association of meaning to sounds through initially identifying the source objects and their behavior, it seems plausible that opportunistic maps are the most economic compromise to subserve such demands. They make no specific assumptions on the prevalence of one or the other sound property and can characterize any sound.

Categorization implies integration across a variety of stimuli without the need for fine discrimination within a category. On the contrary, contrast between such stimuli should be reduced as, for example, for a vowel spoken by different individuals (Peterson and Barney 1952; Ohl and Scheich 1997b). The only contrasts that are required are at category boundaries even though category formation can be developed from discrimination learning of individual exemplars (Ohl et al. 2001; Selezneva et al. 2006; Scheich et al. 2010).

7 The Role of Feedback in Category Formation

Stimulus categorization is formed by feedback-controlled selection of relevant stimulus dimensions. An example is the categorization of sweep direction ("rising" or "falling") in linearly frequency-modulated tones in the Mongolian gerbil (Wetzel et al. 1998; Ohl et al. 2001; Jeschke et al. 2008; Wetzel et al. 2008). Gerbils were trained in a shuttle-box GO/NO-GO procedure to learn the discrimination of one particular pair of a rising and a falling frequency-modulated tone (reversed in time, but otherwise identical). First, they had to learn through reinforcement that the rising tone meant GO and the falling tone NO–GO. At this stage, gerbils showed typical generalization gradients for various physical stimulus parameters such as modulation steepness and frequency boundaries of the FM tone. Next, each new training block presented stimuli with spectral content falling outside the generalization gradients established by the previous block. This initially led to poor discrimination performance for the novel stimuli. After a number of such training blocks, animals suddenly transferred the conditioned responses of the discrimination training to novel FM tone stimuli irrespective of their particular spectral content. Apparently, the FM direction had been identified as the only relevant stimulus parameter in this particular task and the concept of "modulation direction" had been established. Two categories, "rising" and "falling," were formed into which arbitrary novel stimuli could now be sorted. This sudden transition from a discrimination phase of learning – characterized by shallow learning curves and generalization gradients – to a categorization phase – characterized by instantaneous categorization of novel stimuli and steep psychometric functions – occurred after an individually highly variable number of training blocks. It was paralleled by changes in the ongoing neuronal activity in AC: cortical activity states, defined by high-resolution spatial patterns of electrocorticogram power of the beta and gamma band, emerged from the ongoing activity. They co-varied with a subject's current psychophysical scaling of stimuli rather than acoustic stimulus properties (Ohl et al. 2001). This is evident from a dissimilarity analysis of cortical activity over the entire course of the training (Fig. 17.2): Prior to the formation of categories, dissimilarities between cortical activity states associated with stimuli of the same category were of the same order of magnitude than those associated with stimuli from different categories. After the formation of categories, dissimilarities within a category were significantly smaller than between categories. Category learning generated a metric of representation that reflected the assignment of stimuli to the two newly formed equivalence classes of meaning GO and NO-GO. This type of metric is fundamentally different from the tonotopic representation principle which reflects similar relations of physical stimulus parameters, namely spectro-temporal composition.

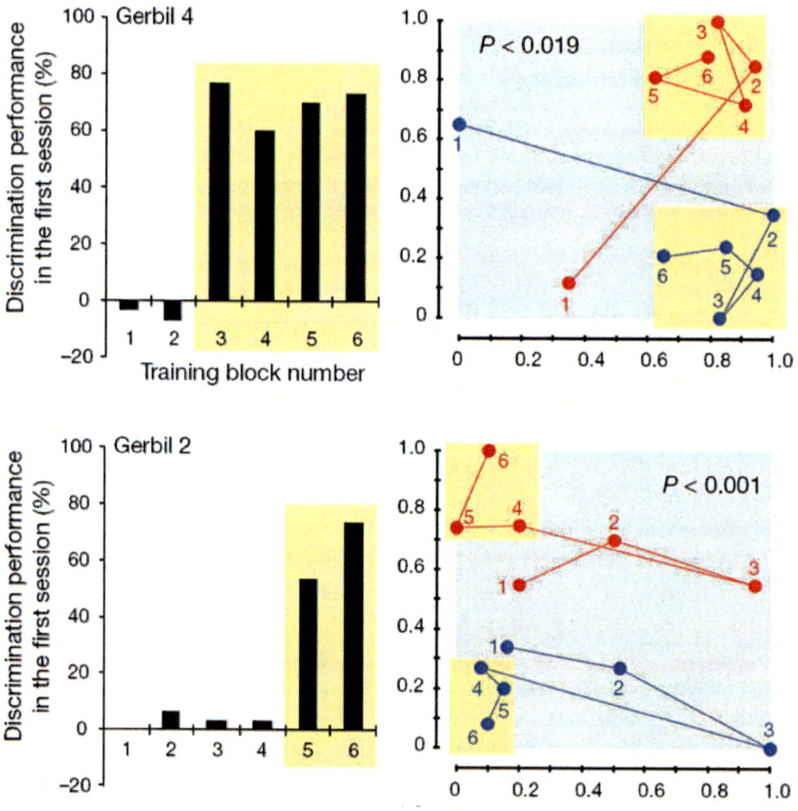

Fig. 17.2 Correspondence between the behavioral transition from the discrimination phase to the categorization phase (*left column*) and the emergence of category-representing mesoscopic activity patterns (*right column*) in two representative animals (*top* and *bottom row*). *Left column*: *Bars* represent discrimination performance for novel stimuli at the beginning of a training block as a function of training block index. Categorization phase is indicated by the abruptly increased discrimination performance (emphasized by *yellow rectangles*) and occurred at different times for different individuals. *Right column*: Similarity–dissimilarity relations between activity patterns found for rising and falling FM tones (*red* and *blue*, respectively) in the sequence of training blocks (*numbers*). The relative dissimilarity between any pair of states is represented by the distance of the corresponding state points in this 2-dimensional display. Note the emergence of point clustering (indicating high state similarity) within categories when transfer from discrimination phase to categorization phase happened

8 Polymodal and Cognitive Contingencies in Categorization

Sound categorization relies on other sensory modalities and on behavioral feedback. This became evident in recordings from monkey AC when different sounds acquired the same meaning (Brosch et al. 2005; Selezneva et al. 2006; Brosch et al. 2010). Monkeys were trained to categorize upward and downward frequency steps in tone sequences irrespective of the absolute tone frequencies. A reward was given upon correct identification of a downward step. The following procedural rule determined the behavioral meaning of the auditory stimuli in a trial: Upon a light signal the monkeys could grasp a holding bar thereby eliciting a tone sequence of unpredictable composition. For any upward step they had to keep holding the bar but needed to release it after a downward step in order to receive the reward. In the fully trained monkeys, auditory cortical neurons (in fields AI and CM) responded to the tone steps in a categorical fashion. Interestingly, the light cue, grasping and releasing of the bar, and the reward delivery, elicited firing in a large proportion of these auditory neurons as well (Fig. 17.3).

As in any learning task that requires deduction of meaning from context and behavioral consequences, a relationship is established between the tones and the poly-modal cues. The intriguing aspect of these results was that the poly-modal cues of the behavioral context were expressed in auditory cortex and that plasticity was not confined to cue-specific modalities of that cortex. This points to the existence of states in AC that include information on the non-auditory, interpretative, side of the sounds. These non-auditory activations were not observed in the same neurons when the monkeys switched to a purely visual discrimination task with the same procedural contingencies. Consequently, non-auditory information is only relevant for AC when it aids the interpretation of an auditory input, i.e., contributing to the solution of a task. Thus, the concept that sound meaning is primarily related to objects that emit the sounds must be extended to the sensory-motor context of producing and reacting to sounds.

These physiological results revealed two other aspects of the special state during categorization. One is that phasically responding neurons increased their response only to the rewarded downward steps. This behavior was independent of best frequency across the neuronal population and signals categorical behavior in that only stimuli with the same meaning were represented in the same fashion. This directional preference was different from naïve animals that showed preferences for steps in either direction (Brosch et al. 1999; Bartlett and Wang 2005). The second finding relates to auditory cortical responses from the monkey's hand. Some neurons slowly increased their firing as soon as the monkey grasped the bar (Selezneva et al. 2006) and their firing leveled off during the tone sequence and showed a sharp decline after the first frequency change. Downward frequency steps resulted in a steep decrease of firing, predicting that the monkey would release the bar. For upward step the decrease of firing was significantly flatter, predicting that the monkey would hold the bar. This association between neuronal and behavioral response held true also for error trials when the monkeys misinterpreted the direction of steps. These results suggest that a categorical state includes category-relevant representation of the stimuli as well as representations of multimodal and cognitive task aspects relevant for the interpretation of stimuli.

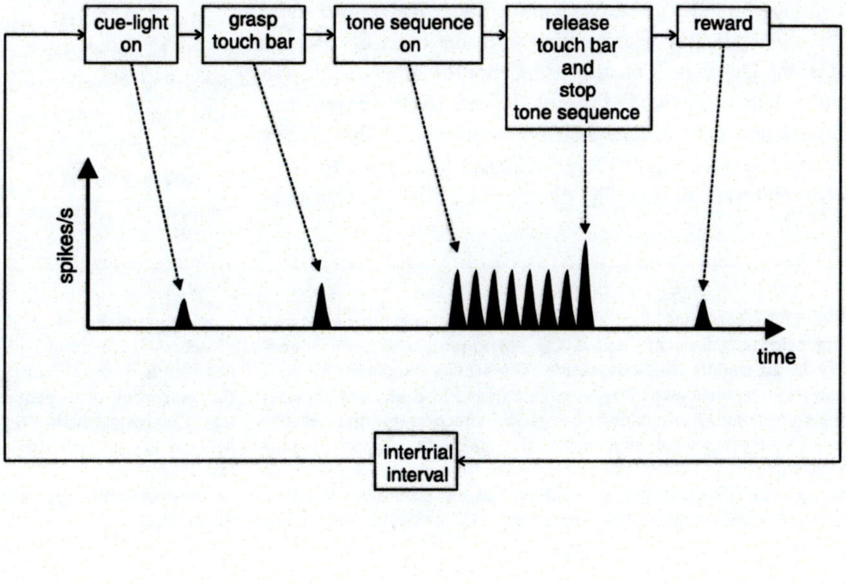

Fig. 17.3 Schematic representation of responses of auditory neurons in field AI to behaviorally relevant events during the tone-sequence categorization task. Note the co-representation of auditory and non-auditory events

9 Potential Influences of Category Formation on Stimulus Representation

The learning processes potentially accounting for category formation and categorization of frequency changes in our studies are outlined as hypothetical changes of stimulus representation in opportunistic maps of auditory cortex (Fig. 17.4). We consider here a stepwise procedure derived from discrimination learning of continuous frequency change (FM) (Ohl et al. 1999; Ohl et al. 2001), or tone sequences with frequency steps (Brosch et al. 2005) that both can be categorized as either rising or falling.

During the discrimination phase a given FM stimulus likely activates all auditory cortical neurons that have receptive field properties sensitive to that stimulus. Many neurons with simple receptive fields will increase their firing when an FM stimulus sweeps into their receptive field towards the best frequency but will decrease their firing when an FM sweeps away from the best frequency (Phillips et al. 1985; Heil et al. 1992, Ohl et al. 2000). These neurons are sensitive to FM direction but not selective for direction because they can change directional sensitivity with change of frequency of stimuli, i.e., variation of other parameters. Other non-specific neurons may respond to start frequencies, stop frequencies or other parameters of the presented FM but are insensitive to FM direction or to a change of direction. Some neurons, however, will distinguish the direction if a rising and a falling FM are in the same frequency range (Nelken 2002) and are considered direction-selective neurons.

The large ensemble of direction-insensitive, sensitive and selective neurons forms the initial representation of an FM stimulus. Thus, two directionally different FM stimuli will generate two representations across neuron populations that may be partially overlapping, depending on how many shared properties they have (Fig. 17.4a). The non-overlapping subpopulation of direction-sensitive and some direction-selective neurons may form the basis of discrimination. Depending on the prevalence of other distinctive properties of any two FM stimuli, it is not likely that discrimination and attribution of different meaning initially depends on the FM directional difference alone. In this case, the specific representation of FM directional difference will only be a portion of the subpopulation that contributes to the discrimination.

In our gerbil experiments, however, we have cued the animals specifically on the FM directional differences by using symmetric pairs of rising and falling FM tones, i.e., all other properties were kept constant. Thus, only in this case we can assume that the category-relevant contrast rising versus falling was already used during the discrimination phase of the experiment. In these gerbils, we found that the attribution of meaning to different stimuli had to be learnt anew with each pair. This suggests the FM directional difference as a common criterion was initially not obvious, i.e., a change of frequency range with a new FM pair recruited two new representations within which the directional contrast had to be established.

But why would the behavioral formation of the two categories emerge as a quasi spontaneous phenomenon after experience with an individually varying number of different FM pairs? This may be understood by hypothetical mechanisms based on reliability of some aspects of the representations emerging after experience with various stimulus pairs.

First, as shown by the scheme in Fig. 17.4b, the exclusive parts of the multiple representations of stimuli with the same direction all overlapped. By using a criterion of "common denominators" a population is selected that most consistently responded to the various stimuli. This selection process covered neurons which, as a common denominator, preferred a given FM direction and neurons that preferred the direction irrespective of change of other stimulus properties such as the FM frequency range. We assume that these direction-selective neurons were reinforced consistently by behavioral feedback – conferring common meaning – in correct trials of footshock avoidance during discrimination learning such that they could form a functionally connected network of "common denominator" neurons.

But even if this distinct network of specialized neurons would be enabled to increase their firing rate as a result of the described learning processes, this selection alone does not fully match our experimental results and would not explain all phenomena of category formation. A second mechanism is required providing suppression of non-category-related neurons, i.e., neurons that are not direction-selective. Once

Fig. 17.4 (continued) from direction-insensitive to categorical direction selectivity for rising and falling FM towards a category boundary in the middle of the *rectangle*. The scheme proposes that in a category-entrained map upon occurrence of an FM-containing stimulus the corresponding ensemble of categorical direction-selective neurons must suppress several other neuronal populations in order to provide unambiguous information in the map: Neurons that are selective to the other direction (1) and neurons that are insensitive to direction (2) whatever their other preferences. Of particular importance is the suppression of all direction-sensitive neurons (3) as they may switch their directional preference with variation of time-frequency parameters of the stimuli. The suppressions seem necessary to obtain the properties of experimental activation patterns described in the text: Categorical activation patterns for stimuli of the same direction become highly "self-similar" independent of diverging other stimulus dimensions; activation patterns for stimuli of different direction become maximally dissimilar independent of converging other stimulus dimensions

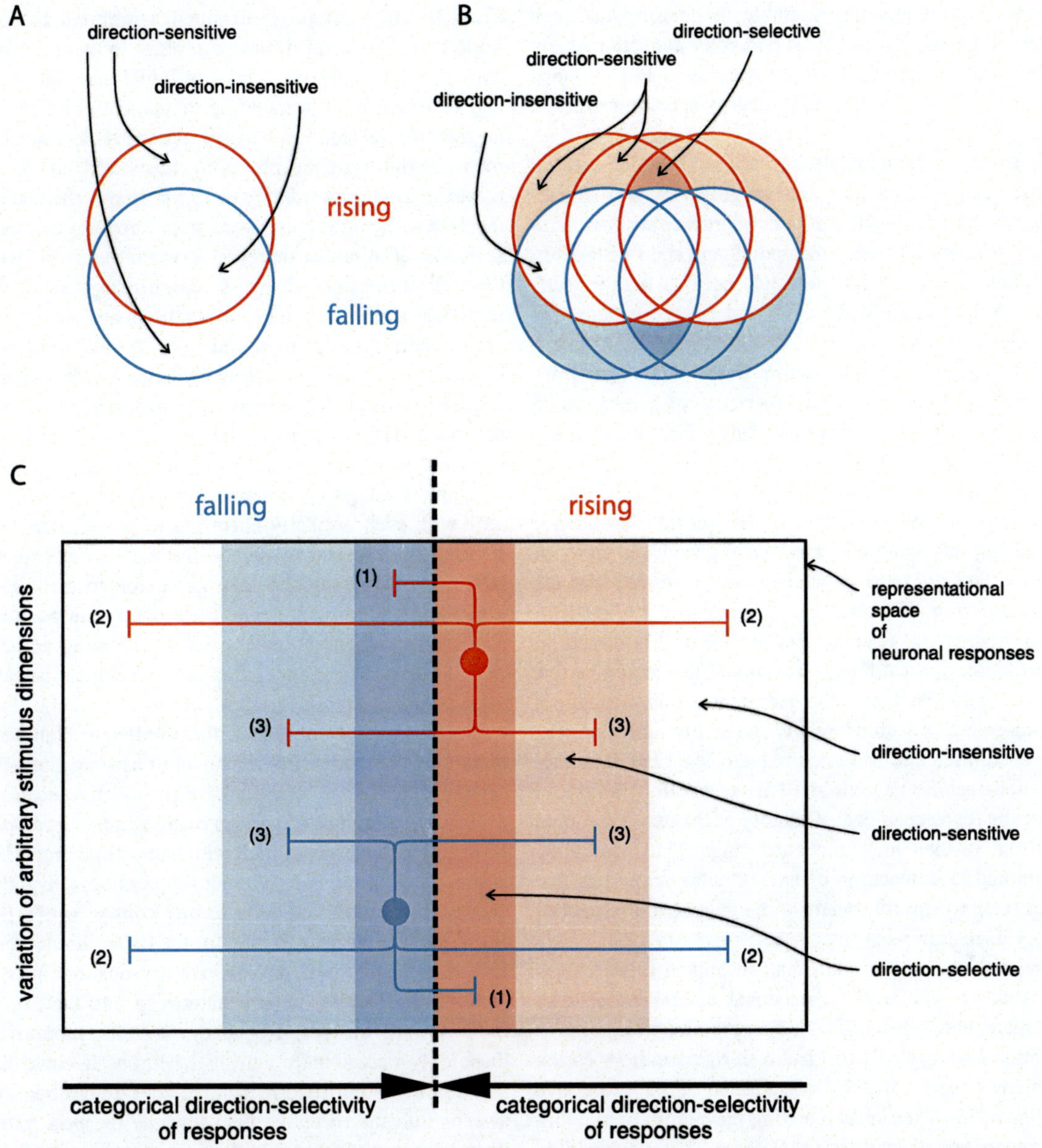

Fig. 17.4 Heuristic scheme of category representation. (**a**) Illustration of neuron ensembles with responses to one individual time-frequency mirror-imaged FM pair. The *two circles* represent the set of neurons responding to the rising (*red*) and the falling (*blue*) FM, respectively (directional sensitivity). The common partial *space* represents the subset of neurons that respond similarly to both directions irrespective of whether they prefer or distinguish other stimulus dimensions (directional insensitivity). In this non-categorical discrimination state, the population of direction-sensitive neurons covertly contains some that with other time-frequency parameters would still prefer one direction (directional selectivity) as well as many non-selective neurons. (**b**) Categorical neuronal representation of various rising and falling FM stimuli. Venn diagrams are multiplied to form two rows with partially overlapping neuronal ensembles. The *upper row* of *circles* represent neurons that all respond to the rising FM stimuli (*red*) but in each circle neurons fall into a different time-frequency range. The *lower row* represents neurons that respond to the corresponding falling direction (*blue*). The unshaded areas in the *middle* represent the subset of direction-insensitive neurons that show similar responses to both directions. The lightly *shaded areas* in the *upper* and *lower row* represent neurons that are direction-sensitive but may lose or switch their directional preference with modifications of time-frequency parameters of stimuli. Only the small subsets of direction-sensitive neurons in the *darkly shaded* areas maintain their directional preference independent of particular time-frequency parameters of FM, i.e., identify category. Training-dependent selectivity may be enhanced by mutual coupling in the ensemble of selective neurons. (**c**) Hypothetical implementation of category-specific neuronal activation patterns in an AC map. The rectangle symbolizes an abstract representational space of neuronal responses in the map to any occurring stimuli that are categorically monitored for any rising or falling FM components. The *vertical axis* represents stimulus dimensions, like frequency, intensity, and time variation, and their complex higher derivatives, to which a mixture of neurons may respond as previously described for a mosaic-like opportunistic map. Along the *horizontal axis* response properties are arbitrarily ordered

the common denominator ensemble of neurons has been selected, this mechanism could suppress the activation of any neurons that are responsive to an occasional FM stimulus but which do not belong to that category-specific ensemble (Fig. 17.4b).

The existence of both selection and suppression mechanisms in category formation was suggested by two characteristic phenomena of gerbil auditory cortex activation. With category formation the spatiotemporal activation patterns in AC for all subsequent FM pairs (1) became highly "self-similar" for a given FM direction and (2) the patterns for the two FM directions became highly dissimilar, even though stimulus properties varied as before (Fig. 17.2). A parsimonious explanation is that with experience of several pairs, the direction-selective neurons covering different frequency ranges with their receptive fields finally formed a common denominator ensemble leading to more unified responses to any stimulus of that direction. The frequency dependence of these responses initially must have introduced most of the dissimilarity of representations between stimuli with the same direction in different pairs.

This process still does not explain the highly dissimilar representation of the different directions. This could be due to a process by which previously direction-sensitive but not direction-selective neurons were excluded from the representation of stimuli that had acquired a categorical meaning. These non-specific neurons must have initially introduced most of the representational similarity of the two directional stimuli due to their joint frequency range. Thus, excluding the neuronal representation of any stimulus properties that are not category-specific seems to be an essential aspect of category formation.

A related phenomenon is found in human fMRI experiments with a similar FM directional categorization task (Brechmann and Scheich 2005). There we found a negative correlation of activated voxel space in right auditory cortex with increase of proficiency of subjects, i.e., a successive reduction of mistakes in categorizing the stimuli by training led to a reduction of activated voxel space. This is compatible with a reduced activation of non-specific neurons presumably in favor of category-specific neurons.

A direct insight into selection and suppression mechanisms of neuronal firing is provided by the primate experiments on categorization of frequency steps in tone sequences (Selezneva et al. 2006). The detection of any downward frequency step led to a reward, while the – false – detection of an upward step or of equal tones (flat steps) entailed no relevant behavioral consequences. Thus, the downward category acquired a specific meaning, while upward steps were no more meaningful than flat steps. This design was different in the gerbil FM directional task where any false response, i.e., no hurdle jump for a rising FM or any hurdle jump for a falling FM, was punished by a footshock. Thus, the two categories acquired a different but equally important behavioral meaning in order to avoid an adverse consequence. In the monkeys, we found that neurons in AI and CM phasically responding to sequential tones showed an increase of responses to the rewarded downward steps over responses to the preceding tones and no change of response to an upward step or to flat steps. The categorical increase of firing to downward steps supports the proposed selection of stimulus representation mechanisms. But it is the lack of response change to upward steps similar to flat steps that sheds more light on the proposed suppression of non-specific stimulus representations. In AC of naïve monkeys selective neurons showing increases of firing for upward steps or for downward steps in tone sequences have been described (Lu and Wang 2000; Brosch et al. 2004). Thus, the lack of neurons with firing rate increases to upward steps in the categorically trained monkeys suggests that neurons with such normally occurring response changes were suppressed. This result demonstrates that the suppressive influence of common denominator neuron ensembles – once a category is formed – not only covers unspecific neurons, but also neurons potentially representing an adjacent or opposite category, and in this way helps to define category boundaries.

To illustrate the solution of the problem of boundary formation by the model (Fig. 17.4), let us assume that only one category, the rising category of FM, was formed from the exclusive experience of various rising stimulus exemplars to be distinguished from stimuli with many other properties but without any falling FM. Without suppression, a hypothetical ensemble of neurons might be formed consisting of relatively few specific neurons selective for the rising direction of FM and those many direction-sensitive neurons that are excited by occasional rising stimuli that sweep into their receptive fields toward the best frequency. Without suppression of these only occasionally activated neurons it could happen that a newly occurring FM stimulus of falling direction that sweeps into the receptive field towards the best frequency activates some of these direction-sensitive but not direction-selective neurons. This would bias the information provided by the selective neurons, namely their lacking responses to falling FM stimuli. Thus, for definition of a sharp category boundary, it seems important to suppress unspecific neurons that did not initially respond to most FM exemplars belonging into the rising category.

This suppression or reduced activation of category-irrelevant neurons in combination with a selection mechanism is prospectively important for later use of a category. Categories are classes across stimulus variations that allow identification of new exemplars. Categories have a predictive function beyond stimuli and mechanisms that have led to their formation (Ohl and Scheich 2001; Scheich et al. 2007; Scheich et al. 2010).

In this perspective, the inclusion criteria of the category for stimulus properties are predictive for a new stimulus because they could be defined by the experiences in the representational space, but the exclusion criteria only to some extent. Exclusion criteria could be innumerable, because they must also cover additional properties of stimuli which belong to the category in question but were not previously experienced, e.g., rising or falling FM that have additional amplitude modulations or consist of harmonic spectra. In this case it would be primarily the matter of an existing network of direction-selective neurons to determine their categorical belongingness and to suppress all neurons newly activated by these stimuli. Only with multiple experiences of such stimuli, some of these neurons may be identified as direction-selective and are incorporated into the category-specific ensemble.

The concept so far suggests that a mosaic-like mixture of various neuronal properties in each auditory cortex map covers sensitivities and selectivities for several acoustic dimensions. This seems appropriate for opportunistic category formation. The required plasticity would primarily target a selection of existing neurons, i.e., the formation of a highly interconnected ensemble and not necessarily change of individual neuronal properties by learning. The latter condition is probably also the case, e.g., for the formation of category-specific sharp boundaries. In our monkey experiments (Selezneva et al. 2006), all neurons irrespective of best frequency responded with an increase of discharge rate to the rewarded downward frequency step and with no change of discharge rate to the non-rewarded stimulus. As this exclusiveness is not a property found across neuronal populations in naive monkeys, it is difficult to assume that the abundance of the neuronal phenomena of direction selectivity in the trained monkeys is just a selection of those neurons that naturally prefer downward steps but also probably involves conditioned changes in receptive field properties of many neurons analogous to findings in discrimination learning.

10 Conclusions

The purpose of this chapter was to provide a comprehensive conceptual framework to substantiate an ongoing debate about learning-induced neuronal plasticity in auditory cortex. To this end it seemed necessary to develop a broader biological view on audition, auditory learning and auditory cortex organization than merely from the perspective of classical conditioning of tones in a laboratory environment. The essence of auditory processing and auditory learning at the level of auditory cortex is to promote behaviorally meaningful conclusions on sounds and not simply auditory pattern analysis. Sounds have a special nature in that they do not form unambiguous descriptions or definitions of sound-generating material objects and sound-eliciting events, unlike light emanating from the same sources, but are rather occasional, fleeting, and highly variable signals. Depending on the causal events, numerous very different sounds can be obtained from the same objects and similar sounds from unrelated objects. But as a sound can have no direct behavioral impact on a listener, it is still the presumed cause of the sound and/or the potential behavioral consequences of the cause in a given context toward which the concern of the listener is ultimately directed and from which its meaning must be derived. Thus any identification and discrimination of sounds typically involve interpretative learning unless they have an inborn significance. But learning strategies for identification (detection) and discrimination, as well as their neuronal correlates, seem to be very different albeit both represent instances of associative learning. Discriminations of multiple sounds with respect to similar or different behavioral consequences can lead to the formation of sound categories. As categories are equivalence classes of meaning, category learning is an indispensable strategy for interpretation in view of sound variabilities. Category learning can develop from discrimination learning but its neuronal correlates in auditory cortex are distinct.

We have argued that the organizational principles of auditory cortex maps with their mix of neurons specialized for various complex acoustic dimensions, unlike separation of different dimensions in visual cortex maps, may serve task-dependent interpretation of variable sounds in an opportunistic fashion. This means that, except for analysis of spectro-temporal coherence of sounds by tonotopic principles, the organization of maps implies little specific assumptions as to prevailing acoustic compositions of naturally occurring sound environments and to rapid changes of the properties of individual sounds over time. Presumably, any sound can be represented descriptively by a mosaic of neurons in any map. But what information must be selectively derived and learned, depending on the type of task, appears to determine which mosaic of neurons in which map and hemisphere (cf. Brechmann et al. 2007; Wetzel et al. 2008) is dominantly engaged and which plastic changes occur at the cellular level.

As sound waves from different sound sources sum up, the formation of simultaneous contrasts to separate sounds cannot be used in the same way as for object delineation in the visual modality. Instead, implicit assumptions on temporal mechanisms of sound generation (e.g., comodulations, onset asynchronies, etc.) allow separating sound components from different sources in opportunistic maps. Also various types of task-dependent contrast formation, including sequential contrast, can be used economically for discrimination and categorization of sounds with multiple properties.

We have shown how physiological correlates of stimulus processing in different learning tasks can be coherently treated in this conceptual framework. The multiplicity of learning-induced changes in neuronal responsiveness for different tasks suggest that the single phenomenon of neuronal best-frequency shifts – central to explaining classical tone conditioning – cannot be the basic building block to account for other types of plasticity such as learning based on sound discrimination.

Acknowledgments This work was supported by grants from the Deutsche Forschungsgemeinschaft (DFG SFB 779, DFG SFB-TR31, DFG SFB-TR62), the German Ministry for Education and Research (BMBF 01GW0621, BMBF BioFuture 0311891, Bernstein grants (01GQ0733, 01GQ0702), the Volkwagen-Stiftung (VW I77356) and the European IST Programme (Project FP6-0027787). We thank Conny Bucks and Kathrin Ohl for technical assistance during the experiments. We also like to thank Drs. Michael Brosch and Peter Heil for critical reading of a previous version of the manuscript.

References

Altmann CF, Bledowski C, Wibral M, and Kaiser J (2007) Processing of location and pattern changes of natural sounds in the human auditory cortex. *Neuroimage* **41**:69–79.

Barbour DL and Wang X (2003) Contrast tuning in auditory cortex. *Science* **299**:1073–1075.

Bartlett EL and Wang X (2005) Long-lasting modulation by stimulus context in primate auditory cortex. *Journal of Neurophysiology* **94**:83–104.

Basole A, Kreft-Kerekes V, White LE, and Fitzpatrick D (2006) Cortical cartography revisited: a frequency perspective on the functional architecture of visual cortex. *Progress in Brain Research* **154**: 121–134.

Bee MA and Klump GM (2004) Primitive auditory stream segregation: a neurophysiological study in the songbird forebrain. *Journal of Neurophysiology* **92**: 1088–1104.

Bee MA and Klump GM (2005) Auditory stream segregation in the songbird forebrain: effects of time intervals on responses to interleaved tone sequences. *Brain, Behavior, and Evolution* **66**:197–214.

Beitel RE, Schreiner CE, Cheung SW, Wang X, and Merzenich MM (2003) Reward-dependent plasticity in the primary auditory cortex of adult monkeys trained to discriminate temporally modulated signals. *Proceedings of the National Academy of Sciences of the United States of America* **100**:11070–11075.

Brechmann A and Scheich H (2005) Hemispheric shifts of sound representation in auditory cortex with conceptual listening. *Cerebral Cortex* **15**:578–587.

Brechmann A, Gaschler-Markefski B, Sohr M, Yoneda K, Kaulisch T, and Scheich H (2007) Working memory-specific activity in auditory cortex: potential correlates of sequential processing and maintenance. *Cerebral Cortex* **17**:2544–2552.

Bregman AS and Rudnicky AI (1975) Auditory segregation: stream or streams? *Journal of Experimental Psychology, Human Perception and Performance* **1**: 263–267.

Bregman AS (1990) *Auditory Scene Analysis: The Perceptual Organization of Sound.* MIT Press, Cambridge.

Brosch M, Schulz A, and Scheich H (1999) Processing of sound sequences in macaque auditory cortex: response enhancement. *Journal of Neurophysiology* **82**:1542–1559.

Brosch M and Schreiner CE (2000) Sequence sensitivity of neurons in cat primary auditory cortex. *Cerebral Cortex* **10**:1155–1167.

Brosch M and Scheich H (2003) Neural representation of sound patterns in the auditory cortex of monkey. In: Ghazanfar AA (ed). *Primate Audition. Ethology and Neurobiology.* CRC Press, Boca Raton, pp.151–175.

Brosch M, Selezneva E, Bucks C, and Scheich H (2004) Macaque monkeys discriminate pitch relationships. *Cognition* **91**:259–272.

Brosch M, Selezneva E, and Scheich H (2005) Nonauditory events of a behavioral procedure activate auditory cortex of highly trained monkeys. *Journal of Neuroscience* **25**:6797–6806.

Brosch M and Scheich H (2008) Tone-sequence analysis in the auditory cortex of awake macaque monkeys. *Experimental Brain Research* **184**:349–361.

Brosch M, Selezneva E, and Scheich H (2010) Formation of associations in auditory cortex by slow changes of tonic firing. *Hearing Research*, doi:10.1016/j.heares.2010.05.003.

Budinger E, Heil P, and Scheich H (2000a) Functional organization of auditory cortex in the Mongolian gerbil (Meriones unguiculatus). III. Anatomical subdivisions and corticocortical connections. *European Journal of Neuroscience* **12**:2425–2451.

Budinger E, Heil P, and Scheich H (2000b) Functional organization of auditory cortex in the Mongolian gerbil (Meriones unguiculatus). IV. Connections with anatomically characterized subcortical structures. *European Journal of Neuroscience* **12**:2452–2474.

Budinger E, Heil P, Hess A, and Scheich H (2006) Multisensory processing via early cortical stages: connections of the primary auditory cortical field with other sensory systems. *Neuroscience* **143**:1065–1083.

Budinger E, Laszcz A, Lison H, Scheich H, and Ohl FW (2008) Non-sensory cortical and subcortical connections of the primary auditory cortex in Mongolian gerbils: bottom-up and top-down processing of neuronal information via field AI. *Brain Research* **1220**:2–32.

Butts DA, Goldman MS (2006) Tuning curves, neuronal variability, and sensory coding. *Public Library of Science Biology* **4**:1–8.

Cahill L, Ohl F, and Scheich H (1996) Alteration of auditory cortex activity with a visual stimulus through conditioning: a 2-deoxyglucose analysis. *Neurobiology of Learning and Memory* **65**:213–222.

Carlyon RP (2004) How the brain separates sounds. *Trends in Cognitive Sciences* **8**:465–471.

Chklovskii DB and Koulakov AA (2004) Maps in the brain: what can we learn from them? *Annual Reviews of Neuroscience* **27**:369–392.

Clarey JC, Barone P, and Imig TJ (1992) Physiology of thalamus and cortex. In: Popper AN and Fay RR (eds). *Springer Handbook of Auditory Research*, volume 2, *The Mammalian Auditory Pathway: Neurophysiology*. Springer-Verlag, New York, pp. 232–334.

Darwin CJ, Turvey MT, and Crowder RG (1972) An auditory analogue of the Sperling partial report procedure: evidence for brief auditory storage. *Cognitive Psychology* **3**:255–267.

Dear SP, Fritz J, Haresign T, Ferragamo M, and Simmons JA (1993) Tonotopic and functional organization in the auditory cortex of the big brown bat, *Eptesicus fuscus*. *Journal of Neurophysiology* **70**:1988–2009.

Deike S, Gaschler-Markefski B, Brechmann A, and Scheich H (2004) Auditory stream segregation relying on timbre involves left auditory cortex. *Neuroreport* **15**:1511–1514.

Eggermont JJ (1998a) Is there a neural code? *Neuroscience and Biobehavioral Reviews* **22**:355–720.

Eggermont JJ (1998b) Representation of spectral and temporal sound features in three cortical fields of the cat. Similarities outweigh differences. *Journal of Neurophysiology* **80**:2743–2764.

Ernst SM and Verhey JL (2008) Peripheral and central aspects of auditory across-frequency processing. *Brain Research* **1220**:246–255.

Estes WK (1996) *Classification and Cognition.* Oxford University Press, Oxford.

Fishbach A, Nelken I, and Yeshurun Y (2001) Auditory edge detection: a neural model for physiological and psychoacoustical responses to amplitude transients. *Journal of Neurophysiology* **85**:2303–2323.

Fishman YI, Arezzo JC, and Steinschneider M (2004) Auditory stream segregation in monkey auditory cortex: effects of frequency separation, presentation rate, and tone duration. *Journal of the Acoustical Society of America* **116**:1656–1670.

Fritz J, Shamma S, Elhiali M, and Klein D (2003) Rapid task-related plasticity of spectrotemporal receptive fields in primary auditory cortex. *Nature Neuroscience* **6**:1216–1223.

Fritz JB, Elhilali M, and Shamma SA (2005) Differential dynamic plasticity of A1 receptive fields during multiple spectral tasks. *Journal of Neuroscience* **25**:7623–7635.

Froemke RC, Merzenich MM, and Schreiner CE (2007) A synaptic memory trace for cortical receptive field plasticity. *Nature* **450**: 425–429.

Ghazanfar AA and Schroeder CE (2006) Is neocortex essentially multisensory? *Trends in Cognitive Science* **10**:278–285.

Ghose GM (2004) Learning in mammalian sensory neocortex. *Current Opinions in Neurobiology* **14**:513–518.

Ghose GM, Yang T, and Maunsell JHR (2001) Physiological correlates of perceptual learning in monkey V1 and V2. *Journal of Neurophysiology* **87**:1867–1888.

Gonzalez-Lima F and Scheich H (1986). Neural substrates for tone-conditioned bradycardia demonstrated with 2-deoxyglucose. II. Auditory cortex plasticity. *Behavioral Brain Research* **20**:281–293.

Griffiths TD and Warren JD (2004) What is an auditory object? *Nature Reviews Neuroscience* **5**:887–892.

Griffiths TD, Warren JD, Scott SK, Nelken I, and King AJ (2004) Cortical processing of complex sound. A way forward. *Trends in Neurosciences* **27**:181–185.

Hall JW, Haggard MP, and Fernandes MA (1984) Detection in noise by spectro-temporal pattern analysis. *Journal of the Acoustical Society of America* **76**:50–56.

Handel S (1989) *Listening: An Introduction to the Perception of Auditory Events*. The MIT Press, Cambridge.

Handel S (2006) *Perceptual Coherence. Hearing and Seeing*. Oxford University Press, Oxford.

Happel MF, Jeschke M, and Ohl FW (2010) Spectral integration in primary auditory cortex attributable to temporally precise convergence of thalamocortical and intracortical input. *Journal of Neuroscience* **30**:11114–111127.

Harper NS and McAlpine D (2004) Optimal neural population coding of an auditory spatial cue. *Nature* **430**:682–686.

Hartline HK (1949) Inhibition of activity of visual receptors by illuminating nearby retinal areas in the Limulus eye. *Federal Proceedings* **8**:69.

Heil P, Rajan R, and Irvine DR (1992) Sensitivity of neurons in cat primary auditory cortex to tones and frequency-modulated stimuli. II: organization of response properties along the 'isofrequency' dimension. *Hearing Research* **63**:135–156.

Heil P, Rajan R, and Irvine DR (1994) Topographic representation of tone intensity along the isofrequency axis of cat primary auditory cortex. *Hearing Research* **76**:188–202.

Hose B, Langner G, and Scheich H (1987) Topographic representation of periodicities in the forebrain of the mynah bird: one map for pitch and rhythm? *Brain Research* **422**:367–373.

Hubel DH and Wiesel TN (1962) Receptive fields, binocular interaction and functional architecture in the cat's visual cortex. *Journal of Physiology* **160**:106–154.

Imig TJ, Irons WA, and Samson FR (1990) Single-unit selectivity to azimuthal direction and sound pressure level of noise bursts in cat high-frequency primary auditory cortex. *Journal of Neurophysiology* **63**:1448–1466.

Irvine DRF and Wright BA (2005) Plasticity of spectral processing. *International Reviews in Neurobiology* **70**:435–472.

Jeschke M, Lenz D, Budinger E, Herrmann CS, and Ohl FW (2008) Gamma oscillations in gerbil auditory cortex during a target-discrimination task reflect matches with short-term memory. *Brain Research* **1220**:70–80.

Joris PX, Schreiner CE, and Rees A (2004) Neural processing of amplitude-modulated sounds. *Physiological Reviews* **84**:541–577.

Kaas JH and Hackett TA (2000) Subdivisions of auditory cortex and processing streams in primates. *Proceedings of the National Academy of Sciences of the United States of America* **97**: 11793–11799.

Kadia SC and Wang X (2003) Spectral integration in A1 of awake primates: neurons with single- and multipeaked tuning characteristics. *Journal of Neurophysiology* **89**:1603–1622.

Komatsu LK (1992) Recent views of conceptual structure. *Psychological Bulletin* **112**:500–526.

Kraus N and Disterhoft JF (1982) Response plasticity of single neurons in rabbit auditory association cortex during tone-signalled learning. *Brain Research* **246**:205–215.

Kuffler SW (1952) Neurons in the retina; organization, inhibition and excitation problems. *Cold Spring Harbor Symposium on Quantitative Biology* **17**:281–292.

Kurt S, Deutscher A, Crook JM, Ohl FW, Budinger E, Moeller CK, Scheich H, and Schulze H (2008) Auditory cortical contrast enhancing by global winner-take-all inhibitory interactions. *Public Library of Science ONE* **3**:e1735.

Lange-Malecki B, Poppinga J, and Creutzfeldt OD (1990) The relative contribution of retinal and cortical mechanisms to simultaneous contrast. *Naturwissenschaften* **77**:394–398.

Langner G, Bonke D, and Scheich H (1981) Neuronal discrimination of natural and synthetic vowels in field L of trained mynah birds. *Experimental Brain Research* **43**:11–24.

Linden JF and Schreiner CE (2003) Columnar transformations in auditory cortex? A comparison to visual and somatosensory cortices. *Cerebral Cortex* **13**:83–89.

Lu T and Wang X (2000) Temporal discharge patterns evoked by rapid sequences of wide- and narrowband clicks in the primary auditory cortex of cat. *Journal of Neurophysiology* **84**:236–246.

Machens CK, Wehr MS, and Zador AM (2004) Linearity of cortical receptive fields measured with natural sounds. *Journal of Neuroscience* **24**:1089–1100.

Mendelson JR, Schreiner CE, Sutter ML, and Grasse KL (1993) Functional topography of cat primary auditory cortex: responses to frequency-modulated sweeps. *Experimental Brain Research* **94**: 65–87.

Middlebrooks JC, Dykes RW, and Merzenich MM (1980) Binaural response-specific bands in primary auditory cortex (AI) of the cat: topographical organization orthogonal to isofrequency contours. *Brain Research* **181**:31–48

Moore BC (1990) Co-modulation masking release: spectro-temporal pattern analysis in hearing. *British Journal of Audiology* **24**: 131–137.

Näätänen R and Winkler I (1999) The concept of auditory stimulus representation in cognitive neuroscience. *Psychological Bulletin* **125**:826–859.

Nelken I (2002) Feature detection by the auditory cortex. In: Oertel D, Fay RR, and Popper AN (eds). *Springer Handbook of Auditory Research*, volume 15, *Integrative Functions in the Mammalian Auditory Pathway*. Springer-Verlag, New York, pp. 358–416.

Nelken I (2004) Processing of complex stimuli and natural scenes in the auditory cortex. *Current Opinions in Neurobiology* **14**:474–480.

Nelken I, Rotman Y, and Bar Yosef O (1999) Responses of auditory-cortex neurons to structural features of natural sounds. *Nature* **397**:154–157.

Nelken I, Fishbach A, Las L, Ulanovsky N, and Farkas D (2003) Primary auditory cortex of cats: feature detection or something else? *Biological Cybernetics* **89**:397–406.

Nelson ME and Bower JM (1990) Brain maps and parallel computers. *Trends in Neurosciences* **13**:403–408.

Ohl FW and Scheich H (1996) Differential frequency conditioning enhances spectral contrast sensitivity of units in the auditory cortex (field AI) of the alert Mongolian gerbil. *European Journal of Neuroscience* **8**:1001–1017.

Ohl FW and Scheich H (1997a) Learning-induced dynamic receptive field changes in primary auditory cortex of the unanaesthetized Mongolian gerbil. *Journal of Comparative Physiology A* **181**: 685–696.

Ohl FW and Scheich H (1997b) Orderly cortical representation of vowels based on formant interaction. *Proceedings of the National Academy of Sciences of the United States of America* **94**:9440–9444.

Ohl FW, Wetzel W, Wagner T, Rech A, and Scheich H (1999) Bilateral ablation of auditory cortex in Mongolian gerbil affects discrimination of frequency modulated tones but not of pure tones. *Learning and Memory* **6**:347–362.

Ohl FW, Schulze H, Scheich H, and Freeman WJ (2000) Spatial representation of frequency-modulated tones in gerbil auditory cortex revealed by epidural electrocorticography. *Journal of Physiology, Paris* **94**:549–554.

Ohl FW, Scheich H, and Freeman WJ (2001) Change in pattern of ongoing cortical activity with auditory category learning. *Nature* **412**:733–736.

Ohl FW and Scheich H (2004) Fallacies in behavioural interpretation of auditory cortex plasticity. *Nature Reviews Neuroscience*, online correspondence doi:10.1038/nrn1366-c1.

Peterson GE and Barney HL (1952) Control methods used in a study of the vowels. *Journal of the Acoustical Society of America* **24**: 175–184.

Phillips DP and Cynader MS (1985) Some neural mechanisms in the cat's auditory cortex underlying sensitivity to combined tone and wide-spectrum noise stimuli. *Hearing Research* **18**:87–102.

Phillips DP, Mendelson JR, Cynader MS, and Douglas RM (1985) Responses of single neurons in cat auditory cortex to time-varying stimuli: frequency-modulated tones of narrow excursion. *Experimental Brain Research* **58**:443–454.

Phillips DP, Semple MN, Calford MB, and Kitzes LM (1994) Level-dependent representation of stimulus frequency in cat primary auditory cortex. *Experimental Brain Research* **102**:210–226.

Plack CJ and Viemeister NF (1993) Suppression and the dynamic range of hearing. *Journal of the Acoustical Society of America* **93**: 976–982.

Priebe NJ and Ferster D (2008) Inhibition, spike threshold, and stimulus selectivity in primary visual cortex. *Neuron* **57**:482–497.

Qin L, Sakai M, Chimoto S, and Sato Y (2004) Spectral-edge sensitivity of primary auditory cortex neurons in alert cats. *Brain Research* **1014**:1–13.

Qin L, Sakai M, Chimoto S, and Sato Y (2005) Interaction of excitatory and inhibitory frequency-receptive fields in determining fundamental frequency sensitivity of primary auditory cortex neurons in awake cats. *Cerebral Cortex* **15**:1371–1383.

Raiguel S, Vogels R, Mysore SG, and Orban GA (2006) Learning to see the difference specifically alters the most informative V4 neurons. *Journal of Neuroscience* **26**:6589–6602.

Rasch RA (1978) The perception of simultaneous notes such as in polyphonic music. *Acustica* **40**: 21–33.

Rauschecker JP, Tian B, Pons T, and Mishkin M (1997) Serial and parallel processing in rhesus monkey auditory cortex. *Journal of Comparative Neurology* **382**:89–103.

Rauschecker JP and Tian B (2004) Processing of band-passed noise in the lateral auditory belt cortex of the rhesus monkey. *Journal of Neurophysiology* **91**:2578–2589.

Recanzone GH, Schreiner CE, and Merzenich MM (1993) Plasticity in the frequency representation of primary auditory cortex following discrimination training in adult owl monkeys. *Journal of Neuroscience* **13**:87–103.

Read HL, Winer JA, and Schreiner CE (2002) Functional architecture of auditory cortex. *Current Opinions in Neurobiology* **12**:433–440.

Rohde WS and Greenberg S (1992) Physiology of the cochlear nuclei. In: Popper AN and Fay RR (eds). *Springer Handbook of Auditory Research*, volume 2: *The Mammalian Auditory Pathway: Neurophysiology*. Springer-Verlag, New York, pp. 94–152.

Riquimaroux H, Gaioni SJ, and Suga N (1991) Cortical computational maps control auditory perception. *Science* **251**:565–568.

Sachs MB and Young ED (1979) Encoding of steady-state vowels in the auditory nerve: representation in terms of discharge rate. *Journal of the Acoustical Society of America* **66**:470–479.

Scheich H, Langner G, and Bonke D (1979) Responsiveness of units in the auditory neostriatum of the guinea fowl (*Numida meleagris*) to species specific calls and synthetic stimuli. II. Discrimination of iambus-like calls. *Journal of Comparative Physiology* **132**:257–276.

Scheich H, Baumgart F, Gaschler-Markefski B, Tegeler C, Tempelmann C, Heinze HJ, Schindler F, and Stiller D (1998) Functional magnetic resonance imaging of a human auditory cortex area involved in foreground-background decomposition. *European Journal of Neuroscience* **10**:803–809.

Scheich H (1991) Auditory cortex: comparative aspects of maps and plasticity. *Current Opinions in Neurobiology* **1**:236–47.

Scheich H, Ohl FW, Schulze H, Hess A, and Brechmann A (2006) What is reflected in sensory cortical activity: external stimuli or what the cortex does with them? In: van Hemmen JL and Sejnowski TJ (eds). *23 Problems in Systems Neuroscience*. Oxford University Press, Oxford, pp. 343–366.

Scheich H, Brechmann A, Brosch M, Budinger E, and Ohl FW (2007) The cognitive auditory cortex: task-specificity of stimulus representations. *Hearing Research* **229**:213–224.

Scheich H, Brechmann A, Brosch M, Budigner E, Ohl FW, Selezneva E, Stark H, Tischmeyer W, and Wetzel W (2010) Behavioral semantics of learning and crossmodal processing in auditory cortex: The semantic processor concept. *Hearing Research* (epub ahead of print).

Schoups A, Vogels RNQ, and Orban G (2001) Practising orientation identification improves orientation coding in V1 neurons. *Nature* **412**:549–553.

Schreiner CE (1995) Order and disorder in auditory cortical maps. *Current Opinions in Neurobiology* **5**:489–496.

Schreiner CE (1998) Spatial distribution of responses to simple and complex sounds in the primary auditory cortex. *Audiology and Neurootology* **3**:104–122.

Schreiner CE and Urbas JV (1988) Representation of amplitude modulation in the auditory cortex of the cat. II. Comparison between cortical fields. *Hearing Research* **32**:49–63.

Schreiner CE and Winer JA (2007) Auditory cortex mapmaking: principles, projections, and plasticity. *Neuron* **56**:356–365.

Schulze H, Hess A, Ohl FW, and Scheich H (2002) Superposition of horseshoe-like periodicity and linear tonotopic maps in auditory cortex of the Mongolian gerbil. *European Journal of Neuroscience* **15**:1077–1084.

Selezneva E, Scheich H, and Brosch M (2006) Dual time scales for categorical decision making in auditory cortex. *Current Biology* **16**:2428–2433.

Shamma SA, Fleshman JW, Wiser PR, and Versnel H (1993) Organization of response areas in ferret primary auditory cortex. *Journal of Neurophysiology* **69**:367–383.

Smith EC and Lewicki MS (2006) Efficient auditory coding. *Nature* **439**:978–982.

Snyder JS, Alain C, and Picton TW (2006) Effects of attention on neuroelectric correlates of auditory stream segregation. *Journal of Cognitive Neuroscience* **18**:1–13.

Sperling G (1960) The information available in brief visual presentations. *Psychological Monographs* **74**:1–29.

Suga N (1990) Cortical computational maps for auditory imaging. *Neural Networks* **3**:3–21.

Suga N and Jen PH (1976) Disproportionate tonotopic representation for processing CF-FM sonar signals in the mustache bat auditory cortex. *Science* **194**:542–544.

Suga N and Ma X (2003) Multiparametric corticofugal modulation and plasticity in the auditory system. *Nature Reviews Neuroscience* **4**:783–794.

Sutter ML and Schreiner CE (1991) Physiology and topography of neurons with multipeaked tuning curves in cat primary auditory cortex. *Journal of Neurophysiology* **65**:1207–1226.

Sutter ML, Schreiner CE, McLean M, O'Connor KN, and Loftus WC (1999) Organization of inhibitory frequency receptive fields in cat primary auditory cortex. *Journal of Neurophysiology* **82**: 2358–2371.

Takagaki K, Lippert MT, Dann B, Wanger T, and Ohl FW (2008) Normalization of voltage-sensitive dye signal with functional activity measures. *Public Library of Science ONE* **3**:e4041.

Tan AY, Zhang LI, Merzenich MM, and Schreiner CE (2004) Tone-evoked excitatory and inhibitory synaptic conductances of primary auditory cortex neurons. *Journal of Neurophysiology* **92**: 630–643.

Thivard L, Belin P, Zilbovicius M, Poline JB, and Samson Y (2000) A cortical region sensitive to auditory spectral motion. *Neuroreport* **11**:2969–2972.

Tian B and Rauschecker JP (2004) Processing of frequency-modulated sounds in the lateral auditory belt cortex of the rhesus monkey. *Journal of Neurophysiology* **92**:2993–3013.

Ulanovsky N, Las L, and Nelken I (2003) Processing of low-probability sounds by cortical neurons. *Nature Neuroscience* **6**:391–398.

Verhey JL, Pressnitzer D, and Winter IM (2003) The psychophysics and physiology of comodulation masking release. *Experimental Brain Research* **153**:405–517.

Wehr M and Zador AM (2003) Balanced inhibition underlies tuning and sharpens spike timing in auditory cortex. *Nature* **426**: 442–446.

Weinberger NM, Hopkins W, and Diamond DM (1984) Physiological plasticity of single neurons in auditory cortex of the cat during acquisition of the pupillary conditioned response: I. Primary field (AI). *Behavioral Neuroscience* **98**:171–188.

Weinberger NM (2004a) Specific long-term memory traces in primary auditory cortex. *Nature Reviews Neuroscience* **5**:279–290.

Weinberger NM (2004b) Correcting misconceptions of tuning shifts in auditory cortex. *Nature Reviews Neuroscience* online correspondence doi:10.1038/nrn1366-c2.

Weinberger NM (2007a) Associative representational plasticity in the auditory cortex: a synthesis of two disciplines. *Learning and Memory* **14**:1–16.

Weinberger NM (2007b) Auditory associative memory and representational plasticity in the primary auditory cortex. *Hearing Research* **229**:54–68.

Weinberger NM (2007c) Associative representational plasticity in the auditory cortex: resolving conceptual and empirical problems. *Debates in Neuroscience* **1**:85–98.

Wetzel W, Wagner T, Ohl FW, and Scheich H (1998) Categorical discrimination of direction in frequency-modulated tones by Mongolian gerbils. *Behavioral Brain Research* **91**:29–39.

Winter Y and Stich KP (2005) Foraging in a complex naturalistic environment: capacity of spatial working memory in flower bats. *Journal of Experimental Biology* **208**:539–548.

von Békésy G (1967) Mach band type lateral inhibition in different sense organs. *Journal of General Physiology* **50**:519–532.

Wetzel W, Ohl FW, and Scheich H (2008) Global vs. local processing of frequency-modulated tones in gerbils: an animal model of lateralized auditory cortex functions in humans. *Proceedings of the National Academy of Sciences of the United States of America* **105**:6753–6758.

Witte RS and Kipke DR (2005) Enhanced contrast sensitivity in auditory cortex as cats learn to discriminate sound frequencies. *Cognitive Brain Research* **23**:171–184.

Chapter 18

Functional Specialization in Primary and Non-primary Auditory Cortex

Stephen G. Lomber and Amee J. McMillan

Abbreviations

2DG	2-deoxyglucose
AAF or A	anterior auditory field
AES	auditory field of the anterior ectosylvian sulcus
AI	primary auditory cortex
AII	second auditory cortical area
D	dorsal
dPE or EPD	dorsal posterior ectosylvian
DZ	dorsal zone of auditory cortex
FAES	auditory field of the anterior ectosylvian sulcus
IN	insular region
iPE or EPI	intermediate posterior ectosylvian
P	posterior
PAF or P	posterior auditory field
PE	posterior ectosylvian
PS	posterior suprasylvian sulcus
SIV	fourth somatotopic cortical representation
SMI-32	monoclonal antibody to subunits of neurofilament proteins
SPL	sound pressure level
T	temporal region
TI	temporal-insular
V	ventral
VPAF or VP	ventral posterior auditory field
vPE or EPV	ventral posterior ectosylvian
SST	somatostatin
STG	superior temporal gyrus

S.G. Lomber (✉)
Cerebral Systems Laboratory, Department of Physiology and Pharmacology, Department of Psychology, The University of Western Ontario, N6A 5C1, London, ON, Canada
e-mail: steve.lomber@uwo.ca

1 Introduction

A long-term goal of auditory neuroscience is to elucidate the behavioral "division of labor" within cat auditory cortex and determine the relative contributions that the different auditory fields make to acoustic behaviors. Here we outline some recent work using reversible cooling deactivation to examine sound localization encoding and the functional cartography of cat auditory cortex. These results, when combined with investigations of underlying cerebral connections and neural function, will constrain hierarchical or network theories that best explain processing in auditory cortex.

Functional specialization is a common characteristic of the cerebral cortex. Globally, regions are specialized to perform particular sensory or motor functions. Extrastriate visual cortex has been extensively studied using behavioral, electrophysiological, and anatomical approaches. Regions specialized for spatial localization and pattern identification have been identified in monkeys (Ungerleider and Mishkin 1982), humans (Ungerleider and Haxby 1994; Courtney et al. 1996), and cats (Lomber et al. 1996a,b). The behavioral correlate for such functional specializations or a "division of labor" within auditory cortex is largely unknown. Anatomical and electrophysiological assessments have indicated that similar specializations may also exist in non-primary auditory cortex of monkeys and humans (Rauschecker 1998a; Rauschecker and Tian 2000). There are also recent findings that regions of cat non-primary auditory cortex are specialized for the spatial localization of acoustic stimuli or the recognition of auditory patterns (Lomber and Malhotra 2008).

In the cat, pioneering work in understanding the functional cartography of auditory cortex focused on examining sound localization (Table 18.1) or the ability of animals to discriminate different acoustic temporal sequences. These latter studies focused on the temporal-insular (TI) region of auditory cortex and showed that TI lesions impair performance of cats at discriminating acoustic temporal sequences (Diamond and Neff 1957; Dewson 1964; Cornwell 1967;

J.A. Winer, C.E. Schreiner (eds.), *The Auditory Cortex*,
DOI 10.1007/978-1-4419-0074-6_18, © Springer Science+Business Media, LLC 2011

Table 18.1 Studies of sound localization abilities following unilateral and/or bilateral deactivation of auditory cortex. Lesion size is indicated as small (generally restricted to AI or a specific non-primary area) or large (AI and multiple other auditory areas). The deactivation method, method of testing, and species studied are also indicated

Study	Unilateral lesions	Bilateral lesions	Lesion size	Deactivation method	Testing method	Species
Beitel and Kaas (1993)	X	X	L	Aspiration	Head orienting	Cat
Casseday and Diamond (1977)	X		L	Ablation	"Y" maze	Cat
Cranford et al. (1971)	X		L	Ablation	"Y" maze	Cat
Neff et al. (1956)	X	X	L	Aspiration	Headset and localization	Cat
Thompson and Welker (1963)		X	L	Aspiration	Head orienting	Cat
Masterton and Diamond (1964)	X	X	S	Aspiration	Discrimination	Cat
Jenkins and Masterton (1982)	X		L	Aspiration	Localization	Cat
Jenkins and Merzenich (1984)	X		S	Ischemically induced	Localization	Cat
Neff (1968)	X	X	L	Ablation	Localization	Cat
Riss (1959)		X	S	Ablation	Head orienting	Cat
Strominger (1969a) Subdivisions		X	L & S	Ablation	Two-choice orienting	Cat
Strominger (1969b) Localization	X	X	L	Ablation	Two-choice orienting	Cat
Whitfield et al. (1972)	X		L	Aspiration	"Y" maze	Cat
Malhotra et al. (2004)	X		S	Cooling	Localization	Cat
Malhotra and Lomber (2007)		X	S	Cooling	Localization	Cat
Malhotra et al. (2008)	X	X	S	Cooling	Localization	Cat
Ravizza and Masterton (1972)		X	L	Aspiration	Detection	Opossum
Girden (1939)	X	X	L	Temporal lobec-tomy	Discrimination	Dog
Heffner (1978)		X	L	Aspiration	Discrimination	Dog
Stepien et al. (1990)		X	L	Aspiration	Discrimination	Dog
Bizley et al. (2007)	X	X	L & S	Muscimol and aspi-ration	Vertical location dis-crimination	Ferret
Kavanagh and Kelly (1987)	X	X	L & S	Aspiration	Localization	Ferret
Smith et al. (2004)		X	S	Muscimol	Localization	Ferret
Heffner and Heffner (1990)		X	L	Aspiration	Discrimination	Japanese macaque
Heffner (1997)	X	X	L	Ablation	Discrimination	Macaque
Wegener (1964)	X	X	L			O.W. monkey
Heffner and Masterton (1975)		X	L & S	Aspiration	Discrimination	Rhesus
Thompson and Cortez (1983)	X		S	Aspiration	Localization	Squirrel monkey
Ravizza and Diamond (1974)		X	L	Ablation	Localization	Hedgehog and bushbaby
Adriani et al. (2003)	X		L	Stroke/tumor	Earphone localization	Human
Clarke et al. (2000)	X		L	Ischemia	Earphone localization	Human
Zatorre and Penhune (2001)	X		L	Excision	Point to location	Human

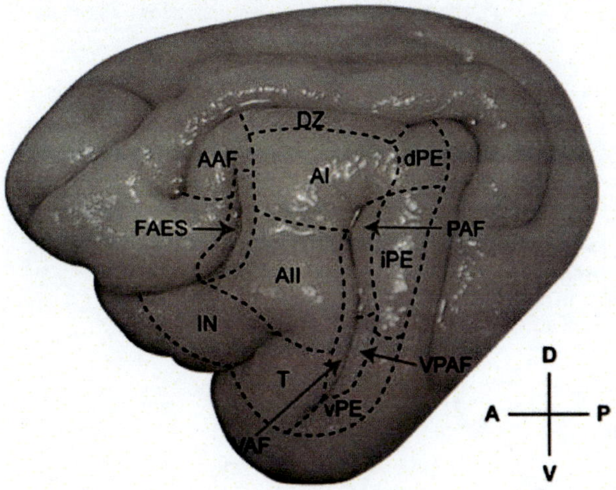

Fig. 18.1 Lateral view of the left hemisphere of cat cerebral cortex showing the generally recognized auditory areas. Auditory cortical areas: AI – primary auditory cortex; AII – second auditory cortex; AAF – anterior auditory field; AES – auditory field of the anterior ectosylvian sulcus; dPE – dorsal posterior ectosylvian gyrus; DZ – dorsal zone of auditory cortex; IN – insular region; iPE – intermediate posterior ectosylvian gyrus; PAF – posterior auditory field; T – temporal region; VAF – ventral auditory field; VPAF – ventral posterior auditory field; and vPE – ventral posterior ectosylvian gyrus. Other abbreviations: A – anterior; D – dorsal; P – posterior; and V – ventral (Compiled from Reale and Imig 1980; Clascá et al. 1997; Ribaupierre 1997; Tian and Rauschecker 1998; Read et al. 2002)

Kelly 1973; Cornwell et al. 1998). Furthermore, lesions of temporal-insular cortex of cats (Cornwell 1967; Colavita 1972; Colavita et al. 1974) were found to impair the discrimination of temporal patterns of sounds as well as the discrimination of complex acoustical signals such as speech sounds that differ mostly in the pattern of their harmonics (Dewson 1964; Dewson et al. 1969) (Fig. 18.1).

Most of the early investigations into functional localization in auditory cortex relied on using lesions of the cerebrum and post-lesion testing (Table 18.1). However, there are many drawbacks to this method including: (1) lesions can only reliably be defined post mortem, (2) comparisons must be made between animals and internal double dissociations are not possible, and (3) the mature cerebrum is plastic and connections can be activated, strengthened, or modified following lesions (Lomber et al. 1999). Therefore, it is desirable to rapidly and reversibly deactivate the cerebrum and assess the functional roles of the individual regions of auditory cortex. We have used reversible cooling deactivation to begin to reveal the functional cartography of the auditory cortex.

2 Reversible Cooling Deactivation

The cooling method to reversibly deactivate neural tissue is an exciting, potent, and appropriate technique for examining cerebral contributions to behavior and has a number of highly beneficial and practical features (Lomber 1999). (1) Limited regions of the cerebral cortex can be selectively and reversibly deactivated in a controlled and reproducible way. Baseline and experimental measures can be made within minutes of each other (Lomber et al. 1996b). (2) Repeated coolings over months or years produce stable, reversible deficits, with little evidence of attenuation or neural compensations (Lomber et al. 1994, 1999). (3) Repeated cooling induces neither local nor distant degenerations that might compromise conclusions (Yang et al. 2006). (4) Compared to traditional ablation studies, fewer animals are needed because within-animal-comparisons and double dissociations are possible, permitting large volumes of high quality data to be acquired from each animal (Lomber et al. 1996b). (5) Finally, as the major effect of cooling is to block synaptic transmission, activity in fibers of passage is not compromised (Jasper et al. 1970; Bénita and Condé 1972). Overall, the technique induces localized hypothermia in a restricted region of the brain. The locus of the deactivation is kept small by the constant perfusion of warm blood into, and around, the cooled region. The cooling disrupts calcium channel function on the pre-synaptic side of the synapse and disrupts normal neurotransmitter release (reviewed by Brooks 1983).

It has been determined that the surgical procedure to implant cryoloops, their presence in contact with the cerebrum, and their operation disrupts neither the normal structural nor functional integrity of cortex (Lomber et al. 1999). In every instance, cell and myelin stains are rich, and the cyto- and myelo-architecture of the region are characteristic of the region investigated, with no signs of pathology, as might be revealed by a marked pale staining of neurons or gliosis or light staining of cytochrome oxidase (Lomber and Payne 1996). However, the lack of damage to the cortex means that it is not possible to use traditional histological techniques to determine the region that was deactivated. A convenient and reliable way to estimate the extent of deactivated cortex is to administer radio-labeled 2-deoxyglucose (2DG) prior to sacrifice (Payne and Lomber 1999). Since cooling silences evoked responses by neurons and depresses metabolic activity, little 2DG is taken up by the silenced neurons compared to regions at normal temperature and with normal levels of high activity. Consequently, the deactivated region is identified as a very pale region surrounded by dark gray, active tissue in 2DG autoradiograms (Payne and Lomber 1999; Lomber and Payne 2000).

It is feasible to use cooling deactivation to block transmission of efferent signals and to test the contributions discrete cortical areas make to behavior. Synaptic transmission in the mammalian brain is blocked by temperatures below 20°C (Brooks 1983; Jasper et al. 1970; Lomber et al. 1999). For a loop cooled to 3°C, the thermocline measures show that the 20°C isotherm lies at the base of layer VI (see Fig. 7 in Lomber and Payne 2000). Moreover, at this same loop

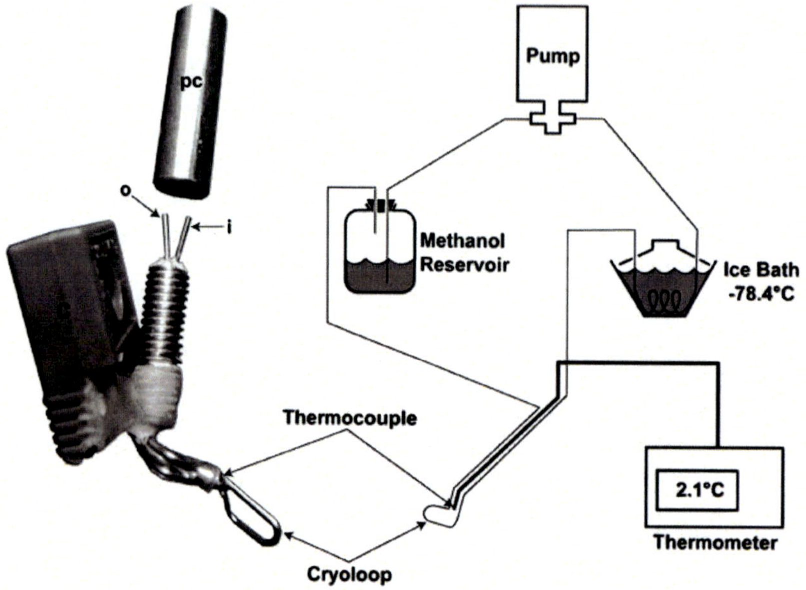

Fig. 18.2 Schematic drawing of the cryoloop cooling deactivation system as described in Lomber et al. (1999). Methanol is used as a coolant and is drawn from a reservoir and pumped through a methanol and dry ice bath. The coolant passes though Teflon tubing and is connected to the implant (*left*) fixed to the animal's skull. After passing through the cryoloop, the coolant returns to reservoir creating a closed system. The cryoloop consists of 23G hypodermic stainless steel tubing that is shaped to conform to the surface of the cortical region of interest. A thermocouple at the union of the loops monitors the loop temperature which is displayed on a digital thermometer. The thermocouple is connected to the thermometer via a commercially available connector (*c*). The flow rate through the systems alters the temperature of the cooling loop. When the implant is not connected to the system, the inflow (*I*) and outflow (*O*) tubes are covered by a stainless steel protective cap (*pc*)

temperature, electrophysiological and 2DG measures show complete silencing of neuronal activity throughout cortical thickness, and behavioral measures reveal substantial impairments (Lomber et al. 1996b; Payne and Lomber 1999). For a loop cooled to 3°C a stable cortical temperature is reached within ~5 min of initiating cooling and normal brain temperature is regained within ~2 min after the cessation of cooling due to the infusion of warm blood (Lomber et al. 1994, 1996b).

After the region of cortex to be cooled is selected, several cryoloops are prepared to conform to different gyral shapes from a collection of fixed and rubber-replica brains. Each cryoloop is fashioned out of 23G hypodermic tubing and has a copper/constantin microthermocouple attached at the union of the inlet and outlet tubes (Fig. 18.2). The loops are designed to fit snugly, after final adjustment, in contact with cortex. The loops resting on the cortical surface are secured to the skull with screws and acrylic, and the dura and bone flaps are replaced. With care, neither the surgical procedures nor repeated coolings alter cortical structure or function (e.g., Lomber et al. 1996a,b, 1999; Vanduffel et al. 1997; Lomber and Payne 2000). The cooling of cortex is effected by pumping chilled methanol through the loop tubing (Fig. 18.2). Loop temperature is monitored and accurately governed within 1°C of the desired value by controlling the rate of methanol flow. For detailed cryoloop procedures, see Lomber et al. (1999).

In humans there is considerable evidence for asymmetries in cerebral functions such as language comprehension, speech production, attentional mechanisms, and spatial representations (Bisiach et al. 1979; Robertson et al. 1988; Robertson 1989; Bradshaw and Rogers 1993; Heilman et al. 1993; Driver and Mattingley 1998; Kolb and Whishaw 1996). The cooling deactivation method is ideally suited to specific testing for asymmetries in cat brain function. This is possible because regions in the left and right hemispheres can be deactivated either independently or in unison, and deficits in performance during the three conditions can be compared to reveal the presence or absence of lateralized operations.

3 Behavioral Double Dissociation in Auditory Cortex

Similar to the visual system, a current model of auditory cortical organization proposes that the auditory system may also contain discernable cortical specializations and separate cortical processing streams specialized for either pattern discrimination or spatial processing (Rauschecker 1997, 1998a,b; Rauschecker et al. 1997). Based on electrophysiological studies in non-human primates, it is proposed that fields rostral to primary auditory cortex (AI) may be specialized for auditory pattern processing and fields caudal to AI may be specialized for accurately determining the spatial location of a sound source. Electrophysiological (Tian et al. 2001) and connectional (Romanski et al. 1999) studies provide evidence buttressing the proposed functional dissociation in the monkey. Recently, we have attempted to provide the critical missing link in the chain of evidence in support of "what" and "where" functional specializations in auditory cortex by performing a behavioral double dissociation similar to that demonstrated in extrastriate visual cortex of monkeys and cats (Ungerleider and Mishkin 1982; Lomber et al. 1996b). To accomplish this, we trained cats to perform a battery of experimental and control behavioral tasks, bilaterally placed cooling loops (Lomber et al. 1999) over both the anterior auditory field (AAF) and the posterior auditory field (PAF) (Fig. 18.3), and then tested the animals while deactivating AAF and PAF individually in order to determine their contributions to the acoustic behaviors (Lomber and Malhotra 2008). This experimental design permitted double dissociations to be performed within the same animal.

The first task (spatial localization) required the cats to accurately localize the spatial position of a broad-band noise burst. The cats were first trained in a semicircular arena to identify the location of a 100 ms broad-band noise burst (20 dB SPL above a background level of 58 dB SPL).

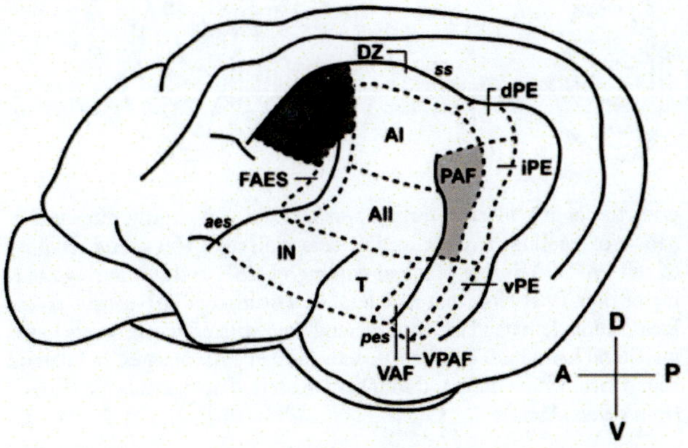

Fig. 18.3 Lateral view of the left hemisphere of the cat cerebrum showing the generally recognized auditory areas. The two areas examined in the "what" and "where" double dissociation experiments are shaded. Sulci are indicated by lower-case lettering (aes – anterior ectosylvian sulcus, pes – posterior ectosylvian sulcus, ss – suprasylvian sulcus). Auditory cortical areas: AI – primary auditory cortex; AII – second auditory cortex; AAF – anterior auditory field; AES – auditory field of the anterior ectosylvian sulcus; dPE – dorsal posterior ectosylvian gyrus; DZ – dorsal zone of auditory cortex; IN – insular region; iPE – intermediate posterior ectosylvian gyrus; PAF – posterior auditory field; T – temporal region; VAF – ventral auditory field; VPAF – ventral posterior auditory field; and vPE – ventral posterior ectosylvian gyrus. A – anterior; D – dorsal; P – posterior; and V – ventral. From Lomber and Malhotra (2008)

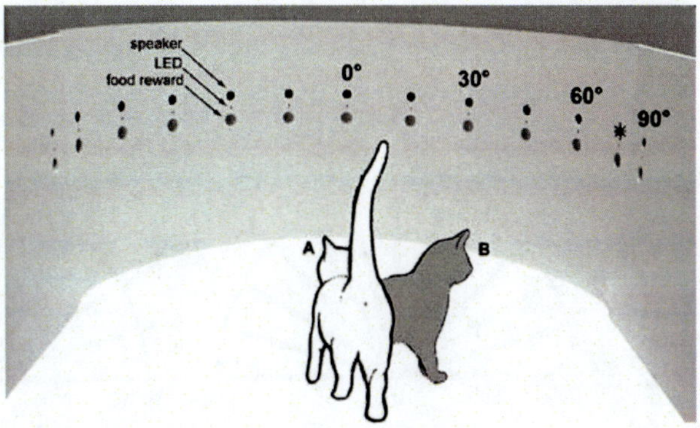

Fig. 18.4 Acoustic and visual orienting arena. A loudspeaker (*top circle*) and a light-emitting diode (LED, *black dot*) were located above a food reward locus (*lower circle*) at each of 13 regularly spaced (15°) intervals (for sake of clarity, only 30° intervals are labeled). **a** The animal was first required to fixate on the central (0°) LED. **b** It then had to orient to, and approach, a secondary acoustic (100 ms broad-band noise) or visual (illumination of an LED) stimulus to receive a food reward

After attending to a central visual stimulus (red LED) the cats had to orient to, and approach, the acoustic stimulus that was emitted randomly from 1 of 13 speakers placed at 15° intervals across 180° of azimuth (Fig. 18.4). Prior to, and following the conclusion of, each cortical deactivation, acoustic spatial localization accuracy and precision was excellent, with performance across all 13 positions for each of the three cats at >85% correct (Fig. 18.5a). In contrast, bilateral deactivation of PAF profoundly impaired the ability of all the cats to accurately and precisely[23] localize the acoustic stimulus (Fig. 18.5b*v*). Regardless of spatial location, performance dropped at all 13 tested positions to levels just above chance (7.7%). On average, sound localization performance fell to about 15% correct. Unilateral deactivation of PAF cortex resulted in deficits restricted to targets presented in the contralateral hemifield (Fig. 18.5b(*iv*,*vi*)). The PAF deactivation-induced impairments were stable, with no evidence of deficit attenuation over the 7 months of testing (Lomber and Malhotra 2008). Neither bilateral, nor unilateral, deactivation of AAF impaired sound localization function.

More recent studies have also revealed similar findings when tonal stimuli are used. Bilateral deactivation of PAF, but not AAF, also impairs the ability of the animals to accurately localize tonal (15 kHz) stimuli presented at the same intensity level and duration as the original noise burst stimuli (Fig. 18.6). The overall ability of the animals to localize the tonal stimuli is significantly less than for burst stimuli (compare Figs. 18.5a and Fig. 18.6a). This finding was in agreement with the earlier work of other investigators (Populin and Yin 1998).

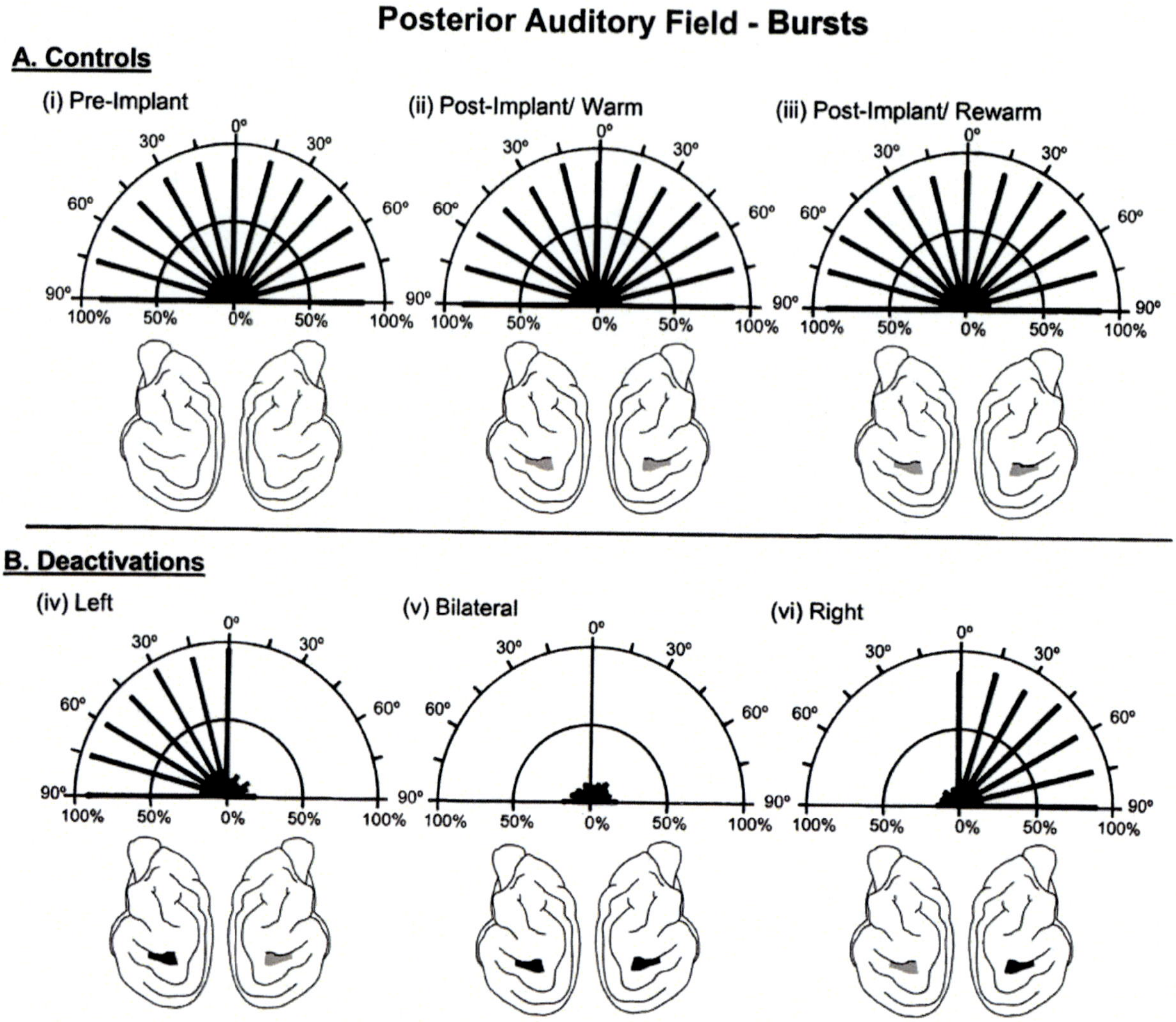

Fig. 18.5 Orienting responses to an acoustic (100 ms broad-band noise burst) stimulus during deactivation of PAF. Lateral view icons of the cat brain indicate the presence and position of a cryoloop (*gray shading*), and its operational status (*black* indicates loop was on and cortex was deactivated). In this and subsequent data graphs, the two concentric semicircles represent 50 and 100% correct response levels and the length of each *bold line* corresponds to the percentage of correct responses at each location tested. **a** Control data collected: (*i*) prior to PAF cryoloop implantation, (*ii*) after PAF cryoloop implantation and prior to cooling in each testing session, and (*iii*) shortly after termination of cooling. **b** Deactivation data collected: (*iv*) during cooling of left PAF, (*v*) during bilateral cooling of PAF, and (*vi*) during cooling of right PAF. Note that unilateral deactivation of PAF caused sound localization deficits in the contralateral field with no impairments in the ipsilateral hemifield. Bilateral deactivation of PAF resulted in bilateral sound localization deficits throughout the tested field. Data summarized from seven animals

The second task (pattern discrimination) required the cats to discriminate between different temporal patterns of acoustic stimuli of the same temporal duration. For this task, we trained the same cats to perform temporal pattern discrimination in a two-choice apparatus utilizing procedures similar to that of a classical delayed match-to-sample task. The stimuli consisted of broad-band noise bursts (825–1525 ms in duration, 78 dB SPL) with an imbedded irregular gap sequence that made the stimuli similar to Morse Code sequences (Lomber and Malhotra 2008). Overall, the cats became very good at discriminating temporal patterns. Normal average performance levels for each animal were >80% (Fig. 18.7a, Pre/Post). During bilateral deactivation of PAF cortex (Fig. 18.7a, PAF), performance was no different from normal control levels (Fig. 18.7a, Pre/Post). However, during bilateral deactivation of AAF cortex in the same cats, performance significantly ($p<0.01$) dropped to levels not different from chance (50%; Fig. 18.7a, AAF). This deficit

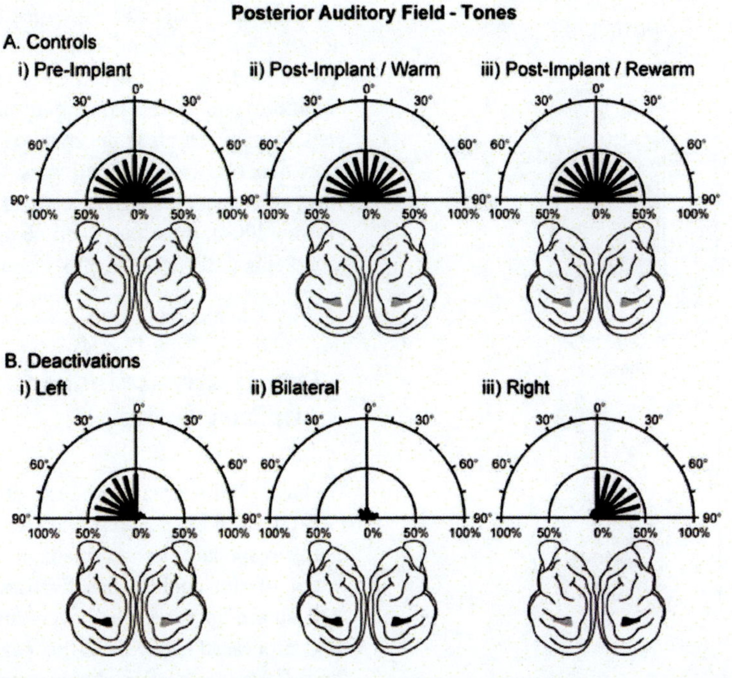

Fig. 18.6 Orienting responses to an acoustic stimulus (100 ms, 15 kHz tone) during deactivation of PAF. For conventions, see Fig. 18.5. **a** Control data collected: (*i*) prior to PAF cryoloop implantation, (*ii*) after PAF cryoloop implantation and prior to cooling in each testing session, and (*iii*) shortly after termination of cooling. **b** Deactivation data collected: (*iv*) during cooling of left PAF, (*v*) during bilateral cooling of PAF, and (*vi*) during cooling of right PAF. Note that unilateral deactivation of PAF caused sound localization deficits in the contralateral field with no impairments in the ipsilateral hemifield. Bilateral deactivation of PAF resulted in bilateral sound localization deficits. Data summarized from five animals

was not localized to one hemisphere as unilateral deactivation of either left or right AAF cortex did not significantly decrease performance. Therefore, from these results we concluded that AAF cortex, but not PAF cortex, is critical for discriminating temporal pattern sequences.

More recent studies have also revealed similar findings when tonal stimuli are used. Bilateral deactivation of AAF, but not PAF, also impairs the ability of the animals to accurately discriminate tonal (15 kHz) gap sequence patterns (Fig. 18.7b) presented at the same intensity level and duration as the original noise burst stimuli (Fig. 18.7a). Therefore, bilateral deactivation of AAF impairs the ability of cats to discriminate different temporal patterns, regardless of whether the patterns consist of burst or tonal stimuli.

The aim of Lomber and Malhotra (2008) was to determine if it is possible to doubly dissociate two fundamentally different cerebral operations in non-primary auditory cortex. Indeed, the results of the present study show that cerebral operations involving the localization of sound and the discrimination of acoustic patterns can be doubly dissociated from each other in posterior and anterior regions of auditory cortex, respectively (Fig. 18.8). Classically, double dissociations are sought by testing two independent groups of subjects, each with a different locus of brain damage (e.g., Petrides 2000; Winters et al. 2004). However, the study of Lomber and Malhotra (2008) did not examine two different populations, but through the use of reversible deactivation was able to demonstrate the dissociations within the same experimental animals.

These results demonstrate a clear division of labor in auditory cortex (Lomber and Malhotra 2008). While a one-to-one relationship might not be expected between functional streams in visual cortex and functional streams in auditory cortex, the results from this study significantly strengthen the notion that functional segregations and processing streams are a common attribute of mammalian cortical sensory systems (Lomber et al. 1996b). Specifically, the proposal that "what" and "where" streams may exist in auditory cortex is significantly supported (Rauschecker et al. 1997; Rauschecker 1998a; Rauschecker and Tian 2000). However, the spatial and pattern processing dichotomy is not the only proposed cortical processing configuration. Goodale and Milner (1992) proposed the existence of visual cortical processing pathways subserving perception (ventral stream) and action (dorsal stream). This theory emphasizes the output requirements of the dorsal and ventral pathways rather than the input or sensory distinctions. Evidence in support of this model has been obtained from double dissociation

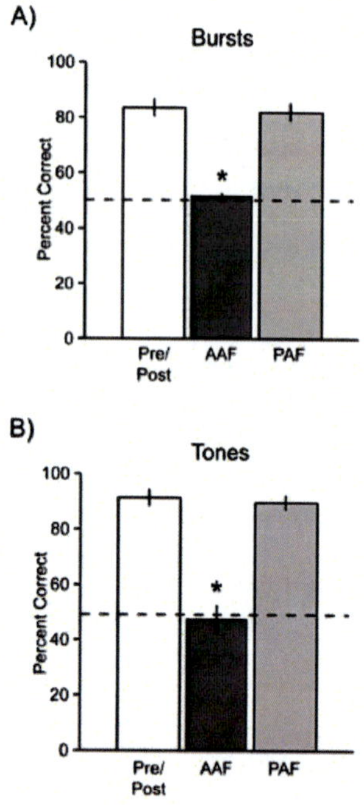

Fig. 18.7 Mean temporal pattern discrimination performance (mean ± SE) prior to (Pre/Post – *no shading*), and during bilateral cooling of PAF cortex (*light gray*), and during bilateral cooling of AAF cortex (*dark gray*). **a** Data from testing with white-noise bursts. **b** Data from testing with 15 kHz tones. Chance = 50%. Asterisk indicates significant difference (p<0.01) from control (Pre/Post). Mean performance for each condition is based on 20 testing sessions of 50 trials each, across the five animals examined on this task

Task	PAF Deactivation	AAF Deactivation
Sound Localization "Where"	DEFICIT	No Deficit
Acoustic Pattern Discrimination "What"	No Deficit	DEFICIT

Fig. 18.8 Diagram summarizing the results of Lomber and Malhotra (2008). PAF and AAF deactivations were bilateral. The sound localization task involved the accurate localization of 100 ms noise bursts, while the pattern discrimination task involved the discrimination of different temporal patterns

studies of neurological patients (Goodale and Milner 1992; Goodale et al. 1991; Goodale and Westwood 2004), as well as functional imaging studies of healthy subjects (Cavina-Pratesi et al. 2007). When applied to our present findings, this proposal would suggest that AAF is involved in perception and PAF is involved with action. Indeed, deactivation of AAF disrupted the perception of the gap sequences and the deactivation of PAF disrupted the action of accurately directing the head and body, and subsequent approach, to the acoustic stimulus. Therefore, although the Lomber and Malhotra (2008) study did not specifically test the perception–action dichotomy, the results do support this cortical segregation as well. Finally, it is also important to consider that there may be more than two processing streams. Given the large number of ways an acoustic pattern/object can be defined (Griffiths and Warren 2004), there may well be more than a single "what" processing pathway in auditory cortex.

4 Deactivation of Individual Regions in Cat Auditory Cortex

We have used reversible cooling deactivation (Lomber et al. 1999) to examine the contributions that 12 regions of acoustically responsive cortex made to sound localization. The extent of the cooling deactivations was determined from 2DG autoradiograms which were matched with adjacent sections processed for SMI-32 that permitted the delineation of the different areas of auditory cortex (Mellott et al. 2010). The positions of these 12 loci, as well as how they relate to the cortical maps of other investigators, are described below.

4.1 Tonotopically Organized Regions

We examined four regions of tonotopically organized auditory cortex: the primary auditory cortex (AI), the posterior auditory field (PAF), the anterior auditory field (AAF), and the ventral posterior auditory field (VPAF). An AI loop was approximately 7 mm long and extended lengthwise across the middle ectosylvian gyrus, from the dorsal tip of the anterior ectosylvian sulcus to just anterior of the posterior ectosylvian sulcus (about A2–A9 Horsley and Clarke (1908) coordinates (stereotaxic coordinates are provided using the Horsley and Clarke (1908) system as described by Reinoso-Suárez (1961)); Reale and Imig (1980); Fig. 18.1). Cryoloops were also placed on PAF (Reale and Imig 1980; Phillips and Orman 1984), located caudal and ventral to A1. Loops were approximately 6 mm long and extended from the anterior one-third of the dorsal posterior ectosylvian gyrus to the fundus of the dorsal half of the posterior ectosylvian (PE) sulcus. A heat shielding compound was also applied to the anterior side of the PAF and VPAF loops to keep the cooling deactivations localized to the posterior bank of the PE sulcus. All deactivations extended down the posterior bank of the PE sulcus to the fundus and did not include the anterior bank. Therefore, the deactivated region included all of area PAF or area P (Fig. 18.1; Imig et al. 1982; Phillips and Orman

1984). The AAF (Knight 1977; Reale and Imig 1980; Phillips and Irvine 1982) cryoloops were approximately 6.5 mm long and were located on the crown of the anterior suprasylvian gyrus between A11 and A17.5. Therefore, the deactivations included all of area AAF or area A (Fig. 18.1), as defined by Knight (1977) and Reale and Imig (1980). Loops approximately 8 mm long were placed on VPAF. These loops extended from the anterior one-third of the ventral PE gyrus and extended to the fundus of the dorsal PE sulcus. In all cases, the deactivated region included all of area VP, as defined by Imig et al. (1982).

4.2 Non-tonotopically Organized Regions

We have also examined eight regions of non-tonotopically organized auditory cortex: the dorsal zone of auditory cortex (DZ), the auditory field of the anterior ectosylvian sulcus (AES), the second auditory cortex (AII), the insular (IN) region, the temporal (T) region, the dorsal posterior ectosylvian (dPE) gyrus, and the ventral posterior ectosylvian (vPE) gyrus. The DZ loops were approximately 8 mm long and extended along the dorsal edge of the middle ectosylvian gyrus along the lip of the middle suprasylvian sulcus (about stereotaxic A2–A10; Paula-Barbosa et al. 1975; Reale and Imig 1980; Fig. 18.1). Only half of the lower limb of the DZ loop came in contact with the cortical surface. The upper limb did not contact the brain. Therefore, the region deactivated by a DZ loop included the dorsal zone, as defined by Middlebrooks and Zook (1983), which has been previously described as the suprasylvian fringe (Woolsey 1961; Paula-Barbosa et al. 1975; Niimi and Matsuoka 1979; Beneyto et al. 1998).

The auditory field contained in the AES occupies a region at the posterior end of the anterior ectosylvian sulcus with the largest portion of the field located on the dorsal bank and fundus (Mucke et al. 1982; Clarey and Irvine 1986; Meredith and Clemo 1989). Loops approximately 6 mm × 3 mm were placed in the posterior two-thirds of the AES where both the auditory and visual representations are located (Rauschecker and Korte 1993). Therefore, cooling of each AES cryoloop silenced the acoustically responsive field of the AES (Clarey and Irvine 1986; Meredith and Clemo 1989; Mucke et al. 1982; Rauschecker and Korte 1993), the visually responsive field (AEV; Olson and Graybiel 1983, 1987), and ventral portions of SIV (Mori et al. 1996).

Loops approximately 6 × 3 mm were placed on AII, which lies ventral to AI and extends between the anterior and posterior ectosylvian sulci (Reale and Imig 1980; Fig. 18.1). The longest dimension of the loop extended anterior to posterior. In all cases the deactivation included all of area AII as defined by Woolsey (1961) and Reale and Imig (1980).

The insular (IN) region occupies a swath of cortex on the anterior sylvian gyrus, between the anterior ectosylvian and sylvian sulci. IN cortex is ventral to AII. Cooling loops (6 × 3 mm) were placed lengthwise over the anterior sylvian gyrus. Therefore, the area IN loops deactivated the majority of the anterior sylvian area as defined by Clascá et al. (1997, 2000). The region of deactivation extended ventrally into the dorsal division of the agranular insular area (Clascá et al. 1997, 2000).

We defined the temporal (T) area as a band across the posterior sylvian gyrus from the sylvian sulcus, anteriorly, to a position approximately 2 mm anterior to the posterior ectosylvian sulcus. The region visible just anterior to the PE sulcus is the ventral auditory field (VAF; Reale and Imig 1980). Temporal area loops deactivated area Te of Clascá et al. (2000), while the VAF loops deactivated the ventral auditory field (VAF or V; Reale and Imig 1980).

On the PE gyrus, anatomical and electrophysiological investigations suggest that the gyrus contains three parallel and vertically oriented "belts" (Woolsey 1960; Reale and Imig 1980; Updyke 1986; Bowman and Olson 1988a,b). The anterior belt contains the two tonotopically organized regions described previously (PAF and VPAF). The middle belt is unimodal and responds to acoustic stimuli, but lacks a tonotopic organization (Bowman and Olson 1988a,b). This middle belt has been further subdivided into dorsal (dPE), intermediate (iPE), and ventral (vPE) subdivisions based on cytoarchitecture and patterns of extrinsic connections (Winer 1992). The posterior belt along the entrance to the posterior suprasylvian sulcus contains both visually and acoustically responsive neurons (Updyke 1986; Bowman and Olson 1988a,b). We have subdivided the central and posterior belts into dorsal and ventral halves and placed vertically oriented 3 × 8 mm cooling loops over each region. The dPE loop deactivations included both the dorsal (dPE) and intermediate (iPE) divisions of the PE gyrus, or EPD and EPI, of Winer (1992). The vPE loop deactivations included the ventral division of the PE gyrus (vPE), or EPV, of Winer (1992) and portions of area PS of Updyke (1986).

5 Contributions of Auditory Cortex to Sound Localization

One of the fundamental functions of the auditory system is to accurately determine the location of a sound source. Therefore, it is curious that while the auditory cortex does not contain any maps of auditory space (Middlebrooks 2002; Middlebrooks et al. 2002), lesions of auditory cortex produce profound sound localization deficits. Behavioral studies in cats, ferrets, dogs, and monkeys have identified that unilateral ablations of the entire auditory cortex produce profound sound localization deficits (Table 18.1; Girden 1939; Neff et al. 1956; Neff 1968; Strominger 1969b; Cranford et al. 1971; Whitfield et al. 1972; Casseday and Diamond 1977; Jenkins and Masterton 1982; Kavanagh and Kelly 1987; Heffner 1997). Large bilateral cortical ablations encompassing most or all of acoustically responsive cortex in cats (Neff et al. 1956; Thompson and Welker 1963; Neff 1968; Strominger 1969a,b), ferrets (Kavanagh and Kelly 1987), opossums (Ravizza and Masterton 1972), dogs (Girden 1939; Heffner 1978), old-world monkeys (Heffner and Heffner 1990; Heffner 1997), and new-world monkeys (Ravizza and Diamond 1974) result in sound localization deficits throughout the entire field. Similar results have been identified when lesions have been restricted to individual areas of auditory cortex. However, at an areal level, virtually all early investigations have only considered the AI contributions. Restricted unilateral lesions, limited to AI, impair an animal's ability to localize brief sounds in the contralesional field (Masterton and Diamond 1964; Wegener 1964; Heffner and Masterton 1975; Jenkins and Merzenich 1984; Thompson and Cortez 1983; Kavanagh and Kelly 1987). Following bilateral AI ablations, sound localization deficits throughout both hemifields have been identified in carnivores and non-human primates (Riss 1959; Strominger 1969a; Masterton and Diamond 1964; Wegener 1964; Heffner and Masterton 1975; Thompson and Cortez 1983; Jenkins and Merzenich 1984; Kavanagh and Kelly 1987).

In all of the studies conducted in our lab (Malhotra et al. 2004, 2008; Malhotra and Lomber 2007; Lomber and Malhotra 2008), we have utilized adult, intact cats that received one or two bilateral pairs of cooling loops over discrete regions of auditory cortex. The cat is an appealing model system for investigations on cerebral networks in auditory cortex because: (1) they can be trained to perform complex auditory tasks; (2) the majority of the auditory areas are easily approachable because they are exposed on the surfaces of gyri (Fig. 18.1), rather than being buried in the depths of a deep sulcus; (3) each area is small enough so that it may be cooled by a single cryoloop (Lomber et al. 1999); and (4) the spatial and non-spatial response properties of units in the auditory cortical fields of the cat have been well characterized by a number of labs.

Prior to our investigations, anatomical and physiological investigations suggested that, in addition to AI, the auditory field of AES, PAF, and DZ would likely play a role in sound localization. Anatomical studies implicated the auditory field in the anterior ectosylvian sulcus (Meredith and Clemo 1989) in auditory spatial processing due to its dense projection to the intermediate and deep layers of the superior colliculus (Meredith and Clemo 1989). Electrophysiological studies of FAES found neurons with spatial selectivity (Korte and Rauschecker 1993; Middlebrooks et al. 1994; Nelken et al. 1997). The neurons in FAES are particularly sensitive to features of sounds that are spatially informative (Nelken et al. 1997). In addition, electrophysiological studies have identified neurons in the posterior auditory field (PAF) that show enhanced specificity for acoustic stimulus location (Stecker et al. 2003). Neurons in the dorsal zone of auditory cortex have many responses properties similar to PAF (Stecker et al. 2005). However, compared with PAF, DZ responses were of shorter overall latency and more DZ units preferred stimulation from ipsilateral azimuths (Stecker et al. 2005). Therefore, both electrophysiological and anatomical investigations suggested that regions outside of AI, such as FAES, PAF, and DZ, would be likely to play a role in sound localization behavior.

To test the accurate localization of a sound source, cats were trained in an orienting arena that allowed for the presentation of either acoustic or visual stimuli. The apparatus (Fig. 18.4), training, and testing procedures are described in detail in Malhotra et al. (2004) and Malhotra and Lomber (2007) and will only be briefly described here. The speakers emitted broad-band noise bursts (100 ms in duration) that were 20 dB(A) above background. For the experimental stimulus, we used broad-band noise bursts rather than pure tones because orienting responses to short broad-band noise bursts have been identified to be much more accurate than responses to tones (Populin and Yin 1998).

Prior to any cooling loop implantations, the cats were highly proficient at accurately orienting to a sound source presented at all locations out to 90°. We found no evidence for any change in sound localization performance following implantation in the uncooled state prior to cooling. Furthermore, following the termination of cooling deactivation, orienting responses to acoustic stimuli returned to normal levels throughout the 180° field examined. Therefore, the similarities in orienting responses across the three control conditions confirmed that neither the presence of the cooling loops, nor their repeated cooling, affected orienting performance.

Data collected during bilateral deactivation of each of 12 auditory areas collected from 33 cats are summarized in Fig. 18.9. Bilateral deactivation of PAF resulted in an almost complete spatial localization impairment through the entire field examined, with performance significantly ($p<0.01$) falling by 83% (Fig. 18.9). Similarly, bilateral deactivation of FAES also resulted in performance dropping by 81% across the entire tested field (Fig. 18.9). Earlier studies had shown that bilateral deactivation of both AI and DZ resulted in decreases in performance as profound as those seen during bilateral deactivation of PAF (Malhotra and Lomber 2007). However, more recent work has identified that bilateral deactivation of neither AI nor DZ results in deficits as profound as those identified during bilateral PAF deactivation (Fig. 18.9; Malhotra et al. 2008). Bilateral deactivation of any of the

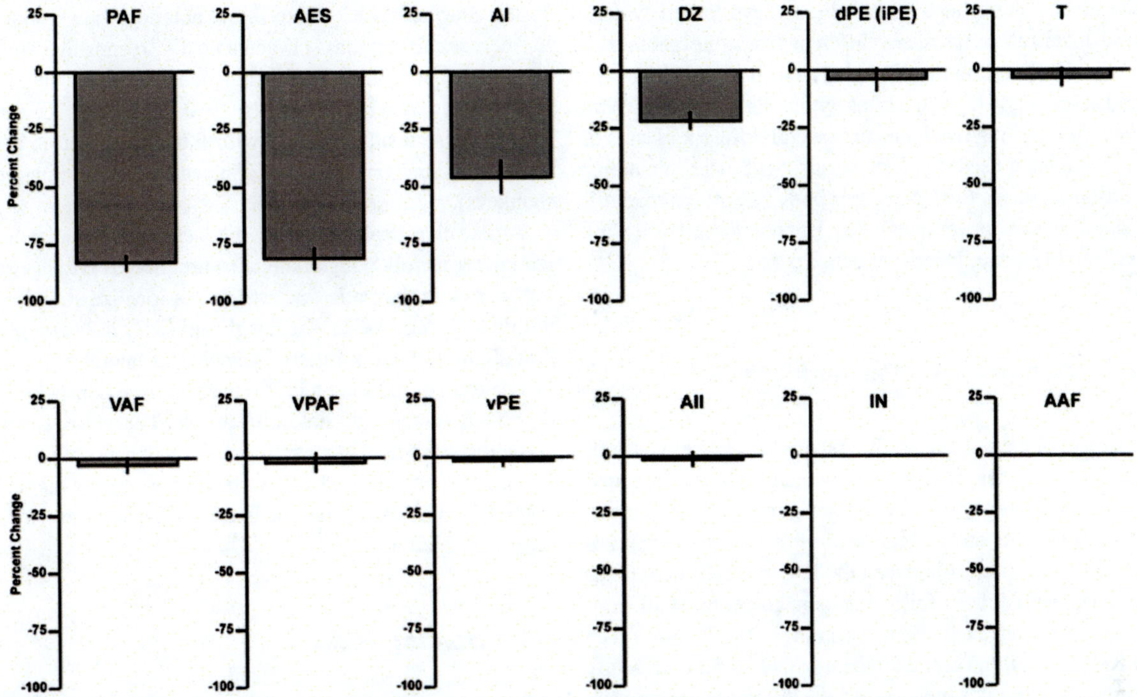

Fig. 18.9 Percent change in acoustic localization, from control levels, during bilateral deactivation of each of the 12 cortical loci examined. Percent change ≤ 1% is not shown. *Error bars* indicate standard error of the mean. Bilateral deactivations of PAF, AES, A1, and DZ result in performance levels that are significantly reduced throughout the tested field

other regions of auditory cortex does not result in any significant decreases in sound localization performance as assessed with acoustic orienting (Fig. 18.9).

During cooling deactivation, when the animals did not respond to the correct sound location, they seldom went to the central position. Instead, we found that the animals made responses to incorrect speaker locations. Bilateral deactivation of PAF or FAES yielded sound localization errors across the entire field (Malhotra and Lomber 2007). The errors identified during bilateral deactivation of either area were comparable. During bilateral deactivation of areas PAF or FAES inaccurate responses tended to stay within the same hemifield as the target location and were comprised of both undershoots and overshoots of the target position with the majority of the errors being undershoots (Malhotra and Lomber 2007). Overall, the errors made to more peripheral locations tended to be of greater magnitude with the range of errors being greatest at the most lateral locations tested.

Bilateral deactivation of AI or DZ also resulted in sound localization deficits throughout the entire field. However, unlike the profound sound localization deficit that occurs when PAF or FAES are bilaterally deactivated (Malhotra and Lomber 2007), deactivation of either AI or DZ alone produced partial and distinct deficits. For AI, bilateral deactivation resulted in sound localization performance falling from ~90% correct to ~45% correct (Malhotra et al. 2008). The errors made during the AI deactivations tended to be within ≤30° of the target and were almost always made to the same hemifield as the target. In contrast, with bilateral deactivation of DZ, sound localization performance dropped from ~90% correct to ~60% correct (Malhotra et al. 2008). The errors made during the DZ deactivations tended to be ≥60° from the target and large numbers of errors were made to the incorrect hemifield (Malhotra et al. 2008). Therefore, individual deactivation of either AI or DZ produced specific and unique sound localization deficits. The results of the present study suggest that the contributions of other cortical regions (PAF and FAES) to sound localization are more significant than either AI or DZ.

6 Task-Specific Sound Localization Deficits with Cortical Deactivations

The vast majority of previous sound localization behavioral studies have examined sound localization in animals following unilateral or bilateral ablation of primary auditory cortex alone, or primary auditory cortex combined with many non-primary areas (Table 18.1). While the regions of cortex investigated have remained fairly constant across species, the types of behavioral testing techniques have varied widely. Sound localization tasks are different from one another in terms of the reporting mechanisms, as well as the number of sound sources and their locations within the hemifields. With respect to reporting mechanisms, several different classes of tasks have been used: (1) Left versus right discriminations; (2) Orienting tasks (which include body orienting, head orienting, and eye movements); (3) Unconditioned orienting tasks; and (4) Conditioned avoidance tasks.

6.1 Left Versus Right Discriminations

Regardless of the position of the sound source within a hemifield, many studies had animals indicate whether a sound source was located within the left or right hemifield. For carnivores (both cats and ferrets) studies that required animals to discriminate sounds between the left and right hemifields found profound deficits following large ablations of all auditory cortices (Neff et al. 1956; Strominger 1969b; Kavanagh and Kelly 1987). In contrast, the animals had little difficulty distinguishing sounds presented in the left versus right hemifield following ablations restricted to primary auditory cortex (Kavanagh and Kelly 1987). Furthermore, when animals had to distinguish sounds within the left or right hemifield, profound sound localization deficits were evident when primary auditory cortex was ablated (Kavanagh and Kelly 1987). Our cooling deactivation results are in agreement with both of these observations. Overall, bilateral deactivation of primary auditory cortex severely impairs the ability to discriminate the spatial location of sounds *within* a hemifield, but not the spatial location of sounds *between* the left and right hemifields.

With respect to individual regions of auditory cortex, nearly all behavioral studies of sound localization have examined the role of primary auditory cortex, regardless of species. The results from these studies are quite consistent. In all mammals with highly developed cerebral cortices, unilateral ablations or deactivations restricted to primary auditory cortex result in spatial localization deficits restricted to the contralateral field. Furthermore, bilateral deactivation of primary auditory cortex results in sound localization deficits throughout the auditory field. These deficits following AI deactivations have been identified in cats (Jenkins and Masterton 1982; Jenkins and Merzenich 1984; Malhotra et al. 2004), ferrets (Kavanagh and Kelly 1987; Smith et al. 2004), and monkeys (Wegener 1964; Heffner and Masterton 1975). Our cooling deactivation results are in agreement with these earlier studies. However, similar results were not obtained in rats. Bilateral destruction of primary auditory cortex in the rat does not result in spatial localization errors (Kelly and Glazier 1978; Kelly 1980; Kelly and Kavanagh 1986). Furthermore, when these ablations are expanded to include all regions of acoustically responsive cortex, no significant sound localization deficits can be identified (Kelly and Glazier 1978; Kelly 1980; Kelly and Kavanagh 1986). These results contrast with those identified in carnivores (cats and ferrets) and primates (old- and new-world monkeys).

A plausible explanation for the lack of any spatial localization errors following bilateral destruction of primary auditory cortex in the rat is that rats do not localize sounds well within the lateral fields. Kavanagh and Kelly (1986) reported that albino rats are virtually incapable of lateral field sound localization of brief sounds and base their responses primarily on left versus right field differences. Therefore, since we argue that the sound localization deficits in cats are primarily within-field deficits (rather than left versus right), the results obtained from rats may not be in such sharp contrast to those for other species.

6.2 Orienting Tasks

In sound localization tasks, the most popular reporting mechanism has been to use orienting of the head or whole body towards the stimulus, which is then often accompanied with an approach to the sound source (Neff et al. 1956; Jenkins and Masterton 1982; Jenkins and Merzenich 1984; Kavanagh and Kelly 1987; Heffner and Heffner 1990; Rauschecker and Kniepert 1994; Malhotra et al. 2004). Consistent in these studies of carnivores and monkeys are the findings that unilateral deactivations of auditory cortex impair sound localization within the hemifield contralateral to the deactivated cortex, and that bilateral ablation of auditory cortex impairs sound localization throughout the entire field. We have been unable to identify any studies that have compared sound localization using head orienting versus whole-body orienting during deactivations of auditory cortex. It would be interesting to determine if the deficits using head orienting as a reporting mechanism would be as profound as studies using whole-body orienting.

6.3 Unconditioned Orienting Tasks

Both the left versus right discrimination tasks and orienting tasks required conditioned responses. However, it is also possible to study unconditioned responses to a sound source locus. For example, cats will reflexively orient their heads to a white-noise sound burst. Beitel and Kaas (1993) examined unconditioned head orienting responses to white-noise bursts following both unilateral and bilateral ablations of auditory cortex in its entirety. Beitel and Kaas (1993) found that bilateral removal of auditory cortex severely impaired the ability of cats to accurately orient to a sound source. However, in cats with unilateral ablations of auditory cortex, there were no deficits in orienting to sounds presented in either the contralateral or ipsilateral hemifields (Beitel and Kaas 1993). This result is contrary to the results using conditioned orienting tasks that found that unilateral destruction of all acoustically responsive cortex resulted in sound localization impairments in the contralateral, but not ipsilateral, hemifield (Neff et al. 1956; Thompson and Welker 1963; Neff 1968; Strominger 1969b; Whitfield et al. 1972; Casseday and Diamond 1977). Therefore, while the conditioned and unconditioned orienting deficits are consistent for bilateral ablations of auditory cortex, profound *conditioned* orienting deficits follow unilateral ablations of auditory cortex while no *unconditioned* orienting deficits follow similar unilateral ablations.

With respect to acoustic orienting work done in cats (Malhotra et al. 2004, 2008; Malhotra and Lomber 2007; Lomber and Malhotra 2008), the study of Beitel and Kaas (1993) differs not only in the behavioral measures employed, but also in the deactivation methods used. In the study of Beitel and Kaas (1993) deficit attenuation following the unilateral lesions may be a possible explanation for the lack of a sound localization deficit. The cats may have actually had a deficit immediately following the lesion. Unfortunately, it is likely that the cats were first tested weeks after the lesion, and after the deficits may have attenuated due to cortical plasticity. One of the great advantages of using reversible deactivation approaches is that the confound of cortical plasticity can be largely avoided (Lomber 1999).

6.4 Conditioned Avoidance Tasks

Conditioned avoidance tasks are also used to examine sound localization following ablations of auditory cortex. In these tasks, animals are typically trained to make or break contact with a water spout to indicate the location of a sound source. The animal learns to break contact with the water spout because it will receive a mild electric shock, through the water spout, if it fails to do so. Using this procedure an animal can be trained to perform two-choice discrimination, such as left–right discrimination or the discrimination of two positions within the same hemifield. The use of these conditioned avoidance procedures with monkeys confirms the left versus right discrimination results described earlier for cats and ferrets. Specifically, bilateral deactivation of primary auditory cortex severely impairs the ability of monkeys to discriminate the spatial location of sounds *within* a hemifield, but not the spatial location of sounds *between* the left and right hemifields (Heffner and Heffner 1990; Heffner 1997).

7 Field-Specific Contributions to Sound Localization

Our studies have shown that unilateral or bilateral deactivations of PAF, AI, DZ, or FAES resulted in profound sound localization deficits (Malhotra et al. 2004, 2008; Malhotra and Lomber 2007). However, these studies also revealed that neither unilateral nor bilateral deactivation of AAF, VPAF, AII, insular region (IN), temporal region (T), VAF, dorsal posterior ectosylvian area (dPE), intermediate posterior ectosylvian area (iPE) nor ventral posterior ectosylvian area (vPE) had any effect on the sound localization task (Malhotra and Lomber 2007). Therefore, one major conclusion that could be drawn from these earlier results is that most of auditory cortex is not necessary for accurate sound localization.

Other conclusions can also be drawn when the results from this present study are compared to earlier studies. First, while AI does play a role in sound localization, its role is not as significant as that described in earlier reports. The earliest reports implicating AI in sound localization involved large physical ablations of AI and much or all of the remaining acoustically responsive cortex (Neff 1968; Neff et al. 1956; Strominger 1969a,b; Thompson and Welker 1963). These studies reported significant sound localization deficits following large lesions in auditory cortex. Subsequent studies made smaller lesions that included AI (Jenkins and Merzenich 1984; Masterton and Diamond 1964; Riss 1959; Strominger 1969b). However, these studies also included portions, if not all, of DZ in their ablations. Even the most recent reversible deactivation studies examining the contributions of AI did not investigate the contributions of AI alone, but examined the contributions of AI together with DZ (Malhotra et al. 2004; Malhotra and Lomber 2007). Both the later ablations, studies and reversible deactivation experiments described profound sound localization deficits following lesion or inactivation of AI/DZ. Therefore, it was impossible to discern the individual contributions of AI or DZ from any previous studies. In the present study, we explicitly examined the individual contributions of both AI and DZ to sound localization behavior. The present results show that deactivations restricted to AI alone do not produce deficits that are as severe as those reported by earlier studies (Jenkins and Merzenich 1984; Malhotra et al. 2004; Malhotra and Lomber 2007; Masterton and Diamond 1964; Riss 1959; Strominger 1969b). Therefore, while primary auditory cortex does play a role in sound localization, its role may not be as significant as that described in earlier reports.

Second, PAF and FAES each play more critical roles in coordinating accurate orienting to an acoustic stimulus that either AI or DZ. Earlier studies have reported that deactivation of AI/DZ, PAF, or FAES results in sound localization deficits that reduce normal performance to chance levels (Malhotra et al. 2004; Malhotra and Lomber 2007). The present study examined AI and DZ individually and revealed that deactivation of neither area results in deficits as severe as those identified during deactivation of PAF or FAES (Malhotra et al. 2004; Malhotra and Lomber 2007). Therefore, the roles of PAF and FAES in sound localization appear to be more significant that either AI or DZ. Considering the positions of FAES and PAF in a proposed sound localization pathway in auditory cortex (Lomber et al. 2007), we hypothesize that PAF is more involved in the perceptual machinery underlying sound localization and the FAES is more involved in the audiomotor execution of sound localization behaviors.

Acknowledgments We thank Erin Woller, Shveta Malhotra, and Jeff Mellott for help with various phases of the work described. This work was supported by grants from the Canadian Institutes of Health Research, the Natural Sciences and Engineering Research Council of Canada, the Canada Foundation for Innovation, and the Hearing Foundation of Canada.

References

Adriani M, Maeder P, Meuli R, Thiran AB, Frischknecht R, Villemure J-G, Mayer J, Annoni J-M, Bogousslavsky J, Fornari E, Thiran J-P, and Clarke S (2003) Sound recognition and localization in man: specialized cortical networks and effects of acute circumscribed lesions. *Experimental Brain Research* **221**:591–604.

Beitel RE and Kaas JH (1993) Effects of bilateral and unilateral ablation of auditory cortex in cats on the unconditioned head orienting response to acoustic stimuli. *Journal of Neurophysiology* **70**:351–369.

Beneyto M, Winer JA, Larue DT, and Prieto JJ (1998) Auditory connections and neurochemistry of the sagulum. *Journal of Comparative Neurology* **401**:329–351.

Bénita M and Condé H (1972) Effects of local cooling upon conduction and synaptic transmission. *Brain Research* **36**:133–151.

Bisiach E, Luzzatti C, and Perani D (1979) Unilateral neglect, representational schema and consciousness. *Brain* **102**:609–618.

Bizley JK, Nodal FR, Parsons CH and King AJ (2007) Role of auditory cortex in sound localization in the midsagittal plane. *Journal of Neurophysiology* **98**:1763–1774.

Bowman EM and Olson CR (1988a) Visual and auditory association areas of the cat's posterior ectosylvian gyrus: cortical afferents. *Journal of Comparative Neurology* **272**:30–42.

Bowman EM and Olson CR (1988b) Visual and auditory association areas of the cat's posterior ectosylvian gyrus: thalamic afferents. *Journal of Comparative Neurology* **272**:15–29.

Bradshaw J and Rogers L (1993) *The evolution of lateral asymmetries, language, tool use, and intellect*. Academic Press, New York.

Brooks VB (1983) Study of brain function by local, reversible cooling. Reviews of Physiology. *Biochemistry & Pharmacology* **95**:1–109.

Casseday JH and Diamond IT (1977) Symmetrical lateralization of function in the auditory system of the cat: Effects of unilateral ablation of cortex. *Annals of the New York Academy of Sciences* **299**:255–263.

Cavina-Pratesi C, Goodale MA, and Culham JC (2007) fMRI reveals a dissociation between grasping and perceiving the size of real 3D objects. *Public Library of Science ONE* **5**:1–14.

Clarey JC and Irvine DRF (1986) Auditory response properties of neurons in the anterior ectosylvian sulcus of the cat. *Brain Research* **386**:12–19.

Clarke S, Bellmann A, Meuli RA, Assal G, and Steck AJ (2000) Auditory agnosia and auditory spatial deficits following left hemispheric lesions: evidence for distinct processing pathways. *Neuropsychologia* **38**:797–807.

Clascá F, Llamas A, and Reinoso-Suárez F (1997) Insular cortex and neighboring fields in the cat: a redefinition based on cortical microarchitecture and connections with the thalamus. *Journal of Comparative Neurology* **384**:456–482.

Clascá F, Llamas A, and Reinoso-Suárez F (2000) Cortical connections of the insular and adjacent parieto-temporal fields in the cat. *Cerebral Cortex* **4**:371–399.

Colavita FB (1972) Auditory cortical lesions and visual pattern discrimination in cat. *Brain Research* **39**:437–447.

Colavita FB, Szeligo FV, and Zimmer SD (1974) Temporal pattern discrimination in cats with insular-temporal lesions. *Brain Research* **79**:153–156.

Cornwell P (1967) Loss of auditory pattern discrimination following insular-temporal lesions in cats. *Journal of Comparative and Physiological Psychology* **63**:165–168.

Cornwell P, Nudo R, Straussfogel D, Lomber SG, and Payne BR (1998) Dissociation of visual and auditory pattern discrimination functions within the cat's temporal cortex. *Behavioral Neuroscience* **112**: 800–811.

Courtney SM, Ungerleider LG, Keil K, and Haxby JV (1996) Object and spatial visual working memory activate separate neural systems in human cortex. *Cerebral Cortex* **6**:39–49.

Cranford J, Ravizza R, Diamond IT, and Whitfield IC (1971) Unilateral ablation of the auditory cortex in the cat impairs complex sound localization. *Science* **172**:286–288.

Dewson J (1964) Speech sound discrimination by cats. *Science* **144**:555–556.

Dewson JH, Pribram KH, and Lynch JC (1969) Effects of ablations of temporal cortex upon speech sound discrimination in the monkey. *Experimental Neurology* **24**:579–591.

Diamond IT and Neff WD (1957) Ablation of temporal cortex and discrimination of auditory patterns. *Journal of Neurophysiology* **20**: 300–315.

Driver J and Mattingley JB (1998) Parietal neglect and visual awareness. *Nature Neuroscience* **1**:17–22.

Girden E (1939) Cerebral determinants of auditory localization. *The American Journal of Psychology* **51**:1–15.

Goodale MA and Milner AD (1992) Separate visual pathways for perception and action. *Trends in Neurosciences* **15**:20–25.

Goodale MA, Milner AD, Jakobson LS, and Carey DP (1991) A neurological dissociation between perceiving objects and grasping them. *Nature* **349**:154–156 (1991).

Goodale MA and Westwood DA (2004) An evolving view of duplex vision: separate but interacting cortical pathways for perception and action. *Current Opinion in Neurobiology* **14**:203–211.

Griffiths TD and Warren JD (2004) What is an auditory object? *Nature Reviews Neuroscience* **5**:887–892.

Heilman KM, Watson RT, and Valenstein E (1993) Neglect and related disorders. In: Heilman KM and Valenstein E (eds). *Clinical Neuropsychology*, 3rd edition. Oxford University Press, New York, pp. 296–346.

Heffner H (1978) Effect of auditory cortex ablation on localization and discrimination of brief sounds. *Journal of Neurophysiology* **41**: 963–976.

Heffner HE (1997) The role of Macaque auditory cortex in sound localization. *Acta Oto-Laryngologica. Supplementum* **532**: 22–27.

Heffner HE and Heffner RS (1990) Effect of bilateral auditory cortex lesions on sound localization in Japanese Macaques. *Journal of Neurophysiology* **64**:915–931.

Heffner H and Masterton B (1975) Contribution of auditory cortex to sound localization in the Monkey (Macaca mulatta). *Journal of Neurophysiology* **38**:1340–1358.

Horsley V and Clarke RH (1908) The structure and function of the cerebellum examined by a new method. *Brain* **31**:45–124.

Imig TJ, Reale RA, and Brugge JF (1982) The auditory cortex: patterns of corticocortical projections related to physiological maps in the cat. In: Woolsey CN (ed). *Cortical Sensory Organization*, volume 3, *Multiple Auditory Areas*. Humana Press, New Jersey, pp. 1–42.

Jasper H, Shacter DG, and Montplaisir J (1970) The effect of local cooling upon spontaneous and evoked electrical activity of cerebral cortex. *Canadian Journal of Physiology and Pharmacology* **48**:640–652.

Jenkins WM and Masterton RB (1982) Sound localization: effects of unilateral lesions in central auditory system. *Journal of Neurophysiology* **47**:987–1016

Jenkins WM and Merzenich MM (1984) Role of cat primary auditory cortex for sound-localization behavior. *Journal of Neurophysiology* **57**:819–847.

Kavanagh GL and Kelly JB (1986) Midline and lateral field sound localization in the albino rat (Rattus norvegicus). *Behavioral Neuroscience* **100**:200–205.

Kavanagh GL and Kelly JB (1987) Contribution of auditory cortex to sound localization by the ferret (*Mustela putorius*). *Journal of Neurophysiology* **57**:1746–1766.

Kelly JB (1973) The effects of insular and temporal lesions in cats on two types of auditory pattern discrimination. *Brain Research* **62**: 71–87.

Kelly JB (1980) Effects of auditory cortical lesions on sound localization by the rat. *Journal of Neurophysiology* **44**:1161–1174.

Kelly JB and Glazier SJ (1978) Auditory cortex lesions and discrimination of spatial location by the rat. *Brain Research* **145**:315–321.

Kelly JB and Kavanagh GL (1986) Effects of auditory cortical lesions on pure-tone sound localization by the albino rat. *Behavioral Neuroscience* **100**:569–575.

Knight PL (1977) Representation of the cochlea within the anterior auditory field (AAF) of the cat. *Brain Research* **130**:447–467.

Kolb B and Whishaw IQ (1996) *Fundamentals of Human Neuropsychology*, 4th edition. W.H. Freeman, New York.

Korte M and Rauschecker JP (1993) Auditory tuning of cortical neurons is sharpened in cats with early blindness. *Journal of Neurophysiology* **70**:1717–1721.

Lomber SG (1999) The advantages and limitations of permanent or reversible deactivation techniques in the assessment of neural function. *Journal of Neuroscience Methods* **86**:109–117.

Lomber SG and Malhotra S (2008) Double dissociation of "what" and "where" processing in auditory cortex. *Nature Neuroscience* **11**:609–616.

Lomber SG and Payne BR (1996) Removal of two halves restores the whole: reversal of visual hemineglect during bilateral cortical or collicular inactivation in the cat. *Visual Neuroscience* **13**: 1143–1156.

Lomber SG and Payne BR (2000) Translaminar differentiation of visually-guided behaviors revealed by restricted cerebral cooling deactivation. *Cerebral Cortex* **10**:1066–1077.

Lomber SG, Payne BR, and Horel JA (1999) The cryoloop: an adaptable reversible cooling deactivation method for behavioral or electrophysiological assessment of neural function. *Journal of Neuroscience Methods* **86**: 179–194.

Lomber SG, Payne BR, and Cornwell P (1996a) Learning and recall of form discriminations during reversible cooling deactivation of ventral-posterior suprasylvian cortex in the cat. *Proceedings of the National Academy of Sciences of the United States of America* **93**:1654–1658.

Lomber SG, Payne BR, Cornwell P, and Long KD (1996b) Perceptual and cognitive visual functions of parietal and temporal cortices in the cat. *Cerebral Cortex* **6**:673–695.

Lomber SG, Cornwell P, Sun JS, MacNeil MA, and Payne BR (1994) Reversible inactivation of visual processing operations in middle suprasylvian cortex of the behaving cat. *Proceedings of the National Academy of Sciences of the United States of America* **91**:2999–3003.

Lomber SG, Malhotra S, and Hall AJ (2007) Functional specialization in non-primary auditory cortex of the cat: areal and laminar contributions to sound localization. *Hearing Research* **229**: 31–45.

Malhotra S and Lomber SG (2007) Sound localization during homotopic and heterotopic bilateral cooling deactivation of primary and nonprimary auditory cortical areas in the cat. *Journal of Neurophysiology* **97**:26–43.

Malhotra S, Hall AJ, and Lomber SG (2004) Cortical control of sound localization in the cat: unilateral cooling deactivation of 19 cerebral areas. *Journal of Neurophysiology* **92**:1625–1643.

Malhotra S, Stecker GC, Middlebrooks JC and Lomber SG (2008) Sound localization deficits during reversible deactivation of primary auditory cortex and/or the dorsal zone. *Journal of Neurophysiology* **99**:1628–1642.

Masterton RB and Diamond IT (1964) Effects of auditory cortex ablation on discrimination of small binaural time differences. *Journal of Neurophysiology* **27**:15–36.

Mellott JG, Van der Gucht E, Lee CC, Carrasco A, Winer JA, and Lomber SG (2010) Areas of the cat auditory cortex as defined by neurofilament proteins expressing SMI-32. *Hearing Research* **267**:119–136.

Meredith MA and Clemo HR (1989) Auditory cortical projection from the anterior ectosylvian sulcus (field AES) to the superior colliculus in the cat: an anatomical and electrophysiological study. *Journal of Comparative Neurology* **289**:687–707.

Middlebrooks JC (2002) Auditory space processing: here, there or everywhere? *Nature Neuroscience* **5**:824–826.

Middlebrooks JC and Zook JM (1983) Intrinsic organization of the cat's medial geniculate body identified by projections to binaural response-specific bands in the primary auditory cortex. *Journal of Neuroscience* **1**:203–224.

Middlebrooks JC, Clock AE, Xu L, and Green DM (1994) A panoramic code for sound localization by cortical neurons. *Science* **264**: 842–844.

Middlebrooks JC, Xu L, Furukawa S, and Macpherson EA (2002) Cortical neurons that localize sounds. *The Neuroscientist* **8**: 73–83.

Mori A, Fuwa T, Kawai A, Yoshimoto T, Hiraba Y, Uchiyama Y, and Minejima T (1996) The ipsilateral and contralateral connections of the fifth somatosensory area (SV) in the cat cerebral cortex. *NeuroReport* **7**:2385–2387.

Mucke L, Norita M, Benedek G, and Creutzfeldt O (1982) Physiologic and anatomic investigation of a visual cortical area situated in the ventral bank of the anterior ectosylvian sulcus of the cat. *Experimental Brain Research* **46**:1–11.

Neff WD (1968) Localization and lateralization of sound in space. In: de Reuch AVS and Knight J (eds). *Ciba Foundation Symposium on Hearing Mechanisms in Vertebrates*. J. and A. Churchill Ltd, London, pp. 207–231.

Neff WD, Fisher JF, Diamond IT, and Yela M (1956) Role of auditory cortex in discrimination requiring localization of sound in space. *Journal of Neurophysiology* **19**:500–512.

Nelken I, Bar-Yosef O, and Young ED (1997) Responses of field AES neurons to virtual space stimuli. In: Palmer AR, Rees A, Summerfield AQ, and Meddis R (eds). *Psychophysical and Physiological Advances in Hearing*. Whurr Publisher, London, pp. 504–512.

Niimi K and Matsuoka H (1979) Thalamocortical organization of the auditory system in the cat studied by retrograde and axonal transport of horseradish peroxidase. *Advances in Anatomy, Embryology and Cell Biology* **57**:1–56.

Olson CR and Graybiel AM (1983) An outlying visual area in the cerebral cortex of the cat. *Progress in Brain Research* **58**:239–245.

Olson CR and Graybiel AM (1987) Ectosylvian visual area of the cat: location, retinotopic organization, and connections. *Journal of Comparative Neurology* **261**:277–294.

Paula-Barbosa MM, Feyo PB, and Sousa-Pinto A (1975) The association connexions of the suprasylvian fringe (SF) and other areas of the cat auditory cortex. *Experimental Brain Research* **23**: 535–554.

Payne BR and Lomber SG (1999) A method to assess the functional impact of cerebral connections on target populations of neurons. *Journal of Neuroscience Methods* **86**:195–208.

Petrides M (2000) Dissociable roles of mid-dorsolateral prefrontal and anterior inferotemporal cortex in visual working memory. *Journal of Neuroscience* **20**:7496–7503.

Phillips DP and Irvine DR (1982) Properties of single neurons in the anterior auditory field (AAF) of cat cerebral cortex. *Brain Research* **248**:237–244.

Phillips DP and Orman SS (1984) Responses of single neurons in posterior field of cat auditory cortex to tonal stimulation. *Journal of Neurophysiology* **51**:147–163.

Populin LC and Yin TCT (1998) Behavioral studies of sound localization in the cat. *Journal of Neuroscience* **6**:2147–2160.

Rauschecker JP (1997) Processing of complex sounds in the auditory cortex of cat, monkey and man. *Acta Otolaryngologica Supplement* **532**:34–38.

Rauschecker JP (1998a) Parallel processing in the auditory cortex of primates. *Audiology and Neurootology* **3**:86–103.

Rauschecker JP (1998b) Cortical processing of complex sounds. *Current Opinion in Neurobiology* **8**:516–521.

Rauschecker JP and Korte M (1993) Auditory compensation for early blindness in cat cerebral cortex. *Journal of Neuroscience* **13**: 4538–4548.

Rauschecker JP and Kniepert U (1994) Auditory localization behaviour in visually deprived cats. *European Journal of Neuroscience* **6**: 149–160.

Rauschecker JP and Tian B (2000) Mechanisms and streams for processing of "what" and "where" in auditory cortex. *Proceedings of the National Academy of Sciences of the United States of America* **97**:11800–11806.

Rauschecker JP, Tian B, Pons T, and Mishkin M (1997) Serial and parallel processing in rhesus monkey auditory cortex. *Journal of Comparative Neurology* **382**:89–103.

Ravizza R and Diamond IT (1974) Role of auditory cortex in sound localization: a comparative ablation study of hedgehog and bush-baby. *Federation Proceedings* **33**:1917–1919.

Ravizza RJ and Masterton B (1972) Contribution of neocortex to sound localization in opossum (*Didelphis virginiana*). *Journal of Neurophysiology* **35**:344–356.

Read HL, Winer JA, and Schreiner CE (2002) Functional architecture of auditory cortex. *Current Opinion in Neurobiology* **12**: 433–440.

Reale RA and Imig TJ (1980) Tonotopic organization in auditory cortex of the cat. *Journal of Comparative Neurology* **192**:265–291.

Reinoso-Suárez F (1961) Topographical atlas of the cat brain for experimental-physiological research. Merck, Darmstadt.

Ribaupierre F de (1997) Acoustical information processing in the auditory thalamus and cerebral cortex. In: Ehret G and Romand R (eds). *The Central Auditory System*. Oxford University Press, New York, pp. 317–397.

Riss W (1959) Effect of Bilateral temporal cortical ablation on discrimination of sound direction. *Journal of Neurophysiology* **22**:374–384.

Robertson LC (1989) Anomalies in the laterality of omissions in left unilateral visual field neglect: implications for an attentional theory of neglect. *Neuropsychologia* **27**:157–165.

Robertson LC, Lamb MR, and Knight RT (1988) Effects of lesions of temporo-parietal junction on perceptual and attentional processing in humans. *Journal of Neuroscience* **8**:3757–3769.

Romanski LM, Tian B, Fritz J, Mishkin M, Goldman-Rakic PS, and Rauschecker JP (1999) Dual streams of auditory afferents target multiple domains in the primate prefrontal cortex. *Nature Neuroscience* **2**:1131–1136.

Smith AL, Parsons CH, Lanyon RG, Bizley JK, Akerman CJ, Baker GE, Dempster AC, Thompson ID, and King AJ (2004) An investigation of the role of auditory cortex in sound localization using muscimol-releasing Elvax. *European Journal of Neuroscience* **19**:3059–3072.

Stepien I, Stepien L, and Lubinska E (1990) Function of dog's auditory cortex in tests involving auditory location cues and directional instrumental response. *Acta Neurobiologiae Experimentalis* **50**: 1–12.

Stecker GC, Mickey BJ, Macpherson EA, and Middlebrooks JC (2003) Spatial sensitivity in field PAF of cat auditory cortex. *Journal of Neurophysiology* **89**:2889–2903.

Stecker GC, Harrington IA, Macpherson EA, and Middlebrooks JC (2005) Spatial sensitivity in the dorsal zone (area DZ) of cat auditory cortex. *Journal of Neurophysiology* **94**:1267–1280.

Strominger NL (1969a) Subdivisions of auditory cortex and their role in localization of sound in space. *Experimental Neurology* **24**: 348–362.

Strominger NL (1969b) Localization of sound in space after unilateral and bilateral ablation of auditory cortex. *Experimental Neurology* **25**:521–533.

Thompson GC and Cortez AM (1983) The inability of squirrel monkeys to localize sound after unilateral ablation of auditory cortex. *Behavioural Brain Research* **8**:211–216.

Thompson RF and Welker WI (1963) Role of auditory cortex in reflex head orientation by cats to auditory stimuli. *Journal of Comparative and Physiological Psychology* **56**:996–1002.

Tian B and Rauschecker JP (1998) Processing of frequency-modulated sounds in the cat's posterior auditory field. *Journal of Neurophysiology* **79**:2629–2642.

Tian B, Reser D, Durham A, Kustov A, and Rauschecker JP (2001) Functional specialization in rhesus monkey auditory cortex. *Science* **292**:290–293.

Ungerleider LG and Haxby JV (1994) "What" and "where" in the human brain. *Current Opinion in Neurobiology* **4**:157–165.

Ungerleider LG and Mishkin M (1982) Two cortical visual systems. In: Ingle DJ, Goodale MA, and Mansfield RJW (eds). *Analysis of Visual Behavior*. MIT Press, Cambridge, pp. 486–549.

Updyke BV (1986) Retinotopic organization within the cat's posterior suprasylvian sulcus and gyrus. *Journal of Comparative Neurology* **246**:265–280.

Vanduffel W, Payne BR, Lomber SG, and Orban GA (1997) Functional impact of cerebral connections. *Proceedings of the National Academy of Sciences of the United States of America* **94**:7617–7620.

Wegener JG (1964) Auditory brain discrimination behavior of brain damaged monkeys. *Journal of Auditory Research* **4**:227–254.

Whitfield IC, Cranford J, Ravizza R, and Diamond IT (1972) Effects of unilateral ablation of auditory cortex in cat on complex sound localization. *Journal of Neurophysiology* **35**:718–731.

Winer JA (1992) The functional architecture of the medial geniculate body and primary auditory cortex. In: Webster DB, Popper AN, and Fay RE (eds). *Springer Handbook of Auditory Research*, volume 1: *The Mammalian Auditory Pathway: Neuroanatomy*. Springer, New York, pp. 222–409.

Winters BD, Forwood SE, Cowell RA, Saksida LM, and Bussey TJ (2004) Double dissociation between the effects of peri-postrhinal cortex and hippocampal lesions on tests of object recognition and spatial memory: heterogeneity of function within the temporal lobe. *Journal of Neuroscience* **24**:5901–5908.

Woolsey CN (1960) Organization of cortical auditory system: A review and synthesis. In: Rasmussen GL and Windle WF (eds). *Neural Mechanisms of the Auditory and Vestibular Systems*. CC Thomas, Springfield, pp. 165–180.

Woolsey CN (1961) Organization of cortical auditory system. In: Rosenblith WA (ed). *Sensory Communication*. MIT Press, Cambridge, pp. 235–257.

Yang XF, Kennedy BR, Lomber SG, Schmidt RE, and Rothman SM (2006) Cooling produces minimal neuropathology in neocortex and hippocampus. *Neurobiology of Disease* **23**:637–643.

Zatorre RJ and Penhune VB (2001) Spatial localization after excision of human auditory cortex. *Journal of Neuroscience* **21**: 6321–6328.

Chapter 19

The Evolution of Auditory Cortex: The Core Areas

Jon H. Kaas

Abbreviations

A	anterior auditory area
AAF	anterior auditory field
AchE	acetylcholinesterase
AI	primary auditory cortex
AII	second auditory cortex; nonprimary auditory cortex
CM	caudal medial field
CO	cytochrome oxidase
DC	caudodorsal field
DZ	dorsal zone
fMRI	functional magnetic resonance imaging
MG	medial geniculate body
MGm	medial division of the medial geniculate body
MGd	dorsal division of the medial geniculate body
MGv	ventral division of the medial geniculate body
P	posterior auditory area
PAF	posterior auditory field
PL	posterior lateral field
PPF	pseudosylvian field
PSF	posterior suprasylvian field
R	rostral area
RT	rostrotemporal area
SI	primary somatic sensory cortex
SSF	suprasylvian fringe
UF	ultrasonic field
VP	ventral posterior area

1 Introduction

An alternative title might be "What, if Anything, is AI?" AI, of course, is primary auditory cortex, an area of cortex that likely all mammals have. Thus, this seems a naive or a puzzling question. Yet, an important issue is hidden in this question. And this type of question was formulated long ago: "What, if anything, is a rabbit?" (Wood 1957). Classification was the issue, and it concluded that rabbits had been mistakenly classified as rodents. That view has prevailed, and rabbits are now considered Lagomorphs. Some time ago, I asked "What, if anything, is S1?" (Kaas 1983). I felt that the term S1 was being used inconsistently to refer to four areas (areas 3a, 3b, 1 and 2) in human and other anthropoid primates, while only one of these areas was considered to be S1 (area 3b) in most mammals. Again, this pertains to the issue of proper identification. All mammals appear to have a region of auditory cortex, but the descriptions of how it is organized vary across species, and even between studies on the same species. Rather than deal with the daunting task of considering the many auditory areas proposed, this review focuses on the so-called core areas, those better-defined areas that have many or most of the characteristics of primary auditory areas, such as AI. In many of the well-studied mammals, where AI has been identified, one or two other areas have been described that have many of the defining features of AI, such as inputs from the ventral division of the medial geniculate complex (MGv), pronounced architectonic features of sensory cortex, and a tonotopic organization. Ambiguous designations of AI could create confusion and misidentification, and the conclusion here is that this has happened. We wish to identify in different species for the same area originally identified as AI in cats, that is, the area in other species that is homologous to cat AI. While homologues originally were defined as features or body structures that were the same in two or more species, a modern definition more specifically requires that the similarity is the result of the retention of the feature or structure from a common ancestor, and that the resemblance is not simply the result of convergent evolution. In addition, homologous structures need not be identical or similar in all ways since they evolve different specializations in branching lines of descent. Thus, AI need not be identical across species, and other non-homologous areas might resemble AI closely because of convergent or parallel evolution. AI is therefore best identified by features

J.H. Kaas (✉)
Department of Psychology, Vanderbilt University, Nashville, TN 37240, USA
e-mail: jon.h.kaas@vanderbilt.edu

J.A. Winer, C.E. Schreiner (eds.), *The Auditory Cortex*,
DOI 10.1007/978-1-4419-0074-6_19, © Springer Science+Business Media, LLC 2011

that are unlikely to be present in other fields. One criterion would feature position relative to other areas. For AI, we add the feature of the orientation of the tonotopic gradient. While other areas may have the same tonotopic gradient as AI, they would not have both the same position relative to other areas and the same tonotopic gradient. Using this line of reasoning, the goal is to identify homologous core areas across taxa. Given the present stage of understanding, the homologues of secondary auditory areas present an even more challenging task best left for another occasion.

2 Cortical Areas Are the Larger Subdivisions of the Cortical Sheet

Cortical areas have been called the organs of the brain (Brodmann 1909). This implies that each cortical area has a unique set of functions. To perform these functions, this often meant some level of specialization of cortical cellular structure within the area. Therefore, early anatomists used histological differences in the appearance of cortical regions to identify subdivisions with presumed functional significance, the cortical areas. As cortical areas mediate function by transforming inputs and redistributing information to their outputs, each area should also be distinguished by a unique pattern of extrinsic connections. Often this includes a systematic arrangement of inputs and outputs so that an orderly map of these arrangements can be revealed within an area. For sensory areas such as the several auditory areas, this suggests an orderly representation of the peripheral receptor array leading to patterns of tonotopic or cochleotopic organization. Neurons in auditory and other sensory areas may also have other response properties that distinguish the areas. Areas are most reliably identified by a congruence of histological, connectional, and physiological distinctions (Kaas 1982). The hypothesis that an area has been identified validly by distinctive traits can be tested by inactivation, ablation, and microstimulation experiments that show that the proposed area is uniquely involved in certain brain functions.

3 The Origin of Auditory Cortex

Components of the mammalian brain stem auditory system can be found in amphibians, reptiles, and birds (Bruce 2007; Sterbing-D'Angelo 2007). However, comparative studies do not reveal how auditory cortex emerged in mammals. While the immediate ancestors of mammals are traditionally called mammal-like reptiles (Colbert and Morales 1991), present day mammals and reptiles are not closely related. Thus, early reptiles are now referred to as stem amniotes (descendants of amphibians which adapted to terrestrial life by developing an amniote egg or amniotic membranes in live-bearers). These stem amniotes formed two major clades some 320 million years ago: the Sauropsida leading to modern reptiles and birds, and the Synapsida, with only mammals surviving. Thus, nothing is known about the forebrain organization of the extinct mammal-like reptiles that preceded mammals. It is a reasonable surmise that the amniote ancestors of mammals had a dorsal cortex much like that of extant reptiles, which is widely regarded as homologous to mammalian neocortex (sometimes called isocortex) (Northcutt and Kaas 1995; Striedter 1997; Kaas 2007; Medina 2007). The dorsal reptilian cortex is a thin, with only one main layer of cells, while neocortex is thick and has six traditionally defined layers. The inputs to dorsal cortex are widespread within it (Ulinski 2007), and there are few functional subdivisions that could define regional areas. More importantly, there is no evidence that any of dorsal cortex is auditory, as most or all of the projections from the auditory thalamus terminate in the striatum or the dorsal ventricular ridge, rather than dorsal cortex (Bruce 2007; Medina 2007). How a structure like dorsal cortex might transform into a much larger, thicker, and laminated neocortex with distinct sensory areas is unknown. Thus, this review is restricted to mammalian auditory cortex. Before considering auditory cortex organization in a phylogenetic distribution of species (Fig. 19.1), we begin by reviewing proposals for how auditory cortex is organized in domestic cats, and then other studied carnivores. The justification for this is that cat auditory cortex has been the focus of many early studies in which key concepts of auditory cortical organization were developed.

4 Auditory Cortex Organization in Cats and Other Carnivores

Current understanding of auditory cortex organization in cats began when electrical stimulation of different sectors of the cochlea was used to activate auditory cortex and identify two systematic representations of the cochlea, areas AI (primary auditory cortex) and AII (nonprimary auditory cortex) (Woolsey and Walzl 1942). This early AI was somewhat larger than AI as presently construed (Fig. 19.2), and included parts of the present day AAF (anterior auditory field). This early AI was larger because the methods then used, brain-surface electrode recordings with electrical stimulation of the cochlea, were not very sensitive to reversals of tonotopic organization and other boundary markers. Early AI represented the cochlea from base to apex in a caudal to rostral direction, corresponding to a low-to-high tone frequency representation repeatedly confirmed in more modern studies of AI (e.g., Merzenich et al. 1973, 1975). Thus, for any caudorostral series of recording sites across AI, the characteristic or best frequency (the frequency of

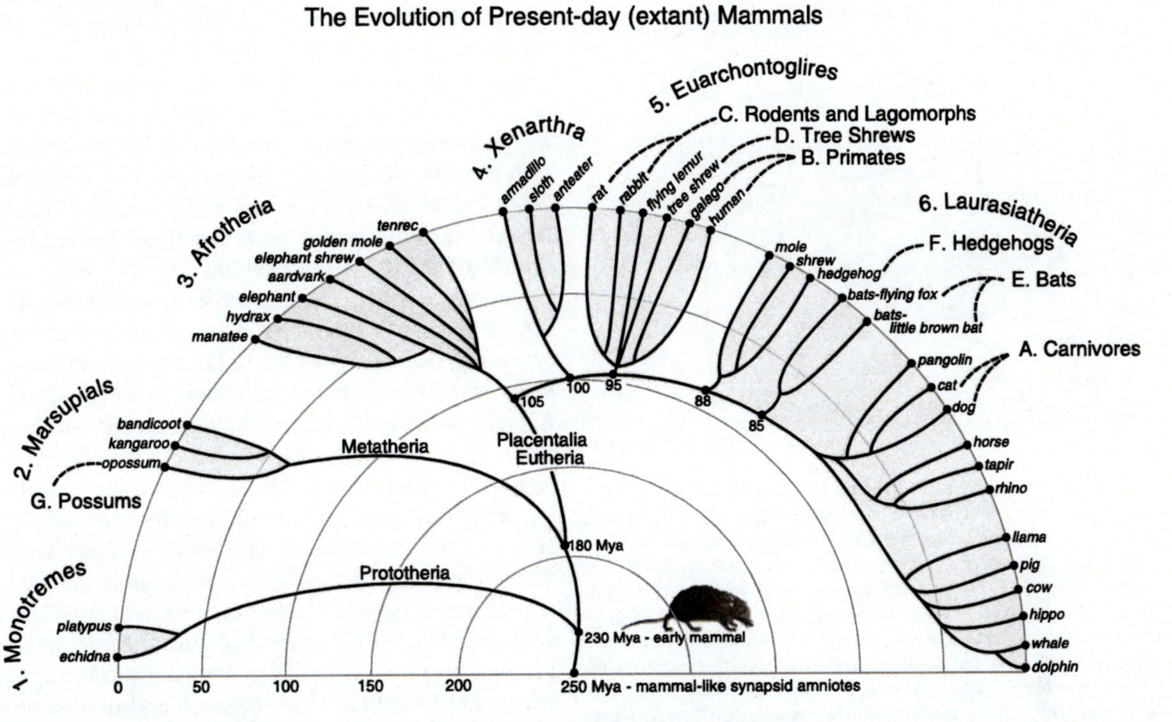

Fig. 19.1 The phyletic distribution of present-day (extant) mammals. Studies of molecular similarities divide extant mammals into six major clades or superorders (Murphy et al. 2004). *Numbers* denote the estimated times of divergence of each clade and several of its major branches from its ancestral common origin from another clade. Prototherian mammals (monotremes) thus diverged from the ancestors that gave rise to other extant mammals ~230 million years ago. The mammal-like reptiles that gave rise to mammals are designated cladistically as synapsid amniotes. Some of the mammals considered in this review are noted *(A–G)*

the tone that would activate the recorded neuron or neuron cluster at the lowest sound intensity) for activating cells progressed from low-to-high tones. In the dorsoventral axis across AI, best frequencies did not change, and this was considered the direction of isorepresentation of tone frequency. The early AI was associated with a region of architectonically distinct cortex (Rose and Woolsey 1949a) and a pattern of thalamic connections with the medial geniculate complex (Rose and Woolsey 1949b). Subsequent studies further defined the features of AI. While the general location of cat AI has been readily established physiologically, determining the precise boundaries can be difficult as adjacent areas have a tonotopic sequence bordering AI (Fig. 19.2). In addition, the neuronal properties of AI cells vary within it (Read et al. 2002), and between AI and other areas. Thus, some investigators even distinguish dorsal, central, and ventral sectors of AI, and central AI neurons have the lowest response thresholds and more regular response properties (Mendelson et al. 1997). AI subregions also vary in the spatial representation of spectral integration (Imaizumi and Schreiner 2007). Because borders of AI and other areas can be difficult to precisely locate with physiological measures, architectonic studies can be useful in delimiting AI and other fields, although AI and adjoining primary-like fields (e.g., AAF in cats) can have a similar architecture. Cytoarchitectonically, cat AI has a thick layer 4 that is densely packed with smaller neurons (Rose 1949; Winer 1984) and the middle layers are more densely myelinated and express more cytochrome oxidase (CO), parvalbumin, and acetylcholinesterase (AChE) than do non-primary cortical areas (Wallace et al. 1991). AI also has a denser staining pattern with the monoclonal antibody (CAT-301) which recognizes a cell-surface proteoglycan. AI receives a dense, topographically organized input from the ventral (principal) division (MGv) of the medial geniculate complex, which is also tonotopically organized (Winer et al. 1977; Morel and Imig 1987; Brandner and Redies 1990; Lee et al. 2004; Lee and Winer 2008a). Ipsilateral cortical connections with adjoining and other auditory areas are widespread, including AAF, AII, and P (the posterior auditory area) (Lee et al. 2004; Lee and Winer 2005, 2008b; Winer and Lee 2007). A second auditory area, AII, was also proposed and was thought to have a tonotopic organization reversed from that in AI, with high tone frequencies represented caudally and low frequencies represented rostrally (Woolsey and Walzl 1942). This erroneous assumption was likely influenced by results from

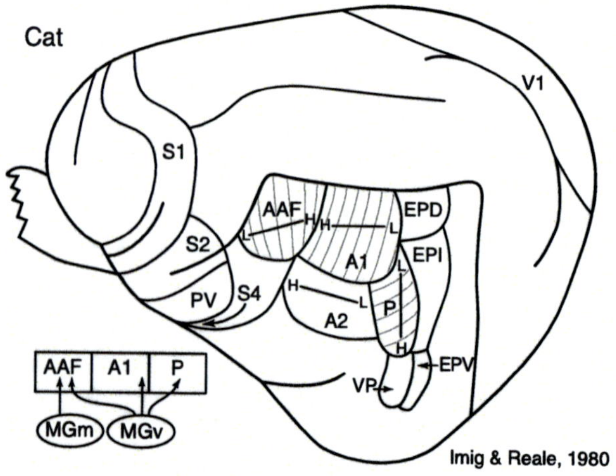

Fig. 19.2 Auditory cortex subdivisions for domestic cats on a dorsolateral view of the left cerebral hemisphere. Auditory areas include the primary area (AI), the second area (AII or A2), the anterior auditory field (AAF), the posterior auditory field (P), the ventroposterior field (VP), and dorsal (EPD), intermediate (EPI) and ventral (EPV) divisions of auditory cortex of the ectosylvian gyrus. For reference, primary visual (V1) and somatic sensory (S1) are outlined, as well as the second (S2), fourth (S4), and parietal ventral (PV) somatic sensory fields. *Boxes* (*lower left*) indicate that AAF, AI, and P all receive inputs from the ventral nucleus of the medial geniculate complex (MGv). AAF may receive more input than AI from the medial nucleus (MGm) (Imig and Reale 1980)

cortex now considered to be outside of AII. The current, smaller extensive AII has a tonotopic organization parallel to that of AI, ranging from low to high frequencies in a caudorostral sequence (Fig. 19.2) (Schreiner and Cynader 1984). AII has a less precise tonotopic organization than AI, neurons with broader frequency tuning and a higher response threshold, and a marked reduction in the architectonic features pronounced in AI and other primary cortical areas. Projections from the ventral nucleus of the medial geniculate complex to AII are sparse (Morel and Imig 1987; Lee and Winer 2008a). Overall, cat AII would be considered as part of the auditory belt (nontonotopic areas) in primates, while AI would be part of the primary or primary-like (tonotopic) core (Kaas and Hackett 2000). Cortex on the dorsal AI border, the poorly defined suprasylvian fringe (SSF) or dorsal auditory zone (DZ) area, in the suprasylvian fissure, also has belt-like physiological and anatomical features (Wallace et al. 1991; He and Hashikawa 1998), and would be considered belt cortex in primates. The part of the SSF immediately adjoining AI has been redefined as the dorsal zone (DZ) (Reale and Imig 1980; Stecker et al. 2005).

In contrast to AI bordering areas AII and SSF, the cortex rostral to AI is primary-like. The anterior auditory field (AAF) (Fig. 19.2) includes cortex that was originally considered part of AI but is distinguished by a tonotopic organization that is a reversed or a mirror image of that in AI (Knight 1977; Imaizumi et al. 2004). Thus, AAF represents tone frequencies from high to low in a caudorostral sequence, and isorepresentation lines course dorsoventrally as in AI. AAF has architectonic features like those of AI, and receives dense projections from MGv, although lighter than those to AI (Imig and Morel 1985; Huang and Winer 2000; Lee et al. 2004; Lee and Winer 2005, 2008a). Both AI and AAF also receive other significant inputs from the dorsal and medial (magnocellular) divisions of the medial geniculate complex. The interconnections between AI and AAF place them at the same hierarchical level of cortical processing (Rouiller et al. 1991). Finally, the response properties of AAF neurons resemble those in AI (Knight 1977; Eggermont 1998; Imaizumi et al. 2004) but are more broadly tuned for frequency, and have shorter response latencies. AI and AAF both are primary areas, processing subcortical auditory inputs in parallel, as originally postulated (Knight 1977). However, the two fields are functionally distinct since the deactivation of AI, but not AAF, results in sound localization deficits in the contralateral auditory field (Malhotra and Lomber 2007). Cortex on the caudoventral border of AI, the posterior area (P) (Fig. 19.2), is also tonotopically organized (Reale and Imig 1980). The tonotopic organization of field P reverses from that in AI, with low tones represented next to the dorsal part of AI and high tones in ventral P. The orientation of P is thus rotated so that isofrequency lines are roughly caudorostral in orientation. The response properties of P neurons are primary-like, but less so than in AI (Phillips and Orman 1984). Their response latencies are longer than those in AI and AAF, and neurons may be more involved in coding stimulus intensity. Thalamic inputs include those from MGv, and from other divisions of the medial geniculate complex (Morel and Imig 1987; Huang and Winer 2000; Lee and Winer 2008a). The architectonic features of P have not been described. P has some of the characteristics of a primary cortical field, and may be part of a primary-like core, but this is less certain than for AI and AAF.

Several other auditory cortical areas have been proposed for cats (Lee et al. 2004), including the ventral posterior area (VP) (Fig. 19.2). VP represents tones from high to low in a dorsoventral sequence (Imig and Reale 1980), and the area receives significant inputs from MGv (Huang and Winer 2000; Lee and Winer 2008a). Thus, VP has some of the features of primary cortex, although it is widely considered to be a secondary area. Auditory cortex subdivisions of the posterior ectosylvian gyrus (EPD, EPI, and FPV) all appear to be secondary or higher-order auditory fields.

In summary, cat auditory cortex consists of a core of two, or possibly three, primary or primary-like fields surrounded by a fringe or belt of secondary or higher-order fields. The question addressed next is how the organization proposed for cats compares to that proposed for auditory cortex in other mammals, beginning with other carnivores.

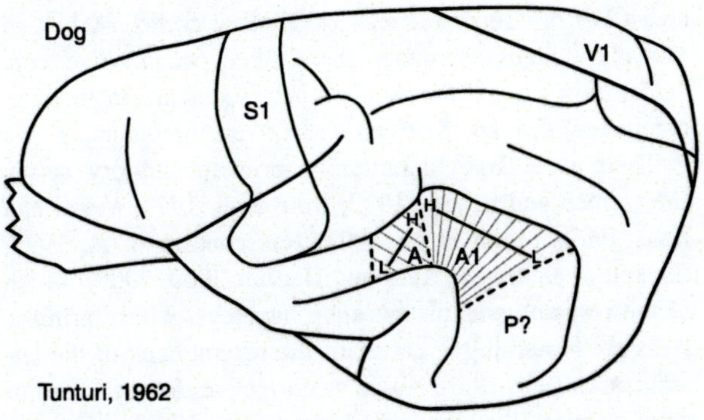

Fig. 19.3 Auditory cortex in dogs. Two areas were defined; one corresponding to AI and the other to AAF of cats (Tunturi 1962). The region (P?) may partly correspond to the posterior area (P) of cats, and was also responsive to sounds. Conventions as in Fig. 19.2 (Tunturi 1962)

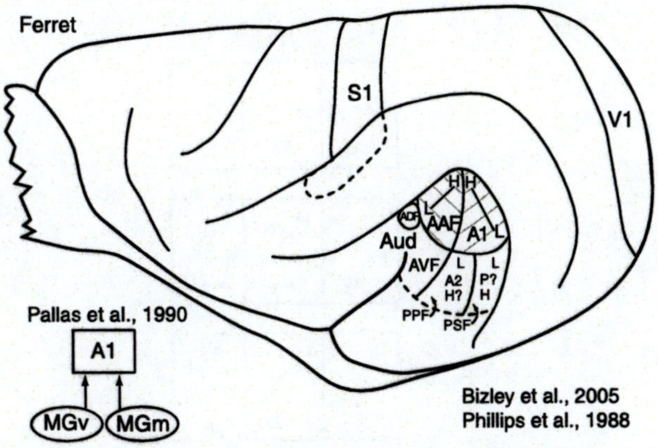

Fig. 19.4 Auditory cortex in ferrets. Both AAF and AI have been identified, and homologues of P and A2 (AII) have been suggested, a posterior pseudosylvian field (PPF) and a posterior suprasylvian field (PSF); the anterior dorsal (ADF) and the anterior ventral (AVF) fields are also noted. Conventions as in Fig. 19.2 (Pallas et al. 1990; Bizley et al. 2005; Phillips et al. 1988)

In addition to cats, auditory cortex organization has been studied in dogs and ferrets, although not to the same extent. Evoked responses to different tone frequencies in dog auditory cortex, recorded with surface electrodes, provided early evidence for areas corresponding to AI and AAF in dogs (Tunturi 1962). As in cats, frequencies were represented from low to high in a caudorostral sequence in AI and in a rostrocaudal sequence in AAF (Fig. 19.3), and there was evidence for an auditory region caudal to AI, which could correspond to the cat posterior area P. Lesions of the medial geniculate complex produced fiber degeneration in AI and AAF (Tunturi 1970). Subsequent studies using injections of retrograde tracers found that AI and the AAF regions receive major inputs from MGv, and that P and AAF have connections with AI (Kosmal 2000; Malinowska and Kosmal 2003). Thus, there is good evidence for AI and AAF in dogs, and other areas, including P, may exist.

The organization of carnivore auditory cortex has also been studied in ferrets (Fig. 19.4) and two primary-like fields, AI and AAF, have been identified (Kelly et al. 1986; Phillips et al. 1988; Shamma et al. 1993; Kowalski et al. 1995; Wallace et al. 1997; Nelken et al. 2004; Bizley et al. 2005). As in cats and dogs, these areas are tonotopically organized, but they are not simple mirror reversals of each other. Unlike cats and dogs, AI in ferrets represents low-to-high frequencies in a ventrodorsal direction with a rostralward slope, while AAF represents low-to-high frequencies in a ventrodorsal direction with a caudalward slope, as if AI and AAF were folded at their dorsal junction and shared a longer common border, with similar tonotopic progressions. Neurons in both fields were primary-like and responded well to pure tones, with narrow tuning curves at characteristic frequency (Bizley et al. 2005). AAF neurons had slightly shorter response latencies, and similar or slightly broader frequency turning curves (Kowalski et al. 1995; Bizley et al. 2005) as in cats. Although the thalamic connections of these areas have not been studied in detail, AI receives inputs from MGv and from other divisions (Pallas et al. 1990; Pallas and Sur 1993). AI and AAF are both densely myelinated in ferrets (Wallace et al. 1997), as are primary areas in other mammals. Layer 4 of AAF and AI also has a koniocellular appearance in Nissl preparations and a dense expression of cytochrome oxidase (Bajo et al. 2007). In addition, AI, AAF, and a posterior area are more metabolically active than other areas, as shown by deoxyglucose utilization (Wallace et al. 1997). Finally, only AAF and AI project to the tonotopically organized central nucleus of the inferior colliculus (Bajo et al. 2007). Thus, AAF and AI have been identified in ferrets, and resemble primary sensory cortex. They form two separate tonotopic gradients that join dorsally and drift apart ventrally.

Other auditory fields besides AAF and AI have been proposed in ferrets (Bizley et al. 2005), with two fields immediately ventral to AI, a posterior pseudosylvian field (PPF), and a posterior suprasylvian field (PSF). PPF was thought to be homologous to cat AII, while PSF to the cat field P. PSF is weakly tonotopic, with neurons having longer response latencies than AI or AAF (Bizley et al. 2005). PSF is also referred to as the ventral posterior area, VP (Wallace et al. 1997), a term that can be confused with the differently located cat ventroposterior field (Fig. 19.2). Ferret PSF expresses less myelin than AI and AAF, but more than other auditory areas. PSF also has a glucose uptake level (deoxyglucose) comparable to that of AAF and AI (Wallace et al. 1997). Both AAF and PSF are reciprocally connected to AI (Wallace and Bajwa 1991). Thus, PSF has primary-like features, but they are not as marked as in AI or AAF. PSF (or VP) is the likely homologue of the cat area P. Other auditory areas lacking the characteristics of primary sensory

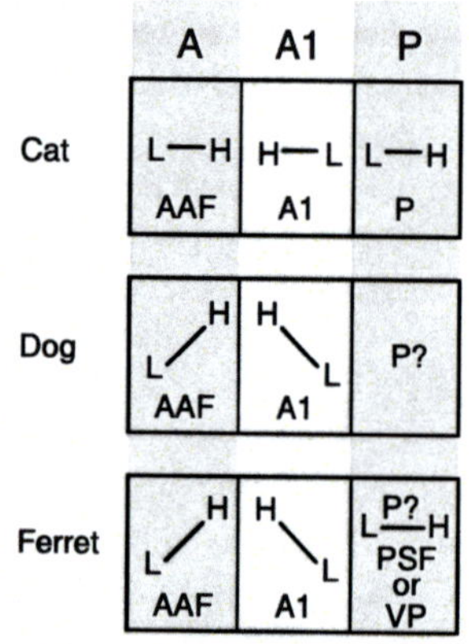

Fig. 19.5 Tonotopic gradients for A (AAF), AI, and P in carnivores. L-H, gradients of tonotopic organization from low-to-high frequencies. Area P in ferrets has been called the posterior suprasylvian field (PSF) or the ventroposterior field (VP). P? reflects a proposed change in nomenclature or uncertainty about the presence of P

cortex have also been proposed for ferrets, including the anterior dorsal (ADF), and the anterior ventral (AVF) fields, both ventral to AAF (Bizley et al. 2005).

Cats, dogs, ferrets, and presumably other carnivores have two primary-like auditory areas, AAF and AI, and a third field with lesser primary-like characteristics, P (Fig. 19.5). These fields form a rostrocaudal sequence of tonotopic representation that reverses at high or low tone boundaries, but this pattern is distorted in ferrets where AI and AAF fold on each other. In these carnivores, P extends ventrally from the caudal margin of AI. Auditory areas surrounding these fields are secondary in nature, and constitute an auditory belt, while AAF, AI, and possibly P form the auditory core. As there are at least two primary fields in cats and other carnivores, criteria for identifying them as homologous across taxa are needed. Clearly not just any tonotopically organized area can be assumed to be AI.

5 Primate Auditory Cortex

It might seem illogical to first compare carnivores (superorder Laurasiatheria) to primates (superorder Euarchontoglires) (Fig. 19.1), but the monkey organization of auditory cortex was a focus of early research that soon followed studies on dogs and cats, so that the concepts of cat cortical organization were applied to monkeys. Early studies in monkeys identified AI, and an adjoining region of cortex was termed AII (Woolsey et al. 1971). Subsequent investigators abandoned the concept of AII and retained a modified AI . As in carnivores, another primary-like area was identified, the rostral area (R), and a further, rostrotemporal area (RT) has some features of primary cortex. A belt of secondary fields surrounds these three primary-like core fields. How do the fields in primates compare to those in carnivores? Can any fields be regarded as homologous?

There is a large literature on primate auditory cortex (Merzenich and Brugge 1973; Imig et al. 1977; Morel and Kaas 1992; Morel et al. 1993; Rauschecker et al. 1995; Hackett et al. 1998a; Kaas and Hackett 2000, 2005). Much of the research was on macaque monkeys, whose primary areas are buried in the cortex of the ventral bank of the lateral sulcus (Fig. 19.6). Auditory cortex consists of a core of three primary-like areas which are tonotopically organized, respond well to pure tones, receive input from the MGv and other divisions of the medial geniculate complex, and resemble primary auditory cortex architectonically (Merzenich and Brugge 1973; Morel et al. 1993; Kosaki et al. 1997; Hackett et al. 1998a,b, 2001). The core areas project to the belt areas, and the belt to the parabelt (Galaburda and Pandya 1983; Morel et al. 1993; Hackett et al. 1998b; Jones 2006). Of the belt areas, the caudomedial area (CM) is unusual in having architectonic features intermediate to those of the core and those of the belt (Hackett et al. 2001; de la Mothe et al. 2006a). However, CM may depend on AI input for its tonotopic organization (Rauschecker et al. 1997), and many CM neurons are responsive to somatic sensory as well as auditory stimuli (Schroeder et al. 2001; Fu et al. 2003). It has been postulated that cortex in the medial belt adjoining AI was AII (Woolsey 1971), but that cortex is now included in the medial belt areas. The organization of the primate auditory core can be considered further by comparing the conclusions of various investigators in different monkeys and prosimian galagos. The proposed organization of the core auditory cortex in four species of monkeys (Fig. 19.7) shows Old World macaque monkeys have three core areas, with AI and R forming mirror reversals of each other in tonotopic organization (Fig. 19.7a). RT may form a third reversal, but this has not been fully established in macaques. New World owl monkeys (Fig. 19.7b) have a similar arrangement of three core areas, and a lateral part of RT represents low tones (Imig et al. 1977; Morel and Kaas 1992; Kaas and Morel 1993; Recanzone et al. 1999). Auditory cortex in New World marmoset monkeys (Fig. 19.7c) shows extensive evidence for a representation of high-to-low tones in a caudorostral direction that conforms to AI (Aitkin et al. 1986; Luethke et al. 1989; Kajikawa et al. 2005; Philibert et al. 2005), with evidence for a rostral area (R), and a rostrotemporal area (RT) (Bendor and Wang 2005). R represents low-to-high tones progressing from the AI border; RT represents high-to-low tones from the RT border. Area CM has been found on the caudomedial AI border (Kajikawa et al. 2005). Although the tonotopic organization of CM in marmosets mirrors that of AI, CM does not have core architectonic features, and its neurons are often bisensory and receive inputs from dorsal

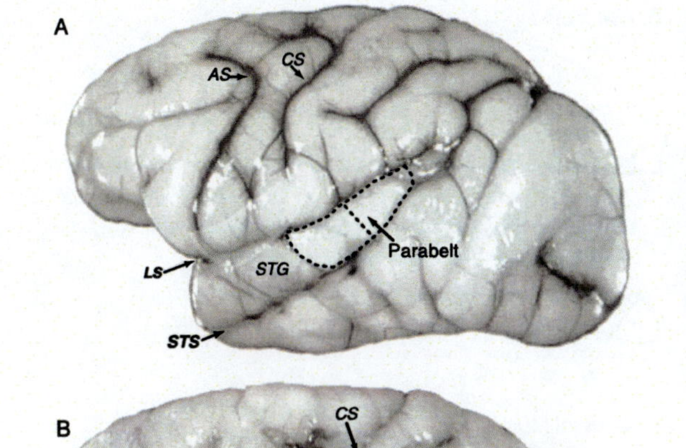

Fig. 19.6 The locations of primary and secondary auditory areas in the cortex of macaque monkeys. **a** The primary areas are within the ventral bank of the lateral sulcus, and are not apparent in this lateral view of the intact brain. Only the parabelt, a third level of auditory processing, is apparent. The lateral sulcus (LS), superior temporal sulcus (STS), and the central sulcus (CS) are indicated for reference. **b** Cortex of the upper bank of the lateral sulcus has been removed (*dashed line*) to reveal the auditory core and belt on the lower bank of the lateral sulcus. The insula (INS) is an island of cortex between the two banks. **c** A schematic of auditory cortex organization. A core of primary-like areas includes AI, a rostral area (R), and a rostrotemporal area (RT). Each of these areas is tonotopically organized from low (L) to high (H) frequencies. *Lines* of isorepresentation are shown for AI and R. The core is surrounded by a belt of secondary areas denoted by location: CL, caudolateral area; CM caudomedial area; ML, middle lateral area; RM, rostromedial area; AL, anterolateral area; RTL, lateral rostrotemporal area; RTM, medial rostrotemporal area. The lateral parabelt, a third level of processing, has been divided into rostral (RPB) and caudal (CPB) zones. Many of the belt areas are at least crudely tonotopically organized (Kaas and Hackett 2000)

and medial divisions of the medial geniculate complex rather than the ventral division.

New World squirrel monkeys have been the subjects of microelectrode mapping studies (Fig. 19.7d), and the area explored in detail was termed AI (Cheung et al. 2001; Cheung 2005; Godey et al. 2005), though the region identified had a pattern of tonotopic organization (low-to-high tones in a caudorostral direction) like that of R rather than AI. There were primary-like areas both rostral and caudal to the proposed AI (Cheung et al. 2001). It seems possible that R was identified as AI, and that squirrel monkeys have AI, R, and RT, as do other monkeys.

An auditory core has been described in prosimian galagos (Brugge 1982), with AI having the tonotopic organization expected for primates, and an area R with a reversed tonotopic organization, as expected (Fig. 19.8). A posterior lateral field (PL) had a mirror reversal tonotopic organization to that in AI and may correspond to the CM field of macaque monkeys, which is intermediate to core and belt in response and architectonic characteristics. The evidence for CM in galagos and in both New World and Old World monkeys suggests that it exists in all or most primates.

Less is known about auditory cortex organization in apes and humans. Architectonically, the chimpanzees' and humans' core has the same elongated shape as that in macaque monkeys (Hackett et al. 2001). This suggests that the same three divisions of the core exist in these primates. Functional imaging (fMRI) studies in humans that reveal cortical regions activated by different frequencies, provide evidence for two tonotopic maps in the architectonic core that form mirror-image representations reversing at a low frequency border. Talavage et al. (2004) proposed that the medial auditory koniocortex defined by others (Galaburda and Sanides 1980) corresponds to macaque area R, while lateral koniocortex corresponds to AI. Both regions had been considered subfields of AI. However, various investigators have delimited human primary auditory cortex (koniocortex) in different ways, usually as a region smaller than original descriptions (Brodmann 1909) of area 41 (Hackett 2002; Talavage et al. 2004; Sweet et al. 2005).

In summary, studies in primates recognize a core of two or three primary areas (Fig. 19.9) that include an AI and a very similar rostral area R. The similarities in neuron response properties in AI and R are so great that it is likely that area R has been mistaken for AI in squirrel monkeys. In other studies, some of R may have been included in AI.

A critical question implicit in the discovery of three core-like primate areas (RT, R, and AI) is how these compare to the core-like areas in carnivores (AAF, AI, and P). Note that monkey AI has a caudorostral tonotopic organization from high to low, while in carnivores the high-to-low tonotopic gradient is rostrocaudal. If tonotopic gradients are stable in evolution, cat AI is more like area R than AI of monkeys.

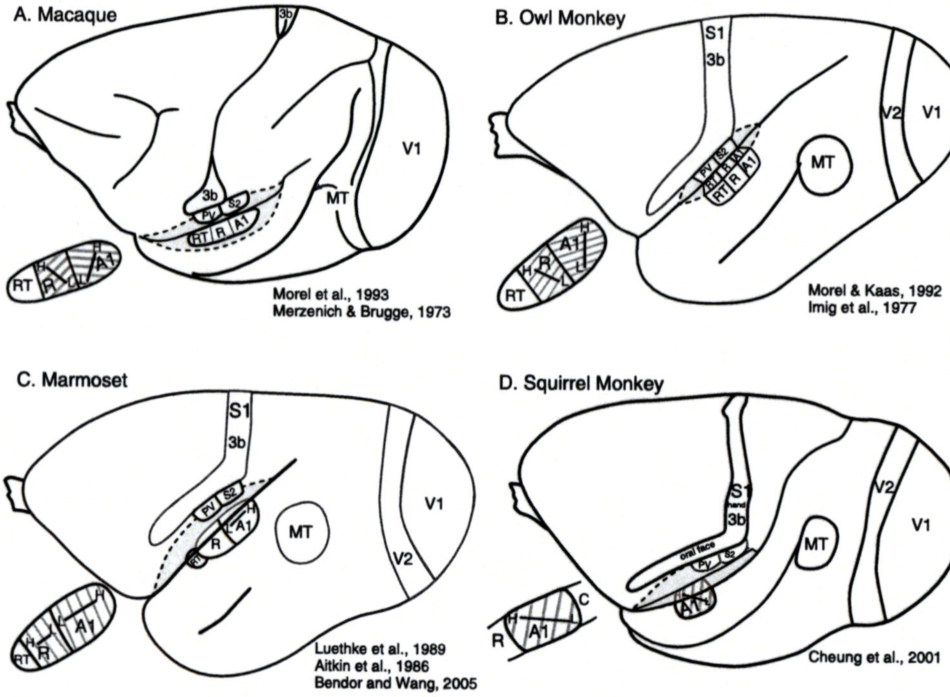

Fig. 19.7 Core auditory areas in monkeys. Areas on the hidden lower bank of the lateral sulcus are shown on dorsolateral views with the sulcus partly opened to reveal them. Figures are based largely on the results of studies cited in the text. **a** Macaque brain showing areas RT, R, and AI. Tonotopic patterns of representation and lines of isorepresentation for AI and R are on the *lower left*. The organization of RT has not been fully determined. Exposed parts of area 3b (S1), which is largely in the central sulcus, somatic sensory areas S2 and PV, visual area MT in the superior temporal sulcus, and V1 are shown for reference. **b** Core areas in owl monkeys. Conventions as in (**a**). **c** Core areas in marmosets. **d** Core areas in squirrel monkeys. The area identified as AI may actually be R, with the rostral region (R) corresponding to RT and the caudal region (c) to AI (see text). Based on: Morel et al. 1993; Merzenich and Brugge 1973; Morel and Kaas 1992; Imig et al. 1977; Luethke et al. 1989; Aitkin et al. 1986; Bendor and Wang 2005; Cheung et al. 2001

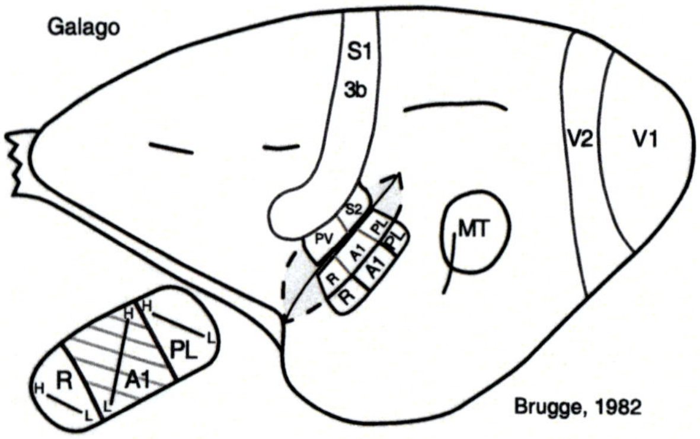

Fig. 19.8 Core auditory areas in galagos, a prosimian primate. PL, posterior lateral area. Other conventions as above (Fig. 19.7). By position and tonotopy, PL is likely to correspond to area CM of other primates (Brugge 1982)

If area R of squirrel monkeys can be misidentified as AI of monkeys, perhaps area R of monkeys is homologous with AI of cats (Kaas 2005). However, an argument against this is that the expansion of the monkey temporal lobe has rotated auditory core nearly 180° reversing the monkey tonotopic relationship of AI to that of cats (Jones 2006).

One way to further evaluate the premise of the monkey AI rotation is to visualize AI and other auditory areas in these species relative to somatic sensory and visual cortex on flattened, surface views of cortex (Fig. 19.10) where owl monkey (Fig. 19.10a) and cat (Fig. 19.10b) neocortex have been flattened manually and histologically processed to identify primary cortical areas. The core is rotated by an expansion of monkey temporal cortex so that the long axis of the core becomes more vertical (mediolateral) than in cats. RT is rotated further forward by the lateral fissure. With the expanded temporal cortex and the presumed rotation of the auditory core, monkey AI attains a high-to-low tonotopy comparable to cat AI, and AI would be the most rostral core field, while RT would be the most caudal. The argument from relative positions suggests that if AI of monkeys is homologous to cat AI, then monkey area R is homologous with cat area P, and possibly monkey area RT of monkeys is homologous with cat VP. This leaves the puzzle of cat area

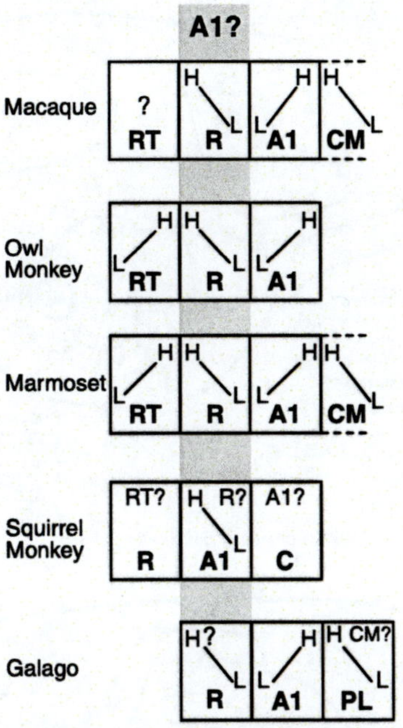

Fig. 19.9 A summary of the proposed organization of core auditory areas in primates. The relative position and the tonotopic organization of squirrel monkey area 'AI' better fits that of R

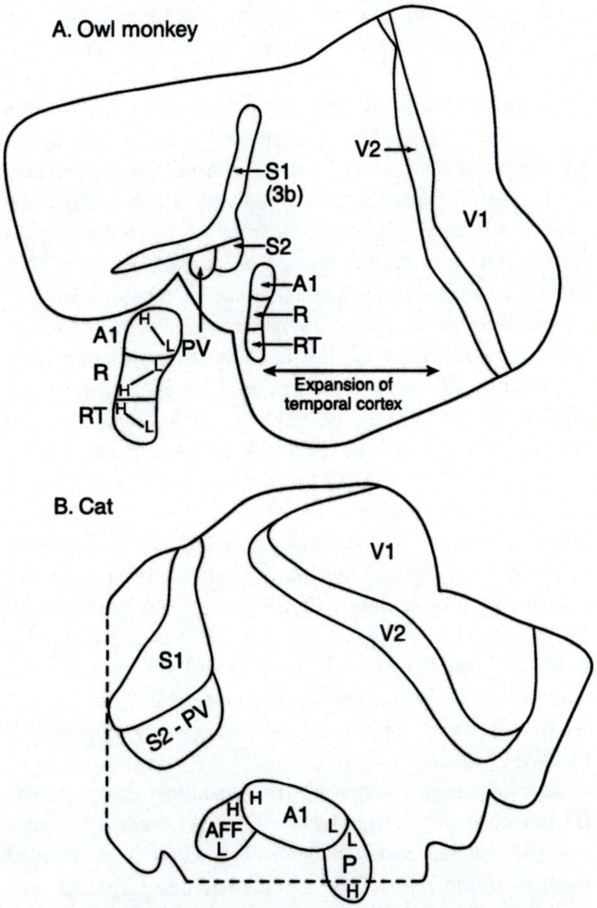

Fig. 19.10 Auditory core areas of owl monkeys (**a**) and cats (**b**) on surface views of the flattened neocortex. **a** In owl monkeys, areas AI, R, and RT are shown in temporal cortex, and again on the *lower left* so that tonotopic patterns of representation from low (L) to high (H) frequencies can be shown. **b** For cats, auditory areas **a** (AAF), AI, and P are shown in a similar manner. The *dashed line* in **b** is where some cortex was removed. The flattened owl monkey cortex was based on prior work (Tootell et al. 1985) as is the flattened cat cortex (Olavarria and Van Sluyters 1985). Somatic sensory areas and visual areas are indicated for reference

AAF, which certainly is a primary field. Is area AAF a homologue of monkey area CM, meaning that AAF and CM have rotated relative to AI and somatic sensory cortex in different directions (counterclockwise while anchored to cat AI and clockwise in monkey). While such rotations seem possible, other major changes would have also occurred, as CM does not have the primary features of AAF. Most notably, CM appears to depend on AI input for its tonotopic organization, while AAF does not, and CM does not receive input from MGv, as primary auditory areas do. Other parallels between CM and AAF would need to be considered. Evidence for homologues depends not only on the similarities between species, but also on the cladistic distribution of the characters (auditory areas) under consideration. Thus, the organization of auditory cortex in the well-studied rodents is considered next, then that of auditory cortex in other mammals.

6 Auditory Cortex in Rodents and Lagomorphs

Lagomorphs (rabbits, hares, and pikas) and rodents are sister orders in the clade Glires (Fig. 19.1), which diverged from other placental mammals (Asher et al. 2005) ~67 million years ago, and lagomorphs diverged from rodents over the next few million years. The rodent radiation includes several distinct groups. There is information on the organization of auditory cortex in South American Caviomorphs (guinea pigs, chinchillas, and degus), Muroides (rats, mice, hamsters, and gerbils), and Sciuromorphs (grey squirrels). Auditory cortex organization has also been studied in domestic rabbits.

The Mongolian gerbil is sensitivity to low frequencies, has an accessible cochlea and central auditory structures, and is robust as a laboratory animal. In microelectrode mapping, 2-deoxyglucose and other experiments, the tonotopic organization of several auditory cortex divisions have been determined (Thomas et al. 1993; Scheich and Zuschratter 1995; Goldschmidt et al. 2004). A primary auditory area (AI), with a tonotopic gradient from low-to-high frequencies in a caudorostral direction and an anterior auditory field (AAF), with a reversed tonotopic organization was defined (Fig. 19.11a), much like the AI and AAF gradients in cat (Fig. 19.2). Two fields were defined caudal to AI, a dorsal posterior field (DP) with a tonotopic organization of concentric rings from a low tone perimeter to a high tone center, and a ventral posterior field (VP) with a low-to-high frequency sequence rostrocaudally from the low frequency AI border. AAF and AI have the densely packed layer 4 of cells that characterized primary auditory cortex, more myelin than surrounding fields, dense immunoreactivity for parvalbumin, and a distinctive laminar banding pattern when reacted for the neurofilament protein labeled by SMI-32 antibody (Budinger et al. 2000a). VP has some of these features, but so less than in AI and AAF. Both AI and AAF receive dense inputs from MGv, while AAF also receives substantial inputs from the medial nucleus (MGm) (Budinger et al. 2000b). Areas DP and VP received input from MGv, MGm, and the dorsal nucleus (MGd). An auditory belt ventral to these did not appear to be tonotopically organized, nor was a dorsal fringe area. The ventromedial field (VM) is in the relative position of cat AIIs. Gerbils have an auditory core of AAF and AI, and perhaps a DP-VP region. Gerbil AAF, AI, and VP have the relative positions and tonotopic organizations of cat AAF, AI, and P.

Rats have been a common target of auditory cortex studies (Polley et al. 2007), and have a large AI flanked by anterior (A) and posterior (P) fields (Fig. 19.11b). A detailed microelectrode map of AI found that this large AI represents low-to-high frequencies caudorostrally (Sally and Kelly 1988). AI and adjoining posterior (P) and anterior (A) fields have been mapped in microelectrode recordings (Doron et al. 2002; Rutkowski et al., 2003; Kalatsky et al., 2005; Polley et al., 2007) and optical imaging (Kilgard and Merzenich 1999; Kalatsky et al. 2005) experiments. Fields A and P have mirror reversals of the AI representation and the three fields are within the architectonically defined auditory cortex core (Doron et al. 2002) and receive input from MGv (Ryugo and Killackey 1974; Horikawa et al. 1988; Roger and Arnault 1989; Clerici and Coleman 1990; Romanski and LeDoux 1993). AI and A cells have short latency responses, while P neurons have longer latencies and less evidence of a tonotopic gradient (Pandya et al. 2008; Polley et al. 2007). Non-primary areas in rat abut the borders of A, AI, and P, but are not well established. They include a ventral secondary belt (Fig. 19.11b), part of which is delineated as an anterior ventral area (Horikawa et al. 1988) or a ventral area (Donishi et al. 2006) ventral to the anterior field, and a posterior dorsal area (PD) dorsal to caudal AI (Horikawa et al. 1988). A supra-rhinal auditory field (Polley et al. 2007) was renamed from earlier work (Kalatsky et al. 2005). The ventral auditory field and the supra-rhinal auditory field appear to be tonotopically organized.

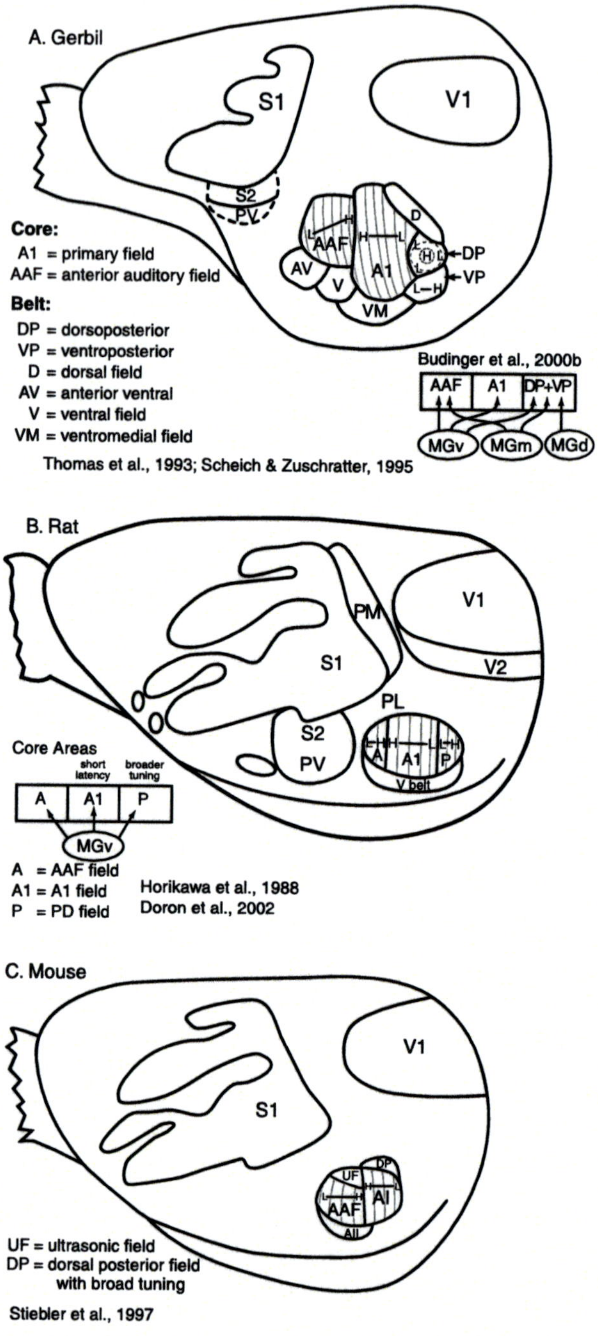

Fig. 19.11 Auditory cortex organization in Muroide rodents (gerbils, rats, and mice). **a** Some auditory areas proposed for gerbils. Gradients of tonotopic organization are indicated for high (H) to low (L) frequencies. Area DP may have a complex tonotopic organization with low tones represented along the periphery and higher frequencies in the center. Areas are identified on the *lower left* and connections with subdivisions of the medial geniculate complex are noted on the *lower right*. Conventions as in previous figures. **b** Auditory areas proposed for rats. **c** Auditory areas proposed or mice. Based on: Budinger et al. 2000b; Thomas et al. 1993; Scheich and Zuschratter 1995; Horikawa et al. 1988; Doron et al. 2002; Stiebler et al. 1997

Rats thus have at least two primary fields, AI and A (or AAF); A third field, P (or PAF) has some features of a core area, but somewhat more broadly tuned cells, longer response latencies, and less pronounced tonotopy.

Mice have been less studied than rats, and two primary areas have been described, AI and AAF (Stiebler et al. 1997), which are tonotopically organized mirror-image representations reversing at the high frequency common border (Fig. 19.11c). Part of this border represents frequencies >45 kHz (Stiebler et al. 1997) and has been called the ultrasonic field (UF). Such a specialization may include parts of both AI and AAF, and may also occur in rats (Polley et al. 2007). In mice, UF receives MGv input, as does AI and AAF (Hofstetter and Ehret 1992). AI and AAF may be in the koniocortical architectonic area 41 (Caviness 1975). Mice also have a cortical zone ventral to AAF with broadly tuned neurons that rapidly habituate and which is designated as AII (Stiebler et al. 1997). A dorsoposterior field (DP) on the dorsocaudal border of AI had no tonotopy, and broadly tuned cells.

Of the Caviomorph South American rodents, auditory cortex has been studied in guinea pigs, chinchillas, and Degus. Guinea pigs have large bulla, accessible cochlea, and breed easily (Wallace et al. 2000). Their core has at least two fields that have been named differently than in gerbils, rats, and mice. Some defined a rostral AI and a caudodorsal field (Wallace et al. 2000), DC (Fig. 19.12a), and others also identified the rostral field as AI and a caudal AII (Fig. 19.12b) (Horikawa et al. 2001). In both schemes, the rostral AI field has the position and tonotopic organization of gerbil AAF and the caudal DC or AII field has the position and tonotopic organization of AI. The two core guinea pig fields were originally called anterior and posterior or dorsocaudal fields (Kayser and Legouix 1963; Hellweg et al. 1977). While it was insightfully speculated that the anterior field corresponded to cat AAF and the posterior field to AI, this identification did not persist (Redies et al. 1989). The tonotopic patterns in these fields have been shown in microelectrode mapping (Hellweg et al. 1977; Redies et al. 1989; Wallace et al. 1997) and optical imaging (Taniguchi et al. 1997; Hosokawa et al. 2004; Nishimura et al. 2007) experiments. Both fields receive significant MGv input (Redies et al. 1989), and both lie within densely myelinated cortex that expresses high levels of cytochrome oxidase (Wallace et al. 2000). Other surrounding belt or secondary fields show some tonotopy (Nishimura et al. 2007). Part of a field has been denoted as rostral (R), and the ventrorostral belt (Wallace et al. 2000) has been subdivided (Nishimura et al. 2007). In brief, guinea pigs have a core of two or three areas resembling those in gerbils, rats and mice, but named differently. More specifically, AI in guinea pigs appears to be AAF, and DC or AII may correspond to AI. Part of the posterior belt may be DCB (dorsocaudal belt in Fig. 19.12a) or

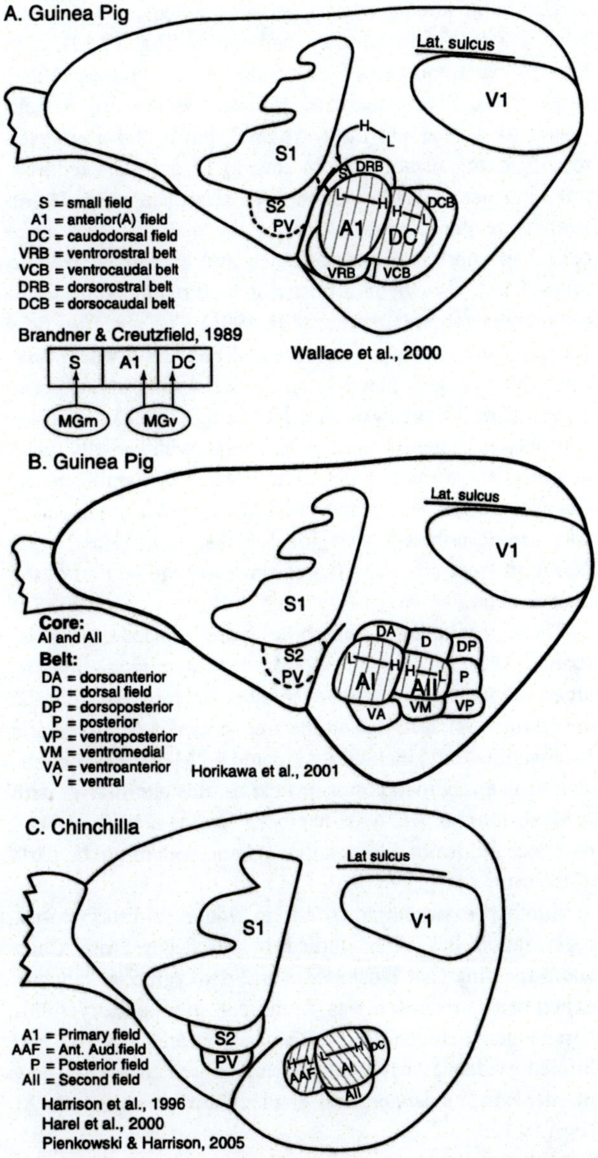

Fig. 19.12 Auditory areas in Caviomorph rodents (guinea pigs and chinchillas). Note that the field in the position of the AAF of Muriode rodents (Fig. 19.11) has been called AI in guinea pigs. In addition, the AAF and AI fields proposed for chinchillas have tonotopic organizations that are reversed from those for AAF and AI of Muriode rodents. Comparisons with other rodents suggest that AI of guinea pigs and chinchillas is AAF and P or DC is AI. **a** Guinea pigs (Wallace et al. 2000). **b** Guinea pigs (Horikawa et al. 2001). **c** Auditory cortex in chinchillas. Conventions as in previous figures. Based on: Brandner and Creutzfield 1989; Wallace et al. 2000; Horikawa et al. 2001; Harrison et al. 1996; Harel et al. 2000; Pienkowski and Harrison 2005

posterior area P (Fig. 19.12b). Auditory cortex organization has been investigated in two other caviomorph rodents, the chinchilla and the degus. Chinchillas have been used extensively in studies of the peripheral auditory system. Two cortical auditory fields include an AI (Fig. 19.12c) (Harrison et al. 1996), although its tonotopic gradient was reversed from that of AI of rats, mice, and gerbils (Fig. 19.11), with high-to-low frequencies in a caudorostral sequence (Harel et al. 2000; Pienkowski and Harrison 2005a, b). A field rostral to AI was not fully explored, but it had a reversed tonotopic organization, from low to high in a caudorostral sequence and was named AAF (Harrison et al. 1996). Both were considered as part of the auditory core since cells had short response latencies and responded well to tones. They also found a posterior field of uncertain tonotopic organization (Harrison et al. 1996). The posterior field was later defined as the dorsocaudal belt (DC) where neurons were broadly tuned to frequencies without tonotopic organization (Pienkowski and Harrison 2005a,b). Area AII was defined along the ventral AI border, with broadly tuned neurons that formed a tonotopic pattern in parallel to AI (Pienkowski and Harrison 2005b), although AII organization was also described as orthogonal to that of AI (Harel et al. 2000). In these studies, AII was considered to be part of the auditory core.

These results from chinchillas pose a puzzle, as areas termed AI and AAF have opposite tonotopic gradients than areas defined as AI and AAF in other rodents. Possibly the area termed AI is AAF, and the area termed AAF is specific to chinchillas or chinchilla AAF could be AI of other rodents, and AI is a modified posterior field. In this alternative, with AAF serving as AI, AAF has been lost or it has not been detected. AII is not different than AI and both might be parts of the same field.

Another caviomorph rodent in which auditory cortex organization has been studied is the degus from Chile and Argentina (not illustrated). In 2-deoxyglucose labeling experiments, evidence was found for five auditory fields (Braun and Scheich 1997). The largest area was AI and limited evidence suggested that it represented high-to-low frequencies in a caudorostral gradient, as does chinchilla AI (Fig. 19.12c).

The final rodent to be considered is the gray squirrel, where visual and somatic sensory areas have been studied more extensively than auditory cortex. A primary area, AI, had high frequencies were represented caudally and low frequencies rostrally (Fig. 19.13d) (Luethke et al. 1988). This AI corresponded well with most of the primary architectonic anterior temporal cortex field (TA) (Kaas et al. 1972) and had densely packed cells in layer 4 and heavy myelination. A more rostral field (R) was not fully explored, but may have a reversed tonotopic organization from that in AI. Subsequent auditory cortex recordings (Luethke et al. 1988) confirmed a reversed tonotopic pattern in R. Area R is within the architectonic field TA, but TA is not uniform in appearance, and the distinctive primary-like features of TA are reduced in R (Merzenich et al. 1976). Cortex caudal to AI, the temporal intermediate field (TI), lacks the characteristics of primary sensory cortex (Kaas et al. 1972), and was unresponsive to auditory stimuli (Merzenich et al. 1976). Cortex rostral and ventral to AI was variably responsive to sound, with area R most consistently responsive. Some neurons in the somatic sensory parietal ventral field (PV) (Krubitzer et al. 1986) responsive to acoustical stimuli (Luethke et al. 1988). The cortical connections of AI included areas R, cortex ventral to AI, and PV (Luethke et al. 1988). Both AI and R receive MGv input.

The results from various rodents present a confusing picture. Considering only the tonotopically organized and most readily characterized fields reveals great variation in the profile of cortical organization across rodent species and between groups of investigators (Fig. 19.14). For the gerbil, there is a rostrocaudal sequence of tonotopically organized areas (AAF, AI, and VP or VP and DP) with tonotopic gradients and reversals that match cat areas AAF, AI, and P. Thus, based on relative position and tonotopic organizations, gerbil AAF and AI at least, may be homologues of cat AAF and AI, and of AAF and AI in other carnivores. Similar patterns exist in rats, where AAF, AI, and P have been identified, and mice,

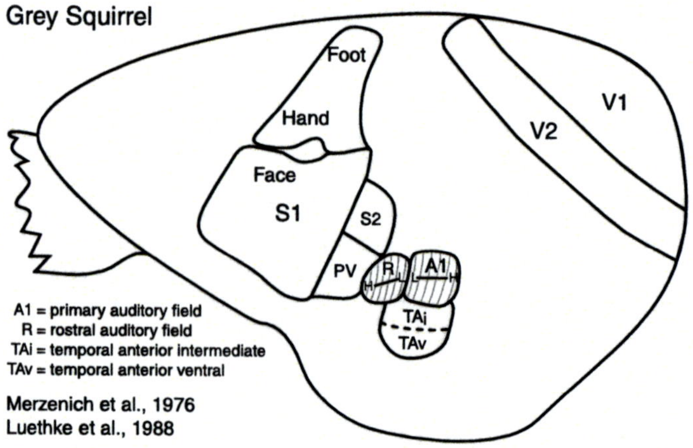

Fig. 19.13 Auditory areas proposed for squirrels. The rostral area, R, has the tonotopic organization of AI in most other rodents, while squirrel AI matches the posterior area (P) (Fig. 19.11). Conventions as in previous figures (Merzenich et al. 1976; Luethke et al. 1988)

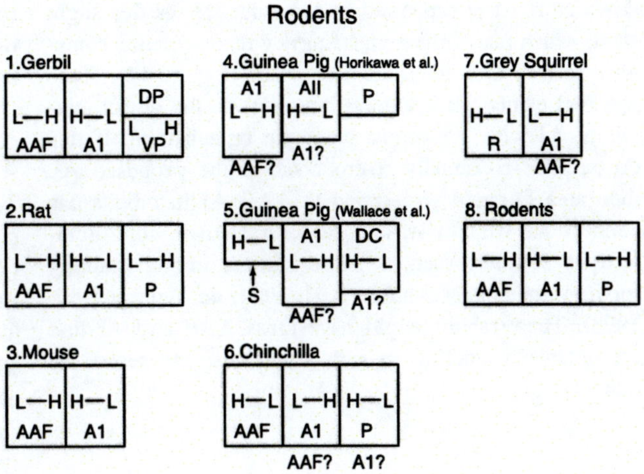

Fig. 19.14 Schematic of auditory core organization in rodents. Areas are designated after the current proposals (Figs. 19.11 and 19.12). To promote consistency in names between studies in rodents, some proposed areas have been renamed below the *boxes*. A proposal for the organization of the auditory core in most or all rodents is shown in schematic 8

where AAF and AI have been described. Guinea pigs fit into this scheme less easily, but one approach identified AI and AII (Fig. 19.12B), except that AI would correspond to AAF and the area designated as AII (Horikawa et al. 2001) would be AI. There is evidence for an area P caudal to this renamed AI. This renaming results in a close correspondence of two or three areas in guinea pigs, gerbils, rats, and mice. Another interpretation of auditory cortex organization in guinea pigs (Wallace et al. 2000) would also fit this rodent scheme with a renaming of areas (Fig. 19.12A). Thus, AI would become AAF, and DC would become AI. Unfortunately, the proposed plan of the chinchilla and squirrel auditory core does not fit this scheme so easily. The most clearly characterized field in both, AI, has a tonotopic gradient that is reversed from that of other rodents. Perhaps AI is actually AAF in both rodents, as the tonotopic gradient corresponds to AAF, but this interpretation leaves no AI in squirrels, and only an ill-defined chinchilla area P for AI. Perhaps squirrels and chinchillas have lost or greatly modified AI, so that only AAF remains as a dominant primary area, which seems unlikely. In addition, this would place primary-like auditory fields rostral to the renamed AAF (area R in squirrels and AAF in chinchillas), and no such fields have been identified in other rodents. Alternatives are that AAF and AI have reversed their tonotopic organization in squirrels and chinchillas, which also seems unlikely, as there are no known examples where a primary sensory area has reversed its internal organization. A reasonable proposal is that auditory cortex in most rodents has two or three core or core-like areas, an AAF with a low-to-high tonotopic organization from rostral to caudal, an AI with a reversed tonotopic organization, and possibly an area P with a reversed tonotopic organization. However, most evidence would exclude area P from the primary core. For the secondary auditory areas, present descriptions are too variable and incomplete to homologize these fields across rodents.

The proposed scheme for rodents would be further supported if a similar pattern of cortical organization could be demonstrated in lagomorphs (rabbits, hares, and picas), which are the closest living relatives of rodents (Fig. 19.1), so that parallels in cortical organization would be much more expected than with distantly related carnivores. The tonotopy of rabbit auditory cortexes has been incompletely studied, and only in dorsoventral microelectrode penetrations coursing parallel to the cortical layers, rather than perpendicular (McMullen and Glaser 1982; Velenovsky et al. 2003). Information on the tonotopic gradient exists in the dorsoventral plane only. An auditory area was described in which high frequencies were represented dorsally and low tones ventrally, with a slight inclination of the isorepresentation lines dorsocaudally to ventrorostrally (Fig. 19.15) (McMullen and Glaser 1982; Velenovsky et al. 2003). That

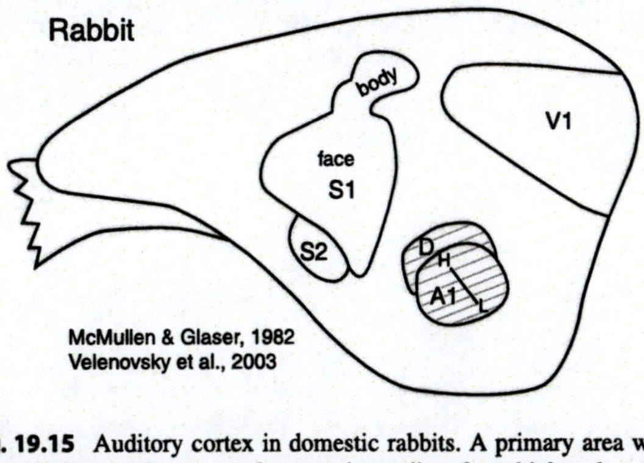

Fig. 19.15 Auditory cortex in domestic rabbits. A primary area with a predominantly dorsoventral tonotopic gradient from high-to-low frequencies with a caudalward slant has been described as AI. As the high frequency representation is displaced rostrally, this organization, allowing for some rotation, is consistent with the area often considered as AI in rodents. A dorsal area (D) may correspond to AAF. Conventions as in previous figures. Based on: McMullen and Glaser 1982; Velenovsky et al. 2003

this area is AI is supported by the evidence for this slight slant, which places the high frequency representation somewhat rostral, as in the rodent schematic (Fig. 19.14). There was less evidence for a dorsal area (D) on the dorsal margin of AI with a reversed tonotopic organization. If this dorsal area is actually rostrodorsal to the proposed AI, then area D could correspond to AAF. AI in rabbits has connections with MGv, and has characteristic core architectonic features, such as dense laminar immunostaining for parvalbumin (McMullen et al. 1994; de Venecia et al. 1998). Thus, rabbits could have areas AAF and AI that are organized much as in gerbils and rats, and secondary areas.

7 Auditory Cortex Organization in Tree Shrews (Scandentia)

The superorder Euarchontoglires includes Glires (rodents and lagomorphs), and Euarchontans, consisting of primates, flying lemurs, and tree shrews. Flying lemurs (Dermoptera) are rare, leaving tree shrews (Scandentia) as the closest living relative of primates available for study. Unfortunately, little is known about the organization of tree shrew auditory cortex. One study (J.H. Kaas, W.C. Hall and M.M. Merzenich, unpublished observations) used microelectrodes to map its tonotopic organization and found evidence for only one tonotopically organized area, a large area designated as AI, with an organization from high-to-low frequency in the rostrocaudal direction, with isorepresentation lines in a dorsoventral axis, inclined slightly rostrally (Fig. 19.16). This organization is consistent with that of the proposed rodent (and possibly rabbit) AI , as well as cats. Tree shrew AI has a primary architectonic appearance, and it receives MGv input (Casseday et al. 1976). The location of tree shrew AI would seem to leave little room for a more anterior area (AAF), and the existence of other fields remains uncertain.

8 Auditory Cortex in Bats

Bats and carnivores have the same (superorder Laurasiatheria). Bats belong to the order Chiroptera, which contains megachiroptera (megabats), the fruit eating bats without echolocation, and microchiroptera (microbats), which echolocate and feed predominantly on insects. Most interest in bats has been on echolocating bats, whose auditory system is highly specialized. The somatic sensory system is also unusual as well, with adaptations related to use of the forelimb as a wing (Calford et al. 1985; Wise et al. 1986) and somatic sensory modulation of flight (Zook 2007).

Recordings from mustached bat auditory cortex provide evidence of specialized cortical auditory areas, most with no apparent homologues in other mammals (Suga 1990, 1994; Fitzpatrick et al. 1998). Only one of these areas is considered here. A primary area, AI has a low-to-high frequency caudorostral gradient (Fig. 19.17A) from 10 to 100 kHz. An expansion in central AI represents biosonar pulse frequencies of 60–62 kHz. It is uncertain if there is an AAF-like area rostral to AI, although other, possibly secondary fields have been described. AI (divided into three sectors) receives input from MGv, as expected (Pearson et al. 2007). An unusual feature of the auditory system of the mustached bat, and perhaps other echolocating bats, is a direct MG projection to frontal cortex, where neurons respond to auditory stimuli (Casseday et al. 1989; Kanwal et al. 2000).

Auditory cortex organization has also been studied in the big brown bat (Fig. 19.17b) that has a large tonotopically organized area, with a caudorostral gradient of low-to-high frequencies, characterized as AI (Dear et al. 1993). A

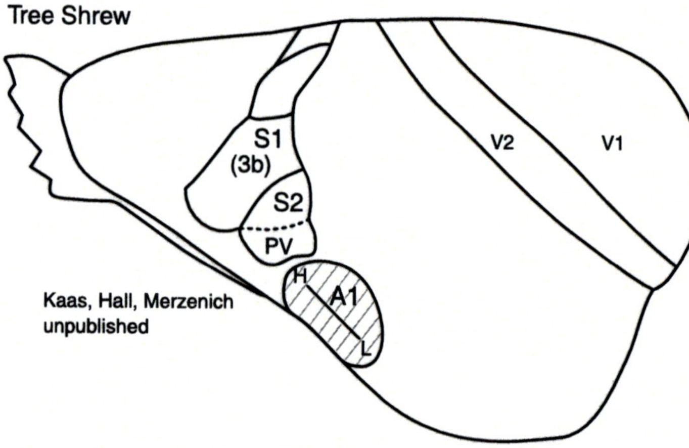

Fig. 19.16 Auditory cortex organization in tree shrews. Only one auditory area has been identified, and was denoted as AI. Conventions as in previous figures

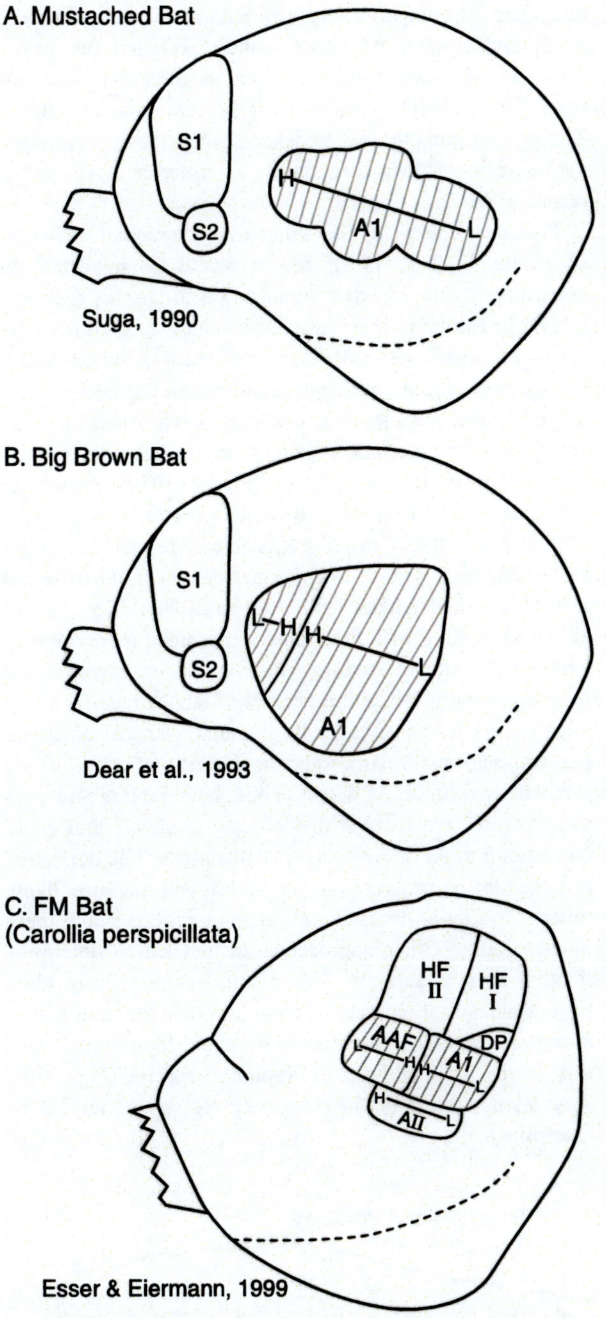

Fig. 19.17 Auditory cortex organization in echolocating bats. **a** The location of the primary area, AI in the mustached bat. Other highly specialized areas have been identified, but none obviously correspond to areas in other mammals. **b** Two tonotopic areas are described in the big brown bat, one with the tonotopic organization of AI and the other matching the relative position and tonotopic organization of AAF. **c** Three tonotopic fields have been proposed for the short-tailed FM fruit bat, two corresponding to AI and AAF, respectively, and a more ventral region proposed as AII. A dorsoposterior area (DP) and two high frequency areas have also been proposed. Conventions as in previous figures

smaller, anterior region had a reversed tonotopic organization, and a relative position and tonotopic organization like that of AAF. The border region between these fields was activated by 60–90 kHz frequencies and was variable among bats.

Other aerial insectivore bats studied include the rufous horseshoe bat (not illustrated), whose AI has a caudorostral low-to-high frequency tonotopic gradient, and tonotopic anterior and posterior fields which have been homologized with cat and rodent AAF and ventroposterior and posterior fields (Radtke-Schuller and Schuller 1995). Both AI and the posterior field receive MGv input, suggesting that they are parts of a primary-like auditory core, together with AAF (Radtke-Schuller 2004). AI has primary-like architectonic features, while AAF has architectonic features intermediate to AI and dorsal secondary auditory fields (Radtke-Schuller 2001). Finally, in a frequency modulating (FM) bat (*Myotis lucifugus*), auditory cortex (not illustrated) has a tonotopically organized AI (with low frequencies caudal), and an anterior field with lower frequencies again represented, possibly corresponding to AAF (Wong and Shannon 1988).

Besides the above insectivorous bats, auditory cortex has been investigated in the short-tailed FM microchiropteran fruit bat (*Carollia perspicillata*), an FM bat with biosonar echolocating capacity that is less specialized than the bats discussed above, as it eats fruits and nectar, and seeks insects opportunistically. Microelectrode mapping defined three tonotopic fields: AI, AAF, and AII (Fig. 19.17c) (Esser and Eiermann 1999). AI and AAF had caudorostral progressions from low-to-high frequencies in AI and high-to-low in AAF, and both fields were considered core. A secondary area, AII, with tonotopic organization lies along the ventral AI and AAF border and these cells were habituated rapidly and were more broadly tuned. A dorsoposterior field above AI had no tonotopic organization and high response thresholds. Much of auditory cortex consisted of high-frequency representations dorsal (HF-I and HF-II) to AI and AAF.

There have been no studies of auditory cortex organization in the non-echolocating fruit bats, the megabats of tropical areas. Their auditory areas may be less specialized than in echolocating bats.

Echolocating bats have a highly specialized auditory system. Nevertheless, a primary area, AI, with a caudorostral frequency representation has been consistently recognized. An AAF with features that suggest that it could be part of the auditory core has also been identified. AI and AAF have tonotopic organizations and other features that suggest that they are homologues of cat AI and AAF. An AII-like area is found in an FM fruit-eating bat, while all bats have highly specialized, dorsally located, secondary auditory fields. In addition, the organization of AI is distorted by having a large auditory foveal region for the echolocation frequencies.

9 Auditory Cortex in Other Mammals

Little is known about auditory cortex organization in other mammals. An auditory region has often been identified architectonically, and is assumed to be AI, but without experimental studies that conclusion is uncertain.

Hedgehogs are insectivores in the Laurasiatherian superorder (Fig. 19.1). They have small brains with little neocortex, and have long been of interest in comparative studies of brain evolution. Auditory cortex investigation of the long-eared hedgehog has found at least two auditory areas (Fig. 19.18), AI and P (posterior) (Batzri-Izraeli et al. 1990). Both were organized tonotopically, with a caudorostral representation of low-to-high frequencies in P, and a reversed pattern in AI. Both areas are in a more densely myelinated region of cortex. Hedgehog cortex is not well differentiated, lacks a koniocellular architectonic appearance and was difficult to distinguish from adjoining cortex. Tracer injections into AI labeled MGv neurons. The connections of P were not determined. The response properties in both fields were similar, although P neuron latencies were slightly shorter than in AI (Batzri-Izraeli and Wollberg 1992). As hedgehog AI compares very well with carnivore and rodent AAF, and P with AI, these authors weighed the possibility that the two fields had been misidentified, and felt that more evidence was needed before renaming them. The present comparative evidence strongly favors the alternative view of two core fields, an anterior AAF and a posterior AI.

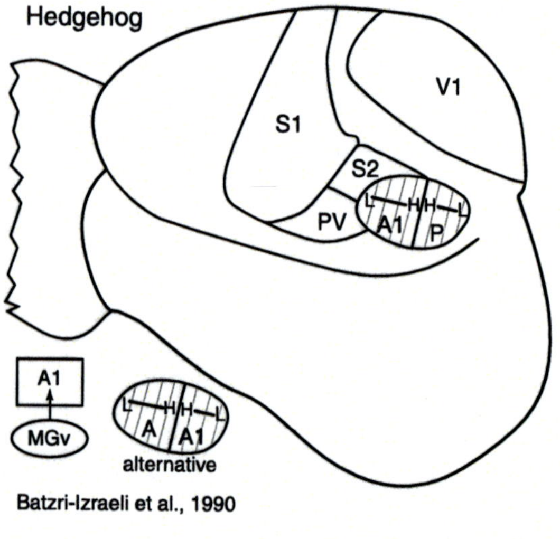

Fig. 19.18 Auditory cortex organization in hedgehogs. Two tonotopically organized fields were described (AI and P (posterior)). AI thalamic connections of were largely from MGv. The comparative evidence favors the alternative interpretation that the anterior field is homologous with AAF of rodents, carnivores, and bats, and the posterior field is AI. The somatic sensory areas (S2, PV and S1) are based on prior work (Catania et al. 2000). Conventions as in previous figures

Possums belong to the impressive marsupial radiation of mammals (Fig. 19.1), and it would be important to determine auditory cortex organization in species distantly related to eutherian mammals. Few studies of auditory cortex organization are available in the brush-tailed opossum. An auditory region with dense myelination has been identified in several marsupials (Beck et al. 1996; Huffman et al. 1999), but the tonotopic organization has been determined only in the brush-tailed possum (Fig. 19.19). Dorsoventral electrode penetrations through auditory cortex recorded neurons at successively lower frequencies. One field, thought to be AI, had high-to-low frequencies in a dorsoventral axis (Gates and Aitkin 1982). As in rabbits (Fig. 19.15), either AI or AAF with high tones represented more dorsally would yield a dorsoventral progression of neurons responsive to progressively lower frequencies. Without precise alignment of recording sites in the parallel, vertical electrode penetrations, any rostrocaudal component of a frequency gradient is difficult to detect. Thus, both rabbits and possums could have areas with tonotopic gradients that could correspond to AI or AAF in other mammals. The dorsoventral frequency progressions in rabbit and possum likely reflect the similar dorsoventral mapping procedures in these investigations. Other approaches are needed to determine if other core areas exist. This would be essential in identifying areas homologous to those in other mammals. It is disappointing that so little is known about Monotreme auditory cortex, or Afrotheria or Xenarthra species (Fig. 19.1). Any organizations of their auditory regions remain to be determined.

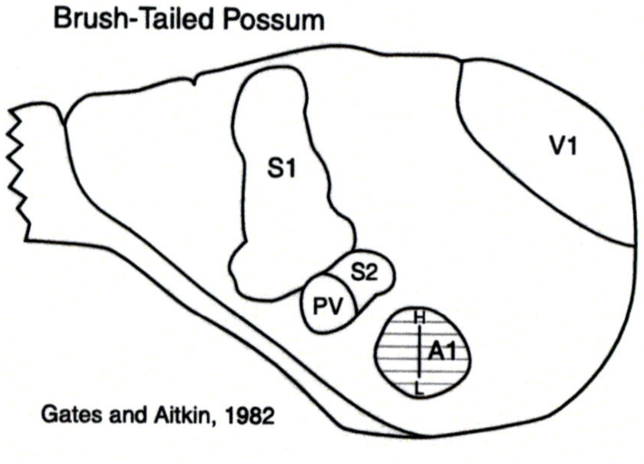

Fig. 19.19 Auditory cortex organization in the brush-tailed possum. Conventions as in previous figures

10 Summary and Conclusions

10.1 Defining the Auditory Cortex

Early studies in cats designated a primary area (AI) and a secondary field (AII). Subsequently, another primary-like field, the anterior auditory field (AAF) was found rostral to AI, and a less primary-like posterior field (P or PAF) caudal to AI. AI and AAF primary-like attributes include tonotopic organization with sharply tuned neurons, direct inputs from the ventral nucleus of the medial geniculate complex and architectonic features of primary sensory cortex. Cats have two, and possibly three, primary areas (AI, AAF, and perhaps P). Adjoining auditory fields in cats are secondary in structure, function, and connections. Comparable but more limited studies in dogs and ferrets suggest that areas AI, AAF, and P likely exist in all carnivores.

10.2 Core Fields of Auditory Cortex

A core of three primary or primary-like areas has also been identified in primates. From caudal to rostral along the lower bank of the lateral sulcus, these areas have been termed AI, the rostral area (R), and the rostrotemporal area (RT). Because both AI and R have pronounced primary-like features, either could be homologous with area AI of cats. As AI in cats has a caudorostral representation of tone frequencies from low to high, and the proposed AI of monkeys has a rostrocaudal progression, these opposite tonotopic gradients do not appear to support the assumption that the two areas termed AI are homologous. However, the possibility that AI of monkeys has been rotated nearly 180° by the expansion of the temporal lobe in primates has been proposed as an explanation for the opposite orientations of the tonotopic gradients. A further consideration of this possibility indicates that a rotation of as much as 90° may have occurred in New World monkeys, and perhaps more in some anthropoids, but the rotation hypothesis is still questionable. In addition, the rotation hypothesis would leave no primary-like area such as AAF on the caudal border of primate AI where CM is located. Thus, the rotation hypothesis seems inconsistent with other observations. Alternatively, R could be the homolog of carnivore AI (and seems to have been mistaken for AI in one detailed study). However, if primate area R is actually AI, the area more caudal to R, now defined as AI, seems too primary-like to correspond to carnivore area P (or PAF). Primate area RT closely resembles carnivore area AAF, in position and tonotopy relative to R redefined as AI, but RT is less primary-like, and smaller than expected for AAF. Thus, the homologies between core areas in primates and carnivores remain uncertain.

10.3 Common and Unique Features in Defining Auditory Cortex

Results from most, but not all rodents, conform to the carnivore pattern of an AI and an anterior auditory field. The two rodent fields have not been consistently named, but their tonotopic gradients, inputs from MGv, and histological features all support homologies between fields AI and AAF in carnivores and most rodents. There is also support for considering the posterior fields in rodents and carnivores as homologues. The apparent differences between results from most rodents, and those from squirrels and chinchillas are difficult to explain, but fundamental differences in the presence or absence of AI and AAF across rodent taxa would be surprising.

10.4 Bats and Other Species

Bats are the only other taxon whose auditory cortex has been well studied in special species and there is good evidence for adjoining core areas homologous to carnivore AAF and AI. The parallels in these two areas in bats, rodents, and carnivores – all different branches of the placental mammal radiation – suggest that early placental mammals had both AAF and AI, and these core areas were widely, perhaps universally, retained by subsequent placentals. The posterior area (P) could be part of this primitive constellation.

10.5 The Future of Comparative Studies of Auditory Cortex

While one tonotopically organized, primary-like area has been demonstrated in a marsupial, there is insufficient information to reach firm conclusions about auditory cortex organization in this mammalian radiation. Nothing is known about the organization of auditory cortex in monotreme mammals. Even our understanding of auditory cortex in placental (eutherian) mammals is highly fragmented, resting largely on two orders (carnivora and chiroptera) of the Laurasiatherian superorder, and on rodentia and primates of the Euarchontoglires superorder, with no species from the Afrotheria and Xenarthra superorders. Bridging this gap in comparative research on auditory cortex should be a major feature of any future agenda.

Acknowledgement Troy Hackett, Correy Camalier, and Peiyan Wong provided helpful comments on the manuscript.

References

Aitkin LM, Merzenich MM, Irvine DRF, Clarey JC, and Nelson JE (1986) Frequency representation in auditory cortex of the common marmoset (*Callithrix jacchus jacchus*). *Journal of Comparative Neurology* **252**:175–185.

Asher RJ, Meng J, Wible JR, McKenna MC, Rougier GW, Dashzeveg D, and Novacek MJ (2005) Stem lagomorpha and the antiquity of Glires. *Science* **307**:1091–1094.

Bajo UM, Nodal FR, Bizley JK, Moore DR, and King AJ (2007) The ferret auditory cortex: descending projections to the inferior colliculus. *Cerebral Cortex* **17**:475–491.

Batzri-Izraeli R, Kelly JB, Glendenning KK, Masterton RB, and Wollberg Z (1990) Auditory cortex of the long-eared hedgehog (*Hemiechinus auritus*). *Brain Behavior and Evolution* **36**:237–248.

Batzri-Izraeli R and Wollberg Z (1992) Auditory cortex of the long-eared hedgehog (*Hemiechinus auritus*): II. Tuning properties. *Brain Behavior and Evolution* **39**:143–152.

Beck PD, Pospichal MW, and Kaas JH (1996) Topography, architecture, and connections of somatosensory cortex in opossums: evidence for five somatosensory areas. *Journal of Comparative Neurology* **366**:109–133.

Bendor D and Wang X (2005) The neuronal representation of pitch in primate auditory cortex. *Nature* **436**:1161–1165.

Bizley JK, Nodal FR, Nelken I, and King AJ (2005) Functional organization of ferret auditory cortex. *Cerebral Cortex* **15**:1637–1653.

Brandner S and Redies H (1990) The projection of the medial geniculate body to field AI: organization in the isofrequency dimension. *Journal of Neuroscience* **10**:50–61.

Braun S and Scheich H (1997) Influence of experience on the representation of the "mothering call" in frontoparietal and auditory cortex of pups of the rodent *Octodon degus*: FDG mapping. *Journal of Comparative Physiology A* **181**:697–709.

Brodmann K (1909) *Vergleichende Lokalisationslehre der Grosshirnrhinde*. J.A. Barth, Leipzig.

Bruce LL (2007) Evolution of the nervous system in reptiles. In: Kaas JH (ed). *Evolution of Nervous Systems*, volume 3. Elsevier, London, pp. 125–156.

Brugge JF (1982) Auditory areas in primates. In: Woolsey CN (ed). *Cortical Sensory Organization* volume 3, *Multiple Auditory Areas*. Humana Press, Clifton, pp. 59–70.

Budinger E, Heil P, and Scheich H (2000a) Functional organization of auditory cortex in the Mongolian gerbil *(Meriones unguiculatus)*. III. Anatomical subdivisions and corticocortical connections. *European Journal of Neuroscience* **12**:2425–2451.

Budinger E, Heil P, and Scheich H (2000b). Functional organization of auditory cortex in the Mongolian gerbil (*Meriones unguiculatus*). IV. Connections with anatomically characterized subcortical structures. *European Journal of Neuroscience* **12**: 2452–2474.

Calford MB, Graydon ML, Huerta MF, Kaas JH, and Pettigrew JD (1985) A variant of the mammalian somatotopic map in a bat. *Nature* **313**:477–479.

Casseday JH, Diamond IT, and Harting JK (1976) Auditory pathways to the cortex in Tupaia glis. *Journal of Comparative Neurology* **166**:303–340.

Casseday JH, Kobler JB, Isbey SF, and Covey E (1989) Central acoustic tract in an echolocating bat: an extralemniscal auditory pathway to the thalamus. *Journal of Comparative Neurology* **287**:247–259.

Caviness VS (1975) Architectonic map of neocortex of the normal mouse. *Journal of Comparative Neurology* **164**:247–264.

Cheung SW, Bedenbaugh PH, Nagarajan SS, and Schreiner CE (2001) Functional organization of squirrel monkey primary auditory cortex: responses to pure tones. *Journal of Neurophysiology* **85**: 1732–1749.

Cheung SW (2005) Frequency map variations in squirrel monkey primary auditory cortex. *Laryngoscope* **115**:1136–1144.

Clerici WJ and Coleman JR (1990) Anatomy of the rat medial geniculate body: I. cytoarchitecture, myeloarchitecture, and neocortical connectivity. *Journal of Comparative Neurology* **297**:14–31.

Colbert EH and Morales M (1991) *Evolution of the Vertebrates*. Wiley-Liss, New York.

Dear SP, Fritz J, Haresign T, Ferragamo M, and Simmons JA (1993) Tonotopic and functional organization in the auditory cortex of the big brown bat, *Eptesicus fuscus*. *Journal of Neurophysiology* **70**:1988–2009.

Donishi T, Kimura A, Okamoto K, and Tamai Y (2006) "Ventral" area in the rat auditory cortex: a major auditory field connected with the dorsal division of the medial geniculate body. *Neuroscience* **141**:1553–1567.

Doron NN, LeDoux JE, and Semple MN (2002) Redefining the tonotopic core of rat auditory cortex: physiological evidence for a posterior field. *Journal of Comparative Neurology* **453**:345–360.

Eggermont JJ (1998) Representation of spectral and temporal sound features in three cortical fields of the cat. Similarities outweigh differences. *Journal of Neurophysiology* **80**:2743–2764.

Esser K-H and Eiermann A (1999) Tonotopic organization and parcellation of auditory cortex in the FM-bat *Carollia perspicillata*. *European Journal of Neuroscience* **11**:3669–3682.

Fitzpatrick DC, Suga N, and Olsen JF (1998) Distribution of response types across entire hemispheres of the mustached bat's auditory cortex. *Journal of Comparative Neurology* **391**:353–365.

Fu KM, Johnston TA, Shah AS, Arnold LL, Smiley JF, Hackett TA, Garraghty PE, and Schroeder CE (2003) Auditory cortical neurons respond to somatosensory stimulation. *Journal of Neuroscience* **23**:7510–7515.

Galaburda A and Sanides F (1980) Cytoarchitectonic organization of the human auditory cortex. *Journal of Comparative Neurology* **190**:597–610.

Galaburda AM and Pandya DN (1983) The intrinsic architectonic and connectional organization of the superior temporal region of the rhesus monkey. *Journal of Comparative Neurology* **221**:169–184.

Gates GR and Aitkin LM (1982) Auditory cortex in the marsupial possum *Trichosurus vulpecula*. *Hearing Research* **7**:1–11.

Godey B, Atencio CA, Bonham BH, Schreiner CE, and Cheung SW (2005) Functional organization of squirrel monkey primary auditory cortex: responses to frequency-modulation sweeps. *Journal of Neurophysiology* **94**:1299–1311.

Goldschmidt J, Zuschratter W, and Scheich H (2004) High-resolution mapping of neuronal activity by thallium autometallography. *NeuroImage* **23**:639–647.

Hackett TA (2002) The comparative anatomy of the primate auditory cortex. In: Ghazanfar AA (ed). *Primate Audition: Ethology and Neurobiology*. CRC Press, Boca Raton, pp. 199–219.

Hackett TA, Preuss TM, and Kaas JH (2001) Architectonic identification of the core region in auditory cortex of macaques, chimpanzees, and humans. *Journal of Comparative Neurology* **441**: 197–222.

Hackett TA, Stepniewska I, and Kaas JH (1998a) Subdivisions of auditory cortex and ipsilateral cortical connections of the parabelt auditory cortex in macaque monkeys. *Journal of Comparative Neurology* **394**:475–495.

Hackett TA, Stepniewska I, and Kaas JH (1998b) Thalamocortical connections of the parabelt auditory cortex in macaque monkeys. *Journal of Comparative Neurology* **400**:271–286.

Harel N, Mori N, Sawada S, Mount RJ and Harrison RV (2000) Three distinct auditory areas of cortex (AI, AII, and AAF) defined by optical imaging of intrinsic signals. *NeuroImage* **11**:302–312.

Harrison RV, Kakigi A, Hirakawa H, Harel N, and Mount RJ (1996) Tonotopic mapping in auditory cortex of the chinchilla. *Hearing Research* **100**:157–163.

He J and Hashikawa T (1998) Connections of the dorsal zone of cat auditory cortex. *Journal of Comparative Neurology* **400**:334–348.

Hellweg FC, Koch R, and Vollrath M (1977) Representation of the cochlea in the neocortex of guinea pigs. *Experimental Brain Research* **29**:467–474.

Hofstetter KM and Ehret G (1992) The auditory cortex of the mouse: connections of the ultrasonic field. *Journal of Comparative Neurology* **323**:370–386.

Horikawa J, Hess A, Nasu M, Hosokawa Y, Scheich H, and Taniguchi I (2001) Optical imaging of neural activity in multiple auditory cortical fields of guinea pigs. *NeuroReport* **12**:3335–3339.

Horikawa J, Ito SI, Hosokawa Y, Homma T, and Murata K (1988) Tonotopic representation in the rat auditory cortex. *Proceedings of the Japanese Academy of Sciences* **64**:260–267.

Hosokawa Y, Sugimoto S, Kubota M, Taniguchi I, and Horikawa J (2004) Optical imaging of binaural interaction in multiple fields of the guinea pig auditory cortex. *Auditory and Vestibular System* **15**:1093–1097.

Huang CL and Winer JA (2000) Auditory thalamocortical projections in the cat: laminar and areal patterns of input. *Journal of Comparative Neurology* **427**:302–331.

Huffman KJ, Nelson J, Clarey J, and Krubitzer L (1999) The organization of somatosensory cortex in three species of marsupials: neural correlates of morphological specializations. *Journal of Comparative Neurology* **403**:5–32.

Imaizumi K, Priebe NJ, Crum PAC, Bedenbaugh PH, Cheung SW, and Schreiner CE (2004) Modular functional organization of cat anterior auditory field. *Journal of Neurophysiology* **92**:444–457.

Imaizumi K and Schreiner CE (2007) Spatial interaction between spectral integration and frequency gradient in primary auditory cortex. *Journal of Neurophysiology* **98**:2933–2942.

Imig TJ and Morel A (1985) Tonotopic organization in ventral nucleus of medial geniculate body in the cat. *Journal of Neurophysiology* **53**:309–340.

Imig TJ and Reale RA (1980) Patterns of cortico-cortical connections related to tonotopic maps in cat auditory cortex. *Journal of Comparative Neurology* **192**:293–332.

Imig TJ, Ruggero MA, Kitzes LM, Javel E, and Brugge JF (1977) Organization of auditory cortex in the owl monkey (*Aotus trivirgatus*). *Journal of Comparative Neurology* **171**:111–128.

Jones EG (2006) *The Thalamus*. Cambridge University Press, New York.

Kaas JH (1982) The segregation of function in the nervous system: why do sensory systems have so many subdivisions? In: Neff WP (ed). *Contributions to Sensory Physiology*. Academic Press, New York, pp. 201–240.

Kaas JH (1983) What, if anything, is S1? Organization of first somatosensory area of cortex. Somatosensory areas of Cortex. *Physiological Reviews* **63**:206–231.

Kaas JH (2005) The future of mapping sensory cortex in primates: three of many remaining issues. *Philosophical Transactions of the Royal Society of London, Series B, Biological Sciences* **360**:653–664.

Kaas JH (2007) Reconstructing the organization of the forebrain of the first mammals. In: Kaas JH (ed). *Evolution of Nervous Systems*, volume 3, Elsevier, London, pp. 27–48.

Kaas JH and Hackett TA. (2000) Subdivisions of auditory cortex and processing streams in primates. *Proceedings of the National Academy of Science of the United States of America* **97**: 11793–11799.

Kaas JH and Hackett TA (2005) Subdivisions and connections of the auditory cortex in primates: a working model. In: Konig R, Heil P, Budinger E, Scheich H (eds). *The Auditory Cortex: A Synthesis of Human and Animal Research*. Lawrence Erlbaum Associates, London, pp. 7–25.

Kaas JH, Hall WC, and Diamond IT (1972) Visual cortex of the grey squirrel (*Sciureus carolinensis*): architectonic subdivisions and connections from the visual thalamus. *Journal of Comparative Neurology* **145**:273–305.

Kaas JH and Morel A (1993) Connections of visual areas of the upper temporal lobe of owl monkeys: the MT crescent and dorsal and ventral subdivisions of FST. *Journal of Neuroscience* **13**: 534–546.

Kajikawa Y, de la Mothe L, Blumell S, and Hackett TA (2005) A comparison of neuron response properties in areas A1 and CM of the marmoset monkey auditory cortex: tones and broadband noise. *Journal of Neurophysiology* **93**:22–34.

Kalatsky VA, Polley DB, Merzenich MM, Schreiner CE, and Stryker MP (2005) Fine functional organization of auditory cortex revealed by Fourier optical imaging. *Proceedings of the National Academy of Science of the United States of America* **102**: 13325–13330.

Kanwal JS, Gordon M, Peng JP, and Heinz-Esser K (2000) Auditory responses from the frontal cortex in the mustached bat, *Pteronotus parnellii*. *NeuroReport* **11**:367–372.

Kayser D and Legouix JP (1963) Tonotopic projections on the guinea pig auditory cortex. *Comptes Rendues Sciences du Societé Biologie et Philosophe* **157**:2161–2164.

Kelly JB, Judge PW, and Phillips DP (1986) Representation of the cochlea in primary auditory cortex of the ferret (*Mustela putorius*). *Hearing Research* **24**:111–115.

Kilgard MP and Merzenich MM (1999) Distributed representation of spectral and temporal information in rat primary auditory cortex. *Hearing Research* **134**:16–28.

Knight PL (1977) Representation of the cochlea within the anterior auditory field (AAF) of the cat. *Brain Research* **130**:447–467.

Kosaki H, Hashikawa T, He J, and Jones EG (1997) Tonotopic organization of auditory cortical fields delineated by parvalbumin immunoreactivity in macaque monkeys. *Journal of Comparative Neurology* **386**:304–316.

Kosmal A (2000) Organization of connections underlying the processing of auditory information in the dog. *Progress in Neuro-Psychopharmacology and Biological Psychiatry* **24**:825–854.

Kowalski N, Versnel H, and Shamma SA (1995) Comparison of responses in the anterior and primary auditory fields of the ferret cortex. *Journal of Neurophysiology* **73**:1513–1523.

Krubitzer LA, Sesma MA, and Kaas JH (1986) Microelectrode maps, myeloarchitecture, and cortical connections of three somatotopically organized representations of the body surface in the parietal cortex of squirrels. *Journal of Comparative Neurology* **250**: 403–430.

Lee CC, Imaizumi K, Schreiner CE, and Winer JA (2004) Concurrent tonotopic processing streams in auditory cortex. *Cerebral Cortex* **14**:441–451.

Lee CC and Winer JA (2005) Principles governing auditory cortex connections. *Cerebral Cortex* **15**:1804–1811.

Lee CC and Winer JA (2008a) Connections of cat auditory cortex: I. Thalamocortical system. *Journal of Comparative Neurology* **507**:1879–1900.

Lee CC and Winer, JA (2008b) Connections of cat auditory cortex: III. Corticocortical system. *Journal of Comparative Neurology* **507**:1920–1943.

Luethke LE, Krubitzer LA, and Kaas JH (1988) Cortical connections of electrophysiologically and architectonically defined subdivisions of auditory cortex in squirrels. *Journal of Comparative Neurology* **268**:181–203.

Luethke LE, Krubitzer LA, and Kaas JH (1989) Connections of primary auditory cortex in the New World monkey, (*Saguinus*). *Journal of Comparative Neurology* **285**:487–513.

Malhotra S and Lomber SG (2007) Sound localization during homotopic and heterotopic bilateral cooling deactivation of primary and nonprimary auditory cortical areas in the cat. *Journal of Neurophysiology* **97**:26–43.

Malinowska M and Kosmal A (2003) Connections of the posterior thalamic region with the auditory ectosylvian cortex in the dog. *Journal of Comparative Neurology* **467**:185–206.

McMullen NT and Glaser EM (1982) Tonotopic organization of rabbit auditory cortex. *Experimental Biology* **75**:208–220.

McMullen NT, Smelser CB, and de Venecia RK (1994) A quantitative analysis of parvalbumin neurons in rabbit auditory neocortex. *Journal of Comparative Neurology* **349**:493–511.

Mendelson JR, Schreiner CE, and Sutter ML (1997) Functional topography of cat primary auditory cortex; response latencies. *Journal of Comparative Physiology A* **181**:615–633.

Medina L (2007) Do birds and reptiles possess homologues of mammalian visual, somatosensory and motor cortices? In: Kaas JH (ed). *The Evolution of Nervous Systems*, volume 2. Elsevier, London, pp. 163–194.

Merzenich MM and Brugge JF (1973) Representation of the cochlear partition on the superior temporal plane of the macaque monkey. *Brain Research* **50**:275–296.

Merzenich MM, Kaas JH, and Roth GL (1976) Auditory cortex in the grey squirrel: tonotopic organization and architectonic fields. *Journal of Comparative Neurology* **166**:387–401.

Merzenich MM, Knight PL, and Roth GL (1973) Cochleotopic organization of primary auditory cortex in the cat. *Brain Research* **63**:343–346.

Merzenich MM, Knight PL, and Roth GL (1975) Representation of cochlea within primary auditory cortex in the cat. *Journal Neurophysiology* **38**:231–249.

Morel A, Garraghty PE, and Kaas JH (1993) Tonotopic organization, architectonic fields, and connections of auditory cortex in macaque monkeys. *Journal of Comparative Neurology* **335**:437–459.

Morel A and Imig TJ (1987) Thalamic projections to fields A, AI, P, and VP in the cat auditory cortex. *Journal of Comparative Neurology* **265**:119–144.

Morel A and Kaas JH (1992) Subdivisions and connections of auditory cortex in owl monkeys. *Journal of Comparative Neurology* **318**: 27–63.

de la Mothe LA, Blumell S, Kajikawa Y, and Hackett TA (2006a) Cortical connections of the auditory cortex in marmoset monkeys: core and medial belt regions. *Journal of Comparative Neurology* **496**:27–71.

Murphy WJ, Pevzner PA, and O'Brien JO (2004) Mammalian phytogenomics comes of age. *Trends in Genetics* **20**:631–639.

Nelken I, Bizley JK, Nodal FR, Ahmed B, Schnupp JWH, and King AJ (2004) Large-scale organization of ferret auditory cortex revealed using continuous acquisition of intrinsic optical signals. *Journal of Neurophysiology* **92**:2574–2588.

Nishimura M, Shirasawa H, Kaizo H, and Song WJ (2007) New field with tonotopic organization in guinea pig auditory cortex. *Journal of Neurophysiology* **97**:927–932.

Northcutt RG and Kaas JH (1995) The emergence and evolution of mammalian neocortex. *Trends in Neuroscience* **18**:373–379.

Olavarria J and Van Sluyters R (1985) Organization and postnatal development of callosal connections in the visual cortex of the rat. *Journal of Comparative Neurology* **239**:1–26.

Pallas SL, Roe AW, and Sur M (1990) Visual projections induced into the auditory pathway of ferrets. I. Novel inputs to primary auditory cortex (AI) from the LP/pulvinar complex and the topography of the MGN-AI projection. *Journal of Comparative Neurology* **298**: 50–68.

Pallas SL and Sur M (1993) Visual projections induced into the auditory pathway of ferrets: II. Corticocortical connections of primary auditory cortex. *Journal of Comparative Neurology* **337**: 317–333.

Pandya PK, Rathbun, DL, Moucha R, Engineer ND, and Kilgard MP (2008) Spectral and temporal processing I rat posterior auditory cortex. *Cerebral Cortex* **18**:301–314.

Pearson JM, Crocker WD, and Fitzpatrick DC (2007) Connections of functional areas in the mustached bat's auditory cortex with the auditory thalamus. *Journal of Comparative Neurology* **500**:401–418.

Philibert B, Beitel RE, Nagarajan SS, Bonham BH, Schreiner CE, and Cheung SW (2005) Functional organization and hemispheric comparison of primary auditory cortex in the common marmoset (Callithrix jacchus). *Journal of Comparative Neurology* **487**: 391–406.

Phillips DP, Judge PW, and Kelly JB (1988) Primary auditory cortex in the ferret (Mustela putorius): neural response properties and topographic organization. *Brain Research* **443**:281–294.

Phillips DP and Orman SS (1984) Responses of single neurons in posterior field of cat auditory cortex to tonal stimulation. *Journal of Neurophysiology* **51**:147–163.

Pienkowski M and Harrison RV (2005a). Tone frequency maps and receptive fields in the developing chinchilla auditory cortex. *Journal of Neurophysiology* **93**:454–466.

Pienkowski M and Harrison RV (2005b) Tone responses in core versus belt auditory cortex in the developing chinchilla. *Journal of Comparative Neurology* **492**:101–109.

Polley DB, Read HL, Storace DA and Merzenich MM (2007) Multiparametric auditory receptive field organization across five cortical fields in the albino rat. *Journal of Neurophysiology* **97**:3621–3638.

Radtke-Schuller S (2001) Neuroarchitecture of the auditory cortex in the rufous horseshoe bat (*Rhinolophus rouxi*). *Anatomy & Embryology (Berlin)* **204**:81–100.

Radtke-Schuller S (2004) Cytoarchitecture of the medial geniculate body and thalamic projections to the auditory cortex in the rufous horseshoe bat (*Rhinolophus rouxi*). I. Temporal fields. *Anatomy & Embryology (Berlin)* **209**:59–76.

Radtke-Schuller S and Schuller G (1995) Auditory cortex of the rufous horseshoe bat: 1. Physiological response properties to acoustic stimuli and vocalizations and the topographical distribution of neurons. *European Journal of Neuroscience* **7**:570–591.

Rauschecker JP, Tian B, and Hauser M (1995) Processing of complex sounds in the macaque nonprimary auditory cortex. *Science* **268**:111–114.

Rauschecker JP, Tian B, Pons TP and Mishkin M (1997) Serial and parallel processing in rhesus monkey auditory cortex. *Journal of Comparative Neurology* **382**:89–103.

Read HL, Winer JA, and Schreiner CE (2002) Functional architecture of auditory cortex. *Current Opinion in Neurobiology* **12**: 433–440.

Reale RA and Imig TJ (1980) Tonotopic organization in auditory cortex of the cat. *Journal of Comparative Neurology* **192**:265–291.

Recanzone GH, Schreiner CE, Sutter ML, Beitel RE, and Merzenich MM (1999) Functional organization of spectral receptive fields in the primary auditory cortex of the owl monkey. *Journal of Comparative Neurology* **415**:460–481.

Redies H, Sieben U, and Creutzfeldt OD (1989) Functional subdivisions in the auditory cortex of the guinea pig. *Journal of Comparative Neurology* **282**:473–488.

Roger M and Arnault P (1989) Anatomical study of the connections of the primary auditory area in the rat. *Journal of Comparative Neurology* **287**:339–356.

Romanski LM and LeDoux JE (1993) Organization of rodent auditory cortex: anterograde transport of PHA-L from MGv to temporal neocortex. *Cerebral Cortex* **3**:499–514.

Rose JE (1949) The cellular structure of the auditory area of the cat. *Journal of Comparative Neurology* **91**:409–439.

Rose JE and Woolsey CN (1949a) Organization of the mammalian thalamus and its relationships to the cerebral cortex. *Electroencephalography and Clinical Neurophysiology* **1**:391–404.

Rose JE and Woolsey CN (1949b) The relations of thalamic connections, cellular structure and evocable electrical activity in the

auditory region of the cat. *Journal of Comparative Neurology* **91**:441–466.
Rouiller EM, Simm GM, Villa AEP, de Ribaupierre Y, and de Ribaupierre F (1991) Auditory corticocortical interconnections in the cat: evidence for parallel and hierarchical arrangement of the auditory cortical areas. *Experimental Brain Research* **86**:483–505.
Rutkowski RG, Miasnikov AA, and Weinberger NM (2003) Characterisation of multiple physiological fields within the anatomical core of rat auditory cortex. *Hearing Research* **181**:116–130.
Ryugo DK and Killackey HP (1974) Differential telencephalic projections of the medial and ventral divisions of the medial geniculate body of the rat. *Brain Research* **82**:173–177.
Sally SL and Kelly JB (1988) Organization of auditory cortex in the albino rat: sound frequency. *Journal of Neurophysiology* **59**: 1627–1638.
Scheich H and Zuschratter W (1995) Mapping of stimulus features and meaning in gerbil auditory cortex with 2-deoxyglucose and c-Fos antibodies. *Behavioral Brain Research* **66**:195–205.
Schreiner CE and Cynader MS (1984) Basic functional organization of second auditory cortical field (AII) of the cat. *Journal of Neurophysiology* **51**:1284–1305.
Schroeder CE, Lindsley RW, Specht C, Marcovici A, Smiley JF, and Javitt DC (2001) Somatosensory input to auditory association cortex in the macaque monkey. *Journal of Neurophysiology* **85**: 1322–1327.
Shamma SA, Fleshman JW, Wiser PR, and Versnel H (1993) Organization of response areas in ferret primary auditory cortex. *Journal of Neurophysiology* **69**:367–383.
Stecker GC, Harrington IA, Macpherson EA, and Middlebrooks JC (2005) Spatial sensitivity in the dorsal zone (area DZ) of cat auditory cortex. *Journal of Neurophysiology* **94**:1267–1280.
Sterbing-D'Angelo S (2007) Evolution of sound localization in mammals. In: Kaas JH (ed). *Evolution of Nervous Systems*, volume 3, Elsevier, London, pp. 253–260.
Stiebler I, Neulist R, Fichtel I, and Ehret G (1997) The auditory cortex of the house mouse: left-right differences, tonotopic organization and quantitative analysis of frequency representation. *Journal of Comparative Neurology* **181**:559–571.
Striedter GF (1997) The telencephalon of tetrapods in evolution. *Brain Behavior and Evolution* **49**:179–213.
Suga N (1990) Cortical computational maps for auditory imaging. *Neurological Networks* **3**:3–21.
Suga N (1994) Multi-function theory for cortical processing of auditory information: implications of single-unit and lesion data for future research. *Journal of Comparative Physiology A* **175**: 135–144.
Sweet RA, Dorph-Petersen KA, and Lewis DA (2005) Mapping auditory core, lateral belt, and parabelt cortices in the human superior temporal gyrus. *Journal of Comparative Neurology* **491**: 270–289.
Talavage TM, Sereno MI, Melcher JR, Ledden PJ, Rosen BR, and Dale AM (2004) Tonotopic organization in human auditory cortex revealed by progressions of frequency sensitivity. *Journal of Neurophysiology* **91**:1282–1296.
Taniguchi I, Horikawa J, Hosokawa Y, and Nasu M (1997) Optical imaging of neural activity in auditory cortex induced by intracochlear electrical stimulation. *Acta Otolaryngologica* **32**:83–88.
Thomas H, Tillein J, Heil P, and Scheich H (1993) Functional organization of auditory cortex in the Mongolian gerbil (*Meriones unguiculatus*). I. Electrophysiological mapping of frequency representation and distinction of fields. *European Journal of Neuroscience* **5**: 882–897.
Tootell RBH, Hamilton SL, and Silverman MS (1985) Topography of cytochrome oxidase activity in Owl Monkey cortex. *Journal of Neuroscience* **10**:2786–2800.
Tunturi AR (1962) Frequency arrangement in anterior ectosylvian auditory cortex of dog. *American Journal of Physiology* **203**: 185–193.
Tunturi AR (1970) The pathway from the medial geniculate body to the ectosylvian auditory cortex in the dog. *Journal of Comparative Neurology* **138**:131–136.
Ulinski PS (2007) Visual cortex of turtles. In: Kaas JH (ed). *Evolution of Nervous Systems*, volume 2. Elsevier, London, pp. 195–203.
Velenovsky DS, Cetas JS, Price RO, Sinex DG, and McMullen NT (2003) Functional subregions in primary auditory cortex defined by thalamocortical terminal arbors: an electrophysiological and anterograde labeling study. *Journal of Neuroscience* **23**: 308–316.
de Venecia RK, Smelser CB, and McMullen NT (1998) Parvalbumin is expressed in a reciprocal circuit linking the medial geniculate body and auditory neocortex in the rabbit. *Journal of Comparative Neurology* **400**:349–362.
Wallace MN and Bajwa S (1991) Patchy intrinsic connections of the ferret primary auditory cortex. *NeuroReport* **2**:417–420.
Wallace MN, Kitzes LM, and Jones EG (1991) Chemoarchitectonic organization of the cat primary auditory cortex. *Experimental Brain Research* **86**:518–526.
Wallace MN, Roeda P, and Harper MS (1997) Deoxyglucose uptake in the ferret auditory cortex. *Experimental Brain Research* **117**: 488–500.
Wallace MN, Rutkowski RG, and Palmer AR (2000) Identification and localization of auditory areas in guinea pig cortex. *Experimental Brain Research* **132**:445–456.
Winer JA (1984) Anatomy of layer IV in cat primary auditory cortex (AI). *Journal of Comparative Neurology* **222**:535–567.
Winer JA, Diamond IT, and Raczkowski D (1977) Subdivisions of the auditory cortex of the cat: the retrograde transport of horseradish peroxidase to the medial geniculate body and posterior thalamic nuclei. *Journal of Comparative Neurology* **176**:387–417.
Winer JA and Lee CC (2007) The distributed auditory cortex. *Hearing Research* **229**:3–13.
Wise LZ, Pettigrew JD, and Calford MB (1986) Somatosensory cortical representation in the Australian ghost bat. *Macroderma gigus*. *Journal Comparative Neurology* **248**:257–262.
Wood AE (1957) What, if anything, is a rabbit? *Evolution* **11**:417–425.
Wong D and Shannon SL (1988) Functional zones in the auditory cortex of the echolocating bat, *Myotis lucifugus*. *Brain Research* **453**: 349–352.
Woolsey CN (1971) Tonotopical organization of the auditory system. In: Sachs MB (ed). *Physiology of the Auditory System*. National Educational Consultants, Baltimore, pp. 271–282.
Woolsey CN, Carlton TG, Kaas JH, and Earls FJ (1971) Projection of visual field on superior colliculus of ground squirrel (*Citellus tridecemlineatus*). *Vision Research* **11**:115–127.
Woolsey CN and Walzl EM (1942) Topical projection of nerve fibers from local regions of the cochlea to the cerebral cortex of the cat. *Bulletin of Johns Hopkins Hospital* **71**:315–344.
Zook JM (2007) Somatosensory adaptations of flying mammals. In: Kaas JH (ed). *Evolution of Nervous Systems*, volume 3. Elsevier, London pp. 215–226.

Chapter 20

The Avian Auditory Pallium

Taffeta M. Elliott and Frédéric E. Theunissen

Abbreviations

AAF	anterior auditory field
AI	primary auditory cortex
AID	dorsomedial part of the intermediate arcopallium
AIV	ventromedial part of the intermediate arcopallium
AIVM	ventromedial nucleus of the intermediate arcopallium
Bas	nucleus basorostralis
BB	broad band
BOS	bird's own song
eMTF	ensemble modulation transfer function
CN	cochlear nuclei
GABA	γ-aminobutyric acid
HVC	letter-based proper name
IC	inferior colliculus
ICx	external nucleus of the inferior colliculus
ICo	intercollicular nucleus
IEG	immediately early gene
L1-L3	subregions of the auditory nidopallium Field L
LLD	dorsal nucleus of the lateral lemniscus
LLI	intermediate nucleus of the lateral lemniscus
LLV	ventral nucleus of the lateral lemniscus
M	mesopallium
MGB	medial geniculate body
MLd	dorsal part of the lateral mesencephalon
MTF	modulation transfer function
NA	angular nucleus
NB	narrow band
NCM	caudal medial nidopallium
Nd	dorsal nidopallium
NFl	lateral frontal nidopallium
NIVL	ventrolateral nidopallium intermedium
NL	laminar nucleus
NM	magnocellular nucleus
nVIII	auditory nerve
Ov	nucleus ovoidalis
PMI	nucleus paramedianus internus thalami
RA	robust nucleus of the arcopallium
SO	superior olive
SPO	nucleus semilunaris parovoidalis
SSA	stimulus-specific adaptation
STRF	spectrotemporal receptive field
TS	torus semicircularis
VMH	ventromedial hypothalamus
WB	wideband

F.E. Theunissen (✉)
Department of Psychology, University of California, Berkeley, CA 94720-1650, USA
e-mail: theunissen@berkeley.edu

1 Introduction

The functional and anatomical similarities between the avian auditory pallium and the mammalian auditory cortex are arguably as striking as their differences. Here, we hope to demonstrate the potential of a comparative approach in auditory physiology. On the one hand, birds and mammals face similar problems in auditory scene analysis and therefore it is not surprising to find evolutionary convergence in the functional strategies of information processing both by individual cells and by circuits. On the other hand, parallel evolution has resulted in similarly radial connections between layers in avian pallium, yet by means of dendrites that span fewer lamina than in the columnar connections within mammalian cortex (Wang 2010). Furthermore, feedback between primary sensory cortex and the thalamus is less extensive and involves indirect routes (Wang 2010). Although these anatomical differences are substantial, their functional significance is still unclear. This experiment of nature provides a conundrum, which rather than obfuscating functional value behind the cloud of circumstantial details may instead reveal the definitive characteristics in either system. We will examine these similarities and differences and explore how far we can go in making use of the opportunity offered by parallel evolution.

J.A. Winer, C.E. Schreiner (eds.), *The Auditory Cortex*,
DOI 10.1007/978-1-4419-0074-6_20, © Springer Science+Business Media, LLC 2011

Remarkably, song birds, parrots, and hummingbirds share the faculty of vocal learning with humans and a small number of mammals (bats, elephants, cetaceans) (Jarvis 2009a). In this ability, humans thus resemble birds more closely than most other mammals. Although the strongest implications of this functional similarity may be for the motor system, vocal learning has led in all these orders to complex signaling system for intra-species communication. These species' auditory systems are therefore also specialized for learning to recognize specific and individual vocal communication signals. Thus, the avian auditory system provides a model of how semantic auditory memories are formed. And, finally, the presence of vocal learning also allows for comparative study of coupling between the auditory and vocal system.

We begin with a brief overview of the evolution of mammalian and avian auditory systems and the auditory behavior of birds before reviewing the known anatomy and physiology of the avian pallium. Throughout the review, we compare aspects of avian and mammalian sound processing, with an eye toward convergent solutions to the challenge of recognizing complex sounds. We will not cover sound localization, although avian research has also played an important role in elucidating the neural underpinnings of that auditory percept (Cohen and Knudsen, 1999; Konishi, 2003; see Chapter 15).

2 Phylogeny of the Avian and Mammalian Auditory Systems

Tympanic hearing evolved separately in archosaurs and mammals since their last common ancestor, a stem amniote (lizard-like creature), which was acquiring terrestrial traits late in the Carboniferous period (Clack 2002; Jarvis 2009b). Thus, existing homologies indicate parallel evolution of solutions to shared problems in the neural coding of sound. Gross organization of the auditory system is conserved among vertebrates generally: most notably, the number of auditory nuclei is similar, as is the pattern of feed-forward connections from the cochlear nucleus to the auditory forebrain. Other anatomical and functional homologies between birds and mammals include: closed middle ear cavities, tonotopic organization (Zaretsky and Konishi 1976; Muller and Leppelsack 1985), the use of midbrain and peripheral binaural comparisons in sound localization, feature detection, convergence of sensory modalities along the ascending pathways, and descending modulation.

Communication systems specialized for vocal learning evolved in parallel in mammals (four groups: humans, bats, elephants, and cetaceans) and birds (three groups: songbirds, parrots, and hummingbirds) (Jarvis 2009a). Vocal abilities in these species are acquired through imitation and individuation, in contrast to the instinct that governs innate vocalizations produced by other animals (see Chapters 16 and 26). For these reasons, bird song learning has provided uniquely tractable opportunities for comparative investigations into the origin of human language. In the service of modeling human speech perception, comparisons of auditory processing in birds and mammals may reveal common principles and specialized adaptations.

3 Auditory Behavior

3.1 Psychophysical Studies in the Laboratory

Psychophysical measurements of auditory sensitivity and resolution largely resemble those in other vertebrates, including humans (Dooling et al. 2000). Specifically, perceptual tasks quantifying loudness threshold and discrimination, pitch discrimination, and temporal discrimination show a performance that is similar to that of humans or slightly worse (Dooling 1982). More importantly, songbirds appear to excel in psychoacoustical tasks involving sounds that have some of the spectral and temporal qualities of their own vocalizations (Lohr and Dooling 1998). Bird perception of song has become a model of complex sound processing (Theunissen and Shaevitz 2006).

3.2 Natural Auditory Behaviors

Understanding the processing of sound in the avian pallium may shed light on an impressive range of auditory and vocal behaviors. Birds use song and other vocalizations for an array of communication tasks in the wild. In males, songs are used for territorial defense and mate attraction (see Catchpole and Slater 1995) and in male–female pairs, songs are used for pair bonding and cooperation (Hile et al. 2000; Marshall-Ball and Slater 2004). While only males produce songs in most songbird species, both males and females produce communication calls that are more varied in function than song. Calls are used to maintain contact (contact call), restore contact (separation call), obtain food (begging call), or advertise danger (alarm call) (see Marler 2004). Male and female birds recognize the distance call of their mate and their response to these calls depends on the social context (Vignal et al. 2004).

The reproductive success of both male and female birds depends on auditory communication. Juvenile songbirds learn the song they use in courtship, first by detecting the song of a tutor, and then by learning the sensorimotor skill required to vocalize a similar song (see Chapter 26). Calls and songs both carry information for the listener about species identity, territorial familiarity, sexual receptivity, and kin relationships. Birds must also passively localize sound sources such as heterospecific prey and predators, as well as environmental sounds that inform them about their surroundings. Cave swiftlets and oil birds can even forage or detect nest locations by echolocating with clicks (Konishi and Knudsen 1979; Coles et al. 1987).

The behavioral discrimination of conspecific and heterospecific calls and song has been well documented in the laboratory (Dooling et al. 1992; Appeltants et al. 2005). Within a species, songbirds have also been shown to use song to discriminate between neighbors and strangers, relatives and non-relatives, mates and non-mates, and familiar and unfamiliar song (Clayton 1988; Searcy and Brenowitz 1988; Sherman et al. 1997; Riebel 2000; Riebel et al. 2002). The discrimination of subtle features of vocalizations has also been demonstrated with physiological measures by evaluating the effect of song features on hormonal levels (Gil et al. 2004; Marshall et al. 2005).

4 The Auditory Pallium: Culmination of the Central Auditory System

The pallium in the telencephalon of birds, like mammalian cortex, receives input from many ascending sensory pathways and is the source of sensory, motor, and modulatory descending pathways. The primary auditory pallium, in particular, receives projections from the auditory thalamus and has targets in secondary sensory areas which in turn project to the vocal motor nuclei of the song system.

4.1 Ascending Auditory Pathways to the Primary Auditory Pallium

Processing stages in the avian auditory system follow a gross anatomical plan quite similar to the mammalian auditory system, all the way from the ear to the secondary auditory areas in the telencephalon. Similarities between birds and mammals include the number of auditory nuclei and the pattern of feed-forward connections from the cochlear nucleus to the auditory forebrain (Fig. 20.1): afferents from the hair cells in the ear branch toward two primary nuclei in the medulla, called nucleus magnocellularis (NM) and nucleus angularis (NA), which are analogous to the anterior ventral and dorsal subdivisions of the mammalian cochlear nucleus, respectively. As in mammals, there are two parallel ascending pathways from these cochlear nuclei which converge in the auditory midbrain: a direct route that processes information about sound level and is monaural, and an indirect route that processes timing information by combining binaural information. In birds, the extra processing stage of the indirect route is from NM to the nucleus laminaris (NL), where sensitivity to interaural time differences emerges (Carr and Christensen-Dalsgaard 2009).

These two pathways pass through the superior olive and the nuclei of the lateral lemniscus before converging in the central nucleus of the midbrain, namely the dorsal lateral nucleus of the mesencephalon (MLd), which is analogous to the inferior colliculus (IC) in mammals (Konishi 2003), and to the torus semicircularis (TS) in reptiles (Carr and Code 2000). From the midbrain, MLd projects bilaterally to a central nucleus in the dorsal thalamus, nucleus Ovoidalis (Ov), just as the IC projects to the medial geniculate body (MGB) in mammals (Karten 1967). Field L, as the primary auditory area in the pallium, is the principal recipient of ascending input from this dorsal auditory thalamic nucleus. In this sense, Field L is analogous to the primary auditory cortical areas A1 and AAF in mammals.

4.2 Subregion Connectivity: Afferent Inputs and Projections

Field L is comprised of three sandwiched layers: L1, L2, and L3 (Bonke et al. 1979a). Subfield L2 receives the primary thalamic input, whereas L1 and L3 are output layers containing the projection neurons of field L. L2 contains a high concentration of cytochrome oxidase and is comprised of L2a and L2b, subdivisions which differ in cytoarchitecture and their reception of parallel ascending projections (Fortune and Margoliash 1992; Wild et al. 1993; Vates et al. 1996). A fifth subregion known simply as "L" is indistinct from subregion L2b in cell morphology and arrangement; L comprises the ventro-caudal extent of L2b (Fortune and Margoliash 1992).

4.2.1 Afferent Inputs to Pallium

Auditory thalamic input to field L goes primarily to subregions L2a and L2b, which in turn project to layers L1 and L3 (Wang et al. 2010). The thalamic nucleus ovoidalis (Ov) comprises the principal part of the input (Karten 1968; Zaretsky and Konishi 1976) and targets subregion L2a only. The remainder of the input, from nucleus semilunaris parovoidalis (SPO) and nucleus ovoidalis shell (OVs), targets L2b mainly, but also L1 and L3 (Fig. 20.1) (Wild et al. 1993; Carr and Code 2000).

4.2.2 Projections from Pallium

Subregions L1 and L3 make bi-directional connections with two secondary auditory areas in the pallium: the nidopallium

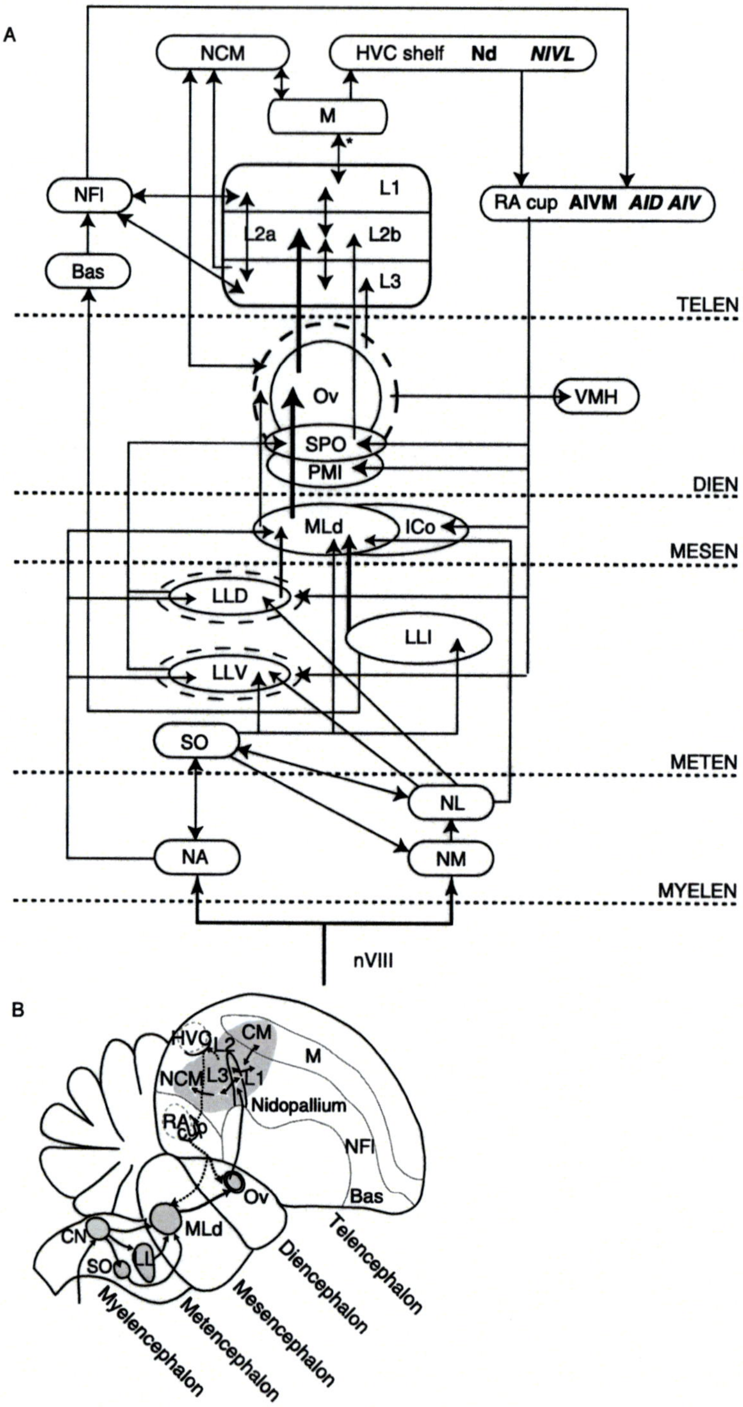

Fig. 20.1 **a** Schematic of the avian auditory pathway, adapted from Fig. 5.9 of Carr and Code (2000) and Fig. 17 of Wild et al. (1993) with modifications (Durand et al. 1993; Vates et al. 1996; Farabaugh and Wild 1997; Wild et al. 2010). Only major projections are depicted. *Thicker arrows* indicate the densest projections. In nuclei with multiple labels, *labels in plain font* apply to oscines, *bold labels* apply to *Columba livia*, and *italic labels* apply to budgerigars. The *asterisk* (*) indicates that oscines and galliforms have a reciprocal connection between Field L and M, whereas *C. livia* does not. *Dashed lines* indicate shell regions of nuclei; *dotted lines* delineate major structures of the brain. **b** Composite cartoon of parasagittal sections through the oscine auditory pathway, with auditory nuclei in *gray*, and *dotted lines* outlining song nuclei targets HVC and RA (adjacent shell and cup regions, respectively, are enclosed in *solid lines*). Only the densest connections are shown. L1 and L3 are layers of field L parallel to the outline of layer L2. *Solid arrows* are feed-forward connections, and feedback connections are drawn as *dotted arrows*. The cochlear nuclei (CN) include NA, NM, and NL from **a**

caudo-medial (NCM) surrounding the medial border of Field L and the adjacent mesopallium (M) (Wild et al. 1993; Wang et al. 2010), particularly in oscines to the caudal mesopallium (CM), which is ventral in the mesopallium (Vates et al. 1996). There is, therefore, a symmetry in the projections from L2: through L1 to M on the one hand, and through L3 to NCM on the other hand. The potentially distinct functional roles of these two parallel auditory pathways are currently explored (see below).

Although there is no direct feedback from the primary auditory pallium to the auditory thalamus or auditory midbrain, there are feedback connections from secondary avian auditory areas that might play a similar role as the corticothalamic connections in mammals. In *Columba livia*, (rock pigeon) L1 and L3 project to the dorsal nidopallium; in oscines, to the dorsal nidopallium area surrounding the song nucleus HVC, the HVC shelf; and in budgerigars, to the ventro-lateral nidopallium (Wild et al. 1993; Metzger et al. 1998). These nidopallium targets in turn project, in oscines, to the caudal arcopallium in the area surrounding the song nucleus RA, known as the RA cup; or, in *C. livia* and budgerigars, to the intermediate arcopallium (Wild et al. 1993). This arcopallial stage then projects to shell regions around the thalamic and midbrain auditory nuclei (Wild et al. 1993; Mello et al. 1998). Uniquely in budgerigars, however, the major auditory input to the vocal nuclei of the arcopallium originates in nucleus basorostralis and passes through the frontal nidopallium (Durand et al. 1997; Striedter 1997). It has been argued that the connectivity of this descending auditory pathway is similar to that in mammals (Jarvis 2009b; Wang et al. 2010) although the physiology and corresponding function in both systems remain relatively unknown.

4.2.3 Internal Organization and Cell Types in Field L, NCM, and CM

The microcircuitry of higher auditory areas, the morphology of neuron types, and their cellular properties have not been examined in great detail. Golgi stains suggest the presence of at least four types of neurons in the auditory forebrain (Saini and Leppelsack 1981; Fortune and Margoliash 1992). There are GABA-ergic stellate neurons packed densely in the auditory pallium (particularly in L1, L2a, L3, NCM, and CMM), and these are presumed to be inhibitory interneurons (Pinaud and Mello 2007). NCM has a particularly high number of inhibitory interneurons with tuning properties distinct from those found in excitatory neurons (Pinaud and Mello 2007; Pinaud et al. 2008).

The four neuronal cell types in field L are categorized by soma size (cell type 1: 12 μm; types 2 and 3: 8–10 μm; type 4: 5–6 μm), as well as by morphology and diameter of dendritic arborization (type 1: >200 μm; type 2: 130–200 μm; type 3: <100 μm; type 4: stumpy) (Fig. 27.2a) (Saini and Leppelsack 1981; Fortune and Margoliash 1992). Furthermore, type 3 can be divided into "unoriented" cells with spherical dendritic arborization and "oriented" cells with dendrites extending parallel to the plane of L2a in any parasagittal section (Fortune and Margoliash 1992). Oriented type 3 cells are found exclusively in L2a and nearby L2b. L1 contains fewer type 1 cells and L3 contains fewer type 4 cells than a random distribution would predict (Fortune and Margoliash 1992).

Interestingly, following passive exposure to song, all the regions of the avian auditory pallium show activation of the immediately early gene (IEG) *zenk*, with the exception of the thalamic recipient area L2 (see Mello 2002). As described below (and see Fig. 1), the gross anatomical organization of the pallium and the heterogeneity of cell types are starting to be investigated in relation to the functional diversity observed in physiological studies.

4.2.4 Comparison to Mammalian Anatomy

Amniotes are diverse in the structure and function of their telencephalic organization. Increased use of sound for communication by vertebrates correlates with large changes in organization. Noticeable differences between the mammalian and the avian auditory system are observed in the feedback and inter-hemispheric connectivity patterns. In mammals, the primary auditory cortex (AI) shows strong feedback projections to the thalamus and more limited ones to the midbrain. In birds this feedback circuitry exists, but it involves two additional processing stages in the forebrain, in the shell regions of song system structures HVC and the robust nucleus of the arcopallium (RA) (Mello et al. 1998). Feedback circuitry lies in similar anatomical locations in non-songbirds (Wild et al. 1993). The direct inter-hemispheric connectivity within primary auditory cortex present in mammals is absent in birds.

Another major difference between the pallium of birds and the neocortex of mammals lies in the cytoarchitectonic organization. However, similar circuit configurations, such as connections between thalamus and pallial areas, may make up for differences in the particular configuration of cytoarchitecture (Butler and Hodos 2005). More specifically, it has been proposed that L2 area could be analogous to cortical layer 4 of AI; L1, L3, (and by extension) NCM, and CM to cortical layers 2 and 3; and HVC shelf and RA cup to layers 5 and 6 (Wang et al. 2010). Whether this connectivity-based analogy holds for other cellular, molecular, and physiological properties remains to be determined (Karten 1991; Medina and Reiner 2000; Jarvis et al. 2005).

4.3 Response Properties of Pallial Auditory Neurons

4.3.1 Tonotopy

As in the mammalian system, neurons in the avian auditory pallium have characteristic frequency responses which together form tonotopic representations. In field L, neighboring cells have overlapping frequency tuning curves, and the shared component of their bandwidth (but not the "best frequencies" which elicit their strongest responses) forms a single tonotopic representation in which isofrequency contours pass through all three subfield lamina. The major cochleotopic gradient represents low frequencies dorso-laterally and higher frequencies ventro-medially (Wild et al. 1993), with increasing frequencies from L2b to L2a (Zaretsky and Konishi 1976; Bonke et al. 1979b; Heil and Scheich 1985; Muller and Leppelsack 1985; Rübsamen and Dorrscheidt 1986; Scheich 1990).

In the mammalian auditory cortex, tonotopy is used to define the boundaries of the two primary fields, AI and AAF. Similarly, in birds, the auditory pallium may be further divided into functional subregions based on a regular organization of frequency selectivity (Gehr et al. 1999; Terleph et al. 2007). However, in order to validate these preliminary findings in songbirds, more detailed mapping studies must still be performed.

4.3.2 Spectrotemporal Tuning

Neurons in the avian auditory pallium exhibit complex neural responses that cannot be explained by a linear frequency tuning. Sensitivity to temporal context (e.g., syllable combinations and longer term memory effects) and complex spectral patterns (e.g., a harmonic feature) have been explored by many studies both in primary auditory area field L (Leppelsack and Vogt 1976; Leppelsack 1978, 1983; Muller and Leppelsack 1985; Lewicki and Arthur 1996; Hausberger et al. 2000; Grace et al. 2003; Amin et al. 2004; Boumans et al. 2007) and in secondary areas NCM and CM (Stripling et al. 1997; Gentner and Margoliash 2003; Phan et al. 2006; Terleph et al. 2006, 2007; Bauer et al. 2008; Pinaud et al. 2008). In all these studies, an attempt is made to relate the complex response tuning properties to the recognition of communication sounds (with a focus on song) or to the memory of sounds. This neuroethological approach is powerful in directly investigating putative functions of high-level sensory areas (see also Section 4.3.3.), but it can only indirectly be related to the results from the classical approach used more extensively in mammalian research in which responses to more simply described synthetic sounds are characterized. The neuroethological approach also faces greater leaps in explaining the mechanisms of how complex responses arise from auditory circuitry.

A third alternative approach is to systematically characterize the joint frequency and temporal tuning of neurons and to relate such tuning to functions of feature extraction and to anatomical structure. Neurons in Field L can be reasonably well described by their spectrotemporal receptive fields (STRFs) (Theunissen et al. 2000; Sen et al. 2001), in particular when the model incorporates compressive static non-linearities and gain control (Gill et al. 2006). STRFs in field L are diverse and show complex tuning properties (Sen et al. 2001; Cousillas et al. 2005; Woolley et al. 2005). The distribution of tuning properties in Field L has been analyzed by classifying neurons into functional groups (Fig. 20.2b) (Nagel and Doupe 2008; Woolley et al. 2009). The clustering of STRF shape properties indicates that groups of auditory cells are specialized to represent distinct spectrotemporal modulation features that cue the fundamental auditory percepts of pitch, timbre, and rhythm (Woolley et al. 2009). This functional clustering suggests the presence of parallel networks associated with different percepts, and these parallel networks can extract distinct information-bearing features in song. The tuning for temporal modulations is correlated with structural properties of neurons, with cells that exhibit selectivity for slower modulations having lower firing rates and wider spike wave forms (Nagel and Doupe 2008). These observations provide initial evidence for a gross anatomical organization of functional properties, in particular that more narrowband neurons are found in region L2.

As in the ascending mammalian pathway, spectrotemporal tuning in the songbird primary auditory pallium (field L) either conforms to that found in the midbrain or else shows more complexity, such as tuning for harmonicity (Woolley et al. 2009). Responses in the secondary auditory area, CM, are less linear than those found in field L in the sense that the STRF model (even with gain control included) predicts a smaller portion of the response variance than in Field L (Gill et al. 2008). Neural sensitivity for spectrotemporal features in CM is better understood in terms of a response to surprising features given expectations based on the sound statistics found in song. This computation results in a highly efficient representation of song that might be useful for memory formation (see below). Neuronal responses in NCM show interesting transient and sustained temporal phases that are shaped by a large inhibitory network. It is the sustained response that is heavily dependent on GABA-ergic inhibition, and these long responses are postulated to play a role in facilitating memory formation (Pinaud et al. 2008).

The studies mentioned above set the stage for a more detailed analysis of the micro-circuitry in the auditory pallium, with the ultimate goal of relating structure to physiological properties and, in turn, physiological properties to feature extraction functions.

4.3.3 Songbird Selectivity for Communication Signals

The ascending auditory pathway in songbirds shows increasing levels of specialization for processing natural sounds (Hsu et al. 2004), and in particular conspecific vocalizations (see Theunissen and Shaevitz 2006). Information theoretic measures provide evidence that natural sounds, particularly vocalizations, are efficiently encoded in the response patterns

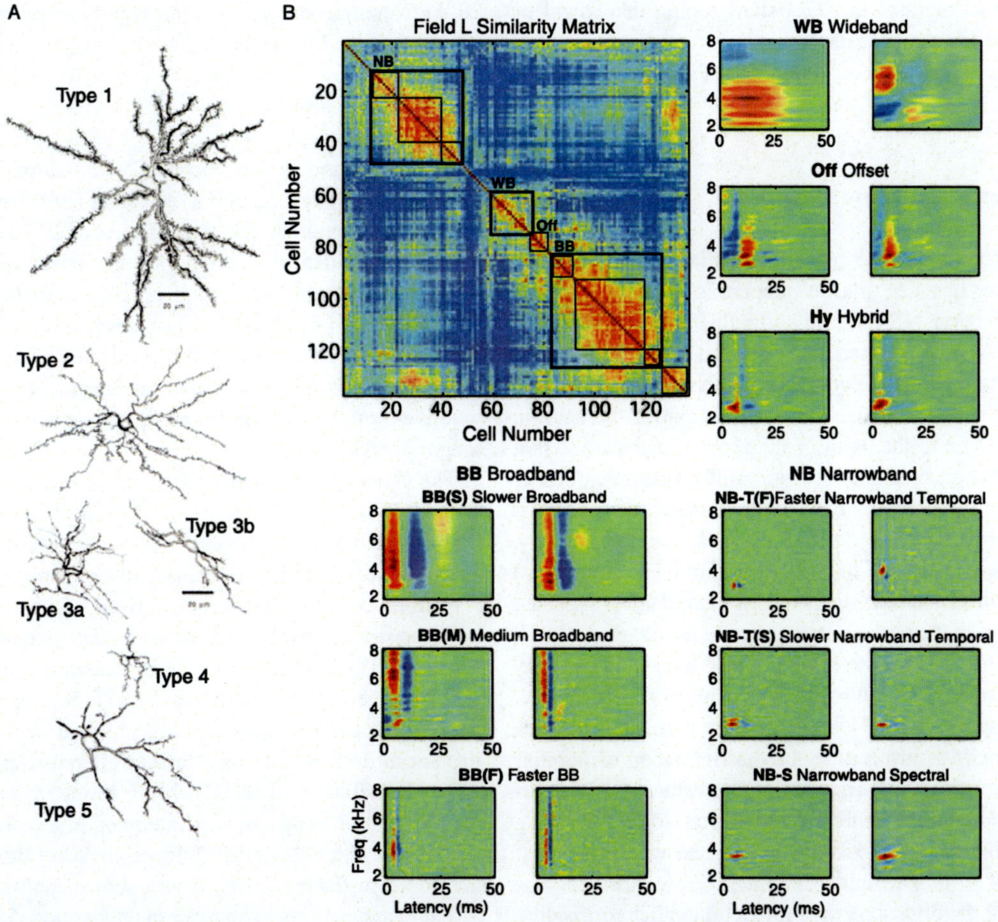

Fig. 20.2 **a** Camera lucida drawings of four cell types in auditory pallial region field L of the zebra finch, with some out-of-plane dendrites omitted. Type 3a cell is unoriented; the 3b cell is the "oriented" type with dendrites extended parallel to L2a in this parasagittal section. Dorsal is up. Axons are labeled with "Ax" or an *arrow*. Scale bars apply to all cells. Reproduced with modifications from Fortune and Margoliash (1992). **b** Functional groups of auditory neurons in field L were defined using the similarity matrix (*top left*) comparing pairs of cells using a genetic algorithm. Clusters comprised five groups: broadband (BB) and narrowband (NB) were the largest; wideband (WB), offset (Off), and hybrid (Hy – not labeled on matrix) groups exhibited receptive field features observed in fewer cells. Example spectrotemporal receptive fields (STRFs, *bottom three rows and right column*) from the five groups illustrate the typical acoustic features preferentially eliciting responses in neurons of each group. Reproduced from Woolley et al. (2009)

of primary auditory neurons (Hsu et al. 2004). The efficiency of the processing is manifested in the matching of ensemble tuning properties to the informative sound features present in song (Woolley et al. 2005).

Field L neurons select for spectrotemporal modulations that are common in song, accurately representing the most prevalent acoustic features (Theunissen et al. 2004). Auditory responses in field L neurons thus resemble those in presynaptic midbrain neurons (MLd), insofar as information rates increase in both field L and MLd when stimuli contain the spectrotemporal statistics of natural sounds. But many field L neurons show greater selectivity because concomitantly their response to simple synthetic sounds is weaker (Theunissen and Shaevitz 2006). Furthermore, selectivity is implicated in vocal learning, because in juveniles the selectivity for conspecific song increases at the developmental stage in which song preferences emerge (Amin et al. 2007).

Similarly, neurons in lateral CM of adult zebra finches are selective for complex natural sounds, and their information rates are disproportionately higher for vocalizations than for complex synthetic sounds. In comparison to field L, NCM and CM neurons have even stronger selectivity for the bird's own song, as well as for familiar songs with behavioral importance (Theunissen et al. 2004) (see Section 4.3.4.).

Immediate early gene (IEG) expression in NCM has been shown to be largest for songs that have greater behavioral significance, such as conspecific song (Mello et al. 1992; Gentner et al. 2001). Similarly, neurophysiological recordings of neurons in NCM show tuning that might be specialized for the acoustic structure found in the calls of conspecifics (Terleph et al. 2006, 2007; Pinaud et al. 2008).

4.3.4 Development and Plasticity

Tuning properties in the secondary auditory areas CM and NCM have revealed plastic properties that require further examination. NCM exhibits stimulus-specific adaptation (SSA). SSA manifests itself as a reduced neural response to repeated stimulation. However, unlike the ubiquitous neural adaptation which is a function of the output of the neuron, SSA is specific to the input of the neuron in the sense that the presentation of a novel (or unfamiliar) stimulus during SSA yields an unadapted response magnitude. SSA can last days and can therefore be considered a form of memory. SSA has been measured in NCM using both IEG studies and neurophysiological recordings. NCM habituates to repeated presentation of the same conspecific song (Mello et al. 1995), and the degree of adaptation is correlated with song familiarity (Chew et al. 1995, 1996; Stripling et al. 1997). These IEG and electrophysiological experiments are consistent with the idea that NCM is involved in the discrimination of familiar songs relative to novel songs. It has also been suggested that this discrimination of familiar songs extends to the tutor song and that therefore NCM could be the site where the neural trace of a tutor template is found. IEG expression was stronger for the tutor song than for an unfamiliar conspecific song and the strength of the response was correlated with the degree of how well the tutor song was learned (Bolhuis et al. 2000; Terpstra et al. 2004). More recently, it was observed that responses to the tutor song show SSA that was characteristic of very familiar songs even when they had not been heard for a prolonged period of time. The familiarity index was correlated with how well the bird was able to copy the song (Phan et al. 2006). Responses in NCM might also be modulated by social context, since the strength of IEG expression in response to calls depends on the presence of other conspecific birds (Vignal et al. 2005).

Auditory neurons in CM, particularly in its more medial extent (CMM), show properties suggestive of a potential role for both memory and song discrimination. Lesion, IEG, and neurophysiological studies have implicated CMM in perception of familiar conspecific song (Gentner et al. 2001). A lesion study in female zebra finches showed that CMM but not HVC was important for song discrimination for mate choice (MacDougall-Shackleton et al. 1998). Similarly, an IEG study in female zebra finches showed that the *zenk* response in CMM to the female birds' father song correlated with the degree of learning measured in behavioral tests of preference for songs like the father's (Terpstra et al. 2006). Neurophysiological recordings have shown that single neurons, and the ensemble of neurons, in CMM become more responsive to conspecific song that is being learned in a perceptual discrimination task (Gentner and Margoliash 2003).

Finally, as mentioned above, CMM neurons appear to show selectivity for the bird's own song (BOS) and are also responsive to auditory feedback (Bauer et al. 2008). On the other hand, evidence is lacking for any selectivity for BOS or tutor song in neuronal responses in the more lateral extent of CM (CLM) (Amin et al. 2004; Shaevitz and Theunissen 2007). Instead of being tuned for conspecific song in general, neurons in CLM are tuned to unexpected features of sounds given expectations about the statistics of conspecific song (Gill et al. 2008).

The avian auditory system has great potential as a model to study the effect of sensory experience (beyond tutor song) on neural development. The first study on this subject observed significant changes in the response of field L neurons of birds that were deprived of normal acoustical experience during early development (Cousillas et al. 2004). Both the selectivity and the organization of frequency tuning properties were altered. In a follow-up study, the same group showed that not only physical deprivation but also social deprivation could lead to altered response properties (Cousillas et al. 2008). More recently, the effect of species-specific exposure was examined in a cross-fostering experiment where Bengalese finches served as surrogate parents to zebra finches. First, it was shown that the auditory midbrain and pallium of Bengalese finches and zebra finches did show selective tuning for species-specific song. However, there were significant differences in auditory responsivity and neural discriminability across the two species: zebra finch auditory neurons had higher response rates to song, and higher discrimination quantified by information metrics, than auditory neurons of Bengalese finches. The cross-fostering experiment then showed that this species difference was in large part influenced by early sensory exposure since the auditory responses in the cross-fostered zebra finches were more similar to Bengalese finches than to normal zebra finches (Woolley et al. 2010).

4.3.5 Auditory Scene Analysis

Auditory scene analysis resolves mixtures of sounds into recognizable descriptions that are distinct from background noise, by segregating the sensory components arising from distinct environmental sources into separate perceptual representations (Micheyl et al. 2007). Avian research has begun to address the neural basis of scene analysis. In one study, the effect of noise on auditory responses to behaviorally relevant signals was examined in field L. It was found that the neural signal for song was degraded by noise and this neural decrease in signal-to-noise ratio was correlated with a decrease in behavioral performance (Narayan et al. 2007). However, the same results also show that field L neurons do not separate signal from noise.

Sound localization helps in the formation of distinct auditory objects. Spatially selective neurons are found in the auditory arcopallium of the barn own (*Tyto alba*), which result from field L input rather than input from the auditory space map in the external nucleus of the inferior colliculus (ICx) (Cohen et al. 1998). Inactivation by pharmacology showed that this pallial pathway can independently compute spatial localizations without input from the midbrain localization pathway.

4.3.6 Functional Comparison to Mammalian Physiology

There are many functional similarities between the physiological properties of the avian pallium and the mammalian auditory cortex. In both groups, there is strong evidence for hierarchical processing of sound features (see Eggermont 2001). This hierarchical processing is evident when neural responses are evaluated either in terms of their tuning properties (Gill et al. 2008; Woolley et al. 2009) or in terms of their information content (Hsu et al. 2004; Chechik et al. 2006).

Neurons in the avian auditory pallium and mammalian auditory cortex (Depireux et al. 2001; Miller et al. 2002) are relatively well described by their STRFs, although there are context effects that cannot be captured in a single STRF model (Theunissen et al. 2000; Ahrens et al. 2008; Gourevitch et al. 2009). The spectrotemporal tuning of mammalian cortical neurons and avian field L neurons can be compared by examining the ensemble modulation transfer function (eMTF) of each population (Escabi and Schreiner 2002; Woolley et al. 2005). The MTF is the gain of an STRF obtained by taking the amplitude component of its 2D Fourier transform. The eMTF can thus be interpreted as the density function for modulation gain. The eMTFs of avian pallial and mammalian cortex neurons are shown in Fig. 20.3. The gain distributions have a similar shape: lower

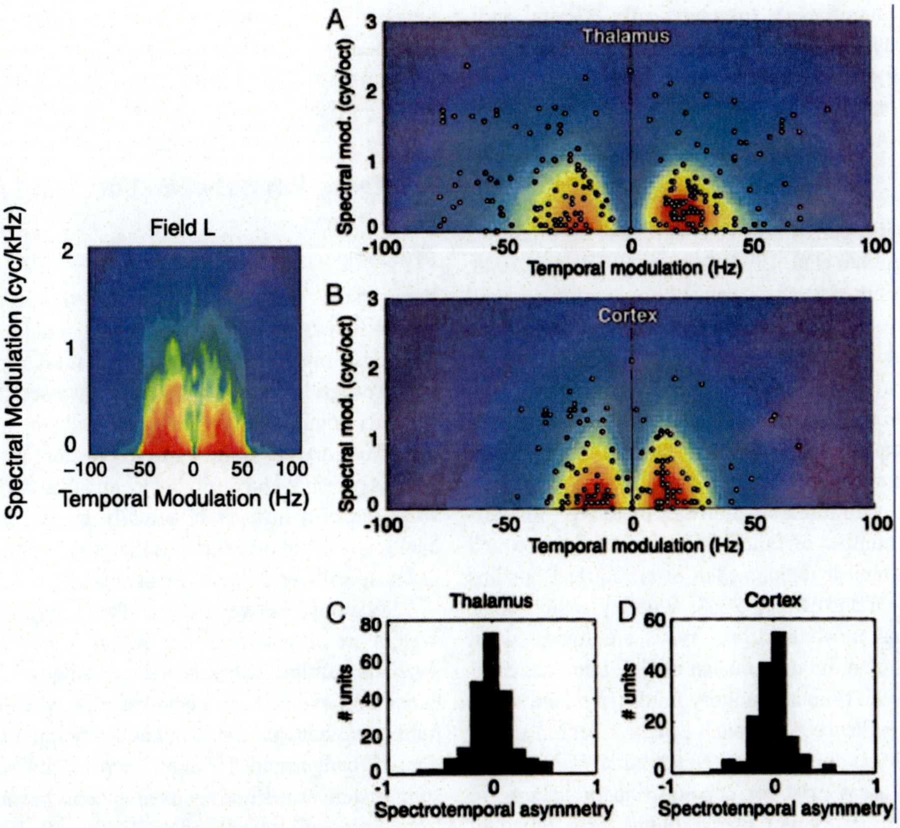

Fig. 20.3. A comparison of the ensemble tuning of neurons in the avian primary auditory pallium, Field L (*left panel*) and the mammalian primary auditory cortex, A1 (*right panel*). The plots show the composite gain of the STRF obtained for single neurons in the space of temporal (*x*-axis) and spectral modulations (*y*-axis). The avian data were obtained in adult male zebra finches, and the mammalian data in adult cats. Note that the spectral modulation axes have a different scale and are in different units (cycles/kHz vs. cycles/octave). Reproduced from Woolley et al. (2005) and from Miller et al. (2002)

spectral modulations and intermediate temporal modulations are emphasized. Woolley et al. (2005) have argued that the overall gain in the avian eMTF is advantageous for processing natural sounds because it emphasizes the sound features that vary the most across natural sounds. The same argument could be made for the mammalian eMTF despite a difference in tuning, namely that the mammalian cortical neurons show less gain for higher temporal modulations. It is possible that avian pallial neurons could be faster than mammalian neurons and that their need for speed might have an ecological origin. However, before this speculation can be generalized it should include evidence from as yet unobtained eMTFs from additional avian and mammalian species. Also, cortical responses of the granular layer and L2 will have to be distinguished from those in other cortical layers, and in L1, L3, CM, and NCM. At the single neuron level, both field L and A1 exhibit a diverse set of simple and more complex STRFs. A cluster analysis of the mammalian STRFs similar to that performed by Woolley et al. (2009) would reveal whether similar functional groups are found in both classes (aves and mammals). This functional analysis should be performed with the behavioral correlates in mind. Does the functional organization correspond to perceptual dimensions or to ecologically relevant auditory tasks? In this respect, avian auditory research might be ahead of mammalian research. But analysis of structural bases for functions of feature processing is further along in the mammalian literature. Even though anatomical specialization remains controversial at the level of cortical areas (see below), it has been demonstrated in the mammalian system that tuning properties differ between neurons in different cortical layers (Sugimoto et al. 1997; Wallace and Palmer 2008; Atencio and Schreiner 2010) as well as between different cell types. Excitatory and inhibitory neurons in the mammalian auditory cortex have distinct STRFs (Atencio and Schreiner 2008).

Both the avian auditory pallium and the mammalian auditory cortex appear to lack a clear functional organization at the level of a map or cytoarchitectural representation. For example, in the avian research, although there are some differences in the number of functional neuronal types found in different anatomical regions (Sen et al. 2001; Boumans et al. 2007; Nagel and Doupe 2008; Woolley et al. 2009), it is striking that most functional types are found in all regions. Similarly, in the mammalian cortex, there are clear differences between primary auditory fields (Imaizumi et al. 2004); there is evidence for a pitch-sensitive region in the primate secondary auditory cortex (Bendor and Wang 2005); and specialized areas exist for echolocation processing in bats (Suga et al. 1978); but many studies have failed to find clearly differentiated parallels in pathways or functional specializations within primary or secondary auditory areas (Nelken et al. 2008). In contrast, neuroimaging studies have suggested specializations in human cortical areas (Hall et al. 2003). Also, neural tracing experiments in animals are consistent with parallels in processing streams (Rouiller et al. 1991). Thus, the search in the mammalian and avian auditory systems for parallel pathways processing the distinct acoustic features that mediate different percepts remains an active and somewhat controversial area of research (Griffiths et al. 2004).

Response plasticity that is enhanced in comparison to the auditory processing at lower brain regions is a trait that has been observed in both the mammalian cortex and the avian pallium. For example, SSA has been found in the secondary auditory pallial area NCM and in auditory cortical areas but not in the thalamus (Ulanovsky et al. 2003), but see Anderson et al. (2009). In avian research, the SSA has been linked to the formation of long-term memories, or familiarity, whereas in the mammalian literature, the function of SSA has been linked to the detection of low-probability events. In both systems, the underlying mechanisms and the actual functions of SSA remain to be elucidated. Plasticity in tuning properties have also been found as a result of learning in both systems (e.g., Gentner and Margoliash 2003). Thus, song birds might offer an advantageous opportunity to study the link between the formation of short-term and long-term auditory memories.

5 Interaction Between Vocal and Auditory Systems

Birds have a specialized set of interconnected motor nuclei, known as the "song system," which control singing. Songbirds must learn the song of a tutor and then learn the motor programs that produce their own song. Vocal learning depends upon auditory perception influencing the song system. Auditory areas must convey distinct information about what is possible and desirable to sing (the tutor's song), about what the bird himself is actually singing (real-time feedback), and about other sounds that trigger singing (e.g., songs of conspecific males and female calls).

Interactions between the auditory system and the vocal system are an active area of research and current evidence supports multiple pathways (see Chapter 26). Species differences exist even among vocal learners, whereas in songbirds, field L provides the principal auditory input to the song system; in budgerigars (a small parrot), auditory input to the song system arises mainly from nucleus basorostralis and the frontal nidopallium (Striedter 1994).

In oscines, the first and best established auditory projection is from secondary auditory area CLM (lateral CM) to the song nucleus NIf, which in turn projects to HVC (Vates et al. 1996). Second, CM appears to project directly to HVC (Shaevitz and Theunissen 2007; Bauer et al. 2008). As to the purpose of these dual pathways, it is hypothesized that CM might be a secondary auditory region that is specialized for processing auditory feedback. Characterizations of neurons in CM that select for the BOS seem to support this hypothesis (Bauer et al. 2008). Tentative observations indicate that sparse connections may exist also between field L and HVC shelf, and HVC shelf and HVC proper, but confirmation is still pending (Fortune and Margoliash 1995). Auditory information might also affect the song system indirectly. For example, Ov projects not only to auditory pallium but also to a cholinergic nucleus in the ventral pallidum (VP) (Li et al. 2000). VP in turn projects to the song nuclei HVC and RA. Thus VP, which could be involved in song learning, may be regulated in part by auditory information originating in Ov (Hall et al. 2003).

The nature of acoustic information entering the song system is another area of active research. Because selectivity for the BOS does not arise solely in field L or CM but increases in HVC and the nucleus interface (NIf) (Bauer et al. 2008), the selectivity of the song system may originate in a single sensorimotor processing step between the auditory and vocal systems (Amin et al. 2004).

Real-time auditory feedback has been recorded in birds both in the song nuclei (Sakata and Brainard 2008) and in the primary auditory pallium (Keller and Hahnloser 2009). In the auditory pallium study, some neurons in field L were shown to be particularly sensitive to perturbed song feedback. These responses are reminiscent of the depressed responses to vocal feedback that are found in primate auditory cortex, which lead to enhanced sensitivity for perturbations (Eliades and Wang 2008). This striking functional similarity between aves and mammals again suggests convergent neural computations for shared problems.

6 Conclusions

The avian auditory pallium is a complex network of primary and secondary regions that show a heterogeneous set of complex and plastic response properties. Although the neural substrates in birds and mammals appear to be quite different, circuit analogies can be made both at the levels of brain regions and cellular processing. Perhaps more remarkably, physiological responses in the two animal classes share similarities that we strongly believe to be the product of convergent evolution selecting for solutions to similar problems in auditory scene analysis.

The potential benefit of further comparative research is underscored by historical reflection. From 1960 to 1980, the avian auditory system was studied with the classical auditory physiological approaches that were also used in mammalian research (e.g., frequency-intensity response curves using pure tones). With the advance of vocal research in songbirds from 1980 to 2000, most avian auditory research took on an ethological bend to focus primarily on the processing of communication sounds. More recently, the two approaches have been combined. Researchers in the avian research have re-embraced more general techniques employing complex synthetic sounds and spectrotemporal receptive field estimation to more systematically probe the auditory pallium. In doing so, the avian research has found once more a common ground for sharing techniques and ideas with mammalian research. In addition, avian researchers have maintained their focus on natural behavior, which has begun to influence mammalian work. This comparative approach is truly synergistic and holds great promise for advancing our understanding of auditory function.

Acknowledgments FET is grateful to have known Prof. Winer and to have benefited from his unwavering support at the earlier stages of his career in auditory science. The two authors thank Prof. Schreiner for his patience and careful editing. The work done by the Theunissen laboratory described here was supported by grants from the NIDCD and the NIMH.

References

Ahrens MB, Linden JF, and Sahani M (2008) Nonlinearities and contextual influences in auditory cortical responses modeled with multilinear spectrotemporal methods. *Journal of Neuroscience* **28**: 1929–1942.

Amin N, Grace JA, and Theunissen FE (2004) Neural response to bird's own song and tutor song in the zebra finch field L and caudal mesopallium. *Journal of Comparative Physiology A* **190**:469–489.

Amin N, Doupe A, and Theunissen FE (2007) Development of selectivity for natural sounds in the songbird auditory forebrain. *Journal of Neurophysiology* **97**:3517–3531.

Anderson LA, Christianson GB, and Linden JF (2009) Stimulus-specific adaptation occurs in the auditory thalamus. *Journal of Neuroscience* **29**:7359–7363.

Appeltants D, Gentner TQ, Hulse SH, Balthazart J, and Ball GF (2005) The effect of auditory distractors on song discrimination in male canaries (*Serinus canaria*). *Behavioral Processes* **69**:331–341.

Atencio CA and Schreiner CE (2008) Spectrotemporal processing differences between auditory cortical fast-spiking and regular-spiking neurons. *Journal of Neuroscience* **28**:3897–3910.

Atencio CA and Schreiner CE (2010) Laminar diversity of dynamic sound processing in cat primary auditory cortex. *Journal of Neurophysiology* **103**:192–205.

Bauer EE, Coleman MJ, Roberts TF, Roy A, Prather JF, and Mooney R (2008) A synaptic basis for auditory-vocal integration in the songbird. *Journal of Neuroscience* **28**:1509–1522.

Bendor D and Wang X (2005) The neuronal representation of pitch in primate auditory cortex. *Nature* **436**:1161–1165.

Bolhuis JJ, Zijlstra GG, den Boer-Visser AM, and Van Der Zee EA (2000) Localized neuronal activation in the zebra finch brain is related to the strength of song learning. *Proceedings of the National Academy of Sciences of the United States of America* **97**:2282–2285.

Bonke BA, Bonke D, and Scheich H (1979a) Connectivity of the auditory forebrain nuclei in the Guinea fowl *Numida-meleagris*. *Cell and Tissue Research* **200**:101–122.

Bonke D, Scheich H, and Langner G (1979b) Responsiveness of units in the auditory neostriatum of the guinea fowl (*Numida meleagris*) to species-specific calls and synthetic stimuli I. Tonotopy and functional zones of field L. *Journal of Comparative Physiology A* **132**:243–255.

Boumans T, Theunissen FE, Poirier C, and Van Der Linden A (2007) Neural representation of spectral and temporal features of song in the auditory forebrain of zebra finches as revealed by functional MRI. *European Journal of Neuroscience* **26**:2613–2626.

Butler AB and Hodos W (2005) *Comparative Vertebrate Neuroanatomy*: Wiley, Hoboken.

Carr CE and Code RA (2000) The central auditory system of reptiles and birds. In: Dooling RJ, Fay RR, and Popper AN (eds). *Springer Handbook of Auditory Research*, volume 13. *Comparative Hearing: Birds and Reptiles* Springer, New York, pp. 197–248.

Carr CE and Christensen-Dalsgaard J (2009) Evolution of the auditory system. In: Binder MD, Hirokawa N, and Windhorst U (eds). *Encyclopedia of Neuroscience, Reptiles and Birds*. Springer, Berlin, p. 3145

Catchpole CK and Slater PJB (1995) *Bird Song: Biological Themes and Variations*. Cambridge University Press, Cambridge.

Chechik G, Anderson MJ, Bar-Yosef O, Young ED, Tishby N, and Nelken I (2006) Reduction of information redundancy in the ascending auditory pathway. *Neuron* **51**:359–368.

Chew SJ, Vicario DS, and Nottebohm F (1996) A large-capacity memory system that recognizes the calls and songs of individual birds. *Proceedings of the National Academy of Sciences of the United States of America* **93**:1950–1955.

Chew SJ, Mello C, Nottebohm F, Jarvis E, and Vicario DS (1995) Decrements in auditory responses to a repeated conspecific song are long-lasting and require two periods of protein synthesis in the songbird forebrain. *Proceedings of the National Academy of Sciences of the United States of America* **92**:3406–3410.

Clack JA (2002) Patterns and processes in the early evolution of the tetrapod ear. *Journal of Neurobiology* **53**:251–264.

Clayton NS (1988) Song tutor choice in zebra finches and Bengalese finches the relative importance of visual and vocal cues. *Behaviour* **104**:281–299.

Cohen YE and Knudsen EI (1999) Maps versus clusters: different representations of auditory space in the midbrain and forebrain. *Trends in Neuroscience* **22**:128–135.

Cohen YE, Miller GL, and Knudsen EI (1998) Forebrain pathway for auditory space processing in the barn owl. *Journal of Neurophysiology* **79**:891–902.

Coles RB, Konishi M, and Pettigrew JD (1987) Hearing and echolocation in the Australian gray swiftlet Collocalia-spodiopygia. *Journal of Experimental Biology* **129**:365–372.

Cousillas H, George I, Henry L, Richard J-P, and Hausberger M (2008) Linking social and vocal brains: could social segregation prevent a proper development of a central auditory area in a female songbird? *Public Library of Science ONE* **3**:e2194.

Cousillas H, Richard J-P, Mathelier M, Henry L, George I, and Hausberger M (2004) Experience-dependent neuronal specialization and functional organization in the central auditory area of a songbird. *European Journal of Neuroscience* **19**:3343–3352.

Cousillas H, Leppelsack HJ, Leppelsack E, Richard JP, Mathelier M, and Hausberger M (2005) Functional organization of the forebrain auditory centres of the European starling: a study based on natural sounds. *Hearing Research* **207**:10–21.

Depireux DA, Simon JZ, Klein DJ, Shamma SA (2001) Spectrotemporal response field characterization with dynamic ripples in ferret primary auditory cortex. *Journal of Neurophysiology* **85**: 1220–1234.

Dooling RJ (1982) Auditory perception in birds. In: Kroodsma DE and Miller EH (eds). *Acoustic Communication in Birds*. Academic Press, New York, pp. 95–130.

Dooling RJ, Lohr B, and Dent ML (2000) Hearing in birds and reptiles. In: Dooling RJ, Fay RR, and Popper AN (eds). *Springer Handbook of Auditory Research*, volume 13, *Comparative Hearing: Birds and Reptiles*. Springer, New York, pp. 308–359.

Dooling RJ, Brown SD, Klump GM, and Okanoya K (1992) Auditory perception of conspecific and heterospecific vocalizations in birds: evidence for special processes. *Journal of Comparative Psychology* **106**:20–28.

Durand SE, Zuo MX, Zhou SL, and Cheng M-F (1993) Avian auditory pathways show met-enkephalin-like immunoreactivity. *Neuroreport* **4**:727–730.

Durand SE, Ehaton JT, Amateau SA, and Brauth SE (1997) Vocal control pathways through the anterior forebrain of a parrot (*Melopsittacus undulatus*). *Journal of Comparative Neurology* **377**:179–206.

Eggermont JJ (2001) Between sound and perception: reviewing the search for a neural code. *Hearing Research* **157**:1–42.

Eliades SJ and Wang X (2008) Neural substrates of vocalization feedback monitoring in primate auditory cortex. *Nature* **453**:1102–1106.

Escabi MA and Schreiner CE (2002) Nonlinear spectrotemporal sound analysis by neurons in the auditory midbrain. *Journal of Neuroscience* **22**:4114–4131.

Farabaugh SM and Wild JM (1997) Reciprocal connections between primary and secondary auditory pathways in the telencephalon of the budgerigar (*Melopsittacus undulatus*). *Brain Research* **747**: 18–25.

Fortune ES and Margoliash D (1992) Cytoarchitectonic organization and morphology of cells of the field L complex in male zebra finches (*Taenopygia guttata*). *Journal of Comparative Neurology* **325**: 388–404.

Fortune ES and Margoliash D (1995) Parallel pathways and convergence onto HVc and adjacent neostriatum of adult zebra finches (*Taeniopygia guttata*). *Journal of Comparative Neurology* **360**: 413–441.

Gehr DD, Capsius B, Grabner P, Gahr M, and Leppelsack HJ (1999) Functional organisation of the field-L-complex of adult male zebra finches. *Neuroreport* **10**:375–380.

Gentner TQ and Margoliash D (2003) Neuronal populations and single cells representing learned auditory objects. *Nature* **424**:669–674.

Gentner TQ, Hulse SH, Duffy D, and Ball GF (2001) Response biases in auditory forebrain regions of female songbirds following exposure to sexually relevant variation in male song. *Journal of Neurobiology* **46**:48–58.

Gil D, Leboucher G, Lacroix A, Cue R, and Kreutzer M (2004) Female canaries produce eggs with greater amounts of testosterone when exposed to preferred male song. *Hormons and Behavior* **45**:64–70.

Gill P, Woolley SMN, Fremouw T, and Theunissen FE (2008) What's that sound? Auditory area CLM encodes stimulus surprise, not intensity or intensity changes. *Journal of Neurophysiology* **99**: 2809–2820.

Gill P, Zhang J, Woolley SM, Fremouw T, and Theunissen FE (2006) Sound representation methods for spectro-temporal receptive field estimation. *Journal of Computational Neuroscience* **21**:5–20.

Gourevitch B, Norena A, Shaw G, and Eggermont JJ (2009) Spectrotemporal receptive fields in anesthetized cat primary auditory cortex are context dependent. *Cerebral Cortex* **19**: 1448–1461.

Grace JA, Amin N, Singh NC, and Theunissen FE (2003) Selectivity for conspecific song in the zebra finch auditory forebrain. *Journal of Neurophysiology* **89**:472–487.

Griffiths TD, Warren JD, Scott SK, Nelken I, and King AJ (2004) Cortical processing of complex sound: a way forward? *Trends in Neurosciences* **27**:181–185.

Hall DA, Hart HC, and Johnsrude IS (2003) Relationships between human auditory cortical structure and function. *Audiology and Neurotology* **8**:1–18.

Hausberger M, Leppelsack E, Richard J, and Leppelsack HJ (2000) Neuronal bases of categorization in starling song. *Behavioral Brain Research* **114**:89–95.

Heil P and Scheich H (1985) Quantitative analysis and two-dimensional reconstruction of the tonotopic organization of the auditory field L in the chick from 2-deoxyglucose data. *Experimental Brain Research* **58**:532–543.

Hile AG, Plummer TK, and Striedter GF (2000) Male vocal imitation produces call convergence during pair bonding in budgerigars, *Melopsittacus undulatus*. *Animal Behaviour* **59**:1209–1218.

Hsu A, Woolley SM, Fremouw TE, and Theunissen FE (2004) Modulation power and phase spectrum of natural sounds enhance neural encoding performed by single auditory neurons. *Journal of Neuroscience* **24**:9201–9211.

Imaizumi K, Priebe NJ, Crum PA, Bedenbaugh PH, Cheung SW, and Schreiner CE (2004) Modular functional organization of cat anterior auditory field. *Journal of Neurophysiology* **92**:444–457.

Jarvis ED (2009a) Bird song systems: evolution. In: Squire LR (ed). *Encyclopedia of Neuroscience*. Academic Press, Oxford. pp. 217–225.

Jarvis ED (2009b) Evolution of the pallium in birds and eptiles. In: Binder MD, Hirokawa N, and Windhorst U (eds). *New Encyclopedia of Neuroscience*. Springer, Berlin, pp. 15–30.

Jarvis ED et al. (2005) Avian brains and a new understanding of vertebrate brain evolution. *Nature Reviews in Neuroscience* **6**:151–159.

Karten HJ (1967) The organization of the ascending auditory pathway in the pigeon (*Columba livia*) I. Diencephalic projections of the inferior colliculus (nucleus mesencephali lateralis, pars dorsalis). *Brain Research* **6**:409–427.

Karten HJ (1968) The ascending auditory pathway in the pigeon *Columba-livia* telencephalic projections of the nucleus ovoidalis thalami mammals. *Brain Research* **11**:134–153.

Karten HJ (1991) Homology and evolutionary origins of the neocortex. *Brain Behavior and Evolution* **38**:264–272.

Keller GB and Hahnloser RHR (2009) Neural processing of auditory feedback during vocal practice in a songbird. *Nature* **457**: 187–190.

Konishi M (2003) Coding of auditory space. *Annual Review of Neuroscience* **26**:31–55.

Konishi M and Knudsen EI (1979) The oilbird: hearing and echolocation. *Science* **204**:425–427.

Leppelsack HJ (1978) Unit responses to species-specific sounds in the auditory forebrain center of birds. *Federal Proceedings* **37**: 2336–2341.

Leppelsack HJ (1983) Analysis of song in the auditory pathway of song birds. In: Ewert JP, Capranica RR, and Ingle DJ (eds). *Advances in Vertebrate Neuroethology*. Plenum, New York, pp 783–799.

Leppelsack HJ and Vogt M (1976) Responses of auditory neurons in the forebrain of a songbird to stimulation with species-specific sounds. *Journal of Comparative Physiology A* **107**:263–274.

Lewicki MS and Arthur BJ (1996) Hierarchical organization of auditory temporal context sensitivity. *Journal of Neuroscience* **16**: 6987–6998.

Li R, Taniguchi I and Sakaguchi H (2000) Auditory-vocal cholinergic pathway in the songbird brain. *Canadian Journal of Physiology and Pharmacology* **78**:1072–1076.

Lohr B and Dooling RJ (1998) Detection of changes in timbre and harmonicity in complex sounds by zebra finches (*Taeniopygia guttata*) and budgerigars (*Melopsittacus undulatus*). *Journal of Comparative Psychology* **112**:36–47.

MacDougall-Shackleton SA, Hulse SH, and Ball GF (1998) Neural bases of song preferences in female zebra finches (*Taeniopygia guttata*). *Neuroreport* **9**:3047–3052.

Marler P (2004) Bird calls: their potential for behavioral neurobiology. *Annals of the New York Academy of Sciences* **1016**:31–44.

Marshall-Ball L and Slater PJB (2004) Duet singing and repertoire use in threat signalling of individuals and pairs. *Proceedings of the Royal Society Biological Sciences Series B* **271**: S440–S443.

Marshall RC, Leisler B, Catchpole CK, and Schwabl H (2005) Male song quality affects circulating but not yolk steroid concentrations in female canaries (*Serinus canaria*). *Journal of Experimental Biology* **208**:4593–4598.

Medina L and Reiner A (2000) Do birds possess homologues of mammalian primary visual, somatosensory and motor cortices? *Trends in Neurosciences* **23**:1–12.

Mello C, Nottebohm F, and Clayton D (1995) Repeated exposure to one song leads to a rapid and persistent decline in an immediate early gene's response to that song in zebra finch telencephalon. *Journal of Neuroscience* **15**:6919–6925.

Mello CV (2002) Mapping vocal communication pathways in birds with inducible gene expression. *Journal of Comparative Physiology A* **188**:943–959.

Mello CV, Vicario DS, and Clayton DF (1992) Song presentation induces gene expression in the songbird forebrain. *Proceedings of the National Academy of Sciences of the United States of America* **89**:6818–6822.

Mello CV, Vates GE, Okuhata S, and Nottebohm F (1998) Descending auditory pathways in the adult male zebra finch (*Taeniopygia guttata*). *Journal of Comparative Neurology* **395**:137–160.

Metzger M, Jiang S, and Braun K (1998) Organization of the dorsocaudal neostriatal complex: a retrograde and anterograde tracing study in the domestic chick with special emphasis on pathways relevant to imprinting. *Journal of Comparative Neurology* **395**:380–404.

Micheyl C, Carlyon RP, Gutschalk A, Melcher JR, Oxenham AJ, Rauschecker JP, Tian B, and Courtenay Wilson E (2007) The role of auditory cortex in the formation of auditory streams. *Hearing Research* **229**:116–131.

Miller LM, Escabi MA, Read HL, and Schreiner CE (2002) Spectrotemporal receptive fields in the lemniscal auditory thalamus and cortex. *Journal of Neurophysiology* **87**:516–527.

Muller CM and Leppelsack HJ (1985) Feature extraction and tonotopic organization in the avian auditory forebrain. *Experimental Brain Research* **59**:587–599.

Nagel KI and Doupe AJ (2008) Organizing principles of spectrotemporal encoding in the avian primary auditory area field L. *Neuron* **58**:938–955.

Narayan R, Best V, Ozmeral E, McClaine E, Dent M, Shinn-Cunningham B, and Sen K (2007) Cortical interference effects in the cocktail party problem. *Nature Neuroscience* **10**:1601–1607.

Nelken I, Bizley JK, Nodal FR, Ahmed B, King AJ, and Schnupp JWH (2008) Responses of auditory cortex to complex stimuli: functional organization revealed using intrinsic optical signals. *Journal of Neurophysiology* **99**:1928–1941.

Phan ML, Pytte CL, and Vicario DS (2006) Early auditory experience generates long-lasting memories that may subserve vocal learning in songbirds. *Proceedings of the National Academy of Sciences of the United States of America* **103**:1088–1093.

Pinaud R and Mello CV (2007) GABA immunoreactivity in auditory and song control brain areas of zebra finches. *Journal of Chemical Neuroanatomy* **34**:1–21.

Pinaud R, Terleph TA, Tremere LA, Phan ML, Dagostin AA, Leao RM, Mello CV, and Vicario DS (2008) Inhibitory network interactions shape the auditory processing of natural communication signals in the songbird auditory forebrain. *Journal of Neurophysiology* **100**:441–455.

Riebel K (2000) Early exposure leads to repeatable preferences for male song in female zebra finches. *Proceedings of the Royal Society Biological Sciences Series B* **267**:2553–2558.

Riebel K, Smallegange IM, Terpstra NJ, and Bolhuis JJ (2002) Sexual equality in zebra finch song preference: evidence for a dissociation between song recognition and production learning. *Proceedings of the Royal Society Biological Sciences Series B* **269**:729–733.

Rouiller EM, Simm GM, Villa AEP, Ribaupierre Y, and Ribaupierre F (1991) Auditory corticocortical interconnections in the cat: evidence for parallel and hierarchical arrangement of the auditory cortical areas. *Experimental Brain Research* **86**:483–505.

Rübsamen R and Dorrscheidt G (1986) Tonotopic organization of the auditory forebrain in a songbird, the European starling. *Journal of Comparative Physiology A* **158**:639–646.

Saini KD and Leppelsack HJ (1981) Cell types of the auditory caudomedial neostriatum of the starling (*Sturnus vulgaris*). *Journal of Comparative Neurology* **198**:209–229.

Sakata JT and Brainard MS (2008) Online contributions of auditory feedback to neural activity in avian song control circuitry. *Journal of Neuroscience* **28**:11378–11390.

Scheich H (1990) Representational geometries of telencephalic auditory maps in birds and mammals. In: Finlay BL, Innocenti G, and Scheich H (eds).*The Neocortex -Ontogeny and Phylogeny*. Plenum, New York, pp. 119–135.

Searcy WA and Brenowitz EA (1988) Sexual differences in species recognition of avian song. *Nature* **332**:152–154.

Sen K, Theunissen FE and Doupe AJ (2001) Feature analysis of natural sounds in the songbird auditory forebrain. *Journal of Neurophysiology* **86**:1445–1458.

Shaevitz SS and Theunissen FE (2007) Functional connectivity between auditory areas Field L and CLM and song system nucleus HVC in anesthetized zebra finches. *Journal of Neurophysiology* **98**: 2747–2764.

Sherman PW, Reeve HK, and Pfennig DW (1997) Recognition systems. In: Krebs JR and Davies NB (eds), *Behavioral Ecology*. Blackwell Science, Malden, pp 69–96.

Striedter GF (1994) The vocal control pathways in budgerigars differ from those in songbirds. *Journal of Comparative Neurology* **343**: 35–56.

Striedter GF (1997) The telencephalon of tetrapods in evolution. *Brain Behavior and Evolution* **49**:179–213.

Stripling R, Volman SF, and Clayton DF (1997) Response modulation in the zebra finch neostriatum: relationship to nuclear gene regulation. *Journal of Neuroscience* **17**:3883–3893.

Suga N, O'Neill WE, and Manabe T (1978) Cortical neurons sensitive to combinations of information-bearing elements of biosonar signals in the mustache bat. *Science* **200**:778–781.

Sugimoto S, Sakurada M, Horikawa J, and Taniguchi I (1997) The columnar and layer-specific response properties of neurons in the primary auditory cortex of Mongolian gerbils. *Hearing Research* **112**:175–185.

Terleph TA, Mello CV, and Vicario DS (2006) Auditory topography and temporal response dynamics of canary caudal telencephalon. *Journal of Neurobiology* **66**:281–292.

Terleph TA, Mello CV, and Vicario DS (2007) Species differences in auditory processing dynamics in songbird auditory telencephalon. *Developmental Neurobiology* **67**:1498–1510.

Terpstra NJ, Bolhuis JJ, and den Boer-Visser AM (2004) An analysis of the neural representation of birdsong memory. *Journal of Neuroscience* **24**:4971–4977.

Terpstra NJ, Bolhuis JJ, Riebel K, van der Burg JM, and den Boer-Visser AM (2006) Localized brain activation specific to auditory memory in a female songbird. *Journal of Comparative Neurology* **494**:784–791.

Theunissen FE and Shaevitz SS (2006) Auditory processing of vocal sounds in birds. *Current Opinion in Neurobiology* **16**: 400–407.

Theunissen FE, Sen K, and Doupe AJ (2000) Spectral-temporal receptive fields of nonlinear auditory neurons obtained using natural sounds. *Journal of Neuroscience* **20**:2315–2331.

Theunissen FE, Amin N, Shaevitz SS, Woolley SM, Fremouw T, and Hauber ME (2004) Song selectivity in the song system and in the auditory forebrain. *Annals of the New York Academy of Sciences* **1016**:222–245.

Ulanovsky N, Las L, and Nelken I (2003) Processing of low-probability sounds by cortical neurons. *Nature Neuroscience* **6**:391–398.

Vates GE, Broome BM, Mello CV, and Nottebohm F (1996) Auditory pathways of caudal telencephalon and their relation to the song system of adult male zebra finches. *Journal of Comparative Neurology* **366**:613–642.

Vignal C, Mathevon N, and Mottin S (2004) Audience drives male songbird response to partner's voice. *Nature* **430**:448–451.

Vignal C, Andru J, and Mathevon N (2005) Social context modulates behavioural and brain immediate early gene responses to sound in male songbird. *European Journal of Neuroscience* **22**: 949–955.

Wallace M and Palmer A (2008) Laminar differences in the response properties of cells in the primary auditory cortex. *Experimental Brain Research* **184**:179–191.

Wang Y, Brzozowska-Prechtl A, and Karten HJ (2010) Laminar and columnar auditory cortex in avian brain. *Proceedings of the National Academy of Sciences* **107**:12676–12681.

Wild JM, Karten HJ, and Frost BJ (1993) Connections of the auditory forebrain in the pigeon (*Columba livia*). *Journal of Comparative Neurology* **337**:32–62.

Wild JM, Krützfeldt NOE, and Kubke MF (2010) Connections of the auditory brainstem in a songbird, *Taeniopygia guttata*. III. Projections of the superior olive and lateral lemniscal nuclei. *Journal of Comparative Neurology* **518**:2149–2167.

Woolley SM, Hauber ME, and Theunissen FE (2010) Developmental experience alters information coding in auditory midbrain and forebrain neurons. *Developmental Neurobiology* **70**:235–252.

Woolley SM, Fremouw TE, Hsu A, and Theunissen FE (2005) Tuning for spectro-temporal modulations as a mechanism for auditory discrimination of natural sounds. *Nature Neuroscience* **8**: 1371–1379.

Woolley SMN, Gill PR, Fremouw T, and Theunissen FE (2009) Functional groups in the avian auditory system. *Journal of Neuroscience* **29**:2780–2793.

Zaretsky MD and Konishi M (1976) Tonotopic organization in the avian telencephalon. *Brain Research* **111**:167–171.

Chapter 21

Development of the Auditory Cortex

Andrej Kral and Sarah L. Pallas

Abbreviations

AC	auditory cortex
AI	primary auditory cortex
AMPA	a-amino-3-hydroxy-5-methyl-4-isoxazole-4-propionic acid
BMP	bone morphogenetic protein
EI	excitatory–inhibitory
EPSP	excitatory postsynaptic potential
GABA	gamma-aminobutyric acid
IC	inferior colliculus
IPSP	inhibitory postsynaptic potential
LTD	long-term depression
LTP	long-term potentiation
MGB	medial geniculate body
MMN	mismatch negativity
NMDA	*N*-methyl-D-aspartic acid
PSP	postsynaptic potential
TCA	thalamocortical afferents
VI	primary visual cortex
VZ	ventricular zone

1 The Ontogenetic Framework

Neuronal development is a progressive series of constructive and reductive events including division of progenitors, their accretion at specific locations, differentiation into neuronal and glial subtypes, and circuit refinement. The final goal is to establish adaptive neuronal circuits controlling the behavior of the organism. The complex architecture of the adult auditory cortex (AC) is thus the consequence of many developmental processes taking place prenatally and postnatally. The end of the developmental period is traditionally defined by sexual maturity; however, substantial adaptations in cortical circuitry continue throughout life. We identify some rules applicable to cortical development in general and to AC in particular, concentrating on the species most common in hearing research. We build on comparative reviews on the structural and functional development of the auditory system (Payne 1992; Cant 1998; Sanes and Walsh 1998; Romand 1997; Yan 2003). We also consider studies on the AC structural and functional plasticity during development. Studies on adult plasticity are beyond the scope of this analysis.

AC requires considerable early plasticity because of the complex behaviors it mediates. Auditory object recognition involves learning, and much of it early in life, given the considerable evolutionary pressure to interpret the meaning of environmental sounds. The exceptional neocortical capacity for adjustment to external conditions has been well known since studies of the of immature monkey cortex found that it was more adaptive in response to damage than mature cortex (Kennard 1938). The ontogenetic period of enhanced adaptability allows the organism to respond optimally to postnatal environmental conditions. The cortex matures substantially as the peripheral sensory and motor organs become functional. The concept of neocortical circuitry as a blank slate (tabula rasa) in newborns has been proposed since some neurons initially connect in an apparently random fashion (Kalisman et al. 2005). From such non-specific early connectivity, meaningful circuits are later selected by experience. However, to what degree the cortical circuits represent a neonatal blank slate, and whether they are predisposed toward input of certain patterns, remains uncertain.

In many species the behavioral repertoire is limited at birth, whereas others are born with a more mature behavioral program. Two primary groups are recognized: altricial species (e.g., mice, rats, rabbits, ferrets, cats) are born relatively early in gestation, with immature sensory organs,

A. Kral (✉)
Institute of Audioneurotechnology and Department of Experimental Otology, ENT Clinics, Medical University Hannover, Feodor-Lynen-Str. 35, D-30625 Hanover, Germany
e-mail: a.kral@uke.de

S.L. Pallas (✉)
Neuroscience Institute, Georgia State University, Atlanta, GA 30302-5030, USA
e-mail: spallas@gsu.edu

J.A. Winer, C.E. Schreiner (eds.), *The Auditory Cortex*,
DOI 10.1007/978-1-4419-0074-6_21, © Springer Science+Business Media, LLC 2011

closed eyelids and external ear canals, and the middle ear filled with a viscous fluid. Their orientation toward stimuli is mainly based on the somatic sensory and olfactory systems, and their motor behavior is rudimentary and dominated by reflexive responses. In contrast, precocial species (e.g., ungulates, guinea pigs, chinchillas) emerge with functional peripheral sensory organs and a more mature motor system. Despite their postnatal brain development, some essential behavioral programs are already available at birth. Primates are in a special position with respect to brain development, with sensory systems that become functional during intrauterine life and are available at birth, and a motor system that is immature. Central circuitry is also immature and the postnatal maturational sequence is extended.

Table 21.1 Gestation times (days) of representative laboratory species

Gestation times	
Rat	21.5
Hamster	15.5
Mouse	18.5
Ferret	41
Cat	65
Macaque	165
Human	270

Due to this developmental diversity (gestation times of common laboratory species are in Table 21.1), it is of dubious value to compare individual developmental events across different species with birth date as a reference. A model enabling cross-species developmental comparisons is available (Clancy et al. 2001). A cardinal milestone for interspecific comparisons of cortical developmental stages is the arrival of thalamic afferents in the cortical plate, where they can directly influence cortical developmental events. A model enabling cross-species developmental comparisons has been proposed (Clancy et al. 2001). AC development can therefore be divided into the time before thalamic afferents arrive and after their arrival (Table 21.2). We define early development as before thalamic afferents have reached the cortex, and late development as after this event. However thalamic afferents activate can also cortical neurons indirectly, before they have entered the cortex (via the cortical subplate; see below).

Table 21.2 Arrival of thalamic afferents in the cortex of representative species in days postconception and the earliest auditory brain stem-evoked responses; to obtain the conceptional age dates, the gestation duration (Table 21.1) must be added. In humans, the date appears as gestational week since brain stem responses have been tested prenatally. In monkeys, acoustic ability develops prenatally, but electrophysiology was performed postnatally. C.a.: conceptional age. For details see the original studies (reviewed in Cant 1998; Clancy et al. 2001)

Cortex: thalamic innervation			
	Subplate	Layer IV	ABRs
Mouse	16.2	21.2	P12
Rat	17.5	25/P3.5	P11
Ferret	37	50.5	P27
Cat	41.5	61.5	P12
Macaque	78	91	Before birth
Human	93.1	130.2	24th week c.a.

Late cortical development can be further subdivided: phase 1 shows an absence of electrically evoked activity due to cochlear immaturity. Thus, this period is after the arrival thalamic afferents in cortex, but before hearing onset (characterized by the first evoked cortical responses). Spontaneous activity likely influences developmental connectivity in this phase. Very early auditory spontaneous activity is characterized by bursts (Friauf and Kandler 1990; Gummer and Mark 1994; Lippe 1994; Tritsch and Bergles, 2010), which may contribute significantly to the interconnection patterns in the immature auditory system (Friauf and Kandler 1990). Phase 2 is characterized by rapid developmental changes immediately after hearing onset. In phase 3 these developmental processes slow down and continue until late childhood, sexual maturity, or beyond.

2 Early Cortical Development

Cerebral cortex develops from the telencephalic vesicle of the embryonic forebrain, a process involving many different molecular signals that specify the three-dimensional patterning of the cells into columns, layers, and areas. These include transcription factors and secreted morphogens. Essential steps include proliferation of neuronal and glial precursors, establishment of regional and, eventually, areal positional information, and migration of postmitotic neurons to the cortical plate and to their final laminar position. Establishing synaptic connections begins at the end of this stage and continues into late cortical development, when it is heavily influenced by neuronal activity. In what follows, we consider the molecular factors controlling each step (Price and Willshaw 2000; Erzurumlu et al. 2006; Rubenstein 2010).

The establishment of positional identity in the nervous system begins before neural tube formation (Shimamura et al. 1997; Lee and Jessell 1999). After induction of the neuroepithelium at gastrulation by noggin and chordin (Sasai and De Robertis 1997), gradients of sonic hedgehog (shh) arising from the floor plate and opposing gradients of bone morphogenetic protein (BMP) from the roof plate interact to establish the dorsoventral central nervous system axis. The anteroposterior axis is defined by retinoic acid, FGF, and antagonists of Wnt and BMP such as Dkk and noggin (Glinka et al. 1997, 1998; Kudoh et al. 2002). Forebrain–hindbrain segregation occurs via antagonistic interactions between Otx2 and Gbx2. Dorsal and ventral telencephalons have differential expression of Pax6 and Emx1/2 dorsally and Dlx ventrally. There is a rostral-to-caudal progression of cortical maturation orchestrated by temporal expression gradients of these molecular factors.

The positional identity of individual cortical neurons, defined in part by their final location, restricts their fate and thus their function. Specification of the fate is a stepwise process directed by cascades of transcription factors controlling aspects of neuronal identity including laminar fate, cellular morphology, neurotransmitter production, and other unique features. Glutamatergic pyramidal cell fate is controlled by a cascade that includes Pax6, Tbr-2/1, and NeuroD (Hevner et al. 2006). The gamma-aminobutyric acid-accumulating (GABAergic) interneurons that eventually reside in cerebral cortex arise not from the ventricular zone of dorsal telencephalon, but from the ventral telencephalic ganglionic eminence, from which they migrate into the dorsal telencephalon and their final positions in the cortical plate (Wonders and Anderson 2006).

Regional cortical patterning is under the control of several genes, and among the most well studied are Pax6 and Emx2 (Manuel and Price 2005). These transcription factors are particularly interesting because they are expressed in opposing gradients in the embryonic cortical epithelium, with Pax6 at high levels rostrolaterally and Emx2 high caudomedially. This arrangement could allow assignment of a unique identity to each topographic location, as do diencephalon and midbrain gradients of ephrins and their Eph receptors (Uziel et al. 2006). Consistent with this idea, a lack of Pax6 or Emx2 retards the formation of areas in which they were normally highly expressed (Bishop et al. 2000, 2003; Mallamaci et al. 2000) and causes mistargeting of thalamocortical afferents (TCAs). Subsequent work found that the TCAs were diverted to the ventral telencephalon rather than the cortical plate (Pratt et al. 2002; Molnar et al. 2003). Pax6 and Emx2 probably work competitively to control the number of cells that exit the cell cycle to become neurons (Heins et al. 2001; Estivill-Torrus et al. 2002). Although it seems unlikely that these genes directly define cortical areal borders, they may indirectly establish restricted gene expression patterns that control subsequent and precise thalamocortical targeting (see below).

The actual parcellation of neocortex into different functional areas occurs through unknown mechanisms. Certainly auditory cortex achieves its auditory identity because it receives information arising from the cochleae, but why it receives that modality of input and not another is not at all clear. Indeed, redirection experiments, in which a normal target of sensory axons in a modality is removed and another target in an alternative modality made available, show cross-modal colonization. We address these questions next.

2.1 Terminal Phase of Early Development: Arrival of Thalamic Afferents

The cortex develops in an inside–out pattern (Fig. 21.1): neurons in deeper layers arrive and differentiate before upper cortical layers (McConnell 1995). Cortical cells derive from stem cell progenitors near the ventricular zone (VZ; His 1874). A web of cells, the preplate, prefigures the future cortex (Rakic 1972). Preplate cells generate molecular factors (e.g., Filamin 1, Doublecortin, LIS1) that influence the migration of the ventricular neuroblasts zone to the cortex (Gleeson and Walsh 2000). The mouse preplate emerges at embryonic day 10–12 (E10–E12) and in humans in embryonic week nine (Meyer et al. 2000). Neuroblasts in the

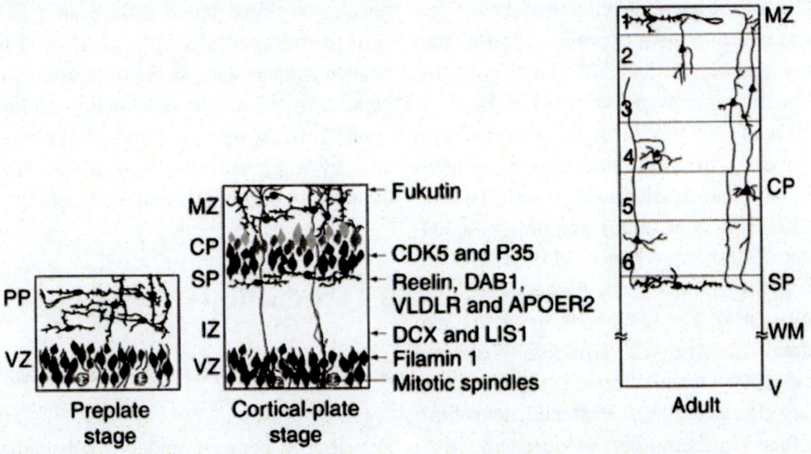

Fig. 21.1 Early stages of mammalian cerebral cortex development. **a** Cells from ventricular zone migrate to the preplate (future cerebral cortex) prenatally. **b** Differentiation of the first cells gives rise to the cortical plate and the subplate. Afferent (thalamic) fibers first enter the subplate and remain transiently within it, before penetrating the cortical plate. **c** Cortical layers arise in an inside-out pattern: with infragranular layers VI and V first. Layer IV and supragranular layers follow. CP: cortical plate; IZ: intermediate zone; MZ: marginal zone; PP: preplate; SP: subplate; VZ: ventricular zone; WM: white matter. Reproduced from the original source with permission (Gleeson and Walsh 2000)

ventricular zone divide and migrate toward the preplate, enter it, and divide it into the subplate and the marginal zone. For targeted thalamic axon ingrowth, additional factors are required (reelin, DAB1, VLDR, APOER2; Gleeson and Walsh 2000). The so-called cortical plate forms between the subplate and the marginal zone and marks the future cortex (Marin-Padilla 1992, 1995). The cortical plate appears at E10–E17 in mice, E30 in the cat, and at 9–18 weeks in humans (for marsupial development, see Aitkin et al. 1991). In the cat, the final migrants into the cortical plate arrive 3–4 weeks after birth (Shatz and Luskin 1986).

The subplate contains differentiated neurons that are the first postmitotic neocortical population (Allendoerfer and Shatz 1994). Beside their intrinsic projections within the subplate, these cells project to the cortical plate, particularly later in development (Valverde and Facal-Valverde 1988). A strong candidate for subplate afferent innervation is the thalamus (Rakic 1977; Friauf et al. 1990). In monkeys (Rakic 1977), cats (Shatz and Luskin 1986), and ferrets (Herrmann et al. 1994), thalamocortical subplate input arrives weeks before their ultimate target neurons in layer IV complete their migration into the cortical plate. The subplate sends reciprocal projections to layer IV before the thalamic afferents arrive (Arber 2004). Thalamic afferents pause in the subplate before they enter the cortical plate. In the rat this happens in the first postnatal week (Shatz 1990; Goodman and Shatz 1993), in the cat, at E46, and by E55 a weak geniculocortical projection reaches the deeper half of the cortical plate (visual cortex: Shatz and Luskin 1986; auditory cortex: Payne et al. 1988b), and even then the major portion of the projection is confined to the subplate. By E57 most cells destined for layer 4 have already migrated to positions above layers 5 and 6. At birth, a substantial geniculocortical projection to cortical layer 4 exists in cats (Shatz and Luskin 1986).

In the human AC, morphological developmental changes in cytoarchitecture and neurofilament expression follow the same time course in areas 41, 42, and 22. The first axons with neurofilament staining appear in the marginal layer at 22 weeks (Moore and Guan 2001). At birth, these axons form a prominent band in layer I. These axons are created by neurons intrinsic to layer I (Cajal-Retzius cells; Marin-Padilla and Marin-Padilla 1982; Ding et al. 2000), ascending medial geniculate body neurons (Hashikawa et al. 1995; Cetas et al. 1999) and descending projections from higher-order AC (Galaburda and Pandya 1983). The first axons arrive in deep cortical layers in humans in week 22 (Krmpotic-Nemanic et al. 1983; Honig et al. 1996), when efferent neurons appear in deep cortical layers (Hevner 2000). Postnatal neurofilament staining is first found in these fibers (Moore and Guan 2001).

The subplate and the marginal zone respond to electrical stimulation of afferent tracts with the shortest latencies in vitro (cat visual cortex, E47–51). At ~E57 in the cat, evoked activity is also found in the cortical plate, though with longer latencies than in the subplate. Subplate evoked activity levels are higher and have shorter latency than those in cortical plate until postnatal life (Friauf and Shatz 1991). The time of arrival of thalamic afferents in the cortex for different species (Table 21.2) (Clancy et al. 2001) marks the switch from experience-independent development to activity-modulated development. We define afferent activity as having both evoked and spontaneous components.

In addition to afferents from the auditory thalamus, input from other subcortical sources arrives in the cortex (Jacobson 1991; Sutor 2002). Monoaminergic locus coereleus projections arrive in the rat at E18, before thalamic afferents. At P1–4 they penetrate the cortical plate and end in layers I and VI, suggesting a holdover from preplate innervation, whereas in adult rats these projections reach layer V. Likewise, rostral mesencephalic dopaminergic projections reach the cortex at E17 and enter the prefrontal and temporal cortical plate at P1–3. Serotoninergic raphe nuclear projections arrive in the cortex in the first postnatal month. In postnatal rat primary AC, acetylcholinesterase is expressed transiently in layers III and IV at P3, peaking at P8–10 and declining to the low, adult pattern by P23 (Robertson et al. 1991). In ferret AC, acetylcholinesterase expression is strongest in layers I, IV, and VI and gradually increases from P21 to adulthood. Noradrenergic fibers are scattered sparsely in cortex but their distribution and density show little change with age. Dopaminergic fibers are densest in layers V and VI, appear at P28, peak at P35, and return to baseline levels 2 weeks later (Harper and Wallace 1995). This transient peak in density does not occur in the adjacent suprasylvian gyrus, confirming interareal dopaminergic innervation differences. Serotoninergic projections into the cat primary AC form a fine, evenly distributed axon system in all layers at P0 and shift to supragranular layers I–III at 3 weeks postnatal. The beaded axon system, in contrast, is far weaker in primary AC, appearing at 3 weeks postnatal in all layers, and confined to layers I–III, where the number of fibers gradually increases and, by week 4, forms pericellular arrays which are unique to auditory cortex (Vu and Törk 1992).

3 Late Cortical Development

3.1 *Cell Death in the Neocortex*

Developing neurons undergo substantial changes, and historic and recent evidence demonstrates that many of them die (Oppenheim 1991). Here we discuss the programmed cell death, so called because it occurs without obvious cause. It is part of the architectural process shaping the macro- and microscopic brain architecture (Finlay 1992), and it is also an adaptive process, contributing to cortical circuit functionality by matching input and target neuron populations.

Cell death during cortical development occurs in two phases (Clarke et al. 1998): during rapid cell proliferation, early in development and later, during synaptogenesis. The causes of cell death at these times are likely different. Cell death during the proliferation period may reflect competition for trophic factors (Voyvodic 1996). Estimates of the proportion of dying cells in this period differ, but may reach 50% (Blaschke et al. 1996). The second phase of apoptosis during late development coincides with synaptogenesis (see below; Ferrer et al. 1992; Clarke et al. 1998). An example is the ontogenetic elimination of subplate and Cajal-Retzius cells during development (ferret AC: Gao et al. 1999, 2000b; cat: Luskin and Shatz 1985; human AC: Krmpotic-Nemanic et al. 1987). Two signals modulate this process: a slow survival signal involving tyrosine kinase receptors (Trk) and a rapid death signal from tumor necrosis factor receptor p75 (Yoon et al. 1998; Majdan and Miller 1999).

One of the proposed functions of the late apoptotic process is to eliminate targeting errors (neurons projecting to incorrect regions). Transient projections connect remote areas in the cortex: in newborn kittens, axons transiently link primary AC to primary visual cortex (Innocenti and Clarke 1984; Dehay et al. 1988; rat: Ding and Elberger 2001). Such heterotopic projections in the cat disappear in the second postnatal month (collateral elimination). Contrary to the central claim of this hypothesis, however, collateral elimination is not caused by apoptosis of the projecting cells, because their cell somata survive (Innocenti et al. 1988).

Another more likely function of cell death is to adjust the number of neurons to the size of their axonal targets (population matching). Because of the trophic interdependence of axons and their targets, failure to acquire sufficient synaptic space often leads to the death of the projecting cell resulting in matching a change in the size of a target or input population (Finlay and Pallas 1989). This process is important not only during development but as an evolutionary substrate (Pallas 2007).

In the rat subplate, the number of dying neurons peaks in the first postnatal week and declines thereafter (Ferrer et al. 1992). In the cortex, cell death rates peak at P7 (Naruse and Keino 1995), before the peak in collateral elimination and decrease substantially during the second week. In mouse and hamster visual cortex, peak late apoptosis is at the end of the first postnatal week (Finlay and Slattery 1983; Pearlman 1985).

The process initiating the late phase of cortical cell death remains obscure. Natural pruning of axons does not initiate it: transitory cortical efferents are eliminated in the second week in the cat, after the peak in apoptosis (Innocenti et al. 1988). The number of dead cells is unaffected by destruction of their targets if alternative targets are available and utilized (Pallas et al. 1988; Windrem et al. 1988). With elimination of afferent connections, contradictory data on its effect on cell death are reported (Ferrer et al. 1992). Neuronal survival may depend critically on the age at which afferents are eliminated, on the method, and on the extent of the lesion.

An open question is whether brain cell division and proliferation occur during late development, and in adults. Stem cell proliferation occurs in the olfactory bulb and hippocampus (Fuchs and Gould 2000; Hastings and Gould 2003). For the AC, data are not yet available.

3.2 Structural Development of the Brain and Neocortex

The brain grows in late development by cellular proliferation, by adding new neuronal branches and synaptic connections, and by axonal myelination. This growth is particularly rapid during the first and second postnatal weeks in rats (Dobbing and Sands 1971), mice (Hahn et al. 1983), and gerbils (Wilkinson 1986). Thereafter, brain growth slows. In cats, the peak growth ends before the end of the first month, then the growth slows down. In ferrets, brain size even decreases before adulthood (Kruska 1993). In humans, brain weight increases by a factor of 4 during the first 3 years after birth, then slows down, and ends at ~15 years (Dekaban 1978).

Myelination apposes layers of glial membranes onto axons, which enhances conduction velocity. In late maturation, axons begin a long period of myelination. In humans, myelination starts in the lower auditory pathway before birth, and by week 23 myelination appears in the brain stem (Moore et al. 1995). Few fibers ascend to the thalamus and none reaching the AC are myelinated at birth (Cant 1998). The human acoustic radiation myelinates further in the first 4 years (Yakovlev and Lecour 1967; Kinney et al. 1988). The myelination of corticocortical connections begins in childhood (Paus et al. 1999) and is incomplete in some projections until adulthood (Yakovlev and Lecour 1967; Giedd 2004). Myelination is experience dependent: bilateral eyelid suturing significantly decreased the number of myelinated fibers in primary visual cortex, particularly in the supragranular layers (Winfield 1983). Similarly, in deaf humans demyelination of thalamocortical projections is seen in AC (Emmorey et al. 2003), although cortical volume is preserved even in prelingual deafness (Penhune et al. 2003).

Whereas layer I in human AC has neurofilament-immunoreactive axons during intrauterine life, deep cortical layers exhibit the first neurofilament positive fibers in the second half of the first postnatal year, and the increase in staining density continues until age five (Moore and Guan 2001). A similar time course was noted in the myelination of these layers. Myelination and neurofilament staining appear later in supragranular layers: the first myelinated axons appear at 6 years and the first true neurofilament-positive fiber plexus appears at five. Morphological fiber maturation in AC continues until years 11–12 (Conel 1939–1967).

Cerebral cortex contains the last auditory cells to differentiate (Morest 1969b). In rats, cortical thickness increases greatly from P0 to P7, and it is mature at P30 (Coleman 1990). The large pyramidal cells mature first, and apical dendrites develop before basal ones (Fox 1968). Deeper cortical layers mature earlier than layers II–IV. In dogs, the dendritic morphology becomes more complex during the first postnatal month (Fox 1968). The most complete study was performed in humans (Conel 1939–1967; Becker et al. 1984; quantitative analysis of Conel's data appears in Shankle et al. 1998). A striking finding is the morphologic change of Golgi-impregnated perinatal cortical neurons (Fig. 21.2). Before birth, they have small dendritic trees and few, sparsely branched dendrites in the primary AC. Postnatally, the dendritic trees become increasingly complex and peak in the AC at ~4 years, after which the complexity decreases.

Similar dendritic developmental changes were seen in the rabbit, though the peak in dendritic complexity appears

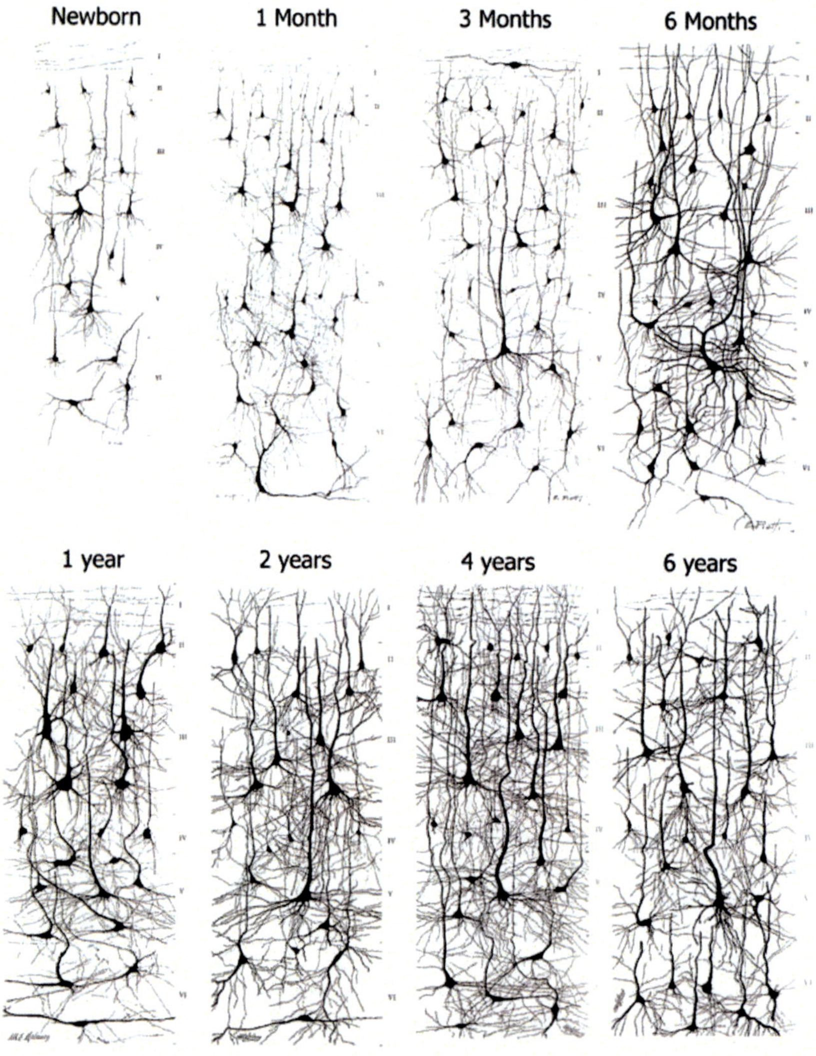

Fig. 21.2 A suite of Golgi-impregnated neurons in the primary auditory cortex (AC) from children at different postnatal ages. The dendritic arbors increase in complexity progressively, peaking at 4 years, with a slight subsequent loss in complexity. The pattern is comparable to other areas; however, in the motor cortex, peak complexity is 2 years and the decrease begins at 4. Compiled and reproduced with permission from the original source (Conel 1939–1967)

near hearing onset (McMullen et al. 1988). This captures the complications of interspecies comparisons, since hearing onset and the pace of development after hearing onset differ. In ferrets, dendritogenesis is also postnatal and dendritic arborization peaks at P21 in layer V and P28 in layers II/III (Zervas and Walkley 1999).

The auditory system develops its interconnections independent of the hierarchical positions of the connected structures in the system (Payne 1992; Cant 1998). For example, projections from the medial geniculate body (MGB) to the AC appear first at E13–14 in rats (Coggeshall 1964), when the MGB is not yet innervated by the inferior colliculus (IC). Corticofugal projections are established before hearing onset: corticofugal input to the cat IC appears between E55 and P4 (Cornwell et al. 1984). Other major cortical afferent and efferent connections with subcortical structures are formed at E56 in cats and auditory nuclei have a topographic organization before hearing onset (Payne et al. 1988b). The corpus callosum develops relatively late. At birth in cats, commissural projections are present, but they are widespread (Feng and Brugge 1983; Payne 1992) and the mature pattern appears at P8 (Feng and Brugge 1983). The last projection to develop is the ipsilateral corticocortical connectivity pattern (Payne et al. 1988a). In ferrets, the clustered projections along the isofrequency axis develop without an initial period of diffuse connectivity, though some terminal clusters are initially mis-located. An adult pattern of clustered terminals along the isofrequency axis emerges by P60 (Gao and Pallas unpublished observations).

3.3 Formation of Cortical Circuits: Synaptogenesis

Dendritic trees thus undergo massive remodeling in development. Dendrites are a primary site of synaptic contact, and their growth and maturation occur in concert with formation of synapses (Morest 1969a,b).

Synaptic maturation can be investigated using dynamic microscopic morphology (presence and number of vesicles, docked vesicles, presence and morphology of postsynaptic density, etc.) (Benson et al. 2001; Vicario-Abejón et al. 2002). The primary changes are as follows:

1. At the first axonal contact with the postsynaptic membrane, there is little pre- and postsynaptic specialization/differentiation (nascent synapse).
2. At stage two, the presynaptic and postsynaptic terminals are distinct morphologically (labile synapse, which can readily be disassembled).
3. The last stage of synapse formation is maturation or stabilization (Goodman and Shatz 1993).

Not all synapses reach the third stage. Some neurons project directly to their targets and do not require stabilization, whereas, promiscuous, neurons project to many targets and their synapses require functional stabilization (Vicario-Abejón et al. 2002).

One view on regulation of pathfinding of axons and synapse formation relies on the hypothesis by Roger Sperry, who proposed dual, perpendicular molecular concentration gradients prospectively guiding the migrating axon. The current view of this process is more complex, but the basic idea appears valid (Sutor 2002; Sur and Rubenstein 2005). Neuroligins and b-neurexin likely initiate neural interactions in vitro at the first stage of synapse formation, whereas ephrins, acting through Eph-receptors, act as repulsive factors in the preceding, axon pathfinding stage. The transformation from a labile to a stable synapse involves neurotrophins and causes changes in adhesion molecule expression. Cadherins, protocadherins, and other junctional proteins stabilize the pre- and postsynaptic elements and couple them to the cytoskeleton. This slow process refines the functional properties of synapses. The extracellular matrix, especially the perineuronal net and its enzymatic degradation (initiated via tissue plasminogen activator), plays an essential role (Oray et al. 2004). Synaptic stabilization is also regulated extrinsically by neural activity.

In the rat AC, early synapses at E16 occur above and below the cortical plate (Konig et al. 1975). In rabbits, on P5 (2 days before hearing onset) synapses can be identified at all cortical layers (Konig and Marty 1974). In human AC extensive studies show that at 8.5–18 weeks of conceptual age, synapses are found only above and below the cortical plate (Molliver et al. 1973). By 12–13 weeks, cortical layers emerge (Krmpotic-Nemanic et al. 1979). The first axodendritic synapses appear between 19 and 23 weeks (Molliver et al. 1973). By 28 weeks, a columnar acetylcholinesterase staining pattern suggests possible thalamic innervation (Krmpotic-Nemanic et al. 1980).

Dendritic spine number and density of in AC might reflect synaptogenesis. Spine density increases postnatally in rats and peaks around P35 (Coleman 1990). In rabbits spine density peaks at P12–P15 and declines until P30 (McMullen et al. 1988).

3.3.1 Synaptic Selection in the Neocortex

There is consensus that the late phase of development includes adjustment to the type of sensory input by selective synaptic stabilization and elimination (Changeux and Danchin 1976). This theory of activity-dependent synaptic stabilization has received considerable interest with regard to alterations in synapse number and density. Much data has come from work in visual cortex.

Synapses are formed in development at various times in different structures. Massive synaptogenesis occurs in cat visual cortex during the first postnatal month (Cragg 1975; Winfield, 1981, 1983; O'Kusky 1985). In newborns, the number of cortical synapses is <10% of the adult value, increases most in the first month, peaks at P70, and then slowly declines by ~30% to adult values after 4 months (Winfield 1983). No comparable structural data are available for the AC, but functional data support a similar timescale of synaptic changes (Fig. 21.3; Kral et al. 2005). Supragranular synaptic currents develop before those in infragranular layers (Kral et al. 2005; visual cortex: Friauf and Shatz 1991). Peak synaptic densities appear in supragranular layers first for asymmetric and later for symmetric synapses (near P110). In other layers the peak is less clear and is between P70 and P110 (Winfield 1983). In the macaque monkey, rapid synaptogenesis occurs 6–9 months postnatally, with a subsequent reduction in synaptic density of 25% (O'Kusky and Colonnier 1982a). Concomitantly, cortical thickness peaks at 6 months postnatally. This increase in cortical thickness is mainly due to the growth of layers II and III.

A similar pattern of synaptic densities development is seen in different animal species and in humans: massive synaptogenesis in late development, a peak in early juvenile stages, and a slow, adult decline (~25–30% of peak densities in carnivores and primates, 15% in rodents). Because brain size also increases concomitantly, a possible explanation of the reduction in synaptic density is that this growth occurs without a parallel change in synapse number. A study in adult macaques found that synaptic elimination was more pronounced in terms of absolute synaptic counts across the entire primary visual cortex, though cortical volume decreased after 6 months (Fig. 21.4; O'Kusky and Colonnier 1982a). This volume decrease reflected a reduced neuropil volume. The topic merits further study because the decrease in cortical volume may not apply in all species.

In human AC, increased synaptic density occurs between postconceptual days 200 and 800, and the decrease between days 1,500 and 4,000. The peak synaptic densities are between 3 months and 3.5 years (Huttenlocher and Dabholkar 1997), with laminar differences such that supragranular layers often develop more slowly than layer IV and infragranular layers (Marin-Padilla 1970), following their time of generation. Methodological problems in counting synapses may also contribute (Guillery 2005). The peak in synaptic densities fits the early data on neuronal structure

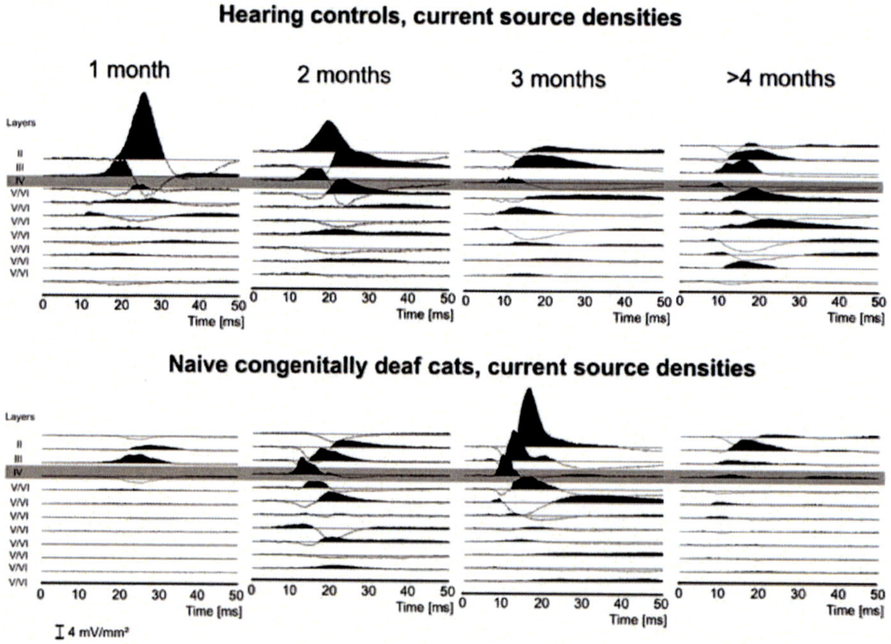

Fig. 21.3 Development of current source density profiles in the primary AC of cochlear-implant stimulated hearing cats and congenitally deaf cats. Current source density signals correspond to extracellular components of gross synaptic currents from many active synapses around the recording depth. Peak synaptic currents in hearing subjects are small before hearing onset, highest 1–2 months postnatally and, later, are smaller and show more fine structure. The functional peak at 1-2 months corresponds to the period when, in visual cortex, the synaptic density is highest. In deaf cats, the postnatal development differs, with the largest synaptic currents at 3 months. This amplified and delayed peak corresponds to the increased and delayed peak in synaptic densities in the primary visual cortex of enucleated animals. In conclusion, developmental alteration in cortical activation occur in congenitally deaf cats. Reproduced with permission from Kral et al. (2005)

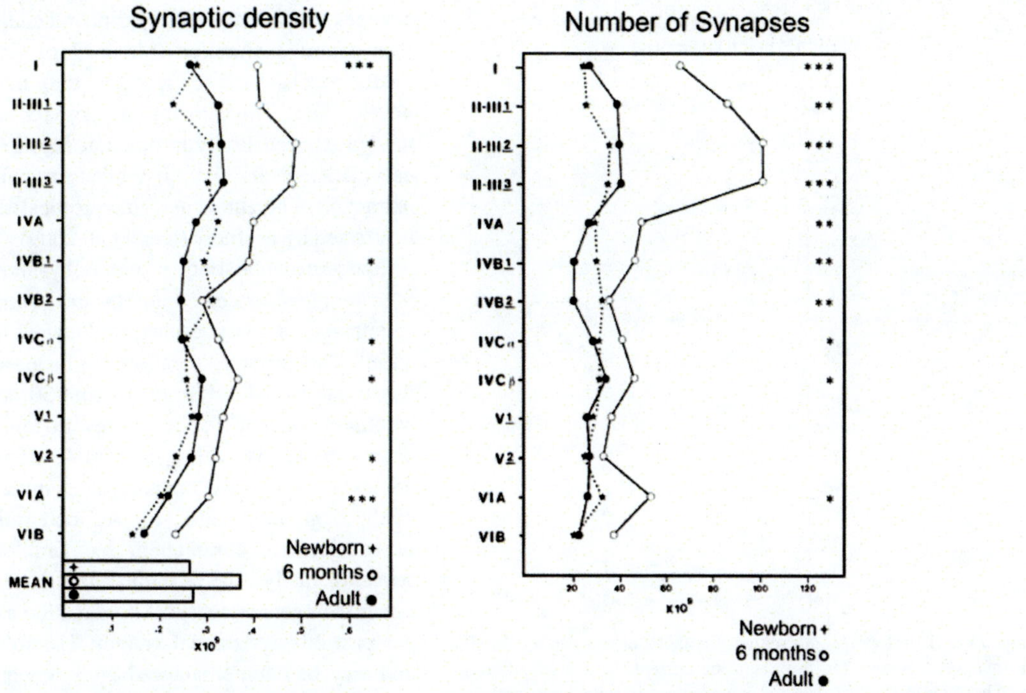

Fig. 21.4 Changes in synaptic densities and synaptic numbers in macaque monkey cerebral cortex. **a** Postnatal development entails an overabundance of synapses (late phase of synaptogenesis) and, after peak synaptic densities are reached, a reduction (synaptic elimination). **b** Corresponding to the decreased synaptic densities, synaptic numbers are also reduced, indicating that the synaptic density decrements are not the consequence of the change in neuropil volume. Modified and reproduced with permission from O'Kusky and Colonnier (1982b)

(Conel 1939–1967), and it is in line with functional data for a sensitive period in AC maturation from 3.5 to 7 years (Sharma et al. 2005).

3.4 Synaptic Properties of Developing Cortical Cells

Synaptic properties have been studied extensively in rat neocortex. The first synapses are of the Gray type I (asymmetric, excitatory) class at E19–P0 (Miller, 1988). Newborn rats show only the NMDA-type of synaptic transmission with no significant AMPA currents (LoTurco et al. 1991); synaptic AMPA receptors first appear at P3 (Carmignoto and Vicini 1992). Inhibition develops later than excitation. Gamma-aminobutyric acid (GABA)-elicited postsynaptic potentials (PSPs) appear between P5 and P8; however, until P10 GABA–receptor binding produces mainly excitatory postsynaptic potentials (EPSPs) rather than inhibitory PSPs (IPSPs) due to a difference in chloride balance in immature versus adult neurons (Ben-Ari 2002; Pallas 2007). The number of inhibitory neurons also changes during development. In ferrets, there are two developmental peaks: a smaller one at P1 and another at P60 (Gao et al. 1999, 2000b). In cats the first, likely prenatal peak, has not been reported.

In sensory cortex before P5, EPSPs have long latencies and long durations and are primarily NMDA receptor based. Stimulus repetition rates >2 Hz cause a substantial decrease in evoked response amplitudes in immature animals (Kim et al. 1995). Adult-like EPSPs with short latencies and short durations are found at the end of the third week in rats (Carmignoto and Vicini 1992). These differences reflect a change in NMDA channels subunit composition and a parallel change in gating properties (van Zundert et al. 2004). In rat AC, long-duration EPSPs in young animals (Fig. 21.5) (Aramakis et al. 2000) decrease with age and correlate with a progressive increase in the levels of NR2A subunit mRNA postnatally (Hsieh et al. 2002). The number of inhibitory neurons and synapses in the rat matures between P12 and P21 (Miller 1988).

The immature cortex expresses electrical synapses that couple the neurons and represent functional units. Such coactive neuronal ensembles in immature cortex span several cortical layers and are 50-μm diameter or larger and are seen with calcium imaging (Yuste et al. 1992). Reductions of gap junction permeability suppress domain formation (Yuste et al. 1995). In adult cortex, domains are absent and gap junctions exist only between glial cells and fast spiking

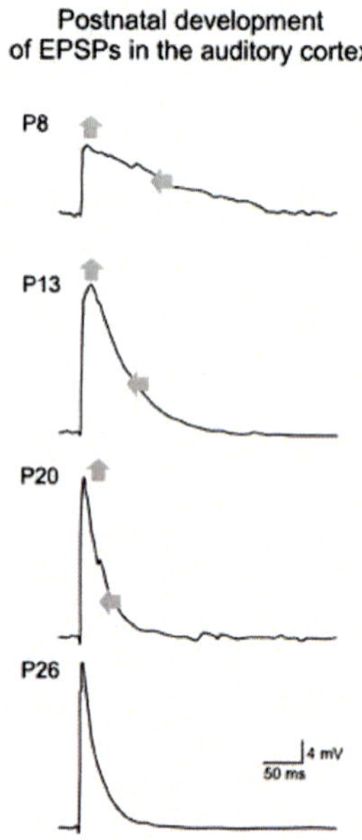

Fig. 21.5 Development of excitatory synaptic currents in postnatal rat primary AC. Soon after birth, synaptic currents have a long duration and low amplitude. With age, the excitatory postsynaptic potential duration decreases and amplitude increases. At the same time, the excitatory postsynaptic potential latency also decreases. Modified and reproduced with permission from Aramakis et al. (2000)

(inhibitory) and low-threshold spiking neurons (Galarreta and Hestrin 1999; Gibson et al. 1999).

3.5 Functional Development in the Auditory Cortex

Investigation of the AC in vivo reveals many ontogenetic processes, as the cells' responses reflect both cortical and subcortical developmental changes.

AC sound sensitivity development (in neuronal thresholds) reflects cochlear sensitivity (Brugge et al. 1988), with low-frequency sensitivity emerging first in many vertebrates (Brugge 1992). In cats before P10, units are insensitive to frequencies >10 kHz. Central mechanisms contribute little, if any, to these changes, and the maturation of central auditory stimulation thresholds closely follows cochlear maturation. In rats, cortical units respond first to higher frequencies, with low-frequency sensitivity emerging at P13–P22 and later (Zhang et al. 2001). Binaural properties in cat AC follow brain stem maturation and basic binaural interactions (e.g., EI interaction: inhibition of a unit response to contralateral stimulation during simultaneous ipsilateral stimulation) mature early in the second week (Brugge et al. 1988; Brugge 1992). In high-frequency ferret neurons, their spatial receptive fields are broader at P33–P39 than in older animals, changes which may reflect peripheral auditory growth rather than central maturation (Mrsic-Flogel et al. 2003). Thus, central mechanisms for spatial coding may emerge early in development. This conclusion is further supported by residual cortical sensitivity to binaural cues in congenitally deaf cats (Tillein et al. 2010). It applies particularly to subcortical extraction of binaural cues, since cortical aural representation is affected by deafness (Kral et al. 2009).

Spontaneous activity in neurophysiological experiments is strongly influenced by anesthesia. In anesthetized cat AC, spontaneous activity reached adult values at P70 (Eggermont 1996). This property parallels the increased synaptic densities in cat visual cortex in the first 30 days (Cragg 1975). Minimum latency for responses to tone pips in cat AC decreases steeply with age, from 40 to 60 ms at P9–12 to 18 ms at P40, when mature values emerge (Brugge et al. 1988; Eggermont 1996; for AC maturation with auditory nerve electrical stimulation, see Kral et al. 2005). This sequence likely reflects the maturation of synaptic currents rather than geniculocortical myelination as the latter continues past this age (visual system: Tsumoto and Suda 1982) and may be counterbalanced by a concomitant increase in projection length (Eggermont 1996).

Many developmental studies concentrated on frequency tuning. Mean bandwidth of cat AC increases with age (Eggermont 1996; Brugge et al. 1988; Bonham et al. 2004) due to the growing proportion of broadly tuned units mainly in the ventral and dorsal parts of adult AI (Schreiner and Mendelson 1990; Heil et al. 1992; Schreiner and Sutter 1992), regions unresponsive in young animals (Bonham et al. 2004). Increasing bandwidth of AI units contrasts with the decreasing tuning curve bandwidth in feline IC at 30–35 days postnatal (Moore and Irvine 1979). Influences shaping tuning curves include inhibition, thalamic divergence, and the type of interaction (corticocortical vs. thalamocortical). Audible frequency range increases early in life, which biases such investigations (rat: Zhang et al. 2001). However, the spread of AC excitation with peripheral stimulation is larger in young rats and in cats (Zhang et al. 2001; Kral et al. 2005) (Fig. 21.6). This implies more thalamocortical divergence in young animals. In rat, units are tuned more broadly at birth (Zhang et al. 2001). Broadened frequency tuning in cats might thus reflect the more complex organization of the AI, including the later development of ventral and the dorsal subregions of broadly tuned cells (Bonham et al. 2004).

AC unit temporal properties have a slow postnatal maturation in cat, with the best modulation frequency reaching mature values by P60, and maximum best modulation frequencies at P150 (Eggermont 1991, 1996). This may reflect postnatal changes in inhibitory function, which suppress spontaneous activity after response onset and cause a poststimulus rebound at 120–150 ms (Eggermont 1992). Rebound responses mature at about P150 (see also Kral et al.

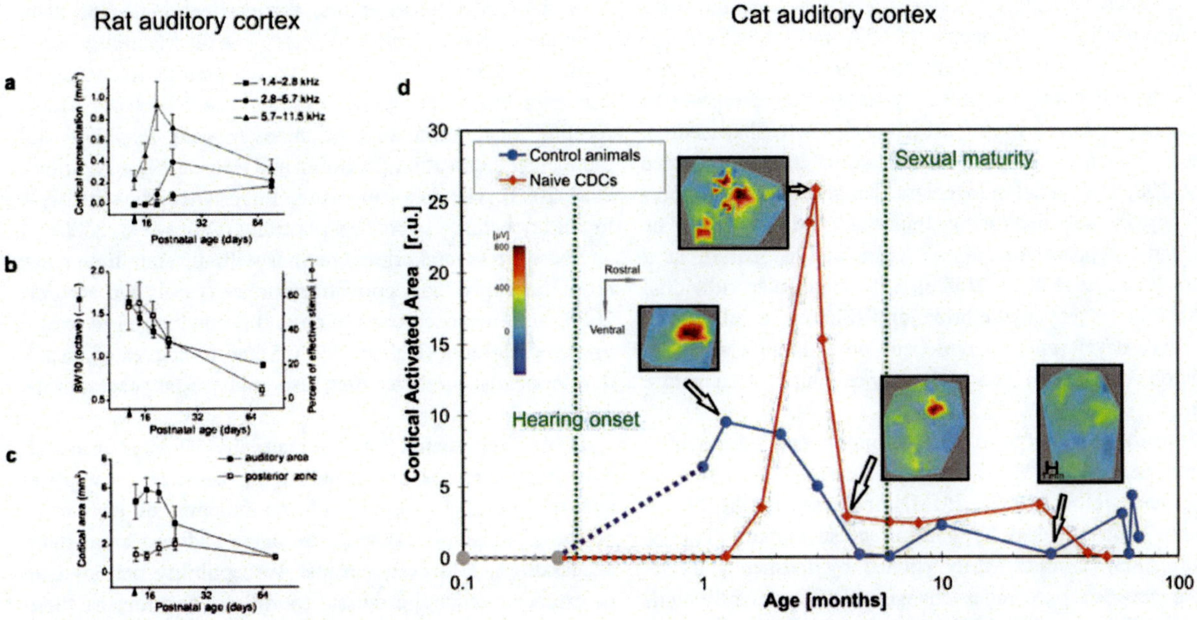

Fig. 21.6 Postnatal development of AC function in rat and cat. Postnatal changes in **a** cortical representation of different frequency bands, **b** in bandwidth of tuning curves, and **c** in cortical areas sensitive to acoustic stimulation in rats. A peak in AC extent occurs at ~P16 (rat), after which the functional cortical area declines with age, despite brain growth. Reproduced with permission from Zhang et al. (2001). **b** Changes in AC area in hearing controls and congenitally deaf cats stimulated electrically via a cochlear implant. *Green dotted lines* denote hearing onset in hearing cats (P10) and sexual maturity (P180). *Grey* areas activated from animals below the age of hearing onset; at P0 and P3 no local field potentials were elicited with cochlear implant stimulation, at P8 small amplitude (<100 μV) local field potentials were recorded. Activated area had responses >300 μV; all animals had cortical responses, however, in some, they were smaller than 300 μV. In both hearing and deaf animals, a peak in such an area responding to the electrical stimulus emerges at 2 months in hearing cats and at 3 months in deaf cats. The functional cortical activated area shrinks, despite slight brain growth. Peak cortical area is significantly larger in deaf cats. Modified from the original from Kral et al. (2005)

2005). Thus, the slowest development in AI in vivo is in the temporal response domain. Precocious species have more mature functional properties than altricial cats and rats (chinchilla: Pienkowski and Harrison 2005; guinea pigs: Sedlacek 1976).

3.6 Human Functional Development

Cochlear functionality (brain stem-evoked responses, behavioral reactions to sounds, etc.) begins 6 months after conception in humans (Granier-Deferre et al. 1985; Moore et al. 1996). However, auditory development, measured by evoked responses, extends long past birth (Eggermont 1989).

Some AC specialization, e.g., sensitivity to the maternal voice and native language can be identified in newborns (psychophysics: DeCasper and Fifer 1980; Mehler et al. 1988; Locke 1997; imaging and electroencephalographic studies: Dehaene-Lambertz 2000; Dehaene-Lambertz et al. 2002; Pena et al. 2003; Winkler et al. 2003; Pallas 2005). Nonetheless, speech discrimination improves markedly after birth. Although children readily discriminate different phonemes early in life, phonetic specialization for the native language is present at 8–12 months (Kuhl 2004; Friederici 2006), when children lose the ability to discriminate foreign language phonemes whose acoustic features are not distinctive in their native language. The massive concomitant increase in synaptic densities and dendritic branching complexity suggests a plausible basis for these processes.

Non-invasive electrophysiological methods find massive changes in cortical function in early infancy. Evoked potential latencies of individual waves of middle- and long-latency potentials decrease with age and the morphology of the individual waves changes significantly. Middle-latency responses (especially P0 and Na) can be recorded reliably in preterm infants (Rotteveel 1992). The next middle-latency response wave, Pa, increases linearly in detectability (from 0 to 50%) from 29 to 52 weeks postconception. Wave Na latency decreases up to 30 weeks postconception, with slower decrements from the third postnatal month. In contrast, Pa latency decreases postnatally, most strongly in the first 3 months (Rotteveel 1992; Kushnerenko et al. 2002). Even more marked changes occur in the long latency range. Response latencies similar to immature N2 and P3/P4 values occur at in 25–29 week premature infants (Schulte et al.

1977; Rotteveel, 1992). Whether these waves are the functional homologues of mature N2, P3, and P4 in adults is unclear. The cortical P1 response, generated by auditory thalamic and cortical sources, systematically decreases in latency up to 12–16 years (Ohlrich et al. 1978; Sharma et al. 1997; Pasman et al. 1999; Ponton et al. 2000; Ceponiene et al. 2002b). Parallel ontogenetic changes in response morphology include decreasing amplitude and duration. The N1 wave appears first at ~3 years and is mature at 9 years (Ceponiene et al. 2002b), with large inter-individual variability in response morphology (Sharma et al. 1997). N1 wave development may depend on auditory experience (Ponton et al. 1996b). Longer-latency responses often mature later.

Sensitivity to acoustic features investigated using mismatch negativity (MMN) can be detected in newborns (Ceponiene et al. 2002a). MMN increases in the first 6 months when evoked by gaps in noise stimuli (Trainor et al. 2003). Human MMN elicited by frequency differences between pure tones changes developmentally until 11 years (Martin et al. 2003), including changes in latency and in the brain generators. Although oscillatory phenomena are much studied, developmental auditory studies are rare. Oscillations in the gamma band and their synchronization for task-specific processes occur in 9–16 year olds. Spontaneous gamma band power did not change in this span, while the activity in an auditory attention task decreased from 9–11 to 16 years. Subtle differences in the spatial distribution of gamma-band responses suggest changes in cortical processing at 12–13 years (Yordanova et al. 2002; see also Engel et al. 2001).

4 Developmental Plasticity and Sensitive Periods

Plasticity is the neural capacity to adapt to environmental influences. Its substrates include changes in cell number, synaptic number, projections, synaptic functional properties receptors and associated ionic channels, and the other processes.

A basic mechanism for plasticity involves changes in synaptic efficacy with repetitive stimulation (long-term potentiation, LTP) and the complementary process (long-term depression; LTD; Citri and Malenka, 2008). LTP and LTD are elicited more easily in immature animals (Crair and Malenka 1995; Kirkwood et al. 1996; Rittenhouse et al. 1999; Sermasi et al. 1999), which is related to developmental changes in the composition of postnatal NMDA receptors and the replacement of NMDA receptors by AMPA receptors (Lu and Constantine-Paton 2004; van Zundert et al. 2004).

Besides NMDA receptors, intracortical inhibition also plays an important role. Inhibitory circuit maturation correlates with the end of the sensitive period. Accentuated excitation blocks the natural developmental switch of NMDA receptor maturation and, in mice, reduces sensitivity to monocular deprivation (Fagiolini and Hensch 2000; Fagiolini et al. 2003). The data support the importance of excitatory–inhibitory balance in cortical plasticity (Pallas et al. 2006).

There are several critical periods with different time windows, both in visual deprivation studies (Lewis and Maurer 2005) and in speech perception and production in hearing-impaired children (Ruben 1997). Some principles of visual developmental plasticity may also apply to language development.

In AC, the search for the neurophysiological basis of critical periods has been challenging because a complete and reversible suppression of the function of an ear is not possible, and because of the many auditory brain stem decussations. However, because susceptibility periods can be extended under appropriate conditions, and because their mechanisms remain under study in the auditory system, we prefer the term sensitive period.

A sensitive period is defined as a period when a specific stimulus is required for normal development, preservation or recovery of neural function, and during which the sensory system is vulnerable to sensory manipulation. Three basic types of sensitive periods are observed (Lewis and Maurer 2005):

1. A developmental sensitive period, the time when sensory stimuli trigger the emergence of a given perceptual capacity.
2. A sensitive period for recovery, during which abnormal development (caused, for example, by the absence of sensory stimuli) is reversible.
3. A sensitive period for damage, when abnormal sensory experience can have a permanent deleterious effect on sensory development.

Psychophysically, sensitive periods in the auditory system include a phonetic specialization in the first year, when disruption of hearing (e.g., severe otitis media) affects phonetic performance long after (Wallace et al. 1988). Semantic performance is affected if hearing is impaired in the first 4 years, with impact on syntax after hearing deficits in the first 15 years (Neville et al. 1992; Ruben 1997). Neurophysiologically, sensitive periods have been investigated mainly in primary AC. After congenital auditory deprivation, electrical stimulation via a cochlear implant found a rudimentary AC cochleotopic gradient, suggesting that coarse topography does not require auditory experience (Hartmann et al. 1997; see also Raggio and Schreiner 2003; however compare Fallon et al. 2009), though it may require spontaneous neural activity. The tonotopic map in normal rats can be modified significantly by early acoustic alterations (Zhang et al. 2001). Presenting pulsed white-noise stimuli during development disrupts AI tonotopic organization from P9 to P28, but not after P30 (Zhang et al. 2002). Such disruption may reflect desynchronization of cortical activity by the incoherence of the white noise stimuli. A constant environmental noise substantially delays and negatively affects cortical development (Chang and Merzenich 2003) and exposure to complex tones within the sensitive period elicits large-scale reorganization of AI with segregation of two frequency regions (Nakahara et al. 2004). Correspondingly, the developmental sequence is aberrant in the first 3 months in congenitally deaf cats, with AC showing enhanced sensitivity to auditory input from a cochlear implant, and lower cortical thresholds (Kral et al. 2005). Electrically evoked local cortical field potentials in deaf cats are delayed and altered developmentally. At 3 months, gross extracellular synaptic currents (current sinks) in supragranular layers are larger in deaf kittens than in hearing controls (Kral et al. 2005), decreasing with age to fall below hearing controls in adult hood (Kral et al. 2000). Weaker source currents (corresponding to outward transmembrane currents) in deep layer III and layer IV from the first month (Kral et al. 2005) suggest a down-regulation of inhibition (generating outward transmembrane currents) in these layers (Hubka et al. 2004; Kral et al. 2006b). Down-regulation of inhibition has been shown at several auditory levels in brain slices from young, binaurally deprived animals (Vale and Sanes 2002; Vale et al. 2003, 2004). The down-regulation of layer III–IV inhibition in congenitally deaf cats is in accord with the effects of visual deprivation on layer IV in V1 (Maffei et al. 2004). The visual effect occurs only if deprivation is early (P14–P17), and a brief delay in the deprivation window (from P18 to P21) causes the opposite effect: feedback inhibition is potentiated in layer IV (Maffei et al. 2006). A down-regulation of inhibition in congenital deafness, however, cannot fully explain the decreased cortical threshold, because inhibition is not effective at threshold intensities. Increased AC EPSPs in gerbils with early noise-induced hearing loss (Kotak et al. 2005) reflect a retention of NR2B NMDA receptor subunits at thalamocortical and corticocortical synapses. Studies on congenitally deaf cats find deficits in the cortical microcircuit functionality, including desynchronization of synaptic activity within a column, causing deficits in infragranular activation which may disable descending, feedback ("top-down") modulation of infragranular layer activity (Kral et al. 2006b).

In congenitally deaf cats chronic electrical stimulation from a cochlear implant caused extensive adaptation to the electrical stimuli (Klinke et al. 1999) and massive changes in cortical circuitry (Klinke et al. 1999; Kral et al. 2006b), with maximal effects in the subarea responding to the stimulus, and a peak fivefold areal increase after months of experience (Kral et al. 2002). Long-latency responses, including the rebound response, were re-introduced after chronic stimulation, and the functional connectivity between different layers could be normalized and infragranular layers reactivated. There was increased response complexity, some units becoming selective to electrical stimuli (Kral et al. 2001, 2006b). With age, there was decreased plasticity in the extent of the areas activated, Pa wave latency, and cortical field potential morphology, extending the sensitive recovery period to 5 postnatal months (Kral et al. 2001, 2002, 2006a,b).

Two sensitive periods occur in prelingually deaf children receiving a cochlear implant: recipients in their teens have poor speech recognition scores long after implantation and do not develop the N1 wave (Ponton and Eggermont 2001). Subjects implanted between 4 and their teens show a developmental delay in the P1 wave correlated with the deafness duration (Ponton et al. 1996a; Eggermont et al. 1997). A second (earlier) sensitive period occurs in children implanted before 3.5 years; they show rapid development of evoked potentials in the normal latency range after a few months of auditory experience (Sharma et al. 2002a,b, 2005), unlike those implanted later.

Development of inhibition and excitation, changes in receptor subunit composition and distribution (Quinlan et al. 1999a,b), and change in neural growth factors and membrane properties, each undoubtedly contributes to the developmental control of synaptic plasticity. A difference between early and adult plasticity in mouse somatosensory cortex showed that extensive plasticity did not change the overall number of synapses (Trachtenberg et al. 2002), although new synapses are formed (O'Kusky and Colonnier 1982b). In adults, plasticity can be elicited by pairing stimuli with some instructional factor such as activation of nucleus basalis (Bakin and Weinberger 1996; Kilgard and Merzenich 2002; Bao et al. 2003), and perhaps top-down modulation (Kral 2007).

5 Cross-modal Developmental Reorganization

Cross-modal reorganization may depend on the level of cortical hierarchy and the experimental design: even normal AC, once considered unimodal, can respond to input from other modalities (Schroeder et al. 2001; Wallace et al. 2004; Bizley and King 2009). These responses usually are weak in the primary AC (Kral et al. 2003; Lakatos et al. 2007) and increasingly robust at higher levels in the cortical hierarchy (Bizley et al. 2007). Nonetheless, even weak non-auditory inputs can, under certain conditions, enhance primary AC responsiveness to auditory stimuli (Lakatos et al. 2007). It has been suggested that nearly all cortical areas are essentially multisensory (Ghazanfar and Schroeder 2006).

Training hearing animals to associate an auditory stimulus with visual stimuli leads to emergence/increases in visually evoked responses in AC if the experimental context is preserved (Brosch et al. 2005). Within VI, training increases unit responses to non-visual inputs in the infragranular layers (Shuler and Bear 2006). Perhaps weak corticocortical influences from other modalities (and nonsensory influences) on sensory cortex can be strengthened under such paradigms (Kral 2007).

The AC in deaf subjects can undergo substantial cross-modal reorganization (Bavelier and Neville 2002), including recruitment for visual and somatic sensory functions. Such reorganizational capacity is modest in primary AC, even in congenital deafness (cat: Stewart and Starr 1970; Hartmann et al. 1997; Kral et al. 2003; human: Nishimura et al. 1999; Petitto et al. 2000). There is some evidence for cross-modal recruitment of primary AC, only in the right hemisphere (Finney et al. 2001). Nonetheless, extensive cross-modal reorganization of higher-order auditory areas was seen in all studies (Levanen et al. 1998; Nishimura et al. 1999; Petitto et al. 2000; Fine et al. 2005). Recently, this concept has been supported by differential visual cross-modal reorganization in different auditory cortical fields in congenitally deaf cats, demonstrating the high degree of specificity of this process (Lomber et al. 2010).

Sensory experience influences the development of sensory cortex, but what are the limits of this influence? Cross-modal plasticity studies show that this impact can be profound and that primary AC can be induced to process visual information (Pallas 2001, 2007) in addition to auditory processing (Mao et al. 2007; Mao and Pallas 2009; and submitted). Visual stimuli can produce evoked potentials in early-deafened human AC (Neville 1990; Finney et al. 2001; Fine et al. 2005), and the converse prevails in blind subjects (Kujala et al. 1997, 2000; Cohen et al. 1999). Animal models suggest that such sensory substitution occurs at the cortex (Rauschecker 1995; Bronchti et al. 2002; Kahn and Krubitzer 2002; Larsen et al. 2009), and whether its basis is the same across species, systems, and experimental conditions is unknown. As the resolution of non-invasive approaches improves, this question can be asked in humans.

Experimentally induced cross-modal plasticity provides insight into the functional capacity of developing cortical circuits by challenging them to process novel sensory input. In neonatal ferrets or hamsters, superficial lesions of the IC and superior colliculus re-route retinal axons to the MGB (Schneider 1973; Sur et al. 1988). Thus, visual stimuli activate AI without altering the geniculocortical pathway (Pallas et al. 1990; Pallas and Sur 1993) and many AC neurons respond to visual or electrical stimulation of the optic chiasm in addition to sound stimulation. Unexpectedly, the visually-responsive AI neurons have tuning properties much like those in normal visual cortex, such as direction and velocity tuning, preference for moving oriented edges, simple and complex spatial arrangements of receptive fields, and end inhibition (Roe et al. 1992). This occurs despite the fact that the retinal–MGB pathway involves W (and not the X- and Y-like) retinal ganglion cells providing most of the input to VI (Roe et al. 1993). Moreover, the retina is organized retinotopically in cross-modal AI (Roe et al. 1990). Perhaps this two-dimensional retinal mapping enables AI to represent visual edges. Such visual information does not reflect input from another visual cortical area; AI receives input from other AC structures (Pallas and Sur 1993). However, some corticocortical connections are substantially rearranged (Gao and Pallas 1999; Pallas et al. 1999) and link neurons with similar visual tuning properties (Sharma et al. 2000). There are also changes in the arrangement and morphology of local inhibitory interneurons (Gao et al. 2000a). Behavioral experiments find that the cross-modal ferrets can perform some rudimentary visual perception using the retinal–MGB–AI pathway (von Melchner et al. 2000). Thus, changing the modality of sensory input, even when the information passes through the same geniculocortical pathway, can induce substantial plasticity in, and even alter, AC function. Findings on cross-modal plasticity can serve as a model for examining sensory substitution in humans and demonstrate how exquisitely sensitive AC is to the pattern of developmental stimulation.

Acknowledgments Space limitations preclude more complete citations of the primary literature; as an alternative, reference to comprehensive reviews was often made. Supported by: grants from Deutsche Forschungsgemeinschaft (Germany) and the National Institutes of Health (USA) (A.K.); and by the Deafness Research Foundation, Fight for Sight, Whitehall Foundation, National Science Foundation, and the National Institutes of Health (USA) (S.L.P.).

References

Aitkin L, Nelson J, Farrington M, and Swann S (1991) Neurogenesis in the brain auditory pathway of a marsupial, the northern native cat (*Dasyurus hallucatus*). *Journal of Comparative Neurology* **309**:250–260.

Allendoerfer KL and Shatz CJ (1994) The subplate, a transient neocortical structure: its role in the development of connections between thalamus and cortex. *Annual Reviews of Neurosciences* **17**:185–218.

Aramakis VB, Hsieh CY, Leslie FM, and Metherate R (2000) A critical period for nicotine-induced disruption of synaptic development in rat auditory cortex. *Journal of Neuroscience* **20**:6106–6116.

Arber S (2004) Subplate neurons: bridging the gap to function in the cortex. *Trends in Neuroscience* **27**:111–113.

Bakin JS and Weinberger NM (1996) Induction of a physiological memory in the cerebral cortex by stimulation of the nucleus basalis.

Proceedings of the National Academy of Sciences of the United States of America **93**:11219–11224.

Bao SW, Chang EF, Davis JD, Gobeske KT, and Merzenich MM (2003) Progressive degradation and subsequent refinement of acoustic representations in the adult auditory cortex. *Journal of Neuroscience* **23**:10765–10775.

Bavelier D and Neville HJ (2002) Cross-modal plasticity: where and how? *Nature Reviews Neuroscience* **3**:443–452.

Becker LE, Armstrong DL, Chan F, and Wood MM (1984) Dendritic development in human occipital cortical neurons. *Brain Research* **315**:117–124.

Ben-Ari Y (2002) Excitatory actions of GABA during development: the nature of the nurture. *Nature Reviews Neuroscience* **3**:728–739.

Benson DL, Colman DR, and Huntley GW (2001) Molecules, maps and synapse specificity. *Nature Reviews Neuroscience* **2**: 899–909.

Bishop KM, Garel S, Nakagawa Y, Rubenstein JL, and O'Leary DD (2003) Emx1 and Emx2 cooperate to regulate cortical size, lamination, neuronal differentiation, development of cortical efferents, and thalamocortical pathfinding. *Journal of Comparative Neurology* **457**:345–360.

Bishop KM, Goudreau G, and O'Leary DD (2000) Regulation of area identity in the mammalian neocortex by Emx2 and Pax6. *Science* **288**:344–349.

Bishop D and Mogford K (1993) Language Development in Exceptional Circumstances Lawrence Erlbaum Assoc., Howe, Hillsdale.

Bizley JK and King AJ (2009) Visual influences on ferret auditory cortex. *Hearing Research* **258**:55–63.

Bizley JK, Nodal FR, Bajo VM, Nelken I, and King AJ (2007) Physiological and anatomical evidence for multisensory interactions in auditory cortex. *Cerebral Cortex* **17**:2172–2189.

Blaschke AJ, Staley K, and Chun J (1996) Widespread programmed cell death in proliferative and postmitotic regions of the fetal cerebral cortex. *Development* **122**:1165–1174.

Bonham BH, Cheung SW, Godey B, and Schreiner CE (2004) Spatial organization of frequency response areas and rate/level functions in the developing AI. *Journal of Neurophysiology* **91**:841–854.

Bronchti G, Heil P, Sadka R, Hess A, Scheich H, and Wollberg Z (2002) Auditory activation of 'visual' cortical areas in the blind mole rat (*Spalax ehrenbergi*). *European Journal of Neuroscience* **16**:311–329.

Brosch M, Selezneva E, and Scheich H (2005) Nonauditory events of a behavioral procedure activate auditory cortex of highly trained monkeys. *Journal of Neuroscience* **25**:6797–6806.

Brugge JF (1992) Development of the lower auditory brainstem of the cat. In: Romand R (ed). *Development of Auditory and Vestibular Systems* 2. Elsevier Science Publishers B.V., Amsterdam, pp. 173–296.

Brugge JF, Reale RA, and Wilson GF (1988) Sensitivity of auditory cortical neurons of kittens to monaural and binaural high frequency sound. *Hearing Research* **34**:127–140.

Cant NB (1998) Structural development of the mammalian auditory pathways. In: Rubel EW, Popper AN, and Fay RR (eds). *Springer Handbook of Auditory Research*, volume 9, *Development of the Auditory System*. Springer, New York, pp. 315–413.

Carmignoto G and Vicini S (1992) Activity-dependent decrease in NMDA receptor responses during development of the visual cortex. *Science* **258**:1007–1011.

Ceponiene R, Kushnerenko E, Fellman V, Renlund M, Suominen K, and Näätänen R (2002a) Event-related potential features indexing central auditory discrimination by newborns. *Cognitive Brain Research* **13**:101–113.

Ceponiene R, Rinne T, and Näätänen R (2002b) Maturation of cortical sound processing as indexed by event-related potentials. *Clinical Neurophysiology* **113**:870–882.

Cetas JS, de Venecia RK, and McMullen NT (1999) Thalamocortical afferents of Lorente de Nó: medial geniculate axons that project to primary auditory cortex have collateral branches to layer I. *Brain Research* **830**:203–208.

Chang EF and Merzenich MM (2003) Environmental noise retards auditory cortical development. *Science* **300**:498–502.

Changeux J-P and Danchin A (1976) Selective stabilisation of developing synapses as a mechanism for the specification of neuronal networks. *Nature* **264**:705–712.

Citri A and Malenka RC (2008) Synaptic plasticity: multiple forms, functions, and mechanisms. *Neuropsychopharmacology* **33**: 18–41.

Clancy B, Darlington RB, and Finlay BL (2001) Translating developmental time across mammalian species. *Neuroscience* **105**:7–17.

Clarke PG, Posada A, Primi MP, and Castagne V (1998) Neuronal death in the central nervous system during development. *Biomedical Pharmacotherapy* **52**:356–362.

Coggeshall RE (1964) A study of diencephalic development in the albino rat. *Journal of Comparative Neurology* **122**:241–269.

Cohen LG, Weeks RA, Sadato N, Celnik P, Ishii K, and Hallett M (1999) Period of susceptibility for cross-modal plasticity in the blind. *Annals of Neurology* **45**: 451–460.

Coleman J (1990) Development of auditory system structures. In: Coleman J (ed). *Development of Sensory Systems in Mammals*. Wiley, New York, pp. 205–247.

Conel JL (1939–1967) *The Postnatal Development of Human Cerebral Cortex. Vol. I–VIII*. Harvard University Press, Cambridge.

Cornwell P, Ravizza R, and Payne B (1984) Extrinsic visual and auditory cortical connections in the 4-day-old kitten. *Journal of Comparative Neurology* **229**:97–120.

Cragg BG (1975) The development of synapses in the visual system of the cat. *Journal of Comparative Neurology* **160**:147–166.

Crair MC and Malenka RC (1995) A critical period for long-term potentiation at thalamocortical synapses. *Nature* **375**: 325–328.

DeCasper AJ and Fifer WP (1980) Of human bonding: newborns prefer their mothers' voices. *Science* **208**:1174–1176.

Dehaene-Lambertz G (2000) Cerebral specialization for speech and non-speech stimuli in infants. *Journal of Cognitive Neuroscience* **12**:449–460.

Dehaene-Lambertz G, Dehaene S, and Hertz-Pannier L (2002) Functional neuroimaging of speech perception in infants. *Science* **298**:2013–2015.

Dehay C, Kennedy H, and Bullier J (1988) Characterization of transient cortical projections from auditory, somatosensory, and motor cortices to visual areas 17, 18, and 19 in the kitten. *Journal of Comparative Neurology* **272**:68–89.

Dekaban AS (1978) Changes in brain weights during the span of human life: relation of brain weights to body heights and body weights. *Annals of Neurology* **4**:345–356.

Ding SL and Elberger AJ (2001) Postnatal development of biotinylated dextran amine-labeled corpus callosum axons projecting from the visual and auditory cortices to the visual cortex of the rat. *Experimental Brain Research* **136**:179–193.

Ding SL, Rockland KS, and Zheng DS (2000) Parvalbumin immunoreactive Cajal-Retzius and non-Cajal-Retzius neurons in layer I of different cortical regions of human newborn. *Anatomy & Embryology (Berlin)* **201**:407–417.

Dobbing J and Sands J (1971) Vulnerability of developing brain. IX. The effect of nutritional growth retardation on the timing of the brain growth-spurt. *Biology of the Neonate* **19**:363–378.

Eggermont JJ (1989) The onset and development of auditory function: contributions of evoked potential studies. *Journal of Speech and Language Pathology and Audiology* **13**:5–27.

Eggermont JJ (1991) Maturational aspects of periodicity coding in cat primary auditory cortex. *Hearing Research* **57**:45–56.

Eggermont JJ (1992) Stimulus induced and spontaneous rhythmic firing of single units in cat primary auditory cortex. *Hearing Research* **61**:1–11.

Eggermont JJ (1996) Differential maturation rates for response parameters in cat primary auditory cortex. *Auditory Neuroscience* **2**: 309–327.

Eggermont JJ, Ponton CW, Don M, Waring MD, and Kwong B (1997) Maturational delays in cortical evoked potentials in cochlear implant users. *Acta Oto-Laryngologica* **117**:161–163.

Emmorey K, Allen JS, Bruss J, Schenker N, and Damasio H (2003) A morphometric analysis of auditory brain regions in congenitally deaf adults. *Proceedings of the National Academy of Sciences of the United States of America* **100**:10049–10054.

Engel AK, Fries P, and Singer W (2001) Dynamic predictions: oscillations and synchrony in top-down processing. *Nature Reviews Neuroscience* **2**:704–716.

Erzurumlu RS, Chen ZF, and Jacquin MF (2006) Molecular determinants of the face map development in the trigeminal brainstem. *Anatomical Record* **288**:121–134.

Estivill-Torrus G, Pearson H, van Meininger V, Price DJ, and Rashbass P (2002) Pax6 is required to regulate the cell cycle and the rate of progression from symmetrical to asymmetrical division in mammalian cortical progenitors. *Development* **129**:455–466.

Fagiolini M and Hensch TK (2000) Inhibitory threshold for critical-period activation in primary visual cortex. *Nature* **404**:183–186.

Fagiolini M, Katagiri H, Miyamoto H, Mori H, Grant SG, Mishina M, and Hensch TK (2003) Separable features of visual cortical plasticity revealed by *N*-methyl-D-aspartate receptor 2A signaling. *Proceedings of the National Academy of Sciences of the United States of America* **100**:2854–2859.

Fallon JB, Irvine DR, and Shepherd RK (2009) Cochlear implant use following neonatal deafness influences the cochleotopic organization of the primary auditory cortex in cats. *Journal of Comparative Neurology* **512**:101–114.

Feng JZ and Brugge JF (1983) Postnatal development of auditory callosal connections in the kitten. *Journal of Comparative Neurology* **214**:416–426.

Ferrer I, Soriano E, Del Rio JA, Alcantara S, and Auladell C (1992) Cell death and removal in the cerebral cortex during development. *Progress in Neurobiology* **39**:1–43.

Fine I, Finney EM, Boynton GM, and Dobkins KR (2005) Comparing the effects of auditory deprivation and sign language within the auditory and visual cortex. *Journal of Cognitive Neuroscience* **17**:1621–1637.

Finlay BL (1992) Cell death and the creation of regional differences in neuronal numbers. *Journal of Neurobiology* **23**:1159–1171.

Finlay BL and Pallas SL (1989) Control of cell number in the developing mammalian visual system. *Progress in Neurobiology* **32**:207–234.

Finlay BL and Slattery M (1983) Local differences in the amount of early cell death in neocortex predict adult local specializations. *Science* **219**:1349–1351.

Finney EM, Fine I, and Dobkins KR (2001) Visual stimuli activate auditory cortex in the deaf. *Nature Neuroscience* **4**:1171–1173.

Fox MW (1968) Neuronal development and ontogeny of evoked potentials in auditory and visual cortex of the dog. *Electroencephalography and Clinical Neurophysiology* **24**:213–226.

Friauf E and Kandler K (1990) Auditory projections to the inferior colliculus of the rat are present by birth. *Neuroscience Letters* **120**:58–61.

Friauf E, McConnell SK, and Shatz CJ (1990) Functional synaptic circuits in the subplate during fetal and early postnatal development of cat visual cortex. *Journal of Neuroscience* **10**:2601–2613.

Friauf E and Shatz CJ (1991) Changing patterns of synaptic input to subplate and cortical plate during development of visual cortex. *Journal of Neurophysiology* **66**:2059–2071.

Friederici AD (2006) The neural basis of language development and its impairment. *Neuron* **52**:941–952.

Fuchs E and Gould E (2000) Mini-review: in vivo neurogenesis in the adult brain: regulation and functional implications. *European Journal of Neuroscience* **12**:2211–2214.

Galaburda AM and Pandya DN (1983) The intrinsic architectonic and connectional organization of the superior temporal region of the rhesus monkey. *Journal of Comparative Neurology* **221**: 169–184.

Galarreta M and Hestrin S (1999) A network of fast-spiking cells in the neocortex connected by electrical synapses. *Nature* **402**:72–75.

Gao WJ, Newman DE, Wormington AB, and Pallas SL (1999) Development of inhibitory circuitry in visual and auditory cortex of postnatal ferrets: immunocytochemical localization of GABAergic neurons. *Journal of Comparative Neurology* **409**:261–273.

Gao WJ and Pallas SL (1999) Cross-modal reorganization of horizontal connectivity in auditory cortex without altering thalamocortical projections. *Journal of Neuroscience* **19**:7940–7950.

Gao W-J, Power JL, Misra V, and Pallas SL (2000a) Cross-modal alteration of inhibitory circuitry in primary auditory cortex. *Society of Neuroscience Abstracts* **26**:1608.

Gao WJ, Wormington AB, Newman DE, and Pallas SL (2000b) Development of inhibitory circuitry in visual and auditory cortex of postnatal ferrets: immunocytochemical localization of calbindin- and parvalbumin-containing neurons. *Journal of Comparative Neurology* **422**:140–157.

Ghazanfar AA and Schroeder CE (2006) Is neocortex essentially multisensory? *Trends in Cognitive Sciences* **10**:278–285.

Gibson JR, Beierlein M, and Connors BW (1999) Two networks of electrically coupled inhibitory neurons in neocortex. *Nature* **402**: 75–79.

Giedd JN (2004) Structural magnetic resonance imaging of the adolescent brain. *Annals of the New York Academy of Sciences* **1021**: 77–85.

Gleeson JG and Walsh CA (2000) Neuronal migration disorders: from genetic diseases to developmental mechanisms. *Trends in Neurosciences* **23**:352–359.

Glinka A, Wu W, Delius H, Monaghan AP, Blumenstock C, and Niehrs C (1998) Dickkopf-1 is a member of a new family of secreted proteins and functions in head induction. *Nature* **391**:357–362.

Glinka A, Wu W, Onichtchouk D, Blumenstock C, and Niehrs C (1997) Head induction by simultaneous repression of Bmp and Wnt signalling in Xenopus. *Nature* **389**:517–519.

Goodman CS and Shatz CJ (1993) Developmental mechanisms that generate precise patterns of neuronal connectivity. *Cell* **72** Supplement:77–98.

Granier-Deferre C, Lecanuet JP, Cohen H, and Busnel MC (1985) Feasibility of prenatal hearing test. *Acta Otolaryngologica (Stockholm) Supplement* **421**:93–101.

Guillery RW (2005) Is postnatal neocortical maturation hierarchical? *Trends in Neurosciences* **28**:512–517.

Gummer AW and Mark RF (1994) Patterned neural activity in brain stem auditory areas of a prehearing mammal, the tammar wallaby (*Macropus eugenii*). *Neuroreport* **5**:685–688.

Hahn ME, Walters JK, Lavooy J, and DeLuca J (1983) Brain growth in young mice: evidence on the theory of phrenoblysis. *Developmental Psychobiology* **16**:377–383.

Harper MS and Wallace MN (1995) Changes in density of brainstem afferents in ferret primary auditory cortex (AI) during postnatal development. *Journal of Anatomy (London)* **186**:373–382.

Hartmann R, Shepherd RK, Heid S, and Klinke R (1997) Response of the primary auditory cortex to electrical stimulation of the auditory nerve in the congenitally deaf white cat. *Hearing Research* **112**: 115–133.

Hashikawa T, Molinari M, Rausell E, and Jones EG (1995) Patchy and laminar terminations of medial geniculate axons in monkey auditory cortex. *Journal of Comparative Neurology* **362**:195–208.

Hastings NB and Gould E (2003) Neurons inhibit neurogenesis. *Nature Medicine* **9**:264–266.

Heil P, Rajan R, and Irvine DRF (1992) Sensitivity of neurons in cat primary auditory cortex to tones and frequency-modulated stimuli. II: Organization of response properties along the 'isofrequency' dimension. *Hearing Research* **63**:135–156.

Heins N, Cremisi F, Malatesta P, Gangemi RM, Corte G, Price J, Goudreau G, Gruss P, and Gotz M (2001) Emx2 promotes symmetric cell divisions and a multipotential fate in precursors from the cerebral cortex. *Molecular and Cell Neuroscience* **18**:485–502.

Herrmann K, Antonini A, and Shatz CJ (1994) Ultrastructural evidence for synaptic interactions between thalamocortical axons and subplate neurons. *European Journal of Neuroscience* **6**:1729–1742.

Hevner RF (2000) Development of connections in the human visual system during fetal mid-gestation: a DiI-tracing study. *Journal of Neuropathology and Experimental Neurology* **59**:385–392.

Hevner RF, Hodge RD, Daza RA, and Englund C (2006) Transcription factors in glutamatergic neurogenesis: conserved programs in neocortex, cerebellum, and adult hippocampus. *Neuroscience Research* **55**:223–233.

His W (1874) *Unsere Körperform und das physiologische Problem innerer Entstehung*. FCW Vogel, Leipzig.

Honig LS, Herrmann K, and Shatz CJ (1996) Developmental changes revealed by immunohistochemical markers in human cerebral cortex. *Cerebral Cortex* **6**:794–806.

Hsieh CY, Chen Y, Leslie FM, and Metherate R (2002) Postnatal development of NR2A and NR2B mRNA expression in rat auditory cortex and thalamus. *Journal of the Association for Research in Otolaryngology* **3**:479–487.

Hubka P, Kral A, and Klinke R (2004) Input desynchronization and impaired columnar activation in deprived auditory cortex revealed by independent component analysis. In: Syka J and Merzenich MM (eds). *Plasticity and Signal Representation in the Auditory System*. Springer, Berlin, pp. 161–165.

Huttenlocher PR and Dabholkar AS (1997) Regional differences in synaptogenesis in human cerebral cortex. *Journal of Comparative Neurology* **387**:167–178.

Innocenti GM, Berbel P, and Clarke S (1988) Development of projections from auditory to visual areas in the cat. *Journal of Comparative Neurology* **272**:242–259.

Innocenti GM and Clarke S (1984) Bilateral transitory projection to visual areas from auditory cortex in kittens. *Brain Research* **316**:143–148.

Innocenti GM, Clarke S, and Kraftsik R (1986) Interchange of callosal and association projections in the developing visual cortex. *Journal of Neuroscience* **6**:1384–1409.

Jacobson M (1991) *Developmental Neurobiology*. Plenum Press, New York.

Kahn DM and Krubitzer L (2002) Massive cross-modal cortical plasticity and the emergence of a new cortical area in developmentally blind mammals. *Proceedings of the National Academy of Sciences of the United States of America* **99**:11429–11434.

Kalisman N, Silberberg G, and Markram H (2005) The neocortical microcircuit as a tabula rasa. *Proceedings of the National Academy of Sciences of the United States of America* **102**:880–885.

Kennard MA (1938) Reorganization of motor function in the cerebral cortex of monkeys deprived of motor and premotor areas in infancy. *Journal of Neurophysiology* **1**:477–496.

Kilgard MP and Merzenich MM (2002) Order-sensitive plasticity in adult primary auditory cortex. *Proceedings of the National Academy of Sciences of the United States of America* **99**:3205–3209.

Kim HG, Fox K, and Connors BW (1995) Properties of excitatory synaptic events in neurons of primary somatosensory cortex of neonatal rats. *Cerebral Cortex* **5**:148–157.

Kinney HC, Brody BA, Kloman AS, and Gilles FH (1988) Sequence of central nervous system myelination in human infancy. II. Patterns of myelination in autopsied infants. *Neuropathology and Experimental Neurology* **47**:217–234.

Kirkwood A, Rioult MC, and Bear MF (1996) Experience-dependent modification of synaptic plasticity in visual cortex. *Nature* **381**: 526–528.

Klinke R, Kral A, Heid S, Tillein J, and Hartmann R (1999) Recruitment of the auditory cortex in congenitally deaf cats by long-term cochlear electrostimulation. *Science* **285**:1729–1733.

Konig N and Marty R (1974) On functions and structure of deep layers of immature auditory cortex. *Journal of Physiology (Paris)* **68**: 145–155.

Konig N, Roch G, Marty R (1975) The onset of synaptogenesis in rat temporal cortex. *Anatomy & Embryology (Berlin)* **148**:73–87.

Kotak VC, Fujisawa S, Lee FA, Karthikeyan O, Aoki C, and Sanes DH (2005) Hearing loss raises excitability in the auditory cortex. *Journal of Neuroscience* **25**:3908–3918.

Kral A (2007) Unimodal and crossmodal plasticity in the "deaf" auditory cortex. *International Journal of Audiology* **46**:479–493.

Kral A, Hartmann R, and Klinke R (2006a) Recruitment of the auditory cortex in congenitally deaf cats. In: Lomber SG and Eggermont JJ (eds). *Reprogramming the Cerebral Cortex*. Oxford University Press, Oxford, pp. 191–210.

Kral A, Hartmann R, Tillein J, Heid S, and Klinke R (2002) Hearing after congenital deafness: central auditory plasticity and sensory deprivation. *Cerebral Cortex* **12**:797–807.

Kral A, Hartmann R, Tillein J, Heid S, and Klinke R (2001) Delayed maturation and sensitive periods in the auditory cortex. *Audiology & Neurootology* **6**:346–362.

Kral A, Hartmann R, Tillein J, Heid S, and Klinke R (2000) Congenital auditory deprivation reduces synaptic activity within the auditory cortex in a layer-specific manner. *Cerebral Cortex* **10**: 714–726.

Kral A, Schroder JH, Klinke R, and Engel AK (2003) Absence of crossmodal reorganization in the primary auditory cortex of congenitally deaf cats. *Experimental Brain Research* **153**:605–613.

Kral A, Tillein J, Hubka P, Schiemann D, Heid S, Hartmann R, and Engel AK (2009). Spatiotemporal patterns of cortical activity with bilateral cochlear implants in congenital deafness. *Journal of Neuroscience* **29**:811–827.

Kral A, Tillein J, Heid S, Hartmann R, and Klinke R (2005) Postnatal cortical development in congenital auditory deprivation. *Cerebral Cortex* **15**:552–562.

Kral A, Tillein J, Heid S, Klinke R, and Hartmann R (2006b) Cochlear implants: cortical plasticity in congenital deprivation. *Progress in Brain Research* **157**:283–313.

Krmpotic-Nemanic J, Kostovic I, Kelovic Z, and Nemanic D (1980) Development of acetylcholinesterase (AChE) staining in human fetal auditory cortex. *Acta Otolaryngologica* **89**:388–392.

Krmpotic-Nemanic J, Kostovic I, Kelovic Z, Nemanic D, and Mrzljak L (1983) Development of the human fetal auditory cortex: growth of afferent fibres. *Acta Anatomica (Basel)* **116**:69–73.

Krmpotic-Nemanic J, Kostovic I, Nemanic D, and Kelovic Z (1979) The laminar organization of the prospective auditory cortex in the human fetus (11–13.5 weeks of gestation). *Acta Otolaryngologica* **87**:241–246.

Krmpotic-Nemanic J, Kostovic I, Vidic Z, Nemanic D, and Kostovic-Knezevic L (1987) Development of Cajal-Retzius cells in the human auditory cortex. *Acta Otolaryngologica* **103**:477–480.

Kruska D (1993) Evidence of decrease in brain size in ranch mink, *Mustela vison f. dom.*, during subadult postnatal ontogenesis. *Brain, Behavior and Evolution* **41**:303–315.

Kudoh T, Wilson SW, and Dawid IB (2002) Distinct roles for Fgf, Wnt and retinoic acid in posteriorizing the neural ectoderm. *Development* **129**:4335–4346.

Kuhl PK (2004) Early language acquisition: cracking the speech code. *Nature Reviews Neuroscience* **5**:831–843.

Kujala T, Alho K, Huotilainen M, Ilmoniemi RJ, Lehtokoski A, Leinonen A, Rinne T, Salonen O, Sinkkonen J, Standertskjold-Nordenstam CG, and Näätänen R (1997) Electrophysiological evidence for cross-modal plasticity in humans with early- and late-onset blindness. *Psychophysiology* **34**:213–216.
Kujala T, Alho K, and Näätänen R (2000) Cross-modal reorganization of human cortical functions. *Trends in Neurosciences* **23**: 115–120.
Kushnerenko E, Ceponiene R, Balan P, Fellman V, Huotilainen M, and Näätänen R (2002) Maturation of the auditory event-related potentials during the first year of life. *Neuroreport* **13**:47–51.
Lakatos P, Chen CM, O'Connell MN, Mills A, and Schroeder CE (2007) Neuronal oscillations and multisensory interaction in primary auditory cortex. *Neuron* **53**:279–292.
Larsen DD, Luu JD, Burns ME, and Krubitzer L (2009) What are the effects of severe visual impairment on the cortical organization and connectivity of primary visual cortex? *Frontiers in Neuroanatomy* **3**:30.
Lee KJ and Jessell TM (1999) The specification of dorsal cell fates in the vertebrate central nervous system. *Annual Reviews of Neuroscience* **22**:261–294.
Levanen S, Jousmaki V, and Hari R (1998) Vibration-induced auditory-cortex activation in a congenitally deaf adult. *Current Biology* **8**:869–872.
Lewis TL and Maurer D (2005) Multiple sensitive periods in human visual development: evidence from visually deprived children. *Developmental Psychobiology* **46**:163–183.
Lippe WR (1994) Rhythmic spontaneous activity in the developing avian auditory system. *Journal of Neuroscience* **14**:1486–1495.
Locke JL (1997) A theory of neurolinguistic development. *Brain and Language* **58**:265–326.
Lomber SG, Meredith A, and Kral A (2010) Cross-modal plasticity in specific auditory areas underlies visual compensations in deaf. *Nature Neuroscience* **13**:1421–1427.
LoTurco JJ, Blanton MG, and Kriegstein AR (1991) Initial expression and endogenous activation of NMDA channels in early neocortical development. *Journal of Neuroscience* **11**:792–799.
Lu W and Constantine-Paton M (2004) Eye opening rapidly induces synaptic potentiation and refinement. *Neuron* **43**:237–249.
Luskin MB and Shatz CJ (1985) Studies of the earliest generated cells of the cat's visual cortex: cogeneration of subplate and marginal zones. *Journal of Neuroscience* **5**:1062–1075.
Maffei A, Nataraj K, Nelson SB, and Turrigiano GG (2006) Potentiation of cortical inhibition by visual deprivation. *Nature* **443**: 81–84.
Maffei A, Nelson SB, and Turrigiano GG (2004) Selective reconfiguration of layer 4 visual cortical circuitry by visual deprivation. *Nature Neuroscience* **7**:1353–1359.
Majdan M and Miller FD (1999) Neuronal life and death decisions: functional antagonism between the Trk and p75 neurotrophin receptors. *International Journal of Developmental Neuroscience* **17**:153–161.
Mallamaci A, Muzio L, Chan CH, Parnavelas J, and Boncinelli E (2000) Area identity shifts in the early cerebral cortex of *Emx2*$^{-/-}$ mutant mice. *Nature Neuroscience* **3**:679–686.
Manuel M and Price DJ (2005) Role of *Pax6* in forebrain regionalization. *Brain Research Bulletin* **66**:387–393.
Mao Y-T, Hua T-M, and Pallas SL (2007) Cross-modal ferret auditory cortex contains both auditory and visual representations. *Neuroscience Meeting Planner*. Society for Neuroscience, San Diego, p. 36.2.
Mao Y-T and Pallas SL (2009) Compensation and compromise of cortical function after neonatal midbrain damage. *Neuroscience Meeting Planner*. Society for Neuroscience, Chicago, p. 810.2.
Marin-Padilla M (1995) Prenatal development of fibrous (white matter), protoplasmic (gray matter), and layer I astrocytes in the human cerebral cortex: a Golgi study. *Journal of Comparative Neurology* **357**:554–572.
Marin-Padilla M (1992) Ontogenesis of the pyramidal cell of the mammalian neocortex and developmental cytoarchitectonics: a unifying theory. *Journal of Comparative Neurology* **321**:223–240.
Marin-Padilla M (1970) Prenatal and early postnatal ontogenesis of the human motor cortex: a Golgi study. I. The sequential development of the cortical layers. *Brain Research* **23**:167–183.
Marin-Padilla M and Marin-Padilla TM (1982) Origin, prenatal development and structural organization of layer I of the human cerebral (motor) cortex. A Golgi study. *Anatomy & Embryology (Berlin)* **164**:161–206.
Martin BA, Shafer VL, Morr ML, Kreuzer JA, and Kurtzberg D (2003) Maturation of mismatch negativity: a scalp current density analysis. *Ear and Hearing* **24**:463–471.
McConnell SK (1995) Strategies for the generation of neuronal diversity in the developing central nervous system. *Journal of Neuroscience* **15**:6987–6998.
McMullen NT, Goldberger B, and Glaser EM (1988) Postnatal development of lamina III/IV nonpyramidal neurons in rabbit auditory cortex: quantitative and spatial analyses of Golgi-impregnated material. *Journal of Comparative Neurology* **278**:139–155.
Mehler J, Jusczyk P, Lambertz G, Halsted N, Bertoncini J, and Amiel-Tison C (1988) A precursor of language acquisition in young infants. *Cognition* **29**:143–178.
Meyer G, Schaaps JP, Moreau L, and Goffinet AM (2000) Embryonic and early fetal development of the human neocortex. *Journal of Neuroscience* **20**:1858–1868.
Miller MW (1988) Development of projection and local circuit neurons in neocortex. In: Peters A and Jones EG (eds). *Cerebral Cortex*, volume 7, *Development and Maturation of Cerebral Cortex*. Plenum Press, New York, pp. 133–175.
Molliver ME, Kostovic I, and Van der Loos H (1973) The development of synapses in cerebral cortex of the human fetus. *Brain Research* **50**:403–407.
Molnar Z, Higashi S, and Lopez-Bendito G (2003) Choreography of early thalamocortical development. *Cerebral Cortex* **13**:661–669.
Moore DR and Irvine DRF (1979) The development of some peripheral and central auditory responses in the neonatal cat. *Brain Research* **163**:49–59.
Moore JK and Guan YL (2001) Cytoarchitectural and axonal maturation in human auditory cortex. *Journal of the Association for Research in Otolaryngology* **2**:297–311.
Moore JK, Perazzo LM, and Braun A (1995) Time course of axonal myelination in the human brainstem auditory pathway. *Hearing Research* **87**:21–31.
Moore JK, Ponton CW, Eggermont JJ, Wu BJ, and Huang JQ (1996) Perinatal maturation of the auditory brain stem response: changes in path length and conduction velocity. *Ear and Hearing* **17**:411–418.
Morest DK (1969a) The differentiation of cerebral dendrites: a study of the post-migratory neuroblast in the medial nucleus of the trapezoid body. *Zeitschrift für Anatomie und Entwicklungsgeschichte* **128**:271–289.
Morest DK (1969b) The growth of dendrites in the mammalian brain. *Zeitschrift für Anatomie und Entwicklungsgeschichte* **128**:290–317.
Mrsic-Flogel TD, Schnupp JWH, and King AJ (2003) Acoustic factors govern developmental sharpening of spatial tuning in the auditory cortex. *Nature Neuroscience* **6**:981–988.
Nakahara H, Zhang LI, and Merzenich MM (2004) Specialization of primary auditory cortex processing by sound exposure in the "critical period". *Proceedings of the National Academy of Sciences of the United States of America* **101**:7170–7174.
Naruse I and Keino H (1995) Apoptosis in the developing CNS. *Progress in Neurobiology* **47**:135–155.
Neville HJ (1990) Intermodal competition and compensation in development. Evidence from studies of the visual system in congenitally

deaf adults. *Annals of the New York Academy of Sciences* **608**: 71–87.

Neville HJ, Mills DL, and Lawson DS (1992) Fractionating language: different neural subsystems with different sensitive periods. *Cerebral Cortex* **2**:244–258.

Nishimura H, Hashikawa K, Doi K, Iwaki T, Watanabe Y, Kusuoka H, Nishimura T, and Kubo T (1999) Sign language 'heard' in the auditory cortex. *Nature* **397**:116.

O'Kusky J and Colonnier M (1982a) Postnatal changes in the number of neurons and synapses in the visual cortex (area 17) of the macaque monkey: a stereological analysis in normal and monocularly deprived animals. *Journal of Comparative Neurology* **210**: 291–306.

O'Kusky J and Colonnier M (1982b) A laminar analysis of the number of neurons, glia, and synapses in the adult cortex (area 17) of adult macaque monkeys. *Journal of Comparative Neurology* **210**:278–290.

O'Kusky JR (1985) Synapse elimination in the developing visual cortex: a morphometric analysis in normal and dark-reared cats. *Brain Research* **354**:81–91.

Ohlrich ES, Barnet AB, Weiss IP, and Shanks BL (1978) Auditory evoked potential development in early childhood: a longitudinal study. *Electroencephalography and Clinical Neurophysiology* **44**:411–423.

Oppenheim RW (1991) Cell death during development of the nervous system. *Annual Reviews in Neuroscience* **14**:453–501.

Oray S, Majewska A, and Sur M (2004) Dendritic spine dynamics are regulated by monocular deprivation and extracellular matrix degradation. *Neuron* **44**:1021–1030.

Pallas SL (2001) Intrinsic and extrinsic factors that shape neocortical specification. *Trends in Neurosciences* **24**:417–423.

Pallas SL (2005) Pre- and postnatal sensory experience shapes functional architecture in the brain. In: Hopkins B and Johnson SP (eds). *Advances in Infancy Research*, volume 14, *Prenatal Development of Postnatal Functions*. Praeger, Westport, pp. 1–30.

Pallas SL (2007) Compensatory innervation in development and evolution. In: Kaas J, Striedter GF, and Rubenstein JLR (eds). *Evolution of Nervous Systems*, volume 1, *Theories, Development, and Invertebrates*. Elsevier Academic Press, Amsterdam, pp. 153–168.

Pallas SL, Gilmour SM, and Finlay BL (1988) Control of cell number in the developing neocortex. I. Effects of early tectal ablation. *Brain Research* **471**:1–11.

Pallas SL, Littman T, and Moore DRF (1999) Cross-modal reorganization of callosal connectivity without altering thalamocortical projections. *Proceedings of the National Academy of Sciences of the United States of America* **96**:8751–8756.

Pallas SL, Roe AW, and Sur M (1990) Visual projections induced into the auditory pathway of ferrets. I. Novel inputs to primary auditory cortex (AI) from the LP/pulvinar complex and the topography of the MGN-AI projection. *Journal of Comparative Neurology* **298**:50–68.

Pallas SL and Sur M (1993) Visual projections induced into the auditory pathway of ferrets: II. Corticocortical connections of primary auditory cortex. *Journal of Comparative Neurology* **337**: 317–333.

Pallas SL, Wenner P, Gonzalez-Islas C, Fagiolini M, Razak KA, Kim G, Sanes D, and Roerig B (2006) Developmental plasticity of inhibitory circuitry. *Journal of Neuroscience* **26**:10358–10361.

Pasman JW, Rotteveel JJ, Maassen B, and Visco YM (1999) The maturation of auditory cortical evoked responses between (preterm) birth and 14 years of age. *European Journal of Paediatric Neurology* **3**:79–82.

Paus T, Zijdenbos A, Worsley K, Collins DL, Blumenthal J, Giedd JN, Rapoport JL, and Evans AC (1999) Structural maturation of neural pathways in children and adolescents: in vivo study. *Science* **283**:1908–1911.

Payne B, Pearson HE, and Cornwell P (1988a) Development of visual and auditory cortical connections in the cat. In: Peters A and Jones EG (eds). *Cerebral Cortex*, volume 7, *Development and Maturation of Cerebral Cortex*. Plenum Press, New York, pp. 309–389.

Payne BR (1992) Development of the auditory cortex. In: Romand R (ed). *Development of Auditory and Vestibular Systems*, volume 2. Elsevier Science Publishers Amsterdam, pp. 357–390.

Payne BR, Pearson HE, and Cornwell P (1988b) Neocortical connections in fetal cats. *Neuroscience Research* **5**, 513–543.

Pearlman AL (1985) The visual cortex of the normal mouse and the reeler mutant. In: Peters A and Jones EG (eds). *Cerebral Cortex*, volume 3, *Visual Cortex*. New York, Plenum, pp. 1–18.

Pena M, Maki A, Kovacic D, Dehaene-Lambertz G, Koizumi H, Bouquet F, and Mehler J (2003) Sounds and silence: an optical topography study of language recognition at birth. *Proceedings of the National Academy of Sciences of the United States of America* **100**:11702–11705.

Penhune VB, Cismaru R, Dorsaint-Pierre R, Petitto LA, and Zatorre RJ (2003) The morphometry of auditory cortex in the congenitally deaf measured using MRI. *NeuroImage* **20**:1215–1225.

Petitto LA, Zatorre RJ, Gauna K, Nikelski EJ, Dostie D, and Evans AC (2000) Speech-like cerebral activity in profoundly deaf people processing signed languages: implications for the neural basis of human language. *Proceedings of the National Academy of Sciences of the United States of America* **97**:13961–13966.

Pienkowski M and Harrison RV (2005) Tone frequency maps and receptive fields in the developing chinchilla auditory cortex. *Journal of Neurophysiology* **93**:454–466.

Ponton CW, Don M, Eggermont JJ, Waring MD, Kwong B, and Masuda A (1996a) Auditory system plasticity in children after long periods of complete deafness. *Neuroreport* **8**:61–65.

Ponton CW, Don M, Eggermont JJ, Waring MD, and Masuda A (1996b) Maturation of human cortical auditory function: differences between normal-hearing children and children with cochlear implants. *Ear and Hearing* **17**:430–437.

Ponton CW and Eggermont JJ (2001) Of kittens and kids: altered cortical maturation following profound deafness and cochlear implant use. *Audiology & Neurootology* **6**:363–380.

Ponton CW, Eggermont JJ, Kwong B, and Don M (2000) Maturation of human central auditory system activity: evidence from multi-channel evoked potentials. *Clinical Neurophysiology* **111**: 220–236.

Pratt T, Quinn JC, Simpson TI, West JD, Mason JO, and Price DJ (2002) Disruption of early events in thalamocortical tract formation in mice lacking the transcription factors Pax6 or Foxg1. *Journal of Neuroscience* **22**:8523–8531.

Price DJ and Willshaw DJ (2000) *Mechanisms of Cortical Development*. Oxford University Press, New York.

Quinlan EM, Olstein DH, and Bear MF (1999a) Bidirectional, experience-dependent regulation of *N*-methyl-D-aspartate receptor subunit composition in the rat visual cortex during postnatal development. *Proceedings of the National Academy of Sciences of the United States of America* **96**:12876–12880.

Quinlan EM, Philpot BD, Huganir RL, and Bear MF (1999b) Rapid, experience-dependent expression of synaptic NMDA receptors in visual cortex in vivo. *Nature Neuroscience* **2**:352–357.

Raggio MW and Schreiner CE (2003) Neuronal responses in cat primary auditory cortex to electrical cochlear stimulation: IV. Activation pattern for sinusoidal stimulation. *Journal of Neurophysiology* **89**:3190–3204.

Rakic P (1977) Prenatal development of the visual system in rhesus monkey. *Philosophical Transactions of the Royal Society London, Series B: Biological Sciences* **278**:245–260.

Rakic P (1972) Mode of cell migration to the superficial layers of fetal monkey neocortex. *Journal of Comparative Neurology* **145**: 61–83.

Rauschecker JP (1995) Compensatory plasticity and sensory substitution in the cerebral cortex. *Trends in Neurosciences* **18**:36–43.
Rittenhouse CD, Shouval HZ, Paradiso MA, and Bear MF (1999) Monocular deprivation induces homosynaptic long-term depression in visual cortex. *Nature* **397**:347–350.
Robertson RT, Mostamand F, Kageyama GH, Gallardo KA, and Yu J (1991) Primary auditory cortex in the rat: transient expression of acetylcholinesterase activity in developing geniculocortical projections. *Developmental Brain Research* **58**:81–95.
Roe AW, Garraghty PE, Esguerra M, and Sur M (1993) Experimentally induced visual projections to the auditory thalamus in ferrets: evidence for a W cell pathway. *Journal of Comparative Neurology* **334**:263–280.
Roe AW, Pallas SL, Hahm JO, and Sur M (1990) A map of visual space induced in primary auditory cortex. *Science* **250**:818–820.
Roe AW, Pallas SL, Kwon YH, and Sur M (1992) Visual projections routed to the auditory pathway in ferrets: receptive fields of visual neurons in primary auditory cortex. *Journal of Neuroscience* **12**:3651–3664.
Romand R (1997) Modification of tonotopic representation in the auditory system during development. *Progress in Neurobiology* **51**:1–17.
Rotteveel JJ (1992) Development of brainstem, middle latency and cortical auditory evoked responses in the human. In: Romand R (ed). *Development of Auditory and Vestibular Systems, Volume 2*. Elsevier Science Publishers, Amsterdam, pp. 321–356.
Ruben RJ (1997) A time frame of critical/sensitive periods of language development. *Acta Oto-Laryngologica* **117**:202–205.
Rubenstein JLR (2010) Development of the cerebral cortex: implications for neurodevelopmental disorders. *Journal of Child Psychology and Psychiatry* DOI: 10.1111/j.1469-7610.2010.02307.x
Sanes DH and Walsh EJ (1998) The Development of Central Auditory Processing. In: Rubel EW, Popper AN, and Fay RR (eds). *Springer Handbook of Auditory Research*, volume 9, *Development of the Auditory System*. Springer, New York, pp. 271–314.
Sasai Y and De Robertis EM (1997) Ectodermal patterning in vertebrate embryos. *Developmental Biology* **182**:5–20.
Schneider GE (1973) Early lesions of superior colliculus: factors affecting the formation of abnormal retinal projections. *Brain, Behavior and Evolution* **8**:73–109.
Schreiner CE and Mendelson JR (1990) Functional topography of cat primary auditory cortex: distribution of integrated excitation. *Journal of Neurophysiology* **64**:1442–1459.
Schreiner CE and Sutter ML (1992) Topography of excitatory bandwidth in cat primary auditory cortex: single-neuron versus multiple-neuron recordings. *Journal of Neurophysiology* **68**:1487–1502.
Schroeder CE, Lindsley RW, Specht C, Marcovici A, Smiley JF, and Javitt DC (2001) Somatosensory input to auditory association cortex in the macaque monkey. *Journal of Neurophysiology* **85**:1322–1327.
Schulte FJ, Stennert E, Wulbrand H, Eichborn W, and Lenard HG (1977) The ontogeny of sensory perception in preterm infants. *European Journal of Pediatrics* **126**:211–224.
Sedlacek J (1976) Foetal and neonatal development of evoked responses in guinea-pig auditory cortex. *Physiologica Bohemoslovaca* **25**: 13–21.
Sermasi E, Tropea D, and Domenici L (1999) Long term depression is expressed during postnatal development in rat visual cortex: a role for visual experience. *Developmental Brain Research* **113**:61–65.
Shankle WR, Romney AK, Landing BH, and Hara J (1998) Developmental patterns in the cytoarchitecture of the human cerebral cortex from birth to 6 years examined by correspondence analysis. *Proceedings of the National Academy of Sciences of the United States of America* **95**:4023–4028.
Sharma A, Dorman MF, and Spahr AJ (2002a) A sensitive period for the development of the central auditory system in children with cochlear implants: implications for age of implantation. *Ear & Hearing* **23**:532–539.
Sharma A, Dorman MF, and Kral A (2005) The influence of a sensitive period on central auditory development in children with unilateral and bilateral cochlear implants. *Hearing Research* **203**:134–143.
Sharma A, Dorman MF, and Spahr AJ (2002b) Rapid development of cortical auditory evoked potentials after early cochlear implantation. *Neuroreport* **13**:1365–1368.
Sharma A, Kraus N, McGee TJ, and Nicol TG (1997) Developmental changes in P1 and N1 central auditory responses elicited by consonant-vowel syllables. *Electroencephalography and Clinical Neurophysiology* **104**:540–545.
Sharma J, Angelucci A, and Sur M (2000) Induction of visual orientation modules in auditory cortex. *Nature* **404**:841–847.
Shatz CJ (1990) Impulse activity and the patterning of connections during CNS development. *Neuron* **5**:745–756.
Shatz CJ and Luskin MB (1986) The relationship between the geniculocortical afferents and their cortical target cells during development of the cat's primary visual cortex. *Journal of Neuroscience* **6**: 3655–3668.
Shimamura K, Martinez S, Puelles L, and Rubenstein JLR (1997) Patterns of gene expression in the neural plate and neural tube subdivide the embryonic forebrain into transverse and longitudinal domains. *Developmental Neuroscience* **19**:88–96.
Shuler MG and Bear MF (2006) Reward timing in the primary visual cortex. *Science* **311**:1606–1609.
Stewart DL and Starr A (1970) Absence of visually influenced cells in auditory cortex of normal and congenitally deaf cats. *Experimental Neurology* **28**:525–528.
Sur M, Garraghty PE, and Roe AW (1988) Experimentally induced visual projections into auditory thalamus and cortex. *Science* **242**:1437–1441.
Sur M and Rubenstein JLR (2005) Patterning and plasticity of the cerebral cortex. *Science* **310**:805–810.
Sutor B (2002) Neurogenesis and maturation of synaptic circuitry in the developing cortex. In: Hohmann CF (ed). *Cortical Development*. Springer, Berlin, pp. 53–73.
Tillein J, Hubka P, Syed E, Hartmann R, Engel AK, and Kral A (2010) Cortical representation of interaural time difference in congenital deafness. *Cerebral Cortex* **20**:492–506.
Trachtenberg JT, Chen BE, Knott GW, Feng G, Sanes JR, Welker E, and Svoboda K (2002) Long-term in vivo imaging of experience-dependent synaptic plasticity in adult cortex. *Nature* **420**:788–794.
Trainor L, McFadden M, Hodgson L, Darragh L, Barlow J, Matsos L, and Sonnadara R (2003) Changes in auditory cortex and the development of mismatch negativity between 2 and 6 months of age. *International Journal of Psychobiology* **51**:5–15.
Tritsch NX and Bergles DE (2010) Developmental regulation of spontaneous activity in the Mammalian cochlea. *Journal of Neuroscience* **30**:1539–550.
Tsumoto T and Suda K (1982) Postnatal development of the corticofugal projection from striate cortex to lateral geniculate nucleus in kittens. *Brain Research* **256**:323–332.
Uziel D, Garcez P, Lent R, Peuckert C, Niehage R, Weth F, and Bolz J (2006) Connecting thalamus and cortex: the role of ephrins. *Anatomical Record* **288**:135–142.
Vale C, Juiz JM, Moore DR, and Sanes DH (2004) Unilateral cochlear ablation produces greater loss of inhibition in the contralateral inferior colliculus. *European Journal of Neuroscience* **20**: 2133–2140.
Vale C and Sanes DH (2002) The effect of bilateral deafness on excitatory and inhibitory synaptic strength in the inferior colliculus. *European Journal of Neuroscience* **16**:2394–2404.
Vale C, Schoorlemmer J, and Sanes DH (2003) Deafness disrupts chloride transporter function and inhibitory synaptic transmission. *Journal of Neuroscience* **23**:7516–7524.

Valverde F and Facal-Valverde MV (1988) Postnatal development of interstitial (subplate) cells in the white matter of the temporal cortex of kittens: a correlated Golgi and electron microscopic study. *Journal of Comparative Neurology* **269**:168–192.

van Zundert B, Yoshii A, and Constantine-Paton M (2004) Receptor compartmentalization and trafficking at glutamate synapses: a developmental proposal. *Trends in Neurosciences* **27**:428–437.

Vicario-Abejón C, Owens D, McKay R, and Segal M (2002) Role of neurotrophins in central synapse formation and stabilization. *Nature Reviews Neuroscience* **3**:965–974.

von Melchner L, Pallas SL, and Sur M (2000) Visual behaviour mediated by retinal projections directed to the auditory pathway. *Nature* **404**:871–876.

Voyvodic JT (1996) Cell death in cortical development: how much? Why? So what? *Neuron* **16**:693–696.

Vu DH and Törk I (1992) Differential development of the dual serotoninergic fiber system in the cerebral cortex of the cat. *Journal of Comparative Neurology* **317**:156–174.

Wallace IF, Gravel JS, McCarton CM, and Ruben RJ (1988) Otitis media and language development at 1 year of age. *Journal of Speech and Hearing Disorders* **53**:245–251.

Wilkinson F (1986) Eye and brain growth in the Mongolian gerbil (*Meriones unguiculatus*). *Behavioral Brain Research* **19**: 59–69.

Wallace MT, Ramachandran R, and Stein BE (2004) A revised view of sensory cortical parcellation. *Proceedings of the National Academy of Science of the United States of America* **101**:2167–2172.

Windrem MS, de Beur SJ, and Finlay BL (1988) Control of cell number in the developing neocortex. 2. Effects of corpus callosum section. *Developmental Brain Research* **43**:13–22.

Winfield DA (1981) The postnatal development of synapses in the visual cortex of the cat and the effects of eyelid closure. *Brain Research* **206**:166–171.

Winfield DA (1983) The postnatal development of synapses in the different laminae of the visual cortex in the normal kitten and in kittens with eyelid suture. *Brain Research* **285**:155–169.

Winkler I, Kushnerenko E, Horvath J, Ceponiene R, Fellman V, Huotilainen M, Näätänen R, and Sussman E (2003) Newborn infants can organize the auditory world. *Proceedings of the National Academy of Sciences of the United States of America* **100**: 11812–11815.

Wonders CP and Anderson SA (2006) The origin and specification of cortical interneurons. *Nature Reviews Neuroscience* **7**: 687–696.

Yakovlev PL and Lecour A-R (1967) The myelogenic cycles of regional maturation of the brain. In: Minkowski A (ed). *Regional Development of the Brain in Early Life*. Blackwell Scientific Publications, Oxford, pp. 3–70.

Yan J (2003) Canadian association of neuroscience review: development and plasticity of the auditory cortex. *Canadian Journal of Neural Science* **30**:189–200.

Yoon SO, Casaccia-Bonnefil P, Carter B, and Chao MV (1998) Competitive signaling between TrkA and p75 nerve growth factor receptors determines cell survival. *Journal of Neuroscience* **18**:3273–3281.

Yordanova J, Kolev V, Heinrich H, Woerner W, Banaschewski T, and Rothenberger A (2002) Developmental event-related gamma oscillations: effects of auditory attention. *European Journal of Neuroscience* **16**:2214–2224.

Yuste R, Nelson DA, Rubin WW, and Katz LC (1995) Neuronal domains in developing neocortex: mechanisms of coactivation. *Neuron* **14**:7–17.

Yuste R, Peinado A, and Katz LC (1992) Neuronal domains in developing neocortex. *Science* **257**:665–669.

Zervas M and Walkley SU (1999) Ferret pyramidal cell dendritogenesis: changes in morphology and ganglioside expression during cortical development. *Journal of Comparative Neurology* **413**: 429–448.

Zhang LI, Bao S, and Merzenich MM (2002) Disruption of primary auditory cortex by synchronous auditory inputs during a critical period. *Proceedings of the National Academy of Sciences of the United States of America* **99**: 2309–2314.

Zhang LI, Bao S, and Merzenich MM (2001) Persistent and specific influences of early acoustic environments on primary auditory cortex. *Nature Neuroscience* **4**:1123–1130.

Chapter 22

Reconceptualizing the Primary Auditory Cortex: Learning, Memory and Specific Plasticity

Norman M. Weinberger

Abbreviations

2-DG	2-deoxyglucose
AC	auditory cortex
AI	primary auditory cortex
AII	second auditory field
APS	auditory problem solver
BF	best frequency
CF	characteristic frequency
CR	conditioned response
CS	conditioned stimulus
EEG	electroencephalogram
ITI	intertrial intervals
LFP	local field potential
MGB	medial geniculate body
RF	receptive field
RM	reference memory
SMI	specific memory trace
SPL	sound pressure level
US	unconditioned stimulus
WM	working memory

1 Introduction

Since 1985, attitudes about the role of the primary auditory cortex (AI) in learning, memory, and adult plasticity have changed from a denial, or studied disinterest, to an acceptance of these roles and, presently, to a new lack of interest. From the traditional assumption that AI is only an acoustic analyzer to the prevalent belief that learning-induced plasticity serves only to facilitate sensory analysis, auditory neuroscientists are expressing (more in private than in publication) a growing boredom with cortical plasticity. One worker wondered: "How much longer must we be subjected to endless demonstrations of plasticity?" From one viewpoint, this attitude is completely understandable, because (almost) every study of plasticity finds plasticity, first for acoustic frequency and, more recently, for any other acoustic parameter that has been used as a signal for reward or punishment. If all that has been gained is the continued compilation of plasticity demonstrations, boredom would be justified. But that is not all there is to it.

From a broader perspective, demonstrations of plasticity have had two major effects. The first provides a foundation for understanding how the contents of auditory experience are acquired, represented, stored, and employed in adaptive behavior. The second compels us to confront the need for a reconceptualization of auditory cortical function.

These two consequences of learning-induced plasticity in AI are of central importance to two disciplines. The auditory cortex is fundamental to auditory neuroscience. Understanding how experiences become memories is crucial to the neurobiology of learning and memory. These two fields have a deep common interest that has been overlooked: each seeks to understand the fate of sensory stimuli (acoustic in the present context). Thus, auditory neurophysiology seeks, at least, to discover how sounds are coded. This is a question of the neural representation of basilar membrane displacements. The neurophysiology of learning and memory, in contrast, seeks to discover how a sound becomes a psychological object. This is a question of the transformation of stimulus representations so that, however sounds are coded, they become signals for some other event, such as a reward or punishment. Clearly, both disciplines are concerned with auditory representations.

In fact, the fundamental paradigms of auditory neurophysiology and the neurophysiology of learning and memory can be seen as complementary. Auditory neurophysiology varies the physical parameters of sounds while holding constant their psychological parameters, i.e., the behavioral relevance or meaning of sounds. Complementing this approach, the neurophysiology of learning and memory holds constant the former while varying the latter (Fig. 22.1).

N.M. Weinberger (✉)
Qureshey Research Laboratory, Department of Neurobiology and Behavior, Center of Neurobiology Learning and Memory, University of California at Irvine, Irvine, CA 92697-3800, USA
e-mail: nmweinbe@uci.edu

J.A. Winer, C.E. Schreiner (eds.), *The Auditory Cortex*,
DOI 10.1007/978-1-4419-0074-6_22, © Springer Science+Business Media, LLC 2011

Discipline	Stimulus Parameters	
	Physical [a]	Psychological [b]
AudPhys	Vary	Constant
L-MPhys	Constant	Vary

[a] E.g., wavelength, kHz, db, etc.
[b] E.g., trials to criterion, # correct responses, etc.

Fig. 22.1 The fundamental paradigms of the disciplines auditory neurophysiology (AudPhys) and the experimental psychology of learning/memory (L-MPhys) are complementary. Stimulus parameters have two basic properties: physical and psychological. Auditory neurophysiology typically manipulates the physical aspects of sound while seeking neural responses, holding constant the psychological meaning of acoustic stimuli. Learning/memory studies do the opposite. Recent years have seen an increase in the combination of these two approaches within the same experiment

Nonetheless, the two fields developed separately with minimal cross-fertilization. This unfortunate situation may be traced to the late nineteenth and early twentieth centuries, when ideas of mental functions and their presumptive cortical substrates asserted that sensory-perceptual analysis and the interpretation-behavioral meaning of perceptions were completely separable. A landmark architectonic monograph on functional localization (Campbell 1905) promulgated this dogma. He labeled the primary auditory cortex 'auditory sensory,' while adjacent auditory regions (now called auditory belt fields) were termed 'auditory psychic' (Diamond 1979; Weinberger 2008a, 2009). A major consequence of this conceptual distinction, based on anatomical grounds, has been that the primary auditory cortex, as other primary sensory cortices, was largely ignored by the study of learning and memory. A comprehensive understanding of auditory cortex requires a synthesis of auditory neurophysiology and neurophysiology of learning and memory (Weinberger 2007a).

2 Scope and Approach

This chapter is an overview of learning, memory, and related plasticity in the primary auditory cortex. Conceptual issues are emphasized since prior reviews have favored empirical studies (Merzenich and Sameshima 1993; Palmer et al. 1998; Weinberger and Bakin 1998; Rauschecker 1999; Kilgard et al. 2002; Edeline 2003; Suga and Ma 2003; Ohl and Scheich 2005). However, experimental findings will be considered and cautions and object lessons discussed, though the thrust is not methodological; advantages and limitations of canonical experimental designs are considered elsewhere (Weinberger 2004b, 2008a, 2009).

Coverage of empirical findings focuses on experimental animal studies, afford precise stimulus control, allow localization of recording sites, and yield various neurophysiological data, including unitary discharges. Moreover, animal studies have yielded most of the information on auditory neurophysiology and neurophysiology of learning and memory. Space limitations preclude considerations of mechanisms of learning-induced plasticity (Weinberger 2009). The final sections attempt a synthesis and anticipate potential research directions as a basis for a new view of the auditory cortex. They emphasize that what we can know depends on the questions that we ask. And many important questions have not yet been addressed.

3 Levels of Analysis and Codes in the Auditory Cortex

An overview of levels of analysis and codes in the primary auditory cortex provides a helpful framework (Fig. 22.2). This schema relates auditory neurophysiology to the neurophysiology of learning and memory.

The relevant levels are psychological, neural systems, and neuronal. At the psychological level, events give rise to percepts, some small fraction of which become the contents of memories. At the neural systems level, sensory stimuli are transduced and then are processed from first to *n*th order within the auditory system, some small fraction of stimulus processing become the substrates of memories, referred to as engrams. The neuronal level is summarized as consisting of generator potentials underlying transduction, action potentials (in this simplified schema) underlying auditory processing, and modified synaptic strengths as a substrate of engrams.

This simplified framework denotes basic relationships between learning/memory and auditory (and other sensory) physiology by linking perception to memory and by separating perception and memory from their underlying neural substrates.

3.1 Levels

Linking memory to perception underscores their organic relationship; there can be no memory without sensory/perceptual events in some form. Only a fraction of all percepts become memories.

The significance of emphasizing the different levels is to reduce conceptual confusions. Unfortunately, there is a widespread conflation of levels. Many equate plasticity with

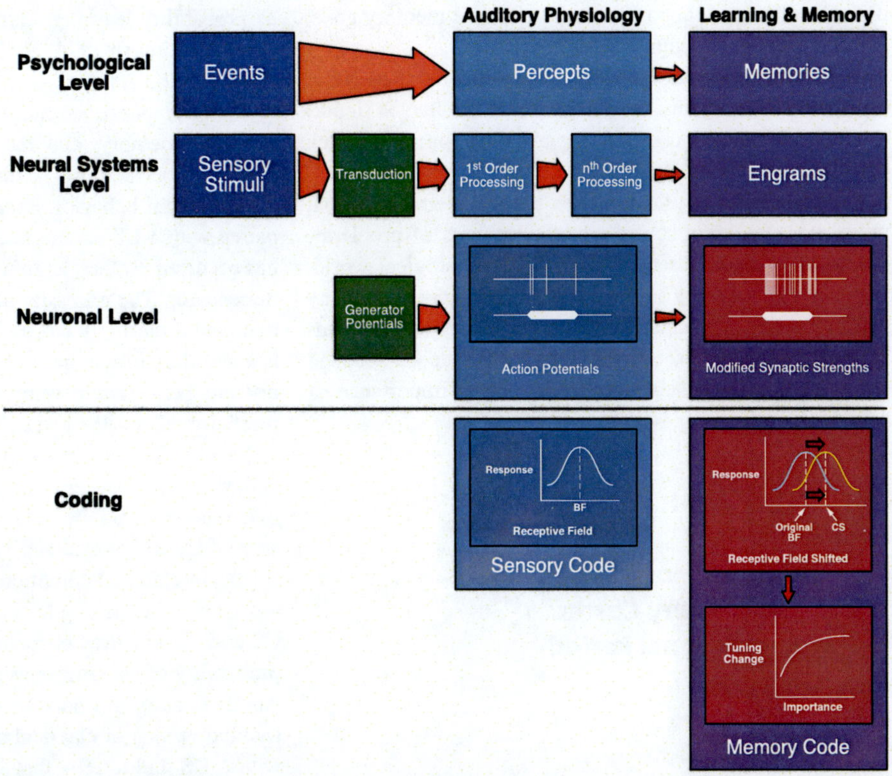

Fig. 22.2 Auditory physiology and learning/memory: levels of analysis and coding (see text for details)

memory. As used here, plasticity refers to any learning-induced change in neuronal activity lasting for minutes to lifetimes, to distinguish it from sensory responses that may last seconds. In this schema, plasticity is a property at the neuronal level, conventionally assigned to changes in synaptic strength, and to the interactions of neurons at the systems level. However, the schema holds regardless of the ultimate cellular mechanisms.

A popular example of fallacious level logic yields the belief that a particular example of plasticity, long-term potentiation, is memory. Particular instances of plasticity can, and should, be tested to determine if they constitute an actual memory trace, that is, a substrate of information storage. This issue will be discussed later, when the specificity of plasticity is considered. Conflating neural plasticity with memory is experimentally confusing and conceptually fallacious. While learning and memory undoubtedly are caused by neural plasticity, equating the two is a category error, i.e., attributing a property of the whole to one of its parts (Ryle 1963).

A consequence of the separation of levels and the relationships between levels is that the development of physiological plasticity during learning cannot be used to conclude that learning or memory have been formed. Rather, such validation must take place at the psychological/behavioral level.

3.2 Sensory Codes and Memory Codes

Coding broadly denotes the algorithm that mediates the transformation of one entity into another form or representation. Thus, coding is not another level of organization nor the neural substrate of percepts nor of memories, but rather an abstraction. A code can denote a specific input–output function. Neuroscientists are familiar with sensory codes. Sensory codes as defined here reveal how the nervous system solves a specific problem, i.e., the representation of the sensory world. For example, a sensory code for stimulus level describes the transform function from the magnitude of basilar membrane displacement (input) to the rate of discharge in VIIIth nerve axons (output).

However, memory codes require explication. A first issue is: Does the primary auditory cortex acquire and store information that is a substrate of auditory memory, i.e., does AI hold specific memory traces? This question is the major subject of the present account. The findings reviewed below strongly support an affirmative answer. A second, more abstract question may pertain to all brain regions that store information non-transiently: Does the primary auditory cortex use memory codes to represent cardinal features of information/memory storage?

In contrast to the extensive research on the first question, the issue of memory codes has been addressed in one study only. The findings support the existence of a memory code. But first, what are memory codes?

A memory code describes the transform function from, e.g., patterns of sensory-derived neuronal discharges (input) to enduring changes in neural organization (output), which constitute an engram (Fig. 22.2). Like sensory codes, memory codes are not physical entities but are abstractions that reveal how the brain solves a particular class of problems, specifically, the representation of the psychological world. Just as there are many sensory codes, there may exist many memory codes. The putative auditory memory code considered here is a key characteristic of all experienced acoustic stimuli: the learned behavioral relevance or importance of sound.

4 Learning, Memory, Plasticity, and the Auditory Cortex: The Foundational Period

4.1 Introduction

Definitions of learning and memory abound. A common definition is that learning is a change of behavior or the potential for behavior unrelated to injury, illness, drugs, exercise, etc. This definition is inconvenient as it is exclusionary and requires both the compilation and the justification of a list of rejected causes that may not be exhaustive. The definition used here is simpler: learning is the acquisition of information; memory is the storage of acquired information. This initially leaves open how to determine which information is acquired and maintained. However, there are ample objective behavioral tests to answer this question; several of these will be noted below.

Almost all studies of the role of AI in learning and memory analyzed associative learning, either the transformation of an acoustic stimulus into a signal for subsequent positive or negative reinforcement (in Pavlovian/classical conditioning) or as a signal for a behavioral response that will yield positive reinforcement or the avoidance of negative reinforcement (in Thorndikian/instrumental conditioning). We use conditioned stimulus (CS) for acoustic signals and unconditioned stimulus (US) for positive and negative (rewarding and punishing) reinforcers, respectively, in both forms of conditioning.

Associative processes have a surprising richness that includes aspects of learning and behavior more complex than one-tone conditioning, although the latter is a solid foundation for understanding complex associative phenomena. Conditioning can account for complex processes such as probability learning, categorization, and concept formation (Mackintosh 1974, 1983; Domjan 1998; Schwartz et al. 2002; Bouton 2007).

Two points are essential to appreciate the literature on learning, memory, and the auditory cortex. First, neither learning nor memory is directly observable and must be inferred from behavior. Even repeated presentation of a CS paired with a US cannot justify the conclusion that learning has occurred. Rather, learning must be inferred from changes in behavior. The necessity of behavioral validation suggests that neural plasticity cannot be used to verify learning. This is a corollary of a point noted above that learning and memory are behavioral/psychological level attributes, whereas plasticity reflects the level of neurons and their systems.

Second, because learning and memory are inferred from behavior, other causes of a behavioral change, besides learning, must be excluded. These include general changes in state of arousal or excitability, or sensitization due to food or shock. In classical conditioning a standard control for non-associative factors employs a second group that receives the CS and US unpaired or randomly, but with the same overall probability of occurrence as the paired (conditioning) group. Another control for non-associative factors is discrimination training, which in classical conditioning uses a CS+ paired with a US and a CS− that is not followed by a US or any stimulus. Successful discrimination training demonstrates an association between the CS+ and the US but not between the CS− and the US. Such associative specificity cannot be explained by factors like sensitization.

An invalid conclusion that subjects have learned a CS–US association can occur if they can accurately estimate the time of presentation of the next conditioning trial. A subject can show temporal conditioning when intertrial intervals (ITIs) are fixed and, thus, become predictable. Pavlov circumvented this confound by using ITIs of variable and unpredictable length. Fixed ITIs are still found in some learning studies (Suga and Ma 2003).

4.2 Is Auditory Learning Actually Perceptual Learning?

It is often assumed that all learning involving sensory systems, including the auditory cortex, must be perceptual learning. Perceptual learning denotes increased sensory acuity in a stimulus dimension, usually in progressively difficult discrimination training (Kellman 2002) that typically requires many trials over many days, e.g., frequency discrimination learning of 4,000–5,000 trials (Irvine et al. 2000), pitch discrimination of >10,000 trials (Demany and Semal 2002), or melodic patterns of ~1,200 different stimuli (Tervaniemi et al. 2001). Perceptual learning can occur in far fewer trials in special circumstances (Hawkey et al. 2004), but even so associative learning can emerge in only five trials (Edeline et al. 1993). Associative learning can include discrimination learning but, unlike perceptual learning, associations can form without discrimination training. Thus, it is more basic than perceptual learning. Therefore, associative learning and perceptual learning differ, although both can affect primary sensory cortex such as AI (Weinberger 2008a).

We can best appreciate the distinction by asking, "After an episode of perceptual learning, what is changed in the auditory cortex?" The answer would seem to be that the "machinery" of the cortex has been altered to enable greater acuity. This is certainly an interesting aspect of learning. However, in contrast to associative learning, perceptual learning does not seem to include "perceptual memory". This absence probably reflects the fact that investigators of perceptual learning are more concerned with sensory/perceptual processes than with learning processes. But of greater import, subjects apparently do not actually remember the specific "contents" of their experience, that is, the particular stimuli or stimulus values given during certain of their multitude of trials. Thus, while perceptual learning alters the gateway to memory, increased acuity by itself is not necessary for memory, as the term is normally understood, i.e., as the "contents" of experience. However, the extant level of acuity can determine the precision with which the information is analyzed, and may then be encoded and stored.

A case can be made that perceptual learning is a subclass of associative learning because subjects must first associate a sound stimulus and a subsequent event, which may be a different sound (classical conditioning) or a reward after a response to one of two discriminative acoustic stimuli (instrumental conditioning). Next, the discriminations become more difficult. Because basic associative learning and its correlated cortical plasticity develop rapidly, they may clarify mechanisms of subsequent perceptual learning.

4.3 Learning and Plasticity in Primary Auditory Cortex

From 1956–1984, learning-induced plasticity in AI was discovered, validated, and well characterized. Nearly all assessments of learning effects were from recordings during training trials. The distinction between recording during training trials versus before and after training trials would emerge in subsequent work. However, in this epoch it was assumed that recording during trials would adequately reveal the features of plasticity, an approach now known to have severe limitations. Habituation and conditioning were studied extensively.

4.3.1 Habituation

Early work on habituation yielded a singularly uniform finding: repeated presentation of the same sound reduced the magnitude of AI-evoked responses, both for local field potentials (LFPs) and neuronal discharges. Spontaneous recovery occurred after minutes of silence. After decremental responses to a sound were established, presenting a novel sound-evoked normal responses. Continued presentation of the novel sound also caused response decrements (Marsh et al. 1961; Dunlop et al. 1966; Wickelgren 1968; Weinberger 2008b). Some studies recorded electroencephalographic (EEG) activation in drowsy or sleeping animals to repeated acoustic stimulation, with similar results, i.e., habituation of EEG activation (Jasper and Sharpless 1956).

These experiments established the development of auditory cortex response plasticity under the simplest possible conditions: presentation of one isolated sound. AI actively suppresses responses to sounds with minimal behavioral salience. Equally important, the habituative decrement is selective since neural and behavioral responses to novel sounds are normal. Thus, the auditory system continually monitors and evaluates current sound with reference to prior experience.

4.3.2 Conditioning

Investigation of learning and the neurophysiology of the cortex began with the discovery of conditioned, EEG desynchronized activation (decreased slow waves, increased fast waves) to a click paired with a flashing light (Durup and Fessard 1935). Conditioned EEG activation was easily found; further study revealed that its cortical distribution during training shrank from widespread cortical activation to the sensory cortex of the CS. As most studies used acoustic CSs, conditioned EEG effects were repeatedly found in the auditory cortex. Controls established that conditioned activation was associative using auditory discrimination training (Gluck and Rowland 1959; Rowland and Gluck 1960). The CS+ produced EEG activation while the CS– did not (John 1961; Morrell 1961; Thomas 1962; Galeano 1963).

The EEG was extremely difficult to quantify as it contains various frequency bands, conventionally delta (1–4 Hz), theta (4–8 Hz), alpha (8–12 Hz), beta1 (12–20 Hz), beta2 (20–32 Hz), and gamma (32 Hz and above). Although the duration of EEG activation could be measured, the EEG could not be dissected into its component bands. Contemporary analysis of conditioned EEG effects to a tone has revealed band-specific associative effects in AI (McLin et al. 2003). But the growing use of LFPs, whose amplitude could be measured precisely, foreshadowed the demise of EEG studies.

In early work on AI learning and LFP plasticity, cats received an auditory (click) CS paired with an immediately following puff of air to the face (US). From this classical conditioning procedure, CS-elicited AI LFPs increased in size and behavioral conditioned responses emerged (Galambos et al.1956). They controlled for CS constancy at the ear by obtaining the same findings with subjects under neuromuscular blockade, thus eliminating head and pinna movements and middle ear muscle contractions. They did not include a non-associative control group that received the CS and US unpaired. However, other studies with proper controls confirmed the associative nature of increased response magnitude of AI LFPs (Marsh et al. 1961; Majkowski and Sobieszek 1975; Molnár et al. 1988).

LFP research was extended to many conditioning tasks and Pavlovian processes, with similar findings of enhanced responses to sounds that became behaviorally important (Weinberger and Diamond 1987). Similar results were seen for instrumental avoidance learning and systematic increases in CS-evoked LFPs were found in the same dogs that alternated between classical conditioning and instrumental avoidance conditioning with the same CS and US, despite learning different responses for each type of conditioning (Cassady et al. 1973).

Some investigators contested the associative nature of increases in LFP magnitude, contending that fear, rather than the learned signaling capability of the acoustic CS, was responsible (Hall and Mark 1967; Mark and Hall 1967). Others argued that arousal caused increased responses to the conditioned stimulus (Kitzes et al. 1978). However, this issue was resolved in favor of associative processes by later studies of CS-specific receptive field shifts. Both sets of critics had assumed that increased responses to the CS due to fear or arousal per se represented the same process as increased responses to the CS during associative learning. However, receptive field analysis found that the former was caused by sensitization, which increased responses to all frequencies, while the latter reflected learning as shown by increased responses to the frequency-specific CS and decreased responses to other frequencies (Weinberger and Diamond 1988). However, this could not be determined at the time because the critiques predated the use of receptive field analysis (see Section 4.5.2).

Besides LFPs, cellular discharges were recorded in AI from clusters of cells (multiple-unit activity). The results were often similar, i.e., increased discharges to an acoustic CS with conditioned responses. This response increase was consistent in one- and in two-tone discrimination training (Buchwald et al. 1966; Saunders and Chabora 1969). Reversal was found when the CS+ and CS– acoustic stimuli were interchanged after initial learning. Moreover, acoustic CS+ stimuli acquired the ability to elicit responses in the sensory cortex of the shock US in primary somatic sensory cortex (Oleson et al. 1975).

Such cluster recordings have the advantage over single unit recordings of yielding good data over many hours or days and the disadvantage of not showing if different single cells develop increased and decreased response plasticity; when increases dominate, the decreased responses would not be detected.

Studies of auditory cortex single units during learning also found plasticity. However, despite the detection of many cells that developed increased responses to the CS, many others decreased their responses or were unchanged (Gasanov and Galashina 1976; Woody et al. 1976; Dumenko and Sachenko 1978, 1979; Weinberger et al. 1984b). Such heterogeneity of unit discharge plasticity was also found in auditory area AII (Diamond and Weinberger 1984). Divergent results were not caused by inadvertent changes in effective acoustic stimulus level in the periphery, undetected movements, muscle contractions, or muscle spindle feedback as they were seen in animals trained under neuromuscular blockade (Weinberger et al. 1984b).

While single-unit plasticity was shown to be associative, the findings of opposite sign made little functional sense. Although recording in AI during training trials provided foundational information, this research approach appeared to give diminishing returns after almost 30 years.

4.4 Disinterest in Learning-Induced Auditory Cortical Plasticity

Although EEG, multiple-units, and single-unit studies established that AI was not just an acoustic analyzer, the findings had little impact on understanding learning/memory and auditory neuroscience. Contributory factors may include dramatic findings that the hippocampus seemed essential for the formation of memory with the discovery of patient H.M. In animal conditioning, attention to model systems, such as conditioned eyeblink, grew. Second, there was no conceptual framework for the findings of auditory cortex associative plasticity, particularly with sensory cortex consigned to the status of stimulus analyzer. Within auditory neuroscience, the indifference may have reflected the use of an impoverished set of sounds in studies of learning (e.g., at most a CS+ and a CS–) inadequate to describe the cell's response properties with so few sounds. Even when learning-induced plasticity was found, it could not be interpreted in terms of auditory coding and processing.

4.5 Limitations of Auditory Cortical Plasticity Obtained During Training Trials

We next consider new approaches and findings. First, however, the limitations of neurophysiological recordings during training trials are evaluated (Weinberger 2004b).

4.5.1 State Factors

During learning trials, both learning processes and non-learning performance factors occur, e.g., arousal level may be high early in training, during acquisition, and lower after performance improves. Arousal level can alter the magnitude of sound-elicited auditory cortex responses (Murata and Kameda 1963; Teas and Kiang 1964; Wickelgren 1968; Molnár et al. 1988). Cortical plasticity obtained during training is associative, given controls for sensitization, etc. That non-learning factors are operative does not weaken the case for associativity. However, they can modify the expression of associative plasticity so that it may be difficult to obtain pure associative effects. Using behavioral data from training trials alone to infer learning strength of memory processes is problematic (Rescorla 1988) and requires appropriate post-training behavior assessments. This is equally applicable to neurophysiological plasticity emerging during training. In short, one cannot assume that the neurophysiological plasticity observed in response to signal acoustic stimuli during training actually represents exclusively the product of learning.

4.5.2 Specificity of Plasticity

A second problem with relying on neurophysiological data obtained during training is that it cannot give adequate information about the specificity of plasticity. Unlike the problem of state factors, which might be controlled by on-line monitoring of arousal level, the limitation on plasticity is endemic to the learning situation. As noted above, the number of different stimulus values used in training is too small to permit determination of the specificity of plasticity, such as changes in frequency receptive fields. For example, in a two-tone discrimination study, increased responses to the CS+ and decreased responses to the CS− yield neurophysiological discriminative plasticity. However, they are insufficient to determine if frequency tuning has shifted, e.g., toward the frequency of the CS. Presenting a sufficiently large set of stimuli cannot solve the problem because they would either constitute additional CS+s, if followed by a US, or additional CS–s, if followed by no reinforcement. Thus, post-training assessments of plasticity are required to determine learning effects.

5 Contemporary Approaches: A Synthesis of Two Disciplines

In the 1980s a new question was posed: instead of asking if learning-induced plasticity involved associative plasticity in AI, attention shifted to the specificity of such plasticity. Does learning cause a re-tuning of AI?

This question altered the research agenda and engendered new experimental paradigms. While this paradigmatic shift took various forms, the essential factor was that the new required integration of integrates auditory neurophysiology and learning/memory.

5.1 Auditory Neurophysiology and Learning

Theories of the history of science sometime contrast the "great-person theory" with the "zeitgeist theory". The great-person theory in science history holds that major advances are the products of a single individual who steers research in new directions with a seminal idea or technique. The zeitgeist theory argues that advances are inevitable, reflecting current need and the availability of methods (Boring 1929). The zeitgeist theory seems well suited to the issue of the representational specificity of learning-induced plasticity in sensory systems (Weinberger 2008a). The new paradigms that revealed representational specificity were first applied to auditory cortex and inaugurated such inquiry for many sensory systems.

The zeitgeist theory may apply since two laboratories independently sought representational specificity simultaneously, with complementary experimental designs. One group attacked the problem with a metabolic technique, 2-deoxyglucose (2-DG) (Gonzalez-Lima and Scheich 1984, 1986). The other laboratory studied receptive field frequency plasticity (Weinberger et al. 1984a; Diamond and Weinberger 1986).

The metabolic approach relies on knowledge of the locus of representation of particular frequencies in the AI tonotopic map and often uses fear conditioning in which a CS sound is paired with an aversive US. After training, animals are exposed for several minutes to the CS before sacrifice and are processed for changes in auditory cortex metabolic response. Increased metabolic activity in the region representing the CS frequency demonstrated AI associative CS-specific representational plasticity.

The receptive field approach obtains AI neuron tuning curves before and after a learning experience. The training may be the same as in metabolic studies, e.g., fear conditioning to a tone, but the assessment approach differs. In receptive field (RF) studies, the pre-training RF is subtracted from the post-training RF; the difference reflects the conditioning, or a control treatment; tuning shifts toward the CS frequency indicate associative CS-specific representational plasticity.

Both approaches are equally valid and use complementary experimental designs. The metabolic approach requires a between-groups design because the 2-DG technique can be performed only once on a subject. Thus, a paired group is compared to an unpaired group to validate associative plasticity. The RF approach can use a within-subjects design because RFs can be obtained before learning, after, and at various retention periods. This permits within-subject tracking of the evolution (consolidation) of plasticity. In a two-tone discrimination protocol, differential CS+ and CS− effects obviate the need for a non-associative control group.

Finally, both approaches ameliorate or solve the state problem. The metabolic studies reduce arousal, attention, and motivational factors by presenting the CS outside of the training situation, after training has been completed. Thus, the absence of a motivational US, such as shock, should reduce changes in arousal, while attention to the CS should be consistent as training ends. The RF methodology also obtains tuning data outside of the training situation, but avoids behavioral responses to the CS when it is given as part of the stimulus set used to obtain RFs. Experimental extinction during post-training determination of receptive fields must be eliminated. This can occur if the subject regards presentations of the CS frequency as one of many tones in the RF stimulus set, as the original CS, and learns that it no longer predicts the US. The metabolic prolonged post-training CS also runs the risk of experimental extinction.

The solution to these problems for RF (and similar) studies is contextual. This issue is both of considerable importance and often poorly understood. Before reviewing the findings, we need to consider it in some detail.

5.2 Contextual Importance: Reduction of State Factors and Extinction

How can state factors and experimental extinction be reduced or eliminated by obtaining RFs before and after training (Weinberger 1998)? The solution is to markedly change the context of the training period from the context of the pre- and post-training assessments of receptive fields. (The term "receptive field" is a proxy for other measures of auditory neuronal response that may exhibit plasticity after learning, e.g., threshold, level or bandwidth.) If the contexts are sufficiently different, then subjects do not treat the same tonal frequency as the CS when it is presented outside of the training period.

Several changes in context are possible and, together, control state factors, and prevent experimental extinction. The absence of a reinforcer (food or shock) and a marked difference in the acoustic environment reduces and can eliminate state changes. Training with a CS tone in a discrete conditioning trial with a 1–5-s duration, a 1–3-min intertrial interval, and a stimulus level far above threshold (60–80 dB SPL) is relatively standard. In contrast, RF determination involves completely different parameters with, e.g., 24 tones at quarter-octave intervals, 100 ms duration, and 400 ms intertone intervals at 0–80 dB SPL to cover the audible range. Test tones are given repeatedly and randomly to obtain statistically reliable RFs. In short, the acoustic context of RF determination must differ from that in conditioning trials. Training and RF testing can also use different laboratory rooms and illumination conditions.

Minimizing similarities between the training context and the testing circumstances reduces state factors. This reduces generalization from the training environment to the testing environment. Extinction is circumvented if the subjects do not initially respond to the CS frequency during RF determination, when it is presented as one of the many brief tone bursts.

Context differences can eliminate behavioral or arousal response to the CS frequency when it is embedded as a brief tone in a series of test tones. Objective measures (pupillary size) show that the CS frequency is not regarded as a conditioned stimulus during RF determination, eliminating experimental extinction (Diamond and Weinberger 1989).

While the contexts between training and assessment of RFs must differ, those during pre-training and post-training RF recordings must be the same. Subtraction of the pre-training from post-training RF data can reveal the intervening training only if there are no other differences between these periods. This state can be achieved by adapting subjects to the RF determination environment and by recording the EEG, heart rate, or other physiological state indices. It is also feasible to eliminate any possibility of arousal confounds by training subjects while they are awake (of course) but obtaining RFs while they are under general anesthesia. CS-specific plasticity is expressed with subjects under general anesthesia (Lennartz and Weinberger 1992; Weinberger et al. 1993).

Of course, responses to the training stimulus in discrete trial presentation can be recorded if the potential state factor confound is kept in mind. It is beneficial to compare plasticity to a CS tone in training trials with RF post-training plasticity. Such a study found little correspondence between training-induced CS changes to responses to that frequency when it was presented in a series of rapidly presented frequencies in the post-training period. In many cases, the sign of change was opposite, e.g., a decrement in response to the CS tone but a specific increase in response to that frequency during RF determination, when tuning shifted toward or to the frequency of the conditioned stimulus. These plasticity differences suggest that changes in the CS response in training trials reflect both associative and state processes (Diamond and Weinberger 1989).

5.3 Habituation

The specificity of AI habituatory response decrements has received little attention. Metabolic studies of acoustic habituation have used noise stimuli and thus provide no evidence for putative frequency-specific reductions in 2-DG uptake at known tonotopic loci (Gonzalez-Lima et al. 1989,b). Habituation to stimuli can develop when a sound is repeated randomly or pseudorandomly to an US, e.g., shock. In such circumstances, response decrements to noise occur in primary (TE1) and other auditory cortical fields (TE2, TE3) during reduced behavioral responses to the noise. Decrements in 2-DG uptake were found in the brain stem reticular arousal system, consistent with an arousal decrement during acoustic habituation i.e., a state confound (Poremba et al. 1998).

The specificity of habituation with a control to preclude state confounds was investigated early in the history of cortical plasticity. Although RF analysis was not used, theses studies were the first to use the basic design of determining auditory responses before and after learning (Westenberg and Weinberger 1976; Westenberg et al. 1976). Two frequencies (A and B) were given as alternating (interleaved) tone bursts (pre-habituation), followed by repetition of one tone (balanced design), then interleaved again. Average LFPs show a selective decrease in response amplitude for the repeated tone. Because the pre and post tones were interleaved, the average responses were obtained for both when the subjects were in the same state so that differences between tone responses could not reflect differences in state either within or between the pre- and post-test periods. These findings were the first to show that repeated acoustic stimulation produces frequency-specific auditory habituation in the auditory system (Westenberg and Weinberger 1976; Westenberg et al. 1976).

RF analysis has been used more recently. After determining the tuning of unit clusters, and insuring their stability, subjects received single tone pips for several minutes. A comparison of pre- and post-repetition frequency RFs revealed a highly specific decrease in response to the repeated stimulus; frequencies 0.125 octaves from the habituated frequency exhibited no response decrement (Fig. 22.3).

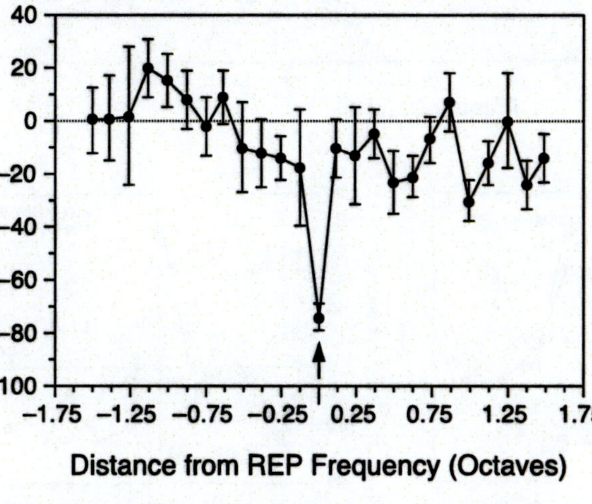

Fig. 22.3 The effects of habituation on frequency receptive fields in the guinea pig primary auditory cortex. Data are normalized to octave distance from the repeated frequency. Habituation produces a frequency-specific decreased response

The extreme degree of specificity is noteworthy and reveals that AI tracks prior sounds with a great deal of precision, even in the absence of any biologically significant events (Condon and Weinberger 1991).

5.4 Conditioning: Initial Studies and Controls for Reactive State Confounds

Metabolic effects of association were studied with 2-DG uptake after training in a terminal treatment with a between-groups design (Gonzalez-Lima and Scheich (1984, 1986). Gerbils received tones paired with strong aversive electrical stimulation of the mesencephalic reticular formation or various controls: CS–US unpaired, CS alone, US alone. The paired group alone developed the behavioral index of learning, conditioned bradycardia. After learning, all groups received continual CS alone presentation during a 2-DG injection in a terminal post-training session. AI 2-DG uptake revealed a CS-specific increase in metabolic activity for appropriate CS frequency. As this necessary design may involve some experimental extinction, the associative findings might be somewhat weakened. The negative outcome in the other groups showed that the CS-specific plasticity was associative.

Subsequent studies assessed Pavlovian phenomena such as compound conditioning (Poremba et al. 1998), blocking (Poremba et al. 1997; Jones and Gonzalez-Lima 2001), differential inhibition (Jones and Gonzalez-Lima 2001), latent inhibition (Puga et al. 2007), conditioned inhibition (McIntosh and Gonzalez-Lima 1995), extinction (Nair et al. 2001a,b; Barrett et al. 2003), and renewal (Bruchey et al. 2007). Innovative analyses using structural equation modeling also have been applied (McIntosh and Gonzalez-Lima 1992, 1993). However, the specificity of auditory cortex associative plasticity apparently has not been pursued.

RF analysis was first applied to classical fear conditioning in the cat. Single-unit discharges were recorded in non-primary auditory fields, secondary (AII) and ventral ectosylvian (VE) cortices (Diamond and Weinberger 1986, 1989). Cats were trained in one brief (20–45 trials) session of tone-shock pairing and behavioral learning was validated by the pupillary dilation conditioned response (CR). CS-specific plasticity was found in the paired group but not for unpaired tone and shock. Extinction (additional CS presentation without the shock US) eliminated RF plasticity. The findings received little notice, probably because these non-primary auditory fields were not well understood compared to AI.

Similar studies were then undertaken in AI of the guinea pig with behavioral validation of associative learning, e.g., conditioned bradycardia. Following determination of frequency RFs (frequency tuning), the frequency to be used as the CS was then selected to not be the best frequency (BF), in order to determine if conditioning caused shifts of tuning toward the CS frequency. Animals then received a single session (30-45 trials) of tone paired with shock. A comparison of post-training with pre-training RFs revealed a dominance of CS-specific increased responses. Moreover, responses to the pre-training BF and other frequencies tended to decrease. These opposing changes were often large enough to produce frank shifts of tuning toward, and even to, the CS frequency, which could become the new best frequency (Fig. 22.4) (Bakin and Weinberger 1990). RF plasticity is associative, as it requires stimulus pairing; sensitization training (no pairing) produces only a general response increase to all frequencies across the RF (Bakin and Weinberger 1990; Bakin et al. 1992).

The independent, largely simultaneous discovery of CS-specific associative AI plasticity in metabolic and RF studies suggested that it might acquire and store specific information and be a site of auditory memory (see Section 5.8). Mnemonic functions are in reality not easy to assign to neural tissue. Neural correlates of learning and memory might arise from many other sources. For example, animals might move closer to an acoustic source that provides sounds, which are becoming behaviorally more important. Also, subjects are likely to pay greater attention to sounds that have become more interesting. Subjects may get more excited or aroused when meaningful sounds are likely to occur.

We have already considered how context differences between the training and the testing environments can prevent behavioral responses to a CS frequency presented during RF determination. But one can argue that contextual control must be verified rather than assumed. Only in the first RF study (Diamond and Weinberger 1986, 1989) was direct behavioral assessments used. However, there is another defense against non-mnemonic confounds.

First, acoustic control can be maintained by keeping constant the relationship between the speaker and the external auditory meatus, e.g., by head-fixation, appropriate sound field construction, or earphones. More subtle controls for arousal and attention are needed as they are endemic to obtaining frequency RFs because many different frequencies

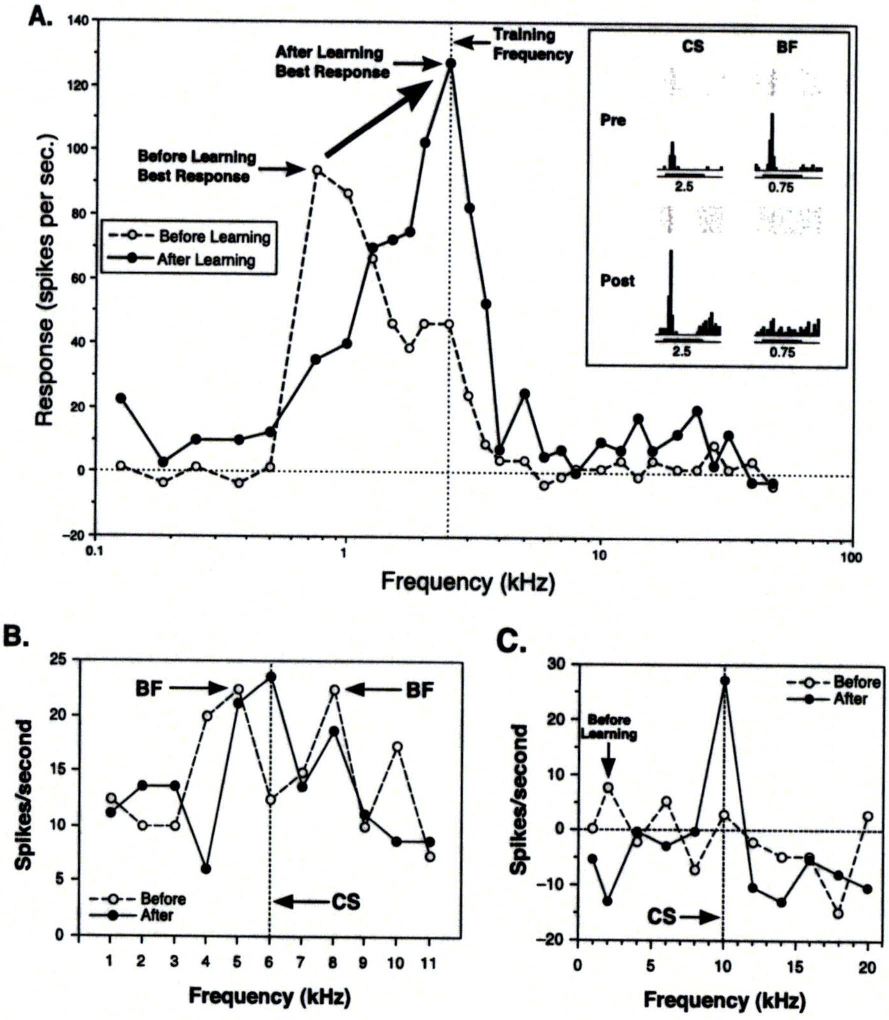

Fig. 22.4 Classical conditioning produces CS-specific facilitation and tuning shifts. **a** An example of a complete shift of frequency tuning of a single cell in guinea pig AI, from a pre-training best frequency (BF) of 0.75 kHz to the CS frequency of 2.5 kHz after 30 conditioning trials. Inset, pre- and post-training post-stimulus time histograms (PSTHs) for the pre-training BF and the CS frequencies. **b** Double-peaked tuning, with pre-training BFs at 5.0 and 8.0 kHz. The CS was selected to be 6.0 kHz, a low point. After conditioning (30 trials), responses to the CS frequency increased to become the peak of tuning. **c** A cell, which exhibited minimal or no response to tones before tuning, developed tuning specifically to the CS frequency after conditioning (30 trials)

are given rapidly (e.g., 2/s) and repeatedly (often in a pseudo-random order) to generate enough responses across the frequency spectrum to enable frequency tuning curve construction. Arousal levels cannot change rapidly enough to track the presentation of different tones. Attention might be paid to the CS frequency during RF determination, but attention would be invoked only after the tone had been processed and identified, too late for attention to affect discharges 10–50 ms after CS onset. If learning-induced tuning plasticity occurred only after hundreds of milliseconds, then reactive changes in arousal, selective attention, or both could be a problem.

5.5 Does the Primary Auditory Cortex Hold Memory Traces?

5.5.1 Introducing Memory Traces

A specific memory trace (SMT) is an enduring neural record of a particular aspect of experience. How can one determine if RF plasticity indexes SMTs? Perhaps destruction of AI should abolish its memory traces, which would then be shown by behavioral tests in a specific loss of auditory memories. This apparently simple and decisive test will prove to be neither. The complexities of lesions merit their own consideration and are deferred.

If (provisionally) not lesions, then what might be done? One approach is to attempt to defeat the proposal that memory traces form in AI. As AI does form associative plasticity, it could be argued that besides such plasticity, SMTs should have the major characteristics of behavioral associative memory. This would impose a second level of criteria not previously demanded of any neurophysiological studies of learning and memory; nonetheless, this is not an unreasonable demand.

Besides being associative, SMTs should also exhibit specificity, fairly rapid formation, long-term retention, and continued strengthening after training without further reinforcement i.e., consolidation. Another feature is that memory can be formed in various learning tasks rather than being confined to, e.g., classical conditioning. Moreover, memory should transcend a particular motivation and develop in both appetitive and aversive tasks, and SMTs should be manifest for any CS or signal stimuli used in training, as for genuine associative memory. That is, SMTs should not be limited to plasticity of frequency representation but should develop for any acoustic parameter that can serve as a signal for reward or punishment. Finally, as for memory, SMTs should be biologically conserved and develop across diverse taxa.

The findings from several laboratories support the conclusion that SMTs develop in AI. Moreover, as this is an active area of inquiry, new acoustic parameters are continually being studied. Although this chapter can never be up-to-date, at least one prediction can be made: If an acoustic parameter can serve as a signal or gain behavioral relevance through learning, then its processing in AI (and perhaps other auditory cortical fields) can develop representational plasticity.

5.5.2 Specificity of Frequency Plasticity

Let us first consider frequency tuning and representation because it has been studied most extensively. We have already noted that RF shifts are directed toward and to the CS frequency and that these are associative. Additionally, RF plasticity is highly specific; the maximum increase in response is at the CS frequency while neighboring frequencies show no change or decreased response. Specificity is also evident in two-tone discrimination learning, in which a reinforced CS+ develops increased responses whereas a non-reinforced CS– has diminished responses (Fig. 22.5). Second, RF plasticity develops very rapidly, in as few as five training trials, as rapidly as the first behavioral (e.g., cardiac) signs of association (Edeline et al. 1993) (Fig. 22.6). Third, RF plasticity shows long-term retention, enduring for the longest periods tested, up to eight weeks after a single 30-trial conditioning session (Weinberger et al. 1993) (Fig. 22.7). Fourth, RF plasticity consolidates, i.e., continues to develop increased responses to the frequency of the CS vs. decreased responses to other frequencies in the absence of further training over hours (Edeline and Weinberger 1993; Weinberger et al. 1993; Galván and Weinberger 2002) and days (Weinberger et al. 1993; Galván and Weinberger 2002).

RF plasticity also has other key features of memory. It develops in all tasks tested to date: in simple instrumental avoidance conditioning (Bakin et al. 1996), simple classical conditioning (above), in two-tone instrumental avoidance conditioning (Bakin et al. 1996), two-tone classical discrimination training (Edeline and Weinberger 1993; Edeline et al. 1990), and in one-tone appetitive classical conditioning in which the US was rewarding ventral tegmental electrical stimulation (Kisley and Gerstein 2001).

CS-specific associative tuning shifts develop in the AI of all species studied: the guinea pig (*Cavia porcellus*) (Bakin and Weinberger 1990), the echolocating big brown bat (*Eptesicus fuscus*) (Gao and Suga 1998, 2000; Ji and Suga 2003; Ji et al. 2001; Suga and Ma 2003), cat (*Felis catus*) (Diamond and Weinberger 1986), and the rat (*Rattus rattus*) (Kisley and Gerstein 2001). CS-specific expanded representations in the AI tonotopic, map which are predicted from CS-directed RF shifts, have been found in the owl monkey (*Aotus trivirgatus boliviensis*) (Recanzone et al. 1993) and rat (Rutkowski and Weinberger 2005).

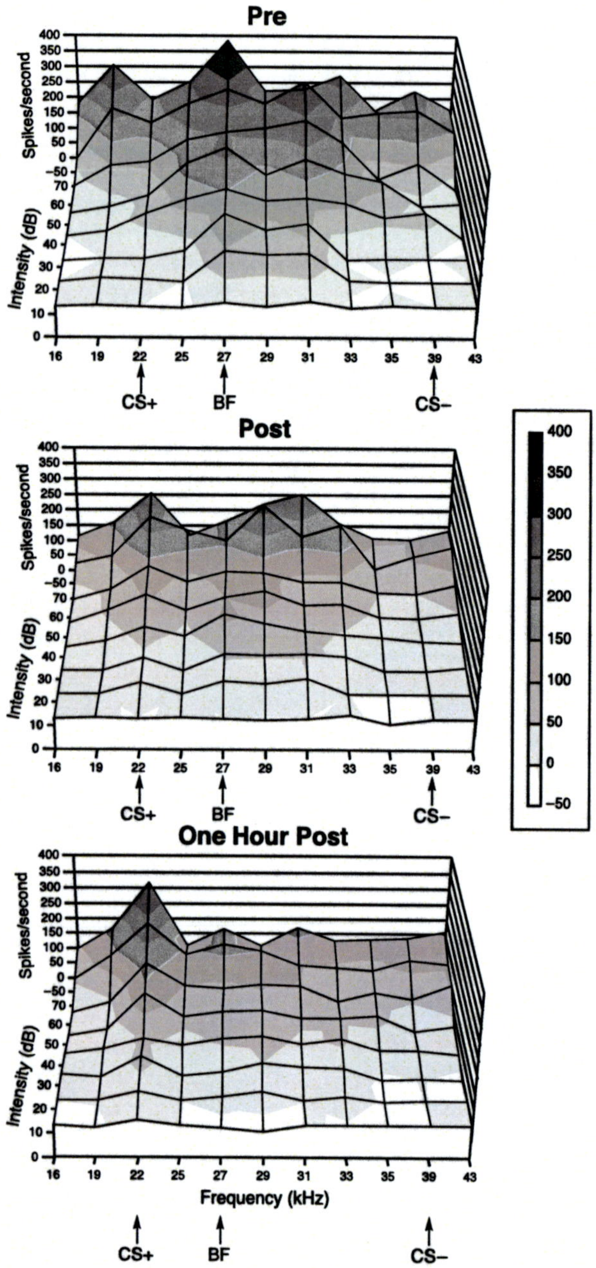

Fig. 22.5 Two-tone discrimination. Representation of neuronal responses in AI before, immediately after, and 1 h after two-tone discrimination training (30 each, CS+ [22.0 kHz] and CS- [39.0 kHz], intermixed trials). Displayed are rates of discharge (*Y*-axis) as a function of tonal frequency (*X*-axis) and level of testing stimuli (10–70 dB). Note that conditioning changed the topography of neuronal response. The pre-training best frequency of 27.0 kHz was reduced as was the CS– frequency, while responses to the CS+ frequency increased. Note consolidation, in the form of a continued development of these changes; after 1 h of silence, the only excitatory response is at the CS+ frequency (Edeline and Weinberger 1993)

Learning-induced tuning plasticity using the same paradigm of classical conditioning (tone paired with a mildly noxious stimulus) produces concordant CS-specific associative changes in the human primary auditory cortex (*Homo sapiens*) (Molchán et al. 1994; Schreurs et al. 1997; Morris et al. 1998).

5.5.3 Arguments to the Contrary

It has been questioned whether tuning shifts embody learning or spontaneous changes observed as tuning changes over days without training (Kisley and Gerstein 1999). This concern is curious as it had already been shown that tuning shifts during learning move toward, not away from, the CS frequency, and develop only in animals receiving paired CS and US, and are discriminative, i.e., toward the CS+ only (Bakin and Weinberger 1990; Bakin et al. 1992; Edeline and Weinberger 1993). Spontaneous shifts of tuning could not account for any of these attributes of learning-induced receptive field plasticity.

The "spontaneous tuning shifts" reported are readily explained as an artifact of their data analysis. In a subsequent experiment a tone was paired with rewarding intracranial self-stimulation (Kisley and Gerstein 2001). Conditioning produced shifts directed toward or to the CS and in agreement with prior studies, this associative plasticity required CS–US pairing. The learning effects exceeded any spontaneous changes in tuning. The evidence for spontaneous drifts of tuning was that entire tuning curves became less correlated over days without conditioning, but they did not determine if the actual tuning of their cells drifted. Moreover, tuning curves consist of strong responses to the BF and some nearby frequencies, but also include weak or even inconsistent responses at more distant frequencies. Therefore, reduced correlations over days could easily have been caused by spontaneous changes in the responses to weak or even inconsistent responses to frequencies distant from the BF, i.e., at the lower and upper limits of the tuning curves. A better way to address the issue of tuning drift is to directly measure tuning, and related parameters, over days. This has been accomplished for a period of 14-21 days. Best frequency did not drift over days. Neither did thresholds or bandwidths (Galván et al. 2001; Galván and Weinberger 2002).

In summary, learning-induced CS-specific shifts of AI frequency tuning are not an artifact of spontaneous changes in tuning nor of state . We have reviewed above both empirical findings and design features of the experiments that rule out all but associative effects. The next issue is whether the effects of learning on the primary auditory cortex are confined to the domain of acoustic frequency or are general to whatever acoustic parameter serves as a signal for positive or negative reinforcement, i.e., food or a nociceptive stimulus.

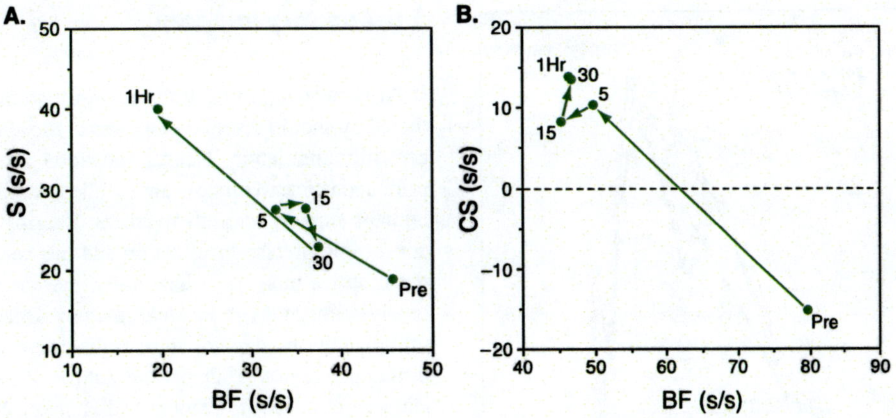

Fig. 22.6 Rapid induction of RF plasticity, shown as vector diagrams of changes in response to the pre-training best frequency (BF) and the CS frequency for two cases. **a** After five trials, responses to the BF decreased while those to the CS increased, changes maintained after 15 and 30 trials, but further change developed after 1 h (consolidation), at which time the CS frequency became the new BF. **b** Sign change in which the CS frequency was inhibitory pre-training but became excitatory after only five training trials; the initial response to the CS was too weak for it to become the new BF or show consolidation in 1 h (Edeline et al. 1993)

5.5.4 Specificity of Plasticity for Other Acoustic Parameters

The studies available show that learning alters the processing of acoustic parameters other than frequency. For example, the preferred repetition rate of noise pulses can be modified by associative learning. Rats were trained in a sound maze in which food reward was contingent upon successful navigation using only auditory cues. The repetition rate of noise pulses grew as the distance between the rat and the target location decreased. After subjects had learned this maze, the neural responses in AI studied in a terminal session had been enhanced for high-rate noise pulses and showed increased phase-locking to the stimuli. This plasticity was due to learning because controls receiving identical sound stimulation with free access to food did not show such temporal processing plasticity and did not differ from naïve subjects (Bao et al. 2004). Similarly, owl monkeys trained to detect an increase in the envelope frequency of a sinusoidally modulated 1 kHz tone show increased sensitivity to small changes in envelope frequency and robust phase-locking to modulation frequencies that usually are only weakly responsive (Beitel et al. 2003).

The processing of sound intensity (level) is also modifiable by learning (Polley et al. 2004). Rats were trained to move to a site in a small arena where sound levels to ongoing sound bursts became maximal. They were guided by stimulus level increases as they approached the otherwise unidentified locus, and levels were reduced as they moved away. Yoked controls received the same acoustic experience, which was not linked to their behavior. In trained animals only did AI responses became selective to more restricted ranges of sound intensities and, as a population, represented a broader range of preferred sound levels. The findings indicate that associative processes could selectively alter the representation of stimulus magnitude (see also Polley et al. 2006).

5.5.5 Working and Reference Memory

Neural correlates of working memory (short term; WM) and reference memory (long-term memory storage; RM) in AI have also been found (Sakurai 1990, 1992). Rats were trained on a continuous non-matching-to-sample task (low and high tones). Unit discharges were recorded during repeated engagement of both types of memory. WM was studied while subjects recalled whether the current tone was the same as the prior tone. RM was studied as they recalled that a low tone required one type of behavioral response, a high tone another type of response. About 20% of AI and medial geniculate body (MGB) single units showed sustained differential activity during the delay period after exposure to the sample tone, suggesting that the thalamocortical auditory system retains auditory information in working memory. Unit recordings from the hippocampal areas CA1, CA3, and dentate gyrus found neural correlates to either WM or RM, but not both. In contrast, AI cells could increase activity for both tasks, indicating enhanced memory task processes (Sakurai 1994). Cross-correlations between pairs of neurons to detect cell assemblies revealed that most correlated pairs in the hippocampal formation occurred during

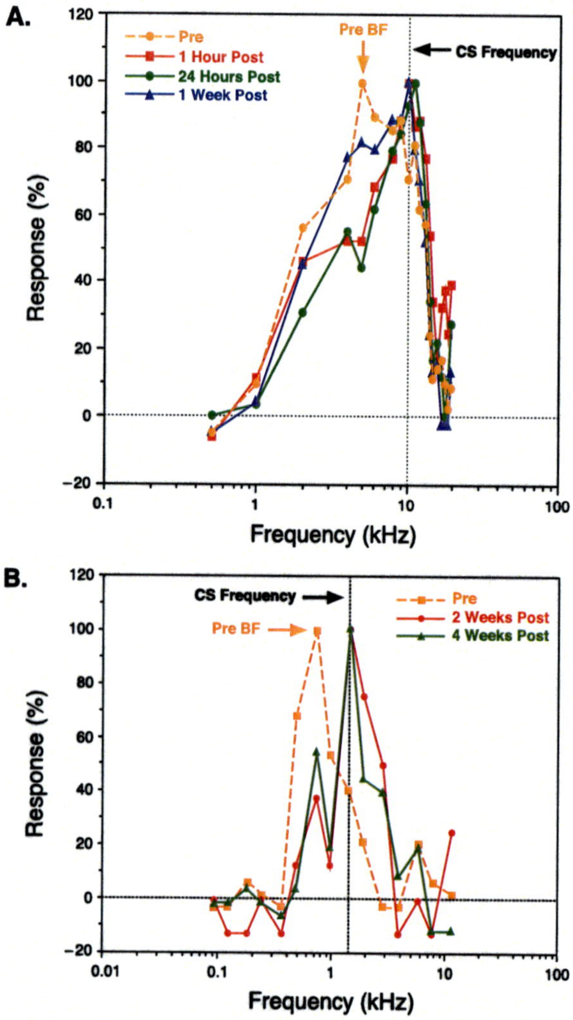

Fig. 22.7 Long-term retention of associative, specific receptive field plasticity in guinea pig AI (multiple-unit recordings). In both examples, conditioning induced a tuning shift to the frequency of the conditioned stimulus (CS). **a** An example of a CS-specific tuning shift over 1 week. The best frequency (BF) shifted from 4.9 kHz to the CS frequency of 10.0 kHz, detected 1 h after completion of training (tone–shock pairing, 30 trials). This shift was maintained at 24 h and 1 week post-training. **b** An example of a CS-specific tuning shift over 4 weeks. The peak of tuning shifted from the pre-training BF of 0.69 kHz to the CS frequency of 1.45 kHz. Data depict tuning at 2 and 4 weeks post-training (tone–shock pairing, 30 trials)

WM, whereas correlated cells in AI could participate equally in WM and RM (Sakurai 1998). Thus, despite the accepted view that the hippocampus has mnemonic functions whereas the primary auditory cortex does not, in fact neurons in AI can exhibit more comprehensive involvement in auditory memory than do cells in the hippocampus.

5.6 Auditory Imagery

If AI networks participate in memory storage and retrieval, then they should reveal themselves in the absence of relevant acoustic stimulation. Neural activation should occur when prior acoustic experiences are recalled. Studies of imagery in humans support such involvement. Bearing in mind caveats concerning precise localization and the need to validate the presumptive imagery behaviorally, there is evidence for the involvement of AI in musical imagery. Imagery for musical timbre activates AI with some right-side asymmetry as does timbre perception (Halpern and Zatorre 1999; Halpern 2001; Halpern et al. 2004; Zatorre and Halpern 2005). Correlates of non-musical imagery in non-musicians have also been reported (Kraemer et al. 2005; Meyer et al. 2006), so this aspect of auditory cortical plasticity may reflect a process that is not limited to a specialized population.

5.7 Interim Summary: Specific Memory Traces in AI

While the fact of associative CS-specific plasticity in the primary auditory cortex is now firmly established, some workers would consider this finding alone to be insufficient to conclude that AI holds specific memory traces. Rather, they may require that plasticity should also satisfy several other criteria. It should (a) exhibit the major attributes of memory and show generality across (b) tasks, (c) motivational valence, (d) acoustic stimulus parameters and (e) species.

The associative plasticity of frequency RFs satisfies all of these criteria. It has the main attributes of associative memory: besides associativity, it is specific, discriminative, rapidly acquired, consolidates over hours and days and shows retention (weeks). Moreover, this plasticity develops in all tasks studied including habituation, both simple and discriminative, classical and instrumental conditioning. RF tuning shifts generalize across positive and negative motivational circumstances and specific plasticity develops for the several acoustic stimulus parameters. Finally, it shows species generality including humans. The conclusion that specific memory traces form and are retained in AI is well-justified. Also, AI also shows correlates of working memory, reference memory, and auditory imagery.

Having survived this gauntlet of criteria, one may ask what other structures in the brain have passed the same level of scrutiny. Remarkably, it seems that none except AI have been evaluated to this extent. The irony seems palpable. Neuroscience, having traditionally excluded primary sensory cortices from both conceptual and empirical legitimacy as loci of memory storage, now finds that the primary auditory cortex is apparently that part of the cerebral cortex for which the storage of specific information is most extensively documented.

5.8 Primary Auditory Cortex Lesions: Rationale, Assumptions, and Limitations

Lesions or ablations of the auditory cortex traditionally have been used to infer the auditory perceptual processes that require its integrity (Neff et al. 1975). Animals were trained to determine if a pre-training lesion impaired learning. On this basis, perceptual functions were often assigned to the auditory cortex. However, that auditory cortical neurons are involved in both perceptual/analytic and mnemonic processes complicates the interpretation of cortical ablations. Deficits in auditory learning might ensue from the disruption of learning processes rather than perceptual deficits and auditory cortical lesions seem to not shed much light on the role of the auditory cortex in learning and memory.

Others express concern about the absence of impairments following cortical lesions. They seem to conclude that unless destruction of AI prevents learning, it cannot hold specific memory traces (Ohl and Scheich 2004; Weinberger 2004a, 2007b). This view reflects an idea of memory as a localized process, at odds to contemporary conceptions of distributed representation of stored experience. Highly localized memory storage typifies only stimulus–response learning of discrete skeletal motor responses in which the conditioned response follows only by a limited and largely stereotyped pattern of muscle actions. While studies of the conditioned eyeblink response have been extraordinarily successful in locating underlying memory traces in the cerebellum (Christian and Thompson 2005), learning more often involves stimulus–stimulus associations, which may be expressed either at the time of learning, at a propitious future occasion, or both. Moreover, the behavioral means of expression of the vast bulk of acquired information are not constrained. Rather, the same knowledge may be communicated in innumerable ways. It matters not whether this sentence was typed by one or many fingers, a nose, or by dictation to an extremely talented monkey.

Further, distributed representation does not imply that parts of the same memory are stored in different locations, so that AI is restricted to storing only a memory fragment. Rather, the auditory cortex likely forms SMTs of the same memory as the subcortical auditory system, even when the latter can completely accommodate an auditory computation, whether memory of a tone or that a tone is followed by shock. However, the auditory cortex, having formed parallel SMTs for even simple situations, can use this information to solve future problems that require cortical participation because it has access to a much wider range of information than the subcortical auditory system. For example, while simple auditory conditioning does not require an intact AI (DiCara et al. 1970; Berntson et al. 1983; Romanski and LeDoux 1992), AI is required for two-tone discrimination (Teich et al. 1988) and for experimental extinction (Teich et al. 1989).

The failure of AI lesions to destroy a behavioral indication of learning cannot refute the evidence that AI forms and holds memory traces. The standard lesion logic is legitimate only for cases in which the entire substrate of a memory is localized to the destroyed tissue. Such localization has not been demonstrated for any auditory memory.

6 Is There an Auditory Memory Code for the Acquired Importance of Sound?

Given that AI acquires and stores information that can underlie specific memory, we can now address the second question posed at the outset. Does learning-induced plasticity in AI also reflect the operation of a memory code for a cardinal feature of auditory memory?

If the acquired behavioral salience of a tone is represented by the number of neurons that become tuned toward or to its frequency, then the greater the importance, the larger should be the cortical representation of the behaviorally relevant frequency. Rats were trained to bar-press for water contingent upon the presence of a 6 kHz tone, each at different levels of water deprivation. Their behavior reflected the amount of water deprivation with thirstier subjects showing better performance. At asymptote, subjects were mapped in a terminal experiment and the AI representation of the octave bands determined. Supporting the hypothesis, the higher the motivation and the behavioral import of the 6 kHz tone, the greater the representation of the 4.0–8.0 kHz octave band (Fig. 22.8) (Rutkowski and Weinberger 2005). Moreover, the greater the increase in representational area, the stronger the memory (Bieszczad and Weinberger 2010a). These findings support the "memory code hypothesis" that the auditory cortex reflects a representational transformation whereby the behavioral significance of a frequency is encoded by the amount of tissue devoted to it. Searching for memory codes is advantageous in sensory systems for which a topographic or other easily studied functional organization exists.

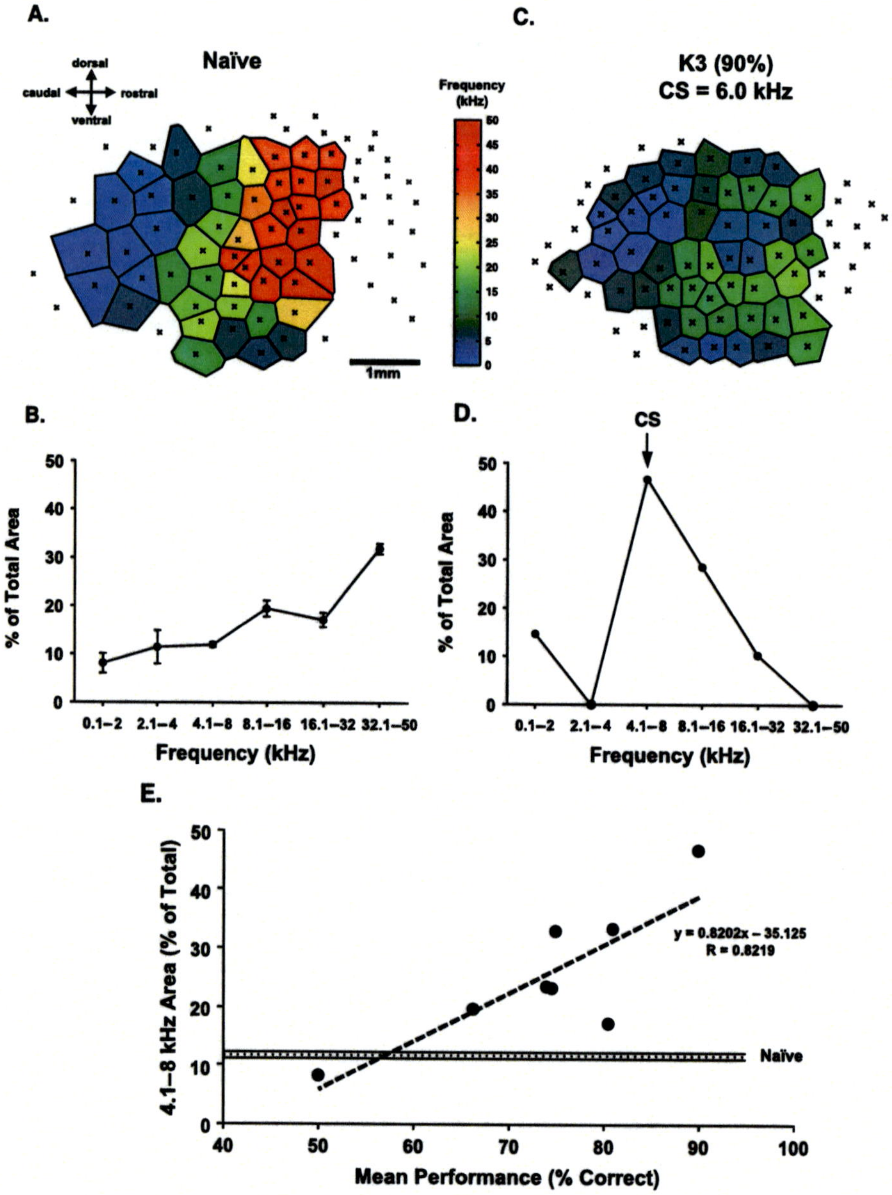

Fig. 22.8 The degree of acquired importance of a tone is correlated directly with the amount of area of frequency representation in the tonotopic map of AI. Trained rats received water reward for bar-presses in the presence of a 6.0 kHz tone. **a–d** Tonotopic maps and quantifications of percent of total area (octave frequency bands) for a naïve rat (*left*) and a rat that attained over 90% correct performance (*right*). Training greatly increased the area of representation for the frequency band containing the 6.0 kHz tone signal. **e** Evidence of a memory code for the acquired behavioral importance of sound. The level of tone importance was controlled by the motivation for water (amount of water deprivation). Asymptotic performance was significantly correlated with motivation level. The area of representation of the frequency band containing the 6.0 kHz tone signal increases as a direct function of the level of behavioral importance of the tone, as operationally indexed by the level of correct performance (Weinberger 2007a)

7 Reconceptualizing the Primary Auditory Cortex

7.1 *Is AI only an Acoustic Analyzer with Adaptive Properties?*

AI is involved in the acquisition and storage of specific information that satisfies the criteria for a substrate of memory, and the magnitude of an expanded CS representation is proportional to the behavioral importence of sound. These characteristics transcend traditional conceptions of AI as a sensory analyzer. They would seem to call for a reconceptualization of the auditory cortex.

An alternative position might maintain the traditional idea of AI by accepting specific learning-induced plasticity but considering it an adjunct to the standard acoustic analytic functions of AI, construing AI as an adaptive analyzer. It still performs its traditional role of acoustic analysis, but biases its responses toward behaviorally important sounds, obviating any need for its reconceptualization.

Let us consider this modified formulation. For the sake of argument, let us grant the premise, and consider whether adaptive analysis constitutes an adequate account of the function of AI. There are two grounds upon which to question this position, theoretical and empirical. We will start with the former.

7.2 Conceptual Problem: Conflation of Analytic and Interpretative Processes

Adaptive analysis must include specific long-term memory traces in or operating on AI; without structural, functional, or both traces, it would be impossible to achieve an experienc-based adaptive bias. But the acceptance of long-term memory traces transforms the concept of acoustic analysis beyond all recognition, as it saves "analysis" by keeping its label but discarding its core role of unambiguous sound identification.

The problem is that the discharges of neurons in AI are affected not merely by the physical parameters of acoustic stimuli, but also by their psychological parameters, i.e., their acquired meaning. Therefore, a cell's response would seem inherently ambiguous. The same discharge might be caused by a loud unimportant sound or a quiet important sound, since "importance" stored in the cortex would produce a larger discharge to a quiet important sound than to a quiet, unimportant sound. Perhaps discharge rate is not used by the auditory system and the detailed temporal pattern of discharge is key. This may be the case, so the issue is by no means settled. Still, abandoning rate coding would render our current understanding of level coding, whether described by monotonic or non-monotonic functions, erroneous.

Another approach to maintaining the traditional role of AI as an acoustic analyzer is to argue that many AI cells are impervious to the behavioral relevance of sound. These serve a purely analytic function and, it can be argued, are grossly under-sampled in studies of learning and memory, as shown by the fact that not all experimental recordings show associative processes in any single study.

This position cannot be refuted given the data available and the lack of morphological identification of cells that develop or fail to show specific associative plasticity. If this position is strongly supported in the future, then a reconceptualization of AI still would be required and might take the form of viewing AI as having two populations of cells with radically different properties. In any event, one still could not claim that AI is purely an acoustic analyzer.

7.3 Empirical Problem: Beyond Learning and Memory in AI

Neurophysiological correlates of other cognitive processes in AI have been documented (Weinberger 2009). Some samples are now considered.

7.3.1 Selective Attention

On-line selective attention to a target frequency rapidly retunes ferret AI cells, as detected in spectrotemporal receptive fields (STRFs). Attention modulates AI by facilitating responses to the target frequency (Fritz et al. 2003) (Fig. 22.9). This rapid, specific retuning, which favors the target stimulus, appears whenever the reinforcement contingencies are switched between two frequencies and also for targets in gap-detection tasks (Fritz et al. 2005,b).

7.3.2 Expectancy

Expectancy of a tone–location combination as a signal for a Go/NoGo appetitive instrumental task elicits specific functional interactions (cross-correlations of spike trains) in the waiting period before signal onset. These patterns had a precise spike discharge repetition with long intervals in the absence of a change in mean rate. This led to the suggestion that AI network activity reflects "... participation of recurrent neuronal networks in processes anticipating the expected sensory input" (Villa et al. 1998).

7.3.3 Concept Formation

Perceptual concept formation involves grouping sensory stimuli by abstract relationships based on physical attributes. Rats were trained to form the categories of rising and falling frequency modulation of tones, independent of their absolute frequencies. EEG recordings from AI show changed dynamics of cortical stimulus representation when the subjects exhibited the acquired abstract concept of the direction of frequency change (Ohl et al. 2001) (Fig. 22.10), showing that AI is involved in the processes of abstract category formation (see also Wetzel et al. 1998).

7.3.4 Cross-Modality Effects

In monkeys trained in a complex auditory discrimination, the cue light that signaled trial availability acquired the ability to elicit responses in AI (Brosch et al. 2004). In humans, the sight of speech without the sound elicits neural activity in AI (Pekkola et al. 2005). An anatomical study in the gerbil found a surprisingly large number of inputs to AI from non-auditory cortical and thalamic regions that might account for some cross-modality effects (Budinger et al. 2006).

7.3.5 Learning Strategy

Specific plasticity in AI is not invariable during learning but is a function of the learning strategy employed rather than the level of asymptotic learning. Rats trained to bar-press for

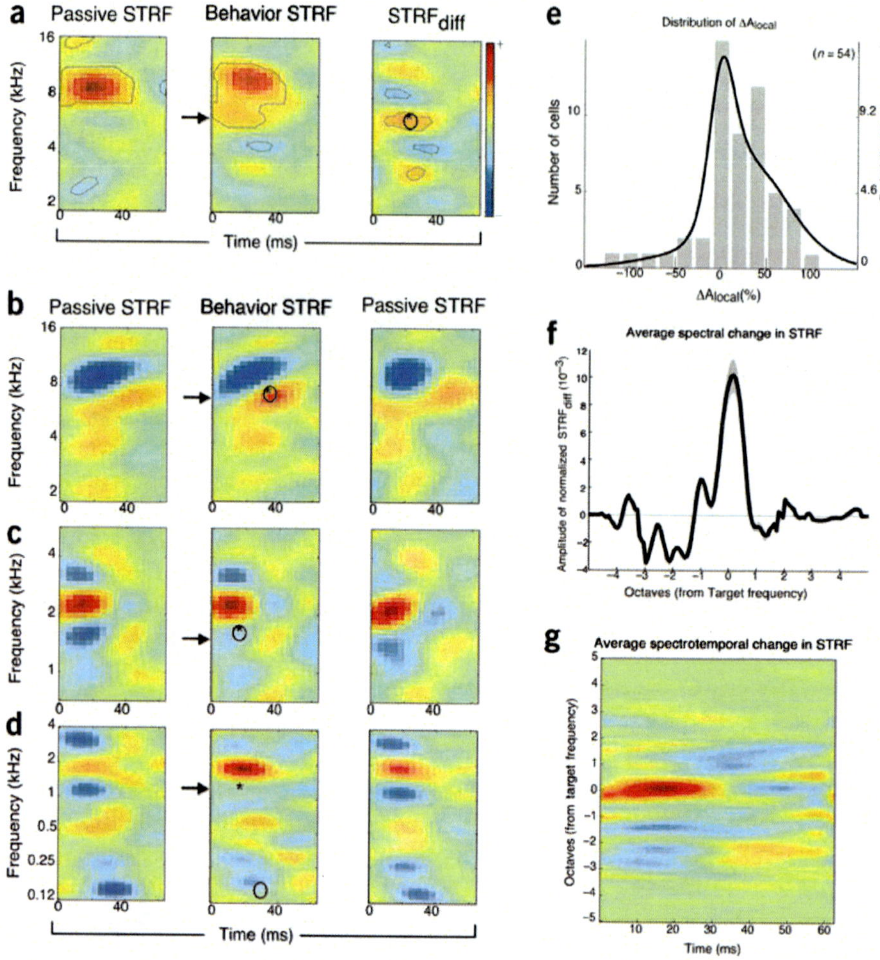

Fig. 22.9 Selective attention for a target tone involves specific modifications of spectrotemporal receptive fields (STRFs). Data from four single units in AI show typical changes observed during performance of the detection task. **a** Comparison of a pre-behavior passive STRF (*left*) and a behavioral STRF (*middle*). *Color scale* represents increased (*red*) to suppressed (*blue*) firing about the mean firing rate (*green*). *Black arrow*, frequency of the target tone during the detection task. *Right*, the difference between the normalized passive and behavior STRF. *Asterisk*, the location of maximal local change. *Circle*, the global change. The local and global maximal changes were both at the target frequency in this case, as in about half of all cells. **b** Localized enhancement of an excitatory region in the STRF during behavior (*left* and *middle*). The post-behavior passive STRF (*right*) reverted immediately to its original shape. **c** Local decrease or elimination of inhibitory sidebands in the behavior STRF. **d** A global weakening of inhibitory fields during behavior. Immediately following behavior, the STRF recovered its pre-behavior shape. **e** Summary histogram and smoothed distribution of local STRF changes from all STRFs. The histogram (*left ordinate*) and distribution (*right ordinate*) are significantly skewed toward positive changes. **f** Average spectral change in the STRF at all frequencies relative to the target frequency. There was facilitation for about one octave around the target and asymmetric suppressive sidebands outside of this range. **g** Average spectrotemporal changes in the STRF derived from all units. The facilitative and suppressive changes near the target frequency, as well as the relatively rapid onset of these STRF changes, can be seen here (Fritz et al. 2003)

water in the presence of a CS tone develop specific plasticity only if they attend to tone onset rather than tone duration (Berlau and Weinberger 2008) (Fig. 22.11). Learning strategy can even be more critical in the formation of specific associative plasticity than a very high level of motivation (Bieszczad and Weinberger 2010b) and the greater the use of a strategy, the greater the signal-specific gain in representational area (Bieszczad and Weinberger 2010c).

7.3.6 Pre-motor Processes

Rats learned in a Go/NoGo task to distinguish combinations of two frequency-modulated sounds and two speaker locations. A pattern detection algorithm revealed reliable spatiotemporal activity patterns predictive of the forthcoming Go or NoGo response (Villa et al. 1999).

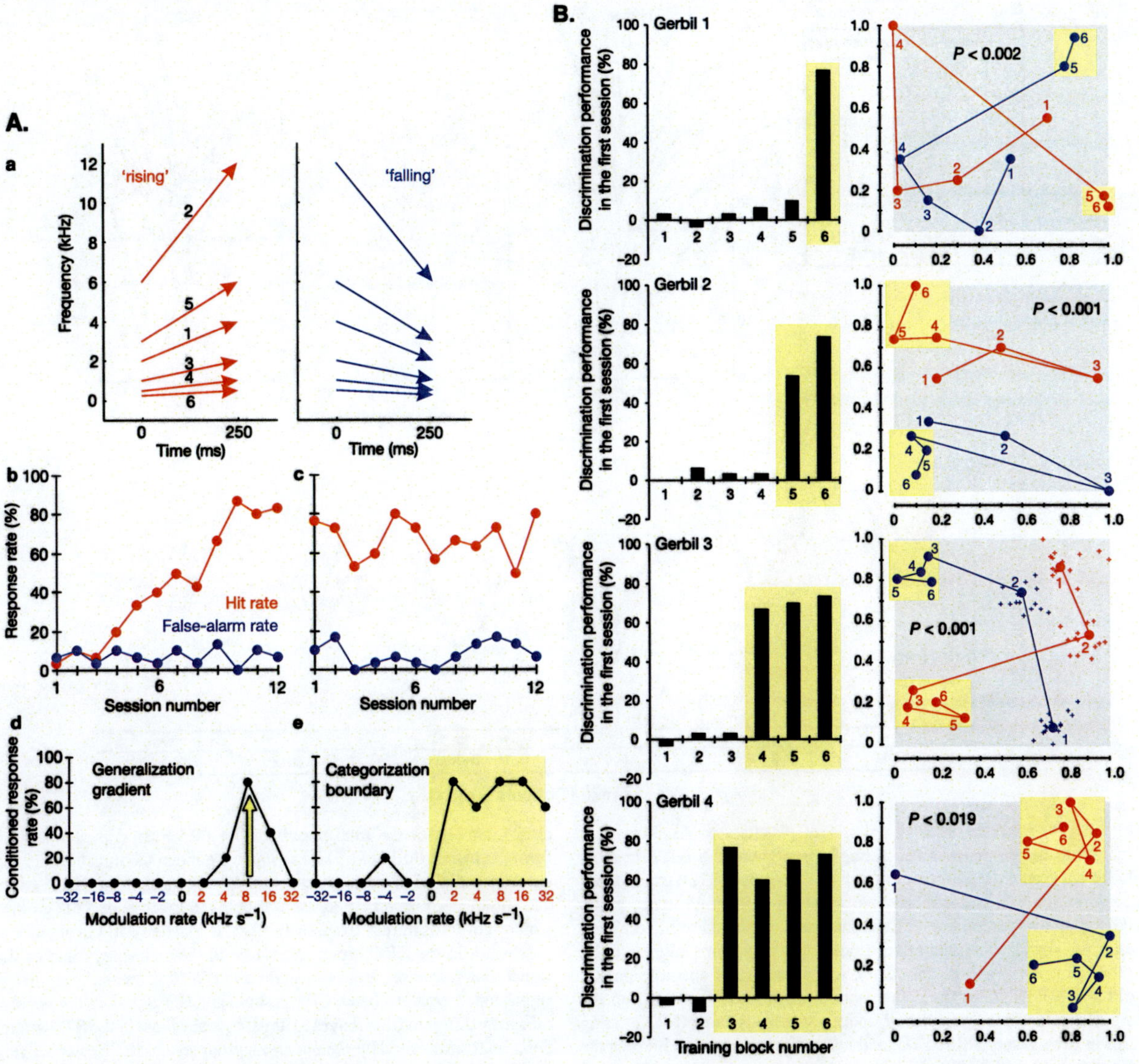

Fig. 22.10 A Stimuli and behavioral measures of category learning. **a** Rising (*red*) and falling (*blue*) frequency-modulated tones used in the six sequential training blocks (*numbers*). **b** Sample learning curve of gerbil 3 before transition to categorization. **c** Sample learning curve of same animal after transition to categorization. **d** Psychometric function for modulation rate obtained after training block shown in **b**. Peak modulation rate of 8 kHz s^{-1} (*arrow*) corresponds to modulation from 2 kHz to 4 kHz in 250 ms used in this block. **e** Sigmoid psychometric function obtained after training block shown in **c**. **B** Behavioral transition to categorization (*left column*) parallels development of cortical spatial activity patterns (*right column*). *Left column*: discrimination performance in the first session of each of the six training blocks. *Yellow areas* indicate categorization phase (Wilcoxon's test, $P < 0.05$). *Right column*: similarity relations between spatial activity patterns during the marked states. Transition to category learning in the behavioral data correlates with clustering (*P*-values of resampling test given) of the marked states within category (*yellow areas*). Only the activity pattern during the marked state that gave rise to the maximum peak value of the dissimilarity function for each category is plotted for each training block (*numbers*). For gerbil 3, marked states of later sessions in blocks 1 and 2 have been included (+) to demonstrate that these point clouds do not fall into the clusters found after the transition to categorization. Absolute coordinates of points have no particular meaning other than scaling relative distances between any pair of points (Ohl et al. 2001)

7.4 Thematic Summary

The conception of AI as dedicated to the analysis of the physical parameters of sound is difficult, if not impossible, to reconcile with the results of many types of behavioral studies. In addition to learning-induced associative specific plasticity that closely resembles specific memory traces, working memory, reference memory, selective attention, concept formation, cross-modality processes, learning

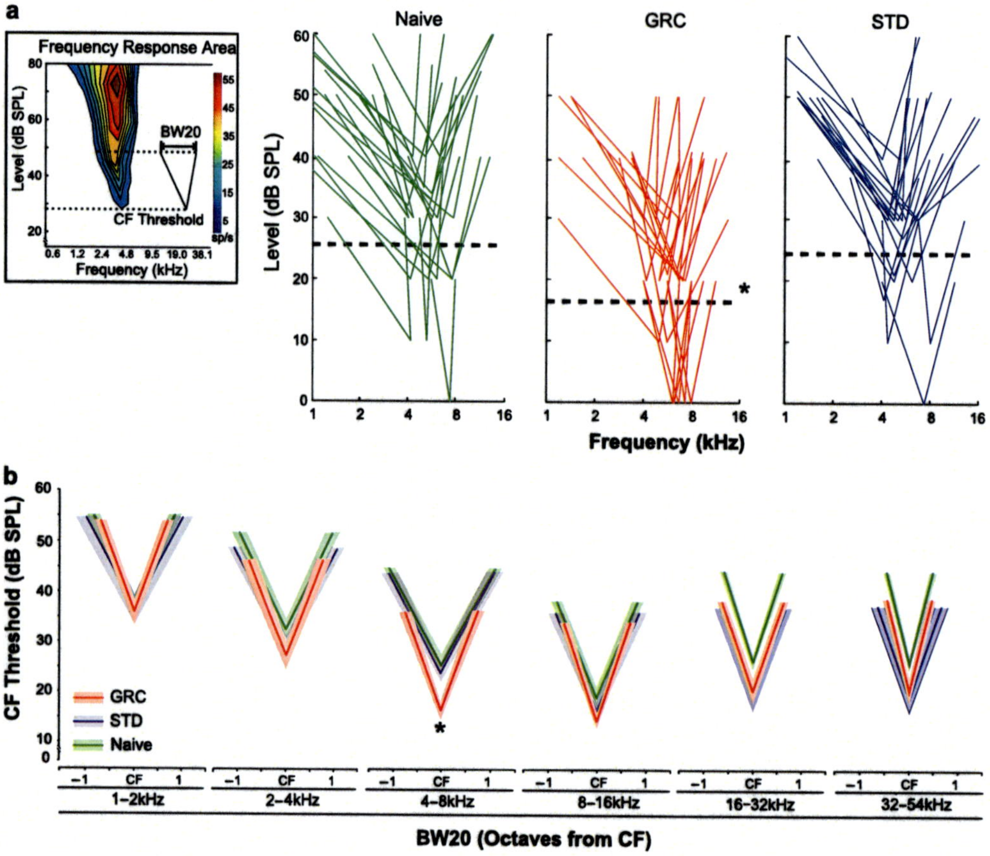

Fig. 22.11 Learning strategy determines plasticity in AI. Rats were trained to bar-press for water during presentation of a 6.0 kHz tone. Different groups used different strategies. One group (STD) used a tone-duration strategy, ceasing bar-pressing at tone offset. Another group (GRC) started responding at tone onset and continued until receiving an error signal after tone offset (tone-onset-to-error). Only the GRC group developed specific plasticity, i.e., reduction of absolute threshold and bandwidth in the octave band (4.0–8.0 kHz) centered on the signal frequency. **a** Examples of frequency-response area (FRA) tuning tips in each group: GRC, STD, and naïve. Each V shape delineates the CF threshold, and BW20 of a recorded FRA (*inset*) that had a CF within the signal-tone frequency band (4–8 kHz) in each respective group. For clarity, subsets of the total population of FRAs are depicted starting from the lowest threshold. *Dashed lines*, the mean CF threshold for the entire population of each group. *Asterisk* shows that the GRC group had significantly lower CF thresholds than either naïve or STD groups, which did not different from each other. **b** Plasticity in threshold and bandwidth in the GRC group is specific to the frequency band of the signal tone (*asterisk*). Both threshold and BW20 decreased only in the signal-tone frequency band. CF threshold and BW20 values are not significantly different from naïves in any frequency band in the STD group. *Solid lines* surrounded by *shaded* areas, group means ± SE, respectively (Berlau and Weinberger 2008)

strategy, and pre-motor processes have each been implicated in AI function.

7.5 A Brief Note on Cerebral Cortex Functional Organization

The traditional view of cerebral cortex functional organization follows a particular sequence of hierarchical functions: sensory analysis, sensory interpretation, and motor performance, executed by anatomically distinct areas. This tripartate schema continues to exert a powerful, if implicit, influence on contemporary neuroscience. However, as AI participates in both analysis and interpretation, this schema is no longer tenable. Moreover, primary auditory cortex, as commonly functionally understood, does not exist, nor do primary somatic sensory or visual cortices. All have learning- and memory-related functions and are involved in other cognitive processes (Weinberger 2008a).

7.6 Toward a New Conception of the Primary Auditory Cortex

It has been argued that theories that no longer account for findings are not discarded until a new theory better explains the results (Kuhn 1970). There are probably two reasons. First, there seems a lack of realization that such a theory is needed. Second, the types of experiments that could produce the foundations for such a theory are rare, probably because many new types of questions have not been asked very often. Some boundary conditions for a theory are next explored. Some new questions will be posed, leading to experiments that should promote the generation of testable theories of the auditory cortex, beginning with AI.

7.6.1 Some Boundary Conditions

We propose that the domain of the functions of AI is greater than now thought, but has limits. Understanding these limits should help circumscribe the borders of AI function.

First, and most obvious, the function of AI is auditory. Despite its involvement in pre-motor functions, AI is nonetheless devoted to sound and hearing.

Second, AI has supra-stimulus functions, i.e., with a role in sound that transcends analysis of the physical features of sound. It can flexibly generalize sounds with similar characteristics, as in concept formation for the direction of frequency change. A myriad of features could be used to form categories, depending upon the reward–punishment contingencies, in both the laboratory and natural settings. The major point is that the domain of AI's concern with sound can be highly abstract.

Third, AI has extra-modal functions. Non-auditory modalities can elicit responses in AI.

Fourth, AI has extra-stimulus functions. This reflects its involvement with processes that are neither auditory nor emanate from any other sensory system, such as the finding that the magnitude of frequency representation expansion is directly related to the level of acquired behavioral importance (Rutkowski and Weinberger 2005). Thus, motivational information that may not be assessed in AI is nonetheless represented there. A similar process was found in primary visual cortex (Shuler and Bear 2006).

7.6.2 The Primary Auditory Cortex as an Auditory Problem Solver

We suggest that the overall function of AI is as an auditory problem solver (APS). This proposal is based largely on its breadth of implicated functions. The diversity of processes involved in AI plasticity simply requires a much broader conceptualization. In considering AI to be an APS, such breadth is honored.

Explication of this construct can only be outlined given limitations of space. So first, let us consider what an APS is not. It is not a high-level acoustic analyzer, although such analysis is no doubt part of its province and the term problem solver can be used to refer to the analysis of complex spectrotemporal patterns of sound.

Also, however AI is conceptualized, an ever-present caveat is that AI almost certainly does not operate in isolation from other auditory cortical fields. It just happens to be that auditory area which is most convenient to study, thanks largely to its organization of tonotopy and other acoustic parameters. But as we presently understand much less about other fields, no more can be said about their roles as APSs.

The core idea advanced here is that whenever the acoustic environment presents a challenge to an organism, AI is involved in meeting that challenge.

Some examples may prove helpful. If a specific motor act is required with respect to acoustic stimulation, then the auditory cortex would integrate (unknown aspects of) motor function with the analysis, acquisition, and storage of information about the relevant sounds. The result would not be confined to auditory information but rather auditory information combined with relevant motor and spatial information for the situation in question. This may explain why selective lesions of frequency bands within the tonotopic map of AI in the cat produce selective impairment of locomotion to the source of the corresponding sound frequency in the cat (Jenkins and Merzenich 1984). But, in no case would AI execute the requisite behavioral act.

If a biological needs state, such as hunger, involves using sound to obtain reward (or in other situations, avoid discomfort or pain), then AI would be involved in the integration of motivational information with relevant sound information. However, in no case would AI itself assess or determine the motivational state, or determine the nature of the goal object, such as food, water or opportunity for sex. Nonetheless, AI would be a major site in which the relevant information would be integrated.

Clearly, this liberalized view of AI is preliminary and really little more than an outline of an idea. The devil is in the details. What constitutes an acoustic challenge? What is integrated in AI and how are the results of such integration used to implement behavior? The list of questions is both long and undelineated at present. But even if the proposed problem solving function of AI is wrong, or worse, fails to lead to testable hypotheses, it is hoped that the need for a reconceptualization of AI has become clear.

8 Some Future Directions

Systematic relevant studies on multi-functional aspects of the auditory cortex do not exist because of the lack of a conceptual framework. Therefore, only a scattered and piecemeal literature exists on functions other than strict acoustic analysis. Fortunately, sufficient investigations of learning and memory provide a basis for new directions.

8.1 Beyond the Documentation of Plasticity

The near and foreseeable futures are likely to see a continuing growth in studies that demonstrate which acoustic parameters are subject to the development of specific cortical plasticity during learning. The major problem with the enumeration of plasticity of acoustic parameters is the enormous parameter space. Therefore, focusing on the most critical and revealing issues is probably more important than which parameters are used. In short, along with demonstrations of plasticity, framing and testing general hypotheses are critical now that the specific has been established.

8.2 The Contents of Auditory Perception and Memory

Auditory learning and memory are always about something, i.e., there must be content. The same holds for auditory perception; there is no such thing as an empty percept of sound. Thus, one new direction would be to determine both what is perceived in a given situation and the contents of that fraction of auditory perception that becomes stored in AI. A first step is to determine what has been learned. For example, training with a 5 kHz CS might lead to learning that 5 kHz predicts food. Alternatively, the learning might be that sound predicts food. Obtaining behavioral frequency generalization gradients after training (during extinction to prevent new learning) will reveal which is the case. A flat gradient indicates that sound, rather than frequency, was learned. A generalization gradient peaked at the CS frequency indicates that the subject learned about the actual acoustic frequency. At the same time, the relationship between responses of AI and the stored aspect of sound must be determined. AI responses differ for the same sound, depending upon whether the frequency is or is not remembered, a new link could be formed between perception and memory. An understanding of the functions of auditory cortex may ultimately depend upon linking the contents of auditory perception, learning and memory to the physiology of the auditory cortex.

8.3 Factors that Determine Plasticity

A frequency-specific increase in the area of AI representation during perceptual learning in the owl monkey (Recanzone et al. 1993) was not found in the cat (Brown et al. 2004).

Several explanations can be offered after a failure to obtain plasticity during learning. First, there are species differences. However, this begs the question of what is critical. Second, a failure to replicate might suggest that the phenomenon is not robust; this restatement of the lack of replication provides no insight. Third, testable hypotheses might explain the different outcomes. We suggest that the development of specific plasticity in AI is governed by multiple factors. Three classes of factors could provide a foundation for such inquiry: stimulus, training, and cognitive.

8.3.1 Stimulus Factors

While AI may be an auditory problem solver, this does not imply that it participates in all auditory problems. Determining the features of auditory situations that induce specific plasticity is an empirical issue, not one of definition.

The acoustic stimulus involved might constitute one factor that affects cortical involvement, such as the number of onset transients in the auditory stimulus. Natural stimuli have many transients, and so may be predisposed to the formation of AI plasticity because it is particularly responsive to transients (Phillips et al. 2002). Natural sounds are often brief, suggesting an acoustic adaptation to extract information from transients (Masterton 1993). Acoustic onsets have a privileged status in both perception and AI discharge. Therefore, the formation of specific learning-induced AI plasticity may be affected by the use of sounds that best exploit cortical proclivities.

8.3.2 Training Factors

Training factors include the amount of training and the asymptotic level of performance. While often correlated, if training continues after asymptote then such overtraining may shift the modes of stimulus representation and behavioral response initiation to more automatic processes (Packard and McGaugh 1996). For example, early in training and while subjects undergo the dynamic phase of learning, relationships between and among sensory stimuli often predominate. Perhaps overtraining reduces or eliminates AI plasticity as the required behavior becomes automatized in the presence of the relevant sound. Thus, the time course of plasticity, and its relationship to the stage of behavioral learning, should be studied.

8.3.3 Learning Strategy

The formation of specific plasticity in AI can depend upon the learning strategy employed, rather than the asymptotic level of performance. Thus, CS-specific decreases in threshold and bandwidth in rat AI developed only if rats focused on the tonal signal onset rather than its duration (Berlau and Weinberger 2008). Moreover, the gain in representational area for a signal frequency is highly correlated with the amount of use of a particular learning strategy (Bieszczad and Weinberger 2010c). The effect of learning strategy may explain an apparent failure to replicate in the cat (Brown et al. 2004) the specific expanded representation found in the owl monkey (Recanzone et al. 1993). The reason may indeed be due to a species difference, but that could be secondary to the use of different learning strategies. For example, given the rich vocal repertoire of these primates, and the fact that acoustic transients are particularly important for such vocalizations, the owl monkeys may have solved the tone discrimination problem by paying particular attention to tonal onset transients whereas the cats did not. This possibility can be tested by determining the strategy employed in future studies.

8.4 Functions of Plasticity

The functions of learning-induced specific plasticity in AI may be the key problem. One might assume that the function of such plasticity is to improve auditory perception, at least for the acoustic parameter employed. Plasticity may confer a perceptual advantage to sounds that become behaviorally important. They may be optimized in a noisy environment, perceived at lower sound pressure levels, and more readily distinguished from other sounds, etc. But not all auditory learning is perceptual.

AI also has associative and mnemonic functions. Thus, specific plasticity appears to bestow advantages to the memory traces stored in the auditory cortex and ultimately could enhance behavior that is dependent upon memory. As memory traces strengthen with plasticity, auditory memories better resist interference (Bieszczad and Weinberger 2010a).

9 Concluding Comments

That the primary auditory cortex is not only an acoustic analyzer, adaptive or otherwise, is a major challenge. If assumptions about the differential cortical localization of stimulus analysis and stimulus interpretation/meaning are no longer tenable, then we understand far less than we think we know about what the auditory system does and how it does it. However, we have virtually unlimited opportunities to make rapid progress. This will require different types of questions and an open mind.

The virtual disappearance of the disciplinary boundaries between auditory neuroscience and the neurobiology of learning, memory, and cognition will be crucial to the future. This may require dual education of new generations of auditory and learning/memory neuroscientists. The strengths of each scientific domain need to be brought to bear because the problem of the primary auditory cortex, and certainly the rest of the auditory cortex, demands no less.

Acknowledgments Sincere thanks to Gabriel K. Hui and Jacquie Weinberger for assistance in the preparation of this chapter. Supported by research grants from the National Institutes of Health/National Institute on Deafness and Other Communication Disorders (NIDCD), DC-02938 and DC-05592 (N.M.W.)

References

Bakin JS, Lepan B, and Weinberger NM (1992) Sensitization induced receptive field plasticity in the auditory cortex is independent of CS-modality. *Brain Research* **577**:226–235.

Bakin JS, South DA, and Weinberger NM (1996) Induction of receptive field plasticity in the auditory cortex of the guinea pig during instrumental avoidance conditioning. *Behavioral Neuroscience* **110**:905–913.

Bakin JS and Weinberger NM (1990) Classical conditioning induces CS-specific receptive field plasticity in the auditory cortex of the guinea pig. *Brain Research* **536**:271–286.

Bao S, Chang EF, Woods J, and Merzenich MM (2004) Temporal plasticity in the primary auditory cortex induced by operant perceptual learning. *Nature Neuroscience* **7**:974–981.

Barrett D, Shumake J, Jones D, and Gonzalez-Lima F (2003) Metabolic mapping of mouse brain activity after extinction of a conditioned emotional response. *Journal of Neuroscience* **23**:5740–5749.

Beitel RE, Schreiner CE, Cheung SW, Wang X, and Merzenich MM (2003) Reward-dependent plasticity in the primary auditory cortex of adult monkeys trained to discriminate temporally modulated signals. *Proceedings of the National Academy of Sciences of the United States of America* **100**:11070–11075.

Berlau KM and Weinberger NM (2008) Learning strategy determines auditory cortical plasticity. *Neurobiology of Learning and Memory* **89**:153–166.

Berntson GG, Tuber DS, Ronca AE, and Bachman DS (1983) The decerebrate human: associative learning. *Experimental Neurology* **81**:77–88.

Bieszczad KM and Weinberger NM (2010a) Representational gain in cortical area underlies increase of memory strength. *Proceedings of the National Academy of Sciences of the United States of America* **107**:3793–3798.

Bieszczad KM and Weinberger NM (2010b) Learning strategy trumps motivational level in determining learning-induced auditory cortical plasticity. *Neurobiology of Learning and Memory* **93**:229–239.

Bieszczad KM and Weinberger NM (2010c) Remodeling the cortex in memory: increased use of a learning strategy increases the representational area of relevant acoustic cues. *Neurobiology of Learning and Memory* **94**:127–144.

Boring EG (1929) *A History of Experimental Psychology*. Appleton-Century, Oxford.

Bouton ME (2007) *Learning and Behavior: A Contemporary Synthesis*. Sinauer Associates, Sunderland.

Brosch M, Selezneva E, Bucks C, and Scheich H (2004) Macaque monkeys discriminate pitch relationships. *Cognition* **91**:259–272.

Brown M, Irvine DRF, and Park VN (2004) Perceptual learning on an auditory frequency discrimination task by cats: association with changes in primary auditory cortex. *Cerebral Cortex* **14**:952–965.

Bruchey AK, Shumake J, and Gonzalez-Lima F (2007) Network model of fear extinction and renewal functional pathways. *Neuroscience* **145**:423–437.

Buchwald JS, Halas ES, and Schramm S (1966) Changes in cortical and subcortical unit activity during behavioral conditioning. *Physiology & Behavior* **1**:11–22.

Budinger E, Heil P, Hess A, and Scheich H (2006) Multisensory processing via early cortical stages: connections of the primary auditory cortical field with other sensory systems. *Neuroscience* **143**:1065–1083.

Campbell AW (1905) *Histological Studies on the Localisation of Cerebral Function*. Cambridge University Press, Cambridge.

Cassady JM, Cole M, Thompson RF, and Weinberger NM (1973) Neural correlates of asymptotic avoidance and classical conditioned leg flexion. *Experimental Neurology* **40**:207–215.

Christian KM and Thompson RF (2005) Long-term storage of an associative memory trace in the cerebellum. *Behavioral Neuroscience* **119**:526–537.

Condon CD and Weinberger NM (1991) Habituation produces frequency-specific plasticity of receptive fields in the auditory cortex. *Behavioral Neuroscience* **105**:416–430.

Demany L and Semal C (2002) Learning to perceive pitch differences. *Journal of the Acoustical Society of America* **111**:1377–1388.

Diamond DM and Weinberger NM (1984) Physiological plasticity of single neurons in auditory cortex of the cat during acquisition of the pupillary conditioned response: II. Secondary field (AII). *Behavioral Neuroscience* **98**:189–210.

Diamond DM and Weinberger NM (1986) Classical conditioning rapidly induces specific changes in frequency receptive fields of single neurons in secondary and ventral ectosylvian auditory cortical fields. *Brain Research* **372**:357–360.

Diamond DM and Weinberger NM (1989) Role of context in the expression of learning-induced plasticity of single neurons in auditory cortex. *Behavioral Neuroscience* **103**:471–494.

Diamond IT (1979) The subdivisions of the neocortex: a proposal to revise the traditional view of sensory, motor, and associational areas. *Progress in Psychobiology and Physiological Psychology* **8**: 1–43.

DiCara LV, Braun JJ, and Pappas BA (1970) Classical conditioning and instrumental learning of cardiac and gastrointestinal responses following removal of neocortex in the rat. *Journal of Comparative and Physiological Psychology* **73**:208–216.

Domjan M (1998) *The Principles of Learning and Behavior*. Brooks-Cole, Pacific Grove.

Dumenko VN and Sachenko VV (1978) Spontaneous activity of auditory cortical neurons of waking cats at rest and during defensive conditioning. *Neurophysiology* **10**:159–168.

Dumenko VN and Sachenko VV (1979) Unit responses of the auditory cortex of waking cats at rest and after defensive conditioning. *Neurophysiology* **11**:17–23.

Dunlop CW, Webster WR, and Rodger RS (1966) Amplitude changes of evoked potentials in the auditory system of unanesthetized cats during acoustic habituation. *Journal of Auditory Research* **6**:47–66.

Durup G and Fessard A (1935) L'électrencéphalogramme de l'homme: observations psycho-physiologiques relatives à l'action des stimuli visuels et auditifs. *L'Année Psychologique* **36**:1–32.

Edeline J-M (2003) The thalamo-cortical auditory receptive fields: regulation by the states of vigilance, learning and the neuromodulatory systems. *Experimental Brain Research* **153**:554–572.

Edeline J-M, Neuenschwander-El Massioui N, and Dutrieux G (1990) Discriminative long-term retention of rapidly induced multiunit changes in the hippocampus, medial geniculate and auditory cortex. *Behavioural Brain Research* **39**:145–155.

Edeline J-M, Pham P, and Weinberger NM (1993) Rapid development of learning-induced receptive field plasticity in the auditory cortex. *Behavioral Neuroscience* **107**:539–551.

Edeline J-M and Weinberger NM (1993) Receptive field plasticity in the auditory cortex during frequency discrimination training: selective retuning independent of task difficulty. *Behavioral Neuroscience* **107**:82–103.

Fritz J, Elhilali M, and Shamma S (2005a) Active listening: task-dependent plasticity of spectrotemporal receptive fields in primary auditory cortex. *Hearing Research* **206**:159–176.

Fritz JB, Elhilali M, and Shamma SA (2005b) Differential dynamic plasticity of A1 receptive fields during multiple spectral tasks. *Journal of Neuroscience* **25**:7623–7635.

Fritz J, Shamma S, Elhilali M, and Klein D (2003) Rapid task-related plasticity of spectrotemporal receptive fields in primary auditory cortex. *Nature Neuroscience* **6**:1216–1223.

Galambos R, Sheatz G, and Vernier V (1956) Electrophysiological correlates of a conditioned response in cats. *Science* **123**: 376–377.

Galeano C (1963) Electrophysiological aspects of brain activity during conditioning: a review. *Acta Neurológica Latinoamericana* **9**: 395–413.

Galván VV, Chen J, and Weinberger NM (2001) Long-term frequency tuning of local field potentials in the auditory cortex of the waking guinea pig. *Journal of the Association for Research in Otolaryngology* **2**:199–215.

Galván VV and Weinberger NM (2002) Long-term consolidation and retention of learning-induced tuning plasticity in the auditory cortex of the guinea pig. *Neurobiology of Learning and Memory* **77**:78–108.

Gao E and Suga N (1998) Experience-dependent corticofugal adjustment of midbrain frequency map in bat auditory system. *Proceedings of the National Academy of Sciences of the United States of America* **95**:12663–12670.

Gao E and Suga N (2000) Experience-dependent plasticity in the auditory cortex and the inferior colliculus of bats: role of the corticofugal system. *Proceedings of the National Academy of Sciences of the United States of America* **97**:8081–8086.

Gasanov UG and Galashina AG (1976) Study of plastic changes in cortical interneuronal connections. *Pavlov Journal of Higher Nervous Activity* **26**:820–827.

Gluck H and Rowland V (1959) Defensive conditioning of electrographic arousal with delayed and differentiated auditory stimuli. *Electroencephalography and Clinical Neurophysiology* **11**: 485–496.

Gonzalez-Lima F, Finkenstädt T, and Ewert JP (1989a) Learning-related activation in the auditory system of the rat produced by long-term habituation: a 2-deoxyglucose study. *Brain Research* **489**:67–79.

Gonzalez-Lima F, Finkenstädt T, and Ewert JP (1989b) Neural substrates for long-term habituation of the acoustic startle reflex in rats: a 2-deoxyglucose study. *Neuroscience Letters* **96**:151–156.

Gonzalez-Lima F and Scheich H (1984) Neural substrates for tone-conditioned bradycardia demonstrated with 2-deoxyglucose: I. activation of auditory nuclei. *Behavioral Brain Research* **14**:213–233.

Gonzalez-Lima F and Scheich H (1986) Neural substrates for tone-conditioned bradycardia demonstration with 2-deoxyglucose: II. Auditory cortex plasticity. *Behavioral Brain Research* **20**:281–293.

Hall RD and Mark RG (1967) Fear and the modification of acoustically evoked potentials during conditioning. *Journal of Neurophysiology* **30**:893–910.

Halpern AR (2001) Cerebral substrates of musical imagery. *Annals of the New York Academy of Sciences* **930**:179–192.

Halpern AR and Zatorre RJ (1999) When that tune runs through your head: a PET investigation of auditory imagery for familiar melodies. *Cerebral Cortex* **9**:697–704.

Halpern AR, Zatorre RJ, Bouffard M, and Johnson JA (2004) Behavioral and neural correlates of perceived and imagined musical timbre. *Neuropsychologia* **42**:1281–1292.

Hawkey DJ, Amitay S, and Moore DR (2004) Early and rapid perceptual learning. *Nature Neuroscience* **7**:1055–1056.

Irvine DR, Martin RL, Klimkeit E, and Smith R (2000) Specificity of perceptual learning in a frequency discrimination task. *Journal of the Acoustical Society of America* **108**:2964–2968.

Jasper H and Sharpless S (1956) Habituation of the arousal reaction. *Brain* **79**:655–680.

Jenkins WM and Merzenich MM (1984) Role of cat primary auditory cortex for sound-localization behavior. *Journal of Neurophysiology* **52**:819–847.

Ji W, Gao E, and Suga N (2001) Effects of acetylcholine and atropine on plasticity of central auditory neurons caused by conditioning in bats. *Journal of Neurophysiology* **86**:211–225.

Ji W and Suga N (2003) Development of reorganization of the auditory cortex caused by fear conditioning: effect of atropine. *Journal of Neurophysiology* **90**:1904–1909.

John ER (1961) High nervous functions: brain functions and learning. *Annual Review of Physiology* **23**:451–484.

Jones D and Gonzalez-Lima F (2001a) Associative effects of Pavlovian differential inhibition of behaviour. *European Journal of Neuroscience* **14**:1915–1927.

Jones D and Gonzalez-Lima F (2001b) Mapping Pavlovian conditioning effects on the brain: Blocking, contiguity, and excitatory effects. *Journal of Neurophysiology* **86**:809–823.

Kellman PJ (2002) Perceptual learning. In: Pashler H and Gallistel R (eds). *Learning, Motivation and Emotion*, 3rd edition. Wiley, Hoboken, pp. 259–299.

Kilgard MP, Pandya PK, Engineer ND, and Moucha R (2002) Cortical network reorganization guided by sensory input features. *Biological Cybernetics* **87**:333–343.

Kisley MA and Gerstein GL (1999) Trial-to-trial variability and state-dependent modulation of auditory-evoked responses in cortex. *Journal of Neuroscience* **19**:10451–10460.

Kisley MA and Gerstein GL (2001) Daily variation and appetitive conditioning-induced plasticity of auditory cortex receptive fields. *European Journal of Neuroscience* **13**:1993–2003.

Kitzes LM, Farley GR, and Starr A (1978) Modulation of auditory cortex unit activity during the performance of a conditioned response. *Experimental Neurology* **62**:678–697.

Kraemer DJ, Macrae CN, Green AE, and Kelley WM (2005) Musical imagery: sound of silence activates auditory cortex. *Nature* **434**:158.

Kuhn TS (1970) *The Structure of Scientific Revolutions*. University of Chicago Press, Chicago.

Lennartz RC and Weinberger NM (1992) Frequency-specific receptive field plasticity in the medial geniculate body induced by Pavlovian fear conditioning is expressed in the anesthetized brain. *Behavioral Neuroscience* **106**:484–497.

Mackintosh NJ (1974) *The Psychology of Animal Learning*. Academic Press, New York.

Mackintosh NJ (1983) *Conditioning and Associative Learning*. Oxford University Press, New York.

Majkowski J and Sobieszek A (1975) Evolution of average evoked potentials in cats during conditioning before and after tegmental lesions. *Physiology & Behavior* **14**:123–131.

Mark RG and Hall RD (1967) Acoustically evoked potentials in the rat during conditioning. *Journal of Neurophysiology* **30**: 875–892.

Marsh JT, McCarthy DA, Sheatz G, and Galambos R (1961) Amplitude changes in evoked auditory potentials during habituation and conditioning. *Electroencephalography and Clinical Neurophysiology* **13**:224–234.

Masterton RB (1993) Central auditory system. *Otorhinolaryngology* **55**:159–163.

McIntosh AR and Gonzalez-Lima F (1992) The application of structural modeling to metabolic mapping of functional neural systems. In: Gonzalez-Lima F, Finkenstadt T, and Scheich H (eds). *Advances in Metabolic Mapping Techniques for Brain Imaging of Behavioral and Learning Functions*. Dordrecht, Boston, pp. 219–255.

McIntosh AR and Gonzalez-Lima F (1993) Network analysis of functional auditory pathways mapped with fluorodeoxyglucose: associative effects of a tone conditioned as a Pavlovian excitor or inhibitor. *Brain Research* **627**:129–140.

McIntosh AR and Gonzalez-Lima F (1995) Functional network interactions between parallel auditory pathways during Pavlovian conditioned inhibition. *Brain Research* **683**:228–241.

McLin DE 3rd, Miasnikov AA, and Weinberger NM (2003) CS-specific *gamma*, *theta*, and *alpha* EEG activity detected in stimulus generalization following induction of behavioral memory by stimulation of the nucleus basalis. *Neurobiology of Learning and Memory* **79**:152–176.

Merzenich MM and Sameshima K (1993) Cortical plasticity and memory. *Current Opinion in Neurobiology* **3**:187–196.

Meyer M, Baumann S, and Jancke L (2006) Electrical brain imaging reveals spatio-temporal dynamics of timbre perception in humans. *NeuroImage* **32**:1510–1523.

Molchán SE, Sunderland T, McIntosh AR, Herscovitch P, and Schreurs BG (1994) A functional anatomical study of associative learning in humans. *Proceedings of the National Academy of Sciences of the United States of America* **91**:8122–8126.

Molnár M, Karmos G, Csépe V, and Winkler I (1988) Intracortical auditory evoked potentials during classical aversive conditioning in cats. *Biological Psychology* **26**:339–350.

Morrell F (1961) Electrophysiological contributions to the neural basis of learning. *Physiological Review* **41**:443–494.

Morris JS, Friston KJ, and Dolan RJ (1998) Experience-dependent modulation of tonotopic neural responses in human auditory cortex. *Proceedings of the Royal Society of London, series B, Biological Sciences* **265**:649–657.

Murata K and Kameda K (1963) The activity of single cortical neurones of unrestrained cats during sleep and wakefulness. *Archives Italiennes de Biologie* **101**:306–331.

Nair HP, Berndt JD, Barrett D, and Gonzalez-Lima F (2001a) Maturation of extinction behavior in infant rats: large-scale regional interactions with medial prefrontal cortex, orbitofrontal cortex, and anterior cingulate cortex. *Journal of Neuroscience* **21**:4400–4407.

Nair HP, Berndt JD, Barrett D, and Gonzalez-Lima F (2001b) Metabolic mapping of brain regions associated with behavioral extinction in preweanling rats. *Brain Research* **903**:141–153.

Neff WD, Diamond IT, and Casseday JH (1975) Behavioral studies of auditory discrimination: central nervous system. In: Keidel WD and Neff WD (eds). *Handbook of Sensory Physiology*, volume 5: *Auditory System, Part 2: Physiology (CNS), Behavioral Studies, Psychoacoustics*. Springer, New York, pp. 307–400.

Ohl FW and Scheich H (2004) Fallacies in behavioural interpretation of auditory cortex plasticity. *Nature Reviews Neuroscience* **5**. Published online at http://www.nature.com/nrn/journal/v5/n4/full/nrn1366-c1.html

Ohl FW and Scheich H (2005) Learning-induced plasticity in animal and human auditory cortex. *Current Opinion in Neurobiology* **15**:470–477.

Ohl FW, Scheich H, and Freeman WJ (2001) Change in pattern of ongoing cortical activity with auditory category learning. *Nature* **412**:733–736.

Oleson TD, Ashe JH, and Weinberger NM (1975) Modification of auditory and somatosensory system activity during pupillary conditioning in the paralyzed cat. *Journal of Neurophysiology* **38**:1114–1139.

Packard MG and McGaugh JL (1996) Inactivation of hippocampus or caudate nucleus with lidocaine differentially affects expression of place and response learning. *Neurobiology of Learning and Memory* **65**:65–72.

Palmer CV, Nelson CT, and Lindley GA 4th (1998) The functionally and physiologically plastic adult auditory system. *Journal of the Acoustical Society of America* **103**:1705–1721.

Pekkola J, Ojanen V, Autti T, Jääskeläinen IP, Möttönen R, Tarkiainen A, and Sams M (2005) Primary auditory cortex activation by visual speech: an fMRI study at 3 T. *NeuroReport* **16**:125–128.

Phillips DP, Hall SE, and Boehnke SE (2002) Central auditory onset responses, and temporal asymmetries in auditory perception. *Hearing Research* **167**:192–205.

Polley DB, Heiser MA, Blake DT, Schreiner CE, and Merzenich MM (2004) Associative learning shapes the neural code for stimulus magnitude in primary auditory cortex. *Proceedings of the National Academy of Sciences of the United States of America* **101**: 16351–16356.

Polley DB, Steinberg EE, and Merzenich MM (2006) Perceptual learning directs auditory cortical map reorganization through top-down influences. *Journal of Neuroscience* **26**:4970–4982.

Poremba A, Jones D, and Gonzalez-Lima F (1997) Metabolic effects of blocking tone conditioning on the rat auditory system. *Neurobiology of Learning and Memory* **68**:154–171.

Poremba A, Jones D, and Gonzalez-Lima F (1998) Classical conditioning modifies cytochrome oxidase activity in the auditory system. *European Journal of Neuroscience* **10**:3035–3043.

Puga F, Barrett DW, Bastida CC, and Gonzalez-Lima F (2007) Functional networks underlying latent inhibition learning in the mouse brain. *NeuroImage* **38**:171–183.

Rauschecker JP (1999) Auditory cortical plasticity: a comparison with other sensory systems. *Trends in Neurosciences* **22**:74–80.

Recanzone GH, Schreiner CE, and Merzenich MM (1993) Plasticity in the frequency representation of primary auditory cortex following discrimination training in adult owl monkeys. *Journal of Neuroscience* **13**:87–103.

Rescorla RA (1988) Behavioral studies of Pavlovian conditioning. *Annual Review of Neuroscience* **11**:329–352.

Romanski LM and LeDoux JE (1992) Equipotentiality of thalamo-amygdala and thalamo-cortico-amygdala circuits in auditory fear conditioning. *Journal of Neuroscience* **12**:4501–4509.

Rowland V and Gluck H (1960) Electrographic arousal and its inhibition as studied by auditory conditioning. In: Wortis J (ed). *Recent Advances in Biological Psychiatry*, volume 2. Grune and Stratton, Oxford, pp. 96–105.

Rutkowski RG and Weinberger NM (2005) Encoding of learned importance of sound by magnitude of representational area in primary auditory cortex. *Proceedings of the National Academy of Sciences of the United States of America* **102**:13664–13669.

Ryle G (1963) *The Concept of Mind*. Hutchinson, London.

Sakurai Y (1990) Cells in the rat auditory system have sensory-delay correlates during the performance of an auditory working memory task. *Behavioral Neuroscience* **104**:856–868.

Sakurai Y (1992) Auditory working and reference memory can be tested in a single situation of stimuli for the rat. *Behavioural Brain Research* **50**:193–195.

Sakurai Y (1994) Involvement of auditory cortical and hippocampal neurons in auditory working memory and reference memory in the rat. *Journal of Neuroscience* **14**:2606–2623.

Sakurai Y (1998) The search for cell assemblies in the working brain. *Behavioural Brain Research* **91**:1–13.

Saunders JC and Chabora JT (1969) Effects of appetitive drive on evoked potentials in cochlear nucleus and auditory cortex in cats. *Journal of Comparative and Physiological Psychology* **69**:355–361.

Schreurs BG, McIntosh AR, Bahro M, Herscovitch P, Sunderland T, and Molchán SE (1997) Lateralization and behavioral correlation of changes in regional cerebral blood flow with classical conditioning of the human eyeblink response. *Journal of Neurophysiology* **77**:2153–2163.

Schwartz B, Wasserman EA, and Robbins SJ (2002) *Psychology of Learning and Behavior*. W.W. Norton and Company, New York.

Shuler MG and Bear MF (2006) Reward timing in the primary visual cortex. *Science* **311**:1606–1609.

Suga N and Ma X (2003) Multiparametric corticofugal modulation and plasticity in the auditory system. *Nature Reviews Neuroscience* **4**:783–794.

Teas DC and Kiang NY (1964) Evoked responses from the auditory cortex. *Experimental Neurology* **10**:91–119.

Teich AH, McCabe PM, Gentile CG, Jarrell TW, Winters RW, Liskowsky DR. and Schneiderman N (1988) Role of auditory cortex in the acquisition of differential heart rate conditioning. *Physiology & Behavior* **44**:405–412.

Teich AH, McCabe PM, Gentile CG, Schneiderman LS, Winters RW, Liskowsky DR, and Schneiderman N (1989) Auditory cortex lesions prevent the extinction of Pavlovian differential heart rate conditioning to tonal stimuli in rabbits. *Brain Research* **480**:210–218.

Tervaniemi M, Rytkönen M, Schröger E, Ilmoniemi RJ, and Näätänen R (2001) Superior formation of cortical memory traces for melodic patterns in musicians. *Learning and Memory* **8**:295–300.

Thomas GJ (1962) Neurophysiology of learning. *Annual Review of Psychology* **13**:71–106.

Villa AEP, Hyland B, Tetko IV, and Najem A (1998) Dynamical cell assemblies in the rat auditory cortex in a reaction-time task. *Biosystems* **48**:269–277.

Villa AEP, Tetko IV, Hyland B, and Najem A (1999) Spatiotemporal activity patterns of rat cortical neurons predict responses in a conditioned task. *Proceedings of the National Academy of Sciences of the United States of America* **96**:1106–1111.

Weinberger NM (1998) Physiological memory in primary auditory cortex: characteristics and mechanisms. *Neurobiology of Learning and Memory* **70**:226–251.

Weinberger NM (2004a) Correcting misconceptions of tuning shifts in auditory cortex. *Nature Reviews Neuroscience* **5**. Published online at http://www.nature.com/nrn/journal/v5/n4/full/nrn1366-c2.html

Weinberger NM (2004b) Experience-dependent response plasticity in the auditory cortex: Issues, characteristics, mechanisms, and functions. In: Parks TN, Rubel EW, Fay RR, and Popper AN (eds). *Springer Handbook of Auditory Research*, volume 23, *Plasticity of the Auditory System*. Springer, New York, pp. 173–227.

Weinberger NM (2007a) Associative representational plasticity in the auditory cortex: A synthesis of two disciplines. *Learning and Memory* **14**:1–16.

Weinberger NM (2007b) Auditory associative memory and representational plasticity in the primary auditory cortex. *Hearing Research* **229**:54–68.

Weinberger NM (2008a) Cortical plasticity in associative learning and memory. In: Byrne JH (ed). *Learning and Memory: A Comprehensive Reference*, volume 3. Academic Press, New York, pp. 187–218.

Weinberger NM (2008b) Retuning the brain by learning, literature, and logic: reply to Suga. *Learning and Memory* **15**:202–207.

Weinberger NM (2009) The cognitive auditory cortex. In: Palmer AR and Rees A (eds). *The Oxford Handbook of Auditory Science: The Auditory Brain*. Oxford University Press, Oxford, pp. 439–475.

Weinberger NM and Bakin JS (1998) Learning-induced physiological memory in adult primary auditory cortex: receptive fields plasticity, model, and mechanisms. *Audiology & Neuro-otology* **3**: 145–167.

Weinberger NM and Diamond DM (1987) Physiological plasticity in auditory cortex: rapid induction by learning. *Progress in Neurobiology* **29**:1–55.

Weinberger NM and Diamond DM (1988) Dynamic modulation of the auditory system by associative learning. In: Edelman GM, Gall WE, and Cowan WM (eds). *Auditory Function: Neurobiological Bases of Hearing*. Wiley, New York, pp. 485–512.

Weinberger NM, Diamond DM, and McKenna TM (1984a) Initial events in conditioning: Plasticity in the pupillomotor and auditory systems. In: Lynch G, McGaugh JL, and Weinberger NM (eds). *Neurobiology of Learning and Memory*. Guilford Press, New York, pp. 197–227.

Weinberger NM, Hopkins W, and Diamond DM (1984b) Physiological plasticity of single neurons in auditory cortex of the cat during acquisition of the pupillary conditioned response: I. Primary field (AI). *Behavioral Neuroscience* **98**:171–188.

Weinberger NM, Javid R, and Lepan B (1993) Long-term retention of learning-induced receptive-field plasticity in the auditory cortex. *Proceedings of the National Academy of Sciences of the United States of America* **90**:2394–2398.

Westenberg IS, Paige G, Golub B, and Weinberger NM (1976) Evoked potential decrements in auditory cortex: I. Discrete-trial and continual stimulation. *Electroencephalography and Clinical Neurophysiology* **40**:337–355.

Westenberg IS and Weinberger NM (1976) Evoked potential decrements in auditory cortex. II: critical test for habituation. *Electroencephalography and Clinical Neurophysiology* **40**: 356–369.

Wetzel W, Wagner T, Ohl FW, and Scheich H (1998) Categorical discrimination of direction in frequency-modulated tones by Mongolian gerbils. *Behavioural Brain Research* **91**:29–39.

Wickelgren WO (1968) Effect of acoustic habituation on click-evoked responses in cats. *Journal of Neurophysiology* **31**: 777–785.

Woody CD, Knispel JD, Crow TJ, and Black-Cleworth PA (1976) Activity and excitability to electrical current of cortical auditory receptive neurons of awake cats as affected by stimulus association. *Journal of Neurophysiology* **39**:1045–1061.

Zatorre RJ and Halpern AR (2005) Mental concerts: musical imagery and auditory cortex. *Neuron* **47**:9–12.

Chapter 23

Cortical Effects of Aging and Hearing Loss

Julie R. Mendelson and Ramesh Rajan

Abbreviations

ABER	auditory brainstem-evoked response
AAF	anterior auditory field
AC	auditory cortex
AI	primary auditory cortex
AM	amplitude modulation
AMFR	amplitude modulation following response
BF	best frequency
CDWC	congenitally deaf white cats
CF	characteristic frequency
CI	cochlear implant
CNS	central nervous system
EFRA	excitatory frequency response area
EP	evoked potential
ERP	event-related potential
FM	frequency modulation
FMFR	frequency modulation following response
FRA	frequency response area
GABA	γ-aminobutyric acid
GAD	glutamic acid decarboxylase
IC	inferior colliculus
IFRA	inhibitory frequency response area
IPD	interaural phase difference
LFP	local field potential
LSO	lateral superior olive
MEG	magnetoencephalography
MGN	medial geniculate nucleus
MLD	masking level differences
MLR	middle-latency response
MMN	mismatch negativity
MRI	magnetic resonance imaging
NAHL	non-aging hearing losses
NECD	normal early-and-chronically deafened
SAM	sinusoidal amplitude modulation
SGC	spiral ganglion cell
SR	spontaneous activity
VOT	voice onset time

J.R. Mendelson (✉)
Research, Toronto Rehabilitation Institute, Toronto, ON, M5G 2A2, Canada
e-mail: j.mendelson@utoronto.ca

R. Rajan (✉)
Department of Physiology, Monash University, Clayton, VIC 3168, Australia
e-mail: ramesh.rajan@med.monash.edu.au

1 Introduction

Aging and hearing loss have profound consequences for the function of auditory cortex (AC). We first discuss the effects of aging on auditory processing followed by surveying consequences of deafness in the AC. In the aging effects, we begin with an overview of the changes that occur in older humans, followed by animal models of aging and cortical function. Next, we discuss possible neural mechanisms underlying aging effects and potential interventions that may retard or even reverse some of these age-related changes.

In the second part of the chapter, we elaborate on the consequences of cochlear hearing losses induced by factors other than aging (non-aging hearing losses, NAHL) for responses from AC in animals and humans.

2 Aging in Human Psychophysics and Physiology

One of the most common age-related changes to occur in the auditory system is an increase in threshold, i.e., a decrease in sensitivity, particularly for higher frequencies. However, one of the more debilitating behavioral changes is a decline in speech discrimination ability. This difficulty can occur under noisy or quiet conditions. It can also occur in individuals with normal audiograms suggesting that some of the difficulties may be due to changes in temporal processing.

J.A. Winer, C.E. Schreiner (eds.), *The Auditory Cortex*,
DOI 10.1007/978-1-4419-0074-6_23, © Springer Science+Business Media, LLC 2011

A primary physiological measure for cortical activity is the N1–P2 waveform of event-related potentials. It is thought to reflect synchronous neural activity in the thalamo-cortical pathway (Näätänen and Picton 1987). In young adults, the N1 wave peaks at approximately100 ms, while the P2 wave peaks around 200 ms following stimulus onset. Increases in N1 and/or P2 latencies are thought to indicate changes in excitatory and inhibitory processes as well as an overall slowing of temporal processing (Harris et al. 2007). One method that is frequently used to record these cortical events is the mismatch negativity (MMN). The MMN is an event-related potential (ERP) that is automatically elicited 100–200 ms after a deviant stimulus is detected within a repetitive series of standard stimuli (Näätänen 2000). A primary advantage of the MMN is that it does not require subjects to attend to the stimuli, thereby reducing any potential confound due to deficits in memory and/or cognitive processing capabilities.

2.1 Temporal Processing

A major age-related change in the central auditory system is a degradation in temporal resolution. This can be expressed in a number of different ways including difficulty in understanding rapid talkers, decreased ability to discriminate speech in noisy environments, and mishearing words. These types of temporal processing can be behaviorally assessed through gap detection, detection of stimulus duration, stimulus presentation rate, speech compression, or by modulating the frequency (FM) or amplitude (AM) of a signal over time.

Speech Stimuli: Tremblay et al. (2003) recorded psychophysical and neural responses to voice onset time (VOT), which provides cues that allow subjects to distinguish between consonants, e.g., along the /ba/–/pa/ continuum. They found that older subjects had greater difficulty discriminating between stimuli with different VOT duration. This difference was also reflected in an increase in the N1 latencies. Tremblay et al. (2003) suggested that longer N1 latencies in elderly subjects may reflect age-related delays in synchronous firing and changes in the neuronal refractory period. In other words, neurons in the aged auditory system may not be able to recover quickly enough before they are required to fire again.

Stimulus Duration: Stimulus duration provides information on vowel duration or consonant transitions (Miller and Liberman 1979). Ostroff et al. (2003) compared the N1 and P2 waveforms of young, middle-aged, and old subjects in response to stimuli of differing durations. While the N1 amplitude increased with increases in stimulus duration for all age groups, the P2 did so only in young- and middle-aged but not in aged subjects. The results suggest that aging does disrupt the auditory system's ability to process small differences in sound duration.

Presentation Rate: Tremblay et al. (2004) recorded longer latencies in the P2 wave of the N1–P2 complex in older adults when speech stimuli were presented at a fast rate. However, if speech stimuli were presented at a slower rate, no age difference in latency responses was observed. This correlates with their behavioral data whereby older subjects had greater difficulty in discriminating the same speech stimuli presented at the faster rates. Interestingly, there were no physiological or functional age-related differences when subjects were asked to attend to simple pure tone as compared to speech stimuli (Tremblay et al. 2004).

Gap Detection: Studies examining gap detection require subjects to indicate when they hear a silent interval between two signals. Researchers have found that elderly subjects require a larger gap in a continuous stimulus before they can detect an interruption (Schneider et al. 1994). This may help to explain why many elderly individuals experience difficulty in discriminating /ba/ from /pa/ because the interval between consonant and vowel onset, the VOT, is too short for the senescent auditory system to encode (Tremblay et al. 2004). However, when subjects are asked to detect a gap within a pure tone stimulus, young and aged subjects perform at the same level (Bertoli et al. 2002; Tremblay et al. 2004). In ERP recordings, older subjects required longer gaps in the stimuli in order to elicit MMNs (Bertoli et al. 2002) and also had reduced MMN peak amplitudes and increased peak latencies compared to younger subjects.

Spectral Content: Dynamic spectral information contained in the second formant transition (F2) provides salient information for listeners to distinguish between speech stimuli. Harkrider et al. (2005) varied the onset frequency of the F2 along the /ba/–/da/–/ga/ continuum. They found that while aging affected the categorical perception of the phonemes, accurate discrimination was dependent upon the position of one phoneme relative to another. The authors also found that the P2 latencies were longer and N1 amplitudes larger for older subjects. However, when the researchers amplified the formant transition cue relative to the rest of the consonant–vowel stimulus, performance and neural processing improved. The authors concluded that while there is an age-related degradation of spectral cue processing, it can be overcome, at least to a certain degree, if relevant information—in this case amplification of the F2—is manipulated.

Amplitude-Modulated and Frequency-Modulated Stimuli: In processing speech, it is essential to discriminate between frequency and amplitude changes over time. For example, the ability to accurately discriminate /ga/ from /da/ depends largely on the rapid rate of frequency change occurring at the beginning of the F2 transition. Consequently, if temporal processing of formant transitions deteriorates in the aged auditory system, then the ability to discriminate /da/ from /ga/ could be impaired because the F2 transition in /da/ may be changing too rapidly for the senescent auditory system to accurately encode. This could then result in mishearing words, such as 'doll' for 'gall.' These types of speech discrimination errors are common in the elderly.

The amplitude modulation following response (AMFR) is a steady-state auditory response that reflects thalamo-cortical and/or auditory cortical processing (Ro et al. 2000) and is thought to provide a physiological measure of intensity discrimination. Boettcher et al. (2001) recorded the AMFR in young and aged subjects and found no differences between the two age groups. Similar to the AMFR, the frequency modulation following response (FMFR) has been used to study frequency discrimination in the aged auditory system. In contrast to AMFR, aging does affect FMFR processing. Boettcher et al. (2002) found that elderly subjects had larger amplitudes and greater modulation depths than young subjects when stimuli were presented at faster modulation rates. The authors suggested that AMFR and FMFR may be generated at different neural sites, thus accounting for the age-related differences observed.

Dimitrijevic et al. (2004) used a unique stimulus that allows for the simultaneous testing of both AM and FM signals. The independent amplitude- and frequency-modulated stimulus consists of a carrier that is simultaneously modulated in amplitude at one rate and in frequency at another rate. They observed that only younger subjects had larger evoked FM responses than AM responses. This lends further support to the suggestion that AM and FM processing are affected by aging in different ways.

2.2 Sound Localization

In order to accurately localize an acoustic signal, the auditory system relies on a number of binaural cues that include interaural level, time, and phase differences. Older listeners require longer interaural time differences between two tones in order to be able to accurately localize their sources (Cranford et al. 1990; Strouse et al. 1998). However, for simple lateralization of an acoustic signal, Grady et al. (2008) found no age differences between subjects. While there are currently no physiological data on sound localization in the senescent human cortex, Grady et al. (2008) observed that older subjects exhibited greater sustained neural activity in the left superior parietal cortex than younger subjects when asked to lateralize a signal.

Interaural Phase Differences (IPDs): Ross et al. (2007) binaurally presented young, middle-aged, and old subjects with sinusoidal amplitude-modulated (SAM) tones that included an IPD (the steady state equivalent of interaural time disparities). They found that behavioral changes began to occur in middle-aged subjects while the physiological differences were only apparent in the aged subjects. They also found that the amplitude of the N1–P2 waveforms in response to IPDs decreased with increasing frequency for older subjects. Finally, they reported that evoked potentials in older subjects exhibited longer latencies than in younger subjects.

Level Discrimination: Level discrimination among the aged population appears to be dependent upon the frequency of the stimulus. One study showed an age-related effect for low but not high frequencies when subjects were asked to indicate if two acoustic signals differed from each other in terms of level (He et al. 1998). The authors' proposal that this was most likely due to deterioration in phase discrimination abilities in the older population received physiological support from Harris et al. (2007) who found that the N1–P2 response thresholds were significantly higher for older subjects, but only for lower frequencies. They also noted that response latencies were longer for older subjects at lower but not higher frequencies. Harris et al. (2007) suggested that these results could indicate an age-related decline in inhibitory control within the central auditory system.

2.3 Signal Segregation

Another difficulty which older listeners experience is in distinguishing between two or more simultaneously occurring conversations. This may be due to an age-related decline in the ability to separate signals into different components (Alain et al. 2001; Grube et al. 2003). Snyder and Alain (2005) examined the ability of young and old subjects to discriminate between two concurrently presented vowels. They found that older subjects had greater difficulty in accomplishing the task than younger subjects. This difference was mirrored by a decrease in neural activity.

Masking Level Differences (MLDs): Masking level differences provide a measure of an individual's ability to segregate a signal from noise. For example, many elderly listeners experience problems in discriminating speech from background noise (Schneider et al. 2000). While there are currently no physiological data on MLDs in the aged auditory cortex, we thought it prudent to include some psychophysical data here. Researchers have found significant differences between young and aged subjects for MLDs that may reflect why many elderly subjects experience difficulty in understanding speech in noisy situations (Pichora-Fuller and Schneider 1991). For example, Cobb et al. (1993), using a backward masking level paradigm, found that the slope of the MLD function was significantly steeper in elderly subjects, suggestive of a decline in temporal resolution.

2.4 Interhemispheric Differences

It appears that the contralateral and ipsilateral auditory pathways are affected differently by aging. Using /ga/ and /da/ stimuli, Bellis et al. (2000) found that the P1–N1 response amplitudes (where P1 latency response occurs 50–100 ms after stimulus onset) were symmetrical in both hemispheres of older but not younger subjects. This corresponded to poorer performance on a speech discrimination task. Other studies have shown that aging delays signal processing in response to changes in interstimulus intervals in the ipsilateral but not contralateral AC (Pekkonen et al. 1995). Finally, physiological recordings have revealed a right-hemisphere dominance for mistuned harmonics in older subjects (Hiraumi et al. 2005; Alain and McDonald 2007).

2.5 Human Anatomical and Morphological Changes

While MRI technology has advanced our understanding of morphological and anatomical changes within the aging human brain, most of the studies do not specifically report on what occurs in the auditory cortex. However, these studies can provide some glimpses of what may transpire in the senescent auditory cortex. For example, Salat et al. (2004) measured changes in cortical thickness as a function of aging. While significant thinning was observed in many cortical regions, including the occipital and prefrontal regions, the temporal cortex exhibited only minor atrophy. Other manifestations of what appear to occur in the aging cortex include changes in volume, white matter, number of neurons, and loss of dendrites and dendritic spines, particularly in pyramidal cells (Anderson and Rutledge 1996; Wong 2002; Sowell et al. 2003).

In summary, perceptually older listeners seem to require longer gaps, slower presentation rates, and longer durations especially for some aspects of speech stimuli such as VOT. They also experience greater difficulty in distinguishing between two or more conversations that are occurring simultaneously. In addition, there appears to be an age-related difference for processing FM but not AM stimuli. These differences appear to be reflected by physiological changes in the auditory cortex. Collectively theses studies suggest that there is a general slowing down of temporal processing in the AC with aging.

3 Aging in Animal Behavior and Physiology

In comparison to studies with human subjects, there is a relative dearth of research on aging in AC of animals. However, as with humans, hearing sensitivity in animals declines with age, particularly for high frequencies (Bennett et al. 1983; Cooper et al. 1990; Proctor et al. 1998; McFadden et al. 1997; Willott 1986; Zheng et al. 1999).

Frequency Distribution: While Willott et al. (1993) found a decreased sensitivity to higher frequencies in the cortex of aging C57 mice, they also observed a change in the distribution of frequency responses. Specifically, they found an increased cortical representation of the intact middle and low frequencies, similar to the plasticity of cochleotopic AC maps that occurs after cochlear damage (see second part of this chapter).

Tuning Response Profiles: Frequency response tuning properties of cortical neurons also appear to change with age. Dunn (1983) reported that response thresholds were 44 dB higher and latencies 3.6 ms longer in older guinea pigs. When examining tuning curve profiles of layer V neurons in aged rats, Turner et al. (2005a) found a reduction in the number of classic V/U-shaped excitatory receptive fields and an increase in the number of neurons showing more variable receptive field properties. These latter "complex" neurons were characterized by poorly defined receptive fields, greater variability in response to repetition of the same stimuli, and increased spontaneous activity. In addition, for both V/U and complex neurons, responses were less reliable upon repetition of the same stimuli and firing rates were altered in the senescent cortex. Turner et al. (2005b) found that the classic V/U neurons were associated with larger pyramidal cells, while the complex neurons were more often found to be smaller pyramidal cells. The authors suggested that these results indicate degraded signal-to-noise processing that is consistent with decreases in GABAergic neurotransmission in the senescent auditory system. As detailed in the next section of this chapter, with cochlear hearing losses induced by factors other than ageing, a common change induced in cortex is a reduction or loss of inhibition and this leads, in those cases, to changes in tuning properties that also mimic the changes described here as a consequence of aging.

Frequency-Modulated Sweeps (FM): As mentioned above, it has been suggested that temporal processing speed slows down with age and that this, in turn, may affect the auditory system's ability to accurately encode rapidly changing acoustic stimuli such as formant transitions (Konkle et al. 1977; Schneider et al. 1994; Mendelson and Ricketts 2001). One stimulus that lends itself well to investigating this ability is the FM sweep that has features in common with formant transitions. FM sweeps are characterized by changes in speed (rate of frequency change) and direction (upward-directed: changing from a low to high frequency, or downward-directed: changing from a high to a low frequency). By using FM sweeps that varied in both speed and direction Mendelson and colleagues reported an age-related change in temporal processing speed in the AC of old rats (Mendelson and Ricketts 2001; Mendelson and Lui 2004; De Rivera et al. 2006). They found that the majority of cells recorded from young rats responded most vigorously to fast and medium speeds of FM sweeps but, in contrast, the majority of units recorded from the aged animals responded best to slow speeds (Fig. 23.1a). For preferred direction of FM sweep, no age differences were observed (Fig. 23.1b). The results demonstrated an age-related difference in the preferred rate of frequency change in AC (Mendelson and Ricketts 2001; De Rivera et al. 2006). This change in temporal processing speed may account, in part, for some of the difficulties that the elderly experience in discriminating speech.

Sound Localization: Directional hearing appears to be degraded by aging (Brown 1984; Heffner et al. 2001). In a behavioral lateralization study, Brown (1984) found that aged rats were less accurate than young rats in pressing a bar on the side from which an acoustic stimulus was presented. McFadden and Willott (1994) found in the inferior colliculus (IC) of aged mice that the pattern of excitatory and inhibitory responses was altered as a stimulus was moved from the contralateral to the ipsilateral side. Preliminary results of sound localization studies in elderly macaque monkeys (Juarez-Salinas et al. 2008) revealed several consequences of aging. A disproportionally high number of neurons was encountered with their best azimuth located in ipsilateral space as compared to younger animals. The bandwidth of the spatial tuning curve was broader in all cortical areas compared to monkeys without hearing loss. The authors concluded that spatial tuning of cortical neurons is broadened in all areas tested and that an enhanced spatial tuning of neurons in the caudal-lateral cortical field seen in younger animals is lost with age.

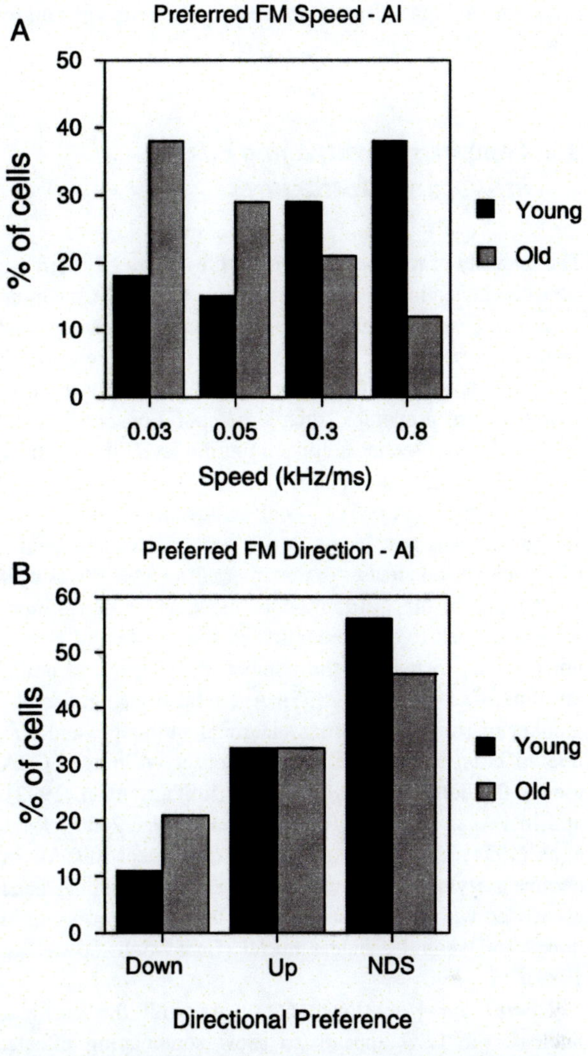

Fig. 23.1 Distribution of preferred speed (**a**) and direction (**b**) selective responses in auditory cortex of young *black* and old *gray* rats. There was a significant age difference for preferred speed with the majority of cells recorded from old animals responding best to slower sweeps while those recorded from young animals responding best to faster sweeps. There was no age difference in preferred direction (Mendelson and Ricketts 2001)

3.1 Anatomical and Neurotransmitter Changes with Aging

Ling et al. (2005) found no significant age-related loss of neurons in the primary AC of the senescent rat confirming what Peters et al. (1997) had previously reported in the cortex of aged monkeys. As mentioned above, it is possible that the difficulty older listeners have in discriminating speech is due in part to a degradation in temporal processing (Fitzgibbons and Gordon-Salant 1994; Ostroff et al. 2003; Tremblay et al. 2003). Changes in inhibitory circuits may alter responses to time-varying stimuli (Walton et al. 1998; Caspary et al. 2002; Liang et al. 2002). GABA (gamma-aminobutyric acid) likely plays a prominent role in this (Wang et al. 2000). Caspary and colleagues found an age-related reduction in GABA content across all cortical layers as well as changes in $GABA_A$ receptor composition (Ling et al. 2005; Caspary et al. 2008). In addition, they examined glutamic acid decarboxylase (GAD) protein levels in the aged cortex and found that all but layer 5 showed significant decreases in the levels.

It is interesting to note that in the senescent visual cortex of monkeys, iontophoretic administration of GABA to cortical cells has been shown to restore both direction and orientation selectivity (Leventhal et al. 2003). It is worth speculating that iontophoretic administration of GABA to neurons in the senescent AC may also restore speed selectivity for rapidly changing FM sweeps.

3.2 Comparison with Aging Effects in Subcortical Structures

The majority of research on aging in the auditory system has been conducted in subcortical structures. Generally, the more basic response properties, such as elevated thresholds, and increase in number of poorly responsive cells, are affected by aging throughout the auditory system. For more complex response properties, such as temporal processing, the effects of aging are most apparent at the level of the cortex (Mendelson and Ricketts 2001).

Cochlear Nucleus: The cochlear nucleus exhibits some age-related changes. In the dorsal cochlear nucleus neural responses show increases in maximum discharge rate as well as changes in the distribution of the number of different response types (e.g., a reduction in the number of pauser units and an increase in the number of chopper and build-up units; Caspary et al. 2005). Age-related changes in the number of surviving neurons, neuronal shrinkage, and volume of brain region have also been observed in aged CBA mouse (Willott et al. 1988), in rat (Keithley et al. 1992), and in rhesus monkey (Torre III and Fowler 2000). Raza et al. (1994) reported no age-related differences in GAD or choline acetyltransferase while changes in glycine have been associated with altered intensity and temporal processing in dorsal and ventral cochlear nuclei (Banay-Schwartz et al. 1989).

Lateral Superior Olive (LSO): As with the cochlear nucleus, the LSO appears to show some aging effects. Finlayson and Caspary (1993) found significant age-related changes in the inhibitory and excitatory responses to pure tone and click stimuli of LSO neurons. However, no age differences were observed for rate-level functions evoked by contralateral stimuli, discharge rate, or conduction latencies.

Inferior Colliculus (IC): While many response properties in the IC of the mouse remain unchanged throughout its lifespan (Willott et al. 1988), changes in the frequency representation comparable to the plasticity of cochleotopic maps seen after partial cochlear damage were observed (Willott 1996). However, the degree of plasticity in the IC was not as great as that observed in the auditory cortex.

There are some age-related changes in temporal processing in the IC. When examining responses to gap stimuli the mean minimum gap threshold response was longer and fewer cells responded to the shortest gap interval in the aged IC (Walton et al. 1998; Barsz et al. 2002). Using FM stimuli, Lee et al. (2002) found no age-related differences in the preferred rate or direction of FM sweeps in the IC. This is consistent with the results of Palombi et al. (2001) who found minimal age-related changes using AM stimuli. However, using SAM noise carriers, some age-related changes became apparent. For example, there were reports of an increase in response rate to SAM noise carriers, a decrease in the median upper cutoff frequency, and a decrease in latency responses (Walton et al. 2002; Simon et al. 2004).

Age-related changes in the IC of the rat include fewer cells containing GABA, a decrease in GAD activity, and changes in GABA receptors at synapses (Caspary et al. 1995). Caspary and colleagues have suggested that the changes in GABA are not due to changes in uptake or degradation of GABA but rather to the activity or amount of GAD present (Milbrandt et al. 1994; Raza et al. 1994; Caspary et al. 1999, 2002).

Medial Geniculate Nucleus (MGN): To our knowledge only one study has examined the effects of aging in MGN. In that study, Mendelson and Lui (2004) found no age-related differences in response to FM sweep rate or direction. Cells recorded from the MGN of aged rats responded best to fast FM sweeps just as they did in the young animals. Mendelson and Lui (2004) compared FM sweep responses in IC, MGN, and AC of young and old animals and found that the responses of cells in IC and MGN did not appear to be affected by aging (Fig. 23.2) and neurons in these two structures appeared to be similar to those recorded from the AC of young animals. This lends greater support to the hypothesis that, at least for temporal processing speed, aging is predominantly a cortical phenomenon.

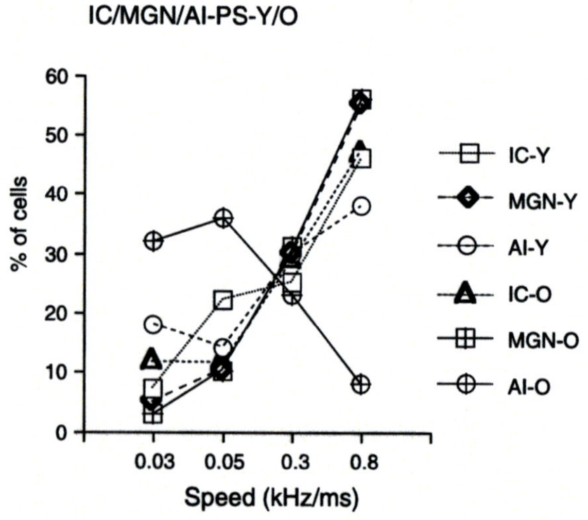

Fig. 23.2 Comparison of FM sweep speed in the inferior colliculus (IC), medial geniculate nucleus (MGN), and primary auditory cortex (AI) of young (Y) and old (O) rats. Only neurons recorded from AI were affected by aging

3.3 Neural Mechanisms

Several possible mechanisms may be at least partially responsible for some of the age-related effects observed in auditory cortex. One mechanism may be changes in dendritic and synaptic structure, abnormal axon terminals, pathological dendrites, and a reduction of dendritic spines (Feldman and Vaughan 1979; Peters and Vaughan 1981; Cha et al. 1997). In humans and rats, synaptic loss has also been associated with age-related functional changes (Jucker and Ingram 1997).

Another mechanism may be a change in calcium homeostatis (Khachaturian 1984). Ca^{2+}-binding proteins have been shown to change with age and are thought to act as protective agents against excitotoxicity (Baimbridge et al. 1992). Wang (1998) presented a model in which he showed that calcium may be involved in modulating cortical pyramidal cells' responses to time-varying inputs. Thus, it is possible that changes in calcium could be reflected in changes in neural responses to stimuli such as FM sweeps.

Finally, changes in GABA may also contribute significantly to changes in temporal processing. The loss or reduction of GABA-mediated inhibition, as well as a decrease in the level of GAD, may alter the balance between excitatory and inhibitory neurotransmitter function thereby compromising the function of neurons in AC. The changes observed in GABA and GAD could result in a down-regulation of neuro-inhibitory transmitters that in turn contribute to the degradation of processing time-varying stimuli.

3.4 Interventions

Many of the effects of aging in general may be due to the insults of oxidative stress and inflammation in the brain and much research has been devoted to studying ways in which the brain can be protected from these insults. Two such approaches involve calorie restriction and specialized diets supplemented with antioxidants. Studies in which caloric restriction has been implemented have yielded mixed results (Casadesus et al. 2002), while research using an antioxidant-enriched diet has resulted in a reversal of some cognitive and motor behaviors (Joseph et al. 1999).

Most studies have shown that a calorie-restricted diet does not affect age-related hearing loss (as assessed by ABER) or the amount of cochlear lesions observed at death. However, Seidman (2000) recorded ABERs in rats that had been placed on a calorie-reduced or antioxidant-enriched diet and found that those rats placed on a calorie-reduced diet had the most acute auditory sensitivity, the lowest quantity of mitochondrial DNA deletions (which is associated with deafness), and the least amount of outer hair cell loss. Rats placed on the antioxidant-enriched diet exhibited some benefits, but not to the same extent as their calorie-restricted cohorts.

In contrast to the results of Seidman (2000), de Rivera et al. (2006) did observe a positive effect when animals were placed on an antioxidant-enriched diet. Responses to both the speed and the direction of FM sweep recorded from aged rats on the diet were similar to those observed in young animals (Fig. 23.3a), i.e., the majority of cells responded more vigorously to the faster sweeps. In contrast, the majority of cells recorded from aged animals that were not on an antioxidant-enriched diet responded more vigorously to slower speeds (Ricketts et al. 1998; Mendelson and Ricketts 2001). For direction selectivity, there was no age difference or effect due to diet (Fig. 23.3b). These results suggest that antioxidants can play a significant role in reversing the deleterious effects of aging on temporal processing speed, at least at the level of the auditory cortex.

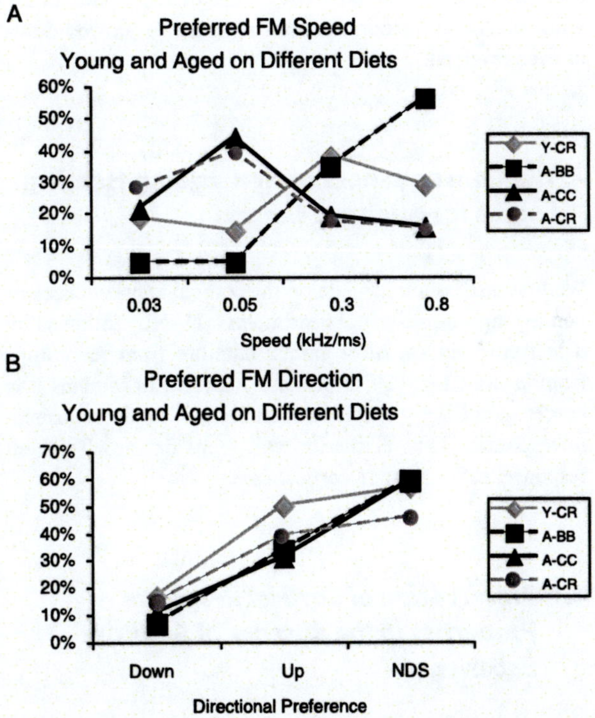

Fig. 23.3 Distribution of preferred speed (**a**) and direction (**b**) responses for young (Y) and aged animals (A). (**a**) Cells recorded from rats in the aged blueberry group (A-BB) and the young control group (Y-CR) preferred the faster sweeps. In contrast, cells recorded from old rats on a control diet (A-CC) or regular diet (A-CR) preferred slower sweeps. (**b**) There were no age or diet effects for preferred direction of FM sweep (De Rivera et al. 2006)

It has been suggested that behavioral training strategies of perception may prevent or reverse age-related decline in hearing ability (e.g.,Tremblay et al. 2003). In a recent study, the effect of intensive behavioral auditory training on properties of auditory cortex in aged rats was examined (de Villers-Sadani et al. 2010). They showed, following perceptual training, a nearly complete reversal of a majority of previously observed functional and structural cortical impairments. This suggests that age-related sensory and cognitive decline is a tightly regulated plastic process, and demonstrates that many of these age-related changes are reversible.

In conclusion, many auditory behaviors and cortical response properties are adversely affected by the aging process. In addition to the ubiquitous decrease in the frequency threshold, there appears to be a degradation in temporal processing particularly for speed of FM sweeps. GABA seems to be a primary candidate responsible for some of the temporally-based response changes observed. When compared to subcortical structures, cortical responses appear to be most affected by aging. Finally, interventions, such as a diet rich in antioxidants or behavioral sensory training, may provide means by which some of the deficits, particularly those involving temporal processing, may be slowed down or even reversed.

4 The Consequences of Non-ageing Hearing Losses for Auditory Cortex

We first discuss animal data on the cortical consequences of non-ageing sensori-neural hearing loss (NAHL) followed by data from humans. Most animal data are from recordings from layers III and IV in primary AC (AI). In humans it is harder to precisely specify which cortical fields contribute to responses. Thus discussion will focus on the effects on responses rather than on cortical area.

4.1 Preservation of Cortical Response Properties in the Absence of Auditory Experience

There are two main animal models of total deafness with no post-partum auditory experience. In the congenitally deaf white cat (CDWC) the Organ of Corti is absent by 3 weeks of age in most animals (hearing in this period is unlikely since hearing in the first few weeks of life in normal cats is at very high thresholds, at > 100 dB SPL; Brugge et al. 1988). There is a very slight loss of spiral ganglion cells (SGCs). Even with complete auditory deprivation, core AC (AI and AAF) responds well to sound from an early age. Cochlear electrical stimulation in CDWCs evokes middle-latency cortical responses (MLRs) albeit smaller than in normal-hearing electrically stimulated animals. A rudimentary cochleotopic organization is seen in evoked potential (EP) and single unit thresholds (Hartmann et al. 1997; Klinke et al. 1999, 2001). Single neuron rate-intensity and latency-intensity functions can be comparable to those in electrically stimulated normal-hearing animals. Monotonic and non-monotonic amplitude/rate intensity functions can be recorded for EPs and single neurons (Hartmann et al. 1997), as well as with cochlear electrical stimulation in normal-hearing acutely deafened animals (Raggio and Schreiner 1994), although with much poorer synchronization. There is even some rudimentary binaural sensitivity in single unit responses (Kral 2007; Kral and Eggermont 2007). Long-latency responses are totally absent (Klinke et al. 1999, 2001).

The other model is the genetically normal cat that is deafened soon after birth ("Normal early-and-chronically deafened", NECD) with ototoxic drugs that cause total destruction of hair cells and a major loss (>50%) of SGCs. Generally similar AI effects to the CDWC are seen in the NECD cat with the exception of little or no cochleotopy (>2 years post-deafening: Raggio and Schreiner 1999; 5–13 months: Fallon et al. 2009) although electrically evoked ABERs and basic AI neural responses are comparable to responses in normally hearing acutely deafened animals (Raggio and Schreiner 1994, 1999; Fallon et al. 2009); note that cochleotopy is seen in acutely deafened cats (Raggio and Schreiner 1999). Absent cochleotopy here is consistent with studies in rats that patterned auditory input before a critical period is required for formation of AI maps (Zhang et al. 2002). The contrast with the crude CDWC AI cochleotopy may reflect differences in number of surviving SGCs or criterion differences. In NECD cats, there is a threefold increase in cortex responding to a stimulus slightly above minimum cortical threshold and only monotonic rate-intensity functions are reported to cochlear stimulation in these cats and in acutely deafened cats (Fallon et al. 2009). Otherwise, the NECD model has reinforced the view derived from CDWCs: cortical neural rate-intensity functions and other properties are comparable to responses in electrically stimulated acutely deafened normal-hearing animals (Raggio and Schreiner 1994, 1999; Fallon et al. 2009) including phase-locking to low-frequency stimuli (in the NECD cat model: Raggio and Schreiner 1999; Middlebrooks et al. 2005). In both models, cortical threshold to electrical stimulation is significantly lower than in hearing controls (Kral et al. 2005; Fallon et al. 2009), and this does not reflect any sub-cortical changes (Fallon et al. 2009). The lower cortical thresholds could indicate decreased inhibition (Kral et al. 2005; Kral and Eggermont 2007) due to incomplete or delayed development of inhibitory synapses in deep layer III and layer IV (Hubka et al. 2004; Kral et al. 2005). In gerbil pups deafened at postnatal day 10 (just before onset of response to airborne sound), layer II/III cortical neurons in brain slices show changes in inhibitory inputs paralleled with changes in monosynaptic thalamic excitatory input, changes in passive membrane properties, and diminished maximum-evoked inhibitory potentials (Kotak et al. 2005, 2008). While all these results indicate a major effect due to changes in inhibition, it has been suggested that inhibition changes alone do not account for the lower thresholds but are conjunctive with changes in excitatory transmission and altered cortical microcircuitry (Kral et al. 2005; Kral and Eggermont 2007; Kral 2007). However, similar lower thresholds (with increased neuronal dynamic range) can be obtained in normally hearing adult animals exposed to electrical stimulation even after only ~2 weeks of profound deafness (Raggio and Schreiner 1999) when it is unlikely that the microcircuitry would be as profoundly altered as likely in the CDWC.

From effects in CDWC cortex it has been suggested that congenital absence of auditory experience leads to "functional decoupling" between primary and higher order cortices (Kral 2007; Kral and Eggermont 2007), with altered information flow from layer IV to supragranular layers and affecting feedback projections to subcortical structures (Kral 2007; Kral et al. 2002, 2005; Kral and Eggermont 2007). Such decoupling may allow other sensory input to take over higher order AC in congenitally absent auditory experience (Sharma et al. 2007).

Generally, even despite the absent cochleotopic organization, an early-deprived cortex appears to respond to electrical stimulation as would normal-hearing acutely deafened adults deafened for at least 2 weeks (Raggio and Schreiner 1994, 1999). However, the CDWC AI shows many processing deficits. Synaptic activity within cortical columns is significantly reduced and layer-specific activity patterns altered (Kral et al. 2000, 2001; Kral and Eggermont 2007), and deficits exist in stimulus activation of thalamo-cortico-thalamic loops. These changes have been attributed to delayed activation and maturation of upper layers relative to layer IV contributing to reduced activity in deep layers, combined with disorganized cortical microcircuitry and reduced descending higher order cortical modulatory activity (Kral 2007; Kral et al. 2005; Kral and Eggermont 2007). Thus, although CDWC AI shows early responsiveness to a cochlear implant (Kral et al. 2005), compared to acutely deafened normal-hearing animals there is a delay of almost 2 months in development of MLRs. Cortical development involves processes such as synapse elimination as well as formation of new synapses both of which critically appear to require peripheral input which is absent without a functional cochlea. It is only surprising that greater differences compared to the normal-hearing acutely deafened case are not seen after chronic absence of auditory experience, possibly due to cortical plasticity.

4.2 Reinstating "Auditory" Input After Development Without Auditory Experience

When input is reinstated through a cochlear implant (CI), both types of deafened animals learn awareness of "sound." CDWCs can be conditioned to respond to "tones" for food rewards and, with behaviorally relevant stimuli, learn to actively search for sound sources, react to voices, and be awakened by sounds (Klinke et al. 1999). Both "early-" (2–2.5 months old) or "late-" implanted (>5 mo) CDWCs can be conditioned though it is faster in the former (Kral et al. 2001, 2002). With continued behaviorally relevant electrical stimulation, many cortical changes occur including reorganization of AI microcircuitry, increasing area of cortical activation, increasing synaptic activity with stimulation duration, increasing synchronization between layers, and recruitment of deep layers (review by Kral 2007). Comparison between early- versus late-implanted versus naïve acutely activated CDWCs established a sensitive period for many effects (Kral, 2007), in which period neural activity and developmental cues interact to effect production of neurotrophic factors important for dendritic growth and synaptic formation.

Although these results suggest that chronic, behaviorally relevant, electrical cochlear stimulation allows experience-dependent maturation of basic AI neuronal responses to be restored, there are caveats about the extent to which this mirror processes in a normal animal. AC in the deaf cats is *not* normal. In normal cortex functional development involves reorganizations and refinements driven by bottom-up processes but complemented by top-down influences (Kral 2007; Kral and Eggermont 2007). In development the influence of bottom-up processes appears to diminish while those of top-down processes increase. When peripheral input is deprived during development, this process is substantially affected: higher order representations cannot be established without auditory experience and so the developmental decrease in "bottom-up"-regulated reorganizations cannot be complemented by top-down modulations and learning is compromised. Unsurprisingly, cortical plasticity in these animals is described as crude and restricted (Kral 2007; Kral and Eggermont 2007).

Chronic electrical stimulation has also been used to reinstate "auditory" experience in the NECD cat (Fallon et al. 2009) in which the absence of auditory input has little effect on basic AI neuronal responses but results in complete loss of cochleotopic organization. The only major effect of stimulation is establishment of a crude cochleotopic map. All other properties are essentially the same between chronically and acutely stimulated animals and acutely deafened and stimulated animals (Fallon et al. 2009). The effect on cochleotopy parallels the observation that chronic stimulation of a single intracochlear location in the NECD cat results in expansion of the AI representation of only that cochlear region (Dinse et al. 2003).

5 Cortical Changes Following NAHL in Adult Animals with Auditory Experience

Cortical effects of peripheral NAHL in animals with auditory experience will be classed according to grade of hearing loss (see Fig. 23.4) to reflect our belief that cortical changes depend on grade of cochlear hearing loss insofar as they influence what will be the dominant mechanistic change in auditory processing.

5.1 Changes in AI After Mild-to-Moderate NAHL

Studies of NAHL in adults show that: (a) many cortical neuronal effects mirror cochlear effects (not necessarily to the same extent or by the same mechanisms) and (b) even a mild cochlear hearing loss alters CNS inhibition, producing effects not directly predicted from the cochlear changes thus accounting for some perceptual effects. For example, changes in auditory nerve response rates alone do not account for loudness recruitment after cochlear hearing loss, suggesting that central changes in representation of level must be involved (Heinz et al. 2005).

Mild-to-moderate hearing losses from traumatic sound to awake (not anesthetized) guinea pigs reduce wide-field responses at cochlea and midbrain but enhance cortical EPs especially to broadband stimuli from levels ~50 dB > pre-trauma threshold (Popelář et al. 1987; Syka and Popelář 1982). The cortical enhancement may result from decreased inhibition (Syka et al. 1994) or changes in heat shock proteins that protect neurons from excitotoxicity (Sun et al. 2008). Enhancement of the cortical MLR is independent of whether the cochlear loss is temporary or permanent (Popelář et al. 2008) and can alter the bilateral balance in AC to input from one ear, but only at higher test frequencies of 4, 8, and 16 kHz. An initial phase of these changes may be related to cessation of [excitatory] neuronal activity while a later phase may result from reduced GABA-mediated inhibition (Popelář et al. 1994).

Recordings of LFPs, multi-units, and single neurons confirm that mild-to-moderate cochlear hearing losses cause cortical effects mirroring peripheral effects and additional effects reflecting changes in CNS inhibition. Kimura and Eggermont (1999) reported that most changes in three cortical fields post loud sound mirrored known cochlear changes, e.g., when characteristic frequency (CF; frequency of greatest sensitivity) in cortical neurons is a frequency at which a cochlear hearing loss is caused, there is a mild loss in CF sensitivity, a small CF shift to an adjacent less-affected frequency, and an increase in bandwidth of the excitatory frequency response area (EFRA). Calford et al. (1993) recorded AI neural EFRAs immediately after loud sounds that produced mainly mild cochlear losses at a frequency that was the CF of the studied neuron. In addition to mild-to-moderate CF desensitization in all neurons, about equal numbers of neurons (~40% each) showed expansion of EFRA boundaries likely reflecting reduced cortical inhibition or contraction of EFRA boundaries or total loss of responsiveness, both likely reflecting increased inhibition. The CF desensitization was linearly related to the cochlear loss and, for a similar amount of cochlear loss, CF desensitization in single AI neurons was inversely related to the neuron's initial sensitivity.

Studies by Rajan (1998, 2001, 2002, 2005) provide the most direct evidence for changes in inhibition after mild cochlear damage, showing that (i) such losses did not change the AI cochleotopic map with a mild CF desensitization in neurons whose CF matched the frequency(ies) with hearing losses; (ii) few neurons had surround inhibition outside the EFRA, but this was not due to a decrease in overall gain of inhibition since many neurons still showed inhibition from within the EFRA; (iii) there was a large decrease in number of neurons showing selectivity in frequency and level dimensions (with a corresponding decrease in number of the specific frequency-and-intensity selective EFRAs believed to be shaped by surround inhibition), a marked increase in the number of neurons responding to broadband sounds, and a marked decrease in the number showed level selectivity for pure tone (CF) and for narrow band sounds.

Loss of neuronal selectivity in frequency and level dimensions is consistent with the role of surround inhibition in shaping neuronal selectivity for narrow and broadband stimuli. This has led to predictions (Rajan 2005) that such neuronal changes cause loss in perceptual selectivity for sounds in background noise. Our unpublished human studies with chronic mild hearing losses show, consistent with these expectations, that mild hearing losses (of $\leq$ 10 dB worse than the normal audiometric range) do not affect factors such as loudness minimum comfort levels or speech discrimination per se, but did significantly reduce discrimination of speech in noise.

Mild-to-moderate unilateral or asymmetric hearing loss in squirrel monkeys evolves from initial gross divergence of

Fig. 23.4 Summary of the effects of cochlear non-ageing hearing loss (NAHL) on cortical responses in adult animals with hearing losses acquired in adulthood. EFRA = excitatory frequency response area. "?" indicates that the effects are unknown

Degree of cochlear hearing loss	Acute effects on cortical neural responses/processes	Chronic effects on cortical neural responses/processes
Mild to low-moderate hearing loss (up to 30 dB cochlear hearing loss)	Small desensitization of cochlear outflow results in small CF desensitization when CF is at a frequency with cochlear loss. Additionally, dramatic changes in excitatory and inhibitory balance in cortical neurons. Idiosyncratic contractions or exapnsions in EFRA because of neuron-specific changes in excitatory and inhibitory balance. Other changes unknown.	Small CF desensitization when CF is at a frequency with cochlear loss. Loss of surround inhibition but not a general loss of inhibition in cortical neurons. Large reduction in number of EFRAs shaped by inhibition, in non-monotonic tone and noise level functions, and in neurons unresponsive to noise. No change in cochleotopic map in A1.
High-moderate to moderate/severe hearing loss (up to about 50-65 dB cochlear hearing loss)	?	Large desensitization of cochlear outflow with major changes in cochlear neuronal bandwidth and in relative contribution of different classes of cochlear neurons to coding of level. Cortical neuronal changes appear to quite faithfully reflect the peripheral changes; changes in inhibition are subsumed in this dominant peripheral effect. Any changes in cochleotopic map in A1 likely to reflect residue of pre-existing inputs.
Severe to profound hearing loss (accompanied by the presence of a cochlear region with large or total deprivation of outflow)	? *Note: Currently-reported effects complicated by non-specific cochlear changes as well as cochlear recovery from deafening procedure*	Plasticity of the cochleotopic map in A1. In cortical regions deprived of their normal CF input, a larger-than-normal number of inputs are expressed at supra-threshold levels (likely reflecting a decrease in strength of inhibition), with a single cochlear "lesion-edge" frequency becoming the new CF. This results in an increase in supra-threshold bandwidth in EFRAs. Other neuronal responses such as latency not different from normal.

auditory cortical inputs from the two ears to near convergence over a 6-month recovery period (Cheung et al. 2009). A large interaural frequency misalignment of >1 octaves at 6 weeks in cortical neurons after overstimulation decreases substantially to ~0.6 octave at 24 weeks. Interaural cortical threshold misalignment faithfully reflects peripheral asymmetric hearing loss at 6 and 12 weeks. However, AI threshold misalignment between inputs from the two ears essentially disappears at 24 weeks, primarily because the cortical thresholds from the normal ear have become unexpectedly elevated relative to peripheral thresholds and match again the cortical threshold from the impaired ear. The findings demonstrate that plastic change in central processing of sound stimuli arriving from the nominally better hearing ear may account for progressive realignment of both interaural frequency and threshold maps (Cheung et al. 2009).

5.2 Changes in AI After High-Moderate and Moderate-to-Severe NAHL

The effects of high-moderate and moderate-severe hearing loss on cortical neurons have been examined after a 1-h long 5- or 6-kHz loud sound that caused asymptotic ABER losses

(Noreña et al. 2003). Examples of EFRAs (apparently in the period before losses had asymptoted) showed changes similar to the effects of altered inhibition reported by Calford et al. (1993). However, when threshold losses had asymptoted, the dominant effect mirrored cochlear changes. These effects were segregated according to neural CF relative to trauma frequency as being below exposure frequency (suffering mild asymptotic cochlear NAHL), at the exposure frequency and up to 1 octave above exposure frequency (the frequency range with increasing cochlear losses up to ~50 dB at 10–12 kHz), and more than 1 octave above exposure frequency (flat loss ~55 dB). Group cortical effects appeared well to reflect the amount of cochlear NAHL and consequent expected change in cochlear outflow. Cortical neural CF threshold desensitization increased systematically across the three bands of increasing ABER losses. There was little evidence for any significant CF or best frequency (BF; evoking the strongest responses) shifts except in the third CF/BF band containing the higher frequencies with the highest ABER losses; then the CF/BF shifts were predominantly to lower frequencies. EFRA bandwidths increased only in the higher frequency bands with moderate and moderate/severe ABER hearing losses. Critically for the notion of changes in inhibition, there was a change in monotonicity of rate-level functions for CFs in which only mild ABER losses were caused.

Thus, the effects of cochlear hearing losses depend on the relationship between AI neural CF and frequency of the traumatic loud sound and this relationship appears very well to be explained by amount of frequency-specific hearing loss. For cochlear losses in the high-moderate to moderate-severe range (~50–65 dB losses at the cochlea), changes in cortical neural coding appeared to mirror quite faithfully changes in cochlear neural responses.

5.3 Changes in AI After Severe and Profound NAHL

Adult AC can show cochleotopic map plasticity after damage that produces severe-to-profound hearing losses from a selected region of the cochlea. The effects on cortical responses of such cochlear losses are detailed in three major reports (Robertson and Irvine 1989; Rajan et al. 1993; Rajan and Irvine 1998) and are only briefly summarized below.

Mechanical lesions to the basal turn basilar membrane have produced a restricted "notch"-type severe/profound hearing loss with normal hearing on either side of this notch, or a broad high-frequency "plateau" loss with total desensitization at high frequencies; in both cases some region of cortex was deprived of normal CF input. Months after the lesion, the deprived AI region contained neurons having a CF at a frequency at the edge of the cochlear hearing-loss-range (a "lesion-edge" frequency mapped in the cochlear region most immediately adjacent to the cochlear lesion). A critical feature establishing that the remapping was plasticity was that neuronal threshold and latency at the new CF(s) in the remapped areas were comparable to CF thresholds in neurons from normal animals with that CF or, if elevated by within 20 dB of normal, reflected an elevation in cochlear sensitivity. Response latency at the new CF inputs was also normal. However, neuronal multi-unit EFRA bandwidth in remapped AI was broader than normal, and this increased with distance from normal AI into the remapped area. This suggested that although a single input became established as the most sensitive (CF) input, the remapped AI received new inputs emanating from a number of normal cochlear regions, not just from the cochlear lesion-edge region. Bandwidths were nearly normal in remapped AI just adjacent to normal AI but increased with increasing distance away (Rajan and Irvine 1998) suggesting that across the expression of a number of new inputs across the entire remapped area but that in cortex closest to normal AI, surround inhibition may have been re-established to narrow the EFRA bandwidths but its strength relative to the new excitation may have decreased with distance into the remapped area.

Plasticity in AI after cochlear lesions in adulthood occurred only for cochlear hearing losses > 20 dB and only when there were very large losses over some cochlear region and a relatively steep slope of hearing sensitivity from the region of normal or near-normal hearing sensitivity to the region of loss. Temporary cochlear desensitization produced by fluid drainage or other non-specific effects, and which resolved over time, did not result in AI map plasticity even if the neural outflow was as desensitized (at least temporarily) as in the case of the lesions, nor did conductive hearing losses due to non-cochlear effects. Plasticity also did not occur immediately after the cochlear lesions. These effects indicate that the lesion had to result in permanent removal of afferent outflow from some part of the cochlea and plasticity required gradual emergent changes in CNS processing. However, it is unclear if the latter was due to CNS requirements or if the need to resolve non-specific cochlear effects of the lesion (which results in temporary hearing losses over a larger range of frequencies) will finally show permanent hearing losses from damage to the hair cells or afferent neurons (see also Su et al. 2008). Such non-specific effects could temporarily mask any immediate plasticity and an appropriate time course study has yet to be done to resolve this issue. Note that these are conditions required for plasticity in cortex when manipulations are carried out in adulthood to produce cochlear damage and some conditions appear to be less stringent when manipulations are carried out in early life.

Finally, profound steeply sloping hearing losses can also result in changes in neurons sensitive to lower frequencies with normal hearing sensitivities, with changes in temporal coding (of gap duration) being reported in AC (Yin et al. 2008).

6 Changes in AI After NAHL in Younger Animals with Auditory Experience

6.1 Changes in AI After Mild-to-Moderate NAHL

Cats exposed to loud sound as juveniles aged 5–7 weeks (Eggermont and Komiya 2000; Seki and Eggermont 2002) or at ~17 weeks (Seki and Eggermont 2002), and tested 2–5 months post-trauma as adults have shown a gradually sloping cochlear hearing loss of 30–50 dB for frequencies > 5 or 6 kHz, with an average of 30 dB hearing loss for frequencies above the trauma tone frequency. Peripheral threshold loss > 20–25 dB in juveniles resulted in cortical map plasticity (Seki and Eggermont 2002) of the type in adults, with thresholds at the new CF(s) similar to normal adult thresholds at those CF(s) (Eggermont and Komiya 2000) or elevated in parallel with cochlear hearing loss at comparable frequencies (Seki and Eggermont 2002). An important difference between the two age groups is that in the adults there was total loss over some cochlear region but in juveniles the loss reached a plateau of average 30 dB loss beyond the trauma frequency; note that in the adult case it was postulated that plasticity would only be evoked when there was total loss over some cochlear region bordering a normal-hearing region (Rajan and Irvine 1998). Another important difference is that in juveniles, bandwidths of multi-neuronal EFRAs (and of inhibitory FRAs) were not different in regions of AI plasticity (Eggermont and Komiya 2000), whereas in adult plasticity EFRAs were broadened. The correlated changes in CF threshold and bandwidth could indicate that the juvenile map changes did not reflect plasticity as in adult-deafened animals but were the residue of pre-existing inputs in neurons deprived of CF inputs by cochlear damage (see Rajan and Irvine 1998; Robertson and Irvine 1989) but this has been argued against Seki and Eggermont (2002). Alternatively, there are studies supporting the hypothesis that the bandwidth difference could indicate that juvenile but not adult plasticity re-established the cortical inhibition required to produce normal excitatory tuning curves (Seki and Eggermont 2002).

In AI map plasticity there is an increase in neural spontaneous rates (SR) (Eggermont and Komiya 2000) possibly from the occurrence of cochlear NAHL or from the noise trauma, not from plasticity itself (Seki and Eggermont 2002). Increases in AI neural SR only occurred from a few hours post-exposure (Noreña and Eggermont 2003; Eggermont and Roberts 2004) unlike tinnitus which is often experienced immediately after sound trauma. A more likely substrate candidate for tinnitus is increase in synchronization of SR of several neurons which is elevated for neurons in affected frequency regions immediately after noise trauma (Noreña and Eggermont 2003) as after application of quinine (Ochi and Eggermont 1997) which also causes tinnitus in humans. Synchrony in affected frequency regions also increases with time (Noreña and Eggermont 2003) and is confined to reorganized AI. Eggermont and Roberts (2004) also note that tinnitus percept ratings appear to be constrained to this frequency region.

Minimum latency of CF neuronal responses (Eggermont and Komiya 2000) increases after noise trauma in juvenile animals but not in adult plasticity (Rajan and Irvine 1998). In juveniles post-trauma changes in cortical signaling of timing occur after even mild cochlear NAHL (which does not produce cortical map plasticity, even in juveniles): such juvenile NAHL affected neuronal ability to signal minimum voice onset time (VOT) for phonemic stimuli and minimum gap duration of noise bursts (Aizawa and Eggermont 2006), due to either decreased inhibition or decreased adaptation. The issue has been raised above that a complication of studies of cortical effects of manipulations in 5–7 week old kittens is how trauma-induced changes interacted with developmental changes. Cats aged 5–7 weeks have mature hearing thresholds and cortical response latencies, but other cortical responses, such as frequency-tuning curve bandwidth and duration of post-activation suppression, remain immature up to 100–120 days of age (Eggermont 1996), as do cortical frequency place maps (Bonham et al. 2004) particularly in the area representing units with CFs between 3 and 15 kHz (Bonham et al. 2004). Spontaneous firing rates take even longer to mature. This consideration raises the question as to whether the reported effects are the outcomes of effects of noise-induced cochlear damage on cortical neural properties or to interference with development processes, an issue that remains unresolved.

6.2 Changes in AI After Severe and Profound NAHL

Limited work has been done on effects of severe-to-profound hearing losses in young animals with some degree of auditory experience (modeling post-lingually deafened children). Harrison and colleagues (1991, 1992; 1993) examined only a cochleotopic map (i.e., no other response properties) in neonatal cats (exact age not specified) with severe-to-profound hearing losses in extensive regions of the middle-

and high-frequency cochlear regions from application of ototoxic drugs. There was plasticity of the cortical cochleotopic map in AI and the adjacent anterior auditory field as seen in adult animals. These effects were also compared to effects reported by Dinse et al. (2003) who found, in neonatally deafened cats, that chronic stimulation initiated at ~2 months of age and restricted to a single intracochlear location caused expansion of the representation of that cochlear region in AI. Plasticity in AI following unilateral hearing losses in early development after the onset of hearing but before the full maturation of the auditory pathway has also been suggested by the absence of a decrease in 2-deoxyglucose uptake to tone burst stimuli to the intact ear (Hutson et al. 2008) but, as noted above, such studies are complicated by the intersection of the maturation of developmental processes and any plasticity evoked by the peripheral manipulation.

7 Effects of Hearing Losses on Human Cortical Responses

7.1 *Effects in Users of Cochlear Implants*

Interpretation of cortical activity imaged in CI users is complicated by the fact that the activity is dependent on a host of factors such as duration of CI use, age of implantation, electrode configuration, duration of deafness, auditory stimulus type, and task (Green et al. 2005; Herzog et al. 1991; Hirano et al. 1997; Ito et al. 1993; Kim et al. 1997; Lee et al. 2007; Naito et al. 1995, 1997; Okazawa et al. 1996; Suarez et al. 1999; Mortensen et al. 2005). With these caveats, these studies have shown that AC in profoundly deaf users exhibits low activity negatively correlated with deafness duration, and activity increases with duration of CI use that is generally greater in cortex contralateral to the implanted cochlea (see Fallon et al. 2008). It must be noted that many imaging studies used positron emission technology that has limitations in relating imaged activation to functional use of a brain area (Truy et al. 1995).

Electrophysiological studies suggest that while AC plasticity can occur in CI children, a critical period influences type of activity recorded and loci from which waveforms originate. Stimulation through the CI in children gradually causes an increase in amplitude and decrease in latency of wave P1 and appearance of wave N1 of long-latency components of cortical ERPs (Sharma et al. 2002, 2005), the latter in prelingually deaf children implanted before the age of 3.5 years but not in children implanted after 7 years (Ponton and Eggermont 2001; Sharma and Dorman 2006). This fits with observations that long-term CI stimulation in post-lingually deaf and in early-implanted prelingually deaf people activated primary and higher order cortex (Naito et al. 1997, Nishimura et al. 1999) and that higher order areas were recruited significantly less by similar long-term auditory experience in late-implanted prelingually deaf people (Naito et al. 1997). Note the interesting parallels with the CDWC animal model where, depending on stimulation duration (i.e., experience), early-implanted cats could show a larger activated A1 area and a shorter latency of the first positive wave. More recently, Gilley et al. (2008) reported that, in response to a /ba/ stimulus, normal-hearing and early-implanted CI children (aged <~4) showed electrical activity emanating from AC, while late-implanted children (aged >~5) showed activity emanating from parieto-temporal cortex. This suggests that if implantation is delayed there is reorganization of auditory pathways. This may account for observations of more successful speech perception in early-implanted children. Sharma et al. (2007) have suggested that the Gilley et al. (2008) results are consistent with the hypothesis (Kral et al. 2005) of a "functional disconnection" between primary and higher order cortex at the end of a sensitive period where the absence of auditory cortical activity in late-implanted children suggests absent or weak connections between primary and association areas and weak feedback activity to thalamic areas.

In contrast to these results are findings in CI children that, regardless of implantation age, the likelihood of detecting the MLR and its amplitude increased with duration of CI use, and immediately post-implantation was actually two times more likely to be found in the oldest than the youngest implant-age group (8–17 years versus 0–2.9 years of age; Gordon et al. 2005). Implantation time had no effect except between the oldest implant-time age group and all others, whereas duration of use had a major effect across all groups. Differences in age-of-implantation effects on middle- and long-latency responses indicate that re-introduction of activity through the CI interacts with complex differential effects of development on different generators of cortical-evoked potentials. Two recent studies (Davids et al. 2008a, b) suggest that, at least in children with ~1 year of CI use, electrophysiological measures of CNS function including from cortex are little affected by the age of deafness onset, the time of implantation, and duration of use. Mortensen et al. (2005) have shown that exactly which sites will be activated in post-lingually implanted CI users depends significantly on the task used to examine cortical activity: stimulation with tone bursts leads to activation "close" to the primary AC bilaterally. This study suggests that the nature and complexity of the task play a significant role in determining exactly what areas of cortex will be activated.

Finally, an unresolved issue beyond our scope is whether, in the absence of auditory experience, the AC is taken over by other sensory inputs. We note only that, in the animal studies where takeover of one sensory cortex by another sense is seen, it requires removal of peripheral input but also of destruction of specific thalamo-cortical pathways and this is not yet been demonstrated to be the case in any auditory model of congenital deafness.

7.2 Hearing Loss in Adults with Auditory Experience

Plasticity in human AC has been reported from the effects of a profound unilateral acquired hearing loss on ipsilateral–contralateral differences, like those examined in guinea pigs (Popelář et al. 1994). A profound unilateral hearing loss in humans reduces differences (Khosla et al. 2003; Moore et al. 2005; Ponton et al. 2001; Vasama and Makela 1995, 1997) similar to those observed in animals (Popelář et al. 1994)—but either only when the damaged ear was the left ear (Khosla et al. 2003) or being more prominent when that was the damaged ear (Moore et al. 2005) and then also affecting the normal left cortex dominance to tonal stimuli (Moore et al. 2005).

Plasticity of the AI cochleotopic map with expanded representation of lesion-edge frequencies has been reported in an MEG study (Dietrich et al. 2001) in humans with a profound steeply-sloping high-frequency hearing loss like that which produces AI map reorganization in animals (Rajan and Irvine 1998). The study showed that frequency discrimination was altered (Thai-Van et al. 2007) in a way consistent with predictions from plasticity in animals. Other evidence for plasticity is the recovery of responses over a long period after unilateral deafness caused by acoustic neuroma removal (see Vasama and Makela 1997). However, evidence for plasticity is not found in other studies of cortical ERPs, using MLRs to tones or clicks (e.g., Museik et al. 1984) or short-latency scalp potentials to sinusoidal amplitude-modulated tones (Kuwada et al. 1986). These studies report that thresholds and/or amplitudes of responses faithfully reflect cochlear sensitivity, even in cases where the audiogram was of the form (e.g., two of the high-frequency hearing loss cases in Kuwada et al. 1986) that appeared to evoke plasticity in the MEG study (Dietrich et al. 2001) as in the animal studies. A complication in interpreting evoked responses in hearing-impaired humans relative to effects seen in recordings in cortical neurons in animals is that the generators of these auditory-evoked responses may change after hearing damage, at least after unilateral sensori-neural hearing loss. This may be further exacerbated when the deafness has occurred in childhood (see Vasama and Makela 1997).

In conclusion, there is diversity in almost every feature of studies of cortical effects of cochlear NAHL. This is complicated by the fact that cochlear NAHL can evoke cortical as well as subcortical plasticity (Robertson and Irvine 1989; Frisina and Rajan 2005) but only for moderate or larger hearing losses. Importantly, this can be reversed (Noreña and Eggermont 2005). It is therefore difficult to always ascertain whether cortical responses post peripheral trauma reflect responses seen with induction of plasticity, interruption of normal developmental processes (when manipulations are done in early life), or simply reflect relayed effects (with sub-cortical transforms) of cochlear changes.

References

Alain C, McDonald KL, Ostroff JM, and Schneider B (2001) Age-related changes in detecting a mistuned harmonic. *Journal of the Acoustical Society of America* **109**:2211–2216.

Alain C and McDonald L (2007) Age-related differences in neuromagnetic brain activity underlying concurrent sound perception. *Journal of Neuroscience* **27**:1308–1314.

Anderson B and Rutledge V (1996) Age and hemisphere effects on dendritic structure. *Brain* **119**:1983–1990.

Aizawa N and Eggermont JJ (2006) Effects of noise-induced hearing loss at young age on voice onset time and gap-in-noise representations in adult cat primary auditory cortex. *Journal of the Association for Otolaryngology* **7**:71–81.

Baimbridge KG, Celio MR, and Roger JH (1992) Calcium-binding proteins in the nervous system. *Trends in Neuroscience* **15**:303–307.

Banay-Schwartz M, Lajitha A, and Palkovits M (1989) Changes with aging in the levels of amino acids in rat CNS structural elements: 1. Glutamate and related amino acids. *Neurochemistry Research* **14**:555–562.

Barsz K, Ison JR, Snell KB, and Walton JP (2002) Behavioral and neural measures of auditory temporal acuity in aging humans and mice. *Neurobiology of Aging* **23**:565–578.

Bellis TJ, Nicol T, and Kraus N (2000) Aging affects hemispheric asymmetry in the neural representation of speech sounds. *Journal of Neuroscience* **20**:791–797.

Bennett CL, Davis RT, and Miller JM (1983) Demonstration of presbycusis across repeated measures in a nonhuman primate species. *Behavioral Neuroscience* **97**:602–607.

Bertoli S, Smurzynski J, and Probst R (2002) Temporal resolution in young and elderly subjects as measured by mismatch negativity and a psychoacoustic gap detection task. *Clinical Neurophysiology* **113**:396–406.

Boettcher FA, Poth EA, Mills JH, and Dubno JR (2001) The amplitude-modulation following response in young and aged human subjects. *Hearing Research* **153**:32–42.

Boettcher FA, Madhorta D, Poth EA, and Mills JH (2002) The frequency-modulation following response in young and aged human subjects. *Hearing Research* **165**:10–18.

Bonham BH, Cheung SW, Godey B, and Schreiner CE (2004) Spatial organization of frequency response areas and rate/level functions in the developing AI. *Journal of Neurophysiology* **91**:841–854.

Brown CH (1984) Directional hearing in aging rats. *Experimental Aging Research* **10**:35–38.

Brugge JF, Reale RA, and Wilson GF (1988) Sensitivity of auditory cortical neurons of kittens to monaural and binaural high frequency sound. *Hearing Research* **34**:127–140.

Calford MB, Rajan R, and Irvine DRF (1993) Rapid changes in the frequency tuning of neurons in cat auditory cortex resulting from pure-tone-induced temporary threshold shift. *Neuroscience* **55**: 953–964.

Casadesus G, Shukitt-Hale B, and Joseph JA (2002) Qualitative versus quantitative caloric restriction: are they equivalent paths to successful aging? *Neurobiology of Aging* **23**:747–769.
Caspary DM, Milbrandt JC, and Helfert RH (1995) Central auditory aging: GABA changes in the inferior colliculus. *Experimental Gerontology* **30**:349–360.
Caspary DM, Holder TM, Hughes LF, Milbrandt JC, McKernan RM, and Naritoku DK (1999) Age-related changes in $GABA_A$ receptor subunit composition and function in rat auditory system. *Neuroscience* **93**:307–312.
Caspary DM, Palombi PS, and Hughes LF (2002) GABAergic inputs shape responses to amplitude-modulated stimuli in the inferior colliculus. *Hearing Research* **168**:163–173.
Caspary DM, Schatterman TA, and Hughes LF (2005) Age-related changes in the inhibitory response properties of dorsal cochlear nucleus output neurons: Role of inhibitory inputs. *Journal of Neuroscience* **25**:10952–10959.
Caspary DM, Ling L, Turner JG, and Hughes LF (2008) Inhibitory neurotransmission, plasticity and aging in the mammalian central auditory system. *The Journal of Experimental Biology* **211**: 1781–1791.
Cha CI, Lee YI, Lee EY, Park KH, and Baik SH (1997) Age-related changes of VIP, NPY and somatostatin-immunoreactive neurons in the cerebral cortex of aged rats. *Brain Research* **753**: 235–244.
Cheung SW, Bonham BH, Schreiner CE, Godey B, and Copenhaver DA (2009) Realignment of interaural cortical maps in asymmetric hearing loss. *Journal of Neuroscience* **29**:7065–7078.
Cobb FE, Jacobson GP, Newman CW, Kretschmer LW, and Donnelly KA (1993) Age-associated degeneration of backward masking task performance: evidence of declining temporal resolution abilities in normal listeners. *Audiology* **32**:260–271.
Cooper WA, Coleman JR, and Newton EH (1990) Auditory brainstem responses to tonal stimuli in young and aging rat. *Hearing Research* **43**:171–180.
Cranford JL, Boose M, and Moore CA (1990) Effects of aging on the precedence effect in sound localization. *Journal of Speech and Hearing Research* **33**:654–659.
Davids T, Valero J, Papsin BC, Harrison RV, and Gordon KA (2008a) Effects of stimulus manipulation on electrophysiological responses in pediatric cochlear implant users. Part I: duration effects. *Hearing Research* **244**:7–14.
Davids T, Valero J, Papsin BC, Harrison RV, and Gordon KA (2008b) Effects of stimulus manipulation on electrophysiological responses of pediatric cochlear implant users. Part II: Rate effects. *Hearing Research* **244**:15–24.
De Rivera C, Shukitt-Hale B, Joseph JA, and Mendelson JR (2006) The effects of an antioxidant-enriched diet on temporal processing speed in the senescent auditory cortex. *Neurobiology of Aging* **27**: 1035–1044.
de Villers-Sadani E, Alzghoul L, Zhou X, Simpson KL, Lin RC, and Merzenich MM (2010) Recovery of functional and structural age-related changes in the rat primary auditory cortex with operant training. Proceedings of the National Academy of Sciences of the United States of America **107**:13900–13905.
Dietrich V, Nieschalk M, Stoll S, Rajan R, and Pantev C (2001) Cortical reorganization in patients with high frequency cochlear hearing loss. *Hearing Research* **158**:95–101.
Dimitrijevic A, John MS, and Picton TW (2004) Auditory steady-state responses and word recognition scores in normal-hearing and hearing-impaired adults. *Ear and Hearing* **25**:68–84.
Dinse HR, Godde B, Reuter G, Cords SM, and Hilger T (2003) Auditory cortical plasticity under operation: reorganization of auditory cortex induced by electrical cochlear stimulation reveals adaptation to altered sensory input statistics. *Speech Communication* **41**: 201–219.
Dunn N (1983) Effects of age upon auditory evoked potentials from the inferior colliculus and cortex in the guinea pig. *Archives of Oto-Rhino-Laryngology* **238**:251–261.
Eggermont JJ (1996) Differential maturation rates for response parameters in cat primary auditory cortex. *Auditory Neuroscience* **2**: 309–327.
Eggermont JJ and Komiya H (2000) Moderate noise trauma in juvenile cats results in profound cortical topographic map changes in adulthood. *Hearing Research* **142**:89–101.
Eggermont JJ and Roberts LE (2004) The neuroscience of tinnitus. *Trends in Neurosciences* **27**:676–682.
Fallon JB, Irvine DRF, and Shepherd RB (2008) Cochlear implants and brain plasticity. *Hearing Research* **238**:110–117.
Fallon JB, Irvine DRF, and Shepherd RB (2009) Cochlear implant use following neonatal deafness influences the cochleotopic organization of the primary auditory cortex in cats. *The Journal of Comparative Neurology* **512**:101–114.
Feldman ML and Vaughan DW (1979) Changes in the auditory pathway with age. In: Hans SS and Coons DH (eds). *Special Senses in Aging*. Institute of Gerontology, Ann Arbor, pp. 143–162.
Finlayson PG and Caspary DM (1993) Response properties in young and old Fischer-344 rat lateral superior olive neurons: a quantitative approach. *Neurobiology of Aging* **14**:127–139.
Fitzgibbons PJ and Gordon-Salant S (1994) Age effects on measures of auditory duration discrimination. *Journal of Speech and Hearing Research* **37**:662–670.
Frisina RD and Rajan R (2005) Inferior colliculus: aging and plasticity. In: Winer JA and Schreiner CE (eds). *The Inferior Colliculus*. Springer-Verlag, New York, pp. 559–584.
Gilley PM, Sharma A, and Dorman MF (2008) Cortical reorganization in children with cochlear implants. *Brain Research* **1239**: 566–565.
Gordon KA, Papsin BC, and Harrison RV (2005) Effects of cochlear implant use on the electrically evoked middle latency response in children. *Hearing Research* **204**:78–89.
Grady CL, Yu H, and Alain C (2008) Age-related differences in brain activity underlying working memory for spatial and nonspatial auditory information. *Cerebral Cortex* **18**:189–199.
Green KMJ, Julyan PJ, Hastings DL, and Ramsden RT (2005) Auditory cortical activation and speech perception in cochlear implant users: effects of implant experience and duration of deafness. *Hearing Research* **205**:184–192.
Grube M, von Crammon DY, and Rubsamen R (2003) Inharmonicity detection. Effects of age and contralateral distractor sounds. *Experimental Brain Research* **153**:637–642.
Harkrider AW, Plyler PN, and Hedrick MA. (2005) Effects of age and spectral shaping on perceptual and neural representation of stop consonant stimuli. *Clinical Neurophysiology* **116**: 2153–2164.
Harris KC, Mills JH, and Dubno JR (2007) Electrophysiologic correlates of intensity discrimination in cortical evoked potentials of younger and older adults. *Hearing Research* **228**:58–68.
Harrison RV, Nagasawa A, Smith DW, Stanton SG, and Mount RJ (1991) Reorganization of auditory cortex after neonatal high frequency cochlear hearing loss. *Hearing Research* **54**:11–19.
Harrison RV, Smith DW, Nagasawa A, Stanton SG, and Mount RJ (1992) Developmental plasticity of AC in cochlear haring loss: physiological and psychophysical findings. In: Cazals Y, Horner K, and Demany L (eds). *Auditory Physiology and Perception*. Pergamon Press, Oxford, pp. 625–633.
Harrison RV, Stanton SG, Ibrahim D, Nagasawa A, and Mount RJ (1993) Effects of chronic cochlear damage on threshold and frequency tuning of neurons in A1 auditory cortex. *Acta Otolaryngology* **519**:30–35.
Hartmann R, Shepherd RK, Heid S, and Klinke R (1997) Response of the primary auditory cortex to electrical stimulation of the auditory

nerve in the congenitally deaf white cat. *Hearing Research* **112**: 115–133.

He N, Dubno JR, and Mills JH (1998) Frequency and intensity discrimination measured in a maximum-likelihood procedure from young and aged normal-hearing subjects. *Journal of the Acoustical Society of America* **103**:553–568.

Heffner RS, Koay G, and Heffner HE (2001) Sound-localization acuity changes with age in C57BL/6 J mice. In: Willott JF (ed). *Handbook of Mouse Auditory Research: From Behavior to Molecular Biology*. CRC Press, Boca Raton, pp. 31–35.

Heinz MG, Issa JB, and Young ED (2005) Auditory-nerve rate responses are inconsistent with common hypotheses for the neural correlates of loudness recruitment. *Journal of the Association for Otolaryngology* **6**:91–105.

Herzog H, Lamprecht A, Kuhn A, Roden W, Vosteen KH, and Feinendegen LE (1991) Cortical activation in profoundly deaf patients during cochlear implant stimulation demonstrated by $H_2(15)O$ PET. *Journal of Computer Assisted Tomography* **15**: 369–375.

Hirano S, Naito Y, Okazawa H, Kojima H, Honjo I, Ishizu K, Yenokura Y, Nagahama Y, Fukuyama H, and Konishi J (1997) Cortical activation by monaural speech sound stimulation demonstrated by positron emission tomography. *Experimental Brain Research* **113**:75–80.

Hiraumi H, Nagamine T, Morita T, Naito Y, Fukuyama H, and Ito J (2005) Right hemispheric predominance in the segregation of mistuned partials. *European Journal of Neuroscience* **22**:1821–1824.

Hubka P, Kral A, and Klinke R (2004) Input desynchronization and impaired columnar activation in deprived AC revealed by independent component analysis. In: Syka J and Merzenich MM (eds). *Plasticity and signal representation in the auditory system*. Springer, Berlin, pp. 161–165.

Hutson K, Durham D, Imig T, and Tucci DL (2008) Consequences of unilateral hearing loss: Cortical adjustment to unilateral deprivation. *Hearing Research* **237**:19–31.

Ito J, Sakakibara J, Iwasaki Y, and Yonekura Y (1993) Positron emission tomography of auditory sensation in deaf patients and patients with cochlear implants. *Annals of Otology Rhinology and Laryngology* **102**:797–801.

Joseph JA, Shukitt-Hale B, Denisova NA, Bielinski D, Martin A, McEwen JJ, and Bickford PC (1999) Reversals of age-related declines in neuronal signal transduction, cognitive, and motor behavioral deficits with blueberry, spinach, or strawberry dietary supplementation. *The Journal of Neuroscience* **19**:8114–8121.

Juarez-Salinas DL, Nyon S, Engle J, Navarro X, and Recanzone GH (2008) Spatial tuning of auditory cortical neurons is degraded in aged macaque monkeys. In: *Neuroscience Meeting Planner*. Society for Neuroscience, Washington, 566.24/JJ8.

Jucker M and Ingram DK (1997) Murine models of brain aging and age-related neurodegenerative diseases. *Behavioral and Brain Research* **85**:1–25.

Keithley EM, Ryan AF, and Feldman ML (1992) Cochlear degeneration in aged rats of four strains. *Hearing Research* **59**:171–178.

Khachaturian ZS (1984) Towards theories of brain aging. In: Kay DW and Burrows GD (eds). *Handbook of Studies on Psychiatry and Old Age*. Elsevier, Amsterdam, pp. 139–165.

Khosla D, Ponton CW, Eggermont JJ, Kwong B, Don M, and Vasama J-P (2003) Differential ear effects of profound unilateral deafness on the adult human central auditory system. *Journal of the Association for Otolaryngology* **4**:235–249.

Kim CS, Oh SH, and Byun SW (1997) Auditory cortical activity in cochlear implantees before and after sound stimulation using positron emission tomography. *Advances in Otorhinolaryngology* **52**:24–26.

Kimura M and Eggermont JJ (1999) Effects of pure tone induced hearing losses on response properties in three auditory cortical field in cat. *Hearing Research* **135**:146–162.

Klinke R, Kral A, Heid S, Tillein J, and Hartmann R (1999) Recruitment of the auditory cortex in chronically deaf cats by long-term electrostimulation through a cochlear implant. *Science* **285**: 1729–1733.

Klinke R, Hartmann R, Heid S, Tillein J, and Kral A (2001) Plastic changes in the auditory cortex of congenitally deaf cats following cochlear implantation. *Audiology & Neuro-Otology* **6**:203–206.

Konkle DF, Beaslet DS, and Bess FH (1977) Intelligibility of time-altered speech in relation to chronological aging. *Journal of Speech and Hearing Research* **20**:108–115.

Kotak VC, Fujisawa S, Lee FA, Karthikeyan O, Aoki C, Sanes DH. 2005. Hearing loss raises excitability in the auditory cortex. *The Journal of Neuroscience* **25**:3908–3918.

Kotak VC, Takesian AE, Sanes DH (2008) Hearing loss prevents the maturation of GABAergic transmission in the auditory cortex. *Cerebral Cortex* **18**:2098–2108.

Kral A (2007) Unimodal and cross-modal plasticity in the 'deaf' auditory cortex. *International Journal of Audiology* **46**:479–493.

Kral A and Eggermont JJ (2007) What's to lose and what's to learn: development under auditory deprivation, cochlear implants and limits of cortical plasticity. *Brain Research Reviews* **56**:259–269.

Kral A, Hartmann R, Tillein J, Heid S, and Klinke R (2000) Congenital auditory deprivation reduces synaptic activity within the auditory cortex in a layer-specific manner. *Cereb Cortex* **10**:714–726.

Kral A, Hartmann R, Tillein J, Heid S, and Klinke R (2001) Delayed maturation and sensitive periods in the auditory cortex. *Audiology & Neuro-Otology.* **6**:346–362.

Kral A, Hartman R, Tillein J, Heid S, and Klinke R (2002) Hearing after congenital deafness. *Cerebral Cortex* **12**:797–807.

Kral A, Tillein J, Heid S, Hartmann R, and Klinke R (2005) Postnatal cortical development in congenital auditory deprivation. *Cerebral Cortex* **15**:552–562.

Kuwada S, Batra R, and Maher VL (1986) Scalp potentials of normal and hearing-impaired subjects in response to sinusoidally amplitude-modulated tones. *Hearing Research* **21**:179–192.

Lee HJ, Wallani T, and Mendelson JR (2002) Effects of aging on temporal processing speed in the inferior colliculus. *Hearing Research* **174**:64–74

Lee HJ, Giraud AL, Kang E, Oh SH, Kang H, Kim CS, and Lee DS (2007) Cortical activity at rest predicts cochlear implantation outcome. *Cerebral Cortex* **17**:909–917.

Leventhal AG, Wang Y, Pu M, Zhou Y, and Ma Y (2003) GABA and its agonists improved visual cortical function in senescent monkeys. *Science* **300**:812–815.

Liang L, Lu T, and Wang X (2002) Neural representations of sinusoidal amplitude and frequency modulations in the primary AC of awake primates. *Journal of Neurophysiology* **87**:2237–2261.

Ling LL, Hughes LF, and Caspary DM (2005) Age-related loss of the GABA synthetic enzyme gultamic acid dcarboxylase in rat primary auditory cortex. *Neuroscience* **132**:1103–1113.

McFadden SL and Willott JF (1994) Responses of inferior colliculus neurons in C57BL/6 J mice with and without sensorineural hearing loss: effects of changing the azimuthal location of an unmasked pure-tone stimulus. *Hearing Research* **78**:115–131.

McFadden SL, Campo P, Quaranta N, and Henderson D (1997) Age-related decline of auditory function in the chinchilla (*Chinichilla laniger*). *Hearing Research* **111**:114–126.

Mendelson JR and Ricketts C (2001) Age-related temporal processing speed deterioration in auditory cortex. *Hearing Research* **158**: 84–94.

Mendelson JR and Lui B (2004) The effects of aging on temporal processing speed in the auditory system: a comparison between the inferior colliculus, medial geniculate nucleus, and auditory cortex. *Hearing Research* **191**:21–33.

Middlebrooks JC, Bierer JA, and Snyder RL (2005) Cochlear implants: the view from the brain. *Current Opinion in Neurobiology* **15**: 488–493.

Milbrandt JC, Albin RL, and Caspary DM (1994) Age-related decrease in GABAB receptor binding in the fischer-344 rat inferior colliculus. *Neurobiology of Aging* **15**:699–703.

Miller JL and Liberman AM (1979) Some effects of later-occurring information on the perception of stop consonant and semivowel. *Perception and Psychophysics* **25**:457–465.

Moore DR, Devlin JT, Raley J, Tunbridge E, Lanary K, Floyer-Lea A, Narain C, Cohen I, Jezzard P, Burton MJ, and Matthews PM (2005) Effects of long-term unilateral hearing loss on the lateralization of fMRI measured activation in human auditory cortex. In: Syka J and Merzenich MM (eds). *Plasticity and Signal Representation in the Auditory System*. Springer, Berlin, pp. 335–346.

Mortensen MV, Madsen S, and Gjedde A (2005) Cortical responses to promontorial stimulation in postlingual deafness. *Hearing Research* **209**:32–41.

Museik FE, Geurkink NA, Weider DJ, and Donnelly K (1984) Past, present, and future applications of the auditory middle latency response. *Laryngoscope* **94**:1545–1553.

Näätänen R (2000) Mismatch negativity (MMN): perspectives for application. *International Journal of Psychophysiology* **37**:3–10.

Näätänen R and Picton T (1987) The N1 wave of the human electric and magnetic response to sound: a review and an analysis of the component structure. *Psychophysiology* **24**:375–425.

Naito Y, Okazawa H, Honjo I, Hirano S, Takahashi H, Shiomi Y, Hoji W, Kawano M, Ishizu K, and Yonekura Y (1995) Cortical activation with sound stimulation in cochlear implant users demonstrated by positron emission tomography. *Brain Research Cognitive Brain Research* **2**:207–214.

Naito Y, Hirano S, Honjo I, Okazawa H, Ishizu K, Takahashi H, Fujiki N, Shiomi Y, Yonekura Y, and Konishi J (1997) Sound induced activation of auditory cortices in cochlear implant users with post- and prelingual deafness demonstrated by positron emission tomography. *Acta Otolaryngology* **117**:490–496.

Nishimura H, Hashikawa K, Doi K, Iwaki T, Watanabe Y, Kusuoka H, Nishimura T, and Kubo T (1999) Sign language "heard" in the auditory cortex. *Nature* **397**:116.

Noreña AJ and Eggermont JJ (2003) Changes in spontaneous neural activity immediately after an acoustic trauma: implications for neural correlates of tinnitus. *Hearing Research* **183**:137–153.

Noreña AJ and Eggermont JJ (2005) Enriched acoustic environment after noise trauma reduces hearing loss and prevents cortical map reorganization. *Journal of Neuroscience* **25**:699–705.

Noreña AJ, Tomita M, and Eggermont JJ (2003) Neural changes in cat auditory cortex after a transient pure-tone trauma. *Journal of Neurophysiology* **90**:2387–2401.

Ochi K and Eggermont JJ (1997) Effects of quinine on neural activity in cat primary auditory cortex. *Hearing Research* **105**:105–118.

Okazawa H, Naito Y, Yonekura Y, Sadato N, Hirano S, Nishizawa S, Magata Y, Ishizu K, Tamaki N, Honjo I, and Konishi J (1996) Cochlear implant efficiency in pre- and post-lingually deaf subjects. A study with $H_2(15)O$ and PET. *Brain* **119**:1297–1306.

Ostroff JM, McDonald KL, Schneider BA, and Alain C (2003) Aging and the processing of sound duration in human auditory cortex. *Hearing Research* **181**:1–7.

Palombi PS, Backoff PM, and Caspary DM (2001) Responses of young and aged Fischer 344 rat inferior colliculus neurons to sinusoidally amplitude modulated tonal stimuli. *Hearing Research* **153**:174–180.

Pekkonen E, Huotilainen M, Virtanen J, Sinkkonen J, Rinne T, Ilmoniemi RJ, and Näätänen R (1995) Age-related functional differences between auditory cortices: a whole head MEG study. *Neuroreport* **6**:1803–1806.

Peters A and Vaughan SW (1981) Central nervous system. In: Johnson JE (ed). *Aging and Cell Structure*, Plenum, New York, pp. 1–34.

Peters A, Nigro NJ, and McNally KJ (1997) A further evaluation of the effect of age on striate cortex of the rhesus monkey. *Neurobiology of Aging* **18**:29–36.

Pichora-Fuller MK and Schneider BA (1991) Masking-level differences in the elderly: a comparison of antiphasic and time-delay dichotic conditions. *Journal of Speech and Hearing Research* **34**: 1410–1422.

Ponton CW and Eggermont JJ (2001) Of kittens and kids: Altered cortical maturation following profound deafness and cochlear implant use. *Audiology and Neurootology* **6**:363–380.

Popelář J, Erre J-P, Aran J-M, and Cazals Y (1994) Plastic changes in ipsi-contralateral differences of auditory cortex and inferior colliculus evoked potentials after in the adult guinea pig. *Hearing Research* **72**:125–134.

Popelář J, Grecova J, Rybalko N, and Syka J (2008) Comparison of noise-induced changes of auditory brainstem and middle latency response amplitudes in rats. *Hearing Research* **245**:82–91.

Ponton CW, Vasama JP, Tremblay K, Khosla D, Kwong B, and Don M (2001) Plasticity in the adult human central auditory system: evidence from late-onset profound unilateral deafness. *Hearing Research* **154**:32–44.

Popelář J, Syka J, and Bemdt H (1987) Effect of noise on auditory evoked responses in awake guinea pigs. *Hearing Research* **26**:239–247.

Proctor TB, Velde TM, Dayal VS, and Bhattacharyya TK (1998) Auditory brainstem response in young and old guinea pigs. *American Journal of Otolaryngology* **19**:226–229.

Raggio MW and Schreiner CE (1994) Neural responses in cat primary auditory cortex to electrical stimulation. I. Intensity dependence of firing rate and response latency. *Journal of Neurophysiology* **72**:2334–2359.

Raggio MW and Schreiner CE (1999). Neuronal responses in cat primary auditory cortex to electrical cochlear stimulation: III. Activation patterns in short- and long-term deafness. *Journal of Neurophysiology* **82**:3506–3526.

Rajan R (1998) Receptor organ damage causes loss of cortical surround inhibition without topographic map plasticity. *Nature Neuroscience* **1**:138–143.

Rajan R (2001) Plasticity of excitation and inhibition in the receptive field of primary auditory cortical neurons after limited receptor organ damage. *Cerebral Cortex* **11**:171–182.

Rajan R (2002) Plasticity in auditory cortex: excitatory and inhibitory processes. *IETE Journal of Research* **49**:157–170.

Rajan R (2005) Small cochlear damage causes unmasking and plasticity in supra-threshold cortical responses. In: Syka J and Merzenich MM (eds). *Plasticity and Signal Representation in the Auditory System*. Springer, Berlin, pp. 153–168.

Rajan R and Irvine DRF (1998) Neuronal responses across cortical field A1 in plasticity induced by peripheral auditory damage. *Audiology and Neurootology* **3**:123–144.

Rajan R, Irvine DRF, Wise LZ, and Heil P (1993) Effect of unilateral partial cochlear lesions on the representation of lesioned and unlesioned cochleas in primary auditory cortex of the cat. *Journal of Comparative Neurology* **338**:17–49.

Raza A, Milbrandt JC, Arneric SP, and Casparay DM (1994) Age-related changes in brainstem auditory neurotransmitters: measures of GABA and acetylcholine function. *Hearing Research* **77**: 221–230.

Ricketts C, Mendelson JR, Anand B, and English R (1998) Responses to time-varying stimuli in rat auditory cortex. *Hear. Res.* **123**:27–30.

Ro B, Borgmann C, Draganova R, Roberts LE, and Pantev C (2000) A high-precision magnetoencephalographic study of human auditory steady-state responses to amplitude-modulated tones. *Journal of the Acoustical Society of America* **108**:579–591.

Robertson D and Irvine DRF (1989) Plasticity of frequency organization in auditory cortex of guinea pigs with partial unilateral deafness. *Journal of Comparative Neurology* **282**:456–471.

Ross B, Fujioka T, Tremblay KL, and Picton TW (2007) Aging in binaural hearing begins in mid-life: evidence from cortical auditory-evoked responses to changes in interaural phase. *Journal of Neuroscience* **27**:11172–11178.

Salat DH, Buckner RL, Snyder AZ, Greve DN, Diskan RSR, Busa E, Morris JC, Dale AM, and Fischl B (2004) Thinning of the cerebral cortex in aging. *Cerebral Cortex* **14**:721–730.

Schneider BA, Pichora-Fuller MK, Kowalchuk D, and Lamb M (1994) Gap detection and the precedence effect in young and old adults. *Journal of the Acoustical Society of America* **95**: 980–991.

Schneider BA, Daneman M, Murphy DR, and Kwong See S (2000) Listening to discourse in distracting settings: the effects of aging. *Psychology and Aging* **15**:110–125.

Seidman MD (2000) Effects of dietary restriction and antioxidants on presbycusis. *Laryngoscope* **110**:727–738.

Seki S and Eggermont JJ (2002) Changes in cat primary auditory cortex after minor-to-moderate pure-tone induced hearing loss. *Hearing Research* **173**:172–186.

Sharma A and Dorman MF (2006) Central auditory development in children with cochlear implants: clinical implications. *Advances in Otorhinolaryngology* **64**:66–88.

Sharma A, Dorman MF, and Kral A (2005) The influence of a sensitive period on central auditory development in children with unilateral and bilateral cochlear implants. *Hearing Research* **203**: 134–143.

Sharma A, Dorman MF, and Spahr AJ (2002) A sensitive period for the development of the central auditory system in children with cochlear implants: Implications for age of implantation. *Ear and Hearing* **23**:532–539.

Sharma A, Gilley PM, Dorman MF, and Baldwin R (2007) Deprivation-induced cortical reorganization in children with cochlear implants. *International Journal of Audiology* **46**:494–499.

Simon H, Frisina RD, and Walton JP (2004) Age reduces response latency of mouse inferior colliculus neurons to AM sounds. *Journal of the Acoustical Society of America* **116**:469–477.

Snyder JS and Alain C (2005) Age-related changes in neural activity associated with concurrent vowel segregation. *Brain Research and Cognitive Brain Research* **24**:492–499.

Sowell ER, Peterson BS, Thompson OM, Welcome SE, Henkenius AL, and Toga AW (2003) Mapping cortical change across the human life span. *Nature Neuroscience* **6**:309–315.

Strouse A, Ashmead DH, Ohde RN, and Grantham DW (1998) Temporal processing in the aging auditory system. *Journal of the Acoustical Society of America* **104**:2385–2397.

Su GL, Colesa DJ, and Pfingst BE (2008) Effects of deafening and cochlear implantation procedures on postimplantation psychophysical electrical detection thresholds. *Hearing Research* **241**: 64–72.

Suarez H, Mut F, Lago G, Silveira A, De Bellis C, Velluti R, Pedemonte M, and Svirsky M (1999) Changes in the cerebral blood flow in postlingual cochlear implant users. *Acta Otolaryngologica* **119**: 239–243.

Sun W, Zhang L, Lu J, Yang G, Laundrie E, and Salvi RJ (2008) Noise exposure–induced enhancement of auditory cortex response and changes in gene expression. *Neuroscience* **156**:374–380.

Syka J and Popelář J (1982) Noise impairment in the guinea pig. I. Changes in electrical evoked activity along the auditory pathway. *Hearing Research* **8**:263–212.

Syka J, Rybalko N, and Popelář J (1994) Enhancement of the auditory cortex evoked responses in awake guinea pigs after noise exposure. *Hearing Research* **78**:158–168.

Thai-Van H., Micheyl C, Noreña A, Veuillet E, Gabriel D, and Collet L (2007) Enhanced frequency discrimination in hearing-impaired individuals: a review of perceptual correlates of central neural plasticity induced by cochlear damage. *Hearing Research* **233**: 14–22.

Torre III P and Fowler CG (2000) Age-related changes in auditory function of rhesus monkeys (Macaca mulatta). *Hearing Research* **142**:131–140.

Tremblay KL, Piskosz M, and Souza P. (2003) Effects of age-related hearing loss on the neural representation of speech cues. *Clinical Neurophysiology* **114**:1332–1343.

Tremblay KL, Billings C, and Rohila N (2004) Speech evoked cortical potentials: effects of age and stimulus presentation rate. *Journal American Academy Audiology* **15**:226–237.

Truy E, Deiber M-P, Cinotti L, Mauguiere F, Froment J-C, and Morgon A (1995) Auditory cortex activity changes in long-term sensorineural deprivation during crude cochlear electrical stimulation: evaluation by positron emission tomography. *Hearing Research* **86**:34–42.

Turner JG, Hughes LF, and Caspary DM (2005a) Affects of aging on receptive fields in rat primary AC layer V neurons. *Journal of Neurophysiology* **94**:2738–2747.

Turner JG, Hughes LF, and Caspary DM (2005b) Divergent response properties of layer V neurons in rat primary auditory cortex. *Hearing Research* **202**:129–140.

Vasama J-P and Makela JP (1995) Auditory pathway plasticity in adult humans after unilateral idiopathic sudden sensorineural hearing loss. *Hearing Research* **87**:132–140.

Vasama J-P and Makela JP (1997) Auditory cortical responses in humans with profound unilateral sensorineural hearing loss from early childhood. *Hearing Research* **104**:183–190.

Walton JP, Frisina RD, and O'Neill WE (1998) Age-related alteration in processing of temporal sound features in the auditory mid-brain of CBA mouse. *Journal of Neuroscience* **18**: 2764–2776.

Walton JP, Simon H, and Frisina RD (2002) Age-related alterations in the neural coding of envelope periodicities. *Journal of Neurophysiology* **79**:565–578.

Wang J, Caspary D, and Salvi RJ (2000) GABA-A antagonist causes dramatic expansion of tuning in primary auditory cortex. *Neuroreport* **11**:1137–1140.

Wang X-J (1998) Calcium coding and adaptive temporal computation in cortical pyramidal neurons. *Journal of Neurophysiology* **79**: 1549–1566.

Willott JF (1986) Effects of aging, hearing loss and anatomical location on thresholds if inferior colliculus neurons in C57BL/6 and CBA mice. *Journal of Neurophysiology* **56**:93–100.

Willott JF, Hunter KP, and Coleman JR (1988) Aging and presycusis: effects on 2-deoxy-D-glucose uptake in the mouse auditory brainstem in quiet. *Experimental Neurology* **99**: 615–621.

Willott JF Aitkin LM, and McFadden SM (1993) Plasticity of auditory cortex associated with sensorineural hearing loss in adult C57BL/6 J mice. *Journal of Comparative Neurology* **329**:402–411.

Willott JF (1996) Anatomic and physiologic aging: a behavioral neuroscience perspective. *Journal of the American Academy of Audiology* **7**:141–151.

Wong TP (2002) Aging of the cerebral cortex. *McGill Journal of Medicine* **6**:104–113.

Yin S-K, Feng Y-M, Chen Z-N, and Wang J (2008) The effect of noise-induced sloping high-frequency hearing loss on the gap-response in the inferior colliculus and auditory cortex of guinea pigs. *Hearing Research* **239**:126–140.

Zhang LI, Bao S, and Merzenich MM (2002) Disruption of primary auditory cortex by synchronous auditory inputs during a critical period. *Proceedings of the National Academy of Sciences of the United States of America* **99**:2309–2314.

Zheng QY, Johnson KR, and Erway LC (1999) Assessment of hearing in 80 inbred strains of mice by ABR threshold analyses. *Hearing Research* **130**:94–107.

Chapter 24

Corticofugal Modulation and Plasticity for Auditory Signal Processing

Nobuo Suga

Abbreviations

AC	auditory cortex
ACh	acetylcholine
AI	primary auditory cortex
APV	2-amino-5-phosphovalerate
BAZ	best azimuth
BDe	best delay
BDu	best duration
BF	best frequency
BMI	bicuculline methiodide
CF	constant frequency
CM	cochlear microphonics
COCB	crossed olivo-cochlear bundle
CS	conditioned stimulus
DPD	dorsoposterior division
DSCF	Doppler-shifted constant frequency
EE	excitatory–excitatory
ES	electric stimulation
FM	frequency modulation
GABA	γ-aminobutyric acid
IC	inferior colliculus
ICc	central nucleus of the inferior colliculus
IE	inhibitory–excitatory
MGB	medial geniculate body
MGBv	ventral division of the medial geniculate body
MGBm	medial division of the medial geniculate body
MT	minimum threshold
NMDA	*N*-methyl-d-aspartic acid
PIN	posterior intralaminar nucleus
US	unconditioned stimulus

1 Introduction

The auditory system has ascending and descending (corticofugal) subsystems. Corticofugal modulation of subcortical neurons for auditory signal processing is an important auditory cortex function. Auditory signals analyzed in the cochlea are sent to the brain by the auditory nerve, then ascend to the auditory cortex through the cochlear nucleus, superior olivary complex, lateral lemniscal nuclei, inferior colliculus, and medial geniculate body. The ascending auditory system has further complexity since divergent and convergent projections occur in multiple levels for parallel and hierarchical signal processing. As a result, many physiologically distinct types of neurons are produced in the subcortical nuclei and physiologically distinct areas exist in the auditory cortex (Fig. 24.1). A classical view of signal processing is based on excitatory and inhibitory interactions occurring between cells in the ascending auditory system and on parallel and hierarchical auditory signal processing (Suga 1984, 1990, 1994; Covey and Casseday 1999). By contrast, the contribution of the auditory corticofugal feedback system on signal processing is a more recent development. Research on corticofugal function shows that it sharpens and shifts subcortical tuning curves in the frequency, amplitude, time, and spatial domains and plays a vital role in the auditory reorganization of normal animals based on experience. For stimulus-specific reorganization in auditory fear conditioning, the auditory cortex, corticofugal system, somatic sensory cortex, and cholinergic basal forebrain all play key roles. The auditory system can be reorganized by expansion and compression of physiological maps. The former is found in many species and sensory systems, the latter thus far only in the mustached bat auditory system, which is specialized for echolocation.

2 Corticofugal Projections

Neurons in the deep layers of the auditory cortex (AC) project to the medial geniculate body, inferior colliculus, or subcollicular auditory nuclei. Corticothalamic fibers project only to the ipsilateral medial geniculate body and thalamic reticular nucleus. However, corticocollicular fibers project

N. Suga (✉)
Department of Biology, Washington University, St. Louis, MO 63130, USA
e-mail: suga@biology.wustl.edu

J.A. Winer, C.E. Schreiner (eds.), *The Auditory Cortex*,
DOI 10.1007/978-1-4419-0074-6_24, © Springer Science+Business Media, LLC 2011

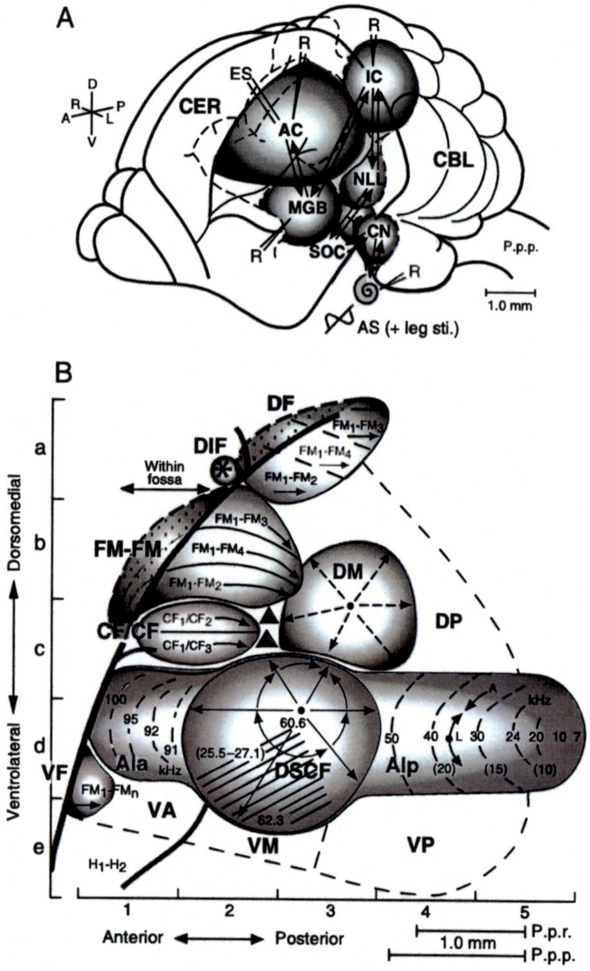

Fig. 24.1 The mustached bat auditory system (**a**) and auditory cortex (**b**). *Arrows*, ascending and descending (corticofugal) auditory systems. ES and R, electric stimulation and recording electrodes, respectively. Acoustic stimuli (AS) with or without electric leg stimulation are given in addition to cortical electric stimulation. (**b**) The physiological map of the auditory cortex (AC). *Numbers* and *lines* in the anterior (AIa) and posterior (AIp) divisions of the primary AC and in the Doppler-shifted constant frequency (DSCF) area denote iso-best-frequency lines. The CF/CF (constant frequency) area responds to CF combinations and has two subdivisions with a Doppler-shift (velocity) axis. The dorsal intrafossa area (DIF) also responds to CF combinations. The FM–FM (frequency modulated), dorsal (DF) and ventral fringe (VF) areas contain three subdivisions with an echo-delay (range) axis. CBL, cerebellum; CER, cerebrum; CN, cochlear nucleus; DM, dorsomedial area; DP, dorsoposterior area; IC, inferior colliculus; MGB, medial geniculate body; NLL, nucleus of the lateral lemniscus; SOC, superior olivary complex; VA, ventroanterior area; VM, ventromedial area; VP, ventroposterior area; P.p.r., *Pteronotus parnellii rubiginosus*; P.p.p., *Pteronotus parnellii parnellii* (**b**: after Suga 1994)

bilaterally. The ipsilateral projection is far larger and more topographically organized than the contralateral projection (Saldaña et al. 1996). Corticofugal projections are bilateral to the superior olivary complex and cochlear nucleus (Feliciano et al. 1995). Corticofugal modulation reaches even the cochlea via olivocochlear neurons in the superior olivary complex (Xiao and Suga 2002a). The central nucleus of the inferior colliculus (ICc) projects to the ventral and medial divisions of the medial geniculate body (MGBv and MGBm) and the superior colliculus and projects to medial olivocochlear neurons, which mostly project to contralateral cochlear outer hair cells. Olivocochlear neurons usually project bilaterally to the cochlea, although there are some species differences. The corticothalamic projection is the shortest feedback loop, and the projection to cochlear hair cells the longest (Kelly and Wong 1981; Huffman and Henson 1990; Ojima 1994; Saldaña et al. 1996).

3 Principles of Corticofugal Modulation

How does the corticofugal auditory system modulate signal processing in the frequency, amplitude, time, and spatial domains? Many studies on corticofugal modulation of MGB and/or IC neurons found that strong activation or inactivation of the primary auditory cortex (AI) evoked subcortical excitation and/or inhibition. These data were controversial, some finding only or mainly inhibitory effects (Desmedt and Mechelse 1958; Massopust and Ordy 1962; Watanabe et al. 1966; Aitkin and Dunlop 1969; Amato et al. 1969; Sun et al. 1996), others only or predominantly excitatory or facilitatory effects (Andersen et al. 1972; Orman and Humphrey 1981; Villa et al. 1991; He et al. 2002), and still others saw equal effects (Ryugo and Weinberger 1976; Syka and Popelar 1984; Jen et al. 1998). These differences might be resolved if the frequency dependence of excitation (facilitation) and inhibition and the relationship in tuning between stimulated and recorded cells were considered.

To this issue, unanesthetized animals were characterized electrophysiologically by measuring the tuning curves of the cortical neurons to be activated or inactivated as well as the cortical and subcortical neurons to be examined. Focally applied electric stimulation or drugs for activation or inactivation of the cortical neurons and its effects on the subcortical or other cortical cells were evaluated as was frequency-dependent facilitation and inhibition. Corticofugal modulation is highly specific and systematic for the improvement and adjustment of auditory signal processing in the frequency and time domains. Studies of anesthetized bats and mice, respectively, extended the findings of corticofugal modulation in the frequency and time domains to modulation in the amplitude and spatial domains.

The corticofugal modulation originates from the changes within AI. That is, cortical changes modulate the subcortical auditory nuclei via the many corticofugal fibers. However, our understanding of corticofugal modulation is still limited to the changes in AI, the subcortical auditory nuclei, and cochlear hair cells evoked by focal AI activation/inactivation.

3.1 Corticofugal Modulation in the Frequency Domain in Bats and Rodents

3.1.1 Frequency-Dependent Facilitation and Inhibition and Best Frequency Shifts

Electric stimulation of AI neurons evokes facilitation and inhibition of the responses of IC and MGB cells and nearby AI neurons. The amount of facilitation and inhibition varies with the frequency of a tone burst to which the neurons respond and with the relationship in frequency tuning between the stimulated and recorded neurons. The response threshold of a neuron is usually lowest at a specific frequency, the neuron's best frequency (BF). When a recorded neuron is matched in BF to the stimulated AI neurons, the matched neuron's response at its BF is augmented and inhibited at frequencies above and/or below the BF, sharpening its frequency tuning. The unmatched neuron is inhibited at its BF response and facilitated at non-BF responses, so that its frequency tuning shifts (Figs. 24.2 and 24.3). The shift of a frequency-tuning curve is always accompanied by this so-called BF shift that improves the input to the stimulated AI neurons as well as that of the matched subcortical and cortical representations of the stimulus parameter values to which the stimulated cortical neurons are tuned (Zhang et al. 1997; Yan and Suga 1998; Zhang and Suga 2000; Chowdhury and Suga 2000; Ma and Suga 2001a; Sakai and Suga 2001, 2002). This corticofugal function is called egocentric selection (Yan and Suga 1996). When strong electric stimulation of AI stimulates many cortical columns, egocentric selection becomes unclear and subcortical neurons are mostly inhibited. Therefore, many studies use 0.2 ms, 100 nA electric pulses, a very weak stimulus (Suga et al. 1995).

Focal inactivation of AI evoked IC BF shifts opposite to those evoked by focal electric stimulation (Fig. 24.3b,d) (Zhang and Suga 2000), whereas non-focal AI inactivation did not, although it evoked large changes in IC auditory responses (Zhang and Suga 1997; Yan and Suga 1999). In

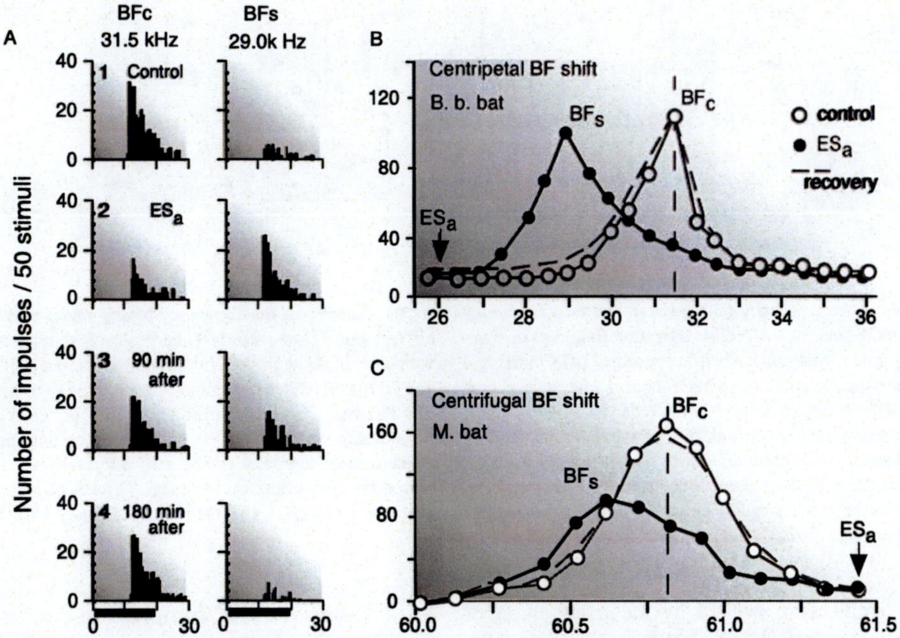

Fig. 24.2 Corticofugal modulation of the auditory responses and frequency-response curves of IC neurons. **(a)** The peristimulus time (PST) histograms with inhibition (*left column*) and facilitation (*right column*) of the responses of a single IC neuron at the best frequencies in the control (BF_c) and shifted (BF_s) conditions. The histograms were obtained before (1), immediately after (2), 90 min after (3), and 180 min after (4) electric stimulation (ES_a) of AI neurons tuned to 26 kHz. The horizontal bars (*bottom*) represent sounds 20-ms long. Centripetal **(b)** and centrifugal **(c)** BF shifts evoked by electric stimulation of AI neurons tuned to 26.0 kHz (**b**, *arrow*) or 61.5 kHz (**c**, *arrow*). *Open* and *filled circles* and *dashed lines*, the frequency-response curves 10 dB above minimum threshold in control, shifted, and recovery conditions, respectively. **a, b** from the big brown bat (Ma and Suga 2001a) and **c** from the mustached bat (Zhang and Suga 2000)

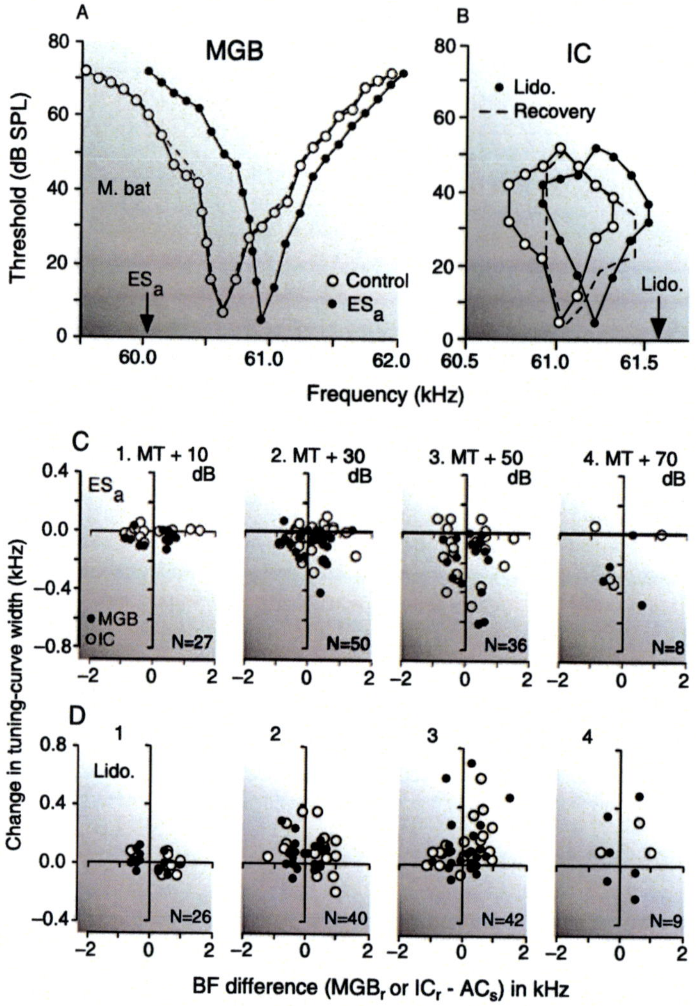

Fig. 24.3 Changes in the frequency-tuning curve and best frequency of MGB and IC DSCF neurons evoked by activation (**a** and **c**) or inactivation (**b** and **d**) of AI DSCF mustached bat neurons. (**a**) Centrifugal shift in the frequency-tuning curve of an MGB neuron after a focal cortical electric stimulation, ES_a (0.2 ms, 100 nA pulse at 5/s for 7 min). (**b**) Centripetal frequency-tuning curve shift of an IC neuron evoked by 90 nl of 0.1% lidocaine hydrochloride (Lido) applied to AI. *Arrows*, the best frequencies (BFs) of the activated/inactivated AI neurons. The *curves* were measured before (control; *open circles*), during (*closed circles*), and after (recovery; *dashed lines*) AI activation or inactivation. (**c**, **d**) Changes in the frequency-tuning curve widths of MGB (*filled circles*) and IC (*open circles*) neurons evoked by activation (**c**) or inactivation (**d**) of AI neurons. Widths were measured at 10 (1), 30 (2), 50 (3), and 70 (4) dB above minimum threshold (MT) of MGB or IC neurons. The abscissae represent differences in BF between recorded MGB_r or IC_r neurons and activated or inactivated cortical neurons (AC_s). The small sample numbers at MT +70 dB reflect the upper threshold or closed frequency-tuning curve (**b**). The BF of stimulated AC neurons averaged 61.2 kHz. **a**, **c** (Zhang and Suga 2000); **b**, **d** (Zhang et al. 1997)

the rat, no BF shifts were evoked by the inactivation of the entire AI (Nwabueze-Ogbo et al. 2002). Eliciting a BF shift requires a focal and uneven distribution of neural activity across AI. Widespread simultaneous activation of AI by strong electric stimulation or by broadband noise does not evoke BF shifts, although it does evoke changes in subcortical auditory responses. BF shifts can be evoked by auditory fear conditioning. The time courses of BF shifts evoked by conditioning or by focal cortical electric stimulation are considered below.

3.1.2 Expanded and Compressed Frequency Map Reorganization

Focal cortical electric stimulation evokes centripetal and centrifugal BF shifts of unmatched neurons. Centripetal BF shifts move toward the BF of the electrically stimulated AI neurons (Fig. 24.2b) or the frequency of a stimulus tone, whereas centrifugal BF shifts move away from the stimulated AI BF (Fig. 24.2c) or tone frequency. These BF shifts underlie reorganization of a frequency map and occur in a specific spatial pattern in the ICc, MGBv, and AI. The spatial distribution pattern of BF shifts is very similar in the ICc and AI (Fig. 24.5 a,c) (Zhang and Suga 1997; Zhang et al. 1997; Gao and Suga 1998; Yan and Suga 1998; Chowdhury and Suga 2000; Ma and Suga 2001a; Xiao and Suga 2002b). BF distribution shifts along a frequency map were studied in AI in the Mongolian gerbil and big brown bat because AI can be easily mapped. In AI in the big brown bat (Fig. 24.5a) (Ma and Suga 2004), Mongolian gerbil (Figs. 24.4a and 24.5d) (Sakai and Suga 2002) and house mouse (Fig. 24.5e) (Yan and Ehret 2002), centripetal BF shifts occur in a large area surrounding the matched neurons and small centrifugal BF shifts are in a narrow zone around this large centripetal area, i.e., a center-surround reorganization (Fig. 24.4). Many neurons in the surrounding zone do not show BF shifts (Fig. 24.4b). The major reorganization in AI is thus due to centripetal BF shifts. However, in a specialized mustached bat AI subarea, the Doppler-shifted constant frequency (DSCF) area (Fig. 24.1), only centrifugal BF shifts occur in a large zone surrounding the matched neurons (Fig. 24.5c).

BF shifts are largest for neurons along the frequency axis crossing stimulated neurons (Fig. 24.4e,f) (Sakai and Suga 2002; Ma and Suga 2004). The relationship between the BF shifts and the BF differences for recorded and stimulated neurons (or stimulus tone) is called the BF shift-difference curve. This curve can be different in AI of different species and between different cortical areas of a species (Fig. 24.5). It may vary with the locus of electric stimulation along the AI frequency axis. The difference in the BF shift is related to the difference in the frequency axis. In an expanded part of the frequency axis, with more sharply tuned neurons, BF shifts are smaller than in other portions of the frequency axis. This is seen in the mustached bat AI (Fig. 24.5c).

Centripetal BF shifts increase the number of neurons responding to the same frequency as the stimulated AI BF or the stimulus tone (Yan and Suga 1998; Gao and Suga 1998, 2000; Ma and Suga 2001a): this is expanded reorganization, while centrifugal BF shifts reduce representation in matched neurons (Zhang et al. 1997). This augments the responses, sharpens the tuning curves and is called compressed reorganization (Suga et al. 2002).

This description is based only on the major change in BF. Expanded reorganization entails overrepresentation of one frequency at the cost of underrepresentation of others. Therefore, for center-surround reorganization, there is a large overrepresentation from centripetal BF shifts at the center, a small underrepresentation from centrifugal BF shifts at the surround, and a small overrepresentation from centrifugal BF shifts just beyond the surround. Likewise, compressed reorganization consists of a large underrepresentation from centrifugal BF shifts near the center and a small overrepresentation outside the centrifugal BF shift area. If the overrepresentation increases sensitivity and/or discrimination, this would occur at the stimulated BF, with minor increases at two frequencies higher and lower than the stimulated BF. Behavioral experiments related to expanded and compressed reorganization remain to be performed, but we speculate that compressed reorganization increases contrast in the auditory signal neural representation, enhancing acoustic signal discrimination better than in an expanded reorganization.

Two groups of IC cells exist: corticofugally plastic (49%) and non-plastic (51%) (Zhou and Jen 2000). BF shifts evoked by AI stimulation depend on the relationship in tuning between stimulated AI and recorded IC or nearby AI neurons (Yan and Suga 1998; Zhang and Suga 2000; Ma and Suga 2001a, 2003), and on the relationship in relative location along an iso-BF line or slab between them (Fig. 24.4) (Sakai and Suga 2002; Ma and Suga 2004). Therefore, it is difficult to evaluate the significance of these percentages.

3.1.3 Role of Facilitation and Inhibition in Producing Two Types of Reorganization

Corticofugal facilitation and inhibition are hypothesized to evoke centripetal and centrifugal BF shifts of unmatched subcortical neurons, respectively. When excitation is stronger and spreads to more nearby unmatched neurons than inhibition, it evokes centripetal BF shifts. In contrast, when stronger inhibition spreads to more neighboring unmatched neurons than excitation, it evokes centrifugal BF shifts (Suga et al. 2000).

In the mustached bat, electric stimulation of AI DSCF neurons evokes the centrifugal BF shifts of IC and AI DSCF neurons (Fig. 24.5c). However, application of the $GABA_A$ receptor antagonist bicuculline methiodide (BMI) to the stimulation site evokes centripetal BF shifts of both IC and AI unmatched neurons. Compressed reorganization becomes expanded reorganization becomes expanded reorganization when AI inhibition is removed or reduced. Electric stimulation of the posterior AI evokes centripetal BF shifts of nearby AI neurons (Fig. 24.5b). BMI applied to the stimulation site augments these centripetal BF shifts (Xiao and Suga 2002b, 2005). In the big brown bat, focal AI electric stimulation evokes center-surround reorganization (Fig. 24.5a), as in the gerbil (Figs. 24.4 and 24.5d). BMI applied to AI changes centrifugal BF shifts at the surround into centripetal BF shifts, whereas the $GABA_A$ receptor agonist (muscimol) applied to AI changes centripetal BF shifts at the center into centrifugal BF shifts (Ma and Suga 2004).

In the AI FM–FM area of the mustached bat, an echo delay (time interval of paired sounds) is systematically mapped (Suga and O'Neill 1979; O'Neill and Suga 1982).

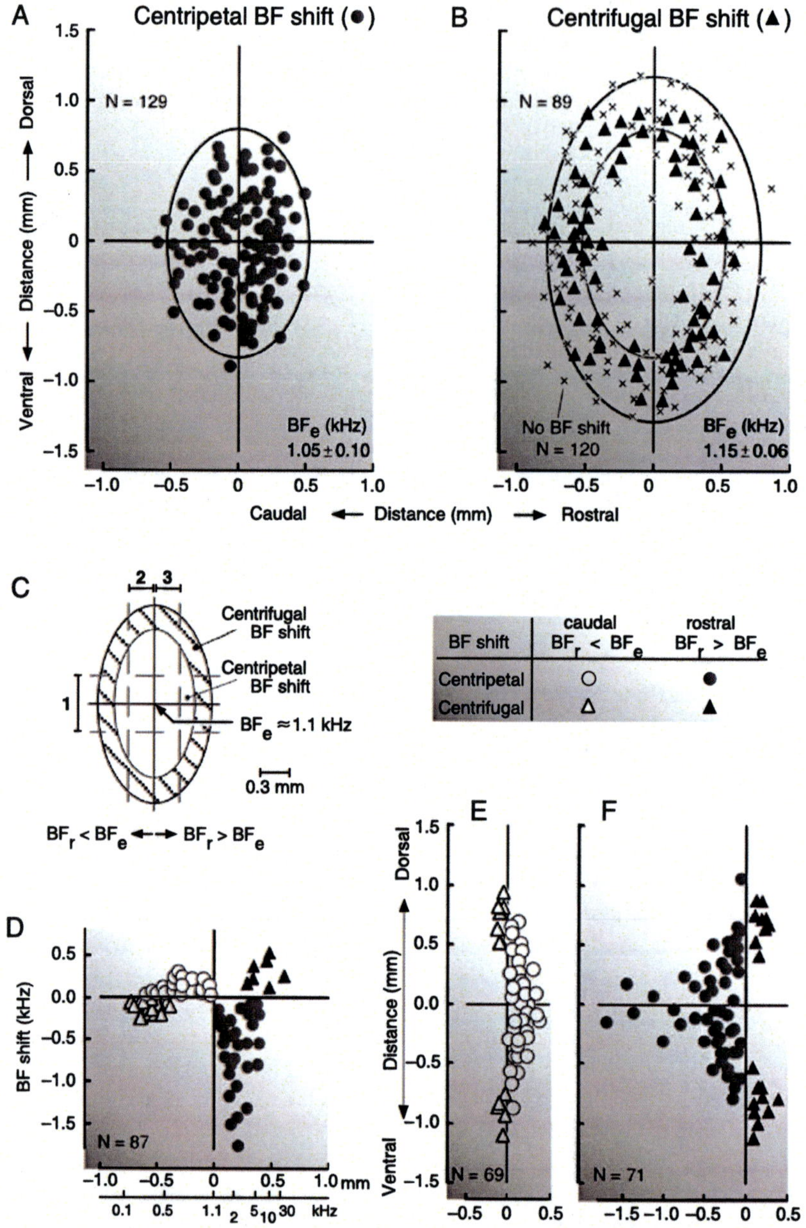

Fig. 24.4 Distributions of centripetal and centrifugal BF shifts in AI after focal AI electric stimulation in the Mongolian gerbil. Electrical stimulation of 1.1 kHz-tuned AI neurons evokes centripetal (**a**, *circles*) or centrifugal (**b**, *triangles*) BF shifts of other AI neurons. Recorded neurons locations on the cortical surface are plotted relative to the stimulated cortical neurons at the coordinate origins. *x* and *y* axes: axes across the frequency map in AI, respectively. X (**b**) neurons without BF shifts. Pooled data are from 16 hemispheres of 11 animals. Confidence ellipses for neurons with centripetal (**a**) or centrifugal (**b**) BF shifts. The BF shifts were measured parallel (*1*) or orthogonal (*2*, *3*) to the AC frequency axis (**c**). The directions and amounts of BF shifts of neurons in the rostrocaudal (C:1) and dorsoventral (C:2, 3) zones plotted (**d–f**) as a function of distance along the AI surface. BF_e: BF of electrically stimulated neurons; BF_r: BF of recorded cortical neurons. *Inset* (*middle right*), *symbols* (Sakai and Suga 2002)

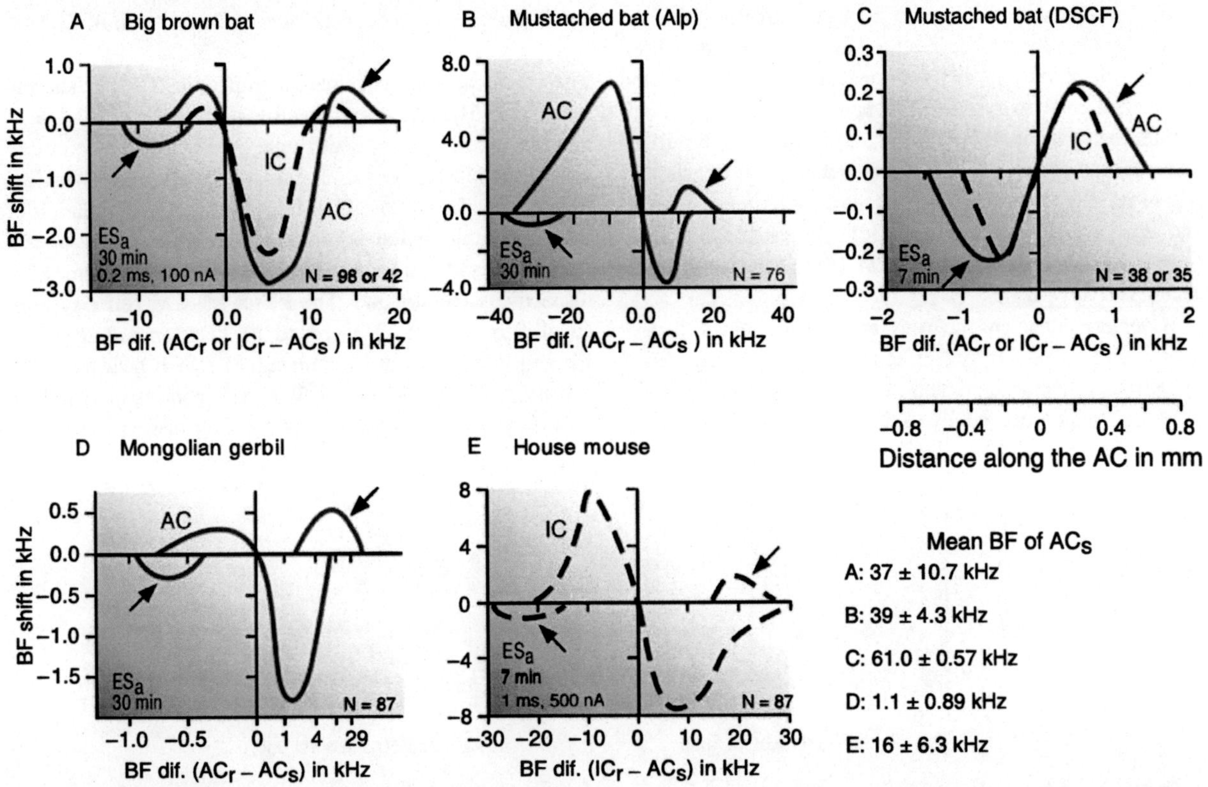

Fig. 24.5 BF shift-difference curves in AC or IC of four mammalian species. BF shift changes as a function of the BF difference between recorded IC (IC_r, *dashed lines*) or AC (AC_r, *solid lines*) neurons and electrically stimulated AC neurons (AC_s). Each BF shift-difference curve encompasses a scatter plot of BF shifts of many neurons studied (*N*). Species and area differences in curves are shown. **a**, **b**, **d** and **e**: centripetal BF shifts, except where indicated by *arrows*. A prominent centripetal BF shift is ~5 kHz higher than the stimulated AI BF in the big brown bat (**a**) and ~1 kHz higher than that in the Mongolian gerbil (**d**). By contrast, the centripetal BF shift is at ~10 kHz lower than the stimulated AC BF in the posterior division of the mustached bat primary AC (AIp) (**b**). In the house mouse, prominent centripetal BF shifts occur at ~9 kHz higher and lower than the AC BF (**e**). (**c**) Centrifugal BF shifts in the mustached bat Doppler-shifted constant frequency (DSCF) area are ~0.5 kHz higher and lower than the stimulated AC BF. The shape of these BF shift-difference curves can change with the mean BF of the stimulated AC neuron (AC_s). The BF mean and standard deviation of stimulated AC neurons in **a**–**e** are shown (*bottom right*). The electrical stimulation (ES_a) was 0.2 ms, 100 nA pulses (**a**–**d**) and 1 ms, 500 nA pulses (**e**). Stronger stimulation presumably increases both BF shifts and the frequency range for shifts. Modified from published data **a**: Chowdhury and Suga (2000) and Ma and Suga (2001a); **b**: Sakai and Suga (2001); **c**: Xiao and Suga (2002b); **d**: Sakai and Suga (2002); **e**: Yan and Ehret (2002))

Focal electric stimulation of the FM–FM area evokes centrifugal best delay (BDe) shifts of IC (Fig. 24.7a) (Yan and Suga 1996) and nearby AI (Xiao and Suga 2004) neurons. BMI applied to the FM–FM area changes centrifugal BDe shifts into centripetal BDe shifts (Xiao and Suga 2004).

The DSCF and FM–FM AI areas are highly specialized for the representation of specific biosonar information (Suga 1984, 1990, 1994). Therefore, these observations indicate that in such areas inhibition is stronger and more widespread than excitation and evokes compressed reorganization, whereas in less-specialized AI, excitation is stronger and more widespread than inhibition and evokes expanded reorganization. Cortical and subcortical reorganization can differ between specialized and non-specialized auditory systems.

3.1.4 Sharpening and Broadening of Frequency-Tuning Curves

In the big brown bat, IC neurons are either corticofugally facilitated (26%) or inhibited (74%) (Jen et al. 1998). For facilitated neurons, AI electric stimulation augmented their auditory responses and broadened their frequency-tuning curves. For inhibited neurons, AI stimulation inhibited their auditory responses and sharpened their frequency-tuning curves. Such corticofugal modulation was observed even in IC neurons with BFs up to 50 kHz different from the AI BF. Their strong electric stimulation presumably stimulated much of AI and many cortical neurons matched and unmatched in BF to a recorded IC neuron. Thus, it is difficult to refer their data to the BF difference between the stimulated and recorded neurons. In the mustached bat, electric stimulation of AI DSCF neurons sharpens the frequency tuning of IC and MGB DSCF neuron (Fig. 24.3c), whereas focal AI inactivation evokes broadening (Fig. 24.3d) (Zhang et al. 1997; Zhang and Suga 2000).

IC external nucleus neurons are broadly tuned in frequency and are inhibitory (Jen et al. 2001). They are excited by corticocollicular fibers and in turn inhibit neurons in the ICc which are corticofugally excited.

3.1.5 Corticofugal Modulation of Cochlear Hair Cells

The mustached bat cochlear microphonic (CM) is sharply tuned to ~61 kHz. Rapid electric stimulation of AI DSCF neurons evokes a brief centrifugal BF shift of the contralateral CM, while slower stimulation, which elicits IC BF shifts, does not evoke a CM BF shift. Surprisingly, stimulation of the dorsal and ventral portions of the DSCF area evokes a centripetal and a centrifugal BF shift of the ipsilateral CM, respectively (see below). The BF of the CM shifts systematically up to 0.25 kHz around 61 kHz, according to the BF and location of the stimulated AI DSCF neurons (Xiao and Suga 2002a). The AI DSCF frequency map is critical for the systematic modulation of cochlear hair cell frequency

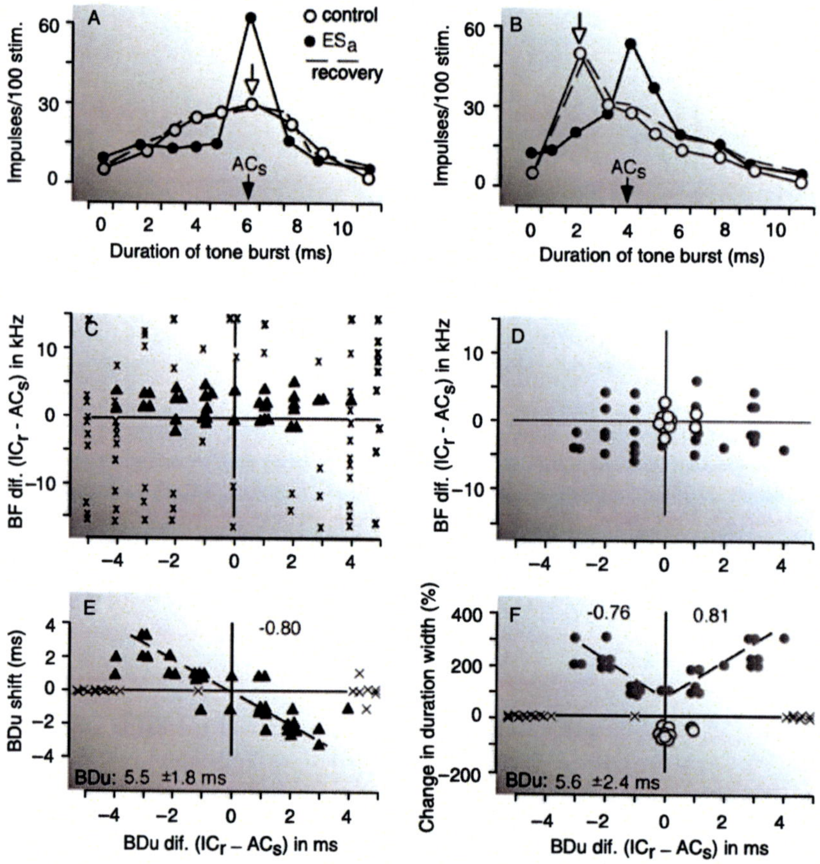

Fig. 24.6 Corticofugal modulation of IC duration-tuned neurons evoked by electrical stimulation of AC duration-tuned neurons. The stimulated AC (AC_s) and recorded IC (IC_r) neurons are matched (**a**) or unmatched (**b**) in best frequency (BF) and best duration (BDu). *Arrows*, BDu of IC_r or AC_s neurons. AC stimulation sharpened (**a**) or shifted (**b**) duration-response curves. (**c,d**) Distributions of three types of changes in duration-response curves: BDu shifts (*triangles*), sharpening (*open circles*), and broadening (*filled circles*). The abscissae and ordinates represent BDu and BF differences between IC_r and AC_s neurons, respectively. Each *triangle* (**c**) is a BDu and a BF difference between paired AC_s and BDu-shifted IC_r neurons. Each data point (**d**) is the relationship between a BDu and a BF difference of paired AC_s and IC_r neurons with sharpened (*open circles*) or broadened (*filled circles*) duration-response curves. Crosses (**c, e, f**) mark neurons unchanged in BDu and duration-response curve width. Changes in duration tuning only occur when the BF and BDu differences are <6 kHz and 4 ms, respectively. (**e,f**) BDu shifts (**e**) and width changes (**f**) in duration-response curves. The change depends on the BDu difference between IC_r and AC_s neurons. The larger the BDu difference, the larger the change. The correlation coefficient (*r*) for each regression line, the mean, and standard deviation of the ACs BDu are shown (Ma and Suga 2001)

tuning. Without electric stimulation, the CM BF in the awake mustached bat changes up to 0.15 kHz in an unpredictable way during biosonar emissions (Goldberg and Henson 1998). Such a change may be evoked by the corticofugal system and related to auditory attention to echoes.

3.1.6 Ipsilateral Versus Contralateral Corticofugal Modulation

In the big brown bat AI and ICc, the contralateral BF shift is similar to, but smaller than, the ipsilateral one (Ma and Suga 2001a). In the mustached bat whose DSCF area is highly specialized to process Doppler-shifted echoes, however, the contralateral shift differs from the ipsilateral one.

The DSCF area, specialized for fine frequency and amplitude tuning, is large (Suga and Jen 1976; Suga et al. 1987), with dorsal (DSCFd) and ventral (DSCFv) parts; DSCFd has ipsilaterally inhibited and contralaterally excited (IE) neurons tuned to intense sounds, whereas DSCFv has bilaterally excited (EE) neurons tuned to weak sounds (Manabe et al. 1978). DSCFv has a commissural projection (Liu and Suga 1997).

Electric stimulation of AI DSCFd/v always evokes centrifugal BF shifts of ipsilateral AI and IC DSCF neurons and of contralateral hair cells. However, stimulation of AI DSCFd/v neurons evokes centripetal and centrifugal BF shifts of contralateral DSCFd neurons and centrifugal and centripetal BF shifts of contralateral DSCFv neurons, respectively. Thus, the shift direction reverses depending on the stimulation sites (Xiao and Suga 2005). IC DSCF neurons are clustered in the dorsoposterior division (DPD) (Zook et al. 1985). DSCFd/v stimulation, respectively, evokes centrifugal and centripetal BF shifts of dorsal contralateral DPD neurons and centripetal and centrifugal BF shifts of ventral contralateral DPD neurons (Fig. 24.6). When BMI is applied to the AI DSCF area, ICc and cochlear BF shifts are all centripetal and bilateral. Thus, centrifugal BF shifts are evoked by inhibition in the stimulated AI DSCF area (Xiao and Suga 2002a, 2005). In DPD, two types of binaural neurons, as well as monaural neurons, are clustered separately (Wenstrup et al. 1986). The AI DSCF neurons probably differently modulate DPD IE and EE neurons both in the frequency and spatial domains.

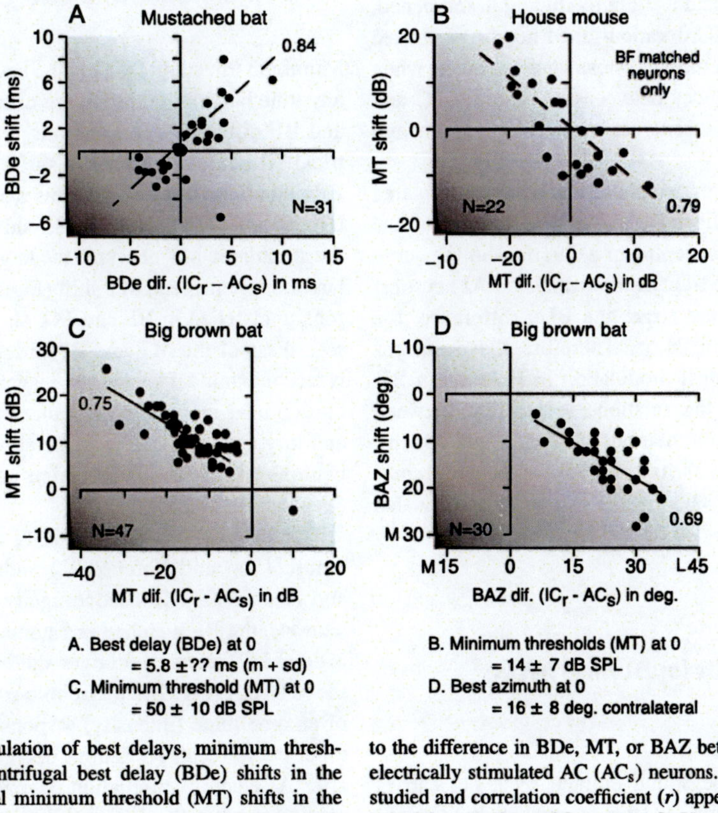

Fig. 24.7 Corticofugal modulation of best delays, minimum thresholds, and best azimuths. Centrifugal best delay (BDe) shifts in the (**a**) mustached bat, centripetal minimum threshold (MT) shifts in the (**b**) house mouse, and (**c**) big brown bat, and centripetal best azimuth (BAZ) shifts in the (**d**) big brown bat. All shifts are linearly related to the difference in BDe, MT, or BAZ between recorded IC (IC_r) and electrically stimulated AC (AC_s) neurons. The number of neurons (*N*) studied and correlation coefficient (*r*) appear in each *panel*. From published data (**a**: Yan and Suga 1996; **b**: Yan and Ehret 2002; **c**, **d**: Jen and Zhou 2003)

4 Multiparametric Corticofugal Modulation

Animal sounds, including human speech sounds, have multiple parameters such as frequency, amplitude (intensity), duration, time interval between sounds, etc. Auditory neurons are tuned to different acoustic parameters besides frequency (Suga 1973, 1984, 1994; Covey and Casseday 1999; Rauschecker and Tian 2000). Corticofugal modulation occurs for different types of subcortical neurons and is multiparametric in the frequency, amplitude, time, and spatial domains (Suga and Ma 2003).

4.1 Modulation of Duration Tuning in the Big Brown Bat

Duration-tuned neurons are sensitive to a particular sound duration and frequency (Pinheiro et al. 1991; Casseday et al. 1994; Ehrlich et al. 1997; Galazyuk and Feng 1997; Ma and Suga 2001b). The maximum neural response at a certain sound duration is the best duration (BDu). When AI duration-tuned neurons are electrically stimulated, an IC duration-tuned neuron matched in BDu and BF with the stimulated AI neuron is augmented and its duration tuning is sharpened, whereas an unmatched IC duration-tuned neuron is shifted or broadened in duration tuning. These changes occur when the BDu and BF differences between the recorded IC and the stimulated cortical neurons are respectively <4 ms and <6 kHz (Fig. 24.6a–d).

The BDu shifts are toward the BDu of the stimulated AI neuron: the larger the BDu difference, the larger the shift (Fig. 24.6e). The broadening of a duration-tuning curve mostly occurs toward the BDu of the stimulated AI neurons: within a certain range, the larger the BDu difference, the larger the broadening (Fig. 24.6f). Therefore, these changes are centripetal. Corticofugal modulation in BDu, as in BF, is specific and reflects the relationship in BDu between the recorded and stimulated neurons (Ma and Suga 2001b). Centripetal BF and BDu shifts can evoke expanded representation of a particular sound in the ICc and logically also in AI.

4.2 Modulation of Delay Tuning in the Mustached Bat

The echo delay from the sound emitted by a bat carries target-distance information. In the mustached bat, delay-tuned neurons are in the AI FM–FM area (Fig. 24.1) (Suga 1994). Electric stimulation of AI delay-tuned neurons augments the response at the cortical and subcortical best delay (BDe) of delay-tuned neurons matched in BDe to the AI neuron, sharpening their delay tuning without shifting their BDe's. It simultaneously suppresses the responses at the BDe of unmatched AI and subcortical delay-tuned neurons and shifts their BDe's away from that of the stimulated AI neuron, evoking centrifugal BDe shifts. The BDe shifts are proportional to the BDe differences in a certain range, between the stimulated and recorded delay-tuned neurons (Fig. 24.7a) (Yan and Suga 1996; Xiao and Suga 2004).

Inactivation of AI delay-tuned neurons with local anesthetic evokes changes in the subcortical delay-tuned neurons opposite to those evoked by AI activation (Yan and Suga 1996). Inactivation studies show that the auditory responses of delay-tuned neurons and the normal delay map are both maintained by the corticofugal system. Bicuculline applied to the FM–FM area transforms BDe shifts from centrifugal to centripetal (Xiao and Suga 2004). Inhibition in the FM–FM as in the DSCF area is apparently strong and widespread and produces centrifugal BDe shifts.

4.3 Modulation of Minimum Threshold in the Mouse and Bat

Minimum threshold (MT) is the neuron's lowest threshold to any tone burst stimulus. In the central auditory system, MT and BF differ between neurons. In the house mouse, electrical AI stimulation evokes shifts in both the MT and the BF of an IC neuron. IC neurons matched to stimulated AI in BF but not in MT show no BF shifts, but MT shifts toward the stimulated MT neurons: the larger the MT difference, the larger the centripetal MT shift (Fig. 24.7b). However, IC neurons unmatched in BF and MT show BF and MT increases regardless of the MT and BF differences. The MT increase is accompanied by a large response decrease. Therefore, corticofugal modulation enhances the neural contrast of an auditory signal by the centripetal MT shifts of BF matched neurons and by the suppression of BF unmatched neurons (Yan and Ehret 2002).

Corticofugal modulation of MT differs between the house mouse (Yan and Ehret 2002) and the big brown bat (Jen and Zhou 2003). In corticofugally inhibited big brown bat neurons, the BF matched and unmatched neurons show centripetal MT shifts: the larger the MT difference, the larger the MT shift (Fig. 24.7c). In both species, the dynamic range of an amplitude (intensity) response (impulses per stimulus tone) curve of an IC neuron decreases when MT increases after AI electric stimulation. Dynamic range and MT were studied only at the BF in the control condition. However, the dynamic ranges and the MTs at the control (original) and shifted (new) BFs of a given IC neuron are most appropriately compared when evaluating corticofugal modulation. If AI stimulation evokes IC BF shifts as in Fig. 24.3a, the dynamic range must increase at the shifted BF and decrease at the control BF.

4.4 Modulation of Spatial Tuning in the Big Brown Bat

The ear's spatial tuning (directional sensitivity) varies with the frequency of a stimulus tone. Binaural interactions in the brain produce neurons whose spatial tuning differs from those determined by the ear. Big brown bat AI stimulation sharpens the spatial tuning curves of corticofugally inhibited cells and broadens those of facilitated neurons (Jen et al. 1998). IC neurons show centripetal best azimuth (BAZ) shifts for AI electric stimulation only when the BF difference between the stimulated and recorded neurons is <6 kHz: the larger the BAZ difference between the stimulated and recorded neurons, the larger the centripetal BAZ shift (Fig. 24.7d) (Zhou and Jen 2005).

Studies on corticofugal modulation in the big brown bat show that tuning curves in the frequency, amplitude, time (duration), and spatial domains each show systematic centripetal shifts for expanded representation of auditory signals frequently stimulating the animal. The IC shifts evoked by electrical stimulation contribute to producing large, long term AI changes that last up to 3.5 h.

4.5 Frequency Map Reorganization After Cochlear Lesions

A partial cochlear lesion causes a permanent partial central auditory inactivation. The central frequency-map reorganization (BF shifts) after the lesion has usually been studied many days postlesion (Harrison et al. 1996; Irvine and Rajan 1996). In contrast, frequency-map reorganization in fear conditioning or AI focal activation/inactivation has been examined usually less than 60 min after the end of the conditioning, activation, or inactivation. Unlike the lesion, however, these evoke reversible reorganization. The neural mechanism for the cochlear lesion-induced reorganization is more like that evoked by focal cortical inactivation than that from conditioning because the cochlear lesion is not related to associative learning but to AI and the corticofugal feedback system. The stimulated BF shifts are brief but last as long as the stimulus, so AI and the corticofugal system continuously participate in postlesion reorganization.

4.6 Non-auditory Augmentation of Corticofugal Modulatory Systems

Does acoustic stimulation evoke corticofugal modulation as cortical electric stimulation does? In the big brown bat only, the BF shifts (plasticity) of IC and AI neurons have been studied with repetitively delivered tone bursts, AI focal electric stimulation (Yan and Suga 1998; Gao and Suga 1998; Chowdhury and Suga 2000; Ma and Suga 2001a), and auditory fear conditioning (Gao and Suga 1998, 2000; Ji et al. 2001). These stimuli elicit the same BF shifts in the ICc, although tone burst stimulation is less effective for eliciting BF shifts. Activation and inactivation experiments (Suga et al. 2000, 2002) show that conditioning shifts IC BFs via the corticofugal system, as does AI stimulation. Therefore, the IC BF shifts evoked by such stimulation are not epiphenomena, and AI stimulation is a valid approach allowing stimulation data to be related to those obtained from conditioning experiments.

4.6.1 Time Course of Frequency Shifts After Electric Stimulation and Fear Conditioning

For 30-min long focal AI electric stimulation, IC and AI BF shifts develop together and peak at stimulus termination. These BF shifts disappear within 3.5-h poststimulation. The AI BF shift tends to last slightly longer than the IC shift (Chowdhury and Suga 2000; Ma and Suga 2001a, 2003). The slow recovery indicates that these tuning shifts are related to auditory plasticity.

In a 30-min long conditioning session with the pairs of conditioning tonal stimulation (CS) and unconditioned leg stimulation (US), the IC BF shift develops and then slowly decays like that evoked by AI stimulation, whereas the AI shift develops slowly and is still present 26 hours later. The long-term AI shift does not reflect BF shift saturation since another conditioning session 3.5 h later causes a short-term IC BF shift and enhances the long-term AI BF shift (Gao and Suga 2000).

When atropine is applied to the ICc 70 min after conditioning onset, the IC shift is transiently reduced but the AI shift which has developed to a plateau is unaffected (Ji et al. 2001). The conditioning-elicited IC shift is largely evoked by the corticofugal feedback transferring conditioning-elicited cortical changes. The IC shift contributes to large long-term AI shifts via feedback loops (Ji et al. 2001). The cholinergic system has an important role in evoking both AI and IC shifts. The AI and IC shifts from AI stimulation are affected by acetylcholine (ACh) and atropine as are those elicited by the conditioning. ACh applied to AI augments short-term shifts evoked by AI stimulation and changes it into a long-term BF shift. ACh applied to the IC augments the short-term IC shift only (Ma and Suga 2005).

Application of the glutamatergic agonist NMDA to AI or ICc augments the auditory responses as ACh does, whereas 2-amino-5-phosphovalerate (APV), an antagonist of NMDA receptors, reduces the auditory responses as atropine does. Although none of these four drugs evokes BF shifts without acoustic stimuli, they influence the development of the long-term cortical and short-term collicular BF shifts elicited by conditioning. Like ACh, NMDA augments both the cortical and collicular BF shifts regardless of whether it is applied to the AI or ICc. Blockade of collicular NMDA receptors by APV abolishes the development of the collicular BF shift and makes the cortical BF shift small and short term. Blockade of cortical NMDA receptors by APV reduces the cortical and collicular BF shifts and renders the cortical BF shift short term. NMDA applied to the AI prior to the conditioning increases the cortical BF shift, however, only for a short while without the presence of ACh. Without NMDA, ACh applied to the AC prior to the conditioning can produce a long-term cortical BF shift, although it is small (Ji et al. 2005).

4.6.2 Circuitry for Frequency Changes in Auditory Fear Conditioning

A conditioned bat shows body movements (Gao and Suga 1998) and a heart rate decrease (Ji and Suga 2007) to the conditioned tone. Behavioral conditioned responses have been well studied and we focus here on the neural mechanisms of BF shifts. The cortical and collicular BF shifts observed in the big brown bat are evoked by auditory fear conditioning, i.e., by short acoustic stimulation (CS) paired with leg stimulation (US) but not by either US or CS alone. Backward conditioning (US–CS) does not evoke these BF shifts which are specific to the frequency of the CS. Unlike the collicular BF shift, the cortical BF shift increases and reaches a stable plateau after the termination of the conditioning (Gao and Suga 2000; Suga et al. 2000; Ji et al. 2001). This long-term cortical BF shift is elicited by an increase in cortical acetylcholine (Suga et al. 2000; Ji et al. 2001; Ji and Suga 2003) and, therefore, is related to physiological associative memory.

Thirteen neurophysiological findings are directly related to the mechanisms for the cortical and collicular BF shifts elicited by auditory fear conditioning.

(1) Auditory fear conditioning elicits the IC BF shift that lasts up to 3.5 h and is specifically related to the frequency of the CS (Gao and Suga 1998, 2000; Ji et al. 2001, 2005; Ji and Suga 2003). In the ascending auditory system, the IC in the midbrain is located below the MGB in the thalamus but shows the short-term BF shift because of corticofugal feedback. Therefore, the MGBm is not the first nucleus in the ascending auditory system where CS–US-associated responses are found.

(2) Focal electric stimulation of AI evokes AI and IC BF shifts which last up to 3.5 h and are specifically related to the BF of the stimulated neurons (Ma and Suga 2003; Chowdhury and Suga 2000; Sakai and Suga 2001, 2002; Xiao and Suga 2002b; Yan and Ehret 2002). The direction of these AI and IC BF shifts changes when an antagonist (Xiao and Suga 2002b; Ma and Suga 2004) or an agonist (Ma and Suga 2004) of GABA-A receptors is applied to AI, the site of the electric stimulation. This indicates that the neural circuit within AI plays an essential role in evoking BF shifts and that CS–US association in the MGBm is not required for the AI BF shift.

(3) Bilateral inactivation of the somatosensory cortex by an agonist of GABA-A receptors does not affect AI and IC auditory responses (Gao and Suga 1998, 2000) and the development of non-specific augmentation of AI auditory neurons elicited by pseudoconditioning (Ji and Suga 2007), but selectively abolishes the development of the conditioning-dependent AI and IC BF shifts (Gao and Suga 1998, 2000), particularly lengthening the duration of the AI BF shift. This does not occur if the cholinergic basal forebrain is lesioned (Ma and Suga 2001a, 2003). This pair of inactivation and activation experiments indicates that the somatosensory cortex, through the cholinergic basal forebrain, plays an essential role in the development of conditioning-dependent BF shifts.

(4) Inactivation of AI by an agonist of GABA-A receptors blocks the development of the conditioning-dependent, short-term IC BF shift without affecting the IC frequency tuning (Gao and Suga 1998). Focal electric stimulation of AI evokes a short-term IC BF shift that is the same as that elicited by the conditioning (Ma and Suga 2001a, 2003; Gao and Suga 2000). This pair of inactivation and activation experiments indicates that the conditioning-dependent IC BF shift is evoked by the corticofugal feedback system and that this short-term IC BF shift contributes to the development of the large long-term AI BF shift (Ji et al. 2001).

(5) Focal electric stimulation of the ICc (Zhang and Suga 2005) or the MGBv (Wu and Yan 2007) evokes the IC BF shift, which does not occur when AI is inactivated. Electric stimulation of the MGBv evokes the AI BF shift (Jafari et al. 2007). These findings indicate that the lemniscal pathway is important in evoking the BF shifts.

(6) Focal electric stimulation of AI evokes short-term AI and IC BF shifts that are nearly identical to each other

in amount and time course (Ma and Suga 2001a). When this AI electric stimulation is accompanied by an ACh application to AI, the large long-term AI BF shift and the large short-term IC BF shifts are evoked, as are those elicited by the conditioning (Ma and Suga 2005). ACh plays an essential role in evoking the large long-term BF shifts. The long-term AI BF shift can be evoked without CS–US association in the multi-sensory thalamic nuclei (MGBm and PIN) and without conditioned behavioral responses.

(7) ACh applied to AI or the IC prior to conditioning augments both AI and IC BF shifts. Atropine (an antagonist of muscarinic ACh receptors) applied to AI prior to conditioning blocks the AI BF shift and reduces the IC BF shift. Atropine applied to the IC prior to that blocks the IC BF shift and reduces the AI BF shift, changing it from long term to short term (Ji et al. 2001). These findings indicate that muscarinic ACh receptors and the feedback loops formed by the descending (corticofugal) and ascending auditory pathways play a role in evoking the cortical and subcortical BF shifts.

(8) Electric stimulation of the cholinergic basal forebrain (nucleus basalis) augments the development of AI and IC BF shifts evoked by either AI electric stimulation or tone burst stimulation (Ma and Suga 2003). It also evokes the large long-lasting AI BF shift when it is delivered together with tone burst stimulation (Bakin and Weinberger 1996; Bjordahl et al. 1998; Kilgard and Merzenich 1998; Yan and Zhang 2005; Zhang et al. 2005). These findings indicate that ACh released in AI by the nucleus basalis augments the AI and IC BF shifts and makes the AI BF shift long term. It also indicates that the BF shifts of AI neurons can be evoked without CS–US association emanating from the multi-sensory MGBm and PIN and without conditioned behavioral responses.

(9) Conditioning-dependent changes in impulse discharges occur in the cholinergic basal forebrain (Quirk et al. 1995, 1997; Maren 2000) and the lateral amygdala (Li et al. 1996; Armony et al. 1998) before AI. The development of long-latency, conditioning-dependent discharges in AI is abolished by a lesion of the amygdala (Armony et al. 1998). (BF shifts in the lateral amygdala have not yet been studied.) Inactivation of the amygdala prevents the development of conditioning-dependent plastic changes in the MGBm (Poremba and Gabriel 2001; Maren et al. 2001). These findings indicate that the origin of conditioning-dependent changes in AI and the lateral amygdala is not located in the MGBm and PIN that project to them. It is possible that the conditioning-dependent changes in AI are evoked via the pathway from the amygdala to the cholinergic basal forebrain and to AI, and that the changes in the MGBm are evoked via the pathway from AI to the IC and to the MGBm.

(10) In general, MGBm neurons have a broad or multi-peaked frequency-tuning curve and habituate after several stimulus presentations (Aitkin 1973; Calford 1983; Bordi and LeDoux 1994a, b). They broadly project to cortical auditory areas including AI (Rose et al. 1958). Therefore, MGBm neurons are not suited for the fine adjustment of the central auditory system for auditory signal processing and for evoking the AI BF shift, but for evoking AI plasticity other than the BF shift.

(11) Electric stimulation of AI facilitates MGBv neurons, but inhibits MGBm neurons through the thalamic reticular nucleus (Yu et al. 2004). If MGBm neurons evoked the AI BF shift, MGBm neurons should not be inhibited by AI.

(12) ACh depolarizes MGBv neurons, but hyperpolarizes MGBm neurons (Mooney et al. 2004). If MGBm neurons evoked the AI BF shift, MGBm neurons should not be suppressed by ACh, because augmentation of the BF shifts in the central auditory system depends on ACh.

(13) For the measurement of a BF shift, the frequency of a tone burst is scanned by, e.g., 0.5-kHz steps. Then, the BF shift less than 0.5 kHz is hardly detected. Such a small BF shift may be defined as a subthreshold BF shift. The presence of a subthreshold BF shift is evident because it is easily changed into a large BF shift by ACh or NMDA applied to AI (Ji et al. 2001, 2005) or by electric stimulation of the nucleus basalis (Bakin and Weinberger 1996; Bjordahl et al. 1998; Kilgard and Merzenich 1998; Ma and Suga 2003; Yan and Zhang 2005; Zhang et al. 2005). A short train of tone bursts such as a CS evokes the subthreshold AI and IC BF shifts (Gao and Suga 1998; Ji et al. 2001). Such subthreshold BF shifts can also be changed into small BF shifts by lengthening the train duration, e.g., by a long train of tone bursts lasting 30 min (Yan and Suga 1998; Gao and Suga 1998, 2000; Chowdhury and Suga 2000; Ma and Suga 2001a, 2003). What is important is that acoustic stimulation alone can evoke small or subthreshold BF shifts and that there are neural pathways via the prefrontal cortex (Zaborszky et al. 1999; Golmayo et al. 2003; Rasmusson et al. 2007) or amygdala through which the cortical ACh level is increased by acoustic, somatosensory, or visual stimulation alone.

Based on these thirteen findings, Gao and Suga (1998) proposed a circuit model to explain the tone-specific cortical and collicular plasticity, BF shifts, excluding the conditioned behavioral responses or discharge rate changes, elicited by auditory fear conditioning (CS–US). The model states that small and short-term cortical and collicular BF shifts specific to tone bursts (CS) are evoked by the AI and corticofugal feedback system activated by CS alone, and that this cortical BF shift is augmented and stabilized by ACh released into the AI from the cholinergic basal forebrain which is activated by the auditory and somatosensory cortices through the association cortex and the amygdala. That is, the signal for the cortical change due to CS–US association arrives at the AI through the amygdala and cholinergic basal forebrain. The collicular BF shift, which needs ACh in addition to corticofugal signals (Ji et al. 2001), is also increased by the augmented cortical BF shift through the corticofugal system and contributes to the development of the large long-term cortical BF shift. According to this model, the minimally necessary neural elements for evoking the large long-term cortical BF shift are AI, the corticofugal system, and neurons releasing ACh into AI. The collicular BF shift evoked by corticocollicular feedback has been discussed here in detail. A thalamic BF shift evoked by corticothalamic feedback would function in a way similar to corticocollicular feedback (Zhang and Suga 2000).

The Gao–Suga model differs from the Weinberger model (1998). In that model, CS–US association occurs in the multi-sensory thalamic nuclei – the medial division of the medial geniculate body (MGBm) and the posterior intralaminar nucleus (PIN). Then, the MGBm and PIN send the "associated" signal to the AC and evokes a small short-term cortical BF shift. At the same time, the MGBm and PIN send the associated signal to the amygdala. The amygdala then sends an associated signal to the cholinergic basal forebrain, which releases ACh in the AC. This augments the small short-term cortical BF shift and stabilizes it for the long term. As described above, the large long-term cortical BF shift can be evoked without the activation of the MGBm by CS–US association. Physiological properties of MGBm neurons (Aitkin 1973; Calford 1983; Bordi and LeDoux 1994a, b) appear not to be suited for the fine adjustment and improvement of the central auditory system for signal processing. At this stage, no electric stimulation or lesion experiments have been performed to show that the MGBm-AC projection is essential in evoking the tone-specific cortical BF shift.

Changes in the impulse discharges of MGBm neurons related to CS–US association have been reported. Such discharge changes and BF shifts are perhaps mutually related, but likely constitute different neurophysiological events, because BF shifts depend not only on frequency-dependent facilitation but also on inhibition. The role of MGBm neurons in evoking the changes in the frequency-tuning curves of central auditory neurons remains to be further examined.

4.6.3 Tone-Specific BF Shifts Versus Conditioned Behavioral Responses

The projections to the amygdala from the MGBm/PIN (LeDoux et al. 1990a, b; Turner and Herkenham 1991) and also from the spino-thalamic and spino-parabrachial tracts (Lanuza et al. 1999) play an essential role in eliciting conditioned behavioral responses (Iwata et al. 1986; LeDoux et al. 1984, 1986). In the Gao–Suga model, the neural pathway for the cortical BF shift is largely separated from that of conditioned behavioral responses.

Decorticated animals (DiCara et al. 1970; Norman et al. 1974; Mauk and Thompson 1987) and humans (Berntson et al. 1983) retain or acquire conditioned behavioral responses. A large lesion of the AC does not prevent the acquisition of conditioned behavioral responses (LeDoux et al. 1990b; Romanski and LeDoux 1993; Armony et al. 1997). This indicates that the cortical BF shift is not directly involved in evoking conditioned behavioral responses.

Tone bursts accompanied with electric stimulation of the cholinergic basal forebrain evoke cortical BF shifts (Bakin and Weinberger 1996; Bjordahl et al. 1998; Kilgard and Merzenich 1998a; Ma and Suga 2003) and induce tone-specific behavioral responses (McLin et al. 2002). This does not imply that the behavioral responses are induced by the activity of the cortical BF-shifted neurons, because decorticated animals show conditioned behavioral responses (DiCara et al. 1970; Norman et al. 1974; Mauk and Thompson 1987), and these behavioral responses are more likely induced by the activation of the amygdala which receives both auditory signals through the MGBm and neural signals from the electrically stimulated basal forebrain.

The cortico-amygdala and MGBm-amygdala projections converge into the same region of the lateral amygdala (LeDoux et al. 1991) and even onto the same neurons (Li et al. 1996). These two inputs must differ in function. The cortico-amygdala projection has an important role in the conditioning-dependent augmentation of the tone-specific BF shift (Gao and Suga 2000) and, presumably, in modulating the activity of the MGBm-amygdala projection for conditioned behavioral responses. The finding that tone-discrimination behavior trained with CS1-US vs. CS2 changes into fear responses to both CS1 and CS2 with a lesioned AC (Jarrell et al. 1987) indicates that the cortico-amygdala projection indeed has an influence on the function of the MGBm-amygdala projection.

4.6.4 Auditory Memory Versus Associative Memory

The cortical BF shift evoked by repetitive acoustic stimulation may be considered a physiological auditory memory trace even if it is augmented by electric stimulation of the cholinergic basal forebrain or by auditory fear conditioning that activates the cholinergic basal forebrain. It is uncertain whether the conditioning-augmented cortical BF shift is a physiological "associative" memory trace as described by Weinberger and Bakin (1998), because it has not yet been demonstrated that the BF-shifted neurons produced by the conditioning carry the information of both auditory and somatosensory stimuli. If conditioned behavioral responses cannot be evoked without the activation of the cortical BF-shifted neurons, the cortical BF shift would be an associative memory trace. By contrast, if conditioned behavioral responses can be evoked without the activation of the cortical BF-shifted neurons, the BF shift itself would not be an associative memory trace. As discussed earlier, the cortical BF shift is a neuronal change that can be separated from conditioned behavioral responses. Therefore, it is likely not a physiological associative memory trace.

An AC lesion abolishes fear conditioning in animals with a disrupted MGBm-amygdala projection in contrast to the finding in decorticated animals (DiCara et al. 1970; Norman et al. 1974; Mauk and Thompson 1987). Therefore, it has been concluded that the AC is not necessary for, but can mediate, simple fear conditioning (Romanski and LeDoux 1993). The neuronal circuit in adult animals for auditory signal processing is plastic (e.g., Heffner and Heffner 1984). When an animal with a disrupted MGBm-amygdala projection is trained for fear conditioning, its function in the normal animal might be taken over by the cortico-amygdala projection. Therefore, the result of the above lesion experiment (Romanski and LeDoux 1993) is not necessarily contradictory to the data obtained from decorticated animals.

4.6.5 Multiparametric Cholingeric and Dopaminergic Modulation

The augmentation of auditory cortical plasticity in the frequency domain (BF shifts) by the cholinergic basal forebrain (Bakin and Weinberger 1996; Kilgard and Merzenich 1998a; Ma and Suga 2003) or the dopaminergic ventral tegmental area (Bao et al. 2001) has been well documented. In rats, responses of cortical neurons to amplitude-modulated sounds (Kilgard and Merzenich 1998b) or a combination of sounds (Kilgard and Merzenich 2002) are augmented by electric stimulation of the basal forebrain. FM–FM neurons of the mustached bat are combination-sensitive neurons that are tuned to specific spectrotemporal patterns of sounds (O'Neill and Suga 1982; Suga et al. 1983). Activation or inactivation of cortical FM–FM neurons modulates subcortical FM–FM neurons (Yan and Suga 1996, 1999). Therefore, multiparametric changes evoked by electric stimulation of the cholinergic or dopaminergic system together with acoustic stimulation at least partially depend on the corticofugal feedback.

5 Functions of the Corticofugal System

5.1 *Egocentric Selection in Different Species and Sensory Systems*

Egocentric selection is based on positive feedback associated with lateral inhibition occurring in AI and the subcortical auditory nuclei (Suga et al. 1995; Yan and Suga 1996; Zhang et al. 1997; Zhang and Suga 1997; He 1997). In terms of reorganization, cortical facilitation carries an attractive message, whereas inhibition implies rejection.

Auditory reorganization in adult animals depends on the balance between these two messages. The effect of egocentric selection on the subcortical nuclei and AI is brief, though short-term plasticity is augmented and becomes long term when an acoustic signal is made behaviorally salient via coactivation of the auditory and other sensory systems.

Egocentric selection occurs in the auditory systems of the big brown bat (Fig. 24.5a) (Yan and Suga 1998; Chowdhury and Suga 2000; Ma and Suga 2001a), mustached bat (Fig. 24.5b) (Sakai and Suga 2001), Mongolian gerbil (Fig. 24.5d) (Sakai and Suga 2001, 2002), house mouse (Fig. 24.5e) (Yan and Ehret 2002) and cat (He 1997, 2003). Tone bursts paired with basal forebrain electric stimulation evoke centripetal BF shifts, egocentric selection, in AI in the rat (Kilgard and Merzenich 1998a), big brown bat (Ma and Suga 2003), and house mouse (Zhang et al. 2005; Yan and Zhang 2005). Auditory fear conditioning causes centripetal BF shifts in the guinea pig AI (Weinberger and Bakin 1998) and big brown bat ICc and AI (Gao and Suga 1998, 2000; Ji et al. 2001).

Egocentric selection occurs in the mustached bat cochlea (Xiao and Suga 2002a) and probably in humans (Khalfa et al. 2001). In guinea pigs, the crossed olivocochlear fibers increase signal-to-noise ratio (Nieder and Nieder 1970; Dolan and Nuttall 1989; Kawase et al. 1993) and in monkeys, they improve complex sound discrimination in noise (Dewson 1968). Such improvement is most likely due to corticofugally mediated egocentric selection.

In the cat visual system, corticothalamic modulation is prominent (Singer 1993; Sillito et al. 1993, 1994) and orientation-sensitive neurons enable egocentric selection (Tsumoto et al. 1978; McClurkin and Marrocco 1984; Marrocco and McClurkin 1985; Murphy et al. 1999). Shifts in visual cortex orientation selectivity are also centripetal (Godde et al. 2002). The intracortical feedback projection from the motion-sensitive area to area 18 augments the visual responses of neurons in area 18 and strongly contributes to the emergence of direction sensitivity (Galuske et al. 2002).

In the somatic sensory system, egocentric selection occurs (Malmierca and Nunez 1998; Canedo and Aguilar 2000;

Ghazanfar et al. 2001) and the monkey and rat corticofugal somatic sensory system plays a role in thalamic plasticity (Ergenzinger et al. 1998; Krupa et al. 1999; Rasmusson 2000; Chowdhury et al. 2004). Receptive field shifts in somatic sensory cortex neurons are centripetal in several species of mammals (Buonomano and Merzenich 1998; Rasmusson 2000).

In humans, corticofugal effects on cochlear hair cells (Khalfa et al. 2001) are comparable to those in the mustached bat (Xiao and Suga 2002a). Therefore, corticofugal functions such as egocentric selection and subcortical plasticity in the big brown bat are apparently seen in the auditory, visual, and somatic sensory systems of non-bat species. In contrast, compressed reorganization has thus far been found only in the mustached bat (Yan and Suga 1996; Zhang et al. 1997; Zhang and Suga 2000; Xiao and Suga 2002a,b).

5.2 Other Corticofugal Contributions

Is the corticofugal function observed in echolocation-specialized bats relevant to other species (He 2003)? Many aspects of the corticofugal function for egocentric selection and plasticity in bats are also found in the auditory system of non-bat species and in the visual and somatic sensory systems. It is important to identify the common and specialized functions between species and to explore common and specialized mechanisms. Only in the mustached bat's specialized Doppler shift or echo delay systems are shifts in frequency and delay tuning all centrifugal, and thus appear to be unique. This difference reflects a quantitative, not qualitative, difference in facilitatory-inhibitory balance in these specialized subsystems and in other auditory systems such as those in the big brown bat and Mongolian gerbil. By strengthening inhibition, the unique mustached bat corticofugal modulation derives from common corticofugal neural mechanisms. Therefore, even specialized bat subsystems share corticofugal functions and mechanisms with non-bat mammals.

Gain control: Despite some differences, egocentric selection may be viewed as selective gain control. Effects of excitatory and/or inhibitory corticofugal modulation found in cats (Watanabe et al. 1966; Massopust and Ordy 1962; Andersen et al. 1972; Villa et al. 1991; Orman and Humphrey 1981; Amato et al. 1969; Ryugo and Weinberger 1976), rats (Syka and Popelar 1984) and bats (Jen et al. 1998; Sun et al. 1996; Zhang and Suga 1997; Yan and Suga 1999) can be interpreted as gain control for auditory signal processing. COCB activity changes the minimum thresholds of cochlear hair cells (guinea pig, Brown and Nuttall 1984) and cochlear nerve fibers (cat, Wiederhold 1970) and changes the dynamic range of intensity coding in auditory nerve fibers (Geisler 1974). Corticofugal modulation of the dynamic range has also been observed in the ICc of mice (Yan and Ehret 2002) and bats (Zhou and Jen 2000).

Attention: The cholinergic basal forebrain may play an essential role in attentional modulation of sensory signal processing through the corticofugal system (Sarter and Bruno 2000; Montero 2000). In cats, visual attention reduces the auditory responses of the dorsal cochlear nucleus (Hernandez-Peon et al. 1956) and a visual discrimination task reduces auditory nerve responses to clicks (Oatman 1971; Oatman and Anderson 1977). In humans, visual attention reduces auditory nerve responses (Lukas 1980) and sound emissions by the cochlea evoked by clicks (Puel et al. 1988). In the mustached bat, cochlear best frequencies randomly fluctuate during the emission of biosonar pulses (Goldberg and Henson 1998), but systematically vary with the location of focal cortical electric stimulation. The corticofugal system likely mediates attentional modulation of auditory signal processing (Xiao and Suga 2002a).

Protection: Cochlear outer hair cells act as an amplifier. If the gain of this cochlear amplifier is significantly reduced by the COCB, cochlear hair cells may be reflexively protected from acoustic injury (Xie and Henson 1998; Maison and Liberman 2000). This reflex may be corticofugally modulated.

Binding: The problem of feature binding has been extensively studied in the visual system. In the cat's visual system, corticofugal activity can evoke feature-linked synchronized discharges in thalamic neurons (Sillito et al. 1994; Gray et al. 1989). It is unknown whether the auditory corticofugal system has a similar effect. In the mustached bat, the responses of thalamic and collicular combination-sensitive neurons are greatly augmented by corticofugal feedback (Yan and Suga 1999). Perhaps creation of combination sensitivity can be considered a local binding phenomenon that is influenced by the auditory corticofugal system.

Adaptive filtering: The thalamic reticular nucleus receives axon collaterals from both ascending thalamo-cortical and descending corticofugal fibers. Corticofugal positive feedback has a high gain (Zhang and Suga 1997; Yan and Suga 1999), evoking long-lasting discharges if the feedback is not modulated by the thalamic reticular nucleus. Improper function of the thalamic reticular nucleus may result in long-lasting discharges, perhaps responsible for tinnitus (Suga et al. 1995). In cats, cooling of the AC had complex effects on the auditory responses of the MGB and reticular nuclear neurons. The reticular nucleus may play a role as an adaptive filter (Villa et al. 1991).

Oscillations: The corticofugal visual system transmits slow oscillatory changes in cortical activity to the thalamic visual nucleus modulating neural excitability, interacting with spindles generated in the thalamus, and producing different brain rhythms characteristic of various behavioral states (Steriade 1999). The corticofugal auditory system may also transmit slow oscillatory changes in cortical activity to the thalamic auditory nucleus (He 2003).

6 Epilogue

The corticofugal auditory system sharpens and shifts subcortical neural tuning curves in the frequency, amplitude, time, and spatial domains. It plays a key role in auditory reorganization in multiparametric domains in adults according to auditory experiences.

Corticofugal functions for egocentric selection and plasticity in bats are shared with the auditory systems of non-bat species as well as with the visual and somatosensory systems. Our goal is to identify common and specialized functions and to explore the underlying mechanisms. The highly specialized processing subsystems for Doppler shifts and echo delays in the mustached bat are unique implementations but are based on principles derived from general corticofugal neural mechanisms.

Cortical and subcortical reorganization by corticofugal modulation occurs by an uneven distribution of neural activity in AI, and it can be evoked by acoustic stimulation, focal electric stimulation, focal drug application, and cochlear lesion. The reorganization is controlled by neuromodulatory systems. For stimulus-specific auditory system reorganization in auditory fear conditioning, key roles are played by the auditory cortex, corticofugal system, somatic sensory cortex, amygdala, and the cholinergic nucleus basalis. Auditory system reorganization can be expansion or compression of the physiological maps. The former is found in many species of animals and sensory systems, whereas the latter has been found so far only in the mustached bat's specialized auditory subsystems (Suga and Ma 2003). For behaviorally unimportant sounds repetitively delivered, AI's neuronal response properties change only slightly; when the sound becomes behaviorally salient, the AI neural network and corticofugal systems implement large and enduring changes through associative learning (Gao and Suga 1998, 2000). A neuroethological approach can reveal the cortical auditory mechanisms enabling this plasticity in a functional context.

Acknowledgments Our work has been supported by a research grant from the National Institute on Deafness and Other Communication Disorders (DC-000175). I thank my collaborators Drs Syed M. Chowdhury, Enquan Gao, Weiqing Ji, Xiaofeng Ma, Masashi Sakai, Zhongju Xiao, Jun Yan, and Yungfeng Zhang for their excellent research and also my laboratory technician, Ms Sally E. Miller, for editing this chapter.

References

Aitkin LM (1973) Medial geniculate body of the cat: responses to tonal stimuli of neurons in medial division. *Journal of Neurophysiology* **36**:275–283.

Aitkin LM and Dunlop CW (1969) Inhibition in the medial geniculate body of the cat. *Experimental Brain Research* **7**:68–83.

Amato G, La Grutta V, and Enia F (1969) The control exerted by the auditory cortex on the activity of the medial geniculate body and inferior colliculus. *Archives of Scientific Biologica* (Bologna) **53**:291–313.

Andersen P, Junge K, and Sveen O (1972) Cortico-fugal facilitation of thalamic transmission. *Brain, Behavior and Evolution* **6**:170–184.

Armony JL, Quirk GJ, and LeDoux JE (1998) Differential effects of amygdala lesions on early and late plastic components of auditory cortex spike trains during fear conditioning. *Journal of Neuroscience* **18**:2592–2601.

Armony JL, Servan-Schreiber D, Romanski LM, Cohen JD, and LeDoux JE (1997) Stimulus generalization of fear responses: effects of auditory cortex lesions in a computational model and in rats. *Cerebral Cortex* **7**:157–65.

Bakin JS and Weinberger NM (1996) Induction of physiological memory in the cerebral cortex by stimulation of the nucleus basalis. *Proceedings of the National Academy of Sciences of the United States of America* **93**:11219–11224.

Bao S, Chan VT, and Merzenich MM (2001) Cortical remodeling induced by activity of ventral tegmental dopamine neurons. *Nature* **412**:79–83.

Berntson GG, Tuber DS, Ronca AE, and Bachman DS (1983) The decerebrate human: associative learning. *Experimental Neurology* **81**:77–88.

Bjordahl TS, Dimyan MA, and Weinberger NM (1998) Induction of long-term receptive field plasticity in the auditory cortex of the waking guinea pig by stimulation of the nucleus basalis. *Behavioral Neuroscience* **112**:467–479.

Bordi F and LeDoux JE (1994a) Response properties of single units in areas of rat auditory thalamus that project to the amygdala. I. Acoustic discharge patterns and frequency receptive fields. *Experimental Brain Research* **98**:261–274.

Bordi F and LeDoux JE (1994b) Response properties of single units in areas of rat auditory thalamus that project to the amygdala. II. Cells receiving convergent auditory and somatosensory inputs and cells antidromically activated by amygdala stimulation. *Experimental Brain Research* **98**:275–286.

Brown MC and Nuttall AL (1984) Efferent control of cochlear inner hair cell responses in the guinea-pig. *Journal of Physiology (London)*. **354**:625–646.

Buonomano DV and Merzenich MM (1998) Cortical plasticity: from synapses to maps. *Annual Review of Neuroscience* **21**:149–186.

Calford MB (1983) The parcellation of the medial geniculate body of the cat defined by the auditory response properties of single units. *Journal of Neuroscience* **3**:2350–2364.

Canedo A and Aguilar J (2000) Spatial and cortical influences exerted on cuneothalamic and thalamocortical neurons of the cat. *European Journal of Neuroscience* **12**:2515–2533.

Casseday JH, Ehrlich D, and Covey E (1994) Neural tuning for sound duration: role of inhibitory mechanisms in the inferior colliculus. *Science* **264**:847–850.

Chowdhury SA, Greek KA, and Rasmusson DD (2004) Changes in corticothalamic modulation of receptive fields during peripheral injury-induced reorganization. *Proceedings of the National Academy of Sciences of the United States of America* **101**:7135–7140.

Chowdhury SA and Suga N (2000) Reorganization of the frequency map of the auditory cortex evoked by cortical electrical stimulation in the big brown bat. *Journal of Neurophysiology* **83**:1856–1863.

Covey E and Casseday JH (1999) Timing in the auditory system of the bat. *Annual Review of Physiology* **61**:457–476.

Desmedt JE and Mechelse K (1958) Suppression of acoustic input by thalamic stimulation. *Proceedings of the Society of Experimental Biological Medicine* **99**:772–775.

Dewson JH (1968) Efferent olivocochlear bundle: some relationships to stimulus discrimination in noise. *Journal of Neurophysiology* **31**:122–130.

DiCara LV, Braun JJ, and Pappas BA (1970) Classical conditioning and instrumental learning of cardiac and gastrointestinal responses following removal of neocortex in the rat. *Journal of Comparative Physiology and Psychology* **73**:208–216.

Dolan DF and Nuttall AL (1989) Inner hair cell responses to tonal stimulation in the presence of broadband noise. *Journal of the Acoustical Society of America* **86**:1007–1012.

Ehrlich D, Casseday JH, and Covey E (1997) Neural tuning to sound duration in the inferior colliculus of the big brown bat, *Eptesicus fuscus*. *Journal of Neurophysio*logy **77**:2360–2372.

Ergenzinger ER, Glasier MM, Hahm JO, and Pons TP (1998) Cortically induced thalamic plasticity in the primate somatosensory system. *Nature Neuroscience* **1**:226–229.

Feliciano M, Saldana E, and Mugnaini E (1995) Direct projections from the rat primary auditory neocortex to nucleus sagulum, paralemniscal regions, superior olivary complex and cochlear nuclei. *Auditory Neuroscience* **1**:287–308.

Galazyuk AV and Feng AS (1997) Encoding of sound duration by neurons in the auditory cortex of the little brown bat, *Myotis lucifugus*. *Journal of Comparative Physiology A* **180**:301–311.

Galuske RA, Schmidt KE, Goebel R, Lomber SG, and Payne BR (2002) The role of feedback in shaping neural representations in cat visual cortex. *Proceedings of the National Academy of Sciences of the United States of America* **99**:17083–17088.

Gao E and Suga N (1998) Plasticity of midbrain auditory frequency map mediated by the corticofugal system in bat. *Proceedings of the National Academy of Sciences of the United States of America* **95**:12663–12670.

Gao E and Suga N (2000) Experience-dependent plasticity in the auditory cortex and the inferior colliculus of bats: role of the corticofugal system. *Proceedings of the National Academy of Sciences of the United States of America* **97**:8081–8086.

Geisler CD (1974) Model of crossed olivocochlear bundle effects. *Journal of the Acoustical Society of America* **56**:1910–1902.

Ghazanfar AA, Krupa DJ, and Nicolelis MA (2001) Role of cortical feedback in the receptive field structure and nonlinear response properties of somatosensory thalamic neurons. *Experimental Brain Research* **141**:88–100.

Godde B, Leonhardt R, Cords SM, and Dinse HR (2002) Plasticity of orientation preference maps in the visual cortex of adult cats. *Proceedings of the National Academy of Sciences of the United States of America* **99**:6352–6357.

Goldberg RL and Henson OW Jr (1998) Changes in cochlear mechanics during vocalization: evidence for a phasic medial efferent effect. *Hearing Research* **122**:71–81.

Golmayo L, Nunez A, and Zaborszky L (2003) Electrophysiological evidence for the existence of a posterior cortical-prefrontal-basal forebrain circuitry in modulating sensory responses in visual and somatosensory rat cortical areas. *Neuroscience* **119**:597–609.

Gray CM, König P, Engel AK, and Singer W (1989) Oscillatory responses in cat visual cortex exhibit inter-columnar synchronization which reflects global stimulus properties. *Nature* **338**, 334–337.

Harrison RV, Ibrahim D, Stanton SG, and Mount RJ (1996) Reorganization of frequency maps in chinchilla auditory midbrain after long-term basal cochlear lesions induced at birth. In: Salvi RJ, Hendersen D, Fiorino F, and Colletti V (eds). *Auditory System Plasticity and Regeneration*, volume 19. Thieme Medical Publishers, New York, pp. 238–255.

He J (1997) Modulatory effects of regional cortical activation on the onset responses of the cat medial geniculate neurons. *Journal of Neurophysiology* **77**:896–908.

He J (2003) Corticofugal modulation of the auditory thalamus. *Experimental Brain Research* **153**:579–590.

He J, Yu YQ, Xiong Y, Hashikawa T, and Chan YS (2002) Modulatory effect of cortical activation on the lemniscal auditory thalamus of the Guinea pig. *Journal of Neurophysiology* **88**:1040–1050.

Heffner HE and Heffner RS (1984) Temporal lobe lesions and perception of species-specific vocalizations by macaques. *Science* **226**:75–76.

Hernandez-Peon R, Scherrer H, and Jouvet M (1956) Modification of electric activity in cochlear nucleus during attention in unanesthetized cats. *Science* **123**:331–332.

Huffman RF and Henson OW Jr (1990) The descending auditory pathway and acousticomotor systems: connections with the inferior colliculus. *Brain Research Brain Research Reviews* **15**: 295–323.

Irvine DRF and Rajan R (1996) Injury- and use-related plasticity in the primary sensory cortex of adult mammals: possible relationship to perceptual learning. *Clinical and Experimental Pharmacology and Physiology* **23**:939–947.

Iwata J, LeDoux JE, Meeley MP, Arneric S, and Reis DJ (1986) Intrinsic neurons in the amygdaloid field projected to by the medial geniculate body mediate emotional responses conditioned to acoustic stimuli. *Brain Research* **383**:195–214.

Jafari MR, Zhang Y, and Yan J (2007) Multiparametric changes in the receptive field of cortical auditory neurons induced by thalamic activation in the mouse. *Cerebral Cortex* **17**:71–80.

Jarrell TW, Gentile CG, Romanski LM, McCabe PM, and Schneiderman N (1987) Involvement of cortical and thalamic auditory regions in retention of differential bradycardiac conditioning to acoustic conditioned stimuli in rabbits. *Brain Research* **412**:285–294.

Jen PH-S, Chen QC, and Sun XD (1998) Corticofugal regulation of auditory sensitivity in the bat inferior colliculus. *Journal of Comparative Physiology-A* **183**:683–697.

Jen PH-S, Sun XD, and Chen QC (2001) An electrophysiological study of neural pathways for corticofugally inhibited neurons in the central nucleus of the inferior colliculus of the big brown bat, *Eptesicus fuscus*. *Experimental Brain Research* **137**:292–302.

Jen PH-S and Zhou X (2003) Corticofugal modulation of amplitude domain processing in the midbrain of the big brown bat, Eptesicus fuscus. *Hearing Research* **184**:91–106.

Ji W, Gao E, and Suga N (2001) Effects of acetylcholine and atropine on plasticity of central auditory neurons caused by conditioning in bats. *Journal of Neurophysiology* **86**:211–225.

Ji W and Suga N (2003) Development of reorganization of the auditory cortex caused by fear conditioning: effect of atropine. *Journal of Neurophysiology* **90**:1904–1909.

Ji W and Suga N (2007). Serotonergic modulation of plasticity of the auditory cortex elicited by fear conditioning. *Journal of Neuroscience* **27**:4910–4918.

Ji W, Suga N, and Gao E (2005) Effects of agonists and antagonists of NMDA and ACh receptors on plasticity of bat auditory system elicited by fear conditioning. *Journal of Neurophysiology* **94**: 1199–1211.

Kawase T, Delgutte B, and Liberman MC (1993) Antimasking effects of the olivocochlear reflex. II. Enhancement of auditory-nerve response to masked tones. *Journal of Neurophysiology* **70**:2533–2549.

Kelly JP and Wong D (1981) Laminar connections of the cat's auditory cortex. *Brain Research* **212**:1–15.

Khalfa S, Bougeard R, Morand N, Veuillet E, Isnard J, Guenot M, Ryvlin P, Fischer C, and Collet L (2001) Evidence of peripheral auditory activity modulation by the auditory cortex in humans. *Neuroscience* **104**:347–358.

Kilgard MP and Merzenich MM (1998a) Cortical map reorganization enabled by nucleus basalis activity. *Science* **279**:1714–1718.

Kilgard MP and Merzenich MM (1998b) Plasticity of temporal information processing in the primary auditory cortex. *Nature Neuroscience* **1**:727–731.

Kilgard MP and Merzenich MM (2002) Order-sensitive plasticity in adult primary auditory cortex. *Proceedings of the National Academy of Sciences of the United States of America* **99**:3205–3209.

Krupa DJ, Ghazanfar AA, and Nicolelis MA (1999) Immediate thalamic sensory plasticity depends on corticothalamic feedback. *Proceedings of the National Academy of Sciences of the United States of America* **96**:8200–8205.

Lanuza E, Nader K, and Ledoux JE (1999) Unconditioned stimulus pathways to the amygdala: effects of posterior thalamic and cortical lesions on fear conditioning. *Neuroscience* **125**:305–315.

LeDoux JE, Cicchetti P, Xagoraris A, and Romanski LM (1990b) The lateral amygdaloid nucleus: sensory interface of the amygdala in fear conditioning. *Journal of Neuroscience* **10**:1062–1069.

LeDoux JE, Farb C, and Ruggiero DA (1990a) Topographic organization of neurons in the acoustic thalamus that project to the amygdala. *Journal of Neuroscience* **10**:1043–1054.

LeDoux JE, Farb CR, and Romanski LM (1991) Overlapping projections to the amygdala and striatum from auditory processing areas of the thalamus and cortex. *Neuroscience Letters* **134**: 139–144.

LeDoux JE, Sakaguchi A, Iwata J, and Reis DJ (1986) Interruption of projections from the medial geniculate body to an archi-neostriatal field disrupts the classical conditioning of emotional responses to acoustic stimuli. *Neuroscience* **17**:615–627.

LeDoux JE, Sakaguchi A, and Reis DJ (1984) Subcortical efferent projections of the medial geniculate nucleus mediate emotional responses conditioned to acoustic stimuli. *Journal of Neuroscience* **4**:683–698.

Li XF, Stutzmann GE, and LeDoux JE (1996) Convergent but temporally separated inputs to lateral amygdala neurons from the auditory thalamus and auditory cortex use different postsynaptic receptors: in vivo intracellular and extracellular recordings in fear conditioning pathways. *Learning and Memory* **3**:229–242.

Liu W and Suga N (1997) Binaural and commissural organization of the primary auditory cortex of the mustached bat. *Journal of Comparative Physiology A* **181**:599–605.

Lukas JH (1980) Human auditory attention: the olivocochlear bundle may function as a peripheral filter. *Psychophysiology* **17**:444–452.

Ma X and Suga N (2001a) Plasticity of bat's central auditory system evoked by focal electric stimulation of auditory and/or somatosensory cortices. *Journal of Neurophysiology* **85**:1078–1087.

Ma X and Suga N (2001b) Corticofugal modulation of duration-tuned neurons in the midbrain auditory nucleus in bats. *Proceedings of the National Academy of Sciences of the United States of America* **98**:14060–14065.

Ma X and Suga N (2003) Augmentation of plasticity of the central auditory system by the basal forebrain and/or somatosensory cortex. *Journal of Neurophysiology* **89**:90–103.

Ma X and Suga N (2004) Lateral inhibition for center-surround reorganization of the frequency map of bat auditory cortex. *Journal of Neurophysiology* **92**:3192–3199.

Ma X and Suga N (2005) Long-term cortical plasticity evoked by electric stimulation and acetylcholine applied to the auditory cortex. *Proceedings of the National Academy of Sciences of the United States of America* **102**:9335–9340.

Maison SF and Liberman MC (2000) Predicting vulnerability to acoustic injury with a noninvasive assay of olivocochlear reflex strength. *Journal of Neuroscience* **20**:4701–4707.

Malmierca E and Nunez A (1998) Corticofugal action on somatosensory response properties of rat nucleus gracilis cells. *Brain Research* **810**:172–180.

Manabe T, Suga N, and Ostwald J (1978) Aural representation in the Doppler-shifted-CF processing area of the auditory cortex of the mustache bat. *Science* **200**:339–342.

Maren S (2000) Auditory fear conditioning increases CS-elicited spike firing in lateral amygdala neurons even after extensive overtraining. *European Journal of Neuroscience* **12**:4047–4054.

Maren S, Yap SA, and Goosens KA (2001) The amygdala is essential for the development of neuronal plasticity in the medial geniculate nucleus during auditory fear conditioning in rats. *Journal of Neuroscience* **21**:RC135, 1–6.

Marrocco RT and McClurkin JW (1985) Evidence for spatial structure in the cortical input to the monkey lateral geniculate nucleus. *Experimental Brain Research* **59**:50–56.

Massopust LC Jr and Ordy JM (1962) Auditory organization of the inferior colliculus in the cat. *Experimental Neurology* **6**:465–477.

Mauk MD and Thompson RF (1987) Retention of classically conditioned eyelid responses following acute decerebration. *Brain Research* **403**:89–95.

McLin DE 3rd, Miasnikov AA, and Weinberger NM (2002) Induction of behavioral associative memory by stimulation of the nucleus basalis. *Proceedings of the National Academy of Sciences of the United States of America* **99**:4002–4007.

McClurkin JW and Marrocco RT (1984) Visual cortical input alters spatial tuning in monkey lateral geniculate nucleus cells. *Journal of Physiology* **348**:135–152.

Montero VM (2000) Attentional activation of the visual thalamic reticular nucleus depends on 'top-down' inputs from the primary visual cortex via corticogeniculate pathways. *Brain Research* **864**:95–104.

Mooney DM, Zhang L, Basile C, Senatorov VV, Ngsee J, Omar A, and Hu B (2004) Distinct forms of cholinergic modulation in parallel thalamic sensory pathways. *Proceedings of the National Academy of Sciences of the United States of America* **101**:320–324.

Murphy PC, Duckett SG, and Sillito AM (1999) Feedback connections to the lateral geniculate nucleus and cortical response properties. *Science* **286**:1552–1554.

Nieder P and Nieder I (1970) Antimasking effect of crossed olivocochlear bundle stimulation with loud clicks in guinea pig. *Experimental Neurology* **28**:179–188.

Norman RJ, Villablanca JR, Brown KA, Schwafel JA, and Buchwald JS (1974) Classical eyeblink conditioning in the bilaterally hemispherectomized cat. *Experimental Neurology* **44**:363–380.

Nwabueze-Ogbo FC, Popelar J, and Syka J (2002) Changes in the acoustically evoked activity in the inferior colliculus of the rat after functional ablation of the auditory cortex. *Physiological Research* **51**:S95–S104.

Oatman LC (1971) Role of visual attention on auditory evoked potentials in unanesthetized cats. *Experimental Neurology* **32**:341–356.

Oatman LC and Anderson BW (1977) Effects of visual attention on tone burst evoked auditory potentials. *Experimental Neurology* **57**: 200–211.

Ojima H (1994) Terminal morphology and distribution of corticothalamic fibers originating from layers 5 and 6 of cat primary auditory cortex. *Cerebral Cortex* **6**:646–663.

Orman SS and Humphrey GL (1981) Effects of changes in cortical arousal and of auditory cortex cooling on neuronal activity in the medial geniculate body. *Experimental Brain Research* **42**: 475–82.

O'Neill WE and Suga N (1982) Encoding of target range and its representation in the auditory cortex of the mustached bat. *Journal of Neuroscience* **2**:17–31.

Pinheiro AD, Wu M, and Jen PH-S (1991) Encoding repetition rate and duration in the inferior colliculus of the big brown bat, *Eptesicus fuscus*. *Journal of Comparative Physiology-A* **169**:69–85.

Poremba A and Gabriel M (2001) Amygdalar efferents initiate auditory thalamic discriminative training-induced neuronal activity. *Journal of Neuroscience* **21**:270–278.

Puel JL, Bonfils P, and Pujol R (1988) Selective attention modifies the active micromechanical properties of the cochlea. *Brain Research* **447**:380–383.

Quirk GJ, Armony JL, and LeDoux JE (1997) Fear conditioning enhances different temporal components of tone-evoked spike trains in auditory cortex and lateral amygdala. *Neuron* **19**:613–624.

Quirk GJ, Repa C, and LeDoux JE (1995) Fear conditioning enhances short-latency auditory responses of lateral amygdala neurons: parallel recordings in the freely behaving rat. *Neuron* **15**:1029–1039.

Rasmusson DD (2000) The role of acetylcholine in cortical synaptic plasticity. *Behavioral Brain Research* **115**:205–218.

Rasmusson DD, Smith SA, and Semba K (2007) Inactivation of prefrontal cortex abolishes cortical acetylcholine release evoked by sensory or sensory pathway stimulation in the rat. *Neuroscience* **149**:232–241.

Rauschecker JP and Tian B (2000) Mechanisms and streams for processing of "what" and "where" in auditory cortex. *Proceedings of the National Academy of Sciences of the United States of America* **97**:11800–11806.

Romanski LM and LeDoux JE (1993) Information cascade from primary auditory cortex to the amygdala: corticocortical and corticoamygdaloid projections of temporal cortex in the rat. *Cerebral Cortex* **3**:515–532.

Rose JE and Woolsey CN (1958) Cortical connections and functional organization of the thalamic auditory system of the cat. In: Harlow HF and Woolsey CN (eds). *Biological and Biochemical Bases of Behavior*. University of Wisconsin Press, Madison, pp. 127–150.

Ryugo DK and Weinberger NM (1976) Corticofugal modulation of the medial geniculate body. *Experimental Neurology* **51**:377–391.

Sakai M and Suga N (2001) Plasticity of the cochleotopic (frequency) map in specialized and nonspecialized auditory cortices. *Proceedings of the National Academy of Sciences of the United States of America* **98**:3507–3512.

Sakai M and Suga N (2002) Centripetal and centrifugal reorganizations of frequency map of the auditory cortex in gerbils. *Proceedings of the National Academy of Sciences of the United States of America* **99**:7108–7112.

Saldaña E, Feliciano M, and Mugnaini E (1996) Distribution of descending projections from primary auditory neocortex to inferior colliculus mimics the topography of intracollicular projections. *Journal of Comparative Neurology* **371**:15–40.

Sarter M and Bruno JP (2000) Cortical cholinergic inputs mediating arousal, attentional processing and dreaming: differential afferent regulation of the basal forebrain by telencephalic and brainstem afferents. *Neuroscience* **95**:933–952.

Sillito AM, Cudeiro J, and Murphy PC (1993) Orientation sensitive elements in the corticofugal influence on center-surround interactions in the dorsal lateral geniculate nucleus. *Experimental Brain Research* **93**:6–16.

Sillito AM, Jones HE, Gerstein GL, and West DC (1994) Feature-linked synchronization of thalamic relay cell firing induced by feedback from the visual cortex. *Nature* **20**:1–2.

Singer W (1993) Synchronization of cortical activity and its putative role in information processing and learning. *Annual Review of Physiology* **55**:349–374.

Steriade M (1999) Coherent oscillations and short-term plasticity in corticothalamic networks. *Trends in Neuroscience* **22**:337–345.

Suga N (1973) Feature extraction in the auditory system of bats. In: Møller AR (ed). *Basic Mechanisms in Hearing*. Academic Press, New York, pp. 675–742.

Suga N (1984) The extent to which biosonar information is represented in the bat auditory cortex. In: Edelman GM, Gall WE, and Cowan WM (eds). *Dynamic Aspects of Neocortical Function*. Wiley, New York, pp. 315–373.

Suga N (1990) Cortical computational maps for auditory imaging. *Neural Networks* **3**:3–21.

Suga N (1994) The processing of auditory information carried by species-specific complex sounds. In: Gazzaniga MS (ed). *The Cognitive Neurosciences*. MIT Press, Cambridge, pp. 295–313.

Suga N, Butman JA, Teng H, Yan J, and Olsen JF (1995) In: Flock A, Ottoson D, and Ulfendahl M (eds). *Active Hearing*. Elsevier Science Publishers, London, pp. 13–30.

Suga N, Gao E, Zhang Y, Ma X, and Olsen JF (2000) The corticofugal system for hearing: recent progress. *Proceedings of the National Academy of Sciences of the United States of America* **97**:11807–11814.

Suga N and Jen PH-S (1976) Disproportionate tonotopic representation for processing CF-FM sonar signals in the mustache bat auditory cortex. *Science* **194**:542–544.

Suga N and Ma X (2003) Multiparametric corticofugal modulation and plasticity in the auditory system. *Nature Reviews. Neuroscience* **4**:783–794.

Suga N, Niwa H, Taniguchi I, and Margoliash D (1987) The personalized auditory cortex of the mustached bat: adaptation for echolocation. *Journal of Neurophysiology* **58**:643–654.

Suga N and O'Neill (1979) WE Neural axis representing target range in the auditory cortex of the mustache bat. *Science* **206**:351–353.

Suga N, O'Neill WE, Kujirai K, and Manabe T (1983) Specificity of combination-sensitive neurons for processing of complex biosonar signals in auditory cortex of the mustached bat. *Journal of Neurophysiology* **49**:1573–1626.

Suga N, Xiao Z, Ma X, and Ji W (2002) Plasticity and corticofugal modulation for hearing in adult animals. *Neuron* **36**:9–18.

Sun X, Chen QC, and Jen PH-S (1996) Corticofugal control of central auditory sensitivity in the big brown bat, *Eptesicus fuscus*. *Neuroscience Letters* **212**:131–134.

Syka J and Popelar J (1984) Inferior colliculus in the rat: neuronal responses to stimulation of the auditory cortex. *Neuroscience Letters* **51**:235–240.

Tsumoto T, Creutzfeldt OD, and Legendy CR (1978) Functional organization of the corticofugal system from visual cortex to lateral geniculate nucleus in the cat (with an appendix on geniculo-cortical mono-synaptic connections) *Experimental Brain Research* **32**: 345–364.

Turner BH and Herkenham M (1991) Thalamoamygdaloid projections in the rat: a test of the amygdala's role in sensory processing. *Journal of Comparative Neurology* **313**:295–325.

Villa AEP, Rouiller EM, Simm GM, Zurita P, de Ribaupierre Y, and de Ribaupierre F (1991) Corticofugal modulation of the information processing in the auditory thalamus of the cat. *Experimental Brain Research* **86**:506–517.

Watanabe T, Yanagisawa K, Kanzaki J, and Katsuki Y (1966) Cortical efferent flow influencing unit responses of medial geniculate body to sound stimulation. *Experimental Brain Research* **2**:302–317.

Weinberger NM (1998) Physiological memory in primary auditory cortex: characteristics and mechanisms. *Neurobiology of Learning and Memory* **70**:226–251.

Weinberger NM and Bakin JS (1998) Learning-induced physiological memory in adult primary auditory cortex: receptive fields plasticity, model, and mechanisms. *Audiology and Neurootology* **3**:145–167.

Wenstrup JJ, Ross LS, and Pollak GD (1986) Binaural response organization within a frequency-band representation of the inferior colliculus: implications for sound localization. *Journal of Neuroscience* **6**:962–973.

Wiederhold ML (1970) Variations in the effects of electric stimulation of the crossed olivocochlear bundle on cat single auditory-nerve-fiber responses to tone bursts. *Journal of the Acoustical Society of America* **48**:966–977.

Wu Y and Yan J (2007) Modulation of the receptive fields of midbrain neurons elicited by thalamic electrical stimulation through corticofugal feedback. *Journal of Neuroscience* **27**:10651–10658.

Xiao Z and Suga N (2002a) Modulation of cochlear hair cells by the auditory cortex in the mustached bat. *Nature Neuroscience* **5**: 57–63.

Xiao Z and Suga N (2002b) Reorganization of the cochleotopic map in the bat's auditory system by inhibition. *Proceedings of the National Academy of Sciences of the United States of America* **99**:15743–15748.

Xiao Z and Suga N (2004) Reorganization of the auditory cortex specialized for echo-delay processing in the mustached bat. *Proceedings of the National Academy of Sciences of the United States of America* **101**:1769–1774.

Xiao Z and Suga N (2005) Asymmetry in corticofugal modulation of frequency tuning in mustached bat auditory system. *Proceedings of the National Academy of Sciences of the United States of America* **102**:19162–19167.

Xie DH and Henson OW Jr (1998) Tonic efferent-induced cochlear damping in roosting and echolocating mustached bats. *Hearing Research* **124**, 60–68.

Yan J and Ehret G (2002) Corticofugal modulation of midbrain sound processing in the house mouse. *European Journal of Neuroscience* **16**:119–128.

Yan J and Suga N (1996) Corticofugal modulation of time-domain processing of biosonar information in bats. *Science* **273**:1100–1103.

Yan J and Suga N (1999) Corticofugal amplification of facilitative auditory responses of subcortical combination-sensitive neurons in the mustached bat. *Journal of Neurophysiology* **81**:817–824.

Yan J and Zhang Y (2005) Sound-guided shaping of the receptive field in the mouse auditory cortex by basal forebrain activation. *European Journal of Neuroscience* **21**:563–576.

Yan W and Suga N (1998) Corticofugal modulation of the midbrain frequency map in the bat auditory system. *Nature Neuroscience* **1**:54–58.

Yu YQ, Xiong Y, Chan YS, and He J (2004) In vivo intracellular responses of the medial geniculate neurons to acoustic stimuli in anaesthetized guinea pigs. *Journal of Physiology* **560**:191–205.

Zaborszky L, Pang K, Somogyi J Nadasdy Z, and Kallo I (1999) The basal forebrain corticopetal system revisited. *Annals of the New York Academy of Sciences* **877**:339–367.

Zhang Y, Hakes JJ, Bonfield SP, and Yan J (2005) Corticofugal feedback for auditory midbrain plasticity elicited by tones and electrical stimulation of basal forebrain in mice. *European Journal of Neuroscience* **22**:871–879.

Zhang Y and Suga N (1997) Corticofugal amplification of subcortical responses to single tone stimuli in the mustached bat. *Journal of Neurophysiology* **78**:3489–3492.

Zhang Y and Suga N (2000) Modulation of responses and frequency tuning of thalamic and collicular neurons by cortical activation in mustached bat. *Journal of Neurophysiology* **84**:325–333.

Zhang Y and Suga N (2005) Corticofugal feedback for collicular plasticity evoked by electric stimulation of the inferior colliculus. *Journal of Neurophysiology* **94**:2676–2682.

Zhang Y, Suga N, and Yan J (1997) Corticofugal modulation of frequency processing in bat auditory system. *Nature* **387**: 900–903.

Zhou X and Jen PH-S (2000) Brief and short-term corticofugal modulation of subcortical auditory responses in the big brown bat, *Eptesicus fuscus*. *Journal of Neurophysiology* **84**:3083–3087.

Zhou X and Jen PH-S (2005) Corticofugal modulation of directional sensitivity in the midbrain of the big brown bat, *Eptesicus fuscus*. *Hearing Research* **203**:201–215.

Zook JM, Winer JA, Pollak GD, and Bodenhamer RD (1985) Topology of the central nucleus of the mustache bat's inferior colliculus: correlation of single unit properties and neuronal architecture. *Journal of Comparative Neurology* **231**:530–546.

Chapter 25

Auditory Evoked Potentials and Their Utility in the Assessment of Complex Sound Processing

Mitchell Steinschneider, Catherine Liégeois-Chauvel, and John F. Brugge

Abbreviations

AEP	auditory evoked potential
AI	primary auditory cortex
AII	secondary auditory cortex
AM	amplitude modulation
BF	best frequency
CSD	current source density
ECP	equivalent current dipole
EEG	electroencephalogram
EPSP	excitatory postsynaptic potential
ERP	event-related potential
fMRI	functional magnetic resonance imaging
GFP	global field power
HG	Heschl's gyrus
ISI	interstimulus interval
LD	Laplacian derivation
LLR	long-latency response
MEG	magnetoencephalography
MLR	middle-latency response
MMN	mismatch negativity
MUA	multi-unit activity
ORN	object-related negativity
PAC	primary auditory cortex
PCP	processing-contingent potential
PET	positron emission tomography
PT	planum temporale
SAC	secondary auditory cortex
SSR	steady-state response
STG	superior temporal gyrus
STP	supratemporal plane
VOT	voice onset time

M. Steinschneider (✉)
Kennedy Center, Albert Einstein Medical College, Bronx, NY 10461, USA
e-mail: steinsch@aecom.yu.edu

1 Studying Human Auditory Cortex

Human auditory cortex is, in the classical sense, composed of multiple fields distributed both on the exposed surface of the superior temporal gyrus (STG) and on the areas buried within the Sylvian fissure on the supratemporal plane (STP). In addition, cortex of the parietal and frontal lobes, while not generally considered part of the classical auditory forebrain, also participates in higher-order operations involving acoustic input (Romanski et al. 1999; Cohen et al. 2004; Gifford and Cohen 2005). Understanding the functions of these various auditory cortical areas requires complementary experimental approaches. This chapter will highlight how event-related potentials (ERPs) and the electroencephalogram (EEG) are important tools in understanding human auditory cortical physiology. All advances in this field cannot be reviewed fully in one chapter. Instead, certain key issues related to the use of these approaches to understanding complex acoustic processing at the cortical level will be discussed, and the relevance of their measures for evolving concepts of auditory cortical function and dysfunction will be highlighted.

We first provide a brief overview of the generally accepted classification of ERPs evoked by acoustic stimulation, auditory evoked potentials (AEPs), and their attentional modulation. The locations of neural generators of the AEP are discussed, and various experimental and modeling approaches are used to obtain this information. Thus, emphasis is placed on the relationships between electrophysiological results obtained in human subjects and those obtained in laboratory animals as they share many processing mechanisms. We describe the postnatal maturation of AEPs, as many studies using AEPs use both normally developing children and those with developmental disabilities.

Human intracranial recordings of AEPs and EEG, with their superior spatial and temporal resolution, have the unique potential to provide the ideal criterion for defining auditory cortical function (Lachaux et al. 2003; Engel et al. 2005). The ability to record simultaneously action potential

J.A. Winer, C.E. Schreiner (eds.), *The Auditory Cortex*,
DOI 10.1007/978-1-4419-0074-6_25, © Springer Science+Business Media, LLC 2011

activity and the AEP enhances the utility of the approach and allows for pointed comparisons of human and animal auditory cortical activity (Creutzfeldt et al. 1989; Howard et al. 1996; Jacobs et al. 2007). These invasive approaches, along with recording potentials from the scalp, complement other non-invasive imaging methods described elsewhere such as magnetoencephalography (MEG), functional magnetic resonance imaging (fMRI), and positron emission tomography (PET).

We discuss the roles of AEP recordings in cognitive neuroscience by examining their contributions in testing hypotheses about auditory scene analysis (Bregman 1990) and the dual-stream hypothesis about the identity and location of acoustic objects (Rauschecker and Tian 2000). We suggest that the definition of unimodal auditory cortex has to be reevaluated based on evidence in humans and monkeys of multisensory input to classically defined temporal lobe auditory cortical areas (Lakatos et al. 2007; Reale et al. 2007). We highlight the role of AEPs in cognition by examining how context modulates auditory cortical activity and how cortical temporal response properties may shape complex acoustic and phonetic perceptions. Normal response patterns are compared with those in people with hearing and language disorders. Dyslexia is the model by which hypotheses of perceptual dysfunction can be assessed by physiological means.

AEPs emphasize time-locked responses at the expense of stimulus-related but not precisely time-locked activity. We conclude by discussing the growing literature revealing the extent of these stimulus-induced responses often seen within high-frequency EEG bands (Freiwald et al. 2001; Ward 2003; Herrmann et al. 2004). These bands exhibit considerable stimulus specificity and sensitivity and thus provide an additional way of assessing auditory cortical activity.

2 Average Evoked Potentials: Definition and Classification

Cortical AEPs can be divided into three main categories based on the timing and polarity of voltage deflections after the onset of an effective acoustic stimulus. These are middle-latency (MLR) and long-latency (LLR) responses, and processing-contingent potentials (PCPs). MLRs and LLRs are true sensory evoked potentials in that they are evoked by physical attributes of sounds. PCPs, in contrast, are a diverse group of responses that are not directly related to the physical characteristics of the stimulus, but instead reflect some additional sound processing; they are often referred to as endogenous potentials. PCPs include both the so-called automatic discriminative responses such as the mismatch negativity (MMN) and those that require active, attention-dependent sound processing. However, these distinctions are often blurred since attention can strongly modulate exogenous and endogenous waveforms (Woldorff and Hillyard 1991; Woldorff et al. 1993; Alain and Woods 1997; Winkler et al. 2006).

3 Middle-Latency Responses

The MLR is a sequence of lower amplitude AEP deflections with latencies from 12 to 50 ms after sound onset (Picton et al. 1974; Borgmann et al. 2001; Yvert et al. 2005). The five waves are conventionally labeled by their voltage polarity and temporal sequence. They include P_0, N_a, P_a, N_b, and P_b. An earlier wave, N_0 (latency ~8 ms), is usually believed to be of subcortical origin. N_a (peak latency 15–25 ms) and P_a (peak latency 25–30 ms) are the most reliably recorded MLRs. P_b (also termed P1) is often considered the first LLR deflection. A variant of the MLR is the steady-state response (SSR), a quasi-sinusoidal response elicited by repetitive, amplitude- or frequency-modulated sounds that match the modulation frequency of the stimulus and may represent a composite of MLR components (Herdman et al. 2002; Poulsen et al. 2007).

4 Long-Latency Responses

The most extensively studied exogenous AEP waveforms contain larger and more reliably recorded P1, N1, and P2 deflections which peak near 50, 100, and 200 ms, respectively. P1 and N1 have voltage maxima over the frontocentral scalp, whereas the P2 maximum is more posteriorly, near the vertex (Wood and Wolpaw 1982; Näätänen and Picton 1987; Crowley and Colrain 2004). All deflections invert in polarity in the mastoid region, ventral to the underlying Sylvian fissure, which is consistent, at least in part, with their neural generators being located on the STP.

Although study of the LLR has been invaluable in probing the physiology of auditory cortex, the significance ascribed to changes in its waveform, especially N1, are often overinterpreted. For instance, it is difficult to ascribe to N1 a crucial role in perception if this wave dissipates at interstimulus intervals (ISIs) typical of speech or music (Snyder and Large 2004). It is also problematic to infer details of auditory cortical organization from changes in amplitude or scalp voltage distribution of each waveform deflection, as these likely have multiple neural generators in distinct cytoarchitectonic fields (Näätänen and Picton 1987; Scherg et al. 1989; Giard et al. 1994). Without information acquired by other means, such as intracranial recordings, we cannot determine the relative voltage contributions of the many fields comprising auditory cortex. More general aspects of auditory cortical function, such as detecting acoustic change, remain amenable to such analysis (Martin and Boothroyd 2000).

Often considered as part of an N1/P2 complex that covaries with changes in stimulus parameters, P2 has its own developmental time course and can be experimentally dissociated from N1 (Ponton et al. 2000; Shahin et al. 2003, 2005; Crowley and Colrain 2004). Many consider P2 as resulting from non-lemniscal pathways, in contrast to N1 and its presumed relationship with the ascending lateral lemniscal system (Crowley and Colrain 2004).

5 Automatic Processing: Contingent Responses

MMN is now one of the best studied PCP deflections. In its simplest form, MMN is generated when a repetitive stimulus (the standard) is occasionally replaced by a different stimulus (the deviant) that varies along some physical dimension. When the AEP evoked by the standard is subtracted from the AEP evoked by the deviant, the difference waveform negative deflection, the MMN, has a latency of ~150–200 ms. This difference wave may be a physiological manifestation of an automatic change-detection process coding a discernable difference between a new sound input and a preceding sound pattern (Näätänen et al. 2005). In essence, MMN is generated when a sound deviates from the sensory-memory representation of the prior acoustic environment. This pre-perceptual process occurs without attentional requirements. Often, the MMN amplitude increases, and its latency decreases, in parallel with the degree of dissimilarity between the standard and deviant stimuli (Friedman et al. 2001). As a pre-perceptual measure, the MMN has become a primary means for evaluating auditory sensory memory and sensory discrimination objectively (Picton et al. 2000; Näätänen et al. 2001). MMN is used to examine the representation and discrimination of simple acoustic attributes, combinations of sound features (Pakarinen et al. 2007), speech (Sharma et al. 1993; Sharma and Dorman 2000), and more complex patterns (Alain et al. 1998; Sussman 2005). Its capacity to examine key aspects of hearing, and its elicitation in passive behavioral states, have made MMN an attractive physiological tool in the assessment of both children and adults with difficulty in task-related performance (Cheour et al. 1998; Ferri et al. 2003; Leppänen et al. 2004; Jing and Benasich 2006).

Another interesting PCPs is P3a (Escera et al. 1998; Friedman et al. 2001), a positive wave, which peaks at about 300 ms after stimulus onset and is classically elicited by a novel sound. It is now known to arise after a large change in the acoustic background and may be an electrophysiological sign of automatic, attentional switching mechanisms. While usually preceded by MMN, P3a does not require MMN for its elicitation, although its emergence may require N1 (Rinne et al. 2006; Sabri et al. 2006). It occurs without attention and has maximal amplitude over frontal regions. As expected for a novelty response, P3a habituates with repeated stimulus presentation. P3a is distinct from a slightly later positive wave, the P3b, which is an attention-dependent positive deflection with maximal amplitude over parietal areas and which is elicited after the subject's detection of a target event.

6 Role of Attention

Attention is a powerful contextual modulator of AEPs. Several waves in the AEP are dependent upon attention and are task related. These attention-related potentials often overlap with non-attention-dependent waves, requiring additional methods to separate activity associated with different processes. Typically, the AEP evoked when a behavioral response is not required is subtracted from the AEP elicited when it is, yielding a difference waveform thought to be associated with attention. A valuable approach uses selective attention tasks, which allow simultaneous acquisition of AEPs evoked by attended and unattended stimuli. One paradigm involves attending to target stimuli embedded in a stream of non-target stimuli presented to one ear only while ignoring similar trains of stimuli presented to the unattended ear. Integration of attentional paradigms with AEP acquisition allows the assessment of whether selective attention is based on early cortical gain control of sound input by comparing the amplitudes of obligatory AEP waveforms in the attended and unattended conditions, or whether early processes are relatively unmodulated and new and later neural events are engaged (Näätänen 1990; Coull 1998; Giard et al. 2000).

Using stimulus paradigms described above, selective attention clearly enhances MLR waves, suggesting that attention can modulate early stages of cortical neural activity (Woldorff and Hillyard 1991; Woldorff et al. 1993). Enhancement of exogenous AEP components includes the N1 and P2 waves, and parallels increase in task difficulty and attentional load (Woods et al. 1994; Neelon et al. 2006; Sabri et al. 2006). These results thus support a gain control theory of attention by showing that MLR and LLR waves, which represent measures of auditory cortical activation elicited by external stimuli, are modulated by attentional constraints.

Attention also induces complex modulatory effects on automatic PCPs. Increased MMN amplitude results when deviant stimuli are presented to an attended ear (Szymanski et al. 1999). However, its amplitude decreases in more demanding tasks and when MMN is elicited by task-irrelevant stimulus changes. P3a modulation is also induced by attentional mechanisms. When attending to a different sensory modality, irrelevant changes in sound deviance evoke a smaller P3a compared to attention directed at auditory stimuli, while more demanding sound discrimination tasks enhance P3a if evoked by changes in task-irrelevant stimulus changes (Sabri et al. 2006).

Processing negativities are also generated by attention (Hillyard and Picton 1987; Näätänen 1990; Woods et al. 1994). Both early and late negativities, termed Nd for negative difference waves, are seen. The early Nd wave overlaps the N1/P2 complex and has shorter latencies when the target is easier to distinguish from the background, making this wave a useful measure of the speed of stimulus discrimination. Later segments of Nd can be distinguished by their differential topography: earlier portions have a frontocentral maximum; later ones a more frontal distribution (Giard et al. 2000). These processing negativities occur to both target and non-target stimuli. In contrast, another processing negativity (N2) is elicited only to target stimuli (Novak et al. 1990; Woods et al. 1994).

7 Neural Bases of Event-Related Potentials

7.1 Non-invasive Measures of Event-Related Potential Generator Localization

Mapping the locations and spatial distributions of ERP generators usually involves, first, creating isopotential plots of the voltage at each recording site serially in time. Fundamental premises include that different scalp topographies reflect different neural generators in the cortex acting over time (Michel et al. 2004). There are problems, however, in identifying ERP generators only from scalp topographies. First, scalp-recorded ERPs have inherently poor spatial resolution, due largely to the spatial blurring effect of the skull (Babiloni et al. 2001; Nunez and Srinivasan 2006). Second, the various voltage deflections need not arise near the electrode sites with maximal response amplitude (Arezzo et al. 1975; Wood and Wolpaw 1982; Gloor1985). Third, reference electrode activity may seriously bias waveform polarity and amplitude, without changing overall potential gradients (Skrandies 1990). Fourth, a sensory stimulus activates multiple brain generators which summate at the scalp to produce complex voltage topographies and waveforms. Scalp-recorded ERPs thus reflect both augmentation and cancellation of neural activity in active tissue subregions. Cortical ERPs reflect mainly the postsynaptic activity in pyramidal neurons that is subject to the largest spatial and temporal summation, with each pyramidal cell neuronal column behaving as an electrical dipole (Speckmann and Elger 1999). Thus, not all active brain regions yield easily detected voltage fields on the scalp, including cortical layered with little or no dipolar cellular architecture, or deep brain nuclei with closed field architecture. These localization issues in scalp-recorded topography highlight difficulties in the well-known inverse problem: the volume conductor of the brain contains, theoretically, an infinite number of potential neural sources that can produce a specific topographic pattern.

AEP components are traditionally characterized by the timing and polarity of positive and negative waveform peaks. Peak latencies may vary as a function of electrode site due to the simultaneous activation of multiple generators (Michel et al. 2004). Examining only those waveforms recorded at a single, user-chosen electrode site may yield inaccurate AEP categorization. Global field power (GFP) is a more objective measure of the AEP waveform, and it is the square root of the mean of the squared voltage differences between all electrode sites (Lehmann and Skrandies 1980; Skrandies 1990; Michel et al. 2004). It is a reference-free, user-independent measure of the net power of the electric field derived from the electrode grid. GFP peaks can be used to characterize objectively AEP waveforms and peak latencies. Additional methods can then be used to clarify generator source identification.

Comparisons can also be made in the field topography across time points in the waveform under one experimental condition or across different conditions at the same time point to assess whether similar generator configurations are present. The global dissimilarity measure is a simple means to compare topographies, and it is the square root of the mean of the squared differences between all corresponding electrodes (Lehmann and Skrandies 1980; Skrandies 1990; Michel et al. 2004). Prior to this calculation, all amplitudes are normalized by dividing activity by its own GFP to minimize topographical changes that might reflect changes in component amplitude rather than generator distribution.

Field topography is sharpened by computing a second spatial derivative (Laplacian derivation, LD) of the raw data (Gevins et al. 1999; Babiloni et al. 2001; Nunez and Srinivasan 2006). From this spatial filtering, contributions to waveforms from the reference electrode or distant sources are reduced or eliminated and the LD estimates the transcranial flow to and from the skull directly beneath the recording electrode. The LD is computed by comparing the activity at an electrode site with the mean of its nearest neighbors (Gevins et al. 1999; Babiloni et al. 2001). More accurate estimates require spline interpolation (Perrin et al. 1987; Gevins et al. 1999; Babiloni et al. 2001). Models better approximating head shape, including the subject's own shape from their MRI, further enhance spatial resolution (Gevins et al. 1999; Babiloni et al. 2001). While scalp AEP distributions are sharpened, the LD is still includes contributions from multiple overlapping current generators and does not unequivocally identify sources of neural activity. For this identification, approximations to solving the inverse problem and direct intracranial recordings are required.

Several sophisticated models provide solutions to the inverse problem. Each makes various assumptions about the neural sources, the conductivity of the brain and its coverings, and head geometry (Baillet 2001; Michel et al. 2004). All methods require the ability to solve first the forward solution, so that an accurate voltage distribution across the scalp can be derived from a series of known generators with given strengths, locations, and orientations (Darvas et al. 2004).

Equivalent current dipole (ECD) models and distributed source models are the two main algorithms used to identify generators in scalp voltage topography. Their principal assumption is that a few dipoles with varying strengths, locations, and orientations identify the underlying generators (Scherg and von Cramon 1985; Cuffin 1998; Ebersole and Wade 1990), each representing the summed activity from a circumscribed brain region. ECD algorithms usually calculate the best fitting dipole locations, strengths, and orientations using a reiterative process to reduce the residual variance between predicted scalp topography derived from a forward solution and the actual, voltage distributions. Dipole parameters are systematically modified to obtain the best-fit solution (Michel et al. 2004). Ultimately, ECD models decompose the evoked potential into a series of source waveforms providing the best statistical fit to the empirical data for an assumed number of dipoles. Advantages include relative ease of use, resistance to noise, and relatively accurate results for focal brain activation (Darvas et al. 2004; Im et al. 2005). Disadvantages are the high dependence upon user-provided decisions and, in its classic application, loose coupling between results and detailed anatomical information (Michel et al. 2004; Im et al. 2005). It is prudent to view an ECD as a center of gravity for activity in a given brain volume, understanding that details of the actual activation are inaccessible and that large activated areas may be mislocalized (Kobayashi et al. 2005).

Attempts have been made to use a physiologically plausible number of dipoles (Im et al. 2005). Known anatomy and physiology of a structure place realistic constraints on their location and orientation and MRI images can help constrain dipoles to the grey matter and suggest accurate cranial models (Babiloni et al. 2001; Michel et al. 2004). fMRI is used to find and estimate the number of equivalent ECD sources required in a paradigm (Mulert et al. 2004; Molholm et al. 2005; Schönwiesner et al. 2007). Thus, the relationship between fMRI and dipole estimation is complex (Logothetis 2003; Ahlfors and Simpson 2004; Benar et al. 2006) and each technique addresses different aspects of neural function. It is unrealistic to assume a direct correspondence between the two measures (Nunez and Silberstein 2000; Devor et al. 2003) since the locations and dimensions of fMRI activation are not always related to functional maps derived from scalp-recorded ERPs, and the latter may not always reflect fMRI changes (Ahlfors and Simpson 2004; Mulert et al. 2004).

Distributed source models do not require a predetermined number of dipoles to arrive at an inverse solution. Instead, brain activity is reconstructed from a three-dimensional grid of solution points distributed uniformly on the cortical surface, each functioning as a dipole of fixed location with varying strength and orientation (Michel et al. 2004). As there are many more unknowns (several thousand dipoles) than data (about 100 measurement points), each algorithm requires a mathematical constraint to have a unique solution. Several distributed source models with various assumptions have emerged and include the minimum norm estimation (Hämäläinen and Ilmoniemi 1994), and the LORETA, and LAURA models (Baillet 2001; Darvas et al. 2004; Michel et al. 2004; Bai et al. 2007). Reviews of distributed source models, and the problems inherent with ECD models, capture the imperfect nature of source localization based solely on indirect means and emphasize the role of more direct methods in supporting or modifying putative generators seen in non-invasive techniques.

8 Invasive Measures of Evoked Potential Generator Localization

Human intracranial recordings promise to help determine contributions made by neural structures to AEP generation (Halgren et al. 1998; Lachaux et al. 2003). Recordings obtained from patients undergoing evaluation for medically intractable epilepsy are an invaluable and unique window into human brain physiology, despite many limitations. The number and locations of recording sites and the time available for data acquisition are determined by clinical constraints. Electrodes may not be optimally oriented to map directly activity from a presumed cortical generator to the head surface, hampering a straightforward interpretation of the relation between surface recordings and their sources. Further, evoked activity within a region may not be uniform for a specific stimulus. Finally, patients with neurological dysfunction in the brain region of interest require caution when extrapolating to the neurologically normal subject (Boatman and Miglioretti 2005; Boatman et al. 2006).

8.1 Generators of Specific Components

Despite its limitations, recording AEPs directly from the cerebral cortex allows relatively precise characterization of functional auditory areas for the stimuli and the tasks studied. When combined with data from non-invasive recordings, clues to the identity of the main generators of the AEP waveform are further strengthened. AEP deflections are often described as waveform components, implying that each deflection represents a discrete underlying neural process near the latency of the maxima or minima of the deflection. Given the uncertain location of sources of the scalp-evoked potential, this interpretation is not strictly tenable for extracranial recording data. It may be more valid for data from intracranial recording, where electrode contacts are very near known sources.

One key to accurate interpretation of intracerebral responses is that electrode polarity inversion between two adjacent recording sites indicates that passage through the dipole generating the component (Vaughan and Arezzo 1988). The higher the amplitude of a component, the closer is the generator to the recording site. This permits distinguishing local field potentials from volume-conducted potentials, as the morphology and timing of the latter change little with distance (Badier and Chauvel 1995).

There is consensus that the most posteromedial parts of Heschl's gyrus (HG) contribute to the early P_0 and N_a components of the MLR (Scherg and Von Cramon 1986; Liégeois-Chauvel et al. 1994; Godey et al. 2001; Yvert et al. 2005). With two HG, a normal anatomical variant, the generator is on the more anterior gyrus but may extend into the intervening sulcus (Yvert et al. 2005). This includes the anatomically defined posteromedial auditory core cortex (Hackett et al. 2001). Posteromedial HG, slightly more anterolateral HG segments, Heschl's sulcus, the planum temporale, and the posterior STG all may contribute to P_a (Liégeois-Chauvel et al. 1994; Steinschneider et al. 1999; Howard et al. 2000; Yvert et al. 2005). The involvement of multiple auditory cortical regions to this cortical wave is not surprising, given that electrical stimulation of posteromedial HG evokes short-latency responses within the posterolateral STG, anterolateral HG, and planum temporale (Liégeois-Chauvel et al. 1991; Howard et al. 2000; Brugge et al. 2003). Non-invasive dipole source localization emphasizes the critical contribution of the HG posteromedial segment to the scalp-recorded P_a (Scherg and Von Cramon 1986; Borgmann et al. 2001; Yvert et al. 2001).

Multifocal generators predominate for the remainder of the AEP, with significant variability in waveform peak latencies, which is likely based on differences across subjects, electrode placements, and stimulus parameters (Howard et al. 2000; Godey et al. 2001). In most studies the evoked waveforms persist for several hundred milliseconds after stimulus onset, including auditory core (Howard et al. 2000; Godey et al. 2001; Brugge et al. 2008). A complex pattern of temporally overlapping waves recorded from diverse auditory cortex regions shows AEPs elicited by a 1-kHz tone burst and recorded simultaneously from electrodes located in auditory core cortex (posteromedial HG, anterior and posterior primary auditory cortex (PAC)) and from non-core auditory cortex (secondary auditory cortex (SAC) and planum temporal (PT)). The earliest components (<30 ms) were from the auditory core (sites P2 and H5), with later activity occurring from multiple regions including the lateral part of HG (site T4) and the PT (site H7) (Fig. 25.1).

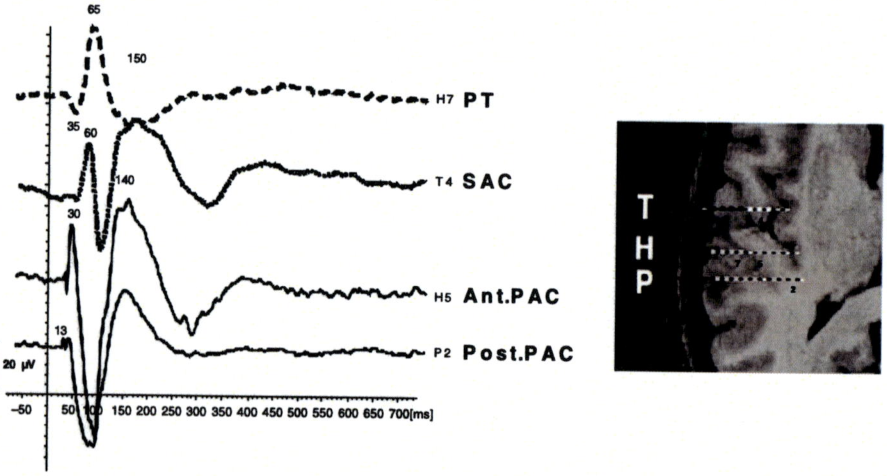

Fig. 25.1 *Left*: AEPs evoked by a 1-kHz tone recorded simultaneously from four intracranial sites in auditory cortex. Shortest onset latencies (<30 ms) are restricted to the auditory core on medial portions of Heschl's gyrus (anterior PAC, posterior PAC). Overlapping activity occurs at later time segments in the core, planum temporale (PT), and the surrounding belt area on more lateral parts of Heschl's gyrus (SAC). *Right*: MRI depicting the location of electrode contacts. Electrodes (T, H, P) and the contact numbers from which AEPs were recorded are to the *right* of the waveforms

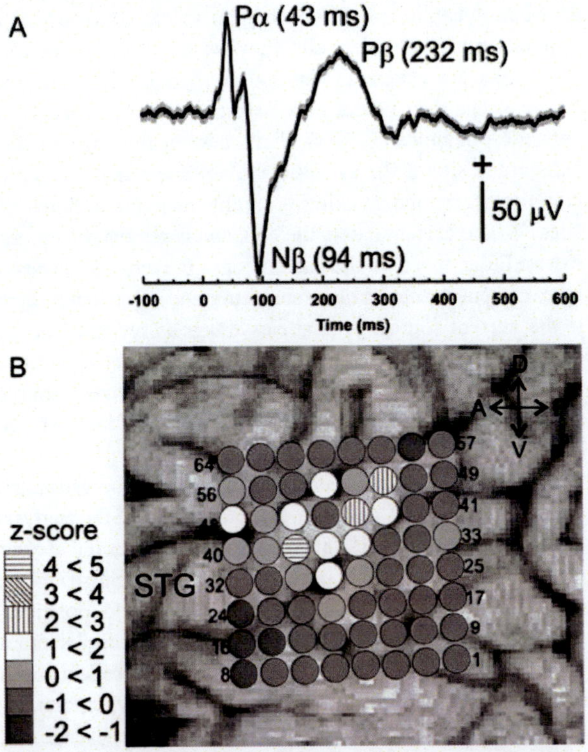

Fig. 25.2 (a) An AEP evoked by brief burst of clicks recorded from channel 38 of a subdural electrode grid overlying the left posterior temporal lobe. Mean (*black line*) and standard error (*grey shading*) are shown. (b) Normalized responses of the P_a AEP component. Maximum responses are restricted to electrodes over the posterior STG

Early and sustained activation of auditory cortex on the lateral surface of posterior STG is also observed. The AEP evoked by a brief burst of clicks (Fig. 25.2a) was recorded from the subdural grid electrode and contained the largest response, with multiple components labeled appropriately (Howard et al. 2000). An early positivity (P_a) 20-ms onset latency and a 43-ms peak latency initiate the response. This is followed by waves (N_b and P_b) which also overlap with activity recorded from the other auditory areas (Fig. 25.1). The normalized distribution of the P_a amplitude on the grid shows maxima distributed along the posterior STG with rapid decline in amplitude at surrounding sites (Fig. 25.2).

While many simultaneously active generators are the rule in forming AEP waveforms, emerging patterns include a lateral spread of activity from medial HG such that P1 has a predominant generator in more intermediate HG sectors (Liégeois-Chauvel et al. 1994; Godey et al. 2001; Yvert et al. 2005), and the largest negativity on the STP in the N1 time range often lies in the planum temporale (Scherg et al. 1989; Liégeois-Chauvel et al. 1994; Yvert et al. 2005). P2 generators for have been less well studied. Dipole source analysis places them in cortex slightly anterior and medial to the N1 center of gravity (Pantev et al. 1996; Shahin et al. 2003). Intracranial recordings find multifocal generators in the planum temporale and posterior STG (Howard et al. 2000; Godey et al. 2001). In summary, 30–50 ms after stimulus onset, auditory cortical activity spreads into multiple regions. Thus, single dipole models of scalp-recorded waveforms for even some of the earliest auditory cortical activity do not provide a full and accurate picture of the neural events underlying it. At various time points in the waveform, however, different generators may predominate.

The question of the MMN generators has become especially controversial. At one extreme is the view that MMN represents differential adaptation of N1 generators anterior and posterior to HG (Jääskeläinen et al. 2004), with N1 generated in more posterior portions of auditory cortex readily adapting to repeated sounds. Anterior generators of N1 peak later than posterior generators and do not readily adapt in part from more narrow sound frequency tuning. When the standard and deviant sound frequency difference is relatively small, the subtraction waveform enhances the relative contribution of the later anterior N1 response. The center of gravity for dipoles shifts anteriorly and peaks later, producing an illusory difference that is ascribed as an MMN. MMN is seen as the differential adaptation of subcomponents of the composite N1 wave rather than as a discrete process representing changes in sound pattern through sensory memory mechanisms.

A contrary argument interprets MMN as a bona fide metric of memory-related activity (Näätänen et al. 2005) based on (1) latency differences between MMN and anterior contributors of N1, (2) MMN elicitation when a stimulus is omitted from a sound sequence and when N1 is absent, (3) elicitation of MMN when feature-specific adaptation cannot occur, as in a change in a tone pattern sequences that steadily rises or falls in frequency, (4) scalp distribution differences and hemispheric asymmetries for N1 and MMN (Picton et al. 2000), and (5) dissociation of MMN and N1 sensitivity to various experimental manipulations and the subjects' experiential background.

Other hypotheses reflect different stimulus paradigms (Haenschel et al. 2005). In one study, trains of tone bursts at a given frequency were presented followed by a train at a different frequency, with the first tone of a new train the deviant and the last tone of the preceding train the standard. The number of tones in a train varied between 2, 6, and 36 repetitions. MMN recorded from frontocentral electrodes was enhanced with longer trains by the development of a positive component (repetition positivity) that was largest in the last tone of the longer trains. Subtracting the more positive standard from the deviant apparently enhanced MMN. MMN recorded from frontocentral electrodes has also been viewed as an effect of stimulus-specific adaptation, whereas MMN recorded over the mastoid was insensitive to train repetition number, did not induce a repetition positivity, and was consistent with a bona fide MMN from a more temporally located source (Jääskeläinen et al. 2004). MMN may not be a unitary phenomenon, but is best appreciated as an electrophysiological measure for many distinct processes that facilitate auditory change detection (Sabri et al. 2006).

Intracranially acquired data also support the multifaceted nature of MMN. Reliably identified MMN was seen at STP loci slightly anterior and lateral to the main N1 generators and far from the most medial HG (Halgren et al. 1995). In a tone sequence of low–medium–high frequency the deviant repeated the preceding tone, minimizing confounding sensory dishabituation, and the response was consistent with a valid MMN wave. A smaller study also failed to find a generator in posteromedial HG, but found an MMN-like response on more lateral STP that appeared to be generated by ISI sensitivity (Kropotov et al. 2000). A more classic MMN best ascribed to deviance from a preceding memory trace was recorded from the posterolateral STG. These studies support the idea that multiple change-detection and context-dependent mechanisms exist on the STP and STG.

AEPs evoked by novel sounds are also recorded in many regions. A negative–positive wave sequence after the MMN occurs in direct recordings on the STP and these overlapped in time with the scalp-recorded P3a (Halgren et al. 1995). Triphasic waves, with a positivity peaking at ~300 ms, were evoked by novel stimuli in supramarginal gyrus, dorsolateral prefrontal cortex, and cingulate cortex (Halgren et al. 1998; Brázdil et al. 2005). Generators for attention-dependent processing negativities have not been well characterized. Early Nd waves likely are generated widely in auditory cortex, while later waves may have a frontal cortex origin (Woods et al. 1994; Kasai et al. 1999; Giard et al. 2000). N2 elicited by target stimuli show hemispheric asymmetry that depends on the phonetic context of the sounds, with major generators prospectively located in temporo-parietal and fronto-temporal areas (Kayser et al. 1998; Celsis et al. 1999).

9 Animal Models

While intracortical and scalp recordings may assist in identifying the locations and timing of AEP generators, they do not address the neural mechanisms that underlie AEP sensitivity and selectivity. For this we turn to experiments in laboratory animals.

AEPs result from synchronized transmembrane current flows within neuronal populations. The cortex has predominantly synaptic currents and associated passive current flows. When these transmembrane currents flow in cells with similar orientation and with similar asymmetrical activation, as when synaptic events occur upon apical dendrites of a population of pyramidal cells, they act as dipolar generators to produce instantaneous volume currents in the brain and its coverings. Net neural depolarization by excitatory postsynaptic potentials (EPSPs) elicits current flowing into the cells at the site of the synaptic activity and passive, circuit-completing, capacitive currents exiting the neurons at nearby sites. Where current enters the cell, and is removed from the extracellular space, is the current sink; the site for extracellular current reentry from transmembrane capacitance flow is the current source. A positive voltage is recorded in the extracellular space at the current source, a negative voltage at the current sink. Polarity inversion occurs between source and sink. Voltages diminish with the square of the distance between the active zone and the recording site.

Extracellular source and sink patterns can be estimated by one-dimensional current source density (CSD) analysis (Vaughan and Arezzo 1988), which approximates the second spatial derivative of the intracortical potential distribution and calculates, at each time point in the field potential, whether a recording site is a source or sink. Sinks often occur at sites of net depolarization and can also reflect zones of passive current return induced by nearby hyperpolarization. Similarly, sources usually reflect sites of passive current return for EPSPs, but can reflect locations of inhibitory activity too. Disentangling these possibilities requires concurrent recordings of action potentials from the same neuronal population. A sink coincident with above baseline multiunit activity (MUA) indicates excitatory activity, while a source colocated with an MUA reduction in suggests hyperpolarization.

Using CSD methods, a characteristic pattern of major sources and sinks is seen in primary auditory cortex (AI) of several mammalian species (Muller-Preuss and Mitzdorf 1984; Steinschneider et al. 1994, 2003, 2008; Metherate and Cruikshank 1999; Cruikshank et al. 2002; Lakatos et al. 2007). The earliest response is an initial sink in layer 4 and lower layer 3, with nearby sources and associated increases in MUA. The short latency of the initial part of the sink (onset <10 ms in the monkey) and of the MUA that can be traced to the white matter suggests that these early potentials are generated, in part, by thalamocortical axonal activity (Steinschneider et al. 1992). A small surface-negative wave is seen sometimes with this early evoked activity, perhaps from a portion of the scalp-recorded N_0 wave. Later parts of the initial sink embody current sources extending to supragranular layers, suggesting excitatory monosynaptic connections within proximal portions of lower layer 3 pyramidal cells layer and passive current returning along their apical dendritic segments. This component is associated with a small surface positive wave with peak latency 12–15 ms in the monkey, consistent with a contribution to the generation of the P_0 MLR (Steinschneider et al. 1992).

A second major sink in upper layer 3 is associated with more superficial current sources (Steinschneider et al. 1994, 2003; Lakatos et al. 2007). It is concurrent with a large amplitude AI surface-positive wave, peaks at 21–28 ms, and inverts in polarity in supragranular layers. This precedes an even more superficial sink with distributed sources in deeper layers and which is associated with a surface-negative wave peaking at ~45–60 ms and, often, a later current sink in deep layer 3 with surrounding sources. A surface-positive wave peaking at 80–110 ms may reflect this source/sink configuration.

This method of analysis is shown by the layer profiles of AEPs, CSD, and MUA concurrently recorded to an 11.5-kHz best frequency (BF) tone in awake monkey AI (Fig. 25.3). Approximate boundaries for middle cortical layer are shown at the far left. The AEP has a prominent surface positivity with peaks at 15 and 22 ms and inverts in polarity within layer 3. This positivity is associated with a layer 4/lower layer 3 initial sink layer and a slightly later, more superficial sink in upper layer 3. The peaks of these sinks in the CSD correspond to the peaks in the superficial AEP positivity. Current sources bracket the sinks. Large amplitude MUA in middle layers is concurrent with the initial sink. Thus, the initial cortical positivity is generated by multilayer events that reflect both mono- and polysynaptic activation of AI. A negative/positive wave complex with peaks at 48 and 80 ms follows the initial positivity in the AEP. These waves are associated with multilayer sources and sinks, though the surface response waveform is dominated by the superficial CSD patterns. Therefore, the layer distribution of sources and sinks that generate field potentials within an auditory cortex slab is complex in time and layer origin. While major sources and sinks have characteristic layer locations, other depths may serve as a source or sink at any time-point in the waveform. Thus, the ensuing field potential will be a complex sum of each source and sink, weighted by its strength and distance from the recording site.

Given these caveats, and assuming that results obtained in non-human AI reflect similar mechanisms in the human core auditory fields, we suggest that a principal generator of the MLR P_a wave in posteromedial HG is a sink in upper layer 3 and a more superficial source, and that negative waves in the N1 time frame are generated by multilayer events that include superficial current sinks and deeper sources. While this scheme might hold for AEP generators in posteromedial HG, the situation becomes complex since, by the time of P_a, many fields are simultaneously active. An activated region will thereby reflect a layer sequence of events that engage sources and sinks and which may differ from those in auditory core. If similar, then large amplitude surface-positive waves such as P1 and P2 partially result from sinks in layer 3 and more superficial sources and that surface-negative waves such as N1 are the partially from superficial layer sinks.

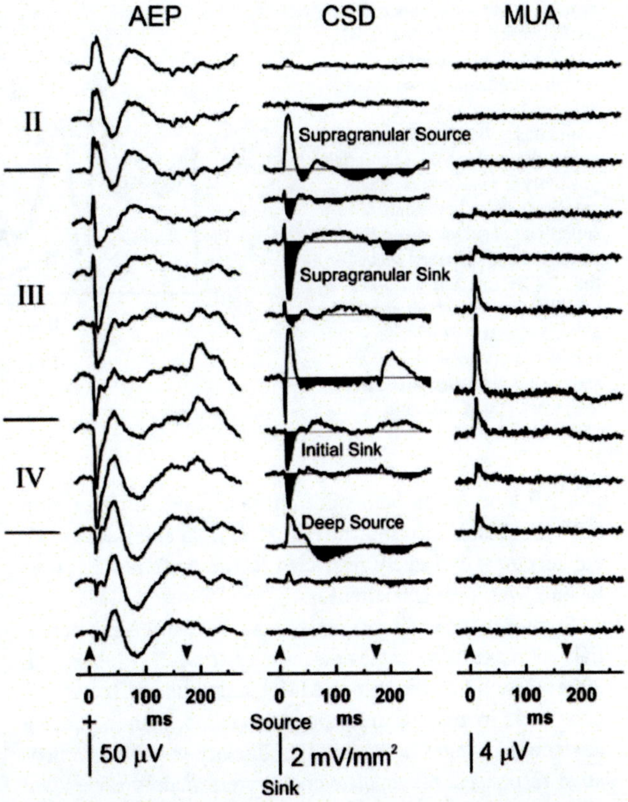

Fig. 25.3 Laminar profiles of AEPs, CSD, and MUA concurrently recorded at 150-μm intervals to a 11.5-kHz BF tone in AI of an awake monkey. Approximate boundaries for middle cortical layers are at the *far left*. See text for details

The neural mechanisms underlying stimulus-change-detection and MMN have been clarified using CSD analysis in monkey AI, which show that deviant click intensity had its most marked effects in supragranular layers, beginning in upper layer 3 sink followed by later current-sink maxima in layers 1 and 2 (Javitt et al. 1994, 1996). The initial sink was not significantly modulated by the stimulus deviance and the enhanced activity could be blocked by infusion of *N*-methyl d-aspartate (NMDA) receptor antagonists (Javitt et al. 1996), in keeping with their presumed role in memory.

Studies in monkey AI also suggest that MMN is associated with enhanced unit activity compared to responses evoked by standard stimuli. Superimposed AEPs evoked by a 12.8-kHz BF tone (determined by the amplitude of the MUA evoked by stimulus onset) when the tone was a standard and when it was a deviant are compared (Fig. 25.4, left side). Enhanced negative waves analogous to both an augmented human N1 and MMN appear to stimuli presented as deviant (Fig. 25.4: arrows). The initial positivity indicative of early cortical processes is relatively unchanged. The amplitude of sustained MUA for the 12.8-BF tone and a 7.6-kHz

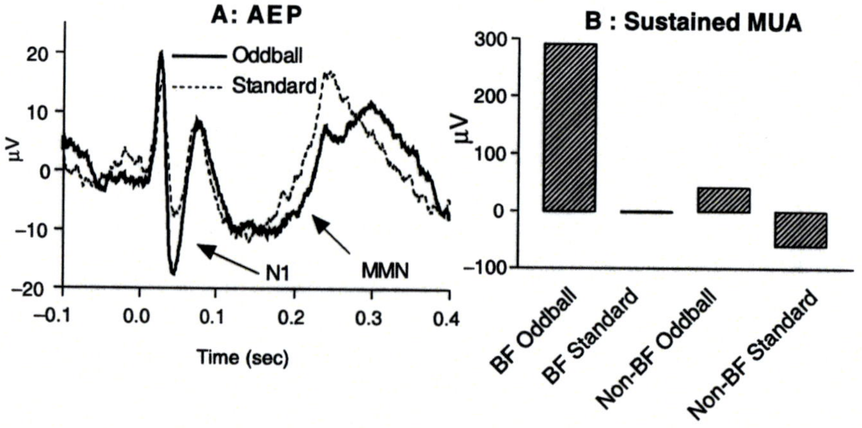

Fig. 25.4 (**a**) Superimposed AEPs evoked by a 12.8-kHz BF tone used as a standard and a deviant stimulus. *Arrows* denote enhanced negative waves analogous to both an augmented human N1 and MMN. The initial positivity is relatively stable in amplitude, despite the enhanced superficial negative waves. (**b**) Amplitude of sustained MUA for the 12.8 BF tone and a 7.6-kHz non-BF tone when both were either a standard or oddball stimulus. Enhancement of the response is restricted to the BF tone when it was a deviant stimulus

non-BF tone, when both were either a standard or an oddball stimulus, is also shown (Fig. 25.4, right side). Response enhancement is largely restricted to the preferred BF tone when it was a deviant stimulus.

Studies in cat auditory cortex show that MMN-like potentials are maximal over the non-primary area AII (Pincze et al. 2001). Much like the human MMN, amplitude of the analogous wave increased, and latency decreased, with increasing deviance. Changes in the initial obligatory positive and negative responses, maximal over AI, were mainly dependent on ISI, whereas MMN-like waves maximal over AII were primarily dependent on deviance and inter-deviance interval, thus mirroring properties of the human MMN (Pincze et al. 2002).

Unit responses in cat AI to pure tones were enhanced with deviant stimuli (Ulanovsky et al. 2003). This increased activity, analogous to that described above in the monkey, reflected the degree of tone frequency difference and the rarity of the sound. As in the monkey, enhanced unit activity was primarily for sustained responses and not the initial response evoked by stimulus onset. These effects were not seen in the medial geniculate nucleus, suggesting that the processes underlying MMN are intracortical. The history of presented stimuli markedly altered the responses of AI neurons, an effect that could persist for seconds (Ulanovsky et al. 2004). Thus, not only was a response enhanced for one stimulus when immediately preceded by a different one (a local effect), but the probability of enhancement was partially determined by the temporally integrated window of sound patterns that preceded the test tone (a global effect). Such findings not only provide information on the underlying change-detection processes in cortex but may explain psychoacoustical phenomena such as perceiving global versus immediate (local) pitch patterns (Sanders and Poeppel 2007) or context-dependent phonetic perceptions based on the spectral characteristics of the preceding acoustic environment (Holt 2006). While MMN elicited by a change in more complex acoustic patterns may not be explained entirely by the activity profiles of AI, similar mechanisms may operate in other auditory cortical fields whose cells integrate activity over wider frequency bands and with longer time constants, resulting in greater sensitivity and/or selectivity to more complex acoustic stimuli.

10 Development

Auditory cortex undergoes profound anatomical changes during development that continue well into adolescence (Huttenlocher and Dabholkar 1997; Moore and Guan 2001). These changes are accompanied by functional changes documented in scalp-recorded AEPs in children whose temporal waveforms and spatial distribution patterns differ from those in adults. Understanding the normal development of AEPs assumes added importance since the mechanisms associated with developmental language disorders may be reflected in these physiologic responses (Bradlow et al. 1999; Nagarajan et al. 1999; Giraud et al. 2005; Kujala et al. 2006).

Immature AEPs bear little resemblance to those in older children (Kurtzberg et al. 1984; Novak et al. 1989; Wunderlich and Cone-Wesson 2006; Wunderlich et al. 2006). Long-latency positive waves or positive–negative wave complexes recorded over the midline and temporal regions characterize the early postnatal AEP. Dipole modeling of the corresponding magnetic response suggests an auditory cortex generator (Huotilainen et al. 2003). Responses recorded from central scalp regions thought to reflect activity from STP auditory cortex, mature more rapidly than activity from temporal electrodes, and are thought to reflect activity in lateral auditory fields. Responses to changes in the acoustic stimulus can also be seen in early infancy, including those evoked by subtle acoustic differences relevant for phonetic perception (Dehaene-Lambertz and Dehaene 1994; Dehaene-Lambertz and Baillet 1998).

MLRs such as P_a can be reliably recorded in children 4–5 years old (Kraus et al. 1989). Latencies are stable from this age range into adulthood, suggesting relatively mature generators for this early response (Ponton et al. 2002). In contrast, long-latency components undergo marked developmental changes. In 5–8 year olds, the AEP is dominated by a broad positivity with several peaks suggestive of P1 and P2, followed by a large-amplitude negativity, N2 (Sharma et al. 1997; Albrecht et al. 2000; Ponton et al. 2000; Gilley et al. 2005; Wunderlich and Cone-Wesson 2006). N1 is usually absent unless very long ISIs are used (Ceponiene et al. 1998; Gilley et al. 2005; Wunderlich and Cone-Wesson 2006). If present, it may represent a relative trough centered in the positive wave complex. N1 first appears clearly at 11–14 years in most recordings (Albrecht et al. 2000; Ponton et al. 2000). P1 and N1 latencies progressively shorten through the teen years (Albrecht et al. 2000; Ponton et al. 2000). In parallel with the emergence of N1, peak latency of P2 moves toward adult values and N2 amplitude decreases (Ponton et al. 2000; Wunderlich and Cone-Wesson 2006). Scalp distributions and dipole sources for all these waves, while consistent with generators that can be ascribed to auditory cortex situated on or near the STP, vary with age and may represent differences in the dominant cortical fields that contribute to the scalp recordings (Albrecht et al. 2000; Wunderlich and Cone-Wesson 2006). This is exemplified by N1, whose absence or near absence in younger children, coupled to data that support a principal generator in the planum temporale, suggests that this region is functionally quite immature through most of childhood (Ceponiene et al. 2002).

Developmental progression is seen in activity recorded from the scalp above lateral STG auditory cortical fields (Tonnquist-Uhlen et al. 2003). While STP cortical activity contributes to scalp recordings from the lateral surface, dipole source modeling shows that local AEPs with a negative–positive–negative wave sequence (T-complex) are clearest at anterior temporal locations, which is less contaminated by volume-conducted STP activity (Albrecht et al. 2000; Ponton et al. 2002). These waves are present in school-aged children at latencies longer than in adults.

MMN can reliably be recorded in children of most ages (Ceponiene et al. 1998; Kushnerenko et al. 2002). Individual subject reliability is not optimal, however, and MMN is best studied with group data (Uwer and von Suchodoletz 2000). MMN can be a fine assay for the profound learning and plasticity changes in sound perception during development. A study of MMN evoked by the vowels /õ/ and /ö/ against the standard stimulus /e/ in Finnish and Estonian infants found that MMN in 6-month-old children of both groups was larger in amplitude for /õ/ than /ö/. This finding is consistent with the acoustic processing of the speech sounds, as /õ/ is more dissimilar than /ö/ from the standard /e/. However, at 1 year, MMN amplitude reversed in size for the two deviant stimuli only for Finnish children (Cheour et al. 1998). In Finnish, /ö/ is phonemic and /õ/ is not used, while in Estonian both vowels are used. This reversal in MMN amplitude, against responses expected based on acoustical differences, can indicate that language-learned phonetic discrimination becomes dominant by 1 year. Since all three sounds are phonemic in Estonian, MMN in this group would be expected to, and did, vary along the persistent acoustic contrasts.

P3a has latencies in young children only slightly longer than in adults (Cycowicz and Friedman, 1997; Kilpeläinen et al. 1999). While less well studied, processing negativities related to selective attention also undergo marked developmental changes. Nd wave amplitude increases and peak latency decreases with age (Berman and Friedman, 1995). This can support the hypothesis that children have more difficulty attending to specific stimuli when confronted with competing inputs. Results from another study support this conclusion by showing that target stimuli do not elicit processing negativities specific to the attended ear in 9 year olds (Määttä et al. 2005).

11 Average Evoked Potentials in Auditory Sensory and Cognitive Neuroscience

11.1 Context Dependence of Auditory Cortical Activity: Electroencephalographic Modulation

There are two classic views of the interaction between the AEP and the on-going EEG (Kruglikov and Schiff 2003). In the first, the AEP is considered as independent of EEG rhythmic activity. In the second, the AEP results from the sensory stimulus disrupting and resetting on-going EEG rhythms. These views are challenged by a third notion derived from emerging evidence that the EEG itself alters the AEP and thus participates in auditory cortex context-dependent processing. Human P_a and P1 amplitudes can be strongly modulated by the phase of the preceding EEG rhythm (Kruglikov and Schiff 2003). In monkeys, the strength of both stimulus-evoked AEPs and MUA in AI was modulated by the phase of delta activity at the time of sound onset, an effect that was maximal in supragranular layers (Lakatos et al. 2005). A hierarchical pattern of activity was seen with activity in a higher EEG frequency band modulated by activity in a lower band, with concurrent ramifications for unit firing. Moreover, somatic sensory stimulation modulated AI EEG rhythms and could enhance or suppress sound-evoked activity (Lakatos et al. 2007). Enhancement or suppression was determined by the temporal relationship between the onsets of the acoustic and somatic stimulations and was related to the periods of delta, theta, and gamma oscillations.

11.2 Modulation by Sound Context

Perhaps no process in hearing exemplifies the importance of context better than auditory scene analysis, wherein intermingled attributes of sounds are grouped or segregated into discrete sound objects (Bregman 1990). Auditory scene analysis can be divided into two: integration or segregation of simultaneous sound components into one or more objects, and integration or segregation of sequential sound components into one or more streams of perceptual objects (i.e., auditory stream segregation). Many of these processes operate independent of a listener's attention. Further, because animals identify sound sources in complicated acoustic environments, what may be interpreted as scene analysis can be studied in non-human species as well (Hulse et al. 1997; Izumi 2002).

With reference to the first category, multiple sound attributes serve as important cues for determining whether sound features constitute one or more objects (Yost 1991). The role of harmonicity in the integration or segregation of simultaneous sound components shows that harmonically related elements are grouped into one object, while those harmonically unrelated are perceptually segregated and assigned to separate sources. For low, resolved harmonics, a mistuned component is perceived as the emergence as a discrete tone percept at thresholds as low as 2–3% (Moore et al. 1985, 1986; Hartmann et al. 1990).

Several AEP studies have identified neural processes that appear to relate to perceptual "pop-out" of a mistuned sound component. The best studied is a negative difference wave, the object-related negativity (ORN) that overlies N1 and P2 (Alain et al. 2002, 2003). It is seen when AEPs evoked by sound complexes with a single mistuned harmonic are compared to their tuned counterpart. The ORN has a frontocentral voltage distribution and inverts in polarity at inferior temporal sites, consistent with a generator on the STP (Alain et al. 2003; Hautus and Johnson 2005). ORN is attention independent, and its amplitude increases with mistuning, which in turn parallels the likelihood of reporting the presence of two acoustic objects (Alain et al. 2003; Alain and McDonald 2007). It is also reliably seen in 8–12 year olds (Alain et al. 2003). A decrease in ORN amplitude in elderly adults parallels a diminished capacity to discriminate two sounds as a function of mistuning (Alain and McDonald 2007). While the ORN may represent a necessary pre-attentive marker for discriminating among simultaneously presented objects, its presence alone is not sufficient for behavioral discrimination. For instance, pre-adolescents have a larger ORN amplitude than adults but are behaviorally less acute in detecting mistuned harmonics (Alain et al. 2003).

While not a sufficient marker for discrimination of acoustic objects, the ORN behaves in a way suggesting its importance in scene analysis. ORN was elicited in subjects asked to identify two simultaneously presented vowels, a scenario mimicking the classic cocktail party effect (Snyder and Alain 2005). ORN amplitude was reduced in older adults and paralleled a decreased ability to identify both vowel sounds. Extending the analogy, ORN could be elicited at very low levels of harmonic mistuning when the harmonic complexes were presented from locations different from the mistuned sound component (McDonald and Alain 2005). This effect paralleled perceptual enhancement of the subjects distinguishing one versus two acoustic objects when sound components were separated.

An AEP wave resembling the ORN occurs in monkeys, suggesting shared neural mechanisms in humans and non-human primates. AEPs were evoked by harmonic complexes (Fig. 25.5, dotted line, left side) whose third harmonic was set at the BF of recording sites and the average AEP evoked by complexes whose third harmonic was mistuned by 8 and 16% both above and below the BF (solid line). Such mistuning precludes changes in the responses from being ascribed to modulation of the AEP by the pure tone-tuning characteristics of the recording sites. The basic AEP waveform is a positive–negative–positive voltage sequence modeling the human P1, N1, and P2. A negative voltage difference between the responses to the tuned and mistuned harmonic complexes parallels the human ORN. Thus, this negative voltage difference overrides the monkey analogue of the N1 and P2 components while other portions of the waveform, including the P1 analog, are similar across stimulus conditions. A further ORN analogue was seen when the third harmonic was kept at the BF while the remainder of the complexes was shifted by 8 and 16% (Fig. 25.5, right-hand side). These enhanced responses at specific tonotopically

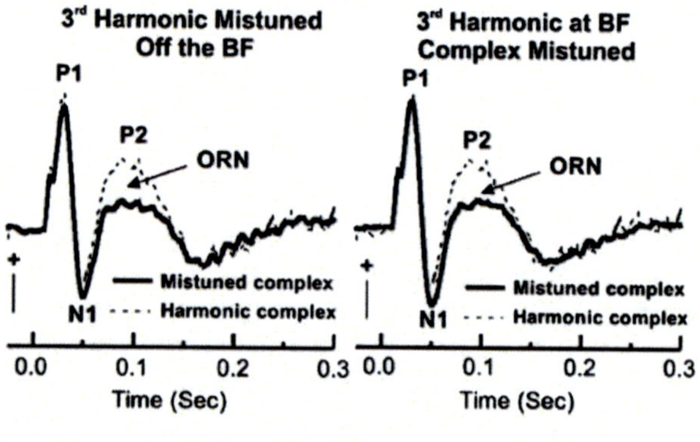

Fig. 25.5 An AEP component recorded from an awake monkey and similar to the human ORN component. The *left half* of the figure illustrates AEPs evoked by harmonic complexes (*dotted line*) whose third harmonic was set at the BF of recording sites and the average AEP evoked by complexes whose third harmonic was mistuned by 8 and 16% both above and below the BF (*solid line*). A negative voltage difference parallels the human ORN. The *right half* of the figure depicts an ORN analogue when the third harmonic was kept at the BF, while the remainder of the complexes was shifted by 8 and 16%

organized locations might facilitate perceptual emergence of the mistuned harmonic.

Auditory stream segregation is the second major category of scene analysis (Bregman 1990). Like simultaneous sound-source discrimination, stream segregation may rely on both attentive and pre-attentive mechanisms. A basic paradigm is the repeated presentation of two sequential tones ("A" and "B") differing in frequency. When frequency separation (ΔF) is small, the tones are heard as one sound stream of alternating tones. As ΔF grows, two streams will begin to be heard that contain only "A" or "B" tones. Increasing sound duration or decreasing the ISI enhances stream segregation (Fishman et al. 2004). This tendency to perceive sound sequences as discrete streams increases over seconds.

Analysis of AEPs has used stream segregation paradigms. P1, N1, P2 and a large negativity recorded from the lateral surface of the STG evoked by "B" tones increase in amplitude as ΔF increases (Snyder et al. 2006). These increases paralleled the subjects' ability to discriminate two segregated streams (see Gutschalk et al. 2005 for related MEG study). A sustained positivity developed early after tone sequence onset, suggesting a relationship with growing stream segregation (Snyder et al. 2006; Snyder and Alain 2007). This voltage increase was enhanced further when subjects attended to the sounds. It was concluded that the findings support both pre-attentive mechanisms for stream segregation, as in the enhanced responses to the "B" tones, and attention-dependent processes in the sustained positivity response.

MMN has also been used to probe acoustic stream organization. Listening to speech in a complex environment suggests that the various sound components must first be segregated into discrete streams before speech can be perceived. This was investigated using AEPs evoked by a cycle of six tones that alternated between high and low frequency and contained internally repeating patterns of three low and three high tones that increased in frequency (i.e., L1, H1, L2, H2, L3, H3). Deviant sounds were the reversal of one internal pattern (L3, H1, L2, H2, L1, H3). At rapid ISIs, which promoted segregation into streams of low and high tones, MMN was elicited by the deviant pattern. However, no MMN was elicited at long ISIs that promoted the perception of alternating low and high tones when the internal pattern changes could not be discerned (Sussman et al. 1999). Thus, detection of detailed patterns within the acoustic environment requires stream segregation as a prerequisite (Sussman 2005).

Additional work documents the impact of attention on stream segregation. Presenting the same sequences of tones described above at a long ISI that promoted perception of one sound stream elicited no MMN in the "ignore" condition (Sussman et al. 1998). When subjects attended to the high tones, the attentional filter generated two perceptual streams. Now, pattern deviations were detected for both the low- and high-frequency deviants. Thus, attention can strongly modulate the organization of sound sequences and facilitate the automatic process of deviance detection in sound streams.

Basic neural mechanisms of stream segregation have also being identified in experimental animals (Micheyl et al. 2007). MUA and CSD response patterns in monkey AI evoked by "A" and "B" tones were presented in an alternating sequence (ABAB) (Fishman et al. 2001a). "A" tones were set at the BF of the recording sites, while "B" tones were displaced from the BF by a variable ΔF. Presentation rate was also varied. At fast rates or large values of ΔF, responses evoked by "B" tones were preferentially suppressed relative to the responses evoked by BF "A" tones. At small values of ΔF or at slow presentation rates, responses were evoked by both tones. These findings paralleled perceptual data and suggest that forward masking targeting non-optimal stimuli was a key component of stream segregation. It was hypothesized that one stream would be heard when population responses evoked by 'A' and 'B' tones overlap significantly across the tonotopic array of AI space. In contrast, stream segregation is facilitated when the responses to the 'A' and 'B' tones spatially segregate within AI. Later work parametrically varied sound features that modulate stream segregation (ΔF, presentation rate, tone duration) and found response patterns supporting psychoacoustical data (Fishman et al. 2004), and this has been replicated in animal models (Kanwal et al. 2003; Bee and Klump 2004; Micheyl et al. 2005) including showing that response patterns modulate over time in a way that parallels perceptual build-up of stream segregation (Micheyl et al. 2005). Thus, responses to 'B' tones decline relative to an 'A' tone responses over a several-second time scale. A likely basis for the enhanced segregation as tone duration increases or ISI decreases is that each parameter change augments sound density, which in turn sharpens AI spectral tuning (Blake and Merzenich, 2002). This interaction between the acoustic environment and the specificity of neuronal activity, emphasized by the sensitivity of AI responses to the past history of sound stimulation (Ulanovsky et al. 2004), can thus have profound impacts on neural activity related to scene analysis.

Context dependence of AEPs is also seen in studies of a dual-stream hypothesis of complex sound processing. Auditory cortical fields analyze complex sound over both serial and parallel pathways. How subcortical auditory information flows in streams to and through auditory cortical fields and beyond, eventually reaching conscious perception, is not understood.

Studies of monkey auditory cortex indicate that complex sound processing is a hierarchical serial and parallel operation (Kaas and Hackett 1998). Activity representing a complex sound ascends the auditory pathway to reach the auditory core fields and is then distributed along two divergent corticocortical pathways, one carrying information on the identity of sound content, the other data on sound location (Rauschecker and Tian 2000; Arnott et al. 2004;

Scott 2005). Both postulated 'what' and 'where' streams reach different regions of prefrontal cortex (Romanski et al. 1999). Human AEP studies of this dual stream hypothesis have compared responses to sound objects to those of sound location (Alain et al. 2001; De Santis et al. 2007), to different sound objects (Murray et al. 2006), and to sound location, sound motion, and to the interaural and timing cues used in sound localization (Ducommun et al. 2002; Tardif et al. 2006).

These and related studies partially support the dual-stream hypothesis. Larger positive wave contributions to AEPs at frontal electrode sites are evident for distinctions based on pitch versus sound location, while the reverse pattern holds for parietal sites (Alain et al. 2001; De Santis et al. 2007). Activity of the right temporo-parietal cortex is greater when the stimulus parameter is sound location rather than pitch (De Santis et al. 2007). Using global dissimilarity measures, the patterns of generators for 'what/where' distinctions for a pitch versus sound location become distinct after ~100 ms (De Santis et al. 2007). Related findings are seen for mechanisms involved in sound localization, which is thought to occur more rapidly than pattern discrimination (Altmann et al. 2007). By 75 ms after stimulus onset, global dissimilarity in AEPs emerges for interaural intensity and time attributes (Tardif et al. 2006). Source analysis indicates that these 'where' attributes evoke AEPs with deflections associated with posterior superior temporal regions, extending into parietal areas. Sound location versus sound movement produces differential scalp distributions of the AEP from ~250 ms after stimulus onset (Ducommun et al. 2002).

Although these results indicate a differential spatial distribution of activity based on sound identity and location, AEP analyses and complementary fMRI imaging have also identified auditory cortex regions activated by aspects of both sound identity and location, including core auditory cortex, posterior STG, and parts of the planum temporale and prefrontal and parietal cortex (Alain et al. 2001; Altmann et al. 2007; De Santis et al. 2007). Taken together, these studies do not support a strict segregation of processing along 'what/where' pathways, but instead suggest a more graded differential pattern of regional activation.

12 Cortical Representation of Temporal Information

12.1 Amplitude Modulation

Auditory cortex encodes both the temporal and spectral structure of natural sounds. In the time domain, humans respond to sounds with temporal modulations ranging from a few Hertz, characteristic of syllable and phoneme repetition in speech, to several hundred Hertz, which approximates the upper limit of pitch perception (Rosen 1992). Amplitude-modulated (AM) sounds are often used to study how central auditory neurons encode temporal information (Joris et al. 2004). Neurons at all levels of the lemniscal auditory pathway can represent modulation frequency by their responses time locked to the modulation envelope, although the upper limit of such temporal coding decreases at higher auditory stations. In primary auditory cortex the upper limit is <200 Hz (Eggermont 1998; Liang et al. 2002). In an intracranial study of human auditory cortex, best phase locking evoked by AM white noise was at AM frequencies of 4–16 Hz, in accord with laboratory animal single unit studies and which correspond to the range of envelope modulation required for accurate speech comprehension (Liégeois-Chauvel et al. 2004).

While best modulation frequency in human auditory cortex may be associated with the encoding of speech temporal modulation, sound roughness may be coded by higher rates of phase-locked activity. In intracranial studies of human and monkey auditory cortex, AM response amplitude was correlated with the degree of consonance and dissonance of musical chords (Fishman et al. 2001b). Dissonant chords such as a minor second had more roughness and elicited significant phase locking to the beat frequencies embedded in the sound envelope. Consonant chords (octave) were perceived as smoother and elicited minimal phase locking. Other work found the best modulation frequency and the maximum modulation frequency of AM tones evoking phase-locked activity in monkey AI closely paralleled human psychoacoustical functions for roughness (Fishman et al. 2000). Together, these findings strongly support the hypothesis that perception of roughness is related to AI phase-locked response patterns. Transformations from a temporal to a rate code may account for the capacity to detect higher modulation frequencies and be a dominant mechanism for the representation of these time-varying signals in AI and other areas (Lu et al. 2001).

Phase-locked responses may also subserve temporal pitch encoding. Classic studies (Flanagan and Guttman 1960a,b; Rosenberg 1965) found that the pitch of click trains with rates <100 Hz was equal to the pulse rate whether the pulses were the same or of alternating polarity (f_0 of alternating polarity pulses equals one-half the rate). Above 200 Hz, the pitch equaled the f_0 and was dependent upon whether the pulses were of the same or alternating polarity (spectral pitch). The 100–200 Hz region is a transition zone. Temporal pitch was mediated by high-frequency auditory channels, while pitch based on f_0 was mediated by low-frequency channels. Both MUA and CSD measures in monkey AI displayed similar characteristics (Steinschneider et al. 1998). Phase locking dominated below 100 Hz regardless of click polarity pattern in high-BF regions of AI and decayed between 100 and 200 Hz. In contrast, low-BF areas showed poor phase locking, and response amplitudes reflected the f_0 of the click trains and the tuning characteristics of the sites. Similar limiting rates of ~100 Hz are seen in AEPs evoked by both the AM of speech sounds (Steinschneider et al. 1999) the click trains (Brugge et al. 2008, Fig. 25.6) in posteromedial HG. These findings are consistent with psychoacoustical data suggesting two pitch mechanisms and support the role of phase-locked activity for encoding temporal pitch (Carlyon and Shackleton 1994; Plack and Carlyon 1995).

12.2 Speech Sounds

The importance of speech perception has engendered a large literature devoted to its underlying neural mechanisms. A well-studied speech parameter is voice onset time (VOT), a feature used by most languages (Lisker and Abramson 1964). VOT is the interval between the consonant release (onset) and the onset of voicing (periodic vocal cord vibrations). A non-overlapping distribution of VOTs occurs in almost all languages: voicing either begins before consonant release (lead), near the time of consonant release, or after the time of release (lag). In American English, voiced stop consonants, such as /b/, /d/, and /g/ contain short VOTs, whereas their unvoiced counterparts /p/, /t/, and /k/ have long lag VOTs. In French, voiced stop consonants have a long lead

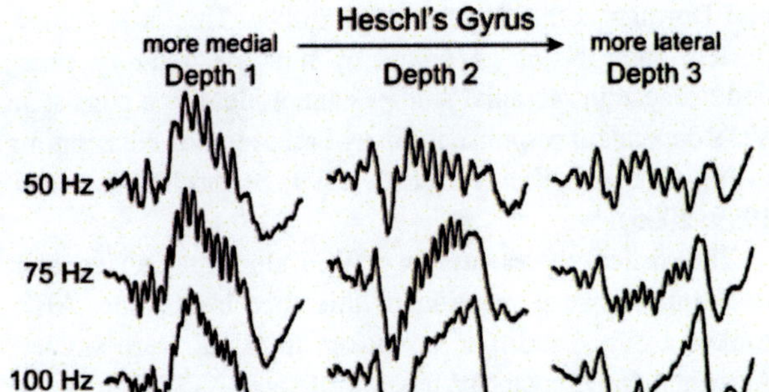

Fig. 25.6 Phase-locked responses to click trains presented at various rates recorded from three electrode sites on the medial portion of Heschl's gyrus. Rates of stimulation are shown at the *far left*. Low amplitude phase-locked responses at the most medial electrode can still be observed at a rate of 125 Hz. Maximal rates that evoke phase-locked responses are lower at more lateral sites

VOT, while voicing onset and consonant release occur almost simultaneously for unvoiced stop consonants.

Human intracranial studies find patterns of neural activity that reflect VOT duration (Liégeois-Chauvel et al. 1999; Steinschneider et al. 1999, 2005; Trébuchon-Da Fonseca et al. 2005). Stop consonant–vowel syllables with short VOTs elicit AEPs in medial HG with a single response complex evoked by stimulus onset. In contrast, syllables with long VOTs, both in lead and lag position, elicit responses with two components, one time locked to consonant release and the other to voicing onset. Perceptually, temporal information with severely limited spectral content suffices to produce excellent discrimination of voiced from unvoiced stop consonants (Shannon et al. 1995). These temporal features are prominent in Heschl's gyrus activity (Fig. 25.7). Intracerebral AEPs from medial HG to French /ba/ and /pa/ in a French patient elicited two response complexes with a first biphasic potential (N40/P80) time locked to voicing onset, then a second (N150/P200) time locked to consonant release. In contrast, the French /pa/, which has voicing onset nearly simultaneous with consonant release, elicited one response complex.

While there is accord from intracranial recordings on the temporal representation of VOT, questions remain as to the laterality of representation. In French subjects, temporal encoding is concentrated in the left hemisphere in patients with left-hemispheric language dominance (Liégeois-Chauvel et al. 1999), and in the right auditory cortex in patients with bilateral or right hemispheric dominance (Trébuchon-Da Fonseca et al. 2005). In American subjects, responses reflecting VOT are bilateral (Steinschneider et al. 1999, 2005, unpublished observations). The VOT distinction used in American English may represent use of a natural psychoacoustical boundary in mammalian hearing, seen in the ability of infants to distinguish VOT contrasts important for the English language when that contrast is phonetically irrelevant for the child's native language (Eilers et al. 1979; Jusczyk et al. 1989). Several American studies indicate an important right hemisphere role in VOT processing (Molfese and Molfese 1988; Simos et al. 1997), and the right hemisphere activity may reflect this non-language-determined boundary. One important feature of phonetic processing regardless of the language is that perception is categorical (Kuhl 1986; Laguitton et al. 2000), e.g., perception of a phoneme is relatively constant around the modal value of its VOT. However, when the VOT changes sufficiently, perception abruptly changes to another phoneme. In American English, a /t/ will be perceived regardless of a VOT of +60 or +40 ms. If the same 20-ms VOT difference ranges from +40 to +20 ms, the percept will be of /d/. Thus, any physiological process presumed to involve VOT in speech perception should reflect its categorical nature. Intracranial AEPs time locked to consonant release and voicing onset appear to

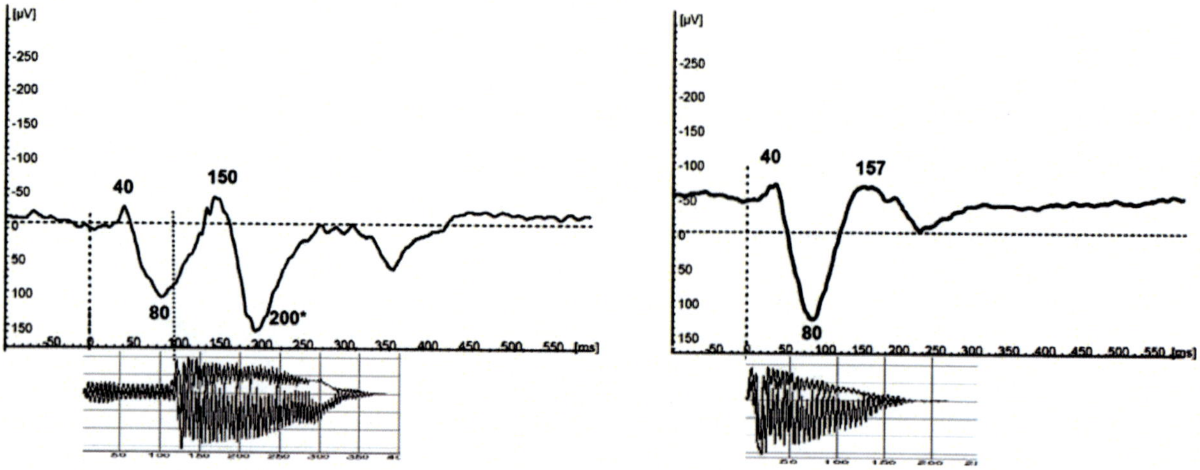

Fig. 25.7 AEPs evoked by the French syllables /ba/ (*left*) and /pa/ (*right*) recorded from an intracranial electrode located in mesial Heschl's gyrus. Stimulus waveforms are depicted below the waveforms. See text for details

conform to this requirement and display categorical-like features (Steinschneider et al. 1999). Thus, /da/ with either a VOT of 0 or 20 ms evokes a single response time locked to consonant release, while /ta/ with VOTs of 40, 60, or 80 ms elicited responses time locked to both consonant release and voicing onset.

A related intracranial study recorded responses in medial HG and found a VOT boundary shift with changes in stop consonant place of articulation (Steinschneider et al. 2005). In American English, perceptual boundaries are shortest for the differential perception of the stop consonants /b/ and /p/ (~20 ms), intermediate for the stops /d/ and /t/ (~30 ms), and longest for the stops /g/ and /k/ (~40 ms) (Lisker and Abramson 1964). Synthetic syllables with varying first formant frequencies mimic the changes that occur with shifts in consonant place of articulation. Temporal response patterns averaged across the HG electrode array paralleled perceptual findings. When the first formant was of low frequency and mimicked the condition when the boundary is longest, the subject heard /t/ for only a VOT of 60 ms. Similarly, AEPs only had time-locked responses to both consonant release and voicing onset for the same VOT. At higher first formant frequencies, the perception of /t/ occurred for VOTs >20 ms. Now, AEPs were time locked to both consonant release and voicing onset for VOTs with values as short as the perceptual findings. A gradient of short-to-long physiological boundaries across more lateral-to-medial HG electrodes was seen. This suggests that this changing pattern reflected the interaction between syllabic spectral characteristics and tonotopic organization of HG (Howard et al. 1996), a pattern supported by multiunit responses in monkeys (Steinschneider et al. 2005). The finding that averaged activity best correlated with perceptual phenomena is supported by work showing that averaged population activity is a determinant for perceptual outcomes (Sanger 2003; Ma et al. 2006).

When intracranial and scalp-recorded AEPs from the same subjects were compared, both showed similar temporal patterns and left hemispheric dominance in VOT for French phonemes (Trébuchon-Da Fonseca et al. 2005). The MEG equivalent of N1 shows a marked amplitude decrement when evoked by syllables with a +40- or +60-ms VOT rather than that for 0- or 20-ms VOT (Simos et al. 1998a). This decrement was also seen in intracranial data low-pass filtered like that in the MEG study (Steinschneider et al. 1999). The smaller N1 could reflect reintroduction of overlapping positive waves evoked by the longer VOT stimuli. Other non-invasive AEP studies suggest the N1 is not a reliable index of the voiced/unvoiced distinction for stop consonants (Sharma and Dorman 2000; Sharma et al. 2000). This is not unexpected, since N1 is generated by multiple auditory areas, and because intracranial studies cannot identify a consistent VOT-dependent response in many brain regions contributing to N1 (Liégeois-Chauvel et al. 1999; Steinschneider et al. 1999; 2005).

The cortical representation of VOT alone may not account fully for human perception of this speech attribute. AEPs evoked by /ba/ and /pa/ as spoken in Hindi were studied. As it had for French, /b/ has a prolonged VOT lead, while the /p/ VOT is nearly simultaneous with consonant release. Two time-locked responses evoked by voicing onset and consonant release were recorded for /ba/ with a –90-ms VOT in both English- and Hindi-native speakers. However, the English speakers could not distinguish the /b/ and /p/, even though their AEPs suggested that the VOT information was represented in their cortex. In contrast, MMN was larger in native language speakers than in non-native subjects (Sharma and Dorman 2000). Perhaps MMN relates more closely to the perceptual attributes of the speech sounds than the time-locked responses evoked by the syllables' acoustic transients.

Several electrophysiological studies address basic psychoacoustic hypotheses on VOT (Hirsh 1959; Pisoni 1977). Seminal work postulated that VOT discrimination is partly determined by whether consonant release and voicing onset occur simultaneously or sequentially. Using two-tone analogs of VOT, categorical perception for this perceptual distinction had a boundary at ~20 ms. In a similar paradigm the magnetic counterpart of N1 decreased in amplitude between the +20- and +40-ms boundary in a way that paralleled perception of the two-tone stimuli (Simos et al. 1998b). Abrupt AEP changes in newborns also had marked response changes between two-tone stimuli with a similar boundary (Simos and Molfese 1997), supporting results in 2-month-old babies seen with a high-amplitude sucking procedure (Jusczyk et al. 1989).

Responses elicited by speech sounds in animals resemble those recorded in humans, with similar categorical-like features and physiological boundaries (Steinschneider et al. 1994, 2003, 2005; Eggermont 1995, 1999; McGee et al. 1996). VOT temporal response patterns include features of gap processing and are seen with two-tone stimuli (Eggermont 1995, 1999, 2000; Steinschneider et al. 2005), and basic cortical circuit mechanisms for such boundaries have been proposed (Eggermont, 2000; Steinschneider et al. 2003). In summary, both perceptual and physiological studies support a temporal processing mechanism as part of VOT perception. This mechanism is partly based on the ability to track acoustic event onsets in time and by fundamental aspects of mammalian physiology modified by language experience.

13 Average Evoked Potentials in Neurological Disorders

AEP recording is a powerful method for examining the pathophysiology of many hearing, speech, and language disorders, including those as diverse as cognitive impairment in aging (Irimajiri et al. 2005), schizophrenia (Light and Braff, 2005; Oades et al. 2006), and dyslexia. Theories for the causes of dyslexia (Rosen 1999; Ramus et al. 2003) range from the phonological hypothesis, which considers dyslexia as a specific impairment in encoding speech sounds, to dyslexia construed as an auditory processing disorder whose primary deficit is the inability to encode rapidly changing sounds (Tallal et al. 1993; Lorenzi et al. 2000; Rocheron et al. 2002). Other suggested impairments in auditory processing include dysfunction of mechanisms for scene analysis and deficiencies in stimulus-specific adaptation (Helenius et al. 1999; Goswami et al. 2002; Foxton et al. 2003; Petkov et al. 2005; Ahissar et al. 2006).

Risk factors implicating auditory processing deficits can be identified in pre-verbal children using perceptual and physiological indices. Infants with family histories of developmental language disorders had deficits in rapid auditory temporal processing compared to control infants, and these deficiencies could predict later language performance (Benasich and Tallal 2002). Aberrant AEPs to deviant speech sounds embedded in a sequence of standard syllables were found in neonates genetically at risk for dyslexia (Leppänen et al. 1999). These early signs, appearing before the emergence of language, suggest a fundamental impairment in central auditory processing that may precede dyslexia.

Studies beyond the early neonatal period further identify auditory processing deficits in dyslexic subjects. Responses reflecting VOT were aberrant in French-speaking adult dyslexics (Giraud et al. 2005). In contrast to the typical of response patterns time locked to voicing onset and consonant release, dyslexics either failed to reveal differential responses to syllables varying in VOT, or had an many more response components with unusually long latencies, and the characteristic left hemisphere response dominance was also absent. Other studies also find aberrant obligatory AEP responses to both speech and non-speech stimuli in dyslexic adults (Helenius et al. 2002; Moisescu-Yifach and Pratt 2005).

Abnormalities in MMN also suggest auditory processing deficiencies as a feature of dyslexia. An abnormally low-amplitude MMN evoked by slow frequency-modulated tones in adult dyslexics occurred even with normal perceptual scores, suggesting that physiological measures may be more sensitive assays than behavior (Stoodley et al. 2006). An abnormal MMN was evoked by a tone-pair reversal when followed by a third tone (Kujala et al. 2003), consistent with both dysfunctional backward masking (Wright et al. 1997) and perceptual grouping abnormalities in theories of deficits in auditory scene analysis. Further support for scene analysis deficits comes from a study of MMN evoked by a tone omission in a stream of tones (Fisher et al. 2006). MMN was not elicited in adult dyslexic subjects when the ISI was 100 ms, suggesting that the dyslexic subjects failed to group the tones as a perceptual stream where omission of one element should have been detected as deviant.

Auditory processing impairment contributory to dyslexia is found in the physiological ramifications of remediation therapy (Kujala et al. 2001). Seven-year-old dyslexics trained with an audio-visual computer game requiring matching of non-verbal sound elements with co-varying visual rectangles had MMN recorded to deviant tone pairs whose order was reversed from the standard before and after training. MMN in the trained subjects increased in amplitude and was correlated with improved reading scores over untrained dyslexic controls. Thus, a non-linguistic audio-visual training protocol enhanced both linguistic skills and physiological measures of auditory cortical function, a finding seen for exogenous AEP components in learning-disabled children trained in an audio-visual program for enhancing phonological skills and auditory processing (Hayes et al. 2003). The evidence favors auditory cortical dysfunction as at least a concomitant, and likely a contributing, factor in dyslexia, despite the lack of a specific core deficit.

Temporal processing impairments as a cause of poor speech sound discrimination have also been found in subjects with cochlear implants (Roman et al. 2004). AEPs recorded to the French speech sounds /ba/ and /pa/ show time-locked deflections to the temporal cues of VOT in control and implanted subjects. However, relatively poor time-locking was associated with relatively poor speech sound discrimination in implant subjects. Like dyslexic subjects, temporal response patterns encoding key acoustic transients of speech normally observed by competent auditors were abnormal in subjects with impaired speech capacities.

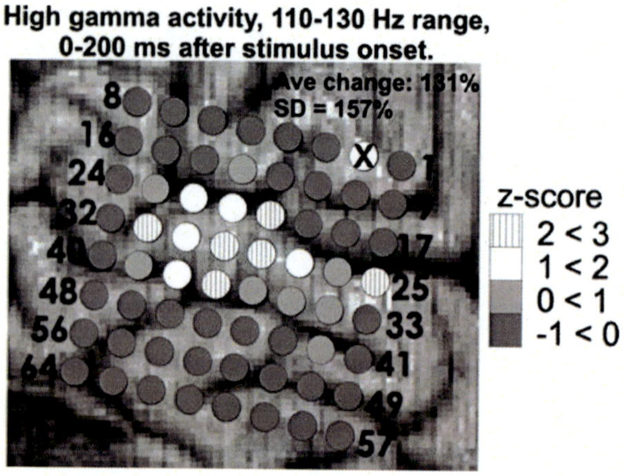

Fig. 25.8 Normalized distribution of high gamma-band activity recorded from subdural grid electrodes overlying the right posterior temporal lobe. Depicted responses were derived from the averaged activity evoked by /ba/, /da/, /ga/, /pa/, /ta/, and /ka/. Maximum responses are distributed along the posterior STG. Average change of activity from baseline levels and the standard deviation across the array are also shown

14 Electroencephalography and Auditory Cortex

EEG rhythms may determine the dynamic genesis of functional cell assemblies (Freiwald et al. 2001; Ward 2003; Herrmann et al. 2004; Kahana 2006). This hypothesis suggests that specific phases of EEG rhythms are associated with neurons in relatively depolarized or hyperpolarized states, which enhances the synchronization of neuronal firing in such populations (Jacobs et al. 2007). Lower EEG frequencies in the theta band (4–8 Hz) may be of special importance for long-range interactions, whereas higher gamma-band frequencies (>30 Hz) may influence local neuronal interactions. Historically, focus on the functional significance of lower gamma frequencies ~40 Hz preceded interest in higher gamma frequencies >100 Hz. High gamma-band activity better correlates with fMRI results than activity indexed by evoked potentials or lower EEG frequencies (Mukamel et al. 2005; Niessing et al. 2005).

In monkey auditory cortex, higher gamma frequencies are more sensitive indicators of activation and more specific indices of tonotopic organization than lower EEG frequencies (Kayser et al. 2007; Steinschneider et al. 2008). Human intracranial recordings (Crone et al. 2001, 2006; Trautner et al. 2006) show that the greatest EEG frequency changes after sound stimulation were at ~100 Hz (Edwards et al. 2005). Specificity is revealed by large power increases in the high gamma band at intracortical sites during language naming tasks, which correlated strongly with disruption of naming by electrical stimulation at the same electrodes (Sinai et al. 2005).

Sensitivity of high gamma activity reveals activation of auditory cortex (Fig. 25.8). Normalized activity in the 110–130 Hz gamma range sampled from grid electrodes over the posterior temporal lobe is shown in response to the averaged activity evoked by six syllables. Maximal responses in the 200 ms after syllable onset occur only along the posterior STG, and these localized increases persisted beyond 200 ms after syllable onset and contrasted with weaker, shorter duration increases for concurrently recorded low frequencies (data not shown).

Non-invasive study of higher frequency gamma-band activity is still in its infancy. Several studies document the feasibility of recording these low-amplitude components and their modulation during sound processing (Kaiser et al. 2002; Palva et al. 2002). Parallel studies have identified correlates between gamma activity and conscious visual perception (Goffaux et al. 2004; Melloni et al. 2007) and changes seen in various neurological disorders (Herrmann and Demiralp 2005; Uhlhass and Singer 2006). Gamma activity may facilitate locally coherent pyramidal cell firing, binding of multiple stimulus attributes, and synaptic plasticity within neural networks (Freiwald et al. 2001; Salinas and Sejnowski 2001; Traub et al. 2005; Sejnowski and Paulsen 2006). These roles, and the abnormalities of gamma activity in neurological disorders, suggest that effects of disease states on complex sound perceptions can be addressed by non-invasively acquired EEG.

15 Future Directions

The refinement and extension of studies using AEPs and EEG will depend on careful integration with multiple methodologies. Human intracranial studies are essential to define auditory cortical physiology in detail, to integrate these findings with those from non-invasive recording techniques, and to reveal cortical auditory processing mechanisms common to humans and non-human mammals. More than 1,500 patients per year undergo invasive EEG monitoring (Kahana 2006). The continued growth of patient recruitment for participation in scientific study while meeting their clinical needs is crucial for the success of this endeavor. Integration with translational animal studies of key physiological processes in AEPs and EEG needs to be further developed in order to enhance the interpretive power of human physiological findings. Finally, the results from studies of the physiology of normal sound processing will enhance understanding of the pathology underlying the many disorders affecting speech and hearing.

Acknowledgments The authors gratefully acknowledge financial support from NIH grants DC00657 (M.S.) and DC04290 (J.F.B. and M.S.), and the French National Institute of Health and Medical Research (INSERM) (C.L.-C.).

References

Ahissar M, Lubin Y, Putter-Katz H, and Banai K (2006) Dyslexia and the failure to form a perceptual anchor. *Nature Neuroscience* **9**:1558–1564.

Ahlfors SP and Simpson GV (2004) Geometrical interpretation of fMRI-guided MEG/EEG inverse estimates. *NeuroImage* **22**: 323–332.

Alain C and Woods DL (1997) Attention modulates auditory pattern memory as indexed by event-related potentials. *Psychophysiology* **34**:534–546.

Alain C, Cortese F, and Picton TW (1998) Event-related brain activity associated with auditory pattern processing. *NeuroReport* **9**: 3537–3541.

Alain C, Arnott SR, Hevenor S, Graham S, and Grady CL (2001) "What" and "where" in the human auditory system. *Proceedings of the National Academy of Sciences of the United States of America* **98**:12301–12306.

Alain C, Schuler BM, and McDonald KL (2002) Neural activity associated with distinguishing concurrent auditory objects. *Journal of the Acoustical Society of America* **111**:990–995.

Alain C, Theunissen EL, Chevalier H, Batty M, and Taylor MJ (2003) Developmental changes in distinguishing concurrent auditory objects. *Cognitive Brain Research* **16**:210–218.

Alain C and McDonald KL (2007) Age-related differences in neuromagnetic brain activity underlying concurrent sound perception. *Journal of Neuroscience* **27**:1308–1314.

Albrecht R, Suchodoletz Wv, and Uwer R (2000) The development of auditory evoked dipole source activity from childhood to adulthood. *Clinical Neurophysiology* **111**:2268–2276.

Altmann CF, Bledowski C, Wibral M, and Kaiser J (2007) Processing of location and pattern changes of natural sounds in the human auditory cortex. *NeuroImage* **35**:1192–1200.

Arezzo JC, Pickoff A, and Vaughan HG Jr (1975) The sources and intracerebral distribution of auditory evoked potentials in the alert rhesus monkey. *Brain Research* **90**:57–73.

Arnott SR, Binns MA, Grady CL, and Alain C (2004) Assessing the auditory dual-pathway model in humans. *NeuroImage* **22**:401–408.

Babiloni F, Cincotti F, Carducci F, Rossini PM, and Babiloni C (2001) Spatial enhancement of EEG data by surface Laplacian estimation: the use of magnetic resonance imaging-based head models. *Clinical Neurophysiology* **112**:724–727.

Badier JM and Chauvel P (1995) Spatio-temporal characteristics of paroxysmal interictal events in human temporal lobe epilepsy. *Journal of Physiology (Paris)* **89**:255–264.

Bai X, Towle VL, He EJ, and Hea B (2007) Evaluation of cortical current density imaging methods using intracranial electrocorticograms and functional MRI. *NeuroImage* **35**:598–608.

Baillet S (2001) Electromagnetic brain mapping. *IEEE Signal Processing Magazine* **18**:14–30.

Bee MA and Klump GM (2004) Primitive auditory stream segregation: a neurophysiological study in the songbird forebrain. *Journal of Neurophysiology* **92**:1088–1104.

Benar CG, Grova C, Kobayashi E, Bagshaw AP, Aghakhani Y, Dubeau F, and Gotman J (2006) EEG-fMRI of epileptic spikes: concordance with EEG source localization and intracranial EEG. *NeuroImage* **30**:1161–1170.

Benasich AA and Tallal P (2002) Infant discrimination of rapid auditory cues predicts later language impairment. *Behavioral Brain Research* **136**:31–49.

Berman S and Friedman D (1995) The development of selective attention as reflected by event-related brain potentials. *Journal of Experimental Child Psychology* **59**:1–31.

Blake DT and Merzenich MM (2002) Changes of A1 receptive fields with sound density. *Journal of Neurophysiology* **88**:3409–3420.

Boatman DF and Miglioretti DL (2005) Cortical sites critical for speech discrimination in normal and impaired listeners. *Journal of Neuroscience* **25**:5475–5480.

Boatman DF, Lesser RP, Crone NE, Krauss G, Lenz FA, and Miglioretti DL (2006) Speech recognition impairments in patients with intractable right temporal lobe epilepsy. *Epilepsia* **47**: 1397–1401.

Borgmann C, Roß B, Draganova R, and Pantev C (2001) Human auditory middle latency responses: influence of stimulus type and intensity. *Hearing Research* **158**:57–64.

Bradlow AR, Kraus N, Nicol TG, McGee TJ, Cunningham J, Zecker SG, and Carrell TD (1999) Effects of lengthened formant transition duration on discrimination and neural representation of synthetic CV syllables by normal and learning-disabled children. *Journal of the Acoustical Society of America* **106**:2086–2096.

Brázdil M, Dobšík M, Mikl M, Hluštík P, Daniel P, Pažourková M, Krupa P, and Rektor I (2005) Combined event-related fMRI and intracerebral ERP study of an auditory oddball task. *NeuroImage* **26**:285–293.

Bregman AS (1990) *Auditory Scene Analysis: The Perceptual Organization of Sound.* Massachusetts Institute of Technology Press, Cambridge.

Brugge JF, Volkov IO, Garell PC, Reale RA, and Howard MA III (2003) Functional connections between auditory cortex on Heschl's gyrus and on the lateral superior temporal gyrus in humans. *Journal of Neurophysiology* **90**:3750–3763.

Brugge JF, Volkov IO, Oya H, Kawasaki H, Reale RA, Fenoy A, Steinschneider M, and Howard MA III (2008). Functional localization of auditory cortical fields of human: click-train stimulation. *Hearing Research* **238**:12–24.

Carlyon RP and Shackleton TM (1994) Comparing the fundamental frequency of resolved and unresolved harmonics: evidence for two pitch mechanisms? *Journal of the Acoustical Society of America* **95**:3541–3554.

Celsis P, Doyon B, Boulanouar K, Pastor J, Demonet J-F, and Nespoulous J-L (1999) ERP correlates of phoneme perception in speech and sound contexts. *NeuroReport* **10**:1523–1527.

Ceponiene R, Cheour M, and Näätänen R (1998) Interstimulus interval and auditory event-related potentials in children: evidence for multiple generators. *Electroencephalography and Clinical Neurophysiology* **108**:345–354.

Ceponiene R, Rinne T, and Näätänen R (2002) Maturation of cortical sound processing as indexed by event-related potentials. *Clinical Neurophysiology* **113**:870–882.

Cheour M, Ceponiene R, Lehtokoski A, Luuk A, Allik J, Alho K, and Näätänen R (1998) Development of language-specific phoneme representations in the infant brain. *Nature Neuroscience* **1**:351–353.

Cohen YE, Russ BE, Gifford GW, Kiringoda R, and MacLean KA (2004) Selectivity for the spatial and nonspatial attributes of auditory stimuli in the ventrolateral prefrontal cortex. *Journal of Neuroscience* **24**:11307–11316.

Coull JT (1998) Neural correlates of attention and arousal: insights from electrophysiology, functional neuroimaging and psychopharmacology. *Progress in Neurobiology* **55**:343–361.

Creutzfeldt O, Ojemann G, and Lettich E (1989) Neuronal activity in the human lateral temporal lobe. I. Responses to speech. *Experimental Brain Research* **77**:451–475.

Crone NE, Boatman D, Gordon B, and Hao L (2001) Induced electrocorticographic gamma activity during auditory perception. *Clinical Neurophysiology* **112**:565–582.

Crone NE, Sinai A, and Korzeniewska A (2006) High-frequency gamma oscillations and human brain mapping with electrocorticography. In: Neuper C and Klimesch W (eds). *Progress in Brain Research*, volume 159: *Event-Related Dynamics of Brain Oscillations*. Elsevier Science Publishers, Amsterdam, pp. 275–295.

Crowley KE and Colrain IM (2004) A review of the evidence for P2 being an independent component process: age, sleep and modality. *Clinical Neurophysiology* **115**:732–744.

Cruikshank SJ, Rose HJ, and Metherate R (2002) Auditory thalamocortical synaptic transmission in vitro. *Journal of Neurophysiology* **87**:361–384.

Cuffin BN (1998) EEG dipole source localization. *IEEE Engineering in Medicine and Biology Magazine: The Quarterly Magazine of the Engineering in Medicine & Biology Society* **17**:118–122.

Cycowicz YM and Friedman D (1997) A developmental study of the effect of temporal order on the ERPs elicited by novel environmental sounds. *Electroencephalography and Clinical Neurophysiology* **103**:304–318.

Darvas F, Pantazis D, Kucukaltun-Yildirim E, and Leahy RM (2004) Mapping human brain function with MEG and EEG: methods and validation. *NeuroImage* **23**:S289–S299.

Dehaene-Lambertz G and Dehaene S (1994). Speed and cerebral correlates of syllable discrimination in infants. *Nature* **370**:292–295.

Dehaene-Lambertz G and Baillet S (1998) A phonological representation in the infant brain. *NeuroReport* **9**:1885–1888.

De Santis L, Clarke S, and Murray MM (2007) Automatic and intrinsic auditory "what" and "where" processing in humans revealed by electrical neuroimaging. *Cerebral Cortex* **17**:9–17.

Devor A, Dunn AK, Andermann ML, Ulbert I, Boas DA, and Dale AM (2003). Coupling of total hemoglobin concentration, oxygenation, and neural activity in rat somatosensory cortex. *Neuron* **39**: 353–359.

Ducommun CY, Murray MM, Thut G, Bellmann A, Viaud-Delmon I, Clarke S, and Michel CM (2002) Segregated processing of auditory motion and auditory location: an ERP mapping study. *NeuroImage* **16**:76–88.

Ebersole JS and Wade PB (1990) Spike voltage topography and equivalent dipole localization in complex partial epilepsy. *Brain Topography* **3**:21–34.

Edwards E, Soltani M, Deouell LY, Berger MS, and Knight RT (2005) High gamma activity in response to deviant auditory stimuli recorded directly from human cortex. *Journal of Neurophysiology* **94**:4269–4280.

Eggermont JJ (1995) Neural correlates of gap detection and auditory fusion in cat auditory cortex. *NeuroReport* **6**:1645–1648.

Eggermont JJ (1998). Representation of spectral and temporal sound features in three cortical fields of the cat. Similarities outweigh differences. *Journal of Neurophysiology* **80**:2743–2764.

Eggermont JJ (1999) Neural correlates of gap detection in three auditory cortical fields in the cat. *Journal of Neurophysiology* **81**:2570–2581.

Eggermont JJ (2000) Neural responses in primary auditory cortex mimic psychophysical, across-frequency-channel, gap-detection thresholds. *Journal of Neurophysiology* **84**:1453–1463.

Eilers RE, Gavin W, and Wilson WR (1979) Linguistic experience and phonemic perception in infancy: a cross-linguistic study. *Child Development* **50**:14–18.

Engel AK, Moll CKE, Fried I, and Ojemann GA (2005) Invasive recordings from the human brain: clinical insights and beyond. *Nature Reviews Neuroscience* **6**:35–47.

Escera C, Alho K, Winkler I, and Näätänen R (1998) Neural mechanisms of involuntary attention to acoustic novelty and change. *Journal of Cognitive Neuroscience* **10**:590–604.

Ferri R, Elia M, Agarwal N, Lanuzza B, Musumeci SA, and Pennisi G (2003) The mismatch negativity and the P3a components of the auditory event-related potentials in autistic low-functioning subjects. *Clinical Neurophysiology* **114**:1671–1680.

Fisher AE, Barnes GR, Hillebrand A, Holliday IE, Witton C, and Richards IL (2006). Abnormality of mismatch negativity in response to tone omission in dyslexic adults. *Brain Research* **1077**:90–98.

Fishman YI, Reser DH, Arezzo JC, and Steinschneider M (2000) Complex tone processing in primary auditory cortex of the awake monkey. I. Neural ensemble correlates of roughness. *Journal of the Acoustical Society of America* **108**:235–246.

Fishman, YI, Reser DH, Arezzo JC, and Steinschneider M (2001a) Neural correlates of auditory stream segregation in primary auditory cortex of the awake monkey. *Hearing Research* **151**:167–187.

Fishman YI, Volkov IO, Noh MD, Garell PC, Bakken H, Arezzo JC, Howard MA 3rd, and Steinschneider M (2001b) Consonance and dissonance of musical chords: neural correlates in auditory cortex of monkeys and humans. *Journal of Neurophysiology* **86**:2761–2788.

Fishman, YI, Arezzo JC, and Steinschneider M (2004) Auditory stream segregation in monkey auditory cortex: effects of frequency separation, presentation rate, and tone duration. *Journal of the Acoustical Society of America* **116**:1656–1670.

Flanagan JL and Guttman N (1960a) On the pitch of periodic pulses. *Journal of the Acoustical Society of America* **32**:1308–1319.

Flanagan JL and Guttman N (1960b) Pitch of periodic pulses without fundamental component. *Journal of the Acoustical Society of America* **32**:1319–1328.

Foxton JM, Talcott JB, Witton C, Brace H, McIntyre F, and Griffiths TD (2003) Reading skills are related to global, but not local, acoustic pattern perception. *Nature Neuroscience* **6**:343–344.

Freiwald WA, Kreiter AK, and Singer W (2001) Synchronization and assembly formation in the visual cortex. In: Nicolelis MAL (ed). *Advances in Neural Population Coding, Progress in Brain Research*, volume 130. Elsevier Science Publishers, Amsterdam, pp. 111–140.

Friedman D, Cycowicz YM, and Gaeta H (2001) The novelty P3: an event-related brain potential (ERP) sign of the brain's evaluation of novelty. *Neuroscience and Biobehavioral Reviews* **25**:355–373.

Gevins A, Smith ME, McEvoy LK, Leong H, and Le J (1999) Electroencephalographic imaging of higher brain function. *Philosophical Transactions of the Royal Society of London, Series B* **354**:1125–1134.

Giard M-H, Perrin F, Echallier JF, Thévenet M, Froment JC, and Pernier J (1994) Dissociation of temporal and frontal components in the human auditory N1 wave: a scalp current density and dipole model analysis. *Electroencephalography and Clinical Neurophysiology* **92**:238–252.

Giard M-H, Fort A, Mouchetant-Rostaing Y, and Pernier J (2000) Neurophysiological mechanisms of auditory selective attention in humans. *Frontiers in Bioscience* **5**:84–94.

Gifford GW and Cohen YE (2005) Spatial and non-spatial auditory processing in the lateral intraparietal area. *Experimental Brain Research* **162**:509–512.

Gilley PM, Sharma A, Dorman M, and Martin K (2005) Developmental changes in refractoriness of the cortical auditory evoked potential. *Clinical Neurophysiology* **116**:648–657.

Giraud K, Démonet JF, Habib M, Marquis P, Chauvel P, and Liégeois-Chauvel C (2005) Auditory evoked potential patterns to voiced and voiceless speech sounds in adult developmental dyslexics with persistent deficits. *Cerebral Cortex* **15**:1524–1534.

Gloor P (1985) Neuronal generators and the problem of localization in electroencephalography: application of volume conductor theory to electroencephalography. *Journal of Clinical Neurophysiology* **2**: 327–354.

Godey B, Schwartz D, de Graaf JB, Chauvel P, and Liégeois-Chauvel C (2001) Neuromagnetic source localization of auditory evoked fields and intracerebral evoked potentials: a comparison of data in the same patients. *Clinical Neurophysiology* **112**:1850–1859.

Goffaux V, Mouraux A, Desmeta S, and Rossion B (2004) Human non-phase-locked gamma oscillations in experience-based perception of visual scenes. *Neuroscience Letters* **354**:14–17.

Goswami U, Thomson J, Richardson U, Stainthorp R, Hughes D, Rosen S, and Scott SK (2002) Amplitude envelope onsets and developmental dyslexia: a new hypothesis. *Proceedings of the National Academy of Sciences of the United States of America* **99**:10911–10916.

Gutschalk A, Micheyl C, Melcher JR, Rupp A, Scherg M, and Oxenham AJ (2005) Neuromagnetic correlates of streaming in human auditory cortex. *Journal of Neuroscience* **25**:5382–5388.

Hackett TA, Preuss TM, and Kaas JH (2001) Architectonic identification of the core region in auditory cortex of macaques, chimpanzees, and humans. *Journal of Comparative Neurology* **441**:197–222.

Haenschel C, Vernon DJ, Dwivedi P, Gruzelier JH, and Baldeweg T (2005) Event-related brain potential correlates of human auditory sensory memory-trace formation. *Journal of Neuroscience* **25**:10494–10501.

Halgren E, Baudena P, Clarke JM, Heit G, Liégeois C, Chauvel P, and Musolino A (1995) Intracerebral potentials to rare target and detractor auditory and visual stimuli. I. Superior temporal plane and parietal lobe. *Electroencephalography and Clinical Neurophysiology* **94**:191–220.

Halgren E, Marinkovic K, and Chauvel P (1998) Generators of the late cognitive potentials in auditory and visual oddball tasks. *Electroencephalography and Clinical Neurophysiology* **106**: 156–164.

Hämäläinen MS and Ilmoniemi RJ (1994) Interpreting magnetic fields of the brain: minimum norm estimates. *Medical and Biological Engineering and Computing* **32**:35–42.

Hartmann WM, McAdams S, and Smith BK (1990) Hearing a mistuned harmonic in an otherwise periodic complex tone. *Journal of the Acoustical Society of America* **88**:1712–1724.

Hautus MJ and Johnson BW (2005) Object-related brain potentials associated with the perceptual segregation of a dichotically embedded pitch. *Journal of the Acoustical Society of America* **117**: 275–280.

Hayes EA, Warrier CM, Nicol TG, Zecker SG, and Kraus N (2003) Neural plasticity following auditory training in children with learning problems. *Clinical Neurophysiology* **114**:673–684.

Helenius P, Uutela K and Hari R (1999) Auditory stream segregation in dyslexic adults. *Brain* **122**:907–913.

Helenius P, Salmelin R, Service E, Connolly JF, Leinonen S, and Lyytinen H (2002) Cortical activation during spoken-word segmentation in nonreading-impaired and dyslexic adults. *Journal of Neuroscience* **22**:2936–2944.

Herdman AT, Lins O, Van Roon P, Stapells DR, Scherg M and Picton TW (2002) Intracerebral sources of human auditory steady-state responses. *Brain Topography* **15**:69–86.

Herrmann CS, Munk MHJ, and Engel AK (2004) Cognitive functions of gamma-band activity: memory match and utilization. *Trends in Cognitive Sciences* **8**:347–355.

Herrmann CS and Demiralp T (2005) Human EEG gamma oscillations in neuropsychiatric disorders. *Clinical Neurophysiology* **116**:2719–2733.

Hillyard SA and Picton TW (1987) Electrophysiology of cognition. In: Mountcastle VB, Plum F, and Geiger SR (eds). *Handbook of Physiology*, Part 2, *Higher Functions of the Brain*. American Physiological Society, Baltimore, pp. 519–584.

Hirsh IJ (1959) Auditory perception of temporal order. *Journal of the Acoustical Society of America* **31**:759–767.

Holt LL (2006) The mean matters: effects of statistically defined non-speech spectral distributions on speech categorization. *Journal of the Acoustical Society of America* **120**:2801–2817.

Howard MA III, Volkov IO, Abbas PJ, Damasio H, Ollendieck MC, and Granner MA (1996) A chronic microelectrode investigation of the tonotopic organization of human auditory cortex. *Brain Research* **724**:260–264.

Howard MA III, Volkov IO, Mirsky R, Garell PC, Noh MD, Granner M, Damasio H, Steinschneider M, Reale RA, Hind JE, and Brugge JF (2000) Auditory cortex on the human posterior superior temporal gyrus. *Journal of Comparative Neurology* **416**:79–92.

Hulse SH, MacDougall-Shackleton SA, and Wisniewski AB (1997) Auditory scene analysis by songbirds: stream segregation of birdsong by European starlings (*Sturnus vulgaris*). *Journal of Comparative Psychology* **111**:3–13.

Huotilainen M, Kujala A, Hotakainen M, Shestakova A, Kushnerenko E, Parkkonen L, Fellman V, and Näätänen R (2003) Auditory magnetic responses of healthy newborns. *NeuroReport* **14**:1871–1875.

Huttenlocher PR and Dabholkar AS (1997) Regional differences in synaptogenesis in human cerebral cortex. *Journal of Comparative Neurology* **387**:167–178.

Im C-H, Jung H-K, and Fujimaki N (2005) Anatomically constrained dipole adjustment (ANACONDA) for accurate MEG/EEG focal source localizations. *Physics in Medicine and Biology* **50**: 4931–4953.

Irimajiri R, Golob EJ, and Starr A (2005) Auditory brain-stem, middle- and long-latency evoked potentials in mild cognitive impairment. *Clinical Neurophysiology* **116**:1918–1929.

Izumi A (2002) Auditory stream segregation in Japanese macaques. *Cognition* **82**:B113–B122.

Jääskeläinen IP, Ahveninen J, Bonmassar G, Dale AM, Ilmoniemi RJ, Levänen S, Lin F-H, May P, Melcher J, Stufflebeam S, Tiitinen H, and Belliveau JW (2004) Human posterior auditory cortex gates novel sounds to consciousness. *Proceedings of the National Academy of Sciences of the United States of America* **101**: 6809–6814.

Jacobs J, Kahana MJ, Ekstrom AD, and Fried I (2007) Brain oscillations control timing of single-neuron activity in humans. *Journal of Neuroscience* **27**:3839–3844.

Javitt DC, Steinschneider M, Schroeder CE, Arezzo JC, and Vaughan HG Jr (1994) Detection of stimulus deviance within primary auditory cortex: intracortical mechanisms of mismatch negativity (MMN) generation. *Brain Research* **667**:192–200.

Javitt DC, Steinschneider M, Schroeder CE, and Arezzo JC (1996) Role of cortical *N*-methyl-D-aspartate receptors in auditory sensory memory and mismatch negativity generation: implications for schizophrenia. *Proceedings of the National Academy of Sciences of the United States of America* **93**:11962–11967.

Jing H and Benasich AA (2006) Brain responses to tonal changes in the first two years of life. *Brain & Development* **28**:247–256.

Joris PX, Schreiner CE, and Rees A (2004) Neural processing of amplitude-modulated sounds. *Physiological Reviews* **84**:541–577.

Jusczyk PW, Rosner BS, Reed MA, and Kennedy LJ (1989) Could temporal order differences under 2-month-olds' discrimination of

English voicing contrasts? *Journal of the Acoustical Society of America* **85**:1741–1749.

Kaas JH and Hackett TA (1998) Subdivisions of auditory cortex and levels of processing in primates. *Audiology & Neurotology* **3**:73–85.

Kahana MJ (2006) The cognitive correlates of human brain oscillations. *Journal of Neuroscience* **26**:1669–1672.

Kaiser J, Birbaumer N, and Lutzenberger W (2002) Magnetic oscillatory responses to lateralization changes of natural and artificial sounds in humans. *European Journal of Neuroscience* **15**:345–354.

Kanwal JS, Medvedev AV, and Micheyl C (2003) Neurodynamics for auditory stream segregation: tracking sounds in the mustached bat's natural environment. *Network* **14**:413–435.

Kasai K, Nakagome K, Itoh K, Koshida I, Fukuda M, Watanabe A, Kamio S, Murakami T, Hata A, Iwanami A, Hiramatsu K-I, and Kato N (1999) Electrophysiological evidence for sequential activation of multiple brain regions during the auditory selective attention process in humans. *NeuroReport* **10**:3837–3842.

Kayser C, Petkov CI, and Logothetis NK (2007) Tuning to sound frequency in auditory evoked potentials. *Journal of Neurophysiology* **98**:1806–1809.

Kayser J, Tenke CE, and Bruder GE (1998) Dissociation of brain ERP topographies for tonal and phonetic oddball tasks. *Psychophysiology* **35**:576–590.

Kilpeläinen R, Koistinen A, Könönen M, Herrgård E, Partanen J, and Karhu J (1999) P300 sequence effects differ between children and adults for auditory stimuli. *Psychophysiology* **36**:343–350.

Kobayashi K, Yoshinaga H, Ohtsuka Y, and Gotman J (2005) Dipole modeling of epileptic spikes can be accurate or misleading. *Epilepsia* **46**:397–408.

Kraus N, McGee T, and Comperatore C (1989) MLRs in children are consistently present during wakefulness, stage 1, and REM sleep. *Ear and Hearing* **10**:339–345.

Kropotov JD, Alho K, Näätänen R, Ponomarev VA, Kropotova OV, Anichkov AD, and Nechaev VB (2000) Human auditory-cortex mechanisms of pre-attentive sound discrimination. *Neuroscience Letters* **280**:87–90.

Kruglikov SY and Schiff SJ (2003) Interplay of electroencephalogram phase and auditory-evoked neural activity. *Journal of Neuroscience* **23**:10122–10127.

Kuhl P (1986) Theoretical contributions of tests on animals to the special-mechanisms debate in speech. *Experimental Biology* **45**:233–265.

Kujala T, Karma K, Ceponiene R, Belitz S, Turkkila P, Tervaniemi M, and Näätänen R (2001) Plastic neural changes and reading improvement caused by audiovisual training in reading-impaired children. *Proceedings of the National Academy of Sciences of the United States of America* **98**:10509–10514.

Kujala T, Belitz S, Tervaniemi M, and Näätänen R (2003) Auditory sensory memory disorder in dyslexic adults as indexed by the mismatch negativity. *European Journal of Neuroscience* **17**:1323–1327.

Kujala T, Halmetoja J, Näätänen R, Alhu P, Lyytinen H, and Sussman E (2006) Speech- and sound-segmentation in dyslexia: evidence for a multiple-level cortical involvement. *European Journal of Neuroscience* **24**:2420–2427.

Kurtzberg D, Hilbert PL, Kreuzer JA, and Vaughan HG Jr (1984) Differential maturation of cortical auditory evoked potentials to speech sounds in normal full term and very low-birthweight infants. *Developmental Medicine and Child Neurology* **26**:466–475.

Kushnerenko E, Ceponiene R, Balan P, Fellman V, and Näätänen R (2002) Maturation of the auditory change detection response in infants: a longitudinal ERP study. *NeuroReport* **13**: 1843–1848.

Lachaux JPh, Rudrauf D, and Kahane P (2003) Intracranial EEG and human brain mapping. *Journal of Physiology, Paris* **97**:613–628.

Laguitton V, De Graaf JB, Chauvel P, and Liégeois-Chauvel C (2000) Identification reaction times of voiced/voiceless continua: a right-ear advantage for VOT values near the phonetic boundary. *Brain and Language* **75**:153–162.

Lakatos P, Shah AS, Knuth KH, Ulbert I, Karmos G, and Schroeder CE (2005) An oscillatory hierarchy controlling neuronal excitability and stimulus processing in the auditory cortex. *Journal of Neurophysiology* **94**:1904–1911.

Lakatos P, Chen C-M, O'Connell MN, Mills A, and Schroeder CE (2007) Neuronal oscillations and multisensory interaction in primary auditory cortex. *Neuron* **53**:279–292.

Lehmann D and Skrandies W (1980) Reference-free identification of components of checkerboard-evoked multichannel potential fields. *Electroencephalography and Clinical Neurophysiology* **48**:609–621.

Leppänen PHT, Pihko E, Eklund KM, and Lyytinen H (1999) Cortical responses of infants with and without a genetic risk for dyslexia: II. Group effects. *NeuroReport* **10**:969–973.

Leppänen PHT, Guttorm TK, Pihko E, Takkinen S, Eklund KM, and Lyytinen H (2004) Maturational effects on newborn ERPs measured in the mismatch negativity paradigm. *Experimental Neurology* **190**:S91–S101.

Liang L, Lu T, and Wang X (2002) Neural representation of sinusoidal amplitude and frequency modulations in the primary auditory cortex of awake primates. *Journal of Neurophysiology* **87**:2237–2261.

Liégeois-Chauvel C, Musolino A, and Chauvel P (1991) Localization of the primary auditory area in man. *Brain* **114**:139–153.

Liégeois-Chauvel C, Musolino A, Badier JM, Marquis P, and Chauvel P (1994) Evoked potentials recorded from the auditory cortex in man: evaluation and topography of the middle latency components. *Electroencephalography and Clinical Neurophysiology* **92**:204–214.

Liégeois-Chauvel C, de Graaf JB, Laguitton V, and Chauvel P (1999) Specialization of left auditory cortex for speech perception in man depends on temporal coding. *Cerebral Cortex* **9**:484–496.

Liégeois-Chauvel C, Lorenzi C, Trébuchon A, Régis J, and Chauvel P (2004) Temporal envelope processing in the human left and right auditory cortices. *Cerebral Cortex* **14**:731–740.

Light GA and Braff DL (2005). Stability of mismatch negativity deficits and their relationship to functional impairments in chronic schizophrenia. *American Journal of Psychiatry* **162**: 1741–1743.

Lisker L and Abramson AS (1964) A cross-language study of voicing in initial stops: acoustical measurements. *Word* **20**:384–422.

Logothetis NK (2003) The underpinnings of the BOLD functional magnetic resonance imaging signal. *Journal of Neuroscience* **23**: 3963–3971.

Lorenzi C, Dumont A, and Füllgrabe C (2000) Use of temporal envelope cues by children with developmental dyslexia. *Journal of Speech, Language, and Hearing Research* **43**:1367–1379.

Lu T, Liang L, and Wang X (2001) Temporal and rate representations of time-varying signals in the auditory cortex of awake primates. *Nature Neuroscience* **4**:1131–1138.

Ma WJ, Beck, JM, Latham PE, and Pouget A (2006) Bayesian inference with probabilistic population codes. *Nature Neuroscience* **9**: 1432–1438.

Määttä S, Pääkkönen A, Saavalainen P, and Partanen J (2005) Selective attention event-related potential effects from auditory novel stimuli in children and adults. *Clinical Neurophysiology* **116**:129–141.

Martin BA and Boothroyd A (2000) Cortical, auditory, evoked potentials in response to changes of spectrum and amplitude. *Journal of the Acoustical Society of America* **107**:2155–2161.

McDonald KL and Alain C (2005) Contribution of harmonicity and location to auditory object formation in free field: evidence from event-related brain potentials. *Journal of the Acoustical Society of America* **118**:1593–1604.

McGee T, Kraus N, King C, and Nicol T (1996) Acoustic elements of speech-like stimuli are reflected in surface recorded responses over the guinea pig temporal lobe. *Journal of the Acoustical Society of America* **99**:3606–3614.

Melloni L, Molina C, Pena M, Torres D, Singer W, and Rodriguez E (2007) Synchronization of neural activity across cortical areas correlates with conscious perception. *Journal of Neuroscience* **27**:2858–2865.

Metherate R and Cruikshank SJ (1999) Thalamocortical inputs trigger a propagating envelope of gamma-band activity in auditory cortex in vitro. *Experimental Brain Research* **126**:160–174.

Michel CM, Murray MM, Lantz G, Gonzalez S, Spinelli L, and Grave de Peralta R (2004) EEG source imaging. *Clinical Neurophysiology* **115**:2195–2222.

Micheyl C, Tian B, Carlyon BP, and Rauschecker JP (2005) Perceptual organization of sound sequences in the auditory cortex of awake macaques. *Neuron* **48**:139–148.

Micheyl C, Carlyon RP, Gutschalk A, Melcher JR, Oxenham AJ, Rauschecker JP, Tian B, and Wilson EC (2007) The role of auditory cortex in the formation of auditory streams. *Hearing Research* **229**:116–131.

Molfese DL and Molfese VJ (1988). Right-hemisphere responses from preschool children to temporal cues to speech and nonspeech materials: electrophysiological evidence. *Brain and Language* **33**: 245–259.

Moisescu-Yiflach T and Pratt H (2005) Auditory event related potentials and source current density estimation in phonologic/auditory dyslexics. *Clinical Neurophysiology* **116**:2632–2647.

Molholm S, Martinez A, Ritter W, Javitt DC, and Foxe JF (2005) The neural circuitry of pre-attentive auditory change-detection: an fMRI study of pitch and duration mismatch negativity generators. *Cerebral Cortex* **15**:545–551.

Moore BCJ, Peters RW, and Glasberg BR (1985) Thresholds for the detection of inharmonicity in complex tones. *Journal of the Acoustical Society of America* **77**:1861–1867.

Moore BCJ, Glasberg BR, and Peters RW (1986) Thresholds for hearing mistuned partials as separate tones in harmonic complexes. *Journal of the Acoustical Society of America* **80**:479–483.

Moore JK and Guan Y-L (2001) Cytoarchitectural and axonal maturation in human auditory cortex. *Journal of the Association for Research in Otolaryngology* **2**:297–311.

Mukamel R, Gelbard H, Arieli A, Hasson U, Fried I, and Malach R (2005) Coupling between neuronal firing, field potentials, and fMRI in human auditory cortex. *Science* **309**:951–954.

Mulert C, Jäger L, Schmitt R, Bussfeld P, Pogarell O, Möller H-J, Juckel G, and Hegerl U (2004) Integration of fMRI and simultaneous EEG: towards a comprehensive understanding of localization and time-course of brain activity in target detection. *NeuroImage* **22**: 83–94.

Muller-Preuss P and Mitzdorf U (1984) Functional anatomy of the inferior colliculus and the auditory cortex: current source density analysis of click-evoked potentials. *Hearing Research* **16**:133–142.

Murray MM, Camen C, Gonzalez Andino SL, Bovet P, and Clarke S (2006) Rapid brain discrimination of sounds of objects. *Journal of Neuroscience* **26**:1293–1302.

Näätänen R and Picton T (1987) The N1 wave of the human electric and magnetic response to sound: a review and an analysis of the component structure. *Psychophysiology* **24**:375–425.

Näätänen R (1990) The role of attention in auditory information processing as revealed by event-related potentials and other brain measures of cognitive function. *Behavioral and Brain Sciences* **13**:201–288.

Näätänen R, Tervaniemi M, Sussman E, Paavilainen P, and Winkler I (2001) 'Primitive intelligence' in the auditory cortex. *Trends in Neurosciences* **24**:283–288.

Näätänen R, Jacobsen T, and Winkler I (2005) Memory-based or afferent processes in mismatch negativity (MMN): a review of the evidence. *Psychophysiology* **45**:25–32.

Nagarajan S, Mahncke H, Salz T, Tallal P, Roberts T, and Merzenich MM (1999) Cortical auditory signal processing in poor readers. *Proceedings of the National Academy of Sciences of the United States of America* **96**:6483–6488.

Niessing J, Ebisch B, Schmidt KE, Niessing M, Singer W, and Galuske RAW (2005) Hemodynamic signals correlate tightly with synchronized gamma oscillations. *Science* **309**:948–951.

Neelon MF, Williams J, and Garell PC (2006) The effects of attentional load on auditory ERPs recorded from human cortex. *Brain Research* **1118**:94–105.

Novak GP, Kurtzberg D, Kreuzer JA, and Vaughan HG Jr (1989) Cortical responses to speech sounds and their formants in normal infants: maturational sequence and spatiotemporal analysis. *Electroencephalography and Clinical Neurophysiology* **73**: 295–305.

Novak GP, Ritter W, Vaughan HG Jr, and Wiznitzer ML (1990) Differentiation of negative event-related potentials in an auditory discrimination task. *Electroencephalography and Clinical Neurophysiology* **75**:255–275.

Nunez PL and Silberstein RB (2000) On the relationship of synaptic activity to macroscopic measurements: does co-registration of EEG with fMRI make sense? *Brain Topography* **13**:79–96.

Nunez PL and Srinivasan R (2006) *Electric Fields of the Brain: The Neurophysics of EEG* (Second Edition). Oxford University Press, New York.

Oades RD, Wild-Wall N, Juran SA, Sachsse J, Oknina LB, and Röpcke B (2006) Auditory change detection in schizophrenia: sources of activity, related neuropsychological function and symptoms in patients with a first episode in adolescence, and patients14 years after an adolescent illness-onset. *BMC Psychiatry* **6**:7.

Pakarinen S, Takegata R, Rinne T, Huotilainen M, and Näätänen R (2007) Measurement of extensive auditory discrimination profiles using the mismatch negativity (MMN) of the auditory event-related potential (ERP). *Clinical Neurophysiology* **118**:177–185.

Palva S, Palva JM, Shtyriov Y, Kujala T, Ilmoniemi RJ, Kaila K, and Näätänen R (2002) Distinct gamma-band evoked responses to speech and non-speech sounds in humans. *Journal of Neuroscience* **22**:RC211:1–5.

Pantev C, Eulitz C, Hampson S, Ross B, and Roberts LE (1996) The auditory evoked "off" response: sources and comparison with the "on" and the "sustained" responses. *Ear and Hearing* **17**:255–265.

Petkov CI, O'Connor KN, Benmoshe G, Baynes K, and Sutter ML (2005) Auditory perceptual grouping and attention in dyslexia. *Cognitive Brain Research* **24**:343–354.

Perrin F, Pernier J, Bertrand O, Giard MH, and Echallier JF (1987) Mapping of scalp potentials by surface spline interpolation. *Electroencephalography and Clinical Neurophysiology* **66**:75–81.

Picton TW, Hillyard SA, Krausz HI, and Galambos R (1974) Human auditory evoked potentials. I: Evaluation of components. *Electroencephalography and Clinical Neurophysiology* **36**: 179–190.

Picton TW, Alain C, Otten L, and Ritter W (2000) Mismatch negativity: different water in the same river. *Audiology & Neurotology* **5**: 111–139.

Pincze Z, Lakatos P, Rajkai C, Ulbert I, and Karmos G (2001) Separation of mismatch negativity and the N1 wave in the auditory cortex of the cat: a topographic study. *Clinical Neurophysiology* **112**:778–784.

Pincze Z, Lakatos P, Rajkai C, Ulbert I, and Karmos G (2002) Effect of deviant probability and interstimulus/interdeviant interval on the auditory N1 and mismatch negativity in the cat auditory cortex. *Cognitive Brain Research* **13**:249–253.

Pisoni DB (1977) Identification and discrimination of the relative onset time of two component tones: implications for voicing perception in stops. *Journal of the Acoustical Society of America* **61**: 1352–1361.

Plack CJ and Carlyon RP (1995) Differences in frequency modulation detection and fundamental frequency discrimination between complex tones consisting of resolved and unresolved harmonics. *Journal of the Acoustical Society of America* **98**:1355–1364.

Ponton C, Eggermont JJ, Kwong B, and Don M (2000) Maturation of human central auditory system activity: evidence from multi-channel evoked potentials. *Clinical Neurophysiology* **111**: 220–236.

Ponton C, Eggermont JJ, Khosla D, Kwong B, and Don M (2002) Maturation of human central auditory system activity: separating auditory evoked potentials by dipole source modeling. *Clinical Neurophysiology* **113**:407–420.

Poulsen C, Picton TW, and Paus T (2007) Age-related changes in transient and oscillatory brain responses to auditory stimulation in healthy adults 19–45 years old. *Cerebral Cortex* **17**: 1454–1467.

Ramus F, Rosen S, Dakin SC, Day BL, Castellote JM, White S, and Frith U (2003) Theories of developmental dyslexia: insights from a multiple case study of dyslexic adults. *Brain* **126**:841–865.

Rauschecker JP and Tian B (2000) Mechanisms and streams for processing of "what" and "where" in auditory cortex. *Proceedings of the National Academy of Sciences of the United States of America* **97**:11800–11806.

Reale RA, Calvert GA, Thesen T, Jenison RL, Kawasaki H, Oya H, Howard MA, and Brugge JF (2007) Auditory-visual processing represented in the human superior temporal gyrus. *Neuroscience* **145**:162–184.

Rinne T, Särkkä A, Degerman A, Schröger E, and Alho K (2006) Two separate mechanisms underlie auditory change detection and involuntary control of attention. *Brain Research* **1077**:135–143.

Rocheron I, Lorenzi C, Füllgrabe C, and Dumont A (2002) Temporal envelope perception in dyslexic children. *NeuroReport* **13**: 1683–1687.

Roman S, Canévet G, Lorenzi C, Triglia J-M, and Liégeois-Chauvel C (2004) Voice onset time encoding in patients with left and right cochlear implants. *NeuroReport* **15**:601–605.

Romanski LM, Tian B, Fritz J, Mishkin M, Goldman-Rakic PS, and Rauschecker JP (1999) Dual streams of auditory afferents target multiple domains in the primate prefrontal cortex. *Nature Neuroscience* **2**:1131–1136.

Rosen S (1992) Temporal information in speech: acoustic, auditory and linguistic aspects. *Philosophical Transactions of the Royal Society London, Series B* **336**:367–373.

Rosen S (1999) Language disorders: a problem with auditory processing? *Current Biology* **9**:R698–R700.

Rosenberg AE (1965) Effect of masking on the pitch of period pulses. *Journal of the Acoustical Society of America* **38**:747–758.

Sabri M, Liebenthal E, Waldron EJ, Medler DA, and Binder JR (2006) Attentional modulation in the detection of irrelevant deviance: a simultaneous ERP/fMRI study. *Journal of Cognitive Neuroscience* **18**:698–700.

Salinas E and Sejnowski TJ (2001) Correlated neuronal activity and the flow of neural information. *Nature Reviews Neuroscience* **2**:539–550.

Sanders LD and Poeppel D (2007) Local and global auditory processing: behavioral and ERP evidence. *Neuropsychologia* **45**: 1172–1186.

Sanger TD (2003) Neural population codes. *Current Opinion in Neurobiology* **13**:238–249.

Scherg M and von Cramon D (1985). Two bilateral sources of the late AEP as identified by a spatio-temporal dipole model. *Electroencephalography and Clinical Neurophysiology* **62**: 32–44.

Scherg M and von Cramon D (1986) Evoked dipole source potentials of the human auditory cortex. *Electroencephalography and Clinical Neurophysiology* **65**:344–360.

Scherg M, Vajsar J, and Picton TW (1989) A source analysis of the late human auditory evoked potentials. *Journal of Cognitive Neuroscience* **1**:336–355.

Schönwiesner M, Novitski N, Pakarinen S, Carlson S, Tervaniemi M, and Näätänen R (2007) Heschl's gyrus, posterior superior temporal gyrus, and mid-ventrolateral prefrontal cortex have different roles in the detection of acoustic changes. *Journal of Neurophysiology* **97**:2075–2082.

Scott SK (2005) Auditory processing – speech, space and auditory objects. *Current Opinion in Neurobiology* **15**:197–201.

Sejnowski TJ and Paulsen O (2006) Network oscillations: emerging computational principles. *Journal of Neuroscience* **26**:1673–1676.

Shahin A, Bosnyak DJ, Trainor LJ, and Roberts LE (2003) Enhancement of neuroplastic P2 and N1c auditory evoked potentials in musicians. *Journal of Neuroscience* **23**:5545–5552.

Shahin A, Roberts LE, Pantev C, Trainor LJ, and Ross B (2005) Modulation of P2 auditory-evoked responses by the spectral complexity of musical sounds. *NeuroReport* **16**:1781–1785.

Shannon RV, Zeng F-G, Kamath V, Wygonski J, and Ekelid M (1995) Speech recognition with primarily temporal cues. *Science* **270**:303–304.

Sharma A, Kraus N, McGee T, Carrell T, and Nicol T (1993) Acoustic versus phonetic representation of speech as reflected by the mismatch negativity event-related potential. *Electroencephalography and Clinical Neurophysiology* **88**:64–71.

Sharma A, Kraus N, McGee TJ, and Nicol TG (1997) Developmental changes in P1 and N1 central auditory responses elicited by consonant-vowel syllables. *Electroencephalography and Clinical Neurophysiology* **104**:540–545.

Sharma A and Dorman MF (2000) Neurophysiological correlates of cross-language phonetic perception. *Journal of the Acoustical Society of America* **107**:2697–2703.

Sharma A, Marsh CM, and Dorman MF (2000) Relationship between N1 evoked potential morphology and the perception of voicing. *Journal of the Acoustical Society of America* **108**:3030–3035.

Simos PG and Molfese DL (1997) Electrophysiological responses from a temporal order continuum in the newborn infant. *Neuropsychologia* **35**:89–98.

Simos PG, Molfese DL, and Brenden RA (1997) Behavioral and electrophysiological indices of voicing-cue discrimination: laterality patterns and development. *Brain and Language* **57**:122–150.

Simos PG, Diehl RL, Breier JI, Molis MR, Zouridakis G, and Papanicolaou AC (1998a) MEG correlates of categorical perception of a voice onset time continuum in humans. *Cognitive Brain Research* **7**:215–219.

Simos PG, Breier JI, Zouridakis G, and Papanicolaou AC (1998b) Magnetic fields elicited by a tone onset time continuum in humans. *Cognitive Brain Research* **6**:285–294.

Sinai A, Bowers CW, Crainiceanu CM, Boatman D, Gordon B, Lesser RP, Lenz FA, and Crone NE (2005) Electrocortigraphic high gamma activity versus electrical cortical stimulation mapping of naming. *Brain* **128**:1556–1570.

Skrandies W (1990) Global field power and topographic similarity. *Brain Topography* **3**:137–141.

Snyder JS and Large EW (2004) Tempo dependence of middle- and long-latency auditory responses: power and phase modulation of the EEG at multiple time-scales. *Clinical Neurophysiology* **115**: 1885–1895.

Snyder JS and Alain C (2005) Age-related changes in neural activity associated with concurrent vowel segregation. *Cognitive Brain Research* **24**:492–499.

Snyder JS, Alain C, and Picton TW (2006) Effects of attention on neuroelectric correlates of auditory stream segregation. *Journal of Cognitive Neuroscience* **18**:1–13.

Snyder JS and Alain C (2007) Sequential auditory scene analysis is preserved in normal aging adults. *Cerebral Cortex* **17**:501–512.

Speckmann EJ and Elger CE (1999) Introduction to the neurophysiological basis of the EEG and DC potentials. In: Niedermeyer E and Lopes Da Silva F (eds). *Electroencephalography: Basic Principles, Clinical Applications, and Related Fields* (Fourth edition). Lippincott, Williams and Wilkins, Baltimore, pp. 15–27.

Steinschneider M, Tenke CE, Schroeder CE, Javitt DC, Simpson GV, Arezzo JC, and Vaughan HG Jr (1992) Cellular generators of the cortical auditory evoked potential initial component. *Electroencephalography and Clinical Neurophysiology* **84**: 196–200, 1992.

Steinschneider M, Schroeder CE, Arezzo JC, and Vaughan HG Jr (1994) Speech-evoked activity in primary auditory cortex: effects of voice onset time. *Electroencephalography and Clinical Neurophysiology* **92**:30–43.

Steinschneider M, Reser D, Fishman YI, Schroeder CE, and Arezzo JC (1998) Click train encoding in primary auditory cortex of the awake monkey: evidence for two mechanisms subserving pitch perception. *Journal of the Acoustical Society of America* **104**: 2935–2955.

Steinschneider M, Volkov IO, Noh MD, Garell PC, and Howard MA III (1999) Temporal encoding of the voice onset time phonetic parameter by field potentials recorded directly from human auditory cortex. *Journal of Neurophysiology* **82**:2346–2357.

Steinschneider M, Fishman YI, and Arezzo JC (2003) Representation of the voice onset time (VOT) speech parameter in population responses within primary auditory cortex of the awake monkey. *Journal of the Acoustical Society of America* **114**:307–321.

Steinschneider M, Volkov IO, Fishman YI, Oya H, Arezzo JC, and Howard MA III (2005) Intracortical responses in human and monkey primary auditory cortex support a temporal processing mechanism for encoding of the voice onset time phonetic parameter. *Cerebral Cortex* **15**:170–186.

Steinschneider M, Fishman YI, and Arezzo JC (2008) Spectrotemporal analysis of evoked and induced electroencephalographic responses in primary auditory cortex (A1) of the awake monkey. *Cerebral Cortex* **18**:610–625.

Stoodley CJ, Hill PR, Stein JF, and Bishop DVM (2006) Auditory event-related potentials differ in dyslexics even when auditory psychophysical performance is normal. *Brain Research* **1121**:190–199.

Sussman ES (2005) Integration and segregation in auditory scene analysis. *Journal of the Acoustical Society of America* **117**:1285–1298.

Sussman E, Ritter W, and Vaughan HG Jr (1998). Attention affects the organization of auditory input associated with the mismatch negativity system. *Brain Research* **789**:130–138.

Sussman E, Ritter W, and Vaughan HG Jr (1999) An investigation of the auditory streaming effect using event-related brain potentials. *Psychophysiology* **36**:22–34.

Szymanski MD, Yund EW, and Woods DL (1999) Phonemes, intensity and attention: differential effects on the mismatch negativity. *Journal of the Acoustical Society of America* **106**:3492–3505.

Tallal P, Miller S, and Fitch RH (1993) Neurobiological basis of speech: a case for the preeminence of temporal processing. *Annals of the New York Academy of Sciences* **682**:27–47.

Tardif E, Murray MM, Meylan R, Spierer L, and Clarke S (2006) The spatio-temporal brain dynamics of processing and integrating sound localization cues in humans. *Brain Research* **1092**:161–176.

Tonnquist-Uhlen I, Ponton CW, Eggermont JJ, Kwong B, and Don M (2003) Maturation of human central auditory system activity: the T-complex. *Clinical Neurophysiology* **114**:685–701.

Traub RD, Bibbig A, LeBeau FEN, Cunningham MO, and Whittington MA (2005) Persistent gamma oscillations in superficial layers of rat auditory neocortex: experiment and model. *Journal of Physiology (London)* **562**:3–8.

Trautner P, Rosburg T, Dietl T, Fell J, Korzyukov OA, Kurthen M, Schaller C, Elger CE, and Boutros NN (2006) Sensory gating of auditory evoked and induced gamma band activity in intracranial recordings. *NeuroImage* **32**:790–798.

Trébuchon-Da Fonseca A, Giraud K, Badier J-M, Chauvel P, and Liégeois-Chauvel C (2005) Hemispheric lateralization of voice onset time (VOT) comparison between depth and scalp EEG recordings. *NeuroImage* **27**:1–14.

Uhlhass PJ and Singer W (2006) Neural synchrony in brain disorders: relevance for cognitive dysfunctions and pathophysiology. *Neuron* **52**:155–168.

Ulanovsky N, Las L, and Nelken I (2003) Processing of low-probability sounds by cortical neurons. *Nature Neuroscience* **6**:391–398.

Ulanovsky N, Las L, Farkas D, and Nelken I (2004) Multiple time scales of adaptation in auditory cortex neurons. *Journal of Neuroscience* **24**:10440–10453.

Uwer R and von Suchodoletz W (2000) Stability of mismatch negativity in children. *Clinical Neurophysiology* **111**:45–52.

Vaughan HG Jr and Arezzo JC (1988) The neural basis of event-related potentials. In: Picton TW (ed). *Handbook of Electroencephalography and Clinical Neurophysiology*, volume 3, *Human Event-Related Potentials*. Elsevier, Amsterdam, pp 45–96.

Ward LM (2003) Synchronous neural oscillations and cognitive processes. *Trends in Cognitive Sciences* **7**:553–559.

Winkler I, van Zuijen T, Sussman E, Horváth J, and Näätänen R (2006) Object representation in the human auditory system. *European Journal of Neuroscience* **24**:625–634.

Woldorff MG and Hillyard SA (1991) Modulation of early auditory processing during selective listening to rapidly presented tones. *Electroencephalography and Clinical Neurophysiology* **79**: 170–191.

Woldorff MG, Gallen CC, Hampson SR, Hillyard SA, Pantev C, Sobel D, and Bloom FE (1993) Modulation of early sensory processing in human auditory cortex during auditory selective attention. *Proceedings of the National Academy of Sciences of the United States of America* **90**:8722–8726.

Wood CC and Wolpaw JR (1982) Scalp distribution of human auditory evoked potentials. II. Evidence for overlapping sources and involvement of auditory cortex. *Electroencephalography and Clinical Neurophysiology* **54**:25–38.

Woods DL, Alho K, and Algazi A (1994) Stages of auditory feature conjunction: an event-related brain potential study. *Journal of Experimental Psychology: Human Perception and Performance* **20**:81–94.

Wright BA, Lombardino LJ, King WM, Puranik CS, Leonard CM, and Merzenich MM (1997) Deficits in auditory temporal and spectral resolution in language-impaired children. *Nature* **387**:176–178.

Wunderlich, JL and Cone-Wesson BK (2006) Maturation of CAEP in infants and children: a review. *Hearing Research* **212**:212–223.

Wunderlich, JL, Cone-Wesson BK, and Shepherd R (2006) Maturation of the cortical auditory evoked potential in infants and young children. *Hearing Research* **212**:185–202.

Yost WA (1991) Auditory image perception and analysis: the basis for hearing. *Hearing Research* **56**: 8–18.

Yvert B, Crouzeix A, Bertrand O, Seither-Preisler A, and Pantev C (2001) Multiple supratemporal sources of magnetic and electric auditory evoked middle latency components in humans. *Cerebral Cortex* **11**:411–423.

Yvert B, Fischer C, Bertrand O, and Pernier J (2005) Localization of human supratemporal auditory areas from intracerebral auditory evoked potentials using distributed source models. *NeuroImage* **28**:140–153.

Chapter 26
Auditory Memories and Feedback Processing for Vocal Learning

Ana Amador and Daniel Margoliash

Abbreviations

AFP	anterior forebrain pathway
BOS	bird's own song
CLM	caudolateral mesopallium
CM	caudal mesopallium
CMM	caudomedial mesopallium
IEG	immediate early gene
IMM	intermediate and medial mesopallium
LMAN	lateral magnocellular nucleus of the anterior nidopallium
MLd	dorsal lateral nucleus of the mesencephalon
NCM	caudomedial nidopallium
NIf	nucleus interfacialis of the nidopallium
nXIIts	tracheosyringeal part of the hypoglossal nucleus
Ov	nucleus ovoidalis
PAm	parambigualis nucleus
RA	robust nucleus of the arcopallium
RAm	retroambigualis nucleus
Uva	nucleus uvaeformis

1 Vocal Learning and Its Evolution

Vocal learning can be defined as the ability to acquire new vocalizations or modify the spectral or temporal structure of existing vocalizations based on environmental cues. This definition can admit a rich set of non-auditory cues that may influence the vocal learning process (e.g., Baptista and Petrinovich 1984; West and King 1988; King et al. 2005; Beecher et al. 2007), but here we focus on processing of auditory cues that are memorized and then drive changes in motor patterns (Janik and Slater 2000). Vocal learning is distinct from auditory perceptual learning because only the former is associated with a change in vocal output. Nevertheless, vocal learning requires specific forms of auditory perceptual learning, typically related to auditory processing of species-specific vocalizations. At the same time, animals exhibit numerous behaviors related to species-specific perceptual processing that do not involve changes in the structure of vocal output (Falls 1982; Kroodsma and Miller 1996), and structures in the brain that may participate in auditory perceptual learning may also participate in vocal learning (see Chapter 20). Thus the choice as to which forms of auditory perceptual learning to include in a discussion of vocal learning is imprecise and rather artificial.

In the vertebrates, a brainstem and midbrain system is recruited during production of "calls", which typically are innately specified vocalizations (Winter et al. 1974; Konishi 1978; Seller 1981; Jürgens 2002). This distinguishes innate calls from learned vocalizations such as songs—and some calls (Mundinger 1979)—that are learned and involve forebrain structures (e.g., Nottebohm et al. 1976; Williams and Vicario 1993). Elements of this innate system for calling apparently were present early in vertebrate evolution, prior to the divergence of the sarcopterygian (lungfish and coelacanths) and the actinopterygian (ray-fined fishes) linages over 400 million years ago (Bass et al. 2008). In contrast to this remarkable example of a conserved trait, vocal learning has evolved multiple times independently in the vertebrates, to date known only in the higher vertebrates (e.g., Nottebohm 1972; Baptista and Schuchmann 1990; Pinker and Bloom 1990; Jarvis et al. 2000; Noad et al. 2000; Wilbrecht and Nottebohm 2003). As exemplified by studies in humans and birds, it may be that all examples of vocal learning in birds and mammals have in common strong forebrain regulation of descending motor pathways arising from non-primary auditory forebrain pathways. Within birds there may be more commonality of pathways, comparing vocal learning in hummingbirds, parrots, and songbirds, than has been traditionally accepted (Jarvis et al. 2000). In addition, recent observations suggest that within songbirds, vocal learning may not be restricted to oscine passerine birds ("true" songbirds),

D. Margoliash (✉)
Department of Organismal Biology and Anatomy, Department of Psychology, University of Chicago, Chicago, IL 60637, USA
e-mail: dan@bigbird.uchicago.edu

J.A. Winer, C.E. Schreiner (eds.), *The Auditory Cortex*,
DOI 10.1007/978-1-4419-0074-6_26, © Springer Science+Business Media, LLC 2011

but that suboscine birds may also exhibit various forms of vocal learning, emphasizing changes in temporal patterning including duetting (Amador et al. 2005; Saranathan et al. 2007).

Within mammals beyond humans, vocal learning is well established by extensive behavioral observations of geographic variation and cultural evolution—if not by the gold standards of isolation rearing and feedback modification—in cetaceans (Noad et al. 2000) and some species of bats (Boughman 1998). There are additionally anecdotal observations of vocal learning that suggest that it may be present more broadly in mammals than currently recognized (e.g., Poole et al. 2005).

The suggestion that vocal learning in hominids arose out of a gestural system (Rizzolatti and Arbib 1998; Corballis 2002) remains controversial, but if this hypothesis obtains, this suggests the existence of different mechanisms for vocal learning in humans compared to birds. Other suggestions include the hypothesis that vocal learning evolved out of duetting implying a more direct role of audition in the evolution of vocal learning in humans. Forebrain influence on the midbrain calling system could provide for additional regulation of calling in a social context and might be the precursor to vocal learning.

Much less is known about vocal learning pathways in mammals outside of humans. In general, the diversity of the pathways and mechanisms, including forebrain auditory regulation, involved in the various lineages in which vocal learning arises remains unresolved. This problem is now being approached in the birds, but remains poorly developed in studies of mammalian species. This remains a major challenge in producing an integrated picture of the auditory mechanisms of vocal learning that spans the vertebrates.

In the spirit of this volume, we attempt to provide a thematic organization identifying similar limitations of our knowledge and posing questions for future studies throughout this chapter.

2 The Sensory Phase of Vocal Learning

The most intensively studied of vocal learners are the oscine passerines ("true" songbirds), and indeed there are many behavioral aspects of vocal learning that are far better studied in birds than in humans, let alone neuronal mechanisms. Thus we focus our attention here on song in songbirds and speech and language in humans.

In the traditional model, vocal learning commences with a sensory phase, where an individual hears appropriate adult vocalizations and commits these to memory. This seemingly simple observation has broad theoretical implications that remain a central focus of research (Konishi 1965, 1978, 2004). The memory of song has been conceived of as an "acquired sensory template," implying that feedback during subsequent sensorimotor practice is compared against the template. While behavioral experiments have further identified some constraints on the functional organization of the template (Rose et al. 2004), it remains unresolved how the acquired sensory template acts and whether it is localized to a single nucleus or is distributed across one or more networks.

The template is acquired during a critical period early in development whose timing varies by species and is sensitive to environmental cues. Failure to experience appropriate songs during a critical period in development results in impoverished adult singing. Whether this ontogenetic effect is the result of direct action on the acquired template (sensory system) during development and/or the result of irreversible development of motor behavior in the absence of appropriate sensory cues remains unresolved. The fact that lack of appropriate song exposure extends the duration of the critical period tends to implicate a sensory locus for critical period mechanism that is driven perhaps by hormonal cues but this is speculative. There is also some evidence consistent with the hypothesis that the acquired template is retained into adulthood, as demonstrated by the ability of adult female zebra finches (*Taeniopygia guttata*) (which do not sing) to discriminate their father's song last heard in early development (Miller 1979; Riebel et al. 2002). Differences in neuronal responses to own song and tutor song that are retained late into song development support this hypothesis, in that they imply that the auditory representations of own and tutor songs are overlapping but separate (Solis and Doupe 1999; Nick and Konishi 2005b).

Birds raised in isolation of tutor songs produce abnormal songs (Marler 1970; Price 1979), but as with normal development, these abnormal "isolate" songs require auditory feedback for their development. The final, adult isolate songs tend to retain species-typical characteristics that are more restrictive than would be expected from motor constraints alone, suggesting that an "innate template" acts in the absence of critical period exposure to conspecific songs. Several different models have been proposed to explain such observations (Konishi 1978; Marler 1997). Analogously, humans with normal hearing but deprived of hearing human speech when growing up also develop highly abnormal vocalizations (Fromkin et al. 1974). Whether the innate template is distinct from the acquired template and whether it is retained if a juvenile bird has normal exposure to tutor song models remains unresolved.

Finally, a juvenile bird does not indiscriminately choose any song as a model. At the onset of the sensory phase of vocal learning, white-crowned sparrows (*Zonotrichia leucophrys*), swamp sparrows (*Melospiza georgiana*), and zebra finches display a stronger response to conspecific song compared to heterospecific song (Dooling and Searcy 1980; Nelson and Marler 1993; Braaten and Reynolds 1999). This is also correlated to the degree of copying of conspecific songs over heterospecific songs. The ability to acquire heterospecific syllables varies across species. For instance, in white-crowned sparrows, the presence of a pure-tone whistle (a universal sound in this species) in the song enables the acquisition of normally rejected sounds (Soha and Marler 2000), suggesting that the whistle could serve as a cue for song learning. These observations identify yet another constraint on the auditory system which exhibits itself as an innate predisposition, a tendency, to select own-species song in a balanced choice experiment. Collectively, these sets of observations represent powerful insights at a behavioral level into constraints on auditory representations associated with song learning (Adret 2004). Still, the insights into the neural mechanisms of these fundamental behavioral observations remain quite limited.

3 Avian Auditory and Motor ("Song System") Pathways

Until recently the organization of the avian brain, especially the forebrain, was confused by misconceptions arising from the work of anatomists early in the 20th century. A modern synthesis with new terminology has been broadly accepted, which we adopt here (Reiner et al. 2004). A history of this development can be found elsewhere (Jarvis et al. 2005). Deep homologies have been described between regions of avian forebrain and layers of mammalian cortex (Karten 1997).

3.1 Auditory Pathways

A sketch of the avian auditory forebrain pathways is shown in Fig. 26.1. The ascending pathways in birds were elaborated starting with the pioneering work of Karten (1967, 1968). Very briefly, the dorsal lateral nucleus of the mesencephalon (MLd, equivalent to the central nucleus of the mammalian inferior colliculus) projects to the thalamus to the nucleus ovoidalis (Ov) in multiple parallel pathways not discussed here. Ov sends multiple parallel projections to the pallium including to Field L, which has been compared to the primary auditory cortex in mammals. Field L is subdivided in several subfields, L1, L2, and L3, with L2 being further subdivided into L2a and L2b, and an additional surrounding region termed L has also been described (Fortune and Margoliash 1992). L2a is composed of a dense granular thalamorecipient cell layer and has reciprocal projections to L1 and L3. Field L in turn has a complex pattern of projections to what can be thought of as a secondary auditory pathway comprising three structures that "bridge" between Field L subdivisions. The caudolateral mesopallium (CLM) has reciprocal projections with all Field L subdivisions. The caudomedial nidopallium (NCM) receives a projection from L3. Thus, traversing the bridge in one direction is distinct from traversing it in the opposite direction. A caveat, however, is

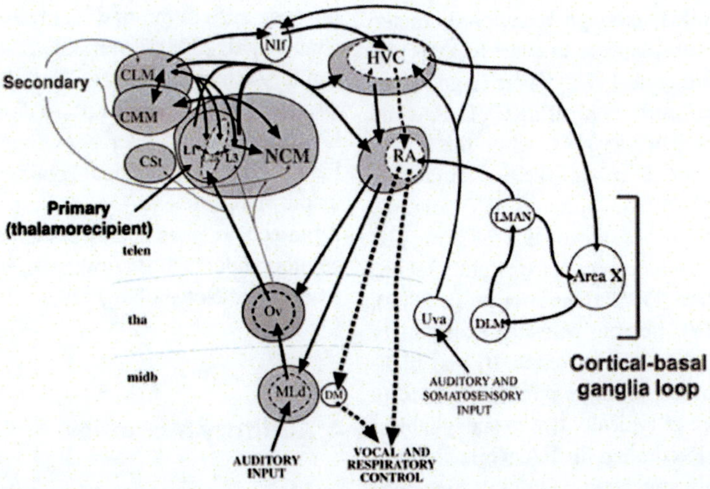

Fig. 26.1 The auditory and song system in a songbird is composed of a network of nuclei that can be loosely organized by functions. The "song system" (*white colored*) includes the AFP or cortical-basal ganglia loop (nuclei Area X, DLM, and LMAN) and a posterior vocal pathway composed by the ventral motor pathway (nuclei Uva and Nif) and the descending motor pathway (*broken arrows*) including HVC, RA, DM, and the nuclei involved in vocal and respiratory control (nXIIts, PAm, RAm). Auditory nuclei (*gray colored*) provide inputs to the song system, at least to HVC, NIf, and Uva. The ascending auditory pathway includes the nuclei MLd in the midbrain, Ov in the thalamus and the telencephalic nuclei Field L, NCM, CMM, CLM, and CSt (see text for abbreviations). Schematic modified from Mello et al. (2004)

that NCM in many species is a particularly large structure in the caudal forebrain that clearly has internal structure but has yet to be extensively described, so its projections are also likely to be poorly described. Both NCM and CLM project bidirectionally to a third structure, the caudomedial mesopallium (CMM). This positions CMM as integrating information from NCM and CLM but not interacting directly with Field L.

The above description of forebrain auditory pathways beyond Field L arises mostly from work on zebra finches (Vates et al. 1996) and pigeons (*Columba livia*) (Wild et al. 1993). There remains a need to broaden the scope of investigation, especially including additional songbird species which receive considerable physiological attention but whose patterns of auditory forebrain connectivity are assumed.

3.2 Song System Pathways

In songbirds a "song system" has been described, which is composed of a posterior vocal pathway and an anterior forebrain pathway (AFP) (see Fig. 26.1.). Here we focus on the posterior vocal pathway, which is thought to receive the principal auditory input to the song system. One subdivision of the posterior vocal pathway consists of the HVC (the proper name), which projects to the robust nucleus of the arcopallium (RA), which in turn innervates the premotor nuclei nXIIts (tracheosyringeal part of the hypoglossal nucleus) and the respiratory nuclei retroambigualis (RAm) and parambigualis (PAm). This pathway is directly related with song production: bilaterally lesioning of HVC or RA impairs song production in adult birds, although behaviorally muted canary males (*Serinus canaria*) appear to court females and attempt to sing (Nottebohm et al. 1976). There is good evidence for functional specialization comparing HVC, RA, and the brainstem. Micro-stimulation in HVC while the bird is singing generates a reset of the song whereas simulating RA affects one syllable selectively. Moreover, HVC chronic recordings in singing zebra finches showed that the firing pattern of interneurons and projection neurons are tightly locked to song (Yu and Margoliash 1996; Hahnloser et al. 2002), with the RA-projecting HVC neurons activated exceedingly sparsely, exhibiting just a single brief (circa 10 ms) high-frequency burst of spike at only a single point each motif (a circa 500–1,000 s sequence of typically two to six syllables) (Hahnloser et al. 2002). Recordings in RA while the bird is singing also show highly structured, very high frequency bursts tightly locked to the syllable production. Each neuron emits multiple bursts per syllable and a total of circa 10–20 bursts per motif (Yu and Margoliash 1996), with the activity of the population of neurons densely representing singing at a "clock" rate of perhaps 10 ms (Leonardo and Fee 2005). It has been shown for RA that variations in the neural firing pattern are correlated with variations in the acoustic features of the syllable (such as pitch and amplitude of the sound wave) (Sober et al. 2008).

3.2.1 Auditory Input to the Song System

Important projections from the auditory system to the song system have been identified (see Fig. 26.1.), arising from brainstem nuclei to the nucleus uvaeformis (Uva), and forebrain nuclei CLM and possibly Field L to the nucleus interfacialis of the nidopallium (NIf) and HVC. Inactivation of HVC suppresses auditory activity in the rest of the forebrain song system (Doupe and Konishi 1991). NIf is a major source of auditory input to HVC (Janata and Margoliash 1999; Coleman and Mooney 2004), as reversibly inactivating NIf suppresses auditory activity in at least some classes of HVC neurons. The thalamic nucleus Uva receives inputs from the auditory system and PAm (Wild 2004) and projects to NIf and to HVC. Bilaterally lesioning Uva does not immediately ablate singing but results in disruption of the normal temporal organization of the adult zebra finch song (Williams and Vicario 1993), and it has been hypothesized that Uva and NIf serve to coordinate sequences of syllables via feedback from the brainstem (Schmidt et al. 2004). Uva also receives a projection from the ventral nucleus of the lateral lemniscus, which could serve to regulate a gating function that has been ascribed to Uva, as high-frequency stimulation in this nucleus simultaneously suppresses auditory activity in HVC and NIf (Coleman et al. 2007).

CLM gives rise to major sources of auditory input to the song system through its projections to NIf and to HVC (Bauer et al. 2008). Reversibly inactivating CLM suppresses auditory activity in NIf and HVC (Bauer et al. 2008). There is also some evidence to support a direct projection of Field L to HVC, and synaptic interactions between NIf and surrounding Field L, but the functional significance of these pathways is unknown (Fortune and Margoliash 1995). Additionally, Field L projects to areas nearby HVC and RA that could provide auditory input to those structures, but this remains unresolved (Kelley and Nottebohm 1979).

4 Auditory Memories

4.1 Representations of Song Memories in the Auditory System

Although the template hypothesis is widely accepted as a concise statement of a broad range of behavioral observations, subsequent behavioral experiments have provided only modest guidance as to the associated neural mechanisms (Margoliash 2002). It remains unclear if the different implied functional structures—guiding innate predisposition for choosing a conspecific song model, guiding song development in the absence of such a model, and in its presence—are distributed or localized, and to what structures. Along the auditory pathway, nuclei express different levels of selectivity, showing a hierarchy of sensory processing for complex auditory stimuli (Theunissen et al. 2004). In the telencephalon, the first level of selectivity is found in Field L, where neurons can discriminate heterospecific song or synthetic sounds over conspecific song. The next level includes neurons in NCM and CM that tend to be tuned more selectively to behaviorally specific complex sounds, such as particular familiar or recently heard sounds. Neurons in Field L, NCM, and CM also respond to the playback of the bird's own song (BOS), but in a non-selective fashion. A special form of song selectivity is found in the song system, one of the targets of the forebrain auditory pathways. In the song system neurons respond selectively to playback of the BOS (see Section 6).

A weakness in such hierarchical descriptions is that they commonly lack associated behavioral studies giving insight into the functional significance of the proposed hierarchy, as is the case here. We have little hard evidence differentiating the functional roles of Field L, NCM, and CM, that is, their contribution to the multitude of auditory behaviors. It is possible for example, but we judge it a dubious proposition, that a single hierarchical scheme will suffice to account for the organization of all auditory memories. Furthermore, electrophysiological results may vary dramatically depending on behavioral state (e.g., Nick and Konishi 2005b). This is rarely evaluated but may have profound implications for processing schemes (see below and Section 7).

In addition to electrophysiological experiments, analyses of immediate early gene (IEG) expression, and to a lesser case lesion studies, are techniques that have been extensively used to study the pathways involved in perception of auditory stimuli. Gene expression has been used as an alternative way to measure activation in neural populations, although the extent to which the electrophysiological and gene expression responses are coupled varies among different areas and neuronal population (Mello and Jarvis 2008). The picture that emerges is that Field L, NCM, and CMM comprise a caudomedial auditory lobule involved in and probably necessary for a broad range of song perceptual processing tasks. Substantial evidence indicates that Field L, NCM, and CM are involved in both processing of perceptual information concerning song complexity and in storage of song memory in songbirds and parrots.

4.1.1 Field L

Early studies explored the responses of Field L neurons to complex, species-specific vocalizations. Briefly, a tonotopy was observed across the extent of Field L and involving other parts of the caudal forebrain (Scheich et al. 1979a; Müller and Leppelsack 1985). Field L neurons exposed to an extensive repertoire of "motif" units of starling song showed a broad range of selectivity, with a few units exhibiting a high degree of selectivity (Leppelsack and Vogt 1976). In white-crowned sparrows, attempts to show the effects of song learning on receptive field properties of Field L neurons met with limited success (Leppelsack 1983; Margoliash 1986). Studies of Field L in the guinea fowl (*Numida meleagris*), a non-vocal learner, also showed some neurons with complex and selective responses to the Iambus call (a complex vocalization associated with individual social roles) mostly restricted to the L1 and L3 subdivisions (Scheich et al. 1979b). Although some insight was gained into the acoustic selectivity underlying those responses, to date quantitative analysis of the receptive field properties of Field L neurons (except for L2 neurons) remains elusive (Nagel and Doupe 2008; see Chapter 20).

4.1.2 Caudomedial Nidopallium

Based on connectivity, the caudomedial nidopallium (NCM) occupies a position comparable to that of superficial (supragranular) layers of the mammalian auditory cortex. It has been found that NCM neurons increase their electrophysiological spike activity in response to complex auditory stimuli (Müller and Leppelsack 1985; Chew et al. 1995). These responses habituate with repeated presentation of the same song, persisting for 48 h (Chew et al. 1995). Auditory responses in NCM fit well into a functional hierarchical scheme. NCM neurons respond more robustly to conspecific song, followed by heterospecific song, and non-song auditory stimuli, where the presentation of tone stimuli or white noise has little or no effect at all (Mello et al. 1992; Chew et al. 1996). Moreover, ZENK expression in canaries indicates that the activation patterns in NCM contain enough information to discriminate natural whistles over synthetic whistles or guitar notes that were matched in intensity and pitch (Ribeiro et al. 1998). A strong influence of the behavioral state has been found in the expression of ZENK: it is abolished if the bird is anesthetized and can be enhanced by shock (or light stimulus) associated with tone or presentation of the image of another bird together with the sound (Vignal et al. 2005).

In zebra finches of both sexes, behavioral studies suggest that song-specific memories are formed at the onset of the sensory period (Clayton 1988; Böhner 1990). Interestingly, NCM song-selectivity also emerges on this developmental time, with electrophysiology selectivity found earlier in development than IEG selectivity (Jin and Clayton 1997; Stripling et al. 2001). During adulthood, the number of cells that express ZENK in response to tutor song playback correlates positively with the extent to which the individual copied the tutor song during development (Bolhuis et al. 2000, 2001). Moreover, electrophysiological recordings indicate that the rate of habituation in NCM neurons upon the presentation of the tutor's song during adulthood correlates with the degree to which the bird has copied the tutor song (Phan et al. 2006). This provides evidence for a possible neural mechanism in NCM contributing to the perceptual memorization of song.

Overall, these findings suggest that circuits including NCM may be involved in some aspect of formation and/or retention of song auditory memories. A particular long-lasting memory that NCM participates in is that of the tutor's song, which is expressed into adulthood (Phan et al. 2006). These exciting results notwithstanding, NCM remains a challenging nucleus to study. It occupies a very large part of the caudal medial forebrain and is poorly defined. It likely contains subdivisions whose location and connectivity are not well described. Resolving these limitations of knowledge is an important goal for future research.

4.1.3 Caudal Mesopallium

The caudal mesopallium (CM) has also been associated with processing behaviorally relevant song stimuli. For example, lesions of CLM (the lateral subdivision of CM) result in a disruption of normal song preferences in female zebra finches (MacDougall-Shackleton et al. 1998). CM neurons in male zebra finches show selectivity for conspecific songs but no selectivity has been found for the tutor's song (Amin et al. 2004). Electrophysiological recordings in European starlings (*Sturnus vulgaris*) have shown that the selectivity properties of CM neurons are strongly dependent on the bird's experience and can be modified by perceptual learning (Gentner and Margoliash 2003).

A revealing study showed that neurons in CLM of awake zebra finches respond robustly to playback of BOS and are also active during singing (Keller and Hahnloser 2009). The two firing patterns were remarkably similar, being the spike patterns in CLM anticipatory to song onset. This premotor activity in an auditory nucleus suggests that in addition to the known auditory input, neurons in CLM are informed regarding premotor activity by a yet-to-be described mechanism. CLM neurons also responded to altered auditory feedback, highlighting their potential role in evaluating auditory feedback, a critical feature of sensorimotor song learning. This interpretation is supported by the observation that birds showed no evidence of modification of vocal output yet almost all the cells responded to feedback or playback perturbation (Keller and Hahnloser 2009). These properties, however, are not limited to CLM neurons but were also observed for Field L neurons (Keller and Hahnloser 2009). These exciting results seemingly implicate large parts of the forebrain auditory axis in being informed of the structure of premotor activity during singing.

The sensitivity of CLM and Field L neurons to auditory feedback distinguishes these cells from neurons in the cortico-basal ganglia loop of the song system which under similar conditions did not show any response to feedback perturbations (Leonardo 2004). This tends to implicate auditory feedback-related processing in the premotor pathway of the song system. Indeed, CM projects directly to the song system nuclei NIf and HVC (Bauer et al. 2008). The feedback-dependent auditory responses are presumed to be conveyed to HVC (Prather et al. 2009).

NCM and CMM have been observed broadly in numerous bird species (Mello and Jarvis 2008). The intermediate and medial mesopallium (IMM), a brain region that partially overlaps with CMM, has also been implicated in visual imprinting in the domestic chick (Horn 1985). Thus, the medial part of the mesopallium may be part of a general recognition system in birds, containing representation of imprinted stimuli, conditioned stimuli, and learned song.

4.2 Representations of Auditory Memories in the Song System

Many neurons in the song system exhibit selective responses to playback of the bird's own song (BOS), which is typically expressed as stronger responses to playback of BOS as compared to conspecific songs or synthetic sounds (e.g., Margoliash 1983; Margoliash and Konishi 1985). Moreover, the pattern of response to BOS can be remarkably similar to the pattern of the same neuron recorded during singing. This has been observed in sleeping adult zebra finches in the premotor nucleus RA (Dave and Margoliash 2000), and in awake swamp sparrows (*M. georgiana*) and Bengalese finches (*Lonchura striata domestica*) in HVC-X projecting neurons (Prather et al. 2008). This suggests that neurons in the adult song system encode a representation of the song that maps between auditory and motor modalities. It is not clear, however, if such a mapping is acting as an auditory memory of the song, as an expression of patterning in the motor system, or both. This pattern could be a representation of a corollary discharge used for sensorimotor planning and learning (Crapse and Sommer 2008; Prather et al.

2008) (Section 6.2) and might emerge from state-dependent processes associated with developmental song learning as observed in juvenile songbirds (Shank and Margoliash 2009; Section 7).

The BOS stimulus that elicits the strongest response changes during development. HVC neurons in anesthetized or sleeping juvenile male zebra finches respond selectivity to BOS as defined by the current vocalizations, even in the early sensorimotor phase when the bird's vocalizations are very variable. In the late sensorimotor phase, the selectivity to BOS recorded earlier in development decays, and neurons become tuned to more complex (but still plastic) vocalizations (Nick and Konishi 2005a). These results suggest that there is a need for motor activity in order to develop selectivity for BOS, and this is presumably coupled to a non-specific role of maturation.

If and how the tutor song is represented in the song system is not as clear. In awake zebra finches, neurons in HVC in a late sensory/early sensorimotor phase respond preferentially to tutor's song playback but this response decays markedly over development (Nick and Konishi 2005b). This suggests the hypothesis that the song system (at least, HVC) expresses a transitory auditory representation during development whereby responses to feedback during (daytime) singing can be modified by a representation of the tutor song. Whether such hypothesized changes occur on-line (during singing) or off-line (spontaneous replay during the day after singing) remains unresolved.

The song system cortico-basal ganglia "anterior forebrain" pathway (AFP) has also been explored as a site of the acquired sensory template. The response properties of AFP neurons (such as song selectivity) also show ontogenetic changes associated with song development (Doupe 1997). During development, the majority of AFP neurons recorded in anesthetized zebra finches exhibit selectivity for BOS, with the strongest response elicited by playback of songs sung by the juveniles in the recent past (Solis and Doupe 1997). Some AFP neurons are also selective for playback of the tutor song, which is particularly evident in cases where a bird fails to accurately copy the tutor song (Solis and Doupe 1999). Experimentally reducing the degree of song copying by manipulating the vocal periphery also resulted in a reduction of selectivity of response to BOS and a change in the distribution of BOS-selective vs. tutor-selective neuronal responses (Solis and Doupe 2000).

One proposal is that auditory feedback is transmitted to the AFP via the HVC—Area X projection. In the AFP this input would be compared with a stored representation of the tutor song, and evaluation of this comparison would influence activity in RA (premotor forebrain output) via LMAN (lateral magnocellular nucleus of the anterior nidopallium; Fig. 26.1). Possibly, this scheme may be obtained during development, but if so its effect presumably changes during development and is largely eliminated in the adult (see also Aronov et al. 2008). In adults, although LMAN responds selectively to BOS in anesthetized or sleeping zebra finches (Doupe and Konishi 1991; Doupe 1997), the response is not observed in awake individuals. AFP activity increases during singing in adults, but it is not clear that this activity carries auditory information. Singing-related discharges in LMAN and Area X in adult birds are premotor, firing before song output, and occurring whether birds can hear or not (Hessler and Doupe 1999). Furthermore, during singing in adults, HVC neurons projecting to Area X carry a motor corollary discharge that is insensitive to modification of auditory feedback (Prather et al. 2008).

Collectively, these results do not admit a simple hypothesis that within the song system an auditory memory for the tutor song is localized to a single nucleus. There appears to be a representation of the tutor song in the song system but it appears across multiple nuclei, is developmentally labile, and varies with behavioral state.

5 Song Processing in Females

Although female songbirds do not sing, they form a memory of an adult male conspecific song. It is possible, although not necessary, that the mechanisms and neural substrate of memory formation in the two sexes are similar. Females have an analogous song system, but in many species, nuclei are generally significantly smaller than in their male counterparts. The degree of sexual dimorphism is well correlated with the amount of singing females engage in (Brenowitz et al. 1985). In species where females sing little or not at all, this presents the opportunity to separate song auditory processing from vocal-motor control.

The relative role of the song system nuclei in female perception varies across species, in a manner consistent with sexual dimorphism in those species. In female canaries, neurons in HVC have different responses to sexually attractive songs (Del Negro et al. 2000) and bilateral lesions in this nucleus disrupt the ability to discriminate between conspecific over heterospecific songs (Brenowitz 1991). In contrast, in female zebra finches, nuclei outside the song system have been implicated in song perception. Electrolytic lesions in CMM and not in HVC disrupt the ability of female zebra finches to discriminate conspecific from heterospecific songs (MacDougall-Shackleton et al. 1998). Adult female canaries can sing spontaneously and respond rapidly with singing to exogenous testosterone, whereas adult female zebra finches do not.

A more uniform pattern of results across species has been observed exploring the role in vocal perception of regions outside the song system. Song-induced IEG expression in

NCM was found in female canaries and European starlings, and in CMM and NCM in zebra finches and white-crowned sparrows (Ribeiro et al. 1998; Gentner et al. 2001; Bailey et al. 2002; Terpstra et al. 2006). Female zebra finches that were reared with their fathers showed significantly increased neuronal activation in CMM but not in NCM, measured through IEG expression (Terpstra et al. 2006). These findings suggest that CMM is part of a neural substrate for the memory of the father's song. Also, female (and male) starlings trained in an operant task to recognize conspecific songs showed memory-related electrophysiological responsiveness in CMM (Gentner and Margoliash 2003).

6 Auditory Processing of Feedback in the Song System

The sensorimotor phase of vocal learning (reviewed in Hultsch and Todt 2004) involves sound production and evaluation of auditory feedback, presumably with the internalized representation of adult songs. Briefly, this period starts with vocalizations that are highly variable in morphology and duration called subsong, comparable with the "babbling" period in humans, even at a time a juvenile bird reliably produces many well-structured calls. The distinction between the production of subsongs and calls helps to emphasize the distinct pathways involved in calling and singing, although there may be more of an influence of auditory feedback on male calls, at least in some species, than has previously been recognized (Liu et al. 2009). During this period the initial vocalizations undergo a transformation and gradually gain similarity with the adult vocalization using mechanisms of auditory feedback evaluation. Human speech learning shows a similar requirement for auditory feedback: subsequent to hearing loss in children, speech deteriorates markedly (Waldstein 1990). This phase ends with the stabilization of song (and speech), when it is said that the song is "crystallized." These similarities between humans and songbirds show a crucial importance in auditory stimuli and feedback for the process of vocal learning.

There is a broad range of variation across species regarding the timing and duration of auditory feedback-regulated developmental vocal learning. Such differences extend into adulthood. A traditional distinction that has not fared well over time has been between "close-ended" and "open-ended" learners. Close-ended learners do not modify their songs in adulthood in the sense of adding vocal material based on newly experienced environmental cues. Nevertheless, there is a much broader range across species of sensitivity to auditory feedback than was originally anticipated. Species such as white-crowned sparrows show minimal effects on the morphology of song elements even after sustained periods of lack of auditory feedback following adult deafening (Konishi 1965). Other species such as zebra finches or Bengalese finches, traditionally thought also to be close-ended learners, show song degradation following adult deafening (Nordeen and Nordeen 1992; Okanoya and Yamaguchi 1997; Fig. 26.2a) and are sensitive to more temporally restricted real-time disruptions of feedback (e.g., Leonardo and Konishi 1999; Sakata and Brainard 2006; Andalman and Fee 2009). Conversely, open-ended learners are not actually open-ended: the evidence to date suggests that species such as canaries, mockingbirds (*Mimus polyglottus*), indigo buntings (*Passerina cyanea*), and starlings incorporate new sounds into their adult songs but tend to do so more commonly earlier in adult life (e.g., Payne 1981).

Collectively, these and other data (e.g., hormonal sensitivity) describe variation in the role of auditory feedback across species and across development. There is as yet little mechanistic insight into the number and action of control elements that might give rise to such variation. Resolving this issue is a central problem motivating research in birdsong learning.

6.1 Maintenance of Adult Song

Once speech and birdsong are learned, they often remain remarkably stable, maintaining the acoustical properties within each syllable, as well as certain "accents" termed dialects, that are characteristic of a given geographical locale (Kroodsma 2005). In some birds as zebra finches or Bengalese finches, the probability of transition from one syllable to another is also well established and maintained throughout the adult life. This stabilization of the behavior could, in principle, be due to loss of plasticity in the neural structures that control vocal motor output, or a reduction in the efficacy of auditory input to those structures that drives vocal plasticity.

A traditional experimental approach has been to examine the effects of deafening adult birds. In zebra finches, this results in a progressive disruption of the song (Nordeen and Nordeen 1992; Lombardino and Nottebohm 2000; Fig. 26.2a). In zebra finches, song degrades over a period of several weeks or more, with minor but measurable changes observed after 3 days (Horita et al. 2008). In Bengalese finches, major effects can be observed after 5 days (Okanoya and Yamaguchi 1997). It has been also shown that age is an important factor for vocal disruption (Lombardino and Nottebohm 2000), relating this effect with neuronal plasticity and neurogenesis (Pytte et al. 2007). Older zebra finches take more time to degrade song, whereas younger zebra finches are more susceptible to auditory feedback disruption. Even so, there is considerable variation across individuals in their response to feedback modification. The source of this

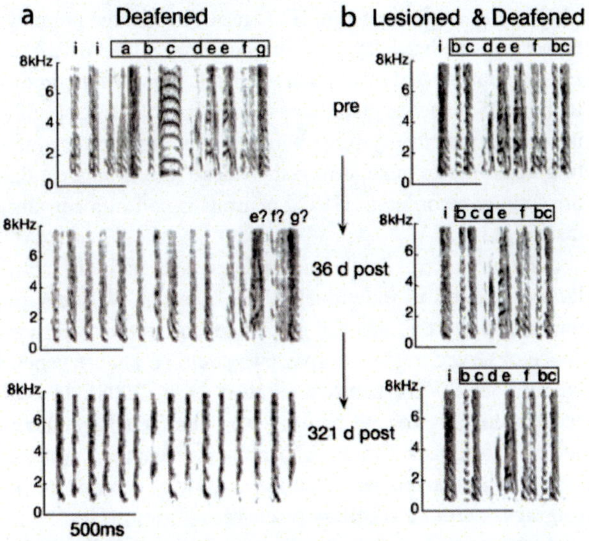

Fig. 26.2 Auditory feedback and the role of the AFP. **a** Adult zebra finch song is disrupted after deafening. The initial syllables (identified by letters in a *box*) cannot be recognized after some days. **b** Lesions in LMAN (output nucleus of the AFP) prevent degradation of song. Data from Brainard and Doupe (2000b)

variation remains unknown. Nevertheless, these experiments demonstrate that adult vocalizations are maintained in many songbird species due to an active process, relaying strongly in auditory feedback. Likewise, in some individuals, speech deteriorates after profound hearing loss in adulthood (Borden 1979; Waldstein 1990; Lane and Webster 1991). As in a given songbird species, there is considerable variation across patients as to the effect of hearing loss on speech production.

Reversible experiments of feedback disruption in birds, such as delayed auditory feedback (Leonardo and Konishi 1999) or regeneration of hair cells (Woolley and Rubel 2002), showed that after deterioration of the song, when auditory feedback is restored, the song gradually recovers its original characteristics. Thus, the internal representation of the adult song remained relatively stable during the period of deafness, implying a long-term memory. Whether, under these experimental conditions, the memory is in the form of motor patterns or an auditory representation that is then used in the mode of feedback correction has yet to be tested (Brainard and Doupe 2000a).

6.2 Role of the AFP in Maintenance of Adult Song

Neurophysiological studies have shown that the AFP is necessary for feedback-dependent song decrystallization. As shown in Fig. 26.2b, lesions in LMAN prevent disruption of adult song that would otherwise occur upon cutting the tracheosyringeal nerve (Williams and Mehta 1999) or upon surgical deafening (Brainard and Doupe 2000b). These experiments show that the AFP is involved in circuitry related in processing auditory feedback, although the specific role remains elusive.

These observations were complemented by neural recordings showing that LMAN neurons exhibit highly selective responses to auditory presentation of BOS in anesthetized birds (Doupe 1997; Doupe and Solis 1997). The BOS selectivity of LMAN neurons in adult birds is maintained by plastic mechanisms. Upon lesioning the tracheosyringeal nerve in adult zebra finches, the acoustic features of the syllables were disrupted whereas the temporal sequence, which is controlled by the respiratory muscles, remained intact. Such birds continued to sing but with abnormal song. Within a few days, LMAN neurons dramatically reduced the response to the original BOS and started responding to the distorted BOS (Roy and Mooney 2007). In order to study auditory processing along this pathway, a lesion study was performed in NIf, one source of auditory input to HVC (Janata and Margoliash 1999; Coleman and Mooney 2004). Bilateral NIf lesions did not prevent the decrystallization process (Roy and Mooney 2009). Moreover, LMAN responses to the new BOS were found in NIf-lesioned birds. A plausible explanation for these results is that other sources of auditory input to the song system were spared by the NIf lesions. In support of this, reversible inactivation of CM suppressed LMAN responses to BOS, showing that CM auditory inputs to the song system are functional and could have a role in vocal plasticity. One possible pathway that could explain these results is the recently discovered direct projection of CM onto HVC (Bauer et al. 2008; Akutagawa and Konishi 2010).

6.3 On-Line Processing of Auditory Feedback in the Song System

Studies of speech production in humans indicate that on-line perturbation of auditory feedback can modulate vocal production very rapidly, suggesting that auditory signals have real-time access to vocal premotor circuitry (Houde and Jordan 1998). In an analogous experiment conducted in Bengalese finches (Sober and Brainard 2009), the pitch of a targeted syllable was modified using custom-designed headphones. Birds compensated for the imposed auditory error by adjusting the pitch of song in the opposite direction. When the perturbation was removed, pitch returned to baseline value. The change in pitch occurred slowly over many days. One possible explanation for the observed time course of plasticity is that birds detected the perceptual difference but were not able to rapidly (or fully) modify motor output. These results indicate that adult Bengalese finches correct vocal errors by comparing auditory feedback to a sensory target. Comparing Bengalese finches with humans, it was proposed that lifelong error correction could be a general principle of learned vocal behavior (Sober and Brainard 2009). The results in these two species could represent two examples along a continuum of vocal learning behavior.

Delayed auditory feedback in Bengalese finches has been studied in detail. In an illustrative study (Sakata and Brainard 2006), a targeted syllable was played back through a loud speaker with a certain time delay. This protocol was sufficient to disrupt the normal sequencing of the song, producing a change in the transition probabilities of singing different chains (sequences) of syllables. Individual syllables could also be dropped, and in addition, a change in syllable timing was also observed. For humans, there is a well-characterized relationship between the timing of delayed auditory feedback and the degree to which that feedback disrupts ongoing speech. Auditory feedback is maximally disruptive when presented at delays of 150–200 ms, the approximate duration of a typical syllable of human speech. A similar result was obtained for Bengalese finches, with the most disruptive delays on order of 45–65 ms. An interval of 64 ms corresponds to the average duration of Bengalese finch syllables (Sakata and Brainard 2006).

Chronic neurophysiological recordings gave additional insight into mechanisms of auditory feedback during vocal behavior. Putative HVC interneurons responded robustly and selectively to BOS in awake silent birds (Sakata and Brainard 2008). In the context of delayed auditory feedback, HVC activity consistently decreased at a short latency after the perturbation of normal feedback during ongoing song (Sakata and Brainard 2008). This represents a neurophysiological demonstration that information derived from auditory feedback is rapidly available to vocal premotor structures during singing, including HVC.

A complementary set of experiments to auditory feedback processing in the song system has been reported in swamp sparrows (Prather et al. 2008). The swamp sparrow song is composed of repetitive notes conforming to a trill. Swamp sparrows countersing in response to song presentation, which can be experimentally useful in evaluating auditory vs. motor neural activity. Only the HVC neurons projecting to Area X (HVC_X) showed robust activity during BOS-playback and during singing. Moreover, the activity in individual neurons was locked to certain syllables in the trill, having the auditory and the premotor response the same phase to the onset of the song. This result, first observed in RA (Dave and Margoliash 2000), is surprising because in principle it was expected that premotor activity recorded during singing should precede the auditory activity recorded during playback. The result in swamp sparrows implicates a delay mechanism that is likely to be implemented in HVC itself, so that the premotor response is being delayed before entering to the AFP. Also, HVC_X neurons that respond to playback showed suppression to the BOS response just before the onset and during singing, indicating that the system switched from an auditory state to a vocal-motor state. Altogether these results suggest that the activity recorded in HVC_X neurons is part of a corollary discharge.

Similar patterns of singing-related and auditory activity were found in individual HVC_X neurons of Bengalese finches, suggesting that the precise sensorimotor correspondence observed in swamp sparrows could be a more general property of HVC_X neurons (Prather et al. 2008). In both species, singing-related activity was unaffected by altered auditory feedback (loud white noise presentation), further supporting the hypothesis that the activity of neurons during singing is part of a corollary discharge.

A theoretical model (Troyer and Doupe 2000a, b) has been developed in order to discuss possible mechanisms for auditory-vocal integration in songbirds. In this model, it is proposed that the motor signal that goes from HVC to RA is also sent as an internal sensory efference copy to the AFP. The two signals, one motor and the other efference copy, would arrive at RA and would be compared there, but auditory feedback from bird's own song would be compared in HVC (in particular, in HVC_X). So, one possibility is that rather than evaluating auditory feedback, the AFP may receive a prediction of expected feedback, perhaps created by the association of premotor signals and auditory feedback in HVC (Brainard and Doupe 2000a; Troyer and Doupe 2000a, b).

7 Sleep and Auditory Processing

In humans, it has been shown that developing motor skills involves both on-line and off-line processing. On-line learning is the skill enhancement that occurs during practice (Newell 1991). Off-line learning is post-practice skill improvement usually associated with sleep cycles (Karni et al. 1998; Walker 2005) and there is a substantial literature linking auditory processing to memory consolidation during sleep (reviewed in Margoliash and Fenn 2009). In principle, off-line processing can occur at any time when the individual is not performing the task, and indeed, neuronal "replay" has been observed in awake animals (Foster and Wilson 2006; Karlsson and Frank 2009). Off-line learning has been linked with processes of memory consolidation in auditory processing tasks in humans (e.g., Fenn et al. 2003) and recently in European starlings (Brawn et al. 2010).

Experience-dependent brain reactivation during sleep has been observed in rodents (Pavlides and Winson 1989; Wilson and McNaughton 1994; Lee and Wilson 2002), nonhuman primates (Hoffman and McNaughton 2002), humans (Maquet et al. 2000), and songbirds (Dave and Margoliash 2000), pointing to a very general biological phenomenon. An example of neuronal replay is shown in Fig. 26.3. Importantly, post-acquisition brain reactivation during sleep has been shown to be proportional to memory acquisition in humans (Peigneux et al. 2003).

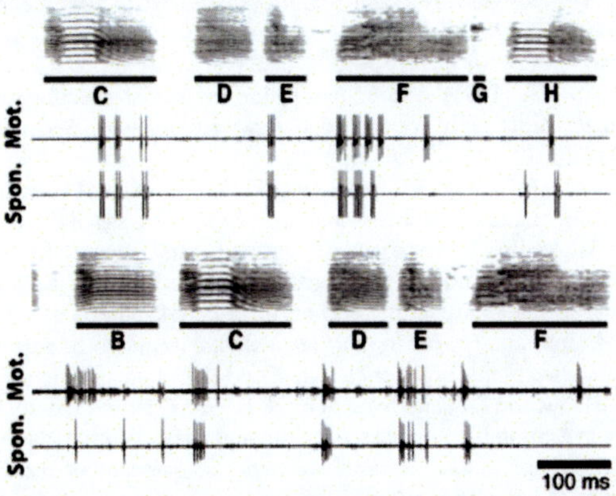

Fig. 26.3 Neuronal replay during sleeping in a songbird. The neuronal activity during sleep in a zebra finch (Spon.) is very similar to the premotor activity while the bird is singing (Mot.). In the spectrograph of the song, each syllable is identified with a letter and the corresponding raw traces of neuronal activity belong to two different neurons from one bird. Data from Dave and Margoliash (2000)

A significant role of sleep in vocal learning has been determined behaviorally (Deregnaucourt et al. 2005). Birds isolated until circa 40 days of age and then given access under an instrumental conditioning paradigm to a tutor song for the first time begin to modify their vocal output, the day after the first day of tutor song exposure (Deregnaucourt et al. 2005; Shank and Margoliash 2009). However, even on the night after the first day of tutor song exposure, RA neurons exhibit robust increase in bursting activity including tutor song-related changes in activity, and this was dynamic, so that changing the tutor song the bird experienced changed the properties of the RA neurons. The results suggest that replay during sleep is related to an auditory percept, at least in the juvenile songbird first acquiring a memory of the tutor song (Shank and Margoliash 2009). This was observed as differences in the average second-order statistics of populations of neurons collapsed across groups of birds exposed to one of the three different tutor songs. To exploit this phenomenon more fully, it will be necessary to explore single-cell correlates of the learning phenomenon. Furthermore, the changes in RA activity were blocked when birds could not hear themselves sing, or could not sing, even if the birds were exposed to the tutor songs (Shank and Margoliash 2009). This implies that birds are actively stimulating structures involved in auditory memories when they sing and that this is necessary to establish the acquired template for song learning. This raises the interesting possibility that the song template is not purely a sensory representation.

8 Conclusions and Future Directions

After many years of research, finally we are at the threshold of expressing the conceptual entities of the template theory for vocal learning in terms of neurophysiological mechanisms. The focus of one research program is on memory representations formed during early song exposure. The evidence to date suggests that the acquired sensory template is represented in multiple nuclei, in a developmentally labile and state-dependent fashion. A localized structure which is the principal site of memory formation and which drives these extensive down-stream changes may yet emerge, but nevertheless it appears that the acquired sensory template is broadly represented in a distributed fashion.

A second research program focuses on processing of feedback during vocalizations. Altered auditory feedback experiments have shown that song learning and adult maintenance is an active process, but little is known about the neural mechanisms of auditory feedback evaluation. It will be particularly valuable in future experiments to alter auditory feedback in real time within a physiological range such that an animal could assimilate the alteration as its own vocal error. More recent studies (Andalman and Fee 2009; Sober and Brainard 2009) go in this direction. This is opening a new line of experiments, including studies of "error signals," which nuclei are involved and how the error is processed, and explorations of the physiological mechanisms involved in error correction of vocal output.

One important distinction is the control mechanisms an animal employs for a given alteration of feedback. This involves not only central nervous system (CNS) regulation, but also arises from the interaction between the periphery and the CNS. More integration is needed in this respect, studying in detail the behavior (e.g., characteristic of the song, Tchernichovski and Mitra 2002), and the physical processes occurring in the peripheral system during singing (e.g., Mindlin and Laje 2005). A deep understanding of the peripheral system would allow the construction of biomimetic devices driven by physiological related parameters, which can be integrated in experiments of altered auditory feedback (Zysman et al. 2005).

Finally, it appears that vocal learning during real-time performance is consolidated off-line, with sleep playing an integral role in consolidating learning, possibly through replay of auditory representations. Whether the role of off-line processing involves sensory and/or sensorimotor learning remains to be resolved. It remains a possibility that the acquired sensory template is modified by sensorimotor processes. If so, the distinction between the sensory and sensorimotor phases may have to be re-evaluated.

Acknowledgments This work was supported in part by NIH grants MH59831 and DC007206 to DM and Human Frontiers Science Program fellowship LT000735 to AA. We are grateful for comments on the manuscript provided by Patrice Adret and Morgan Wirthlin.

References

Adret P (2004) In search of the song template. *Behavioral Neurobiology of Birdsong* **1016**:303–324.

Akutagawa E and Konishi M (2010) New brain pathways found in the vocal control system of a songbird. *Journal of Comparative Neurology* **518**:3086–3100.

Amador A, Trevisan MA, and Mindlin GB (2005) Simple neural substrate predicts complex rhythmic structure in duetting birds. *Physical Review E* **72**:031905.

Amin N, Grace JA, and Theunissen FE (2004) Neural response to bird's own song and tutor song in the zebra finch field L and caudal mesopallium. *Journal of Comparative Physiology A* **190**:469–489.

Andalman AS and Fee MS (2009) A basal ganglia-forebrain circuit in the songbird biases motor output to avoid vocal errors. *Proceedings of the National Academy of Sciences of the United States of America* **106**:12518–12523.

Aronov D, Andalman AS, and Fee MS (2008) A specialized forebrain circuit for vocal babbling in the juvenile songbird. *Science* **320**: 630–634.

Bailey DJ, Rosebush JC, and Wade J (2002) The hippocampus and caudomedial neostriatum show selective responsiveness to conspecific song in the female zebra finch. *Journal of Neurobiology* **52**:43–51.

Baptista LF and Petrinovich L (1984) Social interaction, sensitive phrases and the song template hypothesis in the white-crowned sparrow. *Animal Behaviour* **32**:172–181.

Baptista LF and Schuchmann KL (1990) Song learning in the Anna hummingbird (*Calypte-Anna*). *Ethology* **84**:15–26.

Bass AH, Gilland EH, and Baker R (2008) Evolutionary origins for social vocalization in a vertebrate hindbrain-spinal compartment. *Science* **321**:417–421.

Bauer EE, Coleman MJ, Roberts TF, Roy A, Prather JF, and Mooney R (2008) A synaptic basis for auditory-vocal integration in the songbird. *Journal of Neuroscience* **28**:1509–1522.

Beecher MD, Burt JM, O'Loghlen AL, Templeton CN, and Campbell SE (2007) Bird song learning in an eavesdropping context. *Animal Behaviour* **73**:929–935.

Böhner J (1990) Early Acquisition of song in the zebra finch, Taeniopygia-guttata. *Animal Behaviour* **39**:369–374.

Bolhuis JJ, Hetebrij E, Den Boer-Visser AM, De Groot JH, and Zijlstra GGO (2001) Localized immediate early gene expression related to the strength of song learning in socially reared zebra finches. *European Journal of Neuroscience* **13**:2165–2170.

Bolhuis JJ, Zijlstra GGO, den Boer-Visser AM, and Van der Zee EA (2000) Localized neuronal activation in the zebra finch brain is related to the strength of song learning. *Proceedings of the National Academy of Sciences of the United States of America* **97**:2282–2285.

Boughman JW (1998) Vocal learning by greater spear-nosed bats. *Proceedings of the Royal Society of London Series B-Biological Sciences* **265**:227–233.

Borden GJ (1979) An interpretation of research of feedback interruption in speech. *Brain and Language* **7**:307–319.

Braaten RF and Reynolds K (1999) Auditory preference for conspecific song in isolation-reared zebra finches. *Animal Behaviour* **58**: 105–111.

Brainard MS and Doupe AJ (2000a) Auditory feedback in learning and maintenance of vocal behaviour. *Nature Reviews Neuroscience* **1**:31–40.

Brainard MS and Doupe AJ (2000b) Interruption of a basal ganglia-forebrain circuit prevents plasticity of learned vocalizations. *Nature* **404**:762–766.

Brawn TP, Nusbaum HC, and Margoliash D (2010) Sleep-dependent consolidation of auditory discrimination learning in adult starlings. *Journal of Neuroscience* **30**:609–613.

Brenowitz EA (1991) Altered perception of species-specific song by female birds after lesions of a forebrain nucleus. *Science* **251**: 303–305.

Brenowitz EA, Arnold AP, and Levin RN (1985) Neural correlates of female song in tropical duetting birds. *Brain Research* **343**:104–112.

Chew SJ, Mello C, Nottebohm F, Jarvis E, and Vicario DS (1995) Decrements in auditory responses to a repeated conspecific song are long-lasting and require 2 periods of protein-synthesis in the songbird forebrain. *Proceedings of the National Academy of Sciences of the United States of America* **92**:3406–3410.

Chew SJ, Vicario DS, and Nottebohm F (1996) A large-capacity memory system that recognizes the calls and songs of individual birds. *Proceedings of the National Academy of Sciences of the United States of America* **93**:1950–1955.

Clayton NS (1988) Song discrimination-learning in zebra finches. *Animal Behaviour* **36**:1016–1024.

Coleman MJ and Mooney R (2004) Synaptic transformations underlying highly selective auditory representations of learned birdsong. *Journal of Neuroscience* **24**:7251–7265.

Coleman MJ, Roy A, Wild JM, and Mooney R (2007) Thalamic gating of auditory responses in telencephalic song control nuclei. *Journal of Neuroscience* **27**:10024–10036.

Corballis MC (2002) *From Hand to Mouth: The Origins of Language*. Princeton University Press, Princeton.

Crapse TB and Sommer MA (2008) Corollary discharge across the animal kingdom. *Nature Reviews Neuroscience* **9**:587–600.

Dave AS and Margoliash D (2000) Song replay during sleep and computational rules for sensorimotor vocal learning. *Science* **290**: 812–816.

Del Negro C, Kreutzer M, and Gahr M (2000) Sexually stimulating signals of canary (*Serinus canaria*) songs: evidence for a female-specific auditory representation in the HVc nucleus during the breeding season. *Behavioral Neuroscience* **114**:526–542.

Deregnaucourt S, Mitra PP, Feher O, Pytte C, and Tchernichovski O (2005) How sleep affects the developmental learning of bird song. *Nature* **433**:710–716.

Dooling R and Searcy M (1980) Early perceptual selectivity in the swamp sparrow. *Developmental Psychobiology* **13**:499–506.

Doupe AJ (1997) Song- and order-selective neurons in the songbird anterior forebrain and their emergence during vocal development. *Journal of Neuroscience* **17**:1147–1167.

Doupe AJ and Konishi M (1991) Song-selective auditory circuits in the vocal control system of the zebra finch. *Proceedings of the National Academy of Sciences of the United States of America* **88**:11339–11343.

Doupe AJ and Solis MM (1997) Song- and order-selective neurons develop in the songbird anterior forebrain during vocal learning. *Journal of Neurobiology* **33**:694–709.

Falls JB (1982) Individual recognition by sound in birds. In: Kroodsma DE and Miller EH (eds). *Acoustic Communication in Birds*, volume 2, *Song Learning and Its Consequences*. Academic Press, New York, pp. 237–278.

Fenn KM, Nusbaum HC, and Margoliash D (2003) Consolidation during sleep of perceptual learning of spoken language. *Nature* **425**:614–616.

Fortune ES and Margoliash D (1992) Cytoarchitectonic organization and morphology of cells of the Field L complex in male zebra finches (*Taenopygia guttata*). *Journal of Comparative Neurology* **325**:388–404.

Fortune ES and Margoliash D (1995) Parallel pathways and convergence onto HVC and adjacent neostriatum of adult zebra finches (*Taeniopygia guttata*). *Journal of Comparative Neurology* **360**: 413–441.

Foster DJ and Wilson MA (2006) Reverse replay of behavioural sequences in hippocampal place cells during the awake state. *Nature* **440**:680–683.

Fromkin V, Krashen S, Curtiss S, Rigler D, and Rigler M (1974) The development of language in Genie: a case of language acquisition beyond the "critical period". *Brain and Language* **1**: 81–107.

Gentner TQ, Hulse SH, Duffy D, and Ball GF (2001) Response biases in auditory forebrain regions of female songbirds following exposure to sexually relevant variation in male song. *Journal of Neurobiology* **46**:48–58.

Gentner TQ and Margoliash D (2003) Neuronal populations and single cells representing learned auditory objects. *Nature* **424**:669–674.

Hahnloser RHR, Kozhevnikov AA, and Fee MS (2002) An ultra-sparse code underlies the generation of neural sequences in a songbird. *Nature* **419**:65–70.

Hessler NA and Doupe AJ (1999) Singing-related neural activity in a dorsal forebrain-basal ganglia circuit of adult zebra finches. *Journal of Neuroscience* **19**:10461–10481.

Hoffman KL and McNaughton BL (2002) Coordinated reactivation of distributed memory traces in primate neocortex. *Science* **297**: 2070–2073.

Horita H, Wada K, and Jarvis ED (2008) Early onset of deafening-induced song deterioration and differential requirements of the pallial-basal ganglia vocal pathway. *European Journal of Neuroscience* **28**:2519–2532.

Horn G (1985) *Memory, Imprinting, and the Brain: An Inquiry Into Mechanisms*. Oxford University Press, New York.

Houde JF and Jordan MI (1998) Sensorimotor adaptation in speech production. *Science* **279**:1213–1216.

Hultsch H and Todt D (2004) Learning to sing. In: Marler P and Slabbekoorn H (eds). *Nature's Music: The Science of Birdsong*. Elsevier Academic Press, San Diego, pp. 80–107.

Janata P and Margoliash D (1999) Gradual emergence of song selectivity in sensorimotor structures of the male zebra finch song system. *Journal of Neuroscience* **19**:5108–5118.

Janik VM and Slater PJB (2000) The different roles of social learning in vocal communication. *Animal Behaviour* **60**:1–11.

Jarvis ED, et al. (2005) Avian brains and a new understanding of vertebrate brain evolution. *Nature Reviews Neuroscience* **6**:151–159.

Jarvis ED, Ribeiro S, da Silva ML, Ventrua D, Vielliard J, and Mello CV (2000) Behaviourally driven gene expression reveals song nuclei in hummingbird brain. *Nature* **406**:628–632.

Jin H and Clayton DF (1997) Localized changes in immediate-early gene regulation during sensory and motor learning in zebra finches. *Neuron* **19**:1049–1059.

Jürgens U (2002) Neural pathways underlying vocal control. *Neuroscience and Biobehavioral Review* **26**: 235–258.

Karlsson MP and Frank LM (2009) Awake replay of remote experiences in the hippocampus. *Nature Neuroscience* **12**:913–918.

Karni A, Meyer G, Rey-Hipolito, Jezzard P, Adams MM, Turner R, and Ungerleider LG (1998) The acquisition of skilled motor performance: fast and slow experience-driven changes in primary motor cortex. *Proceedings of the National Academy of Sciences of the United States of America* **95**:861–868.

Karten H (1967) The organization of the ascending auditory pathway in the pigeon (*Columba livia*). I. Diencephalic projections of the inferior colliculus (nucleus mesencephali lateralis, pars dorsalis). *Brain Research* **6**:409–427.

Karten H (1968) The ascending auditory pathway in the pigeon (*Columba livia*). II. Telencephalic projections of the nucleus ovoidalis thalami. *Brain Research* **11**:134–153.

Karten HJ (1997) Evolutionary developmental biology meets the brain: the origins of mammalian cortex. *Proceedings of the National Academy of Sciences of the United States of America* **94**: 2800–2804.

Kelley DB and Nottebohm F (1979) Projections of a telencephalic auditory nucleus -Field L-in the canary. *Journal of Comparative Neurology* **183**:455–469.

Keller GB and Hahnloser RHR (2009) Neural processing of auditory feedback during vocal practice in a songbird. *Nature* **457**:187–190.

King AP, West MJ, and Goldstein MH (2005) Nonvocal shaping of avian song development: parallels to human speech development. *Ethology* **111**:101–117.

Konishi M (1965) The role of auditory feedback in the control of vocalization in the white-crowned sparrow. *Zeitschrift für Tierpsychologie* **22**:770–783.

Konishi M (1978) Auditory environment and vocal development in birds. In: Walk RD and Pick HLJ (eds). *Perception and Experience*. Plenum, New York, pp. 105–118.

Konishi M (2004) The role of auditory feedback in birdsong. *Behavioral Neurobiology of Birdsong* **1016**:463–475.

Kroodsma DE (2005) *The Singing Life of Birds: The Art and Science of Listening to Birdsong*. Houghton Mifflin Publishers, New York.

Kroodsma DE and Miller EH (1996) *Ecology and Evolution of Acoustic Communication in Birds*. Cornell University Press, Ithaca.

Lane H, Webster JW (1991) Speech deterioration in postlingually deafened adults. *Journal of the Acoustical Society of America* **89**:859–866.

Lee AK and Wilson MA (2002) Memory of sequential experience in the hippocampus during slow wave sleep. *Neuron* **36**:1183–1194.

Leonardo A (2004) Experimental test of the birdsong error-correction model. *Proceedings of the National Academy of Sciences of the United States of America* **101**:16935–16940.

Leonardo A and Fee MS (2005) Ensemble coding of vocal control in birdsong. *Journal of Neuroscience* **25**:652–661.

Leonardo A and Konishi M (1999) Decrystallization of adult birdsong by perturbation of auditory feedback. *Nature* **399**:466–470.

Leppelsack H-J (1983) Analysis of song in the auditory pathway of song birds. In: Ewert JP, Capranica RR, and Ingle DJ (eds). *Advances in Vertebrate Neuroethology*. Plenum, New York, pp. 783–799.

Leppelsack HJ and Vogt M (1976) Responses of auditory neurons in the forebrain of a songbird to stimulation with species-specific sounds. *Journal of Comparative Physiology A* **107**:263–274.

Liu W-C, Wada K, and Nottebohm F (2009) Variable food begging calls are harbingers of vocal learning. *Public Library of Science One* **4**:e5929.

Lombardino AJ and Nottebohm F (2000) Age at deafening affects the stability of learned song in adult male zebra finches. *Journal of Neuroscience* **20**:5054–5064.

MacDougall-Shackleton SA, Hulse SH, and Ball GF (1998) Neural bases of song preferences in female zebra finches (*Taeniopygia guttata*). *Neuroreport* **9**:3047–3052.

Maquet P, et al. (2000) Experience-dependent changes in cerebral activation during human REM sleep. *Nature Neuroscience* **3**:831–836.

Margoliash D (1983) Acoustic parameters underlying the responses of song-specific neurons in the white-crowned sparrow. *Journal of Neuroscience* **3**:1039–1057.

Margoliash D (1986) Preference for autogenous song by auditory neurons in a song system nucleus of the white-crowned sparrow. *Journal of Neuroscience* **6**:1643–1661.

Margoliash D (2002) Evaluating theories of bird song learning: implications for future directions. *Journal of Comparative Physiology A* **188**:851–866.

Margoliash D and Fenn KM (2009) Sleep and memory consolidation in audition. In: Dallos P and Oertel D (eds). *The Senses: A Comprehensive Reference*, volume 3, *Audition*. Elsevier, Amsterdam, pp. 895–911.

Margoliash D and Konishi M (1985) Auditory representation of autogenous song in the song system of white-crowned sparrows. *Proceedings of the National Academy of Sciences of the United States of America* **82**:5997–6000.

Marler P (1970) A comparative approach to vocal learning: song development in white-crowned sparrows. *Journal of Comparative and Physiological Psychology* **71**:1–25.

Marler P (1997) Three models of song learning: evidence from behavior. *Journal of Neurobiology* **33**:501–516.

Mello CV and Jarvis ED (2008) Behavior-dependent expression of inducible genes in vocal learning birds. In: Zeigler HP and Marler P (eds). *Neuroscience of Birdsong*. Cambridge University Press, Cambridge, pp. 381–397.

Mello CV, Velho TAF, and Pinaud R (2004) Song-induced gene expression. A window on song auditory processing and perception. *Annals of the New York Academy of Sciences* **1016**:263–281.

Mello CV, Vicario DS, and Clayton DF (1992) Song presentation induces gene expression in the songbird forebrain. *Proceedings of the National Academy of Sciences of the United States of America* **89**:6818–6822.

Miller DB (1979) Long-term recognition of fathers song by female zebra finches. *Nature* **280**:389–391.

Mindlin GB and Laje R (2005) *The Physics of Birdsong*. Springer, Berlin.

Müller CM and Leppelsack HJ (1985) Feature-extraction and tonotopic organization in the avian auditory forebrain. *Experimental Brain Research* **59**:587–599.

Mundinger PC (1979) Call learning in the *Carduelinae*: ethological and systematic considerations. *Systematic Zoology* **28**: 270–283.

Nagel KL and Doupe AJ (2008) Organizing principles of spectro-temporal encoding in the avian primary auditory area Field L. *Neuron* **58**:938–955.

Nelson DA and Marler P (1993) Innate recognition of song in white-crowned sparrows – a role in selective vocal learning. *Animal Behaviour* **46**:806–808.

Newell KM (1991) Motor skill acquisition. *Annual Review of Psychology* **42**:213–237.

Nick TA and Konishi M (2005a) Neural auditory selectivity develops in parallel with song. *Journal of Neurobiology* **62**:469–481.

Nick TA and Konishi M (2005b) Neural song preference during vocal learning in the zebra finch depends on age and state. *Journal of Neurobiology* **62**:231–242.

Noad MJ, Cato DH, Bryden MM, Jenner MN, and Jenner KCS (2000) Cultural revolution in whale songs. *Nature* **408**:537.

Nordeen KW and Nordeen EJ (1992) Auditory feedback is necessary for the maintenance of stereotyped song in adult zebra finches. *Behavioral and Neural Biology* **57**:58–66.

Nottebohm F (1972) Origins of vocal learning. *American Naturalist* **106**:116–140.

Nottebohm F, Stokes TM, and Leonard CM (1976) Central control of song in the canary, Serinus canarius. *Journal of Comparative Neurology* **165**:457–486.

Okanoya K and Yamaguchi A (1997) Adult Bengalese finches (Lonchura striata var. domestica) require real-time auditory feedback to produce normal song syntax. *Journal of Neurobiology* **33**:343–356.

Payne RB (1981) Song learning and social interaction in indigo buntings. *Animal Behaviour* **29**:688–697.

Pavlides C and Winson J (1989) Influences of hippocampal place cell firing in the awake state on the activity of these cells during subsequent sleep episodes. *Journal of Neuroscience* **9**:2907–2918.

Peigneux P, et al. (2003) Learned material content and acquisition level modulate cerebral reactivation during posttraining rapid-eye-movements sleep. *Neuroimage* **20**:125–134.

Phan ML, Pytte CL, and Vicario DS (2006) Early auditory experience generates long-lasting memories that may subserve vocal learning in songbirds. *Proceedings of the National Academy of Sciences of the United States of America* **103**:1088–1093.

Pinker S and Bloom P (1990) Natural language and natural selection. *Behavioral and Brain Sciences* **13**:707–784.

Poole JH, Tyack PL, Stoeger-Horwath AS, and Watwood S (2005) Elephants are capable of vocal learning. *Nature* **434**: 455–456.

Prather JF, Peters S, Nowicki S and Mooney R (2008) Precise auditory-vocal mirroring in neurons for learned vocal communication. *Nature* **451**:305–310.

Prather JF, Nowicki S, Anderson RC, Peters S, and Mooney R (2009) Neural correlates of categorical perception in learned vocal communication. *Nature Neuroscience* **12**:221–228.

Price PH (1979) Developmental determinants of structure in zebra finch song. *Journal of Comparative and Physiological Psychology* **93**:260–277.

Pytte CL, Gerson M, Miller J, and Kirn JR (2007) Increasing stereotypy in adult zebra finch song correlates with a declining rate of adult neurogenesis. *Developmental Neurobiology* **67**:1699–1720.

Reiner A, et al. (2004) Revised nomenclature for avian telencephalon and some related brainstem nuclei. *Journal of Comparative Neurology* **473**:377–414.

Ribeiro S, Cecchi GA, Magnasco MO, and Mello CV (1998) Toward a song code: evidence for a syllabic representation in the canary brain. *Neuron* **21**:359–371.

Riebel K, Smallegange IM, Terpstra NJ, and Bolhuis JJ (2002) Sexual equality in zebra finch song preference: evidence for a dissociation between song recognition and production learning. *Proceedings of the Royal Society of London Series B, Biological Sciences* **269**: 729–733.

Rizzolatti G and Arbib MA (1998) Language within our grasp. *Trends in Neurosciences* **21**:188–94.

Rose GJ, Goller F, Gritton HJ, Plamondon SL, Baugh AT, and Cooper BG (2004) Species-typical songs in white-crowned sparrows tutored with only phrase pairs. *Nature* **432**:753–758.

Roy A and Mooney R (2007) Auditory plasticity in a basal ganglia-forebrain pathway during decrystallization of adult birdsong. *Journal of Neuroscience* **27**:6374–6387.

Roy A and Mooney R (2009) Song decrystallization in adult zebra finches does not require the song nucleus NIf. *Journal of Neurophysiology* **102**:979–991.

Sakata JT and Brainard MS (2006) Real-time contributions of auditory feedback to avian vocal motor control. *Journal of Neuroscience* **26**:9619–9628.

Sakata JT and Brainard MS (2008) Online contributions of auditory feedback to neural activity in avian song control circuitry. *Journal of Neuroscience* **28**:11378–11390.

Saranathan V, Hamilton D, Powell GVN, Kroodsma DE, and Prum RO (2007) Genetic evidence supports song learning in the three-wattled bellbird *Procnias tricarunculata* (*Cotingidae*). *Molecular Ecology* **16**:3689–3702.

Scheich H, Bonke BA, Bonke D, and Langner G (1979a) Functional organization of some auditory nuclei in the guinea fowl demonstrated by the 2-deoxyglucose technique. *Cell and Tissue Research* **204**:17–27.

Scheich H, Langner G, and Bonke D (1979b) Responsiveness of units in the auditory neostriatum of the Guinea fowl (*Numida meleagris*) to species-specific calls and synthetic stimuli. II. Discrimination of iambus-like calls. *Journal of Comparative Physiology A* **132**: 257–276.

Schmidt MF, Ashmore RC, and Vu ET (2004) Bilateral control and interhemispheric coordination in the avian song motor system. *Behavioral Neurobiology of Birdsong* **1016**:171–186.

Seller TJ (1981) Midbrain vocalization centres in birds. *Trends in Neurosciences* **4**:301–303.

Shank SS and Margoliash D (2009) Sleep and sensorimotor integration during early vocal learning in a songbird. *Nature* **458**:73–74.

Sober SJ and Brainard MS (2009) Adult birdsong is actively maintained by error correction. *Nature Neuroscience* **12**:927–931.

Sober SJ, Wohlgemuth MJ, and Brainard MS (2008) Central contributions to acoustic variation in birdsong. *Journal of Neuroscience* **28**:10370–10379.

Soha JA and Marler P (2000) A species-specific acoustic cue for selective song learning in the white-crowned sparrow. *Animal Behaviour* **60**:297–306.

Solis MM and Doupe AJ (1997) Anterior forebrain neurons develop selectivity by an intermediate stage of birdsong learning. *Journal of Neuroscience* **17**:6447–6462.

Solis MM and Doupe AJ (1999) Contributions of tutor and bird's own song experience to neural selectivity in the songbird anterior forebrain. *Journal of Neuroscience* **19**:4559–4584.

Solis MM and Doupe AJ (2000) Compromised neural selectivity for song in birds with impaired sensorimotor learning. *Neuron* **25**: 109–121.

Stripling R, Kruse AA, and Clayton DF (2001) Development of song responses in the zebra finch caudomedial neostriatum: role of genomic and electrophysiological activities. *Journal of Neurobiology* **48**:163–180.

Tchernichovski O and Mitra PP (2002) Towards quantification of vocal imitation in the zebra finch. *Journal of Comparative Physiology A* **188**:867–878.

Terpstra NJ, Bolhuis JJ, Riebel K, van der Burg JMM, and den Boer-Visser AM (2006) Localized brain activation specific to auditory memory in a female songbird. *Journal of Comparative Neurology* **494**:784–791.

Theunissen FE, Amin N, Shaevitz SS, Woolley SM, Fremouw T, and Hauber ME (2004) Song selectivity in the song system and in the auditory forebrain. *Annals of the New York Academy of Sciences* **1016**:222–245.

Troyer TW and Doupe AJ (2000a) An associational model of birdsong sensorimotor learning I. Efference copy and the learning of song syllables. *Journal of Neurophysiology* **84**:1204–1223.

Troyer TW and Doupe AJ (2000b) An associational model of birdsong sensorimotor learning II. Temporal hierarchies and the learning of song sequence. *Journal of Neurophysiology* **84**:1224–1239.

Vates GE, Broome BM, Mello CV, and Nottebohm F (1996) Auditory pathways of caudal telencephalon and their relation to the song system of adult male zebra finches. *Journal of Comparative Neurology* **366**:613–642.

Vignal C, Andru J, and Mathevon N (2005) Social context modulates behavioural and brain immediate early gene responses to sound in male songbird. *European Journal of Neuroscience* **22**: 949–955.

Waldstein RS (1990) Effects of postlingual deafness on speech production: implications for the role of auditory feedback. *Journal of the Acoustical Society of America* **88**:2099–2114.

Walker MP (2005) A refined model of sleep and the time course of memory formation. *Behavioral and Brain Sciences* **28**:51–64.

West MJ and King AP (1988) Female visual displays affect the development of male song in the cowbird. *Nature* **334**:244–246.

Wilbrecht L and Nottebohm F (2003) Vocal learning in birds and humans. *Mental Retardation and Developmental Disabilities Research Reviews* **9**:135–148.

Wild JM (2004) Functional neuroanatomy of the sensorimotor control of singing. *Behavioral Neurobiology of Birdsong* **1016**:438–462.

Wild JM, Karten HJ, and Frost BJ (1993) Connections of the auditory forebrain in the pigeon (*Columba livia*). *Journal of Comparative Neurology* **337**:32–62.

Williams H and Mehta N (1999) Changes in adult zebra finch song require a forebrain nucleus that is not necessary for song production. *Journal of Neurobiology* **39**:14–28.

Williams H and Vicario DS (1993) Temporal patterning of song production: participation of nucleus uvaeformis of the thalamus. *Journal of Neurobiology* **24**:903–912.

Wilson MA and McNaughton BL (1994) Reactivation of hippocampal ensamble memories during sleep. *Science* **265**:676–679.

Winter P, Handley P, Ploog D, and Schott D (1974) Ontogeny of squirrel monkey calls under normal conditions and under acoustic isolation. *Behavior* **47**:230–239.

Woolley SMN and Rubel EW (2002) Vocal memory and learning in adult Bengalese Finches with regenerated hair cells. *Journal of Neuroscience* **22**:7774–7787.

Yu AC and Margoliash D (1996) Temporal hierarchical control of singing in birds. *Science* **273**:1871–1875.

Zysman D, Mendez JM, Pando B, Aliaga J, Goller F, and Mindlin GB (2005) Synthesizing bird song. *Physical Review E* **72**:051926.

Chapter 27

Population Dynamics in Auditory Cortex: Optical Imaging

Hubert R. Dinse and Junsei Horikawa

Abbreviations

AAF	anterior auditory field
AC	auditory cortex
AEG	anterior ectosylvian gyrus
AI	primary auditory cortex
AII	secondary auditory area
AMPA	a-amino-3-hydroxy-5-methyl-4-isoxazolepropionic acid
APV	2-amino-5-phosphono-valerate
CES	cochlear nerve electrical stimulation
CF	characteristic frequency
CI	cochlear implant
CN	cochlear nucleus
CNQX	6-cyano-7-nitroquinozaline-2,3-dione
D	dorsal field
DC	dorsocaudal field
DP	dorsoposterior field
EE	binaural excitation
EI	binaural inhibition
FM	frequency modulation
fMRI	functional magnetic resonance imaging
GABA	g-aminobutyric acid
ICA	independent component analysis
MEG	middle ectosylvian gyrus
MGB	medial geniculate body
NA	numerical aperture
NADH	dihydronicotinamide adenine dinucleotide
NMDA	*N*-methyl-D-aspartate
OI	optical imaging
P	posterior field
PAF	posterior auditory field
PEG	posterior ectosylvian gyrus
PET	positron emission tomography
PSF	point spread function
RF	receptive field
S	small field
SPL	sound pressure level
VAF	ventral auditory field
VAAF	ventral anterior auditory field
VC	ventro-caudal field
VCB	ventrocaudal belt field
VI	primary visual cortex
VM	ventromedial field
VR	ventrorostral field
VRB	ventrorostral belt field
VP	ventroposterior field
VSD	voltage-sensitive dye

H.R. Dinse (✉)
Department of Theoretical Biology, Institute for Neuroinformatics, Ruhr-University Bochum, 44780 Bochum, Germany
e-mail: hubert.dinse@neuroinformatik.ruhr-uni-bochum.de

1 Introduction

An important article, entitled "Single units and sensation: a neuron doctrine for perceptual psychology" proposed that "active high-level neurons directly and simply cause the elements of our perception" (Barlow 1972). This work articulated the conceptual framework at that time and had a great impact on research of sensory information processing. In the 1950s, single neuron recordings, the monitoring of extracellular potential changes, had become routine in the laboratory, boosting the conceptual framework of single cell analysis.

Rapid technical progress in recording technologies now allows for simultaneous recordings from up to 1,000 neurons (Nicolelis et al. 2003), enabling understanding and explaining higher brain processes in terms of dynamics of large populations of neurons (Ghazanfar et al. 2000; Carmena et al. 2005). "As in any good democracy, individual neurons count for little; it is population activity that matters" (Averbeck et al. 2006). It should be stressed, however, that the emphasis on distributed population activity instead that of a single cell does not imply underestimating the performance of single cells. Evidence is growing that axons, passive and excitable dendrites, and spines play a possibly underestimated role in signal transfer and processing (Segev and Rall 1998). Further, the role of single cells in signaling relevant

J.A. Winer, C.E. Schreiner (eds.), *The Auditory Cortex*,
DOI 10.1007/978-1-4419-0074-6_27,

behavioral information is still highly controversial (Quiroga et al. 2005; Houweling and Brecht 2008).

From a phenomenological perspective the use of optical imaging techniques are well suited to capture the response profile when a huge number of broadly tuned neurons are activated, even after the simplest form of sensory stimulation. This mass activity includes both spiking and suprathreshold activity. The widespread patterns of cortical activation evoked by point-like stimuli are referred to as cortical point spread function (PSF), which have been recorded in visual, auditory, and somatic sensory areas (Grinvald et al. 1994; Dinse and Schreiner 2002). It implies that, irrespective of the stimulus, populations of thousands of neurons are invoked. New non-invasive or semi-invasive techniques for recording neural activity or indirect markers of neural activity can measure equivalents of the cortical PSF, as is the case for PET (positron emission tomography), fMRI (functional magnetic resonance imaging), and optical imaging (OI) of intrinsic or dye-coupled signals.

An important constraint for OI data comes from the fact that in order to obtain activity that can be recorded optically the activation must be confined spatially and temporally. In a completely distributed network no optical signal could be recorded: what is required is a given locality and synchronicity in neural discharge.

2 Methodology for Optical Imaging

2.1 Sources of Intrinsic Signals

Optical imaging of intrinsic signals is based upon a close coupling between neural activation and metabolism. Techniques employed include recording of activity-dependent changes in cerebral blood volume or flow or oxygen saturation (Kety and Schmidt 1948; Lassen and Ingvar 1961; Sokoloff 1978; Fox et al. 1988). Neuronal activity produces at least three characteristic types of intrinsic optical changes in brain tissue that affect the intensity of light reflected from the active cortex. Light scattering signals (Hill and Keynes 1949) have multiple origins. A second intrinsic signal originates from changes in the absorption or fluorescence of the transition states of intrinsic chromophores such as hemoglobin, cytochromes, or NADH (Chance et al. 1962; Cohen 1973; Jöbsis and Rosenthal 1978). Of special interest is the well-studied transition from oxyhemoglobin to hemoglobin in response to increased electrical activity (oximetry). A third type of intrinsic signal originates from changes in blood volume affecting the overall light absorption by hemoglobin.

2.2 Wavelength and the Capillary System

The different sources of intrinsic signals can be separated and selected by choosing appropriate wavelength of illumination of the cortical surface (Fig. 27.1a,b). Changes of blood volume due to local capillary recruitment are observed at blue-green light (~450 nm). In contrast, changes of oxygen saturation of hemoglobin reflecting the hemoglobin–oxyhemoglobin ratio can be recorded at orange to red illumination, often at 610 nm. In addition, using near-infrared light, changes in light scattering can be monitored. Studies in the visual cortex revealed that largest signal amplitudes are seen with green light; however, blood vessel artifacts become smaller at longer wavelength. Optical imaging maps obtained in gerbil whisker cortex activity maps obtained at 577 and 605 nm with green light were much broader than those seen with red light, though their focus of activation was almost identical (Hess et al. 2000). For reasons that remain elusive, it is far more difficult to generate maps in auditory cortex using red light than in visual cortex. Most studies applying optical recording of intrinsic signals in auditory cortex used green light, and by that predominantly recorded blood volume changes.

The spatial resolution that limits intrinsic signal imaging depends greatly on the density and fine-structure of the underlying capillary system. Preparing capillary networks by corrosion cast methods reveals that the intrinsic signals associated with auditory cortex activation correlate with discrete capillary beds. The capillary beds in superficial cortical layers are distributed in a non-uniform fashion (Harrison et al. 2002). This study also provided evidence for small-scaled (<10 μm) flow control structures for both the arterial supply and the capillary network.

2.3 Fourier-Based Imaging Techniques

The main complication of recording intrinsic signals comes from the fact that the signal changes are very small, and that metabolically induced changes in the microcirculation are loaded with cardiovascular artifacts, which are all cyclic: heart beat, respiration, and vasomotor signals. A simple way to remove such artifacts is to record signal changes to a stimulus and to a blank condition (i.e., without stimulation), and then to subtract both signals. In this procedure, called episodic averaging, signal accumulation can take hours depending on the artifact. An elegant solution in fMRI research uses periodic stimulus presentations in combination with continuous image acquisition. This is also a useful procedure for optical recording (Kalatsky and Stryker 2003). First, continuous stimulation permits exhaustive coverage

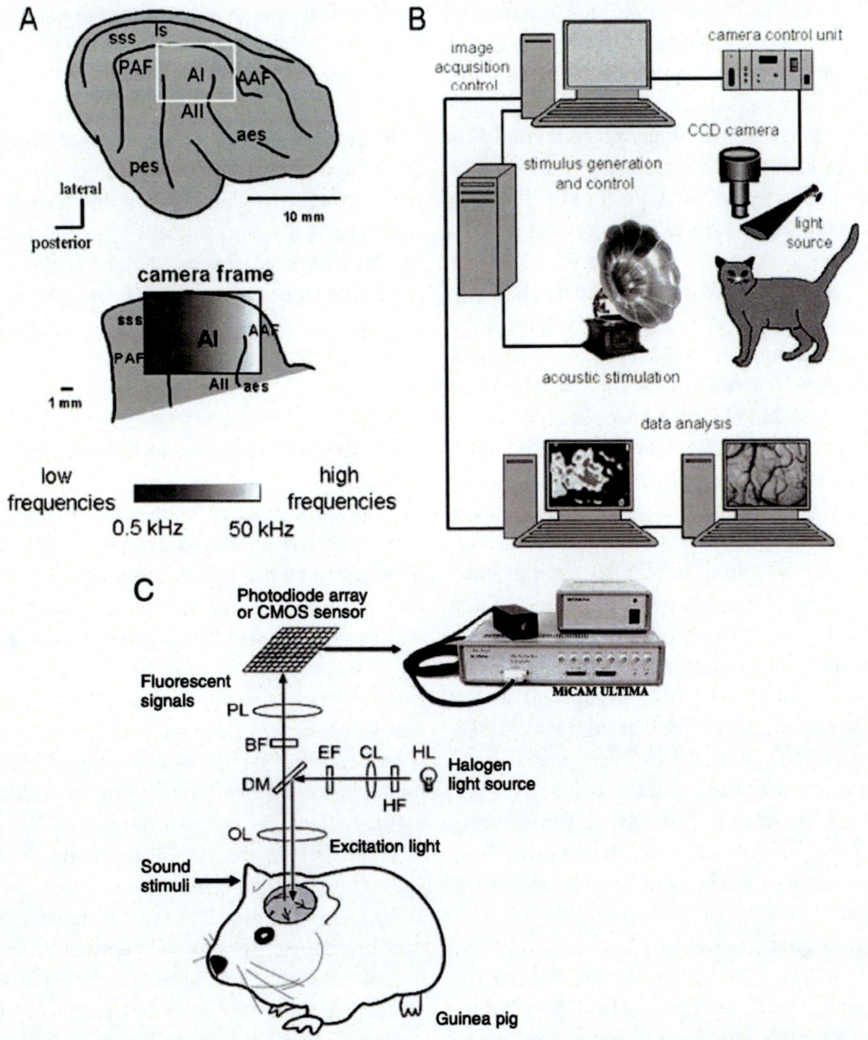

Fig. 27.1 **a** Cat auditory cortical fields (*top*). AAF, anterior auditory field; aes, anterior ectosylvian sulcus; AI, primary auditory cortex; AII, secondary auditory field; ls, lateral sulcus; PAF, posterior auditory field; pes, posterior ectosylvian sulcus sss, suprasylvian sulcus. Location of the frame usually used for optical imaging is indicated (*top* and *bottom*). Enlarged view with a schematic drawing of the rostrocaudal frequency gradient (*bottom*). Modified from the original (Imig and Reale 1980). **b** Schematic arrangement of the optical imaging set-up used for recording intrinsic signals. **c** VSD-OI methods. The auditory cortex stained by a VSD is epiilluminated by excitation light and fluorescent signals emitted from the cortex in response to sound stimuli are recorded by a photodiode array or a CMOS-sensor camera mounted on a microscope. BF, barrier filter; CL, condenser lens; DM, dichroic mirror; EF, excitation filter; HF, heat filter; HL, halogen lamp; OL, object lens; PL, projection lens

of stimulus space: every neuron is exposed to its optimal value of a stimulus parameter that is varied continuously. Second, periodic stimulation allows discrimination of the stimulus-evoked responses from intrinsic noise with Fourier analysis of the continuous data stream. This technique allows reconstruction of functional maps of much higher spatial resolution and lower noise than those obtained by conventional optical imaging methods (Kalatsky and Stryker 2003) and is now widely used in auditory and visual cortex studies (Mrsic-Flogel et al. 2003; Nelken et al. 2004, 2008; Cang et al. 2005; Kalatsky et al. 2005).

2.4 Optical Imaging Using Voltage-Sensitive Dyes

2.4.1 In Vivo and In Vitro Optical Imaging

Optical imaging using voltage-sensitive dyes (VSD) has been developed after the discovery of stimulus-dependent optical signals from stained squid giant axons (Tasaki et al. 1968). This method enabled recording electrical activity of neuron populations optically. Extensive screening of VSD

(Cohen et al. 1974; Cohen and Lesher 1986) led to a discovery of dye groups (merocyanine, oxonol, and styril) having measurable ΔF/*F* ratios (on the order of 10^{-3}) of light absorption or fluorescence emission (ΔF is a change in light intensity caused by change in membrane potential and *F* is the background light intensity). Simultaneous recording of intracellular membrane potentials and VSD optical signals showed that the change of the optical intensity was proportional to the membrane potential in the range of ±100 mV (Ross et al. 1977; Cohen and Lesher 1986). The mechanisms of the voltage dependency of VSD signals are thought to be the conformational changes or displacements of electrons caused by the membrane potential in the light-absorbing or light-emitting radicals of the VSD that attaches on or in the cell membrane (Waggoner 1979; Wolf and Waggoner 1986).

Voltage-sensitive dye optical imaging (VSD-OI) using two-dimensional sensors has been used in various nervous systems in both in vivo and in vitro conditions. In vivo recording can reveal neural activity to natural stimuli and the in vitro recording can reveal neural circuitry in the deeper structures of the brain. In vivo VSD-OI was first conducted on the auditory cortex of guinea pigs with a stylil dye (RH795, fluorescent, fast responsive) and a 12 × 12 pixel photodiode array (Fukunishi et al. 1992; Taniguchi et al. 1992). In vivo VSD-OI was also conducted in the hamster cochlear nucleus (di-2-ANEPEC (stylil), 96 × 64 CCD camera, MiCAM01, Brainvision) (Kaltenbach and Zhang 2004). In vitro auditory VSD-OI was done in chick embryo brain stem slice preparations (NK2761; merocyanine, absorption, fast responsive, 34 × 32 photodiode array, Hamamatsu photonics) (Asako et al. 1999; Sato and Momose-Sat 2003), rat auditory cortex (RH795, di-2-ANEPEC, 128 x 128 MOS-sensor camera, HR deltaron, Fuji Photo Film, NK3630, 464 pixel photodiode array, Neuroplex, Redshirt-Imaging) (Kubota et al. 1997, 1999; Wu et al. 2001) and in the superior olivary nuclei (RH795, Neuroplex) (Srinivasan et al. 2004). In these recordings, the time resolution was sufficiently fast (0.5–1.4 ms/frame) to record postsynaptic membrane potentials but insufficient to record action potentials. The recording area per pixel was 19–250 μm^2, in which summated membrane potentials from several-to-thousands neurons were recorded (Fig. 27.1c).

2.4.2 Depth Dependency of the Optical Signal and Spatial Resolution

The intensity and spatial resolution of optical signals depend on the depth of the signal source in the tissue due to absorption and scattering of light. The effects of focus and scattering on the distribution of light on the sensor array have been measured (Saltzberg et al. 1977; Orback and Cohen 1983). Optical intensity decreased to 50% of the original by moving 150 μm out of focus for 0.6 numerical aperture (NA) objective or 300 μm for 0.4 NA objective in salamander cortex slice preparations (Cohen and Lesher 1986). Spatial resolution also decreased with the depth. A 40-μm light spot (750 nm) spreads to ~200 μm after insertion of a 500-μm thick slice of olfactory bulb into the light path or by moving 500 μm out of focus (Orback and Cohen 1983). A 30-μm light spot (705 nm) spreads to a 50-μm spot after insertion of a *Navanax* buccal ganglia preparation into the light path, when measured at more than 50% intensity. The spread increased to about 100 μm with 510 nm light (Cohen and Lesher 1986; London et al. 1986).

2.4.3 Recording Layers and the Signal Source

The optical signals of the in vivo auditory cortex (Taniguchi et al. 1992; Horikawa et al. 1996; Song et al. 2006) were assumed to be the signals from layer II/III neurons because the microscope of the recording equipment was focused at the depths of 200–300 μm from the pia and optical signals from the deeper layers IV–VI were much attenuated by the tissue. This is also supported by the result that the fluorescent intensity was maximal just beneath the border between layer I and II of in vivo stained cortical slices and it declined rapidly toward deeper layers (Song et al. 2006). Layer I activity may affect the recorded signals, but it is smaller than that in layer II/III in the slice preparation (Kubota et al. 1997, 1999; Song et al. 2006).

VSD optical signal intensity depends on the neuronal membrane area. Because the membrane area of the dendrites is estimated at 1,000 times that of cell somata, the optical signals are thought to represent primarily dendritic potential changes than those in the cell somata (Grinvald et al. 1994).

2.4.4 Noise Canceling

In the in vivo optical recording, canceling of noise from brain movements caused by pulsation and respiration is necessary, whereas it is not in the in vitro optical recording, although it is necessary for both recordings to correct the signal amplitude from signal decays by breaching of the dye and its toxic effect on the neurons during the several-hour long recordings (Grinvald et al. 1986). In the in vivo recording, the respiratory noise is avoided by halting the artificial ventilation under anesthesia and paralysis during each recording for 1–5 s, which is brief enough to cause no significant effect on the brain. Cardiac noise canceling is performed by subtraction between the recordings with and without stimulation synchronized with the heart beat. Independent component analysis (ICA) can remove respiratory and pulsation noise in the optical signals recorded without halting artificial ventilation and heart beat synchronization (Maeda et al. 2001; Inagaki et al. 2003) and it successfully filters noise components from the signals.

3 Analysis of Auditory Representations by Optical Imaging

3.1 Frequency Representations

An early account of OI of guinea pig primary auditory cortex (AI) using suprathreshold pure tones revealed very broad and patchy activation pattern (Bakin et al. 1996). Optical imaging of intrinsic signals largely confirmed the topography of isofrequency maps from electrophysiological studies in cats (Dinse et al. 1997; Spitzer et al. 2001; Tsytsarev et al. 2004; Ojima et al. 2005), rats (Kalatsky et al. 2005), ferrets (Versnel et al. 2002; Nelken et al. 2004), chinchillas (Harrison et al. 1998, 2000; Harel et al. 2000), and gerbils (Hess and Scheich 1996) by demonstrating broad activation pattern with large spatial overlap to single tone bursts. A study in cat AI with tone bursts of 1–24 kHz (40 dB SPL) revealed regions of reflectance changes (delta R) with an average octave separation of ~1.5 mm (Fig. 27.2) (Dinse et al. 2000). The area of the two-dimensional signal distribution increased fairly linearly with amplitude. At 75% of the maximal reflectance changes an average cortical territory of 1.2–3.5 mm^2 was found, and this activated area increased to 7.7–15.8 mm^2 at 50%. From this, a one-dimensional mean space constant can be calculated according to which a 10% delta R results in a spatial spread of 400–450 μm (Dinse et al. 1997). These studies show that, with green light, a topographic distribution of activity maps coincides with preferred frequency tuning in parallel single unit recordings.

VSD-OI visualized the tonotopic organization in the AI and dorsocaudal (DC) field (Fukunishi et al. 1992; Taniguchi et al. 1992; Uno et al. 1993; Horikawa et al. 2001, 2006a; Song et al. 2006) and ventro-rostral (VR) and ventro-caudal (VC) fields (Nishimura et al. 2007) of guinea pig auditory cortex and in hamster cochlear nucleus (CN) (Kaltenbach and Zhang 2004) (Fig. 27.3). The tonotopic organization shown by VSD-OI resembled that seen in microelectrode studies: the center of the activated area shifted along the tonotopic axis as did the characteristic frequencies (CF), the frequency at the lowest threshold measured by microelectrodes. The activated cortical area in the cortex was a spot or band ~2–3 mm along and 0.5–1.5 mm across the isofrequency contour. The activated areas to the adjacent frequencies overlapped (Fig. 27.3).

The breadth of the activation pattern in VSD-OI results from various factors such as neuron population activation at sub- and suprathreshold stimulus levels, polysynaptic spread of activity, broad single neuron dendritic arborizations (over 300–500 μm diameters), and signal blurring by the tissue. As described below (Section 6), the polysynaptic spread of

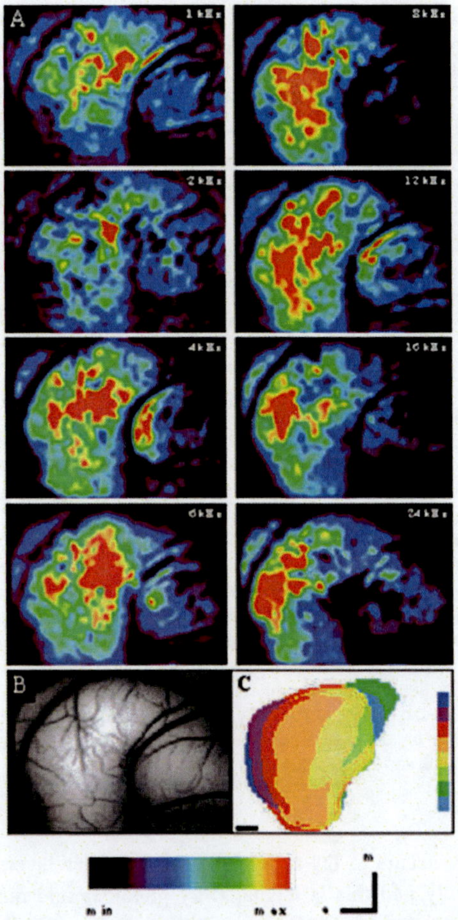

Fig. 27.2 Isofrequency domains in cat auditory cortex as revealed by optical imaging of intrinsic signals. **a** Single condition response maps for stimulation with pure tone bursts of 1–24 kHz at 40 dB sound pressure level as indicated on *bottom left*. Warm colors (see *color plate section*) show regions of reflectance changes, indicating enhanced cortical activation. Each single condition map was individually scaled to its maximal reflectance change. The full *color scale* corresponds to fractional reflectance changes of maximal 4.9 × 10^{-2}. There is a gradual and coherent shift of the areas of reflectance changes from caudal to rostral when stimulation frequency is increased. The single condition maps were computed by subtracting a stimulus from a non-stimulus condition and are the average of 12 trials. **b** Frequency composite map calculated from the single condition maps (**a**). For each pixel comprising the optical map, the frequency preference is color coded according to the *color bar* (*right-hand side*). The composite frequency map shows a smooth and highly ordered representation of frequencies in AI, while AII (at the most ventral aspects of the mapped area) has no comparable consistent topography of frequency representation. Average octave separation in AI is ~1.5 mm. **c** Image of the cortical surface (left hemisphere) shows sulci and pattern of blood vessels. aes, anterior ectosylvian sulcus; AI primary auditory cortex, AII secondary auditory cortex, P, posterior auditory field; pes, posterior ectosylvian sulcus; sss suprasylvian sulcus,. Scale bar is 1 mm. From the original source (Dinse et al. 2000)

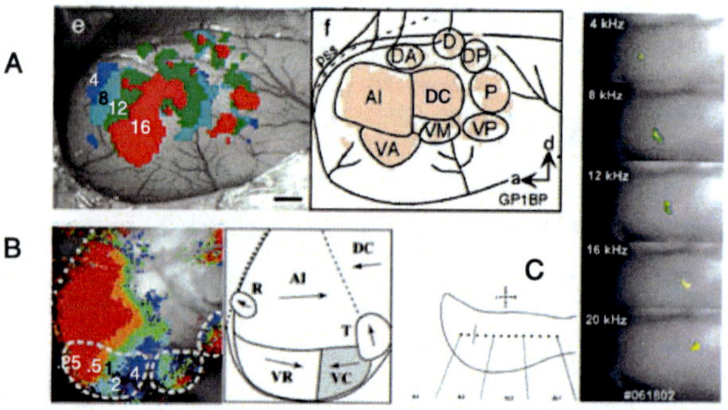

Fig. 27.3 Tonotopic organizations of the guinea pig auditory cortex (**a**, **b**) and the cochlear nucleus of the golden hamster (**c**) revealed by VSD-OI. A *Left panel*, superimposed areas responding to pure tones at 4 (*blue*), 8 (*light blue*), 12 (*green*) and 16 kHz (*red*). *Right panel* shows schematic drawing of AI, DC, and surrounding belt fields. Modified with permission from the original source (Horikawa et al. 2001, Fig. 2). B *Left panel*, tonotopic organization in areas AI, VR, and VC shown by superimposition of responses to 0.25 (*red*), 0.5 (*orange*), 1 (*green*), 2 (*light blue*), and 4 kHz (*blue*) 14 ms after the shortest latency in AI. *Dashed lines*, borders between AI, VR, VC, and T. *Right panel*, schematic drawing of the fields and the direction of tonotopy (*arrows*). Modified from the original with permission (Nishimura et al. 2007, Figs. 1 and 3). C Tonotopic organization of the cochlear nucleus is shown by the response epicenters to tone bursts at 4, 8, 12, 16, and 20 kHz on the images of the DCN. Schematic drawing of the cochlear nucleus is on the *left*. Modified from the original with permission (Kaltenbach and Zhang 2004, Figs. 3 and 6)

activity occurs in the supra- and infragranular layers independently and this is mediated by glutamatergic receptors and regulated by GABAergic inhibition. The GABAergic inhibition greatly attenuates the breadth of the activation, without which it spreads over the auditory cortex.

Both OI approaches converge on the observation of broad activation pattern resulting in substantial representational overlap. Large overlap has also been observed by using intrinsic optical imaging in somatic sensory cortex (Godde et al. 1995) and has therefore been proposed as a general principle of topographic maps (see Section 8.1).

3.2 Multiple Auditory Areas

A general feature of mammalian sensory cortex is multiple representational maps. While this applies to all modalities, the most detailed information is available for primary, first-order cortical maps, and less is known for higher order maps which are smaller in overall size and have less well-defined receptive fields and areal borders. Thus, little is known about their functional role, and in some species there is still uncertainty about the number of areas. Optical imaging provides a unique means to assess simultaneously activity pattern that can provide valuable information to delineate different areas.

Using standard OI of intrinsic signals, in the chinchilla temporal cortex three auditory areas are seen (Harel et al. 2000). The primary auditory cortex (AI), and secondary auditory areas AII and the anterior auditory field (AAF) were identified on the basis of onset-response latencies obtained in parallel single-unit recordings. Using Fourier imaging, a large-scale tonotopic organization of ferret auditory cortex was found. Auditory cortex consists of a low-frequency area at the center of the middle ectosylvian gyrus (MEG), with areas of higher frequency sensitivity surrounding it. In this study, there was no frequency reversal at the AI/AAF border, suggesting that they are continuous. Besides AI/AAF, as-yet uncharacterized, high-frequency areas on anterior (AEG) and posterior ectosylvian gyrus (PEG) may be homologous to higher order auditory fields in cats and other species (Nelken et al. 2004). Frequency reversals in the imaging data suggest that any tonotopic order, if present at all, is less precise than that in AI/AAF.

In rat auditory cortex, at least five fields were revealed by high-resolution optical imaging of intrinsic signals and confirmed by electrophysiology (Fig. 27.4) (Kalatsky et al. 2005): dorsal auditory field or primary auditory cortex (AI), anterior auditory field (AAF), ventral auditory field (VAF), and ventral anterior auditory field (VAAF). So-called belt fields such as posterior-dorsal area in the imaging data appeared as weak non-tonotopic halos on the outer perimeter of the main fields (Fig. 27.4). It was suggested that acoustically responsive cortex may extend further ventrally and that the activation ventral and rostral to VAAF may reveal other auditory fields or may constitute ventral extensions of it. The average cortical linear magnification factors in the central regions of the tonotopic fields perpendicular to isofrequency lines differed between fields (AI, 1.8 octave/mm; AAF, 2.3 octave/mm; VAF, 2.2 octave/mm; and VAAF, 2.9 octave/mm) reflecting areal distinctions in the size and resolution of frequency maps.

An advantage of VSD to explore multiple fields is that the real-time dynamics of cortical activation can be used to assess response latencies. Typically, the time needed to reach a certain processing stage serves as a marker of representational hierarchy, with longer latencies indicating a higher position in the processing stream. In the guinea pig auditory cortex, the core fields AI and DC are surrounded by fields consisting of small (S) dorsocaudal belt (DCB), ventrocaudal belt (VCB), and ventrorostral belt (VRB) fields (Redies et al. 1989; Wallace et al. 2000). In vivo VSD-OI analyses of frequency representation, response latency, and response duration further divided these fields: DCB to D and DP, VRB to P and VP, and VRB to VM, VA, and V (Horikawa et al. 2001). Further tonotopically organized areas (VR and VC) are ventroanterior to guinea pig AI (Nishimura et al. 2007). VA is disputed and it may be included in ventral AI. VSD-OI visualized the sequence

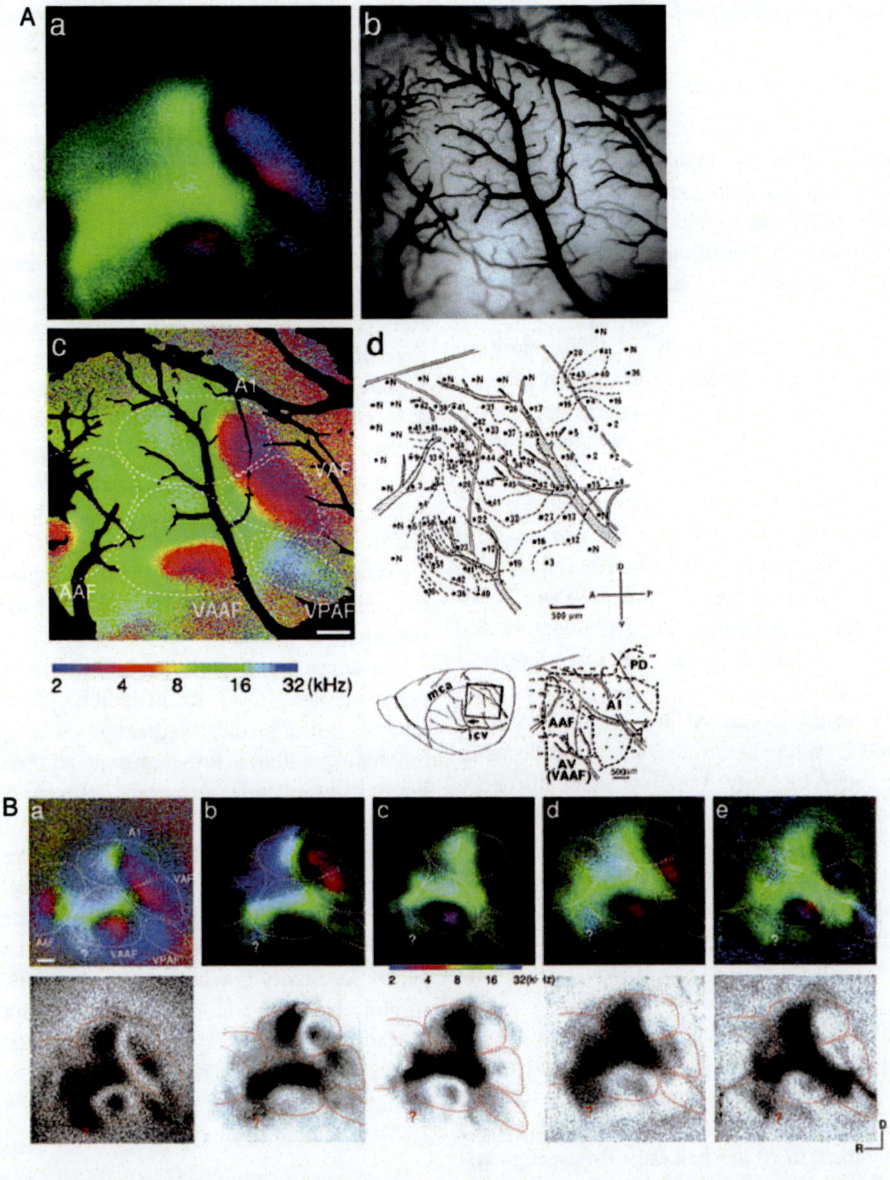

Fig. 27.4 A Maps of multiple topographic areas in rat auditory cortex. (*a*) Polar map (hue encodes response phase and saturation encodes response magnitude) of absolute tonotopy. The hemodynamic delay was removed by the method of the stimulus reversal. (*b*) Surface vessel pattern from the same animal. (*c*) The map in (*a*) plotted as a phase map with overlaid vessel pattern from (*b*). The contours outline identified auditory fields. The *dotted lines*, drawn at ≈16 kHz, outline the high-frequency region. The large uniform region (*green*) has a fine spectral structure that is emphasized on the map of double phase (Fig. g). (*d*) Frequency representation of the rat auditory cortex constructed by using standard electrophysiological mapping technique. A, anterior; D, dorsal; P, posterior; V, ventral; VPAF, ventral posterior auditory field. (Scale bar, 500 μm.). Modified from the original with permission (Kalatsky et al. 2005). (**B**) Stability of multiple fields. Optical polar and magnitude maps, respectively, from five rats 25, 50, 55 (male), 60, and 62 days old are shown. The contours were drawn (Fig. 27.1c), then transferred without distortion to these figures, then translated and rotated until a match was achieved. The fields identified on Fig. 1c for P75 rat almost perfectly match the optically defined fields for all subjects. VPAF, ventral posterior auditory field; "?", possible additional auditory field

of the activation from the core fields to the belt fields (Horikawa et al. 2001, 2006a) and found a functional hierarchy from the core to belt fields in the sound processing stream. However, functional differences among the multiple auditory areas have not been elucidated. Only one VSD-OI study revealed evidence that the caudal belt fields may relate to binaural processing (Hosokawa et al. 2004; see Section 5.3).

4 Beyond Frequency Representation

4.1 Intensity Representation

Auditory cortical neurons are highly sensitive to sound intensity and have complex monotonic and non-monotonic response characteristics. Because frequency maps are in principle one-dimensional, a prevailing question is the systematic representation of intensity along the second axis of the tonotopic maps.

Intrinsic signals evoked in cat AI by a 10-kHz pure-tone stimulus at 40, 50, 55, and 60 dB SPL show a steady increase of the magnitude of intrinsic signal with sound pressure level. Due to the weak stimulus-related intrinsic signal for <40 dB SPL, threshold could not be determined directly, but interpolation suggests a minimum sound pressure level of $\sim$21.5 dB SPL, near the average threshold for the generation of spike activities (Tsytsarev et al. 2004). Similar results were reported for cat AI with tone bursts between 40 and 70 dB SPL. Zones of elongated activation along the dorsoventral axis did not change in size or position at supra-threshold levels (Ojima et al. 2004).

In contrast, a study in cat AI for sound intensities from 10–60 dB SPL reported expanded areas of reflectance with increasing sound intensity. Low intensities elicited a dorsoventrally elongated zone of activation and, at higher levels, a further expansion along the rostrocaudal axis. A reduction in response area indicated global non-monotonic behavior at very high intensities (Dinse et al. 2000). Increasing intensity leads to a complex activity pattern that includes areal recruitment, shifts, and contraction of total activated areas.

A study of intensity coding using VSD-OI in guinea pig AI also found an increase in the activated area and a dorsal shift of the activated area's center at higher intensities (Taniguchi and Nasu 1993). Subsequent VSD-OI studies found no intensity maps in AI and belt areas. Microelectrode studies find cells with different threshold distributed broadly in auditory cortex (Phillips et al. 1994; Linden and Schreiner 2003) and >50% of cat and monkey AI neurons and 98% of cat PAF cells have non-monotonic intensity functions with different best intensities (the value eliciting the maximum response). The VSD-OI results indicate that auditory cortex population activity does not reflect local clustering of neurons with different thresholds or preferred intensity.

4.2 Periodicity Representation

Harmonic sounds, such as voiced speech sounds and many animal communication signals, have a pitch related to their envelope periodicities. While frequency information is extracted by cochlear mechanical filtering, periodicity information is analyzed by brain stem temporal filter mechanisms. Mammalian auditory midbrain envelope periodicity is represented in maps orthogonal to sound frequency representation. However, how periodicity is represented within primary auditory cortex remains controversial.

Optical recording of intrinsic signals found evidence of a periodicity map in cat AI (Langner et al. 2009). While pure tone stimulation replicated the typical rostrocaudal AI frequency gradient, harmonic sound stimulation showed segregated bands of activation, indicating spatially localized preferences for specific periodicities along a dorsoventral axis, nearly orthogonal to the tonotopic gradient. Analysis of the response locations found an average gradient of $100 \pm 10°$ for the periodotopic, and $12 \pm 18°$ for the tonotopic map, with a mean angle difference of 88°. The gradients (0.65 ± 0.08 mm/octave for periodotopy and 1.07 ± 0.16 mm/octave for tonotopy) indicate that more cortical territory is devoted to the octave representation along the tonotopic than the periodotopic gradient.

An unusual periodicity representation seen in gerbil primary auditory cortex are periodicity maps with a circular, horseshoe-like gradient superimposed on the typical, linear tonotopic gradient in low-frequency AI (Schulze et al. 2002). Ferret auditory cortex responses collected during continuous stimulation by sound sequences differing in spectral structure had the same periodicity, and, therefore, evoke the same pitch percept (click trains, sinusoidally amplitude-modulated tones, and iterated ripple noise). These stimuli revealed no periodotopic map across the imaged auditory fields. Rather, period sensitivity gradients differed for the various periodic stimuli (Nelken et al. 2008). There is some evidence for human auditory cortex periodotopic organization (Langner et al. 1997; Hall et al. 2006).

4.3 Binaural Organization

Electrophysiological studies in different species find similar binaural properties clustered across AI that show binaural excitation (EE) or binaural inhibition (EI). It has been suggested that AI has functional maps of binaurality similar to visual cortex ocular dominance maps.

Studies in rat AI show temporal differences in the intrinsic signal after contralateral, ipsilateral, and diotic broadband noise burst presentations, but provided no spatial difference maps (Tsytsarev and Tanaka 2002). Imaging of the developing ferret AI showed that in adults contralateral stimuli evoked stronger responses and activated a larger AI regions than ipsilateral stimuli (Mrsic-Flogel et al. 2006). This was confirmed in ferret AI with binaural interactions to contralateral or ipsilateral single or sequences of broadband noise bursts presented in opposite phase to both ears. Contralateral responses were larger and ipsilateral responses smallest, but the area activated was large and comparable in all configurations. Thus, imaged ferret primary and nonprimary areas do not appear to contain topographic maps of simple binaural properties (Nelken et al. 2008).

Responses in guinea-pig AI to monaural and binaural stimulation presented to each ear separately with earphones were mapped with VSD optical imaging (Hosokawa et al. 1997, 1999). Simultaneous ipsilateral and contralateral stimulation caused inhibition of the AI responses in a wide area of AI with no patch- or band-like structures visible, unlike what has been seen in guinea pig AI with microelectrodes (Rutkowski et al. 2000). This could reflect species and methodological differences between the VSD-OI and the microelectrode recording. No VSD-OI study on the segregation of binaural characteristics has been performed in the cat AI because of problems with staining the cortex due to the cat's blood vessel-rich and relatively thick arachnoid membrane.

The guinea pig caudal belt fields responded with stronger inhibition to ipsilateral stimulation and showed more sensitivity to interaural intensity differences than AI and DC (Hosokawa et al. 2004). This suggests a functional augmentation of binaural processes in higher auditory fields and may support the hypothesis of dual streams of auditory cortex information processing: caudal pathways process sound source and ventral and rostral pathways perform sound object analysis, as in monkey auditory cortex microelectrode studies (Rauschecker and Tian 2000).

4.4 Complex Sounds and Animal Vocalization

The cortical distribution of pure tone responses has a tonotopic organization in several fields. How, then, are dynamic, frequency-modulated (FM) sounds represented within this map? VSD-OI visualized spatiotemporal responses to FM sounds in guinea pig AI auditory cortex (Horikawa et al. 1998, 2006a; Yamaguchi et al. 2001; Sugimoto et al. 2008). The response appeared first at the frequency band of the initial FM sound resembling the response to pure tones, and then a focal activity spot, rather than a band, traversed the tonotopic axis corresponding to the instantaneous frequency of the FM sound. The activity profile traversing the tonotopic axis formed a circumscribed spot, implying neural mechanisms of spatiotemporal inhibition that prevents activation of the whole isofrequency band. For more complex FM sound stimuli, containing 4–16 and 8–32 kHz components swept for 300 ms, AI responses resembled the single-component FM sounds except that focal activity in response to the instantaneous frequency of the upper FM component traversed the isofrequency bands, inhibiting the lower FM response component (Yamaguchi et al. 2001; Horikawa et al. 2006a). The latter component appeared after the inhibition by the upper FM component was reduced.

The mechanisms of spatiotemporal inhibition in the responses to the simple and complex FM sounds resemble those for two-tone sounds with simultaneous and nonsimultaneous masker-probe tone pair. This elicited a localized activity spot in the isofrequency band of the probe frequency in AI. This spatial focusing of the response was induced by inhibition from the masker tone (Sugimoto et al. 2002) and was a function of probe delay and frequency distance between the components (Horikawa et al. 1997; Sugimoto et al. 2002). Spatial focusing was not observed for the simultaneous probes but for probes delayed 15–20 ms from the masker, a value that is consistent with the minimum value of the human onset asynchrony discrimination of two tones (Sugimoto et al. 2002). The AI spatial focusing may contribute to onset asynchrony discrimination by contrasting the neural responses to the complex sound frequency components and may contribute to sound grouping and segregation.

AI responses to guinea pig vocalizations, which have many temporally modulated spectral components, appear as complex, dynamic activity patterns varying spatially and temporally (Horikawa et al. 2006a). The activity in the frequency strips corresponded to prominent instantaneous sound frequency components with spatiotemporal mutual inhibition between frequency strips. Although too complex to be analyzed completely, the spatiotemporal inhibitory activity for the complex tones and FM sounds may also pertain to the responses to animal vocalizations.

4.5 Auditory Cortex Plasticity

4.5.1 Cochlear Implants and Cochlear Implant Plasticity

Cochlear prostheses have become successful tools to restore sound perception in patients with profound sensorineural deafness by electrically stimulating acoustic nerve fibers; the central auditory system can be activated systematically to provide a meaningful hearing experience. While patient-related questions require studies in humans, the issue of how electrical stimulation affects the auditory cortex is salient. Animal models differing in the time of deafening and stimulation onset and duration can provide further insights.

Cochlear implants (CI) provide sound perception to adults and increasingly in prelingual children with profound sensorineural deafness (Zeng et al. 2004). How chronic electrical stimulation of the cochlea affects the developing auditory cortex is pertinent. Optical imaging of intrinsic signals in cat AI evoked by CI stimulation using human multichannel electrodes was mapped (Dinse et al. 2003). In neonatally deafened adult animals, acoustic deprivation severely distorted cochleotopic maps and elevated thresholds (Fig. 27.5). Three months of CI stimulation, using the continuous interleaved sampling approach, conserved responsiveness greatly. Most importantly, the CI stimulation did not restore the normal adult status, but elicited a new topographical organization with large, joint representations of all stimulated electrode sites (Fig. 27.5). It was suggested that the effective CI stimulation might rely primarily on re-learning input pattern arising from artificial sensory inputs via electrical stimulation, thus supporting the importance of learning and training for understanding CI stimulation effects. Furthermore, the ability for acquiring/restoring speech comprehension mediated by CI stimulation might reflect cortical processing strategies to interpret new peripheral patterns. These strategies may emerge from adaptational response capacities to the constraints inherent in the new input statistics that result from the employed stimulation strategy. Conceivably, such reorganizational changes could mediate the variable improvement in open speech comprehension with practice in young CI patients.

Modern CI prostheses have 20–22 channel electrode pairs for cochlear nerve electrical stimulation (CES) and require sound spectrum preprocessing and setting appropriate stimulus intensity for each channel (Zeng et al. 2004). To study the CES parameters, population intensity-response functions were investigated with VSD-OI in the AI. The cortical response to the CES has shorter latency and increases and saturates much faster (has a narrower dynamic range) than that to sound stimulation (Raggio and Schreiner 1994).

VSD-OI showed that the CES at two sites elicited two local responses in guinea pig contralateral AI, at the same distance along the tonotopic axis as those to two pure tones with frequency position estimated from the interval between the cochlear stimulation electrodes (Taniguchi et al. 1997). The responses to CES form a band along the isofrequency contours resembling pure tone stimulation.

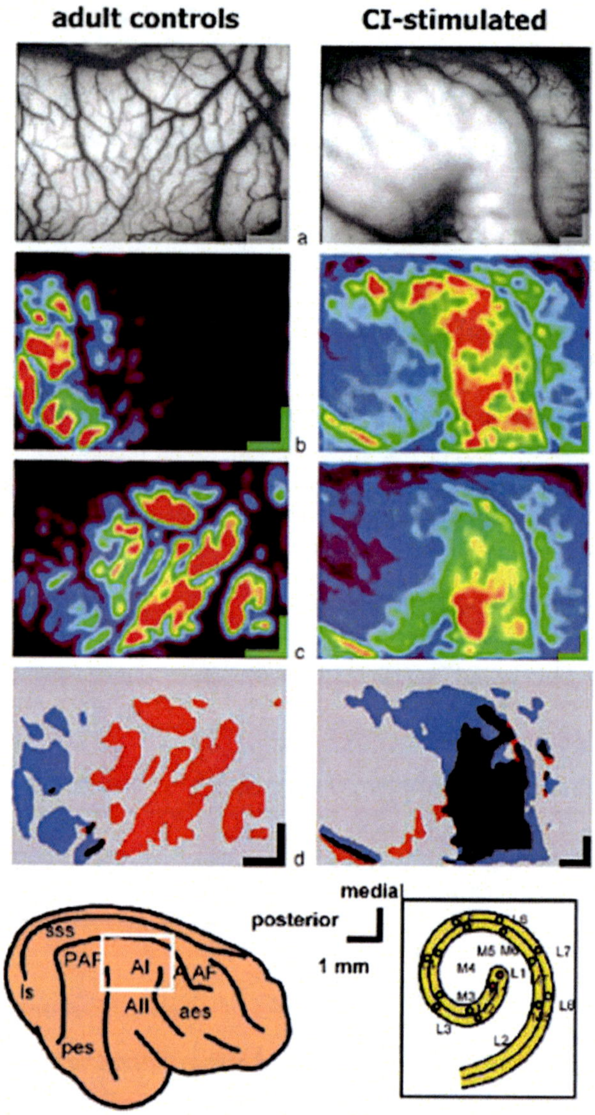

Fig. 27.5 Spatial distribution of reflectance changes of optically recorded intrinsic signals corresponding to neural activity maps recorded in cat AI in response to CI stimulation. Warm colors, regions of maximal reflectance changes, indicating enhanced cortical activation. Each single condition map was individually scaled to its maximal values. Scale bar is 1 mm. Posterior is left, medial up. **a** Image of the exposed brain surface. Location used for optical imaging (*white frame*) is indicated in the schematic drawing of a cat brain to illustrate the layout and parcellation of cat auditory cortical fields (*bottom*). AAF, anterior auditory field; aes, anterior ectosylvian sulcus; AI, primary auditory cortex; AII, secondary auditory field; 1 s, lateral sulcus; PAF, posterior auditory field; pes, posterior ectosylvian sulcus sss, suprasylvian sulcus;. *Left column*: Examples from an adult cat with normal hearing experience that was implanted directly prior to the imaging experiment. Electrical stimulation of electrode pairs 11–1 m (**b**) and 21–2 m (**c**) with 600 lA. The full *color scale* corresponds to fractional reflectance changes of 0.9 and 0.8%. *Right column*: examples from an adult cat deafened neonatally and imaged after 3 months of CI stimulation. Electrical stimulation of electrode pairs 11–1 m (**b**) and 21–2 m (**c**) with 50 lA. The full *color scale* corresponds to fractional reflectance

Fig. 27.5 (continued) changes of 5.4 and 2.4%. **d** Composite maps calculated from the single condition maps in (**b**) and (**c**) displaying selectivity for a preferred electrode pair, which is *color coded* (*blue* indicating electrodes 11–1 m, *red* 21–2 m electrodes). *Black marks* overlap. The computation was performed for activation at the 50% level of maximal signal amplitude. *Bottom right*, a schematic drawing of the multichannel electrode indicating the electrode sites being stimulated (11–1 m in (**b**), and 21–2 m in (**c** Modified from the original with permission (Dinse et al. 2003)

Electrophysiological studies on auditory cortex frequency representation plasticity in animals found that, after local laser spot lesions of cochlear hair cells, the cortical frequency zone affected by the lesion shrank and the areas responding to the adjacent frequencies expanded into the deprived frequency area (Robertson and Irvine 1989). A VSD-OI study of guinea pig auditory cortex tonotopic reorganization after monaural and binaural deafening by cochlear kanamycin injection found that the distance between the AI zones responding to the electrical stimulation of the first and second cochlear turns was 25–33% of the controls after 2 months, indicating post-deprivation changes in AI tonotopic organization (Horikawa et al. 2000).

4.5.2 Cross-Modal Plasticity and Rewiring in AI

The mammalian cortex contains discrete areas that receive, process, and transmit neural signals along functional pathways to form a system of complex networks assembled developmentally and refined connectionally by maturation. To explore cortical pathway formation, maintenance, and experience-dependent plasticity, a model explores the rewiring of visual input to the auditory thalamus and subsequent auditory cortex remodeling to process visual information (Sur and Leamey 2001; Horng and Sur 2006). Primary visual cortex contains groups of neurons with a preferred stimulus orientation that form a orientation module preference map.

Ferret auditory thalamic deafferentation at birth induces retinal axons to innervate the medial geniculate body (MGB) (Angelucci et al. 1997). Retinal input is then relayed via the MGB to primary auditory cortex, which develops with a different pattern of input activity than normal AI. A map of visual space arises in rewired AI, and visually driven cells are orientation selective. OI evidence suggests that the rewired cortical cells form an orientation map with well-defined orientation preference maps and iso-orientation regions whose cells share an orientation preference, and singularities, where adjacent stimulus orientations form pinwheels (Fig. 27.6) (Sharma et al. 2000). Although AI and primary visual cortex (VI) maps both contain a pinwheel organization, the density of rewired AI pinwheel centers was significantly lower, as was preferred the spatial frequency. These data document the profound influence of afferent activity on cortical circuitry, including thalamocortical and local intracortical connections involved in the generation of orientation preference maps.

Cross-modal adult activation plays important roles in sensory integration and multimodal processing. Asymmetric plastic changes in the functional connections between rat auditory cortex (area 41) and the visual cortex (area 18a) slices are seen using Ca^{2+} imaging with rhod-2 and field potential recording techniques. Layer V postsynaptic field potentials were evoked auditory cortex after stimulation of the visual cortex and the reverse. Ca^{2+} imaging showed that alternate electrical stimulation of layer V of the auditory and visual cortex at 1-min intervals under bicuculline (a g-aminobutyric acid $(GABA)_A$ receptor antagonist) potentiated a layer V Ca^{2+} increase after more than five times stimulation of the visual cortex. Such potentiation did not occur in visual cortex layer V after auditory cortex stimulation. The results indicate asymmetric cross-modal activity-dependent auditory cortex plasticity is influenced by visual information (Hishida et al. 2003).

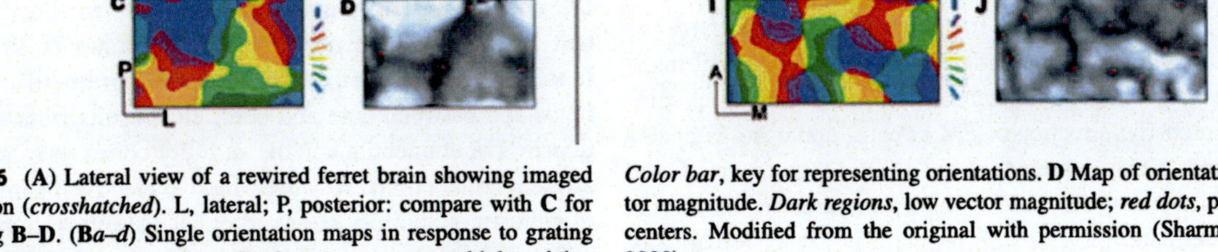

Fig. 27.6 (A) Lateral view of a rewired ferret brain showing imaged A1 region (*crosshatched*). L, lateral; P, posterior: compare with **C** for orienting **B–D**. (**Ba–d**) Single orientation maps in response to grating stimuli of different orientations. *Dark regions* represent high activity. Scale bar, 0.5 mm for **B–D**. **C** Composite map of orientation preference. *Color bar*, key for representing orientations. **D** Map of orientation vector magnitude. *Dark regions*, low vector magnitude; *red dots*, pinwheel centers. Modified from the original with permission (Sharma et al. 2000)

5 Primary Auditory Cortex Organization: Evidence for Multiple Functional Maps?

Early studies on sensory cortex functional organization using OI of intrinsic signals were in cat and monkey visual cortex. While single cell orientation selectivity was known (Hubel and Wiesel 1962), little data on the topography of orientation selectivity was available. The detection of visual cortex pinwheel-like orientation preference maps in OI enabled the simultaneous assessment of two-dimensional activation patterns (Bonhoeffer and Grinvald 1991). A search for further functional maps with OI found maps for ocular dominance, motion direction, and spatial frequency, with complex geometric relationships between them (Hübener et al. 1997; Kim et al. 1999).

Functional OI maps compute the preferred parameter value, e.g., causing the largest change in optical signals at each cortical location. A reconstruction scheme extracts information from the averaged activity distribution and implicitly assumes that thresholded activity reveals pertinent aspects of brain function. From a physiological perspective all locations respond, though in a graded manner. However, thresholding suggests that some locations respond, while others do not.

Some self-organizing models (Kohonen and Hari 1999; Swindale 2000) suggest that topographical gradients and local patches are essential consequences of self-organizing algorithms optimized for representing multiple behaviorally relevant dimensions of environmental scenes. However, there seem to be limitations in the number of functional maps that can be represented according to that scheme. A ceiling of ~9–10 maps may be imposed by the numbers of neurons or minicolumns available to represent a feature in a given cortical microdomain (Swindale 2000). A related question is the relation between the topographic map and the overlay of functional maps, because visual space representation is anisotropic, with the elevation and azimuthal axes having different magnification factors. OI showed that this anisotropy is reflected in the orientation, ocular dominance, and spatial frequency domains, which are elongated such that their directions of rapid change, or high-gradient axes, are orthogonal to the high-gradient axis of the visual map, demonstrating the impact of the visual map on each feature map (Yu et al. 2005).

Are there feature maps in AI? While there is evidence from microelectrode mapping for multiple functional maps representing parameters such as sharpness of tuning, preferred intensity, direction of FM sweeps, and onset latencies (Schreiner 1998; Schreiner et al. 2000), OI so far has failed to provide clear evidence for them compared to visual cortex. What neuronal response properties vary systematically within an auditory cortical column, and, therefore, what stimulus features might be the substrates for columnar processing (Linden and Schreiner 2003)? What distinguishes auditory and visual processing different and constrains the emergence of feature maps? While these views assume that AI contains stimulus features yet undiscovered, an interesting alternative has been proposed (Nelken et al. 2008): auditory cortex neurons show considerable information independence when tested with complex sounds. Under such conditions, average activity of many neurons may not be a good indicator of the interesting processing carried out by these neurons, which therefore cannot be reduced to simple feature detection models and are therefore unlikely to form simple gradient or topographic feature maps.

6 Visualization of Dynamic Response Properties and Synaptic Mechanisms of Primary Auditory Cortex

VSD-OI has the unique advantage of resolving neural activation in the millisecond range. Accordingly, the advantage of recording simultaneously the two-dimensional activity distribution can be combined with the analysis of the temporal and dynamic cortical activation pattern.

6.1 Visualization of Dynamic Response Properties

In vivo VSD-OI studies visualized dynamic response patterns in guinea pig auditory cortex core and belt fields (Fukunishi and Murai 1995; Tokioka et al. 2000; Yamaguchi et al. 2001; Horikawa et al. 2001, 2006a; Song et al. 2006). The response to a pure tone appeared first in dorsal AI ~20 ms after stimulus onset, then spread dorsoventrally along, and to some extent across, the isofrequency contours in 15–20 ms, and was followed by an inhibition of 100–150 ms. The activity spread was more expansive along than across the isofrequency contours, shaping a response band. The velocity of the spread along the isofrequency contour was 0.4–0.6 m/s. The response appeared in the dorsal part of DC and spread dorsoventrally, as in AI, then spread to the belt fields 30–60 ms poststimulus (Horikawa et al. 2001). It is not known whether the response spread in the core fields and between core and belt fields results from the corticocortical connections in the auditory cortex or subcortical afferent connections. However, in vivo electrical stimulation of guinea pig auditory cortex after MGB lesions (Song et al. 2006) and local cortical inhibition (Horikawa et al. 2006b) suggests that the response spread reflects corticocortical connections within AI and those between the core and belt fields, although thalamocortical input may contribute since the speed of the tone stimulation-induced response spread exceeds that of AI electrical stimulation (0.25 m/s) after the MGB lesion (Song et al. 2006).

VSD-OI visualized the dynamic shift of neural activity in AI and the parallel and hierarchical activation of the belt fields. These dynamic activity patterns reflect spatiotemporal activation of many neurons with different spectrotemporal receptive fields responding to sub- and supra-threshold stimuli via polysynaptic corticocortical and thalamocortical connections.

6.2 The Layer-Specific Spread of Activity: Neural Circuitry In and Between Cortical Layers

In vitro VSD-OI in rat auditory cortex slice preparations showed layer-specific spatiotemporal responses (Kubota et al. 1997, 1999) (Figs. 27.7 and 27.8). White matter or layer VI electrical stimulation in an AI slice preparation elicited a spot-like response in layer VI and, 2 ms later, in layers II/III. The response spread horizontally over 2 mm in 20 ms in the layers V/VI and in layers II/III (Figs. 27.7b and 27.8a). Layers I and IV had less activity than other layers and the layer IV response spread little horizontally. The horizontal spread velocity was 0.13 m/s in layers II/III and 0.22 m/s in layers V/VI (Kubota et al. 1999). These responses confirm known connections: MGB-specific thalamocortical input to layer IV, from layer IV to layers II/III, from layers II/III to layers V/VI, from layers V/VI to II/III (vertical interlaminar connections), and from layers II/III to other parts of layers II/III and from layers V/VI to other parts of layers V/VI (horizontal intralaminar connections). VSD-OI also revealed that the horizontal spreads in layers II/III and in layers V/VI are independent (Kubota et al. 1999) (Fig. 27.7). Their activity spread horizontally even after cutting the connections between the supra- and infragranular layers with the same conduction velocity as in the intact slices. The horizontal spread can occur without vertical interactions between layers. Contributions of the vertical connections on the horizontal spread remain to be studied.

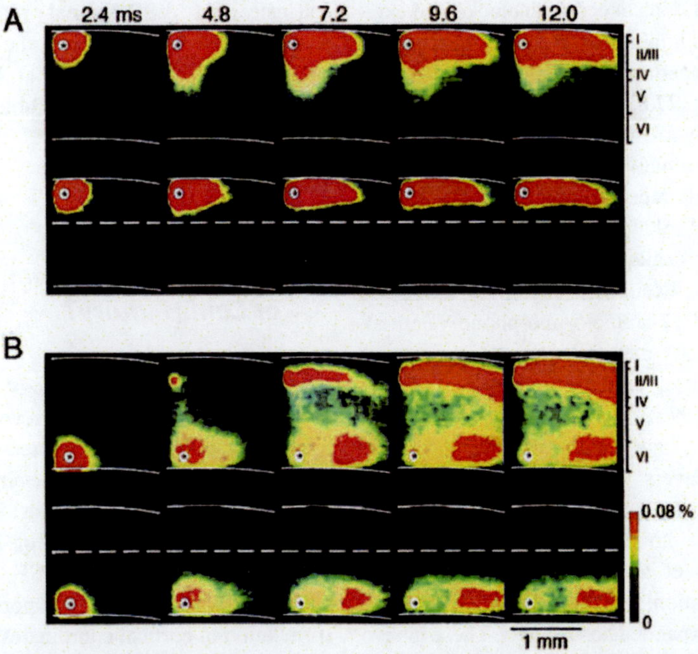

Fig. 27.7 VSD-OI of horizontal propagation of excitation in rat auditory cortex slice preparations. Horizontal propagation can occur without interaction between the supragranular and infragranular layers. (**a**) Optical signals of excitation elicited by electrical stimulation of layers II/III before (*upper panel*) and after (*lower panel*) a horizontal cut through layer IV. (**b**) Optical signals of excitation elicited by electrical stimulation of layer VI before (*upper panel*) and after (*lower panel*) a horizontal cut through layer IV. Time (ms) after stimulation is indicated at the *top*. *Asterisks*, stimulation site. The cortical surface and the border between the *white* matter and layer VI are depicted by *upper* and *lower white lines*, respectively. A horizontal cut is marked by a *dashed line*. Cortical layers are indicated on the *right*. A *color scale bar* shows the linear percentage change in light absorption. Modified from the original with permission (Kubota et al. 1999, Fig. 1)

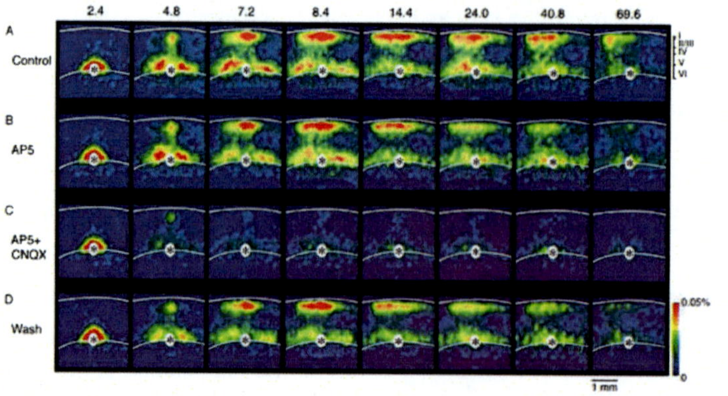

Fig. 27.8 Effects of glutamate receptor antagonists on spatiotemporal patterns of excitation elicited by electrical stimulation of the border between the *white* matter and layer VI in a rat auditory cortex slice preparation . (**a**) Responses in control solution. (**b**) Responses in solution containing D-AP5 (50 μM). The later part of the response was slightly reduced. (**c**) Responses in solution containing both D-AP5 (50 μM) and CNQX (20 μM). Horizontal spread of excitation in layers II/III was blocked and that in layers V/VI was reduced markedly. (**d**) Responses after a 1-h perfusion with the control solution, indicating recovery from the glutamate receptor antagonists. The stimulation site is marked by an *asterisk*. The cortical surface and the border between the *white* matter and layer VI are depicted by *white upper* and *lower lines*, respectively. Labels as in Fig. 27.7. Modified from the original with permission (Kubota et al. 1997, Fig. 2)

6.3 Synaptic Mechanisms of Vertical and Horizontal Spread of Activity

In vivo and in vitro VSD-OI in concert with blockers of excitatory (glutamatergic) and inhibitory (GABAergic) receptors revealed auditory cortex spatiotemporal properties and their synaptic mechanisms (Fig. 27.8). In guinea pig auditory cortex in vivo CNQX (6-cyano-7-nitroquinozaline-2,3-dione), an antagonist of AMPA (alpha-amino-3-hydroxy-5-methyl-4-isoxazolepropionic acid) receptors abolished the excitatory response to tones (Horikawa et al. 1996). Bicuculline enhanced the subsequent excitation and induced considerable horizontal spread of excitation, especially along the tonotopic axis, and APV (2-amino-5-phosphono-valerate), an antagonist of NMDA (*N*-methyl-D-aspartate) receptors abolished the enhanced later excitation and the excitation spread (Horikawa et al. 1996). In an in vitro slice preparation of rat auditory cortex, bath application of CNQX/APV completely suppressed layers II/III and V/VI responses after white matter electrical stimulation, but APV only slightly reduced the late horizontal responses in layers II/III and V/VI (Kubota et al. 1997, 1999) (Fig. 27.8). After bicuculline application, the layers II/III and V/VI responses increased and expanded horizontally. These in vivo and in vitro results show that the vertical and horizontal spreads of excitatory activity are mediated by AMPA and NMDA receptors and the spread of excitation, especially along the tonotopic axis, is regulated by $GABA_A$-mediated inhibition.

7 Comparison Between Intrinsic and Voltage-Sensitive Dye Signals

Intrinsic and VSD signal recordings differ in many aspects, most significantly in the time course of responses. Accordingly, the questions at hand can suggest which method is more appropriate (Table 27.1).

8 Outlook

8.1 Optical Imaging: Complementary or Confirmatory?

Single-neuron electrophysiology and optical imaging are complementary methods that reveal different aspects of cortical processing. Single cell recordings measure receptive fields (RFs) and their properties in parametric space thereby providing a window to the outside world. In contrast, optical imaging, either VSD or of intrinsic signals, provides a measure of the cortical PSF, i.e., the two-dimensional mass activity of a large neural population evoked by sensory stimulation in cortical coordinates.

These differences can create misunderstandings. If a stimulus that activates one neuron, as recorded electrophysiologically, whose RF is constrained in stimulus space, recording the cortical PSF with OI using the same stimulus reveals a much broader cortical activation.

Table 1 Optical signals: morphology and physical parameters

	VSD signals	Intrinsic signals
Origin	Voltage-dependent fluorescence induced by specific membrane-bound dye	Indirect measure reflecting metabolic demand of neural activity
Source	Membrane potentials of somata and dendrites	Subthreshold and spiking activity
Wave length	Absorption and emission of light specific to VSD	Selects signal type: changes in blood volume or oxygenation
Time course	follow the membrane potential within 1–10 μs	Follow neural activation within 0.5–5 s
Signal strength	$\Delta F/F = 10^{-3}$ (ΔF: signal, *F*: background light intensity)	Wavelength dependent
Constraints	Requires staining; this leads to a difficulty of extended stimulation and long recording because of VSD bleaching and toxicity	Does not require staining; therefore, can be done over days
Advantages	Imaging of real-time electrical activity of neuron population	Imaging through the skull opens the possibility for non-invasive, chronic recordings

Several factors are responsible for this perhaps counterintuitive finding.

1. *Cortical magnification*. For the PSF the amount of overlap between cortical RFs is a crucial variable. For a large cortical RF overlap, corresponding to a large cortical magnification, a recording electrode must be moved several millimeters in cortical space to find an RF that does not overlap with the initial one. As a result, the corresponding PSF will be several millimeters wide.
2. *Subthreshold activity*. An impediment to comparing single cell and OI data is that OI does not record action potentials only (as in extracellular neuron recordings), but includes subthreshold activity in a graded and activity-dependent matter. High-resolution AI maps obtained with Fourier imaging find that the optical response strength and the spiking bandwidth at the intensity of stimulation for optical imaging was not correlated, suggesting that the AI optical signal is evoked not only by spiking activity but also by other metabolic-dependent activities such as subthreshold membrane phenomena (Kalatsky et al. 2005).
3. *Spread of activity*. Polysynaptic corticocortical connections contribute to PSF generation as well as the parallel activation of cortical neurons via thalamocortical connections. VSD-OI revealed that the spread of intracortical activity can be elicited by the point stimulation alone after thalamocortical deactivation and that the spread of activity is regulated by intrinsic inhibitory cortical mechanisms.
4. *Cortical dendritic arborizations*. The dendritic fields of cortical neurons overlap with those of adjacent neurons perhaps contributing to the broadness and overlap of activation. As VSD-OI depicts mostly the dendritic membrane potentials of cortical cells, the width of the cortical activation cannot be less than the width of the dendritic arborization and their overlap inevitably enhances PSF overlap.
5. *Optical tissue effects*. Signal blurring is introduced by optical properties of the cortical tissue. This methodological artifact increases the apparent breadth of activity.

Besides the cortical magnification and polysynaptic activity spread, subthreshold activity contributes to OI signals, either intrinsic or VSD based. VSD imaging studies of cat visual cortex showed that the spiking-based PSF was only a fraction of the optical one (Grinvald et al. 1984). More recent visual cortex VSD studies show that, spatiotemporal patterns of subthreshold synaptic potentials have significant implications for cortical processing and for perceptual shaping (Jancke et al. 2004).

8.2 What Electrophysiology Can Learn From Optical Imaging and Vice Versa

Single neuron recordings and OI reflect different levels of processing: single cells and populations of cells, and therefore cannot be commensurate. The available OI data on AI tonotopicity emphasize the broad activation pattern even to circumscribed stimulation as a cortical processing principle, thus providing more than a confirmation of what was known about tonotopic maps. Further, functional maps entail the coactivation of nearby groups of neurons to certain stimulus features, and therefore are particularly suited for OI methods. For reasons to be clarified, no feature maps other than for frequency have been observed in AI to date (see Section 5). Electrophysiological studies emphasize the role of spectrotemporal processing and the dynamic activity pattern arising from spatiotemporal activation of many excitatory and inhibitory neurons. VSD-OI can provide unique insight into global dynamics of three-dimensional auditory cortex neural circuitry processing sound information. Given that single neuron recordings and OI sample different aspects of cortical processing, an ideal solution is to combine them to their mutual advantage.

8.3 Advanced Optical Recording Techniques

Current in vivo VSD-OI technique cannot record (1) depth dependency of cortical activity, (2) individual neuron activity, and (3) selective activity of excitatory and inhibitory neurons. Multiphoton confocal microscopy can visualize in vivo the three-dimensional microscopic architectures (to ~700-μm depth) of cortical neurons expressed by fluorescent protein. For such VSD-OI microscopy, signals 100 times the size of those now available are required. If such a VSD is constructed, the multiphoton confocal microscope could visualize three-dimensional microscopic in vivo neural activity in animals. Recording from individual neurons or each of excitatory and inhibitory neurons requires selective labeling of single cells or excitatory and inhibitory neuron groups. Technical developments on the selective labeling of single or a group of neurons by activity-dependent indicators are under way and will be fruitful for future studies on the three-dimensional dynamics of cortical circuitry.

References

Angelucci A, Clascá F, Bricolo E, Cramer K, and Sur M (1997) Experimentally induced retinal projections to the ferret auditory thalamus: development of clustered eye-specific patterns in a novel target. *Journal of Neuroscience* **17**: 2040–2055.

Asako M, Doi T, Matsumoto A, Yang SM, and Yamashita T (1999) Spatial and temporal patterns of evoked neural activity from auditory nuclei in chick brainstem detected by optical recording. *Acta Oto-Laryngologica* **119**:900–904.

Averbeck BB, Latham PE, and Pouget A (2006) Neural correlations, population coding and computation. *Nature Reviews Neuroscience* **7**:358–366.

Bakin JS, Kwon MC, Masino SA, Weinberger NM, and Frostig RD (1996) Suprathreshold auditory cortex activation visualized by intrinsic signal optical imaging. *Cerebral Cortex* **6**:120–130.

Barlow HB (1972) Single units and sensation: a neuron doctrine for perceptual psychology? *Perception* **1**:371–394.

Bonhoeffer T and Grinvald A (1991) Iso-orientation domains in cat visual cortex are arranged in pinwheel-like patterns. *Nature* **353**:429–431.

Cang J, Kalatsky VA, Löwel S, and Stryker MP (2005) Optical imaging of the intrinsic signal as a measure of cortical plasticity in the mouse. *Visual Neuroscience* **22**:685–691.

Carmena JM, Lebedev MA, Henriquez CS, and Nicolelis MAL (2005) Stable ensemble performance with single-neuron variability during reaching movements in primates. *Journal of Neuroscience* **25**:10712–1076.

Chance B, Cohen P, Jöbsis F, and Schoener B (1962) Intracellular oxidation-reduction states in vivo. *Science* **137**:499–508.

Cohen LB (1973) Changes in neuron structure during action potential propagation and synaptic transmission. *Physiological Reviews* **53**:373–418.

Cohen LB and Lesher S (1986) Optical monitoring of membrane potential: methods of multisite optical measurement. In: Weer PD and Salzberg BM (eds). *Optical Methods in Cell Physiology*. Wiley-Interscience, New York, pp. 71–99.

Cohen LB, Salzberg BM, Davila HV, Ross WN, Landowne D, Waggoner AS, and Wang CH (1974) Changes in axon fluorescence during activity: molecular probes of membrane potential. *Journal of Membrane Biology* **19**:1–36.

Dinse HR and Schreiner CE (2002) Do primary sensory areas play homologous roles in different sensory modalities? In: Schüz A and Miller R (eds). *Cortical Areas: Unity and Diversity: Conceptual Advances in Brain Research*. Taylor & Francis, London, pp. 273–310.

Dinse HR, Godde B, Hilger T, and Schreiner CE (2000) Optical imaging of relationships between functional topographies in cat auditory cortex. *Institut für Neuroinformatik, Ruhr-University Bochum, Internal Report IRINI* 2000-01, pp. 1–11.

Dinse HR, Godde B, Hilger T, Reuter G, Cords SM, Lenarz T, and von Seelen W (1997) Optical imaging of cat auditory cortex cochleotopic selectivity evoked by acute electrical stimulation of a multi-channel cochlear implant. *European Journal of Neuroscience* **9**:113–119.

Dinse HR, Godde B, Reuter G, Cords SM, and Hilger T (2003) Auditory cortical plasticity under operation: reorganization of auditory cortex induced by electric cochlear stimulation reveals adaptation to altered sensory input statistics. *Speech Communication* **41**: 201–219.

Fox PT, Raichle ME, Mintun MA, and Dence C (1988) Nonoxidative glucose consumption during focal physiologic neural activity. *Science* **241**:462–464.

Fukunishi K and Murai N (1995) Temporal coding in the guinea-pig auditory cortex as revealed by optical imaging and its pattern-time-series analysis. *Biological Cybernetics* **72**:463–473.

Fukunishi K, Murai N, and Uno H (1992) Dynamic characteristics of the auditory cortex of guinea pigs observed with multichannel optical recording. *Biological Cybernetics* **67**:501–509.

Ghazanfar AA, Stambaugh CR, and Nicolelis MAL (2000) Encoding of tactile stimulus location by somatosensory thalamocortical ensembles. *Journal of Neuroscience* **20**:3761–3775.

Godde B, Hilger T, von Seelen W, Berkefeld T, and Dinse HR (1995) Optical imaging of rat somatosensory cortex reveals representational overlap as topographic principle. *NeuroReport* **7**:24–28.

Grinvald A, Lieke EE, Frostig RD, and Hildesheim R (1994) Cortical point-spread function and long-range lateral interactions revealed by real-time optical imaging of macaque monkey primary visual cortex. *Journal of Neuroscience* **14**:2545–2568.

Grinvald A, Segal M, Huhnt U, Hildesheim R, Manker A, Anglister L, and Freeman JA (1986) Real-time optical mapping of neuronal activity in vertebrate CNS in vitro and in vivo. In: Weer P and Salzberg MM (eds). *Optical Methods in Cell Physiology*. Wiley-Interscience, New York, pp. 165–197.

Hall DA, Edmondson-Jones AM, and Fridriksson J (2006) Periodicity and frequency coding in human auditory cortex. *European Journal of Neuroscience* **24**:3601–3610.

Harel N, Mori N, Sawada S, Mount RJ, and Harrison RV (2000) Three distinct auditory areas of cortex (AI, AII, and AAF) defined by optical imaging of intrinsic signals. *Neuroimage* **11**:302–312.

Harrison RV, Harel N, Hamrahi H, Panesar J, Mori N, and Mount RJ (2000) Local haemodynamic changes associated with neural activity in auditory cortex. *Acta Otolaryngologica* **120**:255–258.

Harrison RV, Harel N, Kakigi A, Raveh E, and Mount RJ (1998) Optical imaging of intrinsic signals in chinchilla auditory cortex. *Audiology & Neurootology* **3**:214–223.

Harrison RV, Harel N, Panesar J, and Mount RJ (2002) Blood capillary distribution correlates with hemodynamic-based functional imaging in cerebral cortex. *Cerebral Cortex* **12**:225–233.

Hess A and Scheich H (1996) Optical and FDG mapping of frequency-specific activity in auditory cortex. *NeuroReport* **7**:2643–2647.

Hess A, Stiller D, Kaulisch T, Heil P, and Scheich H (2000) New insights into the hemodynamic blood oxygenation level-dependent response through combination of functional magnetic resonance imaging and optical recording in gerbil barrel cortex. *Journal of Neuroscience* **20**:3328–3338.

Hill DK and Keynes RD (1949) Opacity changes in stimulated nerve. *Journal of Physiology (London)* **108**:278–281.

Hishida R, Hoshino K, Kudoh M, Norita M, and Shibuki K (2003) Anisotropic functional connections between the auditory cortex and area 18a in rat cerebral slices. *Neuroscience Research* **46**:171–182.

Horikawa J, Hess A, Hosokawa Y, and Taniguchi I (2006a) Spatiotemporal processing in the guinea pig auditory cortex. In Kanwal JS and Ehret G (eds). *Behavior and Neurodynamics for Auditory Communication*. Cambridge University Press, New York, pp. 189–204.

Horikawa J, Uchiyama D, Matsui T, and Sugimoto S (2006b) Functional connections between the core and belt fields of the guinea-pig auditory cortex observed by optical recording and partial cortical inhibition using muscimol. *Neuroscience Research* Supplement **1**:S153.

Horikawa J, Hess A, Nasu M, Hosokawa Y, Scheich H, and Taniguchi I (2001) Optical imaging of neural activity in multiple auditory cortical fields of guinea pigs. *NeuroReport* **12**:3335–3339.

Horikawa J, Hosokawa Y, Kubota M, Nasu M, and Taniguchi I (1996) Optical imaging of spatiotemporal patterns of glutamatergic excitation and GABAergic inhibition in the guinea-pig auditory cortex in vivo. *Journal of Physiology (London)* **497**:629–638.

Horikawa J, Hosokawa Y, Nasu M, and Taniguchi I (1997) Optical study of spatiotemporal inhibition evoked by two-tone sequences in the guinea pig auditory cortex. *Journal of Comparative Physiology A* **181**:677–684.

Horikawa J, Hosokawa, Y, Nasu M, and Taniguchi I (2000) Optical study of cortical responses induced by electrical stimulation to cochlea of deafened guinea pigs. In: Wada H, Takasaka T, Ikeda K, Ohyama K, and Koike T (eds). *Recent Developments in Auditory Mechanics*. World Scientific, Singapore, pp. 457–462.

Horikawa J, Nasu M, and Taniguchi I (1998) Optical recording of responses to frequency-modulated sounds in the auditory cortex. *NeuroReport* **9**:799–802.

Horng SH and Sur M (2006) Visual activity and cortical rewiring: activity-dependent plasticity of cortical networks. *Progress in Brain Research* **157**:3–11.

Hosokawa Y, Horikawa J, Nasu M, and Taniguchi I (1997) Real-time imaging of neural activity during binaural interaction in the guinea pig auditory cortex. *Journal of Comparative Physiology A* **181**: 607–614.

Hosokawa Y, Horikawa J, Nasu M, and Taniguchi I (1999) Spatiotemporal representation of binaural difference in time and intensity of sound in the guinea pig auditory cortex. *Hearing Research* **134**:123–132.

Hosokawa Y, Sugimoto S, Kubota M, Taniguchi I, and Horikawa J (2004) Optical imaging of binaural interaction in multiple fields of the guinea pig auditory cortex. *NeuroReport* **15**:1093–1097.

Houweling AR and Brecht M (2008) Behavioural report of single neuron stimulation in somatosensory cortex. *Nature* **451**:65–68.

Hubel D and Wiesel T (1962) Receptive fields, binocular interaction and functional architecture in the cat's visual cortex. *Journal of Physiology (London)* **160**:106–154.

Hübener M, Shoham D, Grinvald A, and Bonhoeffer T (1997) Spatial relationships among three columnar systems in cat area 17. *Journal of Neuroscience* **17**:9270–9284.

Inagaki S, Katura T, Kawaguchi H, and Song W-J (2003) Isolation of neural activities from respiratory and heartbeat noises for in vivo optical recording in guinea pigs using independent component analysis. *Neuroscience Letters* **352**:9–12.

Jancke D, Chavane F, Naaman S, and Grinvald A (2004) Imaging cortical correlates of illusion in early visual cortex. *Nature* **428**: 423–426.

Jöbsis FF and Rosenthal M (1978) Cerebral energy consumption and provision: the predominance of neuronal oxidative metabolic processes. In: Purves MJ and Elliott K (eds). *Cerebral Vascular Smooth Muscle and Its Control*, volume 56. *Ciba Foundation Symposium*. Elsevier, Amsterdam, pp. 129–148.

Kalatsky VA and Stryker MP (2003) New paradigm for optical imaging: temporally encoded maps of intrinsic signal. *Neuron* **38**:529–545.

Kalatsky VA, Polley DB, Merzenich MM, Schreiner CE, and Stryker MP (2005) Fine functional organization of auditory cortex revealed by Fourier optical imaging. *Proceedings of the National Academy of Sciences of the United States of America* **102**:13325–13330.

Kaltenbach JA and Zhang JS (2004) In vivo optical imaging of tone-evoked activity in the dorsal cochlear nucleus with a voltage sensitive dye. *Journal of Neuroscience Research* **78**:908–917.

Kety SS and Schmidt CF (1948) The nitrous oxide method for the quantitative determination of cerebral blood flow in man: theory, procedure and normal values. *Journal of Clinical Investigation* **27**:476–483.

Kim DS, Matsuda Y, Ohki K, Ajima A, and Tanaka S (1999) Geometrical and topological relationships between multiple functional maps in cat primary visual cortex. *NeuroReport* **10**: 2515–2522.

Kohonen T and Hari R (1999) Where the abstract feature maps of the brain might come from. *Trends in Neuroscience* **22**:135–139.

Kubota M, Nasu M, and Taniguchi I (1999) Layer-specific horizontal propagation of excitation in the auditory cortex. *NeuroReport* **10**:2865–2867.

Kubota M, Sugimoto S, Horikawa J, Nasu M, and Taniguchi I (1997) Optical imaging of dynamic horizontal spread of excitation in rat auditory cortex slices. *Neuroscience Letters* **237**:77–80.

Langner G, Dinse HR, and Godde B (2009) A map of periodicity orthogonal to frequency representation in the cat auditory cortex. *Frontiers in Integrative Neuroscience* **3**:27.

Langner G, Sams M, Heil P, and Schulze H (1997) Frequency and periodicity are represented in orthogonal maps in the human auditory cortex: evidence from magnetoencephalography. *Journal of Comparative Physiology A* **181**:665–676.

Lassen NA and Ingvar DH (1961) The blood flow of the cerebral cortex determined by radioactive Krypton. *Experientia* **17**:42–43.

Linden JF and Schreiner CE (2003) Columnar transformations in auditory cortex? A comparison to visual and somatosensory cortices. *Cerebral Cortex* **13**:83–89.

London JA, Zecevic D, and Cohen LB (1986) Simultaneous monitoring of activity of many neurons from invertebrate ganglia using a multielement detection system. In: Weer PD and Salzberg BM (eds). *Optical Methods in Cell Physiology*. Wiley-Interscience, New York, pp. 115–131.

Maeda S, Inagaki S, Kawaguchi H, and Song W-J (2001) Separation of signal and noise from in vivo optical recording in Guinea pigs using independent component analysis. *Neuroscience Letters* **302**:137–140.

Mrsic-Flogel T, Hübener M, and Bonhoeffer T (2003) Brain mapping: new wave optical imaging. *Current Biology* **13**: R778–R780.

Mrsic-Flogel TD, Versnel H, and King AJ (2006) Development of contralateral and ipsilateral frequency representations in ferret primary auditory cortex. *European Journal of Neuroscience* **23**: 780–792.

Nelken I, Bizley JK, Nodal FR, Ahmed B, Schnupp JWH, and King AJ (2004) Large-scale organization of ferret auditory cortex revealed using continuous acquisition of intrinsic optical signals. *Journal of Neurophysiology* **92**:2574–2588.

Nelken I, Bizley JK, Nodal FR, Ahmed B, King AJ, and Schnupp JWH (2008) Responses of auditory cortex to complex stimuli: functional organization revealed using intrinsic optical signals. *Journal of Neurophysiology* **99**:1928–1941.

Nicolelis MAL, Dimitrov D, Carmena JM, Crist R, Lehew G, Kralik JD, and Wise SP (2003) Chronic, multisite, multielectrode recordings in macaque monkeys. *Proceedings of the National Academy of Sciences of the United States of America* **100**: 11041–11046.

Nishimura M, Shirasawa H, Kaizo H, and Song W-J (2007) New field with tonotopic organization in guinea pig auditory cortex. *Journal of Neurophysiology* **97**:927–932.

Ojima H, Takayanagi M, Potapov D, and Homma R (2005) Isofrequency band-like zones of activation revealed by optical imaging of intrinsic signals in the cat primary auditory cortex. *Cerebral Cortex* **15**:1497–1509.

Orback HS and Cohen LB (1983) Optical monitoring of activity from many areas of the in vivo and in vitro salamander olfactory bulb: a new method for studying functional organization in the vertebrate central nervous system. *Journal of Neuroscience* **3**: 2251–2262.

Phillips DP, Semple MN, Calford MB, and Kitzes LM (1994) Level-dependent representation of stimulus frequency in cat primary auditory cortex. *Experimental Brain Research* **102**:210–226.

Quiroga RQ, Reddy L, Kreiman G, Koch C, and Fried I (2005) Invariant visual representation by single neurons in the human brain. *Nature* **435**:1102–1107.

Rauschecker JP and Tian B (2000) Mechanisms and streams for processing of "what" and "where" in auditory cortex. *Proceedings of the National Academy of Sciences of the United States of America.* **97**:11800–11806.

Redies H, Sieben U, and Creutzfeldt OD (1989) Functional subdivisions in the auditory cortex of the guinea pig. *Journal of Comparative Neurology* **282**:473–488.

Raggio MW and Schreiner CE (1994) Neuronal responses in cat primary auditory cortex to electrical cochlear stimulation. I. Intensity dependence of firing rate and response latency. *Journal of Neurophysiology* **72**:2334–2359.

Robertson D and Irvine DRF (1989) Plasticity of frequency organization in auditory cortex of the guinea pig with partial unilateral deafness. *Journal of Comparative Neurology* **282**:456–471.

Ross WN, Salzberg BM, Cohen LB, Grinvald A, Davila HV, Waggoner AS, and Wang CH (1977) Changes in absorption, fluorescence, dichroism, and birefringence in stained giant axons: optical measurement of membrane potential. *Journal of Membrane Biology* **33**:141–183.

Rutkowski RG, Wallace MN, Shackleton TN, and Palmer AR (2000) Organisation of binaural interactions in the primary and dorsocaudal fields of the guinea pig auditory cortex. *Hearing Research* **145**: 177–189.

Saltzberg BM, Grinvald A, Cohen LB, Davila HV, and Ross WN (1977) Optical recording of neuronal activity in an invertebrate central nervous system: simultaneous monitoring of several neurons. *Journal of Neurophysiology* **40**:1281–1291.

Sato K and Momose-Sato Y (2003) Optical detection of developmental origin of synaptic function in the embryonic chick vestibulocochlear nuclei. *Journal of Neurophysiology* **89**:3215–3224.

Schreiner CE (1998) Spatial distribution of responses to simple and complex sounds in the primary auditory cortex. *Audiology and Neurotology* **3**:104–122.

Schreiner CE, Read HL, and Sutter MI (2000) Modular organization of frequency integration in primary auditory cortex. *Annual Review of Neuroscience* **23**:501–529.

Schulze H, Hess A, Ohl FW, and Scheich H (2002) Superposition of horseshoe-like periodicity and linear tonotopic maps in auditory cortex of the Mongolian gerbil. *European Journal of Neuroscience* **15**:1077–1084.

Segev I and Rall W (1998) Excitable dendrites and spines: earlier theoretical insights elucidate recent direct observations. *Trends in Neurosciences* **21**:453–460.

Sharma J, Angelucci A, and Sur M (2000) Induction of visual orientation modules in auditory cortex. *Nature* **404**:841–847.

Sokoloff L (1978) Local cerebral energy metabolism: its relationships to local functional activity and blood flow. In: Purves MJ and Elliott K (eds). *Cerebral Vascular Smooth Muscle and Its Control*, volume 56, *Ciba Foundation Symposium*. Elsevier, Amsterdam, pp. 171–197.

Song W-J, Kawaguchi H, Totoki S, Inoue Y, Katura T, Maeda S, Inagaki S, Shirasawa H, and Nishimura M (2006) Cortical intrinsic circuits can support activity propagation through an isofrequency strip of the guinea pig primary auditory cortex. *Cerebral Cortex* **16**: 718–729.

Spitzer MW, Calford MB, Clarey JC, Pettigrew JD, and Roe AW (2001) Spontaneous and stimulus-evoked intrinsic optical signals in primary auditory cortex of the cat. *Journal of Neurophysiology* **85**:1283–1298.

Srinavasan G, Friauf E, and Loehrke S (2004) Functional glutamatergic and glycinergic inputs to several superior olivary nuclei of the rat revealed by optical imaging. *Neuroscience* **128**: 617–634.

Sugimoto S, Hosokawa Y, Horikawa J, Nasu M, and Taniguchi I (2002) Spatial focusing of neuronal responses induced by asynchronous two-tone stimuli in the guinea pig auditory cortex. *Cerebral Cortex* **12**:506–514.

Sugimoto S, Suzuki U, Tanaka H, Kubota M, and Horikawa J (2008) Time-frequency feature extraction from frequency-modulated sounds in a population of auditory cortical neurons. *Acoustical Science & Technology* **29**:36–40.

Sur M and Leamey CA (2001) Development and plasticity of cortical areas and networks. *Nature Reviews Neuroscience* **2**:251–262.

Swindale NV (2000) How many maps are there in visual cortex? *Cerebral Cortex* **10**:633–643.

Taniguchi I, Horikawa J, Hosokawa Y, and Nasu M (1997) Optical imaging of neural activity in auditory cortex induced by intracochlear electrical stimulation. *Acta Oto-Laryngologica* Supplement **532**:83–88.

Taniguchi I, Horikawa J, Moriyama T, and Nasu M (1992) Spatio-temporal pattern of frequency representation in the auditory cortex of guinea pigs. *Neuroscience Letters* **146**:37–40.

Taniguchi I and Nasu M (1993) Spatio-temporal representation of sound intensity in the guinea pig auditory cortex observed by optical recording. *Neuroscience Letters* **151**:178–181.

Tasaki I, Watanabe A, Sandlin R, and Carnay L (1968) Changes in fluorescence, turbidity, and birefringence associated with nerve excitation. *Proceedings of the National Academy of Sciences of the United States of America* **61**:883–888.

Tokioka R, Kawaguchi H, and Fukunishi K (2000) Spatio-temporal analyses of stimulus-evoked and spontaneous stochastic neural activity observed by optical imaging in guinea pig auditory cortex. *Brain Research* **861**:271–280.

Tsytsarev V and Tanaka S (2002) Intrinsic optical signals from rat primary auditory cortex in response to sound stimuli presented to contralateral, ipsilateral and bilateral ears. *NeuroReport* **13**: 1661–1666.

Tsytsarev V, Yamazaki T, Ribot J, and Tanaka S (2004) Sound frequency representation in cat auditory cortex. *Neuroimage* **23**: 1246–1255.

Uno H, Murai N, and Fukunishi K (1993) The tonotopic representation in the auditory cortex of the guinea pig with optical recording. *Neuroscience Letters* **150**:179–182.

Versnel H, Mossop JE, Mrsic-Flogel TD, Ahmed B, and Moore DR (2002) Optical imaging of intrinsic signals in ferret auditory cortex: responses to narrowband sound stimuli. *Journal of Neurophysiology* **88**:1545–1558.

Waggoner AS (1979) Dye indicators of membrane potential. *Annual Reviews of Biophysics and Bioengineering* **8**:47–68.

Wallace MN, Rutkowski RG, and Palmer AR (2000) Identification and localization of auditory areas in guinea pig cortex. *Experimental Brain Research* **132**:445–456.

Wolf BE and Waggoner AS (1986) Optical studies of the mechanism of membrane potential sensitivity of merocyanine 540. In: Weer PD and Salzberg BM (eds). *Optical Methods in Cell Physiology*, Wiley-Interscience, New York, pp. 101–113.

Wu JY, Guan L, Bai L, and Yang Q (2001) Spatiotemporal properties of an evoked population activity in rat sensory cortical slices. *Journal of Neurophysiology* **86**:2461–2474.

Yamaguchi Y, Horikawa J, and Taniguchi I. (2001) Neural dynamics of vocal processing in the auditory cortex. In: *Biophysical Neural Networks*, Mary Ann Liebert Inc., New York, pp. 343–362.

Yu H, Farley BJ, Jin DZ, and Sur M (2005) The coordinated mapping of visual space and response features in visual cortex. *Neuron* **47**: 267–280.

Zeng F-G, Arthur N, Popper AN, Richard R, Fay RR (2004) *Springer Handbook of Auditory Research Cochlear Implants*, volume 20, *Auditory Prostheses and Electric Hearing*, Springer, New York.

Chapter 28

From Tones to Speech: Magnetoencephalographic Studies

Bernd Lütkenhöner and David Poeppel

Abbreviations

AEP	auditory evoked magnetic field
AM	amplitude modulation
CV	consonant-vowel
DC	direct current
EEG	electroencephalography
ERP	event related potential
FM	frequency modulation
fMRI	functional magnetic resonance
HP	Huggins pitch
MEG	magnetoencephalography
MMF	mismatch field
MMN	mismatch negativity
POR	pitch onset response
RIS	regular interval sound
RMS	root-mean-square
SF	sustained field
STG	supratemporal gyrus
TN	tone embedded in noise

1 Basics of Magnetoencephalography

Electrical activity in the brain generates a weak magnetic field in the vicinity of the head. Recording this signal with sensitive detectors is called magnetoencephalography (MEG). The technique may be considered the magnetic counterpart of electroencephalography (EEG), where the signal is recorded from electrodes attached to the scalp. An outstanding feature of MEG (as well as EEG) is that its temporal resolution is virtually unlimited. Thus, MEG is excellently suited for studying brain dynamics. For a magnetic field arising from a single circumscribed brain area, MEG offers high source localization accuracy as well. But this is not the situation in typical experiments. In general, the observed activity is more complex and comprises contributions from several simultaneously active sources. A unique interpretation of the recorded data does not exist under such circumstances, and any conclusion depends on modeling assumptions about the number and configuration of the underlying neuronal sources. A proper understanding of experimental MEG results therefore requires at least an elementary knowledge of the theoretical foundations. Thus, this introduction will briefly explain the basics of MEG. Comprehensive reviews can be found elsewhere (Williamson and Kaufman 1987; Hämäläinen et al. 1993; Baillet et al. 2001; Hämäläinen and Hari 2002; Lütkenhöner and Mosher 2007).

1.1 From Neural Currents to the Magnetic Fields: The Forward Problem

The main sources of the magnetic field recorded by MEG are postsynaptic currents in cortical pyramidal cells. These primary currents cause volume currents in the surrounding conductive medium so that a closed circuit is formed. Microscopic details of the currents are not reflected in the MEG signal. Thus, a relatively coarse, macroscopic model can be used to describe the generation of this signal. It is sufficient for that purpose to imagine a limited brain area as a battery, where the current flowing inside the battery represents the sum of all primary currents in the respective area. At a certain distance from the battery, the strength of the magnetic field is proportional to the product of current and battery length, which is called the dipole moment (typically expressed in nanoamperemeters, nAm). The model is usually simplified even further by assuming a battery of infinitesimal length. In this way, the battery model turns into the model

D. Poeppel (✉)
Department of Psychology, New York University, New York, NY 10003, USA
e-mail: david.poeppel@nyu.edu

J.A. Winer, C.E. Schreiner (eds.), *The Auditory Cortex*,
DOI 10.1007/978-1-4419-0074-6_28, © Springer Science+Business Media, LLC 2011

of a current dipole. Given sufficient knowledge of the primary currents in the brain and the conductivity profile of the head, the MEG signal can be accurately predicted. This issue is often called the forward problem. While the EEG signal crucially depends on the volume currents in brain, skull and scalp, the MEG signal depends, under typical experimental conditions, mainly on the primary currents. Thus, as an approximation, MEG can be used to visualize cortical events directly through the skull (Hämäläinen and Hari 2002). A peculiarity of MEG is that it is mainly sensitive to currents (dipoles) that are oriented tangentially to the inner surface of the skull. By contrast, sources oriented perpendicularly to that surface (radial sources) are generally considered as silent sources. MEG is also relatively insensitive to sources located deep in the brain, so that the observed field is mainly of cortical origin. Figure 28.1 illustrates the recording of a magnetic field (thin arrows) caused by a tangential dipole in the cortex (thick arrow). Radially oriented magnetometers distributed about the head would record the spatial pattern displayed as a contour map in Fig. 28.2, which provides a view of the measurement surface from above.

1.2 Interpretation of Magnetic Fields

To interpret measured data they have to be explained in terms of the underlying sources in the brain. This type of problem is called the inverse problem. If the magnetic field were of a dipolar nature (as in Fig. 28.2), the parameters of the underlying dipole could be easily estimated using an iterative optimization procedure. But typical experimental data are more complicated and the number of contributing sources is usually not obvious. A fundamental dilemma is that the inverse problem does not have a unique solution. Thus, any MEG measurement allows, in principle, an infinite number of interpretations. Although certain types of solutions can generally be excluded based on prior knowledge, assumptions and plausibility considerations, some uncertainty always remains.

Because of these difficulties, many analysis techniques have been developed over the years. Two main classes can be distinguished: parametric and imaging techniques. In the first case, it is typically postulated that the observed magnetic field resulted from a limited number of current dipoles (far fewer than the number of measuring channels). In the second case, a huge number of dipoles with known locations and directions (generally far more than the number of measuring channels) is assumed, and the task is to estimate the activation strengths of these dipoles (the image) from the measured data. Parametric methods have the advantage of a considerable data reduction, but they have to be tailored to the type of experiment, and the choice of an inappropriate model risks

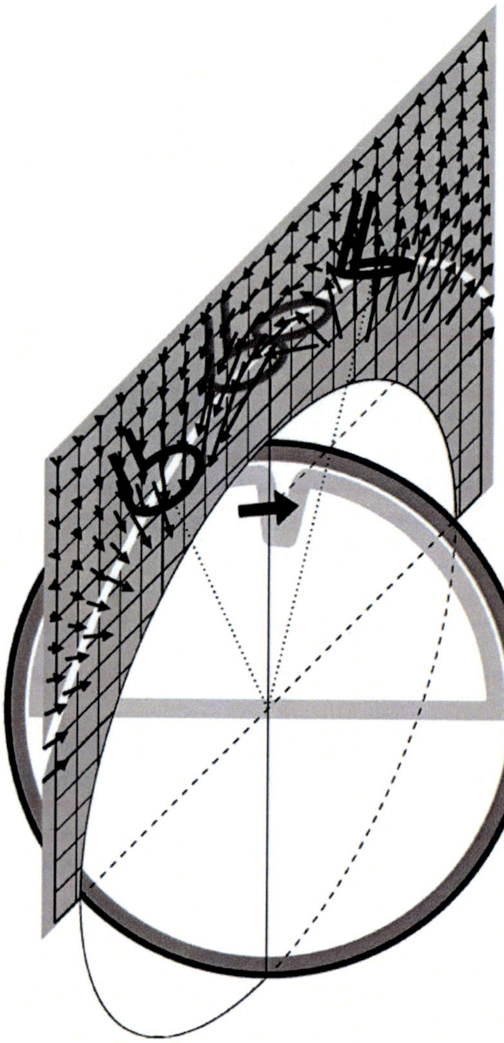

Fig. 28.1 Measurement of the magnetic field caused by a single current dipole in a simplistic model of the head. Scalp and skull are represented by spherical shells (*black* and *dark gray*, respectively). Moreover, a primitive cortex with a single sulcus at the top is plotted (*light gray*). Apical dendrites of cortical pyramidal cells are oriented roughly perpendicular to the cortical surface, and this is also the preferred orientation of the current dipoles considered in MEG. One such dipole is displayed (*thick arrow*). Because of its placement in the depth of a sulcus, it represents a tangential source, which is a favorable condition for MEG. While such a dipole causes a non-zero magnetic field almost everywhere, the strongest field is in a plane perpendicular to the dipole. This plane is represented by the *gray area* with the coordinate grid; *arrows* indicate the strength and the direction of the magnetic field at the respective location. A single magnetometer coil detects only one component of the three-dimensional magnetic field vector, and typically the radial component (roughly perpendicular to the scalp) is chosen. For a given measurement surface (*white curve*), the strength of that component exhibits two maxima (indicated by *dotted lines* emanating from the center of the sphere). A radially oriented magnetometer coil (with two leads) is shown at either location. Directly above the dipole, the magnetic field is purely tangential, i.e. its radial component is zero. A magnetic field here may be recorded with a planar gradiometer consisting of two oppositely wound coils (*dark gray*)

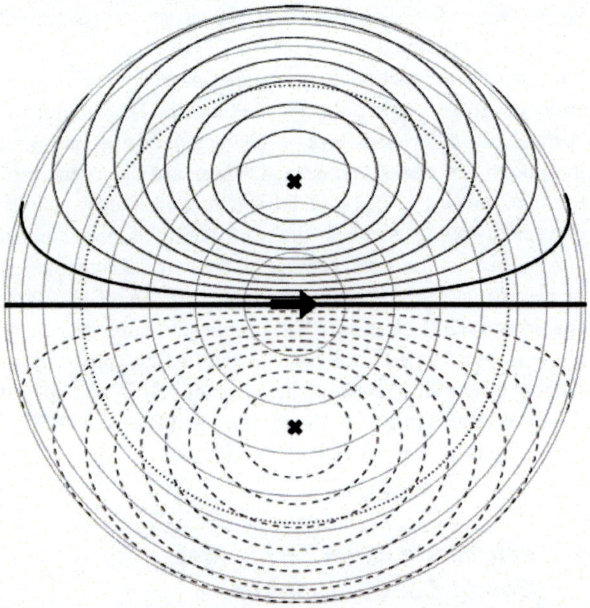

Fig. 28.2 Radial component of the magnetic field caused by a single current dipole. The spherical measurement surface (Fig. 28.1) is viewed from above. *Thin circles*, angular distances of 10, 20,. . ., 90° from the pole of the sphere; *dotted circle*, the outermost contour of the scalp. The contour map shows how the radial component of the magnetic field depends on the measurement location. On the meridian that corresponds to the direction of the dipole (*arrow*), the radial component is zero (*thick solid line*). Magnetic flux directed out of the head (defined as the positive polarity) is shown as the *solid curves* in the *upper* half of the *plot*. A completely symmetric pattern is found in the bottom half of the plot, but here the magnetic flux is directed into the head (*dashed curved lines*). The locations corresponding to maximum and minimum are indicated by x-marks

serious misinterpretations. Images are easiest to calculate, but they may be hardest to interpret, and in the worst case they can yield a misleading impression of the activity in the brain (Lütkenhöner and Mosher 2007).

It is clear from the above that the interpretation of MEG data crucially depends on models and that each methodological approach has specific advantages and drawbacks. Approaches that will be of particular relevance later in this chapter are now considered in more detail.

1.3 Estimation of a Single Current Dipole

Magnetic fields exhibiting an approximately dipolar spatial pattern (Fig. 28.2) are quite common. This is the reason why a data interpretation in terms of a single current dipole still belongs to the most popular approaches. Fitting a dipole to experimental data is analytically straightforward, and unless the data are too noisy—or fundamentally in conflict with the assumption of a single dipolar source—the solution is generally unique and numerically stable. More problematic may be the interpretation of the result. The choice of the dipole model is often justified with the argument that the goodness of fit (percent of the variance explained by the model) is greater than, e.g., 90%. But such an argument is acceptable only in the case of relatively noisy data. If the data exhibit an excellent signal-to-noise ratio, such a goodness-of-fit would basically confirm that more than one source contributed to the observed field. Even goodness-of-fit values above 99% do not exclude the possibility that two or more cortical sources with a distance of several centimeters were active (Lütkenhöner 1998). In the case of multiple sources, the location of the estimated current dipole often corresponds to the center of gravity of the sources. But this is not certain, especially if the primary currents flow in opposite directions so that the associated magnetic fields partially cancel each other (Lütkenhöner and Mosher 2007).

In spite of these problems, the single-current-dipole approach is both useful and powerful. The better a single dipole can explain the data, the more difficult it is to find a convincing alternative model: because of a lack of information in the data, analyses with more complex models tend to be critically dependent on constraints and assumptions. Thus, the above-mentioned problems do not disqualify the dipole model itself, they merely suggest that a careful interpretation of the model parameters are necessary (Lütkenhöner et al. 2006). In any case, the critique of dipole models must acknowledge that the alternative models face with similarly daunting challenges in generating convincing neurophysiological interpretations.

1.4 Multi-dipole Approaches

A natural extension of the single-dipole model is a multi-dipole model. As long as the dipoles are well separated (e.g., one dipole in the auditory cortex of each hemisphere), this approach has basically the same constraints as the single-dipole approach. Additional problems emerge, however, if some of the dipoles are located relatively near one another. Even with prior knowledge of their exact locations it may be difficult to obtain independent estimates of the dipole moments (which may be assumed to reflect the net activities in the respective cortical regions). Moreover, it is not a trivial task to determine the number of dipoles actually needed. Because of such difficulties, it is often better to use a simpler model that imperfectly describes the major features of the data than to add many dipoles until the correspondence between model prediction and data is almost perfect. A frequent shortcoming of the latter approach is that many solutions of similar sophistication exist (Lütkenhöner and Mosher 2007).

1.5 Synthetic Sensors: Beamformers

Synthetic (or virtual) sensors can be realized by linearly combining the signals provided by the actual sensors (Vrba and Robinson 2002). They usually have an improved spatial specificity and are often used to interrogate the activity going on in a specific brain region. An early example is the software lens (Freeman 1980). A general methodological framework is provided by the theory of beamforming (Van Veen and Buckley 1988). An ideal beamformer would correspond to a spatial filter which allows activity from a location of interest pass while blocking other activity. Although such an ideal technique does not exist, many methods can be considered variations of beamforming, even single-dipole modeling (Lütkenhöner 2003). In the latter case, the estimated dipole moment is essentially a linear combination of the signals measured by the individual sensors, with coefficients depending (among other factors) on the location and the orientation of the dipole. Thus, the estimation of a dipole moment may be considered a measurement with a synthetic sensor focusing on the dipole. For two dipoles, the beamformer for the first dipole blocks the signal from the second dipole, and vice versa (Lütkenhöner and Mosher 2007). More general implementations of beamforming require not only knowledge about the noise, but also strong assumptions about source models and statistics (Hillebrand and Barnes 2005). When applied to auditory data elicited by relatively simple acoustic signals, beamforming techniques can yield source reconstructions that are comparable to dipole models—while also incorporating information about time and frequency attributes of the neuronal activity (Sekihara et al. 2001). As our anatomic models of human auditory cortex become more refined, perhaps the distinct modeling approaches will become less similar. For now, there is no principled reason in studies of human auditory cortex to prefer one method over the other without strong prior hypotheses about the source configuration that are more amenable to one or the other data-analytic approach.

1.6 Spectro-Temporal Approaches: Peaks, Peak Activation Sequences, Oscillations, Phase

Much effort is required to perform and justify source analysis, but it is reasonable to assert that the MEG data relating most closely to neurophysiological considerations derive from analyses of the timing and morphology of the neuromagnetic activity elicited by acoustic stimulation. Briefly, four types of analyses are commonly encountered. First, as is also typical for EEG, individual response peaks (described below) are examined for variations in peak amplitudes and latencies as a function of the experimental manipulation (Roberts et al. 2000; Salajegheh et al. 2004). Second, the cortical activation sequence is characterized, i.e., at what time-point are peaks visible and where are they generated (Salmelin et al. 1994). Third, the role that oscillatory activity plays is studied, particularly the contribution of the canonical bands (theta, alpha, beta, gamma) to the cortical construction of perceptual representations in speech (Palva et al. 2002) and nonspeech (Luo et al. 2005) regimes. Finally, the role of phase is investigated, motivated by the fact that phase information is central to encoding auditory signals in non-speech (Patel and Balaban 2004), speech (Luo and Poeppel 2007) or binaural phase (Ross et al. 2007) conditions.

1.7 Relation to Other Techniques: Unique Contributions of Magnetoencephalography

Functional magnetic resonance imaging (fMRI) has emerged as a dominant technique in cognitive neuroscience, and its remarkable spatial resolving power is impressive. Scanners with a field strength of three Tesla or more are now widely available, making it possible to generate functional images with an in-plane resolution of 1 mm or better, thereby enriching the understanding of the functional anatomy of the human auditory cortex. The highly model-dependent spatial resolution of MEG sources is about 5–10 mm. Because the imaging approaches are hemodynamically based (so far), their temporal resolution does not match the rate at which many auditory phenomena occur (milliseconds), and the electromagnetic recording techniques EEG and MEG therefore remain essential to study the dynamics of the neuronal encoding and representation of acoustic signals. In particular, hypotheses that connect insights from human recordings to animal physiology are most powerful at the level of electrophysiological phenomena, and it follows that questions of coding are best addressed by considering electrophysiological data with the appropriate temporal resolution (Luo et al. 2006). In this context, MEG provides a unique contribution to auditory neuroscience, complementing what EEG offers. First, a practical feature of MEG is that preparation time is brief. A subject can be in the scanner within 10–15 min because the time consuming task of applying and checking electrodes is obviated. Second, the anatomic (sulcal) location of large parts of auditory cortex in the human brain (Morosan et al. 2001) make MEG ideally suited for electrophysiological studies. Many neuromagnetic fields originate from the dorsal aspect of the superior temporal gyrus and the planum (Fig. 28.10), and their net activation is optimally sited for capture by MEG. Third, MEG is especially well suited to investigate lateralized phenomena. For biophysical reasons noted above evoked responses measured with EEG/ERP are best visible and quantified most precisely at midline electrodes, making hemispherically asymmetric effects more difficult to characterize. Effects related to speech, language, and pitch processing, which are often lateralized, are effectively captured by MEG, enabling us to build more nuanced models of the neurocomputational principles underlying auditory lateralization.

2 Auditory Evoked Magnetic Fields: Basic Phenomena

Elementary auditory stimuli such as clicks or tone bursts generally elicit an auditory evoked magnetic field (AEF) with a stereotyped, highly reproducible time course. Some aspects of this temporal pattern are conserved in many experimental conditions, and the goal of many MEG studies is to examine how a specific feature (e.g., peak amplitude or latency) depends on certain stimulus properties (e.g., frequency or intensity). The question as to where in the cortex the respective phenomenon originates is generally of secondary importance in such studies. Thus, difficulties inherent to the solution of the inverse problem can be largely ignored. If the main focus is on the time course of the AEF, it may suffice to consider the channel with the strongest signal or to calculate the root-mean-square (RMS) value for an appropriate subset of channels. Alternatively, an elementary source analysis may be performed by representing the auditory cortex of each hemisphere by a single current dipole with invariant location and direction, but time-dependent dipole moment. The estimated dipole moment may then be interpreted as the signal of a synthetic channel (beamformer) focusing on auditory cortex. Regarding the signal-to-noise ratio, the dipole moment is usually superior to single measurement channels.

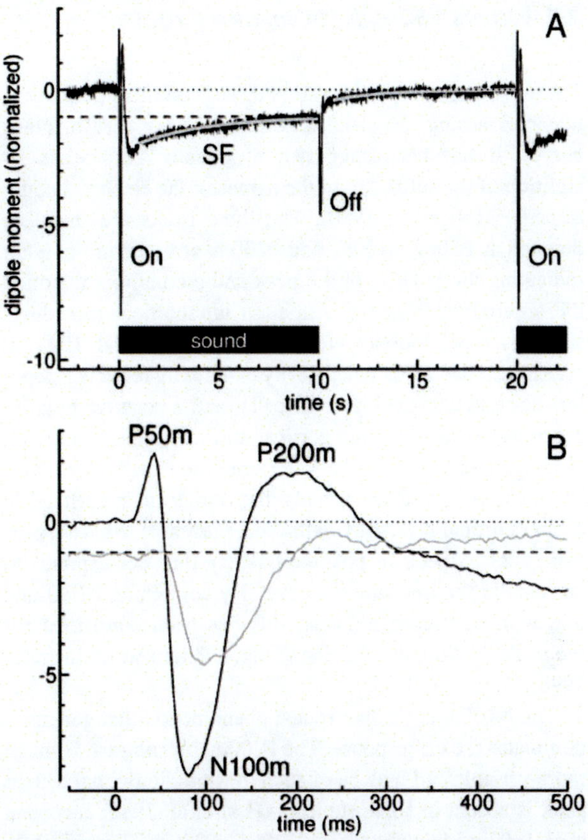

Fig. 28.3 (**a**) Time course of the response elicited by a10-s long tone burst. The *curve* refers to the moment of a dipole representing the entire auditory cortex of one hemisphere. Transitions between silence and sound and vice versa elicit a pronounced on- and a somewhat smaller off-response. Stimulus persistence is reflected in a sustained field (SF). (**b**) On-response (*black curve*) and off-response (*gray curve*) on an enlarged scale. The on-response has deflections with latencies ~50, 100, and 200 ms, termed P50m, N100m, and P200m. The P50m is absent in the off-response. Derived from prior work (Lammertmann and Lütkenhöner 2001, Fig. 9)

2.1 On-Response, Sustained Field, and Off-Response

The transition between sound and silence evokes an on-response at sound onset, a somewhat smaller off-response at sound offset, and a sustained field (SF) which lasts from onset to offset (Fig. 28.3a). These three phenomena are ubiquitous in MEG studies of audition, although they are typically not as distinct as in the present example. The underlying study (Lammertmann and Lütkenhöner 2000) is special in that very long stimuli were presented at a rather low rate. Moreover, experiment and analysis were designed to allow analysis of near-DC components of the response, which normally have a poor signal-to-noise ratio; by this means it was possible to analyze the temporal dynamics of the SF. In the present example, the SF decreases with a time constant of 3.6 s, falls to a much lower level immediately after stimulus offset, and then decays to the baseline (mean potential before the presentation of the next stimulus) with a time constant of 2.7 s. Qualitatively consistent results were obtained using a special direct-current (DC) MEG technique (Mackert et al. 1999). A similar waveshape, consisting of on- and off-response and a sustained component, was found with functional magnetic resonance imaging (fMRI) in both Heschl's gyrus and superior temporal gyrus (Harms and Melcher 2002).

2.2 Waves P50m, N100m, and P200m

The on-response typically shows three prominent peaks with latencies around 50, 100, and 200 ms (Fig. 28.3b, black curve). It has been suggested to denote AEF peaks by addition of the suffix 'm' to the names of the electrical counterparts (Hari et al. 1980). The three peaks are therefore denoted as P50m, N100m, and P200m, with the initial letter indicating the polarity of the peak and the number denoting the approximate latency. Although intracortical recordings suggest a more complex view (Steinschneider et al. 1994), it is assumed that a positive polarity essentially reflects a depolarization in cortical layers III or IV and a negative polarity a depolarization near the cortical surface, perhaps in layer II (Eggermont 2007). The latter condition evidently corresponds to intracellular currents that flow from superficial to deeper cortical layers (by definition, current flows from plus to minus). A peak of positive polarity thus corresponds to currents in the opposite direction. By combining MEG and magnetic resonance imaging, this has been confirmed for the peaks N100m and P200m (Lütkenhöner and Steinsträter 1998).

The N100m is a rather robust phenomenon that generally dominates the on-response. The P200m, by contrast, is much more variable and may be of such low amplitude that a clear peak is absent in some subjects (Hari et al. 1982; Jacobson et al. 1992b; Lütkenhöner et al. 2006). P200m is significantly enlarged in musicians when compared to individuals without musical training (Kuriki et al. 2006).

The earlier finding of a reduced P200m/N100m amplitude ratio in tinnitus patients (Hoke et al. 1989) presumably results from a problematic selection of the normal-hearing reference group since subsequent studies could not replicate that finding (Jacobson et al. 1991; Jacobson and McCaslin 2003). A high interindividual variability also impedes the investigation of the P50m, which is not consistently observed in all subjects (Pantev et al. 1996; Onitsuka et al. 2003; Lütkenhöner et al. 2006).

The off-response (Fig. 28.3a) is typically displayed with time zero referring to the stimulus offset (Fig. 28.3b, gray curve). There is a clear N100m, but no P50m, in accord with earlier studies (Hari et al. 1987; Pantev et al. 1996); the off-counterpart of the P200m is inconspicuous (small peak with respect to the dashed line). In another study (Pantev et al. 1996), an off-P200m was seen in 4 of 10 subjects. Microelectrode recordings from rat auditory cortex suggest that the off-response may be formed by a rebound after inhibitory input (Takahashi et al. 2004). Off responses seem to be more prominent in infants (Wakai et al. 2007).

Both the on– and the off-response are highly dependent on stimulation parameters. The phenomenon studied most systematically is the N100m on-response. The amplitude of this wave is roughly proportional to the square root of sensation level measured in dB (Bak et al. 1985), except near the threshold of hearing, where the amplitude is proportional to sensation level (Lütkenhöner and Klein 2007). Another crucial parameter is the interstimulus interval (Hari et al. 1982). If the interstimulus interval is reduced from 16 to 1 s, the N100m amplitude declines by about a factor of 3–4 (Campbell and Neuvonen 2007). Sequences of six short tone bursts presented at 500 ms intervals, with a 3.4 s silent interval between two sequences, elicited a strong amplitude reduction between the first and the second N100m, but no major difference between the second and the subsequent N100m responses (Lammertmann et al. 2001). Comparable results were found in auditory evoked potentials recorded with scalp electrodes (Fruhstorfer et al. 1970) or intracranially in patients undergoing presurgical evaluation (Rosburg et al. 2004). Although interstimulus intervals as brief as 500 ms (or less) do not usually prevent the development of the N100m, even a 1-s interval does not always ensure that the peak is found (Lütkenhöner et al. 2001). Other factors shaping the N100m response are stimulus duration (Joutsiniemi et al. 1989) and rise time (Biermann and Heil 2000). Moreover, the response significantly depends on the stimulus type, e.g., noise or tone (Lütkenhöner et al. 2006). Aspects such as temporal integration (Forss et al. 1993; Loveless et al. 1996), spectral composition (Jacobson et al. 1992a; Stufflebeam et al. 1998; Roberts et al. 2000; Seither-Preisler et al. 2003), and pitch (Crottaz-Herbette and Ragot 2000; Seither-Preisler et al. 2006b) also influence measurement.

2.3 Transition Responses

An AEF is elicited not only by a transition between silence and sound (on response) and vice versa (off response), but also by a transition from one sound to another, as studies of responses to vowel onsets after voiceless fricative consonants (Kaukoranta et al. 1987) and to noise/square wave transitions (Mäkelä et al. 1988) show. A transition between sounds usually involves a change in stimulus energy. Thus, the elicited response is not necessarily specific to the nature of the transition. A pitch-specific response without contamination by an energy-related component was measured by analyzing a transition from a noise to a regular interval sound (RIS) with the same intensity and bandwidth, but eliciting a sensation of pitch (Krumbholz et al. 2003). The transition from noise to RIS elicited a prominent pitch onset response (POR). The latency and size of the POR were directly related to the pitch value and its salience. Figure 28.4 shows exemplary data from a subsequent study (Seither-Preisler et al. 2006a). The first two seconds, representing the response to noise, qualitatively agree with the response to a tone burst (Fig. 28.3a). Then, however, a noise-RIS transition elicits a POR that is at least as strong as the N100m. There is an approximately linear increase in POR amplitude with the logarithm of noise duration (Seither-Preisler et al. 2004), and RIS induces a larger SF than noise (Fig. 28.4). A similar finding was made by comparing the responses to regular and irregular click trains (Gutschalk et al. 2002, 2004).

Other data, testing binaural pitch, converge with the POR data reported above (Chait et al. 2006). Comparison of the cortical and behavioral responses to Huggins Pitch (HP), a stimulus requiring binaural processing to elicit a pitch percept, with responses to tones embedded in noise (TN) – perceptually quite similar but physically different signals — confirm this idea. As in the above studies, the stimuli were crafted to separate the electrophysiological responses to onset of the pitch percept from the stimulus onset. These data show that, although physically distinct, both HP and TN are mapped onto similar substrates on lateral Heschl's gyrus by 150 ms post-onset. Cumulatively, the data across laboratories provide critical evidence that the pitch-onset response reflects central pitch mechanisms, in agreement with models postulating a single, central pitch extractor sensitive to abstract properties of pitch. A final example of the relevance of transitions lies in how elementary auditory experiences (objects) arise (Chait et al. 2007a,b). The acoustic biotope varies as a consequence of the (dis)appearance of acoustic sources, often manifested as transitions in the pattern of ongoing activity. How does the system detect and process such transitions? MEG data suggest that the dynamics and response morphology of the temporal-edge detection processes depend on the nature of the change. Measurements of auditory cortical responses to transitions between a sequence of random frequency tone pips (disorder) and a constant tone (order) show that these transitions embody key features of auditory edges. Early responses (from 50 ms post-transition) reveal that order-disorder transitions, and vice versa, are mediated by slightly different neural mechanisms. This suggests that cortex optimally adjusts to stimulus statistics—even when this is not required for overt behavior. The response profile (Fig. 28.7) bears a striking similarity to that measured from another order-disorder transition, between interaurally correlated and uncorrelated noise, radically different stimuli (Chait et al. 2005). This parallelism suggests a general mechanism that operates early in the processing stream on the abstract statistics of the auditory input, and is putatively related to the processes of constructing a new representation of the auditory scene.

2.4 Mismatch Negativity

One experimental approach that has been used extensively employs mismatch designs. In mismatch negativity (MMN) studies (in the case of MEG, mismatch field or MMF), a sequence of stimuli is presented such that one stimulus is often repeated and acts as a standard while a second stimulus is interspersed occasionally and is a deviant. The evoked response difference (subtraction) between deviant and standard stimulus is the mismatch response, and is an easily implemented and reliable indicator of change detection in an acoustic sequence. For example, small deviations in frequency, amplitude, timbre, etc. can be tested with the MMN/F design (Näätänen and Alho 1995). Higher-order sequences are also often investigated (e.g., a sequence of speech sounds, or words), highlighting the utility of the mismatch response to assess change detection more generally. The MMN/MMF likely has several cortical generators, including at least one in the superior temporal cortex and one in frontal cortex (Alho 1995). Change detection using MMN is distinct to the transient responses discussed above.

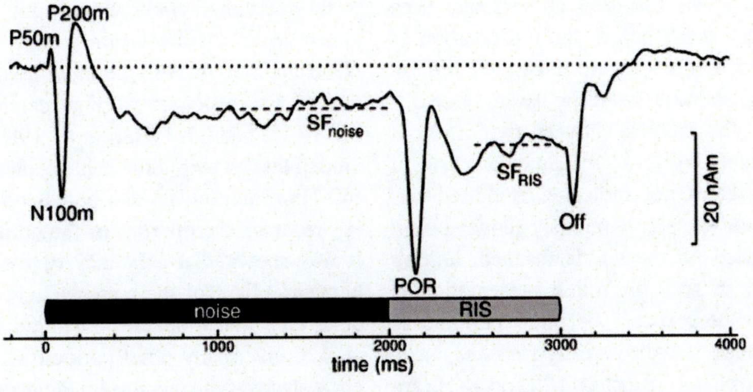

Fig. 28.4 Time course of the response elicited by a 2-s segment of random noise followed, without a gap, by a 1-s segment of regular interval sound (RIS) of identical bandwidth and intensity. The noise onset elicits a prototypical on-response (as in Fig. 28.3), with P50m, N100m, and P200m waves. A strong response is elicited also by the transition from noise to RIS sound; it is called pitch-onset response (POR). RIS sound evokes a stronger sustained field (SF) than noise. Derived from prior work (Seither-Preisler et al. 2006a, Fig. 5b)

2.5 Faster Transient Responses

The existence of a response peak around 50 ms (P50m) implies that there is earlier activity (rising slope of the P50m). However, experiments such as those considered above (Figs. 28.3, 28.4) are unsuited for a more detailed consideration of early activity. To achieve a sufficient enhancement of the signal-to-noise ratio at the beginning of the response, the number of averaged epochs must be increased by at least an order of magnitude, which is practicable only with shorter stimuli presented at a relatively high rate. In the example presented here (Fig. 28.5), a response was elicited by clicks presented at mean intervals of 350 ms. As distinct from the previous figures, not a single waveform (estimated dipole moment) is shown, but the time courses in single magnetometer channels. About 20 ms after click presentation (corresponding to the travel time from the periphery to the cortex), the activity sharply increases to a first maximum around 30 ms, P30m. After a brief reduction in the overall activation level, a second maximum occurs at ~60 ms. The two peaks and the intervening valley are considered counterparts of the Pa, Nb, and Pb (also called P1) waves of the middle-latency auditory evoked potential (Picton et al. 1974; Eggermont and Ponton 2002). But it is appropriate to be cautious since both the magnetic and the electrical responses represent a conglomerate of contributions from various cortical sources, and the mixing ratios may not be entirely the same. This would explain why simultaneous recordings of the two types of responses may show significant differences (Yvert et al. 2001). The N100m and P200m waves are absent, owing to the high stimulus repetition rate (Fig. 28.5).

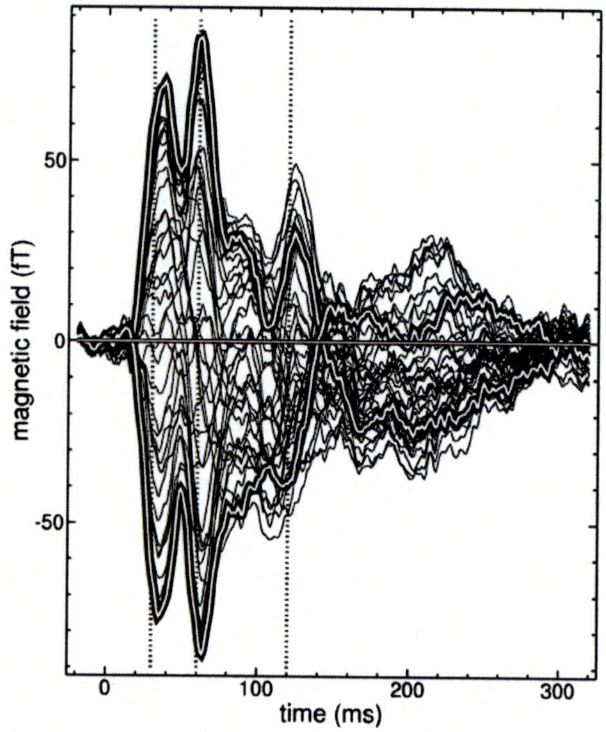

Fig. 28.5 Time course of the response elicited by a click (interstimulus interval 315–385 ms). The response waveforms at 37 locations over one hemisphere were superimposed; the maximum response of each polarity was highlighted (two *white curves* on *black background*). The peaks near 30 and 60 ms (*dotted lines*) and the valley between presumably correspond to the waves Pa, Nb, and Pb of the middle-latency auditory evoked potential. The negative peak is not conspicuous in this example, but the interindividual variability is high, and much more pronounced Nb correlates were found in other experiments (Yoshiura et al. 1996). Later activity has a relatively small amplitude in the present example, and the asymmetry between the two polarities suggests a non-dipolar spatial pattern. Derived from prior work (Lütkenhöner et al. 2003b, Fig. 1a)

2.6 Steady-State Responses

If the stimulus repetition rate is further increased, the situation arises that even faster response components such as the P30m do not fade away before the presentation of the next stimulus. The consequences are illustrated with an example (Fig. 28.6). The curves show responses to three different series of clicks. The interval between two clicks is initially 100 ms, corresponding to an instantaneous rate of 10 Hz (times of click presentation indicated by dotted vertical lines). Then the interval is continuously reduced until the periodic rate indicated on the left is reached. During the time range marked in gray, the click presentation is strictly periodic and so the response becomes periodic as well. This type of response is called steady-state response. Steady-state responses to clicks presented at 20 and 40 Hz are strong, whereas only weak steady-state responses are found at 30 Hz (Fig. 28.6). The effect is qualitatively explained by considering how the major peaks of the transient responses to the individual clicks would sum. The explanation is consistent with a convolution model for steady-state activity (Gutschalk et al. 1999), and with earlier work in which 20- and 40-Hz responses were successfully reconstructed from the 10-Hz response (Hari et al. 1989). The latter study concluded that an amplitude enhancement at a specific frequency (40 Hz in particular) can be explained without hypothesizing resonance properties of the cortical network. Moreover, it was shown that a latency derived from the relationship between phase of the response and stimulation rate (group delay) is of questionable physiological relevance. Apart from click trains, many other periodic stimuli have been used to elicit steady-state responses (Picton et al. 2003). Steady-state responses have been used profitably to test, for example, how simultaneous amplitude modulation (AM) and frequency

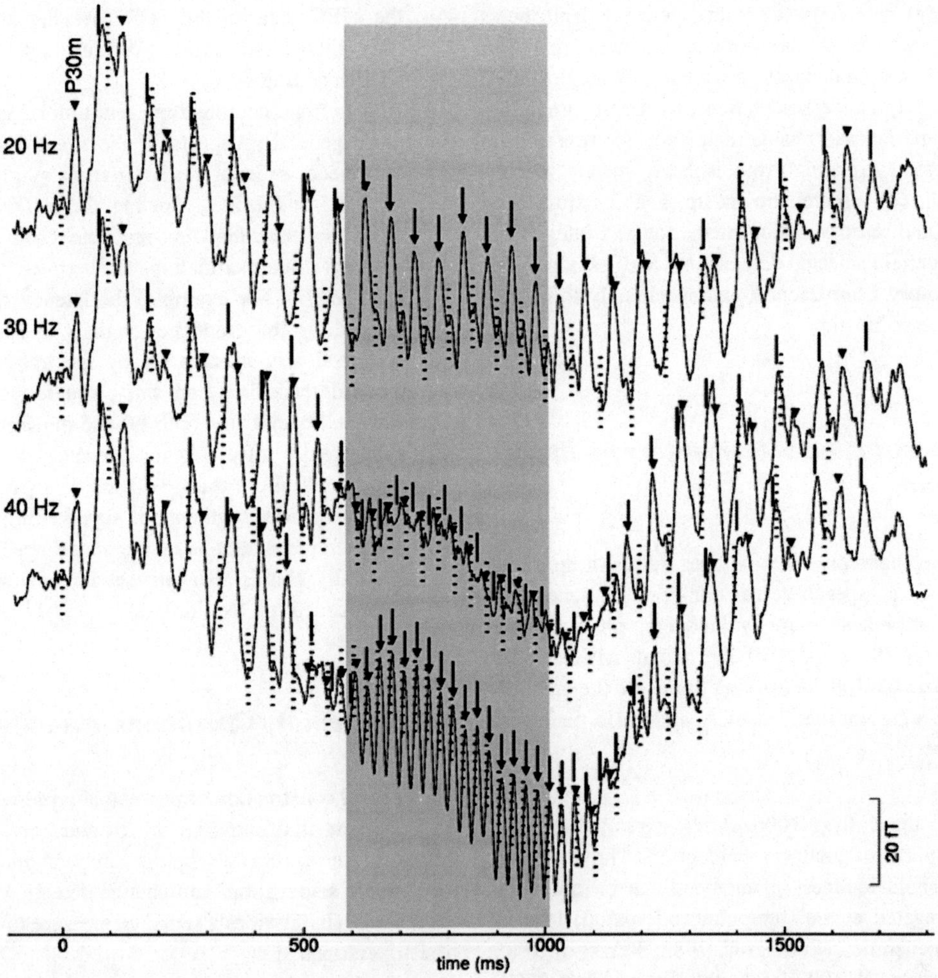

Fig. 28.6 Transition from a sequence of transient responses to a periodic (steady-state) response, and vice versa. In the initial 200 ms, the three response curves are indistinguishable. The first click (click-presentation times marked by *vertical dotted lines*) elicits a P30m and a second positive peak at ~80 ms (*inverted triangle* and a *vertical bar*, respectively). The second click at 100 ms elicits a P30m around 130 ms (*inverted triangle*). This second P30m apparently superimposes on the falling slope of the 80-ms peak elicited by the first click. The intervals between subsequent clicks are next reduced step-by-step until the periodic repetition rate indicated on the left of each curve is reached, which is maintained for about 400 ms (time range marked in gray). The second half of the click series is a mirror image of the first half. The periodic response caused by periodic stimulation is the steady-state response. In this example, the amplitude of the steady-state response is high at click repetition rates of 20 and 40 Hz, and lower at 30 Hz. In the first two cases, P30m peaks likely coincide with 80-ms responses to previous clicks, whereas in the latter case they are assumed to occur in the middle of two 80-ms responses. Based on prior work (Lütkenhöner et al. 2004, Fig. 2)

modulation (FM) are encoded (Luo et al. 2006) and how long acoustic sequences (typical of speech or music) are reflected in the steady-state responses (Patel and Balaban 2004).

3 Domains of Magnetoencephalographic Research in Auditory Cognition

Large portions of the human auditory system are located in sulcal cortex, on dorsal aspects of the superior temporal gyrus (Fig. 28.10). This includes core and belt areas associated with the anatomic structures of Heschl's gyrus (transverse temporal gyrus), the planum temporale, and the planum polare. This anatomic fact—coupled with the millisecond temporal resolution of MEG—renders the technique optimally suited for recording neurophysiological activity noninvasively with remarkably high fidelity, and for investigating how acoustic signals are transformed to yield the auditory representations that form the basis for speech perception, music cognition, and other aspects of auditory cognition.

While many taxonomic schemes are possible, we adopt a simple classification to organize the numerous studies using MEG, identifying several (somewhat overlapping) domains of research: (a) work on elementary perceptual attributes derived from acoustic signals—pitch, loudness, and timbre; (b) work on elementary processing strategies used to generate perceptual representations, including streaming, integration, binding, and change detection; (c) research on speech processing, ranging from isolated vowels to connected speech; (d) research on music; and (e) studies on multisensory and sensory-motor interaction and integration. We briefly highlight selected data on how MEG studies contribute to auditory neuroscience, and specifically to models of auditory cortex function.

3.1 The Construction of Elementary Auditory Attributes

Auditory perceptual representations, regardless of their cognitive identity (e.g., speech versus non-speech), necessarily reflect basic attributes (such as loudness, pitch, timbre) that derive from the physics of the signal. MEG studies have been successful at identifying some of the relations between early neuromagnetic activity and basic perceptual representations.

(i) As noted above, the N100m can be exploited to investigate aspects of loudness perception. Threshold and non-threshold loudness phenomena can be quantified and segregated at the (temporal and spatial) level of N100m generation (Reite et al. 1982; Bak et al. 1985; Stufflebeam et al. 1998; Lütkenhöner and Klein 2007). This is of special interest since psychophysical research shows that a stable percept of loudness is generated at ~200 ms post-stimulus onset (Moore 2003). Models of loudness perception therefore must consider the differing results of a 200-ms time constant, on the one hand, and early sensitivity to intensity evident in the N100m, on the other. Both types of data are highly robust and replicable and require explanation.

(ii) The computational basis of pitch is a vast field. Several recent MEG studies (cf. Section 2.3) make a critical contribution in that regard. Considerable data show that a response generated on the lateral aspect of Heschl's gyrus can be viewed as a pitch-onset-response (POR), regardless of whether the pitch is evoked monotonically or dichotically (Krumbholz et al. 2003; Seither-Preisler et al. 2004, 2006a,b; Chait et al. 2006). Such data thus implicate a local region in lateral Heschl's gyrus in the calculation of pitch at a relatively abstract level, given that the experiments used rather different stimulation that included click trains, iterated rippled noise, dichotic Huggins pitch, and other materials. In this domain, too, the MEG data on the POR (peaking at 100–150 ms after pitch onset) support certain models of pitch and challenge others.

(iii) How cortical neurons represent timbral information, or more generally aspects of the spectral envelope, has become a topic of research from single-unit studies to fMRI and MEG. Viewing the N100m alone, provides evidence that this response covaries in latency with envelope modulations (Roberts et al. 2000; Ritter et al. 2007). For example, the latency of the N100m elicited by low-frequency signals (~100–500 Hz) is systematically affected the by spectral envelope (e.g., the difference between sine-, square-, and saw-tooth waves). Specifically, both F0 and the spectral envelope concurrently affect latency. Because the N100m cannot sample more than ~40 ms of signal (cf. Section 3.2), very brief segments of signal suffice to construct usable representations of the sound spectral envelope. Naturally, this is also relevant for how speech sounds are encoded (cf. Section 3.3).

3.2 Elementary Operations in Auditory Cortex

The cortical construction of perceptual representations relies on processing algorithms that are, by and large, shared across domains. These processes include temporal integration, auditory stream segregation, and change detection. The N100m response again provides a sensitive measure to evaluate such basic operations.

(i) The afferent auditory pathway is subject to temporal summation and integration by the neural substrate subserving the analysis. Neural elements reflect both a temporal integration constant and the temporal resolution afforded by a given response. An N100m temporal integration window of ~25–40 ms must be assumed, i.e., the N100m is affected by acoustic information up to 40 ms, but signals outside this temporal window do not affect its properties (Gage and Roberts 2000; Gage et al. 2006), except at rather low stimulus levels (Lütkenhöner and Klein 2007). Within this brief temporal integration window, acoustic events (e.g., gaps) can be resolved to 2 ms. N100m timing and amplitude show a resolution that is well matched to psychophysical gap detection thresholds while integrating over durations commensurate with the temporal order threshold. Other evoked fields (Fig. 28.3) can be examined in similar studies, forming the basis for a larger-scale model of temporal integration in human auditory cortex.

(ii) Work on stream segregation found auditory cortex displays many aspects of streaming (Micheyl et al. 2007).

Interpretable auditory objects or streams must be assembled from complex input arising from many sources, and here, too, early evoked fields are useful as dependent measures. There is a strong correlation between early cortical activity up to and including the N100m and the representation of separate streams (Gutschalk et al. 2005).

(iii) The detection of change has principally been investigated using the mismatch response (MMN or MMF)—a response pattern that also has significant supratemporal sources. Local changes in stimulus statistics are also reflected in neuromagnetic responses (cf. Section 6.1). The response to change can be used to test where, when, and how changes in stimulus statistics are detected (Fig. 28.7) (Chait et al. 2007a). Stimulus statistics are reflected in change responses by 50–60 ms post change onset.

3.3 Speech Processing: Overview

Investigating the cortical basis of speech processing has been central to MEG research. Because auditory evoked responses exhibit such stereotypical morphology and timing (Fig. 28.3), how these responses are modulated by speech input has been a foundational question. Clicks or brief tone burst elicit the cascade of responses visualized as the P50m, N100m, and P200m. How the responses elicited by single vowels or consonant-vowel (CV) syllables or even single words appear in comparison has been studied extensively, even in nonspeech control experiments (Mäkelä et al. 1988). In this context, it is critical to note that speech perception is not monolithic. The theoretical and neurobiological machinery invoked is quite distinct when studying isolated vowels, isolated CV syllables, isolated

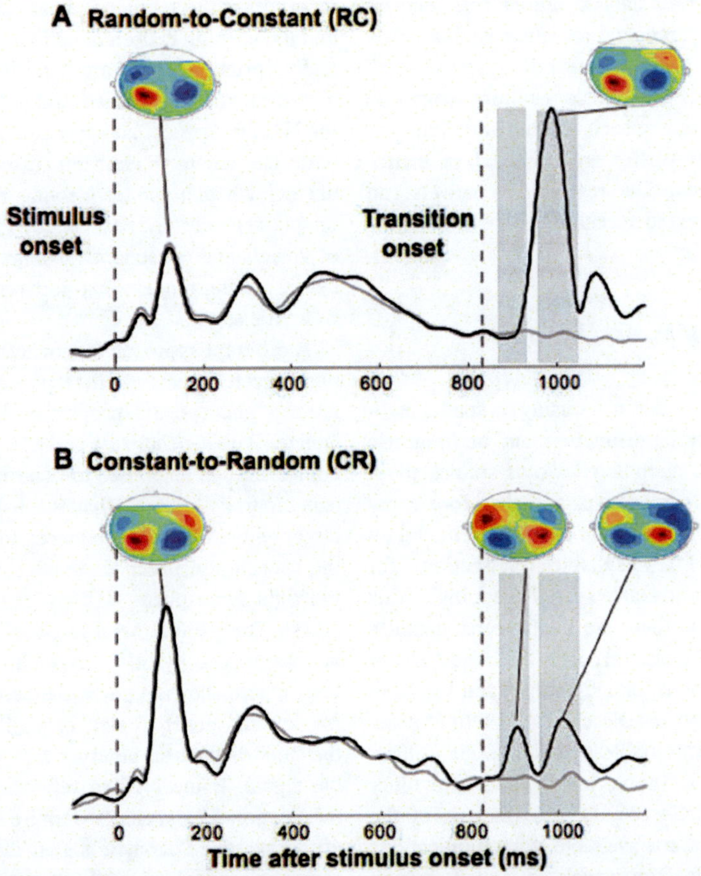

Fig. 28.7 Grand average of the across channels and subjects responses for a stimulus of brief tone pips at randomly changing frequencies between 222 and 2000 Hz (random condition) alternating with a tone (constant condition). *Grey line*, the no-change control condition. Contour maps show the magnetic field distribution at critical time points. Both *panels* illustrate a robust N100m response at the beginning of the stimulus. (**a**) The response profile from random-to-constant shows a single large response after the transition, with a contour resembling the N100m. (**b**) Constant-to-random response with two peaks after the change onset. Because the response profile, timing and distribution differ between these two conditions that are matched along several stimulus dimensions, auditory cortex may maintain an on-line model of the local statistics of the stimulus. The direction of change is critical since in one case a representation is constructed from randomly distributed pips (RC); in the other condition (CR) the representation is destroyed. Adapted from the original (Adapted from Chait et al. 2007b)

words, or connected speech (Hickok and Poeppel 2007; Poeppel et al. 2007). Consequently, generalizations about the neural basis of speech perception should consider that psycholinguistically distinct levels of analysis are associated with varied neurobiological implementation. A further terminological clarification is necessary, since we focus on speech perception but not language comprehension more broadly construed. Language comprehension can be driven by auditory (speech), visual (text, sign), or somatosensory (Braille) information and operates on supramodal linguistic representations. Speech perception is the set of processes transforming acoustic input into a format suitable for language comprehension and further computation (morphology, syntax, etc.). We focus on speech perception, among the many MEG studies on language. Ignoring this fundamental distinction can lead to profound confusion about which computational subroutines are actually at stake.

The study of speech has occurred largely independently from work on basic attributes and operations (cf. Sections 1 and 2). Although basic perceptual attributes such as pitch and primitive operations such as integration apply to all sounds, few MEG studies link basic auditory cognition and speech properties. Many speech studies have inquired whether special signal properties are reflected in neuromagnetic responses. We consider research on vowels and syllables, words, and connected speech.

3.4 Vowels, Consonants, and Syllables

A few distinct approaches are taken, focusing on spatial mapping, on the speech/nonspeech distinction, and on linguistic abstraction (from sound to phonology). Some studies adopt a strongly localizationist perspective, and therefore show extensive source modeling data on the responses elicited by vowels or consonant-vowel (CV) syllables (cf. Section 4 for analysis of dipole localization and spatial mapping). This research often assumes that there are likely to be spatially organized maps in superior temporal cortex that reflect where speech sounds are represented (phonemotopy) and has been used for the analysis of vowels (Diesch et al. 1996; Obleser et al. 2003b) and consonants (Obleser et al. 2003a, 2006). The N100m is subjected to dipole localization as a function of stimulus type (cf. Section 1). For the role of the sustained field, fewer data are available with implications for auditory cortex models. The vowels /a/ and /i/, which are well separated in formant (F1-F2) space, lie far apart on the anterior dorsal aspect of the supratemporal gyrus (STG), the planum polare. In contrast, the spatial position of vowels more closely aligned in formant space is correspondingly closer in auditory cortex. Such work builds on two crucial features: the putative existence of maps in cortex that encode the relevant acoustic features spatially (here corresponding to frequency) and, second, the ability of MEG source modeling to resolve the relevant spatial differences. This relates the systematic representation of speech sounds to cortical maps (speech sound identity is determined by its position on a cortical sheet), and to the results from phonology, raising the possibility that what is mapped may be more abstract than frequency and amplitude. Consonants with conflicting place-of-articulation features are also mapped more distantly in cortical space, which suggests a potential mapping from place of articulation to brain space. That measurements based largely on the N100m stimulate such hypotheses is noteworthy.

It remains controversial whether such maps of speech sounds can be reliably identified in human auditory cortex. The N100m response to speech sounds also varies in systematic ways in time (latency) (Diesch et al. 1996; Roberts et al. 2000; Obleser et al. 2003b). Complementary investigations incorporate the temporal dynamics of the acoustic signals and the ensuing neuromagnetic responses. Perhaps temporal coding principles play a critical role in speech sound representation. F0 as well as spectral peaks are reliably reflected in the N100m latency, an approach that connects more clearly with the auditory elements discussed above. It remains unclear whether models relying more on phonemotopy or on, phonemochrony, best capture the neuronal representation of vowels and consonants. Connecting more explicitly with neural coding models that derive from animal research will be a vital step.

A different approach tests whether speech versus matched non-speech signals elicit responses differing in amplitude, latency, and spectral properties. Measuring the N100m to isolated vowels (/a/, /u/) and CV syllables (/pa/, /ka/) and comparing the responses to materials closely matched spectrotemporally showed that the left N100m is significantly larger and differentiates between the stimulus types, whereas the speech-nonspeech distinction is not robustly visible in the right hemisphere N100m (Fig. 28.8) (Parviainen et al. 2005). The gamma band response differed between speech and nonspeech by 60 ms poststimulus onset, with the right hemisphere showing particular sensitivity to nonspeech and the left to speech (Palva et al. 2002). Such data suggest that this distinction emerges 60–100 ms after the onset of the signal. If the N100m reflects at most 40 ms of signal (cf. Section 6.2), such data require a model that explains how 40 ms worth of acoustic signal can be identified as speech versus nonspeech, given the receptive field properties of neurons in the auditory hierarchy. The data imply that small durations of signal can support subtle distinctions between signal types in the N100m, generated in superior temporal cortex.

The responses to syllables are typically more complex even superficially, encompassing components elicited by the

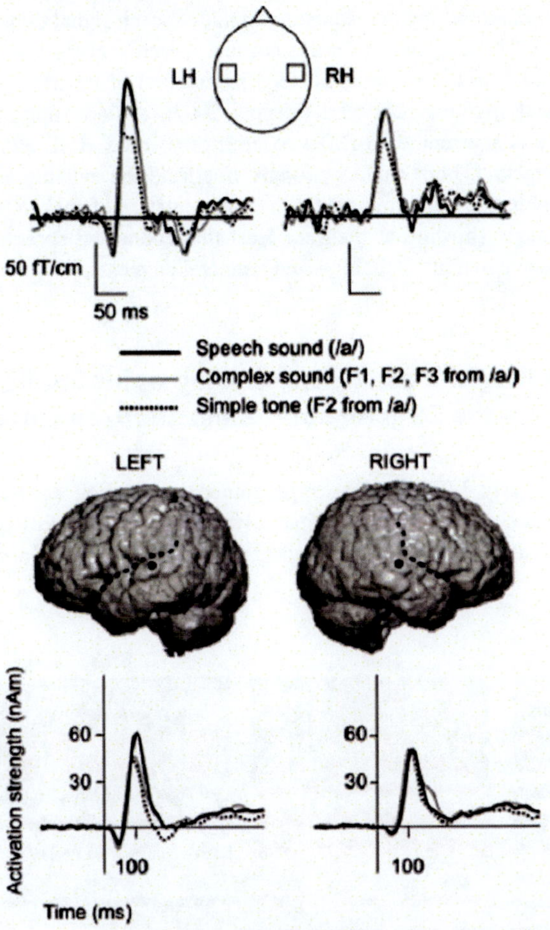

Fig. 28.8 The *top panels* show areal means from several channels for the N100m recorded from both hemispheres. In the left hemisphere, the speech condition has robustly larger responses, discriminating between speech, matched complex sounds, and tones. The *lower panels* show the dipole model fit in the three-dimensional brain (Sylvian fissure marked) and dipole strength over time (*bottom traces*). Again, the speech condition showed the largest response in the *left*, whereas the other two conditions were not differentiated at the N100m. Adapted from the original source (Adapted from Parviainen et al. 2005)

syllable features such as bursts, closure releases, and voicing onsets. The neuromagnetic fields depend on a variety of acoustic-phonetic, phonological, and semantic-contextual factors. Early work capitalizing on MEG's temporal resolution established that intrasyllabic distinctions could be identified. The burst of energy associated with closure release and the onset of voicing, when sufficiently separated in time, as in a voiceless stop such as /t-a/, can be resolved, yielding responses resembling an N100m and N100m, (see Mäkelä et al. 1988 for non-speech controls). The detailed acoustic properties of syllable onsets are also reflected in the N100m, whose amplitude, timing, and lateralization is sensitive to the distinction between stops and continuants (Gage et al. 1998). Such sensitivity transcends acoustic-phonetic factors. Data from (phonological) nasalization restrictions in English show that MEG responses between 60–150 ms poststimulus onset reflect knowledge of the abstract phonological generalizations that a speaker brings to the perceptual task (Flagg et al. 2006). Finally, the response to syllables is conditioned by top-down expectations such that the response to a syllable sharply differs after the N100m, from ~200 ms forward, when presented in isolation versus contexts that facilitate lexical access and other higher-order linguistic subroutines (Bonte et al. 2006).

Mismatch designs (cf. Section 6.4) are tools to study abstract phonological representations in speech by using experimental designs showing auditory cortex sensitivity to a change in loudness or pitch, for example. Psycholinguistic studies use subtle, often crosslinguistic, mismatch designs to test how the native language phonology constrains the early analysis of speech sounds. This approach has established that both language-specific and abstract (phonological) representations can be probed by using mismatch designs, which in turn implies that, by 150 ms, these effects are established (Näätänen et al. 1997; Phillips et al. 2000; Kazanina et al. 2006).

3.5 Words

In contrast to the basic approach exemplified by research on isolated speech sounds, an intermediate level is represented by spoken word recognition. The typical concerns are where, when, and how the recognition process occurs. This implies that acoustic information must interact with long term memory (words), and must therefore be transformed into a usable representation. Experimental design plays a more crucial role in this research. A study using a canonical mismatch design concluded that, by 150 ms after stimulus onset, lexical access has been initiated and phonological and semantic information are evident (Pulvermüller et al. 2006). Because the MMNm originates on dorsal STG and peaks at 150 ms, it is not surprising to see effects at that latency. Another study implemented the mismatch idea by presenting quadruplets of words which either generated semantic or phonological expectations and assessing the evoked fields. Phonological information is reflected reliably by the N100m whereas semantic information appeared from 200 ms on. These data are, thus, more compatible with a view that acoustic-phonetic-phonological analysis executed in superior temporal cortex precedes semantic processing, although both types of information are readily available very early in processing (Uusvuori et al. 2008).

3.6 Connected Speech

Some experiments use connected speech to study how ecologically natural speech is segmented. Listeners are presented sentences at different compression ratios, thereby parameterizing and assessing their intelligibility. A principal component analysis of the MEG showed correlations between auditory cortical phase-locking to the speech envelope and speech intelligibility. Successful phase-locking of the response to the envelope seems a key feature to insure intelligibility. At compression values compromising sentence intelligibility, phase locking was also poor (Ahissar et al. 2001). Analysis of single trials of spoken sentences show that the phase pattern of the cortical theta band (4–8 Hz) response tracks and discriminates between spoken sentences. This discrimination ability is correlated with intelligibility (Fig. 28.9). The data suggest that a 200 ms temporal window (period of theta oscillation) segments the incoming speech signal, resetting and adjusting to track speech signal dynamics. The mechanism for this cortical speech analysis may be based on the stimulus-induced modulation of inherent cortical rhythms (Luo and Poeppel 2007), a view supported by concurrent EEG-fMRI recordings (Giraud et al. 2007). Together, both studies strongly implicate the syllable as a computational primitive for the representation of spoken language, showing at the very least that connected speech is segmented into syllable-sized temporal elements.

4 Magnetoelectroencephalographic Studies on the Structure of Human Auditory Cortex

In early MEG studies, high source localization accuracy was considered the principal advantage compared to EEG. A more reserved view now seems appropriate. Source

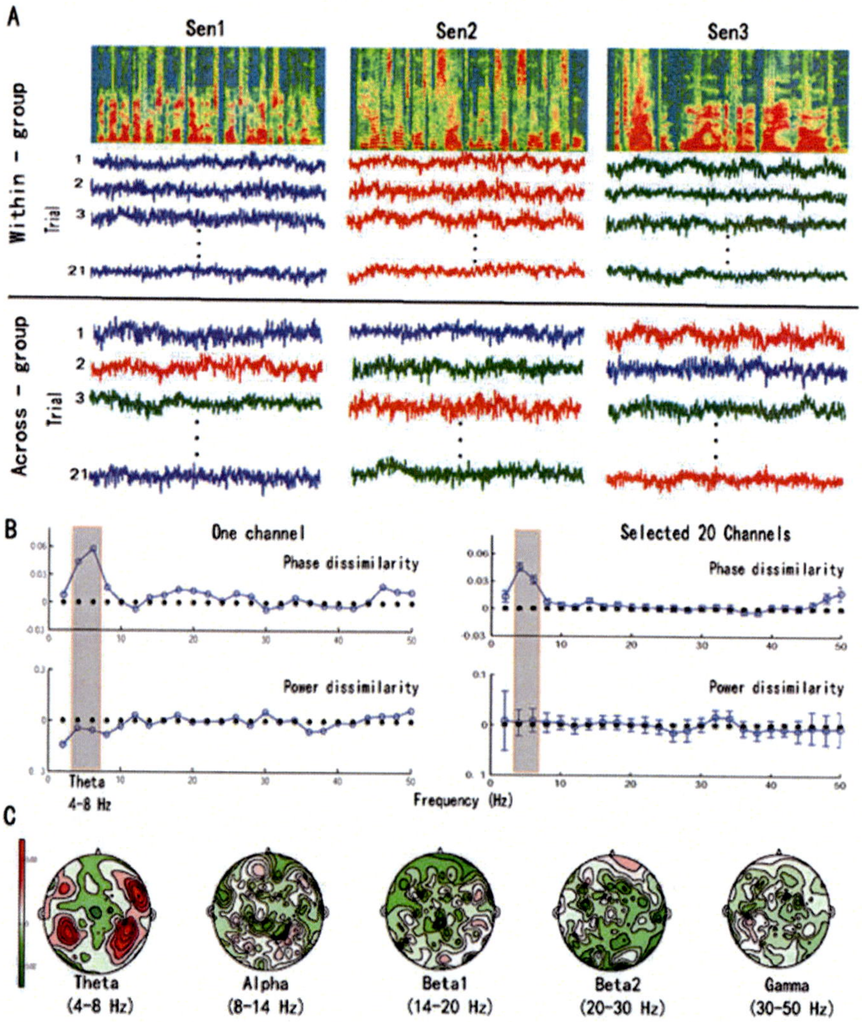

Fig. 28.9 (**a**) Spectrograms of three sentence stimuli and single-trial MEG traces. The analysis evaluated phase coherence across single trials of neuromagnetic responses to the same stimulus (within-group coherence) and compared the response to mixed trials (across-group). (**b**) Phase dissimilarity and power dissimilarity plots for one channel (*left*) and 20 auditory channels (*right*). Strong phase dissimilarity was seen in the theta band, 4–8 Hz, suggesting that phase coherence in that band discriminates the sentence types in single trials. (**c**) Contour plots of the phase dissimilarity plots showing the distribution over auditory cortex (right lateralized). Other responses showed no organized spatial pattern. The recordings suggest that the phase of the theta band response encodes the acoustic envelope of the sentences in single trials of MEG data. Reproduced from the original source (Reproduced with permission from Luo and Poeppel 2007)

localization from MEG is always uncertain, unless the signal arises from one focal source, or a few such well separated sources (Halgren 2004). This prerequisite is not fulfilled in typical MEG studies of audition so that inferences regarding the structure of auditory cortex need to be examined critically.

4.1 Localization of Primary Auditory Cortex

MEG is relatively insensitive to subcortical activity. Thus, supposed that the activation of belt regions of the auditory cortex by nonspecific afferents (Lakatos et al. 2005) can be neglected, the very beginning of the AEF elicited by a short stimulus may be assumed to result essentially from a single focal source: primary auditory cortex. A source analysis of the earliest phase of the response should consequently allow localization of that region. A study of the P30m elicited by clicks, focusing on the initial rise about 20 ms post-click, suggests that this is indeed possible (Lütkenhöner et al. 2003b). Coregistration of the estimated dipoles with magnetic resonance images suggested that primary auditory cortex is near the retroinsular origin of Heschl's gyrus, in good agreement with intracranial recordings (Liégeois-Chauvel et al. 1994).

4.2 Tonotopic Maps

A study of steady-state AEF elicited by amplitude-modulated tones showed that a 4–5 octave frequency change shifted the location of the estimated dipole by about 1 cm, which was interpreted as indicative of a tonotopic map in auditory cortex (Romani et al. 1982). This report triggered numerous investigations with other experimental setups (Lütkenhöner et al. 2003a). Not all authors found evidence of a tonotopic map in their data, but if they did, the typical conclusion was straightforward, with higher frequencies activating more medial regions of auditory cortex than lower frequencies. For the N100m such effects may be highly reproducible in individual subjects (Lütkenhöner and Steinsträter 1998), but the interindividual variability is bewildering (Lütkenhöner et al. 2003a). In most cases, the dipole location either exhibited no significant frequency dependence, or the dipoles for the investigated frequencies were not orderly aligned, or the data disagreed with the single-dipole hypothesis. These results do not support the utility of MEG as a tool for the study of tonotopic maps in auditory cortex. The main obstacle for a successful examination of tonotopy appears to be that the AEF typically arises from multiple sources (Hari 1990; Schreiner 1998; Eggermont and Ponton 2002). With some caveats, the location of the estimated dipole may be considered the center of gravity of the activated cortical areas, where the strength of cortical activation assumes the role of the mass. It is unlikely that stimulus parameters such as frequency will affect the strength and timing of the activities in different cortical areas in precisely the same way. Thus, the center of gravity estimated from multiple sources can, unfortunately, give the illusory impression of a single source with frequency-dependent location, but this would be a pseudotonotopy, without a valid structural correlate in auditory cortex (Lütkenhöner and Mosher 2007).

4.3 Dipoles Representing Multiple Cortical Sources

Even though MEG seems to be unable to reveal structural details of single cortical areas, a cautious interpretation of estimated dipole locations may lead to valuable conclusions. A dipole source analysis may suggest, for example, that one type of activity predominantly originates from more anterior cortical regions than another. But it is crucial to appreciate that each dipole likely represents multiple, spatially distinct sources whose precise locations are unknown. In the case of two sources, for example, the estimated dipole would be expected to be on a line joining them, with the exact location depending on the dipole moments associated with each source. The estimated dipole location is commonly interpreted as the center of gravity of the contributing sources, as noted above But this view requires caution. If the net currents in the contributing sources are in opposite direction (so that the respective magnetic-field contributions partially cancel), the estimated dipole may be outside the activated cortical region (Lütkenhöner and Mosher 2007). Despite this potential pitfall, the idea of a center of gravity is a useful concept if applied with care.

The N100m components of on- and off-response appear to have a similar origin (Hari et al. 1987; Pantev et al. 1996; Noda et al. 1998; Lammertmann and Lütkenhöner 2001). In right-handed subjects, the right-hemispheric N100m dipole was found 6 mm anterior to the left-hemispheric counterpart (Nakasato et al. 1995). While the sources are predominantly in planum temporale (Lütkenhöner and Steinsträter 1998), there seems also to be a component originating in the lateral part of Heschl's gyrus (Godey et al. 2001). Another study (Sams et al. 1993) suggested an anterior and a posterior subcomponent. The posterior subcomponent adapts as sound novelty decreases, and this subcomponent is likely involved in the gating of novel sounds to awareness (Jääskeläinen et al. 2004).

While the P50m dipole was found at a similar location as the N100m dipole (Hari et al. 1987; Mäkelä and Hari 1987; Kanno et al. 2000), the P200m dipole was consistently found at a more anterior location (Hari et al. 1987; Pantev et al. 1996; Lütkenhöner and Steinsträter 1998; Lammertmann and Lütkenhöner 2001). The dipole estimated for the SF either had a similar locus as the N100m dipole (Lammertmann and Lütkenhöner 2001) or was slightly more anterior (Hari et al. 1987; Pantev et al. 1996). Other authors (Gutschalk et al. 2004) distinguished an anterior and a posterior component of the SF. While the anterior component was related to temporal pitch processing, the posterior component was sensitive to stimulus intensity. The localization results obtained for the P30m are not completely consistent. A dipole in the vicinity of the N100m dipole has been seen (Yoshiura et al. 1996), or the source was localized in the dorso-postero-medial part of Heschl's gyrus (Godey et al. 2001), in approximate agreement with other results (Yvert et al. 2001). The POR dipole is described as 12 mm anterior to the N100m dipole, on average (Krumbholz et al. 2003). The POR may represent a source, or sources, on medial Heschl's gyrus (Fig. 28.10), adjacent to a larger region in the anterolateral half of Heschl's gyrus where functional imaging studies find a pattern of activation that is highly correlated with the degree of regularity in RISs (Griffiths et al. 1998; Patterson et al. 2002).

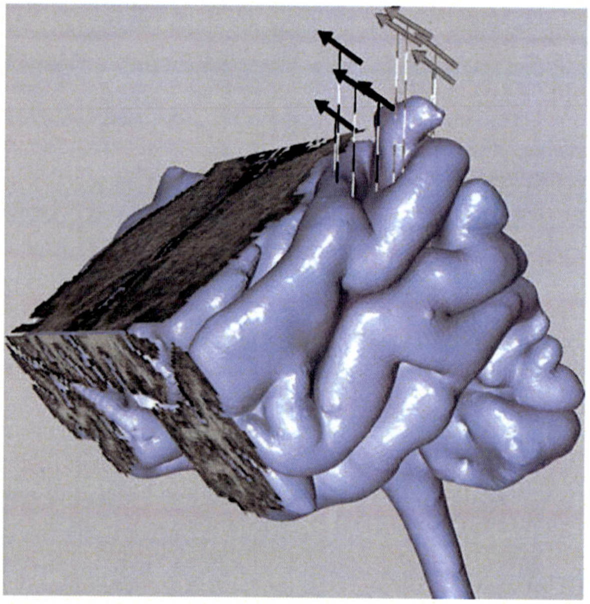

Fig. 28.10 Dipole locations of the POR (*black*) and the N100m (*gray*) for one listener, estimated from four measurement sessions and projected into a three-dimensional reconstruction of the listener's left temporal lobe. The dipoles are shifted upward by 3 cm from their actual position to prevent them from being partially obscured beneath the cortical surface. While the dominant generators of the N100m appear to be in planum temporale, the POR may arise from a source, or sources, on medial Heschl's gyrus. Scenery from the original source (Krumbholz et al. 2003, Fig. 6) modified and viewed from a different perspective

5 Conclusions and Perspectives

Other than invasive (clinically motivated) recording in the human auditory cortices, MEG remains the key method to assess auditory function electrophysiologically. If we accept the view that many of the elementary mechanisms identified in animal preparations will be foundational for human auditory function as well (bandwidth, modulation, spectrotemporal receptive field organization), it is desirable and necessary to understand how human auditory cortex responds to basic attributes of auditory signals. A major advantage of MEG is that it enables concurrent psychophysical and physiological studies. The experimental dimension—both physiologically and psychophysically—should be emphasized more strongly. While a core goal will be to understand those functions that appear to be unique properties of human audition, such as speech perception, building more explicit models from animal work can guide future research. In the animal domain there are serious efforts to correlate hemodynamics with electrophysiology. MEG research can contribute to this effort by developing experimental paradigms of interest to animal physiologists.

On balance, we advocate using MEG largely as an electrophysiological tool (with superior sensitivity in single subject studies at the level of single trials) rather than an imaging method with more modest prospects for precise localization. The power of MEG in localizing functional sources within a reasonable anatomic context such as the gyral and sulcal location can provide a framework for the appropriate electrophysiological analysis of the encoding that particular responses perform. MEG can serve as a functional bridge between human auditory processing and physiology as performed in awake behaving animals.

References

Ahissar E, Nagarajan S, Ahissar M, Protopapas A, Mahncke H, and Merzenich MM (2001) Speech comprehension is correlated with temporal response patterns recorded from auditory cortex. *Proceedings of the National Academy of Sciences of the United States of America* **98**:13367–13372.

Alho K (1995) Cerebral generators of mismatch negativity (MMN) and its magnetic counterpart (MMNm) elicited by sound changes. *Ear and Hearing* **16**:38–51.

Baillet S, Mosher JC, and Leahy RM (2001) Electromagnetic brain mapping. *IEEE Signal Processing Magazine* **18**:14–30.

Bak CK, Lebech J, and Saermark K (1985) Dependence of the auditory evoked magnetic field (100 msec signal) of the human brain on the intensity of the stimulus. *Electroencephalography and Clinical Neurophysiology* **61**:141–149.

Biermann S and Heil P (2000) Parallels between timing of onset responses of single neurons in cat and of evoked magnetic fields in human auditory cortex. *Journal of Neurophysiology* **84**: 2426–2439.

Bonte M, Parviainen T, Hytonen K, and Salmelin R (2006) Time course of top-down and bottom-up influences on syllable processing in the auditory cortex. *Cerebral Cortex* **16**:115–123.

Campbell T and Neuvonen T (2007) Adaptation of neuromagnetic N1 without shifts in dipolar orientation. *NeuroReport* **18**:377–380.

Chait M, Poeppel D, and Simon JZ (2006) Neural response correlates of detection of monaurally and binaurally created pitches in humans. *Cerebral Cortex* **16**:835–848.

Chait M, Poeppel D, and Simon JZ (2007a) Stimulus context affects auditory cortical responses to changes in interaural correlation. *Journal of Neurophysiology* **98**:224–231.

Chait M, Poeppel D, de Cheveigne A, and Simon JZ (2005) Human auditory cortical processing of changes in interaural correlation. *Journal of Neuroscience* **25**:8518–8527.

Chait M, Poeppel D, de Cheveigne A, and Simon JZ (2007b) Processing asymmetry of transitions between order and disorder in human auditory cortex. *Journal of Neuroscience* **27**:5207–5214.

Crottaz-Herbette S and Ragot R (2000) Perception of complex sounds: N1 latency codes pitch and topography codes spectra. *Clinical Neurophysiology* **111**:1759–1766.

Diesch E, Eulitz C, Hampson S, and Ross B (1996) The neurotopography of vowels as mirrored by evoked magnetic field measurements. *Brain & Language* **53**:143–168.

Eggermont JJ (2007) Electric and magnetic fields of synchronous neural activity. In: Burkard RF, Eggermont JJ, and Don M (eds). *Auditory Evoked Potentials. Basic Principles and Clinical Applications*. Lippincott Williams & Wilkins, Philadelphia, pp. 2–21.

Eggermont JJ and Ponton CW (2002) The neurophysiology of auditory perception: from single units to evoked potentials. *Audiology & Neurootology* **7**:71–99.

Flagg EJ, Oram Cardy JE, and Roberts TP (2006) MEG detects neural consequences of anomalous nasalization in vowel-consonant pairs. *Neuroscience Letters* **397**:263–268.

Forss N, Mäkelä JP, McEvoy L, and Hari R (1993) Temporal integration and oscillatory responses of the human auditory cortex revealed by evoked magnetic fields to click trains. *Hearing Research* **68**: 89–96.

Freeman WJ (1980) Use of spatial deconvolution to compensate for distortion of EEG by volume conduction. *IEEE Transactions in Biomedical Engineering* **27**:421–429.

Fruhstorfer H, Soveri P, and Järvilehto T (1970) Short-term habituation of the auditory evoked response in man. *Electroencephalography and Clinical Neurophysiology* **28**:153–161.

Gage N, Roberts TP, and Hickok G (2006) Temporal resolution properties of human auditory cortex: reflections in the neuromagnetic auditory evoked M100 component. *Brain Research* **1069**:166–171.

Gage N, Poeppel D, Roberts TP, and Hickok G (1998) Auditory evoked M100 reflects onset acoustics of speech sounds. *Brain Research* **814**:236–239.

Gage NM and Roberts TP (2000) Temporal integration: reflections in the M100 of the auditory evoked field. *NeuroReport* **11**:2723–2726.

Giraud A, Kleinschmidt A, Poeppel D, Lund T, Frackowiak RSJ, and Laufs H (2007) Endogenous cortical rhythms determine cerebral specialisation for speech perception and production. *Neuron* **56**:1127–1134.

Godey B, Schwartz D, de Graaf JB, Chauvel P, and Liégeois-Chauvel C (2001) Neuromagnetic source localization of auditory evoked fields and intracerebral evoked potentials: a comparison of data in the same patients. *Clinical Neurophysiology* **112**:1850–1859.

Griffiths TD, Büchel C, Frackowiak RSJ, and Patterson RD (1998) Analysis of temporal structure in sound by the human brain. *Nature Neuroscience* **1**:422–427.

Gutschalk A, Patterson RD, Rupp A, Uppenkamp S, and Scherg M (2002) Sustained magnetic fields reveal separate sites for sound level and temporal regularity in human auditory cortex. *NeuroImage* **15**:207–216.

Gutschalk A, Patterson RD, Scherg M, Uppenkamp S, and Rupp A (2004) Temporal dynamics of pitch in human auditory cortex. *NeuroImage* **22**:755–766.

Gutschalk A, Micheyl C, Melcher JR, Rupp A, Scherg M, and Oxenham AJ (2005) Neuromagnetic correlates of streaming in human auditory cortex. *Journal of Neuroscience* **25**:5382–5388.

Gutschalk A, Mase R, Roth R, Ille N, Rupp A, Hahnel S, Picton TW, and Scherg M (1999) Deconvolution of 40 Hz steady-state fields reveals two overlapping source activities of the human auditory cortex. *Clinical Neurophysiology* **110**:856–868.

Halgren E (2004) How can intracranial recordings assist MEG source localization? *Neurology and Clinical Neurophysiology* **2004**:86.

Hämäläinen M and Hari R (2002) Magnetoencephalographic characterization of dynamic brain activation: basic principles and methods of data collection and source analysis. In: Toga AW and Mazziotta JC (eds). *Brain Mapping: The Methods*. Academic Press, Amsterdam, pp. 227–253.

Hämäläinen M, Hari R, Ilmoniemi RJ, Knuutila J, and Lounasmaa OV (1993) Magnetoencephalography – theory, instrumentation, and applications to noninvasive studies of the working human brain. *Reviews of Modern Physics* **65**:413–497.

Hari R (1990) The neuromagnetic method in the study of the human auditory cortex. In: Grandori F, Hoke M, and Romani GL (eds). *Auditory Evoked Magnetic Fields and Electric Potentials. Advances in Audiology*, volume 6. Karger, Basel, pp. 222–282.

Hari R, Hämäläinen M, and Joutsiniemi SL (1989) Neuromagnetic steady-state responses to auditory stimuli. *Journal of the Acoustical Society of America* **86**:1033–1039.

Hari R, Aittoniemi K, Järvinen ML, Katila T, and Varpula T (1980) Auditory evoked transient and sustained magnetic fields of the human brain. Localization of neural generators. *Experimental Brain Research* **40**:237–240.

Hari R, Kaila K, Katila T, Tuomisto T, and Varpula T (1982) Interstimulus interval dependence of the auditory vertex response and its magnetic counterpart: implications for their neural generation. *Electroencephalography and Clinical Neurophysiology* **54**:561–569.

Hari R, Pelizzone M, Mäkelä JP, Hällström J, Leinonen L, and Lounasmaa OV (1987) Neuromagnetic responses of the human auditory cortex to on- and offsets of noise bursts. *Audiology* **26**: 31–43.

Harms MP and Melcher JR (2002) Sound repetition rate in the human auditory pathway: representations in the waveshape and amplitude of fMRI activation. *Journal of Neurophysiology* **88**:1433–1450.

Hickok G and Poeppel D (2007) The cortical organization of speech processing. *Nature Reviews Neurosciences* **8**:393–402.

Hillebrand A and Barnes GR (2005) Beamformer analysis of MEG data. *International Review of Neurobiology* **68**:149–171.

Hoke M, Feldmann H, Pantev C, Lütkenhöner B, and Lehnertz K (1989) Objective evidence of tinnitus in auditory evoked magnetic fields. *Hearing Research* **37**:281–286.

Jääskeläinen IP, Ahveninen J, Bonmassar G, Dale AM, Ilmoniemi RJ, Levänen S, Lin FH, May P, Melcher J, Stufflebeam S, Tiitinen H, and Belliveau JW (2004) Human posterior auditory cortex gates novel sounds to consciousness. *Proceedings of the National Academy of Sciences of the United States of America* **101**: 6809–6814.

Jacobson GP and McCaslin DL (2003) A reexamination of the long latency N1 response in patients with tinnitus. *Journal of the American Academy of Audiology* **14**:393–400.

Jacobson GP, Lombardi DM, Gibbens ND, Ahmad BK, and Newman CW (1992a) The effects of stimulus frequency and recording site on the amplitude and latency of multichannel cortical auditory evoked potential (CAEP) component N1. *Ear and Hearing* **13**:300–306.

Jacobson GP, Ahmad BK, Moran J, Newman CW, Tepley N, and Wharton J (1991) Auditory evoked cortical magnetic field

(M100-M200) measurements in tinnitus and normal groups. *Hearing Research* **56**:44–52.

Jacobson GP, Ahmad BK, Moran J, Newman CW, Wharton J, and Tepley N (1992b) Occurrence of auditory evoked field (AEF) N1m and P2m components in a sample of normal subjects. *Ear and Hearing* **13**:387–395.

Joutsiniemi SL, Hari R, and Vilkman V (1989) Cerebral magnetic responses to noise bursts and pauses of different durations. *Audiology* **28**:325–333.

Kanno A, Nakasato N, Murayama N, and Yoshimoto T (2000) Middle and long latency peak sources in auditory evoked magnetic fields for tone bursts in humans. *Neuroscience Letters* **293**:187–190.

Kaukoranta E, Hari R, and Lounasmaa OV (1987) Responses of the human auditory cortex to vowel onset after fricative consonants. *Experimental Brain Research* **69**:19–23.

Kazanina N, Phillips C, and Idsardi W (2006) The influence of meaning on the perception of speech sounds. *Proceedings of the National Academy of Sciences of the United States of America* **103**: 11381–11386.

Krumbholz K, Patterson RD, Seither-Preisler A, Lammertmann C, and Lütkenhöner B (2003) Neuromagnetic evidence for a pitch processing center in Heschl's gyrus. *Cerebral Cortex* **13**: 765–772.

Kuriki S, Kanda S, and Hirata Y (2006) Effects of musical experience on different components of MEG responses elicited by sequential piano-tones and chords. *Journal of Neuroscience* **26**:4046–4053.

Lakatos P, Pincze Z, Fu KM, Javitt DC, Karmos G, and Schroeder CE (2005) Timing of pure tone and noise-evoked responses in macaque auditory cortex. *NeuroReport* **16**:933–937.

Lammertmann C and Lütkenhöner B (2001) Near-DC magnetic fields following a periodic presentation of long-duration tonebursts. *Clinical Neurophysiology* **112**:499–513.

Lammertmann C, Fujiki N, Lütkenhöner B, and Hari R (2001) Short-term decrement of the auditory N1m response. In: Nenonen J, Ilmoniemi RJ, and Katila T (eds). *Biomag2000, Proceedings of the 12th International Conference on Biomagnetism*. Helsinki University of Technology, Helsinki, pp. 50–53.

Liégeois-Chauvel C, Musolino A, Badier JM, Marquis P, and Chauvel P (1994) Evoked potentials recorded from the auditory cortex in man: evaluation and topography of the middle latency components. *Electroencephalography and Clinical Neurophysiology* **92**:204–214.

Loveless N, Levänen S, Jousmäki V, Sams M, and Hari R (1996) Temporal integration in auditory sensory memory: neuromagnetic evidence. *Electroencephalography and Clinical Neurophysiology* **100**:220–228.

Luo H and Poeppel D (2007) Phase patterns of neuronal responses reliably discriminate speech in human auditory cortex. *Neuron* **54**:1001–1010.

Luo H, Husain FT, Horwitz B, and Poeppel D (2005) Discrimination and categorization of speech and non-speech sounds in an MEG delayed-match-to-sample study. *NeuroImage* **28**:59–71.

Luo H, Wang Y, Poeppel D, and Simon JZ (2006) Concurrent encoding of frequency and amplitude modulation in human auditory cortex: MEG evidence. *Journal of Neurophysiology* **96**: 2712–2723.

Lütkenhöner B (1998) Dipole separability in a neuromagnetic source analysis. *IEEE Transactions of Biomedical Engineering* **45**: 572–581.

Lütkenhöner B (2003) Magnetoencephalography and its Achilles' heel. *Journal of Physiology (Paris)* **97**:641–658.

Lütkenhöner B and Steinsträter O (1998) High-precision neuromagnetic study of the functional organization of the human auditory cortex. *Audiology & Neurootology* **3**:191–213.

Lütkenhöner B and Mosher JC (2007) Source analysis of auditory evoked potentials and fields. In: Burkard RF, Eggermont JJ, and Don M (eds). *Auditory Evoked Potentials. Basic Principles and Clinical Applications*. Lippincott Williams & Wilkins, Philadelphia, pp. 546–569.

Lütkenhöner B and Klein JS (2007) Auditory evoked field at threshold. *Hearing Research* **228**:188–200.

Lütkenhöner B, Lammertmann C, and Knecht S (2001) Latency of auditory evoked field deflection N100m ruled by pitch or spectrum? *Audiology & Neurootology* **6**:263–278.

Lütkenhöner B, Krumbholz K, and Seither-Preisler A (2003a) Studies of tonotopy based on wave N100 of the auditory evoked field are problematic. *Neuroimage* **19**:935–949.

Lütkenhöner B, Seither-Preisler A, and Seither S (2006) Piano tones evoke stronger magnetic fields than pure tones or noise, both in musicians and non-musicians. *NeuroImage* **30**:927–937.

Lütkenhöner B, Borgmann C, Krumbholz K, Seither S, and Seither-Preisler A (2004) Auditory processing at the lower limit of pitch studied by magnetoencephalography. In: Pressnitzer D, de Cheveigné A, McAdams S, and Collet L (eds). *Auditory Signal Processing: Psychophysics, Physiology and Modeling*. Springer, New York, pp. 154–161.

Lütkenhöner B, Krumbholz K, Lammertmann C, Seither-Preisler A, Steinsträter O, and Patterson RD (2003b) Localization of primary auditory cortex in humans by magnetoencephalography. *NeuroImage* **18**:58–66.

Mackert BM, Wübbeler G, Burghoff M, Marx P, Trahms L, and Curio G (1999) Non-invasive long-term recordings of cortical 'direct current' (DC-) activity in humans using magnetoencephalography. *Neuroscience Letters* **273**:159–162.

Mäkelä JP and Hari R (1987) Evidence for cortical origin of the 40 Hz auditory evoked response in man. *Electroencephalography and Clinical Neurophysiology* **66**:539–546.

Mäkelä JP, Hari R, and Leinonen L (1988) Magnetic responses of the human auditory cortex to noise/square wave transitions. *Electroencephalography and Clinical Neurophysiology* **69**: 423–430.

Micheyl C, Carlyon RP, Gutschalk A, Melcher JR, Oxenham AJ, Rauschecker JP, Tian B, and Courtenay Wilson E (2007) The role of auditory cortex in the formation of auditory streams. *Hearing Research* **229**:116–131.

Moore BCJ (2003) *An Introduction to the Psychology of Hearing*. Fifth Edition. Academic Press, San Diego.

Morosan P, Rademacher J, Schleicher A, Amunts K, Schormann T, and Zilles K (2001) Human primary auditory cortex: cytoarchitectonic subdivisions and mapping into a spatial reference system. *NeuroImage* **13**:684–701.

Näätänen R and Alho K (1995) Generators of electrical and magnetic mismatch responses in humans. *Brain Topography* **7**:315–320.

Näätänen R, Lehtokoski A, Lennes M, Cheour M, Huotilainen M, Iivonen A, Vainio M, Alku P, Ilmoniemi RJ, Luuk A, Allik J, Sinkkonen J, and Alho K (1997) Language-specific phoneme representations revealed by electric and magnetic brain responses. *Nature* **385**:432–434.

Nakasato N, Fujita S, Seki K, Kawamura T, Matani A, Tamura I, Fujiwara S, and Yoshimoto T (1995) Functional localization of bilateral auditory cortices using an MRI-linked whole head magnetoencephalography (MEG) system. *Electroencephalography and Clinical Neurophysiology* **94**:183–190.

Noda K, Tonoike M, Doi K, Koizuka I, Yamaguchi M, Seo R, Matsumoto N, Noiri T, Takeda N, and Kubo T (1998) Auditory evoked off-response: its source distribution is different from that of on-response. *NeuroReport* **9**:2621–2625.

Obleser J, Scott SK, and Eulitz C (2006) Now you hear it, now you don't: transient traces of consonants and their nonspeech analogues in the human brain. *Cerebral Cortex* **16**:1069–1076.

Obleser J, Lahiri A, and Eulitz C (2003a) Auditory-evoked magnetic field codes place of articulation in timing and topography around 100 milliseconds post syllable onset. *NeuroImage* **20**: 1839–1847.

Obleser J, Elbert T, Lahiri A, and Eulitz C (2003b) Cortical representation of vowels reflects acoustic dissimilarity determined by formant frequencies. *Brain Research Cognitive Brain Research* **15**: 207–213.

Onitsuka T, Ninomiya H, Sato E, Yamamoto T, and Tashiro N (2003) Differential characteristics of the middle latency auditory evoked magnetic responses to interstimulus intervals. *Clinical Neurophysiology* **114**:1513–1520.

Palva S, Palva JM, Shtyrov Y, Kujala T, Ilmoniemi RJ, Kaila K, and Näätänen R (2002) Distinct gamma-band evoked responses to speech and non-speech sounds in humans. *Journal of Neuroscience* **22**:RC211.

Pantev C, Eulitz C, Hampson S, Ross B, and Roberts LE (1996) The auditory evoked "off" response: sources and comparison with the "on" and the "sustained" responses. *Ear and Hearing* **17**: 255–265.

Parviainen T, Helenius P, Salmelin R (2005) Cortical differentiation of speech and nonspeech sounds at 100 ms: implications for dyslexia. *Cerebral Cortex* **15**:1054–1063.

Patel AD and Balaban E (2004) Human auditory cortical dynamics during perception of long acoustic sequences: phase tracking of carrier frequency by the auditory steady-state response. *Cerebral Cortex* **14**:35–46.

Patterson RD, Uppenkamp S, Johnsrude IS, and Griffiths TD (2002) The processing of temporal pitch and melody information in auditory cortex. *Neuron* **36**:767–776.

Phillips C, Pellathy T, Marantz A, Yellin E, Wexler K, Poeppel D, McGinnis M, and Roberts T (2000) Auditory cortex accesses phonological categories: an MEG mismatch study. *Journal of Cognitive Neuroscience* **12**:1038–1055.

Picton TW, Hillyard SA, Krausz HI, and Galambos R (1974) Human auditory evoked potentials. I. Evaluation of components. *Electroencephalography and Clinical Neurophysiology* **36**: 179–190.

Picton TW, John MS, Dimitrijevic A, and Purcell D (2003) Human auditory steady-state responses. *International Journal of Audiology* **42**:177–219.

Poeppel D, Idsardi W, and van Wassenhove V (2007) Speech perception at the interface of neurobiology and linguistics. *Philosophical Transactions of the Royal Society of London, Series B* **363**: 1071–1086.

Pulvermuller F, Shtyrov Y, Ilmoniemi RJ, and Marslen-Wilson W (2006) Mapping speech comprehension in space and time. *NeuroImage* **31**:1297–1303.

Reite M, Zimmerman JT, Edrich J, and Zimmerman JE (1982) Auditory evoked magnetic fields: response amplitude vs. stimulus intensity. *Electroencephalography and Clinical Neurophysiology* **54**: 147–152.

Ritter S, Dosch HG, Specht HJ, Schneider P, and Rupp A (2007) Latency effect of the pitch response due to variations of frequency and spectral envelope. *Clinical Neurophysiology* **118**:2276–2281

Roberts TPL, Ferrari P, Stufflebeam SM, and Poeppel D (2000) Latency of the auditory evoked neuromagnetic field components: stimulus dependence and insights toward perception. *Journal of Clinical Neurophysiology* **17**:114–129.

Romani GL, Williamson SJ, and Kaufman L (1982) Tonotopic organization of the human auditory cortex. *Science* **216**:1339–1340.

Rosburg T, Trautner P, Korzyukov OA, Boutros NN, Schaller C, Elger CE, and Kurthen M (2004) Short-term habituation of the intracranially recorded auditory evoked potentials P50 and N100. *Neuroscience Letters* **372**:245–249.

Ross B, Tremblay KL, and Picton TW (2007) Physiological detection of interaural phase differences. *Journal of the Acoustical Society of America* **121**:1017–1027.

Salajegheh A, Link A, Elster C, Burghoff M, Sander T, Trahms L, and Poeppel D (2004) Systematic latency variation of the auditory evoked M100: from average to single-trial data. *NeuroImage* **23**:288–295.

Salmelin R, Hari R, Lounasmaa OV, and Sams M (1994) Dynamics of brain activation during picture naming. *Nature* **368**:463–465.

Sams M, Hari R, Rif J, and Knuutila J (1993) The human auditory sensory memory trace persists about 10 sec – neuromagnetic evidence. *Journal of Cognitive Neuroscience* **5**:363–370.

Schreiner CE (1998) Spatial distribution of responses to simple and complex sounds in the primary auditory cortex. *Audiology & Neurootology* **3**:104–122.

Seither-Preisler A, Krumbholz K, and Lütkenhöner B (2003) Sensitivity of the neuromagnetic N100m deflection to spectral bandwidth: a function of the auditory periphery? *Audiology & Neurootology* **8**:322–337.

Seither-Preisler A, Krumbholz K, Patterson RD, Seither S, and Lütkenhöner B (2004) Interaction between the neuromagnetic responses to sound energy onset and pitch onset suggests common generators. *European Journal of Neuroscience* **19**:3073–3080.

Seither-Preisler A, Patterson RD, Krumbholz K, Seither S, and Lütkenhöner B (2006a) From noise to pitch: transient and sustained responses of the auditory evoked field. *Hearing Research* **218**:50–63.

Seither-Preisler A, Patterson R, Krumbholz K, Seither S, and Lütkenhöner B (2006b) Evidence of pitch processing in the N100m component of the auditory evoked field. *Hearing Research* **213**: 88–98.

Sekihara K, Nagarajan SS, Poeppel D, Marantz A, and Miyashita Y (2001) Reconstructing spatio-temporal activities of neural sources using an MEG vector beamformer technique. *IEEE Transactions in Biomedical Engineering* **48**:760–771.

Steinschneider M, Schroeder CE, Arezzo JC, and Vaughan HG, Jr. (1994) Speech-evoked activity in primary auditory cortex: effects of voice onset time. *Electroencephalography and Clinical Neurophysiology* **92**:30–43.

Stufflebeam SM, Poeppel D, Rowley HA, and Roberts TP (1998) Perithreshold encoding of stimulus frequency and intensity in the M100 latency. *NeuroReport* **9**:91–94.

Takahashi H, Nakao M, and Kaga K (2004) Cortical mapping of auditory-evoked offset responses in rats. *NeuroReport* **15**: 1565–1569.

Uusvuori J, Parviainen T, and Salmelin R (2008) Spatiotemporal interaction between sound form and meaning during spoken word perception. *Cerebral Cortex* **18**:456–466.

Van Veen BD and Buckley KM (1988) Beamforming: a versatile approach to spatial filtering. *IEEE Audio, Speech and Language Processing Magazine* **5**:4–24.

Vrba J and Robinson SE (2002) SQUID sensor array configurations for magnetoencephalography applications. *Superconductivity Science and Technology* **15**:R51–R89.

Wakai RT, Lutter WJ, Chen M, and Maier MM (2007) On and off magnetic auditory evoked responses in early infancy: a possible marker of brain immaturity. *Clinical Neurophysiology* **118**:1480–1487.

Williamson SJ and Kaufman L (1987) Analysis of neuromagnetic signals. In: Gevins AS and Remond A (eds). *Methods of Analysis of Brain Electrical and Magnetic Signals. EEG Handbook* (Revised series, volume I). Elsevier, Amsterdam, pp. 405–448.

Yoshiura T, Ueno S, Iramina K, and Masuda K (1996) Human middle latency auditory evoked magnetic fields. *Brain Topography* **8**: 291–296.

Yvert B, Crouzeix A, Bertrand O, Seither-Preisler A, and Pantev C (2001) Multiple supratemporal sources of magnetic and electric auditory evoked middle latency components in humans. *Cerebral Cortex* **11**:411–423.

Chapter 29

The Relationship of Auditory Cortical Activity to Perception and Behavior

Mitchell L. Sutter and Shihab A. Shamma

Abbreviations

AC	auditory cortex
AI	primary auditory cortex
AM	amplitude modulation
BF	best frequency
CV	consonant–vowel
FD	frequency discrimination
FI	frequency irrelevant
FM	frequency modulation
fMRI	functional magnetic resonance imaging
IC	inferior colliculus
LED	light-emitting diode
ITD	interaural time difference
MDTF	modulation detection transfer functions
MEG	magnetoencephalography
MGB	medial geniculate body
MGBd	dorsal nucleus of the medial geniculate body
MGBm	medial nucleus of the medial geniculate body
MGBv	ventral nucleus of the medial geniculate body
MT	middle-temporal field
MTF	modulation transfer function
R	rostral field
ROC	receiver operating characteristics
RT	reaction time
SAM	sinusoidal amplitude modulation
STI	speech transmission index
STMI	spectrotemporal modulation index
STRF	spectrotemporal receptive field
VOT	voice-onset time
VS	vector strength

M.L. Sutter (✉)
Center for Neuroscience, University of California, Davis, CA 95618, USA
e-mail: mlsutter@ucdavis.edu

1 Introduction

A fundamental question in auditory neuroscience is how activity in the brain relates to perception and can a causal link be found? Through the years many approaches have been used. Lesion studies and single unit analysis have led to a better understanding of which areas of the brain are involved in sound processing and how these areas represent important sound features. A growing body of evidence supports a role for primary auditory cortex (AI) not only in simply analyzing sounds, but also in integrating more complex aspects of perception and behavior. In the first section of this chapter we will address how AI activity relates to the perception of stimuli that have been extensively studied behaviorally and psychophysically, and how it relates to well-known psychophysical phenomena associated with perceptually organizing complex 'auditory scenes'. In subsequent sections we address how attributes not directly represented in the stimulus, such as motivation and attention, are potentially represented in auditory cortex, and, finally, how auditory cortical activity is influenced by sensory motor associations, decisions, and rapid adaptive plasticity. Together this provides a picture of auditory cortical activity as not strictly and statically representing the physical attributes of a stimulus, but rather AI activity reflects parameters related to the perception of the stimulus and task-related parameters required to perform the appropriate behavior and motor action in response to the stimulus.

Traditionally, sensory cortices had been conceptualized as primarily extracting and/or representing significant features of simple or complex sensory stimuli. There is good support of this idea from lesion studies that indicate that auditory cortex plays a critical role in perception of most sounds. Cortical lesions impair the ability to localize a sound (e.g., Cranford et al. 1971; Cranford and Oberholtzer 1976; Jenkins and Merzenich 1984; Lomber et al. 2007; Malhotra and Lomber 2007; Malhotra et al. 2008; Neff et al. 1956; Thompson and Cortez 1983), to analyze and process vocal communication including language (Coslett et al. 1984; Graham et al. 1980;

J.A. Winer, C.E. Schreiner (eds.), *The Auditory Cortex*,
DOI 10.1007/978-1-4419-0074-6_29, © Springer Science+Business Media, LLC 2011

Heffner and Heffner 1986a, b; Michel et al. 1980), to process temporal sound features (Cooke et al. 2007; Diamond and Neff 1957; Forrest and Green 1987; Ison et al. 1991; Lomber and Malhotra 2008; Riquimaroux et al. 1991; Samson and Zatorre 1988; Syka et al. 2002), to identify pitch and discriminate frequencies (Cranford et al. 1976; Diamond et al. 1962; Johnsrude et al. 2000; Riquimaroux et al. 1992; Sidtis and Volpe 1988; Stewart et al. 2008; Tramo et al. 2002), and even to detect sounds (Heffner and Heffner 1986a, 1990; Maruyama and Kanno 1961). Frequently, auditory cortical lesions do not produce a lasting deficit, for example in frequency discrimination (Butler et al. 1957; Goldberg and Neff 1961), and these have often been misinterpreted to mean there is no role for auditory cortex in that sound perception task (such as frequency discrimination). However almost always when negative results are obtained, the experiments use non-threshold (easy) tasks with limited or no memory component, and long stimuli, that do not require voluntary movement. The important point to learn from this is that to measure the role of auditory cortex, selection of behavioral paradigm is critical.

The lesion work is supported by years of investigation demonstrating at the neuronal level how auditory cortex extracts and represents important sound features. This has led to some working models of the functional organization of auditory cortex that are well suited for extracting key spectral and temporal composition features of sounds (e.g., Heil et al. 1994; Kusmierek and Rauschecker 2009; Langner et al. 1997; Rauschecker and Tian 2000; Recanzone and Cohen 2009; Schreiner et al. 2000; Schreiner and Sutter 1992; Shamma et al. 1993; Suga 1989; Sutter and Schreiner 1995).

More recently it has become apparent that there is much more to auditory cortex than just statically analyzing a sound's acoustic structure. Auditory cortex has been found to exhibit activity that relates to integrating and enhancing sensory and motor components while performing auditory tasks and behaviors, as well as to enabling learning and adaptation of processing strategies that enhance performance during auditory tasks. In the first section of this chapter, we outline how auditory cortical activity relates to the basic analysis and specific perceptions of sounds. In later parts we describe how non-auditory factors influence auditory cortical activity and processing.

2 Processing of Basic Sound Features

To better understand how auditory cortex contributes to sound perception, numerous studies have investigated the basic response properties of auditory neurons to important stimulus features such as sound frequency and intensity (spectrum), amplitude modulation, frequency modulation, and sound source location. Reviews of the auditory cortical role on spectral (Sutter 2005), temporal (Joris et al. 2004; Recanzone and Sutter 2008), and spatial processing (Recanzone and Sutter 2008) already exist so we will not focus on this basic processing (see also Chapters 13–15). Rather here we describe how activity in AI relates to perception. As such we have chosen a few examples of the relationships of the perception of certain kinds of sounds to auditory cortical activity.

2.1 Methods for Relating Neural Activity to Perception

While it is clear that a goal of neurophysiology is to relate the activity on neurons in the brain to perception, there have been many ways this has been attempted. One tactic is to investigate the activity of single neurons or clusters of neurons and see how this activity relates to different stimuli. By knowing how these stimuli relate to perception we can make inferences about how the activity relates to perception. This class of approach is probably the most prevalent where describing how neuronal responses vary with stimulus parameters is almost synonymous to single neuron sensory physiology. This approach has provided tremendous insights, but has some limitations. One limitation is that it not clear exactly how changes in response rate or timing relate to changes in perception. One manner to improve this is to use similar metrics for both behavior and physiology. Typically this can be done by comparing psychometric functions that deal with trial-to-trial probability functions (e.g., using signal detection theory) to single neuron activity by converting the neural data to neurometric functions. By converting the neural activity into a series of trials and then to probability functions, we can apply the same analysis and statistics to the physiology and behavior and therefore obtain a stronger link. Another way of looking at this is to look at the mean and variance of both neural activity and behavior and see how well they match. An excellent demonstration of this approach has been provided for the visual motion system by Newsome and colleagues (e.g., Britten et al. 1992; Newsome et al. 1989; Tolhurst et al. 1983), where a signal detection approach, the receiver operating characteristics (ROC), was applied to neural and behavioral data to conclude that activity in cortical area MT relates to motion perception. This approach has also been utilized in the auditory system. For example, ROC analysis was applied to auditory nerve fiber responses (Relkin and Pelli 1987; Young and Barta 1986) to see if a single fiber could account for the ability to detect the presence of a tone. Shofner and Dye (1989) used such an approach to see if single cochlear nucleus neurons encode a change of intensity, and Fay and Coombs (1992) tackled a similar problem in goldfish auditory nerve. Zhang et al. (1990) used a d-prime (d') analysis (another type of signal detection analysis based on comparing mean and variance) to determine the ability of auditory nerve fibers to detect a gap in a signal. Pressnitzer et al. (2001) applied a d' analysis to cochlear nucleus neurons to look for a relationship of their activity to perceptual attributes of co-modulation masking release. Only recently has such analysis been extended beyond the cochlear nucleus to more central areas in the auditory system. Takahashi et al. (2003) used a d' analysis to relate sound localization acuity of owl inferior colliculus (IC) neurons to the owl's behaviorally measured ability. Shackleton et al. (2003) used signal detection analysis to determine if single neuron tuning to interaural timing difference (ITD), a critical cue for sound localization, of IC neurons could account for perceptual abilities. Scott et al. (2007) applied these methods to relate responses to auditory spatial perception in the awake monkey. Similar approaches were applied to relate neural firing in the IC to amplitude modulation (AM) detection (Nelson and Carney 2007), to determine if auditory cortical activity in animals electrically stimulated through a cochlear implant have a relationship to perception (Middlebrooks 2004), and to assess the relationship between temporal and rate codes in auditory cortical activity and perceptual vocalization discrimination ability (Walker et al. 2008). In our own work, we are using this approach to determine how well rate vs. phase-locking codes relate to auditory cortical activity (Niwa et al. 2009; Yin et al. 2004). As can be seen from this brief list of studies, signal detection theory approaches appear to be highly promising in establishing a stronger link between neural activity and psychophysically determined perceptual abilities.

While this neurometric approach in non-behaving animals can provide tremendous insights, applying neurometric analysis to data collected while the animal performs perceptual tasks can be even more powerful. With simultaneously performed behavior, neural activity can be compared *directly* to the performance of the animal. One implementation of this is the choice probability analysis, which allows responses of a neuron to be linked with the behavioral choice the animal makes on a trial-to-trial basis. For example, in vision, animals were asked to report the direction of motion as a function of the percentage coherence of dots that move across their field of vision (Britten et al. 1996). In this task a certain percentage of dots on the screen moved in the same direction, while the other dots moved in random directions. The greater the percent coherence (percentage of dots moving in one direction), the easier it is to tell the direction of motion. Near threshold, subjects perform half way between chance and perfect in correctly reporting the direction of motion. Britten et al. (1996) recorded from area MT of visual cortex while the animal was performing such a task. On near threshold trials, where half the time they correctly detect motion direction, one can partition trials based on the animal's choice. This provides information whether differences in neural activity depend on the direction of motion that the animal is reporting. Indeed, Britten et al. (1996) reported that activity was correlated with the animal's perceptual judgment. In a similar study of auditory cortex, Lemus et al. (2009) did not see evidence for such a relationship; however, we recently have found strong evidence for such decision-related activity in AI while monkeys were performing an amplitude modulation detection task (Niwa et al. 2009).

While all of these neurometric analyses provide powerful tools to compare neuronal responses to perceptual ability, one problem is that the most basic form of this analysis is based on the assumption that a single neuron (or small group of neurons) is responsible for the perception. Perception most likely relies on integrating activity from many neurons, and potentially across different stations. How to establish the link from single units to how a population of units relate to perception is not trivial. This can be—at least partially—remedied by modeling how population of neurons can combine to account for threshold performances (Shadlen et al. 1996).

Now, with a better understanding of the techniques used to relate neural activity to perception, we will discuss a few selected examples of how auditory perceptional abilities are related to auditory cortical activity.

2.2 Analysis and Representation of Complex Sounds in the Auditory Cortex

The auditory cortex plays a critical role in the perception and localization of complex sounds. Incoming acoustic stimuli arrive at the ears, are transformed into a neural code in the cochlea, and ascend in an interwoven and bidirectional network of processing centers *en route* to auditory cortex (e.g., Clarey et al. 1992). Despite the rapidly expanding knowledge of the neuroanatomy and connectivity of the auditory cortex (Lee and Winer 2005), relatively little is known about its *functional* organization (Schreiner et al. 2000; Sutter and Schreiner 1995), especially compared to the visual and the motor systems. Nevertheless, a few attributes have been broadly accepted as being vital for auditory behavior, such as sound localization, timbre recognition, and pitch perception. Evidence for the encoding of each of these attributes has been reported or postulated.

The primary auditory cortex (AI) shares with other primary sensory areas' basic neuronal mechanisms and response properties that are thought to be actively involved in a wide range of perceptual processes. For example, the input layers of AI exhibit spectrotemporal receptive fields that are broadly analogous to those of the primary visual cortex (DeValois and DeValois 1990), suggesting common organizational and functional principles underlying these primary areas (O'Leary 1989; Roe et al. 1992; Sur 1988; Sur et al. 1988). One curious finding from *intracellular* recordings in AI is the apparent overlap of the inhibitory and excitatory inputs which may be important for increasing temporal precision and for determining the shape of spectral receptive field (see Chapter 13). This indicates that some of the rich spectral structure of AI receptive fields must be already established in the thalamus (Wehr and Zador 2005; Miller et al. 2001b) and that the thalamocortical input from the lemniscal pathway preferentially confers responses to best frequency of cortical neurons, giving rise to its tonotopic organization. It is also clear that there are horizontal cortico-cortical inputs that shape spectral receptive field properties, especially for away from the neuron's best frequency (Kaur et al. 2004; Metherate et al. 2005) thus creating a much more complex spectral receptive field structure in AI (O'Connor et al. 2005; Sutter 2000; Sutter and Loftus 2003; Sutter et al. 1999).

2.2.1 Amplitude Modulation

Changes of a sound's amplitude envelope over time (amplitude modulations (AM)) are well-suited in relating neural activity to perception because it appears that both rate and temporal codes of the neural activity play an important role in encoding AM (see Joris et al. 2004 and Chapter 14 for more details). AM is an important sound feature acting as an information-bearing parameter in communication sounds (Nelken et al. 1999), such as syllabic features in speech, and playing an important role in segregating sound sources in complex listening environments (Bregman 1990; Grimault et al. 2002; Yost 1991). First we will discuss the perception of amplitude modulation and different ways in which the nervous system encodes AM. Following that, we will address the relationship of auditory activity to the perception of communication sounds and segregating and attending to sound sources in complex listening environments.

The simplest studies of AM perception have used stimuli with sinusoidal modulation. These sine-AM (SAM) stimuli often evoke the perception of a pitch that corresponds to the modulation frequency (Burns and Viemesiter 1976, 1981; Ritsma 1962; Zwicker 1952; Zwicker and Fastl 1999). SAM stimuli often have been used to study the perception of pitch, including how well subjects can determine whether a sound is modulated or not by varying the modulation depth (Bacon and Viemeister 1985; Eddins 1999; Ewert and Dau 2004; Forrest and Green 1987; Viemeister 1979). Modulation depth is a parameter that can continuously vary between unmodulated (flat envelope) to fully modulated (sine envelope varies between 1 and 0); investigation of intermediate modulation depths allows for the determination of a threshold for detecting whether or not the sound is modulated. Varying modulation depth also varies the salience of the associated pitch percept. In general, animal studies find that thresholds for detecting AM match quite closely human thresholds (Dooling and Searcy 1985; Fay 1988; Henderson et al. 1984; Kelly et al. 2006; Langemann and Klump 2007; Moody 1994; O'Connor et al. 2000; Salvi et al. 1982), although temporal integration properties may vary across species (O'Connor et al. 1999).

There are a plethora of studies probing how single units respond to and encode AM, and several different codes for AM have been proposed. In general, the focus of physiological studies has been on two different types of neural codes: temporal and rate codes. Temporal codes can be subdivided into two different types. One type measures how well a neuron's response timing mimics the timing of stimulus events; e.g., how well the neuron's activity phase locks to amplitude modulation cycles. This type of code is often quantified by a measure known as vector strength (VS, Goldberg and Brown 1969). Recently, alternative temporal measures which analyze the timing of spikes and the encoding of stimulus modulation as a spike pattern identifier have also been investigated (Kajikawa and Hackett 2005; Malone et al. 2007). For these measures, changes in temporal pattern across stimuli are more important than temporal following of the sound envelope. Rate codes measure the overall magnitude (e.g., in spikes per second) of a neuron's response to AM without regard to the neuron's phase-locking properties. Although phase-locked responses also exhibit a rate component, a second class of responses to AM that are continuous and sustained without phase locking (a rate code in the absence of a temporal code) is also observed (Lu et al. 2001). This forms two distinct manifestations of rate codes: synchronized and non-synchronized rate codes.

Much neural research has been concentrated on modulation transfer functions (MTFs), which characterize neural responses as a function of modulation frequency. Typically this is done by looking at a rate (for example, total spikes over the stimulus or per AM cycle) and/or a temporal measure of the neural response (usually VS) as a function of modulation frequency. However, the information obtained from MTFs does not easily map onto the large body of psychophysical work. A typical neural MTF shows how well a neuron can distinguish different modulation frequencies when the stimuli are all at 100% modulation depth. Psychophysical studies are focused on determining if a sound is modulated, that is they determine the minimum modulation depth at which the sound is perceived as modulated. These psychophysical modulation detection thresholds are often derived over a range of modulation frequencies to create a psychophysical modulation detection transfer function (MDTF). Comparing single neuron, animal MTFs and human MDTFs is problematic because the former are based solely on responses to 100% modulation depth, while the latter determines perception across different modulation depths. The bias towards collecting MTFs is due to a prevalent opinion of both physiologists and psychophysicists that the nervous system acts as a bank of band-pass filters—each neuron acts as a modulation frequency filter whose properties are well characterized by the MTF. However, recent physiological experiments bring this view into question, raising the possibility that other coding schemes—such as global synchrony or a population spike count code—may play a more relevant role (Malone et al. 2007; Yin et al. 2004, 2010).

Despite the limitations of the MTF approach with respect to comparison to perception, it has revealed some general principles of AM encoding. First, at higher levels of the auditory system the ability of neurons to phase lock to higher modulation frequencies decreases. A second, related principle is that at higher levels of the auditory system, rate codes may become more important in encoding higher modulation frequencies (Lu and Wang 2000). Single neuron responses that do not synchronize to the modulation, but that nonetheless exhibit high firing rates when triggered by a relatively narrow range of modulation frequencies are examples of this type of coding (Bendor and Wang 2008). If cortex is to encode modulation frequencies to which it cannot phase lock, such a rate coding scheme appears to be necessary (Lu and Wang 2000).

There have been a few studies that investigate how neural responses vary as a function of modulation depth (Gleich and Klump 1995; Malone et al. 2007; Middlebrooks 2008; Nelson and Carney 2007). In general, neural thresholds estimated with temporal codes can be quite similar to human psychophysical thresholds, but estimation with rate codes usually results in thresholds somewhat higher than experimentally observed. With modeling, this issue might be resolvable by pooling across more neurons (Johnson and Sutter, unpublished observations). It appears that AI lies in the middle of a temporal to rate transformation and that AM is represented both in the overall degree of synchronized activity and—also to a lesser extent—in the activity of a subset of non-synchronized neurons.

2.2.2 Cortical Spectrotemporal Response Fields, Timbre, and Speech Perception

One of the most important attributes of sound that is thought to emerge in the auditory cortex is that of sound *timbre*, the perceptual attribute responsible for recognizing and classifying complex sounds, such as the distinctions among speech phonemes and different instruments playing at the same pitch and loudness. Over the last decade, extensive data and ideas gained from physiological and psychoacoustical experiments in AI have argued for a distributed representation of *timbre* based on a multiresolution analysis of the cochlear spectrogram. Specifically, much insight has been gained from measurements of the so-called spectrotemporal response fields (STRF) of AI cells (Shamma et al. 1995; Miller et al. 2001). An STRF summarizes the dynamics and sensitivity of a cell or, more precisely, the impulse response of the cell at each frequency. Thus, an STRF displays the excitatory and inhibitory interactions that give the cell its selectivity to spectrotemporal patterns. Some STRFs are responsive (excited or suppressed) over a broad range of frequencies, exceeding an octave, while others are quite narrowly tuned. Dynamically, some STRFs' responses decay rapidly after an impulse, while others last twice as long. Finally, this combined time-frequency sensitivity can take more complex forms that are "inseparable" as in oriented STRFs that are sensitive to frequency modulations (e.g., Klein et al. 2000; Kowalski et al. 1996; see Chapter 13).

STRFs have been measured in many ways including *reverse correlation* with random tone chords or spectrotemporally modulated noise (Chi et al. 2005; deCharms et al. 1998). Another method is the "ripple analysis method" that employs broadband noise with *sinusoidally* modulated spectrotemporal envelopes with different parameters (Klein et al. 2000; Kowalski et al. 1996). Ripples serve the same function as regular sinusoids in measuring the transfer function of linear filters, except that they are two dimensional (spectral and temporal). AI cells respond well to ripples and are usually selective to a narrow range of ripple parameters that reflect details of their receptive fields. By compiling a complete description of the responses of a neuron to all ripple densities and velocities it is possible by an inverse Fourier transform to compute the corresponding STRF.

From a functional perspective, the rich variety of STRFs found in AI implies that each STRF acts as a *modulation selective filter* of its input spectrogram, specifically tuned to a particular range of spectral resolutions (also called *scales*) and a limited range of temporal modulations (or *rates*). The collection of all such STRFs then would constitute a filterbank spanning the broad range of psychoacoustically observed scale and rate sensitivity in humans and animals (Chi et al. 1999; Dau et al. 1997; Green 1986; Klein et al. 2000). Evidence of the importance of spectrotemporal modulations in the perception of complex sounds has come from experiments in which systematic degradations of the speech signal were correlated with the gradual loss of intelligibility (Drullman et al. 1994; Shannon et al. 2004). In fact, the relationship between the temporal modulations and the speech intelligibility has long been codified in the formulation of the widely used Speech Transmission Index (STI) (Houtgast et al. 1980), and the Spectrotemporal Modulation Index (STMI) (Elhilali et al. 2003) which assesses the integrity of both the spectral and temporal modulations in a signal as a measure of intelligibility.

Recently, evidence for these ideas emerged from direct measurements of neuronal responses to continuous speech in the primary auditory cortex of the naive ferrets (Mesgarani et al. 2008). Findings revealed an explicit multidimensional representation made possible by the above-described wide range of spectrotemporal tuning in AI to stimulus frequency, bandwidth, and dynamics. To understand the advantage of such diversity, one should consider the fact that there is always a unique sub-population of neurons that responds well to the distinctive acoustic features of a given phoneme and hence encodes that phoneme in a high-dimensional space.

For example, consider the perception of the plosive consonant /k/ in a consonant–vowel (CV) syllable, which is identified by a conjunction of several acoustic features: an initial silent voice-onset-time (VOT), an onset burst of spectrally broad noise, and the direction of the following formant transitions (Aizawa and Eggermont 2006; Eggermont 1995; Steinschneider et al. 1995, 2003). Each of these features can be encoded in the cortical responses along different dimensions. Thus, neurons selective for broad spectra respond selectively to the noise burst. Fast neurons respond well following the VOT, whereas directional neurons selectively encode the vowel formant transitions. In this manner, /k/ is encoded robustly by a rich pattern of activation that varies in time across the neural population. This neuronal activation pattern constitutes the phoneme representation in AI and presumably forms the input to a set of neural "phoneme classifiers" in higher auditory areas. If one acoustic feature is distorted or absent, the pattern along the other dimensions (and hence the percept) still remains stable.

We have focused here on describing a few prominent features of the response distributions that correspond to well-known distinctive acoustic features of the consonants considered (Stevens 1980). There are clearly many other aspects and more details of the responses that reflect intricate articulatory gestures, contextual effects, or speaker-dependent variability that can only be reliably considered with a much larger sample of responses. One example is the distribution of the directionality index of the responses in the neighborhood of a consonant (reflected by a frequency modulation-like shape of the STRFs), an attribute that would indicate whether the formants are upward or downward sweeping, or if they are converging toward or diverging away from a locus frequency. It should be noted that humans confuse the phonemes of their native tongue when placed in unusual or noisy contexts. Typically, phonemes that share some acoustic features are more confusable than those that do not. This pattern of confusions has been found to mirror that between the cortical responses to the phonemes, leading to the conjecture that human phoneme perception can be explained in large measure by basic auditory representations such as the cortical spectrotemporal analysis common to many mammalian (and also avian) species.

The representation of phonemic features across a population of filters tuned to best frequency, bandwidth, and dynamics suggests a strategy for improved speech recognition systems, and further study may reveal additional strategies for speech processing (Steinschneider et al. 1995). However, many questions about the neural representation of phonemes still remain unclear; for example, how can one extrapolate from such neurophysiological findings to the human perceptual ability to perceive phonemes categorically (also found in monkeys (Steinschneider et al. 2003), cats (Hienz et al. 1996), chinchillas (Kuhl and Miller 1975), birds (Dent et al. 1997; Kluender et al. 1987; Lotto and Kluender 1998), and rat (Pons 2006)), and to shift categorical boundaries arbitrarily between phoneme pairs? In summary, humans' ability to discriminate perfectly their native phonemes is the result of years of training. Naïve animals lack such a history, and hence their perception of clean phonemes is more akin to that of humans listening to noisy phonemes. In both cases, confusion patterns would reflect the acoustic distances between the phonemes. However, if animals are trained to actively discriminate phonemes, it is likely that dimensions useful for this specific discrimination would be emphasized, as discussed below, creating the heightened sensitivity necessary to perform the task through a pattern of behaviorally driven plasticity of AI receptive fields. This is presumably what happens in humans as they learn their phonemes and adapt their neuronal tuning along the dimensions appropriate for the phoneme discrimination task. This same general principle would apply to any complex sound, using additional cortical response dimensions, such as pitch, spatial location, and loudness.

2.2.3 Pitch Perception and Auditory Cortex

Another important attribute of a complex sound is its "pitch," a percept that is directly related to the overall periodicity of a sound. Specifically, a sound complex consisting of several harmonics of a fundamental frequency is heard with a strong pitch at the fundamental frequency of the harmonic series, even if there is no energy at all at that frequency. This percept has been variously called the missing fundamental, virtual pitch or residue pitch (Moore 1989). A large number of psychoacoustical experiments have been carried out to elucidate the nature of this percept, and its relationship to the physical parameters of the stimulus. Basically, all models fall in one of two camps (Stein et al. 2005). The first believes that the pitch is extracted explicitly from the harmonic spectral pattern. This can be accomplished in a variety of ways, for instance by finding the best match between the input pattern and various harmonic templates assumed to be stored in the brain (Goldstein 1973). The second group claims that the pitch is extracted from the periodicities in the time-waveform of responses in the auditory pathway that can be estimated, for example, by computing their autocorrelation functions (Cariani and Delgutte 1996a, b; Moore 1989). The latter is related to the description earlier about the relationship of AM to pitch perception. In these latter models, some form of organized delay lines are assumed to exist in order to do the computations, much like those that seem to exist in the auditory system of the mustached bat for echo-delay processing (Hattori and Suga 1997; Miller et al. 2005; Saitoh and Suga 1995), and in jamming avoidance response in electric fish and sound localization in owls (Carr 1986; Carr and Konishi 1990; Carr et al. 1986), although for jamming avoidance and sound localization the delay lines operate on a much faster timescale. Recordings from awake monkeys (Fishman et al. 1998; Steinschneider et al. 1998) have provided evidence that both mechanisms might operate in AI, depending on the pitch frequency.

In all pitch models, however, it is assumed that the extracted pitch is finally represented as a spatial map in higher auditory centers. This is because many studies have confirmed that neural synchrony to the repetitive features of a stimulus, whether it is the waveform of a tone or its AM, becomes progressively worse toward the cortex (see Joris et al. 2004; Langner 1992; Chapter 14). It is a remarkable aspect of pitch that, despite its fundamental and ubiquitous role in auditory perception, only a few reports exist of physiological evidence of spatial pitch maps, and none have been independently confirmed. One source is human subjects using fMRI and MEG scans of the human auditory cortex (Langner et al. 1997; Penagos et al. 2004). Another is from single-unit and multiunit responses in various pre-cortical auditory structures (Langner and Schreiner 1988; Pressnitzer et al. 2003; Schreiner and Langner 1988), and recently in the auditory cortex (Bendor and Wang 2005; Schulze and Langner 1997). One key difficulty in all experiments seeking to demonstrate physiological correlates of pitch is the cochlear nonlinearity that produces distortion components at the fundamental frequency of the upper harmonics and that unintentionally excites low-BF cells. This "artifact" has cast a shadow of doubt over all discoveries of physiological pitch maps because of the experimental difficulties in avoiding or masking it. Of course, the difficulty in finding a spatial pitch map in the auditory cortex may be due to the fact that it does not exist! This possibility is counter-intuitive given the results of ablation studies that show that bilateral cortical lesions in the auditory cortex severely impair the perception of pitch of complex sounds (Sidtis and Volpe 1988). Another possibility is that the maps sought are not at all as straightforward as we imagine. For example, harmonic complexes may evoke stereotypical patterns that are distributed over large areas in the auditory cortex, and not localized as the simple notion of a pitch map implies. Finally, it is also possible that AI simply functions as one stage that projects sufficient temporal or spectral cues for later cortical stages to extract the pitch.

3 Perception of Stream Segregation

There are several other important properties of cortical responses that have found strong resonance in psychoacoustics. Two in particular are adaptation and synaptic depression which have been hypothesized to be the neural correlates of perceptual phenomena such as "forward masking" (Tan et al. 2004), "buildup of perceptual streams" in auditory scene analysis (Carlyon 2004), and of the "multiple looks hypothesis" for the integration of cues leading to detection (Elhilali et al. 2004; Viemeister and Wakefield 1991). Interesting possible physiological correlates of these perceptual phenomena have been described (Fishman et al. 2001, 2004; Kanwal et al. 2003; Micheyl et al. 2005; Wehr and Zador 2005). One intriguing discovery is that stimulus-specific adaptation takes place on multiple timescales ranging 100-fold from hundreds of milliseconds to tens of seconds (Malone et al. 2002; Malone and Semple 2001; Ulanovsky et al. 2003, 2004), which may play a role in encoding auditory memory. Another important property of auditory cortex is redundancy reduction in the representation of complex spectrotemporal stimuli compared to stimulus-induced redundancy observed in IC and the medial geniculate body (MGB) (Chechik et al. 2006). Neural correlation in spiking patterns between adjacent neurons is more likely when there is overlap of spectrotemporal receptive fields or when the difference in characteristic frequency is small (Eggermont 2006). Finally, rhythmic Gamma oscillations abound in the auditory cortex (Edwards et al. 2005; Jeschke et al. 2008; Lakatos et al. 2004; Palva et al. 2002; Pantev et al. 1991; Steinschneider et al. 2008), much like those found in other cortical areas (Fries et al. 2001). The functional role of oscillatory synchronization of neural activity remains mysterious, but may enhance coincidence detection, could enhance noise tolerance, or play a role in plasticity and attention (Edwards et al. 2005; Fries et al. 2001; Kopell 2005).

4 Perceptual Fill-in in Auditory Cortex

When a sound of interest is interrupted by a loud brief noise, subjects report hearing the complete sound of interest even if the noise was capable of masking a part of it. This mimics the natural condition when we have to focus on one sound in an environment of competing and overlapping sounds. This property can be exploited to create an illusory perception by introducing a silent period into a sound of interest and placing a loud noise into the silent period. When this is done we hear the entire sound of interest as intact, even though any structure of it has been removed in that interval. This phenomena has been observed across every sound of interest studied, from simple tones to complex vocalizations and music. Evidence that this type of filling-in occurs in animals is strong (Braaten and Leary 1999; Miller et al. 2001; Petkov et al. 2003; Seeba and Klump 2009; for review see Petkov and Sutter 2010).

This form of filling-in in the auditory system has been studied at the single-unit level in primary auditory cortex. A conceptually simple correlate of the phenomena has been found (Petkov et al. 2007) in the activity of single neurons in primary auditory cortex. However, the mechanisms involved in creating this mechanism must be quite complex to cover the wide array of interesting sounds the brain must process. In MGB of unanaesthetized guinea pigs, Schreiner (1980) found that onset responses to tones alternated in noise were suppressed when the noise was loud. This correlated with the inability to hear onset and offset transitions of the tone and is consistent with the tone being perceived as continuous. In this study, sustained responses, which should signal the continuation of the tone, were also suppressed—at least partially—which seems inconsistent with illusory fill-in. Sugita (1997) found evidence that response patterns of single units in AI of anesthetized cats were consistent with illusory fill-in. They selected neurons that responded to FM sweeps and placed a gap in the sweep. The neurons seemed to stop responding to the FM, but when the gap was filled with noise the response appeared to return. However, it appears that the narrow-band noise used to fill the gap was offset in frequency from the FM trajectory, and therefore, this stimulus should have been perceived as having a gap. Nonetheless, that study presented behavioral evidence suggesting the cats perceive the target as continuous. It remains difficult to determine if the neuron's response was indeed following the perception.

In awake naïve monkey AI, Petkov et al. (2007) used the same stimuli as in their behavioral study (Petkov et al. 2003) to test a more general model of auditory fill-in. The model predicts that when presented with stimuli that cause fill-in, neurons that respond preferentially to the sound of interest should respond as if the sound is continuous. AI neurons that responded stronger to tones than noise responded to combined tone noise stimuli that causes the continuity illusion as if they were responding to a continuous tone (Petkov et al. 2007). Additionally, the responses were highly nonlinear and inconsistent with inheritance of simple auditory periphery responses such as energetic masking or simple adding of noise energy to the tone. No simple mechanistic model could be found to describe the aggregate of results. It is as if these responses went through a random evolutionary like learning process to modify their responses to correctly respond to the target in the presence of a loud interrupting noise. When one thinks of auditory fill-in occurring for all kinds of sounds of interest it makes sense that no one simple mechanism can explain the response properties. These results yield a simple rule for how neurons might cause the illusion, even though there is no single mechanism to achieve this simple rule and suggests the fill-in phenomenon is heavily selected for by the auditory system.

5 The Effects of Learning, Behavior, and Motivation in Auditory Cortical Responses

So far, we focused on how the auditory cortex responds to, encodes and represents sounds, and how this auditory processing might relate to perception. This has been a focus of auditory cortical research for some time, reflecting a view that the fundamental role of auditory cortex is to encode sounds and that cortical responses should reflect sound perception. This is well based in numerous studies, for example lesion studies, demonstrating a role of auditory cortex in sound perception.

One important finding is that auditory cortex, like most of the brain, is adaptive and plastic. Auditory cortical processing adapts and changes depending on experience and the behavioral and environmental condition the animal finds itself in. Because of this the representation of sound parameters as described above is not static, but constantly changing based on the animals needs. There is also evidence for influences in auditory cortex from other sensory modalities (Bizley and King 2008, 2009; Ghazanfar 2009; Ghazanfar et al. 2005; Hackett et al. 2007; Hackett and Schroeder 2009; Kayser et al. 2008; Lakatos et al. 2007; Schroeder et al. 2001, 2003; Smiley et al. 2007) which are highly contextually dependent. We will look at the adaptive properties of cortical neurons and how stimulus relevance plays an important role in how auditory cortex responds to stimuli.

There is also long history and, more recent, a revival of interest in influences on auditory cortical activity not directly related to the stimulus, but rather to the animal's behavioral condition. One approach that has helped to reveal evidence that activity in auditory cortex does not solely result from stimulus properties is the approach of recording from awake animals while they actively perform a task. One of the earliest studies noting such non-auditory influences was that of Hubel, Galambos, and colleagues (Hubel et al. 1959). They found that ~10% of units in cat auditory cortex only responded when the animal was 'paying attention.' Attention in this context was not strictly defined because not much was known then about it. In the context of Hubel's study, attention could mean when the animal turned its head toward the sounds. While the units are identified as being in auditory cortex, the location of these units across the many auditory fields is not clear. Because it is so common for these combined behavioral and unit recording studies not to define location within auditory cortex, the studies cited here are assumed to have recorded from unspecified areas within the auditory cortex (AC) unless otherwise noted. Over the years the techniques for defining non-auditory influences, such as attention, have improved allowing for a steady increase in our understanding of their impact on auditory cortical activity.

Since Hubel's work evidence has been found that AC responses not only reflect the stimulus, but also are influenced by task engagement, reward value and expectation of reward, attention, stimulus expectation, memory, behavioral training, sensory motor associations, behavioral choices and decisions, motor output, and plasticity and adaptive processes. While this list of types of non-auditory influences on AC is long, many of these non-auditory variables can be interrelated, and future research will need to tease apart how independent these different properties are from each other. We are currently at a critical juncture that could help to increase understanding of the importance of auditory cortex, not just for a machine-like encoding of sounds, but also as a center involved in integrating complex variables involved in listening in situations where animals must learn and behave.

5.1 Long-Term Versus Short-Term Plasticity in Cortex

Long-term auditory experience or learning has been shown to cause profound global effects, such as reshaping of tonotopic maps, and significant local effects by transforming receptive field properties of neurons in the primary auditory cortex (A1) (Fritz et al. 2003; Irvine and Rajan 1996; Irvine et al. 2001; Kilgard et al. 2001; Ma and Suga 2009; Mercado et al. 2001; Percaccio et al. 2007; Recanzone et al. 1993; Robertson and Irvine 1989; Weinberger et al. 1993). Convergent studies of plasticity in the auditory, visual, and motor systems have also demonstrated the capacity for dynamic modulation of representational maps and shown that cortical cells in these systems can undergo rapid, task-dependent, and context-specific changes of their receptive field properties during attentive behavior. The key elements of this form of adaptive plasticity appear to be: (1) directed attention to salient task-related cues, which leads to (2) selective functional reconfiguration of the underlying cortical circuitry that occurs simultaneously with task performance, and causes (3) changes in receptive field properties of individual neurones and the cortical ensemble which may enhance behavioral performance in the current task (Bakin et al. 1996; Crist et al. 2001; Li et al. 2001). These findings, when combined with findings that classical conditioning modifies auditory cortical spectral receptive fields (Diamond and Weinberger 1986; Edeline and Weinberger 1993), suggest that cortical receptive fields are not fixed, but may be constantly adapting and re-organising dynamically to meet the challenges of an ever-changing environment and new behavioral demands and may play an important role in information processing and storage. In this functional model, each primary sensory cortical neuron participates in multiple behavioral contexts, and it is likely that its receptive field properties are differentially modified by top-down influences in each case, and its network connectivity may also be reconfigured in an immediate and reversible manner as the animal switches between behavioral states (Chernyshev and Weinberger 1998; Fritz et al. 2003; Kilgard and Merzenich 1998; Kilgard et al. 2002; Soto et al. 2006; Weinberger et al. 2006). In this way, the same neuronal ensemble can mediate different perceptual functions. The basic adaptive mechanisms that underlie this plasticity may be similar in perceptual and motor learning, and also during optimal performance of a previously learned task. Interestingly, it appears that flexibility in task-dependent processing of similar acoustic stimuli is a fundamental principle, not only at the level of single cells and local networks, but also at the level of hemispheric activation (Brechmann and Scheich 2005).

It is well established that cognition simultaneously involves bottom-up and top-down processes, including interactions between bottom-up, sensory-driven information, and top-down, attentional, memory, and executive processes that modulate bottom-up processing. This bottom-up versus top-down distinction is consistent with the anatomical and physiological evidence of a cortical architecture that abounds with forward and backward axonal projections in the neocortex and associated structures such as the thalamus, the amygdala, the striatum, and the hippocampus. There is increasing evidence that higher brain functions, including the brain's ability to learn from experience, depend on the integration of such forward and feedback signals. Consequently, a complete understanding of how auditory cortical responses encode the acoustic environment must take into account the behavior of the animal within it. For instance, one simply cannot obtain a true understanding of auditory cortical responses to a threatening sound by merely playing it to an anesthetized animal. Instead, when an animal recognizes and escapes threatening sounds, it enters a highly aroused and attentive state in which it categorizes its predators' calls as salient foreground targets to distinguish them from other harmless background sounds. Simultaneously, it likely integrates other sensory cues (visual, olfactory) as well as its stored acoustic memories into its auditory judgment of the nature of the calls. In addition, neuronal correlates of category formations in the prefrontal cortex would likely feedback and adapt the receptive fields of the auditory cortex so as to enhance the perception of the target sounds against a background, and subsequently to generate an appropriate multimodal representation of the scene and plan motor actions to respond to the threat. All these interactions significantly alter auditory responses in the cortex, and a massive descending corticofugal feedback system dynamically reshapes cortical inputs (Winer 2006), perhaps influencing pre-cortical structures all the way down to the cochlea (Suga et al. 2000; Xiao and Suga 2002) and hence must be taken into account when dissecting the nature of auditory cognition.

Finally, an interesting point of intersection between adaptive and representational properties of the auditory cortex is its multimodal responses. Specifically, neuroanatomical and neurophysiological studies have shown a convergence of multisensory (visual and somatosensory) inputs to auditory cortex (Brosch et al. 2005; Budinger et al. 2006; Durif et al. 2003; Fu et al. 2003; Ghazanfar and Schroeder 2006; Lee and Winer 2005; Schroeder et al. 2001). But, interestingly, these influences are strongly modulated by the relevance of these inputs in the auditory behavior, as evidenced by responses in auditory cortex (Brosch et al. 2005) and inferior colliculus (Metzger et al. 2006; Ryan et al. 1984). These results also suggest that rather than a purely unisensory processing stream, responses in the auditory cortex must be understood as an interwoven tapestry of relevant multimodal contextual inputs.

5.2 The Effects of Task Engagement

Several studies have compared activity when an animal performs a task to when the animal sits passively but awake. Most studies (e.g., Gottlieb et al. 1989; Miller et al. 1972; Scott et al. 2007) found increased driven activity when the animal was engaged in a task compared to when awake but not engaged, but some found no effect of engagement (Gilat and Perlman 1984; Hocherman et al. 1976, 1981), or saw decreases in activity (Benson and Hienz 1978; Otazu et al. 2009). Also in all of the studies spontaneous rate either increased or did not change with active engagement, although there was not always a strict relationship between effects on driven and spontaneous activity.

Miller et al. (1972) performed a study designed to look at the effect of training and performance of a task on AC activity. To do this they recorded single-unit activity under three conditions: (1) in monkeys performing the task; (2) in monkeys trained to the task but not performing it during recording and (3) in naïve untrained monkeys. They used a reaction time (RT) task where animals had to depress a telegraph key to begin a trial (after a light cued them it was OK to begin a trial) and release the lever rapidly when any sound was presented. The sounds would be presented between 1 and 4 s after lever press initiated a trial and the animal had to respond in <1 s. Because in the reaction time task the animal is allowed to respond during the stimuli, both stimulus-related and response-related activity are included in reported driven activity. More driven activity was found in the behaving than non-behaving condition and less-driven activity and labile, severely habituating responses were reported in naïve animals.

Several other studies also have found increases in driven activity during task performance compared to non-performance. Benson et al. (1981) found increases in activity when the animals were engaged in a task where they had to press a key next to the perceived sound source location of a noise burst. They also compared active to passive behavior showing significant increases in activity for a location for 22% of neurons and 7% showing significant decreases. These differences between passive and active condition could occur either in or outside of AI. Scott et al. (2007) also found increased activity in the core (fields AI and R) of monkey auditory cortex when recording single neuron responses while monkeys actively discriminated interaural phase (a sound localization cue) compared to passive listening. For active discrimination, peak-driven activity changed in most units (58% of neurons increased, 13% decreased). They found spontaneous rate during active discrimination increased relative to the passive condition for 71% of neurons, and only decreased for 6% of neurons. It appears, however, that the animal may have received its reward during the spontaneous period, so reward-related activity might be included also.

Scott et al. (2007) also performed a neurometric analysis to determine whether the neuron's ability to determine sound location improved in the active condition. This is quite different from studies that asked whether driven activity increased, which could result from a non-stimulus-dependent increase in responses or a gain shift. Neurometric analysis showed neuronal discriminability changes in the behaving condition in more than 50%, the units with 29% showing steeper neurometric functions and 23% showing flatter ones in the behaving condition. This is interpreted as overall neural discriminability not improving, with cells getting either better or worse. It is important to note that interpreting the slopes of neurometric functions depends critically on whether the task is detection, discriminating, or identifying the parameter of interest and on how the neurometric code will be read out (Jazayeri and Movshon 2006). For example, shallow slopes as a function of sound location with low variance can be useful in a single neuron code from an information theory perspective (Jenison 2000; Jenison and Reale 2003; Jenison et al. 1998). In the case of Scott et al., this is less of an issue because about as many slopes get steeper as flatten, suggesting that performance on this discrimination task has no significant net effect on the average neuron's ability to determine interaural phase difference. In this study, sampling of the interaural phase differences in the neurophysiology also might have impacted the results. Very few data points were collected on the slope of the neurometric function. Discrimination thresholds were about 5°, but the tested phase differences were 15, 30, and 60°. The lack of data points near and below thresholds would make slope estimates less accurate. Nevertheless, this is one of the most sophisticated applications of signal detection theory to auditory cortical neuronal data and the analysis supports the interpretation that in this case the main effect of active discrimination behavior seems to be a general non-specific increase in activity.

Recently, evidence has been presented that the ability of AI neurons to discriminate an AM sound from an unmodulated sound increases when the animal is engaged in the AM task (Niwa et al. 2009). In the AM task, the animals pressed a lever to initiate a trial, and two sounds ensued. The first sound was an unmodulated white noise burst. The second sound could be the same as the first or was a sinusoidally modulated noise carrier of various modulation depths (6–100%). If the second sound was an AM signal, the animal was required to release the lever for a reward. If the second sound was unmodulated (0% depth), the animal was required to continue to hold the lever down to receive a reward. Neurometric ROC analysis was used to determine how well each unit discriminated AM from unmodulated sounds by comparing trial-by-trial responses to AM and unmodulated stimuli. Neuronal sensitivity to modulation was improved in the behaving over the passive condition and was independent of a potential general activity increase to sounds because responses increased more to the modulated than to the unmodulated noise. It is interesting that the maximum improvement occurred at the intermediate modulation depths (40–60%). This might be due to a ceiling effect at the highest depths or might also be due to an effect of attention, because the higher modulation depths are easier to discriminate from unmodulated sound than the intermediate depths.

Gottlieb et al. (1989) demonstrated increased driven single activity in posterior belt regions when comparing responses during task performance to responses in passive awake monkeys (*Papio annubis*). Monkeys heard two tones separated by 1 s of silence. If the two tones were the same frequency the monkeys were required to press one button for reward and if not another button. For 65% of units activity during the silence between the tones was significantly higher in performance than in non-performance. Only 2% of units' activity during silence was significantly lower in performance than non-performance.

The general observation of increased activity during task engagement might not be purely cortically derived. In the cochlear nucleus, nuclei of the lateral lemniscus, inferior colliculus, medial geniculate body, and AC (Ryan and Miller 1977; Ryan et al. 1984), increased driven activity was found when comparing responses during a reaction time (RT) task to passive recording. In most of these areas ~43% of units showed changes late in the stimulus (75–200 ms) and ~33% early in the stimulus (0–75 ms). The magnitude of change was larger for late responses (25% increase) than early (10% increase). Late effects were largest above the level of cochlear nucleus and the magnitude of effects and percentage of neurons having significant differences between conditions tended to be larger at higher stations in the auditory system. While spontaneous activity did not change in MGB and AC, sub-thalamic spontaneous rate tended to be higher during task performance. The conclusion of this work (Ryan et al. 1984) with regard to how engagement in this RT task affects neuronal responses is rather important; (1) multiple mechanisms are operating with varying degrees of strengths at different locations in the auditory system and (2) the net effect is to create a more sensitive, low noise signal detector in AC.

Contrasting results showing increases in driven activity when an animal is engaged, several studies have demonstrated equal numbers of units that increase or decrease driven activity when comparing passive and active conditions. Hocherman and colleagues (1976) found in a noise versus tone discrimination task that ~50% of the neurons showed no significant difference between active and passive conditions, ~25% of the neurons had larger responses in the behaving condition, and ~25% of single units had greater responses in the passive condition. When performance on the task was compared to passive stimulation, increases in activity were as likely as decreases. Gilat and Perlman (1984) had monkeys perform a similar task and ~33% of single units increases, ~33% decreased and ~33% did not significantly differ when comparing driven activity during active and passive conditions.

Two studies have predominantly found decreases in driven activity during task performance. Benson and Hienz (1978) compared behaving to non-behaving condition and found 17% of single units in AC increased activity in the performing to non-performing condition, and 27% had more activity in the non-performing condition. The task required the animal to hear sounds presented to both ears but only respond if the sound was presented to the ear they were cued to attend. A block trial design was used, where which ear to attend to remained the same within blocks of 100 trials. A light on the telegraph (response) key indicated which ear to attend. This result differs from Miller et al. (1972) where increases, but not decreases were found in an active versus passive test on a reaction time task. These results, however, are consistent with those of Hocherman et al. (1976) that found about 25% increases and 25% decreases for active versus passive conditions. So it remains unclear whether Benson and Hienz (1978) are tapping in to a propensity to decrease activity during performance or an equal propensity to increase or decrease.

Otazu et al. (2009) report a larger effect of decreasing activity with active engagement. Engagement reduced driven activity but did not affect spontaneous activity in AC. For the active condition, rats put their nose in center of a three-port chamber to start a trial. Rats were required to move to the port on the side from which a target broadband sound was presented to receive a reward. Non-targets were white noise burst trains ranging from 2 to 35 Hz. Task engagement reduced all response component for both target and non-target stimuli in AC. In MGB, engagement had no effect on driven activity but increased spontaneous activity. To explain the cortical decrease the authors propose the hypothesis that engagement decreases responses but selective attention increases them. They also looked at an intermodal auditory versus olfactory selective attention task to investigate this hypothesis and obtained results that are consistent with the hypothesis, with the addition that task difficulty also contributes.

While most data indicate increases in cortical activity with engagement in a task that has higher difficulty and specificity of attention requirements, some contradictory data need to be resolved. The many variables that might contribute to differences of observations relative to all differences accounted for by non-acoustical influences on auditory cortical responses will be discussed throughout the rest of the chapter.

5.3 Reward, Value, and Activity

One factor that might contribute to much of the non-auditory activity in AC might be reward and reward expectation. While effects of reward, reward expectation, and expected value on the responses of parietal cortex (e.g., Platt and Glimcher 1999), frontal/prefrontal cortex (e.g., Leon and Shadlen 1999), and the limbic and modulatory systems (e.g., Schultz et al. 1998) have been relatively extensively studied, little is known about the role of reward on sensory cortex (Pantoja et al. 2007; Serences 2008; Shuler and Bear 2006). In the auditory system, very few studies have manipulated reward which often covaries with other experimental parameters. For example, if rewards are given immediately after the animal's response, activity associated with the animal's response might actually relate to the expectation of reward. Beaton and Miller (1975) varied reward contingencies within a reaction time task to show that some activity might relate to reward expectation. Animals were trained in a task, where they pressed a lever to initiate a trial and had to rapidly release the lever (<1 s after stimulus onset) for target stimuli to receive a reward. In one condition animals were rewarded for rapid lever release to any tone (frequency irrelevant, FI). In the other condition (frequency discrimination, FD) animals were only rewarded for releasing to tones of one frequency (e.g., 500 Hz or 20 kHz). In the FD condition behavioral response time to the same stimuli was slower than in FI. In AC, 25% of the units responded differently to the same tone during FI than FD conditions. Almost always this was an increased onset response in the FD condition. This change was only seen for tones that were unrewarded in FD, and not tones that were rewarded for both. Compared to work with reward from other parts of the brain (reviewed in Sugrue et al. 2005) activity increases for lower reward probabilities is unusual, but this result could be consistent with the model of neurons encoding an error signal of reward (Hollerman and Schultz 1998). Because Beaton and Miller (1975) always limited unit analysis to less than 200 ms after stimulus onset and the animals behavioral response was always > 200 ms, the potential confound of using a reaction time task do not interfere with the proposed interpretation.

Reward-related activity likely is not solely a cortical phenomena in the auditory system as there have been demonstrations of reward-modulated activity in sub-cortical auditory areas. Komura et al. (2001, 2005) specifically manipulated reward expectation when recording from thalamic neurons. An auditory stimulus was presented for 2 s followed by a 1 s delay. Sounds to the left were 'go' sounds and the animal was rewarded if they went to a spout. Sounds to the right were 'no-go' and the animals were not allowed to go to the spout. Both reward size and delay from response to reward were manipulated in a systematic and predictable manner. They found that late (2.5–3.0 s after cue onset), but not early (0.0–0.5 s after cue onset) activity in neurons in the dorsal and medial division of the MGB (MGBd, MGBm) and the posterior intralaminar nucleus was directly related to reward, while no reward-related activity could be found in the ventral division of the MGB (MGBv). This differs from Beaton and Miller (1975) where activity was only analyzed within 200 ms of sound onset.

Metzger et al. (2006) manipulated rewards while recording from the inferior colliculus (IC) of monkeys performing saccades to auditory targets. The flow of the experiment was as follows: first, a light comes on and the monkey must fixate the light; second, after 500–900 ms a noise is turned on from one of the nine locations while the fixation light remains on (the 500 ms before the noise was treated as 'baseline period'); third (overlap period) 500 ms later the fixation light turns off which is a cue that the monkey can move its eyes to the sound which is still on; fourth within 100–500 ms the monkey must initiate a saccade to the sound (the period after light offset before a saccade was initiated, mean 205 ms, is called the 'Pre-saccadic period'); fifth the monkey must maintain fixation at the endpoint of the auditory saccade for 500 ms (the last 200 ms was called the "late period"); and sixth reward was given. In the variable reward task the color of the fixation light cued the reward size and only two target locations were used. Because the overlap period was 500 ms, and the monkey had to wait at least 100 ms to respond following overlap, and had to maintain post-saccade fixation for 500 ms. Only neurons that were more active during the overlap period (auditory + light) than the baseline (just light) were considered auditory and used for further analysis. Most neurons had more activity in the late period than the pre-saccadic period. During both of these periods the sound was on, the light was off and the eyes were stationary (although the eyes were in different positions), but in the later period reward is imminent if the monkey was correct, whereas in the pre-saccadic period saccade initiation is imminent. Analysis by saccade location rules out that the late activity was due to the different eye position. The cue for reward size occurred early in each trial by the color of the fixation light, and all periods (Baseline, pres-saccade, and late) had more activity in high than low reward trials. For many cells recording was also done during a sound only condition where the noise was matched in mean time to the saccade trial but no motor action was required and no reward was given. Here there was no increase in late activity compared to early, further suggesting that the saccade task result was related to reward. These data support reward activity in the IC that starts before the sounds turns on and lasts for more than 1 s. They are different from Komura and colleagues MGB data which were limited to late response components, and different from Beaton and Miller (1975) where activity before the tone did not change and activity was actually lower for rewarded conditions. Nevertheless, the compilation of results suggests reward-related activity can be observed at multiple levels, but might not be simply inherited at each level but both inheritance, feedback, and interacting effects from other sources of reward activity might be present at each level. It also remains possible that the three studies' tasks are tapping into a shared mechanism, but differences are due to the task details and not the area being recorded from.

5.4 Intermodal Selective Attention

There is evidence that responses in auditory cortex change depending on whether the subject is attending to a sound or a stimulus from another modality (e.g., visual or olfactory). Hocherman et al. (1976) did an experiment to determine if the responses to acoustically identical sounds under two different conditions (attend to the sound or attend to a visual stimulus) were the same. During task performance (tone versus noise discrimination), both auditory and visual stimuli were presented simultaneously, but which modality the monkey needed to respond to (attend) differed. For all trials both lights and sounds were presented and for any stimulus block the animal was only rewarded for using the correct cue. For AC neurons, ~1/3 there was no significant difference in response to sounds when the animals performed the auditory or visual task, ~1/3 responded more strongly to sound stimulation during auditory task performance than visual task performance, and the remaining ~1/3 responded stronger to the auditory stimuli during visual than auditory task performance. For units that responded to both tones and noise the auditory attention effect was the same for both. In this task auditory and visual cues signaled similar motor actions, and the behavioral and single-unit responses were influenced by whether the light and sound signaled the same motor action. In the attend-visual task the monkey pushed a button to the left if a light from the left side was flashed and to the right if a light came from the right. For the attend-auditory task the monkeys were trained to push the left button when noise was presented and the right button for a tone. Therefore when a noise was presented and the light was flashed from the right the auditory and visual cues conflicted forcing the animal to correctly attend the correct modality. Similarly if a tone was presented and the light was presented from the left the cues conflicted. Behaviorally animals performed at ~96% for visual and auditory cues signaling the same action, and ~75% when the two signals disagreed. Consistent-cue and inconsistent-cue trials were randomly interspersed, eliminating covariation of cues. For single units ~45% responded stronger for the visual and auditory cues indicating the same motor response, ~45% responded equally well regardless of the pairings, and ~10% of neurons responded stronger when the two cues conflicted (which compelled the animal to attend to the correct modality to obtain a reward). The neurons did not respond to the light presented alone which was tested in occasional light-only behavioral trials. In a non-behaving condition, ~50% of the neurons showed no difference between active and passive conditions, ~25% of the neurons had larger responses in the behaving condition. Two considerations are important for interpreting these results: (1) both these tasks are very easy and this create a low attention demand and (2) the auditory task might have been more difficult than the visual because in the visual task the cue for the location of the button to press is the matched location of the visual stimulus, but in the auditory task the cue for the location of the button to press was the spectrum of the sound (which forced one more level of association). With the above caveats, this study demonstrates that intermodal influences can affect response strength.

As noted above a challenging part of designing intermodal attention experiments is balancing the degree of difficulty of the tasks for each modality. Otazu et al. (2009) looked at intermodal selective attention comparing responses in AC when the animals did a sound versus an olfactory discrimination. Rats were trained to a simple olfactory task and a more difficult (as determined by training time required) tone discrimination task. While performing, stimuli for both modalities were presented, but the animal only was discriminating based on one modality. The auditory trials showed marginally higher sound evoked activity in AC than on olfactory trials. However, the auditory task appeared to be more difficult; therefore the slight increase in sound responses during sound discrimination could be due to attending the sound or could be due to the auditory task being more difficult and demanding more attention. The results of these intermodal studies do not provide conclusive evidence that attending to the auditory modality increases auditory responses in AC. More studies will be needed to assess how intermodal attention influences AC responses.

5.5 Intermodal Predictability and Cortical Responses

Predictability of the auditory stimulus based on co-variation with a visual stimulus also influences AC responses. Hocherman et al. (1981) trained monkeys to push a button to the right for a tone and to the left for a noise. Differing from Hocherman et al. (1976), here a light stimulus usually covaried with the auditory stimulus. A flash of light to the right usually preceded tones, and a flash of light from the left usually preceded noises. In 20–30% of trials false visual cues were given so the expectation of visual stimulus was not matched and relying on the visual cue would result in an incorrect response. To prevent the animals from only using the visual cues, on ~20% of the trials only auditory stimuli were used (blank trials) and no reward was given. Behavioral performance was drastically reduced with presentation of false cues. For single-unit recordings, false cues tended to increase driven activity to tones, but false cueing was as likely to increase as decrease responses to noise. Single-unit physiology showed that ~50% of the tone responses changed with cueing. For false cues to tones, 33% of single units fired significantly greater than for true cues. For true cues to tones, 16% of single units fired significantly greater than for false cues. Cueing also significantly altered noise responses in about 50% of the units. For false cues, 19% of units had significantly greater noise response compared to 24% for true-cue trials. When comparing the results to the baseline response with no visual cue (and no reward), a further simplification can be made. For cells that evoked larger responses for true cues than false cues there was both facilitation relative to the baseline for true cues and suppression below baseline for false cues. For cells that evoked larger responses for false cues than true cues there was facilitation relative to the baseline was seen for false-cues, but no change in response was seen for true cues. These effects relative to baseline appear to be independent of whether the animal's performance was correct or incorrect. So facilitation relative to baseline could occur for true or false cues, but suppression only occurred for false cues. An important consideration with regard to false-cue trials is that the authors noted that when the light came on the monkeys often put their hand on the correct telegraph key likely in anticipation of pressing it when the sound turned on. Therefore for true-cue trials the animal only had to push with the hand already on the key, but for false-cue trials the animal had to move the other hand to the other key before being able to press it. This difference in motor action could have contributed to differences in firing.

Using the same true-cue, false-cue task as Hocherman et al. (1981) and Gilat and Perlman (1984) compared true-cue, false-cue and passive conditions. During the passive conditions no rewards were given, so reward expectation could have contributed to differences between the passive condition and the task performance. Spontaneous rate significantly increased in 40% of units in true-cue blocks when compared to passive blocks. Spontaneous rate significantly decreased for 10% of units. About an equal percentage of units significantly increased (~33%), decreased (~33%), or did not differ (~33%) when comparing driven activity during true-cue and passive. There was no clear relationship between the effect on spontaneous and driven activity. While changes in spontaneous activity differed from cortex, when recording from MGB Gilat and Perlman (1984) found similar percentage of changes (1/3 increased, 1/3 decrease, 1/3 no change) were found for driven activity in MGB. Hocherman and Yirmiya (1990) found in MGB (not broken down by division) that ~6% of neurons had significantly larger responses in true cue than false, 14% had significantly larger responses in false cue than true. In auditory cortex ~13% of neurons had significantly larger responses in true cue than false, 16% had significantly larger responses in false cue than true. There was a slight increase in cortex, particularly for those neurons that fired more in true-cue trials, however it appears that this difference did not reach significance. The average magnitude of increase or decrease in driven activity relative to blank trials is larger in AC than MGB. While there appears to be a slightly larger effect in AC, the similarities between AC and MGB supports one of the three possible interpretations: (1) cortex inherits much (but not all) of its expectation-related properties from MGB; (2) cortex relays most (but not all) of its expectation-related properties back to MGB; (3) cortex and MGB share a common input providing most (but not all) of their expectation-related properties. There are several considerations though about the generalization and interpretation of these results. First, subdivisions of MGB are not considered and AI was targeted in AC, so sampling biases, particularly sampling from different subdivisions of the MGB could greatly skew the results. Second, Hocherman and Yirmiya (1990) report that only very few (4%) of AC neurons had sustained responses, which is extremely low compared to other studies in awake animals (Lu and Wang 2004; Recanzone 2000) and indicates there might be some sampling bias introduced into their cortical recordings. Finally, the number of cortical neurons modulated by expectation is lower than in Hocherman et al. (1981), which suggests that targeting AI (as opposed to all of AC) resulted in a smaller effect in their more recent study.

5.6 Intramodal Selective Attention

There is evidence that responses in auditory cortex change depending on which sound feature the subject is attending. This *intramodal* feature selective attention, which requires attending to a distinct specific sound feature, differs from *intermodal* attention that is directed to different sensory modalities. One paradigm often used is to see if the responses to sounds change depending on whether the subject was attending to the location from which the sound came.

Benson and Hienz (1978) performed an intramodal experiment to determine if the responses to acoustically identical sounds presented to the left or right ear were the same if the animals were attending to one ear of the other. When a sound was presented to the attended ear the animal had to respond; when sounds came from the non-attended ear the animal had to withhold response. This means that motor planning, reward expectation (rewards were only given for response) co-varied with sounds presented to the attended ear. There was no difference in spontaneous activity between conditions. The only differences found were increases in activity for single units when the sound was presented to the attended ear in 18% (14/77). This result is quite different from those from the visual versus auditory attention task of Hocherman and colleagues (1976) where changes were more common and equally likely to be increases as decreases relative to the visual attend condition.

Benson et al. (1981) further investigated whether attending to location (versus detecting a sound or versus passive listening) can change responses as a function of location. In the localization task, the monkey had to press a key next to the location of the sound (five speakers spaced 37.5° apart). In the detection task the animal had to press a button when a sound was detected regardless of its location. During the localization task 8% (15/196) of neurons increased activity for one location, while only 1/196 had decreased activity for one location. It should be noted that significant differences were usually found for only a single location indicating that the effect is location specific. The effects were also restricted to certain parts of auditory cortex. Localization-enhanced activity (comparing driven activity during the localization task to driven activity during the detection task) was only seen in areas outside of AI. The authors interpret their results as indicative of 'a population showing very little change that can be specifically related to localization behavior'.

5.7 Activity Related to Sensory-Motor Associations

Vaadia et al. (1982) designed an experiment to show that sensory-motor associations modulate auditory cortical activity. Monkeys were trained to push a lever to the right for a tone and to the left for a noise. After this initial training, the animals were further trained to switch the contingencies (that is push left for a tone and right for a noise) when cued. In this way, response contingency could be changed while activity was recorded to allow for better disassociation of stimulus- (sensory) versus motor-related activity. From both cortical core regions, including AI, and belt regions they found that ~17% of neurons had activity related to the sensorimotor association where a portion of the apparent stimulus-related activity was altered by the motor contingency. Neurons with sensorimotor association did not appear clustered or localized to any part of cortex and activity appeared to be randomly interspersed with purely sensory neurons in all of auditory cortex.

Durif et al. (2003) tested intramodal effects by training monkeys to change associations between different tone frequencies and different responses (left versus right button press). Within a block one frequency tone indicated press the left button, and another frequency press the right button. A confirmation tone of the same frequency as the instruction tone was presented when the animal performed correctly. For 1/3 of the neurons the response to the instructional and confirmation tone was different even though the sounds were identical (but presented in a different part of the task). It is not specified how many neurons fired more for the instructional tone and vice-versa, but stronger responses to the instructional tone would be supportive of stronger responses when stimulus attention or motor planning was most needed. Stronger responses to the confirmation tone would be supportive of stronger responses when reward expectation was highest. It should be noted that in this study anecdotal examples were given of neurons whose responses to the instructional tone depended on whether the monkey subsequently would be correct or incorrect in his decision.

5.8 Memory, Comparison, or Delay Activity

Gottlieb et al. (1989) demonstrated that in auditory cortex single neurons activity during the period between two sounds depended on whether the first sound needed to be compared the second. The recordings were primarily from the posterior belt regions. Monkeys (*P. annubis*) either sat passively or compared two tones. If the two tones had the same frequency the monkeys were required to press a right button, if they were different subjects needed to press a left button. Fifty percent of units' activity in the time between the two tones depended on first tone during the task and about 33% when there was no response required. For 65% of units activity between the two tones was significantly higher in performance than non-performance. Only 2% of units' activity during the silent interval was significantly lower in performance than non-performance. This suggests that attention, expectation, and/or memory are involved. Because only 3% of units showed behaviorally dependent activity in the silent interval that also depended on the frequency of the first tone, it is unlikely that much of this activity is related to memory about the frequency of the first tone. Also during task performance ~25% of units' responses to the second tone depended on whether its frequency was the same as the first tones. This was never seen during passive recording. This context dependency indicates that the auditory response is modulated by its behavioral meaning (match or non-match) or it could also be preparatory to motor action or reward anticipation. In a related paradigm, Shinba et al. (1995) trained rats to a visual reaction time task. They introduced a predictive warning tone that preceded the visual stimulus. About 25% of AC neurons' firing increased relative to spontaneous during the 1.4 s between the end of the tone and the beginning of the visual stimulus. There was no clustering of neurons with increased delay-period activity. While the task here is operant, the role of the warning tone is somewhat akin to classical conditioning, and it should be noted that responses dependent on interstimulus intervals have also been found in classical conditioning (Kitzes et al. 1978).

5.9 Non-auditory Activity: Sensory and Motor Related

As mentioned previously, Hocherman et al. (1981) linked visual and auditory cues so the visual cue was predictive of the auditory stimulus to which the monkey needed to respond. On false-cue trials monkeys could have high error rates, and neural activity differences were found between false- and true-cue trials. An important consideration with regard to false-cue trials is that when the light cue came on the monkeys often put their hand on the correct telegraph key in anticipation of pressing it when the sound turned on. Therefore, for true-cue trials the animal only had to push with the hand already on the key, but for false-cue trials the animal had to move the other hand to the other key before being able to respond. This difference in motor action could have contributed to differences in firing.

Further evidence of motor-related responses in auditory cortex has been reported. Brosch et al. (2005) trained macaque monkeys on a difficult task to detect stimulus shifts toward lower frequencies. A trial starts with an LED turning on. The monkey then had 3 s to make contact with bar. Once they did, a tone sequence began. The first three tones were of identical frequency. The next three tones could be of lower frequency. Alternatively, 3–6 higher frequency tones could follow which were followed by three low-frequency tones. The monkeys had to release the bar grip when they heard a shift to lower frequency tones. After bar release the cue light was turned off. The initial tone frequency was randomly varied from trial to trial, so the monkeys had to generalize a relative frequency lowering within the sequence (a very difficult task). Recordings during the task were made from auditory cortex including AI. Light-cue-related firing was found in 13% of multi-unit responses. Activity related to bar touch was found in ~62% of recording sites and activity related to bar release was registered in 85% of auditory cortical responses. All of the task-related visual and somatosensory/motor activity seen in AC disappeared when the monkey was performing a visual discrimination task without any sound presentation. This study demonstrated convincingly the presence of a significant amount of non-auditory activity in auditory cortex.

Yin et al. (2008) also found activity that could be related to non-auditory sensory and/or motor task aspects. They trained macaque monkeys on a go-no-go task to detect a 4-note melody and to ignore a variety of non-target sounds. Some ~12% units showed non-acoustic task-related activity linked to bar release. A higher percentage of these units were found in Field R than in AI indicating field differences even among auditory core regions.

5.10 Decision- or Choice-Related Activity

A fundamental question in auditory neuroscience is whether a causal link can be found between brain activity and perception. Recording from animals while they perform perceptual tasks provides a mean to link the animal's perception to brain activity. The work by Newsome and colleagues studying visual motion processing provides an example of the power of this approach. By using statistics that directly relate an animal's judgment regarding perceived motion of a visual stimulus to neural activity a correlation between the animals behavioral choice and single-unit activity in area MT of the visual cortex was found (Britten et al. 1996; Parker and Newsome 1998). The choice probability metrics used in the analysis takes advantage of the fact that, during a threshold discrimination task, near threshold the animal can respond both correctly and incorrectly while the sensory stimulus is exactly the same. If the behavior is done carefully these behavioral responses reflect the animal's perception and a link can be made between single-unit activity and perception. In Britten et al.'s work animals were trained to determine the direction of visual motion and report it in a forced choice task. The motion cue was the percentage of dots moving coherently in the same direction. By partitioning neural data to the same stimulus by the animal's response on trials they could determine whether the neural response contained information about the animals choice and possibly their perception. Decision-based response components were clearly identified even for trials with only random direction motion. It should be noted that such an approach can also be used for bi-stable stimuli such as ambiguous motion (Parker et al. 2002), binocular rivalry (Logothetis and Schall 1989), and illusory contours or streaming.

In most of the early studies mentioned above, a choice probability approach could not be attempted. This is because the employed tasks were generally easy and the animals made very few mistakes, often performing > 90%. In this case it is difficult to obtain enough error trials to have a meaningful statistical comparison between correct and incorrect trials for the same stimulus condition.

In Hocherman et al. (1976) intermodal attention task, animals did make more mistakes when the visual and auditory cues gave conflicting instructions. Here, animal's performance could drop to 75% correct. These authors comment that no difference in activity between misses, correct responses and incorrect responses were reported. However, there is no indication of how this was analyzed. Hocherman et al. (1981) linked visual and auditory cues so the visual cue was predictive of the auditory stimulus to which the monkey needed to respond. On false-cue trials monkeys could have high error rates, so it was possible to compare error and correct trials. Analysis was restricted to false-cue trials where for tones, 34% had significantly greater activity in correct trials and 11% had significantly greater activity in error trials where the animal pushed the incorrect button. For noise, 25% had significantly greater activity in correct trials and 25% had significantly greater activity in error trials where the animal pushed the incorrect button. Misses were recorded when a stimulus was presented but the animal failed to respond. Activity in response to tones for correct trials was greater than misses (both considered only for false-cueing) in 34% of units and more in misses for 11%. For noises the percentage of units were 24 and 29% respectively. Therefore on average it appears as if the activity is slightly higher on correct trials than on incorrect responses or misses, but these differences relate to whether the sound was a noise or a tone which was related to different motor action since the monkey was asked to push different buttons for these two conditions.

The presence of activity that is related to lever release in AC brings to light a shortcoming of the choice probability analysis. Because choice probability relates neural activity to the animal's motor response, it is difficult to disentangle if what is being measured is the animal's decision/perception or pre-motor activity. Activity related to bar release can be intermingled with decision-related activity because the only way to measure the animal's intent to respond is when the animal actually responds (which requires a motor action, in this case bar release). Brosch et al.'s work (2005) indicates that the responses are not purely lever-related since the lever-activity is not observed in AC during performance of a visual task. Therefore the lever-related activity is dependent on the lever touch or release occurring in the context of an auditory task, and not a general somatosensory or motor effect. However more information is needed on the timing of the bar-release to see if what is interpreted as lever-related responses might also relate to perception. It is more clear that the bar-release data of Yin et al. (2008) is not related to the animal's decision because bar-release neurons included neurons whose change in activity only occurred *after* bar release. Such activity after the motor response likely is not caused by either the perception, decision, or planning that leads to the response.

Scott et al. (2007) specifically asked if increased ability of neurons to discriminate interaural phase differences lead to an increased behavioral performance by the monkeys. They did not find a correlation between increased neural discriminability and the percentage of correct trials. It is important to note that conceptually their analysis goes a step beyond that of Britten and Newsome's work. Scott et al. were asking if neural discriminability improved, that is was the neuron more able to tell that two stimuli were different on trials where the animal reports them as different. The previously asked question in the visual system was simply if there was increased neural activity when the motion was judged as going in a specific direction. The latter could simply be increased gain whereas Scott et al.'s approach requires a more specific change as in the temporal response pattern.

Lemus et al. (2009) investigated the activity of AI neurons in monkeys during an auditory 'flutter' discrimination task which was designed analogous to a somatosensory flutter task. They did not identify activity significantly related to the animals' choices.

However Niwa and colleagues (2009) report significant activity related to the animals choice (choice probability) in monkey AI. Here animals had to detect whether or not a sound was amplitude modulated as a function of modulation depth. To analyze the relationship between unit activity and the animals' trial-by-trial decisions, choice probability analysis was used. At each modulation depth, the choice probability analysis statistically compares a unit's firing rate on response trials (animal chooses to respond) and non-response (animal chooses not to respond) trials. Because the stimuli are identical on these trials, any differences in activity must be related to the animal's choice rather than stimulus related. When the activity starting from the beginning of the stimulus up to the animal's behavioral response was compared, ~35% of recording sites and 16% of single units had significantly greater firing rate on response trials than non-response trials. When activity only during the stimulus was used 14% of multiple units and 11% of single units had significantly greater firing rate on response trials than on non-response trials. Using activity timed to the stimulus provides a stronger case for decision-related activity to than using activity linked in time to lever release. To our knowledge this is the first evidence in auditory cortex, and in this case primary auditory cortex (AI) of activity related to the animal's choice.

While only one AC study has found strong evidence for choice-related activity, decision-related activity has been found in the auditory thalamus which suggests it should also be observable in AC. Komura et al. (2005) showed activity related to the animals choice in the auditory thalamus. This activity was observed in suprageniculate nucleus, posterior intralaminar nucleus, medial and dorsal divisions of the MGB, but were not observed in the ventral division of the MGB, the main source of projections to AI. The animals had to go if a 2 s sound was to the left and withhold response if the sound was to the right. The animal had to wait 1 s after sound offset to go. There was a strong tendency for higher late activity (500 ms after sound offset) on trials where the animal responded (hits/false alarms) than when they did not respond (miss/correct rejection). No such trend was seen for early activity (0–100 ms after tone onset). This activity could relate to decision/choice or reward because no reward was given on no-go trials. The work of Niwa et al. (2009) differs in this respect because correct rejections (withholding response for a non-target sounds) were rewarded and no increased activity was seen prior to these rewards, suggesting that the Niwa et al.'s results are not solely driven by reward expectation. These recent results support the presence of choice activity in AI that might be due to different causes than those observed by Komura.

5.11 Conclusion on Non-auditory Influences in Auditory Cortex

Throughout the years unit recordings from animals performing auditory tasks has not been common but, from the studies performed, some unifying themes have been found. The most important general theme is that in auditory cortex, even AI, there is activity that is not solely influenced by the properties of sounds but also by non-auditory sensory, behavioral, and cognitive factors. Often this activity is associated with the motor contingency or actions involved in the task, with reward and non-auditory cues used in the behavioral paradigm, and/or with attention. As can be seen from this section, the results have been varied. Because the number of task and stimulus parameters to control for in these studies is increasing with more knowledge of the involved variables, and because the parameters often co-vary, more studies need to be performed to carefully characterize these influences and resolve the apparent contradictions in the literature.

One property that is important to control for is task difficulty, both with regard to the stimulus and behavioral/cognitive dimension. From the stimulus dimension, this means tasks that require the animal to perform at threshold levels (for example determining the minimum difference in frequency an animal can discern) are far more difficult to perform than asking the animal, for example, to distinguish a tone from a noise. From a behavioral cognitive sense, a task that requires the animal to move in the direction (left/right) of a sound is easier than giving the animal five different buttons to press based on different pitch and timbre properties. Why would this matter? In the visual system consistent results show that attention can increase activity in visual cortex neurons but the ability to observe this depends on how much attention the task demands (Boudreau et al. 2006; Motter 1993; Spitzer et al. 1988). In most of the auditory studies reviewed in this section the task is easy and monkeys performed at very high levels, thereby minimizing the possibility of obtaining interpretable results. As a field, as our ability to train animals on auditory tasks improves, it is likely that more effects of attention on auditory cortical activity will be found.

To properly compare studies, many other factors, such as reward, are important to consider as well. In all these studies properties of the reward and how the reward is administered are critical to know because they might contribute to the observed results. While this has mainly been worked out in association cortex and the limbic system, reward, value, or reward-expectation might contribute to results in sensory cortex of behaving animals, particularly when comparing to a non-behaving condition.

The timing of stimulus and events in the behavioral paradigm also can have large influences on the results. These include delays between the animal initiating a trial (e.g., by grasping a lever) and the beginning of stimulus presentation. Also delays between stimulus onset/offset and when the animal is allowed to respond are important. Adding a delay here can force the animal to hold her decision. This delay can also delay a pre-motor response in neurons, thus allowing separation of decision activity (that could start early) from pre-motor activity (which would be linked to the delayed motor response). The coupling of reward to motor action is also critical. If the reward is given immediately after the motor response, response- and reward-related activity can be confounded. Delaying reward can help to separate these effects, but comes at an expense. Learning is greatly impaired and training made more difficult because delaying the reward dramatically decreases the behavioral association between the response and reward. Finally if the animal is forced to compare stimuli, adding delay between the sounds introduces a more challenging memory component to the task. This has to be considered carefully in audition, where in monkeys at least, it appears that working memory might use fundamentally different mechanisms than visual working memory (Ng et al. 2009). The duration of the stimuli is important also. It appears that longer stimuli are more likely to show non-auditory effects than shorter stimuli because some of these effects have longer latencies. It is also relevant to note if the non-auditory effects are constrained to early or late time periods. Often in behaving studies, a blocked trial design is necessary to keep the behavior feasible. The blocking design, e.g., number in a block and amount of variation per block, can have a large influence on results. Also block designs can lead to entrainment or plasticity effects, which need to be analyzed.

When looking at how different attention-related properties effect performance, task difficulty is critical. For example, in intermodal attention tasks it is critical that the two different modality tasks be matched in difficulty or modality effects can be confounded with attention influences. The same applies for intramodal attention. The amount of training is also an important parameter. The more training an animal receives, the more opportunity there is for plastic changes to occur relative to the task.

The choice of the applied analyses and statistics is critical. Differences in statistics, and in particular what is being compared, can lead to quite different interpretations of results. For example, the methods have to be appropriate for trying to determine if evoked/spontaneous activity or neural discrimination ability is affected. These are two very different questions, yet the difference can be lost when comparing across studies.

The anatomical location of recording and the species studied require careful attention. So far the results obtained in different species vary greatly with respect to non-auditory activity and this needs to be resolved going forward. Confounding this is the lack of detail to date about which cortical areas are recorded from and which might be homologous. To date, the literature on non-auditory activity in auditory cortex mainly looks at the auditory cortex as a single entity. More recent studies (e.g., Scott et al. 2007) have been more careful in defining areas. This care about areas (and anatomical connections) will be essential going forward if the field is to try to meaningfully interpret results from different studies, and how and where non-auditory influences on auditory cortical areas arises.

Acknowledgements Supported by NIH grant DC005779, and the Air Force Office for Scientific Research (to S.A.S.), and NIH grant DC02514, and the McDonnell Foundation (to M.L.S.).

References

Aizawa N and Eggermont JJ (2006) Effects of noise-induced hearing loss at young age on voice onset time and gap-in-noise representations in adult cat primary auditory cortex. *Journal of the Association of Research in Otolaryngology* **7**:71–81.

Bacon SP and Viemeister NF (1985) Temporal modulation transfer functions in normal-hearing and hearing-impaired listeners. *Audiology* **24**:117–134.

Bakin JS, South DA, and Weinberger NM (1996) Induction of receptive field plasticity in the auditory cortex of the guinea pig during instrumental avoidance conditioning. *Behavioral Neuroscience* **110**:905–913.

Beaton R and Miller JM (1975) Single cell activity in the auditory cortex of the unanesthetized, behaving monkey: correlation with stimulus controlled behavior. *Brain Research* **100**:543–562.

Bendor D and Wang X (2005) The neuronal representation of pitch in primate auditory cortex. *Nature* **436**:1161–1165.

Bendor D and Wang X (2008) Neural response properties of primary, rostral, and rostrotemporal core fields in the auditory cortex of marmoset monkeys. *Journal of Neurophysiology* **100**:888–906.

Benson DA and Hienz RD (1978) Single-unit activity in the auditory cortex of monkeys selectively attending left vs. right ear stimuli. *Brain Research* **159**:307–320.

Benson DA, Hienz RD, and Goldstein MH, Jr (1981) Single-unit activity in the auditory cortex of monkeys actively localizing sound sources: spatial tuning and behavioral dependency. *Brain Research* **219**:249–267.

Bizley JK and King AJ (2008) Visual-auditory spatial processing in auditory cortical neurons. *Brain Research* **1242**:24–36.

Bizley JK and King AJ (2009) Visual influences on ferret auditory cortex. *Hearing Research* **258**:55–63.

Boudreau CE, Williford TH, and Maunsell JH (2006) Effects of task difficulty and target likelihood in area V4 of macaque monkeys. *Journal of Neurophysiology* **96**:2377–2387.

Braaten RF and Leary JC (1999) Temporal induction of missing birdsong segments in European Starling. *Psychological Sciences* **10**:162–166.

Brechmann A and Scheich H (2005) Hemispheric shifts of sound representation in auditory cortex with conceptual listening. *Cerebral Cortex* **15**:578–587.

Bregman AS (1990) *Auditory Scene Analysis*. MIT Press, Cambridge, pp. 344–383.

Britten KH, Newsome WT, Shadlen MN, Celebrini S, and Movshon JA (1996) A relationship between behavioral choice and the visual responses of neurons in macaque MT. *Visual Neuroscience* **13**: 87–100.

Britten KH, Shadlen MN, Newsome WT, and Movshon JA (1992) The analysis of visual motion: a comparison of neuronal and psychophysical performance. *Journal of Neuroscience* **12**: 4745–4765.

Brosch M, Selezneva E, and Scheich H (2005) Nonauditory events of a behavioral procedure activate auditory cortex of highly trained monkeys. *Journal of Neuroscience* **25**:6797–6806.

Budinger E, Heil P, Hess A, and Scheich H (2006) Multisensory processing via early cortical stages: connections of the primary auditory cortical field with other sensory systems. *Neuroscience* **143**:1065–1083.

Burns EM and Viemesiter NF (1976) Nonspectral pitch. *Journal of the Acoustical Society of America* **60**:863–869.

Burns EM and Viemesiter NF (1981) Played-again SAM: further observations on the pitch of amplitude-modulated noise. *Journal of the Acoustical Society of America* **70**:1655–1660.

Butler RA, Diamond IT, and Neff WD (1957) Role of auditory cortex in discrimination of changes in frequency. *Journal of Neurophysiology* **20**:108–120.

Cariani PA and Delgutte B (1996a) Neural correlates of the pitch of complex tones. I. Pitch and pitch salience. *Journal of Neurophysiology* **76**:1698–1716.

Cariani PA and Delgutte B (1996b) Neural correlates of the pitch of complex tones. II. Pitch shift, pitch ambiguity, phase invariance, pitch circularity, rate pitch, and the dominance region for pitch. *Journal of Neurophysiology* **76**:1717–1734.

Carlyon RP (2004) How the brain separates sounds. *Trends in Cognitive Sciences* **8**:465–471.

Carr CE (1986) Time coding in electric fish and barn owls. *Brain Behavior and Evolution* **28**:122–133.

Carr CE and Konishi M (1990) A circuit for detection of interaural time differences in the brain stem of the barn owl. *Journal of Neuroscience* **10**:3227–3246.

Carr CE, Maler L, and Taylor B (1986) A time-comparison circuit in the electric fish midbrain. II. Functional morphology. *Journal of Neuroscience* **6**:1372–1383.

Chechik G, Anderson MJ, Bar-Yosef O, Young ED, Tishby N, and Nelken I (2006) Reduction of information redundancy in the ascending auditory pathway. *Neuron* **51**:359–368.

Chernyshev BV and Weinberger NM (1998) Acoustic frequency tuning of neurons in the basal forebrain of the waking guinea pig. *Brain Research* **793**:79–94.

Chi T, Gao Y, Guyton MC, Ru P, and Shamma S (1999) Spectro-temporal modulation transfer functions and speech intelligibility. *Journal of the Acoustical Society of America* **106**:2719–2732.

Chi T, Ru P, and Shamma SA (2005) Multiresolution spectrotemporal analysis of complex sounds. *Journal of the Acoustical Society of America* **118**:887–906.

Clarey JC, Barone P, and Imig TJ (1992) Physiology of thalamus and cortex. In: Popper AN and Fay RR (eds). *Springer Handbook of Auditory Research*, volume 1, *The Mammalian Auditory Pathway: Neurophysiology*. Springer, New York, pp. 232–334.

Cooke JE, Zhang H, and Kelly JB (2007) Detection of sinusoidal amplitude modulated sounds: deficits after bilateral lesions of auditory cortex in the rat. *Hearing Research* **231**:90–99.

Coslett HB, Brashear HR, and Heilman KM (1984) Pure word deafness after bilateral primary auditory cortex infarcts. *Neurology* **34**: 347–352.

Cranford J, Ravizza R, Diamond IT, and Whitfield IC (1971) Unilateral ablation of the auditory cortex in the cat impairs complex sound localization. *Science* **172**:286–288.

Cranford JL, Igarashi M, and Stramler JH (1976) Effect of auditory neocortex ablation on pitch perception in the cat. *Journal of Neurophysiology* **39**:143–152.

Cranford JL and Oberholtzer M (1976) Role of neocortex in binaural hearing in the cat. II. The 'precedence effect' in sound localization. *Brain Research* **111**:225–239.

Crist RE, Li W, and Gilbert CD (2001) Learning to see: experience and attention in primary visual cortex. *Nature Neuroscience* **4**:519–525.

Dau T, Kollmeier B, and Kohlrausch A (1997) Modeling auditory processing of amplitude modulation. II. Spectral and temporal integration. *Journal of the Acoustical Society of America* **102**: 2906–2919.

deCharms RC, Blake DT, and Merzenich MM (1998) Optimizing sound features for cortical neurons. *Science* **280**:1439–1443.

Dent ML, Brittan-Powell EF, Dooling RJ, and Pierce A (1997) Perception of synthetic /ba/-/wa/ speech continuum by budgerigars (Melopsittacus undulatus). *Journal of the Acoustical Society of America* **102**:1891–1897.

DeValois R and DeValois K (1990) *Spatial Vision*. Oxford University Press, New York.

Diamond DM and Weinberger NM (1986) Classical conditioning rapidly induces specific changes in frequency receptive fields of single neurons in secondary and ventral ectosylvian auditory cortical fields. *Brain Research* **372**:357–360.

Diamond IT, Goldberg JM, and Neff WD (1962) Tonal discrimination after ablation of auditory cortex. *Journal of Neurophysiology* **25**:223–235.

Diamond IT and Neff WD (1957) Ablation of temporal cortex and discrimination of auditory patterns. *Journal of Neurophysiology* **20**:300–315.

Dooling RJ and Searcy MH (1985) Temporal integration of acoustic signals by the budgerigar (*Melopsittacus undulatus*). *Journal of the Acoustical Society of America* **77**:1917–1920.

Drullman R, Festen JM, and Plomp R (1994) Effect of reducing slow temporal modulations on speech reception. *Journal of the Acoustical Society of America* **95**:2670–2680.

Durif C, Jouffrais C, and Rouiller EM (2003) Single-unit responses in the auditory cortex of monkeys performing a conditional acousticomotor task. *Experimental Brain Research* **153**:614–627.

Eddins DA (1999) Amplitude-modulation detection at low- and high-audio frequencies. *Journal of the Acoustical Society of America* **105**:829–837.

Edeline JM and Weinberger NM (1993) Receptive field plasticity in the auditory cortex during frequency discrimination training: selective retuning independent of task difficulty. *Behavioral Neuroscience* **107**:82–103.

Edwards E, Soltani M, Deouell LY, Berger MS, and Knight RT (2005) High gamma activity in response to deviant auditory stimuli recorded directly from human cortex. *Journal of Neurophysiology* **94**:4269–4280.

Eggermont JJ (1995) Representation of a voice onset time continuum in primary auditory cortex of the cat. *Journal of the Acoustical Society of America* **98**:911–920.

Eggermont JJ (2006) Properties of correlated neural activity clusters in cat auditory cortex resemble those of neural assemblies. *Journal of Neurophysiology* **96**:746–764.

Elhilali M, Fritz JB, Klein DJ, Simon JZ, and Shamma SA (2004) Dynamics of precise spike timing in primary auditory cortex. *Journal of Neuroscience* **24**:1159–1172.

Elhilali MT, Chi T, and Shamma S (2003) Intelligibility and the spectrotemporal representation of speech in the auditory cortex. *Speech Communication* **41**:331–348.

Ewert SD and Dau T (2004) External and internal limitations in amplitude-modulation processing. *Journal of the Acoustical Society of America* **116**:478–490.

Fay RR (1988) Comparative psychoacoustics. *Hearing Research* **34**:295–305.

Fay RR and Coombs SL (1992) Psychometric functions for level discrimination and the effects of signal duration in the goldfish

(*Carassius auratus*): psychophysics and neurophysiology. *Journal of the Acoustical Society of America* **92**:189–201.

Fishman YI, Arezzo JC, and Steinschneider M (2004) Auditory stream segregation in monkey auditory cortex: effects of frequency separation, presentation rate, and tone duration. *Journal of the Acoustical Society of America* **116**:1656–1670.

Fishman YI, Reser DH, Arezzo JC, and Steinschneider M (1998) Pitch vs. spectral encoding of harmonic complex tones in primary auditory cortex of the awake monkey. *Brain Research* **786**:18–30.

Fishman YI, Reser DH, Arezzo JC, and Steinschneider M (2001) Neural correlates of auditory stream segregation in primary auditory cortex of the awake monkey. *Hearing Research* **151**:167–187.

Forrest TG and Green DM (1987) Detection of partially filled gaps in noise and the temporal modulation transfer function. *Journal of the Acoustical Society of America* **82**:1933–1943.

Fries P, Reynolds JH, Rorie AE, and Desimone R (2001) Modulation of oscillatory neuronal synchronization by selective visual attention. *Science* **291**:1560–1563.

Fritz J, Shamma S, Elhilali M, and Klein D (2003) Rapid task-related plasticity of spectrotemporal receptive fields in primary auditory cortex. *Nature Neuroscience* **6**:1216–1223.

Fu KM, Johnston TA, Shah AS, Arnold L, Smiley J, Hackett TA, Garraghty PE, and Schroeder CE (2003) Auditory cortical neurons respond to somatosensory stimulation. *Journal of Neuroscience* **23**:7510–7515.

Ghazanfar AA (2009) The multisensory roles for auditory cortex in primate vocal communication. *Hearing Research* **258**:113–120.

Ghazanfar AA, Maier JX, Hoffman KL, and Logothetis NK (2005) Multisensory integration of dynamic faces and voices in rhesus monkey auditory cortex. *Journal of Neuroscience* **25**: 5004–5012.

Ghazanfar AA and Schroeder CE (2006) Is neocortex essentially multisensory? *Trends in Cognitive Sciences* **10**:278–285.

Gilat E and Perlman I (1984) Single unit activity in the auditory cortex and the medial geniculate body of the rhesus monkey: behavioral modulation. *Brain Research* **324**:323–333.

Gleich O and Klump GM (1995) Temporal modulation transfer functions in the European Starling (*Sturnus vulgaris*): II. Responses of auditory-nerve fibres. *Hearing Research* **82**:81–92.

Goldberg JM and Brown PB (1969) Response of binaural neurons of dog superior olivary complex to dichotic tonal stimuli: some physiological mechanisms of sound localization. *Journal of Neurophysiology* **32**:613–636.

Goldberg JM and Neff WD (1961) Frequency discrimination after bilateral ablation of cortical auditory areas. *Journal of Neurophysiology* **24**:119–128.

Goldstein JL (1973) An optimum processor theory for the central formation of the pitch of complex tones. *Journal of the Acoustical Society of America* **54**:1496–1516.

Gottlieb Y, Vaadia E, and Abeles M (1989) Single unit activity in the auditory cortex of a monkey performing a short term memory task. *Experimental Brain Research* **74**:139–148.

Graham J, Greenwood R, and Lecky B (1980) Cortical deafness: a case report and review of the literature. *Journal of Neurological Science* **48**:35–49.

Green DM (1986) 'Frequency' and the detection of spectral shape change. In: Moore B (ed). *Auditory Frequency Selectivity*. Plenum Press, New York, pp. 351–359.

Grimault N, Bacon SP, and Micheyl C (2002) Auditory stream segregation on the basis of amplitude-modulation rate. *Journal of the Acoustical Society of America* **111**:1340–1348.

Hackett TA, De La Mothe LA, Ulbert I, Karmos G, Smiley J, and Schroeder CE (2007) Multisensory convergence in auditory cortex. II. Thalamocortical connections of the caudal superior temporal plane. *Journal of Comparative Neurology* **502**:924–952.

Hackett TA and Schroeder CE (2009) Multisensory integration in auditory and auditory-related areas of cortex. *Hearing Research* **258**:1–3.

Hattori T and Suga N (1997) The inferior colliculus of the mustached bat has the frequency-vs-latency coordinates. *Journal of Comparative Physiology A* **180**:271–284.

Heffner HE and Heffner RS (1986a) Hearing loss in Japanese macaques following bilateral auditory cortex lesions. *Journal of Neurophysiology* **55**:256–271.

Heffner HE and Heffner RS (1986b) Effect of unilateral and bilateral auditory cortex lesions on the discrimination of vocalizations by Japanese macaques. *Journal of Neurophysiology* **56**:683–701.

Heffner HE and Heffner RS (1990) Effect of bilateral auditory cortex lesions on absolute thresholds in Japanese macaques. *Journal of Neurophysiology* **64**:191–205.

Heil P, Rajan R, and Irvine DR (1994) Topographic representation of tone intensity along the isofrequency axis of cat primary auditory cortex. *Hearing Research* **76**:188–202.

Henderson D, Salvi R, Pavek G, and Hamernik R (1984) Amplitude modulation thresholds in chinchillas with high-frequency hearing loss. *Journal of the Acoustical Society of America* **75**: 1177–1183.

Hienz RD, Aleszczyk CM, and May BJ (1996) Vowel discrimination in cats: acquisition, effects of stimulus level, and performance in noise. *Journal of the Acoustical Society of America* **99**:3656–3668.

Hocherman S, Benson DA, Goldstein MH, Jr, Heffner HE, and Hienz RD (1976) Evoked unit activity in auditory cortex of monkeys performing a selective attention task. *Brain Research* **117**:51–68.

Hocherman S, Itzhaki A, and Gilat E (1981) The response of single units in the auditory cortex of rhesus monkeys to predicted and to unpredicted sound stimuli. *Brain Research* **230**:65–86.

Hocherman S and Yirmiya R (1990) Neuronal activity in the medial geniculate nucleus and in the auditory cortex of the rhesus monkey reflects signal anticipation. *Brain* **113**:1707–1720.

Hollerman JR and Schultz W (1998) Dopamine neurons report an error in the temporal prediction of reward during learning. *Nature Neuroscience* **1**:304–309.

Houtgast T, Steeneken HJM, and Plomp R (1980) Predicting speech intelligibility in rooms from the modulation transfer function. I. General room acoustics. *Acustica* **46**:60–72.

Hubel DH, Henson CO, Rupert A, and Galambos R (1959) Attention units in the auditory cortex. *Science* **129**:1279–1280.

Irvine DR and Rajan R (1996) Injury- and use-related plasticity in the primary sensory cortex of adult mammals: possible relationship to perceptual learning. *Clinical and Experimental Pharmacology and Physiology* **23**:939–947.

Irvine DR, Rajan R, and Brown M (2001) Injury- and use-related plasticity in adult auditory cortex. *Audiology and Neurootology* **6**:192–195.

Ison JR, O'Connor K, Bowen GP, and Bocirnea A (1991) Temporal resolution of gaps in noise by the rat is lost with functional decortication. *Behavioral Neuroscience* **105**:33–40.

Jazayeri M and Movshon JA (2006) Optimal representation of sensory information by neural populations. *Nature Neuroscience* **9**:690–696.

Jenison RL (2000) Correlated cortical populations can enhance sound localization performance. *Journal of the Acoustical Society of America* **107**:414–421.

Jenison RL and Reale RA (2003) Likelihood approaches to sensory coding in auditory cortex. *Network* **14**:83–102.

Jenison RL, Reale RA, Hind JE, and Brugge JF (1998) Modeling of auditory spatial receptive fields with spherical approximation functions. *Journal of Neurophysiology* **80**:2645–2656.

Jenkins WM and Merzenich MM (1984) Role of cat primary auditory cortex for sound localization behavior. *Journal of Neurophysiology* **52**:819–847.

Jeschke M, Lenz D, Budinger E, Herrmann CS, and Ohl FW (2008) Gamma oscillations in gerbil auditory cortex during a target-discrimination task reflect matches with short-term memory. *Brain Research* **1220**:70–80.

Johnsrude IS, Penhune VB, and Zatorre RJ (2000) Functional specificity in the right human auditory cortex for perceiving pitch direction. *Brain* **123**:155–163.

Joris PX, Schreiner CE, and Rees A (2004) Neural processing of amplitude-modulated sounds. *Physiological Reviews* **84**: 541–577.

Kajikawa Y and Hackett TA (2005) Entropy analysis of neuronal spike train synchrony. *Journal of Neuroscience Methods* **149**:90–93.

Kanwal JS, Medvedev AV, and Micheyl C (2003) Neurodynamics for auditory stream segregation: tracking sounds in the mustached bat's natural environment. *Network* **14**:413–435.

Kaur S, Lazar R, and Metherate R (2004) Intracortical pathways determine breadth of subthreshold frequency receptive fields in primary auditory cortex. *Journal of Neurophysiology* **91**:2551–2567.

Kayser C, Petkov CI, and Logothetis NK (2008) Visual modulation of neurons in auditory cortex. *Cerebral Cortex* **18**:1560–1574.

Kelly JB, Cooke JE, Gilbride PC, Mitchell C, and Zhang H (2006) Behavioral limits of auditory temporal resolution in the rat: amplitude modulation and duration discrimination. *Journal of Comparative Psychology* **120**:98–105.

Kilgard MP and Merzenich MM (1998) Cortical map reorganization enabled by nucleus basalis activity. *Science* **279**:1714–1718.

Kilgard MP, Pandya PK, Engineer ND, and Moucha R (2002) Cortical network reorganization guided by sensory input features. *Biological Cybernetics* **87**:333–343.

Kilgard MP, Pandya PK, Vazquez J, Gehi A, Schreiner CE, and Merzenich MM (2001) Sensory input directs spatial and temporal plasticity in primary auditory cortex. *Journal of Neurophysiology* **86**:326–338.

Kitzes LM, Farley GR, and Starr A (1978) Modulation of auditory cortex unit activity during the performance of a conditioned response. *Experimental Neurology* **62**:678–697.

Klein DJ, Depireux DA, Simon JZ, and Shamma SA (2000) Robust spectrotemporal reverse correlation for the auditory system: optimizing stimulus design. *Journal of Computational Neuroscience* **9**:85–111.

Kluender KR, Diehl RL, and Killeen PR (1987) Japanese quail can learn phonetic categories. *Science* **237**:1195–1197.

Komura Y, Tamura R, Uwano T, Nishijo H, Kaga K, and Ono T (2001) Retrospective and prospective coding for predicted reward in the sensory thalamus. *Nature* **412**:546–549.

Komura Y, Tamura R, Uwano T, Nishijo H, and Ono T (2005) Auditory thalamus integrates visual inputs into behavioral gains. *Nature Neuroscience* **8**:1203–1209.

Kopell N (2005) Does it have to be this complicated? Focus on "Single-column thalamocortical network model exhibiting gamma oscillations, spindles, and epileptogenic bursts". *Journal of Neurophysiology* **93**:1829–1830.

Kowalski N, Depireux DA, and Shamma SA (1996) Analysis of dynamic spectra in ferret primary auditory cortex. II. Prediction of unit responses to arbitrary dynamic spectra. *Journal of Neurophysiology* **76**:3524–3534.

Kraus N, McGee T, Carrell TD, King C, Tremblay K, and Nicol T (1995) Central auditory system plasticity associated with speech discrimination training. *Journal of Cognitive Neuroscience* **7**: 25–32.

Kuhl PK and Miller JD (1975) Speech perception by the chinchilla: voiced-voiceless distinction in alveolar plosive consonants. *Science* **190**:69–72.

Kusmierek P and Rauschecker JP (2009) Functional specialization of medial auditory belt cortex in the alert rhesus monkey. *Journal of Neurophysiology* **102**:1606–1622.

Lakatos P, Chen CM, O'Connell MN, Mills A, and Schroeder CE (2007) Neuronal oscillations and multisensory interaction in primary auditory cortex. *Neuron* **53**:279–292.

Lakatos P, Szilagyi N, Pincze Z, Rajkai C, Ulbert I, and Karmos G (2004) Attention and arousal related modulation of spontaneous gamma-activity in the auditory cortex of the cat. *Brain Research Cognitive Brain Research* **19**:1–9.

Langemann U and Klump GM (2007) Detecting modulated signals in modulated noise: (I) behavioural auditory thresholds in a songbird. *European Journal of Neuroscience* **26**:1969–1978.

Langner G (1992) Periodicity coding in the auditory system. *Hearing Research* **60**:115–142.

Langner G, Sams M, Heil P, and Schulze H (1997) Frequency and periodicity are represented in orthogonal maps in the human auditory cortex: evidence from magnetoencephalography. *Journal of Comparative Physiology A* **181**:665–676.

Langner G and Schreiner CE (1988) Periodicity coding in the inferior colliculus of the cat. I. Neuronal mechanisms. *Journal of Neurophysiology* **60**:1799–1822.

Lee CC and Winer JA (2005) Principles governing auditory cortex connections. *Cerebral Cortex* **15**:1804–1814.

Lemus L, Hernandez A, and Romo R (2009) Neural codes for perceptual discrimination of acoustic flutter in the primate auditory cortex. *Proceedings of the National Academy of Sciences of the United States of America* **106**:9471–9476.

Leon MI and Shadlen MN (1999) Effect of expected reward magnitude on the response of neurons in the dorsolateral prefrontal cortex of the macaque. *Neuron* **24**:415–425.

Li CS, Padoa-Schioppa C, and Bizzi E (2001) Neuronal correlates of motor performance and motor learning in the primary motor cortex of monkeys adapting to an external force field. *Neuron* **30**:593–607.

Logothetis NK and Schall JD (1989) Neuronal correlates of subjective visual perception. *Science* **245**:761–763.

Lomber SG and Malhotra S (2008) Double dissociation of 'what' and 'where' processing in auditory cortex. *Nature Neuroscience* **11**: 609–616.

Lomber SG, Malhotra S, and Hall AJ (2007) Functional specialization in non-primary auditory cortex of the cat: areal and laminar contributions to sound localization. *Hearing Research* **229**: 31–45.

Lotto AJ and Kluender KR (1998) General contrast effects in speech perception: effect of preceding liquid on stop consonant identification. *Perception and Psychophysics* **60**:602–619.

Lu T, Liang L, and Wang X (2001) Temporal and rate representations of time-varying signals in the auditory cortex of awake primates. *Nature Neuroscience* **4**:1131–1138.

Lu T and Wang X (2000) Temporal discharge patterns evoked by rapid sequences of wide- and narrowband clicks in the primary auditory cortex of cat. *Journal of Neurophysiology* **84**:236–246.

Lu T and Wang X (2004) Information content of auditory cortical responses to time-varying acoustic stimuli. *Journal of Neurophysiology* **91**:301–313.

Ma X and Suga N (2009) Specific and nonspecific plasticity of the primary auditory cortex elicited by thalamic auditory neurons. *Journal of Neuroscience* **29**:4888–4896.

Malhotra S and Lomber SG (2007) Sound localization during homotopic and heterotopic bilateral cooling deactivation of primary and nonprimary auditory cortical areas in the cat. *Journal of Neurophysiology* **97**:26–43.

Malhotra S, Stecker GC, Middlebrooks JC, and Lomber SG (2008) Sound localization deficits during reversible deactivation of primary auditory cortex and/or the dorsal zone. *Journal of Neurophysiology* **99**:1628–1642.

Malone BJ, Scott BH, and Semple MN (2002) Context-dependent adaptive coding of interaural phase disparity in the auditory cortex of awake macaques. *Journal of Neuroscience* **22**:4625–4638.

Malone BJ, Scott BH, and Semple MN (2007) Dynamic amplitude coding in the auditory cortex of awake rhesus macaques. *Journal of Neurophysiology* **98**:1451–1474.

Malone BJ and Semple MN (2001) Effects of auditory stimulus context on the representation of frequency in the gerbil inferior colliculus. *Journal of Neurophysiology* **86**:1113–1130.

Maruyama N and Kanno Y (1961) Experimental study of functional compensation of bilateral removal of auditory cortex in cats. *Journal of Neurophysiology* **24**:193–202.

Mercado E, 3rd, Myers CE, and Gluck MA (2001) A computational model of mechanisms controlling experience-dependent reorganization of representational maps in auditory cortex. *Cognitive Affective and Behavioral Neuroscience* **1**:37–55.

Mesgarani N, David SV, Fritz JB, and Shamma SA (2008) Phoneme representation and classification in primary auditory cortex. *Journal of the Acoustical Society of America* **123**:899–909.

Metherate R, Kaur S, Kawai H, Lazar R, Liang K, and Rose HJ (2005) Spectral integration in auditory cortex: mechanisms and modulation. *Hearing Research* **206**:146–158.

Metzger RR, Greene NT, Porter KK, and Groh JM (2006) Effects of reward and behavioral context on neural activity in the primate inferior colliculus. *Journal of Neuroscience* **26**:7468–7476.

Michel F, Peronnet F, and Schott B (1980) A case of cortical deafness: clinical and electrophysiological data. *Brain and Language* **10**: 367–377.

Micheyl C, Tian B, Carlyon RP, and Rauschecker JP (2005) Perceptual organization of tone sequences in the auditory cortex of awake macaques. *Neuron* **48**:139–148.

Middlebrooks JC (2004) Effects of cochlear-implant pulse rate and inter-channel timing on channel interactions and thresholds. *Journal of the Acoustical Society of America* **116**:452–468.

Middlebrooks JC (2008) Auditory cortex phase locking to amplitude-modulated cochlear implant pulse trains. *Journal of Neurophysiology* **100**:76–91.

Miller CT, Dibble E, and Hauser MD (2001a) Amodal completion of acoustic signals by a nonhuman primate. *Nature Neuroscience* **4**:783–784.

Miller LM, Escabi MA, Read HL, and Schreiner CE (2001b) Functional convergence of response properties in the auditory thalamocortical system. *Neuron* **32**:151–160.

Miller JM, Sutton D, Pfingst B, Ryan A, Beaton R, and Gourevitch G (1972) Single cell activity in the auditory cortex of Rhesus monkeys: behavioral dependency. *Science* **177**:449–451.

Miller KE, Casseday JH, and Covey E (2005) Relation between intrinsic connections and isofrequency contours in the inferior colliculus of the big brown bat, Eptesicus fuscus. *Neuroscience* **136**:895–905.

Moody DB (1994) Detection and discrimination of amplitude-modulated signals by macaque monkeys. *Journal of the Acoustical Society of America* **95**:3499–3510.

Moore B (1989) *An Introduction to the Psychology of Hearing*. Academic Press, London.

Motter BC (1993) Focal attention produces spatially selective processing in visual cortical areas V1, V2, and V4 in the presence of competing stimuli. *Journal of Neurophysiology* **70**:909–919.

Neff WD, Fisher JF, Diamond IT, and MY (1956) Role of auditory cortex in discrimination requiring localization of sound in space. *Journal of Neurophysiology* **19**:500–512.

Nelken I, Rotman Y, and Bar Yosef O (1999) Responses of auditory-cortex neurons to structural features of natural sounds. *Nature* **397**:154–157.

Nelson PC and Carney LH (2007) Neural rate and timing cues for detection and discrimination of amplitude-modulated tones in the awake rabbit inferior colliculus. *Journal of Neurophysiology* **97**:522–539.

Newsome WT, Britten KH, and Movshon JA (1989) Neuronal correlates of a perceptual decision. *Nature* **341**:52–54.

Ng CW, Plakke B, and Poremba A (2009) Primate auditory recognition memory performance varies with sound type. *Hearing Research* **256**:64–74.

Niwa M, Johnson JS, Marshall EA, O'Connor KN, and Sutter ML (2009) Decision- and state-related activity in primary auditory cortex (A1) of rhesus macaques performing amplitude modulation (AM) discrimination. *Society for Neuroscience Abstracts* **556**:2.

O'Connor KN, Barruel P, and Sutter ML (2000) Global processing of spectrally complex sounds in macaques (*Macaca mullata*) and humans. *Journal of Comparative Physiology A* **186**:903–912.

O'Connor KN, Petkov CI, and Sutter ML (2005) Adaptive stimulus optimization for auditory cortical neurons. *Journal of Neurophysiology* **94**:4051–4067.

O'Connor KN, Barruel P, Hajalilou R, and Sutter ML (1999) Auditory temporal integration in the rhesus macaque (Macaca mulatta). *Journal of the Acoustical Society of America* **106**:954–965.

O'Leary DD (1989) Do cortical areas emerge from a protocortex? *Trends in Neurosciences* **12**:400–406.

Otazu GH, Tai LH, Yang Y, and Zador AM (2009) Engaging in an auditory task suppresses responses in auditory cortex. *Nature Neuroscience* **12**:646–654.

Palva S, Palva JM, Shtyrov Y, Kujala T, Ilmoniemi RJ, Kaila K, and Naatanen R (2002) Distinct gamma-band evoked responses to speech and non-speech sounds in humans. *Journal of Neuroscience* **22**:RC211.

Pantev C, Makeig S, Hoke M, Galambos R, Hampson S, and Gallen C (1991) Human auditory evoked gamma-band magnetic fields. *Proceedings of the National Academy of Sciences of the United States of America* **88**:8996–9000.

Pantoja J, Ribeiro S, Wiest M, Soares E, Gervasoni D, Lemos NA, and Nicolelis MA (2007) Neuronal activity in the primary somatosensory thalamocortical loop is modulated by reward contingency during tactile discrimination. *Journal of Neuroscience* **27**: 10608–10620.

Parker AJ, Krug K, and Cumming BG (2002) Neuronal activity and its links with the perception of multi-stable figures. *Philosophical Transactions of the Royal Society London, B: Biological Sciences* **357**:1053–1062.

Parker AJ and Newsome WT (1998) Sense and the single neuron: probing the physiology of perception. *Annual Review of Neuroscience* **21**:227–277.

Penagos H, Melcher JR, and Oxenham AJ (2004) A neural representation of pitch salience in nonprimary human auditory cortex revealed with functional magnetic resonance imaging. *Journal of Neuroscience* **24**:6810–6815.

Percaccio CR, Pruette AL, Mistry ST, Chen YH, and Kilgard MP (2007) Sensory experience determines enrichment-induced plasticity in rat auditory cortex. *Brain Research* **1174**:76–91.

Petkov CI, O'Connor KN, and Sutter ML (2003) Illusory sound perception in macaque monkeys. *Journal of Neuroscience* **23**: 9155–9161.

Petkov CI, O'Connor KN, and Sutter ML (2007) Encoding of illusory continuity in primary auditory cortex. *Neuron* **54**:153–165.

Petkov CI, Sutter ML (2010) Evolutionary conservation and neuronal mechanisms of auditory perceptual restoration, *Hearing Research*, doi:10.1016/j.heares.2010.05.011.

Platt ML and Glimcher PW (1999) Neural correlates of decision variables in parietal cortex. *Nature* **400**:233–238.

Pons F (2006) The effects of distributional learning on rats' sensitivity to phonetic information. *Journal of Experimental Psychology Animal Behavior Processes* **32**:97–101.

Pressnitzer D, de Cheveigné A, and Winter IM (2003) Physiological correlates of the perceptual pitch shift for sounds with similar waveform autocorrelation. *Acoustics Research Letters Online* **4**:1–6.

Pressnitzer D, Meddis R, Delahaye R, and Winter IM (2001) Physiological correlates of comodulation masking release in the mammalian ventral cochlear nucleus. *Journal of Neuroscience* **21**:6377–6386.

Rauschecker JP and Tian B (2000) Mechanisms and streams for processing of "what" and "where" in auditory cortex. *Proceedings of the National Academy of Sciences of the United States of America* **97**:11800–11806.

Recanzone GH (2000) Response profiles of auditory cortical neurons to tones and noise in behaving macaque monkeys. *Hearing Research* **150**:104–118.

Recanzone GH and Cohen YE (2009) Serial and parallel processing in the primate auditory cortex revisited. *Behavioral Brain Research* **206**:1–7.

Recanzone GH and Sutter ML (2008) The biological basis of audition. *Annual Reviews in Psychology* **59**:119–142.

Recanzone GH, Schreiner CE, and Mezenich MM (1993) Plasticity in the frequency representation of primary auditory cortex following discrimination training in adult owl monkeys. *Journal of Neuroscience* **13**:87–103.

Relkin EM and Pelli DG (1987) Probe tone thresholds in the auditory nerve measured by two-interval forced-choice procedures. *Journal of the Acoustical Society of America* **82**:1679–1691.

Riquimaroux H, Gaioni SJ, and Suga N (1991) Cortical computational maps control auditory perception. *Science* **251**:565–568.

Riquimaroux H, Gaioni SJ, and Suga N (1992) Inactivation of the DSCF area of the auditory cortex with muscimol disrupts frequency discrimination in the mustached bat. *Journal of Neurophysiology* **68**:1613–1623.

Ritsma RJ (1962) Existence region of the tonal residue. Part I. *Journal of the Acoustical Society of America* **34**:1224–1229.

Robertson D and Irvine DR (1989) Plasticity of frequency organization in auditory cortex of guinea pigs with partial unilateral deafness. *Journal of Comparative Neurology* **282**:456–471.

Roe AW, Pallas SL, Kwon YH, and Sur M (1992) Visual projections routed to the auditory pathway in ferrets: receptive fields of visual neurons in primary auditory cortex. *Journal of Neuroscience* **12**:3651–3664.

Ryan A and Miller J (1977) Effects of behavioral performance on single-unit firing patterns in inferior colliculus of the rhesus monkey. *Journal of Neurophysiology* **40**:943–956.

Ryan AF, Miller JM, Pfingst BE, and Martin GK (1984) Effects of reaction time performance on single-unit activity in the central auditory pathway of the rhesus macaque. *Journal of Neuroscience* **4**:298–308.

Saitoh I and Suga N (1995) Long delay lines for ranging are created by inhibition in the inferior colliculus of the mustached bat. *Journal of Neurophysiology* **74**:1–11.

Salvi RJ, Giraudi DM, Henderson D, and Hamernik RP (1982) Detection of sinusoidally amplitude modulated noise by the chinchilla. *Journal of the Acoustical Society of America* **71**:424–429.

Samson S and Zatorre RJ (1988) Melodic and harmonic discrimination following unilateral cerebral excision. *Brain and Cognition* **7**: 348–360.

Schreiner C (1980) Encoding of alternating acoustical signals in the medial geniculate body of guinea pigs. *Hearing Research* **3**: 265–278.

Schreiner CE and Langner G (1988) Periodicity coding in the inferior colliculus of the cat. II. Topographical organization. *Journal of Neurophysiology* **60**:1823–1840.

Schreiner CE, Read HL, and Sutter ML (2000) Modular organization of frequency integration in primary auditory cortex. *Annual Reviews in Neuroscience* **23**:501–529.

Schreiner CE and Sutter ML (1992) Topography of excitatory bandwidth in cat primary auditory cortex: single-neuron versus multiple-neuron recordings. *Journal of Neurophysiology* **68**:1487–1502.

Schroeder CE, Lindsley RW, Specht C, Marcovici A, Smiley JF, and Javitt DC (2001) Somatosensory input to auditory association cortex in the macaque monkey. *Journal of Neurophysiology* **85**: 1322–1327.

Schroeder CE, Smiley J, Fu KG, McGinnis T, O'Connell MN, and Hackett TA (2003) Anatomical mechanisms and functional implications of multisensory convergence in early cortical processing. *International Journal of Psychophysiology* **50**:5–17.

Schultz W, Tremblay L, and Hollerman JR (1998) Reward prediction in primate basal ganglia and frontal cortex. *Neuropharmacology* **37**:421–429.

Schulze H and Langner G (1997) Representation of periodicity pitch in the primary auditory cortex of the Mongolian gerbil. *Acta Otolaryngologica Supplement* **532**:89–95.

Scott BH, Malone BJ, and Semple MN (2007) Effect of behavioral context on representation of a spatial cue in core auditory cortex of awake macaques. *Journal of Neuroscience* **27**:6489–6499.

Seeba F and Klump GM (2009) Stimulus familiarity affects perceptual restoration in the European starling (*Sturnus vulgaris*). *Public Library of Science One* **4**:e5974.

Serences JT (2008) Value-based modulations in human visual cortex. *Neuron* **60**:1169–1181.

Shackleton TM, Skottun BC, Arnott RH, and Palmer AR (2003) Interaural time difference discrimination thresholds for single neurons in the inferior colliculus of guinea pigs. *Journal of Neuroscience* **23**:716–724.

Shadlen MN, Britten KH, Newsome WT, and Movshon JA (1996) A computational analysis of the relationship between neuronal and behavioral responses to visual motion. *Journal of Neuroscience* **16**:1486–1510.

Shamma S, Versnel H, and Kowalski N (1995) Ripple analysis in the ferret primary auditory cortex. I. Response characteristics of single units to sinusoidally rippled spectra. *Auditory Neuroscience* **1**:233–254.

Shamma SA, Fleshman JW, Wiser PR, and Versnel H (1993) Organization of response areas in ferret primary auditory cortex. *Journal of Neurophysiology* **69**:367–383.

Shannon RV, Fu QJ, and Galvin J, 3rd (2004) The number of spectral channels required for speech recognition depends on the difficulty of the listening situation. *Acta Otolaryngologica Supplement* **552**: 50–54.

Shinba T, Sumi M, Iwanami A, Ozawa N, and Yamamoto K (1995) Increased neuronal firing in the rat auditory cortex associated with preparatory set. *Brain Research Bulletin* **37**:199–204.

Shofner WP and Dye RH, Jr. (1989) Statistical and receiver operating characteristic analysis of empirical spike-count distributions: quantifying the ability of cochlear nucleus units to signal intensity changes. *Journal of the Acoustical Society of America* **86**: 2172–2184.

Shuler MG and Bear MF (2006) Reward timing in the primary visual cortex. *Science* **311**:1606–1609.

Sidtis JJ and Volpe BT (1988) Selective loss of complex-pitch or speech discrimination after unilateral lesion. *Brain and Language* **34**: 235–245.

Smiley JF, Hackett TA, Ulbert I, Karmas G, Lakatos P, Javitt DC, and Schroeder CE (2007) Multisensory convergence in auditory cortex, I. Cortical connections of the caudal superior temporal plane in macaque monkeys. *Journal of Comparative Neurology* **502**:894–923.

Soto G, Kopell N, and Sen K (2006) Network architecture, receptive fields, and neuromodulation: computational and functional implications of cholinergic modulation in primary auditory cortex. *Journal of Neurophysiology* **96**:2972–2983.

Spitzer H, Desimone R, and Moran J (1988) Increased attention enhances both behavioral and neuronal performance. *Science* **240**:338–340.

Stein A, Ewert SD, and Wiegrebe L (2005) Perceptual interaction between carrier periodicity and amplitude modulation in broadband stimuli: a comparison of the autocorrelation and modulation-filterbank model. *Journal of the Acoustical Society of America* **118**:2470–2481.

Steinschneider M, Fishman YI, and Arezzo JC (2003) Representation of the voice onset time (VOT) speech parameter in population responses within primary auditory cortex of the awake monkey. *Journal of the Acoustical Society of America* **114**: 307–321.

Steinschneider M, Fishman YI, and Arezzo JC (2008) Spectrotemporal analysis of evoked and induced electroencephalographic responses in primary auditory cortex (A1) of the awake monkey. *Cerebral Cortex* **18**:610–625.

Steinschneider M, Reser DH, Fishman YI, Schroeder CE, and Arezzo JC (1998) Click train encoding in primary auditory cortex of the awake monkey: evidence for two mechanisms subserving pitch perception. *Journal of the Acoustical Society of America* **104**: 2935–2955.

Steinschneider M, Schroeder CE, Arezzo JC, and Vaughan HG, Jr (1995) Physiologic correlates of the voice onset time boundary in primary auditory cortex (A1) of the awake monkey: temporal response patterns. *Brain and Language* **48**:326–340.

Stevens KN (1980) *Acoustic Phonetics*. The MIT Press, Cambridge.

Stewart L, Overath T, Warren JD, Foxton JM, and Griffiths TD (2008) fMRI evidence for a cortical hierarchy of pitch pattern processing. *Public Library of Science One* **3**:e1470.

Suga N (1989) Principles of auditory information-processing derived from neuroethology. *Journal of Experimental Biology* **146**: 277–286.

Suga N, Gao E, Zhang Y, Ma X, and Olsen JF (2000) The corticofugal system for hearing: recent progress. *Proceedings of the National Academy of Sciences of the United States of America* **97**:11807–11814.

Sugita Y (1997) Neuronal correlates of auditory induction in the cat cortex. *Neuroreport* **8**:1155–1159.

Sugrue LP, Corrado GS, and Newsome WT (2005) Choosing the greater of two goods: neural currencies for valuation and decision making. *Nature Reviews Neuroscience* **6**:363–375.

Sur M (1988) Visual projections induced into auditory thalamus and cortex: implications for thalamic and cortical information processing. *Progress in Brain Research* **75**:129–136.

Sur M, Garraghty PE, and Roe AW (1988) Experimentally induced visual projections into auditory thalamus and cortex. *Science* **242**:1437–1441.

Sutter ML (2000) Shapes and level tolerances of frequency tuning curves in primary auditory cortex: quantitative measures and population codes. *Journal of Neurophysiology* **84**:1012–1025.

Sutter ML (2005) Spectral processing in the auditory cortex. *International Review in Neurobiology* **70**:253–298.

Sutter ML and Loftus WC (2003) Excitatory and inhibitory intensity tuning in auditory cortex: evidence for multiple inhibitory mechanisms. *Journal of Neurophysiology* **90**:2629–2647.

Sutter ML and Schreiner CE (1995) Topography of intensity tuning in cat primary auditory cortex: single-neuron versus multiple-neuron recordings. *Journal of Neurophysiology* **73**:190–204.

Sutter ML, Schreiner CE, McLean M, O'Connor KN, and Loftus WC (1999) Organization of inhibitory frequency receptive fields in cat primary auditory cortex. *Journal of Neurophysiology* **82**: 2358–2371.

Syka J, Rybalko N, Mazelova J, and Druga R (2002) Gap detection threshold in the rat before and after auditory cortex ablation. *Hearing Research* **172**:151–159.

Takahashi TT, Bala AD, Spitzer MW, Euston DR, Spezio ML, and Keller CH (2003) The synthesis and use of the owl's auditory space map. *Biological Cybernetics* **89**:378–387.

Tan AY, Zhang LI, Merzenich MM, and Schreiner CE (2004) Tone-evoked excitatory and inhibitory synaptic conductances of primary auditory cortex neurons. *Journal of Neurophysiology* **92**:630–643.

Thompson GC and Cortez AM (1983) The inability of squirrel monkeys to localize sound after unilateral ablation of auditory cortex. *Behavioral Brain Research* **8**:211–216.

Tolhurst DJ, Movshon JA, and Dean AF (1983) The statistical reliability of signals in single neurons in cat and monkey visual cortex. *Vision Research* **23**:775–785.

Tramo MJ, Shah GD, and Braida LD (2002) Functional role of auditory cortex in frequency processing and pitch perception. *Journal of Neurophysiology* **87**:122–139.

Ulanovsky N, Las L, Farkas D, and Nelken I (2004) Multiple time scales of adaptation in auditory cortex neurons. *Journal of Neuroscience* **24**:10440–10453.

Ulanovsky N, Las L, and Nelken I (2003) Processing of low-probability sounds by cortical neurons. *Nature Neuroscience* **6**:391–398.

Vaadia E, Gottlieb Y, and Abeles M (1982) Single-unit activity related to sensorimotor association in auditory cortex of a monkey. *Journal of Neurophysiology* **48**:1201–1213.

Viemeister NF (1979) Temporal modulation transfer functions based upon modulation thresholds. *Journal of the Acoustical Society of America* **66**:1364–1380.

Viemeister NF and Wakefield GH (1991) Temporal integration and multiple looks. *Journal of the Acoustical Society of America* **90**: 858–865.

Walker KM, Ahmed B, and Schnupp JW (2008) Linking cortical spike pattern codes to auditory perception. *Journal of Cognitive Neuroscience* **20**:135–152.

Wehr M and Zador AM (2005) Synaptic mechanisms of forward suppression in rat auditory cortex. *Neuron* **47**:437–445.

Weinberger NM, Javid R, and Lepan B (1993) Long-term retention of learning-induced receptive-field plasticity in the auditory cortex. *Proceedings of the National Academy of Sciences of the United States of America* **90**:2394–2398.

Weinberger NM, Miasnikov AA, and Chen JC (2006) The level of cholinergic nucleus basalis activation controls the specificity of auditory associative memory. *Neurobiology of Learning and Memory* **86**:270–285.

Winer JA (2006) Decoding the auditory corticofugal systems. *Hearing Research* **212**:1–8.

Xiao Z and Suga N (2002) Modulation of cochlear hair cells by the auditory cortex in the mustached bat. *Nature Neuroscience* **5**:57–63.

Yin P, Johnson JS, and Sutter ML (2010) Coding of amplitude modulation in primary auditory cortex. (in preparation).

Yin P, Mishkin M, Sutter M, and Fritz JB (2008) Early stages of melody processing: stimulus-sequence and task-dependent neuronal activity in monkey auditory cortical fields A1 and R. *Journal of Neurophysiology* **100**:3009–3029.

Yin P, O'Connor KN, Petkov CI, and Sutter ML (2004) Single neuron sensitivity to amplitude-modulation (AM) in auditory cortex. *Society for Neuroscience Abstract* **987**:13.

Yost WA (1991) Auditory image perception and analysis: the basis for hearing. *Hearing Research* **56**:8–18.

Young ED and Barta PE (1986) Rate responses of auditory nerve fibers to tones in noise near masked threshold. *Journal of the Acoustical Society of America* **79**:426–442.

Zhang W, Salvi RJ, and Saunders SS (1990) Neural correlates of gap detection in auditory nerve fibers of the chinchilla. *Hearing Research* **46**:181–200.

Zwicker E (1952) Die Grenzen der Hoerbarkeit der Amplitudenmodulation und der Frequenzmodulation eines Tones. *Acustica* **2**:125–133.

Zwicker E and Fastl H (1999) *Psychoacoustics*. Springer, Berlin.

Chapter 30

Processing Strategies in Auditory Cortex: Comparison with Other Sensory Modalities

Israel Nelken and Mike B. Calford

Abbreviations

AAF	anterior auditory field
AI	primary auditory cortex
BILD	binaural intelligibility level differences
FAES	auditory field in anterior ectosylvian sulcus
fMRI	functional magnetic resonance imaging
IC	inferior colliculus
ILD	interaural level difference
ITD	interaural time difference
LIP	lateral intraparietal area
MT	middle temporal visual area
PAF	posterior auditory field
PET	positron emission tomography
RF	receptive field
RHT	reverse hierarchy theory
RSVP	rapid serial visual presentations
SAM	sinusoidal amplitude modulation
SSA	stimulus-specific adaptation
V2	second visual cortical field
VI	primary visual cortex

1 Common Organizational Themes in Sensory Systems

In discussing parallels between the auditory cortex and the cortical processing of other sensory modalities, we cannot ignore the fact that for most neuroscientists the standard model of sensory processing in cortex is the visual system and that other sensory systems are first and foremost compared to this model. We will therefore begin by briefly recapitulating the main components of this standard model. We will then discuss auditory cortex in the context of this standard model.

The standard model of visual cortex processing has several components, which include five fundamental features: emergent properties, maps, areal multiplicity, functional streams, and the dynamic processing of information. We briefly consider each.

Emergent single-neuron properties: Visual cortex neurons have emergent properties: the sensitivity to features that are not represented in subcortical or lower-order cortical responses. Thus, in primary visual cortex (VI) there are simple and complex cells. In the second visual cortical field (V2), there may be emergent sensitivity to illusory contours, and in macaque monkey area middle temporal visual area (MT) there is an emergent sensitivity to motion (rather than to components of motion). In the standard model, the hierarchy of emergent properties (e.g., simple to complex cells) is often considered to result from an anatomical processing hierarchy (Stone 1983; Nassi and Callaway 2009).

Maps: The second component of the standard model is the concept of local order: nearby neurons share common selectivity to the emergent properties. The local organization of orientation tuning in cat visual cortex was a key early discovery (Hubel and Wiesel 1962) and was refined over the years into a map of orientation domains surrounding pinwheels with little orientation selectivity (Bonhoeffer and Grinvald 1991; Grinvald et al. 1994).

Multiplicity of areas: The third component of the standard model is the existence of multiplicity of anatomical areas that process visual information. The concept of multiple areas preceded the emergence of the standard model in visual physiology and gained its most dramatic expression in the functional anatomical maps of visual cortex (Felleman and Van Essen 1991) which have been much extended since (Van Essen 2005). In particular, neurons in various fields may have different emergent properties, ranging from orientation tuning in area VI to face neurons in inferior-temporal (IT) cortex or to true motion detectors in area MT (Ungerleider and Mishkin 1982).

I. Nelken (✉)
Department of Neurobiology, The Edmond and Lily Safra Center for Brain Sciences and the Silberman Institute of Life Sciences, Hebrew University, Jerusalem 91904, Israel
e-mail: israel@cc.huji.ac.il

J.A. Winer, C.E. Schreiner (eds.), *The Auditory Cortex*,
DOI 10.1007/978-1-4419-0074-6_30, © Springer Science+Business Media, LLC 2011

Functional streams: The multiple visual areas are arranged in (at least) two major processing streams, the so-called spatial localization (where) and object-recognition (what) pathways (Schneider 1969; Ungerleider and Mishkin 1982). The where stream extends into the parietal cortex, and many of its cells are sensitive to the spatial position of a target stimulus, whereas the what pathway has neurons with strong selectivity to the structure of the target stimulus. These pathways are not, however, fully segregated: spatial information is present in the object recognition pathway and vice versa (Woolsey 1981).

Dynamic processing of information: Finally, having specified single-neuron properties, anatomy, local, and global functional anatomy, the standard model posits that information processing through this complex recurrent network is dynamic (e.g., Pettet and Gilbert 1992). By this it is meant that the same neurons may show different response properties depending on a context. Such contexts may include gain fields of eye position, non-classical receptive fields where stimulus components outside the classical receptive field influence neuronal responses, temporal context, and effects of task on neuronal selectivity (Allman et al. 1985).

2 The Standard Model and Auditory Cortex

To what extent does the auditory cortex fit the standard model? The major conclusion of this analysis is that while all of the properties characterizing the standard model are present in auditory cortex, many interesting features of auditory cortex processing have to be understood on their own terms.

2.1 Emergent Properties of Single Neurons in Primary Auditory Cortex

When one asks what are the emergent properties of the processing of sound by the primary auditory cortex, answers are not readily forthcoming. The same question applied to the visual system quickly evokes responses concerned with binocularity and stereopsis, orientation encoding in the phase-dependent (simple cells) and phase-invariant (complex cells) processing regimes (Hubel and Wiesel 1959, 1962). However, visual cortex is unique in this respect. In both, the auditory and the somatic sensory system, it is more difficult to identify fundamental encoding properties which emerge only at the cortical level.

Noting this discrepancy, neuroscientists not involved in audition often lament that auditory cortex physiologists have not yet found the critical stimulus features that will reveal some fundamental aspect of cortical processing, the auditory analogue of orientation encoding (Hubel and Wiesel 1959). This is a surprising statement, for the experimental manipulation of the stimulus space in audition is far simpler than its counterparts in vision (level, contrast, color, movement, two-dimensional Fourier space, etc.) or somatic sensation (all of the physical elements of an auditory stimulus applied to a varying compliant surface at multiple points on or in the body (Knutsen et al. 2006)). Consequently, a basic fundamental encoding element would be difficult to miss entirely, as seems to be the case.

Nevertheless, the question posed is not without merit in that auditory physiology has not presented its data in terms of how basic properties of sound as identified by any listener (in music or speech or space) are encoded. Take for example the following citation: 'the basic elements of any sound are loudness, pitch, contour, duration (or rhythm), tempo, timbre, spatial location, and reverberation' (Levitin 2007, p. 14). While these may be an obvious list of properties to a cognitive psychologist studying music, to the electrophysiologist this is a curious list amalgamating fundamental properties with clear physical correlates (such as loudness, with its relationship to sound pressure level) and pitch (with strong relationship to periodicity) together with highly derived properties (timbre and reverberation), in no specific hierarchical relation or logical sequence. The common feature of the list is the fact that the coding of none of these properties is well understood at the level of primary auditory cortex, and it has barely been explored in the many nonprimary auditory cortex areas.

As a case in point, there is no consensus opinion on how such an elementary property as sound level (or its perceptual counterpart, intensity) is represented in auditory cortex (e.g., Schreiner et al. 1992; Phillips et al. 1994; Zhang et al. 2005). Most auditory physiologists would attempt an account which consists of a description of individual neuron response functions and a distinction between monotonic and non-monotonic forms, and then turn to some form of population response explanation. This question has been addressed by examining the response of single neurons from across the surface of cat primary auditory cortex to a simple tonal stimulus with varying level. Whereas the activation map found is explicable in terms of the tonotopic map and a disparate distribution of monotonic and non-monotonic level response properties, it does not reveal how sound level is encoded at either a single neuron or a population level. Furthermore, it is deeply problematic to attempt to generalize results such as these from pure tones to the activation of auditory cortex by complex sounds (Phillips et al. 1994).

What, then, are the emergent properties at the level of auditory cortex? One conceptualization of auditory cortex is as a multidimensional processor of simple sound properties. Thus, neurons may be sensitive to tone frequency since they respond to sounds that contain energy within a restricted frequency band. They have some temporal response properties as they respond to sounds with a restricted range of repetition rates. They have spatial response properties because they respond to sounds within a restricted (although usually rather wide) sector of space. And the list can be extended. A single neuron would respond to many different sounds, providing weak evidence for the presence of each of these separate features. There may be no correlation between the different properties across neurons, leading to a combinatorial model coding of sound properties. In consequence, the emerging properties are not at the level of the single neurons, but rather at the level of the resulting populations. We will return to the issue of coding by population when considering maps of response properties.

There is, however, something disappointing in this picture when compared with visual cortex. Each of the separate sound properties is analyzed already in subcortical stations, usually with higher resolution. Thus, frequency selectivity of most primary auditory cortex neurons is much wider than in the auditory nerve (Schreiner and Sutter 1992; Suga 1997). The primary auditory cortex responses to repetitive stimuli are much more sluggish than in inferior colliculus (Katsuki et al. 1959; Joris et al. 2004). Spatial response fields are mostly large, sometimes covering full hemifields (Middlebrooks and Pettigrew 1981) and do not necessarily show any improvement relative to similar response fields in the lateral superior olive or in the inferior colliculus (Semple et al. 1983; Irvine 1986). So the search for the definitive coding and visual-like representational response properties of single neurons continues.

Another venerable conceptualization of auditory cortex described classes of combination sensitive neurons, as originally suggested and elegantly shown in the mustached bat (*P. parnellii*) nonprimary auditory cortex (Suga 1984; Fitzpatrick et al. 1993). Relevant natural sounds may be defined by a conjunction of many properties, and therefore cells selective to relevant sounds should be identified by their responses to such combinations of features. For the mustached bat, these combinations are given, for echolocation calls, by a harmonic of the call and a harmonic of the echo, with the difference in onset time directly related to the distance, and the Doppler shift to the velocity, of the target. When tested with the appropriate combinations, neurons respond substantially more strongly than to the individual components of an optimal combination or with non-optimal combinations of features (Kanwal et al. 1999). According to this conceptualization, the rather weak and non-selective responses of auditory cortex neurons reflect the use of multiple stimulus axes, none of which alone is really optimal.

Combination sensitivity has also been described in AI of a terrestrial mammal, the marmoset (Wang et al. 2005): a distinction is drawn between a small set of best stimuli, which have been crafted individually for each neuron by varying frequency, modulation patterns (in amplitude and frequency) and rates, and non-best stimuli, to each of which the neurons responded suboptimally. Notably, best stimuli evoked sustained responses over many seconds of stimulus presentation, whereas suboptimal stimuli typically evoked a transient response which could be as large as that of the best stimulus near stimulus onset, but decayed within a few hundreds of milliseconds to lower rates. Thus, neurons have a best stimulus which corresponds to a conjunction of multiple auditory properties (combination sensitivity), and stimuli remote from this combination induce weaker responses.

Going further, we ask why neurons seem to prefer certain combinations of features and not others? In the bat echolocation system, the relevant combinations can be deduced from the physics of active echolocation. But active echolocation is a special auditory property absent in many terrestrial mammals such as rodents, carnivores, and primates. It has been suggested that neurons in cat primary auditory cortex are really responding to auditory objects, roughly speaking the internal representations of relevant sound sources (Bar-Yosef et al. 2002; Bar-Yosef and Nelken 2007; Nelken 2004). On their face, the results reported in these studies are similar to those in other investigations of auditory cortex: neurons tend to respond to many stimuli, with very strong responses to a subset of the stimuli. However, by playing such sounds alone and in combination, it was shown that in many cases the response to a combination was dominated by one component, and very often by a low-level component. These experiments did not study in any depth the acoustic cues that determined the responses to these components, and the responses may embody a form of combination sensitivity as seen in the marmoset work (Wang et al. 2005). Nevertheless, the striking common feature of these data is that responses to combinations closely resembled the responses to single, low-level components of the combinations, suggesting that the neurons are extracting auditory objects from the soundscape, rather than encoding a complex combination of features.

While the emphasis in the studies reviewed above has been on relatively brief, spectrotemporally complex sounds, another common property of cortical neurons is their sluggish response to repeated stimuli. This is a general property of cortical sensory representation. The visual system is substantially slower than the auditory system beginning at the receptor level, but visual cortical neurons have temporal modulation functions that are slower than those of photoreceptors (Movshon et al. 1978). The somatic sensory system is intermediate, at least at the peripheral level, where Pacinian corpuscles can respond to vibration frequencies of a few hundred Hertz (Sinclair 1981). However, the transfer functions of somatic sensory cortex neurons tend to be much more sluggish, tapering off at ~10 Hz (Hyvärinen et al. 1968).

The temporal sluggishness of auditory cortical neurons is even more striking, not because it is more extreme (in fact, auditory cortex cells can follow repetitive stimuli as well as or even better than somatic sensory neurons), but because it represents an enormous difference relative to the temporal properties of neurons in the auditory nerve or in the inferior colliculus (Irvine 1986; Joris et al. 2004).

This observation begs two questions. First, how are rapid, repetitive events encoded in auditory cortex, if at all? After all, we do perceive such events. One relevant result is the existence of neurons that do code fast events, but with a twist: their responses do not lock to the individual events although their rate increases with the rate of events (Lu et al. 2001). These neurons, present in the somatic sensory cortex of macaque monkeys (Mountcastle et al. 1969) and described in substantial detail in the marmoset auditory cortex (Wang et al. 2005), seem to encode well rates in the flutter region, in which individual events begin to fuse as they produce a pitch percept but still retain some individual quality.

However, the key question raised by the sluggishness of cortical neurons is its role in sensory encoding. It is possible to view sluggishness as a feature rather than as an impediment. In all sensory modalities, the physical events that elicit sensation usually last a certain time, perhaps reflecting their origin in the motion of masses that have to be displaced with limited amount of force. Thus, the vibrissae of a whisking rat may encounter an object once per whisk, which occurs at a few cycles per second (Hartings and Simons 1998). Similarly, whereas fast visual changes tend to fuse into a continuous percept, slow ones remain distinct, presumably because of assumptions about the speed at which objects move. There are several related experimental approaches to these issues in auditory cortex. Probably the most dominant is forward masking, describing the reduction in the response to a stimulus as a function of what happened $\sim$0.1–1 s before (Calford and Semple 1995; Brosch and Schreiner 1997). The general finding is that the response is decreased, but the decrease and recovery are stimulus dependent in non-trivial ways.

The auditory phenomenon of streaming illustrates the relevance of stimulus-specific sluggishness to perception (Bregman 1990; Fishman et al. 2004). In a typical streaming experiment, two tones alternate. As long as the frequency interval between them is not too large, or too small, and the rate of presentation is not too high, the perception is of an alternating melody composed of the two tones. However, at larger frequency intervals and faster presentation rates the two-tone sequence induces two perceptually distinct streams, each containing only one of the frequencies. Physiological studies of the responses to such sequences in macaque auditory cortex suggest a possible neuronal mechanism. The responses to the tone nearest to the neuron's best frequency depress less, and only at faster presentation rates, than the responses to a frequency further away. As a result, the macaque AI neuron response patterns mimic the psychophysical results: a neuron shows response components to both frequencies when the frequencies are near and the presentation rate sufficiently slows, but responds only to one of the frequencies otherwise (Micheyl et al. 2005, 2007).

Another useful consequence of auditory cortex temporal sluggishness, stimulus-specific adaptation, is present even when a stimulus is repeated at $\sim$1/s, far slower than the cut-off frequency of the temporal modulation transfer functions of cortical neurons (Ulanovsky et al. 2004). Nevertheless, responses decrease due to the repetition. A different stimulus, presented rarely, can evoke a substantially larger response even when it is quite similar to the adapting stimulus (Ulanovsky et al. 2003). An adapting stimulus may be a pure tone and the rare stimulus may be another pure tone whose frequency is 10% from that of the adapting stimulus. This frequency difference is substantially smaller than the tuning width of the cortical cell, but nevertheless evokes a substantial difference between the responses to the common and rare stimuli. Stimulus-specific adaptation is strongest at short interstimulus intervals and is present even at inter-stimulus intervals of 2 seconds in cat auditory cortex (Ulanovsky et al. 2003, 2004).

The study of emerging properties in the auditory cortex is impeded by the annoying (at least to cortical physiologists) fact that the subcortical auditory system is very rich and diverse, with a major subcortical station, the inferior colliculus, without a homologue in other sensory systems. A recurring question regarding the complex properties of cortical neurons is whether these properties emerge in cortex, or whether they have been constructed subcortically and are projected to the cortex. This question is rarely raised in the context of the visual cortex, because of the dramatic differences between receptive fields in the visual thalamus and in visual cortex. In the auditory context, the best evidence for specific cortical mechanisms is in fact related to the sluggishness of the cortical neurons. Auditory system sluggishness develops gradually along the ascending pathway, with thalamic responses somewhat less sluggish than cortical ones (Joris et al. 2004). However, cortical sluggishness is special, requiring special mechanisms (Eggermont 2002). Furthermore, it is highly stimulus specific. The stimulus specificity of adaptation, although present to some extent in rat inferior colliculus, might be created in AI (Ulanovsky et al. 2003), at least for very small frequency differences and relatively long interstimulus intervals.

A striking example of emergent properties of single neurons may exist in human auditory cortex. Many very narrowly tuned neurons were seen, and these were far narrower than the expected peripheral filtering of $\sim$1/6 octave (Bitterman et al. 2008). A population of such narrowly tuned neurons has not been described in other species (although there are scattered examples in macaques and marmosets). Whether this narrow tuning emerges in primary auditory cortex (Moshitch et al. 2006) or in subcortical stations is unknown and this question will probably not be resolved soon.

2.2 Maps in Primary Auditory Cortex

The second feature of the standard model is the presence of parameter maps across the cortical surface. This concept, first demonstrated at the single-neuron level in somatic sensory cortex (Mountcastle 1957) and then in visual cortex (Hubel and Wiesel 1959), has been a major driving force in auditory cortex research.

Since the sensory epithelium in the auditory system represents frequency, the basic and most robust auditory cortex maps are for frequency. However, even auditory cortex tonotopic maps are not simple.

Tonotopic maps have traditionally been constructed by combining the tuning properties of multiunit clusters collected separately and in response to different stimuli (Merzenich et al. 1975; Reale and Imig 1980). That is, the frequency to which a cluster responds to at the lowest sound-pressure level or which, at any sound level, evokes the greatest response is entered onto the map. From these data, frequency contours are plotted, and a well-defined tonotopic map emerges, which is how the discipline of auditory physiology presents itself in text books. Imaging of the hemodynamic activation of the cortex (using optical imaging), which has the advantage of including a wide extent of cortex while using a single stimulus, often reveals an activation pattern consistent with such maps (Harrison et al. 1998; Nelken et al. 2004; Kalatsky et al. 2005; Versnel et al. 2002; Zatorre and Belin 2001).

However, these maps can be highly misleading when considering the response of single well-separated neurons, or for suprathreshold stimuli where the properties of individual clusters (bandwidth, threshold, non-monotonicity, etc.) will become apparent. Thus, many inconsistencies are reported between the optical imaging data and those from single neurons recorded in the same preparation (Spitzer et al. 2001). Another, overall rather similar data set emphasized the significant similarity between electrical and optical signals, rather than the marked inconsistencies (Nelken et al. 2008).

Similar considerations apply to the somatotopic maps of the body surface. A detailed analysis of the fine-grained somatotopic representations studied the near-threshold stimulation cutaneous spatial (minimal) RF, the analogue of the best-threshold-defined characteristic auditory frequency and which is often used to derive somatic sensory maps in primary and nonprimary somatic sensory cortex (Favorov and Diamond 1990; Favorov and Whitsel 1988). They also determined the maximal cutaneous extent of effective stimulation using a neuron's optimal stimulus (maximal RF) (Favorov et al. 1987). This two-dimensional concept corresponds in the auditory domain to the one-dimensional width of the frequency response to a stimulus optimized in all other basic acoustic dimensions; no study of auditory cortex has derived a map using such an approach. Maximal RFs were quite large and varied considerably. However, for neurons within a cortical column, all of the maximal RFs overlapped partially and, geometrically, a common intersection could be derived. Further, this common intersection was co-located with the multiunit minimal cortical RF locus. This relationship held within columnar zones, termed segregates, which were 300–400 μm in diameter in cat and macaque. The minimal RF was interpreted as a measure of the central tendency of the distribution of neural firing rate, threshold, feature filtering toward the common intersection of the individual neural response areas. Thus, minimal RFs, and the threshold-field maps constructed from them, should be considered not as physiological measures of neural response properties in the spatial dimension but as statistical constructs (Favorov et al. 1987).

With these cautionary notes in mind, a more refined view of the map concept in AI as a set of multiple, superimposed maps of relatively simple feature detectors has emerged (Favorov et al. 1987; Schreiner 1995; Ehret 1997). In this view, single neurons may have rather simple response properties, and the combination of many neurons constitutes the computational array. In AI, the geometry of the computational array has a special twist. The peripheral sensory epithelium is unidimensional, with frequency mapped across the long but narrow basilar membrane. In contrast, auditory cortex is at least two dimensional. Thus, the issue of feature maps is often rephrased, in auditory cortex, as that of finding the parameter that is mapped along the isofrequency contours (e.g., Schreiner and Mendelson 1990; Heil et al. 1992; Read et al. 2001).

Probably the most densely mapped animal model of auditory cortex is the cat. Features assayed in mapping experiments include the frequency of pure tones (Merzenich et al. 1975), tone intensity (Schreiner et al. 1992), binaural interactions (Imig and Adrián 1977; Middlebrooks et al. 1980), bandwidth of noise bands or transients (Clarey et al. 1995; Read et al. 2001), the direction and speed of frequency-modulated chirps (Poirier et al. 1997; Mendelson et al. 1993), and more (Ehret 1997). With the exception of tone frequency, which is well mapped topographically at near-threshold levels, these studies suggest the presence of local clusters of neurons with similar properties, with only weak overall order. The clusters may have different sizes for different response properties, leading to combinatorial coding of many different components of these properties (Suga 1990; Schreiner 1995).

These conclusions are very different from the regular, almost crystal-like organization of VI (Szentágothai 1975; Rockel et al. 1980; Rose and Dobson 1985), but may still be computationally very useful. For example, using moving ripples, the auditory analogue of moving gratings to study ferret auditory cortex neurons revealed cells tuned to many combinations of spectral and temporal modulation frequencies (Shamma et al. 1993). Although there is no strong order, the existence of these neurons suggests an array of activities that decompose complex incoming sounds into components useful for their further processing. Nevertheless, most such neurons have relatively simple spectrotemporal RFs (Shamma et al. 1993).

In contrast with the extensive information regarding topographic order of simple response properties, there is no accepted description of a topographic map for any of the high-level features of a musical sound (as in Levitin 2007). A case in point is pitch sensitivity (as opposed to sensitivity to the harmonics forming a periodic sound). In the spirit of the search for emerging properties, finding neurons that would generalize pitch across many different physical implementations of the same periodicity would greatly advance understanding the relationships between neural activity and perception. However, most studies of periodicity coding in primary auditory cortex failed to find such neurons. An exception is a study in gerbil auditory cortex using optical imaging of intrinsic signals with sinusoidally amplitude-modulated (SAM) sounds whose carrier frequency is far above the pitch range to produce sounds with no energy within the tuning width of low best-frequency neurons (Schulze et al. 2002). It described an orderly arrangement of pitch selectivity in the low-frequency part of AI, although the periodotopic map did not correspond to the low-frequency pure tone map (Schulze and Langner 1997).

However, two recent studies cast doubt on these conclusions. First, SAM sounds evoked strong cochlear combination tones, and the apparent responses to pitch in auditory cortex could consist of responses to these tones (McAlpine 2004). Second, imaging techniques like those used in ferrets and with a larger set of periodic sounds showed that many sounds had maps of pitch sensitivity, but the maps differed for different families of stimuli (Nelken et al. 2008). Thus, there is no single periodotopic map in primary auditory cortex, overlying the pure frequency map.

2.3 Multiple Auditory Areas and Functional Streams

What are the emergent properties of the processing of sound by secondary auditory cortex? Good answers are, surprisingly, much easier to come by than in primary auditory cortex, despite the fact that secondary auditory cortices are overall much less well-understood than primary auditory cortex. In some sense, this is similar to the situation in visual cortex. Asked what the purpose of VI is, the textbook answer would be that VI represents the world in a general-purpose way. In contrast, extrastriate areas are often assigned to more specific computational tasks, for example, the motion sensitivity of area MT (middle temporal visual field) (Zeki 1980a) and the color sensitivity in area V4 (Zeki 1980b). Specific roles for higher visual areas have been defined in humans as well, using functional magnetic resonance imaging (fMRI). The best known of these higher visual areas is the face area on the parahippocampal gyrus, but there are other examples such as the object area LO in human visual cortex (Bridge and Parker 2007). In the auditory system, there is a model in which the representations are especially well described: the auditory cortex of the mustached bat (Suga 1984, 1990). Detailed mapping found emergent properties: combination sensitivity to components of the bat call and its echo, with separate fields for echo delay (related to the target distance) and Doppler shift (related to the relative speed between the bat and its target). Thus, the mustached bat conforms to at least three properties of the standard model: the emergence of new response properties, the presence of maps of these properties, and the presence of multiple areas in which different aspects of echolocation calls are processed (Suga 1984, 1990).

Remarkably, this may be the only well-studied animal model in which such maps of emergent properties have been uncovered with any consistency. Even in other bat species, whereas auditory cortex combination sensitive neurons are found, they do not necessarily form maps: in the big brown bat (*Eptesicus fuscus*) auditory cortex, delay-tuned neurons abound, and they have some interesting response properties, but they do not form a map (Dear et al. 1993a,b).

In cat auditory cortex, the situation is more complex. Many differences between auditory areas have been described, but the functional relevance of these differences is not entirely clear (Eggermont 1998). A partial list would include differences in temporal sluggishness: AAF cells are faster than AI cells (Imaizumi et al. 2004) which are faster than those in PAF (Schreiner and Urbas 1988; Heil and Irvine 1998). PAF has neurons with frequency-invariant best levels, possibly encoding level in a frequency-independent way (Phillips and Orman 1984). PAF cells have somewhat narrower spatial receptive fields, although the differences between the auditory fields are not large (Stecker et al. 2003). A secondary auditory field (FAES) in the buried region of the anterior ectosylvian sulcus (AES) has a concentration of neurons representing frontal locations (Las et al. 2008).

The situation in primates is as equally complex: in marmosets, there may be a pitch area bordering low-frequency AI (Bendor and Wang 2005). There is a small preponderance of space-sensitive neurons in posterior versus anterior belt areas, and neurons in the lateral intraparietal area (LIP) have spatial selectivity to sounds, although they also encode stimulus identity (Mullette-Gilman et al. 2005). Neurons in the ventral stream areas seem to respond to high-level features such as category. Thus, responses in ventrolateral prefrontal cortex, the endpoint of the ventral, presumed 'what' stream, are not well-predicted by the physical structure of the vocalizations but were predicted better by the outputs of a probabilistic model estimating the likelihood that vocalization belonged to specific classes; these outputs are a complex non-linear transform of the physical properties of the vocalizations (Averbeck and Romanski 2006).

Thinking about the role of higher auditory areas in animal models is dominated by the concept of streams: a where stream from AI to posterior belt areas and from there to parietal cortex, and a what stream arising from the supratemporal plane and ending in ventrolateral prefrontal cortex. As in the macaque visual areas, the data argue against strict segregation of where and what processing: for example, LIP neurons, which are no doubt part of the where pathway, respond to sound identity (and to spatial location) (Gifford and Cohen 2005). The lack of strict segregation is also apparent in human imaging data. For example, a positron emission tomography (PET) study showed activation in posterior auditory areas in a spatial task, but only when subjects were presented with mixtures of spectrally discriminable sounds from multiple locations in space (Zatorre et al. 2002). In cats, behavioral studies found a double dissociation between where and what tasks: cooling anterior auditory areas caused a deficit in what but not in where tasks, whereas cooling posterior auditory areas caused a deficit in what but not in where tasks. The deficits, while highly significant, were nevertheless partial in both cases, suggesting possible cross-talk between the two streams, a possibility that fits better the balance of electrophysiological data (Malhotra et al. 2008).

2.4 Dynamic Processing of Information

Dynamic processing of information occurs in all sensory systems. From the periphery, neurons can adapt to continuous stimuli. In vision, the adaptation to static stimuli is complete: without eye movements, vision disappears (Gerrits et al. 1966). In the somatic sensory system, specific classes of receptors adapt at different rates, encoding different properties of a time-varying stimulus (Catton 1970). Retinal ganglion cells show already quite sophisticated forms of adaptation (Olveczky et al. 2007). For example, during fixational eye movements, the image on the retina is in constant motion. Even in these circumstances ganglion cells are still very sensitive to the onset of a movement of an object in the center of their receptive field. A subset of ganglion cells, both in the retina of the tiger salamander (Schwartz and Berry 2008) and in the mouse (Murphy and Rieke 2006), recognize temporal sequences and respond strongly to a violation of a temporal regularity.

Auditory nerve fibers show a depression in their responses to a probe stimulus immediately following another stimulus (Harris and Dallos 1979). This depression is presumably a neural correlate of forward masking. At the inferior colliculus (IC), a complex form of binaural forward masking may be a correlate of the precedence effect—the dominance of the direct sound in determining the perceived location of a sound source in echoic environments (Litovsky and Yin 1998a,b; Pollak et al. 2003).

Our main interest here is, however, with cortical dynamic processing level. Neurons in VI may respond differently to stimuli in their receptive field (RF) depending on stimulus context, attentional state, and visual experience (Movshon and Lennie 1979; Macaluso et al. 2000). Some of these effects have analogues in auditory cortex as well (Grady et al. 1997).

Contextual effects in VI led to the concept of the non-classical RF: a spatial region in which the neuron does not respond with spiking activity to any stimulus, but which may modify the responses to stimuli within the RF (Allman et al. 1985; Series et al. 2003). Such interactions are often inhibitory (Crook et al. 1996). Intracellular recordings in cat primary visual cortex suggest that membrane potential mechanisms underlie these extended spatial effects, with possible contributions from long-range horizontal connections (Crook et al. 1998).

Similar effects can be demonstrated in auditory cortex. A direct translation of the concept of non-classical RF from the vision to audition would consist of frequency areas outside the tonal excitatory response area that would inhibit the responses to BF tones. Such effects are well known under the name of lateral inhibition. In ferret auditory cortex, a weak topographical asymmetry of lateral inhibition was reflected in directional selectivity for frequency-modulated chirps (Shamma et al. 1993). There are strong effects of wideband backgrounds on the response to bird chirps that consisted of amplitude-modulated, frequency-modulated tones (Bar-Yosef et al. 2002; Bar-Yosef and Nelken 2007). Intracellular recordings in auditory cortex suggest the presence of very wide subthreshold integration fields as well (de Ribaupierre et al. 1972; Wehr and Zador 2003; Kaur et al. 2004; Tan et al. 2004).

The contextual influence that has been studied most in audition is temporal: the responses to the same stimulus depend on stimuli that occurred before at multiple timescales. Studies of in auditory cortex forward masking were described earlier (Calford and Semple 1995; Brosch and Schreiner 1997). Forward masking may be the basic building block of stimulus-specific adaptation: the specific reduction in the responses to a repeated stimulus which does not generalize to other, possibly very similar, stimuli. Stimulus-specific adaptation involves multiple timescales, from seconds to minutes (Nakamoto et al. 2006; Fritz et al. 2007). It transforms auditory cortex into a highly selective processor of novel information.

Stimulus-specific adaptation (SSA) has been studied in visual cortex as well. For example, exposure to high-contrast gratings specifically reduces sensitivity at the spatial frequency of the adapting grating or at its orientation. However, in auditory cortex SSA seems to be much stronger than in visual cortex. Visual studies often test adaptation by shifts approximating the tuning curve width. Analyses of adaptation to orientation used shifts of 22.5° (Dragoi et al. 2000) and a test that used substantially better resolution used shifts of 14° (Muller et al. 1999), about half the tuning width of their neurons. On the other hand, auditory cortex neurons showed significant SSA when the test frequency was 10% or even 4% away from the standard frequency, although the cortical tuning curves are potentially a few octaves wide at the sound levels used in these experiments (King et al. 2007). Further, the size of adaptation in visual cortex is small, and a meta-analysis of visual cortex SSA showed that the largest effects are the same size as the auditory cortex average effect (Ulanovsky 2004). Finally, auditory cortex stimulus-specific adaptation is evoked by stimuli up to two seconds apart (Ulanovsky et al. 2003, 2004). Similar experiments have not been performed in the visual system, but human hyperacuity is absent when stimuli are presented at intervals longer than a few tenths of milliseconds (Westheimer 1981). Thus, auditory cortex stimulus-specific adaptation is more important than in visual cortex. This may be the reason why associated electrophysiological markers in humans, such as mismatch negativity (Molholm et al. 2005), are easily elicited in audition but much more difficult to elicit in vision.

Attention effects in VI are present, though small. The effect of attention when a monkey followed one of the two paths on the screen from one point to another one showed a small, though significant, modulation of the neural responses when the RF was on the attended versus the nonattended path (Roelfsema et al. 1998). Attentional effects have not been convincingly demonstrated in AI, although the few such studies preclude a firm conclusion (Hromadka and Zador 2007).

At longer time constants, visual experience affects VI responses: for example, the responses during a trained task may differ from responses to the same stimuli during a simple fixation task (Gilbert et al. 2000; Crist et al. 2001). Similar effects have been demonstrated in auditory cortex: the frequency selectivity of neurons is task dependent, with improvement in the selectivity to the trained parameter (Fritz et al. 2003, 2005).

Perhaps the largest difference between visual cortex and auditory cortex is the capacity of auditory cortex capacity for experience-dependent reorganization. Visual cortex does not seem to reorganize following learning in the sense that RFs do not change location, size, or orientation selectivity even when the learning procedure presents the stimuli at one location exclusively. On the other hand, auditory cortex maps are highly malleable (Recanzone et al. 1993): classical conditioning strongly shifts the single-neuron RFs (Weinberger 2004), and by mimicking the effects of classical conditioning with stimulation of the basal forebrain (thereby releasing large amounts of acetylcholine in auditory cortex) can dramatically change the representation of auditory cortex spectral and temporal parameters (Kilgard and Merzenich 1998a, b). Somatic sensory cortex may be even more malleable than auditory cortex, with large changes elicited simply by repetitive stimulation, without any behavioral context (Recanzone et al. 1992). Such repeated stimulation has no effect or the opposite effect (Kilgard and Merzenich 2002) in auditory cortex.

2.5 Summary and Interpretation

This overview of the standard model suggests that the auditory system shares significant similarities with the visual system. AI and VI (and presumably higher auditory areas as well) can be described in terms of aggregates of parametric maps of low-level physical features. The hierarchical nature of auditory processing is clearly analogous to the visual system, including perhaps the division between identification or what pathway and localization or where pathways (Romanski et al. 1999; Tian et al. 2001).

However, beyond these superficial similarities lie profound differences. For example, while there is no singular emergent property for AI neurons, there is substantial evidence for highly integrated and complex response properties, possibly beyond the complexity of neurons in VI (Depireux et al. 2001; Miller et al. 2001; Chi et al. 2005). Perhaps these are adequate for assessing and establishing maps, but it must be cautioned that near-threshold stimuli and multiunit or hemodynamic responses do not engage (or average over) the most significant properties of AI neurons, and may not bear strong relationship to their computational capacity. Further, AI has a substantially larger plastic capability, both in the short timescales (stimulus-specific adaptation; Ulanovsky et al. 2003, 2004) and in the longer timescales (malleability of the cortical maps; Recanzone et al. 1993; Kilgard and Merzenich 1998a). This plastic capacity may be the physiological substrate of the complex and rich repertoire of response properties in AI.

A possible interpretation of these differences is that AI is, in fact, higher in the auditory hierarchy than VI. Indeed, VI is the first visual station which integrates binocular information (Hubel and Wiesel 1959, 1962), whereas this occurs at the pontine level in the auditory pathway (Irvine 1986). Furthermore, the auditory system has an important obligatory midbrain station, the IC (Winer and Schreiner 2005), with no direct analogue in other sensory systems. Whereas auditory neurons below the inferior colliculus show relatively uniform and mostly easily categorized response types (e.g., primary-like, chopper and onset response type in the ventral cochlear nucleus; type-II and type-IV response types in the dorsal cochlear nucleus; interaural time difference (ITD) sensitivity in the medial superior olive and interaural level difference (ILD) sensitivity in the lateral superior olive, etc.) (Kiang et al. 1965; Cant and Benson 2003; Irvine 1986; Blauert 1997), in the inferior colliculus the variety of neuronal responses multiplies, as expected from a station that integrates essentially all lower processing streams (Kuwada et al. 1997; Winer and Schreiner 2005). These properties may be organized in parameter maps (Wenstrup et al. 1986; Schreiner and Langner 1988, 1997; Ehret et al. 2003) conceptually very similar to their cortical counterparts. We suggest that in fact the inferior colliculus should be considered as the functional analogue of VI in the auditory system. If so, AI has already a higher order processing task. Finding out what this task is may be the most challenging problem facing research in auditory cortex. Perhaps AI is involved in segregation and grouping of sound components into perceptual objects. According to this hypothesis, the resulting objects may then acquire properties such as pitch, spatial location or phonemic identity in higher auditory areas, which subserve more specialized processing tasks.

3 The Road to Perception

A curious aspect of auditory perception is the distance between the low-level physical characteristics of sounds and their perceptual consequences. Recall the list of elementary properties of a musical sound: loudness, pitch, contour, duration (or rhythm), tempo, timbre, spatial location, and reverberation (Levitin 2007). While we understand much about the coding of some physical correlates of these properties, any electrophysiologist would be hard pressed to explain where each of these properties is encoded and how they are decoded. Thus, space is related to ITD calculations (which are very well understood) and ILD calculations (which have been somewhat less studied), but neither for ITD nor ILD is 'space'-spatial localization depend on the integration of cues across frequency and time (Irvine 1986; Rajan et al. 1990; Stecker and Middlebrooks 2003). In auditory cortex, space is encoded by many neurons distributed throughout all auditory fields (despite clear specializations for encoding space in some fields) (Middlebrooks et al. 1994). The situation for timbre is even more complex: it is implicitly encoded in the spatiotemporal activity pattern of already in the auditory nerve and is also robustly represented in the auditory cortex activity patterns (Shamma et al. 1993; Schwarz and Tomlinson 1990; Mesgarani et al. 2008). However, these are highly distributed representations, and we know of no map of timbre, no brain area where violin neurons respond only to violins (independently of what the violin plays) and not to trumpets.

The same dissociation between physics and perception is strongly at play in speech processing. The categorical flavor of the speech sound perception causes some stimuli that are physically similar to be perceived as distinct sounds. Thus, perception can emphasize some physical differences, but in the context of the high-level category rather than in terms of the low-level physical cues at the ear.

The higher order features of sound all share an important property: they all generalize across many physical axes. Thus many different sounds have the same loudness; pitch can be elicited by pure tones, by click trains, or by aperiodic sounds such as iterated ripple noise; spatial location depends on multiple temporal and spectral cues; and the same speech sound may be realized by different speakers in very different ways. Furthermore, the perceptual quality masks the low-level cues that produce it: we perceive pitch and not a sequence of isolated harmonics, speech and not the spectral peaks and valleys that are associated with a specific vowel. This property suggests that the representation of all of these perceptual qualities is actually relatively high in the processing hierarchy.

Although contrary to much of our introspection, many experiments show that in vision, as in audition, the gist of the scene is perceived very rapidly, while the perception of details is far more difficult (Intraub 1980). In experiments such as rapid serial visual presentations (RSVP) viewers identify substantially above chance pictures that they have seen as part of a very long sequence changing 10 times/s. Effects such as change blindness or repetition blindness (Kanwisher 1987), or even the well-known Stroop effect in which semantic information interferes with color naming, show that high-level features of objects or response selection may strongly influence the perception of low-level features, even in vision (Li et al. 2000, 2004).

Thus, the subordination of details to the higher order properties of the sensory stimuli is not special for audition and is shared with vision. An attempt to account for such relationships is the reverse hierarchy theory (RHT) developed by to account for visual perceptual learning, then extended to visual perception (Ahissar and Hochstein 1997; Hochstein and Ahissar 2006). RHT posits that immediate perception is based on high-level cortical representations, and scene details are incorporated only as needed by the feedback pathways from high- to low-cortical areas. The term 'vision at a glance' referred to visual tasks that are presumably performed at high levels of cortical representations (e.g., discriminating between faces and houses), and 'vision with scrutiny' to refer to visual tasks that are presumably performed at low-cortical representations (requiring discrimination of details or fine within-category discriminations, e.g., the just-noticeable-difference in the orientation of two lines) (Rensink 2000). According to RHT, optimal visual performance with scrutiny is possible only when the low-order representations are accessible. Finding and accessing these levels for the performance of a sensory task is difficult and slow, requiring special conditions (such as those used in classical psychophysical experiments: many repetitions of the same type of stimulus with as little variation as possible).

RHT has been used to account for the ability of listeners to optimally use low-level physical cues (in this case, binaural decorrelation) in order to perform a speech discrimination task in noise (Nahum et al. 2008). Thresholds were measured in diotic and in dichotic conditions, where the noise was identical in the two ears and the words were phase reversed. The use of the decorrelation cues was measured by calculating the differences between the diotic and the dichotic thresholds (binaural intelligibility level differences, BILD). When the words to be discriminated were phonetically similar, optimal use of binaural decorrelation (equivalent to an ideal-observer level of BILD) was limited: subjects achieved ideal-listener performance (which is optimal with respect to low-level information) only when the thresholds were measured in separate diotic and dichotic tracks. If the tracks were interleaved, dichotic thresholds suffered significant degradation. Even when the dichotic thresholds were measured alone, thresholds were degraded if the listeners had to perform a semantic association task. In contrast, when the words to be discriminated had no phonetic overlap, performance was always at the ideal listener level (Nahum et al. 2008).

RHT gives simple account of these results. For phonetically non-overlapping words, the high-level representations (e.g., phonemes) already use all the available acoustic cues. Therefore, performance, which is based on these representations, is optimal independently of the experimental manipulation. For phonetically overlapping words, however, the overlap in the phonemic representation requires listeners to access explicitly the low-level representation of binaural decorrelation. Access to this representation is possible, but only under special circumstances. With interleaved binaural conditions, the search for the right representation was impeded by the constant need to switch between informative representations (spectrotemporal energy in the diotic case and decorrelation in the dichotic case). In the semantic association task, the search for the right low-level representation was impeded by the constant need to use even higher, semantic, representation levels.

For the purpose of this discussion, RHT supplies a natural account for the dissociation between high-order perceptual qualities (timbre, tempo, reverberation, spatial location) and the low-level acoustic cues associated with them. Thus, the speech mode (Liberman and Mattingly 1989) is interpreted simply as perception that occurs so high in the system that it pre-empts the perception of the physical features underlying its construction.

While these are only early steps on the road of relating auditory perception with physiological representations and hierarchies, the fact that a theory developed for vision can account for a complex pattern of results in the auditory domain may indicate that, despite having significant differences, the global architecture of the visual and auditory systems (and presumably of other sensory systems as well) shares common large-scale design principles which are derived from the ethological needs of animals. The study of such principles may be a path to usefully relate visual and auditory physiology.

References

Ahissar M and Hochstein S (1997) Task difficulty and the specificity of perceptual learning. *Nature* **387**:401–406.

Allman J, Miezin F, and McGuiness E (1985) Stimulus specific responses from beyond the classical receptive field: neurophysiological mechanisms for local-global comparisons in visual neurons. *Annual Review of Neuroscience* **8**:407–430.

Averbeck BB and Romanski LM (2006) Probabilistic encoding of vocalizations in macaque ventral lateral prefrontal cortex. *Journal of Neuroscience* **26**:11023–11033.

Bar-Yosef O and Nelken I (2007) The effects of background noise on the neural responses to natural sounds in cat primary auditory cortex. *Frontiers in Computational Neuroscience* **1**:1–14. doi:10.3389/neuro.10.003.2007.

Bar-Yosef O, Rotman Y, and Nelken I (2002) Responses of neurons in cat primary auditory cortex to bird chirps: effects of temporal and spectral context. *Journal of Neuroscience* **22**:8619–8632.

Bendor D, and Wang X (2005) The neuronal representation of pitch in primate auditory cortex. *Nature* **436**:1161–5.

Bitterman Y, Mukamel R, Malach R, Fried I, and Nelken I (2008) Ultra-fine frequency tuning revealed in single neurons of human auditory cortex. *Nature* **451**:197–201.

Blauert J (1997) *Spatial Hearing. The Psychophysics of Human Sound Localization*. MIT Press, Cambridge.

Bonhoeffer T and Grinvald A (1991) Iso-orientation domains in cat visual cortex are arranged in pinwheel-like patterns. *Nature* **353**:429–431.

Bregman AS (1990) *Auditory Scene Analysis: The Perceptual Organization of Sound*. MIT Press, Cambridge.

Bridge H and Parker AJ (2007) Topographical representation of binocular depth in the human visual cortex using fMRI. *Journal of Vision* **7**:1–14.

Brosch M and Schreiner CE (1997) Time course of forward masking tuning curves in cat primary auditory cortex. *Journal of Neurophysiology* **77**:923–943.

Calford MB and Semple MN (1995) Monaural inhibition in cat auditory cortex. *Journal of Neurophysiology* **73**:1876–1891.

Cant NB and Benson CG (2003) Parallel auditory pathways: projection patterns of the different neuronal populations in the dorsal and ventral cochlear nuclei. *Brain Research Bulletin* **60**:457–474.

Catton WT (1970) Mechanoreceptor function. *Physiological Reviews* **50**:297–318.

Chi T, Ru P, and Shamma SA (2005) Multiresolution spectrotemporal analysis of complex sounds. *Journal of the Acoustical Society of America* **118**:887–906.

Clarey JC, Barone P, Irons WA, Samson FK, and Imig TJ (1995) Comparison of noise and tone azimuth tuning of neurons in cat primary auditory cortex and medial geniculate body. *Journal of Neurophysiology* **74**:961–980.

Crist RE, Li W, and Gilbert CD (2001) Learning to see: experience and attention in primary visual cortex. *Nature Neuroscience* **4**: 519–525.

Crook JM, Kisvárday ZF, and Eysel UT (1996) GABA-induced inactivation of functionally characterized sites in cat visual cortex (area 18): effects on direction selectivity. *Journal of Neurophysiology* **75**:2071–2088.

Crook JM, Kisvárday ZF, and Eysel UT (1998) Evidence for a contribution of lateral inhibition to orientation tuning and direction selectivity in cat visual cortex: reversible inactivation of functionally characterized sites combined with neuroanatomical tracing techniques. *European Journal of Neuroscience* **10**:2056–2075.

Dear SP, Fritz J, Haresign T, Ferragamo M, and Simmons JA (1993a) Tonotopic and functional organization in the auditory cortex of the big brown bat, *Eptesicus fuscus*. *Journal of Neurophysiology* **70**:1988–2009.

Dear SP, Simmons JA, and Fritz J (1993b) A possible neuronal basis for representation of acoustic scenes in auditory cortex of the big brown bat. *Nature* **364**:620–623.

Depireux DA, Simon JZ, Klein DJ, and Shamma SA (2001) Spectro-temporal response field characterization with dynamic ripples in ferret primary auditory cortex. *Journal of Neurophysiology* **85**: 1220–1224.

de Ribaupierre F, Goldstein MH, Jr, and Yeni-Komshian G (1972) Intracellular study of the cat's primary auditory cortex. *Brain Research* **48**:185–204.

Dragoi V, Sharma J, and Sur M (2000) Adaptation-induced plasticity of orientation tuning in adult visual cortex. *Neuron* **28**:287–298.

Eggermont JJ (1998) Representation of spectral and temporal sound features in three cortical fields of the cat. Similarities outweigh differences. *Journal of Neurophysiology* **80**:2743–2764.

Eggermont JJ (2002) Temporal transfer functions in cat primary auditory cortex: separating stimulus effects from neural mechanisms. *Journal of Neurophysiology* **87**:305–321.

Ehret G (1997) The auditory cortex. *Journal of Comparative Physiology A* **181**:547–557.

Ehret G, Egorova M, Hage SR, and Muller BA (2003) Spatial map of frequency tuning-curve shapes in the mouse inferior colliculus. *Neuroreport* **14**:1365–1369.

Favorov O and Whitsel BL (1988) Spatial organization of the peripheral input to area 1 cell columns. II. The forelimb representation achieved by a mosaic of segregates. *Brain Research* **472**:43–56.

Favorov OV and Diamond ME (1990) Demonstration of discrete place-defined columns—segregates—in the cat SI. *Journal of Comparative Neurology* **298**:97–112.

Favorov OV, Diamond ME, and Whitsel BL (1987) Evidence for a mosaic representation of the body surface in area 3b of the somatic cortex of cat. *Proceedings of the National Academy of Sciences of the United States of America* **84**:6606–6610.

Felleman DJ and Van Essen DC (1991) Distributed hierarchical processing in the primate cerebral cortex. *Cerebral Cortex* **1**:1–47.

Fishman YI, Arezzo JC, and Steinschneider M (2004) Auditory stream segregation in monkey auditory cortex: effects of frequency separation, presentation rate, and tone duration. *Journal of the Acoustical Society of America* **116**:1656–1670.

Fitzpatrick DC, Kanwal JS, Butman JA, and Suga N (1993) Combination-sensitive neurons in the primary auditory cortex of the mustached bat. *Journal of Neuroscience* **13**:931–940.

Fritz J, Elhilali M, and Shamma S (2005) Active listening: task-dependent plasticity of spectrotemporal receptive fields in primary auditory cortex. *Hearing Research* **206**:159–176.

Fritz J, Shamma S, Elhilali M, and Klein D (2003) Rapid task-related plasticity of spectrotemporal receptive fields in primary auditory cortex. *Nature Neuroscience* **6**:1216–1223.

Fritz JB, Elhilali M, David SV, and Shamma S (2007) Does attention play a role in dynamic receptive field adaptation to changing acoustic salience in A1? *Hearing Research* **229**:186–203.

Gerrits HJ, De Haan B, and Vendrik AJ (1966) Experiments with retinal stabilized imaged. Relations between the observations and neural data. *Vision Research* **6**:427–440.

Gifford GW 3rd and Cohen YE (2005) Spatial and non-spatial auditory processing in the lateral intraparietal area. *Experimental Brain Research* **162**:509–512.

Gilbert C, Ito M, Kapadia M, and Westheimer G (2000) Interactions between attention, context and learning in primary visual cortex. *Vision Research* **40**:1217–1226.

Grady CL, Van Meter JW, Maisog JM, Pietrini P, Krasuski J, and Rauschecker JP (1997) Attention-related modulation of activity in primary and secondary auditory cortex. *NeuroReport* **8**: 2511–2516.

Grinvald A, Lieke EE, Frostig RD, and Hildesheim R (1994) Cortical point-spread function and long-range lateral interactions revealed by real-time optical imaging of macaque monkey primary visual cortex. *Journal of Neuroscience* **14**:2545–2568.

Harris DM and Dallos P (1979) Forward masking of auditory nerve fiber responses. *Journal of Neurophysiology* **42**:1083–1107.

Harrison RV, Harel N, Kakigi A, Raveh E, and Mount RJ (1998) Optical imaging of intrinsic signals in chinchilla auditory cortex. *Audiology & Neurootology* **3**:214–223.

Hartings JA and Simons DJ (1998) Thalamic relay of afferent responses to 1- to 12-Hz whisker stimulation in the rat. *Journal of Neurophysiology* **80**:1016–1019.

Heil P, Rajan R, and Irvine DRF (1992) Sensitivity of neurons in cat primary auditory cortex to tones and frequency-modulated stimuli. II: organization of response properties along the 'isofrequency' dimension. *Hearing Research* **63**:135–156.

Heil P and Irvine DRF (1998) The posterior field P of cat auditory cortex: coding of envelope transients. *Cerebral Cortex* **8**: 125–141.

Hochstein S and Ahissar M (2002) View from the top: hierarchies and reverse hierarchies in the visual system. *Neuron* **36**:791–804.

Hromadka T and Zador AM (2007) Toward the mechanisms of auditory attention. *Hearing Research* **229**:180–185.

Hubel DH and Wiesel TN (1959) Receptive fields of single neurones in the cat's striate cortex. *Journal of Physiology (London)* **148**: 574–591.

Hubel DH and Wiesel TN (1962) Receptive fields, binocular interaction and functional architecture in the cat's visual cortex. *Journal of Physiology (London)* **160**:106–154.

Hyvärinen J, Sakata H, Talbot WH, and Mountcastle VB (1968) Neuronal coding by cortical cells of the frequency of oscillating peripheral stimuli. *Science* **162**:1130–1132.

Imaizumi K, Priebe NJ, Crum PAC, Bedenbaugh PH, Cheung SW, and Schreiner CE (2004) Modular functional organization of cat anterior auditory field. *Journal of Neurophysiology* **90**:444–457.

Imig TJ and Adrián HO (1977) Binaural columns in the primary field (A1) of cat auditory cortex. *Brain Research* **138**:241–257.

Intraub H (1980) Presentation rate and the representation of briefly glimpsed pictures in memory. *Journal of Experimental Psychology: Human Perception & Performance* **6**:1–12.

Irvine DRF (1986) *The Auditory Brainstem. A Review of the Structure and Function of Auditory Brainstem Processing Mechanisms*. In: Autrum H, Ottoson D, Perl ER, Schmidt RF, Shimazu H and Willis WD (eds). *Progress in Sensory Physiology*. Springer, Berlin, pp. 1–279.

Joris PX, Schreiner CE, and Rees A (2004) Neural processing of amplitude-modulated sounds. *Physiological Reviews* **84**:541–577.

Kalatsky VA, Polley DB, Merzenich MM, Schreiner CE, and Stryker MP (2005) Fine functional organization of auditory cortex revealed by Fourier optical imaging. *Proceedings of the National Academy of Sciences of the United States of America* **102**:13325–13330.

Kanwal JS, Fitzpatrick DC, and Suga N (1999) Facilitatory and inhibitory frequency tuning of combination-sensitive neurons in the primary auditory cortex of mustached bats. *Journal of Neurophysiology* **82**:2327–2345.

Kanwisher NG (1987) Repetition blindness: type recognition without token individuation. *Cognition* **27**:117–143.

Katsuki Y, Watanabe T, and Maruyama N (1959) Activity of auditory neurons in upper levels of brain of cat. *Journal of Neurophysiology* **22**:343–359.

Kaur S, Lazar R, and Metherate R (2004) Intracortical pathways determine breadth of subthreshold frequency receptive fields in primary auditory cortex. *Journal of Neurophysiology* **91**:2551–2567.

Kiang NY-S, Watanabe T, Thomas EC, and Clark LF (1965) *Discharge Patterns of Single Fibers in the Cat's Auditory Nerve*. MIT Press, Cambridge.

Kilgard MP and Merzenich MM (1998a) Cortical map reorganization enabled by nucleus basalis activity. *Science* **279**:1714–1718.

Kilgard MP and Merzenich MM (1998b) Plasticity of temporal information processing in the primary auditory cortex. *Nature Neuroscience* **1**:727–731.

Kilgard MP and Merzenich MM (2002) Order-sensitive plasticity in adult primary auditory cortex. *Proceedings of the National Academy of Sciences of the United States of America* **99**:3205–3209.

King AJ, Bajo VM, Bizley JK, Campbell RA, Nodal FR, Schulz AL, and Schnupp JWF (2007) Physiological and behavioral studies of spatial coding in the auditory cortex. *Hearing Research* **229**: 106–115.

Knutsen PM, Pietr M, and Ahissar E (2006) Haptic object localization in the vibrissal system: behavior and performance. *Journal of Neuroscience* **26**:8451–8464.

Kuwada S, Batra R, Yin TCT, Oliver DL, Haberly LB, and Stanford TR (1997) Intracellular recordings in response to monaural and binaural stimulation of neurons in the inferior colliculus of the cat. *Journal of Neuroscience* **17**:7565–7581.

Las L, Shapira AH, and Nelken I (2008) Functional gradients of auditory sensitivity along the anterior ectosylvian sulcus of the cat. *Journal of Neuroscience* **28**:3657–3667.

Levitin DJ (2007) *This Is Your Brain on Music: The Science of a Human Obsession*. Penguin Publisher, New York.

Li W, Piech V, and Gilbert CD (2004) Perceptual learning and top-down influences in primary visual cortex. *Nature Neuroscience* **7**:651–657.

Li W, Thier P, and Wehrhahn C (2000) Contextual influence on orientation discrimination of humans and responses of neurons in V1 of alert monkeys. *Journal of Neurophysiology* **83**:941–954.

Liberman AM and Mattingly IG (1989) A specialization for speech perception. *Science* **243**:489–494.

Litovsky RY and Yin TCT (1998a) Physiological studies of the precedence effect in the inferior colliculus of the cat. I. Correlates of psychophysics. *Journal of Neurophysiology* **80**:1285–1301.

Litovsky RY, and Yin TCT (1998b) Physiological studies of the precedence effect in the inferior colliculus of the cat. II. Neural mechanisms. *Journal of Neurophysiology* **80**:1302–1316.

Lu T, Liang L, and Wang X (2001) Temporal and rate representations of time-varying signals in the auditory cortex of awake primates. *Nature Neuroscience* **4**:1131–1138.

Macaluso E, Frith CD, and Driver J (2000) Modulation of human visual cortex by crossmodal spatial attention. *Science* **289**: 1206–1208.

Malhotra S, Stecker GC, Middlebrooks JC, and Lomber SG (2008) Sound localization deficits during reversible deactivation of primary auditory cortex and/or the dorsal zone. *Journal of Neurophysiology* **99**:1628–1642.

McAlpine D (2004) Neural sensitivity to periodicity in the inferior colliculus: evidence for the role of cochlear distortions. *Journal of Neurophysiology* **92**:1295–1311.

Mendelson JR, Schreiner CE, Sutter ML, and Grasse KL (1993) Functional topography of cat primary auditory cortex: responses to frequency-modulated sweeps. *Experimental Brain Research* **94**:65–87.

Merzenich MM, Knight PL, and Roth GL (1975) Representation of cochlea within primary auditory cortex in the cat. *Journal of Neurophysiology* **38**:231–249.

Mesgarani N, David SV, Fritz JB, and Shamma SA (2008) Phoneme representation and classification in primary auditory cortex. *Journal of the Acoustical Society of America* **123**:899–909.

Micheyl C, Tian B, Carlyon RP, and Rauschecker JP (2005) Perceptual organization of tone sequences in the auditory cortex of awake macaques. *Neuron* **48**:139–48.

Micheyl C, Carlyon RP, Gutschalk A, Melcher JR, Oxenham AJ, Rauschecker JP, Tian B, and Wilson EC (2007) The role of auditory cortex in the formation of auditory streams. *Hearing Research* **229**:116–131.

Middlebrooks JC, Clock AE, Xu L, and Green DM (1994) A panoramic code for sound location by cortical neurons. *Science* **264**: 842–844.

Middlebrooks JC, Dykes RW, and Merzenich MM (1980) Binaural response-specific bands in primary auditory cortex (AI) of the cat: topographical organization orthogonal to isofrequency contours. *Brain Research* **181**:31–48.

Middlebrooks JC and Pettigrew JD (1981) Functional classes of neurons in primary auditory cortex of the cat distinguished by sensitivity to sound location. *Journal of Neuroscience* **1**:107–120.

Miller LM, Escabí MA, Read HL, and Schreiner CE (2001) Functional convergence of response properties in the auditory thalamocortical system. *Neuron* **32**:151–160.

Molholm S, Martinez A, Ritter W, Javitt DC, and Foxe JJ (2005) The neural circuitry of pre-attentive auditory change-detection: an fMRI study of pitch and duration mismatch negativity generators. *Cerebral Cortex* **15**:545–551.

Moshitch D, Las L, Ulanovsky N, Bar-Yosef O, and Nelken I (2006) Responses of neurons in primary auditory cortex (A1) to pure tones in the halothane-anesthetized cat. *Journal of Neurophysiology* **95**:3756–3769.

Movshon JA, Thompson ID, and Tolhurst DJ (1978) Spatial and temporal contrast sensitivity of neurones in areas 17 and 18 of the cat's visual cortex. *Journal of Physiology (London)* **283**:101–120.

Mountcastle VB (1957) Modality and topographic properties of single neurons of cat's somatic sensory cortex. *Journal of Neurophysiology* **20**:408–434.

Mountcastle VB, Talbot WH, Sakata H, and Hyvärinen J (1969) Cortical neuronal mechanisms in flutter-vibration studied in unanesthetized monkeys. Neuronal periodicity and frequency discrimination. *Journal of Neurophysiology* **32**:452–484.

Movshon JA and Lennie P (1979) Pattern-selective adaptation in visual cortical neurones. *Nature* **278**:850–852.

Muller JR, Metha AB, Krauskopf J, and Lennie P (1999) Rapid adaptation in visual cortex to the structure of images. *Science* **285**:1405–1408.

Mullette-Gillman OA, Cohen YE, and Groh JM (2005) Eye-centered, head-centered, and complex coding of visual targets in the intraparietal sulcus. *Journal of Neurophysiology* **94**:2331–2352.

Murphy GJ and Rieke F (2006) Network variability limits stimulus-evoked spike timing precision in retinal ganglion cells. *Neuron* **52**:511–524.

Nahum M, Nelken I, and Ahissar M (2008) Low-level information and high-level perception: the case of speech in noise. *Public Library of Science Biology* **6**:978–991.

Nakamoto KT, Zhang J, and Kitzes LM (2006) Temporal nonlinearity during recovery from sequential inhibition by neurons in the cat primary auditory cortex. *Journal of Neurophysiology* **95**:1897–1907.

Nassi JJ and Callaway EM (2009) Parallel processing strategies of the primate visual system. *Nature Reviews Neuroscience* **10**:360–372.

Nelken I (2004) Processing of complex stimuli and natural scenes in the auditory cortex. *Current Opinions in Neurobiology* **14**:474–480.

Nelken I, Bizley JK, Nodal FR, Ahmed B, King AJ, and Schnupp JWH (2008) Responses of auditory cortex to complex stimuli: functional organization revealed using intrinsic optical signals. *Journal of Neurophysiology* **99**:1928–1941.

Nelken I, Bizley JK, Nodal FR, Ahmed B, Schnupp JW, and King AJ (2004) Large-scale organization of ferret auditory cortex revealed using continuous acquisition of intrinsic optical signals. *Journal of Neurophysiology* **92**:2574–2588.

Olveczky BP, Baccus SA, and Meister M (2007) Retinal adaptation to object motion. *Neuron* **56**:689–700.

Pettet MW and Gilbert CD (1992) Dynamic changes in receptive-field size in cat primary visual cortex. *Proceedings of the National Academy of Sciences of the United States of America* **89**: 8366–8370.

Phillips DP and Orman SS (1984) Responses of single neurons in posterior field of cat auditory cortex to tonal stimulation. *Journal of Neurophysiology* **51**:147–163.

Phillips DP, Semple MN, Calford MB, and Kitzes LM (1994) Level-dependent representation of stimulus frequency in cat primary auditory cortex. *Experimental Brain Research* **102**:210–226.

Poirier P, Jiang H, Lepore F, and Guillemot JP (1997) Positional, directional and speed selectivities in the primary auditory cortex of the cat. *Hearing Research* **113**:1–13.

Pollak GD, Burger RM, and Klug A (2003) Dissecting the circuitry of the auditory system. *Trends in Neurosciences* **26**:33–39.

Rajan R, Aitkin LM, Irvine DRF, and McKay J (1990) Azimuthal sensitivity of neurons in primary auditory cortex of cats. I. Types of sensitivity and the effects of variations in stimulus parameters. *Journal of Neurophysiology* **64**:872–887.

Read HL, Winer JA, and Schreiner CE (2001) Modular organization of intrinsic connections associated with spectral tuning in cat auditory cortex. *Proceedings of the National Academy of Sciences of the United States of America* **98**:8042–8087.

Reale RA, and Imig TJ (1980) Tonotopic organization in auditory cortex of the cat. *Journal of Comparative Neurology* **192**:265–291.

Recanzone GH, Merzenich MM, and Dinse HR (1992) Expansion of the cortical representation of a specific skin field in primary somatosensory cortex by intracortical microstimulation. *Cerebral Cortex* **2**:181–196.

Recanzone GH, Schreiner CE, and Merzenich MM (1993) Plasticity in the frequency representation of primary auditory cortex following discrimination training in adult owl monkeys. *Journal of Neuroscience* **13**:87–103.

Rensink RA (2000) Seeing, sensing, and scrutinizing. *Vision Research* **40**:1469–1487.

Rockel AJ, Hiorns RW and Powell TPS (1980) The basic uniformity in structure of the neocortex. *Brain* **103**:221–244.

Roelfsema PR, Lamme VA, and Spekreijse H (1998) Object-based attention in the primary visual cortex of the macaque monkey. *Nature* **395**:376–381.

Romanski LM, Tian B, Fritz J, Mishkin M, Goldman-Rakic PS, and Rauschecker JP (1999) Dual streams of auditory afferents target multiple domains in the primate prefrontal cortex. *Nature Neuroscience* **2**:1131–1136.

Rose D and Dobson VG (1985) *Models of Visual Cortex*. Wiley, New York.

Schneider GE (1969) Two visual systems. *Science* **163**:895–902.

Schreiner CE (1995) Order and disorder in auditory cortical maps. *Current Opinion in Neurobiology* **5**:489–96.

Schreiner CE and Langner G (1988) Periodicity coding in the inferior colliculus of the cat. II. Topographical organization. *Journal of Neurophysiology* **60**:1823–1840.

Schreiner CE and Langner G (1997) Laminar fine structure of frequency organization in auditory midbrain. *Nature* **388**: 383–3866.

Schreiner CE and Mendelson JR (1990) Functional topography of cat primary auditory cortex: distribution of integrated excitation. *Journal of Neurophysiology* **64**:1442–1459.

Schreiner CE, Mendelson JR, and Sutter ML (1992) Functional topography of cat primary auditory cortex: representation of tone intensity. *Experimental Brain Research* **92**:105–122.

Schreiner CE and Sutter ML (1992) Topography of excitatory bandwidth in cat primary auditory cortex: single-neuron versus multiple-neuron recordings. *Journal of Neurophysiology* **68**: 1487–1502.

Schreiner CE and Urbas JV (1988) Representation of amplitude modulation in the auditory cortex of the cat. II. Comparison between cortical fields. *Hearing Research* **32**:49–63.

Schulze H and Langner G (1997) Representation of periodicity pitch in the primary auditory cortex of the Mongolian gerbil. *Acta Otolaryngologica Supplement* **532**:89–95.

Schulze H, Hess A, Ohl FW, and Scheich H (2002) Superposition of horseshoe-like periodicity and linear tonotopic maps in auditory cortex of the Mongolian gerbil. *European Journal of Neuroscience* **15**:1077–1084.

Schwartz G and Berry MJ 2nd (2008) Sophisticated temporal pattern recognition in retinal ganglion cells. *Journal of Neurophysiology* **99**:1787–1798.

Schwarz DWF and Tomlinson RW (1990) Spectral response patterns of auditory cortex neurons to harmonic complex tones in alert monkey (*Macaca mulatta*). *Journal of Neurophysiology* **64**: 282–298.

Semple MN, Aitkin LM, Calford MB, Pettigrew JD, and Phillips DP (1983) Spatial receptive fields in the cat inferior colliculus. *Hearing Research* **10**:203–215.

Series P, Lorenceau J, and Fregnac Y (2003) The "silent" surround of V1 receptive fields: theory and experiments. *Journal of Physiology (Paris)* **97**:453–474.

Shamma SA, Fleshman JW, Wiser PR, and Versnel H (1993) Organization of response areas in ferret primary auditory cortex. *Journal of Neurophysiology* **69**:367–383.

Sinclair D (1981) *Mechanisms of Cutaneous Sensation*. Oxford University Press, London.

Spitzer MW, Calford MB, Clarey JC, Pettigrew JD, and Roe AW (2001) Spontaneous and stimulus-evoked intrinsic optical signals in primary auditory cortex of the cat. *Journal of Neurophysiology* **85**:1283–1298.

Stecker GC, Mickey BJ, Macpherson EA, and Middlebrooks JC (2003) Spatial sensitivity in field PAF of cat auditory cortex. *Journal of Neurophysiology* **89**:2889–2903.

Stecker GC and Middlebrooks JC (2003) Distributed coding of sound locations in the auditory cortex. *Biological Cybernetics* **89**: 341–349.

Stone J (1983) *Parallel Processing in the Visual System. The Classification of Retinal Ganglion Cells and Its Impact on the Neurobiology of Vision*. Plenum Press, New York and London.

Suga N (1984) The extent to which biosonar information is represented in the bat auditory cortex. In: Edelman GM, Gall WE, and Cowan WM (eds). *Dynamic Aspects of Neocortical Function*. Wiley, New York, pp. 315–373.

Suga N (1990) Cortical computational maps for auditory imaging. *Neural Networks* **3**:3–21.

Suga N (1997) Tribute to Yasuji Katsuki's major findings: sharpening of frequency tuning in the central auditory system. *Acta Otolaryngologica Supplement* **532**:9–12.

Szentágothai J (1975) The "module-concept" in cerebral cortex architecture. *Brain Research* **95**:475–496.

Tan AYY, Zhang LI, Merzenich MM, and Schreiner CE (2004) Tone-evoked excitatory and inhibitory synaptic conductances of primary auditory cortex neurons. *Journal of Neurophysiology* **92**:630–643.

Tian B, Reser D, Durham A, Kustov A, and Rauschecker JP (2001) Functional specialization in rhesus monkey auditory cortex. *Science* **292**:290–293.

Ulanovsky N (2004) Neuronal adaptation in cat auditory cortex. Ph.D. dissertation, Hebrew University, Jerusalem, Israel.

Ulanovsky N, Las L, Farkas D, and Nelken I (2004) Multiple time scales of adaptation in auditory cortex neurons. *Journal of Neuroscience* **24**:10440–10453.

Ulanovsky N, Las L, and Nelken I (2003) Processing of low-probability sounds by cortical neurons. *Nature Neuroscience* **6**:391–398.

Ungerleider LG and Mishkin M (1982) Two cortical visual systems. In: Ingle DJ, Goodale MA, and Mansfield RJW (eds). *Advances in the Analysis of Visual Behavior*. MIT Press, Cambridge, pp. 549–586.

Van Essen DC (2005) Corticocortical and thalamocortical information flow in the primate visual system. *Progress in Brain Research* **69**:215–237.

Versnel H, Mossop JE, Mrsic-Flogel TD, Ahmed B, and Moore DR (2002) Optical imaging of intrinsic signals in ferret auditory cortex: responses to narrowband sound stimuli. *Journal of Neurophysiology* **88**:1545–1558.

Wang X, Lu T, Snider RK, and Liang L (2005) Sustained firing in auditory cortex evoked by preferred stimuli. *Nature* **435**: 341–346.

Weinberger NM (2004) Specific long-term memory traces in primary auditory cortex. *Nature Reviews Neuroscience* **5**:279–290.

Wehr M and Zador AM (2003) Balanced inhibition underlies tuning and sharpens spike timing in auditory cortex. *Nature* **426**: 442–446.

Wenstrup JJ, Ross LS, and Pollak GD (1986) Binaural response organization within a frequency-band representation of the inferior colliculus: implications for sound localization. *Journal of Neuroscience* **6**:962–973.

Westheimer G (1981) Visual hyperacuity. *Progress in Sensory Physiology* **1**:1–30.

Winer JA and Schreiner CE (eds.) (2005) *The Inferior Colliculus*, Springer, New York.

Woolsey CN (ed.) (1981) *Cortical Sensory Organization*, volume 2, *Multiple Visual Areas*. Humana Press, Clifton.

Zatorre RJ and Belin P (2001) Spectral and temporal processing in human auditory cortex. *Cerebral Cortex* **11**:946–953.

Zatorre RJ, Bouffard M, Ahad P, and Belin P (2002) Where is 'where' in the human auditory cortex? *Nature Neuroscience* **5**:905–909.

Zeki SM (1980a) The response properties of cells in the middle temporal area (area MT) of owl monkey visual cortex. *Proceedings of the Royal Society of London, series B* **207**:239–248.

Zeki SM (1980b) The representation of colours in the cerebral cortex. *Nature* **284**:412–418.

Zhang J, Nakamoto KT, and Kitzes LM (2005) Modulation of level response areas and stimulus selectivity of neurons in cat primary auditory cortex. *Journal of Neurophysiology* **94**:2263–2274.

Chapter 31

Cortical Speech and Music Processes Revealed by Functional Neuroimaging

Robert J. Zatorre and Marc Schönwiesner

Abbreviations

AI	primary auditory cortex
BOLD	blood oxygenated level dependent
EEG	electroencephalography
fMRI	functional magnetic resonance imaging
HG	Heschl's gyrus
MEG	magnetoencephalography
MMN	mismatch negativity
PET	positron emission tomography
PT	planum temporale
STG	superior temporal gyrus
STS	superior temporal sulcus

1 Introduction

Economists tell us that wealth is created by trade and exchange. Assuming the same principle holds for intellectual wealth, the interactions between different levels of analysis, and the exchanges across disciplines that characterize contemporary neuroscience should provide us with great riches. Looking at the developments over the past decade in cognitive neuroscience of auditory processing would appear to bear this out. A significant amount of progress has been made, and much of it can be attributed to the possibilities for crossing boundaries afforded by neuroimaging tools. This chapter focuses on recent advances in our understanding of the human auditory cortex in the light of research using functional neuroimaging techniques. We emphasize the processing of music and speech, and how this knowledge complements knowledge drawn from other domains and other species.

M. Schönwiesner (✉)
International Laboratory for Brain, Music and Sound Research, Université de Montréal, Montréal, QC, Canada H2V 4P3
e-mail: marc.schoenwiesner@umontreal.ca

1.1 Human Auditory Cortex: Specializations for Music and Speech

Evolution shapes the nervous system of different species in distinct ways according to what is adaptive. Whereas echolocation is adaptive for bats—who hunt in the dark—communication of cognitive representations is adaptive for humans—who live in complex societies and transmit learned knowledge across generations. This communication yields cultural evolution, the hallmark of the human condition. In the context of the specialization of auditory cortex, therefore, it is not surprising to find that bats have specialized neural systems for echolocation, while humans possess specializations for communication, notably for speech and music. We argue that speech and music are both the most complex and the most characteristically human of acoustical signals, and hence it is these domains that we must explore to understand the uniqueness of our complex brains. The claim is therefore not one of human superiority, but rather a matter of adaptation: since all human societies speak and have music, our auditory system must be adapted for these functions, just as the auditory systems of other species are adapted for other functions.

1.2 Pathways and Hierarchies

Taking the evolutionary argument seriously requires an understanding of how human brain specializations may have evolved from structures and systems already extant in its precursors. Although we argue that the human auditory cortex is uniquely specialized, these functional properties must have evolved from earlier features. Two major insights have emerged from investigations into the connectivity and cytoarchitecture of primate auditory cortex: first is the idea of a hierarchical arrangement, such that information processing proceeds from core to belt and then to parabelt regions (Kaas et al. 1999); second is the concept that multiple parallel

J.A. Winer, C.E. Schreiner (eds.), *The Auditory Cortex*,
DOI 10.1007/978-1-4419-0074-6_31, © Springer Science+Business Media, LLC 2011

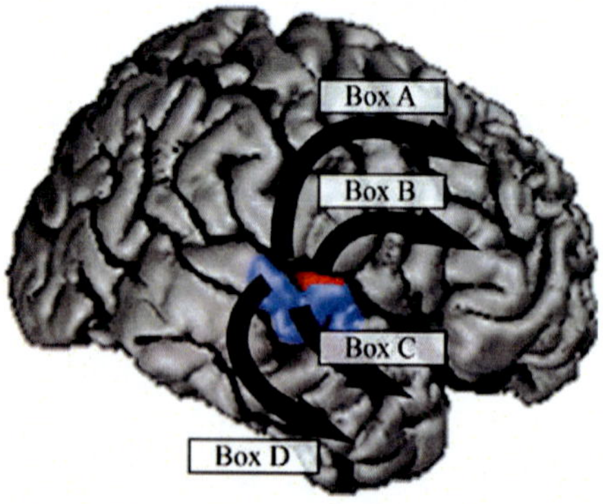

Fig. 31.1 Schematic of putative functional pathways for auditory information processing in human brain. There is evidence for two pathways targeting dorsal (*A*) and ventral (*B*) aspects of premotor and prefrontal cortices. Pathways originating in core auditory areas project outward in a parallel but hierarchical fashion toward belt and parabelt cortices. Subsequently, several distinct functional streams may be identified. (*D*) Ventrally, processing streams progress toward targets in superior temporal sulcus and inferior temporal gyrus. (*C*) An additional ventral stream may exist projecting from core areas in anterior portions of the superior temporal gyrus. Dorsally, projections also follow a hierarchical organization and lead toward distinct targets in parietal and frontal cortices

processing pathways perform computationally different tasks (Rauschecker and Tian 2000), akin to the dorsal and ventral visual pathways (Ungerleider and Haxby 1994). There is growing evidence that each of the two pathways themselves contain dorsal and ventral branches (Kaas et al. 1999) although this concept remains to be worked out in detail (Fig. 31.1). A large body of neuroimaging findings supports a role for a ventral pathway in speech processing (see below) and in nonspeech auditory object processing (Zatorre et al. 2004; Warren et al. 2005a). A spatial processing role for the dorsal pathway has also been supported by many neuroimaging studies (Baumgart et al. 1999; Pavani et al. 2002; Warren et al. 2002); an alternative interpretation is that this pathway may be better characterized as integrating spatially relevant monaural and binaural cues, although spatial processing may be relevant (Griffiths and Warren 2002), with subsequent computations in parietal cortex necessary for transformation of coordinate frames of reference (Zatorre et al. 2002b). This concept would also be in keeping with evidence that the dorsal auditory pathway is closely related to auditory–motor processes (Hickok and Poeppel 2004; Warren et al. 2005b) and offers a parallel to models of the visual system as well (Milner and Goodale 1995). Regardless of the specific functional attribution (for discussion see Belin and Zatorre 2000; Romanski et al. 2000), the twin ideas of hierarchical and parallel processing pathways have proven key in helping to interpret a large body of evidence, and hence provide an important organizational framework.

Primate neurophysiology is thus critical to help build models of human auditory cortical function but, until recently, integrating primate and human data had proven difficult. This task has been facilitated by several recent developments allowing functional magnetic resonance imaging (fMRI) data and neurophysiological data serve as complementary sources of information. First, new insights have clarified the physiological basis for the fMRI blood-oxygenated level-dependent (BOLD) signal (Logothetis and Wandell 2004). This work found a complex, but quantifiable, relationship between neuronal activity and fMRI hemodynamic signal. Second, high rates of correlation between hemodynamic signals from fMRI and single-unit spike activity and local field potentials in human patients undergoing electrophysiological recording in response to natural auditory stimuli (a movie soundtrack) have been demonstrated (Mukamel et al. 2005). These findings hence validate fMRI as a surrogate for direct recordings of neurophysiological activity, considering that fMRI signals pertain to large populations of neurons, without providing information about spike timing or other microtemporal features of neural activity. Third, auditory fMRI data collected from macaque monkeys (Petkov et al. 2006) provide a direct, specific link between monkey single-unit recordings and functional imaging and thus validates conclusions about human auditory responses only available from fMRI studies. Finally, technical advances permit the loud acoustic noise associated with fMRI acquisition to be mitigated by the use of sparse-sampling protocols (Belin et al. 1999; Hall et al. 1999). All of these developments set the stage for investigations of the human auditory cortex using fMRI.

1.3 Pitch and Music: Low-Level Specializations Versus Higher Order Distributed Mechanisms

Pitch is a critical component of both speech and music. Pitch relationships are important for language, since tone languages (e.g., Mandarin, Thai) use pitch contours to signal different meanings at the syllable level, while non-tone languages also use pitch patterns suprasegmentally, to signal syntactic or affective components at the sentence level (for a review of tone-language neuroimaging studies, see Zatorre and Gandour 2008). As for music, it is probably safe to say that it would not exist as we know it without the ability to perceive pitch.

Pitch is often defined psychophysically as a perceptual attribute of a sound that allows ordering on a high-to-low axis. This definition does little justice to the impact of pitch on our experience of sounds, including the tone of a speaker's voice, a musical melody, or environmental sounds like bird songs. Pitch and frequency may be conceptually distinguished: frequency is a physical property of a sound, whereas pitch is a percept, and hence computed in the auditory brain. For sounds with only one spectral component (pure tones) the tone pitch is equivalent to the frequency of the component. Sounds with several components have a more complicated relationship between the pitch and the frequencies of the components. For a harmonic complex, for instance, the pitch is typically heard at the so-called fundamental frequency, i.e., the lowest harmonic of which all other component frequencies are multiples.

1.4 Tonotopy

Because a pure tone involves a direct relation between pitch and frequency, we will briefly review frequency representation in the auditory cortex. Sound frequency is represented along the ascending auditory pathway in an ordered topographic fashion (tonotopy). Electrophysiological recordings in the primary auditory cortex of nonhuman primates find two (Merzenich and Brugge 1973; Morel et al. 1993) or three (Morel and Kaas 1992; Kaas et al. 1999) tonotopic maps with mirror-oriented frequency gradients. Orthogonal to the direction of the frequency gradient run isofrequency bands, within which other response properties vary (Read et al. 2002). The secondary auditory cortex includes seven or more cortical areas, some of which may also be organized tonotopically (Pandya and Sanides 1973; Kaas and Hackett 1998, 2000). Neuroimaging data generally confirm this organization in humans (Fig. 31.1). Several studies reported a large-scale gradient with low frequencies represented laterally along Heschl's gyrus (HG) and high frequencies represented medially (Talavage et al. 2000; Schönwiesner et al. 2002). Mirror-symmetric tonotopic maps exist on and near HG as seen in a high-resolution fMRI study which recorded responses to six different pure tones and fitted a Gaussian function as a model of a frequency tuning curve to the responses. The best frequency, as determined from the maximum of the fitted function, varied systematically across the surface of HG from medial to lateral, and some subjects showed a preferential anterior–posterior gradient, with a gradient reversal midway. Probing discrete frequency points makes it difficult to conclusively demonstrate a frequency gradient: two areas with different frequency selectivity may be part of a continuous map or neighboring functional areas without a gradient between (Formisano et al. 2003). To address this concern long frequency sweeps between 125 Hz and 8 kHz were used to demonstrate multiple frequency gradients on the superior temporal plane within and outside of primary auditory cortex (Talavage et al. 2004).

Despite these promising results, fMRI tonotopic maps are still not as robust as retinotopic maps, and they are not yet used systematically to delineate auditory cortical fields in humans [although this is possible in monkeys (Petkov et al. 2006)].

Tonotopic maps reflect the layout of the ascending subcortical projections. Other low-level sound feature representations may be more indicative of specialized tuning of the auditory cortex to stimuli of importance to humans, like speech and music. In animals, tuning for spectro-temporal modulations has been proposed as a mechanism for discrimination of sounds (Woolley et al. 2005). Using methods adapted from animal neurophysiology Schönwiesner and Zatorre (2009) demonstrate selective tuning to combined spectro-temporal modulations in the primary and secondary auditory cortex. They presented dynamic ripples, complex broadband stimuli with a drifting sinusoidal spectral envelope (Depireux et al. 2001). These sounds combine one spectral modulation rate with one temporal modulation rate. Forty-nine ripple conditions were presented, with spectral and temporal modulation rates systematically varying on a 7-by-7 grid (Fig. 31.2c). Modulation transfer functions (MTF) were obtained from small patches of cortex (single voxels) by arranging the magnitudes of the fMRI responses to all ripples in the same grid. Figure 31.2d shows examples of several types of MTFs, extracting different spectro-temporal features, with a high degree of interaction between spectral and temporal parameters. The mean MTF of all voxels in the auditory cortex shows a low-pass modulation rate preference that matches the modulation content of speech (Fig. 31.2e). Such a match suggests that the human auditory system has increased sensitivity to commonly encountered or highly relevant sound modulations that allow their efficient encoding. The topographic distribution for these features was confined to the superior temporal plane, highly reproducible within listeners, but highly variable across listeners (Fig. 31.2f shows an example of a 'preferred ripple map' on one person's right temporal lobe). This variability may represent a signature of an individual's auditory cortical organization. This approach, unlike tonotopic mapping or any other kind of unidimensional mapping, captures the interactions between spectral and temporal responses in the human cortex and allows a more principled investigation of cortical response properties than has been possible until now. In addition this approach permits a relatively direct comparison to animal studies of auditory response properties that use the same method of dynamic ripple mapping (Depireux et al. 2001; Fritz et al. 2003; Klein et al. 2006; Kowalski et al. 1996; Linden et al. 2003; Sen et al. 2001; Versnel and Shamma 1998; Woolley et al. 2005).

1.5 The Cortical Pitch-Sensitive Area

Human lesion studies suggest a special role for Heschl's gyrus and parts of the superior temporal plane immediately surrounding it in the extraction of complex pitch, in tasks such as perceiving the pitch of sounds with a missing

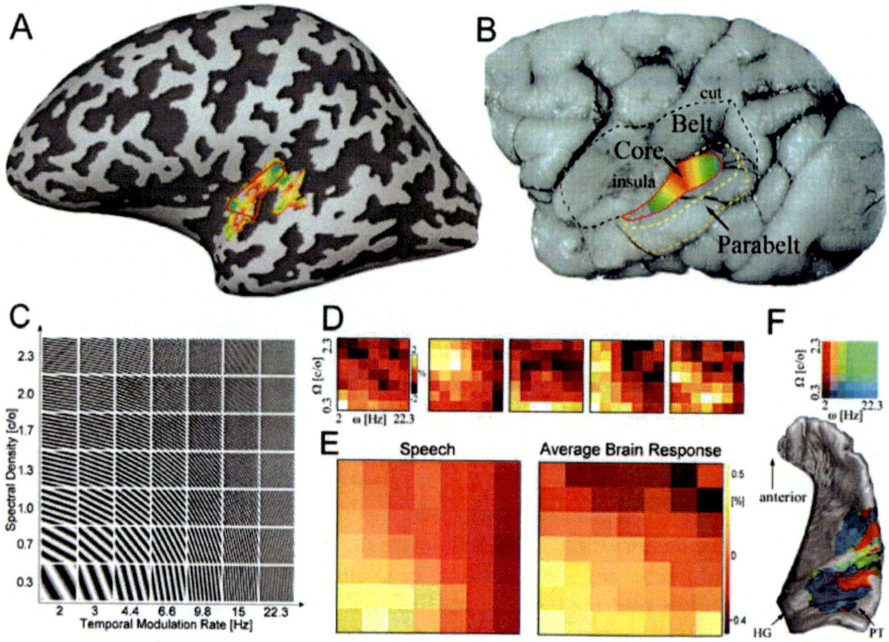

Fig. 31.2 Tonotopic maps in the (**a**) human and (**b**) macaque monkey. Modified from the original sources (**a** Formisano et al. 2003. **b** Kaas and Hackett 2000). The color gradients run from *blue/green* (high frequencies) to *yellow/red* (low frequencies). The *red outline* marks the medial two-thirds of Heschl's gyrus in the inflated representation of the human brain, and the cytoarchitectonic core area in the macaque brain. **c** Each square is a spectrogram of a dynamic ripple, ordered along spectral density and temporal modulation rate axes (49 conditions on a 7-by-7 grid). **d** Color-coded response magnitudes in single voxels to all 49 ripple conditions, ordered according to the stimulus grid. This representation is the 2D modulation transfer function (MTF) of the voxel. The five examples show, from *left* to *right*, two focal responses, a temporal bandpass, a spectral band pass, and a two-peaked MTF. **e** The average modulation content of speech (calculated from random excerpts of radio discussion programs in different languages) matches the average modulation transfer function of the human auditory cortex. Both exhibit a spectral and temporal low-pass characteristic. The color code shows the % BOLD signal change across all active voxels for each ripple condition. **f** A map of the preferred ripple (the ripple condition to which a given voxel responded maximally), color-coded with a logarithmic 2D color map and superimposed on a rendering the left temporal lobe surface. Areas with *reddish* colors respond best to high spectral densities; areas with *bluish* colors respond best to high temporal rates. *Green* corresponds to a conjunction of high spectral and temporal rates (Schönwiesner and Zatorre, 2009)

harmonic fundamental (Zatorre 1988) and in pitch direction discrimination (Johnsrude et al. 2000). A series of neuroimaging studies has provided consistent evidence for an area on the lateral Heschl's gyrus (Fig. 31.3) specific for the extraction of pitch (Griffiths et al. 1998; Patterson et al. 2002; Penagos et al. 2004). When brain activity is compared in response to noise and spectrally matched stimuli containing a pitch based on temporal regularities (this

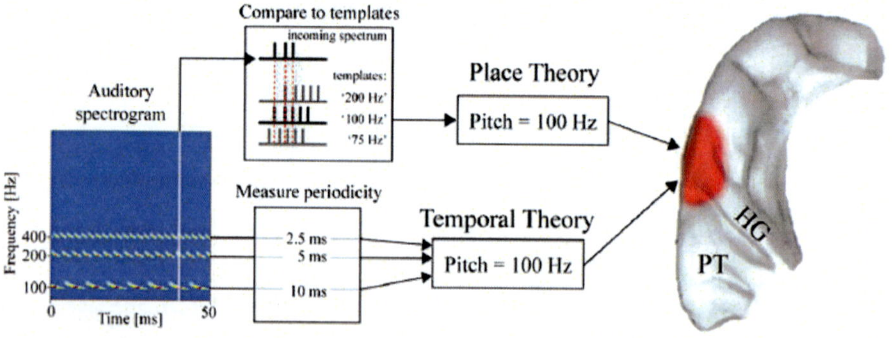

Fig. 31.3 Pitch extraction based on spectral and temporal regularities. The auditory spectrogram represents neural activity in the auditory nerve, i.e., the spectro-temporal information reaching the auditory system. The place theory posits that the distribution of activity along a tonotopic axis is compared with templates for different harmonic tones and that the best match indicates the pitch of the sound. The temporal theory posits that the temporal periodicity within each frequency band is measured. The inverse of the longest cycle duration gives the pitch of the sound. Both mechanisms might operate in parallel with the output converging in the cortical pitch center on lateral Heschl's gyrus (*red highlight*). HG Heschl's gyrus, PT planum temporale. Modified from the original (Shamma 2004)

stimulus is called iterated ripple noise), lateral HG is differentially active in the presence of pitch, but activity in this area is not modulated by the presence of a melody. This suggests that lateral HG is involved in assessing pitch value or strength rather than evaluating pitch changes across a tone sequence (Patterson et al. 2002). Other studies have also implicated this region in computational functions that seem crucial for extracting pitch from complex sounds, such as spectral information integration over time or frequency channels (Zatorre and Belin 2001; Hall et al. 2002; Schönwiesner et al. 2005). A final piece of evidence comes from a neurophysiological study in marmoset monkeys which found that neurons in a small region anterolateral to the medial primary field (AI) represent the missing fundamental of harmonic stimuli (Bendor and Wang 2005). The location of this area relative to other auditory fields is analogous to that of the pitch-sensitive region in the human brain.

Additional evidence concerning the characteristics of pitch-related responses in the human auditory cortex comes from magnetoencephalography (MEG). These recordings have isolated a transient 'pitch onset response' component (Krumbholz et al. 2003) and a sustained pitch response component (Gutschalk et al. 2002, 2004) of the auditory evoked potential that indicates pitch processing. These potentials have longer latencies than typical responses to the sound onset, indicating a hierarchical process in which pitch extraction follows sound onset-related processing, or a longer computation time for pitch extraction. Dipole modeling identified sources of these responses on the lateral Heschl's gyrus, in accord with the results of the functional imaging studies. This convergence of neuroimaging and MEG data provides strong evidence for pitch processing in the lateral Heschl's gyrus.

While the description of pitch as a one-dimensional measure is useful in many respects, musicians recognize pitch as having two dimensions, termed pitch height and pitch chroma (Shepard 1982). This reflects the special status of the octave interval between pitches in music perception. If pitch were perceptually one dimensional, one would expect ratings of pitch similarity to decrease monotonically as the pitch difference increases. However, pitches one octave apart are judged more similar than pitches with smaller separations. This introduces a circularity in the perceptual mapping of sound frequency and pitch similarity that is best expressed using a two-dimensional helical model with pitch height increasing along the helical axis and pitch chroma varying circularly along the perimeter (Krumhansl 1990). These parameters have different cortical representations. A neuroimaging experiment used sequences of notes in which chroma or height were independently varied between notes and found that lateral Heschl's gyrus is activated by changing both pitch chroma and pitch height. From this stage on, the representation of the two parameters diverges. Areas specifically activated by pitch chroma change are anterior in the planum polare, whereas areas specifically activated by pitch height change are located posterior in planum temporale (Warren et al. 2003).

1.6 Mechanism of Pitch Extraction

Finding a cortical area specialized in the extraction of pitch from complex sounds does not solve the question of the computational mechanism of pitch extraction embodied in this area. Two fundamental mechanisms have been proposed: one based on the spectral components of a stimulus and their representation in the locus of activity within the tonotopic map (place theory of pitch), and another based on the phase-locked activity within each frequency band that encodes temporal regularities in the sound (temporal theory of pitch). Mechanisms that rely on the place code might involve harmonic templates that are matched to the incoming spectrum, with the best match determining the pitch value. This template matching would be relatively robust against the absence or mistuning of individual harmonics, including the fundamental. Mechanisms that rely on the temporal code would extract the periods of the prominent temporal regularities in the auditory activity and find their lowest common denominator, which corresponds to the fundamental frequency.

Computational implementations of these algorithms are effective in determining the pitch of complex tones in both cases and mimic many aspects of human pitch perception (Goldstein 1973; Cohen et al. 1995; Patterson et al. 1995). In the brain these two mechanisms are by no means mutually exclusive, since the place and temporal codes work in parallel (Fig. 31.3). The question is therefore probably not which mechanism is implemented, but rather the relative contributions of the two mechanisms to our perception of pitch. The fact that we can hear a pitch in iterated rippled noise, a stimulus that includes temporal regularities but no spectral differences from noise that the auditory periphery can resolve, is strong evidence that the auditory system can extract pitch solely based on temporal regularities (Yost et al. 1996). On the other hand, a correct topographic representation of a sound's frequency content is also necessary for pitch perception. In a very compelling experiment temporal and place information were dissociated with transposed stimuli, in which a high-frequency carrier tone is modulated with a low frequency. In the cochlea this tone excites a location corresponding to the high frequency of the carrier. The tone is demodulated at the auditory nerve, whose activity carries temporal regularity associated with the low-frequency modulation, making it possible to present temporal information of low-frequency sinusoids to locations in the cochlea tuned to high frequencies (Oxenham et al. 2004). A strong version of the temporal theory of pitch would predict that the temporal modulation of auditory nerve activity determines the pitch, and the location of the modulated activity within the tonotopic axis is immaterial. The experimental results show that none of the listeners was able to extract the fundamental frequency from multiple low-frequency harmonics presented to high-frequency cochlear regions. The answer to the question of spectral versus temporal pitch extraction might be, as is often the case in natural systems, that both mechanisms contribute to perception; the dominant mechanism in a given listening situation might depend on which of them provides the more reliable pitch estimate under the circumstances. It remains to be determined precisely how this computation is made in the brain, but the presence of a pitch-sensitive region that responds both to temporal and spectral pitch suggests that this area represents a convergence zone that integrates cues coming from the periphery to create a more stable percept.

1.7 Auditory Stream Segregation

In daily life we rarely hear an isolated pitch. We usually hear sounds from several speakers or instruments simultaneously. These sources must be separated for us to follow individual sources while ignoring others. The auditory system apparently solves the separation problem with ease and permits us to perceive the acoustic world as consisting of coherent objects instead of a confusing mix of sounds. This is a tremendously difficult engineering task. The output of each source may consist of several spectral components, some or all of which can overlap with the output of other sources in time or frequency. The process of grouping and separating these components has been named auditory stream segregation (Bregman 1990). A stream in this sense is a computational stage en route to a mental reconstruction of the acoustic environment. This perceptual segregation of sounds into different streams also improves the extraction of information from a particular stream. A prominent experimental model for the study of stream segregation employs a so-called galloping rhythm sequence of sounds, developed by using two tones of different pitches (A and B) that are presented in a pattern of repeating triplets (i.e., ABA_ABA) (van Noorden 1975). When the pitch difference between the tones is small and the sequence is played slowly, listeners hear a galloping rhythm corresponding to the repeating triplets. When the pitch difference is large and the sequence is played fast, the galloping rhythm disappears and the two pitches separate perceptually into two concurrent streams. At intermediate pitch differences and speeds the perception of the sequence is bistable and flickers randomly between one and two streams every few seconds (Fig. 31.4).

As noted above, most auditory system neurons are frequency selective; hence, two tones of sufficiently different frequencies would activate different neural populations in any tonotopically organized processing stage. This invites speculating about a very basic neural mechanism for the segregation of the two pure tone sequences: when the frequency difference between the tones is large, the neural activity evoked by the two tones is well separated along the tonotopic map, thus representing two streams. When the frequency separation decreases, there is more spatial overlap in neural activity, and neural responses increasingly reflect the galloping rhythm pattern. The effect of the interstimulus interval could arise from short-term adaptation. Consider a neuron that is tuned to the frequency of tone A. At long interstimulus intervals, the neuronal responses to the two tones interact very little, but as the sequence is speeded up and the time interval between the tones shortens, short-term adaptation produced by the strong responses to tone A would reduce the responses to tone B. These effects have been observed in AI of awake monkeys (Fishman et al. 2001, 2004; Micheyl et al. 2005). Psychophysical experiments in humans and single-unit recordings in the macaque auditory cortex using identical stimuli found that the typical curve of the temporal build-up of streaming for different frequency separations can be accurately reproduced by the neural activity in the monkey primary auditory cortex (Fig. 31.4) (Micheyl et al. 2005).

In humans, results from neuroimaging show that the hemodynamic response from the auditory cortex to an ABAB sequences reflects the frequency separation of the tones. The extent and amplitude of the response increased monotonically with frequency separation, and the shape of the response became more sustained (on- and offset peaks diminished) with separation. An analogous change occurs when the physical rate of presentation is lowered. The changes in the response from the auditory cortex are consistent with the perceived decrease in presentation rate when the AB sequence is separated into two interleaved streams (Wilson et al. 2007).

A simple model of auditory stream segregation based on the frequency selectivity of neurons is necessarily incomplete. It does not account for several important effects of stream segregation: although pitch differences appear to be a dominant cue (Singh 1987; Vliegen et al. 1999), sounds without spectral differences can be segregated, for instance, on the basis of intensity differences or differences in the envelope periodicities (Moore and Gockel 2002). As noted above, at intermediate pitch separations and speeds the perception of a galloping rhythm sequence is bistable and alternates between one and two streams. The time intervals between the flips follow a gamma or lognormal distribution, suggesting an underlying random process (Pressnitzer and Hupé 2006),

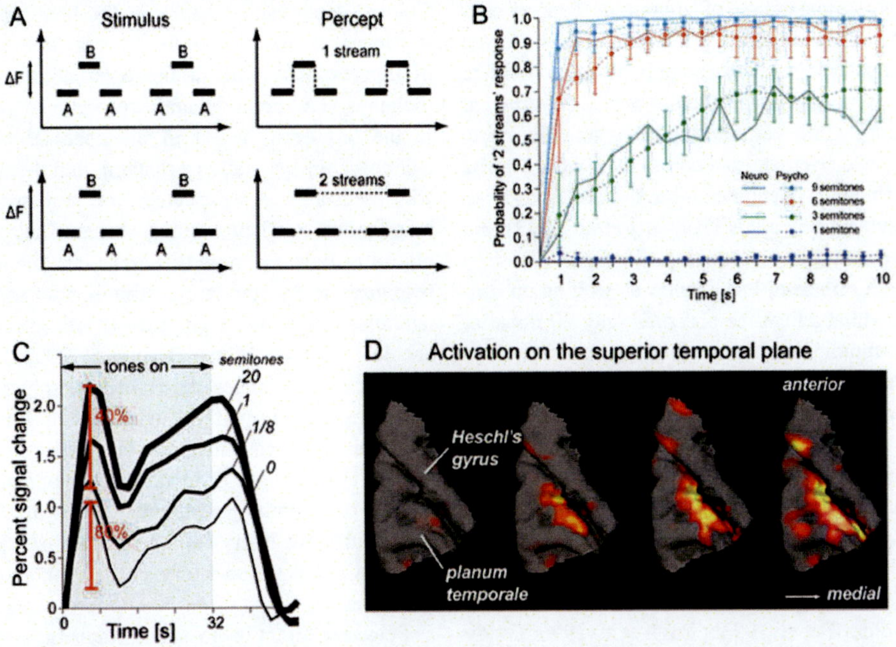

Fig. 31.4 Auditory stream segregation. **a** In a galloping rhythm sequence, ABA tone triplets are repeated. When the A and B tones are close in frequency, the sequence is perceived as one stream. When the frequency separation (ΔF) is large, the A and B tones are perceptually segregated into two streams. **b** Neurometric and psychometric build-up functions of stream segregation. The *dotted lines* represent the varying proportion of 'two streams' responses from sequence onset in humans listening to a galloping rhythm sequence. *Error bars*: 95% confidence intervals around the mean proportions estimated using statistical bootstrap. *Solid lines* show the probability of 'two streams' responses predicted from neural responses measured in the monkey auditory cortex. For different frequency separations the probability of segregated streams increases at different rates in both psychometric and neurometric functions. Reproduced from the original source (Micheyl et al. 2005). **c** Time course of hemodynamic responses from the human auditory cortex to sequences of alternating pure tones with different frequency separations. The response magnitude of the increases with frequency separation, the response becomes more sustained (peristimulus modulation depth decreases from 80 to 40%), and (**d**) the extent of the activation increases (*left* to *right* 0, 1/8, 1, and 20 semitones separation, respectively). Modified from the original source (Wilson et al. 2007)

which, despite its randomness, might have adaptive value (Leopold and Logothetis 1999).

An elegant magnetoencephalographic experiment measured responses to a bistable galloping rhythm sequence of pure tones while listeners indicated whether they perceived one or two streams. Selective averaging based on the listener's perception found that when two streams were perceived the B tones were suppressed more than when one stream was perceived, even though there was no physical change in the stimulus (Gutschalk et al. 2005). Since an attentional effort can influence the perception of a bistable stimulus and this perceptual state is reflected in neuronal activity in the auditory cortex, there must be top-down modulatory pathways mediating the volitional effect descending to the auditory cortex. This is also apparent in the influence of previously learned knowledge on our ability to segregate sounds [(schema-driven selection (Bregman 1990)]. For instance, it is much easier to detect a familiar tune in a mixture of sounds than an unfamiliar one (Dowling 1973; Bey and McAdams 2002). Attentional and other cognitive influences thus shape auditory stream segregation (Snyder and Alain 2007).

1.8 Effects of Attention on Auditory Cortex Activation

Humans have a remarkable ability to direct their attention voluntarily to one of the many competing streams of information, and thereby select certain pieces of sensory information while ignoring others. A seminal work, framed this question as follows: "One of our most important faculties is our ability to listen to, and follow, one speaker in the presence of others. This is such a common experience that we may take it for granted; we may call it "the cocktail party problem. No machine has been constructed to do just this, to filter out one conversation from a number jumbled together" (Cherry 1957). We now briefly discuss evidence from neuroimaging for a modulation of the human auditory cortex by attention and other cognitive factors.

Since the earliest studies of the auditory cortex, it has been clear that attentional selection modulates the responses of single auditory neurons: attention units respond to sound exclusively when the focus of attention coincides with the location of the sound source (Hubel et al. 1959).

Another striking demonstration of attentional control over information flow extending to the brain stem comes showed that electric responses to repetitive sounds from the cochlear nucleus of cats practically vanished when a stimulus of greater significance to the cat than the test sounds was presented: a live mouse (Hernández-Peón et al. 1956). Similar attentional modulation phenomena have been shown in humans using evoked potentials (Hillyard et al. 1973) and have been linked directly to the region of primary auditory cortex, based both on signal localization as well as on timing, since the modulation can be observed only 20 ms after the onset of the stimuli (Woldorff et al. 1993a).

How can attention, in principle, influence the representation of sensory stimuli in the auditory cortex? It would be intuitively satisfying to assume that directing attention toward a stimulus might increase the response gain and thus the dynamic range or contrast of the neural responses, as occurs in the visual system (Desimone 1998). An enhancement of activation of the human auditory cortex due to attention has been repeatedly demonstrated with neuroimaging methods (Woldorff et al. 1993b; Pugh et al. 1996; Grady et al. 1997; Tzourio et al. 1997; Jäncke et al. 1999; Petkov et al. 2004). A particularly stringent study compared auditory cortical activation to attended and unattended sounds. In the unattended condition participants were distracted from listening to the sounds by a visual task; in the attended condition a pitch discrimination task was used to focus attention on the test sounds. The sound-energy-driven response was largest in primary auditory cortex, depended on sound frequency, and showed adaptation to repetitive stimulation. In contrast, the attention-dependent increase in activation was strongest in the lateral superior temporal plane (secondary and higher cortices), independent of sound frequency, and became stronger with repetitive stimulation (Petkov et al. 2004).

1.9 Gain Control for Task-Relevant Areas

The enhancement of neural activity appears to depend on what the listener intends to do with the sound information. The context of an ongoing stream of stimuli can influence the gain of an auditory area and thus increase the response in areas specialized in processing information relevant to the task at hand. Frequency-modulated tones of different durations were played to participants asked to discriminate either pitch direction or duration. Compared to stimulus exposure, categorization of pitch direction increased hemodynamic activity in the right posterior auditory cortex, whereas duration categorization increased activity in the left posterior auditory cortex (Brechmann and Scheich 2005). Context-dependent changes in the hemispheric balance of the response to a stimulus may be due to a modulation of the response gain at the level of the brain stem and thalamus. The hemodynamic response to sounds presented to the left or right ear in the cochlear nuclei, the inferior colliculi, the auditory thalami, and the auditory cortices was measured. As expected, the responses have similar magnitudes at all processing stages. They then introduced blocks of binaural moving sounds into the stimulation. The responses to the monaural stimuli now showed a marked lateralization to the right side in all structures above the cochlear nuclei (Schönwiesner et al. 2007a). The lateralization pattern resembles the right-hemispheric specialization for the processing of acoustic spatial information found in lesion (Zatorre and Penhune 2001) and neuroimaging studies (Krumbholz et al. 2005). The modulation did not depend on attention (participants were involved in a visual task), but more likely resulted from a top-down interaction between low-level auditory cortical areas and subcortical structures.

Selective attention to sound phonetic content modulates response adaptation in the anterior secondary auditory cortex, whereas attending to the location of the same sound modulates responses in the posterior secondary auditory cortex. These findings support results from animal work that showed highly specific effects of attention rather than a simple overall enhancement of activity (Ahveninen et al. 2006). Receptive fields of ferret cortical auditory neurons change their shape in a predictable manner contingent on the frequency content of the task-relevant stimulus (Fritz et al. 2003, 2005, 2007).

The above results collectively show that the context of a sound stimulus may select cortical areas specialized in the relevant type of processing by changing response gain at cortical and subcortical levels. Attention may not only enhance activity related to the attended stimulus, but also suppress activity related to ignored stimuli. For example, participants presented with melodies and shapes either alone or simultaneously were asked to detect changes in either the visual or the auditory information while ignoring the modality not currently monitored. A subsequent memory test showed that attended stimuli were remembered significantly better than ignored stimuli in bimodal conditions (Johnson and Zatorre 2005, 2006). When attention was focused on the auditory stimuli, responses increased in lateral portions of temporal auditory cortex, consistent with previous studies (Lewis et al. 2000; Laurienti et al. 2002). When focusing on the visual stimuli, responses decreased in these same auditory regions. This reciprocal relationship demonstrates that selective attention enhances processing of one modality at the expense of the other in terms of neural responses and memory encoding.

1.10 Auditory Imagery

In all of the above examples, a response to a physically present sound is modulated by attention. An extreme case in the spectrum of top-down modulatory effects of auditory cortex activity is provided by the phenomenon of imagery: when an auditory percept is produced entirely by mental effort in the absence of sound. Several interrelated phenomena are relevant, including auditory rehearsal (Hickok et al. 2003), perceptual expectancies (Voisin et al. 2006), and even cross-modal interactions, such as those involved in silent lip reading (Calvert et al. 1997). In each case, auditory cortex activity cannot be accounted for solely by any stimulus that is present. Musical imagery is perhaps phenomenologically the most obvious instance of imagery, since most people have a clear experience of being able to hear music in their mind. A review of neuroimaging studies concluded that auditory cortex outside of core areas responds during the performance of specific musical imagery tasks in which behavior is controlled (Zatorre and Halpern 2005). That this activity reflects an essential process and not an epiphenomenon is supported by lesion data (Zatorre and Halpern 1993). It remains unclear however which subfields may be active during imagery, and how such activity is initiated. Presumably retrieval functions involving interactions between frontal cortex and auditory areas would be important for volitional imagery; conversely, impairment of these feedback interactions might be related to certain hallucinatory phenomena (Griffiths 2000).

1.11 Involuntary Capture of Auditory Attention

The previous discussion construed attention as a voluntary process that helps to separate acoustic information deemed relevant from other, interfering information. The auditory system also has automatic mechanisms to detect relevant information and trigger attentional switches without volitional control. These mechanisms can provide a basic pre-attentive context dependence of responses to auditory stimuli. Context dependency is defined as the response to a given stimulus based on the immediate or longer-term stimulus history. A typical example of automatic attention capture that a reader of this book might encounter while dozing on the way to a conference is the sudden detection a subtle change in the hum of the plane's jet engine. In such situations, the relevant or interesting event that automatically captures attention is a change in the acoustic environment. Brain mechanisms for the detection of rare sound events have been extensively studied using an experimental paradigm in which infrequent deviant sounds are presented in a stream of repeating standard sounds. The deviant sounds evoke a frontal negative deflection in the auditory event-related potential, the mismatch negativity (MMN) (Näätänen et al. 1978). The MMN can be recorded in response to any discriminable change in the stimulus stream, its occurrence correlates highly with perceptual detection thresholds, and it is largely independent of attention (Näätänen 1995). The MMN is usually interpreted as supporting the existence of a sensory-memory trace in which frequently occurring acoustic features are represented. A new sound that fails to match the stored description triggers a mismatch signal that may, after further evaluation of the sound for sufficient novelty, lead to an involuntary redirection of attention toward the sound.

Many neuroimaging studies have described brain correlates for various stages of this model (Opitz et al. 1999; Muller et al. 2002; Opitz et al. 2002; Doeller et al. 2003; Liebenthal et al. 2003b; Marco-Pallarés et al. 2005; Molholm et al. 2005; Rinne et al. 2005; Schönwiesner et al. 2007b). Hemodynamic and evoked potential responses to rare changes in the duration of sounds in a repetitive sequence were measured while participants attended to an unrelated visual task (Fig. 31.5). Primary and secondary auditory fields, as well as the mid-ventrolateral prefrontal cortex, responded to the acoustic changes, but in different ways. Responses from medial Heschl's gyrus indicate that acoustic changes are first detected at or below the level of the primary auditory cortex. Responses from posterior secondary areas represented the magnitude of the acoustic change most accurately, suggesting that these structures might extract the details of the acoustic change. Activity in the frontal cortex followed the auditory cortex activation with a lag of ~50 ms and was independent of the magnitude of the acoustic change (Schönwiesner et al. 2007b). This region has been associated with memory-based decisions and may signal the novelty value of the acoustic change and thereby determine whether a switch of attention toward the sound is initiated. If a conscious switch in attention is triggered, other frontal and parietal lobe areas are activated (Watkins et al. 2007).

1.12 Pitch Patterns: Melodies

The analysis of pitch information becomes very complex when combinations of periodic sounds that form patterns must be processed. Music is perhaps the paradigmatic case in which pitch combinations are critical; indeed, it is the relationships between pitches, rather than the absolute value of each pitch, which are crucial in encoding and recognizing melodies. Melodies for example are recognized on the basis of the intervals (frequency ratios) between successive pitches

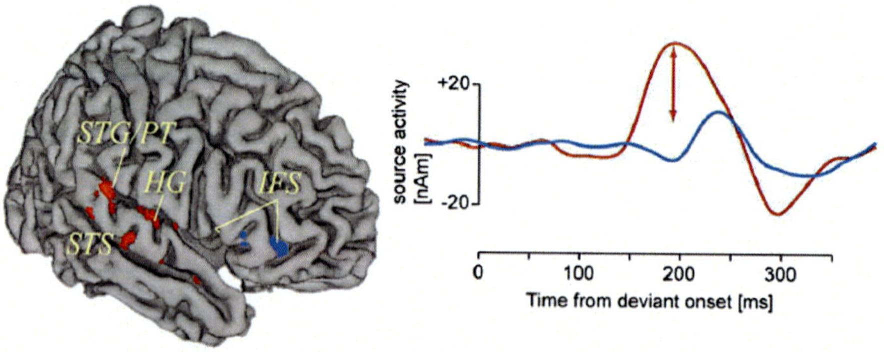

Fig. 31.5 **a** Portions of the temporal (*red*) and frontal lobes (*blue*) respond to rare deviations in a repetitive sound sequence. **b** The time course of activation in those areas obtained from the equivalent current dipole models of electroencephalographic responses. The temporal lobe responses (*red*) vary with the acoustical difference between the deviant and standard sounds (*red arrow*). About 50 ms later the right frontal areas respond (*blue*) independently of the acoustical difference between the deviant and standard sounds. Latency differences in of temporal and frontal responses suggest that change-related frontal lobe activity relies on afferent projections from the perisylvian region of the temporal lobes. HG, Heschl's gyrus; IFS inferior frontal sulcus; PT, planum temporale; STG, superior temporal gyrus; STS, superior temporal sulcus,. Modified from the original source (Schönwiesner et al. 2007b)

(Attneave and Olson 1971), and not on the absolute values of the tone frequencies. Similarly, the interval relationships of simultaneously sounded tones form the basis for harmony, an important element in most musical systems. Besides intervals, other parameters of tonal patterns are also pertinent to perception, including contour, which refers to the pitch directions contained in a melody (Dowling and Harwood 1986).

Early neuroimaging studies on pitch combinations in melodies yielded indications that cortical areas in the superior temporal gyrus (STG) outside HG were active during perception of melodies and compared to various baseline conditions, including acoustically matched noise bursts (Zatorre et al. 1994; Binder et al. 2000) or silence (Griffiths et al. 1999). Areas both anterior and posterior to HG were recruited in these studies (Griffiths and Warren 2002; Zatorre et al. 2002a). The more posterior regions usually within the planum temporale (PT) are sensitive to frequency modulation in general since this area is active when modulated stimuli are compared to static ones (Thivard et al. 2000; Hall et al. 2002; Hart et al. 2003). The relationship of these regions to the pitch-sensitive area described above remains to be determined in detail, but it seems from connectivity information that both anterior STG and the PT receive input from it. In keeping with hierarchical processing, one may therefore hypothesize that these regions perform computations beyond pitch extraction, perhaps related to assembling pitch sequences and determining parameters such as interval size and contour, which as noted above, are crucial for perception. One study (Patterson et al. 2002) is directly relevant here since it dissociated neural activity originating from the lateral portion of HG (during processing of simple pitch) from activity in posterior and anterior STG areas, which was specific to processing of melodies, consistent with the proposal that these distal regions are involved in higher order feature analysis of melodic information.

These investigators also reported the emergence of a right-sided asymmetry within anterior and posterior STG during melodic processing, joining many neuroimaging studies which have observed similar hemispheric differences in a wide variety of tonal processing tasks. These include pitch judgments within melodies (Zatorre et al. 1994) or tones (Binder et al. 1997); maintenance of pitch while singing (Perry et al. 1999); imagery for tunes (Halpern and Zatorre 1999) or for instrument timbres (Halpern et al. 2004); discrimination of pitch and duration in short patterns (Griffiths et al. 1999); reproduction of tonal rhythmic patterns (Penhune et al. 1998); timbre judgments in dichotic stimuli (Hugdahl et al. 1999); and detection of deviant chords (Tervaniemi et al. 2000). These findings are also consistent with a large body of lesion evidence which support similar hemispheric specialization effects (Stewart et al. 2006). Although these studies vary in the specifics of the areas shown to be asymmetric, and the conditions under which the asymmetry emerges, the consensus about right auditory cortex primacy for tonal processes demands an explanatory model.

As discussed below, one hypothesis to explain these data relates to differential hemispheric resolution in the spectral versus temporal domains (Zatorre et al. 2002a); a related concept is that of differential temporal integration windows (Poeppel 2003). Specifically with respect to tonal processing, it is proposed that certain right auditory cortex fields may be more sensitive to fine differences in pitch. Lesion studies showed that excision near the right, but not the left, pitch-sensitive region increased the threshold for discrimination of the pitch of two tones but did not abolish this ability (Johnsrude et al. 2000). Functional imaging

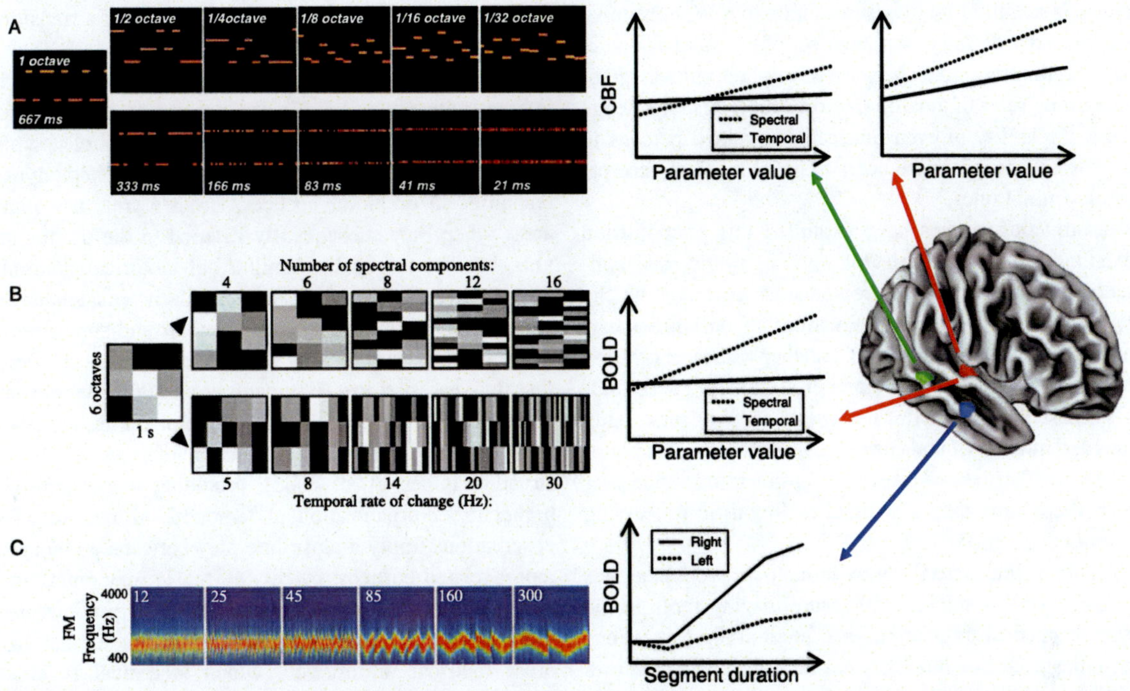

Fig. 31.6 Summary of multiple studies investigating hemispheric asymmetries in auditory cortical response by manipulating spectral and/or temporal features of nonspeech stimuli. The *left side* of the figure shows the stimulus manipulations used for each study; the *right side* of the figure shows a schematic of the obtained results and an illustration of the position of the activation sites within the right auditory cortices. All three studies find consistent evidence for hemispheric differences in response to temporal and/or spectral parameters. **a** A PET study of spectral separation and temporal rate (Zatorre and Belin 2001). Stimuli consisted of a series of pure tones. For the spectral manipulation (*top row*), the frequency separation varied from one octave to 1/32 of an octave, keeping rate constant; for the temporal manipulation, the alternation rate of a quasirandom duty cycle varied from 667 ms (slowest) to 21 ms (fastest), keeping spectral separation constant. Two sites of response to spectral variation were found within the right temporal cortex, one on the lateral portion of Heschl's gyrus (*red*) and one in the superior temporal sulcus (*green*). In both cases the response to the temporal change was low or absent; the reverse pattern was observed for left auditory cortices, with stronger responses to temporal change (not shown). **b** An fMRI study of spectral and intensity changes (Schönwiesner et al. 2005). Stimuli consisted of amplitude-modulated noise bands with spectrograms equivalent to the *square matrices* shown (time on the abscissa; frequency on the ordinate). The number of independently modulated frequency bands (spectral components; *top row*) and the rate of intensity change within each band (*bottom row*) was parametrically varied. Responses from Heschl's gyrus (*red*) and part of the right superior temporal gyrus anterolateral from Heschl's gyrus (not shown) covaried with the spectral parameter, whereas activity in the equivalent region on the left superior temporal gyrus (not shown) covaried with the temporal parameter. **c** An fMRI study of segment duration (Boemio et al. 2005). Stimuli consisted of concatenated band-limited noise. The length of the noise segments was varied, and in each segment the center frequency of the noise could either be stationary (*bottom row*) or sweeping up or down (*top row*). Results show that most of auditory cortex in both hemispheres is sensitive to the temporal rate of the stimulus. A hemispheric asymmetry emerged in the superior temporal sulcus (*blue*), where activity on the right side, but not the left, varied as a function of segment duration

studies have found that auditory cortices on both sides respond to increasingly fine spectral spacing, but with an asymmetry favoring the right hemisphere (Fig. 31.6). Hemodynamic responses to pitch variation in tone sequences emerged near the right pitch-sensitive region even to quite small pitch changes, whereas a response on the left did not emerge until the pitch changes were much larger (Hyde et al. 2008). That the asymmetry does not depend on the presence of periodicity was shown by using stimuli consisting of noise bands that systematically varied in their spectral width and temporal rate of change (Fig. 31.6); hemodynamic responses increased bilaterally in lateral portions of HG as a function of increasing number of noise bands (with correspondingly narrower bandwidths), but the slope of this function was steeper in the right cerebral hemisphere (Schönwiesner et al. 2005), consistent with studies using periodic stimuli (Zatorre and Belin 2001; Jamison et al. 2006).

The studies reviewed to this point focus on auditory areas in processing pitch patterns, but much more complex interactions between auditory cortex and other brain areas are required for the many complex cognitive phenomena associated with musical processing. For example, a simple pitch-memory task for a melody recruits extratemporal areas in frontal and parietal regions (Zatorre et al. 1994; Gaab et al. 2003). Another prominent example arises from the expectancies generated by tonal musical structures, which embody abstract, implicit knowledge acquired from listening to music in a given culture, much like the syntax for native language is acquired by exposure. Thus, in a Western tonal music context, hearing a particular set of notes leads one to expect certain other notes because of implicit knowledge of the governing rules of tonality (Huron 2006); this knowledge explains the ability of even musically untrained persons to readily detect inappropriate notes in novel musical excerpts that follow tonal rules.

Neuroimaging studies have exploited this phenomenon to examine the neural basis for musical syntax and consistently found that inferior frontal areas (stronger on the right) respond to unexpected violations of unfamiliar harmonic sequences (Koelsch 2005), together with temporal and inferior parietal regions (Tillmann et al. 2006). These findings suggest that interactions between sensory processing regions and inferior frontal cortex generate representations of structural regularities and integrate ongoing information over time, perhaps providing a parallel to linguistic processing (Friederici et al. 2003).

The role of the parietal cortex in melodic processing has also recently been studied with neuroimaging tools in the context of musical transformations. Internal representations of melodies exist in a relatively abstract form, which allows them to be manipulated mentally. One such manipulation is transposition, which refers to the situation where all the individual pitches of a melody are changed by the same amount up or down; because of the invariance of perception under this type of transformation (Attneave and Olson 1971), the same melody is perceived, but in a different musical key. The neural substrate of the ability to perceive transposition was recently studied by Foster and Zatorre (2009) who found that activity within the intraparietal sulcus (IPS) was higher when a melody discrimination task required transposition than when it did not. Moreover, the degree of activity in this region independently predicted behavioral performance to a high degree. Recruitment of a similar IPS region was also reported by Zatorre et al. (2010) with a different manipulation task, requiring listeners to make judgments about temporally reversed melodies. What these two experimental tasks have in common is that they both require that the sensory information be transformed from one reference frame into another one. There is widespread evidence for the idea that the IPS is important for mental transformations, for example, in visual cognition (Zacks 2008), and more generally from models of parietal function derived from the visuomotor literature (Culham et al. 2006). A role for the IPS is also reported in visual working memory when the task requires manipulation, rather than monitoring of the information (Champod and Petrides 2007). The findings from the musical studies therefore extend our understanding of the role of the dorsal stream of processing in the auditory domain and fit with the broad idea that this system is organized to perform computations in which precise relationships between elements (pitches, spatial distances, numerical relationships) are maintained in the context of a transformation to a different frame of reference.

Neuroimaging studies also reveal how sensory regions interact with other areas in the context of auditory–motor interactions (Zatorre et al. 2007). For example, subjects trained to play a pattern on a keyboard show hemodynamic responses in premotor cortices, Broca's area, and parietal areas when they subsequently listened to the trained stimulus, but not to equally familiar but motorically untrained melodies (Lahav et al. 2007). Similarly, musicians show evidence of activity in motor (premotor and supplementary motor) and sensory regions when listening to musical pieces that they can perform (Baumann et al. 2005; Bangert et al. 2006). Dorsal portions of the premotor cortex specifically appear to be engaged as a function of metrical rhythmic structure (Chen et al. 2006), providing a mechanism for higher order organization of temporal information. These observations imply a close interplay between auditory- and motor-related cortices: perceiving music may entail activation of motor programs associated with its production, but these motor interactions may also help to extract higher order metrical information, which is critical in creating rhythmic and melodic expectancies. The conclusion that sensory–motor loops are important for music is comparable to concepts developed in models of speech (see below) and can also be related to models of visuomotor integration and action observation (Rizzolatti and Craighero 2004).

A final point in considering pitch processing in music is that musical performance training may substantially affect auditory cortex function (Münte et al. 2002). For example, MEG studies show that brain responses to piano tones are 25% larger in musicians than in nonmusicians (Pantev et al. 1989), an effect more pronounced for tones from a musician's own type of instrument (Pantev et al. 2003) and strongly implying use-dependent plasticity. Neural activity evoked by pure tones is larger in professional musicians than nonmusicians (Schneider 2002) and is accompanied by morphological changes. These findings underscore the importance of understanding the stimulus characteristics as well as the history of interaction between the listener and the class of stimuli, a concept also applicable to speech studies, as explored below.

2 Speech-Related Functions

2.1 Low-Level Specializations Versus Higher Order Distributed Mechanisms

Speech has been the focus of many neuroimaging studies since technical developments allowed scans of human subjects. Many early functional neuroimaging studies were concerned with identifying the pathways associated with speech sound processing. Most of these studies demonstrated that certain portions of the superior temporal gyrus responded to speech sounds, with a left side asymmetry for speech sounds as compared to nonspeech controls, such as tones (Binder et al. 2000), amplitude-modulated noise (Zatorre et al. 1996), or spectrally rotated speech (Scott et al. 2000). It is less clear whether these findings can be interpreted as evidence for a dedicated, specialized speech processing system given the complexity of speech; if speech sounds elicit a certain activity pattern not observed in a control stimulus, the response may be to speech qua speech or to an acoustical feature present in the speech signal but not in the control sound. Conversely, if a nonspeech sound akin to speech elicits auditory cortical activation overlapping with speech, then it may do so because of its similarity to speech.

2.1.1 Speech: Left Auditory Cortex Specialization

A productive approach to this problem is to identify and manipulate systematically the acoustical features which may be relevant to speech, to see to what extent they can explain the neural activity patterns associated with speech. This reductionist approach has largely validated the claim that many aspects of speech specificity be explained as the consequence of low-level feature processing in the temporal domain. However, the findings do not imply that higher order constraints have no influence on speech analysis.

The hypothesis that rapidly changing spectral energy, as found in speech consonants, requires specialized left auditory cortical mechanisms dates back to observations that patients with speech perception impairments also have difficulty in nonspeech temporal processing tasks (Swisher and Hirsh 1972; Phillips and Farmer 1990; Tallal et al. 1993). Neuroimaging has allowed this model to be extended (Zatorre et al. 2002a; Poeppel 2003) to other phenomena, including tonal processing (see above). For example, left auditory cortical blood flow responses were similar to both slower and faster formant transitions in pseudospeech sounds, whereas right auditory cortex responded most to the slower transitions (Belin et al. 1998). Functional imaging studies have tested this general hypothesis using parametric nonspeech stimuli varying systematically in their temporal and spectral characteristics (Fig. 31.6). One study used pure-tone sequences that alternated in pitch by one octave at different temporal rates. As the speed of the alternation increased, so did the neural response in the mid-portion of Heschl's gyrus in both hemispheres, with a significantly greater magnitude on the left than on the right (Zatorre and Belin 2001). The reverse pattern was seen for spectral manipulation. Others replicated these findings and showed individual subject consistency (Jamison et al. 2006). Another fMRI study tested a similar hypothesis using a different stimulus manipulation consisting of noise bands that systematically varied in their spectral width and temporal rate of change (Fig. 31.6) (Schönwiesner et al. 2005). Increasing rate of temporal change again elicited a larger left auditory cortex response and vice-versa for the spectral manipulation. The cortical areas sensitive to temporal change were not identical to those of the prior studies, which may reflect the different stimuli used. The consistency of the lateralization pattern across studies was clear, however.

A sophisticated approach to understanding the role of temporal information to differential activation of left and right auditory cortex comes from a study which varied the segment transition rates parametrically in a set of concatenated narrow-band noise stimuli (Fig. 31.6) so that segment durations varied across a range from rapidly changing (12 ms) to more slowly changing (300 ms) (Boemio et al. 2005). Sensitivity to this parameter was bilateral and symmetrical in primary and adjacent auditory cortices. However, the more slowly modulated signals preferentially drove activity in the right but not the left STS. The authors conclude that two timescales may exist within right and left auditory cortices, such that right and left hemisphere receive afferents carrying information processed on longer and shorter timescales, respectively. They hence support the proposal that left auditory cortex is specialized for high temporal resolution, whereas right auditory cortex is specialized for high-frequency resolution.

Further support for this trend comes from a study (Zaehle et al. 2004) which observed significant overlap in fMRI activation in the left auditory cortex between speech syllables that differed in voice-onset time and that of a nonspeech analog (noises differing in gap duration). This finding indicates that physical cues in the stimuli, as opposed to their linguistic status, sufficed to recruit left auditory cortex. A related study used a factorial design to contrast speech/nonspeech versus slow/fast changes and found a greater left STG response to both speech syllables and nonspeech tone sweeps that contained rapidly changing information (Joanisse and Gati 2003). This effect was not seen to stimuli with more slowly changing temporal information, again supporting the hypothesis of enhanced left auditory cortex temporal resolution.

2.1.2 Speech: Higher Order Constraints

These findings indicate that cortical response asymmetry can be explained on the basis of the single assumption that rapidly varying acoustical information is preferentially processed by left auditory cortex mechanisms. This conclusion is predicated strictly upon the nature of the acoustical input and the resulting neural response. But this approach does not take into account the role of higher order representations in predicting patterns of activity, such as the degree to which the acoustical stimulus matches sound features in the speaker's language. Several neuroimaging studies have shown that portions of the left anterior and posterior temporal cortex respond to intelligible but not to unintelligible speech sentences, whether they are produced naturally (i.e., by a human vocal tract) or generated via computer algorithms which have an unusual timbre (Scott et al. 2000; Davis and Johnsrude 2003; Narain et al. 2003; see also Binder et al. 2000). This result reveals a convergence of processing for different stimulus types at a level of analysis where meaning is decoded. These findings indicate that top-down effects operate in speech processing; the source of such effects, as well as the pathways underlying them, remain to be fully understood.

To understand how abstract knowledge of speech sounds alters the neural response is challenging for another reason: differences in neural activity for a known versus an unknown speech sound may be confounded by acoustic differences in the stimuli. One approach to this problem is to use a stimulus which can be perceived as speech or not under different circumstances. Sine-wave speech is such a stimulus, because it is perceived by naïve subjects as a meaningless sound, but after training can usually be perceived to have linguistic content (Remez et al. 1981). Three fMRI studies have exploited sine-wave speech by comparing how these sounds are perceived before and after such training (Dehaene-Lambertz et al. 2005; Liebenthal et al. 2003a; Möttönen et al. 2006); each found that processing within the left superior temporal cortex was modulated by training. One study saw an enhanced response only in those subjects who learned to identify the stimuli as speech, making the link between perception and brain activity even more explicit (Möttönen et al. 2006). Despite the differences between studies attributable to details of the stimulation and training paradigms, the overall pattern of findings converges to show that otherwise identical physical sounds are processed differently when perceived as speech and that this difference is present in the response pattern of the left auditory cortex neural (Liebenthal et al. 2003a).

Similar conclusions have been reached in studies where categorically perceived speech syllables are compared to stimuli containing the same acoustical cues but which are neither perceived as speech nor perceived categorically. An activation site within the left superior temporal sulcus (STS) exclusively for the categorically perceived stimuli indicated that the response is linked to more than just the acoustical features (Liebenthal et al. 2005). This region likely performs an intermediate stage of processing linking early processing regions with more anteroventral auditory cortical areas containing stored sound representations. Consistent with this conclusion, an fMRI adaptation paradigm found that the left STS shows a larger response when a speech continuum changes from one phonetic category to another, than when an equivalent acoustical change does not produce a change in a phonetic perceptual category (Joanisse et al. 2007). Similarly, left STG/STS areas responded to intelligible isolated stop consonant sounds but not spectrally rotated control sounds, whereas more posterior STG and PT coded the acoustical structure of the sounds (Obleser et al. 2007).

Another recent study explored the phenomenon of neural response modulation in auditory cortex in the context of speech learning (Golestani and Zatorre 2004). Changes in hemodynamic response to speech sounds were studied before and after listeners learned to distinguish a phonetic contrast not present in their native language. A part of the left posterior STG responded more after training; since the stimulus was unchanged, only training could have caused the change in neural activation. Moreover, the region of auditory cortex recruited after training for the foreign speech sound spatially overlapped the response obtained to native speech sounds. Before training this region would respond only to native speech contrasts, whereas after foreign speech contrasts engaged the same region.

These results all show that experience with sounds, and not only the physical cues involved, influences auditory cortex activity patterns. Sounds which are of linguistic significance are treated differentially within the auditory parabelt cortex anterior and ventral to the left AI. This conclusion is consistent with cross-language studies using other methodologies. A study using an event-related response paradigm showed that the size of the mismatch negativity response, presumably originating from left auditory cortex, is affected by a listener's knowledge of native linguistic vowel categories (Näätänen et al. 1997). Also, behavioral training with speech stimuli in adults causes a significant change in the mismatch negativity duration and magnitude (Kraus et al. 1995).

2.1.3 Speech: Interactions Between Auditory and Other Cortices

This discussion has so far focused primarily on modulations of neural activity produced within classically defined auditory cortical areas. But there are also interactions involving top-down mechanisms which include areas well outside traditional auditory cortex. In the domain of speech processing, frontal-lobe areas are among the most clearly documented extratemporal regions to be involved. An early neuroimaging finding was that ventral portions of the left premotor cortex were active in a purely perceptual task (Zatorre et al. 1992, 1996). This region, close to or within what traditionally had been described as Broca's area by aphasiologists was recruited when listeners made phonetic judgments of speech (for example, deciding if two words begin or end with the same consonant). This effect was not seen for other judgments, such as pitch contour identification, which instead recruited right inferior frontal cortex (consistent with the asymmetries described above for pitch). This same left premotor area was subsequently shown to be specifically involved in phonetic segmentation, because it was active only when a speech discrimination task required segmenting a phoneme out of a syllable, but not when the discrimination could be accomplished on the basis of the whole syllable (Burton et al. 2000). Taken together, these findings suggested that interactions between auditory and motor representations are necessary for certain speech processes, a concept related to the motor theory of speech perception, which had been proposed much earlier on the basis of behavioral evidence (Liberman and Mattingly 1985).

This work has been extended to show auditory–premotor interactions under other circumstances. For example, a premotor region cortex dorsal to Broca's area can be active during passive listening to speech (Wilson et al. 2004), and along with superior temporal regions, is differentially sensitive to phoneme pronounceability (Wilson and Iacoboni 2006). Frontal cortex activity as well as that in temporal cortex is greater for speech embedded in noise than when it is presented in silence (Davis and Johnsrude 2003). These and other findings have led to the idea that speech processing may use two hierarchical processing pathways for different processes. Ventral temporal lobe areas may process auditory information to recover phonetic and semantic information, whereas the more dorsal auditory component (Fig. 31.1) may be linked to motor representations (Scott and Johnsrude 2003; Hickok and Poeppel 2004). These streams would permit some redundancy in speech processing, which could partly account for its robust nature and more importantly suggests separate auditory and motor representations of speech. Such a dual nature model of speech better fits much of the behavioral evidence and would provide a substrate for mechanisms for disambiguation of speech sounds, as well as for speech learning via imitation.

The foregoing findings can be taken as globally indicative of an ongoing interplay between neural information processing resulting from stimulus energy decoding as information ascends hierarchical processing pathways from the periphery to the cortex, interacting with corticofugal mechanisms which influence the processing at each stage (Winer 2006). Speech and music processing research have both come to parallel conclusions regarding top-down influences from frontal cortex to auditory areas in general, and auditory–motor interactions in particular (Hickok et al. 2003; Warren et al. 2005b; Zatorre et al. 2007). However, despite the existence of quite detailed knowledge about the extensive afferent and efferent connections in the mammalian auditory neuraxis, the precise pathways and mechanisms mediating these interactions in the human brain remain largely unknown, and hence provide fertile ground for future research.

2.2 Voice

An interesting and challenging aspect of speech is that not only is the signal itself complex, but it arises from an equally complex substrate: the human voice. Findings in functional imaging studies on voice information processing are in general accord with the idea of a ventral processing pathway representing information about auditory objects. For voice, the object in question is not the message content, but the acoustic features associated with the origin of the vocal information, the speaker. Information about speaker identity is both relevant to but independent from the message being conveyed. Identity here can refer to general characteristics of the speaker (gender, size, age), or to a specific individual. Invariant acoustical features in the vocal signature are available to recover identity information, but accessing these requires a process of abstraction, since every vocalization is likely to differ in each instance. Thus, the auditory cognitive system must contain mechanisms able to compute common features across multiple instances of a vocalization from the same speaker, such that one can identify the speaker upon hearing a novel utterance.

The discovery of voice-sensitive regions in the STS was the first step in understanding voice processing (Fig. 31.7). By contrasting stimulus sets of a variety of vocalizations to environmental sounds without voices, it was shown that several regions along the upper bank of the STS in both hemispheres were sensitive to vocal information (Belin et al. 2000; Belin and Zatorre 2003). A high degree of vocal selectivity was also seen in central STS regions, which preferred vocal sounds to matched control stimuli, including scrambled voices and amplitude-modulated noise. This showed that voice information processing in STS areas, particularly in the right hemisphere, could be dissociated from speech processing. A highly convergent fMRI finding was reported in a study which manipulated subjects' attention either to the speech content or to the speaker identity of an utterance. When attention was directed toward the speaker, rather than the linguistic content, activity in the right STS was observed, resembling that noted in the prior studies, whereas the reverse contrast elicited left STS activity (von Kriegstein et al. 2003).

Subsequent studies have repeatedly confirmed the principal finding of a voice-sensitive region in the upper bank of the STS (Belin et al. 2004; Warren et al. 2006). What remains open is how these voice-sensitive regions fit into the larger picture of ventral stream processing. Are they a special adaptation of the auditory cognitive system to voices due to their

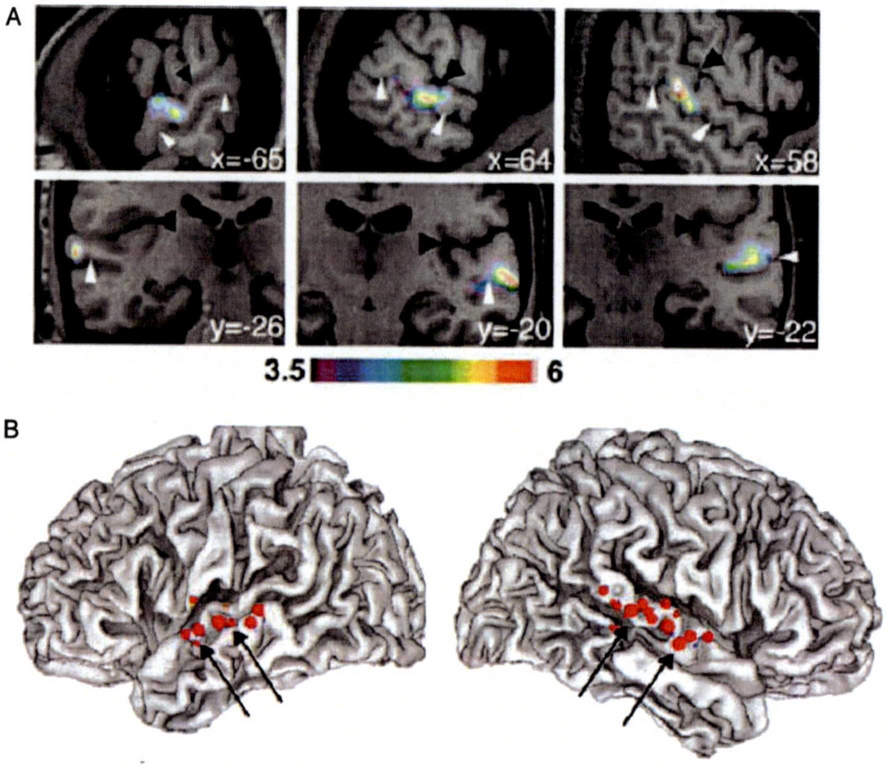

Fig. 31.7 Voice-selective responses. A representative individual fMRI data showing hemodynamic responses to vocal stimuli compared to a variety of nonvocal stimuli (*top row*: sagittal views; *bottom row*: coronal views). **b** Cluster analysis of vocal responses projected onto reconstruction of cortical surface in both hemispheres. In both cases vocal responses often fall within the superior temporal sulcus. Modified from the original source (Belin et al. 2000, 2004)

ecological and evolutionary importance, or are they a product of more general sound-source identification mechanisms, operative for any type of sound category?

3 Summary and Conclusions

The cognitive neuroscience of higher order auditory processing has advanced enormously in a brief time, in large part from neuroimaging approaches. Yet the integration of the many findings remains more a desired goal than an achieved state. Here, we have done no more than outline a series of converging lines of evidence that we hope will help to systematize this large and growing body of information. Among the challenges posed specifically by the study of music and speech is that although the auditory system is their gateway, they clearly involve many other processes that make demands on systems beyond auditory cortex. Hence, we have emphasized the theme that we have much to gain from understanding large-scale interactions between unimodal auditory mechanisms and the rest of the nervous system. The classic approach to studying auditory cortical responses, both in neurophysiology but also in neuroimaging, has typically involved varying some set of stimulus parameters and recording neuronal activity in order to derive a response function. Such an approach has served the field very well and yielded key insights into the functioning of auditory cortex. But we would argue that to understand the neural events involved in speech and music processing it is also necessary to take into account a variety of other factors that do not typically form part of the classic paradigm. The evidence presented in this contribution indicates that cortical responses cannot be predicted simply by knowing all about the acoustical features of a stimulus. Rather, it requires one to know something about the history of the interaction between the listener and that stimulus; about the current status of the stimulus (relative to other stimuli, or relative to internal states); and about the listener's expectations, or future intentions toward the stimulus. These concepts, we feel, will continue to expand and inform more advanced models as the field of cognitive neuroimaging of sound processing continues to grow and evolve.

References

Ahveninen J, Jaaskelainen IP, Raij T, Bonmassar G, Devore S, Hamalainen M, Levanen S, Lin FH, Sams M, Shinn-Cunningham BG, Witzel T, and Belliveau JW (2006) Task-modulated "what" and "where" pathways in human auditory cortex. *Proceedings of the National Academy of Sciences of the United States of America* **103**:14608–14613.

Attneave F and Olson RK (1971) Pitch as a medium: a new approach to psychophysical scaling. *American Journal of Psychology* **84**: 147–166.

Bangert M, Peschel T, Schlaug G, Rotte M, Drescher D, Hinrichs H, Heinze HJ, and Altenmuller E (2006) Shared networks for auditory and motor processing in professional pianists: evidence from fMRI conjunction. *NeuroImage* **30**:917–926.

Baumann S, Koeneke S, Meyer M, Lutz K, and Jancke L (2005) A network for sensory-motor integration: what happens in the auditory cortex during piano playing without acoustic feedback? *Annals of the New York Academy of Sciences* **1060**:186–188.

Baumgart F, Gaschler-Markefski B, Woldorff M, Heinze H-J, and Scheich H (1999) A movement-sensitive area in auditory cortex. *Nature* **400**:724–725.

Belin P and Zatorre RJ (2000) What, where, and how in auditory cortex. *Nature Neuroscience* **3**:965–966.

Belin P and Zatorre RJ (2003) Adaptation to speaker's voice in right anterior temporal lobe. *Neuroreport* **14**:2105–2109.

Belin P, Fecteau S, and Bédard C (2004) Thinking the voice: neural correlates of voice perception *Trends in Cognitive Sciences* **8**: 129–135.

Belin P, Zatorre RJ, Hoge R, Evans AC, and Pike B (1999) Event-related fMRI of the auditory cortex. *NeuroImage* **10**:417–429.

Belin P, Zatorre RJ, Lafaille P, Ahad P, and Pike B (2000) Voice-selective areas in human auditory cortex. *Nature* **403**:309–312.

Belin P, Zilbovicius M, Crozier S, Thivard L, Fontaine A, Masure M-C, and Samson Y (1998) Lateralization of speech and auditory temporal processing. *Journal of Cognitive Neuroscience* **10**: 536–540.

Bendor D and Wang X (2005) The neuronal representation of pitch in primate auditory cortex. *Nature* **436**:1161.

Bey C and McAdams S (2002) Schema-based processing in auditory scene analysis. *Perception & Psychophysics* **64**:844–854.

Binder J, Frost J, Hammeke T, Cox R, Rao S, and Prieto T (1997) Human brain language areas identified by functional magnetic resonance imaging. *Journal of Neuroscience* **17**:353–362.

Binder J, Frost J, Hammeke T, Bellgowan P, Springer J, Kaufman J, and Possing J (2000) Human temporal lobe activation by speech and nonspeech sounds. *Cerebral Cortex* **10**:512–528.

Boemio A, Fromm S, Braun A, and Poeppel D (2005) Hierarchical and asymmetric temporal sensitivity in human auditory cortices. *Nature Neuroscience* **8**:389.

Brechmann A and Scheich H (2005) Hemispheric shifts of sound representation in auditory cortex with conceptual listening. *Cerebral Cortex* **15**:578–587.

Bregman A (1990) *Auditory Scene Analysis*. MIT Press, Cambridge.

Burton MW, Small SL, and Blumstein SE (2000) The role of segmentation in phonological processing: an fMRI investigation. *Journal of Cognitive Neuroscience* **12**:679.

Calvert GA, Bullmore ET, Brammer MJ, Campbell R, Williams SCR, McGuire PK, Woodruff PWR, Iversen SD, and David AS (1997) Activation of auditory cortex during silent lipreading. *Science* **276**:593–596.

Champod AS and Petrides M (2007) Dissociable roles of the posterior parietal and the prefrontal cortex in manipulation and monitoring processes. *Proc Natl Acad Sci U S A* **104**(37): 14837–14842.

Chen JL, Zatorre RJ, and Penhune VB (2006) Interactions between auditory and dorsal premotor cortex during synchronization to musical rhythms. *NeuroImage* **32**:1771–1781.

Cherry C (1957) *Human Communication*. Wiley, London.

Cohen MA, Grossberg S, and Wyse LL (1995) A spectral network model of pitch perception. *Journal of the Acoustical Society of America* **98**:862–879.

Culham JC, Cavina-Pratesi C, and Singhal A (2006) The role of parietal cortex in visuomotor control: what have we learned from neuroimaging?. *Neuropsychologia* **44**(13):2668–2684.

Davis MH and Johnsrude IS (2003) Hierarchical processing in spoken language comprehension. *Journal of Neuroscience*. **23**:3423–3431.

Dehaene-Lambertz G, Pallier C, Serniclaes W, Sprenger-Charolles L, Jobert A, and Dehaene S (2005) Neural correlates of switching from auditory to speech perception. *NeuroImage* **24**:21–33.

Desimone R (1998) Visual attention mediated by biased competition in extrastriate visual cortex. *Philosophical Transactions of the Royal Society of London, Series B, Biological Sciences* **353**: 1245–1255.

Doeller CF, Opitz B, Mecklinger A, Krick C, Reith W, and Schröger E (2003) Prefrontal cortex involvement in preattentive auditory deviance detection: neuroimaging and electrophysiological evidence. *NeuroImage* **20**:1270–1282.

Dowling WJ (1973) Rhythmic groups and subjective chunks in memory for melodies. *Perception & Psychophysics* **14**:37–40.

Dowling WJ and Harwood D (1986) *Music Cognition*. Academic Press, Orlando.

Fishman YI, Arezzo JC, and Steinschneider M (2004) Auditory stream segregation in monkey auditory cortex: effects of frequency separation, presentation rate, and tone duration. *Journal of the Acoustical Society of America* **116**:1656–1670.

Fishman YI, Reser DH, Arezzo JC, and Steinschneider M (2001) Neural correlates of auditory stream segregation in primary auditory cortex of the awake monkey. *Hearing Research* **151**:167–187.

Formisano E, Kim D-S, Di Salle F, van de Moortele P-F, Ugurbil K, and Goebel R (2003) Mirror-symmetric tonotopic maps in human primary auditory cortex. *Neuron* **40**:859–869.

Foster NE and Zatorre RJ (2009) A role for the intraparietal sulcus in transforming musical pitch information. *Cerebral Cortex* (Epub doi:10.1093/cercor/bhp199).

Friederici AD, Ruschemeyer S-A, Hahne A, and Fiebach CJ (2003) The role of left inferior frontal and superior temporal cortex in sentence comprehension: localizing syntactic and semantic processes. *Cerebral Cortex* **13**:170–177.

Fritz J, Elhilali M, and Shamma S (2007) Adaptive changes in cortical receptive fields induced by attention to complex sounds. *Journal of Neurophysiology* **98**:2337–2346

Fritz J, Shamma S, Elhilali M, and Klein D (2003) Rapid task-related plasticity of spectrotemporal receptive fields in primary auditory cortex. *Nature Neuroscience* **6**:1216–1223.

Fritz JB, Elhilali M, and Shamma SA (2005) Differential dynamic plasticity of a1 receptive fields during multiple spectral tasks. *Journal of Neuroscience* **25**:7623–7635.

Gaab N, Gaser C, Zaehle T, Jäncke L, and Schlaug G (2003) Functional anatomy of pitch memory—an fMRI study with sparse temporal sampling. *NeuroImage* **19**:1417–1426.

Goldstein JL (1973) An optimum processor theory for the central formation of the pitch of complex tones. *Journal of the Acoustical Society of America* **54**:1496–1516.

Golestani N and Zatorre RJ (2004) Learning new sounds of speech: reallocation of neural substrates. *NeuroImage* **21**:494–506.

Grady CL, Van Meter JW, Maisog JM, Pietrini P, Krasuski J, and Rauschecker JP (1997) Attention-related modulation of activity in primary and secondary auditory cortex. *Neuroreport* **8**:2511–2516.

Griffiths TD (2000) Musical hallucinosis in acquired deafness. Phenomenology and brain substrate. *Brain* **123**:2065–2076.

Griffiths TD and Warren JD (2002) The planum temporale as a computational hub. *Trends in Neuroscience* **25**:348–353.

Griffiths TD, Buchel C, Frackowiak RJS, and Patterson RD (1998) Analysis of temporal structure in sound by the human brain. *Nature Neuroscience* **1**:422–427.

Griffiths TD, Johnsrude IS, Dean JL, and Green GGR (1999) A common neural substrate for the analysis of pitch and duration pattern in segmented sound? *Neuroreport* **10**:3825–3830.

Gutschalk A, Patterson RD, Rupp A, Uppenkamp S, and Scherg M (2002) Sustained magnetic fields reveal separate sites for sound level and temporal regularity in human auditory cortex. *NeuroImage* **15**:207–216.

Gutschalk A, Patterson RD, Scherg M, Uppenkamp S, and Rupp A (2004) Temporal dynamics of pitch in human auditory cortex. *NeuroImage* **22**:755–766.

Gutschalk A, et al. (2005) Neuromagnetic correlates of streaming in human auditory cortex. *Journal of Neuroscience* **25**:5382–5388.

Hall D, Haggard M, Akeroyd M, Palmer A, Summerfield A, Elliott M, Gurney E, and Bowtell R (1999) "Sparse" temporal sampling in auditory fMRI. *Human Brain Mapping* **7**:213–223.

Hall DA, Johnsrude IS, Haggard MP, Palmer AR, Akeroyd MA, and Summerfield AQ (2002) Spectral and temporal processing in human auditory cortex. *Cerebral Cortex* **12**:140–149.

Halpern AR and Zatorre RJ (1999) When that tune runs through your head: a PET investigation of auditory imagery for familiar melodies. *Cerebral Cortex* **9**:697–704.

Halpern AR, Zatorre RJ, Bouffard M, and Johnson JA (2004) Behavioral and neural correlates of perceived and imagined musical timbre. *Neuropsychologia* **42**:1281–1292.

Hart HC, Palmer AR, and Hall DA (2003) Amplitude and frequency-modulated stimuli activate common regions of human auditory cortex. *Cerebral Cortex* **13**:773–781.

Hernández-Peón R, Scherrer H, and Jouvet M (1956) Modification of electric activity in cochlear nucleus during "attention" in unanesthetized cats. *Science* **123**:331–332.

Hickok G and Poeppel D (2004) Dorsal and ventral streams: a framework for understanding aspects of the functional anatomy of language. *Cognition* **92**:67–99.

Hickok G, Buchsbaum B, Humphries C, and Muftuler T (2003) Auditory-motor interaction revealed by fMRI: speech, music, and working memory in area Spt. *Journal of Cognitive Neuroscience* **15**:673–682.

Hillyard S, Hink R, Schwent V, and Picton T (1973) Electrical signs of selective attention in the human brain. *Science* **182**:177–180.

Hubel DH, Henson CO, Rupert A, and Galambos R (1959) Attention units in the auditory cortex. *Science* **129**:1279–1280.

Hugdahl K, Bronnick K, Kyllingsbaek S, Law I, Gade A, and Paulson O (1999) Brain activation during dichotic presentations of consonant-vowel and musical instrument stimuli: a 15o-pet study. *Neuropsychologia* **37**:431–440.

Huron D (2006) *Sweet Anticipation. Music and the Psychology of Expectation.* MIT Press, Cambridge.

Hyde KL. Peretz I, and Zatorre RJ (2008) Evidence for the role of the right auditory cortex in fine pitch resolution. *Neuropsychologia* **46**:632–639.

Jamison HL, Watkins KE, Bishop DVM, and Matthews PM (2006) Hemispheric specialization for processing auditory nonspeech stimuli. *Cerebral Cortex* **16**:1266–1275.

Jäncke L, Mirzazade S, and Shah NJ (1999) Attention modulates activity in the primary and the secondary auditory cortex: a functional magnetic resonance imaging study in human subjects. *Neuroscience Letters* **266**:125–128.

Joanisse MF and Gati JS (2003) Overlapping neural regions for processing rapid temporal cues in speech and nonspeech signals. *NeuroImage* **19**:64.

Joanisse MF, Zevin JD, and McCandliss BD (2007) Brain mechanisms implicated in the preattentive categorization of speech sounds revealed using fMRI and a short-interval habituation trial paradigm. *Cerebral Cortex* **17**:2084–2093.

Johnson JA and Zatorre RJ (2005) Attention to simultaneous unrelated auditory and visual events: behavioral and neural correlates. *Cerebral Cortex* **15**:1609–1620.

Johnson JA and Zatorre RJ (2006) Neural substrates for dividing and focusing attention between simultaneous auditory and visual events. *NeuroImage* **31**:1673–1681.

Johnsrude IS, Penhune VB, and Zatorre RJ (2000) Functional specificity in the right human auditory cortex for perceiving pitch direction. *Brain* **123**:155–163.

Kaas JH and Hackett TA (1998) Subdivisions of auditory cortex and levels of processing in primates. *Audiology & Neurootology* **3**: 73–85.

Kaas JH and Hackett TA (2000) Subdivisions of auditory cortex and processing streams in primates. *Proceedings of the National Academy of Sciences of the United States of America* **97**: 11793–11799.

Kaas JH, Hackett TA, and Tramo MJ (1999) Auditory processing in primate cerebral cortex. *Current Opinion in Neurobiology* **9**: 164–170.

Klein DJ, Simon JZ, Depireux DA, and Shamma SA (2006) Stimulus-invariant processing and spectrotemporal reverse correlation in primary auditory cortex. *Journal of Computational Neuroscience* **20**:111–136.

Koelsch S (2005) Neural substrates of processing syntax and semantics in music. *Current Opinion in Neurobiology* **15**:207–212.

Kowalski N, Depireux DA, and Shamma SA (1996) Analysis of dynamic spectra in ferret primary auditory cortex. I. Characteristics of single-unit responses to moving ripple spectra. *J Neurophysiol* **76**:3503–3523.

Kraus N, McGee T, Carrell T, King C, Tremblay K, and Nicol T (1995) Central auditory system plasticity associated with speech discrimination training. *Journal of Cognitive Neuroscience* **7**:25–32.

Krumbholz K, Patterson RD, Seither-Preisler A, Lammertmann C, and Lütkenhöner B (2003) Neuromagnetic evidence for a pitch processing center in Heschl's gyrus. *Cerebral Cortex* **13**: 765–772.

Krumbholz K, Schönwiesner M, von Cramon DY, Rübsamen R, Shah NJ, Zilles K, and Fink GR (2005) Representation of interaural temporal information from left and right auditory space in the human planum temporale and inferior parietal lobe. *Cerebral Cortex* **15**:317–324.

Krumhansl CL (1990) *Cognitive Foundations of Musical Pitch.* Oxford University Press, New York.

Lahav A, Saltzman E, and Schlaug G (2007) Action representation of sound: audiomotor recognition network while listening to newly acquired actions. *Journal of Neuroscience* **27**:308–314.

Laurienti PJ, Burdette JH, Wallace MT, Yen Y-F, Field AS, and Stein BE (2002) Deactivation of sensory-specific cortex by cross-modal stimuli. *Journal of Cognitive Neuroscience* **14**:420–429.

Leopold DA and Logothetis NK (1999) Multistable phenomena: changing views in perception. *Trends in Cognitive Sciences* **3**:254–264.

Lewis JW, Beauchamp MS, and DeYoe EA (2000) A comparison of visual and auditory motion processing in human cerebral cortex. *Cerebral Cortex* **10**:873–888.

Liberman AM and Mattingly IG (1985) The motor theory of speech perception revised. *Cognition* **21**:1–36.

Liebenthal E, Binder JR, Piorkowski R, and Remez R (2003a) Short-term reorganization of auditory analysis induced by phonetic experience. *Journal of Cognitive Neuroscience* **15**:549–558.

Liebenthal E, Binder JR, Spitzer SM, Possing ET, and Medler DA (2005) Neural substrates of phonemic perception. *Cerebral Cortex* **15**:1621–1631.

Liebenthal E, Ellingson ML, Spanaki MV, Prieto TE, Ropella KM, and Binder JR (2003b) Simultaneous ERP and fMRI of the auditory cortex in a passive oddball paradigm. *NeuroImage* **19**:1395–1404.

Linden JF, Liu RC, Sahani M, Schreiner CE, and Merzenich MM (2003) Spectrotemporal structure of receptive fields in areas AI and AAF of mouse auditory cortex. *Journal of Neurophysiology* **90**:2660–2675.

Logothetis NK and Wandell BA (2004) Interpreting the bold signal. *Annual Review of Physiology* **66**:735–769.

Marco-Pallarés J, Grau C, and Ruffini G (2005) Combined ICA-LORETA analysis of mismatch negativity. *NeuroImage* **25**:471–477.

Merzenich MM and Brugge JF (1973) Representation of the cochlear partition on the superior temporal plane of the macaque monkey. *Brain Research* **60**:315–333.

Micheyl C, Tian B, Carlyon RP, and Rauschecker JP (2005) Perceptual organization of tone sequences in the auditory cortex of awake macaques. *Neuron* **48**:139–148.

Milner A and Goodale M (1995) *The Visual Brain in Action*. Oxford University Press, Oxford.

Molholm S, Martinez A, Ritter W, Javitt DC, and Foxe JJ (2005) The neural circuitry of pre-attentive auditory change-detection: an FMRI study of pitch and duration mismatch negativity generators. *Cerebral Cortex* **15**:545–551.

Moore BCJ and Gockel H (2002) Factors influencing sequential stream segregation. *Acta Acustica/Acustica* **88**:320–333.

Morel A and Kaas JH (1992) Subdivisions and connections of auditory cortex in owl monkeys. *Journal of Comparative Neurology* **318**: 27–63.

Morel A, Garraghty PE, and Kaas JH (1993) Tonotopic organization, architectonic fields, and connections of auditory cortex in macaque monkeys. *Journal of Comparative Neurology* **335**:437–459.

Möttönen R, Calvert GA, Jääskeläinen IP, Matthews PM, Thesen T, Tuomainen J, and Sams M (2006) Perceiving identical sounds as speech or non-speech modulates activity in the left posterior superior temporal sulcus. *NeuroImage* **30**:563–569.

Mukamel R, Gelbard H, Arieli A, Hasson U, Fried I, and Malach R (2005) Coupling between neuronal firing, field potentials, and fMRI in human auditory cortex. *Science* **309**:951–954.

Muller BW, Juptner M, Jentzen W, and Muller SP (2002) Cortical activation to auditory mismatch elicited by frequency deviant and complex novel sounds: a PET study. *NeuroImage* **17**:231–239.

Münte TF, Altenmüller E, and Jäncke L (2002) The musician's brain as a model of neuroplasticity. *Nature Reviews Neuroscience* **3**: 473–478.

Näätänen R (1995) The mismatch negativity: a powerful tool for cognitive neuroscience. *Ear & Hearing* **16**:6–18.

Näätänen R, Gaillard AW, and Mäntysalo S (1978) Early selective-attention effect on evoked potential reinterpreted. *Acta Psychologica (Amsterdam)* **42**:313–329.

Näätänen R, Lehtokoski A, Lennes M, Cheour M, Huotilainen M, Livonen A, Vainio M, Alku P, Ilmoniemi R, Luuk A, Allik J, Sinkkonen J, and Alho K (1997) Language-specific phoneme representations revealed by electric and magnetic brain responses. *Nature* **385**:432–434.

Narain C, Scott SK, Wise RJS, Rosen S, Leff A, Iversen SD, and Matthews PM (2003) Defining a left-lateralized response specific to intelligible speech using fMRI. *Cerebral Cortex* **13**: 1362–1368.

Obleser J, Zimmermann J, Van Meter J, and Rauschecker JP (2007) Multiple stages of auditory speech perception reflected in event-related FMRI. *Cerebral Cortex* **17**:2251–2257.

Opitz B, Mecklinger A, von Cramon DY, and Kruggel F (1999) Combining electrophysiological and hemodynamic measures of the auditory oddball. *Psychophysiology* **36**:142–147.

Opitz B, Rinne T, Mecklinger A, von Cramon DY, and Schröger E (2002) Differential contribution of frontal and temporal cortices to auditory change detection: fMRI and ERP results. *NeuroImage* **15**:167–174.

Oxenham AJ, Bernstein JG, and Penagos H (2004) Correct tonotopic representation is necessary for complex pitch perception. *Proceedings of the National Academy of Sciences of the United States of America* **101**:1421–1425.

Pandya DN and Sanides F (1973) Architectonic parcellation of the temporal operculum in rhesus monkey and its projection pattern. *Zeitschrift für Anatomie und Entwicklungsgeschichte* **139**:127–161.

Pantev C, Hoke M, Lütkenhöner B, and Lehnertz K (1989) Tonotopic organization of the auditory cortex: pitch versus frequency representation. *Science* **246**:486–488.

Pantev C, Engelien A, Candia V, and Elbert T (2003) Representational cortex in musicians. In: Peretz I and Zatorre R (eds). *The Cognitive Neuroscience of Music*. Oxford University Press, Oxford, pp. 382–395.

Patterson RD, Allerhand MH, and Giguère C (1995) Time-domain modeling of peripheral auditory processing: a modular architecture and a software platform. *Journal of the Acoustical Society of America* **98**:1890–1894.

Patterson RD, Uppenkamp S, Johnsrude IS, and Griffiths TD (2002) The processing of temporal pitch and melody information in auditory cortex. *Neuron* **36**:767–776.

Pavani F, Macaluso E, Warren JD, Driver J, and Griffiths TD (2002) A common cortical substrate activated by horizontal and vertical sound movement in the human brain. *Current Biology* **12**: 1584–1590.

Penagos H, Melcher JR, and Oxenham AJ (2004) A neural representation of pitch salience in nonprimary human auditory cortex revealed with functional magnetic resonance imaging. *Journal of Neuroscience* **24**:6810–6815.

Penhune VB, Zatorre RJ, and Evans AC (1998) Cerebellar contributions to motor timing: a PET study of auditory and visual rhythm reproduction. *Journal of Cognitive Neuroscience* **10**:752–765.

Perry DW, Zatorre RJ, Petrides M, Alivisatos B, Meyer E, and Evans AC (1999) Localization of cerebral activity during simple singing. *Neuroreport* **10**:3979–3984.

Petkov CI, Kayser C, Augath M, and Logothetis NK (2006) Functional imaging reveals numerous fields in the monkey auditory cortex. *Public Library of Science Biology* **4**:1213–1226.

Petkov CI, Kang X, Alho K, Bertrand O, Yund EW, and Woods DL (2004) Attentional modulation of human auditory cortex. *Nature Neuroscience* **7**:658–663.

Phillips DP and Farmer ME (1990) Acquired word deafness and the temporal grain of sound representation in the primary auditory cortex. *Behavioral Brain Research* **40**:84–90.

Poeppel D (2003) The analysis of speech in different temporal integration windows: cerebral lateralization as 'asymmetric sampling in time.' *Speech Communication* **41**:245–255.

Pressnitzer D and Hupé JM (2006) Temporal dynamics of auditory and visual bistability reveal common principles of perceptual organization. *Current Biology* **16**:1351–1357.

Pugh KR, Shaywitz BA, Shaywitz SE, Fulbright RK, Byrd D, Skudlarski P, Shankweiler DP, Katz L, Constable RT, Fletcher J, Lacadie C, Marchione K, and Gore JC (1996) Auditory selective attention: an fMRI investigation. *NeuroImage* **4**: 159–173.

Rauschecker JP and Tian B (2000) Mechanisms and streams for processing of "what" and "where" In auditory cortex. *Proceedings of the National Academy of Sciences of the United States of America* **97**:11800–11806.

Read HL, Winer JA, and Schreiner CE (2002) Functional architecture of auditory cortex. *Current Opinion in Neurobiology* **12**:433–440.

Remez RE, Rubin PE, Pisoni DB, and Carrell TD (1981) Speech perception without traditional speech cues. *Science* **212**: 947–949.

Rinne T, Degerman A, and Alho K (2005) Superior temporal and inferior frontal cortices are activated by infrequent sound duration decrements: an fMRI study. *NeuroImage* **26**:66–72.

Rizzolatti G and Craighero L (2004) The mirror-neuron system. *Annual Reviews of Neuroscience* **27**:169–192.

Romanski L, Tian B, Fritz J, Mishkin M, Goldman-Rakic P, and Rauschecker J (2000) Reply to "'what', 'where' and 'how' in auditory cortex." *Nature Neuroscience* **10**:966.

Schneider P, Scherg, Dosch G, Specht H, and Gutschalk A (2002) Morphology of Heschl's gyrus reflects enhanced activation in the auditory cortex of musicians. *Nature Neuroscience* **5**:688–694.

Schönwiesner M and Zatorre RJ (2009) Spectro-temporal modulation transfer function of single voxels in the human auditory cortex measured with high-resolution fMRI. *Proceedings of the*

National Academy of Sciences of the United States of America **106**: 14611–14616.

Schönwiesner M, von Cramon DY, and Rübsamen R (2002) Is it tonotopy after all? *NeuroImage* **17**:1144–1161.

Schönwiesner M, Rübsamen R, and von Cramon DY (2005) Hemispheric asymmetry for spectral and temporal processing in the human antero-lateral auditory belt cortex. *European Journal of Neuroscience* **22**:1521–1528.

Schönwiesner M, Krumbholz K, Rübsamen R, Fink GR, and von Cramon DY (2007a) Hemispheric asymmetry for auditory processing in the human auditory brain stem, thalamus, and cortex. *Cerebral Cortex* **17**:492–499.

Schönwiesner M, Novitski N, Pakarinen S, Carlson S, Tervaniemi M, and Näätänen R (2007b) Heschl's gyrus, posterior superior temporal gyrus, and mid-ventrolateral prefrontal cortex have different roles in the detection of acoustic changes. *Journal of Neurophysiology* **97**:2075–2082.

Scott SK and Johnsrude IS (2003) The neuroanatomical and functional organization of speech perception. *Trends in Neurosciences* **26**: 100–107.

Scott SK, Blank CC, Rosen S, and Wise RJS (2000) Identification of a pathway for intelligible speech in the left temporal lobe. *Brain* **123**:2400–2406.

Sen K, Theunissen FE, and Doupe AJ (2001) Feature analysis of natural sounds in the songbird auditory forebrain. *Journal of Neurophysiology* **86**:1445–1458.

Shamma S (2004) Topographic organization is essential for pitch perception. *Proceedings of the National Academy of Sciences of the United States of America* **101**:1114–1115.

Shepard RN (1982) Geometrical approximations to the structure of musical pitch. *Psychological Reviews* **89**:305–333.

Singh PG (1987) Perceptual organization of complex-tone sequences: a tradeoff between pitch and timbre?. *Journal of the Acoustical Society of America* **82**:886–899.

Snyder JS and Alain C (2007) Toward a neurophysiological theory of auditory stream segregation. *Psychological Bulletin* **133**:780–799.

Stewart L, von Kriegstein K, Warren JD, and Griffiths TD (2006) Music and the brain: disorders of musical listening. *Brain* **129**: 2533–2553.

Swisher L and Hirsh IJ (1972) Brain damage and the ordering of two temporally successive stimuli. *Neuropsychologia* **10**:137–152.

Talavage T, Sereno M, Melcher J, Ledden P, Rosen B, and Dale A (2004) Tonotopic organization in human auditory cortex revealed by progressions of frequency sensitivity. *Journal of Neurophysiology* **91**:1282–1296.

Talavage TM, Ledden PJ, Benson RR, Rosen BR, and Melcher JR (2000) Frequency-dependent responses exhibited by multiple regions in human auditory cortex. *Hearing Research* **150**: 225–244.

Tallal P, Miller S and Fitch R (1993) Neurobiological basis of speech: a case for the preeminence of temporal processing. *Annals of the New York Academy of Sciences* **682**:27–47.

Tervaniemi M, Medvedev S, Alho K, Pakhomov S, Roudas M, van Zuijen T, and Näätänen R (2000) Lateralized automatic auditory processing of phonetic versus musical information: a pet study. *Human Brain Mapping* **10**:74–79.

Thivard L, Belin P, Zilbovicius M, Poline J, and Samson Y (2000) A cortical region sensitive to auditory spectral motion. *Neuroreport* **11**:2969–2972.

Tillmann B, Koelsch S, Escoffier N, Bigand E, Lalitte P, Friederici AD, and von Cramon DY (2006) Cognitive priming in sung and instrumental music: activation of inferior frontal cortex. *NeuroImage* **31**:1771–1782.

Tzourio N, Massioui FE, Crivello F, Joliot M, Renault B, and Mazoyer B (1997) Functional anatomy of human auditory attention studied with PET. *NeuroImage* **5**:63–77.

Ungerleider L and Haxby J (1994) "What" and "where" in the human brain. *Current Biology* **4**:157–165.

van Noorden LPAS (1975) Temporal coherence in the perception of tone sequences. University of Technology, Eindhoven.

Versnel H and Shamma SA (1998) Spectral-ripple representation of steady-state vowels in primary auditory cortex. *Journal of the Acoustical Society of America* **103**:2502–2514.

Vliegen J and Oxenham AJ (1999) Sequential stream segregation in the absence of spectral cues. *Journal of the Acoustical Society of America* **105**:339–346.

Voisin J, Bidet-Caulet A, Bertrand O, and Fonlupt P (2006) Listening in silence activates auditory areas: a functional magnetic resonance imaging study. *Journal of Neuroscience* **26**:273–278.

von Kriegstein K, Eger E, Kleinschmidt A, and Giraud A-L (2003) Modulation of neural responses to speech by directing attention to voices or verbal content. *Cognitive Brain Research* **17**:48–55.

Warren JD, Uppenkamp S, Patterson RD, and Griffiths TD (2003) Separating pitch chroma and pitch height in the human brain. *Proceedings of the National Academy of Sciences of the United States of America* **100**:10038–10042.

Warren JD, Zielinski BA, Green GG, Rauschecker JP, and Griffiths TD (2002) Perception of sound-source motion by the human brain. *Neuron* **34**:139–148.

Warren JD, Jennings AR, and Griffiths TD (2005a) Analysis of the spectral envelope of sounds by the human brain. *NeuroImage* **24**:1052.

Warren JD, Scott SK, Price CJ, and Griffiths TD (2006) Human brain mechanisms for the early analysis of voices. *NeuroImage* **31**: 1389–1397.

Warren JE, Wise RJ, and Warren JD (2005b) Sounds do-able: auditory-motor transformations and the posterior temporal plane. *Trends in Neuroscience* **28**:636–643.

Watkins S, Dalton P, Lavie N, and Rees G (2007) Brain mechanisms mediating auditory attentional capture in humans. *Cerebral Cortex* **17**:1694–1700.

Wilson EC, Melcher JR, Micheyl C, Gutschalk A, and Oxenham AJ (2007) Cortical fMRI activation to sequences of tones alternating in frequency: relationship to perceived rate and streaming. *Journal of Neurophysiology* **97**:2230–2238.

Wilson SM and Iacoboni M (2006) Neural responses to non-native phonemes varying in producibility: evidence for the sensorimotor nature of speech perception. *NeuroImage* **33**:316–325.

Wilson SM, Saygin AP, Sereno MI, and Iacoboni M (2004) Listening to speech activates motor areas involved in speech production. *Nature Neuroscience* **7**:701–702.

Winer JA (2006) Decoding the auditory corticofugal system. *Hearing Research* **212**:1–8.

Woldorff MG, Gallen CC, Hampson SA, Hillyard SA, Pantev C, Sobel D, and Bloom FE (1993a) Modulation of early sensory processing in human auditory cortex during selective attention. *Proceedings of the National Academy of Sciences of the United States of America* **90**:8722–8726.

Woldorff MG, Gallen CC, Hampson SA, Hillyard SA, Pantev C, Sobel D, and Bloom FE (1993b) Modulation of early sensory processing in human auditory cortex during auditory selective attention. *Proceedings of the National Academy of Sciences of the United States of America* **90**:8722–8726.

Woolley SM, Fremouw TE, Hsu A, and Theunissen FE (2005) Tuning for spectro-temporal modulations as a mechanism for auditory discrimination of natural sounds. *Nature Neuroscience* **8**:1371–1379.

Yost WA, Patterson R, and Sheft S (1996) A time domain description for the pitch strength of iterated rippled noise. *Journal of the Acoustical Society of America* **99**:1066–1078.

Zacks JM (2008) Neuroimaging studies of mental rotation: a meta-analysis and review. *Journal of Cognitive Neuroscience* **20**:1–19.

Zaehle T, Wustenberg T, Meyer M, and Jancke L (2004) Evidence for rapid auditory perception as the foundation of speech processing:

a sparse temporal sampling fMRI study. *European Journal of Neuroscience* **20**:2447–2456.

Zatorre RJ (1988) Pitch perception of complex tones and human temporal-lobe function. *Journal of the Acoustical Society of America* **84**:566–572.

Zatorre RJ and Halpern AR (1993) Effect of unilateral temporal-lobe excision on perception and imagery of songs. *Neuropsychologia* **31**:221–232.

Zatorre RJ and Penhune VB (2001) Spatial localization after excision of human auditory cortex. *Journal of Neuroscience* **21**:6321–6328.

Zatorre RJ and Belin P (2001) Spectral and temporal processing in human auditory cortex. *Cerebral Cortex* **11**:946–953.

Zatorre RJ and Halpern AR (2005) Mental concerts: musical imagery and auditory cortex. *Neuron* **47**:9–12.

Zatorre RJ and Gandour JT (2008) Neural specializations for speech and pitch: moving beyond the dichotomies. *Philosophical Transactions of the Royal Society of London, Series B, Biological Sciences* **363**:1087–1104.

Zatorre RJ, Evans AC, and Meyer E (1994) Neural mechanisms underlying melodic perception and memory for pitch. *Journal of Neuroscience* **14**:1908–1919.

Zatorre RJ, Belin P, and Penhune VB (2002a) Structure and function of auditory cortex: Music and speech. *Trends in Cognitive Sciences* **6**:37–46.

Zatorre RJ, Bouffard M, and Belin P (2004) Sensitivity to auditory object features in human temporal neocortex. *Journal of Neuroscience* **24**:3637–3642.

Zatorre RJ, Chen JL, and Penhune VB (2007) When the brain plays music: Auditory-motor interactions in music perception and production. *Nature Reviews Neuroscience* **8**:547–558.

Zatorre RJ, Halpern AR, Bouffard M (2010) Mental reversal of imagined melodies: a role for the posterior parietal cortex. *Journal of Cognitive Neuroscience* **22**:775–789.

Zatorre RJ, Evans AC, Meyer E, and Gjedde A (1992) Lateralization of phonetic and pitch processing in speech perception. *Science* **256**:846–849.

Zatorre RJ, Meyer E, Gjedde A, and Evans AC (1996) Pet studies of phonetic processing of speech: review, replication, and re-analysis. *Cerebral Cortex* **6**:21–30.

Zatorre RJ, Bouffard M, Ahad P, and Belin P (2002b) Where is 'where' in the human auditory cortex? *Nature Neuroscience* **5**: 905–909.

Chapter 32

Toward a Synthesis of Cellular Auditory Forebrain Functional Organization

Jeffery A. Winer and Christoph E. Schreiner

Abbreviations

AAF	anterior auditory field
AC	auditory cortex
AI	primary auditory cortex
GABA	gamma aminobutyric acid
IC	inferior colliculus
MGB	medial geniculate body
TC	thalamocortical

1 Auditory Forebrain Organization

There is no global theory of auditory forebrain function since the facts available cannot support such an edifice. New technologies, some outlined in the previous chapters, have broadened the issues of functional organization and elevated the discussion to more global perspectives. In the following we are not attempting to provide a global synthesis. We rather address some questions preliminary to such a theory with the explicit view from the cellular level.

1.1 Auditory Forebrain Serial Processing

Why are there so many stations in the central auditory system? A crucial issue is the nature and purpose of information transformations at successive hierarchical stages from cochlea to cortex. This process may seem more amenable to experimental scrutiny in the visual sensory pathway, but that accessibility is more apparent than actual as the complexity of lateral geniculate interactions (Sherman 2004) and the intricate cortical circuitry (Miller et al. 2001a; Ohki et al. 2005) demonstrate, and a similar case can be made for the somatic sensory system (Kaas 1983). The cardinal advantage of the auditory system in general and the forebrain in particular is its distributed architecture, which enables analysis of synaptic traffic and serial transformations. A case in point is an approach in which state control of thalamocortical (TC) transmission (Miller and Schreiner 2000) is used to dissect the classes of serial interaction (Miller et al. 2001c). This reveals three fundamental principles: inheritance entails precise information transfer with minimal transformation; constructive convergence permits cell assemblies to interact in mosaic fashion; and ensemble convergence allows differential enhancement of specific feature elements (Miller et al. 2001b). Analysis of pairs of synaptically coupled thalamic and cortical neurons finds that spectral integration properties were comparable (indicating inheritance), whereas cortical temporal modulation rate was lower and uncorrelated to that in thalamus, and cortical binaural, contralateral excitatory cells were almost twice as numerous (Miller et al. 2002). The clustered TC projections (McMullen and de Venecia 1993) and their precise laminar distribution (Huang and Winer 2000) could underlie inheritance, while the binaural transformation may ensue from thalamic (Middlebrooks and Zook 1983) and commissural (Imig and Adrián 1977; Imig and Brugge 1978) interactions (Winer et al. 2005b). Further challenges to understanding the function of auditory cortex are subsequent intracortical transformations within a module (Szentágothai 1975), between small ensembles (Read et al. 2001), across areas (Chapter 7), and in the descending systems (Winer 2006).

1.2 Topography of Projections

How is the differential distribution of many physiological variables (characteristic frequency, binaurality, amplitopic organization, etc.) scaled in MGB and AC (and in other parts

C.E. Schreiner (✉)
Coleman Memorial Laboratory, Department of Otolaryngology, W.M. Keck Center for Integrative Neuroscience, University of California, School of Medicine, San Francisco, CA 94143-0732, USA
e-mail: chris@phy.ucsf.edu

J.A. Winer, C.E. Schreiner (eds.), *The Auditory Cortex*,
DOI 10.1007/978-1-4419-0074-6_32, © Springer Science+Business Media, LLC 2011

of the auditory system)? Perhaps the question is trivial since the connections between (or within) cortex and thalamus are so 'seamless' that only synaptic delays impede information transfer. The answer is less evident considering that different AC representations of the basilar membrane have a several-fold scaling difference, e.g., area AI is about three times larger than the adjoining AAF (anterior auditory field) (Imaizumi et al. 2005). Sensitive retrograde tracers injected in physiologically defined AI and AAF subregions, or without physiological guidance and in non-tonotopic areas, allow comparison of the topography of TC, corticocortical, and commissural cells of origin to be assessed with three simple metrics (separation, dispersion, and clustering). All connections were highly, and equally, ordered, irrespective of their physiologic arrangement, and independent of any basilar membrane representation, and a similar organization prevailed in prefrontal cortex, suggesting that such topographies may be more general (Lee and Winer 2005).

1.3 Scaling the Projection Systems

What is the contribution of TC, corticocortical, and commissural connections to an AC module? Such knowledge might predict their differential numeric impact in shaping output specificity and dynamics. Studies with retrograde tracers in 13 different AC areas found that 70% of the projection neurons were of cortical origin, and the remainders were divided equally between the MGB and the commissural system (Edeline 2003; Lee et al. 2004; Lee and Winer 2005). Interestingly, the thalamic value matches closely the proportion of synaptic input to layer IV in primary visual cortex (LeVay and Gilbert 1976).

1.4 Parallel Descending Pathways

Do the corticofugal systems mainly provide feedback to the ascending system? If so, the descending systems might have perfectly reciprocal connections with the ascending stream; this is not the case (Colwell 1975). If the descending system was phylogenetically older, it should be present in all species; this is not the case (Wild et al. 1993). If the descending system had primarily a reafferent function then its projections might be morphologically uniform and stereotyped from nucleus to nucleus; however, the projections are specific with regard both to origin (Winer et al. 1999a, b) and target (Winer et al. 1998, 2001). If the descending system were a feedback pathway, it might be smaller than the ascending system; this is not the case (Winer 2006). If the descending system provided feedback primarily, one would predict that activating it should have focal and small effects; this is not the case since extensive reorganization of the frequency domain in the targeted structure is possible in certain regimes (Zhang and Suga 2005). If the corticofugal system was unitary and had one role only, it might arise from one cell type and one layer; this is not the case (Winer et al. 2001).

1.5 Subdivisions of Auditory Cortex

Why do so many AC areas exist? Schemes for subdividing rodent (Shi and Cassell 1997), bat (Fitzpatrick et al. 1998), cat (Lee and Winer 2005), primate (Hackett et al. 2001), and human (Morosan et al. 2001) AC describe more than one primary and non-primary area, and at least in bat (Suga 1978), cat (Reale and Imig 1980), and monkey (Morel and Kaas 1992) multiple tonotopic maps exist. Adjoining subregions whose auditory affiliations are debated, such as insular and temporal cortex, likely have different cortical (Clascá et al. 2000) (and even finer local) and corticofugal (Winer et al. 2001) connections and unique roles (Colavita 1974, 1979). Perhaps the different representations subserve unique functions. It seems counterintuitive that AC would have fewer or less specific functional roles than the cochlear nucleus (Warr 1982) or IC (Casseday et al. 2005).

Does the same area have comparable roles in different species? Lesions of particular tonotopic AI subregions in cat cause equally specific sound localization deficits (Jenkins and Merzenich 1984), while even larger bilateral lesions in rat do not (Kelly and Glazier 1978). Thus, the functions cortical areas (and by extension, in the MGB) may be species specific and otherwise analogous structures may have non-equivalent functions.

Where are the several representations constructed? If the AC maps were merely copies of the cochlear sensory epithelium we might expect that the auditory pathway would be either the same size or even progressively smaller at higher levels. This is not the case. The evidence suggests not merely a progressive increase in size (Kulesza et al. 2002) (with significant exceptions) but concomitant changes in internal organization, with some patterns, e.g., glycinergic caudal brain stem neurons, absent above the midbrain (Winer et al. 1995), whereas others, e.g., the species-specific proportion of MGB interneurons (Winer and Larue 1996), are highly variable and suggest emergent properties that remain to be defined.

2 The Problem of Interneurons

What is the role of Golgi type II cells? The immunocytochemical demonstration of neurotransmitters (Storm-Mathisen 1972) and the repudiation of Dale's principle

(Emson 1983) led to a new picture of the neuron as chemically multivalent, containing many neuroactive compounds, each with a different molecular configuration, intracellular storage regime, specialized mode of release, particular synaptic or postjunctional targets, and independent time courses of kinetics. The classic view that interneurons as arbiters mainly of lateral and recurrent inhibitory events (Windhorst 1990) has evolved to accommodate findings that GABAergic neurons can project remotely (Winer et al. 1996) and transmit impulses rapidly (Peruzzi et al. 1997) to their synaptic targets, while AC interneurons have a layer-specific typology (Prieto et al. 1994) and their chemical subvarieties include cells immunolabeled by antibodies to somatostatin, cholecystokinin, vasoactive intestinal polypeptide, and substance P (Cipolloni and Pandya 1991), as well as nitric oxide (Wakatsuki et al. 1998). When considered from the perspectives of colocalization (Yingcharoen et al. 1989; Jones and Hendry 1986) and receptor subunit distributions (Hsieh et al. 2002), the ensuing neurochemical diversity within a layer becomes imposing. Multiplied by specific constellations of connectivity (Briggs and Callaway 2001), it suggests a new kind of complexity in cortex that is abetted further by sublaminar differences that restrict or amplify extrinsic input spatially with even more precision (Lund 1990; Lund et al. 2001). A major task, then, is to define with more precision in AC the manifold and intricate roles of inhibition, as has been pursued in visual cortex (Crook et al. 1997, 1998; Martinez et al. 2005). Thus, iontophoresis of GABA antagonist in gerbil AC has no effect on sharpness of tuning for pure tones but does alter temporal modulation envelopes (Kurt et al. 2006). It remains to relate such findings to specific classes of neurons and particular patterns of synaptic arrangements, an endeavor well advanced in the cochlear nucleus (Josephson and Morest 1998; Davis 2002). Such data underlie more refined models of the dynamics of AC performance and for assessing system-to-system impact (Emri et al. 2003). It is startling that we have only a very scant picture available for the ordinal flow of information within an AC module (Atencio et al. 2009; Atencio and Schreiner 2010a,b), and the dataset that would permit the prediction of corticocortical transformations from acoustical to conceptually meaningful content remains severely limited (Bar-Yosef and Nelken 2007; Atencio et al. 2009; King and Nelken 2009).

By the same token, in the MGB, the presence of at least two classes of GABAergic neurons that exist in very different proportions raises analogous questions (Huang et al. 1999), and such a pattern exists in the visual thalamus as well (Montero and Zempel 1985). While MGB intrinsic organization may seem less complex than that in AC, any such conclusion must be tempered by the GABAergic inputs arising from the inferior colliculus (Winer et al. 1996), which themselves likely represent different classes of such cells (Oliver et al. 1994), and those from thalamic reticular nucleus projections whose impact on ongoing thalamic sensory processing may be profound (Crabtree et al. 1998).

3 A Case for Comparative Neuroscience

Can one model of the auditory forebrain suffice for all species? Appeals for a comparative perspective on neural function are often couched in cautionary terms as exhortations to search for a mammalian plan, a strategy consonant with the long history of characterizing species differences (Diamond 1973). This strategy has been especially fruitful in the auditory periphery, where the range of variation in the shape, size, and internal configuration of the basilar membrane in reptiles alone (Wever 1978) is to comparative morphology what the form of variations is to music. While many elements are conserved in their particular relations, even more depart from any single metric that the concept 'reptile ear' must be enormously elastic to embrace all of the variants, and the range of microarchitectonic adaptations might seem to exceed the plausible capacity of accommodation of any simple theory. This strategy has also elicited a sense of unease since the link between postulated homologies and function is a tenuous one at best (Striedter 2002) and there is a pervasive sense that comparative questions as such are mainly of theoretical interest. The position taken here, for reasons articulated below, is that there is an essential relation between these matters and the larger question of how the auditory system works.

A crux of this argument is the well-documented variability in the proportion of GABAergic Golgi type II cells in the MGB, which ranges from $<1\%$ in the mustached bat (Winer et al. 1992) to $\sim 1\%$ in rat, to perhaps 30% in the rhesus macaque (Winer and Larue 1996) and $\sim 25\%$ in the cat ventral division (Huang et al. 1999). This species-specific pattern is found also in the expression of such neurons in the ventrobasal complex of the thalamus, while the lateral geniculate body appears to have many more GABAergic neurons than its auditory and somatic sensory counterparts except in non-rodent species (Arcelli et al. 1997). Since there are abundant GABAergic neurons in rodent and bat inferior colliculus and in the auditory cortex (Winer and Larue 1996; Winer et al. 1995; Winer 1992), the thalamus in general and certain specific nuclei in it appear to depart from any common plan since the functional impact of these patterns suggests a difference in kind rather than a mere variation in quantity. Many other elements of thalamic organization are conserved: these species each have bushy tufted neurons (Winer 1992; Winer et al. 1999a) that project to cerebral cortex (Winer et al. 1999c), each receives robust inferior colliculus input (Wenstrup et al. 1994; Wenstrup 2005), and they each have a topographic arrangement of characteristic frequency in the ventral division (Winer 1992), with high Q_{10}dB values (Olsen and Suga 1991) or tectothalamic projections (Malmierca et al. 1997) consonant with a lemniscal role. The questions then become whether this is the same MGB functionally, and whether any single *Bauplan* can capture the range of thalamic GABAergic function on the one hand and the diversity of substrates for intrinsic processing on the other. Whatever their ultimate answers, these questions demand and deserve a considered treatment. Perhaps we should less seek any global pattern than more clearly to capture species differences and their functional ramifications with precision. When that has been accomplished, striking differences have been documented (Zirrinpar and Callaway 2006).

4 Neuropil: A Synaptic Nexus

In higher mammals, the neuropil constitutes the largest functional compartment in the brain (Winer 1984; Peters et al. 1991). Only a comparatively few (albeit powerful) synapses concentrate on the soma, axon hillock, and initial segment of most neurons. The vast majority congregate in the neuropil where most synaptic traffic likewise occurs (Peters et al. 1991), and these inputs represent a challenge to structure–function studies. Undoubtedly, some of the inputs farthest from the spike triggering and electrogenic membrane have roles related to tonic levels of discharge or sleep (Steriade and Timofeev 2003) rather than arousal (Steriade 1996). How each input contributes differentially is a daunting task which can be made more amenable by several strategies. First, depletion of one synaptic transmitter store or another with chemically specific agents can reveal otherwise hidden aspects of functional organization (Persico et al. 2000). Second, antibody-specific destruction of specific classes of neuron might have selective effects (Crabtree et al. 1986) which remain to be explored in the auditory forebrain. Third, slice preparations from avians (Müller 1988) and rodents (Spreafico et al. 1994) with a paucity of local circuit neurons could provide insight into information transfer under such regimes. Third, knockout preparations offer a prospectively exciting way to manipulate the neuropil which might have behavioral significance (Ko et al. 2005). Fourth, an expanded morphological agenda can be expected to reveal novel features of neural organization, such as the possible absence of Golgi type II cells in the inferior colliculus (Oliver et al. 1994), the gap junctions between somata in layer IV of primate auditory cortex (Smith and Moskowitz 1979), autaptic synaptic coupling for visual cortex pyramidal cells (Lübke et al. 1996), and gap junctions from GABAergic cortical interneurons that contribute to a dendrodendritic intracortical network (Fukuda et al. 2006). It is disappointing that two of these arbitrarily selected examples are from the auditory system, that they are the oldest among the group, and that neither seems to have elicited the correlative functional studies that would document their impact on function.

5 Auditory Forebrain Maps: Topography in a State of Flux

There is consensus that multiple independent representations of the basilar membrane are a hallmark of the AC (Reale and Imig 1980), and there are common processing strategies among the maps (Eggermont 1998) as well as differences between them (Imaizumi et al. 2005). However, the majority of AC areas (Schreiner 1995) and of MGB subdivisions (Calford 1983) have just one or a few such representations, or have such coarse topographies of registration (Rouiller et al. 1989) that they do not constitute maps in the strict sense (Tusa et al. 1981). A similar case prevails in the visual (Palmer et al. 1978) and somatic sensory (Kaas 1983) cortex, where multiple and partial representations of the peripheral receptor epithelium are the rule, adjoining a core of a few more or less complete representations. The case that topographic representations are themselves essential computational devices has been scrutinized, sometimes with diametrically opposed conclusions (Kaas 1997; Weinberg 1997). While the core issues in the debate are presently irresolvable, it is now indisputable that activation of the cholinergic nucleus basalis in conjunction with specific sensory experience in the awake, behaving animal can re-tune the AC tonotopic map (Kilgard and Merzenich 1998; Weinberger 1998), and that corticocollicular projections are able likewise to readjust the frequency representation in the inferior colliculus after appropriate regimes of AC stimulation (Zhang and Suga 2005). Several conclusions follow from this finding. It challenges the view that such maps are immutable and suggests that the corticofugal system has a role far from that predicted by theories of its operation in which its principal task is 'feedback.' It proposes to replace that static view with more dynamic concepts such as top-down control (Przybyszewski 1998; Fritz et al. 2007) and signal selection (King 1997).

Such findings are consistent with the complexity and individuation of the corticothalamic (Winer and Prieto 2001), corticocollicular (Winer et al. 1998), corticopontine (Perales et al. 2006), corticoolivary (Schofield and Coomes 2004), and corticocochlear (Weedman and Ryugo 1996) systems, each of which might be expected to be a somewhat different form of descending control of representational plasticity. This is also unsurprising since layers V (Winer et al. 2001) and VI (Prieto and Winer 1999) drive this process and constitute a large segment of AC (Winer 1992). Moreover, these findings help to explain the several otherwise enigmatic instances of proximity and likely interaction between ascending (Huang and Winer 2000) and descending (Hallman et al. 1988; White et al. 1994) systems. Finally, the size and diverse targets of the corticofugal system suggests a relation with behavioral and cognitive complexity since species without neocortex may have fundamentally different behavioral repertoires and capacities for sensorimotor adjustment (Winer 2005; Winer et al. 2005b). The several origins of corticofugal projections could support the proposition that parallel descending auditory pathways complement, and may even be larger than, the classical ascending auditory system (Winer 2005).

6 Conclusions

As pointed out at the beginning and in Chapter 2, a fully developed theory of auditory forebrain function is still beyond our reach. However, sufficient and in some regard astonishing progress has been made that has put such a task in the grasp of the next generation of auditory neuroscientists. Perhaps the collection of thoughts and facts in these pages will encourage, guide, and even inspire our colleagues to refine the rough sketch that has been compiled so far.

Acknowledgments This work was supported by United States Public Health Service grants R01 DC02260-16 (C.E.S.) and DC02319-30 (J.A.W.).

References

Arcelli P, Frassoni C, Regondi MC, De Biasi, and Spreafico R (1997) GABAergic neurons in mammalian thalamus: a marker of thalamic complexity? *Brain Research Bulletin* **42**:27–37.

Atencio CA, Sharpee TO, and Schreiner CE (2009) Hierarchical computation in the canonical auditory cortical circuit. *Proceedings of the National Academy of Sciences of the United States of America* **106**:21894–21899.

Atencio CA and Schreiner CE (2010a) Columnar connectivity and laminar processing in cat primary auditory cortex. *Public Library of Science One* **5**:e9521.

Atencio CA and Schreiner CE (2010b) Laminar diversity of dynamic sound processing in cat primary auditory cortex. *Journal of Neurophysiology* **103**:192–205.

Bar-Yosef O and Nelken I (2007) The effects of background noise on the neural responses to natural sounds in cat primary auditory cortex. *Frontiers in Computational Neuroscience* 1:1:3, doi: 10.3389/neuro.3310/3003.2007.

Briggs F and Callaway EM (2001) Layer-specific input to distinct cell types in layer 6 of monkey primary visual cortex. *Journal of Neuroscience* **21**:3600–3608.

Calford MB (1983) The parcellation of the medial geniculate body of the cat defined by the auditory response properties of single units. *Journal of Neuroscience* **3**:2350–2364.

Casseday JH, Schreiner CE, and Winer JA (2005) The inferior colliculus: past, present, and future. In: Winer JA and Schreiner CE (eds). *The Inferior Colliculus*. Springer, New York, pp. 626–640.

Cipolloni PB and Pandya DN (1991) Golgi, histochemical, and immunocytochemical analyses of the neurons of auditory-related cortices of the rhesus monkey. *Experimental Neurology* **114**: 104–122.

Clascá F, Llamas A, and Reinoso-Suárez F (2000) Cortical connections of the insular and adjacent parieto-temporal fields in the cat. *Cerebral Cortex* **10**:371–399.

Crabtree JW, Collingridge GL, and Issac JTR (1998) A new intrathalamic pathway linking modality-related nuclei in the dorsal thalamus. *Nature Neuroscience* **1**:389–394.

Crabtree JW, Spear PD, McCall MA, Tong L, Jones KR, and Kornguth SE (1986) Dose-response analysis of effects of antibodies to large ganglion cells on the cat's retinogeniculate pathways. *Journal of Neuroscience* **6**:1199–1210.

Crook JM, Kisvárday ZF, and Eysel UT (1997) GABA-induced inactivation of functionally characterized sites in cat striate cortex: effects on orientation tuning and direction selectivity. *Visual Neuroscience* **14**:141–158.

Crook JM, Kisvárday ZF, and Eysel UT (1998) Evidence for a contribution of lateral inhibition to orientation tuning and direction selectivity in cat visual cortex: reversible inactivation of functionally characterized sites combined with neuroanatomical tracing techniques. *European Journal of Neuroscience* **10**:2056–2075.

Colavita F (1979) Temporal pattern discrimination in cats with insular-temporal lesions. *Physiology & Behavior* **18**:513–521.

Colavita FB (1974) Insular-temporal lesions and vibrotactile temporal pattern discrimination in cats. *Physiology & Behavior* **12**: 215–218.

Colwell S (1975) Thalamocortical-corticothalamic reciprocity: a combined anterograde-retrograde tracer technique. *Brain Research* **92**:443–449.

Davis KA (2002) Evidence of a functionally segregated pathway from dorsal cochlear nucleus to inferior colliculus. *Journal of Neurophysiology* **87**:1824–1835.

Diamond IT (1973) The evolution of the tectal-pulvinar system in mammals: structural and behavioural studies of the visual system. *Symposia of the Zoological Society of London* **33**:205–233.

Edeline JM (2003) The thalamo-cortical auditory receptive fields: regulation by the states of vigilance, learning and the neuromodulatory systems. *Experimental Brain Research* **153**:554–572.

Eggermont JJ (1998) Representation of spectral and temporal sound features in three cortical fields of the cat. Similarities outweigh differences. *Journal of Neurophysiology* **80**:2743–2764.

Emri Z, Antal K, and Crunelli V (2003) The impacts of corticothalamic feedback on the output dynamics of a thalamocortical neurone model: the role of synapse location and metabotropic glutamate receptors. *Neuroscience* **117**:229–239.

Emson PC (1983) *Chemical Neuroanatomy*, Raven Press, New York.

Fitzpatrick DC, Olsen JF, and Suga N (1998) Connections among functional areas in the mustached bat auditory cortex. *Journal of Comparative Neurology* **391**:366–396.

Fritz JB, Elhilai M, and Shamma SA (2007) Auditory attention - focusing the searchlight on sound. *Current Opinions in Neurobiology* **17**:437–455.

Fukuda T, Kosaka T, Singer W, and Galuske RA (2006) Gap junctions among dendrites of cortical GABAergic neurons establish a dense and widespread intercolumnar network. *Journal of Neuroscience* **26**:3454–3464.

Hackett TA, Preuss TM, and Kaas JH (2001) Architectonic identification of the core region in auditory cortex of macaques, chimpanzees, and humans. *Journal of Comparative Neurology* **441**:197–222.

Hallman EL, Schofield BR, and Lin C-S (1988) Dendritic morphology and axon collaterals of corticotectal, corticopontine, and callosal neurons in layer V of primary visual cortex of the hooded rat. *Journal of Comparative Neurology* **272**:149–160.

Hsieh CY, Chen Y, Leslie FM, and Metherate R (2002) Postnatal development of NR2A and NR2B mRNA expression in rat auditory cortex and thalamus. *Journal of the Association for Research in Otolaryngology* **3**:479–487.

Huang CL, Larue DT, and Winer JA (1999) GABAergic organization of the cat medial geniculate body. *Journal of Comparative Neurology* **415**:368–392.

Huang CL and Winer JA (2000) Auditory thalamocortical projections in the cat: laminar and areal patterns of input. *Journal of Comparative Neurology* **427**:302–331.

Imaizumi K, Lee CC, Linden JF, Winer JA, and Schreiner CE (2005) The anterior field of auditory cortex: neurophysiological and neuroanatomical organization. In: König R, Heil P, Budinger E, and Scheich H (eds). *The Auditory Cortex. A Synthesis of Human and Animal Research.* Lawrence Erlbaum Associates, New York, pp. 95–110.

Imig TJ and Adrián HO (1977) Binaural columns in the primary auditory field (A1) of cat auditory cortex. *Brain Research* **138**:241–257.

Imig TJ and Brugge JF (1978) Sources and terminations of callosal axons related to binaural and frequency maps in primary auditory cortex of the cat. *Journal of Comparative Neurology* **182**: 637–660.

Jenkins WM and Merzenich MM (1984) Role of cat primary auditory cortex for sound-localization behavior. *Journal of Neurophysiology* **52**:819–847.

Jones EG and Hendry SHC (1986) Colocalization of GABA and neuropeptides in neocortical neurons. *Trends in Neurosciences* **9**:71–76.

Josephson EM and Morest DK (1998) A quantitative profile of the synapses on the stellate cell body and axon in the cochlear nucleus of the chinchilla. *Journal of Neurocytology* **27**:841–864.

Kaas JH (1983) What, if anything is SI? Organization of first somatosensory area of cortex. *Physiological Reviews* **63**:206–231.

Kaas JH (1997) Topographic maps are fundamental to sensory processing. *Brain Research Bulletin* **44**:107–112.

Kelly JB and Glazier SJ (1978) Auditory cortex lesions and discrimination of spatial location by the rat. *Brain Research* **145**:315–321.

Kilgard MP and Merzenich MM (1998) Cortical map reorganization enabled by nucleus basalis activity. *Science* **279**:1714–1718.

King AJ (1997) Signal selection by cortical feedback. *Current Biology* **7**:R85–R88.

King AJ and Nelken I (2009) Unraveling the principles of auditory cortical processing: can we learn from the visual system?. *Nature Neuroscience* **12**:698–701.

Ko S, Zhao MG, Toyoda H, Qiu CS, and Zhuo M (2005) Altered behavioral responses to noxious stimuli and fear in glutamate receptor 5 (GluR5)- or GluR6-deficient mice. *Journal of Neuroscience* **25**:977–984.

Kulesza RJ, Viñuela A, Saldaña E, and Berrebi AS (2002) Unbiased stereological estimates of neuron number in subcortical auditory nuclei of the rat. *Hearing Research* **168**:12–24.

Kurt S, Crook JM, Ohl FW, Scheich H, and Schulze H (2006) Differential effects of iontophoretic in vivo application of GABAA-antagonists bicuculline and gabazine in sensory cortex. *Hearing Research* **212**:224–235.

Lee CC, Schreiner CE, Imaizumi K, and Winer JA (2004) Tonotopic and heterotopic projection systems in physiologically defined auditory cortex. *Neuroscience* **128**:871–887.

Lee CC and Winer JA (2005) Principles governing auditory forebrain connections. *Cerebral Cortex* **15**:1804–1814.

LeVay S and Gilbert CD (1976) Laminar patterns of geniculocortical projection in the cat. *Brain Research* **113**:1–19.

Lübke J, Markram H, Frotscher M, and Sakmann B (1996) Frequency and dendritic distribution of autapses established by layer 5 pyramidal neurons in the developing rat neocortex: comparison with synaptic innervation of adjacent neurons of the same class. *Journal of Neuroscience* **16**:3209–3218.

Lund JS (1990) Excitatory and inhibitory circuiting and laminar mapping strategies in the primary visual cortex of the monkey. In: Edelman GM, Gall WE, and Cowan WM (eds). *Signal and Sense: Local and Global Order in Perceptual Maps.* Wiley-Liss, New York, pp. 51–82.

Lund JS, Griffiths S, Rumberger A, and Levitt JB (2001) Inhibitory synapse cover on the somata of excitatory neurons in macaque monkey visual cortex. *Cerebral Cortex* **11**:783–795.

Malmierca MS, Rees A, and Le Beau FEN (1997) Ascending projections to the medial geniculate body from physiologically identified loci in the inferior colliculus. In: Syka J (ed). *Acoustical Signal Processing in the Central Auditory System.* Plenum, New York, pp. 295–302.

Martinez LM, Wang Q, Reid RC, Pillai C, Alonso J-M, Sommer FT, and Hirsch JA (2005) Receptive field varies with layer in the primary visual cortex. *Nature Neuroscience* **8**:372–379.

McMullen NT and de Venecia RK (1993) Thalamocortical patches in auditory neocortex. *Brain Research* **620**:317–322.

Middlebrooks JC and Zook JM (1983) Intrinsic organization of the cat's medial geniculate body identified by projections to binaural response-specific bands in the primary auditory cortex. *Journal of Neuroscience* **3**:203–225.

Miller KD, Pinto DJ, and Simons DJ (2001a) Processing in layer 4 of the neocortical circuit: new insights from visual and somatosensory cortex. *Current Opinion in Neurobiology* **11**:488–497.

Miller LM, Escabí MA, Read HL, and Schreiner CE (2001b) Functional convergence of response properties in the auditory thalamocortical system. *Neuron* **32**:151–160.

Miller LM, Escabí MA, Read HL, and Schreiner CE (2002) Spectrotemporal receptive fields in the lemniscal auditory thalamus and cortex. *Journal of Neurophysiology* **87**:516–527.

Miller LM, Escabí MA, and Schreiner CE (2001c) Feature selectivity and interneuronal cooperation in the thalamocortical system. *Journal of Neuroscience* **21**:8136–8144.

Miller LM and Schreiner CE (2000) Stimulus based state control in the thalamocortical system. *Journal of Neuroscience* **20**:7011–7016.

Montero VM and Zempel J (1985) Evidence for two types of GABA-containing interneurons in the A-laminae of the cat lateral geniculate nucleus: a double-label HRP and GABA-immunocytochemical study. *Experimental Brain Research* **60**:603–609.

Morel A and Kaas JH (1992) Subdivisions and connections of auditory cortex in owl monkeys. *Journal of Comparative Neurology* **318**: 27–63.

Morosan P, Rademacher J, Schleicher A, Amunts K, Schormann T, and Zilles K (2001) Human primary auditory cortex: cytoarchitectonic subdivisions and mapping into a spatial reference system. *NeuroImage* **13**:694–701.

Müller CM (1988) Distribution of GABAergic perikarya and terminals in the centers of the higher auditory pathway of the chicken. *Cell and Tissue Research* **252**:99–106.

Ohki K, Cheung S, Ch'ng YH, Kara P, Fabene PF, and Reid RC (2005) Functional imaging with cellular resolution reveals precise microarchitecture in visual cortex. *Nature* **433**:597–603.

Oliver DL, Winer JA, Beckius GE, and Saint Marie RL (1994) Morphology of GABAergic cells and axon terminals in the cat inferior colliculus. *Journal of Comparative Neurology* **340**: 27–42.

Olsen JF and Suga N (1991) Combination-sensitive neurons in the medial geniculate body of the mustached bat: encoding of relative velocity information. *Journal of Neurophysiology* **65**: 1254–1274.

Palmer LA, Rosenquist AC, and Tusa RJ (1978) The retinotopic organization of lateral suprasylvian visual areas in the cat. *Journal of Comparative Neurology* **177**:237–256.

Perales M, Winer JA, and Prieto JJ (2006) Focal projections of cat auditory cortex to the pontine nuclei. *Journal of Comparative Neurology* **497**:959–980.

Persico AM, Calia AE, Puglisi-Allegra S, Ventura R, Lucchese F, and Keller F (2000) Serotonin depletion and barrel cortex development: impact of growth impairment vs. serotonin effects on thalamocortical endings. *Cerebral Cortex* **10**:181–191.

Peruzzi D, Bartlett E, Smith PH, and Oliver DL (1997) A monosynaptic GABAergic input from the inferior colliculus to the medial geniculate body in rat. *Journal of Neuroscience* **17**: 3766–3777.

Peters A, Palay SL, and Webster H (1991) *The Fine Structure of the Nervous System. The Neurons and their Supporting Cells*. Oxford University Press, New York.

Prieto JJ, Peterson BA, and Winer JA (1994) Morphology and spatial distribution of GABAergic neurons in cat primary auditory cortex (AI). *Journal of Comparative Neurology* **344**:349–382.

Prieto JJ and Winer JA (1999) Layer VI in cat primary auditory cortex (AI): Golgi study and sublaminar origins of projection neurons. *Journal of Comparative Neurology* **404**:332–358.

Przybyszewski AW (1998) Does top-down processing help us to see? *Current Biology* **8**:R135-R139.

Read HL, Winer JA, and Schreiner CE (2001) Modular organization of intrinsic connections associated with spectral tuning in cat auditory cortex. *Proceedings of the National Academy of Sciences of the United States of America* **98**:8042–8047.

Reale RA and Imig TJ (1980) Tonotopic organization in auditory cortex of the cat. *Journal of Comparative Neurology* **192**:265–291.

Rouiller EM, Rodrigues-Dagaeff C, Simm GM, de Ribaupierre Y, Villa AEP, and de Ribaupierre F (1989) Functional organization of the medial division of the medial geniculate body of the cat: tonotopic organization, spatial distribution of response properties and cortical connections. *Hearing Research* **39**:127–146.

Schofield BR and Coomes DL (2004) Projections from the auditory cortex to the superior olivary complex in guinea pigs. *European Journal of Neuroscience* **19**:2188–2200.

Schreiner CE (1995) Order and disorder in auditory cortical maps. *Current Opinion in Neurobiology* **5**:489–496.

Sherman SM (2004) Interneurons and triadic circuitry of the thalamus. *Trends in Neurosciences* **27**:670–675.

Shi C-J and Cassell MD (1997) Cortical, thalamic, and amygdaloid projections of rat temporal cortex. *Journal of Comparative Neurology* **382**:153–175.

Smith DE and Moskowitz N (1979) Ultrastructure of layer IV of the primary auditory cortex of the squirrel monkey. *Neuroscience* **4**:349–359.

Spreafico R, Frassoni C, Arcelli P, and De Biasi S (1994) GABAergic interneurons in the somatosensory thalamus of the guinea-pig: a light and ultrastructural immunocytochemical investigation. *Neuroscience* **59**:961–973.

Steriade M (1996) Arousal: Revisiting the reticular activating system. *Science* **272**:225–226.

Steriade M and Timofeev I (2003) Neuronal plasticity in thalamocortical networks during sleep and waking oscillations. *Neuron* **37**:563–576.

Storm-Mathisen J (1972) Glutamate decarboxylase in the rat hippocampal region after lesions of the afferent fibre systems: evidence that enzyme is localized in intrinsic neurons. *Brain Research* **40**: 215–235.

Striedter GF (2002) Brain homology and function: An uneasy alliance. *Brain Research Bulletin* **57**:239–242.

Suga N (1978) Specialization of the auditory system for reception and processing of species-specific sounds. *Federation of the American Society for Experimental Biology Proceedings* **37**:2342–2354.

Szentágothai J (1975) The "module-concept" in cerebral cortex architecture. *Brain Research* **95**:475–496.

Tusa RJ, Palmer LA, and Rosenquist AC (1981) Multiple cortical visual areas: visual field topography in the cat. In: Woolsey CN (ed). *Cortical Sensory Organization*, volume 2, *Multiple Visual Areas*. Humana Press, Clifton, pp. 1–31.

Wakatsuki H, Gomi H, Kudoh M, Kimura S, Takahashi K, Takeda M, and Shibuki K (1998) Layer-specific NO dependence of long-term potentiation and biased NO release in layer V in the rat auditory cortex. *Journal of Physiology* **513**:71–81.

Warr WB (1982) Parallel ascending pathways from the cochlear nucleus: neuroanatomical evidence of functional specialization. In: Neff WD (ed). *Contributions to Sensory Physiology*. Academic Press, New York, pp. 1–38.

Weedman DL and Ryugo DK (1996) Projections from auditory cortex to the cochlear nucleus in rats: synapses on granule cell dendrites. *Journal of Comparative Neurology* **371**:311–324.

Weinberg RJ (1997) Are topographic maps fundamental to sensory processing? *Brain Research Bulletin* **44**:113–116.

Weinberger NM (1998) Tuning the brain by learning and by stimulation of the nucleus basalis. *Trends in Cognitive Sciences* **2**:271–273.

Wenstrup JJ (2005) The tectothalamic system. In: Winer JA and Schreiner CE (eds). *The Inferior Colliculus*. Springer, New York, pp. 200–230.

Wenstrup JJ, Larue DT, and Winer JA (1994) Projections of physiologically defined subdivisions of the inferior colliculus in the mustached bat: targets in the medial geniculate body and extrathalamic nuclei. *Journal of Comparative Neurology* **346**:207–236.

Wever EG (1978) *The Reptile Ear. Its Structure and Function*. Princeton University Press, Princeton.

White EL, Amitai Y, and Gutnick MJ (1994) A comparison of synapses onto the somata of intrinsically bursting and regular spiking neurons in layer V of rat SmI cortex. *Journal of Comparative Neurology* **342**:1–14.

Wild JM, Karten HJ, and Frost BJ (1993) Connections of the auditory forebrain in the pigeon (*Columba livia*). *Journal of Comparative Neurology* **337**:32–62.

Windhorst U (1990) Activation of Renshaw cells. *Progress in Neurobiology* **35**:135–179.

Winer JA (1984) Anatomy of layer IV in cat primary auditory cortex (AI). *Journal of Comparative Neurology* **224**:535–567.

Winer JA (1992) The functional architecture of the medial geniculate body and the primary auditory cortex. In: Webster DB, Popper AN, and Fay RR (eds). *Springer Handbook of Auditory Research*, volume 1, *The Mammalian Auditory Pathway: Neuroanatomy*. Springer, New York, pp. 222–409.

Winer JA (2005) Three systems of descending projections to the inferior colliculus. In: Winer JA and Schreiner CE (eds). *The Inferior Colliculus*. Springer, New York, pp. 231–247.

Winer JA (2006) Decoding the auditory corticofugal systems. *Hearing Research* **212**:1–8.

Winer JA, Diehl JJ, and Larue DT (2001) Projections of auditory cortex to the medial geniculate body of the cat. *Journal of Comparative Neurology* **430**:27–55.

Winer JA, Kelly JB, and Larue DT (1999a) Neural architecture of the rat medial geniculate body. *Hearing Research* **130**:19–41.

Winer JA and Larue DT (1996) Evolution of GABAergic circuitry in the mammalian medial geniculate body. *Proceedings of the National Academy of Sciences of the United States of America* **93**:3083–3087.

Winer JA, Larue DT, Diehl JJ, and Hefti BJ (1998) Auditory cortical projections to the cat inferior colliculus. *Journal of Comparative Neurology* **400**:147–174.

Winer JA, Larue DT, and Huang CL (1999b) Two systems of giant axon terminals in the cat medial geniculate body: convergence of cortical and GABAergic inputs. *Journal of Comparative Neurology* **413**:181–197.

Winer JA, Larue DT, and Pollak GD (1995) GABA and glycine in the central auditory system of the mustache bat: structural substrates

for inhibitory neuronal organization. *Journal of Comparative Neurology* **355**:317–353.
Winer JA, Lee CC, Imaizumi K and Schreiner CE (2005a) Challenges to a neuroanatomical theory of forebrain auditory plasticity. In: Syka J and Merzenich MM (eds). *Plasticity of the Central Auditory System and Processing of Complex Acoustic Signals*. Springer, New York, pp. 109–127.
Winer JA, Miller LM, Lee CC, and Schreiner CE (2005b) Auditory thalamocortical transformation: structure and function. *Trends in Neurosciences* **28**:255–263.
Winer JA and Prieto JJ (2001) Layer V in cat primary auditory cortex (AI): cellular architecture and identification of projection neurons. *Journal of Comparative Neurology* **434**:379–412.
Winer JA, Saint Marie RL, Larue DT and Oliver DL (1996) GABAergic feedforward projections from the inferior colliculus to the medial geniculate body. *Proceedings of the National Academy of Sciences of the United States of America* **93**:8005–8010.
Winer JA, Sally SL, Larue DT and Kelly JB (1999c) Origins of medial geniculate body projections to physiologically defined regions of rat auditory cortex. *Hearing Research* **130**:42–61.
Winer JA, Wenstrup JJ and Larue DT (1992) Patterns of GABAergic immunoreactivity define subdivisions of the mustached bat's medial geniculate body. *Journal of Comparative Neurology* **319**:172–190.
Yingcharoen K, Rinvik E, Storm-Mathisen J, and Ottersen OP (1989) GABA, glycine, glutamate, aspartate and taurine in the perihypoglossal nuclei: an immunocytochemical investigation with particular reference to the issue of amino acid colocalization. *Experimental Brain Research* **78**:345–357.
Zhang Y and Suga N (2005) Corticofugal feedback for collicular plasticity evoked by electric stimulation of the inferior colliculus. *Journal of Neurophysiology* **94**:2676–2682.
Zirrinpar A and Callaway EM (2006) Local connections of specific types of layer 6 neurons in the rat visual cortex. *Journal of Neurophysiology* **95**:1751–1761.

Index

J.A. Winer, C.E. Schreiner (eds.), *The Auditory Cortex*,
DOI 10.1007/978-1-4419-0074-6, © Springer Science+Business Media, LLC 2010

C

D

E

I

T